Chemical Process
Design and Integration

Chemical Process
Design and Integration

Second Edition

Robin Smith
School of Chemical Engineering and Analytical Science, The University of Manchester, UK

WILEY

This edition first published 2016

Registered office
John Wiley & Sons Ltd, The Atrium, Southern Gate, Chichester, West Sussex, PO19 8SQ, United Kingdom

For details of our global editorial offices, for customer services and for information about how to apply for permission to reuse the copyright material in this book please see our website at www.wiley.com.

Library of Congress Cataloging-in-Publication Data

Names: Smith, Robin (Chemical engineer)
Title: Chemical process design and integration / Robin Smith.
Description: Second edition. | Chichester, West Sussex, United Kingdom : John Wiley & Sons, Inc., 2016. | Includes index. | Description based on print version record and CIP data provided by publisher; resource not viewed.
Identifiers: LCCN 2015034820 (print) | LCCN 2015032671 (ebook) | ISBN 9781118699089 (ePub) | ISBN 9781118699096 (Adobe PDF) | ISBN 9781119990147 (hardback) | ISBN 9781119990130 (paper)
Subjects: LCSH: Chemical processes. | BISAC: TECHNOLOGY & ENGINEERING / Chemical & Biochemical.
Classification: LCC TP155.7 (print) | LCC TP155.7 .S573 2016 (ebook) | DDC 660/.28–dc23 LC record available at http://lccn.loc.gov/2015034820

A catalogue record for this book is available from the British Library.

ISBN: 9781119990147 (hardback)
ISBN: 9781119990130 (paper)

Set in 9.5/12 pt TimesLTStd-Roman by Thomson Digital, Noida, India

Printed in Singapore by C.O.S. Printers Pte Ltd

2 2017

To the next generation
George, Oliver, Ava and Freya

Contents

Preface to the Second Edition

PREFACE TO THE SECOND EDITION

This book deals with the design and integration of chemical processes. The Second Edition has been rewritten, restructured and updated throughout from the First Edition. At the heart of the book are the conceptual issues that are fundamental to the creation of chemical processes and their integration to form complete manufacturing systems. Compared with the First Edition, this edition includes much greater consideration of equipment and equipment design, including materials of construction, whilst not sacrificing understanding of the overall conceptual design. Greater emphasis has also been placed on physical properties, process simulation and batch processing. Increasing environmental awareness has dictated the necessity of a greater emphasis on environmental sustainability throughout. The main implication of this for process design is greater efficiency in the use of raw materials, energy and water and a greater emphasis on process safety. Consideration of integration has not been restricted to individual processes, but integration across processes has also been emphasized to create environmentally sustainable integrated manufacturing systems. Thus, the text integrates equipment, process and manufacturing system design. This edition has been rewritten to make it more accessible to undergraduate students of chemical engineering than the First Edition, as well as maintaining its usefulness to postgraduate students of chemical engineering and to practicing chemical engineers.

As with the first edition, this edition as much as possible emphasizes understanding of process design methods, as well as their application. Where practical, the derivation of design equations has been included, as this is the best way to understand the limitations of those equations and to ensure their wise application.

The book is intended to provide a practical guide to chemical process design and integration for students of chemical engineering at all levels, practicing process designers and chemical engineers and applied chemists working in process development. For undergraduate studies, the text assumes basic knowledge of material and energy balances and thermodynamics, together with basic spreadsheeting skills. Worked examples have been included throughout the text. Most of these examples do not require specialist software and can be solved either by hand or using spreadsheet software. A suite of Excel spreadsheets has also been made available to allow some of the more complex example calculations to be performed more conveniently. Finally, a number of exercises has been added at the end of each chapter to allow the reader to practice the calculation procedures. A solutions manual is available.

Robin Smith

Acknowledgements

ACKNOWLEDGEMENTS

The author would like to express gratitude to a number of people who have helped in the preparation of the Second Edition.

From the University of Manchester: Mary Akpomiemie, Adisa Azapagic, Stephen Doyle, Victor Manuel Enriquez Gutierrez, Oluwagbemisola Oluleye, Kok Siew Ng, Li Sun and Colin Webb

Interns at the University of Manchester: Rabia Amaaouch, Béatrice Bouchon, Aymeric Cambrillat, Leo Gandrille, Kathrin Holzwarth, Guillemette Nicolas and Matthias Schmid

From National Technical University of Athens: Antonis Kokossis

From the University of Huddersfield: Grant Campbell

From Nova Process Ltd: Stephen Hall

From Norwegian University of Science and Technology: Truls Gundersen

From BP: Paul Oram

Gratitude is also expressed to Lucy Adams, Ellen Gleeson and Tim Mummery for help in the preparation of the figures and the text.

Finally, gratitude is expressed to all of the member companies of the Process Integration Research Consortium, both past and present. Their support has made a considerable contribution to research in the area, and hence to this text.

Nomenclature

NOMENCLATURE

a	Activity (−), or constant in cubic equation of state ($N \cdot m^4 \cdot kmol^{-2}$), or correlating coefficient (units depend on application), or cost law coefficient (\$), or order of reaction (−)
a_{mn}	Group interaction parameter in the UNIFAC Model (K)
a_1, a_2	Profile control parameters in optimization (−)
A	Absorption factor in absorption (−), or annual cash flow (\$), or constant in vapor pressure correlation ($N \cdot m^{-2}$, bar), or heat exchanger area (m^2)
A_C	Cross-sectional area of column (m^2)
A_{CF}	Annual cash flow ($\$ \cdot y^{-1}$)
A_D	Area occupied by distillation downcomer (m^2)
A_{DCF}	Annual discounted cash flow ($\$ \cdot y^{-1}$)
A_{FIN}	Area of fins (m)
A_I	Heat transfer area on the inside of tubes (m^2), or interfacial area (m^2, $m^2 \cdot m^{-3}$)
A_M	Membrane area (m^2)
$A_{NETWORK}$	Heat exchanger network area (m^2)
A_O	Heat transfer area on the outside of tubes (m^2)
A_{ROOT}	Exposed outside root area of a finned tube (m)
A_{SHELL}	Heat exchanger area for an individual shell (m^2)
AF	Annualization factor for capital cost (−) capital cost law coefficient (units depend on cost law), or constant in cubic equation of state ($m^3 \cdot kmol^{-1}$), or correlating coefficient (units depend on application), or order of reaction (−)
b_i	Bottoms flowrate of Component i ($kmol \cdot s^{-1}$, $kmol \cdot h^{-1}$)
B	Baffle spacing in shell-and-tube heat exchangers (m), or Bottoms flowrate in distillation ($kg \cdot s^{-1}$, $kg \cdot h^{-1}$, $kmol \cdot s^{-1}$, $kmol \cdot h^{-1}$), or breadth of device (m), or constant in vapor pressure correlation ($N \cdot K \cdot m^{-2}$, $bar \cdot K$), or moles remaining in batch distillation (kmol)
B_C	Baffle cut for shell-and-tube heat exchangers (−)
BOD	Biological oxygen demand ($kg \cdot m^{-3}$, $mg \cdot l^{-1}$)
c	Capital cost law coefficient (−), or correlating coefficient (units depend on application), or order of reaction (−)
c_D	Drag coefficient (−)
c_f	Fanning friction factor (−)
c_{fS}	Smooth tube Fanning friction factor (−)
c_L	Loss coefficient for pipe or pipe fitting (−)
C	Concentration ($kg \cdot m^{-3}$, $kmol \cdot m^{-3}$, ppm), or constant in vapor pressure correlation (K), or number of components (separate systems) in network design (−)
C_B	Base capital cost of equipment (\$)
Ce	Environmental discharge concentration (ppm)
C_E	Equipment capital cost (\$), or unit cost of energy ($\$ \cdot kW^{-1}$, $\$ \cdot MW^{-1}$)
C_F	Fixed capital cost of complete installation (\$)
C_P	Specific heat capacity at constant pressure ($kJ \cdot kg^{-1} \cdot K^{-1}$, $kJ \cdot kmol^{-1} \cdot K^{-1}$)
$\overline{C_P}$	Mean heat capacity at constant pressure ($kJ \cdot kg^{-1} \cdot K^{-1}$, $kJ \cdot kmol^{-1} \cdot K^{-1}$)
C_S	Corrected superficial velocity in distillation ($m \cdot s^{-1}$)
C_V	Specific heat capacity at constant volume ($kJ \cdot kg^{-1} \cdot K^{-1}$, $kJ \cdot kmol^{-1} \cdot K^{-1}$)
C^*	Solubility of solute in solvent ($kg \cdot kg\ solvent^{-1}$)
CC	Cycles of concentration for a cooling tower (−)
CC_{STEAM}	Cumulative cost ($\$ \cdot t^{-1}$)
COD	Chemical oxygen demand ($kg \cdot m^{-3}$, $mg \cdot l^{-1}$)
COP	Coefficient of performance (−)
COP_{AHP}	Coefficient of performance of an absorption heat pump (−)
COP_{AHT}	Coefficient of performance of an absorption heat transformer (−)
COP_{AR}	Coefficient of performance of absorption refrigeration (−)
COP_{CHP}	Coefficient of performance of a compression heat pump (−)
COP_{HP}	Coefficient of performance of a heat pump (−)
COP_{REF}	Coefficient of performance of a refrigeration system (−)
CP	Capacity parameter in distillation ($m \cdot s^{-1}$) or heat capacity flowrate ($kW \cdot K^{-1}$, $MW \cdot K^{-1}$)

CP_{EX}	Heat capacity flowrate of heat engine exhaust ($kW \cdot K^{-1}$, $MW \cdot K^{-1}$)
CW	Cooling water
d	Diameter (μm, m), or correlating coefficient (units depend on application)
d_C	Column inside diameter (m)
d_i	Distillate flowrate of Component i ($kmol \cdot s^{-1}$, $kmol \cdot h^{-1}$)
d_I	Inside diameter of pipe or tube (m)
d_P	Distillation and absorption packing size (m)
d_R	Outside tube diameter for a finned tube at the root of fins (m)
D	Distillate flowrate ($kg \cdot s^{-1}$, $kg \cdot h^{-1}$, $kmol \cdot s^{-1}$, $kmol \cdot h^{-1}$)
D_B	Tube bundle diameter for shell-and-tube heat exchangers (m)
D_S	Inside shell diameter for shell-and-tube heat exchangers (m)
$DCFRR$	Discounted cash flowrate of return (%)
e	Wire diameter (m)
E	Activation energy of reaction ($kJ \cdot kmol^{-1}$), or entrainer flowrate in azeotropic and extractive distillation ($kg \cdot s^{-1}$, $kmol \cdot s^{-1}$), or exchange factor in radiant heat transfer (–), or extract flowrate in liquid–liquid extraction ($kg \cdot s^{-1}$, $kmol \cdot s^{-1}$), or stage efficiency in separation (–)
E_O	Overall stage efficiency in distillation and absorption (–)
EP	Economic potential ($\$ \cdot y^{-1}$)
f	Fuel-to-air ratio for gas turbine (–)
f_i	Capital cost installation factor for Equipment i (–), or feed flowrate of Component i ($kmol \cdot s^{-1}$, $kmol \cdot h^{-1}$), or fugacity of Component i ($N \cdot m^{-2}$, bar)
f_P	Capital cost factor to allow for design pressure (–)
f_T	Capital cost factor to allow for design temperature (–)
F	Feed flowrate ($kg \cdot s^{-1}$, $kg \cdot h^{-1}$, $kmol \cdot s^{-1}$, $kmol \cdot h^{-1}$), or future worth a sum of money allowing for interest rates (\$), or number of degrees of freedom (–), or volumetric flowrate ($m^3 \cdot s^{-1}$, $m^3 \cdot h^{-1}$)
F_{FOAM}	Foaming factor in distillation (–)
F_{LV}	Liquid–vapor flow parameter in distillation (–)
F_{RAD}	Fraction of heat absorbed in fired heater radiant section (–)
F_{SC}	Correction factor for shell construction in shell-and-tube heat exchangers (–)
F_T	Correction factor for noncountercurrent flow in shell-and-tube heat exchangers (–)
F_{TC}	Correction factor for tube count in shell-and-tube heat exchangers (–)
F_{Tmin}	Minimum acceptable F_T for noncountercurrent heat exchangers (–)
F_{XY}	Factor to allow for inclination in structured packing (–)
F_σ	Factor to allow for inadequate wetting of packing (–)
g	Acceleration due to gravity ($9.81\ m \cdot s^{-2}$)
g_{ij}	Energy of interaction between Molecules i and j in the NRTL equation ($kJ \cdot kmol^{-1}$)
G	Free energy (kJ), or gas flowrate ($kg \cdot s^{-1}$, $kmol \cdot s^{-1}$)
$\overline{G}_i$	Partial molar free energy of Component i ($kJ \cdot kmol^{-1}$)
$\overline{G}_i^O$	Standard partial molar free energy of Component i ($kJ \cdot kmol^{-1}$)
GCV	Gross calorific value of fuel ($J \cdot m^{-3}$, $kJ \cdot m^{-3}$, $J \cdot kg^{-1}$, $kJ \cdot kg^{-1}$)
h	Settling distance of particles (m)
h_B	Boiling heat transfer coefficient for the tube bundle ($W \cdot m^{-2} \cdot K^{-1}$, $kW \cdot m^{-2} \cdot K^{-1}$)
h_C	Condensing film heat transfer coefficient ($W \cdot m^{-2} \cdot K^{-1}$, $kW \cdot m^{-2} \cdot K^{-1}$)
h_I	Film heat transfer coefficient for the inside ($W \cdot m^{-2} \cdot K^{-1}$, $kW \cdot m^{-2} \cdot K^{-1}$)
h_{IF}	Fouling heat transfer coefficient for the inside ($W \cdot m^{-2} \cdot K^{-1}$, $kW \cdot m^{-2} \cdot K^{-1}$)
h_L	Head loss in a pipe or pipe fitting (m)
h_{NB}	Nucleate boiling heat transfer coefficient ($W \cdot m^{-2} \cdot K^{-1}$, $kW \cdot m^{-2} \cdot K^{-1}$)
h_O	Film heat transfer coefficient for the outside ($W \cdot m^{-2} \cdot K^{-1}$, $kW \cdot m^{-2} \cdot K^{-1}$)
h_{OF}	Fouling heat transfer coefficient for the outside ($W \cdot m^{-2} \cdot K^{-1}$, $kW \cdot m^{-2} \cdot K^{-1}$)
h_{RAD}	Radiant heat transfer coefficient ($W \cdot m^{-2} \cdot K^{-1}$, $kW \cdot m^{-2} \cdot K^{-1}$)
h_W	Heat transfer coefficient for the tube wall ($W \cdot m^{-2} \cdot K^{-1}$, $kW \cdot m^{-2} \cdot K^{-1}$)
H	Enthalpy (kJ, $kJ \cdot kg^{-1}$, $kJ \cdot kmol^{-1}$), or height (m), or Henry's Law Constant ($N \cdot m^{-2}$, bar, atm), or stream enthalpy ($kJ \cdot s^{-1}$, $MJ \cdot s^{-1}$)
H_F	Height of fin (m)
H_T	Tray spacing (m)
$\overline{H}_i^O$	Standard heat of formation of Component i ($kJ \cdot kmol^{-1}$)
ΔH^O	Standard heat of reaction (J, kJ)
ΔH_{COMB}	Heat of combustion ($J \cdot kmol^{-1}$, $kJ \cdot kmol^{-1}$)
ΔH_{COMB}^O	Standard heat of combustion at 298 K ($J \cdot kmol^{-1}$, $kJ \cdot kmol^{-1}$)
ΔH_{FUEL}	Heat to bring fuel to standard temperature ($J \cdot kmol^{-1}$, $kJ \cdot kg^{-1}$)
ΔH_{IS}	Isentropic enthalpy change of an expansion ($J \cdot kmol^{-1}$, $kJ \cdot kg^{-1}$)

ΔH_P	Heat to bring products from standard temperature to the final temperature ($J \cdot kmol^{-1}$, $kJ \cdot kg^{-1}$)
ΔH_R	Heat to bring reactants from their initial temperature to standard temperature ($J \cdot kmol^{-1}$, $kJ \cdot kmol^{-1}$)
ΔH_{STEAM}	Enthalpy difference between generated steam and boiler feedwater (kW, MW)
ΔH_{VAP}	Latent heat of vaporization ($kJ \cdot kg^{-1}$, $kJ \cdot kmol^{-1}$)
$HETP$	Height equivalent of a theoretical plate (m)
HP	High pressure
HR	Heat rate for gas turbine ($kJ \cdot kWh^{-1}$)
i	Fractional rate of interest on money (–), or number of ions (–)
I	Total number of hot streams (–)
J	Total number of cold streams (–)
k	Reaction rate constant (units depend on order of reaction), or step number in a numerical calculation (–), or thermal conductivity ($W \cdot m^{-1} \cdot K^{-1}$, $kW \cdot m^{-1} \cdot K^{-1}$)
k_F	Fin thermal conductivity ($W \cdot m^{-1} \cdot K^{-1}$, $kW \cdot m^{-1} \cdot K^{-1}$)
$k_{G,i}$	Mass transfer coefficient in the gas phase ($kmol \cdot m^{-2} \cdot Pa^{-1} \cdot s^{-1}$)
k_{ij}	Interaction parameter between Components i and j in an equation of state (–)
$k_{L,i}$	Mass transfer coefficient of Component i in the liquid phase ($m \cdot s^{-1}$)
k_0	Frequency factor for heat of reaction (units depend on order of reaction)
k_W	Wall thermal conductivity ($W \cdot m^{-1} \cdot K^{-1}$, $kW \cdot m^{-1} \cdot K^{-1}$)
K	Overall mass transfer coefficient ($kmol \cdot Pa^{-1} \cdot m^{-2} \cdot s^{-1}$), or rate constant for fouling ($m^2 \cdot K \cdot W^{-1} \cdot day^{-1}$), or total number of enthalpy intervals in heat exchanger networks (–)
K_a	Equilibrium constant of reaction based on activity (–)
K_i	Ratio of vapor-to-liquid composition at equilibrium for Component i (–)
$K_{M,i}$	Equilibrium partition coefficient of membrane for Component i (–)
K_p	Equilibrium constant of reaction based on partial pressure in the vapor phase (–)
K_T	Parameter for terminal settling velocity ($m \cdot s^{-1}$)
K_x	Equilibrium constant of reaction based on mole fraction in the liquid phase (–)
K_y	Equilibrium constant of reaction based on mole fraction in vapor phase (–)
L	length (m), or liquid flowrate ($kg \cdot s^{-1}$, $kmol \cdot s^{-1}$), or number of independent loops in a network (–)
L_W	Distillation tray weir length (m)
LP	Low pressure
m	Mass flowrate ($kg \cdot s^{-1}$), or molar flowrate ($kmol \cdot s^{-1}$), or number of items (–)
m_C	Mass flowrate of water contaminant ($g \cdot h^{-1}$, $g \cdot d^{-1}$)
m_{COND}	Mass of condensate (kg)
m_{EX}	Mass flowrate of exhaust ($kg \cdot s^{-1}$)
m_{FUEL}	Mass of fuel (kg)
m_{max}	Maximum mass flowrate ($kg \cdot s^{-1}$)
m_{STEAM}	Mass flowrate of steam ($kg \cdot s^{-1}$)
m_W	Mass flowrate of pure water ($t \cdot h^{-1}$, $t \cdot d^{-1}$)
m_{WL}	Limiting mass flowrate of pure water ($t \cdot h^{-1}$, $t \cdot d^{-1}$)
m_{Wmin}	Minimum mass flowrate of fresh water ($t \cdot h^{-1}$, $t \cdot d^{-1}$)
m_{WT}	Target mass flowrate of fresh water ($t \cdot h^{-1}$, $t \cdot d^{-1}$)
m_{WTLOSS}	Target mass flowrate of fresh water involving a water loss ($t \cdot h^{-1}$, $t \cdot d^{-1}$)
M	Constant in capital cost correlations (–), or molar mass ($kg \cdot kmol^{-1}$), or number of variables (–)
MP	Medium pressure
MC_{STEAM}	Marginal cost of steam ($\$ \cdot t^{-1}$)
n	Number of items (–), or number of years (–), or polytropic coefficient (–), or slope of Willans Line ($kJ \cdot kg^{-1}$, $MJ \cdot kg^{-1}$)
N	Number of compression stages (–), or number of independent equations (–), or number of moles (kmol), or number of theoretical stages (–), or rate of transfer of a component ($kmol \cdot s^{-1} \cdot m^{-3}$), or rotational speed ($s^{-1}$, min^{-1})
N_F	Number of fins per unit length (m^{-1})
N_i	Number of moles of Component i (kmol)
N_{i0}	Initial number of moles of Component i (kmol)
N_{min}	Minimum number of theoretical stages (–)
N_P	Number of tube passes (–)
N_R	Number of tube rows (–)
N_{SHELLS}	Number of number of 1–2 shells in shell-and-tube heat exchangers (–)
N_R	Number of tube rows (–)
N_S	Specific speed of centrifugal pump (–)
N_T	Number of tubes (–)
N_{TR}	Number of tubes per row (–)
N_{UNITS}	Number of units in a heat exchanger network (–)
NC	Number of components in a multicomponent mixture (–)
NCV	Net calorific value of fuel ($J \cdot m^{-3}$, $kJ \cdot m^{-3}$, $J \cdot kg^{-1}$, $kJ \cdot kg^{-1}$)
$NPSH$	Net positive suction head (m)
NPV	Net present value ($)
Nu	Nusselt number (–)
p	Partial pressure ($N \cdot m^{-2}$, bar), or helical pitch (m)

p_C Pitch configuration factor for tube layout (–)
p_T Tube pitch (m)
P Present worth of a future sum of money ($), or pressure ($N \cdot m^{-2}$, bar), or probability (–), or thermal effectiveness of 1–2 shell-and-tube heat exchanger (–)
P_C Critical pressure ($N \cdot m^{-2}$, bar)
P_{max} Maximum thermal effectiveness of 1–2 shell-and-tube heat exchangers (–)
$P_{M,i}$ Permeability of Component i for a membrane ($kmol \cdot m \cdot s^{-1} \cdot m^{-2} \cdot bar^{-1}$, kg solvent $\cdot m^{-1} \cdot s^{-1} \cdot bar^{-1}$)
$\overline{P}_{M,i}$ Permeance of Component i for a membrane ($m^3 \cdot m^{-2} \cdot s^{-1} \cdot bar^{-1}$)
P_{N-2N} Thermal effectiveness over N_{SHELLS} number of 1–2 shell-and-tube heat exchangers in series (–)
P_{1-2} Thermal effectiveness over each 1–2 shell-and-tube heat exchanger in series (–)
P^{SAT} Saturated liquid–vapor pressure ($N \cdot m^{-2}$, bar)
Pr Prandtl number (–)
ΔP Pressure drop ($N \cdot m^{-2}$, bar)
ΔP_{FLOOD} Pressure drop under flooding conditions ($N \cdot m^{-2}$, bar)
q Heat flux ($W \cdot m^{-2}$, $kW \cdot m^{-2}$), or thermal condition of the feed in distillation (–), or Wegstein acceleration parameter for the convergence of recycle calculations (–)
q_C Critical heat flux ($W \cdot m^{-2}$, $kW \cdot m^{-2}$)
q_{C1} Critical heat flux for a single tube ($W \cdot m^{-2}$, $kW \cdot m^{-2}$)
q_i Individual stream heat duty for Stream i ($kJ \cdot s^{-1}$), or pure component property measuring the molecular van der Waals surface area for Molecule i in the UNIQUAC Equation of UNIFAC Model (–)
q_{RAD} Radiant heat flux ($W \cdot m^{-2}$, $kW \cdot m^{-2}$)
Q Heat duty (kW, MW)
Q_{ABS} Absorber heat duty (kW, MW)
Q_C Cooling duty (kW, MW)
Q_{Cmin} Target for cold utility (kW, MW)
Q_{COND} Condenser heat duty (kW, MW)
Q_{CONV} Convective heat duty (kW, MW)
Q_{EVAP} Evaporator heat duty (kW, MW)
Q_{EX} Heat duty for heat engine exhaust (kW, MW)
Q_{FEED} Heat duty to the feed (kW, MW)
Q_{FUEL} Heat from fuel in a furnace, boiler, or gas turbine (kW, MW)
Q_{GEN} Heat pump generator heat duty (kW, MW)
Q_H Heating duty (kW, MW)
Q_{Hmin} Target for hot utility (kW, MW)
Q_{HE} Heat engine heat duty (kW, MW)
Q_{HEN} Heat exchanger network heat duty (kW, MW)
Q_{INPUT} Heat input from fuel (kW, MW)
Q_{HP} Heat duty on high-pressure steam (kW, MW), or heat pump heat duty (kW, MW)
Q_{LP} Heat duty on low-pressure steam (kW, MW)
Q_{LOSS} Stack loss from furnace, boiler, or gas turbine (kW, MW)
Q_{OUTPUT} Heat output to steam generation (kW, MW)
Q_{RAD} Radiant heat duty (kW, MW)
Q_{REACT} Reactor heating or cooling duty (kW, MW)
Q_{REB} Reboiler heat duty (kW, MW)
Q_{REC} Heat recovery (kW, MW)
Q_{SITE} Site heating demand (kW, MW)
Q_{STEAM} Heat input for steam generation (kW, MW)
r Molar ratio (–), or pressure ratio (–), or radius (m)
r_i Pure component property measuring the molecular van der Waals volume for Molecule i in the UNIQUAC Equation and UINFAC Model (–), or rate of reaction of Component i ($kmol^{-1} \cdot s^{-1}$), or recovery of Component i in separation (–)
R Fractional recovery of a component in separation (–), or heat capacity ratio of 1–2 shell-and-tube heat exchanger (–), or raffinate flowrate in liquid–liquid extraction ($kg \cdot s^{-1}$, $kmol \cdot s^{-1}$), or ratio of heat capacity flowrates (–), or reflux ratio in distillation (–), or removal ratio in water treatment (–), or residual error (units depend on application), or universal gas constant (8314.5 $N \cdot m \cdot kmol^{-1} \cdot K^{-1} = J \cdot kmol^{-1} \cdot K^{-1}$, 8.3145 $kJ \cdot kmol^{-1} \cdot K^{-1}$)
R_{AF} Mass ratio of air to fuel (–)
R_{min} Minimum reflux ratio (–)
R_F Fouling resistance in heat transfer ($m^{-2} \cdot K \cdot W^{-1}$), or ratio of actual to minimum reflux ratio (–)
R_{SITE} Site power-to-heat ratio (–)
ROI Return on investment (%)
Re Reynolds number (–)
s Reactor space velocity (s^{-1}, min^{-1}, h^{-1}), or steam-to-air ratio for gas turbine (–)
S Entropy ($kJ \cdot K^{-1}$, $kJ \cdot kg^{-1} \cdot K^{-1}$, $kJ \cdot kmol^{-1} \cdot K^{-1}$), or number of streams in a heat exchanger network (–), or reactor selectivity (–), or reboil ratio for distillation (–), or selectivity of a reaction (–), or slack variable in optimization (units depend on application), or solvent flowrate ($kg \cdot s^{-1}$, $kmol \cdot s^{-1}$), or stripping factor in absorption (–)

S_C	Number of cold streams (–)
S_H	Number of hot streams (–)
S_W	Dimensionless swirl parameter (–)
t	Batch time (s, h), or time (s, h)
T	Temperature (°C, K)
T_{ABS}	Absorber temperature (°C, K)
T_{BPT}	Normal boiling point (°C, K)
T_C	Critical temperature (K), or temperature of heat sink (°C, K)
T_{COND}	Condenser temperature (°C, K)
T_E	Equilibrium temperature (°C, K)
T_{EVAP}	Evaporation temperature (°C, K)
T_{FEED}	Feed temperature (°C, K)
T_{GEN}	Heat pump generator temperature (°C, K)
T_H	Temperature of heat source (°C, K)
T_R	Reduced temperature T/T_C(–)
T_{REB}	Reboiler temperature (°C, K)
T_S	Stream supply temperature (°C)
T_{SAT}	Saturation temperature of boiling liquid (°C, K)
T_T	Stream target temperature (°C)
T_{TFT}	Theoretical flame temperature (°C, K)
T_W	Wall temperature (°C, K)
T_{WBT}	Wet bulb temperature (°C)
T^*	Interval temperature (°C)
ΔT_{LM}	Logarithmic mean temperature difference (°C, K)
ΔT_{min}	Minimum temperature difference (°C, K)
$\Delta T_{THRESHOLD}$	Threshold temperature difference (°C, K)
TAC	Total annual cost ($\$ \cdot y^{-1}$)
TOD	Total oxygen demand ($kg \cdot m^{-3}$, $mg \cdot l^{-1}$)
u_{ij}	Interaction parameter between Molecule i and Molecule j in the UNIQUAC Equation ($kJ \cdot kmol^{-1}$)
U	Overall heat transfer coefficient ($W \cdot m^{-2} \cdot K^{-1}$, $kW \cdot m^{-2} \cdot K^{-1}$)
v	Velocity ($m \cdot s^{-1}$)
v_D	Downcomer liquid velocity ($m \cdot h^{-1}$)
v_S	Shell-side fluid velocity ($m \cdot s^{-1}$), or superficial vapor velocity ($m \cdot s^{-1}$)
v_T	Terminal settling velocity ($m \cdot s^{-1}$), or tube-side tube velocity ($m \cdot s^{-1}$)
v_V	Superficial vapor velocity in empty column ($m \cdot s^{-1}$)
V	Molar volume ($m^3 \cdot kmol^{-1}$), or vapor flowrate ($kg \cdot s^{-1}$, $kmol \cdot s^{-1}$), or volume (m^3), or volume of gas or vapor adsorbed ($m^3 \cdot kg^{-1}$)
V_{min}	Minimum vapor flow ($kg \cdot s^{-1}$, $kmol \cdot s^{-1}$)
VF	Vapor fraction (–)
w	Mass of adsorbate per mass of adsorbent (–)
W	Shaft power (kW, MW), or shaft work (kJ, MJ)
W_{GEN}	Power generated (kW, MW)
W_{GT}	Power generated by gas turbine (kW, MW)
W_{INT}	Intercept of Willans Line (kW, MW)
W_{LOSS}	Power loss in gas or steam turbines (kW, MW)
W_{SITE}	Site power demand (kW, MW)
x	Control variable in optimization problem (units depend on application), or liquid-phase mole fraction (–)
x_F	Final value of control variable in optimization problem (units depend on application), or mole fraction in the feed (–)
x_B	Mole fraction in the distillation bottoms (–)
x_D	Mole fraction in the distillate (–)
x_0	Initial value of control variable in optimization problem (units depend on application)
X	Reactor conversion (–), or wetness fraction of steam (–)
X_E	Equilibrium reactor conversion (–)
X_{OPT}	Optimal reactor conversion (–)
X_P	Fraction of maximum thermal effectiveness P_{max} allowed in a 1–2 shell-and-tube heat exchanger (–)
XP	Cross-pinch heat transfer in heat exchanger network (kW, MW)
y	Integer variable in optimization (–), or twist ratio for twisted tape (–), or vapor-phase mole fraction (–)
y_F	Distance between fins (m)
z	Elevation (m), or feed mole fraction (–)
Z	Compressibility of a fluid (–)

Greek Letters

α	Constant in cubic equation of state (–), or constants in vapor pressure correlation (units depend on which constant), or fraction open of a valve (–), or helix angle of wire to the tube axis (degrees), or relative volatility between a binary pair (–)
α_{ij}	Ideal separation factor or selectivity of membrane between Components i and j (–), or parameter characterizing the tendency of Molecule i and Molecule j to be distributed in a random fashion in the NRTL equation (–), or relative volatility between Components i and j (–)
α_{LH}	Relative volatility between light and heavy key components (–)
α_P	Packing surface area ($m^2 \cdot m^3$)
β_{ij}	Separation factor between Components i and j (–)
γ	Logic variable in optimization (–), or ratio of heat capacities for gases and vapors (–)
γ_i	Activity coefficient for Component i (–)
δ	Thickness (m)
δ_F	Fin thickness (m)
δ_M	Membrane thickness (m)
ε	Emissivity (–), or extraction factor in liquid-liquid extraction (–), or

	pipe roughness (mm)
η	Carnot factor (−), or efficiency (−)
η_{AHP}	Absorber heat pump efficiency (−)
η_{AHT}	Absorber heat transformer efficiency (−)
η_{BOILER}	Boiler efficiency (−)
η_C	Carnot efficiency (−)
η_{CHP}	Compression heat pump efficiency (−)
η_{COGEN}	Cogeneration efficiency (−)
η_F	Fin efficiency (−)
η_{GT}	Efficiency of gas turbine (−)
η_{IS}	Isentropic efficiency of compression or expansion (−)
η_{MECH}	Mechanical efficiency of steam turbine (−)
η_P	Polytropic efficiency of compression or expansion (−)
η_{POWER}	Power generation efficiency (−)
η_{ST}	Efficiency of steam turbine (−)
η_W	Weighted fin efficiency (−)
θ	Angle (degrees), or fraction of feed permeated through membrane (−), or root of the Underwood Equation (−)
ϑ	Logic variable in optimization (−)
λ	Ratio of latent heats of vaporization (−)
λ_{ij}	Energy parameter characterizing the interaction of Molecule *i* with Molecule *j* ($kJ \cdot kmol^{-1}$)
μ	Fluid viscosity ($kg \cdot m^{-1} \cdot s^{-1}$, $mN \cdot s \cdot m^{-2} = cP$)
π	Osmotic pressure ($N \cdot m^{-2}$, bar)
ρ	Density ($kg \cdot m^{-3}$, $kmol \cdot m^{-3}$)
σ	Stephan–Boltzmann constant ($W \cdot m^{-2} \cdot K^{-4}$), or surface tension ($mN \cdot m^{-1} = mJ \cdot m^{-2} = dyne \cdot cm^{-1}$)
τ	Reactor space time (s, min, h), or residence time (s, min, h)
τ_W	Wall shear stress ($N \cdot m^{-2}$)
υ_k^i	Number of interaction Groups *k* in Molecule *i* (−)
Φ	Fugacity coefficient (−), or logic variable in optimization (−)
ω	Acentric factor (−)

Subscripts

axial	Axial direction
B	Blowdown, or bottoms in distillation
BFW	Boiler feedwater
BW	Bridgewall
cont	Contribution
C	Cold stream, or contaminant
CN	Condensing
COND	Condensing conditions
CP	Cold plane, or continuous phase
CW	Cooling water
D	Distillate in distillation
DS	De-superheating
e	Enhanced, or end zone on the shell side of a heat exchanger, or environment, or equivalent
E	Evaporation, or extract in liquid–liquid extraction
EVAP	Evaporator conditions
EX	Exhaust
final	Final conditions in a batch
F	Feed, or final, or fluid
FIN	Fin on a finned tube
FG	Flue gas
G	Gas phase
H	Hot stream
HP	Heat pump, or high pressure
i	Component number, or stream number
in	Inlet
I	Inside
IF	Inside fouling
IMP	Impeller
IS	Isentropic
J	Component number, or stream number
K	Enthalpy interval number in heat exchanger networks
L	Liquid phase
LP	Low pressure
M	Stage number in distillation and absorption
max	Maximum
min	Minimum
M	Makeup
MIX	Mixture
N	Stage number in distillation and absorption
out	Outlet
O	Outside, or standard conditions
OF	Outside fouling
p	Stage number in distillation and absorption
prod	Products of reaction
P	Particle, or permeate
PINCH	Pinch conditions
react	Reactants
R	Raffinate in liquid–liquid extraction
REACT	Reaction
ROOT	Root of a finned tube
S	Solvent in liquid–liquid extraction
SAT	Saturated conditions
SF	Supplementary firing
SUP	Superheated conditions
SW	Swirl direction
T	Treatment

Te	Tube side enhanced
TW	Treated water
V	Vapor phase
w	Window section on the shell side of a heat exchanger
W	Conditions at the tube wall, or water
WBT	Wet bulb conditions
WW	Waste water
∞	Conditions at distillate pinch point

Superscripts

I	Phase *I*
II	Phase *II*
III	Phase *III*
IDEAL	Ideal behavior
L	Liquid phase
O	Standard conditions
V	Vapor phase
*	Adjusted parameter

Chapter 1

The Nature of Chemical Process Design and Integration

1.1 Chemical Products

Chemical products are essential to modern living standards. Almost all aspects of everyday life are supported by chemical products in one way or another. However, society tends to take these products for granted, even though a high quality of life fundamentally depends on them.

When considering the design of processes for the manufacture of chemical products, the market into which they are being sold fundamentally influences the objectives and priorities in the design. Chemical products can be divided into three broad classes:

1) *Commodity or bulk chemicals.* These are produced in large volumes and purchased on the basis of chemical composition, purity and price. Examples are sulfuric acid, nitrogen, oxygen, ethylene and chlorine.
2) *Fine chemicals.* These are produced in small volumes and purchased on the basis of chemical composition, purity and price. Examples are chloropropylene oxide (used for the manufacture of epoxy resins, ion-exchange resins and other products), dimethyl formamide (used, for example, as a solvent, reaction medium and intermediate in the manufacture of pharmaceuticals), *n*-butyric acid (used in beverages, flavorings, fragrances and other products) and barium titanate powder (used for the manufacture of electronic capacitors).
3) *Specialty or effect or functional chemicals.* These are purchased because of their effect (or function), rather than their chemical composition. Examples are pharmaceuticals, pesticides, dyestuffs, perfumes and flavorings.

Because commodity and fine chemicals tend to be purchased on the basis of their chemical composition alone, they can be considered to be *undifferentiated*. For example, there is nothing to choose between 99.9% benzene made by one manufacturer and that made by another manufacturer, other than price and delivery issues. On the other hand, specialty chemicals tend to be purchased on the basis of their effect or function and therefore can be considered to be *differentiated*. For example, competitive pharmaceutical products are differentiated according to the efficacy of the product, rather than chemical composition. An adhesive is purchased on the basis of its ability to stick things together, rather than its chemical composition, and so on.

However, in practice few products are completely undifferentiated and few completely differentiated. Commodity and fine chemical products might have impurity specifications as well as purity specifications. Traces of impurities can, in some cases, give some differentiation between different manufacturers of commodity and fine chemicals. For example, 99.9% acrylic acid might be considered to be an undifferentiated product. However, traces of impurities, at concentrations of a few parts per million, can interfere with some of the reactions in which it is used and can have important implications for some of its uses. Such impurities might differ between different manufacturing processes. Not all specialty products are differentiated. For example, pharmaceutical products like aspirin (acetylsalicylic acid) are undifferentiated. Different manufacturers can produce aspirin, and there is nothing to choose between these products, other than the price and differentiation created through marketing of the product. Thus, the terms undifferentiated and differentiated are more relative than absolute terms.

The scale of production also differs between the three classes of chemical products. Fine and specialty chemicals tend to be produced in volumes less than $1000\,t{\cdot}y^{-1}$. By contrast, commodity chemicals tend to be produced in much larger volumes than this. However, the distinction is again not so clear. Polymers are differentiated products because they are purchased on the basis of their mechanical properties, but can be produced in quantities significantly higher than $1000\,t{\cdot}y^{-1}$.

Chemical Process Design and Integration, Second Edition. Robin Smith.

Companion Website: www.wiley.com/go/smith/chemical2e

When a new chemical product is first developed, it can often be protected by a patent in the early years of its commercial exploitation. For a product to be eligible to be patented, it must be novel, useful and unobvious. If patent protection can be obtained, this effectively gives the producer a monopoly for commercial exploitation of the product until the patent expires. Patent protection lasts for 20 years from the filing date of the patent. Once the patent expires, competitors can join in and manufacture the product. If competitors cannot wait until the patent expires, then alternative competing products must be developed.

Another way to protect a competitive edge for a new product is to protect it by secrecy. The formula for Coca-Cola has been kept a secret for over 100 years. Potentially, there is no time limit on such protection. However, for the protection through secrecy to be viable, competitors must not be able to reproduce the product from chemical analysis. This is likely to be the case only for certain classes of specialty chemicals and food products for which the properties of the product depend on both the chemical composition and the method of manufacture.

Figure 1.1 illustrates different product *life cycles* (Sharratt, 1997; Brennan, 1998). The general trend is that when a new product is introduced into the market, the sales grow slowly until the market is established and then more rapidly once the market is established. If there is patent protection, then competitors will not be able to exploit the same product commercially until the patent expires, when competitors can produce the same product and take market share. It is expected that competitive products will cause sales to diminish later in the product life cycle until sales become so low that a company would be expected to withdraw from the market. In Figure 1.1, Product *A* appears to be a poor product that has a short life with low sales volume. It might be that it cannot compete well with other competitive products and alternative products quickly force the company out of that business. However, a low sales volume is not the main criterion to withdraw a product from the market. It might be that a product with low volume finds a market niche and can be sold for a high value. On the other hand, if it were competing with other products with similar functions in the same market sector, which keeps both the sale price and volume low, then it would seem wise to withdraw from the market. Product *B* in Figure 1.1 appears to be a better product, showing a longer life cycle and higher sales volume. This has patent protection but sales decrease rapidly after patent protection is lost, leading to loss of market through competition. Product *C* in Figure 1.1 is an even better product. This shows high sales volume with the life of the product extended through reformulation of the product (Sharratt, 1997). Finally, Product *D* in Figure 1.1 shows a product life cycle that is typical of commodity chemicals. Commodity chemicals tend not to exhibit the same kind of life cycles as fine and specialty chemicals. In the early years of the commercial exploitation, the sales volume grows rapidly to a high volume, but then volume does not decline and enters a mature period of slow growth, or, in some exceptional cases, slow decline. This is because commodity chemicals tend to have a diverse range of uses. Even though competition might take away some end uses, new end uses are introduced, leading to an extended life cycle.

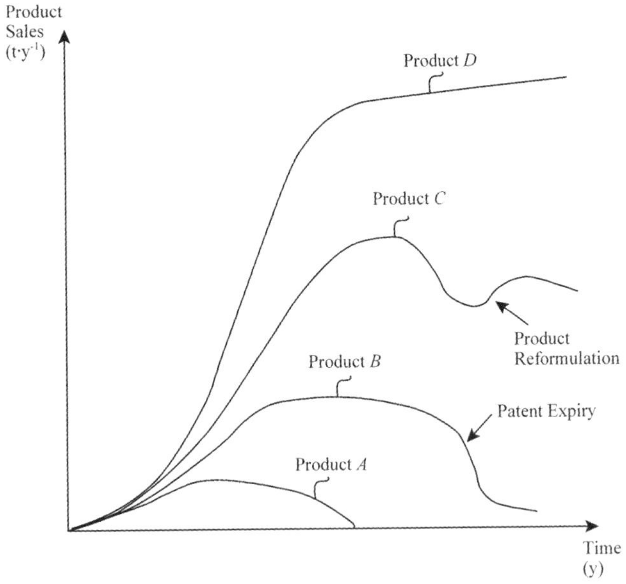

Figure 1.1

Product life cycles. (Adapted from Sharratt PN, 1997, Handbook of Batch Process Design, Chapman & Hall, reproduced by permission.)

The different classes of chemical products will have very different *added value* (the difference between the selling price of the product and the purchase cost of raw materials). Commodity chemicals tend to have low added value, whereas fine and specialty chemicals tend to have high added value. Commodity chemicals tend to be produced in large volumes with low added value, while fine and specialty chemicals tend to be produced in small volumes with high added value.

Because of this, when designing a process for a commodity chemical, it is usually important to keep operating costs as low as possible. The capital cost of the process will tend to be high relative to a process for fine or specialty chemicals because of the scale of production.

When designing a process for specialty chemicals, priority tends to be given to the product, rather than to the process. This is because the unique function of the product must be protected. The process is likely to be small scale and operating costs tend to be less important than with commodity chemical processes. The capital cost of the process will be low relative to commodity chemical processes because of the scale. The time to market for the product is also likely to be important with specialty chemicals, especially if there is patent protection. If this is the case, then anything that shortens the time from basic research, through product testing, pilot plant studies, process design, construction of the plant to product manufacture will have an important influence on the overall project profitability.

All this means that the priorities in process design are likely to differ significantly, depending on whether a process is being designed for the manufacture of a commodity, fine or specialty chemical. In commodity chemicals, there is likely to be relatively little product innovation, but intensive process innovation. Also, equipment will be designed for a specific process step. On the other hand, the manufacture of fine and specialty chemicals might involve:

- selling into a market with low volume;
- a short product life cycle;
- a demand for a short time to market, and therefore less time is available for process development, with product and process development proceeding simultaneously.

As a result, the manufacture of fine and specialty chemicals is often carried out in multipurpose equipment, perhaps with different chemicals being manufactured in the same equipment at different times during the year. The life of the equipment might greatly exceed the life of the product.

The development of pharmaceutical products demands that high-quality products must be manufactured during the development of the process to allow safety and clinical studies to be carried out before full-scale production. Pharmaceutical production represents an extreme case of process design in which the regulatory framework controlling production makes it difficult to make process changes, even during the development stage. Even if significant improvements to processes for pharmaceuticals can be suggested, it might not be feasible to implement them, as such changes might prevent or delay the process from being licensed for production.

1.2 Formulation of Design Problems

Before a process design can be started, the design problem must be formulated. Formulation of the design problem requires a product specification. If a well-defined chemical product is to be manufactured, then the specification of the product might appear straightforward (e.g. a purity specification). However, if a specialty product is to be manufactured, it is the functional properties that are important, rather than the chemical properties, and this might require a *product design* stage in order to specify the product (Seider *et al.*, 2010; Cussler and Moggridge, 2011).

The initial statement of the design problem is often ill defined. For example, the design team could be asked to expand the production capacity of an existing plant that produces a chemical that is a precursor to a polymer product, which is also produced by the company. This results from an increase in the demand for the polymer product and the plant producing the precursor currently being operated at its maximum capacity. The design team might well be given a specification for the expansion. For example, the marketing department might assess that the market could be expanded by 30% over a two-year period, which would justify a 30% expansion in the process for the precursor. However, the 30% projection can easily be wrong. The economic environment can change, leading to the projected increase being either too large or too small. It might also be possible to sell the polymer precursor in the market to other manufacturers of the polymer and justify an expansion even larger than 30%. If the polymer precursor can be sold in the marketplace, is the current purity specification of the company suitable for the marketplace? Perhaps the marketplace demands a higher purity than the current company specification. Perhaps the current specification is acceptable, but if the specification could be improved, the product could be sold for a higher value and/or at a greater volume. An option might be to not expand the production of the polymer precursor to 30%, but instead to purchase it from the market. If it is purchased from the market, is it likely to be up to the company specifications or will it need some purification before it is suitable for the company's polymer process? How reliable will the market source be? All these uncertainties are related more to market supply and demand issues than to specific process design issues.

Closer examination of the current process design might lead to the conclusion that the capacity can be expanded by 10% with a very modest capital investment. A further increase to 20% would require a significant capital investment, but an expansion to 30% would require an extremely large capital investment. This opens up further options. Should the plant be expanded by 10% and a market source identified for the balance? Should the plant be expanded to 20% similarly? If a real expansion in the marketplace is anticipated and expansion to 30% would be very expensive, why not be more aggressive and, instead of expanding the existing process, build an entirely new process? If a new process is to be built, then what should be the process technology? New process technology might have been developed since the original plant was built that enables the same product to be manufactured at a much lower cost. If a new process is to be built, where should it be built? It might make more sense to build it in another country that would allow lower operating costs, and the product could be shipped back to be fed to the existing polymer process. At the same time, this might stimulate the development of new markets in other countries, in which case, what should be the capacity of the new plant?

Thus, from the initial ill-defined problem, the design team must create a series of very specific options and these should then be compared on the basis of a common set of assumptions regarding, for example, raw materials and product prices. Having specified an option, this gives the design team a well-defined problem to which the methods of engineering and economic analysis can be applied.

In examining a design option, the design team should start out by examining the problem at the highest level, in terms of its feasibility with the minimum of detail to ensure the design option is worth progressing (Douglas, 1985). Is there a large difference between the value of the product and the cost of the raw materials? If the overall feasibility looks attractive, then more detail can be added, the option re-evaluated, further detail added, and so on. Byproducts might play a particularly important role in the economics. It might be that the current process produces some byproducts that can be sold in small quantities to the market. However, as the process is expanded, there might be market constraints for the new scale of production. If the byproducts cannot be sold, how does this affect the economics?

In summary, the original problem posed to process design teams is often ill defined, even though it might appear to be well defined in the original design specification. The design team must then formulate a series of plausible design options to be screened by the methods of engineering and economic analysis.

These design options are formulated into very specific design problems. In this way, the design team turns the ill-defined problem into a series of well-defined design options for analysis.

1.3 Synthesis and Simulation

In a chemical process, the transformation of raw materials into desired chemical products usually cannot be achieved in a single step. Instead, the overall transformation is broken down into a number of steps that provide intermediate transformations. These are carried out through reaction, separation, mixing, heating, cooling, pressure change, particle size reduction or enlargement for solids. Once individual steps have been selected, they must be interconnected to carry out the overall transformation (Figure 1.2a). Thus, the *synthesis* of a chemical process involves two broad activities. First, individual transformation steps are selected. Second, these individual transformations are interconnected to form a complete process that achieves the required overall transformation. A *flowsheet* or *process flow diagram* (PFD) is a diagrammatic representation of the process steps with their interconnections.

Once the flowsheet structure has been defined, a *simulation* of the process can be carried out. A simulation is a mathematical model of the process that attempts to predict how the process would behave if it was constructed (Figure 1.2b). Material and energy balances can be formulated to give better definition to the inner workings of the process and a more detailed process design can be developed. Having created a model of the process, the flowrates, compositions, temperatures and pressures of the feeds can be assumed. The simulation model then predicts the flowrates, compositions, temperatures, pressures and properties of the products. It also allows the individual items of equipment in the process to be sized and predicts, for example, how much raw material is being used or how much energy is being consumed. The performance of the design can then be evaluated.

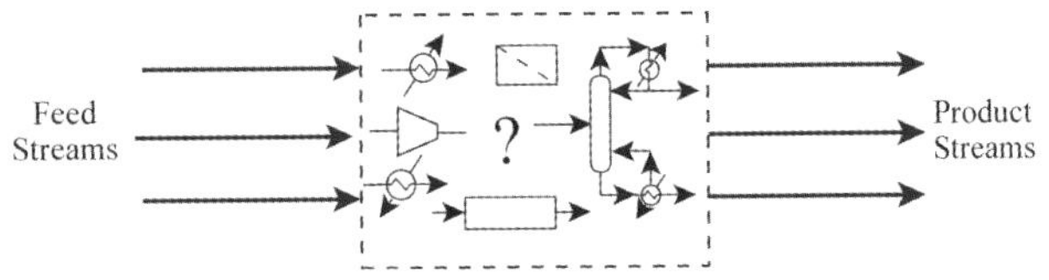

(a) Process design starts with the synthesis of a process to convert raw materials into desired products.

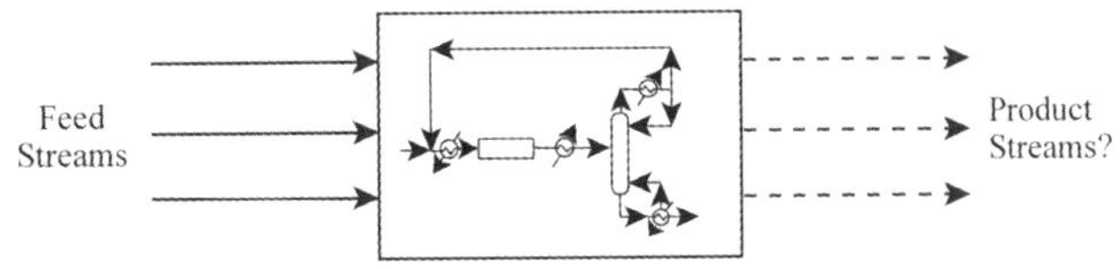

(b) Simulation predicts how a process would behave if it was constructed.

Figure 1.2

Synthesis is the creation of a process to transform feed streams into product streams. Simulation predicts how it would behave if it was constructed.

1) *Accuracy of design calculations.* A simulation adds more detail once a design has been synthesized. The design calculations for this will most often be carried out in a general purpose simulation software package and solved to a high level of precision. However, a high level of precision cannot usually be justified in terms of the operation of the plant after it has been built. The plant will almost never work precisely at its original design flowrates, temperatures, pressures and compositions. This might be because the raw materials are slightly different from what is assumed in the design. The physical properties assumed in the calculations might have been erroneous in some way, or operation at the original design conditions might create corrosion or fouling problems, or perhaps the plant cannot be controlled adequately at the original conditions, and so on, for a multitude of other possible reasons. The instrumentation on the plant will not be able to measure the flowrates, temperatures, pressures and compositions as accurately as the calculations performed. High precision might be required in the calculations for certain specific parts of the design. For example, a polymer precursor might need certain impurities to be very tightly controlled, perhaps down to the level of parts per million, or it might be that some contaminant in a waste stream might be exceptionally environmentally harmful and must be extremely well defined in the design calculations.

   Even though a high level of precision cannot be justified in many cases in terms of the plant operation, the design calculations will normally be carried out to a reasonably high level of precision. The value of precision in design calculations is that the consistency of the calculations can be checked to allow errors or poor assumptions to be identified. It also allows the design options to be compared on a valid like-for-like basis.

   Because of all the uncertainties in carrying out a design, the specifications are often increased beyond those indicated by the design calculations and the plant is *overdesigned*, or *contingency* is added, through the application of *safety factors* to the design. For example, the designer might calculate the number of distillation plates required for a distillation separation using elaborate calculations to a high degree of precision, only to add an arbitrary extra 10% to the number of plates for contingency. This allows for the feed to the unit not being exactly as specified, errors in the physical properties, upset conditions in the plant, control requirements, and so on. If too little contingency is added, the plant might not work. If too much contingency is added, the plant will not only be unnecessarily expensive but too much overdesign might make the plant difficult to operate and might lead to a less efficient plant. For example, the designer might calculate the size of a heat exchanger and then add in a large contingency and significantly oversize the heat exchanger. The lower fluid velocities encountered by the oversized heat exchanger can cause it to have a poorer performance and to foul-up more readily than a smaller heat exchanger.

Too little overdesign might lead to the plant not working. Too much overdesign will lead to the plant becoming unnecessarily expensive, and perhaps difficult to operate and less efficient. A balance must be made between different risks.

2) *Physical properties in process design.* Almost all design calculations require physical properties of the solids, liquids and gases being fed, processed and produced. Physical properties can be critical to obtaining meaningful, economic and safe designs. When carrying out calculations in computer software packages there is most often a choice to be made for the physical property correlations and data. However, if poor decisions are made by the designer regarding physical properties, the design calculations can be meaningless or even dangerous, even though the calculations have been performed to a high level of precision. Using physical property correlations outside the ranges of conditions for which they were intended can be an equally serious problem. Appendix A discusses physical properties in process design in more detail.

3) *Evaluation of performance.* There are many facets to the evaluation of performance. Good economic performance is an obvious first criterion, but it is certainly not the only one. Chemical processes should be designed to maximize the *sustainability* of industrial activity. Maximizing sustainability requires that industrial systems should strive to satisfy human needs in an economically viable, environmentally benign and socially beneficial way (Azapagic, 2014). For chemical process design, this means that processes should make use of materials of construction that deplete the resource as little as practicable. Process raw materials should be used as efficiently as is economic and practicable, both to prevent the production of waste that can be environmentally harmful and to preserve the reserves of manufacturing raw materials as much as possible. Processes should use as little energy as is economic and practicable, both to prevent the build-up of carbon dioxide in the atmosphere from burning fossil fuels and to preserve the reserves of fossil fuels. Water must also be consumed in sustainable quantities that do not cause deterioration in the quality of the water source and the long-term quantity of the reserves. Aqueous and atmospheric emissions must not be environmentally harmful and solid waste to landfill must be avoided. The boundary of consideration should go beyond the immediate boundary of the manufacturing facility to maximize the benefit to society to avoid adverse health effects, unnecessarily high burdens on transportation, odour, noise nuisances, and so on.

The process must also meet required health and safety criteria. Start-up, emergency shutdown and ease of control are other important factors. Flexibility, that is, the ability to operate under different conditions, such as differences in feedstock and product specification, may be important. Availability, that is, the portion of the total time that the process meets its production requirements, might also be critically important. Uncertainty in the design, for example, resulting from poor design data, or uncertainty in the economic data might guide the design away from certain options. Some of these factors, such as economic performance, can be readily quantified; others, such as safety, often cannot. Evaluation of the factors that are not readily quantifiable, the intangibles, requires the judgment of the design team.

4) *Materials of construction.* Choice of materials of construction affects both the mechanical design and the capital cost of equipment. Many factors enter into the choice of the materials of construction. Among the most important are (see Appendix B):

- mechanical properties (particularly yield and tensile strength, compressive strength, ductility, toughness, hardness, fatigue limit and creep resistance);
- effect of temperature on mechanical properties (both low and high temperatures),
- ease of fabrication (machining, welding, and so on);
- corrosion resistance;
- availability of standard equipment in the material;
- cost (e.g. if materials of construction are particularly expensive, it might be desirable to use a cheaper material together with a lining on the process side to reduce the cost).

Estimation of the capital cost and preliminary specification of equipment for the evaluation of performance requires decisions to be made regarding the materials of construction. The discussion of the more commonly used materials of construction is given in Appendix B.

5) *Process safety.* When evaluating a process design, process safety should be the prime consideration. Safety considerations must not be left until the design has been completed. Safety systems need to be added to the design later for the relief of overpressure, to trip the process under dangerous conditions, etc. However, by far the largest impact on process safety can be made early in the design through measures to make the design *inherently safer*. This will be discussed in detail in Chapter 28. Inherently safer design means avoiding the need for hazardous materials if possible, or using less of them, or using them at lower temperatures and pressures or diluting them with inert materials. One of the principal approaches to making a process inherently safer is to limit the inventory of hazardous material. The inventories to be avoided most of all are flashing flammable or toxic liquids, that is, liquids under pressure above their atmospheric boiling points (see Chapter 28).

6) *Optimization.* Once the basic performance of the design has been evaluated, changes can be made to improve the performance; the process is *optimized*. These changes might involve the synthesis of alternative structures, that is, *structural optimization*. Thus, the process is simulated and evaluated again, and so on, optimizing the structure. Each structure can be subjected to *parameter optimization* by changing operating conditions within that structure. This is illustrated in Figure 1.3.

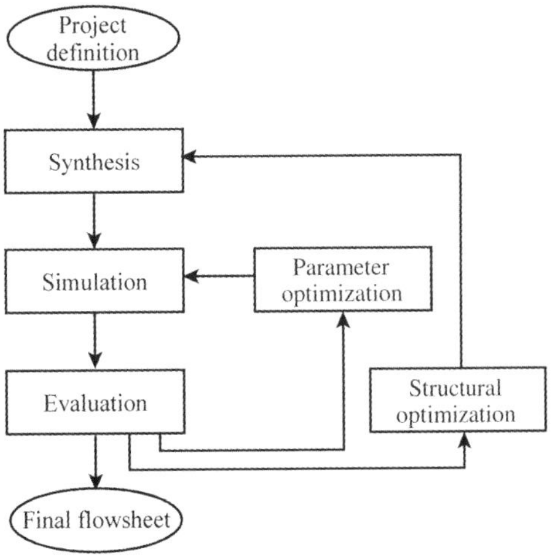

Figure 1.3

Optimization can be carried out as structural or parameter optimization to improve the evaluation of the design.

From the project definition an initial design is synthesized. This can then be simulated and evaluated. Once evaluated, the design can be optimized in a parameter optimization through changing the continuous parameters of flowrate, composition, temperature and pressure to improve the evaluation. However, this parameter optimization only optimizes the initial design configuration, which might not be an optimal configuration. So the design team might return to the synthesis stage to explore other configurations in a structural optimization. Also, if the parameter optimization adjusts the settings of the conditions to be significantly different from the original assumptions, then the design team might return to the synthesis stage to consider other configurations in the structural optimization. The different ways this design process can be followed will be considered later in this chapter.

7) *Keeping design options open.* To develop a design concept requires design options to be first generated and then evaluated. There is a temptation to carry out preliminary evaluation early in the development of a design and eliminate options early that initially appear to be unattractive. However, this temptation must be avoided. In the early stages of a design the uncertainties in the evaluation are often too serious for early elimination of options, unless it is absolutely clear that a design option is not viable. Initial cost estimates can be very misleading and the full safety and environmental implications of early decisions are only clear once detail has been added. If it was possible to foresee everything that lay ahead, decisions made early might well be different. There is a danger in focusing on one option without rechecking the assumptions later for validity when more information is available. The design team must not be boxed in early by preconceived ideas. This means that design options should be left open as long as practicable until it is clear that options can be closed down. All options should be considered, even if they appear unappealing at first.

1.4 The Hierarchy of Chemical Process Design and Integration

Consider the process illustrated in Figure 1.4 (Smith and Linnhoff, 1988). The process requires a reactor to transform the *FEED* into *PRODUCT* (Figure 1.4a). Unfortunately, not all the *FEED* reacts. Also, part of the *FEED* reacts to form *BYPRODUCT* instead of the desired *PRODUCT*. A separation system is needed to isolate the *PRODUCT* at the required purity. Figure 1.4b shows one possible separation system consisting of two distillation columns. The unreacted *FEED* in Figure 1.4b is recycled and the *PRODUCT* and *BYPRODUCT* are removed from the process. Figure 1.4b shows a flowsheet where all heating and cooling is provided by external *utilities* (steam and cooling water in this case). This flowsheet is probably too inefficient in its use of energy and heat should be recovered. Thus, *heat integration* is carried out to exchange heat between those streams that need to be cooled and those that need to be heated. Figure 1.5 (Smith and Linnhoff, 1988) shows two possible designs for the *heat exchanger network*, but many other heat integration arrangements are possible.

The flowsheets shown in Figure 1.5 feature the same reactor design. It could be useful to explore the changes in reactor design. For example, the size of the reactor could be increased to increase the amount of *FEED* that reacts (Smith and Linnhoff, 1988). Now there is not only much less *FEED* in the reactor effluent but also more *PRODUCT* and *BYPRODUCT*. However, the increase in *BYPRODUCT* is larger than the increase in *PRODUCT*. Thus, although the reactor has the same three components in its effluent as the reactor in Figure 1.4a, there is less *FEED*, more *PRODUCT* and significantly more *BYPRODUCT*. This change in reactor design generates a different task for the separation system and it is possible that a separation system different from that shown in Figures 1.4 and 1.5 is now appropriate. Figure 1.6 shows a possible alternative. This also uses two distillation columns, but the separations are carried out in a different order.

Figure 1.6 shows a flowsheet without any heat integration for the different reactor and separation system. As before, this is probably too inefficient in the use of energy, and heat integration schemes can be explored. Figure 1.7 (Smith and Linnhoff, 1988) shows two of the many possible flowsheets involving heat recovery.

Different complete flowsheets can be evaluated by simulation and costing. On this basis, the flowsheet in Figure 1.5b might be more promising than the flowsheets in Figures 1.5a and 1.7a and b. However, the best flowsheet cannot be identified without first optimizing the operating conditions for each. The flowsheet in Figure 1.7b might have greater scope for improvement than that in Figure 1.5b, and so on.

Thus, the complexity of chemical process synthesis is twofold. First, can all possible structures be identified? It might be considered that all the structural options can be found by inspection, at least all of the significant ones. The fact that

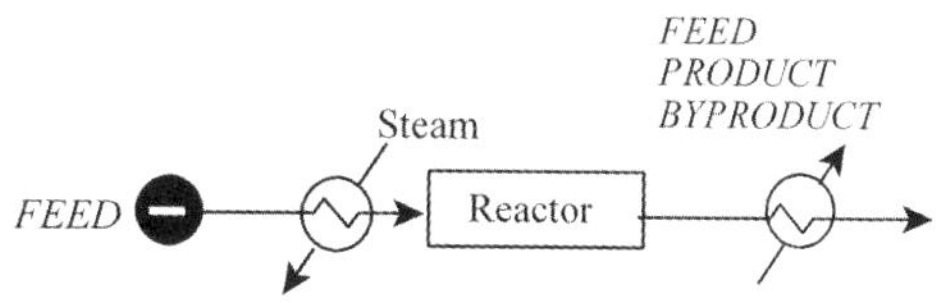

(a) A reactor transforms *FEED* into *PRODUCT* and *BYPRODUCT*.

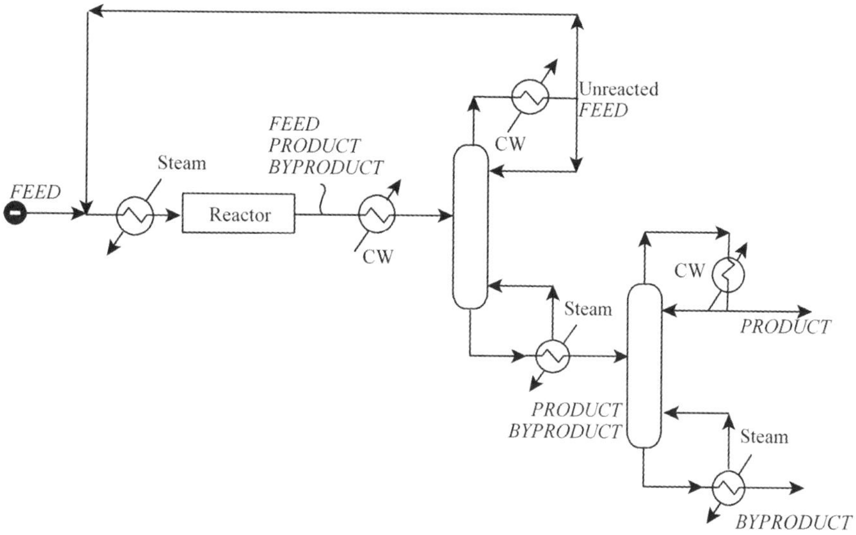

(b) To isolate the *PRODUCT* and recycle unreacted *FEED* we need a separation system.

Figure 1.4

Process design starts with the reactor. The reactor design dictates the separation and recycle problem. (Reproduced from Smith R and Linnhoff B, 1998, *Trans IChemE ChERD*, **66**: 195 by permission of the Institution of Chemical Engineers.)

even long-established processes are still being improved bears evidence to just how difficult this is. Second, can each structure be optimized for a valid comparison? When optimizing the structure, there may be many ways in which each individual task can be performed and many ways in which the individual tasks can be interconnected. This means that the operating conditions for a multitude of structural options must be simulated and optimized. At first sight, this appears to be an overwhelmingly complex problem.

It is helpful when developing a methodology if there is a clear picture of the nature of the problem. If the process requires a reactor, this is where the design starts. This is likely to be the only place in the process where raw material components are converted into components for the products. The chosen reactor design produces a mixture of unreacted feed materials, products and byproducts that need separating. Unreacted feed material is recycled. The reactor design dictates the separation and recycle problem. Thus, design of the separation and recycle system follows the reactor design. The reactor and separation and recycle system designs together define the process for heating and cooling duties. Thus, the heat exchanger network design comes next. Those heating and cooling duties that cannot be satisfied by heat recovery dictate the need for external heating and cooling *utilities* (furnace heating, use of steam, steam generation, cooling water, air cooling or refrigeration). Thus, utility selection and design follows the design of the heat recovery system. The selection and design of the utilities is made more complex by the fact that the process will most likely operate within the context of a site comprising a number of different processes that are all connected to a common utility system. The process and the utility system will both need water, for example, for steam generation, and will also produce aqueous effluents that will have to be brought to a suitable quality for discharge. Thus, the design of the water and aqueous effluent treatment system comes last. Again, the water and effluent treatment system must be considered at the site level as well as the process level.

This hierarchy can be represented symbolically by the layers of the "onion diagram" shown in Figure 1.8 (Linnhoff *et al.*, 1982). The diagram emphasizes the sequential, or hierarchical, nature of process design. Other ways to represent the hierarchy have also been suggested (Douglas, 1985).

Some processes do not require a reactor, for example, some processes just involve separation. Here, the design starts with the separation system and moves outward to the heat exchanger network, utilities, and so on. However, the same basic hierarchy prevails.

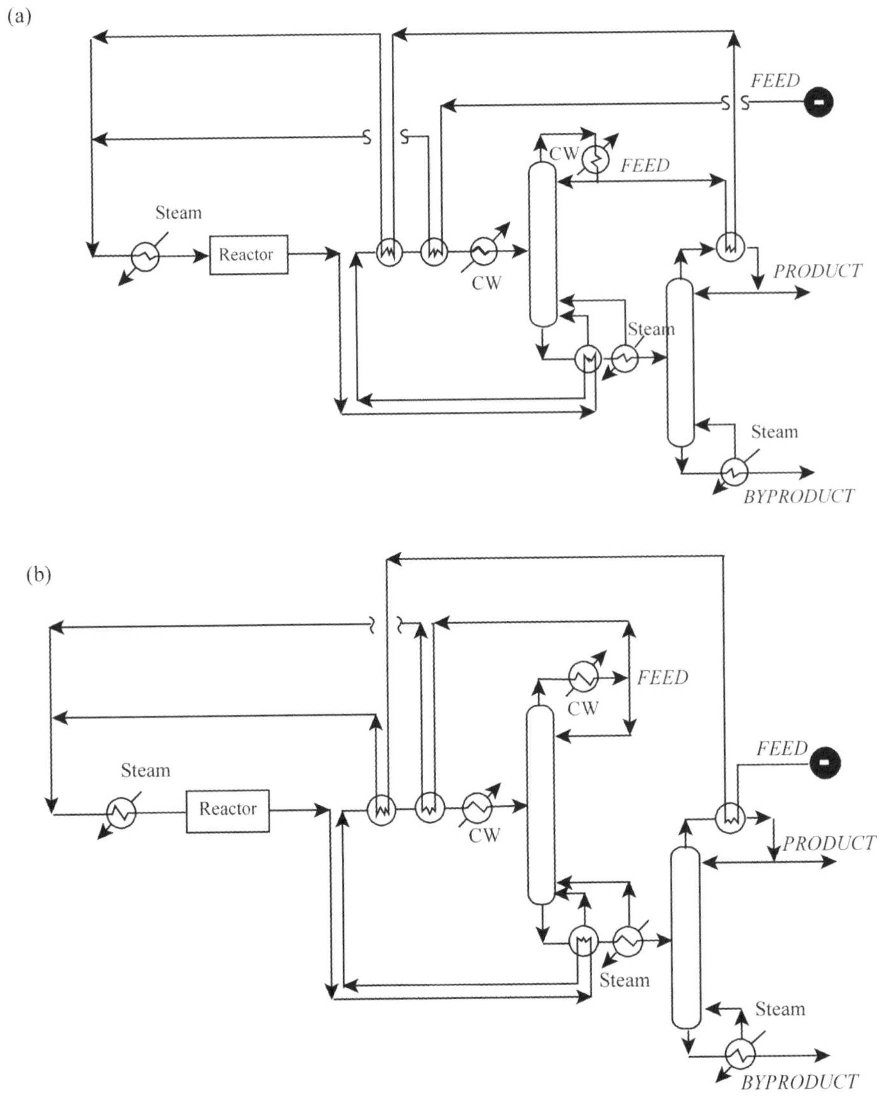

Figure 1.5

For a given reactor and separator design there are different possibilities for heat integration. (Reproduced from Smith R and Linnhoff B, 1998, *Trans IChemE ChERD*, **66**: 195 by permission of the Institution of Chemical Engineers.)

The synthesis of the correct structure and the optimization of parameters in the design of the reaction and separation systems are often the most important tasks of process design. Usually there are many options, and it is impossible to fully evaluate them unless a complete design is furnished for the "outer layers" of the onion model. For example, it is not possible to assess which is better, the basic scheme from Figure 1.4b or that from Figure 1.6, without fully evaluating all possible designs, such as those shown in Figures 1.5a and b and 1.7a and b, all completed, including utilities. Such a complete search is normally too time-consuming to be practical.

Later, in Chapter 17, an approach will be presented in which some early decisions (i.e. decisions regarding reactor and separator options) can be evaluated without a complete design for the "outer layers".

1.5 Continuous and Batch Processes

When considering the processes in Figures 1.4 to 1.6, an implicit assumption was made that the processes operated continuously. However, not all processes operate continuously. In a *batch* process, the main steps operate discontinuously. In contrast with a continuous process, a batch process does not deliver its product continuously but in discrete amounts. This means that heat, mass, temperature, concentration and other properties vary with time. In practice, most batch processes are made up of a series of batch and *semi-continuous* steps. A semi-continuous step runs continuously with periodic start-ups and shutdowns.

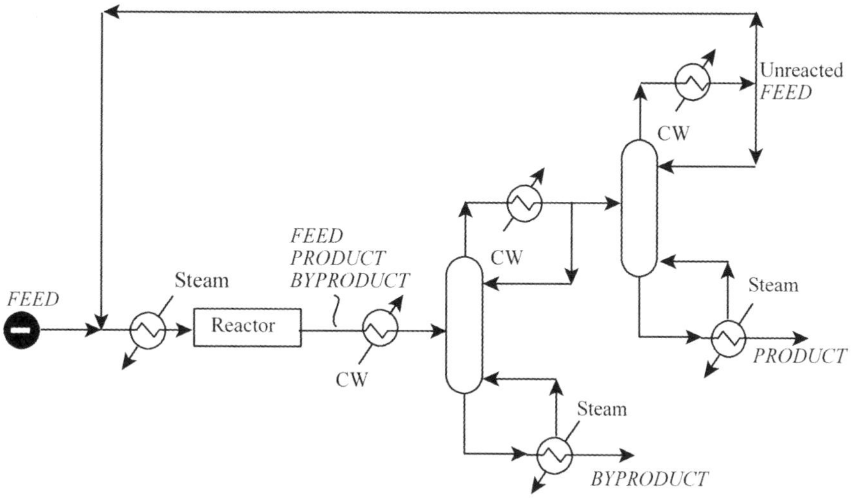

Figure 1.6

Changing the reactor dictates a different separation and recycle problem. (Reproduced from Smith R and Linnhoff B, 1998, *Trans IChemE ChERD*, **66**: 195 by permission of the Institution of Chemical Engineers.)

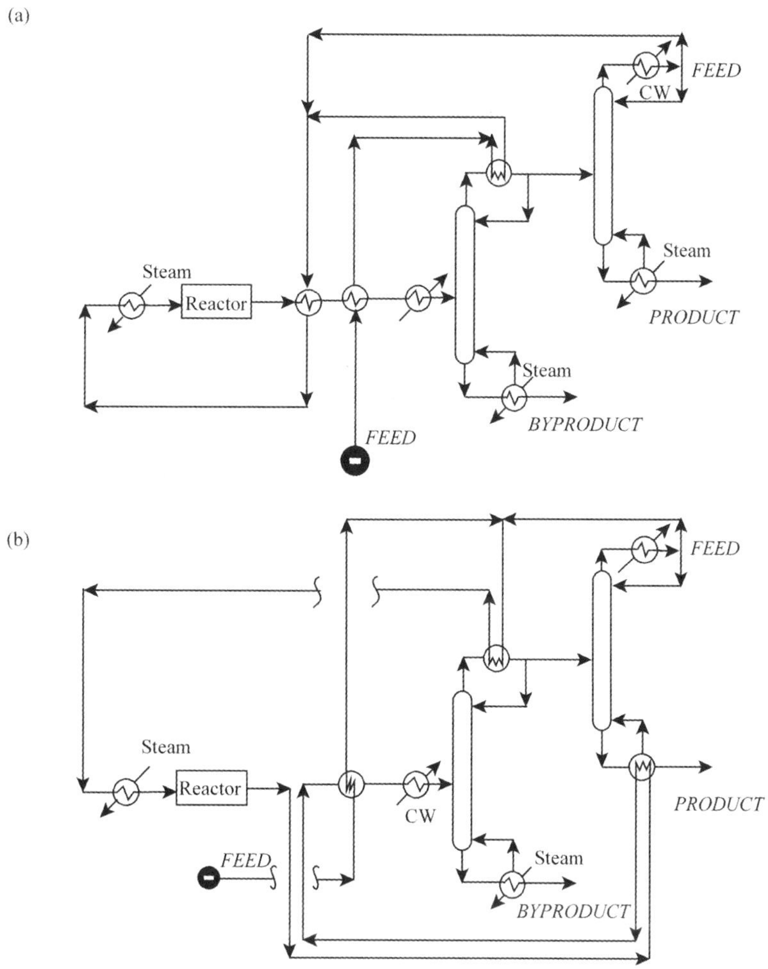

Figure 1.7

A different reactor design not only leads to a different separation system but additional possibilities for heat integration. (Reproduced from Smith R and Linnhoff B, 1998, *Trans IChemE ChERD*, **66**: 195 by permission of the Institution of Chemical Engineers.)

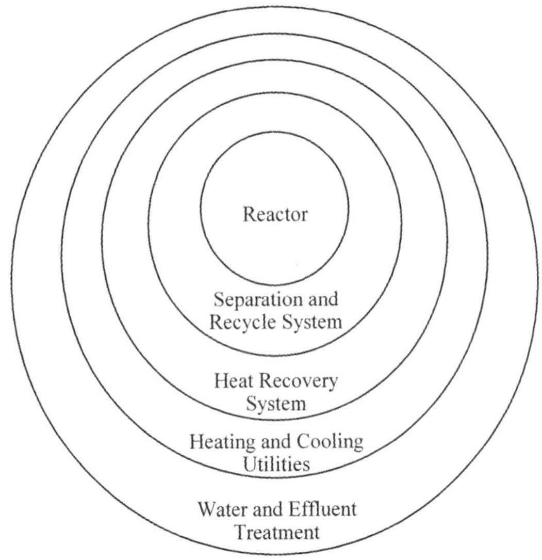

Figure 1.8

The onion model of process design. A reactor is needed before the separation and recycle system can be designed, and so on.

Consider the simple process shown in Figure 1.9. Feed material is withdrawn from storage using a pump. The feed material is preheated in a heat exchanger before being fed to a batch reactor. Once the reactor is full, further heating takes place inside the reactor by passing steam into the reactor jacket before the reaction proceeds. During the later stages of the reaction, cooling water is applied to the reactor jacket. Once the reaction is complete, the reactor product is withdrawn using a pump. The reactor product is cooled in a heat exchanger before going to storage.

The first two steps, pumping for reactor filling and feed preheating, are both semi-continuous. The heating inside the reactor, the reaction itself and the cooling using the reactor jacket are all batch. The pumping to empty the reactor and the product-cooling step are again semi-continuous.

The hierarchy in batch process design is no different from that in continuous processes and the hierarchy illustrated in Figure 1.8 prevails for batch processes also. However, the time dimension brings constraints that do not present a problem in the design of continuous processes. For example, heat recovery might be considered for the process in Figure 1.9. The reactor effluent (which requires cooling) could be used to preheat the incoming feed to the reactor (which requires heating). Unfortunately, even if the reactor effluent is at a high enough temperature to allow this, the reactor feeding and emptying take place at different times, meaning that this will not be possible without some way to store the heat. Such heat storage is possible but usually uneconomic, especially for small-scale processes.

If a batch process manufactures only a single product, then the equipment can be designed and optimized for that product. The dynamic nature of the process creates additional challenges for design and optimization. It might be that the optimization calls for variations in the conditions during the batch through time, according to some *profile*. For example, the temperature in a batch reactor might need to be increased or decreased as the batch progresses.

Multiproduct batch processes, with a number of different products manufactured in the same equipment, present even bigger challenges for design and optimization (Biegler, Grossman and Westerberg, 1997). Different products will demand different designs, different operating conditions and, perhaps, different trajectories for the operating conditions through time. The design of equipment for multiproduct plants will thus require a compromise to be made across the requirements of a number of different products. The more flexible the equipment and the configuration of the equipment, the more it will be able to adapt to the optimum requirements of each product.

Batch processes:

- are economical for small volumes;
- are flexible in accommodating changes in product formulation;
- are flexible in changing the production rate by changing the number of batches made in any period of time;

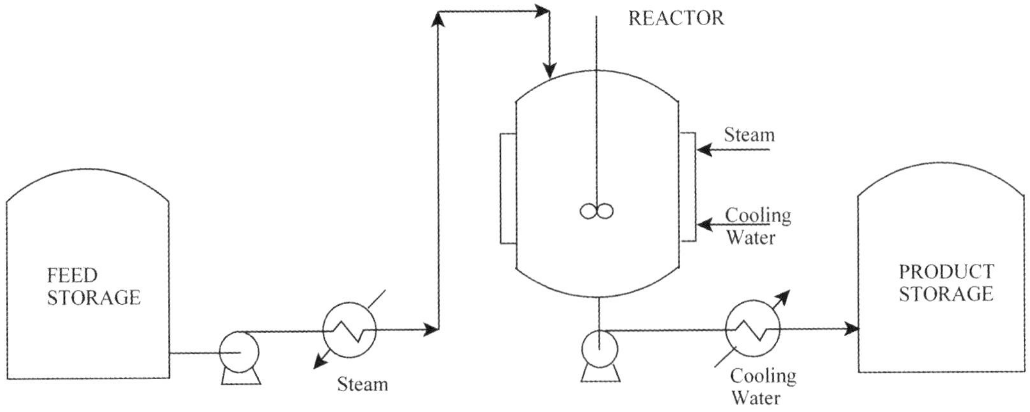

Figure 1.9

A simple batch process.

- allow the use of standardized multipurpose equipment for the production of a variety of products from the same plant;
- are best if equipment needs regular cleaning because of fouling or needs regular sterilization;
- are amenable to direct scale-up from the laboratory and
- allow product identification. Each batch of product can be clearly identified in terms of when it was manufactured, the feeds involved and conditions of processing. This is particularly important in industries such as pharmaceuticals and foodstuffs. If a problem arises with a particular batch, then all the products from that batch can be identified and withdrawn from the market. Otherwise, all the products available in the market would have to be withdrawn.

One of the major problems with batch processing is batch-to-batch conformity. Minor changes to the operation can mean slight changes in the product from batch to batch. Fine and specialty chemicals are usually manufactured in batch processes. However, these products often have very tight tolerances for impurities in the final product and demand batch-to-batch variation to be minimized.

Batch processes will be considered in more detail in Chapter 16.

1.6 New Design and Retrofit

There are two situations that can be encountered in process design. The first is in the design of *new plant* or *grassroot* design. In the second, the design is carried out to modify an existing plant in *retrofit* or *revamp*. The motivation to retrofit an existing plant could be, for example, to increase capacity, allow for different feed or product specifications, reduce operating costs, improve safety or reduce environmental emissions. One of the most common motivations is to increase capacity. When carrying out a retrofit, whatever the motivation, it is desirable to try and make as effective use as possible of the existing equipment. The basic problem with this is that the design of the existing equipment might not be ideally suited to the new role that it will be put to. On the other hand, if equipment is reused, it will avoid unnecessary investment in new equipment, even if it is not ideally suited to the new duty.

When carrying out a retrofit, the connections between the items of equipment can be reconfigured, perhaps adding new equipment where necessary. Alternatively, if the existing equipment differs significantly from what is required in the retrofit, then in addition to reconfiguring the connections between the equipment, the equipment itself can be modified. Generally, the fewer the modifications to both the connections and the equipment, the better.

The most straightforward design situations are those of grassroot design as it has the most freedom to choose the design options and the size of equipment. In retrofit, the design must try to work within the constraints of existing equipment. Because of this, the ultimate goal of the retrofit design is often not clear. For example, a design objective might be given to increase the capacity of a plant by 50%. At the existing capacity limit of the plant, at least one item of equipment must be at its maximum capacity. Most items of equipment are likely to be below their maximum capacity. The differences in the spare capacity of different items of equipment in the existing design arises from different design allowances (or *contingency*) in the original design, changes to the operation of the plant relative to the original design, errors in the original design data, and so on. An item of equipment at its maximum capacity is the *bottleneck* to prevent increased capacity. Thus, to overcome the bottleneck or *debottleneck*, the item of equipment is modified, or replaced with new equipment with increased capacity, or a new item is placed in parallel or series with the existing item, or the connections between existing equipment are reconfigured, or a combination of all these actions is taken. As the capacity of the plant is increased, different items of equipment will reach their maximum capacity. Thus, there will be thresholds in the plant capacity created by the limits in different items of equipment. All equipment with capacity less than the threshold must be modified in some way, or the plant reconfigured, to overcome the threshold. To overcome each threshold requires capital investment. As capacity is increased from the existing limit, ultimately it is likely that it will be prohibitive for the investment to overcome one of the design thresholds. This is likely to become the design limit, as opposed to the original remit of a 50% increase in capacity in the example.

Another important issue in retrofit is the downtime required to make the modifications. The cost of lost production while the plant is shut down to be modified can be prohibitively expensive. Thus, one of the objectives for retrofit is to design modifications that require only a short shutdown. This often means designing modifications that allow the bulk of the work to be carried out while the process is still in operation. For example, new equipment can be installed with final piping connections made when the process is shut down. Decisions whether to replace a major process component completely, or to supplement with a new component working in series or parallel with the existing component, can be critical to the downtime required for retrofit.

1.7 Reliability, Availability and Maintainability

As already discussed, *availability* is often an important issue in process design. Unless the plant is operating in its intended way, it is not productive. Availability measures the portion of the total time that the process meets its production requirements. Availability is related to *reliability* and *maintainability*. *Reliability* is the probability of survival after the unit/system operates for a certain period of time (e.g. a unit has a 95% probability of survival after 8000 hours). Reliability defines the failure frequency and determines the *uptime* patterns. *Maintainability* describes how long it takes for the unit/system to be repaired, which determines the *downtime* patterns. Availability measures the percentage of uptime the process operates at its production requirements over the time horizon, and is determined by reliability and maintainability.

Availability can be improved in many ways. Maintenance policy has a direct influence. *Preventive maintenance* can be used to prevent unnecessary breakdowns. *Condition monitoring* of equipment using techniques such as monitoring vibration of rotating equipment like compressors can be used to detect mechanical problems early, and again prevent unnecessary breakdowns. In design, using *standby* components (sometimes referred to as *spare* or *redundant* components) is a common way to increase system availability. Instead of having one item of equipment on line and vulnerable to breakdown, there may be two, with one on-line and one off-line. These two items of equipment can be sized and operated in many ways:

- 2 × 100% one on-line, one off-line switched off;
- 2 × 100% one on-line, one off-line idling;
- 2 × 50% both on-line, with system capacity reduced to 50% if one fails;
- 2 × 75% both on-line operating at 2/3 capacity when both operating, but with system capacity 75% if one fails;
- and so on.

Over-sizing equipment, particularly rotating equipment like pumps and compressors, can make it more reliable in some cases. Determining the optimum policy for standby equipment involves complex trade-offs that need to consider capital cost, operating cost, maintenance costs and reliability.

1.8 Process Control

Once the basic process configuration has been fixed, a *control system* must be added. The control system compensates for the influence of external *disturbances* such as changes in feed flowrate, feed conditions, feed costs, product demand, product specifications, product prices, ambient temperature, and so on. Ensuring safe operation is the most important task of a control system. This is achieved by monitoring the process conditions and maintaining them within safe operating limits. While maintaining the operation within safe operating limits, the control system should optimize the process performance under the influence of external disturbances. This involves maintaining product specifications, meeting production targets and making efficient use of raw materials and utilities.

A control mechanism is introduced that makes changes to the process in order to cancel out the negative impact of disturbances. In order to achieve this, instruments must be installed to measure the operational performance of the plant. These *measured variables* could include temperature, pressure, flowrate, composition, level, pH, density and particle size. *Primary measurements* may be made to directly represent the control objectives (e.g. measuring the composition that needs to be controlled). If the control objectives are not measurable, then *secondary measurements* of other variables must be made and these secondary measurements related to the control objective. Having measured the variables that need to be controlled, other variables need to be *manipulated* in order to achieve the control objectives. A control system is then designed that responds to variations in the measured variables and manipulates other variables to control the process.

Having designed a process configuration for a continuous process and having optimized it to achieve some objective (e.g. maximize profit) at steady state, is the influence of the control system likely to render the previously optimized process to now be nonoptimal? Even for a continuous process, the process is always likely to be moving from one state to another in response to the influence of disturbances and control objectives. In the steady-state design and optimization of continuous processes, these different states can be allowed for by considering *multiple operating cases*. Each operating case is assumed to operate for a certain proportion of the year. The contribution of the operating case to the overall steady-state design and optimization is weighted according to the proportion of the time under which the plant operates at that state.

While this takes some account of operation under different conditions, it does not account for the dynamic transition from one state to another. Are these transitory states likely to have a significant influence on the optimality? If the transitory states were to have a significant effect on the overall process performance in terms of the objective function being optimized, then the process design and control system design would have to be carried out simultaneously. Simultaneous design of the process and the control system presents an extremely complex problem. It is interesting to note that where steady-state optimization for continuous processes has been compared with simultaneous optimization of the process and control system, the two process designs have been found to be almost identical (Bansal *et al.*, 2000a, 2000b; Kookos and Perkins, 2001).

Industrial practice is to first design and optimize the process configuration (taking into account multiple states, if necessary) and then to add the control system. However, there is no guarantee that design decisions made on the basis of steady-state conditions will not lead to control problems once process dynamics are considered. For example, an item of equipment might be oversized for contingency, because of uncertainty in design data or future debottlenecking prospects, based on steady-state considerations. Once the process dynamics are considered, this oversized equipment might make the process difficult to control, because of the large inventory of process materials in the oversized equipment. The approach to process control should adopt an approach that considers the control of the whole process, rather than just the control of the individual process steps in isolation (Luyben, Tyreus and Luyben, 1999).

The control system arrangement is shown in the *piping and instrumentation diagram* (P & I D) for the process (Sinnott and Towler, 2009). The piping and instrumentation diagram shows all of the process equipment, piping connections, valves, pipe fittings and the control system. All equipment and connections are shown. This includes not only the main items of equipment and connections but also standby equipment, equipment and piping used for start-up, shutdown, maintenance operations and abnormal operation. Figure 1.10a illustrates a very simple *process flow diagram*. This shows only the main items of equipment and the normal process flows. The information shown on such process flow diagrams, and their style, vary according to the practice of different companies. As an example, Figure 1.10a shows the component flowrates and stream temperatures and pressures. By contrast Figure 1.10b shows the corresponding piping and instrumentation diagram. This shows all of the equipment (including the standby pump in this case), all piping connections and fittings, including

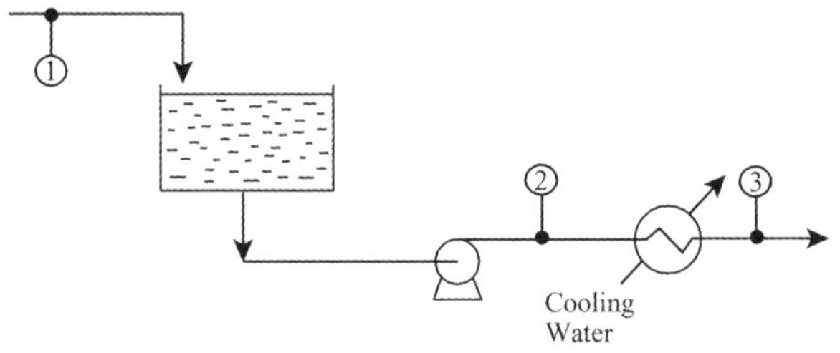

Line	1 Aqueous Recycle	2 Pump Discharge	3 Cooled Recycle
Water ($t \cdot h^{-1}$)	11.6	11.6	11.6
Salts ($t \cdot h^{-1}$)	0.37	0.37	0.37
Sulfuric Acid ($t \cdot h^{-1}$)	-	-	-
Total ($t \cdot h^{-1}$)	11.97	11.97	11.97
Temperature (°C)	60	60	30
Pressure (bar)	1.1	5.0	4.4

(a) A simple process flow diagram.

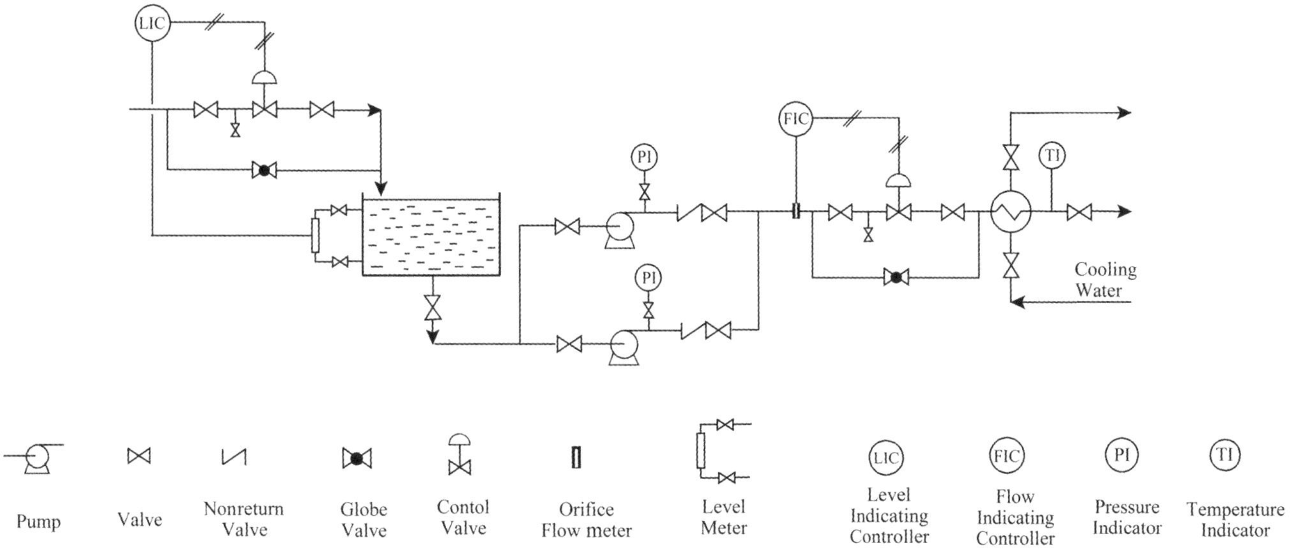

(b) Piping and instrumentation diagram.

Figure 1.10
Process flow diagrams (PFD) and piping and instrumentation diagrams (P&ID).

those used for start-up, shutdown, maintenance and abnormal operation. It also shows the layout of the control system. Additional information normally included would be identification numbers for the equipment, piping connections and control equipment. Information on materials of construction might also be included. But information on process flows and conditions would not normally be shown. As with process flow diagrams, the style of piping and instrumentation diagrams varies according to the practice of different companies.

This text will concentrate on the design and optimization of the process configuration and will not deal with process control. Process control demands expertise in different techniques and will be left to other sources of information (Luyben, Tyreus and Luyben, 1999). Thus, the text will describe how to develop a *flowsheet* or *process flow diagram*, but will not take the final step of adding the instrumentation, control and auxiliary pipes and valves required for the final engineering design in the piping and instrumentation diagram.

Batch processes are, by their nature, always in a transitory state. This requires the dynamics of the process to be optimized, and will be considered in Chapter 16. However, the control systems required to put this into practice will not be considered.

1.9 Approaches to Chemical Process Design and Integration

In broad terms, there are three approaches to chemical process design and integration:

1) *Creating an irreducible structure.* The first approach follows the "onion logic", starting the design by choosing a reactor and then moving outward by adding a separation and recycle system, and so on. At each layer, decisions must be made on the basis of the information available at that stage. The ability to look ahead to the completed design might lead to different decisions. Unfortunately, this is not possible and, instead, decisions must be based on an incomplete picture.

   This approach to creation of the design involves making a series of best local decisions. This might be based on the use of *heuristics* or *rules of thumb* developed from experience (Douglas, 1985) or on a more systematic approach. Equipment is added only if it can be justified economically on the basis of the information available, albeit an incomplete picture. This keeps the structure *irreducible* and features that are technically or economically redundant are not included.

There are two drawbacks to this approach:

a) Different decisions are possible at each stage of the design. To be sure that the best decisions have been made, the other options must be evaluated. However, each option cannot be evaluated properly without completing the design for that option and optimizing the operating conditions. This means that many designs must be completed and optimized in order to find the best.

b) Completing and evaluating many options gives no guarantee of ultimately finding the best possible design, as the search is not exhaustive. Also, complex interactions can occur between different parts of a flowsheet. The effort to keep the system simple and not add features in the early stages of design may result in missing the benefit of interactions between different parts of the flowsheet in a more complex system.

The main advantage of this approach is that the design team can keep control of the basic decisions and interact as the design develops. By staying in control of the basic decisions, the intangibles of the design can be included in the decision making.

2) *Creating and optimizing a superstructure.* In this approach, a *reducible* structure, known as a *superstructure*, is first created that has embedded within it all feasible process options and all feasible interconnections that are candidates for an optimal design structure. Initially, redundant features are built into the superstructure. As an example, consider Figure 1.11 (Kocis and Grossmann, 1988). This shows one possible structure of a process for the manufacture of benzene from the reaction between toluene and hydrogen. In Figure 1.11, the hydrogen enters the process with a small amount of methane as an impurity. Thus, in Figure 1.11, the option of either purifying the hydrogen feed with a membrane or of passing it directly to the process is embedded. The hydrogen and toluene are mixed and preheated to reaction temperature. Only a furnace has been considered feasible in this case because of the high temperature

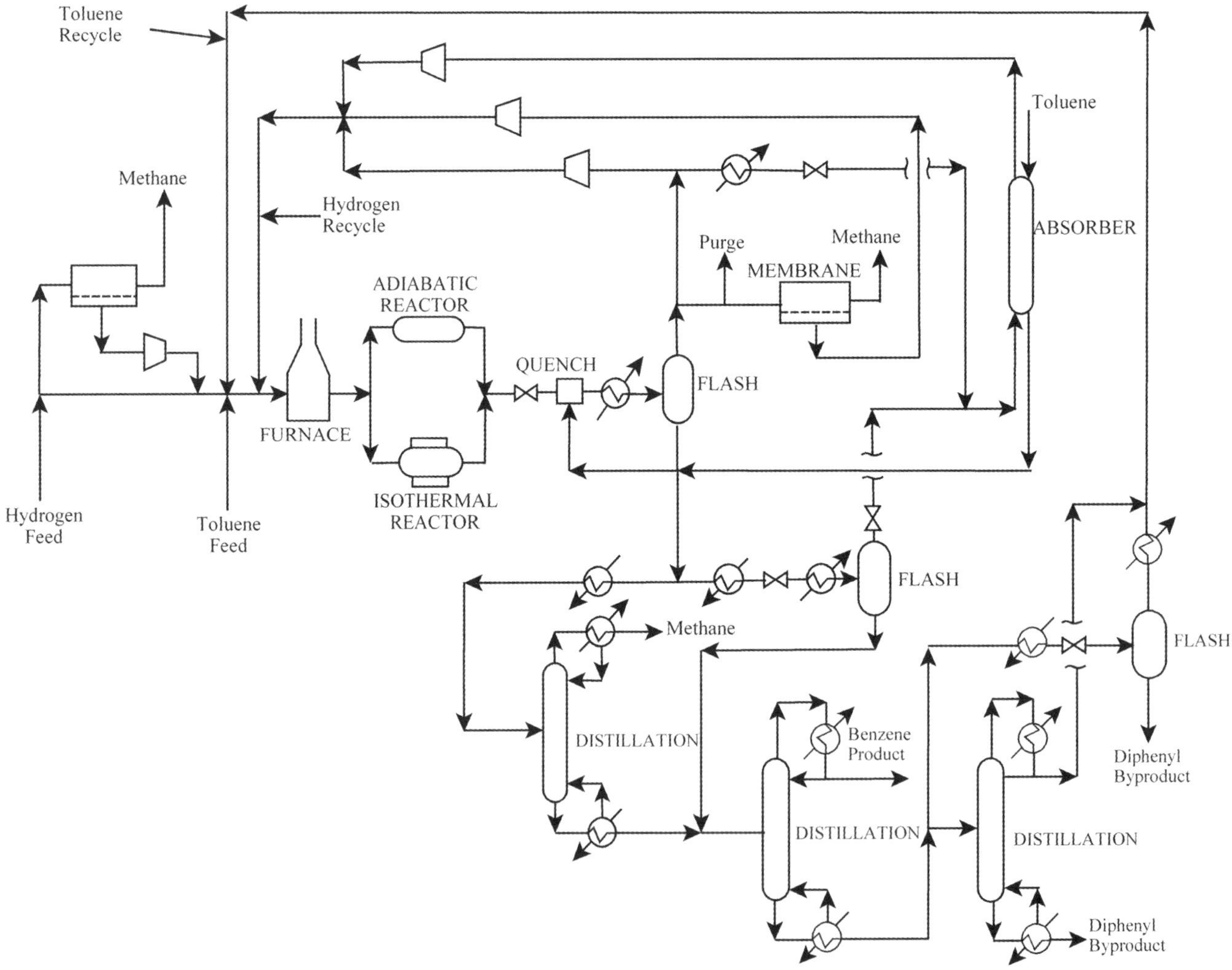

Figure 1.11

A superstructure for the manufacture of benzene from toluene and hydrogen incorporating some redundant features. (Reproduced from Kocis GR and Grossman IE, *Comp Chem Eng*, **13**: 797, with permission from Elsevier.)

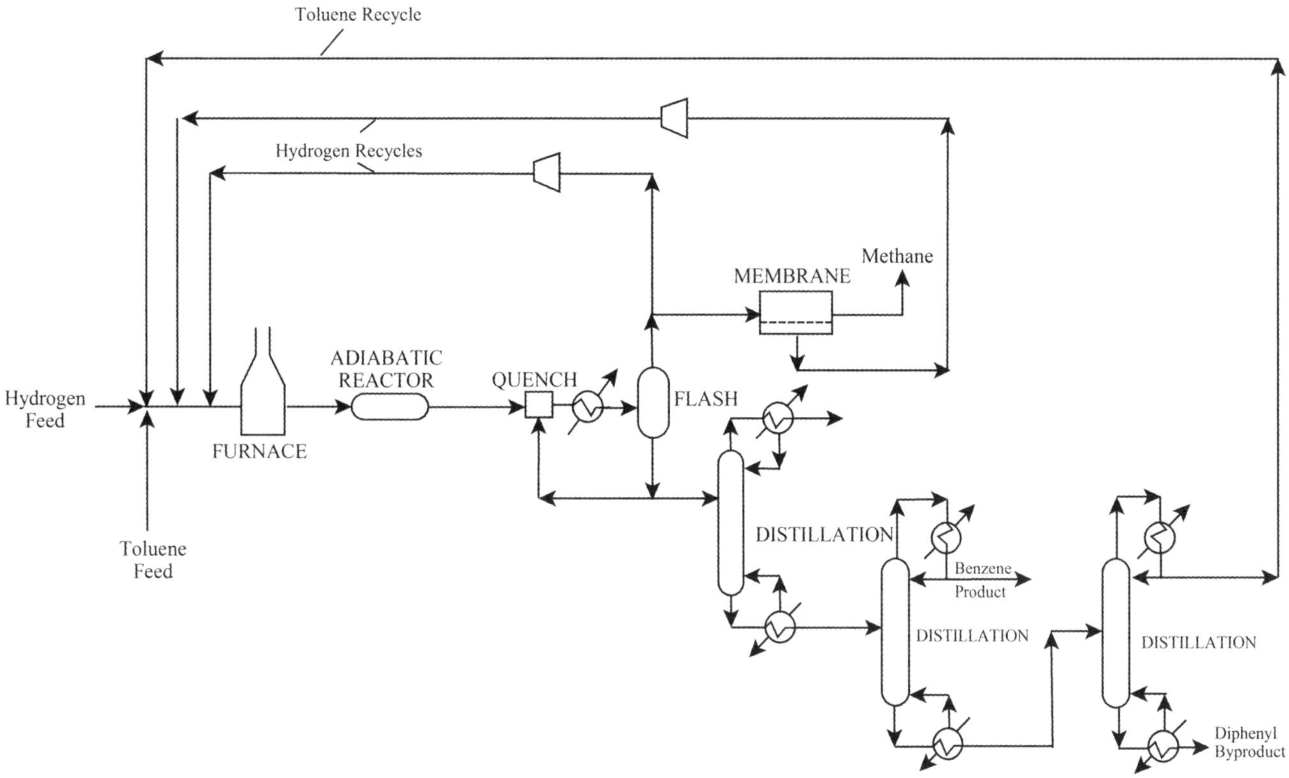

Figure 1.12

Optimization discards many structural features leaving an optimized structure. (Reproduced from Kocis GR and Grossman IE, *Comp Chem Eng*, **13**: 797, with permission from Elsevier.)

required. Then the two alternative reactor options, isothermal and adiabatic reactors, are embedded, and so on. Redundant features have been included in an effort to ensure that all features that could be part of an optimal solution have been included.

The design problem is next formulated as a mathematical model. Some of the design features are continuous, describing the operation of each unit (e.g. flowrate, composition, temperature and pressure) and its size (e.g. volume, heat transfer area, etc.). Other features are discrete (e.g. whether a connection in the flowsheet is included or not, a membrane separator is included or not). Once the problem is formulated mathematically, its solution is carried out through the implementation of an optimization algorithm. An *objective function* is maximized or minimized (e.g. profit is maximized or cost is minimized) in a *structural and parameter* optimization. The optimization justifies the existence of structural features and deletes those features from the structure that cannot be justified economically. In this way, the structure is reduced in complexity. At the same time, the operating conditions and equipment sizes are also optimized. In effect, the discrete decision-making aspects of process design are replaced by a discrete/continuous optimization. Thus, the initial structure in Figure 1.11 is optimized to reduce the structure to the final design shown in Figure 1.12 (Kocis and Grossmann, 1988). In Figure 1.12, the membrane separator on the hydrogen feed has been removed by optimization, as has the isothermal reactor and many other features of the initial structure shown in Figure 1.11.

There are a number of difficulties associated with this approach:

a) The approach will fail to find the optimal structure if the initial structure does not have the optimal structure embedded somewhere within it. The more options included, the more likely it will be that the optimal structure has been included.

b) If the individual unit operations are represented accurately, the resulting mathematical model will be extremely large and the objective function that must be optimized will be extremely irregular. The profile of the objective function can be like the terrain in a range of mountains with many peaks and valleys. If the objective function is to be maximized (e.g. maximize profit), each peak in the mountain range represents a *local optimum* in the objective function. The highest peak represents the *global optimum*. Optimization requires searching around the mountains in a thick fog to find the highest peak, without the benefit of a map and only a compass to tell the direction and an altimeter to show the height. On reaching the top of any peak, there is no way of knowing whether it is the highest peak because of the fog. All peaks must be searched to find the highest. There are crevasses to fall into that might be impossible to climb out of.

Such problems can be overcome in a number of ways. The first way is by changing the model such that the solution space becomes more regular, making the optimization simpler. This most often means simplifying the mathematical model. A second way is by repeating the search many times, but starting each new search from a different initial location. A third way exploits mathematical transformations and bounding techniques for some forms of mathematical expression to allow the global optimum to be found (Floudas, 2000). A fourth way is by allowing the optimization to search the solution space so as to allow the possibility of going downhill, away from an optimum point, as well as uphill. As the search proceeds, the ability of the algorithm to move downhill must be gradually taken away. These problems will be dealt with in more detail in Chapter 3.

c) The most serious drawback of this approach is that the design engineer is removed from the decision making. Thus, the many intangibles in design, such as safety and layout, which are difficult to include in the mathematical formulation, cannot be taken into account satisfactorily.

On the other hand, this approach has a number of advantages. Many different design options can be considered at the same time. The complex multiple trade-offs usually encountered in chemical process design can be handled by this approach. Also, the entire design procedure can be automated and is capable of producing designs quickly and efficiently.

3) *Creating an initial design and evolving through structural and parameter optimization.* The third approach is a variation on creating and optimizing a superstructure. In this approach, an initial design is first created. This is not necessarily intended to be an optimal design and does not necessarily have redundant features, but is simply a starting point. The initial design is then subjected to *evolution* through structural and parameter optimization, using an optimization algorithm (see Chapter 3). As with the superstructure approach, the design problem is formulated as a mathematical model. The design is then evolved one step at a time. Each step is known as a *move*. Each move in the evolution might change one of the continuous variables in the flowsheet (e.g. flowrate, composition, temperature or pressure) or might change the flowsheet structure. Changing the flowsheet structure might mean adding or deleting equipment (together with the appropriate connections), or adding new connections or deleting existing connections.
At each move, the objective function is evaluated (e.g. profit or cost). New moves are then carried out with the aim of improving the objective function. Rules must be created to specify the moves. The same structural options might be allowed as included in the superstructure approach, but in this evolutionary approach the structural moves are carried out to add or delete structural features one at a time. In addition to the structural moves, continuous moves are also carried out to optimize the flowsheet conditions.

The difficulties associated with this approach are similar to those for the creation and optimization of a superstructure:

a) The approach will fail to find the optimal structure if the structural moves have not been defined such that the sequence of moves can lead to the optimal structure from the initial design.

b) The objective function that must be optimized will be extremely irregular. Thus, again there are difficulties finding the global optimum. The optimization methods normally adopted for this approach allow the search to proceed even if the objective function deteriorates after a move. The ability of the algorithm to accept the deteriorating objective function is gradually taken away as the optimization proceeds. This approach will be dealt with in more detail in Chapter 3.

c) Again, this approach has the disadvantage that the design engineer is removed from the decision making.

In summary, the three general approaches to chemical process design have advantages and disadvantages. However, whichever is used in practice, there is no substitute for understanding the problem.

This text concentrates on developing an understanding of the concepts required at each stage of the design. Such an understanding is a vital part of chemical process design and integration, whichever approach is followed.

1.10 The Nature of Chemical Process Design and Integration – Summary

Chemical products can be divided into three broad classes: commodity, fine and specialty chemicals. Commodity chemicals are manufactured in large volumes with low added value. Fine and specialty chemicals tend to be manufactured in low volumes with high added value. The priorities in the design of processes for the manufacture of these three classes of chemical products will differ.

The original design problem posed to the design team is often ill defined, even if it appears on the surface to be well defined. The design team must formulate well-defined design options from the original ill-defined problem, and these must be compared on the basis of consistent criteria.

Design starts by synthesizing design options, followed by simulation and evaluation. The simulation allows further detail to be added to the design. Care should be taken when carrying out simulation to choose appropriate physical property correlations and data. Safety should be considered as early as possible in the development of the design to make the design inherently safe. Both structural and parameter optimization can be carried out to improve the evaluation. Design options should be left open as far as practicable to avoid potentially attractive options being eliminated inappropriately on the basis of uncertain data.

The design might be new or retrofit of an existing process. If the design is a retrofit, then one of the objectives should be to maximize the use of existing equipment, even if it is not ideally suited to its new purpose. Another objective is to minimize the downtime required to carry out the modification.

Both continuous and batch process operations can be used. Batch processes are generally preferred for small-scale and specialty chemicals production.

When developing a chemical process design, there are two basic problems:

- Can all possible structures be identified?
- Can each structure be optimized such that all structures can be compared on a valid basis?

Design starts at the reactor, because it is likely to be the only place in the process where raw materials are converted into the desired chemical products. The reactor design dictates the separation and recycle problem. Together, the reactor design and separation and recycle dictate the heating and cooling duties for the heat exchanger network. Those duties that cannot be satisfied by heat recovery dictate the need for external heating and cooling utilities. The process and the utility system both have a demand for water and create aqueous effluents, giving rise to the water system. This hierarchy is represented by the layers in the "onion diagram" of Figure 1.8. Both continuous and batch process design follow this hierarchy, even though the time dimension in batch processes brings additional constraints in process design.

There are three general approaches to chemical process design:

- creating an irreducible structure;
- creating and optimizing a superstructure;
- creating an initial design and evolving through structural and parameter optimization.

Each of these approaches have advantages and disadvantages.

References

Azapagic A (2014) Sustainability Considerations for Integrated Biorefineries, *Trends in Biotechnology*, **32**: 1.

Bansal V, Perkins JD, Pistikopoulos EN, Ross R and van Shijnedel JMG (2000a) Simultaneous Design and Control Optimisation Under Uncertainty, *Comp Comput Chem Eng*, **24**: 261.

Bansal V, Ross R, Perkins JD, Pistikopoulos EN (2000b) The Interactions of Design and Control: Double Effect Distillation, *J Process Control*, **10**: 219.

Biegler LT, Grossmann IE and Westerberg AW (1997), *Systematic Methods of Chemical Process Design*, Prentice Hall.

Brennan D (1998) *Process Industry Economics*, IChemE, UK.

Cussler EL and Moggridge GD (2011) *Chemical Product Design*, 2nd Edition, Cambridge University Press.

Douglas JM (1985) A Hierarchical Decision Procedure for Process Synthesis, *AIChE J*, **31**: 353.

Floudas CA (2000) *Deterministic Global Optimization: Theory, Methods and Applications*, Kluwer Academic Publishers.

Kocis GR and Grossmann IE (1988) A Modelling and Decomposition Strategy for the MINLP Optimization of Process Flowsheets, *Comp Comput Chem Eng*, **13**: 797.

Kookos IK and Perkins JD (2001) An Algorithm for Simultaneous Process Design and Control, *Ind Eng Chem Res*, **40**: 4079.

Linnhoff B, Townsend DW, Boland D, Hewitt GF, Thomas BEA, Guy AR and Marsland RH (1982) *A User Guide on Process Integration for the Efficient Use of Energy*, IChemE, Rugby, UK.

Luyben WL, Tyreus BD and Luyben ML (1999) *Plant-wide Process Control*, McGraw-Hill.

Seider WD, Seader JD, Lewin DR and Widagdo S (2010) *Product and Process Design Principles*, 3rd Edition, John Wiley & Sons, Inc.

Sharratt PN (1997) *Handbook of Batch Process Design*, Blackie Academic and Professional.

Sinnott R and Towler G (2009) *Chemical Engineering Design*, 5th Edition, Butterworth-Heinemann, Oxford, UK.

Smith R and Linnhoff B (1988) The Design of Separators in the Context of Overall Processes, *Trans IChemE ChERD*, **66**: 195.

Chapter 2

Process Economics

2.1 The Role of Process Economics

The purpose of chemical processes is to make money. An understanding of process economics is therefore critical in process design. Process economics has three basic roles in process design:

1) *Evaluation of design options.* Costs are required to evaluate process design options; for example, should a membrane or an adsorption process be used for a purification?
2) *Process optimization.* The settings of some process variables can have a major influence on the decision making in developing the flowsheet and on the overall profitability of the process. Optimization of such variables is usually required.
3) *Overall project profitability.* The economics of the overall project should be evaluated at different stages during the design to assess whether the project is economically viable.

Before discussing how to use process economics for decision making, the most important costs that will be needed to compare options must first be reviewed.

2.2 Capital Cost for New Design

The total investment required for a new design can be broken down into four main parts that add up to the total investment:

- Battery limits investment
- Utility investment
- Off-site investment
- Working capital.

1) *Battery limits investment.* The battery limit is a geographic boundary that defines the manufacturing area of the process. This is that part of the manufacturing system that converts raw materials into products. It includes process equipment and buildings or structures to house it but excludes boiler-house facilities, site storage, pollution control, site infrastructure, and so on. The term battery limit is sometimes used to define the boundary of responsibility for a given project, especially in retrofit projects.

The battery limits investment required is the purchase of the individual plant items and their installation to form a working process. Equipment costs may be obtained from equipment vendors or from published cost data. Care should be taken as to the basis of such cost data. What is required for cost estimates is delivered cost, but cost is often quoted as FOB (free on board). Free on board means the manufacturer pays for loading charges on to a shipping truck, railcar, barge or ship, but not freight or unloading charges. To obtain a delivered cost requires typically 5 to 10% to be added to the FOB cost. The delivery cost depends on location of the equipment supplier, location of site to be delivered to, size of the equipment, and so on.

The cost of a specific item of equipment will be a function of:

- size,
- materials of construction,
- design pressure,
- design temperature.

Cost data are often presented as cost versus capacity charts, or expressed as a power law of capacity:

$$C_E = C_B\left(\frac{Q}{Q_B}\right)^M \tag{2.1}$$

where C_E = equipment cost with capacity Q
C_B = known base cost for equipment with capacity Q_B
M = constant depending on equipment type

A number of sources of such data are available in the open literature (Guthrie, 1969; Anson, 1977; Hall, Matley and McNaughton, 1982; Ulrich, 1984; Hall, Vatavuk and Matley, 1988; Remer and Chai, 1990; Gerrard, 2000; Peters, Timmerhaus and West, 2003). Published data are often old, sometimes from a variety of sources, with different ages. Such

Chemical Process Design and Integration, Second Edition. Robin Smith.

Companion Website: www.wiley.com/go/smith/chemical2e

data can be brought up-to-date and put on a common basis using cost indexes:

$$\frac{C_1}{C_2} = \frac{INDEX_1}{INDEX_2} \quad (2.2)$$

where C_1 = equipment cost in year 1
C_2 = equipment cost in year 2
$INDEX_1$ = cost index in year 1
$INDEX_2$ = cost index in year 2

Commonly used indexes are the Chemical Engineering Indexes (1957–1959 index = 100) and Marshall and Swift (1926 index = 100), published in *Chemical Engineering* magazine, and the Nelson–Farrar Cost Indexes for refinery construction (1946 index = 100), published in the *Oil and Gas Journal*. The Chemical Engineering (CE) Indexes are particularly useful. CE Indexes are available for equipment covering:

- Heat Exchangers and Tanks
- Pipes, Valves and Fittings
- Process Instruments
- Pumps and Compressors
- Electrical Equipment
- Structural Supports and Miscellaneous.

A combined CE Index of Equipment is available. CE Indexes are also available for:

- Construction and Labor Index
- Buildings Index
- Engineering and Supervision Index.

All of the above indexes are combined to produce a CE Composite Index.

Table 2.1 presents data for a number of equipment items on the basis of January 2000 costs (Gerrard, 2000) (CE Composite Index = 391.1, CE Index of Equipment = 435.8). A guide to the selection of materials of construction can be found in Appendix B.

Cost correlations for vessels are normally expressed in terms of the mass of the vessel. This means that not only a preliminary sizing of the vessel is required but also a preliminary assessment of the mechanical design (Mulet, Corripio and Evans, 1981a, 1981b).

Materials of construction have a significant influence on the capital cost of equipment. Table 2.2 gives some approximate average factors to relate the different materials of construction for equipment capital cost.

It should be emphasized that the factors in Table 2.2 are average and only approximate and will vary, amongst other things, according to the type of equipment. As an example, consider the effect of materials of construction on the capital cost of distillation columns. Table 2.3 gives materials of construction cost factors for distillation columns.

The cost factors for shell-and-tube heat exchangers are made more complex by the ability to construct the different components from different materials of construction. Table 2.4 gives typical materials of construction factors for shell-and-tube heat exchangers.

The operating pressure also influences equipment capital cost as a result of thicker vessel walls to withstand increased pressure. Table 2.5 presents typical factors to account for the pressure rating.

As with materials of construction correction factors, the pressure correction factors in Table 2.5 are average and only approximate and will vary, amongst other things, according to the type of equipment. Finally, its operating temperature also influences equipment capital cost. This is caused by, amongst other factors, a decrease in the allowable stress for materials of construction as the temperature increases. Table 2.6 presents typical factors to account for the operating temperature.

Thus, for a base cost for carbon steel equipment at moderate pressure and temperature, the actual cost can be estimated from:

$$C_E = C_B \left(\frac{Q}{Q_B}\right)^M f_M f_P f_T \quad (2.3)$$

where C_E = equipment cost for carbon steel at moderate pressure and temperature with capacity Q
C_B = known base cost for equipment with capacity Q_B
M = constant depending on equipment type
f_M = correction factor for materials of construction
f_P = correction factor for design pressure
f_T = correction factor for design temperature

In addition to the purchased cost of the equipment, investment is required to install the equipment. Installation costs include:

- cost of installation,
- piping and valves,
- control systems,
- foundations,
- structures,
- insulation,
- fire proofing,
- electrical,
- painting,
- engineering fees,
- contingency.

The total capital cost of the installed battery limits equipment will normally be two to four times the purchased cost of the equipment (Lang, 1947; Hand, 1958).

In addition to the investment within the battery limits, investment is also required for the structures, equipment and services outside the battery limits that are required to make the process function.

2) *Utility investment.* Capital investment in utility plant could include equipment for:

- electricity generation,
- electricity distribution,

Table 2.1

Typical equipment capacity delivered capital cost correlations.

Equipment	Material of construction	Capacity measure	Base size Q_B	Base cost C_B ($)	Size range	Cost exponent M
Agitated reactor	CS	Volume (m^3)	1	1.15×10^4	1–50	0.45
Pressure vessel	SS	Mass (t)	6	9.84×10^4	6–100	0.82
Distillation column (empty shell)	CS	Mass (t)	8	6.56×10^4	8–300	0.89
Sieve trays (10 trays)	CS	Column diameter (m)	0.5	6.56×10^3	0.5–4.0	0.91
Valve trays (10 trays)	CS	Column diameter (m)	0.5	1.80×10^4	0.5–4.0	0.97
Structured packing (5 m height)	SS (low grade)	Column diameter (m)	0.5	1.80×10^4	0.5–4.0	1.70
Scrubber (including random packing)	SS (low grade)	Volume (m^3)	0.1	4.92×10^3	0.1–20	0.53
Cyclone	CS	Diameter (m)	0.4	1.64×10^3	0.4–3.0	1.20
Vacuum filter	CS	Filter area (m^2)	10	8.36×10^4	10–25	0.49
Dryer	SS (low grade)	Evaporation rate (kg $H_2O{\cdot}h^{-1}$)	700	2.30×10^5	700–3000	0.65
Shell-and-tube heat exchanger	CS	Heat transfer area (m^2)	80	3.28×10^4	80–4000	0.68
Air-cooled heat exchanger	CS	Plain tube heat transfer area (m^2)	200	1.56×10^5	200–2000	0.89
Centrifugal pump (small, including motor)	SS (high grade)	Power (kW)	1	1.97×10^3	1–10	0.35
Centrifugal pump (large, including motor)	CS	Power (kW)	4	9.84×10^3	4–700	0.55
Compressor (including motor)		Power (kW)	250	9.84×10^4	250–10,000	0.46
Fan (including motor)	CS	Power (kW)	50	1.23×10^4	50–200	0.76
Vacuum pump (including motor)	CS	Power (kW)	10	1.10×10^4	10–45	0.44
Electric motor		Power (kW)	10	1.48×10^3	10–150	0.85
Storage tank (small atmospheric)	SS (low grade)	Volume (m^3)	0.1	3.28×10^3	0.1–20	0.57
Storage tank (large atmospheric)	CS	Volume (m^3)	5	1.15×10^4	5–200	0.53
Silo	CS	Volume (m^3)	60	1.72×10^4	60–150	0.70
Package steam boiler (fire-tube boiler)	CS	Steam generation ($kg{\cdot}h^{-1}$)	50,000	4.64×10^5	50,000–350,000	0.96
Field erected steam boiler (water-tube boiler)	CS	Steam generation ($kg{\cdot}h^{-1}$)	20,000	3.28×10^5	10,000–800,000	0.81
Cooling tower (forced draft)		Water flowrate ($m^3{\cdot}h^{-1}$)	10	4.43×10^3	10–40	0.63

CS = carbon steel; SS (low grade) = low-grade stainless steel, for example, type 304; SS (high grade) = high-grade stainless steel, for example, type 316.

Table 2.2

Typical average equipment materials of construction capital cost factors.

Material	Correction factor f_M
Carbon steel	1.0
Aluminum	1.3
Stainless steel (low grades)	2.4
Stainless steel (high grades)	3.4
Hastelloy C	3.6
Monel	4.1
Nickel and inconel	4.4
Titanium	5.8

Table 2.3

Typical materials of construction capital cost factors for pressure vessels and distillation columns (Mulet, Corripio and Evans, 1981a, 1981).

Material	Correction factor f_M
Carbon steel	1.0
Stainless steel (low grades)	2.1
Stainless steel (high grades)	3.2
Monel	3.6
Inconel	3.9
Nickel	5.4
Titanium	7.7

- steam generation,
- steam distribution,
- process water,
- cooling water,
- firewater,
- effluent treatment,
- refrigeration,
- compressed air,
- inert gas (nitrogen).

The cost of utilities is considered from their sources within the site to the battery limits of the chemical process served.

3) *Off-site investment.* Off-site investment includes:
- auxiliary buildings such as offices, medical, personnel, locker rooms, guardhouses, warehouses and maintenance shops,
- roads and paths,
- railroads,

Table 2.4

Typical materials of construction capital cost factors for shell-and-tube heat exchangers (Anson, 1977).

Material	Correction factor f_M
CS shell and tubes	1.0
CS shell, aluminum tubes	1.3
CS shell, monel tubes	2.1
CS shell, SS (low grade) tubes	1.7
SS (low grade) shell and tubes	2.9

Table 2.5

Typical equipment pressure capital cost factors.

Design pressure (bar absolute)	Correction factor f_P
0.01	2.0
0.1	1.3
0.5–7	1.0
50	1.5
100	1.9

Table 2.6

Typical equipment temperature capital cost factors.

Design temperature (°C)	Correction factor f_T
0–100	1.0
300	1.6
500	2.1

- fire protection systems,
- communication systems,
- waste disposal systems,
- storage facilities for end product, water and fuel not directly connected with the process,
- plant service vehicles, loading and weighing devices.

The cost of the utilities and off-sites (together sometimes referred to as *services*) ranges typically from 20 to 40% of the total installed cost of the battery limits plant (Bauman, 1955). In general terms, the larger the plant, the larger will tend to be the fraction of the total project cost that goes to utilities and off-sites. In other words, a small project will require typically 20% of the total installed cost for utilities and off-sites. For a large project, the figure can be typically up to 40%.

4) *Working capital,* Working capital is what must be invested to get the plant into productive operation. This is money invested before there is a product to sell and includes:
   - raw materials for plant start-up (including wasted raw materials),
   - raw materials, intermediate and product inventories,
   - cost of transportation of materials for start-up,
   - money to carry accounts receivable (i.e. credit extended to customers) less accounts payable (i.e. credit extended by suppliers),
   - money to meet payroll when starting up.

   Theoretically, in contrast to fixed investment, this money is not lost but can be recovered when the plant is closed down.

   Stocks of raw materials, intermediate and product inventories often have a key influence on the working capital and are under the influence of the designer. Issues relating to storage will be discussed in more detail in Chapters 14 and 16. For an estimate of the working capital requirements, take either (Holland, Watson and Wilkinson, 1983):

   a) 30% of annual sales or

   b) 15% of total capital investment.

5) *Total capital cost.* The total capital cost of the process, services and working capital can be obtained by applying multiplying factors or *installation factors* to the purchase cost of individual items of equipment (Lang, 1947; Hand, 1958):

$$C_F = \sum_i f_i C_{E,i} \tag{2.4}$$

where C_F = fixed capital cost for the complete system
$C_{E,i}$ = cost of equipment i
f_i = installation factor for equipment i

If an average installation factor for all types of equipment is to be used (Lang, 1947),

$$C_F = f_I \sum_i C_{E,i} \tag{2.5}$$

where f_I = overall installation factor for the complete system.

The overall installation factor for new design is broken down in Table 2.7 into component parts according to the dominant phase being processed. The cost of the installation will depend on the balance of gas and liquid processing versus solids processing. If the plant handles only gases and liquids, it can be characterized as fluid processing. A plant can be characterized as solids processing if the bulk of the material handling is solid phase. For example, a solid processing plant could be a coal or an ore preparation plant. Between the two extremes of fluid processing and solids processing are processes that handle a significant amount of both solids and fluids. For example, a shale oil plant involves preparation of the shale oil followed by extraction of fluids from the shale oil and then separation and processing of the fluids. For these types of plant, the contributions to the capital cost can be estimated from the two extreme values in Table 2.7 by interpolation in proportion of the ratio of major processing steps that can be characterized as fluid processing and solid processing.

Table 2.7

Typical factors for capital cost based on delivered equipment costs.

	Type of process	
Item	**Fluid processing**	**Solid processing**
Direct costs		
Equipment delivered cost	1	1
Equipment erection, f_{ER}	0.4	0.5
Piping (installed), f_{PIP}	0.7	0.2
Instrumentation and controls (installed), f_{INST}	0.2	0.1
Electrical (installed), f_{ELEC}	0.1	0.1
Utilities, f_{UTIL}	0.5	0.2
Off-sites, f_{OS}	0.2	0.2
Buildings (including services), f_{BUILD}	0.2	0.3
Site preparation, f_{SP}	0.1	0.1
Total capital cost of installed equipment	3.4	2.7
Indirect costs		
Design, engineering and construction, f_{DEC}	1.0	0.8
Contingency (about 10% of fixed capital costs), f_{CONT}	0.4	0.3
Total fixed capital cost	4.8	3.8
Working capital		
Working capital (15% of total capital cost), f_{WC}	0.7	0.6
Total capital cost, f_I	5.8	4.4

A number of points should be noted about the various contributions to the capital cost in Table 2.7. The values are:

- based on carbon steel, moderate operating pressure and temperature;
- average values for all types of equipment, whereas in practice the values will vary according to the type of equipment;
- only guidelines and the individual components will vary from project to project;
- applicable to new design only.

When equipment uses materials of construction other than carbon steel, or operating temperatures are extreme, the capital cost needs to be adjusted accordingly. Whilst the equipment cost and its associated pipework will be changed, the other installation costs will be largely unchanged, whether the equipment is manufactured from carbon steel or exotic materials of construction. Thus, the application of the factors from Tables 2.2 to 2.6 should only be applied to the equipment and pipework:

$$\begin{aligned} C_F = & \sum_i \left[f_M f_P f_T (1 + f_{PIP})\right]_i C_{E,i} \\ & + (f_{ER} + f_{INST} + f_{ELEC} + f_{UTIL} + f_{OS} + f_{BUILD} \\ & + f_{SP} + f_{DEC} + f_{CONT} + f_{WS}) \sum_i C_{E,i} \end{aligned} \tag{2.6}$$

Thus, to estimate the fixed capital cost:

1) List the main plant items and estimate their size.
2) Estimate the equipment cost of the main plant items.
3) Adjust the equipment costs to a common time basis using a cost index.
4) Convert the cost of the main plant items to carbon steel, moderate pressure and moderate temperature.
5) Select the appropriate installation subfactors from Table 2.7 and adjust for individual circumstances.
6) Select the appropriate materials of construction, operating pressure and operating temperature correction factors for each of the main plant items.
7) Apply Equation 2.6 to estimate the total fixed capital cost.

Equipment cost data used in the early stages of a design will by necessity normally be based on capacity, materials of construction, operating pressure and operating temperature. However, in reality, the equipment cost will depend also on a number of factors that are difficult to quantify (Brennan, 1998):

- multiple purchase discounts,
- buyer–seller relationships,
- capacity utilization in the fabrication shop (i.e. how busy the fabrication shop is),
- required delivery time,
- availability of materials and fabrication labor,
- special terms and conditions of purchase, and so on.

Care should also be taken to the geographic location. Costs to build the same plant can differ significantly between different locations, even within the same country. Such differences will result from variations in climate and its effect on the design requirements and construction conditions, transportation costs, local regulations, local taxes, availability and productivity of construction labor, and so on (Gary and Handwerk, 2001). For example, in the United States of America, Gulf Coast costs tend to be the lowest, with costs in other areas typically 20 to 50% higher, and those in Alaska two or three times higher than the US Gulf Coast (Gary and Handwerk, 2001). In Australia, costs tend to be the lowest in the region of Sydney and the other metropolitan cities, with costs in remote areas such as North Queensland typically 40 to 80% higher (Brennan, 1998). Costs also differ from country to country. For example, relative to costs for a plant located in the US Gulf Coast, costs in India might be expected to be 20% cheaper, in Indonesia 30% cheaper, but in the United Kingdom 15% more expensive, because of labor costs, cost of land, and so on (Brennan, 1998).

It should be emphasized that capital cost estimates using installation factors are at best crude and at worst highly misleading. When preparing such an estimate, the designer spends most of the time on the equipment costs, which represents typically 20 to 40% of the total installed cost. The bulk costs (civil engineering, labor, etc.) are factored costs that lack definition. At best, this type of estimate can be expected to be accurate to ±30%. To obtain greater accuracy requires detailed examination of all aspects of the investment. Thus, for example, to estimate the erection cost accurately requires knowledge of how much concrete will be used for foundations, how much structural steelwork is required, and so on. Such detail can only be included from access to a large database of cost information.

The shortcomings of capital cost estimates using installation factors are less serious in preliminary process design if used to compare options on a common basis. If used to compare options, the errors will tend to be less serious as the errors will tend to be consistent across the options.

Example 2.1 A new heat exchanger is to be installed as part of a large project. Preliminary sizing of the heat exchanger has estimated its heat transfer area to be 250 m^2. Its material of construction is low-grade stainless steel and its pressure rating is 5 bar. Estimate the contribution of the heat exchanger to the total cost of the project (CE Index of Equipment = 682.0).

Solution From Equation 2.1 and Table 2.1, the capital cost of a carbon steel heat exchanger can be estimated from:

$$\begin{aligned} C_E &= C_B \left(\frac{Q}{Q_B}\right)^M \\ &= 3.28 \times 10^4 \left(\frac{250}{80}\right)^{0.68} \\ &= \$7.12 \times 10^4 \end{aligned}$$

The cost can be adjusted to bring it up-to-date using the ratio of cost indexes:

$$\begin{aligned} C_E &= 7.12 \times 10^4 \left(\frac{682.0}{435.8}\right) \\ &= \$1.11 \times 10^5 \end{aligned}$$

The cost of a carbon steel heat exchanger needs to be adjusted for the material of construction. Because of the low pressure rating, no correction for pressure is required (Table 2.5), but the cost needs to be adjusted for the materials of construction. From Table 2.4, $f_M = 2.9$, and the total cost of the installed equipment can be estimated from Equation 2.6 and Table 2.7. If the project is a complete new plant, the contribution of the heat exchanger to the total cost can be estimated to be:

$$\begin{aligned} C_F &= f_M(1 + f_{PIP})C_E + (f_{ER} + f_{INST} + f_{ELEC} + f_{UTIL} + f_{OS} \\ &\quad + f_{BUILD} + f_{SP} + f_{DEC} + f_{CONT} + f_{WS})C_E \\ &= 2.9(1 + 0.7)1.11 \times 10^5 + (0.4 + 0.2 + 0.1 + 0.5 + 0.2 \\ &\quad + 0.2 + 0.1 + 1.0 + 0.4 + 0.7)1.11 \times 10^5 \\ &= 8.73 \times 1.11 \times 10^5 \\ &= \$9.69 \times 10^5 \end{aligned}$$

Had the new heat exchanger been an addition to an existing plant that did not require investment in electrical services, utilities, off-sites, buildings, site preparation or working capital, then the cost would be estimated from:

$$\begin{aligned} C_F &= f_M(1 + f_{PIP})C_E + (f_{ER} + f_{INST} + f_{DEC} + f_{CONT})C_E \\ &= 2.9(1 + 0.7)1.11 \times 10^5 + (0.4 + 0.2 + 1.0 + 0.4)1.11 \times 10^5 \\ &= 6.93 \times 1.11 \times 10^5 \\ &= \$7.69 \times 10^5 \end{aligned}$$

Installing a new heat exchanger into an existing plant might require additional costs over and above those estimated here. Connecting new equipment to existing equipment, modifying or relocating existing equipment to accommodate the new equipment and downtime might all add to the costs.

2.3 Capital Cost for Retrofit

Estimating the capital cost of a retrofit project is much more difficult than for new design. In principle, the cost of individual items of new equipment will be the same, whether it is a grassroot design or a retrofit. By contrast, installation factors for equipment in retrofit can be completely different from grassroot design, and could be higher or lower. If the new equipment can take advantage of existing space, foundations, electrical cabling, and so on, the installation factor might in some cases be lower than in new design. This will especially be the case for small items of equipment. However, most often, retrofit installation factors will tend to be higher than in grassroot design and can be very much higher. This is because existing equipment might need to be modified or moved to allow installation of new equipment. Also, access to the area where the installation is required is likely to be much more restricted in retrofit than in the phased installation of new plant. Smaller projects (as the retrofit is likely to be) tend to bring higher cost of installation per unit of installed equipment than larger projects.

As an example, one very common retrofit situation is the replacement of distillation column internals to improve the performance of the column. The improvement in performance sought is often an increase in the throughput. This calls for existing internals to be removed and then to be replaced with the new internals. Table 2.8 gives typical factors for the removal of old internals and the installation of new ones (Bravo, 1997).

As far as utilities and off-sites are concerned, it is also difficult to generalize. Small retrofit projects are likely not to require any investment in utilities and off-sites. Larger-scale retrofit might demand a major revamp of the utilities and off-sites, which can be particularly expensive because existing equipment might need to be modified or removed to make way for new utilities and off-sites equipment.

Table 2.8

Modification costs for distillation column retrofit (Bravo, 1997).

Column modification	Cost of modification (multiply factor by cost of new hardware)
Removal of trays to install new trays	0.1 for the same tray spacing 0.2 for different tray spacing
Removal of trays to install packing	0.1
Removal of packing to install new trays	0.07
Installation of new trays	1.0–1.4 for the same tray spacing 1.2–1.5 for different tray spacing 1.3–1.6 when replacing packing
Installation of new structured packing	0.5–0.8

Working capital is also difficult to generalize. Most often, there will be no significant working capital associated with a retrofit project. For example, if a few items of equipment are replaced to increase the capacity of a plant, this will not significantly change the raw materials and product inventories, money to carry accounts receivable, money to meet payroll, and so on. On the other hand, if the plant changes function completely, significant new storage capacity is added, and so on, there might be a significant element of working capital.

One of the biggest sources of cost associated with retrofit can be the *downtime* (the period during which the plant will not be productive) required to carry out the modifications. The cost of lost production can be the dominant feature of retrofit projects. The cost of lost production should be added to the capital cost of a retrofit project. To minimize the downtime and cost of lost production requires that as much preparation as possible is carried out whilst the plant is operating. The modifications requiring the plant to be shut down should be minimized. For example, it might be possible for new foundations to be installed and new equipment put into place while the plant is still operating, leaving the final pipework and electrical modifications for the shutdown. Retrofit projects are often arranged such that the preparation is carried out prior to a regular maintenance shutdown, with the final modifications coinciding with the planned maintenance shutdown. Such considerations often dominate the decisions made as to how to modify the process for retrofit.

Because of all of these uncertainties, it is difficult to provide general guidelines for the capital cost of retrofit projects. The basis of the capital cost estimate should be to start with the required investment in new equipment. Installation factors for the installation of equipment for grassroot design from Table 2.7 need to be adjusted according to circumstances (usually increased). If old equipment needs to be modified to take up a new role (e.g. move an existing heat exchanger to a new duty), then an installation cost must be applied without the equipment cost. In the absence of better information, the installation cost can be taken to be that for the equivalent piece of new equipment. Some elements of the total cost breakdown in Table 2.7 will not be relevant and should not be included. In general, for the estimation of capital cost for retrofit, a detailed examination of the individual features of retrofit projects is necessary.

Example 2.2 An existing heat exchanger is to be repiped to a new duty in a retrofit project without moving its location. The only significant investment is piping modifications. The heat transfer area of the existing heat exchanger is 250 m^2. The material of construction is low-grade stainless steel and its design pressure is 5 bar. Estimate the cost of the project (CE Index of Equipment = 682.0).

Solution All retrofit projects have individual characteristics and it is impossible to generalize the costs. The only way to estimate costs with any certainty is to analyze the costs of all of the modifications in detail. However, in the absence of such detail, a very preliminary estimate can be obtained by estimating the retrofit costs from the appropriate installation costs for a new design. In this case, piping costs can be estimated from those for a new heat exchanger of the same specification, but excluding the equipment cost. For Example 2.1, the cost of a new stainless steel heat exchanger with an area of 250 m^2 was estimated to be \$1.11 × 10^5. The piping costs (stainless steel) can therefore be estimated to be:

$$\begin{aligned}\text{Piping cost} &= f_M f_{PIP} C_E \\ &= 2.9 \times 0.7 \times 1.11 \times 10^5 \\ &= 2.03 \times 1.11 \times 10^5 \\ &= \$2.25 \times 10^5\end{aligned}$$

This estimate should not be treated with any confidence. It will give an idea of the costs and might be used to compare retrofit options on a like-for-like basis, but could be very misleading.

Example 2.3 An existing distillation column is to be revamped to increase its capacity by replacing the existing sieve trays with stainless steel structured packing. The column shell is 46 m tall and 1.5 m diameter and currently fitted with 70 sieve trays with a spacing of 0.61 m. The existing trays are to be replaced with stainless steel structured packing with a total height of 30 m. Estimate the cost of the project (CE Index of Equipment = 682.0).

Solution First, estimate the purchase cost of the new structured packing from Equation 2.1 and Table 2.1, which gives costs for a 5 m height of packing:

$$\begin{aligned}C_E &= C_B\left(\frac{Q}{Q_B}\right)^M \\ &= 1.8 \times 10^4 \times \frac{30}{5}\left(\frac{1.5}{0.5}\right)^{1.7} \\ &= \$6.99 \times 10^5\end{aligned}$$

Adjusting the cost to bring it up-to-date using the ratio of cost indexes:

$$\begin{aligned}C_E &= 6.99 \times 10^5\left(\frac{682.0}{435.8}\right) \\ &= \$1.09 \times 10^6\end{aligned}$$

From Table 2.8, the factor for removing the existing trays is 0.1 and that for installing the new packing is 0.5 to 0.8 (say 0.7). Estimated total cost of the project:

$$\begin{aligned}&= (1 + 0.1 + 0.7)1.09 \times 10^6 \\ &= \$1.96 \times 10^6\end{aligned}$$

2.4 Annualized Capital Cost

Capital for new installations may be obtained from:

a) Loans from banks
b) Issue by the company of common stock (ordinary shares), preferred stock (preference shares) or bonds
c) Accumulated net cash flow arising from company profit over time.

Interest on loans from banks, preferred stock and bonds is paid at a fixed rate of interest. A share of the profit of the company is paid as a dividend on common stock and preferred stock (in addition to the interest paid on preferred stock).

The cost of the capital for a project thus depends on its source. The source of the capital often will not be known during the early stages of a project and yet there is a need to select between process options and carry out optimization on the basis of both capital and operating costs. This is difficult to do unless both capital and operating costs can be expressed on a common basis. Capital costs can be expressed on an annual basis if it is assumed that the capital has been borrowed over a fixed period (for large projects, usually 5 to 10 years) at a fixed rate of interest, in which case the capital cost can be annualized according to:

$$\text{Annualized capital cost} = \text{Capital cost} \times \frac{i(1+i)^n}{(1+i)^n - 1} \qquad (2.7)$$

where i = fractional interest rate per year
n = number of years

The derivation of Equation 2.7 is given in Appendix C.

As stated previously, the source of capital is often not known, and hence it is not known whether Equation 2.7 is appropriate to represent the cost of capital. Equation 2.7 is, strictly speaking, only appropriate if the money for capital expenditure is to be borrowed over a fixed period at a fixed rate of interest. Moreover, if Equation 2.7 is accepted, then the number of years over which the capital is to be annualized is known, as is the rate of interest. However, the most important thing is that, even if the source of capital is not known, and uncertain assumptions are necessary, Equation 2.7 provides a common basis for the comparison of competing projects and design alternatives within a project.

Example 2.4 The purchased cost of a new distillation column installation is $1 million. Calculate the annual cost of installed capital if the capital is to be annualized over a five-year period at a fixed rate of interest of 5%.

Solution First calculate the installed capital cost (Table 2.7):

$$\begin{aligned} C_F &= f_i C_E \\ &= 5.8 \times (1,000,000) \\ &= \$5,800,000 \end{aligned}$$

$$\begin{aligned} \text{Annualization factor} &= \frac{i(1+i)^n}{(1+i)^n - 1} \\ &= \frac{0.05(1+0.05)^5}{(1+0.05)^5 - 1} \\ &= 0.2310 \end{aligned}$$

$$\begin{aligned} \text{Annualized capital cost} &= 5,800,000 \times 0.2310 \\ &= \$1,340,000\ \text{y}^{-1} \end{aligned}$$

When using annualized capital cost to carry out optimization, the designer should not lose sight of the uncertainties involved in the capital annualization. In particular, changing the annualization period can lead to very different results when, for example, carrying out a trade-off between energy and capital costs. When carrying out optimization, the sensitivity of the result to changes in the assumptions should be tested.

2.5 Operating Cost

1) *Raw materials cost.* In most processes, the largest individual operating cost is raw materials. The cost of raw materials and the product selling prices tend to have the largest influence on the economic performance of the process. The cost of raw materials and price of products depends on whether the materials in question are being bought and sold under a contractual arrangement (either within the same company or outside the company) or on the open market. Open market prices for some chemical products can fluctuate considerably with time. Raw materials might be purchased and products sold below or above the open market price when under a contractual arrangement, depending on the state of the market. Buying and selling on the open market may give the best purchase and selling prices but give rise to an uncertain economic environment. A long-term contractual agreement may reduce profit per unit of production but gives a degree of certainty over the project life.

   The values of raw materials and products can be found in various on-line sources and trade journals. However, the values reported in such sources will be subject to short-term fluctuations, and long-term forecasts will be required for investment analysis.

2) *Catalysts and chemicals consumed in manufacturing other than raw materials.* Catalysts will need to be replaced or regenerated though the life of a process (see Chapter 6). The replacement of catalysts might be on a continuous basis if homogeneous catalysts are used (see Chapters 5 and 6). Heterogeneous catalysts might also be replaced continuously if they deteriorate rapidly, and regeneration cannot fully reinstate the catalyst activity. More often for heterogeneous catalysts, regeneration or replacement will be carried out on an intermittent basis, depending on the characteristics of the catalyst deactivation.

   In addition to the cost of catalysts, there might be significant costs associated with chemicals consumed in manufacturing that do not form part of the final product. For example, acids and alkalis might be consumed to adjust the pH of streams. Such costs might be significant.

3) *Utility operating cost.* Utility operating cost is usually the most significant variable operating cost after the cost of raw materials. This is especially the case for the production of commodity chemicals. Utility operating cost includes:

   - fuel,
   - electricity,
   - steam,
   - cooling water,
   - refrigeration,
   - compressed air,
   - inert gas.

   Utility costs can vary enormously between different processing sites. This is especially true of fuel and power costs. Not only do fuel costs vary considerably between different fuels (coal, oil, natural gas) and geographic locations but costs also tend to be sensitive to market fluctuations. Contractual relationships also have a significant effect on fuel costs. The price paid for fuel depends very much on how much is purchased and the pattern of usage.

   When electricity is bought from centralized power-generation companies under long-term contract, the price tends to be more stable than fuel costs, since power-generation companies tend to negotiate long-term contracts for fuel supply. However, purchased electricity prices (and sales price if excess electricity is generated and exported) are normally subject to tariff variations. Electricity tariffs can depend on the season of the year (e.g. winter versus summer), the time of the week (e.g. weekend versus weekday) and the time of day (e.g. night versus day). In hot countries, electricity is usually more expensive in the summer than in the winter because of the demand from air-conditioning systems. In cold countries, electricity is usually more expensive in the winter than in the summer because of the demand from space heating. The price structure for electricity can be complex, but should be predictable if based on contractual arrangements. If electricity is purchased from a spot market in those countries that have such arrangements, then prices can vary wildly.

Steam costs vary with the price of fuel and electricity. If steam is only generated at low pressure and not used for power generation in steam turbines, then the cost can be estimated from fuel costs assuming an efficiency of generation and distribution losses. The efficiency of generation depends on the boiler efficiency and the steam consumed in boiler feedwater production (see Chapter 23). Losses from the steam distribution system include heat losses from steam distribution and condensate return pipework to the environment, steam condensate lost to drain and not returned to the boilers and steam leaks. The efficiency of steam generation (including auxiliary boiler-house requirements, see Chapter 23) is typically around 85 to 90% and distribution losses of perhaps another 10%, giving an overall efficiency for steam generation and distribution of typically around 75 to 80% (based on the net calorific value of the fuel). Care should be exercised when considering boiler efficiency and the efficiency of steam generation. These figures are often quoted on the basis of *gross calorific value* of the fuel, which includes the latent heat of the water vapor from combustion. This latent heat is rarely recovered through condensation of the water vapor in the combustion gases. The *net calorific value* of the fuel assumes that the latent heat of the water vapor is not recovered and is therefore the most relevant value, yet figures are often quoted on the basis of gross calorific value.

If high-pressure steam mains are used, then the cost of steam should be related in some way to its capacity to generate power in a steam turbine rather than simply to its additional heating value. The high-pressure steam is generated in the utility boilers and the low-pressure steam is generated by reducing pressure through steam turbines to produce power. This will be discussed in more detail in Chapter 23. One simple way to cost steam is to calculate the cost of the fuel required to generate the high-pressure steam (including any losses), and this fuel cost is then the cost of the high-pressure steam. Although this neglects water costs, labor costs, and so on, it will be a reasonable estimate, as costs are normally dominated by fuel costs. Low-pressure mains have a value equal to that of the high-pressure mains minus the value of power generated by letting the steam down to the low pressure in a steam turbine. To calculate the cost of steam that has been expanded though a steam turbine, the power generated in such an expansion can be calculated. The simplest way to do this is on the basis of a comparison between an ideal and a real expansion though a steam turbine. Figure 2.1 shows a steam turbine expansion on an enthalpy–entropy plot. In an ideal turbine, steam with an initial pressure P_1 and enthalpy H_1 expands isentropically to pressure P_2 and enthalpy $H_{2,IS}$. In such circumstances, the ideal work output is $(H_1 - H_{2,IS})$. Because of the frictional effects in the turbine nozzles and blade passages, the exit enthalpy is greater than it would be in an ideal turbine, and the work output is consequently less, given by H_2 in Figure 2.1. The actual work output is given by $(H_1 - H_2)$. The turbine isentropic efficiency η_{IS} measures the ratio of the actual to ideal work obtained:

$$\eta_{IS} = \frac{H_1 - H_2}{H_1 - H_{2,IS}} \tag{2.8}$$

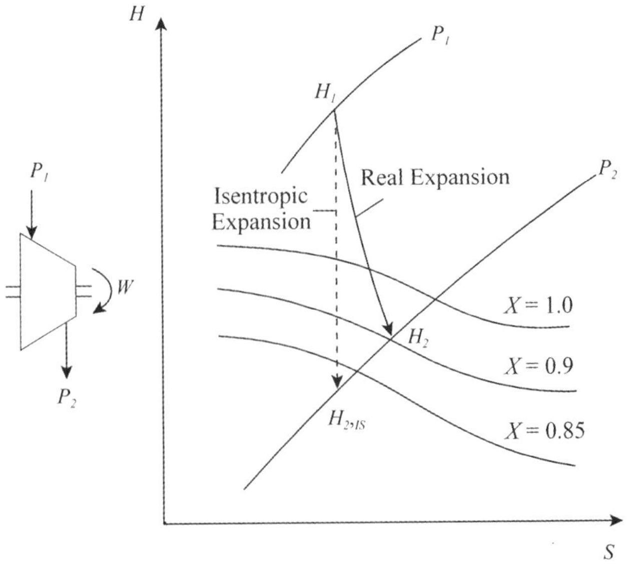

Figure 2.1
Steam turbine expansion.

The output from the turbine might be superheated or partially condensed, as is the case in Figure 2.1. The following example illustrates the approach.

Example 2.5 The pressures of three steam mains have been set to the conditions given in Table 2.9. High-pressure (HP) steam is generated in boilers at 41 barg and superheated to 400 °C. Medium-pressure (MP) and low-pressure (LP) steam are generated by expanding high-pressure steam through a steam turbine with an isentropic efficiency of 80%. The cost of fuel is $\$4.00\,GJ^{-1}$ and the cost of electricity is $\$0.07\,kW^{-1}\cdot h^{-1}$. Boiler feedwater is available at 100 °C with a heat capacity of $4.2\,kJ\cdot kg^{-1}\cdot K^{-1}$. Assuming an efficiency of steam generation of 85% and distribution losses of 10% of the fuel fired, estimate the cost of steam for the three levels.

Table 2.9
Steam mains pressure settings.

Mains	Pressure (barg)
HP	41
MP	10
LP	3

Solution *Cost of 41 barg steam.* From steam tables, for 41 barg steam at 400 °C:

$$\text{Enthalpy} = 3212\,kJ \cdot kg^{-1}$$

For boiler feedwater:

$$\begin{aligned} \text{Enthalpy} &= 4.2(100-0) \quad \text{(relative to water at } 0\,^\circ\text{C)} \\ &= 420\ \text{kJ}\cdot\text{kg}^{-1} \end{aligned}$$

To generate 41 barg steam at 400 °C:

$$\text{Heat duty} = 3212 - 420 = 2792\ \text{kJ}\cdot\text{kg}^{-1}$$

For 41 barg steam:

$$\begin{aligned} \text{Cost} &= 4.00\times10^{-6}\times2792\times\frac{1}{0.75} \\ &= 0.01489\ \$\cdot\text{kg}^{-1} \\ &= 14.89\ \$\cdot\text{t}^{-1} \end{aligned}$$

Cost of 10 barg steam. Here 41 barg steam is now expanded to 10 barg in a steam turbine.

From steam tables, inlet conditions at 41 barg and 400 °C are:

$$\begin{aligned} H_1 &= 3212\ \text{kJ}\cdot\text{kg}^{-1} \\ S_1 &= 6.747\ \text{kJ}\cdot\text{kg}^{-1}\cdot\text{K}^{-1} \end{aligned}$$

Turbine outlet conditions for isentropic expansion to 10 barg are:

$$\begin{aligned} H_{2,IS} &= 2873\ \text{kJ}\cdot\text{kg}^{-1} \\ S_2 &= 6.747\ \text{kJ}\cdot\text{kg}^{-1}\cdot\text{K}^{-1} \end{aligned}$$

For single-stage expansion with isentropic efficiency of 80%:

$$\begin{aligned} H_2 &= H_1 - \eta_{IS}(H_1 - H_{2,IS}) \\ &= 3212 - 0.8(3212 - 2873) \\ &= 2941\ \text{kJ}\cdot\text{kg}^{-1} \end{aligned}$$

From steam tables, the outlet temperature is 251 °C, which corresponds to a superheat of 67 °C. Although steam for process heating is preferred at saturated conditions, it is not desirable in this case to de-superheat by boiler feedwater injection to bring to saturation conditions. If saturated steam is fed to the main, then the heat losses from the main will cause a large amount of condensation in the main, which is undesirable. Hence, it is standard practice to feed steam to the main with a superheat of at least 10 °C to avoid condensation in the main.

$$\begin{aligned} \text{Power generation} &= 3212 - 2941 \\ &= 271\ \text{kJ}\cdot\text{kg}^{-1} \end{aligned}$$

$$\begin{aligned} \text{Value of power generation} &= 271\times\frac{0.07}{3600} \\ &= 0.00527\ \$\cdot\text{kg}^{-1} \end{aligned}$$

$$\begin{aligned} \text{Cost of 10 barg steam} &= 0.01489 - 0.00527 \\ &= 0.00962\ \$\cdot\text{kg}^{-1} \\ &= 9.62\$\cdot\text{t}^{-1} \end{aligned}$$

Cost of 3 barg steam. Here 10 barg steam from the exit of the first turbine is assumed to be expanded to 3 barg in another turbine. From steam tables, inlet conditions of 10 barg and 251 °C are:

$$\begin{aligned} H_1 &= 2941\ \text{kJ}\cdot\text{kg}^{-1} \\ S_1 &= 6.880\ \text{kJ}\cdot\text{kg}^{-1}\cdot\text{K}^{-1} \end{aligned}$$

Turbine outlet conditions for isentropic expansion to 3 barg are:

$$\begin{aligned} H_{2,IS} &= 2732\ \text{kJ}\cdot\text{kg}^{-1} \\ S_2 &= 6.880\ \text{kJ}\cdot\text{kg}^{-1}\cdot\text{K}^{-1} \end{aligned}$$

For a single-stage expansion with isentropic efficiency of 80%:

$$\begin{aligned} H_2 &= H_1 - \eta_{IS}(H_1 - H_{2,IS}) \\ &= 2941 - 0.8(2941 - 2732) \\ &= 2774\ \text{kJ}\cdot\text{kg}^{-1} \end{aligned}$$

From steam tables, the outlet temperature is 160 °C, which is superheated by 16 °C. Again, it is desirable to have some superheat for the steam fed to the low-pressure main.

$$\begin{aligned} \text{Power generation} &= 2941 - 2774 \\ &= 167\ \text{kJ}\cdot\text{kg}^{-1} \end{aligned}$$

$$\begin{aligned} \text{Value of power generation} &= 167\times\frac{0.07}{3600} \\ &= 0.00325\ \$\cdot\text{kg}^{-1} \end{aligned}$$

$$\begin{aligned} \text{Cost of 3 barg steam} &= 0.00962 - 0.00325 \\ &= 0.00637\ \$\cdot\text{kg}^{-1} \\ &= 6.37\ \$\cdot\text{t}^{-1} \end{aligned}$$

If the steam generated in the boilers is at a very high pressure and/or the ratio of power to fuel costs is high, then the value of low-pressure steam can be extremely low or even negative (see Chapter 23). This simplistic approach to costing steam is often unsatisfactory, especially if the utility system already exists. Steam costs will be considered in more detail in Chapter 23.

The operating cost for cooling water tends to be low relative to the value of both fuel and electricity. The principal operating cost associated with the provision of cooling water is the cost of power to drive the cooling tower fans and cooling water circulation pumps. The cost of cooling duty provided by cooling water is in the order of 1% that of the cost of power. For example, if power costs \$0.07 $\text{kW}^{-1}\cdot\text{h}^{-1}$, then cooling water will typically cost $0.07\times0.01/3600 = \$0.19\times10^{-6}\ \text{kJ}^{-1}$ or \$0.19 GJ^{-1}. Cooling water systems will be discussed in more detail in Chapter 24.

The cost of power required for a refrigeration system can be estimated as a multiple of the power required for an ideal system:

$$\frac{W_{IDEAL}}{Q_C} = \frac{T_H - T_C}{T_C} \tag{2.9}$$

where W_{IDEAL} = ideal power required for the refrigeration cycle

Q_C = the cooling duty

T_C = temperature at which heat is taken into the refrigeration cycle (K)

T_H = temperature at which heat is rejected from the refrigeration cycle (K)

The ratio of ideal to actual power is often around 0.6. Thus:

$$W = \frac{Q_C}{0.6}\left(\frac{T_H - T_C}{T_C}\right) \tag{2.10}$$

where W is the actual power required for the refrigeration cycle.

Example 2.6 A process requires 0.5 MW of cooling at −20 °C. A refrigeration cycle is required to remove this heat and reject it to cooling water supplied at 25 °C and returned at 30 °C. Assuming a minimum temperature difference (ΔT_{min}) of 5 °C and both vaporization and condensation of the refrigerant occur isothermally, estimate the annual operating cost of refrigeration for an electrically driven system operating 8000 hours per year. The cost of electricity is $0.07 kW^{-1}·h^{-1}.

Solution

$$W = \frac{Q_C}{0.6}\left(\frac{T_H - T_C}{T_C}\right)$$

$$T_H = 30 + 5 = 35\,^\circ\text{C} = 308\text{ K}$$

$$T_C = -20 - 5 = -25\,^\circ\text{C} = 248\text{ K}$$

$$W = \frac{0.5}{0.6}\left(\frac{308 - 248}{248}\right) = 0.202\text{ MW}$$

$$\text{Cost of electricity} = 0.202 \times 103 \times 0.07 \times 8000 = \$113,120\text{ y}^{-1}$$

More accurate methods to calculate refrigeration costs will be discussed in Chapter 24.

4) *Labor cost.* The cost of labor is difficult to estimate. It depends on whether the process is batch or continuous, the level of automation, the number of processing steps and the level of production. When synthesizing a process, it is usually only necessary to screen process options that have the same basic character (e.g. continuous), have the same level of automation, have a similar number of processing steps and have the same level of production. In this case, labor costs will be common to all options and hence will not affect the comparison.

   If, however, options are to be compared that are very different in nature, such as a comparison between batch and continuous operation, some allowance for the difference in the cost of labor must be made. Also, if the location of the plant has not been fixed, the differences in labor costs between different geographical locations can be important.

5) *Maintenance.* The cost of maintenance depends on whether processing materials are solids on the one hand or gas and liquid on the other. Handling solids tends to increase maintenance costs. Highly corrosive process fluids increase maintenance costs. Average maintenance costs tend to be around 6% of the fixed capital investment (Peters, Timmerhaus and West, 2003).

2.6 Simple Economic Criteria

To evaluate design options and carry out process optimization, simple economic criteria are needed. Consider what happens to the revenue from product sales after the plant has been commissioned. The sales revenue must pay for both fixed costs that are independent of the rate of production and variable costs, which do depend on the rate of production. After this, taxes are deducted to leave the net profit.

Fixed costs independent of the rate of production include:

- Capital cost repayments
- Routine maintenance
- Overheads (e.g. safety services, laboratories, personnel facilities, administrative services)
- Quality control
- Local taxes
- Labor
- Insurance.

Variable costs that depend on the rate of production include:

- Raw materials
- Catalysts and chemicals consumed in manufacturing (other than raw materials)
- Utilities (fuel, steam, electricity, cooling water, process water, compressed air, inert gases, etc.)
- Maintenance costs incurred by operation
- Royalties
- Transportation costs.

There can be an element of maintenance that is a fixed and an element that is variable. Fixed maintenance costs cover routine maintenance, such as statutory maintenance on safety equipment, that must be carried out irrespective of the rate of production. Variable maintenance costs result from certain items of equipment needing more maintenance as the production rate increases. Also, the royalties that cover the cost of purchasing another company's process technology may have different bases. Royalties may be a variable cost, since they can sometimes be paid in proportion to the rate of production or sales revenue. Alternatively, the royalty might be a single-sum payment at the beginning of the project. In this case, the single-sum payment will become part of the project capital investment. As such, it will be included in the annual capital repayment, and this becomes part of the fixed cost.

Two simple economic criteria are useful in process design:

Economic potential (EP):

$$EP = \text{Value of products} - \text{Fixed costs} - \text{Variable costs} - \text{Taxes} \tag{2.11}$$

Total annual cost (TAC):

$$TAC = \text{Fixed costs} + \text{Variable costs} + \text{Taxes} \tag{2.12}$$

When synthesizing a flowsheet, these criteria are applied at various stages when the picture is still incomplete. Hence, it is usually not possible to account for all the fixed and variable costs listed above during the early stages of a project. Also, there is little point in calculating taxes until a complete picture of operating costs and cash flows has been established.

The preceding definitions of economic potential and total annual cost can be simplified if it is accepted that they will be used to compare the relative merits of different structural options in the flowsheet and different settings of the operating parameters. Thus, items that will be common to the options being compared can be neglected.

2.7 Project Cash Flow and Economic Evaluation

As the design progresses, more information is accumulated. The best methods of assessing the profitability of alternatives are based on projections of the cash flows during the project life (Allen, 1980).

Figure 2.2 shows the cash flow pattern for a typical project. The cash flow is a cumulative cash flow. Consider Curve 1 in Figure 2.2. From the start of the project at Point *A*, cash is spent without any immediate return. The early stages of the project consist of development, design and other preliminary work, which causes the cumulative curve to dip to Point *B*. This is followed by the main phase of capital investment in buildings, plant and equipment, and the curve drops more steeply to Point *C*. Working capital is spent to commission the plant between Points *C* and *D*. Production starts at *D*, where revenue from sales begins. Initially, the rate of production is likely to be below design conditions until full production is achieved at *E*. At *F*, the cumulative cash flow is again zero. This is the project breakeven point. Toward the end of the project's life at *G*, the net rate of cash flow may decrease owing to, for example, increasing maintenance costs, a fall in the market price for the product, and so on.

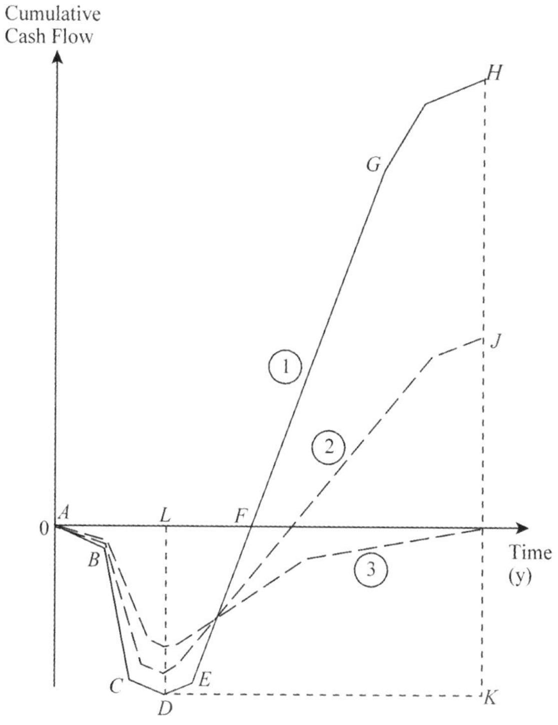

Figure 2.2

Cash flow pattern for a typical project. (Reproduced from Allen DH, (1980) A Guide to the Economic Evaluation of Projects, by permission of the Institution of Chemical Engineers.)

Ultimately, the plant might be permanently shut down or given a major revamp. This marks the end of the project, *H*. If the plant is shut down, working capital is recovered and there may be salvage value, which would create a final cash inflow at the end of the project.

The predicted cumulative cash flow curve for a project throughout its life forms the basis for more detailed evaluation. Many quantitative measures or indices have been proposed. In each case, important features of the cumulative cash flow curve are identified and transformed into a single numerical measure as an index.

1) *Payback time.* Payback time is the time that elapses from the start of the project (*A* in Figure 2.2) to the breakeven point (*F* in Figure 2.2). The shorter the payback time, the more attractive is the project. Payback time is often calculated as the time to recoup the capital investment based on the mean annual cash flow. In retrofit, payback time is usually calculated as the time to recoup the retrofit capital investment from the mean annual improvement in operating costs.
2) *Return on investment (ROI).* Return on investment (ROI) is usually defined as the ratio of average yearly income over the productive life of the project to the total initial investment, expressed as a percentage. Thus, from Figure 2.2:

$$ROI = \frac{KH}{KD} \times \frac{100}{LD} \text{ \% per year} \tag{2.13}$$

Payback and ROI select particular features of the project cumulative cash flow and ignore others. They take no account of the *pattern* of cash flow during a project. The other indices to be described, net present value and discounted cash flow return, are more comprehensive because they take account of the changing pattern of project net cash flow with time. They also take account of the *time value* of money.

3) *Net present value (NPV).* Since money can be invested to earn interest, money received now has a greater value than money if received at some time in the future. The net present value of a project is the sum of the present values of each individual cash flow. In this case, the *present* is taken to be the start of a project.

Time is taken into account by discounting the annual cash flow A_{CF} with the rate of interest to obtain the annual discounted cash flow A_{DCF}. Thus at the end of year 1:

$$A_{DCF1} = \frac{A_{CF1}}{(1+i)}$$

2

at the end of year 2:

$$A_{DCF2} = \frac{A_{CF2}}{(1+i)^2}$$

and at the end of year n:

$$A_{DCFn} = \frac{A_{CFn}}{(1+i)^n} \tag{2.14}$$

The sum of the annual discounted cash flows over n years ΣA_{DCF} is known as the *net present value* (NPV) of the project:

$$NPV = \sum A_{DCF} \tag{2.15}$$

The value of *NPV* is directly dependent on the choice of the fractional interest rate i and project lifetime n.

Returning to the cumulative cash flow curve for a project, the effect of discounting is shown in Figure 2.2. Curve 1 is the original curve with no discounting, that is, $i = 0$, and the project *NPV* is equal to the final net cash position given by *H*. Curve 2 shows the effect of discounting at a fixed rate of interest and the corresponding project *NPV* is given by *J*. Curve 3 in Figure 2.2 shows a larger rate of interest, but it is chosen such that the *NPV* is zero at the end of the project.

The greater the positive *NPV* for a project, the more economically attractive it is. A project with a negative *NPV* is not a profitable proposition.

4) *Discounted cash flow rate of return.* The discounted cash flow rate of return is defined as the discount rate i, which makes the *NPV* of a project equal to zero (Curve 3 in Figure 2.2):

$$NPV = \sum A_{DCF} = 0 \tag{2.16}$$

The value of i given by this equation is known as the *discounted cash flow rate of return* (*DCFRR*). It may be found graphically or by trial and error.

Example 2.7 A company has the alternative of investing in one of two projects, *A* or *B*. The capital cost of both projects is \$10 million. The predicted annual cash flows for both projects are shown in Table 2.10. Capital is restricted, and a choice is to be made on the basis of the discounted cash flow rate of return, based on a five-year lifetime.

Table 2.10

Predicted annual cash flows.

Year	Cash flows ($\$10^6$)	
	Project A	Project B
0	−10	−10
1	1.6	6.5
2	2.8	5.2
3	4.0	4.0
4	5.2	2.8
5	6.4	1.6

Solution Project *A*

Start with an initial guess for *DCFRR* of 20% and increase as detailed in Table 2.11.

Table 2.11

Calculation of *DCFRR* for Project *A*.

Year	A_{CF}	DCF 20%		DCF 30%		DCF 25%	
		A_{DCF}	ΣA_{DCF}	A_{DCF}	ΣA_{DCF}	A_{DCF}	ΣA_{DCF}
0	−10	−10	−10	−10	−10	−10	−10
1	1.6	1.33	−8.67	1.23	−8.77	1.28	−8.72
2	2.8	1.94	−6.73	1.66	−7.11	1.79	−6.93
3	4.0	2.31	−4.42	1.82	−5.29	2.05	−4.88
4	5.2	2.51	−1.91	1.82	−3.47	2.13	−2.75
5	6.4	2.57	0.66	1.72	−1.75	2.10	−0.65

Twenty percent is too low since ΣA_{DCF} is positive at the end of year 5. Thirty percent is too high since ΣA_{DCF} is negative at the end of year 5, as is the case with 25%. The answer must be between 20 and 25%. Interpolating on the basis of ΣA_{DCF} the $DCFRR \approx 23\%$.

Project *B*

Again, start with an initial guess for *DCFRR* of 20% and increase as detailed in Table 2.12.

Table 2.12

Calculation of *DCFRR* for Project *B*.

		DCF 20%		DCF 40%		DCF 35%	
Year	A_{CF}	A_{DCF}	ΣA_{DCF}	A_{DCF}	ΣA_{DCF}	A_{DCF}	ΣA_{DCF}
0	−10	−10	−10	−10	−10	−10	−10
1	6.5	5.42	−4.58	4.64	−5.36	4.81	−5.19
2	5.2	3.61	−0.97	2.65	−2.71	2.85	−2.34
3	4.0	2.31	1.34	1.46	−1.25	1.63	−0.71
4	2.8	1.35	2.69	0.729	−0.521	0.843	0.133
5	1.6	0.643	3.33	0.297	−0.224	0.357	0.490

From ΣA_{DCF} at the end of year 5, 20% is to low, 40% too high and 35% also too low. Interpolating on the basis of ΣA_{DCF}, the $DCFRR \approx 38\%$. Project *B* should therefore be chosen.

2.8 Investment Criteria

Economic analysis should be performed at all stages of an emerging project as more information and detail become available. The decision as to whether to proceed with a project will depend on many factors. There is most often stiff competition within companies for any capital available for investment in projects. However, in an efficient capital market there should be plenty of capital available, providing the returns are high enough. The decision as to where to spend capital on a particular project will, in the first instance, but not exclusively, depend on the economic criteria discussed in the previous section. Criteria that account for the timing of the cash flows (the *NPV* and *DCFRR*) should be the basis of the decision making. The higher the value of the *NPV* and *DCFRR* for a project, the more attractive it is. The absolute minimum acceptable value of the *DCFRR* is the market interest rate. If the *DCFRR* is lower than the market interest rate, it would be better to put the money in the bank. However, since the bank account is less risky than most technical projects, it is prudent to set a return target of at least 5 to 10% higher than the return on a bank account. The essential distinction between *NPV* and *DCFRR* is:

- *Net present value* measures profit but does not indicate how efficiently capital is being used.
- *DCFRR* measures the rate of return that a project might achieve, but gives no indication of how competitive the project would be.

When choosing between competing projects, these will have different cash flow patterns and different capital investments. If the goal is to maximize profit, then projects with the highest *NPV* should be selected. This does not necessarily equate with selecting the project with the highest *DCFRR*. A lower *DCFRR* on a larger investment could yield a higher *NPV* than a higher *DCFRR* on a smaller investment. *NPV* gives a direct cash measure of the attractiveness of a project. For competing projects with different capital investments, a quick way to compare between the projects is the *investment efficiency*:

$$Investment\ efficiency = \frac{NPV}{Capital\ investment} \tag{2.17}$$

Predicting future cash flows for a project is extremely difficult. There are many uncertainties, including the project life. Also, the appropriate interest rate will not be known with certainty. The acceptability of the rate of return will depend on the risks associated with the project and the company investment policy. For example, a *DCFRR* of 20% might be acceptable for a low-risk project. A higher return of, say, 30% might be demanded of a project with some risk, whereas a high-risk project with significant uncertainty might demand a 50% *DCFRR*.

The sensitivity of the economic analysis to the underlying assumptions should always be tested. A sensitivity analysis should be carried out to test the sensitivity of the economic analysis to:

- errors in the capital cost estimate,

- delays in the start-up of the project after the capital has been invested (particularly important for a high capital cost project),
- changes in the cost of raw materials,
- changes in the sales price of the product,
- reduction in the market demand for the product, and so on.

When carrying out an economic evaluation, the magnitude and timing of the cash flows, the project life and interest rate are not known with any certainty. However, providing that consistent assumptions are made for projections of cash flows and the assumed rate of interest, the economic analysis can be used to choose between competing projects. It is important to compare different projects and options within projects on the basis of consistent assumptions. Thus, even though the evaluation will be uncertain in an absolute sense, it can still be meaningful in a relative sense for choosing between options. Because of this, it is important to have a reference against which to judge any project or option within a project.

However, the final decision to proceed with a project will be influenced as much by business strategy as by the economic measures described above. The business strategy might be to gradually withdraw from a particular market, perhaps because of adverse long-term projections of excessive competition, even though there might be short-term attractive investment opportunities. The long-term business strategy might be to move into different business areas, thereby creating investment priorities. Priority might be given to increasing market share in a particular product to establish business dominance in the area and achieve long-term global economies of scale in the business.

2.9 Process Economics — Summary

Process economics is required to evaluate design options, carry out process optimization and evaluate overall project profitability. Two simple criteria can be used:

- economic potential,
- total annual cost.

These criteria can be used at various stages in the design without a complete picture of the process.

The dominant operating cost is usually raw materials. However, other significant operating costs involve catalysts and chemicals consumed other than raw materials, utility costs, labor costs and maintenance.

Capital costs can be estimated by applying installation factors to the purchase costs of individual items of equipment. However, there is considerable uncertainty associated with cost estimates obtained in this way, as equipment costs are typically only 20 to 40% of the total installed costs, with the remainder based on factors. Utility investment, off-site investment and working capital are also needed to complete the capital investment. The capital cost can be annualized by considering it as a loan over a fixed period at a fixed rate of interest.

As a more complete picture of the project emerges, the cash flows through the project life can be projected. This allows more detailed evaluation of project profitability on the basis of cash flows. Net present value can be used to measure the profit, taking into account the time value of money. The discounted cash flow rate of return measures how efficiently the capital is being used.

Overall, there are always considerable uncertainties associated with an economic evaluation. In addition to the errors associated with the estimation of capital and operating costs, the project life or interest rates are not known with any certainty. The important thing is that different projects, and options within projects, are compared on the basis of consistent assumptions. Thus, even though the evaluation will be uncertain in an absolute sense, it will still be meaningful in a relative sense for choosing between options.

2.10 Exercises

1. A new agitated reactor with a new external shell-and-tube heat exchanger and new centrifugal pump are to be installed in an existing facility. The agitated reactor is to be glass-lined, which can be assumed to have an equipment cost of three times the cost of a carbon steel vessel. The heat exchanger, pump and associated piping are all high-grade stainless steel. The equipment is rated for moderate pressure. The reactor has a volume of 9 m^3, the heat exchanger an area of 50 m^2 and the pump has a power of 5 kW. No significant investment is required in utilities, off-sites, buildings, site preparation or working capital. Using Equation 2.1 and Table 2.1 (extrapolating beyond the range of the correlation if necessary), estimate the cost of the project (CE Index of Equipment = 680.0).

2. Steam is distributed on a site via high-pressure and low-pressure steam mains. The high-pressure main is at 40 bar and 350 °C. The low-pressure main is at 4 bar. The high-pressure steam is generated in boilers. The overall efficiency of steam generation and distribution is 75%. The low-pressure steam is generated by expanding the high-pressure stream through steam turbines with an isentropic efficiency of 80%. The cost of fuel in the boilers is 3.5 $\$\cdot GJ^{-1}$ and the cost of electricity is \$0.05 $kW^{-1}\cdot h^{-1}$. The boiler feedwater is available at 100 °C with a heat capacity of 4.2 $kJ\cdot kg^{-1}\cdot K^{-1}$. Estimate the cost of the high-pressure and low-pressure steam.

3. A refrigerated distillation condenser has a cooling duty of 0.75 MW. The condensing stream has a temperature of −10 °C. The heat from a refrigeration circuit can be rejected to cooling water at a temperature of 30 °C. Assuming a temperature difference in the distillation condenser of 5 °C and a temperature difference for heat rejection from refrigeration to cooling water of 10 °C, estimate the power requirements for the refrigeration.

4. Acetone is to be produced by the dehydrogenation of an aqueous solution of isopropanol according to the reaction:

$$\underset{\text{Isopropanol}}{(CH_3)_2CHOH} \rightarrow \underset{\text{Acetone}}{CH_3COCH_3} + \underset{\text{Hydrogen}}{H_2}$$

Table 2.13

Data for Exercise 4.

Component	Flowrate in vapor ($kmol{\cdot}h^{-1}$)	Raw material value ($\${\cdot}kmol^{-1}$)	Fuel value ($\${\cdot}kmol^{-1}$)
Hydrogen	51.1	0	0.99
Acetone	13.5	34.8	6.85

The effluent from the reactor enters a phase separator that separates vapor from liquid. The liquid contains the bulk of the product and the vapor is a waste stream. The vapor stream is at a temperature of 30 °C and an absolute pressure of 1.1 bar. The component flowrates in the vapor stream are given in Table 2.13, together with their raw material values and fuel values. Three options are to be considered:

a) Burn the vapor in a furnace.

b) Recover the acetone by absorption in water recycled from elsewhere in the process with the tail gas being burnt in a furnace. It is expected that 99% will be recovered by this method at a cost of 1.8 $\${\cdot}kmol^{-1}$ acetone recovered.

c) Recover the acetone by condensation using refrigerated coolant with the tail gas being burnt in a furnace. It is anticipated that a temperature of −10 °C will need to be achieved in the condenser. It can be assumed that the hydrogen is an inert gas that will not dissolve in the liquid acetone. The vapor pressure of acetone is given by

$$\ln P = 10.031 - \frac{2940.5}{T - 35.93}$$

where P = pressure (bara)
T = absolute temperature (K)

The cost of refrigerant is $\$11.5\,GJ^{-1}$, the mean molal heat capacity of the vapor is $40\,kJ{\cdot}kmol^{-1}{\cdot}K^{-1}$ and the latent heat of acetone is $29{,}100\,kJ{\cdot}kmol^{-1}$.

Calculate the economic potential of each option given the data in Table 2.13.

5. A process for the production of cellulose acetate fiber produces a waste stream containing mainly air but with a small quantity of acetone vapor. The flowrate of air is $300\,kmol{\cdot}h^{-1}$ and that of acetone is $4.5\,kmol{\cdot}h^{-1}$. It is proposed to recover the acetone from the air by absorption into water followed by distillation of the acetone–water mixture. The absorber requires a flow of water 2.8 times that of the air.

a) Assuming acetone costs 34.8 $\${\cdot}kmol^{-1}$, fresh water is to be used in the absorber at a cost of $\$0.004\,kmol^{-1}$ and the process operates for $8000\,h{\cdot}y^{-1}$, calculate the maximum economic potential assuming complete recovery of the acetone.

Table 2.14

Cash flows for two competing projects.

Year	Cash flows \$1000	
	Project *A*	Project *B*
0	−1000	−1000
1	150	500
2	250	450
3	350	300
4	400	200
5	400	100

b) The absorber and the absorber and distillation column are both assumed to operate at 99% recovery of acetone. If the product acetone overhead from the distillation column must be 99% pure and there is assumed to be no loss of water in the air stream, sketch the flowsheet for the system and calculate the flows of acetone and water to and from the distillation column.

c) Calculate the revised economic potential to allow for incomplete recovery in the absorption and distillation columns. In addition, the effluent from the bottom of the distillation column will cost \$50 for each kmol of acetone plus \$0.004 for each kmol of water to treat before it can be disposed of.

d) How might the costs be decreased for the same recoveries and purities in the separations?

6. A company has the option of investing in one of the two projects, *A* or *B*. The capital cost of both projects is \$1,000,000. The predicted annual cash flows for both projects are shown in Table 2.14. For each project, calculate:

a) The payback time for each project in terms of the average annual cash flow.

b) The return on investment.

c) The discounted cash flow rate of return.

What do you conclude from the result?

7. A company has the option of investing in one of the two projects, *A* or *B*. The capital cost of both projects is \$1,000,000. The predicted annual cash flows for both projects are shown in Table 2.15. For each project, calculate:

a) The payback time for each project in terms of the average annual cash flow.

b) The return on investment.

c) The discounted cash flow rate of return.

What do you conclude from the result?

8. A process has been developed for a new product for which the market is uncertain. A plant to produce $50{,}000\,t{\cdot}y^{-1}$ requires

Table 2.15

Cash flows for two competing projects.

Year	Cash flows $1000	
	Project *A*	Project *B*
0	−1000	−1000
1	150	500
2	250	450
3	350	300
4	400	200
5	400	100

an investment of $10,000,000 and the expected project life is five years. Fixed operating costs are expected to be 750,000 $\$\cdot y^{-1}$ and variable operating costs (excluding raw materials) are expected to be 40 $\$\cdot t^{-1}$ product. The stoichiometric raw material costs are 80 $\$\cdot t^{-1}$ product. The yield of product per ton of raw material is 80%. Tax is paid in the same year as the relevant profit is made at a rate of 20%. Calculate the selling price of the product to give a minimum acceptable discounted cash flowrate of return of 15% year.

9. How can the concept of simple payback be improved to give a more meaningful measure of project profitability?

10. It is proposed to build a plant to produce 170,000 $t\cdot y^{-1}$ of a commodity chemical. A study of the supply and demand projections for the product indicates that current installed capacity in the industry is $6.8\times10^6\ t\cdot y^{-1}$, whereas total production is running at $5.0\times10^6\ t\cdot y^{-1}$. Maximum plant utilization is thought to be around 90%. If the demand for the product is expected to grow at 8% per year and it will take 3 years to commission a new plant from the start of a project, what do you conclude about the prospect for the proposed project?

References

Allen DH (1980) *A Guide to the Economic Evaluation of Projects*, IChemE, Rugby, UK.

Anson HA (1977) *A New Guide to Capital Cost Estimating*, IChemE, UK.

Bauman HC (1955) Estimating Costs of Process Auxiliaries, *Chem Eng Prog*, **51**: 45.

Bravo JL (1997) Select Structured Packings or Trays? *Chem Eng Prog*, **July**: 36.

Brennan D (1998) *Process Industry Economics*, IChemE, UK.

Gary JH and Handwerk GE (2001) *Petroleum Refining Technology and Economics*, 4th Edition, Marcel Dekker.

Gerrard AM (2000) *Guide to Capital Cost Estimating*, 4th Edition, IChemE, UK.

Guthrie KM (1969) Data and Techniques for Preliminary Capital Cost Estimating, *Chem Eng*, **76**: 114.

Hall RS, Matley J and McNaughton KJ (1982) Current Costs of Process Equipment, *Chem Eng*, **89**: 80.

Hall R.S, Vatavuk WM and Matley J (1988) Estimating Process Equipment Costs, *Chem Eng*, **95**: 66.

Hand WE (1958) From Flowsheet to Cost Estimate, *Petrol Refiner*, **37**: 331.

Holland FA, Watson FA and Wilkinson JK (1983) *Introduction to Process Economics*, 2nd Edition, John Wiley & Sons, Inc., New York.

Lang HJ (1947) Cost Relationships in Preliminary Cost Estimation, *Chem Eng*, **54**: 117.

Mulet A, Corripio AB and Evans LB (1981a) Estimate Costs of Pressure Vessels Via Correlations, *Chem Eng*, **Oct**: 145.

Mulet A, Corripio AB and Evans LB (1981b) Estimate Costs of Distillation and Absorption Towers Via Correlations, *Chem Eng*, **Dec**: 77.

Peters MS, Timmerhaus KD and West RE (2003) *Plant Design and Economics for Chemical Engineers*, 5th Edition, McGraw-Hill, New York.

Remer DS and Chai LH (1990) Design Cost Factors for Scaling-up Engineering Equipment, *Chem Eng Prog*, **86**: 77.

Ulrich GD (1984) *A Guide to Chemical Engineering Process Design and Economics*, John Wiley & Sons, Inc., New York.

Chapter 3

3

Optimization

3.1 Objective Functions

Optimization will almost always be required at some stage in a process design. It is usually not necessary for a designer to construct an optimization algorithm in order to carry out an optimization, as general-purpose software is usually available for this. However, it is necessary for the designer to have some understanding of how optimization works in order to avoid the pitfalls that can occur. More detailed accounts of optimization can be found elsewhere (Floudas, 1995; Biegler, Grossmann and Westerberg, 1997; Edgar, Himmelblau and Lasdon, 2001).

Optimization problems in process design are usually concerned with maximizing or minimizing an *objective function*. The objective function is a measure of the quality of the solution and might typically cause economic potential to be maximized or cost to be minimized. For example, consider the recovery of heat from a hot waste stream. A heat exchanger could be installed to recover the waste heat. The heat recovery is illustrated in Figure 3.1a as a plot of temperature versus enthalpy. There is heat available in the hot stream to be recovered to preheat the cold stream. However, how much heat should be recovered? Expressions can be written for the recovered heat as:

$$Q_{REC} = m_H C_{P,H}(T_{H,in} - T_{H,out}) \tag{3.1}$$

$$Q_{REC} = m_C C_{P,C}(T_{C,out} - T_{C,in}) \tag{3.2}$$

where Q_{REC} = recovered heat
m_H, m_C = mass flowrates of the hot and cold streams
$C_{P,H}$, $C_{P,C}$ = specific heat capacity of the hot and cold streams
$T_{H,in}$, $T_{H,out}$ = hot stream inlet and outlet temperatures
$T_{C,in}$, $T_{C,out}$ = cold stream inlet and outlet temperatures

Chemical Process Design and Integration, Second Edition. Robin Smith.

Companion Website: www.wiley.com/go/smith/chemical2e

The effect of the heat recovery is to decrease the energy requirements. Hence, the energy cost of the process:

$$Energy\ cost = (Q_H - Q_{REC})C_{Energy} \tag{3.3}$$

where Q_H = process hot utility requirement prior to heat recovery from the waste stream
C_{Energy} = unit cost of energy

There is no change in cost associated with cooling as the hot stream is a waste stream being sent to the environment. An expression can also be written for the heat transfer area of the recovery exchanger (see Chapter 12):

$$A = \frac{Q_{REC}}{U\,\Delta T_{LM}} \tag{3.4}$$

where A = heat transfer area
U = overall heat transfer coefficient
ΔT_{LM} = logarithmic mean temperature difference
$= \dfrac{(T_{H,in} - T_{C,out}) - (T_{H,out} - T_{C,in})}{\ln\left[\dfrac{T_{H,in} - T_{C,out}}{T_{H,out} - T_{C,in}}\right]}$

In turn, the area of the exchanger can be used to estimate the annualized capital cost:

$$Annualized\ capital\ cost = (a + bA^c)AF \tag{3.5}$$

where a, b, c = cost coefficients
AF = annualization factor (see Chapter 2)

Suppose that the mass flowrates, heat capacities and inlet temperatures of both streams are fixed and the current hot utility requirement, unit cost of energy, overall heat transfer coefficient, cost coefficients and annualization factor are known. The five Equations 3.1 to 3.5, together with the specifications for the 13 variables m_H, m_C, $C_{P,H}$, $C_{P,C}$, $T_{H,in}$, $T_{C,in}$, U, a, b, c, AF, Q_H and C_{Energy}, constitute 18 *equality constraints*. In addition to these 13 variables, there are a further six unknown variables Q_{REC}, $T_{H,out}$, $T_{C,out}$, *energy cost*, *annualized capital cost* and A. Thus, there are 18 equality constraints and 19 variables, and the problem cannot be

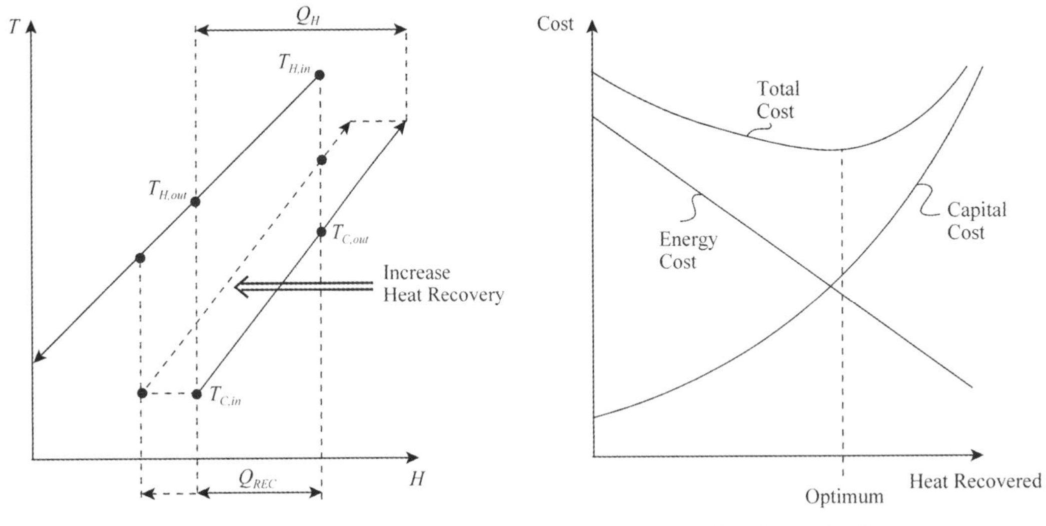

(a) Conditions in heat recovery exchanger. (b) Cost trade-offs.

Figure 3.1

Recovery of heat from a waste steam involves a trade-off between reduced energy cost and increased capital cost of heat exchanger.

solved. For the system of equations (equality constraints) to be solved, the number of variables must be equal to the number of equations (equality constraints). It is underspecified. Another specification (equality constraint) is required to solve the problem; there is one *degree of freedom*. This degree of freedom can be optimized. In this case, it is the sum of the annualized energy and capital costs (i.e. the total cost), as shown in Figure 3.1b.

If the mass flowrate of the cold stream through the exchanger had not been fixed, there would have been one fewer equality constraint, and this would have provided an additional degree of freedom and the optimization would have been a two-dimensional optimization. Each degree of freedom provides an opportunity for optimization.

Figure 3.1b illustrates how the investment in the heat exchanger is optimized. As the amount of recovered heat increases, the cost of energy for the system decreases. On the other hand, the size and capital cost of the heat exchange equipment increase. The increase in size of heat exchanger results from the greater amount of heat to be transferred. Also, because the conditions of the waste stream are fixed, as more heat is recovered from the waste stream, the temperature differences in the heat exchanger become lower, causing a sharp rise in the capital cost. In theory, if the recovery was taken to its limit of zero temperature difference, an infinitely large heat exchanger with infinitely large capital cost would be needed. The costs of energy and capital cost can be expressed on an annual basis, as explained in Chapter 2, and combined as shown in Figure 3.1b to obtain the total cost. The total cost shows a minimum, indicating the optimum size of the heat exchanger.

Given a mathematical model for the objective function in Figure 3.1b, finding the minimum point should be straightforward and various strategies could be adopted for this purpose. The objective function is continuous. Also, there is only one extreme point. Functions with only one extreme point (maximum or minimum) are termed *unimodal*. By contrast, consider the objective functions in Figure 3.2. Figure 3.2a shows an objective function to be minimized that is discontinuous. If the search for the minimum is started at point x_1, it could be easily concluded that the optimum point is at x_2, whereas the true optimum is at x_3. Discontinuities can present problems to optimization algorithms searching for the optimum. Figure 3.2b shows an objective function that has a number of points where the gradient is zero. These points where the gradient is zero are known as *stationary points*. Functions that exhibit a number of stationary points are known as *multimodal*. If the function in Figure 3.2b is to be minimized, it has a *local optimum* at x_2. If the search is started at x_1, it could be concluded that the global optimum is at x_2. However, the *global optimum* is at x_3. The gradient is zero at x_4, which is a *saddle point*. By considering the gradient only, this could be confused with a maximum or minimum. Thus, there is potentially another local optimum at x_4. Like discontinuities, multimodality presents problems to optimization algorithms. It is clear that it is not sufficient to find a point of zero gradient in the objective function in order to ensure that the optimum has been reached. Reaching a point of zero gradient is a *necessary condition* for optimality but not a *sufficient condition*.

To establish whether an optimal point is a local or a global optimum, the concepts of *convexity* and *concavity* must be introduced. Figure 3.3a shows a function to be minimized. In Figure 3.3a, if a straight line is drawn between any two points on the function, then the function is *convex* if all the values of this line lie above the curve. Similarly, if an objective function is to be maximized, as shown in Figure 3.3b, and a straight line drawn between any two points on the function, then if all values on this

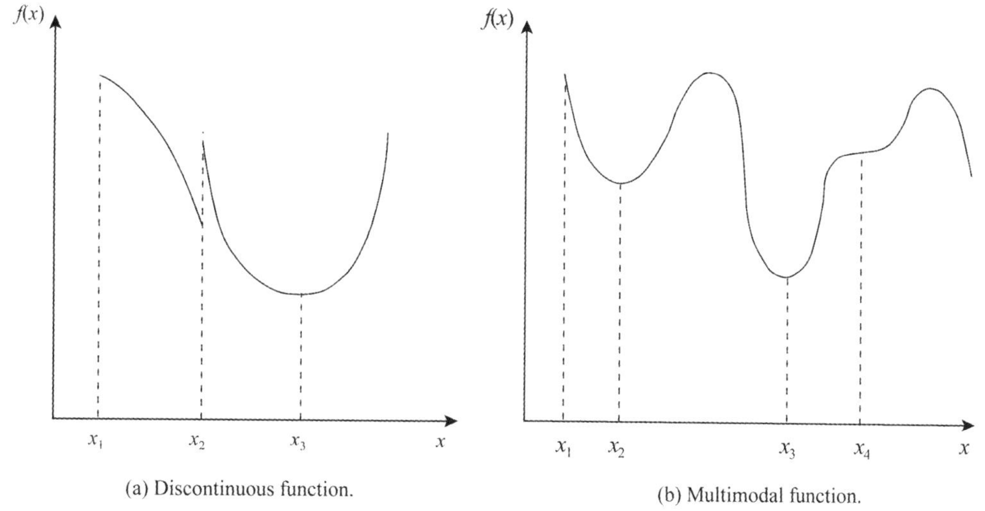

Figure 3.2
Objective functions can exhibit complex behavior.

line lie below the curve, the function is *concave*. A convex or concave function provides a single optimum. Thus, if a minimum is found for a function that is to be minimized and is known to be convex, then it is the global optimum. Similarly, if a maximum is found for a function that is to be maximized and known to be concave, then it is the global optimum. On the other hand, a nonconvex or nonconcave function may have multiple local optima.

Searching for the nonlinear optimum in Figures 3.1 to 3.3 constitutes a one-dimensional search. If the optimization involves two variables, say x_1 and x_2 corresponding to the function $f(x_1, x_2)$, it can be represented as a *contour plot*, as shown in Figure 3.4. Figure 3.4 shows a function $f(x_1, x_2)$ to be minimized. The contours represent lines of uniform values of the objective function. The objective function in Figure 3.4 is multimodal, involving a local optimum and a global optimum. The concepts of convexity and concavity can be extended to problems with more than one dimension (Edgar, Himmelblau and Lasdon, 2001). The objective function in Figure 3.4 is nonconvex. A straight line cannot be drawn between any two points on the surface represented by the contours to ensure that all points on the straight line are below the surface. These concepts can be expressed in a formal way mathematically, but is outside the scope of this text (Floudas, 1995; Biegler, Grossmann and Westerberg, 1997; Edgar, Himmelblau and Lasdon, 2001).

The optimization might well involve more than two variables in a *multivariable optimization*, but in this case it is difficult to visualize the problem.

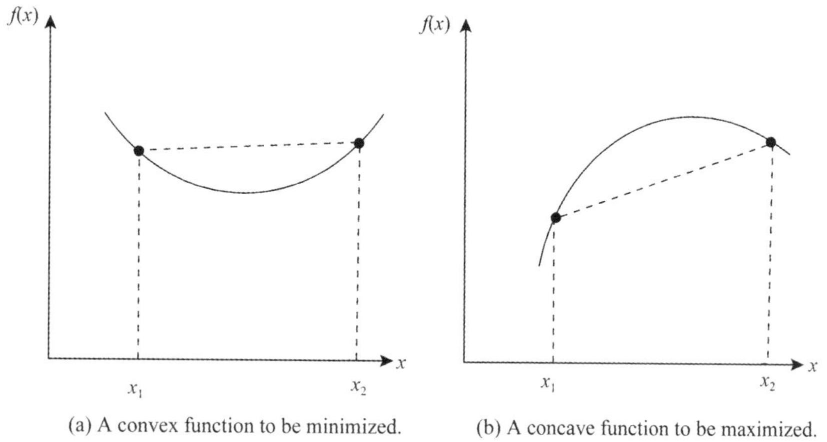

Figure 3.3
Convex and concave functions.

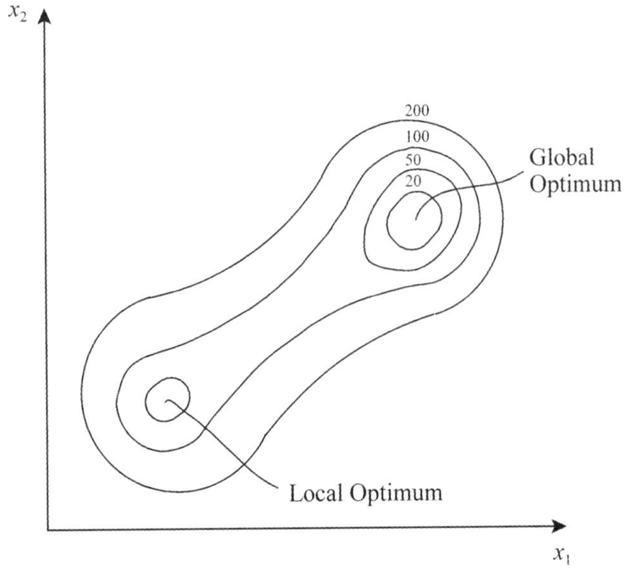

Figure 3.4

A contour plot of a multimodal function to be minimized.

3.2 Single-Variable Optimization

Searching for the optimum (minimum) for the objective function in Figure 3.1b involves a one-dimensional search across a single variable. In the case of Figure 3.1b, a search is made for the amount of heat to be recovered. An example of a method for a single-variable search is *region elimination*. The function is assumed to be unimodal. Figure 3.5 illustrates the approach. In Figure 3.5 the search region is for convenience set between 0 and 1. This is not a restriction on the approach, but for simplicity of explanation. The approach in Figure 3.5 has been to divide the solution space into four equal intervals and to evaluate the function at both the boundaries and the internal points, leading in Figure 3.5a to the minimum amongst these five evaluations at x_1. It cannot yet be determined exactly where the minimum point is. However, if the function is assumed to be unimodal, the minimum point must lie somewhere between 0 and x_2. The region between x_2 and 1 can therefore be eliminated. Figure 3.5b shows a different example with a different outcome from the five function evaluations in which the minimum point is at x_2. In this case, because the function is assumed to be unimodal, the minimum point must lie between x_1 and x_3, with the regions between 0 and x_1 and between x_3 and 1 eliminated. A third example with a different outcome is shown in Figure 3.5c, in which the minimum is located at x_3. In this case, the region between 0 and x_2 can be eliminated. In each of the cases in Figure 3.5, the search region has been halved by the evaluation of the function at five points.

The simple strategy adopted for region elimination in Figure 3.5 can in some cases eliminate more than half of the search region. Figure 3.6a shows a different example in which the minimum of the five points is on the lower boundary. In this case, the minimum must lie somewhere between 0 and x_1 and the region between x_1 and 1 can be eliminated. Figure 3.6b shows an example where the minimum of the five points is on the upper boundary. In this case, the minimum must lie somewhere between x_3 and 1 and the region between 0 and x_3 can be eliminated. In Figure 3.6c the example shows two of the function evaluations being equal and minimum, in this case at x_1 and x_2. This might be an unusual circumstance, but if it happens the minimum can be located between x_1 and x_2. Thus, in the cases shown in Figure 3.6, three-quarters of the search regions have been eliminated by the evaluation of the function at five points.

The strategy shown in Figures 3.5 and 3.6 has allowed part of the search region to be eliminated and the location of the optimum point narrowed down. The location of the optimal point can be narrowed down further by repeating the strategy in the remaining region that has not been eliminated. The location of the optimum is identified once the region has been narrowed down to within the desired tolerance.

For Figures 3.5 and 3.6, intervals were chosen to be equally spaced. In most cases, this allowed the search region to be halved. In special circumstances a greater part of the search region could be eliminated. However, a symmetrical location of the search points leads to a more efficient search method known as *Golden Section*.

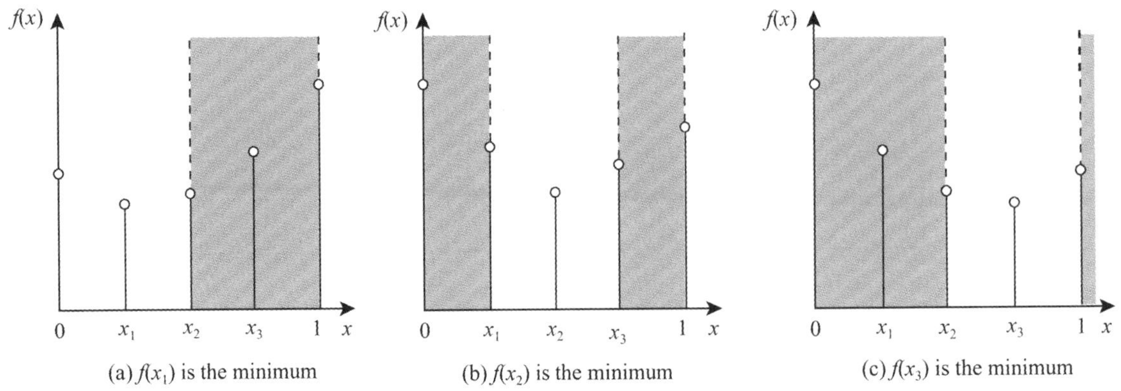

Figure 3.5

Region elimination for the optimization of a single variable.

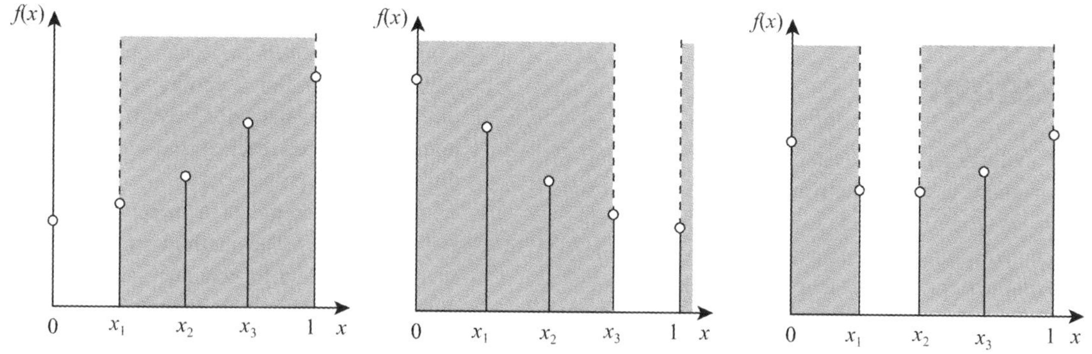

(a) $f(x) = 0$ is the minimum (b) $f(x) = 1$ is the minimum (c) $f(x_1)$ and $f(x_2)$ are both minimum

Figure 3.6

Region elimination can in some cases eliminate much greater ranges.

Consider Figure 3.7a. Again the search region is for convenience set between 0 and 1. In Figure 3.7a two search points have been located in such a way that the ratio of the whole region $[l + (1 - l)]$ to the larger subregion l is the same as the ratio of the ratio of the larger subregion l to the smaller region $(1 - l)$. Thus:

$$\frac{l + (1 - l)}{l} = \frac{l}{1 - l} \tag{3.6}$$

Since:

$$l + (1 - l) = 1 \tag{3.7}$$

substituting and rearranging gives:

$$l^2 + l - 1 = 0 \tag{3.8}$$

This quadratic equation can be solved to give $l = 0.618$. In Figure 3.7b, the function has been evaluated at these two search points. If the function is assumed to be unimodal, then in this case the region between x_2 and 1 can be eliminated. The remaining subregion of length l now has a search point located interior to it such that the symmetry of the search pattern is maintained. Hence in Figure 3.7c the new search point is located at $x_3 = 0.382 \times 0.618 = 0.236$. Evaluation of the function at this search point allows the region between x_1 and x_2 to be further eliminated, assuming a unimodal function. This process can then be repeated in the remaining region in order to identify more precisely the location of the optimum. Each new search point reduces the region of the search space by 0.618. For most problems this is more efficient than the five-point method illustrated in Figure 3.5.

A number of other methods for single-variable optimization is also possible (Edgar, Himmelblau and Lasdon, 2001).

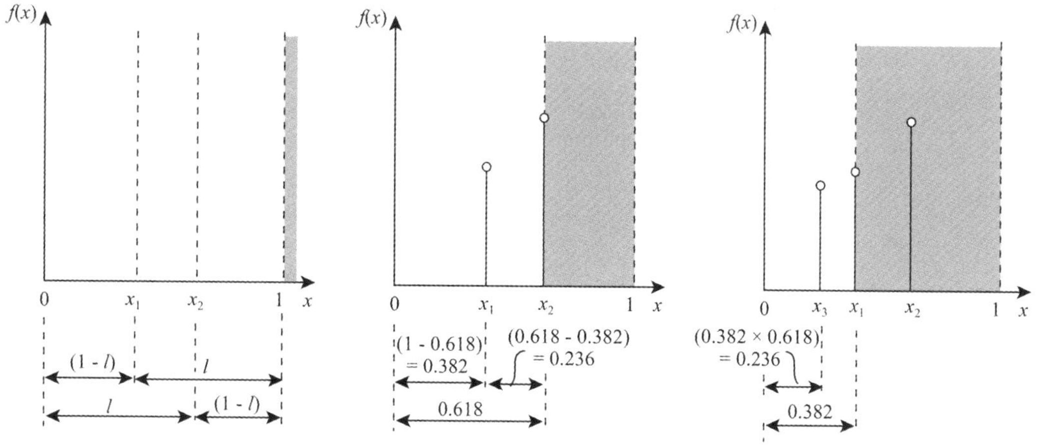

(a) Setting the two initial points symetrically. (b) Eliminating the first sub-region (c) Location of the third inner point to maintain symmetry.

Figure 3.7

Golden section region elimination.

3.3 Multivariable Optimization

The problem of multivariable optimization is illustrated in Figure 3.4. Search methods used for multivariable optimization can be classified as *deterministic* and *stochastic*. Deterministic methods follow a predetermined search pattern and do not involve any guessed or random steps. On the other hand, stochastic search methods use random choice to guide the search. Stochastic search methods generate a randomized path to the solution on the basis of probabilities.

1) *Deterministic methods*. Deterministic methods can be further classified into *direct* and *indirect* search methods. Direct search methods do not require derivatives (gradients) of the function. Indirect methods use derivatives, even though the derivatives might be obtained numerically rather than analytically.

   a) **Direct search methods.** An example of a direct search method is a *univariate search*, as illustrated in Figure 3.8. All of the variables except one are fixed and the remaining variable is optimized. Once a minimum or maximum point has been reached, this variable is fixed and another variable optimized, with the remaining variables being fixed. This is repeated until there is no further improvement in the objective function. Figure 3.8 illustrates a two-dimensional search in which x_1 is first fixed and x_2 optimized. Then x_2 is fixed and x_1 optimized, and so on until no further improvement in the objective function is obtained. In Figure 3.8, the univariate search is able to locate the global optimum. It is easy to see that if the starting point for the search in Figure 3.8 had been at a lower value of x_1, then the search would have located the local optimum, rather than the global optimum. For searching multivariable optimization problems, often the only way to ensure that the global optimum has been reached is to start the optimization from different initial points.

   Another example of a direct search is a *sequential simplex search*. The method uses a regular geometric shape (a *simplex*) to generate search directions. In two dimensions, the simplest shape is an equilateral triangle. In three dimensions, it is a regular tetrahedron. The objective function is evaluated at the vertices of the simplex, as illustrated in Figure 3.9. The objective function must first be evaluated at the Vertices *A*, *B* and *C*. The general direction of search is projected away from the worst vertex (in this case Vertex *A*) through the centroid of the remaining vertices (*B* and *C*) (Figure 3.9a). A new simplex is formed by replacing the worst vertex by a new point that is the mirror image of the simplex (Vertex *D*), as shown in Figure 3.9a. Then Vertex *D* replaces Vertex *A*, as Vertex *A* is an inferior point. The simplex vertices for the next step are *B*, *C* and *D*. This process is repeated for successive moves in a zigzag fashion, as shown in Figure 3.9b. The direction of search can change, as illustrated in Figure 3.9c. When the simplex is close to the optimum, there may be some repetition of simplexes, with the search going around in circles. If this is the case, then the size of the simplex should be reduced.

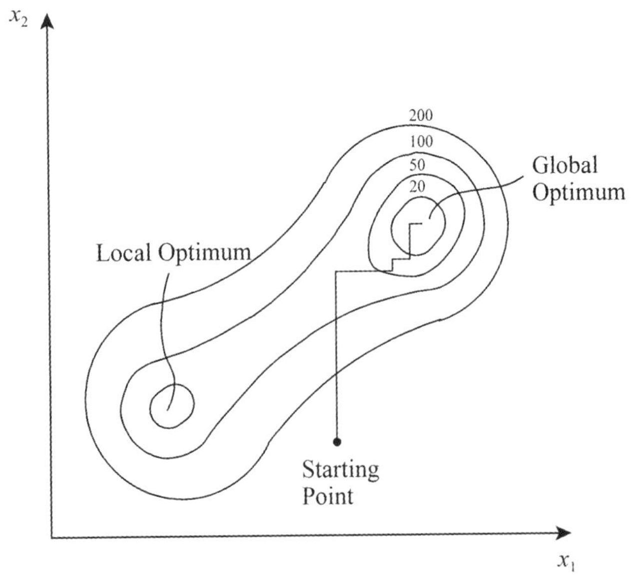

Figure 3.8
A univariate search.

   b) **Indirect search methods.** Indirect search methods use derivatives (gradients) of the objective function. The derivatives may be obtained analytically or numerically. Many methods are available for indirect search. An example is the method of steepest descent in a minimization problem. The direction of steepest descent is the search direction that gives the maximum rate of change for the objective function from the current point. The method is illustrated in Figure 3.10. One problem with this search method is that the appropriate step size is not known and this is under circumstances when the gradient might change significantly during the search. Another problem is that the search can slow down significantly as it reaches the optimum point. If a search is made for the maximum in an objective function, then the analogous search is one of steepest ascent.

   One fundamental practical difficulty with both the direct and indirect search methods is that, depending on the shape of the solution space, the search can locate local optima, rather than the global optimum. Often, the only way to ensure that the global optimum has been reached is to repeat the optimization starting from different initial points and repeat the process.

2) *Stochastic search methods*. In all of the optimization methods discussed so far, the algorithm searches the objective function seeking to improve the objective function at each step, using information such as gradients. Unfortunately, as already noted, this process can mean that the search is attracted towards a local

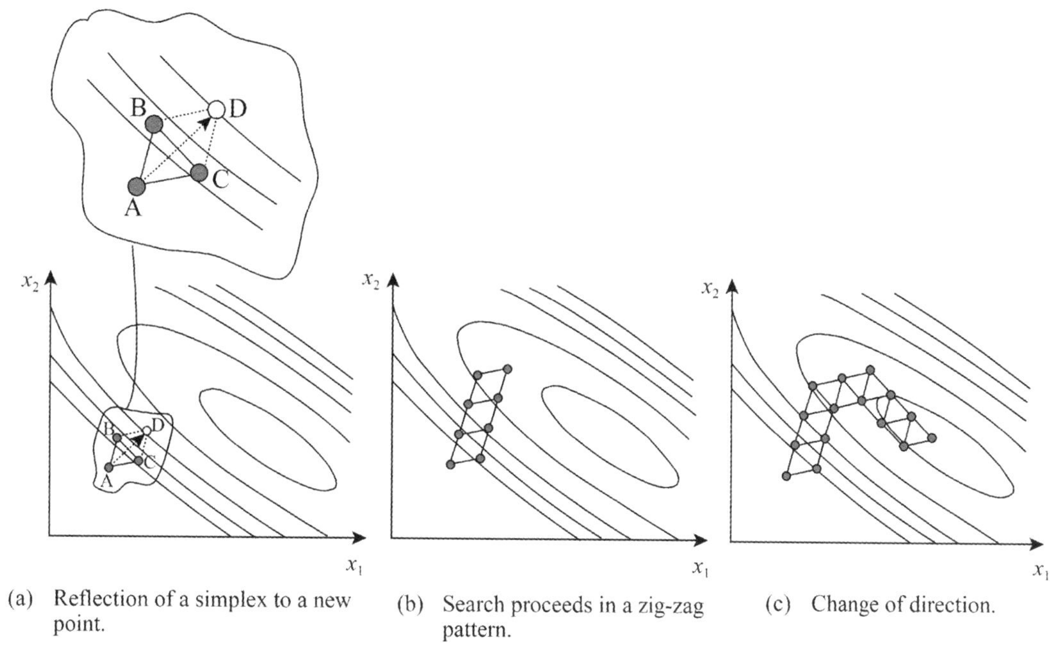

Figure 3.9

The simplex search.

optimum. On the other hand, stochastic search methods use random choice to guide the search and can allow deterioration of the objective function during the search. It is important to recognize that a randomized search does not mean a directionless search. Stochastic search methods generate a randomized path to the solution on the basis of probabilities. Improvement in the objective function becomes the ultimate rather than the immediate goal, and some deterioration in the objective function is tolerated, especially during the early stages of the search. In searching for a minimum in the objective function, rather than the search always attempting to go downhill, stochastic search methods allow the search to also sometimes go uphill. Similarly, if the objective function is to be maximized, stochastic search methods allow the search to sometimes go downhill. This helps to reduce the problem of being trapped in a local optimum.

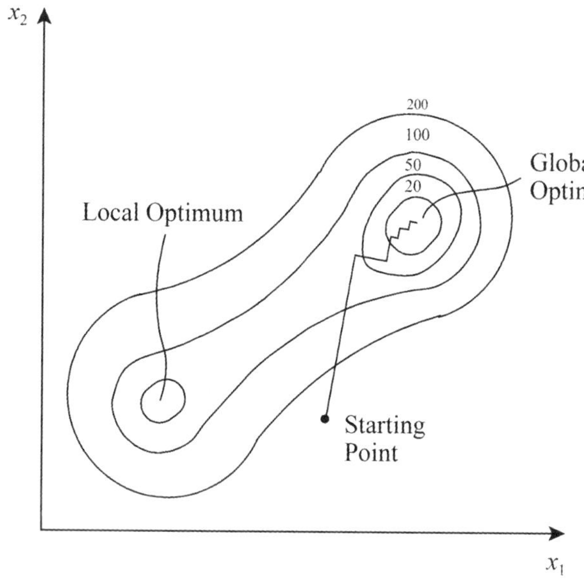

Figure 3.10

Method of steepest descent.

Stochastic search methods do not need auxiliary information, such as derivatives, in order to progress. They only require an objective function for the search. This means that stochastic search methods can handle problems in which the calculation of the derivatives would be complex and cause deterministic methods to fail.

Two of the most popular stochastic search methods are *simulated annealing* and *genetic algorithms*.

a) Simulated annealing. Simulated annealing emulates the physical process of annealing of metals (Metropolis *et al.*, 1953; Kirkpatrick, Gelatt and Vecchi, 1983). In the physical process, at high temperatures, the molecules of the liquid metal move freely with respect to one another. If the liquid is cooled slowly, thermal mobility is lost. The atoms are able to line themselves up and form perfect crystals. This crystal state is one of minimum energy for the system. If the liquid metal is cooled quickly, it does not reach this state but rather ends up in a polycrystalline or amorphous state having higher energy. Therefore the essence of the process is slow cooling, allowing ample time for redistribution of the atoms as they lose mobility to reach a state of minimum energy.

With this physical process in mind, an algorithm can be suggested in which a system *moves* from one point to another. A *move* might be a change in the temperature, pressure, flowrate, etc., in the problem. Rules must be defined to create the moves, for example the size of step change in temperature. The moves are then selected randomly. If the objective function from iteration i to $(i+1)$ changes from E_i to E_{i+1} and the objective function is to be minimized, then a move in which $(E_{i+1} - E_i)$ is negative (i.e. the objective function improves) is accepted. If $(E_{i+1} - E_i)$ is positive, the objective function deteriorates, but this does not mean that the move will be rejected. The probability of accepting a change in which the objective function deteriorates can be assumed to follow a relationship similar to the Boltzmann probability distribution, maintaining the analogy with physical annealing (Metropolis *et al.*, 1953):

$$P = \exp\left[-(E_{i+1} - E_i)/T\right] \tag{3.9}$$

where P is the probability, E is the objective function (the analogy of energy) and T is a control parameter (the analogy of annealing temperature). To determine whether a move is accepted or not according to Equation 3.9, a random number generator creates random numbers between zero and unity. If Equation 3.9 predicts a probability greater than the random number generator, then the move is accepted. If it is less, then the move is rejected and another move is attempted instead. It is evident from Equation 3.9 that when T is a large value, virtually all modifications made to the system are accepted. When T is close to zero, virtually all modifications that yield $(E_{i+1} - E_i) > 0$ are rejected. Thus, Equation 3.9 dictates whether a move is accepted or rejected. In this way, the method will always accept a downhill step when minimizing an objective function, while sometimes it will take an uphill step.

The algorithm starts with a high value of T, allowing a high probability of moves to be accepted that cause deterioration in the objective function. The value of T is gradually decreased as the search progresses. At any value of T a series of random moves is made that constitutes a *Markov chain*. In a Markov chain, the algorithm moves from one state to another in a chain-like manner in a random process in which the next state depends only on the current state and not on the past. It is necessary to specify the length of the Markov chain. A short Markov chain length reduces the likelihood of achieving equilibrium at each value of T. A long Markov chain would make the computation excessively expensive. The appropriate Markov chain length depends on the type of problem to be solved.

A *cooling schedule* needs to be proposed. Aarts and Van Laarhoven (1985) suggested a cooling schedule expressed in the form:

$$T_{k+1} = T_k\left(1 + \frac{\ln(1+\theta)T_k}{3\sigma(T_k)}\right) \tag{3.10}$$

where T_k = annealing temperature setting k for a Markov chain

σ = standard deviation of the objective functions generated at various Markov moves at annealing temperature T_k

θ = cooling parameter that controls how fast the annealing temperature is decreased

The bigger the value of θ, the faster the cooling process, but the greater the likelihood of being trapped into a local optimum. The performance of the algorithm is greatly affected by the initial value of the annealing temperature T_0. Too high a temperature will unnecessarily increase the time required for the algorithm to converge. On the other hand, too low a temperature will limit the number and magnitude of the uphill moves accepted, thus losing the ability of the algorithm to escape from local optima. The optimal initial annealing temperature depends on the nature of the problem and the scale of the objective function.

Thus, the way the algorithm works is to set an initial value for the annealing temperature. At this setting, a series of random moves are made. Equation 3.9 dictates whether an individual move is accepted or rejected. The annealing temperature is lowered and a new series of random moves is made, and so on. As the annealing temperature is lowered, the probability of accepting deterioration in the objective function, as dictated by Equation 3.9, decreases. In this way, the acceptability for the search to move uphill in a minimization or downhill during maximization is gradually withdrawn.

Whilst simulated annealing can be extremely powerful in solving difficult optimization problems with many local optima, it has a number of disadvantages. Initial and final values of the annealing temperature, an annealing schedule and the number of random moves for each Markov chain must be specified, which depend on the class of problem.

b) Genetic algorithms. Genetic algorithms draw their inspiration from biological evolution (Goldberg, 1989). Unlike all of the optimization methods discussed so far, which move from one point to another, a genetic algorithm moves from one set of points (termed a *population*) to another set of points. Populations of strings called *chromosomes* are created to represent an underlying set of parameters (e.g. temperatures, pressures or concentrations). A simple genetic algorithm exploits three basic operators: reproduction (selection), crossover and mutation.

Reproduction or *selection* is a process in which individual members of a population are copied according to the value of the objective function in order to generate new population sets. The operator is inspired by "natural selection" and the "survival of the fittest". The easiest way to understand selection is to make an analogy with a roulette wheel. In a roulette wheel, the probability of selection is proportional to the area of the slots in the wheel. In a genetic algorithm, the probability of selection is proportional to the *fitness* (the objective function). Although the selection procedure is stochastic, fitter chromosomes (with better values of the objective function) are given a better chance of selection (survival). The selection operator can be implemented in a genetic algorithm in many ways (Goldberg, 1989).

Crossover involves the combination of genetic material from two successful *parents* to form two *offspring* (children). Crossover involves cutting two parent chromosomes at random points and combining them differently to form new offspring. The crossover point is generated randomly. Crossover spreads good properties amongst the population. The fraction of new population generated by crossover is generally large (as observed in nature) and is controlled stochastically. The crossover operator can be implemented in a genetic algorithm in many ways (Goldberg, 1989).

Mutation creates new chromosomes by randomly changing (mutating) parts of chromosomes, but (as with nature) with a low probability of occurring. A random change is made in one of the genes in order to preserve diversity. Mutation creates a new solution in the neighborhood of a point undergoing mutation. However, the probability of mutation is usually set to be low. If the probability of mutation is set too high, the search will turn into a primitive random search.

Thus, a genetic algorithm works by first generating an initial population randomly. The population is evaluated according to its fitness (value of the objective function). A selection operator then provides an intermediate population using stochastic selection but biased towards survival of the fittest. Crossover and mutation operators are then applied to the intermediate population to create a new generation of population. The new population is evaluated according to its fitness and the search is continued with further selection, crossover and mutation until the population meets the required convergence criterion (e.g. maximum number of generations or the difference between the average and maximum fitness value).

Consider the example of a function $f(x, y)$ for which x and y must be manipulated to minimize the function, subject to $0.5 \le x \le 7.5$ and $20.3 \le y \le 80.0$. Values of x and y can be normalized such that $0 \le x \le 1$ and $0 \le y \le 1$; for example, for, say, $(x, y) = (3.2, 52.8)$. Normalizing to four digits gives $x = (3.2 - 0.5)/(7.5 - 0.5) = 0.3857$ and $y = (52.8 - 20.3)/(80.0 - 20.3) = 0.5444$. The two normalized values can then be *encoded* as *genes* [3857] and [5444] into the chromosome as {38575444}. This chromosome encodes $(x, y) = (3.2, 52.8)$. The genetic algorithm starts by creating an initial population that is randomly generated, for example {27066372}, {83876194}, {48473693}, and so on. Although this initial population is normally generated randomly, chromosomes can be filtered out, representing undesirable pairs of values. On the other hand, values can be forced to be included in the chromosomes. This initial population is then ranked by fitness according to the function $f(x, y)$. Fit chromosomes, which in this case means low values of the objective function, might be selected for reproduction by copying to the next generation according to, for example, the roulette wheel approach. Fit parents are also subject to crossover that attempts to combine portions of good individuals to possibly create even better individuals. Crossover sites along the chromosomes can be chosen in different ways. For example, the two chromosomes {27|0663|72} and {48|4736|93} can be subject to crossover to give {27|4736|72} and {48|0663|93}. The offspring of these two parents are decoded, evaluated and possibly allowed into the next generation. Mutation alters one or more values in the chromosome from its initial state and helps to prevent the population from stagnating at a local optimum. For example, mutation of the chromosome {270|6|6372} might be to {270|9|6372}. After mutation, chromosomes are decoded, evaluated and possibly allowed into the next generation. The genetic algorithm searches over a specified number of generations to improve the population.

The major strengths of stochastic search methods are that they can tackle the most difficult optimization problems with a high probability of locating solutions in the region of the global optimal. Another major advantage is that, if the solution space is highly irregular, they can produce a range of solutions with close to optimal performance, rather than a single optimal point. This opens up a range of solutions to the designer, rather than having just one option. However, there are also significant disadvantages with stochastic search optimization methods. They can be very slow in converging. The various operations in the stochastic search methods require parameters to be set. The most appropriate values for these parameters vary between different classes of problem and usually need to be adapted to solve particular problems. This means that the methods need tailoring to suit different applications.

3.4 Constrained Optimization

Most optimization problems involve constraints. For example, it might be necessary for a maximum temperature or maximum flowrate not to be exceeded. Thus, the general form of an optimization problem involves three basic elements:

1) An objective function to be optimized (e.g. minimize total cost, maximize economic potential, etc.).
2) Equality constraints, which are equations describing the model of the process or equipment.
3) Inequality constraints, expressing minimum or maximum limits on various parameters.

These three elements of the general optimization problem can be expressed mathematically as:

$$\begin{aligned} &\text{minimize } f(x_1, x_2, \ldots, x_n) \\ &\text{subject to } h_i(x_1, x_2, \ldots, x_n) = 0 \; (i = 1, p) \\ &\qquad\qquad g_j(x_1, x_2, \ldots, x_n) \le 0 \; (j = 1, q) \end{aligned} \tag{3.11}$$

In this case, there are n design variables, with p equality constraints and q inequality constraints. The existence of such constraints can simplify the optimization problem by reducing the size of the problem to be searched or avoiding problematic regions of the objective function. In general, though, the existence of the constraints complicates the problem relative to the problem with no constraints.

Now consider the influence of the inequality constraints on the optimization problem. The effect of inequality constraints is to reduce the size of the solution space that must be searched. However, the way in which the constraints bound the feasible region is important. Figure 3.11 illustrates the concept of convex and nonconvex regions. Figure 3.11a shows a convex region. In a convex region, a straight line can be drawn between any two points A and B located within the feasible region and all points on this straight line will also be located within the feasible region. By contrast, Figure 3.11b shows a nonconvex region. This time, when a straight line is drawn between two points A and B located within the region, some of the points on the straight line can fall outside the feasible region. Figure 3.11c shows a region that is constrained by a set of linear inequality constraints. The region shown in Figure 3.10c is convex, but it is worth noting that a set of linear inequality constraints will always provide a convex region (Edgar, Himmelblau and Lasdon, 2001). These concepts can be represented mathematically for the general problem (Floudas, 1995; Biegler, Grossmann and Westerberg, 1997; Edgar, Himmelblau and Lasdon, 2001).

Now superimpose the constraints on to the objective function. Figure 3.12a shows a contour diagram that has a set of inequality constraints imposed. The feasible region is convex and an appropriate search algorithm should be able to locate the unconstrained optimum. The unconstrained optimum lies inside the feasible region in Figure 3.12a. At the optimum none of the constraints are *active*. By contrast, consider Figure 3.12b. In this case, the region is also convex, but the unconstrained optimum lies outside the feasible region. This time, the optimum point is on the edge of the feasible region and one of the constraints is active. When the inequality constraint is satisfied, as in Figure 3.12b, it

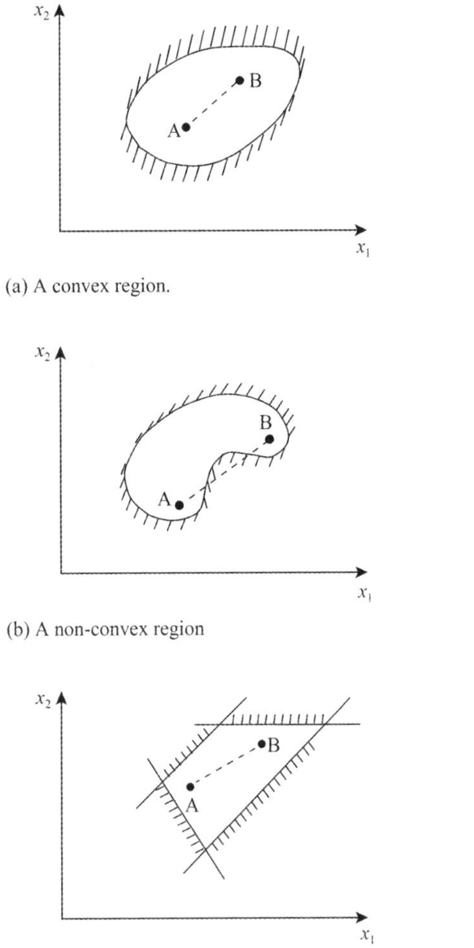

(a) A convex region.

(b) A non-convex region

(c) A set of linear constraints always provides a convex region.

Figure 3.11

Convex and nonconvex regions.

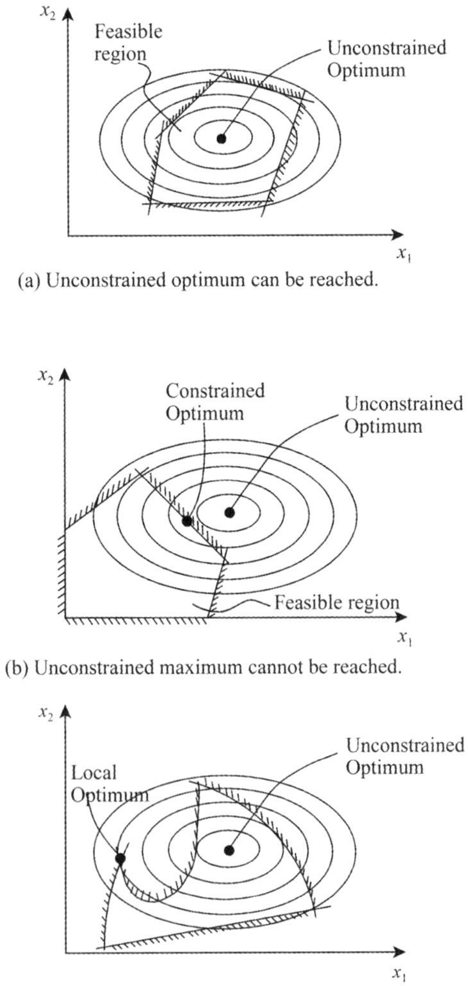

(a) Unconstrained optimum can be reached.

(b) Unconstrained maximum cannot be reached.

(c) A non-convex region might prevent the global optimum from being reached.

Figure 3.12

Effects of constraint and optimization.

becomes an equality constraint. Figure 3.12c illustrates the potential problem of having a nonconvex region. If the search is initiated in the right-hand part of the diagram at high values of x_1, then it is likely that the search will find the global optimum. However, if the search is initiated to the left of the diagram at low values of x_1, it is likely that the search will locate the local optimum (Edgar, Himmelblau and Lasdon, 2001).

It should be noted that it is not sufficient to simply have a convex region in order to ensure that a search can locate the global optimum. The objective function must also be convex if it is to be minimized or concave if it is to be maximized.

The stochastic search optimization methods described previously are readily adapted to the inclusion of constraints. For example, in simulated annealing, if a move suggested at random takes the solution outside the feasible region, then the algorithm can be constrained to prevent this by simply setting the probability of that move to zero. Other methods in stochastic search optimization introduce penalties into the objective function if the constraints are violated.

3.5 Linear Programming

An important class of the constrained optimization problems is one in which the objective function, equality constraints and inequality constraints are all linear. A linear function is one in which the dependent variables appear only to the first power. For example, a linear function of two variables x_1 and x_2 would be of the general form:

$$f(x_1, x_2) = a_0 + a_1 x_1 + a_2 x_2 \tag{3.12}$$

where a_0, a_1 and a_2 are constants. Search methods for such problems are well developed in *linear programming* (LP). Solving such linear optimization problems is best explained through a simple example.

Example 3.1 A company manufactures two products (Product 1 and Product 2) in a batch plant involving two steps (Step I and Step II). The value of Product 1 is 3 $\$\cdot kg^{-1}$ and that of Product 2 is 2 $\$\cdot kg^{-1}$. Each batch has the same capacity of 1000 kg per batch but batch cycle times differ between products. These are given in Table 3.1.

Step I has a maximum operating time of 5000 $h\cdot y^{-1}$ and Step II 6000 $h\cdot y^{-1}$. Determine the operation of the plant to obtain the maximum annual revenue.

Table 3.1

Times for different steps in the batch process.

	Step I (h)	Step II (h)
Product 1	25	10
Product 2	10	20

Solution For Step I, the maximum operating time dictates that:

$$25n_1 + 10n_2 \le 5000$$

where n_1 and n_2 are the number of batches per year manufacturing Products 1 and 2 on Step I. For Step II, the corresponding equation is:

$$10n_1 + 20n_2 \le 6000$$

The feasible solution space can be represented graphically by plotting the above inequality constraints as equality constraints:

$$25n_1 + 10n_2 = 5000$$
$$10n_1 + 20n_2 = 6000$$

This is shown in Figure 3.13. The feasible solution space in Figure 3.13 is given by *ABCD*.

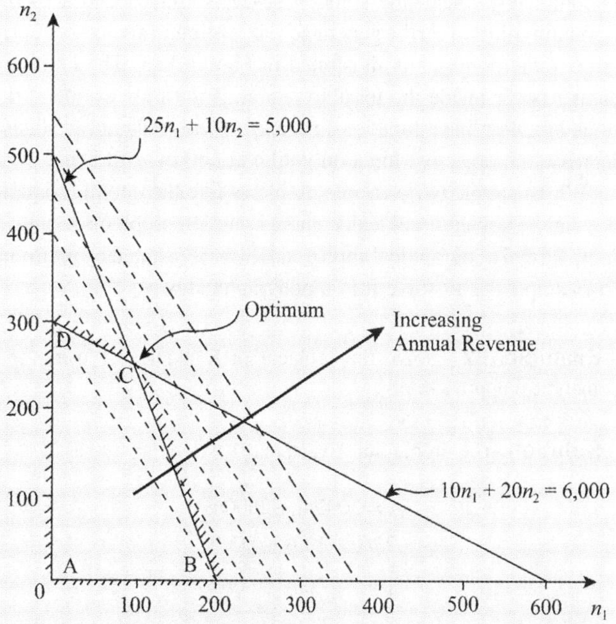

Figure 3.13

Graphical representation of the linear optimization problem from Example 3.1.

The total annual revenue A is given by:

$$A = 3000n_1 + 2000n_2$$

On a plot of n_1 versus n_2 as shown in Figure 3.13, lines of constant annual revenue will follow a straight line given by:

$$n_2 = -\frac{3}{2}n_1 + \frac{A}{2000}$$

Lines of constant annual revenue are shown as dotted lines in Figure 3.13, with revenue increasing with increasing distance from the origin. It is clear from Figure 3.13 that the optimum point corresponds with the extreme point at the intersection of the two equality constraints at Point C.

At the intersection of the two constraints:

$$n_1 = 100$$
$$n_2 = 250$$

As the problem involves discrete batches, it is apparently fortunate that the answer turns out to be two whole numbers. However, had the answer turned out not to be a whole number, then the solution would still have been valid because, even though part batches might not be able to be processed, the remaining part of a batch can be processed the following year.

Thus the maximum annual revenue is given by:

$$\begin{aligned} A &= 3000 \times 100 + 2000 \times 250 \\ &= \$800,000\,\$ \cdot \mathrm{y}^{-1} \end{aligned}$$

Whilst Example 3.1 is an extremely simple example, it illustrates a number of important points. If the optimization problem is completely linear, the solution space is convex and a global optimum solution can be generated. The optimum always occurs at an extreme point, as is illustrated in Figure 3.13. The optimum cannot occur inside the feasible region; it must always be at the boundary. For linear functions, running up the gradient can always increase the objective function until a boundary wall is hit.

Whilst simple two-variable problems like the one in Example 3.1 can be solved graphically, more complex problems require a more formal nongraphical approach. This is illustrated by returning to Example 3.1 to solve it in a nongraphical way.

Example 3.2 Solve the problem in Example 3.1 using an analytical approach.

Solution The problem in Example 3.1 was expressed as:

$$A = 3000n_1 + 2000n_2$$
$$25n_1 + 10n_2 \leq 5000$$
$$10n_1 + 20n_2 \leq 6000$$

To solve these equations algebraically, the inequality signs must first be removed by introducing *slack variables* S_1 and S_2 such that:

$$25n_1 + 10n_2 + S_1 = 5000$$
$$10n_1 + 20n_2 + S_2 = 6000$$

In other words, these equations show that if the production of both products does not absorb the full capacities of both steps, then the slack capacities of these two processes can be represented by the variables S_1 and S_2. Since slack capacity means that a certain amount of process capacity remains unused, it follows that the economic value of slack capacity is zero. Realizing that negative production rates and negative slack variables are infeasible, the problem can be formulated as:

$$3000n_1 + 2000n_2 + 0S_1 + 0S_2 = A \qquad (3.13)$$

$$25n_1 + 10n_2 + 1S_1 + 0S_2 = 5000 \qquad (3.14)$$

$$10n_1 + 20n_2 + 0S_1 + 1S_2 = 6000 \qquad (3.15)$$

where $n_1, n_2, S_1, S_2 \geq 0$

Equations 3.14 and 3.15 involve four variables and can therefore not be solved simultaneously. At this stage, the solution can lie anywhere within the feasible area marked *ABCD* in Figure 3.13. However, providing the values of these variables is not restricted to integer values; two of the four variables will assume zero values at the optimum. In this example, n_1, n_2, S_1 and S_2 are treated as real and not integer variables.

The problem is started with an initial feasible solution that is then improved by a stepwise procedure. The search will be started at the worst possible solution when n_1 and n_2 are both zero. From Equations 3.14 and 3.15:

$$S_1 = 5000 - 25n_1 - 10n_2 \qquad (3.16)$$
$$S_2 = 6000 - 10n_1 - 20n_2 \qquad (3.17)$$

When n_l and n_2 are zero:

$$S_1 = 5000$$
$$S_2 = 6000$$

Substituting in Equation 3.13:

$$\begin{aligned} A &= 3000 \times 0 + 2000 \times 0 + 0 \times 5000 + 0 \times 6000 \\ &= 0 \end{aligned} \qquad (3.18)$$

This is Point *A* in Figure 3.13. To improve this initial solution, the value of n_1 and/or the value of n_2 must be increased, because these are the only variables that possess positive coefficients to increase the annual revenue in Equation 3.13. However, which variable, n_1 or n_2, should be increased first? The obvious strategy is to increase the variable that makes the greatest increase in the annual revenue, which is n_1. According to Equation 3.17, n_1 can be increased by 6000/10 = 600 before S_2 becomes negative. If n_1 is assumed to be 600 in Equation 3.16, then S_1 would be negative. Since negative slack variables are infeasible, Equation 3.16 is the dominant constraint on n_1 and it follows that its maximum value is 5000/25 = 200. Rearranging Equation 3.16:

$$n_1 = 200 - 0.4n_2 - 0.04S_1 \qquad (3.19)$$

which would give a maximum when n_1 and S_1 are zero. Substituting the expression for n_1 in the objective function, Equation 3.13 gives:

$$A = 600,000 + 800n_2 - 120S_1 \qquad (3.20)$$

Since n_2 is initially zero, the greatest improvement in the objective function results from making S_1 zero. This is equivalent to making n_1 equal to 200 from Equation 3.19, given that n_2 is initially zero. For $n_1 = 200$ and $n_2 = 0$, the annual revenue $A = 600{,}000$. This corresponds with Point *B* in Figure 3.13. However, Equation 3.20 also shows that the profit can be improved further by increasing the value of n_2. Substituting n_1 from Equation 3.19 in Equation 3.17 gives:

$$n_2 = 250 + 0.025S_1 - 0.0625S_2 \qquad (3.21)$$

This means that n_2 takes a value of 250 if both S_1 and S_2 are zero. Substituting n_2 from Equation 3.21 in Equation 3.19 gives:

$$n_1 = 100 - 0.05S_1 + 0.025S_2 \qquad (3.22)$$

This means that n_1 takes a value of 100 if both S_1 and S_2 are zero. Finally, substituting the expression for n_1 and n_2 in the objective function, Equation 3.13 gives:

$$A = 800,000 - 100S_1 - 50S_2 \qquad (3.23)$$

Equations 3.21 to 3.23 show that the maximum annual revenue is \$800,000 y^{-1} when $n_1 = 100$ and $n_2 = 250$. This corresponds with Point C in Figure 3.13.

It is also interesting to note that Equation 3.23 provides some insight into the sensitivity of the solution. The annual revenue would decrease by \$100 for each hour of production lost through poor utilization of Step I. The corresponding effect for Step II would be a reduction of \$50 for each hour of production lost. These values are known as *shadow prices*. If S_1 and S_2 are set to their availabilities of 5000 and 6000 hours respectively, then the revenue from Equation 3.23 becomes zero.

While the method used for the solution of Example 3.1 is not suitable for automation, it gives some insights into the way linear programming problems can be automated. The solution is started by turning the inequality constraints into equality constraints by the use of slack variables. Then the equations are solved to obtain an initial feasible solution. This is improved in steps by searching the extreme points of the solution space. It is not necessary to explore all the extreme points in order to identify the optimum. The method usually used to automate the solution of such linear programming problems is the *simplex algorithm* (Biegler, Grossmann and Westerberg, 1997; Edgar, Himmelblau and Lasdon, 2001). Note, however, that the simplex algorithm for linear programming should not be confused with the simplex search described previously, which is quite different. Here the term simplex is used to describe the shape of the solution space, which is a convex polyhedron, or simplex.

If the linear programming problem is not formulated properly, it might not have a unique solution, or even any solution at all. Such linear programming problems are termed *degenerate* (Edgar, Himmelblau and Lasdon, 2001). Figure 3.14 illustrates some degenerate linear programming problems (Edgar, Himmelblau and Lasdon, 2001). In Figure 3.14a, the objective function contours are parallel with one of the boundary constraints. Here there is no unique solution that maximizes the objective function within the feasible region. Figure 3.14b shows a problem in which the feasible region is unbounded. Hence the objective function can increase without bound. A third example is shown in Figure 13.14c, in which there is no feasible region according to the specified constraints.

3

3.6 Nonlinear Programming

When the objective function, equality or inequality constraints of Equation 3.11 are nonlinear, the optimization becomes a *nonlinear programming* (NLP) problem. The worst case is when all three are nonlinear. Direct and indirect methods that can be used for nonlinear optimization have previously been discussed. Whilst it is possible to include some types of constraints, the methods discussed are not well suited to the inclusion of complex sets of constraints. The stochastic search methods discussed previously can readily handle constraints by restricting moves to infeasible solutions, for example in simulated annealing by setting their probability to 0. The other methods discussed are not well suited to the inclusion of complex sets of constraints. It has already been observed in Figure 3.12 that, unlike the linear optimization problem, for the nonlinear optimization problem the optimum may or may not lie on the edge of the feasible region and can, in principle, be anywhere within the feasible region.

One approach that has been adopted for solving the general nonlinear programming problem is *successive linear programming*. These methods linearize the problem and successively apply the linear programming techniques described in the previous section. The procedures involve initializing the problem and linearizing the objective function and all of the constraints about

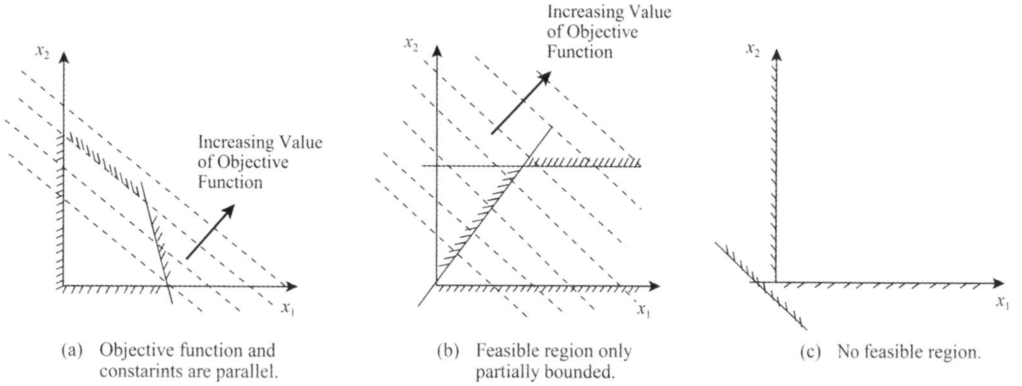

Figure 3.14

Degenerate linear programming problems.

the initial point, so as to fit the linear programming format. Linear programming is then applied to solve the problem. An improved solution is obtained and the procedure repeated. At each successive improved feasible solution, the objective function and constraints are linearized and the linear programming solution repeated, until the objective function does not show any significant improvement. If the solution to the linear programming problem moves to an infeasible point, then the nearest feasible point is located and the procedure applied at this new point.

Another method for solving nonlinear programming problems is based on *quadratic programming* (QP) (Edgar, Himmelblau and Lasdon, 2001). Quadratic programming is an optimization procedure that minimizes a quadratic objective function subject to linear inequality or equality (or both types of) constraints. For example, a quadratic function of two variables x_1 and x_2 would be of the general form:

$$f(x_1, x_2) = a_0 + a_1 x_1 + a_2 x_2 + a_{11} x_1^2 + a_{22} x_2^2 + a_{12} x_1 x_2 \tag{3.24}$$

where a_{ij} are constants. Quadratic programming problems are the simplest form of nonlinear programming with inequality constraints. The techniques used for the solution of quadratic programming problems have many similarities with those used for solving linear programming problems (Edgar, Himmelblau and Lasdon, 2001). Each inequality constraint must either be satisfied as an equality or it is not involved in the solution of the problem. The quadratic programming technique can thus be reduced to a vertex searching procedure, similar to linear programming (Edgar, Himmelblau and Lasdon, 2001). In order to solve the general nonlinear programming problem, quadratic programming can be applied successively, in a similar way to that for successive linear programming, in *successive* (or *sequential*) *quadratic programming* (SQP). In this case, the objective function is approximated locally as a quadratic function. For a function of two variables, the function would be approximated by Equation 3.24. By approximating the function as a quadratic and linearizing the constraints, this takes the form of a quadratic programming problem that is solved in each iteration (Edgar, Himmelblau and Lasdon, 2001). In general, successive quadratic programming tends to perform better than successive linear programming, because a quadratic rather than a linear approximation is used for the objective function.

It is important to note that neither successive linear nor successive quadratic programming are guaranteed to find the global optimum in a general nonlinear programming problem. The fact that the problem is being turned into a linear or quadratic problem, for which global optimality can be guaranteed, does not change the underlying problem that is being optimized. All of the problems associated with local optima are still a feature of the background problem. When using these methods for the general nonlinear programming problem, it is important to recognize this and to test the optimality of the solution by starting the optimization from different initial conditions.

Stochastic search optimization methods described previously, such as simulated annealing, can also be used to solve the general nonlinear programming problem. These have the advantage that the search is sometimes allowed to move uphill in a minimization problem, rather than always searching for a downhill move. Alternatively, in a maximization problem, the search is sometimes allowed to move downhill, rather than always searching for an uphill move. In this way, the technique is less vulnerable to the problems associated with local optima.

3.7 Structural Optimization

1) *Mixed integer linear programming.* In Chapter 1, different ways were discussed that can be used to develop the structure of a flowsheet. In the first way, an irreducible structure is built by successively adding new features if these can be justified technically and economically. The second way to develop the structure of a flowsheet is to first create a superstructure. This superstructure involves redundant features but includes the structural options that should be considered. This superstructure is then subjected to optimization. The optimization varies the settings of the process parameters (e.g. temperature, flowrate) and also optimizes the structural features. Thus, to adopt this approach, both structural and parameter optimization must be carried out. So far the discussion of optimization has been restricted to parameter optimization. Consider now how structural optimization can be carried out.

   The methods discussed for linear and nonlinear programming can be adapted to deal with structural optimization by introducing integer (binary) variables that identify whether a feature exists or not. If a feature exists, its binary variable takes the value 1. If the feature does not exist, then it is set to 0. Consider how different kinds of decisions can be formulated using binary variables (Biegler, Grossman and Westerberg, 1997).

   a) **Multiple choice constraints.** It might be required to select only one item from a number of options. This can be represented mathematically by a constraint:

$$\sum_{j=1}^{J} y_j = 1 \tag{3.25}$$

   where y_j is the binary variable to be set to 0 or 1 and the number of options is J. More generally, it might be required to select only m items from a number of options. This can be represented by:

$$\sum_{j=1}^{J} y_j = m \tag{3.26}$$

   Alternatively, it might be required to select at most m items from a number of options, in which case the constraint can be represented by:

$$\sum_{j=1}^{J} y_j \leq m \tag{3.27}$$

On the other hand, it might be required to select at least m items from a number of options. The constraint can be represented by:

$$\sum_{j=1}^{J} y_j \geq m \tag{3.28}$$

b) **Implication constraints.** Another type of logical constraint might be that if Item k is selected, Item j must be selected, but not vice versa, then this is represented by the constraint:

$$y_k - y_j \leq 0 \tag{3.29}$$

A binary variable can be used to set a continuous variable to 0. If a binary variable y is 0, the associated continuous variable x must also be 0 if a constraint is applied such that:

$$x - Uy \leq 0, x \geq 0 \tag{3.30}$$

where U is an upper limit to x.

c) **Either–or constraints.** Binary variables can also be applied to either–or constraints, known as *disjunctive constraints*. For example, either constraint $g_1(x) \leq 0$ or constraint $g_2(x) \leq 0$ must hold:

$$g_1(x) - M\,y \leq 0 \tag{3.31}$$

$$g_2(x) - M(1 - y) \leq 0 \tag{3.32}$$

where M is a large (arbitrary) value that represents an upper limit $g_1(x)$ and $g_2(x)$. If $y = 0$, then $g_1(x) \leq 0$ must be imposed from Equation 3.31. However, if $y = 0$, the left-hand side of Equation 3.32 becomes a large negative number whatever the value of $g_2(x)$ and, as a result, Equation 3.32 is always satisfied. If $y = 1$, then the left-hand side of Equation 3.31 is a large negative number whatever the value of $g_1(x)$ and Equation 3.31 is always satisfied. However, now $g_2(x) \leq 0$ must be imposed from Equation 3.32.

Some simple examples can be used to illustrate the application of these principles.

Example 3.3 A gaseous waste stream from a process contains valuable hydrogen that can be recovered by separating the hydrogen from the impurities using pressure swing adsorption (PSA), a membrane separator (MS) or a cryogenic condensation (CC). The pressure swing adsorption and membrane separator can in principle be used either individually or in combination. Write a set of integer equations that would allow one from the three options of pressure swing adsorption, membrane separator or cryogenic condensation to be chosen, but also allow the pressure swing adsorption and membrane separator to be chosen in combination.

Solution Let y_{PSA} represent the selection of pressure swing adsorption, y_{MS} the selection of the membrane separator and y_{CC} the selection of cryogenic condensation. First restrict the choice between pressure swing adsorption and cryogenic condensation:

$$y_{PSA} + y_{CC} < 1$$

Now restrict the choice between membrane separator and cryogenic condensation:

$$y_{MS} + y_{CC} < 1$$

These two equations restrict the choices, but still allow the pressure swing adsorption and membrane separator to be chosen together.

Example 3.4 The temperature difference in a heat exchanger between the inlet temperature of the hot stream $T_{H,\,in}$ and the outlet of the cold stream $T_{C,\,out}$ is to be restricted to be greater than a practical minimum value of ΔT_{min}, but only if the option of having the heat exchanger is chosen. Write a disjunctive constraint in the form of an integer equation to represent this constraint.

Solution The temperature approach constraint can be written as:

$$T_{H,in} - T_{C,out} \geq \Delta T_{min}$$

However, this should apply only if the heat exchanger is selected. Let y_{HX} represent the option of choosing the heat exchanger:

$$T_{H,in} - T_{C,out} + M(1 - y_{HX}) \geq \Delta T_{\min}$$

where M is an arbitrary large number. If $y_{HX} = 0$ (i.e. the heat exchanger is not chosen), then the left-hand side of this equation is bound to be greater than ΔT_{min} no matter what the values of $T_{H,\,in}$ and $T_{C,\,out}$ are. If $y_{HX} = 1$ (i.e. the heat exchanger is chosen), then the equation becomes $(T_{H,\,in} - T_{C,\,out}) \geq \Delta T_{min}$ and the constraint must apply.

When a linear programming problem is extended to include integer (binary) variables, it becomes a *mixed integer linear programming problem* (MILP). Correspondingly, when a nonlinear programming problem is extended to include integer (binary) variables, it becomes a *mixed integer nonlinear programming problem* (MINLP).

First consider the general strategy for solving an MILP problem. Initially, the binary variables can be treated as continuous variables, such that $0 \leq y_i \leq 1$. The problem can then be solved as an LP. The solution is known as a *relaxed solution*. The most likely outcome is that some of the binary variables will exhibit noninteger values at the optimum LP solution. Because the relaxed solution is less constrained than the true mixed integer solution in which all of the binary variables have integer values, it will in general give a better value for the objective function than the true mixed integer solution. In general, the noninteger values of the binary variables cannot simply be rounded to the nearest integer value, either because the rounding may lead to an infeasible solution (outside the feasible region) or because the rounding may render the solution nonoptimal (not at the edge of the feasible

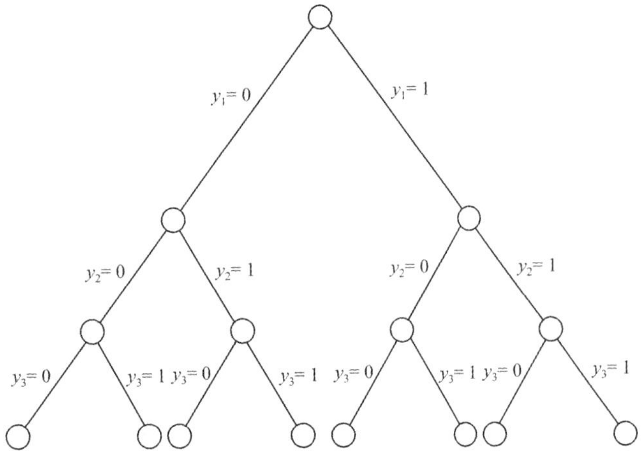

Figure 3.15

Setting the binary variables to zero or one creates a tree structure. (Reproduced from Floudas CA, 1995, Nonlinear and Mixed-Integer Optimization, by permission of Oxford University Press.)

region). However, this relaxed LP solution is useful in providing a *lower bound* to the true mixed integer solution to a minimization problem. For maximization problems, the relaxed LP solutions form the *upper bound* to the solution. The noninteger values can then be set to either 0 or 1 and the LP solution repeated. The setting of the binary variables to be either 0 or 1 creates a solution space in the form of a tree, as shown in Figure 3.15 (Floudas, 1995). As the solution is stepped through, the number of possibilities increases by virtue of the fact that each binary variable can take a value of 0 or 1 (Figure 3.15). At each point in the search, the best relaxed LP solution provides a *lower bound* to the optimum of a minimization problem. Correspondingly, the best true mixed integer solution provides an upper bound. For maximization problems the best relaxed LP solution forms an *upper bound* to the optimum and the best true mixed integer solution provides the *lower bound*. A popular method of solving MILP problems is to use a *branch and bound search* (Mehta and Kokossis, 1988). This will be illustrated by a simple example from Edgar, Himmelblau and Lasdon (2001).

Example 3.5 A problem involving three binary variables y_1, y_2 and y_3 has an objective function to be maximized (Edgar, Himmelblau and Lasdon, 2001):

$$\begin{aligned} \text{maximize}: \quad & f = 86y_1 + 4y_2 + 40y_3 \\ \text{subject to}: \quad & 774y_1 + 76y_2 + 42y_3 \leq 875 \\ & 67y_1 + 27y_2 + 53y_3 \leq 875 \\ & y_1, y_2, y_3 = 0, 1 \end{aligned} \tag{3.33}$$

Solution The solution strategy is illustrated in Figure 3.16a. First the LP problem is solved to obtain the relaxed solution, allowing y_1, y_2 and y_3 to vary continuously between 0 and 1. Both $y_1 = 1$ and $y_3 = 1$ at the optimum of the relaxed solution but the value of y_2 is 0.776, Node 1 in Figure 3.16a. The objective function for this relaxed solution at Node 1 is $f = 129.1$. From this point, y_2 can be set to be either 0 or 1. Various strategies can be adopted to decide which one to choose. A very simple strategy will be adopted here of picking the closest integer to the real number. Given that $y_2 = 0.776$ is closer to 1 than 0, set $y_2 = 1$ and solve the LP at Node 2 in Figure 3.16a. Now $y_2 = 1$ and $y_3 = 1$ at the optimum of the relaxed solution but the value of y_1 is 0.978, Node 2 in Figure 3.16a. Given that $y_1 = 0.978$ is closer to 1 than 0, set $y_1 = 1$ and solve the LP at Node 3. This time y_1 and y_2 are integers, but $y_3 = 0.595$ is a noninteger. Setting $y_3 = 1$ yields an infeasible solution at Node 4 in Figure 3.16a as it violates the first inequality constraint in Equation 3.33. Backtracking to Node 3 and setting $y_3 = 0$ yields the first feasible integer solution at Node 5 for which $y_1 = 1, y_2 = 1, y_3 = 0$ and $f = 90.0$. There is no point in searching further from Node 4 as it is an infeasible solution or from Node 5 as it is a valid integer solution. When the search is terminated at a node for either reason, it is deemed to be *fathomed*. The search now backtracks to Node 2 and sets $y_1 = 0$. This yields the second feasible integer solution at Node 6 for which $y_1 = 0$, $y_2 = 1$, $y_3 = 1$ and $f = 44.0$. Finally, backtrack to Node 1 and set $y_2 = 0$. This yields the third feasible integer solution at Node 7 for which $y_1 = 1$, $y_2 = 0$, $y_3 = 1$ and $f = 126.0$. Since the objective function is being maximized, Node 7 is the optimum for the problem.

Searching the tree in the way done in Figure 3.16a is known as a *depth first* or *backtracking* approach. At each node, the branch was followed that appeared to be more promising to solve. Rather than using a depth first approach, a *breadth first* or *jumptracking* approach can be used, as illustrated in Figure 3.16b. Again start at Node 1 and solve the relaxed problem in Figure 3.16b. This gives an upper bound for the maximization problem of $f = 129.1$. However, this time the search goes across the tree with the initial setting of $y_2 = 0$. This yields a valid integer solution at Node 2 with $y_1 = 1$, $y_2 = 0, y_3 = 1$ and $f = 126.0$. Node 2 now forms a lower bound and is fathomed because it is an integer solution. In this approach, the search now backtracks to Node 1 whether Node 2 is fathomed or not. From Node 1 now set $y_2 = 1$. The solution at Node 3 in Figure 3.16b gives $y_1 = 0.978$, $y_2 = 1$, $y_3 = 1$ and $f = 128.1$, which is the new upper bound. Setting $y_1 = 0$ and branching to Node 4 gives the second valid integer solution. Now backtrack to Node 3 and set $y_1 = 1$. The solution at Node 5 has $y_1 = 1, y_2 = 1, y_3 = 0.595$ and $f = 113.8$. At this point, Node 5 is fathomed, even though it is neither infeasible nor a valid integer solution. The upper bound of this branch at Node 5 has a value of the objective function lower than that of the integer solution at Node 2. Setting the values to be integers from Node 5 can only result in an inferior solution. In this way, the search is bounded.

In this case, the breadth first search yields the optimum with a fewer number of nodes to be searched. Different search strategies than the ones used here can readily be used (Taha, 1975). It is likely that different problems would be suited to different search strategies.

Thus, the solution of the MILP problem is started by solving the first relaxed LP problem. If integer values are obtained for the binary variables, the problem has been solved. However, if integer values are not obtained, the use of bounds is examined to avoid parts of the tree that are known to be suboptimal. The node with the best noninteger solution provides a lower bound for minimization problems and the node with the best feasible

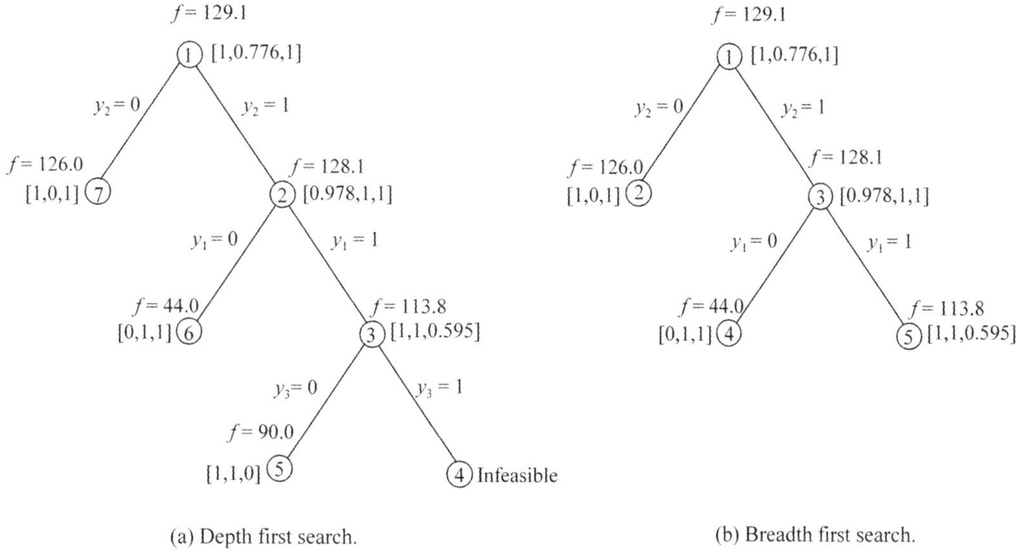

Figure 3.16
Branch and bound search.

mixed integer solution provides an upper bound. In the case of maximization problems, the node with the best noninteger solution provides an upper bound and the node with the best feasible mixed integer solution provides a lower bound. Nodes with noninteger solutions are fathomed when the value of the objective function is inferior to the best integer solution (the lower bound). The tree can be searched by following a depth first approach or a breadth first approach, or a combination of the two. Given a more complex problem than Example 3.5, the search could, for example, set the values of the noninteger variables to be 0 and 1 in turn and carry out an evaluation of the objective function (rather than an optimization). This would then indicate the best direction in which to go for the next optimization. Many strategies are possible (Taha, 1975).

The series of LP solutions required for MILP problems can be solved efficiently by using one LP to initialize the next. An important point to note is that, in principle, a global optimum solution can be guaranteed in the same way as with LP problems.

2) *Mixed Integer Nonlinear Programming.* The general strategy for solving mixed integer nonlinear programming problems is very similar to that for linear problems (Floudas, 1995). The major difference is that each node requires the solution of a nonlinear program, rather than the solution of a linear program. Unfortunately, searching the tree with a succession of nonlinear optimizations can be extremely expensive in terms of the computation time required, as information cannot be readily carried from one NLP to the next, as can be done for LP. Another major problem is that, because a series of nonlinear optimizations is being carried out, there is no guarantee that the optimum will even be close to the global optimum, unless the NLP problem being solved at each node is convex. Of course, different initial points can be tried to overcome this problem, but there can still be no guarantee of global optimality for the general problem.

Another way to deal with such nonlinear problems is to first approximate the solution to be linear and apply MILP, and then apply NLP to the problem. The method then iterates between MILP and NLP (Biegler, Grossmann and Westerberg, 1997).

In some cases, the nonlinearity in a problem can be isolated in a small number of the functions. If this is the case, then one simple way to solve the problem is to linearize the nonlinear function by a series of straight-line segments. Integer logic can then be used to ensure that only one of the straight-line segments is chosen at a time and MILP used to carry out the optimization. For some forms of nonlinear mathematical expressions, deterministic optimization methods can be tailored to find the global optimum through the application of mathematical transformations and bounding techniques (Floudas, 2000).

3) *Stochastic Search Optimization.* Stochastic search optimization can be extremely effective for structural optimization if the optimization is nonlinear in character. Consider again the approach to process design introduced in Chapter 1 in which an initial design is developed and then evolved through structural and parameter optimization. This evolutionary approach is suited to stochastic search optimization. This does not require either a superstructure or a search through a tree as branch and bound methods require. For example, when using simulated annealing, at each setting of the annealing temperature, a series of random moves is performed. These moves can be either step changes to continuous variables or can be changes in structure (either the addition or removal of a structural feature). Thus, the

approach does not require a superstructure to be created, as the approach can add features as well as remove features. However, structural moves must be defined and these must ensure that those moves can somehow create all the structural options that might be candidates for the global optimum structure. In theory, the initialization of the structure does not matter and any initial feasible structure will do to start the optimization, as long as the stochastic search optimization parameters have been set to appropriate values. In practice, it tends to be better for most problems to create an initial design with at least some redundant features, rather like a superstructure, but not necessarily including all possible structural options.

Because stochastic search optimization allows some deterioration in the objective function, it is not as prone to being trapped by a local optimum as MINLP. However, when optimizing a problem involving both continuous variables (e.g. temperature and pressure) and structural changes, stochastic search optimization algorithms take finite steps for the continuous variables and do not necessarily find the exact value for the optimum setting. Because of this, a deterministic method (e.g. SQP) can be applied after stochastic search optimization to fine-tune the answer. This uses the stochastic search method to provide a good initialization for the deterministic method.

Also, it is possible to combine stochastic and deterministic methods as *hybrid* methods. For example, a stochastic search method can be used to control the structural changes and a deterministic method to control the changes in the continuous variables. This can be useful if the problem involves a large number of integer variables, because, for such problems, the tree required for branch and bound methods explodes in size.

3.8 Solution of Equations Using Optimization

It is sometimes convenient to use optimization to solve equations, or sets of simultaneous equations. This arises from the availability of general-purpose optimization software, such as that available in spreadsheets. *Root finding* or *equation solving* is a special case for optimization, where the objective is to reach a value of 0. For example, suppose it is necessary to solve a function $f(x)$ for the value x that satisfies $f(x) = a$, where a is a constant. As illustrated in Figure 3.17a, this can be solved for:

$$f(x) - a = 0 \qquad (3.34)$$

The objective of the optimization would be for the equality constraint given by Equation 3.34 to be satisfied as nearly as possible. There are a number of ways in which this objective can be made specific for optimization. Three possibilities are (Williams, 1997):

(i) minimize $|f(x) - a|$ (3.35)

(ii) minimize $[f(x) - a]^2$ (3.36)

(iii) minimize$(S_1 + S_2)$ (3.37)
subject to

$$f(x) - a + S_1 - S_2 = 0 \qquad (3.38)$$

$$S_1, S_2 \geq 0 \qquad (3.39)$$

where S_1 and S_2 are slack variables

Which of these objectives would be the best to use depends on the nature of the problem, the optimization algorithm being used and the initial point for the solution. For example, minimizing Objective (i) in Equation 3.35 can present problems to optimization methods as a result of the gradient being discontinuous, as illustrated in Figure 3.17b. However, the problem can be transformed such that the objective function for the optimization has no discontinuities in the gradient. One possible transformation is Objective (ii) in Equation 3.36. As illustrated in Figure 3.17c,

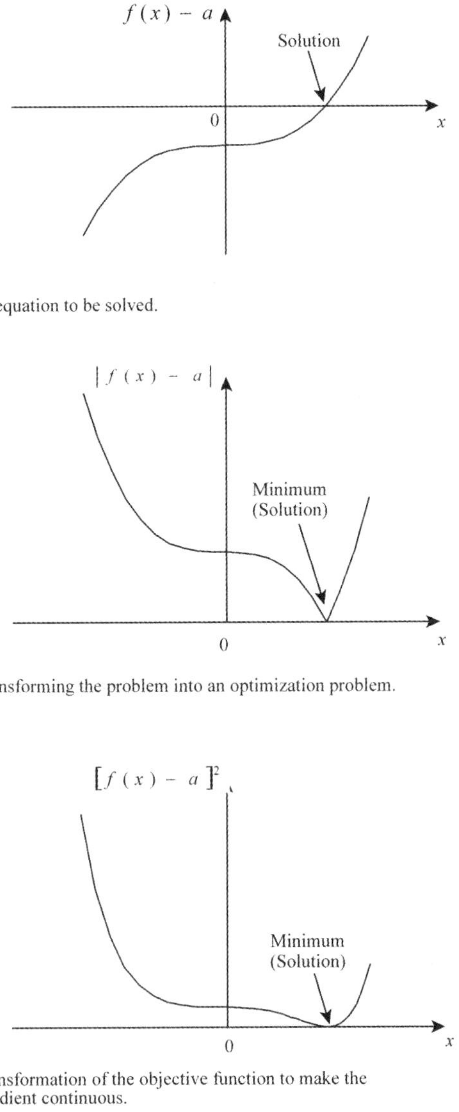

(a) An equation to be solved.

(b) Transforming the problem into an optimization problem.

(c) Transformation of the objective function to make the gradient continuous.

Figure 3.17
Solving equations using optimization.

this now has a continuous gradient. However, a fundamental disadvantage in using the transformation in Equation 3.36 is that if the equations to be solved are linear, then the objective function is transformed from linear to nonlinear. Also, if the equations to be solved are nonlinear, then such transformations will increase the nonlinearity. On the other hand, Objective (iii) in Equation 3.37 avoids both an increase in the nonlinearity and discontinuities in the gradient, but at the expense of introducing slack variables. Note that two slack variables are needed. Slack variable S_1 in Equation 3.38 for $S_1 \geq 0$ ensures that:

$$f(x) - a \leq 0 \tag{3.40}$$

whereas slack variable S_2 in Equation 3.38 for $S_2 \geq 0$ ensures that:

$$f(x) - a \geq 0 \tag{3.41}$$

Equations 3.40 and 3.41 are only satisfied simultaneously by Equation 3.34. Thus, x, S_1 and S_2 can be varied simultaneously to solve Equations 3.37 to 3.39 without increasing the nonlinearity.

The approach can be extended to solve sets of simultaneous equations. For example, suppose a solution is required for x_1 and x_2 such that:

$$f_1(x_1, x_2) = a_1 \quad \text{and} \quad f_2(x_1, x_2) = a_2 \tag{3.42}$$

Three possible ways to formulate the objective for optimization are:

$$\text{(i) minimize } \left\{ |f_1(x_1, x_2) - a_1| + |f_2(x_1, x_2) - a_2| \right\} \tag{3.43}$$

$$\text{(ii) minimize } \left\{ [f_1(x_1, x_2) - a_1]^2 + [f_2(x_1, x_2) - a_2]^2 \right\} \tag{3.44}$$

$$\text{(iii) minimize } (S_{11} + S_{12} + S_{21} + S_{22}) \quad \text{subject to} \tag{3.45}$$

$$f_1(x_1, x_2) - a_1 + S_{11} - S_{12} = 0 \tag{3.46}$$

$$f_2(x_1, x_2) - a_2 + S_{21} - S_{22} = 0 \tag{3.47}$$

$$S_{11}, S_{12}, S_{21}, S_{22} \geq 0 \tag{3.48}$$

where S_{11}, S_{12}, S_{21} and S_{22} are slack variables

As will be seen later, these techniques will prove to be useful when solving design problems in general-purpose software, such as spreadsheets. Many of the numerical problems associated with optimization can be avoided by appropriate formulation of the model. Further details of model building can be found elsewhere (Williams, 1997).

3.9 The Search for Global Optimality

From the discussion in this chapter, it is clear that the difficulties associated with optimizing nonlinear problems are far greater than those for optimizing linear problems. For linear problems, finding the global optimum can, in principle, be guaranteed.

Unfortunately, when optimization is applied to most relatively large design problems, the problem usually involves solving nonlinear optimization. In such situations, standard deterministic optimization methods will find only the first local optimum encountered. Starting from different initializations can allow the optimization to explore different routes through the solution space and might help identify alternative solutions. However, there is no guarantee of finding the global optimum. For some forms of nonlinear mathematical expressions, deterministic optimization methods can be tailored to find the global optimum through the application of mathematical transformations and bounding techniques (Floudas, 2000).

Alternatively, stochastic search optimization (e.g. simulation annealing or genetic algorithms) can be used. These have the advantage of, in principle, being able to locate the global optimum for the most general nonlinear optimization problems. They do not require good initialization and do not require gradients to be defined. However, they involve parameters that are system-dependent and might need to be adjusted for problems that are different in character. Another disadvantage is that they can be extremely slow in solving large complex optimization problems. The relative advantages of deterministic and stochastic search methods can be combined using hybrid methods by using stochastic search methods to provide a good initial point for a deterministic method. As mentioned previously, stochastic and deterministic methods can also be combined to solve structural optimization problems.

In Chapter 1, an objective function for a nonlinear optimization was likened to the terrain in a range of mountains. If the objective function is to be maximized, each peak in the mountain range represents a local optimum in the objective function. The highest peak represents the global optimum. Optimization requires searching around the mountains in a thick fog to find the highest peak, without the benefit of a map and only a compass to tell direction and an altimeter to show height. On reaching the top of any peak, there is no way of knowing whether it is the highest peak because of the fog.

When solving such nonlinear optimization problems, it is not desirable to terminate the search at a peak that is grossly inferior to the highest peak. The solution can be checked by repeating the search but starting from a different initial point.

However, the shape of the optimum for most optimization problems bears a greater resemblance to Table Mountain in South Africa rather than to Mount Everest. In other words, for most optimization problems, the region around the optimum is fairly flat. Although on one hand a grossly inferior solution should be avoided, on the other hand the designer should not be preoccupied with improving the solution by tiny amounts in an attempt to locate exactly the global optimum. There will be uncertainty in the design data, especially economic data. Also, there are many issues to be considered other than simply maximizing economic potential or minimizing cost. There could be many reasons why the solution at the exact location of the global optimum might not be preferred, while a slightly suboptimal solution might be preferred for other reasons, such as safety, ease of control, and so on. Different solutions in the region of the optimum should be examined, rather than considerable effort being expended on finding the solution at the exact location of the global optimum and considering only that solution. In this respect, stochastic search optimization has advantages, as it can provide a range of solutions in the region of the optimum.

3.10 Optimization – Summary

Most design problems will require optimization to be carried out at some stage. The quality of the design is characterized by an objective function to be maximized (e.g. if economic potential is being maximized) or minimized (e.g. if cost is being minimized). The shape of the objective function is critical in determining the optimization strategy. If the objective function is convex in a minimization problem or concave in a maximization problem, then there is a single optimum point. If this is not the case, there can be local optima as well as the global optimum.

Various search strategies can be used to locate the optimum. Indirect search strategies do not use information on gradients, whereas direct search strategies require this information. These methods always seek to improve the objective function in each step in a search. On the other hand, stochastic search methods, such as simulated annealing and genetic algorithms, allow some deterioration in the objective function, especially during the early stages of the search, in order to reduce the danger of being attracted to a local optimum rather than the global optimum. However, stochastic search optimization can be very slow in converging and usually needs to be adapted to solve particular problems.

The addition of inequality constraints complicates the optimization. These inequality constraints can form convex or nonconvex regions. If the region is nonconvex, this means that the search can be attracted to a local optimum, even if the objective function is convex in the case of a minimization problem or concave in the case of a maximization problem. In the case where a set of inequality constraints is linear, the resulting region is always convex.

The general case of optimization in which the objective function, the equality and inequality constraints are all linear can be solved as a linear programming problem. This can be solved efficiently with, in principle, a guarantee of global optimality. However, the corresponding nonlinear programming problem cannot, in general, be solved efficiently and with a guarantee of global optimality. Such problems are solved by successive linear or successive quadratic programming. Stochastic search optimization methods can be very effective in solving nonlinear optimization, because they are less prone to be stuck in a local optimum than deterministic methods.

One of the approaches that can be used in design is to carry out structural and parameter optimization of a superstructure. The structural optimization required can be carried out using mixed integer linear programming in the case of a linear problem or mixed integer nonlinear programming in the case of a nonlinear problem. Stochastic search optimization can also be very effective for structural optimization problems.

3.11 Exercises

1. The cost of closed atmospheric cylindrical storage vessels can be considered to be proportional to the mass of steel required. Derive a simple expression for the dimensions of such a storage tank to give minimum capital cost. Assume the top and bottom are both flat. What are the dimensions for the minimum capital cost if the tank has an open top?

2. The overhead of vapor of a distillation column is to be condensed in a heat exchanger using cooling water. There is a trade-off involving the flowrate of cooling water and the size of the condenser. As the flowrate of cooling water increases, its cost increases. However, as the flowrate increases, the return temperature of the cooling water to the cooling tower decreases. This decreases the temperature differences in the condenser and increases its heat transfer area, and hence its capital cost. The condenser has a duty of 4.1 MW and the vapor condenses at a constant temperature of 80 °C. Cooling water is available at 20 °C with a cost of \$ $0.02\,t^{-1}$. The overall heat transfer coefficient can be assumed to be $500\,W{\cdot}m^{-1}{\cdot}K^{-1}$. The cost of the condenser can be assumed to be $\$2500\,m^{-2}$ with an installation factor of 3.5. Annual capital charges can be assumed to be 20% of the capital costs. The heat capacity of the cooling water can be assumed constant at $4.2\,kJ{\cdot}kg^{-1}{\cdot}K^{-1}$. The distillation column operates for $8000\,h{\cdot}y^{-1}$. Set up an equation for the heat transfer area of the condenser, and hence the annual capital cost of the condenser, in terms of the cooling water return temperature. Using this equation, carry out a trade-off between the cost of the cooling water and the cost of the condenser to determine approximately the optimum cooling water return temperature. The maximum return temperature should be 50 °C. The heat exchange area required by the condenser is given by:

$$A = \frac{Q}{U\,\Delta T_{LM}}$$

where A = heat transfer area (m^2)
Q = heat duty (W)
U = overall heat transfer coefficient ($W{\cdot}m^{-1}{\cdot}K^{-1}$)
ΔT_{LM} = logarithmic mean temperature difference

$$= \frac{(T_{COND} - T_{CW2}) - (T_{COND} - T_{CW1})}{\ln\left(\dfrac{T_{COND} - T_{CW2}}{T_{COND} - T_{CW1}}\right)}$$

T_{COND} = condenser temperature (°C)
T_{CW1} = inlet cooling water temperature (°C)
T_{CW2} = outlet cooling water temperature (°C)

3. A vapor stream leaving a styrene production process contains hydrogen, methane, ethylene, benzene, water, toluene, styrene and ethylbenzene and is to be burnt in a furnace. It is proposed to recover as much of the benzene, toluene, styrene and ethylbenzene as possible from the vapor using low-temperature condensation. The low-temperature condensation requires refrigeration. However, the optimum temperature

Table 3.2

Stream flowrates and component values.

Component	Flowrate ($kmol \cdot s^{-1}$)	Value ($\$ \cdot kmol^{-1}$)
Hydrogen	146.0	0
Methane	3.7	0
Ethylene	3.7	0
Benzene	0.67	21.4
Water	9.4	0
Toluene	0.16	12.2
Ethylbenzene	1.6	40.6
Styrene	2.4	60.1
Total	167.63	

for the condensation needs to be determined. This involves a trade-off in which the amount and value of material recovered increases as the temperature decreases, but the cost of the refrigeration increases as the temperature decreases. The material flows in the vapor leaving the flash drum are given in Table 3.2, together with their values.

The low-temperature condensation requires refrigeration, for which the cost is given by:

$$\text{Refrigeration cost} = 0.033 Q_{COND} \left(\frac{40 - T_{COND}}{T_{COND} + 268} \right)$$

where Q_{COND} = condenser duty (MW)
T_{COND} = condenser temperature (°C)

The fraction of benzene, toluene, styrene and ethylbenzene condensed can be determined from phase equilibrium calculations. The percent of the various components entering the condenser that leave with the vapor are given in Table 3.3 as a function of temperature. The total enthalpy of the flash drum vapor stream as a function of temperature is given in Table 3.4.

Table 3.4

Stream enthalpy data.

Temperature (°C)	Stream enthalpy ($MJ \cdot kmol^{-1}$)
40	0.45
30	−1.44
20	−2.69
10	−3.56
0	−4.21
−10	−4.72
−20	−5.16
−30	−5.56
−40	−5.93
−50	−6.29
−60	−6.63

Calculate the optimum condenser temperature. What practical difficulties do you foresee in using very low temperatures?

4. A tank containing 1500 m^3 of naphtha is to be blended with two other hydrocarbon streams to meet the specifications for gasoline. The final product must have a minimum research octane number (RON) of 95, a maximum Reid vapor pressure (RVP) of 0.6 bar, a maximum benzene content of 2% vol and maximum total aromatics of 25% vol. The properties and costs of the three streams are given in the Table 3.5.

Assuming that the properties of the mixture blend are in proportion to the volume of stream used, how much reformate and alkylate should be blended to minimize cost?

5. A petroleum refinery has two crude oil feeds available. The first crude (Crude 1) is high-quality feed and costs $30 per

Table 3.3

Condenser performance.

	Percent of component entering condenser that leaves with the vapor (%)										
	Condensation temperature (°C)										
	40	30	20	10	0	−10	−20	−30	−40	−50	−60
Benzene	100	93	84	72	58	42	27	15	8	3	1
Toluene	100	80	60	41	25	14	7	3	1	1	0
Ethylbenzene	100	59	33	18	9	4	2	1	0	0	0
Styrene	100	54	29	15	8	4	2	1	0	0	0

Table 3.5
Blending streams.

	RON	RVP (bar)	Benzene (% vol)	Total aromatics (% vol)	Cost $\$\cdot m^{-3}$
Naphtha	92	0.80	1.5	15	275
Reformate	98	0.15	15	50	270
Alkylate	97.5	0.30	0	0	350

Table 3.6
Refinery data.

	Yield (% volume)		Value of product	Maximum production
	Crude 1	Crude 2	$(\$\cdot bbl^{-1})$	$(bbl\cdot day^{-1})$
Gasoline	80	47	75	120,000
Jet fuel	4	8	55	8,000
Diesel	10	30	40	30,000
Fuel oil	6	15	30	–
Processing cost ($\$\ bbl^{-1}$)	1.5	3		

barrel (1 barrel = 42 US gallons). The second crude (Crude 2) is a low-quality feed and costs $20 per barrel. The crude oil is separated into gasoline, diesel, jet fuel and fuel oil. The percent yield of each of these products that can be obtained from Crude 1 and Crude 2 are listed in Table 3.6, together with maximum allowable production flowrates of the products in barrels per day and processing costs.

The economic potential can be taken to be the difference between the selling price of the products and the cost of the crude oil feedstocks. Determine the optimum feed flowrate of the two crude oils from a linear optimization solved graphically.

6. Add a constraint to the specifications for Exercise 5 above such that the production of fuel oil must be greater than $15{,}000\ bbl\cdot day^{-1}$. What happens to the problem? How would you describe the characteristics of the modified linear programming problem?

7. Devise a superstructure for a distillation design involving a single feed, two products, a reboiler and a condenser that will allow the number of plates in the column itself to be varied between 3 and 10 and at the same time vary the location of the feed tray.

8. A reaction is required to be carried out between a gas and a liquid. Two different types of reactor are to be considered: an agitated vessel (AV) and a packed column (PC). Devise a superstructure that will allow one of the two options to be chosen. Then describe this as integer constraints for the gas and liquid feeds and products.

References

Aarts EHL and Van Laarhoven PGM (1985) Statistical Cooling: A General Approach to Combinatorial Optimization Problems, *Philips J Res*, **40**: 193.

Biegler LT, Grossmann IE and Westerberg AW (1997) *Systematic Methods of Chemical Process Design*, Prentice Hall.

Edgar TF, Himmelblau DM and Lasdon LS (2001) *Optimization of Chemical Processes*, 2nd Edition, McGraw-Hill.

Floudas CA (1995) *Nonlinear and Mixed-Integer Optimization*, Oxford University Press.

Floudas CA (2000) *Deterministic Global Optimization: Theory, Methods and Applications*, Kluwer Academic Publishers.

Goldberg DE (1989) *Genetic Algorithms in Search Optimization and Machine Learning*, Addison-Wesley.

Kirkpatrick S, Gelatt CD and Vecchi MP (1983) Optimization by Simulated Annealing, *Science*, **220**: 671.

Mehta VL and Kokossis AC (1988) New Generation Tools for Multiphase Reaction Systems: A Validated Systematic Methodology for Novelty and Design Automation, *Comput Chem Eng*, **22S**: 5119.

Metropolis N, Rosenbluth AW, Rosenbluth MN, Teller AH and Teller E (1953) Equation of State Calculations by Fast Computing Machines, *J Chem Phys*, **21**: 1087.

Taha HA (1975) *Integer Programming Theory and Applications*, Academic Press.

Williams HP (1997) *Model Building in Mathematical Programming*, 3rd Edition, John Wiley & Sons.

Chapter 4

Chemical Reactors I – Reactor Performance

Since process design starts with the reactor, the first decisions are those that lead to the choice of reactor. These decisions are amongst the most important in the whole process design. Good reactor performance is of paramount importance in determining the economic viability of the overall design and fundamentally important to the environmental impact of the process. In addition to the desired products, reactors produce unwanted byproducts. These unwanted byproducts not only lead to a loss of revenue but can also create environmental problems. As will be discussed later, the best solution to environmental problems is not to employ elaborate treatment methods, but to not produce waste in the first place.

Once the product specifications have been fixed, some decisions need to be made regarding the *reaction path*. There are sometimes different paths to the same product. For example, suppose ethanol is to be manufactured. Ethylene could be used as a raw material and reacted with water to produce ethanol. An alternative would be to start with methanol as a raw material and react it with synthesis gas (a mixture of carbon monoxide and hydrogen) to produce the same product. These two paths employ chemical reactor technology. A third path could employ a *biochemical* reaction (or *fermentation*) that exploits the metabolic processes of *microorganisms* in a biochemical reactor. Ethanol could therefore also be manufactured by fermentation of a carbohydrate.

Reactors can be broadly classified as chemical or biochemical. Most reactors, whether chemical or biochemical, are catalyzed. The strategy will be to choose the catalyst, if one is to be used, and the ideal characteristics and operating conditions needed for the reaction system. The issues that must be addressed for reactor design include:

- Reactor type
- Catalyst
- Size (volume)
- Operating conditions (temperature and pressure)
- Phase
- Feed conditions (concentration, temperature and pressure).

Once basic decisions have been made regarding these issues, a practical reactor is selected, approaching as nearly as possible the ideal in order that the design can proceed. However, the reactor design cannot be fixed at this stage since, as will be seen later, it interacts strongly with the rest of the flowsheet. The focus here will be on the choice of reactor and not its detailed sizing. For further details of sizing, see, for example, Levenspiel (1999), Denbigh and Turner (1984) and Rase (1977).

4.1 Reaction Path

As already noted, given that the objective is to manufacture a certain product, there are often a number of alternative reaction paths to that product. Reaction paths that use the cheapest raw materials and produce the smallest quantities of byproducts are to be preferred. Reaction paths that produce significant quantities of unwanted byproducts should especially be avoided, since they can create significant environmental problems.

However, there are many other factors to be considered in the choice of reaction path. Some are commercial, such as uncertainties regarding future prices of raw materials and byproducts. Others are technical, such as safety and energy consumption.

The lack of suitable catalysts is the most common reason preventing the exploitation of novel reaction paths. At the first stage of design, it is impossible to look ahead and see all of the consequences of choosing one reaction path or another, but some things are clear even at this stage. Consider the following example.

Chemical Process Design and Integration, Second Edition. Robin Smith.

Companion Website: www.wiley.com/go/smith/chemical2e

Example 4.1 Given that the objective is to manufacture vinyl chloride, there are at least three reaction paths that can be readily exploited (Rudd, Powers and Siirola, 1973).

Path 1

$$\underset{\text{acetylene}}{C_2H_2} + \underset{\text{hydrogen chloride}}{HCl} \longrightarrow \underset{\text{vinyl chloride}}{C_2H_3Cl}$$

Path 2

$$\underset{\text{ethylene}}{C_2H_4} + \underset{\text{chlorine}}{Cl_2} \longrightarrow \underset{\text{dichloroethane}}{C_2H_4Cl_2}$$

$$\underset{\text{dichloroethane}}{C_2H_4Cl_2} \xrightarrow{\text{heat}} \underset{\text{vinyl chloride}}{C_2H_3Cl} + \underset{\text{hydrogen chloride}}{HCl}$$

Path 3

$$\underset{\text{ethylene}}{C_2H_4} + \underset{\text{oxygen}}{{}^1\!/_2O_2} + \underset{\text{hydrogen chloride}}{2HCl} \longrightarrow \underset{\text{dichloroethane}}{C_2H_4Cl_2} + \underset{\text{water}}{H_2O}$$

$$\underset{\text{dichloroethane}}{C_2H_4Cl_2} \xrightarrow{\text{heat}} \underset{\text{vinyl chloride}}{C_2H_3Cl} + \underset{\text{hydrogen chloride}}{HCl}$$

The market values and molar masses of the materials involved are given in Table 4.1.

Table 4.1

Molar masses and values of materials in Example 4.1.

Material	Molar mass ($kg \cdot kmol^{-1}$)	Value ($\$ \cdot kg^{-1}$)
Acetylene	26	1.0
Chlorine	71	0.23
Ethylene	28	0.58
Hydrogen chloride	36	0.39
Vinyl chloride	62	0.46

Oxygen is considered to be free at this stage, coming from the atmosphere. Which reaction path makes most sense on the basis of raw material costs, product and byproduct values?

Solution Decisions can be made on the basis of the economic potential of the process. At this stage, the best that can be done is to define the economic potential (*EP*) as (see Chapter 2):

$$EP = (\text{Value of products}) - (\text{Raw materials costs})$$

Path 1

$$\begin{aligned} EP &= (62 \times 0.46) - (26 \times 1.0 + 36 \times 0.39) \\ &= -11.52\ \$\cdot kmol^{-1}\ \text{vinyl chloride product} \end{aligned}$$

Path 2

$$\begin{aligned} EP &= (62 \times 0.46 + 36 \times 0.39) - (28 \times 0.58 + 71 \times 0.23) \\ &= 9.99\ \$\cdot kmol^{-1}\ \text{vinyl chloride product} \end{aligned}$$

This assumes the sale of the byproduct HCl. If it cannot be sold, then:

$$\begin{aligned} EP &= (62 \times 0.46) - (28 \times 0.58 + 71 \times 0.23) \\ &= -4.05\ \$\cdot kmol^{-1}\ \text{vinyl chloride product} \end{aligned}$$

Path 3

$$\begin{aligned} EP &= (62 \times 0.46) - (28 \times 0.58 + 2 \times 36 \times 0.39) \\ &= -15.8\ \$\cdot kmol^{-1}\ \text{vinyl chloride product} \end{aligned}$$

Paths 1 and 3 are clearly not viable. Only Path 2 shows a positive economic potential when the byproduct HCl can be sold. In practice, this might be quite difficult, since the market for HCl tends to be limited. In general, projects should not be justified on the basis of the byproduct value.

The preference is for a process based on ethylene rather than the more expensive acetylene, and chlorine, rather than the more expensive hydrogen chloride. Electrolytic cells are a much more convenient and cheaper source of chlorine than hydrogen chloride. In addition, it is preferred to produce no byproducts.

Example 4.2 Devise a process from the three reaction paths in Example 4.1 that uses ethylene and chlorine as raw materials and produces no byproducts other than water (Rudd, Powers and Siirola, 1973). Does the process look economically attractive?

Solution A study of the stoichiometry of the three paths shows that this can be achieved by combining Path 2 and Path 3 to obtain a fourth path.

Paths 2 and 3

$$\underset{\text{ethylene}}{C_2H_4} + \underset{\text{chlorine}}{Cl_2} \longrightarrow \underset{\text{dichloroethane}}{C_2H_4Cl_2}$$

$$\underset{\text{ethylene}}{C_2H_4} + \underset{\text{oxygen}}{{}^1\!/_2O_2} + \underset{\text{hydrogen chloride}}{2HCl} \longrightarrow \underset{\text{dichloroethane}}{C_2H_4Cl_2} + \underset{\text{water}}{H_2O}$$

$$\underset{\text{dichloroethane}}{2C_2H_4Cl_2} \xrightarrow{\text{heat}} \underset{\text{vinyl chloride}}{2C_2H_3Cl} + \underset{\text{hydrogen chloride}}{2HCl}$$

These three reactions can be added to obtain the overall stoichiometry.

Path 4

$$\underset{\text{ethylene}}{2C_2H_4} + \underset{\text{chlorine}}{Cl_2} + \underset{\text{oxygen}}{{}^1\!/_2O_2} \longrightarrow \underset{\text{vinyl chloride}}{2C_2H_3Cl} + \underset{\text{water}}{H_2O}$$

or

$$\underset{\text{ethylene}}{C_2H_4} + \underset{\text{chlorine}}{{}^1\!/_2Cl_2} + \underset{\text{oxygen}}{{}^1\!/_4O_2} \longrightarrow \underset{\text{vinyl chloride}}{C_2H_3Cl} + \underset{\text{water}}{{}^1\!/_2H_2O}$$

Now the economic potential is given by:

$$\begin{aligned} EP &= (62 \times 0.46) - (28 \times 0.58 + 1/2 \times 71 \times 0.23) \\ &= 4.12\ \$\cdot kmol^{-1}\ \text{vinyl chloride product} \end{aligned}$$

In summary, Path 2 from Example 4.1 is the most attractive reaction path if there is a large market for hydrogen chloride. In practice, it tends to be difficult to sell the large quantities of hydrogen chloride produced by such processes. Path 4 is the usual commercial route to vinyl chloride.

4.2 Types of Reaction Systems

Having made a choice of the reaction path, a choice of reactor type must be made, together with some assessment of the conditions in the reactor. This allows assessment of the reactor performance for the chosen reaction path in order for the design to proceed.

Before proceeding to the choice of reactor and operating conditions, some general classifications must be made regarding the types of reaction systems likely to be encountered. Reaction systems can be classified into six broad types:

1) *Single reactions.* Most reaction systems involve multiple reactions. In practice, the *secondary* reactions can sometimes be neglected, leaving a single *primary* reaction to consider. Single reactions are of the type:

$$FEED \longrightarrow PRODUCT \tag{4.1}$$

or

$$FEED \longrightarrow PRODUCT + BYPRODUCT \tag{4.2}$$

or

$$FEED\,1 + FEED\,2 \longrightarrow PRODUCT \tag{4.3}$$

and so on.

An example of this type of reaction that does not produce a byproduct is *isomerization* (the reaction of a feed to a product with the same chemical formula but a different molecular structure). For example, allyl alcohol can be produced from propylene oxide (Waddams, 1978):

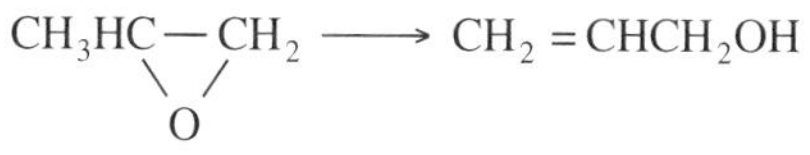

propylene oxide allyl alcohol

An example of a reaction that does produce a byproduct is the production of acetone from isopropyl alcohol, which produces a hydrogen byproduct:

$$\underset{\text{isopropyl alcohol}}{(CH_3)_2CHOH} \longrightarrow \underset{\text{acetone}}{CH_3COCH_3} + H_2$$

2) *Multiple reactions in parallel producing byproducts.* Rather than a single reaction, a system may involve secondary reactions producing (additional) byproducts in *parallel* with the primary reaction. Multiple reactions in parallel are of the type:

$$\begin{aligned} &FEED \longrightarrow PRODUCT \\ &FEED \longrightarrow BYPRODUCT \end{aligned} \tag{4.4}$$

or

$$\begin{aligned} &FEED \longrightarrow PRODUCT + BYPRODUCT\,1 \\ &FEED \longrightarrow BYPRODUCT\,2 + BYPRODUCT\,3 \end{aligned} \tag{4.5}$$

or

$$\begin{aligned} &FEED\,1 + FEED\,2 \longrightarrow PRODUCT \\ &FEED\,1 + FEED\,2 \longrightarrow BYPRODUCT \end{aligned} \tag{4.6}$$

and so on.

An example of a parallel reaction system occurs in the production of ethylene oxide (Waddams, 1978):

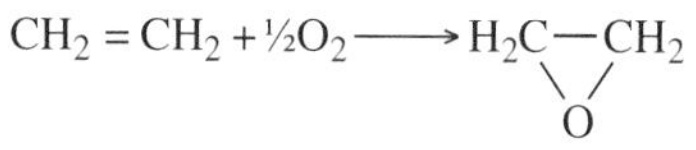

ethylene oxygen ethylene oxide

with the parallel reaction:

$$\underset{\text{ethylene}}{CH_2 = CH_2} + \underset{\text{oxygen}}{3O_2} \longrightarrow \underset{\text{carbon dioxide}}{2CO_2} + \underset{\text{water}}{2H_2O}$$

Multiple reactions might not only lead to a loss of materials and useful product but might also lead to byproducts being deposited on, or poisoning, catalysts (see Chapters 5 and 6).

3) *Multiple reactions in series producing byproducts.* Rather than the primary and secondary reactions being in parallel, they can be in *series*. Multiple reactions in series are of the type:

$$\begin{aligned} &FEED \longrightarrow PRODUCT \\ &PRODUCT \longrightarrow BYPRODUCT \end{aligned} \tag{4.7}$$

or

$$\begin{aligned} &FEED \longrightarrow PRODUCT + BYPRODUCT1 \\ &PRODUCT \longrightarrow BYPRODUCT2 + BYPRODUCT3 \end{aligned} \tag{4.8}$$

or

$$\begin{aligned} &FEED\,1 + FEED\,2 \longrightarrow PRODUCT \\ &PRODUCT \longrightarrow BYPRODUCT\,1 + BYPRODUCT\,2 \end{aligned} \tag{4.9}$$

and so on.

An example of a series reaction system is the production of formaldehyde from methanol:

$$\underset{\text{methanol}}{CH_3OH} + \underset{\text{oxygen}}{{}^1\!/_2O_2} \longrightarrow \underset{\text{formaldehyde}}{HCHO} + \underset{\text{water}}{H_2O}$$

A series reaction of the formaldehyde occurs:

$$\underset{\text{formaldehyde}}{HCHO} \longrightarrow \underset{\text{carbon monoxide}}{CO} + \underset{\text{hydrogen}}{H_2}$$

As with parallel reactions, series reactions might not only lead to a loss of materials and useful products but might also lead to byproducts being deposited on, or poisoning, catalysts (see Chapters 5 and 6).

4) *Mixed parallel and series reactions producing byproducts.* In more complex reaction systems, both parallel and series reactions can occur together. *Mixed parallel and series*

reactions are of the type:

$$\begin{aligned} &FEED \longrightarrow PRODUCT \\ &FEED \longrightarrow BYPRODUCT \\ &PRODUCT \longrightarrow BYPRODUCT \end{aligned} \tag{4.10}$$

or

$$\begin{aligned} &FEED \longrightarrow PRODUCT \\ &FEED \longrightarrow BYPRODUCT\ 1 \\ &PRODUCT \longrightarrow BYPRODUCT\ 2 \end{aligned} \tag{4.11}$$

or

$$\begin{aligned} &FEED\ 1 + FEED\ 2 \longrightarrow PRODUCT \\ &FEED\ 1 + FEED\ 2 \longrightarrow BYPRODUCT\ 1 \\ &PRODUCT \longrightarrow BYPRODUCT\ 2 + BYPRODUCT\ 3 \end{aligned} \tag{4.12}$$

An example of mixed parallel and series reactions is the production of ethanolamines by reaction between ethylene oxide and ammonia (Waddams, 1978):

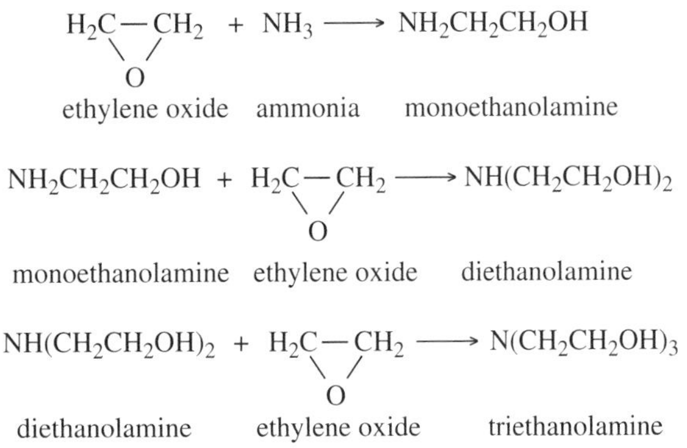

Here the ethylene oxide undergoes parallel reactions, whereas the monoethanolamine undergoes a series reaction to diethanolamine and triethanolamine.

5) *Polymerization reactions.* In polymerization reactions, *monomer* molecules are reacted together to produce a high molar mass *polymer*. Depending on the mechanical properties required of the polymer, a mixture of monomers might be reacted together to produce a high molar mass *copolymer*. There are two broad types of polymerization reactions, those that involve a *termination step* and those that do not (Denbigh and Turner, 1984). An example that involves a termination step is free-radical polymerization of an alkene molecule, known as *addition polymerization*. A free radical is a free atom or fragment of a stable molecule that contains one or more unpaired electrons. The polymerization requires a free radical from an initiator compound such as a peroxide. The initiator breaks down to form a free radical (e.g. $\bullet CH_3$ or $\bullet OH$), which attaches to a molecule of alkene and in so doing generates another free radical. Consider the polymerization of vinyl chloride from a free-radical initiator $\bullet R$. An *initiation step* first occurs:

$$\underset{\text{initiator}}{\dot{R}} + \underset{\text{vinyl chloride}}{CH_2 = CHCl} \longrightarrow \underset{\text{vinyl chloride free radical}}{RCH_2 - \dot{C}HCl}$$

A *propagation step* involving growth around an *active center* follows:

$$\begin{aligned} RCH_2 - \dot{C}HCl + CH_2 = CHCl \longrightarrow \\ RCH_2 - CHCl - CH_2 - \dot{C}HCl \end{aligned}$$

and so on, leading to molecules of the structure:

$$R - (CH_2 - CHCl)_n - CH_2 - \dot{C}HCl$$

Eventually, the chain is terminated by steps such as the union of two radicals that consume but do not generate radicals:

$$\begin{aligned} &R - (CH_2 - CHCl)_n - CH_2 - \dot{C}HCl + \dot{C}HCl - CH_2 \\ &\qquad - (CHCl - CH_2)_m - R \longrightarrow \\ &R - (CH_2 - CHCl)_n - CH_2 - CHCl - CHCl \\ &\qquad - CH_2 - (CHCl - CH_2)_m - R \end{aligned}$$

This termination step stops the subsequent growth of the polymer chain. The period during which the chain length grows, that is before termination, is known as the *active life* of the polymer. Other termination steps are possible.

The orientation of the groups along the carbon chain, its *stereochemistry*, is critical to the properties of the product. The stereochemistry of addition polymerization can be controlled by the use of catalysts. A polymer where repeating units have the same relative orientation is termed *stereoregular*.

An example of a polymerization without a termination step is *polycondensation* (Denbigh and Turner, 1984).

$$\begin{aligned} &HO - (CH_2)_n - COOH + HO - (CH_2)_n - COOH \\ &\longrightarrow HO - (CH_2)_n - COO - (CH_2)_n - COOH + H_2O, \\ &\qquad \text{etc.} \end{aligned}$$

Here the polymer grows by successive esterification with elimination of water and no termination step. Polymers formed by linking monomers with carboxylic acid groups and those that have alcohol groups are known as *polyesters*. Polymers of this type are widely used for the manufacture of artificial fibers. For example, the esterification of terephthalic acid with ethylene glycol produces polyethylene terephthalate.

6) *Biochemical reactions.* Biochemical reactions, often referred to as *fermentations*, can be divided into two broad types. In the first type, the reaction exploits the metabolic pathways in selected microorganisms (especially bacteria, yeasts, moulds and algae) to convert feed material (often called *substrate* in biochemical reactor design) to the required product. The general form of such reactions is:

$$FEED \xrightarrow{\text{microorganisms}} PRODUCT + [\text{More microorganisms}] \tag{4.13}$$

or

$$\begin{aligned} FEED\ 1 + FEED\ 2 \xrightarrow{\text{microorganisms}} PRODUCT \\ + [\text{More microorganisms}] \end{aligned} \tag{4.14}$$

and so on.

In such reactions, the microorganisms reproduce themselves. In addition to the feed material, it is likely that nutrients (e.g. a mixture containing phosphorus, magnesium, potassium, etc.) will need to be added for the survival of the microorganisms. Reactions involving microorganisms include:

- hydrolysis,
- oxidation,
- esterification,
- reduction.

An example of an oxidation reaction is the production of citric acid from glucose:

$$\underset{\text{glucose}}{C_6H_{12}O_6} + {}^3\!/_2\, O_2 \longrightarrow \underset{\text{citric acid}}{HOOCCH_2COH(COOH)CH_2COOH} + 2H_2O$$

In the second group, the reaction is promoted by *enzymes*. Enzymes are the catalyst proteins produced by microorganisms that accelerate chemical reactions in microorganisms. The biochemical reactions employing enzymes are of the general form:

$$FEED \xrightarrow{\text{enzyme}} PRODUCT \qquad (4.15)$$

and so on.

Unlike reactions involving microorganisms, in enzyme reactions the catalytic agent (the enzyme) does not reproduce itself. An example in the use of enzymes is the isomerization of glucose to fructose:

$$\underset{\text{glucose}}{CH_2OH(CHOH)_4CHO} \xrightarrow{\text{enzyme}} \underset{\text{fructose}}{CH_2OHCO(CHOH)_3CH_2OH}$$

Although nature provides many useful enzymes, they can also be engineered for improved performance and new applications. Biochemical reactions have the advantage of operating under mild reaction conditions of temperature and pressure and are usually carried out in an aqueous medium rather than using an organic solvent.

4.3 Measures of Reactor Performance

Before exploring how reactor conditions can be chosen, some measure of reactor performance is required.

For polymerization reactors, the main concern is the characteristics of the product that relate to the mechanical properties. The distribution of molar masses in the polymer product, orientation of groups along the chain, cross-linking of the polymer chains, copolymerization with a mixture of monomers, and so on, are the main considerations. Ultimately, the main concern is the mechanical properties of the polymer product.

For biochemical reactions, the performance of the reactor will normally be dictated by laboratory results, because of the difficulty of predicting such reactions theoretically (Shuler and Kargi, 2002). There are likely to be constraints on the reactor performance dictated by the biochemical processes. For example, in the manufacture of ethanol using microorganisms, as the concentration of ethanol rises, the microorganisms multiply more slowly until at a concentration of around 12% it becomes toxic to the microorganisms.

For other types of reactors, three important parameters are used to describe their performance (Wells and Rose, 1986):

$$\text{Conversion} = \frac{\text{(Reactant consumed in the reactor)}}{\text{(Reactant } fed \text{ to the reactor)}} \qquad (4.16)$$

$$\text{Selectivity} = \frac{\text{(Desired product produced)}}{\text{(Reactant } consumed \text{ in the reactor)} \times \text{Stoichiometric factor}} \qquad (4.17)$$

$$\text{Reactor yield} = \frac{\text{(Desired product produced)}}{\text{(Reactant } fed \text{ to the reactor)} \times \text{Stoichiometric factor}} \qquad (4.18)$$

in which the stoichiometric factor is the stoichiometric moles of reactant required per mole of product. When more than one reactant is required (or more than one desired product produced), Equations 4.16 to 4.18 can be applied to each reactant (or product).

The following example will help clarify the distinctions among these three parameters.

Example 4.3 Benzene is to be produced from toluene according to the reaction (Douglas, 1985):

$$\underset{\text{toluene}}{C_6H_5CH_3} + \underset{\text{hydrogen}}{H_2} \longrightarrow \underset{\text{benzene}}{C_6H_6} + \underset{\text{methane}}{CH_4}$$

Some of the benzene formed undergoes a number of secondary reactions in series to unwanted byproducts that can be represented by the single reaction to diphenyl, according to the reaction:

$$\underset{\text{benzene}}{2C_6H_6} \rightleftharpoons \underset{\text{diphenyl}}{C_{12}H_{10}} + \underset{\text{hydrogen}}{H_2}$$

Table 4.2 gives the compositions of the reactor feed and effluent streams.

Table 4.2

Reactor feed and effluent streams.

Component	Inlet flowrate ($kmol \cdot h^{-1}$)	Outlet flowrate ($kmol \cdot h^{-1}$)
H_2	1858	1583
CH_4	804	1083
C_6H_6	13	282
$C_6H_5CH_3$	372	93
$C_{12}H_{10}$	0	4

Calculate the conversion, selectivity and reactor yield with respect to the:

a) toluene feed
b) hydrogen feed.

Solution

a) Toluene conversion $= \dfrac{\text{(Toluene consumed in the reactor)}}{\text{(Toulene fed to the reactor)}}$

$= \dfrac{372 - 93}{372}$

$= 0.75$

Stoichiometric factor $=$ Stoichiometric moles of toluene required per mole of benzene produced

$= 1$

Benzene selectivity from toluene $= \dfrac{\text{(Benzene produced in the reactor)}}{\text{(Toluene consumed in the reactor)}} \times \text{Stoichiometric factor}$

$= \dfrac{282 - 13}{372 - 93} \times 1$

$= 0.96$

Reactor yield of benzene from toluene $= \dfrac{\text{(Benzene produced in the reactor)}}{\text{(Toluene fed to the reactor)}} \times \text{Stoichiometric factor}$

$= \dfrac{282 - 13}{372} \times 1$

$= 0.72$

b) Hydrogen conversion $= \dfrac{\text{(Hydrogen consumed in the reactor)}}{\text{(Hydrogen fed to the reactor)}}$

$= \dfrac{1858 - 1583}{1858}$

$= 0.15$

Stoichiometric factor $=$ Stoichiometric moles of hydrogen required per mole of benzene produced

$= 1$

Benzene selectivity from hydrogen $= \dfrac{\text{(Benzene produced in the reactor)}}{\text{(Hydrogen consumed in the reactor)}} \times \text{Stoichiometric factor}$

$= \dfrac{282 - 13}{1858 - 1583} \times 1$

$= 0.98$

Reactor yield of benzene from hydrogen $= \dfrac{\text{(Benzene produced in the reactor)}}{\text{(Hydrogen fed to the reactor)}} \times \text{Stoichiometric factor}$

$= \dfrac{282 - 13}{1858} \times 1$

$= 0.14$

Because there are two feeds to this process, the reactor performance can be calculated with respect to both feeds. However, the principal concern is performance with respect to toluene, since it is more expensive than hydrogen.

As will be discussed in the next chapter, if a reaction is reversible there is a maximum conversion, the *equilibrium conversion*, that can be achieved, which is less than 1.0.

In describing reactor performance, selectivity is often a more meaningful parameter than reactor yield. Reactor yield is based on the reactant fed to the reactor rather than on that which is consumed. Part of the reactant fed might be material that has been recycled rather than fresh feed. Reactor yield takes no account of the ability to separate and recycle unconverted raw materials. Reactor yield is only a meaningful parameter when it is not possible for one reason or another to recycle unconverted raw material to the reactor inlet. However, the yield of the overall process is an extremely important parameter when describing the performance of the overall plant, as will be discussed later.

4.4 Rate of Reaction

To define the rate of a reaction, one of the components must be selected and the rate defined in terms of that component. The rate of reaction is the number of moles formed with respect to time, per unit volume of reaction mixture:

$$r_i = \frac{1}{V}\left(\frac{dN_i}{dt}\right) \tag{4.19}$$

where r_i = rate of reaction of Component i ($kmol \cdot m^{-3} \cdot s^{-1}$)
N_i = moles of Component i formed (kmol)
V = reaction volume (m^3)
t = time (s)

If the volume of the reactor is constant ($V =$ constant):

$$r_i = \frac{1}{V}\left(\frac{dN_i}{dt}\right) = \frac{d\,N_i/V}{dt} = \frac{dC_i}{dt} \tag{4.20}$$

where C_i = molar concentration of Component i ($kmol \cdot m^{-3}$)

The rate is negative if the component is a reactant and positive if it is a product. For example, for the general irreversible reaction:

$$bB + cC + \cdots \longrightarrow sS + tT + \cdots \tag{4.21}$$

The rates of reaction are related by:

$$-\frac{r_B}{b} = -\frac{r_C}{c} = -\cdots = \frac{r_S}{s} = \frac{r_T}{t} = \cdots \tag{4.22}$$

If the rate-controlling step in the reaction is the collision of the reacting molecules, then the equation to quantify the reaction rate will often follow the stoichiometry such that:

$$-r_B = k_B C_B^b C_C^c \cdots \tag{4.23}$$

$$-r_C = k_C C_B^b C_C^c \cdots \tag{4.24}$$

$$r_S = k_S C_B^b C_C^c \cdots \tag{4.25}$$

$$r_T = k_T C_B^b C_C^c \cdots \tag{4.26}$$

where r_i = reaction rate for Component i ($\text{kmol}\cdot\text{m}^{-3}\cdot\text{s}^{-1}$)
k_i = reaction rate constant for Component i ($[\text{kmol}\cdot\text{m}^{-3}]^{NC-b-c-\cdots}\,\text{s}^{-1}$)
NC = is the number of components in the rate expression
C_i = molar concentration of Component i ($\text{kmol}\cdot\text{m}^{-3}$)

The exponent for the concentration ($b, c, \ldots$) is known as the *order of reaction*. The reaction rate constant is a function of temperature, as will be discussed in the next chapter.

Thus, from Equations 4.22 to 4.26:

$$\frac{k_B}{b} = \frac{k_C}{c} = \cdots = \frac{k_S}{s} = \frac{k_T}{t} = \cdots \tag{4.27}$$

Reactions for which the rate equations follow the stoichiometry as given in Equations 4.23 to 4.26 are known as *elementary reactions*. If there is no direct correspondence between the reaction stoichiometry and the reaction rate, these are known as *non-elementary reactions* and are often of the form:

$$-r_B = k_B C_B^\beta C_C^\delta \cdots C_S^\varepsilon C_T^\xi \cdots \tag{4.28}$$

$$-r_C = k_C C_B^\beta C_C^\delta \cdots C_S^\varepsilon C_T^\xi \cdots \tag{4.29}$$

$$r_S = k_S C_B^\beta C_C^\delta \cdots C_S^\varepsilon C_T^\xi \cdots \tag{4.30}$$

$$r_T = k_T C_B^\beta C_C^\delta \cdots C_S^\varepsilon C_T^\xi \cdots \tag{4.31}$$

where β, δ, ε, ξ = order of reaction

The reaction rate constant and the order of reaction must be determined experimentally. If the reaction mechanism involves multiple steps involving chemical intermediates, then the form of the reaction rate equations can be of a more complex form than Equations 4.28 to 4.31.

If the reaction is reversible, such that:

$$bB + cC + \cdots \rightleftharpoons sS + tT + \cdots \tag{4.32}$$

then the rate of reaction is the net rate of the forward and reverse reactions. If the forward and reverse reactions are both elementary, then:

$$-r_B = k_B C_B^b C_C^c \cdots - k'_B C_S^s C_T^t \cdots \tag{4.33}$$

$$-r_C = k_C C_B^b C_C^c \cdots - k'_C C_C^s C_T^t \cdots \tag{4.34}$$

$$r_S = k_S C_B^b C_C^c \cdots - k'_S C_S^s C_T^t \cdots \tag{4.35}$$

$$r_T = k_T C_B^b C_C^c \cdots - k'_T C_S^s C_T^t \cdots \tag{4.36}$$

where k_i = reaction rate constant for Component i for the forward reaction
k'_i = reaction rate constant for Component i for the reverse reaction

If the forward and reverse reactions are nonelementary, perhaps involving the formation of chemical intermediates in multiple steps, then the form of the reaction rate equations can be more complex than Equations 4.33 to 4.36.

4.5 Idealized Reactor Models

Three idealized models are used for the design of reactors (Rase, 1977; Denbigh and Turner, 1984; Levenspiel, 1999). In the first (Figure 4.1a), the *ideal-batch* model, the reactants are charged at the beginning of the operation. The contents are subjected to perfect mixing for a certain period, after which the products are discharged. Concentration changes with time, but the perfect mixing ensures that at any instant the composition and temperature throughout the reactor are both uniform.

In the second model (Figure 4.1b), the *mixed-flow* or *continuous well-mixed* or *continuous-stirred-tank reactor* (*CSTR*), feed and product takeoff are both continuous and the reactor contents are assumed to be perfectly mixed. This leads to uniform composition and temperature throughout the reactor. Because of the perfect mixing, a fluid element can leave the instant it enters the reactor or stay for an extended period. The residence time of individual fluid elements in the reactor varies.

In the third model (Figure 4.1c), the *plug-flow* model, a steady uniform movement of the reactants is assumed, with no attempt to

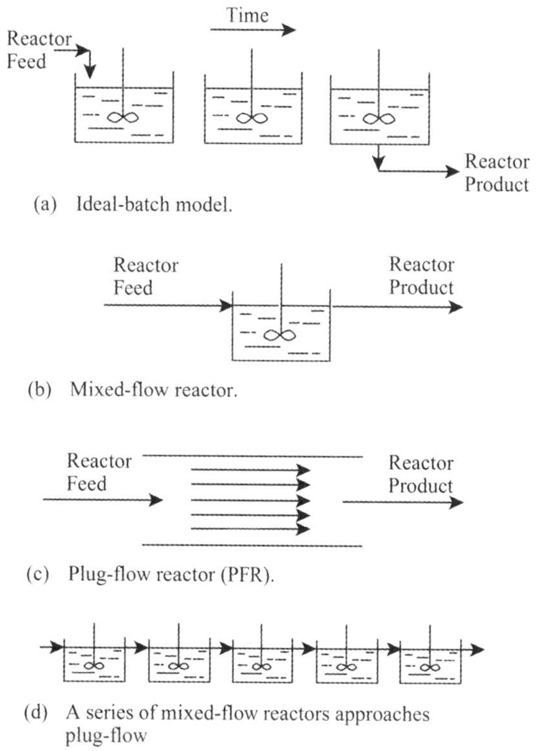

Figure 4.1

The idealized models used for reactor design. (Reproduced from Smith R and Petela EA (1992) Waste Minimization in the Process Industries Part 2 Reactors, Chem Eng, Dec (509–510): 17, by permission of the Institution of Chemical Engineers.)

4

induce mixing along the direction of flow. Like the ideal-batch reactor, the residence time in a plug-flow reactor is the same for all fluid elements. Plug-flow operation can be approached by using a number of mixed-flow reactors in series (Figure 4.1d). The greater the number of mixed-flow reactors in series, the closer is the approach to plug-flow operation.

1) *Ideal-batch reactor.* Consider a batch reactor in which the feed is charged at the beginning of the batch and no product is withdrawn until the batch is complete. It is given that:

$$\begin{bmatrix}\text{Moles of reactant}\\ \text{converted}\end{bmatrix} = -r_i = -\frac{1}{V}\frac{dN_i}{dt} \tag{4.37}$$

Integration of Equation 4.37 gives:

$$t = \int_{N_{i0}}^{N_{it}} \frac{dN_i}{r_i V} \tag{4.38}$$

where t = batch time
N_{i0} = initial moles of Component i
N_{it} = final moles of Component i after time t

Alternatively, Equation 4.37 can be written in terms of reactor conversion X_i:

$$\frac{dN_i}{dt} = \frac{d[N_{i0}(1-X_i)]}{dt} = -N_{i0}\frac{dX_i}{dt} = r_i V \tag{4.39}$$

Integration of Equation 4.39 gives:

$$t = N_{i0}\int_0^{X_i} \frac{dX_i}{-r_i V} \tag{4.40}$$

Also, from the definition of reactor conversion, for the special case of a constant density reaction mixture:

$$X_i = \frac{N_{i0} - N_{it}}{N_{i0}} = \frac{C_{i0} - C_{it}}{C_{i0}} \tag{4.41}$$

where C_i = molar concentration of Component i
C_{i0} = initial molar concentration of Component i
C_{it} = final molar concentration of Component i at time t

Substituting Equation 4.41 into Equation 4.39 and noting that $N_{i0}/V = C_{i0}$ gives:

$$-\frac{dC_i}{dt} = -r_i \tag{4.42}$$

Integration of Equation 4.42 gives:

$$t = -\int_{C_{i0}}^{C_{it}} \frac{dC_i}{-r_i} \tag{4.43}$$

2) *Mixed-flow reactor.* Consider now the mixed-flow reactor in Figure 4.2a in which a feed of Component i is reacting. A material balance for Component i per unit time gives:

$$\begin{bmatrix}\text{Moles of reactant in}\\ \text{feed per unit time}\end{bmatrix} - \begin{bmatrix}\text{Moles of reactant}\\ \text{converted per unit time}\end{bmatrix} = \begin{bmatrix}\text{Moles of reactant}\\ \text{in product per unit time}\end{bmatrix} \tag{4.44}$$

Equation 4.44 can be written per unit time as:

$$N_{i,in} - (-r_i V) = N_{i,out} \tag{4.45}$$

where $N_{i,in}$ = inlet moles of Component i per unit time
$N_{i,out}$ = outlet moles of Component i per unit time

Rearranging Equation 4.45 gives:

$$N_{i,out} = N_{i,in} + r_i V \tag{4.46}$$

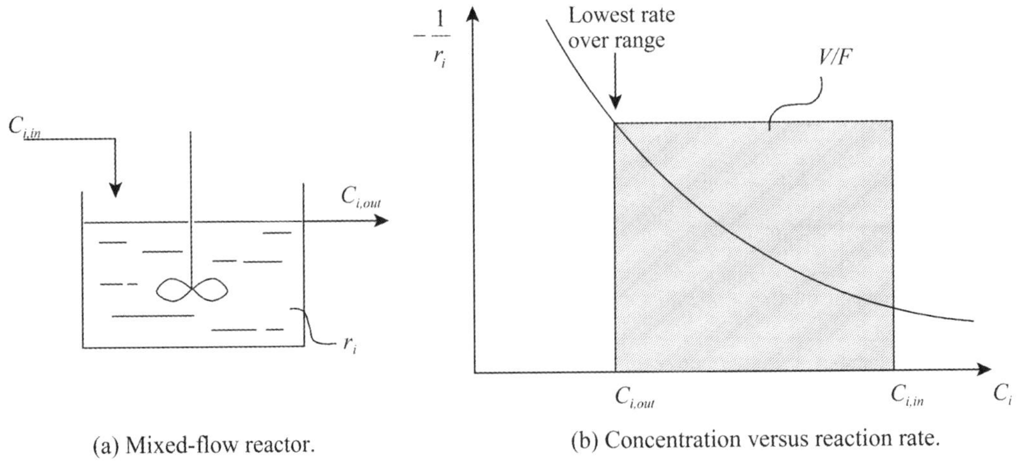

Figure 4.2
Rate of reaction in a mixed-flow reactor.

Substituting $N_{i,out} = N_{i,in}(1 - X_i)$ into Equation 4.46 gives:

$$V = \frac{N_{i,in} X_i}{-r_i} \tag{4.47}$$

For the special case of a constant density system, Equation 4.41 can be substituted to give:

$$V = \frac{N_{i,in}\left(C_{i,in} - C_{i,out}\right)}{-r_i C_{i,in}} \tag{4.48}$$

Analogous to time as a measure of batch process performance, *space–time* (τ) can be defined for a continuous reactor:

$$\tau = [\text{Time required to process one reactor volume of feed}] \tag{4.49}$$

If the space–time is based on the feed conditions:

$$\tau = \frac{V}{F} = \frac{C_{i,in} V}{N_{i,in}} \tag{4.50}$$

where F = volumetric flowrate of the feed ($m^3 \cdot s^{-1}$)

The reciprocal of space–time is *space–velocity* (s):

$$s = \frac{1}{\tau} = [\text{Number of reactor volumes processed in a unit time}] \tag{4.51}$$

Combining Equation 4.48 for the mixed-flow reactor with constant density and Equation 4.50 gives:

$$\tau = \frac{C_{i,in} - C_{i,out}}{-r_i} \tag{4.52}$$

Figure 4.2b is a plot of Equation 4.52. From $C_{i,in}$ to $C_{i,out}$, the rate of reaction decreases to a minimum at $C_{i,out}$. As the reactor is assumed to be perfectly mixed, $C_{i,out}$ is the concentration throughout the reactor; that is, this gives the lowest rate throughout the reactor. The shaded area in Figure 4.2b represents the space–time (V/F).

3) *Plug-flow reactor.* Consider now the plug-flow reactor in Figure 4.3a, in which Component i is reacting. A material balance can be carried out per unit time across the incremental volume dV in Figure 4.3a:

$$\begin{bmatrix}\text{Moles of reactant}\\ \text{entering incremental}\\ \text{volume per unit time}\end{bmatrix} - \begin{bmatrix}\text{Moles of reactant}\\ \text{converted per unit}\\ \text{time}\end{bmatrix} = \begin{bmatrix}\text{Moles of reactant}\\ \text{leaving incremental}\\ \text{volume per unit time}\end{bmatrix} \tag{4.53}$$

Equation 4.53 can be written per unit time as:

$$N_i - (-r_i dV) = N_i + dN_i \tag{4.54}$$

where N_i = moles of Component i per unit time

Rearranging Equation 4.54 gives:

$$dN_i = r_i\, dV \tag{4.55}$$

Substituting the reactor conversion into Equation 4.55 gives:

$$dN_i = d\left[N_{i,in}(1 - X_i)\right] = r_i\, dV \tag{4.56}$$

where $N_{i,in}$ = inlet moles of Component i per unit time

Rearranging Equation 4.56, given $dN_{i,in} = 0$:

$$N_{i,in} dX_i = -r_i\, dV \tag{4.57}$$

Integration of Equation 4.57 gives:

$$V = N_{i,in} \int_0^{X_i} \frac{dX_i}{-r_i} \tag{4.58}$$

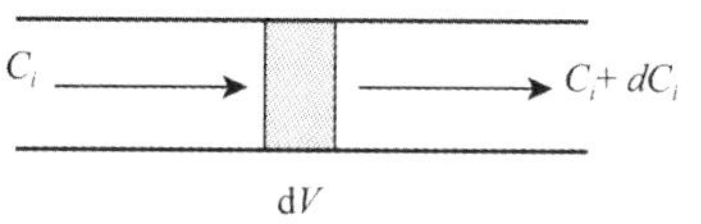

(a) Plug flow reactor.

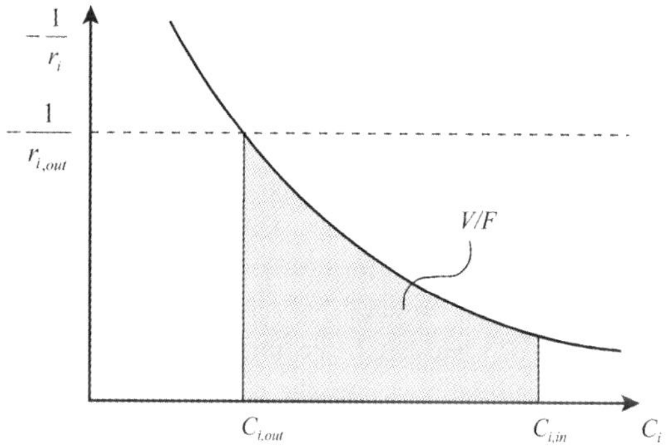

(b) Concentration versus reaction rate.

Figure 4.3

Rate of reaction in a plug-flow reactor.

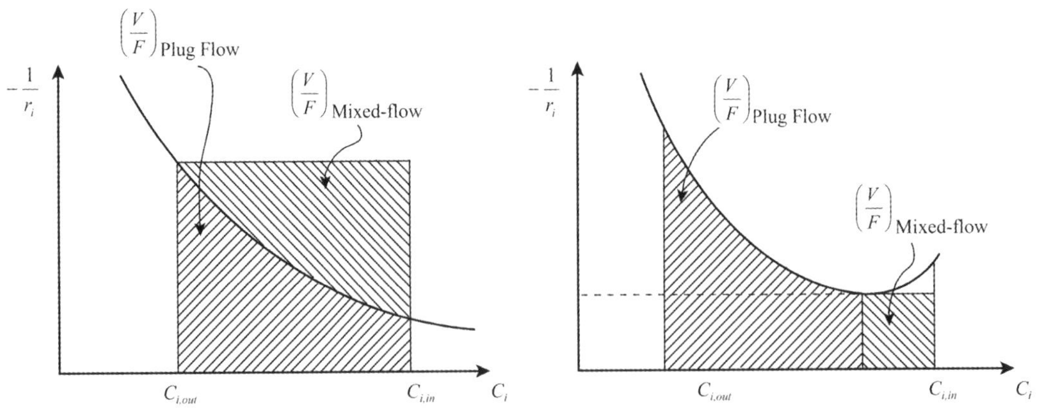

(a) Comparison of mixed-flow and plug-flow reactors. (b) Reaction rate goes through a maximum.

Figure 4.4
Use of mixed-flow and plug-flow reactors.

Writing Equation 4.58 in terms of the space–time:

$$\tau = C_{i,in} \int_0^{X_i} \frac{dX_i}{-r_i} \tag{4.59}$$

For the special case of constant density systems, substitution of Equation 4.50 gives:

$$V = -\frac{N_{i,in}}{C_{i,in}} \int_{C_{i,in}}^{C_{i,out}} \frac{dC_i}{-r_i} \tag{4.60}$$

$$\tau = -\int_{C_{i,in}}^{C_{i,out}} \frac{dC_i}{-r_i} \tag{4.61}$$

Figure 4.3b is a plot of Equation 4.61. The rate of reaction is high at $C_{i,in}$ and decreases to $C_{i,out}$ where it is the lowest. The area under the curve now represents the space–time.

It should be noted that the analysis for an ideal-batch reactor is the same as that for a plug-flow reactor (compare Equations 4.43 and 4.61). All fluid elements have the same residence time in both cases. Thus

$$t_{Ideal-batch} = \tau_{Plug-flow} \tag{4.62}$$

Figure 4.4a compares the profiles for a mixed-flow and plug-flow reactor between the same inlet and outlet concentrations, from which it can be concluded that the mixed-flow reactor requires a larger volume. The rate of reaction in a mixed-flow reactor is uniformly low as the reactant is instantly diluted by the product that has already been formed. In a plug-flow or ideal-batch reactor, the rate of reaction is high initially and decreases as the concentration of reactant decreases.

In an *autocatalytic* reaction, the rate starts low, as little product is present, but increases as product is formed. The rate reaches a maximum and then decreases as the reactant is consumed. This is illustrated in Figure 4.4b. In such a situation, it would be best to use a combination of reactor types, as illustrated in Figure 4.4b, to minimize the volume for a given flowrate. A mixed-flow reactor should be used until the maximum rate is reached and the intermediate product fed to a plug-flow (or ideal-batch) reactor. Alternatively, if separation and recycle of unconverted material is possible, then a mixed-flow reactor could be used up to the point where the maximum rate is achieved, then the unconverted material separated and recycled back to the reactor inlet. Whether this separation and recycle option is cost-effective or not will depend on the cost of separating and recycling material. Combinations of mixed-flow reactors in series and plug-flow reactors with recycle can also be used for autocatalytic reactions, but their performance will always be inferior to either a combination of mixed and plug-flow reactors or a mixed-flow reactor with separation and recycle (Levenspiel, 1999).

Example 4.4 Benzyl acetate is used in perfumes, soaps, cosmetics and household items where it produces a fruity, jasmine-like aroma, and it is used to a minor extent as a flavor. It can be manufactured by the reaction between benzyl chloride and sodium acetate in a solution of xylene in the presence of triethylamine as catalyst (Huang and Dauerman, 1969):

$$C_6H_5CH_2Cl + CH_3COONa \longrightarrow CH_3COOC_6H_5CH_2 + NaCl$$

or

$$A + B \longrightarrow C + D$$

The reaction has been investigated experimentally by Huang and Dauerman (1969) in a batch reaction carried out with initial conditions given in Table 4.3.

The solution volume was 1.321×10^{-3} m^3 and the temperature was maintained to be 102 °C. The measured mole percent benzyl chloride versus time in hours are given in Table 4.4 (Huang and Dauerman, 1969).

Derive a kinetic model for the reaction on the basis of the experimental data. Assume the volume of the reactor to be constant.

Table 4.3

Initial reaction mixture for the production of benzyl acetate.

Component	Moles
Benzyl chloride	1
Sodium acetate	1
Xylene	10
Triethylamine	0.0508

Table 4.4

Experimental data for the production of benzyl acetate.

Reaction time (h)	Benzyl chloride (mole%)
3.0	94.5
6.8	91.2
12.8	84.6
15.2	80.9
19.3	77.9
24.6	73.0
30.4	67.8
35.2	63.8
37.15	61.9
39.1	59.0

Solution The equation for a batch reaction is given by Equation 4.38:

$$t = \int_{N_{A0}}^{N_{Af}} \frac{dN_A}{r_A V}$$

Table 4.5

Expressions for a batch reaction with different kinetic models.

Order of reaction	Kinetic expression	Ideal-batch model
Zero order	$-r_A = k_A$	$\frac{1}{V}(N_{A0} - N_A) = k_A t$
First order	$-r_A = k_A C_A$	$\ln \frac{N_{A0}}{N_A} = k_A t$
Second order	$-r_A = k_A {C_A}^2$	$V\left(\frac{1}{N_A} - \frac{1}{N_{A0}}\right) = k_A t$

Initially, it could be postulated that the reaction could be zero order, first order or second order in the concentration of A and B. However, given that all the reaction stoichiometric coefficients are unity and the initial reaction mixture has equimolar amounts of A and B, it seems sensible to first try to model the kinetics in terms of the concentration of A. This is because, in this case, the reaction proceeds with the same rate of change of moles for the two reactants. Thus, it could be postulated that the reaction could be zero order, first order or second order in the concentration of A. In principle, there are many other possibilities. Substituting the appropriate kinetic expression into Equation 4.47 and integrating gives the expressions in Table 4.5.

The experimental data have been substituted into the three models and presented graphically in Figure 4.5. From Figure 4.5, all three models seem to give a reasonable representation of the data, as all three give a reasonable straight line. It is difficult to tell from the graph which line gives the best fit. The fit can be better judged by carrying out a least squares fit to the data for the three models. The difference between the values calculated from the model and the experimental values are summed according to:

$$\text{minimize } R^2 = \sum_{j}^{10} \left[(N_{A,j})_{calc} - (N_{A,j})_{exp}\right]^2$$

where R = residual
$(N_{A,j})_{calc}$ = calculated moles of A for Measurement j
$(N_{A,j})_{exp}$ = experimentally measured moles of A for Measurement j

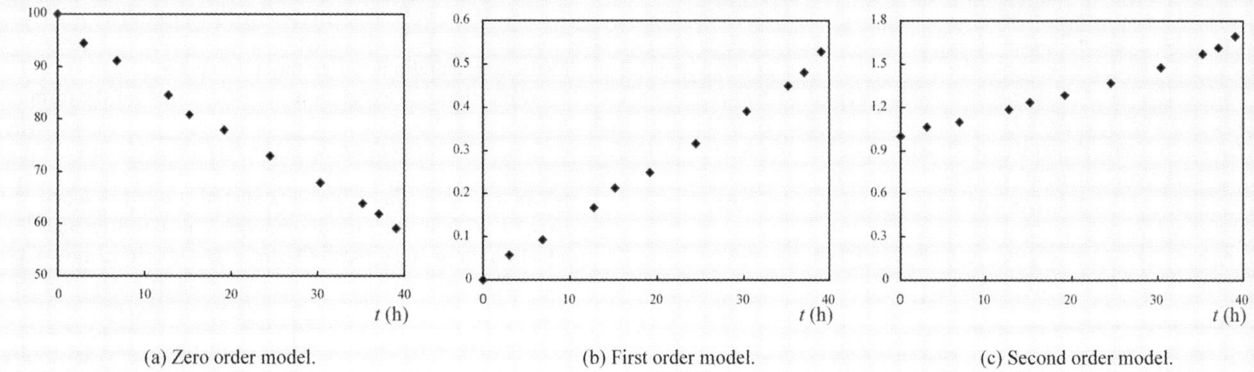

Figure 4.5

Kinetic model for the production of benzyl acetate.

Table 4.6

Results of a least squares fit for the three kinetic models.

Order of reaction	Rate constant	R^2
Zero order	$k_A V = 1.066$	26.62
First order	$k_A = 0.01306$	6.19
Second order	$k_A/V = 0.0001593$	15.65

However, the most appropriate value of the rate constant for each model needs to be determined. This can be determined, for example, in a spreadsheet by setting up a function for R^2 and using the spreadsheet solver to minimize R^2 by manipulating the value of k_A. The results are summarized in Table 4.6.

From Table 4.6, it is clear that the best fit is given by a first-order reaction model: $r_A = k_A C_A$ with $k_A = 0.01306\,h^{-1}$.

Example 4.5 Ethyl acetate is widely used in formulating printing inks, adhesives, lacquers and is used as a solvent in food processing. It can be manufactured from the reaction between ethanol and acetic acid in the liquid phase according to the reaction:

$$\underset{\text{acetic acid}}{CHCOOH} + \underset{\text{ethanol}}{C_2H_5OH} \rightleftharpoons \underset{\text{ethyl acetate}}{CH_3COOC_2H_5} + \underset{\text{water}}{H_2O}$$

or

$$A + B \rightleftharpoons C + D$$

Experimental data are available using an ion-exchange resin catalyst based on batch experiments at 60 °C (Helminen *et al.*, 1998). These data are presented in Table 4.7 (Helminen *et al.*, 1998).

Molar masses and densities for the components are given in Table 4.8.

Initial conditions are such that the reactants are equimolar with product concentrations of initially zero. From the experimental data, assuming a constant density system:

Table 4.7

Experimental data for the production of ethyl acetate.

Sample point	Time (min)	A (mass %)	B (mass %)	C (mass %)	D (mass %)
1	0	56.59	43.41	0.00	0.00
2	5	49.70	38.10	10.00	2.20
3	10	46.30	35.50	15.10	3.10
4	15	42.50	32.50	20.70	4.30
5	30	35.40	27.20	30.90	6.50
6	60	28.10	21.90	41.40	8.60
7	90	24.20	18.60	47.60	9.60
8	120	22.70	17.40	49.80	10.10
9	150	21.20	17.00	51.10	10.70
10	180	20.90	16.50	51.70	10.90
11	210	20.50	16.20	52.30	11.00
12	240	20.30	15.70	52.90	11.10

Table 4.8

Molar masses for the components involved in the manufacture of ethyl acetate.

Component	Molar mass ($kg \cdot kmol^{-1}$)	Density at 60 °C ($kg \cdot m^{-3}$)
Acetic acid	60.05	1018
Ethanol	46.07	754
Ethyl acetate	88.11	847
Water	18.02	980

a) Fit a kinetic model to the experimental data.

b) For a plant producing 10 tons of ethyl acetate per day, calculate the volume required by a mixed-flow reactor and a plug-flow reactor operating at 60 °C. Assume no product is recycled to the reactor and the reactor feed is an equimolar mixture of ethanol and acetic acid. Also, assume the reactor conversion to be 95% of the conversion at equilibrium.

Solution

a) To fit a model to the data, first convert the mass percent data from Table 4.7 into molar concentrations.

Volume of reaction mixture per kg of reaction mixture (assuming no volume change of solution)

$$= \frac{0.566}{1018} + \frac{0.434}{754}$$
$$= 1.1316 \times 10^{-3}\ m^3$$

Molar concentrations are given in Table 4.9.

A reaction rate expression can be assumed of the form:

$$-r_A = k_A C_A^\alpha C_B^\beta - k'_A C_C^\delta C_D^\gamma$$

where α, β, δ, γ can be 0, 1 or 2 such that $n_1 = \alpha + \beta$ and $n_2 = \delta + \gamma$. Many other forms of model could be postulated. The equation for an ideal-batch reactor is given by Equation 4.38:

$$t = -\int_{C_{A0}}^{C_A} \frac{dC_A}{-r_A}$$

Substituting the kinetic model and integrating give the results in Table 4.10 depending on the values of the order of reaction. The integrals can be found from tables of standard integrals (Dwight, 1961).

Table 4.9

Molar concentrations for the manufacture of ethyl acetate.

Time(min)	A (kmol·m^{-3})	B (kmol·m^{-3})	C (kmol·m^{-3})	D (kmol·m^{-3})
0	8.3277	8.3266	0.0000	0.0000
5	7.3138	7.3081	1.0029	1.0789
10	6.8134	6.8094	1.5144	1.5202
15	6.2542	6.2339	2.0761	2.1087
30	5.2094	5.2173	3.0991	3.1875
60	4.1352	4.2007	4.1522	4.2174
90	3.5612	3.5677	4.7740	4.7078
120	3.3405	3.3376	4.9946	4.9530
150	3.1198	3.2608	5.1250	5.2472
180	3.0756	3.1649	5.1852	5.3453
210	3.0167	3.1074	5.2454	5.3943
240	2.9873	3.0115	5.3055	5.4434

Note in Table 4.10 that many of the integrals are common to different kinetic models. This is specific to this reaction where all the stoichiometric coefficients are unity and the initial reaction mixture was equimolar. In other words, the change in the number of moles is the same for all components. Rather than determine the integrals analytically, they could have been determined numerically. Analytical integrals are simply more convenient if they can be obtained, especially if the model is to be fitted in a spreadsheet, rather than being purpose-written software. The least squares fit varies the reaction rate constants to minimize the objective function:

$$\text{minimize} \quad R^2 = \sum_{i=1}^{4}\sum_{j=1}^{11}\left[(C_{i,j})_{calc} - (C_{i,j})_{exp}\right]^2$$

where R = residual

$(C_{i,j})_{calc}$ = calculated molar concentration of Component i for Measurement j

$(C_{i,j})_{exp}$ = experimentally measured molar concentration of Component i for Measurement j

Again, this can be carried out, for example, in spreadsheet software, and the results are given in Table 4.11.

From Table 4.11, there is very little to choose between the best two models. The best fit is given by a second-order model for the forward and a first-order model for the reverse reaction with:

$$k_A = 0.002688\ \text{m}^3\cdot\text{kmol}^{-1}\cdot\text{min}^{-1}$$

$$k'_A = 0.004644\ \text{min}^{-1}$$

However, there is little to choose between this model and a second-order model for both forward and reverse reactions.

b) Now use the kinetic model to size a reactor to produce 10 tons per day of ethyl acetate. First, the conversion at equilibrium needs to be calculated. At equilibrium, the rates of forward and reverse reactions are equal:

$$k_A C_A^2 = k'_A C_C$$

Substituting for the conversion at equilibrium X_E gives:

$$k_A[C_{A0}(1 - X_E)]^2 = k'_A C_{A0} X_E$$

Rearranging gives:

$$\frac{(1 - X_E)^2}{X_E} = \frac{k'_A}{k_A C_{A0}}$$

Substituting $k_A = 0.002688$, $k'_A = 0.004644$ and $C_{A0} = 8.3277$ and solving for X_E by trial and error gives:

$$X_E = 0.6366$$

Assume the actual conversion to be 95% of the equilibrium conversion:

$$X = 0.95 \times 0.6366$$
$$= 0.605$$

A production rate of 10 tons per day equates to 0.0788 kmol·min^{-1}. For a mixed-flow reactor with equimolar feed $C_{A0} = C_{B0} = 8.33$ kmol·m^{-3} at 60 °C:

$$V = \frac{N_{A,in}X_A}{-r_A} = \frac{N_{C,out}}{k_A C_{A0}^2(1 - X_A)^2 - k'_A C_{A0} X_A}$$
$$= \frac{0.0788}{0.002688 \times 8.33^2 \times (1 - 0.605)^2 - 0.004644 \times 8.33 \times 0.605}$$
$$= 13.83\ \text{m}^3$$

Table 4.10

Kinetic models used to fit the data for the manufacture of ethyl acetate.

n_1	n_2	Rate equation	Final concentration from the model
1	1	$-r_A = k_A C_A - k' C_C$	
		$-r_A = k_A C_B - k' C_C$	$C_A = \dfrac{k_A \exp(-(k_A + k'_A)t) + k'_A}{k_A + k'_A} C_{A0}$
		$-r_A = k_A C_A - k'_A C_D$	
		$-r_A = k_A C_B - k'_A C_D$	
2	1	$-r_A = k_A C_A^2 - k'_A C_C$	
		$-r_A = k_A C_B^2 - k'_A C_C$	
		$-r_A = k_A C_A^2 - k'_A C_D$	$C_A = \dfrac{(k'_A - a)(2k_A C_{A0} + k'_A + a) - (k'_A + a)(2k_A C_{A0} + k'_A - a)\exp(-at)}{2k_A(2k_A C_{A0} + k'_A - a)\exp(-at) - 2k_A(2k_A C_{A0} + k'_A + a)}$
		$-r_A = k_A C_B^2 - k'_A C_D$	
		$-r_A = k_A C_A C_B - k'_A C_C$	$a = \sqrt{k'_A + 4k_A k'_A C_{A0}}$
		$-r_A = k_A C_A C_B - k'_A C_D$	
1	2	$-r_A = k_A C_A - k'_A C_C^2$	
		$-r_A = k_A C_B - k'_A C_C^2$	$C_A = \dfrac{(k_A + 2k'_A C_{A0} + b)(k_A - b)\exp(-bt) - (k_A + 2k'_A C_{A0} - b)(k_A + b)}{-2k'_A(k_A + b) + 2k'_A(k_A - b)\exp(-bt)}$
		$-r_A = k_A C_A - k'_A C_D^2$	
		$-r_A = k_A C_B - k'_A C_D^2$	$b = \sqrt{k_A^2 + 4k_A k'_A C_{A0}}$
		$-r_A = k_A C_A - k'_A C_C C_D$	
		$-r_A = k_A C_B - k'_A C_C C_D$	
2	2	$-r_A = k_A C_A^2 - k'_A C_C^2$	
		$-r_A = k_A C_B^2 - k'_A C_C^2$	
		$-r_A = k_A C_A^2 - k'_A C_D^2$	
		$-r_A = k_A C_B^2 - k'_A C_D^2$	$C_A = \dfrac{\sqrt{k_A k'_A}(1 + \exp(2C_{A0}\sqrt{k_A k'_A}t))}{(k_A + \sqrt{k_A k'_A})\exp(2C_{A0}\sqrt{k_A k'_A}t) - (k_A - \sqrt{k_A k'_A})} C_{A0}$
		$-r_A = k_A C_A C_B - k'_A C_C^2$	
		$-r_A = k_A C_A C_B - k'_A C_D^2$	
		$-r_A = k C_A^2 - k' C_C C_D$	
		$-r_A = k_A C_B^2 - k'_A C_C C_D$	
		$-r_A = k_A C_A C_B - k'_A C_C C_D$	

Table 4.11
Results of model fitting for the manufacture of ethyl acetate.

n_1	n_2	k_A	k'_A	R^2
1	1	0.01950	0.01172	1.01108
2	1	0.002688	0.004644	0.3605
1	2	0.01751	0.002080	1.7368
2	2	0.002547	0.0008616	0.5554

For a plug-flow reactor:

$$\tau = \frac{VC_{A0}}{N_{A0}} = \int_0^{X_A} \frac{C_{A0}dX_A}{-k_A C_{A0}^2(1-X_A)^2 + k'_A C_{A0} X_A}$$

From tables of standard integrals (Helminen *et al.*, 1998), this can be integrated to give:

$$\tau = -\frac{1}{a}\ln\frac{(2k_A C_{A0}(1-X_A)+k'_A-a)(2k_A C_{A0}+k'_A+a)}{(2k_A C_{A0}(1-X_A)+k'_A+a)(2k_A C_{A0}+k'_A-a)}$$

where $a = \sqrt{k'_A k'_A + 4k_A k'_A C_{A0}}$. Substituting the values of k_A, k'_A, C_{A0} and X_A gives:

$$\tau = 120.3 \text{ min}$$

The reactor volume is given by:

$$V = \frac{\tau N_{A0}}{C_{A0}} = \frac{\tau N_C}{X_A C_{A0}} = \frac{120.3 \times 0.0788}{0.605 \times 8.33} = 1.88 \text{ m}^3$$

Alternatively, the residence time in the plug-flow reactor could be calculated from the batch equations given in Table 4.10. This results from the residence time being equal for both. Thus the final concentration in a plug-flow reactor is given by (Table 4.10):

$$C_A = \frac{(k'_A - a)(2k_A C_{A0} + k'_A + a) - (k'_A + a)(2k_A C_{A0} + k'_A - a)\exp(-a\tau)}{2k_A(2k_A C_{A0} + k'_A - a)\exp(-a\tau) - 2k_A(2k_A C_{A0} + k'_A + a)}$$

where $a = \sqrt{k'_A k'_A + 4k_A k'_A C_{A0}}$. The final concentration of C_A from the conversion is $8.33 \times 0.395 = 3.29$ kmol·m^{-3} (assuming no volume change). Thus, the above equation can be solved by trial and error to give the residence time of 120.3 min.

As expected, the result shows that the volume required by a mixed-flow reactor is much larger than that for a plug-flow reactor.

A number of points should be noted regarding Examples 4.4 and 4.5:

1) The kinetic models were fitted to experimental data at specific conditions of molar feed ratio and temperature. The models are only valid for these conditions. Use for non-equimolar feeds or at different temperatures will not be valid.
2) Given that kinetic models are only valid for the range of conditions over which they are fitted, it is better that the experimental investigation into the reaction kinetics and the reactor design are carried out in parallel. If this approach is followed, then it can be assured that the range of experimental conditions used in the laboratory cover the range of conditions used in the reactor design. If the experimental program is carried out and completed prior to the reactor design, then there is no guarantee that the kinetic model will be appropriate for the conditions chosen in the final design.
3) Different models often give very similar predictions over a limited range of conditions. However, the differences between different models are likely to become large if used outside the range over which they were fitted to experimental data.

4.6 Choice of Idealized Reactor Model

Consider now which of the idealized models is preferred for the six categories of reaction systems introduced in Section 4.2.

1) *Single reactions.* Consider the single reaction from Equation 4.1:

$$FEED \longrightarrow PRODUCT \quad r = kC_{FEED}^a \tag{4.63}$$

where
r = rate of reaction
k = reaction rate constant
C_{FEED} = molar concentration of *FEED*
a = order of reaction

Clearly, the highest rate of reaction is maintained by the highest concentration of feed (C_{FEED}, kmol·m^{-3}). As already discussed, in the mixed-flow reactor the incoming feed is instantly diluted by the product that has already been formed. The rate of reaction is thus lower in the mixed-flow reactor than in the ideal-batch and plug-flow reactors, since it operates at the low reaction rate corresponding with the outlet concentration of feed. Thus, a mixed-flow reactor requires a greater volume than an ideal-batch or plug-flow reactor. Consequently, for single reactions, an ideal-batch or plug-flow reactor is preferred.

2) *Multiple reactions in parallel producing byproducts.* Consider the system of parallel reactions from Equation 4.4 with the corresponding rate equations (Rase, 1977; Denbigh and Turner, 1984; Levenspiel, 1999):

$$\begin{aligned} FEED &\longrightarrow PRODUCT & r_1 &= k_1 C_{FEED}^{a_1} \\ FEED &\longrightarrow BYPRODUCT & r_2 &= k_2 C_{FEED}^{a_2} \end{aligned} \tag{4.64}$$

where
r_1, r_2 = rates of reaction for primary and secondary reactions
k_1, k_2 = reaction rate constants for primary and secondary reactions
C_{FEED} = molar concentration of *FEED* in the reactor
a_1, a_2 = order of reaction for primary and secondary reactions

4

The ratio of the rates of the secondary and primary reactions gives (Rase, 1977; Denbigh and Turner, 1984; Levenspiel, 1999):

$$\frac{r_2}{r_1} = \frac{k_2}{k_1} C_{FEED}^{a_2 - a_1} \tag{4.65}$$

Maximum selectivity requires a minimum ratio r_2/r_1 in Equation 4.65. A batch or plug-flow reactor maintains higher average concentrations of feed (C_{FEED}) than a mixed-flow reactor, in which the incoming feed is instantly diluted by the *PRODUCT* and *BYPRODUCT*. If $a_1 > a_2$ in Equations 4.64 and 4.65 the primary reaction to *PRODUCT* is favored by a high concentration of *FEED*. If $a_1 < a_2$ the primary reaction to *PRODUCT* is favored by a low concentration of *FEED*. Thus, if:

- $a_2 < a_1$, use a batch or plug-flow reactor,
- $a_2 > a_1$, use a mixed-flow reactor.

In general terms, if the reaction to the desired product has a higher order than the byproduct reaction, use a batch or plug-flow reactor. If the reaction to the desired product has a lower order than the byproduct reaction, use a mixed-flow reactor.

If the reaction involves more than one feed, the picture becomes more complex. Consider the reaction system from Equation 4.6 with the corresponding rate equations:

$$\begin{aligned} &FEED\,1 + FEED\,2 \longrightarrow PRODUCT \qquad && r_1 = k_1\, C_{FEED1}^{a_1}\, C_{FEED2}^{b_1} \\ &FEED\,1 + FEED\,2 \longrightarrow BYPRODUCT \qquad && r_2 = k_2\, C_{FEED1}^{a_2}\, C_{FEED2}^{b_2} \end{aligned} \tag{4.66}$$

where C_{FEED1}, C_{FEED2} = molar concentrations of *FEED 1* and *FEED 2* in the reactor
a_1, b_1 = order of reaction for the primary reaction
a_2, b_2 = order of reaction for the secondary reaction

Now the ratio that needs to be minimized is given by (Rase, 1977; Denbigh and Turner, 1984; Levenspiel, 1999):

$$\frac{r_2}{r_1} = \frac{k_2}{k_1} C_{FEED1}^{a_2 - a_1}\, C_{FEED2}^{b_2 - b_1} \tag{4.67}$$

Given this reaction system, the options are:

- Keep both C_{FEED1} and C_{FEED2} low (i.e. use a mixed-flow reactor).
- Keep both C_{FEED1} and C_{FEED2} high (i.e. use a batch or plug-flow reactor).
- Keep one of the concentrations high while maintaining the other low (this is achieved by charging one of the feeds as the reaction progresses).

Figure 4.6 summarizes these arguments to choose a reactor for systems of multiple reactions in parallel (Smith and Petela, 1992).

3) *Multiple reactions in series producing byproducts.* Consider the system of series reactions from Equation 4.7:

$$\begin{aligned} &FEED \longrightarrow PRODUCT \qquad && r_1 = k_1 C_{FEED}^{a_1} \\ &PRODUCT \longrightarrow BYPRODUCT \qquad && r_2 = k_2 C_{PRODUCT}^{a_2} \end{aligned} \tag{4.68}$$

where r_1, r_2 = rates of reaction for primary and secondary reactions
k_1, k_2 = reaction rate constants
C_{FEED} = molar concentration of *FEED*
$C_{PRODUCT}$ = molar concentration of *PRODUCT*
a_1, a_2 = order of reaction for primary and secondary reactions

For a certain reactor conversion, the *FEED* should have a corresponding residence time in the reactor. In the mixed-flow reactor, *FEED* can leave the instant it enters or remains for an extended period. Similarly, *PRODUCT* can remain for an extended period or leave immediately. Substantial fractions of both *FEED* and *PRODUCT* leave before and after what should be the specific residence time for a given conversion. Thus, the mixed-flow model would be expected to give a poorer selectivity or yield than a batch or plug-flow reactor for a given conversion.

A batch or plug-flow reactor should be used for multiple reactions in series.

4) *Mixed parallel and series reactions producing byproducts.* Consider the mixed parallel and series reaction system from Equation 4.10 with the corresponding kinetic equations:

$$\begin{aligned} &FEED \longrightarrow PRODUCT \qquad && r_1 = k_1\; C_{FEED}^{a_1} \\ &FEED \longrightarrow BYPRODUCT \qquad && r_2 = k_2\; C_{FEED}^{a_2} \\ &PRODUCT \longrightarrow BYPRODUCT \qquad && r_3 = k_3\; C_{PRODUCT}^{a_3} \end{aligned} \tag{4.69}$$

As far as the parallel byproduct reaction is concerned, for high selectivity, if:

- $a_1 > a_2$, use a batch or plug-flow reactor,
- $a_1 < a_2$, use a mixed-flow reactor.

The series byproduct reaction requires a plug-flow reactor. Thus, for the mixed parallel and series system above, if:

- $a_1 > a_2$, use a batch or plug-flow reactor.

However, what is the correct choice if $a_1 < a_2$? Now the parallel byproduct reaction calls for a mixed-flow reactor. On the other hand, the byproduct series reaction calls for a plug-flow reactor. It would seem that, given this situation, some level of mixing between a plug-flow and a mixed-flow reactor will give the best overall selectivity (Smith and Petela, 1992). This could be obtained by:

- a series of mixed-flow reactors (Figure 4.7a),
- a plug-flow reactor with a recycle (Figure 4.7b),
- a series combination of plug-flow and mixed-flow reactors (Figures 4.7c and 4.7d).

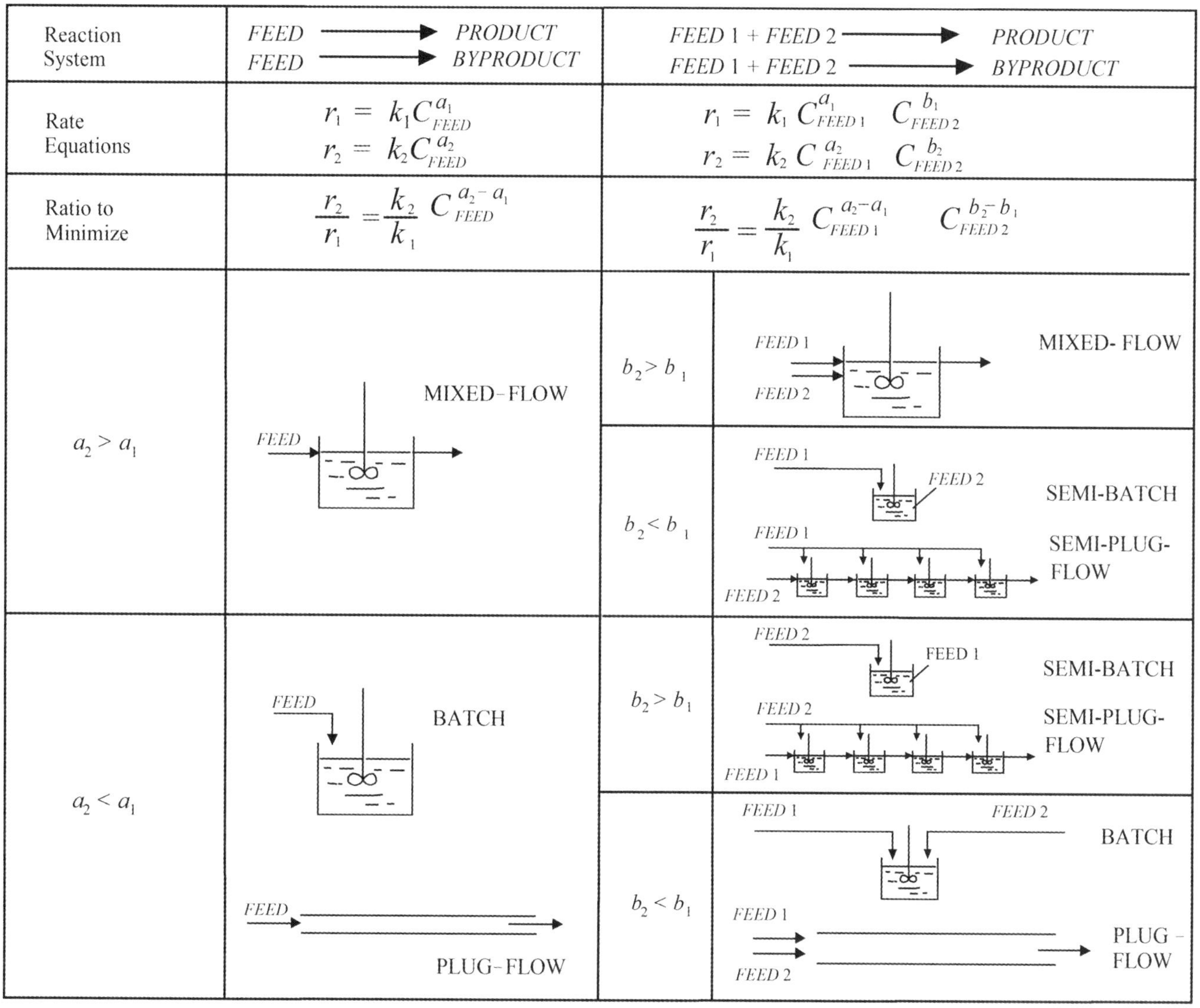

Figure 4.6

Reactor choice for parallel reaction systems. (Reproduced from Smith R and Petela EA (1992) Waste Minimization in the Process Industries Part 2 Reactors, Chem Eng, Dec (509–510): 17, by permission of the Institution of Chemical Engineers.)

The arrangement that gives the highest overall selectivity can only be deduced by a detailed analysis and optimization of the reaction system.

5) *Polymerization reactions.* Polymers are characterized mainly by the distribution of molar mass about the mean as well as by the mean itself. Orientation of groups along chains and cross-linking of the polymer chains also affect the properties. The breadth of distribution of molar mass depends on whether a batch or plug-flow reactor or a mixed-flow reactor is used. The breadth has an important influence on the mechanical and other properties of the polymer, and this is an important factor in the choice of reactor.

Two broad classes of polymerization reactions can be identified (Denbigh and Turner, 1984):

a) In a batch or plug-flow reactor, all molecules have the same residence time, and without the effect of termination (see Section 4.2) all will grow to approximately equal lengths, producing a narrow distribution of molar masses. By contrast, a mixed-flow reactor will cause a wide distribution because of the distribution of residence times in the reactor.

b) When polymerization takes place by mechanisms involving free radicals, the life of these actively growing centers may be extremely short because of termination processes such as the union of two free radicals (see Section 4.2). These termination processes are influenced by free-radical concentration, which in turn is proportional to monomer concentration. In batch or plug-flow reactors, the monomer and free-radical concentrations decline. This produces increasing chain lengths with increasing residence time and thus a broad distribution of molar masses. The mixed-flow reactor

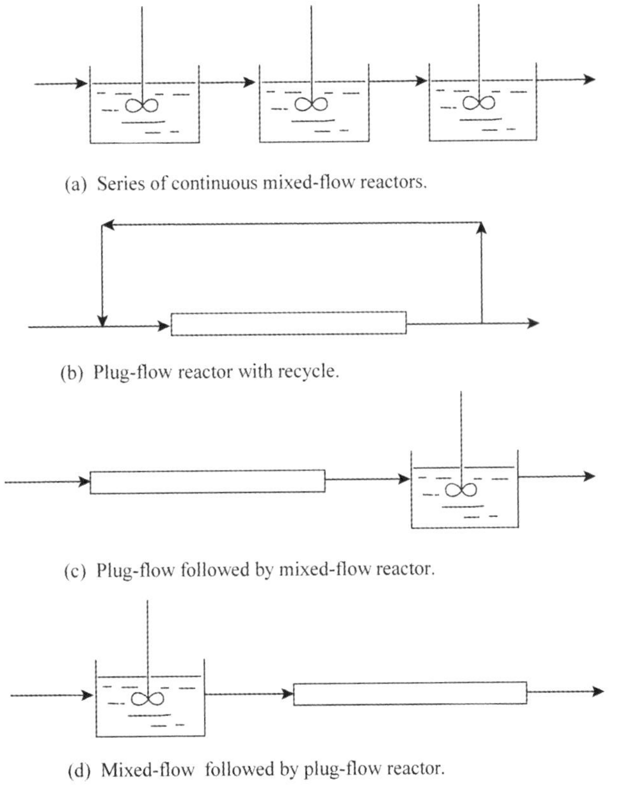

Figure 4.7

Choice of reactor type for mixed parallel and series reactions when the parallel reaction has a higher order than the primary reaction.

maintains a uniform concentration of monomer and thus a constant chain-termination rate. This results in a narrow distribution of molar masses. Because the active life of the polymer is short, the variation in residence time does not have a significant effect.

6) *Biochemical reactions.* As discussed previously, there are two broad classes of biochemical reactions: reactions that exploit the metabolic pathways in selected microorganisms and those catalyzed by enzymes. Consider a microbial biochemical reactive of the type:

$$A \xrightarrow{C} R + C$$

where A is the feed, R is the product and C represents the cells (microorganisms). The kinetics of such reactions can be described by the Monod Equation (Levenspiel, 1999; Shuler and Kargi, 2002):

$$-r_A = k\frac{C_C\,C_A}{C_A + C_M} \tag{4.70}$$

where r_A = rate of reaction
k = rate constant
C_C = concentration of cells (microorganisms)
C_A = concentration of feed
C_M = a constant (Michaeli's constant that is a function of the reaction and conditions)

The rate constant can depend on many factors, such as temperature, the presence of trace elements, vitamins, toxic substances, light intensity, and so on. The rate of reaction depends not only on the availability of feed (food for the microorganisms) but also on the build-up of wastes from the microorganisms that interfere with microorganism multiplication. There comes a point at which their waste inhibits growth. An excess of feed material or microorganisms can slow the rate of reaction. Without poisoning of the kinetics due to the build-up of waste material, Equation 4.70 gives the same characteristic curve as autocatalytic reactions, as shown in Figure 4.4b. Thus, depending on the concentration range, mixed-flow, plug-flow, a combination of mixed-flow and plug-flow or mixed-flow with separation and recycle might be appropriate.

For enzyme-catalyzed biochemical reactions of the form:

$$A \xrightarrow{\text{enzymes}} R$$

The kinetics can be described by the Monod Equation in the form (Levenspiel, 1999; Shuler and Kargi, 2002):

$$-r_A = k\frac{C_E\,C_A}{C_M + C_A} \tag{4.71}$$

where C_E = enzyme concentration

The presence of some substances can cause the reaction to slow down. Such substances are known as *inhibitors*. This is caused either by the inhibitor competing with the feed material for active sites on the enzyme or by the inhibitor attacking an adjacent site and, in so doing, inhibiting the access of the feed material to the active site.

A high reaction rate in Equation 4.71 is favored by a high concentration of enzymes (C_E) and a high concentration of feed (C_A). This means that a plug-flow or ideal-batch reactor is favored if both the feed material and enzymes are to be fed to the reactor.

4.7 Choice of Reactor Performance

It is now possible to define the goals for the choice of the reactor at this stage in the design. Unconverted material usually can be separated and recycled later. Because of this, the reactor conversion cannot be fixed finally until the design has progressed much further than just choosing the reactor. As shall be seen later, the choice of reactor conversion has a major influence on the rest of the process. Nevertheless, some decisions must be made regarding the reactor for the design to proceed. Thus, some initialization for the reactor conversion must be made in the knowledge that this is likely to change once more detail is added to the total system design.

Unwanted byproducts usually cannot be converted back to useful products or raw materials. The reaction to unwanted byproducts creates both raw materials costs due to the raw materials that are wasted in their formation and environmental costs for their disposal. Thus, maximum selectivity is required for the chosen reactor conversion. The objectives at this stage can be summarized as follows.

1) *Single reactions.* In a single reaction, such as Equation 4.2, that produces a byproduct, there can be no influence on the relative amounts of product and byproduct formed. Thus, with single reactions such as Equations 4.1 to 4.3, the goal is to minimize the reactor capital cost, which often (but not always) means minimizing reactor volume, for a given reactor conversion. Increasing the reactor conversion increases size and hence cost of the reactor but, as will be discussed later, decreases the cost of many other parts of the flowsheet. Because of this, the initial setting for reactor conversion for single irreversible reactions is around 95% and that for a single reversible reaction is around 95% of the equilibrium conversion (Smith and Petela, 1992). Equilibrium conversion will be considered in more detail in the next chapter.

   For batch reactors, account must be taken of the time required to achieve a given conversion. Batch cycle time is addressed later.
2) *Multiple reactions in parallel producing byproducts.* Raw materials costs usually will dominate the economics of the process. Because of this, when dealing with multiple reactions, whether parallel, series or mixed, the goal is usually to minimize byproduct formation (maximize selectivity) for a given reactor conversion. Chosen reactor conditions should exploit differences between the kinetics and equilibrium effects in the primary and secondary reactions to favor the formation of the desired product rather than the byproduct, that is, improve the selectivity. Suggesting a reasonable initial setting for conversion is more difficult than with single reactions, since the factors that affect conversion also can have a significant effect on selectivity.

   Consider the system of parallel reactions from Equations 4.64 and 4.65. A high conversion in the reactor tends to decrease C_{FEED}. Thus:

   - $a_2 > a_1$ selectivity increases as conversion increases.
   - $a_2 < a_1$ selectivity decreases as conversion increases.

   If selectivity increases as conversion increases, the initial setting for reactor conversion should be in the order of 95% and that for reversible reactions should be in the order of 95% of the equilibrium conversion. If selectivity decreases with increasing conversion, then it is much more difficult to give guidance. An initial setting of 50% for the conversion for irreversible reactions or 50% of the equilibrium conversion for reversible reactions is as reasonable as can be suggested at this stage. However, these are only initial estimates and will almost certainly be changed later.
3) *Multiple reactions in series producing byproduct.* Consider the system of series reactions from Equation 4.68. Selectivity for series reactions of the types given in Equations 4.7 to 4.9 is increased by low concentrations of reactants involved in the secondary reactions. This means reactor operation with a low concentration of *PRODUCT*, in other words, with low conversion. For series reactions, a significant reduction in selectivity is likely as the conversion increases.

   Again, it is difficult to select the initial setting of the reactor conversion with systems of reactions in series. A conversion of 50% for irreversible reactions or 50% of the equilibrium conversion for reversible reactions is as reasonable as can be suggested at this stage.

   Multiple reactions also can occur with impurities that enter with the feed and undergo reaction. Again, such reactions should be minimized, but the most effective means of dealing with byproduct reactions caused by feed impurities is not to alter reactor conditions but to carry out feed purification.

4.8 Reactor Performance – Summary

Some initialization for the reactor conversion must be made in order that the design can proceed. This is likely to change later in the design because, as will be seen later, there is a strong interaction between the reactor conversion and the rest of the flowsheet.

1) *Single reactions.* For single reactions, a reasonable initial setting is 95% conversion for irreversible reactions and 95% of the equilibrium conversion for reversible reactions.
2) *Multiple reactions.* For multiple reactions, where the byproduct is formed in parallel, the selectivity may increase or decrease as conversion increases. If the byproduct reaction is a higher order than the primary reaction, selectivity increases for increasing reactor conversion. In this case, the same initial setting as single reactions should be used. If the byproduct reaction of the parallel system is a lower order than the primary reaction, a lower conversion than that for single reactions is expected to be appropriate. The best initial suggestion at this stage is to set the conversion to 50% for irreversible reactions or to 50% of the equilibrium conversion for reversible reactions.

   For multiple reactions, where the byproduct is formed in series, the selectivity decreases as conversion increases. In this case, lower conversion than that for single reactions is expected to be appropriate. Again, the best suggestion at this stage is to set the conversion to 50% for irreversible reactions or to 50% of the equilibrium conversion for reversible reactions.

It should be emphasized that these recommendations for the initial settings of the reactor conversion will almost certainly change at a later stage, since reactor conversion is an extremely important optimization variable.

When dealing with multiple reactions, selectivity or reactor yield is maximized for the chosen conversion. The choice of mixing pattern in the reactor and feed addition policy should be chosen to this end.

4.9 Exercises

1. Acetic anhydride is to be produced from acetone and acetic acid. In the first stage of the process, acetone is decomposed at 700 °C and 1.013 bar to ketene via the reaction:

$$\underset{\text{acetone}}{CH_3COCH_3} \longrightarrow \underset{\text{ketene}}{CH_2CO} + \underset{\text{methane}}{CH_4}$$

Unfortunately, some of the ketene formed decomposes further to form unwanted ethylene and carbon monoxide via the reaction:

$$\underset{\text{ketene}}{CH_2CO} \longrightarrow \underset{\text{ethylene}}{{}^1\!/_2\, C_2H_4} + \underset{\text{carbon monoxide}}{CO}$$

Laboratory studies of these reactions indicate that the ketene selectivity S (kmol ketene formed per kmol acetone converted) varies with conversion X (kmol acetone reacted per kmol acetone fed) follows the relationship (Jeffreys, 1964):

$$S = 1 - 1.3X$$

The second stage of the process requires the ketene to be reacted with glacial acetic acid at 80 °C and 1.013 bar to produce acetic anhydride via the reaction:

$$\underset{\text{ketene}}{CH_2CO} + \underset{\text{acetic acid}}{CH_3COOH} \longrightarrow \underset{\text{acetic anhydride}}{CH_3COOCOCH_3}$$

The values of the chemicals involved, together with their molar masses are given in Table 4.12.

Assuming that the plant will produce 15,000 $t{\cdot}y^{-1}$ acetic anhydride:

a) Calculate the economic potential assuming the side reaction can be suppressed and hence obtain 100% yield.

b) Determine the range of acetone conversions (X) over which the plant will be profitable if the side reaction cannot be suppressed.

Table 4.12

Data for acetic anhydride production.

Chemical	Molar mass ($kg{\cdot}kmol^{-1}$)	Value ($\${\cdot}kg^{-1}$)
Acetone	58	0.60
Ketene	42	0
Methane	16	0
Ethylene	28	0
Carbon monoxide	28	0
Acetic acid	60	0.54
Acetic anhydride	102	0.90

Table 4.13

Experimental data for a simple reaction.

Time (min)	Concentration ($kg{\cdot}m^{-3}$)
0	16.0
10	13.2
20	11.1
35	8.8
50	7.1

c) Published data indicates that the capital cost for the project will be at least \$35 m. If annual fixed charges are assumed to be 15% of the capital cost, revise the range of conversions over which the plant will be profitable.

2. Experimental data for a simple reaction showing the rate of change of reactant with time are given to Table 4.13.

Show that the data gives a kinetic equation of order 1.5 and determine the rate constant.

3. Component A reacts to Component B in an irreversible reaction in the liquid phase. The kinetics are first order with respect to A with reaction rate constant $k_A = 0.003\ min^{-1}$. Find the residence times for 95% conversion of A for:

a) a mixed-flow reactor,

b) three mixed-flow reactors in series of equal volume,

c) a plug-flow reactor.

4. Styrene (A) and butadiene (B) are to be polymerized in a series of mixed-flow reactors, each of volume 50 m^3. The rate of reaction of both A and B are given by:

$$r = k\,C_A C_B$$

where $k = 10^{-5}\ m^3{\cdot}kmol^{-1}{\cdot}s^{-1}$

The initial concentration of styrene is 0.8 $kmol{\cdot}m^{-3}$ and of butadiene is 3.6 $kmol{\cdot}m^{-3}$. The feed rate of reactants is 20 $t{\cdot}h^{-1}$. Estimate the total number of reactors required for polymerization of 70% of the styrene. Assume the density of reaction mixture to be 870 $kg{\cdot}m^{-3}$ and the molar mass of styrene is 104 $kg{\cdot}kmol^{-1}$ and that of butadiene is 54 $kg{\cdot}kmol^{-1}$.

5. It is proposed to react 10 $t{\cdot}h^{-1}$ of a pure liquid A to a desired product B. Byproducts C and D are formed through series and parallel reactions:

$$A \xrightarrow{k_1} B \xrightarrow{k_2} C$$
$$B \xrightarrow{k_3} D$$

$$k_1 = k_2 = k_3 = 0.1\ min^{-1}$$

Assume rates of reaction to be first order, the average density to be 800 $kg{\cdot}m^{-3}$ and an inlet concentration of A of

$15\,\mathrm{kmol \cdot m^{-3}}$. Calculate the required residence time, volume and the yield for a reactor that will give the maximum yield of B:

a) a single mixed-flow reactor,

b) three equal-sized mixed-flow reactors in series.

6. What reactor configuration should be used to maximize the selectivity in the following parallel reactions:

$$A + B \longrightarrow R \quad r_R = 15\, C_A^{0.5} C_B$$
$$A + B \longrightarrow S \quad r_S = 15\, C_A C_B$$

where R is the desired product and S is the undesired product.

7. A desired liquid-phase reaction:

$$A + B \xrightarrow{k_1} R \qquad r_R = k_1 C_A^{0.3} C_B$$

is accompanied by a parallel reaction:

$$A + B \xrightarrow{k_2} S \qquad r_S = k_2 C_A^{1.5} C_B^{0.5}$$

a) A number of reactor configurations are possible. Ideal-batch, semi-batch, plug-flow and semi-plug-flow could all be used. In the case of the semi-batch and semi-plug-flow reactors, the order of addition of A and B can be changed. Order the reactor configurations from the least desirable to the most desirable to maximize production of the desired product.

b) The feed stream to a mixed-flow reactor is an equimolar mixture of pure A and B (each has a density of $20\,\mathrm{kmol \cdot m^{-3}}$). Assuming $k_1 = k_2 = 1.0\,\mathrm{kmol \cdot m^{-3} \cdot min^{-1}}$, calculate the composition of the exit stream from the mixed-flow reactor for a conversion of 90%.

8. For a free-radical polymerization and a condensation polymerization process, explain why the molar mass distribution of the polymer product will be different depending on whether a mixed-flow or a plug-flow reactor is used. What will be the difference in the distribution of molar mass?

References

Denbigh KG and Turner JCR (1984) *Chemical Reactor Theory*, 3rd Edition, Cambridge University Press.

Douglas JM (1985) A Hierarchical Decision Procedure for Process Synthesis, *AIChE J*, **31**: 353.

Dwight HB (1961) *Tables of Integrals and Other Mathematical Data*, The Macmillan Company.

Helminen J, Leppamaki M, Paatero E and Minkkinen P (1998) Monitoring the Kinetics of the Ion-Exchange Resin Catalysed Esterification of Acetic Acid with Ethanol Using Near Infrared Spectroscopy with Partial Least Squares (PLS) Model, *Chemometr Intell Lab Syst*, **44**: 341.

Huang I and Dauerman L (1969) Exploratory Process Study, *Ind Eng Chem Prod Res Dev*, **8**(3): 227.

Jeffreys GV (1964) *A Problem in Chemical Engineering Design — The Manufacture of Acetic Acid*, The Institution of Chemical Engineers, UK.

Levenspiel O (1999) *Chemical Reaction Engineering*, 3rd Edition, John Wiley & Sons.

Rase HF (1977) *Chemical Reactor Design for Process Plants*, Vol. 1, John Wiley & Sons.

Rudd DF, Powers GJ and Siirola JJ (1973) *Process Synthesis*, Prentice Hall.

Shuler ML and Kargi F (2002) *Bioprocess Engineering*, 2nd Edition, Prentice Hall.

Smith R and Petela EA (1992) Waste Minimization in the Process Industries. Part 2: Reactors, *Chem Eng*, **Dec**(509–510): 17.

Waddams AL (1978) *Chemicals from Petroleum*, John Murray.

Wells GL and Rose LM (1986) *The Art of Chemical Process Design*, Elsevier.

Chapter 5

Chemical Reactors II – Reactor Conditions

5.1 Reaction Equilibrium

In the preceding chapter, the choice of reactor type was made on the basis of the most appropriate concentration profile as the reaction progressed, in order to minimize reactor volume for single reactions or maximize selectivity (or yield) for multiple reactions for a given conversion. However, there are still important effects regarding reaction conditions of temperature, pressure, phase and concentration to be considered. Before considering reaction conditions, some basic principles of chemical equilibrium need to be reviewed.

Reactions can be considered to be either reversible or essentially irreversible. An example of a reaction that is essentially irreversible is:

$$\underset{\text{ethylene}}{C_2H_4} + \underset{\text{chlorine}}{Cl_2} \rightarrow \underset{\text{dichloroethane}}{C_2H_4Cl_2} \tag{5.1}$$

An example of a reversible reaction is:

$$\underset{\text{hydrogen}}{3H_2} + \underset{\text{nitrogen}}{N_2} \rightleftharpoons \underset{\text{ammonia}}{2NH_3} \tag{5.2}$$

For reversible reactions:

a) For a given mixture of reactants at a given temperature and pressure, there is a maximum conversion (the *equilibrium conversion*) that cannot be exceeded and is independent of the reactor design.

b) The equilibrium conversion can be changed by appropriate changes to the concentrations of reactants, temperature and pressure.

The key to understanding reaction equilibrium is the *Gibbs free energy*, or *free energy*, defined as (Dodge, 1944; Hougen, Watson and Ragatz, 1959; Coull and Stuart, 1964; Smith, 1990):

$$G = H - TS \tag{5.3}$$

where G = free energy (kJ)
H = enthalpy (kJ)
T = absolute temperature (K)
S = entropy ($kJ \cdot K^{-1}$)

Like enthalpy, the absolute value of G for a substance cannot be measured, and only changes in G are meaningful. For a process occurring at constant temperature, the change in free energy from Equation 5.3 is:

$$\Delta G = \Delta H - T\,\Delta S \tag{5.4}$$

A negative value of ΔG implies a spontaneous reaction of reactants to products. A positive value of ΔG implies the reverse reaction is spontaneous. This is illustrated in Figure 5.1. A system is in equilibrium when (Dodge, 1944; Hougen, Watson and Ragatz, 1959; Coull and Stuart, 1964; Smith, 1990):

$$\Delta G = 0 \tag{5.5}$$

The free energy can also be expressed in differential form (Dodge, 1944; Hougen, Watson and Ragatz, 1959; Coull and Stuart, 1964; Smith, 1990):

$$dG = -S\,dT + V\,dP \tag{5.6}$$

where V is the system volume. At constant temperature, $dT = 0$ and Equation 5.6 becomes:

$$dG = V\,dP \tag{5.7}$$

For N moles of an ideal gas, $PV = NRT$, where R is the universal gas constant. Substituting this in Equation 5.7 gives:

$$dG = NRT\frac{dP}{P} \tag{5.8}$$

Chemical Process Design and Integration, Second Edition. Robin Smith.

Companion Website: www.wiley.com/go/smith/chemical2e

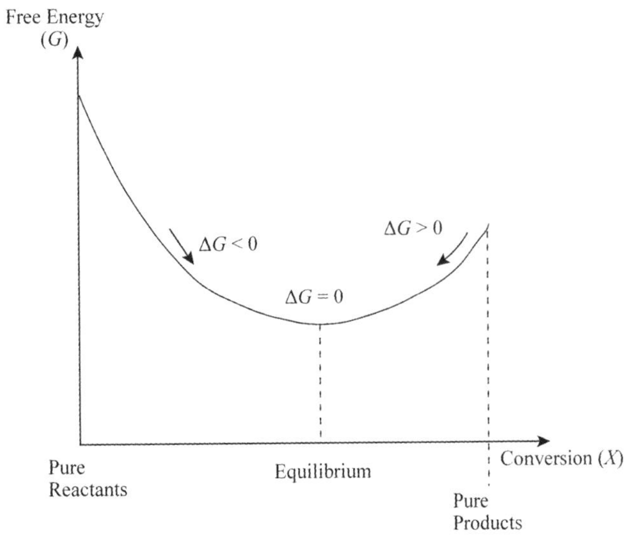

Figure 5.1

Variation of free energy of a reaction mixture.

For a change in pressure from P_1 to P_2 Equation 5.8 can be integrated to give:

$$\Delta G = G_2 - G_1 = NRT \int_{P_1}^{P_2} \frac{dP}{P} \tag{5.9}$$

Values for free energy are usually referred to the standard free energy G^o. The standard state is arbitrary and designates the datum level. A gas is normally considered to be at a standard state if it is at a pressure of 1 bar for the designated temperature of an isothermal process. Thus, integrating Equation 5.9 from standard pressure P_O to pressure P gives:

$$G = G^o + NRT \ln \frac{P}{P_o} \tag{5.10}$$

For a real system, rather than an ideal gas, Equation 5.10 is written as:

$$\begin{aligned} G &= G^o + NRT \ln \frac{f}{f^o} \\ &= G^o + NRT \ln a \end{aligned} \tag{5.11}$$

where f = fugacity (N·m^2 or bar)
f^o = fugacity at standard conditions (N·m^2 or bar)
a = activity (–)
$= f/f^o$

The concepts of *fugacity* and *activity* have no strict physical significance but are introduced to transform equations for ideal systems to real systems. However, it does help to relate them to something that does have physical significance. Fugacity can be regarded as an "effective pressure" and activity can be regarded as an "effective concentration" relative to a standard state (Smith, 1990).

Consider the general reaction:

$$bB + cC + \cdots \rightleftharpoons sS + tT + \cdots \tag{5.12}$$

The free energy change for the reaction is given by:

$$\Delta G = s\overline{G}_S + t\overline{G}_T + \cdots - b\overline{G}_B - c\overline{G}_C - \cdots \tag{5.13}$$

where ΔG = free energy change of reaction (kJ)
$\overline{G}_i$ = partial molar free energy of Component i (kJ·kmol^{-1})

Note that the *partial molar free energy* is used to designate the molar free energy of the individual component *as it exists in the mixture*. This is necessary because, except in ideal systems, the properties of a mixture are not additive properties of the pure components. As a result, $G = \sum_i N_i \overline{G}_i$ for a mixture. The free energy change for reactants and products at standard conditions is given by:

$$\Delta G^O = s\overline{G}_S^O + t\overline{G}_T^O + \cdots - b\overline{G}_B^O - c\overline{G}_C^O - \cdots \tag{5.14}$$

where ΔG^O = free energy change of reaction when all of the reactants and products are at their respective standard conditions (kJ)
$\overline{G}_i^O$ = partial molar free energy of Component i at standard conditions (kJ·kmol^{-1})

Combining Equations 5.13 and 5.14:

$$\begin{aligned} \Delta G - \Delta G^O = {} & s(\overline{G}_S - \overline{G}_S^O) + t(\overline{G}_T - \overline{G}_T^O) + \cdots \\ & - b(\overline{G}_B - \overline{G}_B^O) - c(\overline{G}_c - \overline{G}_C^O) - \cdots \end{aligned} \tag{5.15}$$

Writing Equation 5.11 for the partial molar free energy of Component i in a mixture:

$$\overline{G}_i - \overline{G}_i^O = RT \ln a_i \tag{5.16}$$

where a_i refers to the activity of Component i in the mixture. Substituting Equation 5.16 in Equation 5.15 gives:

$$\begin{aligned} \Delta G - \Delta G^O = {} & sRT \ln a_S + tRT \ln a_T + \cdots \\ & - bRT \ln a_B - cRT \ln a_c - \cdots \end{aligned} \tag{5.17}$$

Since $s \ln a_s = \ln a_S^s$, $t \ln a_T = \ln a_T^t$, . . . , Equation 5.17 can be arranged to give:

$$\Delta G - \Delta G^O = RT \ln \left(\frac{a_S^s a_T^t \cdots}{a_B^b a_C^c \cdots} \right) \tag{5.18}$$

Note that in rearranging $s \ln a_S$ to $\ln a_S^s$, the units of $RT \ln a_S^s$ remain kJ and a_S^s is dimensionless. At equilibrium, $\Delta G = 0$ and Equation 5.18 becomes:

$$\Delta G^O = -RT \ln \left(\frac{a_S^s a_T^t \cdots}{a_B^b a_C^c \cdots} \right) = -RT \ln K_a \tag{5.19}$$

where

$$K_a = \frac{a_S^s a_T^t \cdots}{a_B^b a_C^c \cdots} \tag{5.20}$$

K_a is known as the equilibrium constant. It represents the equilibrium activities for a system under standard conditions and is a constant at constant temperature.

1) *Homogeneous gaseous reactions.* Substituting for the fugacity in Equation 5.20 gives:

$$K_a = \left(\frac{f_S^s f_T^t \cdots}{f_B^b f_C^c \cdots}\right)\left(\frac{f_B^{Ob} f_C^{Oc} \cdots}{f_S^{Os} f_T^{Ot} \cdots}\right) \tag{5.21}$$

For homogeneous gaseous reactions, the standard state fugacities can be considered to be unity, that is, $f_i^O = 1$. If the fugacity is expressed as the product of partial pressure and fugacity coefficient (Dodge, 1944; Hougen, Watson and Ragatz, 1959; Coull and Stuart, 1964; Smith, 1990):

$$f_i = \phi_i p_i \tag{5.22}$$

where f_i = fugacity of Component i
ϕ_i = fugacity coefficient of Component i
p_i = partial pressure of Component i

Substituting Equation 5.22 into Equation 5.21, assuming $f_i^O = 1$:

$$K_a = \left(\frac{\varphi_S^s \varphi_T^t \cdots}{\varphi_B^b \varphi_C^c \cdots}\right)\left(\frac{p_S^s p_T^t \cdots}{p_B^b p_C^c \cdots}\right) = K_\varphi K_P \tag{5.23}$$

Also, by definition (see Appendix A):

$$p_i = y_i P \tag{5.24}$$

where y_i = mole fraction of Component i
P = system pressure

Substituting Equation 5.24 into Equation 5.23 gives:

$$K_a = \left(\frac{\varphi_S^s \varphi_T^t \cdots}{\varphi_B^b \varphi_C^c \cdots}\right)\left(\frac{y_S^s y_T^t \cdots}{y_B^b y_C^c \cdots}\right) P^{\Delta N} = K_\varphi K_y P^{\Delta N} \tag{5.25}$$

where $\Delta N = s + t + \cdots - b - c - \cdots$

For an ideal gas $\phi_i = 1$, thus:

$$K_a = K_p = \left(\frac{p_S^s p_T^t \cdots}{p_B^b p_C^c \cdots}\right) \tag{5.26}$$

$$K_a = K_y P^{\Delta N} = \left(\frac{y_S^s y_T^t \cdots}{y_B^b y_C^c \cdots}\right) P^{\Delta N} \tag{5.27}$$

Note that whilst K_a appears to be dimensional, it is actually dimensionless, since the $\left(f_B^{Ob} f_C^{Oc} \cdots / f_S^{Os} f_T^{Ot} \cdots\right)$ term in Equation 5.21 goes to unity for $f_i^o = 1$ bar, but retains the units of pressure.

2) *Homogeneous liquid reactions.* For homogeneous liquid reactions, the activity can be expressed as (see Appendix A and Dodge, 1944; Hougen, Watson and Ragatz, 1959; Coull and Stuart, 1964; Smith, 1990):

$$a_i = \gamma_i x_i \tag{5.28}$$

where a_i = activity of Component i
γ_i = activity coefficient for Component i
x_i = mole fraction of Component i

The activity coefficient in Equation 5.28 ($\gamma_i = a_i/x_i$) can be considered to be the ratio of the effective to actual concentration. Substituting Equation 5.28 into Equation 5.20 gives:

$$K_a = \left(\frac{\gamma_S^s \gamma_T^t \cdots}{\gamma_B^b \gamma_C^c \cdots}\right)\left(\frac{x_S^s x_T^t \cdots}{x_B^b x_C^c \cdots}\right) = K_\gamma K_x \tag{5.29}$$

For ideal solutions, the activity coefficients are unity, giving:

$$K_a = \left(\frac{x_S^s x_T^t \cdots}{x_B^b x_C^c \cdots}\right) \tag{5.30}$$

3) *Heterogeneous reactions.* For a heterogeneous reaction, the state of all components is not uniform, for example, a reaction between a gas and a liquid. This requires standard states to be defined for each component. The activity of a solid in the equilibrium constant can be taken to be unity.

Consider now an example to illustrate the application of these thermodynamic principles.

Example 5.1 A stoichiometric mixture of nitrogen and hydrogen is to be reacted at 1 bar:

$$\underset{\text{hydrogen}}{3H_2} + \underset{\text{nitrogen}}{N_2} \rightleftharpoons \underset{\text{ammonia}}{2NH_3}$$

Assuming ideal gas behavior ($R = 8.3145\ \text{kJ·K}^{-1}\text{·kmol}^{-1}$), calculate:

a) equilibrium constant
b) equilibrium conversion of hydrogen
c) composition of the reaction products at equilibrium

at 300 K. Standard free energy of formation data for a standard pressure of 1 bar are given in Table 5.1 (Lide, 2010).

Table 5.1

Standard free energy of formation data for ammonia synthesis.

Component	$\overline{G}_{300}^O$ (kJ.kmol^{-1})
H_2	0
N_2	0
NH_3	−16,223

Solution

a) $$3H_2 + N_2 \rightleftharpoons 2NH_3$$

$$\Delta G^O = 2\bar{G}^O_{NH_3} - 3\bar{G}^O_{H_2} - \bar{G}^O_{N_2}$$

$$= -RT \ln K_a$$

$$\ln K_a = \frac{-(2\bar{G}^O_{NH_3} - 3\bar{G}^O_{H_2} - \bar{G}^O_{N_2})}{RT}$$

At 300 K:

$$\ln K_a = \frac{-(-2 \times 16223 - 0 - 0)}{8.3145 \times 300}$$

$$= 13.008$$

$$K_a = 4.4597 \times 10^5$$

Also:

$$K_a = \frac{p^2_{NH_3}}{p^3_{H_2} p_{N_2}}$$

Note again that whilst K_a appears to be dimensional, it is actually dimensionless, since $f_i^O = 1$ bar. Note also that K_a depends on the specification of the number of moles in the stoichiometric equation. For example, if the stoichiometry is written as:

$$\tfrac{3}{2}H_2 + \tfrac{1}{2}N_2 \rightleftharpoons NH_3$$

$$K_a' = \frac{p_{NH_3}}{p^{3/2}_{H_2} p^{1/2}_{N_2}}$$

$$K_a' = \sqrt{K_a}$$

It does not matter which specification is adopted as long as one specification is used consistently.

b) The number of moles and mole fractions, initially and at equilibrium, are given in Table 5.2.

$$K_a = \frac{y^2_{NH_3}}{y^3_{H_2} y_{N_2}} P^{-2}$$

For $P = 1$ bar:

$$K_a = \frac{16X^2(2-X)^2}{27(1-X)^4}$$

At 300 K:

$$4.4597 \times 10^5 = \frac{16X^2(2-X)^2}{27(1-X)^4}$$

This can be solved by trial and error or using spreadsheet software (see Section 3.8):

$$X = 0.97$$

c) To calculate the composition of the reaction products at equilibrium, $X = 0.97$ is substituted in the expressions from Table 5.2 at equilibrium:

$$y_{H_2} = 0.0437$$
$$y_{N_2} = 0.0146$$
$$y_{NH_3} = 0.9418$$

Some points should be noted from this example. Firstly, ideal gas behavior has been assumed. This is an approximation, but it is reasonable for the low pressure assumed in the calculation. Later, the calculation will be repeated at higher pressure when the ideal gas approximation will be poor. Also, it should be clear that the calculation is very sensitive to the thermodynamic data. Errors in the thermodynamic data can lead to a significantly different result. Thermodynamic data, even from reputable sources, should be used with caution.

Table 5.2

Moles and mole fractions for ammonia synthesis.

	H_2	N_2	NH_3
Moles in initial mixture	3	1	0
Moles at equilibrium	$3-3X$	$1-X$	$2X$
Mole fraction at equilibrium	$\frac{3(1-X)}{4-2X}$	$\frac{1-X}{4-2X}$	$\frac{2X}{4-2X}$

Equation 5.19 can be interpreted qualitatively to give guidance on the equilibrium conversion. If ΔG^O is negative, the position of the equilibrium will correspond to the presence of more products than reactants ($\ln K_a > 0$). If ΔG^O is positive ($\ln K_a < 0$), the reaction will not proceed to such an extent and reactants will predominate in the equilibrium mixture. Table 5.3 presents some

Table 5.3

Variation of equilibrium composition with ΔG^O and the equilibrium constant at 298 K.

ΔG^O (kJ)	K_a	Composition of equilibrium mixture
−50,000	6×10^8	Negligible reactants
−10,000	57	Products dominate
−5000	7.5	
0	1.0	
+5000	0.13	
+10,000	0.02	Reactants dominate
+50,000	1.7×10^{-9}	Negligible products

Source: Reproduced from Smith EB, 1990, *Basic Chemical Thermodynamics*, 3rd Edition, Oxford Chemistry Series, by permission of Oxford University Press.

guidelines as to the composition of the equilibrium mixture for various values of ΔG^O and the equilibrium constant (Smith, 1990).

When setting the conditions in chemical reactors, equilibrium conversion will be a major consideration for reversible reactions. The equilibrium constant K_a is only a function of temperature, and Equation 5.19 provides the quantitative relationship. However, pressure change and change in concentration can be used to shift the equilibrium by changing the activities in the equilibrium constant, as will be seen later.

A basic principle that allows the qualitative prediction of the effect of changing reactor conditions on any chemical system in equilibrium is *Le Châtelier's Principle*:

> "If any change in the conditions of a system in equilibrium causes the equilibrium to be displaced, the displacement will be in such a direction as to oppose the effect of the change."

Le Châtelier's Principle allows changes to be directed to increase equilibrium conversion. Now consider the setting of conditions in chemical reactors.

5.2 Reactor Temperature

The choice of reactor temperature depends on many factors. Consider first the effect of temperature on equilibrium conversion. A quantitative relationship can be developed as follows. Start by writing Equation 5.6 at constant pressure:

$$dG = -S\,dT \tag{5.31}$$

Equation 5.31 can be written as:

$$\left(\frac{\partial G}{\partial T}\right)_P = -S \tag{5.32}$$

Substituting Equation 5.31 into Equation 5.3 gives, after rearranging:

$$\left(\frac{\partial G}{\partial T}\right)_P = \frac{G}{T} - \frac{H}{T} \tag{5.33}$$

At standard conditions, G and H are not functions of pressure, by definition. Thus, Equation 5.33 can be written at standard conditions for finite changes in G^O and H^O as:

$$\frac{d\,\Delta G^O}{dT} = \frac{\Delta G^O}{T} - \frac{\Delta H^O}{T} \tag{5.34}$$

Also, because:

$$\begin{aligned}\frac{d}{dT}\left(\frac{\Delta G^O}{T}\right) &= \frac{1}{T}\frac{d\,\Delta G^O}{dT} + \Delta G^O\frac{d(1/T)}{dT}\\ &= \frac{1}{T}\frac{d\,\Delta G^O}{dT} - \frac{\Delta G^O}{T^2}\end{aligned} \tag{5.35}$$

Combining Equations 5.34 and 5.35 gives:

$$\frac{d}{dT}\left(\frac{\Delta G^O}{T}\right) = -\frac{\Delta H^O}{T^2} \tag{5.36}$$

Substituting ΔG^O (kJ) from Equation 5.19 into Equation 5.36 gives:

$$\frac{d\ln K_a}{dT} = \frac{\Delta H^O}{RT^2} \tag{5.37}$$

Equation 5.37 can be integrated to give:

$$\ln K_{a2} - \ln K_{a1} = \frac{1}{R}\int_{T_1}^{T_2}\frac{\Delta H^O}{T^2}\,dT \tag{5.38}$$

If it is assumed that ΔH^O is independent of temperature, Equation 5.38 gives:

$$\ln\frac{K_{a2}}{K_{a1}} = -\frac{\Delta H^O}{R}\left(\frac{1}{T_2} - \frac{1}{T_1}\right) \tag{5.39}$$

In this expression, K_{a1} is the equilibrium constant at T_1 and K_{a2} is the equilibrium constant at T_2. ΔH^O is the standard heat of reaction (kJ) when all the reactants and products are at standard state, given by:

$$\Delta H^O = (s\overline{H}_S^O + t\overline{H}_T^O + \cdots) - (b\overline{H}_B^O + c\overline{H}_C^O + \cdots) \tag{5.40}$$

where $\overline{H}_i^O$ is the molar standard enthalpy of formation of Component i (kJ·kmol^{-1}).

Equation 5.39 implies that a plot of ($\ln K_a$) against ($1/T$) should be a straight line with a slope of ($-\Delta H^O/R$). For an exothermic reaction, $\Delta H^O < 0$ and K_a decreases as temperature increases. For an endothermic reaction, $\Delta H^O > 0$ and K_a increases with increasing temperature.

Equation 5.39 can be used to estimate the equilibrium constant at the required temperature given enthalpy of formation data at some other temperature. Enthalpy of formation data is usually available at standard temperature, and therefore Equation 5.39 can be used to estimate the equilibrium constant at the required temperature given data at standard temperature. However, Equation 5.39 assumes ΔH^O is constant. If data are available for ΔH^O at standard temperature, together with heat capacity data, the equilibrium constant can be calculated more accurately using the thermodynamic path shown in Figure 5.2. Thus (Dodge, 1944; Hougen, Watson and Ragatz, 1959; Coull and Stuart, 1964; Smith, 1990):

$$\Delta H_T^O = \Delta H_{T_o}^O + \int_{T_o}^{T} C_{P,prod}\,dT - \int_{T_o}^{T} C_{P,react}\,dT \tag{5.41}$$

where
- ΔH_T^O = standard enthalpy of formation at T_O
- $C_{P,prod}$ = heat capacity of reaction products as a function of temperature
- $C_{P,react}$ = heat capacity of reactants as a function of temperature

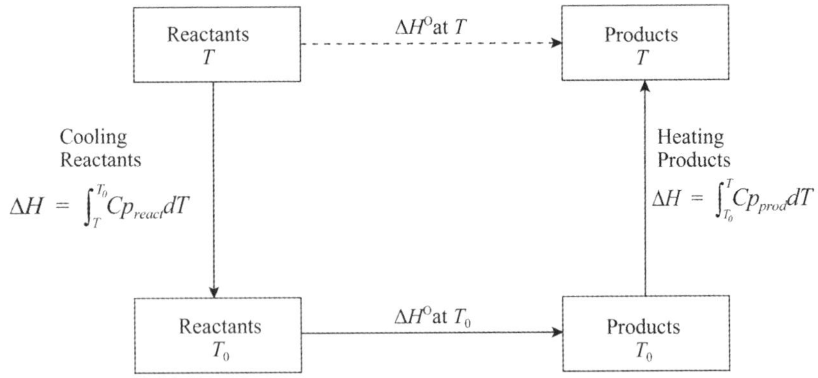

Figure 5.2 Calculation of standard heat of fomation at any temperature from data at standard temperature.

Heat capacity data are often available in some form of polynomial as a function of temperature, for example (Poling, Prausnitz and O'Connell, 2001):

$$\frac{C_P}{R} = \alpha_0 + \alpha_1 T + \alpha_2 T^2 + \alpha_3 T^3 + \alpha_4 T^4 \tag{5.42}$$

where
C_P = heat capacity ($kJ \cdot K^{-1} \cdot kmol^{-1}$)
R = gas constant ($kJ \cdot K^{-1} \cdot kmol^{-1}$)
T = absolute temperature (K)
$\alpha_0, \alpha_1, \alpha_2, \alpha_3, \alpha_4$ = constants determined by fitting experimental data

Thus, Equation 5.41 can be written as:

$$\Delta H_T^O = \Delta H_{T_O}^O + \int_{T_O}^{T} \Delta C_P \, dT \tag{5.43}$$

For the general reaction given in Equation 5.12:

$$\frac{\Delta C_P}{R} = \Delta\alpha_0 + \Delta\alpha_1 T + \Delta\alpha_2 T^2 + \Delta\alpha_3 T^3 + \Delta\alpha_4 T^4 \tag{5.44}$$

From Equations 5.43 and 5.44:

$$\begin{aligned}\Delta H_T^O &= \Delta H_{T_O}^O + R\int_{T_O}^{T} (\Delta\alpha_0 + \Delta\alpha_1 T + \Delta\alpha_2 T^2 + \Delta\alpha_3 T^3 + \Delta\alpha_4 T^4) dT \\ &= \Delta H_{T_O}^O + R\left[\Delta\alpha_0 T + \frac{\Delta\alpha_1 T^2}{2} + \frac{\Delta\alpha_2 T^3}{3} + \frac{\Delta\alpha_3 T^4}{4} + \frac{\Delta\alpha_4 T^5}{5}\right]_{T_O}^{T} \\ &= \Delta H_{T_O}^O + R\left(\Delta\alpha_0 T + \frac{\Delta\alpha_1 T^2}{2} + \frac{\Delta\alpha_2 T^3}{3} + \frac{\Delta\alpha_3 T^4}{4} + \frac{\Delta\alpha_4 T^5}{5}\right) - I\end{aligned} \tag{5.45}$$

where

$$I = R\left(\Delta\alpha_0 T_O + \frac{\Delta\alpha_1 T_O^2}{2} + \frac{\Delta\alpha_2 T_O^3}{3} + \frac{\Delta\alpha_3 T_O^4}{4} + \frac{\Delta\alpha_4 T_O^5}{5}\right) \tag{5.46}$$

Substituting Equation 5.45 into Equation 5.38 gives:

$$\begin{aligned}[\ln K_a]_{K_{aT_O}}^{K_{aT}} &= \frac{1}{R}\int_{T_O}^{T}\left(\frac{\Delta H_{T_O}^O}{T^2} + \frac{R}{T^2}\left[\Delta\alpha_0 T + \frac{\Delta\alpha_1 T^2}{2} + \frac{\Delta\alpha_2 T^3}{3} + \frac{\Delta\alpha_3 T^4}{4} + \frac{\Delta\alpha_4 T^5}{5}\right] - \frac{I}{T^2}\right) dT \\ \ln K_{aT} - \ln K_{aT_O} &= \int_{T_O}^{T}\left(\frac{\Delta H_{T_O}^O}{RT^2} - \frac{I}{RT^2} + \frac{\Delta\alpha_O}{T} + \frac{\Delta\alpha_1}{2} + \frac{\Delta\alpha_2 T}{3} + \frac{\Delta\alpha_3 T^2}{4} + \frac{\Delta\alpha_4 T^3}{5}\right) dT \\ \ln\frac{K_{aT}}{K_{aT_O}} &= \left[-\frac{\Delta H_{T_O}^O}{RT} + \frac{I}{RT} + \Delta\alpha_0 \ln T + \frac{\Delta\alpha_1 T}{2} + \frac{\Delta\alpha_2 T^2}{6} + \frac{\Delta\alpha_3 T^3}{12} + \frac{\Delta\alpha_4 T^4}{20}\right]_{T_O}^{T}\end{aligned} \tag{5.47}$$

Substituting for K_{aT_O}:

$$\ln K_{aT} = \left[-\frac{\Delta H_{T_O}^O}{RT} + \frac{I}{RT} + \Delta\alpha_0 \ln T + \frac{\Delta\alpha_1 T}{2} + \frac{\Delta\alpha_2 T^2}{6} + \frac{\Delta\alpha_3 T^3}{12} + \frac{\Delta\alpha_4 T^4}{20}\right]_{T_O}^{T} - \frac{\Delta G_{T_O}^O}{RT_O} \tag{5.48}$$

where

$$\Delta\alpha_0 = s\alpha_0 + t\alpha_0 + \cdots - b\alpha_0 - c\alpha_0 - \cdots$$

$$\Delta\alpha_1 = s\alpha_1 + t\alpha_1 + \cdots - b\alpha_1 - c\alpha_1 - \cdots$$

and so on.

Given $\Delta G_{T_O}^O$, $\Delta H_{T_O}^O$ and α_0, α_1, α_2, α_3, α_4 for each component, Equation 5.48 is used to calculate K_{aT}.

Alternatively, Equation 5.48 can be written for standard conditions at temperature T:

$$\Delta G_T^O = \Delta H_T^O - T\Delta S_T^O \tag{5.49}$$

Substituting Equation 5.19 into Equation 5.47 gives:

$$\ln K_{aT} = \frac{\Delta S_T^O}{R} - \frac{\Delta H_T^O}{RT} \qquad (5.50)$$

The analogous expression to Equation 5.43 for ΔH_T^O for ΔS_T^O is given by (Dodge, 1944; Hougen, Watson and Ragatz, 1959; Coull and Stuart, 1964; Smith, 1990):

$$\Delta S_T^O = \Delta S_{T_O}^O + \int_{T_O}^{T} \frac{\Delta Cp}{T} dT \qquad (5.51)$$

Substituting Equation 5.44:

$$\begin{aligned}\Delta S_T^O &= \Delta S_{T_O}^O + \int_{T_O}^{T} \frac{R}{T}\left(\Delta\alpha_0 + \Delta\alpha_1 T + \Delta\alpha_2 T^2 + \Delta\alpha_3 T^3 + \Delta\alpha_4 T^4\right) dT \\ &= \Delta S_{T_O}^O + R\int_{T_O}^{T}\left(\frac{\Delta\alpha_0}{T} + \Delta\alpha_1 + \Delta\alpha_2 T + \Delta\alpha_3 T^2 + \Delta\alpha_4 T^3\right) dT \\ &= \Delta S_{T_O}^O + R\left[\Delta\alpha_0 \ln T + \Delta\alpha_1 T + \frac{\Delta\alpha_2 T^2}{2} + \frac{\Delta\alpha_3 T^3}{3} + \frac{\Delta\alpha_4 T^4}{4}\right]_{T_O}^{T}\end{aligned} \qquad (5.52)$$

Thus, ΔH_T^O can be calculated from Equation 5.45 and ΔS_T^O from Equation 5.52 and the results substituted in Equation 5.50.

Example 5.2 Following Example 5.1:

a) Calculate $\ln K_a$ at 300 K, 400 K, 500 K, 600 K and 700 K at 1 bar and test the validity of Equation 5.39. Standard free energy of formation and enthalpy of formation data for NH_3 are given in Table 5.4 (Poling, Prausnitz and O'Connell, 2001). Free energy of formation data for H_2 and N_2 is zero.
b) Calculate the values of $\ln K_a$ from Equations 5.39 and 5.48 from standard data at 298.15 K and compare with values calculated from Table 5.4. Heat capacity coefficients are given in Table 5.5 (Poling, Prausnitz and O'Connell, 2001).
c) Determine the effect of temperature on equilibrium conversion of hydrogen using the data in Table 5.4.
Again assume ideal gas behavior and $R = 8.3145$ kJ·K^{-1}·kmol^{-1}.

Table 5.4

Thermodynamic data for ammonia at various temperatures for a standard pressure of 1 bar (Lide, 2010).

T(K)	$\overline{G}^O_{NH_3}$ (kJ·kmol^{-1})	$\overline{H}^O$ (kJ·kmol^{-1})
298.15	−16,407	−45,940
300	−16,223	−45,981
400	−5980	−48,087
500	4764	−49,908
600	15,846	−51,430
700	27,161	−52,682

Table 5.5

Heat capacity data (Poling, Prausnitz and O'Connell, 200).

	C_P (kJ·kmol^{-1}·K^{-1})				
	α_0	$\alpha_1 \times 10^3$	$\alpha_2 \times 10^5$	$\alpha_3 \times 10^8$	$\alpha_4 \times 10^{11}$
H_2	2.883	3.681	−0.772	0.692	−0.213
N_2	3.539	−0.261	0.007	0.157	−0.099
NH_3	4.238	−4.215	2.041	−2.126	0.761

Solution

a) $\ln K_a = -(2\overline{G}^O_{NH_3} - 3\overline{G}^O_{H_2} - \overline{G}^O_{N_2})/RT$

At	300 K	$\ln K_a = 13.008$ (from Example 5.1)
	400 K	$\ln K_a = 3.5961$
	500 K	$\ln K_a = -2.2919$
	600 K	$\ln K_a = -6.3528$
	700 K	$\ln K_a = -9.3334$

Figure 5.3a shows a plot of $\ln K_a$ versus $1/T$. This is a straight line and appears to be in good agreement with Equation 5.39. From the slope of the graph it can be deduced that ΔH^O has a value of −97,350 kJ. The standard heat of reaction for ammonia synthesis is given by:

$$\begin{aligned}\Delta H^O &= 2\overline{H}^O_{NH_3} - 3\overline{H}^O_{H_2} - \overline{H}^O_{N_2} \\ &= 2\overline{H}^O_{NH_3} - 0 - 0\end{aligned}$$

This implies a standard enthalpy of formation of −48,675 kJ·kmol^{-1}. From standard enthalpy of formation data, ΔH^O varies from −45,981 kJ·kmol^{-1} at 300 K to −52,682 kJ·kmol^{-1} at 700 K, with an average value over the range of −49,332 kJ·kmol^{-1}. This again appears to be in good agreement with the average value from Figure 5.3a. However, this is on the basis of averages across the range of temperature. In the next calculation, the accuracy of Equation 5.37 will be examined in more detail.

b) As a datum, first calculate $\ln K_{aT}$ from the tabulated data for $\overline{G}^O$ listed in Table 5.4:

$$\ln K_{aT} = \frac{\Delta G_T^O}{RT}$$

where $\Delta G_T^O = 2\overline{G}^O_{NH_3,T} - 3\overline{G}^O_{H_2,T} - \overline{G}^O_{N_2,T}$

Substituting values of $\overline{G}^O$ and T leads to the results in Table 5.6. Next calculate K_{aT} for a range of temperatures from Equation 5.37:

$$\begin{aligned}\ln K_{aT} &= -\frac{\Delta H^O_{T_O}}{R}\left(\frac{1}{T}-\frac{1}{T_O}\right)+\ln K_{aT_O}\\ &= -\frac{\Delta H^O_{T_O}}{R}\left(\frac{1}{T}-\frac{1}{T_O}\right)-\frac{\Delta G^O_{T_O}}{RT_O}\end{aligned}$$

$$\begin{aligned}\Delta G^O &= 2\overline{G}^O_{NH_3}-3\overline{G}^O_{H_2}-\overline{G}^O_{N_2}\\ \Delta G_{T_O} &= 2\times(-16,407)-0-0\\ &= -32,814\text{ kJ}\\ \Delta H^O &= 2\overline{H}^O_{NH_3}-3\overline{H}^O_{H_2}-\overline{H}^O_{N_2}\\ \Delta H^O_{T_O} &= 2\times(-45,940)-0-0\\ &= -91,880\ kJ\\ \ln K_{aT} &= -\frac{-91,880}{8.3145}\left(\frac{1}{T}-\frac{1}{298.15}\right)-\frac{-32,814}{8.3145\times 298.15}\end{aligned}$$

Substituting the values of T leads to the results in Table 5.6. Next calculate K_{aT} from Equation 5.46 using the coefficients in Table 5.5:

$$\Delta\alpha_i = 2\alpha_{NH_3}-3\alpha_{H_2}-\alpha_{N_2}$$

Thus

$$\begin{aligned}\Delta\alpha_0 &= -3.7120\\ \Delta\alpha_1 &= -1.9212\times 10^{-2}\\ \Delta\alpha_2 &= 6.3910\times 10^{-5}\\ \Delta\alpha_3 &= -6.4850\times 10^{-8}\\ \Delta\alpha_4 &= 2.2600\times 10^{-11}\end{aligned}$$

Substitute these values in Equation 5.46:

$$I = -12,583\text{ kJ.kmol}^{-1}$$

Substituting the values of temperature into Equation 5.48 leads to the results in Table 5.6. From Table 5.6, it can be seen that calculation of ln K_{aT} from heat capacity data on the basis of $\Delta H^O_{T_O}$ maintains a good agreement with ln K_{aT} calculated from tabulated values of $\overline{G}^O_T$, even for large ranges in temperature. On the other hand, calculation from Equation 5.39 assuming a constant value of $\Delta H^O_{T_O}$ can lead to significant errors when extrapolated over large ranges of temperature.

c)

At	300 K	$K_a = 4.4597\times 10^5$ (from Example 5.1)
	400 K	$K_a = 36.456$
	500 K	$K_a = 0.10107$
	600 K	$K_a = 1.7419\times 10^{-3}$
	700 K	$K_a = 8.8421\times 10^{-5}$

Also from Example 5.1:

$$K_a = \frac{16X^2(2-X)^2}{27(1-X)^4}$$

Substitute K_a and solve for X. This can again be conveniently carried out in spreadsheet software (see Section 3.9):

At	300 K	$X = 0.97$ (from Example 5.1)
	400 K	$X = 0.66$
	500 K	$X = 0.16$
	600 K	$X = 0.026$
	700 K	$X = 0.0061$

Figure 5.3b shows a plot of equilibrium conversion versus temperature. It can be seen that as the temperature increases, the equilibrium conversion decreases (for this reaction). This is consistent with the fact that this is an exothermic reaction. However, it should not necessarily be concluded that the reactor should be operated at low temperature, as the rate of reaction has yet to be considered. Also, catalysts and the deactivation of catalysts have yet to be considered.

It should also be again noted that firstly ideal gas behavior has been assumed, which is reasonable at this pressure, and secondly that small data errors might lead to significant errors in the calculations.

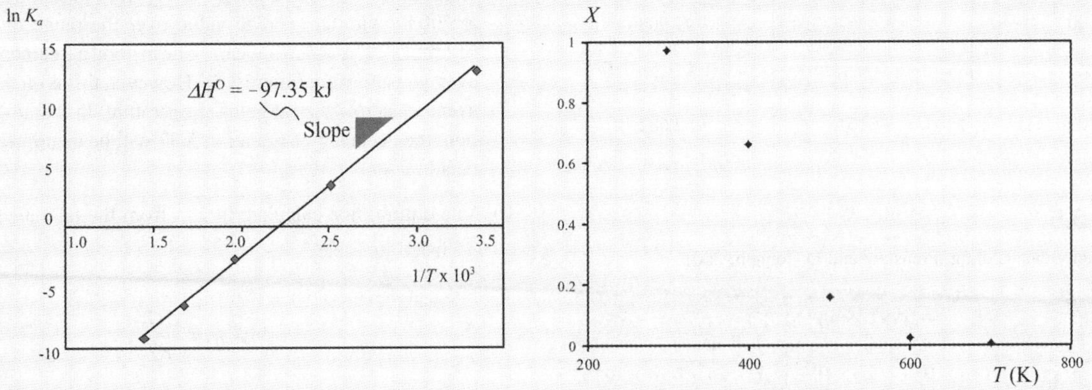

(a) Plot of lnK_a versus $1/T$.

(b) Variation of equilibrium conversion with temperature.

Figure 5.3

The effect of temperature on the ammonia synthesis reaction.

Table 5.6

Comparison of the methods to calculate ln K_{aT} at various temperatures.

T (K)	ln K_{aT}		
	$-\frac{\Delta G_T^O}{RT}$ G_T^O from Table 5.4	$-\frac{\Delta H_{T_O}}{R}\left(\frac{1}{T}-\frac{1}{T_O}\right)-\frac{\Delta G_T^O}{RT_O}$	Equation 5.48
300	13.0078	13.0084	13.0083
400	3.5961	3.7996	3.5977
500	−2.2919	−1.7257	−2.2891
600	−6.3528	−6.4092	−6.3497
700	−9.3334	−8.0403	−9.3300

Example 5.2 shows that for an exothermic reaction, the equilibrium conversion decreases with increasing temperature. This is consistent with Le Châtelier's Principle. If the temperature of an exothermic reaction is decreased, the equilibrium will be displaced in a direction to oppose the effect of the change, that is, increase the conversion.

Now consider the effect of temperature on the rate of reaction. A qualitative observation is that most reactions go faster as the temperature increases. An increase in temperature of 10 °C from room temperature typically doubles the rate of reaction for organic species in solution. It is found in practice that if the logarithm of the reaction rate constant is plotted against the inverse of absolute temperature, it tends to follow a straight line. Thus, at the same concentration, but at different temperatures:

$$\ln k = \text{Intercept} + \text{Slope} \times \frac{1}{T} \tag{5.53}$$

If the intercept is denoted by ln k_O and the slope by $-E/R$, where k_O is called the *frequency factor*, E is the *activation energy* of the reaction and R is the universal gas constant, then:

$$\ln k = \ln k_O - \frac{E}{RT} \tag{5.54}$$

or

$$k = k_O \exp\left[-\frac{E}{RT}\right] \tag{5.55}$$

At the same concentration, but at two different temperatures T_1 and T_2:

$$\ln\frac{r_2}{r_1} = \ln\frac{k_2}{k_1} = \frac{E}{R}\left(\frac{1}{T_1}-\frac{1}{T_2}\right) \tag{5.56}$$

This assumes E to be constant.

Generally, the higher the rate of reaction, the smaller is the reactor volume. Practical upper limits are set by safety considerations, materials-of-construction limitations, maximum operating temperature for the catalyst or catalyst life. Whether the reaction system involves single or multiple reactions, and whether the reactions are reversible, also affects the choice of reactor temperature.

1) *Single reactions*

a) *Endothermic reactions.* If an endothermic reaction is reversible, then Le Châtelier's Principle dictates that operation at a high temperature increases the maximum conversion. Also, operation at high temperature increases the rate of reaction, allowing reduction of reactor volume. Thus, for endothermic reactions, the temperature should be set as high as possible, consistent with safety, materials-of-construction limitations and catalyst life.

Figure 5.4a shows the behavior of an endothermic reaction as a plot of equilibrium conversion against temperature. The plot can be obtained from values of ΔG^O over a range of temperatures and the equilibrium conversion calculated as illustrated in Examples 5.1 and 5.2. If it is assumed that the reactor is operated adiabatically, a heat balance can be carried out to show the change in temperature with reaction conversion. If the mean molar heat capacity of the reactants and products are assumed constant, then for a given starting temperature for the reaction T_{in}, the temperature of the reaction mixture will be proportional to the reactor conversion X for adiabatic operation (Figure 5.4a). As the conversion increases, the temperature decreases because of the reaction endotherm. If the reaction could proceed as far as equilibrium, then it would reach the equilibrium temperature T_E. Figure 5.4b shows how equilibrium conversion can be increased by dividing the reaction into stages and reheating the reactants between stages. Of course, the equilibrium conversion could also have been increased by operating the reactor nonadiabatically and adding heat as the reaction proceeds so as to maximize conversion within the constraints of feasible heat transfer, materials of construction, catalyst life, safety, and so on.

b) *Exothermic reactions.* For single exothermic irreversible reactions, the temperature should be set as high as possible, consistent with materials of construction, catalyst life and safety, in order to minimize reactor volume.

For reversible exothermic reactions, the situation is more complex. Figure 5.5a shows the behavior of an

5

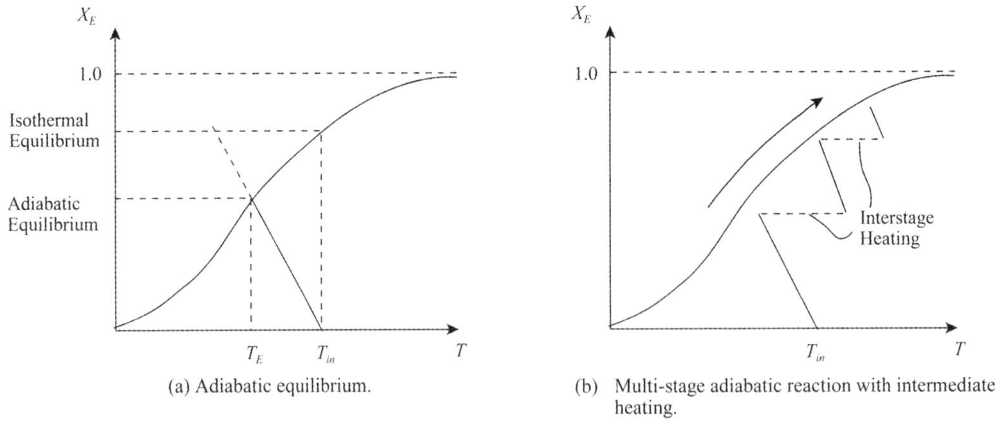

Figure 5.4
Equilibrium behavior with change in temperature for endothermic reactions.

exothermic reaction as a plot of equilibrium conversion against temperature. Again, the plot can be obtained from values of ΔG^O over a range of temperatures and the equilibrium conversion calculated as discussed previously. If it is assumed that the reactor is operated adiabatically, and the mean molar heat capacity of the reactants and products is constant, then for a given starting temperature for the reaction T_{in}, the temperature of the reaction mixture will be proportional to the reactor conversion X for adiabatic operation (Figure 5.5a). As the conversion increases, the temperature rises because of the reaction exotherm. If the reaction could proceed as far as equilibrium, then it would reach the equilibrium temperature T_E (Figure 5.5a). Figure 5.5b shows how equilibrium conversion can be increased by dividing the reaction into stages and cooling the reactants between stages. Rather than adiabatic operation, the equilibrium conversion could also have been increased by operating the reactor non-adiabatically and removing heat as the reaction proceeds so as to maximize conversion within the constraints of feasible heat transfer, materials of construction, catalyst life, safety, and so on.

Thus, if an exothermic reaction is reversible, then Le Châtelier's Principle dictates that operation at a low temperature increases maximum conversion. However, operation at a low temperature decreases the rate of reaction, thereby increasing the reactor volume. Then ideally, when far from equilibrium, it is advantageous to use a high temperature to increase the rate of reaction. As equilibrium is approached, the temperature should be lowered to increase the maximum conversion. For reversible exothermic reactions, the ideal temperature is continuously decreasing as conversion increases.

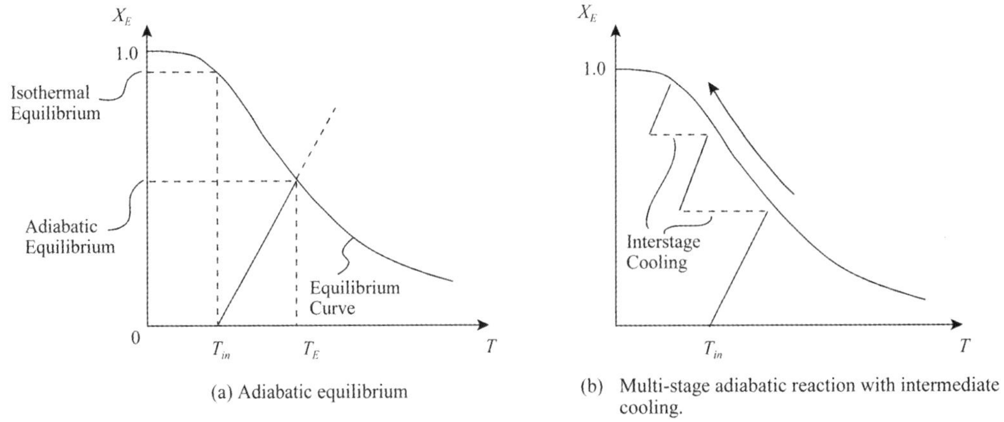

Figure 5.5
Equilibrium behavior with change in temperature for exothermic reactions.

2) *Multiple reactions.* The arguments presented for minimizing reactor volume for single reactions can be used for the primary reaction when dealing with multiple reactions. However, the goal at this stage of the design, when dealing with multiple reactions, is to maximize selectivity or reactor yield, rather than to minimize volume, for a given conversion.

Consider the following equations for parallel reactions:

$$\begin{aligned} &FEED \rightarrow PRODUCT \qquad r_1 = k_1 C_{FEED}^{a_1} \\ &FEED \rightarrow BYPRODUCT \quad r_2 = k_2 C_{FEED}^{a_2} \end{aligned} \tag{4.64}$$

$$\frac{r_2}{r_1} = \frac{k_2}{k_1} C_{FEED}^{a_2 - a_1} \tag{4.65}$$

and for series reactions:

$$\begin{aligned} &FEED \rightarrow PRODUCT \qquad r_1 = k_1 C_{FEED}^{a_1} \\ &PRODUCT \rightarrow BYPRODUCT \quad r_2 = k_2 C_{PRODUCT}^{a_2} \end{aligned} \tag{4.68}$$

The rates of reaction for the primary and secondary reactions both change with temperature, since the reaction rate constants k_1 and k_2 both increase with increasing temperature. The rate of change with temperature might be significantly different for the primary and secondary reactions.

- If k_1 increases faster than k_2, operate at high temperature (but beware of safety, catalyst life and materials-of-construction constraints).
- If k_2 increases faster than k_1, operate at low temperature (but beware of capital cost, since low temperature, although increasing selectivity, also increases reactor size). Here there is an economic trade-off between decreasing byproduct formation and increasing capital cost.

Example 5.3 Example 4.4 developed a kinetic model for the manufacture of benzyl acetate from benzyl chloride and sodium acetate in a solution of xylene in the presence of triethylamine as catalyst, according to:

$$C_6H_5CH_2Cl + CH_3COONa \rightarrow CH_3COOC_6H_5CH_2 + NaCl$$

or

$$A + B \rightarrow C + D$$

Example 4.4 developed a kinetic model for an equimolar feed at a temperature of 102 °C, such that:

$$-r_A = k_A C_A \quad \text{with} \quad k_A = 0.01306\,\text{h}^{-1}$$

Further experimental data are available at 117 °C. The measured mole percent benzyl chloride versus time in hours at 117 °C are given in Table 5.7.

Table 5.7

Experimental data for the production of benzyl acetate at 117 °C.

Reaction time (h)	Benzyl chloride (mole %)
0.00	100.0
3.00	94.5
6.27	88.1
7.23	87.0
9.02	83.1
10.02	81.0
12.23	78.8
13.23	76.9
16.60	69.0
18.00	69.4
23.20	63.0
27.20	57.8

Again, assume the volume of the reactor to be constant. Determine the activation energy for the reaction.

Solution The same three kinetic models as in Example 4.4 can be subjected to a least squares fit given by:

$$\text{minimize } R^2 = \sum_j^{11} \left[(N_{A,j})_{calc} - (N_{A,j})_{exp} \right]^2$$

where R = residual
$(N_{A,j})_{calc}$ = calculated moles of A for Measurement j
$(N_{A,j})_{exp}$ = experimentally measured moles of A for Measurement j

As in Example 4.4, one way this can be carried out is by setting up a function for R^2 in the spreadsheet, and then using the spreadsheet solver to minimize R^2 by manipulating the value of k_A (see Section 3.9). The results are given in Table 5.8.

Table 5.8

Results of a least squares fit for the three kinetic models at 117 °C.

Order of reaction	Rate constant	R^2
Zero order	$k_A V = 1.676$	37.64
First order	$k_A = 0.02034$	7.86
Second order	$k_A/V = 0.0002432$	24.94

Again it is clear from Table 5.8 that the best fit is given by a first-order reaction model:

$$-r_A = k_A C_A \quad \text{with} \quad k_A = 0.02034\,\text{h}^{-1}$$

Next, substitute the results at 102 °C (375 K) and 117 °C (390 K) in Equation 5.56:

$$\ln\frac{r_2}{r_1} = \ln\frac{k_2}{k_1} = \frac{E}{R}\left(\frac{1}{T_1} - \frac{1}{T_2}\right)$$

$$\ln\frac{0.02034}{0.01306} = \frac{E}{8.3145}\left(\frac{1}{375} - \frac{1}{390}\right)$$

$$E = 35,900 \text{ kJ} \times \text{kmol}^{-1}$$

5.3 Reactor Pressure

Now consider the effect of pressure. For reversible reactions, pressure can have a significant effect on the equilibrium conversion. Even though the equilibrium constant is only a function of temperature and not a function of pressure, equilibrium conversion can still be influenced through changing the activities (fugacities) of the reactants and products.

Consider again the ammonia synthesis example from Examples 5.1 and 5.2.

Example 5.4 Following Example 5.2, the reactor temperature will be set to 700 K. Examine the effect of increasing the reactor pressure by calculating the equilibrium conversion of hydrogen at 1 bar, 10 bar, 100 bar and 300 bar. Assume initially ideal gas behavior.

Solution From Equation 5.26 and 5.27:

$$K_a = \frac{p_{NH_3}^2}{p_{H_2}^3 p_{N_2}} = \frac{y_{NH_3}^2}{y_{H_2}^3 y_{N_2}} P^{-2}$$

Note again that K_a is dimensionless and depends on the specification of the number of moles in the stoichiometric equation. From Example 5.2 at 700 K:

$$K_a = 8.8421 \times 10^{-5}$$

Thus:

$$8.8421 \times 10^{-5} = \frac{y_{NH_3}^2}{y_{H_2}^3 y_{N_2}} P^{-2}$$

$$8.8421 \times 10^{-5} = \frac{16X^2(2 - X)^2}{27(1 - X)^4} P^{-2}$$

This can be solved by trial and error, as before.

At	1 bar	$X = 0.0061$ (from Example 5.2)
	10 bar	$X = 0.056$
	100 bar	$X = 0.33$
	300 bar	$X = 0.54$

It is clear that a very significant improvement in the equilibrium conversion for this reaction can be achieved through an increase in pressure.

It should again be noted that ideal gas behavior has been assumed. Carrying out the calculations for real gas behavior using an equation of state and including K_ϕ could change the results significantly, especially at the higher pressures, as will now be investigated.

Example 5.5 Repeat the calculations from Example 5.4 taking into account vapor-phase nonideality. Fugacity coefficients can be calculated from the Peng–Robinson Equation of State (see Poling, Prausnitz and O'Connell (2001) and Appendix A).

Solution The ideal gas equilibrium constants can be corrected for real gas behavior by multiplying the ideal gas equilibrium constant by K_ϕ, as defined by Equation 5.23, which for this problem is:

$$K_\phi = \frac{\phi_{NH_3}^2}{\phi_{H_2}^3 \phi_{N_2}}$$

The fugacity coefficients ϕ_i can be calculated from the Peng–Robinson Equation of State. The values of ϕ_i are functions of temperature, pressure and composition, and the calculations are complex (see Pohling, Prausnitz and O'Connell (2001) and Appendix A). Interaction parameters between components are here assumed to be zero. The results showing the effect of nonideality are given in Table 5.9.

Table 5.9

Equilibrium conversion versus pressure for real gas behavior.

Pressure (bar)	X_{IDEAL}	K_ϕ	X_{REAL}
1	0.0061	0.9990	0.0061
10	0.056	0.9897	0.056
100	0.33	0.8772	0.34
300	0.54	0.6182	0.58

It can be seen in Table 5.9 that as pressure increases, nonideal behavior changes the equilibrium conversion. Note that, like K_a, K_ϕ also depends on the specification of the number of moles in the stoichiometric equation. For example, if the stoichiometry is written as:

$$\tfrac{3}{2}H_2 + \tfrac{1}{2}N_2 \rightleftharpoons NH_3$$

then

$$K'_\phi = \frac{\phi_{NH_3}}{\phi_{N_2}^{1/2}\phi_{H_2}^{3/2}}$$

$$K'_\phi = \sqrt{K_\phi}$$

The selection of reactor pressure for vapor-phase reversible reactions depends on whether there is a decrease or an increase in the number of moles. The value of ΔN in Equation 5.25 dictates whether the equilibrium conversion will increase or decrease with increasing pressure. If ΔN is negative, the equilibrium conversion will increase with increasing pressure. If ΔN is positive, it will decrease. The choice of pressure must also take account of whether the system involves multiple reactions or not.

Increasing the pressure of vapor-phase reactions increases the rate of reaction and hence decreases reactor volume both by decreasing the residence time required for a given reactor conversion and increasing the vapor density. In general, pressure has little effect on the rate of liquid-phase reactions.

1) *Single reactions*

a) *Decrease in the number of moles.* A decrease in the number of moles for vapor-phase reactions decreases the reactor volume as reactants are converted to products. For a fixed reactor volume, this means a decrease in pressure as reactants are converted to products. The effect of an increase in pressure of the system is to cause a shift of the composition of the gaseous mixture toward one occupying a smaller volume. Increasing the reactor pressure increases the equilibrium conversion. Increasing the pressure increases the rate of reaction and also reduces reactor volume. Thus, if the reaction involves a decrease in the number of moles, the pressure should be set as high as practicable, bearing in mind that the high pressure might be costly to obtain through compressor power, mechanical construction might be expensive and high pressure brings safety problems.

b) *Increase in the number of moles.* An increase in the number of moles for vapor-phase reactions increases the volume as reactants are converted to products. Le Châtelier's Principle dictates that a decrease in reactor pressure increases equilibrium conversion. However, operation at a low pressure decreases the rate of reaction in vapor-phase reactions and increases the reactor volume. Thus, initially, when far from equilibrium, it is advantageous to use high pressure to increase the rate of reaction. As equilibrium is approached, the pressure should be lowered to increase the conversion. The ideal pressure would continuously decrease as conversion increases to the desired value. The low pressure required can be obtained by operating the system at reduced absolute pressure or by introducing a diluent to decrease the partial pressure. The diluent is an inert material (e.g. steam) and is simply used to lower the partial pressure in the vapor phase. For example, ethyl benzene can be dehydrogenated to styrene according to the reaction (Waddams, 1978):

$$\underset{\text{ethylbenzene}}{C_6H_5CH_2CH_3} \rightleftharpoons \underset{\text{styrene}}{C_6H_5CH = CH_2} + H_2$$

This is an endothermic reaction accompanied by an increase in the number of moles. High conversion is favored by high temperature and low pressure. The reduction in pressure is achieved in practice by the use of superheated steam as a diluent and by operating the reactor below atmospheric pressure. The steam in this case fulfills a dual purpose by also providing heat for the reaction.

2) *Multiple reactions producing byproducts.* The arguments presented for the effect of pressure on single vapor-phase reactions can be used for the primary reaction when dealing with multiple reactions. Again, selectivity and reactor yield are likely to be more important than reactor volume for a given conversion.

If there is a significant difference between the effect of pressure on the primary and secondary reactions, the pressure should be chosen to reduce as much as possible the rate of the secondary reactions relative to the primary reaction. Improving the selectivity or reactor yield in this way may require changing the system pressure or perhaps introducing a diluent.

For liquid-phase reactions, the effect of pressure on the selectivity and reactor volume is less pronounced, and the pressure is likely to be chosen to:

- prevent vaporization of the products;
- allow vaporization of liquid in the reactor so that it can be condensed and refluxed back to the reactor as a means of removing the heat of reaction;
- allow vaporization of one of the components in a reversible reaction in order that removal increases maximum conversion (to be discussed in Section 5.5);
- allow vaporization to feed the vapor directly into a distillation operation to combine reactions with separation (to be discussed in Chapter 14).

5

5.4 Reactor Phase

Having considered reactor temperature and pressure, the reactor phase can now be considered. The reactor phase can be gas, liquid or multiphase amongst gas, liquid and solid. Given a free choice between gas and liquid-phase reactions, operation in the liquid phase is usually preferred. Consider the single reaction system:

$$FEED \rightarrow PRODUCT \qquad r = kC_{FEED}^{a}$$

Clearly, in the liquid phase much higher concentrations of C_{FEED} ($kmol{\cdot}m^{-3}$) can be maintained than in the gas phase. This makes liquid-phase reactions in general more rapid and hence leads to smaller reactor volumes for liquid-phase reactors.

However, a note of caution should be added. In many multiphase reaction systems, as will be discussed in the next chapter, rates of mass transfer between different phases can be just as important as, or even more important than, reaction kinetics in determining the reactor volume. Mass transfer rates are generally higher in gas-phase than liquid-phase systems. In such situations, it is not so easy to judge whether the gas or liquid phase is preferred.

Very often the choice is not available. For example, if the reactor temperature is above the critical temperature of the chemical species, then the reactor must be in the gas phase. Even if the temperature can be lowered below critical, an extremely high pressure may be required to operate in the liquid phase.

The choice of reactor temperature, pressure and hence phase must, in the first instance, take account of the desired equilibrium and selectivity effects. If there is still freedom to choose between the gas or liquid phase, operation in the liquid phase is preferred.

5.5 Reactor Concentration

When more than one reactant is used, it is often desirable to use an excess of one of the reactants. It is also sometimes desirable to feed an inert material to the reactor or to separate the product partway through the reaction before carrying out a further reaction. Sometimes it is desirable to recycle unwanted byproducts to the reactor. These cases will now be examined.

1) *Single irreversible reactions.* An excess of one feed component can force another component towards complete conversion. As an example, consider the reaction between ethylene and chlorine to dichloroethane:

$$\underset{\text{ethylene}}{C_2H_2} + \underset{\text{chlorine}}{Cl_2} \rightarrow \underset{\text{dichloroethane}}{C_2H_4Cl_2}$$

An excess of ethylene is used to ensure essentially complete conversion of the chlorine, which is thereby eliminated as a problem for the downstream separation system. In a single irreversible reaction (where selectivity is not a problem), the usual choice of excess reactant is to eliminate the component that is more difficult to separate in the downstream separation system. Alternatively, if one of the components is more hazardous (as is chlorine in this example), complete conversion has advantages for safety.

2) *Single reversible reactions.* The maximum conversion in reversible reactions is limited by the equilibrium conversion, and conditions in the reactor are usually chosen to increase the equilibrium conversion:

   a) *Feed ratio.* If to a system in equilibrium, an excess of one of the feeds is added, then the effect is to shift the equilibrium to decrease the feed concentration. In other words, an excess of one feed can be used to increase the equilibrium conversion. Consider the following examples.

Example 5.6 Ethyl acetate can be produced by the esterification of acetic acid with ethanol in the presence of a catalyst such as sulfuric acid or an ion-exchange resin according to the reaction:

$$\underset{\text{acetic acid}}{CH_3COOH} + \underset{\text{ethanol}}{C_2H_5OH} \rightleftharpoons \underset{\text{ethyl acetate}}{CH_3COOC_2H_5} + \underset{\text{water}}{H_2O}$$

Laboratory studies have been carried out to provide design data on the conversion. A stoichiometric mixture of 60 g acetic acid and 46 g ethanol was reacted and held at constant temperature until equilibrium was reached. The reaction products were analyzed and found to contain 63.62 g ethyl acetate.

a) Calculate the equilibrium conversion of acetic acid.
b) Estimate the effect of using a 50% and 100% excess of ethanol.

Assume the liquid mixture to be ideal and the molar masses of acetic acid and ethyl acetate to be 60 $kg \cdot kmol^{-1}$ and 88 $kg \cdot kmol^{-1}$, respectively.

Solution

a) Moles of acetic acid in reaction mixture = 60/60 = 1.0
   Moles of ethyl acetate formed = 63.62/88 = 0.723
   Equilibrium conversion of acetic acid = 0.723
b) Let r be the molar ratio of ethanol to acetic acid. Table 5.10 presents the moles initially and at equilibrium and the mole fractions.

Assuming an ideal solution (i.e. $K_\gamma = 1$ in Equation 5.29):

$$K_a = \frac{X^2}{(1-X)(r-X)}$$

For the stoichiometric mixture, $r = 1$ and $X = 0.723$ from the laboratory measurements:

$$K_a = \frac{0.723^2}{(1-0.723)^2} = 6.813$$

For excess ethanol:

$$6.813 = \frac{X^2}{(1-X)(r-X)}$$

Substitute $r = 1.5$ and 2.0 and solve for X by trial and error. The results are presented in Table 5.11.

Increasing the excess of ethanol increases the conversion of acetic acid to ethyl acetate. To carry out the calculation more accurately would require activity coefficients to be calculated for the mixture (see Poling, Prausnitz and O'Connell (2001) and Appendix A). The activity coefficients depend on correlating coefficients between each binary pair in the mixture, the concentrations and temperature.

Table 5.10

Mole fractions at equilibrium for the production of ethyl acetate.

	CH_3COOH	C_2H_5OH	$CH_3COOC_2H_5$	H_2O
Moles in initial mixture	1	r	0	0
Moles at equilibrium	$1-X$	$r-X$	X	X
Mole fraction at equilibrium	$\frac{1-X}{1+r}$	$\frac{r-X}{1+r}$	$\frac{X}{1+r}$	$\frac{X}{1+r}$

Table 5.11

Variation of equilibrium conversion with feed ratio for the production of ethyl acetate.

r	X
1.0	0.723
1.5	0.842
2.0	0.894

5

Example 5.7 Hydrogen can be manufactured from the reaction between methane and steam over a catalyst. Two principal reactions occur:

$$\underset{\text{methane}}{CH_4} + \underset{\text{water}}{H_2O} \rightleftharpoons \underset{\text{hydrogen}}{3H_2} + \underset{\text{carbon monoxide}}{CO}$$

$$\underset{\text{carbon monoxide}}{CO} + \underset{\text{water}}{H_2O} \rightleftharpoons \underset{\text{hydrogen}}{H_2} + \underset{\text{carbon dioxide}}{CO_2}$$

Assuming the reaction takes place at 1100 K and 20 bar, calculate the equilibrium conversion for a molar ratio of steam to methane in the feed of 3, 4, 5 and 6. Assume ideal gas behavior ($K_\phi = 1$, $R = 8.3145\ kJ{\cdot}K^{-1}{\cdot}kmol^{-1}$). Thermodynamic data for a standard pressure of 1 bar are given in Table 5.12.

Table 5.12

Thermodynamic data for hydrogen manufacture (Lida, 2010).

	$\overline{G}^O_{1100\ K}$ (kJ.kmol^{-1})
CH_4	30,358
H_2O	−187,052
H_2	0.0
CO	−209,084
CO_2	−359,984

Solution For the first reaction:

$$\begin{aligned}
\Delta G_1^O &= 3\Delta\overline{G}^O_{H_2} + \Delta\overline{G}^O_{CO} - \Delta\overline{G}^O_{CH_4} - \Delta\overline{G}^O_{H_2O} \\
&= 3 \times 0 + (-209,084) - 30,358 - (-187,052) \\
&= -52,390\ \text{kJ} \\
-RT \ln K_{a1} &= -52,390\ \text{kJ} \\
K_{a1} &= 307.42
\end{aligned}$$

For the second reaction:

$$\begin{aligned}
\Delta G_2^O &= \Delta\overline{G}^O_{H_2} + \Delta\overline{G}^O_{CO_2} - \Delta\overline{G}^O_{CO} - \Delta\overline{G}^O_{H_2O} \\
&= 0 + (-359,984) - (-209,084) - (-187,052) \\
&= 36,152\ \text{kJ} \\
-RT \ln K_{a2} &= 36,152 \\
K_{a2} &= 1.9201 \times 10^{-2}
\end{aligned}$$

Let r be the molar ratio of water to methane in the feed, X_1 be the conversion of the first reaction and X_2 be the conversion of the second reaction.

$$\underset{1-X_1}{CH_4} + \underset{r-X_1-X_2}{H_2O} \rightleftharpoons \underset{3X_1}{3H_2} + \underset{X_1-X_2}{CO}$$

$$\underset{X_1-X_2}{CO} + \underset{r-X_1-X_2}{H_2O} \rightleftharpoons \underset{X_2}{H_2} + \underset{X_2}{CO_2}$$

$$\begin{aligned}
\text{Total moles at equilibrium} &= (1 - X_1) + (r - X_1 - X_2) + 3X_1 + \\
&\quad (X_1 - X_2) + X_2 + X_2 \\
&= r + 1 + 2X_1
\end{aligned}$$

Mole fractions are presented in Table 5.13.

Table 5.13

Mole fractions at equilibrium for the manufacture of hydrogen.

	CH_4	H_2O	H_2	CO	CO_2
Moles at equilibrium	$1 - X_1$	$r - X_1 - X_2$	$3X_1 + X_2$	$X_1 - X_2$	X_2
Mole fraction at equilibrium	$\dfrac{1 - X_1}{r + 1 + 2X_1}$	$\dfrac{r - X_1 - X_2}{r + 1 + 2X_1}$	$\dfrac{3X_1 + X_2}{r + 1 + 2X_1}$	$\dfrac{X_1 - X_2}{r + 1 + 2X_1}$	$\dfrac{X_2}{r + 1 + 2X_1}$

$$K_{a1} = \frac{y_{H_2}^3 y_{CO}}{y_{CH_4} y_{H_2O}} P^2$$

$$= \frac{\left(\frac{3X_1 + X_2}{r + 1 + 2X_1}\right)^3 \left(\frac{X_1 - X_2}{r + 1 + 2X_1}\right)}{\left(\frac{1 - X_1}{r + 1 + 2X_1}\right)\left(\frac{r - X_1 - X_2}{r + 1 + 2X_1}\right)} P^2 \tag{5.57}$$

$$= \frac{(3X_1 + X_2)^3 (X_1 - X_2)}{(1 - X_1)(r - X_1 - X_2)(r + 1 + 2X_1)^2} P^2$$

$$K_{a2} = \frac{y_{H_2} y_{CO_2}}{y_{CO}\, y_{H_2O}} P^0$$

$$= \frac{\left(\frac{3X_1 + X_2}{r + 1 + 2X_1}\right)\left(\frac{X_2}{r + 1 + 2X_1}\right)}{\left(\frac{X_1 - X_2}{r + 1 + 2X_1}\right)\left(\frac{r - X_1 - X_2}{r + 1 + 2X_1}\right)} P^0 \tag{5.58}$$

$$= \frac{(3X_1 + X_2)X_2}{(X_1 - X_2)(r - X_1 - X_2)} P^0$$

Knowing P, K_{a1} and K_{a2} and setting a value for r, these two equations can be solved simultaneously for X_1 and X_2. However, X_1 or X_2 cannot be eliminated by substitution in this case, and the equations must be solved numerically. This can be done by assuming a value for X_1 and substituting this in both equations, thus yielding values of X_2 from both equations. X_1 is then varied until the values of X_2 predicted by both equations are equal. Alternatively, X_1 and X_2 can be varied simultaneously in a nonlinear optimization algorithm to minimize the errors in the two equations. If spreadsheet software is used, this can be done by taking K_{a1} to the right-hand side of Equation 5.57 and giving the right-hand side a value of, say, *Objective 1*, which must ultimately be zero. Also, K_{a2} can be taken to the right-hand side of Equation 5.58 and the right-hand side given a value of, say, *Objective 2*, which must also ultimately be zero. Values of X_1 and X_2 must then be determined such that *Objective 1* and *Objective 2* are both zero. The spreadsheet solver can then be used to vary X_1 and X_2 simultaneously to search for (see Section 3.8):

$$(Objective\ 1)^2 + (Objective\ 2)^2 = 0 \quad \text{(within a tolerance)}$$

The results are shown in Table 5.14.

From Table 5.14, the mole fraction of CH_4 and H_2 decrease as the molar ratio of H_2O/CH_4 increases. The picture becomes clearer when the results are presented on a dry basis, as shown in Table 5.15.

From Table 5.15, on a dry basis as the molar ratio H_2O/CH_4 increases, the mole fraction of CH_4 decreases and that of H_2 increases.

The next two stages in the process carry out shift conversion at lower temperatures in which the second reaction above is used to convert CO to H_2 to higher conversion.

Table 5.15

Product mole fractions for the manufacture of hydrogen on a dry basis.

H_2O/CH_4	y_{CH_4}	y_{H_2}
3	0.0586	0.7072
4	0.0389	0.7221
5	0.0240	0.7337
6	0.0183	0.7383

Table 5.14

Equilibrium conversions and product mole fractions for the manufacture of hydrogen.

H_2O/CH_4	X_1	X_2	y_{CH_4}	y_{H_2O}	y_{H_2}	y_{CO}	y_{CO_2}
3	0.80	0.015	0.0357	0.3902	0.4313	0.1402	0.0027
4	0.86	0.019	0.0208	0.4644	0.3868	0.1251	0.0028
5	0.91	0.025	0.0115	0.5198	0.3523	0.1132	0.0032
6	0.93	0.031	0.0079	0.5687	0.3184	0.1015	0.0035

b) *Inert concentration.* Sometimes, an inert material is present in the reactor. This might be a solvent in a liquid-phase reaction or an inert gas in a gas-phase reaction. Consider the reaction:

$$A \rightleftharpoons B + C$$

The effect of the increase in moles can be artificially decreased by adding an inert material. Le Châtelier's Principle dictates that this will increase the equilibrium conversion. For example, if the above reaction is in the ideal gas phase:

$$K_a = \frac{p_B p_C}{p_A} = \frac{y_B y_C}{y_A} P = \frac{N_B N_C}{N_A N_T} P \tag{5.59}$$

where N_i = number of moles of Component i
N_T = total number of moles

Thus:

$$\frac{N_B N_C}{N_A} = \frac{K_a N_T}{P} \tag{5.60}$$

Increasing N_T as a result of adding inert material will increase the ratio of products to reactants. Adding an inert material causes the number of moles per unit volume to be decreased, and the equilibrium will be displaced to oppose this by shifting to a higher conversion. If inert material is to be added, then ease of separation is an important consideration. For example, steam is added as an inert to hydrocarbon cracking reactions and is an attractive material in this respect because it is easily separated from the hydrocarbon components by condensation.

Consider the reaction:

$$E + F \rightleftharpoons G$$

For example, if the above reaction is in the ideal gas phase:

$$K_a = \frac{p_G}{p_E p_F} = \frac{y_G}{y_E y_F}\frac{1}{P} = \frac{N_G N_T}{N_E N_F}\frac{1}{P} \tag{5.61}$$

Thus:

$$\frac{N_G}{N_E N_F} = \frac{K_a P}{N_T} \tag{5.62}$$

Decreasing N_T as a result of removing inert material will increase the ratio of products to reactants. Removing inert material causes the number of moles per unit volume to be increased and the equilibrium will be displaced to oppose this by shifting to a higher conversion.

If the reaction does not involve any change in the number of moles, inert material has no effect on equilibrium conversion.

c) *Product removal during reaction.* Sometimes the equilibrium conversion can be increased by removing the product (or one of the products) continuously from the reactor as the reaction progresses, for example, by allowing it to vaporize from a liquid-phase reactor. Another way is to carry out the reaction in stages with intermediate separation of the products. As an example of intermediate separation, consider the production of sulfuric acid, as illustrated in Figure 5.6. Sulfur dioxide is oxidized to sulfur trioxide:

$$\underset{\text{sulfur dioxide}}{2SO_2} + O_2 \rightleftharpoons \underset{\text{sulfur trioxide}}{2SO_3}$$

This reaction can be forced to effectively complete conversion by first carrying out the reaction to approach equilibrium. The sulfur trioxide is then separated (by absorption). Removal of sulfur trioxide shifts the equilibrium and further reaction of the remaining sulfur dioxide and oxygen allows effective complete conversion of the sulfur dioxide (Figure 5.6).

Intermediate separation followed by further reaction is clearly most appropriate when the intermediate separation is straightforward, as in the case of sulfuric acid production.

3) *Multiple reactions in parallel producing byproducts.* After the reactor type is chosen for parallel reaction systems in order to maximize selectivity or reactor yield, conditions can be altered further to improve selectivity. Consider the parallel reaction system from Equation 4.66. To maximize selectivity for this system, the ratio given by Equation 4.67 is minimized:

$$\frac{r_2}{r_1} = \frac{k_2}{k_1} C_{FEED1}^{a_2-a_1} C_{FEED2}^{b_2-b_1} \tag{4.67}$$

Even after the type of reactor is chosen, excess of *FEED 1* or *FEED 2* can be used.

- If $(a_2 - a_1) > (b_2 - b_1)$ use excess *FEED* 2
- If $(a_2 - a_1) < (b_2 - b_1)$ use excess *FEED* 1.

If the secondary reaction is reversible and involves a decrease in the number of moles, such as:

$$\begin{aligned} &FEED\,1 + FEED\,2 \rightarrow PRODUCT \\ &FEED\,1 + FEED\,2 \rightleftharpoons BYPRODUCT \end{aligned} \tag{5.63}$$

then, if inerts are present, increasing the concentration of inert material will decrease byproduct formation. If the secondary reaction is reversible and involves an increase in the number of moles, such as:

$$\begin{aligned} &FEED\,1 + FEED\,2 \rightarrow PRODUCT \\ &FEED\,1 \rightleftharpoons BYPRODUCT\,1 + BYPRODUCT\,2 \end{aligned} \tag{5.64}$$

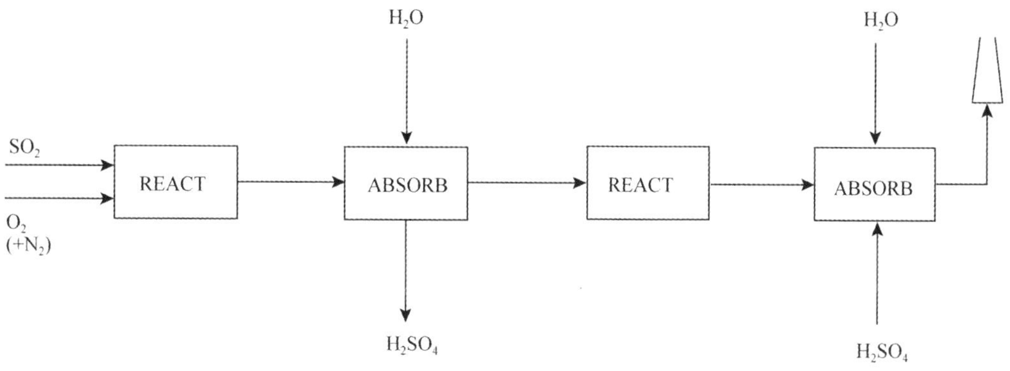

Figure 5.6

The equilibrium conversion for sulfuric acid production can be increased by intermediate separation of the product followed by further reaction.

then, if inert material is present, decreasing the concentration of inert material will decrease byproduct formation. If the secondary reaction has no change in the number of moles, then concentration of inert material does not affect it.

For all reversible secondary reactions, deliberately feeding *BYPRODUCT* to the reactor inhibits its formation at source by shifting the equilibrium of the secondary reaction. This is achieved in practice by separating and recycling *BYPRODUCT* rather than separating and disposing of it directly.

An example of such recycling in a parallel reaction system is in the "Oxo" process for the production of C_4 alcohols. Propylene and synthesis gas (a mixture of carbon monoxide and hydrogen) are first reacted to *n*- and isobutyraldehydes using a cobalt-based catalyst. Two parallel reactions occur (Waddams, 1978):

$$\underset{\text{propylene}}{C_3H_6} + \underset{\substack{\text{carbon}\\\text{monoxide}}}{CO} + \underset{\text{hydrogen}}{H_2} \rightleftharpoons \underset{n\text{-butyraldehyde}}{CH_3CH_2CH_2CHO}$$

$$\underset{\text{propylene}}{C_3H_6} + \underset{\substack{\text{carbon}\\\text{monoxide}}}{CO} + \underset{\text{hydrogen}}{H_2} \rightleftharpoons \underset{\text{isobutyraldehyde}}{CH_3CH(CH_3)CHO}$$

The *n*-isomer is more valuable. Recycling the iso-isomer can be used as a means of suppressing its formation (Waddams, 1978).

4) *Multiple reactions in series producing byproducts.* For the series reaction system in Equation 4.68, the series reaction is inhibited by low concentrations of *PRODUCT*. It has been noted already that this can be achieved by operating with a low conversion.

If the reaction involves more than one feed, it is not necessary to operate with the same low conversion on all the feeds. Using an excess of one of the feeds enables operation with a relatively high conversion of other feed material and still inhibits series reactions. Consider again the series reaction system from Example 4.3:

$$\underset{\text{toluene}}{C_6H_5CH_3} + \underset{\text{hydrogen}}{H_2} \rightarrow \underset{\text{benzene}}{C_6H_6} + \underset{\text{methane}}{CH_4}$$

$$\underset{\text{benzene}}{2C_6H_6} \rightleftharpoons \underset{\text{dyphenyl}}{C_{12}H_{10}} + \underset{\text{hydrogen}}{H_2}$$

It is usual to operate this reactor with a large excess of hydrogen (Waddams, 1978). The molar ratio of hydrogen to toluene entering the reactor is of the order 5:1. The excess of hydrogen encourages the primary reaction directly and discourages the secondary reaction by reducing the concentration of the benzene product. Also, in this case, because hydrogen is a byproduct of the secondary reversible reaction, an excess of hydrogen favors the reverse reaction to benzene. In fact, unless a large excess of hydrogen is used, series reactions that decompose the benzene all the way to carbon become significant, known as *coke formation*.

Another way to keep the concentration of *PRODUCT* low is to remove the product as the reaction progresses, for example, by intermediate separation followed by further reaction. For example, in a reaction system such as Equation 4.68, intermediate separation of the *PRODUCT* followed by further reaction maintains a low concentration of *PRODUCT* as the reaction progresses. Such intermediate separation is most appropriate when separation of the product from the reactants is straightforward.

If the series reaction is also reversible, such as

$$\begin{aligned} &FEED \longrightarrow PRODUCT \\ &PRODUCT \rightleftharpoons BYPRODUCT \end{aligned} \qquad (5.65)$$

then, again, removal of the *PRODUCT* as the reaction progresses, for example, by intermediate separation of the *PRODUCT*, maintains a low concentration of *PRODUCT* and at the same time shifts the equilibrium for the secondary reaction towards *PRODUCT* rather than *BYPRODUCT* formation.

If the secondary reaction is reversible and inert material is present, then to improve the selectivity:

- increase the concentration of inert material if the *BYPRODUCT* reaction involves a decrease in the number of moles;
- decrease the concentration of inert material if the *BYPRODUCT* reaction involves an increase in the number of moles.

An alternative way to improve selectivity for the reaction system in Equation 5.65 is again to deliberately feed *BYPRODUCT* to the reactor to shift the equilibrium of the secondary reaction away from *BYPRODUCT* formation.

An example of where recycling can be effective in improving selectivity or reactor yield is in the production of benzene from toluene. The series reaction is reversible. Hence, recycling diphenyl to the reactor can be used to suppress its formation at the source.

5) *Mixed parallel and series reactions producing byproducts.* As with parallel and series reactions, use of an excess of one of the feeds can be effective in improving selectivity with mixed reactions. As an example, consider the chlorination of methane to produce chloromethane (Waddams, 1978). The primary reaction is:

$$\underset{\text{methane}}{CH_4} + \underset{\text{chlorine}}{Cl_2} \rightarrow \underset{\text{chloro-methane}}{CH_3Cl} + \underset{\substack{\text{hydrogen}\\\text{chloride}}}{HCl}$$

Secondary reactions can occur to higher chlorinated compounds:

$$\underset{\text{chloromethane}}{CH_3Cl} + Cl_2 \rightarrow \underset{\text{dichloro-methane}}{CH_2Cl_2} + HCl$$

$$\underset{\text{dichloromethane}}{CH_2Cl_2} + Cl_2 \rightarrow \underset{\text{chloroform}}{CHCl_3} + HCl$$

$$\underset{\text{chloroform}}{CHCl_3} + Cl_2 \rightarrow \underset{\substack{\text{carbon}\\\text{tetrachloride}}}{CCl_4} + HCl$$

The secondary reactions are series with respect to the chloromethane but parallel with respect to chlorine. A very large excess of methane (molar ratio of methane to chlorine is of the order 10:1) is used to suppress selectivity losses (Waddams, 1978). The excess of methane has two effects. First, because it is only involved in the primary reaction, it encourages the primary reaction. Second, by diluting the product chloromethane, it discourages the secondary reactions, which prefer a high concentration of chloromethane. Removal of the product as the reaction progresses is also effective in suppressing the series element of the byproduct reactions, providing the separation is straightforward.

If the byproduct reaction is reversible and inert material is present, then changing the concentration of inert material if there is a change in the number of moles should be considered, as discussed above. Whether or not there is a change in the number of moles, recycling byproducts can in some cases suppress their formation if the byproduct-forming reaction is reversible. An example is in the production of ethyl benzene from benzene and ethylene (Waddams, 1978):

$$\underset{\text{benzene}}{C_6H_6} + \underset{\text{ethylene}}{C_2H_4} \rightleftharpoons \underset{\text{ethylbenzene}}{C_6H_5CH_2CH_3}$$

Polyethylbenzenes (diethylbenzene, triethylbenzene, etc.) are also formed as unwanted byproducts through reversible reactions in series with respect to ethyl benzene but parallel with respect to ethylene. For example:

$$\underset{\text{ethylbenzene}}{C_6H_5CH_2CH_3} + \underset{\text{ethylene}}{C_2H_4} \rightleftharpoons \underset{\text{diethylbenzene}}{C_6H_4(C_2H_5)_2}$$

These polyethylbenzenes are recycled to the reactor to inhibit formation of fresh polyethylbenzenes (Waddams, 1978). However, it should be noted that recycling of byproducts is by no means always beneficial as byproducts can break down to cause deterioration in catalyst performance.

5.6 Biochemical Reactions

Biochemical reactions must cater for living systems and as a result are carried out in an aqueous medium within a narrow range of conditions. Each species of microorganism grows best under certain conditions. Temperature, pH, oxygen levels, concentrations of reactants and products and possibly nutrient levels must be carefully controlled for optimum operation.

1) *Temperature.* Microorganisms can be classified according to the optimum temperature ranges for their growth (Madigan, Martinko and Parker, 2003):
   - *Psychrophiles* have an optimum growth temperature of around 15 °C and a maximum growth temperature below 20 °C.
   - *Mesophiles* grow best between 20 and 45 °C.
   - *Thermophiles* have optimum growth temperatures between 45 and 80 °C.
   - *Extremphiles* grow under extreme conditions, some species above 100 °C, other species as low as 0 °C.

   Yeasts, moulds and algae typically have a maximum temperature around 60 °C.
2) *pH.* The pH requirements of microorganisms depend on the species. Most microorganisms prefer pH values not far from neutrality. However, some microorganisms can grow in extremes of pH.
3) *Oxygen levels.* Reactions can be carried out under *aerobic* conditions in which free oxygen is required or under *anaerobic* conditions in the absence of free oxygen. Bacteria can operate under aerobic or anaerobic conditions. Yeasts, moulds and algae prefer aerobic conditions but can grow with reduced oxygen levels.
4) *Concentration.* The rate of reaction depends on the concentrations of feed, trace elements, vitamins and toxic substances. The rate also depends on the build-up of wastes from the microorganisms that interfere with microorganism multiplication. There comes a point where their waste inhibits growth.

5

5.7 Catalysts

Most processes are catalyzed where catalysts for the reaction are known. The choice of catalyst is crucially important. Catalysts increase the rate of reaction but are ideally unchanged in quantity and chemical composition at the end of the reaction. If the catalyst is used to accelerate a reversible reaction, it does not by itself alter the position of the equilibrium. However, it should be noted if a porous solid catalyst is used, then different rates of diffusion of different species within a catalyst can change the concentration of the reactants at the point where the reaction takes place, thus influencing the equilibrium indirectly.

When systems of multiple reactions are involved, the catalyst may have different effects on the rates of the different reactions. This allows catalysts to be developed that increase the rate of the desired reactions relative to the undesired reactions. Hence the choice of catalyst can have a major influence on selectivity.

The catalytic process can be homogeneous, heterogeneous or biochemical.

1) *Homogeneous catalysts.* With a homogeneous catalyst, the reaction proceeds entirely in either the vapor or liquid phase. The catalyst may modify the reaction mechanism by participation in the reaction but is regenerated in a subsequent step. The catalyst is then free to promote further reaction. An example of such a homogeneous catalytic reaction is the production of acetic anhydride. In the first stage of the process, acetic acid is pyrolyzed to ketene in the gas phase at 700 °C:

$$\underset{\text{acetic acid}}{CH_3COOH} \rightarrow \underset{\text{ketene}}{CH_2 = C = O} + \underset{\text{water}}{H_2O}$$

   The reaction uses triethyl phosphate as a homogeneous catalyst (Waddams, 1978).

   In general, heterogeneous catalysts are preferred to homogeneous catalysts because the separation and recycling of homogeneous catalysts often can be very difficult. Loss of homogeneous catalyst not only creates a direct expense through loss of material but also creates an environmental problem.
2) *Heterogeneous catalysts.* In heterogeneous catalysis, the catalyst is in a different phase from the reacting species. Most often, the heterogeneous catalyst is a solid, acting on species in

the liquid or gas phase. The solid catalyst can be either of the following:

- *Bulk catalytic materials*, in which the gross composition does not change significantly through the material, such as platinum wire mesh.
- *Supported catalysts*, in which the active catalytic material is dispersed over the surface of a porous solid.

Catalytic gas-phase reactions play an important role in many bulk chemical processes, such as in the production of methanol, ammonia, sulfuric acid and most petroleum refinery processes. In most processes, the effective area of the catalyst is critically important. Industrial catalysts are usually supported on porous materials, since this results in a much larger active area per unit of reactor volume.

As well as depending on catalyst porosity, the reaction rate is some function of the reactant concentrations, temperature and pressure. However, this function may not be as simple as in the case of uncatalyzed reactions. Before a reaction can take place, the reactants must diffuse through the pores to the solid surface. The overall rate of a heterogeneous gas–solid reaction on a supported catalyst is made up of a series of physical steps as well as the chemical reaction. The steps are as follows:

a) Mass transfer of reactant from the bulk gas phase to the external solid surface.

b) Diffusion from the solid surface to the internal active sites.

c) Adsorption on solid surface.

d) Activation of the adsorbed reactants.

e) Chemical reaction.

f) Desorption of products.

g) Internal diffusion of products to the external solid surface.

h) Mass transfer to the bulk gas phase.

All of these steps are rate processes and are temperature dependent. It is important to realize that very large temperature gradients may exist between active sites and the bulk gas phase. Usually, one step is slower than the others, and it is this that is the rate-controlling step. The *effectiveness factor* is the ratio of the observed rate to that which would be obtained if the whole of the internal surface of the pellet were available to the reagents at the same concentrations as they have at the external surface. Generally, the higher the effectiveness factor, the higher the rate of reaction.

The effectiveness factor depends on the size and shape of the catalyst pellet and the distribution of active material within the pellet.

a) *Size of pellet.* If active material is distributed uniformly throughout the pellet, then the smaller the pellet, the higher the effectiveness. However, smaller pellets can produce an unacceptably high pressure drop through packed bed reactors. For gas-phase reactions in packed beds, the pressure drop is usually less than 10% of the inlet pressure (Rase, 1990).

b) *Shape of pellet.* Active material can be distributed on pellets with different shapes. The most commonly used shapes are spheres, cylinders and slabs. If the same amount of active material is distributed uniformly throughout the pellet, then *for the same volume of pellet*, the effectiveness factor is in the order:

$$\text{slab} > \text{cylinder} > \text{sphere}$$

c) *Distribution of active material.* During the catalyst preparation, it is possible to control the distribution of the active material within the catalyst pellet (Rase, 1990). Nonuniform distribution of the active material can increase the conversion, selectivity or resistance to deactivation compared to uniformly active catalysts. It is possible, in principle, to distribute the active material with almost any profile by the use of suitable impregnation techniques (Shyr and Ernst, 1980). Figure 5.7 illustrates some of the possible distributions through the pellet. In addition to uniform distribution (Figure 5.7a), it is possible to distribute as *egg shell*, in which the active material is located towards the outside of the pellet (Figure 5.7b), or *egg yolk*, in which the active material is located towards the core of the pellet (Figure 5.7c). A *middle distribution* locates the active material between the core and the outside of the pellet (Figure 5.7d). For a single reaction involving a fixed amount of active material, it can be shown that for conditions in which there are no catalyst deactivation mechanisms, the effectiveness of a supported catalyst is maximized when the active sites are concentrated at a precise location with zero width as a Dirac Delta Function (Morbidelli, Gavriilidis and Varma, 2001), as illustrated in Figure 5.7e. The optimal performance of the catalyst is obtained by locating the Dirac Delta catalyst distribution so as to maximize the reaction rate by taking advantage of both temperature and concentration gradients within the pellet. The Dirac Delta Function could be located at the surface, in the center or anywhere in between. Unfortunately, it is not possible from a practical point of view to locate the catalyst as a Dirac Delta Function. However, it can be approximated as a step function in a *layered catalyst*, as shown in Figure 5.7f. Providing the thickness of the active layer is less than ~5% of the pellet characteristic dimension (e.g. radius for a spherical pellet), the behavior of the Dirac Delta and step distributions is virtually the same (Morbidelli, Gavriilidis and Varma, 2001). The location of the active material needs to be optimized. However, it should be emphasized that if the catalyst is subject to degradation in its performance because of surface deposits, and so on, the Dirac Delta (and its step function equivalent) can be subject to a sharp deterioration in performance and is not necessarily the best choice under those conditions.

More often than not, solid-catalyzed reactions are multiple reactions. For reactions in parallel, the key to high selectivity is to maintain the appropriate high or low concentration and temperature levels of reactants at the catalyst surface, to encourage the desired reaction and to discourage the byproduct reactions. For reactions in series, the key is to avoid the mixing of fluids of different compositions. These arguments for the gross flow pattern of fluid through any reactor have already been developed.

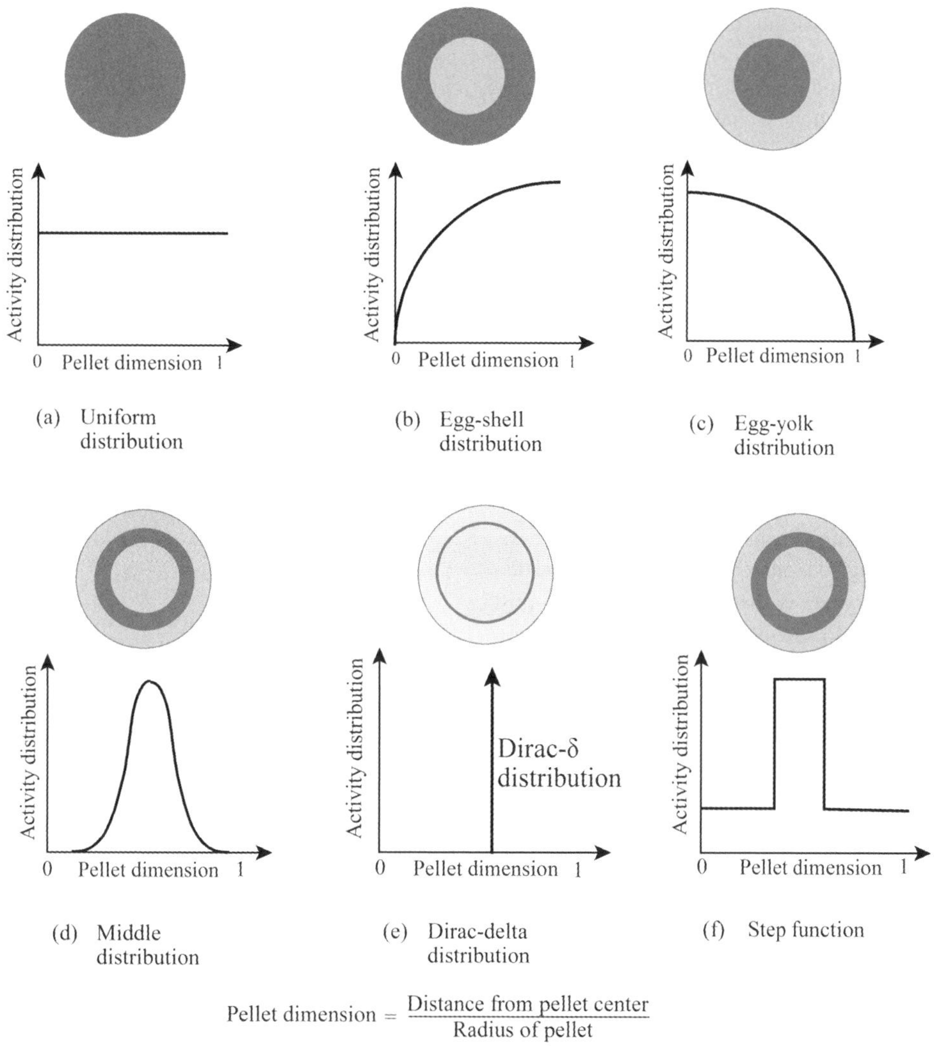

Figure 5.7

Catalyst distribution within a pellet of supported catalyst.

However, before extrapolating the arguments from the gross patterns through the reactor for homogeneous reactions to solid-catalyzed reactions, it must be recognized that in catalytic reactions the concentration and temperature in the interior of catalyst pellets may differ from the main body of the gas. Local nonhomogeneity caused by lowered reactant concentration or change in temperature within the catalyst pellets results in a product distribution different from what would otherwise be observed for a homogeneous system. Consider two extreme cases:

a) *Surface reaction controls.* If the surface reaction is rate controlling, then the concentrations of reactant within the pellets and in the main gas stream are essentially the same. In this situation, the considerations for the gross flow pattern of fluid through the reactor would apply.

b) *Diffusion controls.* If diffusional resistance controls, then the concentration of reactant at the catalyst surface would be lower than in the main gas stream. Referring back to Equation 4.64, for example, lowered reactant concentration favors the reaction of lower order. Hence, if the desired reaction is of lower order, operating under conditions of diffusion control would increase selectivity. If the desired reaction is of higher order, the opposite holds.

For multiple reactions, in cases where there is no catalyst deactivation, as with single reactions, the optimal catalyst distribution within the pellet is a Dirac Delta Function (Shyr and Ernst, 1980). This time the location within the pellet needs to be optimized for maximum selectivity or yield. Again, in practice, a step function approximates the performance of a Dirac Delta Function as long as it is less than ~5% of the pellet characteristic dimension (Morbidelli, Gavriilidis and Varma, 2001). However, a word of caution must again be added. As will be discussed in the next chapter, the performance of supported catalysts deteriorates through time for a variety of reasons. The optimal location of the step function should take account of this deterioration of performance through time. Indeed, as discussed previously, if the catalyst is subject to degradation in its performance because of surface deposits, and so on, the Dirac Delta (and its step function equivalent) might be subject to a sharp deterioration in performance and is not necessarily the best choice when considering the whole of the catalyst life.

Heterogeneous catalysts can thus have a major influence on selectivity. Changing the catalyst can change the relative influence on the primary and byproduct reactions. This might result directly from the reaction mechanisms at the active sites or the relative rates of diffusion in the support material or a combination of both.

3) *Biochemical catalysts.* Some reactions can be catalyzed by enzymes. The attraction in using enzymes, rather than microorganisms, is an enormous rate enhancement that can be obtained in the absence of the microorganisms. This is restricted to situations when the enzyme can be isolated and is also stable. In addition, the chemical reaction does not have to cater for the special requirements of living cells. However, just like microorganisms, enzymes are sensitive, and care must be exercised in the conditions under which they are used. A disadvantage in using enzymes is that use of the isolated enzyme is frequently more expensive on a single-use basis than the use of propagated microorganisms. Another disadvantage in using enzymes is that the enzymes will most likely need to be removed from the product once the reaction has been completed, which might involve an expensive separation. Long reaction times may be necessary if a low concentration of enzyme is used to decrease enzyme cost.

   Some of these difficulties in using enzymes can be overcome by fixing, or *immobilizing*, the enzyme in some way. A number of methods for enzyme immobilization have been developed. These can be classified as follows:

   a) *Adsorption.* The enzyme can be adsorbed onto an ion-exchange resin, insoluble polymer, porous glass or activated carbon.

   b) *Covalent bonding.* Reactions of side groups or cross-linking of enzymes can be used.

   c) *Entrapment.* Enzymes can be entrapped in a gel that is permeable to both the feeds and products, but not to the enzyme. Alternatively, a membrane can be used within which the enzymes have been immobilized, and feed material is made to flow through the membrane by creating a pressure difference across the membrane.

   There are many other possibilities.

Whatever the nature of the reaction, the choice of catalyst and the conditions of reaction can be critical to the performance of the process, because of the resulting influence on the selectivity of the reaction and reactor cost.

5.8 Reactor Conditions – Summary

Chemical equilibrium can be predicted from data for the free energy. There are various sources for data on free energy:

1) Tabulated data are available for standard free energy of formation at different temperatures.
2) Tabulated data are available for ΔH^O and ΔS^O at different temperatures and can be used to calculate ΔG^O from Equation 5.4 written at standard conditions:

$$\Delta G^O = \Delta H^O - T\Delta S^O$$

3) Tabulated data are available for ΔG^O and ΔH^O at standard temperature. This can be extrapolated to other temperatures using heat capacity data.
4) Methods are available to allow the thermodynamic properties of compounds to be estimated from their chemical structure (Poling, Prausnitz and O'Connell, 2001).

The equilibrium conversion can be calculated from knowledge of the free energy, together with physical properties to account for vapor and liquid-phase nonidealities. The equilibrium conversion can be changed by appropriate changes to the reactor temperature, pressure and concentration. The general trends for reaction equilibrium are summarized in Figure 5.8.

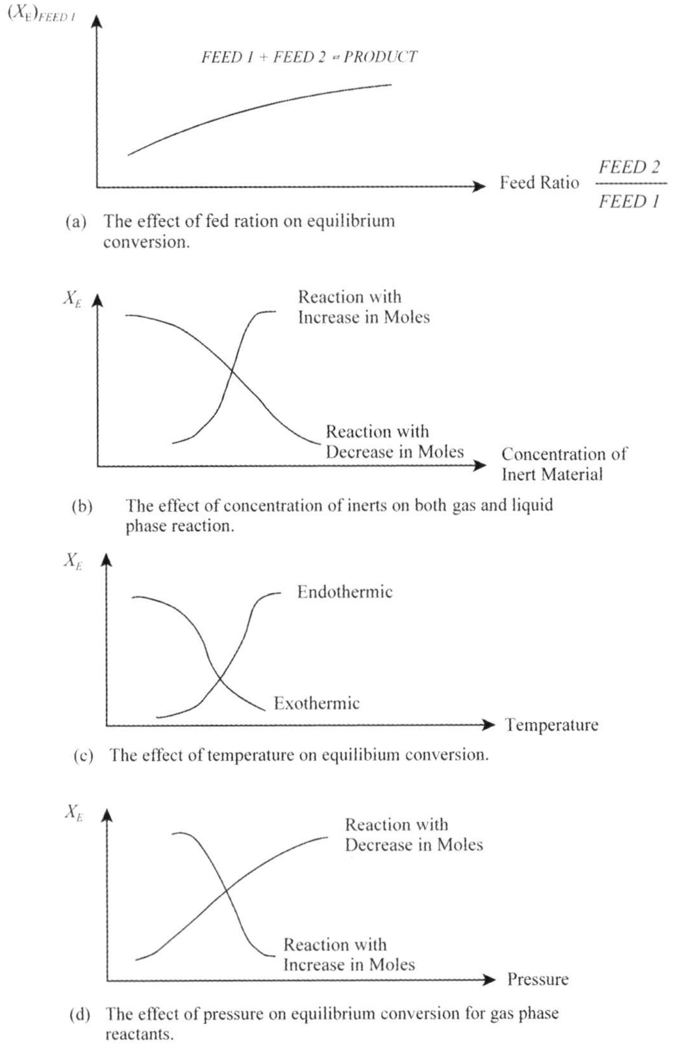

Figure 5.8

Various measures can be taken to increase equilibrium conversion in reversible reactions. (Reproduced from Smith R and Petela EA (1992) Waste Minimization in the Process Industries Part 2 Reactors, Chem Eng, Dec (509–510): 17, by permission of the Institution of Chemical Engineers.)

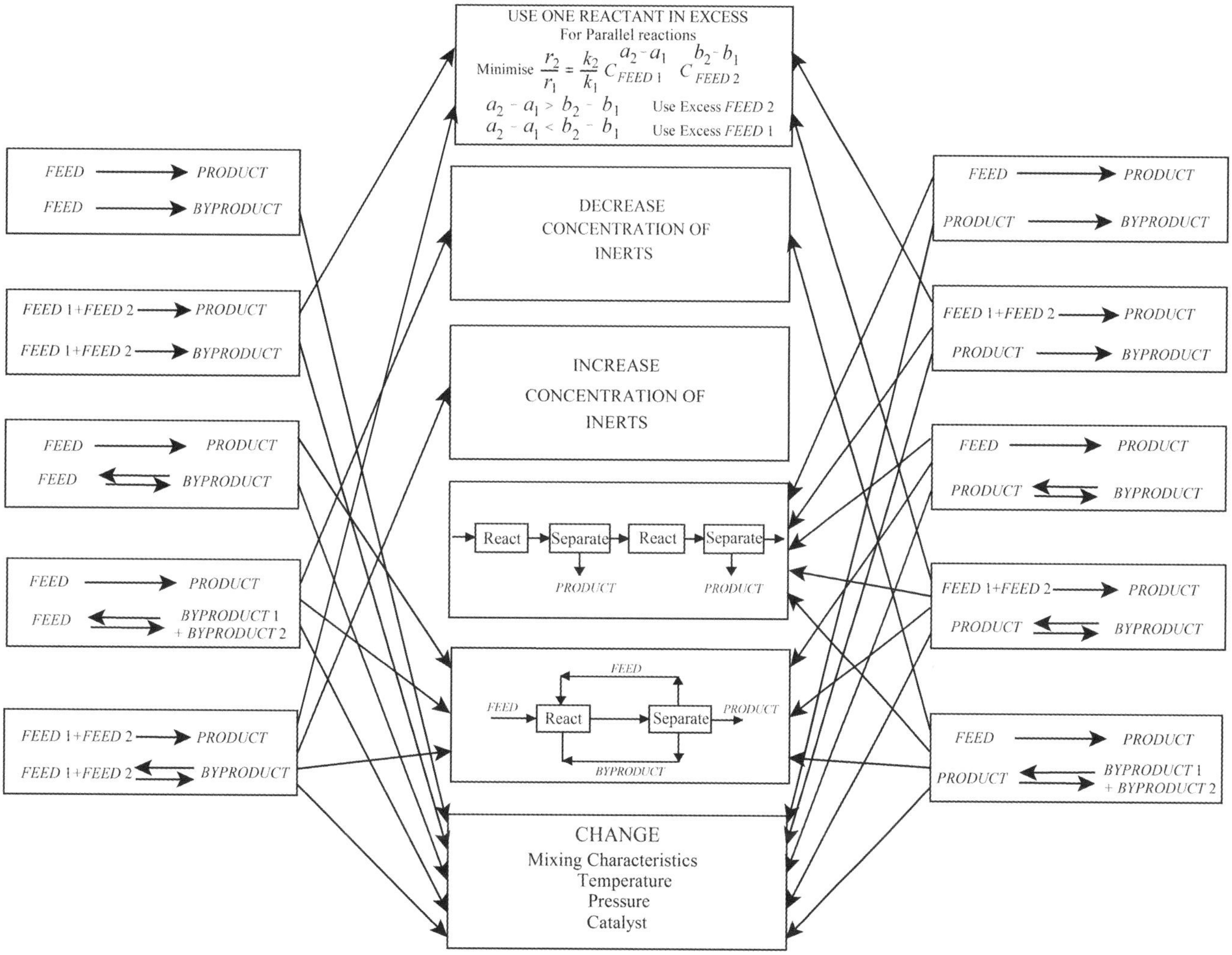

Figure 5.9

Choosing the reactor to maximize security for multiple reactants producing byproducts. (Reproduced from Smith R and Petela EA (1992) Waste Minimization in the Process Industries Part 2 Reactors, Chem Eng, Dec (509–510): 17, by permission of the Institution of Chemical Engineers.)

For reaction systems involving multiple reactions producing byproducts, selectivity and reactor yield can also be enhanced by appropriate changes to the reactor temperature, pressure and concentration. The appropriate choice of catalyst can also influence selectivity and reactor yield. The arguments are summarized in Figure 5.9 (Smith and Petela, 1992).

For supported layered catalysts, optimizing the location of the active sites within the catalyst pellets maximizes the effectiveness or the selectivity or reactor yield.

Reactions can be catalyzed using the metabolic pathways in microorganisms or the direct use of enzymes in biochemical reactors. The use of enzymes directly can have significant advantages if the enzymes can be isolated and immobilized in some way.

5.9 Exercises

1. Ammonia is to be produced by passing a gaseous mixture containing hydrogen, nitrogen and an inert gas over a catalyst at a pressure of 50 bar for the reaction:

$$N_2 + 3H_2 \rightleftharpoons 2NH_3$$

The system can be assumed to follow ideal gas behavior. If $K_P = 0.0125$, estimate the maximum conversion if the feed contains:

a) 60% H_2, 20% N_2 and the remainder inert gas.

b) 65% H_2, 15% N_2 and the remainder inert gas.

2. A gas mixture of 25 mole% CO_2 and 75% CO is mixed with the stoichiometric H_2 for the following reactions:

$$CO_2 + 3H_2 \rightleftharpoons CH_3OH + H_2O \qquad K_P = 2.82 \times 10^{-6}$$

$$CO + 2H_2 \rightleftharpoons CH_3OH \qquad K_P = 2.80 \times 10^{-5}$$

The conversion to equilibrium is carried out in the presence of a catalyst at a pressure of 300 bar. Assuming the behavior to be ideal, calculate the conversion for each reaction and hence the volume composition of the equilibrium mixture.

3. For the reaction:

$$2CO + 4H_2 \rightleftharpoons C_2H_5OH + H_2O$$

The feed consists of r moles of H_2 per mole of CO. The reaction takes place at 20 bar pressure and 573 K for which $\Delta G^O = 8900\ kJ{\cdot}kmol^{-1}$ for a standard pressure of 1 bar.

a) Develop an expression for the fractional conversion of CO in the presence of excess hydrogen (i.e. for $r > 2$).

b) Develop the corresponding expression for the fractional conversion of H_2 when CO is in excess.

4. Table 5.16 presents conversion data obtained from the batch experiments in the gas phase for the isomerization of reactant A to product B.

a) Develop an expression for the reactor conversion for a first-order reversible reaction.

b) Show that the reaction data can be modelled as a first-order reversible reaction by fitting a first-order model at 100 °C and 150 °C.

c) Determine the activation energy for the forward and reverse reactions across the range 100 °C to 150 °C. Assume $R = 8.3145\ kJ{\cdot}K^{-1}{\cdot}kmol^{-1}$.

5. The flow through a plug-flow reactor effecting a first-order forward reaction is increased by 20%, and in order to maintain the fractional conversion at its former value, it has been decided to increase the reactor operating temperature. If the reaction has an activation energy of $18{,}000\ kJ{\cdot}kmol^{-1}$ and the initial temperature is 150 °C, find the new operating temperature. Would the required elevation of temperature be different if the reactor was mixed-flow? Assume $R = 8.3145\ kJ{\cdot}K^{-1}{\cdot}kmol^{-1}$.

6. The flue gas from a combustion process contains oxides of nitrogen NO_x (principally NO and NO_2). The flow of flue gas is $10\ N\ m^3{\cdot}s^{-1}$ and contains 0.1% vol NO_x (expressed as NO_2 at 0 °C and 1 atm) and 3% vol oxygen. There is a reversible reaction in the gas phase between the two principal oxides of nitrogen according to:

$$NO + \tfrac{1}{2}O_2 \rightleftharpoons NO_2$$

The equilibrium relationship for the reaction is given by:

$$K_a = \frac{p_{NO_2}}{p_{NO}\,p_{O_2}^{0.5}}$$

where K_a is the equilibrium constant for the reaction and p the partial pressure. At 25 °C the equilibrium constant is 1.4×10^6 and at 725 °C is 0.14. Assuming ideal gas behaviour, the kilogram molar mass occupies $22.4\ m^3$ at standard conditions and the molar masses of NO and NO_2 are 30 and $46\ kg{\cdot}kmol^{-1}$ respectively, calculate:

a) The molar ratio of NO_2 to NO assuming chemical equilibrium at 25 °C and at 725 °C.

b) The mass flowrate of NO_2 to NO assuming chemical equilibrium at 25 °C and at 725 °C.

7. In the manufacture of sulfuric acid, a mixture of sulfur dioxide and air is passed over a series of catalyst beds, and sulfur trioxide is produced according to the reaction:

$$2SO_2 + O_2 \rightleftharpoons 2SO_3$$

The sulfur dioxide enters the reactor with an initial concentration of 10% by volume, the remainder being air. At the exit of the first bed, the temperature is 620 °C. Assume ideal gas behavior, the reactor operates at 1 bar and $R = 8.314\ kJ{\cdot}K^{-1}{\cdot}kmol^{-1}$. Assume air to be 21% O_2 and 79% N_2. Thermodynamic data at standard conditions at 298.15 K are given in Table 5.17 (Poling, Prausnitz and O'Connell, 2001).

a) Calculate the equilibrium conversion and equilibrium composition for the reactor products after the first catalyst bed at 620 °C assuming ΔH^O is constant over temperature range.

b) Is the reaction exothermic or endothermic?

c) From Le Châtelier's Principle, how would you expect the equilibrium conversion to change as the temperature increases?

d) How would you expect the equilibrium conversion to change as the pressure increases?

Table 5.16

Data for the isomerization reaction.

	Conversion	
Time (s)	**100 °C**	**150 °C**
0	0	0
100	0.05	0.15
500	0.21	0.49
1000	0.35	0.68
2000	0.52	0.81
5000	0.73	0.89

Table 5.17

Thermodynamic data at standard conditions of 298.15 K and 1 bar for sulfuric acid production.

	$\overline{H}^O$ ($kJ{\cdot}kmol^{-1}$)	$\overline{G}^O$ ($kJ{\cdot}kmol^{-1}$)
O_2	0	0
SO_2	−296,800	−300,100
SO_3	−395,700	−371,100

Table 5.18

Heat capacity data for sulfuric acid production.

	C_P (kJ·kmol^{-1}·K^{-1})				
	α_0	$\alpha_1 \times 10^3$	$\alpha_2 \times 10^5$	$\alpha_3 \times 10^8$	$\alpha_4 \times 10^{11}$
O_2	3.630	−1.794	0.658	−0.601	0.179
SO_2	4.417	−2.234	2.344	−3.271	1.393
SO_3	3.426	6.479	1.691	−3.356	1.590

8. Repeat the calculation in Exercise 7 for equilibrium conversion and equilibrium concentration, but taking into account variation of ΔH^O with temperature. Again assume ideal gas behavior. Heat capacity coefficients for Equation 5.42 are given in Table 5.18 (Poling, Prausnitz and O'Connell, 2001). Compare your answer with the result from Exercise 7.

9. In Exercises 7 and 8 it was assumed that the feed to the reactor was 10% SO_2 in air. However, the nitrogen in the air is an inert and takes no part in the reaction. In practice oxygen-enriched air could be fed.

a) From Le Châteliers Principle, what would you expect to happen to the equilibrium conversion, if oxygen enriched air was used for the same O_2 to SO_2 ratio?

b) Repeat the calculation from Exercise 7 assuming the same O_2 to SO_2 ratio in the feed but using an air feed enriched from 21% purity to 50%, 70%, 90% and 99% purity and calculate the component mole fractions at each air purity.

c) Apart from any potential advantages on equilibrium conversion, what advantages would you expect from feeding enriched air?

References

Coull J and Stuart EB (1964) *Equilibrium Thermodynamics*, John Wiley & Sons.

Dodge BF (1944) *Chemical Engineering Thermodynamics*, McGraw-Hill.

Hougen OA, Watson KM and Ragatz RA (1959) *Chemical Process Principles. Part II: Thermodynamics*, John Wiley & Sons.

Lide DR (2010) *CRC Handbook of Chemistry and Physics*, 91st Edition, CRC Press.

Madigan MT, Martinko JM and Parker J (2003) *Biology of Microorganisms*, Prentice Hall.

Morbidelli M, Gavriilidis A and Varma A (2001) *Catalyst Design: Optimal Distribution of Catalyst in Pellets, Reactors, and Membranes*, Cambridge University Press.

Poling BE, Prausnitz JM and O'Connell JP (2001) *The Properties of Gases and Liquids*, 5th Edition, McGraw-Hill.

Rase HF (1990) *Fixed-Bed Reactor Design and Diagnostics – Gas Phase Reactions*, Butterworths.

Shyr Y-S and Ernst WR (1980) Preparation of Nonuniformly Active Catalysts, *J Catal*, **63**: 426.

Smith EB (1990) *Basic Chemical Thermodynamics*, 3rd Edition, Oxford Chemistry Series, Oxford University Press.

Smith R and Petela EA (1992) Waste Minimisation in the Process Industries, IChemE Symposium on Integrated Pollution Control Through Clean Technology, Wilmslow UK, Paper 9: 20.

Waddams AL (1978) *Chemicals from Petroleum*, John Murray.

Chapter 6

Chemical Reactors III – Reactor Configuration

6

In contrast to ideal models, practical reactors must consider many factors other than variations in temperature, concentration and residence time. Consider the temperature control of the reactor first.

6.1 Temperature Control

Before considering how to control the temperature, the heat input or heat output from the reaction needs to be determined. Reactions are *exothermic* when there is a decrease in enthalpy from feed to products, leading to an evolution of heat. Reactions are *endothermic* when there is an increase in enthalpy from feed to products, leading to heat being absorbed. In most cases the reactants and products will be at different temperatures. To calculate the heat released or absorbed from a reaction and the temperature of the products, the thermodynamic path shown in Figure 6.1 can be followed (Hougen *et al.*, 1954). The actual reaction process goes from reactants at temperature T_1 to products at temperature T_2. However, it is more convenient to follow the alternative path from reactants at temperature T_1 that are initially cooled (or heated) to standard temperature of 298 K. The combustion reactions are then carried out at a constant temperature of 298 K. Standard heats of reaction can be calculated from standard heats of formation (Hougen, Watson and Ragatz, 1954). The products of combustion are then heated (or cooled) from 298 K to the final temperature of T_2. The actual heat of combustion is given by (Hougen, Watson and Ragatz, 1954):

$$\Delta H_R = \Delta H_1 + \Delta H^O + \Delta H_2 \tag{6.1}$$

where
ΔH_R = heat of reaction (J·kmol^{-1})
ΔH_1 = heat to bring reactants from their initial temperature to standard temperature (J·kmol^{-1}) $= \int_{T1}^{298} Cp_{react}\, dT$
ΔH^O = standard heat of reaction at 298 K (J·kmol^{-1})
ΔH_2 = heat to bring products from standard temperature to the final temperature (J·kmol^{-1}) $= \int_{298}^{T_2} Cp_{prod}\, dT$
Cp_{react} = heat capacity of reactants (J·kmol^{-1}·K^{-1})
Cp_{prod} = heat capacity of products (J·kmol^{-1}·K^{-1})

For an adiabatic reaction, $\Delta H_R = 0$ and Equation 6.1 becomes:

$$\Delta H_2 = -\Delta H_1 - \Delta H^O \tag{6.2}$$

The following example illustrates the approach.

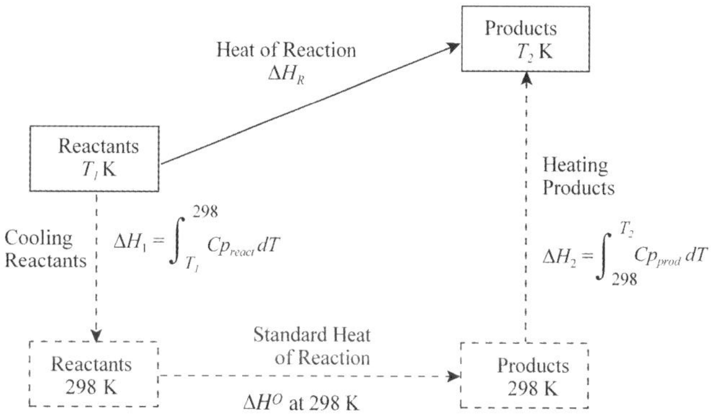

Figure 6.1
The enthalpy change from initial to final state in a reaction process.

Chemical Process Design and Integration, Second Edition. Robin Smith.

Companion Website: www.wiley.com/go/smith/chemical2e

Example 6.1 Ethylene oxide is produced in a reactor at a rate of 100 tons per day from pure ethylene and oxygen by the following reaction:

$$\underset{\text{ethylene}}{CH_2=CH_2} + \underset{\text{oxygen}}{\tfrac{1}{2}O_2} \longrightarrow \underset{\text{ethylene oxide}}{H_2C{-}CH_2 \ (\backslash O /)}$$

$$\Delta H^O = -1.20 \times 10^5 \text{ kJ}\cdot\text{kmol}^{-1}$$

A parallel reaction occurs leading to a selectivity loss:

$$\underset{\text{ethylene}}{CH_2 = CH_2} + \underset{\text{oxygen}}{3O_2} \longrightarrow \underset{\text{carbon dioxide}}{2CO_2} + \underset{\text{water}}{2H_2O}$$

$$\Delta H^O = -1.32 \times 10^6 \text{ kJ}\cdot\text{kmol}^{-1}$$

The reactor conversion can be assumed to be 20% and selectivity 80%. Excess oxygen of 10% above stoichiometric is used.

a) Determine the flowrates of the components entering and leaving the reactor. Molar masses are given in Table 6.1.

b) If the oxygen enters at 150 °C, the ethylene at 200 °C and the products leave at 260 °C find the heat removal rate. Heat capacity data for Equation 5.42 are given in Table 6.2.

Table 6.1

Data for Exercise 1.

Component	Molar mass ($kg\cdot kmol^{-1}$)
C_2H_4	28
O_2	32
C_2H_4O	44
CO_2	44
H_2O	18

Solution

a) Let X designate the reactor conversion, S selectivity, A the kmol of ethylene and B the kmol of oxygen:

Reaction 1	C_2H_4	+	$0.5O_2$	$\longrightarrow$	C_2H_4O
Moles in initial mixture	A		B		0
Moles at outlet condition	$[A - AXS]$		$[B - 0.5AXS]$		$[AXS]$

Reaction 2	C_2H_4	+	$3O_2$	$\longrightarrow$	$2CO_2$	+	$2H_2O$
Moles in initial mixture	A		B		0		0
Moles at outlet conditions	$[A - AX(1-S)]$		$[B - 3AX(1-S)]$		$[2AX(1-S)]$		$[2AX(1-S)]$

Since the reactor is required to produce 100 tons per day of ethylene oxide:

$$AXS = 100\frac{\text{ton}}{\text{d}} \times \frac{1000\text{ kg}}{1\text{ ton}} \times \frac{\text{kmol}}{44\text{ kg}} \times \frac{1\text{ d}}{24\text{ h}} = 94.70\text{ kmol}\cdot\text{h}^{-1}$$

At the outlet conditions:

C_2H_4	O_2	C_2H_4O	CO_2	H_2O
$A - AXS - AX(1-S)$ $= A(1-X)$ $= 473.50$	$B - 0.5AXS - 3AX(1-S)$ $= B - 3AX + 2.5AXS$ $= B - 118.38$	AXS $= 94.70$	$2AX(1-S)$ $= 47.35$	$2AX(1-S)$ $= 47.35$

Table 6.2

Heat capacity data (Poling, 2001).

	C_P ($kJ\cdot kmol^{-1}\cdot K^{-1}$)				
	α_0	$\alpha_1 \times 10^3$	$\alpha_2 \times 10^5$	$\alpha_3 \times 10^8$	$\alpha_4 \times 10^{11}$
C_2H_4	4.221	−8.782	5.795	−6.729	2.511
O_2	3.630	−1.794	0.658	−0.601	0.179
C_2H_4O	−0.9043	26.73	−1.511	0.3117	0.0
CO_2	3.259	1.356	1.502	−2.374	1.056
H_2O	4.395	−4.186	1.405	−1.564	0.632

At the inlet conditions:

$$A = \frac{94.70}{0.2 \times 0.8} = 591.88\ \text{kmol} \cdot \text{h}^{-1}$$

$$B_{react1} = 0.5 \times \frac{94.70}{0.2} = 236.75\ \text{kmol} \cdot \text{h}^{-1}$$

$$B_{react2} = 1.5 \times \frac{47.35}{0.2} = 355.13\ \text{kmol} \cdot \text{h}^{-1}$$

Accounting for O_2 in 10% excess:

$$B = 1.1 \times (236.75 + 355.13) = 651.07\ \text{kmol} \cdot \text{h}^{-1}$$

At outlet conditions:

$$B = 651.07 - 118.38 = 532.69\ \text{kmol} \cdot \text{h}^{-1}$$

Component	Inlet flowrate (kmol·h^{-1})	Outlet flowrate (kmol·h^{-1})
C_2H_4	591.88	473.50
O_2	651.07	532.69
C_2H_4O	0	94.70
CO_2	0	47.35
H_2O	0	47.35

b) Heat balance (see Figure 6.1)

$$\Delta H_R = \Delta H_1 + \Delta H^O + \Delta H_2$$

where ΔH_R = heat of reaction
ΔH_1 = enthalpy change to cool reactants to standard conditions
ΔH^O = standard heat of reaction
ΔH_2 = enthalpy change to heat reactants to final conditions

$$\Delta H_1 = \int_{T1}^{298} Cp_{react}\, dT$$

$$\frac{C_{Pi}}{R} = \alpha_{0i} + \alpha_{1i}T + \alpha_{2i}T^2 + \alpha_{3i}T^3 + \alpha_{4i}T^4$$

$$\Delta H_{1i} = R\left[\alpha_{0i}\,T + \frac{\alpha_{1i}\,T^2}{2} + \frac{\alpha_{2i}\,T^3}{3} + \frac{\alpha_{3i}\,T^4}{4} + \frac{\alpha_{4i}\,T^5}{5}\right]_{T_1}^{298}$$

$$\Delta H_1 = \sum_i m_i \Delta H_{1i}$$

$$\Delta H_2 = \int_{298}^{T_2} Cp_{prod}\, dT$$

$$\Delta H_{2i} = R\left[\alpha_{0i}\,T + \frac{\alpha_{1i}\,T^2}{2} + \frac{\alpha_{2i}\,T^3}{3} + \frac{\alpha_{3i}\,T^4}{4} + \frac{\alpha_{4i}\,T^5}{5}\right]_{298}^{T_2}$$

$$\Delta H_2 = \sum_i m_i \Delta H_{2i}$$

Substituting the heat capacity coefficients and evaluating:

$$\begin{aligned}\Delta H_1 &= m_{C_2H_4}\Delta H_{C_2H_4} + m_{O_2}\Delta H_{O_2}\\ &= 591.88 \times -8954 + 651.07 \times -3730\\ &= -7.728 \times 10^6\ \text{kJ} \cdot \text{h}^{-1}\end{aligned}$$

$$\begin{aligned}\Delta H_2 &= m_{C_2H_4}\Delta H_{C_2H_4} + m_{O_2}\Delta H_{O_2} + m_{C_2H_4O}\Delta H_{C_2H_4O}\\ &\quad + m_{CO_2}\Delta H_{CO_2} + m_{H_2O}\Delta H_{H_2O}\\ &= 473.50 \times 12699 + 532.69 \times 7126 + 94.70 \times 15173\\ &\quad + 47.35 \times 9791 + 47.35 \times 8115\\ &= 12.094 \times 10^6\ \text{kJ} \cdot \text{h}^{-1}\end{aligned}$$

$$\begin{aligned}\Delta H^O &= \Delta H^O_{R1} + \Delta H^O_{R2}\\ &= 94.7 \times -1.2 \times 10^5 + \frac{47.35}{2} \times -1.32 \times 10^6\\ &= -42.615 \times 10^6\ \text{kJ} \cdot \text{h}^{-1}\end{aligned}$$

Therefore the heat removal rate is given by:

$$\begin{aligned}\Delta H_R &= -7.728 \times 10^6 - 42.615 \times 10^6 + 12.094 \times 10^6\\ &= -38.25 \times 10^6\ \text{kJ} \cdot \text{h}^{-1}\end{aligned}$$

a) *Adiabatic operation.* In the first instance, adiabatic operation of the reactor should be considered since this leads to the simplest and cheapest reactor design. If adiabatic operation produces an unacceptable rise in temperature for exothermic reactions or an unacceptable fall in temperature for endothermic reactions, then methods of temperature control need to be considered.

b) *Cold shot and hot shot.* The injection of cold fresh feed directly into the reactor at intermediate points, known as *cold shot*, can be extremely effective for control of temperature in exothermic reactions. This not only controls the temperature by direct contact heat transfer through mixing with cold material, but also controls the rate of reaction by controlling the concentration of feed material. If the reaction is endothermic, then fresh feed that has been preheated can be injected at intermediate points, known as *hot shot*. Again, the temperature control is through a combination of direct contact heat transfer and control of the concentration.

c) *Indirect heat transfer with the reactor.* Indirect heating or cooling can also be considered. This might be by a heat transfer surface inside the reactor, such as carrying out the reaction inside a tube and providing a heating or cooling medium outside of the tube. Alternatively, material could be taken outside of the reactor from an intermediate point to a heat transfer device to provide the heating or cooling and then returned to the reactor. Different arrangements are possible and will be considered in more detail later.

d) *Heat carrier.* An inert material can be introduced with the reactor feed to increase its heat capacity flowrate (i.e. the product of mass flowrate and specific heat capacity) and to reduce the temperature rise for exothermic reactions or reduce temperature fall for endothermic reactions. Where possible, one of the existing process fluids should be used as the heat carrier. For example, an excess of feed material could be used to limit the temperature change, effectively decreasing the conversion, but for temperature-control purposes. Product or byproduct could be recycled to the reactor to limit the temperature change, but care must be taken to ensure that this does not have a detrimental effect on the selectivity or reactor yield. Alternatively, an extraneous inert material such as steam can be used to limit the temperature rise or fall.

e) *Catalyst profiles.* If the reactor uses a heterogeneous catalyst in which the active material is supported on a porous base, the size, shape and distribution of active material within the catalyst pellets can be varied, as discussed in Chapter 5. For a uniform distribution of active material, smaller pellets increase the effectiveness at the expense of an increased pressure drop in packed beds, and the shape for the same pellet volume generally provides effectiveness in the order:

$$\text{Slab} > \text{Cylinder} > \text{Sphere}$$

Cylinders have the advantage that they are cheap to manufacture. In addition to varying the shape, the distribution of the active material within the pellets can be varied, as illustrated in Figure 5.7. For packed-bed reactors, the size and shape of the pellets and the distribution of active material within the pellets can be varied through the length of the reactor to control the rate of heat release (for exothermic reactions) or heat input (for endothermic reactions). This involves creating different zones in the reactor, each with its own catalyst designs.

For example, suppose the temperature of a highly exothermic reaction needs to be controlled by packing the catalyst inside the tubes and passing a cooling medium outside of the tubes. If a uniform distribution of catalyst is used, a high heat release would be expected close to the reactor inlet, where the concentration of feed material is high. The heat release would then gradually decrease through the reactor. This typically manifests itself as an increasing temperature from the reactor inlet, because of a high level of heat release that the cooling medium does not remove completely in the early stages, reaching a peak and then decreasing towards the reactor exit as the rate of heat release decreases. Using a zone with a catalyst design with low effectiveness at the inlet and zones with increasing effectiveness through the reactor would control the rate of reaction to a more even profile through the reactor, allowing better temperature control.

Alternatively, rather than using a different design of pellet in different zones through the reactor, a mixture of catalyst pellets and inert pellets can be used to effectively "dilute" the catalyst. Varying the mixture of active and inert pellets allows the rate of reaction in different parts of the bed to be controlled more easily. Using zones with decreasing amounts of inert pellets through the reactor would control the rate of reaction to a more even profile through the reactor, allowing better temperature control.

As an example, consider the production of ethylene oxide examined in Example 6.1. This uses a silver-supported catalyst (Waddams, 1978). The reaction system is highly exothermic and is carried out inside tubes packed with catalyst, and a coolant is circulated around the exterior of the tubes to remove the heat of reaction. If a uniform catalyst design is used throughout the tubes, a high peak in temperature occurs close to the inlet. The peak in the temperature promotes the secondary reaction, leading to a high selectivity loss. Using a catalyst with lower effectiveness at the reactor inlet can reduce the temperature peak and increase the selectivity. It is desirable to vary the catalyst design in different zones through the reactor to obtain an even temperature profile along the reactor tubes.

As another example, consider the application of a fixed-bed tubular reactor for the production of methanol. Synthesis gas (a mixture of hydrogen, carbon monoxide and carbon dioxide) is reacted over a copper-based catalyst (Supp, 1973). The main reactions are:

$$\underset{\text{carbon monoxide}}{CO} + \underset{\text{hydrogen}}{2H_2} \rightleftharpoons \underset{\text{methanol}}{CH_3OH}$$

$$\underset{\text{carbon dioxide}}{CO_2} + \underset{\text{hydrogen}}{H_2} \rightleftharpoons \underset{\text{monoxide}}{CO} + \underset{\text{water}}{H_2O}$$

The first reaction is exothermic and the second is endothermic. Overall, the reaction evolves considerable heat. Figure 6.2 shows two alternative reactor designs (Supp, 1973). Figure 6.2a shows a shell-and-tube type of device that generates steam on the shell side. The temperature profile shows a peak shortly after the reactor inlet because of a high rate of reaction close to the inlet. The temperature then comes back into control. The temperature profile through the reactor in Figure 6.2a is seen to be relatively smooth. Figure 6.2b shows an alternative reactor design that uses cold-shot cooling. In contrast to the tubular reactor, the cold-shot reactor in Figure 6.2b experiences significant temperature fluctuations. Such fluctuations can, under some circumstances, cause accidental catalyst overheating and shorten catalyst life.

Even if the reactor temperature is controlled within acceptable limits, the reactor effluent may need to be cooled rapidly, or *quenched*, for example, to stop the reaction quickly in order to prevent excessive byproduct formation. This quench can be accomplished by indirect heat transfer using conventional heat transfer equipment or by direct heat transfer by mixing with another fluid. A commonly encountered situation is one in which gaseous products from a reactor need rapid cooling, which is accomplished by mixing with a liquid that evaporates. The heat required to evaporate the liquid causes the gaseous products to cool rapidly. The quench liquid can be a recycled, cooled product or an inert material such as water.

In fact, cooling of the reactor effluent by direct heat transfer can be used for a variety of reasons:

- The reaction is very rapid and must be stopped quickly to prevent excessive byproduct formation.
- The reactor products are so hot or corrosive that if passed directly to a heat exchanger special materials of construction or an expensive mechanical design would be required.

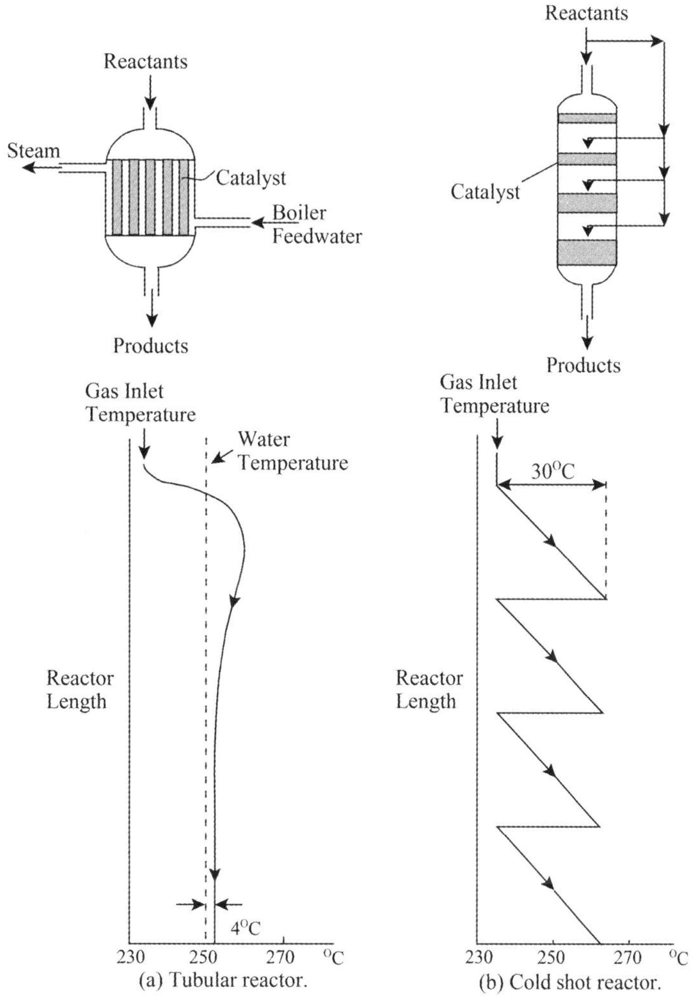

Figure 6.2

Two alternative reactor designs for methanol production give quite different thermal profiles.

- The reactor product cooling would cause excessive fouling in a conventional exchanger.

The liquid used for the direct heat transfer should be chosen such that it can be separated easily from the reactor product and hence recycled with minimum expense. Use of extraneous materials, that is, materials that do not already exist in the process, should be avoided if possible because of the following reasons:

- Extraneous material can create additional separation problems, requiring new separations that do not exist inherently within the process.
- The introduction of extraneous material can create new problems in achieving product purity specifications.
- It is often difficult to separate and recycle extraneous material with high efficiency. Any material not recycled can become an environmental problem. As shall be discussed later, the best way to deal with effluent problems is not to create them in the first place.

6.2 Catalyst Degradation

The performance of most catalysts deteriorates with time (Butt and Petersen, 1988; Rase, 1990; Wijngaarden, Kronberg and Westerterp, 1998). The rate at which the deterioration takes place is not only an important factor in the choice of catalyst and reactor conditions but also the reactor configuration.

Loss of catalyst performance can occur in a number of ways:

a) *Physical loss.* Physical loss is particularly important with homogeneous catalysts, which need to be separated from reaction products and recycled. Unless this can be done with high efficiency, it leads to physical loss (and subsequent environmental problems). However, physical loss, as a problem, is not restricted to homogeneous catalysts. It also can be a problem with heterogeneous catalysts. This is particularly the case when catalytic fluidized-bed reactors (to be discussed later) are employed. Catalyst particles are held in suspension and are mixed by the upward flow of a gas stream that is blown through the catalyst bed. Attrition of the particles causes the catalyst particles to be broken down in size. Particles carried over from the fluidized bed by entrainment are normally separated from the reactor effluent and recycled to the bed. However, the finest particles are not separated and recycled, and are lost.

b) *Surface deposits.* The formation of deposits on the surface of solid catalysts introduces a physical barrier to the reacting species. The deposits are most often insoluble (in liquid-phase reactions) or nonvolatile (in gas-phase reactions) byproducts of the reaction. An example of this is the formation of carbon deposits (known as *coke*) on the surface of catalysts involved in hydrocarbon reactions. Such coke formation can sometimes be suppressed by suitable adjustment of the feed composition. If coke formation occurs, the catalyst can often be regenerated by air oxidation of the carbon deposits at elevated temperatures.

c) *Sintering.* With high-temperature gas-phase reactions that use solid catalysts, *sintering* of the support or the active material can occur. Sintering is a molecular rearrangement that occurs below the melting point of the material and causes a reduction in the effective surface area of the catalyst. This problem is accelerated if poor heat transfer or poor mixing of reactants leads to local hot spots in the catalyst bed. Sintering can also occur during regeneration of catalysts to remove surface deposits of carbon by oxidation at elevated temperatures. Sintering can begin at temperatures as low as half the melting point of the catalyst.

d) *Poisoning. Poisons* are materials that chemically react with, or form strong chemical bonds, with the catalyst. Such reactions degrade the catalyst and reduce its activity. Poisons are usually impurities in the raw materials or products of corrosion. They can either have a reversible or irreversible effect on the catalyst.

e) *Chemical change.* In theory, a catalyst should not undergo chemical change. However, some catalysts can slowly change chemically, with a consequent reduction in activity.

The rate at which the catalyst is lost or degrades has a major influence on the design of the reactor. Deterioration in performance lowers the rate of reaction, which, for a given reactor design, manifests itself as a lowering of conversion with time. An operating policy of gradually increasing the temperature of the reactor through time can often be used to compensate for this deterioration in performance. However, significant increases in temperature can degrade selectivity considerably and can often accelerate the mechanisms that cause catalyst degradation.

If degradation is rapid, the reactor configuration must make provision either by having standby capacity or by removing the catalyst from the bed on a continuous basis. This will be discussed in more detail later, when considering reactor configuration. In addition to the cost implications, there are also environmental implications, since the lost or degraded catalyst represents waste. While it is often possible to recover useful materials from the degraded catalyst and recycle those materials in the manufacture of new catalyst, this still inevitably creates waste since the recovery of material can never be complete.

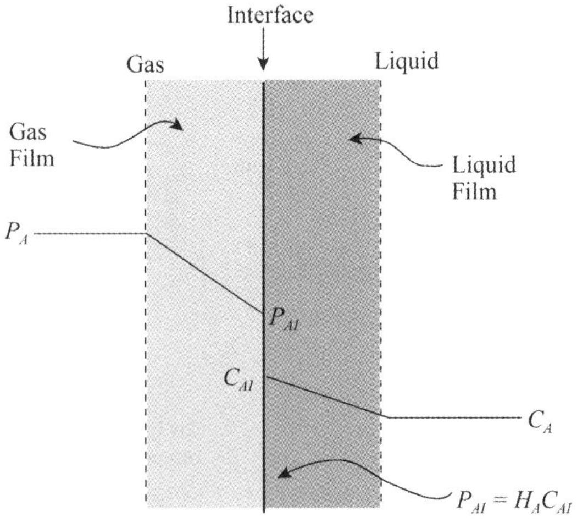

Figure 6.3
The gas–liquid interface.

6.3 Gas–Liquid and Liquid–Liquid Reactors

There are many reactions involving more than one reactant, where the reactants are fed in different phases as gas–liquid or liquid–liquid mixtures. This might be inevitable because the feed material is inherently in different phases at the inlet conditions. Alternatively, it might be desirable to create two-phase behavior in order to remove an unwanted component from one of the phases or to improve the selectivity. If the reaction is two-phase, then it is necessary that the phases be intimately mixed so that mass transfer of the reactants between phases can take place effectively. The overall rate of reaction must take account of the mass transfer resistance in order to bring the reactants together as well as the resistance of the chemical reactions. The three aspects of mixing, mass transfer and reaction can present widely differing difficulties, depending on the problem.

1) *Gas–liquid reactors.* Gas–liquid reactors are quite common. Gas-phase components will normally have a small molar mass. Consider the interface between a gas and a liquid that is assumed to have a flow pattern giving a stagnant film in the liquid and the gas on each side of the interface, as illustrated in Figure 6.3. The bulk of the gas and the liquid are assumed to have a uniform concentration. It will be assumed here that Reactant *A* must transfer from the gas to the liquid for the reaction to occur. There is diffusional resistance in the gas film and the liquid film.

Consider an extreme case in which there is no resistance to reaction and all of the resistance is due to mass transfer. The rate of mass transfer is proportional to the interfacial area and the concentration of the driving force. An expression can be written for the rate of transfer of Component *i* from gas to liquid through the gas film per unit volume of reaction mixture:

$$N_{G,i} = k_{G,i} A_I (p_{G,i} - p_{I,i}) \tag{6.3}$$

where $N_{G,i}$ = rate of transfer of Component i in the gas film ($kmol \cdot s^{-1} \cdot m^{-3}$)
$k_{G,i}$ = mass transfer coefficient in the gas film ($kmol \cdot Pa^{-1} \cdot m^{-2} \cdot s^{-1}$)
A_I = interfacial area per unit volume ($m^2 \cdot m^{-3}$)
$p_{G,i}$ = partial pressure of Component i in the bulk gas phase (Pa)
$p_{I,i}$ = partial pressure of Component i at the interface (Pa)

An expression can also be written for the rate of transfer of Component i through the liquid film, per unit volume of reaction mixture:

$$N_{L,i} = k_{L,i} A_I (C_{I,i} - C_{L,i}) \quad (6.4)$$

where $N_{L,i}$ = rate of transfer of Component i in the liquid film ($kmol \cdot s^{-1} \cdot m^{-3}$)
$k_{L,i}$ = mass transfer coefficient in the liquid film ($m \cdot s^{-1}$)
A_I = interfacial area per unit volume ($m^2 \cdot m^{-3}$)
$C_{I,i}$ = concentration of Component i at the interface ($kmol \cdot m^{-3}$)
$C_{L,i}$ = concentration of Component i in the bulk liquid phase ($kmol \cdot m^{-3}$)

If equilibrium conditions at the interface are assumed to be described by Henry's Law (see Appendix A):

$$p_{I,i} = H_i x_{I,i} = \frac{H_i}{\rho_L} C_{I,i} \quad (6.5)$$

where H_i = Henry's Law constant for Component i (Pa)
$x_{I,i}$ = mole fraction of Component i in the liquid at the interface (–)
$C_{I,i}$ = concentration of Component i in the liquid at the interface ($kmol \cdot m^{-3}$)
ρ_L = molar density of the liquid phase ($kmol \cdot m^{-3}$)

The Henry's Law constant varies between different gases and must be determined experimentally. If steady state is assumed ($N_{G,i} = N_{L,i} = N_i$), then Equations 6.3, 6.4 and 6.5 can be combined to obtain

$$N_i = \frac{1}{\left[\frac{1}{k_{G,i}A_I} + \frac{1}{k_{L,i}A_I}\frac{H_i}{\rho_L}\right]}\left(p_{G,i} - \frac{H_i}{\rho_L}C_{L,i}\right) \quad (6.6)$$

or

$$N_i = K_{GL,i}A_I\left(p_{G,i} - \frac{H_i}{\rho_L}C_{L,i}\right) \quad (6.7)$$

where

$$\frac{1}{K_{GL,i}A_I} = \frac{1}{k_{G,i}A_I} + \frac{1}{k_{L,i}A_I}\frac{H_i}{\rho_L} \quad (6.8)$$

$K_{GL,i}$ = overall mass transfer coefficient ($kmol \cdot Pa^{-1} \cdot m^{-2} \cdot s^{-1}$) and $k_{G,i}$, $k_{L,i}$ and A_I are functions of physical properties and the contacting arrangement. The first term on the right-hand side of Equation 6.8 represents the gas-film resistance and the second term, the liquid-film resistance. If $k_{G,i}$ is large relative to $k_{L,i}/H_i$, the mass transfer is liquid-film controlled. This is the case for low solubility gases (H_i is large). If $k_{L,i}/H_i$ is large relative to $k_{G,i}$, it is gas-film controlled. This is the case for highly soluble gases (H_i is small).

The solubility of gases varies widely. Gases with a low solubility (e.g. N_2, O_2) have large values of the Henry's Law coefficient. This means that the liquid-film resistance in Equation 6.8 is large relative to the gas-film resistance. On the other hand, if the gas is highly soluble (e.g. CO_2, NH_3), the Henry's Law coefficient is small. This leads to the gas-film resistance being large relative to the liquid-film resistance in Equation 6.8. Thus:

- liquid-film resistance controls for gases with low solubility;
- gas-film resistance controls for gases with high solubility.

6

The capacity of a gas to dissolve in a liquid is determined by the solubility of the gas. The capacity of a liquid to dissolve a gas is increased if it reacts with a species in the liquid.

Now consider the effect of chemical reaction. If the reaction is fast, its effect is to reduce the liquid-film resistance. The result is an effective increase in the overall mass transfer coefficient. The capacity of the liquid is also increased.

If the reaction is slow, there is a small effect on the overall mass transfer coefficient. The driving force for mass transfer will be greater than that for physical absorption alone, as a result of the dissolving gas reacting and not building up in the bulk liquid to the same extent as with pure physical absorption.

Figure 6.4 illustrates some of the arrangements that can be used to carry out gas–liquid reactions. The first arrangement in Figure 6.4a shows a countercurrent arrangement where plug-flow is induced in both the gas and the liquid. Packing material or trays can be used to create an interfacial area between the gas and the liquid. In some cases, the packed bed might be a heterogeneous solid catalyst rather than an inert material.

Figure 6.4b shows a packed-bed arrangement where plug flow is induced in both the gas and the liquid, but both phases are flowing cocurrently. This, in general, will give a poorer performance than the countercurrent arrangement in Figure 6.4a. However, the cocurrent arrangement, known as a *trickle-bed reactor*, might be necessary if the flow of gas is much greater than the flow of liquid. This can be the case for some gas–liquid reactions involving a heterogeneous catalyst with a large excess of the gas phase. A continuous gas phase and a liquid phase in the form of films and rivulets characterizes the flow pattern. Countercurrent contacting might well be preferred but a large flow of gas makes this extremely difficult.

Another way to provide cocurrent plug flow for both phases is to feed both phases to a pipe containing an in-line static mixer, as shown in Figure 6.4c. Various designs of static mixer are available, but mixing is usually promoted by repeatedly changing the direction of flow within the device as the liquid and gas flow through. This will give a good approximation to plug flow in both phases with cocurrent flow. Static mixers are particularly suitable when a short residence time is required.

Figure 6.4d shows a spray column. This approximates mixed-flow behavior in the gas phase and plug flow in the

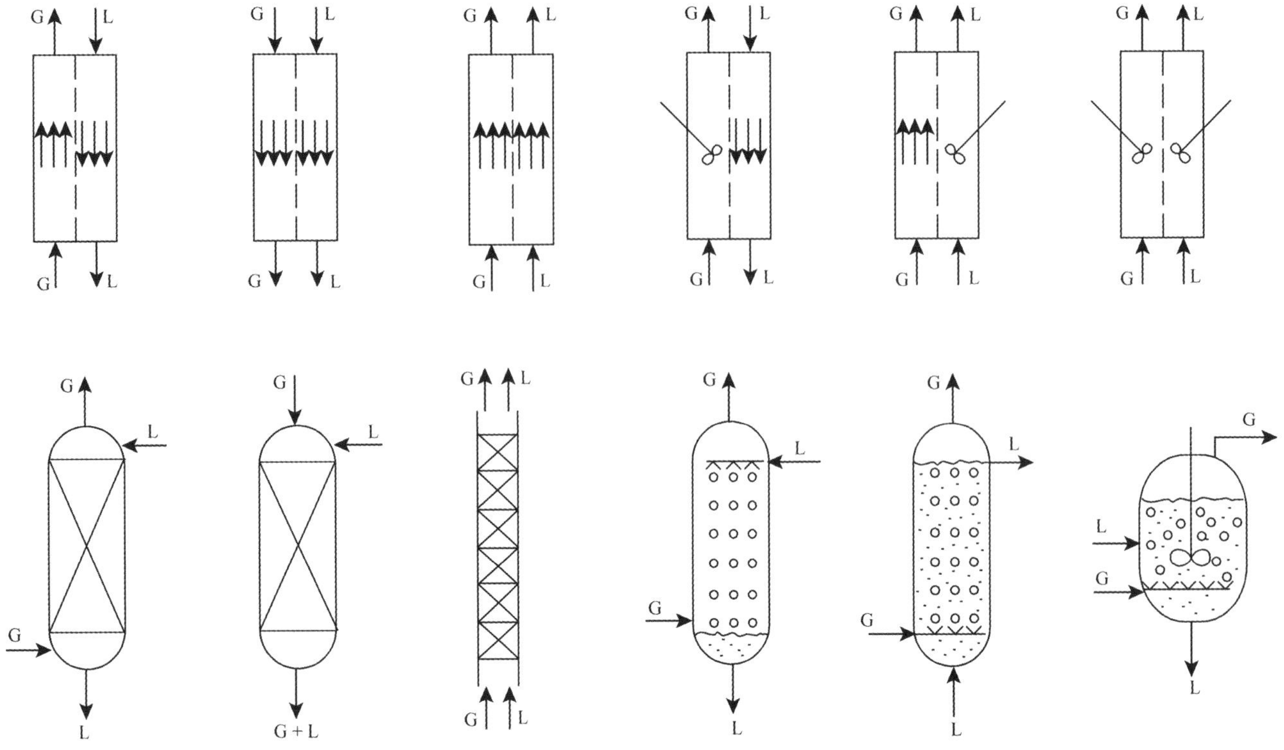

(a) Countercurrent packed bed or plate column. (b) Cocurrent packed bed. (c) In-line static mixer. (d) Spray column. (e) Bubble column. (f) Agitated tanks.

Figure 6.4
Contacting patterns for gas–liquid reactors.

liquid phase. The arrangement tends to induce a high gas-film mass transfer coefficient and a low liquid-film mass transfer coefficient. As the liquid film will tend to be controlling, the arrangement should be avoided when reacting a gas that has a low solubility (large values of Henry's Law coefficient). The spray column will generally give a lower driving force than a countercurrent packed column, but might be necessary if, for example, the liquid contains solids, or solids are formed in the reaction. If the reaction has a tendency to foul a packed bed, a spray column might be preferred for reasons of practicality.

Figure 6.4e shows a bubble column. This approximates plug flow in the gas phase and mixed flow in the liquid phase. The arrangement tends to induce a low gas-film mass transfer coefficient and a high liquid-film mass transfer coefficient. As the gas film will tend to be controlling, the arrangement should be avoided when reacting a gas with a high solubility (small values of Henry's Law coefficient). Although the bubble column will tend to have a lower performance than a countercurrent packed bed, the arrangement has two advantages over a packed bed. First, the liquid hold-up per unit reactor volume is higher than a packed bed, which gives a greater residence time for a slow reaction for a given liquid flowrate. Second, if the liquid contains a dispersed solid (e.g. a biochemical reaction using microorganisms), then a packed bed will rapidly become clogged. A disadvantage is that it will be ineffective if the liquid is highly viscous.

Finally, Figure 6.4f shows an agitated tank in which the gas is sparged through the liquid. This approximates mixed-flow behavior in both phases. The driving force is low relative to a countercurrent packed bed. However, there may be practical reasons to use a sparged agitated vessel. If the liquid is viscous (e.g. a biochemical reaction using microorganisms), the agitator allows the gas to be dispersed as small bubbles and the liquid to be circulated to maintain good contact between the gas and the liquid.

Of the contacting patterns in Figure 6.4, countercurrent packed beds offer the largest mass transfer driving force and agitated tanks the lowest.

The influence of temperature on gas–liquid reactions is more complex than homogeneous reactions. As the temperature increases:

- the rate of reaction increases,
- solubility of the gas in the liquid decreases,
- rates of mass transfer increase,
- volatility of the liquid phase increases, decreasing the partial pressure of the dissolving gas in Equation 6.6.

Some of these effects have an enhancing influence on the overall rate of reaction. Others will have a detrimental effect. The relative magnitude of these effects will depend on the system in question. To make matters worse, if multiple reactions are being considered that are reversible and that also produce byproducts, all of the factors discussed in Chapter 5 regarding the influence of temperature also apply to gas–liquid reactions.

Added to this, the mass transfer can also influence the selectivity. For example, consider a system of two parallel reactions in which the second reaction produces an unwanted byproduct and is slow relative to the primary reaction. The dissolving gas species will tend to react in the liquid film and not reach the bulk liquid in significant quantity for further reaction to occur there to form the byproduct. Thus, in this case, the selectivity would be expected to be enhanced by the mass transfer between the phases. In other cases, little or no influence can be expected.

2) *Liquid–liquid reactors.* Examples of liquid–liquid reactions are the nitration and sulfonation of organic liquids. Much of the discussion for gas–liquid reactions also applies to liquid–liquid reactions. In liquid–liquid reactions, mass needs to be transferred between two immiscible liquids for the reaction to take place. However, rather than gas- and liquid-film resistance, as shown in Figure 6.3, there are two liquid-film resistances. The reaction may occur in one phase or both phases simultaneously. Generally, the solubility relationships are such that the extent of the reactions in one of the phases is so small that it can be neglected.

For the mass transfer (and, hence, reaction) to take place, one liquid phase must be dispersed in the other. A decision must be made as to which phase should be dispersed in a continuous phase of the other. In most cases, the liquid with the smaller volume flowrate will be dispersed in the other. The overall mass transfer coefficient depends on the physical properties of the liquids and the interfacial area. In turn, the size of the liquid droplets and the volume fraction of the dispersed phase in the reactor govern the interfacial area. Dispersion requires the input of power either through an agitator or by pumping of the liquids. The resulting degree of dispersion depends on the power input, interfacial tension between the liquids and their physical properties. While it is generally desirable to have a high interfacial area and, therefore, small droplets, too effective dispersion might lead to the formation of an emulsion that is difficult to separate after the reactor.

Figure 6.5 illustrates some of the arrangements that can be used for liquid–liquid reactors. The first arrangement shown in Figure 6.5a is a packed bed in which the two liquids flow countercurrently. This is similar to Figure 6.4a for gas–liquid reactions. Plates can also be used to create the contact between the two liquid phases, rather than a packed bed. This arrangement will approximate plug-flow in both phases.

Figure 6.5b shows a multistage agitated contactor. A large number of stages and low backmixing will tend to approximate plug flow in both phases. However, the closeness to plug flow will depend on the detailed design.

Figure 6.5c shows an in-line static mixer. Dispersion is usually promoted by repeatedly changing the direction of flow

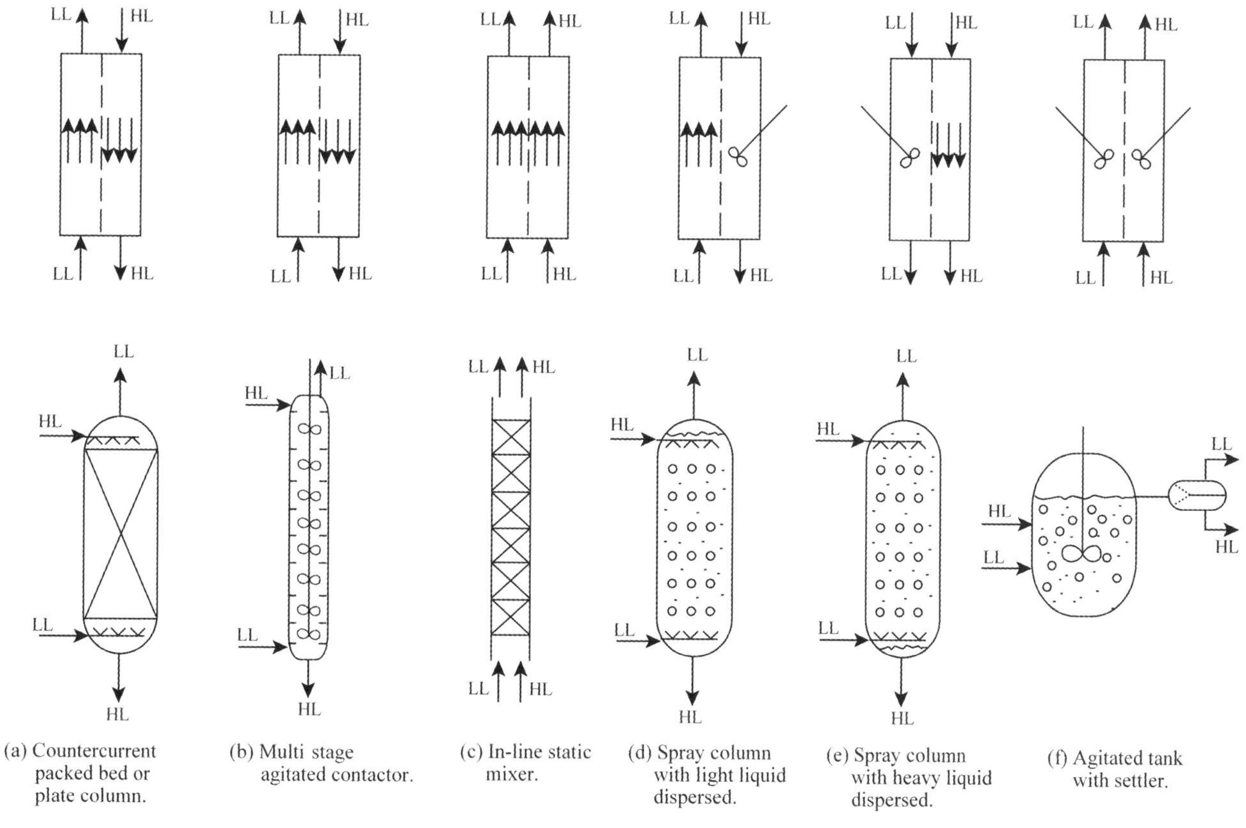

Figure 6.5

Contacting patterns for liquid–liquid reactors.

locally within the mixing device as the liquids are pumped through. This will give a good approximation to plug-flow in both phases in cocurrent flow. As with gas–liquid reactors, static mixers are particularly suitable when a short residence time is required.

Figure 6.5d shows a spray column in which the light liquid is dispersed. This approximates plug-flow in the light liquid phase and mixed-flow behavior in the heavy liquid phase. Figure 6.5e shows a spray column in which the heavy liquid is dispersed. This approximates mixed flow in the light liquid phase and plug-flow behavior in the heavy liquid phase. Spray columns will generally give a lower driving force than countercurrent packed columns, multistage agitated contactors and in-line static mixers.

Figure 6.5f shows an agitated tank followed by a settler in a *mixer–settler* arrangement, which will exhibit mixed-flow in both phases. Although Figure 6.5f shows a single-stage agitated tank and settler, a number of agitated tanks, each followed by a settler, can be connected together. For such a cascade of mixer–settler devices, the two liquid phases can be made to flow countercurrently through the cascade. The more stages that are used, the more the cascade will tend to countercurrent plug-flow behavior. Rather than a countercurrent flow arrangement through the cascade, a cross-flow arrangement can be used in which one of the phases is progressively added and removed at different points through the cascade. Such a flow arrangement can be useful if the reaction is limited by chemical equilibrium. If removal of the liquid removes the product that has formed, the reaction can be forced to a higher conversion than that of a countercurrent arrangement, as discussed in Chapter 5.

6.4 Reactor Configuration

Consider now the more common types of reactor configuration and their use:

1) *Tubular reactors.* Although tubular reactors often take the actual form of a tube, they can be any reactor in which there is steady movement only in one direction. If heat needs to be added or removed as the reaction proceeds, the tubes may be arranged in parallel, in a construction similar to a shell-and-tube heat exchanger. Here, the reactants are fed inside the tubes and a cooling or heating medium is circulated around the outside of the tubes. If a high temperature or high heat flux into the reactor is required, then the tubes are constructed in the radiant zone of a furnace.

   Because tubular reactors approximate plug flow, they are used if careful control of residence time is important, as is the case where there are multiple reactions in series. A high ratio of heat transfer surface area to volume is possible, which is an advantage if high rates of heat transfer are required. It is sometimes possible to approach isothermal conditions or a predetermined temperature profile by careful design of the heat transfer arrangements.

   Tubular reactors can be used for multiphase reactions, as discussed in the previous section. However, it is often difficult to achieve good mixing between phases, unless static mixer tube inserts are used.

   One mechanical advantage tubular devices have is when high pressure is required. Under high-pressure conditions, a small-diameter cylinder requires a thinner wall than a large-diameter cylinder.

2) *Stirred-tank reactors.* Stirred-tank reactors consist simply of an agitated tank and are used for reactions involving a liquid. Applications include:

   - homogeneous liquid-phase reactions,
   - heterogeneous gas–liquid reactions,
   - heterogeneous liquid–liquid reactions,
   - heterogeneous solid–liquid reactions,
   - heterogeneous gas–solid–liquid reactions.

   Stirred-tank reactors can be operated in batch, semi-batch, or continuous mode. In batch or semi-batch mode:

   - operation is more flexible for variable production rates or for manufacture of a variety of similar products in the same equipment;
   - labor costs tend to be higher (although this can be overcome to some extent by use of computer control).

   In continuous operation, automatic control tends to be more straightforward (leading to lower labor costs and greater consistency of operation).

   In practice, it is often possible with stirred-tank reactors to come close to the idealized mixed-flow model, providing the fluid phase is not too viscous. For homogenous reactions, such reactors should be avoided for some types of parallel reaction systems (see Figure 4.6) and for all systems in which byproduct formation is via series reactions.

   Stirred-tank reactors become unfavorable if the reaction must take place at high pressure. Under high-pressure conditions, a small-diameter cylinder requires a thinner wall than a large-diameter cylinder. Under high-pressure conditions, use of a tubular reactor is preferred, although mixing problems with heterogeneous reactions and other factors may prevent this. Another important factor to the disadvantage of the continuous stirred-tank reactor is that, for a given conversion, it requires a large inventory of material relative to a tubular reactor. This is not desirable for safety reasons if the reactants or products are particularly hazardous.

   Heat can be added to or removed from stirred-tank reactors via external jackets (Figure 6.6a), internal coils (Figure 6.6b) or separate heat exchangers by means of a flow loop (Figure 6.6c). Figure 6.6d shows vaporization of the contents being condensed and refluxed to remove heat. A variation on Figure 6.6d would not reflux the evaporated material back to the reactor, but would remove it as a product. Removing evaporated material in this way if it is a product or byproduct of a reversible reaction can be used to increase equilibrium conversion, as discussed in Chapter 5.

   If plug flow is required, but the volume of the reactor is large, then plug-flow operation can be approached by using stirred tanks in series, since large volumes are often more economically arranged in stirred tanks than in tubular devices. This can

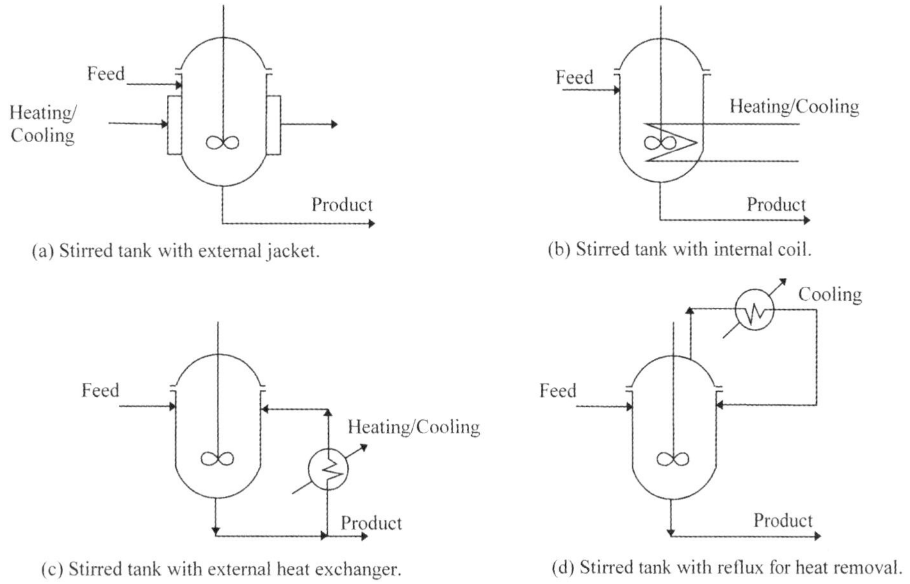

Figure 6.6

Heat transfer to and from stirred tanks.

also offer the advantage of better temperature control than the equivalent tubular reactor arrangement.

3) *Fixed-bed catalytic reactors.* Here, the reactor is packed with particles of solid catalyst. Most designs approximate to plug-flow behavior. The simplest form of fixed-bed catalytic reactor uses an adiabatic arrangement, as shown in Figure 6.7a. If adiabatic operation is not acceptable because of a large temperature rise for an exothermic reaction, or a large decrease for an endothermic reaction, then cold shot or hot shot can be used, as shown in Figure 6.7b. Alternatively, a series of adiabatic beds with intermediate cooling or heating can be used to maintain temperature control, as shown in Figure 6.7c. The heating or cooling can be achieved by internal or external heat exchangers. Tubular reactors similar to a shell-and-tube heat exchanger can be used, in which the tubes are packed with catalyst, as shown in Figure 6.7d. The heating or cooling medium circulates around the outside of the tubes.

Generally, temperature control in fixed beds is difficult because heat loads vary through the bed. The temperature inside catalyst pellets can be significantly different from the bulk temperature of the reactants flowing through the bed, due to the diffusion of reactants through the catalyst pores to the active sites for reaction to occur. In exothermic reactors, the temperature in the catalyst can become locally excessive. Such "hot spots" can cause the onset of undesired reactions or catalyst degradation. In tubular devices such as shown in Figure 6.7d, the smaller the diameter of tube, the better is the temperature control. As discussed in Section 6.1, temperature-control problems also can be overcome by using a profile of catalyst through the reactor to even out the rate of reaction and achieve better temperature control.

If the catalyst degrades (e.g. as a result of coke formation on the surface), then a fixed-bed device will have to be taken off-line to regenerate the catalyst. This can either mean shutting down the plant or using a standby reactor. If a standby reactor is to be used, two reactors are periodically switched, keeping one online while the other is taken off-line to regenerate the catalyst. Several reactors might be used in this way to maintain an overall operation that is close to steady state. However, if frequent regeneration is required, then fixed beds are not suitable, and under these circumstances a moving bed or a fluidized bed is preferred, as will be discussed later.

Gas–liquid mixtures are sometimes reacted in catalytic packed beds. Different contacting methods for gas–liquid reactions have been discussed in Section 6.3.

4) *Fixed-bed noncatalytic reactors.* Fixed-bed noncatalytic reactors can be used to react a gas and a solid. For example, hydrogen sulfide can be removed from fuel gases by reaction with ferric oxide:

$$\underset{\text{ferric oxide}}{Fe_2O_3} + \underset{\substack{\text{hydrogen}\\\text{sulfide}}}{3H_2S} \longrightarrow \underset{\substack{\text{ferric}\\\text{sulfide}}}{Fe_2S_3} + 3H_2O$$

The ferric oxide is regenerated using air:

$$2Fe_2S_3 + 3O_2 \longrightarrow 2Fe_2O_3 + 6S$$

Two fixed-bed reactors can be used in parallel, one reacting and the other regenerating. However, there are many disadvantages in carrying out this type of reaction in a packed bed. The operation is not under steady-state conditions, and this can present control problems. Eventually, the bed must be taken off-line to replace the solid. Fluidized beds (to be

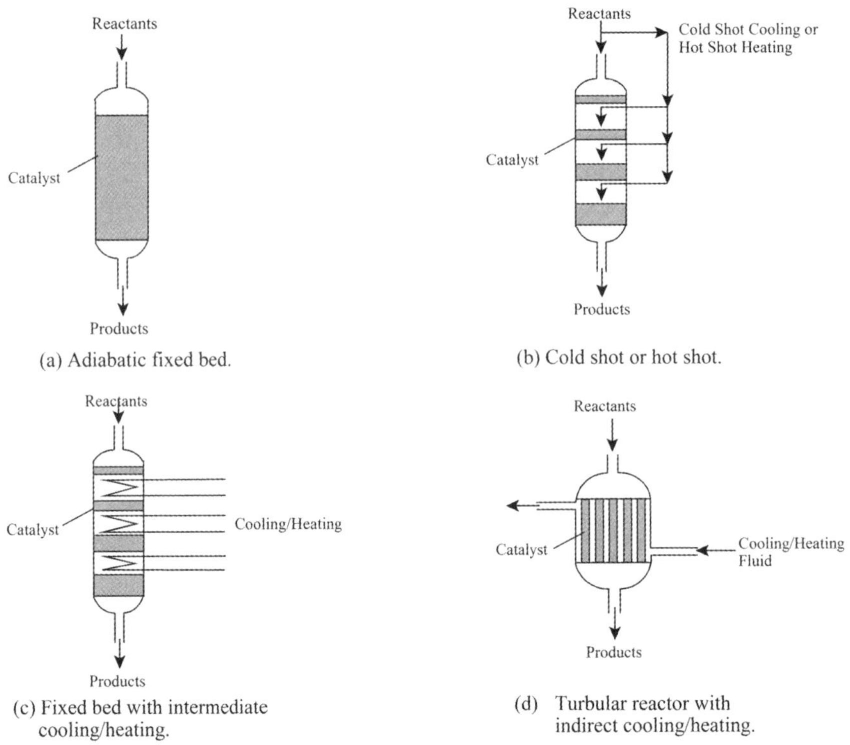

Figure 6.7

Heat transfer arrangements for fixed-bed catalytic reactors.

discussed later) are usually preferred for gas–solid noncatalytic reactions.

Fixed-bed reactors in the form of gas absorption equipment are used commonly for noncatalytic gas–liquid reactions. Here, the packed bed serves only to give good contact between the gas and liquid. Both cocurrent and countercurrent operations are used. Countercurrent operation gives the highest reaction rates. Cocurrent operation is preferred if a short liquid residence time is required or if the gas flowrate is so high that countercurrent operation is difficult.

For example, hydrogen sulfide and carbon dioxide can be removed from natural gas by reaction with monoethanolamine in an absorber, according to the following reactions (Kohl and Riesenfeld, 1979):

$$\underset{\text{monoethanolamine}}{HOCH_2CH_2NH_2} + \underset{\substack{\text{hydrogen}\\ \text{sulfide}}}{H_2S} \rightleftharpoons \underset{\substack{\text{monoethanolamine}\\ \text{hydrogen sulfide}}}{HOCH_2CH_2NH_3HS}$$

$$\underset{\text{monoethanolamine}}{HOCH_2CH_2NH_2} + \underset{\substack{\text{carbon}\\ \text{dioxide}}}{CO_2} + H_2O \rightleftharpoons \underset{\substack{\text{monoethanolamine}\\ \text{hydrogen carbonate}}}{HOCH_2CH_2NH_3HCO_3}$$

These reactions can be reversed in a stripping column. The input of heat in the stripping column releases the hydrogen sulfide and carbon dioxide for further processing. The monoethanolamine can then be recycled.

5) *Moving-bed catalytic reactors.* If a solid catalyst degrades in performance, the rate of degradation in a fixed bed might be unacceptable. In this case, a moving-bed reactor can be used. Here, the catalyst is kept in motion by the feed to the reactor and the product. This makes it possible to remove the catalyst continuously for regeneration. An example of a refinery hydrocracker reactor is illustrated in Figure 6.8a.
6) *Fluidized-bed catalytic reactors.* In fluidized-bed reactors, solid material in the form of fine particles is held in suspension by the upward flow of the reacting fluid. The effect of the rapid motion of the particles is good heat transfer and temperature uniformity. This prevents the formation of the hot spots that can occur with fixed-bed reactors.

The performance of fluidized-bed reactors is not approximated by either the mixed-flow or plug-flow idealized models. The solid phase tends to be in mixed flow, but the bubbles lead to the gas phase behaving more like plug flow. Overall, the performance of a fluidized-bed reactor often lies somewhere between the mixed-flow and plug-flow models.

In addition to the advantage of high heat transfer rates, fluidized beds are also useful in situations where catalyst particles need frequent regeneration. Under these circumstances, particles can be removed continuously from the reactor bed, regenerated and recycled back to the bed. In exothermic reactions, the recycling of catalyst can be used to remove heat from the reactor, or in endothermic reactions it can be used to add heat.

One disadvantage of fluidized beds, as discussed previously, is that attrition of the catalyst can cause the generation of catalyst fines, which are then carried over from the bed and lost

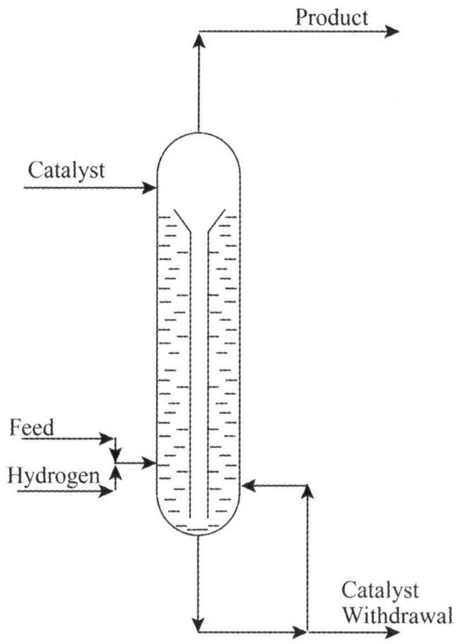

(a) Moving-bed hydrocracker reactor.

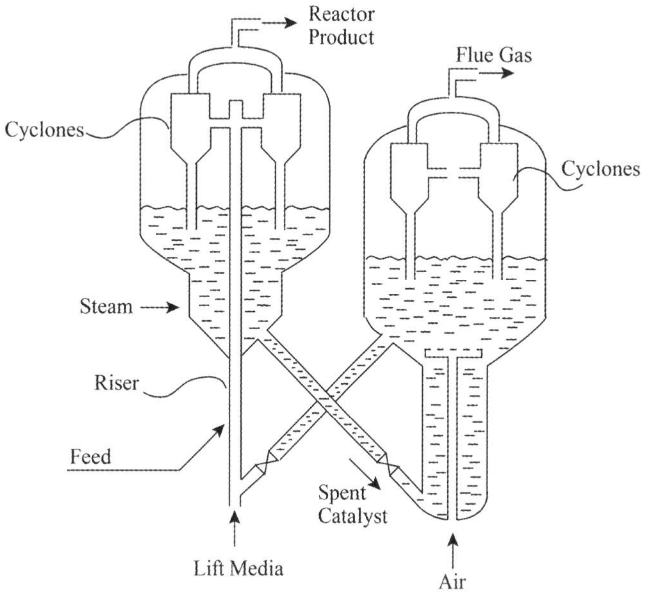

(b) Fluidized-bed catalytic cracker reactor.

Figure 6.8

A moving bed or fluidized bed allows the catalyst to be continuously withdrawn and regenerated.

6

from the system. This carryover of catalyst fines sometimes necessitates cooling the reactor effluent through direct contact heat transfer by mixing with a cold fluid, since the fines tend to foul conventional heat exchangers.

Figure 6.8b shows the essential features of a petroleum refinery catalytic cracker. Large molar mass hydrocarbon molecules are made to *crack* into smaller hydrocarbon molecules in the presence of a solid catalyst. The liquid hydrocarbon feed is atomized as it enters the catalytic cracking reactor and is mixed with the catalyst particles being carried by a flow of steam or light hydrocarbon gas. The mixture is carried up the *riser* and the reaction is essentially complete at the top of the riser. However, the reaction is accompanied by the deposition of carbon (coke) on the surface of the catalyst. The catalyst is separated from the gaseous products at the top of the reactor. The gaseous products leave the reactor and go on for separation. The catalyst flows to a regenerator in which air is contacted with the catalyst in a fluidized bed. The air oxidizes the carbon that has been deposited on the surface of the catalyst, forming carbon dioxide and carbon monoxide. The regenerated catalyst then flows back to the reactor. The catalytic cracking reaction is endothermic and the catalyst regeneration is exothermic. The hot catalyst leaving the regenerator provides the heat of reaction to the endothermic cracking reactions. The catalyst, in this case, provides a dual function of both catalyzing the reaction and exchanging heat between the reactor and regenerator.

7) *Fluidized-bed noncatalytic reactors.* Fluidized beds are also suited to gas–solid noncatalytic reactions. All the advantages described earlier for gas–solid catalytic reactions apply here. As an example, limestone (principally calcium carbonate) can be heated to produce calcium oxide in a fluidized-bed reactor according to the reaction:

$$CaCO_3 \xrightarrow{\text{heat}} CaO + CO_2$$

Air and fuel fluidize the solid particles, which are fed to the bed and burnt to produce the high temperatures necessary for the reaction.

8) *Kilns.* Reactions involving free-flowing solid, paste and slurry materials can be carried out in kilns. In a rotary kiln, a cylindrical shell is mounted with its axis making a small angle to the horizontal and rotated slowly. The solid material to be reacted is fed to the elevated end of the kiln and tumbles down the kiln as a result of the rotation. The behavior of the reactor usually approximates plug flow. High-temperature reactions demand refractory lined steel shells and are usually heated by direct firing. An example of a reaction carried out in such a device is the production of hydrogen fluoride:

$$\underset{\text{calcium fluoride}}{CaF_2} + \underset{\text{sulfuric acid}}{H_2SO_4} \longrightarrow \underset{\text{hydrogen fluoride}}{2HF} + \underset{\text{calcium sulfate}}{CaSO_4}$$

Other designs of kilns use static shells rather than rotating shells and rely on mechanical rakes to move solid material through the reactor.

Having discussed the choice of reactor type and operating conditions, consider two examples.

Example 6.2 Monoethanolamine is required as a product. This can be produced from the reaction between ethylene oxide and ammonia (Waddams, 1978):

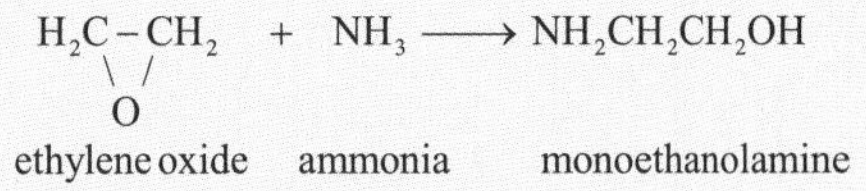

Two principal secondary reactions occur to form diethanolamine and triethanolamine:

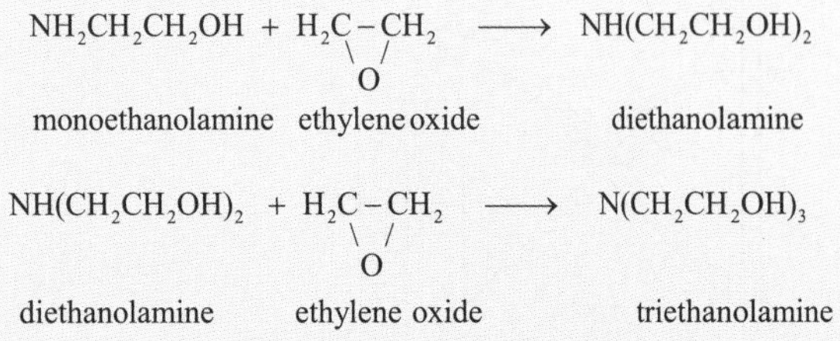

The secondary reactions are parallel with respect to ethylene oxide but are in series with respect to monoethanolamine. Monoethanolamine is more valuable than both the di- and triethanolamine. As a first step in the flowsheet synthesis, make an initial choice of a reactor that will maximize the production of monoethanolamine relative to di- and triethanolamine.

Solution As much as possible, the production of di- and triethanolamine needs to be avoided. These are formed by series reactions with respect to monoethanolamine. In a mixed-flow reactor, part of the monoethanolamine formed in the primary reaction could stay for extended periods, thus increasing its chances of being converted to di- and triethanolamine. The ideal batch or plug-flow arrangement is preferred to carefully control the residence time in the reactor.

Further consideration of the reaction system reveals that the ammonia feed takes part only in the primary reaction and in neither of the secondary reactions. Consider the rate equation for the primary reaction:

$$r_1 = k_1\, C_{EO}^{a_1} C_{NH_3}^{b_1}$$

where r_1 = reaction rate of primary reaction
k_1 = reaction rate constant for the primary reaction
C_{EO} = molar concentration of ethylene oxide in the reactor
C_{NH_3} = molar concentration of ammonia in the reactor
a_1, b_1 = order of primary reaction

Operation with an excess of ammonia in the reactor has the effect of increasing the rate due to the $C_{NH_3}^{b_1}$ term. However, operation with excess ammonia decreases the concentration of ethylene oxide, and the effect is to decrease the rate due to the $C_{EO}^{a_l}$ term. Whether the overall effect is a slight increase or decrease in the reaction rate depends on the relative magnitude of a_1 and b_1. Consider now the rate equations for the byproduct reactions:

$$r_2 = k_2 C_{MEA}^{a_2} C_{EO}^{b_2}$$

$$r_3 = k_3 C_{DEA}^{a_3} C_{EO}^{b_3}$$

where r_2, r_3 = rates of reaction to diethanolamine and triethanolamine respectively
k_2, k_3 = reaction rate constants for the diethanolamine and triethanolamine reactions respectively
C_{MEA} = molar concentration of monoethanolamine
C_{DEA} = molar concentration of diethanolamine
a_2, b_2 = order of reaction for the diethanolamine reaction
a_3, b_3 = order of reaction for the triethanolamine reaction

An excess of ammonia in the reactor decreases the concentrations of monoethanolamine, diethanolamine and ethylene oxide and decreases the rates of reaction for both secondary reactions.

Thus, an excess of ammonia in the reactor has a marginal effect on the primary reaction but significantly decreases the rate of the secondary reactions. Using excess ammonia can also be thought of as operating the reactor with a low conversion with respect to ammonia.

The use of an excess of ammonia is borne out in practice (Waddams, 1978). A molar ratio of ammonia to ethylene oxide of 10:1 yields 75% monoethanolamine, 21% diethanolamine and 4% triethanolamine. Using equimolar proportions, under the same reaction conditions, the respective proportions become 12, 23 and 65%.

Another possibility to improve selectivity is to reduce the concentration of monoethanolamine in the reactor by using more than one reactor with an intermediate separation of the monoethanolamine. Considering the boiling points of the components given in Table 6.3, then separation by distillation is apparently possible. Unfortunately, repeated distillation operations are likely to be very expensive. Also, there is a market to sell both di- and triethanolamine even though their value is lower than monoethanolamine. Thus, in this case, repeated reaction and separation is probably not justified and the choice is a single plug-flow reactor.

An initial setting for the reactor conversion is difficult to make. A high conversion increases the concentration of monoethanolamine and increases the rates of the secondary reactions. A low conversion has the effect of decreasing the reactor capital cost but increasing the capital cost of many other items of equipment in the flowsheet. Thus, an initial value of 50% conversion is probably as good as can be suggested at this stage.

Table 6.3

Normal boiling points of the components.

Component	Normal boiling point (K)
Ammonia	240
Ethylene oxide	284
Monoethanolamine	444
Diethanolamine	542
Triethanolamine	609

Example 6.3 *Tert*-butyl hydrogen sulfate is required as an intermediate in a reaction sequence. This can be produced by the reaction between isobutylene and moderately concentrated sulfuric acid:

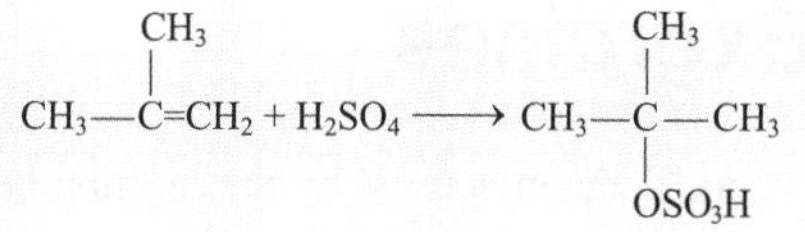

isobutylene sulfuric acid *tert*-butyl hydrogen sulfate

Series reactions occur where the *tert*-butyl hydrogen sulfate reacts to form unwanted *tert*-butyl alcohol:

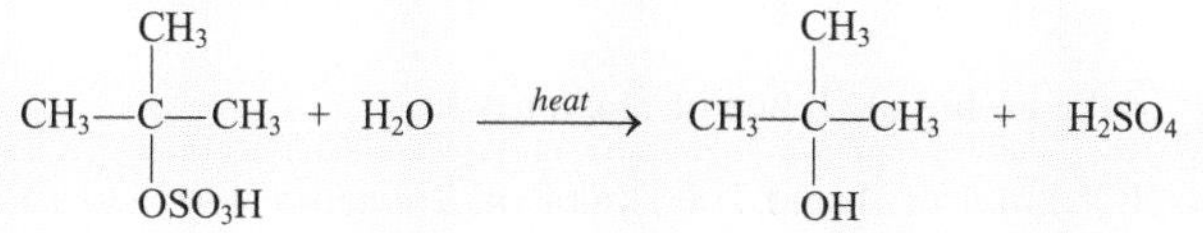

tert-butyl hydrogen sulfate water *tert*-butyl alcohol sulfuric acid

Other series reactions form unwanted polymeric material. Further information on the reaction is:

- The primary reaction is rapid and exothermic.
- Laboratory studies indicate that the reactor yield is maximum when the concentration of sulfuric acid is maintained at 63% (Morrison and Boyd, 1992).
- The temperature should be maintained around 0 °C or excessive byproduct formation occurs (Albright and Goldsby, 1977; Morrison and Boyd, 1992).

Make an initial choice of reactor.

Solution The byproduct reactions to avoid are all series in nature. This suggests that a mixed-flow reactor should not be used; instead either a batch or plug-flow reactor should be used.

However, the laboratory data seem to indicate that maintaining a constant concentration in the reactor to maintain 63% sulfuric acid in the reactor would be beneficial. Careful temperature control is also important. These two factors would suggest that a mixed-flow reactor is appropriate. There is a conflict. How can a well-defined residence time and a constant concentration of sulfuric acid be simultaneously maintained?

Using a batch reactor, a constant concentration of sulfuric acid can be maintained by adding concentrated sulfuric acid as the reaction progresses, that is, semi-batch operation. Good temperature control of such systems can be maintained.

By choosing to use a continuous rather than a batch reactor, plug-flow behavior can be approached using a series of mixed-flow reactors. This again allows concentrated sulfuric acid to be added as the reaction progresses, in a similar way as suggested for some parallel systems in Figure 4.6. Breaking the reactor down into a series of mixed-flow reactors also allows good temperature control.

To make an initial setting for the reactor conversion is again difficult. The series nature of the byproduct reactions suggests that a value of 50% is probably as good as can be suggested at this stage.

6.5 Reactor Configuration For Heterogeneous Solid-Catalyzed Reactions

Heterogeneous reactions involving a solid supported catalyst form an important class of reactors and require special consideration. As discussed in the previous section, such reactors can be configured in different ways:

- fixed-bed adiabatic,
- fixed-bed adiabatic with intermediate cold shot or hot shot,
- tubular with indirect heating or cooling,
- moving bed,
- fluidized bed.

Of these, fixed-bed adiabatic reactors are the cheapest in terms of capital cost. Tubular reactors are more expensive than fixed-bed adiabatic reactors, with the highest capital costs associated with moving and fluidized beds. The choice of reactor configuration for reactions involving a solid supported catalyst is often dominated by the deactivation characteristics of the catalyst.

If deactivation of the catalyst is very short, then moving- or fluidized-bed reactors are required so that the catalyst can be withdrawn continuously, regenerated and returned to the reactor. The example of refinery catalytic cracking was discussed previously, as illustrated in Figure 6.8b, where the catalyst is moved rapidly from the reaction zone in the riser to regeneration. Here, the catalyst deactivates in a few seconds and must be removed from the reactor rapidly and regenerated. If the deactivation is slower, then a moving bed can be used, as illustrated in Figure 6.8a. This still allows the catalyst to be removed continuously, regenerated and returned to the reactor. If the catalyst deactivation is slower, of the order of a year or longer, then a fixed-bed adiabatic or tubular reactor can be used. Such reactors must be taken off-line for the catalyst to be regenerated. Multiple reactors can be used with standby reactors, such that one of the reactors can be taken off-line to regenerate the catalyst, with the process kept running. However, there are significant capital cost implications associated with standby reactors.

Thus, the objective of the designer should be to use a fixed-bed adiabatic reactor if possible. The reactor conditions and the catalyst design can be manipulated to minimize the deactivation. Reactor inlet temperature, pressure, composition of the reactants in the feed, catalyst shape and size, mixtures of inert catalyst, profiles of active material within the catalyst pellets, hot shot, cold shot and the introduction of inert gases in the feed can all be manipulated with this objective. There are often trade-offs to be considered between reactor size, selectivity and catalyst deactivation, as well as interactions with the rest of the process. If the temperature control requires indirect heating or cooling, then a tubular reactor should be considered, with the same variables being manipulated along with the heat transfer characteristics. If everything else fails, then continuous catalytic regeneration needs to be considered (risers, fluidized beds and moving beds).

6.6 Reactor Configuration Summary

In choosing the reactor, the overriding consideration is usually the raw materials efficiency (bearing in mind materials of construction, safety, etc.). Raw materials costs are usually the most important costs in the whole process. Also, any inefficiency in the use of raw materials is likely to create waste streams that become an environmental problem. The reactor creates inefficiency in the use of raw materials in the following ways:

- If low conversion is obtained and unreacted feed material is difficult to separate and recycle.
- Through the formation of unwanted byproducts. Sometimes, the byproduct has value as a product in its own right; sometimes, it simply has value as fuel. Sometimes, it is a liability and requires disposal in expensive waste treatment processes.
- Impurities in the feed can undergo reaction to form additional byproducts. This is best avoided by purification of the feed before reaction.

Temperature control of the reactor can be achieved through:

- cold shot and hot shot,
- indirect heat transfer,
- heat carriers,
- catalyst profiles.

In addition, it is common to have to quench the reactor effluent to stop the reaction quickly or to avoid problems with conventional heat transfer equipment.

Catalyst degradation can be a dominant issue in the choice of reactor configuration, depending on the rate of deactivation. Slow deactivation can be dealt with by periodic shutdown and regeneration or by replacement of the catalyst. If this is not acceptable, then standby reactors can be used to maintain plant operation. If deactivation is rapid, then moving-bed and fluidized-bed reactors, in which catalyst is removed continuously for regeneration, might be the only option.

When dealing with gas–liquid and liquid–liquid reactions, mass transfer can be as equally important a consideration as reaction.

Reactor configurations for conventional designs can be categorized as:

- tubular,
- stirred-tank,
- fixed-bed catalytic,
- fixed-bed noncatalytic,
- moving-bed catalytic,
- fluidized-bed catalytic,
- fluidized-bed noncatalytic,
- kilns.

The decisions made in the reactor design are often the most important in the whole flowsheet. The design of the reactor usually interacts strongly with the rest of the flowsheet. Hence, a return to the decisions for the reactor must be made when the process design has progressed further to understand the full consequences of those decisions.

6.7 Exercises

1. Chlorobenzene is manufactured by the reaction between benzene and chlorine. A number of secondary reactions occur to form undesired byproducts:

$$C_6H_6 + Cl_2 \longrightarrow C_6H_5Cl + HCl$$
$$C_6H_5Cl + Cl_2 \longrightarrow C_6H_4Cl_2 + HCl$$
$$C_6H_5Cl_2 + Cl_2 \longrightarrow C_6H_3Cl_3 + HCl$$

Make an initial choice of reactor type.

2. 1000 kmol of A and 2000 kmol of B react at 400 K to form C and D, according to the reversible reaction:

$$A + B \rightleftharpoons C + D$$

The reaction takes place in a gas phase. Component C is the desired product and the equilibrium constant at 400 K is $K_a = 1$. Assuming ideal gas behavior, calculate the equilibrium conversion and explain what advantage may be gained by employing a reactor in which C is continuously removed from the reaction mixture so as to keep its partial pressure low.

3. Components A and G react by three simultaneous reactions to form three products, one that is desired (D) and two that are undesired (U and W). These gas-phase reactions, together with their corresponding rate laws, are given below.

Desired product:

$$A + G \longrightarrow D$$

$$r_D = 0.0156 \exp\left[18,200\left(\frac{1}{300} - \frac{1}{T}\right)\right] C_A C_G$$

First unwanted byproduct:

$$A + G \longrightarrow U$$

$$r_U = 0.0234 \exp\left[17,850\left(\frac{1}{300} - \frac{1}{T}\right)\right] C_A^{1.5} C_G$$

Second unwanted byproduct:

$$A + G \longrightarrow W$$

$$r_W = 0.0588 \exp\left[3500\left(\frac{1}{300} - \frac{1}{T}\right)\right] C_A^{0.5} C_G$$

T is the temperature (K) and C_A and C_G are the concentrations of A and G.

The objective is to select conditions to maximize the yield of D.

a) Select the type of reactor that appears appropriate for the given reaction kinetics.

b) It has been suggested to introduce inert material. What would be the effect of inerts on the yield?

c) Assess whether the yield can be improved by operating the reactor at high or low temperatures.

d) Assess whether the yield can be improved by operating the reactor at high or low pressures.

4. Pure reactant A is fed at 330 K into an adiabatic reactor where it converts reversibly to useful product B:

$$A \rightleftharpoons B$$

The reactor brings the reaction mixture to equilibrium at the outlet temperature. The reaction is exothermic and the equilibrium constant K_a is given by:

$$K_a = 120,000 \exp\left[-20.0\left(\frac{T-298}{T}\right)\right]$$

where T is the temperature in K. The heat of reaction is $-60{,}000\,\text{kJ}\cdot\text{kmol}^{-1}$. The heat capacities of A and B are $190\,\text{kJ}\cdot\text{kmol}^{-1}\cdot\text{K}^{-1}$.

a) Use an enthalpy balance to calculate the temperature of the reaction mixture as a function of the conversion. Plot the temperature along the reactor length.

b) Calculate the exit temperature and the equilibrium conversion.

c) A choice is required from different reactors of volume V between a single adiabatic reactor, two adiabatic reactors in series or three adiabatic reactors in series. Which of these choices will lead to a better conversion?

5. In the reaction of ethylene to ethanol:

$$\underset{\text{ethylene}}{CH_2 = CH_2} + \underset{\text{water}}{H_2O} \rightleftharpoons \underset{\text{ethanol}}{CH_3 - CH_2 - OH}$$

a side reaction occurs, where diethyl ether is formed:

$$\underset{\text{ethanol}}{2CH_3 - CH_2 - OH} \rightleftharpoons \underset{\text{diethylether}}{CH_3CH_2 - O - CH_2CH_3} + \underset{\text{water}}{H_2O}$$

a) Make an initial choice of reactor as a first step. How would the reactor be able to maximize selectivity to a desired product?

b) What operating pressure would be suitable in this reactor?

c) How would an excess of water (steam) in the reactor feed affect the selectivity of the reactor?

References

Albright LF and Goldsby AR (1977) *Industrial and Laboratory Alkylations*, *ACS Symposium Series No. 55*, ACS, Washington DC.

Butt JB and Petersen EE (1988) *Activation, Deactivation and Poisoning of Catalysts*, Academic Press.

Hougen OA, Watson KM and Ragatz RA (1954) *Chemical Process Principles. Part I: Material and Energy Balances*, 2nd Edition, John Wiley & Sons.

Kohl AL and Riesenfeld FC (1979) *Gas Purification*, Gulf Publishing Company.

Morrison RT and Boyd RN (1992) *Organic Chemistry*, 6th Edition, Prentice-Hall.

Poling BE, Prausnitz JM and O'Connell JP (2001) *The Properties of Gases and Liquids*, 5th Edition, McGraw-Hill.

Rase HF (1990) *Fixed-Bed Reactor Design and Diagnostics – Gas Phase Reactions*, Butterworths.

Supp E (1973) Technology of Lurgi's Low Pressure Methanol Process, *Chem Tech*, **3**: 430.

Waddams AL (1978) *Chemicals from Petroleum*, John Murray.

Wijngaarden RJ, Kronberg A and Westerterp KR (1998) *Industrial Catalysis – Optimizing Catalysts and Processes*, Wiley-VCH.

Chapter 7

Separation of Heterogeneous Mixtures

7.1 Homogeneous and Heterogeneous Separation

Having made an initial specification for the reactor, attention is turned to separation of the reactor effluent. In some circumstances, it might be necessary to carry out separation before the reactor to purify the feed. Whether before or after the reactor, the overall separation task might need to be broken down into a number of intermediate separation tasks. Consider now the choice of separator for the separation tasks. Later in Chapters 10, 11, 14 and 16, consideration will be given to how separation tasks should be connected together and connected to the reactor. As with reactors, emphasis will be placed on the choice of separator, together with its preliminary specifications, rather than its detailed design.

When choosing between different types of reactors, both continuous and batch reactors were considered from the point of view of the performance of the reactor (continuous plug-flow and ideal-batch being equivalent in terms of residence time). If a batch reactor is chosen, it will often lead to a choice of separator for the reactor effluent that also operates in batch mode, although this is not always the case, as intermediate storage can be used to overcome the variations with time. Batch separations will be dealt with in Chapter 16.

If the mixture to be separated is homogeneous, separation can only be performed by the creation of another phase within the system or by the addition of a mass separation agent. For example, if a vapor mixture is leaving a reactor, another phase could be created by partial condensation. The vapor resulting from the partial condensation will be rich in the more volatile components and the liquid will be rich in the less volatile components, achieving a separation. Alternatively, rather than creating another phase, a mass separation agent can be added. Returning to the example of a vapor mixture leaving a reactor, a liquid solvent could be contacted with the vapor mixture to act as a mass separation agent to preferentially dissolve one or more of the components from the mixture. Further separation is required to separate the solvent from the process materials so as to recycle the solvent, and so on. A number of physical properties can be exploited to achieve the separation of homogeneous mixtures (King, 1980; Rousseau, 1987).

If a heterogeneous or multiphase mixture needs to be separated, then separation can be done physically by exploiting the differences in density between the phases. Separation of the different phases of a heterogeneous mixture should be carried out before homogeneous separation, taking advantage of what already exists. Phase separation tends to be easier and should be done first. The phase separations likely to be carried out are:

- Gas–liquid (or vapor–liquid)
- Gas–solid (or vapor–solid)
- Liquid–liquid (immiscible)
- Liquid–solid
- Solid–solid.

A fully comprehensive survey is beyond the scope of this text, and many good surveys are already available (Foust *et al.*, 1980; King, 1980; Rousseau, 1987; Walas, 1988; Coulson *et al.*, 1991; Schweitzer, 1997).

The principal methods for the separation of heterogeneous mixtures are:

- Settling and sedimentation
- Inertial and centrifugal separation
- Electrostatic precipitation
- Filtration
- Scrubbing
- Flotation
- Drying.

Chemical Process Design and Integration, Second Edition. Robin Smith.

Companion Website: www.wiley.com/go/smith/chemical2e

7.2 Settling and Sedimentation

In *settling* processes, particles are separated from a fluid by gravitational forces acting on the particles. The particles can be liquid drops or solid particles. The fluid can be a gas, vapor or liquid.

Figure 7.1a shows a simple device used to separate by gravity a gas–liquid (or vapor–liquid) mixture. The velocity of the gas or vapor through the vessel must be less than the settling velocity of the liquid drops.

When a particle falls under the influence of gravity, it will accelerate until the combination of the frictional drag in the fluid and buoyancy force balances the opposing gravitational force. If the particle is assumed to be a rigid sphere, at this terminal velocity, a force balance gives (Ludwig, 1977; Coulson *et al.*, 1991; Geankopolis, 1993; Schweitzer, 1997):

$$\underbrace{\rho_P \frac{\pi d^3}{6} g}_{\text{gravitational force}} = \underbrace{\rho_F \frac{\pi d^3}{6} g}_{\text{buoyancy force}} + \underbrace{c_D \frac{\pi d^2}{4} \frac{\rho_F v_T^2}{2}}_{\text{drag force}} \tag{7.1}$$

where ρ_P = density of particles ($kg \cdot m^{-3}$)
ρ_F = density of dispersing fluid ($kg \cdot m^{-3}$)
d = particle diameter (m)
g = the gravitational acceleration ($9.81\ m \cdot s^{-2}$)
c_D = drag coefficient (–)
v_T = terminal settling velocity ($m \cdot s^{-1}$)

Rearranging Equation 7.1 gives:

$$v_T = \sqrt{\left(\frac{4gd}{3c_D}\right)\left(\frac{\rho_P - \rho_F}{\rho_F}\right)} \tag{7.2}$$

More generally, Equation 7.2 can be written as:

$$v_T = K_T \sqrt{\frac{\rho_P - \rho_F}{\rho_F}} \tag{7.3}$$

where K_T = parameter for the terminal velocity ($m \cdot s^{-1}$)

If the particles are assumed to be rigid spheres, then from Equations 7.2 and 7.3 (Ludwig, 1977; Coulson *et al.*, 1991; Geankopolis, 1993; Schweitzer, 1997):

$$K_T = \left(\frac{4gd}{3c_D}\right)^{1/2} \tag{7.4}$$

However, more general correlations can be found for K_T in Equation 7.3. If in addition to assuming the particles to be rigid spheres it is also assumed that the flow is in the laminar region, known as the Stoke's Law region, for Reynolds number less than 1 (but can be applied up to a Reynolds number of 2 without much error):

$$c_D = \frac{24}{Re} = \frac{24}{dv_T \rho_F / \mu_F} \qquad 0 < Re < 2 \tag{7.5}$$

where Re = Reynolds Number
μ_F = fluid viscosity ($kg \cdot m^{-1} \cdot s^{-1}$)

Substituting Equation 7.5 into Equation 7.2 gives:

$$v_T = \frac{gd^2(\rho_P - \rho_F)}{18\mu_F} \qquad 0 < Re < 2 \tag{7.6}$$

When applying Equation 7.6, there is a tacit assumption that there is no turbulence in the settler. In practice, any turbulence will mean that a settling device sized on the basis of Equation 7.6 will have a lower efficiency than predicted.

Above a Reynolds number of around 2, Equation 7.5 will underestimate the drag coefficient and hence overestimate the

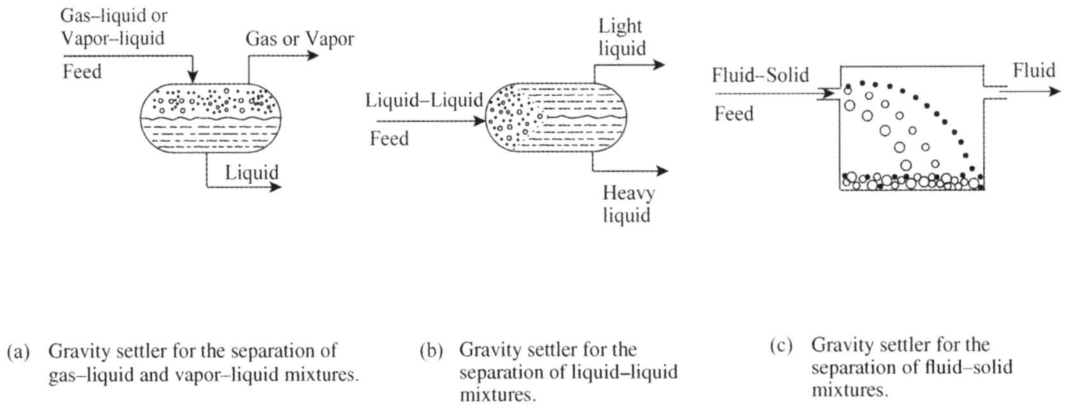

Figure 7.1
Settling processes used for the separation of hetrogeneous mixtures.

settling velocity. Also, for $Re > 2$, an empirical expression must be used (Geankopolis, 1993):

$$c_D = \frac{18.5}{Re^{0.6}} \qquad 2 < Re < 500 \tag{7.7}$$

Substituting Equation 7.7 into Equation 7.2 gives:

$$v_T = \left(\frac{gd^{1.6}(\rho_P - \rho_F)}{13.875\,\rho_F^{0.4}\mu_F^{0.6}}\right)^{0.7143} \qquad 2 < Re < 500 \tag{7.8}$$

For higher values of *Re* (Geankopolis, 1993):

$$c_D = 0.44 \qquad 500 < Re < 200{,}000 \tag{7.9}$$

Substituting Equation 7.9 into Equation 7.2 gives:

$$v_T = \sqrt{\frac{gd(\rho_P - \rho_F)}{3.03\,\rho_F}} \qquad 500 < Re < 200{,}000 \tag{7.10}$$

When designing a settling device of the type in Figure 7.1a, the maximum allowable velocity in the device must be less than the terminal settling velocity. Before Equations 7.6 to 7.10 can be applied, the particle diameter must be known. For gas–liquid and vapor–liquid separations, there will be a range of particle droplet sizes. It is normally not practical to separate droplets less than 100 μm diameter in such a simple device. Thus, the design basis for simple settling devices of the type illustrated in Figure 7.1a is usually taken to be a vessel in which the velocity of the gas (or vapor) is the terminal settling velocity for droplets of 100 μm diameter (Woods, 1995; Schweitzer, 1997).

The separation of gas–liquid (or vapor–liquid) mixtures can be enhanced by installing a mesh pad at the top of the disengagement zone to coalesce the smaller droplets to larger ones. If this is done, then the K_T in Equation 7.3 is normally specified to be $0.11\ m{\cdot}s^{-1}$, although this can take lower values down to $0.06\ m{\cdot}s^{-1}$ for vacuum systems (Ludwig, 1977).

Figure 7.1b shows a simple gravity settler or *decanter* for removing a dispersed liquid phase from another liquid phase. The horizontal velocity must be low enough to allow the low-density droplets to rise from the bottom of the vessel to the interface and coalesce and for the high-density droplets to settle down to the interface and coalesce. The decanter is sized on the basis that the velocity of the continuous phase should be less than the terminal settling velocity of the droplets of the dispersed phase. The velocity of the continuous phase can be estimated from the area of the interface between the settled phases (Ludwig, 1977; Woods, 1995; Schweitzer, 1997):

$$v_{CP} = \frac{F_{CP}}{A_I} \tag{7.11}$$

where v_{CP} = velocity of the continuous phase ($m{\cdot}s^{-1}$)
F_{CP} = volumetric flowrate of the continuous phase ($m^3{\cdot}s^{-1}$)
A_I = decanter area of the interface (m^2)

The terminal settling velocity is given by Equation 7.6 or 7.8. Decanters are normally designed for a droplet size of 150 μm (Woods, 1995; Schweitzer, 1997), but can be designed for droplets down to 100 μm. Dispersions of droplets smaller than 20 μm tend to be very stable. The band of droplets that collect at the interface before coalescing should not extend to the bottom of the vessel. A minimum of 10% of the decanter height is normally taken for this (Schweitzer, 1997).

An empty vessel may be employed, but horizontal baffles can be used to reduce turbulence and assist the coalescence through preferential wetting of the solid surface by the disperse phase. More elaborate methods to assist the coalescence include the use of mesh pads in the vessel or the use of an electric field to promote coalescence. Chemical additives can also be used to promote coalescence.

In Figure 7.1c is a schematic diagram of a gravity settling chamber. A mixture of gas, vapor or liquid and solid particles enters at one end of a large chamber. Particles settle toward the base. Again the device is specified on the basis of the terminal settling velocity of the particles. For gas–solid particle separations, the size of the solid particles is more likely to be known than the other types discussed so far. The efficiency with which the particles of a given size will be collected from the simple settling devices in Figure 7.1c is given by (Dullien, 1989):

7

$$\eta = \frac{h}{H} \tag{7.12}$$

where η = efficiency of collection (–)
h = settling distance of the particles during the residence time in the device (m)
H = height of the settling zone in the device (m)

When high concentrations of particles are to be settled, the surrounding particles interfere with individual particles. This is particularly important when settling high concentrations of solid particles in liquids. For such *hindered* settling, the viscosity and fluid density terms in Equation 7.6 can be modified to allow for this. The walls of the vessel can also interfere with settling (Coulson *et al.*, 1991; Woods, 1995).

When separating a mixture of water and fine solid particles in a gravity settling device, such as the one shown in Figure 7.1c, it is common in such operations to add a *flocculating agent* to the mixture to assist the settling process. This agent has the effect of neutralizing electric charges on the particles that cause them to repel each other and remain dispersed. The effect is to form *aggregates* or *flocs*, which, because they are larger in size, settle more rapidly.

The separation of suspended solid particles from a liquid by gravity settling into a clear fluid and a slurry of higher solids content is called *sedimentation*. Figure 7.2 shows a sedimentation device known as a *thickener*, the prime function of which is to produce a more concentrated slurry. The feed slurry in Figure 7.2 is fed at the center of the tank below the surface of the liquid. Clear liquid overflows from the top edge of the tank. A slowly revolving rake serves to scrape the thickened slurry sludge toward the center of the base for removal and at the same time assists the formation of a more concentrated sludge. Again, it is common in such operations to add a flocculating agent to the mixture to assist the settling

Figure 7.2
A thickener for liquid–solid separation.

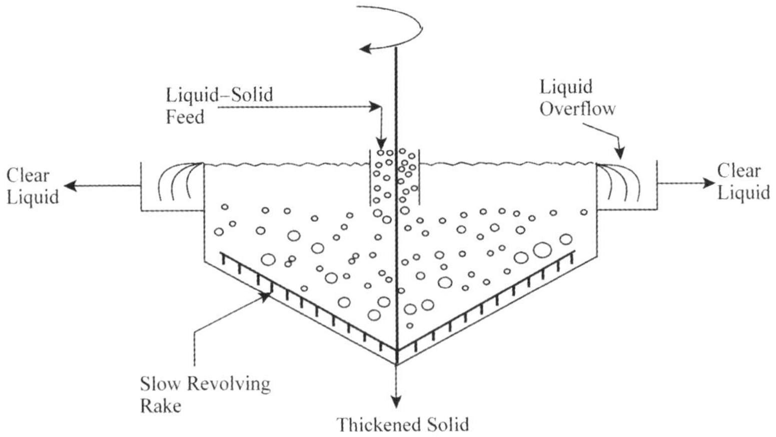

Figure 7.3
Simple gravity settling classifier.

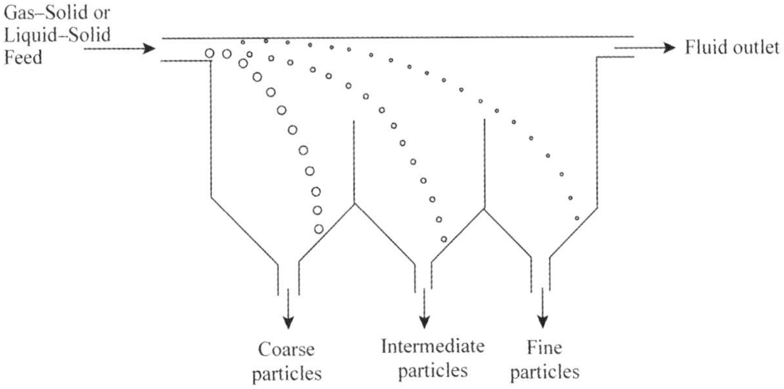

process. When the prime function of the sedimentation is to remove solids from a liquid rather than to produce a more concentrated solid–liquid mixture, the device is known as a *clarifier*. Clarifiers are often similar in design to thickeners.

Figure 7.3 shows a simple type of *classifier*. In the device in Figure 7.3, a large tank is subdivided into several sections. A size range of solid particles suspended in gas, vapor or liquid enters the tank. The larger, faster-settling particles settle to the bottom close to the entrance and the slower-settling particles settle to the bottom close to the exit. The vertical baffles in the tank allow the collection of several fractions.

This type of classification device can be used to carry out solid–solid separation in mixtures of different solids. The mixture of particles is first suspended in a fluid and then separated into fractions of different size or density in a device similar to that in Figure 7.3.

Example 7.1 Solid particles with a size greater than 100 μm are to be separated from larger particles in a settling chamber. The flowrate of gas is 8.5 $m^3 \cdot s^{-1}$. The density of the gas is 0.94 $kg \cdot m^{-3}$ and its viscosity is 2.18×10^{-5} $kg \cdot m^{-1} \cdot s^{-1}$. The density of the particles is 2780 $kg \cdot m^{-3}$.

a) Calculate the settling velocity, assuming the particles are spherical.

b) The settling chamber is to be box shaped, with a rectangular cross-section for the gas flow. If the length and breadth of the settling chamber are equal, what should the dimensions of the chamber be for 100% removal of particles greater than 100 μm?

Solution

a) Assume initially that the settling is in the Stoke's Law region:

$$\begin{aligned} v_T &= \frac{gd^2(\rho_P - \rho_F)}{18\mu_F} \\ &= \frac{9.81 \times (100 \times 10^{-6})^2 \times (2780 - 0.94)}{18 \times 2.18 \times 10^{-5}} \\ &= 0.69\ m \cdot s^{-1} \end{aligned}$$

Check the Reynolds number:

$$\begin{aligned} Re &= \frac{d v_T \rho_F}{\mu_F} \\ &= \frac{100 \times 10^{-6} \times 0.69 \times 0.94}{2.18 \times 10^{-5}} \\ &= 3.0 \end{aligned}$$

This is just outside the range of validity of Stoke's Law. Whilst the errors would not be expected to be too serious, compare with the prediction of Equation 7.8:

$$\begin{aligned} v_T &= \left(\frac{gd^{1.6}(\rho_P - \rho_F)}{13.875\,\rho_F^{0.4}\mu_F^{0.6}}\right)^{0.7143} \\ &= \left(\frac{9.81 \times (100 \times 10^{-6})^{1.6} \times (2780 - 0.94)}{13.875 \times 0.94^{0.4} \times (2.18 \times 10^{-5})^{0.6}}\right)^{0.7143} \\ &= 0.61\ m \cdot s^{-1} \end{aligned}$$

b) For 100% separation of particles:

$$\tau = \frac{H}{v_T}$$

where τ = mean residence time in the settler
H = height of the settling chamber

Also :

$$\tau = \frac{V}{F} = \frac{LBH}{F}$$

where V = volume of settling chamber
F = volumetric flowrate
L = length of settling chamber
B = breadth of settling chamber

Assuming $L = B$, then:

$$\frac{H}{v_T} = \frac{L^2 H}{F} \tag{7.13}$$

where L

$$= \sqrt{\frac{F}{v_T}}$$

$$= \sqrt{\frac{7.5}{0.61}}$$

$$= 3.51 \text{ m}$$

From Equation 7.13, in principle, any height can be chosen. If a large height is chosen, then the particles will have further to settle, but the residence time will be long. If a small height is chosen, then the particles will have a shorter distance to travel, but have a shorter residence time. To keep down the capital cost, a small height should be chosen. However, the bulk velocity through the settling chamber should not be too high; otherwise reentrainment of settled particles will start to occur. The maximum bulk mean velocity of the gas would normally be kept below around 5 m·s^{-1} (Dullien, 1989). Also, for maintenance and access, a minimum height of around 1 m will be needed. If the height is taken to be 1 m, then the bulk mean velocity is $F/LH = 2.1$ m·s^{-1}, which seems reasonable.

Example 7.2 A liquid–liquid mixture containing 5 kg·s^{-1} hydrocarbon and 0.5 kg·s^{-1} water is to be separated in a decanter. The physical properties are given in Table 7.1.

The water can be assumed to be dispersed in the hydrocarbon. Estimate the size of decanter required to separate the mixture in a horizontal drum with a length to diameter ratio of 3 to 1 and an interface across the center of the drum.

Solution Assuming a droplet size of 150 μm and the flow to be in the Stoke's Law region:

$$v_T = \frac{9.81 \times (150 \times 10^{-6})^2 \times (993 - 730)}{18 \times 8.1 \times 10^{-4}}$$

$$= 0.0040 \text{ m} \cdot \text{s}^{-1}$$

Table 7.1

Physical property data for Example 7.2.

	Density (kg·m^{-3})	Viscosity (kg·m^{-1}·s^{-1})
Hydrocarbon	730	8.1×10^{-4}
Water	993	8.0×10^{-4}

Check the Reynolds Number:

$$Re = \frac{730 \times 150 \times 10^{-6} \times 0.0040}{8.1 \times 10^{-4}}$$

$$= 0.54$$

From Equation 7.10:

$$v_{CP} = \frac{F_{CP}}{A_I} = v_T$$

$$0.0040 = \frac{5.0}{730 \times A_I}$$

$$A_I = 1.712 \text{ m}^2$$

Also : $A_I = DL$

where D = diameter of decanter
L = length of decanter

Assume $3D = L$:

$$A_I = \frac{L^2}{3}$$

$$L = \sqrt{3A_I}$$

$$= 2.27 \text{ m}$$

$$D = L/3$$

$$= 0.76 \text{ m}$$

Check the continuous phase (hydrocarbon) droplets that could be entrained in the dispersed phase (water). Then:

$$\text{Velocity of the water phase} = \frac{0.5}{993 \times 1.712}$$

$$= 2.94 \times 10^{-4} \text{ m·s}^{-1}$$

The diameter of hydrocarbon droplets entrained by the velocity of the water phase:

$$v_T = -2.94 \times 10^{-4} \text{ m} \cdot \text{s}^{-1} \text{(i.e. rising)}$$

is:

$$d = \sqrt{\frac{18 v_T \mu_F}{g(\rho_P - \rho_F)}}$$

$$= \sqrt{\frac{18 \times -2.94 \times 10^{-4} \times 8.0 \times 10^{-4}}{9.81 \times (730 - 993)}}$$

$$= 0.4 \times 10^{-6} \text{ m}$$

Only hydrocarbon droplets smaller than 0.4 μm will be entrained.

7

7.3 Inertial and Centrifugal Separation

In the preceding processes, the particles were separated from the fluid by gravitational forces acting on the particles. Sometimes gravity separation may be too slow because of the closeness of the densities of the particles and the fluid, because of small particle size leading to low settling velocity or, in the case of liquid–liquid separations, because of the formation of a stable emulsion.

Inertial or *momentum* separators improve the efficiency of gas–solid settling devices by giving the particles downward momentum, in addition to the gravitational force. Figure 7.4 illustrates three possible types of inertial separator. Many other arrangements are possible (Svarovsky, 1981; Dullien, 1989). The design of inertial separators for the separation of gas–solid separations is usually based on collection efficiency curves, as illustrated schematically in Figure 7.5. The curve is obtained experimentally and shows what proportion of particles of a given size is expected to be collected by the device. As the particle size decreases, the collection efficiency decreases. Collection efficiency curves for standard designs are published (Ludwig, 1977), but it is preferable to use data provided by the equipment supplier.

Centrifugal separators take the idea of an inertial separator a step further and make use of the principle that an object whirled about an axis at a constant radial distance from the point is acted on by a force. Use of centrifugal forces increases the force acting on the particles. Particles that do not settle readily in gravity settlers often can be separated from fluids by centrifugal force.

The simplest type of centrifugal device is the *cyclone* separator for the separation of solid particles or liquid droplets from a gas or vapor (Figure 7.6). This consists of a vertical cylinder with a conical bottom. Centrifugal force is generated by the motion of the fluid. The mixture enters through a tangential inlet near the top, and the rotating motion so created develops a centrifugal force that throws the dense particles radially toward the wall. The entering fluid flows downward in a spiral adjacent to the wall. When the fluid reaches the bottom of the cone, it spirals upward in a smaller spiral at the center of the cone and cylinder. The downward and upward spirals are in the same direction. The particles of dense material are thrown toward the wall and fall downward, leaving the bottom of the cone.

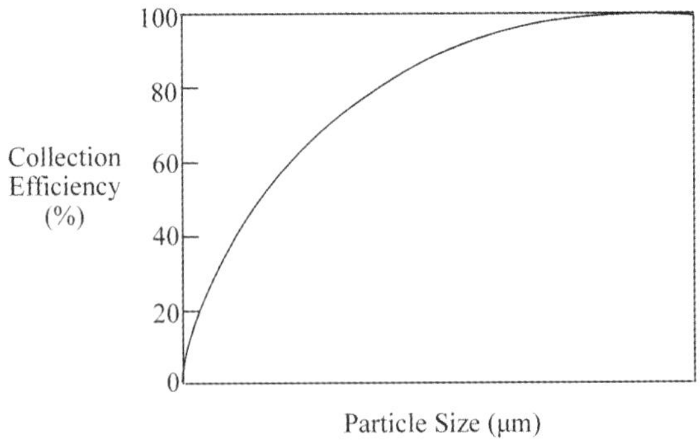

Figure 7.5

A collection efficiency curve shows the fraction of particles of a given particle size that will be collected.

The design of cyclones is normally based on collection efficiency curves, such as the one shown in Figure 7.5 (Svarovsky, 1981: Dullien, 1989). Curves for cyclones with standard dimensions are published that can be scaled to different dimensions using scaling parameters (Svarovsky, 1981; Dullien, 1989). Again, it is preferable to use curves supplied by equipment manufacturers. Standard designs tend to be used in practice and units placed in parallel to process large flowrates.

The same principle can be used for the separation of solids from a liquid in a *hydrocyclone*. Although the principle is the same, whether a gas or vapor is being separated from a liquid, the geometry of the cyclone will change accordingly. Hydrocyclones can also be used to separate mixtures of immiscible liquids, such as mixtures of oil and water. For the separation of oil and water, the

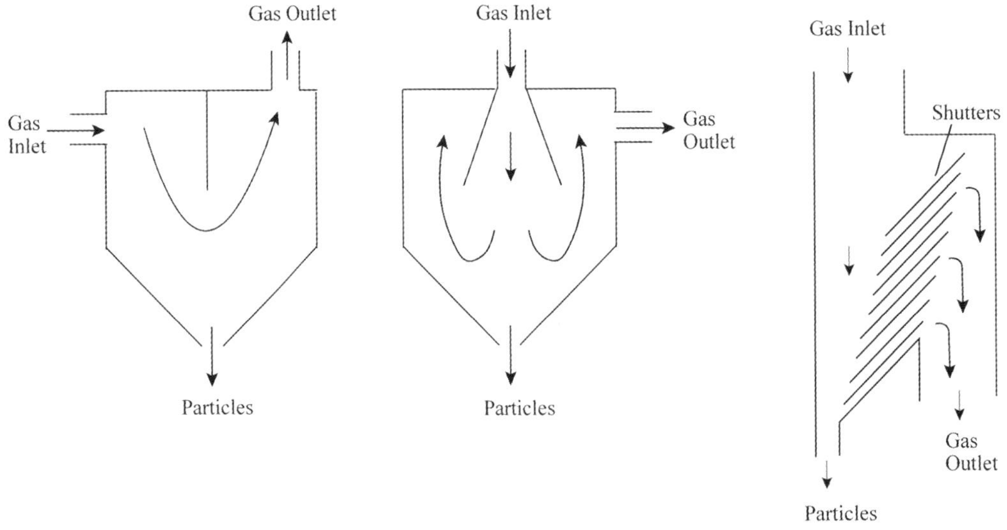

Figure 7.4

Inertial separators increase the efficiency of separation by giving the particles downward momentum.

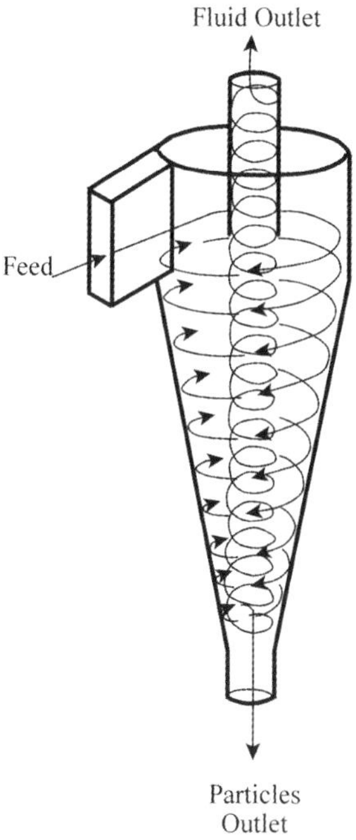

Figure 7.6

A cyclone generates centrifugal force by the fluid motion.

water is denser than the oil and is thrown to the wall by the centrifugal force leaving from the conical base. The oil leaves from the top. Again, the design of hydrocyclones is normally based on collection efficiency curves, such as the one shown in Figure 7.5, with standard designs used and units placed in parallel to process large flowrates.

Figure 7.7 shows *centrifuges*, in which a cylindrical bowl is rotated to produce the centrifugal force. In Figure 7.7a, the cylindrical bowl is shown rotating with a feed consisting of a liquid–solid mixture fed at the center. The feed is thrown outward to the walls of the container. The particles settle horizontally outward. Different arrangements are possible to remove the solids from the bowl. In Figure 7.7b, two liquids having different densities are separated by the centrifuge. The more dense fluid occupies the outer periphery, since the centrifugal force is greater on the more dense fluid.

7.4 Electrostatic Precipitation

Electrostatic precipitators are generally used to separate particulate matter that is easily ionized from a gas stream (Ludwig, 1977; Dullien, 1989; Schweitzer, 1997). This is accomplished by an electrostatic field produced between wires or grids and collection plates by applying a high voltage between the two, as illustrated in Figure 7.9. A corona is established around the negatively charged electrode. As the particulate-laden gas stream passes through the space, the corona ionizes molecules of gases such as O_2 and CO_2 present in the gas stream. These charged molecules attach themselves to particulate matter, thereby charging the particles. The oppositely charged collection plates attract the particles. Particles collect on the plates and are removed by vibrating the collection plates mechanically, thereby dislodging particles that drop to the bottom of the device. Electrostatic precipitation is most effective when separating particles with a high resistivity. The operating voltage typically varies between 25 and 45 kV or more, depending on the design and the operating temperature. The application of electrostatic precipitators is normally restricted to the separation of fine particles of solid or liquid from a large volume of gas. Preliminary designs can be based on collection efficiency curves, as illustrated in Figure 7.5 (Ludwig, 1977).

7

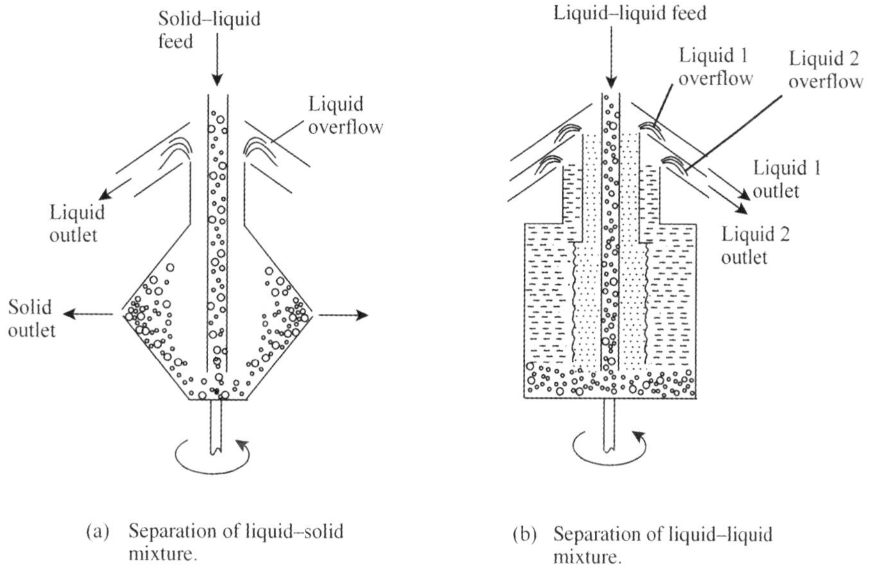

Figure 7.7

A centrifuge uses rotating cylindrical bowl to produce centrifugal force.

Example 7.3 A dryer vent is to be cleaned using a bank of cyclones. The gas flowrate is 60 $m^3 \cdot s^{-1}$, density of the solids is 2700 $kg \cdot m^{-3}$ and the concentration of solids is 10 $g \cdot m^{-3}$. The size distribution of the solids is given in Table 7.2.

The collection efficiency curve for the design of cyclone used is given in Figure 7.8. Calculate the solids removal and the final outlet concentration.

Table 7.2

Particle size distribution for Example 7.3.

Particle size (μm)	Percentage by mass less than
50	90
40	86
30	80
20	70
10	45
5	25
2	10

Solution The particles are first divided into size ranges and the collection efficiency for the average size applied with the size range is shown in Table 7.3.

The overall collection efficiency is 69.5% and the concentration of solids at exit is 3.05 $g \cdot m^{-3}$.

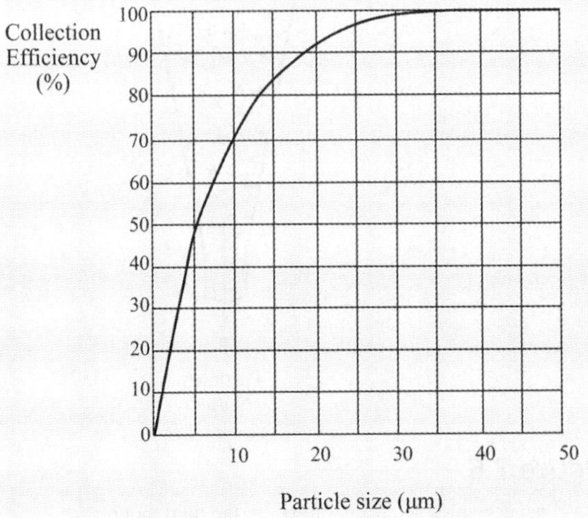

Figure 7.8

Collection efficiency curve for the cyclone design in Example 7.3.

Table 7.3

Collection efficiency for Example 7.3.

Particle size range (μm)	Percent in range (%)	Efficiency at mean size (%)	Overall collection (%)	Outlet	
				Fraction in range	Outlet %
>50	10	100	10	0	0
40–50	4	100	4	0	0
30–40	6	99	5.9	0.1	0.3
20–30	10	96	9.6	0.4	1.3
10–20	25	83	20.8	4.2	13.8
5–10	2	60	12.0	8.0	26.2
2–5	15	41	6.2	8.8	28.9
0–2	10	10	1.0	9.0	29.5
			69.5	30.5	100.0

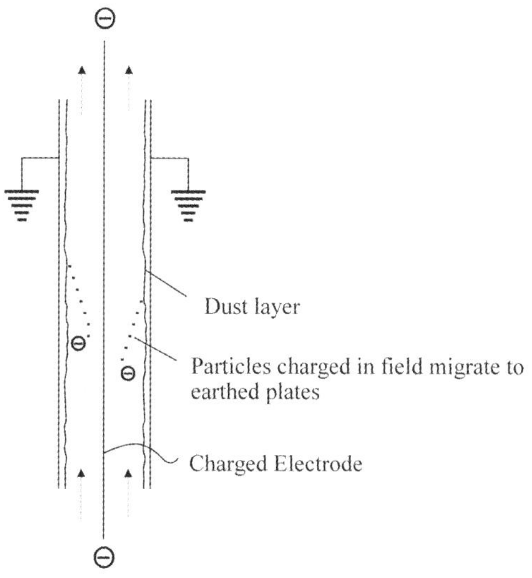

Figure 7.9

Electrostatic precipitation. (Reproduced from Stenhouse JIT, 1981, in Teja (ed) Chemical Engineering and the Environment, with permission from Blackwell Scientific Publications.)

7.5 Filtration

In filtration, suspended solid particles in a gas, vapor or liquid are removed by passing the mixture through a porous medium that retains the particles and passes the fluid (filtrate). The solid can be retained on the surface of the filter medium, which is *cake filtration*, or captured within the filter medium, which is *depth filtration*. The filter medium can be arranged in many ways.

Figure 7.10 shows four examples of cake filtration in which the filter medium is a *cloth* of natural or artificial fibers. In different arrangements, the filter medium might even be ceramic or metal. Figure 7.10a shows the filter cloth arranged between plates in an enclosure in a *plate-and-frame filter* for the separation of solid particles from liquids. Figure 7.10b shows the cloth arranged as a *thimble* or *candle*. This arrangement is common for the separation of solid particles from a gas and is known as a *bag filter*. As the particles build up on the inside of the thimble, the unit is periodically taken off-line and the flow reversed to recover the filtered particles. For the separation of solid particles from gases, conventional filter media can be used up to a temperature of around 250 °C. Higher temperatures require ceramic or metallic (e.g. stainless steel sintered fleece) filter media. These can be used for temperatures of 250 to 1000 °C and higher. Cleaning of high-temperature media requires the unit to be taken off-line and pressure pulses applied to recover the filtered particles. Figure 7.10c shows a rotating belt for the separation of a slurry of solid particles in a liquid and Figure 7.10d shows a rotating drum in which the drum rotates through the slurry. In both cases, the flow of liquid is induced by the creation of a vacuum. When filtering solids from liquids, if the purity of the filter cake is not important, *filter aids*, which are particles of porous solid, can be added to the mixture to aid the filtration process. When filtering solids from a liquid, a thickener is often used upstream of filtration to concentrate the mixture prior to filtration.

7

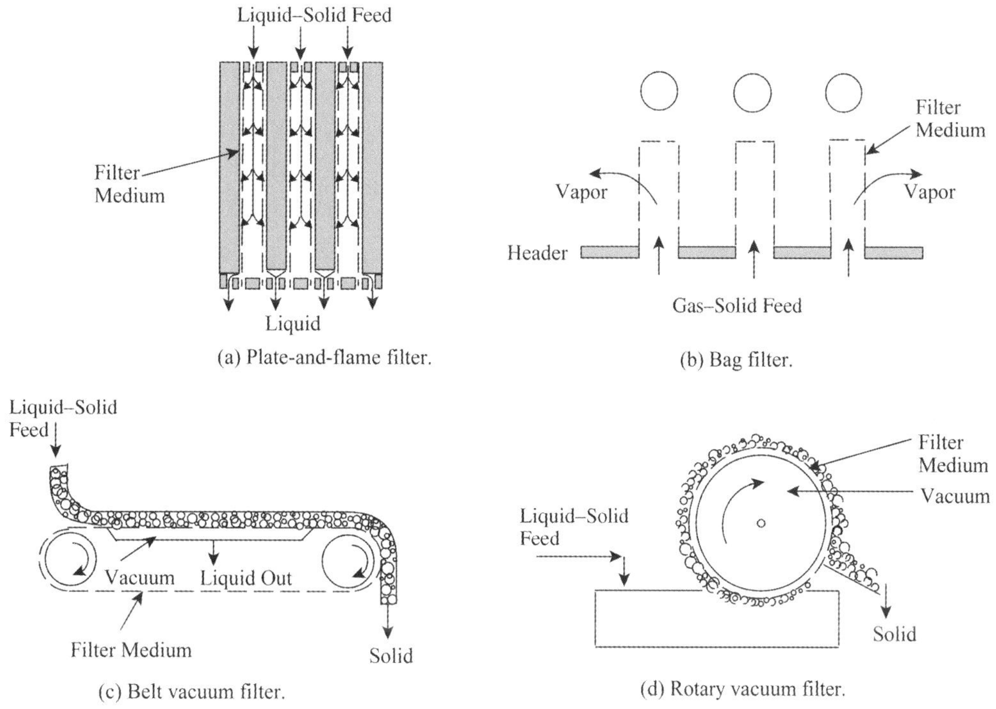

Figure 7.10

Filtration can be arranged in many ways.

When separating solid particles from a liquid filtrate, if the solid filter cake is a product, rather than a waste, then it is usual to wash the filter cake to remove the residual filtrate from the filter cake. The washing of the filter cake after filtration takes place by displacement of the filtrate and by diffusion.

Rather than using a cloth, a *granular* medium consisting of layers of particulate solids on a support grid can be used. Downward flow of the mixture causes the solid particles to be captured within the medium. Such *deep-bed* filters are most commonly used to remove small quantities of solids from large quantities of liquids. To release the solid particles captured within the bed, the flow is periodically reversed, causing the bed to expand and release the particles that have been captured. Around 3% of the throughput is needed for this backwashing.

Whereas the liquid–solid filtration processes described so far can separate particles down to a size of around 10 μm, for smaller particles that need to be separated, a porous polymer membrane can be used. This process, known as *microfiltration*, retains particles down to a size of around 0.05 μm. A pressure difference across the membrane of 1 to 5 bar is used. The two most common practical arrangements are spiral wound and hollow fiber. In the spiral wound arrangement, flat membrane sheets separated by spacers for the flow of feed and filtrate are wound into a spiral and inserted in a pressure vessel. Hollow fiber arrangements, as the name implies, are cylindrical membranes arranged in series in a pressure vessel in an arrangement similar to a shell-and-tube heat exchanger. The feed enters the shell side and permeate passes through the membrane to the center of the hollow fibers. The main factor affecting the flux of the filtrate through the membrane is the accumulation of deposits on the surface of the filter. Of course, this is usual in cake filtration systems when the particles are directed normally toward the surface of the filter medium. However, in the case of microfiltration, the surface deposits cause the flux of filtrate to decrease with time. To ameliorate this effect in microfiltration, *cross-flow* can be used, in which a high rate of shear is induced across the surface of the membrane from the feed. The higher the velocity across the surface of the filter medium, the lower is the deterioration in flux of the filtrate. Even with a significant velocity across the surface of the membrane, there is still likely to be deterioration in the flux caused by fouling of the membrane. Periodic flushing or cleaning of the membrane will be required to correct this. The method of cleaning depends on the type of foulant, the membrane configuration and the membrane's resistance to cleaning agents. In the simplest case, a reversal of the flow (*backflush*) might be able to remove the surface deposits. In the worst case, cleaning agents such as sodium hydroxide might be required. Microfiltration is used for the recovery of paint from coating processes, oil–water separations, separation of biological cells from a liquid, and so on.

7.6 Scrubbing

Scrubbing with liquid (usually water) can enhance the collection of particles when separating gas–solid mixtures. Figure 7.11 shows three of the many possible designs for scrubbers. Figure 7.11a illustrates a packed tower, similar to an absorption tower. Whilst this can be effective, it suffers from the problem that the packing can become clogged with solid particles. Towers using perforated plates similar to a distillation or absorption column can also be used. As with packed columns, plate columns can also encounter

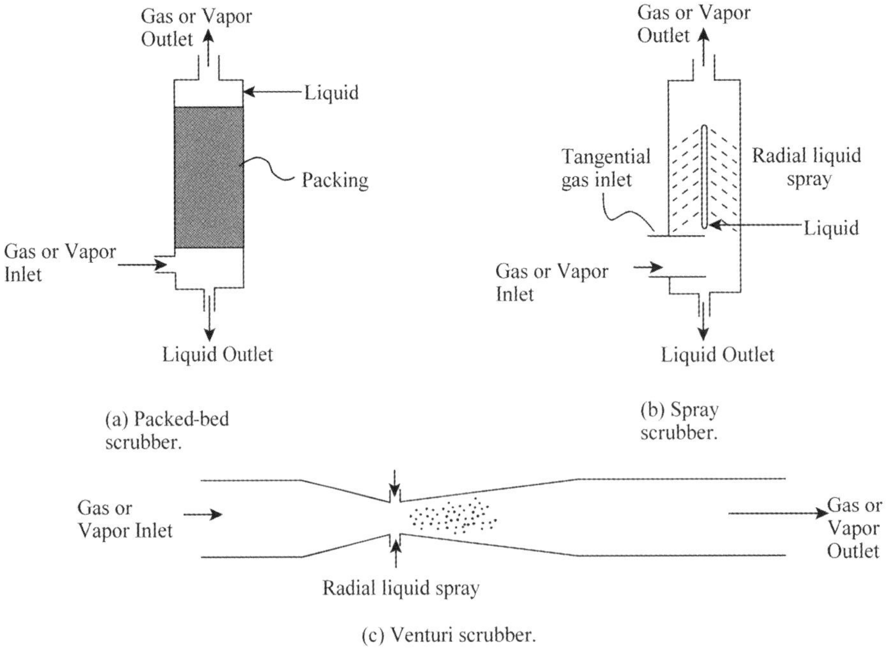

Figure 7.11
Various scrubber designs can be used to separate solid from gas or vapor. (Reproduced from Stenhouse JIT, 1981, in Teja (ed.) Chemical Engineering and the Environment, with permission from Blackwell Scientific Publications.)

problems of clogging. By contrast, Figure 7.11b uses a spray system that will be less prone to fouling. The design in Figure 7.11b uses a tangential inlet to create a swirl to enhance the separation. The design shown in Figure 7.11c uses a *venturi*. Liquid is injected into the throat of the venturi, where the velocity of the gas is highest. The gas accelerates the injected water to the gas velocity and breaks up the liquid droplets into a relatively fine spray. The particles are then captured by the fine droplets. Very high collection efficiencies are possible with venturi scrubbers. The main problem with venturi scrubbers is the high pressure loss across the device. As with other solids separation devices, preliminary design can be carried out using collection efficiency curves like the one illustrated in Figure 7.5 (Ludwig, 1977).

7.7 Flotation

Flotation is a gravity separation process that exploits the differences in the surface properties of particles. Gas bubbles are generated in a liquid and become attached to solid particles or immiscible liquid droplets, causing the particles or droplets to rise to the surface. This is used to separate mixtures of solid–solid particles after dispersion in a liquid, solid particles already dispersed in a liquid or liquid–liquid mixtures of finely divided immiscible droplets. The liquid used is normally water and the particles of solid or immiscible liquid will attach themselves to the gas bubbles if they are *hydrophobic* (e.g. oil droplets dispersed in water).

The bubbles of gas can be generated by three methods:

a) *dispersion*, in which the bubbles are injected directly by some form of sparging system;
b) *dissolution* in the liquid under pressure and then liberation in the flotation cell by reducing the pressure;
c) *electrolysis* of the liquid.

Flotation is an important technique in mineral processing, where it is used to separate different types of ores. When used to separate solid–solid mixtures, the material is ground to a particle size small enough to *liberate* particles of the chemical species to be recovered. The mixture of solid particles is then dispersed in the flotation medium, which is usually water. The mixture is then fed to a flotation cell, as illustrated in Figure 7.12a. Here gas is also fed to the cell where gas bubbles become attached to the solid particles, thereby allowing them to float to the surface of the liquid. The solid particles are collected from the surface by an overflow weir or mechanical scraper. The separation of the solid particles depends on the different species having different surface properties such that one species is preferentially attached to the bubbles. A number of chemicals can be added to the flotation medium to meet the various requirements of the flotation process:

a) *Modifiers* are added to control the pH of the separation. These could be acids, lime, sodium hydroxide, and so on.
b) *Collectors* are water-repellent reagents that are added to preferentially adsorb on to the surface of one of the solids. Coating or partially coating the surface of one of the solids renders the solid to be more hydrophobic and increases its tendency to attach to the gas bubbles.
c) *Activators* are used to "activate" the mineral surface for the collector.
d) *Depressants* are used to preferentially attach to one of the solids to make it less hydrophobic and decrease its tendency to attach to the gas bubbles.
e) *Frothers* are surface-active agents added to the flotation medium to create a stable froth and assist the separation.

Flotation is also used in applications such as the separation of low-density solid particles (e.g. paper pulp) from water and oil droplets from oil–water mixtures. It is not necessary to add reagents if the particles are naturally hydrophobic, as is the case,

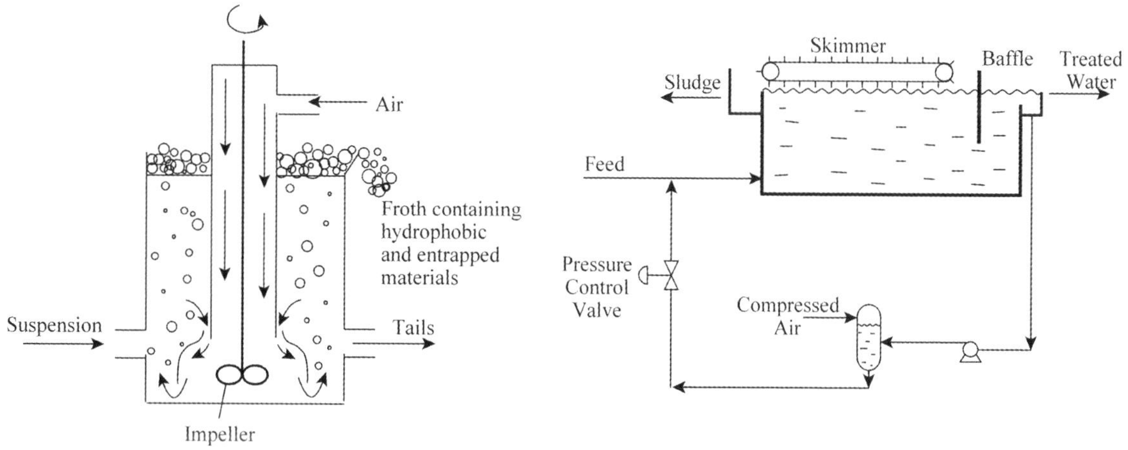

(a) A typical flotation cell for solid separation.

(b) Dissolved air flotation (DAF).

Figure 7.12
Flotation arrangements.

for example, with oil–water mixtures, as the oil is naturally hydrophobic.

When separating low-density solid particles or oil droplets from water, the most common method used is *dissolved-air flotation*. A typical arrangement is shown in Figure 7.12b. This shows some of the effluent water from the unit being recycled and air being dissolved in the recycle under pressure. The pressure of the recycle is then reduced, releasing the air from solution as a mist of fine bubbles. This is then mixed with the incoming feed that enters the cell. Low-density material floats to the surface with the assistance of the air bubbles and is removed.

7.8 Drying

Drying refers to the removal of water from a substance through a whole range of processes, including distillation, evaporation and even physical separations such as centrifuges. Here consideration is restricted to the removal of moisture from solids into a gas stream (usually air) by heat, namely, *thermal drying*. Some of the types of equipment for removal of water also can be used for removal of organic liquids from solids.

Four of the more common types of thermal dryers used in the process industries are illustrated in Figure 7.13.

1) *Tunnel dryers* are shown in Figure 7.13a. Wet material on trays or a conveyor belt is passed through a tunnel, and drying takes place by hot air. The airflow can be countercurrent, cocurrent or a mixture of both. This method is usually used when the product is not free flowing.
2) *Rotary dryers* are shown in Figure 7.13b. Here, a cylindrical shell mounted at a small angle to the horizontal is rotated at low speed. Wet material is fed at the higher end and flows under gravity. Drying takes place from a flow of air, which can be countercurrent or cocurrent. The heating may be direct to the dryer gas or indirect through the dryer shell. This method is usually used when the material is free flowing. Rotary dryers are not well suited to materials that are particularly heat sensitive because of the long residence time in the dryer.
3) *Drum dryers* are shown in Figure 7.13c. This consists of a heated metal roll. As the roll rotates, a layer of liquid or slurry is dried. The final dry solid is scraped off the roll. The product comes off in flaked form. Drum dryers are suitable for handling slurries or pastes of solids in fine suspension and are limited to low and moderate throughput.
4) *Spray dryers* are shown in Figure 7.13d. Here a liquid or slurry solution is sprayed as fine droplets into a hot gas stream. The feed to the dryer must be pumpable to obtain the high pressures required by the atomizer. The product tends to be light, porous particles. An important advantage of the spray dryer is that the product is exposed to the hot gas for a short period. Also, evaporation of the liquid from the spray keeps the product temperature low, even in the presence of hot gases. Spray dryers are thus particularly suited to products that are sensitive to thermal decomposition, such as food products.

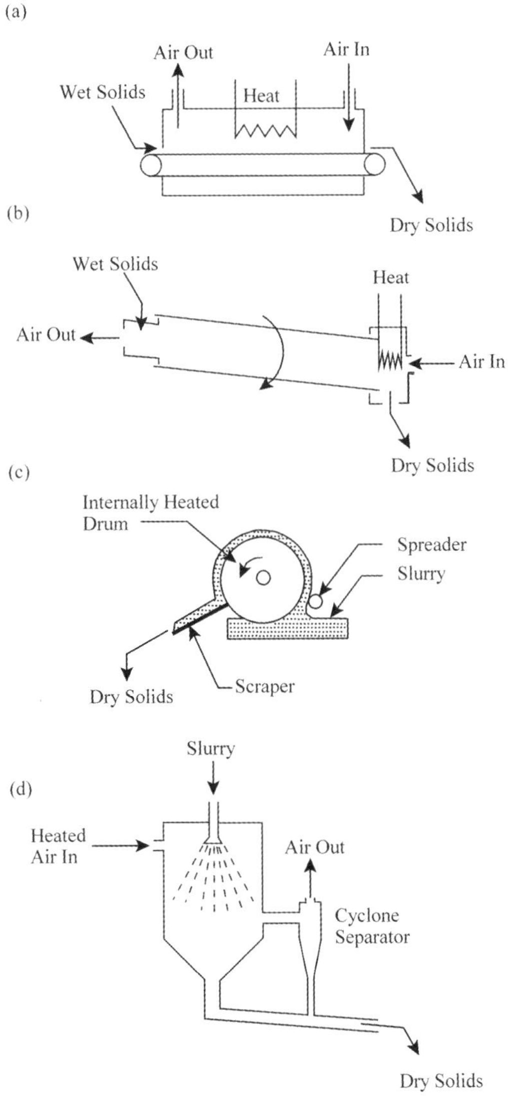

Figure 7.13
Four of the more common types of thermal dryer.

Another important class of dryers is the *fluidized-bed* dryer. Some designs combine spray and fluidized-bed dryers. Choice between dryers is usually based on practicalities such as the materials' handling characteristics, product decomposition, product physical form (e.g. if a porous granular material is required), and so on. Also, dryer efficiency can be used to compare the performance of different dryer designs. This is usually defined as follows:

$$\text{Dryer efficiency} = \frac{\text{Heat of vaporization}}{\text{Total heat consumed}} \tag{7.14}$$

If the total heat consumed is from an external utility (e.g. mains steam), then a high efficiency is desirable, even perhaps at the expense of a high capital cost. However, if the heat consumed is by recovery from elsewhere in the process, as is discussed in Chapter 22, then comparison based on dryer efficiency becomes less meaningful.

7.9 Separation of Heterogeneous Mixtures – Summary

For a heterogeneous or multiphase mixture, separation usually can be achieved by phase separation. Such phase separation should be carried out before any homogeneous separation. Phase separation tends to be easier and should be done first.

The simplest devices for separating heterogeneous mixtures exploit gravitational force in settling and sedimentation devices. The velocity of the fluid through such a device must be less than the terminal settling velocity of the particles to be separated. When high concentrations of particles are to be settled, the surrounding particles interfere with individual particles and the settling becomes hindered. Separating a dispersed liquid phase from another liquid phase can be carried out in a decanter. The decanter is sized on the basis that the velocity of the continuous phase should be less than the terminal settling velocity of the droplets of the dispersed phase.

If gravitational forces are inadequate to achieve the separation, the separation can be enhanced by exploiting inertial, centrifugal or electrostatic forces.

In filtration, suspended solid particles in a gas, vapor or liquid are removed by passing the mixture through a porous medium that retains the particles and passes the fluid.

Scrubbing with liquid (usually water) can enhance the collection of particles when separating gas–solid mixtures. Flotation is a gravity separation process that exploits differences in the surface properties of particles. Gas bubbles are generated in a liquid and become attached to solid particles or immiscible liquid droplets, causing the particles or droplets to rise to the surface.

Thermal drying can be used to remove moisture from solids into a gas stream (usually air) by heat. Many types of dryers are available and can be compared on the basis of their thermal efficiency.

For the separation of gas–solid mixtures, preliminary design of inertial separators, cyclones, electrostatic precipitators, scrubbers and some types of filters can be carried out on the basis of collection efficiency curves derived from experimental performance.

No attempt should be made to carry out any optimization at this stage in the design.

7.10 Exercises

1. A gravity settler is to be used to separate particles less than 75 μm from a flowrate of gas of $1.6\ m^3 \cdot s^{-1}$. The density of the particles is $2100\ kg \cdot m^{-3}$. The density of the gas is $1.18\ kg \cdot m^{-3}$ and its viscosity is $1.85 \times 10^{-5}\ kg \cdot m^{-1} \cdot s^{-1}$. Estimate the dimensions of the settling chamber assuming a rectangular cross-section with the length to be twice that of the breadth.

2. When separating a liquid–liquid mixture in a cylindrical drum, why is it usually better to mount the drum horizontally than vertically?

Table 7.4

Fluid properties for Exercise 7.3.

	Density ($kg \cdot m^{-3}$)	Viscosity ($kg \cdot m^{-1} \cdot s^{-1}$)
Water	993	0.8×10^{-3}
Oil	890	3.5×10^{-3}

3. A liquid–liquid mixture containing $6\ kg \cdot s^{-1}$ of water with $0.5\ kg \cdot s^{-1}$ of entrained oil is to be separated in a decanter. The properties of the fluids are given in Table 7.4.

   Assuming the water to be the continuous phase, the decanter to be a horizontal drum with a length to diameter ratio of 3 to 1 and an interface across the center of the drum, estimate the dimensions of the decanter to separate water droplets less than 150 μm.

4. In Exercise 3, it was assumed that the interface would be across the center of the drum. How would the design change if the interface was lower such that the length of the interface across the drum was 50% of the diameter?

5. A mixture of vapor and liquid ammonia is to be separated in a cylindrical vessel mounted vertically with a mesh pad to assist separation. The flowrate of vapor is $0.3\ m^3 \cdot s^{-1}$. The density of the liquid is $648\ kg \cdot m^{-3}$ and that of the vapor $2.71\ kg \cdot m^{-3}$. Assuming $K_T = 0.11\ m \cdot s^{-1}$, estimate the diameter of vessel required.

6. A gravity settler has dimensions of 1 m breadth, 1 m height and length of 3 m. It is to be used to separate solid particles from a gas with a flowrate of $1.6\ m^3 \cdot s^{-1}$. The density of the particles is $2100\ kg \cdot m^{-3}$. The density of the gas is $1.18\ kg \cdot m^{-3}$ and its viscosity is $1.85 \times 10^{-5}\ kg \cdot m^{-1} \cdot s^{-1}$. Construct an approximate collection efficiency curve for the settling of particles ranging from 0 to 50 μm.

7. Table 7.5 shows the size distribution of solid particles in a gas stream.

   The flowrate of the gas is $10\ m^3 \cdot s^{-1}$ and the concentration of solids is $12\ g \cdot m^{-3}$. The particles are to be separated in a scrubber with a collection efficiency curve as shown in Figure 7.14. Estimate the overall collection efficiency and the concentration of solids in the exit gas.

Table 7.5

Particle size distribution.

Particle size (μm)	Percentage by mass less than
30	95
20	90
10	70
5	30
2	20

7

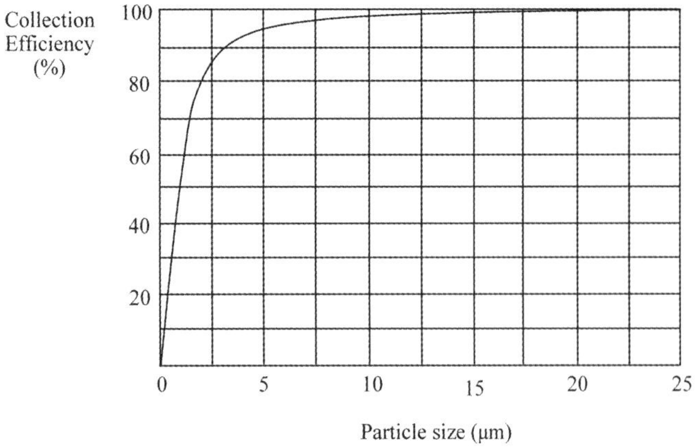

Figure 7.14

Collection efficiency curve for scubber.

References

Coulson JM and Richardson JF with Backhurst JR and Harker JH (1991) *Chemical Engineering*, Vol. **2**, 4th Edition, Butterworth Heinemann.

Dullien FAL (1989) *Introduction to Industrial Gas Cleaning*, Academic Press.

Foust AS, Wenzel LA, Clump CW, Maus L and Anderson LB (1980) *Principles of Unit Operations*, John Wiley & Sons, Inc., New York.

Geankopolis CJ (1993) *Transport Processes and Unit Operations*, 3rd Edition, Prentice Hall.

King CJ (1980) *Separation Processes*, 2nd Edition, McGraw-Hill, New York.

Ludwig EE (1977) *Applied Process Design for Chemical and Petrochemical Plants*, 2nd Edition, Gulf Publishing Company, Houston.

Rousseau RW (1987) *Handbook of Separation Process Technology*, John Wiley & Sons, Inc., New York.

Schweitzer PA (1997) *Handbook of Separation Process Techniques for Chemical Engineers*, 3rd Edition, McGraw-Hill, New York.

Svarovsky L (1981) *Solid–Gas Separation*, Elsevier Scientific, New York.

Walas SM (1988) *Chemical Process Equipment Selection and Design*, Butterworths.

Woods DR (1995) *Process Design and Engineering Practice*, Prentice Hall, New Jersey.

Chapter 8

Separation of Homogeneous Fluid Mixtures I – Distillation

As pointed out in the previous chapter, the separation of a homogeneous fluid mixture requires the creation of another phase or the addition of a mass separation agent. Consider a homogeneous liquid mixture. If this liquid mixture is partially vaporized, then another phase is created and the vapor becomes richer in the more volatile components (i.e. those with the lower boiling points) than the liquid phase. The liquid becomes richer in the less-volatile components (i.e. those with the higher boiling points). If the system is allowed to come to equilibrium conditions, then the distribution of the components between the vapor and liquid phases is dictated by vapor–liquid equilibrium considerations. In principle, all components can appear in both phases.

On the other hand, rather than partially vaporize a liquid, the starting point could have been a homogeneous mixture of components in the vapor phase and the vapor partially condensed. There would still be a separation, as the liquid that was formed would be richer in the less-volatile components, while the vapor would have become depleted in the less-volatile components. Again, the distribution of components between the vapor and liquid is dictated by vapor–liquid equilibrium considerations if the system is allowed to come to equilibrium.

Distillation involves the repeated vaporization and condensation of a homogeneous liquid mixture. In some cases, a single vaporization or condensation can bring an effective separation. However, first consider vapor–liquid equilibrium.

8.1 Vapor–Liquid Equilibrium

For a liquid and vapor in contact and in equilibrium, the K-value relates the vapor and liquid mole fractions (see Appendix A):

$$K_i = \frac{y_i}{x_i} = \frac{\phi_i^L}{\phi_i^V} \tag{8.1}$$

Chemical Process Design and Integration, Second Edition. Robin Smith.

Companion Website: www.wiley.com/go/smith/chemical2e

where K_i = K-value of Component i
y_i = mole fraction of Component i in the vapor phase
x_i = mole fraction of Component i in the liquid phase
ϕ_i^L = liquid-phase fugacity coefficient
ϕ_i^V = vapor-phase fugacity coefficient

Equation 8.1 defines the relationship between the vapor and liquid mole fractions and provides the basis for vapor–liquid equilibrium calculations on the basis of *equations of state*. Thermodynamic models are required for ϕ_i^V and ϕ_i^L from an equation of state (see Appendix A). Alternatively, a *liquid-phase activity coefficient model* can provide the basis for vapor–liquid equilibrium calculations (see Appendix A):

$$K_i = \frac{y_i}{x_i} = \frac{\gamma_i P_i^{SAT}}{\phi_i^V P} \tag{8.2}$$

where γ_i = liquid-phase activity coefficient
P_i^{SAT} = saturated liquid vapor pressure

In Equation 8.2, thermodynamic models are required for ϕ_i^V (from an equation of state) and γ_i from a liquid-phase activity coefficient model. Some of the more important models are given in Appendix A. At moderate pressures, the vapor phase becomes ideal and $\phi_i^V = 1$. For an ideal vapor phase, Equation 8.2 simplifies to:

$$K_i = \frac{y_i}{x_i} = \frac{\gamma_i P_i^{SAT}}{P} \tag{8.3}$$

When the liquid phase behaves as an ideal solution:

- all molecules have the same size;
- all intermolecular forces are equal.

The properties of the mixture depend only on the properties of the pure components comprising the mixture. Mixtures of isomers, such as *o*-, *m*- and *p*-xylene mixtures, and adjacent members of

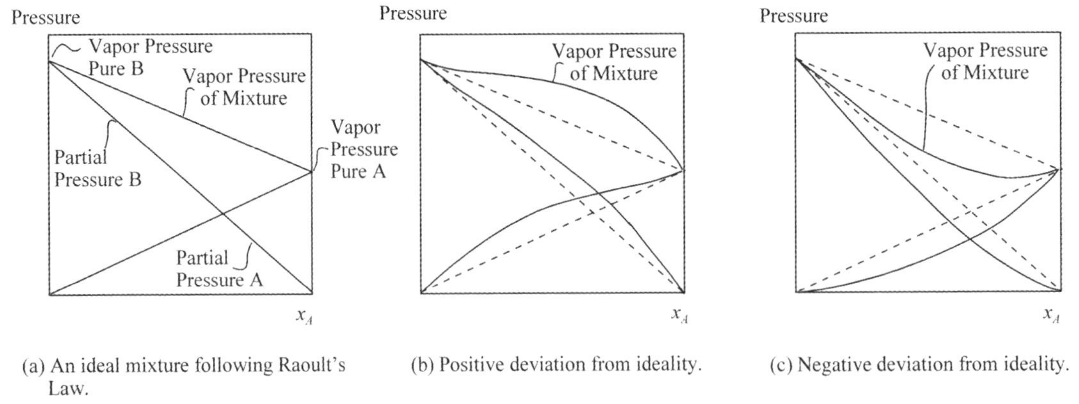

Figure 8.1
Deviations from Raoult's Law.

homologous series, such as *n*-hexane–*n*-heptane and benzene–toluene mixtures, give close to ideal liquid-phase behavior. For this case, $\gamma_i = 1$ and Equation 8.3 simplifies to:

$$K_i = \frac{y_i}{x_i} = \frac{P_i^{SAT}}{P} \tag{8.4}$$

This is Raoult's Law and represents both ideal vapor- and liquid-phase behavior.

Correlations are available to relate component vapor pressure to temperature. One of the most common equations used is the Antoine Equation (see Appendix A):

$$\ln P^{SAT} = A - \frac{B}{C + T} \tag{8.5}$$

where *A*, *B* and *C* are constants determined by correlating experimental data. Extended forms of the Antoine Equation have also been proposed (Poling, Prausnitz and O'Connell, 2001). Great care must be taken when using correlated vapor pressure data not to use the correlation coefficients outside the temperature range over which the data have been correlated, otherwise serious errors can occur.

Comparing Equations 8.3 and 8.4, the liquid-phase nonideality is characterized by the activity coefficient γ_i. When $\gamma_i = 1$, the behavior is ideal. If $\gamma_i \neq 1$, then the value of γ_i can be used to characterize the nonideality:

- $\gamma_i > 1$ represents positive deviations from Raoult's Law;
- $\gamma_i < 1$ represents negative deviations from Raoult's Law.

These different classes of nonideality are illustrated in Figure 8.1 for a binary mixture. Figure 8.1a illustrates Raoult's Law behavior. Raoult's Law states that the equilibrium partial pressure that is exhibited by a component in a solution is proportional to the mole fraction of that component. Thus in Figure 8.1a, as the mole fraction of Component *A* is increased, the partial pressure of Component *A* increases linearly. At the same time, the partial pressure of Component *B* decreases linearly. The combined partial pressures result in the vapor pressure of the mixture to vary linearly with the mole fraction. Deviations from Raoult's Law are characterized by the activity coefficient not being unity. When the activity coefficient is greater than unity, giving a positive deviation from Raoult's Law, the molecules of the components in the system repel each other and exert a higher partial pressure than if their behavior were ideal. This is illustrated in Figure 8.1b. When the activity coefficient is less than unity, the molecules of the components in the system attract each other through chemical interactions and exert a lower partial pressure than if their behavior were ideal. This is illustrated in Figure 8.1c. The components of such solutions occur typically when the separate components show wide differences in polarity.

The ratio of equilibrium K-values for two components measures their *relative volatility*:

$$\alpha_{ij} = \frac{K_i}{K_j} \tag{8.6}$$

where α_{ij} = volatility of Component *i* relative to Component *j*

These expressions form the basis for two alternative approaches to vapor–liquid equilibrium calculations:

a) $K_i = \phi_i^L / \phi_i^V$ forms the basis for calculations based entirely on equations of state. Using an equation of state for both the liquid and vapor phase has a number of advantages. In principle, continuity at the critical point can be guaranteed with all thermodynamic properties derived from the same model. The presence of noncondensable gases, in principle, causes no additional complications. However, the application of equations of state is largely restricted to nonpolar components. The approach is described in more detail in Appendix A together with the models to calculate ϕ_i^L and ϕ_i^V.

b) $K_i = \gamma_i P_i^{SAT} / \phi_i^V P$ forms the basis for calculations based on liquid-phase activity coefficient models. It is used when polar molecules are present. For most systems at low pressures, ϕ_i^V can be assumed to be unity. If high pressures are involved, then ϕ_i^V must be calculated from an equation of state. However, care should be taken when mixing and matching different models for γ_i and ϕ_i^V for high-pressure systems to ensure that appropriate combinations are taken. Again the approach is described in more detail in Appendix A with some of the more important models for γ_i described in detail.

8.2 Calculation of Vapor-Liquid Equilibrium

Consider a single equilibrium stage in which a multicomponent feed is allowed to separate into a vapor and a liquid phase with the phases coming to equilibrium, as shown in Figure 8.2. An overall material balance and component material balances can be written as:

$$F = V + L \tag{8.7}$$

$$Fz_i = Vy_i + Lx_i \tag{8.8}$$

where F = feed flowrate (kmol·s^{-1})
V = vapor flowrate from the separator (kmol·s^{-1})
L = liquid flowrate from the separator (kmol·s^{-1})
z_i = mole fraction of Component i in the feed (–)
y_i = mole fraction of Component i in vapor (–)
x_i = mole fraction of Component i in liquid (–)

The vapor–liquid equilibrium relationship can be defined in terms of K-values from Equation 8.1 by:

$$y_i = K_i x_i \tag{8.9}$$

Equations 8.7 to 8.9 can now be solved to give expressions for the vapor- and liquid-phase compositions leaving the separator:

$$y_i = \frac{z_i}{\dfrac{V}{F} + \left(1 - \dfrac{V}{F}\right)\dfrac{1}{K_i}} \tag{8.10}$$

$$x_i = \frac{z_i}{(K_i - 1)\dfrac{V}{F} + 1} \tag{8.11}$$

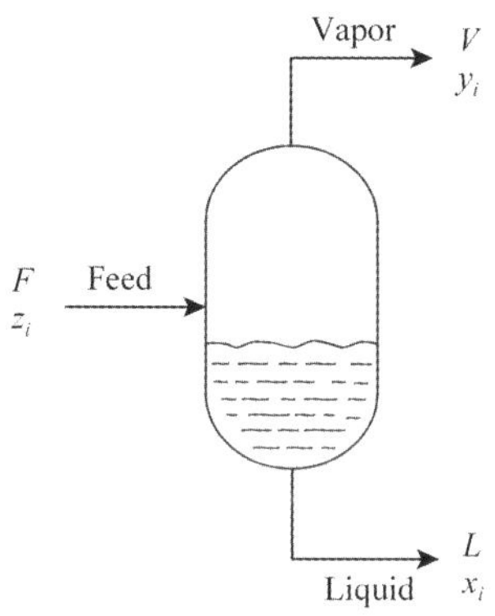

Figure 8.2
Single-stage separation.

The vapor fraction (V/F) in Equations 8.10 and 8.11 lies in the range $0 < V/F < 1$.

For a specified temperature and pressure, Equations 8.10 and 8.11 need to be solved by trial and error. Given that:

$$\sum_i^{NC} y_i = \sum_i^{NC} x_i = 1 \tag{8.12}$$

where NC is the number of components, then:

$$\sum_i^{NC} y_i - \sum_i^{NC} x_i = 0 \tag{8.13}$$

Substituting Equations 8.10 and 8.11 into Equation 8.13, after rearrangement gives (Rachford and Rice, 1952):

$$\sum_i^{NC} \frac{z_i(K_i - 1)}{\dfrac{V}{F}(K_i - 1) + 1} = 0 = f(V/F) \tag{8.14}$$

Equation 8.14 is known as the Rachford–Rice Equation (Rachford and Rice, 1952). To solve Equation 8.14, start by assuming a value of V/F and calculate $f(V/F)$ and search for a value of V/F until the function equals zero.

Many variations are possible around the basic *flash* calculation. Pressure and V/F can be specified and T calculated, and so on (King, 1980; Walas, 1985). However, two special cases are of particular interest. If it is necessary to calculate the *bubble point*, then $V/F = 0$ in Equation 8.14, which simplifies to:

$$\sum_i^{NC} z_i(K_i - 1) = 0 \tag{8.15}$$

and given:

$$\sum_i^{NC} z_i = 1 \tag{8.16}$$

This simplifies to the expression for the bubble point:

$$\sum_i^{NC} z_i K_i = 1 \tag{8.17}$$

Thus, to calculate the bubble point for a given mixture and at a specified pressure, a search is made for a temperature to satisfy Equation 8.17. Alternatively, temperature can be specified and a search made for a pressure, the *bubble pressure*, to satisfy Equation 8.17.

Another special case is when it is necessary to calculate the *dew point*. In this case, $V/F = 1$ in Equation 8.14, which simplifies to:

$$\sum_i^{NC} \frac{z_i}{K_i} = 1 \tag{8.18}$$

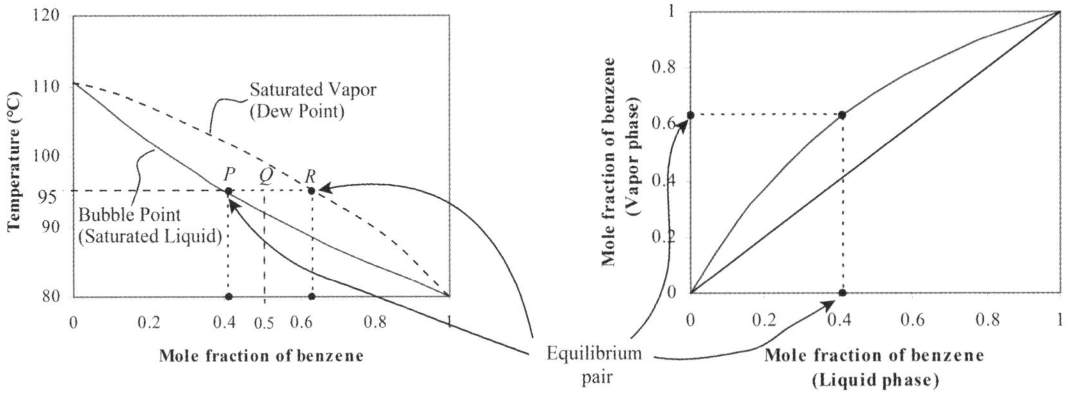

(a) Temperature-composition behavior. (b) x-y representation.

Figure 8.3

Vapor–liquid equilibrium for a binary mixture of benzene and toluene at a pressure of 1 atm. (Reproduced from Smith R and Jobson M (2000) Distillation, Encyclopedia of Separation Science, Academic Press, with permission from Elsevier.)

Again, for a given mixture and pressure, temperature is searched to satisfy Equation 8.18. Alternatively, temperature is specified and pressure searched for the *dew pressure*.

If the K-value requires the composition of both phases to be known, then this introduces additional complications into the calculations. For example, suppose a bubble-point calculation is to be performed on a liquid of known composition using an equation of state for the vapor–liquid equilibrium (see Appendix A). To start the calculation, a temperature is assumed. Then, calculation of K-values requires knowledge of the vapor composition to calculate the vapor-phase fugacity coefficient and that of the liquid composition to calculate the liquid-phase fugacity coefficient (see Appendix A). While the liquid composition is known, the vapor composition is unknown and an initial estimate is required for the calculation to proceed. Once the K-value has been estimated from an initial estimate of the vapor composition, the composition of the vapor can be reestimated, and so on.

Figure 8.3 shows, as an example, the vapor–liquid equilibrium behavior for a binary mixture of benzene and toluene (Smith and Jobson, 2000). Figure 8.3a shows the behavior of temperature of the saturated liquid and saturated vapor (i.e. equilibrium pairs) as the mole fraction of benzene is varied (the balance being toluene). This can be constructed by calculating the bubble and dew points for different concentrations. Figure 8.3b shows an alternative way of representing the vapor–liquid equilibrium in a composition or *x–y diagram*. The *x–y* diagram can be constructed from the relative volatility (Equation 8.6). From the definition of relative volatility for a binary mixture of Components *A* and *B*:

$$\alpha_{AB} = \frac{y_A/x_A}{y_B/x_B} = \frac{y_A/x_A}{(1-y_A)/(1-x_A)} \tag{8.19}$$

Rearranging gives:

$$y_A = \frac{x_A \alpha_{AB}}{1 + x_A(\alpha_{AB} - 1)} \tag{8.20}$$

Thus, by knowing α_{AB} from the vapor–liquid equilibrium and by specifying x_A, y_A can be calculated. Figure 8.3a also shows a typical vapor–liquid equilibrium pair, where the mole fraction of benzene in the liquid phase is 0.4 and that in the vapor phase is 0.62. A diagonal line across the *x–y* diagram in Figure 8.3b represents equal vapor and liquid compositions. The phase equilibrium behavior in Figure 8.3b shows a curve above the diagonal line. This indicates that benzene has a higher concentration in the vapor phase than toluene, that is, benzene is the more volatile component. Figure 8.3b shows the same vapor–liquid equilibrium pair as that shown in Figure 8.3a (Smith and Jobson, 2000).

Figure 8.3a can be used to predict the separation in a single equilibrium stage, given a specified feed to the stage and a stage temperature. For example, suppose the feed is a mixture with equal mole fractions of benzene and toluene of 0.5 and this is brought to equilibrium at 95 °C (Point *Q* in Figure 8.3a). Then the resulting liquid will have a mole fraction of benzene of 0.4 and the vapor a mole fraction of 0.62. In addition, the quantity of each phase formed can be determined from the lengths of the lines *PQ* and *QR* in Figure 8.3a. An overall material balance across the separator gives:

$$m_Q = m_P + m_R \tag{8.21}$$

A material balance for Component *i* gives:

$$m_Q x_{i,Q} = m_p x_{i,P} + m_R x_{i,R} \tag{8.22}$$

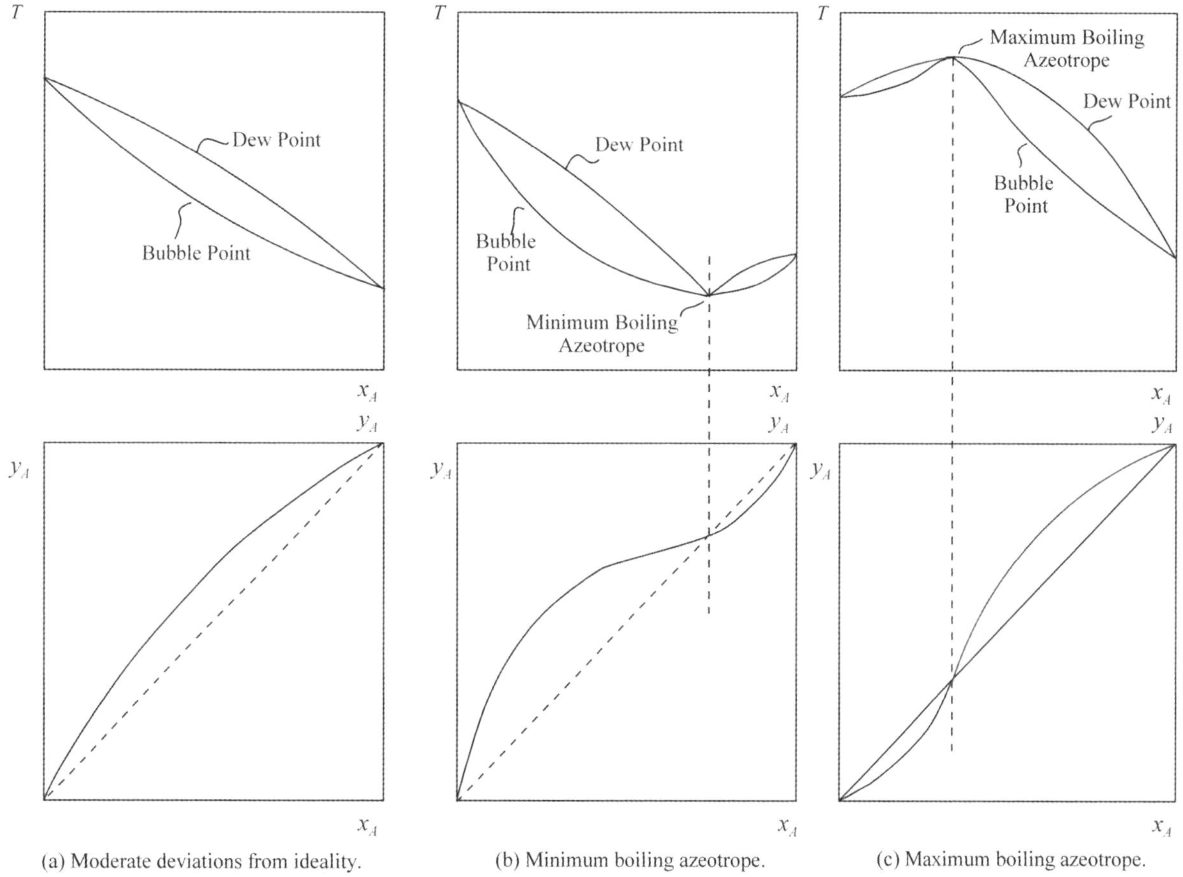

Figure 8.4
Binary-liquid equilibrium behavior.

Substituting Equation 8.21 into Equation 8.22 and rearranging gives:

$$\frac{m_P}{m_R} = \frac{x_{i,R} - x_{i,Q}}{x_{i,Q} - x_{i,P}} \tag{8.23}$$

Thus, in Figure 8.3:

$$\frac{m_P}{m_R} = \frac{QR}{PQ} \tag{8.24}$$

The ratio of molar flowrates of the vapor and liquid phases is thus given by the ratio of the opposite line segments. This is known as the *Lever Rule*, after the analogy with a lever and fulcrum (King, 1980).

Consider now a binary mixture of two Components *A* and *B*; the vapor–liquid equilibrium exhibits only a moderate deviation from ideality, as represented in Figure 8.4a. In this case, as pure *A* boils at a lower temperature than pure *B* in the temperature–composition diagram in Figure 8.4a, Component *A* is more volatile than Component *B*. This is also evident from the vapor–liquid composition diagram (*x*–*y* diagram), as it is above the line of $y_A = x_A$. In addition, it is also clear from Figure 8.4a that the order of volatility does not change as the composition changes. By contrast, Figure 8.4b shows a more highly nonideal behavior in which $\gamma_i > 1$ (positive deviation from Raoult's Law) forms a *minimum-boiling azeotrope*. At the azeotropic composition, the vapor and liquid are both at the same composition for the mixture. The lowest boiling temperature is below that of either of the pure components and is at the minimum-boiling azeotrope. It is clear from Figure 8.4b that the order of volatility of Components *A* and *B* changes, depending on the composition. Figure 8.4c also shows azeotropic behavior. This time, the mixture shows a behavior in which $\gamma_i < 1$ (negative deviation from Raoult's Law) forms a *maximum-boiling azeotrope*. This maximum-boiling azeotrope boils at a higher temperature than either of the pure components and would be the last fraction to be distilled, rather than the least volatile component, which would be the case with nonazeotropic behavior. Again, from Figure 8.4c, it can be observed that the order of volatility of Components *A* and *B* changes depending on the composition. Minimum-boiling azeotropes are much more common than maximum-boiling azeotropes. The formulation of azeotropes presents special problems for the separation of such a mixture, owing to the change in the order of volatility. This situation will be dealt with in detail in Chapter 11. The remainder of the chapter will deal only with separations in which no azeotropes are formed.

Some general guidelines for vapor–liquid mixtures in terms of their nonideality are:

a) Mixtures of isomers usually form ideal solutions.
b) Mixtures of close-boiling aliphatic hydrocarbons are nearly ideal below 10 bar.
c) Mixtures of compounds close in molar mass and structure frequently do not deviate greatly from ideality (e.g. ring compounds, unsaturated compounds, naphthenes, etc.).
d) Mixtures of simple aliphatics with aromatic compounds deviate modestly from ideality.
e) Noncondensables such as CO_2, H_2S, H_2, N_2, and so on, that are present in mixtures involving heavier components tend to behave nonideally with respect to the other compounds.
f) Mixtures of polar and nonpolar compounds are always strongly nonideal.
g) Azeotropes and phase separation into liquid–liquid mixtures represent the ultimate in nonideality.

Moving down the list, the nonideality of the system increases.

Example 8.1 A mixture of ethane, propane, *n*-butane, *n*-pentane and *n*-hexane is given in Table 8.1. For this calculation, it can be assumed that the K-values are ideal and follow Raoult's Law. For the mixture in Table 8.1, an equation of state method might have been a more appropriate choice (see Appendix A). However, this makes the calculation of the K-values much more complex. The ideal K-values for the mixture can be expressed in terms of the Antoine Equation as:

$$K_i = \frac{1}{P}\exp\left[A_i - \frac{B_i}{T + C_i}\right] \tag{8.25}$$

where P is the pressure (bar), T the absolute temperature (K) and A_i, B_i and C_i are constants given in the Table 8.1:

a) For a pressure of 5 bar, calculate the bubble point.
b) For a pressure of 5 bar, calculate the dew point.
c) Calculate the pressure needed for total condensation at 313 K.
d) At a pressure of 6 bar and a temperature of 313 K, how much liquid will be condensed?

Solution

a) The bubble point can be calculated from Equation 8.17 or from Equation 8.14 by specifying $V/F = 0$. The bubble point is calculated from Equation 8.17 in Table 8.2 to be 296.4 K. The search can be readily automated in spreadsheet software.
b) The dew point can be calculated from Equation 8.18 or 8.14 by specifying $V/F = 1$. In Table 8.3, the dew point is calculated from Equation 8.18 to be 359.1 K.
c) To calculate the pressure needed for total condensation at 313 K, the bubble pressure is calculated. The calculation is essentially the same as that in Part a above, with the temperature fixed at 313 K and the pressure varied, instead, in an iterative fashion until Equation 8.17 is satisfied. The resulting bubble pressure at 313 K is 7.4 bar.

   If distillation was carried out giving an overhead composition, as given in Table 8.1, and if cooling water was used in the condenser, then total condensation at 313 K (40 °C) would require an operating pressure of 7.4 bar.
d) Equation 8.14 is used to determine how much liquid will be condensed at a pressure of 6 bar and 313 K. The calculation is detailed in Table 8.4, giving $V/F = 0.0951$; thus, 90.49% of the feed will be condensed.

Simple phase equilibrium calculations, like the one illustrated here, can be readily implemented in spreadsheet software and automated. In practice, the calculations will most often be carried out in commercial physical property packages, allowing more elaborate methods for calculating the equilibrium K-values to be used.

Table 8.1

Feed and physical property for components.

Component	Formula	Feed (kmol)	A_i	B_i	C_i
1. Ethane	C_2H_6	5	9.0435	1511.4	−17.16
2. Propane	C_3H_8	25	9.1058	1872.5	−25.16
3. *n*-Butane	C_4H_{10}	30	9.0580	2154.9	−34.42
4. *n*-Pentane	C_5H_{12}	20	9.2131	2477.1	−39.94
5. *n*-Hexane	C_6H_{14}	20	9.2164	2697.6	−48.78

Table 8.2

Bubble-point calculation.

i	z_i	$T = 275$ K		$T = 350$ K		$T = 300$ K		$T = 290$ K		$T = 296.366$ K	
		K_i	z_iK_i	K_i	z_iK_i	K_i	z_iK_i	K_i	z_iK_i	K_i	z_iK_i
1	0.05	4.8177	0.2409	18.050	0.9025	8.0882	0.4044	6.6496	0.3325	7.5448	0.3772
2	0.25	1.0016	0.2504	5.6519	1.4130	1.9804	0.4951	1.5312	0.3828	1.8076	0.4519
3	0.30	0.2212	0.0664	1.8593	0.5578	0.5141	0.1542	0.3742	0.1123	0.4594	0.1378
4	0.20	0.0532	0.0106	0.6802	0.1360	0.1464	0.0293	0.1000	0.0200	0.1279	0.0256
5	0.20	0.0133	0.0027	0.2596	0.0519	0.0437	0.0087	0.0280	0.0056	0.0373	0.0075
	1.00		0.5710		3.0612		1.0917		0.8532		1.0000

Table 8.3

Dew-point calculation.

i	z_i	$T = 325$ K		$T = 375$ K		$T = 350$ K		$T = 360$ K		$T = 359.105$ K	
		K_i	z_i/K_i	K_i	z_i/K_i	K_i	z_i/K_i	K_i	z_i/K_i	K_i	z_i/K_i
1	0.05	12.483	0.0040	24.789	0.0020	18.050	0.0028	20.606	0.0024	20.370	0.0025
2	0.25	3.4951	0.0715	8.5328	0.0293	5.6519	0.0442	6.7136	0.0372	6.6138	0.0378
3	0.30	1.0332	0.2903	3.0692	0.0977	1.8593	0.1614	2.2931	0.1308	2.2517	0.1332
4	0.20	0.3375	0.5925	1.2345	0.1620	0.6802	0.2941	0.8730	0.2291	0.8543	0.2341
5	0.20	0.1154	1.7328	0.5157	0.3879	0.2596	0.7704	0.3462	0.5778	0.3376	0.5924
	1.00		2.6911		0.6789		1.2729		0.9773		1.0000

8

Table 8.4

Flash calculation.

i	z_i	K_i	$V/F = 0$	$V/F = 0.5$	$V/F = 0.1$	$V/F = 0.05$	$V/F = 0.0951$
			$\frac{z_i(K_i-1)}{\frac{V}{F}(K_i-1)+1}$	$\frac{z_i(K_i-1)}{\frac{V}{F}(K_i-1)+1}$	$\frac{z_i(K_i-1)}{\frac{V}{F}(K_i-1)+1}$	$\frac{z_i(K_i-1)}{\frac{V}{F}(K_i-1)+1}$	$\frac{z_i(K_i-1)}{\frac{V}{F}(K_i-1)+1}$
1	0.05	8.5241	0.3762	0.0790	0.2147	0.2734	0.2193
2	0.25	2.2450	0.3112	0.1918	0.2768	0.2930	0.2783
3	0.30	0.6256	−0.1123	−0.1382	−0.1167	−0.1145	−0.1165
4	0.20	0.1920	−0.1616	−0.2711	−0.1758	−0.1684	−0.1751
5	0.20	0.0617	−0.1877	−0.3535	−0.2071	−0.1969	−0.2060
	1.00		0.2258	−0.4920	−0.0081	0.0866	0.0000

8.3 Single-Stage Separation

When a mixture contains components with large relative volatilities, either a partial condensation from the vapor phase or a partial vaporization from the liquid phase followed by a simple phase split can often produce an effective separation (King, 1980).

Figure 8.2 shows a feed being separated into a vapor and liquid phase and being allowed to come to equilibrium. If the feed to the separator and the vapor and liquid products are continuous, then the material balance is described by Equations 8.10, 8.11 and 8.14 (King, 1980). If K_i is large relative to V/F (typically $K_i > 10$) in Equation 8.10, then (Douglas, 1988):

$$y_i \approx z_i/(V/F) \quad \text{that is,} \quad Vy_i \approx Fz_i \tag{8.26}$$

This means that all of Component i entering with the feed, Fz_i, leaves in the vapor phase as Vy_i. Thus, if a component is required to leave in the vapor phase, its K-value should be large relative to V/F.

On the other hand, if K_i is small relative to V/F (typically $K_i < 0.1$) in Equation 8.11, then (Douglas, 1988):

$$x_i \approx Fz_i/L \quad \text{that is,} \quad Lx_i \approx Fz_i \tag{8.27}$$

This means that all of Component i entering with the feed, Fz_i, leaves with the liquid phase as Lx_i. Thus, if a component is required to leave in the liquid phase, its K-value should be small relative to V/F.

Ideally, the K-value for the light key component in the phase separation should be greater than 10 and, at the same time, the K-value for the heavy key less than 0.1. Such circumstances can lead to a good separation in a single stage. However, use of phase separators might still be effective in the flow sheet if the K-values for the key components are not so extreme. Under such circumstances, a more crude separation must be accepted.

8.4 Distillation

A single equilibrium stage can only achieve a limited amount of separation. However, the process can be repeated by taking the vapor from the single-stage separation to another separation stage and partially condensing it and taking the liquid to another separation stage and partially vaporizing it, and so on. With each repeated condensation and vaporization, a greater degree of separation will be achieved. In practice, the separation to multiple stages is extended by creating a cascade of stages as shown in Figure 8.5. It is assumed in the cascade that liquid and vapor streams leaving each stage are in equilibrium, forming an *equilibrium stage*. Using a cascade of stages in this way allows the more-volatile components to be transferred to the vapor phase and the less-volatile components to be transferred to the liquid phase. In principle, by creating a large enough cascade, an almost complete separation can be carried out.

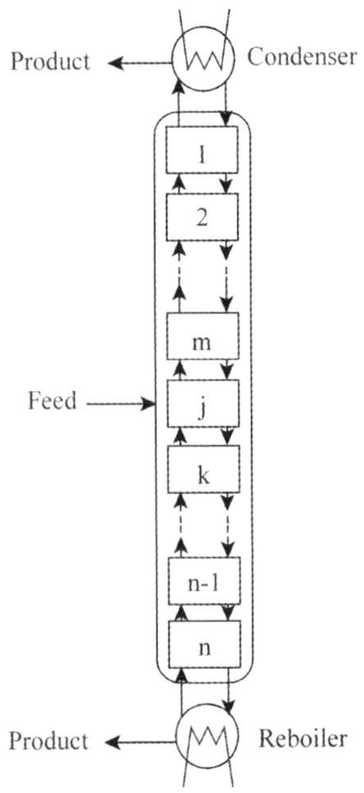

Figure 8.5

A cascade of equilibrium stages with refluxing and reboiling. (Reproduced from Smith R and Jobson M (2000) Distillation, Encyclopedia of Separation Science, Academic Press, with permission from Elsevier.)

The section of the column above the feed is known as the *rectifying section* and that below the feed the *stripping section*. At the top of the cascade in Figure 8.5, liquid is needed to feed the cascade in the rectifying section. This is produced by condensing vapor that leaves the top stage and returning this liquid to the first stage of the cascade as *reflux*. All of the vapor leaving the top stage can be condensed in a *total condenser* to produce a liquid top product. Alternatively, only enough of the vapor to provide the reflux can be condensed in a *partial condenser* to produce a vapor top product if a liquid top product is not desired. Vapor is also needed to feed the cascade at the bottom of the column in the stripping section. This is produced by vaporizing some of the liquid leaving the bottom stage and returning the vapor to the bottom stage of the cascade in a *reboiler*. The feed to the process is introduced at an intermediate stage and products are removed from the condenser and the reboiler.

The methods by which the vapor and liquid are contacted in each stage of the cascade in distillation fall into the two broad categories of tray or packed columns.

1) *Tray columns.* The first arrangement illustrated in Figure 8.6a shows a *plate*, or *tray*, column. Liquid reflux enters the first tray at the top of the column and flows across what is shown in Figure 8.6a as a perforated plate (sieve tray). Vapor bubbles through the liquid flowing across the tray and a froth might be formed. This contacting allows mass transfer to take place. Liquid is prevented from *weeping* through the holes in the tray by the up-flowing vapor. Weeping occurs when the vapor flow is too low to maintain the liquid level on the tray. The liquid from each tray flows over a *weir* and down a *downcomer* to the

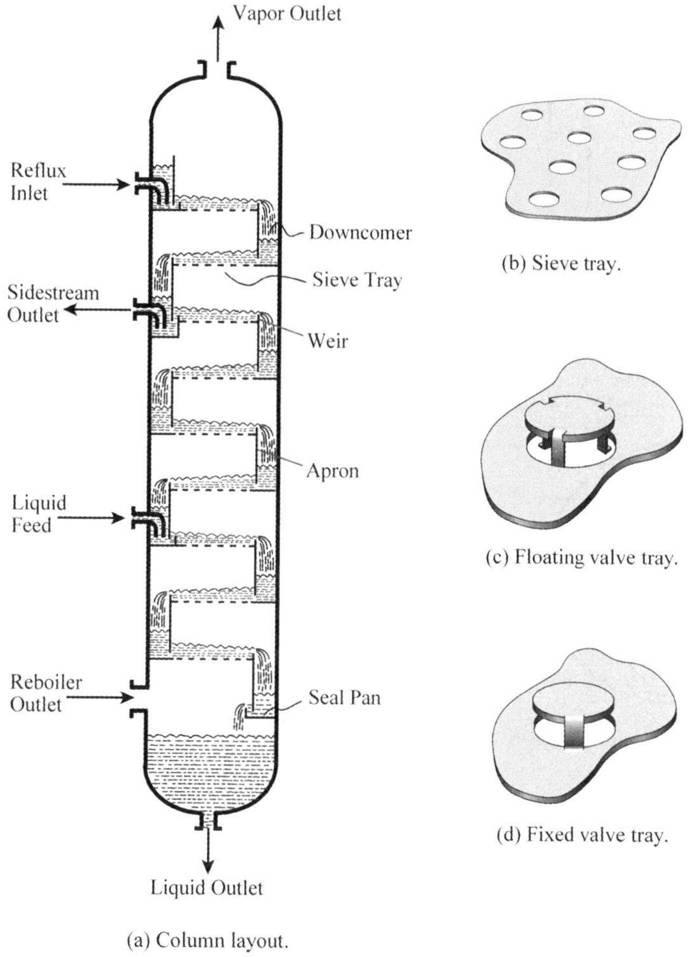

Figure 8.6
A distillation column fitted with trays.

next tray, and so on. Figure 8.7 illustrates further the downcomer design. Most downcomer designs follow the *segmental* or *chord* design, illustrated in Figures 8.6 and 8.7. This is a vertical flat plate that extends down from the outlet weir. Other designs are possible. The flow of vapor up the column leads to a pressure drop across each tray of typically 0.007 bar. The downcomer provides a liquid seal and must provide enough head of liquid to overcome the tray pressure drop across the tray. The three most commonly used types of tray (although there are many designs available) are:

a) *Sieve tray.* The simplest and most commonly used type of tray is the sieve tray, as illustrated in Figure 8.6b. This is a perforated plate with holes drilled or punched through the tray. Hole sizes vary typically between 3 mm and 25 mm, with larger holes used for fouling duties. Sieve trays are cheap, simple and well understood in terms of performance. One major disadvantage of sieve trays is their lack of flexibility. Flexibility is measured by the turndown ratio. For sieve trays this is of the order 2:1, which means that the flowrate can be decreased to ½ while maintaining the design separation.

b) *Floating valve.* Floating valve trays use small moveable metal flaps to adjust the size of the openings of the holes in the tray. The flap rises or lowers in the hole as the vapor rate increases or decreases. A typical design is illustrated in Figure 8.6c. In the arrangement in Figure 8.6c, upward movement is guided and restrained by three metal legs. Other designs mount the metal flap in a cage to guide and restrain movement. The principle advantage of floating valve trays is that they improve the flexibility of operation to be able to cope with a wider variety of liquid and vapor flowrates in the column. By using floating valve trays the turndown ratio can be increased from a typical value for sieve trays of 2:1 to 5:1 or greater for floating valve trays. However, for fouling services, fouling can affect the opening and closing of the valve.

c) *Fixed valve.* Fixed valve trays use metal flaps punched from the tray at a fixed distance from the tray. A typical design is illustrated in Figure 8.6d. The principle advantage of fixed valves compared with sieve trays is that the horizontal vapor velocity created by the valve promotes intense radial mixing on the tray. The fixed position of the valve implies an

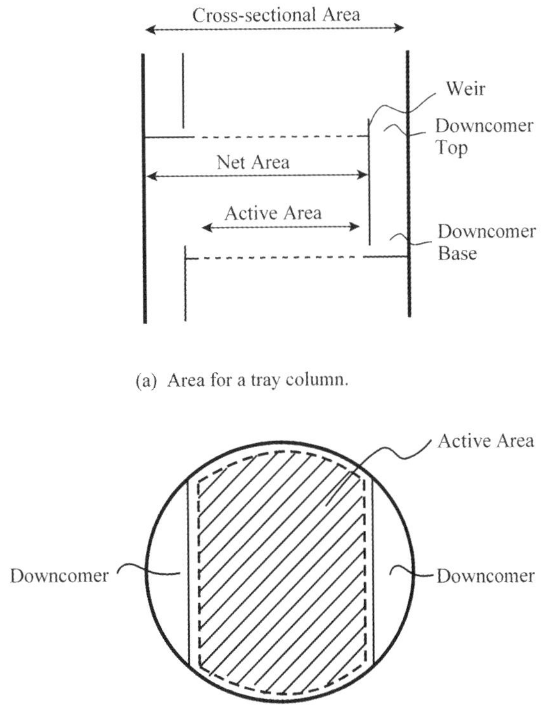

(a) Area for a tray column.

(b) Active area in the column is restricted by the downcomers.

Figure 8.7

Areas of distillation tray layout.

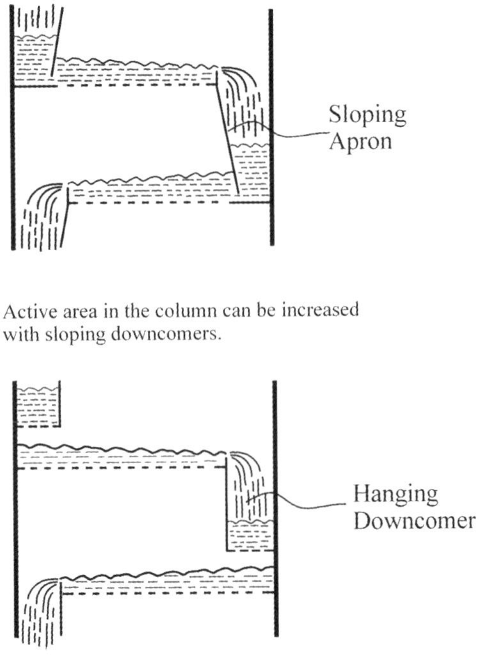

(a) Active area in the column can be increased with sloping downcomers.

(b) Hanging downcomers restrict the flow in the downcomer to allow an increase in active area.

Figure 8.8

Alternative downcomer designs.

operating flexibility similar to that of a sieve tray. However, turndown rates and efficiency are higher than sieve trays but lower than floating valve trays. This radial flow of the vapor also reduces fouling.

Many other designs of tray are available. Other features of tray column design relating to entry of reflux and feed are also illustrated in Figure 8.6a. Equipment for vapor entry at the base of the column varies and depends on whether the exit from the reboiler is vapor or a mixture of liquid and vapor.

One particular disadvantage of distillation trays is that the downcomer arrangement allowing liquid to flow down the column makes a significant proportion of the area within the column shell not available for contacting liquid and vapor. Figure 8.7a defines the areas within the column. The column cross-sectional area is the areas of the empty column without the trays and downcomers. The *net area* or *free area* is the cross-sectional area minus the area at the top of the downcomer (see Figure 8.7). This is the smallest area available for vapor flow between the trays. The *active area* or *bubbling area* is the cross-sectional area minus the downcomer top and base areas and any additional nonperforated areas. This represents the area available for vapor flow close to the tray. Each downcomer typically takes up 10 to 15% of the cross-sectional area. If the required area for the downcomers exceeds this range, or an existing column requires capacity to be increased, then different methods can be adopted to increase the active area. Figure 8.8a shows a *sloped* downcomer design with a sloped apron. This allows more area at the top of the downcomer for vapor and liquid disengagement, allowing a greater area on the tray below to be used as active area. The ratio of the top to bottom downcomer area is normally between 1.5 and 2.0. The sloped downcomer might be a flat plate, or in more elaborate designs can be a semi-conical shape to follow the contour of the column shell. Another possibility is illustrated in Figure 8.8b, which shows a *hanging downcomer* arrangement. In this case, the flow of liquid from the downcomer is restricted, allowing the area below the downcomer to be exploited as an active area. Downcomer arrangements such as hanging downcomers are generally reserved for existing columns when an increase in capacity is required.

Weir loading is an important parameter for tray design. It is calculated as the clear liquid volume divided by the length of the tray outlet weir. Weir loading has a direct influence on the froth height on the tray as higher weir loadings increase the fluid crest over the weir. Increased crest over the weir raises the liquid level on the tray deck, which increases the pressure drop across the tray. This creates higher downcomer backup and *downcomer backup flooding* can occur, preventing the tray functioning. Crests can also become large enough so as not to fit into downcomer opening, and *downcomer choke flooding* can occur, again preventing the tray functioning. If the weir loading is too high, leading to an excessively high weir crest, then the number of downcomers can be increased by introducing multiple passes. Figure 8.9 contrasts single-, two-, three- and

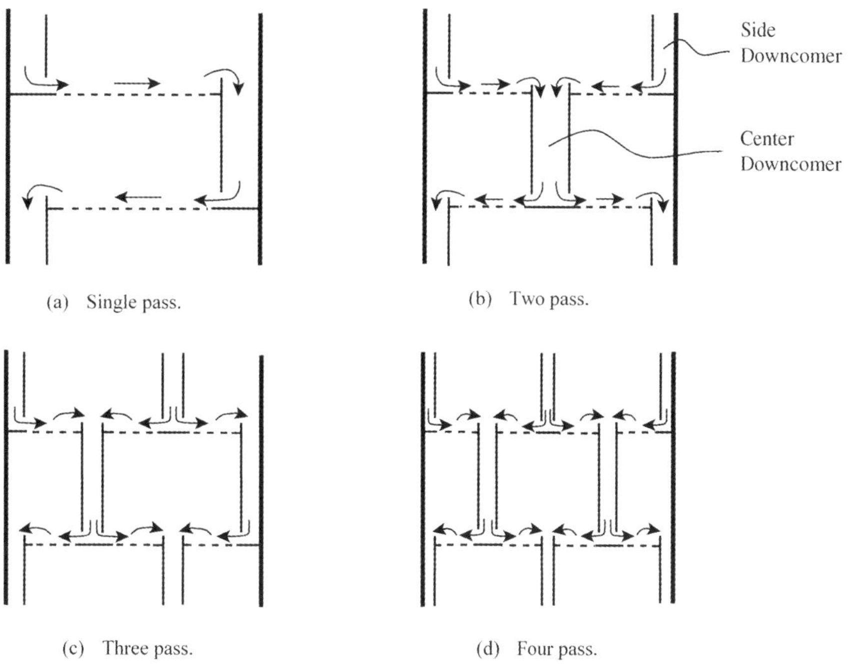

Figure 8.9
Multipass tray layouts.

four-pass designs. Increasing the number of passes brings additional complexities in the tray design to ensure the correct distribution of liquid on the trays (Pilling, 2005). Three-pass designs are rarely used because of the difficulties associated with the asymmetry of the design (Kister and Olsson, 2010). A shorter liquid flow path and possible maldistribution of liquid and vapor streams can result in lower tray efficiency. Rather than use vertical aprons as illustrated in Figure 8.9, sloped aprons can also be used with multipass designs.

If the liquid and vapor being contacted on a tray were able to come to equilibrium, this would constitute an equilibrium stage. In practice, the column will need more trays than the number of equilibrium stages, as mass transfer limitations and poor contacting efficiency prevent equilibrium being achieved on a tray.

2) *Packed columns*. The other broad class of contacting arrangement for the cascade in a distillation column is that of *packed columns*. Here the column is filled with a solid material that has a high surface area. Liquid trickles across the surfaces of the packing and vapor flows upward through the voids in the packing, contacting the liquid on its way up the column. Many different designs of packing are available. Figure 8.10a illustrates a column arrangement that uses *structured packing*. Structured packing is most often manufactured from corrugated sheets with the corrugations inclined to the horizontal in alternating layers, as illustrated in Figure 8.10b. Different angles of inclination are used by different designs. These preformed sheets of metal are joined together to produce preformed cylindrical packing slabs with inclined flow channels and built up as alternating slabs within the distillation column. The intersection of the slabs creates mixing points. Designs can feature additional crimping of the metal sheets and holes in the sheets. The structured packing slabs are often fitted with *wall wipers* to redirect liquid away from the column wall to prevent bypassing of the liquid along the wall. Fabrication materials can be metal alloys, plastics or ceramic material. A variation on structured packaging is *grid* packing. This uses a performed open-lattice structure manufactured in slabs. They have a high open area, leading to high capacity, low pressure drop and a higher tolerance to fouling than structured packing.

An alternative design of packing is *random* or *dumped* packing. The random packing is pieces of preformed metal, plastic or ceramic, which when dumped in the column produce a body with a high surface area. Figure 8.10c shows the simplest form of random packing, a Raschig Ring. Figure 8.10d shows a metal Pall Ring®, Figure 8.10e a metal Intalox Saddle® and Figure 8.10f a Berl Saddle. Many designs of random packing are available. For random packing, the pieces of packing are dumped in beds, as in Figure 8.10a instead of the structured packing. Although structured packings have a high surface area per unit volume, the liquid loading for structured packings is only slightly higher than for random packings. As with structured packing, liquid distribution, liquid collection and vapor distribution are still required. Structured packing generally has a lower pressure drop, higher efficiency and higher capacity for a given column diameter compared with a random packing. Packed designs in general have a lower pressure drop than tray designs for the same separation. Structured packing normally has a pressure drop less than 0.003 $bar \cdot m^{-1}$.

The packed column requires column internals for the distribution of vapor and liquid and the collection of liquid. A liquid distribution device is required for the reflux. A liquid collector

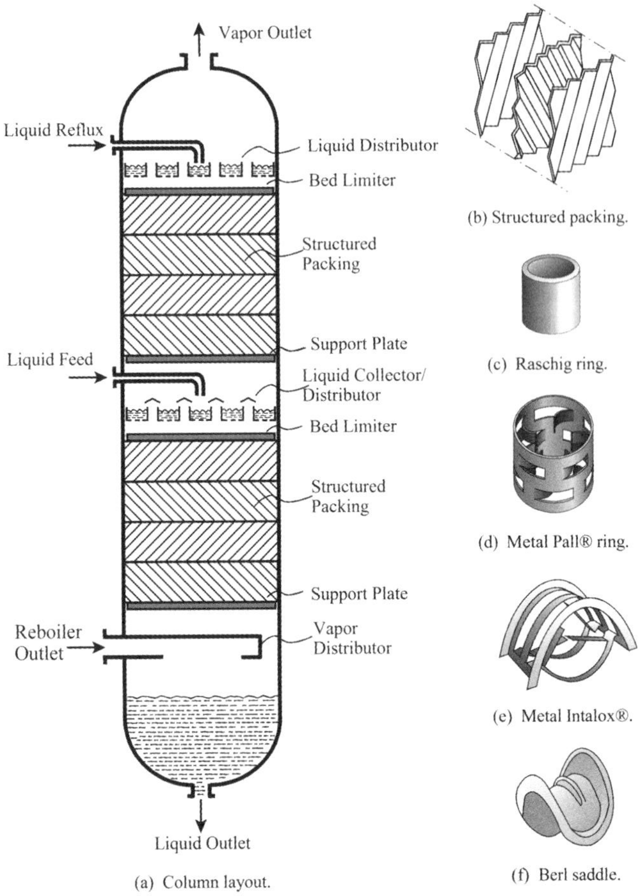

Figure 8.10
A packed distillation column.

is required at the base of the rectifying section. The liquid cascading down from the rectifying section then needs to be redistributed for the stripping section, along with the liquid feed. Vapor distribution devices are required at the base of the rectifying and stripping sections. Many different designs of liquid distribution, liquid collection and vapor distribution are available. For large packing heights, the liquid flowing down the column must be collected and redistributed to ensure efficient contacting of vapor and liquid throughout the bed height. Whereas the change in composition in a plate column is finite from stage to stage, the change in composition in a packed column is continuous.

Before detailed design can be carried out a number of important decisions must be made about the distillation column. The thermal condition of the feed, the number of equilibrium stages, feed location, operating pressure, amount of reflux, and so on, all must be chosen before a simulation can be carried out.

To explore these decisions in a systematic way, shortcut design methods can be used. These exploit simplifying assumptions to allow many more design options to be explored than would be possible with detailed simulation and allow conceptual insights to be gained. Once the major decisions have been made, a detailed simulation needs to be carried out.

Conceptual insights into the design of distillation are best developed by first considering distillation of binary mixtures.

8.5 Binary Distillation

Consider the material balance for a simple binary distillation column. A simple column has one feed, two products, one reboiler and one condenser. Such a column is shown in Figure 8.11. A method of graphical analysis of binary distillation known as the *McCabe–Thiele Method* (McCabe and Thiele, 1925) will now be developed. Although it involves too many simplifications to be used for a final design in distillation, it is a key conceptual tool in understanding distillation.

An overall material balance can be written as:

$$F = D + B \tag{8.28}$$

A material balance can also be written for Component i as:

$$Fx_{i,F} = Dx_{i,D} + Bx_{i,B} \tag{8.29}$$

However, to fully understand the design of the column, the material balance must be followed through the column. To simplify the analysis, it can be assumed that the molar vapor and liquid flowrates are constant in each column section, which is termed *constant molar overflow*. This is strictly true only if the component molar latent heats of vaporization are the same, there is no heat of mixing between the components, heat capacities are constant and there is no external addition or removal of heat. In fact, this turns out to be a reasonable assumption for many mixtures of organic compounds that exhibit reasonably ideal behavior. However, it can also turn out to be a poor assumption for many mixtures, such as many mixtures of alcohol and water.

Start by considering the material balance for the part of the column above the feed in the rectifying section. Figure 8.12 shows the rectifying section of a column and the flows and compositions

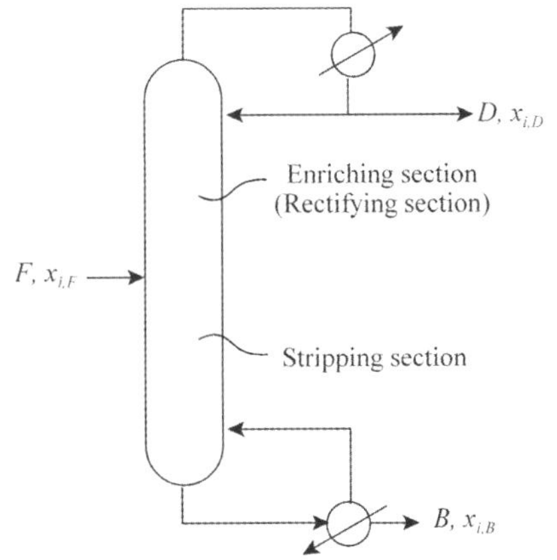

Figure 8.11
Mass balance on a simple disillation column.

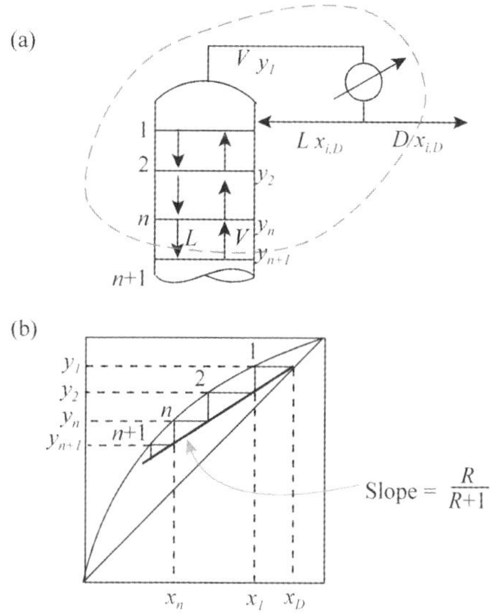

Figure 8.12

Mass balance for the rectifying section. (Reproduced from Smith R and Jobson M (2000) Distillation, Encyclopedia of Separation Science, Academic Press, with permission from Elsevier.)

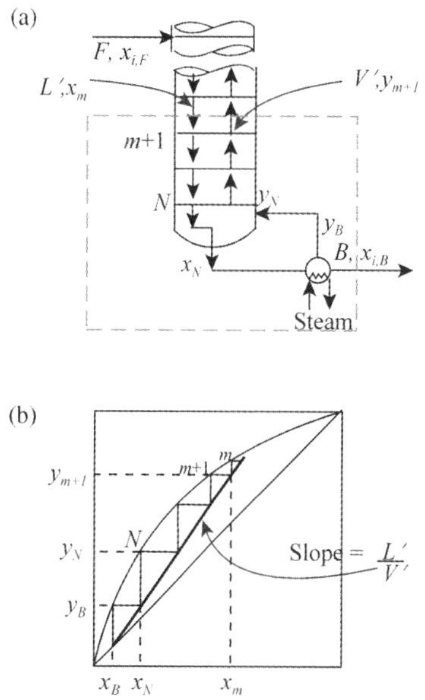

Figure 8.13

Mass balance for the stripping section. (Reproduced from Smith R and Jobson M (2000) Distillation, Encyclopedia of Separation Science, Academic Press, with permission from Elsevier.)

of the liquid and vapor in the rectifying section. First, an overall balance is written for the rectifying section (assuming L and V are constant, i.e. constant molar overflow):

$$V = L + D \tag{8.30}$$

A material balance can also be written for Component i:

$$Vy_{i,n+1} = Lx_{i,n} + Dx_{i,D} \tag{8.31}$$

The reflux ratio, R, is defined to be:

$$R = L/D \tag{8.32}$$

Given the reflux ratio, the vapor flow can be expressed in terms of R:

$$V = (R + 1)D \tag{8.33}$$

These expressions can be combined to give an equation that relates the vapor entering and liquid flows leaving Stage n:

$$y_{i,n+1} = \frac{R}{R+1}x_{i,n} + \frac{1}{R+1}x_{i,D} \tag{8.34}$$

On an x–y diagram for Component i this is a straight line starting at the distillate composition with slope $R/(R + 1)$ and which intersects the diagonal line at $x_{i,D}$.

Starting at the distillate composition $x_{i,D}$ in Figure 8.12b, a horizontal line across to the equilibrium line gives the composition of the vapor in equilibrium with the distillate (y_1). A vertical step down gives the liquid composition leaving Stage 1, x_1. Another horizontal line across to the equilibrium line gives the composition of the vapor leaving Stage 2 (y_2). A vertical line to the operating line gives the composition of the liquid leaving Stage 2 (x_2), and so on. Thus, stepping between the operating line and equilibrium line in Figure 8.12b, the change in vapor and liquid composition is followed through the rectifying section of the column.

Now consider the corresponding mass balance for the column below the feed in the stripping section. Figure 8.13a shows the vapor and liquid flows and compositions through the stripping section of a column. An overall mass balance for the stripping section around Stage $m + 1$ gives:

$$L' = V' + B \tag{8.35}$$

A component balance gives:

$$L'x_{i,m} = V'y_{i,m+1} + Bx_{i,B} \tag{8.36}$$

Again, assuming constant molar overflow (L' and V' are constant), these expressions can be combined to give an equation relating the liquid entering and the vapor leaving Stage $m + 1$:

$$y_{i,m+1} = \frac{L'}{V'}x_{i,m} - \frac{B}{V'}x_{i,B} \tag{8.37}$$

This line can be plotted in an x–y plot as shown in Figure 8.13b. It is a straight line with slope L'/V', which intersects the diagonal line at $x_{i,B}$. Starting from the bottom composition, $x_{i,B}$, a vertical line up to the equilibrium line gives the composition of the vapor leaving the reboiler (y_B). A horizontal line across to the operating line gives the composition of the liquid leaving Stage N (x_N). A vertical line up to the equilibrium line then gives the vapor leaving Stage N (y_N), and so on.

Now the rectifying and stripping sections can be brought together at the feed stage. Consider the point of intersection of the operating lines for the rectifying and stripping sections. From Equations 8.31 and 8.36:

$$Vy_i = Lx_i + Dx_{i,D} \tag{8.38}$$

$$V'y_i = L'x_i - Bx_{i,B} \tag{8.39}$$

where y_i and x_i are the intersection of the operating lines. Subtracting Equations 8.38 and 8.39 gives:

$$(V - V')y_i = (L - L')x_i + Dx_{i,D} + Bx_{i,B} \tag{8.40}$$

Substituting the overall mass balance, Equation 8.28, gives:

$$(V - V')y_i = (L - L')x_i + Fx_{i,F} \tag{8.41}$$

Now it is necessary to determine how the vapor and liquid flowrates change at the feed stage.

What happens at the feed stage depends on the condition of the feed, whether it is subcooled liquid, saturated liquid, partially vaporized, saturated vapor or superheated vapor. To define the condition of the feed, the variable q is introduced, defined as:

$$q = \frac{\text{Heat required to vaporize 1 mole of feed}}{\text{Molar latent heat of vaporization of feed}} \tag{8.42}$$

For a saturated liquid feed $q = 1$ and for a saturated vapor feed $q = 0$. If the feed is subcooled liquid, q is greater than 1. If the feed is superheated vapor, q is less than 0. The flowrate of feed entering the column as liquid is qF. The flowrate of feed entering the column as vapor is $(1 - q)F$. Figure 8.14 shows the feed stage. An overall mass balance on the feed stage for the vapor gives:

$$V = V' + (1 - q)F \tag{8.43}$$

An overall mass balance for the liquid on the feed stage gives:

$$L' = L + qF \tag{8.44}$$

Combining Equations 8.41, 8.43 and 8.44 together gives a relationship between the compositions of the feed and the vapor and liquid leaving the feed tray:

$$y_i = \frac{q}{q-1}x_i - \frac{1}{q-1}x_{i,F} \tag{8.45}$$

This equation is known as the *q-line*. On the x–y plot, it is a straight line with slope $q/(q - 1)$ and intersects the diagonal line at $x_{i,F}$. It is plotted in Figure 8.14 for various values of q.

The material balances for the rectifying and stripping sections can now be brought together. Figure 8.15a shows the complete design. The construction is started by plotting the operating lines for the rectifying and stripping sections. The q-line intercepts the operating lines at their intersection. The construction steps off between the operating lines and the equilibrium lines. The intersection of the operating lines is the correct feed stage, that is, the feed stage necessary to minimize the overall number of theoretical stages. The stepping procedure changes from one operating line to the other at the intersection with the q-line. The construction can be started either from the overhead composition working down or from the bottom composition working up. This method of design is known as the *McCabe–Thiele Method* (McCabe and Thiele, 1925).

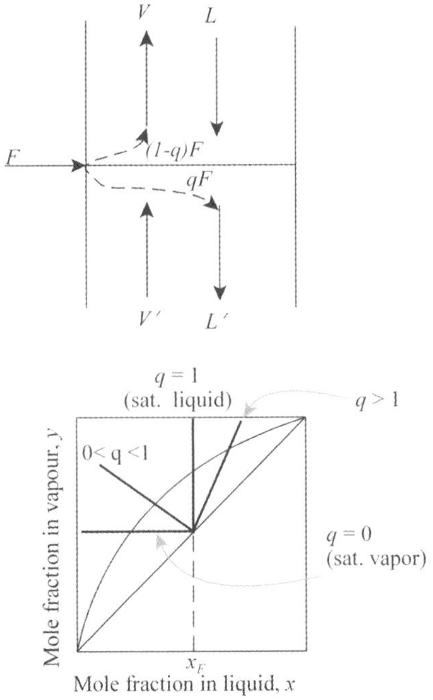

Figure 8.14

Mass balance for the feed stage. (Reproduced from Smith R and Jobson M (2000) Distillation, Encyclopedia of Separation Science, Academic Press, with permission from Elsevier.)

Figure 8.15b shows an alternative stepping procedure in which the change from the rectifying to the stripping operating lines leads to a feed stage location below the optimum feed stage. The result is an increase in the number of theoretical stages for the same separation. Figure 8.15c shows a stepping procedure in which the feed stage location is above the optimum, again resulting in an increase in the number of theoretical stages.

Figure 8.16 contrasts a total condenser (Figure 8.16a) and partial condenser (Figure 8.16b) in the McCabe–Thiele Diagram. Although a partial condenser can provide a theoretical stage in principle, in practice the performance of a condenser will not achieve the performance of a theoretical stage.

Figure 8.17 contrasts the two general classes of reboiler. Figure 8.17a is fed with a liquid. This is partially vaporized and discharges liquid and vapor streams that are in equilibrium. This is termed a *partial reboiler* or *once-through reboiler* or *kettle reboiler* and, as shown in the McCabe–Thiele diagram in Figure 8.17a, acts as a theoretical stage. In the other class of

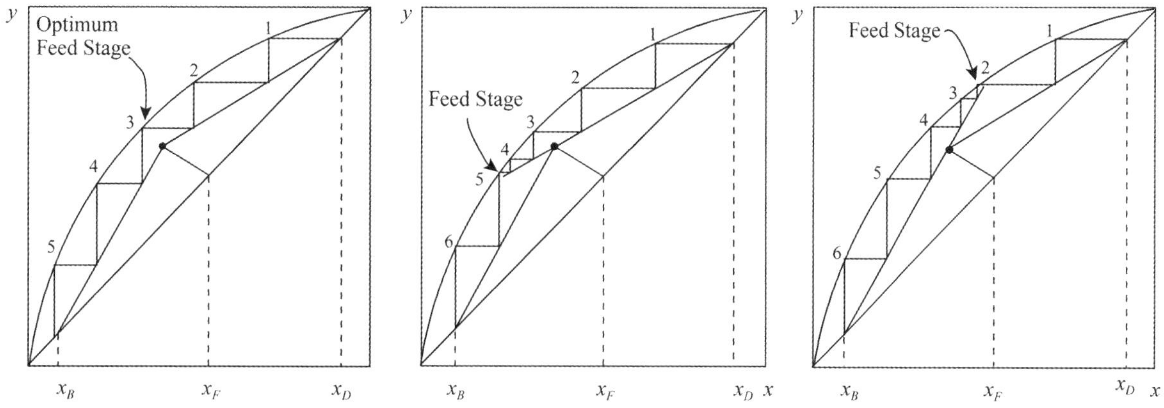

(a) The optimum feed location.

(b) Feed location below the optimum stage.

(c) Feed location above the optimum stage.

Figure 8.15

Combining the rectifying and stripping sections.

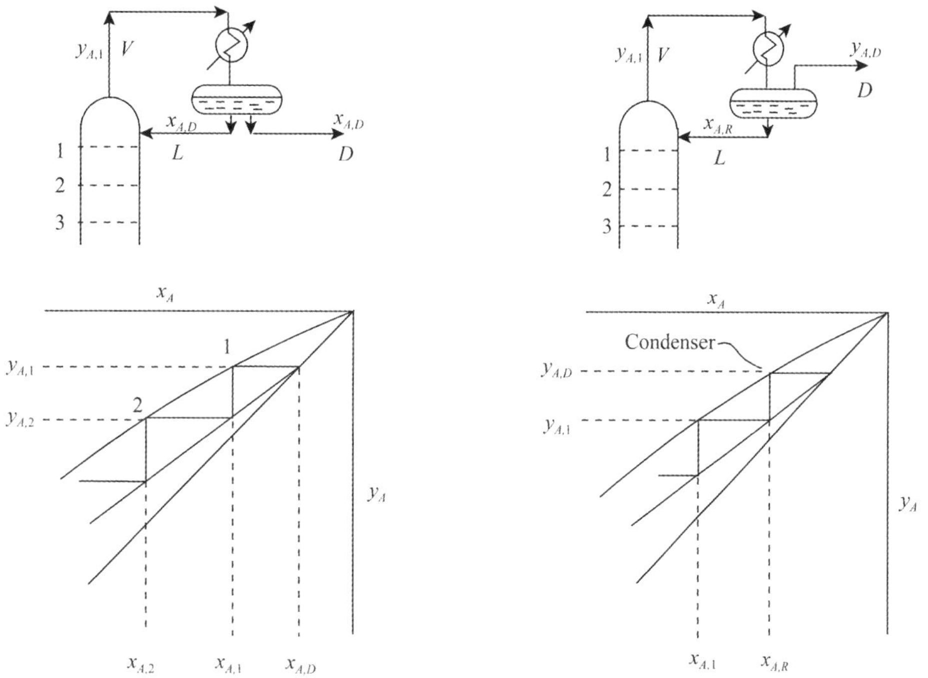

(a) Total condenser.

(b) Partial condenser.

Figure 8.16

Total and partial condensers in the McCabe–Thiele Diagram.

reboilers shown in Figure 8.17b, known as *thermosyphon reboilers*, part of the liquid flow from the bottom of the column is taken and partially vaporized. Whilst some separation obviously occurs in the thermosyphon reboiler, this will be less than that in a theoretical stage. It is safest to assume that the necessary stages are provided in the column itself, as illustrated in the McCabe–Thiele Diagram in Figure 8.17b.

There are two important limits that need to be considered for distillation. The first is illustrated in Figure 8.18a. This is total reflux, in which no products are taken and there is no feed. At total reflux, the entire overhead vapor is refluxed and all the bottoms liquid reboiled. Figure 8.18a also shows total reflux on an x–y plot. This corresponds with the minimum number of stages required for the separation. The other limiting case,

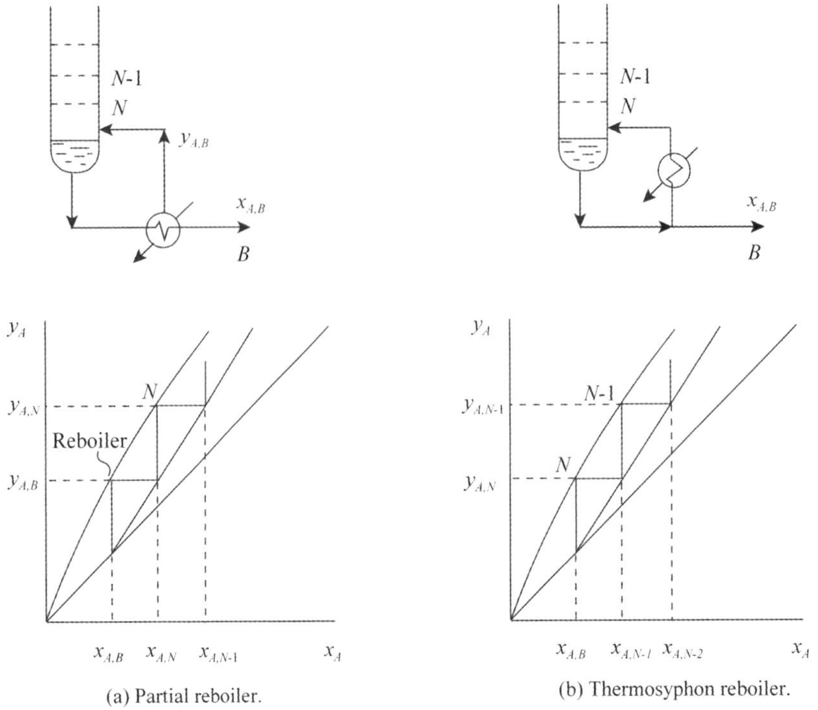

Figure 8.17
Partial and thermosyphon reboilers.

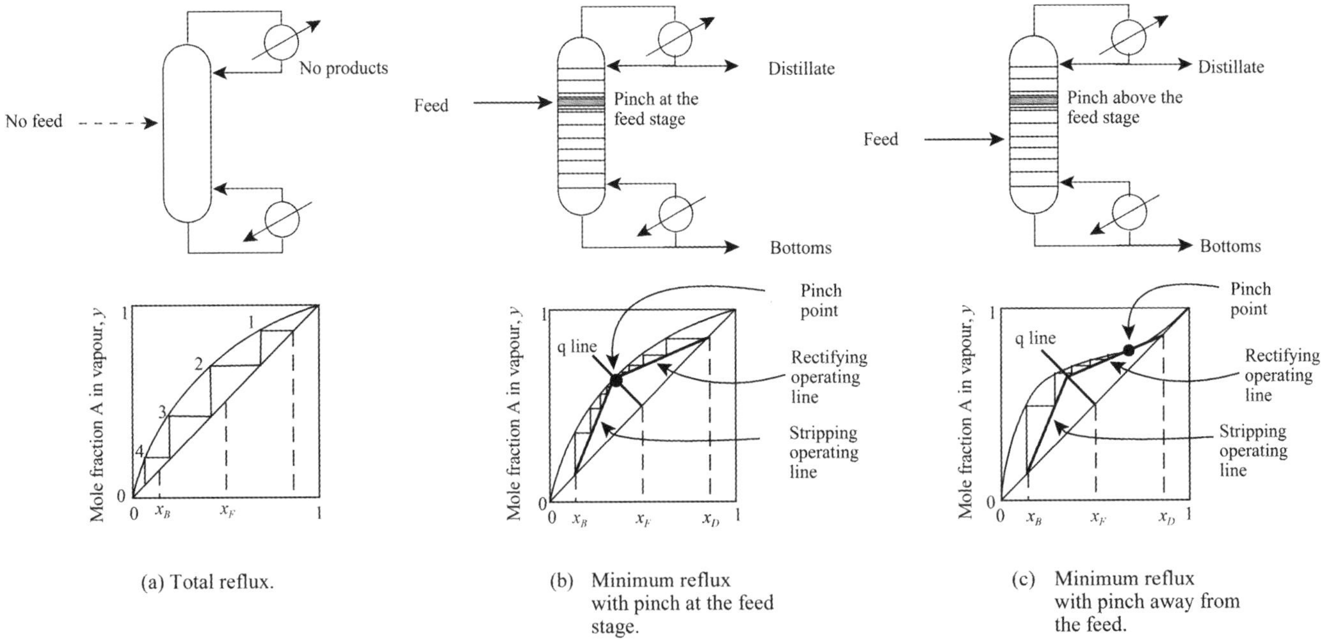

Figure 8.18
Total and maximum reflux in binary distillation. (Adapted from Smith R and Jobson M (2000) Distillation, Encyclopedia of Separation Science, Academic Press, with permission from Elsevier.)

shown in Figure 8.18b, is where the reflux ratio is chosen such that the operating lines intersect at the equilibrium line. As this stepping procedure approaches the q-line from both sides, an infinite number of steps are required to approach the q-line. This is the minimum reflux condition in which there are zones of constant composition (*pinches*) above and below the feed. For binary distillation, the zones of constant composition are usually located at the feed stage, as illustrated in Figure 8.18b. The pinch can also be away from the feed, as illustrated in Figure 8.18c. This time, a *tangent pinch* occurs above the feed stage. A tangent pinch can also be obtained below the feed stage, depending on the shape of the x–y diagram.

The McCabe–Thiele Method is restricted in its application because it only applies to binary systems and involves the simplifying assumption of constant molar overflow. However, it is an important method to understand as it gives important conceptual insights into distillation that cannot be obtained in any other way.

8.6 Total and Minimum Reflux Conditions for Multicomponent Mixtures

Just as with binary distillation, it is important to understand the operating limits for multicomponent distillation. The two extreme conditions of total and minimum reflux will therefore be considered. However, before a multicomponent distillation column can be designed, a decision must be made as to the two *key components* between which it is desired to make the separation. The *light key component* is the one to be kept out of the bottom product according to some specifications. The *heavy key component* is the one to be kept out of the top product according to some specifications. The separation of the nonkey components cannot be specified. Lighter than light key components will tend to go predominantly with the overhead product and heavier than heavy key components will tend to go predominantly with the bottoms product. Intermediate-boiling components between the keys will distribute between the products. In preliminary design, the recovery of the light key or the concentration of the light key in the overhead product must be specified, as must the recovery of the heavy key or the concentration of the heavy key in the bottom product.

Consider first total reflux conditions, corresponding with the minimum number of theoretical stages. The bottom of a distillation column at total reflux is illustrated in Figure 8.19:

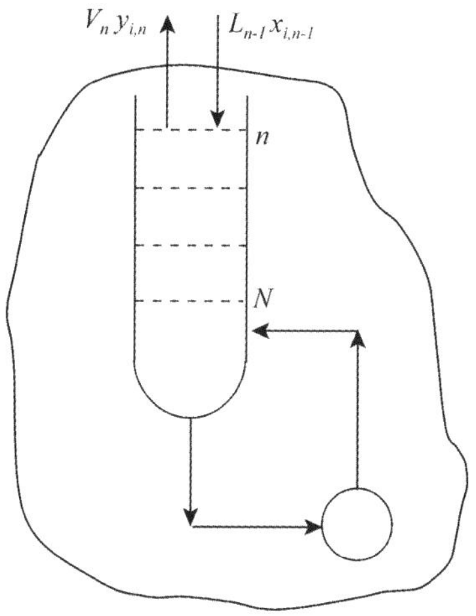

Figure 8.19

Stripping section of a column under total reflux conditions.

$$V_n = L_{n-1} \tag{8.46}$$

A component mass balance gives:

$$V_n y_{i,n} = L_{n-1} x_{i,n-1} \tag{8.47}$$

Combining Equations 8.46 and 8.47 gives:

$$y_{i,n} = x_{i,n-1} \tag{8.48}$$

Consider a binary separation. The relative volatility is defined by Equation 8.19. Applying this equation to the bottom of the column, and because reflux conditions are total, the liquid composition in the reboiler is the same as the bottom composition in this case:

$$\left(\frac{y_A}{y_B}\right)_R = \alpha_R \left(\frac{x_A}{x_B}\right)_R = \alpha_R \left(\frac{x_A}{x_B}\right)_B \tag{8.49}$$

where subscript R refers to the reboiler and subscript B to the bottoms. Combining Equation 8.48 with Equation 8.49 gives:

$$\left(\frac{x_A}{x_B}\right)_N = \alpha_R \left(\frac{x_A}{x_B}\right)_B \tag{8.50}$$

Similarly, for Stage N:

$$\left(\frac{y_A}{y_B}\right)_N = \left(\frac{x_A}{x_B}\right)_{N-1} = \alpha_N \left(\frac{x_A}{x_B}\right)_N \tag{8.51}$$

Combining this with Equation 8.50 gives:

$$\left(\frac{x_A}{x_B}\right)_{N-1} = \alpha_N \alpha_R \left(\frac{x_A}{x_B}\right)_B \tag{8.52}$$

Continuing up the column to the top stage:

$$\left(\frac{x_A}{x_B}\right)_1 = \alpha_2 \cdots \alpha_N \alpha_R \left(\frac{x_A}{x_B}\right)_B \tag{8.53}$$

Combining with Equation 8.19 from the definition of α gives:

$$\left(\frac{y_A}{y_B}\right)_1 = \alpha_1 \alpha_2 \cdots \alpha_N \alpha_R \left(\frac{x_A}{x_B}\right)_B \tag{8.54}$$

Assuming a total condenser and the relative volatility to be constant gives:

$$\left(\frac{x_A}{x_B}\right)_D = \alpha_{AB}^{N_{min}} \left(\frac{x_A}{x_B}\right)_B \tag{8.55}$$

where N_{min} = minimum number of theoretical stages (including the reboiler as an equilibrium stage)
α_{AB} = relative volatility between A and B

Subscript D refers to the distillate. Equation 8.55 predicts the number of theoretical stages for a specified binary separation at total reflux and is known as the *Fenske Equation* (Fenske, 1932).

In fact, the development of the equation does not include any assumptions limiting the number of components in the system. It can also be written for a multicomponent system for any two reference Components i and j (King, 1980; Treybal, 1980; Geankopolis, 1993; Seader, Henley and Roper, 2011):

$$\frac{x_{i,D}}{x_{j,D}} = \alpha_{ij}^{N_{min}} \frac{x_{i,B}}{x_{j,B}} \tag{8.56}$$

where $x_{i,D}$ = mole fraction of Component i in the distillate
$x_{i,B}$ = mole fraction of Component i in the bottoms
$x_{j,D}$ = mole fraction of Component j in the distillate
$x_{j,B}$ = mole fraction of Component j in the bottoms
α_{ij} = relative volatility between Components i and j
N_{min} = minimum number of stages

Equation 8.56 can also be written in terms of the molar flowrates of the products:

$$\frac{d_i}{d_j} = \alpha_{ij}^{N_{min}} \frac{b_i}{b_j} \tag{8.57}$$

where d_i = molar distillate flow of Component i
b_i = molar bottoms flow of Component i

Alternatively, Equation 8.56 can be written in terms of recoveries of the products:

$$\frac{r_{i,D}}{r_{j,D}} = \alpha_{ij}^{N_{min}} \frac{r_{i,B}}{r_{j,B}} \tag{8.58}$$

where $r_{i,D}$ = recovery of Component i in the distillate
$r_{i,B}$ = recovery of Component i in the bottoms
$r_{j,D}$ = recovery of Component j in the distillate
$r_{j,B}$ = recovery of Component j in the bottoms

When Component i is the light key component L, and j is the heavy key component H, this becomes (King, 1980; Treybal, 1980; Geankopolis, 1993; Seader, Henley and Roper, 2011):

$$N_{min} = \frac{\log\left[\dfrac{d_L}{d_H}\dfrac{b_H}{b_L}\right]}{\log \alpha_{LH}} \tag{8.59}$$

$$N_{min} = \frac{\log\left[\dfrac{x_{L,D}}{x_{H,D}}\dfrac{x_{H,B}}{x_{L,B}}\right]}{\log \alpha_{LH}} \tag{8.60}$$

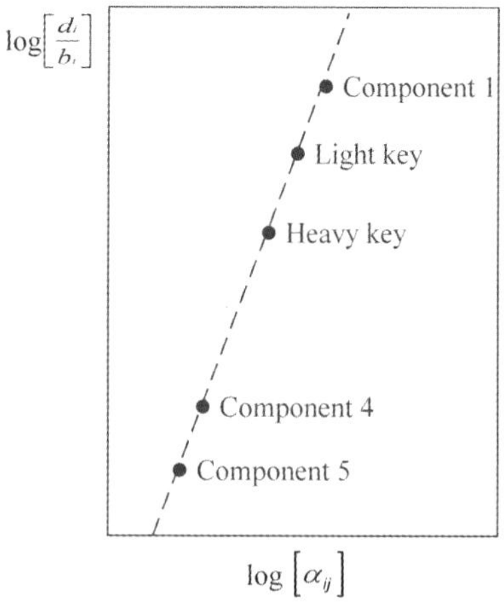

Figure 8.20

A plot of d_i/b_i versus α_{ij} tends to follow a straight line when plotted on logorithmic axes. (Reproduced from Smith R and Jobson M (2000) Distillation, Encyclopedia of Separation Science, Academic Press, with permission from Elsevier.)

$$N_{min} = \frac{\log\left[\dfrac{r_{L,D}}{1-r_{L,D}}\dfrac{r_{H,B}}{1-r_{H,B}}\right]}{\log \alpha_{LH}} \tag{8.61}$$

The Fenske Equation can be used to estimate the composition of the products. Equation 8.57 can be written in the form:

$$\log\left[\frac{d_i}{b_i}\right] = N_{min} \log \alpha_{ij} + \log\left[\frac{d_j}{b_j}\right] \tag{8.62}$$

Equation 8.62 indicates that a plot of log $[d_i/b_i]$ will be a linear function of log α_{ij} with a gradient of N_{min}. This can be rewritten as:

$$\log\left[\frac{d_i}{b_i}\right] = A \log \alpha_{ij} + B \tag{8.63}$$

where A and B are constants. The parameters A and B are obtained by applying the relationship to the light and heavy key components. This allows the compositions of the nonkey components to be estimated, which is illustrated in Figure 8.20. Having specified the distribution of the light and heavy key components, knowing the relative volatilities of the other components allows their compositions to be estimated. The method is based on total reflux conditions. It assumes that the component distributions do not depend on the reflux ratio.

Equation 8.57 can be written in a more convenient form than Equations 8.62 and 8.63. By combining an overall component balance:

$$f_i = d_i + b_i \tag{8.64}$$

with Equation 8.57, the resulting equation is (Seader, Henley and Roper, 2011):

$$d_i = \frac{\alpha_{ij}^{N_{min}} f_i \left(\dfrac{d_j}{b_j}\right)}{1 + \alpha_{ij}^{N_{min}}\left(\dfrac{d_j}{b_j}\right)} \tag{8.65}$$

and

$$b_i = \frac{f_i}{1 + \alpha_{ij}^{N_{min}}\left(\frac{d_j}{b_j}\right)} \tag{8.66}$$

The Fenske Equation assumes that relative volatility is constant. The relative volatility can be calculated from the feed composition, but this might not be characteristic of the overall column (King, 1980). The relative volatility will in practice vary through the distillation column, and an average needs to be taken. The average would best be taken at the average temperature of the column. To estimate the average, first assume that $\ln P_i^{SAT}$ varies linearly as $1/T$ (see Appendix A). If it is assumed that the two key components both vary linearly in this way, then the difference $(\ln P_i^{SAT} - \ln P_j^{SAT})$ also varies as $1/T$, which in turn means $\ln(P_i^{SAT}/P_j^{SAT})$ varies as $1/T$. If it is now assumed that Raoult's Law applies (i.e. ideal vapor–liquid equilibrium behavior, see Appendix A), then the relative volatility α_{ij} is the ratio of the saturated vapor pressures P_i^{SAT}/P_j^{SAT}. Therefore, $\ln \alpha_{ij}$ varies as $1/T$. The mean temperature in the column is given by:

$$T_{mean} = \frac{1}{2}(T_{top} + T_{bottom}) \tag{8.67}$$

Assuming $\ln(\alpha_{ij})$ is proportional to $1/T$:

$$\frac{1}{\ln(\alpha_{ij})_{mean}} = \frac{1}{2}\left(\frac{1}{\ln(\alpha_{ij})_{top}} + \frac{1}{\ln(\alpha_{ij})_{bottom}}\right) \tag{8.68}$$

which on rearranging gives:

$$(\alpha_{ij})_{mean} = \exp\left[\frac{2\ln(\alpha_{ij})_{top}\ln(\alpha_{ij})_{bottom}}{\ln(\alpha_{ij})_{top} + \ln(\alpha_{ij})_{bottom}}\right] \tag{8.69}$$

If, instead of assuming α_{ij} is proportional to $1/T$, it is assumed that α_{ij} is proportional to T, at the mean temperature:

$$\ln(\alpha_{ij})_{mean} = \frac{1}{2}\left[\ln(\alpha_{ij})_{top} + \ln(\alpha_{ij})_{bottom}\right] \tag{8.70}$$

which can be rearranged to give what is the geometric mean:

$$(\alpha_{ij})_{mean} = \sqrt{(\alpha_{ij})_{top}(\alpha_{ij})_{bottom}} \tag{8.71}$$

Either Equation 8.69 or 8.71 can be used to obtain a more realistic relative volatility for the Fenske Equation. Generally, Equation 8.69 gives a better prediction than Equation 8.71. However, the greater the variation of relative volatility across the column, the greater the error in using any averaging equation. By making some assumption of the product compositions, the relative volatility at the top and bottom of the column can be calculated and a mean taken. Iteration can be carried out for greater accuracy. Start by calculating the relative volatility from the feed conditions. Then the product compositions can be estimated (say from the Fenske Equation). The average relative volatility can then be estimated from the top and bottom products. However, this new relative volatility will lead to revised estimates of the product compositions, and hence the calculation should iterate to converge. How good the approximations turn out to be largely depends on the ideality (or nonideality) of the mixture being distilled. Generally, the more nonideal the mixture, the greater the errors.

In addition to the changes in relative volatility resulting from changes in composition and temperature through the column, there is also a change in pressure through the column due to the pressure drop. The change in pressure affects the relative volatility. In preliminary process design, the effect of the pressure drop through the column is often neglected.

Total reflux is one extreme condition for distillation and can be approximated by the Fenske Equation. The other extreme condition for distillation is minimum reflux. At minimum reflux, there must be at least one zone of constant composition across a number of stages for all components. If this is not the case, then additional stages would change the separation. In binary distillation, the zones of constant composition are usually adjacent to the feed (Figure 8.21a). If there are no heavy nonkey components in a multicomponent distillation, the zone of constant composition above the feed will still be adjacent to the feed. Similarly, if there are no light nonkey components, the zone of constant composition below the feed will still be adjacent to the feed (Figure 8.21b). If one or more of the heavy nonkey components do not appear in the distillate, the zone of constant composition above the feed will move to a higher position in the column, so that the nondistributing heavy nonkey can diminish to zero mole fraction in the stages immediately above the feed (Figure 8.21c). If one or more of the light nonkey components do not appear in the bottoms, the zone of constant composition below the feed will move to a lower position in the column so that the nondistributing light key components can diminish to zero mole fraction in the stages immediately below the feed (Figure 8.21d). Generally, there will be a zone of constant composition (pinch) above the feed in the rectifying section and one in the stripping section below the feed (Figure 8.21e).

The Underwood Equations can be used to predict the minimum reflux for multicomponent distillation (Underwood, 1946). The derivation of the equations is lengthy, and the reader is referred to other sources for details of the derivation (Underwood, 1946; King, 1980; Geankopolis, 1993). The equations assume that the relative volatility and molar overflow are constant between the zones of constant composition. There are two equations. The first is given by (Underwood, 1946):

$$\sum_{i=1}^{NC} \frac{\alpha_{ij} x_{i,F}}{\alpha_{ij} - \theta} = 1 - q \tag{8.72}$$

where
α_{ij} = relative volatility
$x_{i,F}$ = mole fraction of Component i in the feed
θ = root of the equation
q = feed condition
= $\dfrac{\text{heat required to vaporize one mole of feed}}{\text{molar latent heat of vaporization of feed}}$
= 1 for a saturated liquid feed, 0 for a saturated vapor feed
NC = number of components

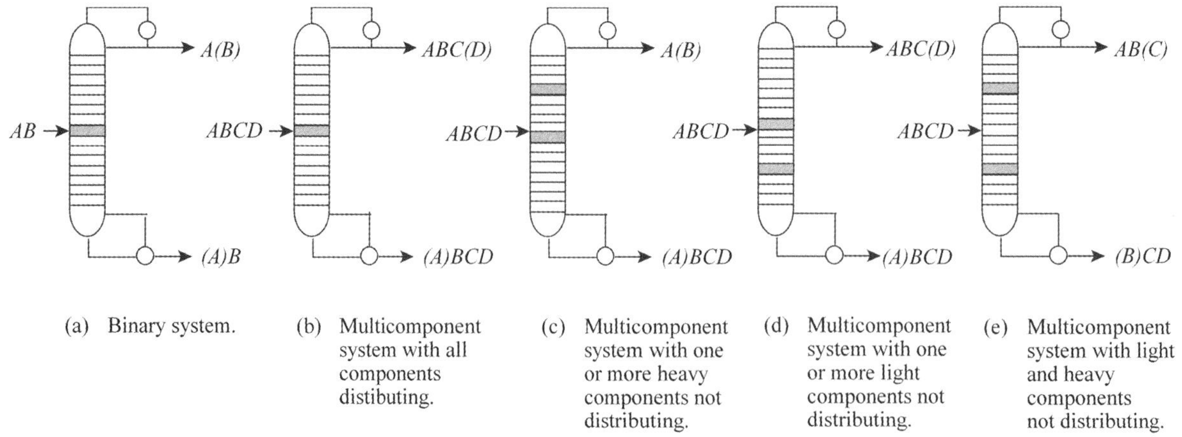

Figure 8.21

Pinch location (zones of constant composition) for binay and multicomponent systems. Brackets indicate key components remaining in a product stream due to incomplete recovery.

To solve Equation 8.72, start by assuming a feed condition such that q can be fixed. Saturated liquid feed (i.e. $q = 1$) is normally assumed in an initial design as it tends to decrease the minimum reflux ratio relative to a vaporized feed. Liquid feeds are also preferred because the pressure at which the column operates can easily be increased if required by pumping the liquid to a higher pressure. Increasing the pressure of a vapor feed is much more expensive as it requires a compressor rather than a pump. Feeding a subcooled liquid or a superheated vapor brings inefficiency to the separation as the feed material must first return to saturated conditions before it can participate in the distillation process.

Equation 8.72 can be written for all NC components of the feed and solved for the necessary values of θ. There are $(NC - 1)$ real positive values of θ that satisfy Equation 8.72, and each lies between the values of α of the components. The second equation is then written for each value of θ obtained to determine the minimum reflux ratio, R_{min} (Underwood, 1946):

$$R_{min} + 1 = \sum_{i=1}^{NC} \frac{\alpha_{ij} x_{i,D}}{\alpha_{ij} - \theta} \tag{8.73}$$

To solve Equation 8.73, it is necessary to know the values of not only α_{ij} and θ but also $x_{i,D}$. The values of $x_{i,D}$ for each component in the distillate in Equation 8.73 are the values at the minimum reflux and are unknown. Rigorous solution of the Underwood Equations, without assumptions of component distribution, thus requires Equation 8.72 to be solved for $(NC - 1)$ values of θ lying between the values of α_{ij} of the different components. Equation 8.73 is then written $(NC - 1)$ times to give a set of equations in which the unknowns are R_{min} and $(NC - 2)$ values of $x_{i,D}$ for the nonkey components. These equations can then be solved simultaneously. In this way, in addition to the calculation of R_{min}, the Underwood Equations can also be used to estimate the distribution of nonkey components at minimum reflux conditions from a specification of the key component separation. This is analogous to the use of the Fenske Equation to determine the distribution at total reflux. Although there is often not too much difference between the estimates at total and minimum reflux, the true distribution is more likely to be between the two estimates.

However, this calculation can be simplified significantly by making some reasonable assumptions regarding the component distributions to approximate $x_{i,D}$. It is often a good approximation to assume all of the lighter than light key components go to the overheads and all of the heavier than heavy key components go to the column bottoms. Also, if the light and heavy key components are adjacent in volatility, there are no components between the keys, and all $x_{i,D}$ are known, as the key component split is specified by definition. Equation 8.72 can then be solved by trial and error for the single value of θ required that lies between the relative volatilities of the key components. This value of θ can then be substituted in Equation 8.73 to solve for R_{min} directly, as all $x_{i,D}$ are known.

If the assumption is maintained that all of the lighter than light key components go to the overheads and all of the heavier than heavy key components go to the column bottoms, for cases where the light and heavy key components are not adjacent in volatility, one more value of θ is required than there are components between the keys. For this case, $x_{i,D}$ are unknown for the components between the keys. The values of θ are obtained from Equation 8.72, where each θ lies between an adjacent pair of relative volatilities. Once the values of θ have been determined, Equation 8.73 can be written for each value of θ with $x_{i,D}$ of the components between the keys as unknowns. This set of equations is then solved simultaneously for R_{min} and $x_{i,D}$ for the components between the keys.

Another approximation that can be made to simplify the solution of the Underwood Equations is to use the Fenske Equation to approximate $x_{i,D}$. These values of $x_{i,D}$ will thus correspond with the total reflux rather than the minimum reflux.

Example 8.2 A distillation column operating at 14 bar with a saturated liquid feed of 1000 kmol·h^{-1} with a composition given in Table 8.5 is to be separated into an overhead product that recovers 99% of the n-butane overhead and 95% of the i-pentane in the bottoms. Relative volatilities are also given in Table 8.5.

a) Calculate the minimum number of stages using the Fenske Equation.
b) Estimate the compositions of the overhead and bottoms products using the Fenske Equation.
c) Calculate the minimum reflux ratio using the Underwood Equations.

Table 8.5

Distillation column feed and relative volatilities.

Component	f_i (kmol·h^{-1})	α_{ij}
Propane	30.3	16.5
i-Butane	90.7	10.5
n-Butane (LK)	151.2	9.04
i-Pentane (HK)	120.9	5.74
n-Pentane	211.7	5.10
n-Hexane	119.3	2.92
n-Heptane	156.3	1.70
n-Octane	119.6	1.00

Solution

a) Substitute $r_{L,D} = 0.99$, $r_{H,B} = 0.95$, $\alpha_{LH} = 1.5749$ in Equation 8.61:

$$N_{min} = \frac{\log\left[\frac{0.99}{(1-0.99)}\frac{0.95}{1-0.95}\right]}{\log 1.5749}$$

$$= 16.6$$

b) Using the heavy key component as the reference:

$$\frac{d_H}{b_H} = \frac{f_H(1-r_{H,B})}{f_H r_{H,B}} = \frac{1-r_{H,B}}{r_{H,B}} = \frac{1-0.95}{0.95} = 0.05263$$

Substitute N_{min},$\alpha_{i,H}$,f_i and (d_H/b_H) in Equation 8.65 to obtain d_i and determine b_i by mass balance. For the first component (propane):

$$d_i = \frac{2.875^{16.6} \times 30.3 \times 0.05263}{1 + 2.875^{16.6} \times 0.05263}$$

$$= 30.30\ \text{kmol·h}^{-1}$$

$$b_i = f_i - d_i$$

$$= 30.30 - 30.30$$

$$= 0\ \text{kmol·h}^{-1}$$

and so on, for the other components to obtain the results in Table 8.6.

Table 8.6

Distribution of components for Example 8.2.

Component	d_i	b_i	$x_{i,D}$	$x_{i,B}$
Propane	30.30	0.0	0.1089	0.0
i-Butane	90.62	0.08	0.3257	0.0001
n-Butane	149.69	1.51	0.5380	0.0021
i-Pentane	6.05	114.86	0.0217	0.1591
n-Pentane	1.55	210.15	0.0056	0.2911
n-Hexane	0.0	119.30	0.0	0.1653
n-Heptane	0.0	156.30	0.0	0.2165
n-Octane	0.0	119.60	0.0	0.1657
Total	278.21	721.80	0.9999	0.9999

Table 8.7

Solution for the root of the Underwood Equation.

			$\frac{\alpha_{ij}x_{i,F}}{\alpha_{ij}-\theta}$			
$x_{F,i}$	α_{ij}	$\alpha_{ij}x_{i,F}$	$\theta=7.0$	$\theta=7.3$	$\theta=7.2$	$\theta=7.2487$
0.0303	16.5	0.5000	0.0526	0.0543	0.0538	0.0540
0.0907	10.5	0.9524	0.2721	0.2796	0.2886	0.2929
0.1512	9.04	1.3668	0.6700	0.7855	0.7429	0.7630
0.1209	5.74	0.6940	−0.5508	−0.4449	−0.4753	−0.4600
0.2117	5.10	1.0797	−0.5682	−0.4908	−0.5141	−0.5025
0.1193	2.92	0.3484	−0.0854	−0.0795	−0.0814	−0.0805
0.1563	1.70	0.2657	−0.0501	−0.0474	−0.0483	−0.0479
0.1196	1.00	0.1196	−0.0199	−0.0190	−0.0193	−0.0191
			−0.2797	0.0559	−0.0532	0.0000

c) To calculate minimum reflux ratio, first solve Equation 8.72. A search must be carried out for the root θ that satisfies Equation 8.72. Since there are no components between the key components, there is only one root, and the root will have a value between α_L and α_H. This involves trial and error to satisfy the summation to be equal to zero, as summarized in Table 8.7.

Now substitute $\theta = 7.2487$ in Equation 8.73, as summarized in Table 8.8:

$$R_{min} + 1 = 3.866$$

$$R_{min} = 2.866$$

The calculation in this example can be conveniently carried out in spreadsheet software. However, many implementations are available in commercial flowsheet simulation software.

Table 8.8

Solution of the second Underwood Equation.

$x_{D,i}$	α_{ij}	$\alpha_{ij}x_{i,D}$	$\frac{\alpha_{ij}x_{i,D}}{\alpha_{ij}-\theta}$
0.1089	16.5	1.7970	0.1942
0.3257	10.5	3.4202	1.0519
0.5380	9.04	4.8639	2.7153
0.0217	5.74	0.1247	−0.0827
0.0056	5.10	0.0285	−0.0133
0.0	2.92	0.0	0.0
0.0	1.70	0.0	0.0
0.0	1.00	0.0	0.0
			3.8655

The Underwood Equation is based on the assumption that the relative volatilities and molar overflow are constant between the pinches. Given that the relative volatilities change throughout the column, which are the most appropriate values to use in the Underwood Equations? The relative volatilities could be averaged according to Equations 8.69 or 8.71. However, it is generally better to use the ones based on the feed conditions rather than the average values based on the distillate and bottoms compositions. This is because the location of the pinches is often close to the feed.

The Underwood Equations tend to underestimate the true value of the minimum reflux ratio. The most important reason for this is the assumption of constant molar overflow. As mentioned previously, the Underwood Equations assumed constant molar overflow between the pinches. So far, in order to determine the reflux ratio of the column, this assumption has been extended to the whole column. However, some compensation can be made for the variation in molar overflow by carrying out an energy balance around the top pinch for the column, as shown in Figure 8.22. Thus:

$$Q_{COND} = V(H_V - H_L) \tag{8.74}$$

where Q_{COND} = heat rejected in the condenser
V = vapor flowrate at the top of the column
H_V, H_L = molar enthalpy of the vapor and liquid at the top of the column

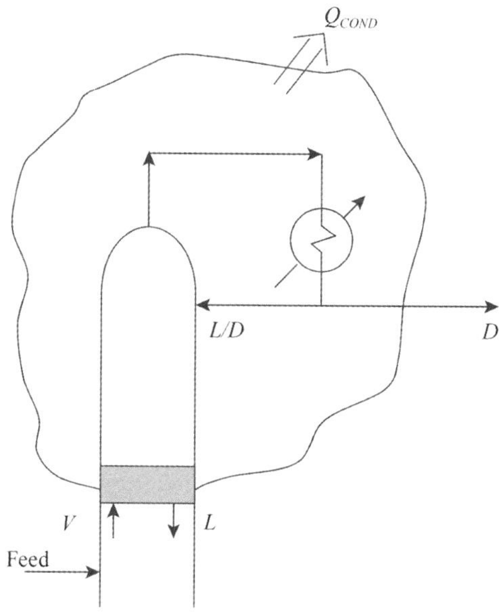

Figure 8.22

Energy balance around the rectifying pinch under minimum reflux conditions.

An energy balance gives:

$$V_\infty H_{V\infty} = L_\infty H_{L\infty} + Dh_L + Q_{COND} \tag{8.75}$$

where V_∞, L_∞ = vapor and liquid flowrates at the rectifying pinch
$H_{V\infty}$, $H_{L\infty}$ = molar enthalpies of the vapor and liquid at the rectifying pinch
D = distillate rate

Also:

$$V_\infty = L_\infty + D \tag{8.76}$$

$$V = L + D \tag{8.77}$$

Combining Equations 8.74 to 8.77 and rearranging gives (Seader, Henley and Roper, 2011):

$$\frac{L}{D} = \frac{(L_\infty/D)(H_{V\infty} - H_{L\infty}) + H_{V\infty} - H_V}{H_V - H_L} \tag{8.78}$$

where (L_∞/D) is the reflux ratio given by the Underwood Equations and (L/D) is the reflux ratio compensated for the variation in molar overflow. One problem remains before Equation 8.78 can be applied. The calculation of $H_{V\infty}$ and $H_{L\infty}$ is required, for which the vapor and liquid composition at the top pinch is required. However, these are given by the Underwood Equations (Underwood, 1946; King, 1980; Seader, Henley and Roper, 2011):

$$x_{i,\infty} = \frac{x_{i,D}}{\dfrac{L_\infty}{D}\left(\dfrac{\alpha_{ij}}{\theta} - 1\right)} \tag{8.79}$$

where $x_{i,\infty}$ = liquid mole fraction of Component i at the rectifying pinch
α_{ij} = relative volatility
θ = root of the Underwood Equation that satisfies the second equation (Equation 8.73) at the known minimum reflux ratio. The appropriate root is the one associated with the heavy key, which is the one below α_{HK}. This is given by $\alpha_{HNK} < \theta < \alpha_{HK}$, where α_{HNK} is the relative volatility of the component below the heavy key. If there is no component heavier than the heavy key $0 < \theta < \alpha_{HK}$.

A material balance around the rectifying pinch gives the mole fraction in the vapor phase:

$$y_{i,\infty} = \frac{x_{i,\infty}(L_\infty/D) + x_{i,D}}{(L_\infty/D) + 1} \tag{8.80}$$

where $y_{i,\infty}$ = vapor mole fraction of Component i at the rectifying pinch

8

Example 8.3 Apply the enthalpy correction to the prediction of the minimum reflux ratio for Example 8.2 to obtain a more accurate estimate of the minimum reflux ratio.

Solution Calculate the liquid composition at the rectifying pinch using Equation 8.79, but first the root of the second Underwood Equation associated with the heavy key component must be calculated. From Table 8.5, this lies in the range:

$$5.10 < \theta < 5.74$$

Assume that components heavier than the heavy key component do not appear in the distillate. The root is determined in Table 8.9.

Now substitute $\theta = 5.6598$ in Equation 8.79 to calculate $x_{i,\infty}$ and solve for $y_{i,\infty}$ from Equation 8.80. The results are given in Table 8.10.

Now calculate the molar enthalpies of the vapor and liquid streams. Enthalpies were calculated here from ideal gas enthalpy data corrected using the Peng–Robinson Equation of State (see Appendix A):

$$\begin{aligned} H_V &= -122{,}980\ \text{kJ·kmol}^{-1} \\ H_L &= -138{,}380\ \text{kJ·kmol}^{-1} \\ H_{V\infty} &= -131{,}210\ \text{kJ·kmol}^{-1} \\ H_{L\infty} &= -151{,}110\ \text{kJ·kmol}^{-1} \end{aligned}$$

Substitute these values and $L_\infty/D = 2.866$ in Equation 8.78:

$$\frac{L}{D} = \frac{2.866(-131{,}210 - (-151{,}110)) - 131{,}210 + 122{,}980}{-122{,}980 - (-138{,}380)}$$

$$= 3.170$$

This is compared with an uncorrected reflux ratio of 2.866. To obtain a rigorous value for the minimum reflux ratio to assess the results of the Underwood Equation, a rigorous simulation of the column is needed. A rigorous simulation model is set up with a large number of stages (say 200 or more) that carry out the separation. The reflux ratio is then gradually turned down and resimulated with each setting. This is repeated until zones of constant composition appear (pinches). For this separation, using the Peng–Robinson Equation of State, it turns out to be $R_{min} = 3.545$. It should be emphasized that this value of 3.545 accounts for both variation in the relative volatility and molar overflow.

Table 8.10

Vapor and liquid mole fractions at the rectifying pinch.

Component	$x_{i,\infty}$	$y_{i,\infty}$
Propane	0.0198	0.0429
i-Butane	0.1329	0.1828
n-Butane	0.3144	0.3722
i-Pentane	0.5348	0.4021
	1.0019	1.0000

Table 8.9

Root of the Underwood Equation.

				$\sum_{i=1}^{NC} \frac{\alpha_{ij} x_{i,D}}{\alpha_{ij} - \theta} - R_{min} - 1$			
Component	$x_{i,D}$	α_{ij}	$\alpha_{ij}\, x_{i,D}$	$\theta = 5.7$	$\theta = 5.6$	$\theta = 5.65$	$\theta = 5.6598$
Propane	0.1089	16.5	1.7970	0.1664	0.1649	0.1656	0.1658
i-Butane	0.3257	10.5	3.4202	0.7126	0.6980	0.7052	0.7066
n-Butane	0.5380	9.04	4.8639	1.4562	1.4139	1.4348	1.4389
i-Pentane	0.0217	5.74	0.1247	3.1180	0.8909	1.3858	1.5551
	0.9943			1.5882	−0.6973	−0.1736	0.0014

8.7 Finite Reflux Conditions for Multicomponent Mixtures

There are a number of important design parameters that need to be determined before a design can be evaluated. The operating costs require the reboiler and condenser duties to be evaluated before the cost of the utilities to supply them can be calculated. The actual number of trays in the column, their spacing and the column diameter need to be known to size the column and estimate the capital cost. Consider first the reboiler and condenser duties.

It has been shown earlier how the Underwood Equations can be used to calculate the minimum reflux ratio. A simple mass balance around the top of the column for constant molar overflow, as shown in Figure 8.23, at minimum reflux gives:

$$V_{min} = D(1 + R_{min}) \tag{8.81}$$

where V_{min} = minimum vapor flowrate above the feed ($kmol{\cdot}s^{-1}$)
R_{min} = minimum reflux ratio (–)
D = distillate flowrate ($kmol{\cdot}s^{-1}$)

Equation 8.81 can also be written at finite reflux (Figure 8.23). Defining R_F to be the ratio R/R_{min} (typically $R/R_{min} = 1.1$):

$$V = D(1 + R_F R_{min}) \tag{8.82}$$

where V = vapor flowrate above the feed ($kmol{\cdot}s^{-1}$)

Thus, the condenser duty for a total condenser is given by:

$$Q_{COND} = \Delta H_{VAP} V = \Delta H_{VAP} D(1 + R_F R_{min}) \tag{8.83}$$

where Q_{COND} = condenser duty ($kJ{\cdot}s^{-1}$, kW)
ΔH_{VAP} = heat of vaporization of overhead vapor ($kJ{\cdot}kmol^{-1}$)

For a partial condenser, the condenser duty is given by:

$$Q_{COND} = \Delta H_{VAP} D R_F R_{min} \tag{8.84}$$

The vapor rate below the feed depends on the feed condition, assuming constant molar overflow:

$$V' = V - (1 - q)F \tag{8.85}$$

where V' = vapor flowrate below the feed ($kmol{\cdot}s^{-1}$)
V = vapor flowrate above the feed ($kmol{\cdot}s^{-1}$)
q = feed condition
$= \dfrac{\text{heat required to vaporize one mole of feed}}{\text{molar latent heat of vaporization of feed}}$
F = feed flowrate ($kmol{\cdot}s^{-1}$)

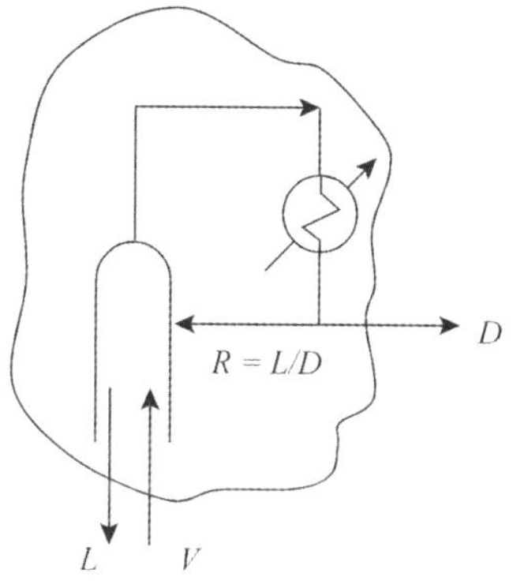

Figure 8.23

Mass balance around top of a distillation column.

Thus, the reboiler duty for a total condenser is given by:

$$Q_{REB} = \Delta H'_{VAP} V' = \Delta H'_{VAP}(V - F + qF) \tag{8.86}$$

where Q_{REB} = reboiler duty ($kJ{\cdot}s^{-1}$, kW)
$\Delta H'_{VAP}$ = heat of vaporization of the distillation bottoms liquid ($kJ{\cdot}kmol^{-1}$)

Substituting Equation 8.82 gives:

$$Q_{REB} = \Delta H'_{VAP}(D + DR_F R_{min} - F + qF) \tag{8.87}$$

Now consider how to estimate the number of trays in the column. Having obtained the minimum number of theoretical stages from the Fenske Equation and minimum reflux ratio from the Underwood Equations, the empirical relationship of Gilliland (1940) can be used to determine the actual number of theoretical stages. The original correlation was presented in graphical form (Gilliland, 1940). Two parameters (X and Y) were used to correlate the data:

$$Y = \frac{N - N_{min}}{N + 1}, \quad X = \frac{R - R_{min}}{R + 1} \tag{8.88}$$

where X, Y = correlating parameters
N = actual number of theoretical stages
N_{min} = minimum number of theoretical stages
R = actual reflux ratio
R_{min} = minimum reflux ratio

Various attempts have been made to represent the correlation algebraically. For example (Rusche, 1999):

$$Y = 0.2788 - 1.3154X + 0.4114X^{0.2910} + 0.8268 \ln X + 0.9020 \ln\left(X + \frac{1}{X}\right) \tag{8.89}$$

Equation 8.89 allows an estimate of the number of theoretical stages the column requires.

Whilst Equation 8.89 provides an estimate of the number of theoretical stages, it does not provide the distribution of the stages between the rectifying and stripping sections. This can be estimated by the empirical equation of Kirkbride (1944):

$$\frac{N_{REC}}{N_{STRIP}} = \left[\frac{B}{D}\left(\frac{x_{H,F}}{x_{L,F}}\right)\left(\frac{x_{L,B}}{x_{H,D}}\right)^2\right]^{0.206} \tag{8.90}$$

where N_{REC} = number of stages above the feed, including if appropriate a partial condenser (–)
N_{STRIP} = number of stages below the feed, including if appropriate a partial reboiler (–)
B = bottoms flowrate ($kmol{\cdot}s^{-1}$, $kmol{\cdot}h^{-1}$)
D = distillate flowrate ($kmol{\cdot}s^{-1}$, $kmol{\cdot}h^{-1}$)
$x_{H,F}$ = mole fraction of heavy key in the feed (–)
$x_{L,F}$ = mole fraction of light key in the feed (–)
$x_{L,B}$ = mole fraction of light key in the bottoms (–)
$x_{H,D}$ = mole fraction of heavy key in the distillate (–)

It should be noted that N_{REC} and N_{STRIP} do not include the feed stage and that $N = N_{REC} + N_{STRIP} + 1$.

Example 8.4 Assuming that the distillation column in Examples 8.2 and 8.3 uses a total condenser and a thermosyphon reboiler, estimate the number of theoretical stages required if $R/R_{min} = 1.1$ for the values of R_{min} obtained in Examples 8.2 and 8.3 and the distribution of the stages between the rectifying and stripping sections.

Solution

$$R_{min} = 2.866$$

Given $R/R_{min} = 1.1$:

$$R = 3.153$$

From Equation 8.88:

$$X = \frac{R - R_{min}}{R+1} = \frac{3.152 - 2.865}{3.152 + 1} = 0.0690$$

Substitute $X = 0.0691$ in Equation 8.89:

$$Y = 0.2788 - 1.3154 \times 0.069 + 0.4114 \times 0.069^{0.2910} + 0.8268 \ln 0.069 + 0.9020 \ln\left[0.069 + \frac{1}{0.069}\right] = 0.5823$$

Substitute $Y = 0.5822$ in Equation 8.88:

$$0.5823 = \frac{N - 16.6}{N+1}$$

$$N = 41.1$$

Thus, 42 theoretical stages are needed in the column. Had a partial reboiler been assumed, in principle this would have meant 41 theoretical stages would be needed in the column.

To calculate the distribution of stages between the rectifying and stripping sections:

$$B = 721.80\ \text{kmol}\cdot\text{h}^{-1}$$
$$D = 278.21\ \text{kmol}\cdot\text{h}^{-1}$$
$$x_{H,F} = 0.1209$$
$$x_{L,F} = 0.1512$$
$$x_{L,B} = 0.0021$$
$$x_{H,D} = 0.0217$$

From Equation 8.90:

$$\frac{N_{REC}}{N_{STRIP}} = \left[\frac{B}{D}\left(\frac{x_{H,F}}{x_{L,F}}\right)\left(\frac{x_{L,B}}{x_{H,D}}\right)^2\right]^{0.206} = \left[\frac{721.80}{278.21}\left(\frac{0.1209}{0.1512}\right)\left(\frac{0.0021}{0.0217}\right)^2\right]^{0.206} = 0.44$$

$$N = N_{REC} + N_{STRIP} + 1$$
$$42 = N_{REC} + N_{STRIP} + 1$$
$$42 = 0.44 N_{STRIP} + N_{STRIP} + 1$$
$$N_{STRIP} = 28.47$$
$$N_{REC} = 41 - 28.47 = 12.53$$
$$\text{say } N_{REC} = 13, N_{STRIP} = 28$$

Repeat the calculation for $R_{min} = 3.170$:

$$R = 3.487$$
$$X = 0.0706$$
$$Y = 0.5799$$
$$N = 41.0$$
$$N_{STRIP} = 27$$
$$N_{REC} = 12$$

Thus, the error in the prediction of the minimum reflux ratio has little effect on the estimate of the number of theoretical stages for this example. The error in the prediction of the energy consumption is likely to be more serious.

The actual number of trays required in the column will be greater than the number of theoretical stages, as mass transfer limitations will prevent equilibrium being achieved on each tray. To estimate the actual number of trays, the number of theoretical stages must be divided by the overall stage efficiency. This is most often in the range between 0.7 and 0.9, depending on the separation being carried out and the design of the distillation tray used. Efficiencies significantly below 0.7 can be encountered (perhaps as low as 0.4 in extreme cases), as well as efficiencies in excess of 0.9. O'Connell (1946) produced a simple graphical plot that can be used to obtain a first estimate of the overall stage efficiency. This graphical plot can be represented by (Kessler and Wankat, 1988):

$$E_O = -0.3143 - 0.285 \log(\alpha_{LH}\,\mu_L) \tag{8.91}$$

where
E_O = overall stage efficiency ($0 < E_O < 1$)
α_{LH} = relative volatility between the key components
μ_L = viscosity of the feed at average column conditions ($\text{kg}\cdot\text{s}^{-1}\cdot\text{m}^{-1} = \text{N}\cdot\text{s}\cdot\text{m}^{-2}$)

Equation 8.91 is only approximate at best. For distillation trays with long flow paths across the active plate area, Equation 8.91 tends to underestimate the overall efficiency.

It should be emphasized that the overall stage efficiency from Equation 8.91 should only be used to derive a first estimate of the actual number of distillation trays. More elaborate methods are available but are outside the scope of this text. In practice, the stage efficiency varies from component to component, and more accurate calculations require much more information on tray type and geometry and physical properties of the fluids (Kister, 1992).

In practice, the number of trays is often increased by 5 to 10% to allow for uncertainties in the design.

8.8 Column Dimensions

1) *Column height for tray columns.* The height of the column can be estimated by multiplying the actual number of trays by the

tray spacing. Tray spacing can vary between 0.15 m and 1.0 m, depending on the system, column diameter and the access requirements. If access is required for maintenance, cleaning or inspection purposes, 0.45 m is a practical minimum spacing. For preliminary design, a value of 0.45 m is usually a reasonable assumption for small columns with less than 1 m diameter and 0.6 m spacing for larger diameters. Additional height of 0.5 m to 1 m needs to be added at the top of the column for vapor disengagement and 1 m to 2 m at the bottom of the column for vapor–liquid disengagement for the reboiler return and a liquid sump to maintain a steady liquid head for the pump for the column bottoms. There is a maximum height for a column, beyond which the column must be split into multiple shells. Large columns are designed to be free standing. This means that the column must be able to mechanically withstand the wind load, as must the foundations. The maximum height depends on the local weather (especially susceptibility to hurricanes and typhoons), ground conditions for foundations and susceptibility to earthquakes. The maximum height is normally restricted to be below 100 m. However, this would be a special design. Most columns will be smaller than 60 m, especially for adverse conditions.

2) *Column height for packed columns.* To relate the continuous change in composition to that of an equilibrium stage requires the concept of the *height equivalent of a theoretical plate* (*HETP*) to be introduced. This is the height of packing required to bring about the same concentration change as an equilibrium stage. The height of packing can be estimated from:

$$H = N \times HETP \tag{8.92}$$

where H = packing height
N = number of theoretical stages
$HETP$ = height equivalent of a theoretical plate

Various correlations are available for *HETP*, but the designer should use them with great caution, as reliable values can only be obtained from experimental data or packing manufacturers. Data for a number of common structured packings are given in Table 8.11. Structured packings are mostly available in two different inclination angles (Figure 8.10b), termed Type 'Y' and Type 'X'. Type 'Y' packings have an inclination angle of 45° from the horizontal axis and are the most widely used. Type 'X' packings have an inclination angle of 60° from the horizontal axis. Type 'Y' packings provide higher efficiency over their corresponding type 'X' counterpart, but at the cost of a higher pressure drop and lower capacity. Type 'X' packings are used in high capacity and low pressure drop applications. The packing factor in Table 8.11 is an empirical factor characteristic of the packing geometry used to correlate the pressure drop through the packing.

For structured packing an approximate *HETP* can be obtained from (Kister, 1992; Perry and Green, 2007):

$$HETP = \frac{100 F_{XY} F_{\sigma}}{a_P} + 0.1 \tag{8.93}$$

Table 8.11

Data for some common structured packings (Perry and Green, 2007).

Packing	Number	Surface area ($m^2 \cdot m^{-3}$)	Packing factor (m^{-1})
Metal corrugated sheets			
Flexipac	1Y	420	98
	1.6Y	290	59
	2Y	220	49
	3.5Y	80	30
	4Y	55	23
	1X	420	52
	1.6X	290	33
	2X	220	23
	3X	110	16
Flexipac high capacity	700	710	223
	1Y	420	82
	1.6Y	290	56
	2Y	220	43
Intalox	1T	310	66
	2T	215	56
	3T	170	43
Mellapak	125Y	125	33
	170Y	170	39
	2Y	223	46
	250Y	250	66
	350Y	350	75
	500Y	500	112
	125X	125	16
	170X	170	20
	2X	223	23
	250X	250	26
	500X	500	82
Mellapak Plus	252Y	250	39
	452Y	350	69
	752Y	500	131

(*continued*)

8

Table 8.11 *(Continued)*

Packing	Number	Surface area ($m^2 \cdot m^{-3}$)	Packing factor (m^{-1})
Montz-Pak	B1-250	250	66
	B1-250M	250	43
Wire mesh			
Sulzer	BX	492	69
Ceramic			
Flexeramic	28	260	131
	48	160	79
	88	100	49
Plastic			
Mellapak	250Y	250	72

where $HETP$ = height equivalent of a theoretical plate (m)
a_P = packing surface area ($m^2 \cdot m^{-3}$)
F_{XY} = factor to allow for angle of inclination (–) ($F_{XY} = 1$ for Type Y and high capacity packings, $F_{XY} = 1.45$ for Type X packings with $a_P \leq 300\ m^2 \cdot m^{-3}$)
F_σ = factor to allow for high surface tension causing inadequate 'wetting' of the packing (–) ($F_\sigma = 1$ for $\sigma \leq 25\ mN \cdot m^{-1}$, $F_\sigma = 2$ for $\sigma \geq 70\ mN \cdot m^{-1}$, for $25\ mN \cdot m^{-1} \leq \sigma \leq 70\ mN \cdot m^{-1}$ use linear interpolation)
σ = surface tension ($mN \cdot m^{-1} = mJ \cdot m^{-2} = dyne \cdot cm^{-1}$)

For random packing an approximate *HETP* can be obtained from (Kister, 1992; Perry and Green, 2007):

$$HETP = 18 d_P F_\sigma$$
$$HETP \geq d_C F_\sigma \text{ for } d_C \leq 0.7\ m \qquad (8.94)$$

where d_P = packing size (m)
d_C = column inside diameter (m)

Data for a number of common random packings are given in Table 8.12.

The height of the column for a packed column needs to allow for liquid distribution and redistribution. As the feed enters the column, the liquid above the feed needs to be collected, combined with the feed liquid and distributed across the packing below the feed using troughs, weirs, drip-pipes, and so on. This requires a height of typically 0.5 to 1 m. Also, as the liquid flows down through the packing, flow gradually becomes less effectively distributed over the packing, mainly due to interaction with the column walls. This requires the

Table 8.12

Data for some common random packings (Perry and Green, 2007).

Packing	Nominal size (mm)	Surface area ($m^2 \cdot m^{-3}$)	Packing factor (m^{-1})
Metal			
Raschig rings	19	245	722
	25	185	472
	50	95	187
	75	66	105
Pall rings	16	360	256
	25	205	183
	38	130	131
	50	105	89
	90	66	59
Metal Intalox	25	207	134
	40	151	79
	50	98	59
	70	60	39
Ceramic			
Raschig rings	13	370	1900
	25	190	587
	50	92	213
	75	62	121
Pall rings	25	220	350
	38	164	180
	50	121	142
	80	82	85
Berl saddles	13	465	790
	25	250	360
	38	150	215
	50	105	150
Plastic			
Ralu-rings	15	320	230
	25	190	135
	38	150	80
	50	110	55

Table 8.12 *(Continued)*

Packing	Nominal size (mm)	Surface area ($m^2 \cdot m^{-3}$)	Packing factor (m^{-1})
	90	75	38
	125	60	30
Pall rings	15	350	320
	25	206	180
	40	131	131
	50	102	85
	90	85	56

liquid to be periodically collected and redistributed over the packing below. Packing manufacturers generally recommend that a single packed bed be limited to no more than 20 theoretical stages. This height should also be no greater than 15 column diameters. Liquid collection and distribution between beds typically require a height of 0.5 m. As with tray columns, additional height of 0.5 to 1 m needs to be added at the top of the column for vapor disengagement and 1 to 2 m at the bottom of the column for vapor–liquid disengagement for the reboiler return and a liquid sump to maintain a steady liquid head for the pump for the column bottoms.

3) *Column diameter for tray columns.* Figure 8.24 illustrates the envelope of satisfactory operation for a tray (Kister, 1992).

- *Weeping* sets the lower limit of operation when liquid descends through the tray perforations. This is caused by low vapor flow when the pressure exerted by the vapor is insufficient to hold the liquid on the tray. Excessive weeping will lead to *dumping*, when all of the liquid will descend to the bottom of the column.

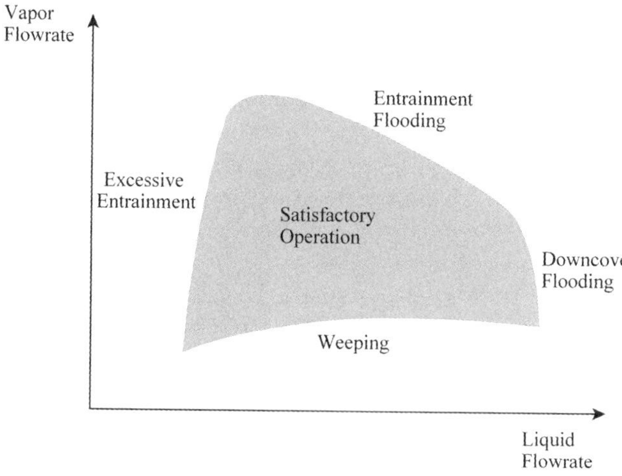

Figure 8.24
Operating range for a plate.

- *Flooding* sets the upper limit and results from excessive vapor flow. There are two basic mechanisms by which this can happen. In *entrainment flooding*, liquid drops are entrained into the vapor flow and carried to the tray above, causing liquid to accumulate on the tray above instead of flowing to the tray below. In *froth entrainment flooding*, a froth forms on the tray, which increases in height as the vapor velocity is increased. For small tray spacing the froth might expand to reach the tray above. For larger tray spacing the froth can turn to spray as vapor velocity increases. Either way, liquid will accumulate on the tray above instead of flowing to the tray below. Spray entrainment flooding is by far the most common. At low liquid flowrates the entrainment can become excessive.
- *Downcomer flooding* occurs at high liquid flowrates. There are two mechanisms for downcomer flooding. In *downcomer backup flooding*, froth backs up in the downcomer. This occurs when the pressure available from the height of liquid and froth in the downcomer cannot overcome the total pressure drop across the tray. This pressure imbalance causes the froth in the downcomer to start backing up until it reaches the tray above, causing an increased accumulation of liquid on it. When the froth level exceeds the downcomer height, the tray floods. The second type of downcomer flooding is *velocity* or *choke flooding*. This occurs when the frictional pressure losses in the downcomer become excessive. Also, the vapor carried into the downcomer must separate from the liquid and then flow countercurrently to the liquid entering the downcomer. When the combination of vapor leaving and liquid entering the downcomer becomes excessive, the downcomer entrance is choked, causing the liquid to back up on the tray.

The liquid entrainment in the column that leads to flooding depends on the vapor velocity in the column. To prevent liquid entrainment, the vapor velocity for tray columns is usually in the range 0.5 to 2.5 $m \cdot s^{-1}$. The entrainment of liquid droplets can be predicted using Equation 7.3 to calculate the settling velocity (Souders and Brown, 1934):

$$v_T = K_T \left(\frac{\rho_L - \rho_V}{\rho_V} \right)^{0.5} \qquad (8.95)$$

where
v_T = terminal settling velocity ($m \cdot s^{-1}$)
K_T = parameter for the terminal velocity ($m \cdot s^{-1}$)
ρ_L = liquid density ($kg \cdot m^{-3}$)
ρ_V = vapor density ($kg \cdot m^{-3}$)

The terminal settling velocity is based on the net area of the column, as illustrated in Figure 8.7. To apply Equation 8.95 requires the parameter K_T to be specified. For distillation using tray columns, K_T is correlated in terms of a liquid–vapor flow parameter F_{LV}, defined by:

$$F_{LV} = \left(\frac{M_L L}{M_V V} \right) \left(\frac{\rho_V}{\rho_L} \right)^{0.5} \qquad (8.96)$$

where
F_{LV} = liquid–vapor flow parameter (–)
L = liquid molar flowrate ($kmol \cdot s^{-1}$)

8

V = vapor molar flowrate (kmol·s^{-1})
M_L = liquid molar mass (kg·kmol^{-1})
M_V = vapor molar mass (kg·kmol^{-1})
ρ_V = vapor density (kg·m^{-3})
ρ_L = liquid density (kg·m^{-3})

The liquid–vapor flow parameter is related to operating pressure, with low values corresponding to vacuum distillation and high values corresponding to high-pressure distillation. Generally, for values of F_{LV} lower than 0.1, packings are preferred. Fair (1961) presented a graphical correlation for K_T that can be used for preliminary design. The original graphical correlation for K_T can be expressed as:

$$K_T = \left(\frac{\sigma}{20}\right)^{0.2} \exp\left[-2.979 - 0.717 \ln F_{LV} - 0.0865(\ln F_{LV})^2 + 0.997 \ln H_T - 0.07973 \ln F_{LV} \ln H_T + 0.256(\ln H_T)^2\right] \quad (8.97)$$

where K_T = parameter for terminal velocity (m·s^{-1})
σ = surface tension (mN·m^{-1} = mJ·m^{-2} = dyne·cm^{-1})
H_T = tray spacing (m)

The correlation is valid in the range $0.25\,\text{m} < H_T < 0.6\,\text{m}$ and should be used with caution outside of this range. The entrainment equations tend to predict higher flooding velocities than are actually experienced. This is to some extent related to the tendency of the system to foam. However, the prediction of higher flooding velocities is not uniquely related to foaming. The problem is mostly associated with high-pressure systems. To account for this, a *derating factor* or *system factor* or *foaming factor* is introduced. The value of K_T from Equation 8.97 is multiplied by a factor that is less than unity. The factors are typically (Kister, 1992; Bahadori and Vuthaluru, 2010):

- Distillation involving light gases typically 0.8 to 0.9
- Crude oil distillation typically 0.85
- Absorbers typically 0.7 to 0.85
- Strippers typically 0.6 to 0.8.

Such foaming factors should be applied in high-pressure separations when the vapor density is greater than 28 kg·m^{-3} and can be correlated approximately by (Kister, 1992; Bahadori and Vuthaluru, 2010):

$$F_{FOAM} = \frac{2.94}{\rho_V^{0.32}} \quad 0 < F_{FOAM} \leq 1 \quad (8.98)$$

where F_{FOAM} = foaming factor (–)
ρ_V = vapor density (kg·m^{-3})

The correlation in Equation 8.97 requires the spacing between the trays to be specified. Tray spacing can vary between 0.15 m and 1 m, depending on the diameter of the column, tray design, vapor and liquid flowrates and physical properties of the fluids. As indicated previously, tray spacing is most often in the range 0.45 m to 0.6 m. If access is required for maintenance, cleaning or inspection purposes, 0.45 m is a practical minimum spacing.

Having determined the flooding velocity from Equation 8.95, the operation of the column is fixed at some proportion of the flooding velocity. Designs usually lie in the range of 70 to 90% of the flooding velocity. In the preliminary design, a value of 80 to 85% of the flooding velocity is usually reasonable. The net area of the column can then be estimated from the vapor flowrate up the column.

From Equation 8.97, K_T increases with increasing tray spacing. Thus, there is a trade-off between tray spacing and column diameter. Increasing the tray spacing increases the vapor flooding velocity, allowing a smaller column diameter. In turn, increasing the tray spacing increases the column height. In practice, the two changes in cost tend to balance each other. However, the trade-off can be exploited to allow simpler mechanical design by maintaining a more consistent diameter requirement through the height of the column.

If conventional trays are to be used, then an allowance must be made for the area of the column cross-section that will be taken up by the downcomers. The size of the downcomer depends on the tray geometry, vapor and liquid flowrates, operating pressure and physical properties of the fluids. The area of the column cross-section taken up by downcomers is typically in the range of 10 to 15% of the cross-sectional area of the column. Generally, the higher the column operating pressure, the smaller is the downcomer area as a proportion of the total cross-sectional area.

The downcomer area must be large enough to allow the liquid to flow freely from the upper to the lower tray, whilst allowing disengagement of vapor and liquid, with only liquid flowing to the lower tray and creating a liquid seal. This will depend on the density difference between the liquid and vapor, the tray spacing and the pressure drop across the tray. Methods for detailed hydraulic design have been presented by Kister (1992). However, for conceptual design, Bahadori and Vuthaluru (2010) presented a correlation for preliminary sizing of downcomer area based on the maximum liquid velocity in the downcomer:

$$v_D = \exp\left[a + \frac{b}{(\rho_L - \rho_V)} + \frac{c}{(\rho_L - \rho_V)^2} + \frac{d}{(\rho_L - \rho_V)^3}\right] \quad (8.99)$$

where v_D = downcomer liquid velocity (m·h^{-1})

$$a = a_1 + \frac{a_2}{H_T} + \frac{a_3}{H_T^2} + \frac{a_4}{H_T^3} \quad (8.100)$$

$$b = b_1 + \frac{b_2}{H_T} + \frac{b_3}{H_T^2} + \frac{b_4}{H_T^3} \quad (8.101)$$

$$c = c_1 + \frac{c_2}{H_T} + \frac{c_3}{H_T^2} + \frac{c_4}{H_T^3} \quad (8.102)$$

$$d = d_1 + \frac{d_2}{H_T} + \frac{d_3}{H_T^2} + \frac{d_4}{H_T^3} \quad (8.103)$$

H_T = tray spacing (m)
ρ_L = liquid density (kg·m^{-3})
ρ_V = vapor density (kg·m^{-3})

The coefficients for Equations 8.99 to 8.103 are given in Table 8.13. For high-pressure systems where the vapor density is greater than 28 kg·m^{-3} the downcomer liquid velocity needs to be multiplied by the foaming factor given in Equation 8.98.

4) *Number of downcomers for tray columns.* As discussed previously, if the liquid loading for the downcomer weirs becomes excessive, then the downcomers can suffer from excessive backup or can choke. The weir loading is calculated as the clear liquid volume divided by the length of the tray outlet weir. The liquid flowrate should not exceed 90 m^3·m^{-1}·h^{-1} (Kister, 1992; Pilling and Holden, 2009). It is useful to be able to relate the weir length to the area of the downcomer and the diameter of the column. Figure 8.25a shows the angle subtended by a chord to represent the downcomer. Some simple geometry gives:

$$\frac{A_D}{A_C} = \frac{\theta}{360} - \frac{1}{\pi}\sin\left(\frac{\theta}{2}\right)\cos\left(\frac{\theta}{2}\right) \tag{8.104}$$

$$\frac{L_W}{d_C} = \sin\left(\frac{\theta}{2}\right) \tag{8.105}$$

where A_D = area occupied by the downcomer (m^2)
A_C = cross-sectional area of the column (m^2)
θ = central angle (degrees)
L_W = weir length (m)
d_C = column diameter (m)

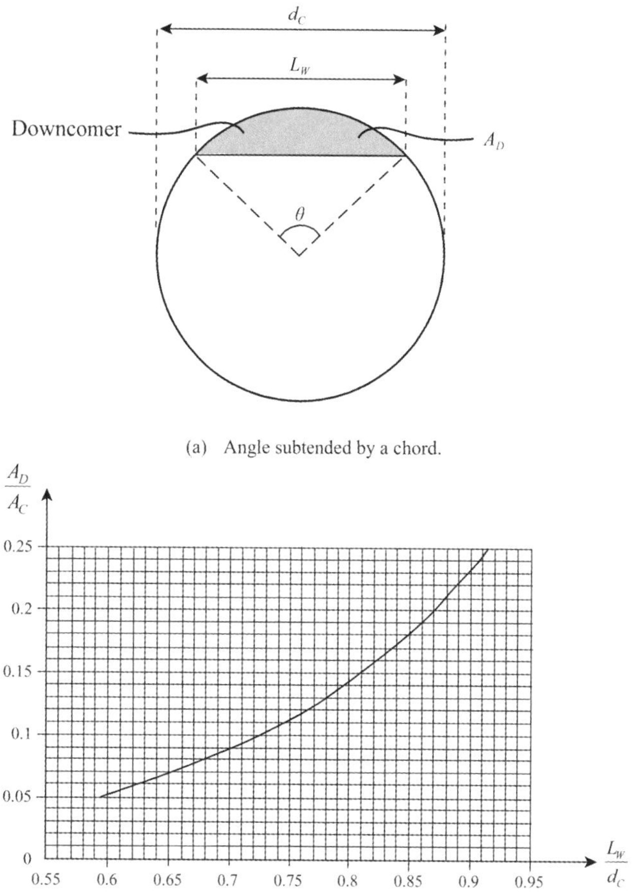

(a) Angle subtended by a chord.

(b) Area of the downcomer versus the wier lenth.

Figure 8.25

Relationship between column areas and downcomer weir length.

Table 8.13

Coefficients for downcomer area (Bahadori and Vuthaluru, 2010).

Coefficient	Value
a_1	6.93401
a_2	6.85621×10^{-1}
a_3	-4.80489×10^{-1}
a_4	7.90633×10^{-2}
b_1	1.06359×10^{3}
b_2	-2.47091×10^{3}
b_3	1.12853×10^{3}
b_4	-1.64928×10^{2}
c_1	-7.17957×10^{5}
c_2	1.21897×10^{6}
c_3	-5.57757×10^{5}
c_4	8.13018×10^{4}
d_1	1.08133×10^{8}
d_2	-1.71306×10^{8}
d_3	7.85777×10^{7}
d_4	-1.14496×10^{7}

Figure 8.25b shows the relationship between A_D/A_C and L_W/d_C. Given that the downcomer area is normally in the range 10 to 15% of the column cross-sectional area, a value of 12.5% corresponding with a value of L_W/d_C of 0.77 is often a reasonable initial assumption. The weir loading can then be checked to see if it is a reasonable value. If it is greater than 90 m^3·h^{-1}·m^{-1}, then the number of tray passes can be increased.

Although increasing the number of passes is normally driven by the necessity to reduce the weir loading, another benefit is that the liquid loading at any point on the tray is also decreased. From Figure 8.9 it can be seen that the liquid loading for a multipass tray is the liquid loading for a single-pass tray divided by the number of passes. This has the benefit of increasing the flooding velocity and decreasing the column diameter.

5) *Column diameter for packing.* The diameter for packed columns is again taken to be that giving a vapor velocity to be some proportion of the flooding velocity. In preliminary design, a value of 80 to 85% of the flooding velocity is usually reasonable. As with plate columns, flooding occurs in packed columns when heavy liquid entrainment occurs and the liquid can no longer flow down the column in such a way as to allow efficient operation of the column (Kister, 1992). This is characterized by a condition in which the pressure drop through the packed bed

starts to increase very rapidly as the vapor flowrates are increased, with simultaneous loss of mass transfer efficiency. Thus packing pressure drop characteristics can be used to determine flooding conditions. An empirical relationship has been suggested by Kister *et al.* (2007) to relate pressure drop at flooding conditions to the packing factor introduced earlier:

$$\Delta P_{FLOOD} = 40.91 F_P^{0.7} \tag{8.106}$$

where ΔP_{FLOOD} = pressure drop under flooding conditions ($N{\cdot}m^{-2}$)
F_P = packing factor (m^{-1})

Tables 8.11 and 8.12 present packing factors for some common types of structured and random packing. Thus specifying the packing factor specifies the pressure drop at flooding from Equation 8.106. The problem then is to relate the ΔP_{FLOOD} to the vapor velocity. Generalized pressure drop correlations are available for packing to predict the pressure drop under any conditions (Kister *et al.*, 2007). These are normally presented in graphical form as a plot of a *capacity parameter* (CP) plotted against the liquid–vapor flow parameter F_{LV} (Equation 8.96). The capacity parameter for the generalized pressure drop plot is defined by (Kister *et al.*, 2007):

$$CP = C_S F_P^{0.5} \left(\frac{\mu_L}{\rho_L}\right)^{0.05} \tag{8.107}$$

where CP = capacity parameter
C_S = C-factor, which is the superficial vapor velocity corrected for vapor and liquid densities ($m{\cdot}s^{-1}$)
$= v_S \left(\frac{\rho_V}{\rho_L - \rho_V}\right)^{0.5}$
v_S = superficial vapor velocity, which is the vapor velocity in the empty column ($m{\cdot}s^{-1}$)
ρ_L = liquid density ($kg{\cdot}m^{-3}$)
ρ_V = vapor density ($kg{\cdot}m^{-3}$)
μ_L = liquid viscosity ($kg{\cdot}m^{-1}{\cdot}s^{-1}$)

The original graphical plot of the capacity factor against the liquid–vapor flow parameter can be represented by (Enríquez-Gutiérrez *et al.*, 2014):

$$CP = A \ln(F_{LV}) + B \tag{8.108}$$

where F_{LV} = liquid–vapor flow parameter (Equation 8.96)

For structured packing:

$$A = -7.31 \times 10^{-11} \Delta P^3 + 2.18 \times 10^{-7} \Delta P^2 - 2.19 \times 10^{-4} \Delta P - 0.0124 \tag{8.109}$$

$$B = 1.28 \times 10^{-10} \Delta P^3 - 3.15 \times 10^{-7} \Delta P^2 + 2.62 \times 10^{-4} \Delta P + 0.0826 \tag{8.110}$$

For random packing:

$$A = 6.80 \times 10^{-8} \Delta P^2 - 1.48 \times 10^{-4} \Delta P - 0.00629 \tag{8.111}$$

$$B = 2.55 \times 10^{-10} \Delta P^3 - 6.08 \times 10^{-7} \Delta P^2 + 4.69 \times 10^{-4} \Delta P + 0.08816 \tag{8.112}$$

where ΔP = packing pressure drop ($Pa{\cdot}m^{-1}$)

At flooding conditions, C_S in Equation 8.107 becomes the parameter for flooding conditions K_T by analogy with Equation 8.95. Thus at flooding conditions Equation 8.107 can be rearranged to give:

$$K_T = \frac{CP_{FLOOD}}{F_P^{0.5} \left(\frac{\mu_L}{\rho_L}\right)^{0.05}} \tag{8.113}$$

where K_T = parameter for the terminal velocity ($m{\cdot}s^{-1}$)
CP_{FLOOD} = capacity factor at flooding conditions

To determine the flooding velocity, the packing is first chosen, which fixes the packing factor F_P. This in turn fixes the flooding pressure drop ΔP_{FLOOD} from Equation 8.106. Substituting ΔP_{FLOOD} into Equations 8.108 to 8.112 gives CP_{FLOOD}, which can then be substituted into Equation 8.113 to give K_T. Substituting K_T into Equation 8.95 gives the flooding velocity v_T.

Example 8.5 For the distillation column from Examples 8.2, 8.3 and 8.4, assuming $R_{min} = 3.170$ and $R/R_{min} = 1.1$, estimate:

a) The actual number of trays for a tray column.
b) Height of the tray column assuming a tray spacing of 0.6 m and a combined allowance at the top and bottom of the column for vapor–liquid disengagement and the bottom for a sump of 4 m.
c) Diameter of the tray column based on conditions at the top and bottom of the column. The system should be assumed to be susceptible to foaming. Assume vapor velocity to be 80% of the flooding velocity.
d) Number of tray passes above and below the feed.

The relative volatility between the key components is 1.57 and the viscosity of the feed is $9.21 \times 10^{-5}\, kg{\cdot}m^{-1}{\cdot}s^{-1}$. The physical properties for the distillate and bottoms compositions are given in Table 8.14.

Table 8.14

Physical properties of distillation and bottoms compositions.

	Distillate	Bottoms
M_L ($kg{\cdot}kmol^{-1}$)	57.0	87.5
M_V ($kg{\cdot}kmol^{-1}$)	55.6	80.3
ρ_L ($kg{\cdot}m^{-3}$)	476	483
ρ_V ($kg{\cdot}m^{-3}$)	34.9	41.2
μ_L ($kg{\cdot}m^{-1}{\cdot}s^{-1}$)	9.39×10^{-5}	9.03×10^{-5}
σ ($mN{\cdot}m^{-1}$)	4.6	3.7

Solution

a) From Example 8.4, for $R_{min} = 3.170$, the number of theoretical stages $N = 41$. The overall plate efficiency can be estimated from Equation 8.91:

$$E_O = -0.3143 - 0.2853 \log(\alpha_{LH} \mu_L)$$

Taking the mean viscosity of $9.21 \times 10^{-5}\,kg{\cdot}m^{-1}{\cdot}s^{-1}$:

$$E_O = -0.3143 - 0.2853 \times \log(1.57 \times 9.21 \times 10^{-5})$$
$$= 0.78$$

$$\text{Number of real trays} = \frac{41}{0.78}$$
$$= 52.6$$
$$\text{say } 53$$

b) Assuming a plate spacing of 0.6 m and 4 m allowance at the top of the column for vapor–liquid disengagement and the bottom for a sump:

$$\text{Height} = 0.6(53 - 1) + 4$$
$$= 35.2\text{ m}$$

c) The diameter of the column will be based on the flooding velocity, which is correlated in terms of the liquid–vapor flow parameter F_{LV}. F_{LV} requires the liquid and vapor rates, which change through the column. Here the assessment will be based on the flow at the top and bottom of the column assuming constant molar overflow. From Example 8.2, the distillate flow is:

$$D = 278.21\text{ kmol}{\cdot}h^{-1}$$

A mass balance around the top of the column gives:

$$V = D + L$$
$$= D(R + 1)$$

where V and L are the vapor and liquid molar flowrates at the top of the column.

$$V = 278.21(3.170 \times 1.1 + 1)$$
$$= 1248.3\text{ kmol}{\cdot}h^{-1}$$
$$L = RD$$
$$= 3.170 \times 1.1 \times 278.21$$
$$= 970.1\text{ kmol}{\cdot}h^{-1}$$

The feed of $1000\,kmol{\cdot}h^{-1}$ is saturated liquid. Thus the liquid flowrate below the feed is:

$$L' = 970.1 + 1000$$
$$= 1970.1\text{ kmol}{\cdot}h^{-1}$$

From Example 8.2, the bottoms flowrate is:

$$B = 721.8\text{ kmol}{\cdot}h^{-1}$$

The vapor flowrate below the feed is:

$$V' = 1970.1 - 721.8$$
$$= 1248.3\text{ kmol}{\cdot}h^{-1}$$

The liquid–vapor flow parameter is given by Equation 8.96:

$$F_{LV} = \left(\frac{M_L L}{M_V V}\right)\left(\frac{\rho_V}{\rho_L}\right)^{0.5}$$

At the top of the column:

$$F_{LV} = \left(\frac{57.0 \times 970.1}{55.6 \times 1248.3}\right)\left(\frac{34.9}{476}\right)^{0.5}$$
$$= 0.2157$$

At the bottom of the column:

$$F'_{LV} = \left(\frac{87.5 \times 1970.1}{80.3 \times 1248.3}\right)\left(\frac{41.2}{483}\right)^{0.5}$$
$$= 0.5023$$

The terminal velocity parameter K_T is given by Equation 8.97:

$$K_T = \left(\frac{\sigma}{20}\right)^{0.2} \exp(-2.979 - 0.717 \ln F_{LV} - 0.0865(\ln F_{LV})^2$$
$$+ 0.997 \ln H_T - 0.07973 \ln F_{LV} \ln H_T + 0.256 (\ln H_T)^2$$

At the top of the column:

$$\sigma = 4.6\text{ mN}{\cdot}m^{-1}$$
$$F_{LV} = 0.2157$$
$$H_T = 0.6\text{ m}$$

Substituting in Equation 8.97:

$$K_T = 0.0560\text{ m}{\cdot}s^{-1}$$

At the bottom of the column

$$\sigma' = 3.7\text{ mN}{\cdot}m^{-1}$$
$$F'_{LV} = 0.5023$$
$$H_T = 0.6\text{ m}$$

Substituting in Equation 8.97:

$$K'_T = 0.0356\text{ m}{\cdot}s^{-1}$$

Vapor density is greater than $28\,kg{\cdot}m^{-3}$. Calculate the foaming factor from Equation 8.98.

Above the feed:

$$F_{FOAM} = \frac{2.94}{\rho_V^{0.32}} \qquad 0 < F_{FOAM} \leq 1$$
$$= \frac{2.94}{34.9^{0.32}}$$
$$= 0.94$$

Below the feed:

$$F'_{FOAM} = \frac{2.94}{41.2^{0.32}}$$
$$= 0.89$$

The vapor flooding velocity can now be calculated from Equation 8.95 with correction for the foaming factor:

$$v_T = F_{FOAM} K_T \left(\frac{\rho_L - \rho_V}{\rho_V}\right)^{0.5}$$

8

At the top of the column:

$$v_T = 0.94 \times 0.0560\left(\frac{476 - 34.9}{34.9}\right)^{0.5}$$
$$= 0.188\ m{\cdot}s^{-1}$$

At the bottom of the column:

$$v'_T = 0.89 \times 0.0356\left(\frac{483 - 41.2}{41.2}\right)^{0.5}$$
$$= 0.104\ m{\cdot}s^{-1}$$

It should be noted that the calculated flooding velocities are lower than most distillation systems. This is due mainly to the relatively low difference in density between the liquid and the vapor in this case. Knowing the flooding velocity allows the column net area to be calculated. Above the feed, for 80% of the flooding velocity:

$$A_N = \frac{VM_V}{0.8v_T\rho_V}$$
$$= \frac{1248.3 \times 55.6}{0.8 \times 0.188 \times 3600 \times 34.9}$$
$$= 3.674\ m^2$$

Below the feed, for 80% of the flooding velocity:

$$A'_N = \frac{1248.3 \times 80.3}{0.8 \times 0.104 \times 3600 \times 41.2}$$
$$= 8.092\ m^2$$

Downcomer liquid velocity above the feed is given by Equations 8.99 to 8.103:

$$a = 7.10805$$
$$b = -6.83344 \times 10^2$$
$$c = 1.40738 \times 10^5$$
$$d = -1.21126 \times 10^7$$
$$v_D = \exp\left[a + \frac{b}{(\rho_L - \rho_V)} + \frac{c}{(\rho_L - \rho_V)^2} + \frac{d}{(\rho_L - \rho_V)^3}\right]$$
$$= \exp\left[7.10805 - \frac{6.83344 \times 10^{-2}}{(476 - 34.9)} + \frac{1.40738 \times 10^5}{(476 - 34.9)^2} - \frac{1.21126 \times 10^7}{(476 - 34.9)^3}\right]$$
$$= 464.6\ m{\cdot}h^{-1}$$
$$A_D = \frac{LM_L}{0.8v_D\rho_L F_{FOAM}}$$
$$= \frac{970.1 \times 57.0}{0.8 \times 464.6 \times 476 \times 0.94}$$
$$= 0.3314\ m^2$$

Below the feed the values of a, b, c and d are the same as above the feed, as the tray spacing is the same. From Equation 8.99:

$$v'_D = \exp\left[7.10805 - \frac{6.83344 \times 10^{-2}}{(483 - 41.2)} + \frac{1.40738 \times 10^5}{(483 - 41.2)^2} - \frac{1.21126 \times 10^7}{(483 - 41.2)^3}\right]$$
$$= 464.9\ m{\cdot}h^{-1}$$
$$A'_D = \frac{1970.1 \times 87.5}{0.8 \times 464.9 \times 483 \times 0.89}$$
$$= 1.073\ m^2$$

Thus, the diameter of the column above the feed is given by:

$$d_C = \sqrt{\frac{4(A_N + A_D)}{\pi}}$$
$$= \sqrt{\frac{4(3.674 + 0.3314)}{\pi}}$$
$$= 2.26\ m$$

Below the feed:

$$d'_C = \sqrt{\frac{4(8.092 + 1.073)}{\pi}}$$
$$= 3.42\ m$$

d) Now check the weir loading. Above the feed:

$$\frac{A'_D}{A'_C} = \frac{0.3314}{3.674 + 0.3314}$$
$$= 0.0827$$

From Figure 8.25:

$$\frac{L_W}{d_C} = 0.69$$
$$L_W = 2.26 \times 0.69$$
$$= 1.56\ m$$
$$\text{Weir loading} = \frac{970.12 \times 57.0}{476 \times 1.56}$$
$$= 74.6\ m^3{\cdot}h^{-1}{\cdot}m^{-1}$$

This is below the maximum of 90 $m^3{\cdot}h^{-1}{\cdot}m^{-1}$ and thus single-pass trays will be adequate. Below the feed:

$$\frac{A'_D}{A_C} = \frac{1.073}{8.092 + 1.073}$$
$$= 0.117$$

From Figure 8.25:

$$\frac{L'_W}{d'_C} = 0.76$$
$$L'_W = 3.42 \times 0.76$$
$$= 2.60\ m$$
$$\text{Weir loading} = \frac{1970.1 \times 87.5}{483 \times 2.60}$$
$$= 137\ m^3{\cdot}h^{-1}{\cdot}m^{-1}$$

This loading is too high for a single-pass tray. Increase the number of passes below the feed to 2. This decreases the liquid load and the diameter below the feed needs to be recalculated:

$$L' = 1970.1/2$$
$$= 985.1\ kmol{\cdot}h^{-1}$$
$$V' = 1248.3\ kmol{\cdot}h^{-1}$$
$$F'_{LV} = 0.2511$$
$$K'_T = 0.05030$$
$$F_{FOAM} = 0.89$$

$$v'_T = 0.147 \text{ m·s}^{-1}$$

$$A'_N = \frac{1248.3 \times 80.3}{0.8 \times 0.147 \times 3600 \times 41.2} = 5.733 \text{ m}^2$$

$$d'_C = \sqrt{\frac{4(5.733 + 1.073)}{\pi}} = 2.94 \text{ m}$$

Assume that the total downcomer area remains the same and the limiting downcomer length will be the side downcomers, rather than the center downcomers (see Figure 8.9). Check the weir loading for the two-pass tray by splitting the downcomer area into two and calculating the new weir length and loading for the side downcomers:

$$A'_D = \frac{1.073}{2} = 0.5365 \text{ m}^2$$

where A'_D now refers to the area of each of the two side downcomers:

$$\frac{A'_D}{A'_C} = \frac{0.5365}{5.733 + 0.5365} = 0.0856$$

From Figure 8.25:

$$\frac{L'_W}{d'_C} = 0.70$$

$$L'_W = 2.94 \times 0.70 = 2.06 \text{ m}$$

Side downcomer weir loading for a two-pass tray

$$= \frac{1970.1 \times 87.5}{483 \times 2 \times 2.06} = 87 \text{ m}^3\text{·h}^{-1}\text{·m}^{-1}$$

This loading is acceptable.

The calculation shows a significant difference between the diameter at the top and bottom. In conceptual design, it is reasonable to take the largest value. Later, when the design is considered in more detail, if different sections of a column require diameters that differ by greater than 20%, it would usually be engineered with different diameters (Kister, 1992). Such decisions can only be made after a much more detailed analysis of the column design. Also, the design here is based on the flooding velocity at the top and bottom of the column only. In practice, the flooding velocity changes throughout the column, and intermediate points might need to be considered in a more detailed analysis.

Example 8.6 For the distillation column from Example 8.5, assuming Flexipac 1.6Y structured packing is used rather than trays, estimate:

a) Height of a packed column using structured packing, assuming a total combined disengagement height at the top and bottom of the column and sump for the column of 4 m, 1 m for liquid collection and distribution at the feed and 0.5 m for liquid collection and redistribution in the rectifying and stripping sections.

b) Diameter of the column with structured packing. The system should be assumed susceptible to foaming. Assume vapor velocity to be 80% of flooding.

Physical properties can be taken from Table 8.14.

Solution

a) First estimate the *HETP* for the structured packing. From Equation 8.93:

$$HETP = \frac{100 F_{XY} F_\sigma}{\alpha_P} + 0.1$$

For Flexipac 1.6Y, $F_{XY} = 1$ and given $\sigma \le 25$ mN·m^{-1}, $F_\sigma = 1$. Also, α_P can be taken from Table 8.11:

$$HETP = \frac{100 \times 1 \times 1}{290} + 0.1 = 0.44 \text{ m}$$

From Example 8.4, above the feed there are 12 theoretical stages:

$$H = N \times HETP = 12 \times 0.44 = 5.28 \text{ m}$$

Below the feed, including the feed stage, there are (27 + 1) theoretical stages:

$$H' = 28 \times 0.44 = 12.32 \text{ m}$$

For a maximum of 20 stages per bed, above the feed can be a single bed, but below the feed two beds are required. Allowing 4 m at the top and bottom of the column, 1 m for liquid collection and distribution at the feed and 0.5 m for liquid collection and redistribution in the stripping section:

$$H = 5.28 + 12.32 + 4 + 1 + 0.5 = 23.10 \text{ m}$$

b) For Flexipac 1.6Y structured packing, from Table 8.11, the packing factor is 59 m^{-1}; thus from Equation 8.106:

$$\Delta P_{FLOOD} = 40.91 \times 59^{0.7} = 710.3 \text{ Pa·m}^{-1}$$

From Equation 8.109 and 8.110:

$$A = -7.31 \times 10^{-11} \times 710.3^3 + 2.18 \times 10^{-7} \times 710.3^2 - 2.19 \times 10^{-4} \times 710.3 - 0.0124 = -0.08416$$

$$B = 1.28 \times 10^{-10} \times 710.3 - 3.15 \times 10^{-7} \times 710.3^2 + 2.62 \times 10^{-4} \times 710.3 + 0.0826 = 0.1556$$

From Equation 8.96 above the feed:

$$F_{LV} = \left(\frac{M_L L}{M_V V}\right)\left(\frac{\rho_V}{\rho_L}\right)^{0.5} = \left(\frac{57.0 \times 947.2}{55.6 \times 1225.4}\right)\left(\frac{34.9}{476}\right)^{0.5} = 0.2146$$

Thus, from Equation 8.108:

$$CP_{FLOOD} = A \ln(F_{LV}) + B$$
$$= -0.08416 \times \ln(0.2146) + 0.1556$$
$$= 0.2852$$
$$K_T = \frac{CP_{FLOOD}}{F_P^{0.5}\left(\frac{\mu_L}{\rho_L}\right)^{0.05}}$$
$$= \frac{0.2852}{59^{0.5}\left(\frac{9.39 \times 10^{-5}}{476}\right)^{0.05}}$$
$$= 0.0803 \text{ m·s}^{-1}$$
$$v_T = K_T\left(\frac{\rho_L - \rho_V}{\rho_V}\right)^{0.5}$$
$$= 0.0803\left(\frac{476 - 34.9}{34.9}\right)^{0.5}$$
$$= 0.286 \text{ m·s}^{-1}$$

Assuming the same foaming factor for the packing as for trays and the design is for 80% of flooding:

$$d_C = \left(\frac{4M_V V}{0.8 \times 0.94 \times \pi \rho_V v_T}\right)^{0.5}$$
$$= \left(\frac{4 \times 55.6 \times 1225.4}{0.8 \times 0.94 \times \pi \times 34.9 \times 0.286 \times 3600}\right)^{0.5}$$
$$= 1.79 \text{ m}$$

Below the feed:

$$F'_{LV} = \left(\frac{87.5 \times 1947.2}{80.3 \times 1225.4}\right)\left(\frac{41.2}{483}\right)^{0.5}$$
$$= 0.5057$$
$$CP'_{FLOOD} = -0.08416 \times \ln(0.5057) + 0.1556$$
$$= 0.2130$$
$$K'_T = \frac{0.2130}{59^{0.5}\left(\frac{9.03 \times 10^{-5}}{483}\right)^{0.05}}$$
$$= 0.0602 \text{ m·s}^{-1}$$
$$v'_T = 0.0602\left(\frac{483 - 41.2}{41.2}\right)^{0.5}$$
$$= 0.1971 \text{ m·s}^{-1}$$
$$d'_C = \left(\frac{4 \times 80.3 \times 1225.4}{0.8 \times 0.89 \times \pi \times 41.2 \times 0.1971 \times 3600}\right)^{0.5}$$
$$= 2.45 \text{ m}$$

In the case of structured packing, as with a tray column, there is again a difference between the diameters at the top and bottom of the column. In the case of a tray column, the number of tray passes can be increased to decrease the difference. This option is not available in the case of a packed column. The calculation in this case results in a smaller column height and diameter for the structured packing design.

If random packing is to be used, the approach is basically the same, but using Equations 8.94 for *HETP* and Equations 8.111 and 8.112 for the flooding calculation.

8.9 Conceptual Design of Distillation

The feed composition and flowrate to the distillation are usually specified. Also, the specifications of the products are usually known, although there may be some uncertainty in product specifications. The product specifications may be expressed in terms of product purities or recoveries of certain components. The conceptual design requires the designer to specify:

- operating pressure,
- reflux ratio,
- feed condition,
- type of condenser,
- type of reboiler,
- preliminary distribution of equilibrium stages,
- preliminary specification of column internals.

1) *Pressure.* The first decision is operating pressure. As pressure is raised:
   - separation becomes more difficult (relative volatility decreases), that is, more stages or reflux are required;
   - latent heat of vaporization decreases, that is, reboiler and condenser duties become lower;
   - vapor density increases, giving a smaller column diameter;
   - reboiler temperature increases with a limit often set by thermal decomposition of the material being vaporized, causing excessive fouling;
   - condenser temperature increases.

   As pressure is lowered, these effects reverse. The lower limit is often set by the desire to avoid:
   - vacuum operation;
   - refrigeration in the condenser.

   Both vacuum operation and the use of refrigeration incur capital and operating cost penalties and increase the complexity of the design. They should be avoided if possible, but it is sometimes not possible.

   For a first pass through the design, it is usually adequate, if process constraints permit, to set distillation pressure to as low a pressure above ambient as allows cooling water or air cooling to be used in the condenser. If a total condenser is to be used and a liquid top product taken, the pressure should be fixed such that:
   - if cooling water is to be used, the bubble point of the overhead product should be typically 10 °C above the summer cooling water temperature or
   - if air cooling is to be used, the bubble point of the overhead product should be typically 20 °C above the summer air temperature or

- the pressure should be set to atmospheric pressure if either of these conditions would lead to vacuum operation.

If a partial condenser is to be used and a vapor top product taken, then the above criteria should be applied to the dew point of the vapor top product, rather than the bubble point of the liquid top product. Also, if a vapor top product is to be taken, then the operating pressure of the destination for the product might determine the column pressure (e.g. the overhead top product being sent to the fuel gas system).

There are two major exceptions to these guidelines:

- If the operating pressure of the distillation column becomes excessive as a result of trying to operate the condenser against cooling water or air cooling, then a combination of high operating pressure and low-temperature condensation using refrigeration should be used. This is usually the case when separating gases and light hydrocarbons.
- If process constraints restrict the maximum temperature of the distillation, then vacuum operation must be used in order to reduce the boiling temperature of the material to below a value at which product decomposition occurs. This tends to be the case when distilling high molar mass material.

2) *Reflux ratio.* Another variable that needs to be set for distillation is the reflux ratio. For a stand-alone distillation column (i.e. utility used for both reboiling and condensing), there is a capital–energy trade-off, as illustrated in Figure 8.26. As the reflux ratio is increased from its minimum, the capital cost decreases initially as the number of plates reduces from infinity, but the utility costs increase as more reboiling and condensation are required (Figure 8.26). If the capital costs of the column, reboiler and condenser are annualized (see Chapter 2) and combined with the annual cost of utilities (see Chapter 2), the optimal reflux ratio is obtained. The optimal ratio of actual to minimum reflux is often less than 1.1. However, most designers are reluctant to design columns closer to the minimum reflux than 1.1, except in special circumstances, since a small error in design data or a small change in operating conditions might lead to an infeasible design. Also, the total cost curve is often relatively flat over a range of relative volatility around the optimum. No attempt should be made to do the optimization illustrated in Figure 8.26 until a picture of the overall process design has been established. Later, when heat integration of the column with the rest of the process is considered, the nature of the trade-off changes, and the optimal reflux ratio for the heat-integrated column can be very different from that for a stand-alone column. Any reasonable assumption is adequate in the initial stages of a design, say, a ratio of actual to minimum reflux of 1.1.

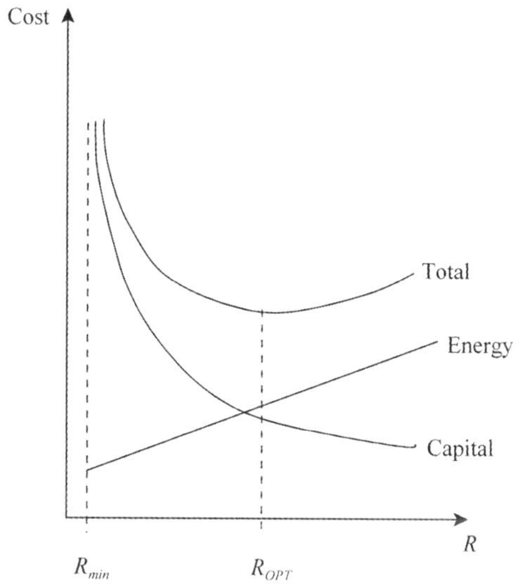

Figure 8.26

The capital–energy trade-off stand-alone distillation columns.

3) *Feed condition.* Another variable that needs to be fixed is the feed condition. For the distillation itself, the optimum feed point should minimize any mixing between the feed and the flows within the column. In theory, for a binary distillation it is possible to achieve an exact match between a liquid feed and the liquid on a feed stage. In practice, this might not be possible to achieve, as changes from stage to stage are finite. For a multicomponent system, in general it is not possible to achieve an exact match, except under special circumstances. If the feed is partially vaporized, the degree of vaporization provides a degree of freedom to obtain a better match between the composition of the liquid and vapor feed and the liquid and vapor flows in the column. However, minimizing feed mixing does not necessarily minimize the operating costs.

Heating the feed most often:

- increases trays in the rectifying section, but decreases trays in the stripping section;
- requires less heat in the reboiler but more cooling in the condenser.

Cooling the feed most often reverses these effects.

Heat added to provide feed preheating may not substitute heat added to the reboiler on an equal basis (Liebert, 1993). The ratio of heat added to preheat the feed divided by the heat saved in the reboiler tends to be less than unity. Although heat added to the feed may not substitute heat added in the reboiler, it changes the minimum reflux ratio as a result of change to the feed condition (q in Equation 8.50). As the condition of the feed is changed from saturated liquid feed ($q = 1$) to saturated vapor feed ($q = 0$), the minimum reflux ratio tends to increase. Thus the ratio of heat added to preheat the feed divided by the heat saved in the reboiler depends on the change in q, the relative volatility between the key components, feed concentration and ratio of actual to minimum reflux. In some circumstances, particularly at high values of actual to minimum reflux ratio, heating the feed can increase the reboiler duty (Liebert, 1993).

For a given separation, the feed condition can be optimized. No attempt should be made to do this in the early stages of an overall design, since heat integration is likely to change the optimal settings later in the design. It is usually adequate to set the feed to saturated liquid conditions. This tends to equalize the vapor rate below and above the feed, and having a liquid feed allows the column pressure to be increased if necessary through the feed pump. Of course if a vaporized feed is required, a liquid feed can be pressurized first with a pump before being vaporized and fed to the column.

4) *Type of condenser.* Either a total or partial condenser can be chosen. Most designs use a total condenser. A total condenser is necessary if the top product needs to be sent to intermediate or final product storage. Also, a total condenser is best if the top product is to be fed to another distillation at a higher pressure as the liquid pressure can readily be increased using a pump.

   If a partial condenser is chosen, then the partial condenser in theory acts as an additional stage, although in practice the performance tends to be less than a theoretical stage. A partial condenser reduces the condenser duty, which is important if the cooling service to the condenser is expensive, such as low-temperature refrigeration. It is often necessary to use a partial condenser when distilling mixtures with low-boiling components that would require very low-temperature (and expensive) refrigeration for a total condenser. Also, in these circumstances a *mixed condenser* might be used. This condenses what is possible as liquid for reflux and a liquid top product is taken as well as a vapor top product. In such designs, the uncondensed material often goes to a gas collection system, or *fuel header*, to be used, for example, as fuel gas. If the uncondensed material from a partial condenser is to be sent to a downstream distillation column for further processing, as has already been noted under the discussion on feed condition, there will be implications for the reboiler and condenser duties of the downstream column. A vapor feed will generally increase the condenser duty and decrease the reboiler duty of the downstream column. Whether this is good or bad for the operating costs of the downstream column depends on whether the cold utility used for the condenser or the hot utility used for the reboiler is more expensive. Also, when heat integration is discussed later, there might be important implications resulting from using partial condensers in the design of the overall process.

5) *Type of reboiler.* The major decision to be made is whether a kettle or thermosyphon reboiler is to be chosen. A kettle reboiler provides a theoretical stage. Even though some separation will occur in a thermosyphon, it will always be less than a theoretical stage. The choice of reboiler depends on:
   - the nature of the process fluid (particularly its viscosity and fouling tendencies);
   - sensitivity of the bottoms product to thermal degradation;
   - operating pressure;
   - temperature difference between the process and heating medium;
   - equipment layout (particularly the space available for headroom).

   Thermosyphon reboilers are usually cheapest but not suitable for high-viscosity liquids or vacuum operation. Further discussion of reboilers will be deferred until Chapter 12. In the preliminary phases of a design, it is best to assume that enough stages are available in the column itself to achieve the required separation.

6) *Preliminary distribution of equilibrium stages.* A preliminary estimate of the number of theoretical stages can be obtained from the Gilliland Correlation and the distribution of stages from the Kirkbride Correlation.

7) *Preliminary specification of column internals.* The first major decision to be made regarding the column internals is whether to use trays or packing. Generally, trays should be used when:
   - the liquid flowrate is high relative to the vapor flowrate (occurs when the separation is difficult);
   - the diameter of the column is large (packings can suffer from maldistribution of the liquid over the packing);
   - there is variation in the feed composition (trays can be more flexible to variations in operating conditions);
   - the column requires multiple feeds or multiple products (tray columns are simpler to design for multiple feeds or products).

   Packings should be used when:
   - the column diameter is small (the relative cost for fabrication of small-diameter trays is high);
   - vacuum conditions are used (packings offer a lower pressure drop and reduced entrainment tendencies);
   - a low pressure drop is required (packings have a lower pressure drop than trays);
   - the system is corrosive (a greater variety of corrosion-resistant materials is available for packings);
   - the system is prone to foaming (packings have a lower tendency to promote foaming);
   - a low liquid hold-up in the column is required (liquid hold-up in packings is lower than that for trays).

   If trays are chosen, then the most common options are sieve, floating valve or fixed valve trays. Sieve trays are the cheapest and most common but have a much poorer turn-down ratio than valve trays. The number of tray passes must also be determined.

   If packing is chosen, then the choice must be made between structured and random packing. Structured packing requires a smaller volume than random packing for the same separation, but is more expensive.

8.10 Detailed Design of Distillation

Once the major decisions have been made in the development of the conceptual design, then a detailed simulation needs to be carried out. To develop a rigorous approach to distillation design, consider Figure 8.27. This shows a general equilibrium stage in the distillation column. It is a general stage and allows for many design options other than simple columns with one feed and two products. Vapor and liquid enter this stage. In principle, an additional product can be withdrawn from the distillation column at an intermediate stage as a liquid or vapor *sidestream*. Feed can enter and heat can be transferred to or from the stage. By using a general representation of a stage, as shown in Figure 8.27, designs with multiple feeds, multiple products and intermediate heat exchange

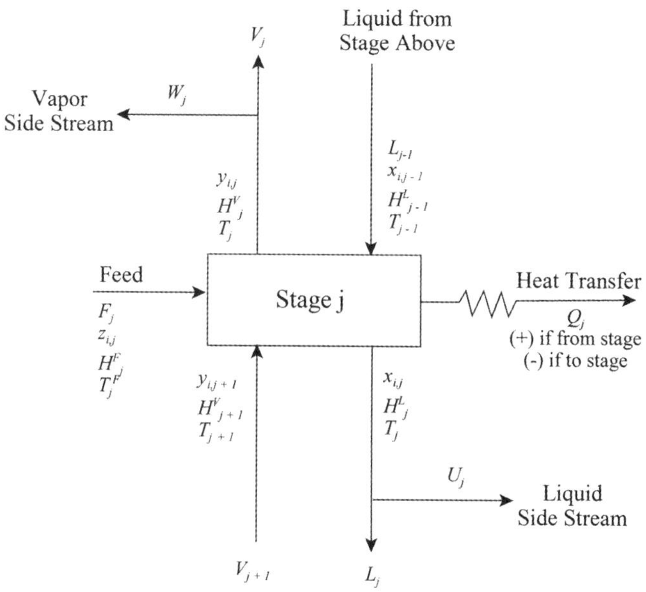

Figure 8.27

A general equilibrium stage for distillation. (Reproduced from Smith R and Jobson M (2000) Distillation, Encyclopedia of Separation Science, Academic Press, with permission from Elsevier.)

are possible. Equations can be written to describe the material and energy balance for each stage:

1) Material balance for Component i and Stage j (NC equations for each stage):

$$L_{j-1}x_{i,j-1} + V_{j+1}y_{i,j+1} + F_j z_{i,j} - (L_j + U_j)x_{i,j} - (V_j + W_j)y_{i,j} = 0 \tag{8.114}$$

2) Equilibrium relation for each Component i (NC equations for each stage):

$$y_{i,j} - K_{i,j}x_{i,j} = 0 \tag{8.115}$$

3) Summation equations (one for each Stage j):

$$\sum_{i=1}^{NC} y_{i,j} - 1.0 = 0, \sum_{i=1}^{NC} x_{i,j} - 1.0 = 0 \tag{8.116}$$

4) Energy balance (one for each Stage j):

$$L_{j-1}H_{j-1}^L + V_{j+1}H_{j+1}^V + F_j H_j^F - (L_j + U_j)H_j^H - (V_j + W_j)H_j^V - Q_j = 0 \tag{8.117}$$

where
$z_{i,j}, y_{i,j}, x_{i,j}$ = feed, vapor and liquid mole fractions for Stage j
F_j, V_j, L_j = feed, vapor and liquid molar flowrates for Stage j
W_j, U_j = vapor and liquid sidestream molar flowrates for Stage j
H_F = molar enthalpy of the feed
H_j^V, H_j^L = vapor and liquid molar enthalpies for Stage j
Q_j = heat transfer from Stage j (negative for heat transfer to the stage)
$K_{i,j}$ = vapor–liquid equilibrium constant between x_i and y_i for Stage j
NC = number of components

Equations 8.114 to 8.117 for the material balances (M), equilibrium relationships (E), summation of compositions (S) and enthalpy balances (H) are known as MESH equations. They require physical property data for vapor–liquid equilibrium and enthalpies, and the set of equations must be solved simultaneously. The calculations are complex, but many methods are available to solve the equations (King, 1980; Kister, 1992; Seader, Henley and Roper, 2011). The details of the methods used are outside the scope of this text. In practice, designers most often use commercial computer simulation packages.

In order to carry out a detailed simulation, the following need to be specified:

a) *Feed composition and flowrate.* The composition of the feed and its flowrate must be specified.
b) *Operating pressure at the top of the column.* Initially, the operating pressure is normally set by the desire to be able to use cooling water or air cooling in the overhead condenser, but vacuum operation should be avoided if possible. If a very high operating pressure is required as a result of trying to operate the condenser against cooling water or air cooling, a combination of high operating pressure and low-temperature condensation using refrigeration should be used. Process constraints might restrict the maximum temperature of the distillation to avoid product decomposition. In these circumstances, vacuum operation might be necessary to reduce the boiling temperature.
c) *Pressure drop.* The pressure drop across the column must be specified or the pressure drop per plate (typically 0.01 bar per plate).
d) *Feed condition.* Specify two from temperature, pressure, q (for saturated liquid feed $q = 1$, saturated vapor feed $q = 0$).
e) *Stage layout.* The number of theoretical stages and the feed stage must be specified. For more complex columns, the layout of the stages needs to be specified.
f) *Material and energy balance.* Specify two from distillate flowrate, bottoms flowrate, a component recovery for one of the products (up to two can be chosen), reflux ratio, reboil S ($S = V / B$), condenser duty, reboiler duty. Other specifications are possible but less common.

Once the detailed design has been developed based on theoretical stages, the stage efficiency can be incorporated if using trays. This might be a fixed overall stage efficiency or using a stage efficiency model (King, 1980; Kister, 1992). For a fixed overall stage efficiency, all components are assumed to have the same stage efficiency. However, in practice, each component has a different stage efficiency. This can only be allowed for by using detailed stage efficiency models, which are outside the scope of this text.

Having determined the layout of the actual trays and the material and energy balance for the column, the column internals can be investigated in more detail (Kister, 1992). If trays are to be used then this requires hydraulic calculations to determine the

8

detailed layout of the trays. A detailed simulation also allows the hydraulic design to be considered on a tray by tray basis. This means that the flooding characteristics (and weeping characteristics) can vary through the rectifying and stripping sections. In turn, this means that in principle the tray design can vary through the rectifying and stripping sections. If packing is to be used, then the column design can also be refined. In principle, the packing design can change through the rectifying and stripping sections. The final hydraulic design is best left to equipment vendors.

Fine tuning the design will involve fixing the feed composition and flowrate, and the separation specification (but not necessarily fixing the separation rigidly as long as the required separation specifications are achieved). The following can then be fine-tuned:

- pressure,
- feed condition,
- reflux ratio,
- material balance (within the constraints of the separation specification),
- stage layout,
- tray type, number of passes, spacing, tray design and downcomer design in different sections of the column in the case of tray columns,
- packing type in different sections of the column in the case of packed columns.

In addition to checking the column design at design conditions, the turndown of the column to reduced capacity needs also to be considered. At turndown conditions, the column design must be checked for weeping (Kister, 1992).

Whilst it is important to establish a preliminary design in order to establish the overall process design, significant effort on detailed design and optimization should be avoided until the context of the separation in the overall process has been fully understood.

Example 8.7 Carry out a rigorous simulation of the distillation from Examples 8.2 to 8.6 and determine if the conceptual design achieves the required recovery specifications of an overhead product that recovers 99% of the *n*-butane overhead and 95% of the *i*-pentane in the bottoms. For the purpose of comparison, assume the column operates at 14 bar with no pressure drop across the column. Assume, as determined from Exercise 8.4, that the column has 41 theoretical stages with a saturated liquid feed fed to Stage 13 (counting from the top of the column) with a total condenser and a reboiler that does not provide a theoretical stage (i.e. all stages are within the column). The energy balance is to be specified by the reflux ratio with $R_{min} = 3.170$ and $R/R_{min} = 1.1$, giving $R = 3.487$, taken from Example 8.3. The material balance is to be specified by the distillate flowrate with a value of 278.21 kmol·h^{-1} taken from Exercise 8.2. Calculate the vapor–liquid equilibrium and liquid and vapor enthalpies from the Peng Robinson equation of state.

Solution Many commercial software packages are available to carry out this calculation. Figure 8.28 shows the resulting material

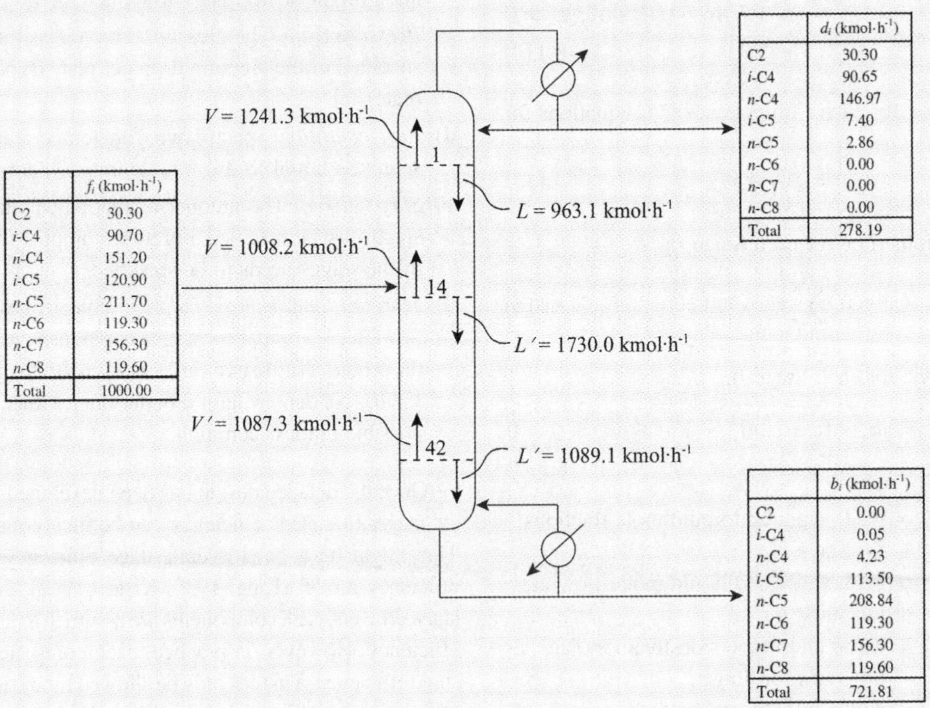

	f_i (kmol·h^{-1})
C2	30.30
i-C4	90.70
n-C4	151.20
i-C5	120.90
n-C5	211.70
n-C6	119.30
n-C7	156.30
n-C8	119.60
Total	1000.00

	d_i (kmol·h^{-1})
C2	30.30
i-C4	90.65
n-C4	146.97
i-C5	7.40
n-C5	2.86
n-C6	0.00
n-C7	0.00
n-C8	0.00
Total	278.19

	b_i (kmol·h^{-1})
C2	0.00
i-C4	0.05
n-C4	4.23
i-C5	113.50
n-C5	208.84
n-C6	119.30
n-C7	156.30
n-C8	119.60
Total	721.81

Figure 8.28
Mass balance for the rigorous simulation from Example 8.7.

balance from the rigorous simulation. From the data in Figure 8.28 the recovery of n-butane in the overhead is 97% compared with the specification of 99%. The recovery of i-pentane in the bottoms is 94%, compared with the specification of 95%. To achieve the required recoveries needs some adjustments to the design.

i) The number of stages in the rectifying and stripping sections can be increased for the same reflux ratio, adjusting the total number of stages and keeping the ratio of the number of stages in the rectifying and stripping stages. This does not allow the required recoveries to be readily achieved in this case. The total number of stages and the ratio of the number of stages in the rectifying and stripping stages need to be adjusted. It should be recalled in Example 8.3 that the minimum reflux ratio predicted by the Underwood Equations tends to be low.
ii) The reflux ratio can be adjusted for the same number of stages. If the reflux ratio is increased from 3.478 to 3.730 for the same number of stages, the recovery of n-butane in the overhead increases to 99%, and the recovery of i-pentane in the bottoms increases to 95.5%, compared with the specification of 95%. Adjusting the reflux ratio and the reboil ratio independently allows both specifications to be achieved simultaneously for the same number of stages. If the reflux ratio and the reboil ratio are adjusted independently, then the specification for the distillate rate needs to be relaxed.
iii) The reflux ratio, reboil ratio and number of stages in the rectifying and stripping sections can be varied simultaneously, Additionally, the feed condition can also be varied simultaneously. This requires significant trial and error.

Figure 8.28 also shows the vapor and liquid flowrates at the top, bottom and feed stages. The vapor and liquid flowrates vary through the rectifying and stripping sections of the column, which means that the hydraulic design of the column internals will vary through the rectifying and stripping sections.

To complete the design requires an allowance for a pressure drop to be included. A pressure drop across each stage of typically 0.01 bar needs to be included. Then the sensitivity of the design to the number of stages, location of the feed, reflux ratio, reboil ratio, feed condition and distillate rate needs to be tested.

8.11 Limitations of Distillation

The most common method for the separation of homogeneous fluid mixtures with fluid products is distillation. Distillation allows virtually complete separation of most homogeneous fluid mixtures. It is no accident that distillation is the most common method used for the separation of homogeneous fluid mixtures with fluid products. Distillation has the following three principal advantages relative to competitive separation processes.

1) The ability to separate mixtures with a wide range of throughputs; many of the alternatives to distillation can only handle low throughput.
2) The ability to separate mixtures with a wide range of feed concentrations; many of the alternatives to distillation can only handle relatively pure feeds.
3) The ability to produce high-purity products; many of the alternatives to distillation only carry out a partial separation and cannot produce pure products.

However, distillation does have limitations. The principal cases where distillation is not well suited for the separation are as follows.

1) *Separation of materials with low molar mass.* Low molar mass materials are distilled at high pressure to increase their condensing temperature and to allow, if possible, the use of cooling water or air cooling in the column condenser. Very low molar mass materials often require refrigeration in the condenser in conjunction with high pressure. This significantly increases the cost of the separation since refrigeration is expensive. Absorption, adsorption and membrane gas separators are the most commonly used alternatives to distillation for the separation of low molar mass materials.
2) *Separation of heat-sensitive materials.* High molar mass material is often heat sensitive and will decompose if distilled at high temperature. Low molar mass material can in some cases also be heat sensitive, particularly when its nature is highly reactive (e.g. organic compounds with double and triple chemical bonds). Such material will normally be distilled under vacuum to reduce the boiling temperature. Crystallization and liquid–liquid extraction can be used as alternatives to the separation of high molar mass heat-sensitive materials.
3) *Separation of components with a low concentration.* Distillation is not well suited to the separation of products that form a low concentration in the feed mixture. Adsorption and absorption are both effective alternative means of separation in this case.
4) *Separation of classes of components.* If a class of components is to be separated (e.g. a mixture of aromatic components from a mixture of aliphatic components), then distillation can only separate according to boiling points, irrespective of the class of component. In a complex mixture where classes of components need to be separated, this might mean isolating many components unnecessarily. Liquid–liquid extraction and adsorption can be applied to the separation of classes of components.
5) *Mixtures with low relative volatility or which exhibit azeotropic behavior.* Some homogeneous liquid mixtures exhibit highly nonideal behavior that form constant boiling azeotropes. At an azeotropic composition, the vapor and liquid are both at the same composition for the mixture. Thus, separation cannot be carried out beyond an azeotropic composition using conventional distillation. The most common method used to deal with such problems is to add a mass separation agent to the distillation to alter the relative volatility of the key separation in a favorable way and make the separation feasible. If the separation is possible but extremely difficult because of low

relative volatility, then a mass separation agent can also be used in these circumstances, in a similar way to that used for azeotropic systems. These processes are considered in detail later in Chapter 11. Crystallization, liquid–liquid extraction and membrane processes can be used as alternatives to distillation for the separation of mixtures with low relative volatility or which exhibit azeotropic behavior.

6) *Separation of mixtures of condensable and noncondensable components.* If a vapor mixture contains both condensable and noncondensable components, then a partial condensation followed by a simple phase separator can often give a good separation. This is essentially a single-stage distillation operation. It is a special case that again deserves attention in some detail later in Chapter 14.

In summary, distillation is not well suited for separating either low molar mass materials or high molar mass heat-sensitive materials. However, distillation might still be the best method for these cases, since the basic advantages of distillation (potential for high throughput, any feed concentration and high purity) still prevail. The next chapter will consider other methods for the separation of homogeneous liquid mixtures.

8.12 Separation of Homogeneous Fluid Mixtures by Distillation – Summary

The design of a distillation separation involves a number of steps:

a) Set the separation and product specifications.
b) Set the operating pressure.
c) Determine the number of theoretical stages required and the energy requirements.
d) Determine the actual number of trays or height of packing needed and the column diameter.
e) Design the column internals, which involves determining the dimensions of the trays, packing, liquid and vapor distribution systems, and so on.

Once the process design is complete, mechanical design is necessary to determine the thickness of the vessel walls, internal fittings, and so on.

Even though the conceptual design of distillation must be carried out in the early stages of the development of a process design or retrofit, the assessment of distillation processes ideally should be done in the context of the total system. As will be discussed later, separators such as distillation that use an input of heat to carry out the separation can often be run at effectively zero energy cost if they are appropriately heat integrated with the rest of the process. Although energy intensive, heat-driven separators can be energy efficient in terms of the overall process if they are properly heat integrated. It is not worth expending significant effort optimizing pressure, feed condition, reflux ratio and number of stages until the overall heat integration picture has been established. These parameters might change later in the design of the overall system.

Table 8.15

Data for a mixture of aromatics.

Component	Feed (kmol)	A_i	B_i	C_i
Benzene	1	9.2806	2789.51	−52.36
Toluene	26	9.3935	3096.52	−53.67
Ethylbenzene	6	9.3993	3279.47	−59.95
Xylene	23	9.5188	3366.99	−59.04

8.13 Exercises

1. For the mixture of aromatics in Table 8.15, determine:
   a) The bubble point at a pressure of 1 bar.
   b) The bubble point at a pressure of 5 bar.
   c) The pressure needed for total condensation at a temperature of 313 K.
   d) At a pressure of 1 bar and a temperature of 400 K, how much liquid will be condensed?

   Assume that K-values can be correlated by Equation 8.25 with pressure in bar, temperature in Kelvin and constants A_i, B_i and C_i given Table 8.15.

2. Acetone is to be produced by the dehydrogenation of 2-propanol. In this process, the product from the reactor contains hydrogen, acetone, 2-propanol and water, and is cooled before it enters a flash drum. The purpose of the flash drum is to separate hydrogen from the other components. Hydrogen is removed in the vapor stream and sent to a furnace to be burnt. Some acetone is, however, carried over in the vapor steam. The minimum temperature that can be achieved in the flash drum using cooling water is 35 °C. The operating pressure is 1.1 bar absolute. The component flowrates to the flash drum are given in Table 8.16. Assume that the K-values can be correlated by Equation 8.25 with pressure in bar, temperature

Table 8.16

Flowrate and vapor–liquid equilibrium data for acetone production.

Component	Flowrate ($kmol\cdot h^{-1}$)	A_i	B_i	C_i
Hydrogen	76.95	7.0131	164.90	3.19
Acetone	76.95	10.0311	2940.46	−35.93
2-propanol	9.55	12.0727	3640.20	−53.54
Water	36.6	11.6834	3816.44	−46.13

in Kelvin and constants A_i, B_i and C_i given in Table 8.16 for the individual components.

a) Estimate the flow of acetone and 2-propanol in the vapor stream.

b) Estimate the flow of hydrogen in the liquid stream.

c) Would another method for calculating K-values have been more appropriate?

3. A mixture of benzene and toluene has a relative volatility of 2.34. Sketch the x–y diagram for the mixture, assuming the relative volatility to be constant.

4. In a mixture of methanol and water, methanol is the most volatile component. At a pressure of 1 atm, the relative volatility can be assumed to be constant and equal to 3.60. Construct the x–y diagram.

5. A feed mixture of methanol and water containing a mole fraction of methanol of 0.4 is to be separated by distillation at a pressure of 1 atm. The overhead product should achieve a purity of 95 mole% methanol and the bottoms product a purity of 95 mole% water. Assume the feed to be saturated liquid. Using the x–y diagram constructed in Exercise 4 and the McCabe–Thiele construction:

a) determine the minimum reflux ratio;

b) for a reflux ratio of 1.1 times the minimum reflux, determine the number of theoretical stages for the separation.

6. A distillation calculation is to be performed on a multicomponent mixture. The vapor–liquid equilibrium for this mixture is likely to exhibit significant departures from ideality, and an activity coefficient model is to be used to model the vapor–liquid equilibrium (see Appendix A). Unfortunately, a complete set of binary interaction parameters is not available. What factors would you consider in assessing whether the missing interaction parameters are likely to have an important effect on the calculations?

7. The two components, A and B, are to be separated in a distillation column for which the physical properties are largely unknown. The mole fractions of A and B in the overheads are 0.96 and 0.04 respectively. The overhead vapor can be assumed to be condensed at a uniform temperature corresponding with the bubble point temperature of the overhead mixture. Cooling water at an assumed temperature of 30 °C is to be used in the condenser. It can be assumed also that the minimum allowable temperature difference in the condenser is 10 °C. No vapor–liquid equilibrium data or vapor pressure data are available for the two components. Only measured normal boiling points are available, together with critical temperatures and pressures that have been estimated from the structure of the components. Vapor pressures can be estimated from these data assuming the behavior follows the Clausius–Clapeyron Equation from the normal boiling point to the critical point:

$$\ln P_i^{SAT} = \alpha_i + \frac{\beta_i}{T} \qquad (8.118)$$

Table 8.17

Physical properties of Components A and B.

Component	Critical temp (K)	Critical pressure (bar)	Normal boiling point (K)
A	369.8	42.5	231.0
B	425.2	39.0	272.6

where P_i^{SAT} is the vapor pressure, T is the absolute temperature and α_i and β_i are constants for each component. The physical property data are summarized in Table 8.17.

Assuming the vapor–liquid equilibrium to be ideal, at what pressure would the distillation column have to operate on the basis of the temperature in the condenser?

8. A saturated liquid mixture of ethane, propane, n-butane, n-pentane and n-hexane given in Table 8.18 is to be separated by distillation such that 95% of the propane is recovered in the distillate and 90% of the butane is recovered in the bottoms. The operating pressure of the column is 10 bar. Assume that the K-values can be correlated by Equation 8.25 where T is the absolute temperature (K), P the pressure (bar) and constants A_i, B_i and C_i are given in Table 8.18.

a) The bubble point of the feed at 10 bar.

b) The distribution of the non-key components in the distillation.

c) Comment on the vapor-liquid equilibrium method used.

9. The second column in the distillation train of an aromatics plant is required to split toluene and ethylbenzene. The recovery of toluene in the overheads must be 95%, and 90% of the ethylbenzene must be recovered in the bottoms. In addition to toluene and ethylbenzene, the feed also contains benzene and xylene. The feed enters the column under saturated conditions at a temperature of 170 °C, with component flowrates given in Table 8.19. Estimate the mass balance around the column using the Fenske Equation. Assume that

Table 8.18

Feed and physical property for constants.

Component	Formula	Feed (kmol·h^{-1})	A_i	B_i	C_i
Ethane	C_2H_6	5	9.0435	1511.4	−17.16
Propane	C_3H_8	25	9.1058	1872.5	−25.16
n-Butane	C_4H_{10}	30	9.0580	2154.9	−34.42
n-Pentane	C_5H_{12}	20	9.2131	2477.1	−39.94
n-Hexane	C_6H_{14}	20	9.2164	2697.6	−49.78

8

Table 8.19

Data for mixture of aromatics.

Component	Feed ($kmol \cdot h^{-1}$)	A_i	B_i	C_i
Benzene	1	9.2806	2789.51	−52.36
Toluene	26	9.3935	3096.52	−53.67
Ethylbenzene	6	9.3993	3279.47	−59.95
Xylene	23	9.5188	3366.99	−59.04

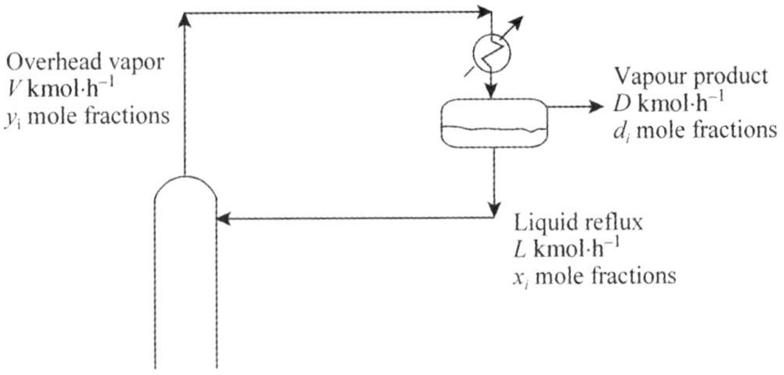

Figure 8.29

Partial condenser for a distillation column.

the K-values can be correlated by Equation 8.25 with constants A_i, B_i and C_i given in Table 8.19.

10. A distillation column separating 150 $kmol \cdot h^{-1}$ of a four-component mixture is estimated to have a minimum reflux ratio of 3.5. The composition and the relative volatility of the feed are shown in Table 8.20. The column recovers 95% of Component *B* to the distillate and 95% of Component *C* to the bottoms product. The feed to the column is a saturated liquid.

    a) Calculate the molar flowrate and composition of the distillate and bottoms products for this column. State any assumptions you need to make.

    b) Estimate the vapor load of the reboiler to this column if the reflux ratio is 25% greater than the minimum reflux ratio. Constant molar overflow can be assumed in the column and that a total condenser is used.

    c) Use the Gilliland Correlation to estimate the minimum number of theoretical stages required to carry out this separation.

11. A distillation column uses a partial condenser as shown in Figure 8.29. Assume that the reflux ratio and the overhead product composition and flowrate and the operating pressure are known and that the behavior of the liquid and vapor phases in the column is ideal (i.e. Raoult's Law holds). How can the flowrate and composition of the vapor feed to the condenser and its products be estimated, given the vapor pressure data for the pure components? Set up the equations that need to be solved.

12. A mixture of ethane, propane, *n*-butane, *n*-pentane and *n*-hexane given in Table 8.21 is to be separated by distillation such that 95% of the propane is recovered in the distillate and 90% of the butane is recovered in the bottoms. The column will operate at 18 bar. Data for the feed and relative volatility are given in Table 8.21.

    For the separation, calculate:

    a) The distribution of the nonkey components using the Fenske Equation.

    b) The minimum reflux ratio from the Underwood Equations.

    c) The actual number of stages at $R/R_{min} = 1.1$ from the Gilliland Correlation.

13. For the same feed, operating pressure and relative volatility as Exercise 12, the heavy key component is changed to pentane. Now 95% of the propane is recovered in the overheads and 90% of the pentane in the bottoms. Assuming that all lighter than light key components go to the overheads and all the heavier than heavy key components go to the bottoms, estimate the distribution of the butane and the minimum reflux ratio using the Underwood Equations.

Table 8.20

Feed characteristics of a four-component mixture.

Component	Mole fraction in the feed	Relative volatility in the feed (relative to component *D*)
A	0.10	3.9
B	0.35	2.5
C	0.30	1.6
D	0.25	1.0

Table 8.21

Data for a five-component system.

Component	Formula	Feed ($kmol \cdot h^{-1}$)	α_{ij}
Ethane	C_2H_6	5	16.0
Propane	C_3H_8	25	7.81
n-Butane	C_4H_{10}	30	3.83
n-Pentane	C_5H_{12}	20	1.94
n-Hexane	C_6H_{14}	20	1.00

References

Bahadori A and Vuthaluru HB (2010) Predictive Tools for the Estimation of Downcomer Velocity and Vapor Capacity Factor in Fractionators, *Applied Energy*, **87**: 2615.

Douglas JM (1988) *Conceptual Design of Chemical Processes*, McGraw-Hill.

Enríquez-Gutiérrez VM, Jobson M and Smith R (2014) A Design Methodology for Retrofit of Crude Oil Distillation Systems, *24th European Symposium on Computer Aided Process Engineering – ESCAPE 24*, June 15–18, Budapest, Hungary.

Fair JR (1961) How to Predict Sieve Tray Entrainment and Flooding, *Petrol Chem Eng*, **33**: 45.

Fenske MR (1932) Fractionation of Straight-Run Pennsylvania Gasoline, *Ind Eng Chem*, **24**: 482.

Geankopolis CJ (1993) *Transport Processes and Unit Operations*, 3rd Edition, Prentice Hall.

Gilliland ER (1940) Multicomponent Rectification – Estimation of the Number of Theoretical Plates as a Function of the Reflux Ratio, *Ind Eng Chem*, **32**: 1220.

Kessler DP and Wankat PC (1988) Correlations for Column Parameters, *Chem Eng*, **Sept**: 72.

King CJ (1980) *Separation Processes*, 2nd Edition, McGraw-Hill.

Kirkbride CG (1944) Process Design Procedure for Multicomponent Fractionators, *Petroleum Refiner*, **23** (9): 87.

Kister HZ (1992) *Distillation Design*, McGraw-Hill.

Kister HZ, Scherffius J, Afshar K and Abkar E (2007) Realistically Predict Capacity and Pressure Drop for Packed Columns, *Chem Eng Progr*, **July**: 28.

Kister HZ and Olsson M (2010) Understanding Maldistribution in 3-Pass Trays, *Distillation Absorption 2010*, 12–15 Sept 2010, Eindhoven, The Netherlands, p. 617.

Liebert TC (1993) Distillation Feed Preheat – Is It Energy Efficient? *Hydrocarbon Process*, **Oct**: 37.

McCabe WL and Thiele EW (1925) Graphical Design of Fractionating Columns, *Ind Eng Chem*, **17**: 605.

O'Connell HE (1946) Plate Efficiency of Fractionating Columns and Absorbers, *Trans AIChE*, **42**: 741.

Perry RH and Green DW (2007) *Perry's Chemical Engineers' Handbook*, 8th Edition, McGraw-Hill.

Pilling M (2005) Ensure Proper Design and Operation of Multi-pass Trays, *Chem Eng Progr*, **June**: 22.

Pilling M and Holden BS (2009) Choosing Trays and Packings for Distillation, *Chem Eng Progr*, **Sept**: 44.

Poling BE, Prausnitz JM and O'Connell JP (2001) *The Properties of Gases and Liquids*, 5th Edition, McGraw-Hill.

Rachford HH and Rice JD (1952) Procedure for Use in Electrical Digital Computers in Calculating Flash Vaporization Hydrocarbon Equilibrium, *Journal of Petroleum Technology*, Sec. 1 **Oct**: 19.

Rusche FA (1999) Gilliland Plot Revisited, *Hydrocarbon Process*, **Feb**: 79.

Seader JD, Henley EJ and Roper DR (2011) *Separation Process Principles*, 3rd Edition, John Wiley & Sons.

Smith R and Jobson M (2000) *Distillation, Encyclopedia of Separation Science*, Academic Press.

Souders M. and Brown GG (1934) Design of Fractionating Columns, Entrainment and Capacity, *Ind Eng Chem*, **38**: 98.

Treybal RE (1980) *Mass Transfer Operations*, 3rd Edition, McGraw-Hill.

Underwood AJV (1946) Fractional Distillation of Multicomponent Mixtures – Calculation of Minimum Reflux Ratio, *J Inst Petrol*, **32**: 614.

Walas SM (1985) *Phase Equilibria in Chemical Engineering*, Butterworth Publishers.

Chapter 9

Separation of Homogeneous Fluid Mixtures II – Other Methods

9.1 Absorption and Stripping

In absorption, a gas or vapor mixture is contacted with a liquid *solvent* that preferentially dissolves or reacts with one or more components of the vapor. Absorption processes can be divided into two broad classes:

1) *Physical absorption*, in which the solvent does not react with the absorbed species. Examples include removal of the volatile organic compounds (VOCs) using water or organic solvents, removal of H_2S or CO_2 using methanol and recovery of heavy hydrocarbons from a vapor mixture of hydrocarbons from using naphtha as a solvent.
2) *Chemical absorption*, in which the solvent and absorbed species undergo a reaction. This can enhance the rate of absorption. Examples include removal of H_2S and CO_2 using amine and alkanolamine solvents and scrubbing of SO_2 using solutions of NaOH.

Absorption is a common alternative to distillation for the separation of low molar mass materials. Absorption processes often require an extraneous material to be introduced into the process to act as a liquid solvent. Both aqueous and organic solvents can be used and the solvent can in some cases be regenerated in a stripper and recycled. If it is possible to use one of the materials already in the process, this should be done in preference to introducing an extraneous material.

Consider first the physical absorption. Liquid flowrate, temperature and pressure are important variables to be set. The vapor–liquid equilibrium for such systems can often be approximated by Henry's Law (see Appendix A):

$$p_i = H_i x_i \tag{9.1}$$

where p_i = partial pressure of Component i
H_i = Henry's Law constant (determined experimentally)
x_i = mole fraction of Component i in the liquid phase

Assuming ideal gas behavior ($p_i = y_i P$):

$$y_i = \frac{H_i x_i}{P} \tag{9.2}$$

Thus, the K-value is $K_i = H_i/P$, and a straight line would be expected on a plot of y_i against x_i. The analysis of absorption processes is greatly simplified if it can be assumed that the vapor and liquid flowrates are constant. This is analogous to the assumption of constant molar overflow in distillation. If constant molar overflow can be assumed in distillation, then a simple graphical method, the McCabe–Thiele method can be developed. For absorption, a mass balance can be carried out around a part of the absorber to obtain the operating line, as shown in Figure 9.1. This assumes there is no change in the vapor and liquid flowrates as a result of the absorption:

$$y = \frac{L}{V}x + \left(y_{in} - \frac{L}{V}\right)x_{out} \tag{9.3}$$

where y, x = vapor and liquid mole fractions
y_{in} = mole fraction of the vapor inlet
x_{out} = mole fraction of liquid outlet
L = liquid flowrate
V = vapor flowrate

This is shown in Figure 9.1 to be a straight line with slope L/V. If it is assumed that the vapor flowrate is fixed, the solvent flowrate can be varied to obtain the minimum solvent flowrate, as shown in Figure 9.2.

Chemical Process Design and Integration, Second Edition. Robin Smith.

Companion Website: www.wiley.com/go/smith/chemical2e

Figure 9.1
Equilibrium and operating lines for an absorber.

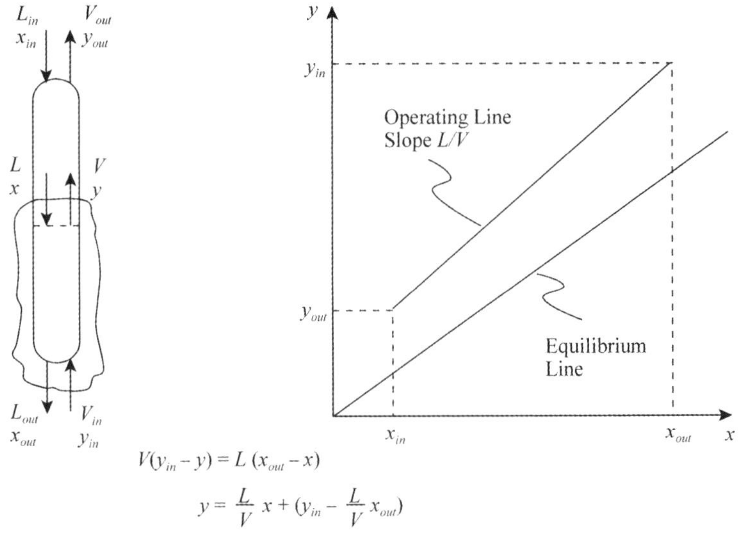

Figure 9.2
Minimum liquid–vapor ratio for absorbers.

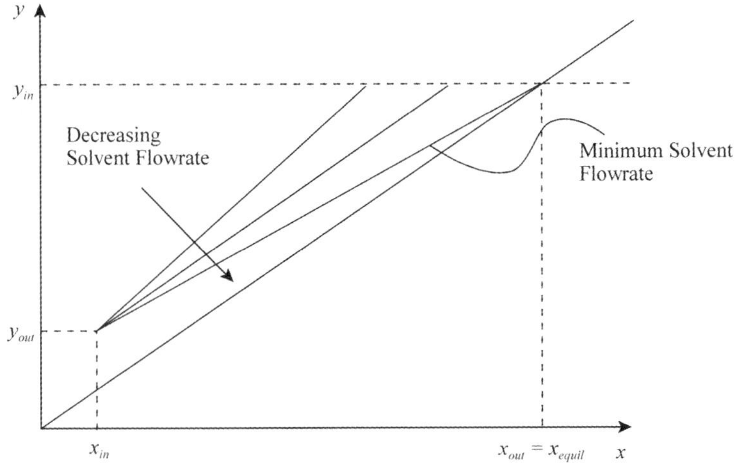

A graphical construction allows evaluation of the number of stages in the absorber analogous to the McCabe–Theile construction in distillation, as shown in Figure 9.3. In Figure 9.3, solvent enters the column on Stage 1. The operating line relates to compositions of passing streams and hence y_1 (y_{out}) and x_{in} are the start of the operating line. Solvent in equilibrium with y_1 is x_1 and is located horizontally across on the equilibrium line. The passing stream for x_1 is y_2 and located vertically up from x_1 on the operating line. The solvent in equilibrium with y_2 is x_2 and is located across horizontally on the equilibrium line, and so on. In this way the changes in liquid and vapor composition can be tracked through the equilibrium stages in the column.

Rather than using a graphical construction, the approach can be expressed analytically as the Kremser Equation (Treybal, 1980;

Figure 9.3
Equilibrium countercurrent stage operation for absorbers.

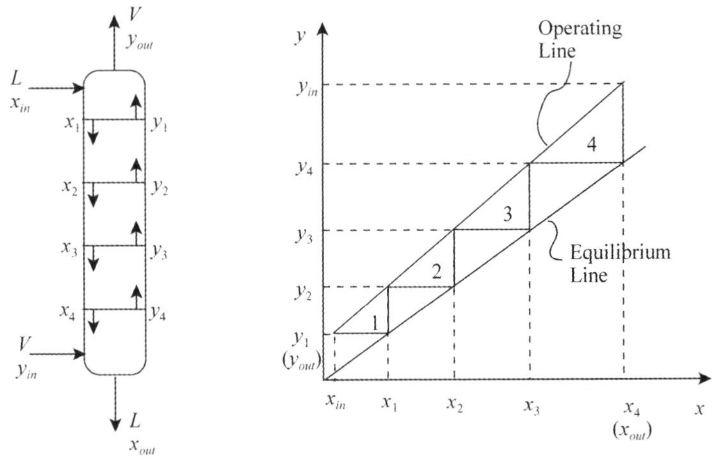

King, 1980; Geankopolis, 1993; Seader, Henley and Roper, 2011). The Kremser Equation assumes that the equilibrium line is straight and intersects the origin of the x–y diagram. The operating line is also assumed to be straight. It provides an analytical expression for the stepping construction shown in Figure 9.3. The derivation of the equation is lengthy and the reader is referred to other sources (King, 1980; Geankopolis, 1993; Seader, Henley and Roper, 2011). If concentrations are known, then the theoretical stages N can be calculated from (Treybal, 1980; King, 1980; Geankopolis, 1993):

$$N = \frac{\ln\left[\left(\frac{A-1}{A}\right)\left(\frac{y_{in} - Kx_{in}}{y_{out} - Kx_{in}}\right) + \frac{1}{A}\right]}{\ln A} \tag{9.4}$$

where $A = L/KV$
K = vapor–liquid equilibrium K-value (y/x)

A is known as the *absorption factor*. For $A = 1$, the equation takes the form:

$$N = \frac{y_{in} - y_{out}}{y_{out} - Kx_{in}} \tag{9.5}$$

If the number of theoretical stages is known, the concentrations can be calculated from:

$$\frac{y_{in} - y_{out}}{y_{in} - Kx_{in}} = \frac{A^{N+1} - A}{A^{N+1} - 1} \tag{9.6}$$

For multicomponent systems, Equation 9.4 can be written for the limiting component, that is, the component with the highest K_i. Having determined the number of stages, the concentrations of the other components can be determined from Equation 9.6.

The overall stage efficiency for absorption and stripping is significantly lower than that for distillation and often in the range 0.1 to 0.2. A first estimate of the overall stage efficiency can be obtained from the empirical correlation (Seader, Henley and Roper, 2011):

$$\log_{10} E_O = -0.773 - 0.415 \log_{10} X - 0.0896(\log_{10} X)^2 \tag{9.7}$$

where E_O = overall stage efficiency ($0 < E_O < 1$)
$X = \frac{KM_L\mu_L}{\rho_L}$
M_L = liquid molar mass (kg·kmol^{-1})
μ_L = liquid (solvent) viscosity (mN·s·m^{-2} = cP)
ρ_L = liquid (solvent) density (kg·m^3)

To calculate the overall stage efficiency for absorbers, the liquid viscosity and density need to be specified at average conditions.

As with distillation, the correlation for overall tray efficiency for absorbers, given in Equation 9.7, should only be used to derive a first estimate of the actual number of trays. More elaborate and reliable methods are available, but these require much more information on tray type and geometry and physical properties. If the column is to be packed, then the height of the packing is determined from Equation 8.92. The height equivalent of a theoretical plate (HETP) can vary significantly according to the packing type and the separation. The performance of a given packing can differ significantly between distillation and absorption applications, and reliable data can only be obtained from packing manufacturers.

Stripping is the reverse of absorption and involves the transfer of solute from the liquid to the vapor phase. This is illustrated in Figure 9.4. Note in Figure 9.4 that the operating line is below the equilibrium line. Again, if it is assumed that K_i, L and V are constant, as shown in Figure 9.4, the graphical construction can be expressed by the Kremser Equation (Treybal, 1980; King, 1980; Geankopolis, 1993; Seader, Henley and Roper, 2011):

$$N = \frac{\ln\left[(1-A)\left(\frac{x_{in} - y_{in}/K}{x_{out} - y_{in}/K}\right) + A\right]}{\ln(1/A)} \tag{9.8}$$

For $A = 1$:

$$N = \frac{x_{in} - x_{out}}{x_{out} - y_{in}/K} \tag{9.9}$$

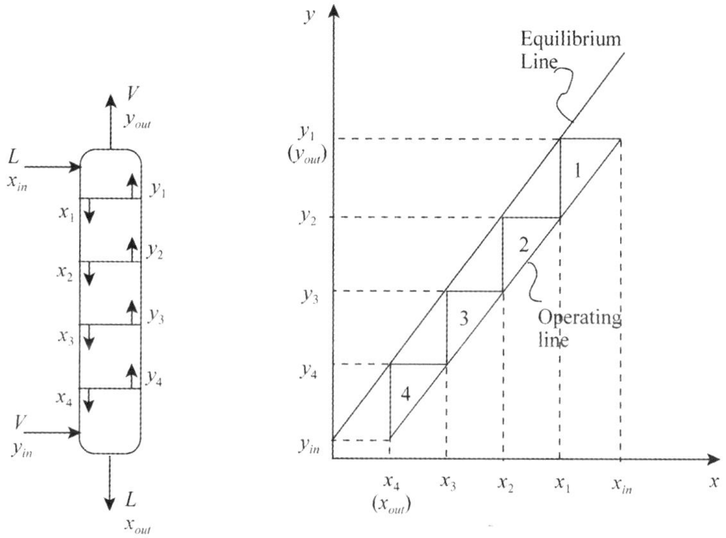

Figure 9.4
Equilibrium countercurrent stage operation for stripping.

If the number of stages is known, then the concentrations can be calculated from:

$$\frac{x_{in} - x_{out}}{x_{in} - y_{in}/K} = \frac{(1/A)^{N+1} - (1/A)}{(1/A)^{N+1} - 1} \quad (9.10)$$

Equations 9.8 to 9.10 can be written in terms of a *stripping factor* (S), where $S = 1/A$. For multicomponent systems, Equations 9.8 and 9.9 can be used for the limiting component (i.e. the component with the lowest K_i) and then Equation 9.10 can be used to determine the distribution of the other components.

Equations 9.4 to 9.6 and 9.8 to 9.10 can be written in terms of mole fractions and molar flowrates. Alternatively, mass fractions and mass flowrates can be used instead, as long as a consistent set of units is used.

If absorption is the choice of separation, then some preliminary selection of the major design variables must be made to allow the design to proceed:

1) *Liquid flowrate.* The liquid flowrate is determined by the absorption factor (L/K_iV) for the key separation. The absorption factor determines how readily Component i will absorb into the liquid phase. When the absorption factor is large, Component i will be absorbed more readily into the liquid phase. The absorption factor must be greater than 1 for a high degree of solute removal; otherwise removal will be limited to a low value by the liquid flowrate. Since L/K_iV is increased by increasing the liquid flowrate, the number of plates required to achieve a given separation decreases. However, at high values of L/K_iV, the increase in the liquid flowrate brings diminishing returns. This leads to an economic optimum in the range $1.2 < L/K_iV < 2.0$. An absorption factor around 1.4 is often used (Douglas, 1988).
2) *Temperature.* Decreasing temperature increases the solubility of the solute. In an absorber, the transfer of solute from gas or vapor to liquid brings about a heating effect. This usually will lead to temperatures increasing down the column. If the component being separated is dilute, the heat of absorption will be small and the temperature rise down the column will also be small. Otherwise, the temperature rise down the column will be large, which is undesirable since solubility decreases with increasing temperature. To counteract the temperature rise in absorbers, the liquid is sometimes cooled at intermediate points as it passes down the column. The cooling is usually down to temperatures that can be achieved with cooling water, except in special circumstances where refrigeration is used.

   If the solvent is volatile, there will be some loss of the gas or vapor. This should be avoided if the solvent is expensive and/or environmentally harmful by using a condenser (refrigerated if necessary) on the vapor leaving the absorber.
3) *Pressure.* High pressure gives greater solubility of solute in the liquid. However, high pressure tends to be expensive to create since this can require a gas compressor. Thus, there is an optimal pressure.

   As with distillation, no attempt should be made to carry out any optimization of liquid flowrate, temperature or pressure at this stage in the design.

   Having dissolved the solute in the liquid, it is often necessary to then separate the solute from the liquid in a stripping operation so as to recycle the liquid to the absorber. Now the stripping factor for Component i (K_iV/L) should be large to concentrate it in the vapor phase and thus be stripped out of the liquid phase. For a stripping column, the stripping factor should be in the range $1.2 < K_iV/L < 2.0$ and is often around 1.4 (Douglas, 1988). As with absorbers, there can be a significant change in temperature through the column. This time, however, the liquid decreases in temperature down the column for reasons analogous to those developed for absorbers. Increasing temperature and decreasing pressure will enhance the stripping.

Example 9.1 A hydrocarbon gas stream containing benzene vapor is to have the benzene separated by absorption in a heavy liquid hydrocarbon stream with an average molar mass of $200\,kg\cdot kmol^{-1}$. The concentration of benzene in the gas stream is 2% by volume and the liquid contains 0.2% benzene by mass. The flowrate of the gas stream is $850\,m^3\cdot h^{-1}$, its pressure is 1.07 bar and its temperature is 25 °C. It can be assumed that the gas and liquid flowrates are constant, the kilogram molecular volume occupies $22.4\,m^3$ at standard conditions, the molar mass of benzene is $78\,kg\cdot kmol^{-1}$ and that the vapor–liquid equilibrium obeys Raoult's Law. At the temperature of the separation, the saturated liquid vapor pressure of benzene can be taken to be 0.128 bar. If a 95% removal of the benzene is required, estimate the liquid flowrate and the number of theoretical stages.

Solution First, estimate the flowrate of vapor in $kmol\cdot h^{-1}$:

$$\begin{aligned} V &= 850 \times \frac{1}{22.4\left(\frac{298}{273}\right)\left(\frac{1.013}{1.07}\right)} \\ &= 36.6\,kmol\cdot h^{-1} \end{aligned}$$

From Raoult's Law:

$$\begin{aligned} K &= \frac{p^{sat}}{P} \\ &= \frac{0.128}{1.07} \\ &= 0.12 \end{aligned}$$

$$\begin{aligned} \text{Assuming } A &= 1.4 \\ L &= 1.4 \times 0.12 \times 36.6 \\ &= 6.15\,kmol\cdot h^{-1} \\ &= 6.15 \times 200\,kg\cdot h^{-1} \\ &= 1230\,kg\cdot h^{-1} \\ x_{in} &= 0.002 \times \frac{200}{78} \\ &= 0.0051 \\ y_{out} &= 0.02(1 - 0.95) \text{ assuming } V \text{ constant} \\ &= 0.001 \end{aligned}$$

$$\begin{aligned} N &= \frac{\ln\left[\left(\frac{A-1}{A}\right)\left(\frac{y_{in} - Kx_{in}}{y_{out} - Kx_{in}} + \frac{1}{A}\right)\right]}{\ln A} \\ &= \frac{\ln\left[\left(\frac{1.4-1}{1.4}\right)\left(\frac{0.02 - 0.12 \times 0.0051}{0.001 - 0.12 \times 0.0051} + \frac{1}{1.4}\right)\right]}{\ln 1.4} \\ &= 8 \text{ theoretical stages} \end{aligned}$$

Separation in absorption is sometimes enhanced by adding a component to the liquid that reacts with the solute. The discussion regarding absorption has so far been restricted to physical absorption. In chemical absorption, chemical reactions are used to enhance absorption. Both irreversible and reversible reactions can be used. An example of an irreversible reaction is the removal of SO_2 from gas streams using sodium hydroxide solution:

$$\underset{\text{sodium hydroxide}}{2NaOH} + SO_2 \longrightarrow \underset{\text{sodium sulphite}}{Na_2SO_3} + H_2O$$

$$\underset{\text{sodium sulfite}}{Na_2SO_3} + {}^1\!/_2O_2 \longrightarrow \underset{\text{sodium sulfate}}{Na_2SO_4}$$

Another example of an irreversible reaction is the removal of oxides of nitrogen from gas streams using hydrogen peroxide:

$$\underset{\text{nitric oxide}}{2NO} + \underset{\text{hydrogen peroxide}}{3H_2O_2} \longrightarrow \underset{\text{nitric acid}}{2HNO_3} + 2H_2O$$

$$\underset{\text{nitrogen dioxide}}{2NO_2} + \underset{\text{hydrogen peroxide}}{H_2O_2} \longrightarrow \underset{\text{nitric acid}}{2HNO_3}$$

Examples of reversible reactions are the removal of hydrogen sulfide and carbon dioxide from gas streams using a solution of monoethanolamine:

$$\underset{\text{monoethanolamine}}{HOCH_2CH_2NH_2} + H_2S \underset{\text{Regeneration}}{\overset{\text{Absorption}}{\rightleftharpoons}} \underset{\text{monoethanolamine hydrogen sulphide}}{HOCH_2CH_2NH_3HS}$$

$$\underset{\text{monoethanolamine}}{HOCH_2CH_2NH_2} + CO_2 + H_2O \underset{\text{Regeneration}}{\overset{\text{Absorption}}{\rightleftharpoons}} \underset{\text{monoethanolamine hydrogen carbonate}}{HOCH_2CH_2NH_3HCO_3}$$

In this case, the reactions can be reversed in a regeneration stage in a stripping column by the input of heat in a reboiler. If the solvent is to be recovered by stripping the solute from the solvent, the chemical absorption requires more energy than physical absorption. This is because the energy input for chemical absorption must overcome the heat of reaction, as well as the heat of solution. However, chemical absorption involves smaller solvent flowrates than physical absorption.

Unfortunately, the design of chemical absorption is far more complex than physical absorption. The vapor–liquid equilibrium behavior cannot be represented by Henry's Law or any of the methods described in Appendix A. Also, different chemical compounds in the gas mixture can become involved in competing reactions. This means that simple methods like the Kremser equation no longer apply and complex simulation software is required to model chemical absorption systems such as the absorption of H_2S and CO_2 in monoethanolamine. This is outside the scope of this text.

9.2 Liquid–Liquid Extraction

Like gas absorption, liquid–liquid extraction separates a homogeneous mixture by the addition of another phase, in this case an immiscible liquid. Liquid–liquid extraction carries out separation by contacting a liquid feed with another immiscible liquid. The equipment used for liquid–liquid extraction is the same as that used for the liquid–liquid reactions illustrated in Figure 6.5. The separation occurs as a result of components in the feed distributing themselves differently between the two liquid phases. The liquid with which the feed is contacted is known as the *extraction solvent*. The extraction solvent extracts *solute* from the feed. The solute in the feed is dissolved in the *feed solvent* or *carrier*. The solvent-rich stream obtained from the separation is known as the *extract* and the residual feed from which the solute has been extracted is known as the *raffinate*. The stagewise contacting is illustrated in Figure 9.5a.

The distribution of solute between the two phases at equilibrium can be quantified by the K-value or *distribution coefficient* (see Appendix A):

$$K_i = \frac{x_{E,i}}{x_{R,i}} = \frac{\gamma_{R,i}}{\gamma_{E,i}} \tag{9.11}$$

where $x_{E,i}, x_{R,i}$ = mole fractions of Component i in the extract and in the raffinate

$\gamma_{E,i}, \gamma_{R,I}$ = activity coefficients of Component i in the extract and in the raffinate

By taking the ratio of the distribution coefficients for the two Components i and j, the *separation factor* can be defined, which is analogous to relative volatility in distillation:

$$\beta_{ij} = \frac{K_i}{K_j} = \frac{x_{E,i}/x_{R,i}}{x_{E,j}/x_{R,j}} = \frac{\gamma_{R,i}/\gamma_{E,i}}{\gamma_{R,j}/\gamma_{E,j}} \tag{9.12}$$

The separation factor (or selectivity) indicates the tendency for Component i to be extracted more readily from the raffinate to the extract than Component j.

The stagewise calculation of liquid–liquid extraction has much in common with the stagewise calculations for distillation, absorbers and strippers. The McCabe–Thiele graphical method assumes constant molar vapor and liquid flowrates and allows convenient stepwise calculation with straight operating lines and a curved equilibrium line. A similar concept can be achieved in liquid–liquid extraction by expressing flowrates on a solute-free basis to give a constant flow rate of *feed solvent* F and a constant flowrate of *extraction solvent* S through the extractor. Because the flowrate is being expressed on a solute-free basis, concentrations must be adjusted to be the ratio of solute to solvent. These assumptions help to keep operating lines straight and simplify the calculations. Three simplifying cases can be identified for the calculations (Schweitzer, 1997):

1) In most cases liquid–liquid extraction systems can be treated as having immiscible solvents. A material balance for the solute around the extractor from Stages 1 to n in Figure 9.5a, assuming $F = R$ and $S = E$, gives:

$$Fx_F + Sx_{E,n+1} = Fx_{R,n} + Ex_{E,out} \tag{9.13}$$

Rearranging gives:

$$x_{E,n+1} = \frac{F}{S}x_{R,n} + \frac{Ex_{E,out} - Fx_F}{S} \tag{9.14}$$

9

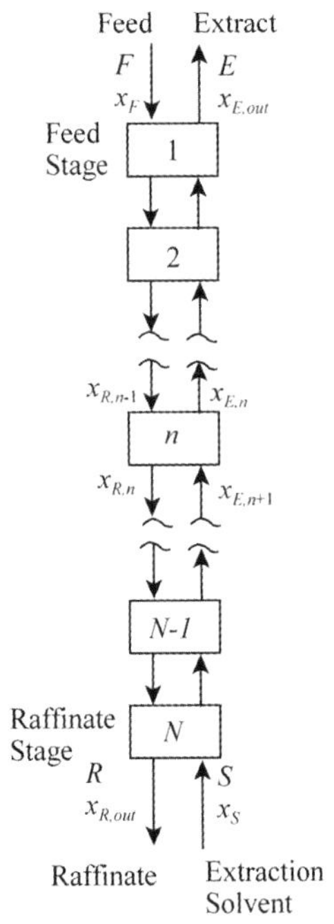

(a) Mass balance for liquid–liquid extraction.

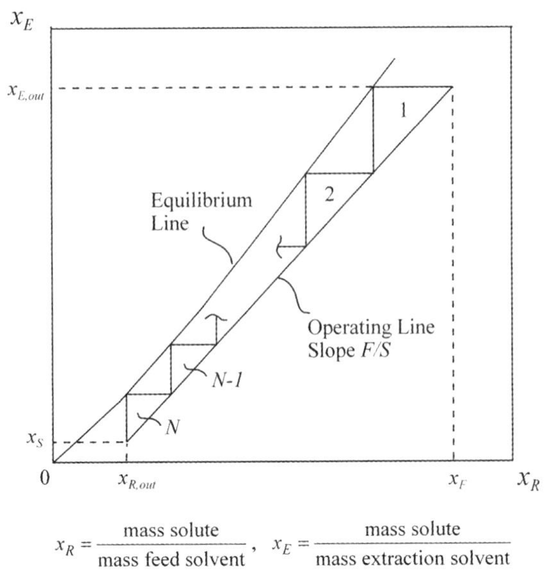

(b) Stepping off for the number of theoretical stages.

Figure 9.5

Counter current liquid–liquid extraction.

This is the equation of a straight line slope F/S, as shown in Figure 9.5b. The same operating line can be obtained from a material balance around the raffinate end of the extractor from Stages N to n, assuming $F = R$ and $S = E$, gives:

$$x_{E,n} = \frac{F}{S} x_{R,n-1} + \frac{Sx_s - Rx_{R,out}}{S} \tag{9.15}$$

An overall balance around the extractor gives:

$$x_{E,out} = \frac{Fx_F + Sx_S - Rx_{R,out}}{E} \tag{9.16}$$

Thus the operating line has end points corresponding to ($x_{R,out}$, x_S) and (x_F, $x_{E,out}$). Figure 9.5b shows a plot of this operating line, together with the equilibrium line. Figure 9.5b also shows the stepping off for liquid–liquid extraction analogous to distillation, absorption and stripping. Mass fractions and mass flowrates or mole fractions and mole flowrates can also be used, as long as a consistent set of units is used.

If the distribution coefficient is constant and the liquid flowrates are also constant on a solute-free basis, the same analysis as that used for stripping in Section 9.1 based on the Kremser Equation can be applied. In liquid–liquid extraction, like stripping, solute is transferred from the feed for $\varepsilon \neq 1$:

$$N = \frac{\ln\left[\left(\frac{\varepsilon - 1}{\varepsilon}\right)\left(\frac{x_F - x_S/K}{x_R - x_S/K}\right) + \frac{1}{\varepsilon}\right]}{\ln \varepsilon} \tag{9.17}$$

where N = number of theoretical stages
x_F = mole fraction of solute in the feed based on solute-free feed
x_S = mole fraction of solute in the solvent inlet based on solute-free solvent
x_R = mole fraction of solute in raffinate based on solute-free raffinate
K = slope of the equilibrium line
ε = extraction factor
= KS/F
S = flowrate of solute-free solvent (kmol·s^{-1})
F = flowrate of solute-free feed (kmol·s^{-1})

For $\varepsilon = 1$:

$$N = \frac{x_F - x_R}{x_R - x_S/K} \tag{9.18}$$

When N is known, the composition can be calculated from:

$$\frac{x_F - x_R}{x_F - x_S/K} = \frac{\varepsilon^{N+1} - \varepsilon}{\varepsilon^{N+1} - 1} \tag{9.19}$$

For multicomponent systems, Equations 9.17 and 9.18 can be used to determine the number of stages for the limiting component (i.e. the component with the lowest K_i). Equation 9.19 can then be applied to determine the compositions of the other components.

Again, Equations 9.17 to 9.19 are written in terms of mole fractions and molar flowrates. However, mass fractions and mass flowrates can also be used, as long as a consistent set of units is used. If the K-value varies across the extractor, then a geometric mean of the value leaving the feed stage and at the raffinate K_1 concentration leaving the raffinate stage K_N can be used (Treybal, 1980):

$$K = \sqrt{K_1 K_N} \tag{9.20}$$

2) In the second case the solvents are partially miscible and often occurs when all solute concentrations are relatively low. To simplify the calculations it is assumed that the extraction solvent dissolves in the feed solvent only in the feed stage and this miscibility remains constant through the extractor. This means that the extraction solvent in Figure 9.5a has a constant flowrate from the inlet of Stage N to the inlet of Stage 1, and experiences a decrease in flowrate in Stage 1 through the mutual solubility in the feed stage. Similarly, feed solvent is assumed to dissolve in the extraction solvent only in the raffinate stage (Stage N) and this miscibility remains constant through the extractor. This means that the feed solvent in Figure 9.5a has a constant flowrate from the inlet of Stage 1 to the inlet of Stage N and experiences a decrease in flowrate in Stage N through the mutual solubility in the raffinate stage (Schweitzer, 1997). Thus the extraction solvent and feed solvent flowrates can be assumed to be constant through the extractor. However, the extract flowrate E is less than S and the raffinate flowrate R is less than F because of the mutual solubilities. The slope of the operating line in Equations 9.14 and 9.15 remains F/S, but only from Stages 2 to $N-1$. Stages 1 and N will not be located on the operating line. For the feed stage, when there is a change in F because of mutual solubility, for passing streams entering Stage 1 to be on the operating line (Equation 9.14) the feed composition must be adjusted to x'_F by substituting $x_{E,out} = x_{E,n+1}$ in Equation 9.14 (Schweitzer, 1997):

$$x_{E,out} = x'_F + \frac{Ex_{E,out} - Fx_F}{S} \tag{9.21}$$

Rearranging gives:

$$x'_F = x_F + \left(\frac{S-E}{F}\right)x_{E,out} \tag{9.22}$$

For the raffinate stage, when there is a change in S because of mutual solubility, for passing streams entering Stage N to be on the operating line (Equation 9.15) the solvent composition must be adjusted to x'_S by substituting $x_{R,n-1} = x_{R,out}$ in Equation 9.15:

$$x'_S = \frac{F}{S}x_{R,out} + \frac{Sx_S - Rx_{R,out}}{S} \tag{9.23}$$

Rearranging gives:

$$x'_S = x_S + \left(\frac{F-R}{S}\right)x_{R,out} \tag{9.24}$$

The operating line must go through points $(x_{R,out}, x'_S)$ and $(x'_F, x_{E,out})$. The stepping off procedure is the same as with the case for immiscible liquids once the adjustment of the solvent and feed concentrations has been made. If both the equilibrium line and operating lines are straight, the Kremser Equation can also be applied to the partially miscible case by using the adjusted compositions x'_F and x'_S, the actual value of $x_{R,out}$ and $\varepsilon = KS/F$. If the K-value varies across the extractor, then a geometric mean value can be used according to Equation 9.20.

3) In the third case the solvents are also partially miscible and often occurs when the solute concentration is high in the feed and extract, but low in the raffinate. This time to simplify the calculations, as in Case 2, it is assumed that the extraction solvent dissolves in the feed solvent only in the feed stage (Stage 1) and this miscibility remains constant through the extractor. However, feed solvent is assumed to dissolve in the extraction solvent only in the feed stage and this miscibility remains constant through the extractor. This means that the feed solvent in Figure 9.5a decreases in the feed stage through the mutual solubility in the raffinate stage and then has a constant flowrate from the inlet of Stage 2 to the outlet of Stage N (Schweitzer, 1997). Thus the extraction solvent and feed solvent flowrates can be assumed to be constant through the extractor. Again the extract flowrate E is less than S, and the raffinate flowrate R is less than F because of the mutual solubilities. The slope of the operating line in Equations 9.14 and 9.15 is now R/S rather than F/S, but only from Stages 2 to N. Stages 1 (feed stage) will not be located on the operating line. For the feed stage when there is a change in F because of mutual solubility, for passing streams entering Stage 1 to be on the operating line (Equation 9.14), the feed composition must be adjusted to x'_F by substituting $x_{E,out} = x_{E,n+1}$ in Equation 9.14, but with the slope R/S instead of F/S (Schweitzer, 1997):

$$x_{E,out} = \frac{R}{S}x'_F + \frac{Ex_{E,out} - Fx_F}{S} \tag{9.25}$$

Rearranging gives:

$$x'_F = \frac{F}{R}x_F + \left(\frac{S-E}{R}\right)x_{E,out} \tag{9.26}$$

The operating line must go through points $(x_{R,out}, x_S)$ and $(x'_F, x_{E,out})$. The stepping off procedure is the same as with the case for immiscible liquids once the adjustment of the feed concentration has been made. If both the equilibrium line and operating lines are straight, the Kremser Equation can also be applied to the partially miscible case by using the adjusted compositions x'_F and the actual values of x_S, $x_{R,out}$ and $\varepsilon = KS/R$. If the K-value varies across the extractor, then a geometric mean value can be used according to Equation 9.20.

Example 9.2 An organic product with a flowrate of 1000 kg·h^{-1} contains a water-soluble impurity with a concentration of 6% by mass. A laboratory test indicates that if the product is extracted with an equal mass of water, then 90% of the impurity is extracted. Assume that water and the organic product are immiscible.

a) For the same separation of 90% removal, estimate how much water would be needed if a two-stage countercurrent extraction is used.

b) If an equal mass flowrate of water to feed is maintained for a two-stage countercurrent extraction, estimate the fraction of impurity extracted.

Solution

a)

$$\begin{aligned}\text{Mass of impurity in feed} &= 1000 \times 0.06\\ &= 60\ \text{kg}\cdot\text{h}^{-1}\end{aligned}$$

$$\begin{aligned}\text{Feed flowrate on a solute-free basis} &= 1000 - 60\\ &= 940\ \text{kg}\cdot\text{h}^{-1}\\ x_F &= \frac{60}{940}\\ &= 0.06383\end{aligned}$$

$$\begin{aligned}\text{Mass of impurity in raffinate} &= 60 \times 0.1\\ &= 6\ \text{kg}\cdot\text{h}^{-1}\end{aligned}$$

$$\begin{aligned}\text{Mass of impurity in extract} &= 60 - 6\\ &= 54\ \text{kg}\cdot\text{h}^{-1}\end{aligned}$$

If the water and organic product are assumed to be immiscible, it can be assumed that $F = R$ and $S = E$:

$$\begin{aligned}x_R &= \frac{6}{940}\\ &= 6.383 \times 10^{-3}\\ x_E &= \frac{54}{1000}\\ &= 0.054\\ K &= \frac{x_E}{x_R}\\ &= \frac{0.054}{6.383 \times 10^{-3}}\\ &= 8.460\end{aligned}$$

From Equation 9.15:

$$\frac{x_F - x_R}{x_F - x_S/K} = \frac{\varepsilon^{N+1} - \varepsilon}{\varepsilon^{N+1} - 1}$$

$$\frac{0.06383 - 6.383 \times 10^{-3}}{0.06383 - 0/8.460} = \frac{\varepsilon^3 - \varepsilon}{\varepsilon^3 - 1}$$

$$0.9 = \frac{\varepsilon^3 - \varepsilon}{\varepsilon^3 - 1}$$

$$0 = 0.1\varepsilon^3 - \varepsilon + 0.9$$

Solving for ε by trial and error:

$$\begin{aligned}\varepsilon &= 2.541\\ S &= \frac{\varepsilon F}{K}\\ &= \frac{2.541 \times 940}{8.460}\\ &= 282.3\ \text{kg}\cdot\text{h}^{-1}\end{aligned}$$

b)

$$\begin{aligned}\varepsilon &= \frac{8.460 \times 1000}{940}\\ &= 10.0\end{aligned}$$

From Equation 9.15:

$$\frac{x_F - x_R}{x_F - x_S/K} = \frac{\varepsilon^{N+1} - \varepsilon}{\varepsilon^{N+1} - 1}$$

$$\frac{0.06383 - x_R}{0.06383 - 0/8.460} = \frac{9.0^3 - 9.0}{9.0^3 - 1}$$

$$x_R = 7.0 \times 10^{-4}$$

$$\begin{aligned}\text{Mass of impurity in raffinate} &= 940 \times 7 \times 10^{-4}\\ &= 0.658\ \text{kg}\cdot\text{h}^{-1}\end{aligned}$$

$$\begin{aligned}\text{Fraction of impurity extracted} &= \frac{60 - 0.658}{60}\\ &= 0.989\end{aligned}$$

Example 9.3 A feed with a flowrate of 1000 kg·h^{-1} contains 30% acetic acid by mass in aqueous solution. The acetic acid (AA) is to be extracted with pure isopropyl ether to produce a raffinate with 2% by mass on a solvent-free basis. Equilibrium data are given in Table 9.1 (Campbell, 1940; Treybal, 1980). It can be seen from Table 9.1 that the water and ether have significant mutual solubility and this must be accounted for.

a) From an initial estimate of the feed solvent transferred to the extraction solvent in the feed and raffinate stages, determine which of the simplified cases is most appropriate for the calculations.

b) Estimate the minimum flowrate of ether for the separation.

c) Estimate graphically the number of theoretical extraction stages if a flowrate of 2500 kg·h^{-1} of ether is used.

d) Estimate the number of theoretical extraction stages for the separation using the Kremser Equation.

Solution

a) To maintain a straight operating line, the concentrations must be expressed as ratios of solute to feed solvent (water) and ratios of solute to extraction solvent (isopropyl ether). The data from Table 9.1 are expressed on this basis in Table 9.2.

To determine which of the simplifying cases is most appropriate, the transfer of feed solvent (water) to extraction solvent (ether) needs to be compared at the feed stage and the extraction stage. First estimate the transfer at the feed stage conditions for which $x_{E,out}$ needs to be known.

Table 9.1

Equilibrium data for acetic acid–water–isopropyl ether (Campbell, 1940; Treybal, 1980). (From Cambell H, 1940, *Trans AIChE*, **36**: 628, reproduced by permission of the American Institute of Chemical Engineers.)

Mass fraction in water phase			Mass fraction ether phase		
Acetic acid	Water	Isopropyl ether	Acetic acid	Water	Isopropyl ether
0.0069	0.981	0.012	0.0018	0.005	0.993
0.0141	0.971	0.015	0.0037	0.007	0.989
0.0289	0.955	0.016	0.0079	0.008	0.984
0.0642	0.917	0.019	0.0193	0.010	0.971
0.1330	0.844	0.023	0.0482	0.019	0.933
0.2550	0.711	0.034	0.1140	0.039	0.847
0.3670	0.589	0.044	0.2160	0.069	0.715

The equilibrium data from Table 9.2 show distribution coefficients that vary significantly. Figure 9.6a shows graphically that the equilibrium line is not straight. Figure 9.6a shows the operating line starting from $x_{R,out} = 0.02/0.98 = 0.0204$ with a maximum slope that touches the equilibrium line at $x_R = 0.1576$ kg AA/kg water. In Figure 9.6a, the operating line for the minimum solvent flowrate touches the equilibrium line at $x_R = 0.1576$ and $x_E = 0.05166$. If there is assumed to be no mutual solubility, the slope of this line is the ratio of feed to solvent flowrates *F/S*. Thus:

$$\frac{F}{S} = \frac{0.05166 - 0}{0.1576 - 0.0204}$$

$$= 0.377$$

$$F = 1000 \times 0.7 = 700\ \text{kg}\cdot\text{h}^{-1}$$

Thus, from the slope:

$$S = \frac{700}{0.377}$$

$$= 1857\ \text{kg}\cdot\text{h}^{-1}$$

A material balance around the extractor on the acetic acid gives:

$$Ex_{E,out} = Fx_F + Sx_S - Rx_{R,out} \tag{9.27}$$

Assuming initially that $S = E$:

$$1857x_{E,out} = 700 \times \frac{0.3}{0.7} + S \times 0 - 700 \times 0.0204$$

$$x_{E,out} = 0.1539$$

9

Table 9.2

Equilibrium data expressed as ratios of feed solvent and extraction solvent.

Mass ratios in water phase		Mass ratios in ether phase		$K = \frac{x_E}{x_R}$
x_{AA} (x_R)	x_{Ether}	x_{AA} (x_E)	x_{Water}	
7.034×10^{-3}	0.01223	1.813×10^{-3}	5.035×10^{-3}	0.2577
0.01452	0.01545	3.741×10^{-3}	7.078×10^{-3}	0.2576
0.03026	0.01675	8.028×10^{-3}	8.130×10^{-3}	0.2653
0.07001	0.02072	0.01988	0.01030	0.2840
0.1576	0.02725	0.05166	0.02036	0.3278
0.3586	0.04782	0.1346	0.04604	0.3753
0.6231	0.07470	0.3021	0.09650	0.4848

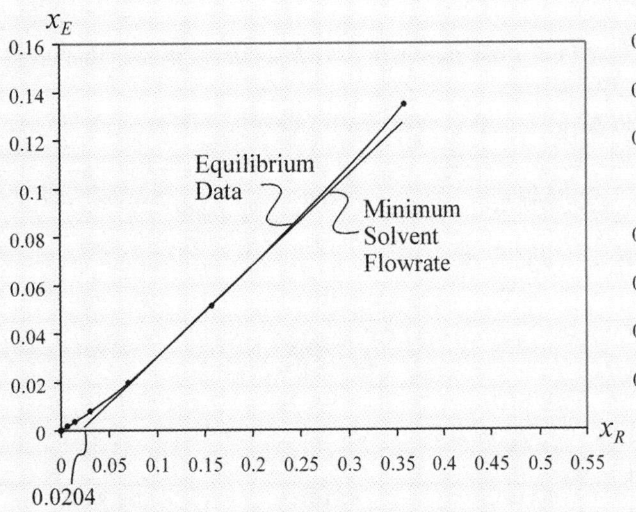

(a) Minimum flowrate of solvent.

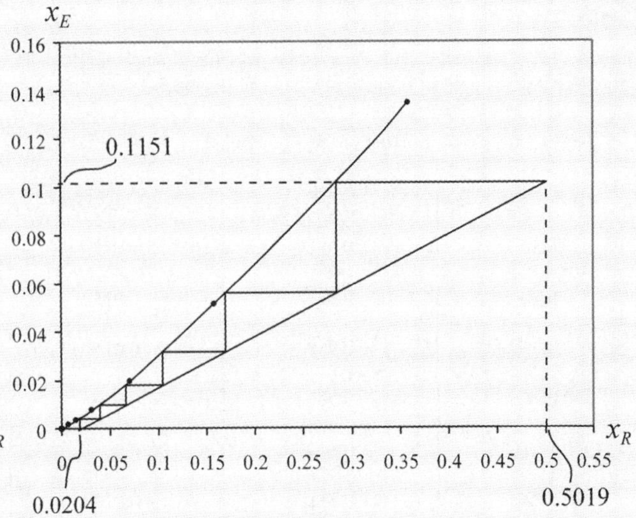

(b) Stepping off reveals the number of theoretical stages.

Figure 9.6

Extraction of acetic acid from aqueous solution using isopropyl ether.

For the ether phase, from Table 9.2, for $x_E = 0.1539$ kg AA/kg ether, the ether phase contains 0.05185 kg water/kg ether (by interpolation).

$$\begin{aligned}\text{Water in extract} &= 0.05185 \times 1857\\ &= 96.3\ \text{kg}\cdot\text{h}^{-1}\end{aligned}$$

$$\begin{aligned}\text{Water in raffinate leaving feed stage} &= 700 - 96.3\\ &= 603.7\ \text{kg}\cdot\text{h}^{-1}\end{aligned}$$

For the raffinate stage, from Table 9.2, at $x_R = 0.0204$ kg AA/kg water, the ether phase contains 0.007471 kg water/kg ether (by interpolation):

$$\begin{aligned}\text{Water in extract leaving the raffinate stage} &= 0.007471 \times 1857\\ &= 13.9\ \text{kg}\cdot\text{h}^{-1}\end{aligned}$$

Thus, in this case the change in flowrate of the raffinate is much greater at the feed stage than the raffinate stage. This means that Case 3, assuming a large amount of feed solvent is dissolved in the extract in the feed stage with an operating line slope of R/S, is the most appropriate.

b) From Part a above the operating line is as shown in Figure 9.6a going through (0.0204, 0) and touching the equilibrium line with a slope of $R/S = 0.377$. Now the extraction solvent flowrate can be reestimated taking into account the mutual solubility from the water flowrate leaving the raffinate stage above:

$$\begin{aligned}S &= \frac{603.7}{0.377}\\ &= 1601\ \text{kg}\cdot\text{h}^{-1}\end{aligned}$$

Now reestimate $x_{E,out}$ from Equation 9.27:

$$\begin{aligned}1601 x_{E,out} &= \frac{0.3}{0.7} \times 700 + 1601 \times 0 - 603.7 \times \frac{0.02}{0.98}\\ &= 0.1797\ \text{kg AA/kg ether}\end{aligned}$$

From Table 9.2 at 0.1797 kg AA/kg ether, the ether phase contains 0.05963 kg water/kg ether (by interpolation):

$$\begin{aligned}\text{Water in extract} &= 0.05963 \times 1601\\ &= 95.5\ \text{kg}\cdot\text{h}^{-1}\end{aligned}$$

$$\begin{aligned}\text{Water in raffinate}(R) &= 700 - 95.5\\ &= 604.5\ \text{kg}\cdot\text{h}^{-1}\end{aligned}$$

Reestimate the extraction solvent flowrate:

$$\begin{aligned}S &= \frac{604.5}{0.377}\\ &= 1604\ \text{kg}\cdot\text{h}^{-1}\end{aligned}$$

The previous calculation is repeated for convergence:

$$x_{E,out} = 0.1779\ \text{kg AA/kg ether}$$

$$\text{Water in extract} = 94.8\ \text{kg}\cdot\text{h}^{-1}$$

$$\text{Water in raffinate}(R) = 605.2\ \text{kg}\cdot\text{h}^{-1}$$

$$S = 1605\ \text{kg}\cdot\text{h}^{-1}$$

Thus the minimum extraction solvent flowrate is estimated to be 1605 kg·h^{-1}.

c) As before:

$$\begin{aligned}F &= 700\ \text{kg}\cdot\text{h}^{-1}\\ x_F &= 0.4286\\ x_R &= 0.0204\end{aligned}$$

Now the extraction solvent flowrate is fixed:

$$S = 2500\ \text{kg}\cdot\text{h}^{-1}$$

Assume initially that $F = R$ and $S = E$ and perform an overall mass balance on acetic acid from Equation 9.27:

$$\begin{aligned} x_E &= \frac{0.4286 \times 700 + 0 \times 2500 - 0.0204 \times 700}{2500} \\ &= 0.1143 \end{aligned}$$

For the ether phase, from Table 9.2, at $x_E = 0.1143$ kg AA/kg water the ether phase contains 0.03976 kg water/kg ether:

$$\begin{aligned} \text{Water in extract} &= 0.03979 \times 2500 \\ &= 99.4\ \text{kg}\cdot\text{h}^{-1} \end{aligned}$$

$$\begin{aligned} \text{Water in raffinate } (R) &= 700 - 99.4 \\ &= 600.6\ \text{kg}\cdot\text{h}^{-1} \end{aligned}$$

For the raffinate, at $x_R = 0.0204$ kg AA/kg water, the water phase contains 0.01594 kg ether/kg water:

$$\begin{aligned} \text{Ether in raffinate} &= 0.01594 \times 600.6 \\ &= 9.6\ \text{kg}\cdot\text{h}^{-1} \end{aligned}$$

$$\begin{aligned} \text{Ether in extract } (E) &= 2500 - 9.6 \\ &= 2490.4\ \text{kg}\cdot\text{h}^{-1} \end{aligned}$$

Now return to Equation 9.27 and iterate to converge:

$$\begin{aligned} E &= 2490.4\ \text{kg}\cdot\text{h}^{-1} \\ R &= 600.0\ \text{kg}\cdot\text{h}^{-1} \\ x_{E,out} &= 0.1151 \end{aligned}$$

Following the assumption that the feed stream only dissolves extraction solvent in the feed stage, then the apparent feed concentration after the mass transfer in the feed stage can be calculated from Equation 9.26 (Schweitzer, 1997):

$$\begin{aligned} x'_F &= \frac{F}{R} x_F + \left(\frac{S-E}{R}\right) x_{E,out} \\ &= \frac{700 \times 0.4286}{600.0} + \left(\frac{2500 - 2490.4}{600.0}\right) \times 0.1151 \\ &= 0.5019 \end{aligned}$$

In Figure 9.6b, the operating line for this is drawn between:

$$\begin{aligned} x'_F &= 0.5019, \quad x_E = 0.1151 \\ x_R &= 0.0204, \quad x_S = 0.0 \end{aligned}$$

Although the operating line is straight as a result of the assumptions made, the equilibrium line is not straight. Figure 9.6b shows that a graphical stepping off between the operating and equilibrium lines requires approximately six equilibrium stages.

d) The Kremser Equation cannot be applied directly, as the equilibrium line is not straight:

$$\begin{aligned} x'_F &= 0.5019 \\ x_R &= 0.0204 \\ x_S &= 0.0 \end{aligned}$$

At $x_{E,out} = 0.1151$, $K_1 = 0.3641$ and at $x_{R,out} = 0.0204$, $K_N = 0.2605$ (from Table 9.2 by interpolation). Thus:

$$\begin{aligned} K &= \sqrt{K_1 \cdot K_N} \\ &= \sqrt{0.3641 \times 0.2605} \\ &= 0.3078 \\ \varepsilon &= \frac{KS}{R} \\ &= \frac{0.3048 \times 2490.4}{600.0} \\ &= 1.278 \\ N &= \frac{\ln\left[\left(\frac{1.278 - 1}{1.278}\right)\left(\frac{0.5019 - 0.0/0.3048}{0.0204 - 0.0/0.3048}\right)\right]}{\ln[1.278]} \\ &= 6.8 \end{aligned}$$

This is close to the answer obtained by stepping off. However, great caution should be exercised when applying the Kremser Equation in situations where the K-value varies significantly over the extractor. The greater the variation in the K-value over the extractor, the greater is the uncertainty in the accuracy of the Kremser Equation.

Having obtained an estimate of the number of theoretical stages, this must be related to the height of the actual equipment or the number of stages in the actual equipment. The equipment used for liquid–liquid extraction is the same as that described for liquid–liquid reactions illustrated in Figure 6.5. For the mixer–settler arrangement shown in Figure 6.5, these can be combined in multiple stages countercurrently, in which each mixing and settling stage represents a theoretical stage. Much less straightforward is the relationship between the stages, or height of the contactor, in the other arrangements in Figure 6.5 and the number of theoretical stages. Typical HETPs for various designs of contactor are (Humphrey and Keller, 1997):

Contactor	HETP (m)
Sieve tray	0.5–3.5
Random packing	0.5–2.0
Structured packing	0.2–2.0
Mechanically agitated	0.1–0.3

The relationship between the number of theoretical and actual stages or contactor height depends on many factors, such as geometry, rate of agitation, flowrates of the liquids, physical properties of the liquids, the presence of impurities affecting the surface properties at the interface, and so on. The only reliable way to relate the actual stages to the theoretical stages in liquid–liquid extraction equipment is to scale from the performance of similar equipment carrying out similar separation duties, or to carry out pilot plant experiments.

When choosing a solvent for an extraction process, there are many issues to consider:

1) *Distribution coefficient.* It is desirable for the distribution coefficient, defined in Equation 9.11, to be large. Large values will mean that less solvent is required for the separation. A useful guide when selecting a solvent is that the solvent should be chemically similar to the solute. In other words, like dissolves like. A polar liquid like water is generally best suited for ionic and polar compounds. Nonpolar compounds like hexane

are better for nonpolar compounds. When a solute dissolves in a solvent, some attractions between solvent molecules must be replaced by solute–solvent attractions when a solution forms. If the new attractions are similar to those replaced, very little energy is required to form the solution. This holds true for many systems, but the molecular interactions that contribute to the solubility of one compound in another are many and varied. Hence, it is only a general guide.

2) *Separation factor.* The separation factor, defined by Equation 9.12, measures the tendency of one component to be extracted more readily than another. If the separation needs to separate one component from a feed relative to another, then it is necessary to have a separation factor greater than unity, and preferably as high as possible.
3) *Insolubility of the solvent.* The solubility of the solvent in the raffinate should be as low as possible.
4) *Ease of recovery.* It is always desirable to recover the solvent for reuse. This is often done by distillation. If this is the case, then the solvent should be thermally stable and not form azeotropes with the solute. Also, for the distillation to be straightforward, the relative volatility should be large and the latent heat of vaporization small.
5) *Density difference.* The density difference between the extract and the raffinate should be as large as possible to allow the liquid phases to coalesce more readily.
6) *Interfacial tension.* The larger the interfacial tension between the two liquids, the more readily coalescence will occur. However, on the other hand, the higher the interfacial tension, the more difficult will be the dispersion in the extraction.
7) *Side reactions.* The solvent should be chemically stable and not undergo any side reactions with the components in the feed (including impurities).
8) *Vapor pressure.* The vapor pressure at working conditions should preferably be low if an organic solvent is to be used. High vapor pressure for an organic solvent will lead to the emission of volatile organic compounds (VOCs) from the process, potentially leading to environmental problems. VOCs will be discussed in more depth later when environmental issues are considered.
9) *General properties.* The solvent should be nontoxic for applications such as the manufacture of foodstuffs. Even for the manufacture of general chemicals, the solvent should be preferably nontoxic for safety reasons. Safety also dictates that the solvent should preferably be nonflammable. Low viscosity and a high freezing point will also be advantageous.

It will rarely be the case that a solvent can be chosen to satisfy all of the above criteria, and hence some compromise will almost always be necessary.

Like absorption, separation is sometimes enhanced by using a solvent that reacts with the solute. The discussion, so far, regarding liquid–liquid extraction has been restricted to physical extraction. Both irreversible and reversible reactions can be used in chemical extraction. For example, acetic acid can be extracted from water using organic bases that take advantage of the acidity of the acetic acid. These organic bases can be regenerated and recycled. Unfortunately, the analysis of chemical extraction is far more complex than that of physical extraction. The phase equilibrium behavior cannot be approximated by constant distribution coefficients. This means that simple methods like the Kremser Equation no longer apply and complex simulation software is required. This is outside the scope of this text.

9.3 Adsorption

Adsorption is a process in which molecules of *adsorbate* become attached to the surface of a solid *adsorbent*. Adsorption processes can be divided into two broad classes:

1) *Physical adsorption*, in which physical bonds form between the adsorbent and the adsorbate.
2) *Chemical adsorption*, in which chemical bonds form between the adsorbent and the adsorbate.

An example of chemical adsorption is the reaction between hydrogen sulfide and ferric oxide:

$$\underset{\text{hydrogen sulphide}}{6H_2S} + \underset{\text{ferric oxide}}{2Fe_2O_3} \longrightarrow \underset{\text{ferric sulfide}}{2Fe_2S_3} + 6H_2O$$

The ferric oxide adsorbent, once it has been transformed chemically, can be regenerated in an oxidation step:

$$\underset{\text{ferric sulfide}}{2Fe_2S_3} + \underset{\text{oxygen}}{3O_2} \longrightarrow \underset{\text{sulphur}}{6S} + \underset{\text{ferric oxide}}{2Fe_2O_3}$$

The adsorbent, having been regenerated, is then available for reuse. By contrast, physical adsorption does not involve chemical bonds between the adsorbent and the adsorbate. Table 9.3 compares physical and chemical adsorption in broad terms (Hougen, Watson and Ragatz, 1959).

Here, attention will focus on physical adsorption. This is a commonly used method for the separation of gases, but is also used for the removal of small quantities of organic components from liquid streams.

A number of different adsorbents are used for physical adsorption processes. All are highly porous in nature. The main types can be categorized as follows:

1) *Activated carbon.* Activated carbon is a form of carbon that has been processed to develop a solid with high internal porosity. Almost any carbonaceous material can be used to manufacture activated carbon. Coal, petroleum residue, wood or shells of nuts (especially coconut) can be used. The most common method used for the manufacture of activated carbon starts by forming and heating the solid up to 600 to 900 °C, usually in an inert atmosphere. This is followed by controlled oxidation (or *activation*) by heating to a higher temperature (up to 1200 °C) in the presence of oxygen, steam or carbon monoxide to develop the porosity and surface activity. Other methods of preparation exploit chemical activation in which carbonaceous material is mixed with phosphoric acid, sodium hydroxide, zinc chloride or another agent, followed by heating up to 500 to 900 °C. Adsorption using activated carbon is the

Table 9.3

Physical and chemical adsorption (Hougen, Watson and Ragatz, 1959).

	Physical adsorption	Chemical adsorption
Heat of adsorption	Small, same order as heat of vaporization (condensation)	Large, many times greater than the heat of vaporization (condensation)
Rate of adsorption	Controlled by resistance to mass transfer	Controlled by resistance to surface reaction
	Rapid rate at low temperatures	Low rate at low temperatures
Specificity	Low, entire surface availability for physical adsorption	High, chemical adsorption limited to active sites on the surface
Surface coverage	Complete and extendable to multiple molecular layers	Incomplete and limited to a layer, one molecule thick
Activation energy	Low	High, corresponding to a chemical reaction
Quantity adsorbed per unit mass	High	Low

most commonly used method for the separation of organic vapors from gases. It is also used for liquid-phase separations. A common liquid-phase application is for decolorizing or deodorizing aqueous solutions. The activated carbon for liquid separations has a different pore structure from that used for vapors and gases.

2) *Silica gel.* Silica gel is a porous amorphous form of silica (SiO_2) and is manufactured by acid treatment of sodium silicate solution and then dried. The silica gel surface has an affinity for water and organic material. It is primarily used to dehydrate gases and liquids.
3) *Activated aluminas.* Activated alumina is a porous form of aluminum oxide (Al_2O_3) with a high surface area, manufactured by heating hydrated aluminum oxide to around 400 °C in air. Activated aluminas are mainly used to dry gases and liquids, but can be used to adsorb gases and liquids other than water.
4) *Molecular sieve zeolites.* Zeolites are crystalline aluminosilicates. They differ from the other three major adsorbents in that they are crystalline and the adsorption takes place inside the crystals. This results in a pore structure different from other adsorbents in that the pore sizes are more uniform. Access to the adsorption sites inside the crystalline structure is limited by the pore size, and hence zeolites can be used to absorb small molecules and separate them from larger molecules, as "*molecular sieves*". Zeolites selectively adsorb or reject molecules on the basis of differences in molecular size, shape and other properties, such as polarity. Applications include a variety of gaseous and liquid separations. Typical applications are the removal of hydrogen sulfide from natural gas, separation of hydrogen from other gases, removal of carbon dioxide from air before cryogenic processing, separation of *p*-xylene from mixed aromatic streams, separation of fructose from sugar mixtures, and so on.

Data used for the design of adsorption processes are normally derived from experimental measurements. The capacity of an adsorbent to adsorb an adsorbate depends on the compound being adsorbed, the type and preparation of the adsorbate, inlet concentration, temperature and pressure. In addition, adsorption can be a competitive process in which different molecules can compete for the adsorption sites. For example, if a mixture of toluene and acetone vapor is being adsorbed from a gas stream on to activated carbon, then toluene will be adsorbed preferentially relative to acetone and will displace the acetone that has already been adsorbed.

The capacity of an adsorbent to adsorb an adsorbate can be represented by adsorption isotherms, as shown in Figure 9.7a, or adsorption isobars, as shown in Figure 9.7b. The general trend can be seen that adsorption increases with decreasing temperature and increases with increasing pressure.

Data for adsorption isotherms can often be correlated by the Freundlich Isotherm Equation. For adsorption from liquids, this takes the form:

$$w = kC^n \tag{9.28}$$

where w = mass of adsorbate per mass of adsorbent at equilibrium
C = concentration
k, n = constants determined experimentally

Although Equation 9.28 is written in terms of concentration of a specific component, it can also be used if the identity of the solute is unknown. For example, adsorption is used to remove color from liquids. In such cases, the concentration of solute can be measured, for example, by a colorimeter and Equation 9.28 expressed in terms of arbitrary units of color intensity, providing that the color scale varies linearly with the concentration of the solute responsible for the color.

Data from the adsorption of gases and vapors are usually correlated in terms of volume adsorbed (at standard conditions of 0 °C and 1 atm pressure) and partial pressure. The adsorption can then be represented by:

$$V = k'p^{n'} \tag{9.29}$$

where V = volume of gas or vapor adsorbed at standard conditions ($m^3 \cdot kg^{-1}$)
p = partial pressure (Pa, bar)
k', n' = constants determined experimentally

9

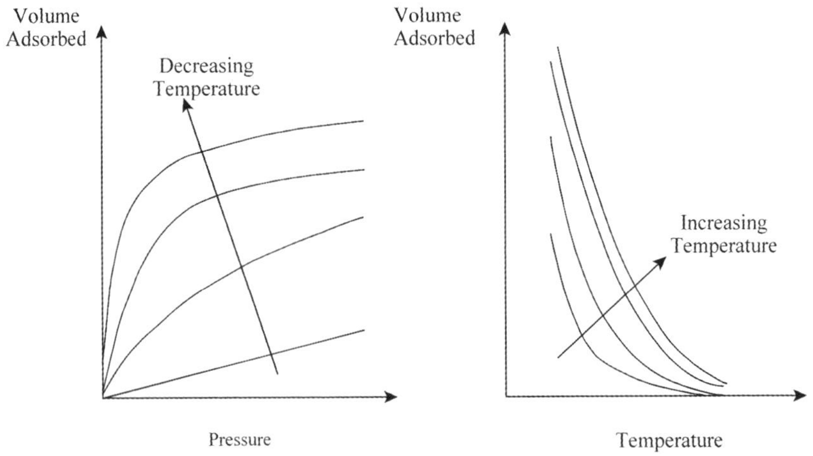

(a) Adsorption isotherms. (b) Adsorption bars.

Figure 9.7

Adsorption of gases and vapors on solids.

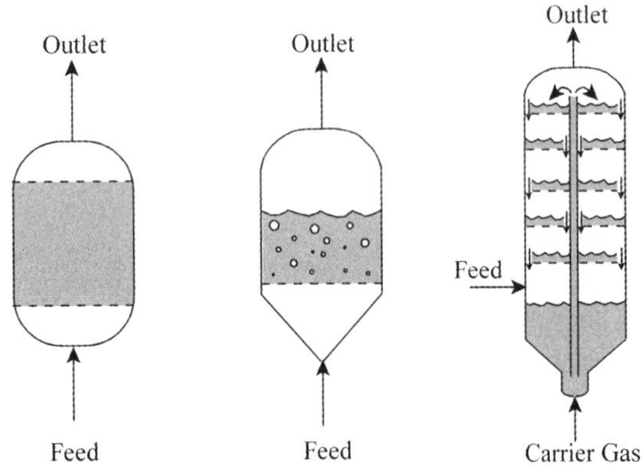

(a) Fixed bed. (b) Fluidized bed. (c) Travelling bed.

Figure 9.8

Different contacting arrangements for adsorber design.

This means that representing the equilibrium adsorbate mass (or volume) adsorbed versus concentration (or partial pressure) should be a straight line if plotted on logarithmic scales. Other theoretically based correlating equations are available for the adsorption of gases and vapors (Hougen, Watson and Ragatz, 1959), but all equations have their limitations.

Adsorption can be carried out in fixed-bed arrangements, as illustrated in Figure 9.8a. This is the most common arrangement. However, fluidized beds and traveling beds can also be used, as illustrated in Figure 9.8b and c. Adsorption in a fixed bed is nonuniform and a *front* moves through the bed with time, as illustrated in Figure 9.9. The point at which the concentration in the outlet begins to rise is known as *breakthrough*. Once breakthrough has occurred, or preferably before, the bed on-line must be taken off-line and regenerated. This can be done by having several beds operating in parallel, one or more on-line, with one being regenerated. The regeneration options are:

1) *Steam.* This is the most commonly used method for recovery of organic material from activated carbon. Low-pressure steam is passed through the bed countercurrently. The steam is condensed along with any recovered organic material.
2) *Hot gas.* Hot gas can be used when the regeneration gas is going to be fed to thermal oxidation. Air is normally used outside the flammability range (see process safety later); otherwise nitrogen can be used.
3) *Pressure swing.* The desorption into a gas stream at a pressure lower than that for adsorption can also be used.
4) *Off-site regeneration.* Off-site regeneration is used when regeneration is difficult or infrequent. For example, organic material can polymerize on the adsorbent, making it difficult to regenerate. If activated carbon is being used, the carbon can

Figure 9.9

Concentration profiles through an adsorption bed exhibit a moving front.

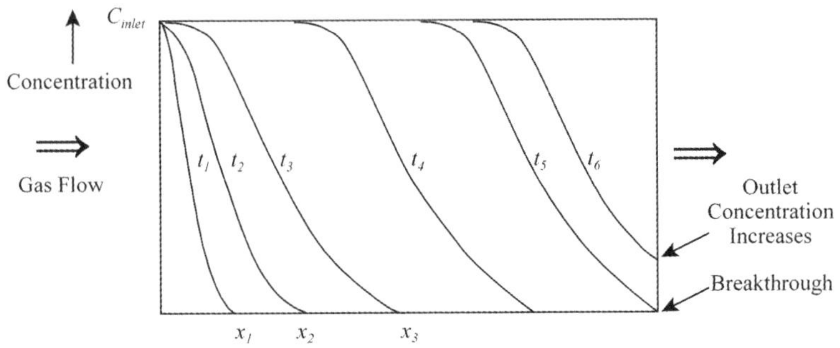

be regenerated by heat treating in a furnace and activating with steam.

When using fixed beds, there are various arrangements used to switch between adsorption and regeneration. Cycles involving two, three and four beds are used. Clearly, the more beds that are used, the more complex is the cycle of switching between the beds, but the more effective is the overall system.

In actual operation, the adsorption bed is not at equilibrium conditions. Also, there is loss of bed capacity due to:

- heat of adsorption;
- other components in the inlet flow competing for the adsorption sites, for example, moisture in an inlet gas stream competing for adsorption sites;
- for gas adsorption, any moisture in the bed after regeneration will block the adsorption sites.

In practice, two to three times the equilibrium capacity tends to be used.

Example 9.4 A gas mixture with a flowrate of $0.1\ m^3 \cdot s^{-1}$ contains $0.203\ kg \cdot m^{-3}$ of benzene. The temperature is 10 °C and the pressure 1 atm (1.013 bar). Benzene needs to be separated to give a gas stream with a benzene concentration of less than $5\ mg \cdot m^{-3}$. It is proposed to achieve this by adsorption using activated carbon in a fixed bed. The activated carbon is to be regenerated using superheated steam. The experimental adsorption isotherms cannot be adequately represented by Freundlich isotherms and, instead, can be correlated at 10 °C by the empirical relationship:

$$\ln V = -0.0113(\ln p)^2 + 0.2071 \ln p - 3.0872 \qquad (9.30)$$

where V = volume of benzene adsorbed ($m^3 \cdot kg^{-1}$)
p = partial pressure (Pa)

It can be assumed that the gas mixture follows ideal gas behavior and that the kilogram molar mass of a gas occupies $22.4\ m^3$ at standard conditions of 0 °C and 1 atm (1.013 bar).

a) Estimate the mass of benzene that can be adsorbed per kg of activated carbon.
b) Estimate the volume of activated carbon required, assuming a cycle time of 2 hours (minimum cycle time is usually around 1.5 hours). Assume that the actual volume is three times the equilibrium volume and the bulk density of activated carbon is $450\ kg \cdot m^{-3}$.
c) The concentration of benzene in the gas stream must be less than $5\ mg \cdot m^{-3}$ after the bed has been regenerated with steam at 200 °C and brought back on-line at 10 °C. What fraction of benzene must be recovered from the bed by the regeneration to achieve this if the bed is assumed to be saturated before regeneration?

Solution

a) Assuming ideal gas behavior, first calculate the partial pressure of benzene at 10 °C:

$$y = 0.203 \times \frac{1}{78} \times 22.4 \times \frac{283}{273}$$
$$= 0.0604$$
$$p = yP = 0.0604 \times 1.013 = 0.0612 \text{ bar}$$

Substitute $p = 6120$ Pa in Equation 9.30:

$$\ln V = -0.0113(\ln 6120)^2 + 0.2071 \ln (6120) - 3.0872$$
$$V = 0.1176\ m^3 \cdot kg^{-1}$$

$$\text{Mass of benzene adsorbed} = 0.1176 \times \frac{1}{22.4} \times 78$$
$$= 0.410\ \text{kg benzene} \cdot \text{kg carbon}^{-1}$$

b) Carbon required for a 2-hour cycle, assuming equilibrium

$$= 0.1 \times 0.203 \times 2 \times 3{,}600 \times \frac{1}{0.410}$$
$$= 356.5\ \text{kg}$$

Assuming a design factor of 3:

$$= 1069.5\ \text{kg}$$

$$\text{Volume of bed} = \frac{1069.5}{450} = 2.38\ m^3$$

c) Concentration of benzene when the regenerated bed is brought back on-line must be less than $5\ mg \cdot m^{-3}$. This results from the partial pressure of any benzene left on the bed after regeneration, until there is a breakthrough from the bed. Thus, for a concentration of $5\ mg \cdot m^{-3}$ at 10 °C:

$$y = 5 \times 10^{-6} \times \frac{1}{78} \times 22.4 \times \frac{283}{273} = 1.488 \times 10^{-6}$$
$$p = 1.488 \times 10^{-6} \times 1.013 = 1.507 \times 10^{-6} \text{ bar}$$
$$= 0.1507 \text{ Pa}$$

Volume of benzene on the bed at 10 °C is given by:

$$\ln V = -0.0113(\ln 0.1507)^2 + 0.2071(\ln 0.1507) - 3.0872$$
$$V = 0.0296\ m^3 \cdot kg^{-1}$$

Benzene recovered from the bed during regeneration, assuming saturation before regeneration

$$= \frac{0.1176 - 0.0296}{0.1176} = 0.75$$

During regeneration, 75% of the benzene must be recovered from the bed if the outlet concentration is to be $5\ mg \cdot m^{-3}$.

9.4 Membranes

Membranes act as a *semi-permeable* barrier between two phases to create a separation by controlling the rate of movement of species across the membrane. The separation can involve two gas (vapor) phases, two liquid phases or a vapor and a liquid phase. The feed mixture is separated into a *retentate*, which is the part of the feed that does not pass through the membrane, and a *permeate*, which is that part of the feed that passes through the membrane. The driving force for separation using a membrane is partial pressure in the case of a gas or vapor and concentration in the case of a liquid. Differences in partial pressure and concentration across the membrane are usually created by the imposition of a pressure

9

differential across the membrane. However, the driving force for liquid separations can also be created by the use of a solvent on the permeate side of the membrane to create a concentration difference, or an electrical field when the solute is ionic.

To be effective for separation, a membrane must possess high *permeance* and a high *permeance ratio* for the two species being separated. The permeance for a given species diffusing through a membrane of given thickness is analogous to a mass transfer coefficient, that is, the flowrate of that species per unit area of membrane per unit driving force (partial pressure or concentration). The flux (flowrate per unit area) of Component *i* across a membrane can be written as (Hwang and Kammermeyer, 1975; Winston and Sirkar, 1992; Mulder, 1996; Seader, Henley and Roper, 2011):

$$N_i = \overline{P}_{M,i} \times (\text{driving force}) = \frac{P_{M,i}}{\delta_M} \times (\text{driving force}) \qquad (9.31)$$

where N_i = flux of Component *i*
$\overline{P}_{M,i}$ = permeance of Component *i*
$P_{M,i}$ = permeability of Component *i*
δ_M = membrane thickness

The flux, and hence the permeance and permeability, can be defined on the basis of volume, mass or molar flowrates. The accurate prediction of permeabilities is generally not possible and experimental values must be used. Permeability generally increases with increasing temperature. Taking a ratio of two permeabilities defines a *separation factor* or *selectivity* α_{ij}, which is defined as:

$$\alpha_{ij} = \frac{P_{M,i}}{P_{M,j}} \qquad (9.32)$$

Another important variable that needs to be defined is the cut or fraction of feed permeated θ, which is defined as:

$$\theta = \frac{F_P}{F_F} \qquad (9.33)$$

where θ = fraction of feed permeated
F_P = volumetric flowrate of permeate
F_F = volumetric flowrate of feed

Membrane materials can be divided into two broad categories:

1) *Microporous*. *Microporous* membranes are characterized by interconnected pores, which are small, but large in comparison to the size of small molecules. If the pores are of the order of size of the molecules for at least some of the components in the feed mixture, the diffusion of those components will be hindered, resulting in a separation. Molecules of size larger than the pores will be prevented from diffusing through the pores by virtue of a *sieving* effect.

   For microporous membranes, the partial pressure profiles, in the case of gas (vapor) systems, and concentration profiles are continuous from the bulk feed to the bulk permeate, as illustrated in Figure 9.10a. Resistance to mass transfer by films adjacent to the upstream and downstream membrane interfaces create partial pressure and concentration differences between the bulk concentration and the concentration adjacent to the membrane interface. Permeability for microporous membranes is high but selectivity is low for small molecules.

2) *Dense*. Nonporous *dense* solid membranes are also used. Separation in dense membranes occurs by components in a gas or liquid feed diffusing to the surface of the membrane, dissolving into the membrane material, diffusing through the solid and desorbing at the downstream interface. The permeability depends on both the solubility and the diffusivity of the permeate in the membrane material. Diffusion through the membrane can be slow, but highly selective. Thus, for the separation of small molecules, a high permeability or high separation factor can be achieved, but not both. This problem is solved by the formation of *composite* or *asymmetric* membranes involving a thin dense layer, called the *permselective* layer, supported on a much thicker microporous layer of *substrate* that provides support for the dense layer. A microporous substrate can be used on one side of the dense layer or on both sides of the dense layer. Supporting the dense layer on both sides has advantages if the flow through the membrane needs to be reversed for cleaning purposes. The flux rate of a species is controlled by the permeance of the thin permselective layer.

   The permeability of dense membranes is low because of the absence of pores, but the permeance of Component *i* in Equation 9.31 can be high if δ_M is very small, even though the

Figure 9.10

Partial pressure and concentration profiles across membranes.

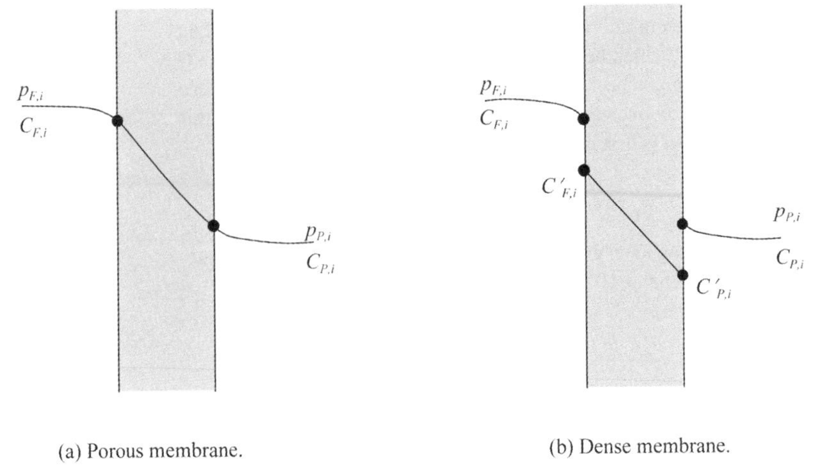

permeability is low. Thickness of the permselective layer is typically in the range 0.1 to 10 μm for gas separations. The porous support is much thicker than this and typically more than 100 μm. When large differences in P_M exist among species, both high permeance and high selectivity can be achieved in asymmetric membranes.

In Figure 9.10a, it can be seen that for porous membranes, the partial pressure and concentration profiles vary continuously from the bulk feed to the bulk permeate. This is not the case with nonporous dense membranes, as illustrated in Figure 9.10b. Partial pressure or concentration of the feed liquid just adjacent to the upstream membrane interface is higher than the partial pressure or concentration at the upstream interface. Also, the partial pressure or concentration is higher just downstream of the membrane interface than in the permeate at the interface. The concentrations at the membrane interface and just adjacent to the membrane interface can be related according to an equilibrium partition coefficient $K_{M,i}$. This can be defined as (see Figure 9.10b):

$$K_{M,i} = \frac{C'_{F,i}}{C_{F,i}} = \frac{C'_{P,i}}{C_{P,i}} \qquad (9.34)$$

Most membranes are manufactured from synthetic polymers. The application of such membranes is generally limited to temperatures below 100 °C and to the separation of chemically inert species. When operation at high temperatures is required, or the species are not chemically inert, microporous membranes can be made of ceramics and dense membranes from metals such as palladium.

Four idealized flow patterns can be conceptualized for membranes. These are shown in Figure 9.11. In Figure 9.11a, both the feed and permeate sides of the membrane are well mixed. Figure 9.11b shows a cocurrent flow pattern in which the fluid on the feed or retentate side flows along and parallel to the upstream surface of the membrane. The permeate fluid on the downstream side of the membrane consists of fluid that has just passed through the membrane at that location plus the permeate flowing to that location. The cross-flow case is shown in Figure 9.11c. In this case, there is no flow of permeate fluid along the membrane surface. Finally, Figure 9.11d shows countercurrent flow in which the feed flows along, and parallel to, the upstream of the membrane and the permeate fluid is the fluid that has just passed through the membrane at that location plus the permeate fluid flowing to that location in a countercurrent arrangement.

For these idealized flow patterns, parametric studies show that, in general, for the same operating conditions, countercurrent flow patterns yield the best separation and require the lowest membrane area. The next best performance is given by cross flow, then by cocurrent flow and the well-mixed arrangement shows the poorest performance. In practice, it is not always obvious as to which idealized flow pattern is the best to assume. The flow pattern not only depends on the geometry of the membrane arrangement but also on the permeation rate and, therefore, the cut fraction. The two most common practical arrangements are spiral wound and hollow fiber. In the spiral wound arrangement, flat membrane sheets separated by spacers for the flow of feed and permeate are wound into a spiral and inserted in a pressure vessel. Hollow fiber arrangements, as the name implies, are cylindrical membranes arranged in series in a pressure vessel in an arrangement similar to a shell-and-tube heat exchanger. The permselective layer is on the outside of the fibers. The feed enters the shell side and permeate passes through the membrane to the center of the hollow fibers. Other arrangements are possible, but less common. For example, arrangements similar to the plate-and-frame filter press illustrated in Figure 7.10a are used for some membrane separations. Membrane separators are usually modular in construction, with many parallel units required for large-scale applications.

With all membrane processes, the condition of the feed has a significant influence on the performance of the unit. This often means that some feed pretreatment is necessary to minimize fouling and the potential to damage the membrane.

9

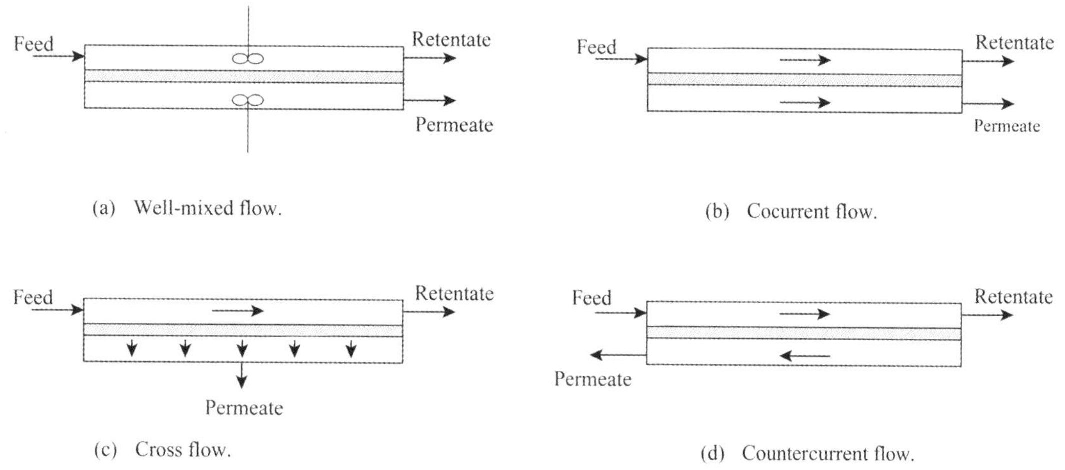

Figure 9.11
Idealized flow patterns in membrane separation.

Now consider the most important membrane separations.

1) *Gas permeation*. In *gas permeation* applications of membranes, the feed is at high pressure and usually contains some low molar mass species (typically less than 50 kg·kmol^{-1}) to be separated from higher molar mass species. The other side of the membrane is maintained at a low pressure, giving a high-pressure gradient across the membrane, typically in the range of 20 to 40 bar. It is also possible to separate organic vapor from a gas (e.g. nitrogen, hydrogen), using a membrane that is more permeable to the organic species than to the gas. Typical applications of gas permeation include:

- separation of hydrogen from methane,
- air separation,
- removal of CO_2 and H_2S from natural gas,
- helium recovery from natural gas,
- adjustment of H_2 to CO ratio in synthesis gas,
- dehydration of natural gas and air,
- removal of organic vapor from air,
- recovery of heavier hydrocarbons (C_3 +) from methane in natural gas, and so on.

The membrane is usually dense but sometimes microporous. If the external resistances to mass transfer are neglected in Figure 9.10, then $p_{F,i} = p'_{F,i}$ and $p_{P,i} = p'_{P,i}$ and Equation 9.31 can be written in terms of the volumetric flux as:

$$N_i = \frac{P_{M,i}}{\delta_M}(p_{F,i} - p_{P,i}) \tag{9.35}$$

where N_i = molar flux of Component i (kmol·m^{-2}·s^{-1})
$P_{M,i}$ = permeability of Component i (kmol·m·s^{-1}·m^{-2}·bar^{-1})
δ_M = membrane thickness (m)
$p_{F,i}$ = partial pressure of Component i in the feed (bar)
$p_{P,i}$ = partial pressure of Component i in the permeate (bar)

Low molar mass gases and strongly polar gases have high permeabilities and are known as *fast gases*. *Slow gases* have high molar mass and symmetric molecules. The ratio of permeabilities (separation factor) of the fast gas and slow gas should normally be greater than 2 for an effective membrane separation. Thus, membrane gas separators are effective when the gas to be separated is already at a high pressure and only a partial separation is required. A near-perfect separation is generally not achievable. Improved performance, overall, can be achieved by creating membrane networks in series, perhaps with recycles. Figure 9.12 illustrates some common examples of membrane networks.

There are important trade-offs to be considered when designing a gas separation using a membrane. The cost of a membrane separation is dominated by the capital cost of the membrane, which is proportional to its area, the capital cost of any compression equipment required and the operating costs of the compression equipment. If the feed is at a low pressure,

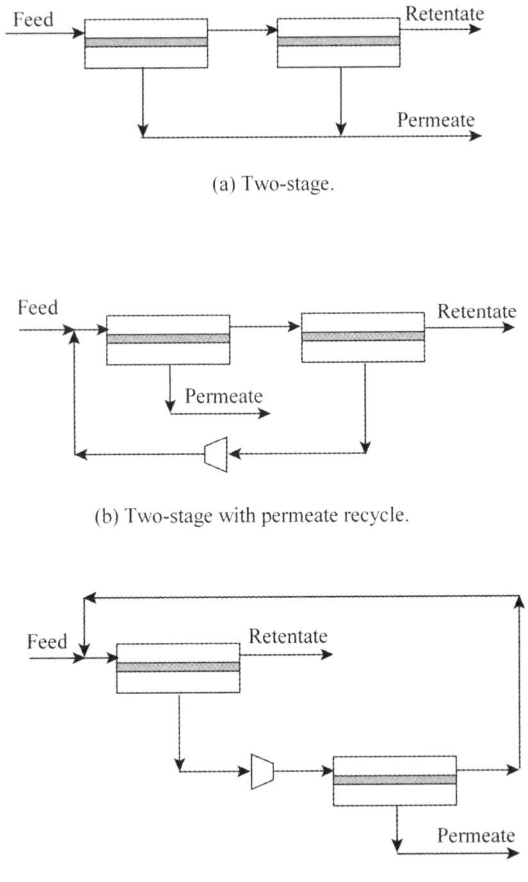

Figure 9.12
Examples of membrane networks.

then its pressure will need to be increased. The low-pressure permeate might need to be recompressed for further processing. Also, if a network of membranes has been used, then compressors might be needed within the membrane network.

In Equation 9.35, if a certain total molar flowrate of a component through the membrane is specified, then choosing a membrane material with a high permeability will decrease the required membrane area and hence the capital cost. Gas permeability increases with increasing temperature. Thus, increasing the temperature for a given flux will decrease the membrane area. This would suggest that the feed to membrane systems should be heated to reduce the area requirements. Unfortunately, the polymer membranes most often used in practice do not allow high-temperature feeds and feed temperatures are normally restricted to below 100 °C. It is common practice to heat the feed to gas membranes to avoid the possibility of condensation, which can damage the membrane. It is also clear from Equation 9.35 that, for a given membrane material, the thinner the membrane and the higher the pressure difference across the membrane, the lower the area requirement for a given component flux. However, the trade-offs are even more complex when the relative flux of components through the membrane and the product purity

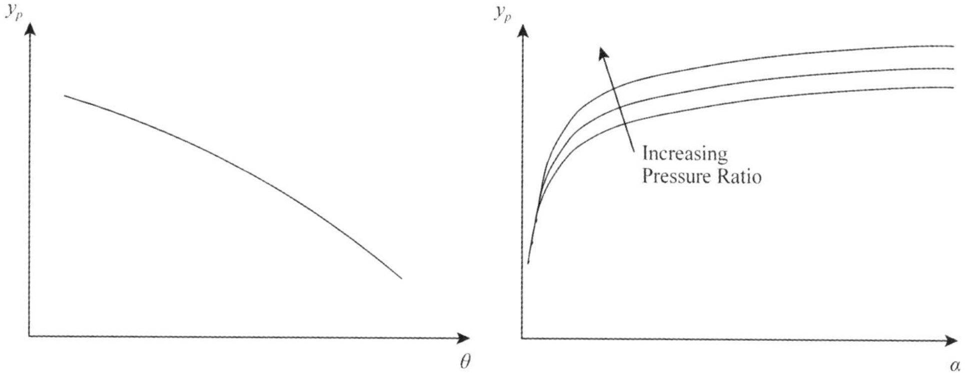

(a) Variation of permeate purity with stage cut.

(b) Variation of permeate purity with separation factor.

Figure 9.13

Trade-offs in membrane design for gas separation.

are considered. Figure 9.13a shows that, as the stage cut increases, the purity of the permeate decreases. In other words, if a component is being recovered as permeate, the more that is recovered, the less pure will be the product that is recovered. This is another important degree of freedom. Figure 9.13b shows the variation of permeate purity with separation factor. As expected, the higher the separation factor, as defined in Equation 9.32, the higher the product purity. Also, in Figure 9.13b, the higher the pressure ratio across the membrane, the higher the permeate purity. However, Figure 9.13b also shows that above a certain separation factor, the product purity is not greatly affected by an increase in the separation factor.

For the designer, given a required separation, there are three major contributions to the total cost to be traded off against each other:

- capital cost of the membrane unit (membrane modules and pressure housings);
- capital and operating costs of compression equipment (for compression of feed, recompression of the permeate or recycling within the membrane network);
- raw material losses.

2) *Reverse osmosis.* In *reverse osmosis*, a solvent permeates through a dense asymmetric membrane that is permeable to the solvent but not to the solute. The solvent is usually water and the solutes are usually dissolved salts. The principle of reverse osmosis is illustrated in Figure 9.14. In Figure 9.14a, a solute dissolved in a solvent in a concentrated form is separated from the same solvent in a dilute form by a dense membrane. Given the difference in concentration across the membrane, a natural process known as *osmosis* occurs, in which the solvent permeates across the membrane to dilute the more concentrated solution. The osmosis continues until equilibrium is established, as illustrated in Figure 9.14b. At equilibrium, the flow of solvent in both directions is equal and a difference in pressure is established between the two sides of the membrane,

9

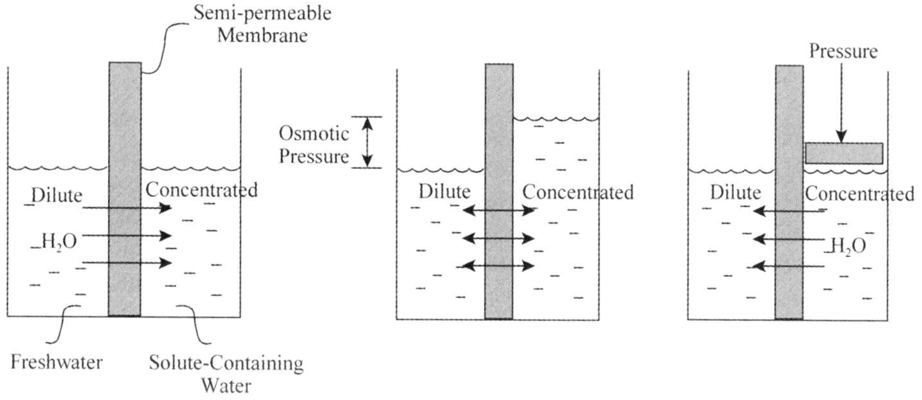

(a) Osmotic flow.

(b) No flow at equilibrium.

(c) Reverse osmosis.

Figure 9.14

Reverse osmosis.

the *osmotic pressure*. Although a separation has occurred as a result of the presence of the membrane, the osmosis is not useful because the solvent is transferred in the wrong direction, resulting in mixing rather than separation. However, applying a pressure to the concentrated solution, as shown in Figure 9.14c, can reverse the direction of transfer of solvent through the membrane. This causes the solvent to permeate through the membrane from a concentrated solution to the dilute solution. This separation process, known as *reverse osmosis*, can be used to separate a solvent from a solute–solvent mixture.

The flux through the membrane can be written as

$$N_i = \frac{P_{M,i}}{\delta_M}(\Delta P - \Delta\pi) \tag{9.36}$$

where N_i = solvent (water) flux through the membrane ($kg \cdot m^{-2} \cdot s^{-1}$)

$P_{M,i}$ = solvent membrane permeability ($kg\ solvent \cdot m^{-1} \cdot bar^{-1} \cdot s^{-1}$)

δ_M = membrane thickness (m)

ΔP = pressure difference across the membranes (bar)

$\Delta\pi$ = difference between the osmotic pressures of the feed and permeate solutions (bar)

Hence, as the pressure difference is increased, the solvent flow increases. The pressure difference used varies according to the membrane and the application, but is usually in the range 10 to 80 bar and can also be up to 100 bar. The osmotic pressure in Equation 9.36 for dilute solutions can be approximated by the Van't Hoff Equation:

$$\pi = iRT\frac{N_S}{V} \tag{9.37}$$

where π = osmotic pressure (bar)

i = number of ions formed if the solute molecule dissociates (e.g. for NaCl, $i = 2$, for $BaCl_2$, $i = 3$)

R = universal gas constant ($0.083145\ bar \cdot m^3 \cdot K^{-1} \cdot kmol^{-1}$)

T = absolute temperature (K)

N_S = number of moles of solute (kmol)

V = volume of pure solvent (m^3)

A phenomenon that is particularly important in the design of reverse osmosis units is that of *concentration polarization*. This occurs on the feed side (concentrated side) of the reverse osmosis membrane. Because the solute cannot permeate through the membrane, the concentration of the solute in the liquid adjacent to the surface of the membrane is greater than that in the bulk of the fluid. This difference causes mass transfer of solute by diffusion from the membrane surface back to the bulk liquid. The rate of diffusion back into the bulk fluid depends on the mass transfer coefficient for the boundary layer on the feed side. Concentration polarization is the ratio of the solute concentration at the membrane surface to the solute concentration in the bulk stream. Concentration polarization causes the flux of solvent to decrease since the osmotic pressure increases as the boundary layer concentration increases and the overall driving force ($\Delta P - \Delta\pi$) decreases.

Figure 9.15a illustrates the separation achieved by reverse osmosis. Reverse osmosis is now widely applied to the desalination of water to produce potable water. Other applications include:

- dewatering/concentrating foodstuffs,
- concentration of blood cells,
- treatment of industrial wastewater to remove heavy metal ions,
- treatment of liquids from electroplating processes to obtain a metal ion concentrate and a permeate that can be used as rinse water,
- separation of sulfates and bisulfates from effluents in the pulp and paper industry,
- treatment of wastewater in dying processes,
- treatment of wastewater inorganic salts, and so on.

Reverse osmosis is particularly useful when it is necessary to separate ionic material from an aqueous solution. A wide range of ionic species is capable of being removed with an efficiency of 90% or greater in a single stage. Multiple stages can increase the separation.

Applications of reverse osmosis are normally restricted to below 50 °C. When used in practice, the membranes for reverse osmosis must be protected by pretreatment of the feed to reduce membrane fouling and degradation (Hwang and Kammermeyer, 1975; Winston and Sirkar, 1992; Mulder, 1996). Spiral wound modules should be treated to remove particles down to 20 to 50 μm, while hollow fiber modules require particles down to 5 μm to be removed. If necessary, pH should be adjusted to avoid extremes of pH. Also, oxidizing agents such as free chlorine must be removed. Other pretreatments are also used depending on the application, such as removal of calcium and magnesium ions to prevent scaling of the membrane. Even with elaborate pretreatment, the membrane may still need to be cleaned regularly.

3) *Nanofiltration*. *Nanofiltration* is a pressure-driven membrane process similar to reverse osmosis, using asymmetric membranes but ones that are more porous. It can be considered to be a "coarse" reverse osmosis and is used to separate covalent ions and larger univalent ions such as heavy metals. Small univalent ions (e.g. Na^+, K^+ and Cl^-) pass through the membrane to a major extent. For example, a nanofiltration membrane might remove 50% NaCl and 90% $CaSO_4$. Nanofiltration membranes use less fine membranes (typically in the range 0.5 to 10 nm) than reverse osmosis and the pressure drop across the membrane is correspondingly lower. Pressure drops are usually in the range of 5 to 30 bar. Also, the fouling rate of nanofiltration membranes tends to be lower than that of reverse osmosis membranes. Figure 9.15b illustrates the separation achieved by nanofiltration. Typical applications include:
   - water softening (removal of calcium and magnesium ions),
   - water pretreatment prior to ion exchange or electrodialysis,
   - removal of heavy metals to allow reuse of water,

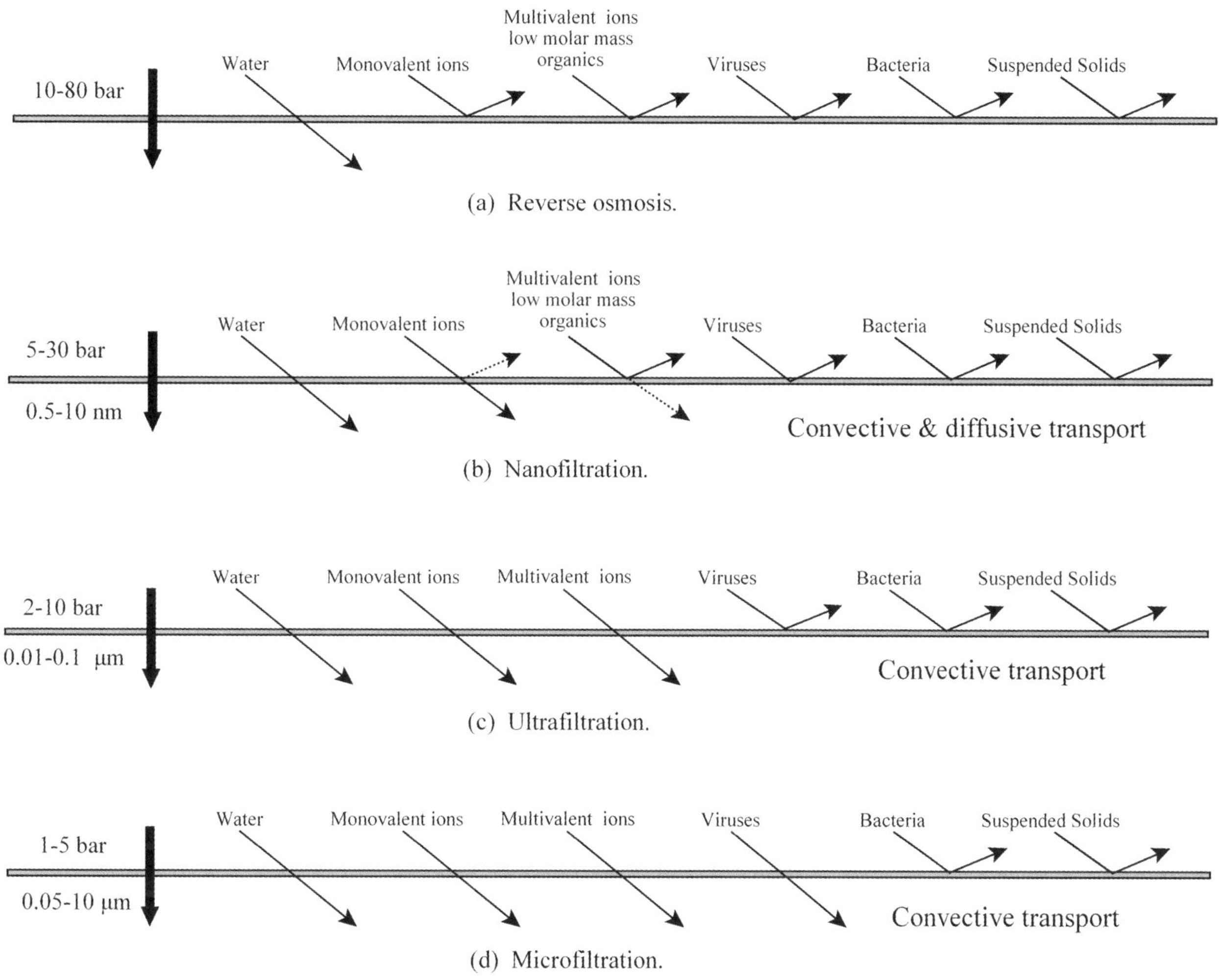

Figure 9.15

Liquid membrane separation. (From Jevons K, 2010, The Chemical Engineer, June, Issue 828: 39–41, reproduced by permission of the Institution of Chemical Engineers.)

9

- concentration of foodstuffs,
- desalination of foodstuffs, and so on.

As with reverse osmosis, feed pretreatment can be used to minimize membrane fouling and degradation, and regular cleaning is necessary.

4) *Ultrafiltration. Ultrafiltration* is again a pressure-driven membrane process similar to reverse osmosis, using asymmetric membranes but ones that are significantly more porous. Particles and large molecules do not pass through the membrane and are recovered as a concentrated solution. Solvent and small solute molecules pass through the membrane and are collected as permeate. Ultrafiltration is used to separate very fine particles (typically in the range 0.01 to 0.1 μm), microorganisms and organic components with high molar mass. The flux through the membrane is described by Equation 9.36. However, the ultrafiltration does not retain species for which the bulk solution osmotic pressure is significant. Pressure drops are usually in the range 2 to 10 bar.

Figure 9.15c illustrates the separation achieved by ultrafiltration. Typical applications of ultrafiltration are:

- clarification of fruit juice,
- recovery of vaccines and antibiotics from fermentation broths,
- oil–water separation,
- color removal, and so on.

Temperatures are restricted to below 70 °C. Again, feed pretreatment can be used to minimize membrane fouling and degradation, and regular cleaning is necessary.

5) *Microfiltration. Microfiltration* is a pressure-driven membrane filtration process and has already been discussed in Chapter 7 for the separation of heterogeneous mixtures. Microfiltration retains particles down to a size of around 0.05 μm. Salts and large molecules pass through the membrane, but particles of the size of bacteria and fat globules are rejected. A pressure difference of 1 to 5 bar is used across the membrane.

Figure 9.15d illustrates the separation achieved by microfiltration. Typical applications include:

- clarification of fruit juice,
- removal of bacteria from foodstuffs,
- removal of fat from foodstuffs, and so on.

6) *Dialysis*. *Dialysis* uses a membrane to separate species by virtue of their different diffusion rates in a microporous membrane. Both plate-and-frame and hollow fiber membrane arrangements are used. The feed solution, or *dialysate* containing the solutes to be separated, flows on one side of the membrane. A solvent, or *diffusate* stream, flows on the other side of the membrane. Some solvent may also diffuse across the membrane in the opposite direction, which reduces the performance by diluting the dialysate. Dialysis is used to separate species that differ appreciably in size and have reasonably large differences in diffusion rates. The driving force for separation is the concentration gradient across the membrane. Hence, dialysis is characterized by low flux rates when compared with other membrane processes such as reverse osmosis, microfiltration and ultrafiltration that depend on applied pressure. Dialysis is generally used when the solutions on both sides of the membrane are aqueous. Applications include recovery of sodium hydroxide, recovery of acids from metallurgical process liquors, purification of pharmaceuticals and separation of foodstuffs. An important application of dialysis is as an artificial kidney in the biomedical field, where it is used for the purification of human blood. Urea, uric acid and other components that have elevated concentrations in the blood diffuse across the membrane to an aqueous dialyzing solution, without removing essential high molar mass materials and blood cells.

7) *Electrodialysis*. *Electrodialysis* enhances the dialysis process with the aid of an electrical field and ion-selective membranes to separate ionic species from solution. It is used to separate an aqueous electrolyte solution into a concentrated and a dilute solution. Figure 9.16 illustrates the principle. The cation-selective membrane passes only cations (positively charged ions). The anion-selective membrane passes anions (negatively charged ions). The electrodes are chemically neutral. When a direct current charge is applied to the cell, the cations are attracted to the cathode (negatively charged) and anions are attracted to the anode (positively charged). Ions in the feed will pass through the appropriate membranes according to their charge, thus separating the ionic species.

   Since electrodialysis is suited only for the removal or concentration of ionic species, it is suited to recovery of metals from solution, recovery of ions from organic compounds, recovery of organic compounds from their salts, and so on.

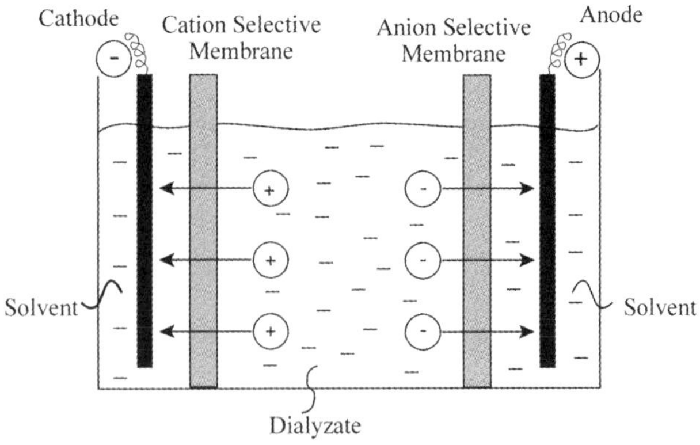

Figure 9.16
Electrodialysis.

8) *Pervaporation*. *Pervaporation* differs from the other membrane processes described so far in that the phase state on one side of the membrane is different from that on the other side. The term pervaporation is a combination of the words *perm*-selective and *evaporation*. The feed to the membrane module is a mixture (e.g. ethanol–water mixture) at a pressure high enough to maintain it in the liquid phase. The liquid mixture is contacted with a dense membrane. The other side of the membrane is maintained at a pressure at or below the dew point of the permeate, thus maintaining it in the vapor phase. The permeate side is often held under vacuum conditions. Pervaporation is potentially useful when separating mixtures that form azeotropes (e.g. ethanol–water mixture). One of the ways to change the vapor–liquid equilibrium to overcome azeotropic behavior is to place a membrane between the vapor and liquid phases. Temperatures are restricted to below 100 °C and, as with other liquid membrane processes, feed pretreatment and membrane cleaning are necessary.

With all membrane processes, there is a potential fouling problem that must be addressed when specifying the unit. With liquid feeds in particular, this usually means pretreating the feed to remove solids, potentially down to very fine particle sizes, as well as other pretreatments. Membranes used for liquid separations often need regular cleaning, perhaps on a daily basis or even more regularly. Cleaning can be carried out by reversal of flow (*backflushing*) and chemical treatment. The system used for cleaning is usually cleaning-in-place, whereby the membrane is taken off-line and connected to a cleaning circuit. However, membranes are easily damaged – chemically, by aggressive components, mechanically, in cleaning cycles and by excessive temperatures.

Example 9.5 A gaseous purge stream from a process has a mole fraction of hydrogen of 0.7. The balance can be assumed to be methane. It is proposed to recover the hydrogen using a membrane. The flowrate of the purge gas is 0.2 kmol·s^{-1}. The pressure is 20 bar and temperature is 30 °C. The permeate will be assumed to be 1 bar. Assume that the gas is well mixed on each side of the membrane and that there is no pressure drop along the membrane surface. The permeance ($P_{M,i}/\delta_M$) of hydrogen and methane for the membrane are given in Table 9.4.

Assume that 1 kmol of gas occupies 22.4 m^3 at standard temperature and pressure (STP). For stage-cut fractions from 0.1 to 0.9, calculate the purity of hydrogen in the permeate, the

Table 9.4

Permeance data for Example 9.5.

	Permeance (m^3 STP·m^{-2}·s^{-1}·bar^{-1})
H_2	3.75×10^{-4}
CH_4	3.00×10^{-6}

membrane area and the fractional hydrogen recovery for a single-stage membrane.

Solution If the gas is assumed to be well mixed, then on the high-pressure (feed-side) side of the membrane, the mole fraction is that of the retentate leaving the membrane. Assuming no pressure drop along the membrane and a binary separation, Equation 9.35 can be written for Component A as:

$$\frac{F_P y_{P,A}}{22.4 A_M} = \frac{P_{M,A}}{22.4\delta_M}(P_F y_{R,A} - P_P y_{p,A}) \tag{9.38}$$

where F_P = volumetric flowrate of the permeate ($m^3 \cdot s^{-1}$)
A_M = membrane area (m^2)
$P_{M,A}$ = permeability of Component A ($m^3 \cdot m \cdot s^{-1} \cdot m^{-2} \cdot bar^{-1}$)
δ_M = membrane thickness (m)
P_F = pressure of feed (bar)
P_P = pressure of permeate (bar)
$y_{P,A}$ = mole fraction of Component A in the permeate
$y_{R,A}$ = mole fraction of Component A in the retentate

Similarly, for Component B:

$$\frac{F_P y_{P,B}}{22.4 A_M} = \frac{P_{M,B}}{22.4\delta_M}(P_F y_{R,B} - P_P y_{p,B}) \tag{9.39}$$

where $y_{P,B}$ = mole fraction of Component B in the permeate
$y_{R,B}$ = mole fraction of Component B in the retentate

For a binary mixture:

$$\begin{aligned} y_{P,B} &= 1 - y_{P,A} \\ y_{R,B} &= 1 - y_{R,A} \end{aligned} \tag{9.40}$$

Substituting Equation 9.40 in Equation 9.39 gives:

$$\frac{F_P(1 - y_{P,A})}{A_M} = \frac{P_{M,A}}{\delta_M}[P_F(1 - y_{R,A}) - P_P(1 - y_{p,A})] \tag{9.41}$$

Dividing Equation 9.38 by 9.41:

$$\frac{y_{P,A}}{1 - y_{P,A}} = \frac{\alpha\left[y_{R,A} - \left(\frac{P_P}{P_F}\right) y_{P,A}\right]}{(1 - y_{R,A}) - \left(\frac{P_P}{P_F}\right)[1 - y_{P,A}]} \tag{9.42}$$

where $\alpha = \dfrac{P_{M,A}}{P_{M,B}}$

The value of $y_{R,A}$ is usually unknown, but it can be eliminated from Equation 9.42 by making an overall balance:

$$F_F = F_R + F_P \tag{9.43}$$

and an overall balance for Component A:

$$F_F y_{F,A} = F_R y_{R,A} + F_P y_{P,A} \tag{9.44}$$

where F_F = volumetric flowrate of the feed
F_R = volumetric flowrate of the retentate
F_P = volumetric flowrate of the permeate

Equations 9.43 and 9.44 can be combined to give:

$$y_{R,A} = \frac{y_{F,A} - \theta y_{P,A}}{(1 - \theta)} \tag{9.45}$$

where $\theta = \dfrac{F_P}{F_F}$

Substituting Equation 9.45 in Equation 9.42 and rearranging gives:

$$0 = a_o + a_1 y_{P,A} + a_2 y_{P,A}^2 \tag{9.46}$$

where
$$\begin{aligned} a_0 &= -\alpha y_{F,A} \\ a_1 &= 1 - (1 - \alpha)(\theta + y_{F,A}) - \frac{P_P}{P_F}(1 - \theta)(1 - \alpha) \\ a_2 &= \theta(1 - \alpha) + \frac{P_P}{P_F}(1 - \theta)(1 - \alpha) \end{aligned}$$

Thus, given α, θ, P_P, P_F and y_F, Equation 9.46 can be solved to give y_P. This can be done numerically or using the general analytical solution to quadratic equations (Hwang and Kammermeyer, 1975; Mulder, 1996):

$$y_{P,A} = \frac{-a_1 + \sqrt{a_1^2 - 4a_2 a_0}}{2a_2} \tag{9.47}$$

Once y_P has been determined, A_M can be determined by substituting Equation 9.45 and $F_p = \theta F_F$ in Equation 9.38 to obtain:

$$A_M = \frac{\theta(1 - \theta) F_F y_{P,A} \delta_M}{P_{M,A}\left[P_F(y_{F,A} - \theta y_{P,A}) - P_P y_{P,A}(1 - \theta)\right]} \tag{9.48}$$

Also, the recovery can be defined as:

$$\begin{aligned} R &= \frac{F_P y_{P,A}}{F_F y_{F,A}} \\ &= \theta \frac{y_{P,A}}{y_{F,A}} \end{aligned} \tag{9.49}$$

From the feed data at STP:

$$\begin{aligned} F_F &= 0.2 \times 22.4 \\ &= 4.48\ m^3 \cdot s^{-1} \end{aligned}$$

Values of θ can now be substituted in Equations 9.46, 9.48 and 9.49 to obtain the results in Table 9.5.

The values of $y_{P,A}$, A_M and R are plotted against θ in Figure 9.17. It should be noted from Figure 9.17 that as the stage cut increases, the permeate concentration approaches that of the inlet gas. Also, as the stage cut increases the membrane area requirements rise steeply above $\theta = 0.6$. The choice of stage cut to be adopted for the design is a trade-off between the required purity, required recovery and membrane area, but is most likely to be less than 0.6 in this case.

Table 9.5

Results for a range of values of stage cut for Example 9.5.

θ	a_0	a_1	a_2	$y_{P,A}$	A_M	R
0.1	−87.5	105.8	−18.0	0.996	96.3	0.142
0.2	−87.5	117.6	−29.8	0.995	206.1	0.284
0.3	−87.5	129.3	−41.5	0.994	339.4	0.426
0.4	−87.5	141.1	−53.3	0.991	519.3	0.567
0.5	−87.5	152.9	−65.1	0.987	811.4	0.705
0.6	−87.5	164.7	−76.9	0.976	1479.6	0.837
0.7	−87.5	176.5	−88.7	0.937	3890.6	0.937
0.8	−87.5	188.2	−100.4	0.854	9626.3	0.976
0.9	−87.5	200.0	−112.2	0.771	16653.9	0.991

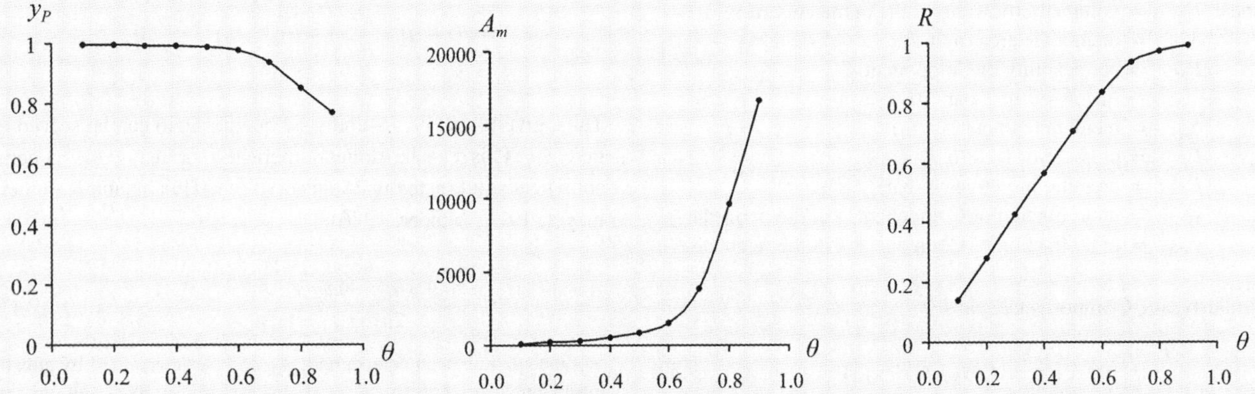

(a) Permeate purity. (b) Membrane area. (c) Recovery.

Figure 9.17

Trade-offs for permeate concentration, membrane area and hydrogen recovery for Example 9.5.

Example 9.5 illustrates the trade-offs for gas permeation. The feed was assumed to be a binary mixture, which simplifies the calculations. For multicomponent mixtures, the same basic equations can be written for each component and solved simultaneously (Hwang and Kammermeyer, 1975). The approach is basically the same, but numerically more complex.

It was also assumed that the gas on both sides of the membrane was well mixed. Again, this simplifies the calculations. In practice, cross-flow is more likely to represent the actual flow pattern. This is again more numerically complex as the separation varies along the membrane. The assumption of well-mixed feed and permeate will tend to overestimate the membrane area requirements. To perform the calculations assuming cross-flow, the model illustrated in Figure 9.18 can be used. If the simplifying assumptions are made that the pressure drop across the membrane is fixed, that there is no pressure drop along the membrane and that the selectivity is fixed, the equations can be solved either analytically or numerically. A simple numerical approach is to assume that the membrane area is divided into incremental areas, as illustrated in Figure 9.18. Each incremental area can be modeled by the well-mixed model developed in Example 9.5. The calculation starts from the feed side of the membrane for the first incremental area. A value of the stage cut θ needs to be assumed for the first incremental area. This allows $y_{P,A}$ to be calculated from Equation 9.47 and the incremental area dA_M to be calculated from Equation 9.48. The retentate concentration $y_{R,A}$ can then be calculated from Equation 9.45. The flow through the membrane dF_P can be calculated from the stage cut θ, and, hence, F_R from a balance around the incremental area. For the next incremental area across the membrane, the feed flowrate is F_R from the first increment and

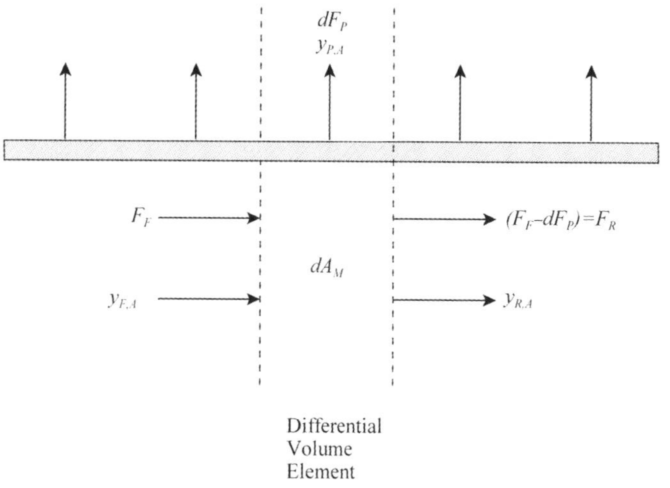

Figure 9.18
Cross-flow model for membranes.

the feed concentration is the retentate concentration from the first increment, which is $y_{R,A}$. The well-mixed model is then applied to the next incremental area, and so on, across the membrane. To ensure that the numerical approach is accurate, the assumed stage cut must be small enough to allow the well-mixed model for the incremental area to be representative of cross-flow. The procedure is carried on across the membrane until the recovery or product purity meets the required specification. Alternatively, if the size of the membrane module is known, the integration across the membrane is continued until the calculated membrane area equals the specified area. The equations for cross-flow can also be solved analytically (Hwang and Kammermeyer, 1975).

Example 9.6 Reverse osmosis is to be used to separate sodium chloride (NaCl) from water to produce 45 $m^3 \cdot h^{-1}$ of water with a concentration of less than 250 ppm NaCl. The initial concentration of the feed is 5000 ppm. A membrane is available for which the following test results have been reported:

Feed concentration	2000 ppm NaCl
Pressure difference	16 bar
Temperature	25 °C
Solute rejection	99%
Flux	$10.7 \times 10^{-6}\ m^3 \cdot m^{-2} \cdot s^{-1}$

The following assumptions can be made:

- Feed and permeate sides of the membrane are both well mixed.
- Solute rejection is constant for different feeds.
- There is no pressure drop along the membrane surfaces.
- Test conditions are such that the cut fraction of water recovered is low enough for the retentate concentration to be equal to the feed concentration.
- Solute concentrations are low enough for the osmotic pressure to be represented by the Van't Hoff Equation ($R = 0.083145\ bar \cdot m^3 \cdot K^{-1} \cdot kmol^{-1}$).
- Density of all solutions is 997 $kg \cdot m^{-3}$ and molar mass of NaCl is 58.5.

For a pressure difference of 40 bar across the membrane, estimate the permeate and retentate concentrations and membrane area for cut fractions ranging from 0.1 to 0.5.

Solution Start by writing an overall balance and a balance for the solute:

$$F_F = F_R + F_P \tag{9.50}$$

$$C_F F_F = C_R F_R + C_P F_P \tag{9.51}$$

where F_F = flowrate of feed ($m^3 \cdot s^{-1}$)
F_R = flowrate of retentate ($m^3 \cdot s^{-1}$)
F_P = flowrate of permeate ($m^3 \cdot s^{-1}$)
C_F = concentration of feed ($kg \cdot m^{-3}$)
C_R = concentration of retentate ($kg \cdot m^{-3}$)
C_P = concentration of permeate ($kg \cdot m^{-3}$)

Combining Equations 9.50 and 9.51 with the definition of the cut fraction, $\theta = F_P/F_F$, after rearranging gives:

$$C_F = C_R(1 - \theta) + C_P\theta \tag{9.52}$$

The solute rejection is defined as the ratio of the concentration difference across the membrane to the bulk concentration on the feed-side of the membrane. If it is assumed that both sides are well-mixed:

$$SR = \frac{C_R - C_P}{C_R} \tag{9.53}$$

where SR = solution rejection (–)

Combining Equations 9.52 and 9.53 gives:

$$C_R = \frac{C_F}{1 - \theta \cdot SR} \tag{9.54}$$

$$C_P = \frac{C_F(1 - SR)}{1 - \theta \cdot SR} \tag{9.55}$$

First, calculate the osmotic pressure on each side of the membrane under test conditions:

$$\begin{aligned} C_R &= 2000 \text{ ppm} \\ &= 0.002 \text{ kg solute} \cdot \text{kg solution}^{-1} \\ &= \frac{0.002}{0.998} \text{ kg solute} \cdot \text{kg solvent}^{-1} \end{aligned}$$

$$\begin{aligned}
\frac{N_R}{V_R} &= \frac{0.002}{0.998} \times \frac{1}{58.5} \times 997 \\
&= 0.0342 \text{ kmol solute} \cdot \text{m}^3 \cdot \text{solvent}^{-1} \\
\pi_R &= i\,RT\frac{N_R}{V_R} \\
&= 2 \times 0.083145 \times 298 \times 0.342 \\
&= 1.69 \text{ bar}
\end{aligned}$$

Similarly for the permeate side:

$$\begin{aligned}
\pi_P &= 0.0169 \text{ bar} \\
\Delta\pi &= 1.69 - 0.0169 \\
&= 1.68 \text{ bar} \\
\overline{P}_M &= \frac{10.7 \times 10^{-6}}{(16 - 1.68)} \\
&= 7.470 \times 10^{-7} \text{ m}^3 \cdot \text{bar}^{-1} \cdot \text{m}^{-2} \cdot \text{s}^{-1}
\end{aligned}$$

The required permeate flux is:

$$\begin{aligned}
F_P &= 45 \text{ m}^3 \cdot \text{h}^{-1} \\
&= 0.0125 \text{ m}^3 \cdot \text{s}^{-1}
\end{aligned}$$

For $\theta = 0.1$:

$$\begin{aligned}
C_R &= \frac{5000}{1 - 0.1 \times 0.99} \\
&= 5549 \text{ ppm} \\
C_P &= \frac{5000(1 - 0.99)}{1 - 0.1 \times 0.99} \\
&= 55.5 \text{ ppm} \\
\pi_R &= 4.71 \text{ bar} \\
\pi_P &= 0.047 \text{ bar} \\
A_M &= \frac{Q_P}{\overline{P}_M(\Delta P - \Delta\pi)} \\
&= \frac{0.0125}{7.470 \times 10^{-7}(40 - 4.71 + 0.047)} \\
&= 473.5 \text{ m}^2
\end{aligned}$$

Table 9.6 gives the results of these calculations for other values of θ.

These trends are plotted in Figure 9.19. It can be seen that as θ increases, Q_F decreases and C_R, C_P and A_M all increase.

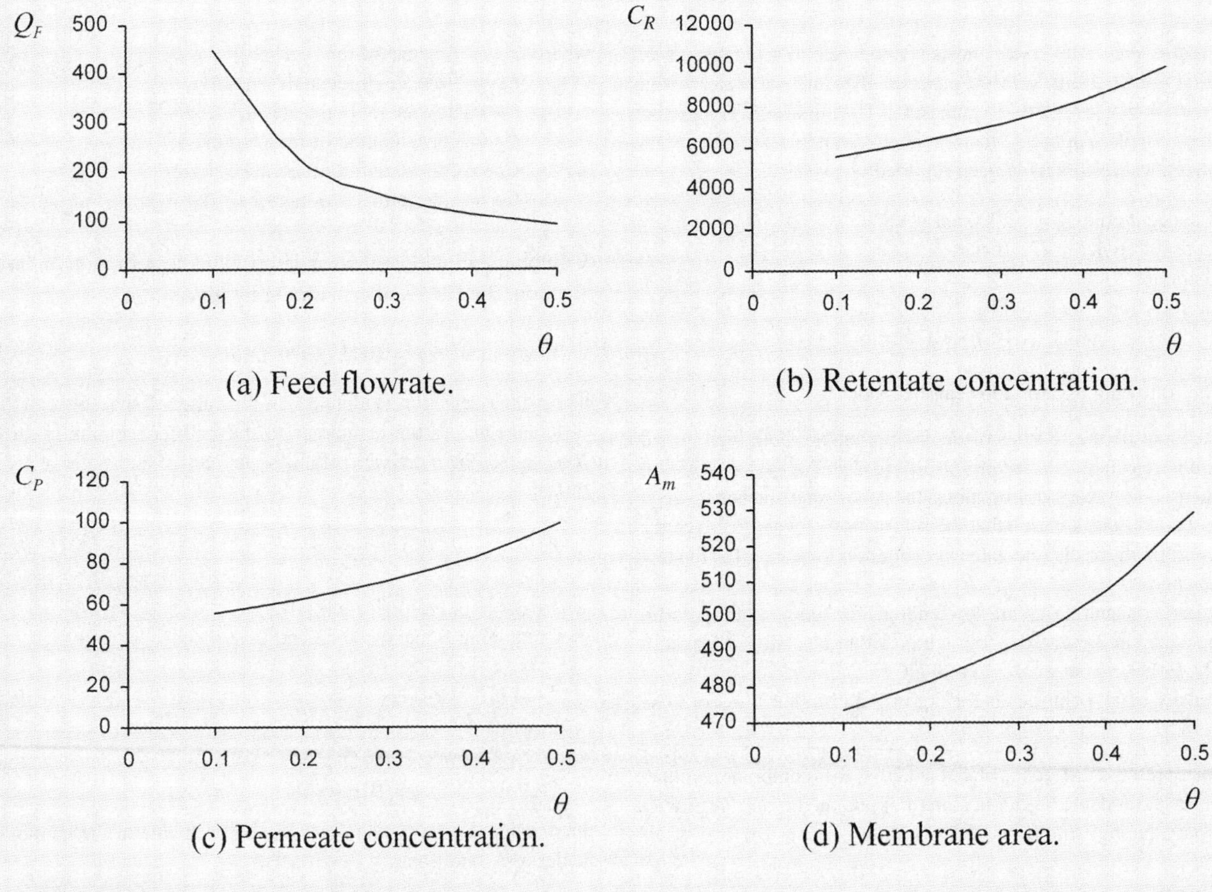

Figure 9.19

Trade-offs for stage cut, retentate concentration, permeate concentration and membrane area for the reverse osmosis separation in Example 9.6.

Table 9.6

Results for a range of values of stage cut for Example 9.6.

θ	F_F ($m^3 \cdot h^{-1}$)	C_R (ppm)	C_P (ppm)	A_M (m^2)
0.1	450	5549	55.5	473.5
0.2	225	6234	62.3	481.5
0.3	150	7112	71.1	492.0
0.4	112.5	8278	82.8	506.8
0.5	90	9901	99.0	528.9

A number of points need to be noted regarding these calculations:

1) The basic assumption of well-mixed fluid on the feed side of the membrane does not reflect the flow patterns for the configurations used in practice. The assumption simplifies the calculations and allows the basic trends to be demonstrated. Cross-flow is a better reflection of practical configurations. The well-mixed assumption is only reasonable for low concentrations and low values of θ. As a comparison, seawater desalination involves a feed with a concentration of the order of 35,000 ppm.
2) The membrane test data assumed that the value of θ was effectively zero. In practice, measurements are usually taken at low values of θ and the solute recovery is based on the average of the feed and retentate concentrations.
3) The basic assumption was made that the solute recovery was constant and independent of both the feed concentration and θ. This is only a reasonable assumption for high values of solute recovery where the flux of solute is inherently very low.
4) The volume of the equipment for a given area requirement depends on the chosen membrane configuration. For example, spiral wound membranes have a typical packing density of around 800 $m^2 \cdot m^{-3}$, whereas the packing density for hollow fiber membranes is much higher, at around 6000 $m^2 \cdot m^{-3}$.

9.5 Crystallization

Crystallization involves formation of a solid product from a homogeneous liquid mixture. Often, crystallization is required as the product is in solid form. The reverse process of crystallization is dispersion of a solid in a *solvent*, termed *dissolution*. The dispersed solid that goes into solution is the *solute*. As dissolution proceeds, the concentration of the solute increases. Given enough time at fixed conditions, the solute will eventually dissolve up to a maximum solubility where the rate of dissolution equals the rate of crystallization. Under these conditions, the solution is saturated with solute and is incapable of dissolving further solute under equilibrium conditions.

Figure 9.20a shows the solubility of a typical binary system of two Components *A* and *B*. The Line *CED* in Figure 9.20a represents the conditions of concentration and temperature that correspond to a saturated solution. If a mixture along the Line *CE* is cooled, then crystals of pure *B* are formed, leaving a residual solution. This continues along Line *CE* until Point *E* is reached, which is the *eutectic point*. At the eutectic point, both components crystallize and further separation is not possible. If a mixture along the Line *DE* is cooled, then crystals of pure *A* are formed, leaving a residual solution. Again, this continues along Line *DE* until Point *E* is reached, the eutectic point, and further separation is not possible. Point *C* in Figure 9.20a is the melting point of pure *B* and Point *D* is the melting point of pure *A*. Below the eutectic temperature, a solid mixture of *A* and *B* forms. Examples of binary mixtures that exhibit the kind of behavior illustrated in Figure 9.20a are benzene–naphthalene and acetic acid–water. Not all binary systems follow the behavior shown in Figure 9.20a. There are other forms of behavior, some of which resemble vapor–liquid equilibrium behavior. Solid–liquid equilibrium can be predicted by thermodynamic methods. The equilibrium behavior can be extremely complex, especially when more than two components are involved (Walas, 1985).

Complexity is increased by *polymorphism* in which solid material can exist in more than one form or crystal structure. For example, glycine (an amino acid commonly found in proteins) is able to form monoclinic and hexagonal crystals. Different polymorphs, although possessing the same chemical composition, may have very different properties. Polymorphism is important in the development of pharmaceutical ingredients, as many drugs receive regulatory approval for only a single crystal form or polymorph. Some solutes can form compounds with their solvents. Such compounds with definite proportions between solutes and solvents are termed *solvates*. If the solvent is water, the compounds formed are termed *hydrates*.

Figure 9.20b shows the equilibrium solubility of various salts in water. Usually, the solubility increases as temperature increases. The solubility of copper sulfate increases significantly with increasing temperature. The solubility of sodium chloride increases with increasing temperature, but the effect of temperature on solubility is small. The solubility of sodium sulfate decreases with increasing temperature. Such reverse solubility behavior is unusual.

In general, solubility is mainly a function of temperature, generally increasing with increasing temperature. Pressure has a negligible effect on solubility.

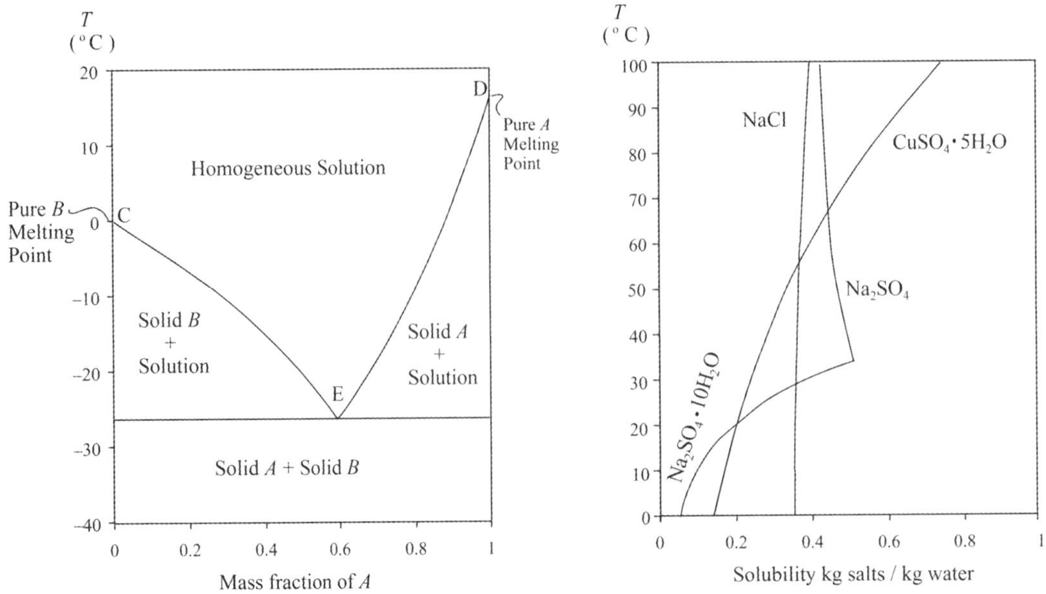

(a) Binary phase diagram for an *A*–*B* mixture (*A* = acetic acid, *B* = water in this example).

(b) Solubility of various salts in water.

Figure 9.20

Equilibrium solubility of solutes versus temperature.

It might be expected that if a solute is dissolved in a solvent at a fixed temperature until the solution achieves saturation and any excess solute is removed and the saturated solution is then cooled, the solute would immediately start to crystallize from solution. However, solutions can often contain more solute than is present at saturation. Such *supersaturated* solutions are thermodynamically metastable and can remain unaltered indefinitely. This is because crystallization first involves *nucleus formation* or *nucleation* and then *crystal growth* around the nucleus. If the solution is free of all solid particles, foreign or of the crystallizing substance, then nucleus formation must first occur before crystal growth starts. *Primary nucleation* occurs in the absence of suspended product crystals. Homogeneous primary nucleation occurs when molecules of solute come together to form clusters in an ordered arrangement in the absence of impurities or foreign particles. The growing clusters become crystals as further solute is transferred from solution. As the solution becomes more supersaturated, more nuclei are formed. This is illustrated in Figure 9.21. The curve *AB* in Figure 9.21 represents the equilibrium solubility curve. Starting at Point *a* in the unsaturated region and cooling the solution without any loss of solvent, the equilibrium solubility curve is crossed horizontally into the *metastable* region. Crystallization will not start until it has been subcooled to Point *c* on the *supersolubility* curve. Crystallization begins at Point *c*, continues

Figure 9.21

Supersaturation in crystallization processes.

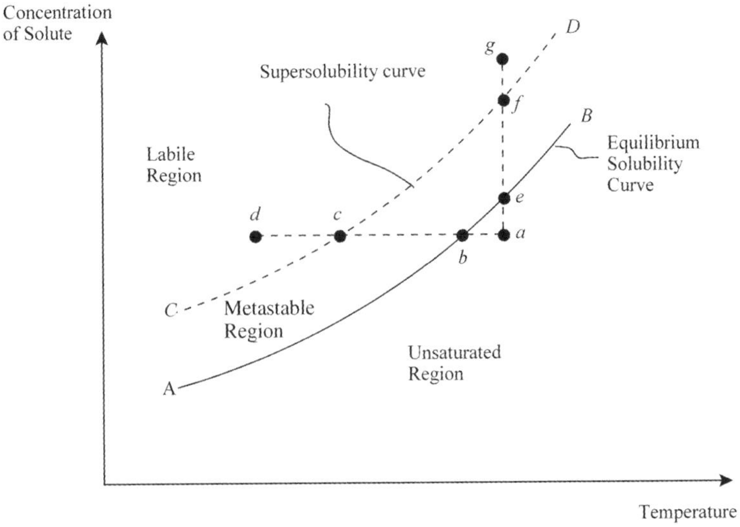

to Point d in the *labile* region and onwards. The curve CD, called the *supersolubility curve*, represents where nucleus formation appears spontaneously and, hence, where crystallization can start. The supersolubility curve is more correctly thought of as a region where the nucleation rate increases rapidly, rather than a sharp boundary. Primary nucleation can also occur heterogeneously on solid surfaces such as foreign particles.

Figure 9.21 also shows another way to create supersaturation instead of reducing the temperature. Starting again at Point a in the unsaturated region, the temperature is kept constant and the concentration is increased by removing the solvent (for example, by evaporation). The equilibrium solubility curve is now crossed vertically at Point e and the metastable region is entered. Crystallization is initiated at Point f and continues to Point g in the labile region and onwards.

Secondary nucleation requires the presence of crystalline product. Nuclei can be formed through attrition either between crystals or between crystals and solid walls. Such attrition can be created either by agitation or by pumping. The greater the intensity of agitation, the greater the rate of nucleation. Referring back to Figure 9.21, the equilibrium solubility curve AB is unaffected by the intensity of agitation. However, the supersolubility curve CD moves closer to the equilibrium solubility curve AB as the intensity of agitation becomes greater. Another way to create secondary nucleation is by adding *seed crystals* to start crystal growth in the supersaturated solution. These seeds should be a pure product. As solids build up in the crystallization, the source of new nuclei is often a combination of primary and secondary nucleation, although secondary nucleation is normally the main source of nuclei. Secondary nucleation is likely to vary with position in the crystallization vessel, depending on the geometry and the method of agitation.

Crystal growth can be expressed in a number of ways, such as the change in a characteristic dimension or the rate of change of mass of a crystal. The different measures are related through crystal geometry. The growth is measured as the increase in length in the linear dimension of the crystals. This increase in length is for geometrically corresponding distances on the crystals.

The size of the crystals and the size distribution of the final crystals are both important in determining the quality of the product. Generally, it is desirable to produce large crystals. Large crystals are much easier to wash and filter and therefore allow a purer final product to be achieved. Because of this, when operating a crystallizer, operation in the labile region is generally avoided. Operation in the labile region, where primary nucleation mechanisms will dominate, will produce a large number of fine crystals. The two phenomena of nucleation and crystal growth compete for the crystallizing solute.

In addition to crystal size and size distribution, the shape of the crystal product might also be important. The term *crystal habit* is used to describe the development of faces of the crystal. For example, sodium chloride crystallizes from aqueous solution with cubic faces. On the other hand, if sodium chloride is crystallized from an aqueous solution containing a small amount of urea, the crystals will have octahedral faces. Both crystals differ in habit.

The choice of crystallizer for a given separation will depend on the method used to bring about supersaturation. Batch and continuous crystallizers can be used. Continuous crystallizers are generally preferred, but special circumstances often dictate the use of batch operation, as will be discussed further in Chapter 16. The methods used to bring about supersaturation can be classified as:

1) *Cooling* a solution through indirect heat exchange. This is most effective when the solubility of the solute decreases significantly with temperature. Rapid cooling will cause the crystallization to enter the labile region. Controlled cooling, perhaps with seeding, can be used to keep the process in the metastable region. Care must be taken to prevent fouling of the cooling surfaces by maintaining a low temperature difference between the process and the coolant. Scraped surface heat exchange equipment might be necessary.
2) *Evaporation* of the solvent can be used to generate supersaturation. This can be used if the solute has a weak dependency of solubility on temperature.
3) *Vacuum* can be used to assist evaporation of the solvent and reduce the temperature of the operation. One arrangement is for the hot solution to be introduced into a vacuum, where the solvent evaporates and cools the solution as a result of the evaporation.
4) *Salting* or *knock-out* or *drown-out* involves adding an extraneous substance, sometimes call a *nonsolvent*, which induces crystallization. The nonsolvent must be miscible with the solvent and must change the solubility of the solute in the solvent. The nonsolvent will usually have a polarity different from that of the solvent. For example, if the solvent is water, the nonsolvent might be acetone, or if the solvent is ethanol, the nonsolvent might be water. This method has the advantage that fouling of heat exchange surfaces is minimized. On the other hand, an additional (extraneous) component is introduced into the system that must be separated and recycled.
5) *Reaction* can create metastable conditions directly. This is an attractive option when the reaction to produce the desired product and the separation can be carried out simultaneously.
6) *pH switch* can be used to adjust the solubility of sparingly soluble salts in aqueous solution.

Given these various methods of creating supersaturation, which is preferred?

- Reaction is preferable if the situation permits. It requires low solubility of the solute formed, but can produce tiny crystals if the solubility is too low.
- Cooling crystallization is also preferred. Here, there are options. The mixture can be crystallized directly in *melt crystallization* or a solvent can be used in *solution crystallization*. If crystallization is carried out without an extraneous solvent in melt crystallization, then a high temperature might be needed for the mixture to melt in order to carry out the separation. A high temperature might lead to product decomposition. If this is the case, there might be no choice other than to use an extraneous solvent. It might also be the case that a solvent is forced on the design. For example, a prior step, such as reaction, might require a certain solvent, and crystallization must be carried out from this solvent. Otherwise, there might be freedom to choose

the solvent. The initial criteria for the choice of solvent will be related to the solubility characteristics of the solute in the solvent. A solvent is preferred that exhibits a high solubility at high temperature but a low solubility at low temperature. In other words, it should have a steep solubility curve. The reason is that it is desirable to use as little solvent as possible, and hence high solubility at high temperature is required, but it is necessary to recover as much solute as possible, and hence low solubility at low temperature is desirable. In the pursuit of a suitable solvent, a pure solvent or a mixed solvent might be used. In addition to the solvent having suitable solubility characteristics, it should preferably have low toxicity, low flammability, low environmental impact, low cost, it should be easily recovered and recycled and have ease of handling, such as suitable viscosity characteristics.

- pH switch is preferred if water can be used as the solvent and if the solute has a solubility in water that is sensitive to changes in pH.
- Evaporative crystallization is not preferred if the product needs to be of high purity. In addition to evaporation concentrating the solute, it also concentrates impurities. Such impurities might form crystals to contaminate the product or might be present in the residual liquid occluded within the solid product.
- Knock-out or drown-out is generally not preferred as it involves adding a further extraneous material to the process. If it is to be successful, it requires a steep solubility curve versus the fraction of nonsolvent added.

Although the crystals are likely to be pure, the mass of crystals will retain some liquid when the solid crystals are separated from the residual liquid. If the adhering liquid is dried on the crystals, this will contaminate the product. In practice, the crystals will be separated from the residual liquid by filtration or centrifuging. Large uniform crystals separated from a low-viscosity liquid will retain the smallest proportion of liquid. Nonuniform crystals separated from a viscous liquid will retain a higher proportion of liquid. It is common practice to wash the crystals in the filter or centrifuge. This might be with fresh solvent or, in the case of melt crystallization, with a portion of melted product.

Example 9.7 A solution of sucrose in water is to be separated by crystallization in a continuous operation. The solubility of sucrose in water can be represented by the expression:

$$C^* = 1.524 \times 10^{-4}T^2 + 8.729 \times 10^{-3}T + 1.795 \qquad (9.56)$$

where C^* = solubility at the operating temperature (kg sucrose·kg H_2O^{-1})
T = temperature (°C)

The feed to the crystallizer is saturated at 60 °C ($C^* = 2.867$ kg sucrose·kg H_2O^{-1}). Compare cooling and evaporative crystallization for the separation of sucrose from water.

a) For cooling crystallization, calculate the yield of sucrose crystals as a function of the temperature of the operation for the crystallizer.

b) For evaporative crystallization, calculate the energy requirement as a function of crystal yield.

Solution

a) First, it is necessary to define the yield. Since there is no change in the volume of water, the yield can be defined as:

$$\text{Yield} = \frac{C_{in} - C_{out}}{C_{in}} \times 100(\%) \qquad (9.57)$$

The operating conditions in the crystallizer will be under supersaturated conditions. Calculation of the supersolubility curve is possible, but complex. The crystallizer will be designed to operate under supersaturated conditions in the metastable region. The degree of supersaturation is an important degree of freedom in the crystallizer design. Detailed design is required to define this with any certainty. Therefore, the yield will be defined here assuming the outlet concentration to be saturated. Then assuming a temperature, the outlet concentration can be calculated from Equation 9.56 and the yield from Equation 9.57. The results are presented in Table 9.7.

Table 9.7 shows that the temperature must be decreased to low values to obtain a reasonable yield from the crystallization process. It should also be noted that cooling to 40 °C should be possible against cooling water, and perhaps even down to 30 °C. Any cooler than this and refrigeration of some kind is required. This increases the cost of the cooling significantly.

b) A mass balance on the solvent gives:

$$F_{in} = F_{out} + F_V \qquad (9.58)$$

where F_{in} = inlet flowrate of liquid solvent
F_{out} = outlet flowrate of liquid solvent
F_V = flowrate of vaporized solvent

A mass balance on the solute gives:

$$C_{in}F_{in} = C_{out}F_{out} + m_{out}F_{out} \qquad (9.59)$$

where C_{in} = inlet solute concentration
C_{out} = outlet solute concentration
m_{out} = mass of crystals leaving the crystallizer

Table 9.7

Yield versus temperature for Example 9.7.

Temperature (°C)	Yield (%)
60	0.0
50	8.9
40	16.7
30	23.5
20	29.7
10	33.8
5	35.7

$$
\begin{aligned}
\text{Yield} &= \frac{m_{out}F_{out}}{C_{in}F_{in}} \\
&= \frac{C_{in}F_{in} - C_{out}F_{out}}{C_{in}F_{in}} \\
&= \frac{C_{in}F_{in} - C_{out}(F_{in} - F_V)}{C_{in}F_{in}} \\
&= 1 - \frac{C_{out}}{C_{in}} + \frac{C_{out}}{C_{in}}\frac{F_V}{F_{in}}
\end{aligned} \tag{9.60}
$$

Equation 9.60 indicates that as the rate of evaporation F_V increases, the yield increases. However, the energy input must also increase. If the simplifying assumption is made that the outlet concentration is the saturated equilibrium concentration C^* at 60 °C, then:

$$C_{in} = C_{out} = C^*$$

and from Equation 9.60:

$$\text{Yield} = \frac{F_V}{F_{in}}$$

Thus, if 10% of the solvent is vaporized, this will lead to a yield of 10%, and so on. Energy input is required to vaporize the solvent at 60 °C. The latent heat of water is 2350 kJ·kg^{-1}. The product of the mass of evaporation and latent heat gives the energy input:

$$\text{Energy input} = 2350\, F_V$$

9.6 Evaporation

Evaporation separates a volatile solvent from a solid. Single-stage evaporators tend to be used only when the capacity needed is small. For larger capacity, it is more usual to employ multistage systems that recover and reuse the latent heat of the vaporized material. Three different arrangements for a three-stage evaporator are illustrated in Figure 9.22.

1) *Forward-feed* operation is shown in Figure 9.22a. The fresh feed is added to the first stage and flows to the next stage in the same direction as the vapor flow. The boiling temperature decreases from stage to stage, and this arrangement is thus used when the concentrated product is subject to decomposition at higher temperatures. It also has the advantage that it is possible to design the system without pumps to transfer the solutions from one stage to the next.
2) *Backward-feed* operation is shown in Figure 9.22b. Here, the fresh feed enters the last and coldest stage and leaves the first stage as concentrated product. This method is used when the concentrated product is highly viscous. The high temperatures in the early stages reduce viscosity and give higher heat transfer coefficients. Because the solutions flow against the pressure gradient between stages, pumps must be used to transfer solutions between stages.
3) *Parallel-feed* operation is illustrated in Figure 9.22c. Fresh feed is added to each stage and product is withdrawn from each stage. The vapor from each stage is still used to heat the next stage. This arrangement is used mainly when the feed is almost saturated, particularly when solid crystals are the product.

Many other *mixed-feed* arrangements are possible, which combine the individual advantages of each type of arrangement. Figure 9.23 shows a three-stage evaporator in temperature–enthalpy terms, assuming that inlet and outlet solutions are at saturated conditions and that all evaporation and condensation duties are at constant temperature.

The three principal degrees of freedom in the design of stand-alone evaporators are:

1) Temperature levels can be changed by manipulating the operating pressure. Figure 9.23a shows the effect of a decrease in pressure.
2) The temperature difference between stages can be manipulated by changing the heat transfer area. Figure 9.23b shows the effect of a decrease in heat transfer area.
3) The heat flow through the system can be manipulated by changing the number of stages. Figure 9.23c shows the effect of an increase from three to six stages.

Given these degrees of freedom, how can an initialization be made for the design? The most significant degree of freedom is the choice of number of stages. If the evaporator is operated using a hot and cold utility, as the number of stages is increased a trade-off might be expected, as shown in Figure 9.24. Here, starting with a single stage, it has a low capital cost but requires a large energy cost. Increasing the stages to two decreases the energy cost in return for a small increase in capital cost, and the total cost decreases. However, as the stages are increased, the increase in capital cost at some point no longer compensates for the corresponding decrease in energy cost, and the total cost increases. Hence, there is an optimal number of stages. However, no attempt should be made to carry out this optimization in the early stages of a design, since the design is almost certain to change significantly when heat integration is considered later.

Another commonly used arrangement is to operate a heat pump across the evaporator. This recovers the heat from the vapor leaving the evaporator by increasing its temperature through compression and then using this compressed vapor as heat input for the evaporation. Heat pumps will be discussed in more detail later. Choosing heat pumping is also inappropriate in the early stages of a design, since this decision should only be made in the context of the full heat integration problem later in the design.

All that can be done is to make a reasonable initial assessment of the number of stages. Having made a decision for the number of stages, the heat flow through the system is temporarily fixed so that the design can proceed. Generally, the maximum temperature in evaporators is set by product decomposition and fouling. Therefore, the highest-pressure stage is operated at a pressure low enough to be below this maximum temperature. The pressure of the lowest-pressure stage is normally chosen to allow heat rejection to cooling water or air cooling. If decomposition and fouling are not a problem, then the stage pressures should be

Figure 9.22
Three possible arrangements for a three-stage evaporator.

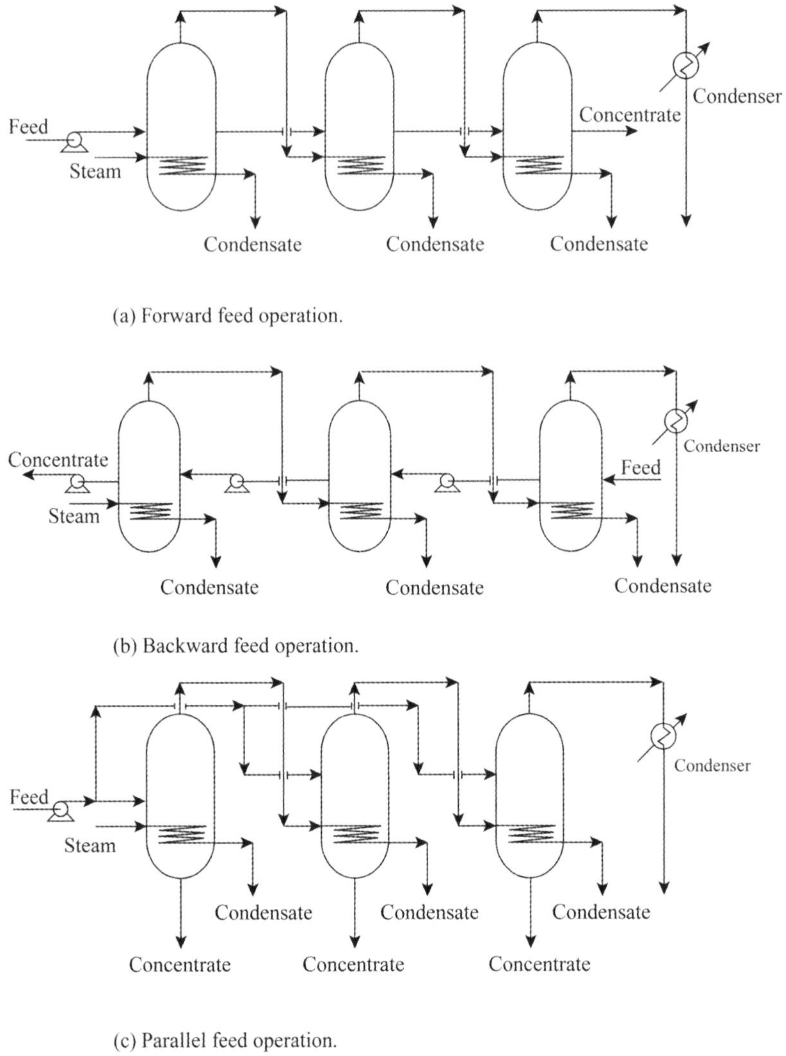

Figure 9.23
The principal degrees-of-freedom in evaporator design.

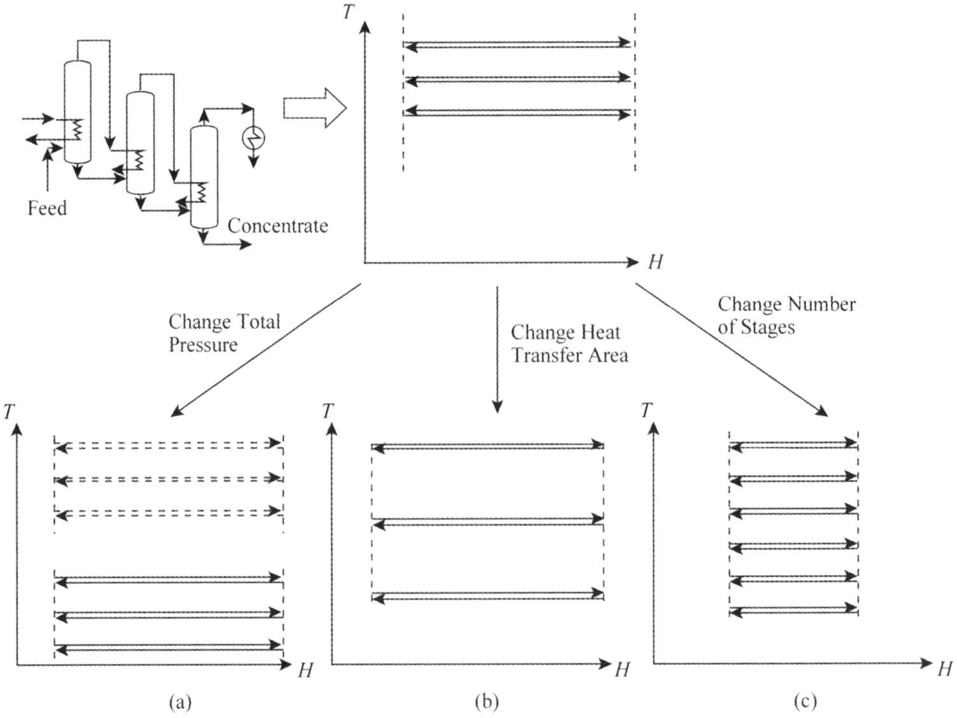

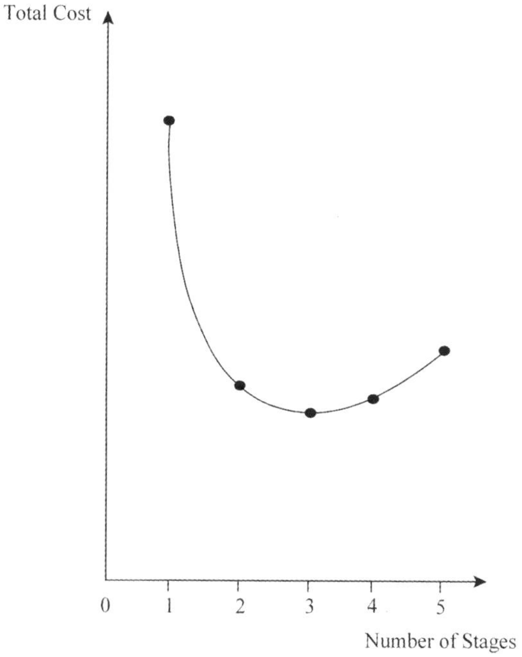

Figure 9.24

Variation of total cost with number of stages indicates that three stages is the optimum number for a stand-alone system in this case.

chosen such that the highest-pressure stage is below steam temperature and the lowest-pressure stage is above cooling water or air-cooling temperature.

For a given number of stages, if:

- all heat transfer coefficients are equal,
- all evaporation and condensation duties are at constant temperature,
- boiling point rise of the evaporating mixture is negligible,
- latent heat is constant through the system,

then the minimum capital cost is given when all temperature differences are equal (Smith and Jones, 1990). If evaporator pressure is not limited by the steam temperature but by product decomposition and fouling, then the temperature differences should be spread out equally between the upper practical temperature limit and the cold utility. This is usually a good enough initialization for most purposes, given that the design might change drastically later when heat integration is considered.

Another factor that can be important in the design of evaporators is the condition of the feed. If the feed is cold, then the backward-feed arrangement has the advantage that a smaller amount of liquid must be heated to the higher temperatures of the second and first stages. However, factors such as this should not be allowed to dictate design options at the early stages of flowsheet design because preheating the cold feed by heat integration with the rest of the process might be possible.

If the evaporator design is considered against a background process and heat integration with the background process is possible, then very different designs can emerge. When making an initial choice of separator, a simple, low-capital-cost design of evaporator should be chosen.

9.7 Separation of Homogeneous Fluid Mixtures by Other Methods – Summary

A common alternative to distillation for the separation of low molar mass materials is absorption. Liquid flowrate, temperature and pressure are important variables to be set, but no attempt should be made to carry out any optimization in the early stages of a design.

Adsoption can be effective for the separation of gases and vapor. Generally, physical adsorption increases with decreasing temperature and increases with increasing pressure. The adsorption bed requires periodic regeneration.

Membranes can be effective both for the separation of gasses and vapor and for the separation of liquids. If the ratio of the permeabilities of two gases through a semi-permeable membrane is significantly greater than 2, then a gas membrane separation can be effective. However, a high pressure drop across the membrane is often required. Membrane processes can also be used for liquid separations. Pressure drop across the membrane is the usual driving force for the separation. The required pressure drop varies significantly, depending on the nature of the separation.

9

As with distillation, when evaporators are chosen, no attempt should be made to carry out any optimization of individual operations in the early stages of a design.

When choosing a separation technique (absorption, stripping, liquid–liquid extraction, etc.), the use of extraneous mass-separating agents should be avoided for the following reasons:

- The introduction of extraneous material can create new problems in achieving product purity specifications throughout the process.
- It is often difficult to separate and recycle extraneous material with high efficiency. Any material not recycled can become an environmental problem. As will be discussed later, the best way to deal with effluent problems is not to create them in the first place.
- Extraneous material can create additional safety and storage problems.

Occasionally, a component that already exists in the process can be used as the mass separation agent, thus avoiding the introduction of extraneous material. However, clearly in many instances, practical difficulties and excessive cost might force the use of extraneous material.

9.8 Exercises

1. A gas stream with a flowrate of 8 N $m^3 \cdot s^{-1}$ contains 0.1% SO_2 by volume. It is necessary to remove 95% of the SO_2 by absorption in water at 10 °C and 1.01 bar. Henry's Law constant

for SO_2 in water at 10 °C is 22.3 bar. It can be assumed that the liquid and vapor flowrates are unchanged through the column as a result of the absorption, the gas behavior is ideal and the molar mass in kilograms occupies 22.4 m^3. For a concentration of water at the exit of the absorber to be 70% of equilibrium, calculate:

a) How much water will be required?

b) How many theoretical stages will be required in the absorber?

c) How might the absorption process be enhanced?

2. A process produces 5 $t{\cdot}h^{-1}$ of aqueous waste containing 25% by mass acetic acid. The acetic acid is to be recovered by extraction with 10 $t{\cdot}h^{-1}$ pure isopropyl ether. Equilibrium data are given in Table 9.1. The relationship between the concentration of acetic acid in the extract (x_E) to that in the raffinate (x_R) can be represented by:

$$x_E = 0.3098x_R^3 + 0.104x_R^2 + 0.3007x_R - 0.0006 \qquad (9.61)$$

Assuming that a single stage mixer settler unit is used for the extraction, and this achieves equilibrium:

a) Sketch the flowsheet.

b) If the mutual solubility is neglected, calculate the fractional recovery of acetic acid in the extract.

c) Repeat the calculation accounting for the mutual solubility to calculate the fractional recovery of acetic acid in the extract.

3. For the same process in Exercise 2 above, the process waste is to be extracted with the same flowrate of pure isopropyl ether but in a three-stage mixer settler arrangement in countercurrent flow.

a) Sketch the flowsheet.

b) If the mutual solubility is neglected, calculate the fractional recovery of acetic acid in the extract using the Kremser Equation assuming equilibrium is achieved in each stage. Equilibrium data are given in Table 9.2.

4. A vent with a flowrate of 40 $m^3{\cdot}h^{-1}$ and pressure of 1.2 bar is saturated with nitrobenzene at 25 °C and is to be treated with carbon adsorption to reduce the nitrobenzene concentration to ppm levels. The duty is small, and therefore it is anticipated that disposable carbon cartridges will be used. Assume that the vapor pressure of nitrobenzene at 25 °C is 0.00026 bar and its molar mass is 123 $kg{\cdot}kmol^{-1}$. It can also be assumed that the vent behaves as an ideal gas in which the molar mass in kilograms occupies 22.4 m^3 at standard conditions. If the carbon is capable of adsorbing 10% of its own mass of nitrobenzene at 25 °C (including a safety factor), calculate how long a disposable cartridge would last if it was loaded with 90 kg of carbon.

5. A vessel with a temperature of 10 °C operating at a pressure of 1.1 bar contains toluene and is purged with nitrogen with a flowrate of 300 $m^3{\cdot}h^{-1}$. Assuming that the nitrogen is saturated with toluene, estimate the size of a carbon adsorption system using two beds with in situ regeneration to reduce the toluene concentration to ppm levels. Assume that toluene vapor pressure at 10 °C is 0.0164 bar and that carbon is capable of absorbing 15% of its own mass of toluene at 10 °C (including a safety allowance). Assume that the superficial velocity of the gas in the bed is 0.2 $m{\cdot}s^{-1}$ (i.e. velocity in an empty bed) and that a cylindrical vessel with a height to bed diameter ratio of 3:1 is to be used. Assume that the density of activated carbon is 450 $kg{\cdot}m^{-3}$. The molar mass of toluene is 92 $kg{\cdot}kmol^{-1}$. It can be assumed that the vent behaves as an ideal gas and that the molar mass in kilograms occupies 22.4 m^3 at standard conditions.

a) Calculate the time of the cycle between regenerations.

b) If instead of fixing the superficial velocity, the duration of bed on stream is fixed to be 2 hours, calculate the volume of the bed.

c) After the cycles of operation and regeneration have become established, what dictates the concentration of the vent from the adsorption bed?

d) What are the steps involved with the regeneration of the bed if regenerated with steam and what effects can a poor regeneration procedure have?

6. Air containing 21% by volume oxygen and 79% by volume nitrogen is to be enriched to provide a gas with a higher oxygen content. A membrane is available for the separation with permeabilities shown in Table 9.8.

The flowrate of air to the membrane is 0.1 $kmol{\cdot}s^{-1}$. The air is to be introduced by a blower with a pressure at the membrane surface of 1.7 bar and the pressure differential maintained across the membrane by reducing the permeate pressure by means of a vacuum pump to 0.25 bar. Examine the relationship between permeate purity, membrane area and stage cut by varying the stage cut between 0.1 and 0.9. To simplify the calculations, assume the gas to be well-mixed on both sides of the membrane. Assume that 1 kmol of gas occupies 22.4 m^3 at standard temperature and pressure.

7. A well-mixed continuous crystallizer is to be used to separate potassium sulfate from an aqueous solution by cooling crystallization. The solubility of potassium sulfate can be represented by the expression:

$$C^* = 0.0666 + 0.0023T - 6 \times 10^{-6}T^2$$

where C^* = solubility at the operating temperature (kg solute per kg solvent)

T = temperature (°C)

Table 9.8

Permeability data for Exercise 6.

Component	Permeability (m^3 STP$\cdot m^{-2}\cdot s^{-1}\cdot bar^{-1}$)
Oxygen	1.80×10^{-3}
Nitrogen	6.93×10^{-4}

The feed to the crystallizer is saturated at 80 °C.

a) Calculate the yield of potassium sulfate crystals for cooling crystallization down to 40 °C.

b) Calculate the cooling temperature required to obtain a crystal yield of 50%.

References

Campbell H (1940) Report on the Estimated Costs of Doubling the Production of the Acetic Acid Concentrating Department, *Trans AIChE*, **36**: 628.

Douglas JM (1988) *Conceptual Design of Chemical Processes*, McGraw-Hill, New York.

Geankopolis CJ (1993) *Transport Processes and Unit Operations*, 3rd Edition, Prentice Hall.

Hougen OA, Watson KM and Ragatz RA (1959) *Chemical Process Principles. Part I: Material and Energy Balances*, John Wiley & Sons.

Humphrey JL and Keller GE (1997) *Separation Process Technology*, McGraw-Hill.

Hwang S-T and Kammermeyer K (1975) *Membranes in Separations*, John Wiley Interscience.

King CJ (1980) *Separation Processes*, 2nd Edition, McGraw-Hill.

Mulder M (1996) *Basic Principles of Membrane Technology*, Kluwer Academic Publishers.

Schweitzer PA (1997) *Handbook of Separation Process Techniques for Chemical Engineers*, 3rd Edition, McGraw-Hill, New York.

Seader JD, Henley EJ and Roper DR (2011) *Separation Process Principles*, 3rd Edition, John Wiley & Sons.

Smith R and Jones PS (1990) The Optimal Design of Integrated Evaporation Systems, *Heat Recovery Systems and CHP*, **10**: 341.

Treybal RE (1980) *Mass Transfer Operations*, 3rd Edition, McGraw-Hill.

Walas SM (1985) *Phase Equilibrium in Chemical Engineering*, Butterworth Publishers.

Winston WS and Sirkar KK (1992) *Membrane Handbook*, Chapman & Hall.

Chapter 10

Distillation Sequencing

Consider now the particular case in which a homogeneous multicomponent fluid mixture needs to be separated into a number of products, rather than just two products. As noted previously, distillation is the most common method of separating homogeneous fluid mixtures and in this chapter the choice of separation will be restricted such that all separations are carried out using distillation only. If this is the case, generally there is a choice of order in which the products are separated, that is, the choice of *distillation sequence*.

10.1 Distillation Sequencing using Simple Columns

Consider first the design of distillation systems comprising only simple columns. These simple columns employ:

- one feed split into two products;
- key components adjacent in volatility, or any components that exist in small quantities between the keys will become impurities in the products;
- a reboiler and a condenser.

If there is a three-component mixture to be separated into three relatively pure products and simple columns are employed, then the decision is between two sequences, as illustrated in Figure 10.1. The sequence shown in Figure 10.1a is known as the *direct sequence* in which the lightest component is taken overhead in each column. The *indirect sequence,* as shown in Figure 10.1b, takes the heaviest component as bottom product in each column.

If the distillation columns have both reboiling and condensation supplied by utilities, then the direct sequence in Figure 10.1a often requires less energy than the indirect sequence in Figure 10.1b. This is because the light material (Component A) is only vaporized once in the direct sequence. However, the indirect sequence can be more energy efficient if the feed to the sequence has a low flowrate of the light material (Component A) and a high flowrate of the heavy material (Component C). In this case, vaporizing the light material twice in the indirect sequence is less important than feeding a high flowrate of heavy material to both of the columns in the direct sequence.

For a three-component mixture to be split into three relatively pure products, there are only two alternative sequences. The complexity increases significantly as the number of products increases. Figure 10.2 shows the alternative sequences for a four-product mixture. Table 10.1 shows the relationship between the number of products and the number of possible sequences for simple columns (King, 1980).

Thus, there may be many ways in which the separation can be carried out to produce the same products. The problem is that there may be significant differences in the capital and operating costs between different distillation sequences that can produce the same products. Operating costs are usually the dominant cost in choosing between different sequences. This might be the cost of providing heat input to reboiling through steam or fired heaters or the cost of heat removal in low-temperature separation from the cost of providing refrigeration to low-temperature condensers.

10.2 Practical Constraints Restricting Options

Process constraints often reduce the number of options that can be considered. Examples of constraints of this type are:

1) Safety considerations might dictate that a particularly hazardous component be removed from the sequence as early as possible to minimize the inventory of that material.
2) Reactive and heat-sensitive components must be removed early to avoid problems of product degradation.
3) Corrosion problems often dictate that a particularly corrosive component be removed early to minimize the use of expensive materials of construction.
4) If thermal decomposition in the reboilers contaminates the product, then this dictates that finished products cannot be taken from the bottoms of columns.

Chemical Process Design and Integration, Second Edition. Robin Smith.

Companion Website: www.wiley.com/go/smith/chemical2e

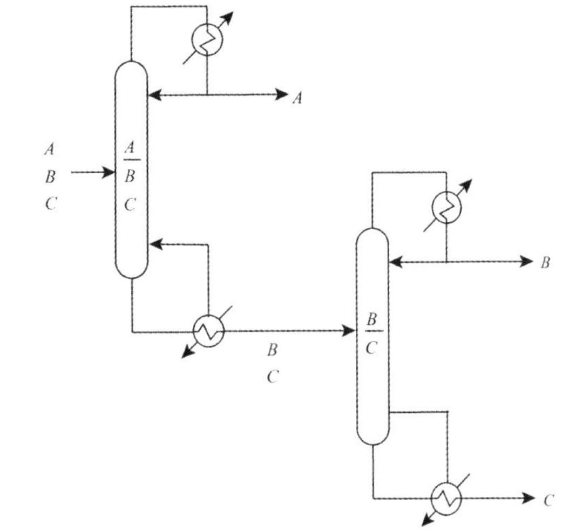

(a) Direct sequence.

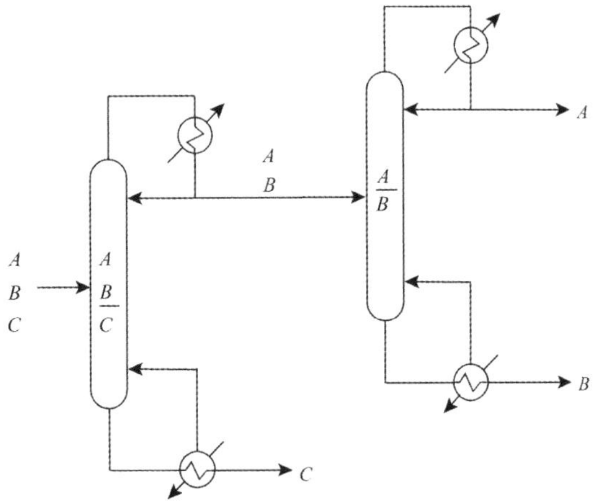

(b) Indirect sequence.

Figure 10.1

The direct and indirect sequences of simple distillation columns for a three-product separation. (Reproduced from Smith R and Linnhoff B, 1998, *Trans IChemE ChERD*, **66**: 195 by permission of the Institution of Chemical Engineers.)

5) Some compounds tend to polymerize when distilled unless chemicals are added to inhibit polymerization. These polymerization inhibitors tend to be nonvolatile, ending up in the column bottoms. If this is the case, it normally prevents finished products being taken from column bottoms.
6) There might be components in the feed to a distillation sequence that are difficult to condense. Total condensation of these components might require low-temperature condensation using refrigeration and/or high operating pressures. Condensation using both refrigeration and operation at high pressure increases operating costs significantly. Under these circumstances, the light components are normally removed from the top of the first column to minimize the use of refrigeration and high pressures in the sequence as a whole.

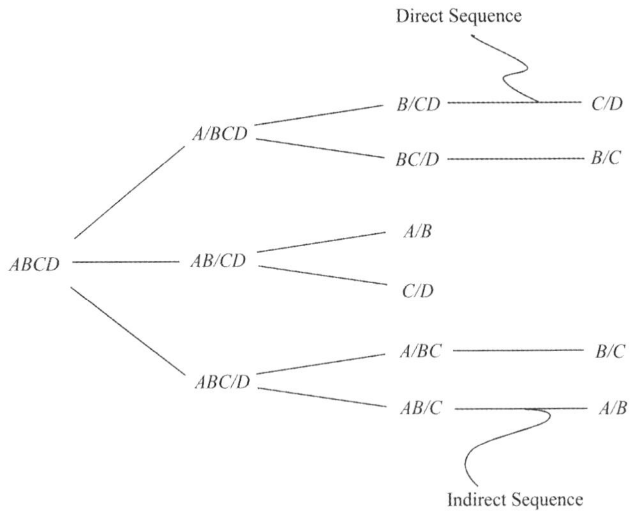

Figure 10.2

Alternative sequences for the separation of a four-product mixture.

10.3 Choice of Sequence for Simple Nonintegrated Distillation Columns

Heuristics have been proposed for the selection of the sequence of simple nonintegrated distillation columns to minimize energy costs (King, 1980). These heuristics attempt to generalize the observations made in many problems, mainly above ambient temperature distillation, where the cost of heat input to reboilers is the dominant cost. Although many heuristics have been proposed to minimize energy costs, they can be summarized by the following four types (Stephanopoulos, Linnhoff and Sophos, 1982):

Heuristic 1. Separations where the relative volatility of the key components is close to unity or that exhibit azeotropic behavior should be performed in the absence of nonkey components. In other words, do the most difficult separation last.

Table 10.1

Number of possible distillation sequences using simple columns.

Number of products	Number of possible sequences
2	1
3	2
4	5
5	14
6	42
7	132
8	429

Heuristic 2. Sequences that remove the lightest components alone one by one in column overheads should be favored. In other words, favor the direct sequence.

Heuristic 3. A component composing a large fraction of the feed should be removed first.

Heuristic 4. Favor splits in which the molar flow between top and bottom products in individual columns is as near equal as possible.

In addition to being restricted to simple columns, the observations are based on no heat integration (i.e. all reboilers and condensers are serviced by utilities). Difficulties can arise when the heuristics are in conflict with each other. This is illustrated in Example 10.1.

The conflicts that can arise with the use of heuristics, as illustrated in Example 10.1, might in principle be avoided by ranking in order of importance. However, the rank order might change from problem to problem. Although in the above example the heuristics do not give a clear indication of good candidate sequences, in some problems they might. It does seem though that a more general method than the heuristics is needed.

Rather than relying on heuristics that are qualitative, and can be in conflict, a quantitative measure of the relative performance of different sequences would be preferred. Ultimately, the choice will be based on capital and operating costs. For most problems, the costs are dominated by the operating costs. Heat integration might have a significant impact later in the design, but at this stage the nonintegrated performance is assessed:

- *Above ambient temperature processes.* The operating costs are dominated by the cost of supplying heat to the reboilers. Multiple heating utilities might be involved. In this case, lower temperature heating utilities should be preferred to higher temperatures. Very large reboilers might be fired heaters, but the majority will be steam heated. Cooling costs can be added to heating costs, but the operating costs are likely to be dominated by the heating costs. Methods to estimate the operating costs have been discussed in Chapter 2.
- *Below ambient temperature processes.* The operating costs are dominated by the cost of providing power to the refrigeration system required to supply cooling to the condensers. Multiple cooling utilities are likely to be involved. These might include multiple refrigeration levels, cooling water and perhaps air cooling. In this case, higher temperature cooling utilities should be preferred to lower temperature ones. The lower the temperature of a refrigeration utility, the greater the power requirement and cost. Cooling water is significantly cheaper than refrigeration. Heating costs can be added to cooling costs, but the operating costs are likely to be dominated by the power costs associated with the provision of cooling by refrigeration. Again, methods to estimate the costs have been discussed in Chapter 2.

Before the costs of heating and cooling can be assessed, the reboiler and condenser duties need to be calculated. As a first level of assessment, for above ambient temperature processes, the combined reboiler duties can be used to screen options. For below ambient temperature processes, the combined condenser duties can be used. In Chapter 8, it was shown how the condenser duty for an individual total condenser is given by:

$$Q_{COND} = \Delta H_{VAP} D(1 + R_F R_{\min}) \tag{10.1}$$

where Q_{COND} = condenser duty ($kJ \cdot s^{-1}$, kW)
ΔH_{VAP} = heat of vaporization of overhead vapor ($kJ \cdot kmol^{-1}$)
D = distillate flowrate ($kmol \cdot s^{-1}$)
R_F = ratio R/R_{min}
R_{min} = minimum reflux ratio (–)

For an individual partial condenser, the condenser duty is given by:

$$Q_{COND} = \Delta H_{VAP} D R_F R_{min} \tag{10.2}$$

Thus, the reboiler duty for an individual total condenser is given by:

$$Q_{REB} = \Delta H'_{VAP}(D + DR_F R_{min} - F + qF) \tag{10.3}$$

where Q_{REB} = condenser duty ($kJ \cdot s^{-1}$, kW)
$\Delta H'_{VAP}$ = heat of vaporization of the distillation bottoms liquid ($kJ \cdot kmol^{-1}$)
F = feed flowrate ($kmol \cdot s^{-1}$)
q = feed condition

$$= \frac{\text{heat required to vaporize one mole of feed}}{\text{molar latent heat of vaporization of feed}}$$

In most cases, for reasons discussed in Chapter 8, a saturated liquid feed is normally preferred ($q = 1$).

The calculation is repeated for all columns in the sequence and, depending on whether the process is above or below ambient temperature, either the reboiler heat loads summed to obtain the overall heating load for the sequence or the condenser heat loads summed to obtain the overall cooling load for the sequence. Different sequences can then be compared on the basis of heating duty or cooling. Alternatively, if the utility temperatures and costs are known, the individual reboiler and condenser utility costs can be added to obtain the total cost. The use of the Underwood equations in sequencing is illustrated in Examples 10.2 and 10.3.

The errors associated with the Underwood Equations used to calculate the minimum reflux ratio were discussed in Chapter 8. The Underwood Equations tend to underpredict the minimum reflux ratio. This introduces uncertainty in the way that the calculations were carried out in Examples 10.2 and 10.3. The differences in the total heat load between different sequences are in some cases small and the differences comparable with the errors associated with the prediction of the minimum reflux ratio using the Underwood Equations. However, as long as the errors are consistently in the same direction for all of the distillation calculations, the approach can still be used to screen between options. Nevertheless, the predictions should be used with caution and options not ruled out because of some small difference in the assessment of different sequences.

Example 10.1 Each component for the mixture of alkanes in Table 10.2 is to be separated into relatively pure products. Table 10.2 shows normal boiling points and relative volatilities to indicate the order of volatility and the relative difficulty of the separations. The relative volatilities have been calculated on the basis of the feed composition to the sequence, assuming a pressure of 6 barg. Vapor–liquid equilibrium has been calculated using the Peng–Robinson Equation of State with interaction parameters set to zero (see Appendix A). Different pressures can, in practice, be used for different columns in the sequence and if a single set of relative volatilities is to be used, the pressure at which the relative volatilities are calculated needs, as much as possible, to be chosen to represent the overall system.

Use the heuristics to identify potentially good sequences that are candidates for further evaluation.

Solution

Heuristic 1. Do *D*/*E* split last since this separation has the smallest relative volatility.

Heuristic 2. Favor the direct sequence and do the *A*/*B* split first.

$$A/BCDE$$

Heuristic 3. Remove the most plentiful component first.

$$ABCD/E$$

Heuristic 4. Favor near-equimolar splits between top and bottom products.

$$ABC/DE$$

$$408.3\ \text{kmol} \times \text{h}^{-1}/498.9\ \text{kmol} \times \text{h}^{-1}$$

All four heuristics are in conflict. Heuristic 1 suggests doing the *D*/*E* split last, whereas Heuristic 3 suggests it should be done first. Heuristic 2 suggests the *A*/*B* split first and Heuristic 4 the *C*/*D* split first.

Take one of the candidates and accept, say, the *A*/*B* split first. Then for the remainder:

Heuristic 1: Do *D*/*E* split last.

Heuristic 2:

$$B/CDE$$

Heuristic 3:

$$BCD/E$$

Heuristic 4:

$$BC/DE$$

$$362.9\ \text{kmol} \times \text{h}^{-1}/498.9\ \text{kmol} \times \text{h}^{-1}$$

Again the heuristics are in conflict. Heuristic 1 again suggests doing the *D*/*E* split last, whereas again Heuristic 3 suggests it should be done first. Heuristic 2 suggests the *B*/*C* split first and Heuristic 4 the *C*/*D* split first.

This process could be continued and possible sequences identified for further consideration. Some possible sequences would be eliminated, narrowing the number down, suggested by Table 10.1.

Table 10.2

Data for a mixture of alkanes to be separated by distillation.

Component	Flowrate ($\text{kmol}\cdot\text{h}^{-1}$)	Normal boiling point (K)	Relative volatility	Relative volatility between adjacent components
A. Propane	45.4	231	5.77	
				1.93
B. *i*-Butane	136.1	261	2.99	
				1.27
C. *n*-Butane	226.8	273	2.36	
				1.95
D. *i*-Pentane	181.4	301	1.21	
				1.21
E. *n*-Pentane	317.5	309	1.00	

Example 10.2 Using the Underwood Equations to predict the minimum reflux ratio, determine the best distillation sequence in terms of overall reboiler load to separate the mixture of alkanes in Table 10.2 into relatively pure products. The recoveries are assumed to be 100%. Assume the ratio of actual to minimum reflux to be 1.1 and all columns are fed with a saturated liquid and total condensers used throughout. Neglect pressure drop across each column. Relative volatilities and latent heats of vaporization can be calculated from the Peng–Robinson Equation of State with interaction parameters assumed to be zero (see Appendix A). Relative volatilities are to be calculated for each column based on the feed composition of each column. Determine the rank order of the distillation sequences on the basis of total reboiler load for:

a. All column pressures fixed to 6 barg.

b. Pressures allowed to vary through the sequence, such that the pressures of each column are minimized to give either the bubble point of the overhead product to be 10 °C above the cooling water return temperature of 35 °C (i.e. 45 °C) or a minimum of atmospheric pressure. This will avoid the necessity for refrigeration.

Solution The results for the two cases are shown in Tables 10.3 and 10.4.

It can be seen from Tables 10.3 and 10.4 that for each case there is little difference between the best few sequences in terms of the total reboiler heat load. For this problem, there is not even too much difference between the best and the worst sequences. Also, when the results of Cases a and b are compared, it should be noted that the rank order changes. The best three sequences for the two cases are illustrated in Figure 10.3. This sensitivity of the rank order to changes in the assumptions is not surprising given the small differences between the various sequences. All of the columns in all sequences are above atmospheric pressure, varying from 0.7 to 14.4 barg.

Table 10.3

Sequences for the separation of the mixture of alkanes, with pressure fixed at 6 barg.

Rank order	Total reboiler duty (MW)	% Best	Sequence			
1	32.40	100	*ABCD/E*	*ABC/D*	*AB/C*	*A/B*
2	32.62	100.7	*ABCD/E*	*AB/CD*	*A/B*	*C/D*
3	33.39	103.1	*ABCD/E*	*ABC/D*	*A/BC*	*B/C*
4	33.76	104.2	*ABCD/E*	*A/BCD*	*BC/D*	*B/C*
5	34.14	105.4	*ABCD/E*	*A/BCD*	*B/CD*	*C/D*
6	35.10	108.3	*A/BCDE*	*BCD/E*	*BC/D*	*B/C*
7	35.21	108.7	*AB/CDE*	*A/B*	*CD/E*	*C/D*
8	35.37	109.2	*ABC/DE*	*AB/C*	*D/E*	*A/B*
9	35.48	109.5	*A/BCDE*	*BCD/E*	*B/CD*	*C/D*
10	36.36	112.2	*ABC/DE*	*A/BC*	*D/E*	*B/C*
11	37.54	115.9	*A/BCDE*	*B/CDE*	*CD/E*	*C/D*
12	37.59	116.0	*A/BCDE*	*BC/DE*	*B/C*	*D/E*
13	37.73	116.5	*AB/CDE*	*A/B*	*C/DE*	*D/E*
14	40.06	123.7	*A/BCDE*	*B/CDE*	*C/DE*	*D/E*

Table 10.4

Sequences for the separation of the mixture of alkanes, with pressure fixed for cooling water in condensers.

Rank order	Total reboiler duty (MW)	% Best	Sequence			
1	31.05	100	*ABC/DE*	*AB/C*	*D/E*	*A/B*
2	31.44	101.2	*ABCD/E*	*ABC/D*	*AB/C*	*A/B*
3	31.60	101.8	*ABCD/E*	*AB/CD*	*A/B*	*C/D*

(continued)

10

Table 10.4 *(Continued)*

Rank order	Total reboiler duty (MW)	% Best	Sequence			
4	31.86	102.6	ABC/DE	A/BC	D/E	B/C
5	32.25	103.9	ABCD/E	ABC/D	A/BC	B/C
6	32.75	105.5	ABCD/E	A/BCD	BC/D	B/C
7	32.99	106.2	AB/CDE	A/B	CD/E	C/D
8	33.02	106.3	ABCD/E	A/BCD	B/CD	C/D
9	33.29	107.2	AB/CDE	A/B	C/DE	D/E
10	33.48	107.8	A/BCDE	BC/DE	B/C	D/E
11	33.82	108.9	A/BCDE	BCD/E	BC/D	B/C
12	34.09	109.8	A/BCDE	BCD/E	B/CD	C/D
13	35.39	114.0	A/BCDE	B/CDE	CD/E	C/D
14	35.69	114.9	A/BCDE	B/CDE	C/DE	D/E

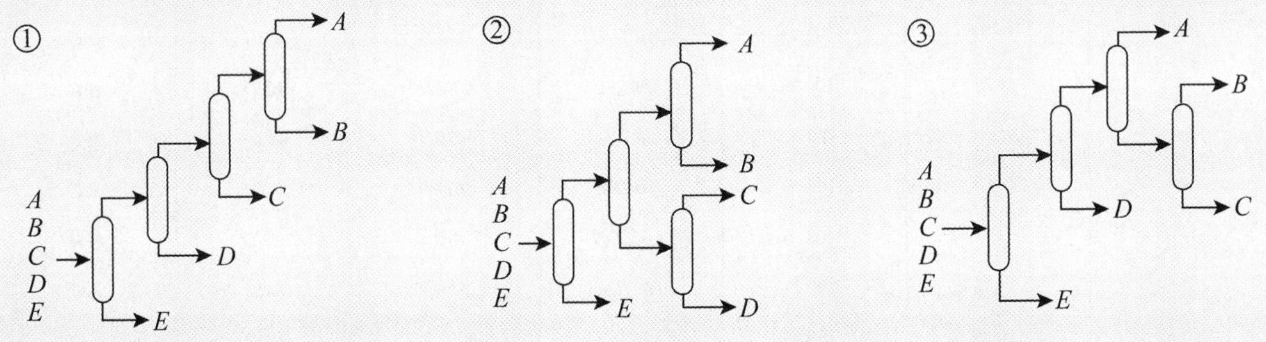

(a) Pressures fixed to 6 barg with relative volatilities recalculated for each column.

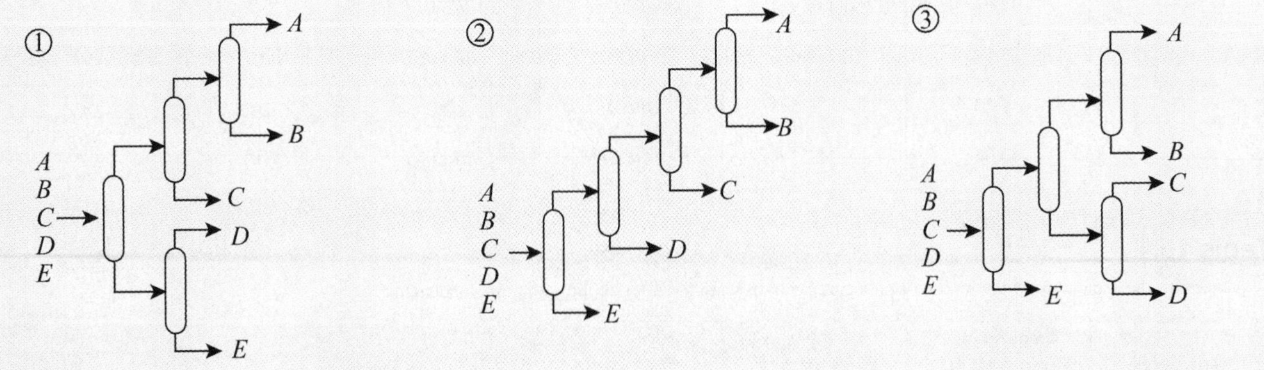

(b) Column pressures fixed for cooling water in condensers.

Figure 10.3

The best sequences in terms of vapor load for the separation of the mixture of alkanes from Example 10.2.

Example 10.3 The mixture of aromatics in Table 10.5 is to be separated into five products. The xylenes are to be taken as a mixed xylenes product. The C9s in Table 10.5 are to be characterized as C_9H_{12} (1-methylethylbenzene). The recoveries are to be assumed to be 100%. Relative volatilities and latent heats of vaporization are to be calculated from the Peng–Robinson Equation of State, assuming that all interaction parameters are zero (see Appendix A). Pressures of each column are to be minimized such that either the bubble point of the overhead product is 10 °C above the cooling water return temperature of 35 °C (i.e. 45 °C) or a minimum of atmospheric pressure. Assuming the ratio of actual to minimum reflux to be 1.1, all columns are fed with a saturated liquid and total condensers are used throughout. Neglect the pressure drop across each column. Determine the rank order of the distillation sequences on the basis of total reboiler load with minimum reflux ratio calculated from the Underwood Equations.

Solution The relative volatilities are recalculated for each column. However, Table 10.6 shows the relative volatilities of the feed mixture to the sequence at a pressure of 1 atm. This shows clearly that the ethyl benzene/xylenes separation is by far the most difficult with relative volatilities for the xylenes close to unity. The volatilities of the components are such that all separations can be carried out at atmospheric pressure and at the same time allow the use of cooling water in the condensers. Thus, column pressures are fixed to atmospheric pressure with the relative volatilities recalculated for the feed composition at this pressure as the concentration changes through the sequence.

Table 10.7 gives the total vapor flow for different sequences in rank order.

Again it can be noted from Table 10.7 that there is not too much difference between the best sequences in terms of the overall reboiler load. The three sequences with the lowest overall reboiler loads are shown in Figure 10.4. At first sight, the structure of the best sequences seems to be surprising. In each case, the most difficult separation (*C*/*D*) is carried out in the presence of other components. Heuristic 1 would suggest that such a difficult separation should be carried out in isolation from other components. Further investigation reveals that the relative volatilities for this most difficult separation are sensitive to the presence of other components. The presence of benzene and toluene increases the relative volatility of the *C*/*D* separation slightly, making it beneficial to carry out the difficult separation in the presence of nonkey components. This problem illustrates some of the dangers faced when dealing with separations that are very close in relative volatility. Special care should be taken in such problems to make sure that the vapor–liquid equilibrium data are specified as accurately as possible. The assumption of zero for the interaction parameters in the Peng–Robinson Equation of State is questionable. In fact, the inclusion of interaction parameters in this case does not change the order of the best few sequences. It does, however, change the absolute values of the calculated reboiler loads. The other questionable assumption is that of complete recovery for the products. Again, changing the assumed recovery to be less than complete does not change the order of the best few sequences in this case, but does change the value of the total reboiler load. Finally, the assumption at this stage is that utilities will be used to satisfy all of the heating and cooling requirements. This leads to all columns operating at atmospheric pressure. Later, in Chapter 21, it will be seen that it is beneficial to operate some of the columns at higher pressures to allow heat recovery to take place between some of the reboilers and condensers in the sequences.

Table 10.5

Data for a five-product mixture of aromatics to be separated by distillation.

Component	Flowrate ($kmol \cdot h^{-1}$)
A. Benzene	269
B. Toluene	282
C. Ethyl benzene	57
D. *p*-Xylene	47
m-Xylene	110
o-Xylene	58
E. C9s	42

Table 10.6

Relative volatilities of the feed to the sequence at 1 atm.

Component	Relative volatility	Relative volatility between adjacent components
Benzene	7.570	
		2.33
Toluene	3.243	
		2.07
Ethyl benzene	1.564	
		1.07
p-Xylene	1.467	
		1.04
m-Xylene	1.417	
		1.16
o-Xylene	1.220	
		1.22
C9s	1.000	

10

Table 10.7

Sequences for the separation of the mixture of aromatics in Example 10.3.

Rank order	Total reboiler duty (MW)	% Best	Sequence			
1	57.40	100	*ABC/DE*	*A/BC*	*D/E*	*B/C*
2	60.04	104.6	*ABC/DE*	*AB/C*	*D/E*	*A/B*
3	61.40	107.0	*A/BCDE*	*BC/DE*	*B/C*	*D/E*
4	64.81	112.9	*ABCD/E*	*ABC/D*	*A/BC*	*B/C*
5	65.98	115.0	*A/BCDE*	*B/CDE*	*C/DE*	*D/E*
6	66.02	115.0	*A/BCDE*	*BCD/E*	*BC/D*	*B/C*
7	66.63	116.1	*A/BCDE*	*B/CDE*	*CD/E*	*C/D*
8	67.02	116.8	*AB/CDE*	*A/B*	*C/DE*	*D/E*
9	67.46	117.5	*ABCD/E*	*ABC/D*	*AB/C*	*A/B*
10	67.67	117.9	*AB/CDE*	*A/B*	*CD/E*	*C/D*
11	68.50	119.3	*ABCD/E*	*A/BCD*	*BC/D*	*B/C*
12	70.26	122.4	*A/BCDE*	*BCD/E*	*B/CD*	*C/D*
13	72.73	126.7	*ABCD/E*	*A/BCD*	*B/CD*	*C/D*
14	73.97	128.9	*ABCD/E*	*AB/CD*	*A/B*	*C/D*

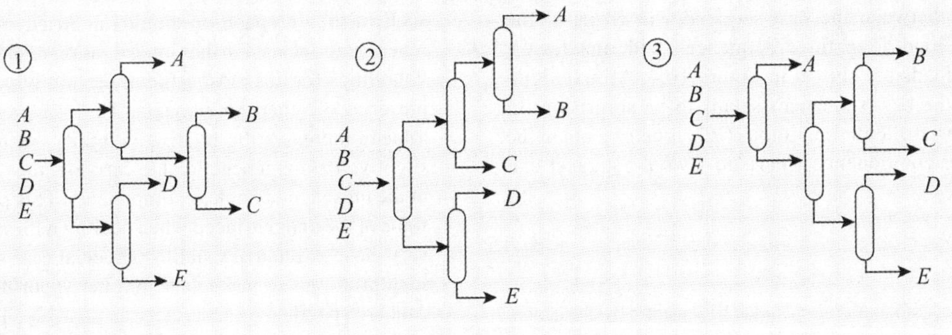

Figure 10.4

The best sequences in terms of vapor load for the separation of the mixture of aromatics from Example 10.3.

The use of total heating or cooling duty, even without any calculation errors, is still only a guide and might not give the correct rank order. Had the utility temperatures and costs been known, the calculation could be taken one step further to calculate the energy costs. In fact, given some computational aid, it is straightforward to size and cost all of the possible sequences using a shortcut sizing calculation, such as the Fenske–Gilliland–Underwood approach discussed in Chapter 8, together with cost correlations discussed in Chapter 2.

It should also be noted that there is likely to be a link between high heat loads in the distillation and high capital costs. Higher heat loads lead to larger and hence more expensive reboilers and condensers. Higher heat loads also lead to higher vapor flowrates in distillation columns, larger diameters and hence more expensive columns. Consequently, sequences with higher heat loads will also tend to have higher capital costs.

Whatever the method used to screen possible sequences, it is important not to give exclusive attention to the one sequence that appears to have the lowest total reboiler, condenser heat load, lowest operating cost or lowest total cost. There is often little to choose in this respect between the best few sequences, particularly when the number of possible sequences is large. Other considerations such as heat integration, operability, safety, and so on, might also have an important bearing on the final decision. Thus, the screening of sequences should focus on the best few sequences rather than exclusively on the single best sequence. It should be recalled from Chapter 1 that it is important to keep design options open until enough information is available to screen design options with confidence.

10.4 Distillation Sequencing using Columns With More Than Two Products

When separating a three-product mixture using simple columns, there are only two possible sequences (Figure 10.1). Consider the first characteristic of simple columns. A single feed is split into two products. As a first alternative to two simple columns, the possibilities shown in Figure 10.5 can be considered. Here three products are taken from one column. The designs are in fact both feasible and cost-effective when compared to simple arrangements on a stand-alone basis (i.e. reboilers and condensers operating on utilities) for certain ranges of conditions. If the feed is dominated by the middle product (typically more than 50% of the feed) and the heaviest product is present in small quantities (typically less than 5%), then the arrangements shown in Figure 10.5a can be an attractive option (Tedder and Rudd, 1978). The heavy product must find its way down the column past the sidestream. Unless the heavy product has a small flow and the middle product a high flow, a reasonably pure middle product cannot be achieved. In these circumstances, the sidestream is usually taken as a vapor product to obtain a reasonably pure sidestream. A large relative volatility between the sidestream Product *B* and the bottom Product *C* is also necessary to obtain a high-purity sidestream.

When it is required to take a vapor sidestream, it should be noted that this is more problematic than taking a liquid sidestream. Whilst it is straightforward to split a liquid flow down a column, it is less straightforward to split a vapor flow up a column to take a vapor sidestream. The flow of vapor up the column must have enough pressure to overcome the pressure drop in the piping and other equipment associated with the sidestream flow. There might need to be some kind of flowrate control for the sidestream. Equipment might also be necessary to prevent the carryover of liquid droplets into the sidestream (e.g. a separation drum with a mist eliminator). To overcome the pressure drop in the sidestream equipment might require additional pressure drop to be created in the column by installing a single tray above the sidestream with a high pressure drop.

If the feed to the column is dominated by the middle product (typically more than 50%) and the lightest product is present in small quantities (typically less than 5%), then the arrangement shown in Figure 10.5b can be an attractive option (Tedder and Rudd, 1978). This time the light product must find its way up the column past the sidestream. Again, unless the light product is a small flow and the middle product a high flow, a reasonably pure middle product cannot be achieved. This time the sidestream is taken as a liquid product to obtain a reasonably pure sidestream. A large relative volatility between the sidestream Product *B* and the overhead Product *A* is also necessary to obtain a high-purity sidestream.

In summary, single-column sidestream arrangements can be attractive when the middle product is in excess and one of the other components is present in only minor quantities. Thus, the sidestream column only applies to special circumstances for the feed composition. More generally applicable arrangements are possible by relaxing the restriction that separations must be between adjacent key components.

Consider a three-product separation as in Figure 10.6a in which the lightest and heaviest components are chosen to be the key separation in the first column. Two further columns are required to produce pure products (Figure 10.6a). This arrangement is known as *distributed distillation* or *sloppy distillation*. The distillation sequence provides parallel flow paths for the separation of a product. At first sight, the arrangement in Figure 10.6a seems to be inefficient in the use of equipment in

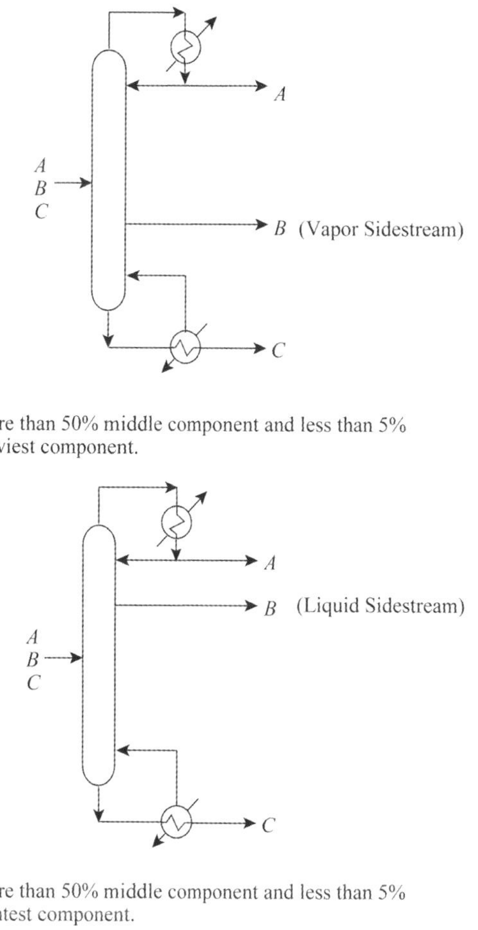

Figure 10.5

Distillation columns with three products. (Reproduced from Smith R and Linnhoff B, 1998, *Trans IChemE ChERD*, **66**: 195 by permission of the Institution of Chemical Engineers.)

10

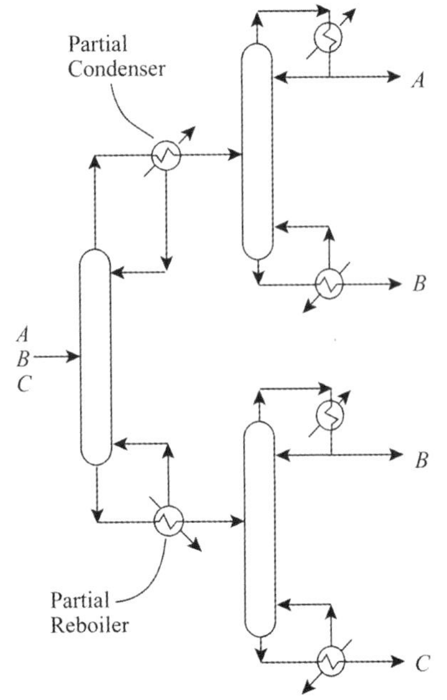

(a) Distributed distillation or sloppy distillation separates nonadjacent key components.

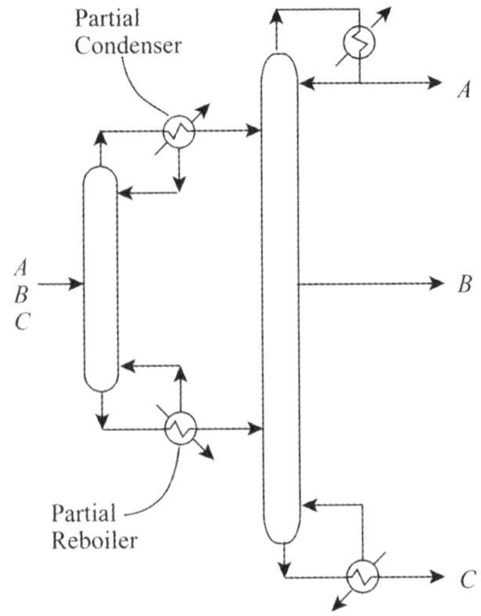

(b) Prefractionator arrangement.

Figure 10.6

Choosing nonadjacent keys leads to the prefractionator arrangement.

that it requires three columns instead of two, with the bottoms and overheads of the second and third columns both producing pure *B*. However, it can be a useful arrangement in some circumstances. In the new design, the three columns can be operated at different pressures. Also, the distribution of the middle Product *B* between the second and third columns is an additional degree of freedom in the design. The additional freedom to vary the pressures and the distribution of the middle product gives significant extra freedom to vary the loads and levels at which the heat is added to or rejected from the distillation. This might mean that the reboilers and condensers can be matched more cost-effectively against utilities, or heat integrated more effectively.

If the second and third columns in Figure 10.6a are operated at the same pressure, then the second and third columns could simply be connected and the middle product taken as a sidestream, as shown in Figure 10.6b. The arrangement in Figure 10.6b is known as a *prefractionator* arrangement. Note that the first column in Figure 10.6b, the prefractionator, has a partial condenser to reduce the overall energy consumption.

Comparing the distributed (sloppy) distillation in Figure 10.6a and the prefractionator arrangement in Figure 10.6b with the conventional arrangements in Figure 10.1, the distributed and prefractionator arrangements typically require 20 to 30% less energy than the conventional arrangements for the same separation duty. The reason for this difference is rooted in the fact that the distributed distillation and prefractionator arrangements are fundamentally thermodynamically more efficient than a simple arrangement. Consider why this is the case.

Consider the sequence of simple columns shown in Figure 10.7. In the direct sequence shown in Figure 10.7, the composition of Component *B* in the first column increases below the feed as the more volatile Component *A* decreases. However, moving further down the column, the composition of Component *B* decreases again as the composition of the less-volatile Component *C* increases. Thus the composition of *B* reaches a peak only to be remixed (Triantafyllou and Smith, 1992).

Similarly, with the first column in the indirect sequence, the composition of Component *B* first increases above the feed as the less-volatile Component *C* decreases. It reaches a maximum only to decrease as the more volatile Component *A* increases. Again, the composition of Component *B* reaches a peak only to be remixed.

This remixing that occurs in both sequences of simple distillation columns is a source of inefficiency in the separation. By contrast, consider the prefractionator arrangement shown in Figure 10.8. In the prefractionator, a crude split is performed so that Component *B* is distributed between the top and bottom of the column. The upper section of the prefractionator separates *AB* from *C*, whilst the lower section separates *BC* from *A*. Thus, both sections remove only one component from the product of that column section and this is also true for all four sections of the main column. In this way, the remixing effects that are a feature of both simple column sequences are avoided (Triantafyllou and Smith, 1992).

In addition, another feature of the prefractionator arrangement is important in reducing mixing effects. Losses occur in distillation operations due to mismatches between the composition of the column feed and the composition at the feed stage. In theory, for binary distillation in a simple column, a good match can be found between the feed composition and the feed stage. However, because the changes from stage to stage are finite, an exact match is not always possible. For multicomponent distillation in a simple

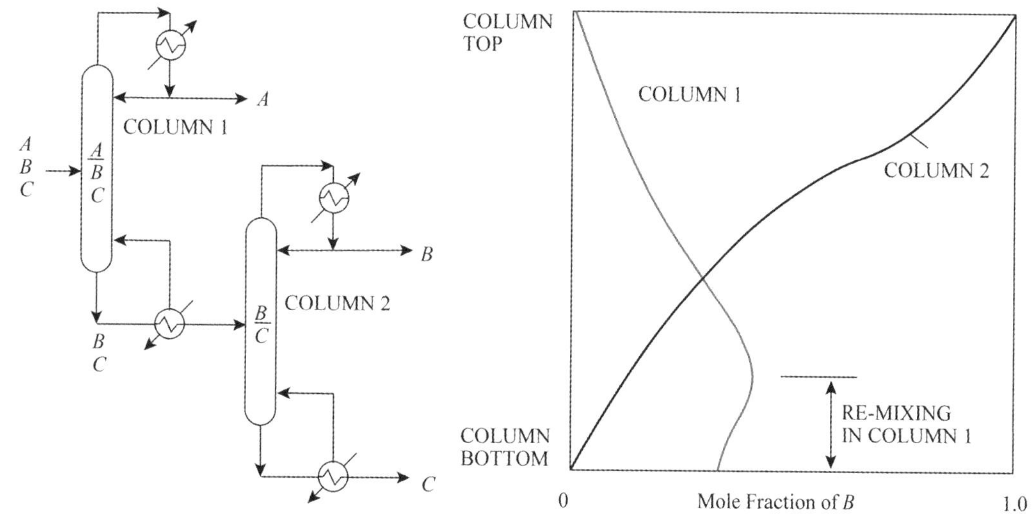

Figure 10.7

Composition profiles for the middle product in the columns of the direct sequence show re-mixing effects. (Reproduced from Triantafyllou C and Smith R (1992) The Design and Optimization of Fully Thermally-coupled Distillation Columns, *Trans IChemE*, **70A**: 118, by permission of the Institution of Chemical Engineers.)

column, except under special circumstances, it is not possible to match the feed composition and the feed stage. In a prefractionator arrangement, because the prefractionator distributes Component B between top and bottom, this allows greater freedom to match the feed composition with one of the trays in the column to reduce mixing losses at the feed tray.

The elimination of mixing losses in the prefractionator arrangement means that it is inherently more efficient than an arrangement using simple columns. The same basic arguments apply to both distributed distillation and prefractionator arrangements, with the additional degree of freedom in the case of distributed distillation to vary the pressures of the second and third columns independently.

10.5 Distillation Sequencing using Thermal Coupling

The final restriction of simple columns stated earlier was that they should have a reboiler and a condenser. It is possible to use material

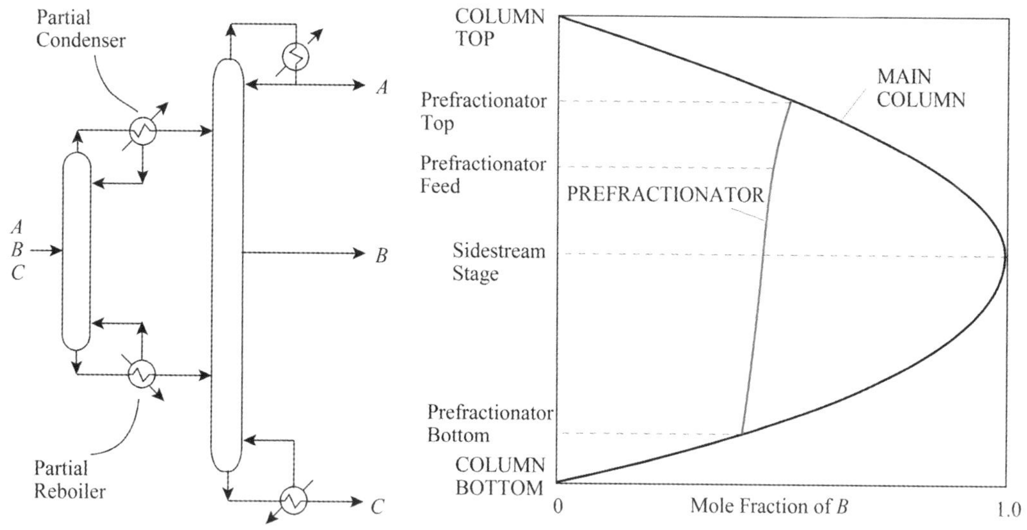

Figure 10.8

Composition profiles for the middle product in the prefractionator arrangement show that there are no re-mixing effects. (Reproduced from Triantafyllou C and Smith R (1992) The Design and Optimization of Fully Thermally-coupled Distillation Columns, *Trans IChemE*, **70A**: 118, by permission of the Institution of Chemical Engineers.)

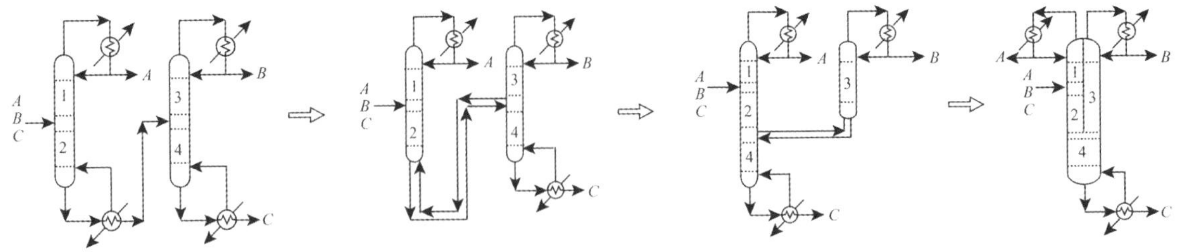

(a) Direct sequence. (b) Thermally coupled direct sequence. (c) Side-rectifier arrangement. (d) Partitioned side-rectifier arrangement.

Figure 10.9

Thermal coupling of the direct sequence.

flows to provide some of the necessary heat transfer by direct contact. This transfer of heat by direct contact is known as *thermal coupling*.

First consider thermal coupling of the simple sequences in Figure 10.9a. Figure 10.9b shows a thermally coupled direct sequence. The reboiler of the first column is replaced by a thermal coupling. Liquid from the bottom of the first column is transferred to the second as before, but now the vapor required by the first column is supplied by the second column, instead of a reboiler on the first column. The four column sections are marked as 1, 2, 3 and 4 in Figure 10.9b. In Figure 10.9c, the four column sections from Figure 10.9b are rearranged to form a *side-rectifier* arrangement (Calberg and Westerberg, 1989). There are practical difficulties in engineering a side-rectifier arrangement associated with taking a vapor sidestream from the main column, as already noted for vapor sidestream columns. Such problems can be avoided by constructing the side-rectifier in a single shell with a *partition wall* as shown in Figure 10.9d to give a *partitioned side-rectifier*. The partition wall in Figure 10.9d should be insulated to avoid heat transfer across the wall as different separations are carried out on each side of the wall and the temperatures on each side will differ. Heat transfer across the wall will have an overall detrimental effect on column performance (Lestak, Smith and Dhole, 1994).

Similarly, Figure 10.10a shows an indirect sequence and Figure 10.10b the corresponding thermally coupled indirect sequence. The condenser of the first column is replaced by thermal coupling. The four column sections are again marked as 1, 2, 3 and 4 in Figure 10.10b. In Figure 10.10c, the four column sections are rearranged to form a *side-stripper* arrangement (Calberg and Westerberg, 1989). As with the side-rectifier, the side-stripper can be arranged in a single shell with a partition wall, as shown in Figure 10.10d, to give a *partitioned side-stripper*. Again the partition wall should be insulated, otherwise heat transfer across the wall will have an overall detrimental effect on the separation (Lestak, Smith and Dhole, 1994).

Both the side-rectifier and side-stripper arrangements have been shown to reduce the energy consumption compared to simple two-column arrangements (Tedder and Rudd, 1978; Glinos and Malone, 1988). This results from reduced mixing losses in the first (main) column. As with the first column of the simple sequence, a peak in composition occurs with the middle product. Now, however, the advantage of the peak is taken by transferring material to the side-rectifier or side-stripper.

The side-rectifier and side-stripper arrangements have some important degrees of freedom for optimization. In these arrangements, there are four column sections. For the side-rectifier, the degrees of freedom to be optimized are:

- number of stages in each of the four column sections;
- reflux ratios (ratio of liquid reflux to overhead product flowrates) in the main column and sidestream column;

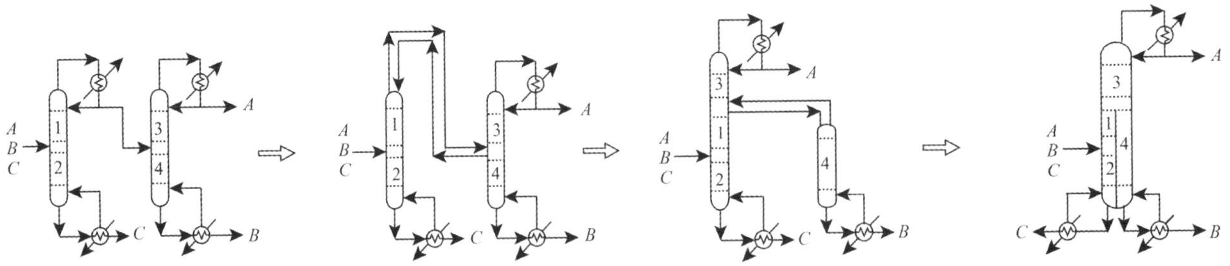

(a) Indirect sequence. (b) Thermally coupled indirect sequence. (c) Side-stripper arrangement. (d) Partitioned side-stripper arrangement.

Figure 10.10

Thermal coupling of the indirect sequence.

- vapor split between the main column and sidestream column;
- feed condition.

For the side-stripper, the degrees of freedom to be optimized are:

- number of stages in each of the four column sections;
- reboil ratios (ratio of stripping vapor to bottom product flowrates) in the main column and sidestream column;
- liquid split between the main column and sidestream column;
- feed condition.

All of these variables must be optimized simultaneously to obtain the best design. Some of the variables are continuous and some are discrete (the number of stages in each column section). Such optimizations are far from straightforward if carried out using detailed simulation. It is often beneficial to carry out some optimization using shortcut methods before proceeding to detailed simulation where the optimization can be fine-tuned.

A simple model for side-rectifiers suitable for shortcut calculation is shown in Figure 10.11. The side-rectifier can be modeled as two columns in the thermally coupled direct sequence. The first column is a conventional column with a condenser and partial reboiler. The second column is modeled as a sidestream column, with a vapor sidestream one stage below the feed stage (Triantafyllou and Smith, 1992). The liquid entering the reboiler and vapor leaving can be calculated from the vapor–liquid equilibrium (see Exercise 11 of Chapter 8). The vapor and liquid streams at the bottom of the first column can then be matched with the feed and sidestream of the second column to allow the calculations for the second column to be carried out.

The side-stripper can be modeled as two columns in the thermally coupled indirect sequence, as shown in Figure 10.12. The first column is modeled as a conventional column with a partial condenser and partial reboiler. The second column is modeled as a sidestream column with a liquid sidestream on stage above the feed stage (Triantafyllou and Smith, 1992). The vapor entering the condenser and liquid leaving can be calculated from the vapor–liquid equilibrium (see Exercise 11 of Chapter 8). This allows the two columns to be linked and the calculations for the second column to be carried out.

In both the cases of the side-rectifier and the side-stripper, the first column in the two-column model can be modeled using the Fenske–Underwood–Gilliland Equations, as described in Chapter 8. The second column is modeled as a sidestream column that can also be modeled using the Fenske–Underwood–Gilliland Equations (Triantafyllou and Smith, 1992). The minimum reflux ratio for a liquid sidestream column one stage above the feed stage can be estimated using the Underwood Equations by combining the sidestream and overhead product as a net overhead product (King, 1980). The minimum reflux ratio for a vapor sidestream column one stage below the feed stage can also be estimated using the Underwood Equations, but this time by combining the sidestream and bottoms product as a net bottoms product (King, 1980).

The optimization can be carried out using nonlinear optimization techniques such as SQP (see Chapter 3). The nonlinear optimization has the problem of local optima if techniques such as SQP are used for the optimization. Constraints need to be added to the optimization in order that a mass balance can be maintained and the product specifications achieved. The optimization of the side-rectifier and side-stripper in a capital-energy trade-off determines the distribution of plates, the reflux ratios in the main and sidestream columns and the condition of the feed. If a partitioned side-rectifier (Figure 10.9d) or partitioned side-stripper (Figure 10.10d) is to be used, then the ratio of the vapor flowrates on each side of the partition can be used to fix the location of the partition across the column. The partition is located such that the ratio of areas on each side of the partition is the same as the optimized ratio of vapor flowrates on each side of the partition. However, the vapor split for the side-rectifier will only follow this ratio if the pressure drop on each side of the partition is the same. Rather than locate the partition in this way, some deterioration in the design performance might be accepted by locating the partition across the diameter of the column (i.e. equal areas on each side of the partition) for

10

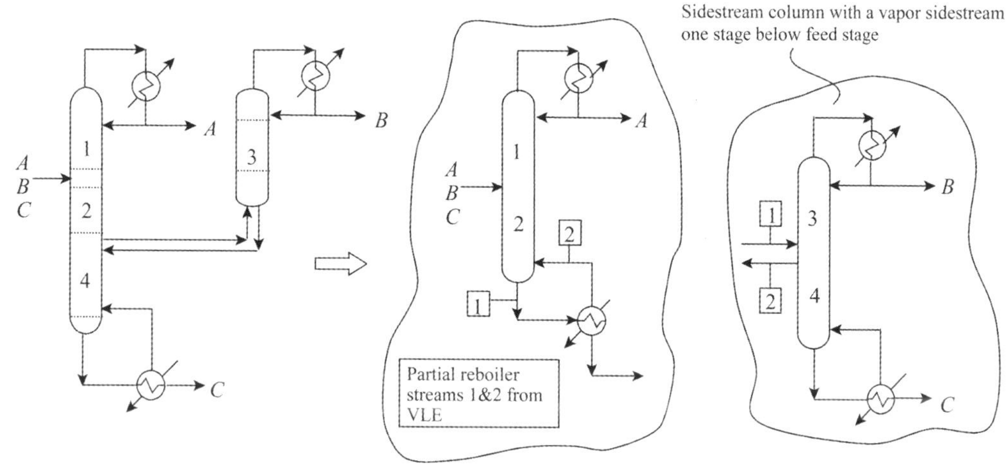

Figure 10.11

A side-rectifier can be modelled as a sequence of two simple columns in the direct sequence.

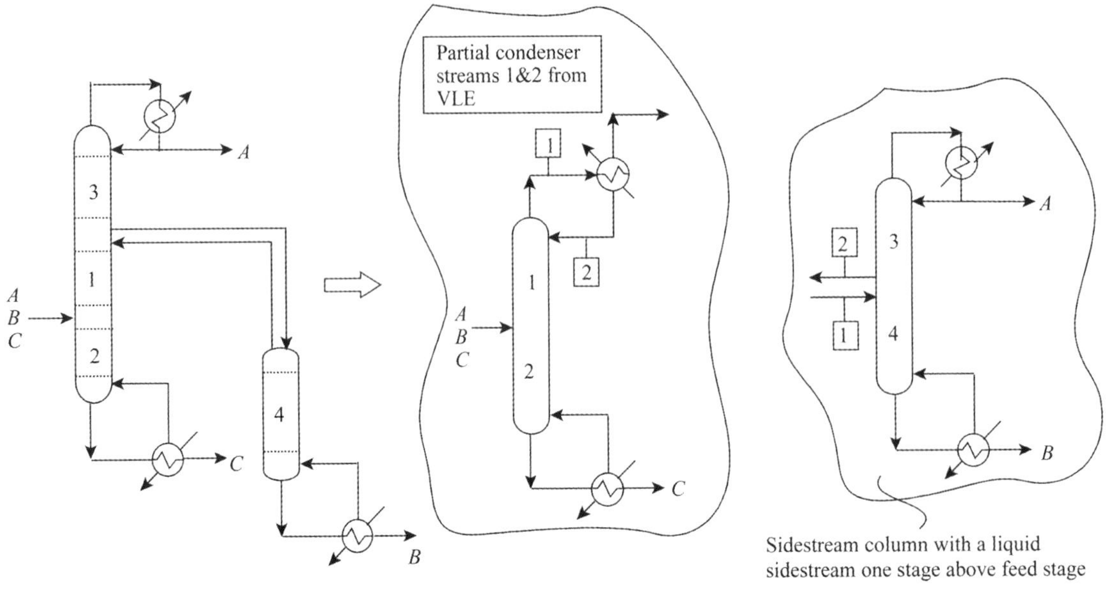

Figure 10.12

A side-stripper can be modelled as a sequence of two simple columns in the indirect sequence.

mechanical simplicity. The sensitivity of the design performance to the location of the partition should be explored before the location is finalized.

Now consider thermal coupling of the prefractionator arrangement from Figure 10.6b. Figure 10.13a shows a prefractionator arrangement with partial condenser and reboiler on the prefractionator. Figure 10.13b shows the equivalent thermally coupled prefractionator arrangement, sometimes known as the Petlyuk column. To make the two arrangements in Figures 10.13a and 13b equivalent, the thermally coupled prefractionator requires extra plates to substitute for the prefractionator condenser and reboiler (Aichele, 1992). The prefractionator arrangement in Figure 10.13a and the thermally coupled prefractionator (Petlyuk Column) in Figure 10.13b are almost the same in terms of total heating and cooling duties (Aichele, 1992). There are differences between the vapor and liquid flows in the top and bottom sections

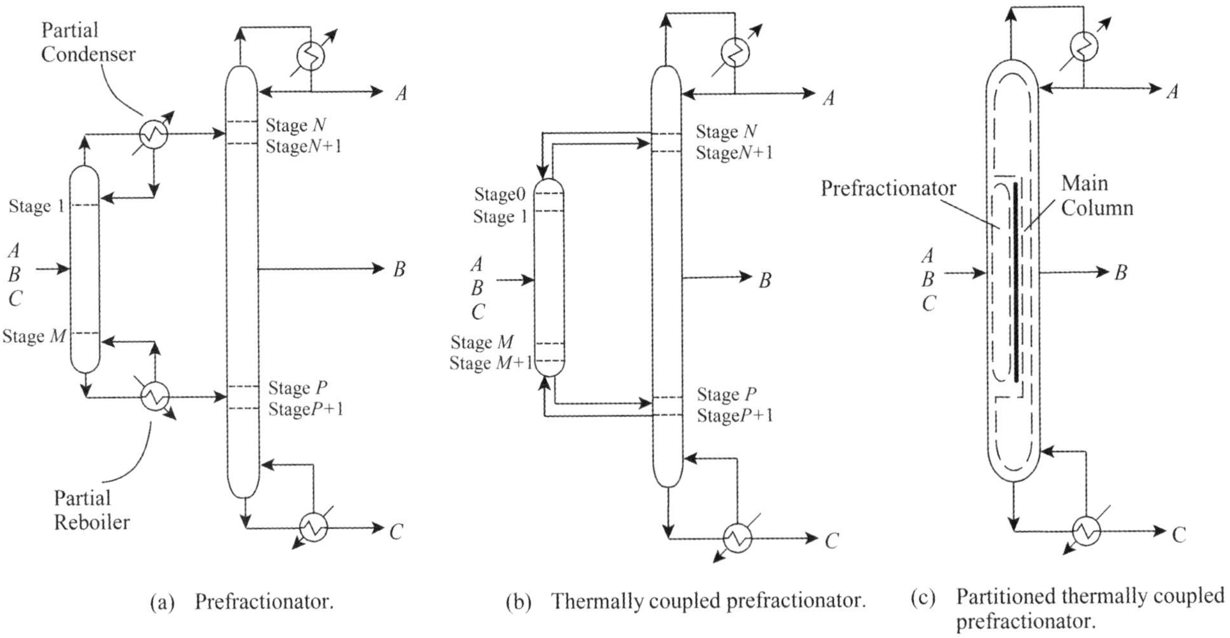

(a) Prefractionator.

(b) Thermally coupled prefractionator.

(c) Partitioned thermally coupled prefractionator.

Figure 10.13

Thermal coupling of the prefractionator arrangement.

of the main column of the designs in Figures 10.13a and 10.13b, resulting from the presence of the partial reboiler and condenser in Figure 10.13a. However, although the total heating and cooling duties are almost the same, there are differences in the temperatures at which the heat is supplied and rejected. In the case of the prefractionator in Figure 10.13a, the heat load is supplied at two points and two different temperatures and rejected from two points and at two different temperatures. Figure 10.13c shows an alternative configuration for the thermally coupled prefractionator that uses a single shell with a vertical partition dividing the central section of the shell into two parts, known as a *dividing-wall column* or *partition prefractionator column*. The arrangements in Figures 10.13b and 10.13c are equivalent if there is no heat transfer across the partition. As with side-rectifiers and side-strippers, the partition wall should be insulated to avoid heat transfer across the wall as different separations are carried out on each side of the wall and the temperatures on each side will differ. Heat transfer across the wall will have an overall detrimental effect on column performance (Lestak, Smith and Dhole, 1994).

Partition prefractionator columns offer a number of advantages over conventional arrangements:

1) Various studies (Petlyuk, Olatonov and Slavinskii, 1965; Stupin and Lockhart, 1972; Kaibel, 1987, 1988; Glinos and Malone, 1988; Mutalib and Smith, 1998; Mutalib, Zeglam and Smith, 1998; Becker *et al.*, 2001; Shultz *et al.*, 2002; Asprion and Kaibel, 2010) have compared the thermally coupled arrangement in Figures 10.13b and 10.13c with a conventional arrangement using simple columns on a stand-alone basis. These studies show that the prefractionator arrangement in Figure 10.13 requires typically 20 to 30% less energy than the best conventional arrangement using simple columns (Figure 10.1). The prefractionator column also requires less energy than the side-rectifier and side-stripper arrangements, for the same separation (Glinos and Malone, 1988). The energy saving for the thermally coupled arrangement is the same as that for the prefractionator, with reduced mixing losses as illustrated in Figures 10.7 and 10.8.
2) In addition, the partition prefractionator column in Figure 10.13c requires typically 20 to 30% less capital cost than a two-column arrangement of simple columns.
3) The partition prefractionator column has one further advantage over the conventional arrangements in Figure 10.1. In partitioned prefractionator columns, the material is only reboiled once and its residence time in the high-temperature zones is minimized. This can be important if distilling heat-sensitive materials.

Standard distillation equipment can be used for the fabrication. Either packing or plates can be used, although packing is more commonly used in practice. Also, the control of partitioned prefractionator columns is straightforward (Mutalib and Smith, 1998; Mutalib, Zeglam and Smith, 1998).

The partition prefractionator column does have a number of disadvantages relative to simple column arrangements:

1) Even though the arrangement might require less energy than a conventional arrangement, all of the heat must be supplied at the highest temperature and all of the heat rejected at the lowest temperature of the separation. This can be particularly important if the distillation is at a low temperature using refrigeration for condensation. In such circumstances, minimizing the amount of condensation at the lowest temperatures can be very important. If differences in the temperature at which the heat is supplied or rejected are particularly important, then distributed distillation or a prefractionator might be better options. These offer the advantage of being able to supply and reject heat at different temperatures. This issue will be addressed again later, in Chapter 21, when dealing with heat integration of distillation.
2) If it is appropriate for the two columns in a simple column arrangement to operate at very different pressures, then this can pose problems for a partition prefractionator column to replace them, as the partition prefractionator column must carry out the whole separation at effectively the same pressure (usually the highest pressure). In general, partition prefractionator columns are not suited to replace sequences of two simple columns that operate at very different pressures.
3) Another disadvantage of partitioned prefractionator columns might arise from materials of construction. If the two columns of a conventional arrangement require two different materials of construction, one being much more expensive than the other, then any savings in capital cost arising from using a partitioned prefractionator column will be diminished. This is because the whole partitioned column will have to be fabricated from the more expensive material.
4) The hydraulic design of the partitioned prefractionator column is such that the pressure must be balanced on either side of the partition. This is usually achieved by designing with the same number of stages on each side of the partition. This constrains the design. Alternatively, a different number of stages can be used on each side of the partition, and different column internals with a different pressure drop per stage used to balance the pressure drop. If foaming is more likely to occur on one side of the partition than the other, this can lead to a hydraulic imbalance and the vapor split changing from design conditions.

The partitioned prefractionator in Figure 10.13c can be simulated using the arrangement in Figure 10.13b as the basis of the simulation. However, like side-rectifiers and side-strippers, fully thermally coupled columns have some important degrees of freedom for optimization. In the fully thermally coupled column, there are six column sections (above and below the partition, above and below the feed in the prefractionator and above and below the sidestream from the main column side of the partition). The degrees of freedom to be optimized in partitioned columns are:

- number of stages in the six column sections;
- reflux ratio;
- liquid split on each side of the partition flowing down the column;
- vapor split on each side of the partition flowing up the column;
- feed condition.

These basic degrees of freedom can be represented in different ways, but all must be optimized simultaneously to obtain the best design. Again, some of the variables are continuous and some are discrete (the number of stages in each column section). As with side-rectifier and side-stripper arrangements, it can be beneficial to carry out some optimization using shortcut methods before proceeding to detailed simulation and fine-tuning of the optimization. Like side-rectifiers and side-strippers, the fully thermally coupled column can be represented as simple columns. This time three simple columns are needed to represent the fully thermally coupled column, as shown in Figure 10.14 (Triantafyllou and Smith, 1992). The prefractionator is modeled as Column 1 in Figure 10.14, which has a partial condenser and partial reboiler initially. Column 2 in Figure 10.14 is the top section of the main column and is modeled as a liquid sidestream column with a liquid sidestream one stage above the vapor feed. A vapor–liquid equilibrium calculation around the partial condenser in Column 1 allows the vapor entering the condenser and liquid leaving the condenser to be determined. These can then be connected to the sidestream and feed for Column 2. Column 3 in Figure 10.14 is the lower part of the main column and can be modeled as a vapor sidestream column with the sidestream one stage below the liquid feed. A vapor–liquid equilibrium calculation around the partial reboiler in the prefractionator allows the vapor leaving the reboiler and liquid entering the reboiler to be determined and can then be connected with the sidestream and feed to Column 3. The three columns can then be represented by the shortcut calculations described above for side-rectifiers and side-strippers on the basis of the Fenske–Gilliland–Underwood Correlations. The optimization can again be carried out using nonlinear optimization techniques such as SQP (see Chapter 3). Constraints need to be added to the optimization in order that it can fulfill the required objectives. As illustrated in Figure 10.14, Columns 2 and 3 together represent the main column, and it is necessary to ensure that the vapor flowrate in Columns 2 and 3 are the same. Also, the bottom product from Column 2 and top product from Column 3 represent the same sidestream from the main column of the thermally coupled configuration. Thus, the bottom product of Column 2 and the top product of Column 3 must be constrained to be the same. Additional constraints must also be added to ensure that the mass balance is maintained and product specifications are achieved. After the design concept has been established and some preliminary optimization carried out, the design can be verified by detailed simulation based on the model in Figure 10.13b.

10.6 Retrofit of Distillation Sequences

It might be the case that an existing plant needs to be modified, rather than a new design carried out. For example, a retrofit study might require the capacity of the unit to be increased. When such a revamp is carried out, it is essential to try and make as effective use as possible of existing equipment. For example, consider a simple two-column sequence like the ones illustrated in Figure 10.1. If the capacity needs to be increased, then, instead of replacing the two existing columns with new ones, a new column might be added to the two existing ones and reconfigured to the distributed distillation arrangement in Figure 10.6a. The vapor and liquid loads for the increased capacity are now spread between three columns, instead of two, taking advantage of the existing two-column shells. Of course, the design of the existing shells might not be ideally suited to the new role to which they will be put, but at least it will allow existing equipment to carry on being used and avoid some

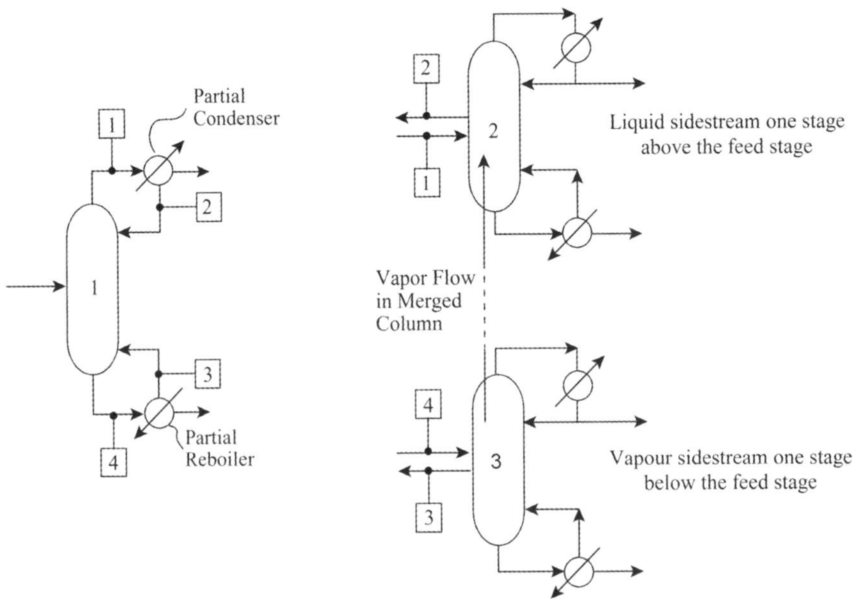

Figure 10.14

Three column model of partitioned thermally coupled prefractionator.

unnecessary investment in new equipment. Of the three shells shown in Figure 10.6a, any of the three could be the new shell. The arrangement of new shell and existing shells will depend on how the existing shells can best be adapted to the new role.

If the existing shells differ significantly from what is required in the retrofitted sequence, then, in addition to reconfiguring the distillation sequence, the existing shells can also be modified. For example, the number of theoretical stages can be increased in the columns by changing the design of the column internals. Generally, however, the fewer the number of modifications carried out the better.

Rather than retrofit a two-column sequence to the distributed distillation arrangement in Figure 10.6a, it can be retrofitted to the prefractionator arrangement in Figure 10.6b. This time, rather than have two-column shells, as shown in Figure 10.6b, three columns are used with the second and third connected directly by vapor and liquid flows. Thus, there is no reboiler in the second column and no condenser in the third. The liquid flow between the two shells that function as the main column is split to provide the reflux for the lower part of the main column and the middle product. Two-column shells are thus used to mimic the main column of the prefractionator. As with the distributed distillation retrofit arrangement, which of the two existing shells and the new shell are used in which role depends on the details of the design of the existing shells and how they can best be fitted to the new role. As with the distributed distillation retrofit arrangement, the existing shells can also be modified.

The retrofit of more complex distillation sequences can also be considered. Within a larger, more complex sequence, any pair of columns that are together in a sequence can be considered as candidates for the same retrofit modifications as those discussed for the two-column retrofit.

10.7 Crude Oil Distillation

One particularly important case of distillation sequencing is worthy of special consideration. This is the case of crude oil distillation, which is the fundamental process underlying the petroleum and petrochemicals industry. Crude oil is an extremely complex mixture of hydrocarbons containing small quantities of sulfur, oxygen, nitrogen and metals. A typical crude oil contains many millions of compounds, most of which cannot be identified. Only the lightest compounds, for example, methane, ethane, propane, benzene, and so on, can normally be identified. In such circumstances, the modeling of vapor–liquid equilibrium for distillation presents a challenge, as the composition of the crude oil feed is unknown. The crude oil feed and the products from the distillation are modelled using *pseudo-components*. These pseudo-components are fictitious components with attributes of molar mass, critical temperature, critical pressure and accentric factor, invented and mixed in a composition so as to reproduce the boiling characteristics of the feed (the amount of feed that vaporizes at a given temperature).

In the first stage of processing crude oil, it is distilled under conditions slightly above atmospheric pressure (typically 1 barg where the feed enters the column). A range of products are taken from the crude oil distillation, based on their boiling temperatures. Designs are normally thermally coupled. Most configurations follow the thermally coupled indirect sequence as shown in Figure 10.15a. However, rather than build the configuration in Figure 10.15a, a configuration based on Figure 10.15b is normally constructed. The two arrangements are equivalent and the

10

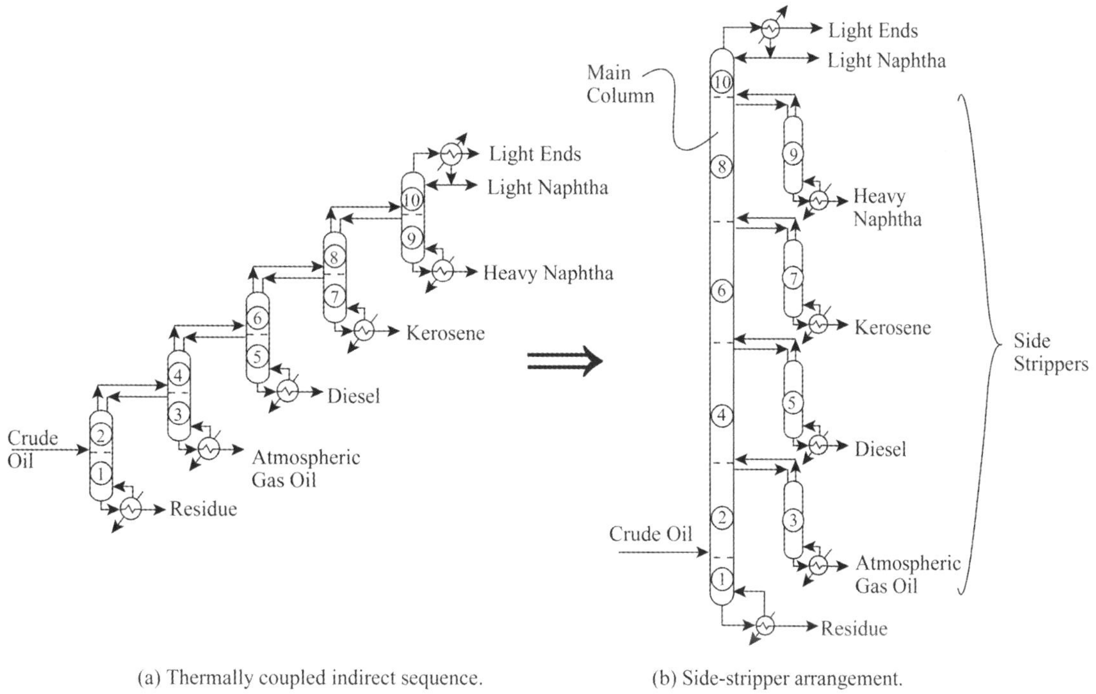

Figure 10.15
The thermally-coupled indirect sequence for crude oil distillation.

corresponding column sections are shown in Figure 10.15. Unfortunately, a practical crude oil distillation cannot be operated in quite the way shown in Figure 10.15b. The first problem is that if high-temperature reboiling is used, as would be required for the higher boiling products in the lower part of the column in Figure 10.15b, extremely high-temperature sources of heat would be required. Steam is usually not available for process heating at such high temperatures. Also, high temperatures in the reboilers would result in significant fouling of the reboilers from decomposition of the hydrocarbons to form coke. The maximum temperature for using reboiling in petroleum refining is usually of the order of 300 °C. Therefore, in practice, some or all of the reboiling is substituted by the direct injection of steam into the distillation. The steam has two functions:

- It provides some of the heat required for the distillation.
- It lowers the partial pressure of the boiling components, making them more volatile.

Because steam is injected into the operation in various places in Figure 10.16, the steam is condensed overhead and is separated in a decanter from the condensing hydrocarbons and the hydrocarbons that do not condense.

Another problem with the arrangement in Figure 10.15b is that as the vapor rises up the main column, its flowrate increases significantly. In Figure 10.16, heat is being removed from the main column at intermediate points. This not only controls the vapor flow up the column but introduces additional control at intermediate points to maintain product quality. Introducing condensation at intermediate points corresponds with introducing some condensation of the vapor at the top of intermediate columns in the arrangement shown in Figure 10.15a. It should be noted that the introduction of some condensation between the columns does not necessarily eliminate all thermal coupling between the columns in Figure 10.15a, but can leave partial thermal coupling, rather than full thermal coupling. How much condensation and how much thermal coupling are to be used at each point are important degrees of freedom to be optimized.

In crude oil distillation, the heat is removed from the column by taking a liquid sidestream, subcooling the liquid and returning it to the column. This provides the condensation by direct contact, but introduces inefficiency into the design. Returning subcooled liquid to the column is inefficient in distillation, as subcooled liquid cannot take part in the distillation until it has returned to saturation conditions.

The heat is removed from the main column in one of two ways. These are shown in Figure 10.17. The first, in Figure 10.17a, is a *pumparound*. Liquid is taken from the column, subcooled and returned to the column at a higher point. Another arrangement shown in Figure 10.17b is a *pumpback*. This takes liquid from the column and subcools it, but this time returns it to a lower point in the column. The problem with the pumpback is that the flowrate that can be withdrawn must be significantly less than the liquid flowing down the column. It is therefore restricted in its capacity to remove heat. The pumparound, on the other hand, is not restricted by the flow of liquid down the column and can recirculate as much liquid as required around a section of the column. By choosing the most appropriate flowrate and temperature for the pumparound, the heat load to be removed can be adjusted to whatever is desired. The trays between the liquid draw and return in a pumparound are functioning more to carry out heat transfer than mass transfer. In addition to returning a subcooled liquid to the column, mixing occurs as material is introduced to a higher point in the column.

One final point needs to be made about the arrangement shown in Figure 10.16. The crude oil entering the main column needs to be preheated. This is done initially by heat recovery to a temperature in the range 90 to 150 °C, and the crude oil is extracted with water to remove salt. The desalting process mixes with water and then separates into two layers, the salt dissolving in the water. The desalted crude oil is then heated further by heat recovery to a temperature usually around 280 °C, or higher if possible, and then by a furnace (fired heater) to around 360 °C before entering the column. Note that this temperature is higher than that previously

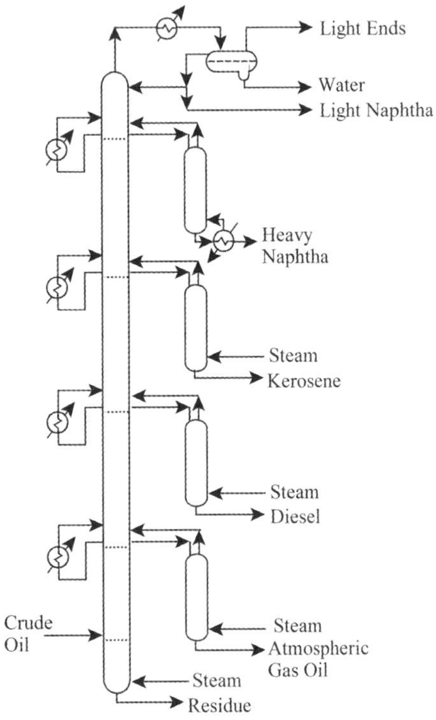

Figure 10.16

Substitute some (or all) of the reboiling with direct steam injection and introduce intermediate condensation.

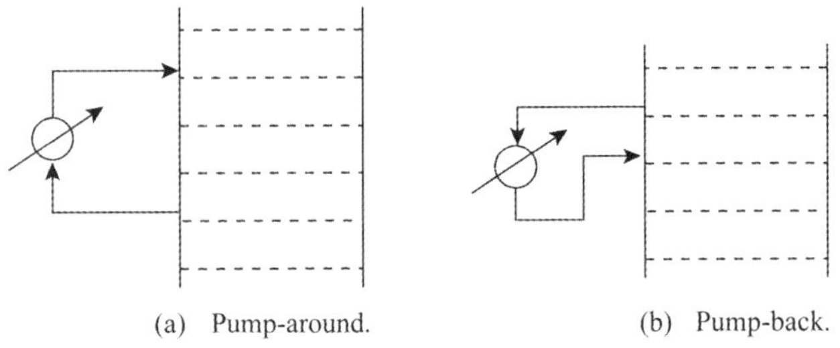

Figure 10.17

Partial condensation can be achived by subcooling liquid for the column.

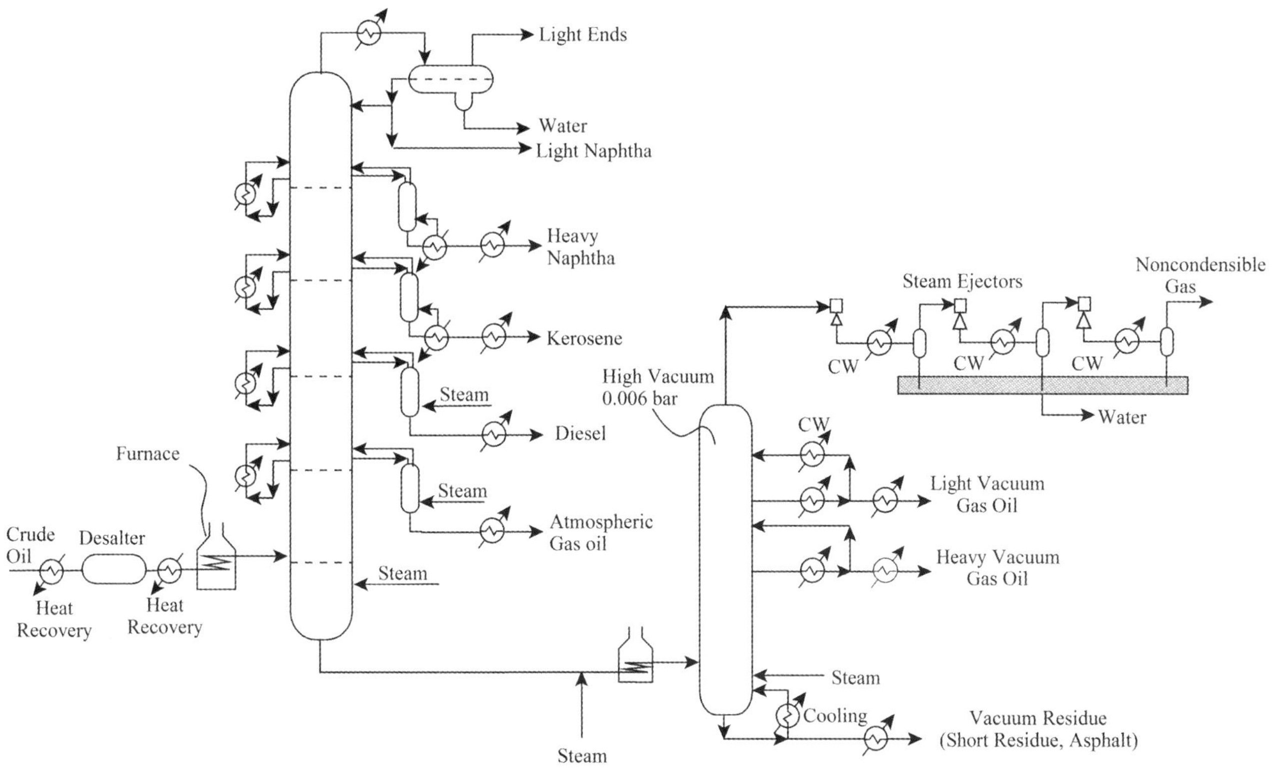

Figure 10.18

A typical complete crude oil distillation system.

quoted as the maximum that can be supplied in a reboiler. However, the decomposition depends on both temperature and residence time, and a high temperature can be tolerated in the furnace if it is only for a short residence time.

All of the material that needs to leave as product above the feed point must vaporize as it enters the column. In addition to this, some extra vapor over and above this flowrate must be created that will be condensed and flow back down through the column as reflux. This extra vaporization to create reflux is known as *overflash*.

The majority of atmospheric crude oil distillation columns follow the configuration shown in Figure 10.16, which is basically the partially thermally coupled indirect sequence. However, there are many other possibilities for different column arrangements that have the potential to reduce the energy consumption. One common variation is to introduce a flash separation partway through the feed preheating, the vapor from which is fed directly to the column. This allows part of the vaporization to be carried out at a lower temperature and reduces the amount of material that needs to be raised to a higher temperature for vaporization.

The distillation of crude oil under conditions slightly above atmospheric pressure is limited by the maximum temperature that can be tolerated by the materials being distilled, otherwise there would be decomposition. Further separation of the bottoms of the column (the *atmospheric residue*) would require a higher temperature, and therefore bring about the decomposition of material. However, this leaves a significant amount of valuable material that could still be recovered from the atmospheric residue. Therefore, the residue from the atmospheric crude oil distillation is usually reheated to a temperature around 400 °C or slightly higher and fed to a second column, the *vacuum column*, which operates under a high vacuum (very low pressure) to allow further recovery of material from the atmospheric residue, as shown in Figure 10.18. The pressure of the vacuum column is typically 0.006 bar at the top of the column where the vacuum is created. The design of the vacuum column is simpler than the atmospheric column with typically two sidestream products being taken, leaving a heavy *vacuum residue* at the bottom of the column.

10.8 Structural Optimization of Distillation Sequences

Consider now ways in which the best arrangement of a distillation sequence can be determined more systematically. Given the possibilities for changing the sequence of simple columns or the introduction of distributed distillation, prefractionators, side-strippers, side-rectifiers and fully thermally coupled arrangements, the problem is complex with many structural options. The problem can be addressed using structural optimization.

Different approaches can be used for structural optimization. Figure 10.19 illustrates one approach using the example of separation of a four-component mixture into relatively pure products. Five basic sequences for the separation of the mixture using simple

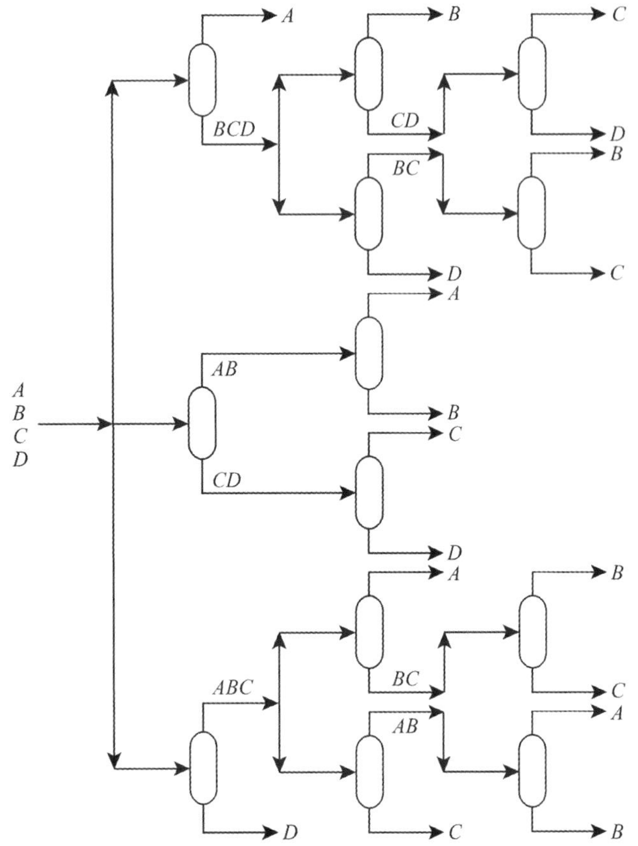

Figure 10.19

Superstructure of simple columns for five product distillation sequence.

columns have been included in a superstructure in Figure 10.19. The superstructure contains all of the options for sequences of simple columns. This superstructure of simple columns contains redundant features that need to be removed by the optimization. The optimization involves not only the distillation configuration but also the operating pressure, reflux ratio and feed condition of each column and choice of partial or total condenser for each column. It should be noted that there are interactions between columns in series as far as the feed condition is concerned. For example, if the bottom product of an upstream distillation column leaves as a saturated liquid and is fed to a downstream column at the same pressure, then the feed to the downstream column will be a saturated liquid. However, if the pressure of the downstream column is increased, the feed to the downstream column will be subcooled. On the other hand, if the pressure of the downstream column is decreased, the feed to the downstream column will become superheated.

Figure 10.19 only includes options for simple columns. Figure 10.20 shows that any two columns connected in series can in principle be replaced by different complex column arrangements. The appropriate complex column arrangement options depend on whether the two simple columns in series to be replaced are in the direct or indirect sequence. In principle, any two columns in series can be merged together.

Structural optimization can be carried out using mixed integer optimization of the superstructure (see Chapter 3), or stochastic search optimization (e.g. simulated annealing algorithm, see Chapter 3). Stochastic search optimization starts with any feasible distillation sequence (e.g. one of the simple column sequences). The design is then evolved in the optimization through a series of moves to improve the design. The moves can include:

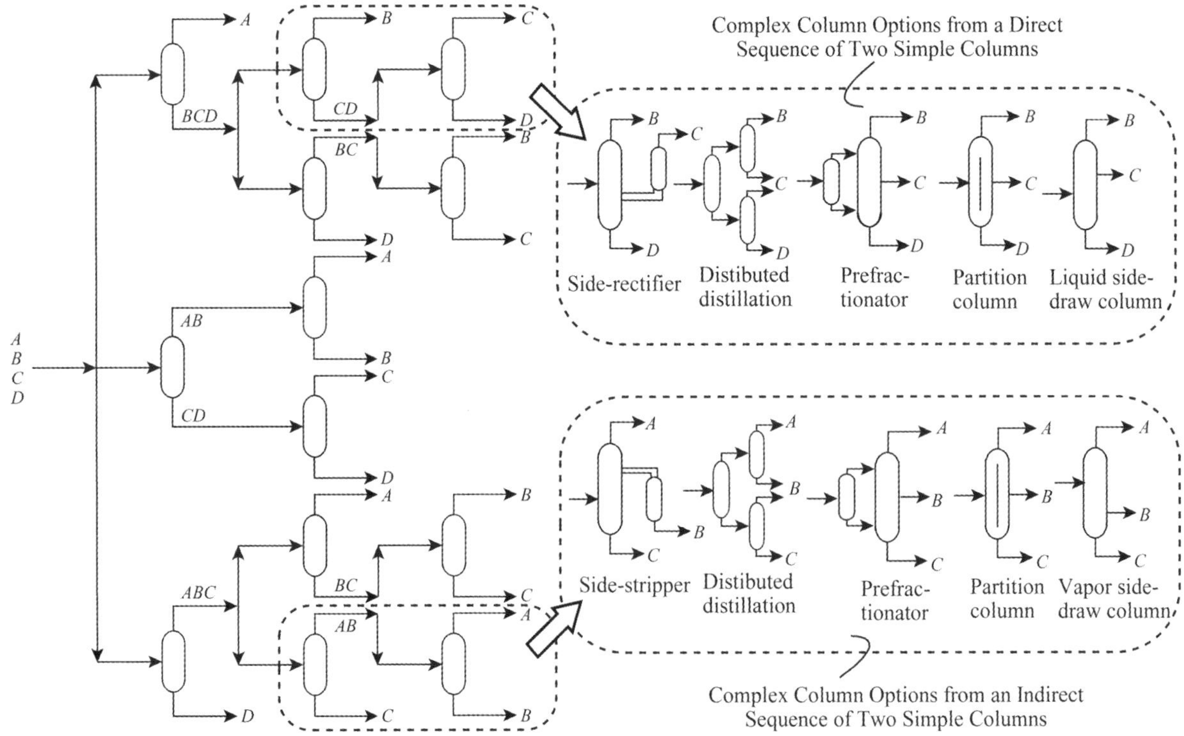

Figure 10.20

Simple column pairs in a superstructure of simple columns can be replaced by complex column arrangements.

- change the base sequence of simple columns;
- merge a sequence of two simple columns in series in the direct sequence to form a complex column with options of side-rectifier, prefractionator, partition column, or liquid side-draw column;
- merge a sequence of two simple columns in series in the indirect sequence to form a complex column with options of side-stripper, prefractionator, partition column or vapor side-draw column;
- split two simple columns in series into a distributed distillation arrangement (as illustrated in Figure 10.6a);
- demerge a complex column into simple columns (note that the complex column arrangement might be demerged to a direct or indirect sequence, depending on the complex column arrangement);
- merge a three-column distributed distillation arrangement into a direct sequence of simple columns;
- merge a three-column distributed distillation arrangement into an indirect sequence of simple columns;
- change the distribution of components in the prefrationation of distributed distillation, prefractionators and partition prefractionator columns;
- change the pressure of any column;
- change the reflux ratio of any column;
- change the feed condition of any column;
- change a total condenser on a column to a partial condenser;
- change a partial condenser on a column to a total condenser.

In this way, the optimization evolves the design by simulating the design after each move and calculating the objective function, which might be total reboiler load, total condenser load, operating cost or total annualized cost. Each simple column or complex column can be evaluated by designing and costing using either shortcut calculations (described in Chapter 8 for simple columns and earlier in this chapter for complex columns) or using detailed simulation and costing calculations. Although the approach commonly uses shortcut calculations, such as the Fenske–Underwood–Gilliland approach, the approach is not restricted to use of shortcut methods.

Rather than using stochastic search optimization and moves from an initial design, a comprehensive superstructure can be developed by embedding all possible combinations of simple and complex columns into a single superstructure. As discussed in Chapter 3, this superstructure can be represented as an MINLP optimization problem and optimized to provide a design. However, the nonlinear nature of the optimization creates special problems associated with being trapped in local optima if deterministic optimization methods are used (see Chapter 3).

Another point to be noted is that, so far, heat integration has not been considered. All reboiling and condensing duties will be satisfied by hot and cold utilities. In principle, there could be multiple utilities. For example, different steam pressures might be available for reboiling and cooling for condensers could be supplied by steam generation, cooling water or different refrigeration levels. The influence of heat integration on distillation sequencing will be considered later in Chapter 20.

Component	Recovery to Products		
	Product A	Product B	Product C
1	1.0	0	0
2	0.2	0.8	0
3	0.05	0.95	0
4	0	0.5	0.5
5	0	0	1
6	0	0	1

Figure 10.21

A typical product recovery matrix.

Before the optimization can begin, the overall material balance needs to be specified. This can be done via a product recovery matrix, as illustrated in the example in Figure 10.21. Components are arranged in order of their volatility. It follows that components in a product must all be adjacent, and a component can only distribute between adjacent products. Also, the lightest component of Product $(i+1)$ must be at least as heavy as the lightest component in Product i and the heaviest component of Product $(i+1)$ must be at least as heavy as the heaviest component in Product i. In this way, the material balance for the sequence can be specified. However, it might not be possible to achieve all of the specifications in the product recovery matrix, as in any single column in the system, only the recoveries of the light and heavy key components can be specified. Nonkey components distribute according to the separation and relative volatilities. Thus the product recovery matrix represents what is desired, rather than what can be achieved in practice.

Example 10.4 The mixture of alkanes given in Table 10.2 is to be separated into relatively pure products. Using the Underwood Equations for minimum reflux ratio, determine sequences of simple and complex columns that minimize the overall reboiler load. The recoveries will be assumed to be 100%. Assume the actual to minimum reflux ratio to be 1.1 and all the columns, with the exception of thermal coupling and prefractionator links, are fed with saturated liquid. Total condensers are to be used throughout. Neglect pressure drop through columns. Relative volatilities and latent heats of vaporization can be calculated from the Peng–Robinson Equation of State with interaction parameters set to zero. Pressures are allowed to vary through the sequence with relative volatilities recalculated on the basis of the feed composition for each column. Pressures of each column are allowed to vary to a minimum such that the bubble point of the overhead product is 10 °C above the cooling water return temperature of 35 °C (i.e. 45 °C) or a minimum of atmospheric pressure.

Solution Figure 10.22 shows three sequences that reduce the total reboiler load compared with the best sequence of simple columns. The performance of the three sequences is effectively the same, given the assumptions made for the calculations. The total reboiler load is around 30% lower than that of the best sequence of simple columns from Table 10.4. The differences between networks of simple and complex columns in terms of the reboiler load vary according to the problem.

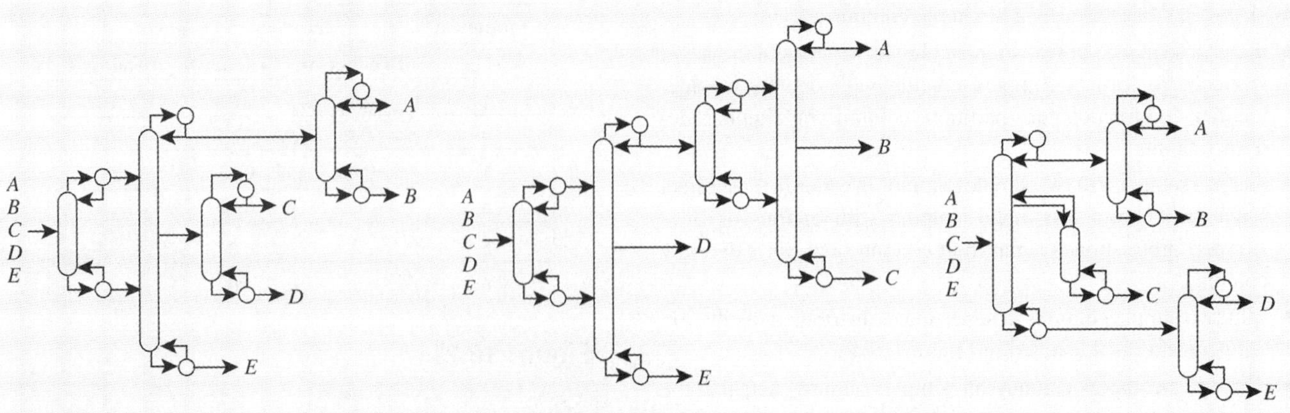

Figure 10.22
Complex column arrangements for the separation of alkanes from Example 10.4.

10.9 Distillation Sequencing – Summary

The best few non-heat-integrated sequences can be identified using the total reboiler load for above ambient temperature distillation or total condenser load for below ambient temperature distillation as a criterion. Alternatively, operating cost can be used. If this is not satisfactory, then the alternative sequences can be sized and the capital cost estimated using shortcut techniques.

Complex column arrangements, such as the prefractionator and thermally coupled arrangements, offer large potential savings in energy compared to sequences of simple columns. Partitioned prefractionator columns (dividing-wall columns) also offer large potential savings in capital cost. However, caution should be exercised in fixing the design at this stage, as the optimum sequence can change later in the design once heat integration is considered.

Crude oil distillation is carried out in a complex column sequence in which live steam is injected into the separation to provide some of the heat required and to reduce the partial pressure of the components to be distilled. The design most often used is the equivalent of the partially thermally coupled indirect sequence. However, other design configurations can also be used.

The design of sequences of simple and complex columns can be carried out using structural optimization. Structural options can be explored using stochastic search optimization or the optimization of a superstructure. The operating pressure, reflux ratio and feed condition of each column and choice of partial or total condenser for each column should also be optimized. It is assumed at this stage that all reboiling and condensing duties will be satisfied by hot and cold utilities. In principle, there could be multiple utilities.

10.10 Exercises

1. Table 10.8 shows the composition of a four-component mixture to be separated by distillation. The K-values for each component at the bubble point temperature of this mixture are given. The liquid- and vapor-phase mixtures of these components may be assumed to behave ideally.

   Figure 10.23 shows a sequence that has been proposed to separate the mixture shown in Table 10.8 into pure component products. The minimum reflux ratio for each column in this sequence is shown. The feed and all distillation products may be assumed to be liquids at their bubble point temperatures.

   a) Calculate the flowrates and composition of all streams shown in Figure 10.23. Assume that the recovery of the light and heavy key components in each column is complete.

   b) Calculate the vapor load (the total flow of vapor produced in all reboilers in the sequence) for this sequence assuming a ratio of actual to minimum reflux ratio of 1:1.

   c) What is the significance of the total vapor load?

2. Apply the heuristics for the sequencing of simple distillation columns to the problem in Table 10.8. Is the sequence shown in Figure 10.23 likely to be a good or bad sequence based on the heuristics?

Table 10.8
Data for the four-component mixture to be separated.

Component	Mole fraction in feed	K-value	Normal boiling point (°C)
n-Pentane	0.05	3.726	36.3
n-Hexane	0.30	1.5373	69.0
n-Heptane	0.45	0.6571	98.47
n-Octane	0.20	0.284	125.7
Total flowrate ($kmol \cdot h^{-1}$)	150		

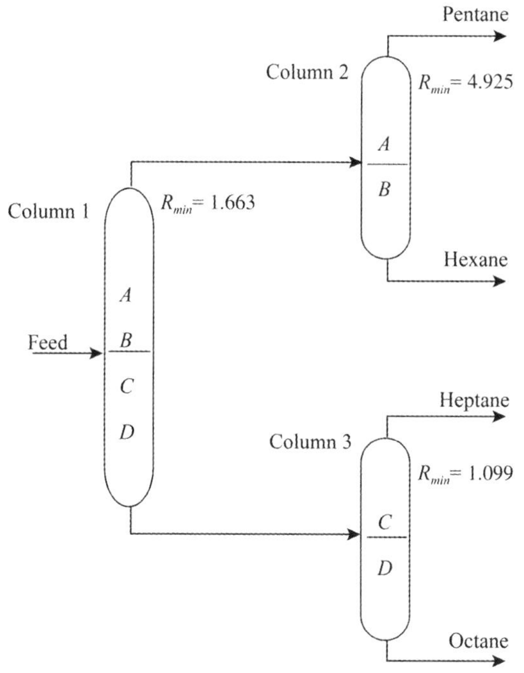

Figure 10.23
Sequence of simple distillation columns.

3. Table 10.9 presents some heuristics for using complex distillation columns to separate a ternary mixture into its pure component products. On the basis of these heuristics, suggest two sequences containing complex columns that can be used to separate the mixture described in Table 10.8 into relatively pure products.

4. A mixture of 180 kmol·h^{-1} of benzene (B), toluene (T), p-xylene (pX) and o-xylene (oX) with the feed composition in Table 10.10 is to be separated into relatively pure products. Assume that recovery is complete.

Table 10.9

Heuristics for separating a mixture of components A, B and C using complex distillation columns. Component A is the most volatile and Component C is the least volatile (Glinos and Malone, 1988).

Complex column	Heuristic for using columns
Sidestream column (sidestream below feed)	$B > 50\%$ of feed, $C < 5\%$ of feed; $\alpha_{BC} \gg \alpha_{AB}$
Sidestream column (sidestream above feed)	$B > 50\%$ of feed, $A < 5\%$ of feed; $\alpha_{AB} \gg \alpha_{BC}$
Side-rectifier	$B < 30\%$ of feed; less C than A in the feed
Side-stripper	$B < 30\%$ of feed; less A than C in the feed
Prefractionator arrangement	Fraction of B in the feed is large; α_{AB} is similar to α_{BC}

Table 10.10

Mixture of aromatics.

Component	Mole fraction	Relative volatility	Relative volatility between adjacent components
Benzene	0.40	6.215	
			2.33
Toluene	0.35	2.673	
			2.33
p-Xylene	0.20	1.48	
			1.148
o-Xylene	0.05	1.00	

a) Using heuristics, suggest a sequence of simple columns for the separation.

b) Suggest two sequences involving complex columns using the heuristics in Table 10.9.

5. When separating a mixture into relatively pure components using simple distillation columns, the different sequences can be decomposed into a series of separation *tasks* (Shah and Kokossis, 2002). Some tasks are common to different sequences. For the separation of the mixture of aromatics into relatively pure products in Exercise 4, the tasks are listed in Table 10.11. For each task the values of R_{min}, N_{min}, ΔH_{VAP}, T_{COND} and T_{REB} are listed in Table 10.11. Assume that the ratio of the actual to minimum reflux ratio is 1.1 and all column feeds are saturated liquid.

a) List the sequences of simple column tasks that can be used for the separation.

b) Calculate the flowrate of the distillate for all simple column tasks in all possible sequences. From the distillate flowrates calculate the total vapor load and total reboiler heating demand for all sequences.

6. Figure 10.24 shows an existing distillation arrangement for the separation of light hydrocarbons (Lestak and Collins, 1997). The system is to be retrofitted to reduce its energy consumption by changing to a complex distillation arrangement. As much of the existing equipment as possible is to be retained.

a) For two of the existing simple distillation columns, suggest different ways the two columns can be merged in a retrofit to form complex column arrangements.

b) For two of the existing simple distillation columns, suggest how the two columns might be re-used in a retrofit to form complex column arrangement by adding a new column shell and the three columns connected to form complex columns. Emphasis should be given to maximum reuse of existing equipment.

Table 10.11

Data for separation tasks in the aromatics separation.

Separation	R_{min}	N_{min}	ΔH_{VAP} (kJ kmol^{-1})	T_{COND} (°C)	T_{REB} (°C)
B/T pX oX	1.49	19.7	35170	80.2	120.1
B T/pX oX	0.63	19.6	35170	91.2	139.8
B T pX/oX	1.96	111.3	35170	97.4	143.6
B/T pX	1.5	19.8	34764	80.2	118.6
B T/pX	0.61	19.5	34764	91.2	139.0
T/pX oX	1.43	22.1	35476	110.3	139.8
T pX/oX	3.74	129.3	35476	118.6	143.6
B/T	1.36	20.0	32599	80.2	110.3
T/pX	1.34	22.8	35182	110.3	139.0
pX/oX	10.31	151.2	36349	138.9	143.6

B: benzene; *T*: toluene; *pX*: *p*-xylene; *oX*: *o*-xylene.

c) If two of the three columns in Figure 10.24 are to be transformed into a complex column arrangement, which two columns would be the most appropriate ones to merge?

d) Devise a number of complex column arrangements involving merging of two of the existing columns in Figure 10.24 that reuse the existing column shells and are likely to bring a significant reduction in the energy consumption.

7. Vapor flowrate in kmol·h^{-1} for each simple distillation task for the separation of a four-component mixture are:

A/BCD	100	*B/CD*	90	*A/B*	70
AB/CD	120	*BC/D*	250	*B/C*	100
ABC/D	240	*A/BC*	130	*C/D*	220
		AB/C	140		

Determine the best distillation sequence for the minimum total vapor flowrate.

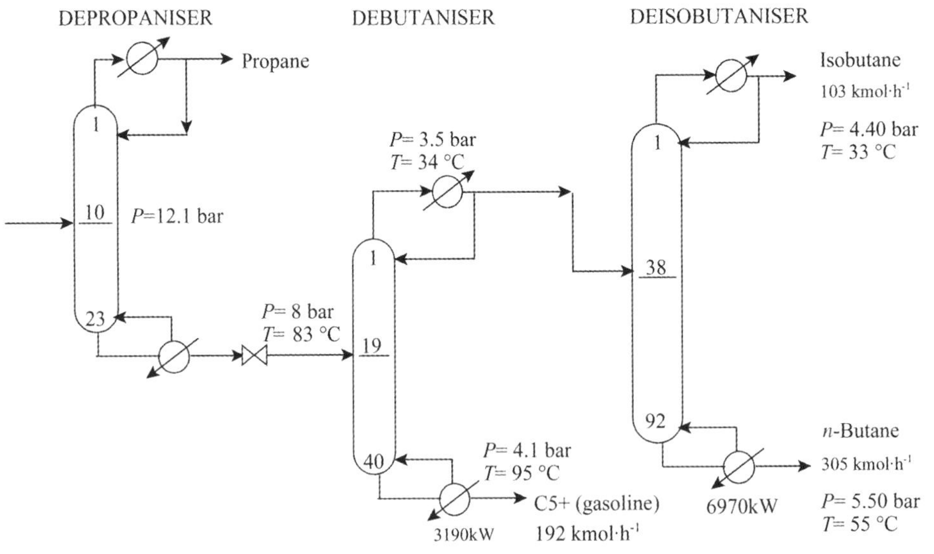

Figure 10.24

Existing arrangement of columns for the separation of light hydrocarbons. (From Lestak F and Collins C, 1997, *Chem Engg*, July: 72, reproduced by permission of McGraw Hill.)

Table 10.12

Vapor flowrates for Exercise 8.

Task number	Task	Vapor flow rate (kmol·h^{-1})
1	A/BCD	100
2	AB/CD	120
3	ABC/D	240
4	B/CD	90
5	BC/D	250
6	A/BC	130
7	AB/C	140
8	C/D	220
9	B/C	100
10	A/B	80

8. Vapor flowrates (kmol·h^{-1}) for each simple distillation task separating a four-component mixture are given in Table 10.12.

a) Determine how the tasks can be combined together for all possible sequences of simple distillation columns.

b) Calculate the vapor flowrate of all possible sequences of simple column assuming there are no interactions between the simple distillation tasks.

c) Which distillation sequence performs the best in terms of vapor flowrate?

9. By using the concept of breaking a sequence of simple distillation columns into distillation column tasks, derive a mixed integer model for the problem in Exercise 7 to determine the best sequence for the minimum vapor flowrate using simple distillation columns.

a) Develop a task-based network superstructure for the mixture *ABCD*.

b) Assign a binary variable (y_i) for each task and write an objective function for vapor load in terms of the binary variables, assuming the loads for each task given in Exercise 7.

c) Write logic constraints for the selection of tasks. For example, if Task 4 is selected ($y_4 = 1$), then Task 8 is also selected ($y_8 = 1$), but if Task 4 is not selected ($y_4 = 0$), then Task 8 may or may not be selected ($y_8 = 0$ or 1); therefore ($y_4 – y_8$) ≤ 0.

d) Formulate the problem as an MILP optimization problem.

References

Aichele P (1992) Sequencing Distillation Operations Being Serviced by Multiple Utilities, MSc Dissertation, UMIST, UK.

Asprion N and Kaibel G (2010) Dividing Wall Columns: Fundamentals and Recent Advances, *Chem Eng and Processing*, **49**: 139–146.

Becker H, Godorr S, Kreis H and Vaughan J (2001) Partitioned Distillation Columns – Why, When and How, *Chem Eng*, **Jan**: 68.

Calberg NA and Westerberg AW (1989) Temperature-Heat Diagrams for Complex Columns 2, Underwood's Method for Side-Strippers and Enrichers, *Ind Eng Chem Res*, **28** (9): 1379.

Glinos K and Malone MF (1988) Optimality Regions for Complex Column Alternatives in Distillation Columns, *Trans IChemE, Chem Eng Res Dev*, **66**: 229.

Kaibel G. (1987) Distillation Columns with Vertical Partitions, *Chem Eng Technol*, **10**: 92.

Kaibel G. (1988) Distillation Column Arrangements with Low Energy Consumption, *IChemE Symp Ser*, **109**: 43.

King CJ (1980) *Separation Processes*, 2nd Edition, McGraw-Hill.

Lestak F and Collins C (1997) Advanced Distillation Saves Energy and Capital, *Chem Eng*, **July**: 72.

Lestak F, Smith R and Dhole VR (1994) Heat Transfer Across the Wall of Dividing Wall Columns, *Trans IChemE*, **49**: 3127.

Mutalib MIA and Smith R (1998) Operation and Control of Dividing Wall Distillation Columns. Part I: Degrees of Freedom and Dynamic Simulation, *Trans IChemE*, **76A**: 308.

Mutalib MIA, Zeglam AO and Smith R (1998) Operation and Control of Dividing Wall Distillation Columns. Part II: Simulation and Pilot Plant Studies Using Temperature Control, *Trans IChemE*, **76A**: 319.

Petlyuk FB, Platonov VM and Slavinskii DM (1965) Thermodynamically Optimal Method for Separating Multicomponent Mixtures, *Int Chem Eng*, **5**: 555.

Shah PB and Kokossis AC (2002) New Synthesis Framework for the Optimization of Complex Distillation Systems, *AIChE J*, **48**: 527.

Shultz MA, Stewart DG, Harris JM, Rosenblum SP, Shakur MS and O'Brien DE (2002) Reduce Costs with Dividing-Wall Columns, *Chem Eng Prog*, **May**: 64.

Stephanopoulos G, Linnhoff B and Sophos A (1982) Synthesis of Heat Integrated Distillation Sequences, *IChemE Symp Ser*, **74**: 111.

Stupin WJ and Lockhart FJ (1972) Thermally-coupled Distillation – A Case History, *Chem Eng Prog*, **68**: 71.

Tedder DW and Rudd DF (1978) Parametric Studies in Industrial Distillation, *AIChE J*, **24**: 303.

Triantafyllou C. and Smith R (1992) The Design and Optimization of Fully Thermally-Coupled Distillation Columns, *Trans IChemE*, **70A**: 118

Chapter 11

Distillation Sequencing for Azeotropic Distillation

11.1 Azeotropic Systems

In Chapter 10, the distillation sequence for the separation of homogeneous liquid mixtures into a number of products was considered. The mixtures considered in Chapter 10 did not involve the kind of highly nonideal vapor–liquid equilibrium behavior that would result in azeotrope formation. If a multicomponent mixture needs to be separated, in which some of the components form azeotropes, one approach is to treat the azeotrope-forming components as if they were a single pseudocomponent and these isolated in a distillation sequence as addressed in Chapter 10. The separation of the azeotropic components can then be carried out in the absence of other components. However, caution should be exercised when following such an approach as it will not always lead to the best solution. Complex interactions can occur between components and the presence of components other than those involved with the azeotrope can sometimes make the azeotropic separation easier. Indeed, if a mass separation agent is used to achieve the separation, as will be discussed later, it is preferable to use a component that already exists in the process.

In many cases, the azeotropic behavior is between two components, involving binary azeotropes. In other cases, azeotropes can involve more than two components, involving multicomponent azeotropes.

The problem with azeotropic behavior, as discussed in Chapter 8, is that for highly nonideal mixtures a constant boiling mixture can form at a specific composition that has the same composition for the vapor and liquid at an intermediate composition. This depends on the vapor–liquid equilibrium physical properties. Distillation can be used to separate to a composition approaching the azeotropic composition, but the azeotropic composition cannot be approached closely in a finite number of distillation stages, and the composition cannot be crossed, even with an infinite number of stages. Thus, if the light and heavy key components form an azeotrope, then something more sophisticated than simple distillation is required. If a mixture that forms an azeotrope is to be separated using distillation, then the vapor–liquid equilibrium behavior must somehow be changed to allow the azeotrope to be crossed. There are three ways in which this can be achieved.

1) *Change in pressure.* The first option to consider when separating a mixture that forms an azeotrope is exploiting change in azeotropic composition with pressure. If the composition of the azeotrope is sensitive to pressure and it is possible to operate the distillation over a range of pressures without any material decomposition occurring, then this property can be used to carry out a separation. A change in azeotropic composition of at least 5% with a change in pressure is usually required (Holland, Gallum and Lockett, 1981).
2) *Add an entrainer to the distillation.* A mass separation agent, known as an *entrainer*, can be added to the distillation. The separation becomes possible because the entrainer interacts more strongly with one of the azeotrope-forming components than the other. This can in turn alter in a favorable way the relative volatility between the key components.
3) *Use a membrane.* If a semi-permeable membrane is placed between the vapor and liquid phases, it can alter the vapor–liquid equilibrium and allow the separation to be achieved. This technique is known as *pervaporation*.

11.2 Change in Pressure

The first option to consider when separating an azeotropic mixture is exploiting change in azeotropic composition with pressure. Azeotropes are often insensitive to change in pressure. However, if the composition of the azeotrope is sensitive to pressure and it is possible to operate the distillation over a range of pressures without any decomposition of material occurring, then this property can be used to carry out a separation. A change in azeotropic composition of at least 5% with a change in pressure is usually required (Holland, Gallun and Lockett, 1981).

Chemical Process Design and Integration, Second Edition. Robin Smith

Companion Website: www.wiley.com/go/smith/chemical2e

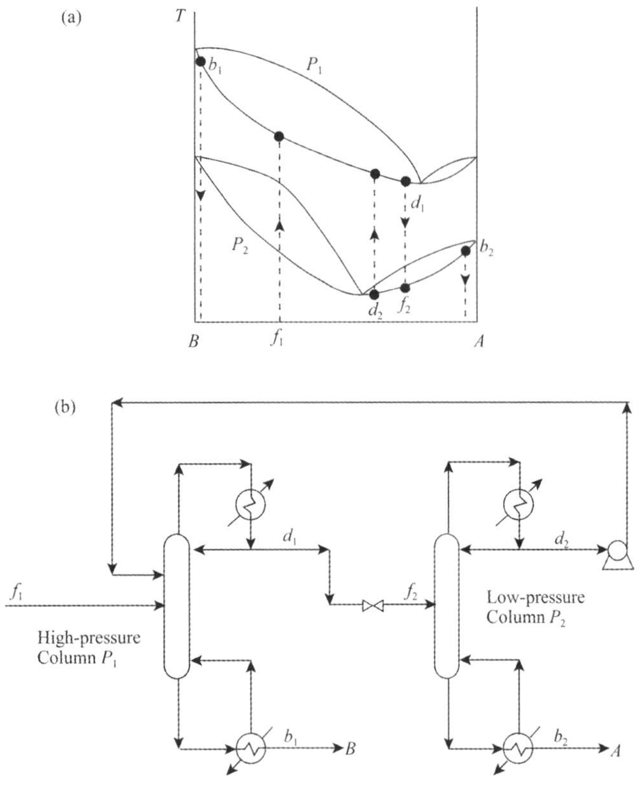

Figure 11.1

Separation of minimum boiling azeotrope by pressure change.

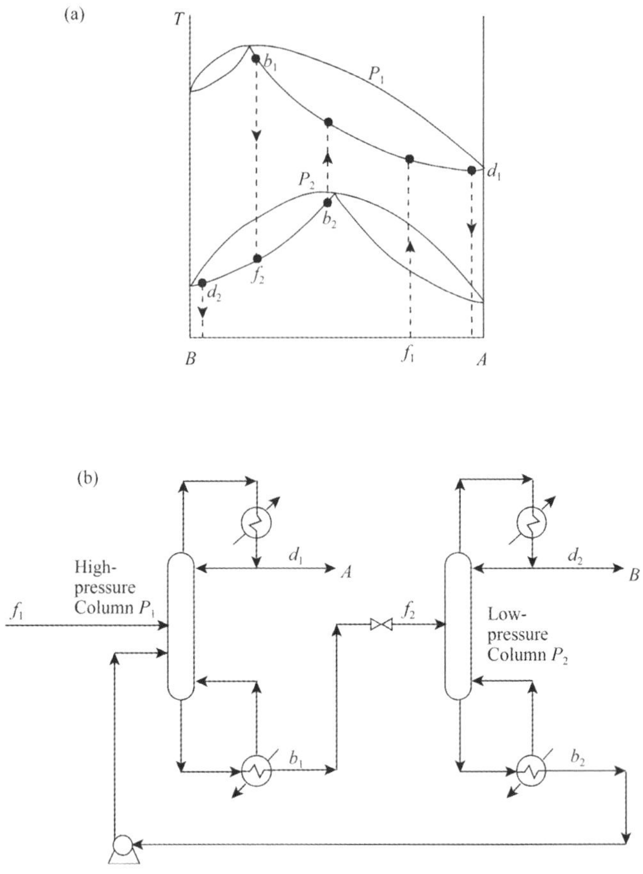

Figure 11.2

Separation of a maximum boiling azeotrope by pressure change.

Figure 11.1a shows the temperature-composition diagram for a mixture that forms a minimum-boiling azeotrope and is sensitive to change in pressure (Holland, Gallun and Lockett, 1981). This mixture can be separated using two columns operating at different pressures, as shown in Figure 11.1b. Feed with composition f_1 is fed to the high-pressure column. The bottom product from this high-pressure column b_1 is relatively pure *B*, whereas the overhead product d_1 approaches the high-pressure azeotropic composition. This distillate d_1 is fed to the low-pressure column as f_2. The low-pressure column produces a bottom product b_2 that is relatively pure *A* and an overhead product d_2 that approaches the high-pressure azeotropic composition. This distillate d_2 is recycled to the high-pressure column.

Figure 11.2a shows the temperature–composition diagram for a mixture that forms a maximum-boiling azeotrope and is sensitive to change in pressure (Holland, Gallun and Lockett, 1981). Again, this mixture can be separated using two columns operating at different pressures, as shown in Figure 11.2b. Feed with composition f_1 is fed to the high-pressure column. The overhead product from this high-pressure column d_1 is relatively pure *A*, whereas the bottom product b_1 approaches the high-pressure azeotropic composition. This bottom product b_1 is fed to the low-pressure column as f_2. The low-pressure column produces an overhead product d_2 that is relatively pure *B* and a bottom product b_2 that approaches the low-pressure azeotropic composition. This bottom product b_2 is recycled to the high-pressure column.

The problem with using a pressure change is that the smaller the change in azeotropic composition, the larger is the recycle in Figures 11.1b and 11.2b. A large recycle means a high energy consumption and high capital cost for the columns, resulting from large feeds to the columns.

If the azeotrope is not sensitive to changes in pressure, then an entrainer can be added to the distillation to alter in a favorable way the relative volatility of the key components. Before the separation of an azeotropic mixture using an entrainer is considered, the representation of azeotropic distillation in ternary diagrams needs to be introduced.

11.3 Representation of Azeotropic Distillation

To gain a conceptual understanding of the separation of such systems, they can be represented graphically on triangular diagrams (Doherty and Malone, 2001). Figure 11.3 shows a triangular diagram for three components, *A*, *B* and *C*. Different forms of triangular diagrams can be used, such as equilateral triangular and right triangular (Hougen, Watson and Ragatz, 1954). The principles are the same for each form of triangular diagram, and the form

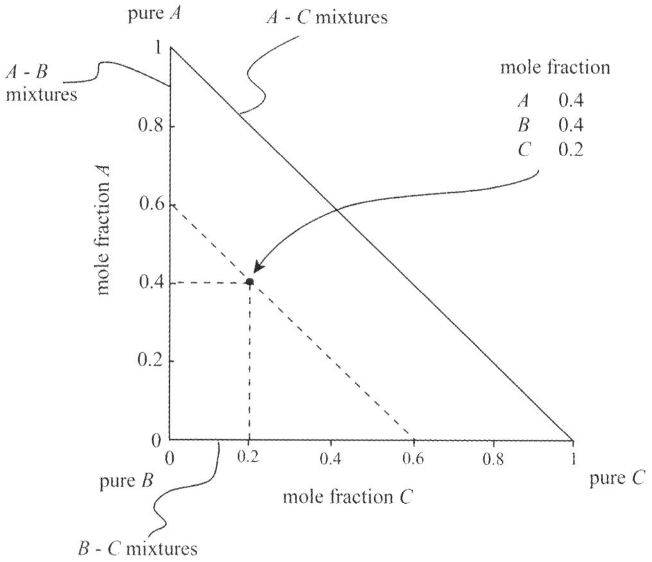

Figure 11.3

The triangular diagram.

of diagram used is only a matter of preference. Here, the right-triangular diagram will be adopted, as illustrated in Figure 11.3. The three edges of the triangle represent *A–B, A–C* and *B–C* mixtures. Anywhere within the triangle represents an *A–B–C* mixture. An example is shown in Figure 11.3 involving a mixture with a mole fraction of $A = 0.4$, $B = 0.4$ and $C = 0.2$. The location of the mixture is at the intersection of a horizontal line corresponding with $A = 0.4$ from the axis on the *A–B* edge of the triangle with a vertical line corresponding with $C = 0.2$ from the axis on the *B–C* edge of the triangle. The concentration of *B* is not read from any scale but is obtained by difference and is 0.4. Lines of constant mole fraction of *B* run parallel with the *A-C* edge of the triangle.

First, consider the representation of a mixer on the triangular diagram (Hougen, Watson and Ragatz, 1954). Figure 11.4a shows

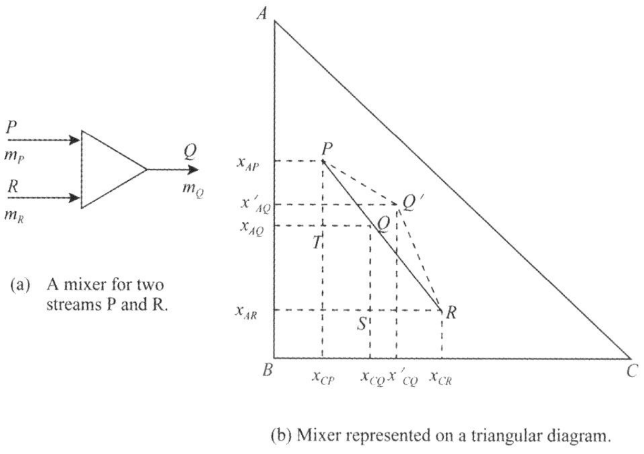

Figure 11.4

The Lever Rule.

a mixer for two streams *P* and *R* producing a mixed product *Q*. An overall mass balance gives:

$$m_Q = m_P + m_R \tag{11.1}$$

A mass balance on Component *A* gives:

$$m_Q\, x_{A,Q} = m_P\, x_{A,P} + m_R\, x_{A,R} \tag{11.2}$$

A mass balance on Component *C* gives:

$$m_Q\, x_{C,Q} = m_P\, x_{C,P} + m_R\, x_{C,R} \tag{11.3}$$

Substituting m_Q from Equation 11.1 into Equation 11.2 gives:

$$\frac{m_R}{m_P} = \frac{x_{A,P} - x_{A,Q}}{x_{A,Q} - x_{A,R}} \tag{11.4}$$

Substituting m_Q from Equation 11.1 into Equation 11.3 gives:

$$\frac{m_R}{m_P} = \frac{x_{C,Q} - x_{C,P}}{x_{C,R} - x_{C,Q}} \tag{11.5}$$

Equating Equations 11.4 and 11.5 gives:

$$\frac{x_{A,P} - x_{A,Q}}{x_{A,Q} - x_{A,R}} = \frac{x_{C,Q} - x_{C,P}}{x_{C,R} - x_{C,Q}} \tag{11.6}$$

Equation 11.6 was developed purely from mass balance arguments, without any reference to triangular diagrams. Thus, point *Q* must be located in a triangular diagram so that Equation 11.6 is satisfied. If *Q* lies on a straight line between *P* and *R*, as shown in Figure 11.4b, the triangles *PQT* and *QRS* in Figure 11.4b are similar triangles from geometry, and from the properties of similar triangles Equation 11.6 is satisfied. Had *Q* not been on a straight line between *P* and *R*, at say *Q′* in Figure 11.4b, the corresponding triangles would not have been similar, and Equation 11.6 would not have been satisfied. Thus, the composition of the mixture *Q* from Figure 11.4a when represented in a triangular diagram must lie on a straight line between the compositions of the two feeds to the mixer. From Figure 11.4b:

$$\frac{x_{AP} - x_{AQ}}{x_{AQ} - x_{AR}} = \frac{PQ}{QR} = \frac{x_{CQ} - x_{CP}}{x_{CR} - x_{CQ}} \tag{11.7}$$

Substituting Equation 11.7 in Equations 11.4 and 11.5 gives

$$\frac{m_R}{m_P} = \frac{PQ}{QR} \tag{11.8}$$

Equation 11.8 indicates that the ratio of molar flowrates of two streams being mixed are given by the ratio of the opposite line segments on a straight line, when represented in a triangular diagram (Hougen, Watson and Ragatz, 1954). As with the relationship developed in Chapter 8 for a single vapor–liquid equilibrium stage, this is again known as the *Lever Rule* from the analogy with a mechanical level and fulcrum.

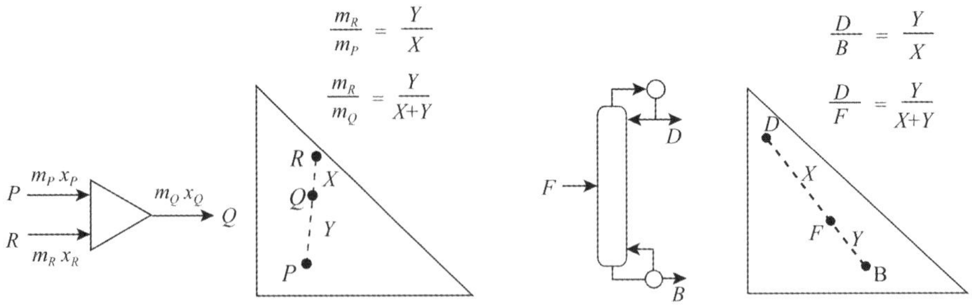

(a) Representing mixers.

(b) Distillation columns are reverse mixers.

Figure 11.5

Mixers and separators on the triangular diagram.

Figure 11.6

Representation of distillation systems on triangular diagrams.

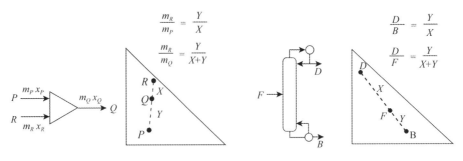

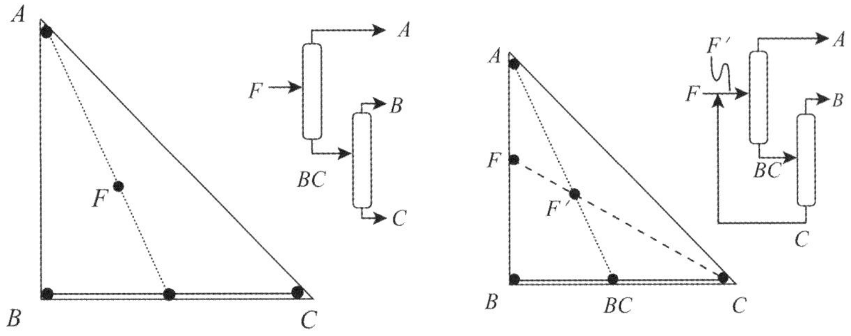

(a) A distillation sequence.

(b) A distillation sequence with recycle.

Figure 11.5a shows again a mixer and how the mixed stream lies on a straight line between the two streams being mixed at a point, located by the Lever Rule. Figure 11.5b shows a distillation column represented on the triangular diagram. The representation is basically the same, as distillation columns are effectively reverse mixers. The feed distillate and bottoms compositions all lie on a straight line with the relative distances again dictated by the Lever Rule.

Figure 11.6a now shows a sequence of two distillation columns represented on a triangular diagram. The separation is assumed to be straightforward with no azeotropes forming in this case. In Figure 11.6a, Component *A* is the most volatile and Component *C* the least volatile. A mixture of *A*, *B* and *C* at Point *F* is first separated into pure *A* and a *B–C* mixture, as represented by the intersection of a line between pure *A*, *F* and the *B–C* edge of the triangle in Figure 11.6a. This *B–C* mixture is then separated in a second distillation column to produce pure *B* and *C*. Again, the relative flowrates are dictated by the Lever Rule. Figure 11.6b shows a slightly more complex arrangement that involves a recycle and a mixer. The feed mixture *F* is a mixture of *A* and *B*. This mixes with pure *C* to form feed *F*′, which forms the feed to the first column. *F*′ is now therefore a mixture of *A*, *B* and *C*. This is separated in the first column into pure *A* overhead and bottoms product of *B–C*. In the second column, *B* and *C* are separated into pure products, with the *C* being recycled to mix with the original feed *F*.

11.4 Distillation at Total Reflux Conditions

Like conventional distillation, when studying azeotropic distillation, it is helpful to consider the two limiting cases of distillation under total reflux conditions and minimum reflux conditions (Doherty and Malone, 2001). Consider distillation at total reflux. Both staged columns and packed columns need to be considered at total reflux. Consider first staged distillation columns at total reflux conditions. The stripping section of a staged distillation column under total reflux conditions is illustrated in Figure 8.19.

In Chapter 8, a relationship was developed for total reflux conditions at any Stage *n*:

$$y_{i,n} = x_{i,n-1} \tag{8.48}$$

Starting from an assumed bottoms composition, the liquid entering and vapor leaving the reboiler are the same at total reflux conditions. This is the same as the vapor entering and liquid leaving the bottom stage. Knowing the composition of the liquid leaving the bottom stage, the composition of the vapor leaving the bottom stage can be calculated from vapor–liquid equilibrium. Then, from Equation 8.48 the composition of the liquid entering the bottom stage is equal to the composition of the vapor leaving the bottom stage at total reflux conditions. Now, knowing the composition of

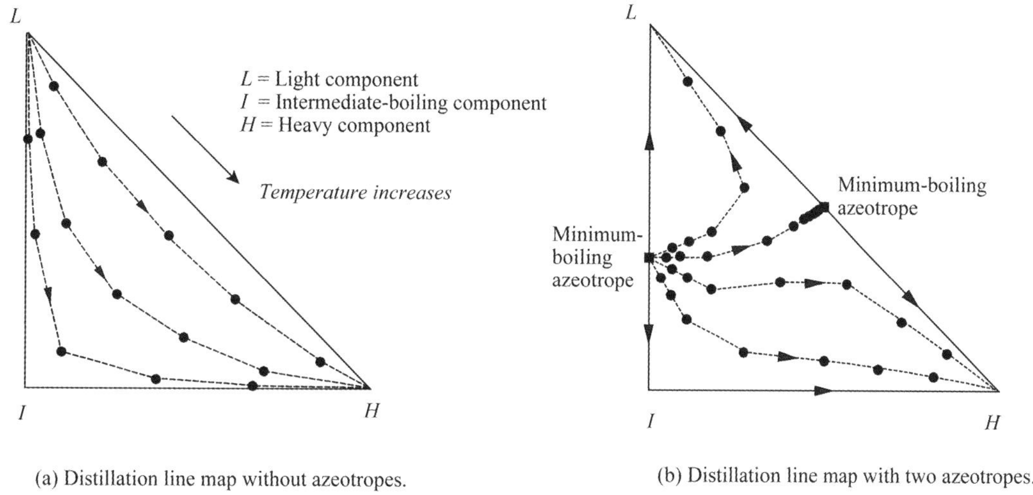

Figure 11.7
Distillation line maps.

the liquid leaving the second from bottom stage, the composition of the vapor leaving the second from the bottom stage can be calculated from the vapor–liquid equilibrium. Applying Equation 8.48 then allows the composition of the liquid leaving the third stage from bottom to be calculated, and so on up the column. In stepping up the column, the pressure is usually assumed to be constant. If the liquid composition at each stage is plotted on a triangular diagram, a *distillation line* is obtained (Zharov, 1968). The distillation line is unique for any given ternary mixture and depends only on vapor–liquid equilibrium data, pressure and the composition of the starting point. Repeating this for different assumed bottoms compositions allows us to plot a *distillation line map*, as illustrated in Figure 11.7 (Petlyuk and Avetyan, 1971; Petlyuk, Kievskii and Serafimov, 1975a). The distillation line map in Figure 11.7a involves a simple system with no azeotropes. Distillation lines in a distillation line map can never cross each other, otherwise the vapor–liquid equilibrium would not be unique for the system. The points on the distillation lines represent the discrete liquid compositions at each stage in a staged column. The *tie lines* between the points have no real physical significance as the changes in a staged column are discrete from stage to stage. The tie lines between the points simply help clarify the overall picture. Arrows can also be assigned to the tie lines to assist in the interpretation of the diagram. Here, arrows will be assigned in the direction of increasing temperature.

A more complex distillation line map is shown in Figure 11.7b. This involves two binary azeotropes. The closeness of the dots on a distillation line is indicative of the difficulty of separation. As an azeotrope is approached, the points become closer together, indicating a smaller change in composition from stage to stage.

The distillation lines in the distillation line map were in this case developed by carrying out a balance around the bottom of the column, as indicated in Figure 8.13. Equally well, the distillation line could have been developed by drawing an envelope around the top of the column at total reflux, and the calculation developed down the column in the direction of increasing temperature.

The distillation lines in Figure 11.7 show the change in composition through a staged column at total reflux. The analogous principle for the packed column at total reflux could also be developed (Doherty and Perkins, 1979a). To begin, consider the mass balance around the rectifying section of a staged column, but this time not at total reflux, as illustrated in Figure 11.8a. To simplify the development of the equations, constant molar overflow will be assumed. A mass balance around the top of the column for Component i gives:

$$Vy_{i,n+1} = Lx_{i,n} + Dx_{i,D} \tag{11.9}$$

Also, by definition of the reflux ratio R:

$$L = RD \tag{11.10}$$

An overall balance around the top of the column gives:

$$V = (R+1)D \tag{11.11}$$

Combining Equations 11.9 to 11.11 gives:

$$y_{i,n+1} = \frac{Rx_{i,n}}{R+1} + \frac{x_{i,D}}{R+1} \tag{11.12}$$

Then carrying out a balance across Stage n for Component i, as illustrated in Figure 11.8b, gives:

$$Lx_{i,n-1} + Vy_{i,n+1} = Lx_{i,n} + Vy_{i,n} \tag{11.13}$$

Rearranging Equation 11.13 gives:

$$\begin{aligned} x_{i,n} - x_{i,n-1} &= \frac{V}{L}(y_{i,n+1} - y_{i,n}) \\ &= \frac{R+1}{R}(y_{i,n+1} - y_{i,n}) \end{aligned} \tag{11.14}$$

11

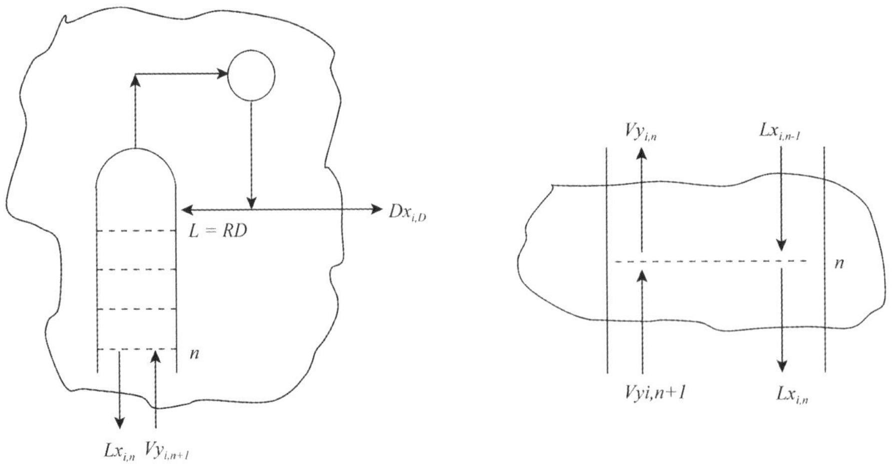

(a) Mass balance around rectifying section.

(b) Mass balance around Stage *n*.

Figure 11.8

Mass balances for rectifying section at finite reflux conditions.

Now combining Equations 11.12 and 11.14 gives:

$$x_{i,n} - x_{i,n-1} = x_{i,n} + \frac{x_{i,D}}{R} - \frac{(R+1)y_{i,n}}{R} \tag{11.15}$$

Equation 11.15 is a difference equation that, for small changes, can be approximated by a differential equation:

$$\frac{dx_i}{dh} = x_i + \frac{x_i}{R} - \frac{(R+1)y_i}{R} \tag{11.16}$$

where h = dimensionless height in a packed column

Equation 11.16 can be considered to represent the change of concentration with height within a packed column, in which change in composition is continuous. At total reflux, $R = \infty$ and Equation 11.16 simplifies to:

$$\frac{dx_i}{dh} = x_i - y_i \tag{11.17}$$

Substituting the vapor–liquid equilibrium K-value:

$$\frac{dx_i}{dh} = x_i(1 - K_i) \tag{11.18}$$

This equation can be integrated at constant pressure from an initial composition $x_{i,0}$ through the dimensionless height h. However, the integration must be carried out numerically. Since K_i is a function of x_i, Equation 11.18 is nonlinear. Also, the equations for the individual components are coupled. Hence, the integration must be carried out numerically, typically using a fourth-order Runge–Kutta integration technique (Doherty and Perkins, 1979a).

Equations 11.17 and 11.18 for packed columns at total reflux could also have been developed by considering a mass balance around the bottom of the column. The same result would have been obtained.

Equation 11.18 allows us to predict the concentration profile through a packed column operating at total reflux conditions. The resulting profile is known as a *residue curve*, since they were first developed by considering the batch vaporization of a mixture through time (Schreinemakers, 1901; Doherty and Perkins, 1978). Such batch vaporization is often termed *simple distillation*. An analysis of a simple distillation process results in the same form of expression as Equations 11.17 and 11.18 (Doherty and Perkins, 1978). However, in the case of simple distillation, the expressions feature dimensionless time instead of the dimensionless packing height. The significance of the residue curve in simple distillation is therefore a plot of the residual liquid composition through time as material is vaporized, and hence the name residue curve. The arrows assigned to residue curves then follow increasing time as well as temperature. Within the present context, the interpretation of the residue curve as a change of composition through height in a packed column at total reflux conditions is more useful. The residue curve is unique for any given ternary mixture and depends only on vapor–liquid equilibrium data, pressure and the composition of the starting point.

Given a number of different starting points, a number of residue curves can be plotted on the same triangular diagram, giving a *residue curve map* (Doherty and Perkins, 1979b). Figure 11.9a shows a residue curve map that does not exhibit any azeotropes. Residue curves in a residue curve map can never cross each other, otherwise vapor–liquid equilibrium would not be unique for the system. Unlike distillation lines, residue curves are continuous. Arrows can again be assigned to help in the interpretation. Here, the arrows will be assigned in the direction of increasing temperature. A slightly more complex residue curve map is shown in Figure 11.9b. This residue curve shows a binary minimum-boiling azeotrope between the light and intermediate-boiling components.

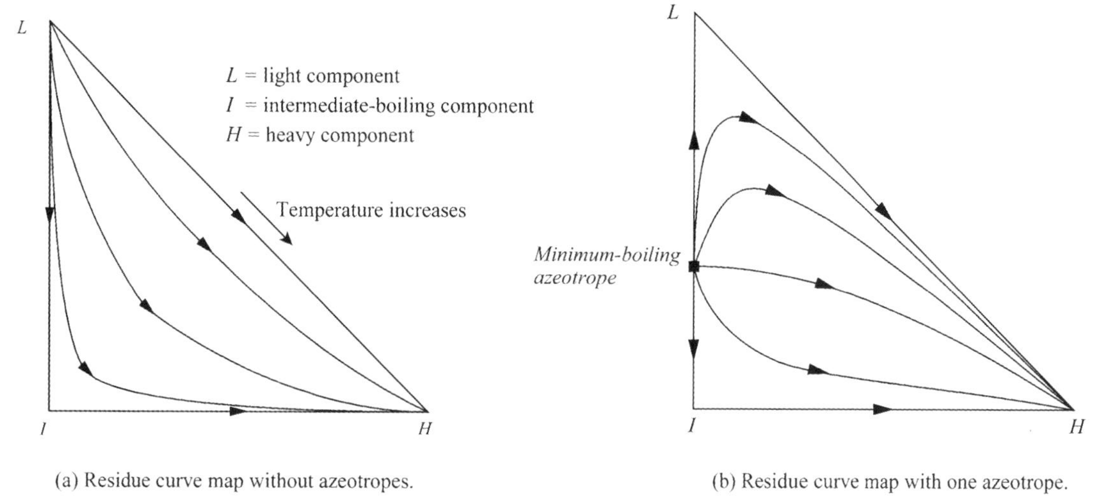

Figure 11.9
Residue curve maps.

An even more complex residue curve map is shown in Figure 11.10. This has binary azeotropes between each of the three components. It also has a ternary minimum-boiling azeotrope involving all three components.

Figure 11.11 superimposes distillation lines and residue curves for the same ternary systems. Figure 11.11a shows the system n-pentane, n-hexane and n-heptane, which is a relatively wide boiling mixture. It can be observed in Figure 11.11a that there are significant differences between the paths of the distillation lines and the residue curves. By contrast, the ternary system ethanol, isopropanol and water shown in Figure 11.11b is a more close-boiling mixture. This time, the distillation lines and residue curves follow each other fairly closely because the difficult separation

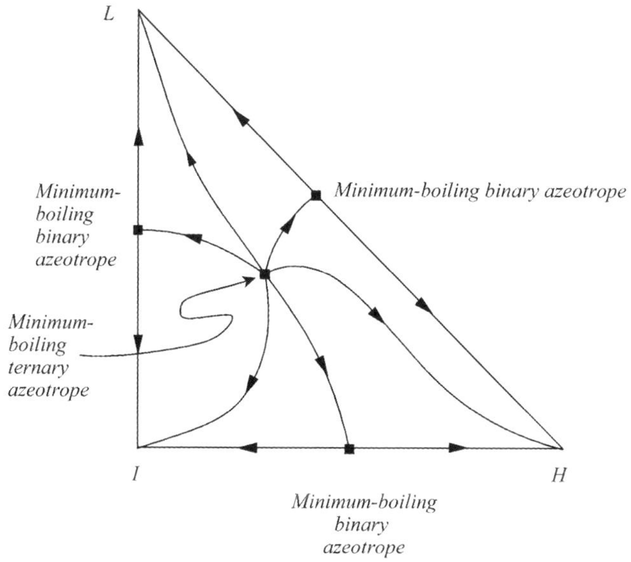

Figure 11.10
A residue curve map with three binary azeotropes and one ternary azeotrope.

means that the changes from stage to stage in a staged column become smaller and approach the continuous changes in a packed column. It is important to note that distillation lines and residue curves have the same properties at fixed points (when the distillation lines and residue curves converge to a pure product or an azeotrope).

Both distillation lines and residue curves can be used to make an initial assessment of the feasibility of an azeotropic separation at high reflux rates. Both will give a similar picture. Figure 11.12a shows a residue curve map involving a binary azeotrope between the light and intermediate-boiling components. Also shown in Figure 11.12a are two suggested distillation separations drawn as straight lines. At total reflux, for a packed column to be feasible, both the distillate and bottoms product should be located on the same residue curve, as well as the feed and products being located on the same straight line. The upper separation in Figure 11.12a shows this to be the case and might in principle be expected to be feasible. The lower separation in Figure 11.12a shows a separation in which the products do not quite lie on the same residue curve. However, total reflux will not be used in the final design and there may be slight changes to the feed and product specifications. All that can be determined from the two separations shown in Figure 11.12a is that they might be feasible. However, there is no guarantee of this. Later, more rigorous assessment of the feasibility of an individual column will be considered. By contrast, two different separations are shown for the same residue curve map in Figure 11.12b. This time, the suggested separations run across the residue curves. In this case, the separations will not be feasible.

Another important feature of residue curve maps (and distillation line maps) is illustrated in Figure 11.13. The residue curve map in Figure 11.13a shows two binary azeotropes. Starting anywhere in the residue curve map will always terminate at the same point, in this case at the binary azeotrope between the intermediate-boiling and heavy components in Figure 11.13a. The residue curve map in Figure 11.13b shows a quite different behavior. Again, there are two binary azeotropes. This time, however, the residue curve map

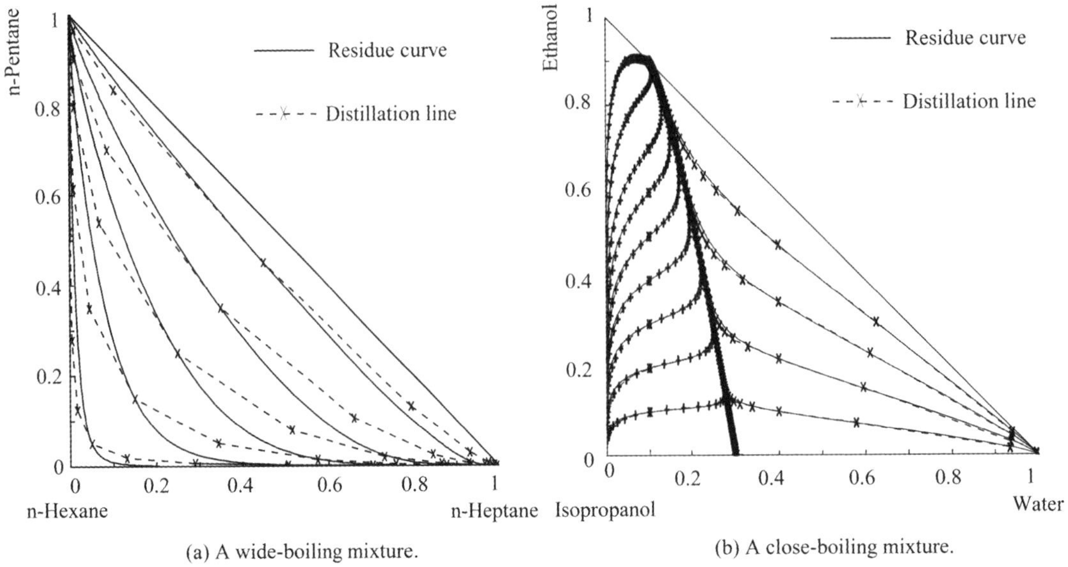

Figure 11.11
Comparison of distillation lines and residue curves.

can be divided into two distinct *distillation regions* separated by a *distillation boundary* (Petlyuk, Kievskii and Serafimov, 1975b; Doherty and Perkins, 1979b). Starting at any point to the left of the distillation boundary in Figure 11.13b will always terminate at the intermediate-boiling component. Starting anywhere to the right of the distillation boundary will always terminate at the heavy boiling component. The residue curve connecting the two azeotropes divides the residue curve map into two distinct regions. The significance of this is critical. Starting with a composition anywhere in Region *I* in Figure 11.13b, a composition in Region *II* cannot be accessed, and vice versa. This means that distillation problems involving distillation boundaries are likely to be extremely problematic. For more complex systems, the residue curve map can be divided into more than two regions. For example, referring back to Figure 11.10, this residue curve map is divided into three regions. The three distillation boundaries are created by the residue curves connecting the binary azeotropes with the ternary azeotrope.

Distillation line maps also exhibit distillation boundaries. For example, referring back to Figure 11.7b, this is divided into two regions. The boundary is formed by the distillation line connecting the two binary azeotropes. The boundary for a residue curve on the system in Figure 11.7b would start and end at the same point but might follow a slightly different path because the residue curve follows a continuous profile rather than the discrete profile of the distillation line boundary in Figure 11.7b. Figure 11.11b also has two regions. The distillation lines and residue curves are superimposed in Figure 11.11b, and it can be seen that the distillation line and residue curve distillation boundaries follow each other closely in this case.

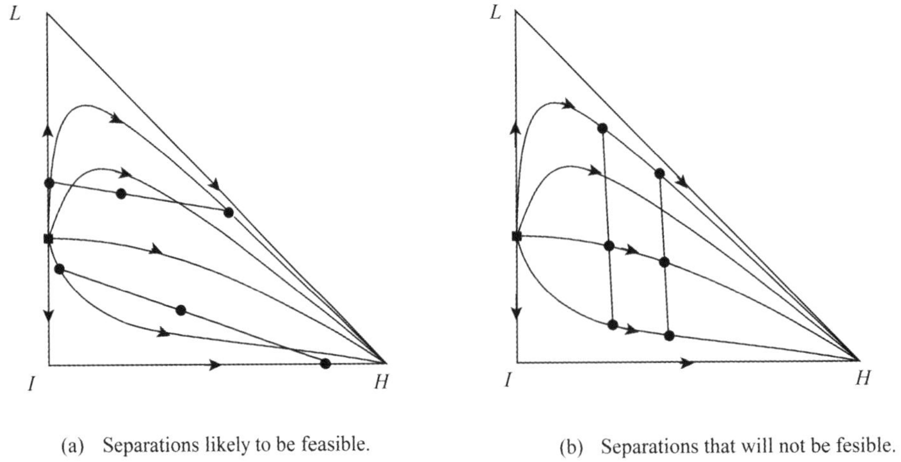

Figure 11.12
Residue curves give an indication of what separations are likely to be feasible.

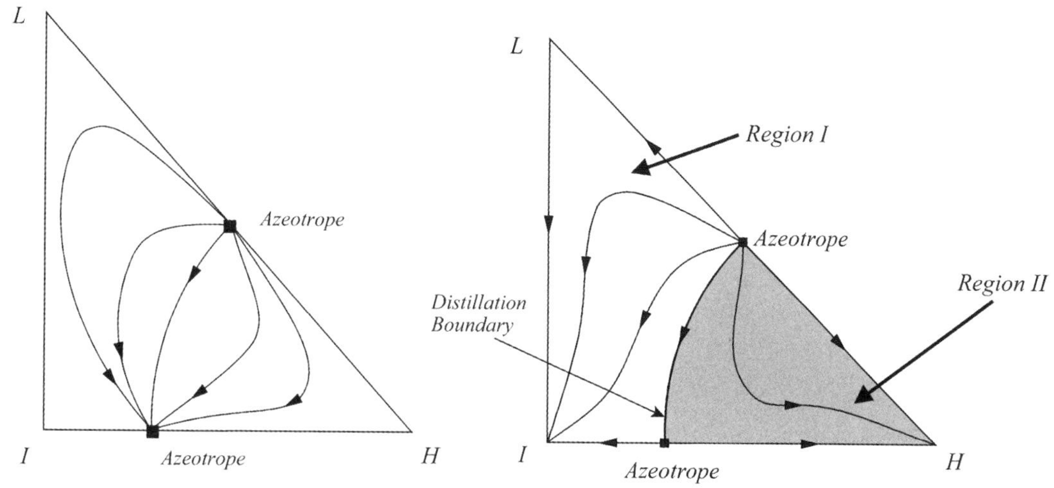

(a) Residue curves giving the same final composition for any intermediate starting point.

(b) Residue curves in which the final compositions are different divide the map into different regions.

Figure 11.13

The residue curve maps can be divided into regions by distillation boundries.

11.5 Distillation at Minimum Reflux Conditions

So far, only the limiting case of total reflux condition has been investigated in developing distillation lines and residue curves for staged and packed columns. The other limiting condition of minimum reflux will now be explored (Wahnschafft *et al.*, 1992; Doherty and Malone, 2001). Consider the rectifying section of a column as illustrated in Figure 11.14a. This is operating at minimum reflux, as there is a pinch in the column in which the changes in composition are incrementally small. This could either be an incrementally small change from stage to stage in a staged column or an incrementally small change with height in a packed column. A mass balance around the top of the column indicates that the liquid composition leaving the top section of the column, the liquid composition of the distillate and the vapor composition of the vapor entering the top section must all be in a straight line on a triangular diagram, as illustrated in Figure 11.14b. Also, at a pinch condition the change in composition is incremental. For an incremental change in composition with height in the residue curve, as illustrated in Figure 11.14b, both incremental points

11

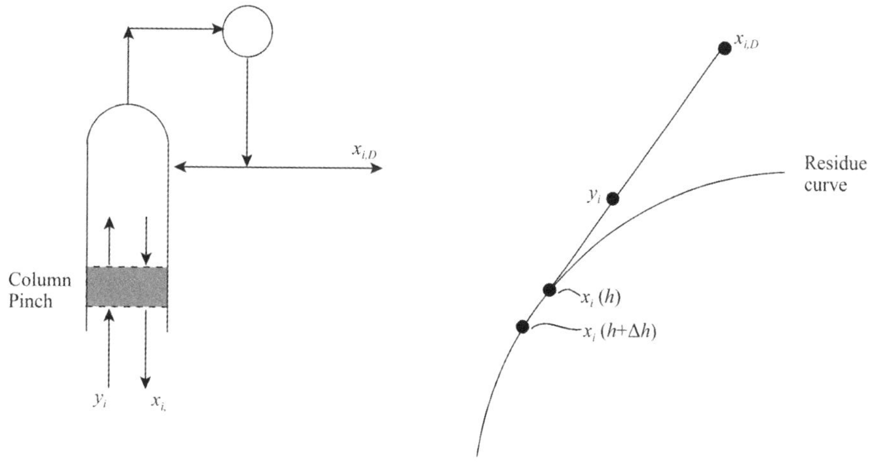

(a) Rectifying section of column under minimum reflux conditions.

(b) The tangent to the residue curve must pass through the distillate composition at minimum reflux conditions.

Figure 11.14

The minimum reflux condition can be identified from the tangent to the residue curve.

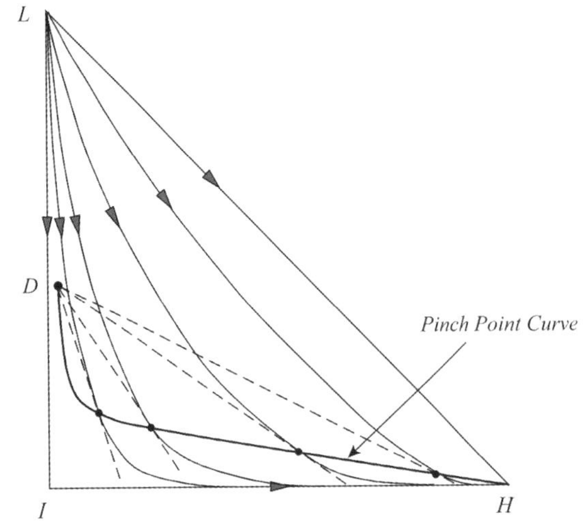

Figure 11.15
The locus of tangents to the residue curves gives the pinch point curve.

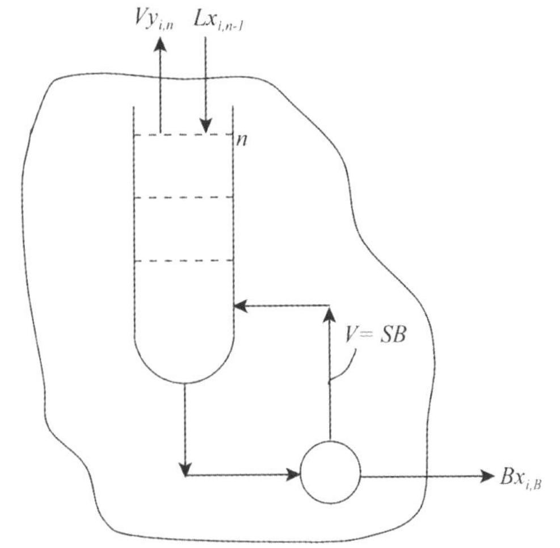

Figure 11.16
Mass balance for the stripping section at finite reflux conditions.

of the pinch must be on the residue curve, and an incremental change defines the line to be a tangent to the residue curve (Wahnschafft *et al.*, 1992). Thus, it follows that the tangent to the residue curve at a pinch point x_i must pass through the distillate composition $x_{i,D}$. This principle gives a very simple way to follow the path of pinch conditions for any given distillate composition. Figure 11.15 shows a distillate composition *D* and a series of tangents drawn to the residue curves. The locus of points joining the tangents to the residue curves defines the *pinch point curve* (Wahnschafft *et al.*, 1992). The same principle can also be applied to the bottom of the column. The same pinch point curve applies to both staged and packed columns, as by definition at a pinch condition the changes are incrementally small.

11.6 Distillation at Finite Reflux Conditions

Consider again the column rectifying section shown in Figure 11.8a at finite reflux conditions. Equation 11.12 relates the composition of passing vapor and liquid streams at any stage, given the reflux ratio and distillate composition. Specifying the distillate composition from the rectifying section in Figure 11.8a allows the vapor composition leaving the top stage of the distillation column to be calculated from the vapor–liquid equilibrium. This would allow the composition of the vapor leaving the first stage and the liquid composition entering the first stage (being the same as the distillate composition) to be determined. The liquid composition leaving the first stage will be at equilibrium with the vapor composition leaving the first stage. Thus, from a vapor–liquid equilibrium calculation, the liquid leaving the first stage can be calculated from the composition of the vapor leaving the first stage. The composition of the vapor entering the first stage can then be calculated from Equation 11.12. This is the vapor leaving the second stage. Thus, the composition of the liquid leaving the second stage can be calculated from the vapor–liquid equilibrium. The vapor entering the second stage can then be calculated from Equation 11.12, and so on down the column. In this way, the composition through the rectifying section can be calculated by specifying the reflux ratio and distillate composition, based on the assumption of constant molar overflow (Levy, Van Dongen and Doherty, 1985).

The corresponding mass balance can also be carried out around the stripping section of a column, as illustrated in Figure 11.16. A mass balance around the bottom of the column for Component *i* gives:

$$Lx_{i,n-1} = V_{yi,n} + Bx_{i,B} \tag{11.19}$$

Also, by definition of the reboil ratio *S*:

$$V = SB \tag{11.20}$$

An overall balance around the bottom of the column gives:

$$L = (S + 1)B \tag{11.21}$$

Combining Equations 11.19 to 11.21 gives:

$$x_{i,n-1} = \frac{Sy_{i,n}}{S+1} + \frac{x_{i,B}}{S+1} \tag{11.22}$$

Equation 11.22 relates the compositions of vapor and liquid streams passing each other in the stripping section of a column. Equation 11.22 can be used together with vapor–liquid equilibrium calculations to calculate a composition profile in the stripping section of the column, similar to that of the rectifying section of the column as described above. The calculation is started with an assumed bottoms composition and Equation 11.22 applied repeatedly with vapor–liquid equilibrium calculations working up the column.

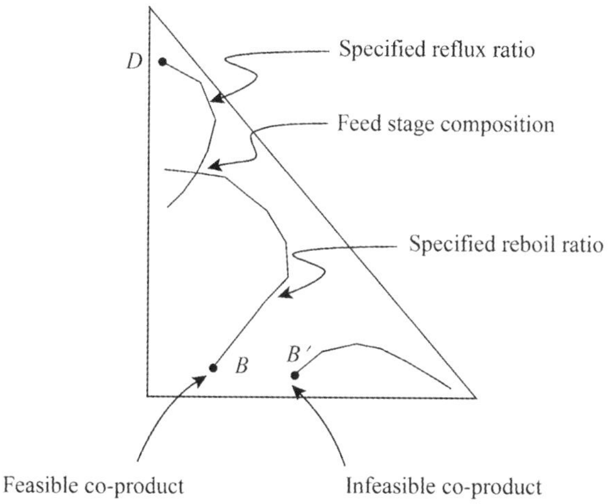

Figure 11.17
Section profiles.

Figure 11.17 shows a plot of *section profiles* developed in this way (Levy, Van Dongen and Doherty, 1985). Starting from an assumed distillate composition *D*, the calculation works down the column with a given reflux ratio. Working down the column, the changes from stage to stage gradually decrease as the lighter components become depleted. The distillate section profile approaches a pinch condition towards its termination as the lighter components become depleted. Also shown in Figure 11.17 is a section profile for the stripping section of a column. Starting at bottoms composition *B* with a given reboil ratio, the section profile works up the column. Towards the end of the section profile for the stripping section, again the changes from stage to stage become smaller as the section becomes depleted in the heavier components and again approaches a pinch condition. The section profiles starting from *D* and *B* in Figure 11.17 intersect each other before reaching a pinch condition. The intersection would correspond with the feed condition. Figure 11.17 indicates that because the section profiles for the rectifying and stripping sections intersect, these settings would, in principle, lead to a feasible column.

Two issues should be noted. First, the section profiles are actually discrete changes from stage to stage, and strictly speaking the compositions of discrete points corresponding with stages for the rectifying and stripping sections need to intersect for feasibility. If the discrete points do not intersect, then usually some small adjustments to the product compositions, reflux ratio or reboil ratio can be used to fine-tune for feasibility. Second, the section profiles were developed by assuming constant molar overflow, which is an approximation.

Also shown in Figure 11.17 is a section profile for a stripping section starting at *B*′. The stripping section profile starting at *B*′ follows a path different from the profile starting at *B* and does not intersect the rectifying section profile. This means that the combination of distillate compositions *D* and *B*′ could not lead to a feasible column design.

The intersection of section profiles as illustrated in Figure 11.17 can be used to test the feasibility of given product compositions and settings for the reflux ratio and reboil ratio. However, the profiles would change as the reflux ratio and reboil ratios change, and trial and error will be required.

It should be noted that although the assumption of constant molar overflow has been used here to develop the section profiles, it is possible to develop rigorous profiles. A rigorous profile requires the material and energy balances to be solved simultaneously. Thus, the rigorous profile would step through the column one stage at a time, solving the material and energy balance for each stage, rather than just the material balance. Obviously, the calculations become more complex.

Figure 11.18 shows a product composition for the system chloroform, benzene and acetone with a series of section profiles for the stripping section of a column starting from bottoms 11

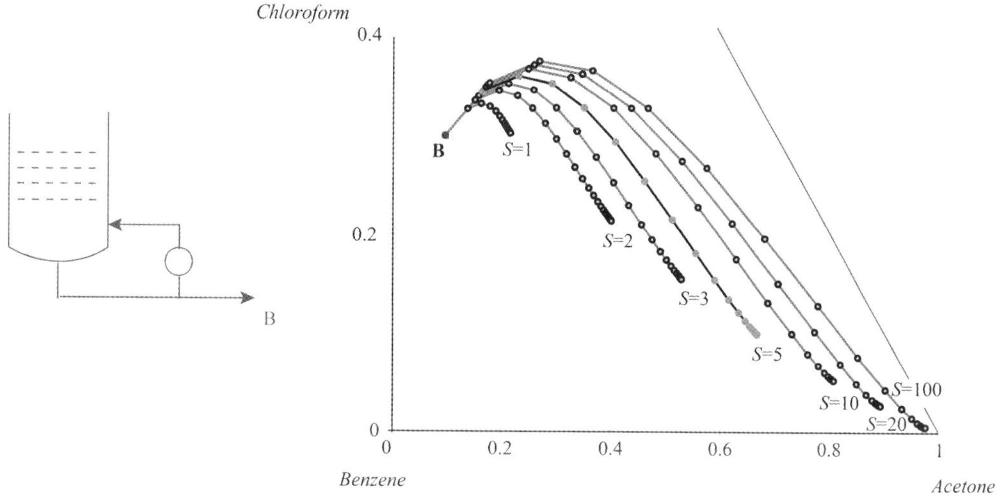

Figure 11.18
For a given bottoms composition, a range of section profiles can be generated by varying the reboil ratio. (Reproduced from Castillo F.J.L, Thong D.Y.C and Towler G.P (1998) Homogeneous Azeotropic Distillation 1. Design Procedure for Single-Feed Columns at Non-total Reflux, *Ind Eng Chem Res*, **37**: 987. Copyright © 1998, American Chemical Society.)

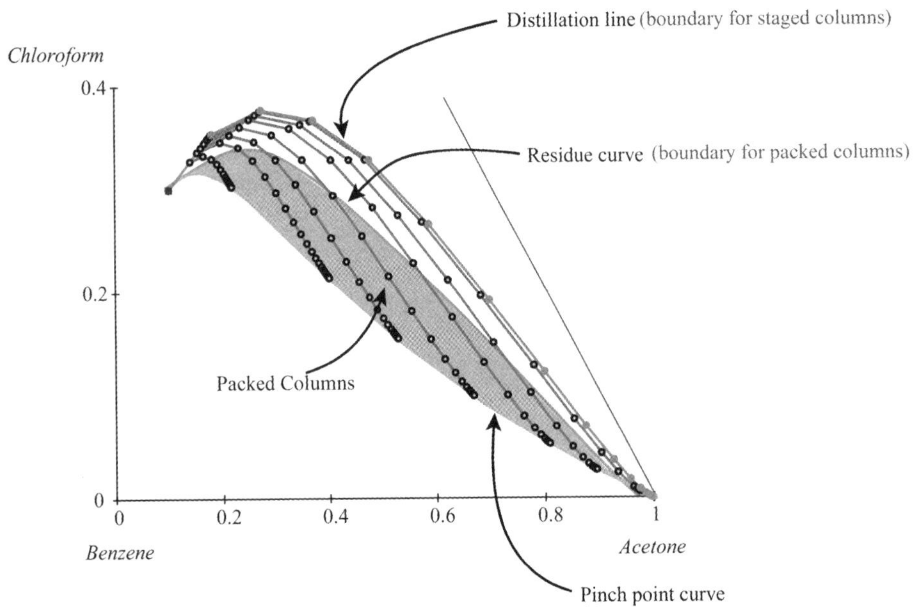

Figure 11.19

The operation leaf is bounded by the distillation line or residue curve and the pinch point curve. (Reproduced from Castillo F.J.L, Thong D.Y.C and Towler G.P (1998) Homogeneous Azeotropic Distillation 1. Design Procedure for Single-Feed Columns at Non-total Reflux, *Ind Eng Chem Res*, **37**: 987. Copyright © 1998, American Chemical Society.)

composition *B*. For a defined product composition, the section profiles can be projected for different settings of the reboil ratio (or the reflux ratio for a rectifying section). The reboil ratio is shown to gradually be increased from $S = 1$ to $S = 100$. The profile changes according to the setting of the reboil ratio. Each of the section profiles terminates in a pinch condition. This shows that there is a large area of the triangular diagram that can be accessed from a starting composition *B* by a column stripping section with the appropriate setting of the reboil ratio. However, it would be convenient to determine the region of all possible composition profiles for a given product without having to plot all of the section profiles, as illustrated in Figure 11.18.

The rectifying or stripping section of a column must operate somewhere between total reflux and minimum reflux conditions. The range of feasible operations of a column section can thus be defined for a given product composition. It can be seen in Figure 11.19 that these section profiles are bounded for a stage column by the distillation line and the pinch point curve. As noted previously, the pinch point curve provides a minimum reflux boundary for both staged and packed columns, as by definition at a pinch condition the changes are incrementally small.

Also shown in Figure 11.19 is the residue curve projected from the same product composition. The area enclosed within the residue curve and the pinch point curve thus provides the feasible compositions that can be obtained by a packed column section from a given product composition. For any given product composition, the *operation leaf* of feasible operation for a column section can be defined by plotting the distillation line (or residue curve) and the pinch point curve (Levy, Van Dongen and Doherty, 1985; Castillo, Thong and Towler, 1998). The column section must operate somewhere between the total and minimum reflux conditions.

For the stripping and rectifying sections to become a feasible column, the two operation leaves must overlap. Figure 11.20 shows the system for chloroform, benzene and acetone. The operation leaf for a distillate composition *D* intersects with the operation leaf for a bottoms composition B_1 in Figure 11.20. This means that for these two products there is some combination of settings for the reflux ratio and reboil ratio that will allow the section profiles to intersect and become a feasible column design. By contrast, the bottoms composition B_2 shows an operation leaf that does not intersect with the operation leaf

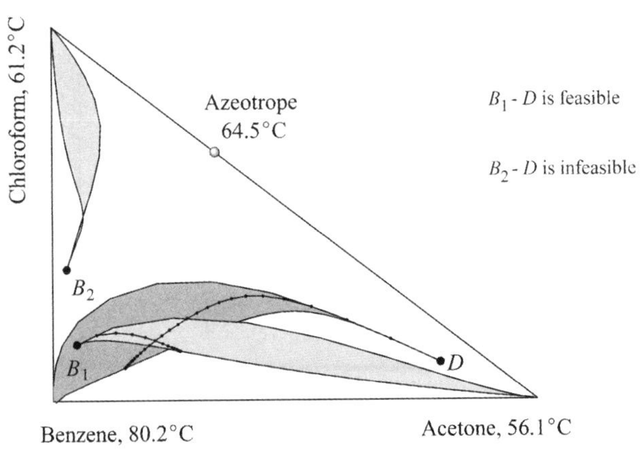

Figure 11.20

Overlap of column section operating leaves is a necessary condition for feasible separation. (Reproduced from Castillo F.J.L, Thong D.Y.C and Towler G.P (1998) Homogeneous Azeotropic Distillation 1. Design Procedure for Single-Feed Columns at Non-total Reflux, *Ind Eng Chem Res*, **37**: 987. Copyright © 1998, American Chemical Society.)

of distillate D. This means that the two products D and B_2 cannot be produced in the same column, and the design is infeasible. No settings of reboil ratio or reflux ratio can make the combination of B_2 and D a feasible design.

The operation leaves are particularly interesting for cases when separations are not feasible at high reflux ratios but can be feasible at lower reflux ratios. The reason that this kind of behavior can happen will be discussed later.

11.7 Distillation Sequencing Using an Entrainer

Consider the separation of a mixture containing a mole fraction of acetone of 0.7 and a mole fraction of heptane of 0.3. This needs to be separated to produce pure acetone. Unfortunately, there is an azeotrope between acetone and heptane at a mole fraction of acetone of 0.95. Figure 11.21a shows the mass balance for the separation of this mixture into pure acetone and heptane using benzene as an entrainer. The feed is first mixed with benzene. Figure 11.21a shows three different mass balances corresponding with different amounts of benzene being mixed with the feed, producing feed mixtures at A, B and C in Figure 11.21a. From the Lever Rule, Point A in Figure 11.21a mixes the largest amount of benzene and Point C the lowest amount of benzene. The first column $C1$ separates the ternary mixture into pure heptane and a mixture of acetone and benzene, with a composition that depends on the amount of benzene mixed with the original feed. A second column $C2$ then separates pure acetone and benzene, with the benzene being recycled. Figure 11.21b shows the distillation sequence that would correspond with this mass balance.

Even though a feasible mass balance has been set up, there is no guarantee that vapor–liquid equilibrium will allow the separation. To test whether the columns are feasible, the operation leaves corresponding with the products suggested by the mass balance need to be plotted. Take the benzene flowrate corresponding with Point C in Figure 11.21a. This would correspond with the lowest flowrate of benzene of the three settings. Figure 11.21c shows the operation leaves for Column $C1$. The stripping section leaf is very narrow and follows the acetone–heptane edge of the triangular diagram from b_1. The rectifying leaf starts at d_1 and is long and narrow along the acetone–benzene edge, after which it opens up and converges towards pure heptane, where it intersects with the

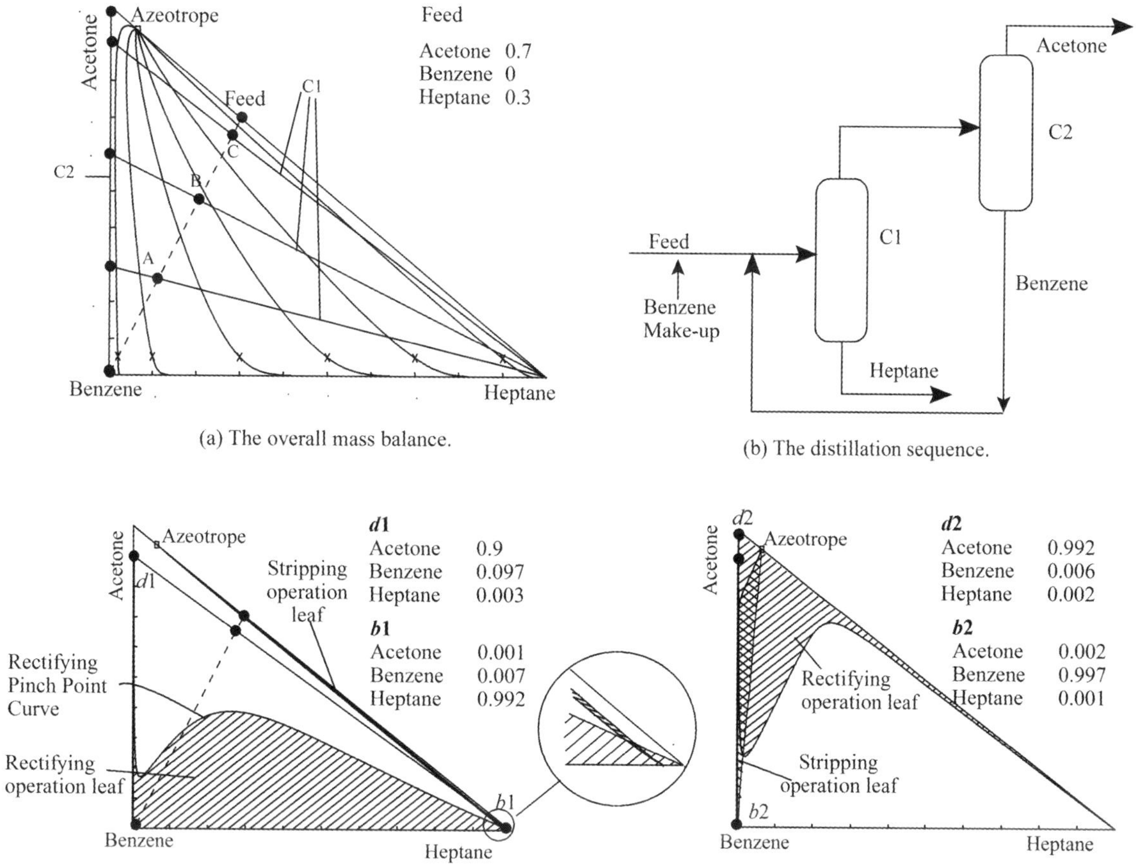

(a) The overall mass balance. (b) The distillation sequence. (c) Operation leaves for Column C1. (d) Operation leaves for Column C2.

Figure 11.21

Distillation sequence for the separation of an acetone–heptane mixture using benzene as entrainer.

stripping operation leaf. The rectifying operation leaf is bounded by the rectifying pinch point curve, as shown in Figure 11.21c, and the residue curve from d_1 that follows the acetone–benzene and benzene–heptane edges of the triangular diagram. Because the stripping and rectifying leaves for Column *C*1 intersect, in principle it is therefore a feasible column. One further point should be noted from the rectifying section leaf. From the shape of the leaf, it would be expected that the concentration of benzene in the column would be low at each end of the column and be higher in the middle of the column. This follows from the fact that benzene is an *intermediate-boiling entrainer*, with a volatility between that of acetone and heptane. A detailed simulation of Column *C*1 would show a benzene concentration through the column following these trends. Figure 11.21d shows the operation leaves for the second Column *C*2, which intersect. Thus, the distillation sequence would be expected to be feasible, and optimization of the design can be considered.

In fact, it is possible to carry out the separation of acetone and heptane using benzene as an entrainer in a different sequence to that shown in Figure 11.21 by separating the acetone in the first column as an overhead product. The heptane is separated in the second column as the bottom product with the overhead of benzene from the second column being recycled.

In Figure 11.21a, different settings were possible for the mass balance, depending on the benzene recycle. At first sight, it might be expected that the lower the benzene recycle, the better. In other words, Point *C* in Figure 11.21a will be better than Point *A*. However, it is not so straightforward. A higher concentration of benzene in Column *C*1 will help the separation. Setting *C* in Figure 11.21a requires a very high reflux ratio for Column *C*1, in turn requiring a large amount of energy for the reboiler. As the benzene recycle increases, moving towards Point *A* in Figure 11.21a, the reflux ratio for Column *C*1 decreases considerably. It would be expected that as the benzene recycle is increased, the improvements in the design of Column *C*1 would reach diminishing returns, and thereafter an excessive load on the system would be created as the recycle of benzene increases. Thus, the flowrate of benzene in the recycle is an important optimization parameter that affects the sizes of both Columns *C*1 and *C*2 in Figure 11.21.

Consider a second example involving the separation of a mixture of ethanol and water that forms an azeotrope at around a mole fraction of ethanol of 0.88. It is proposed to use ethylene glycol as the entrainer. An overall mass balance for the separation is shown in Figure 11.22a. As with the previous example in Figure 11.21, the feed is mixed with the entrainer in varying amounts to give Points *A*, *B* and *C*, and the resulting mixture is first distilled to produce pure ethanol (Figure 11.22b). The water and ethylene glycol are then separated in the second column, with the ethylene glycol being recycled. Operation leaves for the first

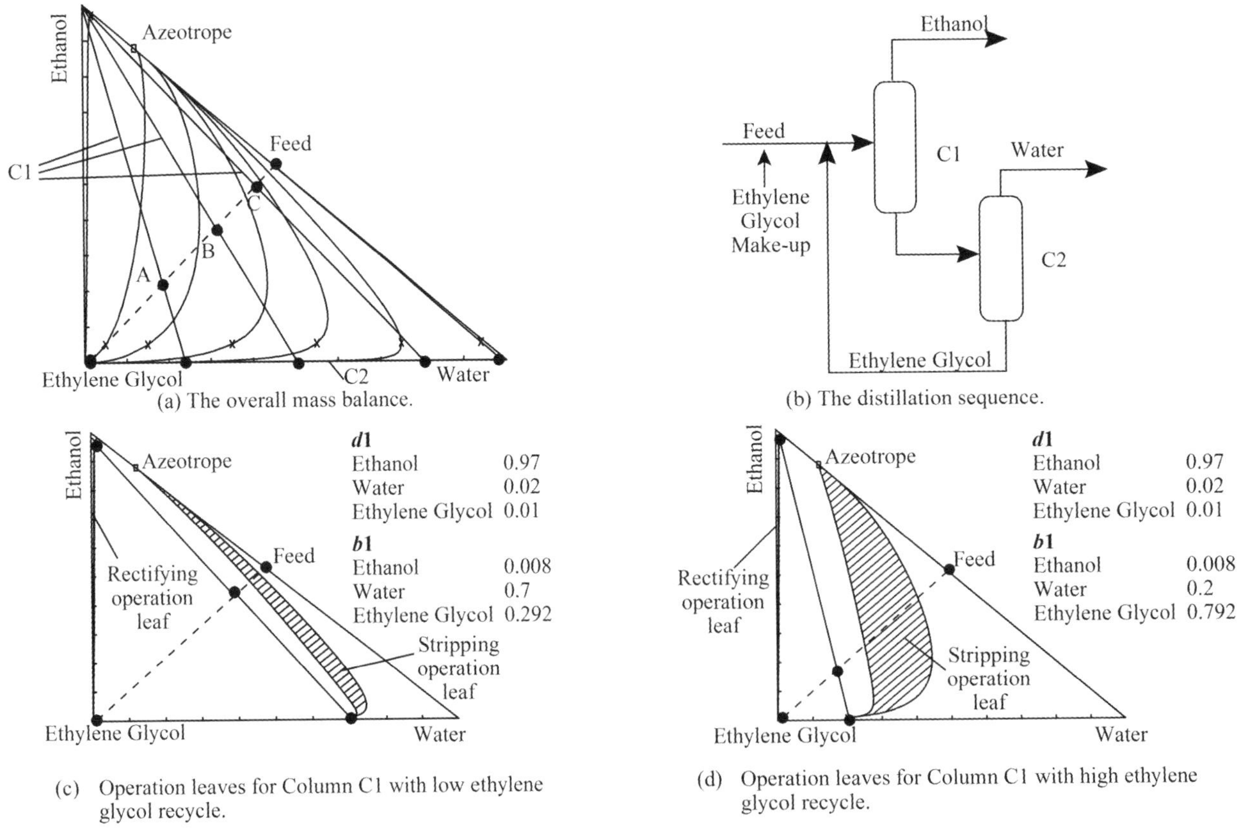

Figure 11.22

Distillation sequence for the separation of an ethanol–water mixture using ethylene glycol as entrainer is infeasible using single feed column.

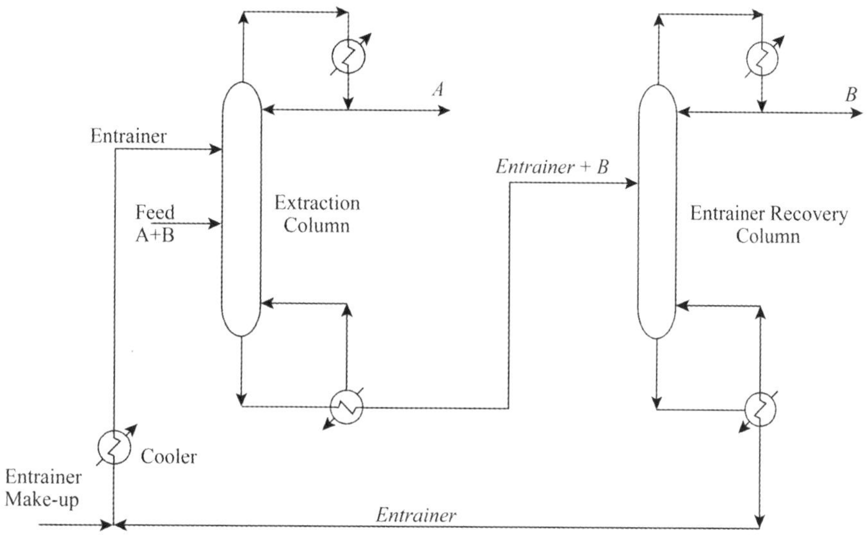

Figure 11.23

Extractive distillation flow sheet.

column can now be constructed to test the feasibility with a low ethylene glycol recycle, as shown in Figure 11.22c. The operation leaves do not overlap, and therefore the column is not feasible. Figure 11.22d also tests the feasibility of Column $C1$, but this time with a higher flowrate of ethylene glycol. Unfortunately, this is also an infeasible design, as the section leaves do not overlap.

One feature common to both designs in Figures 11.21 and 11.22 is that single-feed columns were used with the entrainer being mixed with the feed. In the case of the ethanol–water–ethylene glycol, the operation leaves for the top and bottom sections of the column do not meet and there is a gap. In some systems, it is possible to bridge the gap between the operation leaves between the top and bottom sections by creating a middle section in the column. This is achieved by using a two-feed column, as shown in Figure 11.23, in which a heavy entrainer (sometimes termed a *solvent*) is fed to the column at a point above the feed point for the feed mixture (Doherty and Malone, 2001). The entrainer must not form any new azeotropes in the system. This arrangement, shown in Figure 11.23, is known as *extractive distillation*. In the first column, the *extraction column*, a heavy entrainer is fed to the column above the feed point for the feed mixture. This entrainer flows down the column and creates a bridge between the top and bottom sections. One of the components is distilled overhead as pure A. The other component and the entrainer leave the bottom of the extraction column and are fed to the *entrainer recovery column*, which separates pure B from the entrainer, and the entrainer is recycled. The critical part of the design in Figure 11.23 is the two-feed extraction column. The design of the entrainer recovery column is straightforward, as it is a standard design. For the two-feed column, the middle section operation leaf needs to be constructed to see whether it intersects with the operation leaves for the top section (above the entrainer feed) and bottom section (below the feed point for the feed mixture).

Figure 11.24 shows the mass balance around the top of the column including the entrainer feed. A *difference point* can be defined to be the composition of the net overhead product, that is, the difference between the distillate product and entrainer feed (Wahnschafft and Westerberg, 1993). Therefore:

$$\Delta = D - E \tag{11.23}$$

Also, a stripping section mass balance gives:

$$\Delta = F - B \tag{11.24}$$

Thus, the difference point (net overhead product) is given by the intersection of a straight line joining the Feed F, Bottom Product B with a straight line joining the Distillate D and Entrainer Feed E, as shown in Figure 11.24. Pinch point curves for the middle section can now be constructed by drawing tangents to the residue curves from the difference point (net overhead product). This is shown in Figure 11.25 for the ethanol–water–ethylene glycol system. The area bounded by the pinch point curves defines the middle section operation leaf.

As long as this middle section operation leaf intersects with those for the top section (above the entrainer feed) and the bottom section (below the feed point for the feed mixture), the column design will be feasible. Note that there will always be a maximum reflux ratio, above which the separation will not be feasible because the profiles in the top and bottom sections will tend to follow residue curves, which cannot intersect. Also, the separation becomes poorer at high reflux ratios as a result of the entrainer being diluted by the reflux of lower boiling components.

The size and shape of the middle section operation leaf depends on the location of the difference point, which in turn depends on the flowrate of the entrainer. There will be a minimum flowrate of entrainer for a feasible design. Above the minimum flowrate, the actual flowrate of the entrainer is an important degree of freedom for optimization.

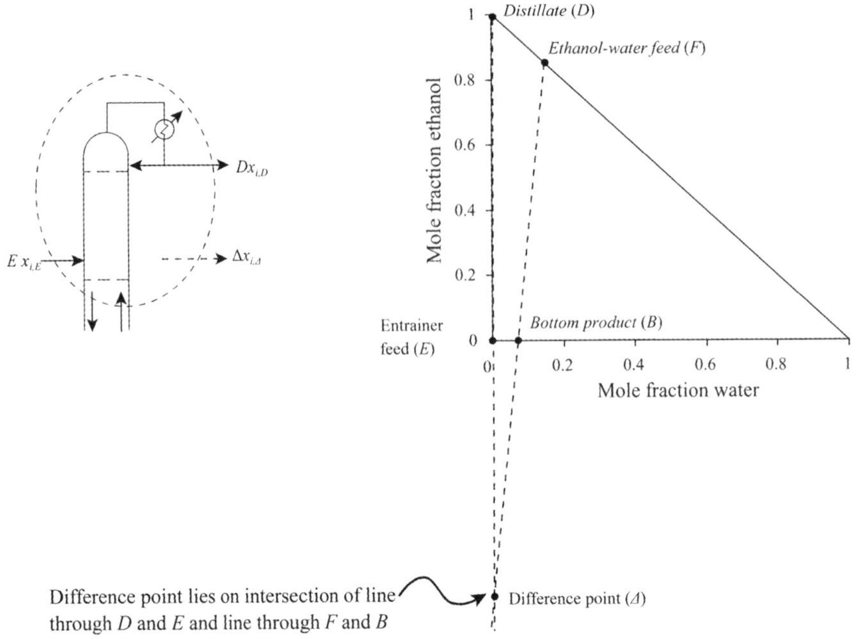

Figure 11.24
The difference point for a two-feed column.

Section profiles can also be developed for a two-feed column in a similar way to the section profiles for a single-feed column (Laroche *et al.*, 1992). The section profiles will be the same as a single-feed column above the entrainer feed (Equation 11.12) and below the feed point for the feed mixture (Equation 11.22). Figure 11.26 shows middle section mass balances. The mass balance can be created either around the top of the column, as shown in Figure 11.26a, or around the bottom of the column, as shown in Figure 11.26b. A mass balance around the top of the column from Figure 11.26a gives:

$$Vy_{i,m+1} + Ex_{i,E} = Lx_{i,m+}Dx_{i,D} \tag{11.25}$$

Since:

$$V = (R+1)D \tag{11.26}$$

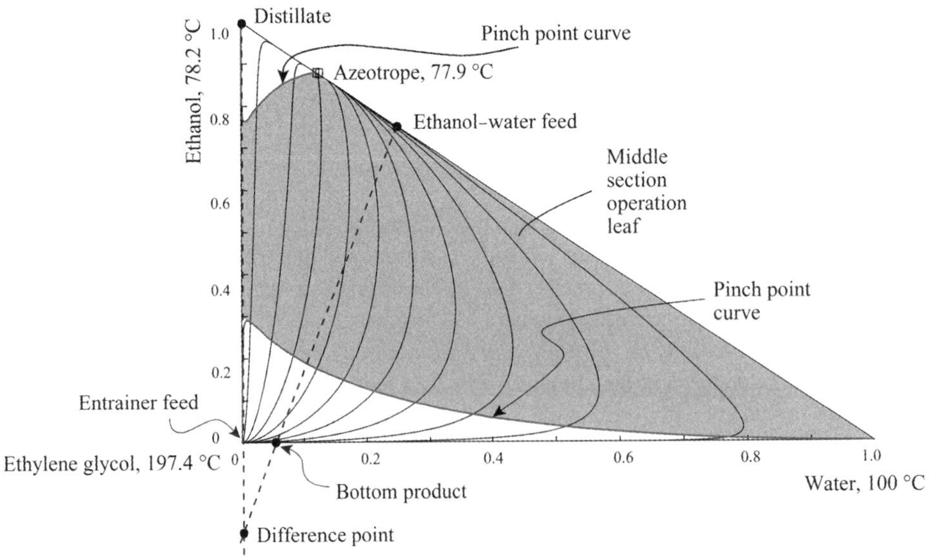

Figure 11.25
The difference point allows the pinch point curves for the middle section operation leaf to be constructed.

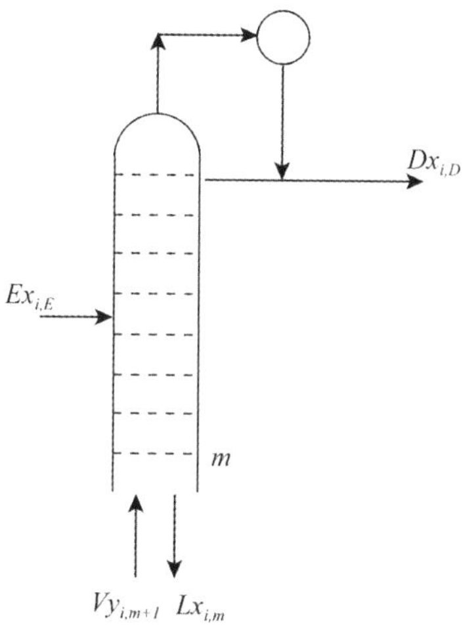

(a) Middle section mass balance around the top of the column.

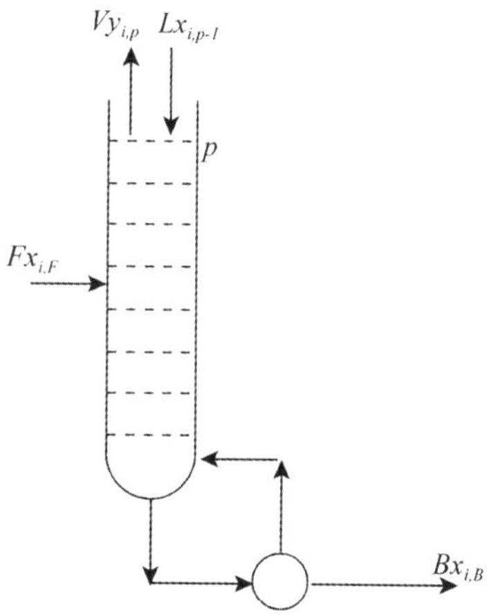

(b) Middle section mass balance around the bottom of the column.

Figure 11.26

Middle section mass balances for a two feed column.

and

$$L = RD + E \tag{11.27}$$

Equation 11.25 can be combined with Equations 11.26 and 11.27 to give:

$$y_{i,m+1} = \frac{(RD + E)x_{i,m} + Dx_{i,D} - Ex_{i,E}}{(R + 1)D} \tag{11.28}$$

Alternatively, a mass balance can be carried out around the bottom of the column from Figure 11.26b to give:

$$Lx_{,p-1} + Fx_{i,F} = Vy_{i,p} + Bx_{i,B} \tag{11.29}$$

Since

$$V = SB \tag{11.30}$$

and

$$\begin{aligned} L &= V + B - F \\ &= SB + B - F \end{aligned} \tag{11.31}$$

Equation 11.29 can be combined with Equations 11.30 and 11.31 to give:

$$x_{i,p-1} = \frac{SBy_{i,p} + Bx_{i,B} - Fx_{i,F}}{SB + B - F} \tag{11.32}$$

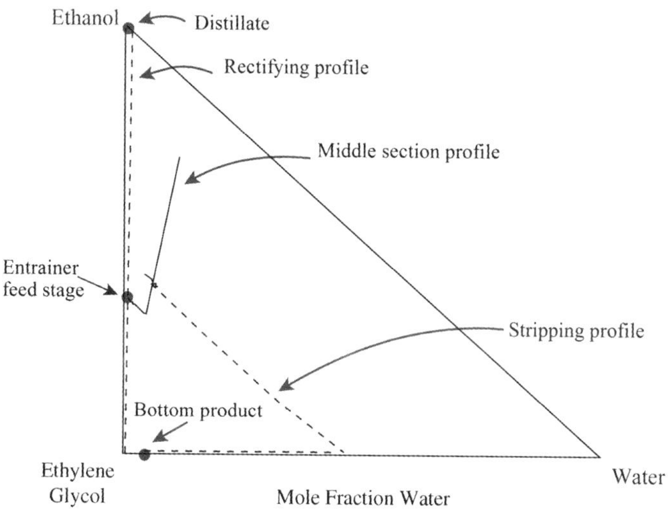

Figure 11.27

Section curves for the three sections of a two-feed column must intersect for a feasible two-feed column design.

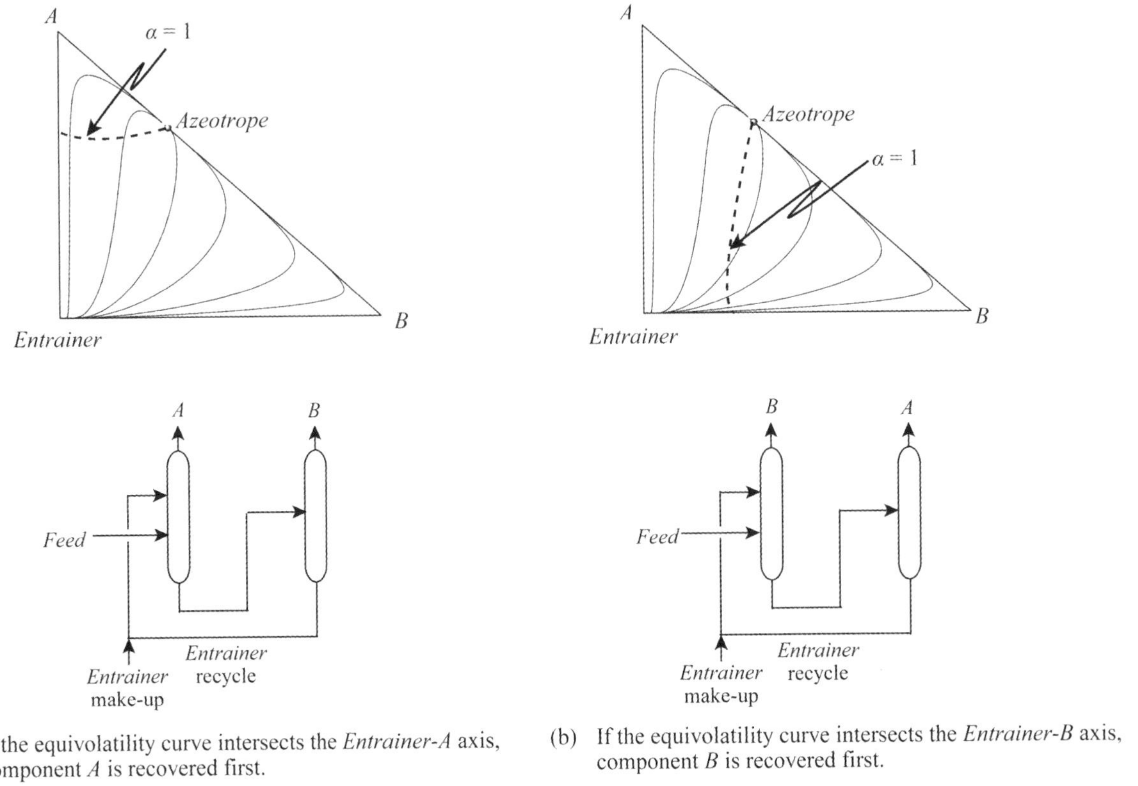
A
α = 1
Azeotrope
Entrainer
B
A
B
Feed
Entrainer
Entrainer recycle
Entrainer make-up
(a) If the equivolatility curve intersects the Entrainer-A axis, component A is recovered first.
A
Azeotrope
α = 1
Entrainer
B
B
A
Feed
Entrainer
Entrainer recycle
Entrainer make-up
(b) If the equivolatility curve intersects the Entrainer-B axis, component B is recovered first.

Figure 11.28
The order of separation in extractive distillation.

Either Equation 11.28 or Equation 11.32 can be used in conjunction with vapor–liquid equilibrium calculations to calculate the section profile for the middle section of the two-feed column.

Figure 11.27 shows the section profiles for the three sections of a two-feed column. The three profiles intersect in Figure 11.27, and the column will, in principle, be feasible.

Thus, the first step in the design of an extractive two-feed distillation column is to set up a system of intersecting operating leaves for the three sections of the column by choosing appropriate products and entrainer feed flowrate. Then by selecting a reflux ratio, the rectifying section profile can be calculated, as described previously. From the feasible entrainer flowrate, the section profile for the middle section can be calculated using Equation 11.28 to calculate the middle section profile. Calculation of the stripping section profile requires first the assumption of a reboil ratio. If three section profiles intersect, then the design is feasible. Some trial and error may be required for the reflux ratio and reboil ratio to obtain an intersection of the three profiles and feasible design. Figure 11.27 shows the section profiles for a two-feed column successfully intersecting to provide a feasible design. The number of stages in each of the column sections can also be obtained from the section profiles. This can then provide all the information required for a rigorous simulation.

One final point regarding extractive distillation is illustrated in Figure 11.28. The order in which the separation occurs depends on the change in relative volatility between the two components to be separated. Figure 11.28 shows both the residue curves and the equivolatility curve for the system *A–B–entrainer*. This equivolatility curve shows where the relative volatility between Components *A* and *B* is unity. On either side of the equivolatility curve, the order of volatility of *A* and *B* changes. In Figure 11.28a, if the equivolatility curve intersects the *A–entrainer* axis, then Component *A* should be recovered first. However, if the equivolatility curve intersects the *B–entrainer* axis, then Component *B* should be recovered first, as shown in Figure 11.28b.

All of the systems discussed so far involve homogeneous separation, that is, a single-phase liquid throughout. Consider now distillation systems involving two-liquid phases.

11.8 Heterogeneous Azeotropic Distillation

As the components in a liquid mixture become more chemically dissimilar, their mutual solubility decreases. This is characterized by an increase in their activity coefficients (for positive deviation from Raoult's Law). If the chemical dissimilarity, and the corresponding increase in activity coefficients, become large enough, the solution can separate into two-liquid phases. For liquid–liquid equilibrium, the fugacity of each component in each phase must be equal: (see Appendix A)

$$(x_i \gamma_i)^I = (x_i \gamma_i)^{II} \tag{11.33}$$

where x_i = mole fraction of Component i in the liquid phase
γ_i = liquid-phase activity coefficient

I and II represent the two-liquid phases in equilibrium.

To calculate the compositions of the two coexisting liquid phases for a binary system, the two equations for phase equilibrium need to be solved:

$$(x_1\gamma_1)^I = (x_1\gamma_1)^{II} \quad \text{and} \quad (x_2\gamma_2)^I = (x_2\gamma_2)^{II} \qquad (11.34)$$

where

$$x_1^I + x_2^I = 1, x_1^{II} + x_2^{II} = 1 \qquad (11.35)$$

Given a prediction of the liquid-phase activity coefficients, from say the NRTL or UNIQUAC equations (see Appendix A), then Equations 11.34 and 11.35 can be solved simultaneously for x_1^I and x_1^{II}. There are a number of solutions to these equations, including a trivial solution corresponding with $x_1^I = x_1^{II}$. For a solution to be meaningful:

$$0 < x_1^I < 1, \qquad 0 < x_1^{II} < 1, \qquad x_1^I \neq x_1^{II} \qquad (11.36)$$

For a ternary system, the corresponding equations to be solved are:

$$(x_1\gamma_1)^I = (x_1\gamma_1)^{II} \text{ and } (x_2\gamma_2)^I = (x_2\gamma_2)^{II} \text{ and } (x_3\gamma_3)^I = (x_3\gamma_3)^{II} \qquad (11.37)$$

These equations can be solved simultaneously with the material balance equations to obtain x_1^I, x_2^I, x_1^{II} and x_2^{II}. Further details of the calculation of liquid–liquid equilibrium and vapor–liquid–liquid equilibrium can be found in Appendix A.

Figure 11.29 shows a triangular diagram for a ternary system that forms two-liquid phases at certain compositions. From Figure 11.29, a binary I–L system is completely miscible, and a binary I–H system is also completely miscible. However, a binary L–H system shows partial miscibility in a range of concentrations. For L–I–H mixtures, there is a region of immiscibility in which two-liquid phases form. A mixture of composition F in the two-phase region in Figure 11.29 will separate into two phases of composition E and R. Phases in equilibrium with each other are termed *conjugate phases*, and the Line E–R connecting the conjugate phases E and R is termed a *tie line*. Any number of tie lines may be constructed in the two-phase region, as shown in Figure 11.29. The ratio of the two phases resulting in the separation of F into E and R is given by the Lever Rule. Thus, the flowrate of E is given by the length of Line F–R in Figure 11.29 and the flowrate of R is given by the length of the Line F–E in Figure 11.29. Figure 11.29 illustrates how to design decanters using a ternary diagram.

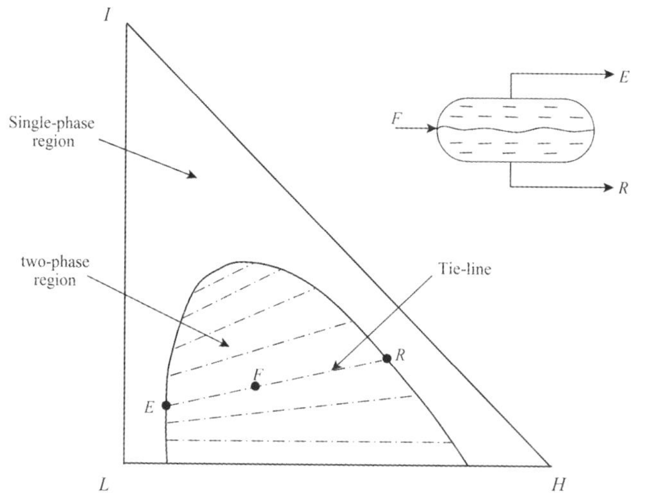

Figure 11.29

Phase splitting in a ternary system.

Now consider the distillation with the decanter. Different arrangements are possible when using a decanter in conjunction with distillation. The first is shown in Figure 11.30 shows a stand-alone decanter where the overhead from the column is condensed to form a two-liquid phase mixture. Part is refluxed and part goes to a decanter that separates two layers E and R. One problem with the arrangement shown in Figure 11.30 is that the reflux to the column at D is a two-phase mixture. It is preferable not to have two-liquid phases inside the column unless this cannot be avoided. Another arrangement is shown in Figure 11.31. This time the decanter is partially integrated with the column and the overhead vapor is condensed to form two-liquid phases D and R. D is taken as a top product and R (a single-phase lower layer) is refluxed to the column. The overall separation is between F, D and B to give the column mass balance as shown in Figure 11.31. Figure 11.32 shows a fully integrated decanter in which the overhead vapor is condensed and fed to the decanter. The decanter separates into a light decanter liquid DL and a heavy decanter liquid DH. This time part of the heavy and light decanter liquids are partially mixed to provide both the reflux and overhead product. The extent of mixing after the decanter for both the reflux and distillate are degrees of freedom for the design. In the triangular diagram the extent of mixing can be reflected by the Lever Rule. 11

One extremely powerful feature of heterogeneous distillation is the ability to cross distillation boundaries. It was noted previously that distillation boundaries divide the compositions into two regions that cannot be accessed from each other. Decanters allow distillation boundaries to be crossed, as illustrated in Figure 11.33. The feed to the decanter at F is on one side of the distillation boundary. This splits in the decanter to two-liquid phases E and R. These two-liquid phases are now on opposite sides of the distillation boundary. Phase splitting in this way is not constrained by a distillation boundary, and exploiting a two-phase separation in this way is an extremely effective way to cross distillation boundaries.

An example is shown in Figure 11.34 in which a feed containing a mole fraction of isopropyl alcohol 0.6 and a mole fraction of water 0.4 needs to be split into relatively pure isopropyl alcohol and water. There is an azeotrope between isopropyl alcohol (IPA) and water with a mole fraction of isopropyl alcohol around 0.68. It is proposed to use di-isopropyl ether (DIPE) as an entrainer to separate the mixture. The ternary diagram in Figure 11.34a shows that the system IPA–DIPE–water forms a complex behavior. New azeotropes are formed between the IPA and DIPE and between DIPE and water. A ternary azeotrope is also formed. The distillation boundaries shown in Figure 11.34a would make this

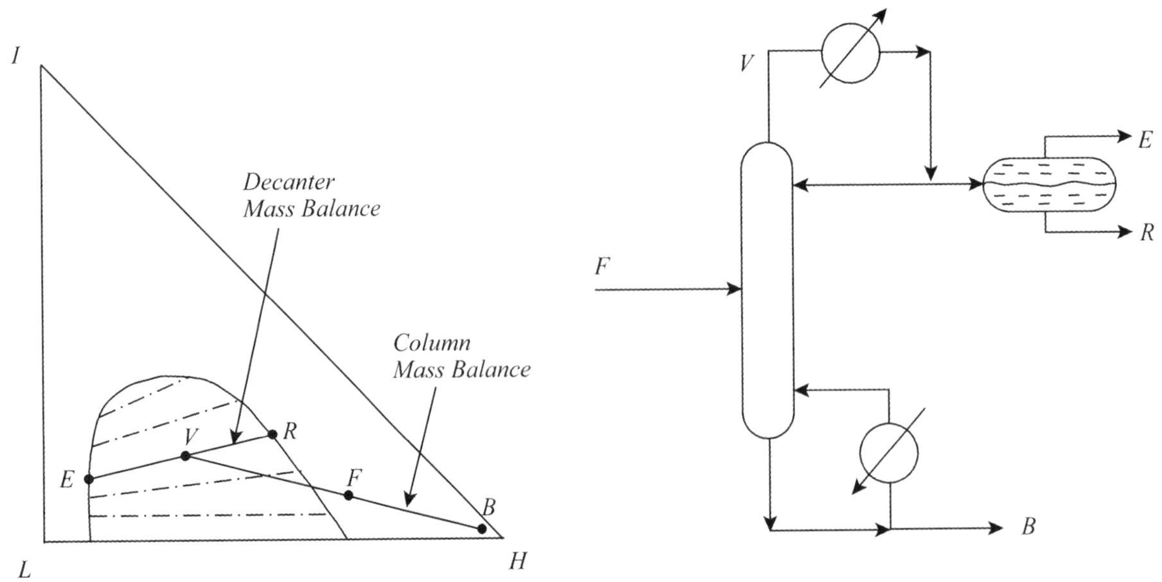

Figure 11.30
Heterogeneous distillation with a stand-alone decanter.

separation extremely difficult. However, the two-phase region shown in Figure 11.34a allows the distillation boundaries to be crossed. The synthesis of the separation system can be started by placing a decanter at the ternary azeotrope, as shown in Figure 11.34a. This follows the tie line and separates the ternary azeotrope into a DIPE-rich layer and a water-rich layer. Next, a distillation column is placed to produce IPA and the ternary azeotrope, as shown in Figure 11.34b. Finally, the DIPE-rich layer is recycled from the decanter to the feed for the column. This mixes with the incoming feed mixture to provide the feed for the column, as shown in Figure 11.34c. The water product from the decanter in Figure 11.34c is not pure and may require further separation. This judgment is based on the prediction of the design of the decanter from Figure 11.34.

The phase equilibrium in the triangular diagrams in Figure 11.34 was predicted using the NRTL equation (see Appendix A). The NRTL equation is capable of predicting vapor–liquid equilibrium, liquid–liquid equilibrium and vapor–liquid–liquid equilibrium. However, it is difficult to find a single set of interaction parameters to represent all of these kinds of behavior well. The parameters for the triangular diagrams in Figure 11.34 were correlated from vapor–liquid equilibrium behavior. Figure 11.35a shows the two-phase region again based on NRTL parameters correlated from vapor–liquid data. Figure 11.35b shows the

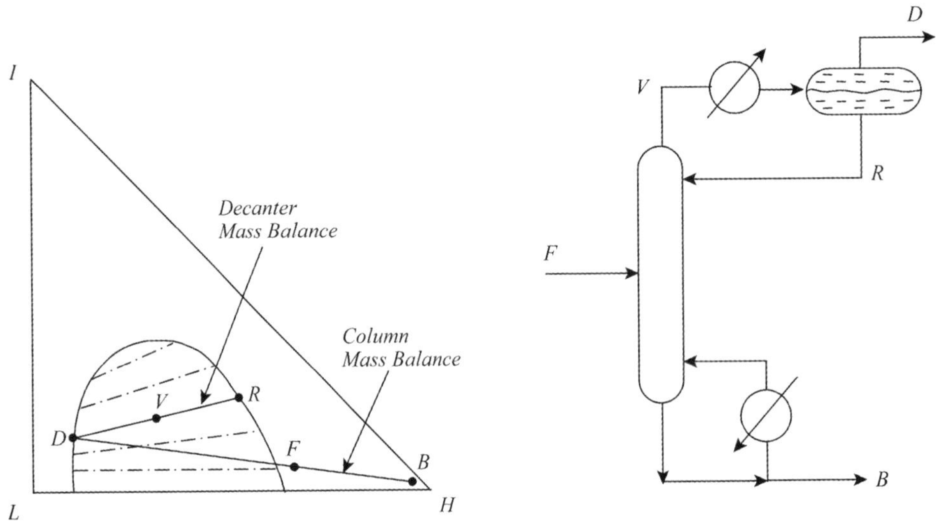

Figure 11.31
Heterogeneous distillation with a partially integrated decanter.

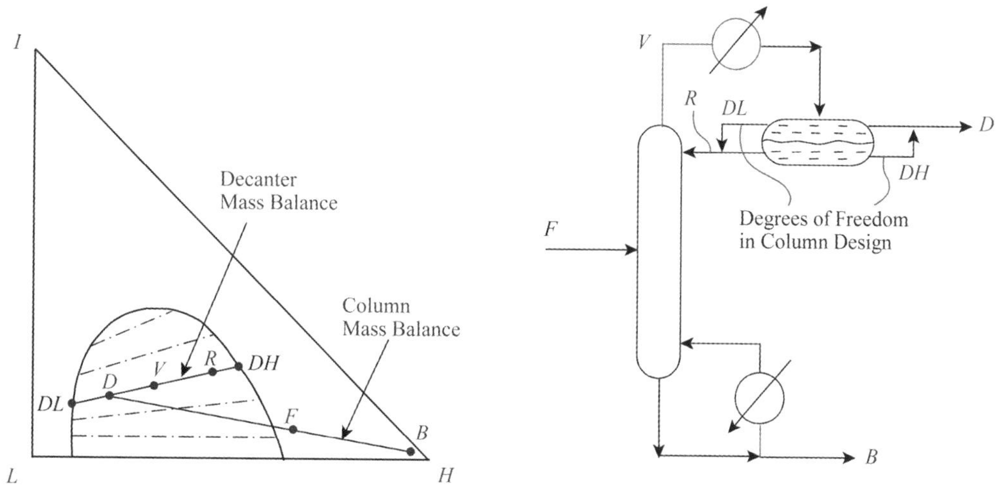

Figure 11.32
Heterogeneous distillation with an integrated decanter.

triangular diagram for the same system, but with the two-phase region calculated from the NRTL equation based on parameters correlated from liquid–liquid equilibrium data. The two-phase region is much larger, and the decanter is capable of producing almost pure water from this prediction.

Even further caution should be exhibited regarding the two-phase region in such diagrams. In Figures 11.34, 11.35a and 11.35b, the phase equilibrium is based on saturated conditions throughout. This is useful in judging the design of the distillation system and where two-liquid phase behavior will occur, as distillation by definition operates under saturated conditions. However, this is not necessarily the case in the decanter. The temperature in the decanter can be fixed, as it is outside of the column. Figure 11.35c shows the two-phase region again on the basis of NRTL parameters correlated from liquid–liquid equilibrium data, but this time plotted at a fixed temperature of 30 °C. The two-phase region is slightly larger at 30 °C when compared with saturated conditions. Generally, the lower the temperature, the larger will be the two-phase region. Lowering the temperature lowers the mutual solubility of the two-liquid phases. This is an important degree of freedom when designing a decanter. A better separation in the decanter can be obtained by condensing the distillation overheads and subcooling before the two-liquid phase separation.

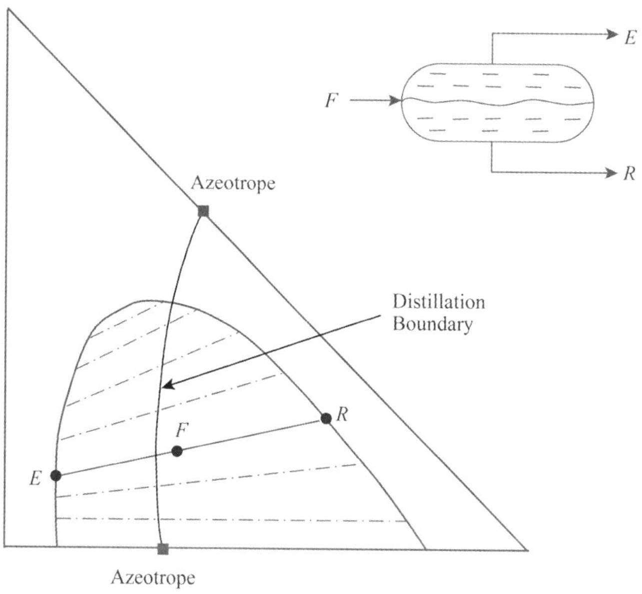

Figure 11.33
Phase splitting can be used to cross distillation boundaries.

It is important to understand whether there will be two-liquid phases present in the column. If two-liquid phases form in a large part of the column, it can make the column difficult to operate. The formation of two-liquid phases also affects the hydraulic design and mass transfer in the distillation (and hence stage efficiency). If it is possible to avoid the formation of two-liquid phases inside the column, then such behavior should be avoided. Unfortunately, there will be many instances when two-liquid phases on some plates cannot be avoided. The formation of two-liquid phases can also be sensitive to changes in the reflux ratio.

11

11.9 Entrainer Selection

When separating azeotropic mixtures, if possible, changes in the azeotropic composition with pressure should be exploited rather than using an extraneous mass-separating agent, since:

- The introduction of extraneous material can create new problems in achieving product purity specifications throughout the process.
- It is often difficult to separate and recycle extraneous material with high efficiency. Any material not recycled can become

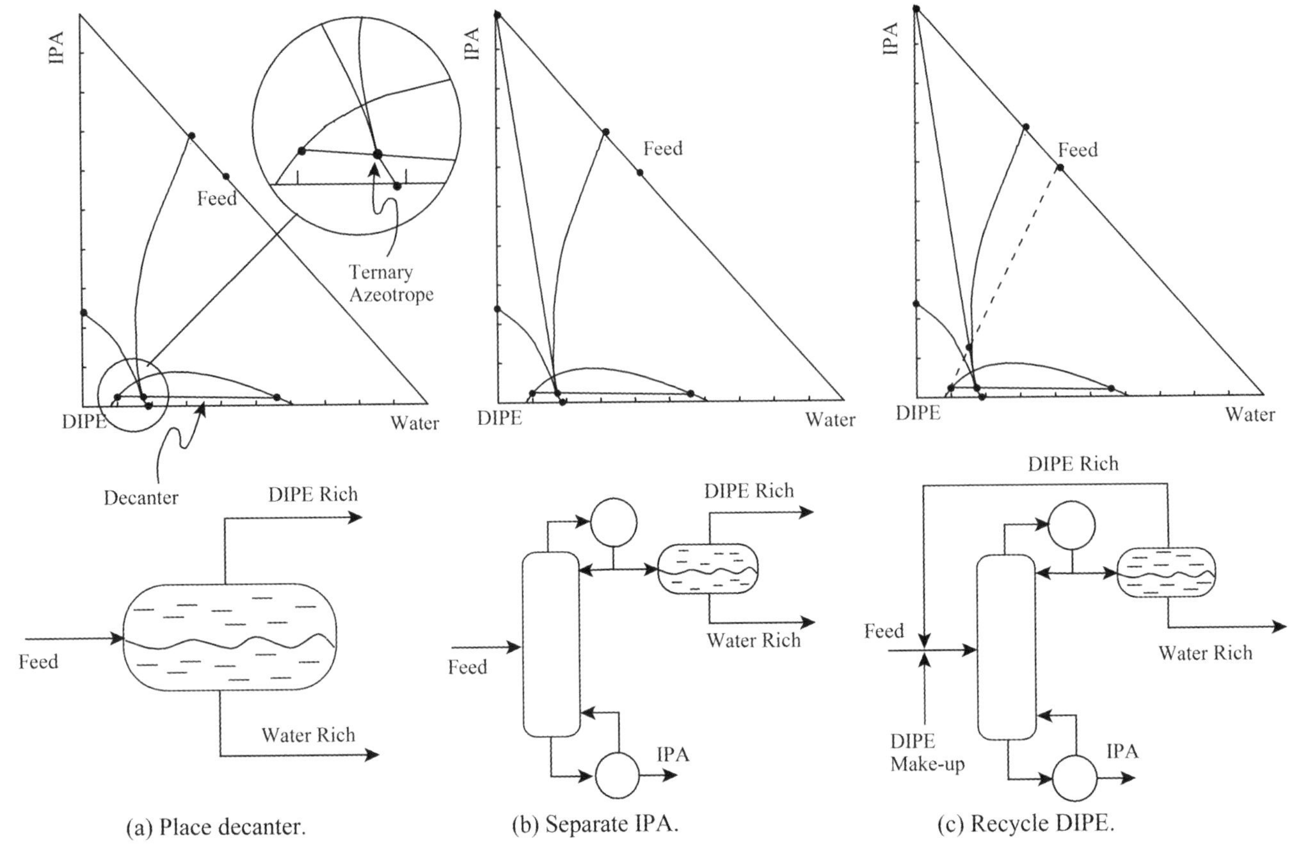

Figure 11.34

Separation of isopropyl alcohol (IPA) and water mixture using diisopropyl ether (DIPE) as entrainer in heterogeneous azeotropic distillation.

an environmental problem. As will be discussed later, the best way of dealing with effluent problems is not to create them in the first place.

- Extraneous material can create additional safety and storage problems.

Occasionally, a component that already exists in the process can be used as the entrainer, thus avoiding the introduction of extraneous materials for azeotropic distillation. However, in many instances practical difficulties and excessive cost might force the use of extraneous material. Whether a component is used

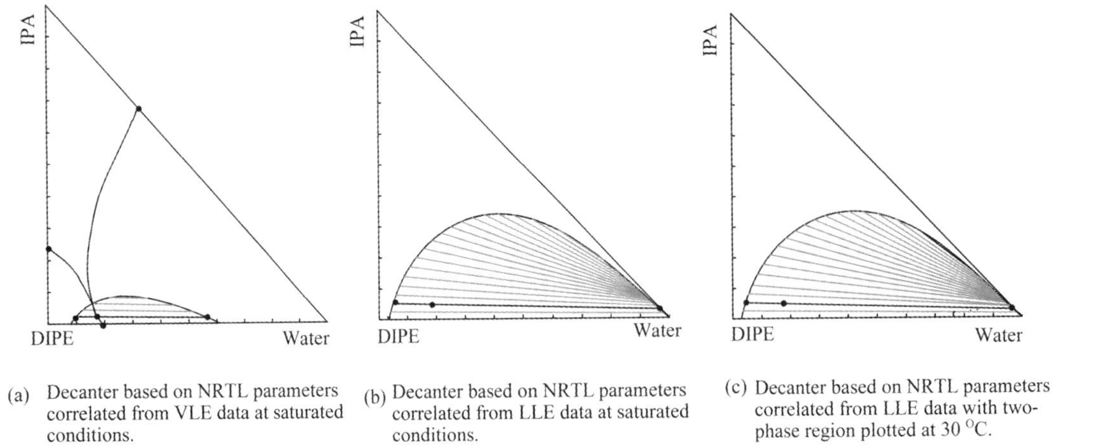

Figure 11.35

The two-phase region can be plotted from correlations fitted to VLE or LLE data, and can be plotted at saturated conditions at a fixed temperature.

that already exists within the process as an entrainer or an extraneous material is used as entrainer, it is necessary to be able to select between different entrainers. Distillation line maps and residue curve maps are particularly useful for entrainer selection, as the difficulty of separation can be judged from the configuration of the map. For example, it has been noted how distillation boundaries divide distillation line and residue curve maps into different regions and that distillation cannot access one region from another. As discussed above, for heterogeneous systems, distillation boundaries can be crossed using liquid–liquid separation. Another way to cross boundaries in a sequence is to mix a stream on one side of the boundary with another on the opposite side of the boundary. Depending on the ratio of the flowrates of the streams being mixed, the resulting stream after the mixer can be on one side of the boundary or the other.

In theory, it is possible to cross a distillation boundary for homogeneous systems, as illustrated in Figure 11.36 (Laroche *et al.*, 1992). The distillation boundary in Figure 11.36 has a marked curvature, and a column can be placed as shown such that the feed is on one side of the boundary and the products are on the other side of the boundary. Depending on the shape of the distillation lines or residue curves, the products *D* and *B* can be on the same distillation line or residue curve. In this way, a distillation boundary can be crossed using distillation, rather than relying on a liquid–liquid separation to cross distillation boundaries. Whilst arrangements like those in Figure 11.36 are possible in theory, there are many potential dangers associated with this, as follows (Laroche *et al.*, 1992).

1) The distillation boundary must be curved as shown in Figure 11.36. However, even if there is very significant curvature across the boundary, a column placed like the one in Figure 11.36 is going to be highly constrained in its operation.
2) There is always uncertainty and inaccuracy with vapor–liquid equilibrium data and correlations. Any errors in these data could mean an incorrect prediction of the location and shape of the boundary.

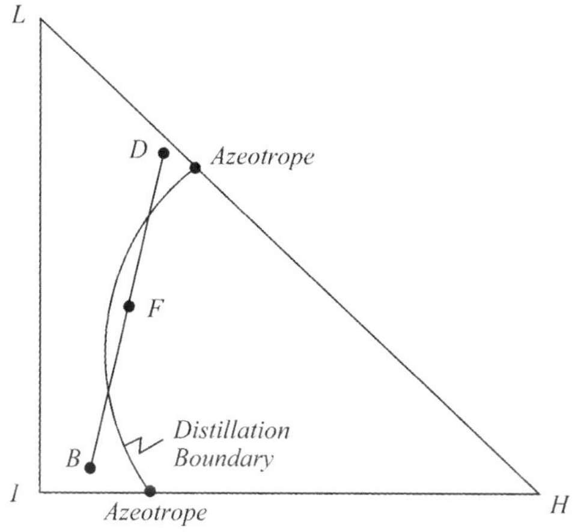

Figure 11.36
Crossing distillation boundaries.

3) All of the discussions so far regarding distillation lines, residue curves and distillation boundaries have assumed equilibrium behavior. Real columns do not work at equilibrium, and stage efficiency must be accounted for. Each component will have its own stage efficiency, which means that each composition will deviate from equilibrium behavior differently. This means that if nonequilibrium behavior is taken into account, the shape of the distillation lines, residue curves and distillation boundaries will change (Castillo and Towler, 1998). Thus, the shape of the distillation boundary will be different in a real column when compared with equilibrium predictions. If a system is designed assuming equilibrium, there is no guarantee that it will still work in a real column with nonideal stages. These nonequilibrium effects can, in principle, be included in the analysis, but there is also considerable uncertainty regarding stage efficiency calculations and the calculations require a considerable amount of information on the geometry of the column and distillation internals (Castillo and Towler, 1998).
4) Even with reassurance regarding uncertainties in the vapor–liquid equilibrium data and nonequilibrium effects, there is often a significant difference between the way a column is required to operate in practice compared with its design. The operation of the overall plant can often be different from the design because of a whole range of reasons. If the design is highly constrained and cannot be flexible to accommodate changes in operation, then it might not be able to function.

Thus, while it is possible in theory to cross a curved distillation boundary as shown in Figure 11.36, it is generally more straightforward to follow designs that will be feasible over a wide range of reflux ratios and in the presence of uncertainties. Such designs can be readily developed using distillation line and residue curve maps.

When introducing an entrainer, it will need to have a significant effect on the relative volatility between the azeotropic components to be separated, and it must be possible to separate the entrainer relatively easily. One way of making sure the entrainer can be easily separated is to choose a component that will introduce a two-liquid phase separation. Such entrainers typically introduce additional distillation boundaries, but the overall separation can be efficient if the two-liquid separation produces mixtures with compositions in the different distillation regions (Doherty and Perkins, 1979b).

When using an entrainer for separation of a homogeneous mixture, it is best to select components that do not introduce any additional azeotropes. The classical method for the separation of homogeneous mixtures separation is extractive distillation, which relies on the effect of a high-boiling entrainer on the relative volatility in the column sections below the entrainer feed. Such columns can work well, but they sometimes exhibit counter-intuitive behavior, in particular, with regard to the detrimental effect of high reflux diluting the entrainer composition. However, in most cases high-boiling entrainers will be the best choice for homogeneous distillation. Another possibility can be to choose an intermediate-boiling entrainer that does not introduce azeotropes, since this will lead to a residue curve map with no distillation boundaries. However, intermediate-boiling entrainers can only be practical for breaking azeotropes of components with large boiling point differences; otherwise an intermediate-boiling entrainer will lead to very difficult, energy-consuming separations of close-boiling mixtures. Finally, using low-boiling entrainers that do not introduce azeotropes is generally not practical, since such components would not tend to accumulate sufficiently in the liquid phase to make a difference on the relative volatility between the components to be separated.

Thus, distillation line and residue curve maps are excellent tools to evaluate feasibility of azeotropic separations, with just one exception, namely, the use of high-boiling entrainers for separation. In such cases, the equivolatility curves discussed in this chapter are a better way of determining separation feasibility.

11.10 Multicomponent Systems

All of the discussion in this chapter has so far related to binary or ternary systems. It will most often be the case that systems involving azeotropic behavior will also be multicomponent. The concepts developed here for ternary systems are readily extended to quaternary systems. The difference is that this cannot be represented on a ternary diagram but must be represented in three dimensions as a pyramid. Lines in the ternary diagram become surfaces in the quaternary diagram. It simply becomes more difficult to represent graphically and to interpret. The concepts can be extended beyond quaternary systems but cannot be represented graphically at all, unless three or four components are picked out and represented to the exclusion of the other components.

When dealing with multicomponent systems, one possible approach is to simplify the problem by lumping components together and representing it in a ternary analysis. Such an approach should be exercised with great caution. Even trace amounts of components can change the analysis very significantly in azeotropic systems. For example, the designer might consider that the system has been represented reasonably well if 99% of the mixture can be accounted for as a ternary mixture and the influence of the other 1% neglected. However, in some systems, varying the makeup of the 1% can significantly change the shape of operation leaves, and their equivalent in multidimensional space, and render a design infeasible that appears to be feasible on the basis of an analysis of the behavior of 99% of the mixture (Thong and Jobson, 2001). Thus, the residue curves, distillation lines, pinch point curves and operation leaves for multicomponent mixtures should be constructed from a full multicomponent calculation, even if the dominant components are to be represented on a ternary diagram.

Once a design has been synthesized, it should be checked carefully with the most detailed simulation possible. Even if the design is confirmed to be feasible by this simulation, the sensitivity of the design should be checked carefully by simulation for:

- errors in the phase equilibrium behavior by perturbing the phase equilibrium data;
- changes in the feed composition.

11.11 Trade-Offs in Azeotropic Distillation

For a simple binary distillation column, once the feed composition has been fixed, only two product component compositions can be specified independently, one in each product. For a simple distillation column separating a ternary system, once the feed composition has been fixed, three product component compositions can be specified, with at least one component composition for each product. The remaining compositions will be determined by colinearity in the ternary diagram. Once the mass balance has been specified, the column pressure, reflux (or reboil ratio) and feed condition must also be specified.

By contrast with nonazeotropic systems, for azeotropic systems there is a maximum reflux ratio above which the separation deteriorates (Laroche *et al.*, 1992). This is because an increase in the reflux ratio results in two competing effects. First, as in nonazeotropic distillation, the relative position of the operating surface relative to the equilibrium surface changes to improve the separation. This is countered by a reduction in the entrainer concentration, owing to dilution by the increased reflux, which results in a reduction in the relative volatility between the azeotropic components, leading to a poorer separation (Laroche *et al.*, 1992).

However, this so far assumes that the feed to the column is fixed. Even if the overall feed to the separation system is fixed, the feed to each column can be changed by changing the amount of entrainer recycled. Such a trade-off has already been seen in Figure 11.21. As the amount of entrainer recycled is increased, this helps the azeotropic separation. This allows the reflux ratio to be decreased. However, as the entrainer recycle increases, it creates an excessive load on the overall system. The amount of entrainer recycled is therefore an important degree of freedom to be optimized.

11.12 Membrane Separation

So far, the separation of azeotropic systems has been restricted to the use of pressure shift and the use of entrainers. The third method is to use a membrane to alter the vapor–liquid equilibrium behavior. *Pervaporation* differs from other membrane processes in that the phase state on one side of the membrane is different from the other side. The feed to the membrane is a liquid mixture at a high-enough pressure to maintain it in the liquid phase. The other side of the membrane is maintained at a pressure at or below the dew point of the permeate, maintaining it in the vapor phase. Dense membranes are used for pervaporation and selectivity results from chemical affinity. Most pervaporation membranes in commercial use are hydrophyllic (Wynn, 2001). This means that they preferentially allow water to permeate and are therefore suitable for the dehydration of organics. Typical applications include the dehydration of ethanol–water and isopropanol–water mixtures, both of which form azeotropes (Wynn, 2001). A flowsheet for ethanol

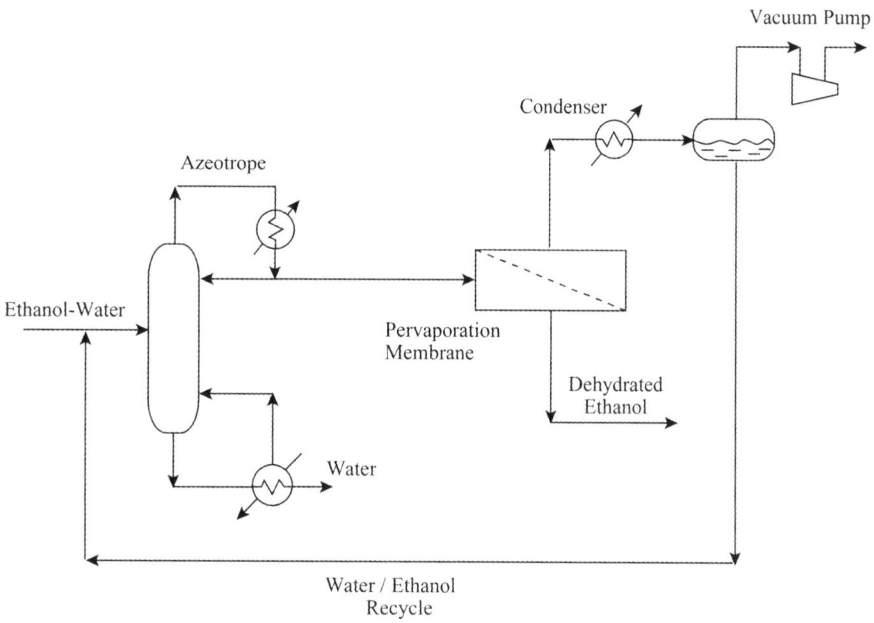

Figure 11.37

Flowsheet for dehydration of ethanol using pervaporation.

dehydration is shown in Figure 11.37. An ethanol–water mixture is fed to a standard distillation column that separates excess water at the bottom from a mixture in the overhead approaching the azeotropic composition. This is then fed to a pervaporation membrane that dehydrates the ethanol past the azeotrope by allowing water to permeate through the membrane. The low-pressure side of the membrane in Figure 11.37 is maintained under vacuum to ensure that the water leaves in the vapor phase. This is condensed and recycled to the distillation column as there is still a significant amount of ethanol that permeates through the membrane and should be recovered.

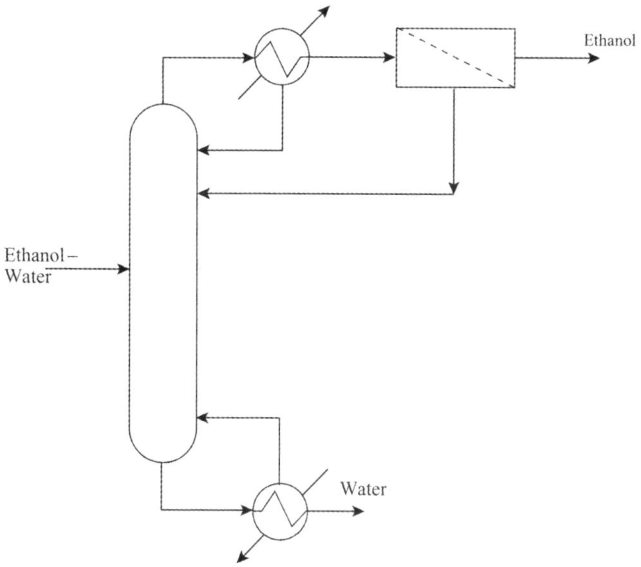

Figure 11.38

Flowsheet for dehydration of ethanol using vapor permeation.

Figure 11.38 shows a flowsheet for the separation of an azeotropic mixture using a membrane, but this time using vapor permeation. The mixture is first distilled to approach the azeotrope using a distillation column with a partial condenser. The uncondensed vapor is fed to a vapor permeation membrane that preferentially permeates the organic material. The retentate vapor is passed back to the distillation column. In this way, the membrane allows the azeotrope to be crossed.

11

11.13 Distillation Sequencing for Azeotropic Distillation – Summary

When liquid mixtures exhibit azeotropic behavior, it presents special challenges for distillation sequencing. At the azeotropic composition, the vapor and liquid are both at the same composition for the mixture. The order of volatility of components changes, depending on which side of the azeotrope the composition occurs. There are three ways of overcoming the constraints imposed by an azeotrope.

- Pressure shift
- Use of an entrainer
- Membrane separation.

Pressure shift should always be explored as the first option when separating an azeotropic system. Adding extraneous components to a separation should always be avoided if possible. Unfortunately, most azeotropes are insensitive to change in pressure, and at least a 5% change in composition with pressure is required for a feasible separation using pressure shift (Holland, Gallun and Lockett, 1981).

If pressure shift cannot be exploited, then the next option is to add an entrainer to the mixture that interacts differently with the components in the mixture to alter the vapor–liquid equilibrium behavior in a favorable way. When dealing with ternary systems, the mass balance and vapor–liquid equilibrium behavior can be represented on a triangular diagram. The two limiting cases of distillation at total reflux and minimum reflux conditions can be used to understand the system. For staged columns at total reflux conditions, distillation lines can be plotted. Residue curves represent the behavior of packed columns at total reflux. The feasibility of a column section can be represented by the area between the total reflux and pinch point lines, as an operating leaf. If the operating leaves for the rectifying and stripping sections of a column intersect, in principle the column will be feasible.

Some systems form two-liquid phases for certain compositions and this can be exploited in heterogeneous azeotropic distillation. The use of liquid–liquid separation in a decanter can be extremely effective and can be used to cross distillation boundaries.

When selecting entrainers for homogeneous mixture separation, the entrainer should preferably not introduce any new azeotropes; otherwise it will be difficult to separate the entrainer from the components to be separated. When separating multicomponent mixtures, the first thing to check generally is if there are components in the feed that can facilitate the separation of azeotrope-forming components, because using such components will typically lead to more cost-effective designs than processes in which the azeotropic separations are left to the end of a sequence and extraneous separating agents are chosen. Using components that are not already in the feed will generally require dedicated additional recovery steps.

Membranes can also be used to alter the vapor–liquid equilibrium behavior and allow separation of azeotropes. The liquid mixture is fed to one side of the membrane and the permeate is held under conditions to maintain it in the vapor phase. Most separations use hydrophyllic membranes that preferentially pass water rather than organic material. Thus, pervaporation is commonly used for the dehydration of organic components.

11.14 Exercises

1. An equimolar mixture of ethanol and ethyl acetate is to be separated by distillation into relatively pure products. The mixture forms a minimum-boiling azeotrope, as detailed in Table 11.1. However, the composition of the azeotrope is sensitive to pressure, showing a significant increase in the mole fraction of ethanol with increasing pressure, as indicated in Table 11.1. Sketch a flow scheme for the separation of the binary mixture that exploits a change in pressure.

2. An equimolar mixture of methanol and ethyl acetate is to be separated by distillation into relatively pure products. As with the ethanol–ethyl acetate system, the mixture forms a minimum-boiling azeotrope, as detailed in Table 11.2. Again, the composition of the azeotrope is sensitive to pressure, showing a significant increase in mole fraction of methanol with increasing pressure, as indicated in Table 11.1. Sketch a flow scheme for the separation of the binary mixture that exploits change in pressure.

3. Figure 11.39 shows a distillation sequence, together with its mass balance. Sketch a representation of the mass balance in a triangular diagram.

4. From the data in Tables 11.1 and 11.2, sketch the distillation line map (residue curve map) for the ethanol–ethyl acetate–

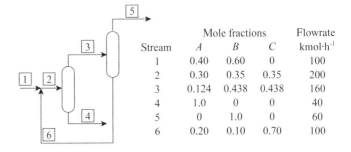

Table 11.1

Data for the ethanol–ethyl acetate system.

Component	Boiling temperature (°C) and azeotrope composition at 1 atm	Boiling temperature (°C) and azeotrope composition at 5 atm
Ethanol	78.2	125.6
Ethyl acetate	77.1	135.8
Ethanol–ethyl acetate azeotrope	72.2 °C, 0.465 mole fraction of ethanol	122.7 °C, 0.677 mole fraction of ethanol

Table 11.2

Data for the methanol-ethyl acetate system.

Component	Boiling temperature (°C) and azeotrope composition at 1 atm	Boiling temperature (°C) and azeotrope composition at 5 atm
Methanol	64.5	111.8
Ethyl acetate	77.1	135.8
Methanol–ethyl acetate azeotrope	62.3 °C, 0.709 mole fraction of methanol	110.6 °C, 0.82 mole fraction of methanol

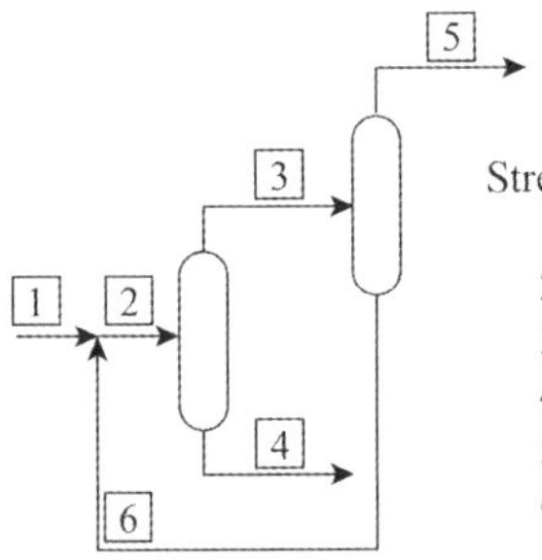

Stream	Mole fractions A	B	C	Flowrate kmol·h^{-1}
1	0.40	0.60	0	100
2	0.30	0.35	0.35	200
3	0.124	0.438	0.438	160
4	1.0	0	0	40
5	0	1.0	0	60
6	0.20	0.10	0.70	100

Figure 11.39

Sequence representation in a triangular diagram.

methanol system at 1 atm and 5 atm. Does the system have a distillation boundary? Is the position of the boundary sensitive to pressure?

5. A ternary mixture of the mole fraction of ethanol of 0.15, ethyl acetate of 0.6 and methanol of 0.25 is to be separated into relatively pure products. A first distillation split can be made to separate the ternary mixture into two products with ethyl acetate and methanol in the overhead product and ethyl acetate and ethanol in the bottom product. These two binary mixtures can then be separated using the flowsheets from Exercises 1 and 2. Sketch a system of distillation columns and mixer arrangements in the triangular diagram to carry out the separation by exploiting the shift in the distillation boundary with pressure. Sketch the flowsheet corresponding with this mass balance.

6. The vapor–liquid equilibrium for a ternary system of Components A, B and C can be represented by:

$$\begin{aligned} y_A &= 0.2x_A \\ y_B &= 2.0x_B \\ y_C &= 1 - y_A - y_C \end{aligned}$$

Starting from a bottoms composition $x_A = 0.95$, $x_B = 0.04$ and $x_C = 0.01$, calculate the stripping section profile for the reboiler and bottom five stages of the column for a reboil ratio of 1. Sketch the profile on a ternary diagram.

7. The separation in Figure 11.21 can be carried out in an alternative sequence by mixing the n-heptane–acetone feed with the benzene entrainer and separating pure acetone in the first column. Sketch the mass balance for the alternative sequence in a triangular diagram and the resulting flowsheet.

References

Castillo FJL and Towler GP (1998) Influence of Multi-component Mass Transfer on Homogeneous Azeotropic Distillation, *Chem Eng Sci*, **53**: 963.

Castillo FJL, Thong DYC and Towler GP (1998) Homogeneous Azeotropic Distillation 1. Design Procedure for Single-Feed Columns at Non-total Reflux, *Ind Eng Chem Res*, **37**: 987.

Doherty MF and Malone MF (2001) *Conceptual Design of Distillation Systems*, McGraw-Hill.

Doherty MF and Perkins JD (1978) On the Dynamics of Distillation Processes: I. The Simple Distillation of Multi-component Non-reacting Homogeneous Liquid Mixtures, *Chem Eng Sci*, **33**: 281.

Doherty MF and Perkins JD (1979a) The Behaviour of Multi-component Azeotropic Distillation Processes, *IChemE Symp Ser*, **56**: 4. 2/21.

Doherty MF and Perkins JD (1979b) On the Dynamics of Distillation Process: III. Topological Structure of Ternary Residue Curve Maps, *Chem Eng Sci*, **34**: 1401.

Holland CD, Gallun SE and Lockett MJ (1981) Modeling Azeotropic and Extractive Distillations, *Chem Eng*, **88** (March 23): 185.

Hougen OA, Watson KM and Ragatz RA (1954) *Chemical Process Principles. Part I: Material and Energy Balances*, 2nd Edition, John Wiley & Sons.

Laroche L, Bekiaris N, Andersen HW and Morari M (1992) Homogeneous Azeotropic Distillation: Separability and Flowsheet Synthesis, *Ind Eng Chem Res*, **31**: 2190.

Levy SG, Van Dongen DB and Doherty MF (1985) Design and Synthesis of Homogeneous Azeotropic Distillation: 2. Minimum Reflux Calculations for Non-ideal and Azeotropic Columns, *Ind Eng Chem Fund*, **24**: 463.

Petlyuk FB and Avetyan VS (1971) Investigation of Three Component Distillation at Infinite Reflux, *Theor Found Chem Eng*, **5**: 499.

Petlyuk FB, Kievskii VY and Serafimov LA (1975a) Thermodynamic and Topological Analysis of the Phase Diagrams of Polyazeotropic Mixtures: I. Definition of Distillation Regions Using a Computer, *Russ J Phys Chem*, **49**: 1834.

Petlyuk FB, Kievskii VY and Serafimov LA (1975b) Thermodynamic and Topological Analysis of the Phase Diagrams of Polyazeotropic Mixtures: II. Algorithm for Construction of Structural Graphs for Azeotropic Ternary Mixtures, *Russ J Phys Chem*, **49**: 1836.

Schreinemakers FAH (1901) Dampfdrueke im System: Wasser Aceton and Phenol, *Z Phys Chem*, **39**: 440.

Thong DYC and Jobson M (2001) Multi-component Azeotropic Distillation 1. Assessing Product Feasibility, *Chem Eng Sci*, **56**: 4369.

Wahnschafft OM and Westerberg AW (1993) The Product Composition Regions of Azeotropic Distillation Columns: II. Separability in Two-Feed Columns and Entrainer Selection, *Ind Eng Chem Res*, **32**: 1108.

Wahnschafft OM, Koehler JW, Blass E and Westerberg AW (1992) The Product Composition Regions of Single-Feed Azeotropic Distillation Columns, *Ind Eng Chem Res*, **31**: 2345.

Wynn N (2001) Pervaporation Comes of Age, *Chem Eng Prog*, **97**(10): 66.

Zharov VT (1968) Phase Representations and Rectification of Multi-component Solutions, *J Appl Chem USSR*, **41**: 2530.

Chapter 12

Heat Exchange

Many types of heat transfer equipment are used in the process industries. Figure 12.1 illustrates a number of different *tubular* designs. The first shown in Figure 12.1a illustrates a *double-pipe* or *hairpin* heat exchanger. Heat transfer tubes are mounted concentrically within pipes and connected by tees and bends. The arrangement in Figure 12.1a shows a single tube mounted within the pipe with the cold fluid flowing through the inside of the tube and the hot fluid flowing outside of the tube in the annular space. In Figure 12.1a, the hot fluid flows downwards. This is normal practice since a hot liquid will become denser as it is cooled, and therefore less buoyant, and would tend to naturally flow downwards as a result of the buoyancy forces. Also, if some condensation of vapor was occurring, this would also tend to flow naturally downwards. The cold fluid on the tube-side of the heat exchanger flows upwards. This is because a cold liquid being heated up would become less dense and therefore more buoyant, and would tend to naturally flow upwards as a result of the buoyancy forces. Alternatively, if a liquid was being partially vaporized, any vapor would tend to naturally flow upwards. An alternative flow arrangement with the hot fluid flowing through the inside of the tube and cold fluid outside with the appropriate flow direction imposed would also be possible. The flow arrangement in Figure 12.1a exhibits true countercurrent behavior. The size of such units can be increased by mounting multiple tubes within the pipe or by stacking many units together. However, the construction of these units limits their capacity.

By far the most commonly used type of heat exchanger is the shell-and-tube design, as illustrated in Figure 12.1b. Figure 12.1b shows the cold fluid flowing through the inside of the tubes on the *tube-side* and the hot fluid flowing around the outside of the tubes on the *shell-side*. The flow arrangements for the hot and cold fluids could have been swapped with the appropriate flow direction imposed. The fluid flowing on the shell-side is made to flow repeatedly across the outside of the tubes by the use of *baffles*. The most common design uses a disc plate with a segment of the disc removed, known as a *segmental baffle*. In Figure 12.1b the segments are arranged to give an up and down cross flow. The baffles can also be rotated to give side to side cross flow.

Chemical Process Design and Integration, Second Edition. Robin Smith.

Companion Website: www.wiley.com/go/smith/chemical2e

Figure 12.1c shows a *kettle reboiler*. A hot fluid flows through the inside of tubes immersed in a pool of boiling liquid. A large-diameter shell allows disengagement of liquid droplets as the liquid boils. The level of liquid is maintained by a weir, over which flows the liquid that is not vaporized. This design of heat exchanger can be used for a variety of vaporization duties, as well as being used for distillation column reboiling. Other designs of distillation reboiler can be used, as discussed in Chapter 8.

Figure 12.1d illustrates one possible design of condenser. The device is mounted horizontally in this case with the cold fluid flowing through the inside of the tubes. Vapor condenses on the outside of the tubes. The design in Figure 12.1d features a horizontal, longitudinal baffle to direct the condensing vapor around the inside of the shell. It also features vertically mounted segmental baffles to direct the condensing vapor to move from side to side across the tubes.

Many other tubular designs of heat exchanger are possible. Figure 12.2 illustrates a number of different shell arrangements for shell-and-tube heat exchangers and their typical application. Also shown in Figure 12.2 is the TEMA (Tubular Exchanger Manufacturers Association) classification for the various shell designs (TEMA, 2007). Although tubular designs are the most common, many designs of heat exchanger other than tubular are available. These will be discussed later in Section 12.12.

12.1 Overall Heat Transfer Coefficients

Consider first the resistance to heat transfer across the wall of the tubes in tubular heat exchangers. Figure 12.3 illustrates the resistance to heat transfer across the wall of the tube. There are five resistances to heat transfer that combine to give the total resistance to heat transfer. Each resistance can be characterized by a *heat transfer coefficient*.

1) *Shell-side film coefficient.* The heat transfer through the resistance created by the fluid on the outside (shell-side) of the tubes is given by:

$$Q = h_S A_O \Delta T_S \tag{12.1}$$

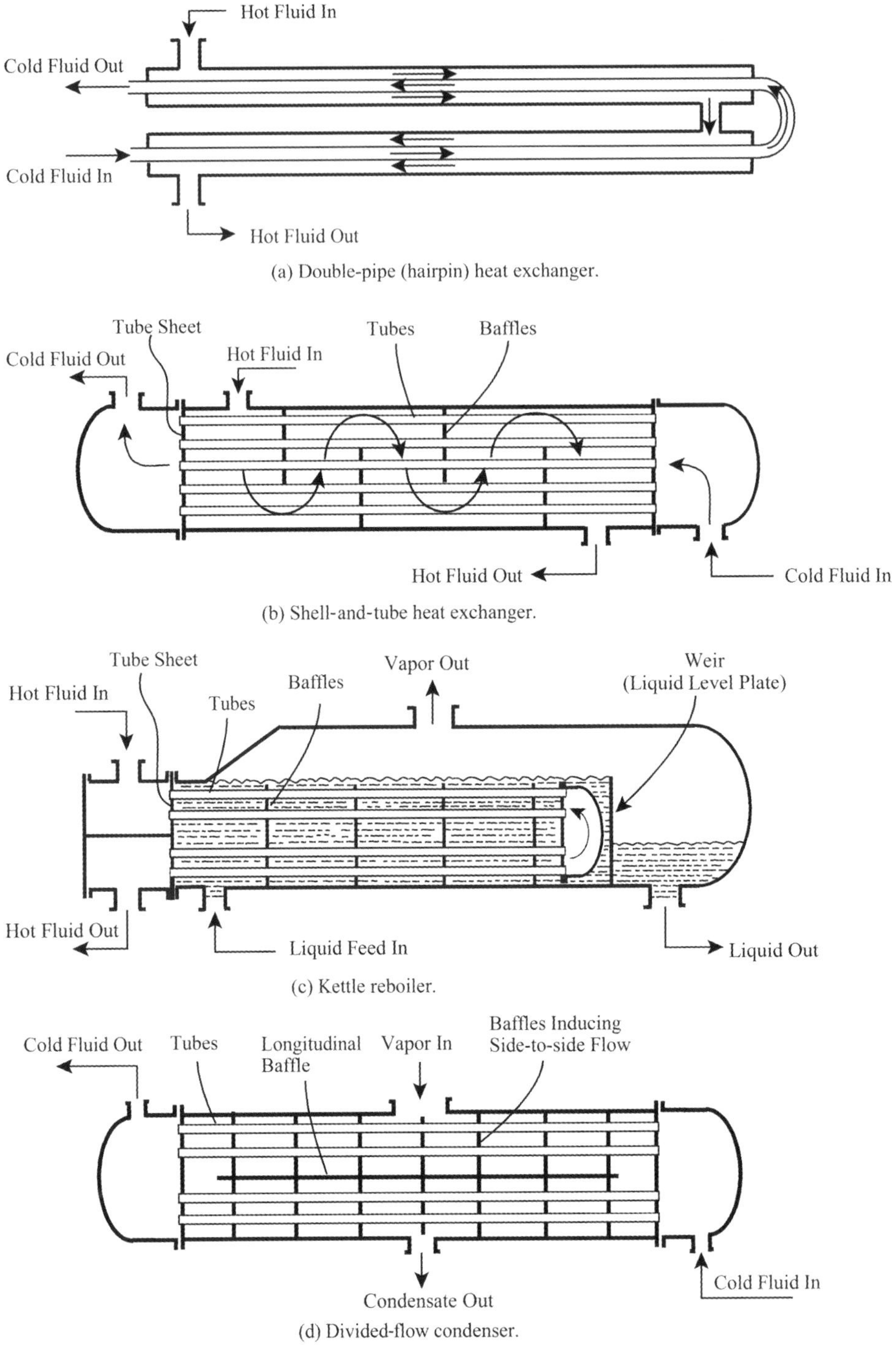

Figure 12.1

Some common tubular heat echangers.

where Q = heat transferred per unit time ($J \cdot s^{-1} = W$)
h_S = film heat transfer coefficient on the outside (shell-side) of the tubes ($W \cdot m^{-2} \cdot K^{-1}$)
A_O = heat transfer area outside (shell-side) of the tubes (m^2)
ΔT_S = temperature difference across the outside (shell-side) film (K)

2) *Shell-side fouling coefficient.* Heat transfer is usually impeded by surface deposits on the heat transfer surface (*fouling*). The material deposited as fouling usually has a low thermal conductivity. Fouling is time dependent and depends on the fluid velocity, temperature and many other factors. Fouling is difficult to predict and allowances are usually based on experience. Design is based on an assumed value to

	TEMA Classification	Description	Applications
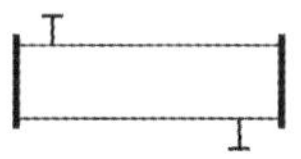	E Shell	One-pass Shell	Suitable for most process heating or cooling applications – the most common shell design
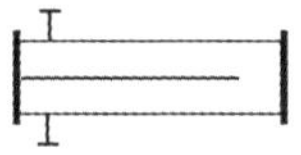	F Shell	Two-pass Shell with Longitudinal Baffle	Countercurrent flow pattern suitable for close approach temperature
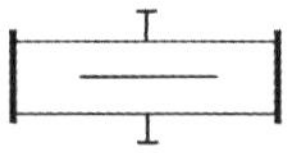	G Shell	Split Flow	Phase change applications on the shell side
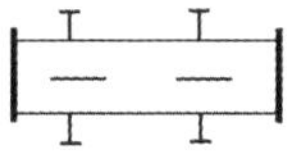	H Shell	Double Split Flow	Phase change applications on the shell side
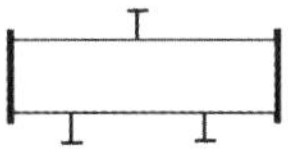	J Shell	Divided Flow	Phase change applications on the shell side where low pressure drop is required
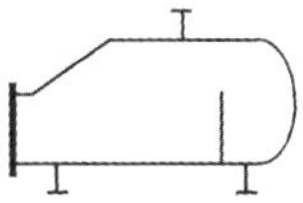	K Shell	Kettle Reboiler	Shell side stream undergoes vaporization with vapor disengagement
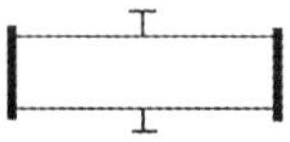	X Shell	Cross Flow	Phase change applications on the shell-side

Figure 12.2

Different shell arrangements for shell-and-tube heat exchangers and their application, classified according to TEMA standards (TEMA, 2007).

be expected after a reasonable period of time before the exchanger is cleaned.

The heat transfer through the resistance created by the outside (shell-side) fouling is quantified by a *fouling coefficient* given by:

$$Q = h_{SF} A_O \Delta T_{SF} \tag{12.2}$$

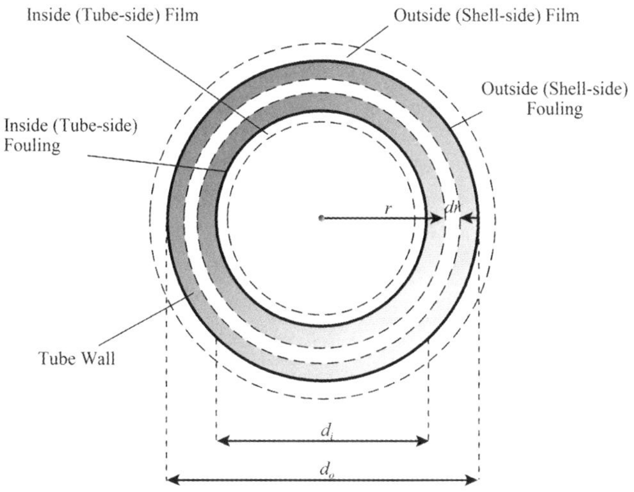

Figure 12.3

Resistance to heat transfer across the tube.

where h_{SF} = outside (shell-side) fouling coefficient ($W \cdot m^{-2} \cdot K^{-1}$)

ΔT_{SF} = temperature difference across the outside (shell-side) fouling resistance (K)

3) *Tube-wall coefficient.* Heat transfer across the tube wall is described by the Fourier Equation (Kern, 1950):

$$Q = -kA\frac{dT}{dr} \tag{12.3}$$

where k = thermal conductivity of the tube wall material ($W \cdot m^{-1} \cdot K^{-1}$)

r = radial distance (m)

A = heat transfer area at radial distance r(m)

Consider an incremental thickness of tube wall dr with radius r, as illustrated in Figure 12.3.

$$A = 2\pi r L \tag{12.4}$$

where L = tube length (m)

Substituting Equation 12.4 into Equation 12.3 and integrating:

$$-\frac{Q}{2\pi kL}\int_{r_I}^{r_O}\frac{dr}{r} = \int_{T_I}^{T_O} dT \tag{12.5}$$

where r_O = outside tube radius (m)
r_I = inside tube radius (m)
T_O = outside surface temperature of the tube (°C)
T_I = inside surface temperature of the tube (°C)

Integrating Equation 12.5 gives:

$$-\frac{Q}{2\pi kL}\ln\left(\frac{r_O}{r_I}\right) = T_O - T_I \tag{12.6}$$

Thus:

$$Q = \frac{2\pi kL}{\ln\left(\frac{d_O}{d_I}\right)}\Delta T_W \tag{12.7}$$

where d_O, d_I = outside and inside tube diameters (m)
ΔT_W = temperature difference across the wall (K)

4) *Tube-side fouling coefficient.* The heat transfer through the resistance created by the inside (tube-side) fouling is given by:

$$Q = h_{TF}A_I\Delta T_{TF} \tag{12.8}$$

where h_{TF} = inside (tube-side) fouling coefficient ($W \cdot m^{-2} \cdot K^{-1}$)
A_I = inside (tube-side) heat transfer area of tubes (m^2)
ΔT_{TF} = temperature difference across the tube-side fouling resistance (K)

5) *Tube-side film coefficient.* The heat transfer through the resistance created by the fluid on the inside (tube-side) of the tubes is given by:

$$Q = h_TA_I\Delta T_T \tag{12.9}$$

where h_T = inside (tube-side) film heat transfer coefficient ($W \cdot m^{-2} \cdot K^{-1}$)
ΔT_T = temperature difference across the inside (tube-side) film (K)

The five resistances can be added. If ΔT represents the temperature difference between the bulk temperature of the fluid on the outside and inside of the tubes, then the temperature differences across the individual resistances can be added to give:

$$\begin{aligned}\Delta T &= \Delta T_S + \Delta T_{SF} + \Delta T_W + \Delta T_{TF} + \Delta T_T \\ &= \frac{Q}{h_SA_O} + \frac{Q}{h_{SF}A_O} + \frac{Q}{2\pi kL}\ln\left(\frac{d_O}{d_I}\right) + \frac{Q}{h_{TF}A_I} + \frac{Q}{h_TA_I}\end{aligned} \tag{12.10}$$

Rearranging Equation 12.10 gives:

$$\Delta T = \frac{Q}{A_O}\left[\frac{1}{h_S} + \frac{1}{h_{SF}} + \frac{d_O}{2k}\ln\left(\frac{d_O}{d_I}\right) + \frac{d_O}{d_I}\frac{1}{h_{TF}} + \frac{d_O}{d_I}\frac{1}{h_T}\right] \tag{12.11}$$

If the overall heat transfer is written as:

$$\Delta T = \frac{Q}{A_O}\frac{1}{U} \tag{12.12}$$

where U = overall heat transfer coefficient based on the outside area of the tube ($W \cdot m^{-2} \cdot K^{-1}$), then comparing Equations 12.11 and 12.12 gives:

$$\frac{1}{U} = \frac{1}{h_S} + \frac{1}{h_{SF}} + \frac{d_O}{2k}\ln\left(\frac{d_O}{d_I}\right) + \frac{d_O}{d_I}\frac{1}{h_{TF}} + \frac{d_O}{d_I}\frac{1}{h_T} \tag{12.13}$$

Alternatively, Equation 12.13 can be written in terms of *fouling resistances*. The fouling resistance is simply the reciprocal of the fouling coefficient:

$$\frac{1}{U} = \frac{1}{h_S} + R_{SF} + \frac{d_O}{2k}\ln\left(\frac{d_O}{d_I}\right) + \frac{d_O}{d_I}R_{TF} + \frac{d_O}{d_I}\frac{1}{h_T} \tag{12.14}$$

where R_{SF} = outside (shell-side) fouling resistance ($m^2 \cdot K \cdot W^{-1}$)
R_{TF} = inside (tube-side) fouling resistance ($m^2 \cdot K \cdot W^{-1}$)

The overall heat transfer coefficient as defined in Equations 12.13 and 12.14 refers to the outside area of the tubes. Alternatively, the inside area could have been chosen as the reference. By convention, the outside area is normally used.

Table 12.1 lists typical values for the film transfer coefficients (Kern, 1950; Hewitt, Shires and Bott, 1994; Hewitt, 2008; Towler and Sinnott, 2013).

Table 12.1

Typical values for film transfer coefficients.

	h_S or h_T ($W \cdot m^{-2} \cdot K^{-1}$)
No change of state	
Water	2000–6000
Gases	10–500
Organic liquid (low viscosity)	1000–3000
Organic liquid (high viscosity)	100–1000
Condensing	
Steam	5000–15,000
Organic (low viscosity)	1000–2500
Organic (high viscosity)	500–1000
Ammonia	3000–6000
Evaporation	
Water	2000–10,000
Organic (low viscosity)	500–2000
Organic (high viscosity)	100–500
Ammonia	1000–2500

Table 12.2

Tube wall coefficients based on the outer diameter for a variety of metals at 100 °C.

		$h_W = \dfrac{2k}{d_O \ln(d_O/d_I)}$ $(W \cdot m^{-2} \cdot K^{-1})$			
		$d_O = 20$ mm		$d_O = 25$ mm	
Metal	k $(W \cdot m^{-1} \cdot K^{-1})$	$d_I = 16.8$ mm	$d_I = 16$ mm	$d_I = 21$ mm	$d_I = 19.8$ mm
Aluminum	240	137,700	107,600	110,100	82,340
Copper	395	226,600	177,000	181,200	135,500
Hastelloy	11.7	6,710	5,240	5,370	4,010
Monel	24	13,770	10,760	11,010	8,230
Nickel	83	47,600	37,200	38,080	28,470
Stainless steel 304	16.5	9,460	7,390	7,570	5,660
Stainless steel 316	15	8,600	6,720	6,880	5,150
Steel	45	25,810	20,170	20,650	15,440
Titanium	21	12,050	9,410	9,640	7,200

Table 12.2 tabulates the tube wall coefficients for a variety of materials for some common tube sizes. It should be noted that the thermal conductivity varies with temperature and the purity of the metal.

From Table 12.2, the heat transfer coefficient for the tube wall in many cases is so high that its contribution to the overall heat transfer coefficient can be neglected.

12.2 Heat Exchanger Fouling

Fouling is the accumulation of undesired material at the heat transfer surfaces and can be classified as:

- *particulate*, in which solid particles suspended in the process stream are carried to the heat transfer surface and accumulate;
- *scaling*, in which solid material is precipitated from solution on to the heat transfer surface through inverse solubility (e.g. calcium carbonate deposit from hardness in water);
- *crystallization*, in which solid material is crystallized from solution as a result of the change in temperature and/or the presence of nucleation sites at the surface;
- *freezing*, in which material is decreased in temperature locally to below its freezing point;
- *chemical reaction*, which results from reactions at the heat transfer surface (e.g. polymerization and cracking reactions);
- *corrosion*, in which the heat transfer surface is exposed to a corrosive fluid that reacts to produce byproducts of corrosion on the surface, which typically are oxides with low thermal conductivity;
- *biological fouling* or *biofouling*, in which a layer of microorganisms grows on the heat transfer surface producing *slime*.

A number of these fouling mechanisms can occur at the same time. Depending on conditions and the fouling mechanism, the change in the fouling resistance can follow different patterns, as illustrated in Figure 12.4 (Bott, 1995; Müller-Steinhagen, 2000). There can be an initiation period for a new heat exchanger, or one that has been cleaned, during which the heat transfer remains unchanged. During the initiation period nuclei are formed for the fouling to begin or nutrients deposited for biological growth. There is no initiation period for particulate fouling. The initiation period might last for seconds or days. It can be seen in Figure 12.4 that the rate of increase of fouling resistance might be constant, giving a linear increase, or exhibit a falling rate, or a rate that falls to approach a constant fouling resistance. Figure 12.4 also shows a saw-tooth increase in fouling resistance caused by the periodic

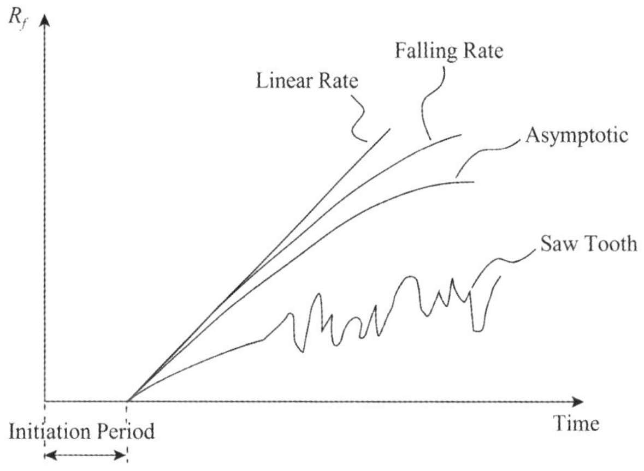

Figure 12.4

Fouling resistance through time.

12

removal of fouling by the flow of the process fluid. For most fouling mechanisms, the fouling resistance increases exponentially with the tube wall temperature according to (Müller-Steinhagen, 2000):

$$\frac{dR_F}{dt} = K \exp\left[-\frac{E}{RT_W}\right] \tag{12.15}$$

where R_F = fouling resistance ($m^2 \cdot K \cdot W^{-1}$)
t = time (day)
K = rate constant that depends on the fouling mechanism and the fluid properties ($m^2 \cdot K \cdot W^{-1} \cdot day^{-1}$)
E = activation energy ($kJ \cdot kmol^{-1}$)
R = universal gas constant ($8.3145\, kJ \cdot kmol^{-1} \cdot K^{-1}$)
T_W = tube wall temperature (K)

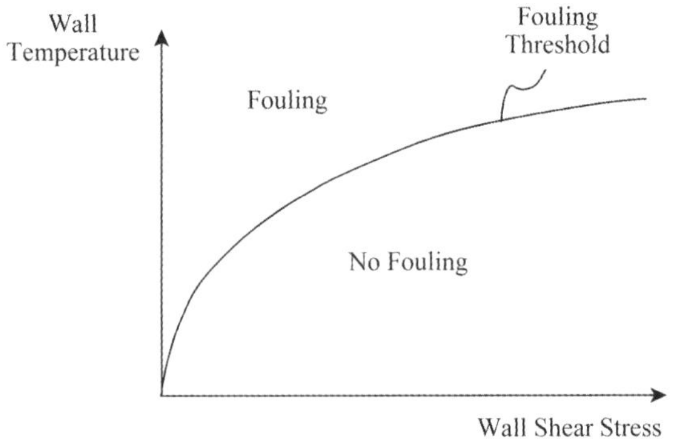

Figure 12.5
Fouling threshold.

For most fouling mechanisms, fouling resistance decreases with increasing wall shear stress (Bott, 1995; Müller-Steinhagen, 2000). This results from wall shear stress removing fouling.

The rate at which fouling occurs is normally a balance between deposition and removal:

$$[\text{Rate of fouling}] = \begin{bmatrix}\text{Rate of fouling}\\ \text{deposition}\end{bmatrix} - \begin{bmatrix}\text{Rate of fouling}\\ \text{removal}\end{bmatrix} \tag{12.16}$$

The rate of fouling deposition depends on many factors, but temperature is usually the most important. The rate of fouling removal depends mainly on the wall shear stress. For flow inside of tubes, the wall shear stress is determined by:

$$\tau_w = c_f \frac{\rho v^2}{2} \tag{12.17}$$

where τ_w = wall shear stress ($N \cdot m^2$)
c_f = Fanning friction factor (–)
ρ = fluid density ($kg \cdot m^{-3}$)
v = mean velocity in the tube ($m \cdot s^{-1}$)

The wall shear stress on the shell-side is much more complex to calculate. Fouling resistance is typically proportional to flow velocity raised to the power minus 1.5 (Müller-Steinhagen, 2000).

If the rate of fouling deposition exceeds the rate of fouling removal then there is a net accumulation. If the rate of fouling removal exceeds the rate of fouling deposition, then the surface will not foul. For some types of fouling, there is a fouling threshold illustrated by Figure 12.5. This is typical of the fouling that occurs, for example, with crude oil. The deposition is dominated by the effect of temperature and the removal by the wall shear stress. At conditions below the fouling threshold, no fouling occurs in Figure 12.5.

Antifouling chemical agents can be used to mitigate the effects of fouling. On the tube-side, *enhanced tubes* can be used rather than plain tubes. These have irregularities on the internal surface. Plain tubes can also be fitted on the inside with *tube inserts*. Heat transfer enhancement promotes additional turbulence and pressure drop and reduces the surface temperature of the tube to mitigate fouling. Heat transfer enhancement will be dealt with in Section 12.8.

The effect of fouling on the shell-side is more complex. On the shell-side, the heat transfer and pressure drop are also affected by the fouling changing the flow pattern. Bypassing and leakage reduce both the shell-side heat transfer coefficient and pressure drop. Fouling will tend to block the clearances, increasing the amount of cross flow. This will have the effect of increasing the shell-side heat transfer coefficient and pressure drop. However, this increase will be countered by fouling resistances of the heat transfer surfaces. Initially, the fouling might even increase the heat transfer coefficient by promoting better cross flow, only to decrease it later as the film resistance increases.

The segmented baffles normally used in shell-and-tube heat exchangers are not the best arrangement for fouling fluids placed on the shell-side, as stagnant zones are created by the flow pattern. Rather than use segmented baffles, helical flow baffles can be used, which induce a spiral flow pattern along the exchanger on the shell-side, eliminating the stagnant zones.

Thus, fouling is a transient process that begins with a clean heat transfer surface and continues until the point where the surface becomes fouled to the point where the heat exchanger can no longer be used effectively. At this point, the heat exchanger must be taken out of service and cleaned. Cleaning can be accomplished by mechanical or chemical processes. Mechanical cleaning involves the use of high-pressure water jets. Chemical cleaning exploits cleaning fluids that react with and/or dissolve surface deposits. This usually involves creating a cleaning circuit involving the heat exchanger and a pump in which the cleaning fluid is recirculated through the heat exchanger at a relatively high velocity. In some instances (e.g. cooling water circuits), chemicals can be added to the heat transfer fluids to inhibit fouling. The chemicals added to inhibit fouling and the means of cleaning depend on the nature of the fouling.

When designing a heat exchanger, it is usual to assume a constant fouling coefficient with a value to be expected after a reasonable period of time before the heat exchanger is cleaned. Table 12.3 gives typical values of fouling coefficients for design (Kern, 1950; Hewitt, Shires and Bott, 1994; TEMA, 2007; Hewitt, 2008; Towler and Sinnott, 2013).

Table 12.3

Typical values of fouling coefficients.

	h_{SF} or h_{TF} ($W \cdot m^{-2} \cdot K^{-1}$)
Water	
Distilled	10,000
Boiler feedwater	5000–10,000
Steam condensate	1500–5000
Potable water	2000–5000
Bore hole water	1000–3000
Clear river	2000–6000
Good-quality cooling water	3000–6000
Poor-quality cooling water	1000–2000
Sea	2000–6000
Boiler blowdown	3000
Liquids	
Aqueous salt solutions	3000–6000
Organic (low viscosity)	3000–11,000
Organic (high viscosity)	1000–3000
Machinery oil	6000
Fuel oils	1000
Tars	500–1000
Vegetable oils	2000
Gases	
Air	5000–10,000
Organic vapor	5000–10,000
Flue gases	2000–5000
Boiling liquids	
Hydrocarbons	2500–10,000
Polymerizing hydrocarbons	2000–4000
Condensing vapors	
Good quality steam	4000–10,000
Contaminated steam	2000–5000
Organics	5000–20,000

However, it must be remembered that the values in Table 12.3 are only design guidelines. Fouling is a dynamic process that starts with a clean heat exchanger and at some point the heat exchanger will be taken off-line and cleaned. Caution must be exercised when designing for fouling. If an excessive allowance is made for fouling, then the heat exchanger will be oversized. This in turn will lead to larger equipment and lower velocities in the exchanger, which will accelerate the fouling.

12.3 Temperature Differences in Shell-and-Tube Heat Exchangers

Consider the heat exchange process in Figure 12.6a. The heat transfer arrangement shows a *concentric pipe* heat exchanger. The flow of the fluids in such a device is truly countercurrent. To understand the heat transfer, assume that it is a steady-state process in which all of the fluid properties and overall heat transfer coefficient U are constant. It is also assumed that there is no phase change and that there is no heat loss from the system. The heat transfer process is shown in Figure 12.6b in a plot of heat transfer

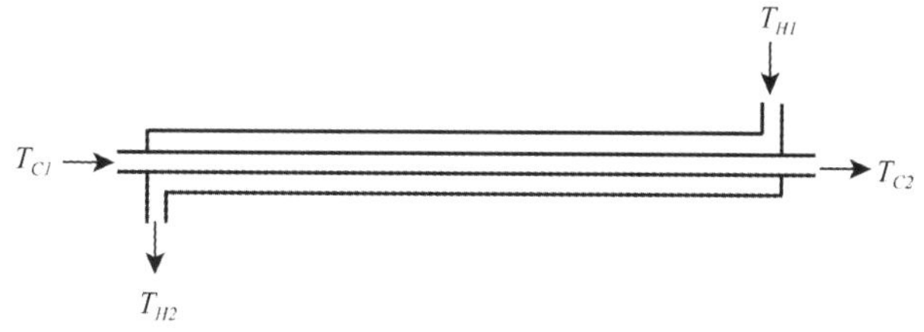

(a) Counter-current heat exchange.

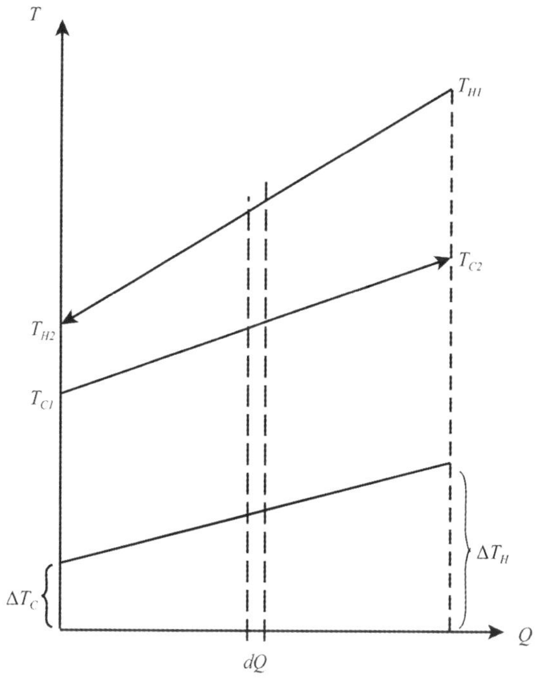

(b) Heat transfer versus temperature.

Figure 12.6

Heat exchange temperature difference.

12

versus temperature. The slope of the ΔT line in Figure 12.6b is given by:

$$\frac{d(\Delta T)}{dQ} = \frac{\Delta T_H - \Delta T_C}{Q} \quad (12.18)$$

Applying Equation 12.12 across the differential element in Figure 12.6b:

$$dQ = U\,dA\,\Delta T \quad (12.19)$$

Combining Equations 12.18 and 12.19 gives:

$$\frac{\Delta T_H - \Delta T_C}{Q} = \frac{d(\Delta T)}{U\,dA\,\Delta T} \quad (12.20)$$

Rearranging and integrating gives:

$$\int_o^A dA = \frac{Q}{\Delta T_H - \Delta T_C}\frac{1}{U}\int_{\Delta T_C}^{\Delta T_H}\frac{d(\Delta T)}{\Delta T} \quad (12.21)$$

Thus:

$$A = \frac{Q}{\Delta T_H - \Delta T_C}\frac{1}{U}\ln\frac{\Delta T_H}{\Delta T_C} \quad (12.22)$$

$$Q = UA\left[\frac{\Delta T_H - \Delta T_C}{\ln\dfrac{\Delta T_H}{\Delta T_C}}\right] \quad (12.23)$$

$$= UA\Delta T_{LM} \quad (12.24)$$

where ΔT_{LM} = logarithmic mean temperature difference

Note that this result would have been obtained if:

a) the fluid positions inside and outside of the tube in Figure 12.6b had been reversed;

b) the direction of one of the fluids had been reversed giving cocurrent flow;

c) either fluid had been isothermal.

If the slope of the hot and cold streams had been equal, then $\Delta T_H = \Delta T_C = \Delta T$ and the logarithmic temperature difference would be replaced by ΔT in Equation 12.24. This is most likely to happen if both fluids are isothermal. It should also be noted that:

$$\frac{\Delta T_H - \Delta T_C}{\ln\dfrac{\Delta T_H}{\Delta T_C}} = \frac{\Delta T_C - \Delta T_H}{\ln\dfrac{\Delta T_C}{\Delta T_H}} \quad (12.25)$$

Although the result in Equation 12.24 applies to both countercurrent and cocurrent flow, in practice, cocurrent flow is almost never used as, given fixed fluid inlet and outlet temperatures, the logarithmic mean temperature difference for countercurrent flow is always larger. This in turn leads to smaller surface area requirements. Also, as shown in Figure 12.7a for countercurrent flow, the final temperature of the hot fluid can be lower than the final temperature of the cold fluid (sometimes known as *temperature cross*), whereas in Figure 12.7b, it is clear that there can never be a temperature cross.

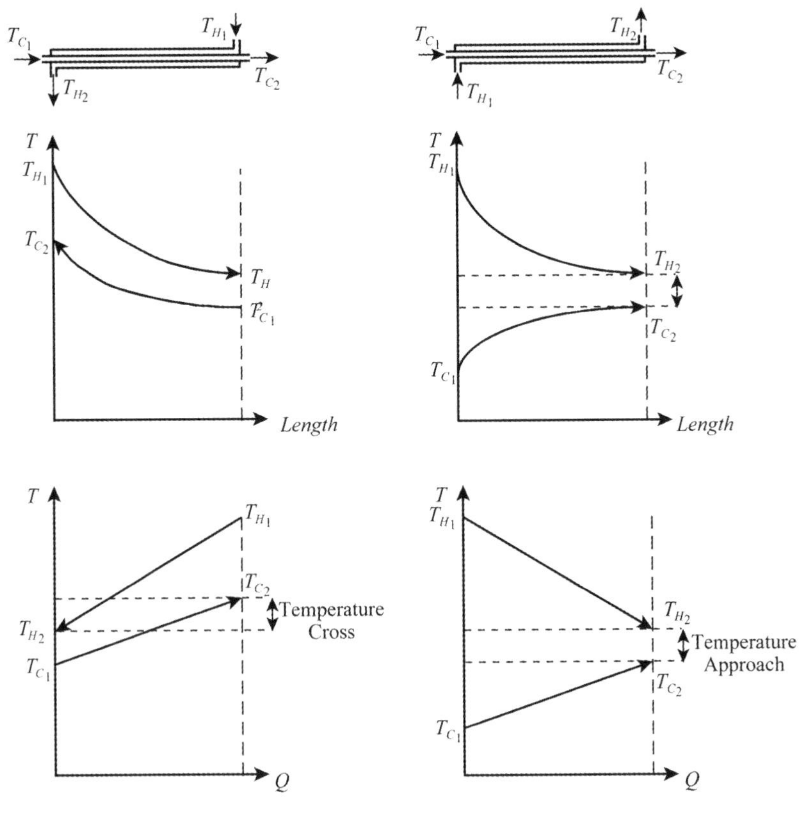

Figure 12.7

Fluid temperatures can never cross in a cocurrent heat exchanger.

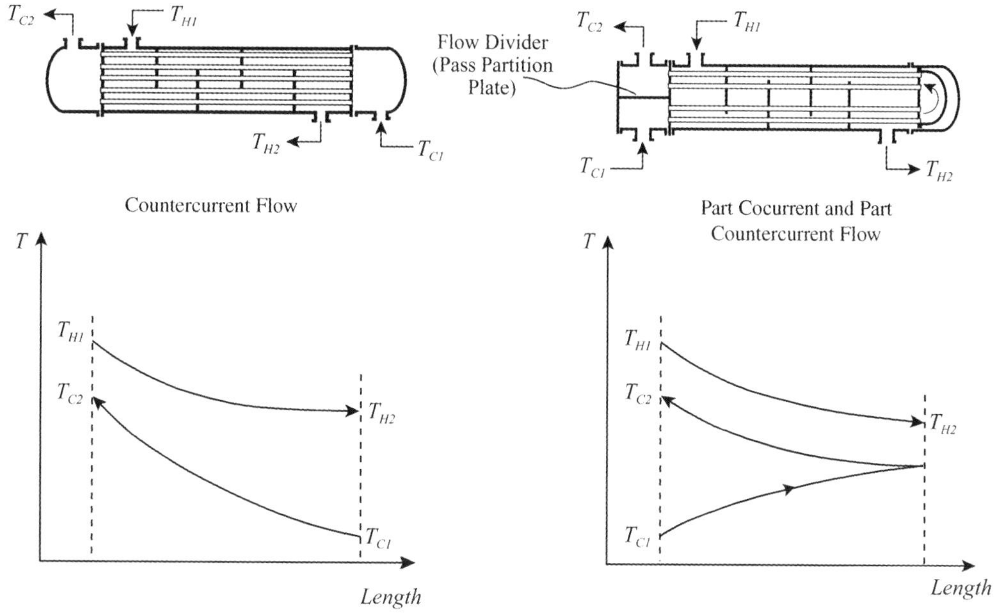

(a) 1–1 Heat Exchanger.

(b) 1–2 Heat Exchanger.

Figure 12.8

1–1 shells approach pure countercurrent flow, whereas 1–2 shells exhibit partial countercurrent and partial cocurrent flow.

For a given heat duty and overall heat transfer coefficient, the fluid flowing through the tubes in a *single pass*, in a 1–1 design (1 shell pass–1 tube pass), as illustrated in Figure 12.8a, offers the lowest requirement for surface area for shell-and-tube heat exchangers. Figure 12.8b shows a design in which the fluid flows in *two passes* through the tube-side in a 1–2 design (1 shell pass–2 tube passes). The tube-side flow is made to change direction by a *flow divider* or *pass partition plate* mounted in the heat exchanger head. The increase in the tube passes increases the tube velocity and tube-side heat transfer coefficient, and hence the overall heat transfer coefficient (as discussed later).

Figure 12.9 shows four different design arrangements to achieve this. Figure 12.9a shows a fixed tube sheet design. This has the tubes secured at both ends to tube sheets fixed to the shell. The outside of the tubes cannot be cleaned mechanically and the design is limited to clean services on the shell-side. Another disadvantage is that a large temperature difference between tubes and the shell creates a differential stress that might need an expansion joint to be incorporated. It is a low cost design, providing no expansion joint is required. An alternative *U-tube* design with only one tube sheet is shown in Figure 12.9b. Additional costs are incurred for bending tubes, and an increased shell diameter is required due to the minimum bend radius allowed. The tube bundle can expand and contract in response to temperature-induced stresses. Also, the bundle can be removed to clean the outside of the tubes. The inside of the tubes cannot be cleaned effectively due to the bends, so application is restricted to clean fluids on the tube-side. A *split ring floating head* design is shown in Figure 12.9c. This has one tube sheet fixed relative to the shell and the other is free to "float", permitting free expansion of the tube bundle. A dished cover is bolted to the floating head tube sheet with a *split backing ring*, which is formed in two halves. The bundle can be removed by removing the *channel* end of the heat exchanger for the tube-side inlet and outlets, the shell cover at the floating head end, the floating head cover and the split ring. The tube bundle can then be pulled through the shell from the channel end, as the diameter of the floating tube sheet is slightly smaller than the shell diameter. Once removed, the outside of the tubes can be cleaned mechanically for certain tube layouts. An alternative, less common floating head design is shown in Figure 12.9d. This time a dished cover is bolted directly to the floating head tube sheet. The tube bundle can be removed directly by removing the channel end of the heat exchanger for the tube-side inlet and outlets and the tube bundle pulled through the shell from the channel end. Again, once removed, the outside of the tubes can be cleaned mechanically for certain tube layouts (see later). However, the simplicity of removal of the tube bundle for pull-through floating head designs is accomplished at the expense of a larger shell diameter. Floating head designs can be used when the fluids on both sides are dirty.

The 1–2 design in Figure 12.8b shows that the flow arrangement involves part countercurrent and part cocurrent flow. This reduces the effective temperature difference for heat exchange compared with a pure countercurrent device. This is accounted for, in design, by the introduction of the F_T factor into the basic heat exchanger design equation (Underwood, 1934; Bowman, 1936; Bowman, Mueller and Nagle, 1940):

$$Q = UA\Delta T_{LM}F_T \quad \text{where} \quad 0 < F_T < 1 \tag{12.26}$$

Thus, for a given exchanger duty and overall heat transfer coefficient, the 1–2 design needs a larger area than the 1–1 design. However, the 1–2 design offers many practical advantages. These

Figure 12.9

Two tube passes can be engineered in different ways.

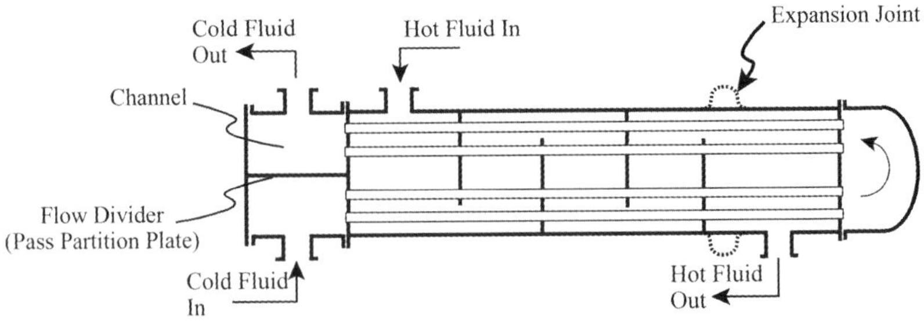

(a) Two tube passes with fixed tube sheets.

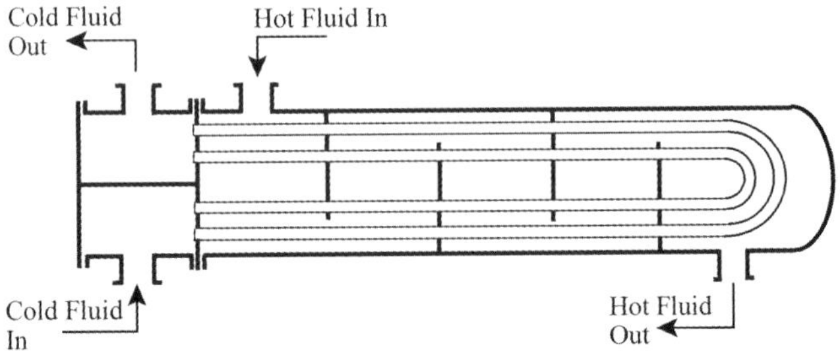

(b) Two tube passes with U-tubes.

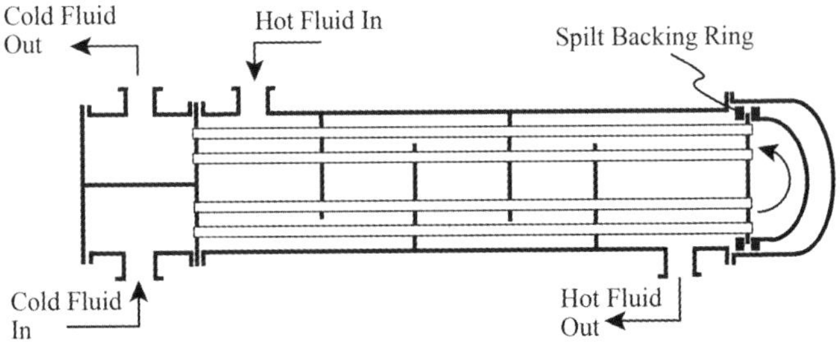

(c) Two tube passes with a split ringfloating head.

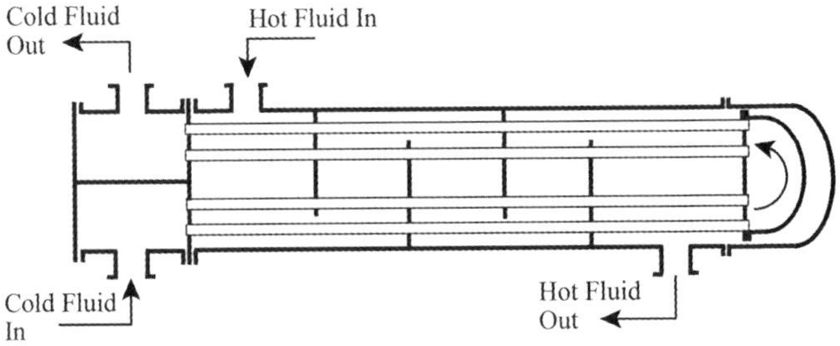

(d) Two tube passes with a pull through floating head.

include, in particular, good heat transfer coefficients on the tube-side (due to higher velocity) and allowance for thermal expansion and easy mechanical cleaning in some designs.

The F_T correction factor is usually correlated in terms of two dimensionless ratios, the ratio of the two heat capacity flowrates (R) and the thermal effectiveness of the exchanger (P) (Bowman, Mueller and Nagle, 1940):

$$F_T = f(R, P) \tag{12.27}$$

where

$$R = CP_C/CP_H = (T_{H1} - T_{H2})/(T_{C2} - T_{C1}) \tag{12.28}$$

and

$$P = (T_{C2} - T_{C1})/(T_{H1} - T_{C1}) \tag{12.29}$$

Note therefore that F_T depends only on the inlet and outlet temperatures of the streams in a 1–2 heat exchanger. An expression can be developed for the F_T in a 1–2 heat exchanger as a function of R and P. The proof is lengthy and can be found elsewhere (Underwood, 1934; Bowman, 1936; Bowman, Mueller and Nagle, 1940; Kern, 1950).

For $R \neq 1$:

$$F_T = \frac{\sqrt{R^2+1}\ln\left[\frac{1-P}{1-RP}\right]}{(R-1)\ln\left[\frac{2-P\left(R+1-\sqrt{R^2+1}\right)}{2-P\left(R+1+\sqrt{R^2+1}\right)}\right]} \tag{12.30}$$

For $R = 1$:

$$F_T = \frac{\left[\frac{\sqrt{2}P}{1-P}\right]}{\ln\left[\frac{2-P\left(2-\sqrt{2}\right)}{2-P\left(2+\sqrt{2}\right)}\right]} \tag{12.31}$$

Rather than use two tube passes, the number of tube passes can be increased to increase the tube-side velocity and hence the tube-side heat transfer coefficient. Strictly, F_T depends on the number of tube-side passes:

$$F_{T1-2} < F_{T1-4} < F_{T1-6} < F_{T1-8} < \cdots < 1 \tag{12.32}$$

However, these values of F_T are very close, with a maximum error of around 2% across all values of P and R (Kern, 1950). Hence F_{T1-2} is used across all F_{T1-2+} cases:

$$F_{T1-2} = F_{T1-2}, F_{T1-4}, F_{T1-6}, F_{T1-8}, \cdots \tag{12.33}$$

The shape of the function from Equations 12.30 and 12.31 is illustrated in Figure 12.10. It can be seen that the slope of the F_T curve for a given R becomes very steep and approaches an asymptote as the thermal effectiveness P increases.

Three basic situations can be encountered when using 1–2 exchangers (Figure 12.10):

1) The final temperature of the hot stream is higher than the final temperature of the cold, as illustrated in Figure 12.10a. This is a so-called *temperature approach*. This situation is straightforward to design for, since it can always be accommodated in a single 1–2 shell.
2) The final temperature of the hot stream is slightly lower than the final temperature of the cold stream, as illustrated in Figure 12.10b. This is known as a *temperature cross*. This situation is usually straightforward to design for, provided the temperature cross is small, because again it can probably be accommodated in a single shell. However, the decrease in F_T increases the heat transfer area requirements significantly.
3) As the amount of temperature cross increases, however, problems are encountered, as illustrated in Figure 12.10c. The F_T decreases significantly, causing a dramatic increase in the heat

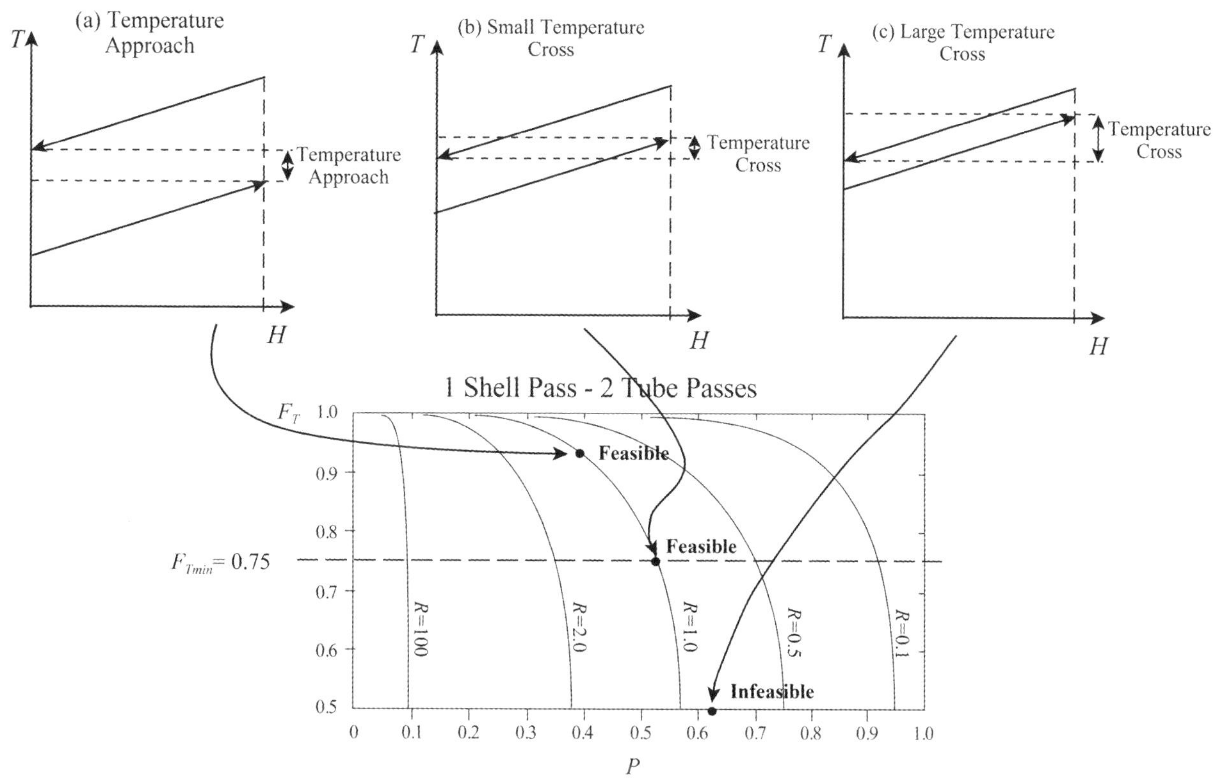

Figure 12.10

Designs with a temperature approach or small temperature cross can be accommodated in a single 1-2 shell, whereas designs with a large temperature cross become infeasible. (From Ahmad S, Linnhoff B and Smith R, 1988, *Trans ASME J Heat Transfer*, **110**: 304, reproduced by permission of the American Society of Mechanical Engineers.)

transfer area requirements. Local reversal of heat flow may also be encountered, which is wasteful in a heat transfer area. The design may even become infeasible. Thus, for a given R, the design of the heat exchanger becomes less and less efficient as the asymptote for the F_T curve is approached.

The maximum temperature cross that can be tolerated is often set by rules of thumb, for example, $F_T > 0.75$ (Kern, 1950). It is important to avoid low values of F_T because:

1) Low values of F_T indicate inefficient use of the heat transfer area.
2) Any violation of the simplifying assumptions used in the approach tends to have a particularly significant effect in areas of the F_T chart where slopes are particularly steep.
3) Any uncertainties or inaccuracies in design data also have a more significant effect when slopes are steep.

Consequently, to be confident in a design, those parts of the F_T chart where slopes are steep should be avoided, irrespective of $F_T > 0.75$ (Ahmad, Linnhoff and Smith, 1988). A simple method to achieve this is based upon the fact that for any value of R there is a maximum asymptotic value for P, say P_{max}, which is given as F_T tends to $-\infty$, and is given by (Ahmad, Linnhoff and Smith, 1988):

$$P_{max} = \frac{2}{R + 1 + \sqrt{R^2 + 1}} \tag{12.34}$$

Equation 12.34 is derived in Appendix D. Practical designs will be limited to some fraction of P_{max}, that is (Ahmad, Linnhoff and Smith, 1988):

$$P = X_P P_{max} \quad 0 < X_P < 1 \tag{12.35}$$

where X_P is a constant defined by the designer.

A line of constant X_P is compared with a line of constant F_T in Figure 12.11 (Ahmad, Linnhoff and Smith, 1990). It can be seen that the line of constant X_P avoids the regions of steep slope.

Situations are often encountered where the design is infeasible in a single 1–2 shell, because the F_T is too low or the F_T slope too large. If this happens, either different types of shell or multiple shell arrangements must be considered (Kern, 1950; Hewitt, Shires and Bott, 1994; Serth, 2007; Hewitt, 2008). Here, consideration will be restricted to multiple shell arrangements of the 1–2 type. By using two 1–2 shells in series (Figure 12.12), the temperature cross in each individual shell is reduced below that for a single 1–2 shell for the same duty. The profiles shown in Figure 12.12 could in principle be achieved either by two 1–2 shells in series or by a single 2–4 shell.

For a number of 1–2 shells in series, a transformation can be developed based on the fact that for N_{SHELLS} in series, each shell pass has the same value of F_T, which also equals the F_T across all N_{SHELLS} passes (Bowman, Mueller and Nagle, 1940). Also, all values of the effectiveness factor of each shell pass (P_{1-2}) are equal, but not equal to the value of the effectiveness factor across all N_{SHELLS} (P). Of course, R is constant across all shells and the overall design.

For $R \neq 1$ (Bowman, Mueller and Nagle, 1940):

$$P = \frac{1 - \left(\frac{1 - P_{1-2}R}{1 - P_{1-2}}\right)^{N_{SHELLS}}}{R - \left(\frac{1 - P_{1-2}R}{1 - P_{1-2}}\right)^{N_{SHELLS}}} \tag{12.36}$$

For $R = 1$ (Bowman, Mueller and Nagle, 1940):

$$P = \frac{P_{1-2}N_{SHELLS}}{P_{1-2}N_{SHELLS} - P_{1-2} + 1} \tag{12.37}$$

Thus, given an overall value of P for the duty involving N_{SHELLS}, Equations 12.36 and 12.37 allow the value of P_{1-2} to be calculated for each shell. To do this, first define a variable Z:

$$Z = \left(\frac{1 - P_{1-2}R}{1 - P_{1-2}}\right)^{N_{SHELLS}} \tag{12.38}$$

Substituting Z into Equation 12.36 and rearranging gives:

$$Z = \frac{1 - PR}{1 - P} \tag{12.39}$$

Thus, starting with the overall P for the duty, Z is first calculated from Equation 12.39. Then P_{1-2} is calculated for each shell by

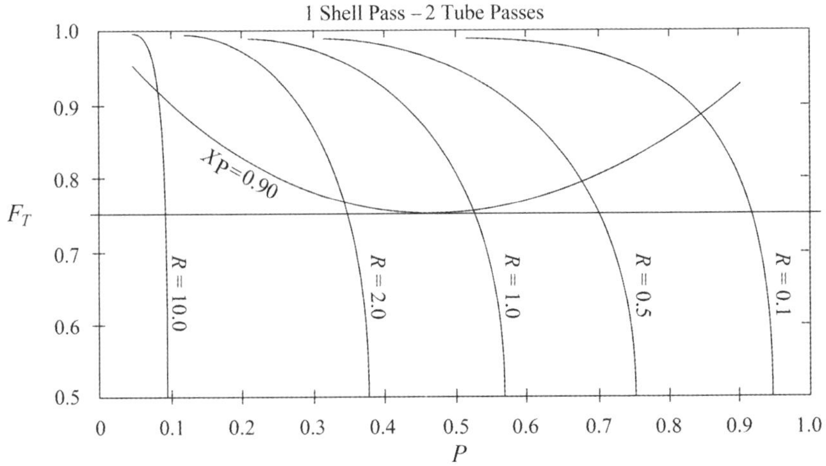

Figure 12.11

The X_P parameter avoids steep slopes on the F_T curves, whereas minimum F_T does not. (From Ahmad S, Linnhoff B and Smith R, 1990, *Computers and Chem Eng*, **7**: 751, reproduced by permission.)

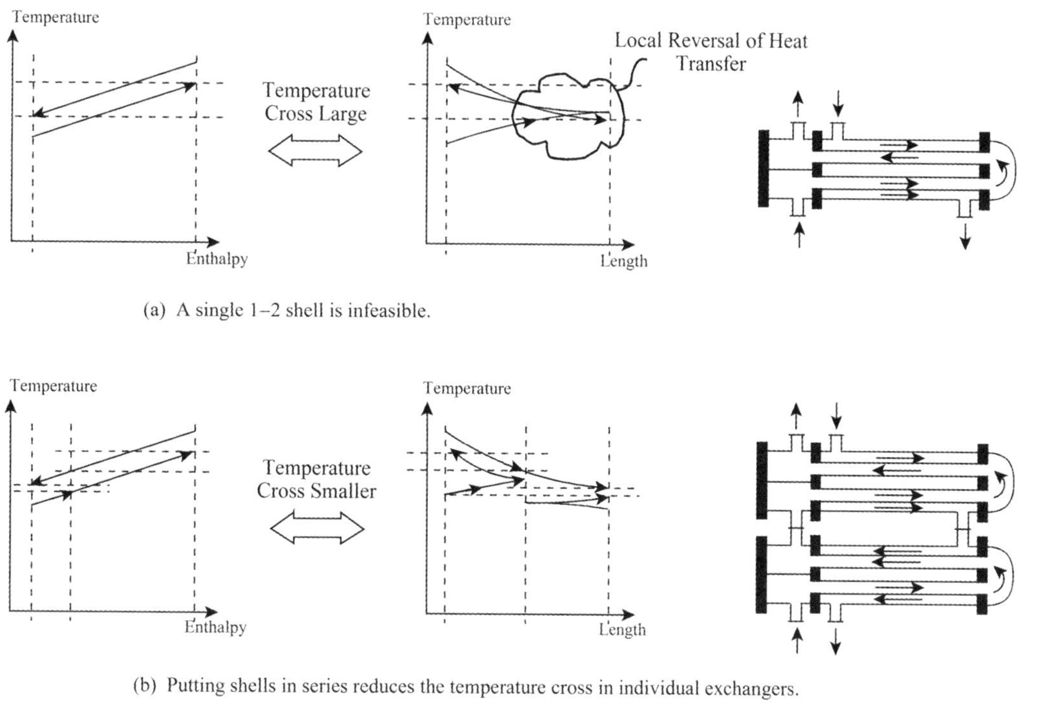

Figure 12.12

A large overall temperature cross requires shells in series to reduce the cross in individual exchangers. (From Ahmad S, Linnhoff B and Smith R,1988, *Trans ASME J Heat Transfer*, **110**: 304, reproduced by permission of the American Society of Mechanical Engineers.)

inverting Equation 12.38 and substituting the value of Z. For $R \neq 1$:

$$P_{1\text{-}2} = \frac{Z^{1/N_{SHELLS}} - 1}{Z^{1/N_{SHELLS}} - R} \quad \text{for } R \neq 1 \qquad (12.40)$$

For $R = 1$, a simple rearrangement of Equation 12.37 gives:

$$P_{1\text{-}2} = \frac{P}{P - PN_{SHELLS} + N_{SHELLS}} \quad \text{for } R = 1 \qquad (12.41)$$

Thus, for N_{SHELLS} in series for $R \neq 1$:

$$F_T = \frac{\sqrt{R^2+1}\ln\left[\dfrac{1 - P_{1\text{-}2}}{1 - RP_{1\text{-}2}}\right]}{(R-1)\ln\left[\dfrac{2 - P_{1\text{-}2}\left(R + 1 - \sqrt{R^2+1}\right)}{2 - P_{1\text{-}2}\left(R + 1 + \sqrt{R^2+1}\right)}\right]} \qquad (12.42)$$

where $P_{1\text{-}2}$ is given by Equation 12.40. For N_{SHELLS} in series for $R = 1$:

$$F_T = \frac{\left[\dfrac{\sqrt{2}P_{1\text{-}2}}{1 - P_{1\text{-}2}}\right]}{\ln\left[\dfrac{2 - P_{1\text{-}2}\left(2 - \sqrt{2}\right)}{2 - P_{1\text{-}2}\left(2 + \sqrt{2}\right)}\right]} \qquad (12.43)$$

where $P_{1\text{-}2}$ is given by Equation 12.41.

Traditionally, the designer would approach a design for an individual unit by trial and error. Starting by assuming one shell, the F_T can be evaluated. If the F_T is not acceptable, then the number of shells in series is progressively increased until a satisfactory value of F_T is obtained for each shell. For a number of 1–2 shells in series:

$$F_{T1\text{-}2} < F_{T2\text{-}4} < F_{T3\text{-}6} < F_{T4\text{-}8} < \cdots < 1 \qquad (12.44)$$

Adopting the design criterion given by Equation 12.35 as the basis, any need for trial and error can be eliminated, since an explicit expression for the number of shells for a given unit is derived in Appendix E (Ahmad, Linnhoff and Smith, 1988). For $R \neq 1$:

$$N_{SHELLS} = \frac{\ln\left(\dfrac{1 - RP}{1 - P}\right)}{\ln W} \qquad (12.45)$$

where

$$W = \frac{R + 1 + \sqrt{R^2 + 1} - 2RX_P}{R + 1 + \sqrt{R^2 + 1} - 2X_P} \qquad (12.46)$$

For $R = 1$:

$$N_{SHELLS} = \frac{\left(\dfrac{P}{1 - P}\right)\left(1 + \dfrac{\sqrt{2}}{2} - X_P\right)}{X_P} \qquad (12.47)$$

12

X_P is chosen to satisfy the minimum allowable F_T (for example, for $F_{Tmin} > 0.75$, $X_P = 0.9$ is used). Once the real (noninteger) number of shells is calculated from Equation 12.45 or 12.47, this is rounded up to the next largest number to obtain the number of shells. Generally, the smaller the number of shells for a given overall duty, the cheaper will be the design. The higher the value of X_P chosen, the larger will be the number of shells, but the safer the design. Thus, a compromise is required. A value of $X_P = 0.9$ is reasonable for most cases.

It should be noted that this approach can be used for 1–4, 1–6, . . . , etc., shells in series with little error. This is because of the small difference between $F_{T1\text{–}2}$ and $F_{T1\text{–}2+}$.

In addition to the F_T limiting the temperature cross in each shell and dividing the overall duty into a number of shells, there is a maximum physical size that can be fabricated in a single shell. For shell-and-tube heat exchangers with removable tube bundles, the maximum size of shell is around 1000 m^2. However, a significantly lower figure might well be preferred for maintenance and cleaning purposes. Fixed bundle heat exchangers can be much larger, typically up to 4500 m^2.

Example 12.1 A hot stream is to be cooled from 300 to 100 °C by exchange with a cold stream being heated from 60 to 200 °C in a single unit. 1–2 shell-and-tube heat exchangers are to be used subject to $X_P = 0.9$. The duty for the exchanger is 3.5 MW and the overall heat transfer coefficient is estimated to be 100 W · m^{-2} · K^{-1}. Calculate:

a) the number of shells required;
b) $P_{1\text{–}2}$ for each shell;
c) F_T for the shells in series;
d) the heat transfer area.

Solution

a)

$$R = \frac{T_{H1} - T_{H2}}{T_{C2} - T_{C1}} = \frac{300 - 100}{200 - 60} = 1.4286$$

$$P = \frac{T_{C2} - T_{C1}}{T_{H1} - T_{C1}} = \frac{200 - 60}{300 - 60} = 0.5833$$

$$W = \frac{R + 1 + \sqrt{R^2 + 1} - 2RX_P}{R + 1 + \sqrt{R^2 + 1} - 2X_P} = 0.6748$$

$$N_{SHELLS} = \frac{\ln\left[\dfrac{(1 - RP)}{(1 - P)}\right]}{\ln W} = 2.33$$

Thus, the unit requires three shells.

b)

$$Z = \frac{1 - PR}{1 - P} = \frac{1 - 0.5833 \times 1.4286}{1 - 0.5833} = 0.4$$

$$P_{1\text{–}2} = \frac{Z^{1/N_{SHELLS}} - 1}{Z^{1/N_{SHELLS}} - R} = \frac{(0.4)^{1/3} - 1}{(0.4)^{1/3} - 1.4286} = 0.3805$$

c)

$$F_T = \frac{\sqrt{R^2 + 1}\ln\left[\dfrac{(1 - P_{1\text{–}2})}{(1 - RP_{1\text{–}2})}\right]}{(R - 1)\ln\left[\dfrac{2 - P_{1\text{–}2}\left(R + 1 - \sqrt{R^2 + 1}\right)}{2 - P_{1\text{–}2}\left(R + 1 + \sqrt{R^2 + 1}\right)}\right]}$$

Substituting $R = 1.4286$ and $P_{1\text{–}2} = 0.3805$:

$$F_T = 0.86$$

d)

$$\Delta T_{LM} = \frac{(T_{H1} - T_{C2}) - (T_{H2} - T_{C1})}{\ln\left[\dfrac{T_{H1} - T_{C2}}{T_{H2} - T_{C1}}\right]} = \frac{(300 - 200) - (100 - 60)}{\ln\left[\dfrac{300 - 200}{100 - 60}\right]} = 65.48\ °\text{C}$$

$$A = \frac{Q}{U\Delta T_{LM} F_T} = \frac{3.5 \times 10^6}{100 \times 65.48 \times 0.86} = 619\ \text{m}^2$$

12.4 Heat Exchanger Geometry

Calculation of the overall heat transfer coefficient from Equation 12.13 requires knowledge of the film transfer coefficients. Although Table 12.1 presents typical values, the actual values depend on the velocities (flowrates) of the fluids, fluid physical properties and exchanger geometry. Methods are needed to calculate the film transfer coefficients for both the tube-side and the shell-side. In addition to heat transfer coefficients, it is also necessary to be able to predict the pressure drops.

Before considering the calculation of heat transfer coefficients and pressure drops, consider details of the exchanger geometry:

1) *Tube diameter.* Tube sizes are specified by an outside diameter and a wall thickness. The TEMA standards (TEMA, 2007) are based on imperial units in which the outside diameters are

Table 12.4

Commonly available tube sizes based on imperial sizes.

Wall thickness (mm)	0.889	1.245	1.651	2.108	2.769	3.404	4.191
Birmingham wire gage (BWG)	20	18	16	14	12	10	8
Outside diameter (mm)	Inside diameter (mm)						
15.88	14.10	13.39	12.57	11.66	10.34	9.07	
19.05	17.27	16.56	15.75	14.83	13.51	12.24	
25.40	23.62	22.91	22.10	21.18	19.86	18.59	17.02
31.75	29.97	29.26	28.45	27.53	26.21	24.94	23.37
38.10			34.80	33.88	32.56	31.29	
76.20				71.98	70.66	69.39	

specified in inches with wall thickness specified in terms of the Birmingham wire gage (BWG). The smaller the BWG, the thicker the wall. TEMA also specify the equivalent metric dimensions (TEMA, 2007). Some of the more commonly used sizes are summarized in Table 12.4. From Table 12.4, the most commonly used tube sizes in shell-and-tube heat exchangers are 19.05 mm and 25.5 mm outside diameter, and to a lesser extent 15.55 mm outside diameter. Tubes of 31.75 mm and 38.10 mm outside diameter are used in reboilers, evaporators and steam boilers. Larger diameter tubes of 76.2 mm to 101.6 mm are used in fired heaters (see later). The wall thickness in the first instance must be able to withstand the maximum pressure difference across the tube wall, including a corrosion allowance. However, this is not the only concern. Tube vibration and also the consequent wearing of the tube against the baffles are also a concern. The usual tube thicknesses used from Table 12.4 are 1.651 mm (16BWG) to 2.769 mm (12BWG). Thinner tube walls are used when expensive materials of construction are used. In preliminary design, when using steel tubes, a wall thickness of 2.108 mm (14BWG) can be assumed.

In addition to the standards based on imperial units, many other standards are used. Table 12.5 gives a sample of commonly used metric sizes available. Common sizes from

Table 12.5

Commonly available tube sizes based on metric sizes.

Wall thickness (mm)	1.6	2	2.6	3.2	3.6	4	4.5
Outside diameter (mm)	Inside diameter (mm)						
16	12.8	12.0	10.8	9.6			
18	14.8	14.0	12.8	11.6	10.8		
20	16.8	16.0	14.8	13.6	12.8	12.0	
22	18.8	18.0	16.8	15.6	14.8	14.0	13.0
25	21.8	21.0	19.8	18.6	17.8	17.0	16.0
30	26.8	26.0	24.8	23.6	22.8	22.0	21.0
32	28.8	28.0	26.8	25.6	24.8	24.0	23.0
38	34.8	34.0	32.8	31.6	30.8	30.0	29.0
40	36.8	36.0	34.8	33.6	32.8	32.0	31.0
70	66.8	66.0	64.8	63.6	62.8	62.0	61.0

Table 12.5 are $d_O = 20$ mm with $d_I = 16$ mm and $d_O = 25$ mm with $d_I = 19.8$ mm.

2) *Tube length.* Preferred tube lengths are quoted by TEMA (2007). Commonly used lengths based on imperial units are 1.83 m, 2.44 m, 3.05 m, 3.66 m, 4.88 m, 6.10 m and 7.32 m. A length of 4.88 m is commonly used in petroleum refinery applications. However, in principle, any length up to the maximum available from the tube manufacturer can be used. Metric lengths of 2.5, 3, 4, 5 and 6 m are also used. In U-tube designs (see Figures 12.9b) the length of the tubes varies according to the position in the tube bundle. Tubes closer to the inside of the tube bundle will have a smaller bend radius and will be shorter in overall length than those at the outside of the bundle. There will be a characteristic mean length across the bundle. The working length of the tube is slightly shorter than its end-to-end length from the length taken up by the tube sheets on which it is mounted. Tube sheets vary in thickness typically between 2 and 4 cm. Figure 12.13 shows how tubes are most commonly joined to tube sheets. Grooves are ground into the holes in the tube sheet and the tubes expanded by hydraulic pressure or roller expansion to create a seal. If the tubes and tube sheet can be welded, the joint can be strengthened by welding. The choice of tube length is a degree of freedom at the discretion of the designer. The same heat transfer area can be made available either with a small number of long tubes in a small diameter but long shell or a large number of short tubes in a large diameter but short shell. The ratio of tube length to shell diameter is usually in the range 5 to 15.

3) *Tube pitch.* Tube *pitch* (p_T) is the center-to-center distance between adjacent tubes. This varies between $1.25d_O$ and $1.5d_O$. Tube pitch should be a minimum of $1.25d_O$ and is often set to this value.

4) *Tube configuration.* Tubes can be arranged in either a *square* or *triangular configuration*, as illustrated in Figure 12.14. Rotated pitches tend to give higher shell-side heat transfer coefficients than in-line pitches. For the same tube pitch and shell-side flowrate the heat transfer coefficient is normally in the order $h_{S,30°} > h_{S,45°} > h_{S,60°} > h_{S,90°}$. A 90° layout has the lowest heat transfer coefficient, but the lowest pressure drop. A square configuration (45° or 90°), as shown in Figures 12.14a and 12.14b, is used for fouling fluids to provide *lanes* between the rows for mechanical cleaning. Cleaning lanes should be continuous through the entire tube bundle. Triangular configurations, as shown in Figures 12.14c and 12.14d, are restricted to nonfouling fluids on the outside of the tubes because they are more difficult to clean mechanically. However, for a given tube pitch, a triangular configuration allows a greater number of tubes in a given shell diameter. In Figure 12.14a, for a square pitch each tube is contained in an area of p_T^2. As shown in Figure 12.14c, for a triangular pitch each triangular pitch with sides p_T, having an area of $0.5p_T^2\sqrt{3}/2$, contains half a tube. Thus, a single tube in a triangular pitch is contained in an area of $p_T^2\sqrt{3}/2 = 0.866p_T^2$. This means that for the same pitch, a triangular configuration can allow a greater number of tubes than a square configuration in the same size of shell. The assumption of a square configuration in a new design will be conservative.

5) *Baffle configuration.* As shown in Figure 12.15a, baffles are used on the shell-side to direct the fluid stream across the tubes. Cross flow gives higher rates of heat transfer than axial flow at the expense of a higher pressure drop. Figure 12.15a shows segmental baffles, which are the most commonly used type. These are circular plates with a segment removed. The baffle not only increases the rate of heat transfer but also provides structural support to the tubes to prevent tubes sagging and to prevent tube vibration that can occur as a result of cyclic forces. The arrangement in Figure 12.15a shows the segmental baffles arranged for up and down flow on the shell-side. The baffles can be rotated through 90° to provide side-to-side flow, as in Figure 12.1d. This is desirable for vapor–liquid mixtures, such as in condensation. Double segmental baffles can also be used, as illustrated in Figure 12.15b. The pressure drop for double segmental baffles is typically one-third to one-half that for single segmental baffles (Bouhairie, 2012). However, this is at the expense of lower cross-flow heat transfer. Double segmental baffles are typically used when gases are on the shell-side and a low-pressure drop is necessary. Baffles are normally supported by tie rods and spacers, as illustrated in Figure 12.15c. A number of other baffle designs are possible (Bouhairie, 2012). In the methods and examples to be used later, it will be assumed that the most common arrangement of single segmental baffles is used.

Figure 12.16a shows an ideal shell-side flow pattern using segmental baffles with combinations of ideal cross flow and ideal axial flow. Cross flow gives both higher rates of heat transfer and higher pressure drops than axial flow. In practice, the flow pattern is not ideal, as illustrated in Figure 12.16a, as leakage occurs through the tube-to-baffle clearance, as illustrated in Figure 12.16b. Also, bypassing occurs between the tube bundle and the shell and is a function of the shell-to-baffle clearance, as illustrated in Figure 12.16b. Both leakage and bypassing act to reduce the rate of heat transfer on the shell-side.

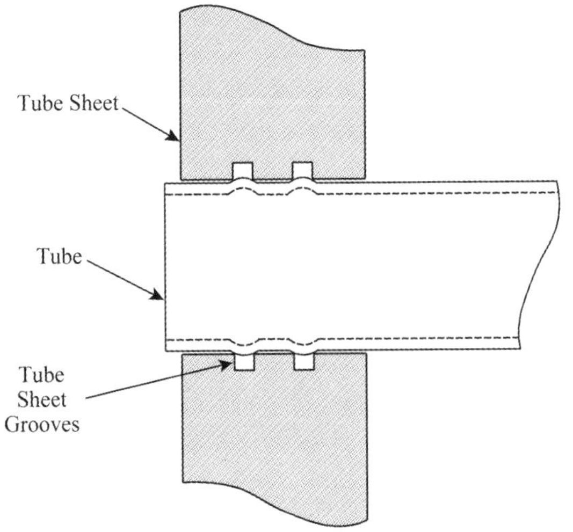

Figure 12.13
Fixing the tubes to the tubesheet.

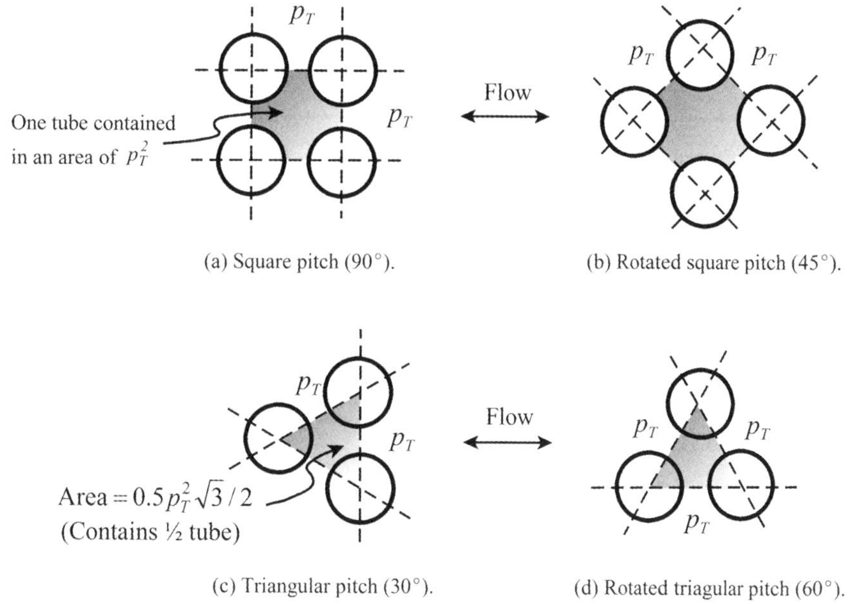

Figure 12.14
Tube configurations.

Longitudinal metal *sealing strips* can be mounted in notches at the edges of the baffles to minimize bypassing of the crossflow at the edge of the tube bundle (see Figure 12.15). These seals extend towards the inside of the shell to divert flow from the shell wall back into the tube bundle. Sealing strips are particularly important where there is a large clearance between the bundle and the shell, typically greater than 30 mm. Sealing strips are not normally fitted to fixed tube-sheet

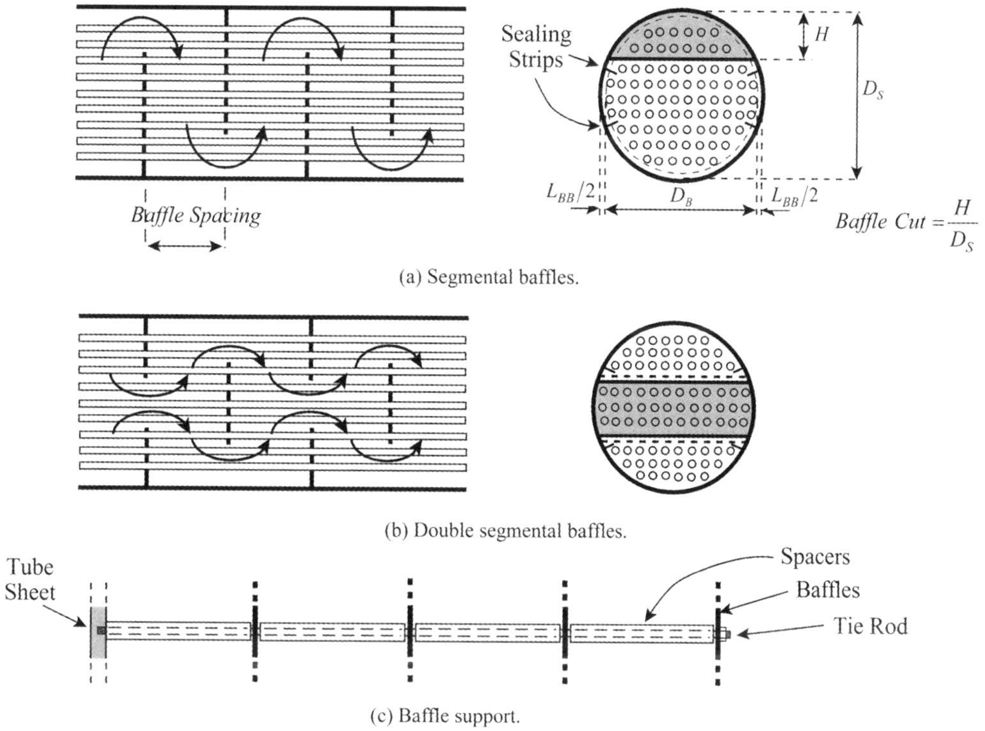

Figure 12.15
Shell-side baffles.

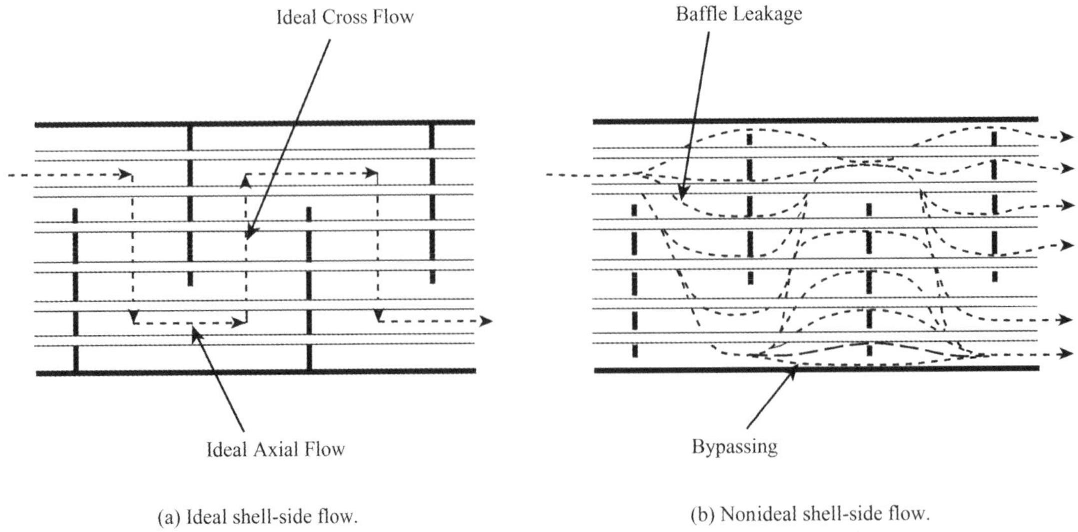

Figure 12.16
Shell-side flow patterns.

and U-tube designs, but split-ring and pull-through floating head designs usually require sealing strips. Bypassing of the bundle can also occur in the gaps in the core of the tube bundle left for the pass partition plates. Such internal bypassing can also be prevented by longitudinal sealing strips.

When using segmental baffles, some important parameters need to be fixed. The minimum spacing between segmental baffles should be larger than $0.2D_S$, where D_S is the shell inside diameter (TEMA, 2007). The optimum ratio of baffle spacing to shell diameter is that which results in the highest conversion of pressure drop to heat transfer and is in the range $0.3D_S <$ *baffle spacing* $< 0.6D_S$ (Bouhairie, 2012). In some cases, larger baffle spacing is used in the inlet and outlet compartments compared with the central baffle spacing. This is to allow for placement of shell-side nozzles without interference with the heat exchanger flanges and without overlapping the first or last baffle. Also, for segmental baffles, the number of baffles and the location of the shell-side nozzles depend on whether the number of baffles is an even or odd number. An even number of baffles implies the shell-side inlet and outlet nozzles both being on the same side of the shell (e.g. see Figure 12.1b). An odd number of baffles implies that the shell-side inlet and outlet nozzles are on opposite sides of the shell.

The *baffle cut* is the height of the segment removed from the baffle as a fraction of the baffle disc diameter, as illustrated in Figure 12.15a. The optimum baffle cut is typically 0.25, although baffle cuts from 0.15 to 0.45 are used in special circumstances. As much as possible, the baffle spacing and baffle cut should match to provide a similar velocity in cross flow and axial flow in the baffle window (Bouhairie, 2012). This avoids repeated acceleration and deceleration of the fluid.

6) *Bundle clearance.* A clearance must be allowed between the tube bundle and the inside shell wall. This clearance L_{BB} is defined as the distance between the inside shell wall and the circle circumscribing the outermost tubes in the bundle (see Figure 12.15a). The bundle diameter D_B is the diameter of the circle circumscribing the outmost tubes (see Figure 12.15a). The value of L_{BB} depends mainly on the design of bundle to be used. Fixed tube sheet and U-tube designs require the smallest clearances. Split-ring floating-head designs require much larger clearances. Pull-through floating-head designs require larger clearances, and are also a function of the shell-side design pressure. The clearance is important in the design performance, as it increases the flow area on the shell-side. Table 12.6 gives typical values for different types of shell design.

7) *Shell diameter.* For a square pitch, each tube is contained in an area of p_T^2 and for a triangular pitch each tube is contained in an area of $0.866p_T^2$. Thus, for heat exchangers with a single tube pass the number of tubes that can in principle fit into a shell with diameter D_S is given by:

$$N_T = \frac{\frac{\pi}{4}D_S^2}{p_C p_T^2} \tag{12.48}$$

Table 12.6
Typical inside shell diameter to bundle clearances (L_{BB} and D_S in m).

Design type	Clearance
Fixed tube sheet and U-tube	$L_{BB} = 0.0048\,D_S + 0.0133$
Split-ring and packed floating heads	$L_{BB} = 0.0169\,D_S + 0.0257$
Pull-through floating heads, 1000 kPa	$L_{BB} = 5 \times 10^{-3}\,D_S^2 + 0.0179\,D_S + 0.0822$
Pull-through floating heads, 2000 kPa	$L_{BB} = 3 \times 10^{-3}\,D_S^2 + 0.0332\,D_S + 0.0827$

where N_T = total number of tubes (–)
D_S = internal diameter of the shell (m)
p_T = tube pitch (m)
p_C = pitch configuration factor (–)
= 1 for a square pitch
= 0.866 for a triangular pitch

In practice, some allowance needs to be made for the clearance between the tube bundle and the shell wall. In addition, if U-tubes are used, the shell diameter needs to be increased to allow for the minimum bend radius. If multiple tube passes are required, the pass partition plates do not allow complete coverage of the shell diameter by tubes, as there needs to be a seal between the pass partition plates and the tube sheet. Also, if floating head designs are used, the shell diameter needs to be increased to allow for withdrawal of the floating head arrangement. Thus, two factors need to be added to Equation 12.48 to allow for these effects:

$$N_T = \frac{\pi D_S^2/4}{F_{TC}F_{SC}p_C p_T^2} \tag{12.49}$$

where F_{TC} = tube count constant that accounts for the incomplete coverage of the shell diameter by the tubes, due to necessary clearances between the shell and the tube bundle and tube omissions due to the location of pass partition plates for multiple pass designs; F_{TC} is given in Table 12.7
F_{SC} = correction factor for the shell construction as given in Table 12.8

Equation 12.49 gives an approximation for the tube count for a given shell diameter. A more accurate tube count can be obtained from tube count tables (TEMA, 2007). Equation 12.49 can be rearranged to give the shell diameter:

$$D_S = \left(\frac{4N_T F_{TC}F_{SC}p_C p_T^2}{\pi}\right)^{1/2} \tag{12.50}$$

The number of tubes can be eliminated from Equation 12.50 by introducing the heat transfer area:

$$A = N_T \pi d_O L \tag{12.51}$$

Table 12.7

F_{TC} for various tube passes (for $D_S > 0.337$ m).

Tube passes	F_{TC}
$N_P = 1$	1.08
$N_P = 2$	1.11
$N_P = 4, 6$	1.45 for $D_s \le 635$ mm 1.18 for $D_s > 635$ mm

Table 12.8

F_{SC} for various tube bundle geometries (for $D_S > 0.337$ m).

Head type	F_{SC}
Fixed head	1.0
Floating head (split backing ring or outside packed loating head)	1.15
U-tube	1.05 1.09 for 25 mm outside diameter tubes on $1.25d_O$

where A = heat transfer area based on the tube outside surface (m^2)
N_T = total number of tubes (–)
d_O = tube outside diameter (m)
L = tube length (m)

Combining Equations 12.50 and 12.51 to eliminate N_T:

$$D_S = \left(\frac{4F_{TC}F_{SC}p_C p_T^2 A}{\pi d_O L}\right)^{1/2} \tag{12.52}$$

Equation 12.52 can be used to approximate the shell diameter if the tube length L is specified. Alternatively, the tube length to shell diameter ratio (L/D_S) can be specified. If this is specified, then Equation 12.52 can be rearranged to give:

$$D_S = \left(\frac{4F_{TC}F_{SC}p_C p_T^2 A}{\pi d_O (L/D_S)}\right)^{1/3} \tag{12.53}$$

The ratio of tube length to shell diameter is usually in the range 5 to 10. For smaller shell diameters, the shell is most often manufactured from standard pipes and for larger diameters manufactured from plate by rolling and welding the plate. It is left to the discretion of the manufacturer to establish a standard system of shell diameters, but tolerances and shell thickness must conform to standard codes (TEMA, 2007).

8) *Inlet and outlet nozzles.* The tube-side and shell-side flows need to be connected to outside pipework via nozzles. The appropriate size for the nozzles depends on both the pressure drop in the nozzle and the pipework to which the heat exchanger is to be connected. Excessive velocity through the shell-side inlet nozzle can cause erosion to the tubes or induce vibration. To avoid vibration, TEMA (2007) recommend:

$\rho v_N^2 \le 2200\ kg \cdot m^{-1} \cdot s^{-2}$ for nonabrasive single-phase fluids

$\rho v_N^2 \le 740\ kg \cdot m^{-1} \cdot s^{-2}$ for all other liquids, including saturated liquids

For velocities in excess of these conditions an *impingement plate* can be installed between the shell-side inlet nozzle and the tubes to protect the tubes.

12

12.5 Allocation of Fluids in Shell-and-Tube Heat Exchangers

In a new design, the allocation of streams to the tube-side or shell-side will need to be defined. The issues to be considered for the allocation include:

1) *Heat transfer coefficients.* The fluid with the lower heat transfer coefficient can be allocated to the shell-side. The heat transfer coefficients will depend on the physical properties of the fluids and the flowrates. If the flowrate of one of the fluids is lower than the other, allocating to the shell-side can allow a higher overall heat transfer coefficient to be obtained. However, allocating the fluid with the lower flowrate to the tube-side allows the velocity to be manipulated by increasing the number of tube passes. Thus, the designer can address the potential problem of a lower heat transfer coefficient in different ways.
2) *Fouling.* The fluid that has the greatest tendency to foul the heat transfer surfaces is often allocated to the tube-side. This will give better control over the design fluid velocity and the higher allowable velocity in the tubes will reduce fouling. Stagnant zones and zones with low velocity can occur on the shell-side, leading to accelerated fouling. Also, the inside of the tubes is easier to clean than the shell-side. However, the fluid with the greater tendency to foul will also often be the one with the lower heat transfer coefficient. Thus, it might still be better to allocate the fluid with the greater fouling tendency to the shell-side. However, if this is done, the shell-side design must reflect this by the appropriate choice of tube layout and baffle spacing.
3) *Materials of construction.* If expensive materials of construction are required for one of the fluids because of its corrosive nature or high temperature, then this fluid should normally be allocated to the tube-side. This can reduce the cost of expensive materials of construction.
4) *Operating pressures.* The stream with the higher pressure should be allocated to the tube-side. The smaller the diameter of a tube, the thinner the wall needed to contain the same pressure. The tubes are therefore more effective at containing high pressure than the shell.
5) *Pressure drop.* For the same pressure drop, higher heat transfer coefficients will be obtained on the tube-side than on the shell-side. The fluid with the lower allowable pressure drop should normally be allocated to the tube-side.
6) *Viscosity.* Generally, a higher heat transfer coefficient will be obtained by allocating the more viscous material to the shell-side, provided the flow is turbulent. The critical Reynolds number for turbulent flow on the shell-side is in the region of 200.
7) *Fluid temperatures.* Placing the hotter fluid in the tubes will reduce the shell surface temperature, and hence the need for lagging to reduce heat loss, and might be desirable for safety reasons.

These general guidelines can contradict each other. If this is the case, some compromise must be made.

12.6 Heat Transfer Coefficients and Pressure Drops in Shell-and-Tube Heat Exchangers

Simple shell-and-tube heat exchanger models are developed in Appendix F in which heat transfer coefficient and pressure drop are both related to velocity. The basic correlations used in the model (Wang *et al.*, 2012; Jiang, Shelley and Smith, 2014) are suitable for the following conditions:

i. single-phase heat transfer in a shell-and-tube heat exchanger;
ii. plain tubes;
iii. single segmental baffles with 20–50% cut;
iv. physical properties are assumed constant, based on an average between the inlet and outlet conditions.

1) *Tube-side heat transfer coefficient*
   a) *For laminar flow* $Re \leq 2100$ and $L \leq 0.05 Re \cdot Pr \cdot d_I$:

$$h_T = K_{hT1} v_T^{1/3} \tag{12.54}$$

where

$$K_{hT1} = 1.86 \frac{k}{d_I} \left[\left(\frac{\rho d_I}{\mu} \right) Pr \left(\frac{d_I}{L} \right) \right]^{1/3} \tag{12.55}$$

   b) *For transition flow* $2100 < Re < 10^4$:

$$h_T = K_{hT2} v_T^{2/3} - K_{hT3} \tag{12.56}$$

where

$$K_{hT2} = 0.116 \frac{k}{d_I} \left(\frac{\rho d_I}{\mu} \right)^{2/3} Pr^{1/3} \left[1 + \left(\frac{d_I}{L} \right)^{2/3} \right] \tag{12.57}$$

$$K_{hT3} = 14.5 \frac{k}{d_I} Pr^{1/3} \left[1 + \left(\frac{d_I}{L} \right)^{2/3} \right] \tag{12.58}$$

   c) *For fully developed turbulent flow* $Re \geq 10^4$:

$$h_T = K_{hT4} v_T^{0.8} \tag{12.59}$$

where

$$K_{hT4} = C \frac{k}{d_I} Pr^{0.4} \left(\frac{\rho d_I}{\mu} \right)^{0.8} \tag{12.60}$$

C = constant (0.024 for heating, 0.023 for cooling)

h_T = tube-side heat transfer coefficient

$Re = \dfrac{\rho v_T d_I}{\mu}$, the tube-side Reynolds number

$Pr = \dfrac{C_P \mu}{k}$, the tube-side Prandtl number

d_I = tube inner diameter

L = tube length

ρ = tube-side fluid density

μ = tube-side fluid viscosity

C_P = tube-side heat capacity

k = tube-side fluid thermal conductivity

v_T = mean fluid velocity inside the tubes $= \dfrac{m_T(N_P/N_T)}{\rho(\pi d_I^2/4)}$

m_T = mass flowrate on the tube-side

N_P = number of tube passes

N_T = number of tubes

Fluid physical properties are taken at the average bulk fluid temperature between the tube-side flow inlet and outlet.

2) *Tube-side pressure drop.* The total tube-side pressure drop ΔP_T for a single shell comprises the pressure drop in the straight tubes (ΔP_{TT}), the pressure drop in the tube entrances, exits and reversals (ΔP_{TE}), and the pressure drop in nozzles (ΔP_{TN}) (Serth, 2007).

$$\begin{aligned}\Delta P_T &= \Delta P_{TT} + \Delta P_{TE} + \Delta P_{TN} \\ &= K_{PT1} N_P L v_T^{2+m_f} + K_{PT2} v_T^2 + K_{PT3}\end{aligned} \tag{12.61}$$

where

$$K_{PT1} = \frac{2F_C \left(\dfrac{\rho d_I}{\mu}\right)^{m_f} \rho}{d_I} \tag{12.62}$$

$$K_{PT2} = 0.5\alpha_R \rho \tag{12.63}$$

$$K_{PT3} = \rho\left(C_{TN,inlet} v_{TN,inlet}^2 + C_{TN,outlet} v_{TN,outlet}^2\right) \tag{12.64}$$

where
$F_C = 16$, $m_f = -1$ for $Re \le 2100$
$F_C = 5.36 \times 10^{-6}$, $m_f = 0.949$ for $2100 > Re > 3000$
$F_C = 0.0791$, $m_f = -0.25$ for $Re \ge 3000$
$\alpha_R = 3.25 N_P - 1.5$ for $500 \le Re \le 2100$
$\alpha_R = 2N_P - 1.5$ for $Re > 2100$
$C_{TN,inlet} = 0.75$ for $100 \le Re_{TN,inlet} \le 2100$
$C_{TN,inlet} = 0.375$ for $Re_{TN,inlet} > 2100$

$$Re_{TN,inlet} = \frac{\rho v_{TN,inlet} d_{TN,inlet}}{\mu}$$

$$v_{TN,inlet} = \frac{m_T}{\rho(\pi d_{TN,inlet}^2/4)}$$

$d_{TN,inlet}$ = inner diameter of the inlet nozzle for the tube-side fluid

$C_{TN,outlet} = 0.75$ for $100 \le Re_{TN,outlet} \le 2100$

$C_{TN,outlet} = 0.375$ for $Re_{TN,outlet} > 2100$

$$Re_{TN,inlet} = \frac{\rho v_{TN,inlet} d_{TN,inlet}}{\mu}$$

$$v_{TN,inlet} = \frac{m_T}{\rho(\pi d_{TN,inlet}^2/4)}$$

$d_{TN,outlet}$ = inner diameter of the outlet nozzle for the tube-side fluid

For heat exchangers with multiple shells connected in series, the total pressure drop on the tube-side fluid is estimated by the product of the pressure drop per shell and shell number in series:

$$\Delta P_{T,N_{SHELLS}} = N_{SHELLS}\, \Delta P_T \tag{12.65}$$

where $\Delta P_{T,N_{SHELLS}}$ = total pressure drop on the tube-side across N_{SHELLS}
ΔP_T = pressure drop per shell, given by Equation 12.61
N_{SHELLS} = number of shells connected in series

For heat exchangers with multiple parallel shells, the total pressure drop on the tube-side fluid is equal to the pressure drop in a single shell ΔP_T.

3) *Shell-side heat transfer coefficient.* The shell-side heat transfer coefficient is calculated from (Ayub, 2005; Wang, *et al.*, 2012):

a) *For $Re_S \le 250$:*

$$h_S = K_{hS1} v_S^2 + K_{hS2} v_S + K_{hS3} \tag{12.66}$$

where

$$K_{hS1} = -3.722 \times 10^{-5} \frac{F_P F_L J_S k^{2/3} c_P^{1/3} \rho^2 d_O}{\mu^{5/3}} \tag{12.67}$$

$$K_{hS2} = 0.03843 \frac{F_P F_L J_S k^{2/3} c_P^{1/3} \rho}{\mu^{2/3}} \tag{12.68}$$

$$K_{PS3} = \rho\left(C_{NS,inlet} v_{NS,inlet}^2 + C_{NS,outlet} v_{S,outlet}^2\right) \tag{12.69}$$

b) *For $250 < Re_S \le 250{,}000$:*

$$h_S = K_{hS4} v_S^{0.6633} \tag{12.70}$$

where

$$K_{hS4} = 0.08747 \frac{F_P F_L J_S k^{2/3} c_P^{1/3} \rho^{0.6633}}{\mu^{0.33} d_O^{0.3367} B_C^{0.5053}} \tag{12.71}$$

h_S = shell-side heat transfer coefficient
d_O = tube outer diameter
ρ = shell-side fluid density
k = shell-side fluid thermal conductivity
c_P = shell-side fluid specific heat capacity
μ = shell-side fluid viscosity
F_P = the pitch factor, which depends on the tube layout of the bundle
= 1.0 for triangular and diagonal square pitch
= 0.85 for in-line square pitch
F_L = the leakage factor to allow for all the stream leakage, which is a function of bundle configuration
= 0.9 for straight tube bundle
= 0.85 for U-tube bundle
= 0.8 for floating head bundle
J_S = correction factor for unequal baffle spacing (Taborek in Hewitt, 2008; Serth, 2007)

$$= \frac{(N_B - 1) + (B_{in}/B)^{2/3} + (B_{out}/B)^{2/3}}{(N_B - 1) + (B_{in}/B) + (B_{out}/B)} \quad \text{for } Re_S < 100$$
$$= \frac{(N_B - 1) + (B_{in}/B)^{0.4} + (B_{out}/B)^{0.4}}{(N_B - 1) + (B_{in}/B) + (B_{out}/B)} \quad \text{for } Re_S \geq 100 \qquad (12.72)$$

B = central baffle spacing
B_{in} = inlet baffle spacing
B_{out} = outlet baffle spacing
B_C = baffle cut (–)
Re_S = Reynolds Number based on tube outside diameter
$= \dfrac{\rho d_O v_S}{\mu}$
v_S = shell-side fluid velocity

$$= \frac{m_S}{\rho B\left[(D_S - D_B) + \dfrac{(D_B - d_O)(p_T - d_O)}{p_{CF} p_T}\right]} \qquad (12.73)$$

D_S = shell inside diameter
D_B = outside diameter of the tube bundle
p_T = tube pitch
p_{CF} = pitch correction factor for flow direction
= 1 for 90° and 30° layouts
= $\sqrt{2}/2$ for 45° layouts
= $\sqrt{3}/2$ for 60° layouts

c) *Shell-side pressure drop.* The total pressure drop ΔP_S for the shell-side in one shell includes the pressure drop in the straight section of shell (ΔP_{SS}) and pressure drop in nozzles (ΔP_{NS}) (Kern and Kraus, 1972). The total pressure drop for the shell-side per shell is:

$$\Delta P_S = \Delta P_{SS} + \Delta P_{NS} = K_{PS1} v_S^{1.875} + K_{PS2} v_S^{1.843} + K_{PS3} \qquad (12.74)$$

where

$$K_{PS1} = 18\left(5\frac{B}{D_S} - 1\right)(N_B - 1 + R_S)\frac{a D_S \rho}{d_e}\left(\frac{B_C}{0.2}\right)^{m_{fo}}\left(\frac{\rho d_e}{\mu}\right)^{-0.125} \qquad (12.75)$$

$$K_{PS2} = 90\left(1 - \frac{B}{D_S}\right)(N_B - 1 + R_S)\frac{b D_S \rho}{d_e}\left(\frac{B_C}{0.2}\right)^{m_{fo}}\left(\frac{\rho d_e}{\mu}\right)^{-0.157} \qquad (12.76)$$

$$K_{PS3} = \rho\left(C_{NS,inlet} v_{NS,inlet}^2 + C_{NS,outlet} v_{S,outlet}^2\right) \qquad (12.77)$$

D_S = shell inner diameter
N_B = number of baffles
R_S = correction factor for unequal baffle spacing
$$= (B/B_{in})^{1.8} + (B/B_{out})^{1.8} \qquad (12.78)$$
B = central baffle spacing
B_{in} = inlet baffle spacing
B_{out} = outlet baffle spacing
d_e = shell-side equivalent diameter
$= C_{De}\dfrac{p_T^2}{d_O} - d_O$
C_{De} = pitch configuration factor
= $4/\pi$ for square pitch
= $2\sqrt{3}/\pi$ for triangular pitch
a = 0.008190 for $D_S \leq 0.9$ m
= 0.01166 for $D_S > 0.9$ m
b = 0.004049 for $D_S \leq 0.9$ m
= 0.002935 for $D_S > 0.9$ m
m_{fo} = −0.26765 for $20\% < B_C \leq 30\%$
m_{fo} = −0.36106 for $30\% \leq B_C < 40\%$
m_{fo} = −0.58171 for $40\% \leq B_C \leq 50\%$
m_S = mass flowrate on the shell-side
$C_{NS,inlet}$ = 0.375 for $Re_{NS,inlet} > 2100$
$C_{NS,inlet}$ = 0.75 for $100 \leq Re_{NS,inlet} \leq 2100$

$$Re_{NS,inlet} = \frac{\rho v_{NS,inlet} d_{NS,inlet}}{\mu}$$

$$v_{NS,inlet} = \frac{m_S}{\rho(\pi d_{NS,inlet}^2/4)}$$

$d_{NS,inlet}$ = inner diameter of the inlet nozzle for the shell-side fluid
$C_{NS,outlet}$ = 0.375 for $Re_{NS,outlet} > 2100$
$C_{NS,outlet}$ = 0.75 for $100 \leq Re_{NS,outlet} \leq 2100$

$$Re_{NS,outlet} = \frac{\rho v_{NS,outlet} d_{NS,outlet}}{\mu}$$

$$v_{NS,outlet} = \frac{m_S}{\rho\left(\pi d_{NS,outlet}^2/4\right)}$$

$d_{NS,outlet}$ = inner diameter of the outlet nozzle for the shell-side fluid

For heat exchangers with multiple shells connected in series, the total pressure drop on the shell-side fluid is estimated by the product of the pressure drop per shell and shell number in series:

$$\Delta P_{S,N_{SHELLS}} = N_{SHELLS} \Delta P_S \qquad (12.79)$$

where $\Delta P_{S,N_{SHELLS}}$ = total pressure drop on the shell-side
ΔP_S = pressure drop per shell, given by Equation 12.74
N_{SHELLS} = number of shells connected in series

For heat exchangers with multiple parallel shells, the total pressure drop on the shell-side fluid is equal to the pressure drop in a single shell ΔP_S.

These shell-and-tube heat exchanger models have been compared with commercial heat exchanger simulation software that uses more complex models and found to be in close agreement (Jiang, Shelley and Smith, 2014).

Before these equations can be applied for design, a number of issues need to be considered:

1) *Fluid velocity*. Equations 12.54, 12.56, 12.59, 12.61, 12.66, 12.70 and 12.74 require knowledge of the fluid velocity. Liquid velocity on the tube-side is usually of the order of 1 to 3 $m \cdot s^{-1}$. Whilst velocity on the tube-side is unambiguous, velocity on the shell-side can be confusing. Velocity changes continuously across the bundle, since the fluid experiences a differing cross-sectional area as it travels across the shell. There is also a difference between the cross-flow velocity across the bundle and that the axial-flow velocity in the baffle window. The basis for the shell-side velocity is normally taken to be the velocity in the cross flow at the widest point in the shell (see Appendix F). However, there is still a potential for confusion when quoting the velocity, as the velocity can be quoted on the basis of no leakage, or can include the leakage caused by the gap between the tube bundle and the shell and possible gaps within the bundle for location of pass partition plates. Liquid velocity on the shell-side is normally lower than that on the tube-side. This is because the geometry on the shell-side interrupts and changes the direction of the flow, inducing turbulence. The shell-side velocity is often quoted on the basis of no leakage and is usually of the order of 0.5 to 1.5 $m \cdot s^{-1}$. Leakage will reduce the velocity. Gas velocities on both the tube-side and the shell-side typically vary in the range 50 to 70 $m{\cdot}s^{-1}$ for gases under vacuum, 10 to 30 $m{\cdot}s^{-1}$ for gases at atmospheric pressure and 5 to 10 $m{\cdot}s^{-1}$ for gases under pressure.

   High velocities will result in high heat transfer coefficients and will tend to reduce fouling. However, high velocities also result in a high pressure drop. Also, if solids are present in the fluids, then high velocities can result in erosion. High velocities on the shell-side for liquids (typically greater than 3 $m \cdot s^{-1}$) can also induce vibration within the heat exchanger that can cause it long-term damage.

   The velocity is a critical parameter for the heat exchanger design. For the tube-side, a higher velocity for the same duty will require a smaller number of, but longer, tubes and a higher tube-side pressure drop. On the shell-side the velocity is predominantly dictated by the shell diameter and the baffle spacing. For a given tube velocity (and therefore a given number of tubes and shell diameter) a higher shell-side velocity will require a smaller baffle spacing and a higher shell-side pressure drop.

2) *Pressure drop*. Rather than specify a fluid velocity, it is often preferred to specify the pressure drop across the heat exchanger. In retrofit situations, where a new heat exchanger is to be installed in an existing plant, the allowable pressure drop is often highly constrained. This is because in a retrofit situation it is often desirable to avoid the installation of new pumps or retrofit the existing pumps. Pumps will be discussed in detail in the next chapter. In the absence of a specific retrofit constraint for pressure drop, the value for liquids is normally in the range 0.35 to 0.7 bar. For low-viscosity liquids (less than 1 $mN \cdot s \cdot m^{-2}$), the pressure drop is likely to be less than 0.35 bar. For gases, the maximum allowable pressure drop varies typically between 1 bar for high-pressure gases (10 bar and above) down to 0.01 bar for gases under vacuum conditions. For gases, the cost associated with the pressure drop is more sensitive to aspects of the design outside the immediate considerations for the heat exchanger design than is the case for liquids.

As pointed out in Section 12.1, fouling adds additional resistances to heat transfer on both the tube-side and the shell-side. If a fluid has a particularly high tendency to foul (e.g. crude oil), then designing for a low-pressure drop can be a false economy. The fouling fluid is often placed on the tube-side. As the fluid fouls the tube-side, not only will the overall heat transfer coefficient decrease significantly but the pressure drop will also increase significantly as a direct result of the fouling. If, for this highly fouling fluid, the exchanger had been designed for an initial high-pressure drop on the tube-side, the resulting high shear stress would not only have increased the initial rate of heat transfer but also decreased the rate of deterioration of the rate of heat transfer and the rate of increase of pressure drop. An initial design for a tube-side pressure drop of 0.7 bar for a highly fouling fluid (e.g. crude oil) might result in a fouled pressure drop of two to three times the clean pressure drop (Nesta and Coutinho, 2011). Had the initial design been for, say, 1.5 bar, this could have led to a lower pressure drop after fouling and an overall better performance.

If a pressure drop on the tube-side needs to be decreased, then the following measures can be considered:

- decrease the number of tube passes;
- decrease the tube length and thereby increase the shell diameter;
- increase the tube diameter.

If the pressure drop on the shell-side needs to be decreased, then the following measures can be considered:

- increase the baffle spacing;
- increase the baffle cut;
- increase the tube pitch;

- change the tube pitch configuration from triangular to square or rotated to in-line, or both;
- use an alternative baffle design.

3) *Final design.* Applying the equations described here will result in certain specifications for the heat transfer area, tube length and shell diameter. The preliminary design would then need to be checked against a detailed simulation of the heat exchanger. One particularly important factor is the variation of physical properties that takes place within the heat exchanger. This can only be accounted for by more detailed models that divide the heat exchanger into zones.

Example 12.2 A crude oil stream is to be preheated by recovering heat from a kerosene product in a shell-and-tube heat exchanger. The flowrates, temperatures and physical properties (at the mean temperatures) are given in Table 12.9.

The exchanger type to be used is 1 shell pass, 2 tube passes with a split-ring floating head. Because crude oil will have a greater tendency to foul than kerosene, crude oil will be allocated to the tube-side. Steel tubes are to be used with a thermal conductivity of $45\ W \cdot m^{-1} \cdot K^{-1}$ and the following assumptions can be made regarding the heat exchanger geometry:

$$\begin{aligned} d_O &= 20\,mm \\ d_I &= 16\,mm \\ p_T &= 1.25 d_O \\ \text{Tube layout} &= 90° \\ B_C &= 0.25 \\ d_{NT,inlet} &= 0.3048\,m \\ d_{NT,outlet} &= 0.3048\,m \\ d_{NS,inlet} &= 0.2027\,m \\ d_{NS,outlet} &= 0.2027\,m \end{aligned}$$

Table 12.9

Data for a heat recovery problem.

	Kerosene	Crude oil
Flowrate ($kg \cdot s^{-1}$)	25	79.07
Initial temperature (°C)	200	35
Final temperature (°C)	95	75
Density ($kg \cdot m^{-3}$)	730	830
Heat capacity ($J \cdot kg^{-1} \cdot K^{-1}$)	2470	2050
Viscosity ($N \cdot s \cdot m^{-2}$)	4.0×10^{-4}	3.6×10^{-3}
Thermal conductivity ($W \cdot m^{-1} \cdot K^{-1}$)	0.132	0.133
Fouling coefficient ($W \cdot m^{-2} \cdot K^{-1}$)	5000	2000

a) Calculate the number of shells required and the F_T based on $X_P = 0.9$.
b) Calculate the overall heat transfer coefficient, heat transfer area, number of tubes, tube length and shell diameter assuming a fluid velocity of $1\ m \cdot s^{-1}$ on both the tube-side and $0.5\ m \cdot s^{-1}$ on the shell-side.
c) Calculate the pressure drop for the tube-side and shell-side assuming a fluid velocity of $1\ m \cdot s^{-1}$ on the tube-side and $0.5\ m \cdot s^{-1}$ on the shell-side.
d) Calculate the overall heat transfer coefficient and heat transfer area for a tube-side pressure drop of 0.15 bar and a shell-side pressure drop of 0.15 bar.

Solution

a) First calculate the values of R, P and F_T:

$$R = \frac{T_{H1} - T_{H2}}{T_{C2} - T_{C1}} = \frac{200 - 95}{75 - 35} = 2.625$$

$$P = \frac{T_{C2} - T_{C1}}{T_{H1} - T_{C1}} = \frac{75 - 35}{200 - 35} = 0.2424$$

$$W = \frac{R + 1 + \sqrt{R^2 + 1} - 2RX_P}{R + 1 + \sqrt{R^2 + 1} - 2X_P} = 0.3688$$

$$N_{SHELLS} = \frac{\ln\left[\frac{1 - RP}{1 - P}\right]}{\ln W} = 0.74$$

Thus, the unit requires 1 shell.

$$F_T = \frac{\sqrt{R^2 + 1}\ \ln\left[\frac{(1 - P)}{(1 - RP)}\right]}{(R - 1)\ln\left[\frac{2 - P\left(R + 1 - \sqrt{R^2 + 1}\right)}{2 - P\left(R + 1 + \sqrt{R^2 + 1}\right)}\right]}$$

Substituting $R = 2.625$ and $P = 0.2424$:

$$F_T = 0.90$$

b) For sizing the heat exchanger, the velocity on both the tube-side and shell-side will be used to guide the design. Velocities must be assumed. For the tube-side, with an assumed velocity of $1\ m \cdot s^{-1}$, calculate the Reynolds number:

$$Re = \frac{\rho v_T d_I}{\mu} = \frac{830 \times 1 \times 0.016}{3.6 \times 10^{-3}} = 3689$$

This is a transitional flow and the tube-side heat transfer coefficient can be calculated from Equation 12.56. However, for transitional (or laminar flow) the calculation of the heat transfer coefficient requires the tube length to be known. Hence, in this case an initial tube length must be assumed, say 6 m:

$$K_{hT1} = 0.116\frac{k}{d_I}\left(\frac{\rho d_I}{\mu}\right)^{2/3} Pr^{1/3}\left[1+\left(\frac{d_I}{L}\right)^{2/3}\right]$$

$$= 0.116 \times \frac{0.133}{0.016} \times \left(\frac{830 \times 0.016}{3.6 \times 10^{-3}}\right)^{2/3}\left(\frac{2050 \times 3.6 \times 10^{-3}}{0.133}\right)^{1/3}$$

$$\times \left[1+\left(\frac{0.016}{6.0}\right)^{2/3}\right]$$

$$= 895.0$$

$$K_{hT2} = 14.5\frac{k}{d_I} Pr^{1/3}\left[1+\left(\frac{d_I}{L}\right)^{2/3}\right]$$

$$= 14.5 \times \frac{0.133}{0.016} \times \left(\frac{2050 \times 3.6 \times 10^{-3}}{0.133}\right)^{1/3} \times \left[1+\left(\frac{0.016}{6.0}\right)^{2/3}\right]$$

$$= 468.6$$

$$h_T = K_{hT1}v_T^{2/3} - K_{hT2}$$
$$= 895.0 v_T^{2/3} - 468.6$$
$$= 426.4\ \text{W}\cdot\text{m}^2\cdot\text{K}^{-1}$$

For the shell-side, with an assumed velocity of $0.5\,\text{m}\cdot\text{s}^{-1}$, calculate the Reynolds number:

$$Re_S = \frac{\rho d_O v_S}{\mu}$$
$$= \frac{730 \times 0.020 \times 0.5}{4.0 \times 10^{-4}}$$
$$= 18{,}250$$

Calculate the shell-side heat transfer coefficient from Equation 12.70:

$$K_{hS4} = 0.08747\frac{F_P F_L J_S k^{2/3} c_P^{1/3} \rho^{0.6633}}{\mu^{0.33} d_O^{0.3367} B_C^{0.5053}}$$

$$= 0.08747 \times \frac{0.85 \times 0.8 \times 1 \times 0.132^{2/3} \times 2470^{1/3} \times 730^{0.6633}}{(4 \times 10^{-4})^{0.33} \times 0.02^{0.3367} \times 0.25^{0.5053}}$$

$$= 1643$$

$$h_S = K_{hS4} v_S^{0.6633}$$
$$= 1643 \times 0.5^{0.6633}$$
$$= 1038\ \text{W}\cdot\text{m}^{-2}\cdot\text{K}^{-1}$$

Now calculate the overall heat transfer coefficient:

$$\frac{1}{U} = \frac{1}{h_S} + \frac{1}{h_{SF}} + \frac{d_O}{2k}\ln\left(\frac{d_O}{d_I}\right) + \frac{d_O}{d_I}\frac{1}{h_{TF}} + \frac{d_O}{d_I}\frac{1}{h_T}$$

$$= \frac{1}{1038} + \frac{1}{5000} + \frac{0.02}{2 \times 45}\ln\left(\frac{0.02}{0.016}\right) + \frac{0.02}{0.016}\left(\frac{1}{2000} + \frac{1}{426.4}\right)$$

$$U = 209.7\ \text{W}\cdot\text{m}^{-2}\cdot\text{K}^{-1}$$

$$\Delta T_{LM} = \frac{(200-75)-(95-35)}{\ln\left[\frac{200-75}{95-35}\right]}$$

$$= 88.6\ ^\circ\text{C}$$

Calculate the first estimate of the required area:

$$A = \frac{Q}{U\,\Delta T_{LM} F_T}$$
$$= \frac{79.07 \times 2050(75-35)}{209.7 \times 88.6 \times 0.9}$$
$$= 387.7\ \text{m}^2$$

Calculate the first estimate of the number of tubes:

$$N_T = \frac{m_T N_P}{\rho(\pi d_I^2/4)v_T}$$
$$= \frac{79.07 \times 2}{830 \times (\pi \times 0.016^2/4) \times 1}$$
$$= 947.6$$

The number of tubes must be an integer and be able to be allocated evenly across the number of tube passes. For 2 tube passes round up to the nearest number divisible by 2:

$$N_T = 948$$

Calculate the tube length:

$$L = \frac{A}{\pi d_O N_T}$$
$$= \frac{387.7}{\pi \times 0.02 \times 948}$$
$$= 6.51\ \text{m}$$

Calculate the shell diameter from Equation 12.52:

$$D_S = \left(\frac{4F_{TC}F_{SC}p_{CP}p_T^2 A}{\pi^2 d_o L}\right)^{1/2}$$

F_{TC} and F_{SC} are defined in Tables 12.7 and 12.8:

$$D_S = \left(\frac{4 \times 1.11 \times 1.15 \times 1 \times 0.025^2 \times 387.7}{\pi^2 \times 0.02 \times 6.51}\right)^{1/2}$$
$$= 0.981\ \text{m}$$

The tube length is slightly different from the initial estimate. Repeating the calculation with a revised initial tube length leads to only a small change to 6.52 m. In practice, the designer might prefer to adjust the length to a specified length (e.g. 6 m) and adjust the number of tubes accordingly.

c) For the tube-side pressure drop $Re = 3689$, which can be calculated from Equation 12.61:

$$K_{PT1} = \frac{2F_C\left(\frac{\rho d_I}{\mu}\right)^{m_f}\rho}{d_I}$$

$$= \frac{2 \times 0.0791\left(\frac{830 \times 0.016}{3.6 \times 10^{-3}}\right)^{-0.25} \times 830}{0.016}$$

$$= 1053.0$$

$$K_{PT2} = 0.5\alpha_R \rho$$
$$= 0.5 \times (2 \times 2 - 1.5) \times 830$$
$$= 1037.5$$

$$v_{NT,inlet} = \frac{m_T}{\rho(\pi d_{NT,inlet}^2/4)} = v_{NT,outlet}$$
$$= \frac{79.07}{830(\pi \times 0.3048^2/4)}$$
$$= 1.31\ \text{m}\cdot\text{s}^{-1}$$

$$Re_{TN,inlet} = Re_{TN,outlet} = \frac{830 \times 1.31 \times 0.3048}{3.6 \times 10^{-3}}$$
$$= 91{,}750$$

12

$$K_{PT3} = \rho\left(C_{TN,inlet}v_{TN,inlet}^2 + C_{TN,outlet}v_{TN,outlet}^2\right)$$
$$= 830(0.375 \times 1.31^2 + 0.375 \times 1.31^2)$$
$$= 1061.1$$

$$\Delta P_T = \Delta P_{TT} + \Delta P_{TE} + \Delta P_{TN}$$
$$= K_{PT1}N_P L v_T^{2+m_f} + K_{PT2}v_T^2 + K_{PT3}$$
$$= 1053.0 \times 2 \times 6.51 \times 1^{2-0.25} + 1037.5 \times 1^2 + 1061.1$$
$$= 15,815\ \text{N}\cdot\text{m}^{-2}$$

For the shell-side pressure drop $Re_S = 18,250$ and can be calculated from Equation 12.74. Assuming equally spaced baffles throughout, calculate the baffle spacing for the assumed velocity. First, however, the bundle clearance must be estimated from Table 12.6:

$$L_{BB} = 0.0169D_S + 0.0257$$
$$= 0.0169 \times 0.981 + 0.0257$$
$$= 0.0423\ \text{m}$$

Calculate the preliminary baffle spacing from Equation 12.73:

$$B = \frac{m_S}{\rho v_S\left[(D_S - D_B) + \dfrac{(D_B - d_O)(p_T - d_O)}{p_{CF}p_T}\right]}$$
$$= \frac{25}{730 \times 0.5\left[0.0423 + \dfrac{(0.981 - 0.0423 - 0.020)(0.025 - 0.020)}{1 \times 0.025}\right]}$$
$$= 0.303\ \text{m}$$

This cannot be the final baffle spacing, as the number of baffles must be an integer:

$$N_B = \frac{6.51}{0.303} - 1$$
$$= 20.48$$

The number of baffles needs to be rounded. Rounding up is always conservative. Rounding to 21 baffles, the actual baffle spacing can be calculated:

$$B = \frac{6.51}{21 + 1}$$
$$= 0.296\ \text{m}$$

$$R_S = (B/B_{in})^{1.8} + (B/B_{out})^{1.8}$$
$$= (0.296/0.296)^{1.8} + (0.296/0.296)^{1.8}$$
$$= 2$$

$$d_e = C_{De}\frac{p_T^2}{d_O} - d_O$$
$$= \frac{4}{\pi}\frac{0.025^2}{0.02} - 0.02$$
$$= 0.01979\ \text{m}$$

$$K_{PS1} = 18\left(5\frac{B}{D_S} - 1\right)(N_B - 1 + R_S)\frac{aD_S\rho}{d_e}\left(\frac{B_C}{0.2}\right)^{m_{fo}}\left(\frac{\rho d_e}{\mu}\right)^{-0.125}$$
$$= 18\left(\frac{5 \times 0.296}{0.981} - 1\right)(21 - 1 + 2)\left(\frac{0.01166 \times 0.981 \times 730}{0.01979}\right)$$
$$\times\left(\frac{0.25}{0.2}\right)^{-0.26765}\left(\frac{730 \times 0.01979}{4.0 \times 10^{-4}}\right)^{-0.125}$$
$$= 21,560$$

$$K_{PS2} = 90\left(1 - \frac{B}{D_S}\right)(N_B - 1 + R_S)\frac{bD_S\rho}{d_e}\left(\frac{B_C}{0.2}\right)^{m_{fo}}\left(\frac{\rho d_e}{\mu}\right)^{-0.157}$$
$$= 90\left(1 - \frac{0.296}{0.981}\right)(21 - 1 + 2)\left(\frac{0.002935 \times 0.981 \times 730}{0.01979}\right)$$
$$\times\left(\frac{0.25}{0.2}\right)^{-0.26765}\left(\frac{730 \times 0.01979}{4.0 \times 10^{-4}}\right)^{-0.157}$$
$$= 26,641$$

Now check the inlet nozzle velocity:

$$v_{NS,inlet} = \frac{m_S}{\rho\left(\pi d_{NS,inlet}^2/4\right)}$$
$$= \frac{25}{730(\pi \times 0.2027^2)/4}$$
$$= 1.06\ \text{m}\cdot\text{s}^{-1}$$

Check the allowable velocity:

$$v_{NS,inlet} \le \sqrt{\frac{2200}{730}} = 1.74\ \text{m}\cdot\text{s}^{-1}$$

The actual velocity is below the allowable. Now calculate the nozzle Reynolds number:

$$Re_{NS} = \frac{\rho v_{NS} d_{NS}}{\mu}$$
$$= \frac{730 \times 1.06 \times 0.2027}{4 \times 10^{-4}}$$
$$= 392,587$$

$$K_{PS3} = \rho\left(C_{NS,inlet}v_{NS,inlet}^2 + C_{NS,outlet}v_{S,outlet}^2\right)$$
$$= 730(0.375 \times 1.06^2 + 0.375 \times 1.06^2)$$
$$= 617$$

$$\Delta P_S = \Delta P_{SS} + \Delta P_{NS}$$
$$= K_{PS1}v_S^{1.875} + K_{PS2}v_S^{1.843} + K_{PS3}$$
$$= 21,560 \times 0.5^{1.875} + 26,641 \times 0.5^{1.843} + 617$$
$$= 13,920\ \text{N}\cdot\text{m}^{-2}$$

d) The tube-side and shell-side pressure drops are now fixed to be:

$$\Delta P_T = 15,000 = K_{PT1}N_P L v_T^{2+m_f} + K_{PT2}v_T^2 + K_{PT3}$$

$$\Delta P_S = 15,000 = K_{PS1}v_S^{1.875} + K_{PS2}v_S^{1.843} + K_{PS3}$$

Thus, v_T and v_S can be varied by trial and error to satisfy these two equations. At each new value of v_T and v_S, the U, A and exchanger geometry need to be calculated for fixed Q, ΔT_{LM} and F_T. The calculation is iterative and can be conveniently carried out using a solver in a spreadsheet to satisfy the equations for ΔP_T and ΔP_S simultaneously. To reach the two variables simultaneously, the objective can be set up such that the difference between the left- and right-hand sides of the equations are given the values, say *Objective* 1 and *Objective* 2, respectively. Then the spreadsheet solver is used to search for:

$$(Objective\ 1)^2 + (Objective\ 2)^2 = 0 \quad \text{(within a tolerance)}$$

The result is:

Tube-side	Shell-side
$v_T = 0.97\ m{\cdot}s^{-1}$	$v_S = 0.53\ m \cdot s^{-1}$
$h_T = 409.7\ W \cdot m^{-1} \cdot K^{-1}$	$h_S = 1072.4\ W \cdot m^{-2} \cdot K^{-1}$
$\Delta P_T = 15{,}000\ N \cdot m^{-2}$	$\Delta P_S = 15{,}000\ N \cdot m^{-2}$
$U = 205.8\ W \cdot m^{-2} \cdot K^{-1}$	
$A = 395.1\ m^2$	
$N_T = 974$	
$D_S = 0.995\ m$	
$L = 6.46\ m$	
$L/D_S = 6.49$	
$N_B = 22$	
$B = 0.281\ m$	
$B/D_S = 0.28$	
$L_{BB} = 0.0425\ m$	

Given freedom to change the pressure drop, the size of the heat exchanger can be traded off against the pressure drop.

The final design needs to consider the inlet and outlet baffle spacing. In this case, the baffle spacing is relatively small, to maintain the velocity on the shell-side. Although the baffle spacing is slightly larger than the shell-side internal nozzle diameter, the mechanical design will require a larger spacing at the inlet and outlet to allow for placement of shell-side nozzles without interference with the heat exchanger flanges to avoid the nozzle overlapping the first or last baffle.

12.7 Rating and Simulation of Heat Exchangers

When designing a heat exchanger, both the inlet and the outlet conditions are normally specified. Knowing the fluid physical properties also means that the heat duty is specified. The task is to design a unit that satisfies the required process duty. Another situation that arises is when a heat exchanger that is already fully specified is to be evaluated for its performance in a required duty.

Rating is the assessment of performance of a heat exchanger design for specified process inlet and outlet conditions. Rating calculates the overall heat transfer coefficient and area of a given design for fixed inlet and outlet conditions. This calculated area is then compared with the area of the actual design to see if it matches. The ratio of the areas might be such that the heat exchanger is overdesigned or underdesigned for the duty.

Whilst rating determines whether a particular design of heat exchanger would be suitable for a process duty, it does not predict how it would perform in practice. *Simulation* calculates the outlet temperatures and heat duty for a specified heat exchanger design for specified process inlet conditions.

Rating or simulation might need to be performed for a number of reasons – to determine:

- the performance of a given heat exchanger design in a given process duty at steady-state conditions to determine whether it will perform to process requirements;
- the performance of a given heat exchanger design in a given process duty through time as the heat exchanger fouls according to one of the fouling models in Figure 12.4;
- the performance of an existing heat exchanger in service in an existing location in an existing heat exchanger network to be relocated to a new duty in the network as part of a retrofit study;
- the overall heat transfer coefficient of an existing heat exchanger in service in an existing heat exchanger network to determine its condition for possible cleaning. In this case the heat transfer area, inlet and outlet temperatures are known and the overall heat transfer coefficient is calculated by trial and error.

Example 12.3 A kerosene stream is to be cooled by cooling water. Details of the heat exchange duty are given in Table 12.10. A heat exchanger is available for this duty with the geometry given in Table 12.11.

Table 12.10

Data for a heat recovery problem.

	Cooling water	Kerosene
Flowrate ($kg \cdot s^{-1}$)	–	25
Initial temperature (°C)	25	95
Final temperature (°C)	35	45
Density ($kg \cdot m^{-3}$)	993	749
Heat capacity ($J \cdot kg^{-1} \cdot K^{-1}$)	4190	2131
Viscosity ($N \cdot s \cdot m^{-2}$)	0.8×10^{-3}	1.1×10^{-3}
Thermal conductivity ($W \cdot m^{-1} \cdot K^{-1}$)	0.62	0.13
Fouling resistance ($W^{-1} \cdot m^2 \cdot K$)	0.00018	0.00033

It is proposed to place the kerosene on the tube-side and cooling water on the shell-side. Carry out a rating of the specified heat exchanger to assess whether it can meet the requirements of the process duty.

Solution Before the tube-side heat transfer coefficient can be calculated, the tube-side velocity must be calculated:

$$v_T = \frac{m_T(N_P/N_T)}{\rho(\pi d_I^2/4)}$$

$$= \frac{25(4/600)}{749(\pi \times 0.016^2/4)}$$

$$= 1.107\ m \cdot s^{-1}$$

Now calculate the Reynolds number for the tube-side, which is given by:

12

Table 12.11
Heat exchanger geometry.

Tube pitch p_T (m)	0.02
Number of tubes N_T	600
Number of tube passes N_P	4
Tube length L (m)	5.5
Tube thermal conductivity ($W \cdot m^{-1} \cdot K^{-1}$)	45
Tube pattern (tube layout angle)	90°
Tube inner diameter d_I (mm)	16
Tube outer diameter d_O (mm)	20
Shell inner diameter D_S (m)	0.8
Baffle spacing B (m)	0.367
Baffle cut B_C	0.2
Shell-bundle diametric clearance L_{BB} (m)	0.0392

$$\begin{aligned} Re &= \frac{\rho v_T d_I}{\mu} \\ &= \frac{749 \times 1.107 \times 0.016}{1.1 \times 10^{-3}} \\ &= 12,060 \end{aligned}$$

This is a turbulent flow and the tube-side heat transfer coefficient can be calculated from Equation 12.59:

$$\begin{aligned} K_{hT4} &= 0.024 \frac{k}{d_I} Pr^{0.4} \left(\frac{\rho d_I}{\mu}\right)^{0.8} \\ &= 0.024 \times \frac{0.13}{0.016} \times \left(\frac{2131 \times 1.1 \times 10^{-3}}{0.13}\right)^{0.4} \times \left(\frac{749 \times 0.016}{1.1 \times 10^{-3}}\right)^{0.8} \\ &= 1008.6 \end{aligned}$$

$$\begin{aligned} h_T &= K_{hT4} v_T^{0.8} \\ &= 1008.6 \times 1.107^{0.8} \\ &= 1094.0 \text{ W} \cdot \text{m}^2 \cdot \text{K}^{-1} \end{aligned}$$

Before the shell-side heat transfer coefficient can be calculated, the shell-side flowrate and velocity must be calculated:

$$\begin{aligned} Q &= m_T C_{PT} \Delta T_T \\ &= 25 \times 2131 \times (95 - 45) \\ &= 2.664 \times 10^6 W \\ &= \frac{2.664 \times 10^6}{4190 \times (35 - 25)} \text{kg} \cdot \text{s}^{-1} \\ &= 63.58 \text{ kg} \cdot \text{s}^{-1} \end{aligned}$$

Flowrate of cooling water

$$\begin{aligned} v_S &= \frac{m_S}{\rho B \left[(D_S - D_B) + \dfrac{(D_B - d_O)(p_T - d_O)}{p_{CF} p_T}\right]} \\ &= \frac{63.58}{993 \times 0.36 \left[0.0392 + \dfrac{(0.8 - 0.0392 - 0.02)(0.025 - 0.02)}{1 \times 0.025}\right]} \\ &= 0.949 \text{ m} \cdot \text{s}^{-1} \end{aligned}$$

Calculate the Reynolds number for the shell-side:

$$\begin{aligned} Re_S &= \frac{\rho d_O v_S}{\mu} \\ &= \frac{993 \times 0.020 \times 0.949}{0.8 \times 10^{-3}} \\ &= 23,560 \end{aligned}$$

Calculate the shell-side heat transfer coefficient from Equation 12.70:

$$\begin{aligned} K_{hS4} &= 0.08747 \frac{F_P F_L J_S k^{2/3} c_P^{1/3} \rho^{0.6633}}{\mu^{0.33} d_O^{0.3367} B_C^{0.5053}} \\ &= 0.08747 \times \frac{0.85 \times 0.8 \times 1 \times 0.62^{2/3} \times 4190^{1/3} \times 993^{0.6633}}{(0.8 \times 10^{-3})^{0.33} \times 0.02^{0.3367} \times 0.20^{0.5053}} \\ &= 6004.2 \end{aligned}$$

$$\begin{aligned} h_S &= K_{hS4} v_S^{0.6633} \\ &= 6004.2 \times 0.949^{0.6633} \\ &= 5799 \text{ W} \cdot \text{m}^{-2} \cdot \text{K}^{-1} \end{aligned}$$

Now calculate the overall heat transfer coefficient:

$$\begin{aligned} \frac{1}{U} &= \frac{1}{h_S} + \frac{1}{h_{SF}} + \frac{d_O}{2k} \ln\left(\frac{d_O}{d_I}\right) + \frac{d_O}{d_I} \frac{1}{h_{TF}} + \frac{d_O}{d_I} \frac{1}{h_T} \\ &= \frac{1}{5799} + 0.00018 + \frac{0.020}{2 \times 45} \ln\left(\frac{0.020}{0.016}\right) + \frac{0.02}{0.016}\left[0.00033 + \frac{1}{1094.0}\right] \end{aligned}$$

$$U = 511.0 \text{ W} \cdot \text{m}^{-2} \cdot \text{K}^{-1}$$

$$\begin{aligned} \Delta T_{LM} &= \frac{(95 - 35) - (45 - 25)}{\ln\left[\dfrac{95 - 35}{35 - 25}\right]} \\ &= 36.41 \text{ °C} \end{aligned}$$

Calculate the F_T correction factor:

$$R = \frac{T_{H1} - T_{H2}}{T_{C2} - T_{C1}} = \frac{95 - 45}{35 - 25} = 5.0$$

$$P = \frac{T_{C2} - T_{C1}}{T_{H1} - T_{C1}} = \frac{35 - 25}{95 - 25} = 0.1429$$

$$F_T = \frac{\sqrt{R^2 + 1} \ln\left[\dfrac{(1 - P)}{(1 - RP)}\right]}{(R - 1) \ln\left[\dfrac{2 - P\left(R + 1 - \sqrt{R^2 + 1}\right)}{2 - P\left(R + 1 + \sqrt{R^2 + 1}\right)}\right]}$$

Substituting $R = 5.0$ and $P = 0.1429$:

$$F_T = 0.93$$

Now calculate the required area:

$$\begin{aligned} A &= \frac{Q}{U \Delta T_{LM} F_T} \\ &= \frac{2.664 \times 10^6}{511.0 \times 36.41 \times 0.93} \\ &= 154.0 \text{ m}^2 \end{aligned}$$

The actual area of the heat exchanger is given by:

$$\begin{aligned} A &= N_T \pi d_O L \\ &= 600 \times \pi \times 0.020 \times 5.5 \\ &= 207.3\ \text{m}^2 \end{aligned}$$

Now compare the existing and required area:

$$\begin{aligned} \text{Overdesign} &= \frac{207.3 - 154.0}{154.0} \times 100\% \\ &= 31\% \end{aligned}$$

To complete the rating, the predicted pressure drop and pressure drop available should also be compared.

Consider now the simulation of a heat exchanger. In this case, the inlet conditions are specified and the outlet conditions need to be calculated. If the outlet temperatures are unknown then ΔT_{LM}, P, F_T and heat transfer area A cannot be calculated. One approach is therefore to assume either the hot or cold stream outlet temperature. This allows the other outlet temperature to be calculated from the energy balance. In turn, this allows ΔT_{LM}, P, F_T and therefore heat transfer area A to be calculated. The calculated area can then be compared with the actual area and the outlet temperature adjusted by trial and error until the calculated and actual areas are equal within a calculation tolerance. However, an alternative approach will now be developed that avoids the need for iteration. In Chapter 18 the approach will be extended to the simulation of networks of heat exchangers.

Consider first the simulation of a countercurrent heat exchanger. The equations describing simulation of a countercurrent heat exchanger are given by (Kotjabasakis and Linnhoff, 1986):

$$Q_H = CP_H(T_{H1} - T_{H2}) \tag{12.80}$$

$$Q_C = CP_C(T_{C2} - T_{C1}) \tag{12.81}$$

$$\begin{aligned} Q_H &= Q_C = UA\,\Delta T_{LM} \\ &= UA\frac{(T_{H1} - T_{C2}) - (T_{H2} - T_{C1})}{\ln\left(\dfrac{T_{H1} - T_{C2}}{T_{H2} - T_{C2}}\right)} \end{aligned} \tag{12.82}$$

where
Q_H = heat duty on the hot stream
Q_C = heat duty on the cold stream
CP_H = heat capacity flowrate for the hot stream (product of mass flowrate and specific heat capacity)
CP_C = heat capacity flowrate for the cold stream (product of mass flowrate and specific heat capacity)
T_{H1} = inlet temperature of the hot stream
T_{H2} = outlet temperature of the hot stream
T_{C1} = inlet temperature of the cold stream
T_{C2} = outlet temperature of the cold stream
U = overall heat transfer coefficient
A = heat transfer area

Also, if the heat capacity is constant:

$$R = \frac{CP_C}{CP_H} = \frac{T_{H1} - T_{H2}}{T_{C2} - T_{C1}} \tag{12.83}$$

Combining Equation 12.81 and 12.82 gives:

$$\frac{T_{H1} - T_{C2}}{T_{H2} - T_{C1}} = \exp\left[\frac{UA}{CP_C} \times \frac{(T_{H1} - T_{H2}) - (T_{C2} - T_{C1})}{(T_{C2} - T_{C1})}\right] \tag{12.84}$$

Combining Equations 12.83 and 12.84 gives:

$$\frac{T_{H1} - T_{C2}}{T_{H2} - T_{C1}} = \exp\left[\frac{UA(R-1)}{CP_C}\right] \tag{12.85}$$

Eliminating T_{C2} between Equations 12.83 and 12.85 gives (Kotjabaskis and Linnhoff, 1986):

$$(R-1)T_{H1} + R(X-1)T_{C1} + (1-RX)T_{H2} = 0 \qquad R \neq 1 \tag{12.86}$$

where

$$X = \exp\left[\frac{UA(R-1)}{CP_C}\right] \tag{12.87}$$

Eliminating T_{H2} between Equations 12.83 and 12.85 gives (Kotjabaskis and Linnhoff, 1986):

$$(X-1)T_{H1} + X(R-1)T_{C1} + (1-RX)T_{C2} = 0 \qquad R \neq 1 \tag{12.88}$$

If the inlet temperatures T_{H1} and T_{C1} are known, along with U, A, CP_H and CP_C, then Equations 12.86 and 12.88 constitute two equations with two unknowns (the outlet temperatures T_{H2} and T_{C2}). Equations 12.86 and 12.88 can be rearranged to give:

$$T_{H2} = \frac{(R-1)T_{H1} + R(X-1)T_{C1}}{(RX-1)} \qquad R \neq 1 \tag{12.89}$$

$$T_{C2} = \frac{(X-1)T_{H1} + X(R-1)T_{C1}}{(RX-1)} \qquad R \neq 1 \tag{12.90}$$

For the special case $R = 1$:

$$Q_C = CP_C(T_{C2} - T_{C1}) = UA(T_{H2} - T_{C1}) \tag{12.91}$$

Also, for $R = 1$:

$$(T_{H1} - T_{H2}) = (T_{C2} - T_{C1}) \tag{12.92}$$

Combining Equations 12.91 and 12.92 gives:

$$T_{H1} + YT_{C1} - (Y+1)T_{H2} = 0 \qquad R = 1 \tag{12.93}$$

where

$$Y = \frac{UA}{CP_C} \tag{12.94}$$

Substituting T_{H2} from Equation 12.92 into Equation 12.91 gives:

$$YT_{H1} + T_{C1} - (Y+1)T_{C2} = 0 \qquad R = 1 \tag{12.95}$$

Thus, for the special case of $R = 1$, Equations 12.93 and 12.95 replace Equations 12.86 and 12.88. For a single heat exchanger where $R = 1$:

$$T_{H2} = \frac{T_{H1} + YT_{C1}}{(Y+1)} \qquad R = 1 \tag{12.96}$$

$$T_{C2} = \frac{YT_{H1} + T_{C1}}{(Y+1)} \qquad R = 1 \tag{12.97}$$

Consider now the simulation of a noncountercurrent heat exchanger. The equation describing such a heat exchanger is given by:

$$Q = UA\Delta T_{LM}F_T \tag{12.98}$$

The F_T parameter creates a potential problem for the simulation of heat exchangers, as the outlet temperatures of the heat exchanger are unknown and F_T depends on the outlet temperatures. At first sight this would seem to require iteration. However, manipulation of the equations for F_T can avoid iteration (Herkenhoff, 1981). This will prove particularly important later in Chapter 18 when simulating heat exchanger networks. To manipulate the equations, consider the basic definition of F_T:

$$F_T = \frac{(UA/CP_C)_{CC}}{(UA/CP_C)} \tag{12.99}$$

The numerator of Equation 12.99 is the countercurrent heat duty and the denominator the actual noncountercurrent duty. Equating Equations 12.81 and 12.82 and rearranging for a countercurrent heat exchanger gives:

$$\left(\frac{UA}{CP_C}\right)_{CC} = \frac{\ln\left[\dfrac{T_{H1} - T_{C2}}{T_{H2} - T_{C1}}\right]}{\left[\dfrac{T_{H1} - T_{H2}}{T_{C2} - T_{C1}}\right] - 1} = \frac{\ln\left[\dfrac{1-P}{1-RP}\right]}{R-1} \tag{12.100}$$

For a series of N_{SHELLS} of 1–2 heat exchangers, Equations 12.99, 12.100 and 12.30 can be combined to give:

$$\frac{\ln\left[\dfrac{1-P}{1-RP}\right]}{(R-1)(UA/CP_C)} = \frac{\sqrt{R^2+1}\,\ln\left[\dfrac{1-P_{1-2}}{1-RP_{1-2}}\right]}{(R-1)\ln\left[\dfrac{2-P_{1-2}\left(R+1-\sqrt{R^2+1}\right)}{2-P_{1-2}\left(R+1+\sqrt{R^2+1}\right)}\right]} \tag{12.101}$$

Equation 12.40 can be rearranged to give:

$$\ln\left[\frac{1-P}{1-RP}\right] = N_{SHELLS}\ln\left[\frac{1-P_{1-2}}{1-RP_{1-2}}\right] \tag{12.102}$$

Combining Equations 12.101 and 12.102 and rearranging gives:

$$P_{1-2} = \frac{2G-2}{G\left(R+1+\sqrt{R^2+1}\right) - \left(R+1-\sqrt{R^2+1}\right)} \tag{12.103}$$

where

$$G = \exp\left[\frac{UA\sqrt{R^2+1}}{CP_C N_{SHELLS}}\right] \tag{12.104}$$

P can then be determined from Equation 12.36. Equation 12.103 is valid for $R = 1$, but P must be determined from Equation 12.37. Thus, if CP_H, CP_C, U, A and N_{SHELLS} are known, then P_{1-2} can be determined from Equation 12.103 and substituted into Equation 12.42 or 12.43 to obtain F_T without knowing the outlet temperatures. This allows a series of N_{SHELLS} of 1–2 heat exchangers to be simulated without iteration (Herkenhoff, 1981).

For noncountercurrent heat exchangers Equation 12.87 becomes:

$$X = \exp\left[\frac{UA(R-1)F_T}{CP_C}\right] \qquad R \neq 1 \tag{12.105}$$

and Equation 12.94 for noncountercurrent heat exchangers becomes:

$$Y = \frac{UAF_T}{CP_C} \qquad R = 1 \tag{12.106}$$

Thus Equations 12.89, 12.90 and 12.96 and 12.97 can be solved for the outlet temperatures if CP_H, CP_C, U, A, T_{H1}, T_{C1} and N_{SHELLS} are fixed, for both the countercurrent and noncountercurrent heat exchangers.

It should be noted that Equation 12.103 only works for noncountercurrent behavior as quantified by Equation 12.30 and cannot be used for other noncountercurrent arrangements, such as cross flow.

Exercise 12.4

A heat exchange unit consist of three 1–2 shells in series where a total heat exchange area of 619 m^2 is to be used to cool a hot steam with an inlet temperature of 300 °C by exchanging heat with a cold stream with an inlet temperature of 60 °C. The overall heat transfer coefficient has been estimated to be 100 W · m^{2} · K^{-1}. The heat capacity flowrates of the streams are CP_H = 17.5 kW · K^{-1} and CP_C = 25.0 kW · K^{-1}.

a) Calculate the outlet temperature of the hot and cold streams.
b) Calculate the F_T and the heat transfer duty for the unit and compare with the design calculation in Example 12.1.

Solution

a) From Equation 12.83:

$$R = \frac{CP_C}{CP_H} = \frac{25.0}{17.5} = 1.429$$

From Equation 12.104:

$$G = \exp\left[\frac{UA\sqrt{R^2+1}}{CP_C N_{SHELLS}}\right] = \exp\left[\frac{100 \times 619\sqrt{1.429^2+1}}{25.0 \times 10^3 \times 3}\right] = 4.217$$

From Equation 12.103:

$$P_{1-2} = \frac{2G-2}{G\left(R+1+\sqrt{R^2+1}\right)-\left(R+1-\sqrt{R^2+1}\right)}$$

$$= \frac{2 \times 4.217 - 2}{4.217\left(1.429+1+\sqrt{1.429^2+1}\right)-\left(1.429+1-\sqrt{1.429^2+1}\right)}$$

$$= 0.3805$$

Now calculate P from Equation 12.36:

$$P = \frac{1-\left(\dfrac{1-P_{1-2}R}{1-P_{1-2}}\right)^{N_{SHELLS}}}{R-\left(\dfrac{1-P_{1-2}R}{1-P_{1-2}}\right)^{N_{SHELLS}}}$$

$$= \frac{1-\left(\dfrac{1-0.3805 \times 4.429}{1-0.3805}\right)^3}{1.429\left(\dfrac{1-0.3805 \times 1.429}{1-0.3805}\right)^3}$$

$$= 0.5834$$

Now calculate T_{C2} from the definition of P in Equation 12.29:

$$T_{C2} = T_{C1} + P(T_{H1} - T_{C1}) = 60 + 0.5834(300 - 60) = 200.0\ ^\circ\text{C}$$

T_{H2} can be calculated from Equation 12.83:

$$T_{H2} = T_{H1} - R(T_{C2} - T_{C1}) = 300 - 1.429(200.0 - 60.0) = 100.0\ ^\circ\text{C}$$

b) F_T can be calculated from Equation 12.42:

$$F_T = \frac{\sqrt{R^2+1}\ln\left(\dfrac{1-P_{1-2}}{1-RP_{1-2}}\right)}{(R-1)\ln\left[\dfrac{2-P_{1-2}(R+1-\sqrt{R^2+1})}{2-P_{1-2}(R+1+\sqrt{R^2+1})}\right]}$$

$$= \frac{\sqrt{1.429^2+1}\ln\left(\dfrac{1-1.03805}{1-1.429 \times 0.3805}\right)}{(1.429-1)\ln\left[\dfrac{2-0.3805\left(1.429+1-\sqrt{1.429^2+1}\right)}{2-0.3805\left(1.429+1+\sqrt{1.429^2+1}\right)}\right]}$$

$$= 0.86$$

$$Q_H = CP_H(T_{H1} - T_{H2}) = 17.5(300.0 - 100.0) = 3500\ \text{kW}$$

$$Q_C = CP_c(T_{C2} - T_{C1}) = 25.0(200.0 - 60.0) = 3500\ \text{kW}$$

Thus, the simulation calculation confirms the design calculation in Example 12.1.

Example 12.5 Simulate the design from Example 12.2d with the heat exchanger geometry given in Table 12.12. For an inlet temperature of kerosene of 200 °C and that of crude oil of 35 °C:

Table 12.12

Heat exchanger geometry.

Tube pitch p_T (mm)	25
Number of tubes N_T	974
Number of tube passes N_P	2
Tube length L (m)	6.46
Tube thermal conductivity (W · m^{-1} · K^{-1})	45
Tube pattern (tube layout angle)	90°
Tube inner diameter d_I (mm)	16
Tube outer diameter d_O (mm)	20
Shell inner diameter D_S (m)	0.995
Number of baffles	22
Baffle spacing B (m)	0.281
Baffle cut B_C	0.25
Shell-bundle diametric clearance L_{BB} (m)	0.0425

a) Calculate the outlet temperature and heat duty and compare with the design calculation in Example 12.2.

b) Enlarge the inlet and outlet baffle spacing to avoid interference between the shell-side nozzles and the inlet and outlet baffles, but keeping the same central baffle spacing. Calculate the adjusted outlet temperatures and heat duty.

Solution

a) Calculate the tube-side velocity and heat transfer coefficient:

$$v_T = \frac{m_T(N_P/N_T)}{\rho(\pi d_I^2/4)}$$

$$= \frac{79.07(2/974)}{830(\pi \times 0.016^2/4)}$$

$$= 0.973 \text{ m} \cdot \text{s}^{-1}$$

$$Re = \frac{\rho v_T d_I}{\mu}$$

$$= \frac{830 \times 0.973 \times 0.016}{3.6 \times 10^{-3}}$$

$$= 3589$$

Calculate the tube-side heat transfer coefficient from Equation 12.56:

$$K_{hT1} = 0.116 \frac{k}{d_I}\left(\frac{\rho d_I}{\mu}\right)^{2/3} Pr^{1/3}\left[1+\left(\frac{d_I}{L}\right)^{2/3}\right]$$

$$= 0.116 \times \frac{0.133}{0.016} \times \left(\frac{830 \times 0.016}{3.6 \times 10^{-3}}\right)^{2/3}\left(\frac{2050 \times 3.6 \times 10^{-3}}{0.133}\right)^{1/3} \times \left[1+\left(\frac{0.016}{6.46}\right)^{2/3}\right]$$

$$= 894.1$$

$$K_{hT2} = 14.5 \frac{k}{d_I} Pr^{1/3}\left[1+\left(\frac{d_I}{L}\right)^{2/3}\right]$$

$$= 14.5 \times \frac{0.133}{0.016} \times \left(\frac{2050 \times 3.6 \times 10^{-3}}{0.133}\right)^{1/3} \times \left[1+\left(\frac{0.016}{6.46}\right)^{2/3}\right]$$

$$= 468.1$$

$$Re_S = \frac{\rho d_O v_S}{\mu}$$

$$= \frac{730 \times 0.020 \times 0.532}{4.0 \times 10^{-4}}$$

$$= 19{,}418$$

Calculate the shell-side heat transfer coefficient from Equation 12.70:

$$K_{hS4} = 0.08747 \frac{F_P F_L J_S k^{2/3} c_P^{1/3} \rho^{0.6633}}{\mu^{0.33} d_O^{0.3367} B_C^{0.5053}}$$

$$= 0.08747 \times \frac{0.85 \times 0.8 \times 1 \times 0.132^{2/3} \times 2470^{1/3} \times 730^{0.6633}}{(4 \times 10^{-4})^{0.33} \times 0.02^{0.3367} \times 0.25^{0.5053}}$$

$$= 1643$$

$$h_S = K_{hS4} v_S^{0.6633}$$

$$= 1643 \times 0.532^{0.6633}$$

$$= 1081 \text{ W} \cdot \text{m}^{-2} \cdot \text{K}^{-1}$$

Now calculate the overall heat transfer coefficient:

$$\frac{1}{U} = \frac{1}{h_S} + \frac{1}{h_{SF}} + \frac{d_O}{2k}\ln\left(\frac{d_O}{d_I}\right) + \frac{d_O}{d_I}\frac{1}{h_{TF}} + \frac{d_O}{d_I}\frac{1}{h_T}$$

$$= \frac{1}{1081} + \frac{1}{5000} + \frac{0.02}{2 \times 45}\ln\left(\frac{0.02}{0.016}\right) + \frac{0.02}{0.016}\left(\frac{1}{2000} + \frac{1}{409.8}\right)$$

$$U = 206.2 \text{ W} \cdot \text{m}^{-2} \cdot \text{K}^{-1}$$

The heat transfer area is given by:

$$A = N_T \pi d_O L$$

$$= 974 \times \pi \times 0.020 \times 6.46$$

$$= 395.3 \text{ m}^2$$

From Equation 12.83:

$$R = \frac{CP_C}{CP_H}$$

$$= \frac{79.07 \times 2050}{25 \times 2470}$$

$$= 2.625$$

From Equation 12.104:

$$G = \exp\left[\frac{UA\sqrt{R^2+1}}{CP_C N_{SHELLS}}\right]$$

$$= \exp\left[\frac{206.2 \times 395.3\sqrt{2.625^2+1}}{79.07 \times 2050 \times 1}\right]$$

$$= 4.106$$

From Equation 12.103:

$$P_{1-2} = \frac{2G-2}{G\left(R+1+\sqrt{R^2+1}\right) - \left(R+1-\sqrt{R^2+1}\right)}$$

$$= \frac{2 \times 4.106 - 2}{4.106\left(2.625+1+\sqrt{2.625^2+1}\right) - \left(2.625+1-\sqrt{2.625^2+1}\right)}$$

$$= 0.2427$$

Given that there is a single shell:

$$P = P_{1-2} = 0.2427$$

Now calculate T_{C2} from the definition of P in Equation 12.29:

$$
\begin{aligned}
T_{C2} &= T_{C1} + P(T_{H1} - T_{C1}) \\
&= 35 + 0.2427(200 - 35) \\
&= 75.0\,^{\circ}\mathrm{C}
\end{aligned}
$$

T_{H2} can be calculated from Equation 12.83:

$$
\begin{aligned}
T_{H2} &= T_{H1} - R(T_{C2} - T_{C1}) \\
&= 200 - 2.625(75.0 - 35) \\
&= 95.0\,^{\circ}\mathrm{C}
\end{aligned}
$$

$$
\begin{aligned}
Q_H &= CP_H(T_{H1} - T_{H2}) \\
&= 25 \times 2470 \times (200.0 - 95.0) \\
&= 6.484 \times 106\ \mathrm{W}
\end{aligned}
$$

$$
\begin{aligned}
Q_C &= CP_C(T_{C2} - T_{C1}) \\
&= 79.07 \times 2050 \times (75.0 - 35.0) \\
&= 6.484 \times 10^6\ W
\end{aligned}
$$

Thus, the simulation calculation agrees with the design calculation in Example 12.2.

b) To avoid interference between the inlet and exit nozzles and baffles, decrease the number of baffles by one. The central baffle spacing is maintained, but the inlet and outlet baffle spacing is increased to $1.5 \times 0.281 = 0.422$ m.

From Equation 12.72:

$$
\begin{aligned}
J_S &= \frac{(N_B - 1) + (B_{in}/B)^{0.4} + (B_{out}/B)^{0.4}}{(N_B - 1) + (B_{in}/B) + (B_{out}/B)} \\
&= \frac{(21 - 1) + (1.5)^{0.4} + (1.5)^{0.4}}{(21 - 1) + (1.5) + (1.5)} \\
&= 0.97 \\
h_S &= 1081 \times 0.97 \\
&= 1049\ \mathrm{W \cdot m^{-2} \cdot K^{-1}}
\end{aligned}
$$

The new overall heat transfer coefficient can be calculated to be $205.1\ \mathrm{W \cdot m^{-2} \cdot K^{-1}}$. Then repeating the simulation from Part a, with $U = 205.1\ \mathrm{W \cdot m^{-2} \cdot K^{-1}}$ and $A = 295.3\ \mathrm{m^2}$ gives:

$$T_{C2} = 74.94\,^{\circ}\mathrm{C}$$

$$T_{H2} = 95.16\,^{\circ}\mathrm{C}$$

$$Q_H = 6.474 \times 10^6\ \mathrm{W}$$

$$Q_C = 6.474 \times 10^6\ \mathrm{W}$$

The performance for unequal baffle spacing is only marginally worse than for equal baffle spacing. This is because of the relatively large number of baffles in this design.

12.8 Heat Transfer Enhancement

The overall heat transfer coefficient is made up of resistances to heat transfer from the outside film, outside fouling, tube wall, inside fouling and inside film. In some situations the outside film or the inside film resistance is significantly higher than the other resistances. In other words, the outside film transfer coefficient or the inside film transfer coefficient is significantly lower than all of the other coefficients. If this is the case, then the outside film resistance or the inside film resistance dominates the overall resistance and can be considered to be the *controlling resistance*. Consider a simple example in which the heat transfer coefficients are as follows:

h_S	= film heat transfer coefficient on the outside of the tubes	$= 1000\ \mathrm{W \cdot m^{-2} \cdot K^{-1}}$
h_{SF}	= outside fouling coefficient	$= 5000\ \mathrm{W \cdot m^{-2} \cdot K^{-1}}$
h_W	= tube wall coefficient	$= 20{,}000\ \mathrm{W \cdot m^{-2} \cdot K^{-1}}$
h_{TF}	= inside fouling coefficient	$= 5000\ \mathrm{W \cdot m^{-2} \cdot K^{-1}}$
h_T	= inside film heat transfer coefficient	$= 200\ \mathrm{W \cdot m^{-2} \cdot K^{-1}}$

This leads to an overall heat transfer coefficient of $155\ \mathrm{W \cdot m^{-2} \cdot K^{-1}}$. Now suppose it is desired to increase this value by manipulating the inside or outside film coefficient. Doubling the outside heat transfer coefficient to $2000\ \mathrm{W \cdot m^{-2} \cdot K^{-1}}$ leads to only a marginal increase in the overall heat transfer coefficient to $168\ \mathrm{W \cdot m^{-2} \cdot K^{-1}}$, an increase of 8%. Doubling the inside heat transfer coefficient to $400\ \mathrm{W \cdot m^{-2} \cdot K^{-1}}$ leads to a much bigger increase in the overall heat transfer coefficient to $253\ \mathrm{W \cdot m^{-2} \cdot K^{-1}}$, an increase of 63%. In this case, the inside film transfer coefficient is controlling. For the inside or outside film heat transfer coefficient to be limiting it must constitute the major contribution to the overall resistance to heat transfer. In the above example, summing the reciprocal of the heat transfer coefficients gives an overall resistance of $0.00645\ \mathrm{m^2 \cdot K \cdot W^{-1}}$. The resistance for the inside film heat transfer coefficient is $0.0050\ \mathrm{m^2 \cdot K \cdot W^{-1}}$, which is 76% of the overall resistance. For the inside or outside film heat transfer coefficient to be limiting, it normally constitutes at least 50% of the overall resistance to heat transfer. If it constitutes more than 60–70% of the overall resistance to heat transfer it dominates. A film transfer coefficient is likely to be controlling when heating or cooling viscous liquids or gases. 12

If the inside film heat transfer coefficient is controlling, then the first thing that can be considered is increasing the tube-side velocity by increasing the number of tube passes. If the outside film heat transfer coefficient is controlling then changing the baffle design can be considered. For segmented baffles, this might be as simple as decreasing the space between the baffles. However, if these measures do not bring the desired benefit with an acceptable pressure drop, then heat transfer enhancement can be considered. It must be emphasized that there will be no significant benefit from enhancing a film transfer coefficient that is not controlling. If one of the film coefficients is controlling, a more compact unit will result if:

- a greater surface area is presented to the controlling fluid by the use of extended-surface tubes or
- the controlling film transfer coefficient is increased through the use of an enhancement technique

1) *Extended surfaces.* Extended surfaces increase the rate of heat transfer per unit length of tube and the resulting exchanger can be much smaller and cheaper than the corresponding plain-tube exchanger. The external tube surface can be extended in one of two general ways:

- *Integrally formed tubes*, in which the extended surface is produced by cold-forming fins on to the surface of the tube by extrusion of the parent tube. Such extended surfaces can be formed on the inside or outside of the tubes.
- *Nonintegrally formed fins*, in which the surface is extended by attaching pieces of metal to the outside of tubes in the form of longitudinal or transverse strips, wire or spines by welding, brazing, grooving and peening, or shrink fitting the extended surface to the tube. The method of fixing the extended surface to the parent tube can create a resistance to heat transfer.

Extended surfaces can be created on the inside of tubes by integrally forming spiral or longitudinal fins. By far the most common design of extended surface tube uses "high" transverse fins on the outside of the tubes, as illustrated in Figure 12.17a. The high transverse fins can be formed integrally or nonintegrally. Nonintegral fins can be attached in a variety of ways. Such high transverse fins on the outside can increase the surface area by a factor of up to 16 relative to plain tubes. Integrally formed "low" transverse fins can be used on the outside of tubes and can be used in conventional shell-and-tube heat exchangers to enhance the outside film transfer coefficient. Low transverse fins can increase the surface area by a factor of around 2.5 relative to plain tubes. Longitudinal fins can also be used on the outside of tubes, as illustrated in Figure 12.17b. However, their application is mostly restricted to small heat exchangers in the form of concentric pipe heat exchangers, similar to the design in Figure 12.1a. In this arrangement, the inner tube would be the extended surface tube with the fins in the annular space to enhance the heat transfer. Longitudinal fins can increase the surface area by a factor of 14 to 20 relative to plain tubes.

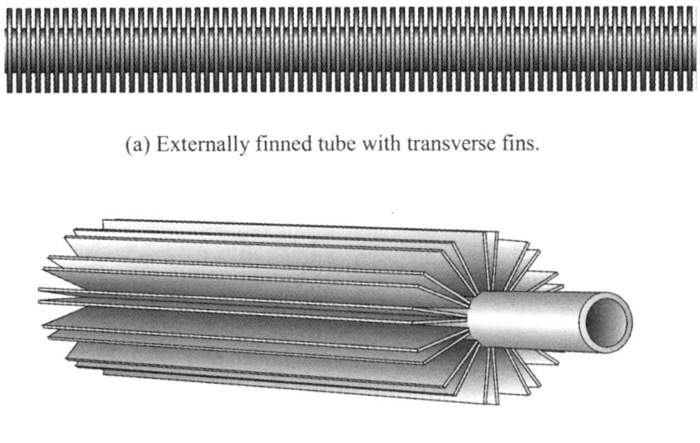

(a) Externally finned tube with transverse fins.

(b) Externally finned tube with longitudinal fins.

Figure 12.17
Tube designs for enhanced heat transfer.

High transverse finned tubes are commonly used in banks to transfer heat between a liquid flowing through the inside of the tubes and a gas flowing on the outside. For these applications, the tubes are mounted on a triangular pitch in a chamber with a rectangular cross-section and parallel sides. The direction of flow through the tubes is across the flow of the gas (Figure 12.18). One common application is for cooling duties where hot fluid is passed through the inside of tubes with ambient air from a fan flowing across the outside in *air coolers* (see Chapter 24). Another application is heat recovery from furnace flue gases (see later in this chapter). Such *cross-flow* arrangements require F_T correction factors different from the case of shell-and-tube heat exchangers. Some assumptions must be made regarding the mixing characteristics of the fluids. For the tube-side, it is reasonable to assume that there is no mixing as the liquid transfers heat to or from the gas. Of course, there is mixing within each tube, but the flow through each tube is independent. The gas side is more complex. If the duct through which the gas is flowing is fitted with baffles in line with the flow of gas to prevent mixing normal to the direction of the flow, then the gas can be considered to be unmixed. If there are no baffles and the tubes are short, then the gas can be considered to be mixed normal to the direction of flow. However, if there are no baffles and the tubes are long, then there will be some mixing normal to the flow, but the mixing will not be perfect across the whole cross-section of the gas flow. A conservative assumption is to consider the gas to be completely mixed normal to the flow. As illustrated in Figure 12.18a for cross flow, the gas side is considered to be mixed normal to its flow but the tube-side flows through tubes in a single pass and is unmixed (Bowman, Mueller and Nagle, 1940):

$$F_T = \frac{\ln\left[\dfrac{1-P}{1-RP}\right]}{(1-R)\ln\left[1+\dfrac{1}{R}\ln(1-RP)\right]} \qquad R \neq 1 \qquad (12.107)$$

$$F_T = \frac{P}{(P-1)\ln[1+\ln(1-P)]} \qquad R = 1 \qquad (12.108)$$

where

$P = \dfrac{T_{P2}-T_{P1}}{T_{G1}-T_{P1}}$ for a hot gas stream

$P = \dfrac{T_{G2}-T_{G1}}{T_{P1}-T_{G1}}$ for a hot process stream

$R = \dfrac{T_{G1}-T_{G2}}{T_{P2}-T_{P1}}$ for a hot gas stream

$R = \dfrac{T_{P1}-T_{P2}}{T_{G2}-T_{G1}}$ for a hot process stream

T_{G1} = inlet gas temperature
T_{G2} = outlet gas temperature
T_{P1} = inlet process temperature
T_{P2} = outlet process temperature

As illustrated in Figure 12.18b, for cross flow in which the gas side is considered to be mixed normal to the flow but the tube-side flows through tubes in a two-pass countercurrent

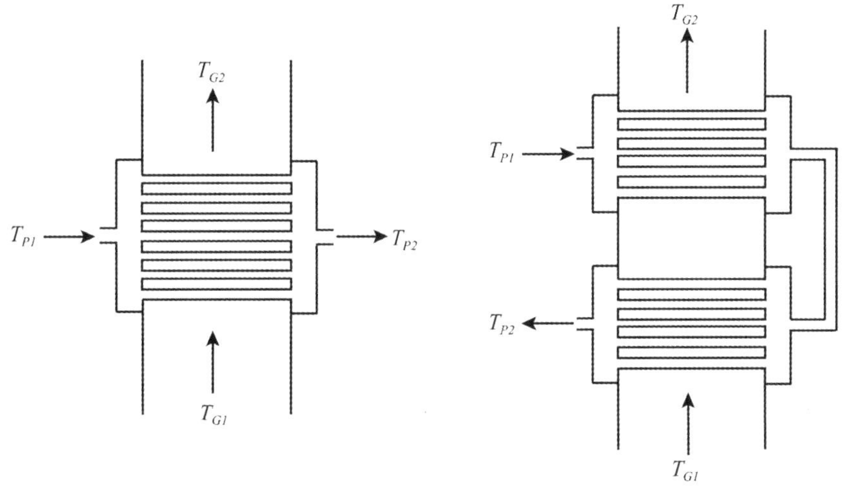

(a) Single-pass cross-flow arrangement.

(b) Two-pass cross-flow countercurrent arrangement.

Figure 12.18
Cross-flow arrangements.

arrangement and is unmixed when flowing though the tubes but mixed to a uniform temperature between the two passes (Bowman, Mueller and Nagle, 1940):

$$F_T = \frac{\ln\left[\dfrac{1-P}{1-RP}\right]}{2(1-R)\ln\left[1-\dfrac{1}{R}\ln\left(\dfrac{\sqrt{\dfrac{1-P}{1-RP}}-\dfrac{1}{R}}{1-\dfrac{1}{R}}\right)\right]} \qquad R \neq 1 \tag{12.109}$$

$$F_T = \frac{P}{2(P-1)\ln\left[1-\ln\left(\dfrac{P}{2(1-P)}+1\right)\right]} \qquad R = 1 \tag{12.110}$$

Cross-flow arrangements will be considered again in Chapter 24 for air coolers.

The outside heat transfer coefficient for banks of finned tubes on a triangular pitch can be calculated from (Briggs and Young, 1963):

$$Nu = 0.134Re^{0.681}Pr^{1/3}\left(\frac{y_F}{H_F}\right)^{0.2}\left(\frac{y_F}{\delta_F}\right)^{0.1134} \tag{12.111}$$

where $Nu = \dfrac{h_O d_R}{k}$

$Re = \dfrac{\rho_O d_R v_{max}}{\mu}$

$Pr = \dfrac{c_P \mu}{k_O}$

h_O = outside heat transfer coefficient $(\mathrm{W \cdot m^{-2} \cdot K^{-1}})$

d_R = outside tube diameter base on the root of the fins (m)

k_O = thermal conductivity of outside fluid $(\mathrm{W \cdot m^{-1} \cdot K^{-1}})$

ρ_O = density of the outside fluid $(\mathrm{kg \cdot m^{-3}})$

v_{max} = maximum velocity based on the minimum flow area $(\mathrm{m \cdot s^{-1}})$

μ = viscosity of the outside fluid $(\mathrm{N \cdot s \cdot m^{-2}})$

c_P = heat capcity of the outside fluid $(\mathrm{J \cdot kg^{-1} \cdot K^{-1}})$

y_F = distance between fins (m)

H_F = fin height (m)

δ_F = fin thickness (m)

Finned tubes are usually specified in terms of a number of fins per unit length. Thus:

$$y_F = \frac{1}{N_F} - \delta_F \tag{12.112}$$

where N_F = number of fins per unit length $(\mathrm{m^{-1}})$

Substituting in Equation 12.111 gives:

$$Nu = 0.134Re^{0.681}Pr^{1/3}\left(\frac{1}{H_F N_F} - \frac{\delta_F}{H_F}\right)^{0.2}\left(\frac{1}{\delta_F N_F} - 1\right)^{0.1134} \tag{12.113}$$

The definition of v_{max} is based on the minimum flow area, which for a triangular pitch is the open area between adjacent tubes and the outermost tubes and the chamber walls. If it is assumed that the tubes are contained within a rectangular chamber, the same spacing is maintained between the tubes and the outermost tubes and the walls of the chamber, and the walls are parallel, then the width of the chamber to accommodate N_{TR} rows on a staggered equilateral triangular pitch

$$\begin{aligned} &= (N_{TR} - 1)p_T + 2\left(p_T - \frac{d_R}{2}\right) + \frac{p_T}{2} \\ &= N_{TR}p_T + \frac{3p_T}{2} - d_R \end{aligned} \tag{12.114}$$

where N_{TR} = number of tubes per row (–)
d_R = tube diameter at the root of the fins (m)
p_T = tube pitch (m)

An additional $p_T/2$ is required for the triangular pitch compared with an in-line pitch. For a large number of tubes in a row for an in-line pitch or triangular pitch, the width can be approximated by $p_T\,(N_{TR} + 1)$. For finned tubes, in addition to the tube cylinder, the fins on each tube take up an area at the narrowest point of $2H_F\delta_F N_F L$, where L is the tube length. Thus, the minimum flow area

$$\begin{aligned} &= L\left[\left(N_{TR}p_T + \frac{3p_T}{2} - d_R\right) - (N_{TR}d_R + 2N_{TR}H_F\delta_F N_F)\right] \\ &= L\left[N_{TR}(p_T - d_R - 2H_F\delta_F N_F) + \left(\frac{3p_T}{2} - d_R\right)\right] \end{aligned} \tag{12.115}$$

where L = tube length (m)

From this flow area, the maximum velocity is given by:

$$v_{max} = \frac{m_O}{\rho L\left[N_{TR}(p_T - d_R - 2H_F\delta_F N_F) + \dfrac{3p_T}{2} - d_R\right]} \tag{12.116}$$

where m_O = mass flowrate of the outside fluid ($kg \cdot s^{-1}$)

The fins on the surface provide additional heat transfer area, but also an additional resistance to heat transfer. In order to account for this additional resistance, a *fin efficiency* is introduced. This is the ratio of the heat transferred through the actual fin to the heat that would be transferred if the fin was isothermal at its root temperature. For a cooling operation, in practice the temperature will decrease continuously from the root to the tip. The rate of decrease depends on the shape of the fin, thermal conductivity of the fin and the outside heat transfer coefficient. To determine the fin efficiency, it is assumed that heat transfer is by convection only and no radiation, that there is no temperature difference across the thickness of the fin and that there is no heat transfer from the fin tip. An allowance can be made for heat transfer from the fin tip by assuming the fin radius to be slightly larger than it is in practice. If the curvature of the outer fin tip is neglected, the tip area is equivalent to an extra $\delta_F/2$ on the fin radius, accounting for both sides of the fin. The fin efficiency of a circumferential fin with uniform thickness is given by (McQuiston and Tree, 1972):

$$\eta_F = \frac{\tanh(\kappa\psi)}{\kappa\psi} \tag{12.117}$$

where η_F = fin efficiency (–)

$$\kappa = \left(\frac{2h_O}{k_F\delta_F}\right)^{1/2}$$

$$\psi = \left(H_F + \frac{\delta_F}{2}\right)\left[1 + 0.35\ln\left(\frac{d_R + 2H_F + \delta_F}{d_R}\right)\right]$$

k_F = thermal conductivity of the fin ($W \cdot m^{-1} \cdot K^{-1}$)

H_F in this equation is extended by $\delta_F/2$ to allow for heat transfer from the fin tip. The heat transfer from the outside of the finned tube is from both the root of the tube and the fin, given by:

$$\begin{aligned} Q &= \left(\frac{1}{\dfrac{1}{h_O} + R_{OF}}\right)A_{ROOT}(T_{ROOT} - T_O) + \left(\frac{1}{\dfrac{1}{h_O} + R_{OF}}\right)\eta_F A_{FIN}(T_{ROOT} - T_O) \\ &= \left(\frac{1}{\dfrac{1}{h_O} + R_{OF}}\right)(A_{ROOT} + \eta_F A_{FIN})(T_{ROOT} - T_O) \\ &= \left(\frac{1}{\dfrac{1}{h_O} + R_{OF}}\right)\left(\frac{A_{ROOT} + \eta_F A_{FIN}}{A_{ROOT}}\right)A_{ROOT}(T_{ROOT} - T_O) \\ &= \left(\frac{1}{\dfrac{1}{h_O} + R_{OF}}\right)\eta_W A_O(T_{ROOT} - T_O) \end{aligned} \tag{12.118}$$

where Q = heat transfer from the outside of the tube (W)
h_O = outside heat transfer coefficient ($W \cdot m^{-2} \cdot K^{-1}$)
R_{OF} = outside fouling resistance ($W^{-1} \cdot m^2 \cdot K$)
A_{ROOT} = exposed root area (m^2)
$= \pi d_R L(1 - N_F\delta_F)$
A_{FIN} = area of fins (m^2)
$= \dfrac{\pi N_F L}{2}\left[(d_R + 2H_F)^2 - d_R^2 + 2\delta_F(d_R + 2H_F)\right]$
A_O = total outside area of tube (m^2)
$= A_{ROOT} + A_{FIN}$
T_{ROOT} = temperature at the root of the tube (°C)
T_O = bulk temperature of the outside fluid (°C)
η_W = weighted fin efficiency (–)

$$= \frac{A_{ROOT} + \eta_F A_{FIN}}{A_{ROOT} + A_{FIN}} \tag{12.119}$$

The overall heat transfer coefficient for a finned tube can now be defined. However, first the area basis must be defined. The overall heat transfer coefficient for a plain tube as defined in Equations 12.13 and 12.14 refers to the outside area of plain tubes. If the total outside area, including the fin area, is chosen as the reference, then the overall heat transfer coefficient is defined as:

$$\begin{aligned} \frac{1}{U} &= \frac{1}{\eta_W h_O} + \frac{R_{OF}}{\eta_W} + \frac{A_O}{\pi d_R L}\frac{d_R}{2k_W}\ln\left(\frac{d_R}{d_I}\right) \\ &\quad + \left(\frac{A_O}{\pi d_I L}\right)\left(R_{IF} + \frac{1}{h_I}\right) \end{aligned} \tag{12.120}$$

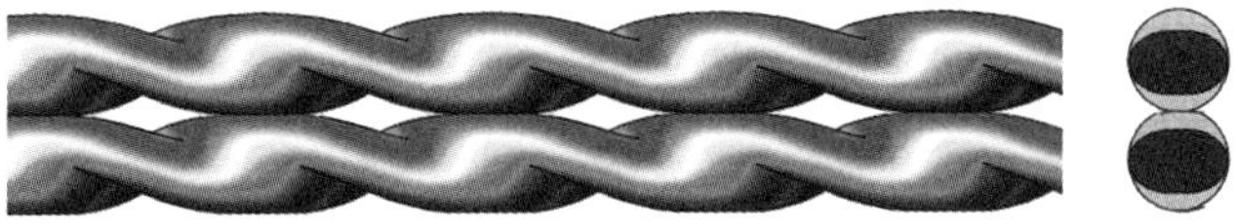

Figure 12.19

Twisted tubes for enhanced heat transfer.

where h_O = outside heat transfer coefficient $(W \cdot m^{-2} \cdot K^{-1})$
h_I = inside heat transfer coefficient $(W \cdot m^{-2} \cdot K^{-1})$
R_{OF} = outside fouling resistance $(W^{-1} \cdot m^2 \cdot K)$
R_{IF} = inside fouling resistance $(W^{-1} \cdot m^2 \cdot K)$
k_W = thermal conductivity of the wall $(W \cdot m^{-1} \cdot K^{-1})$

If the construction of the finned tube is nonintegral, then in theory an additional resistance should be added to Equation 12.120 to allow for the contact resistance between the fin and the tube wall. However, this is most often difficult to quantify.

2) *Twisted tubes. Twisted tubes* enhance the heat transfer coefficients on both the inside and outside of tubes by twisting the tubes to a helical shape, as illustrated Figure 12.19. The corrugations formed by the twisting pattern create additional turbulence both inside and outside of the tube. The tubes are arranged on a triangular pitch and each tube is supported by surrounding tubes, as illustrated in Figure 12.19, which eliminates the need for baffles for tube support. However, the fluid outside of the tubes is free to swirl around the outside of the tubes.

3) *Tube inserts.* Rather than changing the form of the tube to create enhancement, *tube inserts* can be mounted inside plain tubes. Three commonly used designs are illustrated in Figure 12.20. The advantage of tube inserts is that a conventional design, or an existing heat exchanger, can be readily enhanced at low cost. The application in retrofit of existing heat exchangers will be discussed later.

- *Twisted tapes.* These are illustrated in Figure 12.20a and consist of a thin strip of twisted metal with the same width as the tube inside diameter that is slid into the tube. The flow is caused to spiral along the tube length. The helical flow path that is induced creates a higher velocity along the tube wall and creates turbulent flow at a lower Reynolds number than would be the case in plain tubes. The pitch of the twist can be varied according to the application.
- *Coiled wire.* A coiled wire insert is illustrated in Figure 12.20b. The coil is manufactured by tightly wrapping wire on to a rod such that the outside diameter of the coil is slightly larger than the inside diameter of the tube. This ensures that when it is fitted into the tube there is no movement when in service. The coil functions by interrupting the boundary layer adjacent to the tube wall, thus increasing the heat transfer coefficient.
- *Mesh inserts.* Mesh inserts are manufactured by weaving wire coils together to form a matrix. The matrix consists of small loops offset in a helical arrangement. Figure 12.20c illustrates a HiTran© wire matrix insert. The diameter of the wire mesh is slightly larger than the tube diameter to give close contact with the tube wall. The mesh causes random mixing throughout the tube, but most importantly breaks up the fluid boundary layer at the tube wall. Different "densities" of wire mesh can be formed by weaving the wire together more or less frequently per unit length. This allows the enhancement to be tailored to a particular application.

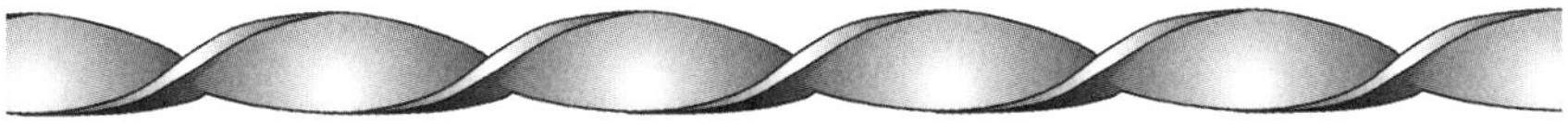

(a) Twisted tape insert.

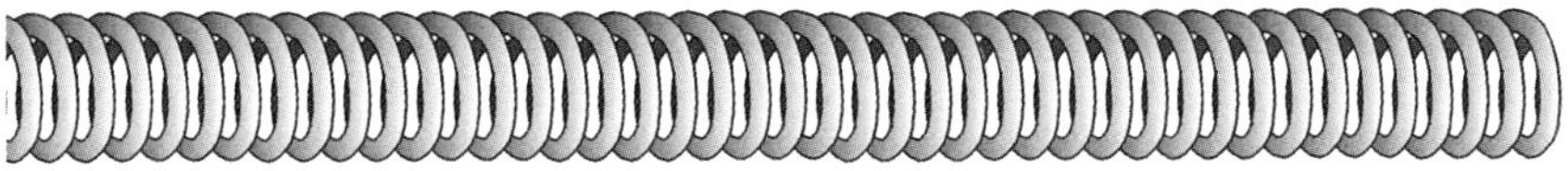

(b) Coiled wire insert.

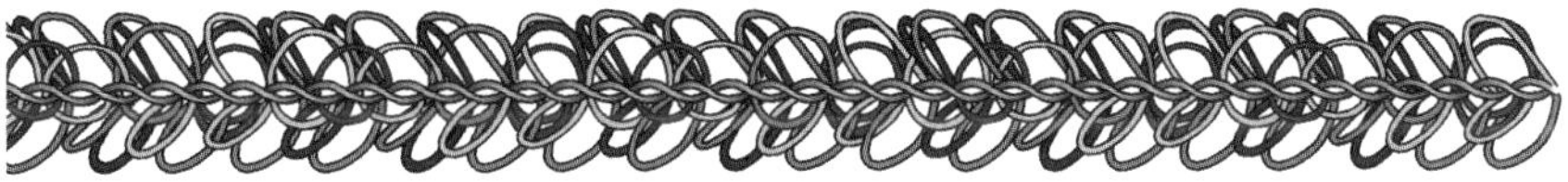

(c) HiTran® wire matrix insert.

Figure 12.20

Tube inserts for enhanced heat transfer.

Tube inserts are most effective when applied to laminar and transitional flow. Although enhancement occurs when applied to fully turbulent flow, the level of enhancement usually does not often justify its application and are normally applied at Reynolds numbers typically below 20,000. However, used in an appropriate way, tube inserts are capable of increasing film heat transfer coefficients by 10 times or greater, especially when applied to laminar flow. Another important advantage of tube inserts is that they can be very effective in reducing fouling. By disturbing the fluid boundary layer at the tube wall, the wall surface temperature can be reduced, decreasing fouling. The inserts also increase the wall shear stress, which assists fouling removal. Both effects act to decrease fouling. The disadvantage of inserts is that they increase the pressure drop for a single tube pass. If the increase in pressure drop is unacceptable and the heat exchanger features multiple tube passes, then the use of inserts coupled with a decrease in the number of tube passes can result in a significantly higher overall heat transfer coefficient with a small increase in the pressure drop across the heat exchanger, or in some cases no increase at all.

No published correlations are available for the performance of wire mesh inserts. However, published correlations are available for twisted tapes and coiled wire inserts.

- *Twisted tape heat transfer.* Twisted tape heat transfer for laminar flow for $Sw < 2000$ and $Re < 10{,}000$ (Manglik and Bergles, 1993a; Jiang et al., 2014Jiang, Shelley and Smith, 2014):

$$Nu = \frac{h_{Te} d_I}{k} = 0.106 Sw^{0.767} Pr^{0.3} \tag{12.121}$$

where

Sw = dimensionless swirl parameter

$$= \frac{Re}{\sqrt{y}} \frac{\pi}{\pi - 4(\delta/d_I)} \left[1 + \left(\frac{\pi}{2y} \right)^2 \right]^{1/2} \tag{12.122}$$

Re = Reynolds number based on bare tubes $= \dfrac{\rho d_I v}{\mu}$

h_{Te} = enhanced tube-side film heat transfer coefficient $(\mathrm{W \cdot m^{-2} \cdot K^{-1}})$

$$v = \frac{4m}{\pi d_I^2 \rho} (\mathrm{m \cdot s^{-1}}) \tag{12.123}$$

m = mass flowrate $(\mathrm{kg \cdot s^{-1}})$
δ = thickness of tape (m)
y = twist ratio, axial length for a 180° turn of the tape divided by the internal diameter of the tube
d_I = internal diameter of the tube (m)

For turbulent flow for $S_W \geq 2000$ and $Re \geq 10{,}000$ (Manglik and Bergles, 1993b; Jiang, Shelley and Smith, 2014):

$$Nu = \frac{h_{Te} d_I}{k} = 0.023\, Re^{0.8}\, Pr^{0.4} \left(1 + \frac{0.769}{y} \right) \left[\frac{\pi}{\pi - 4(\delta/d_I)} \right]^{0.8} \left[\frac{\pi + 2 - 2(\delta/d_I)}{\pi - 4(\delta/d_I)} \right]^{0.2} \tag{12.124}$$

$S_W > 2000$ and $Re < 10{,}000$ as an approximation take the Nusselt number to be the mean of the predictions of Equations 12.121 and 12.124.

- *Twisted tape pressure drop.* Pressure drop for twisted tape laminar flow for $Sw < 2000$ (Manglik and Bergles, 1993a; Jiang, Shelley and Smith, 2014):

$$\Delta P_{Te} = 4 c_f \frac{L_{sw}}{d_I} \left(\frac{\rho v_{sw}^2}{2} \right) \tag{12.125}$$

$$c_f = \frac{15.767}{Re_{sw}} \left[\frac{\pi + 2 - 2(\delta/d_I)}{\pi - 4(\delta/d_I)} \right]^2 \left(1 + 10^{-6} Sw^{2.55} \right)^{1/6} \tag{12.126}$$

where ΔP_{Te} = pressure drop in straight tubes for tube inserts

$$Re_{sw} = \frac{\rho v_{sw} d_I}{\mu}$$

v_{sw} = swirl velocity $(\mathrm{m \cdot s^{-1}})$

$$= v \frac{\pi}{\pi - 4(\delta/d_I)} \left[1 + \left(\frac{\pi}{2y} \right)^2 \right]^{1/2} \tag{12.127}$$

L_{sw} = swirl length (m)

$$= L \left[1 + \left(\frac{\pi}{2y} \right)^2 \right]^{1/2} \tag{12.128}$$

L = tube length (m)

Pressure drop for twisted tape turbulent flow for $Sw \geq 2000$ (Manglik and Bergles, 1993b; Jiang, Shelley and Smith, 2014):

$$\Delta P_{Te} = 4 c_f \frac{L}{d_I} \left(\frac{\rho v^2}{2} \right) \tag{12.129}$$

$$c_f = \frac{0.0791}{Re^{0.25}} \left(1 + \frac{2.752}{y^{1.29}} \right) \left[\frac{\pi}{\pi - 4(\delta/d_I)} \right]^{1.75} \left[\frac{\pi + 2 - 2(\delta/d_I)}{\pi - 4(\delta/d_I)} \right]^{1.25} \tag{12.130}$$

- *Coiled wire heat transfer.* Coiled wire laminar flow heat transfer for $Re \leq 1000$ (Jiang, Shelley and Smith, 2014):

$$Nu = 1.86 Re^{1/3} Pr^{1/3} \left(\frac{p}{d_I} \right)^{-1/3} \left[\frac{\cos\alpha - (e/d_I)^2}{\cos\alpha + (e/d_I)} \right]^{-1/3} \tag{12.131}$$

where

$$\cos\alpha = \sqrt{\frac{1}{\left[\frac{\pi}{(p/d_I)}\right]^2 + 1}}$$

Coiled wire transitional and turbulent flow heat transfer for $1000 < Re < 80{,}000$, $0.07 < e/d_I < 0.1$ and $1.17 < p/d_I < 2.68$ (Garcia, Xicenti and Viedma, 2005; Jiang, Shelley and Smith, 2014):

$$Nu = \frac{h_{Te} d_I}{k} = 0.132\left(\frac{p}{d_I}\right)^{-0.372} Re^{0.72} Pr^{0.37} \tag{12.132}$$

Coiled wire turbulent flow heat transfer for $80{,}000 \le Re \le 250{,}000$, $0.01 < e/d_I < 0.2$, $0.1 < p/d_I < 7$ and $0.3 < \alpha/90 < 1$ (Ravigururajan and Bergles, 1996; Jiang, Shelley and Smith, 2014):

$$Nu = \frac{h_{Te} d_I}{k} = Nu_S\left\{1 + \left[2.64 Re^{0.036}\left(\frac{e}{d_I}\right)^{0.212}\left(\frac{p}{d_I}\right)^{-0.21}\left(\frac{\alpha}{90}\right)^{0.29} Pr^{-0.024}\right]^7\right\}^{1/7} \tag{12.133}$$

where

Nu_S = smooth tube Nusselt number

$$= \frac{h_T d_I}{k} = \frac{RePr(c_{fS}/2)}{1 + 12.7\sqrt{c_{fS}/2}(\mathrm{Pr}^{2/3} - 1)} \tag{12.134}$$

h_T = smooth tube-side film heat transfer coefficient ($\mathrm{W \cdot m^{-2} \cdot K^{-1}}$)

c_{fS} = smooth tube Fanning friction factor (–)

e = wire diameter (m)

p = helical pitch (m)

α = helix angle of wire to the tube axis (degrees)

where

$$\frac{\alpha}{90} = \frac{2}{\pi}\tan^{-1}\left[\frac{\pi}{(p/d_I)}\right]$$

- *Coiled wire pressure drop.* The pressure drop for coiled wire (Ravigururajan and Bergles, 1996; Garcia, Vicenti and Viedma, 2005; Jiang, Shelley and Smith, 2014):

$$\Delta P_{Te} = 4c_f \frac{L}{d_I}\left(\frac{\rho v^2}{2}\right) \tag{12.135}$$

For $Re \le 310$:

$$c_f = \frac{16}{Re} \tag{12.136}$$

For $310 < Re < 30{,}000$, $0.07 < e/d_I < 0.1$, $1.17 < p/d_I < 2.68$ and $16.7 < p/e < 26.8$:

$$c_f = 9.35\left(\frac{p}{e}\right)^{-1.16} Re^{-0.217} \quad \text{for} \quad 310 < Re < 30{,}000 \tag{12.137}$$

For $30{,}000 \le Re \le 250{,}000$, $0.01 < e/d_I < 0.2$, $0.1 < p/d_I < 7$ and $0.3 < \alpha/90 < 1$:

$$c_f = c_{fS}\left\{1.036 + \left[30.15 Re^{a1}\left(\frac{e}{d}\right)^{a2}\left(\frac{p}{d}\right)^{a3}\left(\frac{\alpha}{90}\right)^{a4}\right]^{15/16}\right\}^{16/15} \tag{12.138}$$

where

$$\begin{aligned} a_1 &= 0.67 - 0.06\left(\frac{p}{d}\right) - 0.49\left(\frac{\alpha}{90}\right) \\ a_2 &= 1.37 - 0.157\left(\frac{p}{d}\right) \\ a_3 &= -1.66 \times 10^{-6} Re^{-0.33}\left(\frac{\alpha}{90}\right) \\ a_4 &= 4.59 + 4.11 \times 10^{-6} Re^{-0.15}\left(\frac{p}{d}\right) \end{aligned} \tag{12.139}$$

The total tube-side pressure drop ΔP_T for a single shell comprises the pressure drop in the straight tubes (ΔP_{Te}), pressure drop in the tube entrances, exits and reversals (ΔP_{TE}) and pressure drop in nozzles (ΔP_{TN}):

$$\begin{aligned} \Delta P_T &= \Delta P_{Te} + \Delta P_{TE} + \Delta P_{TN} \\ &= \Delta P_{Te} + K_{PT2} v_T^2 + K_{PT3} \end{aligned} \tag{12.140}$$

where K_{PT2} and K_{PT3} are defined by Equations 12.62 and 12.63. To specify an insert, dimensions are first chosen for the insert. For twisted tapes, this would be the tape thickness and twist ratio. For wire coils, this would be the wire thickness and pitch. The enhanced heat transfer coefficient is then calculated based on the flow regime. The enhanced pressure drop is then checked to ensure the pressure drop is below the allowable limit. If the heat transfer enhancement requirements are not met or the pressure drop constraint exceeded, then another enhancement design is chosen and the heat transfer and pressure drop again checked. If an acceptable pressure drop cannot be achieved, then the number of tube passes can be reduced for a multipass heat exchanger.

12.9 Retrofit of Heat Exchangers

In retrofit situations, an increase in the performance of an existing heat exchanger might be required for a variety of reasons. The existing heat transfer duty might need to be maintained or increased as a result of changes to the inlet flowrates or temperatures or both. Consider the basic equation governing the heat transfer:

$$Q = UA\Delta T_{LM} F_T \tag{12.141}$$

If Q is required to be increased for the same $\Delta T_{LM} F_T$, then this can be accomplished either by an increase in the value of U or A. If the value of $\Delta T_{LM} F_T$ is decreased for the same Q, then this can also be accomplished either by an increase in the value of U or A. Thus, the retrofit of heat exchangers can in principle be accomplished either by manipulating U or A.

Consider possible options for retrofit.

1) *Increase in heat transfer area* If shell-and-tube heat exchangers are being used on a heat transfer match and additional heat transfer area is required, it might be possible to install new tube bundles into the existing shells if the additional area requirement is small. This might be possible if the tube pitch can be decreased or the pitch configuration changed from a square to a triangular pitch. If a significant amount of additional area is required, then the existing area can be supplemented by adding a new shell (or more than one shell if there is a large area requirement). New heat exchanger shells can be added to an existing match in one of two ways:
   - *Series* When new exchanger shells are added in series to an existing match, then the full flowrate goes through the existing exchangers of the match. The pressure drop and heat transfer coefficients of the existing exchangers of the match will only change significantly if the changes lead to significant changes in the operating conditions of the existing match (e.g. increase in flowrate in a debottlenecking study). The addition of new exchangers in series to the existing match will lead to an increase in the overall pressure drop across the match. This might be important if the pump (or compressor) is close to its maximum capacity.
   - *Parallel* If new exchanger shells are added in parallel to an existing match, then the flowrate through the existing exchangers is decreased. This will decrease the flowrate and pressure drop through the existing exchangers of the match. However, it will also decrease the heat transfer coefficient, decreasing its performance. The addition of new exchangers will therefore leave the pressure difference largely unchanged. The pressure drop across the match will be the largest between that of the existing exchangers under the new conditions and the new exchangers installed in parallel.

   Increasing the heat transfer area, either by replacing tube bundles or by adding new shells, is likely to be expensive. In some other designs of heat exchanger, increasing the heat transfer area can be much more straightforward (see Section 12.12).
2) *Heat transfer enhancement using conventional designs* Rather than install additional heat transfer area to cater for the new operational requirements, the overall heat transfer coefficient of the heat exchanger can be increased. Using conventional designs, it might be possible to modify the tube-side flow pattern to increase the number of tube passes, thereby increasing the tube-side velocity and film transfer coefficient. However, this is not as simple as installing new pass partition plates in the heat exchanger heads and will require a new tube bundle to allow for the pass partition plates in the tube layout. On the shell-side, it might be possible to modify the baffle arrangement to increase the shell-side film transfer coefficient by decreasing the baffle spacing. Again, this will require a new tube bundle, as the existing tubes would need to be cut to install the new baffles.
3) *Extended surfaces* Changing the heat transfer surface from a plain surface to a finned, ribbed or nonuniform surface can increase the rate of heat transfer of the tube surfaces. This can be applied to the inside or outside surface of the tube, or both, but requires the tubes to be changed.
4) *Tube inserts* As discussed previously, devices can be inserted into the inside of existing plain tubes to enhance the inside heat transfer coefficient, at the expense of increased pressure drop. Insertion devices include twisted tapes, coiled wire and mesh inserts. The major disadvantage with increasing the heat transfer coefficient using tube inserts is the resulting increase in pressure drop on the tube-side. This problem can be overcome in multipass shell-and-tube heat exchangers by installing tube inserts and at the same time decreasing the number of tube passes. The number of tube passes can be decreased by removing pass partition plates. No change is required to the tube bundle. The result can be a significant increase in the heat transfer coefficient without an increase in pressure drop. However, changes might be required to the inlet and outlet nozzles and the piping. In some cases, the number of tube passes can be decreased without any change to the inlet and outlet nozzles. Table 12.13 lists the changes required to the construction of the

Table 12.13

Reduction of tube passes in shell-and-tube-heat exchangers.

Tube passes	Options to decrease tube passes	Retrofit required
2	1	i. Remove all the partitions ii. Replace the existing heads with two new heads iii. Rearrange or repipe the flows if necessary
4, 8, 12	2	i. Remove all the partitions ii. Turn around the shell or change the position of two heads iii. Install a new partition in the middle of the front head iv. Rearrange or repipe the flows if necessary
	1	i. Remove all the partitions ii. Replace the existing heads with two new heads iii. Rearrange or repipe the flows if necessary
6, 10	2	i. Remove all the partitions except the middle one in the front head
	1	i. Remove all the partitions ii. Replace the existing heads with two new heads iii. Rearrange or repipe the flows if necessary
12	4	i. Remove some partitions and retain those in the typical four-tube-pass layout

heat exchanger for a change in the number of tube passes. In Table 12.13, two underlying criteria have been adopted: the tubes for each pass should be the same after decreasing the tube passes and no new partition plates can be installed to cut existing tubesheets.

It can be seen that the most straightforward reduction of tube passes is from 6 and 10 to 2 or 12 to 4.

For most applications of tube inserts in retrofit, three conditions are required to make tube inserts effective:

- The overall heat transfer coefficient must be limited by the tube-side heat transfer coefficient. In other words, the tube-side coefficient must be the lowest of the five resistances to heat transfer across the tube. If the shell-side is limiting, or the fouling coefficients dominate, then nothing done to the inside coefficient will make a significant impact on the overall coefficient.
- The tube-side heat transfer coefficient should have a Reynolds number of typically less than 10,000 to 20,000 for heat transfer enhancement to have a significant effect on the tube-side heat transfer coefficient.
- The heat exchanger should preferably be multi-pass, allowing a decrease in the number of tube passes.

Example 12.6 Assuming the design developed in Example 12.2d and simulated in Example 12.5b with inlet and outlet baffle spacing of 0.422 m is an existing heat exchanger (see Table 12.12 for the geometry), its performance now needs to be improved through retrofit. The heat exchanger is to have its duty increased by the installation of tube inserts. For an inlet temperature of kerosene of 200 °C and that of crude oil of 35 °C:

a) Examine whether tube inserts are likely to be effective.
b) Calculate the outlet temperatures of the streams, the heat exchanger duty and tube-side pressure drop using twisted tape inserts with a twist ratio of 3 and thickness 0.0016 m.
c) Calculate the outlet temperatures of the streams, the heat exchanger duty and tube-side pressure drop using coiled wire inserts with a pitch of 0.04 m and thickness of 0.0016 m.

Solution

a) As discussed in the previous section, for tube inserts to be effective, the tube-side resistance should be at least 50% of the total resistance and the tube-side Reynolds number should preferably be less than 10,000 to 20,000. From Example 12.5 the tube-side Reynolds number is 3589. Also, from Example 12.5, the heat transfer coefficients and resistances are given by:

	h_S	h_{SF}	h_W	h_{TF}	h_T
Coefficient ($W \cdot m^{-2} \cdot K^{-1}$)	1081	5000	20,166	2000	409.8
% of overall resistance	23.0	4.8	1.2	12.1	58.9

Thus the problem is tube-side controlled and a potential candidate for heat transfer enhancement.

b) For the twisted tape, from Equation 12.122:

$$S_W = \frac{Re}{\sqrt{y}}\frac{\pi}{\pi - 4(\delta/d_I)}\left[1 + \left(\frac{\pi}{2y}\right)^2\right]^{1/2}$$

$$= \frac{3589}{\sqrt{3}}\frac{\pi}{\pi - 4(0.0016/0.016)}\left[1 + \left(\frac{\pi}{2 \times 3}\right)^2\right]^{1/2}$$

$$= 2680.2$$

For the empty tube:

$$v_T = \frac{m_T(N_P/N_T)}{\rho(\pi d_I^2/4)}$$

$$= \frac{79.07(2/974)}{830(\pi \times 0.016^2/4)}$$

$$= 0.973\ m \cdot s^{-1}$$

$$Re = \frac{\rho v_T d_I}{\mu}$$

$$= \frac{830 \times 0.973 \times 0.016}{3.6 \times 10^{-3}}$$

$$= 3589$$

For S_W greater than 2000 and Reynolds number less than 10,000 use a mean of the Nusselt number predicted by Equations 12.121 and 12.124:

$$Nu = \frac{h_{Te} d_I}{k} = 0.106 S_W^{0.767} Pr^{0.3}$$

$$h_{Te} = \frac{0.133}{0.016} \times 0.106 \times 2680.2^{0.767} \times \left(\frac{2050 \times 3.6 \times 10^{-3}}{0.133}\right)^{0.3}$$

$$= 1252.3\ W \cdot m^{-2} \cdot K^{-1}$$

$$Nu = \frac{h_{Te} d_I}{k} = 0.023 Re^{0.8} Pr^{0.4}\left(1 + \frac{0.769}{y}\right)\left[\frac{\pi}{\pi - 4(\delta/d_I)}\right]^{0.8}\left[\frac{\pi + 2 - 2(\delta/d_I)}{\pi - 4(\delta/d_I)}\right]^{0.2}$$

$$h_{Te} = \frac{0.133}{0.016} \times 0.023 \times 3589^{0.8} \times \left(\frac{2050 \times 3.6 \times 10^{-3}}{0.133}\right)^{0.4}$$

$$\times \left[1 + \frac{0.769}{3}\right]\left[\frac{\pi}{\pi - 4(0.0016/0.016)}\right]^{0.8}\left[\frac{\pi + 2 - 2(0.0016/0.016)}{\pi - 4(0.0016/0.016)}\right]^{0.2}$$

$$= 1048.9\ W \cdot m^{-2} \cdot K^{-1}$$

Taking the mean of the two equations gives $h_{Te} = 1150.6\ W \cdot m^{-2} \cdot K^{-1}$:

$$\frac{1}{U} = \frac{1}{h_S} + \frac{1}{h_{SF}} + \frac{d_O}{2k}\ln\left(\frac{d_O}{d_I}\right) + \frac{d_O}{d_I}\left(\frac{1}{h_{Te}} + \frac{1}{h_{TF}}\right)$$

$$= \frac{1}{1,049} + \frac{1}{5,000} + \frac{0.02}{2 \times 45}\ln\left(\frac{0.02}{0.016}\right) + \frac{0.02}{0.016}\left(\frac{1}{1150.6} + \frac{1}{2000}\right)$$

$$U = 343.4\ W \cdot m^{-2} \cdot K^{-1}$$

12

As for Example 12.5:

$$A = 395.3\ m^2$$
$$R = 2.625$$
$$G = \exp\left[\frac{UA\sqrt{R^2+1}}{CP_C N_{SHELLS}}\right]$$
$$= 10.496$$
$$P_{1-2} = \frac{2G-2}{G\left(R+1+\sqrt{R^2+1}\right)-\left(R+1-\sqrt{R^2+1}\right)}$$
$$= 0.2847 = P$$
$$T_{C2} = T_{C1} + P(T_{H1} - T_{C1})$$
$$= 35 + 0.2847 \times (200 - 35)$$
$$= 81.98\ °C$$
$$T_{H2} = T_{H1} - R(T_{C2} - T_{C1})$$
$$= 200 - 2.625 \times (81.98 - 35)$$
$$= 76.68\ °C$$
$$Q_H = CP_H(T_{H1} - T_{H2})$$
$$= 25 \times 2470 \times (200 - 76.68)$$
$$= 7.615 \times 10^6\ W$$

The heat duty has thus increased by $(7.615 - 6.484) \times 10^6$ W $= 1.131 \times 10^6$ W. The friction factor for the tube-side is given by:

$$c_f = \frac{0.0791}{Re^{0.25}}\left(1+\frac{2.752}{y^{1.29}}\right)\left[\frac{\pi}{\pi-4(\delta/d_I)}\right]^{1.75}\left[\frac{\pi+2-2(\delta/d_I)}{\pi-4(\delta/d_I)}\right]^{1.25}$$
$$= 0.0452$$
$$\Delta P_T = \Delta P_{Te} + \Delta P_{TE} + \Delta P_{TN}$$
$$= 4cf\frac{L}{d_I}\left(\frac{\rho v_T^2}{2}\right) + K_{PT2}v_T^2 + K_{PT3}$$
$$= 30,691\ N \cdot m^{-2}$$

This pressure drop is increased from 15,000 $N \cdot m^{-2}$ to 30,691 $N \cdot m^{-2}$ for the enhanced case.

c) For the coiled wire insert, given $Re_T = 3589$, the heat transfer coefficient can be calculated from Equation 12.132:

$$Nu = \frac{h_{Te}d_I}{k} = 0.132\left(\frac{p}{d_I}\right)^{-0.372} Re^{0.72}Pr^{0.37}$$
$$h_{Te} = \frac{0.133}{0.016} \times 0.132 \times \left(\frac{0.04}{0.016}\right)^{-0.372} 3589^{0.72} 55.489^{0.37}$$
$$= 1250.9\ W \cdot m^{-2} \cdot K^{-1}$$
$$U = 354.0\ W \cdot m^{-2} \cdot K^{-1}$$
$$A = 395.3\ m^2$$
$$R = 2.625$$
$$G = 11.302$$
$$P_{1-2} = 0.2866 = P$$
$$T_{C2} = 82.28\ °C$$
$$T_{H2} = 75.88\ °C$$
$$Q_H = 7.664 \times 10^6\ W$$

The heat duty has increased by $(7.664 - 6.484) \times 10^6$ $= 1.18 \times 10^6$ W. The friction factor for the tubeside is given by:

$$c_f = 9.35\left(\frac{p}{e}\right)^{-1.16} Re^{-0.217}$$
$$= 0.0378$$
$$\Delta P_T = 4c_f\frac{L}{d_R}\left(\frac{\rho v_T^2}{2}\right) + K_{PT2}v_T^2 + K_{PT3}$$
$$= 26,039\ N \cdot m^{-2}$$

This pressure drop is increased from 15,000 $N \cdot m^{-2}$ to 26,039 $N \cdot m^{-2}$ for the enhanced case.

12.10 Condensers

The construction of shell-and-tube condensers is very similar to that of shell-and-tube heat exchangers for duties that do not involve a change of phase. Condensers can be horizontally or vertically oriented with the condensation on the tube-side or the shell-side. The magnitude of the condensing film coefficient for a given quantity of vapor condensation on a given surface is significantly different depending on the orientation of the condenser. The condensation normally takes place on the shell-side of horizontal exchangers and the tube-side of vertical exchangers. Horizontal shell-side condensation is normally preferred, as the condensing film transfer coefficients are higher. Condensation on the tube-side of horizontal condensers is normally restricted to the use of condensing steam as a heating medium.

Condensation can take place by one of two mechanisms:

a) *Filmwise condensation*, in which the condensing vapor wets the surface of the tube, forming a continuous film (see Figure 12.21a).

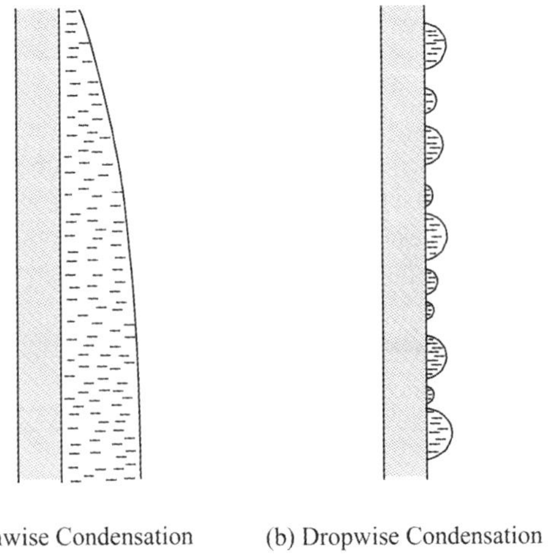

Figure 12.21

Condesation mechanisms.

b) *Dropwise condensation*, in which droplets of condensation do not wet the surface and, after growing, fall from the tube to expose a fresh condensing surface without forming a continuous film (Figure 12.21b).

Although dropwise condensation can produce much higher condensing film transfer coefficients, it is unpredictable, and design is carried out on the basis of filmwise condensation.

The basic equations describing filmwise condensation were developed by Nusselt (1916). The derivation of the equations has been given by Kern (1950), Serth (2007) and others. A number of assumptions are made in the derivation:

- The liquid film flows smoothly and steadily by gravitational forces.
- The liquid film is in laminar flow.
- No noncondensable gases are present in the vapor phase.
- No vapor shear force acts on the liquid–vapor interface.
- Momentum terms are negligible.
- Temperature distribution in the condensate film is linear.
- The only heat transferred across the liquid film is the latent heat of condensation released at the liquid–vapor interface (transfer of sensible heat in the liquid film is negligible).
- The temperature of the liquid–vapor interface is equal to the saturation temperature.
- Physical properties of the liquid film are constant.

For the falling film illustrated in Figure 12.21a, the condensing film heat transfer coefficient is given by:

$$h_C = 0.943\left(\frac{k_L^3\rho_L(\rho_L-\rho_V)\Delta H_{VAP}g}{L\mu_L\Delta T}\right)^{1/4} \qquad (12.142)$$

where
h_C = condensing film coefficient ($W \cdot m^{-2} \cdot K^{-1}$)
k_L = thermal conductivity of the liquid ($W \cdot m^{-1} \cdot K^{-1}$)
ρ_L = density of the liquid ($kg \cdot m^{-3}$)
ρ_V = density of the vapor ($kg \cdot m^{-3}$)
ΔH_{VAP} = latent heat ($J \cdot kg^{-1}$)
g = gravitational constant ($9.81\ m \cdot s^{-2}$)
L = length of the wall (m)
μ_L = viscosity of the liquid ($N \cdot s \cdot m^{-2}$ or $kg \cdot m^{-1} \cdot s^{-1}$)
ΔT = temperature difference across the condensate film (K)

It should be noted that the original form of the Nusselt Equation featured the term ρ_L^2 instead of the term $\rho_L(\rho_L-\rho_V)$. The latter was introduced to account for buoyancy forces (Rosenow, 1956; Rosenow, Webber and Ling, 1956). Such buoyancy forces are usually only important close to the critical point. Also, the constant in the original form of the Nusselt Equation was slightly different due to errors in the way the required integration was performed. In most cases, the two most important factors that cause a significant deviation from Equation 12.142 are the presence of vapor shear forces and noncondensable gases in the vapor. Vapor shear forces act to increase the heat transfer coefficient, whereas noncondensable gases act to decrease it.

Since the ΔT across the film is unknown, it is best eliminated from Equation 12.142. By definition of the condensing film coefficient:

$$m\Delta H_{VAP} = h_C LW\Delta T \qquad (12.143)$$

where
m = flowrate of condensate ($kg \cdot s^{-1}$)
L = length of wall (m)
W = width of wall (m)

Substituting Equation 12.143 into Equation 12.142 and rearranging gives:

$$h_C = 0.925k_L\left(\frac{\rho_L(\rho_L-\rho_V)gW}{\mu_L m}\right)^{1/3} \qquad (12.144)$$

The Reynolds number for the flow can be defined as:

$$Re = \frac{\rho d_e v}{\mu_L}$$

where
d_e = equivalent diameter
$= 4 \times \dfrac{\text{flow area}}{\text{wetted perimeter}}$
$= 4\delta$
δ = film thickness (m)
v = velocity of condensate ($m \cdot s^{-1}$)
$= \dfrac{m}{\rho W\delta}$

Thus the Reynolds number becomes:

$$Re = \frac{4m}{\mu_L W} \qquad (12.145)$$

Equation 12.144, the Nusselt Equation for a falling film, is valid for $Re < 30$.

For condensation inside vertical tubes, neglecting the curvature of the tube wall, Equation 12.143 becomes:

$$m\Delta H_{VAP} = h_C\pi d_I LN_T\Delta T \qquad (12.146)$$

where
d_I = inside diameter of tube (m)
L = length of tube (m)
N_T = number of tubes (m)

Substituting Equation 12.146 in Equation 12.142 gives:

$$h_C = 1.354k_L\left(\frac{\rho_L(\rho_L-\rho_V)d_I N_T g}{\mu_L m}\right)^{1/3} \qquad (12.147)$$

The Reynolds number is defined by:

$$Re = \frac{\rho d_e v}{\mu_L}$$

where

$$d_e = 4 \times \frac{\text{flow area}}{\text{wetted perimeter}}$$

Neglecting curvature:

$$d_e = 4\delta$$
$$v = \frac{m}{\rho \pi d_I N_T \delta}$$

Thus, the Reynolds number becomes:

$$Re = \frac{4m}{\pi d_I N_T \mu_L} \tag{12.148}$$

Equation 12.147 is valid for $Re < 30$. For condensation on the outside of vertical tubes, the equations become:

$$h_C = 1.354 k_L \left(\frac{\rho_L(\rho_L - \rho_V) d_O N_T g}{\mu_L m} \right)^{1/3} \tag{12.149}$$

which is again valid for $Re < 30$ where:

$$Re = \frac{4m}{\pi d_O N_T \mu_L} \tag{12.150}$$

For condensation on the outside of a single horizontal tube the corresponding form of the Nusselt Equation is (Kern, 1950; Rose, 1999):

$$h_C = 0.728 \left(\frac{k_L^3 \rho_L(\rho_L - \rho_V) \Delta H_{VAP} g}{d_O \mu_L \Delta T} \right)^{1/4} \tag{12.151}$$

By definition of the condensing film coefficient:

$$m \Delta H_{VAP} = h_C \pi d_O L N_T \Delta T \tag{12.152}$$

Combining Equations 12.151 and 12.152:

$$h_C = 0.959 k_L \left(\frac{\rho_L(\rho_L - \rho_V) L N_T g}{\mu_L m} \right)^{1/3} \tag{12.153}$$

which is valid for $Re < 30$ where:

$$Re = \frac{4m}{\mu_L L N_T} \tag{12.154}$$

For the shell-side of a horizontal tube bundle, dripping of condensate over successive rows acts to decrease the condensing coefficient. This can be accounted for by multiplying the condensing coefficient for a single tube by an empirical correction involving the number of tubes in a vertical row. Kern (1950) proposed the correction:

$$h_C = 0.959 k_L \left(\frac{\rho_L(\rho_L - \rho_V) L g N_T}{\mu_L m} \right)^{1/3} N_R^{-1/6} \tag{12.155}$$

where N_R = number of tubes in a vertical row (–)

This correction works well for a single vertical row. However, in a circular tube bundle, the number of tubes in a vertical row varies according to the position in the bundle. This can be accounted for by modifying N_T in Equation 12.153 to be $N_T^{2/3}$ (Kern, 1950):

$$\begin{aligned} h_C &= 0.959 k_L \left(\frac{\rho_L(\rho_L - \rho_V) L g N_T^{2/3}}{\mu_L m} \right)^{1/3} \\ &= 0.959 k_L \left(\frac{\rho_L(\rho_L - \rho_V) L g}{\mu_L m} \right)^{1/3} N_T^{2/9} \end{aligned} \tag{12.156}$$

Condensation inside horizontal tubes is complex to analyse on a fundamental basis. An approximate condensing coefficient can be obtained by using a simple correction to the Nusselt Equation for horizontal condensation to account for the accumulation of condensation along the base of the tube. The correction often applied is 0.8 (Butterworth, 1977). No correction for the number of tubes is required. Thus, condensation inside horizontal tubes can be approximated by:

$$h_C = 0.767 k_L \left(\frac{\rho_L(\rho_L - \rho_V) L g N_T}{\mu_L m} \right)^{1/3} \tag{12.157}$$

In the above equations, the film thickness and hence the condensing coefficient varies across the surface. The correlations give an average coefficient applicable to the entire surface. The condensing coefficients are independent of shell-side geometry (e.g. baffle cut, distance, etc.). The Nusselt Equations give reasonably good agreement with experimental data for laminar flow of the condensate film in the absence of vapor shear forces and non-condensable gases in the vapor. In the absence of noncondensable gases, the Nusselt Equations will tend to give a conservative prediction of the condensing coefficient. Vapor shear and turbulence in the film can lead to considerably higher values than those predicted by the Nusselt Equations. Turbulence in the film occurs when condensation occurs on a long vertical surface or the rate of condensation is high. The laminar film goes through a transition to wavy and then turbulent flow. The turbulence acts to increase the heat transfer coefficient, but the thickening of the film acts to decrease it.

For a simple total condenser involving isothermal condensation:

$$Q = m \Delta H_{VAP} = U A \Delta T_{LM} \tag{12.158}$$

where U is defined by Equation 12.13. Note that if the condensing fluid is pure and therefore isothermal, no F_T correction factor is required if multipass exchangers are used, that is, $F_T = 1$.

If the heat exchange involves desuperheating as well as condensation, then the exchanger can be divided into zones with linear temperature–enthalpy profiles in each zone. Figure 12.22a illustrates desuperheating and condensation on the shell-side of a horizontal condenser. The total heat transfer area is the sum of the values for each zone:

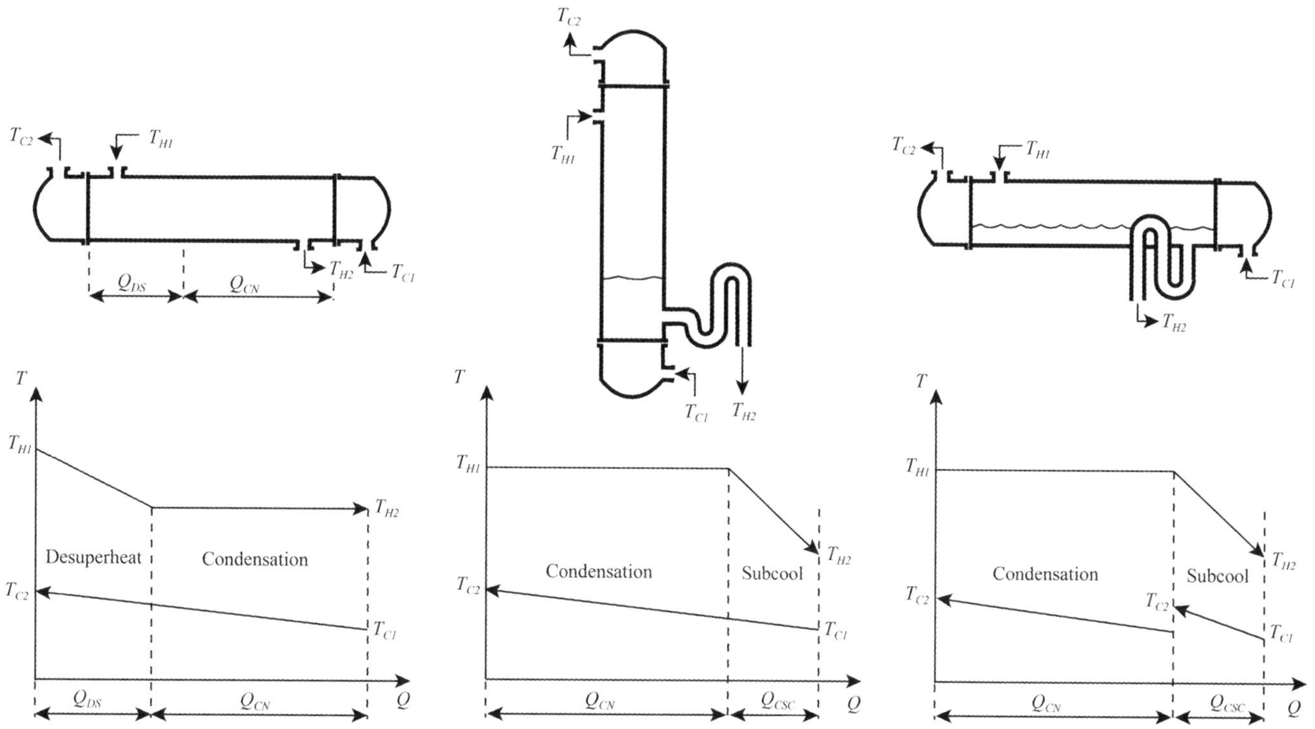

(a) Desuperheating in a horizontal condenser.

(b) Subcooling in a vertical condenser.

(c) Subcooling in a horizontal condenser.

Figure 12.22

Condensation with desuperheating and subcooling.

$$\begin{aligned} A &= A_{DS} + A_{CN} \\ &= \frac{Q_{DS}}{U_{DS}\Delta T_{LM,DS}} + \frac{Q_{CN}}{U_{CN}\Delta T_{LM,CN}} \end{aligned} \qquad (12.159)$$

where
- A = total heat transfer area
- A_{DS} = heat transfer area for the desuperheating zone
- ACN = heat transfer area for the condensing zone
- Q_{DS} = heat transfer duty for the desuperheating zone
- Q_{CN} = heat transfer duty for the condensing zone
- U_{DS} = overall heat transfer coefficient for the desuperheating zone
- U_{CN} = overall heat transfer coefficient for the condensing zone
- $\Delta T_{LM,DS}$ = logarithmic mean temperature difference for the desuperheating zone
- $\Delta T_{LM,CN}$ = logarithmic mean temperature difference for the condensing zone

To calculate the condensing heat transfer coefficient requires the length of the condensing zone L to be specified. Thus, a value of L must be estimated before the calculation can be made. For the value of L to be correct, it must comply with:

$$L = \frac{A_{CN}}{A} \times Tube\ length \qquad (12.160)$$

12

The value of L is then varied until there is agreement with Equation 12.160.

It might also be necessary to subcool the condensate. As with desuperheating, if subcooling is required, the heat exchanger can be divided into zones. Figure 12.22b illustrates subcooling on the shell-side of a vertical condenser. The subcooling arrangement in Figure 12.22b is achieved by using a loop seal to create a partially submerged tube bundle (Kern, 1950). For subcooling, the heat transfer area is given by:

$$\begin{aligned} A &= A_{CN} + A_{SC} \\ &= \frac{Q_{CN}}{U_{CN}\Delta T_{LM,CN}} + \frac{Q_{SC}}{U_{SC}\Delta T_{LM,SC}} \end{aligned} \qquad (12.161)$$

where
- A_{SC} = heat transfer area for the subcooling zone
- Q_{SC} = heat transfer duty for the subcooling zone
- U_{SC} = overall heat transfer coefficient for the subcooling zone
- $\Delta T_{LM,SC}$ = logarithmic mean temperature difference for the subcooling zone

To calculate the condensing heat transfer coefficient again requires the length of the condensing zone L to be specified. Thus, a value of L must be estimated and adjusted until it complies with Equation 12.160.

Figure 12.22c illustrates subcooling on the shell-side of a horizontal condenser. The subcooling arrangement in Figure 12.22c is again achieved by using a loop seal to create a partially submerged tube bundle (Kern, 1950). Rather than use a loop seal, a dam baffle can be used to partially submerge the bundle (Kern, 1950). Figure 12.22c shows the zones this time represented in parallel, rather than the series arrangements in Figures 12.22a and 12.22b. Calculation of the condensing heat transfer coefficient for a horizontal exchanger requires the number of tubes in the condensing zone $N_{T,CN}$ to be specified. Thus, a value of $N_{T,CN}$ must be estimated before the calculation can be made. For the value of $N_{T,CN}$ to be correct, it must comply with:

$$N_{T,CN} = \frac{A_{CN}}{A} \times Number\ of\ tubes \qquad (12.162)$$

The value of $N_{T,CN}$ is then varied until there is agreement with Equation 12.162.

For multicomponent condensation, the condensation will not be isothermal, leading to a nonlinear temperature–enthalpy profile for the condensation. If this is the case, then the exchanger can be divided into a number of zones with the temperature–enthalpy profiles linearized in each zone. Each zone is then modeled separately and zones summed to obtain the overall area requirement (Kern, 1950).

Particular care needs to be adopted if a vapor to be condensed has noncondensable gases present. Here the vapor diffuses through the gas to the cold surface where it condenses. However, as the condensation proceeds, the concentration of the noncondensable gas increases, which increases the diffusional resistance and decreases the condensing coefficient. To take this into account requires complex models, which is outside the scope of this text.

Pressure drop during condensation results essentially from the vapor flow. As condensation proceeds, the vapor flowrate decreases. The equations described previously for pressure drop in shell-and-tube heat exchangers are only applicable under constant flow conditions. Again, the exchanger can be divided into zones. However, in preliminary design, a reasonable estimate of the pressure drop can usually be obtained by basing the calculation on the mean of the inlet and outlet vapor flowrates.

Example 12.7 A flowrate of 0.1 kmol · s^{-1} of essentially pure acetone vapor from the overhead of a distillation column is to be condensed without any condensate subcooling. The condensation is to take place on the shell-side of a horizontal shell-and-tube heat exchanger against cooling water flowing in two passes on the tube-side with a split ring head. The operating pressure of the condenser is 1.52 bar. At this pressure, the acetone condenses at 67 °C. The cooling water can be assumed to be at 25 °C and to be returned to the cooling tower at 35 °C. The condenser can be assumed to be steel with tubes with 20 mm outside diameter and 2 mm wall thickness. The tube pitch can be assumed to be 1.25d_O and a square configuration. The physical property data are given in Table 12.14.

Table 12.14

Physical property data for acetone and water.

Property	Acetone (67 °C)	Water (30 °C)
Liquid density (kg · m^{-3})	736	996
Vapor density (kg · m^{-3})	3.12	–
Heat capacity (J · kg^{-1} · K^{-1})	2320	4180
Liquid viscosity (N · s · m^{-2})	0.213×10^{-3}	0.797×10^{-3}
Thermal conductivity (W · m^{-1} · K^{-1})	0.137	0.618
Heat of vaporization (J · kg^{-1})	494,000	–

The properties of acetone are at 67 °C. Although the average film temperature will be lower than this, the value of $k(\rho^2/\mu)^{1/3}$ tends not to be very sensitive to temperature. The molar mass of acetone can be assumed to be 58 kg · kmol^{-1}. Assume the fouling coefficients to be 11,000 W · m^{-2} · K^{-1} and 5000 W · m^{-2} · K^{-1} for the shell-side and tube-side respectively. For an allowable tube-side velocity of 2 m · s^{-1}, estimate the heat transfer area.

Solution

$$\text{Flowrate of acetone} = 0.1 \times 58 = 5.8\ \text{kg} \cdot \text{s}^{-1}$$

$$\text{Duty on condenser} = 5.8 \times 494{,}000 = 2.8652 \times 10^6\ \text{W}$$

$$\text{Flowrate of cooling water} = \frac{2.8652 \times 10^6}{4180(35-25)} = 68.55\ \text{kg} \cdot \text{s}^{-1} = 0.06882\ \text{m}^3 \cdot \text{s}^{-1}$$

To determine the condensing film coefficient using Equation 12.156 requires the number of tubes to be known. Thus, the solution must be iterative. Assume an initial heat transfer area (say 100 m^2). If the tube-side velocity is 2 m · s^{-1}:

$$N_T = \frac{m_T N_P}{\rho(\pi d_I^2/4)v_T} = \frac{68.55 \times 2}{996(\pi \times 0.016^2/4)2} = 342.3$$

Rounding to the nearest integer that can be configured in two passes:

$$N_T = 342$$

Now calculate the tube length:

$$\begin{aligned}L &= \frac{A}{\pi d_O N_T} \\ &= \frac{100}{\pi \times 0.020 \times 342} \\ &= 4.654\ \text{m}\end{aligned}$$

The shell diameter can be calculated from Equation 12.52:

$$\begin{aligned}D_S &= \left(\frac{4F_{TC}F_{SC}P_C P_T^2 A}{\pi^2 d_O L}\right)^{1/2} \\ &= \left(\frac{4 \times 1.11 \times 1.15 \times 0.025^2 \times 100}{\pi^2 \times 0.02 \times 4.654}\right)^{1/2} \\ &= 0.589\ \text{m}\end{aligned}$$

The condensing coefficient can be calculated for Equation 12.156:

$$\begin{aligned}h_C &= 0.959 k_L \left(\frac{\rho_L(\rho_L - \rho_V)Lg}{\mu_L m}\right)^{1/3} N_T^{2/9} \\ &= 0.959 \times 0.137 \left(\frac{736 \times (736 - 3.12) \times 4.654 \times 9.81}{0.213 \times 10^{-3} \times 5.8}\right)^{1/3} 342^{2/9} \\ &= 1302.7\ \text{W} \cdot \text{m} \cdot \text{K}^{-1}\end{aligned}$$

Calculate the tube-side Reynolds number:

$$\begin{aligned}Re &= \frac{\rho d_I v_T}{\mu} \\ &= \frac{996 \times 0.016 \times 2}{0.797 \times 10^{-3}} \\ &= 39{,}990\end{aligned}$$

The tube-side heat transfer coefficient can be calculated from Equation 12.59:

$$\begin{aligned}K_{hT4} &= 0.024 \frac{k}{d_I}\left(\frac{C_P \mu}{k}\right)^{0.4}\left(\frac{\rho d_I}{\mu}\right)^{0.8} \\ &= 0.024 \times \frac{0.618}{0.016}\left(\frac{4180 \times 0.797 \times 10^{-3}}{0.618}\right)^{0.4}\left(\frac{996 \times 0.016}{0.797 \times 10^{-3}}\right)^{0.8} \\ &= 5017.4 \\ h_T &= K_{hT4} v_T^{0.8} \\ &= 5017.4 \times 2^{0.8} \\ &= 8735.8\ \text{W} \cdot \text{m}^{-2} \cdot \text{K}^{-1}\end{aligned}$$

Now the overall heat transfer coefficient can be calculated:

$$\begin{aligned}\frac{1}{U} &= \frac{1}{h_C} + \frac{1}{h_{SF}} + \frac{d_O}{2k}\ln\frac{d_O}{d_I} + \frac{d_O}{d_I}\frac{1}{h_{TF}} + \frac{d_O}{d_I}\frac{1}{h_T} \\ &= \frac{1}{1302.7} + \frac{1}{11{,}000} + \frac{0.020}{2 \times 45}\ln\frac{0.020}{0.016} + \frac{0.020}{0.016}\left[\frac{1}{5000} + \frac{1}{8735.8}\right] \\ U &= 768.5\ \text{W} \cdot \text{m}^{-2} \cdot \text{K}^{-1} \\ \Delta T_{LM} &= \frac{(67 - 35) + (67 - 25)}{\ln\left[\dfrac{67 - 35}{67 - 25}\right]} \\ &= 36.8\ \text{K}\end{aligned}$$

Now the duty can be calculated from:

$$\begin{aligned}Q &= UA\,\Delta T_{LM} \\ &= 754.6 \times 100 \times 36.8 \\ &= 2.775 \times 10^6\ \text{W}\end{aligned}$$

This does not agree with the specified duty of 2.8652×10^6 W. To make the heat duty balanced requires the heat transfer area A to be adjusted by trial and error. At each value of A, the number of tubes and h_C must be calculated. This can be readily done using a spreadsheet solver. The result is:

$$\begin{aligned}Q &= 2.865 \times 10^6\ \text{W} \\ h_C &= 1307.7\ \text{W} \cdot \text{m}^{-2} \cdot \text{K}^{-1} \\ h_T &= 8735.7\ \text{W} \cdot \text{m}^{-2} \cdot \text{K}^{-1} \\ U &= 770.2\ \text{W} \cdot \text{m}^{-2} \cdot \text{K}^{-1} \\ A &= 101.2\ \text{m}^2 \\ D_S &= 0.589\ \text{m} \\ N_T &= 342\end{aligned}$$

Rather than specify the tube-side velocity, the tube-side pressure drop could have been specified (e.g. $\Delta P_T = 30{,}000$ $\text{N} \cdot \text{m}^{-2}$). Had this been the case, then the calculation would have required an iterative solution, similar to the solution of Example 12.2d.

12.11 Reboilers and Vaporizers

Reboilers are required for distillation columns to vaporize a fraction of the bottom product, as discussed in Chapter 8. It may also be the case that a liquid needs to be vaporized for other purposes; for example, a liquid feed needs to be vaporized before entering a reactor. The discussion here will focus on reboiling a distillation column, but the same basic principles apply to other types of vaporizers. 12

Three common designs of the reboiler are illustrated in Figure 12.23. The first shown in Figure 12.23a is a *kettle reboiler*. Vaporization takes place on the outside of tubes immersed in a pool of liquid. The bottom product is taken from an overflow from the liquid pool and there is no recirculation between the reboiler and the column. In some designs, the tube bundle can be installed in the base of the column as an internal reboiler. The kettle reboiler incorporates a volume above the liquid pool and tube bundle for vapor and liquid disengagement. The shell diameter is typically 40% greater than the bundle diameter to allow for this. The second type of reboiler shown in Figure 12.23b, the *horizontal thermosyphon*, also features vaporization on the outside of the tubes. However, in this case, there is recirculation around the base of the column. A mixture of vapor and liquid leaves the reboiler and enters the base of the column, where it separates. The column

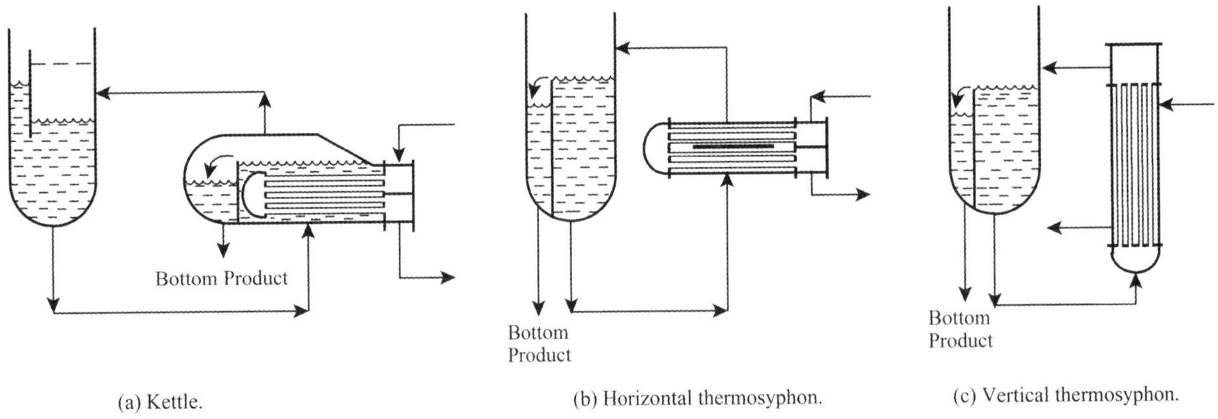

Figure 12.23

Reboiler designs.

internals are arranged to provide a steady hydrostatic head for the circulation. The third type of reboiler, the *vertical thermosyphon*, is illustrated in Figure 12.23c. Again, there is a recirculation around the base of the column, but this time the vaporization takes place inside the tubes. Boiling on the tube-side will normally be carried out in a 1–1 exchanger.

The three reboilers in Figure 12.23 are shown under *natural circulation*. The flow of liquid from the column to the reboiler is created by the difference in hydrostatic head between the column of liquid feeding the reboiler and the vapor–liquid mixture created by the reboiler.

The amount of liquid vaporized in the reboiler should not be more than 80%, otherwise this will tend to lead to excessive fouling of the reboiler. For kettle reboilers, there is no recirculation, but for thermosyphon reboilers, a *recirculation ratio* can be defined as:

$$\text{Recirculation ratio} = \frac{\text{Flowrate of liquid at reboiler outlet}}{\text{Flowrate of vapor at reboiler outlet}}$$

This usually lies between 0.25 and 6. The greater the value of recirculation ratio, the less fouling there is in the reboiler. Lower values tend to be used in horizontal thermosyphons and higher values (greater than 4) used in vertical thermosyphons. The recirculation ratio is a degree of freedom at the discretion of the designer. This should be fixed later when the detailed design is carried out.

The kettle reboiler has the advantage that it is equivalent to a theoretical stage for the distillation but is relatively expensive due to the extra volume required for vapor disengagement. Also, the liquid has a high residence time in the heating zone, which can be a problem if the material is prone to thermal decomposition. If the reboiler must operate at a high pressure, then the large diameter of the kettle shell is a disadvantage. A large-diameter cylindrical shell requires a thicker wall to withstand a given pressure than a small-diameter shell. The horizontal and vertical thermosyphons both have the disadvantage of not providing a complete theoretical stage for the distillation. However, the thermosyphon reboilers are less prone to fouling than kettle reboilers and have a lower residence time in the heating zone. Thermosyphon reboilers require additional height inside the column shell than kettle reboilers to allow for vapor disengagement as the vapor–liquid mixture enters the column. Vertical thermosyphon reboilers require the column to be at a higher elevation than kettle and horizontal thermosyphons. Horizontal thermosyphons tend to be preferred to vertical ones if the heat transfer area is large, as horizontal arrangements are easier to maintain. Although thermosyphon reboilers can be used under vacuum conditions, care must be exercised as the effect of pressure on the boiling point of the fluid entering the reboiler must be considered. When reboiling multicomponent systems, the vaporization can take place over a range of temperatures. The forced flow in a thermosyphon can give a higher mean temperature difference than a kettle for the same percentage of vaporization, as the kettle boiling temperature is uniform. Generally the heat flux (heat transferred per unit area) and heat transfer coefficients are in the order

$$\text{Kettle} < \text{Horizontal thermosyphon} < \text{Vertical thermosyphon}$$

Given these arguments, it is not surprising that the most common design of reboiler is the vertical thermosyphon.

The basic characteristics of the boiling process are illustrated in Figure 12.24. This shows a plot of heat flux versus the temperature difference between the vaporizing surface and the bulk liquid plotted on logarithmic axes. Initially, between Points A and B in Figure 12.24, heat transfer is by natural convection. Superheated liquid rises to the liquid surface where evaporation takes place. As the temperature difference increases beyond Point B in Figure 12.24, nucleate boiling occurs in which vapor bubbles are formed at the heating surface and released from the surface. Nucleate boiling, as its name implies, depends on the presence of nuclei. In boiling, the nuclei are preexisting inclusions of non-condensable gas or vapor in cavities on the heat transfer surface. It thus depends on the character of the heat transfer surface.

Point C in Figure 12.24 is termed the *critical heat flux* or *maximum boiling flux* or *peak boiling flux* as bubbles coalesce on the surface creating a vapor blanket. Critical heat flux occurs because insufficient liquid is able to reach the heat transfer surface due to the rate at which vapor is leaving. Beyond Point D, the

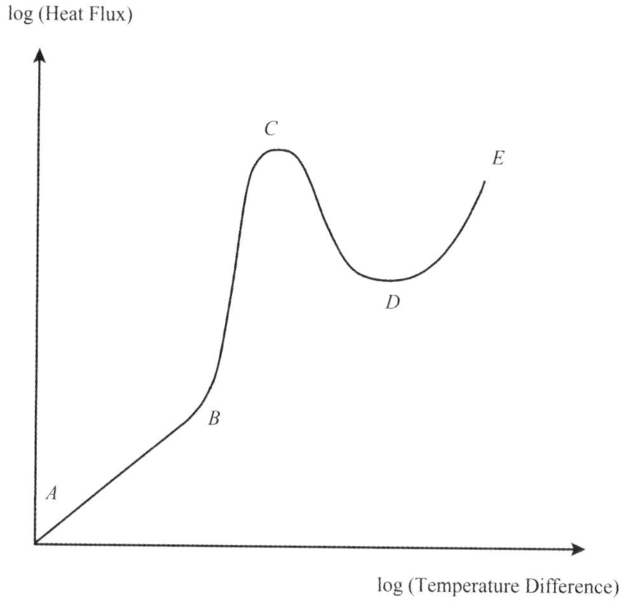

Figure 12.24

Heat transfer characteristics of boiling.

surface is dry and entirely blanketed by vapor and heat is transferred by conduction and radiation.

Reboilers are designed to operate below the peak flux, as beyond it either the heat flux would be lower or a much higher temperature difference would be required. The design is normally restricted to have a heat flux less than 70% of the critical flux. The preliminary design of the kettle and horizontal thermosyphon reboilers can be based on *pool boiling*. In pool boiling, the heating surface is surrounded by a large body of fluid in which the fluid motion is only induced by natural convection currents and the motion of bubbles. A simple correlation for nucleate boiling on a single tube is that due to Palen (in Hewitt, 2008):

$$h_{NB} = 0.18 P_C^{0.69} q^{0.7} \left(\frac{P}{P_C}\right)^{0.17} \quad (12.163)$$

where
h_{NB} = nucleate boiling coefficient ($W \cdot m^{-2} \cdot K^{-1}$)
P_C = liquid critical pressure (bar)
P = operating pressure (bar)
q = heat flux ($W \cdot m^{-2}$)
= $h_{NB}(T_W - T_{SAT})$
T_W = wall temperature of the heating surface (°C)
T_{SAT} = saturation temperature of the boiling liquid (°C)

Equation 12.163 neglects natural convection effects, which are only important for a temperature difference less than 4 °C. For very small temperature differences, a value of 250 $W \cdot m^{-2} \cdot K^{-1}$ for hydrocarbons and 1000 $W \cdot m^{-2} \cdot K^{-1}$ for water can be added to the value calculated from Equation 12.157 to compensate for natural convection (Palen in Hewitt, 2008).

Equation 12.163 applies to vaporization of single components, but can be used for close boiling mixtures without too much error. Coefficients for wide boiling mixtures will be overestimated. This can be corrected for using an empirical correction factor given by Palen (in Hewitt, 2008):

$$h_{NB} = 0.18 P_C^{0.69} q^{0.7} \left(\frac{P}{P_C}\right)^{0.17} \left[\frac{1}{1 + 0.023 q^{0.15} (T_{DEW} - T_{BUB})^{0.75}}\right] \quad (12.164)$$

where
T_{DEW} = dew point of the mixture (°C, K)
T_{BUB} = bubble point of the mixture (°C, K)

Boiling heat transfer coefficients for tube bundles are generally higher than those for single tubes under the same conditions. Palen (in Hewitt, 2008) presented an approximate method that can be used for the preliminary design of kettle and horizontal thermosyphon reboilers:

$$h_B = h_{NB}\left[1.0 + 0.1\left(\frac{0.785 D_B}{p_C (p_T/d_O)^2 d_O} - 1.0\right)^{0.75}\right] + h_{NC} \quad (12.165)$$

where
h_B = boiling heat transfer coefficient for the tube bundle ($W \cdot m^{-2} \cdot K^{-1}$)
h_{NB} = nucleate boiling heat transfer coefficient given by Equations 12.163 and 12.164 ($W \cdot m^{-2} \cdot K^{-1}$)
h_{NC} = natural convection heat transfer coefficient
D_B = tube bundle diameter (m)
p_T = tube pitch (m)
d_O = outside diameter of the tubes (m)
p_C = pitch correction factor ($p_C = 1$ for square pitch, $p_C = 0.866$ for triangular pitch)

Natural convection makes a relatively small contribution to the total heat transfer, except when the temperature difference between the wall and the boiling liquid is less the 4 °C. For larger temperature differences, Palen (in Hewitt, 2008) suggests using a value of h_{NC} of 250 $W \cdot m^{-2} \cdot K^{-1}$ for hydrocarbons and 1000 $W \cdot m^{-2} \cdot K^{-1}$ for water and aqueous solutions.

Mostinski (1963) gives a correlation for the estimation of the critical heat flux for single tubes:

$$q_{C1} = 3.67 \times 10^4 P_C \left(\frac{P}{P_C}\right)^{0.35} \left[1 - \frac{P}{P_C}\right]^{0.9} \quad (12.166)$$

where q_{C1} = critical heat flux for a single tube ($W \cdot m^{-2}$)

This can be corrected for tube bundles by (Palen in Hewitt, 2008):

$$q_C = q_{C1} \phi_B \quad \text{for} \quad \phi_B < 1.0 \quad (12.167)$$

otherwise

$$\phi_B = 1$$

where q_C = critical heat flux for the tube bundle ($W \cdot m^{-2} \cdot K^{-1}$)

12

$$\phi_B = \frac{3.1\pi D_B L}{A}$$

D_B = tube bundle diameter (m)
L = tube length (m)
A = heat transfer area (m^2)

The design of vertical thermosyphon reboilers requires iterative calculations in which the exchanger needs to be divided into zones. The energy and pressure balances need to be performed simultaneously. Frank and Prickett (1973) performed a range of detailed simulations and presented the results graphically. This can be used as the basis of preliminary design. The graphical data can be correlated as:

For *aqueous solutions*:

$$q = 384.52\Delta T + 130.07\Delta T^2 - 2.4204\Delta T^3 \tag{12.168}$$

where q = heat flux ($W \cdot m^{-2}$)
ΔT = mean temperature difference between the heat transfer surface and the bulk fluid (K)

For *organic liquids*:

$$q = -100.98 + 1705.9\Delta T + 26.37\Delta T^2 - 0.288\Delta T^3 - 5902.8T_R\Delta T + 6031.3T_R^2\Delta T \tag{12.169}$$

where T_R = reduced temperature of fluid
= T/T_C
T = temperature (K)
T_C = critical temperature (K)

For the simulations, the heating was assumed to be supplied using saturated steam on the shell-side with a combined condensation and fouling coefficient of 5700 $W \cdot m^{-2} \cdot K^{-1}$. The fouling coefficient on the tube-side was assumed to be 5700 $W \cdot m^{-2} \cdot K^{-1}$.

If these values are appropriate, then the overall heat transfer coefficient can be calculated from:

$$U = \frac{q}{\Delta T} \tag{12.170}$$

The correlation should be used with caution outside the range $0.6 < T_R < 0.8$ and should not be used below a pressure of 0.3 bar. When dealing with a clean, nondegrading material, the process fouling coefficient should be increased to around 11,000 $W \cdot m^{-2} \cdot K^{-1}$, but should be reduced to 1400 to 1900 $W \cdot m^{-2} \cdot K^{-1}$ for material that has a tendency to polymerize (Frank and Prickett, 1973). If a shell-side process fouling coefficient different from 5700 $W \cdot m^{-2} \cdot K^{-1}$ is required, the corrected overall heat transfer coefficient can be calculated from (Frank and Prickett, 1973):

$$\frac{1}{U'} = \frac{1}{U} - \frac{1}{5700} + \frac{1}{h_S} - \frac{1}{5700} + \frac{1}{h_{TF}} \tag{12.171}$$

where U' = corrected overall heat transfer coefficient ($W \cdot m^{-2} \cdot K^{-1}$)
h_S = required shell-side coefficient including fouling ($W \cdot m^{-2} \cdot K^{-1}$)
h_{TF} = required process fouling coefficient ($W \cdot m^{-2} \cdot K^{-1}$)

The mean temperature difference should be less than 35 to 55 °C. This will avoid excessive fouling and excessive vaporization per pass (i.e. a low recirculation ratio), leading to poor heat transfer in the upper parts of the tubes as heat transfer to a liquid annulus is replaced by heat transfer to a mist.

Great caution should be exercised regarding correlations for boiling: they are notoriously unreliable. Unlike other heat transfer phenomena, the goal of reliable prediction of boiling rates has proved to be elusive. Many correlations other than those given here are available in the literature. Their predictions of boiling coefficients for the same duty can differ by an order of magnitude. Even the predictions from detailed simulations should be treated with great caution.

Rather than use natural circulation, as in the designs in Figure 12.23, the liquid can be fed to the reboiler by a pump using *forced circulation*. Boiling normally takes place on the tube-side of a 1–1 heat exchanger, which can be vertical or horizontal. Vaporization in forced convection reboilers is usually limited to be less than 1 to 5% (Palen in Hewitt, 2008). In some cases, it might be desirable to suppress boiling inside the heat exchanger completely by installing a control valve at the exchanger outlet to increase the pressure in the exchanger and suppress boiling. The liquid leaving the exchanger will then partially vaporize as the liquid pressure is decreased across the valve. Suppression of boiling inside the heat exchanger might be desirable if it leads to excessive fouling. In preliminary design, forced convection reboilers can be based on the assumption that heat transfer is by forced convection only. This will give a conservative design if some boiling is allowed in the exchanger and will overestimate the heat transfer area. High tube velocities of the order of 3 to 5 $m \cdot s^{-1}$ or higher are used to reduce fouling.

If forced-convective boiling is to be carried out, the design is essentially the same as any 1–1 exchanger.

Forced convection suppresses nucleate boiling but introduces a significant component of forced convection heat transfer. In fact, in most practical situations, including natural circulation, both forced convection and nucleate boiling are important. In forced-convective boiling, the boiling heat transfer coefficient can be estimated by a combination of convective and nucleate boiling heat transfer with the nucleate boiling component reduced by a suppression factor. There is significant uncertainty associated with the prediction of such components. Because the amount of vapor changes through the exchanger, the exchanger needs to be divided into zones and the correlations applied in each zone.

Natural circulation will lead to cheaper designs than forced circulation, but forced circulation can lead to lower fouling than the corresponding natural circulation designs.

Finally, if vaporization of a liquid is required for applications other than distillation, then the same principles and methods can be applied. The one distinctive difference is that with all designs, with the exception of kettle designs, a vapor–liquid separation device will be required at the vaporizer outlet, as illustrated in Figure 12.25.

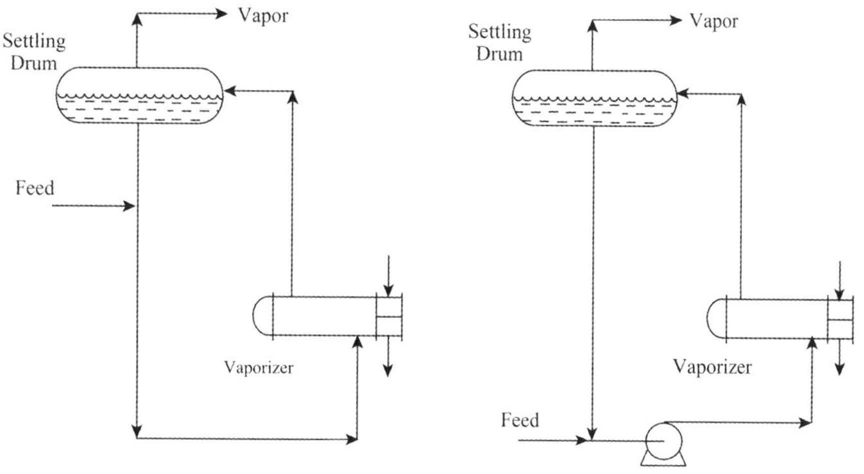

(a) Natural circulation. (b) Forced circulation.

Figure 12.25
Process vaporizer arrangements.

Example 12.8 A reboiler is required to supply 0.1 $\text{kmol} \cdot \text{s}^{-1}$ of vapor to a distillation column. The column bottom product is almost pure butane. The column operates with a pressure at the bottom of the column of 19.25 bar. At this pressure, the butane vaporizes at a temperature of 112 °C. The vaporization can be assumed to be essentially isothermal and is to be carried out using steam with a condensing temperature of 140 °C. The heat of vaporization for butane is 233,000 $\text{J} \cdot \text{kg}^{-1}$, its critical pressure 38 bar, critical temperature 425.2 K and molar mass 58 $\text{kg} \cdot \text{kmol}^{-1}$. The film coefficient (including fouling) for the condensing steam can be assumed to be 5700 $\text{W} \cdot \text{m}^{-2} \cdot \text{K}^{-1}$. Calculate the heat transfer area for:

a) A kettle reboiler assuming steel tubes with 30 mm outside diameter, 2 mm wall thickness and length 2.95 m are to be used, with thermal conductivity of the tube wall of 45 $\text{W} \cdot \text{m}^{-1} \cdot \text{K}^{-1}$ and the fouling coefficient for vaporization to be 2500 $\text{W} \cdot \text{m}^{-2} \cdot \text{K}^{-1}$.

b) A vertical thermosyphon reboiler.

Solution

a)

$$\begin{aligned}\text{Heat load} &= 0.1 \times 58 \times 233{,}000 \\ &= 1.3514 \times 106\ \text{W}\end{aligned}$$

The boiling film coefficient for a kettle reboiler can be estimated from the correlation for pool boiling. Equation 12.163 gives one such method due to Palen (in Hewitt, 2008). However, the correlation requires the heat flux to be known, and therefore the heat transfer area to be known. Hence the calculation will need to be iterative. Assume an initial heat transfer area of, say, 40 m^2.

$$\begin{aligned}q &= \frac{Q}{A} \\ &= \frac{1.3514 \times 10^6}{40} \\ &= 33{,}785\ \text{W·m}^{-2}\end{aligned}$$

Now the boiling film coefficient can be calculated:

$$\begin{aligned}h_{NB} &= 0.18 P_C^{0.67} q^{0.7} \left(\frac{P}{P_C}\right)^{0.17} \\ &= 0.18 \times 38^{0.67} \times 33{,}785^{0.7} \left(\frac{19.25}{38}\right)^{0.17} \\ &= 2714\ \text{W·m}^{-2}\text{·K}^{-1}\end{aligned}$$

The overall heat transfer coefficient can be calculated from:

$$\begin{aligned}\frac{1}{U} &= \frac{1}{h_{NB}} + \frac{1}{h_{SF}} + \frac{d_O}{2k} \ln \frac{d_O}{d_I} + \frac{d_O}{d_I}\frac{1}{h_{TF}} + \frac{d_O}{d_I}\frac{1}{h_T} \\ &= \frac{1}{2714} + \frac{1}{2500} + \frac{0.03}{2 \times 45} \ln \frac{0.03}{0.026} + \frac{0.03}{0.026}\left(\frac{1}{5700}\right) \\ U &= 1125\ \text{W·m}^{-2}\text{·K}^{-1} \\ Q &= UA\Delta T \\ &= 1125 \times 40 \times 28 \\ &= 1.260 \times 106\ \text{W}\end{aligned}$$

The calculated heat duty does not agree with the specified duty of 1.351×10^6 W. To make the duty balance requires the heat transfer area to be adjusted by trial and error. At each value of A, a new heat flux and boiling film coefficient is calculated. This allows the new overall heat transfer coefficient to be calculated, and so on, until the calculated heat duty agrees with a specified duty. This can be readily done using a spreadsheet solver. The result is:

$$
\begin{aligned}
Q &= 1.3514 \times 10^6\ \text{W} \\
h_{NB} &= 2,564\ \text{W}\cdot\text{m}^{-2}\cdot\text{K}^{-1} \\
U &= 1,112\ \text{W}\cdot\text{m}^{-2}\cdot\text{K}^{-1} \\
A &= 43.40\ \text{m}^2 \\
q &= 31,142\ \text{W}\cdot\text{m}^{-2}
\end{aligned}
$$

This heat flux must be checked to see if it is below the critical heat flux. The critical heat flux for a single tube is given by the Mostinski Equation (Equation 12.166):

$$
\begin{aligned}
q_{C1} &= 3.67 \times 10^4 P_C \left(\frac{P}{P_C}\right)^{0.35} \left[1 - \frac{P}{P_C}\right]^{0.9} \\
&= 5.82 \times 10^5\ \text{W}\cdot\text{m}^{-2}
\end{aligned}
$$

To correct this for the bundle requires the bundle diameter to be calculated from Equation 12.52 assuming F_{TC} and F_{SC} to both be unity:

$$
D_B = \left(\frac{4 p_C p_T^2 A}{\pi^2 d_O L}\right)^{1/2}
$$

Assume a pitch of 0.0375 m and square configuration with a tube length of 2.95 m:

$$
\begin{aligned}
D_B &= \left(\frac{4 \times 1 \times 0.0375^2 \times 43.40}{\pi^2 \times 0.03 \times 2.95}\right)^{1/2} \\
&= 0.53\ \text{m} \\
\phi_B &= \frac{3.1 \pi D_B L}{A} \\
&= \frac{3.1 \times \pi \times 0.53 \times 2.95}{43.40} \\
&= 0.35 \\
q_C &= q_{C1} \phi_B \\
&= 5.82 \times 10^5 \times 0.35 \\
&= 2.04 \times 10^5\ \text{W}\cdot\text{m}^{-2}
\end{aligned}
$$

Thus, the flux is well below the maximum predicted by the Mostinski Equation.

It should be noted that the shell diameter will be around 40% greater than that of the tube bundle to allow for vapor disengagement.

b) For a vertical thermosyphon reboiler, the heat flux can be approximated using Equation 12.169 for organic liquids.

$$
\begin{aligned}
T_R &= \frac{T}{T_C} \\
&= \frac{112 + 273.15}{425.2} \\
&= 0.905
\end{aligned}
$$

This is outside the range over which the original data were correlated and hence the results should be treated with caution.

$$
\begin{aligned}
q &= -100.98 + 1705.9 \times 28 + 26.37 \times 28^2 - 0.288 \times 28^3 \\
&\quad - 5902.8 \times 0.91 \times 28 + 6031.3 \times 0.912 \times 28 \\
&= 50,817\ \text{W}\cdot\text{m}^{-2}
\end{aligned}
$$

$$
\begin{aligned}
U &= \frac{q}{\Delta T} \\
&= \frac{50,817}{28} \\
&= 1289\ \text{W}\cdot\text{m}^{-2}\cdot\text{K}^{-1}
\end{aligned}
$$

This does not need to be corrected using Equation 12.171 as the steam condensing film coefficient and process fouling coefficient agree with the assumptions on which the correlation is based.

$$
\begin{aligned}
A &= \frac{Q}{q} \\
&= \frac{1.351 \times 10^6}{50,817} \\
&= 26.6\ \text{m}^2
\end{aligned}
$$

On the basis of these calculations, the vertical thermosyphon would appear to be the better option. However, the correlations to predict boiling have an extremely low reliability. Predicted minor variations in heat transfer area should not be used to choose options, as the predictions for boiling are so unreliable.

12.12 Other Types of Heat Exchangers

While the shell-and-tube heat exchanger is the most common type of heat exchanger in the process industries, it does have some significant limitations:

a) The flow is not truly countercurrent. Even the 1–1 exchanger is not truly countercurrent as the flow on the shell-side is partially cross-flow. This means that the shell-and-tube exchanger is mostly limited to transfer heat with a minimum temperature difference of 10 °C. Designs can achieve a smaller temperature difference (perhaps down to 5 °C), but care is needed in the application. Some heat transfer duties demand very small temperature differences.

b) The *area density* (heat transfer area per unit of volume of exchanger) is relatively low. A conventional shell-and-tube heat exchanger has an area density of the order of $100\ \text{m}^2 \cdot \text{m}^{-3}$. Other designs of heat exchanger can achieve area densities of 300 to $700\ \text{m}^2 \cdot \text{m}^{-3}$ and in some designs greater than $1000\ \text{m}^2 \cdot \text{m}^{-3}$.

The most important alternatives to shell-and-tube designs are:

1) *Gasketed plate heat exchanger* After the shell-and-tube heat exchanger, the most commonly used alternative is the *gasketed plate heat exchanger*. This consists of a series of parallel plates with gaskets between the plates to provide a fluid seal. The plates are corrugated both to increase the turbulence (and hence the film transfer coefficients) and to give mechanical rigidity. The enhancement of heat transfer coefficients from the

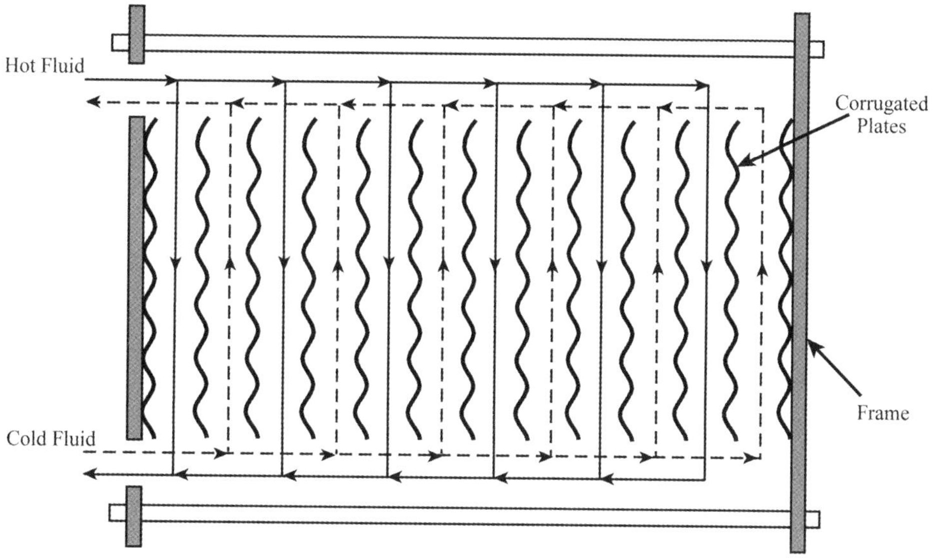

Figure 12.26
Plate-and-frame heat exchanger.

corrugations is a particular advantage when heating and cooling more viscous materials. The turbulence promoted by the corrugations also helps to reduce fouling relative to plain surfaces. The plates are held together and compressed in a frame by the use of lateral bolts. The basic arrangement is illustrated in Figure 12.26. Each corrugated plate is provided with four ports. The gasket arrangement around the four ports forces the hot and cold fluids to flow down alternate channels. This provides countercurrent flow, allowing temperature differences between hot and cold fluids down to around 1 °C. Most applications for gasketed plate heat exchangers are for liquid–liquid duties, but can also be applied to condensing and evaporating duties. The limitations of the gasket seals mean that applications are normally restricted to be between –30 and 200 °C with pressures up to 20 bar.

A gasketed plate heat exchanger is usually significantly cheaper than a shell-and-tube heat exchanger for the same duty. This is especially the case if the shell-and-tube heat exchanger must be fabricated in more expensive materials such as stainless steel.

Another advantage in retrofit is that an existing frame can often accommodate additional plates if a higher capacity is required. For the same flowrate, an increase in the number of plates decreases the flowrate through the channels and therefore the heat transfer coefficients. However, if the flowrate increases, the number of plates can be increased to accommodate a higher duty (at the expense of increased pressure drop).

The flow arrangement shown in Figure 12.26 involves a single pass for each fluid. More complex flow arrangements can also be used.

2) *Welded plate heat exchangers* The limitations of the gaskets in gasketed plate heat exchangers can be overcome by welding the plates together. This eliminates both the gaskets and the frame from the design. Elimination of the gaskets extends the range of application to a wider range of temperatures and pressures. The highest pressures for welded plate heat exchangers can be achieved by mounting the plates within a pressure vessel. A fluid is maintained in the pressure vessel with a higher pressure than the fluids flowing between the plates. Heat transfer only takes place between fluids flowing between the plates. This means that the heat exchanger plates are only exposed to the pressure difference between the fluids flowing between the plates. Welded plate heat exchangers can achieve an area density of up to 300 $m^2 \cdot m^{-3}$. The operating temperature range varies between –200 °C and 900 °C. Pressures up to 300 bar can be accommodated.

   The costs for this form of exchanger are higher than those for gasketed plate heat exchangers. An important limitation is that they can only be cleaned chemically and not mechanically.

3) *Plate-fin heat exchangers.* Another type of plate heat exchanger is the *plate-fin heat exchanger*. This is illustrated in Figure 12.27. The plate-fin heat exchanger consists of a series of flat plates, between which a matrix is formed from corrugated metal that provides a large extended heat transfer area. The components are bonded together either by brazing or diffusion bonding. Many different surface geometries are available to promote heat transfer. Figure 12.27 illustrates just three example surfaces. Heights of surfaces range from 4 and 12 mm. The space between the plates and the surface geometry chosen for each fluid are important degrees of freedom in the design of such units. The area density of such units is typically in the range 850 to 1500 $m^2 \cdot m^{-3}$. The operating range for such units depends both on the bonding technique used and the material of construction. Aluminum-brazed plate-fin heat exchangers are used for cryogenic applications, but can also be used up to temperatures of around

12

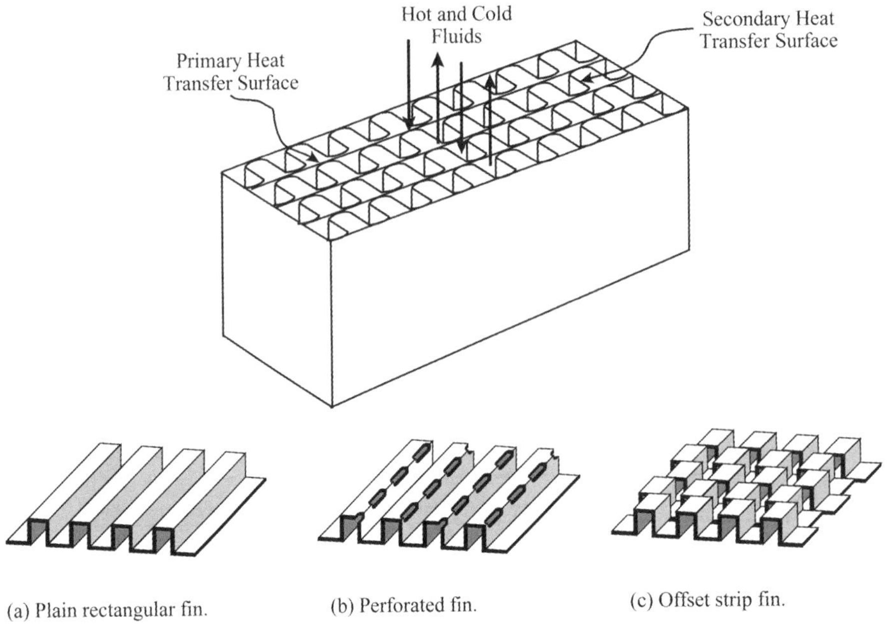

Figure 12.27
Plate fin heat exchangers.

100 °C. Stainless steel plate-fin heat exchangers are able to operate up to 650 °C and titanium units up to 550 °C. Aluminum-brazed units can operate up to 100 bar, stainless steel units up to 50 bar and higher. Higher pressures require diffusion-bonded units. Plate-fin heat exchangers not only have the advantages of high area density and high heat transfer coefficients but also have a number of other advantages that make them overwhelmingly attractive for certain applications. Temperature differences of 1 °C or less can be tolerated in such units. Also, the units can be designed to handle multiple streams through the use of complex header arrangements. This allows, in effect, for a heat exchanger network to be accommodated within a single unit.

4) *Spiral heat exchangers. Spiral heat exchangers* can be thought of as plate heat exchangers in which the plates are formed into a spiral, as illustrated in Figure 12.28. The channels are closed by gasketed end-plates. The hot fluid enters at the center of the unit and flows from the inside outwards. The cold fluid enters at the periphery and flows towards the center in a countercurrent flow arrangement. The gap between the plates can be adjusted to suit the application. Operating temperatures up to 400 °C and operating pressures up to 20 bar are possible. The units have a low tendency to foul, but are easily cleaned by removing the end-plates. Again, true countercurrent flow is possible and therefore lower temperature driving forces can be tolerated. Spiral heat exchangers are suited to small heat transfer duties that have a tendency to foul.

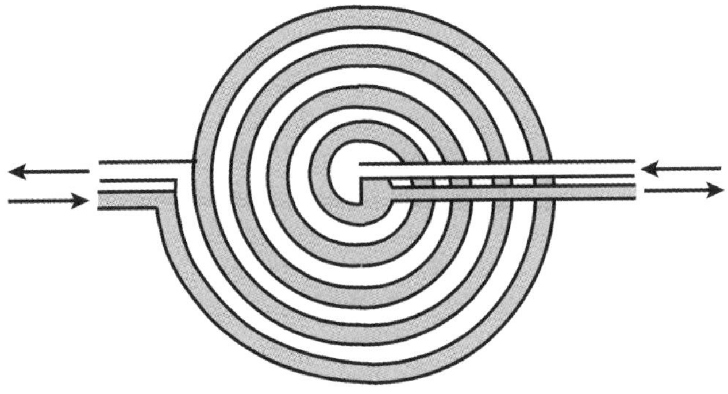

Figure 12.28
Spiral heat exchanger.

12.13 Fired Heaters

In some situations, process heat needs to be supplied:

a) with a high heat duty (e.g. a very large reboiler);
b) at a high temperature (e.g. at a temperature above which heat can be supplied by steam);
c) with a high heat flux (e.g. heat of reaction in situations where a short residence time in the reactor is required).

In such cases, radiant heat transfer is used from the combustion of fuel in a *fired heater* or *furnace*. Sometimes the function is to purely provide heat; sometimes the fired heater is also a reactor and provides heat of reaction. The special case of steam generation in a fired heater (a steam boiler) is dealt with in Chapter 23.

Fired heater designs vary according to the function, heating duty, type of fuel and the method of introducing combustion air. Figure 12.29 shows some typical fired heater designs. Combustion

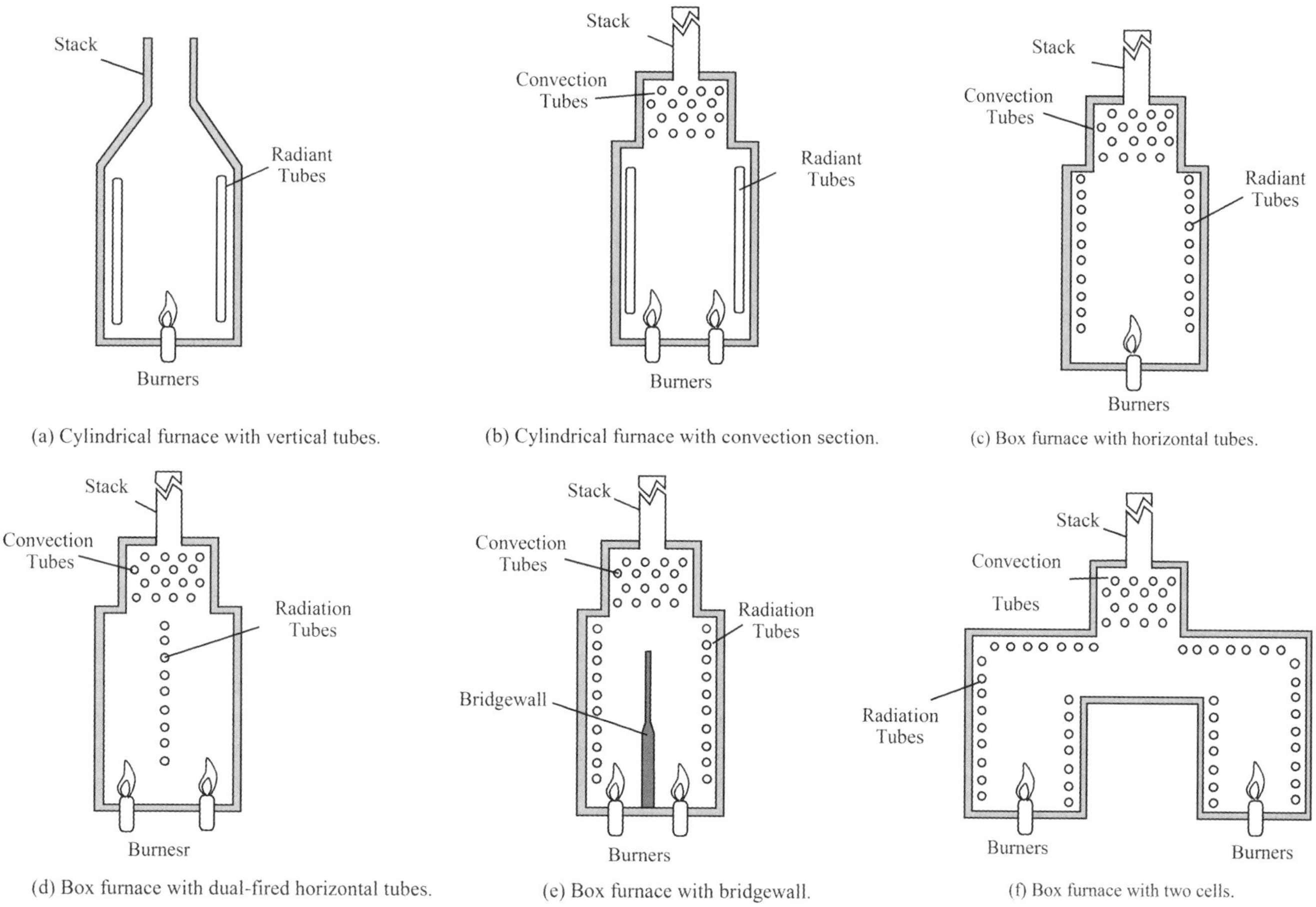

Figure 12.29
Different furnace configurations.

of fuel takes place in the *radiant section* or *firebox*, which is refractory lined. Most often, the walls are lined with one or two rows of tubes mounted horizontally or vertically through which passes the fluid that needs to be heated. Tube dimensions are specified as a nominal diameter (DN) and a schedule that specifies the wall thickness. Table 13.5 gives dimensions for a number of commonly used pipe sizes. The tube pitch is typically 2 × nominal diameter. If two rows of tubes are used, then the tubes are mounted in a staggered arrangement as an equilateral triangle. The gap between the tubes and the refractory walls is typically equal to the nominal diameter. Heat transfer in the radiant section is mainly by radiation, with a small contribution by convection. After the flue gas leaves the radiant section, most furnace designs extract further heat from the flue gas in horizontal banks of tubes in a *convection section*, before the flue gas is vented to the atmosphere through the *stack*. Heat transfer in the convection section is by both radiation and convection, but convection effects dominate.

The simplest design of fired heater is shown in Figure 12.29a, featuring a cylindrical design with the tubes mounted vertically in a circle around the inner walls of the combustion chamber. The design in Figure 12.29a does not feature a convection section and would be an inefficient design, appropriate only for small duties. Figure 12.29b shows the corresponding cylindrical design but with a convection section. The convection section consists of a chamber with a square or rectangular cross section located on top of the cylindrical radiant section. Cylindrical designs are used for duties less than 60 MW. Larger duties are suited to *box* designs. Such fired heaters have a rectangular cross section. The design in Figure 12.29c shows a box furnace with horizontal tubes in the combustion chamber. The convection section consists of a chamber with a rectangular cross section located on top of the cylindrical radiant section. The length of the convection section in box designs is usually the same as the radiant section, but narrower than the radiant section. Figure 12.29d shows a box furnace with dual firing. In this case, burners are located on either side of the radiation tubes. Dual firing allows a higher and more uniform heat flux and is therefore suited to the requirements of endothermic reactions. In this case, burners could be in principle mounted on the floor or the walls of the radiant section. Figure 12.29e shows a box furnace with dual firing, but with a refractory *bridgewall* dividing the combustion chamber into two parts. Such a design allows some independence of the heat transfer arrangements on either side of the bridgewall. A more extreme version is shown in Figure 12.29f, which features a box furnace with two *cells*. Rather than the horizontal tube arrangement in Figure 12.29f, the tubes can be mounted vertically in the center of the cells with burners mounted on the walls in a dual-fired arrangement. Such arrangements are often used to carry out endothermic reactions within the tubes.

Flow of combustion air to the fired heater is created in one of three ways:

1) *Natural draft.* In natural draft designs the air flows into the fired heater as a result of the density difference between the hot gases in the stack and that of the surrounding air. Thus, the height of the stack provides both the driving force for air flow and the mechanism to disperse the waste combustion gases.
2) *Forced draft.* In forced draft designs a fan is used to blow air into the fired heater. The fired heater is thus at a pressure slightly above atmospheric. The height of the stack is dictated purely by the dispersion requirements of the waste combustion gases.
3) *Induced draft.* In induced draft designs a fan is located in the duct for the waste combustion gases after the fired heater before entering the stack. This creates a pressure in the fired heater slightly below atmospheric. As with forced draft arrangements, the height of the stack is dictated purely by the dispersion requirements of the waste combustion gases.

For large fired heaters requiring a significant amount of equipment in the flue to treat the exhaust gases to remove oxides of sulfur and/or oxides of nitrogen, as well as equipment for heat recovery, a combination of forced and induced draft might need to be used in a *balanced draft* design.

Figure 12.30 shows a typical fired heater configuration for a process heating duty. The feed is split into parallel flows, known as *passes*. The example in Figure 12.30 features four passes. Many different pass arrangements are possible (Nelson, 1958; Martin, 1998). Figure 12.30 simply shows one possible example. The feed is first preheated in the *convection section* where the tubes usually have extended surfaces to increase the rate of heat transfer from the flue gas. The feed then passes to the *shield section*, which comprises plain tubes, known as *shock tubes* or *shield tubes*. These tubes need to be robust enough to be able to withstand high temperatures and receive significant radiant heat from the radiant section. The tubes in the convection section and shield section are horizontal and there are typically between 4 and 12 tubes in each row. After the shield section, the fluid passes to the tubes of the *radiant section* where the heat transfer is completed. The radiant tubes in Figure 12.30 are horizontal in this example and feature a single row. For process heating duties the tubes are formed into loops, as shown in Figure 12.30. A *header box* encloses the ends of the loops. This is an insulated compartment separated from the flue gas. Expansion of the furnace tubes is facilitated using pipe sleeves in the furnace wall to guide expansion. In other designs the tubes are oriented vertically. Vertically mounted tubes are supported from the roof of the furnace by heat-resistant metal alloy brackets. The bottoms of the tubes are provided with guides to allow expansion, but preventing lateral movement. For either horizontal or vertical orientation, two tube rows on a triangular pitch can also be used. After combustion of the fuel in the radiant section, the hot flue gases pass through the shield and convective sections before being collected in *breeching* for release through the stack to atmosphere. The flowrate of the gases is controlled by a *damper* in the stack. If the fired heater is required

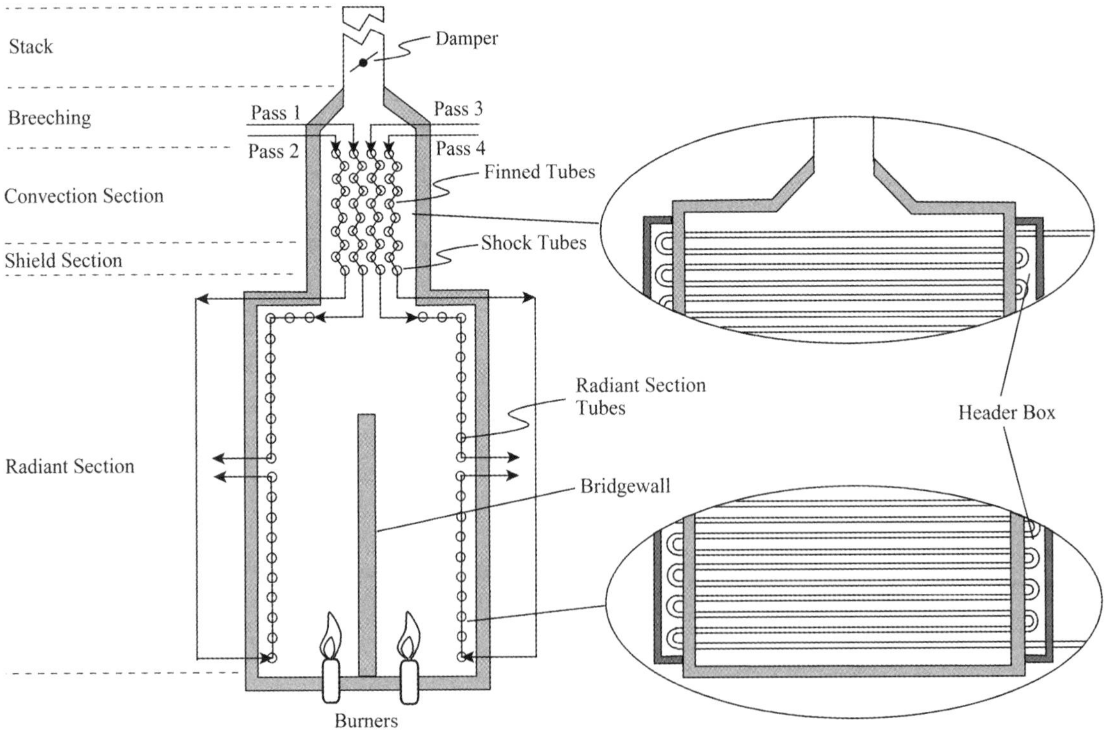

Figure 12.30
A typical furnace arrangement.

to carry out a reaction, it might be necessary to have an arrangement with many more passes than used for process heating only, using parallel flows connected to inlet and outlet manifolds.

Heat input to the fired heater is from three sources:

a) net heat of combustion (Q_{COMB});
b) sensible heat of combustion air (Q_{AIR});
c) sensible heat of the fuel, together with heat from atomizing steam if it is used for heavy fuel oil combustion (Q_{FUEL}).

The heat output from the fired heater is to four sinks:

a) heat transferred to the radiant section tubes ($Q_{RADIANT}$);
b) heat transferred in the convection section (Q_{CONV});
c) casing losses (Q_{CASING});
d) sensible heat of exit flue gas (Q_{LOSS}).

Overall, there must be an energy balance such that:

$$Q_{COMB} = Q_{RADIANT} + Q_{CONV} + Q_{CASING} + Q_{LOSS} - Q_{AIR} - Q_{FUEL} \qquad (12.172)$$

The split of heat duty between the radiant and convection section varies according to the design. Casing losses are usually between 1 and 3% of the heat release from combustion. The temperature of the flue gases at the exit of the radiant section is known as the *bridgewall temperature* and is usually in the range 800 to 1000 °C. Recovery of heat in the convection section to minimize the loss from the stack is constrained by the desire to avoid any condensation of water vapor in the convection section. If there is any sulfur present in the fuel, then the condensate will be corrosive. The temperature at which the flue gas starts to condense is the *acid dew point*. For sulfur-bearing fuels, the temperature of the flue gas is normally kept above 150 to 160 °C. For combustion of sulfur-free gaseous fuels, the temperature can in principle be decreased to below 100 °C.

1) *Fuel combustion*. The fuel for process fired heaters is usually gaseous or liquid. As discussed in Chapter 23, steam boilers can also be fired with coal. A combustion process takes place that reacts carbon, hydrogen, sulfur and nitrogen in the fuel with oxygen to produce carbon dioxide, carbon monoxide, water, sulfur dioxide, sulfur trioxide and oxides of nitrogen (NO_X). Nitrogen in combustion air or elemental nitrogen in a gaseous fuel can also react with oxygen to form oxides of nitrogen at high temperatures. Only small quantities of NO_X are formed. However, even small quantities can be environmentally harmful, as discussed in Chapter 25. In a well-designed and operated combustion device, there should be virtually no carbon monoxide formed, but virtually all carbon should be oxidized to carbon dioxide. Any sulfur present in the fuel can be assumed to react to form sulfur dioxide. In practice, some formation of sulfur trioxide will occur, but measurements on combustion systems indicate that sulfur trioxide is only formed in small quantities and less than what would be predicted by thermodynamic equilibrium with the sulfur dioxide. Thus, it can be assumed that all sulfur in the fuel reacts to sulfur dioxide.

Combustion data for fuels are given in terms of different conditions for the water that enters with the fuel and air and the water that is formed in the combustion. The *gross calorific value* or *higher heating value* of the fuel is the quantity of heat released from the complete combustion of a specified amount of the fuel at standard temperature (25 °C) and standard pressure (1.01325 bar) with water initially present in the fuel and that from the combustion products in the liquid state. The latent heat from the condensation of the water is rarely recovered in combustion equipment. The *net calorific value* or *lower heating value* of the fuel assumes that the latent heat of water vapor is not recovered and is defined as the quantity of heat released from the complete combustion of a specified amount of the fuel at standard temperature (25 °C) and standard pressure (1.01325 bar), with water initially present in the fuel and that from the combustion products in the vapor state at standard temperature. The precombustion reactants are initially at standard temperature and the combustion products are cooled back to standard temperature. The standard temperature for the net and gross calorific value is often taken to be 25 °C. However, some standards use 15 °C (or 60 °F). This is a potential source of confusion when carrying out combustion calculations. Data can be converted from net to gross calorific value and vice versa from knowledge of how many molecules of water are present and the heat of vaporization of water.

Data for some typical gaseous and liquid fuels are given in Tables 12.15 and 12.16.

All combustion processes work with an excess of air or oxygen to ensure complete combustion of the fuel. Excess air typically ranges between 5 and 25% depending on the fuel, burner design and furnace design. For natural draft furnaces typical excess air is 10% for gaseous fuels and 15% for liquid

Table 12.15

Combustion data for some typical natural gases.

	Composition (% by volume)								Net calorific value ($MJ \cdot m^{-3}$)	
Gas	CO_2	N_2	CH_4	C_2H_6	C_3H_8	$C_{10}H_{14}$	C_5H_{12}	C_5H_{14}	Gross	Net
North Sea	0.2	1.5	94.4	3.0	0.5	0.2	0.1	0.1	38.62	34.82
Groningen	0.9	15.0	81.8	2.7	0.4	0.1	0.1	–	33.28	30.00
Algeria	0.2	5.5	83.8	7.1	2.1	0.9	0.4	–	39.1	

Table 12.16

Combustion data for some typical liquid fuels.

Fuel	Composition (% by mass)				Net calorific value ($MJ \cdot kg^{-1}$)	
	C	H	S	O + N + Ash	Gross	Net
Light fuel oil	85.6	11.7	2.5	0.2	43.5	41.1
Medium fuel oil	85.6	11.5	2.6	0.3	43.1	40.8
Heavy fuel oil	85.4	11.4	2.8	0.4	42.9	40.5

fuels. For forced draft and induced draft designs, typical excess air is 5% for gaseous fuels and 10% for liquid fuels. The excess oxygen in the flue gas after combustion should normally be around 3%. However, in large furnaces with the appropriate design, the excess oxygen might be significantly less (e.g. 2% for fuel oil, 1% for natural gas).

2) *Flame temperature.* The actual flame temperature in a burner is extremely difficult to predict. However, the *theoretical flame temperature* or *adiabatic combustion temperature* can be readily predicted. This forms a useful reference point and allows a simple conceptual model to be developed for the fired heater. Theoretical flame temperature is the temperature attained when a fuel is burnt in air or oxygen without loss or gain of heat.

To calculate the heat release from combustion and the temperature of the products of combustion, the thermodynamic path shown in Figure 12.31 can be followed (Hougen, Watson and Ragatz, 1954). The actual combustion process goes directly from reactants at temperature T_1 to products at temperature T_2. However, it is more convenient to follow the alternative path from reactants at temperature T_1 that are initially cooled (or heated) to standard temperature of 298 K. The combustion reactions are then carried out at a constant temperature of 298 K. Standard heats of combustion are available for this. The products of combustion are then heated from 298 K to the final temperature of T_2. The actual heat of combustion is given by (Hougen, Watson and Ragatz, 1954):

$$\Delta H_{COMB} = \Delta H_{FUEL} + \Delta H_{AIR} + \Delta H^O_{COMB} + \Delta H_{FG} \tag{12.173}$$

where ΔH_{COMB} = heat released by the combustion ($J \cdot kmol^{-1}$)

ΔH_{FUEL} = heat to bring the fuel from its initial temperature to standard temperature ($J \cdot kmol^{-1}$) $= \int_{T_1}^{298} C_{P\,FUEL} dT$

ΔH_{AIR} = heat to bring the combustion air from its initial temperature to standard temperature ($J \cdot kmol^{-1}$) $= \int_{T_1}^{298} C_{P\,AIR} dT$

ΔH^O_{COMB} = standard heat of combustion at 298 K ($J \cdot kmol^{-1}$)

ΔH_{FG} = heat to bring products from standard temperature to the final temperature ($J \cdot kmol^{-1}$) $= \int_{298}^{T_2} C_{PFG} dT$

C_{PFUEL} = heat capacity of fuel ($J \cdot kmol^{-1} \cdot K^{-1}$)

C_{PAIR} = heat capacity of combustion air ($J \cdot kmol^{-1} \cdot K^{-1}$)

C_{PFG} = heat capacity of products ($J \cdot kmol^{-1} \cdot K^{-1}$)

For adiabatic combustion, $\Delta H_{COMB} = 0$ and Equation 12.173 becomes:

$$\Delta H_{FG} = -\Delta H_{FUEL} - \Delta H_{AIR} - \Delta H^O_{COMB} \tag{12.174}$$

Standard heats of combustion are widely available. Table 12.17 provides a list of some combustion reactions.

When using data for standard heats of combustion, care should be taken regarding the initial state of the reactants and the final state of the products. If these do not correspond with the conditions for the actual combustion, then errors can arise. Note in Table 12.17 that the final state of water was taken in all cases to be vapor rather than liquid. This relates to the condition

Table 12.17

Standard heats of combustion.

Reaction	ΔH^O_{COMB} at 298 K ($MJ \cdot kmol^{-1}$)
$C\ (sol) + O_2 \rightarrow CO_2$	−393.5
$S\ (sol) + O_2 \rightarrow SO_2$	−297.1
$CO + \frac{1}{2} O_2 \rightarrow CO_2$	−283.0
$H_2 + \frac{1}{2} O_2 \rightarrow H_2O\ (vap)$	−241.8
$CH_4\ (vap) + 2O_2 \rightarrow CO_2 + 2H_2O\ (vap)$	−802.8
$C_2H_6 (vap) + 3\frac{1}{2} O_2 \rightarrow 2CO_2 + 3H_2O\ (vap)$	−1428.7
$C_3H_8\ (vap) + 5O_2 \rightarrow 3CO_2 + 4H_2O\ (vap)$	−2043.2
n-$C_4H_{10}(vap) + 6\frac{1}{2} O_2 \rightarrow 4CO_2 + 5H_2O\ (vap)$	−2657.7
i-$C_4H_{10}(vap) + 6\frac{1}{2} O_2 \rightarrow 4CO_2 + 5H_2O\ (vap)$	−2649.1

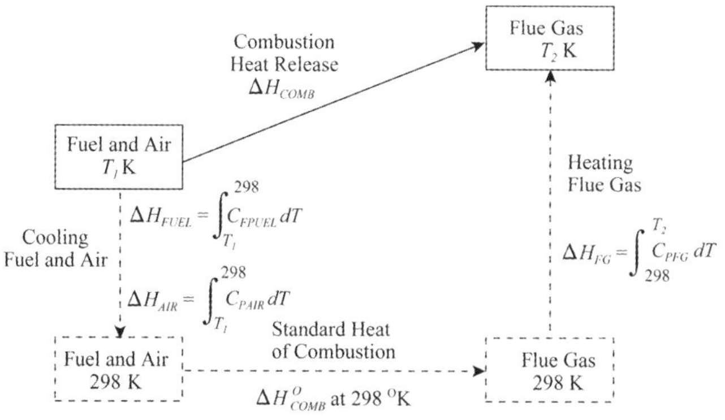

Figure 12.31

The enthalpy change from initial to final state in a combustion process is independent of path.

of the combustion products as they would normally be in practice. One further problem needs to be considered before a calculation can be attempted. The heat balance as illustrated in Figure 12.31 involves extremely large temperature changes for the combustion process. This means that the heat capacity will vary significantly, especially for the products of combustion. The variation in heat capacity can be expressed as a polynomial in temperature, as discussed in Appendix A. However, with combustion processes, care should be taken that the data have been correlated over a wide range of temperature. Thus, for example (Hougen, Watson and Ragatz, 1954):

$$C_P = \alpha_0 + \alpha_1 T + \alpha_2 T^2 + \alpha_3 T^3 \qquad (12.175)$$

where C_P = heat capacity ($J \cdot kmol^{-1} \cdot K^{-1}$ or $kJ \cdot kmol^{-1} \cdot K^{-1}$)
$\alpha_0, \alpha_1, \alpha_2, \alpha_3$ = constants
T = absolute temperature (K)

Thus,

$$\begin{aligned} \Delta H &= \int_{T_1}^{T_2} C_P dT \\ &= \int_{T_1}^{T_2} (\alpha_0 + \alpha_1 T + \alpha_2 T^2 + \alpha_3 T^3) dT \\ &= \left[\alpha_0 T + \frac{\alpha_1 T^2}{2} + \frac{\alpha_2 T^3}{3} + \frac{\alpha_3 T^4}{4} \right]_{T_1}^{T_2} \end{aligned} \qquad (12.176)$$

where ΔH = enthalpy change from T_1 to T_2 ($J \cdot kmol^{-1}$, $kJ \cdot kmol^{-1}$)

Equation 12.176 and Table 12.18 can be used to calculate the enthalpy changes. Alternatively, the heat capacity data can be used to derive mean heat capacities. The mean heat capacity can be defined as (Hougen, Watson and Ragatz, 1954):

$$\overline{C_P} = \frac{\int_{T_1}^{T_2} C_P dT}{T_2 - T_1} \qquad (12.177)$$

where $\overline{C_P}$ = mean heat capacity between temperatures T_1 and T_2 ($J \cdot kmol^{-1} \cdot K^{-1}$ or $kJ \cdot kmol^{-1} \cdot K^{-1}$)

Table 12.19 presents mean heat capacity data between 25 °C and a given temperature for a range of temperatures.

It should be emphasized that the theoretical and real flame temperatures will be significantly different. The real flame temperature will be lower than the theoretical flame temperature because, in practice, heat is lost from the flame (mainly due

Table 12.18

Heat capacity constants.

12

Component	C_P ($kJ \cdot kmol^{-1} \cdot K^{-1}$)			
	α_0	$\alpha_1 \times 10^2$	$\alpha_2 \times 10^5$	$\alpha_3 \times 10^9$
O_2	25.4767	1.5202	−0.7155	1.3117
N_2	28.9015	−0.1571	0.8081	−2.8726
H_2O	32.2384	0.1923	1.0555	−3.5952
CO_2	22.2570	5.9808	−3.5010	7.4693
SO_2	25.7781	5.7945	−3.8112	8.6122
CO	28.3111	0.1675	0.5372	−2.2219
H_2	29.1066	−0.1916	0.4004	−0.8704
CH_4	19.8873	5.0242	1.2686	−11.0113
C_2H_6	6.8998	17.2664	−6.4058	7.2850
C_3H_8	−4.0444	30.4757	−15.7214	31.7359
n-C_4H_{10}	3.9565	37.1495	−18.3382	35.0016
i-C_4H_{10}	−7.9131	41.6000	−23.0065	49.9067

Table 12.19

Mean heat capacity data.

T(°C)	$\overline{C_P}$ ($kJ \cdot kmol^{-1} \cdot K^{-1}$)											
	O_2	N_2	H_2O	CO_2	SO_2	CO	H_2	CH_4	C_2H_6	C_3H_8	n-C_4H_{10}	i-C_4H_{10}
25	29.41	29.07	33.65	37.17	39.89	29.12	28.87	35.69	52.86	73.65	99.30	96.95
100	29.82	29.18	33.94	38.65	41.24	29.40	28.88	37.76	57.87	81.65	109.20	107.56
200	30.33	29.34	34.36	40.47	42.87	29.64	28.92	40.51	64.22	91.59	121.56	120.68
300	30.81	29.54	34.82	42.12	44.33	29.89	28.98	43.25	70.20	100.75	132.99	132.70
400	31.26	29.76	35.31	43.61	45.61	30.16	29.05	45.97	75.84	109.19	143.55	143.69
500	31.67	30.00	35.83	44.96	46.74	30.44	29.14	48.64	81.13	116.95	153.29	153.73
600	32.05	30.26	36.38	46.17	47.74	30.73	29.25	51.26	86.08	124.07	162.27	162.88
700	32.41	30.53	36.94	47.25	48.60	31.02	29.37	53.80	90.72	130.61	170.53	171.23
800	32.73	30.80	37.52	48.23	49.34	31.31	29.51	56.24	95.05	136.61	178.14	178.84
900	33.04	31.09	38.10	49.10	49.99	31.59	29.65	58.58	99.08	142.12	185.14	185.80
1000	33.32	31.36	38.68	49.88	50.54	31.88	29.80	60.79	102.82	147.19	191.59	192.18
1100	33.58	31.64	39.27	50.58	51.02	32.15	29.97	62.86	106.28	151.86	197.53	198.05
1200	33.82	31.90	39.84	51.21	51.43	32.40	30.14	64.77	109.48	156.19	203.03	203.49
1300	34.04	32.15	40.39	51.78	51.79	32.64	30.32	66.51	112.42	160.22	208.12	208.58
1400	34.25	32.39	40.93	52.31	52.12	32.86	30.50	68.05	115.12	163.99	212.88	213.38
1500	34.44	32.60	41.44	52.80	52.42	33.06	30.69	69.39	117.58	167.57	217.34	217.98
1600	34.63	32.78	41.93	53.27	52.70	33.23	30.87	70.50	119.83	170.98	221.56	222.44
1700	34.81	32.93	42.37	53.73	52.99	33.37	31.06	71.37	121.86	174.28	225.59	226.84
1800	34.97	33.05	42.78	54.18	53.29	33.48	31.25	71.98	123.69	177.53	229.49	231.27
1900	35.14	33.13	43.13	54.65	53.62	33.55	31.44	72.32	125.33	180.76	233.31	235.78
2000	35.30	33.16	43.44	55.14	53.99	33.58	31.62	72.37	126.79	184.02	237.09	240.47
2100	35.46	33.14	43.69	55.66	54.41	33.60	31.80	72.11	128.08	187.36	240.91	245.39
2200	35.62	33.07	43.87	56.22	54.89	33.51	31.98	71.52	129.22	190.84	244.79	250.63

to radiation). Also, part of the heat released provides heat for a variety of endothermic dissociation reactions that occur at high temperatures, such as:

$$CO_2 \rightleftharpoons CO + O$$
$$H_2O \rightleftharpoons H_2 + O$$
$$H_2O \rightleftharpoons H + OH$$

However, as the temperature of the flue gas decreases, as heat is extracted, the dissociation reactions reverse and the heat is released. Thus, although the theoretical flame temperature does not reflect the true temperature of the flame, it does provide a convenient reference to indicate how much heat is actually released by combustion as the flue gas is cooled. Figure 12.32 shows the flue gas starting from the theoretical flame temperature. The temperature at the exit of the convection section is the stack temperature. The cooling from the stack temperature to the ambient temperature is the *stack loss*.

In Figure 12.32, the temperatures in the radiant section (above 800 to 1000 °C) are thus not accurately represented by the profile based on the theoretical flame temperature. Also, as the flue gas is cooled and passes through the convection section

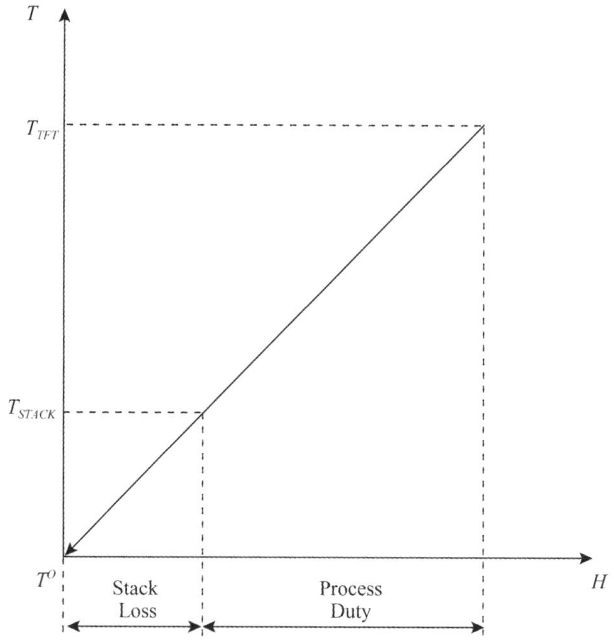

Figure 12.32
The flue gas profile for a fired heater.

of the furnace, the temperatures are more representative of what they would be in practice. However, it should be emphasized again that the simple model in Figure 12.32 does allow the correct heat duty to be represented. Furnace efficiency can be defined as:

$$\text{Furnace efficiency} = \frac{\text{Heat to the process}}{\text{Heat released from combustion}} \tag{12.178}$$

The profile shown in Figure 12.32 represents the furnace efficiency, if there is no heat input from the sensible heat in the fuel or combustion air and the casing heat losses are neglected. Making these assumptions, the process duty plus the stack loss represents the heat released by the fuel. As excess air is reduced, the theoretical flame temperature increases. If the stack temperature is maintained, this has the effect of reducing the stack loss and increasing the thermal efficiency of the furnace for a given process heating duty.

Example 12.9 A gas that can be considered to be pure methane is to be used as fuel in a furnace. Both the fuel gas and combustion air are at 25 °C. Calculate the theoretical flame temperature and furnace efficiency assuming a stack temperature of 150 °C and casing losses of 2% if the methane is burnt in:

a) its stoichiometric ratio in dry air;
b) 15% excess dry air;
c) 15% excess air with a relative humidity of 60%.

Solution

a) Stoichiometric air requirements are defined by

$$\underset{1\ \text{kmol}}{CH_4} + \underset{2\ \text{kmol}}{2O_2} \rightarrow \underset{1\ \text{kmol}}{CO_2} + \underset{2\ \text{kmol}}{2H_2O}$$

Here 2 kmol of oxygen are required per kmol of methane burnt. If air is assumed to be 21% oxygen and nitrogen is assumed inert, then combustion products are:

$$N_2 = \frac{2 \times 0.79}{0.21} = 7.52\ \text{kmol}$$
$$H_2O = 2\ \text{kmol}$$
$$CO_2 = 1\ \text{kmol}$$

Since the fuel and combustion air are at the standard temperature of 25 °C, $\Delta H_{FUEL} = \Delta H_{AIR} = 0$. To calculate ΔH_{FG}, start by estimating the theoretical flame temperature to be 2000 °C. From Table 12.19:

$$\overline{C_{P_{O_2}}} = 35.30\ \text{kJ·kmol}^{-1}\text{·K}^{-1}$$

$$\overline{C_{P_{N_2}}} = 33.16\ \text{kJ·kmol}^{-1}\text{·K}^{-1}$$

$$\overline{C_{P_{H_2O}}} = 43.44\ \text{kJ·kmol}^{-1}\text{·K}^{-1}$$

$$\overline{C_{P_{CO_2}}} = 55.14\ \text{kJ·kmol}^{-1}\text{·K}^{-1}$$

$$\Delta H_{FG} = (7.52 \times 33.16 + 2 \times 43.44 + 1 \times 55.14)(T - 25)$$
$$= 391.38\,(T - 25)$$

From Table 12.17, $\Delta H^o_{COMB} = -802.8 \times 10^3\ \text{kJ.kmol}^{-1}$. Thus, from an energy balance:

$$\Delta H_{FG} = -\Delta H_{FUEL} - \Delta H_{AIR} - \Delta H^o_{COMB}$$

$$391.38\,(T_{TFT} - 25) = 0 - 0 - (-802.8 \times 10^3)$$

$$T_{TFT} = 2076\ ^\circ\text{C}$$

This agrees well with the initial estimate. Had there been significant disagreement, then revised mean heat capacities would need to be taken and the calculation repeated.

$$\text{Furnace efficiency} = \frac{\text{Heat to process}}{\text{Heat released by fuel combustion}}$$
$$= \frac{391.38\,(2076 - 150)}{802.8 \times 10^3 \times 1.02}$$
$$= 0.921$$

b) If 15% excess air is used, then combustion products are:

$$O_2 = 2 \times 0.15 = 0.3\ \text{kmol}$$
$$N_2 = \frac{2 \times 0.79}{0.21} \times 1.15 = 8.65\ \text{kmol}$$
$$H_2O = 2\ \text{kmol}$$
$$CO_2 = 1\ \text{kmol}$$

12

Again, estimate the theoretical flame temperature to be 2000 °C. The mean heat capacities are as before and the heat balance is now:

$$\Delta H_{FG} = (0.3 \times 35.3 + 8.65 \times 33.16 + 2 \times 43.44 + 1 \times 55.14)(T_{TFT} - 25)$$
$$= 439.44(T_{TFT} - 25)$$

$$\Delta H_{FG} = -\Delta H_{FUEL} - \Delta H_{AIR} - \Delta H^o_{COMB}$$

$$439.44\,(T_{TFT} - 25) = 0 - 0 - (-802.8 \times 10^3)$$

$$T_{TFT} = 1852\,°C$$

$$\text{Furnace efficiency} = \frac{439.44\,(1852 - 150)}{802.8 \times 10^3 \times 1.02} = 0.913$$

c) So far the combustion air has been assumed to be dry. If the combustion air is humid, then the water vapor will act as another inert in the combustion. The relative humidity of air is the percentage relative to saturation.

Saturated vapor pressure of water at 25 °C (from steam tables)

$$= 0.03166 \text{ bar}$$

For 60% relative humidity, the partial pressure of water vapor is given by:

$$p_{H_2O} = 0.03166 \times 0.6 = 0.0190 \text{ bar}$$

Thus, the mole fraction of water vapor in the combustion air for a pressure of 1 atm (1.013 bar) is given by:

$$y_{H_2O} = \frac{p_{H_2O}}{P} = \frac{0.0190}{1.013} = 0.0188$$

The combustion air for 15% excess

$$= (2 + 7.52)\,1.15 = 10.95 \text{ kmol}$$

Water from combustion air

$$= 10.95 \times \frac{y_{H_2O}}{1 - y_{H_2O}} = 10.95 \times \frac{0.0188}{1 - 0.0188} = 0.21 \text{ kmol}$$

The total water vapor in the combustion products

$$= 2 + 0.21 = 2.21 \text{ kmol}$$

Again, estimate the theoretical flame temperature to be 2000 °C. The mean heat capacities are as before and the heat balance is:

$$\Delta H_{FG} = (0.3 \times 35.3 + 8.65 \times 33.16 + 2.21 \times 43.44 + 1 \times 55.14) \times (T_{TFT} - 25)$$
$$= 448.57(T_{TFT} - 25)$$
$$448.57(T_{TFT} - 25) = 0 - 0 - (-802.8 \times 10^3)$$

$$T_{TFT} = 1815\,°C$$

Again, iterate with revised heat capacities for greater accuracy. It can be seen that excess air and humidity in the combustion air both act to reduce the theoretical flame temperature.

$$\text{Furnace efficiency} = \frac{448.57(1815 - 150)}{802.8 \times 10^3 \times 1.02} = 0.912$$

In these calculations, the fuel and combustion air were both at the standard temperature of 25 °C. If the temperature of either had been below 25 °C, then ΔH_{FUEL} and ΔH_{AIR} would have acted to decrease the theoretical flame temperature. If either had been above 25 °C, the effect would have been to increase the theoretical flame temperature. For a given stack temperature, the higher the theoretical flame temperature, the higher the furnace efficiency. However, there is a minimum excess air required to ensure that the combustion is efficient.

3) *Heat transfer in the radiant section.* The heat transferred in the radiant section depends on the furnace design and the number of passes, will be between 50 and 85% of the total heat transferred to the process, and is typically 75%. Of the heat transfer in the radiant section, typically 80 to 90% is by radiation and the balance by convection.

Heat transfer by radiation is in accordance with the Stefan–Boltzmann Law, which quantifies the energy radiated from a black body in terms of its temperature:

$$\frac{Q_{RAD}}{A} = \sigma T^4 \qquad (12.179)$$

where Q_{RAD} = radiant heat transfer (W)
A = Area of radiating surface (m^2)
σ = Stefan–Boltzmann Constant = $5.6704 \times 10^{-8}\ W \cdot m^{-2} \cdot K^{-4}$
T = absolute temperature (K)

In practice, the heat radiated is some proportion of the black body radiation defined by the emissivity:

$$\frac{Q_{RAD}}{A} = \varepsilon \sigma T^4 \qquad (12.180)$$

where ε = emissivity (–) $0 < \varepsilon < 1$

The emissivity depends on the material and temperature, and in the case of solid materials, the surface finish.

In a furnace there is an exchange of radiant heat between the hot combustion gases and the cold solid surfaces. Part of the heat is absorbed directly at the front of the tubes. Part passes between the tubes, is absorbed by the walls and reradiated to the rear surface of the tubes. The cold surface is taken to be a cold plane surface equal to the number of tubes multiplied by the tube length and the center-to-center tube spacing. This is known as the *cold plane area*, A_{CP}. However, the tubes will not absorb heat as effectively as a plane surface, so an *absorption effectiveness factor* α_{CP} is introduced to correct for the tubes being represented by a plane surface. The product $\alpha_{CP}A_{CP}$ represents the area of an ideal black plane that has the same absorption as the bank of tubes and is known as the *equivalent cold plane area*. The value of α_{CP} depends on the tube diameter, center-to-center distance between the tubes and whether one or two tube rows are used. Finally, to allow for the nonblack emissivities of the hot gases and cold surfaces, an *exchange factor*, *E*, is introduced. This depends of the emissivity of the gas, the total refractory area (walls, roof and floor) and the equivalent cold plane area. The radiant heat transfer can thus be written as:

$$Q_{RAD} = \sigma\alpha_{CP}A_{CP}E\left(T_G^4 - T_W^4\right) \qquad (12.181)$$

where α_{CP} = absorption effectiveness factor (–)
A_{CP} = cold plane area (m^2)
E = exchange factor (–)
T_G = temperature of the gases (K)
T_W = temperature of the tube walls (K)

In addition to the radiant heat transfer, a minor proportion of the heat is transferred by convection:

$$Q_{CONV} = h_{CONV}(T_G - T_W) \qquad (12.182)$$

where h_{CONV} = convective film transfer coefficient in the radiant zone ($W \cdot m^{-2} \cdot K^{-1}$)

Application of Equations 12.181 and 12.182 requires knowledge of the combustion conditions, geometry of the fired heater and emissivity data. Methods have been presented by Lobo and Evans (1939), Kern (1950) and Tucker and Truelove (in Hewitt, 2008). The key parameter that needs to be determined for the radiant section is the fraction of heat released in the combustion that is absorbed in the radiant section, F_{RAD}. This depends on the combustion conditions, the *mean radiant heat flux to the tubes*, and q_{RAD}. Although q_{RAD} is an output from the detailed design, in a conceptual design a value can be assumed from experience. The fraction of heat released in the combustion that is absorbed in the radiant section, F_{RAD}, varies for process heating applications by typically between 45 and 55%, although it is possible to design outside this range. F_{RAD} decreases with increasing ratio of air to fuel, R_{AF}, as a result of decreasing flame temperature. F_{RAD} also decreases with increasing radiant heat flux to the tubes, q_{RAD}. This is because a high radiant flux requires a lower heat transfer area and a shorter residence time for the combustion gases, decreasing the fraction of heat absorbed. Bahadori and Vuthaluru (2010) presented a correlation for F_{RAD}, which is based on the equations of Wilson, Lobo and Hottel (1932). This correlates the fraction of heat released that is absorbed in the radiant section, F_{RAD}, as a function of the air to fuel mass ratio, R_{AF}, and the mean radiant heat flux to the tubes, q_{RAD}:

$$F_{RAD} = a + bR_{AF} + cR_{AF}^2 + dR_{AF}^3 \qquad (12.183)$$

where F_{RAD} = fraction of heat absorbed in the radiant section (–)
R_{AF} = mass ratio of air to fuel (–)
q_{RAD} = radiant heat flux ($W \cdot m^{-2}$)

$$a = a_1 + b_1 q_{RAD} + c_1 q_{RAD}^2 + d_1 q_{RAD}^3 \qquad (12.184)$$

$$b = a_2 + b_2 q_{RAD} + c_2 q_{RAD}^2 + d_2 q_{RAD}^3 \qquad (12.185)$$

$$c = a_3 + b_3 q_{RAD} + c_3 q_{RAD}^2 + d_3 q_{RAD}^3 \qquad (12.186)$$

$$d = a_4 + b_4 q_{RAD} + c_4 q_{RAD}^2 + d_4 q_{RAD}^3 \qquad (12.187)$$

The coefficients are given in Table 12.20.

Equation 12.183 applies to designs with one row of 200 mm nominal diameter pipes (see Table 13.5) on a pitch of 2 × nominal diameter. For other sizes, a correction is applied to R_{AF}

Table 12.20

Coefficients for heat transfer in the fired heater radiant section.

Variable symbol	Coefficients
a_1	1.77185
b_1	-1.00192×10^{-4}
c_1	3.75347×10^{-9}
d_1	-4.19104×10^{-14}
a_2	-1.36692×10^{-1}
b_2	1.53116×10^{-5}
c_2	-5.96386×10^{-10}
d_2	6.68455×10^{-15}
a_3	6.51975×10^{-3}
b_3	-8.13214×10^{-7}
c_3	3.08515×10^{-11}
d_3	-3.43559×10^{-16}
a_4	-1.10851×10^{-4}
b_4	1.38376×10^{-8}
c_4	-5.17240×10^{-13}
d_4	5.74804×10^{-18}

Table 12.21

Correction factors for tube spacing and number of tube rows.

Number of rows	$2d_N$	$3d_N$
1	1	0.9
2	1.34	1.14

according to (Bahadori and Vuthaluru, 2010):

$$C = 1.09266 - 0.99501d_N + 3.49407d_N^2 - 4.17804d_N^3 \quad (12.188)$$

where d_N = nominal outside diameter of radiant tubes (Table 13.5) (m)

The factor C is multiplied by R_{AF} to correct for diameters different from 200 mm. A correction can also be applied to allow for tube spacing and a number of tubes. This correction is given in Table 12.21 and is also applied to the air to fuel ratio R_{AF}.

Equation 12.183 with the appropriate corrections can be used to calculate the fraction of the heat released that is extracted in the radiant zone. The heat released includes the heat of combustion, sensible heat in the inlet fuel and air, atomizing steam in the case of fuel oil and any recirculated flue gas relative to a reference temperature of 25 °C. Note that the original reference temperature used by Bahadori and Vuthaluru (2010) was 15 °C (288 K).

To apply Equation 12.183 requires the radiant heat flux to be specified. For simple heating, radiant fluxes typically range from 25,000 to 60,000 $W \cdot m^{-2}$. Duties with no vaporization are typically 47,000 $W \cdot m^{-2}$. Duties with a small amount of vaporization are typically 32,000 to 47,000 $W \cdot m^{-2}$. Reboilers with a large, low-temperature vaporization duty are typically 63,000 $W \cdot m^{-2}$. For furnace reactors heat flux can vary typically from 60,000 to 100,000 $W \cdot m^{-2}$.

If the heat released is known, then the fraction of heat absorbed in the radiant section can be calculated from Equation 12.183 (with its correction factors). If the radiant section is considered to be well mixed, then the *bridgewall temperature* (i.e. the temperature of the gases leaving the radiant section) can be calculated by an energy balance. $Q_{RELEASE}$ is defined to be the heat released by the combustion, which is the sum of the standard heat of combustion at 298 K, plus the enthalpy change to bring the fuel and combustion air from its initial temperature to 298 K. From Figure 12.31 and Equation 12.173:

$$F_{RAD}Q_{RELEASE} = -m_{FUEL}\Delta H^O_{COMB} - m_{FG}\Delta H_{FG} \quad (12.189)$$

where $Q_{RELEASE} = -m_{FUEL}\Delta H^O_{COMB} - m_{FUEL}\Delta H_{FUEL} - m_{AIR}\Delta H_{AIR}(W)$

ΔH^O_{COMB} = standard heat of combustion at 298 K ($J \cdot kmol^{-1}$, $J \cdot kg^{-1}$)

ΔH_{FUEL} = enthalpy difference to bring the fuel to 298 K ($J \cdot kmol^{-1}$, $J \cdot kg^{-1}$)

$= \int_{T_{inlet}}^{298} C_{PFUEL}dT$

ΔH_{AIR} = enthalpy difference to bring the inlet air to 298K ($J \cdot kmol^{-1}$, $J \cdot kg^{-1}$)

$= \int_{T_{inlet}}^{298} C_{PAIR}dT$

ΔH_{FG} = heat to bring combustion products from standard temperature to the final temperature ($J \cdot kmol^{-1}$, $J \cdot kg^{-1}$)

$= \int_{298}^{T_{BW}} C_{PFG}dT$

T_{inlet} = inlet temperature of the fuel (K)

T_{BW} = bridgewall temperature (K)

m_{FUEL} = flowrate of fuel ($kmol \cdot s^{-1}$, $kg \cdot s^{-1}$)

m_{AIR} = flowrate of fuel ($kmol \cdot s^{-1}$, $kg \cdot s^{-1}$)

If fuel oil is being burnt that requires atomizing steam then that enthalpy relative to 25 °C would need to be added to $Q_{RELEASE}$. If flue gas was being recirculated, then that enthalpy relative to 25 °C would also need to be added. Equation 12.189 can be solved for the bridgewall temperature in the same way as the theoretical flame temperature. The only difference is that in the case of the theoretical flame temperature, the left-hand side of Equation 12.189 is zero.

4) *Heat transfer in the shield section.* At least the first two rows of tubes after the radiant section will be plain shock tubes. The radiant flux for the first row can be assumed to be that for the radiant tubes. The radiant flux for the second row will be 56% of the radiant tubes (Nelson, 1958). Heat transfer is also by convection.

Convective heat transfer to a bank of plain tubes in a staggered geometry can be predicted by (Butterworth, 1977):

$$Nu = 1.309\, Re^{0.36}\, Pr^{0.34} F_N \qquad 10 < Re < 300 \quad (12.190)$$

$$Nu = 0.273\, Re^{0.635}\, Pr^{0.34} F_N \qquad 30 \le Re \le 200,000 \quad (12.191)$$

$$Nu = 0.124\, Re^{0.70}\, Pr^{0.34} F_N \qquad 200,000 < Re < 2 \times 10^6 \quad (12.192)$$

where Nu = Nusselt Number $= \dfrac{h_{CONV} d_O}{k}$

Re = Reynolds Number $= \dfrac{\rho v_{max} d_O}{\mu}$

Pr = Prandtl Number $= \dfrac{C_P \mu}{k}$

$$F_N = -0.04375(\ln N_R)^2 + 0.27 \ln N_R + 0.61 \quad (12.193)$$

h_{CONV} = convective heat transfer coefficient ($W \cdot m^{-2} \cdot K^{-1}$)

d_O = outside diameter of plain tube (m)

k = thermal conductivity of the flue gas ($W \cdot m^{-2} \cdot K^{-1}$)

v_{max} = fluid velocity at the minimum cross-flow area, from Equation 12.116 ($m \cdot s^{-1}$)

$$= \frac{m_{FG}}{\rho L\left[N_{TR}(p_T - d_O) + \frac{3p_T}{2} - d_O\right]} \quad (12.194)$$

m_{FG} = mass flowrate of flue gas ($kg \cdot s^{-1}$)
ρ = density of flue gas ($kg \cdot s^{-1}$)
L = tube length (m)
N_{TR} = number of tubes in a row (–)
N_R = number of tube rows (–)

Equations 12.190 to 12.192 can be used to predict the convective heat transfer coefficient for the shock tubes.

The physical properties of the flue gas will depend on the fuel burnt and excess air, but most importantly the temperature. The physical properties of the flue gas can be approximated as a function of temperature by (Isachenko, Osipova and Sukomel, 1977):

$$\rho = \frac{1}{4 \times 10^{-9}T^2 + 0.0028T + 0.7701} \quad (12.195)$$

$$C_P = -9.546 \times 10^{-11}T^3 + 1.114 \times 10^{-4}T^2 + 0.2508T + 1043 \quad (12.196)$$

$$\mu = -9.614 \times 10^{-12}T^2 + 4.189 \times 10^{-8}T + 1.625 \times 10^{-5} \quad (12.197)$$

$$k = 9.3 \times 10^{-10}T^2 + 8.51 \times 10^{-5}T + 0.0229 \quad (12.198)$$

where ρ = density of flue gas ($kg \cdot m^{-3}$)
C_P = heat capacity of flue gas ($J \cdot kg^{-1} \cdot K^{-1}$)
μ = viscosity of flue gas ($kg \cdot m^{-1} \cdot s^{-1}$)
k = thermal conductivity of flue gas ($W \cdot m^{-2} \cdot K^{-1}$)
T = temperature of flue gas (°C)

5) *Heat transfer in the convection section.* The radiant component of heat transfer in the convective section can be approximated by (Nelson, 1958):

$$h_{RAD} = 0.02555T - 2.385 \quad (12.199)$$

where h_{RAD} = radiant heat transfer coefficient ($W \cdot m^{-2} \cdot K^{-1}$)
T = mean flue gas temperature in the convective section (°C)

Equations 12.190 to 12.192 can be used to predict the convective heat transfer coefficient of the tubes in the convection section if they are of a plain design. For a bank of finned tubes, in a staggered geometry, Equation 12.111 can be used to predict the convective heat transfer coefficient.

An allowance must also be made for reradiation of heat from the walls of the convective section. This varies between 6 and 15% and an allowance of 10% is normally made (Nelson, 1958). Thus:

$$h_{FG} = 1.1(h_{CONV} + h_{RAD}) \quad (12.200)$$

where h_{FG} = combined convective and radiant heat transfer coefficient in the convective section ($W \cdot m^{-2} \cdot K^{-1}$)

6) *Additional heat recovery from the flue gas.* The minimum temperature for the flue gas that can be achieved by the arrangement in Figure 12.30 is limited both by the area installed in the convection section and the feed temperature of the cold fluid. If the feed temperature is relatively high, then it will not be possible to decrease the flue gas to the minimum temperature of, say, 150 °C and achieve high furnace efficiency. One way to overcome this is to install an additional bank of tubes in the convection section with another lower temperature cold fluid on the inside of the tubes, such as boiler feedwater preheating.

Another technique that can be used to extract additional heat from the flue gas is air preheating. This is illustrated in Figure 12.33. If the combustion air is preheated by heat recovery, then the theoretical flame temperature increases, reducing the stack loss. Although higher flame temperatures reduce the fuel consumption for a given process heating duty, there is one significant disadvantage. Higher flame temperatures increase the formation of oxides of nitrogen, which are environmentally harmful.

7) *Fired heater dimensions.* For cylindrical designs, the radiant tube length to tube circle diameter is normally less than 2.7. For box designs the height to width is normally less than 2.75. Tube diameters vary between DN50 and DN250 (see Table 13.5). DN100 and DN150 (see Table 13.5) are the most commonly used sizes. Tube length varies from 9 to 18 m, with 12 m being a commonly used length. Tube pitch is normally 2 × nominal diameter, with the spacing between the tubes and the refractory walls approximately the tube nominal diameter. Tube velocities for liquid flow are typically 1.5 to 2.5 $m \cdot s^{-1}$.

Example 12.10 Crude oil with a flowrate of 185.4 $kg \cdot s^{-1}$ is to be heated in a fired heater from 280 °C to 360 °C. The mean heat capacity of the crude oil can be assumed to be 3282 $J \cdot kg^{-1} \cdot K^{-1}$. In this case, the mean heat capacity does not just allow for the heat capacity of the liquid but also includes the duty for the partial vaporization necessary as the crude oil leaves the furnace and enters the distillation (Nelson, 1958). The furnace configuration is to be a 6 pass box design with a single row of tubes in the radiant section, 2 rows of bare tubes after the radiant zone in the shield section and 12 rows of fined tubes in the convection section with 6 tubes per row in the shield and convection sections. The tube size to be used has a nominal diameter of 150 mm with d_O = 168.3 mm and d_I = 154.1 mm (see Table 13.5). The tube length is 12 m and it can be assumed that the exposed length is 11.2 m in the radiant, shield and convection sections. The pitch is 300 mm for all tubes in the radiant, shield and convective zones. For the finned tubes assume d_R = 168.3 mm and d_I = 154.1 mm. Take the fin height to

Figure 12.33

Fired heater with air preheat

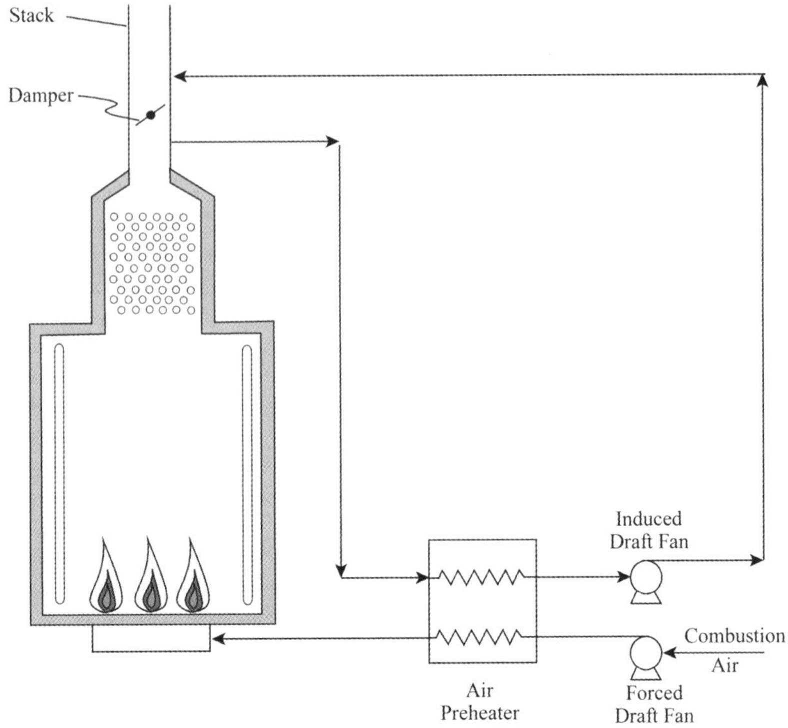

be 55 mm and thickness to be 2 mm and the number of fins per meter to be 120. The conductivity for the tube walls and the fins can be assumed to be 35 $W \cdot m^{-1} \cdot K^{-1}$. Assume the inside heat transfer coefficient to be 500 $W \cdot m^{-2} \cdot K^{-1}$, the inside fouling resistance to be 0.0005 $W^{-1} \cdot m^2 \cdot K$ and the fouling resistance on the outside of the tubes in the shield and convection sections to be 0.0002 $W^{-1} \cdot m^2 \cdot K$.

a) Assuming gaseous fuel with 90% methane and 10% nitrogen with 10% excess air, calculate the air to fuel ratio and the flowrate of the flue gas.

b) Assuming the fuel and air are both fed at 25 °C make an initial estimate of the fuel flowrate and calculate the fuel heat release.

c) Assuming q_{RAD} to be 32,000 $W \cdot m^{-2}$, calculate the fraction of heat absorbed in the radiant section and the inlet temperature of the crude oil to the radiant section, assuming the required outlet temperature.

d) Calculate the number of tubes required in the radiant section.

e) Assuming the radiant section to be well mixed, calculate the bridgewall temperature.

f) Calculate the temperature of the flue gas at the exit of the shield section and temperature of the crude oil at the inlet to the shield section.

g) Calculate the temperature of the flue gas at the exit of the convective section and temperature of the crude oil at the inlet to the convection section.

h) Calculate the flowrate of fuel required to satisfy the process duty.

i) Calculate the furnace efficiency assuming the casing losses to be 2% of the heat release.

Solution

a) For the fuel gas inlet, assuming ideal gas behavior:

$$\begin{aligned}
m_{CH4} &= y_{CH4} \times \frac{P}{RT} \\
&= 0.9 \times \frac{1.013 \times 10^5}{8314 \times 298.15} \\
&= 0.03678 \text{ kmol} \cdot m^{-3} \text{ fuel} \\
&= 0.03678 \times 16 \\
&= 0.5885 \text{ kg} \cdot m^{-3} \text{ fuel} \\
m_{N2} &= 0.1 \times \frac{1.013 \times 10^5}{8314 \times 298.15} \\
&= 0.004087 \text{ kmol} \cdot m^{-3} \text{ fuel} \\
&= 0.004087 \times 28 \\
&= 0.1144 \text{ kg} \cdot m^{-3} \text{ fuel} \\
m_{FUEL} &= 0.5885 + 0.1144 \\
&= 0.7029 \text{ kg} \cdot m^{-3} \text{ fuel}
\end{aligned}$$

Calculate the air requirements for combustion. Each kmol CH_4 requires 2 kmol O_2 for combustion, plus 10% excess air. Again, assuming ideal gas behavior:

$$\begin{aligned}
m_{O_2} &= 2 \times y_{CH_4} \times \frac{P}{RT}(1 + \text{Excess Air}) \\
&= 2 \times 0.9 \times \frac{1.013 \times 10^5}{8314 \times 298.15} \times (1 + 0.10) \\
&= 0.08092 \text{ kmol} \cdot m^{-3} \text{ fuel} \\
&= 0.08092 \times 32 \\
&= 2.5893 \text{ kg} \cdot m^{-3} \text{ fuel}
\end{aligned}$$

Assuming air is 21% O_2 and 79% N_2:

$$\begin{aligned} m_{N_2} &= 0.08092 \times \frac{79}{21} \\ &= 0.3044\ \text{kmol} \cdot \text{m}^{-3}\ \text{fuel} \\ &= 0.3044 \times 28 \\ &= 8.5231\ \text{kg} \cdot \text{m}^{-3}\ \text{fuel} \\ m_{AIR} &= 2.5893 + 8.5231 \\ &= 11.1124\ \text{kg} \cdot \text{m}^{-3}\ \text{fuel} \\ R_{AF} &= \frac{11.1124}{0.7029} \\ &= 15.81 \end{aligned}$$

For the flue gas, assuming ideal gas behavior:

$$\begin{aligned} m_{O_2} &= 2 \times 0.9 \times \frac{1.013 \times 10^5}{8314 \times 298.15} \times 0.1 \\ &= 0.007356\ \text{kmol} \cdot \text{m}^{-3}\ \text{fuel} \\ &= 0.007356 \times 32 \\ &= 0.2354\ \text{kg} \cdot \text{m}^{-3}\ \text{fuel} \\ m_{N_2} &= 0.1144 + 8.5231 \\ &= 8.6375\ \text{kg} \cdot \text{m}^{-3}\ \text{fuel} \end{aligned}$$

For the flue gas, 2 kmol H_2O and 1 kmol CO_2 are formed per kmol CH_4 combusted:

$$\begin{aligned} m_{H_2O} &= 2 \times 0.9 \times \frac{1.013 \times 10^5}{8314 \times 298.15} \\ &= 0.007356\ \text{kmol} \cdot \text{m}^{-3}\ \text{fuel} \\ &= 0.007356 \times 18 \\ &= 1.3241\ \text{kg} \cdot \text{m}^{-3}\ \text{fuel} \\ m_{CO_2} &= 0.9 \times \frac{1.013 \times 10^5}{8314 \times 298.15} \\ &= 0.03678\ \text{kmol} \cdot \text{m}^{-3}\ \text{fuel} \\ &= 0.03678 \times 44 \\ &= 1.6183\ \text{kg} \cdot \text{m}^{-3}\ \text{fuel} \end{aligned}$$

Total flowrate of flue gas

$$\begin{aligned} &= 0.2354 + 8.6375 + 1.3241 + 1.6183 \\ &= 11.8153\ \text{kg} \cdot \text{m}^{-3}\ \text{fuel} \end{aligned}$$

b) Make an initial estimate of the fuel flowrate. Later this figure will be revised. The fuel required will depend on the furnace process duty, stack losses and casing losses. The furnace process duty can be calculated and the casing losses assumed to be 2% of the heat release. However, the stack losses are unknown. The efficiency of such fired heaters is often in the range of 80 to 90%. Assume, as an initial estimate, the furnace efficiency to be 85%. Given that the fuel air and combustion air are both at standard temperature, the heat release is given by:

$$\begin{aligned} Q_{RELEASE} &= -m_{CH_4}\Delta H^0_{COMB} - m_{FUEL}\Delta H_{FUEL} - m_{AIR}\Delta H_{AIR} \\ &= -m_{FUEL} \times \frac{0.9}{16} \times (-802.3 \times 10^6) - 0 - 0 \\ &= 4.5129 \times 10^7 m_{FUEL} \end{aligned}$$

The process duty is given by:

$$\begin{aligned} Q_P &= m_P C_{PP}(T_{Poutlet} - T_{Pinlet}) \\ &= 185.4 \times 3,282(360 - 280) \\ &= 4.8679 \times 10^7\ \text{W} \\ \eta &= 0.85 = \frac{4.8679 \times 10^7}{4.5129 \times 10^7 \times m_{FUEL} \times 1.02} \\ m_{FUEL} &= 1.24\ \text{kg} \cdot \text{s}^{-1} \end{aligned}$$

where m_P = mass flowrate of the process fluid ($\text{kg} \cdot \text{s}^{-1}$)
m_{FUEL} = mass flowrate of fuel ($\text{kg} \cdot \text{s}^{-1}$)
C_{PP} = heat capacity of the process fluid ($\text{J} \cdot \text{kg}^{-1} \cdot \text{K}^{-1}$)
$T_{Poutlet}$ = outlet temperature of the process fluid (°C)
T_{Pinlet} = inlet temperature of the process fluid (°C)
η = furnace efficiency (–)

Thus:

$$\begin{aligned} Q_{RELEASE} &= 4.5129 \times 10^7 \times 1.24 \\ &= 5.5960 \times 10^7\ \text{W} \end{aligned}$$

This represents the first estimate of the total heat release.

c) Calculate the fraction of the heat release absorbed in the radiant section from Equation 12.183:

$$F_{RAD} = a + bR_{AF} + cR_{AF}^2 + dR_{AF}^3$$

For $q_{RAD} = 32{,}000\ \text{W} \cdot \text{m}^{-2}$, from Equation 12.184 to 12.187 and Table 12.20:

$$\begin{aligned} a &= 1.0359 \\ b &= -3.8382 \times 10^{-2} \\ c &= 8.3106 \times 10^{-4} \\ d &= -9.3509 \times 10^{-6} \end{aligned}$$

From Equation 12.188:

$$\begin{aligned} C &= 1.09266 - 0.99501 d_N + 3.49407 d_N^2 - 4.17804 d_N^3 \\ &= 1.09266 - 0.99501 \times 0.15 + 3.49407 \times 0.15^2 \\ &\quad - 4.17804 \times 0.15^3 \\ &= 1.0075 \end{aligned}$$

No correction is needed for the tube spacing or the number of tube rows:

$$\begin{aligned} F_{RAD} &= 1.0359 - 3.8382 \times 10^{-2} \times (1.0075 \times 15.81) \\ &\quad + 8.3106 \times 10^{-4} \times (1.0075 \times 15.81)^2 \\ &\quad -9.3509 \times 10^{-6} \times (1.0075 \times 15.81)^3 \\ &= 0.5976 \end{aligned}$$

Heat duty in the radiant section, Q_{RADSEC}, is given by:

$$\begin{aligned} Q_{RADSEC} &= F_{RAD} Q_{RELEASE} \\ &= 0.5976 \times 5.5960 \times 10^7 \\ &= 3.3442 \times 10^7\ \text{W} \end{aligned}$$

12

Calculate the process inlet temperature to the radiant section assuming an outlet temperature of 360 °C.

$$Q_{RADSEC} = m_P C_{PP}(360 - T_{Pinlet})$$

$$3.3442 \times 10^7 = 185.4 \times 3282 \times (360 - T_{Pinlet})$$

$$T_{Pinlet} = 305.0\ °C$$

d) Calculate the number of tubes requires in the radiant section. The outside diameter of the tubes is given in Table 13.5:

$$N_T = \frac{Q_{RADSEC}}{q_{RAD}\pi d_O L} = \frac{3.3442 \times 10^7}{32,000 \times \pi \times 0.1683 \times 11.2} = 176.5$$

For a six-pass configuration, this should be rounded up to 180 tubes.

e) The bridgewall temperature, T_{BW}, can be determined from Equation 12.189:

$$F_{RAD} Q_{RELEASE} = -m_{FUEL} \Delta H^0_{COMB} - m_{FG} \Delta H_{FG}$$

where m_{FG} = mass flowrate of the flue gas $(kg \cdot s^{-1})$

$$m_{FUEL} = 1.24\ kg \cdot s^{-1}$$

$$m_{FG} = 1.24 \times \frac{11.8153}{0.7029} = 20.84\ kg \cdot s^{-1}$$

ΔH_{FG} can be calculated from enthalpy data from Equation 12.176 and Table 12.18. Alternatively, rather than use the coefficients in Table 12.18, the coefficients in Equation 12.196 can be used to give an approximate measure:

$$\Delta H_{FG} = \left[-9.546 \times 10^{-11}\frac{T^4}{4} + 1.114 \times 10^{-4}\frac{T^3}{3} + 0.2508\frac{T^2}{2} + 1043T\right]_{25}^{T_{BW}}$$

$$= \begin{bmatrix} -2.3865 \times 10^{-11} T_{BW}^4 + 3.7133 \times 10^{-5} T_{BW}^3 \\ 3 + 0.1254 T_{BW}^2 + 1043 T_{BW} - 2.6154 \times 10^4 \end{bmatrix}$$

Substituting into Equation 12.189 gives:

$$0.5976 \times 5.5960 \times 10^7 = \frac{-1.24}{16} \times 0.9 \times (-802.3 \times 10^6) - 20.84 \begin{bmatrix} -2.3865 \times 10^{-11} T_{BW}^4 + 3.7133 \times 10^{-5} T_{BW}^3 \\ + 0.1254 T_{BW}^2 + 1043 T_{BW} - 2.6154 \times 10^4 \end{bmatrix}$$

$$1.1067 \times 10^6 = -2.3865 \times 10^{-11} T_{BW}^4 + 3.7133 \times 10^{-5} T_{BW}^3 + 0.1254 T_{BW}^2 + 1043 T_{BW}$$

Solving by trial and error:

$$T_{BW} = 928.8\ °C$$

f) Heat transfer in the shield section is both by radiation and convection. The radiative contribution is known from the assumption of the radiative flux. Assuming the first tube row has the same heat flux as the radiative section and the second row 56% of that value:

$$Q_{RAD} = (1 + 0.56) \times q_{RAD} \times 2 \times \pi d_O L N_{TR} = 1.56 \times 32,000 \times 2 \times \pi \times 0.1683 \times 11.2 \times 6 = 3.5474 \times 10^6\ W$$

where Q_{RAD} = heat transferred by radiation (W)

The convective heat transfer Q_{CONV} can be calculated from:

$$Q_{CONV} = UA\,\Delta T_{LM}$$

Calculation of ΔT_{LM} requires that the inlet and outlet temperatures of both the flue gas and process must be known. The inlet temperature of the flue gas is the bridgewall temperature with a value of 928.8 °C. The outlet temperature of the shield section is the inlet temperature to the radiative section of 305.0 °C. Both the outlet temperature of the flue gas and the inlet temperature of the process are unknown. If the mechanism for heat transfer could have been defined as only convective, then iteration could be avoided using the approach shown earlier for shell-and-tube heat exchangers. In this case, the radiative heat transfer has a different basis from the convective heat transfer and some iteration cannot be avoided.

Start by making an initial estimate of the outlet temperature of the flue gas from the shield section, say 750 °C. Later, the value will be refined. Calculate the resulting mean temperature in the shield section to calculate the physical properties:

$$\overline{T_{FG}} = \frac{928.8 + 750}{2} = 839.4\ °C$$

At this mean temperature, the corresponding flue gas physical properties can be calculated from Equations 12.195 to 12.198:

$$\rho_{FG} = 0.3203\ kg \cdot m^{-3}$$

$$\mu_{FG} = 4.464 \times 10^{-5}\ kg \cdot m^{-1} \cdot s^{-1}$$

$$C_{PFG} = 1332.0\ J \cdot kg^{-1} \cdot K^{-1}$$

$$k_{FG} = 0.0950\ W \cdot m^{-1} \cdot K^{-1}$$

From the initial estimate of the flue gas outlet temperature, the process inlet temperature, T_{Pinlet}, can be calculated from an energy balance:

$$T_{Pinlet} = 305.0 - \frac{m_{FG} C_{PFG}(T_{BW} - T_{FGoutlet})}{m_P C_{PP}} = 305.0 - \frac{20.84 \times 1332.0(928.8 - 750)}{185.4 \times 3282} = 296.8\ °C$$

The log mean temperature difference can now be calculated:

$$\Delta T_{LM} = \frac{(928.8 - 305.0) - (750 - 296.8)}{\ln\left[\dfrac{928.8 - 305.0}{750 - 296.8}\right]} = 534.0\ °C$$

The calculation of the overall heat transfer coefficient requires the calculation of the flue gas convective heat transfer coefficient from Equation 12.190, 12.191 or 12.192, depending on the Reynolds Number. Calculate v_{max} for the flue gas from Equation 12.194:

$$\begin{aligned} v_{max} &= \frac{m_{FG}}{\rho_{FG}L\left[N_{TR}(p_T - d_O) + \dfrac{3p_T}{2} - d_O\right]} \\ &= \frac{20.84}{0.3202 \times 11.2\left[6(0.3 - 0.1683) + \dfrac{3 \times 0.3}{2} - 0.1683\right]} \\ &= 5.42\ \text{m} \cdot \text{s}^{-1} \end{aligned}$$

$$\begin{aligned} Re &= \frac{\rho_{FG}v_{\max}d_O}{\mu_{FG}} \\ &= \frac{0.3202 \times 5.42 \times 0.1683}{4.464 \times 10^{-5}} \\ &= 6545 \end{aligned}$$

Thus, Equation 12.191 can be used to calculate the outside heat transfer coefficient:

$$\begin{aligned} Pr &= \frac{C_{PFG}\mu_{FG}}{k_{FG}} \\ &= \frac{1332.0 \times 4.464 \times 10^{-5}}{0.0950} \\ &= 0.6259 \end{aligned}$$

Calculate F_N from Equation 12.193:

$$\begin{aligned} F_N &= -0.04375(\ln N_R)^2 + 0.27 \ln N_R + 0.61 \\ &= -0.04375(\ln 2)^2 + 0.27(\ln 2) + 0.61 \\ &= 0.78 \end{aligned}$$

Thus, from Equation 12.191 the convective film coefficient for the flue gas, h_{FG}, is given by:

$$\begin{aligned} h_{FG} &= 0.273\frac{k_{FG}}{d_O}Re^{0.635}\,Pr^{0.34}\,F_N \\ &= 0.273 \times \frac{0.095}{0.1683} \times 6545^{0.635} \times 0.6259^{0.34} \times 0.78 \\ &= 27.1\ \text{W} \cdot \text{m}^{-2} \cdot \text{K}^{-1} \end{aligned}$$

The overall heat transfer coefficient can now be calculated:

$$\frac{1}{U} = \frac{1}{h_{FG}} + R_{FG} + \frac{d_O}{2k_W}\ln\left(\frac{d_O}{d_I}\right) + \frac{d_O}{d_I}R_P + \frac{d_O}{d_I}\cdot\frac{1}{h_P}$$

$$U = \left[\frac{1}{27.1} + 0.0002 + \frac{0.1683}{2 \times 35}\ln\frac{0.1683}{0.1541} + \frac{0.1683}{0.1541} \times 0.0005 + \frac{0.1683}{0.1541} \times \frac{1}{500}\right]^{-1}$$

$$= 24.90\ \text{W} \cdot \text{m}^{-2}\ \text{K}^{-1}$$

Strictly, for convective heat transfer the ΔT_{LM} should be corrected for a noncountercurrent flow arrangement. It is a cross-flow arrangement with multiple tube passes. However, the large temperature differences and the multiple tube passes will make the F_T correction factor high enough to assume countercurrent flow. Calculate the convective heat duty:

$$\begin{aligned} Q_{CONV} &= UA\Delta T_{LM} \\ &= U(2\pi d_O L N_{TR})\Delta T_{LM} \\ &= 24.90(2 \times \pi \times 0.1683 \times 11.2 \times 6)534.0 \\ &= 9.4482 \times 10^5\ \text{W} \end{aligned}$$

The value of the flue gas outlet temperature, $T_{FGoutlet}$, can now be calculated from an energy balance:

$$\begin{aligned} T_{FGoutlet} &= T_{BW} - \frac{Q_{RAD} + Q_{CONV}}{m_{FG}C_{PFG}} \\ &= 928.8 - \frac{3.5474 \times 10^6 + 9.4482 \times 10^5}{20.84 \times 1332.0} \\ &= 767.0\,^{\circ}\text{C} \end{aligned}$$

This compares with the initial estimate of 750 °C. From this new value of the flue gas outlet temperature new values of the physical properties, T_{Pinlet}, ΔT_{LM}, U and Q_{CONV} can be calculated to give a further value of $F_{FGoutlet}$. It is convenient to carry out this iteration by varying $T_{FGoutlet}$ to find the root of:

$$0 = \frac{Q_{CONV}}{U\Delta T_{LM}} - 2\pi d_O L N_{TR}$$

In this case, iteration brings virtually no change with:

$$\begin{aligned} T_{FGoutlet} &= 766.5\,^{\circ}C \\ T_{Pinlet} &= 297.6\,^{\circ}C \end{aligned}$$

g) For the convective section, heat transfer is again by radiation and convection. However, in this case a radiation allowance can be added to the convection heat transfer coefficient. This allows calculation of the performance in principle without iteration. However, iteration might be required for the variable physical properties. In this case the tubes are finned in the convection section. Had there been more rows of plain tubes than two after the radiant section, then the following procedure can be applied to those tubes, but with the convective heat transfer coefficient being calculated from Equations 12.190 to 12.192. For finned tubes, the convective heat transfer coefficient can be calculated from Equation 12.111. Start by making an initial estimate of the temperature of the flue gas at the outlet of the convection section, to calculate the physical properties. Assuming an outlet temperature of the flue gas of 300 °C, the mean temperature in the convection section is given by:

12

$$\begin{aligned} \overline{T_{FG}} &= \frac{766.5 + 300}{2} \\ &= 533.3\,^{\circ}\text{C} \end{aligned}$$

At this mean temperature, the corresponding flue gas physical properties can be calculated from Equations 12.195 to 12.198:

$$\begin{aligned} \rho_{FG} &= 0.4416\ \text{kg} \cdot \text{m}^{-3} \\ \mu_{FG} &= 3.585 \times 10^{-5}\ \text{kg} \cdot \text{m}^{-1} \cdot \text{s}^{-1} \\ C_{PFG} &= 1208.4\ \text{J} \cdot \text{kg}^{-1} \cdot \text{K}^{-1} \\ k_{FG} &= 0.0685\ \text{W} \cdot \text{m}^{-1} \cdot \text{K}^{-1} \end{aligned}$$

The convective heat transfer coefficient can be calculated from Equation 12.113. First, calculate the maximum velocity of the flue gas from Equation 12.116:

$$\begin{aligned} v_{max} &= \frac{m_{FG}}{\rho L\left[N_{TR}(p_T - d_R - 2H_F\delta_F N_F) + \dfrac{3p_T}{2} - d_R\right]} \\ &= \frac{20.84}{0.4416 \times 11.2\left[6(0.3 - 0.1683 - 2 \times 0.055 \times 0.002 \times 120) + \dfrac{3 \times 0.3}{2} - 0.1683\right]} \\ &= 4.61\ \text{m} \cdot \text{s}^{-1} \end{aligned}$$

$$Re = \frac{0.4416 \times 4.61 \times 0.1683}{3.585 \times 10^{-6}}$$
$$= 9,581$$
$$Pr = \frac{1208.4 \times 3.585 \times 10^{-6}}{0.0685}$$
$$= 0.06324$$

$$Nu = 0.134\, Re^{0.681}\, Pr^{1/3} \left(\frac{1}{H_F N_F} - \frac{\delta_F}{H_F}\right)^{0.2} \left(\frac{1}{\delta_F N_F} - 1\right)^{0.1134}$$

The convective film coefficient is given by:

$$h_{CONV} = \frac{0.0685}{0.1683}\left[0.134 \times 9561^{0.681} \times 0.06324^{1/3} \times \left(\frac{1}{0.055 \times 120} - \frac{0.002}{0.055}\right)^{0.2} \times \left(\frac{1}{0.002 \times 120} - 1\right)^{0.1134}\right]$$
$$= 17.80\ \mathrm{W \cdot m^{-2} \cdot K^{-1}}$$

From Equation 12.199 and 12.200, the combined convective and radiative film coefficient is given by:

$$h_{FG} = 1.1(h_{CONV} + 0.02555T - 2.385)$$
$$= 1.1(17.80 + 0.02555 \times 533.3 - 2.385)$$
$$= 31.94\ \mathrm{W \cdot m^{-2} \cdot K^{-1}}$$

The fin efficiency now needs to be calculated from Equation 12.117:

$$\kappa = \left(\frac{2h_{FG}}{k_F \delta_F}\right)^{1/2}$$
$$= \left(\frac{2 \times 31.94}{35 \times 0.002}\right)^{1/2}$$
$$= 30.21$$
$$\psi = \left(H_F + \frac{\delta_F}{2}\right)\left[1 + 0.35 \ln\left(\frac{d_R + 2H_F + \delta_F}{d_R}\right)\right]$$
$$= \left(0.055 + \frac{0.002}{2}\right) \times \left[1 + 0.35 \ln\left(\frac{0.1683 + 2 \times 0.055 + 0.002}{0.1683}\right)\right]$$
$$= 0.0660$$
$$\eta_F = \frac{\tanh(\kappa\psi)}{\kappa\psi}$$
$$= \frac{\tanh(30.21 \times 0.0660)}{30.21 \times 0.0660}$$
$$= 0.4833$$

Calculation of the weighted efficiency from Equation 12.119 requires A_{ROOT} and A_{FIN} to be first calculated:

$$A_{ROOT} = \pi d_R L(1 - N_F \delta_F)$$
$$= \pi \times 0.1683 \times 11.2 \times (1 - 120 \times 0.002)$$
$$= 4.50\ \mathrm{m^2}$$

$$A_{FIN} = \frac{\pi N_F L}{2}\left[(d_R + 2H_F)^2 - d_R^2 + 2\delta_F(d_R + 2H_F)\right]$$
$$= \frac{\pi \times 120 \times 11.2}{2}\left[(0.1683 + 2 \times 0.055)^2 - 0.1683^2 + 2 \times 0.002 \times (0.1683 + 2 \times 0.055)\right]$$
$$= 106.6\ \mathrm{m^2}$$

$$\eta_W = \frac{A_{ROOT} + \eta_F A_{FIN}}{A_{ROOT} + A_{FIN}}$$
$$= \frac{4.500 + 0.4833 \times 106.06}{4.500 + 106.06}$$
$$= 0.5043$$

The total outside area is given by

$$A_O = 4.500 + 106.06$$
$$= 110.56\ \mathrm{m^2}$$

The overall heat transfer coefficient is now calculated from Equation 12.120:

$$\frac{1}{U} = \frac{1}{\eta_W h_O} + \frac{R_{OF}}{\eta_W} + \frac{A_O}{\pi d_R L} \frac{d_R}{2k_W} \ln\left(\frac{d_R}{d_I}\right) + \left(\frac{A_O}{\pi d_I L}\right)\left(R_{IF} + \frac{1}{h_I}\right)$$

$$U = \left[\frac{1}{0.5043 \times 31.94} + \frac{0.0002}{0.5043} + \frac{110.56}{\pi \times 0.1683 \times 11.2} \times \frac{0.1683}{2 \times 35} \ln\left(\frac{0.1683}{0.1541}\right) + \left(\frac{110.56}{\pi \times 0.1541 \times 11.2}\right)\left(0.0005 + \frac{1}{500}\right)\right]^{-1}$$
$$= 8.24\ \mathrm{W \cdot m^{-2} \cdot K^{-1}}$$

Again, the ΔT_{LM} should be corrected for a noncountercurrent flow arrangement by the application of an F_T correction factor for cross flow. However, the large temperature differences and the multiple tube passes will make the F_T correction factor high enough to assume $F_T = 1$. Thus, calculate R and X from Equations 12.83 and 12.87:

$$R = \frac{CP_C}{CP_H} = \frac{T_{H1} - T_{H2}}{T_{C2} - T_{C1}}$$
$$= \frac{185.4 \times 3282}{20.84 \times 0.4416}$$
$$= 24.16$$

$$X = \exp\left[\frac{UA(R-1)}{CP_C}\right]$$
$$= \exp\left[\frac{8.24 \times 110.56(24.16 - 1)}{185.4 \times 3282}\right]$$
$$= 12.1431$$

The process inlet to the convection section can be calculated from Equation 12.90:

$$T_{C2} = \frac{(X-1)T_{H1} + X(R-1)T_{C1}}{(RX-1)} \qquad R \neq 1$$

Rearranging this equation gives:

$$T_{C1} = \frac{(RX-1)T_{C2} - (X-1)T_{H1}}{X(R-1)} \qquad R \neq 1$$
$$= \frac{(24.16 \times 12.1431 - 1)297.6 - (12.1431 - 1)766.5}{12.1431(24.16 - 1)}$$
$$= 279.1\ ^{\circ}\mathrm{C}$$

The flue gas outlet temperature can be calculated from Equation 12.89:

$$T_{H2} = \frac{(R-1)T_{H1} + R(X-1)T_{C1}}{(RX-1)} \qquad R \neq 1$$
$$= \frac{(24.16-1)766.5 + 24.16(12.1431-1)279.1}{(24.16 \times 12.1431 - 1)}$$
$$= 317.7\,°C$$

h) The calculated inlet temperature for the process is 279.1 °C, whereas the specified inlet is 280 °C. This is close to the required temperature. If there is a large difference, the whole calculation needs to be repeated by varying the fuel flowrate until the process inlet temperature is 280 °C. This is a tedious calculation by hand, but easily programmed in a spreadsheet. If the iteration is carried out, the result is:

$$m_{FUEL} = 1.22\ kg \cdot s^{-1}$$
$$m_{FG} = 20.58\ kg \cdot s^{-1}$$
$$Q_{RELEASE} = 5.5253 \times 10^7\ W$$
$$F_{RAD} = 0.5977$$
$$T_{BW} = 928.6\,°C$$
$$T_{FGoutlet,SHIELD} = 764.8\,°C$$
$$T_{FGoutlet,CONVSEC} = 317.3\,°C$$
$$T_{Poulet,CONVSEC} = 298.3\,°C$$
$$T_{Pinlet,RAD} = 305.7\,°C$$

i) Furnace efficiency for casing losses of 2% of heat release is given by:

$$\eta = \frac{185.4 \times 3282 \times (360-280)}{1.22 \times \frac{0.9}{16} \times 802.3 \times 10^6 + 5.5253 \times 10^7 \times 0.02}$$
$$= 0.867$$

It is worth making a number of observations regarding the calculations for furnace design. All of the methods for the design of the radiant zone have considerable uncertainty, especially the empirically based methods used here. For the process side, the heat duty was characterized by a heat capacity, which was assumed constant. In this particular example, an allowance needed to be made both for the sensible heat to the crude oil and the energy required for partial vaporization of the crude oil in the distillation to follow (Nelson, 1958). The inside heat transfer coefficient was assumed, rather than calculated from heat transfer correlations. For the flue gas, the enthalpy was calculated from an average for the flue gas, rather than from a breakdown of the individual components. Also for the flue gas, the temperature has a large range, leading to large variations in the physical properties. The above calculation averaged the physical properties over the shield and convection sections, despite large temperature differences over each.

12.14 Heat Exchange – Summary

Of the many types of heat transfer equipment used in the process industries, the shell-and-tube heat exchanger is by far the most common. A number of different flow arrangements are possible with shell-and-tube heat exchangers, but the one shell pass–one tube pass and the one shell pass–two tube pass arrangements are the most common.

Resistance to heat transfer across the tube wall for shell-and-tube heat exchangers is made up of five individual resistances to heat transfer:

- shell-side film coefficient,
- shell-side fouling coefficient,
- tube-wall coefficient,
- tube-side fouling coefficient,
- tube-side film coefficient.

These are combined to give an overall heat transfer coefficient in which one of the individual coefficients might be controlling.

Simple models are available for shell-and-tube heat exchangers in which both heat transfer coefficients and pressure drops can be related to the fluid velocity. The 1–1 heat exchanger is designed as a countercurrent device. For heat exchangers with more than a single tube pass, there is an element of cocurrent flow as well as the countercurrent flow. This must be corrected by the introduction of a correction to the logarithmic mean temperature difference. This can be taken into account when designing, rating or simulating a heat exchanger.

In retrofit situations, existing heat exchangers might be subjected to changes in flowrate, heat transfer duty, temperature differences or fouling characteristics. Heat transfer coefficients and pressure drops can be approximated from their original values from the change in flowrate. Increased duties can be accommodated by changing the tube bundle to incorporate a greater heat transfer area in the same shell, or by heat transfer enhancement devices, or by changes to the baffle arrangements.

Shell-and-tube heat exchangers are also used extensively for condensing duties. Condensers can be horizontally or vertically mounted with the condensation on the tube-side or the shell-side. Condensation normally takes place on the shell-side of horizontal exchangers and the tube-side of vertical exchangers.

Reboilers are required on distillation columns to vaporize a fraction of the bottom product. Three common designs of reboiler are used: the kettle reboiler and the horizontal and vertical thermosyphon reboilers. The preliminary design of kettle and horizontal thermosyphon reboilers can be based on correlations for pool boiling. The design of vertical thermosyphon reboilers requires the hydraulic and thermal designs to be carried out simultaneously. Preliminary design of such units can be based on correlations derived from a large number of detailed designs. Great care must be exercised in the preliminary design of reboilers as the predictions of the correlations are extremely unreliable. 12

While the shell-and-tube heat exchanger is the most commonly used in the process industries, it has the disadvantages that the flow is not truly countercurrent, which limits the minimum temperature difference that can be accommodated, and the area density is relatively low. Commonly used alternatives for shell-and-tube heat exchangers are:

- gasketed plate heat exchangers,
- welded plate heat exchangers,
- plate-fin heat exchangers,
- spiral heat exchangers.

Some heat transfer operations demand a high heat duty, high temperature and/or high heat flux. In such cases, radiant heat transfer is used from the combustion of fuel in a fired heater. Fired heater designs vary according to the function, heating duty, type of fuel and method of introducing combustion air. In conceptual design, knowledge of the heat duty in the various sections of the furnace is more important than the heat transfer area, as the cost in preliminary design is based on the heat duty. The theoretical flame temperature provides a simple model to allow the heat duty to be determined.

12.15 Exercises

1. A distillation operation separating a low-viscosity hydrocarbon mixture requires three shell-and-tube heat exchangers. The liquid feed is to be preheated to saturated liquid by heat recovery from another low-viscosity hydrocarbon stream. The reboiler is to be a vertical thermosyphon using steam heating. The condenser is to be a horizontal exchanger with the condensing stream on the shell-side and the cooling serviced by cooling water. What would be the expected rank order of the three heat exchangers in terms of their overall heat transfer coefficient?

2. For the three heat exchangers from Exercise 1, make a first estimate of the order of magnitude of the overall heat coefficients from tabulated values of film transfer coefficients and fouling coefficients. Neglect the resistance from the tube walls.

3. Under what circumstances might a thermosyphon reboiler be orientated vertically or horizontally?

4. Under what circumstances might a distillation condenser be orientated vertically or horizontally?

5. The demethanizer distillation column of an ethylene process works at extremely low temperatures. The feed is cooled with extremely small temperature differences of the order of 1 °C to minimize the refrigeration costs associated with the cooling. What type of heat exchanger would you expect to be used for this duty?

6. Gasketed plate heat exchangers are commonly used in food processing. What advantages does the design offer in such applications?

7. A hot stream is to be cooled from 210 to 80 °C by heating a cold stream from 60 to 150 °C with a duty of 1.7 MW. A 1–1 shell-and-tube heat exchanger is to be used and the overall heat transfer coefficient has been estimated to be $120 \text{ W} \cdot \text{m}^{-2} \cdot \text{K}^{-1}$. Calculate the heat transfer area of the unit.

8. Instead of using a 1–1 design in Exercise 7, a 1–2 design is to be used subject to $X_P = 0.9$. Assume that the overall heat transfer coefficient is unchanged. (In practice, it would be expected to increase). Calculate:

   a) the number of shells required;

   b) P_{1-2} for each shell;

Table 12.22

Physical property data for *n*-butanol and water.

	n-Butanol	Water
Density ($\text{kg} \cdot \text{m}^{-3}$)	712	993
Heat capacity ($\text{J} \cdot \text{kg}^{-1} \cdot \text{K}^{-1}$)	3200	4190
Viscosity ($\text{N} \cdot \text{s} \cdot \text{m}^{-2}$)	4.0×10^{-4}	8.0×10^{-4}
Thermal conductivity ($\text{W} \cdot \text{m}^{-1} \cdot \text{K}^{-1}$)	0.127	0.616

   c) F_T for the shells in series;

   d) the heat transfer area.

9. Liquid *n*-butanol at 115 °C is to be cooled to 45 °C against cooling water between 25 and 35 °C. The flowrate of *n*-butanol is $10 \text{ kg} \cdot \text{s}^{-1}$. A 4 pass split ring floating head shell-and-tube heat exchanger is to be used. The *n*-butanol will be allocated to the tube-side and the cooling water to the shell-side. Physical property data for the fluids are given in Table 12.22.

   Assume that the fouling coefficient for *n*-butanol is $10{,}000 \text{ W} \cdot \text{m}^{-2} \cdot \text{K}^{-1}$ and that for cooling water is $3000 \text{ W} \cdot \text{m}^{-2} \cdot \text{K}^{-1}$. Steel tubes are to be used with the heat exchanger data given in Table 12.23.

   a) Calculate the number of shells required and the F_T based on $X_P = 0.9$.

   b) Calculate the overall heat transfer coefficient, heat transfer area, number of tubes, tube length and shell diameter assuming a fluid velocity of $1 \text{ m} \cdot \text{s}^{-1}$ on both the tube-side and on the shell-side.

   c) Calculate the pressure drop for the tube-side and shell-side assuming a fluid velocity of $1 \text{ m} \cdot \text{s}^{-1}$ on the tube-side and $1 \text{ m} \cdot \text{s}^{-1}$ on the shell-side.

Table 12.23

Heat exchanger geometry.

Tube inner diameter d_I (mm)	16
Tube outer diameter d_O (mm)	20
Tube pitch p_T (mm)	25
Tube pattern (tube layout angle)	90°
Number of tube passes N_P	2
Tube thermal conductivity ($\text{W} \cdot \text{m}^{-1} \cdot \text{K}^{-1}$)	45
Baffle cut B_C	0.25
Tube-side inlet nozzle diameter $d_{TN,inlet}$ (m)	0.1023
Tube-side outlet nozzle diameter $d_{TN,outlet}$ (m)	0.1023
Shell-side inlet nozzle diameter $d_{TS,inlet}$ (m)	0.2545
Shell-side outlet nozzle diameter $d_{TS,outlet}$ (m)	0.2545

Table 12.24

Heat exchanger geometry.

Tube inner diameter d_I (mm)	16
Tube outer diameter d_O (mm)	20
Tube pitch p_T (mm)	25
Tube pattern (tube layout angle)	90°
Tube length L (m)	4.96
Number of tubes N_T	280
Number of tube passes N_P	4
Tube thermal conductivity ($W \cdot m^{-1} \cdot K^{-1}$)	45
Shell inner diameter D_S (m)	0.610
Number of baffles	13
Baffle spacing B (m)	0.354
Baffle cut B_C	0.25
Shell-bundle diametric clearance L_{BB} (m)	0.036
Tube-side inlet nozzle diameter $d_{TN,inlet}$ (m)	0.1023
Tube-side outlet nozzle diameter $d_{TN,outlet}$ (m)	0.1023
Shell-side inlet nozzle diameter $d_{TS,inlet}$ (m)	0.2545
Shell-side outlet nozzle diameter $d_{TS,outlet}$ (m)	0.2545
Tube-side fouling coefficient ($W \cdot m^{-2} \cdot K^{-1}$)	10,000
Shell-side fouling coefficient ($W \cdot m^{-2} \cdot K^{-1}$)	3000

10. The heat exchanger in Exercise 12.9 can now be considered to be an existing heat exchanger with a feed to the tube-side of $10\ kg \cdot s^{-1}$ of liquid n-butanol at 115 °C. The feed to the shell-side is $53.46\ kg \cdot s^{-1}$ of cooling water at 25 °C. Physical property data for the fluids are given in Table 12.22. The heat exchanger geometry data are given in Table 12.24.

    Calculate the outlet temperatures and heat duty and compare with the design calculation in Exercise 12.9.

11. A condenser is required to condense a flowrate of $7\ kg \cdot s^{-1}$ of isopropanol at an operating pressure of 1.0 bar. The condensation takes place isothermally at 83 °C without subcooling of the condensate. The cooling is provided by cooling water between 25 and 35 °C. The condenser can be assumed to be a single-pass fixed tube-sheet design manufactured from steel with 20 mm tubes with 2 mm wall thickness having a thermal conductivity of $45\ W \cdot m^{-1} \cdot K^{-1}$. The tube pitch can be assumed to be $1.25d_O$ with a square configuration. The physical property data are given in Table 12.25.

    Assume the fouling coefficients to be $10{,}000\ W \cdot m^{-2} \cdot K^{-1}$ and $5000\ W \cdot m^{-2} \cdot K^{-1}$ for isopropanol and cooling water respectively. Carry out a preliminary design for:

Table 12.25

Physical property data for isopropanol and water.

Property	Isopropanol (83 °C)	Water (30 °C)
Density ($kg \cdot m^{-3}$)	722	996
Vapor density ($kg \cdot m^{-3}$)	2.0	–
Liquid heat capacity ($J \cdot kg^{-1} \cdot K^{-1}$)	3370	4180
Liquid viscosity ($N \cdot s \cdot m^{-2}$)	0.492×10^{-3}	0.797×10^{-3}
Thermal conductivity ($W \cdot m^{-1} \cdot K^{-1}$)	0.131	0.618
Heat of vaporization ($J \cdot kg^{-1}$)	678,000	–
Molar mass ($kg \cdot kmol^{-1}$)	60	–

   a) a horizontal condenser with shell-side condensation with a tube-side velocity of $1.5\ m \cdot s^{-1}$;
   b) a vertical condenser with tube-side condensation with a length to shell diameter ratio of 5 and shell-side velocity of $1\ m \cdot s^{-1}$.

12. A reboiler of a distillation column is required to supply $10\ kg \cdot s^{-1}$ of toluene vapor. The column operating pressure at the bottom of the column is 1.6 bar. At this pressure, the toluene vaporizes at 127 °C and can be assumed to be isothermal. Steam at 160 °C is to be used for the vaporization. The latent heat of vaporization of toluene is $344{,}000\ J \cdot kg^{-1}$, the critical pressure is 40.5 bar and critical temperature is 594 K. Steel tubes with 30 mm outside diameter, 2 mm wall thickness, tube pitch of 0.0375 m and length 3.95 m are to be used. The tube layout angle is 90°. The fouling coefficient for reboiling can be assumed to be $2500\ W \cdot m^{-2} \cdot K^{-1}$. The film coefficient (including fouling) for the condensing steam can be assumed to be $5700\ W \cdot m^{-2} \cdot K^{-1}$. Estimate the heat transfer area for: 12
   a) a kettle reboiler;
   b) a vertical thermosyphon reboiler.

13. The purge gas from a petrochemical process is at 25 °C and contains a mole fraction of methane of 0.6, the balance being hydrogen. This purge gas is to be burnt in a furnace to provide heat to a process with a cold stream pinch temperature of 150 °C ($\Delta T_{min} = 50$ °C). Ambient temperature is 10 °C.
   a) Calculate the theoretical flame temperature if 15% excess air is used in the combustion. Standard heats of combustion are given in Table 12.17 and mean molar heat capacities in Table 12.19.
   b) Calculate the furnace efficiency.
   c) Suggest ways in which the furnace efficiency could be improved.

References

Ahmad S, Linnhoff B and Smith R (1988) Design of Multi-pass Heat Exchangers: An Alternative Approach, *Trans ASME J Heat Transfer*, **110**: 304.

Ahmad S, Linnhoff B and Smith R (1990) Cost Optimum Heat Exchanger Networks, II: Targets and Design for Detailed Capital Cost Models, *Comp Chem Eng*, **7**: 751.

Ayub ZH (2005) A New Chart Method for Evaluating Single-phase Shell-Side Heat Transfer Coefficient in a Single Segmental Shell-and-Tube Heat Exchanger, *Appl Therm Eng*, **25**: 2412.

Bahadori A and Vuthaluru HB (2010) Novel Predictive Tools for Design of Radiant and Convective Sections of Direct Fired Heaters, *Applied Energy*, **87**: 2194.

Bott TR (1995) *Fouling of Heat Exchangers*, Elsevier.

Bouhairie S (2012) Selecting Baffles for Shell and Tube Heat Exchanges, *Chem Engg Progr*, **Feb**: 27.

Bowman RA (1936) Mean Temperature Difference Correction in Multipass Exchangers, *Ind Eng Chem*, **28**: 541.

Bowman RA, Mueller AC and Nagle WM (1940) Mean Temperature Differences in Design, *Trans ASME*, **62**: 283.

Briggs DE and Young EH (1963) Convection Heat Transfer and Pressure Drop of Air Flowing Across Triangular Pitch Banks of Finned Tubes, *Chem Eng Progr Syp Ser*, **59** (41): 1.

Butterworth D. (1977) *Introduction to Heat Transfer*, Oxford University Press.

Frank O and Prickett RD (1973) Design of Vertical Thermosyphon Reboilers, *Chem Eng*, **3**: 107.

García A, Vicente PG and Viedma A (2005) Experimental Study of Heat Transfer Enhancement with Wire Coil Inserts in Laminar-Transition-Turbulent Regimes at Different Prandtl Numbers, *Int J Heat and Mass Transfer*, **48**: 4640.

Herkenhoff RG (1981) A New Way to Rate an Existing Heat Exchanger, *Chem Engg*, **March 23**: 213.

Hewitt GF (2008) *Handbook of Heat Exchangers Design*, Begell House Inc.

Hewitt GF, Shires GL and Bott TR (1994) *Process Heat Transfer*, CRC Press Inc.

Hougen OA, Watson KM and Ragatz RA (1954) *Chemical Process Principles. Part I: Material and Energy Balances*, 2nd Edition, John Wiley & Sons.

Isachenko VP, Osipova VA and Sukomel AS (1977) *Heat Transfer*, 3rd Edition, translated by Semyonov S, Mir Publishers, Moscow.

Jiang N, Shelley JD and Smith R (2014) New Models for Conventional and Enhanced Heat Exchangers for Heat Exchanger Network Retrofit, *Applied Thermal Engg*, **70**: 944.

Kern DQ (1950) *Process Heat Transfer*, McGraw-Hill.

Kern DQ and Kraus AD (1972) *Extended Surface Heat Transfer*, McGraw-Hill, New York.

Kotjabasakis M and Linnhoff B (1986) Sensitivity Tables for the Design of Flexible Processes (I) – How Much Contingency in Heat Exchanger Networks is Cost Effective?, *Chem Eng Res Des*, **64**: 198.

Lobo WE and Evans JE (1939) Heat Transfer in the Radiant Section of Petroleum Heaters, *AIChEJ*, **35**: 743.

Manglik RM and Bergles AE (1993a) Heat Transfer and Pressure Drop Correlations for Twisted-Tape Inserts in Isothermal Tubes, I: Laminar Flows, *J Heat Transfer*, **115**: 881.

Manglik RM and Bergles AE (1993b) Heat Transfer and Pressure Drop Correlations for Twisted-Tape Inserts in Isothermal Tubes, II: Transition and Turbulent Flows, *J Heat Transfer*, **115**: 890.

Martin GR (1998) Heat-Flux Imbalances in Fired Heaters Cause Operating Problems, *Hydrocarbon Processing*, **May**: 103.

McQuiston FC and Tree DR (1972) Optimum Surface Envelopes of the Finned Tube Heat Transfer Surface, ASHRAE Annual Meeting, Bahamas, p. 144.

Mostinski IL (1963) Calculation of Boiling Heat Transfer Coefficients, Based on the Law of Corresponding States, *Br Chem Eng*, **8**: 580.

Müller-Steinhagen H (2000) *Heat Exchanger Fouling – Mitigation and Cleaning Technologies*, Institution of Chemical Engineers, UK.

Nelson WL (1958) *Petroleum Refinery Engineering*, 4th Edition, McGraw-Hill.

Nesta JM and Coutinho CA (2011) Update on Designing for High-Fouling Liquids, *Hydrocarbon Processing*, **May**: 83.

Nusselt W. (1916) Die Oberflachenkondensation des Wasserdampfes, *Z Ver Deut Ing*, **60**: 541, 569.

Ravigururajan TS and Bergles AE (1996) Development and Verification of General Correlations for Pressure Drop and Heat Transfer in Single-Phase Turbulent Flow in Enhanced Tubes, *Exp Therm Fluid Sci*, **13**: 55.

Rose J (1999) Laminar Film Condensation of Pure Vapors, in *Handbook of Phase Change: Boiling and Condensation*, Kandlikar SG, Shoji M and Dhir VK (eds), Taylor Francis.

Rosenow WM (1956) Heat Transfer and Temperature Distribution in Laminar Film Condensation, *Trans ASME*, **78**: 1645.

Rosenow WM, Webber JH and Ling AT (1956) Effect of Velocity on Laminar and Turbulent Film Condensation, *Trans ASME*, **78**: 1645.

Serth RW (2007) *Process Heat Transfer: Principles and Applications*, Academic Press.

TEMA, (Tubular Exchanger Manufacturers Association Inc.) (2007) *Standards of the Tubular Exchanger Manufacturers Association, 25 North Broadway, Tarrytown, New York.*

Towler G and Sinnott RK (2013) *Chemical Engineering Design*, 2nd Edition, Butterworth-Heinemann.

Underwood, AJV (1934) The Calculation of the Mean Temperature Difference in Multi-pass Heat Exchangers, *J Inst Petroleum Tech*, **20**: 145.

Wang YF, Pan M, Bulatov I, Smith R and Kim JK (2012) Application of Intensified Heat Transfer for the Retrofit of Heat Exchanger Network, *Appl Energy*, **89**: 45.

Wilson DW, Lobo WE and Hottel HC (1932) Heat Transmission in Radiant Sections of Tube Stills, *Ind Eng Chem*, **24**: 486.

Chapter 13

Pumping and Compression

Liquids and gases need to be pressurized for transfer to achieve the required pressure of the destination, overcome the pressure drops in equipment and pipelines through which the material needs to be transferred to reach its destination and to overcome the change in elevation. The cost of a pressure increase for liquids in *pumps* is usually small relative to that for gases in *compressors*. On the other hand, to increase the pressure of material in the gas phase in a compressor tends to have a high capital cost and large power requirements, leading to high operating costs.

13.1 Pressure Drops in Process Operations

When transporting material through the process, the pressure drop through reactors, separators, heat transfer equipment, control valves, and other devices must be overcome.

For gas-phase reactions, the pressure drop through the reactor is usually less than 10% of the inlet pressure (Rase, 1990). The pressure drop through trickle-bed reactors is usually less than 1 bar. A value of 0.5 bar is often a reasonable first estimate for packed and trickle-bed reactors, although pressure drops can be higher. The pressure drop through fluidized-bed reactors is usually between 0.02 and 0.1 bar.

A pressure drop across distillation involves a pressure drop across each tray of typically 0.007 bar. A pressure drop across distillation packing is normally lower than for distillation trays for the same separation. Structured packing normally has a pressure drop less than 0.003 bar·m^{-1}.

Heat exchangers for liquids normally have a pressure drop in the range 0.35 to 0.7 bar (see Chapter 12). For gases, heat exchangers have a pressure drop typically between 1 bar for high-pressure gases (10 bar and above), down to 0.01 bar for gases under vacuum conditions (see Chapter 12).

Chemical Process Design and Integration, Second Edition. Robin Smith.

Companion Website: www.wiley.com/go/smith/chemical2e

13.2 Pressure Drops in Piping Systems

Consider the pumping of a liquid between two vessels, as illustrated in Figure 13.1a. If the fluid is assumed to be incompressible and the change in kinetic energy from inlet to outlet is neglected, then the pressure change required to be delivered by the pump is given by:

$$\text{Pump head} = \frac{\Delta P}{\rho g} = \frac{(P_2 - P_1)}{\rho g} + \Delta z + \sum h_L \tag{13.1}$$

where ΔP = pressure increase across the pump (N·m^{-2} = kg·m^{-1}·s^{-2})
ρ = fluid density (kg·m^{-3})
g = gravitational acceleration (9.81 m·s^{-2})
P_2 = pressure at the liquid surface in the discharge vessel (N·m^{-2})
P_1 = pressure at the liquid surface in the feed vessel (N·m^{-2})
Δz = change in elevation (m)
$\sum h_L$ = sum of the frictional head losses for straight pipes and pipe fittings (m)

Pipe fittings include bends, isolation valves, control valves, orifice plates, expansions and reductions. If the material to be transferred is a gas, as illustrated in Figure 13.1b, the change in elevation can normally be neglected. If the gas is assumed to be incompressible and the change in kinetic energy from inlet to outlet is neglected, then the pressure change required to be delivered by the compressor is given by:

$$\text{Compressor head} = \frac{\Delta P}{\rho g} = \frac{(P_2 - P_1)}{\rho g} + \sum h_L \tag{13.2}$$

For straight pipes (Coulson and Richardson, 1999):

$$h_L = 2c_f \frac{L}{d_I} \frac{v^2}{g} \tag{13.3}$$

where c_f = Fanning friction factor
L = pipe length (m)
d_I = internal diameter of pipe (m)
v = mean velocity in the pipe (m·s^{-1})

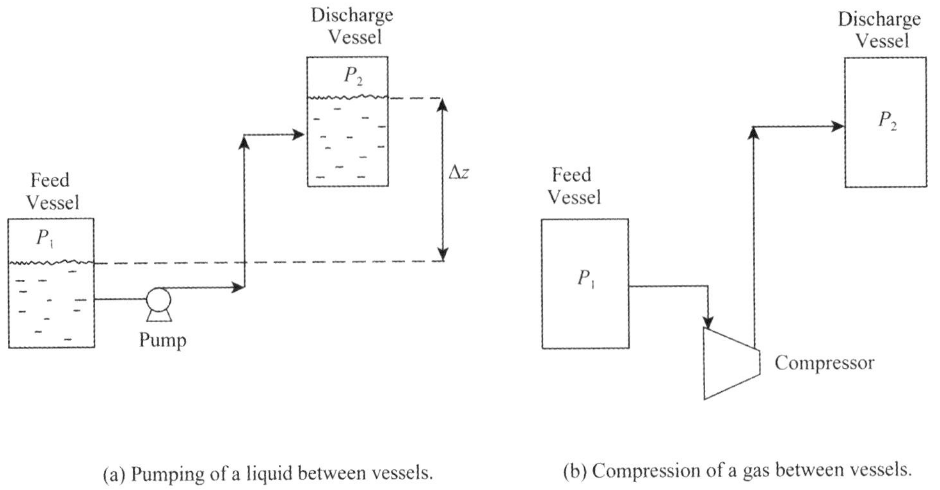

Figure 13.1

Pressure change for a piping system for a liquid transfer.

The Fanning friction factor is given by (Hewitt, Shires and Bott, 1994):

$$c_f = \frac{16}{Re} \qquad Re < 2000 \tag{13.4}$$

$$c_f = 0.046Re^{-0.2} \qquad 2000 < Re < 20,000 \tag{13.5}$$

$$c_f = 0.079Re^{-0.25} \qquad Re > 20,000 \tag{13.6}$$

where $Re = \frac{\rho d_I v}{\mu}$

μ = fluid viscosity ($N{\cdot}s{\cdot}m^{-2}$)

It should be noted that Equations 13.5 and 13.6 apply to smooth pipes, whereas the pipes used for transmission of fluids usually have some surface roughness, which increases the friction factor. However, for short fluid transmission pipes, the overall frictional pressure drop is usually dominated by the frictional pressure drop in the pipe fittings (valves, bends, etc.). Thus, for short transmission pipes, there is little point in calculating the straight pipe frictional pressure drop accurately. If the transmission pipe is long (>100 m) and straight, then the Fanning friction factor can be correlated as (Coulson and Richardson, 1999):

$$\frac{1}{\sqrt{c_f}} = -1.77 \ln\left[0.27\frac{\varepsilon}{d_I} + \frac{1.25}{Re\sqrt{c_f}}\right] \tag{13.7}$$

where ε = surface roughness (m)

Table 13.1 gives some typical values of surface roughness (Coulson and Richardson, 1999).

Many different types of fittings are used in process pipelines. When transporting fluids, valves are required for:

- starting and stopping flow (*block* valve),
- regulating the flow (*control valve*),
- preventing reversal of flow (*nonreturn* or *check* valve).

Figure 13.2 shows some of the more common types of valve used in the process industry.

1) *Globe valve.* A typical globe valve is illustrated in Figure 13.2a. A screw mechanism is used to move a *plug* or *disc* against a *seat*. Figure 13.2a shows both a *plug disc* and a *composite disc* design. The composite disc design uses a seating material incorporated into the disc and has some advantages in the way the disc is guided to the seat. Many different designs of seating arrangements are available. Globe valves can be used to regulate the flow, as well as starting and stopping the flow.
2) *Butterfly valve.* A typical butterfly valve is illustrated in Figure 13.2b. The angle of a disc is manipulated relative to the flow. When the disc is perpendicular to the flow, a seal is created between the disc and the lining of the valve to prevent flow. When the disc is rotated through a quarter turn, the disc is in line with the flow and the valve is open. Butterfly valves can be used to regulate the flow, as well as starting and stopping the flow. Their use is normally restricted to larger diameter pipelines.
3) *Gate valve.* A typical gate valve is illustrated in Figure 13.2c. This operates by a screw mechanism raising and lowering a circular gate in the path of the fluid. The gate faces can form a wedge shape or can be parallel. Gate valves are primarily used for starting and stopping flow, rather than regulating the flow.

Table 13.1

Surface roughness of pipes.

Pipe	ε (mm)
Drawn tubing	0.0015
Commercial steel	0.046
Cast iron	0.26
Concrete	0.3–3.0

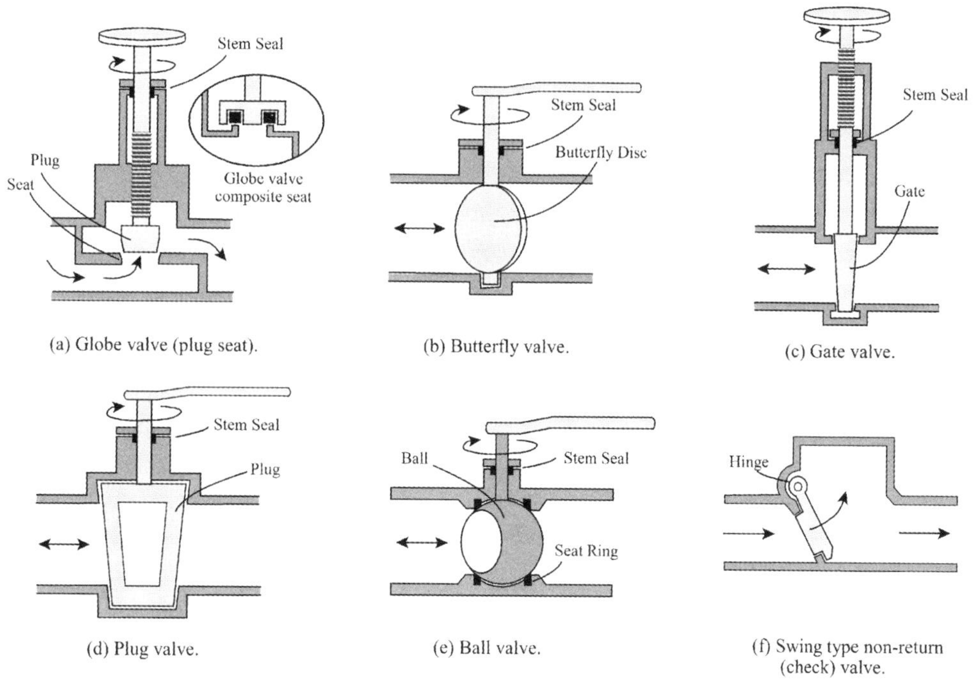

Figure 13.2
Valve types.

4) *Plug valve.* A typical plug valve is illustrated in Figure 13.2d. A cylindrical or more often conically tapered plug with a hollow port is rotated through a quarter turn to allow or prevent flow. The installation space is smaller than a gate valve and operation to open and close is simple and rapid. The tapered plug allows a tight shut-off.

5) *Ball valve.* A typical ball valve is illustrated in Figure 13.2e. A sphere with a cylindrical hollow port is rotated through a quarter turn to allow or prevent flow. Conventional ball valves have a port smaller than the pipe diameter. Full port valves with the same diameter of the pipe are also available, but are less common. Tight shut-off can be achieved by virtue of the seat rings, but the seat material creates temperature limitations. As with plug valves, ball valves require an installation space smaller than a gate valve and operation to open and close is simple and rapid.

6) *Nonreturn (check) valve.* Nonreturn valves are used purely to prevent reversal of flow, for example to prevent liquid from siphoning back from a tank if a pump is switched off. A swing-type nonreturn valve is illustrated in Figure 13.2f. The disc swings on a hinge either on to the seat to prevent reverse flow or swings to allow forward flow. Other designs of nonreturn valves use mechanisms such as forward flow moving a ball mounted in a slot moving away from a circular seat to open the valve with reversal of flow moving the ball in the reverse direction to seal with the seat to prevent reverse flow.

Control valves are used to control flow, pressure, temperature and liquid level by opening and closing in response to signals received from controllers. The opening or closing of control valves occurs remotely by an *actuator*. Actuators can be pneumatic or electrical. Most control valves in the process industry use a globe valve design. The plug and seat are shaped to vary the opening according to the desired control characteristics. Figure 13.3 illustrates a pneumatically operated control valve. An air pressure signal that varies between 0.2 and 1 bar causes the valve to open or close.

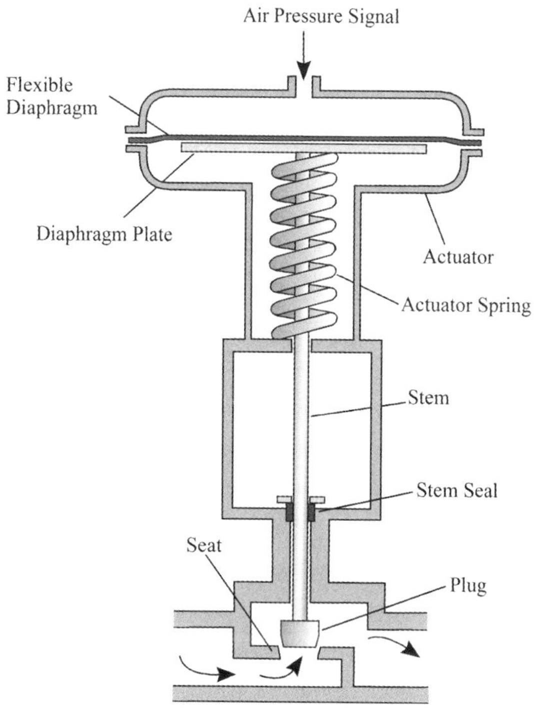

Figure 13.3
Schematic arrangement of a pneumatically operated control value.

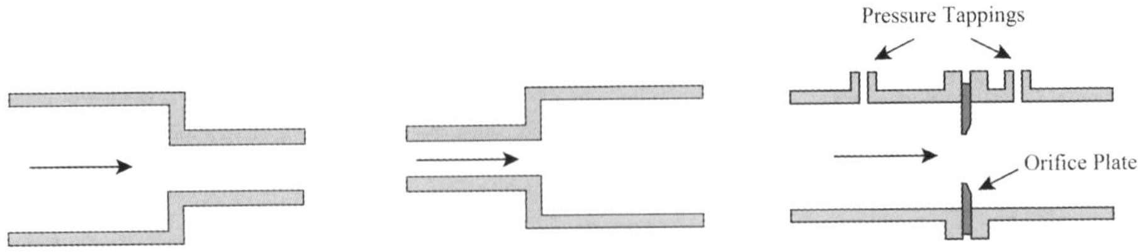

(a) Sudden contraction. (b) Sudden expansion. (c) Orifice plate.

Figure 13.4
Pipe fittings.

For an electrical actuator using an electric motor, an electrical signal between 0 and 10 V or between 4 and 20 mA causes the electric motor to open or close the valve. If there is failure, the valve must be designed such that the most probable failure mode results in the safest condition, known as the *fail safe* condition. The fail safe condition might be fully closed, fully open or to maintain the current valve position to maintain the current operation. The valve design in Figure 13.3 will open if the air signal fails. A different arrangement with the spring and diaphragm inverted can create a design that closes if the air signal fails. Many other control valve designs are available with different actuators.

In addition to valves, other fittings include sudden contraction, sudden enlargement, bends and orifice plates used for flow measurement (Figure 13.4). The head loss in the pipe fittings can be correlated as (Coulson and Richardson, 1999):

$$h_L = c_L \frac{v^2}{2g} \qquad (13.8)$$

where c_L = loss coefficient (–)

For laminar flow:

$$c_L = f(Re, \text{geometry of fitting})$$

For turbulent flow:

$$c_L = f(\text{geometry of fitting})$$

Table 13.2 gives some typical values of the loss coefficient for valves and other fittings (Perry, 1997). It should be noted that values for loss coefficient for the same fitting, but from different manufacturers, will vary as a result of differences in geometry. Table 13.3 gives head losses for sudden contractions, sudden expansions and orifice plates. Note that the formula for orifice plates in Table 13.3 relates to the overall pressure drop and not the pressure drop between the pressure tappings used to measure the flowrate.

In preliminary design, the fluid transmission lines can be designed on the basis of an assumed fluid velocity. For nonviscous liquids ($\mu < 10\,\text{mN·s·m}^{-2} = \text{cP}$), a pipe velocity of 1 to 2 m·s^{-1} is normally used. For viscous fluids, the velocity may be

Table 13.2
Loss coefficients for various pipe fittings.

Pipe fitting	Laminar flow c_L	Turbulent flow c_L	Correction for partial closure of valve in turbulent flow (α = fraction open)
Bend (standard)	$\frac{840}{Re}(Re < 1100)$	0.8	–
Globe valve (plug disc)	$\frac{70}{Re^{0.26}}(Re < 2700)$	9.0	$\frac{c_L}{\alpha^{1.84}}$
Globe valve (composite seat)	$\frac{100}{Re^{0.33}}(Re < 5000)$	6.0	$\frac{c_L}{\alpha^{0.5}}$
Gate valve	$\frac{1200}{Re}(Re < 6000)$	0.2	$\frac{c_L}{\alpha^{3.7}}$
Plug valve	–	0.05	–
Ball valve	–	0.05	–
Butterfly valve	–	0.24	–
Nonreturn valve (swing)	$\frac{1200}{Re^{0.86}}(Re < 1700)$	2.0	–

Table 13.3

Head losses in sudden contractions, sudden expansions and orifice plates.

Fitting	h_L
Sudden contraction	$\left[0.5\left(1-\frac{A_2}{A_1}\right)\right]\left(\frac{v_2^2}{2g}\right)$
Sudden expansion	$\left[1-\frac{A_1}{A_2}\right]^2\left(\frac{v_1^2}{2g}\right)$
Orifice plate	$\frac{1}{c_D^2}\left[1-\frac{A_O}{A}\right]\left[\left(\frac{A}{A_O}\right)^2-1\right]\left(\frac{v^2}{2g}\right)$

where v_1, v_2 = upstream and downstream velocities ($m{\cdot}s^{-1}$)
A_1, A_2 = upstream and downstream pipe areas (m^2)
g = acceleration due to gravity ($9.81\ m{\cdot}s^{-2}$)
A, A_O = areas of pipe and orifice (m^2)
v = velocity in the pipe ($m{\cdot}s^{-1}$)
c_D = 0.62 for preliminary design

constrained by the allowable pressure drop or shear degradation of the fluid (e.g. large molecules breaking down into smaller molecules from high shear rates). Typical values are given in Table 13.4.

For gases and vapors, typical fluid velocities are in the range of 15 to 30 $m{\cdot}s^{-1}$.

Table 13.4

Typical fluid velocities for viscous liquids.

Viscosity ($mN{\cdot}s{\cdot}m^{-2}$ = cP)	Velocity ($m{\cdot}s^{-1}$)
50	0.5–1.0
100	0.3–0.6
1000	0.1–0.3

The fluid velocity must take account of standard pipe sizes. Pipe sizes are specified by a nominal pipe size (NPS), or nominal diameter (DN), and a schedule. The larger the schedule number, the thicker the pipe wall. Table 13.5 gives dimensions for a number of commonly used pipe sizes. The schedule of pipe required depends on the maximum internal pressure and the appropriate corrosion allowance. For steam systems, the smallest schedule normally used is Schedule 40.

If the piping layout is known, then the above correlations can be used to estimate the pressure drop through the pipes. This might be the case, for example, in a retrofit situation. In preliminary design, it might be necessary to make some allowance for the cost of pumping and compression without knowledge of the piping layout. If this is the case, then it is not difficult to make a first estimate of the distances required for transportation of the fluid. What is uncertain

Table 13.5

Commonly used standard pipe sizes for steel pipes.

DN	Outside diameter (mm)	Inside diameter (mm)					
		Schedule					
		5	10	30	40	80	160
15	21.34	18.04	17.12	16.51	15.80	13.87	11.79
20	26.67	23.37	22.45	21.84	20.93	18.85	15.54
25	33.40	30.10	27.86	27.61	26.64	24.31	20.70
32	42.16	38.86	36.62	36.22	35.05	32.46	29.46
40	48.26	44.96	42.72	41.91	40.89	38.10	33.99
50	60.33	57.03	54.79	53.98	52.51	49.26	42.91
65	73.03	68.81	66.93	63.48	62.72	59.01	53.98
80	88.90	84.68	82.80	79.35	77.93	73.66	66.65
100	114.30	110.1	108.2	104.8	102.3	97.18	87.33
150	168.28	162.7	161.5		154.1	146.3	131.8
200	219.08	213.5	211.6	205.0	202.7	193.7	173.1
250	273.05	266.2	264.7	257.5	254.5	242.9	215.9
300	323.85	315.5	314.7	307.1	304.8	288.9	257.2

is the pipe fittings that will be involved. Some general guidelines are therefore required in order to make a first estimate of the pressure drop to take account of the pipe fittings:

- Vessels will often have isolation valves (but this varies between different sectors of the industry).
- Equipment that needs to be taken out of service for maintenance will normally have an isolation valve on each side. This will include pumps, compressors and control valves.
- In some situations, for the sake of process safety, a more secure isolation is required on inlets and outlets. In this case, double block and bleed can be used. This involves two isolation valves with a vent valve between them. By closing the two isolation valves and opening the vent valve between them, any leakage past the upstream valve will go through the vent instead of potentially leaking through the downstream valve.
- Pumps will normally have a nonreturn (check) valve to prevent reversal of flow.
- Flow control will usually be based on the measurement of pressure drop across an orifice plate.
- A line going between vessels or between a vessel and a pipe junction will typically have at least three bends when the pipe design is finalised.

For example, suppose a liquid is being pumped from one vessel into another vessel using a pump and under the action of flow control using an orifice plate to measure the flowrate. The head losses involved will typically be:

- a sudden contraction from the feed vessel into the discharge line;
- an isolation valve for the vessel;
- two isolation valves and a check valve for the pump;
- an orifice plate for flow measurement;
- a control valve;
- two isolation valves for the control valve;
- typically three pipe bends from changes in direction of the pipes;
- an isolation valve for the discharge vessel;
- a sudden expansion for the fluid entering the discharge vessel.

Example 13.1 In a new design, water is to be pumped between two vessels both at atmospheric pressure separated by an estimated distance of 30 m under the action of flow control. An increase in elevation of 5 m is also estimated. The flowrate of water is $100\ m^3 \cdot h^{-1}$, its viscosity is $0.8\ mN \cdot s \cdot m^{-2}$ (equal to centipoise) and its density is $993\ kg \cdot m^{-3}$. Estimate the pressure drop required to be produced by the pump.

Solution First determine the pipe diameter from an assumed velocity of say $2\ m \cdot s^{-1}$. The area of the pipe (A) is given by:

$$A = 100 \times \frac{1}{3600} \times \frac{1}{2}$$
$$= 0.01389\ m^2$$

The internal diameter (d_I) is given by

$$d_I = \sqrt{\frac{4A}{\pi}}$$
$$= \sqrt{\frac{4 \times 0.01389}{\pi}}$$
$$= 0.133\ m$$

This needs to be rounded to the next largest standard pipe diameter of internal diameter. Assuming a Schedule 40 pipe this would be 0.154 m. The actual fluid velocity is therefore:

$$v = 100 \times \frac{1}{3600} \times \frac{4}{\pi \times 0.154^2}$$
$$= 1.49\ m \cdot s^{-1}$$

Reynolds number for the straight pipes is:

$$Re = \frac{\rho d_I v}{\mu}$$
$$= \frac{993 \times 0.154 \times 1.49}{0.8 \times 10^{-3}}$$
$$= 2.85 \times 10^5$$

Head loss in the straight pipe sections:

$$h_L = 2c_f \frac{L}{d_I} \frac{v^2}{g}$$
$$= 2 \times \frac{0.079}{Re^{0.25}} \times \frac{30}{0.154} \times \frac{1.49^2}{9.81}$$
$$= 0.30\ m$$

For the isolation valves, take gate valves fully open, one for each vessel, two for the pump and two for the control valve:

$$h_L = 6 \times c_L \frac{v^2}{2g}$$
$$= 6 \times 0.2 \times \frac{1.49^2}{2 \times 9.81}$$
$$= 0.14\ m$$

Assume a nonreturn (check) valve for the pump:

$$h_L = c_L \frac{v^2}{2g}$$
$$= 2 \times \frac{1.49^2}{2 \times 9.81}$$
$$= 0.23\ m$$

To estimate the control valve, take a globe valve to be half open:

$$h_L = \frac{c_L}{\alpha^{1.84}} \frac{v^2}{2g}$$
$$= \frac{9}{0.5^{1.84}} \frac{1.49^2}{2 \times 9.81}$$
$$= 3.65\ m$$

Assume three bends:

$$h_L = 3 \times 0.8 \times \frac{1.49^2}{2 \times 9.81}$$
$$= 0.27\ m$$

Assume an orifice plate to measure the flowrate with a diameter ratio of 0.4 and discharge coefficient of 0.62:

$$
\begin{aligned}
h_L &= \frac{1}{c_D^2}\left[1 - \frac{A_O}{A}\right]\left[\left(\frac{A}{A_O}\right)^2 - 1\right]\frac{v^2}{2g} \\
&= \frac{1}{0.62^2}\left[1 - 0.4^2\right]\left[\left(\frac{1}{0.4}\right)^4 - 1\right] \times \frac{1.49^2}{2 \times 9.81} \\
&= 9.41 \text{ m}
\end{aligned}
$$

The entrance loss from the feed vessel is given by:

$$
\begin{aligned}
h_L &= 0.5\left[1 - \frac{A_2}{A_1}\right]\frac{v_2^2}{2g} \\
&= 0.5(1 - 0)\frac{1.49^2}{2 \times 9.81} \\
&= 0.06 \text{ m}
\end{aligned}
$$

The exit loss into the receiving vessel is given by:

$$
\begin{aligned}
h_L &= \left[1 - \frac{A_1}{A_2}\right]^2\frac{v_1^2}{2g} \\
&= [1 - 0]^2\frac{1.49^2}{2 \times 9.81} \\
&= 0.11 \text{ m}
\end{aligned}
$$

$$
\frac{\Delta P}{\rho g} = \frac{(P_2 - P_1)}{\rho g} + \Delta z + \sum h_L
$$

Because both the feed and discharge tanks are at atmospheric pressure $(P_2 - P_1) = 0$:

$$
\begin{aligned}
\frac{\Delta P}{\rho g} &= [0+5+(0.30+0.14+0.23+3.65+0.27+9.41+0.06+0.11)] \\
&= 19.17 \text{m} \\
\Delta P &= 993 \times 9.81 \times 19.17 \\
&= 1.87 \times 10^5 \text{N} \cdot \text{m}^{-2}
\end{aligned}
$$

13.3 Pump Types

Pumps can be classified into two general types:

1) *Positive displacement.* In positive displacement pumping energy is transferred to the liquid by trapping a fixed volume and forcing the trapped volume to the pump discharge in arrangements such as reciprocating pistons, or rotary motion of gears, screws or vanes. Positive displacement pumps deliver a definite quantity for each stroke or partial rotation of the device. They are used when the liquid has a high viscosity, low flowrate, or a combination of the two.
2) *Dynamic.* In dynamic pumps energy is transferred to the liquid by means of vanes mounted on a rotating shaft. The liquid enters along or near the rotating shaft and is accelerated by the rotation of the vanes. This imparts kinetic energy to the liquid, which is transformed to pressure energy.

The most common type of pump used in chemical processes is *centrifugal*. This is a form of dynamic pump in which an *impeller* consisting of curved vanes rotates on a shaft. The arrangement is illustrated in Figure 13.5a. The liquid enters near the rotating axis,

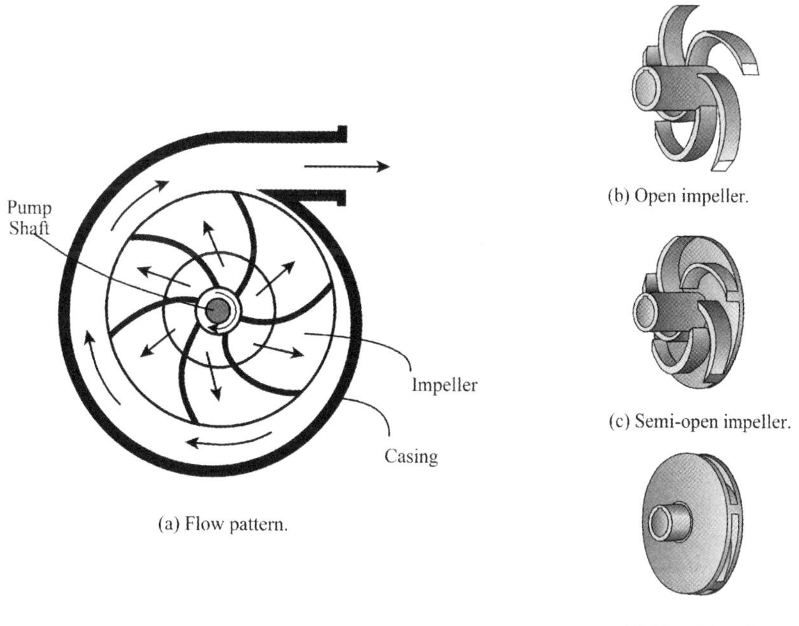

(a) Flow pattern.

(b) Open impeller.

(c) Semi-open impeller.

(d) Closed impeller.

Figure 13.5
Centrifugal pump features.

is accelerated and flows radially outwards into a *diffuser* or *casing* from which it exits. The kinetic energy imparted to the liquid by the impeller is transformed to pressure when the fluid reaches the pump casing. Centrifugal pumps deliver a volume that is dependent on the discharge pressure and the energy added. Figure 13.5b shows an *open* impeller design. A *semi-open* impeller is illustrated in Figure 13.5c, which uses a *shroud* to protect the impeller, vanes. A *closed* impeller using shrouds on both sides is illustrated in Figure 13.5d. Open and semi-open impellers can operate with a higher efficiency than a closed impeller, but can be more susceptible to abrasive wear. Open impellers tend to be used for pumping liquids with solids, where a closed impeller might be susceptible to blockage. The vanes can have a backward curvature (as in Figure 13.5) for high flowrates or a forward curve for a high liquid pressure head or straight radial for either service. Multiple impellers can be used in *multistage pumps*. When starting up, the pump must be filled with liquid, or *primed*, in order to operate. To assist in priming the pump, centrifugal pumps are, where possible, located below the level of the feed to the pump. If this is not possible, an auxiliary pump might be required to supply liquid to the pump inlet for start-up.

Most industrial applications favor the use of centrifugal pumps as they can handle a wide range of fluids and a wide range of pumping conditions at low cost compared with positive displacement devices. The remainder of the discussion about pumps will be restricted to centrifugal pumps, as this is by far the most common type used.

13.4 Centrifugal Pump Performance

The performance of the pump can be characterized by the head delivered, power required and efficiency. Efficiency is a measure of the useful work transferred to the liquid and is defined as (neglecting any change in kinetic energy):

$$\eta = \frac{\text{Power transferred to the liquid}}{\text{Shaft power}} = \frac{F\Delta P}{W} \tag{13.9}$$

where W = shaft power input ($N{\cdot}m{\cdot}s^{-1} = J{\cdot}s^{-1} = W$)
F = volumetric flowrate ($m^3{\cdot}s^{-1}$)
ΔP = pressure drop across the pump ($N{\cdot}m^{-2}$)

The efficiency is normally quoted by pump manufacturers in terms of the performance on water.

Figure 13.6 shows typical performance curves for a centrifugal pump. The speed of the impeller is fixed. Figure 13.6a shows that the head delivered by the pump decreases as the flowrate increases, with the head decreasing more rapidly as the flowrate increases. Figure 13.6b shows that the power increases as the flowrate increases. Figure 13.6c shows that the efficiency of the pumps goes through a maximum (best efficiency point, or BEP) as the flowrate increases.

How a given pump performs in practice depends on the match with the system pressure drop. Figure 13.7 shows the variation of system head with flowrate. At zero flowrate the system head is the sum of the pressure difference between the feed and discharge vessels and the difference in elevation. As the flowrate increases, the system head increases with the square of the velocity.

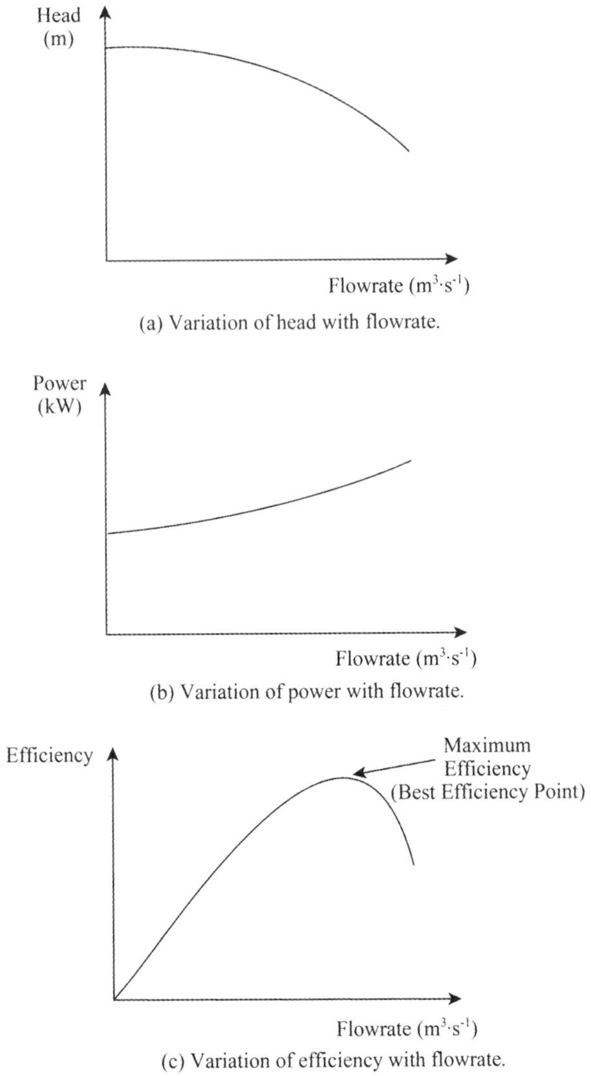

Figure 13.6
Performance characteristics of centrifugal pumps.

The operating point will be at the intersection between the curves for the system head and pump head. If the system is designed well, this point of intersection will correspond with the best efficiency point for the pump. In practice, the correct flowrate is ensured by manipulating the system head curve using a control valve. Changing the opening of the control valve causes the system head curve to pivot around the zero flow point.

Another point that needs attention is the potential for the pressure in the pump to fall below the vapor pressure of the liquid at the pumping temperature. If the pressure in the pump falls locally below the vapor pressure then bubbles, or cavities, of vapor will be formed. These bubbles will collapse when reaching a region of higher pressure. The phenomenon is known as *cavitation*.

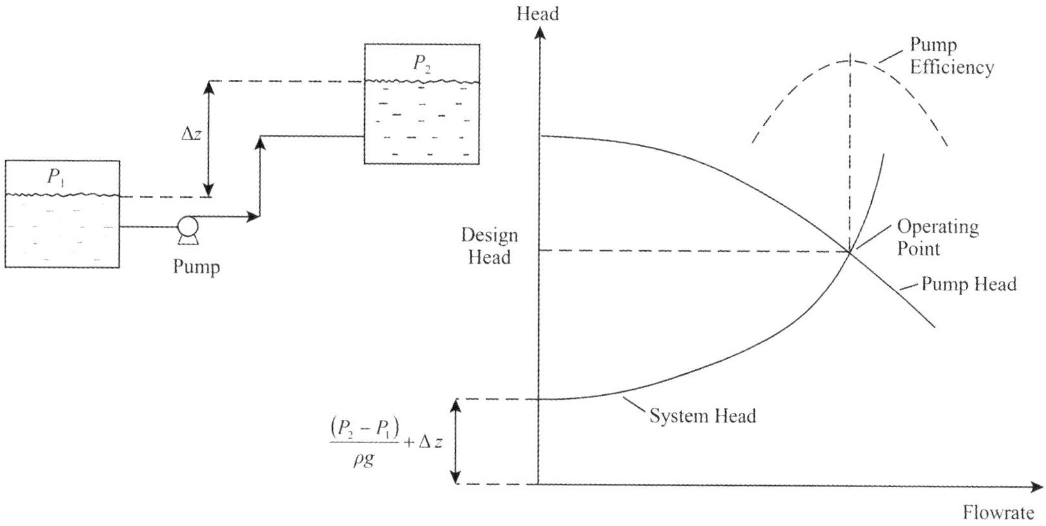

Figure 13.7

Operating point for a centrifugal pump.

Cavitation causes a deterioration in the pump performance, but, more importantly, the collapse of the bubbles (cavities) can cause damage to the pump, particularly to the impeller.

The net positive suction head (NPSH) is the pressure at the pump suction above the vapor pressure of the liquid. Referring to Figure 13.8, the net positive suction head available ($NPSH_A$) can be expressed as a head and defined as:

$$NPSH_A = \left(\frac{P_1}{\rho g} + z_1 - h_L\right) - \frac{P^{SAT}}{\rho g} \tag{13.10}$$

where $NPSH_A$ = net positive suction head available (m)
P_1 = pressure above the liquid in the feed vessel ($N{\cdot}m^{-2}$)
ρ = liquid density ($kg{\cdot}m^{-3}$)
g = gravitational constant ($9.81\ m{\cdot}s^{-2}$)
z_1 = elevation of the liquid in the feed vessel above the pump (m)
h_L = friction head loss in the suction piping (m)
P^{SAT} = saturated liquid vapor pressure ($N{\cdot}m^{-2}$)

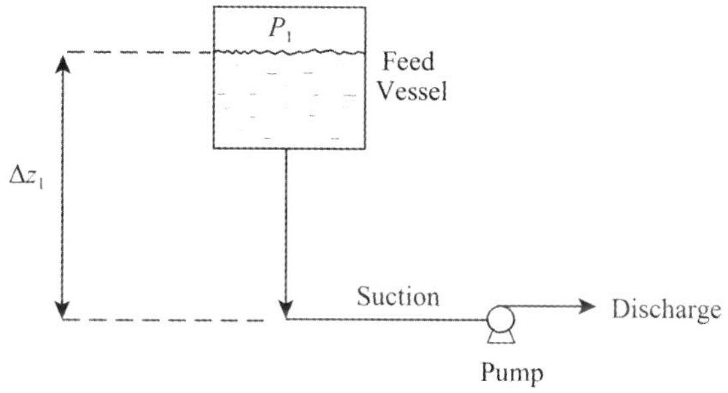

Figure 13.8

Net positive suction head available.

The net positive suction head required $NPSH_R$ by a particular pump is defined by the manufacturer and depends on the design of the pump. Thus, the $NPSH_A$ calculated from Equation 13.10 should be above the value stated to be required by the pump manufacturer. $NPSH_A$ is most often a problem when pumping boiling liquids (e.g. in distillation operations) and pumping liquids from sumps at a low elevation.

If the $NPSH_A$ available from the system is lower than that specified by the pump manufacturer, then this is most often overcome by changing the layout (particularly the elevation). The pressure of the feed vessel might also be increased. Alternatively, a different pump design can be chosen, which is often a slower running pump.

Now consider the main parameters in the design of centrifugal pumps and how they relate to the selection of a particular pump. The power W of a centrifugal pump depends on the density ρ of the liquid, the rotational speed of the impeller N_{IMP}, the impeller diameter D_{IMP}, the change in head across the pump Δh, the volumetric flowrate F and the gravitational acceleration g. Taking Δh and g together:

$$W = f(\rho, N_{IMP}, D_{IMP}, g\Delta h, F) \tag{13.11}$$

A dimensional analysis can be performed on the variables (Coulson and Richardson, 1999). Given the six variables in Equation 13.11 and given that there are three fundamental variables (mass, length, time), then three independent dimensionless groups can be formed. First form a dimensionless group with F and $\rho N_{IMP} D_{IMP}$:

$$F = f(\rho, N_{IMP}, D_{IMP}) = \Pi_1 \rho^a N_{IMP}^b D_{IMP}^c \tag{13.12}$$

where Π_1 = first dimensionless group
a, b, c = integer exponents

13

Writing Equation 13.12 in terms of mass M, length L and time T:

$$M^3T^{-1} = \left(ML^{-3}\right)^a\left(T^{-1}\right)^b\left(L^1\right)^c \tag{13.13}$$

From Equation 13.13, $a=0$, $b=1$ and $c=3$, giving:

$$F = \Pi_1 \rho N_{IMP} D_{IMP}^3 \tag{13.14}$$

Thus:

$$\Pi_1 = \frac{F}{N_{IMP}D_{IMP}^3} = \text{Flow coefficient} \tag{13.15}$$

Repeating the procedure between $g\Delta h$ and $\rho N_{IMP} D_{IMP}$ gives:

$$\Pi_2 = \frac{g\Delta h}{N_{IMP}^2 D_{IMP}^2} = \text{Head coefficient} \tag{13.16}$$

Repeating the procedure between W and $\rho N_{IMP} D_{IMP}$ gives:

$$\Pi_3 = \frac{W}{\rho N_{IMP}^3 D_{IMP}^5} = \text{Power coefficient} \tag{13.17}$$

These dimensional groups are useful to compare different designs.

To find the theoretical speed for a unit discharge under a unit head it is necessary to scale down the operating volume of the pump. This is done by assuming that, when scaling down, the pump is kept geometrically similar with the linear dimensions kept proportional to the impeller diameter. To develop the required relationship the flow coefficient in Equation 13.15 and the head coefficient in Equation 13.16 are rearranged to be explicit in diameter and then equated:

$$\left(\frac{F}{N_{IMP}\Pi_2}\right)^{1/3} = \left(\frac{g\Delta h}{\Pi_3 N_{IMP}^2}\right)^{1/2} \tag{13.18}$$

Rearranging Equation 13.18:

$$\frac{\Delta h^{1/2}}{F^{1/3}N_{IMP}^{2/3}} = \frac{\Pi_3^{1/2}}{\Pi_2^{1/3} g^{1/2}} = \text{constant} \tag{13.19}$$

Thus:

$$\frac{F^{1/3}N_{IMP}^{2/3}}{\Delta h^{1/2}} = \text{constant} \tag{13.20}$$

Normalizing to reach power of unity for N_{IMP}:

$$\frac{F^{1/2}N_{IMP}}{\Delta h^{3/4}} = \text{constant} = N_S \tag{13.21}$$

where N_S = specific speed

N_S is a constant for geometrically similar pumps. It should be noted that N_S defined in this way is a dimensional quantity. Manufacturers quote specific speed data at the best efficiency point (see Figure 13.6). Values for centrifugal pumps for N_S in rpm, F in $m^3 \cdot h^{-1}$ and Δh in m are typically in the range 600 to 4500.

From the definition of specific speed, pumps with high specific speed develop high flows and low heads. Pumps with low specific speeds develop high heads and low flows. In general, higher speed pumps tend to be more susceptible to damage by abrasive material, more susceptible to cavitation and tend to be less reliable. On the other hand, higher-speed pumps tend to be physically smaller for the same duty, which tends to lower the capital cost.

The three dimensionless groups defined in Equations 13.15 to 13.17 can also be used to estimate the performance changes between two pumps from a family of geometrically similar pumps. From Equation 13.15:

$$\frac{F_2}{F_1} = \frac{N_{IMP2}}{N_{IMP1}}\left(\frac{D_{IMP2}}{D_{IMP1}}\right)^3 \tag{13.22}$$

This can be used to predict the new flowrate for changes in pump speed and impeller diameter. From Equation 13.16:

$$\frac{\Delta h_2}{\Delta h_1} = \left(\frac{N_{IMP2}}{N_{IMP1}}\right)^2\left(\frac{D_{IMP2}}{D_{IMP1}}\right)^2 \tag{13.23}$$

This can be used to predict the new pump head for design changes in pump speed and impeller diameter. From Equation 13:17:

$$\frac{W_2}{W_1} = \frac{\rho_2}{\rho_1}\left(\frac{N_{IMP2}}{N_{IMP1}}\right)^3\left(\frac{D_{IMP2}}{D_{IMP1}}\right)^5 \tag{13.24}$$

This can be used to predict the new pump power for changes in fluid, pump speed and impeller diameter.

Equations 13.22 to 13.24 can be used to compare the performance of a family of geometrically similar pumps. The scaling laws for a specific pump will not necessarily be the same as the scaling laws for a family of geometrically similar pumps. Consider first a change in speed for a specific pump. The vanes in the impeller mean that if the speed of the impeller is changed, the flow is changed by the same amount. Thus, for an existing pump with a fixed impeller diameter:

$$\frac{F_2}{F_1} = \frac{N_{IMP2}}{N_{IMP1}} \tag{13.25}$$

Next consider the influence of change of speed on the head delivered. The friction head loss is proportional to the velocity squared. If the pipe system is unchanged, this means the frictional loss is proportional to the volume squared:

$$\frac{\Delta h_2}{\Delta h_1} = \left(\frac{F_2}{F_1}\right)^2 \tag{13.26}$$

Combining with Equation 13.25 gives:

$$\frac{\Delta h_2}{\Delta h_1} = \left(\frac{N_{IMP2}}{N_{IMP1}}\right)^2 \tag{13.27}$$

Next consider the influence of change in speed on power. Because the pressure loss in the existing pipework is proportional to velocity squared:

$$\frac{\Delta P_2}{\Delta P_1} = \left(\frac{F_2}{F_1}\right)^2 \tag{13.28}$$

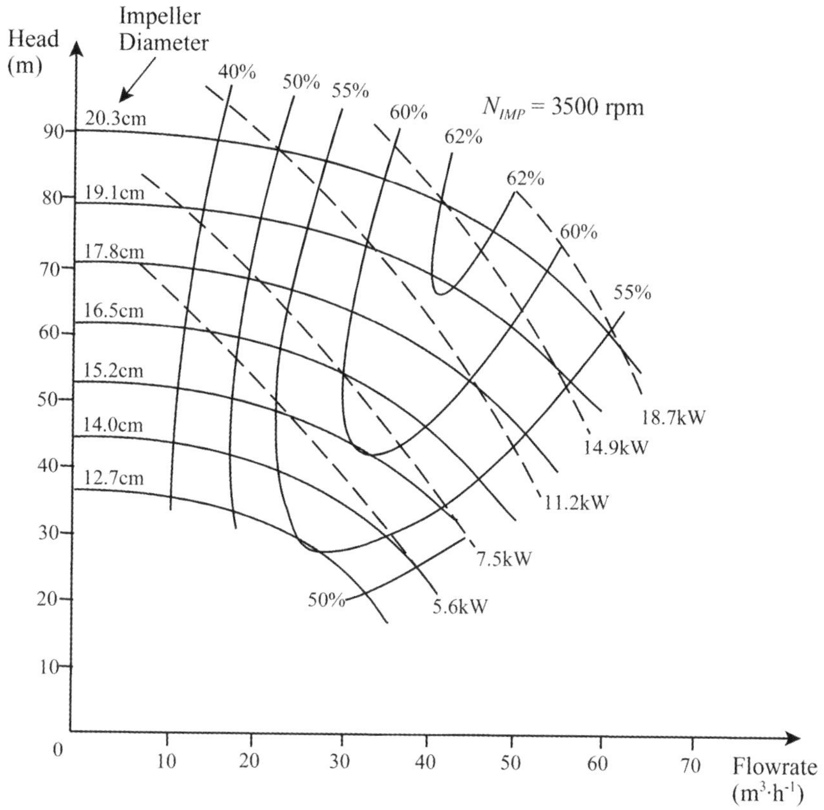

Figure 13.9

Typical performance curves for a design of centrifugal pump.

Thus:

$$\frac{F_2 \Delta P_2}{F_1 \Delta P_1} = \frac{W_2}{W_1} = \left(\frac{F_2}{F_1}\right)^3 = \left(\frac{N_{IMP2}}{N_{IMP1}}\right)^3 \tag{13.29}$$

The scaling of a specific pump has so far considered change in speed. Suppose now that the speed is held constant and the impeller diameter is changed. Changing the impeller diameter means that the velocity at the impeller tip changes by $D_{IMP}/2$. Thus, the change in impeller diameter has the same effect as a change in speed. These arguments can be summarized as follows:

For a specific pump with impeller diameter held constant:

$$\frac{F_2}{F_1} = \frac{N_{IMP2}}{N_{IMP1}} \tag{13.30}$$

$$\frac{\Delta h_2}{\Delta h_1} = \left(\frac{N_{IMP2}}{N_{IMP1}}\right)^2 \tag{13.31}$$

$$\frac{W_2}{W_1} = \left(\frac{N_{IMP2}}{N_{IMP1}}\right)^3 \tag{13.32}$$

For a specific pump with impeller speed held constant:

$$\frac{F_2}{F_1} = \frac{D_{IMP2}}{D_{IMP1}} \tag{13.33}$$

$$\frac{\Delta h_2}{\Delta h_1} = \left(\frac{D_{IMP2}}{D_{IMP1}}\right)^2 \tag{13.34}$$

$$\frac{W_2}{W_1} = \left(\frac{D_{IMP2}}{D_{IMP1}}\right)^3 \tag{13.35}$$

These *similarity laws* or *affinity laws* allow design changes for a family of geometrically similar pumps or design changes to a specific pump to be explored.

Figure 13.9 shows performance curves for a particular design of centrifugal pump at a fixed speed. Different pump head curves are plotted at different impeller diameters. Power and efficiency curves are plotted for different conditions. The NPSH requirements might also be plotted. Pump efficiency is the highest when the largest impeller possible is installed in the pump casing. Pump efficiency falls when a smaller impeller is installed because of increased slippage between the impeller tip and the casing. However, it is not necessarily good practice to install the largest impeller when specifying a new pump. If the demands of the pump increase because of some future change in operation, one method to increase the capacity of an existing pump is to install a larger impeller. Therefore, choosing a combination of casing and impeller that allows a possible future increase in capacity might be desirable.

In the absence of a manufacturer's data, the pump efficiency can be estimated from (Branan, 1999):

$$\eta = 0.8 - 9.367\times10^{-3}\Delta h + 5.461\times10^{-5}\Delta h F - 1.514\times10^{-7}\Delta h F^2 + 5.820\times10^{-5}\Delta h^2 - 3.029\times10^{-7}\Delta h^2 F + 8.348\times10^{-10}\Delta h^2 F^2 \tag{13.36}$$

where η = pump efficiency
Δh = head developed by the pump (m) $15 > \Delta h > 90$
F = flowrate ($m^3 \cdot h^{-1}$) $20 > F > 230$

For flows below $20\,m^3 \cdot h^{-1}$ down to $5\,m^3 \cdot h^{-1}$ the efficiency can be approximated by taking the efficiency at $20\,m^3 \cdot h^{-1}$ and then subtracting the product of 7.9×10^{-4} and the difference between $20\,m^3 \cdot h^{-1}$ and the required flowrate (Branan, 1999).

13

Example 13.2 A centrifugal pump is required to produce a flowrate of 47 $m^3 \cdot h^{-1}$ with a head of 50 m. The performance of a pump is given in Table 13.6. The pump has a speed of 1250 rpm and an impeller diameter of 15 cm. Calculate the speed and impeller diameter of a geometrically similar pump to deliver the required flowrate and head at the best efficiency point.

Table 13.6

Performance of a centrifugal pump.

Flowrate ($m^3 \cdot h^{-1}$)	Head (m)	Efficiency (–)
0	104.4	0
25	104.0	28.0
50	103.2	47.8
75	99.9	59.9
100	92.4	65.0
125	78.8	63.6
150	57.2	56.3
175	25.7	43.5

Solution The pump characteristics are plotted in Figure 13.10. From the efficiency curve, the best efficiency point is at a flowrate of 107 $m^3 \cdot h^{-1}$ and a head of 89 m. Calculate the specific speed at this point:

$$N_S = \frac{N_{IMP} F^{1/2}}{\Delta h^{3/4}}$$
$$= \frac{1250 \times 107^{1/2}}{89^{3/4}}$$
$$= 446.2$$

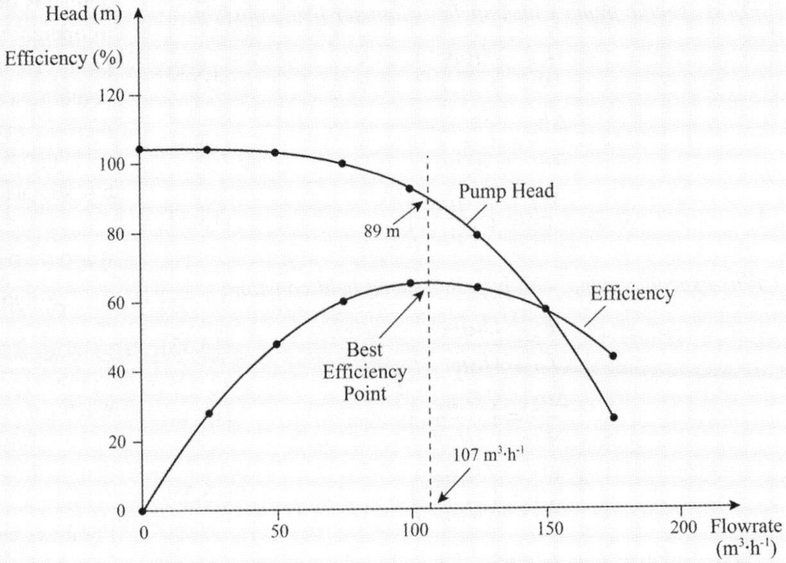

Figure 13.10

Pump head and efficiency curves for Example 13.3.

For a geometrically similar pump:

$$N_{IMP} = \frac{N_S \Delta h^{3/4}}{F^{1/2}}$$
$$= \frac{446.2 \times 50^{3/4}}{47^{1/2}}$$
$$= 1224 \text{ rpm}$$

The impeller diameter for a geometrically similar pump can be determined from Equation 13.22:

$$D_{IMP2} = D_{IMP1} \left(\frac{F_2}{F_1} \frac{N_{IMP1}}{N_{IMP2}} \right)^{1/3}$$
$$= 15 \left(\frac{47}{107} \times \frac{1250}{1224} \right)^{1/3}$$
$$= 11.5 \text{ cm}$$

Alternatively, from Equation 13.23:

$$D_{IMP2} = D_{IMP1} \frac{N_{IMP1}}{N_{IMP2}} \left(\frac{\Delta h_2}{\Delta h_1} \right)^{1/2}$$
$$= 15 \left(\frac{1250}{1224} \right) \left(\frac{50}{89} \right)^{1/2}$$
$$= 11.5 \text{ cm}$$

Example 13.3 A distillation column bottoms is essentially pure toluene and needs to be pumped out of the base of the column using a centrifugal pump at a flowrate of 20 t·h^{-1}. The pressure of the distillation column is 1.5 bar and the bottoms temperature is 125 °C. At this temperature the density of toluene is 764 kg·m^{-3} and viscosity is 0.00025 N·s·m^{-2}. The column has a diameter of 1.25 m. The pump suction pipe is DN80 Schedule 40 pipe with a straight pipe length of 3.5 m, two plug valves and one bend. The level of the liquid surface in the column bottom is 2.75 m above the pump. Calculate the net positive suction head available.

Solution First calculate the frictional loss in the suction pipework. The internal diameter of the pipe is given in Table 13.5 to be 0.0779 m.

$$\begin{aligned} v &= \frac{20 \times 1000}{3600} \times \frac{1}{764} \times \frac{4}{\pi \times 0.0779^2} \\ &= 1.53\ \text{m} \cdot \text{s}^{-1} \end{aligned}$$

Reynolds number for the straight pipes:

$$\begin{aligned} Re &= \frac{\rho d_I v}{\mu} \\ &= \frac{764 \times 0.0779 \times 1.53}{0.00025} \\ &= 3.64 \times 10^5 \end{aligned}$$

Head loss in the straight pipe sections:

$$\begin{aligned} h_L &= 2c_f \frac{L}{d_I} \frac{v^2}{g} \\ &= 2 \times \frac{0.079}{(3.64 \times 10^5)^{0.25}} \times \frac{3.5}{0.0779} \times \frac{1.53^2}{9.81} \\ &= 0.07\ \text{m} \end{aligned}$$

Head loss for two plug values:

$$\begin{aligned} h_L &= 2 \times c_L \frac{v^2}{2g} \\ &= 2 \times 0.05 \times \frac{1.53^2}{2 \times 9.81} \\ &= 0.01\ \text{m} \end{aligned}$$

Head loss for one bend:

$$\begin{aligned} h_L &= c_L \frac{v^2}{2g} \\ &= 0.8 \times \frac{1.53^2}{2 \times 9.81} \\ &= 0.10\ \text{m} \end{aligned}$$

Entrance loss from the distillation:

$$h_L = 0.5\left[1 - \frac{A_2}{A_1}\right] \frac{v_2^2}{2g}$$

Assume $A_2/A_1 = 0$:

$$\begin{aligned} h_L &= 0.5[1 - 0] \frac{1.53^2}{2 \times 9.81} \\ &= 0.06\ \text{m} \end{aligned}$$

If the bottom is at its boiling point at the pressure of the column, then the saturated liquid vapor pressure is the same as the column pressure. From Equation 13.10:

$$\begin{aligned} NPSH_A &= \left(\frac{P_1}{\rho g} + z_1 - h_L\right) - \frac{P^{SAT}}{\rho g} \\ &= \frac{1.5}{764 \times 9.81} + 2.75 - (0.07 + 0.01 + 0.10 + 0.06) - \frac{1.5}{764 \times 9.81} \\ &= 2.51\ \text{m} \end{aligned}$$

This needs to be checked against the net positive suction head required by the pump.

Example 13.4 The transfer in Example 13.1 is to be carried out with a centrifugal pump with the characteristics given in Table 13.7. The pump has an impeller diameter of 12.5 cm and a speed of 1500 rpm.

a) For the piping system given in Example 13.1 develop a system head curve to represent the variation of head with flowrate, assuming the control valve setting is fixed. Determine the flowrate that would occur if the pump was used in this system.

b) Suggest a different pump speed with the same impeller diameter for the specified pump to allow the system to match the required flowrate of 100 m^3·h^{-1}.

c) Suggest a different impeller diameter with the same pump speed for the specified pump to allow the system to match the required flowrate of 100 m^3·h^{-1}.

d) Estimate the pump efficiency and the annual power cost at the desired operating point assuming the pump is driven by an electric motor with an efficiency of 90%, electricity costs \$0.06 kW·h^{-1}, and operates for 8300 hours per year.

Table 13.7

Centrifugal pump characteristic curve.

Flowrate (m^3·h^{-1})	Head (m)
0	72.5
25	72.3
50	71.5
75	68.4
100	61.0
125	47.5
150	26.0

Solution

a) From Example 13.1 the head loss in the fittings can be represented as a function of the velocity. The friction factor for loss in straight pipes is approximated to be independent of velocity:

Straight pipe	$0.135v^2$
Gate valves	$0.063v^2$
Check valve	$0.104v^2$
Globe valve	$1.644v^2$
Bends	$0.122v^2$
Orifice	$4.239v^2$
Entrance loss	$0.027v^2$
Exit loss	$0.050v^2$

From this the total head loss h_L is given by:

$$h_L = 6.384v^2$$

The flowrate is given by:

$$F = \frac{\pi \times 0.154^2}{4} \times 3600 \times v$$
$$= 67.06v$$

Thus the system head curve is given by:

$$h_L = 5 + \frac{6.384}{67.06^2}F^2$$
$$= 5 + 1.420 \times 10^{-3}F^2$$

Figure 13.11 shows the pump head curve matched against the system head curve. The curves intersect at 142.3 $m^3 \cdot h^{-1}$ with a head of 33.4 m. The pump is over size.

b) To scale the pump to the required operating conditions by adjusting the speed, from Equation 13.26:

$$\Delta h_1 = \Delta h_2 \left(\frac{F_1}{F_2}\right)^2$$

where subscript 1 refers to the existing design and subscript 2 to required design. To achieve the desired flowrate of 100 $m^3 \cdot h^{-1}$ with a head of 19.17 m:

$$\Delta h_1 = 19.17 \left(\frac{F_1}{100}\right)^2$$
$$= 0.001917F_1^2$$

Representing this in Figure 13.11, it is a quadratic curve starting at the origin that intersects the existing pump head curve at a flowrate of 138.7 $m^3 \cdot h^{-1}$ with a head of 36.9 m. From Equation 13.30:

$$N_{IMP2} = 1500 \times \frac{100}{138.7}$$
$$= 1081 \text{ rpm}$$

Alternatively, the scaling can use Equation 13.31:

$$N_{IMP2} = 1500\sqrt{\frac{19.17}{36.9}}$$
$$= 1081 \text{ rpm}$$

which agrees. The original values of F and Δh for the pump can be scaled for change in speed using Equations 13.30 and 13.31. The pump head curve after scaling to 1081 rpm is superimposed in Figure 13.11.

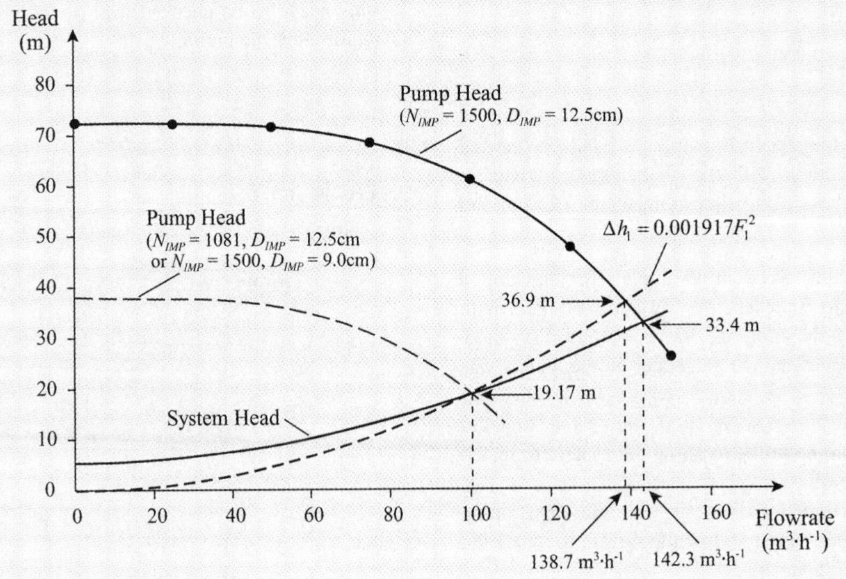

Figure 13.11

Pump head and system curves for Example 13.4.

c) Now fix the speed and change the impeller diameter. Combining Equations 13.33 and 13.34 and writing to achieve the desired flowrate of 100 $m^3 \cdot h^{-1}$ with a head of 19.17 m gives:

$$\Delta h_1 = 19.17\left(\frac{F_1}{100}\right)^2$$
$$= 0.00917F_1^2$$

This is the same curve as Part b and will again intersect the existing pump head curve at 138.7 $m^3 \cdot h^{-1}$ and 36.9 m. From Equation 13.33:

$$D_{IMP2} = 12.5\left(\frac{100}{138.7}\right)$$
$$= 9.0\ \text{cm}$$

Alternatively, the scaling can use Equation 13.34:

$$D_{IMP2} = 12.5\sqrt{\frac{19.17}{36.9}}$$
$$= 9.0\ \text{cm}$$

which agrees. Again, the original values of F and Δh can be scaled, but this time for a change in the impeller diameter, using Equations 13.33 and 13.34. The pump head curve after scaling to a 9.0 cm diameter is the same as the curve for reduced speed with a 12.5 cm diameter, as shown in Figure 13.11.

d) The pump efficiency at the desired operating point can be estimated from Equation 13.36:

$$\begin{aligned}
\eta &= 0.8 - 9.367 \times 10^{-3} \times 19.17 \\
&\quad +5.461 \times 10^{-5} \times 19.17 \times 100 \\
&\quad -1.514 \times 10^{-7} \times 19.17 \times 100^2 \\
&\quad +5.820 \times 10^{-5} \times 19.17^2 \\
&\quad -3.029 \times 10^{-7} \times 19.17^2 \times 100 \\
&\quad +8.348 \times 10^{-10} \times 19.17^2 \times 100^2 \\
&= 0.71
\end{aligned}$$

$$\begin{aligned}
W &= \frac{F\Delta P}{\eta} \times \frac{1}{0.9} \\
&= \frac{F\rho g\Delta h}{\eta} \times \frac{1}{0.9} \\
&= \frac{100 \times 19.17 \times 993 \times 9.81}{3600 \times 0.71 \times 0.9} \\
&= 8118\ \text{W}
\end{aligned}$$

$$\begin{aligned}
\text{Annual cost} &= \frac{8118}{10^3} \times 0.06 \times 8300 \\
&= 4042\ \$ \cdot y^{-1}
\end{aligned}$$

13.5 Compressor Types

Gas compressors can generally be classified as:

1) *Positive displacement compressors* confine successive volumes of fluid within a closed space in which the pressure of the fluid is increased as the volume of the closed space is decreased.
2) *Dynamic compressors or turbocompressors* transfer energy to the gas by dynamic means from a rotating impeller or blades. The kinetic energy of the gas is increased, which is then converted to pressure energy. The direction of gas flow in the centrifugal compressor is radial with respect to the axis of rotation. In an axial-flow compressor, the gas flow is parallel to the axis of rotation. A dynamic compressor with a low-pressure ratio is normally termed a *fan* (pressure ratio up to 1.11) or *blower* (pressure ratio up 1.11 to 1.2).
3) *Ejectors* in which the kinetic energy of a high-velocity *working fluid* or *motive fluid* (steam or a gas) entrains and compresses a second fluid stream. The device has no moving parts. They are inefficient devices and their normal usage is for vacuum service, where small quantities of gas are handled.

The most common types of compressor used for gas compression in the process industries are:

1) *Reciprocating*. Reciprocating compressors incorporate a piston moving backwards and forwards in a cylinder. Inlet and exhaust valves are set to open and close at certain pressures. The compressor can be *single acting* or *double acting*. Single-acting compressors use only one side of the piston, whereas double-acting machines use both sides of the piston. Figure 13.12 illustrates a double-acting reciprocating compressor. A significant disadvantage of reciprocating compressors is that the compressed gas is not delivered continuously. The resulting pulsations in flow and pressure can lead to vibrations (and, in extreme cases, mechanical failure) and poor compressor efficiency (to overcome high-pressure peaks). *Surge drums* and *acoustic filters* are required to dampen the pulsations.

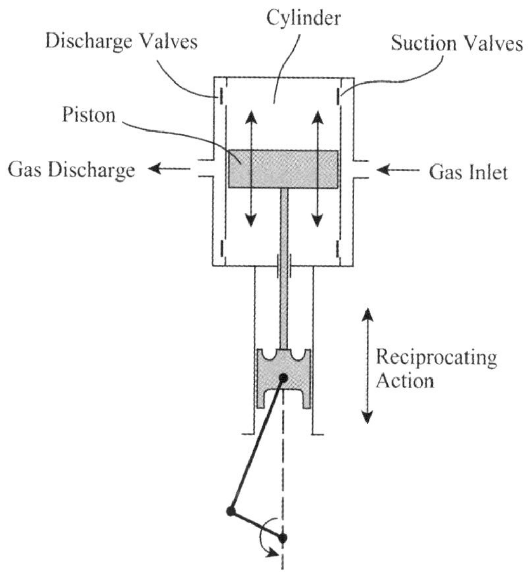

Figure 13.12
Reciprocating compressor.

13

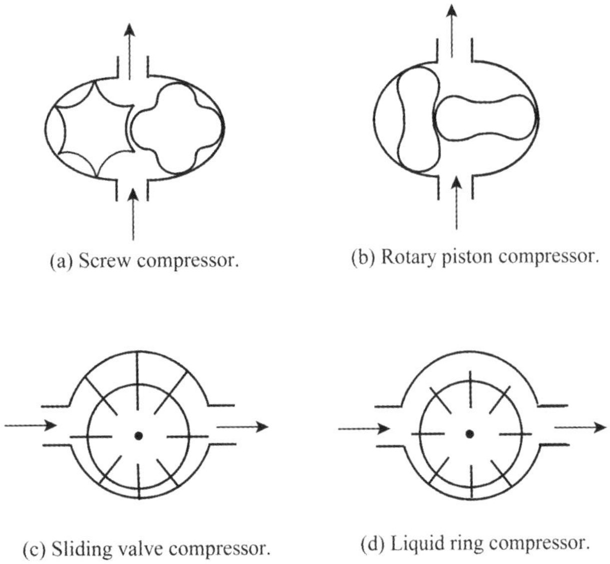

Figure 13.13
Positive displacement rotary compressors.

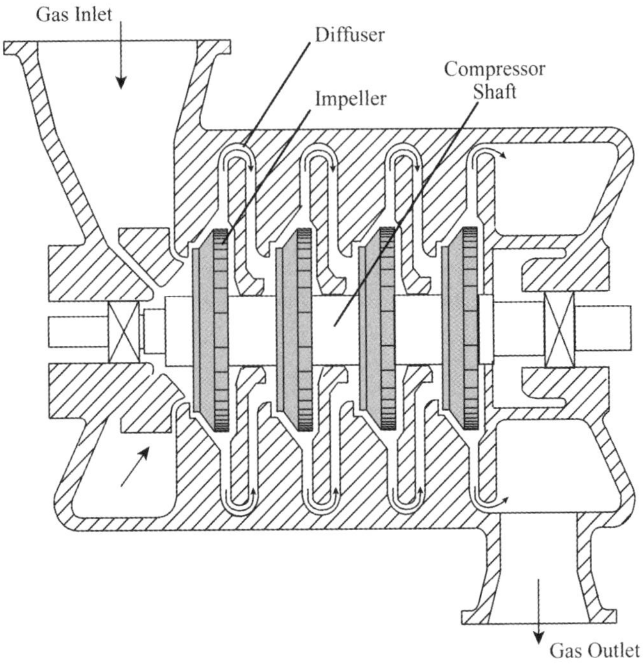

Figure 13.14
Centrifugal compressor.

2) *Positive displacement rotary.* Rotary machines, as their name implies, involve one or two rotating shafts to create closed spaces with decreasing size from inlet to outlet to increase the pressure. There are four broad classes of positive displacement rotary compressor:
   - *Screw* compressors use two counterrotating screw-like shafts (Figure 13.13a). Lubricating oil is required in some designs to lubricate the rotors, seal the gaps between the rotors and reduce the temperature rise of the gas during compression. These have the disadvantage of contaminating the gas with lubricating oil. On the other hand, oil-free machines do not feature mixing of the gas with lubricating oil. In these designs, contact between the rotors is prevented by timing gears outside the working chamber. However, they are more expensive than oil-injected machines.
   - *Rotary piston* or *rotary lobe* or *roots* compressors use two counterrotating matching lobe-shaped rotors (Figure 13.13b). Each revolution of the lobes delivers four pulses of gas.
   - *Sliding vane* compressors use a single rotating shaft with an eccentrically located rotor in the compressor casing with sliding vanes in the rotor (Figure 13.13c).
   - *Liquid ring* compressors use a single rotating shaft with an eccentrically located rotor in the compressor casing with static vanes (Figure 13.13d). A flow of low-viscosity liquid (usually water) through the casing draws the gas into the cells between the vanes, where it is compressed by the movement of the rotor. A settling drum after the compressor separates the liquid from the gas and the liquid is usually recycled.
3) *Centrifugal.* Centrifugal compressors increase gas pressure by accelerating the gas radially outwards through an *impeller* or *wheel*. The increased kinetic energy from the impeller is then converted to pressure energy in a *diffuser* as the gas leaves the impeller. Designs vary considerably, but Figure 13.14 shows a typical arrangement involving four impellers in a single casing.
4) *Axial.* Axial compressors increase gas pressure by accelerating it in the direction of the flow using blades mounted on a rotating shaft that rotate between stationary blades mounted on the compressor casing. This compressor has rotors and stators. The rotors are moving. They increase the kinetic energy of the gas. The stators are not moving. They are used to turn the kinetic energy into an increase in pressure. This is illustrated in Figure 13.15.

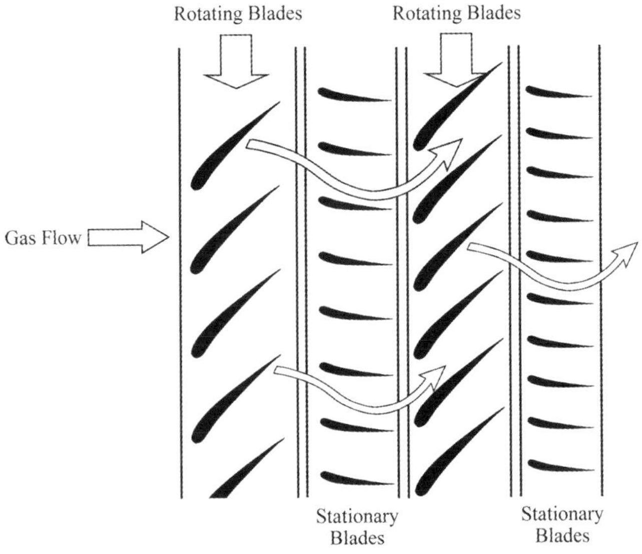

Figure 13.15
Axial flow compressor.

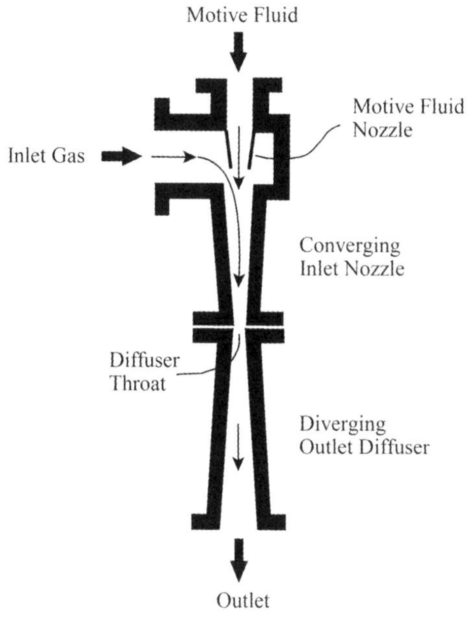

Figure 13.16

An ejector.

5) *Ejector.* Figure 13.16 illustrates an ejector. A high-pressure motive fluid, most often steam, enters the ejector through a nozzle. Fluid under high pressure is converted into a high-velocity jet at the throat of a convergent–divergent nozzle. The increase in velocity at the throat of the nozzle creates a low pressure at that point, drawing the gas to be compressed into the convergent–divergent nozzle where it mixes with the motive fluid. The mixture then passes through a diverging outlet diffuser, where the mixed fluid expands, its velocity decreases, converting the kinetic energy back into pressure energy, and its pressure increases, resulting in compression of the inlet gas. The mixture of steam and compressed gas then passes to a condenser, where the steam is condensed and the gases vented. Ejectors are only suited to compression of low gas volumes and moderate increases in pressure and are used most often for the production of vacuum. For example, in a vacuum distillation, vapor is condensed in the overhead condenser of the column and the uncondensed gases from the distillation condenser are compressed in a steam ejector, creating a vacuum in the distillation condenser. The steam is then condensed and the gases vented. For all but moderate vacuum, multiple stages of ejectors are used in series with condensers between the stages and a condenser after the final stage. Figure 13.17 illustrates a typical three-stage steam ejector arrangement. Steam is condensed between the ejector stages and the condensate is collected in a *hotwell*. A head of liquid condensate held in a *barometric leg* is required between the steam condenser and the hotwell in order to balance the pressure between the condenser and the hotwell. Figure 13.17 illustrates the use of shell-and-tube heat exchangers for the condensation. Condensation is often carried out between stages by direct contact by spraying water into the steam in an empty drum. This is a lower capital cost arrangement, but should be avoided, where possible, as it leads to an unnecessary additional consumption of water. Such direct contact condensation should be avoided unless the condensing duty would cause excessive fouling of a shell-and-tube heat exchanger. Steam ejectors have no moving parts, unlike liquid ring compressors, which are also used for vacuum services.

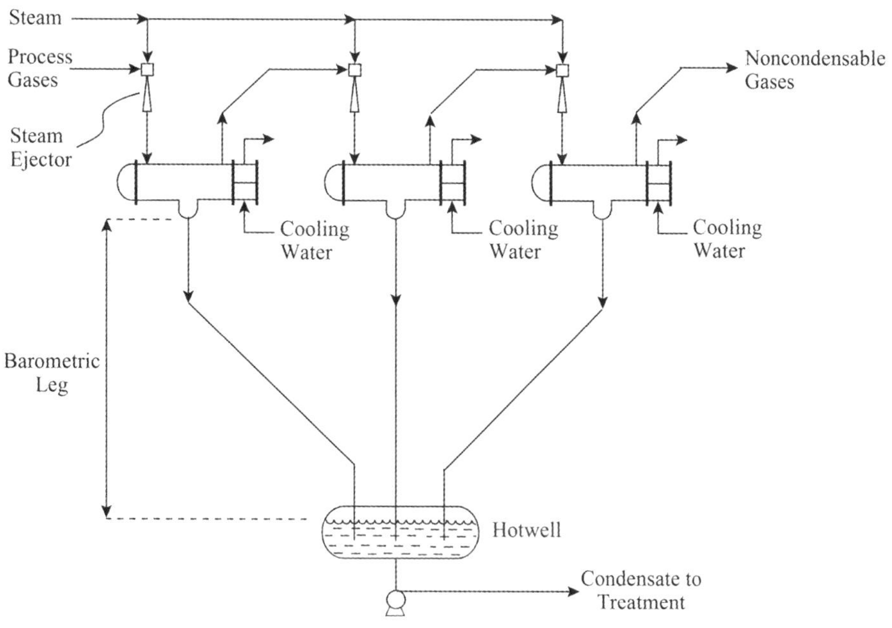

Figure 13.17

A typical three-stage steam ejector arrangement.

13

13.6 Reciprocating Compressors

Most compression processes are carried out close to adiabatic conditions. Adiabatic compression of an ideal gas along a thermodynamically reversible (isentropic) path can be expressed as (see Appendix G):

$$PV^{\gamma} = \text{constant} \tag{13.37}$$

where $\gamma = C_P/C_V$
$= \dfrac{C_P}{(C_P - R)}$ for an ideal gas
C_P = heat capacity at constant pressure
C_V = heat capacity at constant volume

Table 13.8 gives the ratio of heat capacities C_P/C_V for a number of common gases.

Alternatively, the value of C_P/C_V can be calculated from C_P data from the relationship (Hougen, Watson and Ragatz, 1959):

$$\gamma = \frac{C_P}{C_V} = \frac{C_P}{C_P - R} \tag{13.38}$$

where R = universal gas constant

However, it should be noted that the values of C_P/C_V are both temperature and pressure dependent, but can be calculated from an equation of state. Given:

$$\gamma = \frac{C_P}{C_V} = \frac{C_P}{C_P - (C_P - C_V)} \tag{13.39}$$

where $C_P - C_V = -T\left(\dfrac{\partial P}{\partial T}\right)_V \left(\dfrac{\partial V}{\partial P}\right)_T$

Table 13.8

Heat capacity ratio for various gases at 20 °C and 1 bar.

	Typical value		C_P/C_V
Monatomic gases	1.67	He	1.67
		Ar	1.67
Diatomic gases	1.40	H_2	1.41
		N_2	1.40
		CO	1.40
		Air	1.40
		NO	1.39
		O_2	1.40
		Cl_2	1.34
		HCl	1.41
Polyatomic gases	1.30	CH_4	1.30
		NH_3	1.31
		H_2O	1.31
		C_2H_4	1.24
		C_2H_6	1.19
		H_2S	1.32
		CO_2	1.29
		C_3H_6	1.15
		C_3H_8	1.13
		C_4H_{10}	1.09

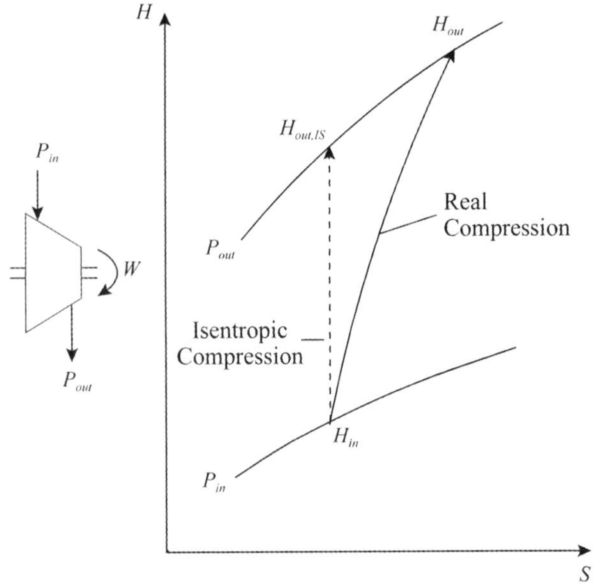

Figure 13.18

Compressor efficiency.

Then the value of $(\partial P/\partial T)_V$ and $(\partial V/\partial P)_T$ can be determined from an equation of state (e.g. that of Peng–Robinson). From Equation 13.37 for an adiabatic ideal gas (isentropic) compression:

$$\frac{P_1}{P_2} = \left(\frac{V_2}{V_1}\right)^{\gamma} \tag{13.40}$$

where P_1, P_2 = initial and final pressures
V_1, V_2 = initial and final volumes

Figure 13.18 illustrates an ideal gas compression. This follows a vertical path on an enthalpy–entropy diagram for the fluid. Also shown in Figure 13.18 is a real compression path. The real compression path can be defined by an isentropic efficiency:

$$\eta_{IS} = \frac{H_{in} - H_{out,IS}}{H_{in} - H_{out}} \tag{13.41}$$

The work required to compress an ideal gas is given by (see Appendix G):

$$W = \left(\frac{\gamma}{\gamma - 1}\right)\frac{P_{in}V_{in}}{\eta_{IS}}\left[1 - \left(\frac{P_{out}}{P_{in}}\right)^{(\gamma-1)/\gamma}\right] \tag{13.42}$$

where W = work required for gas compression (J)
P_{in}, P_{out} = inlet and outlet pressures ($N{\cdot}m^2$)
V_{in} = inlet gas volume (m^3)
η_{IS} = isentropic efficiency (–)
γ = heat capacity ratio C_P/C_V (–)

Expressing V_{in} as a volumetric flowrate returns a value of W in W. Alternatively, the work can be expressed in terms of the molar flowrate by introducing the compressibility factor, noting $PV = ZRT$ (see Appendix A):

$$W = \left(\frac{\gamma}{\gamma - 1}\right)\frac{ZRT_{in}}{\eta_{IS}}\left[1 - \left(\frac{P_{out}}{P_{in}}\right)^{(\gamma-1)/\gamma}\right] \qquad (13.43)$$

where W = actual work produced ($J \cdot kmol^{-1}$)
R = universal gas constant ($8314.5\ J \cdot kmol^{-1} \cdot K^{-1}$)
Z = compressibility factor (–)

The compressibility factor Z can be determined from an equation of state (see Appendix A). It should be noted that the introduction of a compressibility factor from an equation of state in Equation 13.43 corrects for the nonideality in the specification of the compressor inlet volume, but does not compensate for the nonideality through the compression path. If necessary a mechanical efficiency can also be introduced to allow for mechanical losses in the bearings and seals. However, this is usually 98 or 99% and can be neglected.

From thermodynamic convention, the work of compression from Equations 13.42 and 13.43 is negative.

It is also necessary to calculate the outlet temperature from the compression. By definition, the isentropic efficiency η_{IS} is given by Equation 13.41. If the heat capacity is assumed to be constant:

$$\eta_{IS} = \frac{mC_P(T_{out,IS} - T_{in})}{mC_P(T_{out} - T_{in})} = \frac{T_{out,IS} - T_{in}}{T_{out} - T_{in}} \qquad (13.44)$$

where m = mass flowrate
C_P = specific heat capacity
$T_{out,IS}$ = temperature of the outlet stream for an isentropic compression
T_{in} = temperature of the inlet stream
T_{out} = temperature of the outlet stream for a real compression

To obtain the temperature rise for an ideal gas isentropic compression, substitute $V = RT/P$ in Equation 13.37 to give:

$$\frac{T_{out,IS}}{T_{in}} = \left(\frac{P_{out}}{P_{in}}\right)^{(\gamma-1)/\gamma} \qquad (13.45)$$

Equation 13.45 assumes the compression to be an adiabatic ideal gas (isentropic) compression. Equations 13.44 and the 13.45 can be combined to give:

$$\eta_{IS} = \frac{\left(\frac{P_{out}}{P_{in}}\right)^{(\gamma-1)/\gamma} - 1}{\left(\frac{T_{out}}{T_{in}}\right) - 1} \qquad (13.46)$$

If η_{IS} is known, Equation 13.46 can be rearranged to calculate T_{out}:

$$T_{out} = \left(\frac{T_{in}}{\eta_{IS}}\right)\left[\left(\frac{P_{out}}{P_{in}}\right)^{(\gamma-1)/\gamma} + \eta_{IS} - 1\right] \qquad (13.47)$$

Equations 13.42, 13.43 and 13.47 can be used to model gas compression. However, these equations need to be used with caution, as they are based on ideal gas behavior and heat capacity is assumed to be constant. Providing the pressure is not high and the gas does not deviate significantly from ideal behavior, then the equations can be used to model the compression. In addition, the value of γ can vary significantly through the compression process.

An alternative way to calculate the work for an adiabatic compression is from the difference between the total enthalpy of the outlet and inlet flows:

$$W = H_{in} - H_{out} = \frac{H_{in} - H_{out,IS}}{\eta_{IS}} \qquad (13.48)$$

where H_{in} = total enthalpy of the inlet stream ($kJ \cdot s^{-1}$)
$H_{out,IS}$ = total enthalpy of the outlet stream for an isentropic compression ($kJ \cdot s^{-1}$)
H_{out} = total enthalpy of the outlet stream for a real compression ($kJ \cdot s^{-1}$)

The calculation starts with the inlet enthalpy and, given the inlet enthalpy and the pressure, calculates the entropy of the inlet. Then assuming the outlet entropy is the same as the inlet (isentropic), given the outlet pressure, calculate the outlet enthalpy. From the isentropic enthalpy change, the real outlet enthalpy can then be calculated by dividing the isentropic enthalpy change by the isentropic efficiency to get the real enthalpy change. Neglecting any mechanical losses, this gives the overall power requirements. Knowing the outlet enthalpy and pressure allows the outlet temperature to be calculated from physical properties. This is the method normally used in simulation software, but is inconvenient to carry out using hand calculations. The two approaches based on Equations 13.42 and 13.43 or Equation 13.48 should yield the same result as long as the physical properties are correct and consistent. The two approaches are simply different ways to follow an isentropic compression based on physical properties. The problem with using Equations 13.42, 13.43 and 13.47 is that the gas might exhibit significant deviation from ideal behavior and γ might change through the compression path. The approach based on Equation 13.48 will be more reliable, but requires simulation software.

13.7 Dynamic Compressors

Dynamic compressors increase gas pressure by accelerating the gas as it flows, either radially out from a rotating impeller, or axially from blades mounted on a rotating shaft. The gas then flows through a *diffuser* with increasing cross-sectional area where kinetic energy created by the increase in velocity is converted into pressure energy and pressure is increased. Unlike reciprocating compressors, dynamic compressors involve a constant flow through the compressor.

1) *Adiabatic compression.* Consider first adiabatic compression of an ideal gas in a dynamic compressor. The work of compression is also given by Equations 13.42 and 13.43 (see Appendix G).

An alternative way to calculate the work for an adiabatic compression is from the difference between the total enthalpy of the outlet and inlet flows from Equation 13.48. Again, the approach based on Equation 13.48 will be more reliable, but requires simulation software.

2) *Polytropic compression.* Rather than assume that the compression is carried out adiabatically for an ideal gas, a more general process can be defined that depends on how the process is conducted. A general polytropic compression is defined that is neither adiabatic nor isothermal, but specific to the physical properties of the gas and the design of the compressor. A *polytropic* compression is defined as:

$$PV^n = \text{constant} \tag{13.49}$$

where n = polytropic coefficient (–)

If $n = 1$, Equation 13.49 corresponds with an isothermal ideal gas compression. For $n = \gamma$, Equation 13.49 corresponds with the expression for an adiabatic ideal gas. A polytropic compression between two states is given by:

$$\frac{P_1}{P_2} = \left(\frac{V_2}{V_1}\right)^n \tag{13.50}$$

If the initial and final conditions for a given compression process are known, then n can be determined from a rearrangement of Equation 13.50:

$$n = \frac{\ln(P_2/P_1)}{\ln(V_1/V_2)} \tag{13.51}$$

If ideal gas behavior is assumed, then Equation 13.50 becomes:

$$\frac{T_{out}}{T_{in}} = \left(\frac{P_{out}}{P_{in}}\right)^{(n-1)/n} \tag{13.52}$$

Whilst the overall isentropic efficiency, as illustrated in Figure 13.18, is useful as a measure of the overall performance of the compressor it does not give any indication regarding internal losses. This is overcome by introducing the *polytropic efficiency*. To define the polytropic efficiency, the compression is assumed to occur in a series of incremental isentropic steps. Figure 13.19 illustrates the principle. An isentropic compression represents a vertical line from the inlet to the outlet. To define the polytropic efficiency, the real compression is broken down into a series of isentropic steps. Figure 13.19 shows a compression broken down into three isentropic steps for illustration. It should be noted that the sum of the power for the three isentropic steps is greater than the single isentropic step from the inlet to the outlet. This is because the lines of constant pressure diverge as the entropy increases. In practice, rather than use the small number of steps, as illustrated in Figure 13.19, incrementally small steps are used and integrated across the process. The polytropic efficiency is defined as:

$$\eta_P = \frac{dH_{out,IS}}{dH} \tag{13.53}$$

From the thermodynamics of differential energy properties (Hougen, Watson and Ragatz, 1959; Coull and Stuart, 1964):

$$dH = T\,dS + V\,dP \tag{13.54}$$

For an isentropic process $T\,dS = 0$, combining Equations 13.53 and 13.54 thus gives:

$$\eta_P = \frac{V\,dP}{dH} \tag{13.55}$$

For an ideal gas (Hougen, Watson and Ragatz, 1959; Coull and Stuart, 1964):

$$V = \frac{RT}{P}, \quad dH = C_P\,dT \quad \text{and} \quad C_P = \frac{\gamma}{\gamma - 1}R \tag{13.56}$$

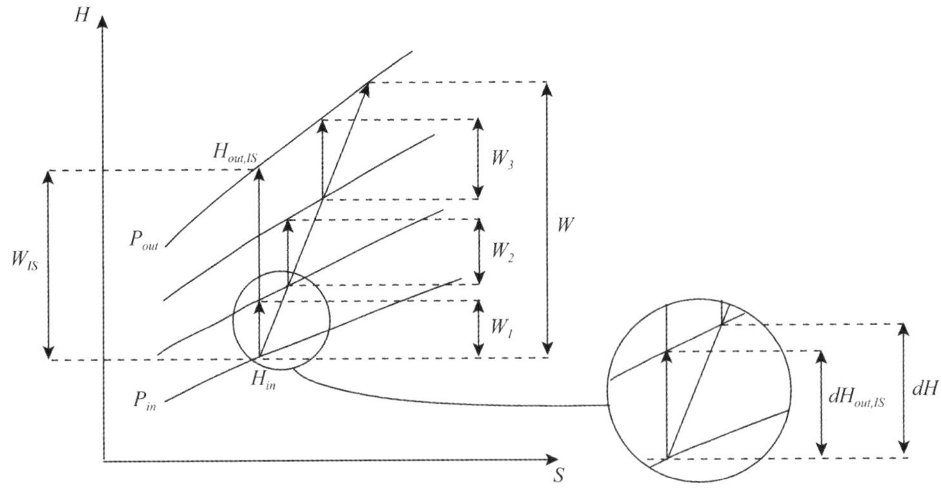

Figure 13.19
Polytropic efficiency.

Substituting Equation 13.56 into Equation 13.55 and rearranging gives:

$$\frac{dT}{T} = \left(\frac{\gamma - 1}{\gamma \eta_P}\right)\frac{dP}{P} \qquad (13.57)$$

Integrating Equation 13.57 between the inlet and outlet conditions assuming η_P to be constant gives:

$$\ln\left(\frac{T_{out}}{T_{in}}\right) = \frac{\gamma - 1}{\gamma \eta_P}\ln\left(\frac{P_{out}}{P_{in}}\right) \qquad (13.58)$$

Substituting Equation 13.52 into Equation 13.58 and rearranging gives:

$$\eta_P = \frac{\left(\dfrac{\gamma - 1}{\gamma}\right)}{\left(\dfrac{n - 1}{n}\right)} \qquad (13.59)$$

Rearranging Equation 13.59 gives:

$$n = \frac{\gamma \eta_P}{\gamma \eta_P - \gamma + 1} \qquad (13.60)$$

Equation 13.60 is a useful expression to determine the polytropic coefficient. Given the polytropic efficiency for a compressor (say from a manufacturer's data), the polytropic coefficient can be obtained from Equation 13.60. It is evident from Equation 13.59 and 13.60 that for normal values of γ and η_P, where $1.1 > \gamma > 1.7$ and $0.5 > \eta_P > 1$, then $n > \gamma$.

Another useful expression can be obtained by combining Equations 13.46, 13.52 and 13.60:

$$\eta_{IS} = \frac{\left(\dfrac{P_{out}}{P_{in}}\right)^{(\gamma-1)/\gamma} - 1}{\left(\dfrac{P_{out}}{P_{in}}\right)^{(\gamma-1)/\gamma\eta_P} - 1} \qquad (13.61)$$

This provides a relationship between the isentropic efficiency and of the polytropic efficiency. From Equation 13.61, the value of the isentropic efficiency is lower than the polytropic efficiency. For example, if it is assumed that for air $\gamma = 1.4$ and for the air compressor $\eta_P = 0.7$, as P_{out}/P_{in} approaches 1.0 then η_{IS} approaches 0.7, but for $P_{out}/P_{in} = 2$, $\eta_{IS} = 0.67$ and for $P_{out}/P_{in} = 4$, $\eta_{IS} = 0.64$.

It should be noted that the isentropic efficiency in Equations 13.46 and 13.61 is a function of the pressure ratio, whereas the polytropic efficiency in Equation 13.59 is not a function of the pressure ratio. The isentropic efficiency, although fundamentally valid, can be misleading if used for comparing the efficiencies of different compressors with differing pressure ratios. For two compressors with different pressure ratios, the higher pressure ratio compressor will have a lower isentropic efficiency because of thermodynamic losses. This property can make it difficult to comparatively evaluate different compressor designs. Polytropic efficiency is better than the isentropic efficiency for comparison between compressors, since the value of the polytropic efficiency does not change as much as the value of the isentropic efficiency. In practice, the polytropic efficiency is a function of the flowrate, physical properties of the gas and the machine design.

To obtain the work for a polytropic compression, Equations 13.59 and 13.61 can be combined with Equation 13.42 to give:

$$W = \frac{n}{n - 1}\frac{P_{in}V_{in}}{\eta_P}\left[1 - \left(\frac{P_{out}}{P_{in}}\right)^{(n-1)/n}\right] \qquad (13.62)$$

where W = actual work produced (J)
η_P = polytropic efficiency (ratio of polytropic work to actual work)

Expressing V_{in} as a volumetric flowrate returns a value of W in W. Alternatively, the work can be expressed in terms of the molar flowrate by introducing the compressibility factor, noting $PV = ZRT$ (see Appendix A):

$$W = \frac{n}{n - 1}\frac{ZRT_{in}}{\eta_P}\left[1 - \left(\frac{P_{out}}{P_{in}}\right)^{(n-1)/n}\right] \qquad (13.63)$$

where W = actual work produced ($J{\cdot}kmol^{-1}$)

Again, it should be noted that the introduction of a compressibility factor from an equation of state in Equation 13.63 corrects for the nonideality in the specification of the compressor inlet volume, but does not compensate for the nonideality through the compression path.

The temperature rise for the polytropic compression can be calculated from a rearrangement of Equation 13.58 to give:

$$\frac{T_{out}}{T_{in}} = \left(\frac{P_{out}}{P_{in}}\right)^{(\gamma-1)/\gamma\eta_P} \qquad (13.64)$$

13.8 Staged Compression

The temperature rise accompanying single-stage gas compression might be unacceptably high because of the properties of the gas, materials of construction of the compressor or the properties of the lubricating oil used in the machine. If this is the case, the overall compression can be broken down into a number of stages with intermediate cooling. Also, intermediate cooling will reduce the volume of gas between stages and reduce the work of compression for the next stage. On the other hand, the intercoolers will have a pressure drop that will increase the work, but this effect is usually small compared with the reduction in work from gas cooling.

Consider a multistage compression in which the intermediate gas is cooled down to the initial temperature. The minimum work for a multistage adiabatic gas compression of an ideal gas is given by (see Appendix G):

$$W = \frac{\gamma}{\gamma - 1}\frac{P_{in}V_{in}N}{\eta_{IS}}\left[1 - (r)^{(\gamma-1)/\gamma}\right] \qquad (13.65)$$

where N = number of compression stages (–)
r = pressure ratio for each stage (–)

13

The pressure ratio for minimum work for N stages is given by (see Appendix G):

$$r = \sqrt[N]{\frac{P_{out}}{P_{in}}} \tag{13.66}$$

In principle, the isentropic efficiency might change from stage to stage. However, if the isentropic efficiency for a reciprocating compressor is assumed to be only a function of the pressure ratio and the pressure ratio is constant between stages, then it is legitimate to use a single value as in Equation 13.65. It should be noted that these results for staged compression are based on adiabatic ideal gas compression and are therefore not strictly valid for real gas compression. It is also assumed that intermediate cooling is back to inlet conditions, which might not be the case with real intercoolers. For fixed inlet conditions and outlet pressure, the overall power consumption is usually not sensitive to minor changes in the intercooler temperature.

The corresponding equation for a polytropic compression is given by (see Appendix G):

$$W = \frac{n}{n-1}\frac{P_{in}V_{in}N}{\eta_P}\left[1-(r)^{(n-1)/n}\right] \tag{13.67}$$

If the polytropic efficiency of a centrifugal or axial compressor is assumed to be a function of volumetric flowrate, then the efficiency, in principle, will change from stage to stage. This is because the density changes between stages, even if the gas is cooled back to the same temperature as a result of the pressure increase. However, such effects are not likely to have a significant influence on the predicted power.

It is important to note that if the intercooling is not back to the original inlet temperature and that there is a pressure drop in the intercooler, then Equation 13.66 does not predict the pressure ratio for minimum power consumption. Consider a two-stage compression in which the intermediate gas is cooled to a defined temperature T_2, different from the inlet temperature T_1, and there is a pressure drop across the intercooler ΔP_{INT}. Let T_2 and P_2 be the outlet temperature and pressure of the intercooler. The conditions for minimum power are given by (see Appendix G):

$$\frac{P_2^{2\gamma-1}}{(P_2+\Delta P_{INT})} = \left(\frac{T_2}{T_1}\right)^{\gamma}(P_1P_3)^{\gamma-1} \tag{13.68}$$

If $\Delta P_{INT}=0$ and $T_2=T_1$, Equation 13.68 simplifies to Equation 13.66. Equation 13.68 predicts the intermediate pressure for minimum shaft work for compression of an ideal gas in a two-stage compression. The corresponding expression for a polytropic compression is given by replacing γ by n in Equation 13.68. The important thing to note is that if the intercooling is not back to the original inlet temperature and that there is a pressure drop in the intercooler, then Equation 13.66 does not predict the pressure ratio for minimum power consumption. For more than two stages the calculation can be broken down into different compressor stages with the appropriate allowance for change in temperature and pressure drop between stages.

13.9 Compressor Performance

The maximum compression ratio (ratio of outlet to inlet pressure) for compressors depends on the design of the machine, the properties of the lubricating oil used in the machine, the ratio of heat capacities of the gas ($C_P/C_V=\gamma$), other properties of the gas (e.g. tendency to polymerize when heated) and the inlet temperature. There are always mechanical losses for the machine from friction in the bearings, etc. However, the mechanical efficiency is typically 98 to 99% and can therefore often be neglected.

1) *Reciprocating compressors.* Reciprocating compressors tend to be used when high compression ratios are required without high flowrates. Compression ratios of up to 10 are used and flowates up to 7000 $m^3 \cdot h^{-1}$ (Dimoplon, 1978). Diatomic gases with $\gamma = 1.4$ can have a pressure ratio per cylinder of up to 4 and hydrocarbon gases with $\gamma = 1.2$ up to 9. For reciprocating compressors the isentropic efficiency may typically be in the range 60–80%, depending on the compression ratio, flowrate, machine design and properties of the gas. A first estimate of the isentropic efficiency of a reciprocating compressor can be obtained from:

$$\eta_{IS} = 0.1091(\ln r)^3 - 0.5247(\ln r)^2 + 0.8577\ln r + 0.3727 \qquad 1.1 < r < 5 \tag{13.69}$$

where r = pressure ratio P_{out}/P_{in}

2) *Screw compressors.* Compression ratios of up to 20 are possible. Diatomic gases with $\gamma = 1.4$ can have a pressure ratio per casing of up to 4.5 and hydrocarbon gases with $\gamma = 1.2$ can have a pressure ratio per casing of up to 10. Maximum discharge pressures are up to 30 bar. Suction flowrates of up to 15 $m^3 \cdot s^{-1}$ and greater are possible.
3) *Rotary piston* or *rotary lobe* or *roots compressors.* Maximum discharge pressures in a single stage are up to 2.5 bar. Suction flowrates of up to 3 $m^3 \cdot s^{-1}$ are possible.
4) *Sliding vane compressors.* The maximum pressure ratio for each casing is restricted to around 3.5. Maximum discharge pressures are up to 10 bar. Suction flowrates of up to 3 $m^3 \cdot s^{-1}$ are possible.
5) *Liquid ring compressors.* Maximum discharge pressures can be greater than 5 bar. Flowrates of up to 3 $m^3 \cdot s^{-1}$ are possible. Liquid ring compressors are most commonly used for vacuum service.
6) *Centrifugal compressors.* Compression ratios of up to 3 are used in a single stage. Centrifugal compressors are used for flowrates typically in the range 1000 to 350,000 $m^3 \cdot h^{-1}$, depending on the pressure ratio (Dimoplon, 1978). For centrifugal compressors the isentropic efficiency may typically be in the range 70 to 90%, depending on the compression ratio, flowrate, machine design and properties of the gas. A first estimate of the polytropic efficiency of a centrifugal compressor can be obtained from:

$$\eta_P = 0.017\ln F + 0.7 \tag{13.70}$$

where F = inlet flowrate of gas ($m^3 \cdot s^{-1}$)

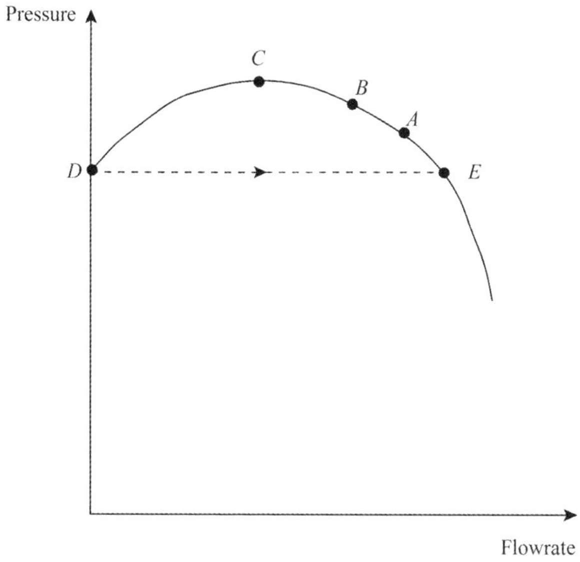

Figure 13.20
Surging in centrifugal compressors.

A significant disadvantage of centrifugal compressors is their limited turndown capacity. It is frequently necessary to operate a compressor at flows below the design capacity. If the flow is reduced far enough, the compressor enters an unstable region known as the *surge region*. Figure 13.20 shows the typical behavior of pressure versus flowrate for a centrifugal compressor. If the compressor is operating at Point *A* in Figure 13.20 and the flow is reduced to Point *B* by restricting the outlet flow, the increased pressure drop caused by the restriction is met by the compressor. Reducing the flow beyond the peak Point *C* results in a discharge mains pressure higher than the compressor pressure. The flow then reverses with the operating point moving to zero flow at Point *D*. Once the discharge main clears, the operating point moves to Point *E*. The excessive flow then moves the operation back to beyond Point *C*. The cycle is then repeated. The surge limit depends on the speed of the compressor, decreasing with decreasing speed. Unstable operations can start at 50 to 70% of the rated flow. Surge is a problem for low flowrates. For high flowrates in Figure 13.20, the performance curve becomes very steep as the *choke point* is reached. This is where the flow in the compressor reaches Mach 1 and no more flow can pass through the compressor. It is often known as "*stone walling*".

The performance of a given centrifugal compressor can be mapped in a *performance map*, as illustrated in Figure 13.21. The compressor can be operated at different speeds, each of which has a performance curve. Operation is restricted to an area between the surge line, capacity limit (choking), maximum speed and minimum speed. The efficiency of the compressor varies within this operating envelope. For a given compressor, the isentropic efficiency can be correlated by the general expression:

$$\eta_{IS} = a_1 F_{in}^{b1} + a_2 \left(\frac{P_{out}}{P_{in}}\right)^{b2} + a_3 \left(F_{in}\frac{P_{out}}{P_{in}}\right)^{b3} \qquad (13.71)$$

where F_{in} = inlet volumetric flowrate ($m^3 \cdot s^{-1}$)
a_1, a_2, a_3 b_1, b_2, b_3 = correlating constants to be determined by regression of data from the machine performance

7) *Axial compressors.* Axial compressors are used for very high flowrates and moderate pressure drops. Flowrates are typically in the range 130,000 $m^3 \cdot h^{-1}$ to 1,300,000 $m^3 \cdot h^{-1}$, depending on the pressure ratio (Dimoplon, 1978). Axial flow compressors have polytropic efficiencies up to 10% higher than centrifugal compressors for the same pressure ratio. A reasonable estimate is for the efficiency to be 5% higher than that of the corresponding centrifugal machine given by Equation 13.70. As with centrifugal compressors, axial compressors can also suffer from surge and choking and must operate within a certain envelope similar to that in Figure 13.21.

For low-pressure difference and large flows, the capital cost of a reciprocating compressor can be twice that of a centrifugal machine with the same flowrate. If the pressure difference is high and the flowrate low, then the costs tend to be more equal. Also, centrifugal machines tend to be more reliable than reciprocating machines. Standby machines might need to be installed if reciprocating machines are used. The capital cost of axial compressors tends to be of the same order as centrifugal machines. At flowrates typically less than 30 $m^3 \cdot s^{-1}$, the capital cost of axial compressors tends to be higher than centrifugal machines. Rotary screw compressors tend to be cheaper than both reciprocating and centrifugal machines in their operating range. Sliding vane compressors tend to be cheaper than reciprocating machines in their operating range.

For high demand compression duties, the overall compression can be broken down into stages and different types of compressor used in the different casings. For example, a large flowrate with a large pressure differential

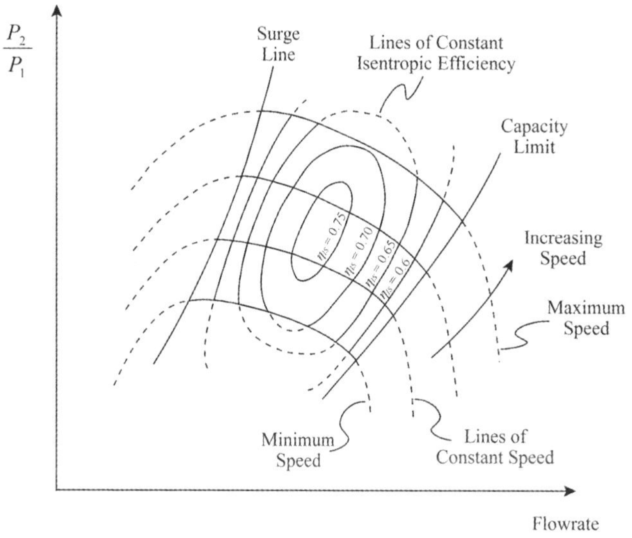

Figure 13.21
Performance map for centrifugal compressors.

13

might use an axial compressor followed by a centrifugal compressor.

Example 13.5 A recycle gas stream containing 88% hydrogen and 12% methane is to be increased in pressure from 81 bar to 98 bar. The inlet temperature is 40 °C and the flowrate is 170,000 $Nm^3 \cdot h^{-1}$ (Nm^3 = normal m^3). Estimate the power requirements for a centrifugal compressor for this duty assuming the polytropic efficiency to be 0.75.

Solution

$$\text{Pressure ratio} = \frac{98}{81} = 1.21$$

Thus, compression in a single stage will be acceptable.

Suction volume of gas at design conditions of 40 °C and 81 bar (assuming an ideal gas) is given by:

$$F_{in} = 170{,}000 \times \frac{313}{273} \times \frac{1.013}{81} = 2438\ m^3 \cdot h^{-1} = 0.677\ m^3 \cdot s^{-1}$$

The ratio of heat capacities for the mixture can be taken as a weighted mean of the values in Table 13.8:

$$\gamma = 0.88 \times 1.40 + 0.12 \times 1.30 = 1.39$$

The polytropic coefficient can then be calculated from Equation 13.60:

$$n = \frac{\gamma \eta_P}{\gamma \eta_P - \gamma + 1} = \frac{1.39 \times 0.75}{1.39 \times 0.75 - 1.39 + 1} = 1.60$$

Now calculate the power from Equation 13.62 with the inlet volume V_{in} replaced by the inlet volumetric flowrate F_{in} to give the compressor power:

$$W = \frac{n}{n-1}\frac{P_{in}F_{in}}{\eta_P}\left[1 - \left(\frac{P_{out}}{P_{in}}\right)^{(n-1)/n}\right] = \frac{1.60}{1.60-1}\frac{81 \times 10^5 \times 0.677}{0.75}\left[1 - \left(\frac{98}{81}\right)^{(1.60-1)/1.60}\right] = -1.44 \times 10^6\ W$$

This calculation assumed the gas to be ideal. For comparison, the calculation can be based on the Peng–Robinson Equation of State (see Appendix A). A number of commercial physical property software packages allow the prediction of gas density and γ for a mixture of hydrogen and methane using the Peng–Robinson Equation of State. Using this, the gas density at normal conditions is 0.1651 $kg \cdot m^{-3}$. At 40 °C and 81 bar, the density is 11.2101 $kg \cdot m^{-3}$. Thus, the suction volume of gas

$$F_{in} = 170{,}000 \times \frac{0.1651}{11.2101} = 2504\ m^3 \cdot h^{-1} = 0.695\ m^3 \cdot s^{-1}$$

At suction conditions:

$$\gamma = 1.38$$

This is close to the estimated value. Greater accuracy again could be obtained from γ at the average compression conditions. From Equation 13.60:

$$n = 1.58$$

From Equation 13.62:

$$W = -1.48 \times 10^6\ W$$

This is a relatively small error given the high pressure of the compression.

13.10 Process Expanders

Process expanders are effectively process compressors working in reverse. If a gas stream is under pressure and needs to be let down in pressure, then a process expander can be used to recover power. The power can be used to drive another machine (e.g. a compressor) or generate electricity. An example of such an application is associated with a petroleum refinery catalytic cracker, as illustrated in Figure 6.8b. In the catalyst regenerator, carbon deposited on the catalyst is burnt off to regenerate the catalyst. The exhaust gases are at a pressure of 1.5 to 2 barg and a temperature of 620 to 650 °C for a partial combustion process or 670 to 750 °C for a complete combustion process. These exhaust gases can be expanded in a process expander to recover power before being vented. Although the pressure is low, the volume is large and the temperature high, which makes power recovery an attractive option. The expander can be used to drive the air compressor for the regenerator or generate electricity.

The modeling of expanders has the same basis as compressors. An ideal expansion is the reverse of an ideal compression. Expanders are most often single-stage machines. For an adiabatic expansion, Equation 13.42 becomes:

$$W = \left(\frac{\gamma}{\gamma - 1}\right)\eta_{IS,E}P_{in}V_{in}\left[1 - \left(\frac{P_{out}}{P_{in}}\right)^{(\gamma-1)/\gamma}\right] \tag{13.72}$$

where W = actual work produced (J)

$\eta_{IS,E}$ = isentropic efficiency for the expansion process (–)

The efficiency term in Equation 13.72 reduces power generation, in comparison with the efficiency term in Equation 13.42, which increases the power input for compression. The power prediction from Equation 13.72 is positive, indicating power generation. Expressing V_{in} as a volumetric flowrate returns a value of W in W. Alternatively, the work can be expressed in terms of the molar flowrate by introducing the compressibility factor and Equation 13.43 becomes:

$$W = \left(\frac{\gamma}{\gamma - 1}\right)\eta_{IS,E} ZRT_{in}\left[1 - \left(\frac{P_{out}}{P_{in}}\right)^{(\gamma-1)/\gamma}\right] \tag{13.73}$$

where R = universal gas constant (8314.5 J·K^{-1}·kmol^{-1})
Z = compressibility factor (–)

By definition, the isentropic efficiency $\eta_{IS,E}$ is given by (see Figure 2.1):

$$\eta_{IS,E} = \frac{H_{in} - H_{out}}{H_{in} - H_{out,IS}} \tag{13.74}$$

For a polytropic expansion, Equations 13.72 and 13.73 become:

$$W = \left(\frac{n}{n - 1}\right)\eta_{P,E} P_{in} V_{in}\left[1 - \left(\frac{P_{out}}{P_{in}}\right)^{(n-1)/n}\right] \tag{13.75}$$

$$W = \left(\frac{n}{n - 1}\right)\eta_{P,E} ZRT_{in}\left[1 - \left(\frac{P_{out}}{P_{in}}\right)^{(n-1)/n}\right] \tag{13.76}$$

where $\eta_{P,E}$ = polytropic efficiency for the expansion process (–)

The compressibility factor Z can be determined from an equation of state (see Appendix A). It is also necessary to calculate the outlet temperature. If the heat capacity is assumed to be constant, Equation 13.74 can be written as:

$$\begin{aligned}\eta_{IS,E} &= \frac{CP(T_{in} - T_{out})}{CP(T_{in} - T_{out,IS})} \\ &= \frac{T_{out} - T_{in}}{T_{out,IS} - T_{in}}\end{aligned} \tag{13.77}$$

where CP = total heat capacity (product of mass flowrate and specific heat capacity)
T_{out} = temperature of the outlet stream for a real expansion
T_{in} = temperature of the inlet stream
$T_{out,IS}$ = temperature of the outlet stream for an isentropic expansion

For an isentropic process, from Equation 13.45:

$$\frac{T_{out,IS}}{T_{in}} = \left(\frac{P_{out}}{P_{in}}\right)^{(\gamma-1)/\gamma} \tag{13.78}$$

Combining Equations 13.77 and 13.78 gives:

$$\eta_{IS,E} = \frac{\left(\dfrac{T_{out}}{T_{in}}\right) - 1}{\left(\dfrac{P_{out}}{P_{in}}\right)^{(\gamma-1)/\gamma} - 1} \tag{13.79}$$

If $\eta_{IS,E}$ is known, Equation 13.79 can be rearranged to calculate T_{out}:

$$T_{out} = T_{in}\eta_{IS,E}\left[\left(\frac{P_{out}}{P_{in}}\right)^{(\gamma-1)/\gamma} + \frac{1}{\eta_{IS,E}} - 1\right] \tag{13.80}$$

Equations 13.72, 13.76 and 13.80 can be used to model gas expansion. However, these equations need to be used with caution, as they are based on ideal gas behavior and heat capacity is assumed to be constant. Providing the pressure is not high and the gas does not deviate significantly from ideal behavior, then the equations can be used to model the expansion. In addition, the value of γ can vary significantly through the expansion process.

An alternative way to calculate the work for an adiabatic expansion is from the difference between the total enthalpy of the outlet and inlet flows:

$$\begin{aligned}W &= H_{in} - H_{out} \\ &= \eta_{IS,E}\left(H_{in} - H_{out,IS}\right)\end{aligned} \tag{13.81}$$

where W = power produced from expansion (N·m·s^{-1} = J·s^{-1} = W)
H_{in} = total enthalpy of the inlet stream (kJ·s^{-1})
$H_{out,IS}$ = total enthalpy of the outlet stream for an isentropic expansion (kJ·s^{-1})
H_{out} = total enthalpy of the outlet stream for a real expansion (kJ·s^{-1})
$\eta_{IS,E}$ = isentropic efficiency for the expansion process (–)

The calculation starts with the inlet enthalpy and, given the inlet enthalpy and the pressure, calculates the entropy of the inlet. Then, assuming the outlet entropy is the same as the inlet entropy (isentropic), given the outlet pressure, the outlet enthalpy is determined. From the isentropic enthalpy change, the real outlet enthalpy can then be calculated by multiplying the isentropic enthalpy change by the isentropic efficiency to get the real enthalpy change. Neglecting any mechanical losses, this gives the overall power production. Knowing the outlet enthalpy and pressure allows the outlet temperature to be calculated from physical properties. This is the method normally used in simulation software, but is inconvenient to carry out using hand calculations. The two approaches based on Equations 13.72 and 13.73 or Equation 13.81 should yield the same result as long as the physical properties are correct and consistent. The problem with using Equations 13.72, 13.73 and 13.80 is that the gas might exhibit significant deviation from ideal behavior and γ might change through the expansion path. The approach based on Equation 13.81 will be more reliable, but requires simulation software.

Example 13.6 The flue gas stream from a fluid catalytic cracker regenerator with a flowrate of 1.25 kmol s^{-1} is at a temperature of 700 °C and a pressure of 2.0 barg. The flue gas contains mainly nitrogen, but with significant quantities of carbon monoxide, carbon dioxide, water vapor and a small amount of oxygen. The gas is to be combusted in a boiler to generate steam before release to atmosphere. The inlet pressure required for the boiler is 0.2 barg. Even though the gas from the regenerator is not at a high pressure, its high flowrate and high temperature mean that power recovery is economic. Assuming the expander has an isentropic efficiency of 0.8, γ for the gas is 1.32 and a compressibility factor of 1.0, calculate the power recovery from the expander and the expander outlet temperature.

Solution From Equation 13.73:

$$W = \left(\frac{\gamma}{\gamma-1}\right)\eta_{IS,E} ZRT_{in}\left[1-\left(\frac{P_{out}}{P_{in}}\right)^{(\gamma-1)/\gamma}\right]$$

$$= 1.25\left(\frac{1.32}{1.32-1}\right)\times 0.8\times 1.0\times 8314.5\times 973.15$$

$$\times\left[1-\left(\frac{0.2+1.013}{2.0+1.013}\right)^{(1.32-1)/1.32}\right]$$

$$= 6.61\times 10^6\ \text{W}$$

$$= 6.61\ \text{MW}$$

The outlet temperature can be calculated from Equation 24.37:

$$T_{out} = T_{in}\eta_{IS,E}\left[\left(\frac{P_{out}}{P_{in}}\right)^{(\gamma-1)/\gamma}+\frac{1}{\eta_{IS,E}}-1\right]$$

$$= 973.15\times 0.8\left[\left(\frac{0.2+1.013}{2.0+1.013}\right)^{(1.32-1)/1.32}+\frac{1}{0.8}-1\right]$$

$$= 819.1\ \text{K}$$

$$= 545.9\ ^\circ\text{C}$$

Most applications of process expanders are in low temperature systems where the expander is used to produce a low temperature gas or vapor stream, in addition to power recovery. Thus, process expanders will be discussed again in detail in Chapter 24.

13.11 Pumping and Compression – Summary

Pumps and compressors are required to achieve the required pressure of the destination, overcome pressure drops in equipment and pipelines for the transfer of gases and liquids and to overcome changes in elevation. Pressure losses in both sections of straight pipe and pipe fittings contribute to the overall pressure drop, with the pressure drop in the pipe fittings normally making the largest contribution to the overall pressure drop. Before a design is available, the greatest uncertainty in calculating pressure drops is regarding the pipe fittings ultimately needed. Simple guidelines for the use of pipe fittings can be used to make a first estimate of the pressure drop.

For liquids, positive displacement and dynamic pumps are used. The most common type of pump used in the process industries is centrifugal. It is important to ensure that the pressure in the pump does not fall below the vapor pressure of the liquid at the pumping temperature, otherwise cavitation will occur, deteriorating the pump performance and possibly causing damage. The net positive suction head is the pressure at the pump suction above the vapor pressure of the liquid. The power required for a centrifugal pump depends on the density of the liquid, the rotational speed of the impeller, the impeller diameter, the change in head across the pump and the volumetric flowrate. Similarity laws can be used to scale from one design of pump to another.

Compressors can be classified as positive displacement, dynamic or ejectors. The most common types of compressor used in the process industries are reciprocating, centrifugal, axial and ejectors. Ejectors differ from the other compressor designs in that they need a high pressure motive fluid that is contacted with the process fluid in a convergent–divergent nozzle. Except in the case of ejectors, compression processes are modelled as either adiabatic or polytropic. If the temperature rise associated with a gas compression is unacceptably high, the compression can be broken down into a number of stages with intermediate cooling. Multistage compression reduces the overall power requirement, but at the price of increased complexity. A significant disadvantage of centrifugal compressors is that if the flow is reduced far enough, the compressor can enter a surge region and become unstable.

Process expanders are effectively process compressors working in reverse and can be used for power recovery from a high-pressure process stream.

13.12 Exercises

1. Toluene is to be pumped between a feed vessel with a pressure of 1.5 bar to another vessel with a pressure of 5 bar, 3 m in elevation above the feed vessel using a centrifugal pump with a flowrate of $30\ \text{t·h}^{-1}$. The pipe internal diameter is 77.93 mm. The pipeline is 35 m long, with four isolation valves (plug), a nonreturn (check) valve and 5 bends. The density of toluene is $778\ \text{kg·m}^{-3}$ and viscosity is $0.251\times 10^{-3}\ \text{N·s·m}^{-2}$.
   a) Estimate the pressure drop through the pipeline.
   b) Estimate the power consumed by the centrifugal pump.
2. Toluene is to be pumped between two vessels both at atmospheric pressure under flow control at a rate of $30\ \text{t·h}^{-1}$. The piping arrangement has yet to be laid out in detail but the approximate distance between the two tanks in plan is 50 m. The density of the toluene is $778\ \text{kg·m}^{-3}$ and viscosity is $0.251\times 10^{-3}\ \text{N·s·m}^{-2}$.
   a) Determine an appropriate size of pipe for a liquid velocity of around $2\ \text{m·s}^{-1}$.
   b) Estimate the pressure drop through the pipeline and the power required if the pumping is accomplished using a centrifugal pump and the tanks are at the same elevation.
   c) How high would the feed tank have to be elevated if the flow was to be accomplished by gravity?
3. The operating conditions for a pump design pumping water are to be changed. The pump at the best efficiency point delivers $30\ \text{m}^3\text{·h}^{-1}$ with a head of 43 m, which requires a power of

6.42 kW. The pump has an impeller diameter of 15.2 cm and a speed of 3500 rpm. The operating conditions are to be decreased to 3000 rpm and the impeller diameter increased to 16.5 cm. Calculate the flowrate, pump head and power for a geometrically similar pump.

4. For the problem in Exercise 1:

a) Develop a system head curve to give the relationship between the flowrate and head.

b) Match the system head curve against the pump head curve from Table 13.7 and determine the actual flowrate that will be delivered.

c) Calculate the speed required from the specific pump to deliver the specified 30 $t \cdot h^{-1}$.

5. A two-stage compression is from ambient pressure (1.013 bar) to a final pressure of 10 bar. Assuming intercooling to the inlet temperature and $\gamma = 1.3$, what should be the pressures across each compression stage for:

a) no pressure drop between stages;

b) a pressure drop of 0.2 bar between stages in the intercooler and associated piping.

6. A compressor is required to compress natural gas with a flowrate of 100,000 $Nm^3 \cdot h^{-1}$ (measured at 15 °C and 1.013 bar) from 1.013 bar to 10 bar. The inlet temperature of the gas can be assumed to be 20 °C and $\gamma = 1.3$. Assuming a two-stage compressor with an isentropic efficiency of 0.85 is to be used with intercooling to 40 °C, estimate the power requirements for:

a) no pressure drop in the intercooler;

b) a pressure drop of 0.3 bar in the intercooler;

c) a pressure drop of 0.3 bar in the intercooler and intercooling to 30 °C.

7. For the compression in Exercise 6a, repeat the calculation based on a polytropic compression assuming no pressure drop in the intercooler, intercooling to 40 °C, $\gamma = 1.3$, $\eta_{IS} = 0.85$ and the same compression ratios as those for the adiabatic compression from Exercise 6. Compare the answers with those from Exercise 6. Calculate the polytropic coefficient, polytropic efficiency and power:

a) for the first stage;

b) for the second stage.

References

Branan CR (1999) *Pocket Guide to Chemical Engineering*, Gulf Publishing Company.

Coull J and Stuart EB (1964) *Equilibrium Thermodynamics*, John Wiley & Sons.

Coulson JM and Richardson JF (1999) *Chemical Engineering, Volume 1: Fluid Flow, Heat Transfer and Mass Transfer*, 6th Edition, Butterworth-Heinemann.

Dimoplon W (1978) What Process Engineers Need to Know About Compressors, *Hydrocarbon Processing*, **57** (May): 221.

Hewitt GF, Shires GL and Bott TR (1994) *Process Heat Transfer*, CRC Press Inc.

Hougen OA, Watson KM and Ragatz RA (1959) *Chemical Process Principles. Part II: Thermodynamics*, John Wiley & Sons.

Perry RH (1997) *Chemical Engineers Handbook*, 7th Edition, McGraw-Hill.

Rase HF (1990) *Fixed-Bed Reactor Design and Diagnostics – Gas Phase Reactions*, Butterworth.

Chapter 14

Continuous Process Recycle Structure

After making some basic decisions regarding the choice of reactor and the resulting separation system, these two systems need to be brought together. Raw materials need to be brought from storage, purified or treated if necessary, and fed to the reaction system. The effluent from the reactor is passed to the separation system and the product isolated, along with byproducts and unreacted feed material. The product will most likely go forward to product storage. However, the direction of material flow is not just forward and some material, especially unreacted feed material, might be *recycled* to earlier steps in the process. Byproducts and unreacted feed material might also require storage. Completing the structure of the reactor, separation and recycle system allows a material and energy balance for the basic process to be completed. A first evaluation of the storage requirements can also be carried out.

14.1 The Function of Process Recycles

The recycling of material is an essential feature of most chemical processes. Thus, it is necessary to consider the main factors that dictate the recycle structure of a process. Start by restricting consideration to continuous processes.

1) *Reactor conversion.* Earlier in Chapters 4 to 6, the initial choice of reactor type, operating conditions and conversion was discussed. The initial assumption for the conversion varies according to whether there are single reactions or multiple reactions producing byproducts and whether reactions are reversible. Consider the simple reaction:

$$FEED \longrightarrow PRODUCT \tag{14.1}$$

To achieve complete conversion of *FEED* to *PRODUCT* in the reactor might require an extremely long residence time, which is normally uneconomic. Thus, if there is no byproduct formation, the initial reactor conversion might be set to be around 95%, as discussed in Chapter 4. The reactor effluent thus contains unreacted *FEED* and *PRODUCT* (Figure 14.1).

Because a pure product is required, a separator is needed. The unreacted *FEED* is usually too valuable to be disposed of and is therefore recycled to the reactor inlet via a pump or compressor (Figure 14.1). In addition, disposal of unreacted *FEED*, rather than recycling, could create an environmental problem.

2) *Byproduct formation.* Consider now the case where a byproduct is formed either by the primary reaction such as:

$$FEED \longrightarrow PRODUCT + BYPRODUCT \tag{14.2}$$

or via a secondary reaction such as:

$$\begin{array}{lcl} FEED & \longrightarrow & PRODUCT \\ PRODUCT & \longrightarrow & BYPRODUCT \end{array} \tag{14.3}$$

An additional separator is now required (Figure 14.2a). Again the unreacted *FEED* is normally recycled but the *BYPRODUCT* must be removed to maintain the overall material balance. An additional complication now arises with two separators because the separation sequence can be changed (Figure 14.2b).

Also, instead of using two separators, a purge can be used (Figure 14.2c). Using a purge saves the cost of a separator but incurs raw material losses, and possibly incurs waste treatment and disposal costs. This might be worthwhile if the *FEED-BYPRODUCT* separation is expensive. To use a purge, the *FEED* and *BYPRODUCT* must be adjacent to each other in the order of separation (e.g. adjacent to each other in order of volatility if differences in volatility are used as the means of separation). Care should be taken to ensure that the resulting increase in concentration of *BYPRODUCT* in the reactor does not have an adverse effect on reactor performance. Too much *BYPRODUCT* might, for example, cause deterioration in the performance of the reactor catalyst.

Chemical Process Design and Integration, Second Edition. Robin Smith.

Companion Website: www.wiley.com/go/smith/chemical2e

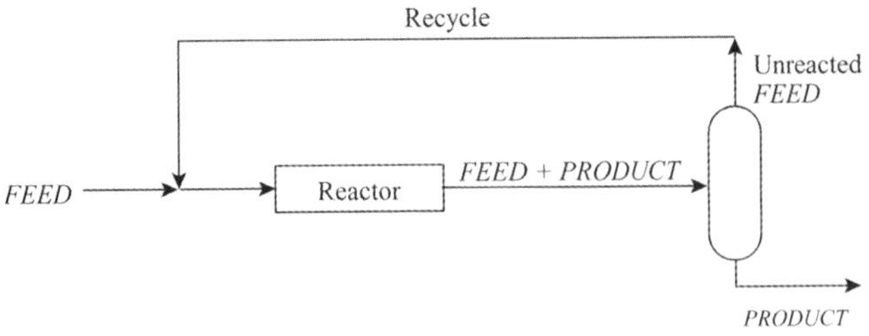

Figure 14.1

Incomplete conversion in the reactor requires a recycle for unconverted feed material.

Clearly, the separation configurations shown in Figure 14.2 change between different processes if the properties on which the separation is based change the order of separation. For example, if distillation is to be used for the separation and the order of volatility between the components changes, then the order of the separation will also change from that shown in Figure 14.2.

3) *Recycling byproducts for improved selectivity.* In systems of multiple reactions, byproducts are sometimes formed in secondary reactions that are reversible, such as:

$$\begin{aligned} &FEED \longrightarrow PRODUCT \\ &FEED \rightleftharpoons BYPRODUCT \end{aligned} \tag{14.4}$$

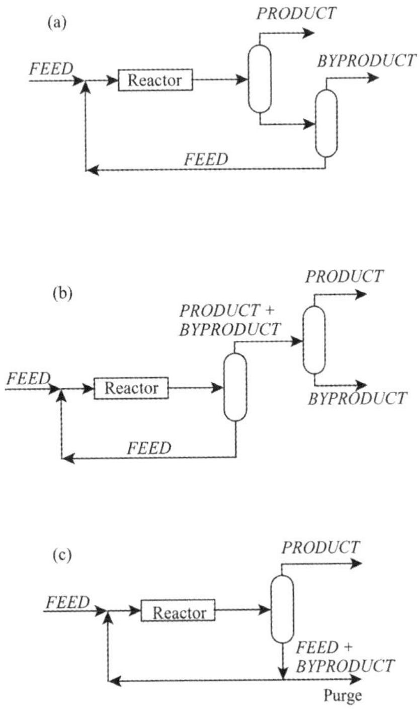

Figure 14.2

If a byproduct is formed in the reactor, then different recycle structures are possible.

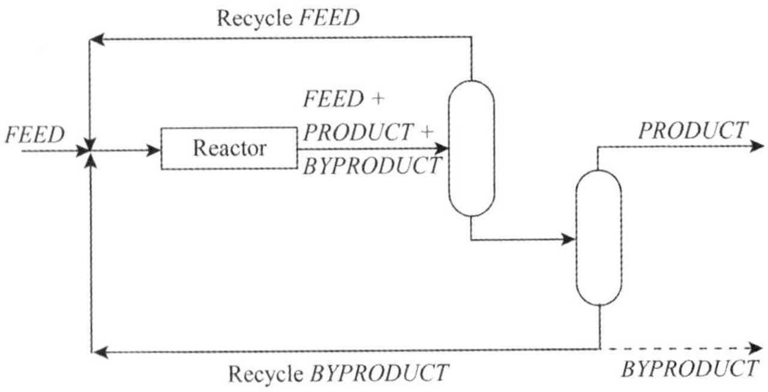

Figure 14.3

If a byproduct is formed via a reversible secondary reaction then recycling the byproduct can inhibit its formation at source.

The three recycle structures shown in Figure 14.2 can also be used with this case. However, because the *BYPRODUCT* is now being formed by a secondary reaction that is reversible, its formation at source can be inhibited by recycling *BYPRODUCT*, as shown in Figure 14.3. In Figure 14.3, the *BYPRODUCT* formation might be inhibited to the extent that it is effectively stopped, or it may be only reduced. If the formation is only reduced, then the net *BYPRODUCT* formation must be removed, as shown in Figure 14.3. Again, the separation configuration will change between different processes as the physical properties on which the separation is based change the order of separation.

4) *Recycling byproducts or contaminants that damage the reactor.* When recycling unconverted feed material, it is possible that some byproducts or contaminants, such as products of corrosion, can poison the catalyst in the reactor. Even trace quantities can sometimes be damaging to the catalyst. It is clearly desirable to remove such damaging components from the recycle in arrangements similar to Figure 14.2.

5) *Feed impurities.* So far it has been assumed that the feed material is pure. An impurity in the feed affects the recycle structure and opens up further options. The first option in Figure 14.4a shows the impurity being separated before entering the process. If the impurity has an adverse effect on the reaction or poisons the catalyst, this is the obvious solution. However, if the impurity does not have a significant effect on the reaction, then it could perhaps be passed through the reactor and be separated as shown in Figure 14.4b. Alternatively, the separation sequence could be changed as shown in Figure 14.4c (Smith and Linnhoff, 1988).

The fourth option shown in Figure 14.4d uses a purge (Smith and Linnhoff, 1988). As with its use to separate byproducts, the purge saves the cost of a separation, but incurs raw material losses. This might be worthwhile if the *FEED-IMPURITY* separation is expensive. To use a purge, the *FEED* and *IMPURITY* must be adjacent to each other in the order of separation (e.g. adjacent in the order of volatility if a single-stage flash or distillation is used for the

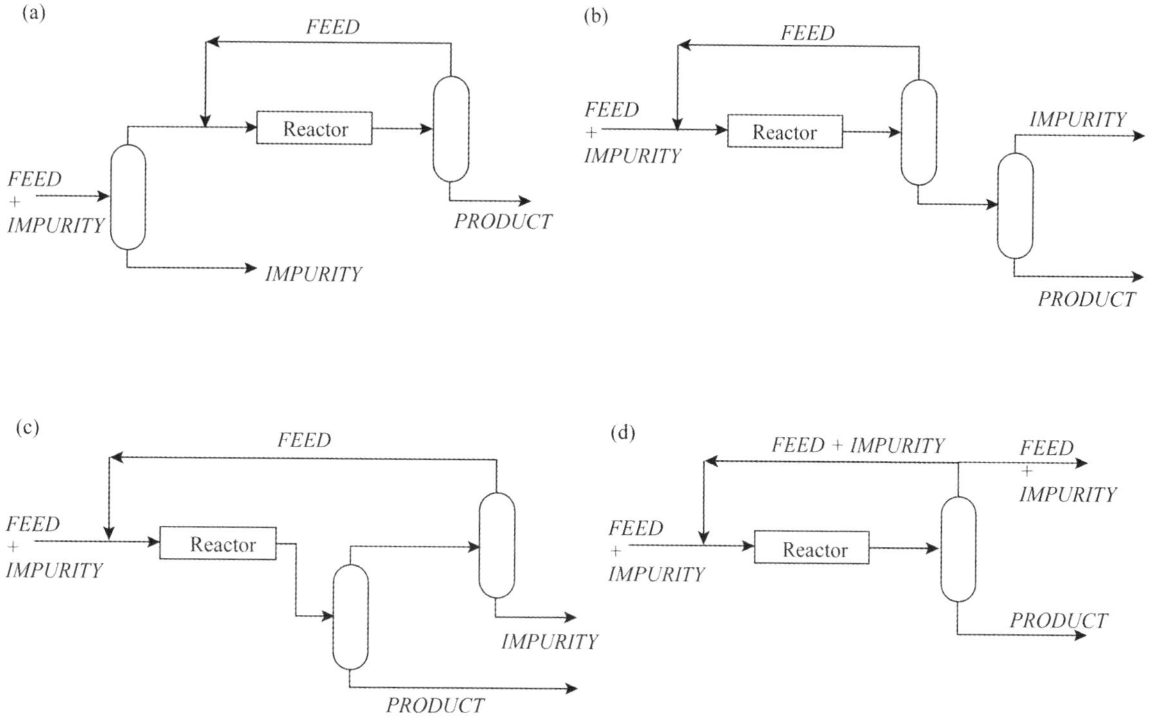

Figure 14.4

Introduction of an impurity in the feed creates further options for recycle structures.

separation). Care should be taken to ensure that the resulting increase in concentration of *IMPURITY* in the reactor does not have an adverse effect on reactor performance.

Again, the separation configuration will change between different processes as the properties on which the separation is based change the order of separation, when compared with Figure 14.4.

6) *Reactor diluents and solvents.* As pointed out in Chapter 5, an inert diluent such as steam is sometimes needed in the reactor to lower the partial pressure of reactants in the vapor phase. Diluents are normally recycled. An example is shown in Figure 14.5. The actual configuration used depends on the order of separation.

   Many liquid-phase reactions are carried out in a solvent. If this is the case, then the solvent should initially be assumed to be separated and recycled in arrangements similar to that in Figure 14.5. In some cases, the solvent will be contaminated with byproducts of reaction that will need to be separated and disposed of (e.g. by thermal oxidation). In this case, it might be cheaper to dispose of the entire solvent, rather than separate and recycle the solvent. However, for the efficient use of materials, every effort should be made to recycle solvents, as illustrated in Figure 14.5.

7) *Reactor heat carrier.* As pointed out in Chapter 6, if adiabatic operation is not possible and it is not possible to control temperature by indirect heat transfer, then an inert material can be introduced to the reactor to increase its heat capacity flowrate (i.e. product of mass flowrate and specific heat capacity). This will reduce a temperature rise for exothermic reactions or reduce a temperature decrease for endothermic reactions. The introduction of an extraneous component as a heat carrier affects the recycle structure of the flowsheet. Figure 14.6a shows an example of the recycle structure for just such a process.

   Where possible, introducing extraneous materials into the process should be avoided, and material already present in the process should be used. Figure 14.6b illustrates the use of the product as the heat carrier. This simplifies the recycle structure of the flowsheet and removes the need for one of the separators (Figure 14.6b). The use of the product as heat carrier is obviously restricted to situations where the product does not undergo secondary reactions to unwanted byproducts. Unconverted feed that is recycled also acts as a heat

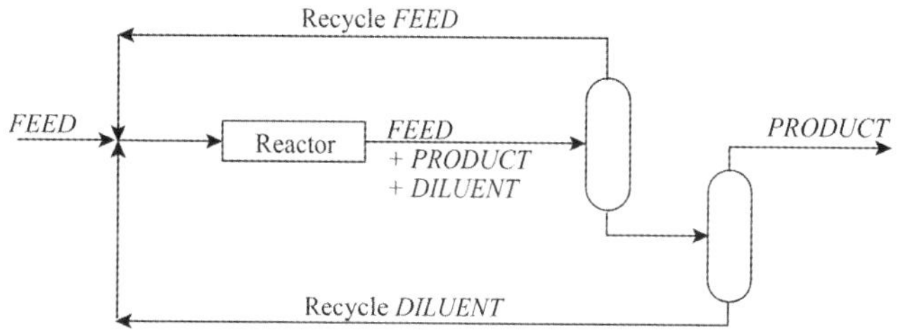

Figure 14.5

Diluents and solvents are normally recycled.

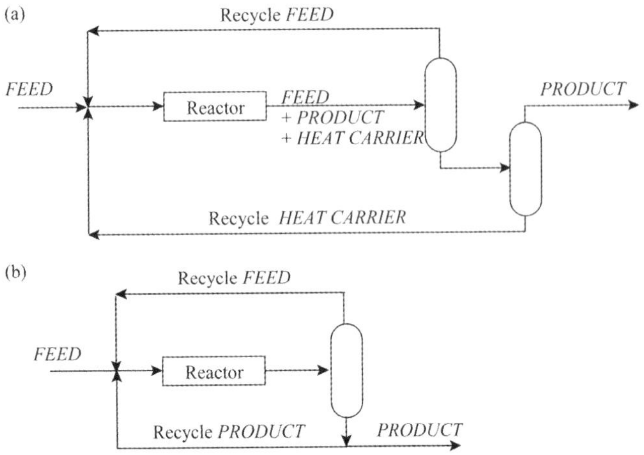

Figure 14.6

Heat carriers are normally recycled.

carrier. Thus, rather than relying on recycled product to limit the temperature rise (fall), simply opt for a low conversion, a high recycle of feed and a resulting small temperature change in the reactor.

Other options are possible. If the process produces a byproduct of reaction, then this can be recycled, provided it does not have an adverse effect on the reactor performance. If the feed enters with an impurity, then the impurity could also be recycled as a heat carrier, provided it does not have an adverse effect on the reactor performance.

Whether an extraneous component, product, feed, byproduct or feed impurity is used as a heat carrier, as before, the actual configuration of the separation configuration will change between different processes as the properties on which the separation is based change the order of separation.

Having considered the main factors that determine the need for recycles, care should be taken if a flowsheet requires multiple recycles. It is counterproductive to separate two components adjacent in the order of separation that are to be recycled to the reactor, since they would only be remixed at some point before entering the reactor. The designer should always be on guard to avoid unnecessary separation and unnecessary mixing.

Example 14.1 Monochlorodecane (MCD) is to be produced from decane (DEC) and chlorine via the reaction (Powers, 1972; Rudd, Powers and Siirota, 1973):

$$\underset{\text{DEC}}{C_{10}H_{22}} + \underset{\text{chlorine}}{Cl_2} \longrightarrow \underset{\text{MCD}}{C_{10}H_{21}Cl} + \underset{\text{hydrogen chloride}}{HCl}$$

A side reaction occurs in which dichlorodecane (DCD) is produced:

$$\underset{\text{MCD}}{C_{10}H_{21}Cl} + \underset{\text{chlorine}}{Cl_2} \longrightarrow \underset{\text{DCD}}{C_{10}H_{20}Cl_2} + \underset{\text{hydrogen chloride}}{HCl}$$

The byproduct DCD is not required. Hydrogen chloride can be sold to a neighboring plant. Assume at this stage that all separations can be carried out by distillation. The normal boiling points are given in the Table 14.1.

Table 14.1

Data for the materials involved with the production of monochlorodecane.

Material	Molar mass ($kg \cdot kmol^{-1}$)	Normal boiling point (K)	Value ($\$ \cdot kg^{-1}$)
Hydrogen chloride	36	188	0.35
Chlorine	71	239	0.21
Decane	142	447	0.27
Monochlorodecane	176	488	0.45
Dichlorodecane	211	514	0

a) Determine alternative recycle structures for the process by assuming different levels of conversion of raw materials and different excesses of reactants.

b) Which structure is most effective in suppressing the side reaction?

c) What is the minimum selectivity of decane that must be achieved for profitable operation? The values of the materials involved, together with their molar mass, are given in the Table 14.1.

Solution

a) Four possible arrangements can be considered:

i) *Complete conversion of both feeds.* Figure 14.7a shows the most desirable arrangement; complete conversion of the decane and chlorine in the reactor. The absence of reactants in the reactor effluent means that no recycles are needed.

Although the flowsheet shown in Figure 14.7a is very attractive, it is not practical. This would require careful control of the stoichiometric ratio of decane to chlorine, taking into account both the requirements of the primary and byproduct reactions. Even if it was possible to balance out the reactants exactly, a small upset in process conditions would create an excess of either decane or chlorine and these would then appear as components in the reactor effluent. If these components appear in the reactor effluent of the flowsheet in Figure 14.7a, there are no separators to deal with their presence and no means of recycling unconverted raw materials.

Also, although there are no selectivity data for the reaction, the selectivity losses would be expected to increase with increasing conversion. Complete conversion would tend to produce high byproduct formation and poor selectivity. Finally, the reactor volume required to give complete conversion would be extremely large.

ii) *Incomplete conversion of both feeds.* If complete conversion is not practical, consider incomplete conversion. This is shown in Figure 14.7b. In this case, all components are present in the reactor effluent, and one additional separator and a recycle are required. Thus the complexity of the flowsheet is significantly increased, compared with that for complete conversion.

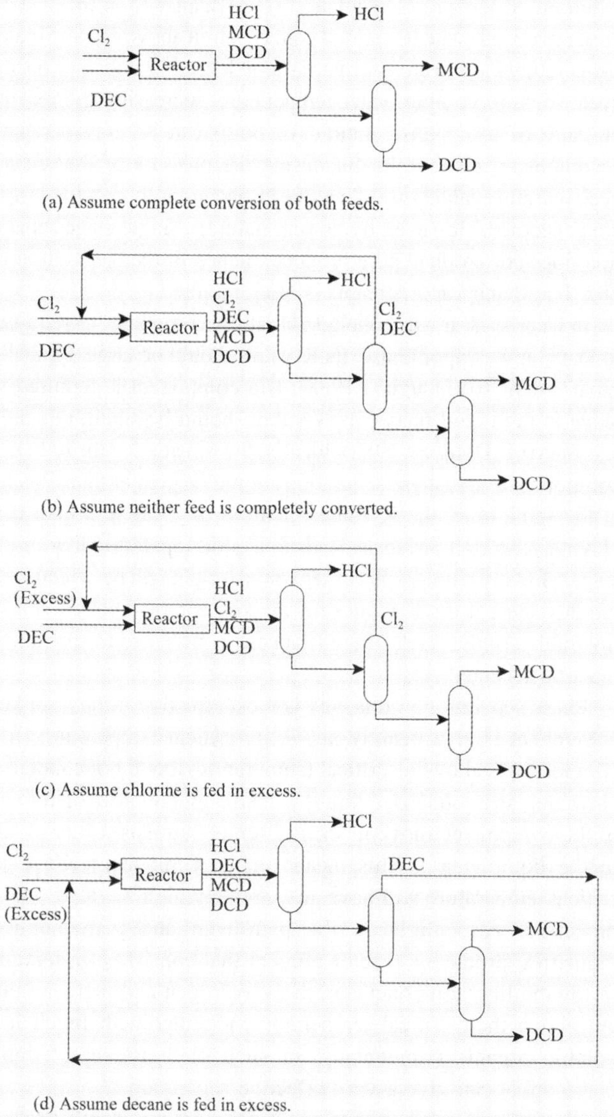

Figure 14.7

Different recycle structures for the production of monochlorodecane.

Note that no attempt has been made to separate the chlorine and decane since they are remixed after recycling to the reactor.

iii) *Excess chlorine.* Use of excess chlorine in the reactor can force the decane to effectively complete conversion (Figure 14.7c). Now there is effectively no decane in the reactor effluent and again three separators and a recycle are required.

In practice, there is likely to be a trace of decane in the reactor effluent. However, this should not be a problem since it can either be recycled with the unreacted chlorine or it can leave with the product monochlorodecane (providing it can still meet product specifications).

At this stage, how large the excess of chlorine should be for Figure 14.7c to be feasible cannot be specified. Experimental data on the reaction chemistry would be required in order to establish this. However, the size of the excess does not change the basic structure.

iv) *Excess decane.* Use of excess decane in the reactor forces the chlorine to effectively complete conversion (Figure 14.7d). Now there is effectively no chlorine in the reactor effluent. Again, three separators are required and a recycle of unconverted raw material.

In practice, there is likely to be a trace of chlorine in the reactor effluent. This can be recycled to the reactor with the unreacted decane or allowed to leave with the hydrogen chloride byproduct (providing this meets with the byproduct specifications).

Again at this stage, it cannot be determined exactly how large an excess of decane would be required in order to make Figure 14.7d feasible. This would have to be established from experimental data, but the size of the excess does not change the basic flowsheet structure.

b) An arrangement is to be chosen to inhibit the side reaction, that is, to give high selectivity. The side reaction is suppressed by starving the reactor of either monochlorodecane or chlorine. Since the reactor is designed to produce monochlorodecane, the former option is not sensible. However, it is practical to use an excess of decane.

The last of the four flowsheet options generated above, which features excess decane in the reactor, is therefore preferred as shown in Figure 14.7d.

c) The selectivity (S) is defined by

$$S = \frac{\text{MCD produced in the reactor}}{\text{DEC consumed in the reactor}} \times \text{Stoichiometric factor}$$

In this case, the stoichiometric factor is one. This is a measure of the MCD obtained from the DEC consumed. To assess the selectivity losses, the MCD produced in the primary reaction is split into that fraction that will become the final product and that which will become the byproduct. Thus the reaction stoichiometry is:

$$C_{10}H_{22} + Cl_2 \longrightarrow SC_{10}H_{21}Cl + (1-S)C_{10}H_{21}Cl + HCl$$

and, for the byproduct reaction:

$$(1-S)C_{10}H_{21}Cl + (1-S)Cl_2 \longrightarrow (1-S)C_{10}H_{20}Cl_2 + (1-S)HCl$$

Adding the two reactions gives overall:

$$\begin{aligned} C_{10}H_{22} + (2-S)Cl_2 \longrightarrow\ & SC_{10}H_{21}Cl + (1-S)C_{10}H_{20}Cl_2 \\ & +(2-S)HCl \end{aligned}$$

Considering raw materials costs only, the economic potential, EP, of the process is defined as:

$$\begin{aligned} EP &= \text{Value of products} - \text{Raw materials costs} \\ &= [176 \times S \times 0.45 + 36 \times (2-S) \times 0.35] \\ &\quad - [142 \times 1 \times 0.27 + 71 \times (2-S) \times 0.21] \\ &= 79.2S - 2.31(2-S) - 38.34\ (\$ \cdot \text{kmol}^{-1}\ \text{decane reacted}) \end{aligned}$$

The minimum selectivity that can be tolerated is given when the economic potential is just zero:

$$\begin{aligned} 0 &= 79.2S - 2.31(2-S) - 38.34 \\ S &= 0.53 \end{aligned}$$

In other words, the process must convert at least 53% of the decane that reacts to monochlorodecane rather than to dichlorodecane for the process to be economic. This figure assumes that the hydrogen chloride is sold to a neighboring process. If this is not the case, there is no value associated with the hydrogen chloride. Assuming there are no treatment and disposal costs for the now waste hydrogen chloride, the minimum economic potential is given by:

$$\begin{aligned} 0 &= [176 \times S \times 0.45] - [142 \times 1 \times 0.27 + 71 \times (2 - S) \times 0.21] \\ &= 79.2S - 14.91(2 - S) - 38.4 \\ S &= 0.72 \end{aligned}$$

Now the process must convert at least 72% of the decane to monochlorodecane.

If the hydrogen chloride cannot be sold, it must be disposed of somehow. Alternatively, it could be converted back to chlorine via the reaction:

$$2HCl + \frac{1}{2}O_2 \rightleftharpoons Cl_2 + H_2O$$

and then recycled to the MCD reactor. Now the overall stoichiometry changes since the $(2 - S)$ moles of HCl that were being produced as byproduct are now being recycled to substitute fresh chlorine feed:

$$(2 - S)HCl + \frac{1}{4}(2 - S)O_2 \longrightarrow \frac{1}{2}(2 - S)Cl_2 + \frac{1}{2}(2 - S)H_2O$$

Thus the overall reaction now becomes:

$$\begin{aligned} &C_{10}H_{22} + \frac{1}{2}(2 - S)Cl_2 + \frac{1}{4}(2 - S)O_2 \longrightarrow \\ &SC_{10}H_{21}Cl + (1 - S)C_{10}H_{20}Cl_2 + \frac{1}{2}(2 - S)H_2O \end{aligned}$$

The economic potential is now given by:

$$\begin{aligned} 0 &= [176 \times S \times 0.45] - [142 \times 1 \times 0.27 + 71 \times 1/2(2 - S) \times 0.21] \\ &= 79.2S - 7.455(2 - S) - 38.34 \\ S &= 0.61 \end{aligned}$$

The minimum selectivity that can now be tolerated becomes 61%.

14.2 Recycles with Purges

As discussed in the previous section, a purge can be used to avoid the cost of separating a component from a recycle. Purges can, in principle, be used either with liquid or vapor (gas) recycles. However, purges are most often used to remove low-boiling components from vapor (gas) recycles.

A common situation is encountered when the effluent from a chemical reactor contains components with large relative volatilities. As discussed in Chapter 8, a partial condensation of the mixture from the vapor phase followed by a simple phase split can often produce an effective separation. Cooling below cooling water temperature is not preferred, otherwise refrigeration is required. A dew-point calculation (see Chapter 8) at the system pressure reveals whether partial condensation above cooling water temperatures is effective. If partial condensation does not occur, even down to cooling water temperature, increasing the reactor pressure or using refrigeration (or both) can be considered to accomplish a phase split. Increasing the pressure of the reactor needs careful evaluation as far as the implications for reactor design are concerned. However, by its very nature, a single-stage separation does not produce pure products; hence further separation of both liquid and vapor streams is often required.

In Chapter 8, it was concluded that if a component is required to leave in the vapor phase, its K-value should be large, typically greater than 10 (Douglas, 1988). If a component is required to leave in the liquid phase, its K-value should be small, typically less than 0.1 (Douglas, 1988). Ideally, the K-value for the light key component in the phase separation should be greater than 10 and, at the same time, the K-value for the heavy key less than 0.1. Having such circumstances leads to a good separation in a single stage. However, use of phase separators might still be effective in the flowsheet if the K-values for the key components are not so favorable. Under such circumstances, a more crude separation must be accepted.

Phase separation in this way is most effective if the light key component is significantly above its critical temperature. If a component is above its critical temperature, it will not condense. However, any condensed liquid will still contain a small amount of the component above its critical temperature as it 'dissolves' in the liquid phase. This means that a component above its critical temperature will have an extremely high K-value. Many processes, particularly in the petroleum and petrochemical industries, produce a reactor effluent that consists of a mixture of low-boiling components such as hydrogen and methane that are above their critical temperature, together with much less volatile organic components. In such circumstances, simple phase splits can give a very effective separation.

If the vapor from the phase split is either predominantly product or predominantly byproduct, then it can be removed from the process. If the vapor contains predominantly unconverted feed material, it is normally recycled to the reactor. If the vapor stream consists of a mixture of unconverted feed material, products and byproducts, then some separation of the vapor may be needed. The vapor from the phase split will be difficult to condense if the feed to the phase split has been cooled to cooling water temperature. If separation of the vapor is needed in such circumstances, one of the following methods can be used:

1) *Refrigerated condensation.* Separation by condensation relies on differences in volatility between the condensing components. Refrigeration, or a combination of high pressure and refrigeration, is needed. If a single-stage separation using refrigerated condensation does not give an adequate separation, the process can be repeated in a *refluxed condenser* or *dephlegmator*. This is a plate-fin heat exchanger (as will be discussed in Chapter 12). The vapor flows up through the vertical condenser and the condensed liquid flows down over the heat exchange surface countercurrently. Mass is exchanged between the upward flowing

vapor and the downward flowing condensed liquid. A separation of typically up to eight theoretical stages can be achieved in such a device.

2) *Low-temperature/high-pressure distillation.* Rather than use a low-temperature single-stage condensation or refluxed condenser, a conventional distillation can be used. To carry out the separation under these circumstances will require a low-temperature condenser for the column, or operation at high pressure, or a combination of both.

3) *Absorption.* Absorption was discussed in Chapter 9. If possible, a component that already exists in the flowsheet should be used as a solvent. Introducing an extraneous component into the flowsheet introduces additional complexity and the possibility of increased environmental and safety problems later in the design.

4) *Adsorption.* Adsorption involves the transfer of a component on to a solid surface (Chapter 9). The adsorbent will need to be regenerated by a gas or vapor when the bed approaches saturation. As discussed in Chapter 9, a vapor or gas can be used for regeneration. However, this can introduce extraneous material into the process, with the regeneration stream needing further separation. Thus, for such applications, regeneration by a change in pressure (pressure swing adsorption) would normally be preferred.

5) *Membrane separation.* Membranes, as discussed in Chapter 9, separate gases by means of a pressure gradient across a membrane, typically 40 bar or greater. Some gases permeate through the membrane faster than others and concentrate on the low-pressure side. Low molar mass gases and strongly polar gases have high permeabilities and are known as *fast gases*. *Slow gases* have higher molar mass and symmetrical molecules. Thus, membrane gas separators are effective when the gas to be separated is already at a high pressure and only a partial separation is required.

In situations where a large vapor (gas) flow having a dew point below cooling water temperature is to be recycled back to the reactor, it is often expensive to carry out such separations on the recycle. This is especially true when relatively small amounts of material need to be separated from the recycle. Rather than carry out a separation on the recycle vapor (gas) stream, a purge from the recycle stream can allow such material to be removed without the need to carry out a separation. Although the purge removes the need for a separator, it incurs raw material losses. Not only can these material losses be expensive, but they can also create environmental problems. However, another option is to use a combination of a purge with a separator on the purge.

As an example, consider ammonia synthesis. In an ammonia synthesis loop, hydrogen and nitrogen are reacted to ammonia. The reactor effluent is partially condensed to separate ammonia as a liquid. Unreacted gaseous hydrogen and nitrogen are recycled to the reactor. A purge on the recycle prevents the build-up of argon and methane that enter the system as feed impurities. Although the purge can be burnt as fuel, considerable quantities of hydrogen are lost and therefore recovery of this hydrogen is usually economic.

For such hydrogen recovery systems, adsorption, a membrane or cryogenic condensation could be used. For hydrogen recovery from ammonia purge gas, a membrane is usually the most economic option. The membrane allows fast gases such as hydrogen to be separated from slow gases such as methane. The driving force for the permeation of the fast gas (and hence the separation of the fast gas from the other slower components) is the difference in partial pressure from one side of the membrane to the other. Hence, for recovery of hydrogen, the product stream must be at a substantially lower pressure than the feed stream.

If the liquid from the phase split requires separation, then this can normally be accomplished by distillation, except under special circumstances. A distillation sequence is most often required with products and byproducts removed from the process and unreacted feed materials recycled. In some situations, byproducts might be recycled for reasons discussed in the previous section.

The following example illustrates the quantitative relationships involving the use of a purge on a gas recycle stream.

Example 14.2 Benzene is to be produced from toluene according to the reaction (Douglas, 1985):

$$\underset{\text{toluene}}{C_6H_5CH_3} + \underset{\text{hydrogen}}{H_2} \longrightarrow \underset{\text{benzene}}{C_6H_6} + \underset{\text{methane}}{CH_4}$$

The reaction is carried out in the gas phase and normally operates at around 700 °C and 40 bar. Some of the benzene formed undergoes a series of secondary reactions. These are characterized here by the single secondary reversible reaction to an unwanted byproduct, diphenyl, according to the reaction:

$$\underset{\text{benzene}}{2C_6H_6} \rightleftharpoons \underset{\text{diphenyl}}{C_{12}H_{10}} + \underset{\text{hydrogen}}{H_2}$$

Laboratory studies indicate that a hydrogen/toluene ratio of 5 at the reactor inlet is required to prevent excessive coke formation in the reactor. Even with a large excess of hydrogen, the toluene cannot be forced to complete conversion. The laboratory studies indicate that the selectivity (i.e. fraction of toluene reacted that is converted to benzene) is related to the conversion (i.e. fraction of toluene fed that is reacted) according to (Douglas, 1985):

$$S = 1 - \frac{0.0036}{(1 - X)^{1.544}}$$

where S = selectivity
X = conversion

The reactor effluent is thus likely to contain hydrogen, methane, benzene, toluene and diphenyl. Because of the large differences in volatility of these components, it seems likely that partial condensation will allow the effluent to be split into a vapor stream containing predominantly hydrogen and methane, and a liquid stream containing predominantly benzene, toluene and diphenyl.

The hydrogen in the vapor stream is a reactant and hence should be recycled to the reactor inlet (Figure 14.8). The methane

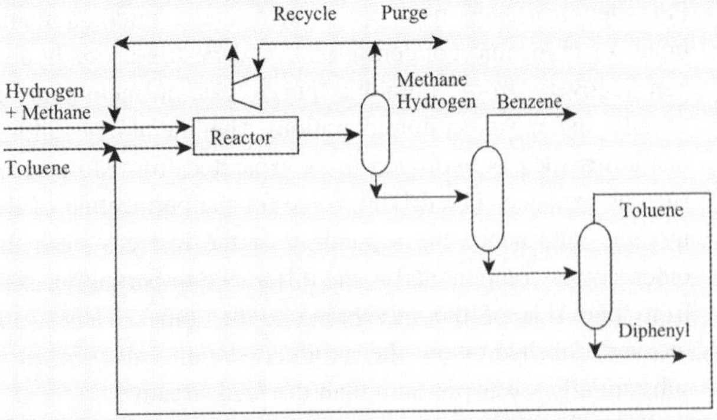

Figure 14.8
A flowsheet for the production of benzene uses purge to remove the methane that enters as a feed impurity and is also formed as a byproduct.

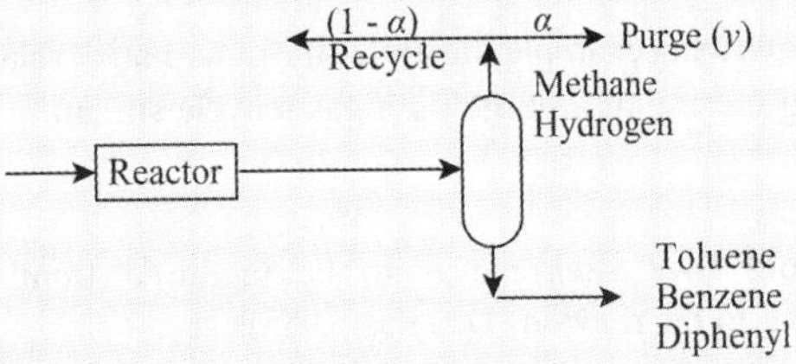

Figure 14.9
Purge fraction for the recycle.

enters the process as a feed impurity and is also a byproduct from the primary reaction and must be removed from the process. The hydrogen–methane separation is likely to be expensive but the methane can be removed from the process by means of a purge (Figure 14.8).

The liquid stream can readily be separated into relatively pure components by distillation, the benzene taken off as product, diphenyl as an unwanted byproduct and the toluene recycled. It is possible to recycle the diphenyl to improve selectivity, but it will be assumed that is not done here. The hydrogen feed contains methane as an impurity at a mole fraction of 0.05. The production rate of benzene required is 265 kmol·h^{-1}. Assume initially that a phase split can separate the reactor effluent into a vapor stream containing only hydrogen and methane, and a liquid containing only benzene, toluene and diphenyl, and that it can be separated to produce essentially pure products. For a conversion in the reactor of 0.75:

a) Determine the relation between the fraction of vapor from the phase split sent to purge (α) and the fraction of methane in the recycle and purge (y), as shown in Figure 14.9.

b) Given the assumptions, estimate the composition of the reactor effluent for the fraction of methane in the recycle and purge of 0.4.

Solution

a) Following Douglas (1985) with the benzene production rate to be P_B, the benzene produced by the primary reaction is split into two parts, one part forming the final product and the other reacting to byproduct:

$$\begin{array}{ccc} C_6H_5CH_3 + H_2 & \longrightarrow & C_6H_6 + C_6H_6 + CH_4 \\ \frac{P_B}{S} + \frac{P_B}{S} & & P_B + P_B\left(\frac{1}{S}-1\right) + \frac{P_B}{S} \end{array}$$

$$\begin{array}{ccc} 2C_6H_6 & \rightleftharpoons & C_{12}H_{10} + H_2 \\ P_B\left(\frac{1}{S}-1\right) & & \frac{P_B}{2}\left(\frac{1}{S}-1\right) + \frac{P_B}{2}\left(\frac{1}{S}-1\right) \end{array}$$

For $X = 0.75$:

$$S = 1 - \frac{0.0036}{(1-0.75)^{1.544}} = 0.9694$$

Toluene balance

$$\text{Fresh toluene feed} = \frac{P_B}{S}$$

$$\text{Toluene recycle} = R_T$$

$$\text{Toluene entering the reactor} = \frac{P_B}{S} + R_T$$

$$\text{Toluene in reactor effluent} = \left(\frac{P_B}{S} + R_T\right)(1-X) = R_T$$

For $P_B = 265$ kmol·h^{-1}, $X = 0.75$ and $S = 0.9694$:

$$R_T = 91.12\ \text{kmol}\cdot\text{h}^{-1}$$

$$\text{Toluene entering the reactor} = \frac{265}{0.9694} + 91.12$$

$$= 364.5\ \text{kmol}\cdot\text{h}^{-1}$$

Hydrogen balance

$$\text{Hydrogen entering the reactor} = 5 \times 364.5$$

$$= 1823\ \text{kmol}\cdot\text{h}^{-1}$$

$$\text{Net hydrogen consumed in reaction} = \frac{P_B}{S} - \frac{P_B}{2}\left(\frac{1}{S}-1\right)$$

$$= \frac{P_B}{S}\left(1 - \frac{1-S}{2}\right)$$

$$= 269.2\ \text{kmol}\cdot\text{h}^{-1}$$

$$\text{Hydrogen in reactor effluent} = 1823 - 269.2$$

$$= 1554\ \text{kmol}\cdot\text{h}^{-1}$$

$$\text{Hydrogen lost in purge} = 1554\alpha$$

$$\text{Hydrogen feed to the process} = 1554\alpha + 269.2$$

Methane balance

$$\text{Methane feed to process as impurity} = (1554\alpha + 269.2)\frac{0.05}{0.95}$$

$$\text{Methane produced by reactor} = \frac{P_B}{S}$$

$$\text{Methane in purge} = \frac{P_B}{S} + (1554\alpha + 269.2)\frac{0.05}{0.95}$$

$$= 81.79\alpha + 287.5$$

$$\text{Total flowrate of purge} = 1554\alpha + 81.79\alpha + 287.5$$

$$= 1636\alpha + 287.5$$

Fraction of methane in the purge (and recycle):

$$y = \frac{81.79\alpha + 287.5}{1636\alpha + 287.5} \tag{14.5}$$

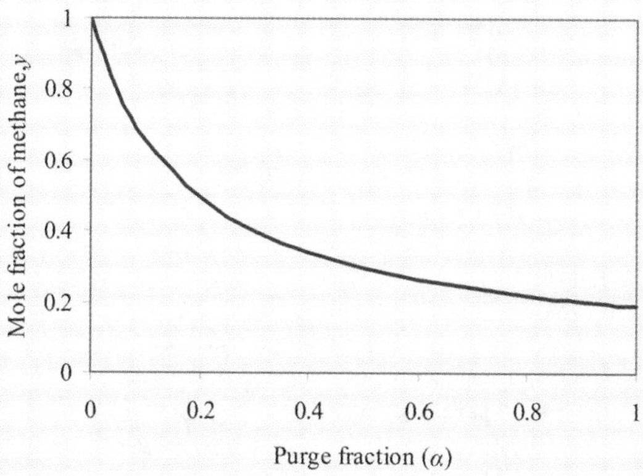

Figure 14.10
Variation of vapor composition with purge fraction.

Figure 14.10 shows a plot of Equation 14.5. As the purge fraction α is increased, the flowrate of purge increases but the concentration of methane in the purge and recycle decreases. This variation (along with reactor conversion) is an important degree of freedom in the optimization of reaction and separation systems as will be discussed later.

b) Methane balance

Mole fraction of methane in vapor from phase separator = 0.4

$$\text{Methane in reactor effluent} = \frac{0.4}{0.6} \times 1554$$
$$= 1036\ \text{kmol} \cdot \text{h}^{-1}$$

Diphenyl balance

$$\text{Diphenyl in reactor effluent} = \frac{P_B}{2}\left(\frac{1}{S} - 1\right)$$
$$= 4\ \text{kmol} \cdot \text{h}^{-1}$$

The estimated composition of the reactor effluent are given in Table 14.2. This calculation assumes that all separations in the phase split are sharp.

Table 14.2
Composition of reactor effluent for Example 14.2.

Component	Flowrate (kmol·h^{-1})
Hydrogen	1554
Methane	1036
Benzene	265
Toluene	91
Diphenyl	4

The above example illustrates some important principles for the design of recycles with purges:

1) There must be an overall mass balance in which the mass of the component to be separated that enters with the feed, or is formed in the reactor, must equal the mass of that component leaving with the purge gas, plus that which leaves with the liquid. If the mass that leaves with the liquid is extremely small relative to the purge, then effectively all of this mass must leave with the purge.
2) The concentration of the recycle can be controlled by varying the purge fraction. Decreasing the purge fraction causes the concentration of the component being purged to increase and vice versa.
3) A given mass of material can be purged with a low flowrate and high concentration, leading to a low loss of useful material in the purge. Alternatively, a given mass of material can be purged with a high flowrate and low concentration, leading to a high loss of useful material in the purge.
4) Purging with a low flowrate and high concentration leads to a relatively high recycle flowrate, and hence high costs for the recycle. On the other hand, purging with a high flowrate and low concentration leads to a relatively low recycle flowrate, and hence low costs for the recycle.
5) There are important costs to be traded off and will be considered in Chapter 15.

14.3 Hybrid Reaction and Separation

So far, it has been assumed that the reaction and separation would be carried out consecutively and connected with recycle streams if appropriate. Consider a liquid-phase exothermic equilibrium reaction, such as:

$$FEED1 + FEED2 \rightleftharpoons PRODUCT + BYPRODUCT \tag{14.6}$$

If the reaction is carried out in a reactor, such as that shown in Figure 14.11a, in which the product is allowed to vaporize and leave the reactor, then the equilibrium is shifted to higher conversion, as discussed in Chapter 5. However, the vaporization would be expected to be such that not just the product would vaporize but also perhaps the feed material and byproduct. A distillation rectifying section could be added to the reactor as shown in Figure 14.11b to carry out rectification and produce a pure *PRODUCT* from the top of the distillation. Of course, for this to be possible, the relative volatilities of the various components must be appropriate both in their order and their magnitude. Also, the temperature at which the reaction and distillation take place needs to be similar. The arrangement shown in Figure 14.11b, in addition to having advantages for the reactor, is also energy efficient in that the vapor supplied to the distillation comes from the heat of reaction. This idea is carried a step further by adding a distillation stripping section, as shown in Figure 14.11c,

14

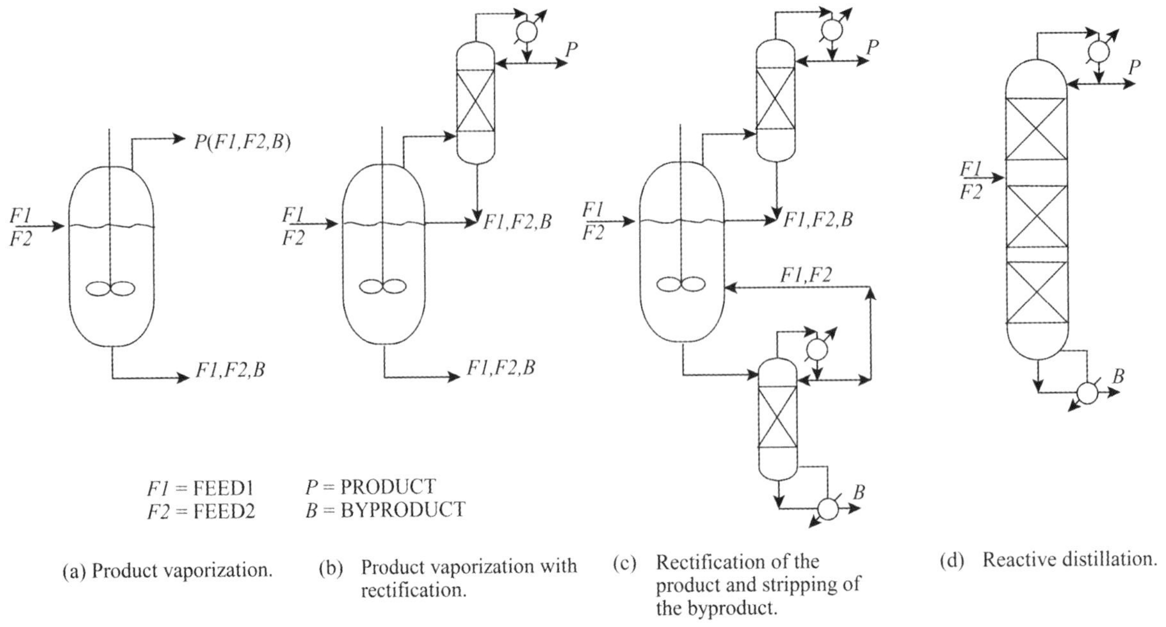

Figure 14.11
Hybrid reaction and distillation – reactive distillation.

to separate the pure *BYPRODUCT* from the liquid leaving the reactor and return the feed material to the reactor. Again, the relative volatilities of the components must be appropriate both in their order and their magnitude to allow this to be achieved. Finally, the whole system is fitted into a single shell and the result is a *reactive distillation*, as shown in Figure 14.11d.

It may also be useful to carry out such hybrid reaction and separation when byproducts are formed from competing reactions, such as:

$$\begin{aligned} FEED1 + FEED2 &\longrightarrow PRODUCT \\ PRODUCT &\longrightarrow BYPRODUCT \end{aligned} \tag{14.7}$$

It is desirable to remove product from the reaction as soon as it is formed to prevent it from reacting further to byproduct. An arrangement such as that shown in Figure 14.11d would allow this to happen if the relative volatilities between the components have the appropriate order and magnitude.

Thus, there are a number of potential advantages for reactive distillation:

- reaction equilibrium can be shifted to higher (or even complete) conversion;
- side reactions can be suppressed and selectivity increased;
- capital investment can be reduced;
- exothermic reaction heat used to provide the heat for the separation and reduce operating costs.

In some fortunate circumstances, azeotropes can be eliminated from the separation that would need to be dealt with if reaction and separation are carried out consecutively.

The disadvantages of reactive distillation are:

- favorable relative volatilities are needed;
- distillation conditions must give an adequate reaction rate;
- thorough research, testing and even pilot plant trials are required.

Figure 14.12 shows another example of hybrid reaction and separation. This shows an esterification reaction in which water is removed between the reaction stages using pervaporation as discussed in Chapter 11 (Wynn, 2001). The esterification uses a heterogeneous catalyst and the process in Figure 14.12 runs in four stages. Each stage includes a reactor where the components are brought close to equilibrium and then the mixture flows through a pervaporation stage where the water generated in the reaction step is removed. This shifts the equilibrium conversion favorably. In the next reaction step, equilibrium is reestablished and again the reaction water is removed, and so on.

14.4 The Process Yield

Having considered the feed, reaction, separation and recycling of material, the streams entering and leaving the process can be established. Figure 14.13 illustrates typical input and output streams. Feed streams enter the process and product, byproduct and purge streams leave after the separation and recycle system has been established.

Raw materials costs dominate the operating costs of most processes (see Chapter 2). Also, if raw materials are not used efficiently, this creates waste that becomes an environmental

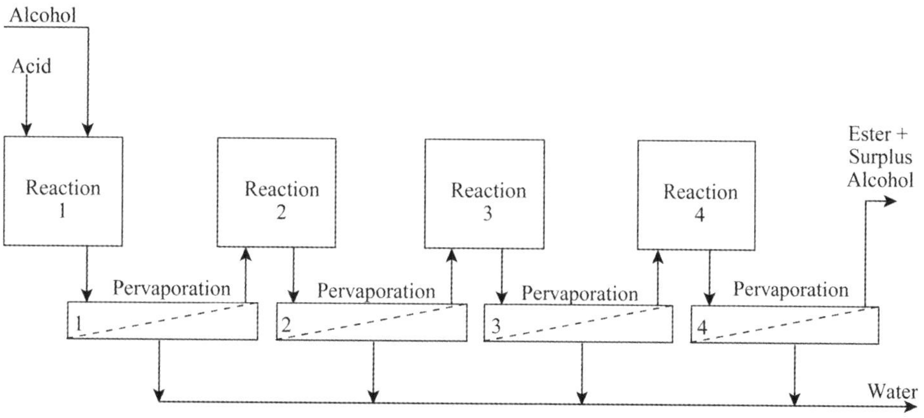

Figure 14.12

Hybrid reaction and pervaporation. (From Wynn N, 2001, *Chem Eng Progr*, Oct: 66, reproduced by permission of AiChE.)

problem. It is therefore important to have a measure of the efficiency of raw materials usage. The process yield is defined as:

$$\text{Process yield} = \frac{\text{Desired product produced}}{\text{Reactant fed to the process}} \times \text{Stoichiometric factor} \tag{14.8}$$

where the stoichiometric factor is the stoichiometric moles of reactant required per mole of product. When more than one reactant is used (or more than one desired product produced) Equation 14.8 can be applied to each reactant (or product).

In broad terms, there are two sources of yield loss in the process:

- losses in the reactor due to byproduct formation (selectivity losses) or unconverted feed material if recycling is not possible;
- losses from the separation and recycle system.

Addressing the streams entering and leaving the process in Figure 14.13, there are material losses in the byproducts and purges that should be reduced if possible. Thus, before proceeding further, a number of questions should be considered:

1) Can byproduct formation be avoided or reduced by recycling? This is sometimes possible when the byproduct is formed by secondary reversible reactions.
2) If a byproduct is formed by reaction involving feed impurities, can this be avoided or reduced by purification of the feed?
3) Can the byproduct be subjected to further reaction and its value upgraded? For example, most organic chlorination reactions produce hydrogen chloride as a byproduct. If this cannot be sold it must be disposed of. An alternative as discussed in Example 14.1 is to convert the hydrogen chloride back to chlorine via the reaction:

$$2HCl + \frac{1}{2}O_2 \rightleftharpoons Cl_2 + H_2O$$

The chlorine can then be recycled.
4) Can the loss of useful material in the purge streams be avoided or reduced by feed purification?
5) Can the loss of useful material in the purge be avoided or reduced by additional separation on the purge? The roles of refrigerated condensation, low-temperature distillation, absorption, adsorption and membranes in this respect have already been discussed.
6) Can the useful material lost in the purge streams be reduced by additional reaction to useful products? If the purge stream contains significant quantities of reactants, then placing a reactor and additional separation on the purge can sometimes be justified. This technique is used in some designs of ethylene oxide processes.

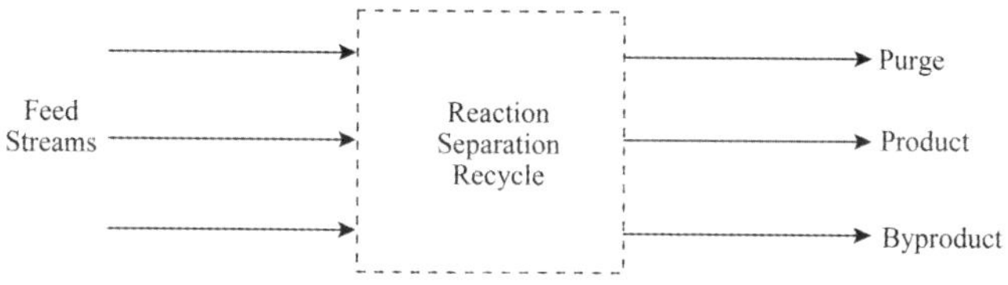

Figure 14.13

The overall process material balance for the process yield.

Example 14.3 Calculate the process yield of benzene from toluene and benzene from hydrogen for the approximate phase split in Example 14.2.

Solution

$$\text{Benzene yield} = \frac{\text{Benzene produced}}{\text{Toluene fed to the process}} \times \text{Stoichiometric factor}$$

$$\text{Stoichiometric factor} = \text{Stoichiometric moles of toluene required per mole of benzene produced} = 1$$

$$\text{Benzene yield from toluene} = \frac{P_B}{P_B/S} \times 1 = S = 0.97$$

14

In this case, because there are no raw materials losses in the separation and recycle system, the only yield loss is in the reactor and the process yield equals the reactor selectivity:

$$\text{Benzene yield from hydrogen} = \frac{\text{Benzene produced}}{\text{Hydrogen fed to the process} \times \text{Stoichiometric factor}}$$

$$\text{Stoichiometric factor} = \text{Stoichiometric moles of hydrogen required per mole of benzene produced} = 1$$

For $y = 0.4$, $\alpha = 0.3013$:

$$\text{Benzene yield from hydrogen} = \frac{P_B}{1554\alpha + 269.2} \times 1 = \frac{265}{1554 \times 0.3013 + 269.2} = 0.36$$

14.5 Feed, Product and Intermediate Storage

Most processes require storage for the feed and product. Storage of feed is required if the delivery of the feed is in batches (e.g. barge, rail car, road truck). Even if the feed is being delivered continuously via pipelines for gases and liquids or conveyors in the case of solids, there will be no guarantee that feed will be free from interruptions in supply. For example, the upstream plant providing the continuous feed will need to be shut down for various reasons and there might be unexpected failures of the delivery, for example, as a result of breakdowns.

Whilst solids and liquids are straightforward to store, gases are difficult. Relatively small quantities of gas can be stored in the gaseous state at ambient temperature in pressurized vessels. Larger quantities of gas storage require the gas to be liquefied. This can be achieved by decreasing the temperature using refrigeration, or increasing the pressure, or a combination of both. High-pressure storage has a high capital cost, as it requires thick-walled vessels. Low-temperature storage also has a high capital cost, as it requires capital investment in refrigeration equipment. Low-temperature storage also has a significant operating cost for power to run the refrigeration. The most appropriate method of storage for gases depends on a number of factors and involves safety, as well as capital and operating cost considerations.

If the feed is delivered in batches and used continuously, there is a fluctuating amount of storage, known as the *active stock*. For example, suppose a plant operating with a liquid feed is at steady state. The liquid feed tank after a delivery might be perhaps 80% full. As the plant operates at steady state, the liquid level falls continuously to, say, 20% when the next delivery arrives and the level returns to 80% as a result of the delivery. The amount of liquid between the 20% and 80% levels is the active stock and 20% is *inactive*. Storage tanks for liquids should not be designed to operate with less than 10% inactive stack at the minimum, as this would create difficulties in operation. On the other hand, tanks for liquids should not be designed to operate more than 90% full at the maximum. An empty space above the liquid (known as *ullage*) is required when the tank is full to allow for safety and expansion. If the flowrate of feed material is m_{FEED} tons per year and the maximum active stock is m_{STOCK}, the number of deliveries per year will be the ratio m_{FEED}/m_{STOCK}.

The capital cost of the storage equipment will be approximately proportional to the storage capacity. This involves the capital cost of storage tanks in the case of gases and liquids, and silos in the case of solids, as well as the capital cost of materials handling equipment (e.g. pumps, conveyors, etc.) and the capital cost of refrigeration equipment. In addition to the capital cost of the equipment, there is also the working capital associated with the value of the material being stored. The greater the value of the material in storage the greater the disincentive to store large quantities of material.

The feed storage:

- provides a supply of feed to the plant between feed deliveries;
- compensates for interruptions in feed delivery due to unforeseen circumstances (e.g. breakdown in the plant manufacturing the feed material);
- allows short-term increases in production if the market for the product is favorable;
- compensates for interruptions in feed delivery due to holiday periods;
- compensates for seasonal variations in feed supply;
- allows feed to be bought under favorable market conditions when it is cheaper and stored for later use;
- dampens out variations in the feed properties.

The amount of feed storage will depend on:

- the frequency of deliveries;
- the size of deliveries;
- the reliability of deliveries;
- the capacity of the plant;
- the phase of the feed (gas, liquid or solid);
- the hazardous nature of the feed material (the inventory of hazardous feed should be kept to a minimum);
- the capital and operating costs (e.g. refrigeration system for the storage of liquefied gas) associated with the feed storage equipment;
- the working capital locked up in the stored feed;
- the economic benefit to be gained from being able to take advantage of market fluctuations in the purchase cost of the raw materials.

The product must also be stored, for similar reasons to those for feed storage. The product delivery will often not be continuous. Also, the product will often be delivered to different customers. If the product is being delivered via a pipeline in the case of a liquid or gas, or continuous conveyer in the case of a solid, then product storage can be minimized.

The product storage:

- balances the difference between the rates of production and dispatch;
- maintains product delivery during plant shutdown for maintenance;
- maintains product delivery during unforeseen plant shutdown;
- compensates for peak and seasonal demands;
- holds materials when holidays prevent dispatch;
- allows materials to be held up for later sale if short-term market conditions are unfavorable, leading to a short-term decrease in the sales price;
- allows variations in product quality to be dampened out.

The amount of product storage will depend on:

- the frequency of product dispatches;
- the size of dispatches;
- the reliability of the dispatches;
- the capacity of the plant;
- the phase of the product (gas, liquid or solid);
- the hazardous nature of the product (the inventory of hazardous products should be kept to a minimum);
- the capital and operating costs (e.g. refrigeration system for the storage of liquefied gas) associated with product storage equipment;
- the working capital locked up in the stored product;
- the economic benefit to be gained from being able to take advantage of market fluctuations in the sales price of the product.

In addition to the storage of the feed and product, chemical intermediates are also often stored within the process. *Intermediate storage* for chemical intermediates is required particularly when the process requires a number of transformation steps between the feed and the product. It creates flexibility in the operation of the plant. For example, consider a process involving a complex reaction system, followed by a complex separation system. The start-up and control of the two sections can be simplified if they can be decoupled. This can be achieved by introducing intermediate storage between the reaction and separation sections. For start-up, the reaction section can be started up independently of the separation section. When starting up, the reaction section produces an intermediate chemical that is accumulated in the intermediate storage. When the reaction section is producing material of a suitable quality to be fed to the separation section, then the separation section can be started up by feeding from intermediate storage. Off-specification material can be kept separate for reworking at a later time, or disposal. The intermediate storage allows the two sections to be operated independently of each other. This is not only important for start-up and shutdown, but allows one of the sections to be operated even if the other breaks down for a short period. The intermediate storage also decouples the control of the two sections.

The intermediate storage between the reaction and separation system can also help dampen out variations in composition, temperature and flowrate between the two sections (for gases and nonviscous liquids, but not solids). Variations in the outlet properties from the storage are reduced compared with variations in the inlet properties.

The greater the amount of intermediate storage the greater the flexibility created in the operation of the process and the simpler will be the control. However, like feed and product storage there are significant costs associated with intermediate storage, involving capital, working and operating costs. In addition, intermediate storage will bring additional safety problems if the material being stored is of a hazardous nature.

In summary, the amount of feed, product and intermediate storage will depend on capital, working and operating costs, together with operability, control and safety considerations.

14.6 Continuous Process Recycle Structure – Summary

The use of excess reactants, diluents or heat carriers in the reactor design has a significant effect on the flowsheet recycle structure. Sometimes the recycling of unwanted byproduct to the reactor can inhibit its formation at source. If this can be achieved, it improves the overall usage of raw materials and reduces effluent disposal problems. However, the recycling results in an increase of some costs.

When a mixture in a reactor effluent contains components with a wide range of volatilities, then a partial condensation from the vapor phase followed by a simple phase split can often produce a good separation. If the vapor from such a phase split is difficult to condense, then further separation needs to be carried out in a vapor separation process such as a membrane. The liquid from the phase split can be sent to a liquid separation unit such as distillation.

The process yield is an important measure of both raw materials efficiency and environmental impact.

Feed, product and intermediate storage can be very significant cost components.

14.7 Exercises

1. Ethylene is to be converted by catalytic air oxidation to ethylene oxide according to the reaction:

$$\underset{\text{ethylene}}{C_2H_4} + \underset{\text{oxygen}}{\frac{1}{2}O_2} \longrightarrow \underset{\text{ethylene oxide}}{C_2H_4O}$$

A parallel reaction occurs leading to a selectivity loss:

$$\underset{\text{ethylene}}{C_2H_4} + \underset{\text{oxygen}}{3O_2} \longrightarrow \underset{\text{carbon dioxide}}{2CO_2} + \underset{\text{water}}{2H_2O}$$

The air (21% O_2, 79% N_2) and ethylene (assumed pure) are mixed in the ratio 10:1 by volume. This mixture is combined with a recycle stream and the two streams are fed to the reactor. Of the ethylene entering the reactor, 40% is converted to

14

ethylene oxide, 20% is converted to carbon dioxide and water, and the rest does not react. The exit gases from the reactor are treated to remove substantially all of the ethylene oxide and water, and the residue recycled. Purging of the recycle is required to avoid accumulation of carbon dioxide and hence maintain a constant feed to the reactor.

a) Sketch the basic flowsheet.

b) Calculate the ratio of purge to recycle if not more than 8% of the ethylene fed is lost in the purge.

c) Calculate the composition of the corresponding reactor feed gas.

2. Benzene is to be produced by the hydrodealkylation of toluene according to the reaction:

$$\underset{\text{toluene}}{C_6H_5CH_3} + \underset{\text{hydrogen}}{H_2} \longrightarrow \underset{\text{benzene}}{C_6H_6} + \underset{\text{methane}}{CH_4}$$

Some of the benzene formed undergoes secondary reactions in series to unwanted byproducts that can be characterized by the reaction to diphenyl, according to the reaction:

$$\underset{\text{benzene}}{2C_6H_6} \rightleftharpoons \underset{\text{diphenyl}}{C_{12}H_{10}} + \underset{\text{hydrogen}}{H_2}$$

Laboratory studies have established that the selectivity (i.e. fraction of toluene reacted that is converted to benzene) is related to the conversion (i.e. fraction of toluene fed that is reacted) according to:

$$S = 1 - \frac{0.0036}{(1 - X)^{1.544}}$$

where S = selectivity
X = conversion

The hydrogen feed to the plant contains methane as an impurity at a mole fraction of 0.05. In the first instance, it can be assumed that the reactor effluent will contain hydrogen, methane, benzene, toluene and diphenyl, and that a simple phase split will produce vapor stream containing all of the hydrogen and methane and a liquid stream containing all of the aromatics. The hydrogen and methane will be recycled to the reactor with a purge to prevent the build-up of methane. The liquid stream containing the aromatics will be separated into pure products and the toluene recycled. The values of the feeds and products are given in Table 14.3.

The purge stream containing hydrogen and methane and the byproduct diphenyl will be burned in a furnace and can be attributed with their fuel values given in Table 14.4.

a) For a production rate of benzene of 300 kmol·h^{-1} and mole fraction of hydrogen in the purge of 0.35, determine the flowrate of hydrogen as a function of the selectivity S.

b) Determine the range of reactor conversions over which the plant is profitable.

Table 14.3

Values of feeds and products for Exercise 2.

	Molar mass (kg·kmol^{-1})	Value (\$·kg^{-1})
Hydrogen	2	1.06
Toluene	92	0.21
Benzene	78	0.34

Table 14.4

Fuel values of waste streams for Exercise 2.

	Molar mass (kg·kmol^{-1})	Fuel value (\$·kg^{-1})
Hydrogen	2	0.53
Methane	16	0.22
Diphenyl	154	0.17

3. An outline process flowsheet for the hydrodealkylation process from Exercise 2 for the production of benzene from toluene is shown in Figure 14.8.

a) Suggest alternative recycle structures for the process to improve the yield of benzene from toluene.

b) Suggest two ways in which the yield of benzene from hydrogen can be improved.

4. Ethylene oxide (EO) reacts with ammonia to form monoethanolamine (MEA) by the following reaction:

$$\underset{\text{ethylene oxide}}{C_2H_4O} + \underset{\text{ammonia}}{NH_3} \rightleftharpoons \underset{\text{monoethanolamine}}{NH_2{-}CH_2{-}CH_2OH}$$

Two principal secondary reactions occur to form diethanolamine (DEA) and triethanolamine (TEA):

$$\underset{\text{monoethanolamine}}{NH_2{-}CH_2{-}CH_2OH} + \underset{\text{ethylene oxide}}{C_2H_4O} \rightleftharpoons \underset{\text{diethanolamine}}{NH(CH_2CH_2OH)_2}$$

$$\underset{\text{diethanolamine}}{NH(CH_2CH_2OH)_2} + \underset{\text{ethylene oxide}}{C_2H_4O} \rightleftharpoons \underset{\text{triethanolamine}}{N(CH_2CH_2OH)_3}$$

All the above reactions are reversible and exothermic. The presence of DEA or TEA in the reactor feed virtually eliminates production of that amine.

The normal boiling points of the components are given below:

Component	Boiling point (°C)
Ammonia	−33
Ethylene oxide	11
Monoethanolamine	170
Diethanolamine	269
Triethanolamine	335

There is market demand for all three amines. Assume that the objective is to design a flexible plant, that is, one that can produce any specified combination of the amines. Typically, the reaction bubbles ethylene oxide through aqueous ammonia. All components are water soluble.

a) How would the product be affected by the ratio of ammonia to EO, the recycling of intermediate products and removal of intermediate products from the reactor?
b) Suggest a recycle–separation scheme that will allow flexible production of high-purity amine products.

5. Consider a process in which each delivery of raw material provides a 10 day supply and is stored in a tank. Delivery from the supplier takes between 5 and 15 days. The minimum inventory in the tank is to be 20 days supply. For what period should the storage tank be sized?

References

Douglas JM (1985) A Hierarchical Decision Procedure for Process Synthesis, *AIChE J*, **31**: 353.
Douglas JM (1988) *Conceptual Design of Chemical Processes*, McGraw-Hill.
Powers GJ (1972) Heuristics Synthesis in Process Development, *Chem Eng Prog*, **68**: 88.
Rudd DF, Powers GJ and Siirola JJ (1973) *Process Synthesis*, Prentice-Hall, New Jersey.
Smith R and Linnhoff B (1988) The Design of Separators in the Context of Overall Processes, *Trans IChemE ChERD*, **66**: 195.
Wynn N (2001) Pervaporation Comes of Age, *Chem Eng Prog*, **97**: 66.

Chapter 15

Continuous Process Simulation and Optimization

Having created an initial design for a flowsheet, an evaluation of the design requires the material and energy balance to be evaluated with greater accuracy. A preliminary sizing of equipment and an economic evaluation of the design is also required. In order to do this, a simulation model of the flowsheet is required. The overall simulation model requires unit models for the various transformation steps (chemical reaction, separation, pressure change, etc.) to be developed and connected by process streams. A key element in these models is the ability to describe the behavior of the fluids involved, which requires the use of thermodynamic equations and physical property models. This overall simulation model is most often developed in general purpose flowsheet simulation software. The overall flowsheet simulation model will allow the feasibility of a proposed design to be assessed. The effect of important design decisions on the performance of the flowsheet can be determined. Model outputs may be related to process economics, process yield, plant capacity, safety, environmental impact, operability, etc. Once a simulation model is developed it can be subjected to optimization to improve the design performance.

15.1 Physical Property Models for Process Simulation

The first important decision when simulating a flowsheet is how to represent the physical properties (see Appendix A). Generally, when using commercial simulation software, a variety of physical property methods will be available. Choosing the most appropriate method from a number of options can be critical to obtaining a reliable design. Phase equilibrium (vapor–liquid, liquid–liquid, solid–liquid, etc.) is usually the most critical physical property. Thermodynamic properties (molar volume, density, enthalpy, enthalpy of vaporization, heat capacity, entropy) and transport properties (viscosity, thermal conductivity, diffusivity) are also required. The prediction of the physical properties of mixtures of components often starts with the prediction of the physical properties of the individual components and then these are combined according to mixing rules (see Appendix A). Such mixing rules introduce errors depending on the physical property and the reliability of the mixing rule.

Correlated experimental data should be used whenever possible. If no such data are available, then the designer must resort to estimation methods. There are two broad categories of estimation methods:

1) Methods whereby known properties of a compound are used to estimate the unknown properties. For example, Riedel (1954) proposed an equation to predict vapor pressure of the form:

$$\ln P^{SAT} = A + \frac{B}{T} + C \ln T + DT^6 \qquad (15.1)$$

where P^{SAT} = saturated liquid vapour pressure
T = absolute temperature
A, B, C, D = constants calculated from critical constants and normal boiling points (see Poling, Prausnitz and O'Connell, 2001)

This allows the prediction of vapor pressure from knowledge of the critical properties and normal boiling point.

2) Group contribution methods, which are based on the concept that a particular physical property of a compound can be considered to be made up of the contributions from the constituent chemical groups and chemical bonds (for example, ethylene C_2H_4 is considered to be constituted of two $=CH_2$ groups, while ethyl chloride CH_3CH_2Cl is considered as one $-CH_3$, one $-CH_2-$ and one $-Cl$). An example of a group contribution method is the prediction of critical properties

Chemical Process Design and Integration, Second Edition. Robin Smith.

Companion Website: www.wiley.com/go/smith/chemical2e

using the method of Joback and Reid (1987). The critical properties can be predicted from:

$$\begin{aligned} T_C &= T_{BPT}\left[0.584 + 0.965\sum\Delta_T - \left(\sum\Delta_T\right)^2\right]^{-1} \\ P_C &= \left(0.113 + 0.0032n_A - \sum\Delta_P\right)^{-2} \\ V_C &= 17.5 + \sum\Delta_V \end{aligned} \tag{15.2}$$

where
T_C = critical temperature (K)
T_{BPT} = normal boiling point (K)
P_C = critical pressure (bar)
V_C = critical volume ($cm^3 \cdot mol^{-1}$)
Δ_T, Δ_P, Δ_V = constants that depend on the atomic group and chemical bonds (see Poling, Prausnitz and O'Connell, 2001)
n_A = number of atoms in the molecule

The accuracy required of physical property data depends on the use to which the data will be put. There are three general considerations:

1) *Design stage.* Exploratory calculations carried out to evaluate process options at a high level require less accurate physical property data than final equipment design calculations.
2) *Reliability of design methods.* If there is significant uncertainty in the design method, there might be little point in using highly accurate physical property data.
3) *Mass and energy transfer with small driving forces.* If mass is being transferred in a process with small driving forces, then phase equilibrium data with high precision are required. This might be, for example, when distillation is carried out between components with very low relative volatility. As an example, consider the separation by distillation of propylene and propane in ethylene production, which has a small relative volatility. An error of +1% in the prediction of the K-value for propane can lead to the requirement for 26% more trays or 25% more reflux (Streich and Kirstenmacher, 1979). Alternatively, if distillation is to be carried out to a high purity, leading to small driving forces in each stage (as seen in a McCabe–Thiele diagram when the equilibrium and operating lines become very close), then again phase equilibrium data with high precision are required. If heat is being transferred between streams with small temperature differences, then accurate physical property data for the enthalpies (heat capacities) of the streams are required to ensure the calculations reflect the change in temperature accurately.

Design calculations can be sensitive or insensitive to errors in physical property data. For example, errors in viscosity often have little effect on calculated pressure drops in turbulent flow. On the other hand, distillation calculations when the relative volatility is small will be sensitive to errors in physical property data. The only way to be sure about the effect of errors in physical property data is to carry out a sensitivity check by repeating the calculation with slight changes in the physical property data.

Physical property design data should be validated wherever possible. This is particularly true of phase equilibrium data. For example, suppose vapor–liquid equilibrium data are required for a mixture of benzene and cyclohexane. This is a hydrocarbon mixture and an equation of state method such as that of Peng–Robinson or Soave–Redlich–Kwong would normally be used (see Appendix A). In fact the mixture forms an azeotrope in this case and the equation of state methods would give spurious results. On the other hand, a mixture of benzene and toluene would be adequately represented by an equation of state method. No single vapor–liquid equilibrium model can be used on all systems. Different mixtures require different models. A simulation model based on inappropriate physical properties is likely to give results of some kind, but those results will be in the worst case meaningless. In the case of vapor–liquid equilibrium, the binary data can be displayed on an *x*–*y* diagram, temperature–composition plot or pressure–composition plot and compared with experimental binary data (Gmehling, Onken and Arlt, 1977; Oellrich *et al.*, 1981) to ensure the model is representative of at least the binary data.

15.2 Unit Models for Process Simulation

As pointed out previously, the simulation of the process must start by creating models for the individual process steps, before connecting them together to form a model of the overall process. The first basic principle that must apply to each unit model is the conservation of mass:

$$\begin{aligned} [\text{Total material out}] &= [\text{Total material in}] \\ &\quad - [\text{Accumulation of material}] \end{aligned} \tag{15.3}$$

For an individual Component *i* in a reacting system:

$$\begin{aligned} [\text{Material out}]_i &= [\text{Material in}]_i + [\text{Generation}]_i \\ &\quad - [\text{Consumption}]_i - [\text{Accumulation}]_i \end{aligned} \tag{15.4}$$

For batch processes there can be an accumulation, but for continuous processes at steady state:

$$[\text{Total material out}] = [\text{Total material in}] \tag{15.5}$$

$$\begin{aligned} [\text{Material out}]_i &= [\text{Material in}]_i + [\text{Generation}]_i \\ &\quad - [\text{Consumption}]_i \end{aligned} \tag{15.6}$$

Either mass or molar units can be used, but for reacting systems it is generally preferred to use molar units. There must also be conservation of energy for each unit:

$$[\text{Energy out}] = [\text{Energy in}] - \begin{bmatrix}\text{Accumulation of energy} \\ \text{within the system}\end{bmatrix} \tag{15.7}$$

For batch processes there can be an accumulation of energy within the system, but for continuous processes at steady state:

$$[\text{Energy out}] = [\text{Energy in}] \tag{15.8}$$

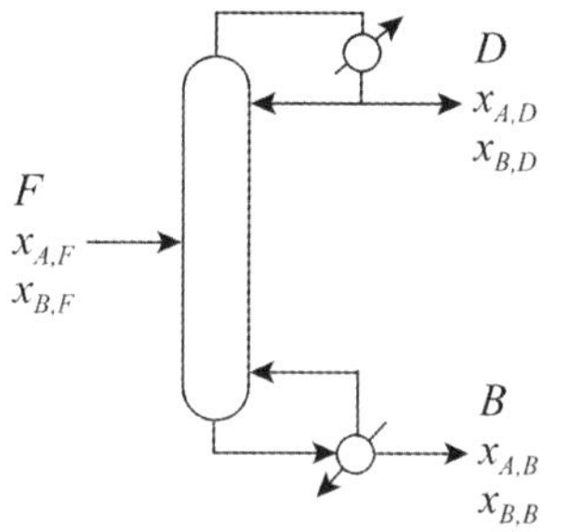

Figure 15.1
Binary distillation material balance.

In its most general form, the energy input and output should account for heat, work, kinetic energy and potential energy. In most process energy balance calculations, change in kinetic and potential energy can be neglected. For chemical reactions there can be a generation of heat (exothermic) or a consumption of heat (endothermic), but overall energy is neither generated nor consumed by the reacting system. There is a difference in the energy associated with the chemical bonds in reactants and products.

The basic specification of unit models is often not straightforward, even when using commercial simulation software. It is both possible to underspecify or overspecify a model. Consider the example of specifying the material balance for a simple binary distillation. This is illustrated in Figure 15.1. Equations can be set up to model the distillation in Figure 15.1.

Material balance equations

1. $F = D + B$ (15.9)

2. $Fx_{A,F} = Dx_{A,D} + Bx_{A,B}$ (15.10)

3. $Fx_{B,F} = Dx_{B,D} + Bx_{B,B}$ (15.11)

Summation equations

4. $x_{A,F} + x_{B,F} = 1$ (15.12)

5. $x_{A,D} + x_{B,D} = 1$ (15.13)

6. $x_{A,B} + x_{B,B} = 1$ (15.14)

Product recovery

7. $r_{A,D} = \dfrac{Dx_{A,D}}{Fx_{A,F}}$ (15.15)

Flowrates and purities

8. $F = 1000\ \text{kmol}\cdot\text{h}^{-1}$ (15.16)

9. $D = 500\ \text{kmol}\cdot\text{h}^{-1}$ (15.17)

10. $x_{A,F} = 0.5$ (15.18)

11. $x_{A,D} = 0.99$ (15.19)

12. $R_{A,D} = 0.98$ (15.20)

To determine whether this is a viable specification the number of variables and equations need to correspond. The number of variables is 10 (F, $x_{A,F}$, $x_{B,F}$, D, $x_{A,D}$, $x_{B,D}$, B, $x_{A,B}$, $x_{B,B}$, $R_{A,D}$), but the number of equations is 12. For there to be a unique solution to this model, the number of variables must be equal to the number of equations. Thus there is no unique solution to the problem as specified. However, before dropping equations, it must be clear that the equations constitute independent sources of information. In the above equations, Equations 2, 3, 4, 5, and 6 can be combined to obtain Equation 1. Alternatively, Equations 1, 2, 4, 5, and 6 can be combined to obtain Equation 3, and so on. Thus, the first six equations are not independent. One equation must be dropped, for example drop Equation 1. This leaves 5 independent equations instead of the original 6, but the problem is still overspecified. One further equation must be relaxed. This could be $D = 500\ \text{kmol}\cdot\text{h}^{-1}$ or $x_{A,D} = 0.99$ or $R_{A,D} = 0.98$. Then the system can be solved. For example, drop the equation $D = 500\ \text{kmol}\cdot\text{h}^{-1}$. Solving the equations gives $D = 494.9\ \text{kmol}\cdot\text{h}^{-1}$, $B = 505.1\ \text{kmol}\cdot\text{h}^{-1}$, $x_{A,B} = 0.0199$, $x_{B,B} = 0.9801$. If a further equation is relaxed, then there is no unique solution.

Consider now the simple flowsheet in Figure 15.2. To simulate this flowsheet requires models for the following units:

1) *Feed streams.* For the feed, specifications are required for the total flowrate and composition (or component flowrates), temperature and pressure. Solid feeds need additional information.
2) *Mixers.* Outlet conditions are fixed by the material and energy balance. Outlet pressure would typically be the minimum of the two feeds minus the specified pressure drop.
3) *Heaters and coolers.* A simple heater does not specify the source of heat for heating. Similarly, a simple cooler does not specify the sink of heat for cooling. It might be that the heating and cooling will be satisfied by utilities, or it might be a preliminary stage in the modeling of the flowsheet that will

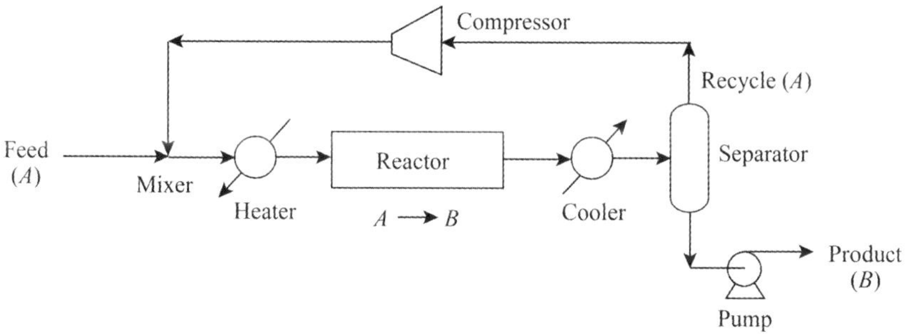

Figure 15.2
A simple process.

ultimately be connected to another heat source or sink. The modeling equation for simple heaters and coolers is:

$$Q = m \int_{T_1}^{T_2} C_P \, dT \qquad (15.21)$$

where Q = heating or cooling duty (kW)
m = stream flowrate ($kg \cdot s^{-1}$)
C_P = stream specific heat capacity as a function of temperature ($kJ \cdot kg^{-1} \cdot K^{-1}$)
T_1 = stream inlet temperature (°C, K)
T_2 = stream outlet temperature (°C, K)

The inlet conditions will be specified by the upstream unit, which might be a feed stream or another unit. The duty on the heater or cooler can be specified in different ways. The outlet temperature might be specified by the heating or cooling heat duty.

4) *Heat exchangers.* A heat exchanger is more complex than a simple heater or cooler, as both the heat source and heat sink need to be specified in the model. Again, the inlet conditions will be specified by the upstream units. Heat exchangers have been considered in more detail in Chapter 12. However, simple modeling equations for countercurrent heat transfer are given by:

$$Q = UA \, \Delta T_{LM} = UA \left[\frac{(T_{H1} - T_{C2}) - (T_{H2} - T_{C1})}{\ln\left(\dfrac{T_{H1} - T_{C2}}{T_{H2} - T_{C1}} \right)} \right] \qquad (15.22)$$

$$Q = m_H C_{P,H} (T_{H1} - T_{H2}) \qquad (15.23)$$

$$Q = m_C C_{P,C} (T_{C2} - T_{C1}) \qquad (15.24)$$

where Q = heat exchanger duty (kW)
U = overall heat transfer coefficient ($kW \cdot m^{-2} \cdot K^{-1}$)
= U (m_H, m_C, physical properties, exchanger geometry)
A = heat transfer area (m^2)
ΔT_{LM} = logarithmic temperature difference (°C, K)
m_H, m_C = flowrates for the hot and cold streams ($kg \cdot s^{-1}$)
$C_{P,H}$, $C_{P,C}$ = specific heat capacities of the hot and cold streams ($kJ \cdot kg^{-1} \cdot K^{-1}$)
T_{H1}, T_{C1} = inlet temperatures of the hot and cold streams (°C, K)
T_{H2}, T_{C2} = outlet temperatures of the hot and cold streams (°C, K)

Equations 15.22 to 15.24 assume that the physical properties are constant through the heat exchanger. Additionally, Equation 15.22 assumes countercurrent heat transfer. More complex models are given in Chapter 12. There are many ways in which the heat exchanger model can be specified. For example, the heat duty can be specified, and, given the inlet conditions and flowrates from upstream units, the outlet temperatures and UA calculated. Alternatively, the UA can be specified and the outlet temperatures and heat duty calculated. Another option might be to specify the value of the overall heat transfer coefficient U and the heat duty Q and calculate the heat transfer area A and the outlet temperatures. Other combinations are possible. If the overall heat transfer coefficient U is to be calculated, details of the heat exchanger geometry need to be supplied. Such calculations have been discussed in more detail in Chapter 12. In addition to the heat duty calculation, the change in pressure across the heat exchanger must either be specified or calculated. Calculation of the pressure drop requires details of the exchanger geometry to be supplied. The most common approach in flowsheet simulation is to specify a reasonable pressure drop for the streams.

5) *Reactors.* Three broad classes of reactor models are used in flowsheet simulation:

a) *Conversion reactor.* Extent of reaction defined by the reactor conversion:

$$\text{Conversion} = X = \frac{\text{Reactant consumed in the reactor}}{\text{Reactant } \textit{fed} \text{ to the reactor}} \qquad (15.25)$$

This reactor model is therefore a simple material balance based on the specified conversion for each of the reactions.

b) *Equilibrium reactor.* For an equilibrium reactor, with reactions such as:

$$A \rightleftharpoons B \qquad (15.26)$$

The equilibrium constants define the equilibrium and the equilibrium conversion, as discussed in Chapter 5:

$$K_a = \frac{y_B}{y_A} \text{ (gas phase reactions)} \qquad (15.27)$$

$$K_a = \frac{x_B}{x_A} \text{ (liquid phase reactions)} \qquad (15.28)$$

The equilibrium constants might be fixed or a function of temperature, for example:

$$\ln(K_a) = a + \frac{b}{T} + c \ln(T) + d\,T \qquad (15.29)$$

where T = reaction temperature
a, b, c, d = constants correlated from experimental or thermodynamic data

The equilibrium constants can be calculated from thermodynamic data, as detailed in Chapter 5. Another approach that is useful when the reactions occurring are not known or are high in number due to many components participating in the reactions, is often known as the "Gibbs reactor". The Gibbs reactor models the system by finding the equilibrium state with the lowest Gibbs Free Energy. The approach effectively finds all the possible equilibrium reactions and allows them all to equilibrate. There is no need to know individual equilibrium constants. Only components listed as reacting in the reaction undergo reaction. Setting components to be reactive and inert is an important aspect in modeling using this approach.

c) *Kinetic reactor.* The third approach is to create a kinetic model, as discussed in Chapters 4 to 6. A reactor flow model must be chosen. For example, for the reaction from Equation 15.26 in plug flow, the kinetic model is:

$$\tau = -\int_{C_{A,in}}^{C_{A,out}} \frac{dC_A}{-r_A} \qquad (15.30)$$

$$-r_A = k_A C_A - k'_A C_B \qquad (15.31)$$

where
τ = reactor space–time (s)
r_A = rate of reaction of Component A ($kmol \cdot m^{-3} \cdot s^{-1}$)
k_A, k'_A = reaction rate constants for the forward and reverse reactions (s^{-1})
C_A = molar concentration of Component A ($kmol \cdot m^{-3}$)
C_B = molar concentration of Component B ($kmol \cdot m^{-3}$)

The corresponding model for a mixed flow reactor is given by:

$$\tau = \frac{C_{A,in} - C_{A,out}}{-r_A} \qquad (15.32)$$

$$-r_A = k_A C_A - k'_A C_B \qquad (15.33)$$

Thus for a basic model the flow pattern and kinetic expressions need to be specified.

6) *Separators.* The flowsheet in Figure 15.2 features a phase separator. There are many possibilities for modeling separators:

a) *Simple splitter.* The simplest option for modeling the separator is a simple flow splitter in which the inlet flow is split in a specified ratio. For this model there is no separation of components and the outlet streams have the same composition. A pressure drop can also be specified.

b) *Component splitter.* In a component splitter the split of each *component* is specified as a ratio. This might be useful for initial calculations for flash or distillation separation in which an estimate of the split on each component is specified. It is also useful when separators are difficult to model (e.g. pressure swing adsorption, chromatography, membranes, etc.). A pressure drop can also be specified.

c) *Flash separator.* A single-stage vaporization or condensation might be used if the relative volatilities are large between some of the components. Vaporization might be through heat input or pressure reduction. However, only a limited separation is possible. If the vapor liquid equilibrium K_i is very large (typically $K_i > 10$), then component i will mostly go with the vapor. If K_i is very small (typically $K_i < 0.1$), then component i will mostly go with the liquid (Douglas, 1985).

d) *Distillation.* If distillation is to be modeled for the separation, then the separation needs to be defined. Summarizing from Chapter 8, two important decisions must be made, whichever way the column is to be modeled. These are:

i. *Operating pressure.* Initially, the operating pressure at the top of the column is normally set by the desire to be able to use cooling water or air cooling in the overhead condenser, but vacuum operation should be avoided if possible. If a very high operating pressure is required as a result of trying to operate the condenser against cooling water or air cooling, a combination of high operating pressure and low temperature condensation using refrigeration should be used. Process constraints might restrict the maximum temperature of the distillation to avoid product decomposition. In these circumstances, vacuum operation might be necessary to reduce the boiling temperature.

ii. *Pressure drop.* The pressure drop across the column must be specified, or the pressure drop per plate.

If the column is to be modelled by shortcut calculations (e.g. those of Fenske–Underwood–Gilliland, see Chapter 8), then the following must be specified additionally:

i. *Feed condition.* The feed condition is specified by q. For saturated liquid feed $q = 1$ and for saturated vapor feed $q = 0$.

ii. *Material balance.* The light and heavy key components and their recovery in the overhead and bottoms need to be defined. The key components are often adjacent in volatility, but do not have to be. There is no control over the split of the nonkey components. The split happens according to the material and energy balance and physical properties. The Fenske equation can be used to estimate the composition of the products (at total reflux).

iii. *Energy balance.* The reflux ratio or ratio of actual to minimum reflux (R/R_{min}) must be specified.

If the column is to be modelled by rigorous calculations, then the following must be specified in addition to operating pressure and pressure drop:

i. *Feed composition and flowrate.* The composition of the feed and flowrate must be specified or calculated from an upstream unit model.

ii. *Feed condition.* The feed condition can be taken to be that from an upstream unit, or specified by two from temperature, pressure and q.

iii. *Number of stages and stage layout.* The number of theoretical stages and the feed stage must be specified. For more complex columns, the layout of the stages needs to be specified.

iv. *Material and energy balance.* Specify two from distillate flowrate, bottoms flowrate, a component recovery in one of the products (up to two can be specified) reflux ratio, reboil ratio, condenser duty, or reboiler duty. Other specifications are possible but less common.

7) *Pumps.* The power required for a given pumping duty can be calculated from:

$$W = \frac{F\,\Delta P}{\eta} \tag{15.34}$$

where W = power required for pumping ($N \cdot m \cdot s^{-1} = J \cdot s^{-1} = W$)
F = volumetric flowrate ($m^3 \cdot s^{-1}$)
ΔP = pressure drop across pump ($N \cdot m^{-2}$)
η = pump efficiency (–)

The efficiency of a pump is a function of both its design and its capacity. The efficiency is a strong function of capacity and might be as high as 90% for a large pump and as low as 30% for a small one. For centrifugal pumps, a first estimate of the pump efficiency can be obtained from (Branan, 1999):

$$\begin{aligned}\eta = 0.8 - 9.367 \times 10^{-3}\Delta h + 5.461 \times 10^{-5}\Delta hF - 1.514 \\ \times 10^{-7}\Delta hF^2 + 5.820 \times 10^{-5}\Delta h^2 - 3.029 \\ \times 10^{-7}\Delta h^2 F + 8.348 \times 10^{-10}\Delta h^2 F^2\end{aligned} \tag{15.35}$$

where η = pump efficiency
Δh = head developed by the pump (m) $\quad 15 < \Delta h < 90$
F = flowrate ($m^3 \cdot h^{-1}$) $\quad 20 < F < 230$

For flows below 20 $m^3 \cdot h^{-1}$ down to 5 $m^3 \cdot h^{-1}$ the efficiency can be approximated by taking the efficiency at 20 $m^3 \cdot h^{-1}$ and then subtracting the product of 7.9×10^{-4} and the difference between 20 $m^3 \cdot h^{-1}$ and the required flowrate (Branan, 1999).

8) *Compressors.* Two basic approaches are used for compressor modeling.

a) *Adiabatic compression.* Figure 13.18 shows a compression process on a plot of enthalpy versus entropy. An ideal compression would follow a vertical (isentropic) path from the inlet pressure to the outlet pressure. A real compression would involve an increase in entropy, as shown in Figure 13.18. By definition, the isentropic efficiency η_{IS} is given by:

$$\eta_{IS} = \frac{H_{in} - H_{out,IS}}{H_{in} - H_{out}} \tag{15.36}$$

where η_{IS} = isentropic compressor efficiency (–)
H_{in} = total enthalpy of the inlet stream ($kJ \cdot s^{-1}$)
$H_{out,IS}$ = total enthalpy of the outlet stream for an isentropic compression ($kJ \cdot s^{-1}$)
H_{out} = total enthalpy of the outlet stream for a real compression ($kJ \cdot s^{-1}$)

The power for a real adiabatic compression can also be calculated from the difference between the total enthalpy of the outlet and inlet flows:

$$\begin{aligned} W &= H_{in} - H_{out} \\ &= \frac{H_{in} - H_{out,IS}}{\eta_{IS}} \end{aligned} \tag{15.37}$$

where W = power for compressor ($kJ \cdot s^{-1}$)

A mechanical efficiency for the machine can also be included:

$$W = \frac{H_{in} - H_{out,IS}}{\eta_{IS}\eta_{MECH}} \tag{15.38}$$

where η_{MECH} = compressor mechanical efficiency (–)

The mechanical efficiency is typically 98 to 99% and can therefore often be neglected. The calculation starts with the inlet enthalpy and, given the inlet enthalpy and the pressure, calculates the entropy of the inlet. Then assuming the outlet entropy is the same as the inlet (isentropic), given the outlet pressure, calculate the outlet enthalpy. From the isentropic enthalpy change, the real outlet enthalpy can then be calculated by dividing the isentropic enthalpy change by the isentropic efficiency to get the real enthalpy change. The power requirements can be calculated by dividing the isentropic enthalpy change by the isentropic efficiency and mechanical efficiency to give the overall power requirements. The outlet temperature can then be calculated from the real outlet enthalpy and the outlet pressure using physical properties predicted by an equation of state (see Appendix A).

The temperature rise accompanying gas compression might be unacceptably high because of the properties of the gas (e.g. decomposition, polymerization, etc.), the materials of construction of the compressor or the properties of the lubricating oil used in the machine. The temperature must be below the flash point of the lubricating oil (i.e. the temperature at which it gives off enough vapor to form an ignitable mixture). If this is the case, the overall compression can be broken down into a number of stages with intermediate cooling. Also, intermediate cooling will reduce the volume of gas between stages and reduce the power for compression of the next stage. On the other hand, the intercoolers will have a pressure drop that will increase the power requirements, but this effect is usually small compared with the reduction in power from gas cooling. The power for the staged compression of a gas can initially be set by (see Appendix G):

$$r = \sqrt[N]{\frac{P_{out}}{P_{in}}} \tag{15.39}$$

where N = number of compression stages
r = stage compression ratio

Equation 15.39 sets the compression ratio to minimize the overall compression power for an N-stage compression.

However, the basis is for adiabatic ideal gas compression and therefore not strictly valid for real gas compression (see Appendix G). It is also assumed that intermediate cooling is back to initial conditions and there is no pressure drop for the gas in the intercoolers, which might not be the case with real intercoolers. Whilst compression ratios for staged compression of 7 or greater can be used for certain types of machine, the maximum per stage is normally taken to be 3 or 4. If the maximum temperature is known, then the maximum pressure ratio can be calculated. For multistage process compressors, each stage can have its own adiabatic efficiency. Between each stage the pressure drop across each intercooler and the outlet temperature of the gas from each intercooler must be specified. Any cooling after the final stage will be dealt with via an external cooler.

Staged compression should not be confused with compressor stages. A centrifugal compressor will often have multiple impellers (or wheels) mounted on the shaft to form a multistage machine without cooling. Staged compression is where the overall compression is broken down into intermediate stages with intercooling.

The isentropic efficiency is a function of the machine design and pressure ratio (P_{out}/P_{in}). For reciprocating compressors the isentropic efficiency may typically be in the range 60–80%, depending on the compression ratio, flowrate, machine design and properties of the gas. For centrifugal compressors the isentropic efficiency may typically be in the range 70–90%, depending on the compression ratio, flowrate, machine design and properties of the gas. The mechanical efficiency, accounting for losses in bearings, etc., is generally assumed to be 98–99% and is often neglected.

The approach to modeling for adiabatic compression discussed here is dependent on calculation of the physical properties (enthalpy and entropy) from an equation of state, and is thus not suited to hand calculations, but is suited to calculations using a software package. Alternative approaches are possible and further details on compressor design and operation are given in Appendix G.

b) *Polytropic compression.* The basis of the adiabatic model was compression along a thermodynamically reversible (isentropic) path. Adiabatic compression of an ideal gas along an isentropic path can be expressed as:

$$PV^{\gamma} = \text{constant} \tag{15.40}$$

where
$\gamma = C_P/C_V$
$= \dfrac{C_P}{(C_P - R)}$ for an ideal gas
C_P = heat capacity at constant pressure
C_V = heat capacity at constant volume

In practice, the compression will be neither adiabatic nor reversible. Instead, the gas compression can be assumed to follow a *polytropic* compression represented by the expression (see Appendix G):

$$PV^{n} = \text{constant} \tag{15.41}$$

where n = polytropic coefficient

The power required for a polytropic compression can be expressed as (see Appendix G):

$$W = \frac{n}{n-1}\,\frac{P_{in}F_{in}N}{\eta_P}\left[1 - (r)^{(n-1)/n}\right] \tag{15.42}$$

where
W = power required for gas compression ($N{\cdot}m{\cdot}s^{-1} = J{\cdot}s^{-1} = W$)
n = polytropic coefficient (–)
P_{in} = inlet pressure ($N{\cdot}m^{-2}$)
F_{in} = inlet volumetric flowrate ($m^3{\cdot}s^{-1}$)
N = number of compression stages (–)
η_P = polytropic efficiency, i.e. ratio of polytropic power to actual power (–)
r = stage compression ratio (–)
$= \sqrt[N]{\dfrac{P_{out}}{P_{in}}}$

The polytropic efficiency is a function of the flowrate, physical properties of the gas and the machine design. Polytropic efficiency is always greater than isentropic efficiency for the same compression in the same machine, how much greater being dependent on the heat capacity ratio and pressure ratio (see Chapter 13). The isentropic efficiency, although fundamentally valid, can be misleading if used for comparing the efficiencies of different compressors with differing pressure ratios. For two compressors with different pressure ratios, the higher pressure ratio compressor will have a lower isentropic efficiency because of thermodynamic losses. This property can make it difficult to comparatively evaluate different compressor designs. Polytropic efficiency is better than isentropic efficiency for comparison between compressors, since the value of the polytropic efficiency does not change as much as the value of the isentropic efficiency.

The isentropic and polytropic efficiencies can be related for the same compression process by (see Chapter 13):

$$\eta_{IS} = \frac{\left(\dfrac{P_{out}}{P_{in}}\right)^{(\gamma-1)/\gamma} - 1}{\left(\dfrac{P_{out}}{P_{in}}\right)^{(\gamma-1)/\gamma\eta_P} - 1} \tag{15.43}$$

The polytropic coefficient can be estimated from the heat capacity ratio and the polytropic efficiency (say from an equipment manufacturer) from the relationship (see Chapter 13):

$$n = \frac{\gamma \eta_P}{\gamma \eta_P - \gamma + 1} \tag{15.44}$$

The value of n is always greater than γ.

9) *Product streams.* When using simulation software product streams must be defined as such.

15.3 Flowsheet Models

The previous section discussed the individual unit models that comprise the flowsheet. These must now be connected to form a processing system. Just as individual unit models can be overspecified or underspecified, the same is true for flowsheets when the unit models are connected together. It was pointed out above that if there are M variables and N independent equations, then if $M = N$ there is a unique solution. If $M < N$ there is no unique solution. If $M > N$, then $(M - N)$ variables must receive their values from other sources. The difference $(M - N)$ is the number of degrees of freedom DF. The degrees of freedom can either be removed by additional specifications or can be exploited for optimization. It should be again noted that the equations must constitute independent sources of information (any relation that can be derived from the others is not independent). When considering flowsheets, both the degrees of freedom for the unit models and the *connecting relations* need to be considered. Degrees of freedom DF for the system are related to the number of degrees of freedom of each unit model DF_i and number of connecting relations:

$$DF = \sum_i DF_i - \begin{bmatrix} \text{Number of} \\ \text{connecting} \\ \text{relations} \end{bmatrix} \tag{15.45}$$

As an example, consider two steam heaters in series providing heat to a feed stream, the first serviced by low-pressure steam and the second by high-pressure steam. Some of the degrees of freedom for the downstream heater are specified by the outlet of the upstream heater. The temperature between the two heaters can be chosen. However, if the outlet temperature of the upsteam low-pressure steam heater is specified, the inlet temperature to the downstream high-pressure steam heater is fixed. Alternatively, if the inlet temperature of the high-pressure steam heater is specified, the outlet temperature to the upstream low-pressure steam heater is fixed. An underspecified problem cannot be solved. An overspecified problem cannot be solved, or at least cannot be solved consistently. Any degrees of freedom provide design variables that can be varied or optimized.

As with unit models, the flowsheet must also comply with material and energy balances, as specified in Equations 15.3 to 15.8.

15.4 Simulation of Recycles

Computer simulation packages are normally used to evaluate the material and energy balance once the recycle structure has been established.

To understand how such computer packages function, consider the simple flowsheet in Figure 15.3a. This involves an isomerization of Component A to Component B. The mixture of A and B from the reactor is separated into relatively pure A, which is recycled, and relatively pure B, which is the product. No byproducts are formed and the reactor performance can be characterized by its conversion. The performance of the separator is to be characterized by the recovery of A to the recycle stream (r_A) and recovery of B to the product (r_B).

In this case, only the material balance will be solved in order to keep the problem simple. If the material balance is to be solved, then a series of material balance equations can be written for the flowsheet in Figure 15.3a:

Mixer

$$m_{A,2} = m_{A,1} + m_{A,5} \tag{15.46}$$

$$m_{B,2} = m_{B,1} + m_{B,5} \tag{15.47}$$

Reactor

$$m_{A,3} = m_{A,2}(1 - X) \tag{15.48}$$

$$m_{B,3} = m_{B,2} + X m_{A,2} \tag{15.49}$$

Separator

$$m_{A,4} = m_{A,3}(1 - r_A) \tag{15.50}$$

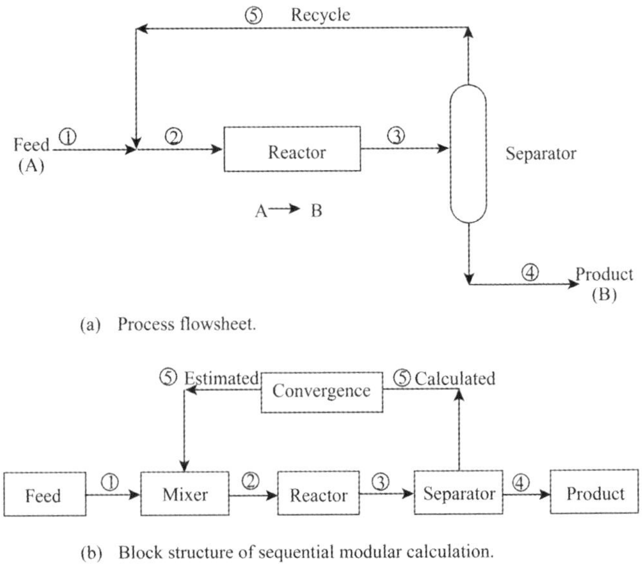

Figure 15.3
A simple process with recycle.

$$m_{A,5} = r_A m_{A,3} \tag{15.51}$$

$$m_{B,4} = r_B m_{B,3} \tag{15.52}$$

$$m_{B,5} = m_{B,3}(1 - r_B) \tag{15.53}$$

where $m_{i,j}$ = molar flowrate of Component i in Stream j
X = reactor conversion
r_i = fractional recovery of Component i

Equations 15.46 to 15.53 form a set of 8 equations and 13 variables (m_A and m_B for each stream, X, r_A, r_B). Specifying the feed stream $m_{A,1}$ and m_{B1}, and X, r_A and r_B allows the set of equations to be solved. There are two basic approaches that could be adopted.

1) *Equation-oriented.* The *equation-oriented* or *equation-based* approach solves the set of equations simultaneously. If the problem involves n design variables, with p equations (equality constraints) and q inequality constraints, the problem becomes one of:

$$\begin{array}{ll} \text{solve} & h_i(x_1, x_2, \ldots, x_n) = 0 \ (i = 1,\ p) \\ \text{subject to} & g_j(x_1, x_2, \ldots, x_n) \le 0 \ (j = 1,\ q) \end{array} \tag{15.54}$$

Whilst this approach seems straightforward for the simple mass balance above, for more complex recycle systems with energy balance equations and phase equilibrium equations, it is not straightforward. Equations describing the flowsheet connectivity are combined with equations describing the various unit models in the flowsheet and, if possible, the physical property correlations into one large equation set (Biegler, Grossman and Westerberg, 1997). The solution of the set of equations can be performed by a general-purpose nonlinear equation solver. Because of the difficulties of including the physical property equations, these are often formulated as distinct procedures and kept separate from equations describing the flowsheet connectivity and unit models (Biegler, Grossman and Westerberg, 1997).

For example, in the simple problem above, if the values are set to:

$$\begin{aligned} m_{A,1} &= 100\ \text{kmol} \\ m_{B,1} &= 0\ \text{kmol} \\ X &= 0.7 \\ r_A &= 0.95 \\ r_B &= 0.95 \end{aligned} \tag{15.55}$$

then the equations can be solved simultaneously to give:

$$\begin{aligned} m_{A,5} &= 39.8601\ \text{kmol} \\ m_{B,5} &= 5.1527\ \text{kmol} \\ m_{A,4} &= 2.0979\ \text{kmol} \\ m_{B,4} &= 97.9021\ \text{kmol} \end{aligned}$$

2) *Sequential modular.* In the *sequential modular* approach, the process equations are grouped within *unit operation blocks*. Each unit operation block contains the equations that relate the outlet stream and the performance variables for the block to the inlet stream variables and specified parameters. Each unit operation block is then solved one at a time in sequence (Biegler, Grossman and Westerberg, 1997). The output calculated from each block becomes the feed to the next block, and so on. Figure 15.3b shows the block structure for the flowsheet in Figure 15.3a. The direction of information flow usually follows that of the material flow. First the feed stream must be created. This then goes to a mixer where the fresh feed is mixed with the recycle stream. Here a problem is encountered, as the flowrate and composition of the recycle are unknown. The sequential modular solution technique is to *tear* one of the streams in the recycle loop. In Figure 15.3b, the recycle stream itself has been torn. In general, tearing the recycle stream itself is only one option for tearing a stream in a loop. Tearing determines those streams or information flows that must be torn to render the system (or subsystem) to be acyclic. A *recycle convergence unit* or *solver* is inserted in the tear stream (Figure 15.3b). To start the calculation of the material balance in Figure 15.3b, values for the component molar flowrates for the recycle stream (tear stream) must be estimated. This allows the material balance in the reactor and separator to be solved. In turn, this allows the molar flowrates for the recycle stream to be calculated. The calculated and estimated values can then be compared to test whether errors are within a specified tolerance. It is usual to specify a scaled error, typically of the form:

$$-Tolerance \le \frac{f(x) - x}{x} \le Tolerance \tag{15.56}$$

where x = estimate of the variable
$f(x)$ = resulting calculated value of the variable

Care needs to be taken if some components are present in trace quantities. If an estimated concentration is 0.5 ppm and the calculated value is 1 ppm, the scaled error is 100%. This is much too large an error for most variables and yet the absolute error might be acceptable for a trace component. In other situations, it might be necessary to define trace components with a high precision. A trace component threshold can be set, below which the convergence criterion is ignored. It is unlikely that the estimated values for the recycle stream will be within tolerance for the initial estimate. If the convergence criteria are not met, then the convergence block needs to update the value of the recycle stream.

The equation-oriented approach and the sequential modular approach each have their relative advantages and disadvantages. The sequential modular approach is intuitive and easy to understand. It allows the designer to interact with the solution as it develops and errors tend to be more straightforward to understand than with the equation-oriented approach. However, large problems may be difficult to converge with the sequential modular approach. On the other hand, the equation-oriented approach can make it difficult to diagnose errors. It is generally not as robust as the sequential modular approach and generally requires a good initialization to solve. One major advantage of the equation-oriented approach is the ability to formulate the problem as an optimization problem, as design problems almost invariably involve some optimization. Thus, the formulation in Equation 15.54 can be readily transformed to the corresponding optimization problem and the material and energy balance solved simultaneously with the optimization problem:

$$\begin{aligned}\text{minimize } & f(x_1, x_2, \ldots, x_n) \\ \text{subject to } & h_i(x_1, x_2, \ldots, x_n) = 0 \;\; (i = 1, p) \\ & g_j(x_1, x_2, \ldots, x_n) \le 0 \;\; (j = 1, q)\end{aligned} \tag{15.57}$$

Of course, the two approaches can be combined and the sequential modular approach used to provide an initialization for the equation-oriented approach.

15.5 Convergence of Recycles

For the sequential modular approach, there are a number of ways that the convergence of the recycle can be achieved.

1) *Direct substitution.* The simplest approach to this is *direct substitution* or *repeated substitution* or *successive substitution* (Biegler, Grossman and Westerberg, 1997, Seider, Seader and Lewin, 2010). This approach is illustrated in Figure 15.4a. The sequence is calculated from an initial estimate of the recycle stream x_1. The calculated value from the *flowsheet response* $f(x_1)$ then becomes x_2, the value for the next iteration. Thus:

$$x_{k+1} = f(x_k) \tag{15.58}$$

This is repeated until all convergence criteria are met, when:

$$x_{k+1} \approx f(x_{k+1}) \tag{15.59}$$

In practice this means iterating until a convergence tolerance is achieved:

$$-Tolerance \le \frac{f(x_{k+1}) - x_k}{x_k} \le Tolerance \tag{15.60}$$

Figure 15.4a shows a schematic representation of the direct substitution strategy that is convergent. By contrast Figure 15.4b shows a direct substitution strategy that is divergent. Fortunately, unstable cases with direct substitution are rare.

For illustration, consider the example from Figure 15.3. Assuming the values in Equation 15.55, the initial estimate could be:

$$\begin{aligned} m_{A,5} &= 50 \text{ kmol} \\ m_{B,5} &= 5 \text{ kmol} \end{aligned}$$

Table 15.1 follows the iterations using direct substitution until convergence is achieved.

Most recycle problems are multivariable in nature. If a material balance is to be solved, then the convergence variables can be taken to be the component molar flowrates or mole fractions. When a material and energy balance is to be solved, the additional convergence variables are usually taken to be pressure and enthalpy. For multivariable systems using direct substitution, for each variable i on iteration k:

$$x_{i,k+1} = f(x_{i,k+1}) \tag{15.61}$$

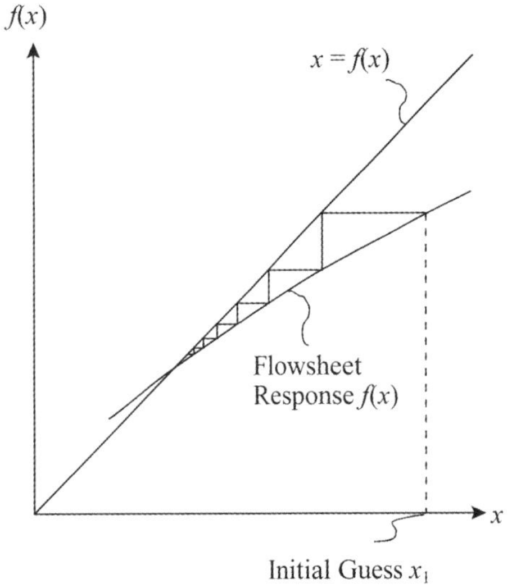

(a) Convergent solution using direct substitution.

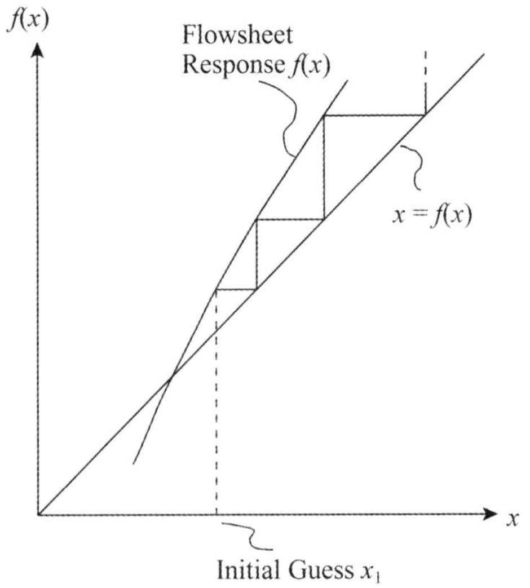

(b) Divergent solution using direct substitution.

Figure 15.4
Solution of recycles by direct substitution.

Table 15.1

Solution of material balance by direct substitution.

Iteration	Assumed		Calculated		Scaled residual	
	$m_{A,5}$ (kmol)	$m_{B,5}$ (kmol)	$m_{A,5}$ (kmol)	$m_{B,5}$ (kmol)	$m_{A,5}$	$m_{B,5}$
1	50	5	42.7500	5.5000	−0.1450	0.1000
2	42.7500	5.500	40.6838	5.2713	−0.0483	−0.0416
3	40.6838	5.2713	40.0949	5.1875	−0.0145	−0.0159
4	40.0949	5.1875	39.9270	5.1627	−0.0042	−0.0048
5	39.9270	5.1627	39.8792	5.1556	−0.0012	−0.0014
6	39.8792	5.1556	39.8656	5.1536	−0.0003	−0.0004
7	39.8656	5.1536	39.8617	5.1530	−0.0001	−0.0001
8	39.8617	5.1530	39.8606	5.1528	0.0000	0.0000

For multidimensional problems, the root mean square of the error (RMS) can be used as a convergence criterion:

$$RMS = \sqrt{\frac{1}{n}\sum_{i=1}^{n}\left(\frac{x_{k+1,i} - x_{k,i}}{x_{k,i}}\right)^2} \leq Tolerance \qquad (15.62)$$

Using direct substitution, the variables are treated as if independent of each other. Yet this is not the case. Despite this, the approach tends to work well for many cases. However, convergence might require many iterations and some problems might fail to converge to the required tolerance. Rather than use direct substitution, the convergence unit can *accelerate* the convergence.

2) *The Wegstein Method.* The most commonly used method to accelerate convergence is the Wegstein Method (1958), illustrated in Figure 15.5. The direct substitution iterations are linearized. A straight-line equation can be written for the two iterations as:

$$f(x) = ax + b \qquad (15.63)$$

where

$$a = \text{slope of the line} = \frac{f(x_k) - f(x_{k-1})}{x_k - x_{k-1}}$$

$f(x_k), f(x_{k-1})$ = calculated values of variables for iterations k and $k-1$

x_k, x_{k-1} = estimated values of variables for iterations k and $k-1$

For iteration k, Equation 15.63 can be written to define the intercept:

$$b = f(x_k) - ax_k \qquad (15.64)$$

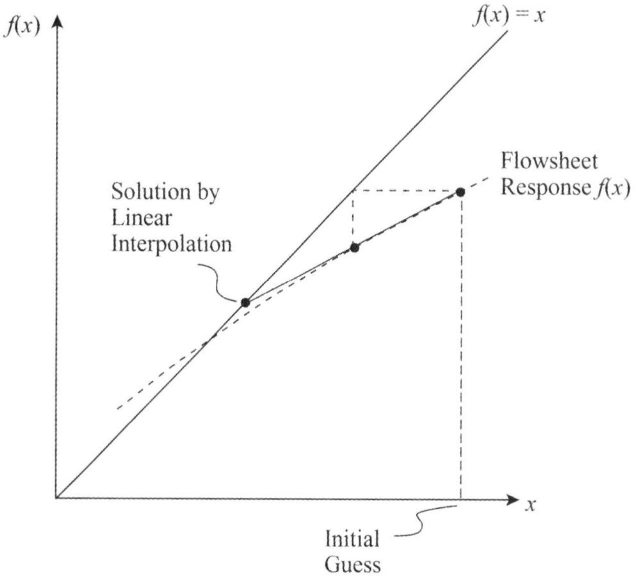

Figure 15.5
The Wegstein Method.

Substituting Equation 15.64 into Equation 15.63 gives:

$$f(x_{k+1}) = ax_{k+1} + [f(x_k) - ax_k] \qquad (15.65)$$

The intersection is required for Equation 15.65 with the equation:

$$f(x_{k+1}) = x_{k+1} \qquad (15.66)$$

Substituting Equation 15.66 into Equation 15.65 gives:

$$x_{k+1} = ax_{k+1} + [f(x_k) - ax_k] \tag{15.67}$$

Rearranging Equation 15.67 gives:

$$x_{k+1} = \left(\frac{a}{a-1}\right)x_k - \left(\frac{1}{a-1}\right)f(x_k) \tag{15.68}$$

Substituting $q = a/(a-1)$ gives:

$$x_{k+1} = qx_k + (1-q)f(x_k) \tag{15.69}$$

Thus, Equation 15.69 can be used to accelerate the convergence and is known as the Wegstein Method (Wengstein, 1958). Bounds are normally set for the value of Wegstein acceleration parameter q to prevent unstable behavior. If $q = 0$ in Equation 15.69, the method becomes direct substitution. If $q < 0$, acceleration of the solution occurs. If q is bounded such that $0 < q < 1$, then slow but stable convergence can occur. Bounding in the range $-5 < q < 0$ is usual (Towler and Sinnott, 2013).

Returning to the example from Figure 15.3, the solution by direct substitution is followed in Table 15.1. If the Wegstein Method is applied after the first two iterations:

$$\begin{aligned} a &= \frac{40.6838 - 42.7500}{42.7500 - 50} \\ &= 0.2850 \\ q &= -0.3986 \end{aligned}$$

Substituting in Equation 15.69 gives:

$$\begin{aligned} x_{k+1} &= -0.3986 \times 42.7500 + (1 - (-0.3986)) \times 40.6838 \\ &= 39.8602 \text{ kmol} \end{aligned}$$

For multivariable problems, Equation 15.69 is written for each variable i on iteration k as:

$$x_{i,k+1} = q_i x_{i,k} + (1-q_i)f(x_{i,k}) \tag{15.70}$$

where $q_i = a_i/(a_i - 1)$

As with the direct substitution approach, the Wegstein Method treats the variables as if independent of each other, but this is not the case. Application of the method starts with a number of direct substitution steps. The convergence is then accelerated. Further direct substitution steps are followed by acceleration, until convergence is achieved. Compared with direct substitution only, this approaches the solution much more rapidly.

3) *Newton–Raphson methods.* More sophisticated methods for convergence are Newton–Raphson methods, which exploit gradients to solve the recycle system. To understand these methods, start with the simplest form of the Newton-Raphson method. To solve the recycle in Figures 15.4 requires the solution of:

$$f(x) = x \tag{15.71}$$

To solve this equation using the Newton–Raphson Method, the root of the following equation is determined:

$$g(x) = f(x) - x = 0 \tag{15.72}$$

Figure 15.6a shows $g'(x_1)$ to be the derivative (slope) of $g(x)$ at x_1. The slope is defined as:

$$g'(x_1) = \frac{g(x_1) - 0}{x_1 - x_2} \tag{15.73}$$

Rearranging gives:

$$x_2 = x_1 - \frac{g(x_1)}{g'(x_1)} \tag{15.74}$$

Thus starting with an initial guess at x_1, the next guess x_2 is given by Equation 15.74. The process is repeated according to:

$$x_{k+1} = x_k - \frac{g(x_k)}{g'(x_k)} \quad \text{for} \quad g'(x_k) \neq 0 \tag{15.75}$$

where $g'(x_k)$ is the derivative of $g(x)$ at x_k on iteration k. Equation 15.75 is applied repeatedly until successive new approximations to the solution change by less than an acceptable tolerance.

This procedure is illustrated in Figure 15.6a. The equivalent illustration in $f(x)$ is shown in Figure 15.6b. A significant disadvantage of this method is the need for gradient information (derivatives). For explicit functions, the derivatives can be obtained analytically. For process simulation in general, and recycle convergence in particular, such analytical derivatives are not available and gradients must be approximated by numerical perturbation by a small value δx. Using this approach, which introduces some errors, then Equation 15.75 becomes:

$$x_{k+1} = x_k - \frac{g(x_k)}{g(x_k + \delta x) - g(x_k)}\delta x \tag{15.76}$$

As an example, apply the Newton–Raphson acceleration to the problem in Figure 15.3a after the first two iterations in Table 15.1:

$$x_k = 42.75 \text{ kmol}, \; f(x_k) = 40.6838 \text{ kmol}$$

$$g(x_k) = 40.6838 - 42.7500 = -2.0663 \text{ kmol}$$

Now perturbate x_k by, say, 0.1:

$$x_k + \delta x = 42.7500 + 0.1 = 42.8500 \text{ kmol}$$

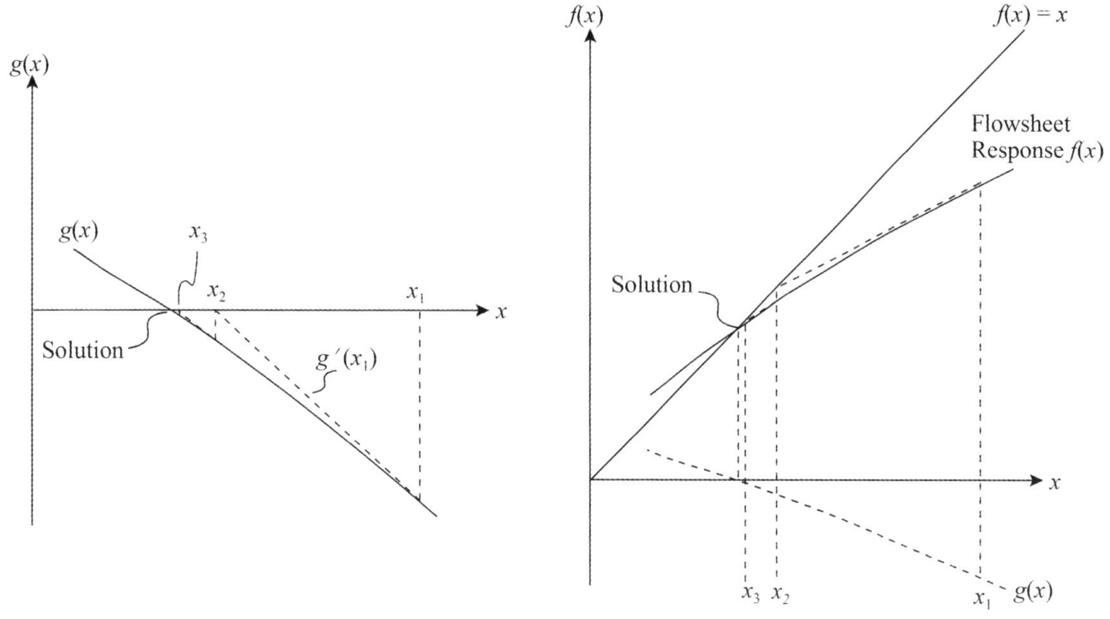

(a) Newton-Raphson Method solving for the root of $g(x)$.

(b) Newton-Raphson Method solving for $f(x) = x$.

Figure 15.6

The Newton–Raphson Method.

Now solving for $f(x + \delta x)$ from the unit operation blocks:

$$f(x_k + \delta x) = 40.7123 \text{ kmol}$$
$$g(x_k + \delta x) = 40.7123 - 42.8500 = -2.1378 \text{ kmol}$$

Now apply the Newton–Raphson acceleration from Equation 15.76:

$$x_{k+1} = 42.7500 - \frac{-2.0663}{-2.1378 - (-2.0663)} \times 0.1$$
$$= 39.8601 \text{ kmol}$$

This is clearly much better than direct substitution. For multivariable problems with n recycle variables, a set of n equations and n unknowns can be written for the tear stream variables:

$$\begin{aligned} g_1(x_1, x_2, \ldots x_n) &= 0 \\ g_2(x_1, x_2, \ldots x_n) &= 0 \\ &\cdot \\ &\cdot \\ &\cdot \\ g_n(x_1, x_2, \ldots x_n) &= 0 \end{aligned} \tag{15.77}$$

Note that each of the n equations is a function of the n unknowns associated with the tear stream. In vector notation, this becomes:

$$\mathbf{g(x) = 0} \tag{15.78}$$

The Newton–Raphson Method becomes:

$$\mathbf{x_{k+1} = x_k - J_k^{-1} g(x_k)} \tag{15.79}$$

where $\mathbf{x_k}$ and $\mathbf{x_{k+1}}$ are the vectors of variable x_i for iterations k and $k+1$, $\mathbf{g(x_k)}$ is the vector of functions g_i for iteration k and $\mathbf{J_k}$ is the Jacobian matrix of partial derivatives:

$$\mathbf{J_k} = \begin{bmatrix} \frac{\partial g_1}{\partial x_1} & \frac{\partial g_1}{\partial x_2} & \cdots & \frac{\partial g_1}{\partial x_n} \\ \frac{\partial g_2}{\partial x_1} & \frac{\partial g_2}{\partial x_2} & \cdots & \frac{\partial g_2}{\partial x_n} \\ \cdot & \cdot & & \cdot \\ \cdot & \cdot & & \cdot \\ \frac{\partial g_n}{\partial x_1} & \frac{\partial g_n}{\partial x_2} & \cdots & \frac{\partial g_n}{\partial x_n} \end{bmatrix}_k \tag{15.80}$$

The Jacobian can in some rare cases be evaluated directly by using analytical partial derivatives for each variable. However, in most cases the Jacobian must be determined numerically using small perturbations of the recycle variables, as given by:

$$
\mathbf{J_k} = \begin{bmatrix}
\frac{g_1(x_1+\delta x_1)-g_1(x_1)}{\delta x_1} & \frac{g_1(x_2+\delta x_2)-g_1(x_2)}{\delta x_2} & \cdots & \frac{g_1(x_n+\delta x_n)-g_1(x_n)}{\delta x_n} \\
\frac{g_2(x_1+\delta x_1)-g_2(x_1)}{\delta x_1} & \frac{g_2(x_2+\delta x_2)-g_2(x_2)}{\delta x_2} & \cdots & \frac{g_2(x_n+\delta x_n)-g_2(x_n)}{\delta x_n} \\
\cdot & \cdot & & \cdot \\
\cdot & \cdot & & \cdot \\
\frac{g_n(x_1+\delta x_1)-g_n(x_1)}{\delta x_1} & \frac{g_n(x_2+\delta x_2)-g_n(x_2)}{\delta x_2} & \cdots & \frac{g_n(x_n+\delta x_n)-g_n(x_n)}{\delta x_n}
\end{bmatrix}_k \tag{15.81}
$$

Using this approach, the variables in the recycle x_i are first guessed. The unit models in the recycle loop are then evaluated using these initial guesses to calculate new values of x_i. To calculate the Jacobian numerically, each value of x_i is perturbed in turn one at a time. For each perturbation the flowsheet response is determined by solving the unit models in the recycle loop. The value of x_i is then re-set and the next variable perturbed, and so on. In this way, the Jacobian of partial derivatives is populated. New values of the recycle variables are then determined from Equation 15.79. The Newton–Raphson approach starts with direct substitution. The convergence is then accelerated. Further direct substitution steps are followed by acceleration and the process repeated until convergence is achieved within a specified tolerance. The convergence criterion is usually the root mean square error. The approach is computationally very expensive. More efficient methods are the *Secant* or *Quasi-Newton* Methods. The Newton–Raphson Method is useful with complex recycle structures and when design specifications are imposed (see Section 15.6).

4) *Secant Methods.* In the Secant Method, instead of using the first derivative to determine the root of $g(x)$, successive iteration values are used to determine the location of the root. The first derivative in the Newton–Raphson Method is approximated as:

$$g'(x_k) \approx \frac{g(x_k) - g(x_{k-1})}{x_k - x_{k-1}} \tag{15.82}$$

Thus the Secant Method for a single-variable problem becomes:

$$x_{k+1} = x_k - g(x_k)\frac{x_k - x_{k-1}}{g(x_k) - g(x_{k-1})} \tag{15.83}$$

The Secant Method is similar to using the Newton–Raphson Method with numerical derivatives, but using a large perturbation to estimate the slope, rather than a small perturbation. Figure 15.7 contrasts the Newton–Raphson and Secant Methods for the first iteration. The advantage of the Secant Method is that the Newton–Raphson Method requires the evaluation of $g(x)$ and $g'(x)$ on each iteration, whereas the Secant Method only requires the evaluation of $g(x)$ on each iteration. The maximum step size needs to be specified. As an example, apply secant acceleration to the problem in Figure 15.3a after the first two iterations in Table 15.1:

$$x_{k-1} = 50.00\ \text{kmol},\ f(x_k) = 42.7500\ \text{kmol},$$
$$g(x_k) = 42.75 - 50.00 = -7.25\ \text{kmol}$$

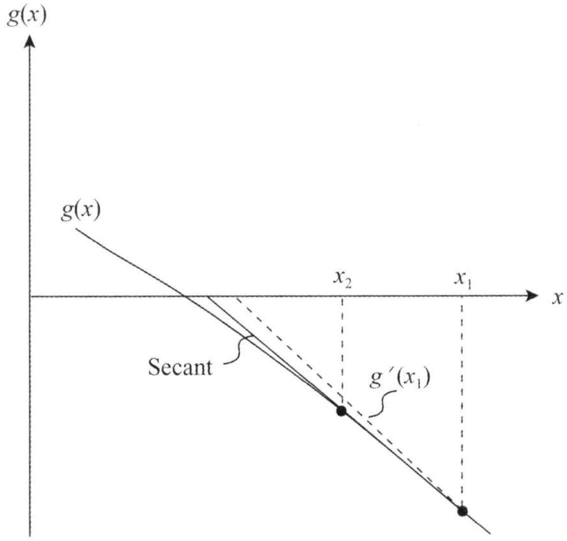

Figure 15.7
The Secant Method.

$$x_k = 42.75\ \text{kmol},\ f(x_k) = 40.6838\ \text{kmol},\ g(x_k) = 40.6838 - 42.7500 = -2.0663\ \text{kmol}$$

Now apply the secant acceleration from Equation 15.83:

$$x_{k+1} = 42.7500 - (7.25)\left(\frac{42.75 - 50.00}{-2.0663 - (-7.2500)}\right) = 39.8601\ \text{kmol}$$

Again, this is clearly much better than direct substitution. It should be noted that the results for the Wegstein, Newton–Raphson and secant accelerations are the same in this case. This results from the linear nature of the flowsheet response curve in this simple example. For multivariable problems, the method of Broyden (1965) is normally used. Broyden's Method is similar to the multivariable Newton–Raphson method, but uses an approximation of the Jacobian. The single variable Secant Equation (Equation 15.83) is generalized and used to determine the approximate Jacobian:

$$\mathbf{J_k}(\mathbf{x_k} - \mathbf{x_{k-1}}) \approx \mathbf{g}(\mathbf{x_k}) - \mathbf{g}(\mathbf{x_{k-1}}) \tag{15.84}$$

However, this does not provide enough information to populate the whole Jacobian. The method uses the current estimate $\mathbf{J_{k-1}}$ with the least change that satisfies the Secant

Formula (Equation 15.84). Different variations are possible for the approach to determination of the approximate Jacobian and the details of the method are outside the scope of this discussion. Having determined the approximate Jacobian, the method proceeds in the Newton–Raphson direction (Equation 15.79) with the approximate Jacobian updated as the iteration progresses. As with the Newton–Raphson Method, the Broyden Method starts with direct substitution steps before acceleration. Broyden's Method is faster than that of Newton–Raphson, but can be less reliable. Like the Newton–Raphson Method, it is useful with complex recycle structures and when design specifications are imposed (see Section 15.6).

Example 15.1

a) Given the estimate of the reactor effluent from Example 14.2 for fraction of methane in the recycle and purge of 0.4, calculate the actual separation in the phase separator assuming the temperature to be 40 °C. Phase equilibrium for this mixture can be represented by the Peng–Robinson Equation of State with binary interaction parameters assumed to be zero. Many computer simulation programs are available commercially to carry out such calculations.

b) Repeat the calculation from Example 14.2 with actual phase equilibrium data in the phase separation instead of assuming a sharp split. Direct substitution can be used to converge the recycle using simulation software.

Solution

a) If such a phase separation is carried out assuming the feed in Table 14.2, the results are given in Table 15.2.

The phase separation at 40 °C gives a good separation of the hydrogen and methane into the vapor phase and benzene, toluene and diphenyl into the liquid phase. Under these conditions, the hydrogen and methane are above their critical temperatures and are effectively noncondensibles. However, some hydrogen and methane dissolve in the liquid phase. Also, some aromatics are carried with the vapor. An important consequence of this is that the flowsheet in Figure 14.8 would need to be modified to separate the hydrogen and methane carried forward with the liquid from the phase separation. This will require another distillation column to separate the hydrogen and the methane from the aromatics before separating the aromatics.

The reader might like to check that as the temperature of the phase separation is increased or its pressure decreased, the separation between the hydrogen, methane and the other components becomes worse.

b) Assuming the phase split operates at 40 bar and 40 °C, a rigorous solution of the phase equilibrium using the Peng–Robinson Equation of State and the recycle equations using flowsheet simulation software gives a composition of the reactor effluent, given in Table 15.3.

Comparing this solution with that based on a sharp phase separation in Example 14.2, the errors are surprisingly small. However, studying the K-values in the phase separator given in Table 15.4, it is not so surprising.

The temperature of the phase split is well above the critical temperatures of both hydrogen and methane, leading to large K-values. On the other hand, the K-values of the benzene, toluene and diphenyl are very low; hence the assumption of a sharp split in Example 14.2 was a good one in this case.

Table 15.2

Phase separation calculated using the Peng–Robinson Equation of State.

Component	Reactor effluent flowrate (kmol·h^{-1})	Vapor flowrate from phase separator (kmol·h^{-1})	Liquid flowrate from phase separator (kmol·h^{-1})
Hydrogen	1554	1550	4
Methane	1036	1020	16
Benzene	265	17.8	247.2
Toluene	91	2.3	88.7
Diphenyl	4	0	4

Table 15.3

Composition of the reactor effluent and phase separation calculated using the Peng–Robinson Equation of State and solving the recycle for Example 15.1.

Component	Reactor effluent flowrate (kmol·h^{-1})	Vapor flowrate from phase separator (kmol·h^{-1})	Liquid flowrate from phase separator (kmol·h^{-1})
Hydrogen	1536	1532	4
Methane	1053	1036	17
Benzene	283	18	265
Toluene	93	3	90
Diphenyl	4	0	4

Table 15.4

K-values for the phase separation based on the Peng–Robinson Equation of State.

Component	K_i
Hydrogen	54
Methane	8.9
Benzene	0.010
Toluene	0.0037
Diphenyl	1.2×10^{-5}

15.6 Design Specifications

For sequential modular simulations, design specifications can be used to set values in the flowsheet that would otherwise involve repeated simulations and trial and error to achieve the required value. For example, the feed flowrate to the flowsheet must be specified, but it may be desirable to specify the flowrate of the product exiting the flowsheet. There is often not a clear link between the feed and product flowrates due to material losses in various places. Specifying the product flowrate can be done by adding a design specification unit to the simulation model that acts like a process controller. The desired value from the flowsheet is specified and the design specification manipulates a variable in order to achieve the desired value. In the case of the product flowrate specification, the design specification unit manipulates the feed flowrate to achieve the desired product flowrate. Another example might be the need to specify the concentration of one of the components in a gas recycle stream to its maximum acceptable value. A unit input variable, process feed variable or other specified input must first be identified that will be used to manipulate the recycle composition. This could be the split fraction for the purge splitter. The variable to be controlled is chosen. This could be the composition of the recycle or purge. A design specification unit is then used to link the two and manipulate the purge split fraction to achieve the desired recycle concentration. In this way, design specifications simulate the steady-state effect of a feedback controller. A design specification can only manipulate the value of one variable. Adding a design specification creates a loop that requires a convergence block to be inserted between the manipulated and controlled variable. A convergence block is required for each design specification. The manipulated variable is adjusted until convergence is achieved:

$$|Specified\ value - Calculated\ value| \leq Tolerance \qquad (15.85)$$

When comparing recycle convergence with design specifications, in the case of design specifications the solution is specified and values during iteration are compared with the correct value. In the case of recycle convergence comparison is made only between neighboring iteration values. Supplying a good estimate for the manipulated variable will help the design specification converge in fewer iterations. This is especially important for large flowsheets with a number of recycles and design specifications.

15.7 Flowsheet Sequencing

When dealing with more complex flowsheets than the one in Figure 15.3, the order in which the calculations take place in the sequential modular approach is important. The first consideration is to tear streams for which a good initial estimate can be provided. Thereafter the choice of tear streams should be to reduce the complexity of the calculation of the solution. Consider the block flowsheet in Figure 15.8a. At first sight there appears to be five tear streams. Figure 15.8b shows a reordered calculation sequence. This calculation order requires tearing of only two streams rather than five. This will greatly simplify the calculations. Also in Figure 15.8b, the calculation sequence has been *partitioned* into two sets of blocks. Partitioning identifies the sets of blocks that must be solved together. In Figure 15.8b, there is little point solving the second partition until the first partition has been solved. If there are multiple tear streams in a partition, then the tear streams can either be converged sequentially or simultaneously. Various algorithms are available for the systematic *partitioning and tearing* of flowsheets (Biegler, Grossman and Westerberg, 1997). Rather than tearing for the minimum number of tear streams, it might be better to give priority to tearing streams for which good estimates can be provided, irrespective of the compatibility with the minimum tear set. Some software packages identify automatically the tear steams for the minimum number. However, the designer might override the automatic choice if good estimates can be provided for certain streams.

15.8 Model Validation

All flowsheet simulation models should be subject to some kind of validation. There is a temptation to simply accept whatever output comes from a computer simulation. However, inappropriate choice of physical property models or unit models can lead to the output being not representative of the real process or, in the worst case, being completely wrong. The best way to validate a flowsheet simulation model is to attempt to model an existing operation or process that is similar to the one being designed. The most important issues in judging whether a process is similar or not are the components being processed, pressure and temperature. If the objective of the simulation is to help to understand or modify an existing process, then validation will occur naturally by comparing the model with the existing operation. However, if it is a new design, validation can be less straightforward. The designer should attempt to identify operating data that have been reported for the key unit models and attempt to reproduce the reported performance in a simulation model before adapting it to the new circumstances. Reported data for a whole process can also be used to help the validation, if this is available.

15.9 Process Optimization

Once the structure of the reaction, separation and recycle system has been established and a simulation model developed, degrees of freedom that have a significant effect on the overall process economics can be optimized. Such optimization will most often be performed automatically. However, because of the difficulties when optimizing, discussed in Chapter 3, particularly when optimizing nonlinear problems, it is important to understand the trends that are to be expected. Consideration here will be restricted to the

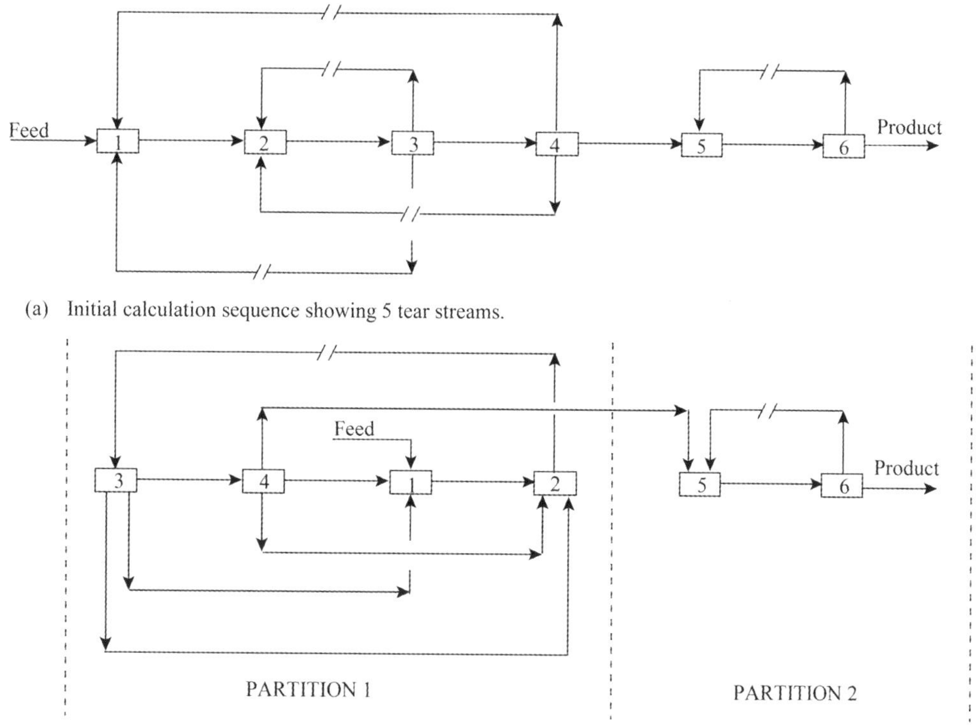

(a) Initial calculation sequence showing 5 tear streams.

(b) Re-ordered calculation sequence showing 2 tear streams.

Figure 15.8

Partitioning and tearing a complex flowsheet.

trends to be expected when optimizing reactor conversion and recycle systems involving a purge.

1) *Optimization of reactor conversion.* Some change in the reactor conversion might be possible. If the reactor conversion is changed so as to optimize its value, then not only is the reactor affected in size and performance but also the separation system, since it now has a different separation task. The size of the recycle will also change. If the recycle requires a compressor, then the capital and operating costs of the recycle compressor will change. In addition, the heating and cooling duties associated with the reactor and the separation and recycle system change. As the reactor conversion increases, the reactor volume increases and hence the reactor capital cost increases. At the same time, the amount of unconverted feed needing to be separated decreases and hence the cost of recycling unconverted feed decreases.

   Consider a simple process in which *FEED* is reacted to *PRODUCT* via the reaction:

$$FEED \rightarrow PRODUCT \tag{15.86}$$

   The flowsheet synthesis is started at the reactor. The effluent from the reactor contains both *PRODUCT* and unreacted *FEED* that must be separated. Unreacted *FEED* is recycled to the reactor via a pump if the recycle is liquid or a compressor if the recycle is vapor.

Optimization of the system can be carried out by minimizing cost or maximizing economic potential (EP), as discussed in Chapter 2. Costs for the process to carry out the reaction in Equation 15.86 are illustrated in Figure 15.9, decomposed according to the layers of the onion model (Smith and Linnhoff, 1988). In Figure 15.9, the annualized reactor cost (capital only) increases since high conversion requires a larger volume and hence a higher capital cost. The annualized separation and recycle cost (capital only in this case) decreases with increasing reactor conversion, since the amount of unreacted *FEED* to separate and recycle decreases. If the recycle had required a compressor, the capital and operating costs of the compressor would have been included in the separation and recycle cost. The cost of the heat exchanger network and utilities is a combination of annualized energy cost and annualized capital cost of all exchangers, heaters and coolers. Later, it will be explained how to estimate the energy cost of the heat exchanger network without having to carry out its detailed design. Figure 15.9 shows that the cost of the heat exchanger network and utilities decreases with increasing conversion, since the separation duty is decreased and also the heating and cooling in the recycle. Combining the reactor, separation and recycle and heat exchanger network costs into a total annual cost (energy and capital) reveals that there is an optimum reactor conversion. From Figure 15.9, for this example, heat integration and the cost of the heat exchanger network and utilities have a

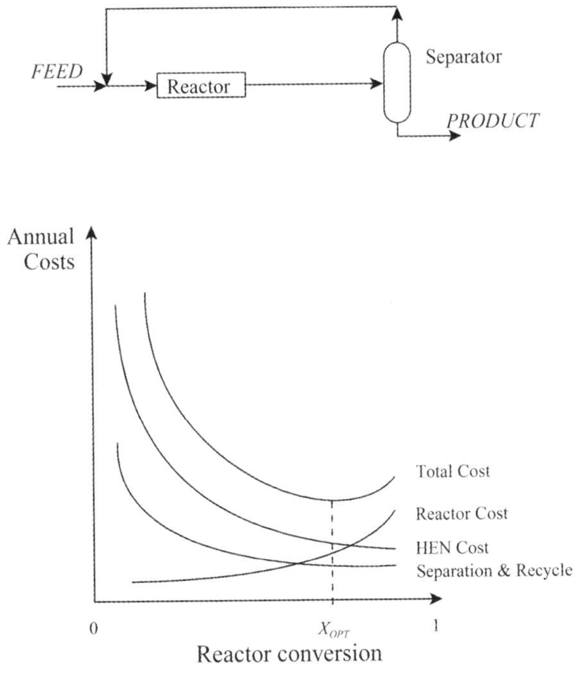

Figure 15.9

Overall cost trade-offs for a simple process as a function of reactor conversion.

significant influence on the optimum conversion. In other cases, the relative importance of the component costs will be different.

If the cost of the heat exchanger network changes, perhaps through a change in energy cost, then the optimum reactor conversion will change. This change will likely dictate a different optimum reactor conversion and hence different separator design and process flowrates.

In Figure 15.9, the only cost forcing the optimum conversion back from high values is that of the reactor. Hence, for such simple reaction systems, a high optimum conversion would be expected. This was the reason in Chapter 5 that an initial value of reactor conversion of 0.95 of the maximum conversion was chosen for simple reaction systems.

In Figure 15.9, the curves are limited by a maximum reactor conversion of 1.0. If the reaction had been reversible, then a similar picture would have been obtained. However, instead of being limited by a reactor conversion of 1.0, the curves would have been limited by the equilibrium conversion (see Chapter 5).

Consider the example of a process that involves the multiple reactions:

$$\begin{aligned} &FEED \longrightarrow PRODUCT \\ &PRODUCT \longrightarrow BYPRODUCT \end{aligned} \tag{15.87}$$

Because there is a mixture of *FEED, PRODUCT* and *BYPRODUCT* in the reactor effluent, an additional separator is required. The economic trade-offs now become more complex and a new cost must be added to the trade-offs. This is a raw materials inefficiency cost due to byproduct formation. If the *PRODUCT* formation is kept constant, despite varying levels of *BYPRODUCT* formation, then the cost can be defined to be (Smith and Omidkhah Nasrin, 1993a):

$$\begin{aligned} &\text{Cost due to } BYPRODUCT \text{ formation} \\ &\quad = \text{cost of } FEED \text{ lost to } BYPRODUCT \\ &\quad\quad - \text{value of } BYPRODUCT \end{aligned} \tag{15.88}$$

The value of *PRODUCT* formation and the raw materials cost of *FEED* that reacts to *PRODUCT* are constant. Alternatively, if the byproduct has no value, the cost of disposal should be included as:

$$\begin{aligned} &\text{Cost due to } BYPRODUCT \text{ formation} \\ &\quad = \text{cost of } FEED \text{ lost to } BYPRODUCT \\ &\quad\quad + \text{cost of disposal of } BYPRODUCT \end{aligned} \tag{15.89}$$

By considering only those raw materials that undergo reaction to an undesired byproduct, only the raw materials costs that are in principle avoidable are considered. Those raw materials costs that are inevitable, that is, the stoichiometric requirements for *FEED* that converts into the desired *PRODUCT*, are not included. Raw materials costs that are in principle avoidable are distinguished from those that are inevitable from the stoichiometric requirements of the reaction (Smith and Omidkhah Nasrin, 1993a, 1993b).

Figure 15.10 shows typical cost trade-offs for this case. At high conversion, the raw materials cost due to byproduct formation is dominant. This is because the reaction to the undesired *BYPRODUCT* is series in nature, which results in the selectivity becoming very low at high conversions. In Chapter 4, the initial setting for reactor conversion was set to be 0.5 of the maximum conversion for such reaction systems. Figure 15.10 shows clearly why a high setting for reactor conversion would be inappropriate. The byproduct formation cost forces the optimum to lower values of conversion. Again, if the primary reaction had been reversible, then a similar picture would have been obtained. However, instead of being limited by a reactor conversion of 1.0, the curves would have been limited by the equilibrium conversion (see Chapter 5).

Also, if there are two separators, the order of separation can change. The trade-offs for these two alternative flowsheets will be different. The choice between different separation sequences can be made using the methods described in Chapters 10 and 11. However, as the reactor conversion changes, the most appropriate sequence can also change. In other words, different separation system structures can become appropriate for different reactor conversions.

2) *Optimization of processes involving a purge.* If an impurity entering with the feed, or a byproduct of reaction, needs to be removed via a purge, the concentration of impurity in the recycle can be varied as a degree of freedom. If the impurity is allowed to build up to a high concentration, then this reduces the loss of valuable raw materials in the purge. However, this decrease in raw materials cost is offset by an increase in the cost of recycling the additional impurity, together with an increased

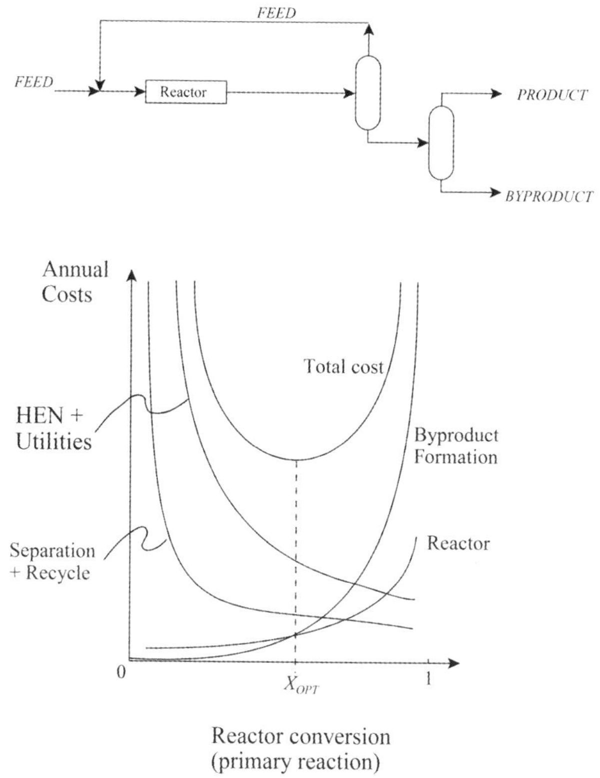

Figure 15.10

Cost trade-offs for a process with byproduct formation from secondary reactions.

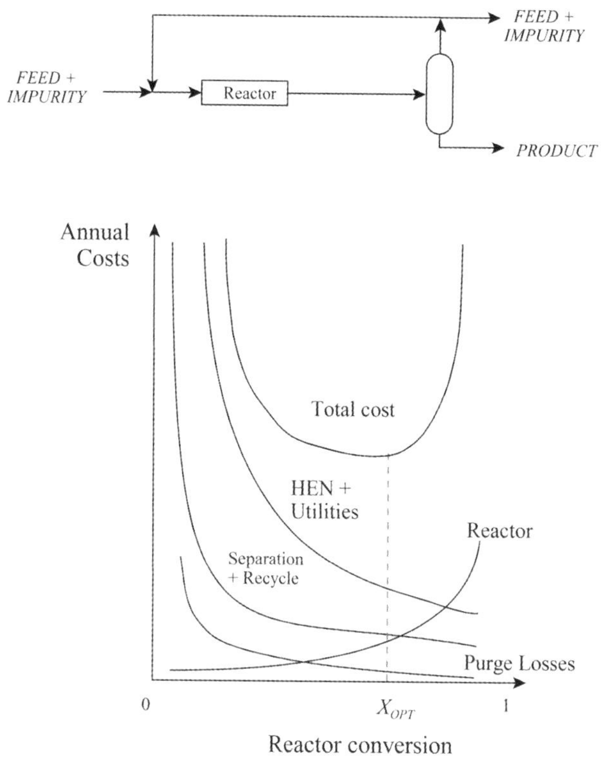

Figure 15.11

Cost trade-offs for a process with a purge for a fixed concentration of impurity in the recycle.

capital cost of equipment in the recycle. Changes in the recycle concentration again make changes throughout the flowsheet.

As with the case of byproduct losses, another cost needs to be added to the trade-offs when there is a purge. This is a raw materials efficiency cost due to purge losses. If the *PRODUCT* formation is constant, this cost can be defined to be (Smith and Omidkhah Nasrin, 1993a, 1993b):

$$\text{Cost of purge losses} = \text{Cost of } FEED \text{ lost to purge} - \text{Value of purge} \qquad (15.90)$$

The purge by its nature is a mixture and usually only has value in terms of its fuel value. Alternatively, if the purge must be disposed of by effluent treatment:

$$\text{Cost of purge losses} = \text{Cost of } FEED \text{ lost to purge} + \text{Cost of disposal of purge} \qquad (15.91)$$

Again, as with the byproduct case, those raw materials costs that are in principle avoidable (i.e. the purge losses) are distinguished from those that are inevitable (i.e. the stoichiometric requirements for *FEED* entering the process that converts to the desired *PRODUCT*). Consider the trade-offs for the reaction in Equation 15.86, but now with *IMPURITY* entering with the *FEED*.

Now there are two variables in the optimization. These are the reactor conversion (as before) but now also the concentration of *IMPURITY* in the recycle. For each setting of the *IMPURITY* concentration in the recycle, a set of trade-offs can be produced analogous to those shown in Figures 15.9 and 15.10. Figure 15.11 shows the trade-offs for the feed impurity case and a purge with fixed concentration of impurity in the recycle (Smith and Omidkhah Nasrin, 1993a, 1993b).

As the concentration of impurity in the recycle is varied, each component cost shows a family of curves when plotted against reactor conversion. The reactor cost (capital only) increases as before with increasing conversion (Figure 15.12a). Separation and recycle costs decrease as before (Figure 15.12b). Figure 15.12c shows the cost of the heat exchanger network and utilities to again decrease with increasing conversion. Each of the component costs in Figure 15.12 increase with increasing concentration of impurity in the recycle. 15

Figure 15.13 shows the corresponding component costs for the purge losses. Figure 15.13a shows component costs that are typical when the value of the raw materials lost in the purge is high relative to the fuel value of the purge. Figure 15.13b shows component costs that are typical when the value of the raw materials lost in the purge is low relative to the fuel value of the purge. If the process produces a byproduct to be purged and the purge has a value, then the component cost of the purge can

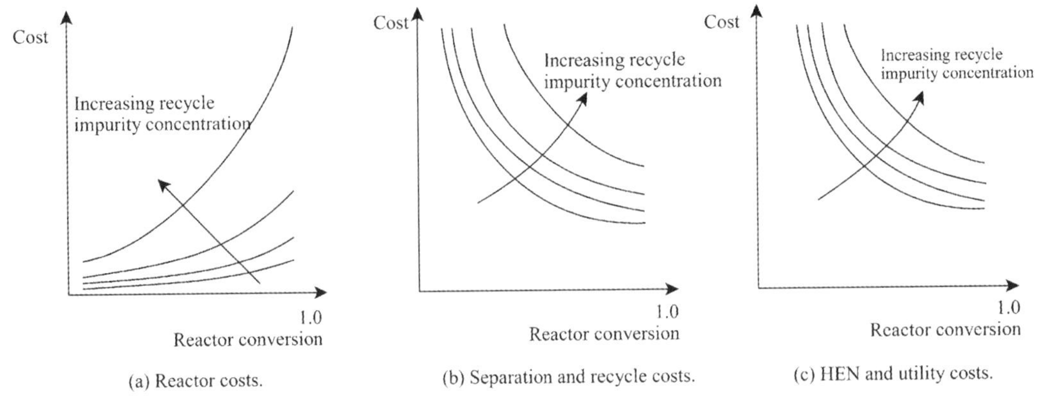

Figure 15.12

Cost trade-offs for processes with a purge as recycle inert concentration changes.

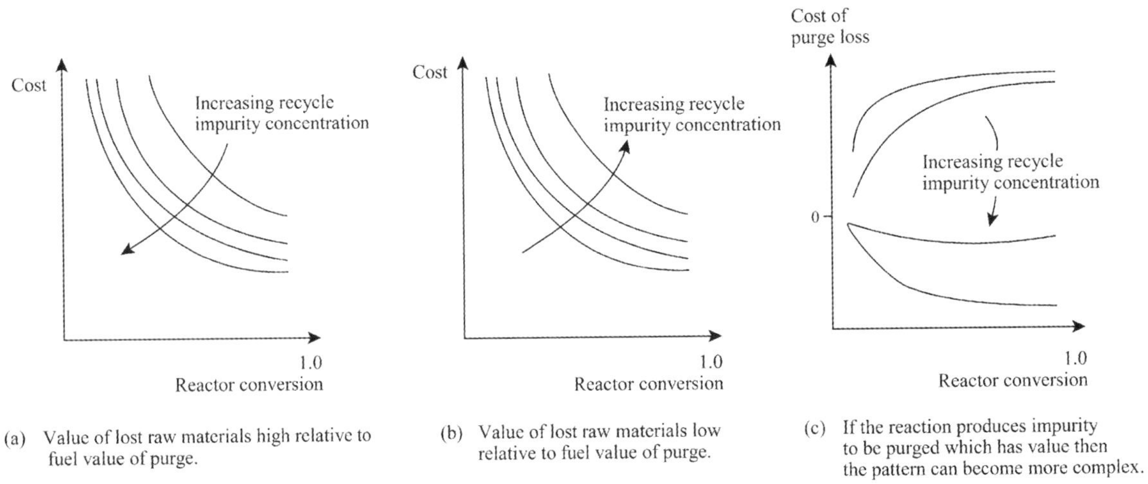

Figure 15.13

Cost trade-offs for purge losses as reactor conversion and recycle impurity concentration change.

show more complex behavior, as shown in Figure 15.13c. This shows a complex pattern that depends on the relative costs of raw materials and fuel value of the purge (Smith and Omidkhah Nasrin, 1993a, 1993b).

Figure 15.14 shows the component costs combined to give a total cost that varies with both reactor conversion and recycle concentration of impurity. Each setting of the recycle impurity concentration shows a cost profile with an optimum reactor conversion. As the recycle impurity concentration is increased, the total cost initially decreases but then increases for larger values of recycle concentration. The optimum conditions in Figure 15.14 are in the region of $y = 0.6$ and $X = 0.5$, although the optimum is quite flat.

An alternative way to view the trade-offs shown in Figure 15.15 is as a contour diagram. The contours in Figure 15.15a are lines of constant total cost. The objective of the optimization is to find the lowest point. A simple strategy

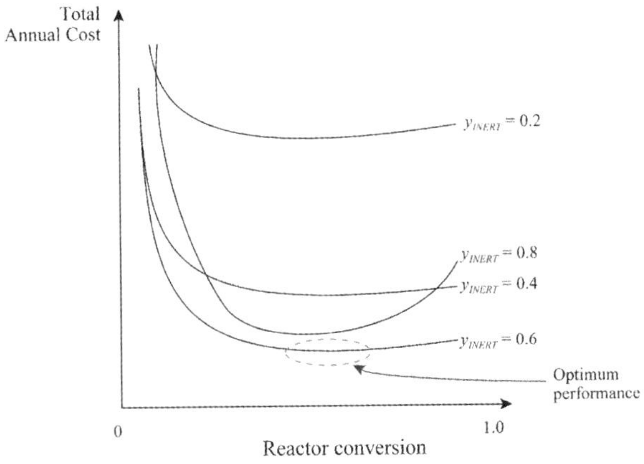

Figure 15.14

Putting all the costs together allows the optimum conversion and recycle inert concentration to be determined.

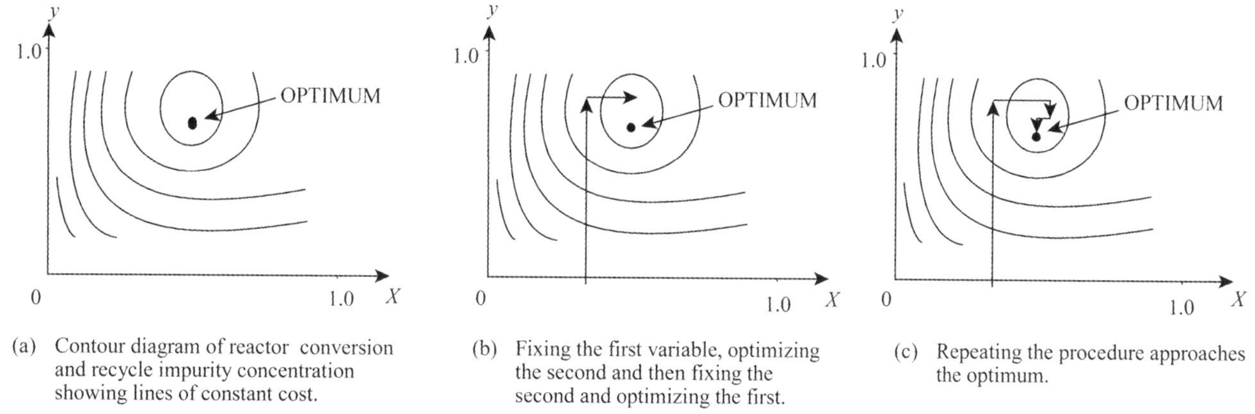
y
1.0
OPTIMUM
0
1.0
X
(a) Contour diagram of reactor conversion and recycle impurity concentration showing lines of constant cost.
y
1.0
OPTIMUM
0
1.0
X
(b) Fixing the first variable, optimizing the second and then fixing the second and optimizing the first.
y
1.0
OPTIMUM
0
1.0
X
(c) Repeating the procedure approaches the optimum.

Figure 15.15
Optimization of reactor conversion and recycle impurity concentration using a univariate search.

would fix the first variable, then optimize the second variable, and then fix the second variable and optimize the first, and so on in a univariate search, as discussed in Chapter 3. This is illustrated in Figure 15.15b, where reactor conversion (X) is first fixed and impurity concentration (y) optimized. Impurity concentration is then fixed and conversion optimized. In this case, after two searches the solution is close to the optimum (Figure 15.15c). Whether this is adequate depends on how flat the solution space is in the region of the optimum. Whether such a strategy will find the actual optimum depends on the shape of the solution space and the initialization for the optimization, as discussed in Chapter 3. Other strategies for the optimization have been discussed in Chapter 3.

Obviously, the use of purges is not restricted to dealing with impurities. Purges can also be used to deal with byproducts. As with the optimization of reactor conversion, changes in the recycle concentration of impurity might change the most appropriate separation sequence.

15.10 Continuous Process Simulation and Optimization – Summary

Once the basic flowsheet has been synthesized, it needs more detailed evaluation. The material and energy balances need to be established and preliminary sizing carried out for the major items of equipment. This requires a detailed simulation model to be created. Such simulation models are normally created with general purpose process simulation packages. The choice of physical properties can be critical to ensure that the simulation model is representative of the process to be constructed. Phase equilibrium models can be critical in this respect. A series of unit models need to be created to model the various transformation steps in the process. These unit models then need to be linked together to form a simulation model of the overall process.

There are two basic approaches to modeling process flowsheets. The first approach is the equation-based approach that solves a set of equations simultaneously. In the second approach, the sequential modular approach, the process equations are grouped within unit operation blocks. Each unit operation block is then solved one at a time in sequence. The sequential modular approach creates problems when solving flowsheets with recycles. The sequential modular solution technique is to tear one of the streams in the recycle loop. A recycle convergence unit is inserted in the tear stream. The recycle convergence unit might use direct substitution as the means to solve the recycle or the calculation can be accelerated using a variety of techniques. Design specifications can also be added using the sequential modular approach. The calculation sequence can often be changed, which can have a significant influence on the convergence of the simulation model.

Flowsheet models should be validated as much as possible by comparing the predictions with actual operation of similar plant or equipment. Once a validated flowsheet model has been developed, it can be subjected to optimization. Reactor conversion and recycle composition for flowsheets involving a purge can be important optimization variables.

15.11 Exercises

1. A process is to be simulated involving a simple reaction, separation and recycle system, as illustrated in Figure 15.3. The reaction is:

$$A \longrightarrow B$$

The feed contains 95 kmol of A and 5 kmol of B. The reactor operates at a conversion of 70%. The performance of the separator gives a recovery of A in the recycle stream of 95% and a recovery of B in the product stream of 98%. In a spreadsheet:

a) Solve the material balance using a sequential modular approach by tearing the recycle stream and using direct (repeated) substitution to a scaled residual convergence of 0.0001 for both components.

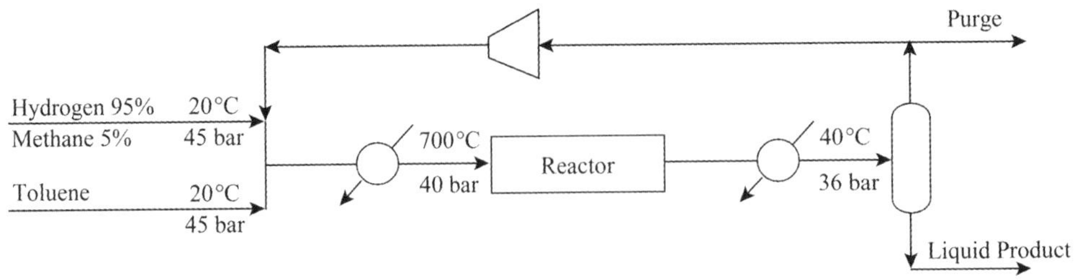

Figure 15.16

Front end of the benzene process from Exercise 3.

b) Solve the material balance by accelerating the solution using the Wegstein Method after the second direct substitution iteration.

c) Check the solutions by solving the equations simultaneously.

d) From the material balance, calculate the process yield for the manufacture of *B*.

2. Create a flowsheet simulation model for the process in Example 15.1 in a flowsheeting package. Compare the answer with the spreadsheet results from Exercise 1. The flowsheet simulation could be set up by introducing two new components to the simulation software. Given that physical properties of the Components *A* and *B* are not necessary for the simple material balance, the simplest approach is to define *A* and *B* as being two components already in the physical property database, e.g.:

$$\text{butane} \longrightarrow \text{isobutane}$$

This can be simulated without introducing any new components.

Assume:

i. The feed enters at 20 °C and 45 bar.
ii. All operations after the feed operate at 40 bar.
iii. The feed is heated to 500 °C before entering the reactor.
iv. The reactor is modelled by a conversion reactor with a conversion of 70%.
v. There is no heat of reaction.
vi. The reactor effluent is cooled to 150 °C before entering the separator.
vii. The separator is modelled using a splitter with a recovery of *A* of 95% in the recycle and a recovery of *B* of 98% in the product.
viii. The recycle convergence is by direct substitution to a scaled residual of 0.0001.

3. Create a flowsheet simulation model for the toluene hydrodealkylation process (Douglas, 1985). Benzene is to be produced from toluene according to the reaction:

$$\underset{\text{toluene}}{C_6H_5CH_3} + \underset{\text{hydrogen}}{H_2} \longrightarrow \underset{\text{benzene}}{C_6H_6} + \underset{\text{methane}}{CH_4}$$

The reaction is carried out in the gas phase and operates at 700 °C and 40 bar. Some of the benzene formed undergoes a series of secondary reactions that can be characterized by the single secondary reversible reaction to an unwanted byproduct, diphenyl:

$$\underset{\text{benzene}}{2C_6H_6} \rightleftharpoons \underset{\text{diphenyl}}{C_{12}H_{10}} + \underset{\text{hydrogen}}{H_2}$$

A hydrogen/toluene ratio of 5 is required at the reactor inlet to prevent excessive coke formation in the reactor. The fraction of toluene reacted that is converted to benzene is related to the conversion (i.e. fraction of toluene fed that is reacted) according to (Douglas, 1985):

$$S = 1 - \frac{0.0036}{(1 - X)^{1.544}}$$

where S = selectivity
X = conversion

The flowsheet for the process front end, without the distillation system and toluene recycle, is shown in Figure 15.16. The toluene feed of 265 kmol · h^{-1} can be assumed to be pure. The hydrogen feed contains methane as an impurity at a mole fraction of 0.05. Methane is also a byproduct from the primary reaction and the methane is to be removed from the process by means of a purge.

Assume:

i. Physical properties can be modelled by the Peng–Robinson Equation of State.
ii. The toluene and hydrogen feeds enter at 25 °C and 45 bar.
iii. The reactor operates at 700 °C and 40 bar.
iv. The reactor can be modelled by a conversion reactor with a toluene conversion of 0.75.
v. The reactor effluent is at 700 °C and 40 bar.
vi. The flash separator after the reactor operates at 40 °C and 36 bar.
vii. The purge fraction (fraction of purge to recycle) is 0.3.
viii. The recycle gas compressor can be modelled as an adiabatic compression with an isentropic efficiency of 0.8.

Using a flowsheeting package, develop a flowsheet simulation model using direct substitution for recycle convergence with a scaled residual of 0.0001.

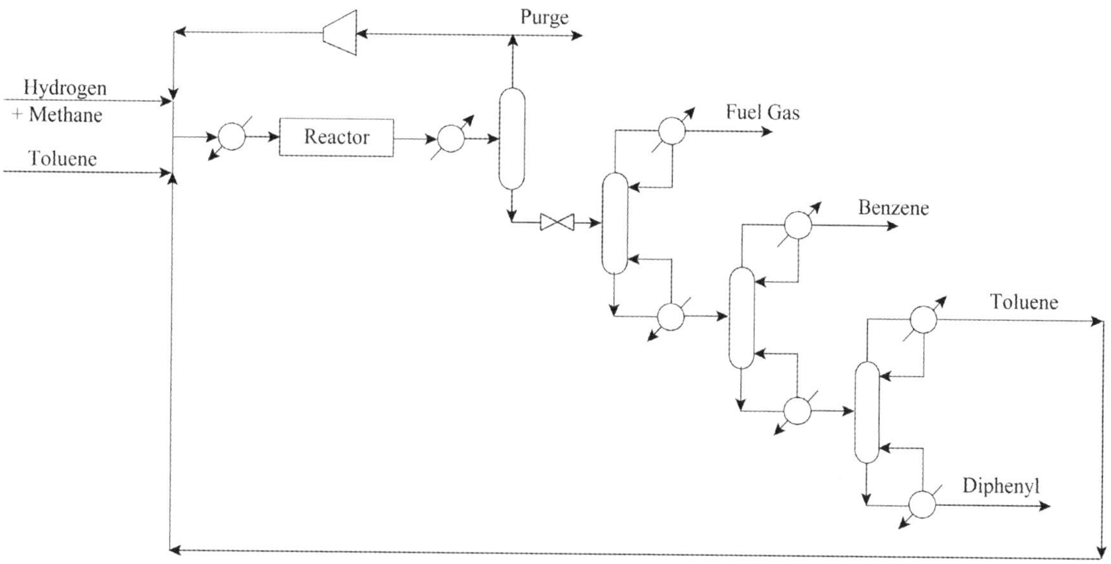

Figure 15.17

Benzene process from Exercise 4.

Table 15.5

Distillation column details for Exercise 3.

Column	Condenser type	Pressure (bar)	Number of stages	Reflux ratio	Distillate rate (kmol·h^{-1})
Stabilizer	Partial	20	4	2	50.5
Benzene column	Total	1.1	15	1	260
Toluene column	Total	1.1	7	1	56

a) Assume a feed of hydrogen with 1400 kmol·h^{-1}.

b) Use a design constraint to impose a hydrogen toluene ratio of 5 by varying the flowrate of the hydrogen feed.

4. The liquid stream product for Figure 15.16 can readily be separated into relatively pure components by distillation, the benzene taken off as product, diphenyl as an unwanted byproduct and the toluene recycled. The flowsheet is shown in

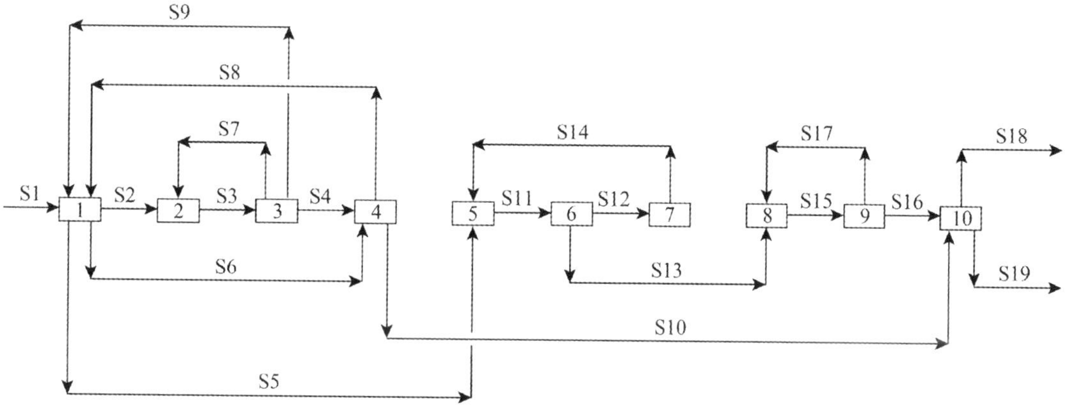

Figure 15.18

Block flowsheet for Exercise 6.

15

Figure 15.17. The distillation columns are to be modelled as rigorous equilibrium models with the details given in Table 15.5. Assume that each of the distillation columns has a saturated liquid feed. Using a flowsheeting package, develop a flowsheet simulation model to solve the material and energy balance by adding the distillation and toluene recycle to the model from Exercise 3b. The flowsheet model should be converged to a scaled residual convergence of 0.0001 using direct substitution.

5. Adapt the flowsheet model from Exercise 4 by adding a design specification to achieve a production rate of benzene of 265 $kmol \cdot h^{-1}$ and a design specification to maintain a recycle concentration of hydrogen in the recycle of 0.6.

6. Figure 15.18 shows the block diagram of a flowsheet simulation model.

   a) Separate the flowsheet model into four partitions such that streams only go forward from one partition to the next and there is no recycling between partitions.

   b) Within each partition, identify the tear streams

   c) For the partitions with more than one tear stream suggest a different flowsheet sequence to decrease the number of tear streams.

References

Biegler LT, Grossman IE and Westerberg AW (1997) *Systematic Methods of Chemical Process Design*, Prentice Hall.

Branan CR (1999) *Pocket Guide to Chemical Engineering*, Gulf Publishing Company.

Broyden CG (1965) A Class of Methods for Solving Nonlinear Simultaneous Equations, *Mathematics of Computation*, **19**: 577.

Douglas JM (1985) A Hierarchical Decision Procedure for Process Synthesis, *AIChE J*, **31**: 353.

Gmehling J, Onken U and Arlt W (1977–1980) *Vapor–Liquid Equilibrium Data Collection*, Dechema Chemistry Data Series.

Joback KG and Reid RC (1987) Estimation of Pure-Component Properties from Group-Contributions, *Chem Eng Comm*, **57**: 233.

Oellrich L, Plöcker U, Prausnitz JM and Knapp H (1981) Equation-of-State Methods for Computing Phase Equilibria and Enthalpies, *Int Chem Eng*, **21**(4): 1.

Poling BE, Prausnitz JM and O'Connell JP (2001) *The Properties of Gases and Liquids*, McGraw-Hill.

Riedel L (1954) Kritischer Koeffizient, Dichte des Gesättigten Dampfes und Verdampfungswärme, *Chem Ing Tech*, **26**: 679.

Seider WD, Seader JD, Lewin DR and Widago S (2010) *Product and Process Design Principles: Synthesis, Analysis and Design*, 3rd Edition, John Wiley & Sons.

Smith R and Linnhoff B (1988) The Design of Separators in the Context of Overall Processes, *Trans IChemE ChERD*, **66**: 195.

Smith R and Omidkhah Nasrin M (1993a) Trade-offs and Interactions in Reaction and Separation Systems, Part I: Reactions with No Selectivity Losses, *Trans IChemE ChERD*, **A5**: 467.

Smith R and Omidkhah Nasrin M (1993b) Trade-offs and Interactions in Reaction and Separation Systems, Part II: Reactions with Selectivity Losses, *Trans IChemE ChERD*, **A5**: 474.

Streich M and Kistenmacher H (1979) Property Inaccuaracies Influence Low Temperature Designs, *Hydrocarbon Process*, **58**: 237.

Towler G and Sinnott R (2013) *Chemical Engineering Design*, 2nd Edition, Butterworth-Heinemann imprint of Elsevier.

Wegstein JH (1958) Accelerating Convergence of Iterative Processes, *Commun Assoc Comp Mach*, **1**: 9.

Chapter 16

Batch Processes

16.1 Characteristics of Batch Processes

As pointed out in Chapter 1, in a batch process the main steps operate discontinuously. This means that temperature, concentration, mass and other properties vary with time. Also as pointed out in Chapter 1, most batch processes are made up of a series of batch and semicontinuous steps. A semicontinuous step operates continuously with periodic start-ups and shutdowns.

Many batch processes are designed on the basis of a scale-up from the laboratory, particularly for the manufacture of specialty chemicals and pharmaceuticals. If this is the case, the process development will produce a *recipe* for the manufacturing process. The recipe is not unlike a recipe used in cookery. It is a step-by-step procedure that resembles the laboratory procedure, but scaled to the quantities required for manufacturing. It provides information on the quantities of material to be used in any step in the manufacturing, the conditions of temperature, pressure, and so on, at any time, and the times over which the various steps take place. The recipe can be thought of as the equivalent of the material and energy balance in a continuous process. However, care should be taken to avoid taking artificial constraints from the laboratory to the manufacturing process (i.e. those constraints imposed by the laboratory procedures that do not apply to industrial plant).

As noted in Chapter 1, the priorities in batch processes are often quite different from those in large-scale continuous processes. Particularly when manufacturing specialty chemicals and pharmaceuticals, the shortest time possible to get a new product to market is often the biggest priority (accepting that the product must meet the specifications and regulations demanded and the process must meet the required safety and environmental standards). This is particularly true if the product is protected by patent. The period over which the product is protected by patent must be exploited to its full. This means that product development, testing, pilot plant work, process design and construction should be *fast tracked* and carried out as much as possible in parallel.

Before a system of batch reaction and separation processes is considered, the main operations that will be used in batch processes need to be reviewed, but with the emphasis on how they will differ from the corresponding operations in continuous processes.

16.2 Batch Reactors

As with continuous processes, the heart of a batch chemical process is its reactor. Idealized reactor models were considered in Chapter 4. In an ideal-batch reactor, all fluid elements have the same residence time. There is thus an analogy between ideal-batch reactors and continuous plug-flow reactors. There are four major factors that affect batch reactor performance:

- Contacting pattern
- Operating conditions
- Agitation for agitated vessel reactors
- Solvent selection.

1) *Contacting pattern.* Most batch reactors are agitated vessels with a standard configuration. However, this is only one way that such reactions can be accomplished. If the reaction is multiphase, then any of the contacting patterns in Figures 6.4 and 6.5 might be considered to enhance the mass transfer relative to that which can be achieved using an agitated vessel. For example, a recirculation system could be used in which material was taken from the reactor and returned to the reactor via a pump in the case of a liquid or a compressor in the case of a gas. If such possibilities are considered, then there is no reason why the kind of equipment used for continuous processes could not be applied, with recirculation of the materials used to create the batch mode. Indeed, the reactor could simply be operated in a semicontinuous mode. If the reaction is rapid and the temperature can be controlled, whether single-phase or multiphase, then a simple static mixer arrangement operated in semicontinuous mode might be the best solution (Figure 6.4 and 6.5).

   Figure 16.1 shows various modes of operation for agitated vessel batch and semibatch reactors. In Figure 16.1a, the feeds are loaded into the reactor at the beginning of the batch, the reaction then proceeds for a specified time, after which the products are removed. By contrast, Figure 16.1b shows

Chemical Process Design and Integration, Second Edition. Robin Smith.

Companion Website: www.wiley.com/go/smith/chemical2e

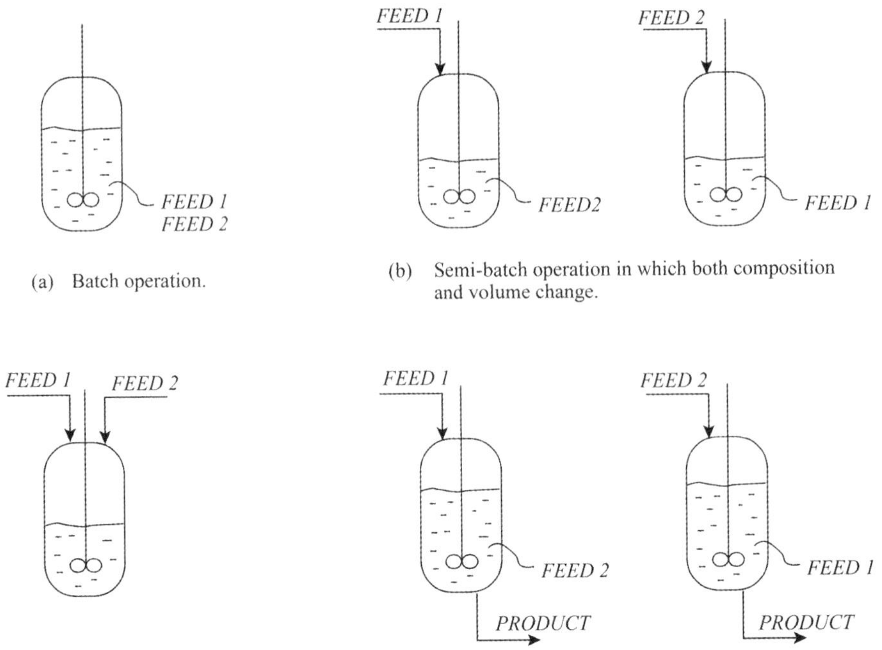

(a) Batch operation.

(b) Semi-batch operation in which both composition and volume change.

(c) Semi-batch operation in which composition is maintained constant, but volume changes.

(d) Semi-batch operation in which composition changes, but volume is maintained constant.

Figure 16.1
Various modes of operation for batch and semi-batch reactors.

semibatch operation in which one of the feeds is initially charged to the reactor and the other feed is charged progressively. This mode of operation was discussed in Chapter 4. The semibatch operation in Figure 16.1b is one in which both the composition and the volume in the reactor change with time. However, batch reactors offer additional degrees of freedom. Figure 16.1c shows a semibatch operation in which the feeds are charged to the reactor as the batch progresses. In this way, composition can in principle be maintained constant, but volume changes. Finally, Figure 16.1d shows semibatch operation in which the product is withdrawn before the end of a batch reaction. In the arrangements shown in 16.1d, the composition changes but the volume can in principle be maintained constant. In Chapter 4, some simple heuristics were developed to help decide the contacting pattern if some knowledge of the reaction kinetics is available. For example, for the parallel reaction:

$$FEED1 + FEED2 \rightarrow PRODUCT \qquad r_1 = k_1 C_{FEED1}^{a_1} C_{FEED2}^{b_1}$$
$$FEED1 + FEED2 \rightarrow BYPRODUCT \quad r_2 = k_2 C_{FEED1}^{a_2} C_{FEED2}^{b_2} \qquad (16.1)$$

Taking the ratio of the reaction rates:

$$\frac{r_2}{r_1} = \frac{k_2}{k_1} C_{FEED1}^{a_2 - a_1} C_{FEED2}^{b_2 - b_1} \qquad (16.2)$$

In order to minimize the formation of the unwanted *BYPRODUCT*, the ratio of the reaction rates in Equation 16.2 should be minimized. Thus (Smith and Petela, 1992; Sharratt, 1997):

- $a_2 > a_1$ and $b_2 > b_1$. The concentration of both feeds should be minimized and each added progressively as the reaction proceeds. Predilution of the feeds might be considered.
- $a_2 > a_1$ and $b_2 < b_1$. The concentration of *FEED*1 should be minimized by charging *FEED*2 at the beginning of the batch and adding *FEED*1 progressively as the reaction proceeds. Predilution of *FEED*1 might be considered.
- $a_2 < a_1$ and $b_2 > b_1$. The concentration of *FEED*2 should be minimized by charging *FEED*1 at the beginning of the batch and adding *FEED*2 progressively as the reaction proceeds. Predilution of *FEED*2 might be considered.
- $a_2 < a_1$ and $b_2 < b_1$. The concentration of *FEED*1 and *FEED*2 should be maximized by rapid addition and mixing.

Whilst heuristics like these are helpful, they have also severe limitations:

- Knowledge of the reaction chemistry and the kinetics is needed.
- They only are helpful when dealing with simple reaction systems.
- They are only qualitative.

Even if it is decided to use semibatch operation in the above example in which *FEED*2 is charged to the reactor initially and then *FEED*1 as the reaction progresses, it is not known how fast the feed should be added to obtain the optimum performance. Feed addition rate and product takeoff rate are degrees of freedom that need to be optimized.

There are many practical issues that need to be considered when choosing mixing equipment and mixing patterns, in addition to those for maximizing yield, selectivity or conversion (Harnby, Nienow and Edwards, 1997). This is especially the case when dealing with multiphase reactions (Harnby, Nienow and Edwards, 1997). A superstructure approach can be used for the optimization of the reaction mixing pattern but this requires a detailed model of the reaction system to be available (Zhang and Smith, 2004).

2) *Operating conditions.* Once batch cycle time and total amount of reactants have been fixed for a given batch reactor system, variables such as temperature, pressure, feed addition rates and product takeoff rates are dynamic variables that change through the batch cycle time. These values form profiles for each variable across the batch cycle time.

    If the rate of feed addition, rate of product takeoff, temperature and pressure are known in each time interval, a simulation of the reactor can be carried out in that time interval. The problem is that the conditions will change from one time interval to subsequent time intervals. The *profile* of the dynamic variables (feed addition, product takeoff, temperature and pressure) needs to be known through time. An approach to profile optimization will be discussed later.

    The resulting optimal profiles for the operating conditions will need to be evaluated in terms of their practicality from the point of view of control and safety. If a complex profile offers only a marginal benefit relative to a fixed value, then simplicity (and possibly safety) will dictate a fixed value to be maintained. However, an optimized profile might offer a significant increase in the performance, in which the complex control problem will be worth addressing.

3) *Agitation.* Different designs of agitated vessel are available. The main differences are related to the way heat is added or removed and the design of the agitator. Figure 16.2 shows three designs of agitated vessel. Figure 16.2a shows a design with an internal coil for the addition or removal of heat. Figure 16.2b shows a design with an external jacket and Figure 16.2c shows a design with half-pipes welded to the exterior for the addition or removal of heat. Figure 16.3 shows four of the more common designs of agitator. Many other designs of agitator are available. Multiple agitators can be mounted on the same shaft. Different agitator designs create different flow patterns in the agitated vessel. Processing duties are central to the selection of the agitator. The duties of the agitator are:

- Mixing
- Solids suspension (suspending sinking solids or incorporating floating solids)
- Dispersion of two liquid phases
- Gas dispersion
- Heat transfer.

Selection of the most appropriate design of agitator is a specialized subject outside of the scope of this text (see, for example, Harnby, Nienow and Edwards, 1997).

Within the context of reaction systems, it is important to note that the selectivity of the system can be effected by agitation when competing reactions are involved that produce byproducts. For example, consider the reaction system:

$$\begin{array}{lcl} FEED1 + FEED2 & \longrightarrow & PRODUCT \\ PRODUCT + FEED2 & \longrightarrow & BYPRODUCT \end{array} \qquad (16.3)$$

Where mixing is imperfect, then regions of excess *FEED2* produce excessive *BYPRODUCT*. Regions low in *FEED2* give lower rates for both the *PRODUCT* and *BYPRODUCT*. This means that imperfect mixing favors the production of the *BYPRODUCT*.

Often the reaction kinetics is not known when trying to design a batch reactor. Indeed, the reaction chemistry is often not known. In such circumstances, scale-up from small-scale experiments to production scale is used. Two broad scales of

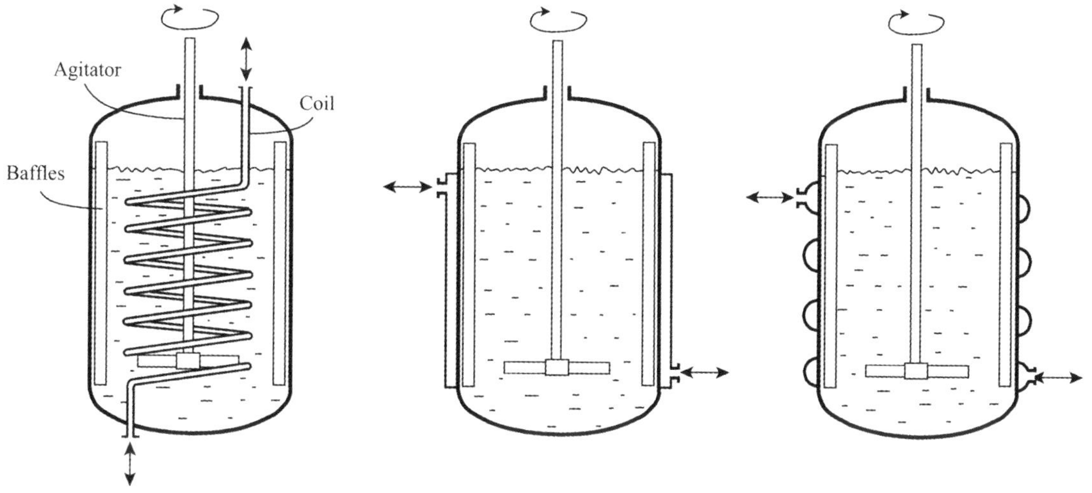

(a) Helical coil. (b) Jacket. (c) Half pipe.

Figure 16.2

Heat transfer in agitated vessels.

16

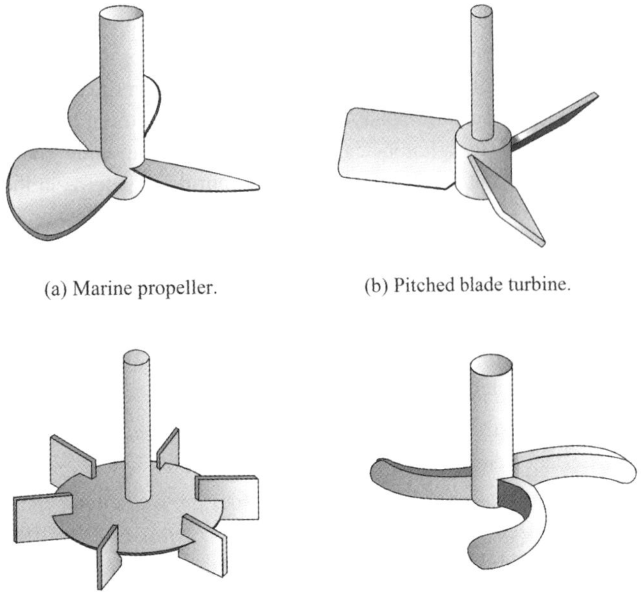

(a) Marine propeller. (b) Pitched blade turbine.

(c) Rushton disk turbine. (d) Pfaudler impeller.

Figure 16.3
Examples of agitator design.

mixing can be distinguished. Mixing through the bulk flow is termed *macromixing*. Mixing at the molecular level through diffusion is termed *micromixing*. The rate of micromixing depends on the operating conditions, viscosity and the energy dissipation rate. If the energy dissipation rate from the agitator in the reaction zone is held constant, the product distribution for the same operating conditions should be independent of the scale. If the reaction is very slow, then the reaction will proceed throughout the whole reactor volume. On the other hand, if the reaction is very fast, the reaction will proceed in only a small fraction of the whole reactor volume. The fraction within which the reaction takes place is scale-dependent and reaction zones spread out to different extents at different scales. This introduces complexities in the scale-up of reaction systems.

4) *Solvent selection.* Solvents are very common in batch reaction systems. The solvent has a number of purposes (Sharratt, 1997):
   - mobilize solid reactants or solid products with high melting points;
   - allow reaction and mass transfer;
   - modify the reaction pathway;
   - act as a heat sink;
   - provide temperature control when the solvent is allowed to evaporate from the reactor, is condensed and returned to the reactor;
   - allow material transfer to other units.

   There are many issues to consider in solvent selection (Sharratt, 1997):
   - a large working range between the melting and boiling points is desirable;
   - viscosity should allow good mixing;
   - interfacial tension should be considered for liquid–liquid reactions;
   - the solvent should be easily recovered for recycling (i.e. no azeotropes if distillation is to be used, low latent heat, etc.);
   - low boiling point can create environmental problems through the release of volatile organic compounds (VOCs) that create environmental problems;
   - the solvent should be easily disposed of if waste is formed (e.g. chlorinated solvents should be avoided);
   - effect on reactions (solvent polarity, bond type, effect on reaction rate, activation energy, equilibrium, changes in reaction mechanism).

16.3 Batch Distillation

Batch distillation has a number of advantages when compared with continuous distillation:

- The same equipment can be used to process many different feeds and produce different products.
- There is flexibility to meet different product specifications.
- One distillation column can separate a multicomponent mixture into relatively pure products.

The disadvantages of batch distillation are:

- High-purity products require the careful control of the column because of its dynamic state.
- The mixture is exposed to a high temperature for extended periods.
- Energy requirements are generally higher.

1) *Batch distillation without reflux.* The simplest batch distillation involves charging a feed to a batch *distillation pot*, as shown in Figure 16.4. There are no trays or packing materials, and no reflux is returned to the pot. The vapor formed is removed from the system. Since this vapor is richer in the more volatile components than the liquid in the pot, the liquid remaining in the pot becomes steadily leaner in these components. The result of this vaporization is that the composition of the product changes progressively through time. The vapor leaving the pot at any time is assumed to be in equilibrium with the liquid in the pot, but since the vapor is richer in the more volatile components, the composition of the liquid and vapor are not constant. To show how the compositions change with time, consider vaporization of an initial charge to the distillation pot in Figure 16.4.

   Initially, restrict considerations to binary mixtures. Let F be the initial binary feed to the pot (kmol) and x_F be the mole fraction of the more volatile component (A) in the feed. Let B be the amount of material remaining in the batch pot at time t, x_B be the mole fraction of Component A in the pot, and x_D the mole fraction of Component A in the vapor phase. If a small amount of liquid dB with a vapor mole fraction x_D is vaporized, a

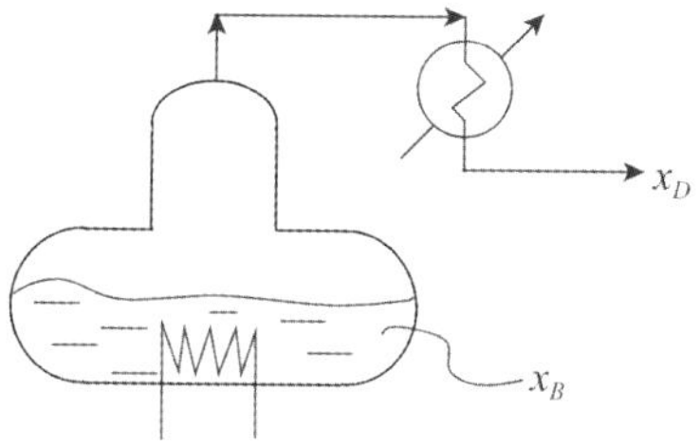

Figure 16.4
Simple distillation from a batch pot.

material balance for Component A gives:

$$\begin{aligned} x_D dB &= d(Bx_B) \\ &= Bdx_B + x_B dB \end{aligned} \tag{16.4}$$

Rearranging Equation 16.4 gives:

$$\frac{dB}{dx_B} = \frac{B}{x_D - x_B} \tag{16.5}$$

Integrating Equation 16.5:

$$\int_F^B \frac{dB}{B} = \int_{x_F}^{x_B} \frac{dx_B}{x_D - x_B} \tag{16.6}$$

or

$$\ln\frac{B}{F} = \int_{x_F}^{x_B} \frac{dx_B}{x_D - x_B} \tag{16.7}$$

where B = total moles in the batch pot at time t
F = total moles in the batch pot at the beginning of the operation
x_B = mole fraction of Component A in the liquid in the batch pot at time t
x_F = mole fraction of Component A in the liquid in the batch pot at the beginning of the operation
x_D = mole fraction of Component A in the vaporized material leaving the batch pot at time t

Equation 16.7 is known as the Rayleigh Equation and describes the material balance around the distillation pot. Equation 16.7 can be integrated analytically for some special cases. If the vapor composition leaving the pot can be related to the composition of the liquid in the pot according to $y = Kx$ where K is assumed to be constant, then substituting $x_D = Kx_B$ in Equation 16.7 and integrating gives:

$$\ln\frac{B}{F} = \frac{1}{K-1}\ln\left[\frac{x_B}{x_F}\right] \tag{16.8}$$

Another simple case is when the relative volatility α can be assumed to be constant. From Equation 8.20:

$$x_D = \frac{\alpha x_B}{1 - x_B(1-\alpha)} \tag{16.9}$$

Substituting in Equation 16.7 gives:

$$\begin{aligned} \ln\frac{B}{F} &= \int_{x_F}^{x_B} \frac{dx_B}{\dfrac{\alpha x_B}{1 + x_B(\alpha - 1)} - x_B} \\ &= \int_{x_F}^{x_B} \left[\frac{1 + x_B(\alpha - 1)}{\alpha_B(\alpha - 1)(1 - x)}\right] dx_B \\ &= \int_{x_F}^{x_B} \left[\frac{1}{x_B(\alpha - 1)(1 - x_B)} + \frac{1}{(1 - x_B)}\right] dx_B \\ &= \left[\frac{1}{(\alpha - 1)}\left(-\ln\frac{1 - x_B}{x_B}\right) + (-\ln(1 - x_B))\right]_{x_F}^{x_B} \\ &= \frac{1}{(\alpha - 1)}\ln\left[\frac{x_B(1 - x_F)}{x_F(1 - x_B)}\right] + \ln\left[\frac{1 - x_F}{1 - x_B}\right] \\ \ln\frac{B}{F} &= \frac{1}{(\alpha - 1)}\left[\ln\left(\frac{x_B}{x_F}\right) + \alpha\ln\left(\frac{1 - x_F}{1 - x_B}\right)\right] \end{aligned} \tag{16.10}$$

If the K-value or relative volatility is not constant, then Equation 16.7 must be integrated numerically. Numerical integration is illustrated in Figure 16.5, in which $1/(x_D - x_B)$ is plotted versus the bottoms mole fraction x_B. Integrating to find the area under the graph (e.g. using the Trapezoidal Rule or Simpson's Rule), the amount of liquid left in the pot at the end of the batch is given by:

$$B_{final} = F\exp(-Area) \tag{16.11}$$

where B_{final} = total moles in the batch pot at the end of the batch

The amount of distillate is by mass balance:

$$D_{final} = F[1 - \exp(-Area)] \tag{16.12}$$

The amount of A in the final product

$$= x_F F - x_{B,final} B_{final}$$

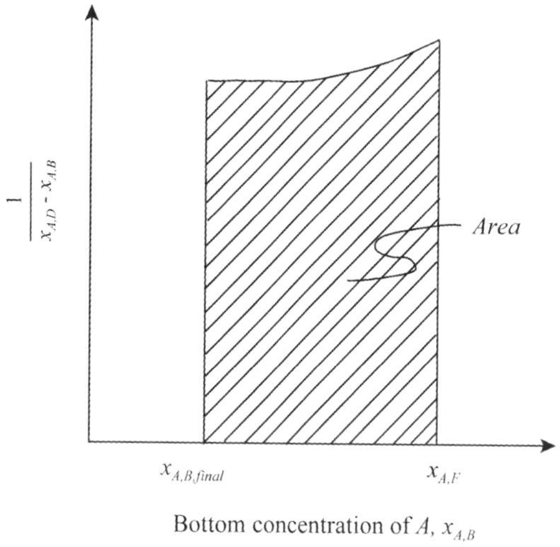

Figure 16.5
Integration of the Rayleigh Equation for constant reflux ratio.

16

Thus the average overhead composition of $\bar{x}_{A,D}$ is given by:

$$\bar{x}_D = \frac{x_F F - x_{B,final} B_{final}}{F - B_{final}} \tag{16.13}$$

Although the equations developed here were for binary mixtures, the Rayleigh Equation (Equation 16.7) can also be written for each component in a muticomponent mixture.

2) *Batch distillation with reflux.* Batch distillation from a pot will not provide a good separation unless the relative volatility is very high. In most cases, a rectifying column with reflux is added to the pot, as shown in Figure 16.6. The operation of a batch distillation can be analyzed for binary systems at a given instant in time using a McCabe–Thiele diagram. The operating line is the same as that for the rectifying section of a continuous distillation (see Chapter 8):

$$y_{i,n+1} = \frac{R}{R+1} x_{i,n} + \frac{1}{R+1} x_{i,D} \tag{8.34}$$

This is shown in Figure 16.6, where the slope of the operating line is given by $R/(R+1)$. The feed is charged to the distillation pot at the beginning of the batch and is subjected to continuous vaporization. This vapor would then flow upward through trays or packing to the condenser and reflux would be returned, as with continuous distillation. However, unlike continuous distillation, the overhead product will change with time. The first material to be distilled will be the more volatile components. As the vaporization proceeds and product is withdrawn overhead, the product will become gradually richer in the less-volatile components.

Figure 16.6 illustrates what happens if the reflux ratio is fixed for a binary separation. Because the reflux ratio is fixed, the slope of the operating line is fixed, but its position varies through time. For a fixed number of stages, the distillate and the bottoms gradually become more concentrated in the heavier component. This means that the product quality deteriorates with time and blending is required to meet the appropriate purity specification. The Rayleigh Equation (Equation 16.7) can be integrated graphically, as illustrated in Figure 16.5 for the separation of an *AB* mixture. The initial charge to the distillation pot is *F* moles with a mole fraction of x_F. The McCabe–Thiele construction can be used to determine the distillate mole fraction x_D for various bottoms mole fractions x_B for a fixed reflux ratio (fixed slope of operating line) and a given number of stages (four stages in the case of Figure 16.6). Thus, the Rayleigh Equation can be integrated numerically as in Figure 16.5.

Alternatively, by careful control of the reflux ratio, it is possible to hold the composition of the distillate constant for a time until the required reflux ratio becomes intolerably large, as illustrated in Figure 16.7. Again, the Rayleigh Equation can be integrated numerically as in Figure 16.5.

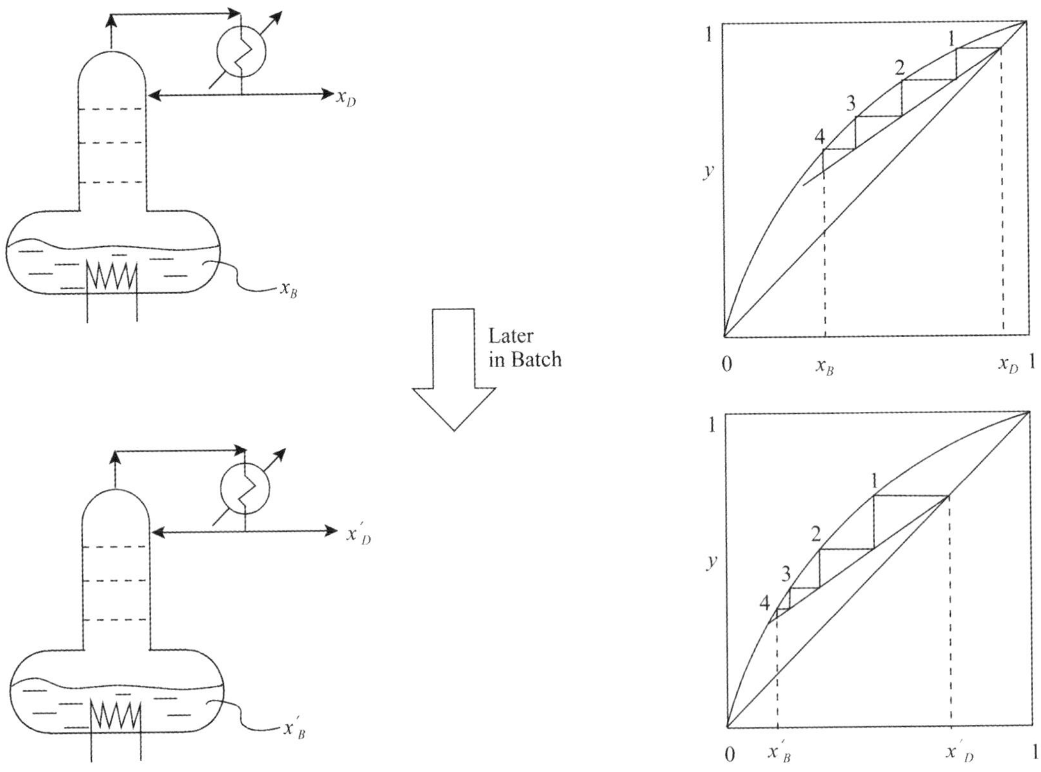

Figure 16.6

If reflux ratio is fixed in batch distillation the overhead product purity decreases.

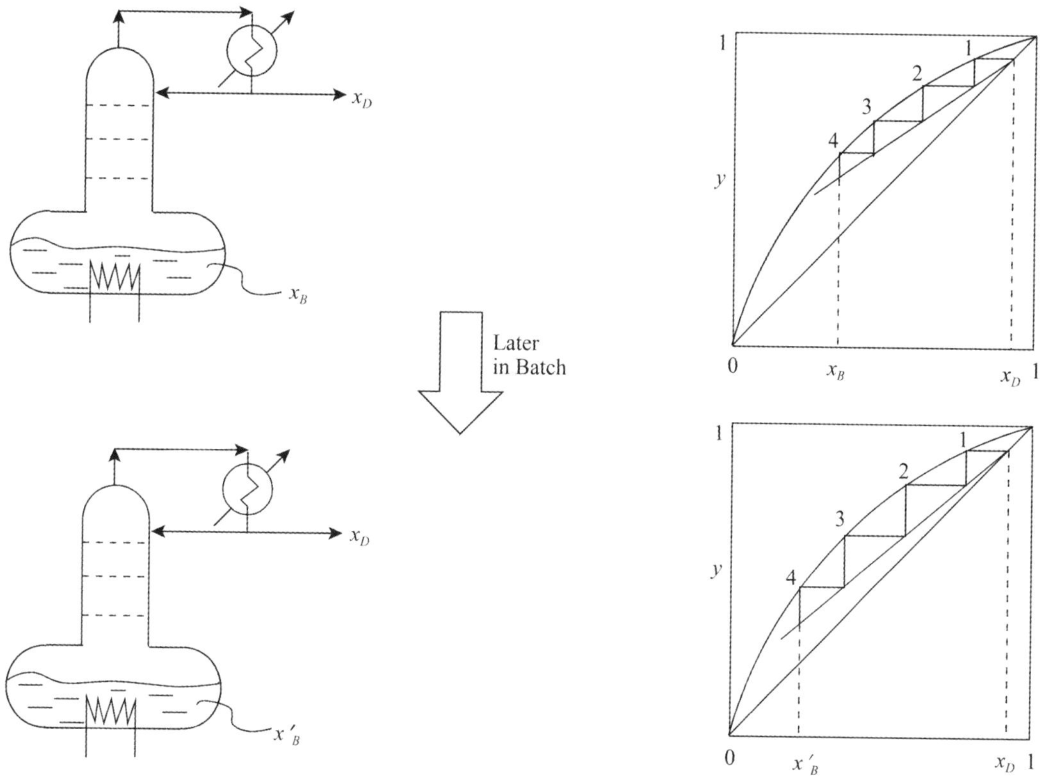

Figure 16.7

Reflux ratio can be varied in batch distillation to maintain overhead product purity.

Operation at constant reflux ratio is better than operation with constant distillate composition for high-yield batch separations. However, operation with constant distillate composition might be necessary if high product purity is required. In fact, it is not necessary to operate in one of these two special cases of constant reflux ratio or constant distillate composition. Given the appropriate control scheme, the reflux ratio can be varied through the batch with an objective different from that of maintaining constant distillate composition, perhaps with a constraint on the final mean distillate composition. Optimization of the profile of the reflux ratio through the batch can be performed, as will be discussed later.

3) *Modeling binary batch distillation.* Batch distillation can be modeled by dividing the batch cycle into time increments and steady-state continuous distillation assumed in each time increment. For a binary distillation with constant relative volatility and constant molar overflow the separation performance is predicted by the McCabe–Thiele diagram in each time increment. However, this requires repeated application of the graphical approach, which is not convenient. An analytical solution of the McCabe–Thiele construction was developed by Smoker (1938). The operation line for the rectifying section can be represented by:

$$y = \frac{R}{R+1}x + \frac{1}{R+1}x_D \tag{8.45}$$

or

$$y = ux + v \tag{16.14}$$

where

y = vapor mole fraction
x = liquid mole fraction
x_D = distillate mole fraction
R = reflux ratio

$$u = \frac{R}{R+1}$$

$$v = \frac{x_D}{R+1}$$

For constant relative volatility:

$$y = \frac{x\alpha}{1 + x(\alpha - 1)} \tag{8.27}$$

where α = relative volatility

y can be eliminated from these equations to give a quadratic in x:

$$u(\alpha - 1)x^2 + [u + (\alpha - 1)v - \alpha]x + v = 0 \tag{16.15}$$

There is only one real root k to this equation, where $0 < k < 1$. Representing the root in this equation:

$$u(\alpha - 1)k^2 + [u + (\alpha - 1)v - \alpha]k + v = 0 \tag{16.16}$$

16

Smoker (1938) showed that the number of theoretical stages for the rectifying section is given by:

$$N = \ln\left[\frac{x_D^*\left(1-\beta x_F^*\right)}{x_F^*\left(1-\beta x_D^*\right)}\right] \Big/ \ln\left[\frac{\alpha}{uv^2}\right] \tag{16.17}$$

where

N = number of theoretical stages

$x_F^* = x_F - k$

$x_D^* = x_D - k$

$\beta = \dfrac{uw(\alpha-1)}{\alpha - uw^2}$

$w = 1 + k(\alpha - 1)$

Rearranging Equation 16.17 gives:

$$x_F^* = \frac{x_D^*}{\left(\dfrac{\alpha}{uw^2}\right)^N \left(1-\beta x_D^*\right) + \beta x_D^*} \tag{16.18}$$

Thus:

$$x_F = \frac{x_D - k}{\left(\dfrac{\alpha}{uw^2}\right)^N [1-\beta(x_D-k)] + \beta(x_D-k)} + k \tag{16.19}$$

It should be noted that Equation 16.17 cannot be rearranged to be explicit in terms of x_D, but can be made explicit in terms of x_F (x_B in a batch rectification process). Thus, it is convenient to set values of x_D in each time increment and the corresponding value of x_B calculated in order to integrate the Rayleigh Equation. The reflux ratio might be held constant across time intervals or varied between intervals. Illustrative examples are presented later.

4) *Modeling multicomponent batch distillation.* A shortcut model for a multicomponent batch rectifier can also be developed (Diwekar and Madhavan, 1991). As with the binary model, the basis of the multicomponent model is to consider the batch distillation column as a continuous column which is at steady state within any time interval, but with a feed that varies between time intervals. In this way, it is possible to apply the Fenske–Underwood–Gilliland (FUG) method (see Chapter 8) to each interval of time. The reflux ratio might be held constant across time intervals or varied. Zero hold-up on the plates and in the condenser is assumed, together with constant relative volatility and constant molar overflow.

First it is necessary to develop the Rayleigh Equation for the multicomponent case. Assuming that the vapor flowrate V is constant, the overall mass balance equation is:

$$\frac{dB}{dt} = -\frac{V}{R+1} \tag{16.20}$$

The mass balance for the component taken as reference is:

$$\frac{d\left(x_{r,B}B\right)}{dt} = -x_{r,D}\frac{V}{R+1} \tag{16.21}$$

where $x_{r,B}$ = mole fraction of the reference component r in the bottoms (pot) at any time t

$x_{r,D}$ = mole fraction of the reference component r in the distillate at any time t

Expanding the left-hand side of Equation (16.21) and combining with Equations 16.20 and 16.21 gives:

$$x_{r,B}\frac{dB}{dt} + B\frac{dx_{r,B}}{dt} = x_{r,D}\frac{dB}{dt} \tag{16.22}$$

or

$$\frac{dB}{B} = \frac{dx_{r,B}}{x_{r,D} - x_{r,B}} \tag{16.23}$$

Writing Equation 16.23 for the other components i for $i = 1$ to NC ($i \neq r$):

$$\frac{dx_{r,B}}{x_{r,D} - x_{r,B}} = \frac{dx_{i,B}}{x_{i,D} - x_{i,B}} \tag{16.24}$$

This equation needs to be integrated across the batch. If the batch cycle is divided into small increments, then for a small defined step in the reference component $dx_{r,B}$ the values of $x_{r,D}$, $x_{r,B}$, $x_{i,D}$ and $x_{i,B}$ at any step $k+1$ can be assumed to be the same as the previous step k for the calculation of the component increments for each component $dx_{i,B}$. Thus, from Equation 16.24:

$$dx_{i,B}^{k+1} = \frac{dx_{r,B}}{x_{r,D}^k - x_{r,B}^k} \times \left(x_{i,D}^k - x_{i,B}^k\right) \tag{16.25}$$

The compositions for Component i in step $k+1$ are given by:

$$\begin{aligned} x_{i,B}^{k+1} &= x_{i,B}^k + dx_{i,B}^{k+1} \\ &= x_{i,B}^k + \frac{dx_{r,B}}{\left(x_{r,D}^k - x_{r,B}^k\right)} \times \left(x_{i,D}^k - x_{i,B}^k\right) \end{aligned} \tag{16.26}$$

It must be assured that the sum of the mole fractions for $x_{i,B}^{k+1}$ must add to 1.0. From Equation 16.26:

$$\begin{aligned} \sum_i^{NC} x_{i,B}^{k+1} &= \sum_i^{NC}\left[x_{i,B}^k + \frac{dx_{r,B}}{\left(x_{r,D}^k - x_{r,B}^k\right)} \times \left(x_{i,D}^k - x_{i,B}^k\right)\right] \\ &= \sum_i^{NC} x_{i,B}^k + \frac{dx_{r,B}}{\left(x_{r,D}^k - x_{r,B}^k\right)}\left[\sum_i^{NC} x_{i,D}^k - \sum_i^{NC} x_{i,B}^k\right] \end{aligned} \tag{16.27}$$

As long as the summation for the previous step k adds to unity, that is:

$$\sum_i^{NC} x_{i,D}^k = 1 \quad \text{and} \quad \sum_i^{NC} x_{i,B}^k = 1 \tag{16.28}$$

then Equation 16.27 becomes:

$$\sum_i^{NC} x_{i,B}^{k+1} = 1 \tag{16.29}$$

Thus Equation 16.26 provides a basis to integrate Equation 16.24. However, Equation 16.26 requires knowledge of the distillate concentrations. The Fenske Equation can be used to relate the distillate and bottoms compositions, as demonstrated in Chapter 8 for continuous distillation:

$$x_{i,D}^{k} = \alpha_{i,r}^{N\,min}\left(\frac{x_{r,D}^{k}}{x_{r,B}^{k}}\right)x_{i,B}^{k} \qquad (16.30)$$

However, the distillate composition of the reference component is needed to complete the calculations. Since the individual overhead compositions must add up to unity, for each interval k:

$$\sum_{i=1}^{NC} x_{i,D} = 1 \qquad (16.31)$$

Combining Equation 16.30 with 16.31:

$$1 = \frac{x_{r,D}^{k}}{x_{r,B}^{k}}\sum_{i=1}^{NC}\alpha_{i,r}^{N\,min}x_{i,B}^{k} \qquad (16.32)$$

or

$$x_{r,D}^{k} = \frac{x_{r,B}^{k}}{\sum \alpha_{i,r}^{N\,min}x_{i,B}^{k}} \qquad (16.33)$$

At the beginning of the batch, the operation is assumed to be at total reflux. If it is assumed there is no hold-up on the trays and in the condenser, then $x_{i,B}^{0}$ are the same as the feed concentration, and N_{min} equals the number of stages in the column. Thus, the distillate compositions can be calculated from Equation 16.33 for Component r and Equation 16.30 for all other components with N_{min}. If N_{min} is known for the subsequent intervals then new bottoms concentrations can be calculated from Equation 16.27 and distillate compositions from Equations 16.33 and 16.30. However, once the distillate product begins to be taken after Interval 0, then N_{min} is unknown and must be calculated. In principle, N_{min} can be calculated from the Gilliland Correlation if the actual reflux ratio R, the minimum reflux ratio R_{min}, and the actual number of stages N are known (see Chapter 8). An explicit equation was presented by Eduljee (1975) to represent the Gilliland Correlation:

$$R_{min}^{G} = R - (R+1)\left[1 - \frac{4}{3}\left(\frac{N - N_{min}}{N+1}\right)\right]^{1/0.5668} \qquad (16.34)$$

where R_{min}^{G} = minimum reflux ratio calculated by the Eduljee Equation
R = actual reflux ratio
N = actual number of theoretical stages
N_{min} = minimum number of theoretical stages

The minimum reflux ratio can be calculated using the Underwood Equations (see Chapter 8):

$$\sum_{i=1}^{NC}\frac{\alpha_i x_{i,B}}{\alpha_i - \theta} = 1 - q \qquad (16.35)$$

$$R_{min}^{U} = \sum_{i=1}^{NC}\frac{\alpha_i x_{i,D}}{\alpha_i - \theta} - 1 \qquad (16.36)$$

where α_i = volatility of Component i relative to the reference Component r
$x_{i,B}$ = mole fraction of Component i in the bottoms
$x_{i,D}$ = mole fraction of Component i in the distillate
θ = root of the equation
q = feed condition
= $\dfrac{\text{heat required to vaporize one mole of feed}}{\text{molar latent heat of vaporization of feed}}$
= 1 for a saturated liquid feed, 0 for a saturated vapor feed
NC = number of components
R_{min}^{U} = minimum reflux ratio calculated by the Underwood Equation

At each interval an initial value of N_{min} is guessed, which enables the calculation of R_{min}^{G} and R_{min}^{U}. The difference $(R_{min}^{G} - R_{min}^{U})$ can then be taken as an objective to solve those equations and find the value of N_{min} iteratively. The moles of material in pot B is also needed. The Rayleigh Equation 16.23 can be integrated, resulting in:

$$\ln\left(\frac{B^k}{B^0}\right) = \int_{x_{r,B}^{0}}^{x_{r,B}^{k}}\frac{dx_{r,B}}{x_{r,D} - x_{r,B}} \qquad (16.37)$$

Equation 16.37 can be solved numerically by a series of incremental changes, using, for example, the Trapezoidal Rule. In the case of a binary model it is more convenient to set values of x_D in each time increment and the corresponding value of x_B calculated in order to integrate the Rayleigh Equation. However, the equations for multicomponent distillation require the value of x_B to be set and the value of x_D calculated in each time increment. The calculation will be illustrated later.

5) *Batch cycle time.* It is clear that Equations 16.7, 16.8, 16.10, 16.13 and 16.23 are independent of time. Yet the batch time is an important consideration. F and B represent the moles of material, but now let V represent the rate at which vapor is leaving the pot. For continuous distillation, it can be seen from Figure 8.21 that $D = V/(R+1)$. Thus for batch distillation at any instant:

$$\frac{dB}{dt} = -\frac{V}{R+1} \qquad (16.38)$$

or

$$dt = -\frac{R+1}{V}dB \qquad (16.39)$$

If an operating policy is adopted of maintaining the reflux ratio constant, then in Equation 16.39 both V and R are constant and the equation can be integrated:

$$\int_0^{t_B} dt = -\frac{R+1}{V}\int_F^B dB \qquad (16.40)$$

$$t_B = \frac{R+1}{V}(F - B) \qquad (16.41)$$

where t_B = batch cycle time.

16

For the general case in which the reflux ratio is allowed to vary, then the Rayleigh Equation for a binary mixture is given by:

$$dB = \frac{Bdx_B}{(x_D - x_B)} \tag{16.42}$$

Substituting Equation 16.40 into Equation 16.37 gives:

$$dt = -\frac{B}{V}\frac{R+1}{(x_D - x_B)}dx_B \tag{16.43}$$

Integrating Equation 16.43 gives:

$$t_B = \int_{x_B}^{x_F} \frac{B}{V}\frac{R+1}{(x_D - x_B)}dx_B \tag{16.44}$$

For the multicomponent systems, the corresponding equation is:

$$t_B = \int_{x_{r,B}^k}^{x_{r,B}^0} \frac{B}{V}\frac{R+1}{(x_{r,D} - x_{r,B})}dx_{r,B} \tag{16.45}$$

6) *Modelling errors.* Approximate models for batch distillation have been developed in this chapter. Using approximate models for steady state might not present errors that are too serious. However, in batch distillation if the Rayleigh Equation needs to be integrated numerically a continuous model is used for a series of small increments. Errors in each increment will accumulate through the calculation. The models also neglect hold-up in the column and the condenser. Extra caution should be applied when interpreting the results of approximate batch distillation calculations. Such calculations can be useful to explore trends, but more rigorous calculations should be applied later when the design is finalized (Diwekar, 2011).

Example 16.1 A solvent *S* is to be recovered from a heavier residue *HR* by distillation in a batch rectifier. The initial charge to the pot is 150 kmol with a mole fraction *S* of 0.6. The relative volatility between *S* and *HR* is 3.5. It can be assumed that the boil-up rate is 100 kmol·h^{-1}.

a) If it is assumed that the column is operated without reflux, calculate the average distillate composition for a recovery of solvent of 90%.
b) If the column is refluxed and the trays and condenser provide the equivalent of four theoretical stages, calculate the recovery for a mean distillate purity of 98% and minimum recovery of solvent 90% for a reflux ratio maintained to be constant at 9.
c) For the refluxed distillation in Part b, calculate the batch cycle time and total energy consumed if the enthalpy of vaporization is 33,000 kJ·kmol^{-1}.

Solution

a) If the solvent recovery is assumed to be 90%:

$$\begin{aligned} Bx_B &= 0.1 \times Fx_F \\ &= 0.1 \times 150 \times 0.6 \\ &= 9\ \text{kmol} \\ x_B &= \frac{9}{B} \end{aligned}$$

From Equation 16.10:

$$\ln\frac{B}{F} = \frac{1}{\alpha - 1}\left[\ln\left(\frac{x_B}{x_F}\right) + \alpha\ln\left(\frac{1-x_F}{1-x_B}\right)\right]$$

$$\ln\left(\frac{B}{150}\right) = \frac{1}{(3.5-1)}\left[\ln\left(\frac{9}{0.6B}\right) + 3.5\ln\left(\frac{1-0.6}{1-9/B}\right)\right]$$

Solving by trial and error:

$$\begin{aligned} B &= 40.08\ \text{kmol} \\ x_B &= 0.2246 \\ Dx_D &= Fx_F - Bx_B \\ &= 150 \times 0.6 - 40.08 \times 0.2246 \\ &= 81.00\ \text{kmol} \\ D &= 150 - 40.08 \\ &= 109.92\ \text{kmol} \\ x_D &= \frac{81.00}{109.92} \\ &= 0.74 \end{aligned}$$

b) For the refluxed column, Equation 16.7 must be solved numerically to obtain the material balance. The variation of bottoms composition is required for the numerical integration. However, the McCabe–Theile Method calculated by the Smoker Equation requires x_B to be calculated from x_D. Thus, x_D will be varied from an initial concentration by incremental reduction and x_B calculated from the Smoker Equation (Equation 16.19). It can be assumed that the distillation will be started at total reflux to establish the separation and then the separation commenced by fixing the reflux ratio to a value of 9. The distillate composition at total reflux for time 0 can be calculated from the Fenske Equation (Equation 8.56):

$$\frac{x_D}{1-x_D} = \alpha^N\left(\frac{x_B}{1-x_B}\right)$$

$$\begin{aligned} x_D &= \frac{\alpha^N\left(\frac{x_B}{1-x_B}\right)}{1+\alpha^N\left(\frac{x_B}{1-x_B}\right)} \\ &= \frac{3.5^4\left(\frac{0.6}{1-0.6}\right)}{1+3.5^4\left(\frac{0.6}{1-0.6}\right)} \\ &= 0.9956 \end{aligned}$$

Table 16.1

Batch distillation with fixed reflux ratio of 9.

Interval	x_D	K	W	β	x_B	$\dfrac{1}{x_D - x_B}$	$\Delta Area$	ln(B/F)	B	$\bar{x}_D$	Bx_B
0	0.9956	0.044216	1.110540	1.045474	0.6000	2.527952	0.000000	0.000000	150.00	0.9956	90.00
1	0.9906	0.043958	1.109896	1.044305	0.4894	1.995445	0.250063	0.250063	116.81	0.9892	57.17
2	0.9856	0.043701	1.109252	1.043139	0.3890	1.676138	0.184438	0.434502	97.14	0.9878	37.78
3	0.9806	0.043444	1.108610	1.041976	0.3244	1.524036	0.103275	0.537776	87.61	0.9869	28.42
4	0.9756	0.043187	1.107969	1.040818	0.2794	1.436478	0.066603	0.604379	81.96	0.9862	22.90
5	0.9706	0.042931	1.107329	1.039662	0.2462	1.380590	0.046737	0.651116	78.22	0.9855	19.26
6	0.9656	0.042676	1.106690	1.038511	0.2208	1.342596	0.034717	0.685833	75.55	0.9849	16.68
7	0.9606	0.042421	1.106052	1.037362	0.2005	1.315711	0.026875	0.712709	73.55	0.9843	14.75
8	0.9556	0.042166	1.105415	1.036217	0.1841	1.296203	0.021468	0.734177	71.99	0.9838	13.25
9	0.9506	0.041912	1.104779	1.035076	0.1705	1.281853	0.017578	0.751755	70.73	0.9833	12.06
10	0.9456	0.041658	1.104144	1.033938	0.1590	1.271257	0.014684	0.766438	69.70	0.9828	11.08
11	0.9406	0.041404	1.103510	1.032803	0.1491	1.263483	0.012471	0.778909	68.84	0.9824	10.26
12	0.9356	0.041151	1.102878	1.031672	0.1406	1.257889	0.010740	0.789650	68.10	0.9820	9.57
13	0.9306	0.040898	1.102246	1.030544	0.1331	1.254020	0.009361	0.799010	67.47	0.9816	8.98
14	0.9256	0.040646	1.101616	1.029420	0.1266	1.251541	0.008243	0.807253	66.91	0.9813	8.47
15	0.9206	0.040394	1.100986	1.028298	0.1207	1.250203	0.007324	0.814577	66.42	0.9809	8.02
16	0.9156	0.040143	1.100358	1.027181	0.1155	1.249816	0.006560	0.821137	65.99	0.9806	7.62
17	0.9106	0.039892	1.099730	1.026066	0.1107	1.250233	0.005916	0.827053	65.60	0.9803	7.26
18	0.9056	0.039642	1.099104	1.024955	0.1064	1.251338	0.005370	0.832424	65.25	0.9800	6.94

This gives the initial distillate composition. Then for the first increment of the batch, the initial distillate composition is decreased by a small increment and the bottoms composition calculated from the Smoker Equation. A series of incremental changes allows Equation 16.7 to be solved numerically, in this case for simplicity by the Trapezoidal Rule. The calculation can be conveniently performed in a spreadsheet. Table 16.1 shows the details of the calculation for an incremental change in x_D of 0.005. From Table 16.1, for a distillate composition of 0.98 the recovery of S in the distillate

$$
\begin{aligned}
&= Fx_F - Bx_B \\
&= 150 \times 0.6 - 6.94 \\
&= 83.06 \text{ kmol} \\
&= 92.2\%
\end{aligned}
$$

This result should be contrasted with that from Part a above where no reflux was used.

c) The batch time for a constant reflux ratio is given by Equation 16.41:

$$
\begin{aligned}
t_B &= \frac{R+1}{V}(F - B) \\
&= \frac{9+1}{100}(150 - 65.25) \\
&= 8.48 \text{ h}
\end{aligned}
$$

The energy required for the separation

$$
\begin{aligned}
&= 100 \times 33{,}000 \times 8.48 \\
&= 27.98 \times 10^6 \text{ kJ}
\end{aligned}
$$

It should be noted that decreasing the size of the increment, and thus increasing the number of increments, changes the answer slightly. For example, decreasing the change in x_D to 0.0025 for a mean distillate purity of 0.98 leads to a recovery of 93.7% and a batch cycle time of 8.61 h. Numerical accuracy could also be improved by using more accurate numerical integration techniques.

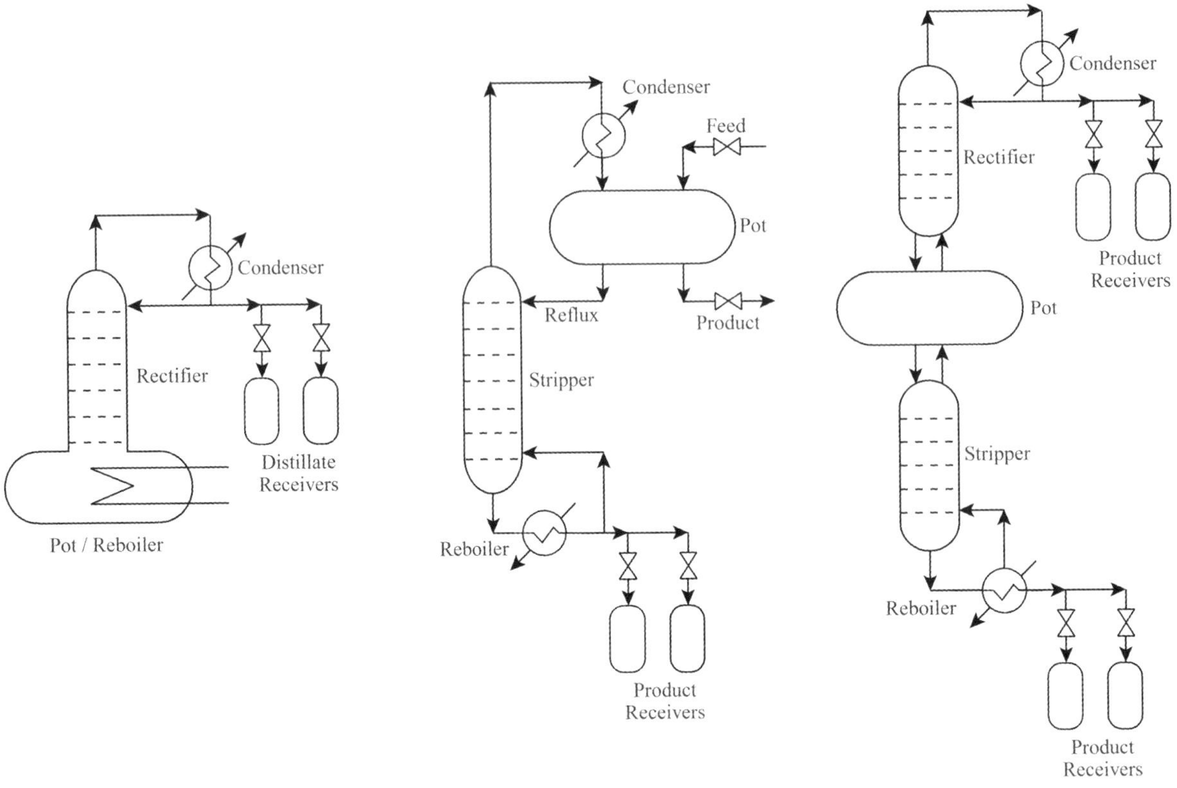

(a) Batch rectifier. (b) Batch stripper. (c) Middle vessel column.

Figure 16.8
Different batch distillation arrangements for multiple products.

Whilst the methods presented here for batch distillation modeling are useful for systems in which the vapor liquid equilibrium is relatively ideal, the vapor–liquid equilibrium most often requires a model more complex than the constant relative volatility illustrated in the previous examples. In such circumstances a more rigorous tray to tray model is required (Diwekar, 2011).

Most batch distillation operations use a batch rectifier arrangement to purify a single product overhead, as illustrated in Figures 16.6 and 16.7. However, the separation can require multiple products, in which case other configurations are sometimes required, as illustrated in Figure 16.8 (Diwekar, 2011). The batch rectifier illustrated in Figure 16.8a allows the possibility of multiple products being distilled overhead from the same feed mixture. Figure 16.8b illustrates a batch stripper. In a batch stripper the pot is located above the stripping column. High boiling components are separated from the liquid in the pot and collected in the bottom product receivers. This arrangement tends to be used if the mixture to be distilled forms a minimum boiling azeotrope with a low boiling entrainer used for the separation. Figure 16.8c illustrates a batch operation involving both rectifier and stripper sections. This

Figure 16.9
Batch distillation of a multiproduct mixture.

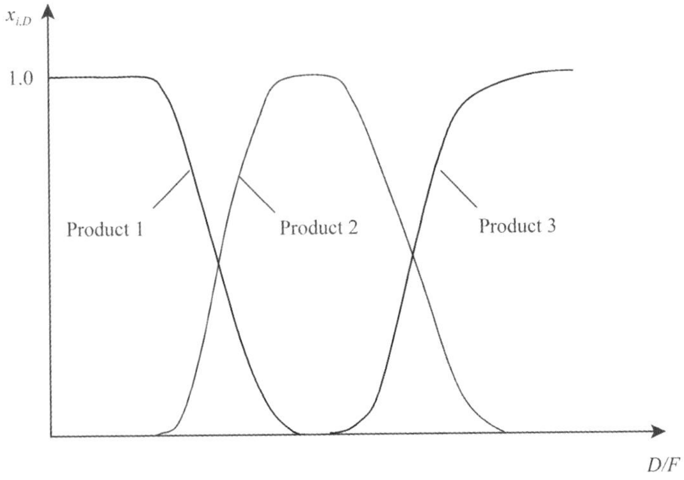

offers more flexibility to accommodate different separations in the same equipment. Each of the arrangements in Figure 16.8 allow multiple products to be distilled from the same feed mixture (Diwekar, 2011). Different product receivers are used to collect the separate products. Figure 16.9 illustrates a profile of distillate composition versus the ratio of distillate to feed flowrate for a variety of products distilled overhead in a batch rectifier. Initially, Product 1 is distilled overhead. Once Product 1 has largely been distilled overhead, its concentration begins to fall in the distillate and the composition of Product 2 begins to rise. Then Product 2 is distilled overhead for a period, after which its concentration begins to fall as the concentration of Product 3 increases in the distillate. The distillation is finished by collecting Product 3 overhead.

When evaluating a batch distillation, either through simulation or from scale-up of a pilot plant, one thing that needs special attention is the hold-up in the column. Liquid hold-up on the trays (or within the packing) and in the condensing and reflux system reduces the sharpness of separation and will require increased reflux or increased number of stages. If hold-up in the column is not accounted for, products might be off specification (Diwekar, 2011). The problem is worst when the feed contains low concentrations of components that need to be separated.

Example 16.2 A charge of 150 kmol of the ternary mixture detailed in Table 16.2 is to be separated in a batch rectifier. Two different distillate receivers will collect the two lightest components as they distil overhead in turn, leaving the heaviest component as a residue in the pot. The column and condenser are equivalent to five theoretical stages and is refluxed with a reflux ratio of 11, which is maintained to be constant. Simulate the recovery of the lightest component, which is to be recovered with a minimum concentration of 99%. If the boil-up rate from the pot is to be maintained at 100 kmol·h^{-1}, determine the time required to recover the first product.

Table 16.2

Data for the ternary batch separation.

Component	x_F	α_i
L	0.45	8
I	0.25	3
H	0.3	1

Solution The heaviest component *H* is to be taken as the reference component. Equation 16.24 must be solved numerically to obtain the material balance. This can be stepped through incrementally in a spreadsheet. However, at each step two iterations are required. For each step a solver can be used first to vary θ for the Underwood Equations to determine the value of R^U_{min}; then N_{min} is varied to minimize the difference $(R^G_{min} - R^U_{min})$. Knowing the value of N_{min} allows the distillate composition to be determined and the material balance for each increment. Table 16.3 shows the results of the calculation for an assumed value of $dx_{r,B} = 0.01$. A series of incremental changes allows Equation 16.24 to be solved numerically.

From Table 16.3 it can be seen that in the first increment when $k = 0$, $x_{i,B}$ are the feed concentrations and are known. N_{min} is also known, as the operation is on total reflux. This allows the distillate concentrations to be determined from the Fenske Equation (assuming no hold-up in the column and the condenser), with $x_{3,D}$ determined from Equation 16.33 and $x_{2,D}$ and $x_{1,D}$ from Equation 16.30. For the next increment when $k = 1$, the heaviest component is the reference component and its concentration is defined by $dx_{r,B} = 0.01$. Then $x_{2,B}$ and $x_{1,B}$ are determined from Equation 16.26. Initial values of $x_{i,D}$ for $k = 1$ can be determined from the Fenske Equation using an initial estimate of N_{min} (say the value from the previous increment), with $x_{3,D}$ determined from Equation 16.33 and $x_{2,D}$ and $x_{1,D}$ from Equation 16.30. However, the value of N_{min} needs to be refined later. The value of θ for the Underwood Equations can then be determined from Equation 16.35 from the bottoms concentrations $x_{i,B}$ and $\alpha_{i,r}$ by iteration using a solver. In turn, this allows an initial value of R^U_{min} to be determined from the initial values of $x_{i,D}$ using Equation 16.36. The value of R^G_{min} can then be determined from Equation 16.34 from the specified R and the initial estimate of N_{min}. Now the value of N_{min} needs to be refined by iteration using a solver to minimize $(R^G_{min} - R^U_{min})$. This allows the distillate concentrations for iteration $k = 1$ to be refined. Now the integral in Equation 16.37 needs to be evaluated for that increment using, say, the Trapezoidal Rule and thus the value of B can be determined. Finally, the mean distillate concentration for increment $k = 1$ can be determined from Equation 16.13. The whole procedure is then repeated for increment $k = 2$ and so on, until the mean concentration of the distillate falls below the specified $x_{1,D} = 0.99$. Details of the calculation are given in Table 16.3.

The batch time to distil the lightest product with a constant reflux ratio is given by Equation 16.41:

$$t_B = \frac{R+1}{V}(F - B)$$

$$= \frac{11+1}{100}(150 - 95.73)$$

$$= 6.5 \text{ h}$$

It should again be noted that decreasing the size of the increment needs to be explored. The simulation is tedious to perform by hand and routine calculations require commercial software. Another point to note, especially when using approximate models for distillation, is the danger of accumulating errors. The model creates an error in each increment, which in the individual increment might not be serious, but the errors will accumulate increment by increment.

Table 16.3

Batch distillation with fixed reflux ratio of 11.

K	$x_{3,B} = x_{r,B}$	$x_{2,B}$	$x_{1,B}$	$x_{3,D} = x_{r,D}$	$x_{2,D}$	$x_{1,D}$	N_{min}	θ	$\sum_{i=1}^{NC} \frac{\alpha_i x_{i,B}}{\alpha_i - \theta}$	R_{min}^U	R_{min}^G	$R_{min}^G - R_{min}^U$	$\frac{1}{x_{r,D} - x_{r,B}}$	$\Delta Area$	$\ln(B/F)$	B	$\bar{x}_D$
0	0.3000	0.2500	0.4500	2.03×10^{-5}	0.0041	0.9959	5.0000						−3.3336	0.0000	0.0000	150.00	
1	0.3100	0.2582	0.4318	3.75×10^{-5}	0.0057	0.9943	4.7391	1.3172	3.69×10^{-8}	0.20028	0.20028	-4.52×10^{-11}	−3.2262	−0.0328	−0.0328	145.16	0.9958
2	0.3200	0.2663	0.4137	4.04×10^{-5}	0.0061	0.9938	4.7386	1.3285	2.95×10^{-7}	0.20262	0.20262	-2.39×10^{-10}	−3.1254	−0.0318	−0.0646	140.62	0.9950
3	0.3300	0.2745	0.3955	4.36×10^{-5}	0.0066	0.9933	4.7380	1.3398	9.93×10^{-7}	0.20499	0.20499	-6.50×10^{-10}	−3.0307	−0.0308	−0.0953	136.36	0.9946
4	0.3400	0.2826	0.3774	4.71×10^{-5}	0.0071	0.9928	4.7375	1.3511	-6.87×10^{-8}	0.20742	0.20742	-1.37×10^{-9}	−2.9416	−0.0299	−0.1252	132.35	0.9943
5	0.3500	0.2907	0.3593	5.10×10^{-5}	0.0077	0.9922	4.7369	1.3625	-1.67×10^{-7}	0.20989	0.20989	-2.53×10^{-9}	−2.8576	−0.0290	−0.1542	128.57	0.9940
6	0.3600	0.2988	0.3412	5.52×10^{-5}	0.0083	0.9916	4.7364	1.3738	-3.41×10^{-7}	0.21243	0.21243	-4.29×10^{-9}	−2.7782	−0.0282	−0.1824	124.99	0.9938
7	0.3700	0.3069	0.3231	6.00×10^{-5}	0.0090	0.9909	4.7358	1.3851	-6.17×10^{-7}	0.21502	0.21502	-6.90×10^{-9}	−2.7031	−0.0274	−0.2098	121.61	0.9935
8	0.3800	0.3149	0.3051	6.53×10^{-5}	0.0098	0.9901	4.7352	1.3964	6.16×10^{-8}	0.21769	0.21769	-1.07×10^{-8}	−2.6320	−0.0267	−0.2365	118.41	0.9932
9	0.3900	0.3229	0.2871	7.12×10^{-5}	0.0107	0.9892	4.7346	1.4076	1.07×10^{-7}	0.22044	0.22044	-1.63×10^{-8}	−2.5646	−0.0260	−0.2624	115.38	0.9929
10	0.4000	0.3309	0.2691	7.80×10^{-5}	0.0117	0.9882	4.7340	1.4189	1.71×10^{-7}	0.22329	0.22329	-1.29×10^{-8}	−2.5005	−0.0253	−0.2878	112.49	0.9926
11	0.4100	0.3389	0.2511	8.57×10^{-5}	0.0128	0.9871	4.7333	1.4301	2.58×10^{-7}	0.22626	0.22626	-3.66×10^{-8}	−2.4395	−0.0247	−0.3125	109.75	0.9923
12	0.4200	0.3469	0.2331	9.45×10^{-5}	0.0141	0.9858	4.7326	1.4412	3.67×10^{-7}	0.22937	0.22937	-5.53×10^{-8}	−2.3815	−0.0241	−0.3366	107.13	0.9920
13	0.4300	0.3548	0.2152	1.05×10^{-4}	0.0157	0.9842	4.7319	1.4523	4.97×10^{-7}	0.23266	0.23266	-8.53×10^{-8}	−2.3261	−0.0235	−0.3601	104.64	0.9917
14	0.4400	0.3627	0.1973	1.17×10^{-4}	0.0174	0.9824	4.7311	1.4634	6.44×10^{-7}	0.23618	0.23618	-1.37×10^{-7}	−2.2733	−0.0230	−0.3831	102.26	0.9913
15	0.4500	0.3705	0.1795	1.31×10^{-4}	0.0196	0.9803	4.7302	1.4744	7.98×10^{-7}	0.23998	0.23998	-2.32×10^{-7}	−2.2229	−0.0225	−0.4056	99.99	0.9909
16	0.4600	0.3783	0.1617	1.49×10^{-4}	0.0221	0.9777	4.7293	1.4853	9.51×10^{-7}	0.24416	0.24416	-4.30×10^{-7}	−2.1746	−0.0220	−0.4276	97.81	0.9904
17	0.4700	0.3861	0.1439	1.71×10^{-4}	0.0253	0.9745	4.7283	1.4962	-1.33×10^{-7}	0.24885	0.24885	-9.27×10^{-7}	−2.1284	−0.0215	−0.4491	95.73	0.9899

16.4 Batch Crystallization

Crystallization is very common in the production of fine and specialty chemicals. Many chemical products are in the form of solid crystals. Also, crystallization has the advantage that it can produce a product with a high purity and can be more effective than distillation from the separation of heat-sensitive materials. Crystallization has already been discussed in Chapter 9 and has two main steps. First, the solute to be crystallized is dissolved in a suitable solvent, unless it is already dissolved, for example, solute dissolved in a solvent from a previous reaction step. Second, the solid is then deposited in the form of crystals from the solution by cooling, evaporation and so on.

Two of the main objectives in crystallization are to maximize the average crystal size and to minimize the coefficient of variation of crystal size. As pointed out in Chapter 9, large crystals are easier to filter and wash in order to produce a high purity product. Table 16.4 compares batch and continuous crystallization.

From Table 16.4, it can be concluded that both batch and continuous crystallization have their relative advantages and disadvantages. The biggest single advantage of batch crystallization is its flexibility. Batch crystallization operations are most often carried out in an agitated vessel. However, the vessel might be fitted with baffles or a *draft tube*. A draft tube is a vertical cylinder placed inside the crystallizer with a diameter typically 70% of the vessel diameter. The agitator induces a vertical flow through the inside of the draft tube and circulation vertically in the opposite direction on the outside of the draft tube. This can help maintain good circulation rates for the slurry.

Table 16.4

Batch versus continuous crystallization.

Batch	Continuous
Flexible	Not flexible
Low capital investment	High capital investment
Small process development requirements	Large process development requirements
Poor reproducibility	Good reproducibility

Cooling crystallization (see Chapter 9) is the most common method of achieving supersaturation. A solvent is preferred that exhibits a high solubility at high temperature, but a low solubility at low temperature. The initial temperature is reduced gradually to the final temperature. Figure 16.10 illustrates a batch cooling crystallization. Starting at Point *A* in the unsaturated region, this can be cooled and the metastable region entered. A cooling profile can then be followed within the metastable region to the final temperature. As pointed out in Chapter 9, it is desirable to stay out of the labile region, otherwise too many fine crystals are created. In cooling from the initial to the final temperature, different cooling profiles can be followed, as illustrated in Figure 16.11. Figure 16.11a shows *natural cooling*, whereby there is a large temperature decrease initially, but as the temperature difference between the crystallization and the ambient temperature decreases, the rate of cooling decreases. Natural cooling is uncontrolled and tends to lead to poor crystal quality. Rather than relying on natural cooling, a coolant such as cooling water could be exploited. Figure 16.11b illustrates *linear cooling* in which the coolant flowrate is used to maintain a constant rate of cooling. This tends to lead to better crystal quality than natural cooling. Finally, Figure 16.11c illustrates controlled cooling in which the flowrate of the coolant is varied in order to follow a specified profile from the initial to the final temperature. The cooling profile is a degree of freedom for optimization and can be optimized using profile optimization (Choong and Smith, 2004a). This will be discussed later.

Although cooling crystallization is the most common method of inducing supersaturation in batch crystallization processes, other methods can be used, as discussed in Chapter 9. For example, evaporation can be used, in which case the profile of the rate of evaporation through the batch can also be optimized (Choong and Smith, 2004b). Indeed, the profiles of both temperature and rate of

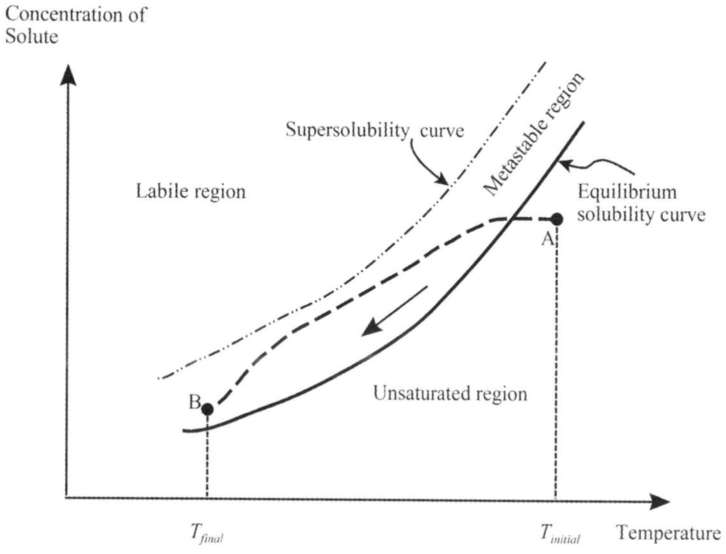

Figure 16.10

Batch cooling crystallization.

16

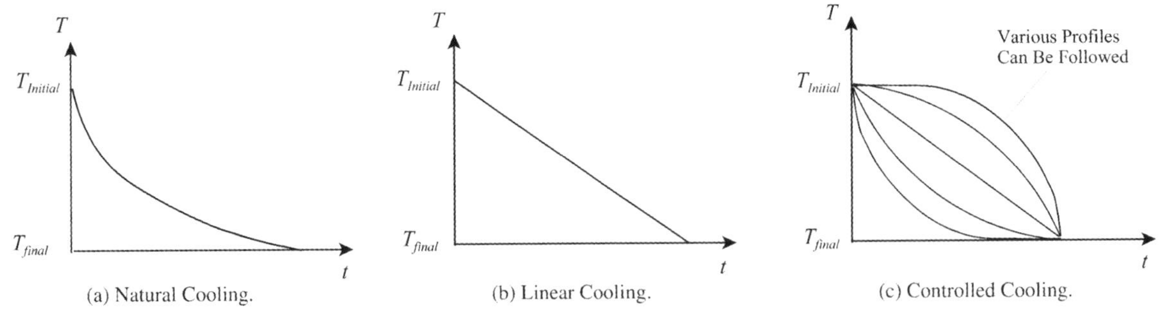

Figure 16.11
Types of batch cooling crystallization.

evaporation can be controlled simultaneously to obtain greater control over the level of supersaturation as the batch proceeds (Choong and Smith, 2004b). However, it should be noted that there is often reluctance to use evaporation in the production of fine, specialty and pharmaceutical products, as evaporation can concentrate any impurities and increase the level of contamination of the final product.

The agitator is a key component of crystallizer design and is required to:

- generate appropriate levels of secondary nucleation throughout the vessel;
- prevent local excessive levels of supersaturation;
- provide adequate temperature control throughout the vessel for cooling crystallization;
- provide adequate rates of heat transfer for evaporation if evaporative crystallization is used;
- keep crystals from settling on the bottom of the crystallizer to maintain crystal growth and a low coefficient of variation of crystal size.

Increasing the power input and fluid shear from agitation increases both crystal breakage and secondary nucleation. As the speed of an agitator is increased, this generally reduces the size of the metastable region. Scale-up of crystallizers from laboratory size to full-scale operation is far from straightforward. If a crystallization operation is scaled up for a constant power input per unit volume of slurry based on geometric similarity, the maximum shear rate at the agitator increases, but the average shear rate decreases. This can result in a greater variation of crystal size from scale-up. Secondary nucleation is likely to vary significantly with location in the crystallizer. Changes in the agitation and vessel geometry can change the local rates of energy dissipation and result in significant changes in total secondary nucleation.

Table 16.5
Optimization variables for batch and semibatch cooling crystallization.

Batch	Semibatch
1. Temperature profile	1. Solute/feed flowrate
2. Number of seeds	2. Temperature profile
3. Size of seeds	3. Number of seeds
4. Agitation	4. Seed addition profile
5. Initial level of supersaturation	5. Size of seeds
6. Batch time	6. Agitation
	7. Initial level of supersaturation
	8. Batch Time

As pointed out in Chapter 9, another way to create secondary nucleation is to add seed crystals to start the crystals growing in the supersaturated solution. These seeds should be pure product.

Just as a reactor can be operated in batch or semibatch mode, a crystallizer can also be operated in batch or semibatch mode. Table 16.5 contrasts the optimization variables for batch and semibatch operation of cooling crystallization.

As mentioned previously, scale-up of crystallization processes from the laboratory is far from straightforward. Various parameters need to be maintained to be as close to those used in the laboratory as possible in order to reproduce the results from the laboratory. For scale-up, supersolubility, agitation (and its effect on secondary nucleation throughout the vessel), fraction of solids in the slurry, seed number and sizes, contact time between growing crystals and liquid all need to be maintained.

16.5 Batch Filtration

Batch filtration involves the separation of suspended solids from a slurry of associated liquid. The required product could be either the solid particles or the liquid filtrate. In batch filtration, the filter medium presents an initial resistance to the fluid flow that will change as particles are deposited. The driving forces used in batch filtration are (Sharratt, 1997):

- gravity,
- vacuum,
- pressure,
- centrifugal.

If a constant pressure difference is maintained across a filter medium and filter cake, then for the filtration of volume of liquid F, filtration time is proportional to the square of F.

As discussed in Chapter 7, filter aids can be added to the slurry to reduce the filter cake resistance. These are materials that have high porosity. Their application is normally restricted to cases where the filtrate is valuable and the solid cake is a waste. In cases where the solid is valuable, the filter aid should be readily separable from the filter cake. Sometimes the filter aid is precoated on to the filter medium. If the solid filter cake is the product, then the cake is normally washed to remove dissolved solutes and solvents present in the original feed. Liquid may be retained within the cake within clusters of particles or at points of contact between particles. In an ideal situation, perfect washing would require only one volume of wash liquid equivalent to the void in the solid bed to wash away unwanted solute and solvent. In practice, perfect washing is not achieved and the actual washing efficiency depends on mixing and mass transfer. After washing the filter cake, the product would normally be dried to remove any wash liquid remaining within the solid.

Experimental results from the laboratory or pilot plant may often be scaled up by a factor of 100 times or more. However, to reduce errors in scale-up, a similar filter, the same slurry mixture, the same filter aid and approximately the same pressure drop should be used.

A detailed discussion of filtration can be found in Rushton and Griffiths (1987) and Svarovsky (1981).

16.6 Batch Heating and Cooling

Consider now the problem of batch heating and cooling. The case assumed here is of an agitated vessel containing a liquid that is perfectly mixed, with no feed to or product removal from the vessel (Figure 16.2). Heat transfer to or from the liquid is accomplished by the flow of a heating or cooling medium through a coil mounted within the vessel (Figure 16.2a), a jacket on the outside of the vessel (Figure 16.2b) or a half-pipe jacket welded to the outside of the vessel (Figure 16.2c). Figure 16.3 shows some examples of different agitator designs. Within the context of heat transfer, different agitator designs will induce different flow patterns that will influence the heat transfer coefficient.

The overall heat transfer coefficient for coil heating (Figure 16.2a) is usually in the range 400–600 $W \cdot m^{-2} \cdot K^{-1}$ (Carpenter, 1997). Typical, overall heat transfer coefficients for jacketed agitated vessels are given in Table 16.6 (Carpenter, 1997).

The overall heat transfer coefficient is usually controlled by the process side. For the inside vessel wall surface, the process side film transfer coefficient can be calculated from (Carpenter, 1997):

$$Nu = a\,Re^{2/3} Pr^{2/3} \left(\frac{\mu}{\mu_w}\right)^{0.14} \tag{16.46}$$

For a coil, the outside overall heat transfer coefficient is given by (Carpenter, 1997):

Table 16.6

Typical overall heat transfer coefficients for jacketed agitated vessels.

	U ($W \cdot m^{-2} \cdot K^{-1}$)
Heating (stainless steel vessel)	400
Heating (glass-lined steel vessel)	310
Cooling (stainless steel vessel)	350
Cooling (glass-lined steel vessel)	200

$$Nu = 1.4 Re^{0.62} Pr^{1/3} \left(\frac{\mu}{\mu_w}\right)^{0.14} \tag{16.47}$$

where
$Nu = \dfrac{hD_{VI}}{k}$
$Re = \dfrac{ND_{IMP}^2 \rho}{\mu}$
$Pr = C_p \mu / k$
C_p = liquid specific heat capactiy ($J \cdot kg^{-1} \cdot K^{-1}$)
k = liquid thermal conductivity ($W \cdot m^{-1} \cdot K^{-1}$)
h = heat transfer coefficient for the process side ($W \cdot m^{-2} \cdot K^{-1}$)
D_{VI} = internal diameter of the vessel (m)
k = liquid thermal conductivity ($W \cdot m^{-1} \cdot K^{-1}$)
N = impeller rotational speed (s^{-1})
D_{IMP} = impeller diameter (m)
ρ = liquid density ($kg \cdot m^{-3}$)
μ = liquid viscosity in the bulk fluid ($N \cdot s \cdot m^{-2}$)
μ_W = liquid viscosity at the heat transfer surface ($N \cdot s \cdot m^{-2}$)
a = constant given in Table 16.7

Heat transfer coefficients for a liquid used in a jacket on the utility side are given by (Carpenter, 1997; Lehrer, 1970):

$$Nu = \frac{0.03 Re^{3/4} Pr}{1 + 1.74 Re^{-1/8}(Pr - 1)} \left(\frac{\mu}{\mu_W}\right)^{0.14} \tag{16.48}$$

where
$Nu = \dfrac{hd_e}{k}$
$Re = \dfrac{\rho d_e \left(\sqrt{\nu_i \nu_A} + \nu_B\right)}{\mu}$

Table 16.7

Constants for agitated vessel heat transfer coefficients (Carpenter, 1997).

Impeller type	a
Propeller	0.46
45° Turbine	0.61
Rushton Turbine	0.87

$$d_e = 0.816(D_J - D_{VO})$$
$$\nu_i = \frac{4Q}{\pi D_{INLET}^2}$$
$$\nu_B = 0.5\sqrt{2z\beta g \Delta T}$$

D_J = diameter of the jacket (m)
D_{VO} = outside diameter of the vessel (m)
Q = liquid flowrate of the cooling liquid ($m^3 \cdot s^{-1}$)
D_{INLET} = inside diameter of the inlet nozzle (m)
z = wetted jacket height (m)
β = coefficient of thermal expansion of the cooling liquid (–)
g = acceleration due to gravity ($m \cdot s^{-2}$)
ΔT = temperature rise of the cooling liquid (K)

$$\nu_A = \frac{4Q}{\pi\left(D_J^2 - D_{VO}^2\right)} \text{ (radial inlet)}$$
$$\nu_A = \frac{2Q}{(D_J - D_{VO})z} \text{ (tangential inlet)}$$

For a liquid used in a half-pipe external jacket:

$$Nu = 0.023Re^{0.8}Pr^{1/3}\left(\frac{\mu}{\mu_W}\right)^{0.14} \tag{16.49}$$

The area for heat transfer should be the whole jacketed area and the diameter should be the hydraulic diameter given by (Carpenter, 1997):

$$d_e = \frac{\pi d_I}{2}$$

where d_I is the diameter of the half-pipe.

For the analysis of batch process heating and cooling, the following assumptions are made:

- Heat transfer area is specified (A).
- Overall heat transfer coefficient (U) is specified and constant across the whole surface and through time.
- Agitation produces a uniform temperature in the vessel.
- Specific heat capacities are assumed constant.
- Heat losses are negligible.

A number of different cases will be considered (Bowman, Mueller and Nagle, 1940; Kern, 1950).

1) *Heating with an isothermal heating medium.* This case corresponds with a condensing pure component being used as a heating medium. This could be the use of steam with no significant superheat and no condensate subcooling. The batch temperature is T at any time t. A differential heat balance gives:

$$dQ = mC_P\frac{dT}{dt} = UA(T_{COND} - T) \tag{16.50}$$

where
dQ = increment of heat transferred ($J \cdot s^{-1}$)
m = mass of liquid in the vessel (kg)
C_P = specific heat capacity of the liquid ($J \cdot kg^{-1} \cdot K^{-1}$)
T = temperature of the liquid in the vessel (°C, K)
t = time (s)
U = overall heat transfer coefficient ($W \cdot m^{-2} \cdot K^{-1}$)
A = heat transfer area (m^2)
T_{COND} = temperature of the heating medium (°C, K)

Rearranging and integrating Equation 16.50 gives:

$$\int_0^{t_B} dt = \frac{mC_P}{UA}\int_{T_1}^{T_2}\frac{dT}{(T_{COND} - T)} \tag{16.51}$$

where T_1 = initial temperature of the batch
T_2 = final temperature of the batch

From Equation 16.51:

$$t_B = \frac{mC_P}{UA}\ln\left[\frac{T_{COND} - T_1}{T_{COND} - T_2}\right] \tag{16.52}$$

2) *Cooling with an isothermal cooling medium.* This case corresponds to a vaporizing pure component being used as the cooling medium. This could be a vaporizing refrigerant. A differential heat balance gives:

$$dQ = -mC_P\frac{dT}{dt} = UA(T - T_{EVAP}) \tag{16.53}$$

where T_{EVAP} = temperature of the cooling medium (°C, K)

Rearranging and integrating Equation 16.53 gives:

$$\int_0^{t_B} dt = -\frac{mC_P}{UA}\int_{T_1}^{T_2}\frac{dT}{(T - T_{EVAP})} \tag{16.54}$$

From Equation 16.54:

$$t_B = \frac{mC_P}{UA}\ln\left[\frac{T_1 - T_{EVAP}}{T_2 - T_{EVAP}}\right] \tag{16.55}$$

3) *Heating with a nonisothermal heating medium.* Assume that the heating medium has a constant flowrate with a fixed inlet temperature and an outlet temperature that varies through the operation. A differential heat balance gives:

$$dQ = m_H C_{PH}(T_{H1} - T_{H2}) = UA\Delta T_{LM} \tag{16.56}$$

where

$$\Delta T_{LM} = \frac{(T_{H1} - T) - (T_{H2} - T)}{\ln\left[\frac{T_{H1} - T}{T_{H2} - T}\right]} = \frac{T_{H1} - T_{H2}}{\ln\left[\frac{T_{H1} - T}{T_{H2} - T}\right]} \tag{16.57}$$

m_H = mass flowrate of the heating medium ($kg \cdot s^{-1}$)

C_{PH} = specific heat capacity of heating medium ($J \cdot kg^{-1} \cdot K^{-1}$)
T_{H1} = inlet temperature of heating medium (°C, K)
T_{H2} = outlet temperature of heating medium (°C, K)

Rearranging Equation 16.56:

$$T_{H2} = T + \frac{T_{H1} - T}{K_1} \quad (16.58)$$

where $K_1 = \exp\left[\frac{UA}{m_H C_{PH}}\right]$

The differential heat balance can be written as:

$$dQ = mC_P \frac{dT}{dt} = m_H C_{PH}(T_{H1} - T_{H2}) \quad (16.59)$$

Substituting Equation 16.58 into Equation 16.59 and rearranging gives:

$$\frac{dT}{dt} = \frac{m_H C_{PH}}{mC_P}\left(\frac{K_1 - 1}{K_1}\right)(T_{H1} - T) \quad (16.60)$$

Rearranging and integrating Equation 16.60 gives:

$$\int_0^{t_B} dt = \frac{mC_P}{m_H C_{PH}}\left(\frac{K_1}{K_1 - 1}\right)\int_{T_1}^{T_2} \frac{dT}{(T_{H1} - T)} \quad (16.61)$$

From Equation 16.61:

$$t_B = \frac{mC_P}{m_H C_{PH}}\left(\frac{K_1}{K_1 - 1}\right)\ln\left[\frac{T_{H1} - T_1}{T_{H1} - T_2}\right] \quad (16.62)$$

4) *Cooling with a nonisothermal cooling medium.* Again assume a fixed flowrate of cooling medium with a fixed inlet temperature. A differential energy balance gives:

$$dQ = m_C C_{PC}(T_{C2} - T_{C1}) = UA\Delta T_{LM} \quad (16.63)$$

where

$$\Delta T_{LM} = \frac{(T - T_{C1}) - (T - T_{C2})}{\ln\left[\frac{T - T_{C1}}{T - T_{C2}}\right]} = \frac{T_{C2} - T_{C1}}{\ln\left[\frac{T - T_{C1}}{T - T_{C2}}\right]} \quad (16.64)$$

m_c = mass flowrate of the cooling medium ($kg \cdot s^{-1}$)
C_{PC} = specific heat capacity of cooling medium ($J \cdot kg^{-1} \cdot K^{-1}$)
T_{C1}, T_{C2} = inlet and outlet temperatures of cooling medium (°C, K)

Rearranging Equation 16.63 gives:

$$T_{C2} = T - \frac{T - T_{C1}}{K_2} \quad (16.65)$$

where $K_2 = \exp\left[\frac{UA}{m_C C_{PC}}\right]$

The differential heat balance can be written as:

$$dQ = -mC_P \frac{dT}{dt} = m_C C_{PC}(T_{C2} - T_{C1}) \quad (16.66)$$

Substituting Equation 16.65 into Equation 16.66 gives:

$$\frac{dT}{dt} = -\frac{m_C C_{PC}}{mC_P}\left(\frac{K_2 - 1}{K_2}\right)(T - T_{C2}) \quad (16.67)$$

Rearranging and integrating Equation 16.67 gives:

$$\int_0^{t_B} dt = -\frac{mC_P}{m_C C_{PC}}\left(\frac{K_2}{K_2 - 1}\right)\int_{T_1}^{T_2} \frac{dT}{(T - T_{C2})} \quad (16.68)$$

From Equation 16.68:

$$t_B = \frac{mC_P}{m_C C_{PC}}\left(\frac{K_2}{K_2 - 1}\right)\ln\left[\frac{T_1 - T_{C2}}{T_2 - T_{C2}}\right] \quad (16.69)$$

The analysis can be extended to heating and cooling recirculation loops using external heat exchangers (Fisher, 1944; Kern, 1950).

Example 16.3 A jacketed vessel contains 10 t of solvent at 100 °C with a heat capacity of $1.72\ kJ \cdot kg^{-1} \cdot K^{-1}$. The contents of the vessel are to be cooled in a batch cooling operation to 40 °C using cooling water at 25 °C. The cooling water has a heat capacity of $4.2\ kJ \cdot kg^{-1} \cdot K^{-1}$ and the maximum return temperature is 40 °C. It can be assumed that the jacket has a heat transfer area of $18\ m^2$ with an overall heat transfer coefficient of $350\ W \cdot m^{-2} \cdot K^{-1}$. Calculate the flowrate of cooling water required and the time required for the operation.

Solution First calculate the cooling water flowrate for a maximum return temperature of 40 °C. The maximum return temperature will occur at the beginning of the batch. Equation 16.65 gives the cooling water return temperature:

$$T_{C2} = T - \frac{T - T_{C1}}{K_2}$$

where $K_2 = \exp\left[\frac{UA}{m_C C_{PC}}\right]$

$$K_2 = \exp\left[\frac{0.350 \times 18}{4.2 m_C}\right]$$

$$T_{C2} = 100 - \frac{100 - 25}{K_2}$$

16

Solve by varying m_C until $T_{C2} = 40\,°C$ to give $m_C = 6.72\ kg \cdot s^{-1}$. Now the batch cooling time can be calculated:

$$K_2 = \exp\left[\frac{0.350 \times 18}{6.72 \times 4.2}\right]$$

$$= 1.25$$

$$t_B = \frac{mC_P}{m_C C_{PC}}\left(\frac{K_2}{K_2 - 1}\right)\ln\left[\frac{T_1 - T_{C2}}{T_2 - T_{C2}}\right]$$

$$t_B = \frac{10{,}000 \times 1.72}{6.72 \times 4.2}\left(\frac{1.25}{1.25 - 1}\right)\ln\left[\frac{100 - 25}{40 - 25}\right]$$

$$= 4902.5\ s$$

$$= 1.36\ h$$

This is a long operation to cool the batch. Such equipment is not well suited to heating and cooling operations.

16.7 Optimization of Batch Operations

Through a batch operation such as a batch reaction or batch distillation, the process conditions vary through time. In order to improve the performance of batch operations, the conditions through the batch can be directed to change through time to improve the overall batch performance. In the batch reaction, the temperature profile through time might be controlled to increase the conversion or yield. For a batch distillation, the reflux ratio profile through time might be varied to increase product recovery or decrease the energy. Should the conditions be maintained constant through the batch? Should they increase or decrease? Should the increase or decrease be linear, exponential, and so on? Should the profile go through a maximum or minimum? This presents a *dynamic optimization*.

A profile generator algorithm can be developed to generate various families of curves that are continuous functions through time. In principle, this can be achieved in many ways (Mehta and Kokossis, 1988; Choong and Smith, 2004a). One way is to exploit two different profiles to generate a wide variety of shapes. Although many mathematical expressions could be used, two types of profile described by the following mathematical equations can be exploited to give a wide variety of profiles, the shape of which can be easily controlled in an optimization (Choong and Smith, 2004a).

Type I

$$x = x_F - (x_F - x_0)\left[1 - \frac{t}{t_F}\right]^{a_1} \tag{16.70}$$

Type II

$$x = x_0 - (x_0 - x_F)\left[\frac{t}{t_F}\right]^{a_2} \tag{16.71}$$

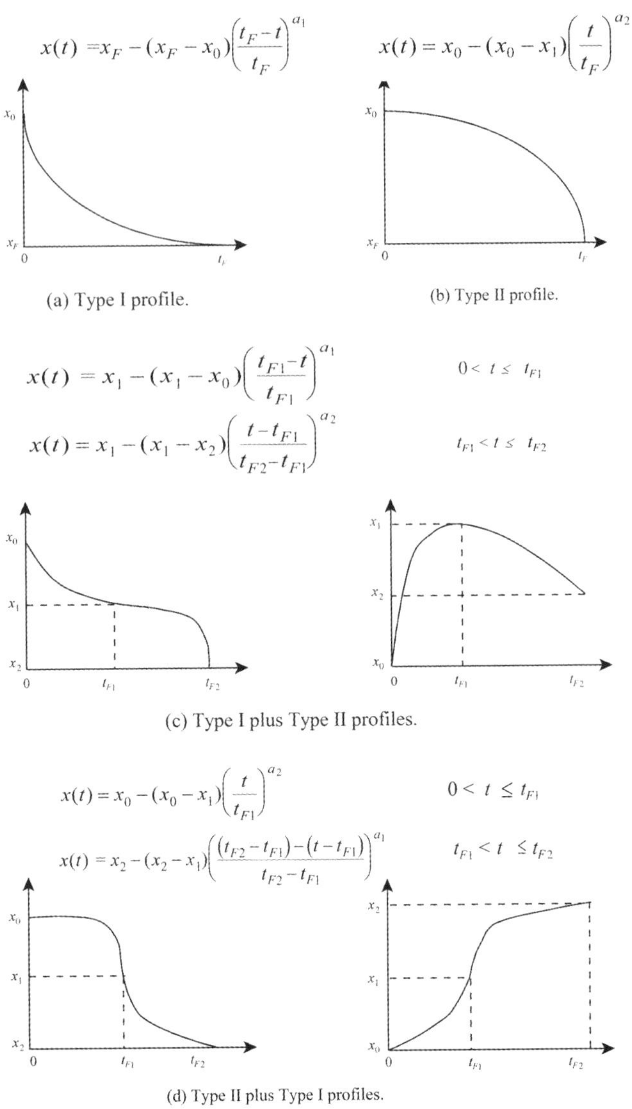

Figure 16.12

Two basic profiles can be combined in different ways.

In these equations, x is the instantaneous value of any control variable at any time t, x_0 is the initial value and x_F is the final value of the control variable. In principle, x can be any control variable such as temperature, reactant feed rate, evaporation rate, heat removed or supplied, and so on; t_F is the final time for the profile. The convexity and concavity of the profiles are governed by the values of a_1 and a_2. Figures 16.12a and 16.12b illustrate typical forms of each curve.

Combining these two profiles across the time horizon allows virtually all types of continuous curves to be produced that can be implemented in a practical design. When the two profiles are combined, two additional variables are needed. The value of t_{F1} indicates the point where the two curves meet and x_1 is the corresponding value of the control variable where the curves meet. Figure 16.12c illustrates the form of Type I followed by Type II and Figure 16.12d the form of Type II followed by Type I.

By combining the two curves together, only six variables are needed to generate the various profiles. These six variables are the initial and final values of the control variable x_0 and x_{F2}, two exponential constants a_1 and a_2, the intermediate point in time where the two profiles converge t_{F1} and the corresponding intermediate value of the control variable where the profiles converge x_1. In limiting cases, only one type of profile will be used, rather than two.

In formulating profiles, emphasis should be placed on searching for profiles that are continuous and easily implemented in practice. Therefore, curves that include serious discontinuities should normally be avoided. It is meaningless to have a global optimum solution with complex and practically unrealizable profiles. Curve combinations from the above equations such as Type I + Type I or Type II + Type II should not normally be considered as there can be a prominent discontinuity at the intermediate point. The profile generator can be easily extended to combine three or more curves across the time horizon instead of two. However, there is little practical use to employ more than two different curves for the majority of problems. The complexity of the profiles increases with the number of curves generated. It should not be forgotten that a way must be found to realize the profile in practice. In a dynamic problem varying through time, a control system must be designed that will allow the profile to be followed through time.

At each point along the profile, the process will have a different performance, depending on the value of the control variable. The process can be modeled in different ways along the profile. The profile can be divided into increments in time, and a model developed for each time increment. The size of the increments must be assessed in such an approach to make sure that the increments are small enough to follow the changes in the profile adequately. Alternatively, the rate of change through time can be modeled by differential equations. The profile functions (Equations 16.70 and 16.71) are readily differentiated to obtain gradients for the solution of the differential equations.

Having evaluated the system performance for each setting of the six variables, the variables are optimized simultaneously in a multidimensional optimization to maximize or minimize an objective function evaluated at each setting of the variables. In practice, many models tend to be nonlinear and hence, for example, SQP or a stochastic search method can be used.

In order to generate a profile through time that involves both Type I and Type II profiles, a more complex mathematical formulation is required than simply Equations 16.70 and 16.71. Type I and II profiles need to be combined and the shape optimized by optimizing both the shape of the individual Type I and II profiles and the order in which they are imposed. In order to switch the order of the profiles, two logic variables γ and ϑ are introduced. A further logic variable ϕ is introduced to prevent potential "divide by zero" problems during optimization.

Type I

$$x = \gamma\left\{\vartheta x_1 + (1-\vartheta)x_2 - [\vartheta(x_1 - x_0) + (1-\vartheta)(x_2 - x_1)] \times \left[\frac{\mathrm{ABS}((1-\vartheta)(t_{F1}+t_{F2}) + \vartheta t_{F1} - (1-\vartheta)t_{F1} - t)}{\vartheta t_{F1} + (1-\vartheta)(t_{F2} - t_{F1}) + \phi_{\mathrm{I}}}\right]^{a_1}\right\} \tag{16.72}$$

Type II

$$x = (\gamma - 1)\left\{\vartheta x_1 + (1-\vartheta)x_0 - [\vartheta(x_1 - x_2) + (1-\vartheta)(x_0 - x_1)] \times \left[\frac{\mathrm{ABS}(t - \vartheta t_{F1})}{\vartheta(t_{F2} - t_{F1}) + (1-\vartheta)t_{F1} + \phi_{\mathrm{II}}}\right]^{a_2}\right\} \tag{16.73}$$

where t = time
t_{F1} = intermediate time where the profile changes from Type I to Type II or vice versa
t_{F2} = final batch time
x = instantaneous value of x at time t
x_0 = initial value of x at time 0
x_1 = intermediate value of x at time t_{F1} where the profile changes from Type I to Type II or vice versa
x_2 = final value of x at time t_{F2}
ϑ = logic variable used to control the order of the profiles:
$\vartheta = 1$ gives Type I + Type II
$\vartheta = 0$ gives Type II + Type I
γ = logic variable used to control the order of the profiles:
IF $\vartheta = 1$ AND $t \le t_{F1}$, $\gamma = 1$
IF $\vartheta = 1$ AND $t \ge t_{F1}$, $\gamma = 0$
IF $\vartheta = 0$ AND $t \le t_{F1}$, $\gamma = 0$
IF $\vartheta = 0$ AND $t \ge t_{F1}$, $\gamma = 1$
ϕ = logic variable used to avoid "divide by zero" numerical problems during optimization:

IF$[\vartheta t_{F1} + (1-\vartheta)(t_{F2} - t_{F1})] = 0$, ϕ_{I} = arbitary value (say 1)
IF$[\vartheta t_{F1} + (1-\vartheta)(t_{F2} - t_{F1})] \neq 0$, $\phi_{\mathrm{I}} = 0$
IF$[\vartheta(t_{F2} - t_{F1}) + (1-\vartheta)t_{F1}] = 0$, ϕ_{II} = arbitary value (say 1)
IF$[\vartheta(t_{F2} - t_{F1}) + (1-\vartheta)t_{F1}] \neq 0$, $\phi_{\mathrm{II}} = 0$

In Equations 16.72 and 16.73 the ABS (absolute value) function avoids numerical problems during optimization caused by attempting to raise negative values to a power of a_1 or a_2. Such negative values do not have a physical significance but occur when a calculation is made on a Type I or Type II function that is to be eliminated by the variable γ. Equations 16.72 and 16.73 can be combined to create a rich variety of shapes that can, in principle, be implemented in a batch control system. The optimization variables are:

$$x_0, x_1, x_2, t_{F1}, a_1, a_2, \vartheta$$

Additionally, the batch time t_{F2} can also be optimized if required. All of the above variables are continuous, with the exception of ϑ, which is integer. This would make the optimization to be typically a mixed integer nonlinear program. This can be done by either a mixed integer deterministic method or stochastic search optimization (see Chapter 3). However, mixed integer optimization can be avoided by simply using a nonlinear program with ϑ set 1 (Type I followed by Type II, as illustrated in Figure 16.12c) and then repeating with ϑ set to zero (Type II followed by Type I, as illustrated in Figure 16.12d).

The control variables can be constrained to fixed values (e.g. fixed initial temperature in a temperature profile) or constrained to be between certain limits. In addition to the six variables dictating the shape of the profile, t_{F2} can also be optimized if required. For example, this can be important in batch processes to optimize the batch cycle time in a batch process, in addition to the other variables.

The approach is readily applied to single profiles, such as the optimization of batch distillation (Jain, Kim and Smith, 2012). It can also be readily extended to problems involving multiple profiles. For example, in a batch crystallization process, the temperature profile and evaporation profile can be optimized simultaneously (Choong and Smith, 2004b). Each profile optimization would involve six variables using the above profile equations.

Once the optimum profile(s) has been established, its practicality for implementation must be assessed. For a continuous process, the equipment must be able to be designed such that the profile can be followed through space by adjusting rates of reaction, mass transfer, heat transfer, and so on. In a dynamic problem, a control system must be designed that will allow the profile to be followed through time. If the profile is not practical, then the optimization must be repeated with additional constraints added to avoid the impractical features.

Example 16.4 The following isomerization reaction is to be carried out in a batch reactor:

$$A \underset{k_2}{\overset{k_1}{\rightleftharpoons}} B \tag{16.74}$$

The forward and reverse reaction rates are given by:

$$k_1 = \exp\left[13.25 - \frac{49{,}900}{RT}\right] \tag{16.75}$$

$$k_2 = \exp\left[38.25 - \frac{121{,}000}{RT}\right] \tag{16.76}$$

where k_1 = rate of forward reaction (min^{-1})
k_2 = rate of reverse reaction (min^{-1})
R = gas constant (8.3145 kJ·K^{-1}·kmol^{-1})
T = reaction temperature (K)

The batch reaction time is fixed to be 6 h.

a) Determine an expression for the variation of concentration of A (C_A) through time from an initial concentration C_{A_0}, which can be applied to incremental changes through time.
b) For an initial concentration of $C_{A_0} = 0$ determine the optimum temperature to maximize conversion in the batch time if the temperature is maintained constant through the batch.
c) For an initial concentration of $C_{A_0} = 0$ determine the optimum temperature profile to maximize conversion in the batch time.

Solution

a) For a batch reaction:

$$\begin{aligned} t &= -\int_{C_{A_1}}^{C_{A_2}} \frac{dC_A}{-r_A} \\ &= -\int_{C_{A_1}}^{C_{A_2}} \frac{dC_A}{k_1C_A - k_2C_B} \\ &= -\int_{C_{A_1}}^{C_{A_2}} \frac{dC_A}{k_1C_A - k_2\left(C_{A_0} - C_A\right)} \\ &= -\int_{C_{A_1}}^{C_{A_2}} \frac{dC_A}{(k_1 + k_2)C_A - k_2C_{A_0}} \\ &= -\frac{1}{k_1 + k_2}\left[\ln\left((k_1 + k_2)C_A - k_2C_{A_0}\right)\right]_{C_{A_1}}^{C_{A_2}} \end{aligned}$$

Substituting and rearranging gives:

$$C_{A_2} = \frac{\left[(k_1 + k_2)C_{A_1} - k_2C_{A_0}\right]\left[\exp - (k_1 + k_2)t\right] - k_2C_{A_0}}{k_1 + k_2} \tag{16.77}$$

By definition of conversion:

$$X = \frac{C_{A_0} - C_{A_2}}{C_{A_0}} \tag{16.78}$$

b) For a constant temperature through the batch, Equations 16.77 and 16.78 define the conversion through the batch with k_1 and k_2 defined by Equations 16.75 and 16.76. The model can be set up in a spreadsheet with the temperature T varied using a solver to maximize the conversion X. A temperature of 324.9 K gives a conversion of 0.722.
c) To determine the optimum temperature profile, the batch time of 260 min is divided into increments of, say, 20 min and Equations 16.77 to 16.78 used to calculate the outlet concentration and conversion for each time increment. The temperature profile through the batch can be defined using Equations 16.72 and 16.73. The optimization can be carried out in a spreadsheet with the variables T_0, T_1, T_2, t_{F1}, a_1 and a_2 varied using the spreadsheet solver. Repeating the calculation for $\vartheta = 1$ and $\vartheta = 0$ allows the optimum profile to be determined. Figure 16.13 shows the optimum temperature profile with:

$$\begin{aligned} T_0 &= 367.3\ \text{K} \\ T_{F1} &= 333.7\ \text{K} \\ T_{F2} &= 316.3\ \text{K} \\ t_{F1} &= 83.5\ \text{min} \\ a_1 &= 2.398 \\ a_2 &= 0.548 \\ \vartheta &= 1 \end{aligned}$$

This gives an optimum conversion of 0.760, an increase of 4%. It should be noted that this is a nonlinear optimization with many local optima around the solution. Thus different combinations of settings allow similar optimum conversions to be achieved. However, all have the same basic shape, as shown in Figure 16.13.

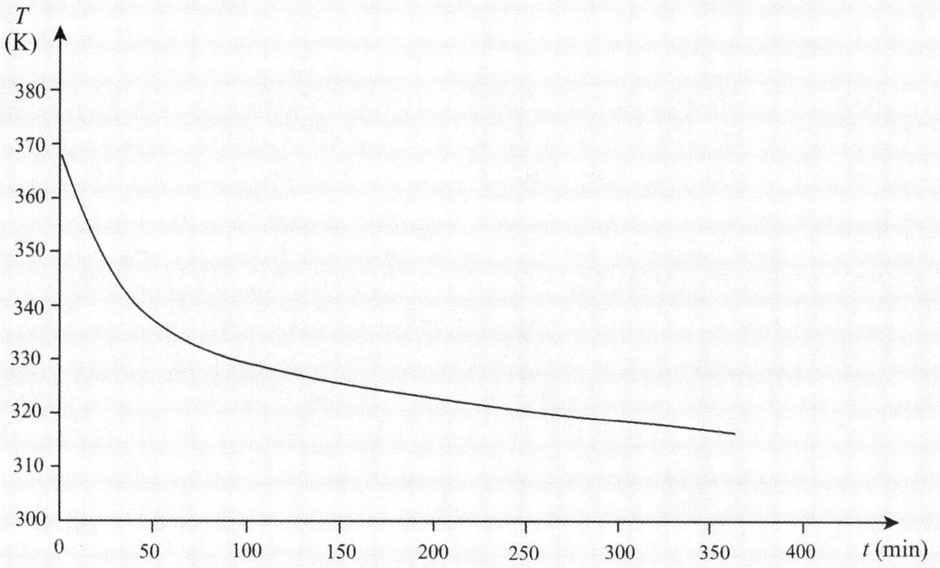

Figure 16.13

Optimum temperature profile for Example 16.4.

Example 16.5 For the separation in Example 16.1, achieving a mean distillate purity of 0.98 and a recovery of S of 90%:

a) Optimize the reflux ratio to minimize the batch cycle time, maintaining the reflux ratio to be constant across the batch cycle.
b) Optimize the reflux ratio profile across the batch cycle time to minimize the batch cycle time.

Solution

a) As with Example 16.1, the material balance must be solved by solving Equation 16.7 numerically. If the reflux ratio is maintained constant across the cycle, then the batch cycle time can be calculated using Equation 16.41. Assuming the distillation is first established at total reflux, then, as in Example 16.1, $x_D = 0.9956$ at $t = 0$. The batch time is to be minimized subject to the constraint $\bar{x}_D = 0.98$, $Bx_B = 9.0$. To achieve this, the reflux ratio must be varied, but also with the increment in x_D (Δx_D) varied simultaneously. This can be set up in a spreadsheet using the spreadsheet solver.

Table 16.8 shows the results from an optimization that divides the batch cycle into 20 increments:

$$R = 6.1712$$
$$\Delta x_D = 0.004136$$

From Equation 16.41:

$$t = 5.93\ \text{h}$$

b) In order to impose and optimize a reflux ratio profile, Equations 16.72 and 16.73 need to be applied across the batch cycle. However, unlike the batch reactor optimization in Example 16.3, the batch cycle time is calculated, rather than set. Also, because the reflux ratio varies across the cycle, the batch cycle time must be calculated by numerically integrating Equation 16.44. Thus, the cycle is divided into increments across an arbitrary scale when applying Equations 16.72 and 16.73 (say 0 to 100). As before, total reflux is assumed at time 0, giving $x_D = 0.9956$. The distillation concentration is decreased in increments through the batch cycle. The batch cycle time is minimized subject to $\bar{x}_D = 0.98$ and $Bx_B = 9.0$, but now optimizing:

$$x_0, x_1, x_2, t_{F1}, a_1, a_2, \vartheta, \Delta x_D$$

The calculations can be performed in a spreadsheet using the spreadsheet solver. The results are given in Table 16.9 based on dividing the batch cycle into 20 intervals. The resulting optimization variables are:

$$x_0 = 10.89$$
$$x_1 = 3.44$$
$$x_2 = 15.02$$
$$t_{F1} = 14.0 (\text{out of } 100)$$
$$a_1 = 2.783$$
$$a_2 = 1.482$$
$$\vartheta = 1$$
$$\Delta x_D = 0.00259$$

The minimized batch cycle time is 4.98 h (compared with 5.93 h when the reflux ratio is optimized, but held constant through the cycle). It should be noted that there are inevitably numerical errors as a result of the numerical integration. Increasing the number of time intervals decreases the errors. Also, more accurate numerical integration can be performed using Simpson's Rule. However, the errors are still small in this case. It should also be noted that different optimal points can be found with similar performance, due to the nonlinear nature of the optimization. The reflux ratio profile is shown in Figure 16.14.

16

Table 16.8

Batch distillation for the optimized reflux ratio held constant over the cycle.

Interval	x_D	k	W	B	x_B	$\frac{1}{x_D - x_B}$	$\Delta Area$	$\ln(B/F)$	B	$\bar{x}_D$	Bx_B
0	0.9956	0.064461	1.161153	1.067679	0.6000	2.527952	0.000000	0.000000	150.00	0.9956	90.00
1	0.9914	0.064129	1.160323	1.066160	0.5439	2.234469	0.133568	0.133568	131.25	0.9925	71.39
2	0.9873	0.063798	1.159495	1.064647	0.4503	1.862121	0.191770	0.325338	108.34	0.9894	48.78
3	0.9832	0.063467	1.158669	1.063140	0.3862	1.675007	0.113411	0.438750	96.73	0.9883	37.35
4	0.9790	0.063138	1.157844	1.061639	0.3395	1.563531	0.075623	0.514372	89.68	0.9874	30.44
5	0.9749	0.062808	1.157021	1.060144	0.3039	1.490325	0.054286	0.568658	84.94	0.9866	25.81
6	0.9708	0.062480	1.156200	1.058655	0.2759	1.439166	0.040996	0.609654	81.53	0.9859	22.50
7	0.9666	0.062152	1.155380	1.057171	0.2533	1.401872	0.032133	0.641788	78.95	0.9853	20.00
8	0.9625	0.061825	1.154563	1.055693	0.2346	1.373870	0.025918	0.667706	76.93	0.9847	18.05
9	0.9584	0.061499	1.153747	1.054220	0.2189	1.352406	0.021385	0.689091	75.30	0.9842	16.49
10	0.9542	0.061173	1.152932	1.052753	0.2056	1.335717	0.017976	0.707067	73.96	0.9837	15.20
11	0.9501	0.060848	1.152120	1.051292	0.1940	1.322629	0.015345	0.722412	72.84	0.9832	14.13
12	0.9459	0.060523	1.151309	1.049836	0.1839	1.312324	0.013271	0.735682	71.88	0.9828	13.22
13	0.9418	0.060200	1.150499	1.048386	0.1751	1.304220	0.011606	0.747288	71.05	0.9824	12.44
14	0.9377	0.059877	1.149692	1.046942	0.1672	1.297886	0.010249	0.757537	70.32	0.9820	11.76
15	0.9335	0.059554	1.148886	1.045502	0.1601	1.293001	0.009129	0.766666	69.68	0.9816	11.16
16	0.9294	0.059233	1.148082	1.044069	0.1538	1.289319	0.008192	0.774858	69.12	0.9813	10.63
17	0.9253	0.058912	1.147279	1.042640	0.1481	1.286648	0.007401	0.782259	68.61	0.9809	10.16
18	0.9211	0.058591	1.146478	1.041217	0.1428	1.284837	0.006727	0.788985	68.15	0.9806	9.73
19	0.9170	0.058272	1.145679	1.039799	0.1380	1.283764	0.006147	0.795132	67.73	0.9803	9.35
20	0.9129	0.057953	1.144882	1.038387	0.1336	1.283332	0.005645	0.800778	67.35	0.9800	9.00

Figure 16.14

Optimum reflux ratio profile for Example 16.5.

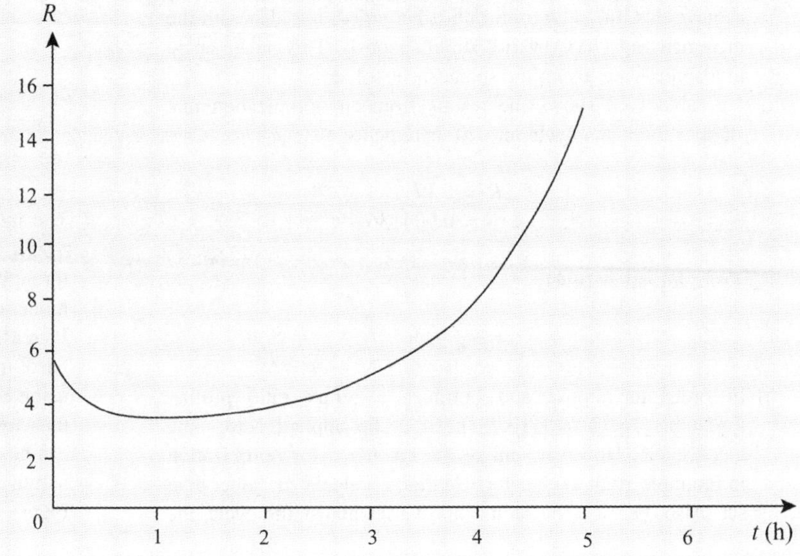

Table 16.9

Batch distillation for the optimized reflux ratio profile.

Interval	t	RR	x_D	x_B	$\frac{1}{x_D - x_B}$	$\Delta Area$	$\ln(B/F)$	B	$\bar{x}_D$	Bx_B	$\frac{B}{V}\frac{R+1}{(x_D - x_B)}$	Δt	t
0	0	10.8899	0.9956	0.6000	2.527952	0.000000	0.000000	150.00	0.9956	90.00	45.0857	0.0000	0.0000
1	5	5.6067	0.9930	0.6000	2.544624	0.000003	0.000003	150.00	0.9943	90.00	25.2175	0.0000	0.0000
2	10	3.6611	0.9904	0.5772	2.420233	0.056576	0.056579	141.75	0.9915	81.82	15.9906	0.4696	0.4696
3	15	3.4543	0.9878	0.5305	2.186889	0.107530	0.164109	127.30	0.9895	67.53	12.4002	0.6626	1.1323
4	20	3.6636	0.9852	0.4778	1.970775	0.109631	0.273740	114.08	0.9881	54.51	10.4850	0.6034	1.7357
5	25	3.9900	0.9826	0.4296	1.808131	0.091139	0.364879	104.14	0.9871	44.73	9.3963	0.4795	2.2152
6	30	4.3984	0.9800	0.3866	1.685285	0.074946	0.439825	96.62	0.9862	37.36	8.7906	0.3902	2.6054
7	35	4.8740	0.9774	0.3490	1.591333	0.061641	0.501466	90.85	0.9854	31.71	8.4918	0.3251	2.9305
8	40	5.4077	0.9748	0.3163	1.518493	0.050903	0.552369	86.34	0.9848	27.31	8.4007	0.2765	3.2070
9	45	5.9934	0.9722	0.2879	1.461248	0.042300	0.594669	82.76	0.9842	23.83	8.4575	0.2393	3.4463
10	50	6.6265	0.9696	0.2633	1.415674	0.035420	0.630089	79.88	0.9836	21.03	8.6246	0.2103	3.6566
11	55	7.3036	0.9671	0.2419	1.378957	0.029904	0.659993	77.53	0.9831	18.75	8.8772	0.1873	3.8439
12	60	8.0216	0.9645	0.2232	1.349054	0.025462	0.685455	75.58	0.9827	16.87	9.1985	0.1687	4.0126
13	65	8.7784	0.9619	0.2069	1.324468	0.021860	0.707315	73.94	0.9822	15.30	9.5767	0.1535	4.1661
14	70	9.5717	0.9593	0.1925	1.304084	0.018918	0.726233	72.56	0.9819	13.96	10.0033	0.1409	4.3070
15	75	10.3999	0.9567	0.1797	1.287064	0.016497	0.742729	71.37	0.9815	12.83	10.4720	0.1304	4.4374
16	80	11.2615	0.9541	0.1684	1.272769	0.014488	0.757217	70.35	0.9812	11.85	10.9782	0.1214	4.5588
17	85	12.1552	0.9515	0.1583	1.260705	0.012808	0.770025	69.45	0.9808	10.99	11.5182	0.1137	4.6725
18	90	13.0797	0.9489	0.1492	1.250488	0.011393	0.781418	68.66	0.9805	10.25	12.0892	0.1071	4.7796
19	95	14.0340	0.9463	0.1410	1.241813	0.010193	0.791611	67.97	0.9803	9.59	12.6890	0.1013	4.8810
20	100	15.0171	0.9437	0.1336	1.234437	0.009167	0.800778	67.35	0.9800	9.00	13.3159	0.0963	4.9772

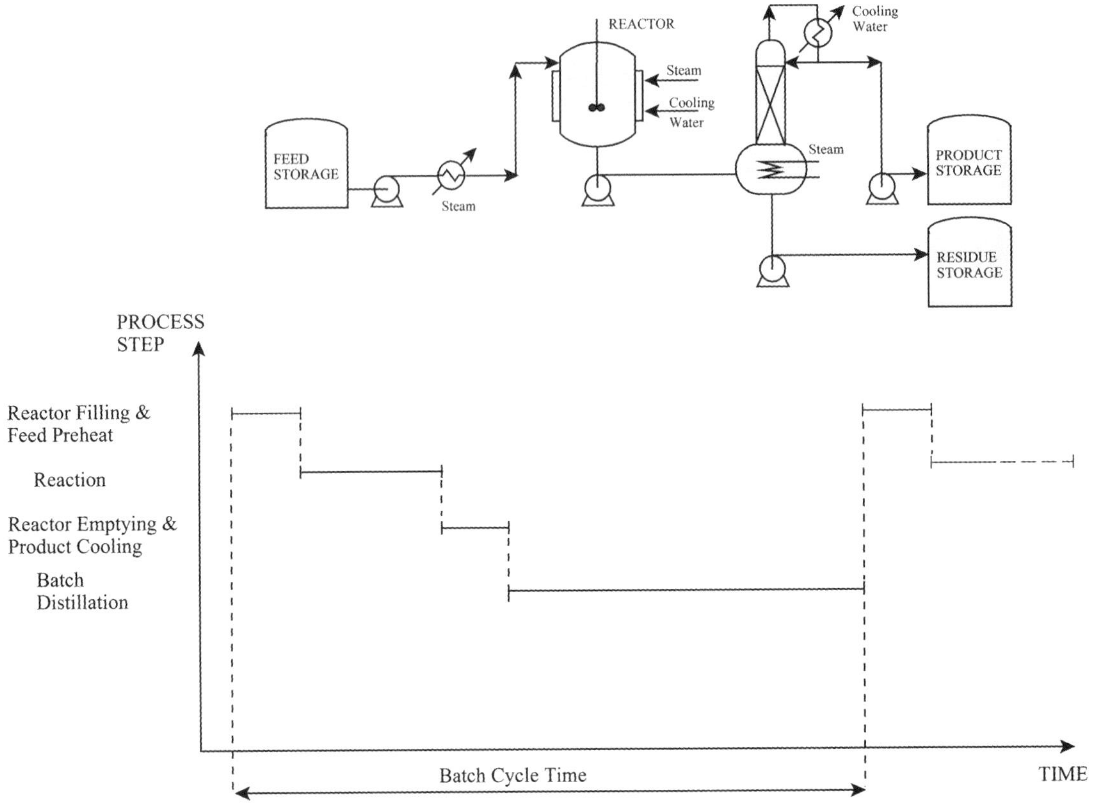

Figure 16.15
Gantt chart for a simple batch process.

16.8 Gantt Charts

Now consider the complete batch process. Figure 16.15 shows a simple process. Feed material is withdrawn from storage using a pump. The feed material is preheated in a heat exchanger before being fed to a batch reactor. Once the reactor is full, further heating takes place inside the reactor using steam to the reactor jacket, before the reaction proceeds. During the later stages of the reaction, cooling water is applied to the reactor jacket. Once the reaction is complete, the reactor product is withdrawn using a pump. The reactor product is passed to a batch distillation that produces a finished product in the overhead and a residue left in the distillation. The product and residue are sent to storage.

The process is also shown in Figure 16.15 as a *Gantt* or *time event chart* (Mah, 1990; Biegler, Grossman and Westerberg, 1997). The first two steps, pumping for reactor filling and feed preheat are both semicontinuous. The heating inside the reactor, the reaction itself and the cooling using the reactor jacket are all batch. The pumping to empty the reactor and charge to the batch distillation is again semicontinuous. The distillation step is batch. It can be seen from the Gantt chart in Figure 16.15 that there is very poor utilization of equipment. There are considerable periods over which the equipment is standing idle, sometimes termed *dead time*. The batch *cycle time* is the time interval between successive batches of product being produced.

High utilization of equipment is one of the goals of batch process design. This can be achieved by *overlapping* batches. Overlapping means that more than one batch, at different stages, resides in the process at any time, as shown in Figure 16.16. This allows the batch cycle time, that is, the time interval between producing successive batches of product, to be decreased considerably. The step with the longest time limits the cycle time. Alternatively, if more than one step is carried out in the same equipment, the cycle time is limited by the longest series of steps in the same equipment. The batch cycle time must be at least as long as the longest step. The rest of the equipment other than the limiting step is then idle for some fraction of the batch cycle.

16.9 Production Schedules for Single Products

Batch processes can be dedicated to the production of a single product or can produce multiple products. Start by considering the simplest case in which the process produces only a single product. Consider the process shown in Figure 16.17 involving three steps (Step *A*, Step *B* and Step *C*) in which Step *A* takes 10 h, Step *B* takes 5 h and Step *C* also takes 5 h. Figure 16.17a shows a *sequential* production schedule. Subsequent batches are only started once the

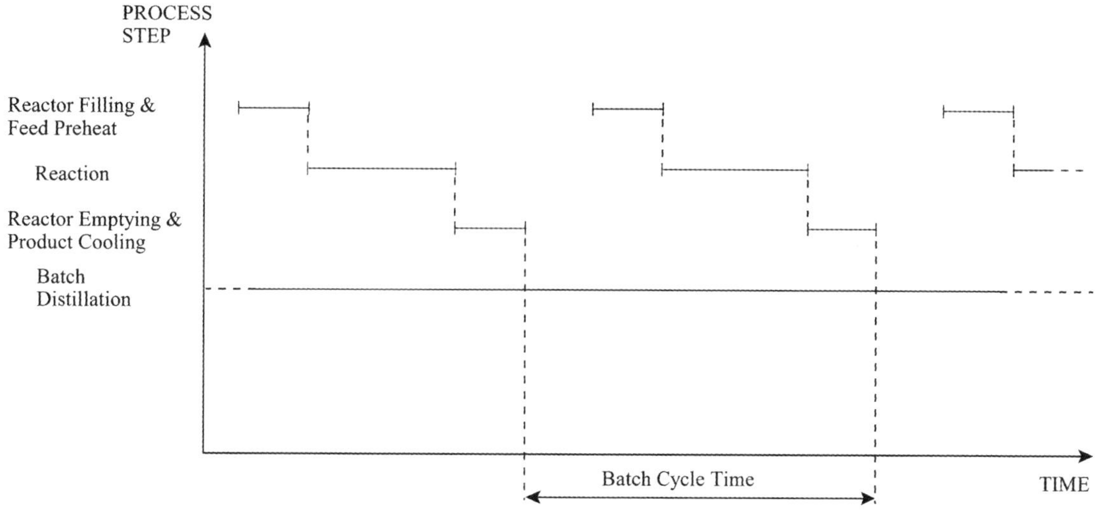

Figure 16.16

Overlapping batches allows the batch cycle time to be decreased.

Figure 16.17

Production schedules for a three-step process.

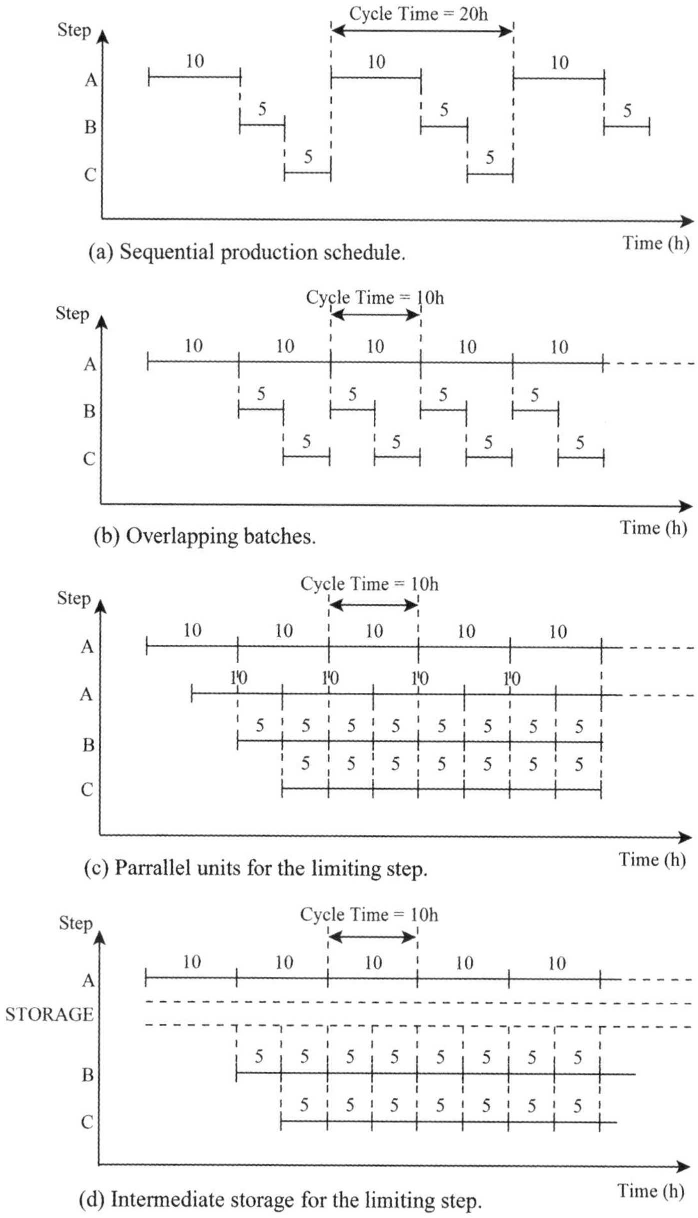

previous batch has been completely finished. For this sequential production schedule, the cycle time is 20 h. This clearly leads to very poor utilization of equipment. It has already been noted that *overlapping* batches can reduce the cycle time. This is illustrated in Figure 16.17b, where subsequent batches are started as soon as the appropriate equipment becomes available. Cycle time in Figure 16.17b decreases to 10 h for overlapping batches (the length of the longest step). If a specified volume of production needs to be achieved over a given period of time, then the equipment in the process that uses overlapping batches in Figure 16.17b can in principle be half the size of the equipment for sequential production in Figure 16.17a.

Even with overlapping batches in Figure 16.17b, Steps *B* and *C* are underutilized. Step *A* is fully utilized and this is the limiting step. Figure 16.17c shows a design in which there are two items of equipment operating Step *A*, but in parallel. This allows both Step *B* and Step *C* to be carried out with complete utilization. If the sizes of the equipment are compared to the sequential production schedule, then each of the two Steps *A*1 and *A*2 in Figure 16.17c can in principle be one-quarter the size of the equipment for Step *A* for sequential production in Figure 16.17a. The size of the equipment for Steps *B* and *C* in Figure 16.17c will also be one-quarter the size of those in the sequential production schedule in Figure 16.17a for the same production rate.

The final option shown in Figure 16.17d is to use *intermediate storage* for the limiting step. Material from Step *A* is sent to storage, from which Step *B* draws its feed. Material is still passed directly from Step *B* to Step *C*. Now all three steps are fully utilized. For the same rate of production over a period of time, the size of Step *A* can in principle be half that relative to the sequential production in Figure 16.17a and the sizes of Steps *B* and *C* can in principle be one-quarter those for sequential production. However, this is at the cost of introducing intermediate storage.

16.10 Production Schedules for Multiple Products

So far plants have been considered involving a single product. However, batch processes often produce multiple products in the same equipment. Here two broad types of process can be distinguished. In *flowshop* or *multiproduct* plants, all products produced require all steps in the process and follow the same sequence of operations. In *jobshop* or *multipurpose* processes, not all products require all steps and/or might follow a different sequence of steps (Biegler, Grossman and Westerberg, 1997).

Figure 16.18 shows a process that produces two products, Product 1 and 2, in a flowshop process. Figure 16.18a shows a production cycle involving a sequential production schedule. Production alternates between Product 1 and Product 2. The cycle time to produce a batch each of Product 1 and 2 is 28 h.

The first thing that can be considered in order to reduce the cycle time and increase equipment utilization is to overlap the batches as shown in Figure 16.18b. This reduces the cycle time to 18 h.

All of the schedules considered so far involved transferring material from one step to another, from a step to storage or from storage to a step without any time delay. This is known as *zero-wait transfer*. An alternative is to exploit the equipment in which a production step has taken place to provide *hold-up*. In this situation, material is held in the equipment until it is required by the production schedule. A schedule using equipment hold-up is shown in Figure 16.18c. This reduces the cycle time to 15 h.

Finally, Figure 16.18d shows the use of intermediate storage. The use of storage is only necessary for Product 2. Use of intermediate storage in this way reduces the cycle time to 14 h.

Consider now another problem involving the production of two products (Product 1 and 2) each involving two steps (Step *A* and *B*) in a flowshop plant. Figure 16.19a shows the production cycle for three batches each of Product 1 and Product 2. It can be

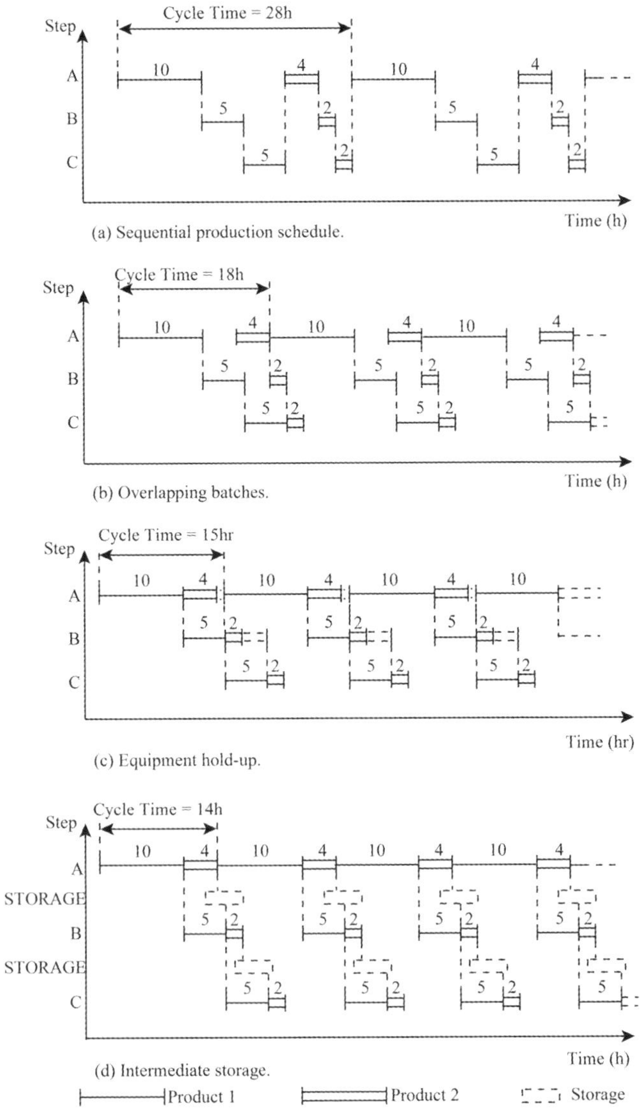

Figure 16.18
Production schedule for two products with a three-step process.

seen from Figure 16.19a that the batches have been overlapped to increase equipment utilization. In order to produce three products each of Product 1 and Product 2, the schedule in Figure 16.19a involves single-product *campaigns*. Three batches of Product 1 and three batches of Product 2 follow directly from each other. For this production schedule, the cycle time is 47 h. The total time required to produce a given number of batches, in this case three batches of each Product 1 and Product 2, is known as the *makespan*. From Figure 16.19a, for single-product *campaigns* the makespan is 53 h.

An alternative production schedule can be suggested by following a mixed-product campaign, as illustrated in Figure 16.19b. Alternating between batches of Product 1 and Product 2 in Figure 16.19b allows the cycle time to be reduced to 45 h and the makespan to be reduced to 51 h.

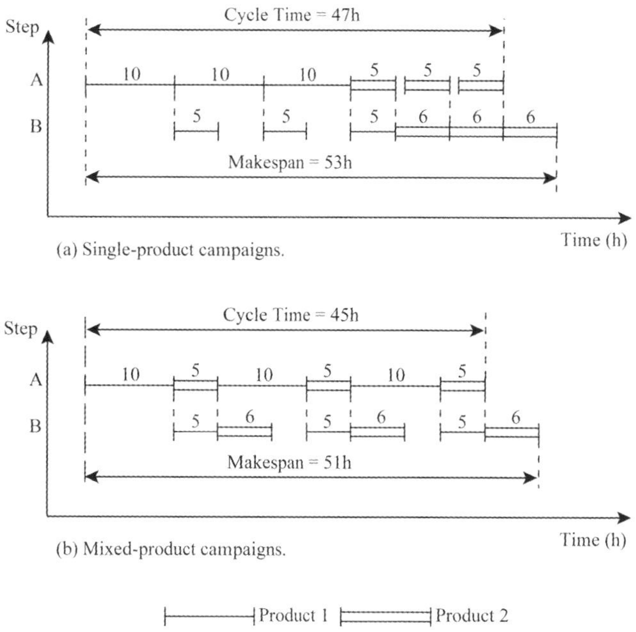

Figure 16.19

Single versus mixed product campaigns for three batches each of two products.

16.11 Equipment Cleaning and Material Transfer

So far, in the discussion of production schedules, some practical issues have been neglected. Two practical issues can be encountered in production scheduling that can have a significant effect on the cycle time and the makespan (Biegler, Grssman and Westerberg, 1997).

Consider first the changeover between two different products. It is usual for the equipment to be cleaned when changing from one product to another. Figure 16.20 shows the single-product campaigns and mixed-product campaigns from Figure 16.19, but with cleaning between product changes. The cleaning increases both the cycle time and the makespan. If the single-product campaign without cleaning in Figure 16.19a is compared with the single-product campaign with cleaning in Figure 16.20a, then the cycle time increases from 47 to 49 h and the makespan from 53 to 55 h. However, when the mixed-product campaign in Figure 16.19a is compared with the mixed-product campaign with cleaning in Figure 16.20b, it can be seen that there is a more significant increase in both the cycle time and the makespan. Cleaning increases the cycle time from 45 to 55 h and the makespan from 51 to 61 h.

Without cleaning the mixed-product campaign would have been considered to be more efficient than the single-product campaigns. Once cleaning is introduced, the mixed-product campaigns are seen to be less efficient than single-product campaigns.

Whether cleaning introduces such a decrease in the overall equipment utilization as that presented in Figures 16.19 and 16.20 depends on the problem. However, it is something that must be taken into account when planning production schedules. Another issue to be addressed later is that when products are changed in a batch production system and equipment needs to be cleaned, this can produce a significant amount of waste from the process that can create a significant environmental problem. This will be discussed again later under cleaner production.

Another important issue that has been neglected so far in the production schedules is that of transfer times between different steps. Figure 16.21 shows again the production schedule from Figure 16.17, but this time introducing an allowance of 1 h to transfer material between storage and the production steps, from one production step to another and from the outlet of a production step to storage. If material is being transferred from one step to another step, then the emptying of one step and filling the next step can be carried out simultaneously; hence the transfer from Step *A* to Step *B* and from Step *B* to

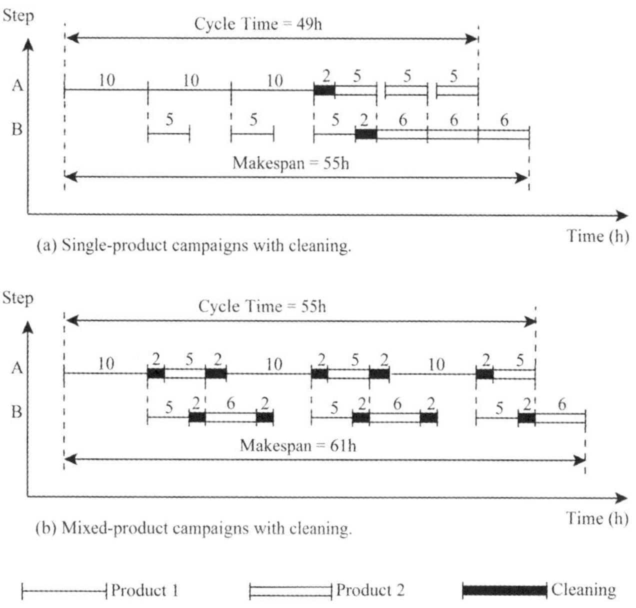

Figure 16.20

Cleaning between product changes extends the cycle times.

16

Figure 16.21
Transfer times extend the cycle times.

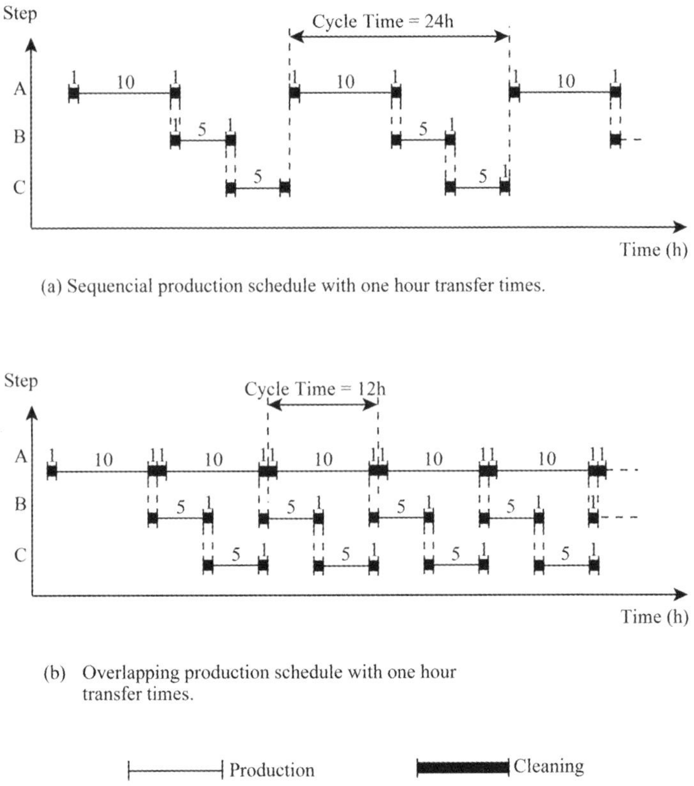

Step C overlaps. The cycle time for sequential production as shown in Figure 16.21a increases from 20 h to 24 h for one-hour transfer times. Figure 16.21b shows overlapping production with 1 h transfer times. In this case, the cycle time increases from 10 h to 12 h.

16.12 Synthesis of Reaction and Separation Systems for Batch Processes

Now consider how to synthesize the reaction and separation system for a batch process. Start by assuming the process to be continuous and then, if choosing to use batch operation, the continuous steps are replaced by batch steps (Myriantheos, 1986). It is simpler to start with continuous process operation because the time dependency of batch operation adds additional constraints over and above those for continuous operation.

However, there is one very significant difference between batch and continuous processes as far as the synthesis of reaction and separation systems is concerned. Continuous processes involve connections in space between processing steps. Batch processes also have connections in space between processing steps. In addition, batch processes have connections in time between processing steps. In batch processes, the connections in space can sometimes be substituted by connections in time. Consider the sequential production schedule in Figure 16.17a. Suppose that Step B and Step C could be carried out in the same equipment as Step A. This would mean that only one piece of equipment would be needed rather than three, but the production schedule would look the same as shown in Figure 16.17a. This would serve to reduce the capital cost of the equipment. It would also give advantages in terms of material transfer. Thus, the transfer times between the steps in Figure 16.21 would be eliminated. Another advantage in terms of cleaning is that there is less equipment to clean and less waste resulting from cleaning. Of course, the option to overlap batches would no longer be available if steps were merged. An example of how steps can be merged is a reaction followed by cooling crystallization. In principle, both steps can be carried out in the same equipment.

Before steps are merged into a single piece of equipment, it must be ensured that the equipment is suitable for multiple purposes in terms of its function, size, materials of construction and pressure rating. Also, it is clear that merging will affect the production schedule and the schedule needs to be considered when merging.

Finally, recycling of materials is difficult in batch processes because the connection in time cannot usually be made between the steps involved in the recycling. This is because different steps take place in different time periods. However, time can be bridged through the use of intermediate storage for the recycle.

The approach is illustrated by the following example.

Example 16.6 Butadiene sulfone (or 3-sulfolene) is an intermediate used for the production of solvents. It can be produced from butadiene and sulfur dioxide according to the reaction (McKetta, 1977; Myriantheos, 1986):

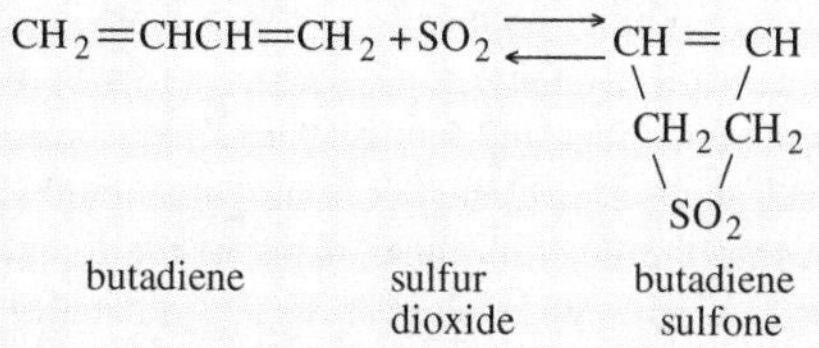

This is an exothermic, reversible, homogeneous reaction that takes place in a single liquid phase. The liquid butadiene feed contains 0.5% *n*-butane as an impurity. The sulfur dioxide is essentially pure. The mole ratio of sulfur dioxide to butadiene must be kept above 1 to prevent unwanted polymerization reactions. A value of 1.2 will be assumed. The temperature in the process must be kept above 65 °C to prevent crystallization of the butadiene sulfone but below 100 °C to prevent its decomposition. The product must contain less than 0.5 wt% butadiene and less than 0.3 wt% sulfur dioxide.

The normal boiling points of the materials are given in the Table 16.10.

Synthesize a continuous reaction, separation and recycle system for the process, with a view that the process will later become batch.

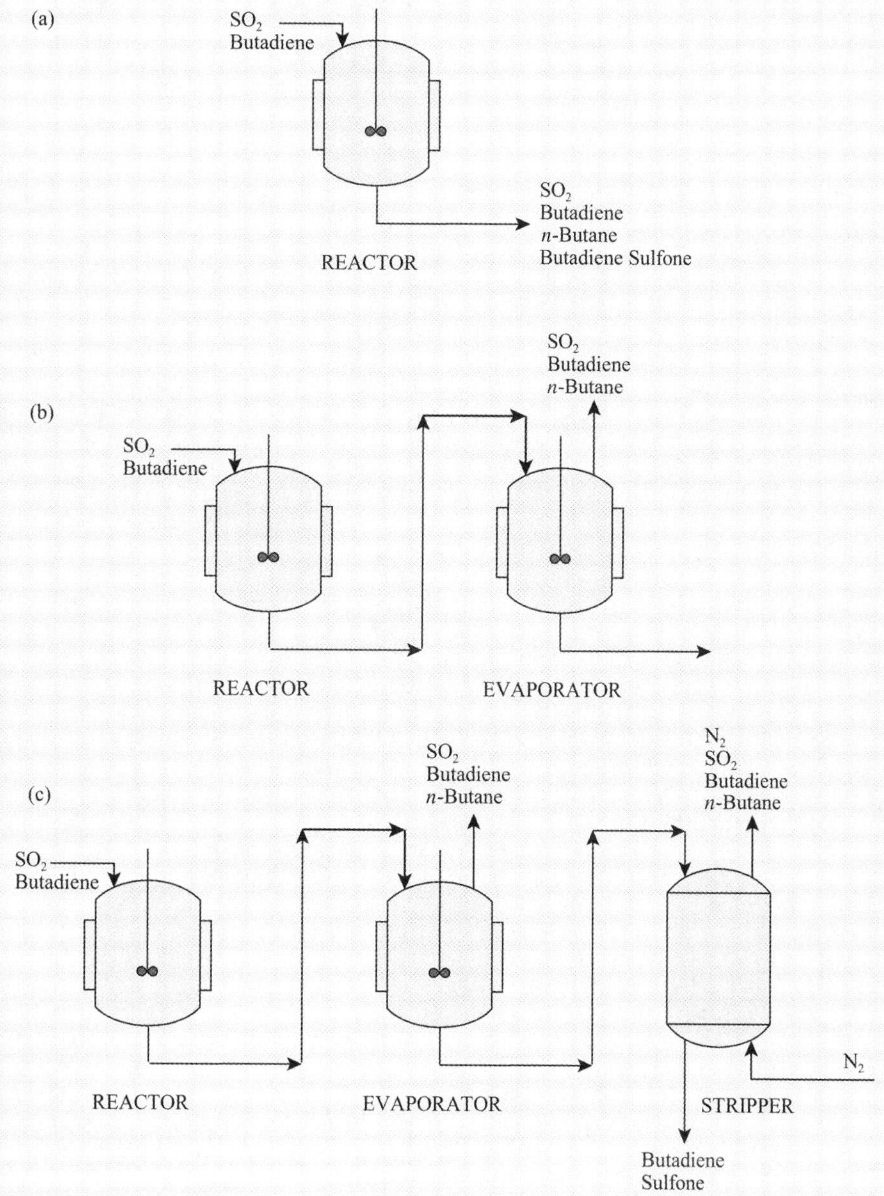

Figure 16.22

Reaction and separation system for the production of butadiene sulfone.

Table 16.10

Normal boiling points for the components.

Material	Normal boiling point (°C)
Sulfur dioxide	−10
Butadiene	−4
n-Butane	−1
Butadiene sulfone	151

Solution The reversible nature of the reaction means that neither of the feed materials can be forced to complete conversion. The reactor design in Figure 16.22a shows that the reactor product contains a mixture of both feed and product materials together with the *n*-butane impurity. These must be separated, but how?

If the relative boiling points of the components in the reactor product are considered, there is a wide range of volatilities. The sulfur dioxide, butadiene and *n*-butane are all low boilers and the butadiene sulfone is a much higher boiling material by comparison. Given that the reaction takes place in the liquid phase, a partial vaporization might well give a good separation between the butadiene sulfone and the other components (Figure 16.22b).

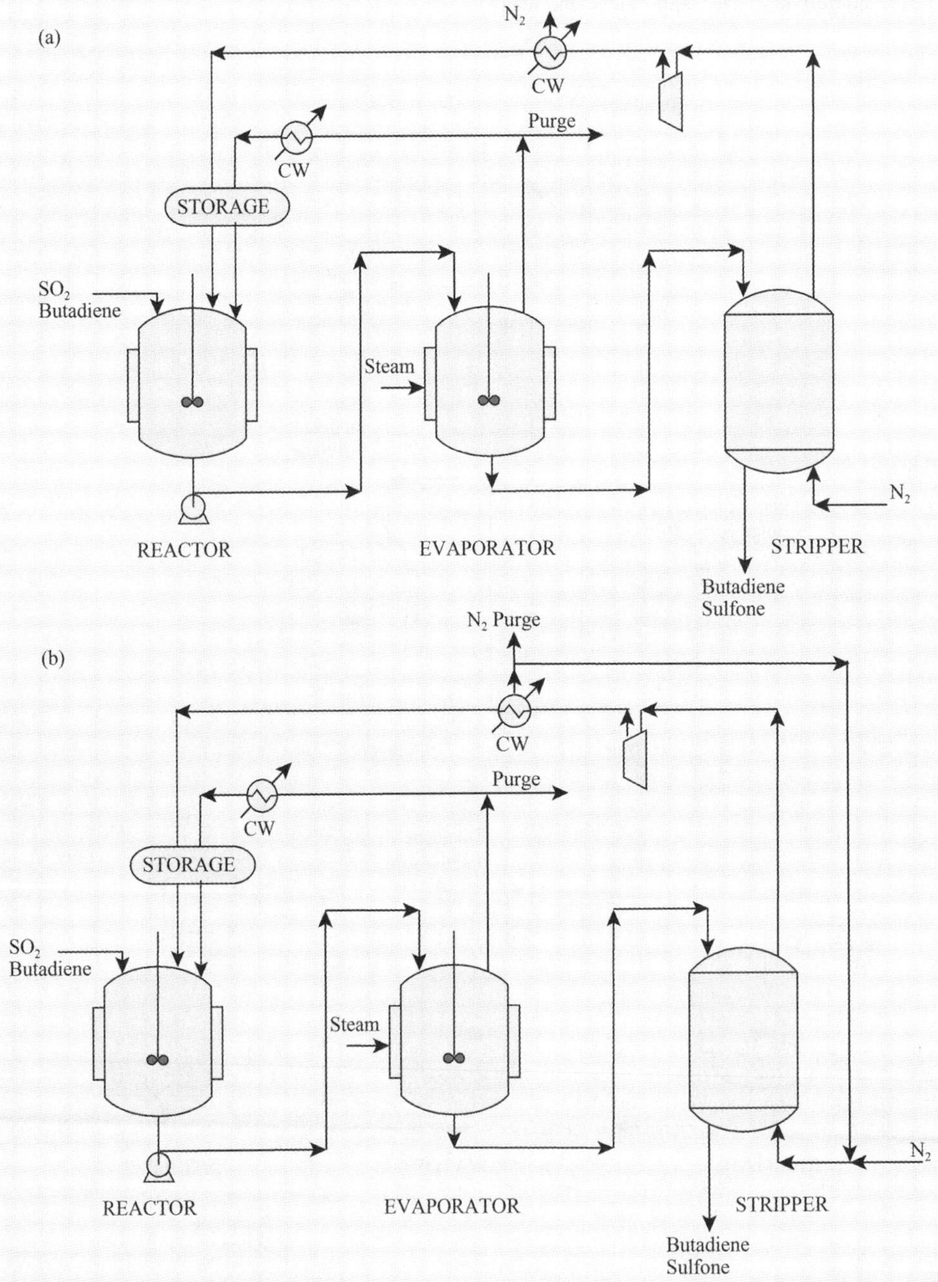

Figure 16.23

The recycle system for the production of butadiene sulfone.

A vapor–liquid equilibrium calculation shows that a good separation is obtained but the required product purity of butadiene < 0.5 wt% and sulfur dioxide < 0.3 wt% is not obtained. Further separation of the liquid is needed. Distillation of the liquid is difficult because of the narrow temperature limits between which the distillation must operate. However, the liquid can be stripped using nitrogen (Figure 16.22c).

The type of equipment illustrated in Figure 16.22 is more typical of batch operation than continuous operation, even though continuous operation is being contemplated at the moment. For example, the evaporator is a stirred tank with a heating jacket. In continuous plant, a more elaborate design with tubular heating of some type would probably have been used, perhaps with multiple stages.

Now consider recycling unconverted feed material to the reactor. Figure 16.23a shows recycles for unconverted feed material. The recycle from the evaporator to the reactor has been made possible by pressurizing the evaporator with the evaporator feed pump. Had this not been done, the vapor recycle would have required a compressor. The stripper works at a lower pressure to allow the unconverted material to be stripped. Thus, the recycle from the stripper requires a compressor. It is then condensed and fed back to the reactor.

Another problem is that most of the *n*-butane impurity that enters with the feed enters the vapor phase from the evaporator. Thus the *n*-butane builds up in the recycle unless a purge is provided (Figure 16.23a). Finally, the possibility of a nitrogen recycle should be considered to minimize the use of fresh nitrogen (Figure 16.23b).

Example 16.7 Convert the continuous process from Example 16.6 into a batch process. Preliminary sizing of the equipment indicates that the duration of the processing steps are given in Table 16.11 (Myriantheos, 1986).

Table 16.11

Duration of processing steps.

Processing step	Duration (h)
Reaction	2.1
Evaporation	0.45
Stripping	0.65
Vessel filling	0.25
Vessel emptying	0.25

Solution Having synthesized the continuous flowsheet shown in Figure 16.23b, now convert this into batch operation. The reactor now becomes batch, requiring the reaction to be completed before the separation can take place. Figure 16.24 shows the production schedule for a sequential batch schedule. Note in Figure 16.24 that there is a small overlap between the process steps. This is to allow for the fact that emptying of one step and filling the following step take place during the same time period.

The Gantt chart shown in Figure 16.24 indicates that individual items of equipment have a poor utilization. To improve the equipment utilization, overlapping batches are shown in Figure 16.25. Clearly, it is not possible to recycle directly from the separators to the reactor since the reactor is fed at a time different from that at which the separation is carried out. A storage tank is needed to hold the recycle material. This material is then used to provide part of the feed for the next batch. The final flowsheet for batch operation is shown in Figure 16.26.

In Figure 16.25, the reactor limits the batch cycle time, that is, it has no dead time. On the other hand, the evaporator and stripper both have significant dead time. Figure 16.27 shows the schedule for an arrangement with two reactors operating in parallel. With parallel operation, the reaction operations can overlap, allowing the evaporation and stripping operations to be carried out more frequently. This improves the overall utilization of equipment, and in principle allows the size of equipment to be reduced.

The batch cycle time has been reduced from 2.6 to 1.3 h. This means that a greater number of batches can be processed and hence, if there are two reactors each with the original capacity, the process capacity has increased. However, the increase in capacity has been achieved at the expense of an increased capital cost for the second reactor. An economic assessment is required before it can judge whether the trade-off is justified.

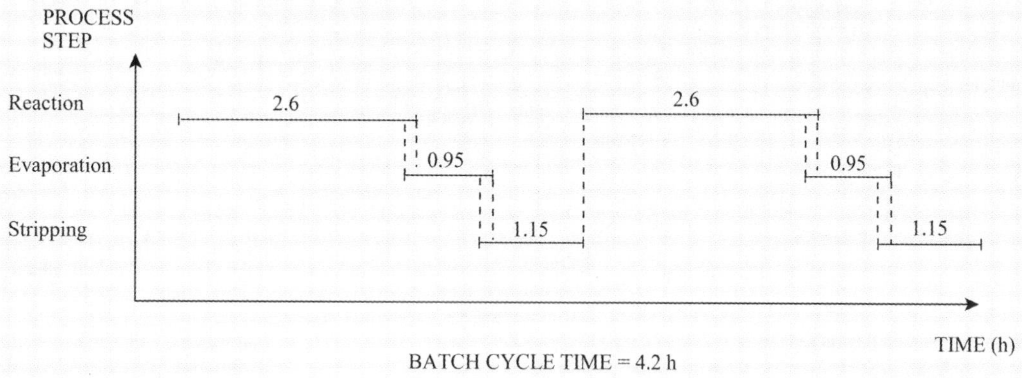

Figure 16.24

Gantt chart for a repeated batch cycle for Example 16.7.

16

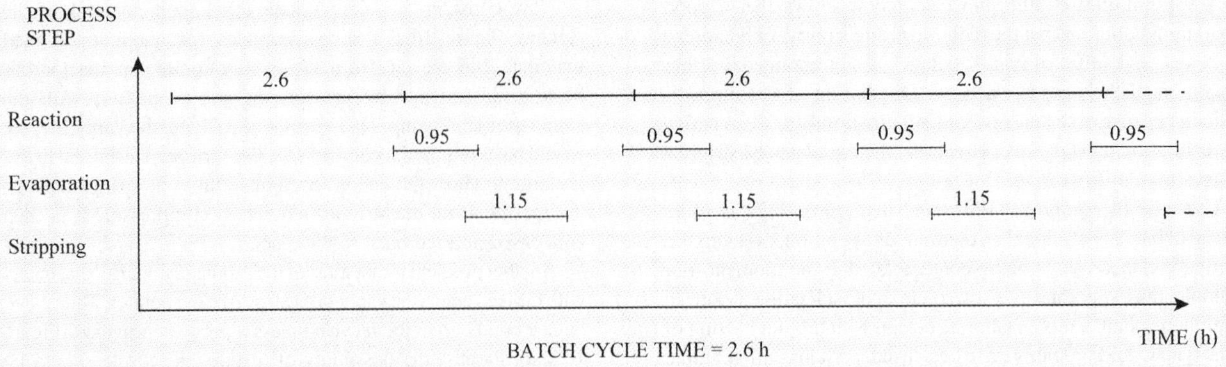

Figure 16.25

Overlapping batches for Example 16.7 reduces the batch cycle time.

Perhaps the additional capacity might not be needed. If it is not needed then the size of the reactors, evaporator and stripper can be reduced. Keeping the original process capacity with parallel operation of the reactors would mean a trade-off between the increased capital cost of two (smaller) reactors versus reduced capital cost of the evaporator and stripper. An economic comparison would be required to judge whether this would be beneficial.

Another option to improve utilization of equipment is, instead of adding a reactor in parallel, installing intermediate storage. Figure 16.28 shows a production schedule with intermediate storage between the reactor and evaporator and between the reactor and stripper. The evaporation step is no longer constrained to start on completion of the reaction step and start the stripping step on completion of the evaporation step. The individual steps can be decoupled via the intermediate storage. This maintains the original batch cycle time of 2.6 h but allows, as shown in Figure 16.28, the elimination of dead time in the evaporation and stripping steps. Now more evaporation and stripping steps can be carried out and the size of the evaporator and stripper reduced accordingly. This time the capital cost of intermediate storage is traded off against reduced capital cost of the evaporator and stripper. In Figure 16.28, the intermediate storage between the reactor and evaporator has a significant effect on equipment utilization. The intermediate storage between the evaporator and stripper has a less pronounced effect and would be more difficult to justify economically.

Finally, merging of operations into the same equipment could be considered to replace connections through space by those through time. For example, it might be possible to carry out the reaction and evaporation in the same equipment. Overlapping of the reaction and evaporation would no longer be possible, but there would be savings in capital cost.

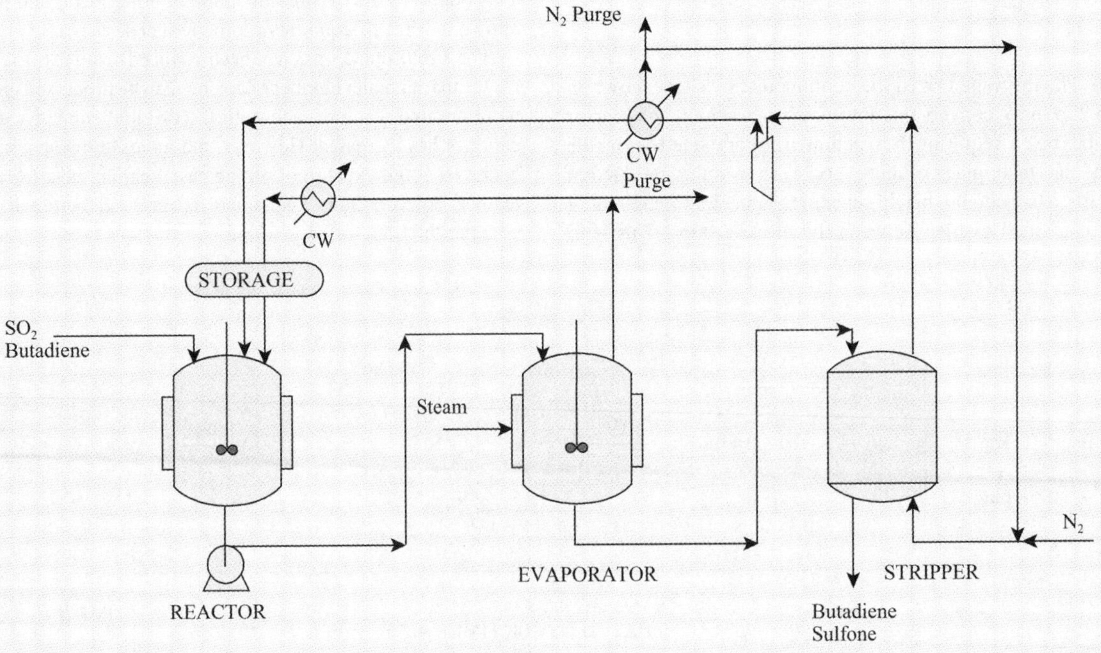

Figure 16.26

Flowsheet for the production of butadiene sulfone in a batch process.

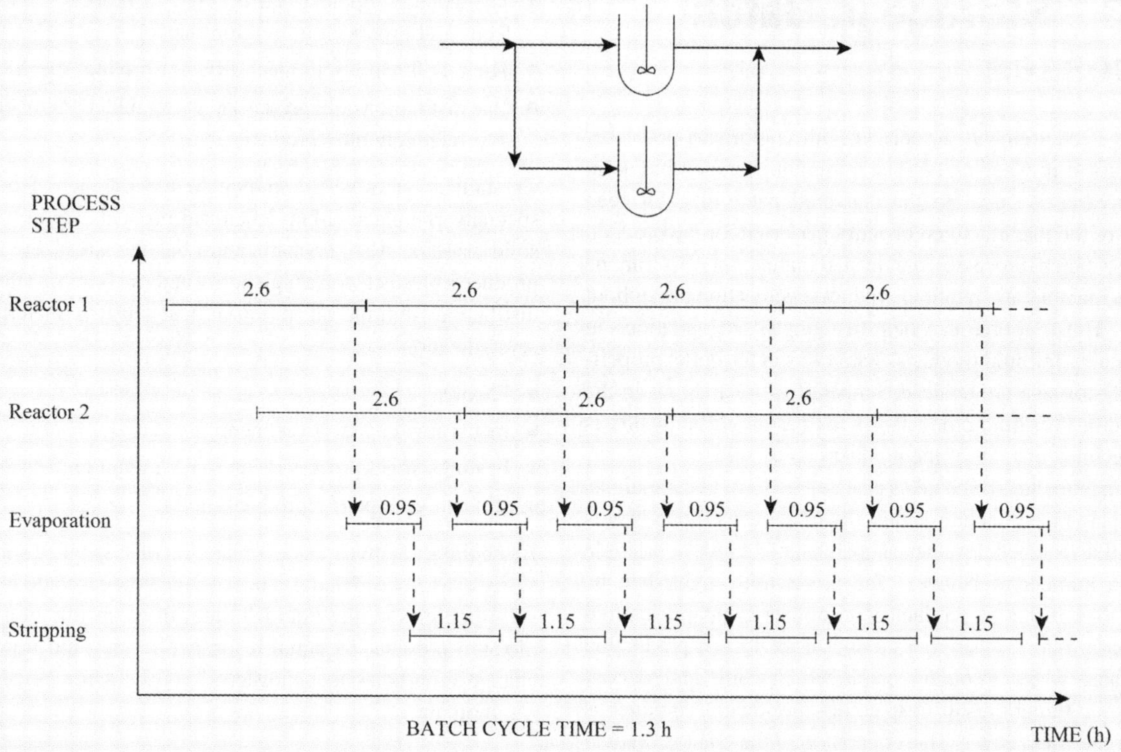

Figure 16.27

Placing units in parallel for the limiting step reduces the batch cycle time for Example 16.7.

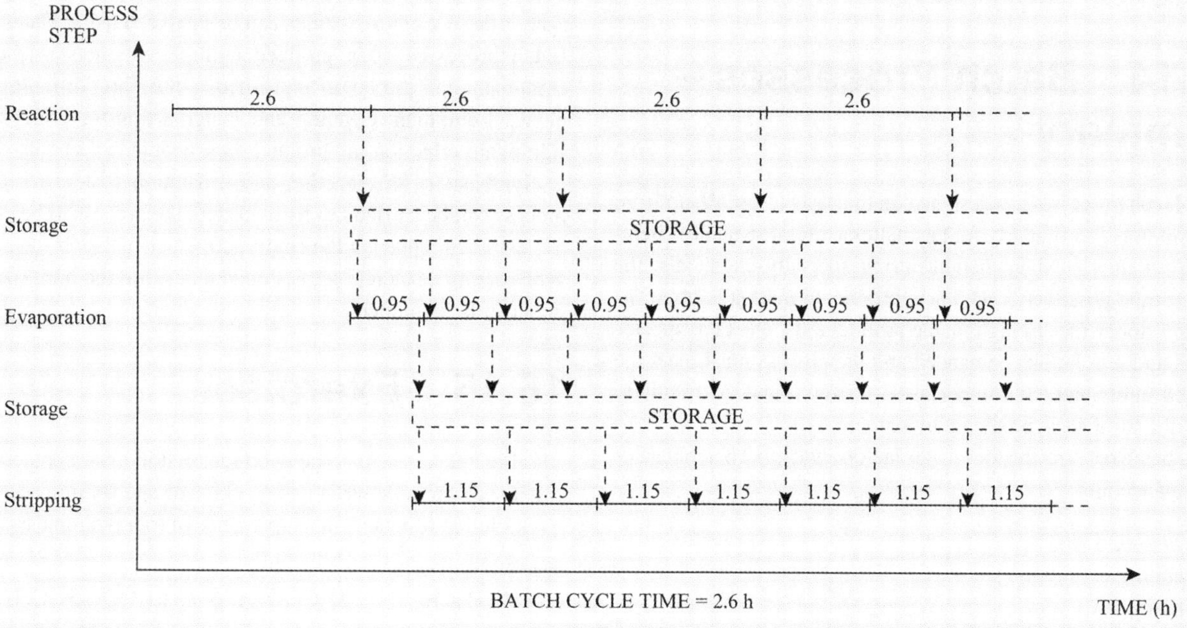

Figure 16.28

Gantt chart for Example 16.7 with intermediate storage.

16

16.13 Storage in Batch Processes

The storage requirements associated with continuous processes were discussed in Chapter 14. The minimum storage volume in batch processes is equal to the delivered batch size for feed materials, or the manufactured quantity per batch for the product.

Because batch processes are in a dynamic state, it can be difficult to maintain the required product specification throughout the batch. Thus, storage can help to dampen out variations in product quality for fluid products. Variations in the outlet properties from the storage are reduced compared with variations in the inlet properties for fluid products. As long as the mean quality is within specifications, all of the product can be sold.

Batch manufacture often takes place in the form of campaigns. Each campaign manufactures a quantity of a given product that is stored until dispatched. Product campaigns might involve a single batch or a number of batches. The production might then be switched to a different product and another campaign carried out for that product. Changing product leads to lost production time, creates waste through equipment cleaning and decontamination and might result in some off-specification product as a result of the changeover. Thus, operation of the plant is favored by large batches and long production campaigns, but this increases storage costs (capital, working and operating).

As with storage for continuous processes, storage tanks for liquids should not be designed to operate with less that 10% inventory at the minimum or more than 90% full at the maximum. As with storage for continuous processes, the amount of feed product and intermediate storage will depend on capital, working and operating costs, together with operability, control and safety considerations.

16.14 Batch Processes – Summary

Many batch processes are designed on the basis of a scale-up from the laboratory, particularly for the manufacture of specialty chemicals. Particularly when manufacturing specialty chemicals, the shortest time possible to get a new product to market is often the biggest priority (accepting that the product must meet the specifications and regulations demanded and the process must meet the required safety and environmental standards). This is particularly true if the product is protected by patent.

For batch reactors the four major factors that affect batch reactor performance are:

- Contacting pattern
- Operating conditions
- Agitation
- Solvent selection.

Batch distillation has a number of advantages when compared with continuous distillation:

- The same equipment can be used to process many different feeds and produce different products.
- There is flexibility to meet different product specifications.
- One distillation column can separate a multicomponent mixture into relatively pure products.

Crystallization is very common in the production of fine and specialty chemicals. Many chemical products are in the form of solid crystals. Also, crystallization has the advantage that it can produce a product with a high purity and can be more effective than distillation from the separation of heat-sensitive materials. Batch crystallization is:

- Flexible
- Incurs low capital investment
- Features small process development requirements
- But can give poor reproducibility.

Batch filtration involves the separation of suspended solids from a slurry of associated liquid. The required product could be either the solid particles or the liquid filtrate. In batch filtration, the filter medium presents an initial resistance to the fluid flow that will change as particles are deposited. The driving forces used in batch filtration are:

- gravity,
- vacuum,
- pressure,
- centrifugal.

Batch heating and cooling can be analyzed for agitated vessels for both constant temperature and variable temperature heating and cooling media.

Production schedules for batch processes can be sequential, overlapping, parallel, use intermediate storage, or use a combination of these. Such schedules can be analyzed using Gantt charts. Batch processes often produce multiple products in the same equipment and can be distinguished as flowshop or multiproduct plants. Equipment cleaning and material transfer policy have a significant effect on the production schedule.

Synthesis of the reaction and separation system for a batch process can be carried out by assuming the process to be continuous and then replacing the continuous steps by batch steps. When synthesizing a batch process, connections through both space and time can be exploited.

16.15 Exercises

1. Benzyl acetate is to be manufactured in a batch reactor from the reaction between benzyl chloride and sodium acetate in a solution of xylene in the presence of triethylamine as catalyst according to the reaction:

$$C_6H_5CH_2Cl + CH_3COONa \rightarrow CH_3COOC_6H_5CH_2 + NaCl$$

or

$$A + B \rightarrow C + D$$

The kinetic model for the reaction is given by

$$-r_A = k_A C_A$$

where r_A = rate of reaction of benzyl chloride
k_A = reaction rate constant
= $0.01306\ h^{-1}$
C_A = molar concentration of benzyl chloride

Assuming no products are present in the reactor feed, calculate the residence time required for a conversion of 40% and 60%.

2. A product C is to be produced in a batch reaction according to:

$$A + B \rightleftharpoons C + D$$

The reactor is to be fed with a large excess of B. The feed ratio to the reactor is to be 5 kmol B per kmol A. Under these conditions the reaction rate can be described by:

$$r_A = -kC_A^2$$

where r_A = rate of reaction
k = reaction rate constant
= $0.0174\ m^3 \cdot kmol^{-1} \cdot min^{-1}$
C_A = concentration of A ($kmol \cdot m^{-3}$)

The density of the reaction mixture can be assumed constant at $10.75\ kmol \cdot m^{-3}$.

a) Calculate the time required to reach 50% conversion of A.

b) If the reactor is shut down for 30 min between batches, calculate the reaction volume required to produce $10\ kmol \cdot h^{-1}$ of C.

3. Ethyl acetate is to be manufactured in a batch reactor from the reaction between ethanol and acetic acid in the liquid phase according to the reaction:

$$\underset{\text{acetic acid}}{CH_3COOH} + \underset{\text{ethanol}}{C_2H_5OH} \rightleftharpoons \underset{\text{ethyl acetate}}{CH_3COOC_2H_5} + \underset{\text{water}}{H_2O}$$

or

$$A + B \rightleftharpoons C + D$$

The reaction rate expression is of the form

$$-r_A = k_A C_A^2 - k_A' C_C$$

where $k_A = 0.002688\ m^3 \cdot kmol^{-1} \cdot min^{-1}$
$k_A' = 0.004644\ min^{-1}$

Molar masses and densities for the components are given in Table 16.12.

The reactants are equimolar with product concentrations of initially zero and reactor volume is constant.

a) Calculate the residence time required for a conversion of 60%.

b) The operating schedule for the reactor requires it to be shut down between batch reactions for 1 h. Calculate the volume of the reactor required to produce $10\ t \cdot d^{-1}$ based on a conversion of 60% and operation of 24 h per day based on the volume of the feed.

Table 16.12

Molar masses for the components.

Component	Molar mass ($kg \cdot kmol^{-1}$)	Density at 60 °C ($kg \cdot m^{-3}$)
Acetic acid	60.05	1018
Ethanol	46.07	754
Ethyl acetate	88.11	847
Water	18.02	980

4. A binary mixture of 125 kmol of components A and B is to be separated by batch distillation. The feed mixture contains a mole fraction of A of 65%. The relative volatility between A and B is 2.75. The distillation is required to recover 90% of A. If only a batch distillation pot is used (i.e. no distillation trays and no reflux), calculate the composition of the recovered A.

a) Using an analytical solution of the Rayleigh Equation.

b) Using a numerical solution of the Rayleigh Equation.

5. The mixture from Exercise 4 is to be separated in a batch rectifier with the equivalent of six theoretical stages. The vapor is generated in the pot at a rate of $95\ kmol \cdot h^{-1}$.

a) Calculate the recovery of A and the batch cycle time required to recover A with a purity of 98% using a constant reflux ratio of 6.

b) Calculate the reflux ratio to minimize the batch cycle time to recover 90% of A with a purity of 98%.

6. A batch process consists of the four steps given in Table 16.13. For repeated batch cycles of the same process, calculate the cycle time for:

a) a sequential production schedule;

b) overlapping batches.

7. A batch process consists of three steps given in Table 16.14. For repeated cycles of the same process, calculate the cycle time for:

a) a sequential production schedule;

b) overlapping batches;

Table 16.13

Batch steps for Exercise 6.

Step	Duration (h)
A	0.75
B	3
C	0.75
D	6

Table 16.14

Batch steps for Exercise 7.

Step	Duration (h)
A	12
B	5
C	5

Table 16.15

Batch steps for Exercise 8.

Step	Product 1 (h)	Product 2 (h)
A	12	4
B	5	3
C	5	3

c) two parallel units for the limiting step;
d) intermediate storage for the limiting step.

8. A batch process manufactures Product 1 and Product 2 in the same process. The manufacture of both products involves three steps with durations given in Table 16.15. Calculate the cycle time and makespan for one batch each of Product 1 and Product 2 with no delay between the two batches for:

a) a sequential production schedule with zero-wait transfer;
b) overlapping batches with zero-wait transfer;
c) overlapping batches with hold-up;
d) overlapping batches with intermediate storage.

9. The following process is proposed for the production of Product C. For the scheduling of the production campaign, overlapping is allowed and a zero-wait transfer is applied for storage policy. The process is represented in Figure 16.29.

Stage I

- Liquid raw materials *A* and *B* are reacted in a batch reactor for 6 h.
- 1 kg of *A* and 1 kg of *B* are mixed to produce *C*.
- A reactor is operated with 70% of yield by mass.
- The mixture density is 800 kg·m^{-3}.

Stage II

- The liquid mixture of *A*, *B* and *C* is fed into a tank and mixed with a solvent for 3 h. The product *C* is converted into solids.
- 1 kg of solvent is added and mixed with *A*, *B* and *C*.
- The mixture density is 950 kg·m^{-3}.

Stage III

- The product *C* is separated by centrifuge for 4 h.
- 90% recovery by mass for product *C* is obtained.
- The mixture density is 900 kg·m^{-3}.

a) Calculate the cycle time for the production of Product *C* when only one unit is used for each stage. Use a Gantt chart to show at least two production batches.
b) The current production campaign must produce 100,000 kg·y^{-1} of product *C*. The plant is operated for 7200 h·y^{-1}. What is the size of the reactor (Stage I) required? (Use the cycle time calculated in Part a).
c) The addition of parallel units can increase the efficiency of equipment utilization. When two reactors are operated for Stage I, one tank for Stage II, and two centrifuges for Stage III, how is the cycle time changed? Use a Gantt chart to show at least four production batches.

10. The plant from Exercise 9 is modified to produce Products *D* and *E* according to the schedule in Table 16.16.

a) When no intermediate storage is applied as the storage policy, calculate the cycle time for the production of *D* and *E* in a sequence *DEDEDE* . . . . Use a Gantt chart to show at least two cycles of production (i.e. *DEDE*).
b) When unlimited intermediate storage is applied for storage policy, calculate the cycle time for the production of *D* and *E* in a sequence *DEEDEEDEE* . . . . Use a Gantt chart to show at least two cycles of production (i.e. *DEEDEE*).

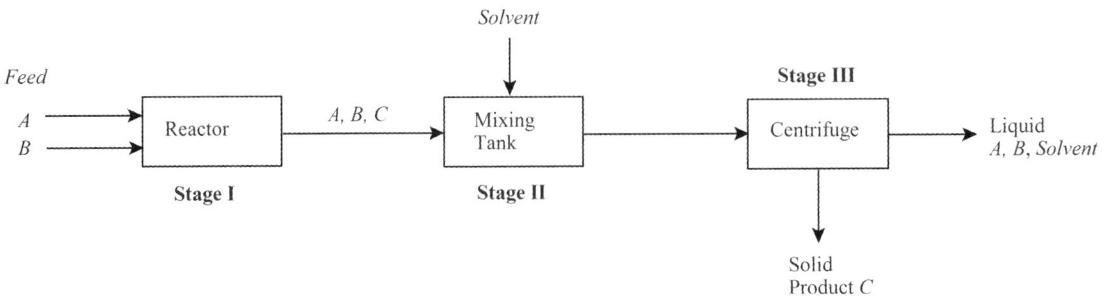

Figure 16.29

A simple batch process involving three steps.

Table 16.16

Modified schedule.

	Product	Stage I	Stage II	Stage III
Processing times (h)	*D*	8	3	5
	E	4	2	4

References

Biegler LT, Grossmann IE and Westerberg AW (1997) *Systematic Methods of Chemical Process Design*, Prentice Hall.

Bowman RA, Mueller AC and Nagle WM (1940) Mean Temperature Difference in Design, *Trans ASME*, **62**: 283.

Carpenter KJ (1997) Agitation, Chapter 4, in Sharratt PN (ed.), *Handbook of Batch Process Design*, Chapman and Hall.

Choong KL and Smith R (2004a) Optimisation of Batch Cooling Crystallization, *Chem Eng Sci*, **59**: 313.

Choong KL and Smith R (2004b) Novel Strategies for Optimization of Batch, Semi-batch and Heating/Cooling Evaporative Crystallization, *Chem Eng Sci*, **59**: 329.

Diweka UM (2011) *Batch Distillation – Simulation, Optimal Design and Control*, 2nd Edition, CRC Publishers.

Diwekar UM and Madhavan KP (1991) Multicomponent Batch Distillation Column Design, *Ind Eng Chem Res*, **30**: 713.

Eduljee HE (1975) Equations Replace Gilliland's Plot, *Hydrocarbon Processing*, **54**: 120.

Fisher RC (1944) Heating and Cooling in Circulating Systems, *Ind Eng Chem*, **36**: 939.

Harnby N, Nienow AW and Edwards MF (1997) *Mixing in the Process Industries*, Butterworth-Heineman.

Jain S, Kim J-K and Smith R (2012) Operational Optimization of Batch Distillation Systems, *Ind Eng Chem Res*, **51**: 5749–5761.

Kern DQ (1950) *Process Heat Transfer*, McGraw-Hill, New York.

Lehrer IH (1970) Jacket Side Nusselt Number, *Ind Eng Chem Proc Dec*, **9**: 533.

Mah RSH (1990) *Chemical Process Structures and Information Flows*, Butterworth.

McKetta JJ (1977) *Encyclopedia of Chemical Processing and Design*, Vol. **5**, Marcel Dekker Inc, New York.

Mehta VL and Kokossis AC (1988) New Generation Tools for Multiphase Reaction Systems: A Validated Systematic Methodology for Novelty and Design Automation, *Comp Chem Eng*, **22S**: 5119.

Myriantheos CM (1986) *Flexibility Targets for Batch Process Design*, MS Thesis, University of Massachusetts, Amherst.

Rushton A and Griffiths P (1987) *Filtration Principles and Practice*, Matteson MJ (ed), 2nd Edition, Marcel Dekker, New York.

Sharratt PN (1997) *Handbook of Batch Process Design*, Blackie Academic and Professional.

Smith R and Petela EA (1992) Waste Minimisation in the Process Industries: Part 2 – Reactors, *Chem Eng*, **12**, 509510:12.

Smoker EH (1938) Analytical Determination of Plates in a Fractionating Column, *Trans AIChE*, **34**: 165.

Svarovsky L (1981) *Solid–Liquid Separation*, 2nd Edition, Butterworth, London.

Zhang J and Smith R (2004) Design and Optimisation of Batch and Semi-batch Reactors, *Chem Eng Sci*, **59**: 459.

Chapter 17

Heat Exchanger Networks I – Network Targets

Fixing the major processing steps (reactors, separators and recycles) fixes the material and energy balance. Thus, the heating and cooling duties for the heat recovery system are known. However, completing the design of the heat exchanger network is not necessary in order to assess energy performance of the completed design. *Targets* can be set for the heat exchanger network to assess the energy performance of the complete process design without actually having to carry out the network design. Moreover, the targets allow the designer to suggest process changes for the reactor and separation and recycle systems to improve the targets for energy cost.

Using targets for the heat exchanger network, rather than designs, allows many design options for the overall process to be screened quickly and conveniently. Screening many design options by completed designs is often not practical in terms of the time and effort required. In Chapters 19 to 22, energy targets will be used to suggest design improvements to the reaction, separation and recycle systems.

17.1 Composite Curves

The analysis of the heat exchanger network first identifies sources of heat (termed *hot streams*) and sinks (termed *cold streams*) from the material and energy balance. Consider first a very simple problem with just one hot stream (heat source) and one cold stream (heat sink). The initial temperature (termed *supply temperature*), final temperature (termed *target temperature*) and enthalpy change of both streams are given in Table 17.1.

Steam is available at 180 °C and cooling water at 20 °C. Clearly, it is possible to heat the cold stream in Table 17.1 using steam and cool the hot stream using cooling water. However, this would incur excessive energy cost. It is also incompatible with the goals of sustainable industrial activity, which call for use of the minimum energy consumption. Instead, it is preferable to try to recover the heat between process streams, if this is possible. The scope for heat recovery can be determined by plotting both streams from Table 17.1 on temperature–enthalpy axes. For feasible heat exchange between the two streams, the hot stream must be at all points hotter than the cold stream. Figure 17.1a shows the temperature–enthalpy plot for this problem with a minimum temperature difference (ΔT_{min}) of 10 °C. The region of overlap between the two streams in Figure 17.1a determines the amount of heat recovery possible (for $\Delta T_{min} = 10$ °C). For this problem, the heat recovery (Q_{REC}) is 11 MW. The part of the cold stream that extends beyond the start of the hot stream in Figure 17.1a cannot be heated by recovery and requires steam. This is the *hot utility target* or *energy target* (Q_{Hmin}), which for this problem is 3 MW. The part of the hot stream that extends beyond the start of the cold stream in Figure 17.1a cannot be cooled by heat recovery and requires cooling water. This is the minimum cold utility (Q_{Cmin}), which for this problem is 1 MW. Also shown at the bottom of Figure 17.1a is the arrangement of heat exchangers corresponding with the temperature–enthalpy plot.

The temperatures or enthalpy change for the streams (and hence their slope) cannot be changed, but the relative position of the two streams can be changed by moving them horizontally relative to each other. This is possible since the reference enthalpy for the hot stream can be changed independently from the reference enthalpy for the cold stream, but the enthalpy difference across the streams cannot be changed. Figure 17.1b shows the same two streams moved to a different relative position such that ΔT_{min} is now 20 °C. The amount of overlap between the streams is reduced (and hence heat recovery is reduced) to 10 MW. A greater amount of the cold stream now extends beyond the start of the hot stream, and hence the amount of steam is increased to 4 MW. Also, more of the hot stream extends beyond the start of the cold stream, increasing the cooling water demand to 2 MW. Thus, the approach of plotting a hot and a cold stream on the same temperature–enthalpy axes can determine hot and cold utility for a given value of ΔT_{min}.

The importance of ΔT_{min} is that it sets the relative location of the hot and cold streams in this two-stream problem, and therefore the amount of heat recovery. Setting the value of ΔT_{min} or Q_{Hmin} or Q_{Cmin} sets the relative location and the amount of heat recovery.

Chemical Process Design and Integration, Second Edition. Robin Smith.

Companion Website: www.wiley.com/go/smith/chemical2e

Table 17.1

Two-stream heat recovery problem.

Stream	Type	Supply temperature T_S (°C)	Target temperature T_T (°C)	ΔH (MW)
1	Cold	40	110	14
2	Hot	160	40	−12

Consider now the extension of this approach to several hot and cold streams. Figure 17.2 shows a simple flowsheet. Temperatures and heat duties for each stream are shown. Two of the streams in Figure 17.2 are sources of heat (hot streams) and two are sinks for heat (cold streams). Assuming that the heat capacities are constant, the data for the hot and cold streams can be extracted as given in Table 17.2. Note that the heat capacities (CP) are total heat capacities, being the product of mass flowrate and specific heat capacity ($CP = m\ C_P$). Had the heat capacities varied significantly, the nonlinear temperature–enthalpy behavior could have been represented by a series of linear *segments* (see Chapter 19).

Instead of dealing with individual streams as given in Table 17.1, an overview of the process is needed. Figure 17.3a shows the two hot streams individually on temperature–enthalpy axes. How these hot streams behave overall can be quantified by combining them in the given temperature ranges (Hohman, 1971; Huang and Elshout, 1976; Linnhoff, Mason and Wardle, 1979). The temperature ranges in question are defined by changes in the overall rate of change of enthalpy with temperature. If heat capacities are constant, then changes will occur only when streams start or finish. Thus, in Figure 17.3, the temperature axis is divided into ranges defined by the supply and target temperatures of the streams.

Within each temperature range, the streams are combined to produce a composite hot stream. This composite hot stream has a CP in any temperature range that is the sum of the individual streams. Also, in any temperature range, the enthalpy change of the composite stream is the sum of the enthalpy changes of the individual streams. Figure 17.3b shows the *composite curve* of the hot streams (Hohman, 1971; Huang and Elshout, 1976; Linnhoff, Mason and Wardle, 1979). The composite hot stream is a single stream that is equivalent to the individual hot streams in terms of temperature and enthalpy. Similarly, the composite curve

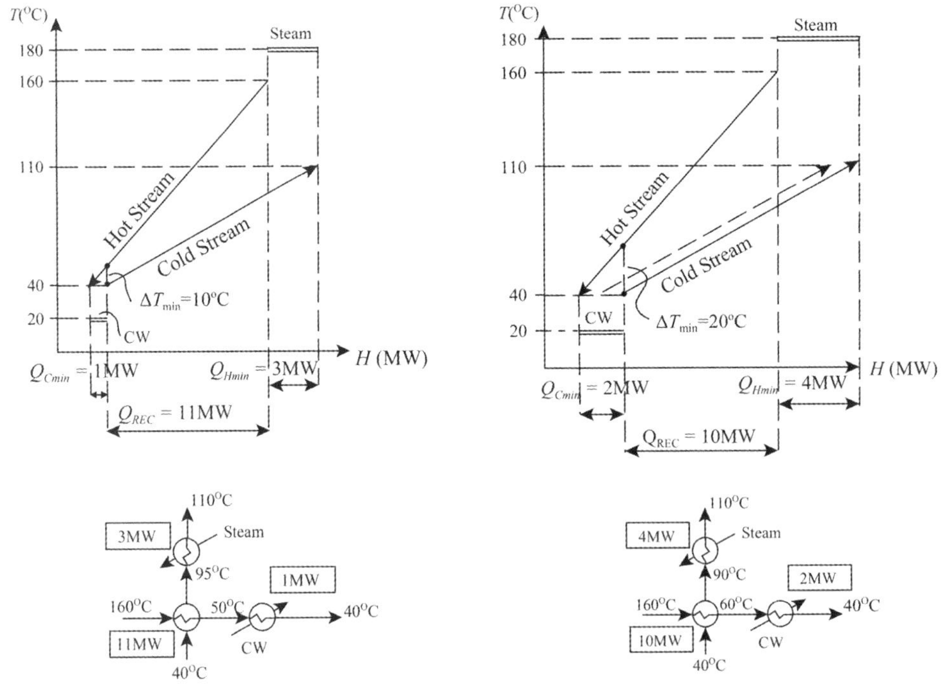

(a) Recovery for ΔT_{min} = 10°C. (b) Recovery for ΔT_{min} = 20°C.

Figure 17.1

A simple heat recovery problem with one hot stream and one cold stream.

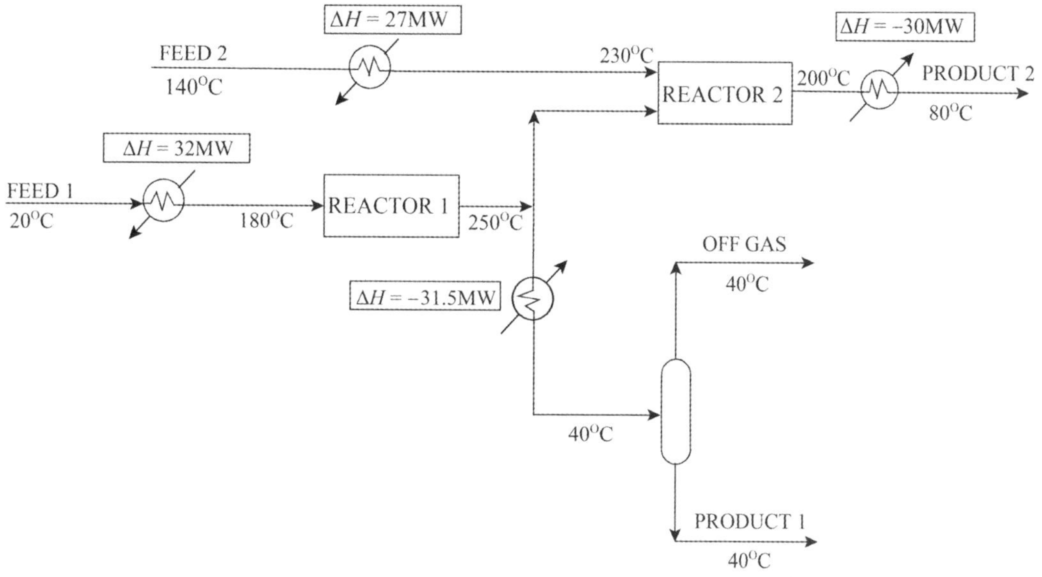

Figure 17.2

A simple flowsheet with two hot streams and two cold streams.

Table 17.2

Heat exchange stream data for the flowsheet in Figure 17.2.

Stream	Type	Supply temperature T_S (°C)	Target temperature T_T (°C)	ΔH (MW)	Heat capacity flowrate CP ($MW \cdot K^{-1}$)
1. Reactor 1 feed	Cold	20	180	32.0	0.2
2. Reactor 1 product	Hot	250	40	−31.5	0.15
3. Reactor 2 feed	Cold	140	230	27.0	0.3
4. Reactor 2 product	Hot	200	80	−30.0	0.25

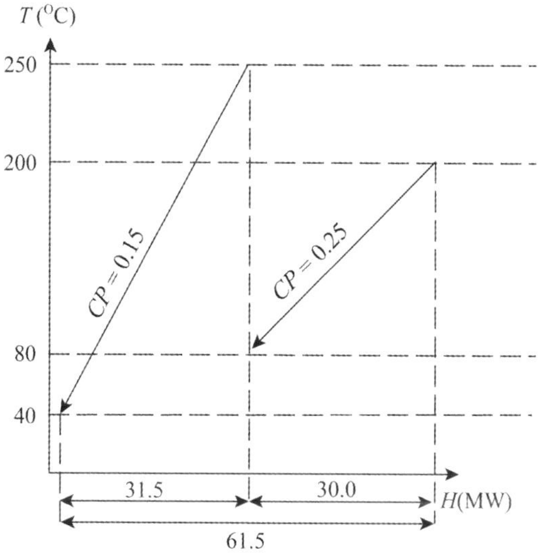

(a) The hot streams plotted separately.

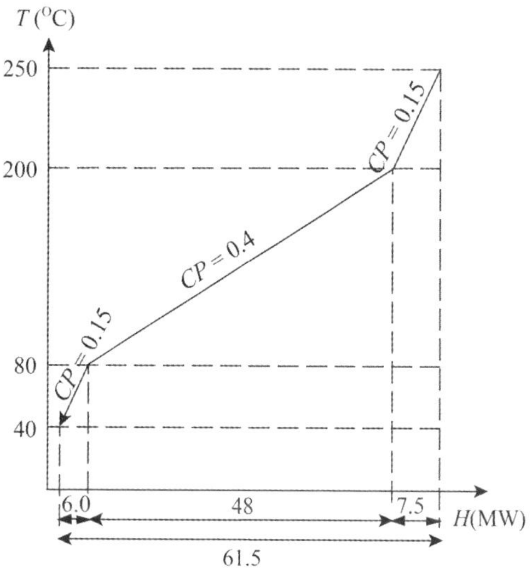

(b) The composite hot stream.

Figure 17.3

The hot streams can be combined to obtain a composite hot stream.

17

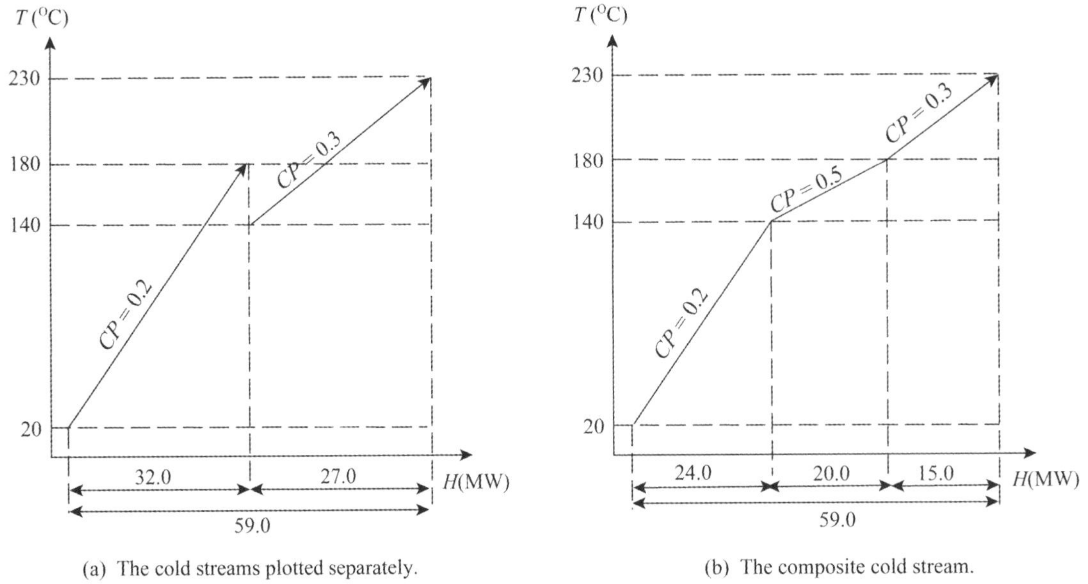

Figure 17.4

The cold streams can be combined to obtain a composite cold stream.

of the cold streams for the problem can be produced, as shown in Figure 17.4. Again, the composite cold stream is a single stream that is equivalent to the individual cold streams in terms of temperature and enthalpy.

The composite hot and cold curves can now be plotted on the same axes, as in Figure 17.5. Plotting the composite hot and cold curves is analogous to plotting the single hot and cold streams in Figure 17.1. The composite curves in Figure 17.5a are set to have a minimum temperature difference (ΔT_{min}) of 10 °C. Where the curves overlap in Figure 17.5a, heat can be recovered vertically from the hot streams that comprise the hot composite curve into the cold streams that comprise the cold composite curve. The way in which the composite curves are constructed (i.e. monotonically decreasing hot composite curve and monotonically increasing cold composite curve) allows maximum overlap between the curves and hence maximum heat recovery. Maximizing the energy recovery thereby minimizes the external requirements for heating and cooling duties and minimizes the energy consumption. In this problem, for $\Delta T_{min} = 10$ °C, the maximum heat recovery (Q_{REC}) is 51.5 MW.

Where the cold composite curve extends beyond the start of the hot composite curve in Figure 17.5a, heat recovery is not possible, and the cold composite curve must be supplied with an external hot utility such as steam. This represents the target for hot utility (Q_{Hmin}). For this problem, with $\Delta T_{min} = 10$ °C, $Q_{Hmin} = 7.5$ MW. Where the hot composite curve extends beyond the start of the cold composite curve in Figure 17.5a, heat recovery is again not possible and the hot composite curve must be supplied with an external cold utility such as cooling water. This represents the target for cold utility (Q_{Cmin}). For this problem, setting $\Delta T_{min} =$ 10 °C gives $Q_{Cmin} = 10.0$ MW.

Specifying the hot utility or cold utility heat duty or ΔT_{min} fixes the relative position of the two curves. As with the simple problem in Figure 17.1, the relative position of the two curves is a degree of freedom (Linnhoff *et al.*, 1982). Again, the relative position of the two curves can be changed by moving them horizontally relative to each other. Clearly, to consider heat recovery from hot streams into cold streams, the hot composite curve must be in a position such that it is always above the cold composite curve for feasible heat transfer. Thereafter, the relative position of the curves can be chosen. Figure 17.5b shows the curves with $\Delta T_{min} = 20$ °C. The hot and cold utility targets are now increased to 11.5 MW and 14 MW respectively.

If utility consumption is to be minimized, the composite curves can be set to their practical minimum, which will be dictated by the type of heat transfer equipment used for the heat transfer in the region of the pinch. To achieve a small ΔT_{min} in a design requires heat exchangers that exhibit pure countercurrent flow. With shell-and-tube heat exchangers this is not possible, even if single-shell pass and single-tube pass designs are used, because the shell-side stream takes periodic cross-flow. Consequently, operating with a ΔT_{min} less than 10 °C should be avoided, unless under special circumstances (Polley, 1993). A smaller value of 5 °C or less can be achieved with plate heat exchangers, and the value can go as low as 1 °C with plate-fin designs (Polley, 1993). For high-temperature processes, typically involving heat transfer with a flue gas, ΔT_{min} less than 20 °C should be avoided. It should be noted, however, that such constraints only apply to the heat transfer equipment that operates around the point of closest approach between the composite curves. Additional constraints apply if vaporization or condensation is occurring at the point of closest approach (see Chapter 12).

Figure 17.6 illustrates what happens to the annualized cost of the system as the relative position of the composite curves is changed over a range of values of ΔT_{min}. When the curves just touch, there is no driving force for heat transfer at one point in the

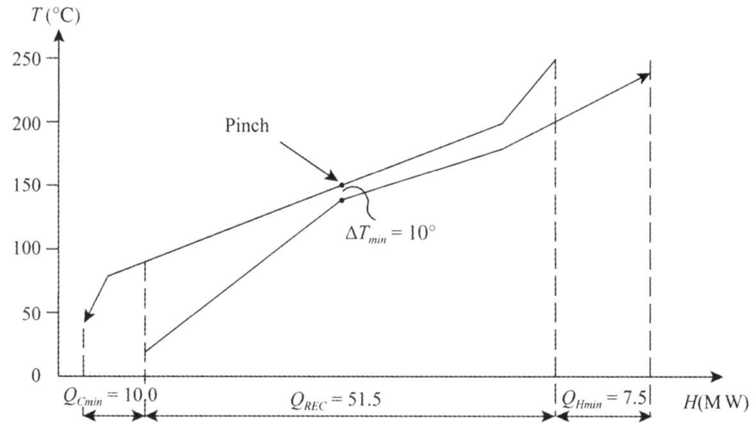

(a) The hot and cold composite curves plotted together at ΔT_{min}=10°C.

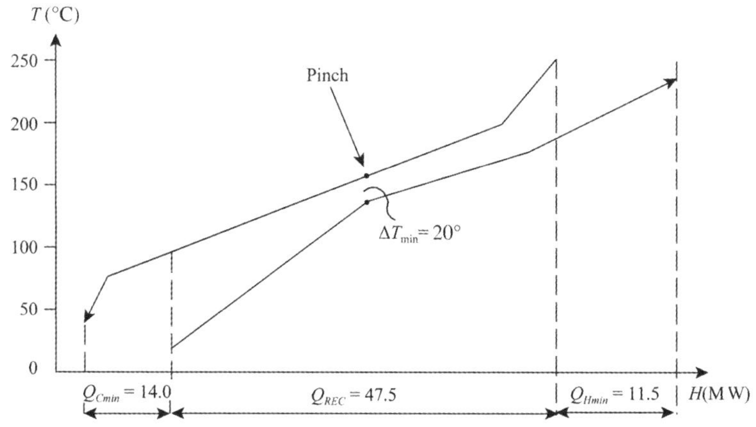

(b) Increasing ΔT_{min} from 10°C to 20°C increases the hot and cold utility targets.

Figure 17.5

Plotting the hot and cold composite curves together allows the targets for hot and cold utility to be obtained.

process, which would require infinite heat transfer area and hence infinite capital cost. As the energy target (and hence ΔT_{min} between the curves) is increased, the annualized capital cost decreases. This results from increased temperature differences throughout the process, decreasing the heat transfer area. On the other hand, the cost of utilities increases as ΔT_{min} increases. There is a trade-off between utility cost and annualized capital cost and an economic amount of energy recovery. It should be noted that the cost of either heating or cooling can dominate the utility cost. For high-temperature processes it tends to be the cost of supplying heat that dominates the trade-off. In low-temperature processes where cooling is required below ambient temperature to be serviced by expensive refrigeration, it tends to be the cost of supplying cooling that dominates the trade-off. The utilities versus capital cost trade-off is also very dependent on the annualization of the capital cost. If a long period is chosen for the capital cost annualization, this forces the optimum point to a lower value of ΔT_{min}.

In practice, the shape of the economic trade-off is in most cases quite flat in the region of the optimum. This means that in most cases, as long as a reasonable value is taken somewhere in the flat region of the optimum, no serious errors will result. Also, the network design to be developed later will be subjected to optimization (see Chapter 18). A typical value appropriate for most above ambient temperature processes is 10 °C. In some cases, especially where heat transfer coefficients are low, values as high as 20 °C are sometimes used. For low-temperature processes requiring moderate temperature refrigeration, a value of 5 °C is reasonable, decreasing to 1 °C for processes requiring very low-temperature refrigeration.

17.2 The Heat Recovery Pinch

Once a value of ΔT_{min} has been chosen, this fixes the relative position of the composite curves and hence the energy target. The value of ΔT_{min} and its location between the composite curves have important implications for design, if the energy target is to be achieved in the design of a heat exchanger network. The ΔT_{min} is normally observed at only one point between the hot and the cold composite curves, called the *heat recovery pinch* (Umeda, Itoh and

17

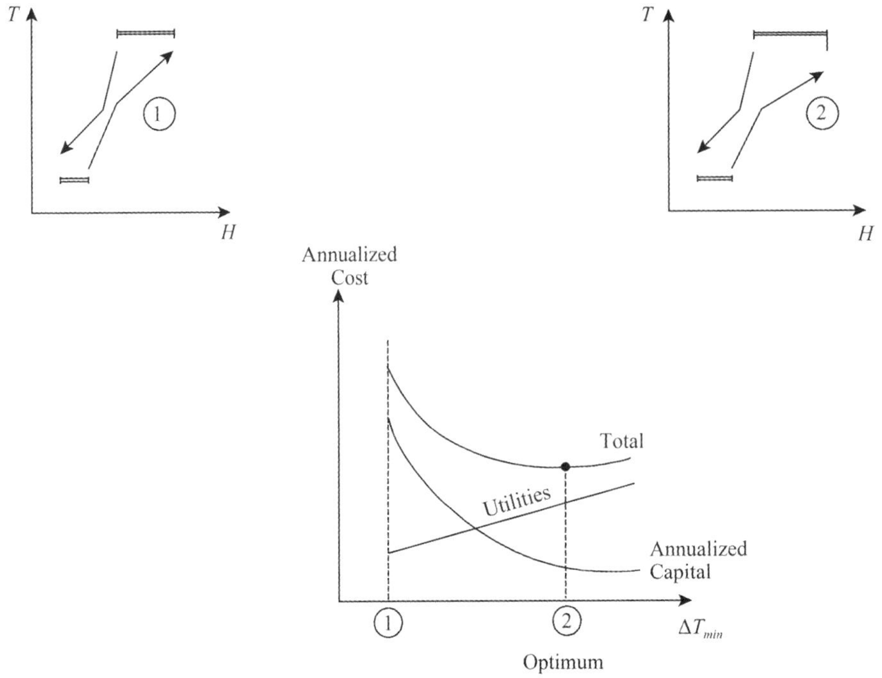

Figure 17.6
The correct setting for ΔT_{min} is fixed by economic trade-offs.

Shiroko, 1978; Umeda, Harada and Shiroko, 1979; Umeda, Niida and Shiroko, 1979; Linnhoff, Mason and Wardle, 1979). The pinch point occurs between the composite curves where there is either a change in slope on the hot composite curve or a change in slope on the cold composite curve. If the pinch occurs at a change in slope on the hot composite curve, the slope moving down the curve becomes less steep at the pinch. This is where a hot stream starts. Alternatively, if the pinch occurs at a change in slope on the cold composite curve, the slope moving up the curve at the pinch also becomes less steep. This is where a cold stream starts. So the pinch point occurs where a hot stream starts or a cold stream starts. The pinch point has a special significance.

The trade-off between energy and capital in the composite curves suggests that, "on average", individual exchangers should have a temperature difference no smaller than ΔT_{min}. A good initialization in heat exchanger network design is to assume that no individual heat exchanger has a temperature difference smaller than the ΔT_{min} between the composite curves.

With this rule in mind, divide the process at the pinch as shown in Figure 17.7a. Above the pinch (in temperature terms) the process is in heat balance with the minimum hot utility, Q_{Hmin}. Heat is received from hot utility and no heat is rejected. The process acts as a heat sink. Below the pinch (in temperature terms), the process is in heat balance with the minimum cold utility, Q_{Cmin}. No heat is received but heat is rejected to cold utility. The process acts as a heat source.

Consider now the possibility of transferring heat between these two systems. Figure 17.7b shows that it is possible to transfer heat from hot streams above the pinch to cold streams below it. The pinch temperature for hot streams for the problem is 150 °C and for cold streams 140 °C. Transfer of heat from above the pinch to below, as shown in Figure 17.7b, means transfer of heat from hot streams with a temperature of 150 °C or greater into cold streams with a temperature of 140 °C or less. This is clearly possible. By contrast, Figure 17.7c shows that heat transfer from hot streams below the pinch to cold streams above is not possible. Such transfer requires heat being transferred from hot streams with a temperature of 150 °C or less into cold streams with a temperature of 140 °C or greater. This is clearly not possible (without violating the ΔT_{min} constraint).

If an amount of heat *XP* is transferred from the system above the pinch to the system below the pinch, as in Figure 17.8a, this will create a deficit of heat *XP* above the pinch and an additional surplus of heat *XP* below the pinch. The only way this can be corrected is by importing an extra *XP* amount of heat from hot utility and exporting an extra *XP* amount of heat to cold utility (Linnhoff, Mason and Wardle, 1979; Linnhoff *et al.*, 1982).

Analogous effects are caused by the inappropriate use of utilities. Utilities are appropriate if they are necessary to satisfy the enthalpy imbalance in that part of the process. Above the pinch, hot utility (in this case, steam) is needed to satisfy the enthalpy imbalance. Figure 17.8b illustrates what happens if inappropriate use of utilities is made. If cooling to cold utility *XP* is used to cool hot streams above the pinch, this creates an enthalpy imbalance in the system above the pinch. To satisfy the enthalpy imbalance above the pinch, an import of ($Q_{Hmin} + XP$) heat from hot utility is required. Overall, ($Q_{Cmin} + XP$) of cold utility is used (Linnhoff, Mason and Wardle, 1979; Linnhoff *et al.*, 1982).

Another inappropriate use of utilities involves heating of some of the cold streams below the pinch by hot utility (steam in this case). Below the pinch, cold utility is needed to satisfy the enthalpy

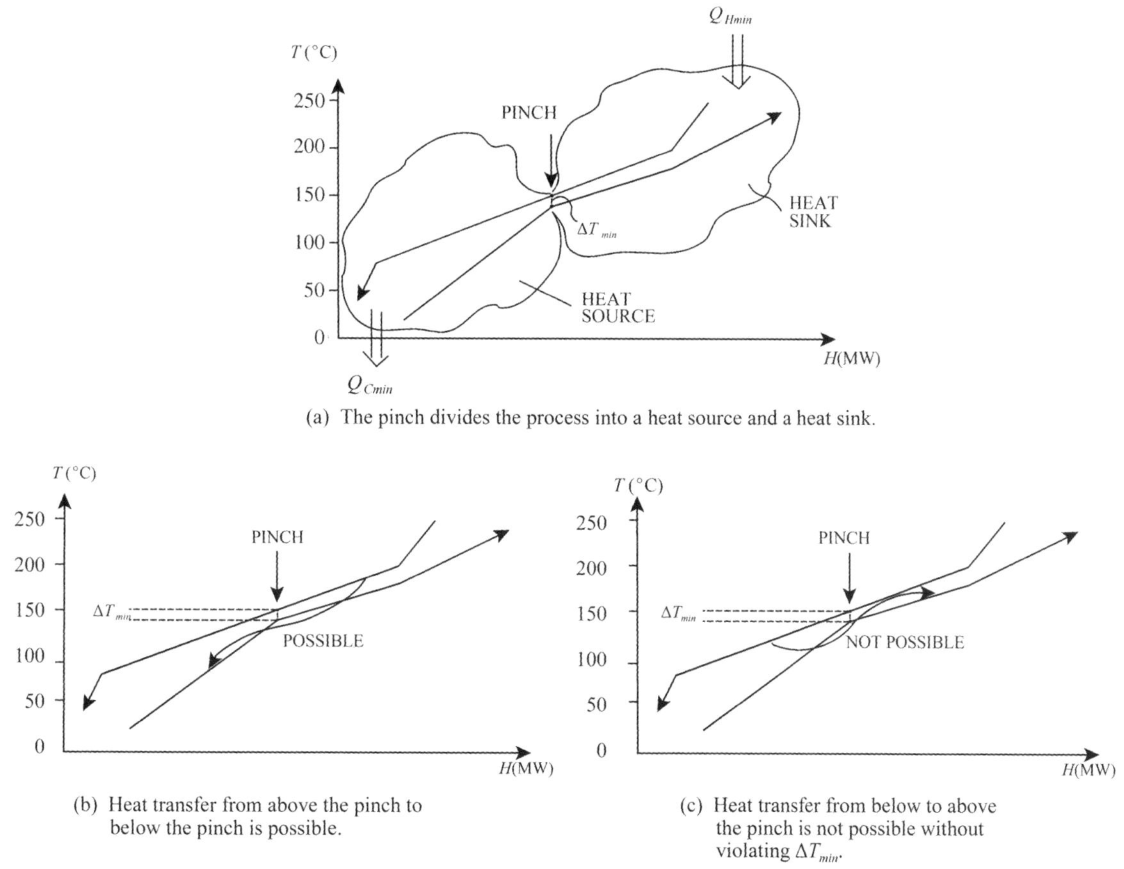

Figure 17.7

The composite curves set the energy target and the location of the pinch.

imbalance. Figure 17.8c illustrates what happens if an amount of heat XP from the hot utility is used below the pinch. Q_{Hmin} must still be supplied above the pinch to satisfy the enthalpy imbalance above the pinch. Overall, $(Q_{Hmin} + XP)$ of steam is used and $(Q_{Cmin} + XP)$ of cooling water (Linnhoff, Mason and Wardle, 1979; Linnhoff *et al.*, 1982).

In other words, to achieve the energy target set by the composite curves, the designer must not transfer heat across the pinch by (Linnhoff, Mason and Wardle, 1979; Linnhoff *et al.*, 1982):

a) process-to-process heat transfer,

b) inappropriate use of utilities.

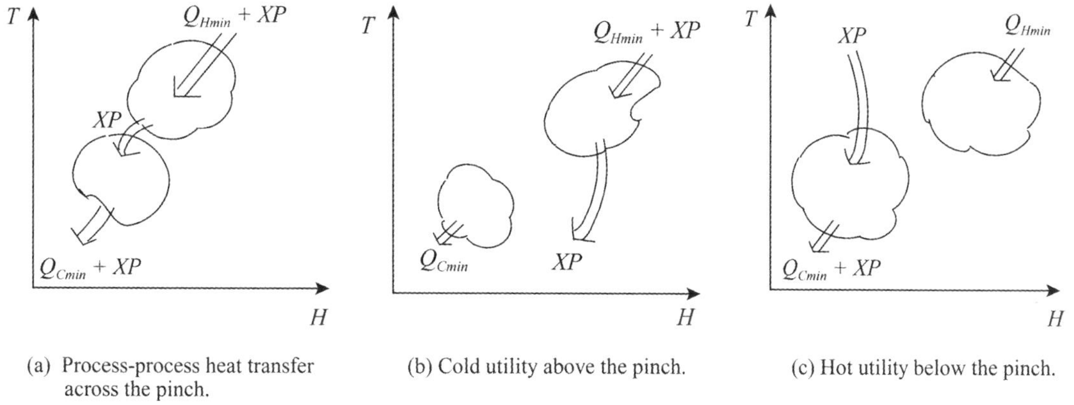

Figure 17.8

Three forms of cross-pinch heat transfer.

17

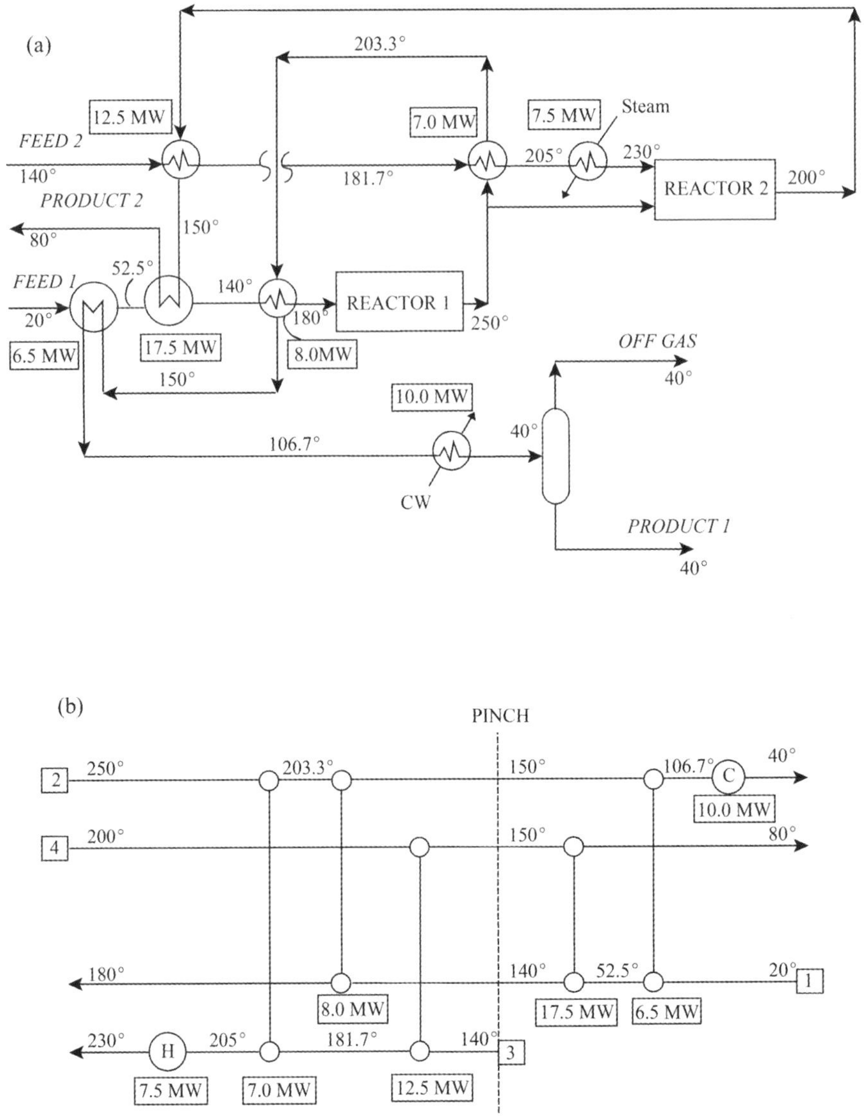

Figure 17.9
A design that achieves the energy target.

These rules are both necessary and sufficient to ensure that the target is achieved, providing that the initialization rule is adhered to that no individual heat exchanger should have a temperature difference smaller than ΔT_{min}.

Figure 17.9a shows a design corresponding to the flowsheet in Figure 17.2, which achieves the target of $Q_{Hmin} = 7.5$ MW and $Q_{Cmin} = 10$ MW for $\Delta T_{min} = 10$ °C. Figure 17.9b shows an alternative representation of the flowsheet in Figure 17.9a, known as the *grid diagram* (Linnhoff and Flower, 1978). The grid diagram shows only heat transfer operations. Hot streams are at the top running left to right. Cold streams are at the bottom running right to left. A heat exchange match is represented by a vertical line joining two circles on the two streams being matched. An exchanger using hot utility is represented by a circle with an "H". An exchanger using cold utility is represented by a circle with a "C". The importance of the grid diagram is clear in Figure 17.9b, since the pinch, and how it divides the process into two parts, is easily accommodated. Dividing the process into two parts on a conventional diagram such as shown in Figure 17.9a is both difficult and extremely cumbersome.

Details of how the design in Figure 17.9 was developed are explained in Chapter 18. For now, simply take note that the targets set by the composite curves are achievable in design, providing that the pinch is recognized and there is no transfer of heat across the pinch, either process-to-process or through inappropriate use of utilities. However, the insight of the pinch is needed to analyze some of the important decisions still to be made before the network design is addressed.

17.3 Threshold Problems

Not all problems have a pinch to divide the process into two parts (Linnhoff *et al.*, 1982). Consider the composite curves in Figure 17.10a. At this setting, both steam and cooling water are required. As the composite curves are moved closer together, both

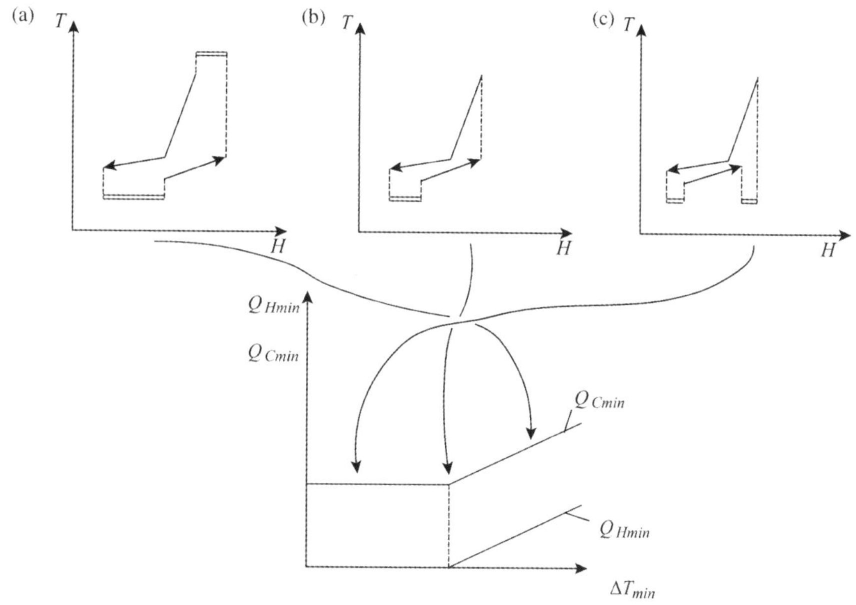

Figure 17.10

As ΔT_{min} is varied, some problems require only cold utility below a threshold value.

the steam and cooling water requirements decrease until the setting shown in Figure 17.10b is reached. At this setting, the composite curves are in alignment at the hot end, indicating that there is no longer a demand for hot utility. Moving the curves closer together as shown in Figure 17.10c, decreases the cold utility demand at the cold end but opens up a demand for cold utility at the hot end corresponding with the decrease at the cold end. In other words, as the curves are moved closer together, beyond the setting in Figure 17.10b, the utility demand is constant and set by the energy balance between the hot and cold streams. The setting shown in Figure 17.10b marks a threshold, and problems that exhibit this feature are known as *threshold problems* (Linnhoff *et al.*, 1982). In some threshold problems, the hot utility requirement disappears as in Figure 17.10. In others, the cold utility disappears as shown in Figure 17.11.

Considering the capital–energy trade-off for threshold problems, there are two possible outcomes, as shown in Figure 17.12. Below the threshold ΔT_{min}, utility costs are constant, since utility demand is constant. Figure 17.12a shows a situation where the optimum occurs at the threshold ΔT_{min}. Figure 17.12b shows a

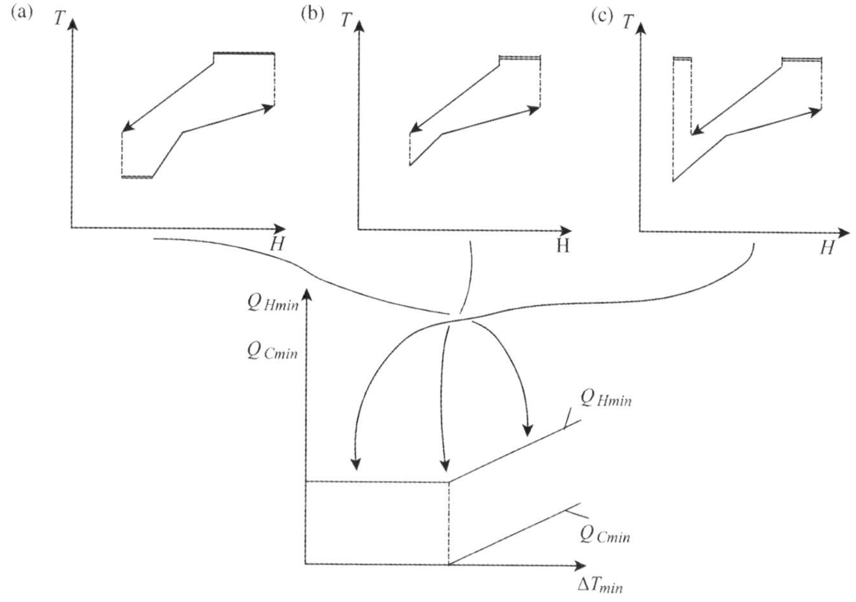

Figure 17.11

In some threshold problems only hot utility is required below the threshold value of ΔT_{min}.

17

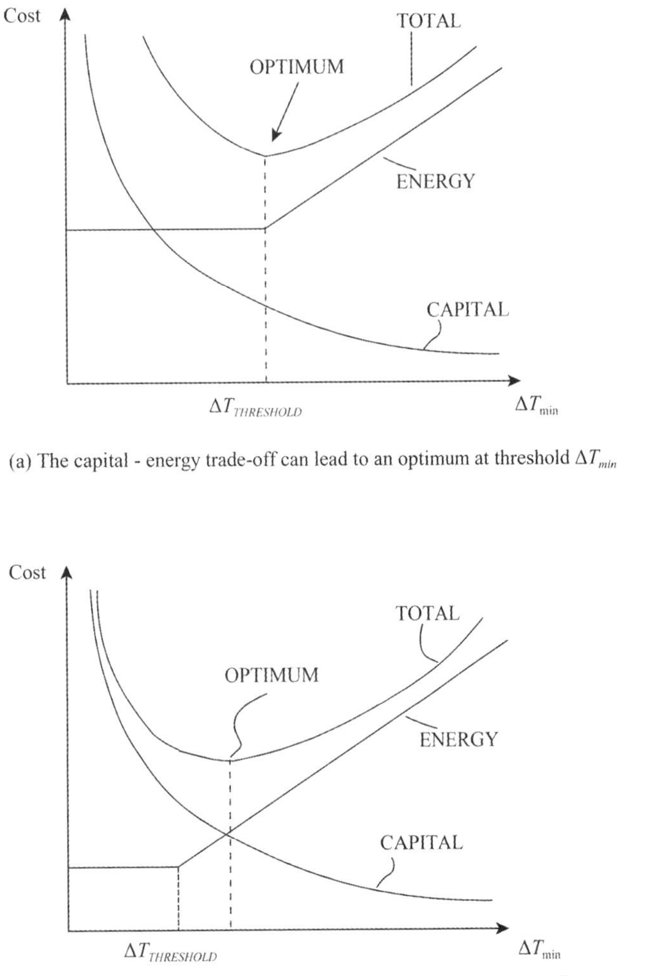

Figure 17.12

The optimum setting of the capital–energy trade-off for threshold problems.

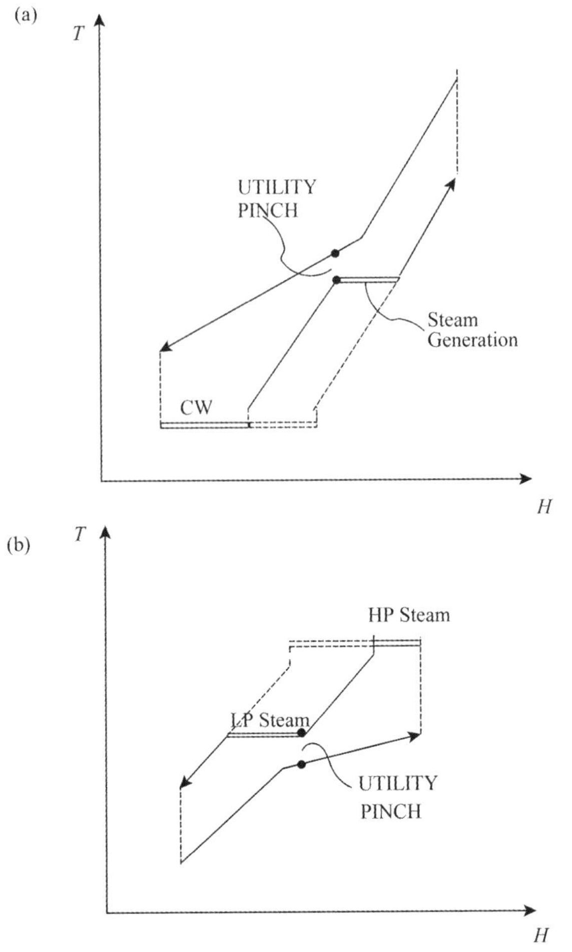

Figure 17.13

Threshold problems are turned into pinched problems when additional utilities are added.

situation where the optimum occurs above the threshold ΔT_{min}. The flat profile of utility costs below the threshold ΔT_{min} means that the optimum can never occur below the threshold value. It can only be at or above the threshold value.

In a situation, as shown in Figure 17.12a, with the optimum ΔT_{min} at the threshold, there is no pinch. On the other hand, in a situation as shown in Figure 17.12b with the optimum above the threshold value, there is a demand for both utilities and there is a pinch.

It is interesting to note that threshold problems are quite common in practice and although they do not have a process pinch, pinches are introduced into the design when multiple utilities are added. Figure 17.13a shows composite curves similar to the composite curves from Figure 17.10 but with two levels of cold utility used instead of one. In this case, the second cold utility is steam generation. The introduction of this second utility causes a pinch. This is known as a *utility pinch* since it is caused by the introduction of an additional utility (Linnhoff *et al.*, 1982).

Figure 17.13b shows composite curves similar to those from Figure 17.11, but with two levels of steam used. Again, the introduction of a second steam level causes a utility pinch.

In design, the same rules must be obeyed around a utility pinch as a process pinch. Heat should not be transferred across it by process-to-process transfer and there should be no inappropriate use of utilities. In Figure 17.13a, this means that the only utility to be used above the utility pinch is steam generation and only cooling water below. In Figure 17.13b, this means that the only utility to be used above the utility pinch is high-pressure steam and only low-pressure steam below.

17.4 The Problem Table Algorithm

Although composite curves can be used to set energy targets, they are inconvenient since they are based on a graphical construction. A method of calculating energy targets directly without the necessity of graphical construction can be developed (Hohman, 1971; Linnhoff and Flower, 1978; Bandyopadhyay and Sahu, 2010). The process is first divided into temperature intervals in the same way as was done for construction of the composite curves.

Figure 17.14a shows composite curves divided into temperature intervals with one temperature interval highlighted. It is not possible to recover all of the heat in each temperature interval since temperature differences are not feasible throughout the interval. Figure 17.14b shows that some heat recovery is possible. The amount that can be recovered is difficult to quantify and depends on both ΔT_{min} and the relative slopes of the two curves in the temperature interval. A simple transformation allows the heat recovery within temperature intervals to be readily quantified. Instead of working in the actual temperature scale, the temperatures are *shifted*. Purely for the purposes of construction, the hot composite is shifted to be $\Delta T_{min}/2$ colder than it is in practice and that the cold composite is shifted to be $\Delta T_{min}/2$ hotter than it is in practice, as shown in Figure 17.14c. The *shifted composite curves* now touch at the pinch. Carrying out a heat balance between the shifted composite curves within a *shifted temperature interval*, as shown in Figure 17.14c, shows that heat transfer is feasible throughout each shifted temperature interval, since hot streams are in practice actually $\Delta T_{min}/2$ hotter and cold streams $\Delta T_{min}/2$ colder. Thus, within each shifted interval, the hot streams are in reality hotter than the cold streams by ΔT_{min} and heat transfer is feasible throughout the shifted temperature interval.

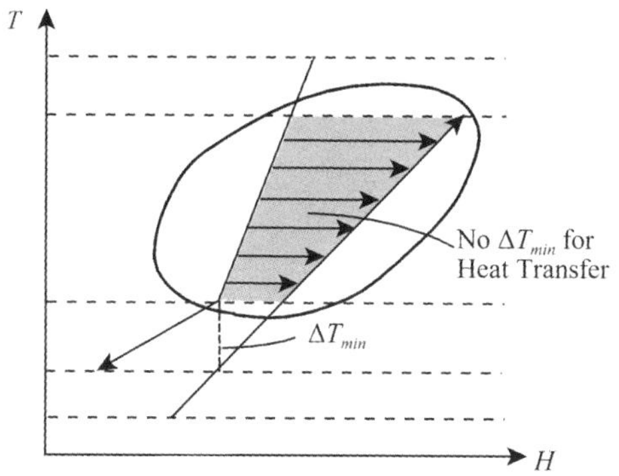

(a) Temperature differences are not feasible within a temperature interval.

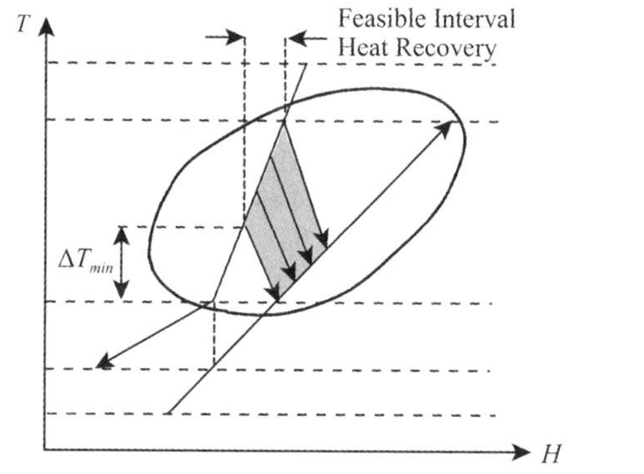

(b) Some heat can be recovered, but is difficult to quantify.

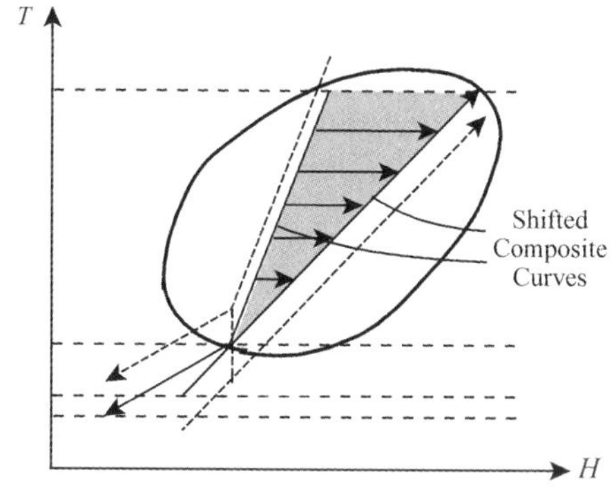

(c) Heat transfer within shifted temperature intervals is feasible.

Figure 17.14

Shifting the composite curves in temperature allows complete heat recovery within temperature intervals.

It is important to note that shifting the curves vertically does not alter the horizontal overlap between the curves. It therefore does not alter the amount by which the cold composite curve extends beyond the start of the hot composite curve at the hot end of the problem. Also, it does not alter the amount by which the hot composite curve extends beyond the start of the cold composite curve at the cold end. The shift simply removes the problem of ensuring temperature feasibility within temperature intervals.

This shifting technique can be used to develop a strategy to calculate the energy targets without having to construct composite curves (Hohman, 1971; Linnhoff and Flower, 1978; Bandyopadhyay and Sahu, 2010):

1) Set up shifted temperature intervals from the stream supply and target temperatures by subtracting $\Delta T_{min}/2$ from the hot streams and adding $\Delta T_{min}/2$ to the cold streams (as in Figure 17.14c).
2) In each shifted temperature interval, calculate a simple energy balance from:

$$\Delta H_i = \left[\sum_{Cold\ Streams} CP_C - \sum_{Hot\ Streams} CP_H \right] \Delta T_i \qquad (17.1)$$

where ΔH_i is the heat balance for shifted temperature interval i and ΔT_i is the temperature difference across it. If the cold streams dominate the hot streams in a temperature interval, then the interval has a net deficit of heat, and ΔH is positive. If hot streams dominate cold streams, the interval has a net surplus of heat, and ΔH is negative. This is consistent with standard thermodynamic convention; for example, for an exothermic reaction, ΔH is negative. If no hot utility is used, this is equivalent to constructing the shifted composite curves shown in Figure 17.15a.
3) The overlap in the shifted curves as shown in Figure 17.15a means that heat transfer is infeasible. At some point, this overlap is a maximum. This maximum overlap is added as hot utility to correct the overlap. The shifted curves now touch at the pinch, as shown in Figure 17.15b. Since the shifted curves just touch, the actual curves are separated by ΔT_{min} at this point (Figure 17.15b).

This basic approach can be developed into a formal algorithm known as the *problem table algorithm* (Linnhoff and Flower, 1978). The algorithm will be explained using the data from Figure 17.2 given in Table 17.2 for $\Delta T_{min} = 10\,°C$.

First determine the shifted temperature intervals (T^*) from actual supply and target temperatures. Hot streams are shifted

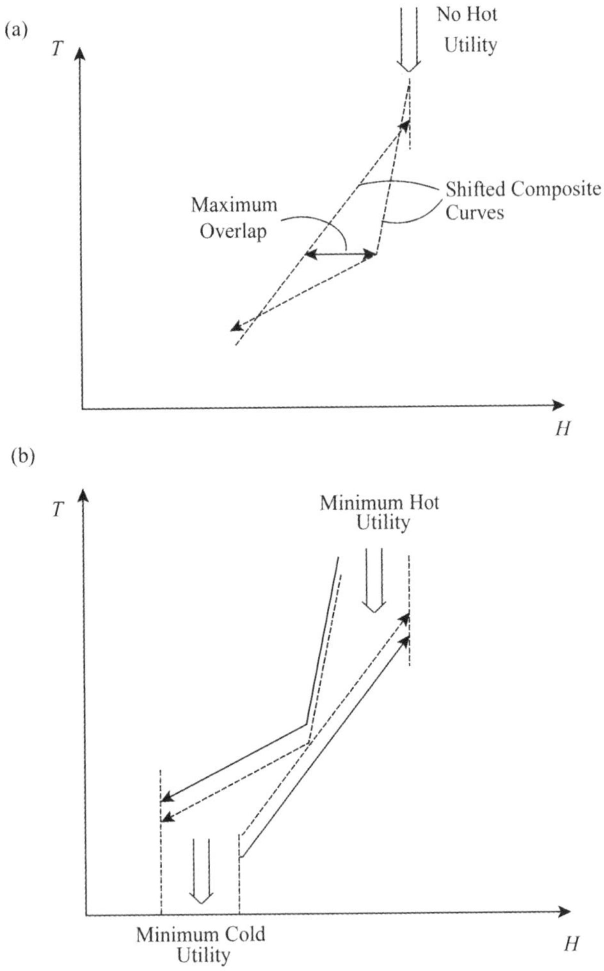

Figure 17.15

The utility target can be determined from the maximum overlap between the shifted composite curves.

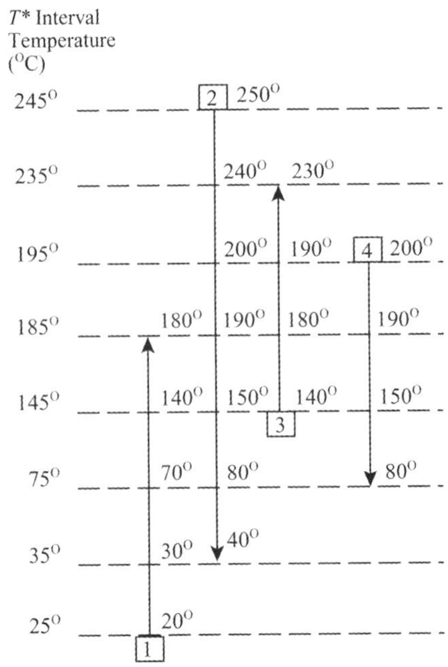

Figure 17.16

The stream population for the data from Figure 17.2.

down in temperature by $\Delta T_{min}/2$ and cold streams up by $\Delta T_{min}/2$, as detailed in Table 17.3.

The stream population is shown in Figure 17.16 with a vertical temperature scale. The interval temperatures shown in Figure 17.16 are set to $\Delta T_{min}/2$ below hot stream temperatures and $\Delta T_{min}/2$ above cold stream temperatures.

Next, a heat balance is carried out within each shifted temperature interval according to Equation 17.1. Summing the interval heat balances in Figure 17.17 provides a cross check on the calculation, as the summation must agree with the overall enthalpy balance across the four streams in Table 17.2, −2.5 MW in this case (Bandyopadhyay and Sahu, 2010). In Figure 17.17, some of the shifted intervals are seen to have a surplus of heat and some have a deficit. The heat balance within each shifted interval allows maximum heat recovery within each interval. However, recovery must also be allowed between intervals.

Now, *cascade* any surplus heat down the temperature scale from interval to interval. This is possible because any excess heat available from the hot streams in an interval is hot enough to supply a deficit in the cold streams in the next interval down. Figure 17.18 shows the cascade for the problem. First, assume no heat is supplied to the first interval from hot utility (Figure 17.18a). The first interval has a surplus of 1.5 MW, which is cascaded to the next interval. This second interval has a deficit of 6 MW, which leaves the heat cascaded from this interval to be −4.5 MW. In the third interval, the process has a surplus of 1 MW, which leaves −3.5 MW, to be cascaded to the next interval, and so on.

Table 17.3

Shifted temperatures for the data from Table 17.2.

Stream	Type	T_S	T_T	T_S*	T_T*
1	Cold	20	180	25	185
2	Hot	250	40	245	35
3	Cold	140	230	145	235
4	Hot	200	80	195	75

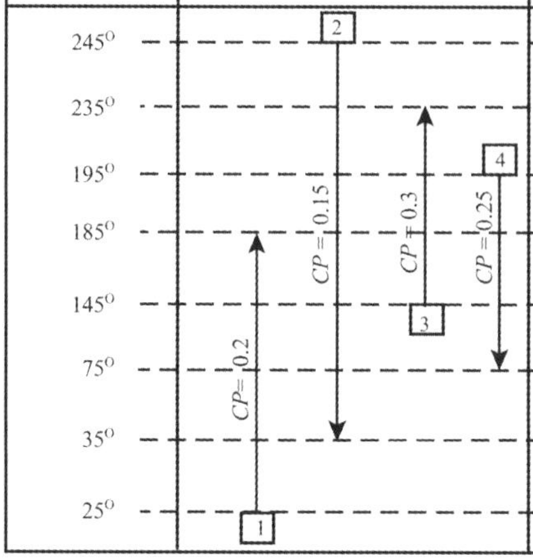

Interval Temperature	$\Delta T_{INTERVAL}$ (°C)	$\sum CP_C - \sum CP_H$ (MW/°C)	$\Delta H_{INTERVAL}$ (MW)	Surplus/Deficit
245°–235°	10	-0.15	-1.5	Surplus
235°–195°	40	0.15	6.0	Deficit
195°–185°	10	-0.1	-1.0	Surplus
185°–145°	40	0.1	4.0	Deficit
145°–75°	70	-0.2	-14.0	Surplus
75°–35°	40	0.05	2.0	Deficit
35°–25°	10	0.2	2.0	Deficit
		$\sum \Delta H_{INTERVAL}$ (MW)	-2.5	Surplus

Figure 17.17

The temperature interval heat balances.

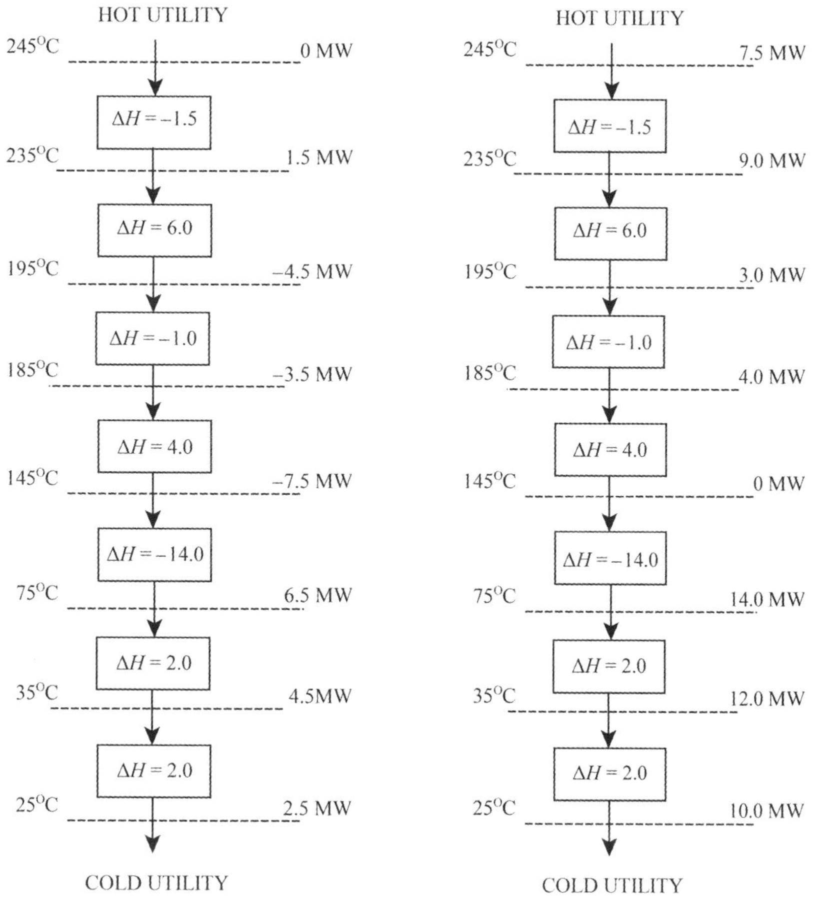

(a) Cascade surplus heat from high to low temperature.

(b) Add heat from hot utility to make all heat flows zero or positive.

Figure 17.18

The problem table cascade.

17

Some of the heat flows in Figure 17.18a are negative, which is infeasible. Heat cannot be transferred up the temperature scale. To make the cascade feasible, sufficient heat must be added from hot utility to make the heat flows to be at least zero. The smallest amount of heat needed from hot utility is the largest negative heat flow from Figure 17.18a, that is 7.5 MW. In Figure 17.18b, 7.5 MW is added from hot utility to the first interval. This does not change the heat balance within each interval, but increases all of the heat flows between intervals by 7.5 MW, giving one heat flow of just zero at an interval temperature of 145 °C.

More than 7.5 MW could be added from hot utility to the first interval, but the objective is to find the minimum hot and cold utility, hence only the minimum is added. Thus from Figure 17.18b, Q_{Hmin} = 7.5 MW and Q_{Cmin} = 10 MW. This corresponds with the values obtained from the composite curves in Figure 17.5. One further important piece of information can be deduced from the cascade in Figure 17.18b. The point where the heat flow goes to zero at T^* = 145 °C corresponds to the pinch. It was noted in Section 17.2 that the pinch must correspond with either the start of a hot stream, or the start of the cold stream. This provides another cross check for the calculation. From Figure 17.17, for the interval pinch temperature to be feasible, it must be one of 254 °C, or 195 °C, or 145 °C, or 25 °C. Thus, for this example the actual hot and cold stream pinch temperatures are 150 °C and 140 °C respectively. Again, this agrees with the result from the composite curves in Figure 17.5a.

The initial setting for the heat cascade in Figure 17.18a corresponds to the setting for the shifted composite curves in Figure 17.15a where there is an overlap. The setting of the heat cascade for zero or positive heat flows in Figure 17.18b corresponds to the shifted composite curve setting in Figure 17.15b.

The composite curves are useful in providing conceptual understanding of the process but the problem table algorithm is a more convenient calculation tool.

Example 17.1 The flowsheet for a low-temperature distillation process is shown in Figure 17.19. Calculate the minimum hot and cold utility requirements and the location of the pinch assuming ΔT_{min} = 5 °C.

Solution First extract the stream data from the flowsheet. This is given in Table 17.4 below.

Next, calculate the shifted interval temperatures. Hot stream temperatures are shifted down by 2.5 °C and cold stream temperatures shifted up by 2.5 °C, as shown in Table 17.5.

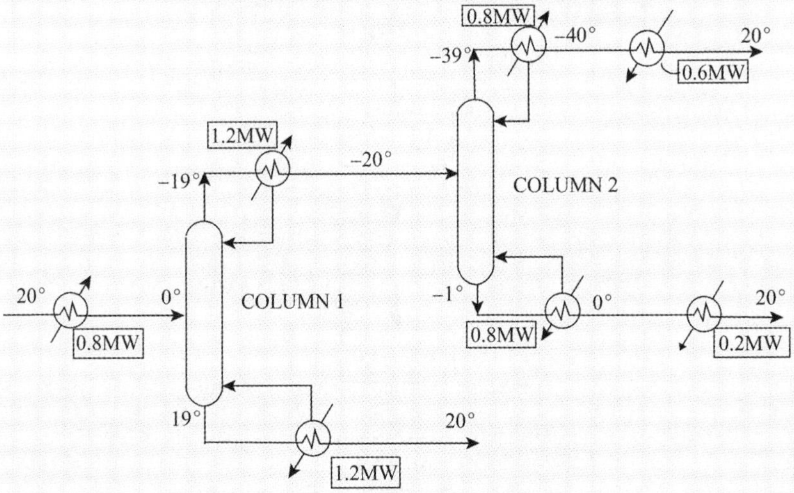

Figure 17.19
A low-temperature distillation process.

Table 17.4
Stream data for a low-temperature distillation process.

Stream	Type	Supply temperature T_S (°C)	Target temperature T_T (°C)	ΔH (MW)	CP (MW·K^{-1})
1. Feed to column 1	Hot	20	0	−0.8	0.04
2. Column 1 condenser	Hot	−19	−20	−1.2	1.2
3. Column 2 condenser	Hot	−39	−40	−0.8	0.8
4. Column 1 reboiler	Cold	19	20	1.2	1.2
5. Column 2 reboiler	Cold	−1	0	0.8	0.8
6. Column 2 bottoms	Cold	0	20	0.2	0.01
7. Column 2 overheads	Cold	−40	20	0.6	0.01

Table 17.5

Shifted temperatures for the data in Table 17.4.

Stream	Type	T_S	T_T	T_S^*	T_T^*
1	Hot	20	0	17.5	−2.5
2	Hot	−19	−20	−21.5	−22.5
3	Hot	−39	−40	−41.5	−42.5
4	Cold	19	20	21.5	22.5
5	Cold	−1	0	1.5	2.5
6	Cold	0	20	2.5	22.5
7	Cold	−40	20	−37.5	22.5

Now carry out a heat balance within each shifted temperature interval, as shown in Figure 17.20. The sum of the interval heat balances is 0.0 MW, which agrees with the sum of the energy balance for the seven streams in Table 17.4. Finally, the heat cascade is shown in Figure 17.21. Figure 17.21a shows the cascade with zero hot utility. This leads to negative heat flows, the largest of which is −1.84 MW. Adding 1.84 MW from hot utility as shown in Figure 17.21b gives $Q_{Hmin} = 1.84$ MW, $Q_{Cmin} = 1.84$ MW. The interval pinch temperature is −21.5 °C, corresponding with the start of Stream 2 in Figure 17.20. The hot stream pinch temperature is −19 °C and cold stream pinch temperature is −24 °C.

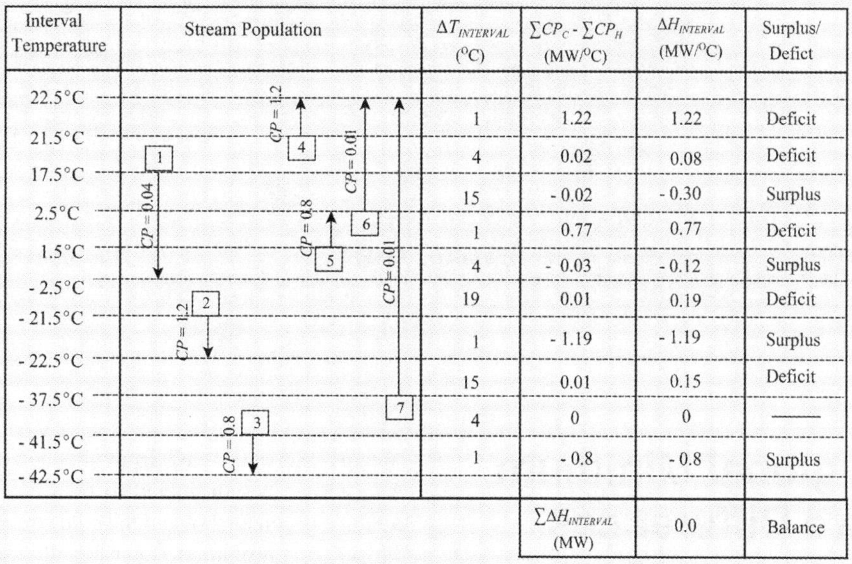

Interval Temperature
Stream Population
$\Delta T_{INTERVAL}$ (°C)
$\sum CP_C - \sum CP_H$ (MW/°C)
$\Delta H_{INTERVAL}$ (MW/°C)
Surplus/ Deficit
22.5°C
21.5°C
17.5°C
2.5°C
1.5°C
- 2.5°C
- 21.5°C
- 22.5°C
- 37.5°C
- 41.5°C
- 42.5°C
1
2
3
4
5
6
7
CP = 1.2
CP = 0.01
CP = 0.04
CP = 0.8
CP = 0.01
CP = 1.2
CP = 0.8
1 1.22 1.22 Deficit
4 0.02 0.08 Deficit
15 - 0.02 - 0.30 Surplus
1 0.77 0.77 Deficit
4 - 0.03 - 0.12 Surplus
19 0.01 0.19 Deficit
1 - 1.19 - 1.19 Surplus
15 0.01 0.15 Deficit
4 0 0
1 - 0.8 - 0.8 Surplus
$\sum \Delta H_{INTERVAL}$ (MW) 0.0 Balance

Figure 17.20

Temperature interval heat balances for Example 17.1.

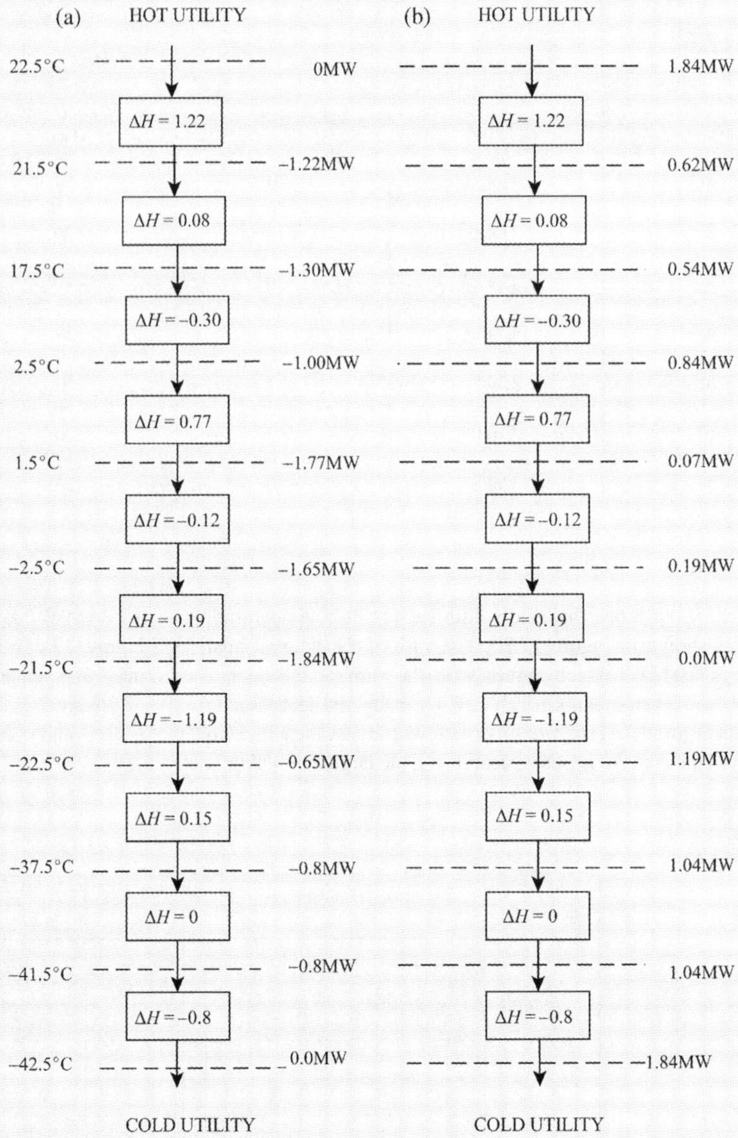

Figure 17.21

The problem table cascade for Example 17.1.

17.5 Non-global Minimum Temperature Differences

So far, it has been assumed that the minimum temperature difference for a heat exchanger network applies globally between all streams in the network. However, there are occasions when non-global minimum temperature differences might be required. For example, suppose a heat exchanger network is servicing some streams that are liquid and some that are gaseous. For the liquid–liquid heat transfer matches, a value of perhaps $\Delta T_{min} = 10\,°C$ is appropriate. However, for the gas–gas matches, a larger temperature minimum temperature difference is required, say $\Delta T_{min} =$ 20 °C. How can this be accommodated in the targeting?

When carrying out the problem table algorithm, the temperatures were shifted according to $\Delta T_{min}/2$ being added to the cold streams and subtracted from the hot streams. This value of $\Delta T_{min}/2$ can be considered to be a *contribution* to the overall ΔT_{min} between the hot and the cold streams. Rather than making the ΔT_{min} contribution equal for all streams, it could be made stream-specific:

$$\begin{aligned} T^*_{H,i} &= T_{H,i} - \Delta T_{min,cont,i} \\ T^*_{C,j} &= T_{C,j} + \Delta T_{min,cont,j} \end{aligned}$$

where $T^*_{H,i}$, $T_{H,i}$ are the shifted and actual temperatures for Hot Stream i, $T^*_{C,j}$, $T_{C,j}$ are the shifted and actual temperatures for Cold Stream j, and $\Delta T_{min,cont,i}$ and $\Delta T_{min,cont,j}$ are the contributions to ΔT_{min} for Hot Stream i and Cold Stream j. Thus, for the above example, if the ΔT_{min} contribution for liquid streams is taken to be 5 °C and for gas streams 10 °C, then a liquid–liquid match would lead to $\Delta T_{min} = 10\,°C$, a gas–gas match would lead to $\Delta T_{min} = 20$

°C and a liquid–gas match would lead to $\Delta T_{min} = 15$ °C (Linnhoff *et al.*, 1982). To include this in the problem table algorithm is straightforward. All that needs to be done is that the appropriate ΔT_{min} contribution is to be allocated to each stream and then that ΔT_{min} contribution is subtracted in the case of hot streams and added in the case of cold streams (rather than just subtracting or adding $\Delta T_{min}/2$). This would lead to different interval temperatures compared with a global minimum temperature difference. The remainder of the problem table algorithm would be the same. Once the interval temperatures based on ΔT_{min} contributions have been established, the interval heat balances can be performed and the cascade set up in the same way as for a global ΔT_{min}.

From the point of view of the composite curves, the location of the pinch and the ΔT_{min} at the pinch would depend on which kind of streams were located in the region of the point of closest approach between the composite curves. If only liquid streams were present around the point of closest approach of the composite curves, then in the above example, $\Delta T_{min} = 10$ °C will apply. If there were only gas streams in the region around the point of closest approach, then in the above example, $\Delta T_{min} = 20$ °C would apply. If there was a mixture of liquid and gas streams at the point of closest approach, then $\Delta T_{min} = 15$ °C would apply.

17.6 Process Constraints

So far it has been assumed that any hot stream could, in principle, be matched with any cold stream providing there is a feasible temperature difference between them. Often, practical constraints prevent this. For example, it might be the case that if two streams are matched in a heat exchanger and a leak develops, such that the two streams come into contact, this might produce an unacceptably hazardous situation. If this were the case, then no doubt a constraint would be imposed to prevent the two streams being matched. Another reason for a constraint might be that two streams are expected to be geographically very distant from each other, leading to unacceptably long pipe runs. Potential control and start-up problems might also call for constraints. There are many reasons why constraints might be imposed.

One common reason for imposing constraints results from *areas of integrity* (Ahmad and Hui, 1991). A process is often normally designed to have logically identifiable sections or areas. An example might be the "reaction area" and "separation area" of the process. These areas might need to be kept separate for reasons such as start-up, shutdown, operational flexibility, safety, and so on. The areas are often made operationally independent by use of intermediate storage between the areas. Such independent areas are generally described as areas of integrity and impose constraints on the ability to transfer heat. Clearly, to maintain operational independence, two areas cannot be dependent on each other for heating and cooling by recovery.

The question now is, given that there are often constraints to deal with, how to evaluate the effect of these constraints on the system performance? The problem table algorithm cannot be used directly if constraints are imposed. However, the effect of constraints on the energy performance can often be evaluated using the problem table algorithm, together with a little common sense. The following example illustrates how (Ahmad and Hui, 1991).

Example 17.2 A process is to be divided into two operationally independent areas of integrity, Area *A* and Area *B*. The stream data for the two areas are given in Table 17.6 (Ahmad and Hui, 1991).

Calculate the penalty in utility consumption to maintain the two areas of integrity for $\Delta T_{min} = 20$ °C.

Solution To identify the penalty, first calculate the utility consumption of the two areas separate from each other, as shown in Figure 17.22a. Next, combine all of the streams from both areas and again calculate the utility consumption (Figure 17.22b). Figure 17.23a shows the problem table cascade for Area *A*, the cascade for Area *B* is shown in Figure 17.23b, and that for Areas *A* and *B* combined is shown in Figure 17.23c.

With Areas *A* and *B* separate, the total hot utility consumption is $(1400 + 0) = 1400$ kW and the total cold utility consumption is $(0 + 1350) = 1350$ kW. With Areas *A* and *B* combined, the total utility consumption is 950 kW of hot utility and 900 kW of cold utility.

The penalty for maintaining the areas of integrity is thus $(1400 - 950) = 450$ kW of hot utility and $(1350 - 900) = 450$ kW of cold utility.

Table 17.6

Stream data for heat recovery between two areas of integrity.

Area *A*				Area *B*			
Stream	T_S (°C)	T_T (°C)	CP (kW·K^{-1})	Stream	T_S (°C)	T_T (°C)	CP (kW·K^{-1})
1	190	110	2.5	3	140	50	20.0
2	90	170	20.0	4	30	120	5.0

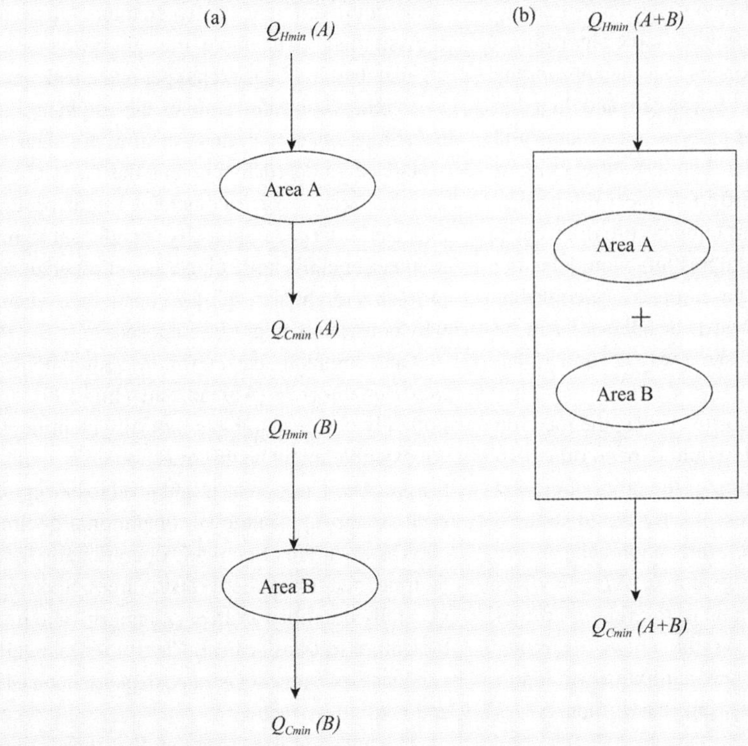

Figure 17.22

The areas of integrity can be targeted separately and then the combination targeted.

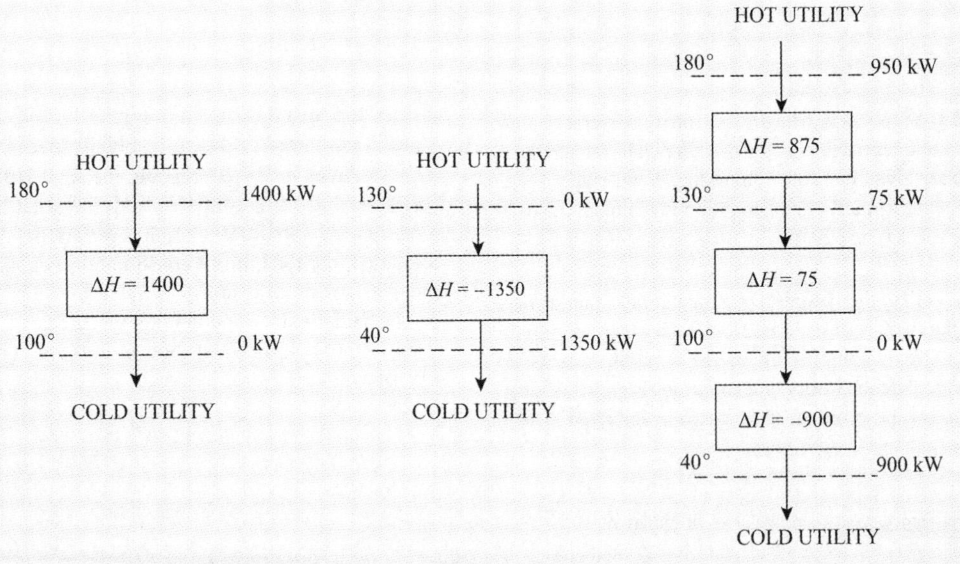

(a) Area A. (b) Area B. (c) Area A+B.

Figure 17.23

Problem table cascade for the separate and combined areas of integrity.

Having quantified the penalty as a result of imposing constraints, the designer can exercise judgment as to whether it is acceptable or too expensive. If it is too expensive, there is a choice between two options:

a) Reject the constraints and operate the process as a single system.

b) Find a way to overcome the constraint. This is often possible by using a heat transfer fluid. The simplest option is via the existing utility system. For example, rather than have a direct match between two streams, it might be possible for the heat source to generate steam to be fed into the steam mains and the heat sink to use steam from the same mains. The utility system then acts as a buffer between the heat sources and sinks. Another possibility might be to use a heat transfer fluid such as hot oil or hot water, which circulates between the two streams being matched. To maintain operational independence, a standby heater and cooler supplied by utilities can be provided in the hot oil circuit or hot water circuit, so that if either the heat source or sink is not operational, utilities could substitute heat recovery for short periods.

Many constraints can be evaluated by scoping the problem with different boundaries. In Example 17.2, the sets of streams that were constrained to be separate were collected to be within each boundary for targeting. Comparing the targets for the streams within each boundary with that for all the streams put together allows the penalty of the constraint to be evaluated. The approach is more widely applicable than just areas of integrity. Whenever a stream, or set of streams, is to be maintained separate from any other set of streams, the same approach can be used. However, this approach of scoping out the problem with different boundaries has limitations in the evaluation of constraints. More complex constraints require linear programming to obtain the energy target (Cerda *et al.*, 1983; Papoulias and Grossman, 1983).

17.7 Utility Selection

After maximizing heat recovery in the heat exchanger network, those heating and cooling duties not serviced by heat recovery must be provided by external utilities. The most common hot utility is steam. It is usually available at several levels. High-temperature heating duties require furnace flue gas or a hot oil circuit. Cold utilities might be refrigeration, cooling water, air cooling, furnace air preheating, boiler feedwater preheating or even steam generation if heat needs to be rejected at high temperatures.

Although the composite curves can be used to set energy targets, they are not a suitable tool for the selection of utilities. The *grand composite curve*, to be developed next, is a more appropriate tool for understanding the interface between the process and the utility system (Itoh, Shiroko and Umeda, 1982; Linnhoff *et al.*, 1982; Townsend and Linnhoff, 1983). It will also be shown in Chapters 20 to 22 to be a useful tool to study the interaction between heat-integrated reactors and separators and the rest of the process.

The grand composite curve is obtained by plotting the problem table cascade. A typical grand composite curve is shown in Figure 17.24. It shows the heat flow through the process against temperature. It should be noted that the temperature plotted here is *shifted temperature* (T^*) and not actual temperature. Hot streams are represented $\Delta T_{min}/2$ colder and cold streams $\Delta T_{min}/2$ hotter than they are in practice. Thus, an allowance for ΔT_{min} is built into the construction.

The point of zero heat flow in the grand composite curve in Figure 17.24 is the heat recovery pinch. The open "jaws" at the top and bottom are Q_{Hmin} and Q_{Cmin} respectively. Thus, the heat sink above the pinch and heat source below the pinch can be identified as shown in Figure 17.24. The shaded areas in Figure 17.24, known as *pockets*, represent areas of additional process-to-process heat transfer. Remember that the profile of the grand composite curve represents residual heating and cooling demands after recovering heat within the shifted temperature intervals in the problem table algorithm. In the pockets in Figure 17.24, a local surplus of heat in the process is used at temperature differences in excess of ΔT_{min} to satisfy a local deficit. This reflects the cascading of excess heat from high-temperature intervals to lower temperature intervals in the problem table algorithm.

Figure 17.25a shows the same grand composite curve with two levels of steam used as hot utility. Figure 17.25b shows again the same grand composite curve but with hot oil used as hot utility.

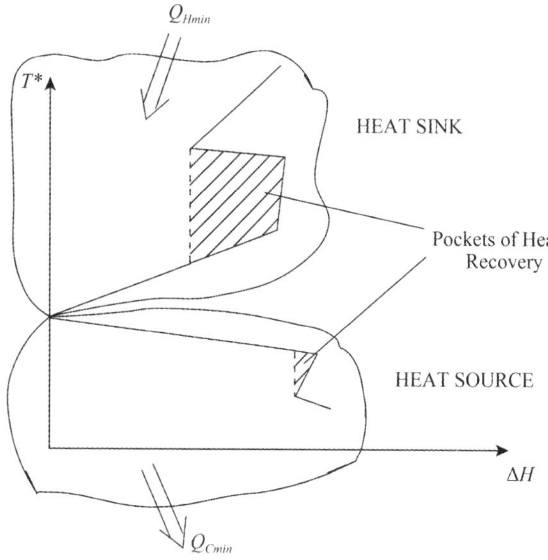

Figure 17.24

The grand composite curve shows the utility requirements both in enthalpy and temperature terms.

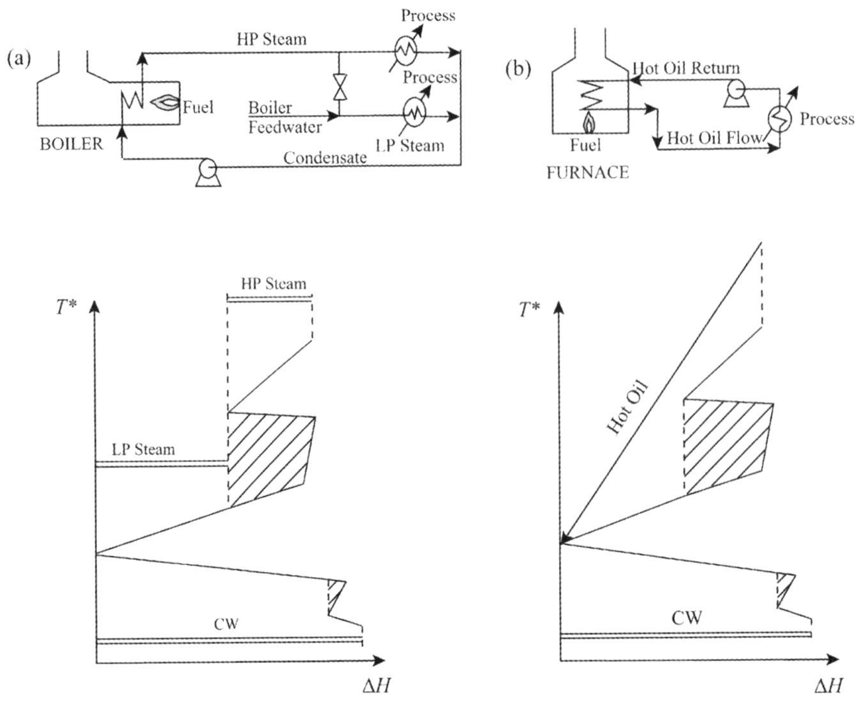

Figure 17.25

The grand composite curve allows alternative utility mixes to be evaluated.

Example 17.3 The problem table cascade for the process in Figure 17.2 is given in Figure 17.18. Using the grand composite curve:

a) For two levels of steam at saturation conditions and temperatures of 240 °C and 180 °C, determine the loads on the two steam levels that maximizes the use of the lower pressure steam.

b) Instead of using steam, a hot oil circuit is to be used with a supply temperature of 280 °C and $C_P = 2.1\ kJ \cdot kg^{-1} \cdot K^{-1}$. Calculate the minimum flowrate of hot oil.

Solution

a) For $\Delta T_{min} = 10$ °C, the two steam levels are plotted on the grand composite curve at temperatures of 235 °C and 175 °C. Figure 17.26a shows the loads that maximize the use of the lower pressure steam. Calculate the load on the low-pressure steam by interpolation of the cascade heat flows. At $T^* = 175$ °C:

$$\text{Load of 180 °C steam} = \frac{175 - 145}{185 - 145} \times 4$$
$$= 3\ \text{MW}$$
$$\text{Load of 240 °C steam} = 7.5 - 3$$
$$= 4.5\ \text{MW}$$

b) Figure 17.26b shows the grand composite curve with hot oil providing the hot utility requirements. If the minimum flowrate is required, then this corresponds to the steepest slope and minimum return temperature. For this problem, the minimum return temperature for the hot oil is the pinch temperature ($T^* = 145$ °C, $T = 150$ °C for hot streams). Check that the hot oil line can fit in the grand composite curve at $T^* = 195$ °C:

$$\text{Load on hot oil from } T^* = 195\text{ °C to } T^* = 145\text{ °C}$$
$$= \frac{195 - 145}{275 - 145} \times 7.5$$
$$= 2.9\ \text{MW}$$

Thus, the hot oil line fits in the grand composite curve:

$$\text{Minimum flowrate} = 7.5 \times 10^3 \times \frac{1}{2.1} \times \frac{1}{(280 - 150)}$$
$$= 27.5\ kg \cdot s^{-1}$$

In other problems, the shape of the grand composite curve away from the pinch could have limited the flowrate.

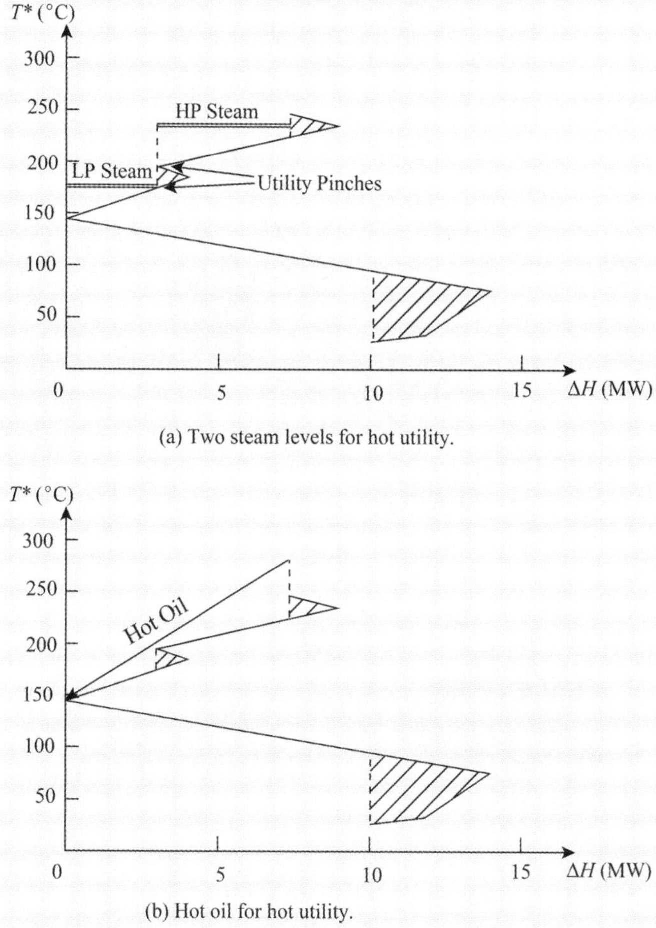

Figure 17.26

Alternative utility mixes for the process in Figure 17.2.

17.8 Furnaces

When a hot utility needs to be at a high temperature and/or provide high heat fluxes, radiant heat transfer is used from combustion of fuel in a furnace. Furnace designs vary according to the function of the furnace, heating duty, type of fuel and the method of introducing combustion air (see Chapter 12). Sometimes the function is to purely provide heat, while sometimes the furnace is also a reactor and provides heat for an endothermic reaction. However, process furnaces have a number of features in common. In the chamber in which combustion takes place, heat is transferred mainly by radiation to tubes around the walls of the chamber, through which passes the fluid to be heated. After the flue gas leaves the combustion chamber, most furnace designs extract further heat from the flue gas in a convection section before the flue gas is vented to atmosphere (see Chapter 12).

Figure 17.27 shows a grand composite curve with a flue gas matched against it to provide hot utility (Linnhoff and de Leur, 1988). The flue gas starts at its theoretical flame temperature (see Chapter 12) shifted for ΔT_{min} on the grand composite curve and presents a sloping profile because it is giving up sensible heat. The theoretical flame temperature is the temperature attained when a fuel is burnt in air or oxygen without loss or gain of heat, as explained in Chapter 12.

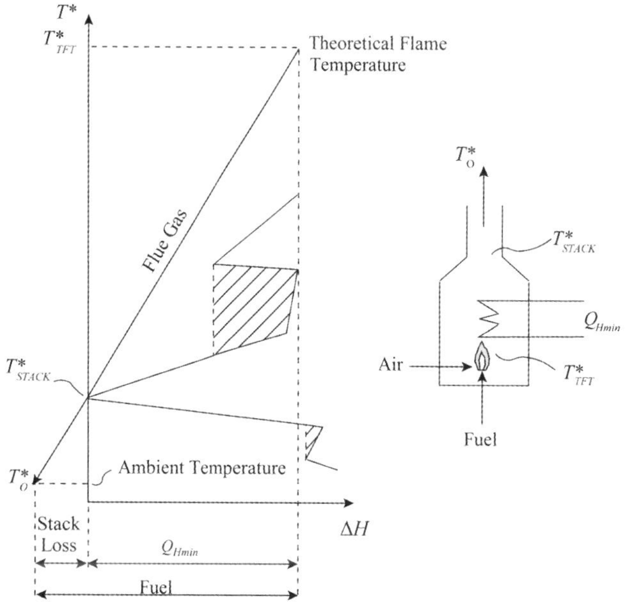

Figure 17.27

Simple furnace model.

17

In Figure 17.27, the flue gas is cooled to the pinch temperature before being released to atmosphere. The heat released from the flue gas between the pinch temperature and ambient is the stack loss. Thus in Figure 17.27, for a given grand composite curve and theoretical flame temperature, the heat from fuel and stack loss can be determined.

All combustion processes work with an excess of air or oxygen to ensure complete combustion of the fuel. Excess air typically ranges between 5 and 20% depending on the fuel, burner design and furnace design. As excess air is reduced, the theoretical flame temperature increases as shown in Figure 17.28. This has the effect of reducing the stack loss and increasing the thermal efficiency of the furnace for a given process heating duty. Alternatively, if the combustion air is preheated (typically by heat recovery), then again the theoretical flame temperature increases as shown in Figure 17.28, reducing the stack loss.

Although, the higher flame temperatures in Figure 17.28 reduce the fuel consumption for a given process heating duty, there is one significant disadvantage. Higher flame temperatures increase the formation of oxides of nitrogen, which are environmentally harmful. This point will be returned to in Chapter 25.

In Figures 17.27 and 17.28, the flue gas is capable of being cooled to the pinch temperature before being released to atmosphere. This is not always the case. Figure 17.29a shows a situation in which the flue gas is released to atmosphere above the pinch temperature for practical reasons. There is a practical minimum, the

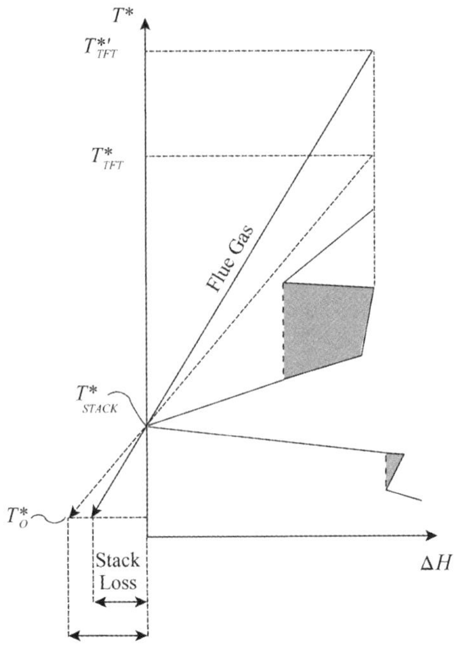

Figure 17.28

Increasing the theoretical flame temperature by reducing excess air or preheating combustion air reduces the stack loss.

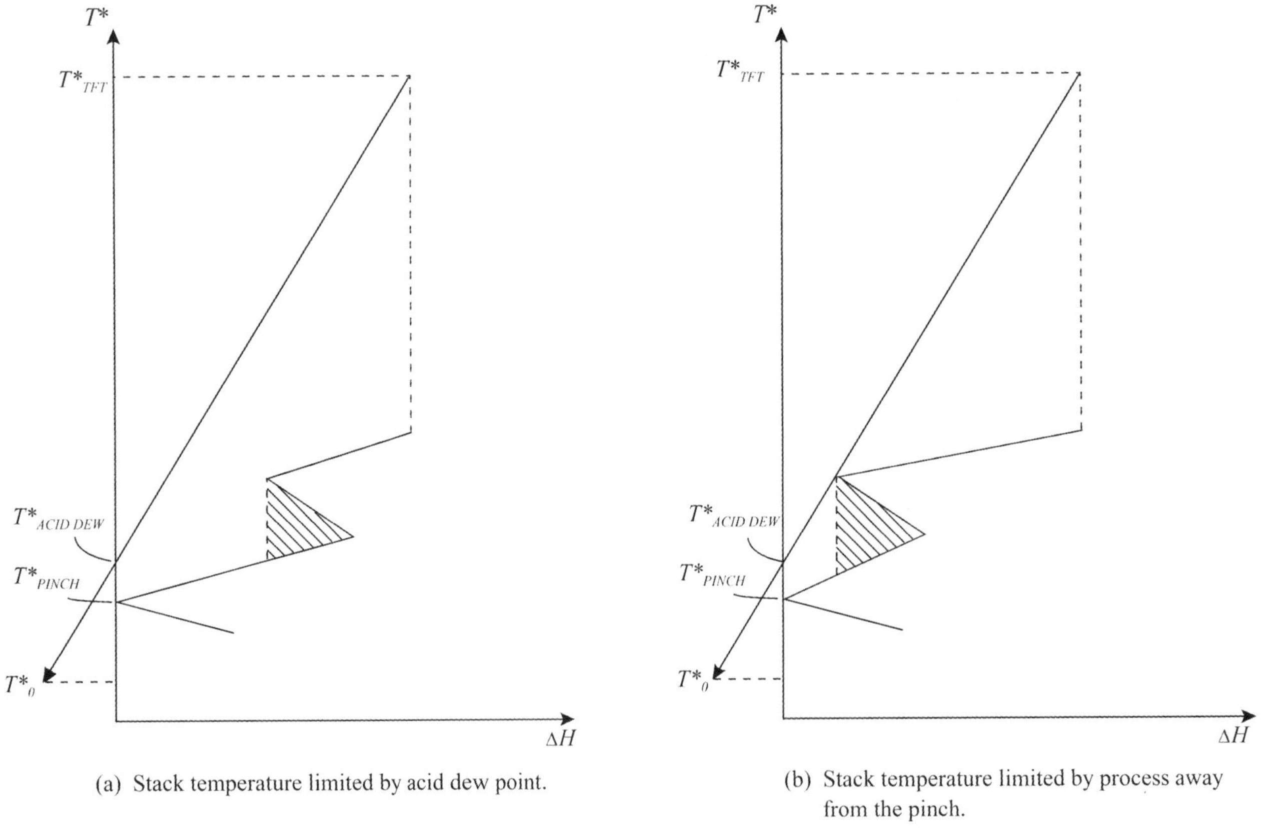

Figure 17.29

Furnace stack temperature can be limited by other factors than pinch temperature.

acid dew point, to which a flue gas can be cooled without condensation causing corrosion in the stack (see Chapter 12). The minimum stack temperature in Figure 17.29a is fixed by the acid dew point. Another case is shown in Figure 17.29b where the process away from the pinch limits the slope of the flue gas line and hence the stack loss.

Example 17.4 The process in Figure 17.2 is to have its hot utility supplied by a furnace. The theoretical flame temperature for combustion is 1800 °C and the acid dew point for the flue gas is 160 °C. The ambient temperature is 10 °C. Assume $\Delta T_{min} = 10$ °C for process-to-process heat transfer but $\Delta T_{min} = 30$ °C for flue gas-to-process heat transfer. A high value for ΔT_{min} for flue gas-to-process heat transfer has been chosen because of poor heat transfer coefficients in the convection bank of the furnace. Calculate the fuel required, stack loss and furnace efficiency.

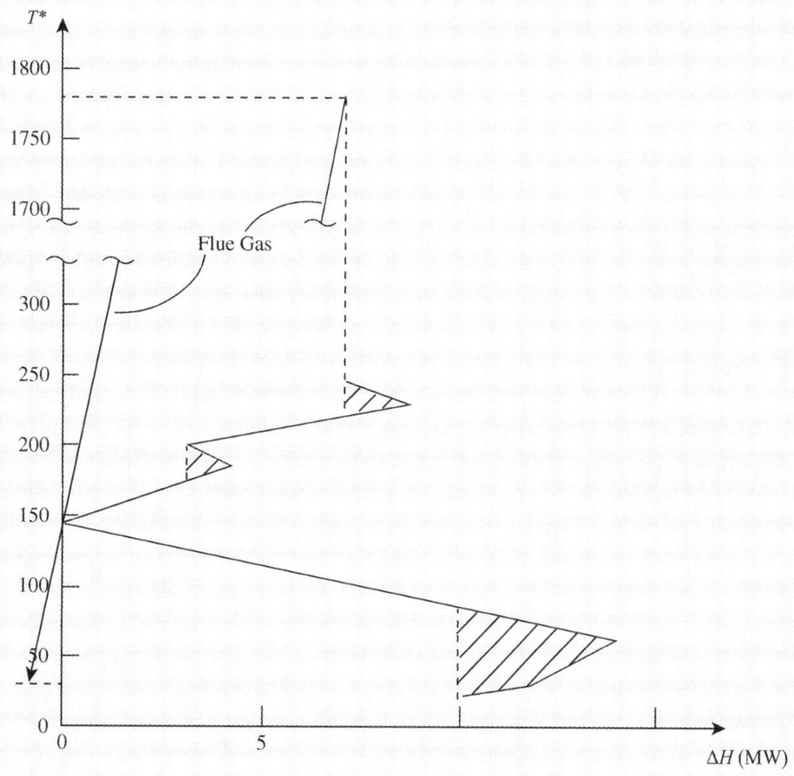

Figure 17.30

Flue gas matched against the grand composite curve of the process in Figure 17.2.

Solution The first problem is that a different value of ΔT_{min} is required for different matches. The problem table algorithm is easily adapted to accommodate this. This is achieved by assigning ΔT_{min} contributions to streams. If the process streams are assigned a contribution of 5 °C and the flue gas a contribution of 25 °C, then a process-to-process match has a ΔT_{min} of $(5+5) = 10$ °C and a process-to-flue gas match has a ΔT_{min} of $(5+25) = 30$ °C. When setting up the interval temperatures in the problem table algorithm, the interval boundaries are now set at hot stream temperatures minus their ΔT_{min} contribution, rather than half the global ΔT_{min}. Similarly, boundaries are now set on the basis of cold stream temperatures plus their ΔT_{min} contribution.

Figure 17.30 shows the grand composite curve plotted from the problem table cascade in Figure 17.18b. The starting point for the flue gas is an actual temperature of 1800 °C, which corresponds to a shifted temperature of $(1800 - 25) = 1775$ °C on the grand composite curve. The flue gas profile is not restricted above the pinch and can be cooled to the pinch temperature corresponding with a shifted temperature of 145 °C before venting to atmosphere. The actual stack temperature is thus $(145 + 25) = 170$ °C. This is just above the acid dew point of 160 °C. Now calculate the fuel consumption:

$$Q_{Hmin} = 7.5 \text{ MW}$$

$$CP_{FLUE\ GAS} = \frac{7.5}{1775 - 145} = 0.0046 \text{ MW} \cdot \text{K}^{-1}$$

The fuel consumption is now calculated by taking the flue gas from the theoretical flame temperature to ambient temperature:

$$\text{Fuel required} = 0.0046\,(1800 - 10) = 8.23 \text{ MW}$$

$$\text{Stack loss} = 0.0046\,(170 - 10) = 0.74 \text{ MW}$$

$$\text{Furnace efficiency} = \frac{Q_{H\,min}}{\text{Fuel required}} \times 100 = \frac{7.5}{8.23} \times 100 = 91\%$$

17

17.9 Cogeneration (Combined Heat and Power Generation)

More complex utility options are encountered when combined heat and power generation (or *cogeneration*) is exploited. Here the heat rejected by a heat engine such as a steam turbine, gas turbine or diesel engine is used as hot utility.

Fundamentally, there are two possible ways to integrate a heat engine exhaust (Townsend and Linnhoff, 1983). In Figure 17.31, the process is represented as a heat sink and heat source separated by the pinch. Integration of the heat engine across the pinch as shown in Figure 17.31a is counterproductive. The process still requires Q_{Hmin} and the heat engine performs no better than operated stand-alone. There is no saving by integrating a heat engine across the pinch (Townsend and Linnhoff, 1983).

Figure 17.31b shows the heat engine integrated above the pinch. In rejecting heat above the pinch, it is rejecting heat into the part of the process, which is overall a heat sink. In so doing, it is exploiting the temperature difference that exists between the utility source and the process sink, producing power at high efficiency. The net effect in Figure 17.31b is the import of extra energy W from heat sources to produce W power. Owing to the process and heat engine acting as one system, apparently conversion of heat to power at 100% efficiency is achieved (Townsend and Linnhoff, 1983).

Now consider the two most commonly used heat engines (steam and gas turbines) in more detail to see whether they achieve this in practice. To make a quantitative assessment of any combined heat and power scheme, the grand composite curve should be used and the heat engine exhaust treated like any other utility.

Figure 17.32 shows a steam turbine integrated with the process above the pinch. A steam turbine is conceptually like a centrifugal compressor, but working in reverse. Steam is expanded from high to low pressure in the machine, producing power. In Figure 17.32 heat Q_{HP} is taken into the process from high-pressure steam. The balance of the hot utility demand Q_{LP} is taken from the steam turbine exhaust. In Figure 17.32a, heat Q_{FUEL} is taken into the boiler from fuel. An overall energy balance gives:

$$Q_{FUEL} = Q_{HP} + Q_{LP} + W + Q_{LOSS} \tag{17.2}$$

The process requires ($Q_{HP} + Q_{LP}$) to satisfy its enthalpy imbalance above the pinch. If there were no losses from the boiler, then fuel W would be converted to power W at 100% efficiency. However, the boiler losses Q_{LOSS} reduce this to below 100% conversion. In practice, in addition to the boiler losses, there can also be significant losses from the steam distribution system. Figure 17.32b shows how the grand composite curve can be used to size steam turbine cycles (Townsend and Linnhoff, 1983). Steam turbines will be dealt with in more detail in Chapter 23.

Figure 17.33a shows a schematic of a simple gas turbine. The machine is essentially a rotary compressor mounted on the same shaft as a turbine. Air enters the compressor where it is compressed before entering a combustion chamber. Here the combustion of fuel increases its temperature. The mixture of air and combustion

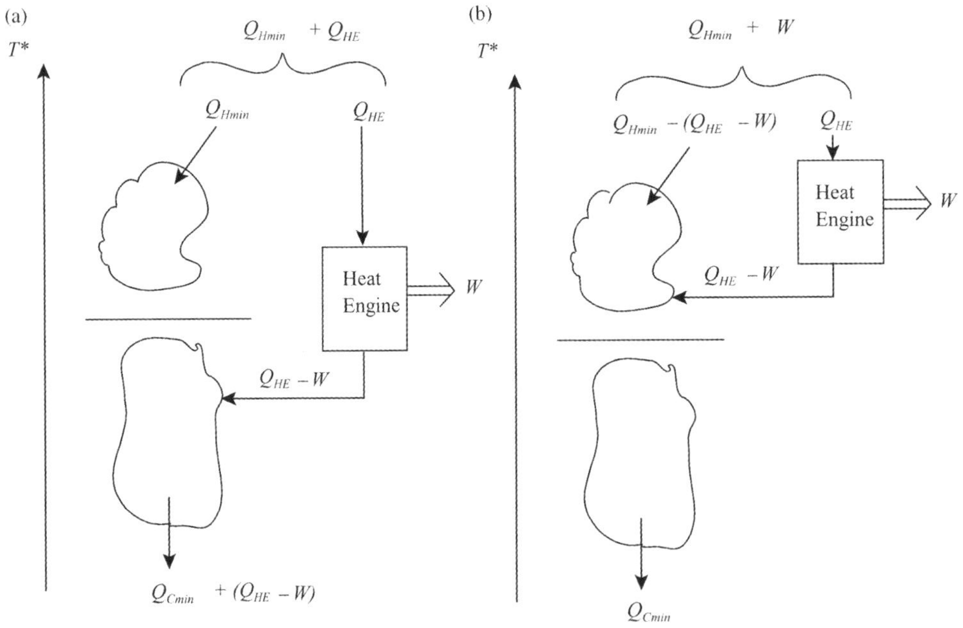

Figure 17.31
Heat engine exhaust can be integrated either across or not across the pinch.

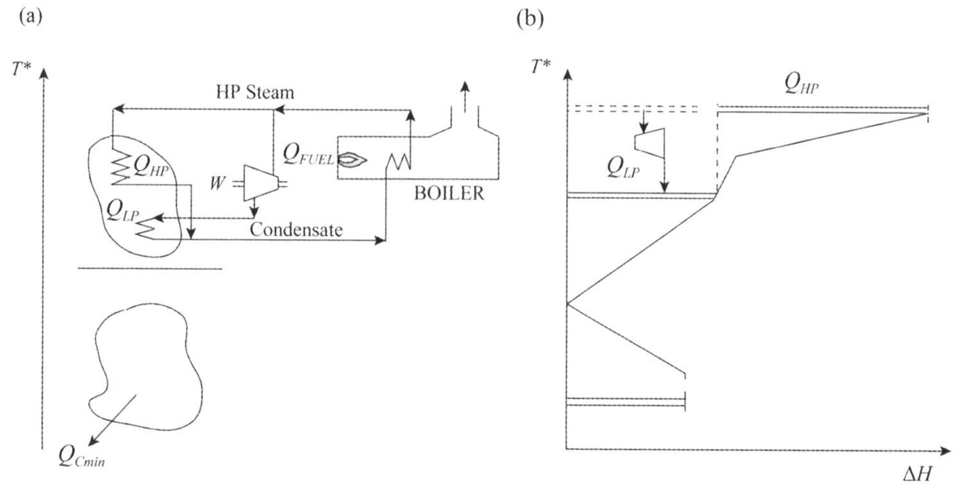

Figure 17.32

Integration of a steam turbine with the process.

gases is expanded in the turbine. The input of energy to the combustion chamber allows enough power to be developed in the turbine to both drive the compressor and provide useful power. The performance of the machine is specified in terms of the power output, airflow rate through the machine, efficiency of conversion of heat to power and the temperature of the exhaust. Gas turbines are normally used only for relatively large-scale applications, and will be dealt with in more detail in Chapter 23.

Figure 17.33b shows a gas turbine matched against the grand composite curve (Townsend and Linnhoff, 1983). As with the steam turbine, if there was no stack loss to atmosphere (i.e. if Q_{LOSS} was zero), then W heat would be turned into W power. The stack losses in Figure 17.33 reduce the efficiency of conversion of heat to power. The overall efficiency of conversion of heat to power depends on the turbine exhaust profile, the pinch temperature and the shape of the process grand composite curve.

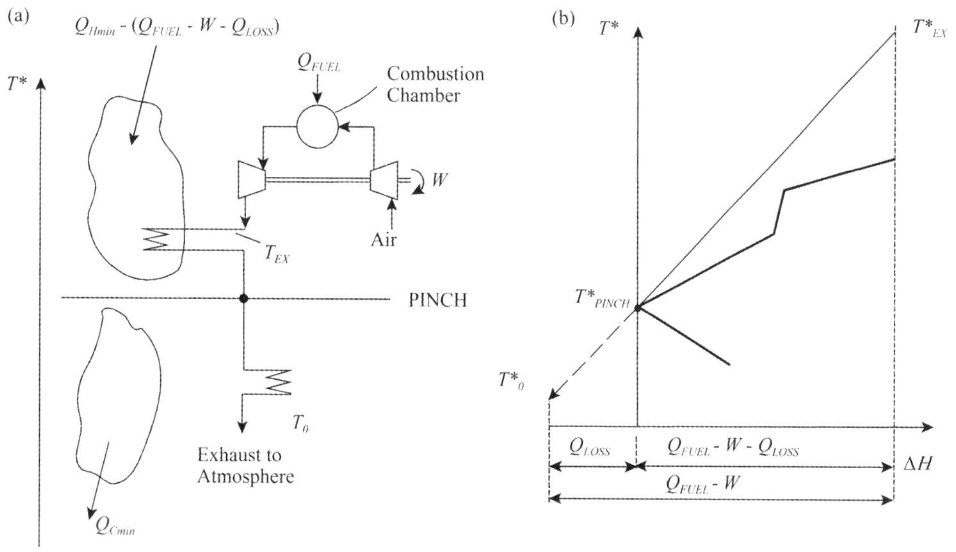

Figure 17.33

A gas turbine exhaust matched against the process (same as a flue gas).

Table 17.7

Stream data for Example 17.5.

Stream		T_S (°C)	T_T (°C)	Heat capacity flowrate ($MW \cdot K^{-1}$)
No.	Type			
1	Hot	450	50	0.25
2	Hot	50	40	1.5
3	Cold	30	400	0.22
4	Cold	30	400	0.05
5	Cold	120	121	22.0

Example 17.5 The stream data for a heat recovery problem are given in Table 17.7.

A problem table analysis for $\Delta T_{min} = 20\,°C$ results in the heat cascade given in Table 17.8.

Table 17.8

Problem table cascade for Example 17.5.

T^* (°C)	Cascade heat flow (MW)
440	21.9
410	29.4
131	23.82
130	1.8
40	0
30	15

The process also has a requirement for 7 MW of power. Two alternative cogeneration schemes are to be compared economically.

a) A steam turbine with its exhaust saturated at 150 °C used for process heating is one of the options to be considered. Superheated steam is generated in the central boiler house at 41 bar with a temperature of 300 °C. This superheated steam can be expanded in a single-stage turbine with an isentropic efficiency of 85%. Calculate the maximum generation of power possible by matching the exhaust steam against the process.

b) A second possible scheme uses a gas turbine with a flowrate of air of $97\,kg \cdot s^{-1}$, which has an exhaust temperature of 400 °C. Calculate the power generation if the turbine has an efficiency of 30% under these conditions of an ambient temperature of 10 °C.

c) The cost of heat from fuel for the gas turbine is $\$4.5\,GW^{-1}$. The cost of imported electricity is $\$19.2\,GW^{-1}$. Electricity can be exported with a value of $\$14.4\,GW^{-1}$. The cost of fuel for steam generation is $\$3.2\,GW^{-1}$. The overall efficiency of steam generation and distribution is 80%. Which scheme is most cost-effective, the steam turbine or the gas turbine?

Solution

a) This is shown in Figure 17.34a. The steam condensing interval temperature is 140 °C.

Heat flow required from the turbine exhaust = 21.9 MW

From steam tables, inlet conditions at $T_1 = 300\,°C$ and $P_1 = 41$ bar are:

$$H_1 = 2959\,kJ \cdot kg^{-1}$$
$$S_1 = 6.349\,kJ \cdot kg^{-1} \cdot K^{-1}$$

Turbine outlet conditions for isentropic expansion to 150 °C from steam tables are:

$$P_2 = 4.77\,bar$$
$$S_2 = 6.349\,kJ \cdot kg^{-1} \cdot K^{-1}$$

The wetness fraction (X) can be calculated from:

$$S_2 = X\,S_L + (1 - X)S_V$$

where S_L and S_V are the saturated liquid and vapor entropies. Taking saturated liquid and vapor entropies from steam tables at 150 °C and 4.77 bar:

$$6.349 = 1.842X + 6.838\,(1 - X)$$
$$X = 0.098$$

The turbine outlet enthalpy for an isentropic expansion can now be calculated from:

$$H_2 = X\,H_L + (1 - X)H_V$$

where H_L and H_V are the saturated liquid and vapor enthalpies. Taking saturated liquid and vapor enthalpies from steam tables at 150 °C and 4.77 bar:

$$H_2 = 0.098 \times 632 + (1 - 0.098)\,2747$$
$$= 2540\,kJ \cdot kg^{-1}$$

For a single-stage expansion with isentropic efficiency of 85%:

$$H_2' = H_1 - \eta_{IS}(H_1 - H_2)$$
$$= 2959 - 0.85(2959 - 2540)$$
$$= 2603\,kJ \cdot kg^{-1}$$

The actual wetness fraction (X') can be calculated from:

$$H_2' = X'\,H_L + (1 - X')H_V$$

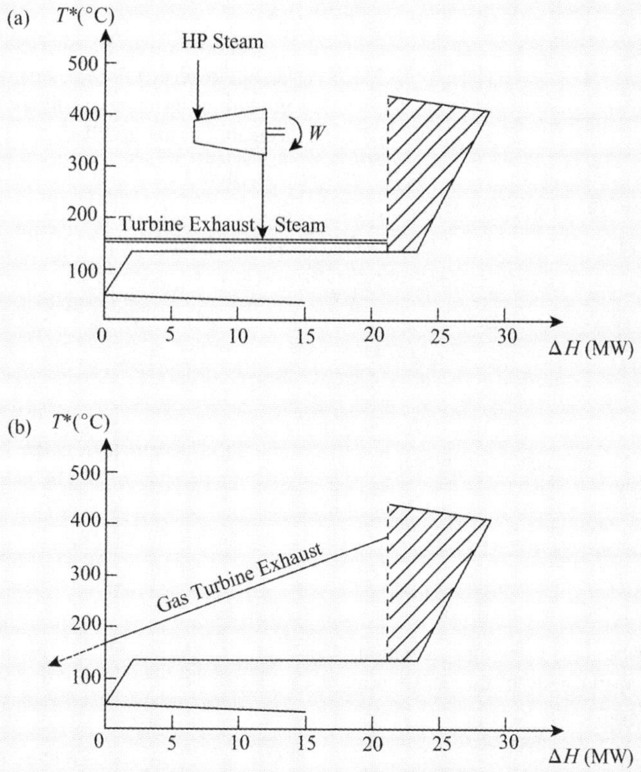

Figure 17.34
Alternative combined heat and power schemes for Example 17.5.

where H_L and H_V are the saturated liquid and vapor enthalpies:

$$H_2' = 2603 = 632X' + 2747\,(1 - X')$$
$$X' = 0.068$$

Assume in this case that the saturated steam and condensate are separated after the turbine and only the saturated steam is used for process heating:

$$\text{Steam flow to process} = \frac{21.9 \times 10^3}{2747 - 632} = 10.35\ \text{kg}\cdot\text{s}^{-1}$$

$$\text{Steam flow through turbine} = \frac{10.35}{(1 - 0.068)} = 11.11\ \text{kg}\cdot\text{s}^{-1}$$

$$\text{Power generated } W = 11.11\,(2959 - 2603) \times 10^{-3} = 3.96\ \text{MW}$$

b) The exhaust from the gas turbine is primarily air with a small amount of combustion gases. Hence, the *CP* of the exhaust can be approximated to be that of the airflow. Assuming C_P for air = 1.03 kJ·kg^{-1}·K^{-1}:

$$CP_{EX} = 97 \times 1.03 = 100\ \text{kW}\cdot\text{K}^{-1}$$

The gas turbine option is shown in Figure 17.34b:

$$Q_{EX} = CP_{EX}(T_{EX} - T_0) = 0.1 \times (400 - 10) = 39\ \text{MW}$$

$$Q_{FUEL} = \frac{Q_{EX}}{(1 - \eta_{GT})} = \frac{39}{(1 - 0.3)} = 55.71\ \text{MW}$$

$$W = Q_{FUEL} - Q_{EX} = 16.71\ \text{MW}$$

c) Steam turbine economics:

$$\text{Cost of fuel} = (21.9 + 3.96) \times \frac{3.2 \times 10^{-3}}{0.8} = \$0.10\ \text{s}^{-1}$$

$$\text{Cost of imported electricity} = (7 - 3.96) \times 19.2 \times 10^{-3} = \$0.06\ \text{s}^{-1}$$

$$\text{Net cost} = \$0.16\ \text{s}^{-1}$$

17

Gas turbine economics:

$$\text{Cost of fuel} = 55.71 \times 4.5 \times 10^{-3}$$
$$= \$0.25\ s^{-1}$$
$$\text{Electricity credit} = (16.71 - 7) \times 14.4 \times 10^{-3}$$
$$= \$0.14\ s^{-1}$$
$$\text{Net cost} = \$0.11\ s^{-1}$$

Hence the gas turbine is the most profitable in terms of energy costs. However, this is only part of the story since the capital cost of a gas turbine installation is likely to be significantly higher than that of a steam turbine installation.

Example 17.6 The problem table cascade for a process is given in Table 17.9 for $\Delta T_{min} = 10$ °C.

It is proposed to provide process cooling by steam generation from boiler feedwater with a temperature of 100 °C.

a) Determine how much steam can be generated at a saturation temperature of 230 °C.

b) Determine how much steam can be generated with a saturation temperature of 230 °C and superheated to the maximum temperature possible against the process.

c) Calculate how much power can be generated from the superheated steam from Part b, assuming a single-stage condensing steam turbine is to be used with an isentropic efficiency of 85%. Cooling water is available at 20 °C and is to be returned to the cooling tower at 30 °C.

Solution

a) Heat available for steam generation at 235 °C interval temperature

$$= 12.0\ \text{MW}$$

Table 17.9

Problem table cascade.

Interval temperature (°C)	Heat flow (MW)
495	3.6
455	9.2
415	10.8
305	4.2
285	0
215	16.8
195	17.6
185	16.6
125	16.6
95	21.1
85	18.1

From steam tables, the latent heat of water at a saturated temperature of 230 °C is 1812 kJ·kg^{-1}.

$$\text{Steam production} = \frac{12.0 \times 10^3}{1812}$$
$$= 6.62\ kg \cdot s^{-1}$$

Taking the heat capacity of water to be 4.3 kJ·kg^{-1}·K^{-1}, heat duty on boiler feedwater preheating

$$= 6.62 \times 4.3 \times 10^{-3}(230 - 100)$$
$$= 3.70\ \text{MW}$$

The profile of steam generation is shown against the grand composite curve in Figure 17.35a. The process can support both boiler feedwater preheat and steam generation.

b) Maximum superheat temperature

$$= 285\ ^\circ\text{C interval}$$
$$= 280\ ^\circ\text{C actual}$$

The profile is shown against the grand composite curve in Figure 17.35b.

From steam tables, enthalpy of superheated steam at 280 °C and 28 bar

$$= 2947\ kJ \cdot kg^{-1}$$

and enthalpy of saturated water at 230 °C and 28 bar

$$= 991\ kJ \cdot kg^{-1}$$
$$\text{Steam production} = \frac{12.0 \times 10^3}{(2947 - 991)}$$
$$= 6.13\ kg \cdot s^{-1}$$

c) In a condensing turbine, the exhaust from the turbine is condensed under vacuum against cooling water. The lower the condensing temperature, the greater the power generation. The lowest condensing temperature for this problem is cooling water temperature plus ΔT_{min}, that is, 30 + 10 = 40 °C.

From steam tables, inlet conditions at $T_1 = 280$ °C and $P_1 = 28$ bar are:

$$H_1 = 2947\ kJ \cdot kg^{-1}$$
$$S_1 = 6.488\ kJ \cdot kg^{-1} \cdot K^{-1}$$

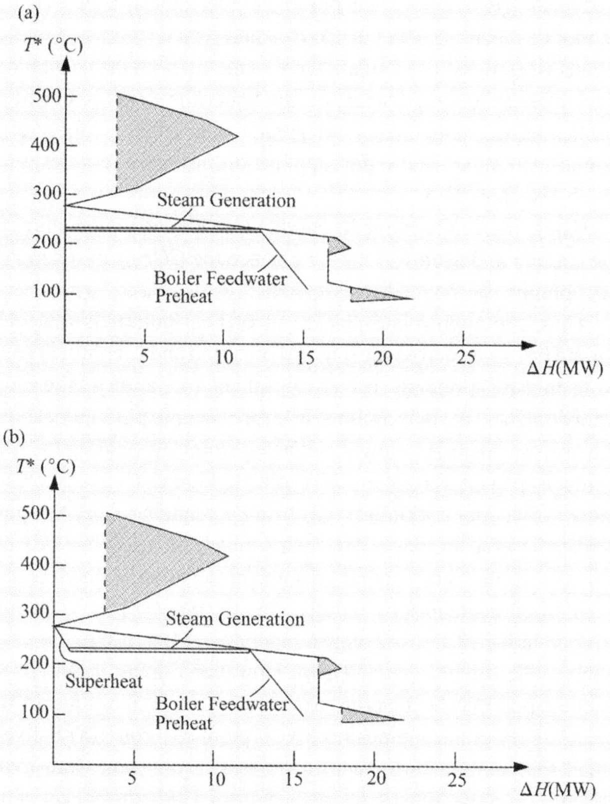

Figure 17.35

Alternative cold utilities for Example 17.6.

Turbine outlet conditions for isentropic expansion to 40 °C from steam tables are:

$$P_2 = 0.074 \text{ bar}$$

For $S_2 = 6.488 \text{ kJ}\cdot\text{kg}^{-1}\cdot\text{K}^{-1}$, the wetness fraction ($X$) and outlet enthalpy H_2 can be calculated as shown in Example 17.5:

$$\begin{aligned} X &= 0.23 \\ H_2 &= 2020 \text{ kJ} \cdot \text{kg}^{-1} \end{aligned}$$

For a single-stage expansion with isentropic efficiency of 85%:

$$\begin{aligned} H_2' &= 2947 - 0.85\,(2947 - 2020) \\ &= 2159 \text{ kJ} \cdot \text{kg}^{-1} \end{aligned}$$

The power generation (W) is given by:

$$\begin{aligned} W &= 6.13\,(2947 - 2159) \times 10^{-3} \\ &= 4.8 \text{ MW} \end{aligned}$$

The wetness fraction for the real expansion is given by:

$$\begin{aligned} H_2' = 2159 &= X' H_L + (1 - X')H_V \\ &= 167.5X' + 2574(1 - X') \\ X' &= 0.17 \end{aligned}$$

This wetness fraction is possibly too high, since high levels of wetness can cause damage to the turbine. To allow a lower wetness fraction of, say, $X = 0.15$, the outlet pressure of the turbine must be raised to 0.2 bar, corresponding to a condensing temperature of 60 °C. However, in so doing, the power generation decreases to 4.2 MW.

17.10 Integration of Heat Pumps

A heat pump is a device that takes in low-temperature heat and upgrades it to a higher temperature to provide process heat. A schematic of a simple vapor compression heat pump is shown in Figure 17.36. In Figure 17.36, the heat pump absorbs heat at a low temperature in the evaporator, consumes power when the working fluid is compressed and rejects heat at a higher temperature in the condenser. The condensed working fluid is expanded and partially vaporizes. The cycle then repeats. The working fluid is usually a pure component, which means that the evaporation and condensation take place isothermally. When considering the integration of a heat pump with the process, there are appropriate and inappropriate ways to integrate heat pumps.

There are two fundamental ways in which a heat pump can be integrated with the process; across and not across the pinch (Townsend and Linnhoff, 1983). Integration not across (above) the pinch is illustrated in Figure 17.37a. This arrangement imports W power and saves W hot utility. In other words, the system converts power into heat, which is not normally worthwhile economically. Another integration not across (below) the pinch is shown in Figure 17.37b. The result is worse economically. Power is turned into waste heat (Townsend and Linnhoff, 1983).

Integration across the pinch is illustrated in Figure 17.37c. This arrangement brings a genuine saving. It also makes overall sense since heat is pumped from the part of the process that is overall a heat source to the part that is overall a heat sink.

Figure 17.38 shows a heat pump appropriately integrated against a process. Figure 17.38a shows the overall balance. Figure 17.38b illustrates how the grand composite curve can be used to size the heat pump. If the cold side of the heat pump is fixed in Figure 17.38b then power is needed to provide the temperature lift. The greater the temperature lift, the greater the power requirement and the greater the heat to be rejected. In Figure 17.38b this follows the *A-B* profile. Figure 17.38b shows the minimum temperature at which the heat can be rejected above the pinch. How the heat pump performs determines its coefficient of performance. The coefficient of performance for a heat pump can

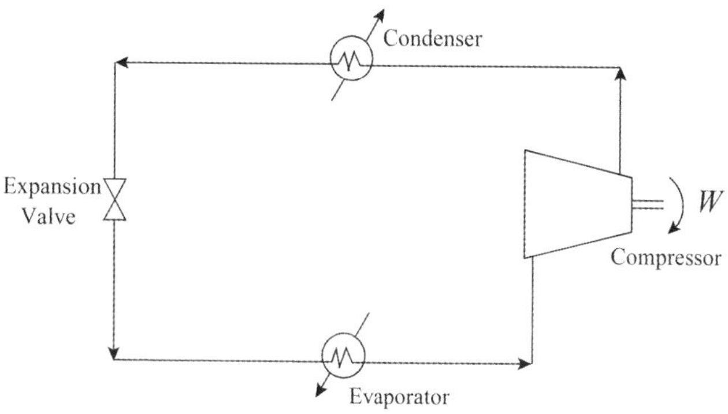

Figure 17.36

Schematic of a simple vapor compression heat pump.

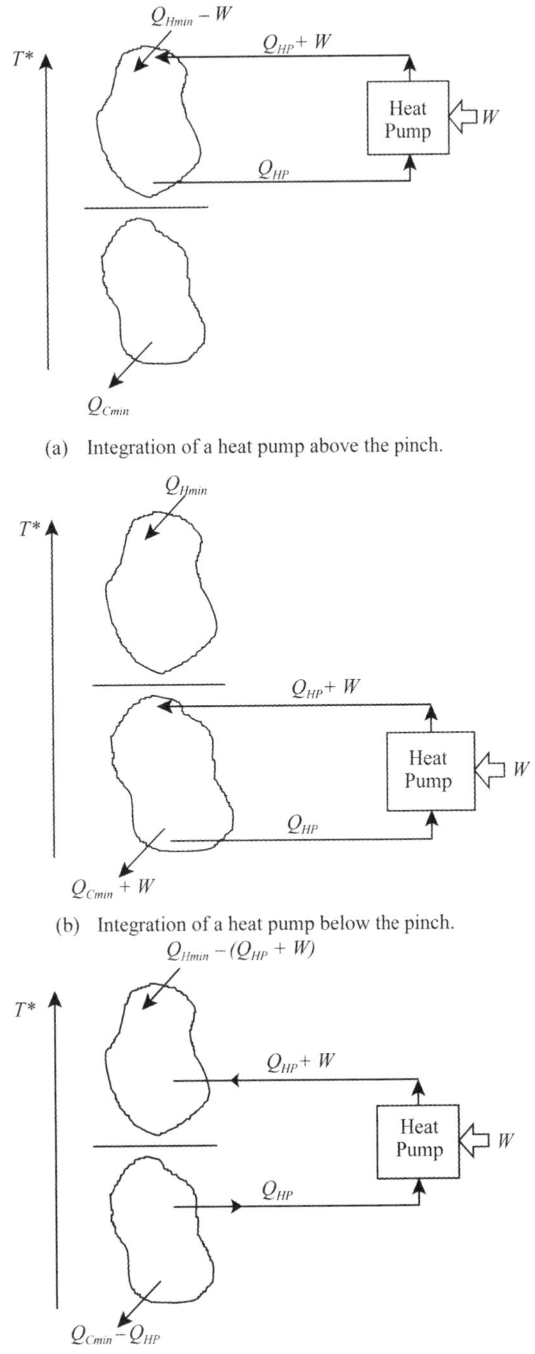

(a) Integration of a heat pump above the pinch.

(b) Integration of a heat pump below the pinch.

(c) Integration of a heat pump across the pinch.

Figure 17.37

Integration of a heat pump with the process.

generally be defined as the useful energy delivered to the process divided by the power expended to produce this useful energy:

$$COP_{HP} = \frac{Q_{HP} + W}{W} \tag{17.3}$$

where COP_{HP} is the heat pump coefficient of performance, Q_{HP} the heat absorbed at low temperature and W the power consumed.

For any given type of heat pump, a higher COP_{HP} leads to better economics. Having a better COP_{HP} and hence better economics means working across a small temperature lift with the heat pump. The smaller the temperature lift, the better the COP_{HP}. For most applications, a temperature lift greater than 25 °C is rarely economic. Attractive heat pump application normally requires a lift much less than 25 °C.

Using the grand composite curve, the loads and temperatures of the cooling and heating duties and hence the COP_{HP} of integrated heat pumps can be readily assessed.

Thus, the appropriate placement of heat pumps is that they should be placed across the pinch (Townsend and Linnhoff, 1983). Note that the principle needs careful interpretation if there are utility pinches. In such circumstances, heat pump placement above the process pinch or below it can be economic, providing the heat pump is placed across a utility pinch. Figure 17.39a shows a process that does not show a significant potential for heat pumping across the process pinch. The heat source just below the process pinch is small. The heat sink just above the process pinch is also small. Significant heat pumping across the process pinch can only be accomplished across a large temperature difference in this case, which will result in a poor COP_{HP}. However, Figure 17.39b shows another possible heat pump arrangement. A heat source in the pocket of the grand composite curve is pumped through a relatively small temperature difference to replace medium-pressure (MP) steam. To compensate for taking heat from within the pocket of the grand composite curve, additional low-pressure (LP) steam is used for process heating to maintain the heat balance. Thus, the heat pump results in a saving in MP steam for a sacrifice of extra LP steam usage and the power required for the heat pump. This might be economic if there is a large cost difference between MP and LP steam. Although the heat pump does not operate across the process pinch, it does not violate the principle of the appropriate placement of heat pumps. The heat pump in Figure 17.39b operates across a utility pinch.

17.11 Number of Heat Exchange Units

In the design to follow the targeting stage, heat exchange *matches* are to be made between hot and cold streams in heat exchange *units*. Consider now how to set a target for the minimum number of units. It should be noted that a unit will not necessarily be a single heat exchanger, but might involve multiple heat exchangers in series or parallel on the same stream. The unit on both sides of the match will have a single inlet and outlet on the same stream, but might constitute multiple heat exchangers in different configurations on either side of the match.

To understand the minimum number of matches or units in a heat exchanger network, some basic results of *graph theory* can be used (Hohman, 1971; Linnhoff, Mason and Wardle, 1979). A *graph* is any collection of points in which some pairs of points are connected by lines. Figures 17.40a and 17.40b give two examples of graphs. Note that the lines such as *BG* and *CE* and *CF* in Figure 17.40a are not supposed to cross, that is, the diagram should

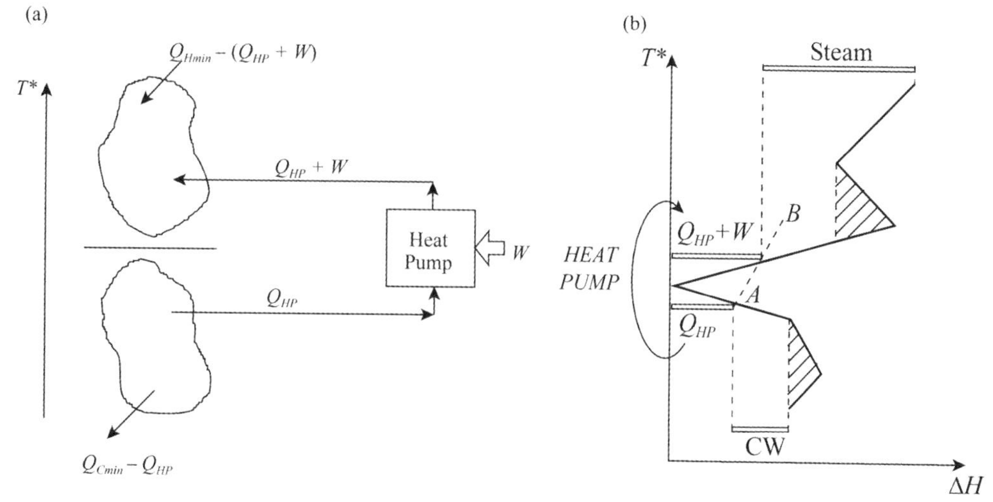

Figure 17.38
The grand composite curve allows heat pump cycles to be sized.

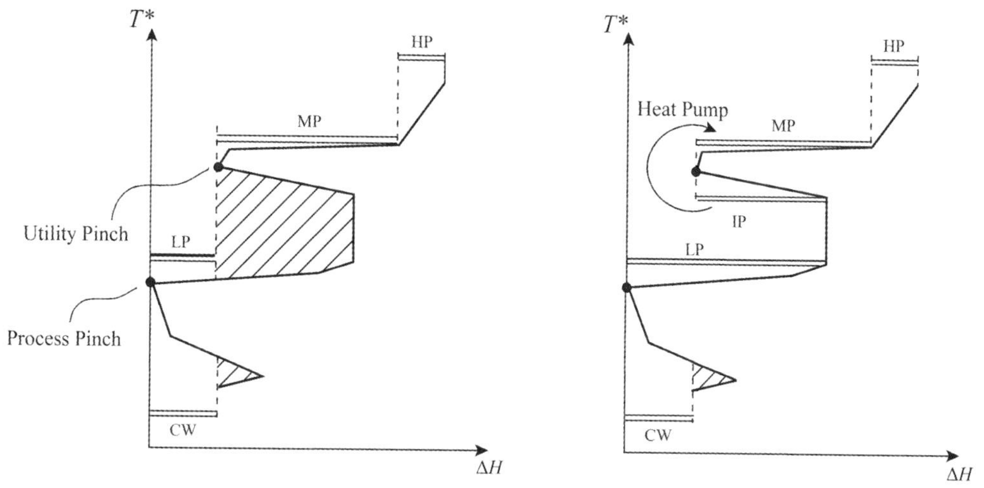

(a) A process that shows little scope for heat pumping across the process pinch.
(b) Heat pump placed across a utility pinch.

Figure 17.39
Heat pumping can be applied across utility pinches as well as the process pinch.

be drawn in three dimensions. This is true for the other lines in Figure 17.40 that appear to cross.

In this context, the points correspond to process and utility streams, and the lines to heat exchange matches between the heat sources and heat sinks.

A *path* is a sequence of distinct lines that are connected to each other. For example, in Figure 17.40a *AECGD* is a path. A graph forms a single *component* (sometimes called a *separate system*) if any two points are joined by a path. Thus, Figure 17.40b has two components (or two separate systems) and Figure 17.40a has only one.

A *loop* is a path that begins and ends at the same point, like *CGDHC* in Figure 17.40a. If two loops have a line in common, they can be linked to form a third loop by deleting the common line. In Figure 17.40a, for example, *BGCEB* and *CGDHC* can be linked to give *BGDHCEB*. In this case, this last loop is said to be *dependent* on the other two.

From graph theory, the main result needed in the present context is that the number of independent loops for a graph is given by (Hohman, 1971; Linnhoff, Mason and Wardle, 1979):

$$N_{UNITS} = S + L - C \tag{17.4}$$

17

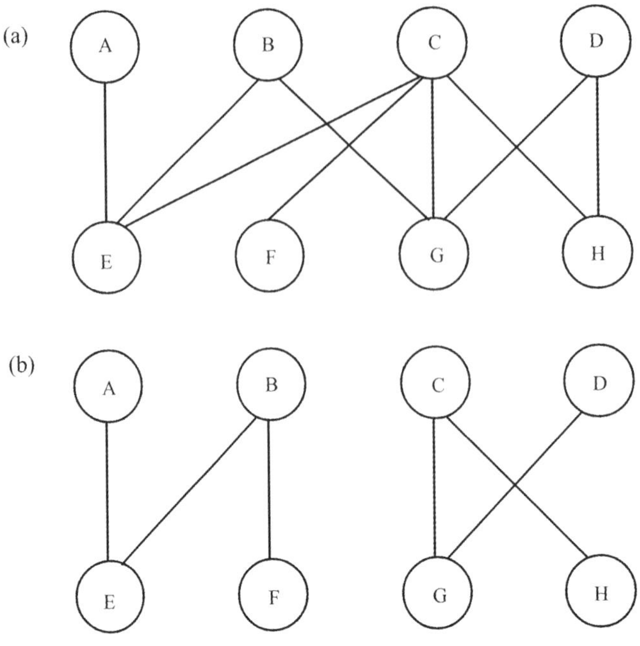

Figure 17.40
Two alternative graphs. (Reproduced from Linnhoff B, Mason D and Wardle I, 1979, *Comp and Chem Engg*, **3**: 279, with permission from Elseiver.)

where N_{UNITS} = number of matches or units (lines in graph theory)
S = number of streams including utilities (points in graph theory)
L = number of independent loops
C = number of components

In general, if the final network design is achieved in the minimum number of units it will keep down the capital cost (although this is not the only consideration to keep down the capital cost). To minimize the number of units in Equation 17.4, L should be zero and C should be a maximum. Assuming L to be zero in the final design is a reasonable assumption. However, what should be assumed about C? Consider the network in Figure 17.40b that has two components. For there to be two components, the heat duties for streams A and B must exactly balance the duties for streams E and F. Also, the heat duties for streams C and D must exactly balance the duties for streams G and H. Such balances are likely to be unusual and not easy to predict.

The safest assumption for C thus appears to be that there will be one component only, that is, $C = 1$. This leads to an important special case when the network has a single component and is loop-free. In this case (Hohman, 1971; Linnhoff, Mason and Wardle, 1979):

$$N_{UNITS} = S - 1 \tag{17.5}$$

Equation 17.5 put in words states that the minimum number of units required is one less than the number of streams (including utility streams).

This is a useful result since, if the network is assumed to be loop-free and has a single component, the minimum number of units can be predicted simply by knowing the number of streams. If the problem does not have a pinch, then Equation 17.5 predicts the minimum number of units. If the problem has a pinch, then Equation 17.5 is applied on each side of the pinch separately (Linnhoff, Mason and Wardle, 1979):

$$N_{UNITS} = [S_{ABOVE\ PINCH} - 1] + [S_{BELOW\ PINCH} - 1] \tag{17.6}$$

Example 17.7 For the process in Figure 17.41, calculate the minimum number of units given that the pinch is at 150 °C for the hot streams and 140 °C for the cold streams.

Solution Figure 17.41 shows the stream grid with the pinch in place dividing the process into two parts. Above the pinch, there are five streams, including the steam. Below the pinch, there are four streams, including the cooling water. Applying Equation 17.6:

$$\begin{aligned} N_{UNITS} &= (5 - 1) + (4 - 1) \\ &= 7 \end{aligned}$$

Looking back at the design presented for this problem in Figure 17.9, it does in fact use the minimum number of units of 7. In the next chapter, design for the minimum number of units will be addressed.

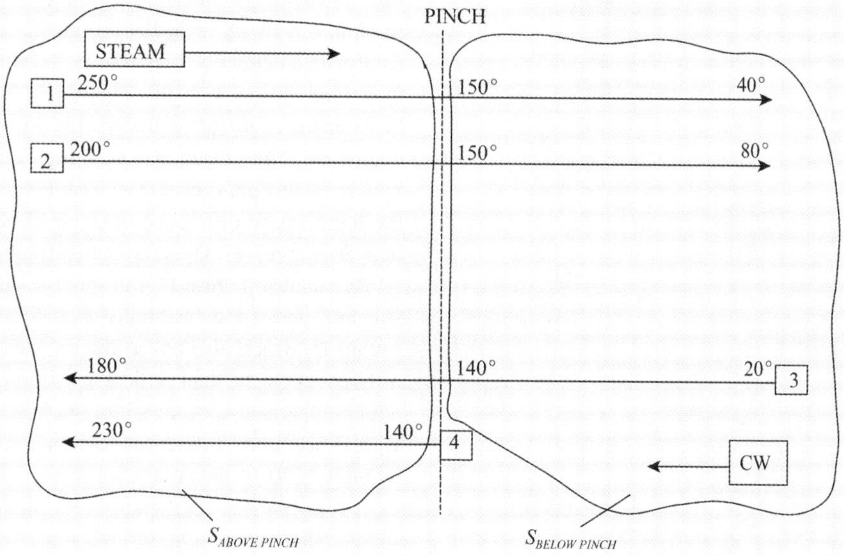

Figure 17.41

To target the number of units for pinched problems the streams above and below the pinch must be counted separately with the appropriate utilities included.

17.12 Heat Exchange Area Targets

In addition to giving the necessary information to predict energy targets, the composite curves also contain the necessary information to predict the network heat transfer area. To calculate the network area from the composite curves, utility streams must be included with the process streams in the composite curves to obtain the *balanced composite curves* (Townsend and Linnhoff, 1984), going through the same procedure as illustrated in Figures 17.3 and 17.4 but including the utility streams. The resulting balanced composite curves should have no residual demand for utilities. The balanced composite curves are divided into vertical *enthalpy intervals*, as shown in Figure 17.42. Assume initially that the overall heat transfer coefficient U is constant throughout the process. Assuming true countercurrent heat transfer, the area requirement for enthalpy interval k for this *vertical heat transfer* is given by (Hohman, 1971; Townsend and Linnhoff, 1984):

$$A_{NETWORK\,k} = \frac{\Delta H_k}{U\,\Delta T_{LMk}} \tag{17.7}$$

where $A_{NETWORK\,k}$ = heat exchange area for vertical heat transfer required by interval k

ΔH_k = enthalpy change over interval k

ΔT_{LMk} = log mean temperature difference for interval k

U = overall heat transfer coefficient

To obtain the network area, Equation 17.7 is applied to all enthalpy intervals (Hohman, 1971; Townsend and Linnhoff, 1984):

$$A_{NETWORK} = \frac{1}{U}\sum_{k}^{INTERVALS\ K} \frac{\Delta H_k}{\Delta T_{LMk}} \tag{17.8}$$

where $A_{NETWORK}$ = heat exchange area for vertical heat transfer for the whole network

K = total number of enthalpy intervals

The problem with Equation 17.8 is that the overall heat transfer coefficient is not constant throughout the process. Is there some way to extend this model to deal with the individual heat transfer coefficients?

The effect of individual stream film transfer coefficients can be included using the following expression, which is derived in Appendix H (Townsend and Linnhoff, 1984; Linnhoff and Ahmad, 1990):

$$A_{NETWORK} = \sum_{k}^{INTERVALS\ K} \frac{1}{\Delta T_{LMk}} \times \left[\sum_{i}^{HOT\ STREAMS\ I} \frac{q_{i,k}}{h_i} + \sum_{j}^{COLD\ STREAMS\ J} \frac{q_{j,k}}{h_j}\right] \tag{17.9}$$

where $q_{i,k}$ = stream duty on hot stream i in enthalpy interval k

$q_{j,k}$ = stream duty on cold stream j in enthalpy interval k

h_i, h_j = film transfer coefficients for hot stream i and cold stream j (including wall and fouling resistances)

I = total number of hot streams in enthalpy interval k

17

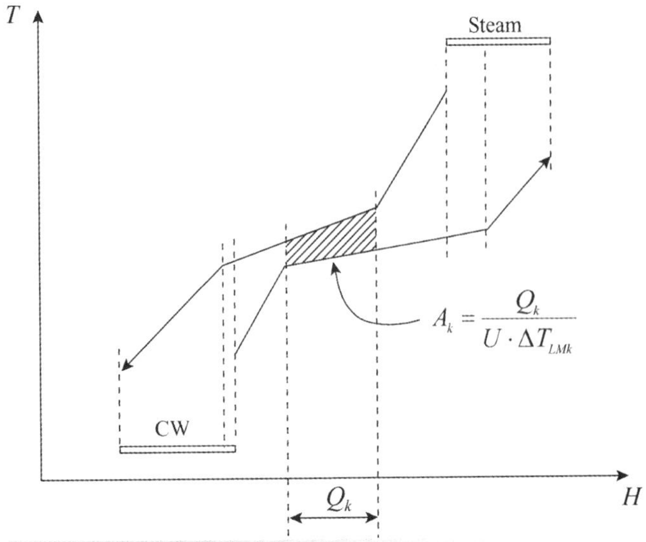

Figure 17.42

To determine the network area the balanced composite curves are divided into enthalpy intervals.

J = total number of cold streams in enthalpy interval k

K = total number of enthalpy intervals

This simple formula allows the network area to be targeted, on the basis of a vertical heat exchange model if film transfer coefficients vary from stream to stream. However, if there are large variations in film transfer coefficients, Equation 17.9 does not predict the true minimum network area. If film transfer coefficients vary significantly, then deliberate nonvertical matching is required to achieve the minimum area. Consider Figure 17.43a. Hot stream A with a low heat transfer coefficient is matched against cold stream C with a high coefficient. Hot stream B with a high heat transfer coefficient is matched with cold stream D with a low coefficient. In both matches, the temperature difference is taken to be the vertical separation between the curves. This arrangement requires 1616 m^2 area overall.

By contrast, Figure 17.43b shows a different arrangement. Hot stream A with a low heat transfer coefficient is matched with cold stream D, which also has a low coefficient but uses temperature differences greater than vertical separation. Hot stream B is matched with cold stream C, both with high heat transfer coefficients, but with temperature differences less than vertical. This arrangement requires 1250 m^2 area overall, less than the vertical arrangement.

If film transfer coefficients vary significantly from stream to stream, the true minimum area must be predicted using linear programming (Saboo, Morari and Colberg, 1986; Ahmad, Linnhoff and Smith, 1990). However, Equation 17.9 is still a useful basis to calculate the network area for the purposes of capital cost estimation, for the following reasons.

1) Providing film coefficients vary by less than one order of magnitude, then Equation 17.9 has been found to predict the network area to within 10% of the actual minimum (Ahmad, Linnhoff and Smith, 1990).
2) Network designs tend *not* to approach the true minimum in practice, since a minimum area design is usually too complex to be practical. Putting the argument the other way around,

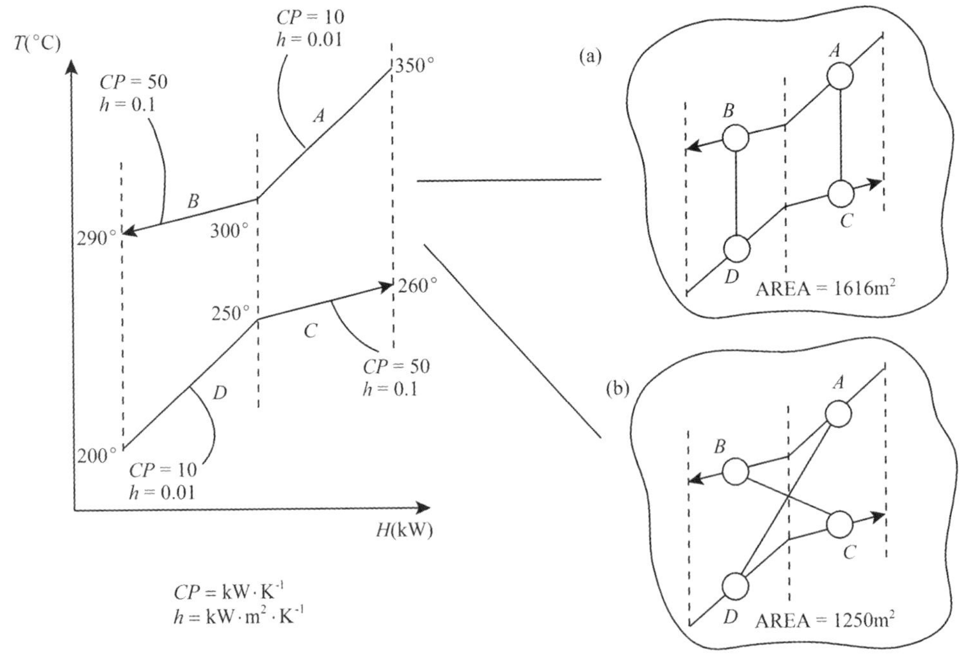

Figure 17.43

If film transfer coefficients differ significantly then nonvertical heat transfer is necessary to achieve the minimum area. (Reproduced from Linnhoff B and Ahmad S (1990) Cost Optimum Heat Exchanger Networks—1. Minimum Energy and Capital Using Simple Models for Capital Cost, *Comp Chem Eng*, **14**: 729., with permission from Elseiver.)

starting with the complex design required to achieve minimum area, then a significant reduction in complexity usually only requires a small penalty in area.

One significant problem remains – where to get the film transfer coefficients h_i and h_j from. There are the following three possibilities.

1) Tabulated experience values (see Chapter 12).
2) By assuming a reasonable fluid velocity, together with fluid physical properties, heat transfer correlations can be used (see Chapter 12).
3) If the pressure drop available for the stream is known, the expressions from Chapter 12 can be used.

The detailed allocation of fluids to the tube side or the shell side can only be made later in the heat exchanger network design. Also, the area targeting formula does not recognize fluids to be allocated to the tube side or the shell side. Area targeting only recognizes the individual film heat transfer coefficients. All that can be done in network area targeting is to make a preliminary estimate of the film heat transfer coefficient on the basis of an initial assessment as to whether the fluid is likely to be suited to tube-side or shell-side allocation in the final design. Thus, in addition to the approximations inherent within the area targeting formula, there is uncertainty regarding the preliminary assessment of the film heat transfer coefficients.

However, one other issue that needs to be included in the assessment often helps mitigate the uncertainties in the assessment of the film heat transfer coefficient. A fouling allowance needs to be added to the film transfer coefficient according to Equations 12.13 and 12.14.

Example 17.8 For the process in Figure 17.2, calculate the target for network heat transfer area for $\Delta T_{min} = 10\,°C$. Steam at 240 °C and condensing to 239 °C is to be used for hot utility. Cooling water at 20 °C and returning to the cooling tower at 30 °C is to be used for cold utility. Table 17.10 presents the stream data, together with utility data and stream heat transfer coefficients.

Calculate the heat exchange area target for the network.

Figure 17.45 now shows the stream population for each enthalpy interval together with the hot and cold stream temperatures. Set up a table to compute Equation 17.9. This is shown in Table 17.11.

Thus, the network area target for this problem for $\Delta T_{min} = 10\,°C$ is 7410 m^2.

Table 17.10

Complete stream and utility data for the process from Figure 17.2.

Stream	Supply temperature T_S (°C)	Target temperature T_T (°C)	ΔH (MW)	Heat capacity flowrate CP ($MW \cdot K^{-1}$)	Film heat transfer coefficient h ($MW \cdot m^{-2} \cdot K^{-1}$)
1. Reactor 1 feed	20	180	32.0	0.2	0.0006
2. Reactor 1 product	250	40	−31.5	0.15	0.0010
3. Reactor 2 feed	140	230	27.0	0.3	0.0008
4. Reactor 2 product	200	80	−30.0	0.25	0.0008
5. Steam	240	239	7.5	7.5	0.0030
6. Cooling water	20	30	10.0	1.0	0.0010

Solution First, the balanced composite curves must be constructed using the complete set of data from Table 17.10. Figure 17.44 shows the balanced composite curves. Note that the steam has been incorporated within the construction of the hot composite curve to maintain the monotonic nature of composite curves. The same is true of the cooling water in the cold composite curve. Figure 17.45 also shows the streams divided into enthalpy intervals according to the location in the composite curves. Where there is a change of slope either on the hot composite curve or on the cold composite curve there is an enthalpy interval boundary.

The network design in Figure 17.9 already achieves minimum energy consumption and it is now possible to judge how close the area target is to design if the area for the individual units in Figure 17.9 is calculated. Using the same heat transfer coefficients as given in Table 17.10, the design in Figure 17.9 requires some 8341 m^2, which is 13% above target. Remember that no attempt was made to steer the design in Figure 17.9 towards the minimum area. Instead, the design achieved the energy target in the minimum number of units and that tends to lead to simple designs.

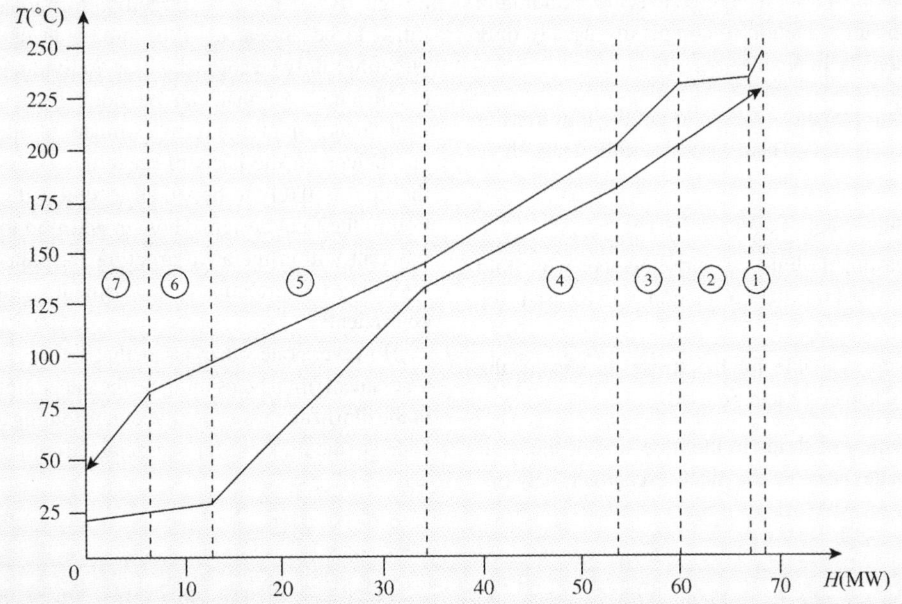

Figure 17.44

The enthalpy intervals for the balanced composite curves of Example 17.8.

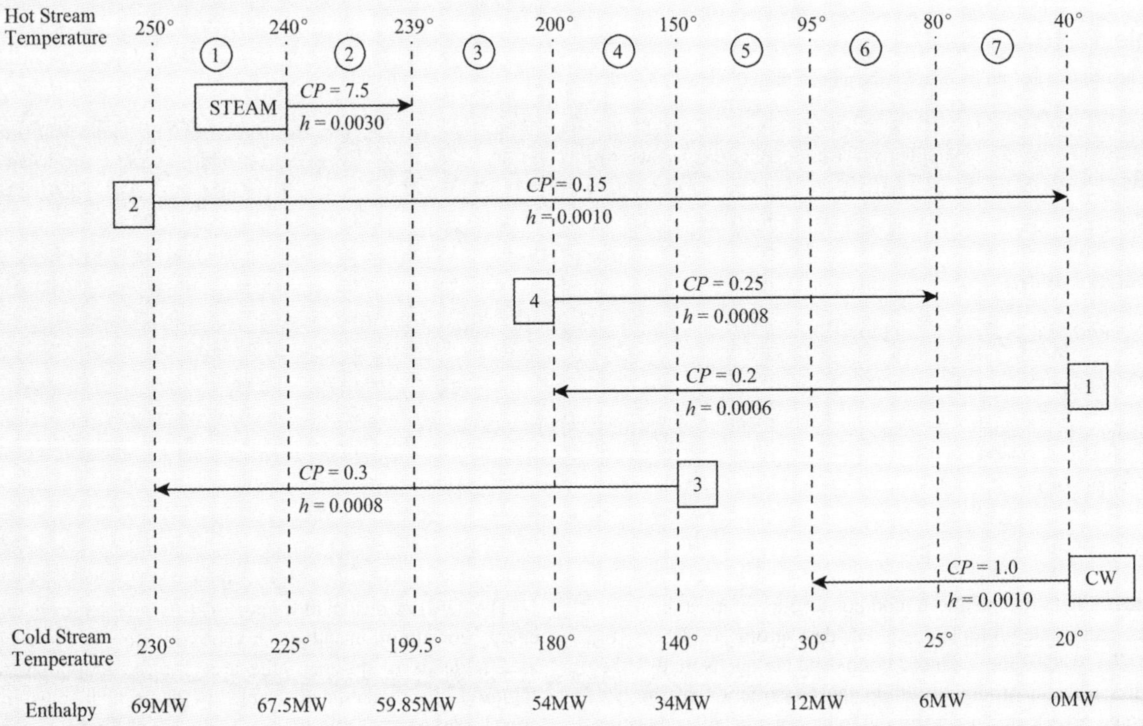

Figure 17.45

The enthalpy interval stream population for Example 17.8.

Table 17.11

Network area target for the process from Figure 17.2.

Enthalpy interval	ΔT_{LMk}	Hot streams $\Sigma(q_i/h_i)_k$	Cold streams $\Sigma(q_j/h_j)_k$		A_k
1	17.38	1500	1875		194.2
2	25.30	2650	9562.5		482.7
3	28.65	5850	7312.5		459.4
4	14.43	23,125	28,333.3		3566.1
5	29.38	25,437.5	36,666.7		2113.8
6	59.86	6937.5	6666.7		227.3
7	34.60	6000	6666.7		366.1
				ΣA_k	7409.6

17.13 Sensitivity of Targets

The targets for minimum energy, minimum number of units and network area are all based on an assumption for ΔT_{min}. It is important whenever carrying out process design to test the sensitivity of the design to the assumptions made. Thus, it is important to understand the sensitivity of the targets to the value of ΔT_{min}. Figure 17.46 illustrates how the three targets vary as the value of ΔT_{min} changes from a low to a high value for the flowsheet in Figure 17.2. Figure 17.46a shows the variation targets for hot and cold utilities with ΔT_{min}. As expected, both increase with increasing value of ΔT_{min}. With more complex problems, the sensitivity of this graph can change as ΔT_{min} varies, depending on the stream population for the problem. In this case the problem is simple and the slopes for the hot and cold utilities are both constant. Figure 17.46b shows the variation of number of units target with ΔT_{min}. In this case, again because it is a simple problem, the number of units target is constant. However, this is not true of all problems and number of units can vary as ΔT_{min} changes. Finally, Figure 17.46c shows the variation of the area target with ΔT_{min}. The area target generally decreases as ΔT_{min} increases. In this case, there are no sudden changes in the slope of the graph. However, there can be discontinuities in the curve with complex problems.

17.14 Capital and Total Cost Targets

To predict the capital cost of a network, it can be assumed that a heat exchange unit with surface area A can be costed according to a simple relationship such as:

$$\text{Installed capital cost of unit} = a + bA^c \qquad (17.10)$$

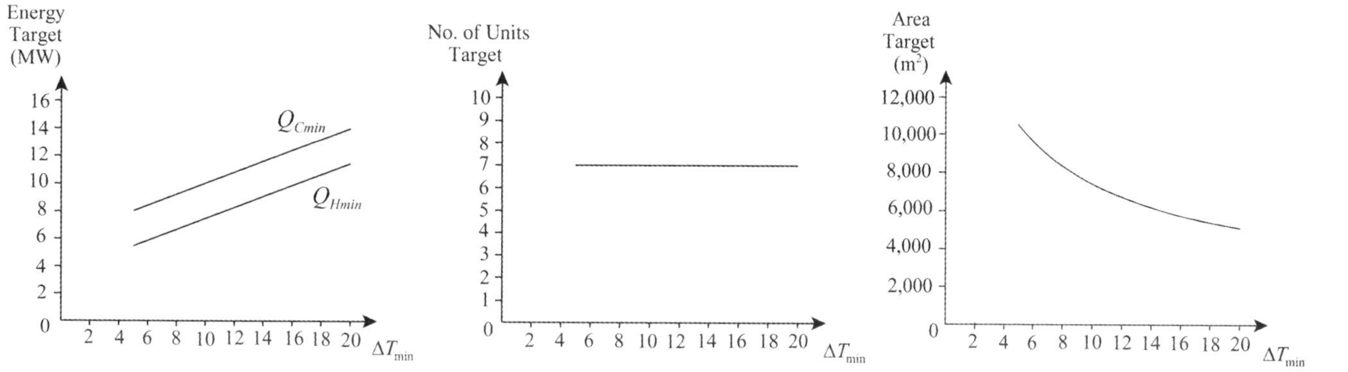

(a) Sensitivity of energy targets.

(b) Sensitivity of number of units target.

(c) Sensitivity of area target.

Figure 17.46

Sensitivity of targets for the process in Figure 17.2.

where a, b, c = cost law constants that vary according to materials of construction, pressure rating and type of exchanger

When cost targeting, the distribution of the targeted area between the units is unknown. Thus, to cost a network using Equation 17.10, some area distribution must be assumed across the units, the simplest being that all units have the same area:

$$\text{Network capital cost} = N_{UNITS}\,[a + b\,(A_{NETWORK}/N_{UNITS})^c] \qquad (17.11)$$

where N_{UNITS} = number of units

At the targeting stage, no given distribution can be judged consistently better than another, since the network is not yet known. Also, increasing the chosen value of process energy consumption increases temperature differences available for heat recovery and hence decreases the necessary heat exchanger surface area, Figure 17.46. The network area can be distributed over the targeted number of units to obtain a capital cost using Equation 17.11. This capital cost can be annualized as detailed in Chapter 2. The annualized capital cost can then be traded off against the annual utility cost, as shown in Figure 17.6. The total cost shows a minimum at the optimum energy consumption.

Example 17.9 For the process in Figure 17.2, determine the value of ΔT_{min} and the total cost of the heat exchanger network at the optimum setting of the capital–energy trade-off. The stream and utility data are given in Table 17.10. The utility costs are:

$$\text{Steam cost} = 120{,}000\ (\$\cdot \text{MW}^{-1}\cdot \text{y}^{-1})$$

$$\text{Cooling water cost} = 10{,}000\ (\$\cdot \text{MW}^{-1}\cdot \text{y}^{-1})$$

The heat exchange units (matches) can be considered to be single heat exchanger shells with an installed capital cost given by:

$$\text{Heat exchanger capital cost} = 40{,}000 + 500A\ (\$)$$

where A is the heat transfer area in m^2. The capital cost is to be paid back over five years at 10% interest.

Solution From Equation 2.7 from Chapter 2:

Annualized heat exchanger capital cost

$$= \text{Capital cost} \times \frac{i(1+i)^n}{(1+i)^n - 1}$$

where i = fractional interest rate per year
n = number of years

Annualized heat exchanger capital cost

$$= [40{,}000 + 500A] \times \frac{0.1(1+0.1)^5}{(1+0.1)^5 - 1}$$

$$= [40{,}000 + 500A]0.2638$$

$$= 10{,}552 + 131.9A$$

Annualized network capital cost

$$= N_{UNITS}\left[10{,}552 + \frac{131.9A_{NETWORK}}{N_{UNITS}}\right]$$

Now scan a range of values of ΔT_{min} and calculate the targets for energy, number of units and network area and combine these into a total cost. The results are given in Table 17.12. The data from Table 17.12 are presented graphically in Figure 17.47. The optimum ΔT_{min} is at 10 °C, confirming the initial value used for this problem. The total annualized cost at the optimum setting of the capital–energy trade-off is 2.05×10^6 \$·y^{-1}.

Table 17.12

Variation of annualized costs with ΔT_{min}.

ΔT_{min}	Q_{Hmin} (MW)	Annual hot utility cost (10^6 \$·y^{-1})	Q_{Cmin} (MW)	Annual cold utility cost (10^6 \$·y^{-1})	$A_{NETWORK}$ (m^2)	N_{UNITS}	Annualized capital cost (10^6 \$·y^{-1})	Annualized total cost (10^6 \$·y^{-1})
2	4.3	0.516	6.8	0.068	15,519	7	2.121	2.705
4	5.1	0.612	7.6	0.076	11,677	7	1.614	2.302
6	5.9	0.708	8.4	0.084	9645	7	1.346	2.138
8	6.7	0.804	9.2	0.092	8336	7	1.173	2.069
10	7.5	0.900	10.0	0.100	7410	7	1.051	2.051
12	8.3	0.996	10.8	0.108	6716	7	0.960	2.064
14	9.1	1.092	11.6	0.116	6174	7	0.888	2.096

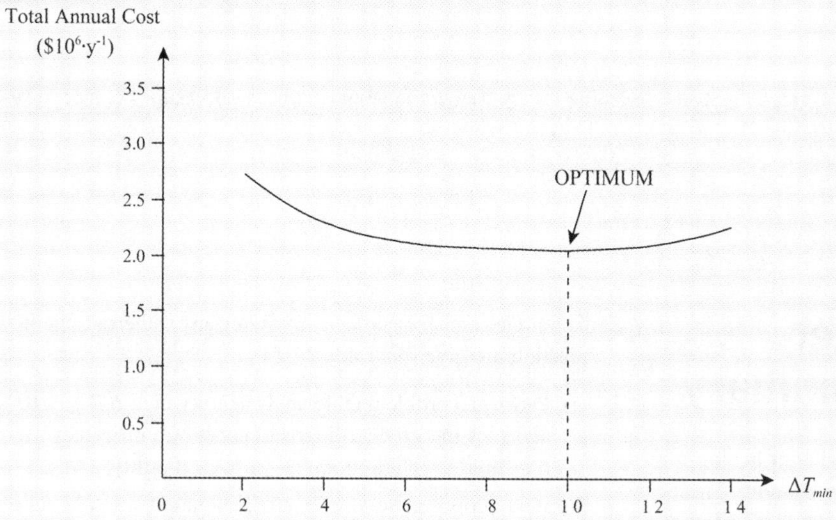

Figure 17.47

Optimization of the capital–energy trade-off for Example 17.9.

So far it has been assumed that each unit in the network can be costed according to Equation 17.10. This assumes that each unit comprises a single countercurrent heat exchanger, all heat exchangers are constructed from the same materials of construction and of the same type, and all heat exchangers are subject to the same pressure rating. However, the capital cost of the heat exchanger network in general depends on a number of factors:

1) *Number of units (matches between hot and cold streams).* This can be targeted by Equation 17.5 for networks without a pinch and by Equation 17.6 for networks with a pinch.
2) *Number of shells.* If multipass shell-and-tube heat exchangers are used, then individual units can comprise a number of shells. It is possible at the targeting stage to predict the number of shells given certain assumptions (Ahmad and Smith, 1989).
3) *Network heat transfer area.* If stream heat transfer coefficients do not vary significantly, then Equation 17.9 can be used to target for the network heat transfer area. However, once stream heat transfer coefficients vary significantly, then preferential matches can be arranged such that larger temperature differences in the network can be exploited to overcome poor heat transfer coefficients (Ahmad, Linnhoff and Smith, 1990). This requires an approach to targeting the network heat transfer area based on linear programming (Saboo, Morari and Colberg, 1986; Ahmad, Linnhoff and Smith, 1990). In addition, if multipass shell-and-tube heat exchangers are used, then the target for network heat transfer area needs to be adjusted for noncountercurrent heat transfer (Ahmad and Smith, 1989).
4) *Materials of construction.* If the same materials of construction can be used for all heat exchangers, then Equation 17.10 can be adjusted for the appropriate materials. However, it can often be the case that different materials of construction are required for different process streams. This can be allowed for in the network heat transfer area target by artificially weighting the heat transfer coefficients associated with a given stream. For example, if a stream requires a more expensive material, then its heat transfer coefficient is weighted to be lower than the real heat transfer coefficient in order to artificially increase the network area and hence the capital cost (Hall, Ahmad and Smith, 1990).
5) *Heat exchanger type.* If the same exchanger type can be used for all heat exchangers, then Equation 17.10 can be adjusted for the appropriate type. However, it can often be the case that different types of heat exchanger are required for different process streams. If this is the case, then, as with mixing materials of construction in the same network, the heat transfer coefficient for the streams requiring different heat exchanger types can be weighted to artificially change the heat transfer area, and therefore reflect the change in capital cost (Hall, Ahmad and Smith, 1990).
6) *Pressure rating.* Different pressure ratings for different streams require different mechanical construction of heat exchangers. The higher the pressure, the more expensive the heat exchanger as a result of the requirements for stronger mechanical design. If the same pressure rating is applied throughout the network, then this can be reflected simply in Equation 17.10. However, different pressure ratings can be required in different parts of the network. Again, this can be allowed for by weighting individual stream heat transfer coefficients in the network area targeting (Hall, Ahmad and Smith, 1990).

There are many uncertainties in trying to predict the capital cost from the stream data before the network has been designed. Because of this, capital and total cost targets should not be used to predict performance. However, such targets can be of value when trying to set a reasonable value for ΔT_{min} for targeting purposes when no experience data are available from previous designs.

Table 17.13

Stream data for Exercise 1.

Stream	Type	Supply temperature (°C)	Target temperature (°C)	Enthalpy change (MW)
1	Hot	100	40	12
2	Cold	10	150	7

17.15 Heat Exchanger Network Targets – Summary

The energy targets for the process can be set without having to design the heat exchanger network and utility system. These energy targets can be calculated directly from the material and energy balance. Thus, energy costs can be established without design for the outer layers of the process onion model. Using the grand composite curve, different utility scenarios can be screened quickly and conveniently, including cogeneration and heat pumps.

In addition to setting energy targets, targets can also be set for the minimum number of units and the network area. The sensitivity of the targets to the assumption of ΔT_{min} should also be examined.

17.16 Exercises

1. A heat recovery problem consists of two streams given in Table 17.13:

Steam is available at 180 °C and cooling water at 20 °C.

a) For a minimum permissible temperature difference (ΔT_{min}) of 10 °C, calculate the minimum hot and cold utility requirements.

b) What are the hot and cold stream pinch temperatures?

c) If the steam is supplied as the exhaust from a steam turbine, is the heat engine appropriately placed?

d) If the (ΔT_{min}) is increased to 20 °C, what will happen to the utility requirements?

2. The stream data for a process involving an exothermic chemical reaction are given in Table 17.14.

a) Perform a problem table analysis for which $\Delta T_{min} = 10$ °C and plot the grand composite curve. Confirm that it is a threshold problem requiring only cold utility.

b) It is proposed to use steam generation as cold utility. Calculate how much steam can be generated by the process at a pressure of 41 bar with a saturated temperature of 252 °C and latent heat 1706 kJ·kg^{-1}. Assume that saturated boiler feedwater is available and that the steam generated is saturated.

c) If the steam is superheated to a temperature of 350 °C, calculate how much steam can be generated at 41 bar. Assume the heat capacity of steam is 4.0 kJ·kg^{-1}·K^{-1}.

d) What would happen if the steam in Part b was generated from boiler feedwater at 100 °C with a heat capacity of 4.2 kJ·kg^{-1}·K^{-1}? How would the steam generation be calculated under these circumstances?

3. The stream data for a process are given in Table 17.15.

a) Sketch the composite curves for $\Delta T_{min} = 10$ °C and determine the hot and cold utility targets.

b) Determine the targets for the hot and cold utilities for $\Delta T_{min} = 10$ °C using the problem table algorithm and compare with the composite curves.

4. The problem table cascade for a process is given in the Table 17.16 for $\Delta T_{min} = 10$ °C.

Table 17.14

Stream data for Exercise 2.

Stream		Enthalpy change (kW)	T_S (°C)	T_T (°C)
No.	Type			
1	Hot	−7000	377	375
2	Hot	−3600	376	180
3	Hot	−2400	180	70
4	Cold	2400	60	160
5	Cold	200	20	130
6	Cold	200	160	260

Table 17.15

Stream data for Exercise 3.

Stream				
No.	Type	T_S (°C)	T_T (°C)	Heat duty (MW)
1	Hot	150	30	7.2
2	Hot	40	40	10
3	Hot	130	100	3
4	Cold	150	150	10
5	Cold	50	140	3.6

The minimum permissible temperature differences for various options are given in Table 17.17.

a) Calculate the amount of fuel required to satisfy the heating requirements for a furnace used to supply hot utility for a theoretical flame temperature of 1800 °C and acid dew point of 150 °C. Ambient temperature is 25 °C.

b) If the furnace from Part a is used in conjunction with high-pressure saturated steam at 300 °C for heating, such that the heat duty for the steam is maximized, what is the heat duty on the steam and the furnace heat duty?

c) If the furnace from Part a is used in conjunction with low-pressure saturated steam at 210 °C for heating, such that the heat duty for the steam is maximized, what is the heat duty on the steam and the furnace heat duty?

d) Instead of the furnace it is proposed to match a gas turbine exhaust with an exhaust temperature of 445 °C. Assume that the flowrate through the gas turbine can be varied to match the process requirements. If the efficiency of the gas turbine is 30% under these conditions, how much power can be generated?

Table 17.16

Problem table cascade for Exercise 4.

Interval temperature (°C)	Cascade heat flow (MW)
360	9.0
310	7.6
230	8.0
210	4.8
190	5.8
190	0
170	1.0
150	7.6
60	3.0

Table 17.17

ΔT_{min} for different matches.

Type of match	ΔT_{min} (°C)
Process/process	10
Process/steam	10
Process/flue gas (furnace or gas turbine)	50

5. The problem table cascade for a process is given in Table 17.18 for $\Delta T_{min} = 10$ °C.

   It is proposed to provide process cooling by steam generation from boiler feedwater with a temperature of 80 °C.

   a) Determine how much saturated steam can be generated at a temperature of 230 °C. The latent heat of steam under these conditions is 1812 kJ·kg^{-1}. The heat capacity of water can be taken as 4.2 kJ·kg^{-1}·K^{-1}.

   b) Determine how much steam can be generated with a saturation temperature of 230 °C and superheated to the maximum temperature possible against the process. The heat capacity of steam can be taken as 3.45 kJ·kg^{-1}·K^{-1}.

   c) Calculate how much power can be generated from the superheated steam from Part b, assuming the exhaust steam is saturated at a pressure of 4 bar with an enthalpy of 2738 kJ·kg^{-1}. Assume the reference temperature for enthalpy is 0 °C.

Table 17.18

Problem table cascade for Exercise 5.

Interval temperature (°C)	Heat flow (MW)
495	3.6
455	9.2
415	10.8
305	4.2
285	0
215	16.8
195	17.6
185	16.6
125	16.6
95	21.1
85	18.1

Table 17.19

Problem table cascade for Exercise 6.

Interval temperature (°C)	Heat flow (MW)
440	21.9
410	29.4
130	23.82
130	1.8
100	0
95	15

6. The problem table cascade for a process is given in Table 17.19 for $\Delta T_{min} = 20\,°C$.

   The process also has a requirement for 7 MW of power. Three alternative utility schemes are to be compared economically:

   a) A steam turbine with its exhaust saturated at 150 °C used for process heating is one of the options to be considered. Steam is raised in the central boiler house at 41 bar with an enthalpy of 3137 kJ·kg^{-1} from boiler feedwater with an enthalpy of 418 kJ·kg^{-1}. The enthalpy of saturated condensate at 41 bar is 1095 kJ·kg^{-1}. The overall efficiency of the steam system is 85%. The enthalpy of saturated steam at 150 °C is 2747 kJ·kg^{-1} and its latent heat is 2115 kJ·kg^{-1}. Calculate the maximum power generation possible by matching the exhaust steam against the process.

   b) A second possible scheme uses a gas turbine with an exhaust temperature of 400 °C and heat capacity flowrate of 0.1 MW·K^{-1}. The minimum stack temperature should be taken to be 100 °C. Calculate the power generation if the turbine converts heat to power with an efficiency of 30% under these conditions. Ambient temperature is 10 °C.

Table 17.20

Utility costs for Exercise 6.

Utility	Cost ($·MW^{-1})
Fuel for gas turbine	0.0042
Fuel for steam generation	0.0030
Imported power	0.018
Credit for exported power	0.014
Cooling water	0.00018

Table 17.21

Problem table cascade for Exercise 7.

Interval temperature (°C)	Heat flow (kW)
160	1000
150	0
130	1100
110	1400
100	900
80	1300
40	1400
10	1800
–10	1900
–30	2200

   c) A third possible scheme integrates a heat pump with the process. The power required by the heat pump is given by:

$$W = \frac{Q_H}{0.6}\,\frac{T_H - T_C}{T_H}$$

   where Q_H is the heat rejected by the heat pump at T_H (K). Heat is absorbed into the heat pump at T_C (K). Calculate the power required by the heat pump. The cost utilities are given in Table 17.20.

   The overall efficiency of steam generation and distribution is 60%. Which scheme is most cost-effective, the steam turbine, the gas turbine or the heat pump?

7. Table 17.21 represents a problem table cascade ($\Delta T_{min} =$ 10 °C).

   The following utilities are available:

   i. Saturated MP steam at 200 °C.
   ii. Saturated LP steam at 130 °C with a latent heat of vaporization of 2174 kJ·kg^{-1} raised from boiler feed water at 90 °C and the specific heat capacity is 4.2 kJ·kg·K^{-1}.
   iii. Cooling water (20 to 30 °C).
   iv. Refrigeration at 0 °C.
   v. Refrigeration at –40 °C.

   For matches between process and refrigeration, $\Delta T_{min} =$ 10 °C. Draw the process grand composite curve and set the targets for the utilities. Below the pinch use of higher temperatures, cold utilities should be maximized.

References

Ahmad S and Hui DCW (1991) Heat Recovery Between Areas of Integrity, *Comp Chem Eng*, **15**: 809.

Ahmad S, Linnhoff B and Smith R (1990) Cost Optimum Heat Exchanger Networks – 2. Targets and Design for Detailed Capital Cost Models, *Comp Chem Eng*, **14**: 751.

Ahmad S and Smith R (1989) Targets and Design for Minimum Number of Shells in Heat Exchanger Networks, *Trans IChemE ChERD*, **67**: 481.

Bandyopadhyay S and Sahu GC (2010) Modified Problem Table Algorithm for Energy Targeting, *Ind Eng Chem Res*, **49**: 11557.

Cerda J, Westerberg AW, Mason D and Linnhoff B (1983) Minimum Utility Usage in Heat Exchanger Network Synthesis – A Transportation Problem, *Chem Eng Sci*, **38**: 373.

Hall SG, Ahmad S and Smith R (1990) Capital Cost Targets for Heat Exchanger Networks Comprising Mixed Materials of Construction, Pressure Ratings and Exchanger Types, *Comp Chem Eng*, **14**: 319.

Hohman EC (1971) Optimum Networks of Heat Exchange, PhD Thesis, University of Southern California.

Huang F and Elshout RV (1976) Optimizing the Heat Recover of Crude Units, *Chem Eng Prog*, **72**: 68.

Itoh J, Shiroko K and Umeda T (1982) Extensive Application of the T–Q Diagram to Heat Integrated System Synthesis, International Conference on Proceedings Systems Engineering (PSE-82) Kyoto, 92.

Linnhoff B and Ahmad S (1990) Cost Optimum Heat Exchanger Networks – 1. Minimum Energy and Capital Using Simple Models for Capital Cost, *Comp Chem Eng*, **14**: 729.

Linnhoff B and Flower JR (1978) Synthesis of Heat Exchanger Networks, *AIChE J*, **24**: 633.

Linnhoff B and de Leur J (1988) Appropriate Placement of Furnaces in the Integrated Process, *IChemE Symp Ser*, **109**: 259.

Linnhoff B, Mason DR and Wardle I (1979) Understanding Heat Exchanger Networks, *Comp Chem Eng*, **3**: 295.

Linnhoff B, Townsend DW, Boland D, Hewitt GF, Thomas BEA, Guy AR and Marsland RH (1982) *A User Guide on Process Integration for the Efficient Use of Energy*, IChemE, UK.

Papoulias SA and Grossmann IE (1983) A Structural Optimization Approach in Process Synthesis – II. Heat Recovery Networks, *Comp Chem Eng*, **7**: 707.

Polley GT (1993) Heat Exchanger Design and Process Integration, *Chem Eng*, **8**: 16.

Saboo AK, Morari M and Colberg RD (1986) RESHEX – An Interactive Software Package for the Synthesis and Analysis of Resilient Heat Exchanger Networks – II Discussion of Area Targeting and Network Synthesis Algorithms, *Comp Chem Eng*, **10**: 591.

Townsend DW and Linnhoff B (1983) Heat and Power Networks in Process Design, *AIChE J*, **29**: 742.

Townsend DW and Linnhoff B (1984) *Surface Area Targets for Heat Exchanger Networks*, IChemE Annual Research Meeting, Bath, UK.

Umeda T, Harada T and Shiroko K (1979) A Thermodynamic Approach to the Synthesis of Heat Integration Systems in Chemical Processes, *Comp Chem Eng*, **3**: 273.

Umeda T, Itoh J and Shiroko K (1978) Heat Exchange System Synthesis, *Chem Eng Prog*, **74**: 70.

Umeda T, Niida K and Shiroko K (1979) A Thermodynamic Approach to Heat Integration in Distillation Systems, *AIChE J*, **25**: 423.

Chapter 18

Heat Exchanger Networks II – Network Design

Having explored the targets for the heat exchanger network, it now remains to develop the design for the network. Both manual and automated methods for network design are possible. Start by developing a manual method that will allow user interaction.

18.1 The Pinch Design Method

The capital–energy trade-off in the heat exchanger networks was discussed in Chapter 17. Varying ΔT_{min}, as shown in Figure 17.6, changed the relative position of the process composite curves. As ΔT_{min} is changed from a small to a large value, the capital cost decreases, but the energy cost increases. When the two costs are combined to obtain a total cost, the optimum point in the capital–energy trade-off is identified, corresponding with an optimum value of ΔT_{min} (Figure 17.6). Such a trade-off can only be carried out with any certainty once a design has been established. Thus, in order to establish a first design, the initial setting of ΔT_{min} must be approximated from heuristics or past experience for similar processes. As pointed out in Chapter 17, the trade-off between energy and capital suggests that individual exchangers should have a temperature difference no smaller than the ΔT_{min} between the composite curves. Although ΔT_{min} is not known with certainty, if it is assumed fixed for each individual exchanger, this allows some rules to be set that simplify the design procedure.

Having made the assumption that no individual heat exchanger should have a temperature difference smaller than ΔT_{min}, two rules were deduced in Chapter 17. If the energy target set by the composite curves (or the problem table algorithm) is to be achieved, the design must not transfer heat across the pinch by:

- process-to-process heat transfer;
- inappropriate use of utilities.

Chemical Process Design and Integration, Second Edition. Robin Smith.

Companion Website: www.wiley.com/go/smith/chemical2e

These rules are necessary for the design to achieve the energy target, assuming that no individual exchanger should have a temperature difference smaller than ΔT_{min}. To comply with these two rules, the process should be divided at the pinch. As pointed out in Chapter 17, this is most clearly done by representing the stream data in the grid diagram. Figure 18.1 shows the stream data from Table 17.2 in grid form with the pinch marked. As pointed out in Chapter 17, either a hot stream or a cold stream will start at the pinch. Above the pinch, steam can be used (up to Q_{Hmin}) and below the pinch cooling water can be used (up to Q_{Cmin}). However, what strategy should be adopted for the design? A number of simple criteria can be developed to help (Linnhoff and Hindmarsh, 1983).

1) *Start at the pinch.* The pinch is the most constrained region of the problem. At the pinch, ΔT_{min} exists between all hot and cold streams. As a result, the number of feasible matches in this region is severely restricted. Often there are essential matches to be made. If such matches are not made, the result will be either the use of temperature differences smaller than ΔT_{min} or excessive use of utilities resulting from heat transfer across the pinch. If the design was started away from the pinch at the hot end or cold end of the problem, then initial matches are likely to need follow-up matches that violate the pinch or the ΔT_{min} criterion as the pinch is approached. Thus, if the design is started at the pinch, then initial decisions are made in the most constrained part of the problem. This is much less likely to lead to difficulties later.
2) *The CP inequality for individual matches.* Figure 18.2a shows the temperature profiles for an individual exchanger at the pinch, above the pinch (Linnhoff *et al.*, 1982; Linnhoff and Hindmarsh, 1983). Moving away from the pinch, temperature differences must increase. Figure 18.2a shows a match between a hot stream and a cold stream that has a CP smaller than the hot stream. At the pinch, the match starts with a temperature difference equal to ΔT_{min}. The relative slopes of the temperature–enthalpy profiles of the two streams mean that the temperature differences become smaller moving away from the pinch, which is infeasible. Alternatively, Figure 18.2b shows a

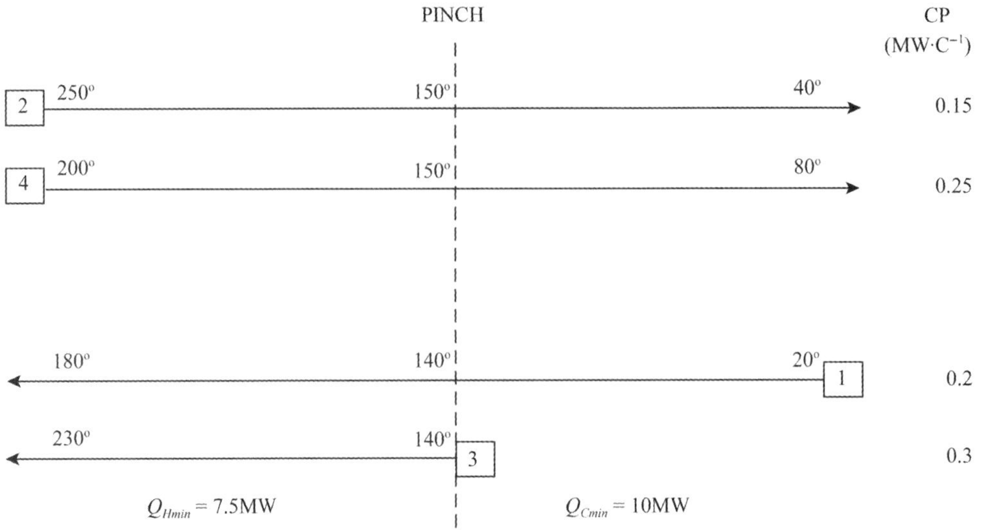

Figure 18.1
The grid diagram for the data from the Table 17.2.

match involving the same hot stream but with a cold stream that has a larger *CP*. The relative slopes of the temperature–enthalpy profiles now cause the temperature differences to become larger moving away from the pinch, which is feasible. Thus, starting with ΔT_{min} at the pinch, for temperature differences to increase moving away from the pinch (Linnhoff *et al.*, 1982; Linnhoff and Hindmarsh, 1983):

$$CP_H \leq CP_C \quad \text{(above pinch)} \tag{18.1}$$

Figure 18.3 shows the situation below the pinch at the pinch. If a cold stream is matched with a hot stream with smaller *CP*, as

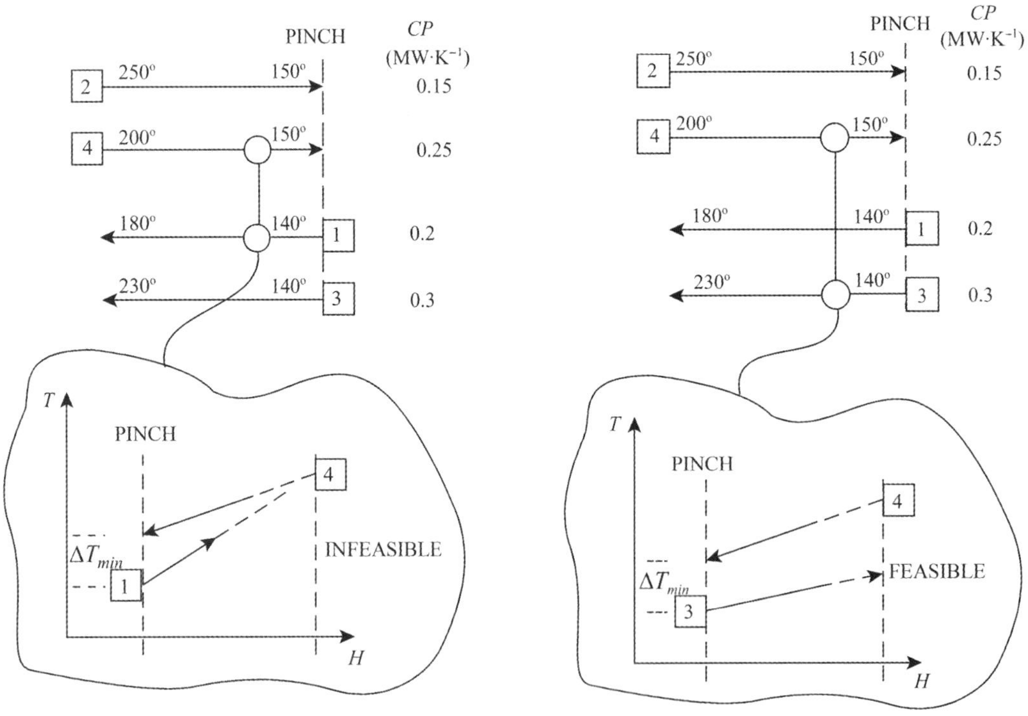

(a) Match is infeasible.

(b) Match is feasible.

Figure 18.2
Criteria for the pinch matches above the pinch.

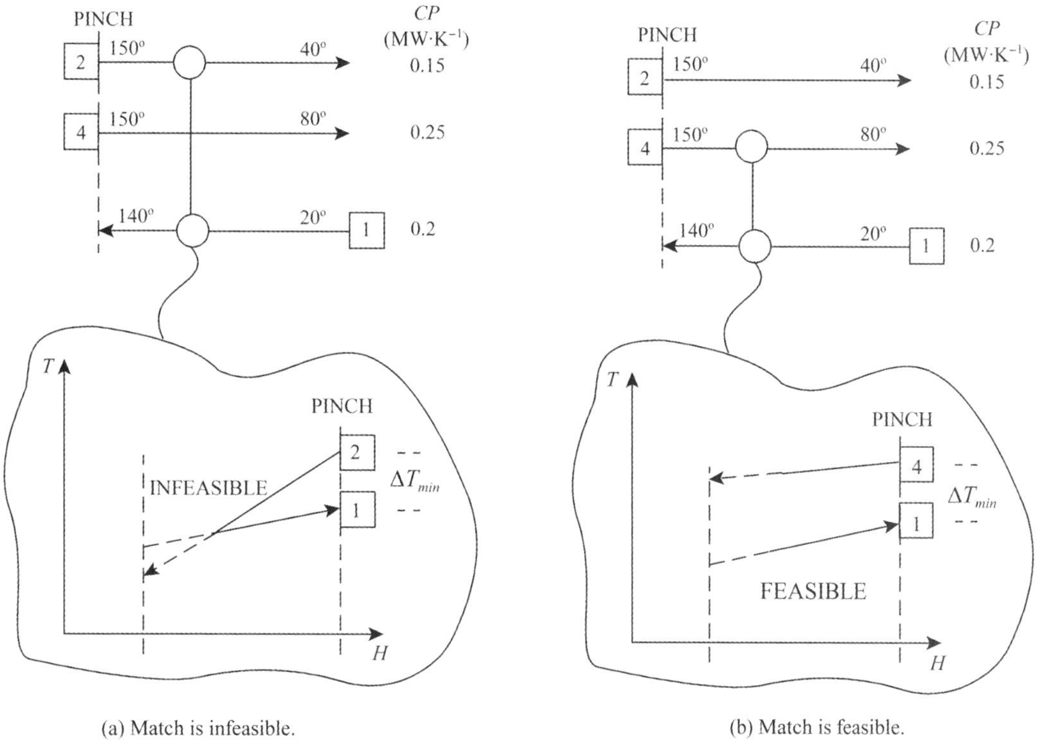

Figure 18.3

Criteria for pinch matches below the pinch.

shown in Figure 18.3a (i.e. a steeper slope), then the temperature differences become smaller (which is infeasible). If the same cold stream is matched with a hot stream with a larger *CP* (i.e. a less steep slope), as shown in Figure 18.3b, then temperature differences become larger, which is feasible. Thus, starting with ΔT_{min} at the pinch, for temperature differences to increase moving away from the pinch (Linnhoff *et al.*, 1982; Linnhoff and Hindmarsh, 1983):

$$CP_H \geq CP_C \quad \text{(below pinch)} \tag{18.2}$$

Note that the *CP* inequalities given by Equations 18.1 and 18.2 only apply at the pinch and when both ends of the match are at pinch conditions.

3) *The CP table.* Identification of the essential matches in the region of the pinch is clarified by use of the *CP table* (Linnhoff *et al.*, 1982; Linnhoff and Hindmarsh, 1983). In the CP table, the *CP* values of the hot and cold streams for the streams at the pinch are listed in descending order.

Figure 18.4a shows the grid diagram with the CP table for the design above the pinch. Cold utility must not be used above the pinch, which means that hot streams must be cooled to pinch temperature by heat recovery. Hot utility can be used, if necessary, on the cold streams above the pinch. Thus, it is essential to match hot streams above the pinch with a cold partner. In addition, if the hot stream is at pinch conditions, the cold stream it is to be matched with must also be at pinch conditions, otherwise the ΔT_{min} constraint will be violated. Figure 18.4a shows a feasible design arrangement above the pinch that does not use temperature differences smaller than ΔT_{min}. Note again that the *CP* inequality only applies when a match is made between two streams that are both at the pinch. Away from the pinch, temperature differences increase and it is no longer essential to obey the *CP* inequalities.

Figure 18.4b shows the grid diagram with the CP table for the design below the pinch. Hot utility must not be used below the pinch, which means that cold streams must be heated to pinch temperature by heat recovery. Cold utility can be used, if necessary, on the hot streams below the pinch. Thus, it is essential to match cold streams below the pinch with a hot partner. In addition, if the cold stream is at pinch conditions, the hot stream it is to be matched with must also be at pinch conditions, otherwise the ΔT_{min} constraint will be violated. Figure 18.4b shows a design arrangement below the pinch that does not use temperature differences smaller than ΔT_{min}.

Having decided that some essential matches need to be made around the pinch, the next question is how big should the matches be?

4) *The "tick-off" heuristic.* Once the matches around the pinch have been chosen to satisfy the criteria for minimum energy, the design should be continued in such a manner as to keep capital costs to a minimum. One important criterion in the capital cost is the number of units (there are others, of course, which shall be addressed later). Keeping the number of units to a minimum can be achieved using the *tick-off heuristic*. To tick off a stream, individual units are made as large as

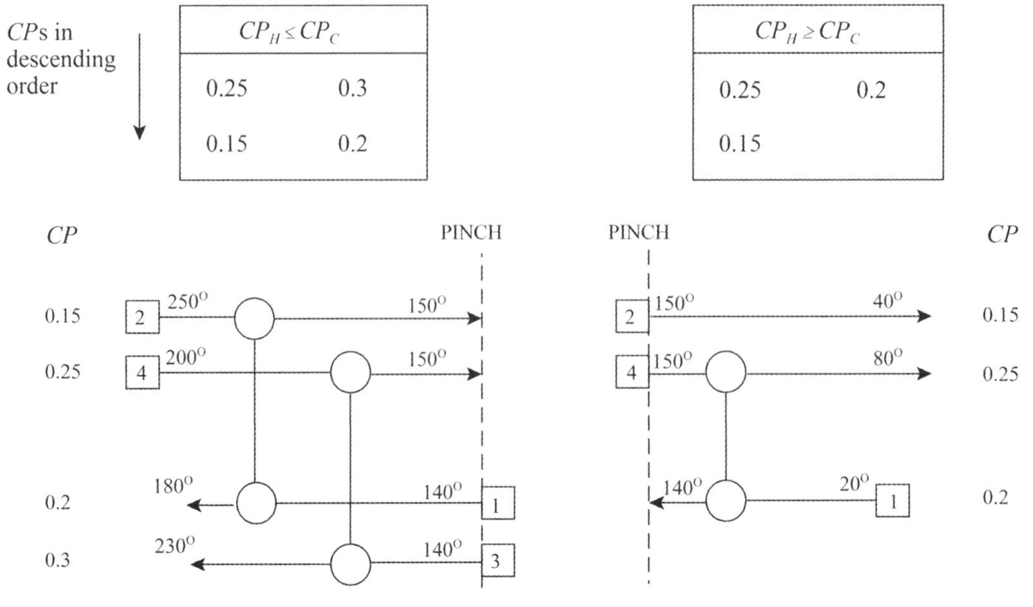

(a) Above pinch. (b) Below pinch.

Figure 18.4
The CP table for the designs above and below the pinch for the problem from Table 17.2.

possible, that is, the smaller of the two heat duties on the streams being matched.

Figure 18.5a shows the matches around the pinch from Figure 18.4a with their duties maximized to tick-off streams. It should be emphasized that the tick-off heuristic is only a heuristic and can in some cases penalize the design. Methods will be developed later that allow such penalties to be identified as the design proceeds.

The design in Figure 18.5a can now be completed by satisfying the heating and cooling duties away from the pinch. Cooling water must not be used above the pinch. Therefore, if there are hot streams above the pinch for which the pinch matches do not satisfy the duties, additional process-to-process heat recovery is required. Figure 18.5b shows an additional match to satisfy the residual cooling of the hot streams above the pinch. Again, the duty on the unit is maximized. Finally, above the pinch, the residual heating duty on the cold streams must be satisfied. Since there are no hot streams left above the pinch, hot utility must be used as shown in Figure 18.5c.

Turning now to the cold end design, Figure 18.6a shows the pinch design with the streams ticked off. If there are any cold streams below the pinch for which the pinch matches do not satisfy the duties, then additional process-to-process heat recovery is required, since hot utility must not be used. Figure 18.6b shows an additional match to satisfy the residual heating of the cold streams below the pinch. Again, the duty on the unit is maximized. Finally, below the pinch, the residual cooling duty on the hot streams must be satisfied. Since there are no cold streams left below the pinch, cold utility must be used (Figure 18.6c).

The final design shown in Figure 18.7 amalgamates the hot end design from Figure 18.5c and the cold end design from Figure 18.6c. The duty on hot utility of 7.5 MW agrees with Q_{Hmin} and the duty on cold utility of 10.0 MW agrees with Q_{Cmin} predicted by the composite curves and the problem table algorithm.

Note one further point from Figure 18.7. The number of units is seven in total (including the heater and cooler). Referring back to Example 17.7, the target for the minimum number of units was predicted to be seven. It therefore appears that there was something in the procedure that naturally steered the design to achieve the target for the minimum number of units.

It is in fact the tick-off heuristic that steered the design toward the minimum number of units (Linnhoff *et al.*, 1982; Linnhoff and Hindmarsh, 1983). The target for the minimum number of units was given by:

$$N_{UNITS} = S - 1 \tag{17.5}$$

Before any matches are placed, the target indicates that the number of units required is equal to the number of streams (including utilities) minus one. The tick-off heuristic satisfied the heat duty on one stream every time one of the units was used. The stream that has been ticked off is no longer part of the remaining design problem. The tick-off heuristic ensures that having placed a unit (and used up one of the available units), a stream is removed from the problem. Thus, Equation 17.5 is satisfied if every match satisfies the heat duty on a stream or a utility.

This design procedure is known as the *pinch design method* and can be summarized in five steps (Linnhoff and Hindmarsh, 1983).

- Divide the problem at the pinch into separate problems.
- The design for the separate problems is started at the pinch, moving away.

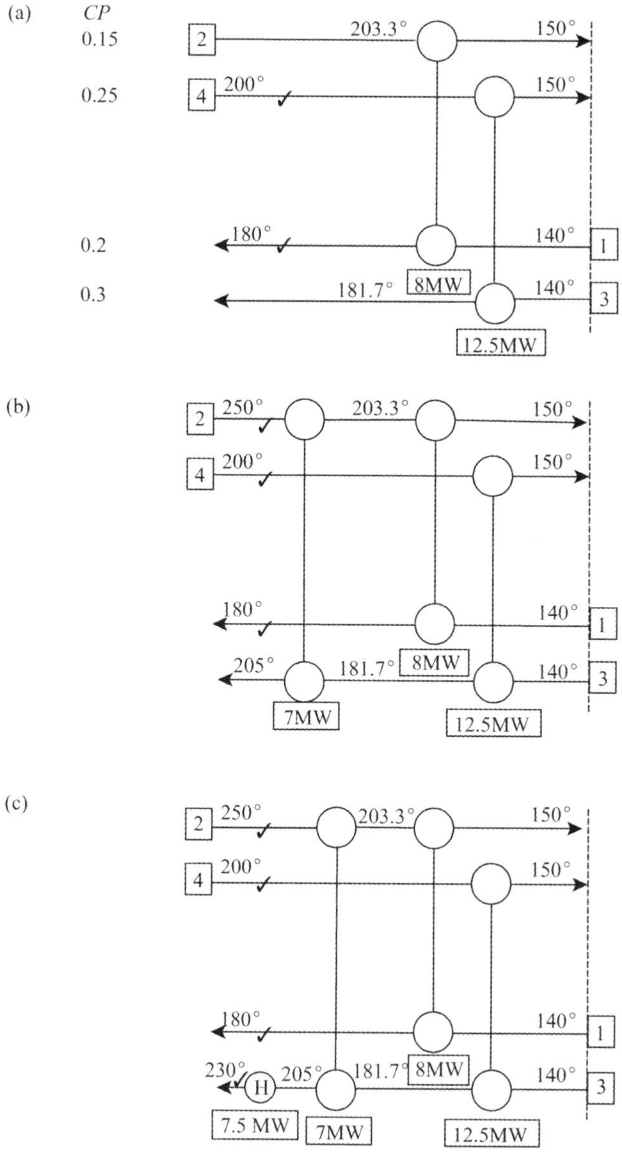

Figure 18.5
Sizing the units above the pinch using the tick-off heuristic.

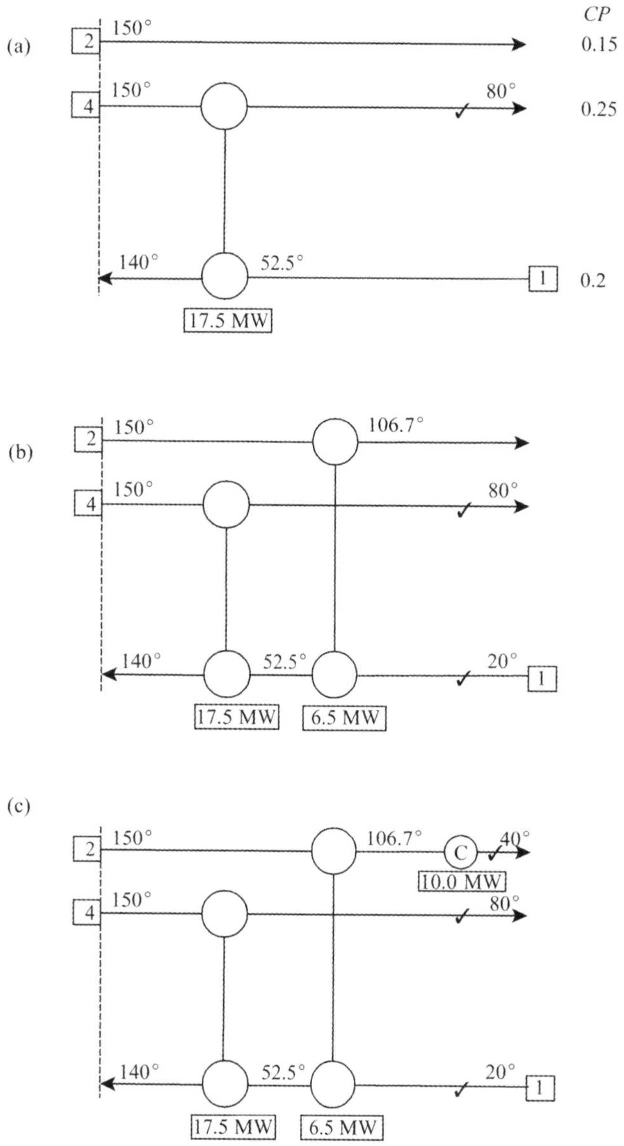

Figure 18.6
Sizing the units below the pinch using the tick-off heuristic.

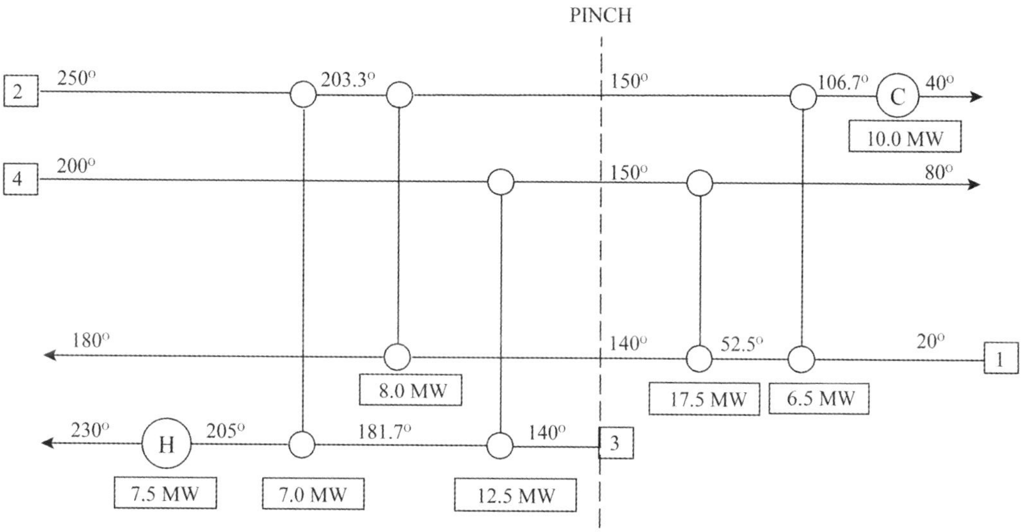

Figure 18.7
The completed design for the data from Table 17.2.

18

- Temperature feasibility requires constraints on the CP values to be satisfied for matches between the streams at the pinch.
- The loads on individual units are determined using the tick-off heuristic to minimize the number of units. Occasionally, the heuristic causes problems.
- Away from the pinch, there is usually more freedom in the choice of matches. In this case, the designer can discriminate on the basis of operability, plant layout and so on.

Example 18.1 The process stream data for a heat recovery network problem are given in Table 18.1.

A problem table analysis on these data reveals that the minimum hot utility requirement for the process is 15 MW and the minimum cold utility requirement is 26 MW for a minimum allowable temperature difference of 20 °C. The analysis also reveals that the pinch is located at a temperature of 120 °C for hot streams and 100 °C for cold streams. Design a heat exchanger network for maximum energy recovery featuring the minimum number of units.

Solution Figure 18.8a shows the hot end design with the CP table. Above the pinch, adjacent to the pinch, $CP_H \leq CP_C$. The duty on the units has been maximized according to the tick-off heuristic.

Figure 18.8b shows the cold end design with the CP table. Below the pinch, adjacent to the pinch, $CP_H \geq CP_C$. Again the duty on units has been maximized according to the tick-off heuristic.

The completed design is shown in Figure 18.8c. The minimum number of units for this problem is given by

$$\begin{aligned} N_{UNITS} &= (S-1)_{ABOVE\ PINCH} + (S-1)_{BELOW\ PINCH} \\ &= (5-1) \qquad\qquad + (4-1) \\ &= 7 \end{aligned}$$

The design in Figure 18.8 is seen to achieve the minimum number of units target.

Table 18.1

Stream data for Example 18.1.

Stream		Supply temperature (°C)	Target temperature (°C)	Heat capacity flowrate ($MW \cdot K^{-1}$)
No.	Type			
1	Hot	400	60	0.3
2	Hot	210	40	0.5
3	Cold	20	160	0.4
4	Cold	100	300	0.6

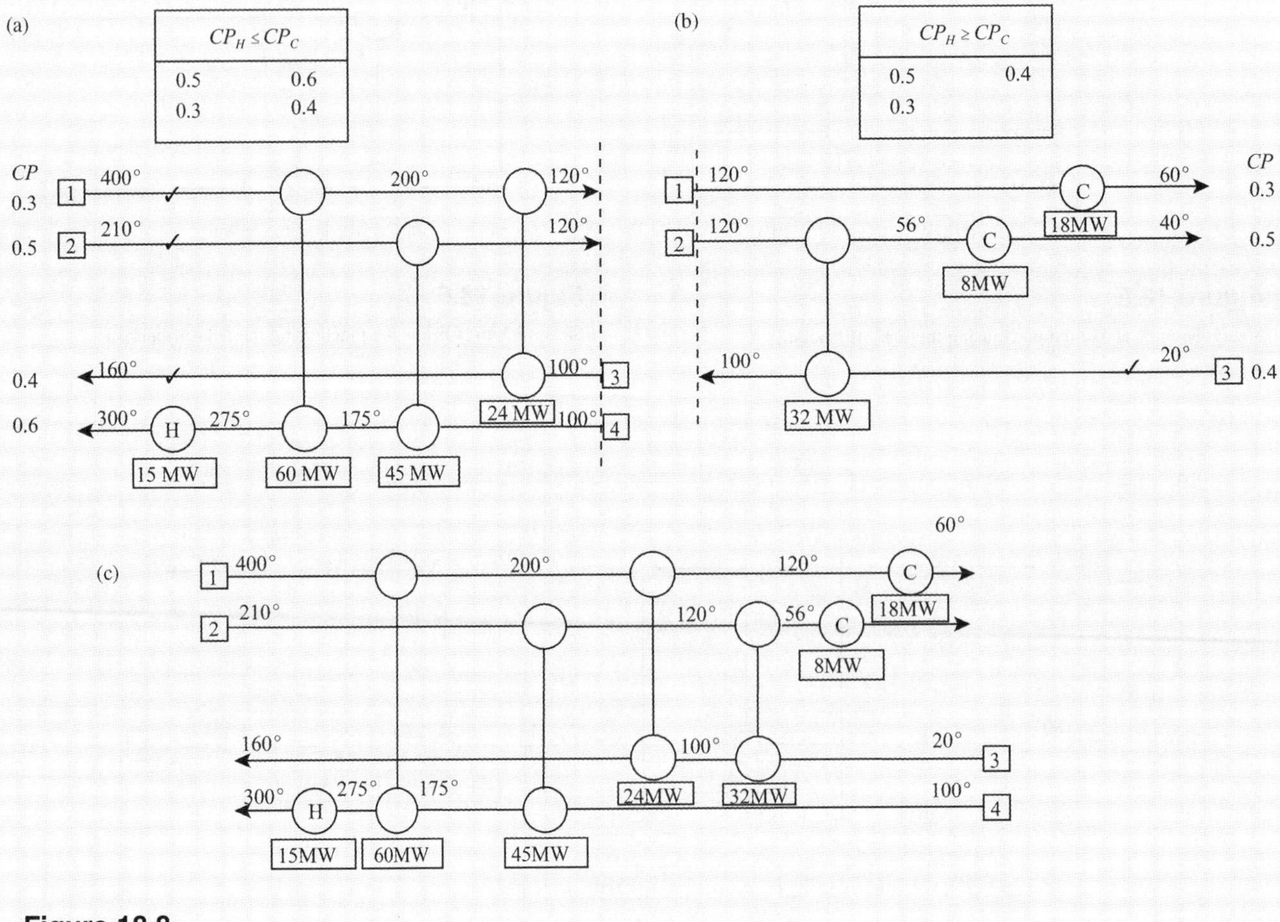

Figure 18.8

Maximium energy recovery design for Example 18.1.

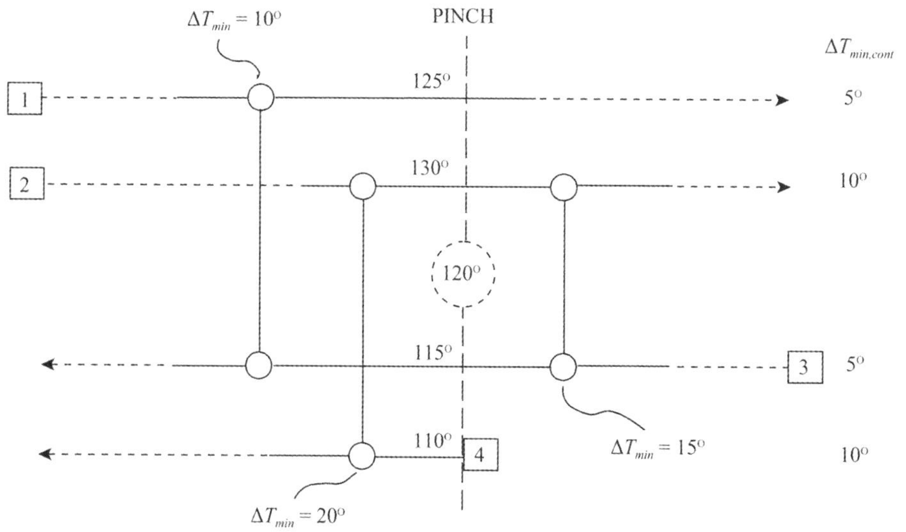

Figure 18.9

The design grid for a problem stream-specific with ΔT_{min} contributions.

The pinch design method, as discussed so far, has assumed the same ΔT_{min} applied between all stream matches. In Chapter 17, it was discussed how the basic targeting methods for the composite curves and the problem table algorithm can be modified to allow stream-specific values of ΔT_{min}. The example was quoted in which liquid streams were required to have a ΔT_{min} contribution of 5 °C and gas streams a ΔT_{min} contribution of 10 °C. For liquid–liquid matches, this would lead to a $\Delta T_{min} = 10$ °C. For gas–gas matches, this would lead to a $\Delta T_{min} = 20$ °C. For liquid–gas matches, it will lead to a $\Delta T_{min} = 15$ °C (Linnhoff *et al.*, 1982). Modifying the problem table and the composite curves to account for these stream-specific values of ΔT_{min} is straightforward, but how is the pinch design method modified to take account of such ΔT_{min} contributions? Figure 18.9 illustrates the approach. Suppose the interval pinch temperature from the problem table is 120 °C. This would correspond with hot stream pinch temperatures of 125 °C and 130 °C for hot streams with ΔT_{min} contributions of 5 °C and 10 °C respectively. For an interval pinch temperature of 120 °C, the corresponding cold stream pinch temperatures would be 115 °C and 110 °C for cold streams with ΔT_{min} contributions of 5 °C and 10 °C respectively. The stream grid is set up as shown in Figure 18.9 with the pinch matches being made and given their appropriate value of ΔT_{min} depending on the streams being matched. The constraints regarding the *CP* inequalities apply in this case, just as the case with a global value of ΔT_{min}.

18.2 Design for Threshold Problems

In Section 17.3, it was discussed that some problems, known as threshold problems, do not have a pinch. They need either hot utility or cold utility, but not both. How should the approach be modified to deal with the design of threshold problems?

The philosophy in the pinch design method was to start the design where it was most constrained. If the design is pinched, the problem is most constrained at the pinch. If there is no pinch, where is the design most constrained? Figure 18.10a shows one typical threshold problem that requires no hot utility, just cold utility. The most constrained part of this problem is the no-utility end (Linnhoff *et al.*, 1982). This is where temperature differences are smallest and there may be constraints, as shown in Figure 18.10b, where the target temperatures on some of the hot streams can only be satisfied by specific matches. Also, if individual matches are required to have a temperature difference no smaller than the threshold ΔT_{min}, the *CP* inequalities described in the pinch design method must be applied. For the most part, problems similar to those in Figure 18.10a are treated as one half of a pinched problem.

Figure 18.11 shows another threshold problem that requires only hot utility. This problem is different in characteristic from the one in Figure 18.10. Now the minimum temperature difference is in the middle of the problem causing a pseudo-pinch. The best strategy to deal with this type of threshold problem is to treat it as a pinched problem. In Figure 18.11, the problem is divided into two parts at the pseudo-pinch and the pinch design method is followed. The only complication in applying the pinch design method for such problems is that one half of the problem (the cold end in Figure 18.11) will not feature the flexibility offered by matching against utility.

18.3 Stream Splitting

The pinch design method developed earlier followed several rules and guidelines to allow design for minimum utility (or maximum energy recovery) in the minimum number of units. Occasionally, it appears not to be possible to create the appropriate matches because one or other of the design criteria cannot be satisfied.

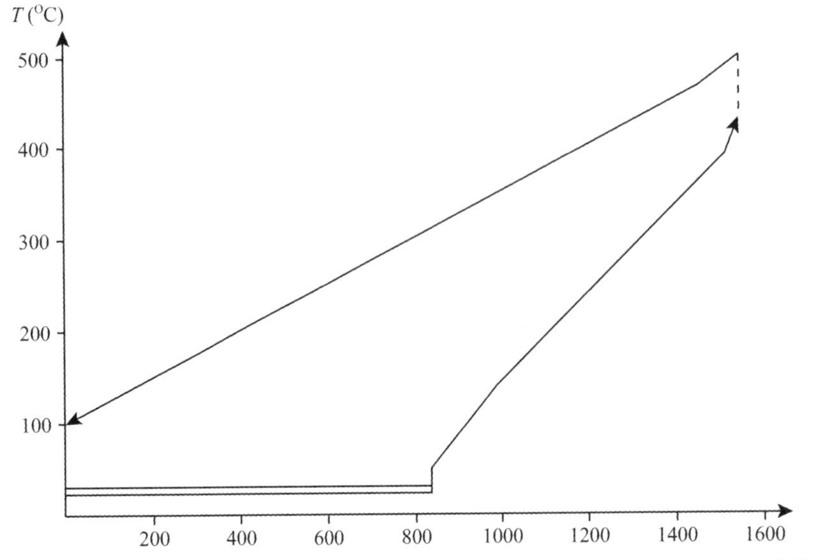

(a) Composite curves for a threshold problem.

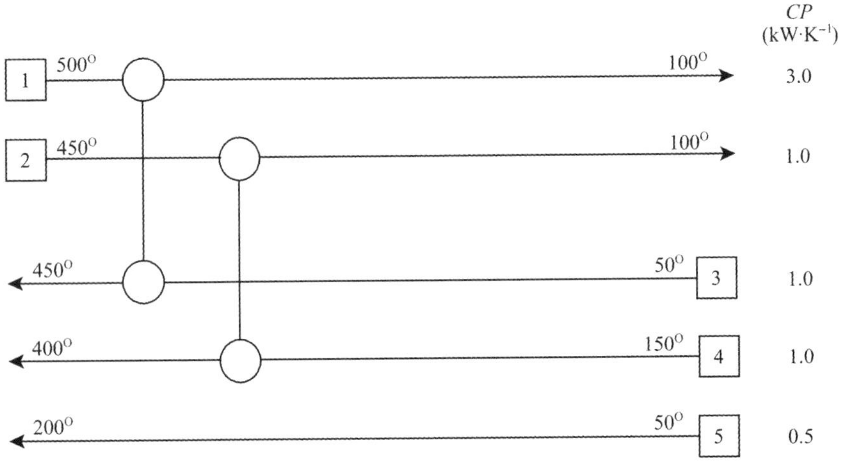

(b) Essential matches for the threshold problem.

Figure 18.10

Even though threshold problems have large driving forces, there are still often essential matches to be made, especially at the no utility end.

Consider Figure 18.12a, which shows the above-pinch part of a design. Cold utility must not be used above the pinch, which means that all hot streams must be cooled to pinch temperature by heat recovery. There are three hot streams and two cold streams in Figure 18.12a. Thus, regardless of the *CP* values of the streams, one of the hot streams cannot be cooled to pinch temperature without some violation of the ΔT_{min} constraint. The problem can only be resolved by splitting a cold stream into two parallel branches, as shown in Figure 18.12b. Now each hot stream has a cold steam with which to match, capable of cooling it to pinch temperature. Thus, in addition to the *CP* inequality criteria introduced earlier, there is a stream number criterion above the pinch such that (Linnhoff *et al.*, 1982; Linnhoff and Hindmarsh, 1983):

$$S_H \leq S_C \quad \text{(above pinch)} \qquad (18.3)$$

where S_H = number of hot streams at the pinch (including branches)

S_C = number of cold streams at the pinch (including branches)

If there had been a greater number of cold streams than hot streams in the design above the pinch, this would not have created a problem, since hot utility can be used above the pinch.

By contrast, now consider part of a design below the pinch, as shown in Figure 18.13a. Here hot utility must not be used, which means that all cold streams must be heated to pinch temperature by heat recovery. There are now three cold streams and two hot streams in Figure 18.13a. Again, regardless of *CP* values, one of the cold streams cannot be heated to pinch temperature without some violation of the ΔT_{min} constraint. The problem can only be

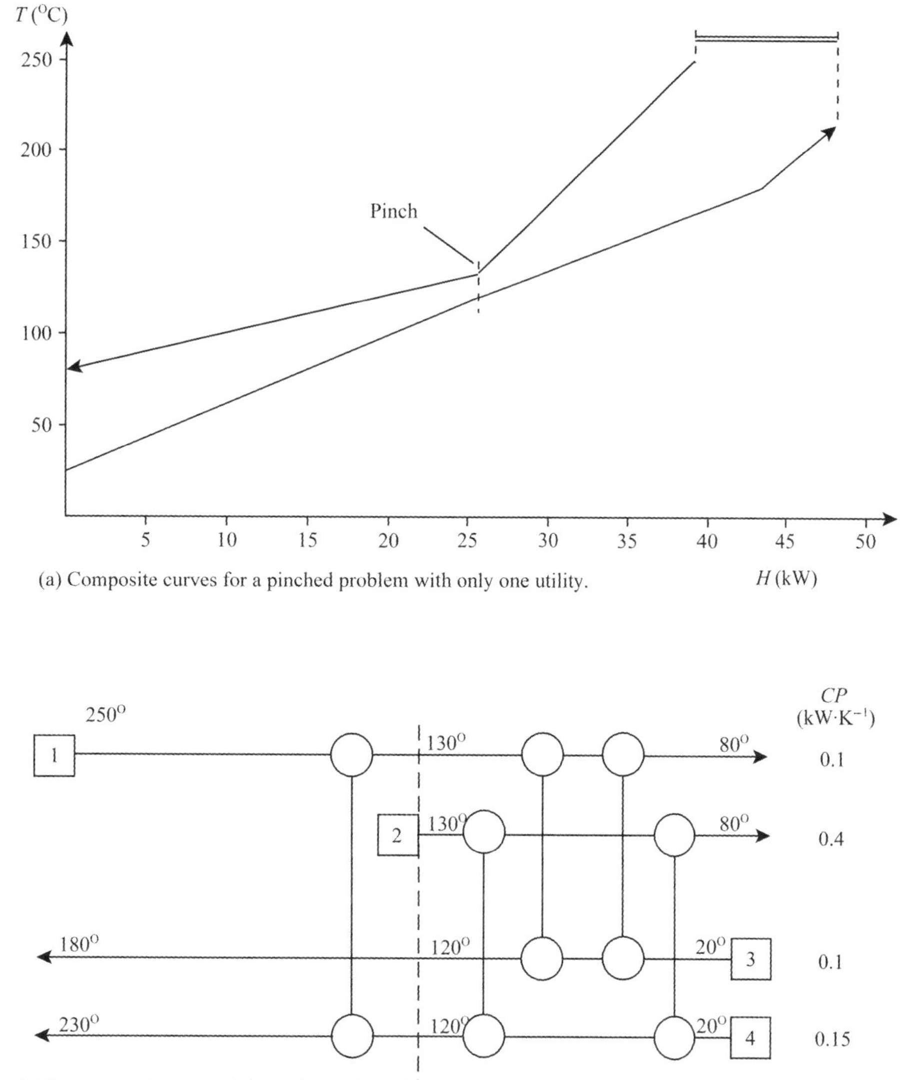

(a) Composite curves for a pinched problem with only one utility.

(b) Design requires essential matches at both the no utility end and the pinch.

Figure 18.11

Some threshold problems must be treated as pinched problems.

resolved by splitting a hot stream into two parallel branches, as shown in Figure 18.13b. Now each cold stream has a hot stream with which to match and be capable of heating it to pinch temperature. Thus there is a stream number criterion below the pinch, such that (Linnhoff *et al.*, 1982; Linnhoff and Hindmarsh, 1983):

$$S_H \geq S_C \quad \text{(below pinch)} \qquad (18.4)$$

Had there been more hot streams than cold below the pinch, this would not have created a problem since cold utility can be used below the pinch.

It is not only the number of streams that creates the need to split streams at the pinch. Sometimes the *CP* inequality criteria, Equations 18.1 and 18.2, cannot be met at the pinch without a stream split. Consider the above-pinch part of a problem in Figure 18.14a. The number of hot streams is less than the number of cold streams, and hence Equation 18.3 is satisfied. However, the *CP* inequality, Equation 18.1, must be satisfied. Neither of the two cold streams has a large enough *CP*. The hot stream can be made smaller by splitting it into two parallel branches (Figure 18.14b).

Figure 18.15a shows the below-pinch part of a problem. The number of hot streams is greater than the number of cold streams and hence Equation 18.4 is satisfied. However, neither of the two hot streams has a large enough *CP* to satisfy the *CP* inequality of Equation 18.2. The cold stream can be made smaller by splitting it into two parallel branches (Figure 18.15b).

Clearly, in designs different from those in Figures 18.14 and 18.15, when streams are split to satisfy the *CP* inequality, this might create a problem with the number of streams at the pinch such that Equations 18.3 and 18.4 are no longer satisfied. This would then require further stream splits to satisfy the stream number criterion. Figures 18.16a and 18.16b present algorithms

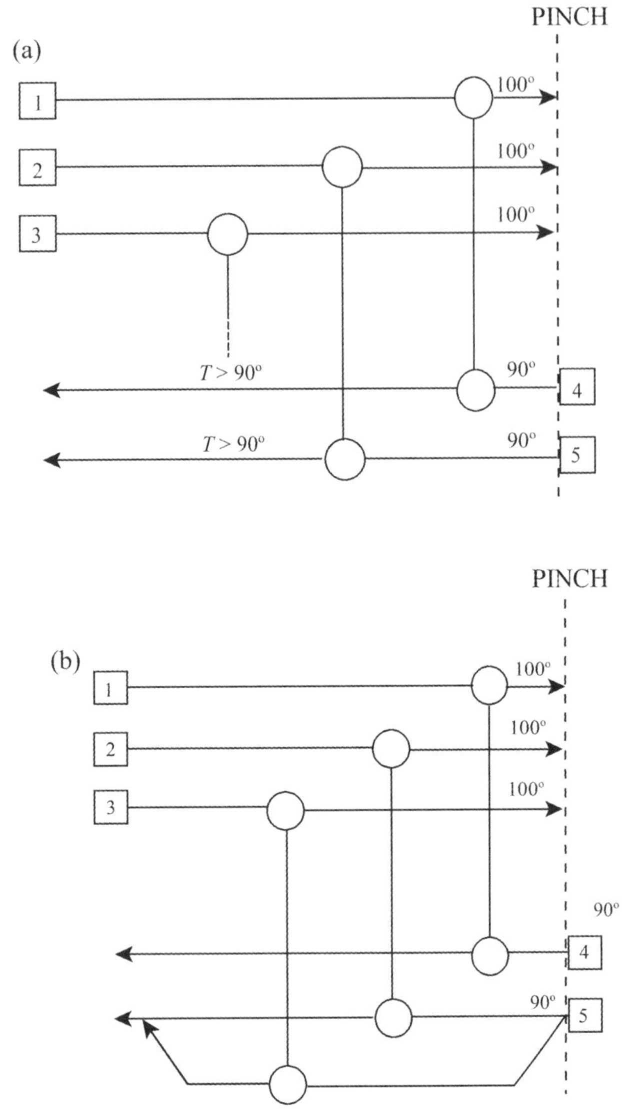

Figure 18.12

If the number of hot streams at the pinch, above the pinch, is greater than the number of cold streams, then stream splitting of the cold streams is required.

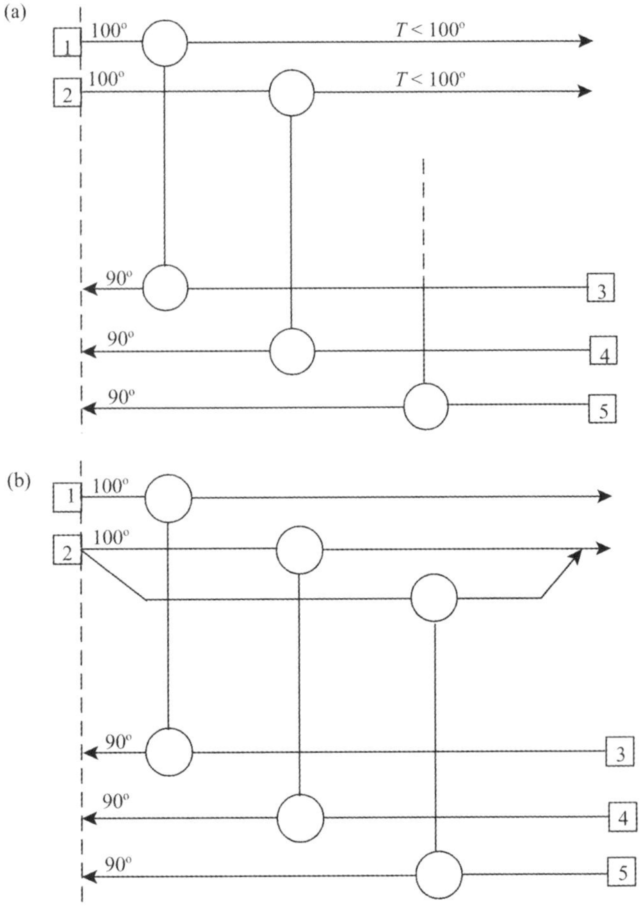

Figure 18.13

If the number of cold streams below the pinch, at the pinch, is greater than the pinch number of hot streams, then stream splitting of the hot steam is required.

for the overall approach (Linnhoff *et al.*, 1982; Linnhoff and Hindmarsh, 1983).

One further important point needs to be made regarding stream splitting. In Figure 18.14, the hot stream is split into two branches with *CP* values of 3 and 2 to satisfy the *CP* inequality criteria. However, a different split could have been chosen. For example, the split could have been into branch *CP* values of 4 and 1, or 2.5 and 2.5, or 2 and 3 (or any setting between 4 and 1 and 2 and 3). Each of these would also have satisfied the *CP* inequalities. Thus, there is a degree of freedom in the design to choose the branch flowrates. By fixing the heat duties on the two units in Figure 18.14b and changing the branch flowrates, the temperature differences across each unit are changed. The best choice can only be made by sizing and costing the various units in the completed network for different branch flowrates. This is an important degree of freedom when the network is optimized. Similar arguments could be made regarding the cold end design in Figure 18.15b.

Example 18.2 A problem table analysis for part of a high-temperature process reveals that for $\Delta T_{min} = 20\,°C$ the process requires 9.2 MW of hot utility, 6.4 MW of cold utility and the pinch is located at 520 °C for hot streams and 500 °C for cold streams. The process stream data are given in Table 18.2.

Table 18.2

Stream data.

Stream		Supply temperature (°C)	Target temperature (°C)	Heat capacity flowrate ($MW \cdot K^{-1}$)
No.	Type			
1	Hot	720	320	0.045
2	Hot	520	220	0.04
3	Cold	300	900	0.043
4	Cold	200	550	0.02

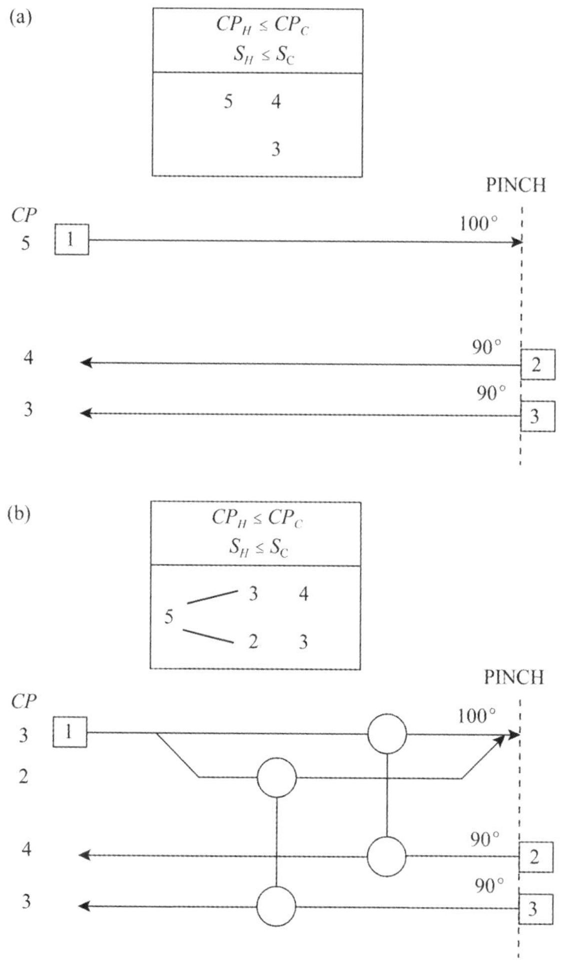

Figure 18.14
The CP inequality rules can necessiate stream splitting above the pinch.

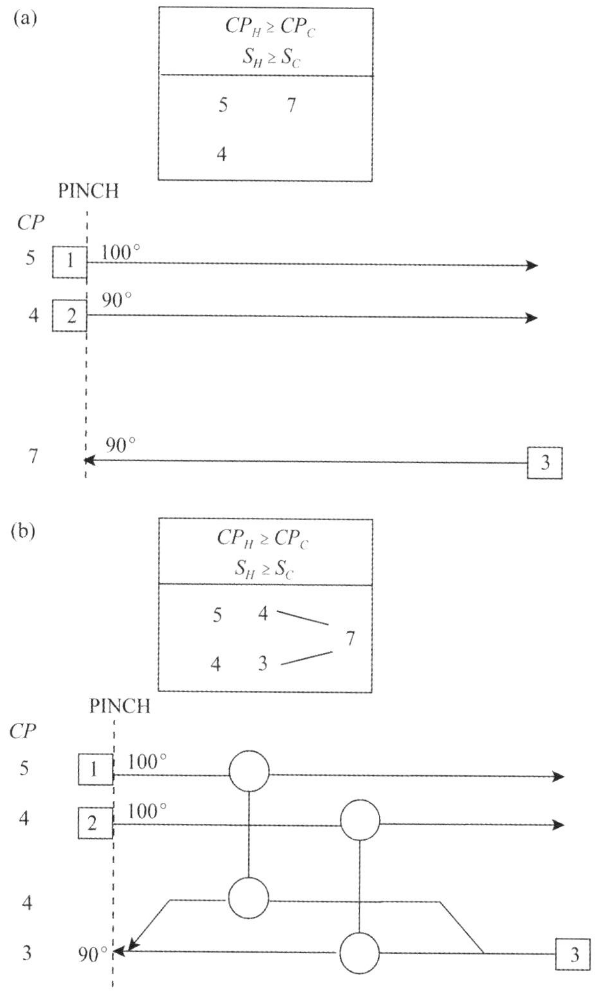

Figure 18.15
The CP inequality rules can necessiate stream splitting below the pinch.

Design a heat exchanger network for maximum energy recovery that features the minimum number of units.

Solution Figure 18.17a shows the stream grid with the CP tables for the above-pinch and below-pinch designs. Following the algorithm in Figure 18.16a, a hot stream must be split above the pinch to satisfy the *CP* inequality, as shown in Figure 18.17b. Thereafter the design is straightforward and the final design is shown in Figure 18.17c.

The target for the minimum number of units is given by:

$$\begin{aligned} N_{UNITS} &= (S-1)_{ABOVE\ PINCH} + (S-1)_{BELOW\ PINCH} \\ &= (4-1) \qquad\qquad + (5-1) \\ &= 7 \end{aligned}$$

The design in Figure 18.17c is seen to achieve the minimum units target.

18.4 Design for Multiple Pinches

In Chapter 17, it was discussed as to how the use of multiple utilities could give rise to multiple pinches. For example, the design for the process in Figure 17.2 could have used either a single level of hot utility or two steam levels (Figure 17.26a). The targeting indicated that instead of using 7.5 MW of high-pressure steam at 240 °C, 3 MW of this could be substituted with low-pressure steam at 180 °C. Where the low-pressure steam touches the grand composite curve in Figure 17.26a results in a utility pinch. The high-pressure steam also causes a utility pinch below the temperature of the high-pressure steam. Figure 18.18a shows the grid diagram when two steam levels are used with the two utility pinches dividing the process into four parts.

Following the pinch rules, heat should not be transferred across either the process pinch or either of the utility pinches by process-

18

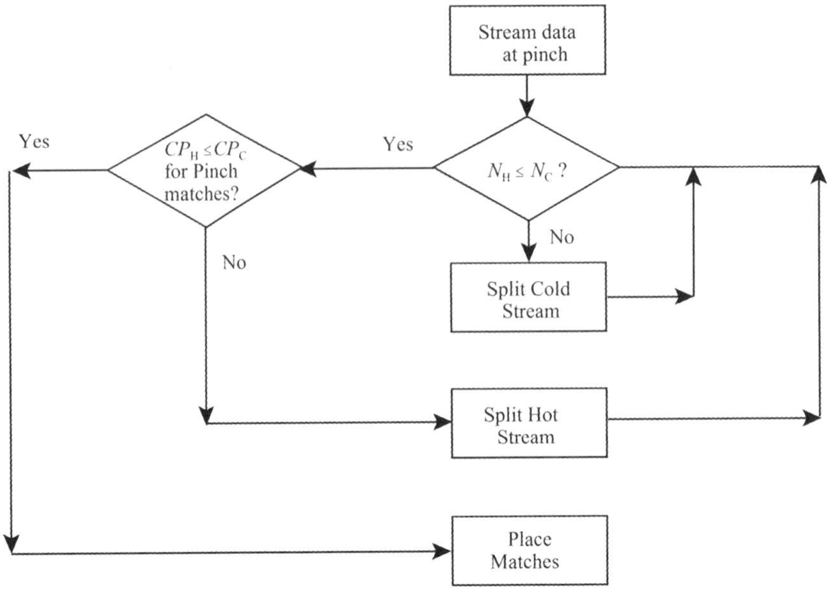

(a) Above the pinch.

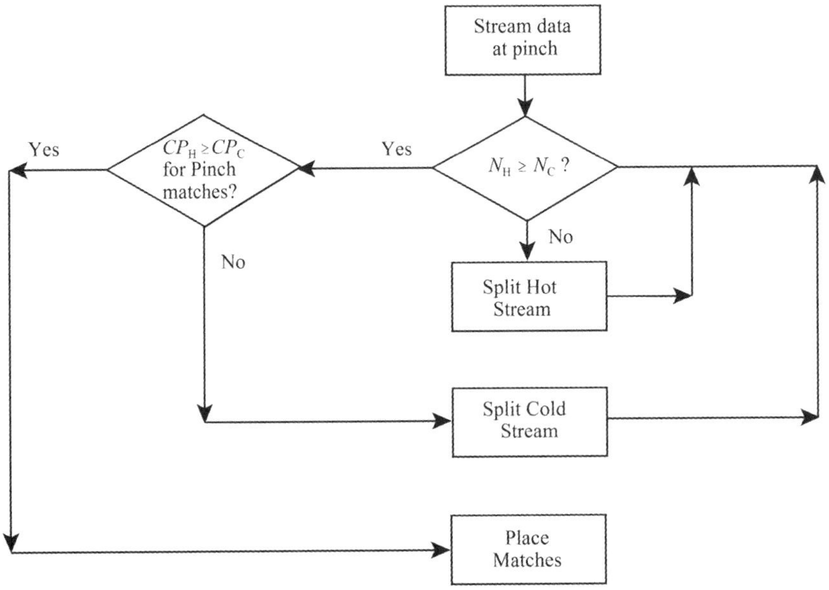

(b) Below the pinch.

Figure 18.16

Stream splitting algorithims.

to-process heat exchange. Also, there must be no inappropriate use of utilities. This means that above the high-temperature utility pinch shown in Figure 18.18a, high-pressure steam should be used and no low-pressure steam or cooling water. Between the utility pinches, no utility at all can be used and the heating and cooling must be satisfied by only heat recovery. Between the low-temperature utility pinch and the process pinch low-pressure steam should be used and no high-pressure steam or cooling water. Below the process pinch in Figure 18.18a only cooling water should be used. The appropriate utility streams have been included with the process streams in Figure 18.18a.

The network can now be designed using the pinch design method (Linnhoff *et al.*, 1982, Linnhoff and Hindmarsh, 1983). The philosophy of the pinch design method is to start at the pinch,

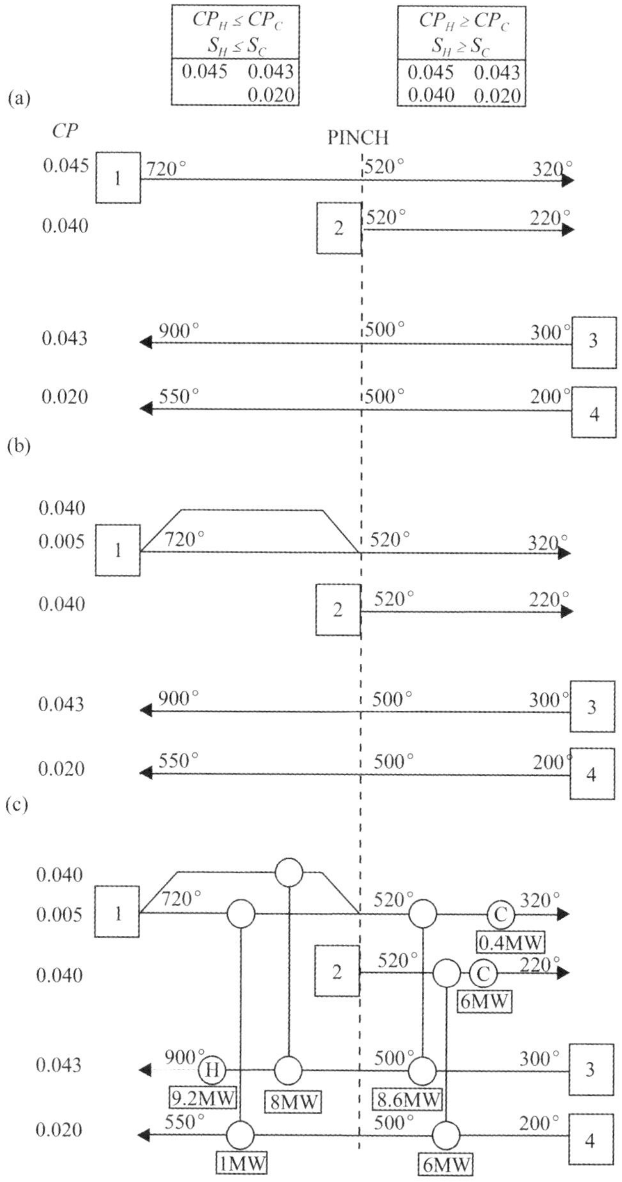

Figure 18.17

Maximum energy recovery design for Example 18.2.

and move away. At the pinch, the rules for the CP inequality and the number of streams must be obeyed. Above the high temperature utility pinch in Figure 18.18a and below the process pinch in Figure 18.18a there is clearly no problem in applying this philosophy. However, between the pinches there is a problem, since designing away from both pinches could lead to a clash.

More careful examination of Figure 18.18a reveals that, between the low-temperature utility pinch and the process pinch, one is more constrained than the other. Below the utility pinch, $CP_H \ge CP_C$ is required and low-pressure steam is available as a hot stream with an extremely large CP. In fact, if steam is assumed to condense or vaporize isothermally, it will have a CP that is infinite.

Thus, following the philosophy of starting the design in the most constrained region, the design between the pinches in Figure 18.18a should be started at the most constrained pinch, which is the process pinch.

Finally, the design between the two utility pinches must satisfy the constraint $CP_H \ge CP_C$ below the high-temperature utility pinch and $CP_H \le CP_C$ above the low-temperature utility pinch without the use of utilities. This part of the design is the most constrained. The design must start from each utility pinch and meet feasibly in the middle. Figure 18.18b shows one possible design.

Following this approach, the final design is shown in Figure 18.18b. It achieves the energy target set in Example 17.3 and in the minimum number of units. Remember that, in this case, to calculate the minimum number of units, the stream count must be performed separately in the four parts of the problem. Note that the stream split on the low-pressure steam in Figure 18.18b is not strictly necessary, but is made for practical reasons. Without the stream split, steam would have to partially condense in one unit and the steam–condensate mixture transferred to the next unit. The stream split allows two conventional steam heaters on low-pressure steam to be used. It is clear from Figure 18.18b that the use of two steam levels has increased the complexity of the design considerably. However, the complexity of the design can be reduced later when the structure is subjected to optimization. The optimization can remove units that are uneconomic.

It is rare for there to be two process pinches in a problem. Multiple pinches usually arise from the introduction of additional utilities causing utility pinches.

Example 18.3 The stream data for a process are given in Table 18.3.

It has been decided to integrate a gas turbine exhaust with the process. The exhaust temperature of the gas turbine is 400 °C with $CP = 0.05\,MW{\cdot}K^{-1}$. Ambient temperature is 10 °C.

a) Calculate the problem table cascade for $\Delta T_{min} = 20\,°C$.

b) Saturated steam is to be generated by the process at a high-pressure level of 250 °C and low-pressure level of 140 °C, each from saturated boiler feedwater. The generation of the higher-pressure steam is to be maximized. How much steam can be generated at the two levels assuming boiler feedwater and final steam conditions are both saturated?

c) Design a network for maximum energy recovery for $\Delta T_{min} = 20\,°C$ that generates steam at these two levels.

d) What is the residual cooling demand?

Solution

a) The problem table cascade is shown in Table 18.4 for $\Delta T_{min} = 20\,°C$.

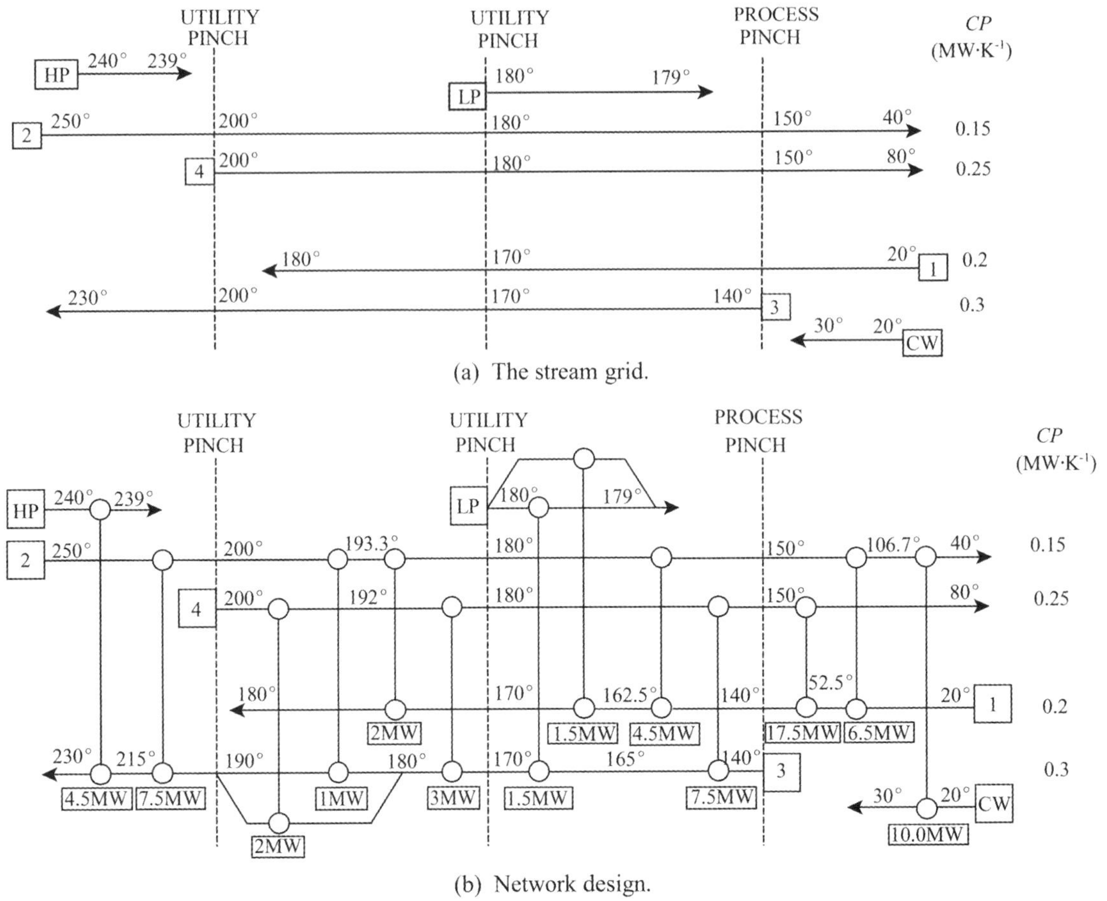

Figure 18.18

Network design for the process from Figure 18.2 using two steam levels.

Table 18.3

Stream data.

Stream No.	Stream Type	Supply temperature (°C)	Target temperature (°C)	Heat capacity flowrate ($MW \cdot K^{-1}$)
1	Hot	635	155	0.044
2	Cold	10	615	0.023
3	Cold	85	250	0.020
4	Cold	250	615	0.020

b) For high-pressure steam, $T^* = 260\,°C$ and for low-pressure steam, $T^* = 150\,°C$. Figure 18.19 shows the grand composite curve plotted from the problem table cascade. The two levels of steam generation are shown.

Duty on high-pressure steam generation

$$= 6.865\ \text{MW}$$

Table 18.4

Problem table cascade.

Interval temperature (°C)	Cascade heat flow (MW)
625	0
390	0.235
260	6.865
145	12.73
95	13.08
20	15.105
0	16.105

By interpolation from the problem table cascade, heat flow at $T* = 150\,°C$

$$= 6.865 + \frac{(260 - 150)}{(260 - 145)} \times (12.73 - 6.865)$$

$$= 12.475\ \text{MW}$$

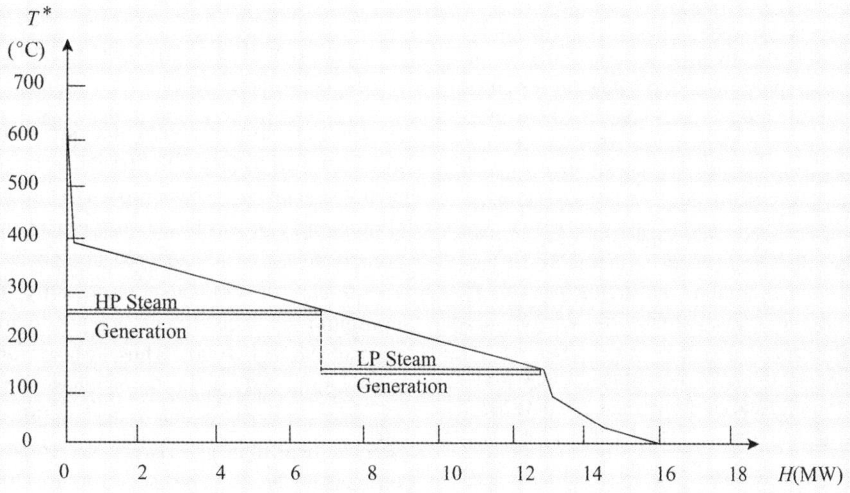

Figure 18.19
Grand composite curve for Example 18.3 showing two levels of steam generation.

Duty on low-pressure steam generation = 12.475 – 6.865 = 5.61 MW

c) The use of two levels of steam generation in Figure 18.19 creates two utility pinches. Thus, the stream grid needs to be divided into three parts. Figure 18.20 shows the final design, which achieves the targets set for both high-pressure (HP) and low-pressure (LP) steam generation. In Figure 18.20 above the HP pinch, the *CP* of Stream 1 and the gas turbine stream are too large to match directly against Streams 3 and 4. This is overcome in Figure 18.20 by splitting Stream 1 and exploiting the infinite *CP* of the HP steam generation. Between the pinches, the design is started below the HP pinch and developed toward the LP pinch, where the infinite *CP* of the LP steam generation can be exploited to satisfy the *CP* inequalities. Below the LP pinch, although the *CP* inequalities can be satisfied by direct matches, the heat duty on Stream 1 is small compared with the other streams. If Stream 1 did not exist below the LP pinch, then this would call for the gas turbine

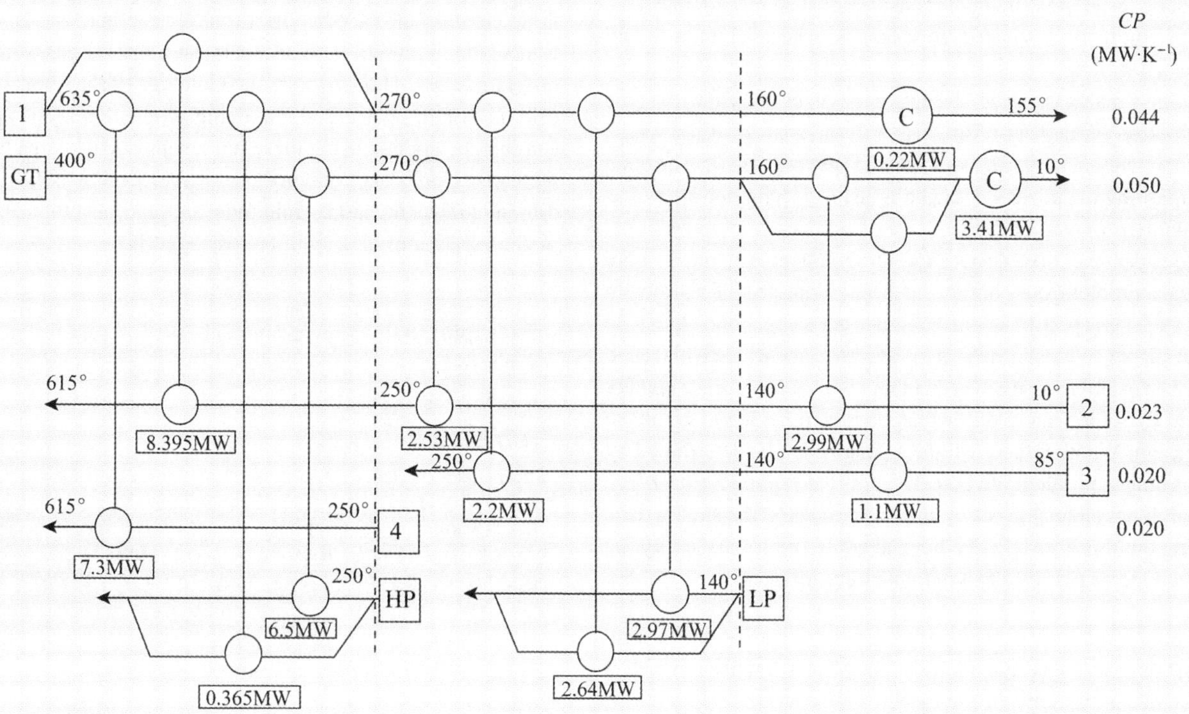

Figure 18.20
Network design for Example 18.3.

stream to be split. This has been done in Figure 18.20 because of the small duty on Stream 1. Although the steam generation for both high-pressure and low-pressure steam is shown with steam splits in Figure 18.20, in practice these units would be individual steam generators, each fed with boiler feedwater. Also, the stream split for the gas turbine exhaust could be accommodated by splitting the gas turbine flow into two ducts or by placing two sets of tubes in the same gas turbine exhaust located in parallel. The design in Figure 18.20 has significant scope for simplification, but at the penalty of reduced energy efficiency. Such trade-offs are at the discretion of the designer.

d) There is a cooling demand of 0.22 MW on Stream 1 that needs to be satisfied by cold utility and cooling demand of 3.41 MW required by the gas turbine exhaust. The cooling of the gas turbine exhaust is satisfied by simply venting it to atmosphere after heat recovery has been completed.

18.5 Remaining Problem Analysis

The considerations addressed so far in network design have been restricted to those of energy performance and number of units. In addition, the problems have all been straightforward to design for maximum energy recovery in the minimum number of units by ticking-off streams. Not all problems are so straightforward. Also, the heat transfer area should be considered when placing matches. Here, a more sophisticated approach is needed (Linnhoff and Ahmad, 1990).

When a match is placed, the duty needs to be chosen with some quantitative assessment of the match in the context of the whole network, without having to complete the network. This can be done by exploiting the powers of targeting using a technique known as *remaining problem analysis*.

Consider the first design for minimum energy in a more complex problem than has so far been addressed. If a problem table analysis is performed on the stream data, Q_{Hmin} and Q_{Cmin} can be calculated. When the network is designed and a match placed, it would be useful to assess whether there will be any energy penalty caused by some feature of the match without having to complete the design. Whether there will be a penalty can be determined by performing a problem table analysis on the *remaining problem*. The problem table analysis is simply repeated on the stream data, leaving out those parts of the hot and cold stream satisfied by the match. One of two results would then occur:

1) The algorithm may calculate Q_{Hmin} and Q_{Cmin} to be unchanged. In this case, the designer knows that the match will not penalize the design in terms of increased utility usage.
2) The algorithm may calculate an increase in Q_{Hmin} and Q_{Cmin}. This means that the match is transferring heat across the pinch or that there is some feature of the design that will cause cross-pinch heat transfer if the design was completed. If the match is not transferring heat across the pinch directly, then the increase in utility will result from the match being too big as a result of the tick-off heuristic.

The remaining problem analysis technique can be applied to any feature of the network that can be targeted, such as a minimum area. In Chapter 17, the approach to targeting for heat transfer area (Equation 17.9) was based on "vertical" heat transfer from the hot composite curve to the cold composite curve. If heat transfer coefficients do not vary significantly, this model predicts the minimum area requirements adequately for most purposes (Linnhoff and Ahmad, 1990). Thus, if heat transfer coefficients do not vary significantly, then the matches created in the design should come as close as possible to the conditions that would correspond with vertical transfer between the composite curves. Remaining problem analysis can be used to approach the area target, as closely as a practical design permits, using a minimum (or near-minimum) number of units. Suppose a match is placed; then its area requirement can be calculated. A remaining problem analysis can be carried out by calculating the area target for the stream data, leaving out those parts of the data satisfied by the match. The area of the match is now added to the area target for the remaining problem. Subtraction of the original area target for the whole-stream data $A_{NETWORK}$ gives the area penalty incurred.

If heat transfer coefficients vary significantly, then the "vertical" heat transfer model adopted in Equation 17.9 predicts a network area that is higher than the true minimum, as illustrated in Figure 17.43. Under these circumstances, a careful pattern of "nonvertical" matching is required to approach the minimum network area. However, the remaining problem analysis approach can still be used to steer the design toward a minimum area under these circumstances. When heat transfer coefficients vary significantly, the minimum network area can be predicted using linear programming (Saboo, Morari and Colberg, 1986; Ahmad, Linnhoff and Smith, 1990). The remaining problem analysis approach can then be applied using these more sophisticated area targeting methods. Under such circumstances, the design is likely to be difficult to steer toward the minimum area and an automated design method can be used, as will be discussed later.

Example 18.4 The stream data for a process are given in Table 18.5.

Table 18.5

Stream data.

Stream		Supply temperature (°C)	Target temperature (°C)	Heat capacity flowrate ($MW \cdot K^{-1}$)
No.	Type			
1	Hot	150	50	0.2
2	Hot	170	40	0.1
3	Cold	50	120	0.3
4	Cold	80	110	0.5

Steam is available condensing between 180 and 179 °C and cooling water between 20 and 30 °C. All film transfer coefficients are 200 $W \cdot m^{-2} \cdot K^{-1}$. For $\Delta T_{min} = 10$ °C, the minimum hot and cold utility duties are 7 MW and 4 MW respectively, corresponding with a pinch at 90 °C on the hot streams and 80 °C on the cold streams.

a) Develop a maximum energy recovery design above the pinch that comes close to the area target in the minimum number of units.

b) Develop a maximum energy recovery design below the pinch that comes as close as possible to the minimum number of units.

Solution

a) The area target for the above-pinch problem shown in Figure 18.21 is 8859 m^2. If the design is started at the pinch with Stream 1, then Figure 18.21a shows a feasible match that obeys the *CP* inequality. Maximizing its duty to 12 MW allows two streams to be ticked off simultaneously. This results from a coincidence in the stream data, the duties for Streams 1 and 3 being equal above the pinch. The area of the match is 6592 m^2 and the target for the remaining problem above the pinch is 3419 m^2, giving a total of 10,011 m^2. Thus, the match in Figure 18.21a causes the overall target to be exceeded by 1152 m^2 (13%). This does not seem to be a good match.

Figure 18.21b shows an alternative match for Stream 1 that also obeys the *CP* inequality. The tick-off heuristic also fixes its duty to be 12 MW. The area for this match is 5087 m^2 and the target for the remaining problem above the pinch is 3788 m^2, giving a total of 8875 m^2. Thus, the match in Figure 18.21b causes the overall target to be exceeded by 16 m^2 (0.2%). This seems to be a better match and therefore is accepted.

Placing the next match above the pinch as shown in Figure 18.21c also allows the *CP* inequality to be obeyed. The area for both matches in Figure 18.21c is 7856 m^2 and the target for the remaining problem is 1020 m^2, giving a total of 8876 m^2. Accepting both matches causes the overall area target to be exceeded by 17 m^2 (0.2%). This seems to be reasonable and both matches are accepted. No further process-to-process matches are possible and it remains to place hot utility.

b) The cold utility target for the problem shown in Figure 18.22 is 4 MW. If the design is started at the pinch with Stream 3, then Stream 3 must be split to satisfy the CP inequality (Figure. 18.22a). Matching one of the branches against Stream 1 and ticking off Stream 1 results in a duty of 8 MW. This is a case in which the tick-off heuristic has caused problems. The match is infeasible, because the temperature difference between the streams at the cold end of the match is infeasible. Its duty must be reduced to 6 MW to be feasible without either stream being ticked off (Figure 18.22b).

Figure 18.22c shows an additional match placed on the other branch for Stream 3 with its duty maximized to 3 MW to tick off Stream 3. No further process-to-process matches are possible and it remains to place cold utility.

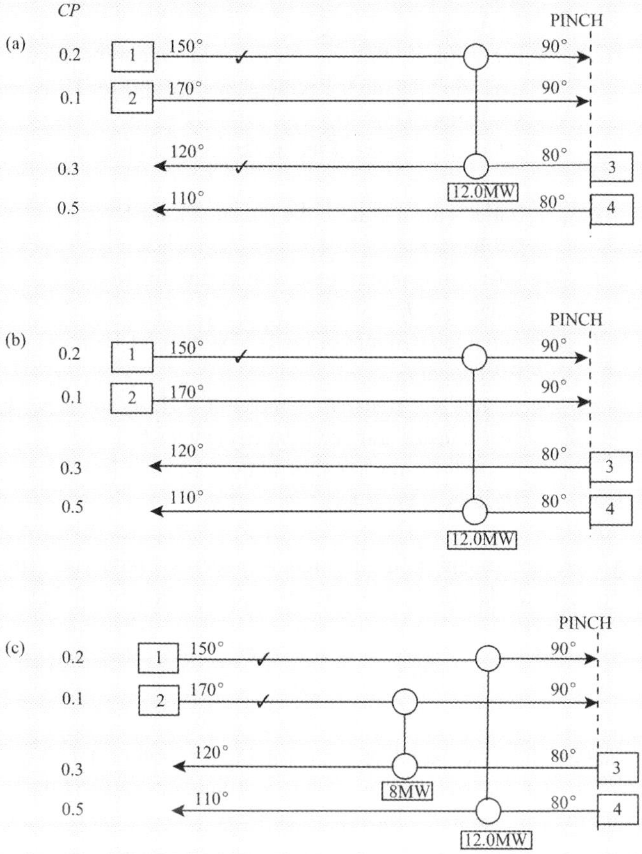

Figure 18.21

Above the pinch design for Example 18.4.

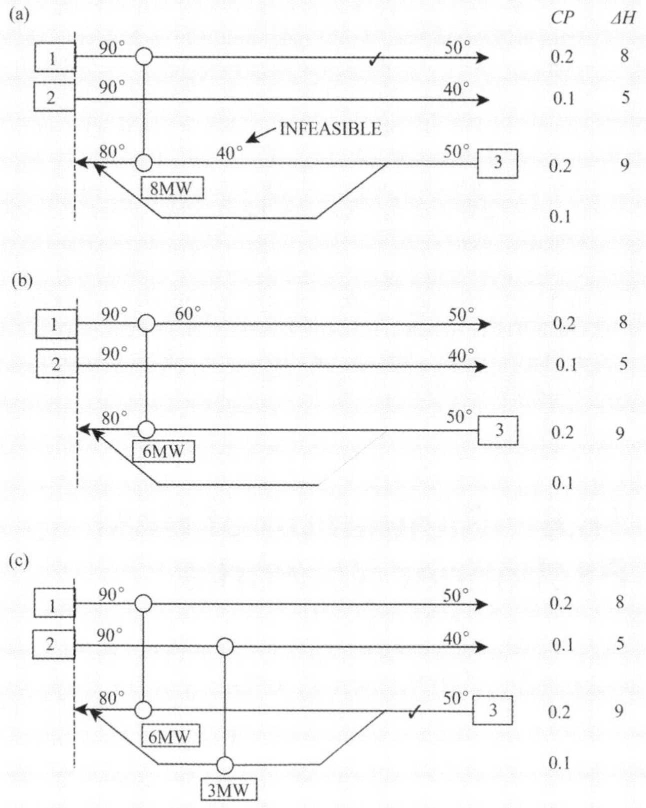

Figure 18.22

Below the pinch design for Example 18.4.

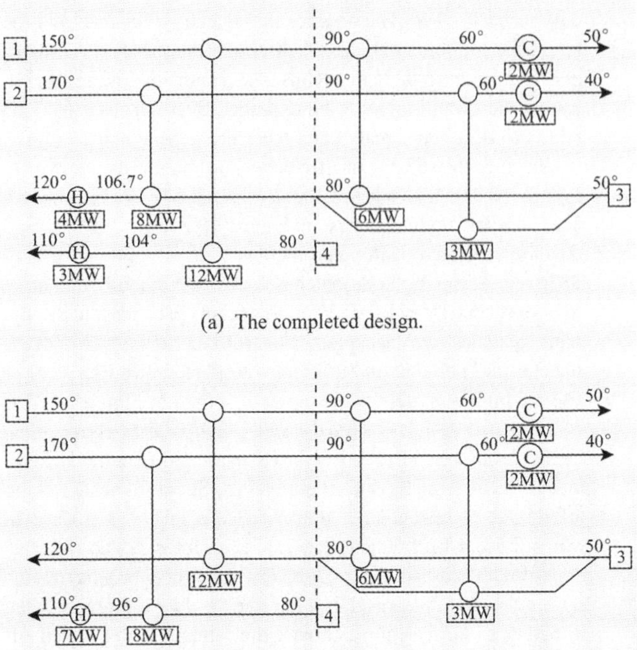

(a) The completed design.

(b) An alternative design.

Figure 18.23
Alternative designs for Example 18.4.

Figure 18.23a shows the complete design, achieving maximum energy recovery in one more unit than the target minimum, due to the inability to tick off streams below the pinch. If the match in Figure 18.21a had been accepted and the design completed, then the design in Figure 18.23b would have been obtained. This achieves the target for the minimum number of units of 7 (at the expense of excessive area). This results from the coincidence of data mentioned earlier in Figure 18.21a, which allowed two streams to be ticked off simultaneously. The result is that the design above the pinch uses one fewer unit than the target, owing to the formation of two components above the pinch. The design below the pinch uses one more than the target and the net result is that the overall design achieves the target for the minimum number of units.

18.6 Simulation of Heat Exchanger Networks

Once a network structure has been created, further design detail can be added. The whole network performance can then be simulated. In the simplest case, the values of U for each heat exchanger are estimated or calculated and then fixed. The ΔT_{LM} for each heat exchanger is known from the preliminary design, which allows the heat transfer area A for each heat exchanger to be determined. If the values of U and A are fixed and the CP for each stream is constant, then the simulation equations for an individual heat exchanger (Equations 12.83, 12.86, 12.88, 12.93, 12.95, 12.105 and 12.106) can be written for each heat exchanger in the network:

For $R \neq 1$:

$$(R_k - 1)T_{H1k} + R_k(X-1)T_{C1k} + (1 - R_kX)T_{H2k} = 0 \qquad R_k \neq 1 \tag{18.5}$$

$$(X-1)T_{H1k} + X(R_k - 1)T_{C1k} + (1 - R_kX)T_{C2k} = 0 \qquad R_k \neq 1 \tag{18.6}$$

For $R = 1$:

$$T_{H1k} + YT_{C1k} - (Y+1)T_{H2k} = 0 \qquad R_k = 1 \tag{18.7}$$

$$YT_{H1k} + T_{C1k} - (Y+1)T_{C2k} = 0 \qquad R_k = 1 \tag{18.8}$$

where

$$R_k = \frac{CP_{Ck}}{CP_{Hk}} \tag{18.9}$$

$$X = \exp\left[\frac{U_kA_k(R_k - 1)F_{Tk}}{CP_{Ck}}\right] \tag{18.10}$$

$$Y = \frac{U_kA_kF_{Tk}}{CP_{Ck}} \tag{18.11}$$

CP_{Hk} = heat capacity flowrate for the hot stream (product of mass flowrate and specific heat capacity) for Heat Exchanger k
CP_{Ck} = heat capacity flowrate for the cold stream (product of mass flowrate and specific heat capacity) for Heat Exchanger k
T_{H1k} = inlet temperature of the hot stream for Heat Exchanger k
T_{H2k} = outlet temperature of the hot stream for Heat Exchanger k
T_{C1k} = inlet temperature of the cold stream for Heat Exchanger k
T_{C2k} = outlet temperature of the cold stream for Heat Exchanger k
U_k = overall heat transfer coefficient for Heat Exchanger k
A_k = heat transfer area for Heat Exchanger k
F_{Tk} = logarithmic mean temperature correction factor for Heat Exchanger k

If there are stream splits, the mixer associated with each split requires an additional equation. Assuming no heat of mixing:

$$T_{MIX} = \frac{1}{\sum_i CP_i}\sum_i CP_i T_i \tag{18.12}$$

where T_{MIX} = temperature of the mixing junction (°C)
CP_i = heat capacity flowrate of Stream i entering the mixing junction ($kW{\cdot}K^{-1}$)
T_i = temperature of Stream i entering the mixing junction (°C)

Consider the application of these equations to a heat exchanger network. The first thing that needs to be understood is the number

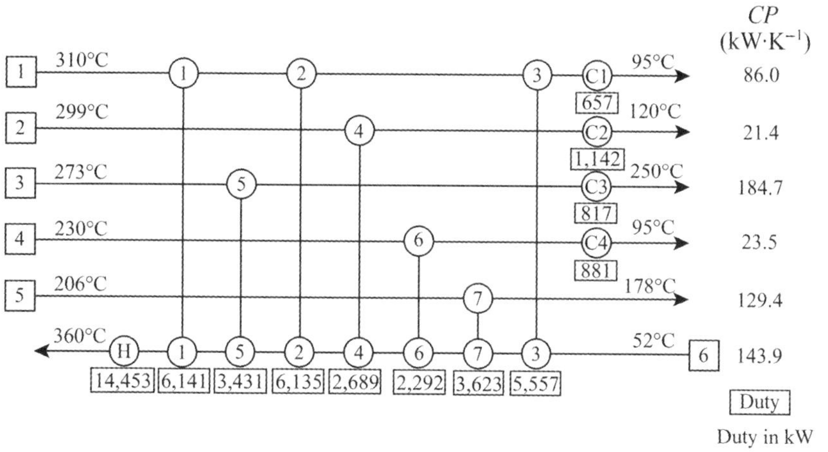

Figure 18.24

An example heat exchanger network. (Reproduced from Ning Jiang, Jacob David Shelley, Steve Doyle, Robin Smith (2014) Heat exchanger network retrofit with a fixed network structure, *Applied Energy*, **127**: 25–33, with permission from Elsevier.)

of unknown temperatures that need to be determined. The number of unique temperature nodes in a network is given by:

$$\begin{bmatrix}\text{Number of}\\ \text{temperature nodes}\end{bmatrix} = \begin{bmatrix}\text{Number of streams}\\ \text{including utilities}\end{bmatrix} + 2\times\begin{bmatrix}\text{Number of}\\ \text{heat exchangers}\end{bmatrix} + \begin{bmatrix}\text{Number of}\\ \text{mixers}\end{bmatrix} \qquad (18.13)$$

As an example, consider the number of unique temperature nodes in the network in Figure 18.24:

$$\begin{bmatrix}\text{Number of}\\ \text{temperature nodes}\end{bmatrix} = [6+2] + 2\times[12] + [0] = 32 \qquad (18.14)$$

Thus there are 32 temperature nodes, some of which are known. The initial temperatures of the process steams and utility streams are known, of which there are eight in total. That leaves 24 temperatures unknown. Two equations can be written for each heat exchanger, which leaves 24 equations and 24 unknown temperatures.

If U_k, A_k, R_K, CP_{Ck}, CP_{Hk} and N_{SHELLS} are fixed then Equations 18.5 to 18.8 and 18.12 are a set of linear equations. Equations 18.5 and 18.8 are linear because Equations 12.42 and 12.43 show that F_T is a function of R and $P_{1\text{-}2}$. In turn, Equation 12.103 shows that $P_{1\text{-}2}$ is a function of U_k, A_k, R_K, CP_{Ck} and N_{SHELLS}. Thus if U_k, A_k, R_K, CP_{Ck}, CP_{Hk} and N_{SHELLS} are fixed then X and Y above are also fixed and if CP_i are fixed in Equation 18.12 then Equations 18.5 to 18.8 and 18.12 are a set of linear equations. This set of simultaneous linear equations can be solved efficiently using, for example, the LU Decomposition Method (Press *et al.*, 2007). It should be noted that the specified target temperatures of the streams will not necessarily be achieved, as this is a simulation calculation in which the final network temperatures are to be calculated.

The approach can be adapted to streams with nonlinear heat capacity flowrates. Figure 18.25 illustrates the approach for nonlinear stream heat capacity, showing how the enthalpy–temperature profile can be linearized between the inlet and outlet temperatures. A similar approach can be adopted as used for the linear CP case. This time, however, the enthalpy of the stream must be defined as a function of temperature, as shown in Figure 18.25. Knowing enthalpy as a function of temperature, each time the temperatures are set at an intermediate point in the solution of Equations 18.5 to 18.11, the inlet and outlet temperatures for each heat exchanger define points on the temperature–enthalpy profile for the stream, as shown in Figure 18.25. This defines the enthalpy change for the stream across the heat exchanger, from which CP_{Hk} and CP_{Ck} can be calculated. However, the nonlinear temperature–enthalpy profile makes the equation set nonlinear. Different methods can be used to solve the set of nonlinear equations (Press *et al.*, 2007). One way is to use a nonlinear equation solver to manipulate the temperatures and use the LU Decomposition Method to solve the network for each iteration.

Finally, it is possible to go one step further and introduce the models for U_k developed in Chapter 12. This allows a detailed simulation of the whole network in a nonlinear model.

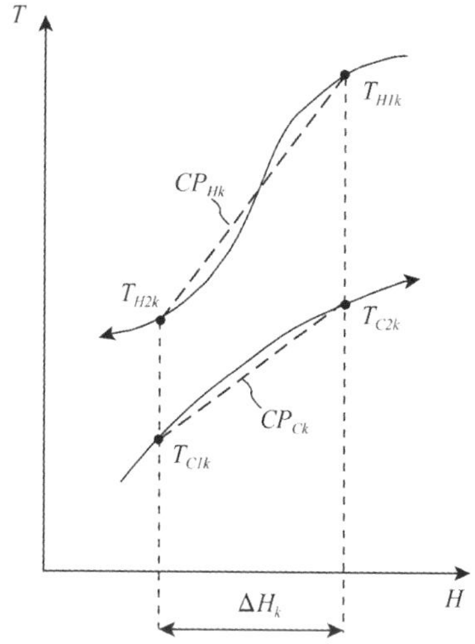

Figure 18.25

Nonlinear stream heat capacity linearization for Heat Exchanger k. (Reproduced from Ning Jiang, Jacob David Shelley, Steve Doyle, Robin Smith (2014) Heat exchanger network retrofit with a fixed network structure, *Applied Energy*, **127**: 25–33, with permission from Elsevier.)

18.7 Optimization of a Fixed Network Structure

The pinch design method creates a network structure based on the assumption that no heat exchanger should have a temperature difference smaller than ΔT_{min}. Having created a structure for the heat exchanger network, the structure can be subjected to continuous optimization. The constraint that no exchanger should have a temperature difference smaller than ΔT_{min} can be relaxed during the optimization. The continuous optimization of heat exchanger networks is based on the redistribution of the exchanger duties. Some exchangers should perhaps be larger, some smaller and some perhaps removed from the design altogether. Exchangers are removed from the design if the optimization sets their duty to zero.

Given a network structure, it is possible to identify loops and paths for it, as in Section 17.11 (Linnhoff *et al.*, 1982; Linnhoff and Hindmarsh, 1983). Within the context of optimization, only those paths that connect two different utilities need to be considered. This could be a path from steam to cooling water or a path from high-pressure steam used as hot utility to low-pressure steam also used as hot utility. These paths between two different utilities will be termed *utility paths*. Loops and utility paths both provide degrees of freedom in the optimization.

Consider Figure 18.26a, which shows the network design from Figure 18.7, but with a loop highlighted. Heat can be shifted around loops. Figure 18.26a shows the effect of shifting heat duty u around the loop. In this loop, heat duty u is simply moved from Unit E to Unit B. The change in heat duties around the loop maintains the network heat balance and the supply and target temperatures of the streams. However, the temperatures around the loop change and hence the temperature differences of the exchangers in the loop change in addition to their duties. The magnitude of u could be changed to different values and the network costed at each value to find the optimum setting for u. If the optimum setting for u turns out to be 6.5 MW (the original duty on Unit E), then the duty on one of the exchangers is zero and should be removed from the design.

Figure 18.26b shows the network with another loop marked. Figure 18.26b shows the effect of shifting heat duty v around the loop. Again the heat balance is maintained, but the temperatures as well as the duties around the loop change. As before, the value of v can be optimized by costing the network at different settings of v. If v is optimized to 7.0 MW (the original duty on Unit A), then the duty on A becomes zero and this unit is removed from the design. Note again that once optimization is started, there is no longer a constraint to maintain temperature differences to be larger than ΔT_{min}.

Figure 18.27a shows the network with a utility path highlighted. Heat duty can be shifted along utility paths in a similar way to that for loops. Figure 18.27a shows the effect of shifting heat duty w along the path. This time the heat balance changes because the loads imported from hot utility and exported to cold utility both change by w. The supply and target temperatures are maintained. If w is optimized to 7.0 MW, this will result in Unit A being removed from the design. Different values of w can be taken and the network sized and costed at each value to find the optimal setting for w. Figure 18.27 shows other utility paths that can be exploited for optimization.

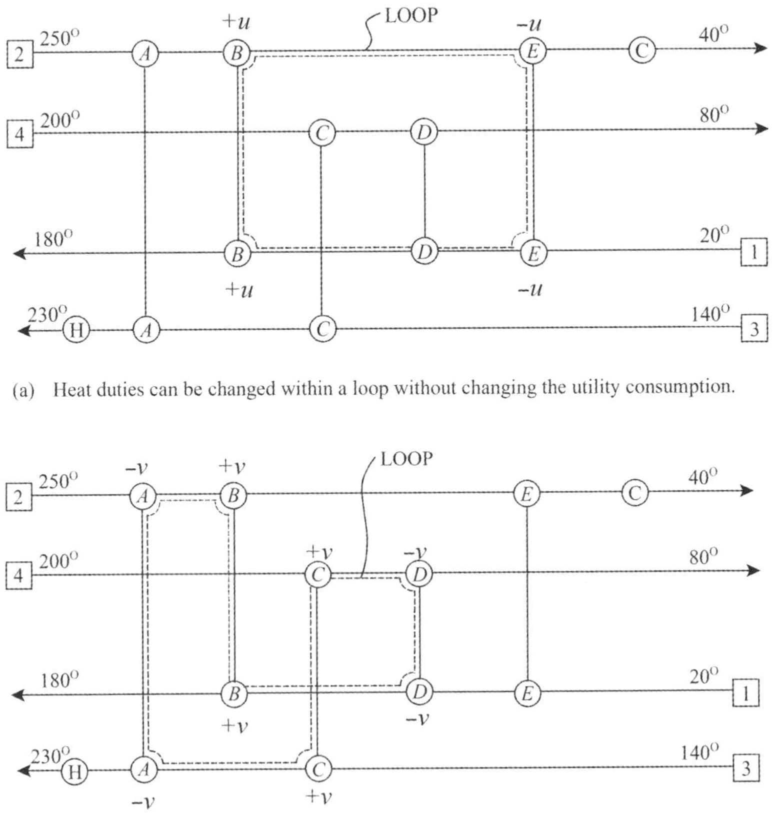

Figure 18.26

The loops that can be exploited for the optimization of the design for Figure 18.7.

(a) Heat duties can be changed within a loop without changing the utility consumption.

(b) Another loop allowing heat duties to be changed without changing the utility consumption.

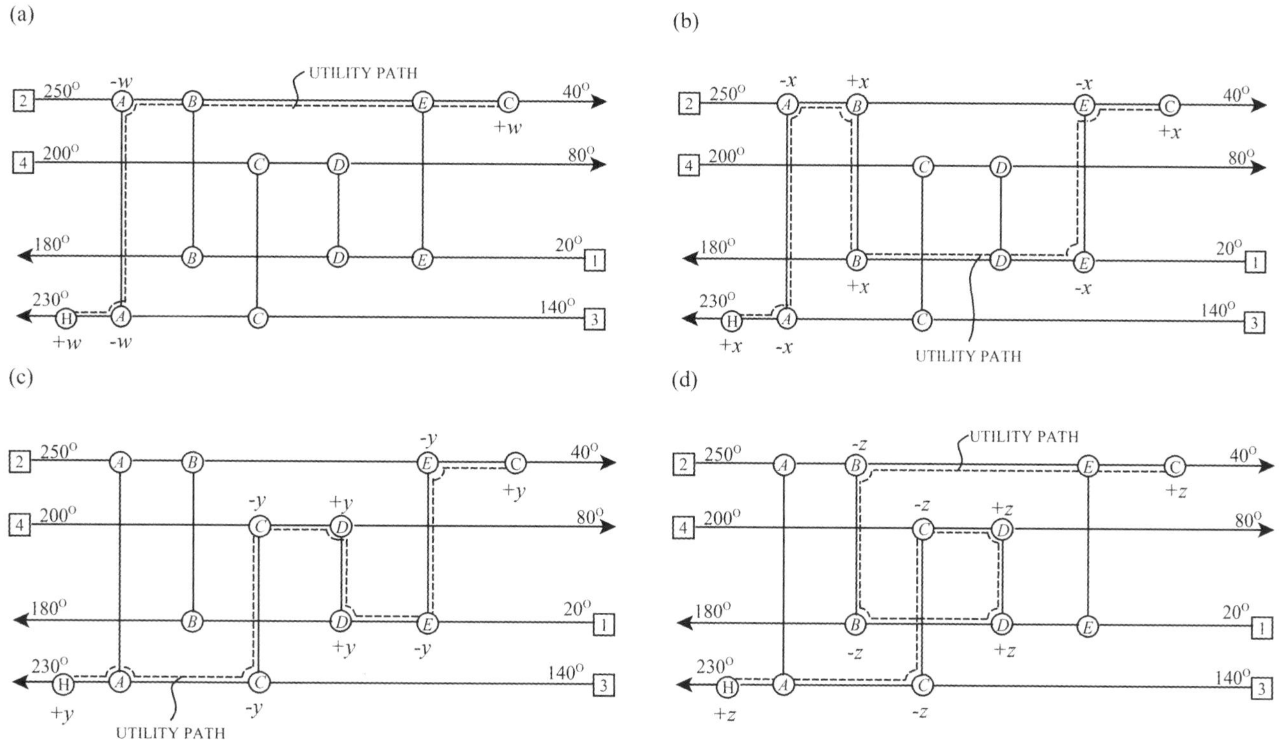

Figure 18.27

The utility paths that can be exploited for the optimisation of the design from Figure 18.7.

In fact, the optimization of the network requires that u, v, w, x, y and z in Figures 18.26 and 18.27 must be optimized simultaneously. Furthermore, stream splits may exist in the design and variations of their branch flowrates can be superimposed on the exploitation of loops and paths in the optimization. During this optimization, the design is no longer constrained to have temperature differences larger than ΔT_{min} (although very small values in individual heat exchangers should be avoided for practical reasons). Also, pinches no longer divide the design into independent thermodynamic regions and there is no longer any concern about cross-pinch heat transfer. The objective now is simply to minimize cost.

Thus, loops, utility paths and stream splits offer the degrees of freedom for manipulating the network cost. The objective function in new design is usually to minimize total cost, that is, combined operating and annualized capital cost. The annualization period chosen for the capital cost will have a direct influence on the optimization. A longer annualization period will lead to more energy-efficient designs.

In practice, rather than manipulate loops and paths explicitly, the optimization is normally formulated such that the individual duties on each match are varied in the multivariable optimization, subject to:

- the total enthalpy change on each stream being within a specified tolerance of the original stream data,
- non-negative heat duty for each match,
- positive temperature difference for each exchanger to be greater than a practical minimum value for a given type of heat exchanger,
- for stream splits, branch flowrates must be positive and above a practical minimum flowrate.

In a network, some of the duties on the matches will not be able to be varied because they are not in a loop or a utility path. This simplifies the optimization. The problem is one of multivariable nonlinear continuous optimization (Gunderson and Naess, 1988).

If the network is optimized at fixed energy consumption, then only loops and stream splits are exploited. When energy consumption is allowed to vary, utility paths must also be included. As the network energy consumption increases, the overall capital cost tends to decrease and vice versa.

Example 18.5 Evolve the heat exchanger network in Figure 18.7 to simplify its structure.

a) Remove the smallest heat recovery unit from the network by exploiting the degree of freedom in a loop.
b) Recalculate the network temperatures and identify any violations of the $\Delta T_{min} = 10\,°C$ constraint.
c) Restore the original $\Delta T_{min} = 10\,°C$ throughout the network by exploiting a utility path.

Solution

a) Figure 18.28a shows the network from Figure 18.7 with a loop highlighted involving the unit with the smallest heat duty (6.5 MW). A heat duty of 6.5 MW has been shifted around the loop to adjust the smallest unit to a duty of zero. This will change the temperatures around the loop.

b) Heat balances around the units for the new heat duties allow the new temperatures in the network to be calculated, as shown in Figure 18.28b. Also highlighted in Figure 18.28b is a unit with an infeasible temperature difference. Not only is it less than the original ΔT_{min} of 10 °C, it is actually negative. Removing a unit in this way will always create a temperature difference smaller than the original ΔT_{min}, if the hot and cold utility remain fixed. There is a minimum number of units to satisfy the problem for hot

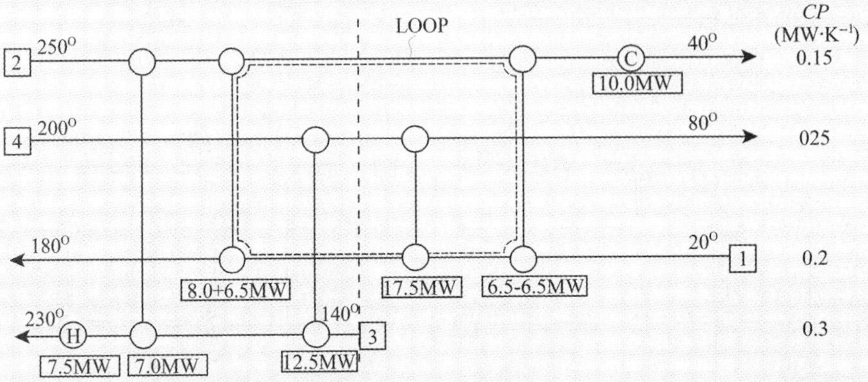

(a) The network with 6.5 MW of heat shifted around a loop.

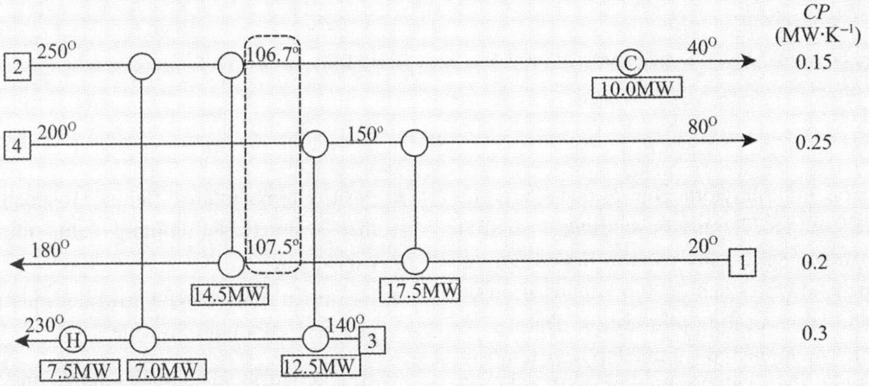

(b) Calculating intermediate network temperatures reveals on infesible temperature difference.

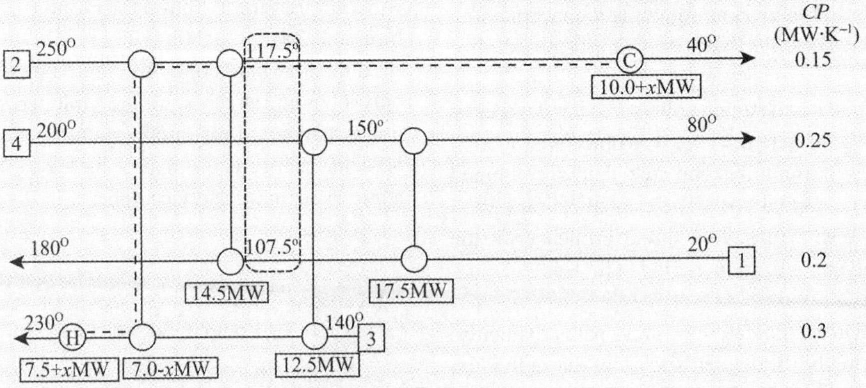

(c) Increasing the heat flow along a utility path allows the feasible temperature difference to be restored.

Figure 18.28

Evolution of a network to remove a unit.

utility consumption of Q_{Hmin} and a cold utility consumption of Q_{Cmin}, subject to the ΔT_{min} constraint. If a unit is then removed, something must be violated. By its nature, shifting heat load around a loop does not change the energy consumption, but does change internal temperatures.

c) Given the infeasible temperature difference in Figure 18.28b, this can be corrected by exploiting a utility path to change the temperatures in the network at the expense of increased energy consumption. Figure 18.28c shows the network with a utility path highlighted. The utility path allows one of the infeasible temperatures to be adjusted (Stream 2 in this case). If the original ΔT_{min} of 10 °C is to be restored, then the intermediate temperature of Stream 2 needs to be adjusted to 117.5 °C, as shown in Figure 18.28c. The unknown is how much additional heat duty (x MW) needs to be shifted along the utility path to restore the temperature to 117.5 °C. This can be determined by a simple heat balance around the cooler:

$$10.0 + x = 0.15(117.5 - 40)$$
$$x = 1.6\,\text{MW}$$

Thus the hot and cold utility consumption both need to be increased by 1.6 MW to restore the ΔT_{min} to the original 10 °C. In fact there is no justification to restore the ΔT_{min} back to the original 10 °C. The amount of additional energy shifted along the utility path is a degree of freedom that should be set by cost optimization. However, the example illustrates how the degrees of freedom can be manipulated in network optimization.

18.8 Automated Methods of Heat Exchanger Network Design

Having considered a manual method for the design of heat exchanger networks, now consider automated approaches. Manual methods for network design are adequate for smaller problems. However, larger and more complex problems demand greater automation. Also, it is often necessary to account for more details of equipment design when creating the network. Two basic approaches are possible for the automation of network design.

1) *Optimization of a superstructure.* Manual methods for heat exchanger network design are based on the creation of an irreducible structure. No redundant features are included. However, after the structure has been created, some features might be removed later by optimization. Any scope for simplification of the network by optimization to remove features is a consequence of the assumptions made during the creation of the initial structure. However, no attempt is made to deliberately include redundant features.

   The first approach to automated network design uses deterministic optimization to optimize a superstructure (Floudas, Ciric and Grossman, 1986). A superstructure is created that deliberately includes redundant features. This is then subjected to deterministic optimization. Redundant features are removed by the optimization, leaving the final network design. Floudas, Ciric and Grossmann (1986) showed how a heat exchanger network superstructure could be set up with all structural features included. The basic network superstructure is illustrated in Figure 18.29a. This shows a single stream split into three branches, with a match on each branch. A number of additional connections are included, such that the basic structure in Figure 18.29a can be evolved to any of the structures in Figure 18.29b by the removal of the appropriate connections from Figure 18.29a. Figure 18.29 illustrates how the superstructure could be evolved to different parallel, series, series–parallel and parallel–series configurations by the removal of the appropriate connections from the superstructure. Figure 18.30 illustrates the approach as applied to part of a heat exchanger network problem involving two hot streams, two cold streams and steam. Figure 18.30a shows a superstructure for that part of the network. All structural features have been included. The approach is to then optimize the superstructure in order to remove the unnecessary features and minimize the cost. Figure 18.30b shows the evolution to one possible outcome.

   Whilst this approach seems simple in principle, the optimization required is a mixed integer nonlinear programming problem (MINLP, see Chapter 3). This is a difficult optimization problem with all of the issues associated with local optima.

   Another issue is how to set up the initial superstructure. The problem could be divided at the pinch, as done in the pinch design method, and a superstructure set up on each side of the pinch, like the one in Figure 18.30. However, for large complex problems this would not be comprehensive enough in terms of the number of structural options. The overall problem could therefore be divided into enthalpy intervals, as shown in Figure 18.31, in which a superstructure is created within each enthalpy interval (Yee, Grossmann and Kravanja, 1990). This brings the potential danger of an overcomplex superstructure with the danger of the optimization finding an unnecessarily complex final design. Rather than dividing the composite curves into enthalpy intervals, as shown in Figure 18.31, with a superstructure within each enthalpy interval, some enthalpy intervals can be merged together into *blocks* (Zhu *et al.*, 1995). Blocks can be created by merging together enthalpy intervals such that there is at least a feasible minimum temperature difference between all hot and cold streams throughout the block. A superstructure is then created within each block, rather than each enthalpy interval. This simplifies the complete superstructure for the whole network.

   The major advantage of the superstructure approach to heat exchanger network design is that, in principle, it is capable of designing large networks with complex constraints and accounting for equipment details. One major disadvantage of the approach is that the optimization is a difficult MINLP, particularly for large problems. Another significant disadvantage is that, because a computer carries out the optimization, the designer is removed from the decision making.

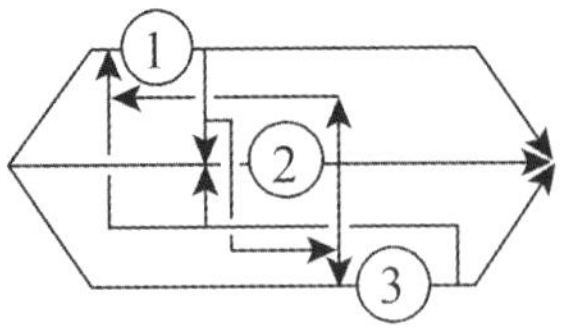

(a) Basic network superstructure.

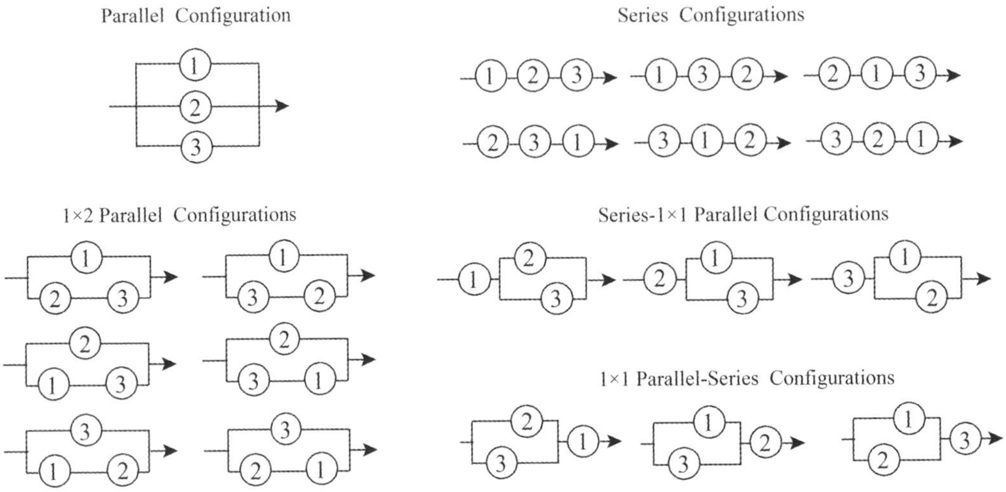

(b) Possible evolutions from the basic superstructure.

Figure 18.29

Basic superstructure allows evolution to all structual options.

2) *Automated design using stochastic search optimization.* Rather than using deterministic optimization to optimize a superstructure, an automated approach based on stochastic search optimization can be developed (Dolan, Cummings and Le Van, 1990). The most commonly used approach exploiting stochastic search optimization is simulated annealing (see Chapter 3). In this approach, an initial design is first created. The initial design can be in principle any feasible design and does not need to include all features that are candidates for the final design. This is in contrast with the superstructure approach. The initial design is then subject to an evolutionary development using *simulated annealing moves*.

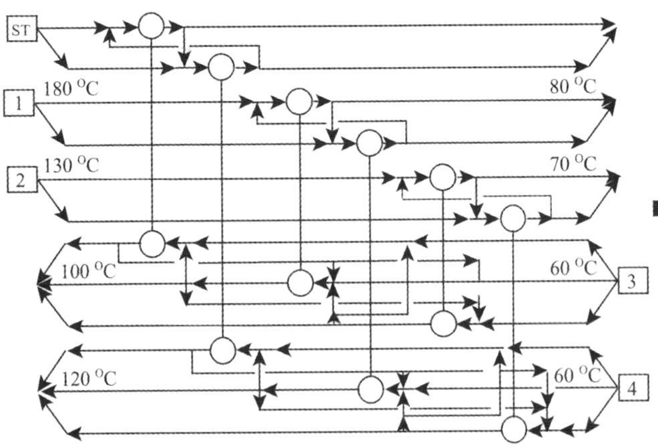

(a) A superstructure for part of a heat exchanger network.

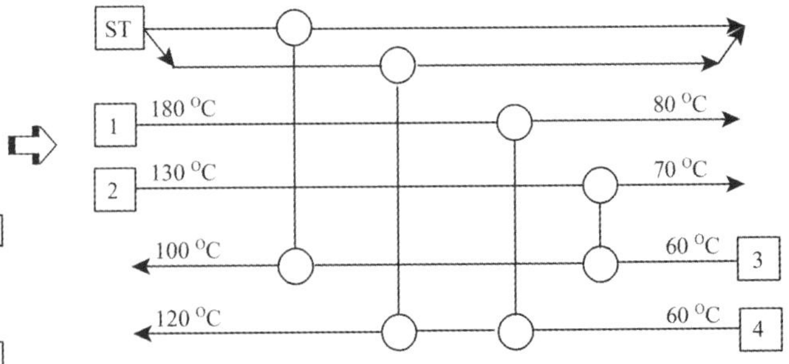

(b) The optimized design.

Figure 18.30

Heat exchanger network design from the optimization of a superstructure.

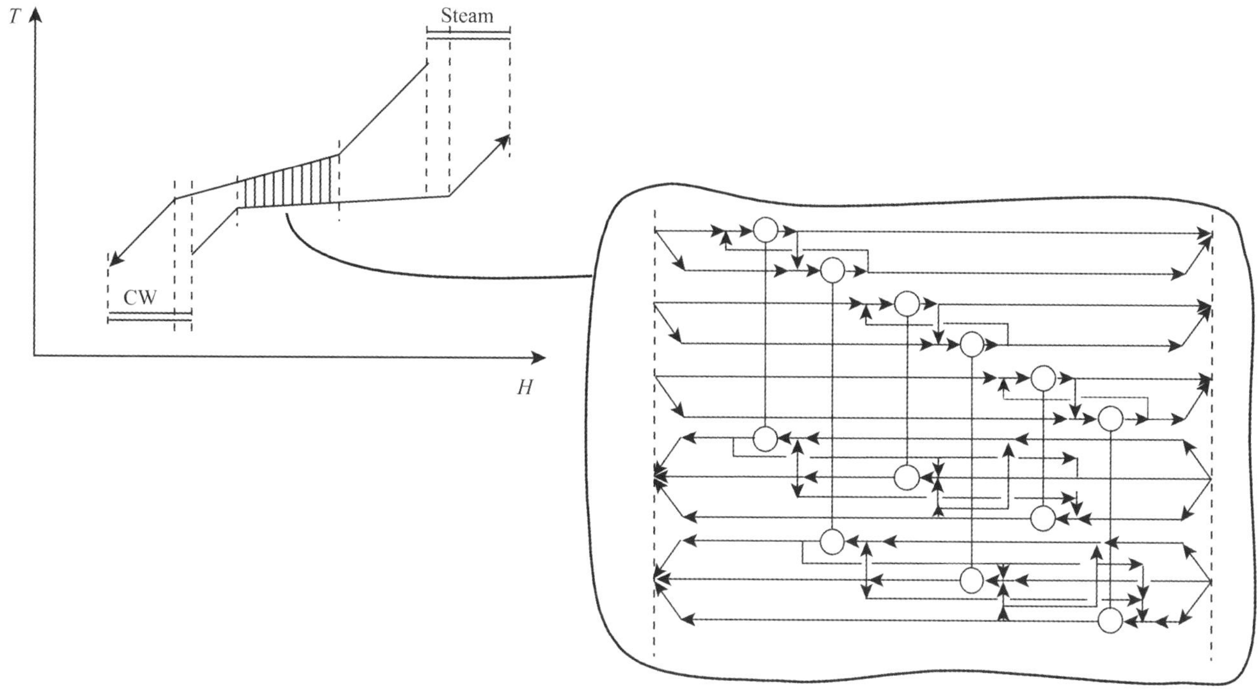

Figure 18.31

Create a superstructure for each enthalpy interval.

The moves can be continuous moves, changing, for example, heat exchanger duties. The moves can also be structural moves, for example, adding a new heat exchanger to the design. The various moves in the design can be categorized as:

Continuous moves:

- Change a heat exchanger duty.
- Change a flow fraction of a stream split.

Discrete moves to change the network configuration:

- Add a heat exchanger.
- Delete a heat exchanger.
- Move a heat exchanger to a new location.
- Add a stream split.
- Delete a stream split.

After each move, the network energy balance needs to be restored, the network simulated and the objective function evaluated. Different mechanisms are possible to restore the energy balance. The stochastic search optimization then follows an evolution as described in Chapter 3. The optimization does not depend on gradients and is not trapped by local optima. In principle, the optimization can start from any initial feasible design and, allowed enough time, will find the region of the global optimum. In practice, because the number of moves is limited by computational constraints and the moves themselves are subject to rules programmed into the optimization, the initial design can make a difference. It is generally better to start with a more complex initial design than a simpler initial design. Another point to note is that even continuous moves, such as changing a heat exchanger duty, take finite steps. Because of this, the final design can be fine-tuned using nonlinear programming continuous optimization. It is straightforward to include constraints in the approach based on stochastic search optimization. For example, if it is necessary to forbid a particular match between two streams, then the probability of that move in simulated annealing can be set to zero (see Chapter 3).

18.9 Heat Exchanger Network Retrofit with a Fixed Network Structure

So far, all considerations regarding the targeting and design of heat exchanger networks have related to new, or grassroot, design. It is often necessary to retrofit an existing heat exchanger network. The need to retrofit might arise from a desire to reduce the utility consumption of the existing network, the need to increase the throughput, modification of the feed to the process or a modification to the product specifications. All of these objectives might require heat duties within the network to be changed. Consider first the case when the network structure is fixed. Whilst this seems to be unnecessarily constrained, it can be desirable to maintain a low capital cost for the retrofit. Changing the network structure of heat exchanger networks to allow new equipment is often extremely expensive because of the piping and civil engineering costs. Even further, modifying existing plant to change the network structure might not be possible as a result of congestion in the plant layout. Cost-effective retrofit most often involves the fewest modifications to the existing network. Thus, retrofitting the network with a fixed structure is an extreme case of the retrofit problem. Later, retrofit allowing change in the network structure will be explored.

If the network overall heat recovery performance is to be increased to allow increased overall heat recovery or to accommodate increased throughput, then this will require increased heat duties for the network matches at various places in the network. There are two general ways that allow the duty of a match to be increased. Either the heat transfer area can be increased or the overall heat transfer coefficient can be increased. Either of these changes on individual matches might enhance the performance of the overall network to a greater or lesser extent, depending on the details of the design of the heat exchangers and the network structure.

1) *Increase in the heat transfer area.* If shell-and-tube heat exchangers are being used, and additional heat transfer area is required, it might be possible to install a new tube bundle with a more compact tube layout into an existing shell if the additional area requirement is small. If this is not feasible, then the existing area can be supplemented by adding a new shell (or more than one shell if there is a large area requirement) in:
   a) *Series.* The addition of new exchangers in series to the existing match will lead to an increase in the overall pressure drop across the match. This might be important if the pump (or compressor) is close to its maximum capacity.
   b) *Parallel.* The addition of new exchangers in parallel will leave the pressure difference largely unchanged. The pressure drop across the match will be the largest between that of the existing exchangers under the new conditions and new exchangers installed in parallel.

   Rather than shell-and-tube heat exchangers, the network might involve plate-and-frame heat exchangers (see Chapter 12). Adding additional area to plate-and-frame heat exchangers is generally more straightforward than shell-and-tube exchangers. Plate-and-frame heat exchangers can often have their area increased by adding additional plates to the existing frame. If an additional frame needs to be added to provide the additional area, then the new frame can be added in series or parallel with the existing frame. Series arrangements increase the pressure drop. Parallel arrangements decrease the flowrate through each frame and decrease the heat transfer coefficient in the existing frame. The pressure drop across the match will be the largest between that of the existing exchangers under the new conditions and new exchangers installed in parallel.
2) *Increase in the overall heat transfer coefficient.* Increasing the overall heat transfer coefficient is equivalent to increasing the heat transfer area. Changes to the number of tube passes or the baffle arrangement might allow the heat transfer coefficient to be enhanced. Alternatively, tube inserts can be used, as discussed in Chapter 12. A major disadvantage in using tube inserts is that it will increase the pressure drop in each tube pass. In retrofit this can be important, as the pumps driving the flow might be limited in their capacity to meet the required increase in pressure drop (see Chapter 13). However, as discussed in Chapter 12, an increase in the pressure drop is not inevitable when using tube inserts for heat transfer enhancement. Although tube-side enhancement increases the pressure drop per tube pass, the number of tube passes might be decreased by changes to the tube-side partition plates.

Given that the network simulation calculations are extremely rapid once set up, it is straightforward to explore changes of a fixed network structure through a *sensitivity analysis*. This is especially the case when stream heat capacities are constant. If the objective of a retrofit is to reduce the load on a particular utility exchanger, then the retrofit is guided by the sensitivity of the inlet temperature to that utility exchanger to changes in the UA of existing heat exchangers in the network. In such a sensitivity analysis, different values of U and A are assumed for a particular match and a simulation calculation carried out for each setting. This is also carried out for other heat exchangers that might be candidates to improve the network performance. The sensitivity of network performance to changes to each candidate heat exchanger is then compared. In this way, the heat exchangers within the network that are most influential to the inlet temperatures of utility exchangers can be identified.

The approach involves the following steps:

Step 1. If changes to the network structure are not allowed, then reduction in the duty on a particular utility exchanger can only be brought about by modifying heat exchangers located on a utility path involving that utility exchanger. Thus the network structure needs to be analyzed to identify those exchangers located on a utility path involving the utility exchanger being studied.

Step 2. For those heat exchangers located on utility paths for the utility exchanger being studied, a sensitivity analysis is carried out to identify the most influential heat exchanger to bring about a reduction in the consumption of that utility.

Step 3. The most influential heat exchanger identified in Step 2 is increased in performance either by heat transfer enhancement or addition of heat transfer area, depending on the feasibility and preferences. If the retrofit objective is to reduce the utility consumption, then the feasibility of the network retrofit will be violated when the addition of duty to a process heat exchanger exceeds the duty on one of the utility exchangers on the utility path. In other words, as the duty on a process heat exchanger is increased, the corresponding reduction on the utility exchangers will mean that at some point the duty on one of the utility exchangers becomes zero and it is essentially removed from the network. Otherwise, the stream target temperatures will not be able to be maintained. In addition to checking that the changes in the duties are feasible, it is also advisable to check whether the changes in the duties result in infeasible temperature differences in the network.

Step 4. Having enhanced one of the heat exchangers, the target temperatures will no longer be achieved. Stream target temperatures are then corrected to bring them within acceptable bounds by adjusting the loads on utility exchangers, and if necessary enhancing additional process heat exchangers.

Table 18.6

Heat exchanger data.

Exchanger	Duty (kW)	UA (kW·K^{-1})	N_{SHELLS}	F_T
1	6141	205	2	0.878
2	6135	259	2	0.820
3	5557	100	1	0.882
4	2689	34.2	1	0.937
5	3431	59.7	1	0.978
6	2292	59.9	1	0.831
7	3623	41.4	1	0.985
C1	657	–	–	–
C2	1142	–	–	–
C3	817	–	–	–
C4	881	–	–	–
H	14,453	–	–	–

Step 5. Having enhanced the network performance and brought the target temperatures back to within acceptable bounds, the new network design becomes a new base case for further possible improvement. Steps 2 to 4 are repeated until the retrofit objective is satisfied or the retrofit scope is exhausted.

The approach can be demonstrated on the network shown in Figure 18.24. The duties, UA values and other details are given in Table 18.6.

Suppose the objective is to retrofit in order to decrease the energy consumption of the heater. This can be interpreted as an objective to increase the process stream temperature entering the heater. From Figure 18.24, all heat exchangers except Heat Exchanger 7 are located on a utility path to the heater. A sensitivity analysis is carried out by varying the UA for Heat Exchangers 1 to 6 in increments and resimulating the network to calculate the change in the inlet temperature to the heater. The simulations are very rapid and robust, allowing this to be done quickly using software. Figure 18.32 shows the results graphically, in which the $\Delta T_{OBJECTIVE}$ is the change in temperature at the inlet to the heater as a result of changes to the UA in each heat exchanger in turn. The change in UA can be interpreted as a change in the U through heat transfer enhancement, or a change in the area A through the addition of new shells, or a combination of both. It is clear from Figure 18.32 that Heat Exchanger 5 is by far the most influential to increase the heater inlet temperature. An examination of the duties along the utility path for Heat Exchanger 5 reveals that the maximum increase in the duty is 817 kW. This is the duty that removes Cooler C3 from the network. Increasing the duty on Heat Exchanger 5 by 817 kW corresponds with a temperature change at the inlet of the heater of 5.7 °C. In turn, from Figure 18.32 this corresponds to an increase of almost 50% in the UA for Heat Exchanger 5. If the duties are changed along the utility path, then a check of the temperatures reveals that there are no small temperature differences, and the proposed retrofit appears feasible in the context of the network.

Assuming an enhanced UA of 49% for Exchanger 5 with all other exchangers having the original UA from Table 18.6, the network is simulated in Figure 18.33a. The basis of the simulation is that the process heat exchangers have a fixed UA, but the utility heat exchangers are assumed to have fixed duty. The required target temperatures are no longer achieved. The network then needs to be balanced to achieve the required target temperatures. For this problem, this can be achieved by adjusting the duties on the heater and the coolers. To balance the network, the network is simulated for different settings of the heater and cooler duties and keeping the UA of the process heat exchangers fixed until the target temperatures are achieved. The result after balancing is shown in

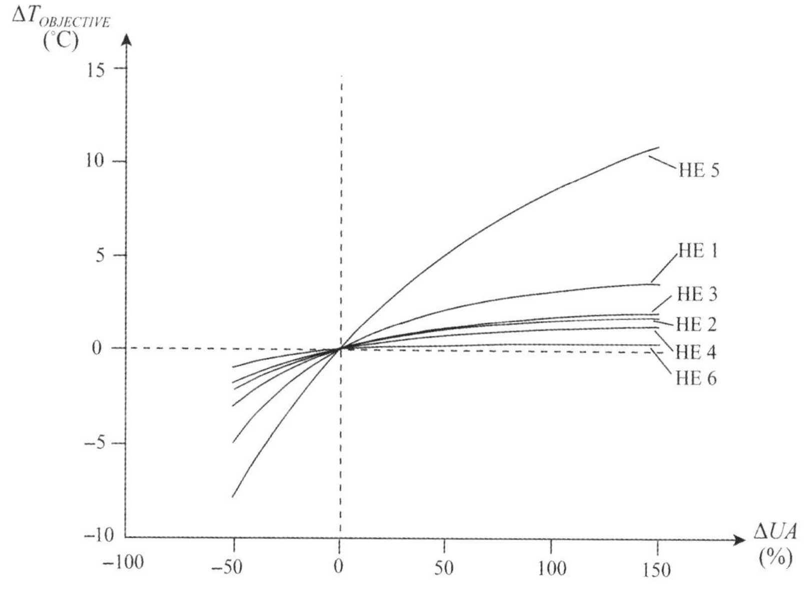

Figure 18.32

Network sensitivity analysis. (Reproduced from Ning Jiang, Jacob David Shelley, Steve Doyle, Robin Smith (2014) Heat exchanger network retrofit with a fixed network structure, *Applied Energy*, **127**: 25–33, with permission from Elsevier.)

Figure 18.33

Network changes after heat transfer enhancement. (Reproduced from Ning Jiang, Jacob David Shelley, Steve Doyle, Robin Smith (2014) Heat exchanger network retrofit with a fixed network structure, *Applied Energy*, **127**: 25–33, with permission from Elsevier.)

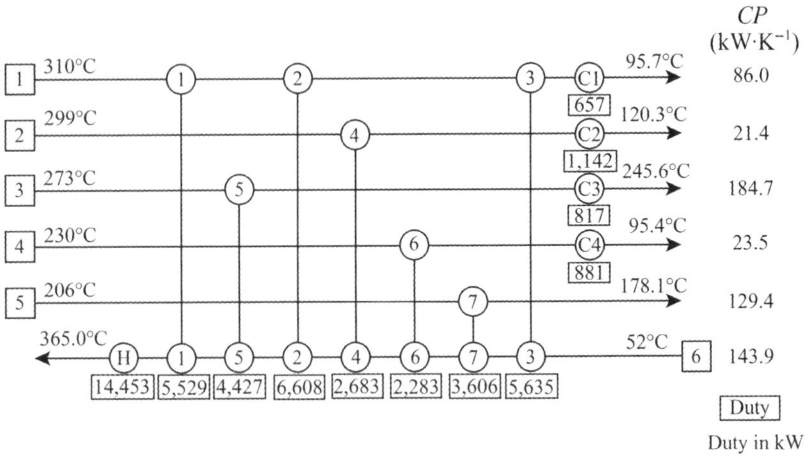

(a) Network after enhancement of Heat Exchanger 5.

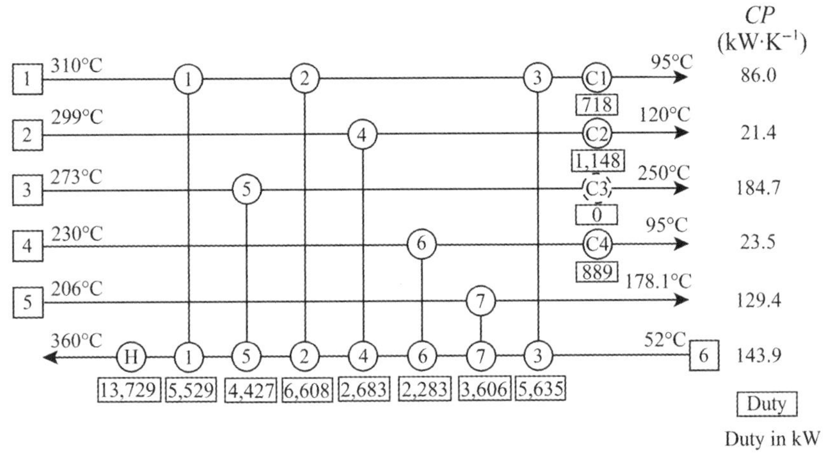

(b) Network after enhancement of Heat Exchanger 5 and network balancing.

Figure 18.33b. The final heater duty is 13,729 kW. This represents a saving of 724 kW of hot utility. The saving in cold utility is 742 kW. This discrepancy between the hot and cold utility saving figures is caused by minor differences between the original target temperatures and those achieved in Figure 18.33b. In practice, there is always some tolerance on the target temperatures, depending on the destination of the stream. The design in Figure 18.33b achieves a saving in hot utility of 5.1%. It should be noted that the change in hot utility is not 817 kW, due to the downstream effects of the enhancement of Heat Exchanger 5. In the overall picture, the heat load on any given process exchanger after balancing depends on the *UA* of the heat exchangers.

In this example the balancing of the network was achieved through manipulation of the utility exchanger duties. In more complex examples, the balancing might require adjustment of the *UA* of some of the process heat exchangers in addition to manipulation of the utility exchanger duties.

Now knowing how much equipment enhancement is required, detailed modification of Heat Exchanger 5 can be explored to achieve the required enhancement. The enhancement of existing heat exchangers can be made by adding a new area or enhancing heat transfer. Two commonly used heat transfer enhancement techniques are considered here: twisted tapes and coiled wires (see Section 12.8). The details of Heat Exchanger 5 are given in Table 18.7.

Heat Exchanger 5 is tube-side controlled and a potential candidate for tube insert enhancement (see Section 12.8). Thus, tube inserts are considered for this case. The performance models for twisted tapes and coiled wires given in Section 12.8 are used to calculate the enhanced heat exchanger performance.

Given the required duty enhancement of 817 kW for Heat Exchanger 5, different enhancement designs can be obtained. The first option is a new unit installed in series with the existing unit with an area of 113 m^2, assuming the heat transfer coefficients for the new heat exchanger are the same as the original heat exchanger. Rather than install a new heat exchanger, tube inserts can be used in the existing heat exchanger. The inside heat transfer film coefficient creates 55% of the total resistance to heat transfer, and hence enhancing the tube side should be effective. Twisted-tape tube inserts with a twist ratio of 6.57 and thickness of 1.6 mm and coiled-wire tube inserts with a pitch of 82 mm and thickness of 1.6 mm can be used. The performances of the tube inserts are compared in Table 18.8 based on the models in Section 12.8. However, the results for wire coils should be treated with caution, as the correlations are out of range for the p/d_I used.

Table 18.7

Heat exchanger data for Heat Exchanger 5.

Streams	Shell side	Tube side
Heat capacity C_P (J·kg^{-1}·K^{-1})	2718.5	2325
Thermal conductivity k (W·m^{-1}·K^{-1})	0.215	0.106
Viscosity μ (N·s·m^{-2})	2.1×10^{-4}	2.38×10^{-3}
Density ρ (kg·m^{-3})	776	544.5
Flowrate m (kg·s^{-1})	67.93	61.87
Inlet temperature T_{in} (°C)	273	193.04
Final temperature (°C)	254.4	216.9
Fouling resistance (m^2·K·W^{-1})	6.90×10^{-4}	5.35×10^{-4}
Film heat transfer coefficient (W·m^{-2}·K^{-1})	2692	505.0
Pressure drop (Pa)	22,988	14,740
Geometry of heat exchanger		
Tube pitch p_T (m)	0.025	
Number of tubes N_T	808	
Number of tube passes N_P	2	
Tube effective length L_{eff} (m)	5.0	
Tube conductivity k_{tube} (W·m^{-1}·K^{-1})	51.91	
Tube pattern (tube layout angle)	90°	
Tube inner diameter d_I (m)	0.016	
Tube outer diameter d_O (m)	0.020	
Shell inner diameter D_S (m)	0.9	
Number of baffles N_B	9	
Baffle spacing B (m)	0.5	
Inlet baffle spacing B_{in} (m)	0.5	
Outlet baffle spacing B_{out} (m)	0.5	
Baffle cut B_C	25%	
Inner diameter of tube-side inlet nozzle $d_{TN,inlet}$ (m)	0.3048	
Inner diameter of tube-side outlet nozzle $d_{TN,outlet}$ (m)	0.3048	
Inner diameter of shell-side inlet nozzle $d_{NS,inlet}$ (m)	0.3048	
Inner diameter of shell-side outlet nozzle $d_{NS,outlet}$ (m)	0.3048	
Shell-bundle diametric clearance (m)	0.041	
Area (m^2)	253.8	
Overall heat transfer coefficient (W·m^{-2}·K^{-1})	235.4	
Duty (kW)	3431	

18

Table 18.8

Enhancement options for Heat Exchanger 5.

Options	h_T (W·m^{-2}·K^{-1})	hs (W·m^{-2}·K^{-1})	U (W·m^{-2}·K^{-1})	Q (kW)	ΔP_T (kPa)	ΔP_S (kPa)
Twisted tapes	963.8	2692	325.2	4247	22.8	26.4
Coiled wires	963.8	2692	325.2	4247	12.5	26.4

Increasing the duty on Heat Exchanger 5 by 817 kW does not lead to a reduction of 817 kW in hot utility. This is because the *UA* of the other heat exchangers are fixed, but their duties are not fixed. The heat transfer enhancement options are likely to be more attractive than new area, as long as the increase in pressure drop can be accommodated.

Further retrofit could be carried out by enhancing the next most influential heat exchanger, which is Heat Exchanger 1 from Figure 18.32. If it is assumed that the *UA* of Exchanger 1 is enhanced by, say, 40%, this allows an additional 557 kW of heat exchange, leading to a reduction of 108 kW of hot utility. To balance the network with Heat Exchanger 1 enhanced requires the level of enhancement on Heat Exchanger 5 to be decreased. The resulting cumulative energy saving is 5.7% (Jiang *et al.*, 2014). Thus the benefits of the second step in the retrofit are marginal and possibly not worth pursuing.

18.10 Heat Exchanger Network Retrofit through Structural Changes

Allowing structural changes opens further options for retrofit. One approach to retrofit allowing structural changes would be to try and evolve the network toward an ideal grassroot design. Following this approach, stream data would be targeted using the composite curves or the problem table to determine the location of the pinch for an assumed ΔT_{min}. Knowing the location of the pinch, the existing units in the network could then be located relative to the pinch. Any cross-pinch heat transfer could then be identified and eliminated by disconnecting the units that are transferring heat across the pinch. These could be process-to-process heat transfer or the inappropriate use of utilities (e.g. use of steam below the pinch). The network could then be reconnected, with as many features of the existing network retained as possible (Tjoe and Linnhoff, 1986). This might improve the energy performance of an existing heat exchanger network, but it has a number of fundamental problems:

- Which ΔT_{min} should be used? As the value of ΔT_{min} changes, the location of the pinch changes and therefore the assessment of which unit transfers heat across the pinch also changes.
- Existing equipment is only reused in an ad hoc way.
- The retrofit is likely to involve a large number of modifications to the existing network. One of the overriding priorities in retrofit is to carry out as few modifications as possible.
- Constraints associated with the existing network are not readily included.

The fundamental problem with this approach is that it is attempting to change the network to a grassroot design, rather than accepting the features that already exist. For these reasons, approaches based on energy targeting and the pinch design method are fundamentally unsuitable for retrofit. Energy targeting can be used to give an idea of the scope to improve the network, but it is most likely that realizing the scope indicated by the target will be uneconomic. A better approach is to evolve the network from the existing structure in order to identify only the most critical, and therefore cost-effective, changes to the network structure (Asante and Zhu, 1997).

Suppose that it is desired to reduce the energy consumption of the existing heat exchanger network shown in Figure 18.34a. Figure 18.34b shows the composite curves for the basic stream data. The composite curves have been adjusted such that the overlap between the composite curves corresponds with the actual existing heat recovery duty of 200 MW. For the composite curves, this corresponds with a ΔT_{min} of 22.5 °C. On the other hand, it can be seen that the existing heat exchanger network has minimum temperature differences of 20 °C. First consider how the energy consumption of the existing network might be decreased *without changing the network structure.* As discussed previously, this is only possible by the exploitation of a utility path. The existing network in Figure 18.34a has only one degree of freedom that can be exploited for network optimization. This is highlighted within the bubble superimposed on the network. It involves a utility path between the heater and the cooler. The matches in the network outside of this bubble are all constrained by the heat duties on individual streams. The only way, therefore, that the utility consumption of the existing network can be decreased is by shifting the heat load along the path between the heater and the cooler.

Figure 18.35a shows the network evolved to reduce the utility consumption to the point where the temperature difference in the existing network is now 0 °C (Asante and Zhu, 1997). This is not a limit that could be achieved in practice, but is taken here for the sake of illustration. For a minimum temperature difference of 0 °C in the existing network, the energy recovery must increase to 220 MW. Compare the composite curves for increased heat recovery. These are shown in Figure 18.35b for their maximum overlap, which is 250 MW. At this setting, the composite curves still have a ΔT_{min} of 6 °C. This reveals that the maximum heat recovery within the existing heat exchanger network structure is different from that theoretically possible from the composite curves. The difference is caused by the fact that the existing heat exchanger network

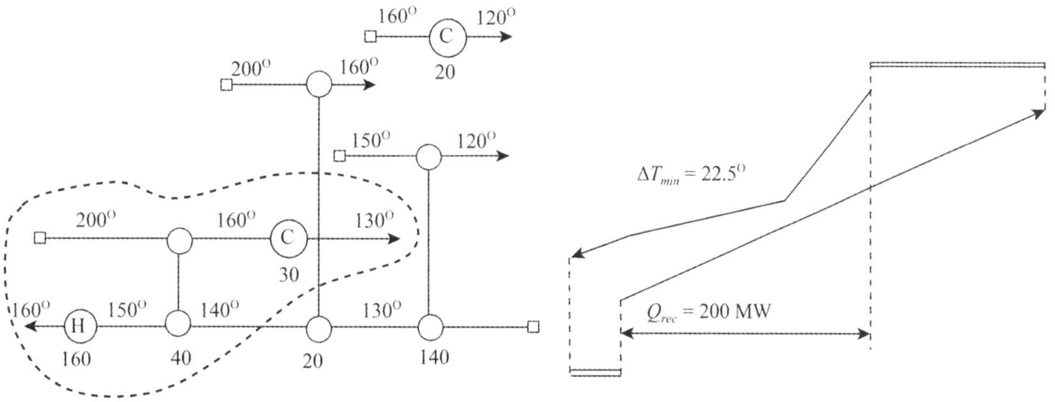

(a) Existing network with Q_{REC} = 200 MW.

(b) Composite curves with Q_{REC} = 200 MW.

Figure 18.34

An existing heat exchanger network. (Reproduced from Asante NDK and Zhu XX (1997) An Automated and Interactive Approach for Heat Exchanger Network Retrofit, *Trans IChemE*, **75**: 349, by permission of Elsevier.)

structure is not appropriate for maximum energy recovery. How can the existing network structure be modified to improve its performance?

The existing heat exchanger network is shown again in Figure 18.36 with its recovery increased to an absolute maximum. Figure 18.36a highlights what is limiting the existing heat exchanger network in terms of its heat recovery. One of the existing units features the minimum temperature difference (0 °C in this case for the sake of illustration). The composite curves are also shown in Figure 18.36b, but set to the same overall heat recovery load as that featuring the existing heat exchanger network at its limit (220 MW). Superimposed on the composite curves are the temperature profiles for the hot streams in each of the existing units. One of the matches limits the heat recovery, in this case the one featuring a temperature difference of 0 °C. This match that limits the heat recovery is known as the *pinching match* (Asante and Zhu, 1997). The point at which this occurs is known as the *network pinch* (Asante and Zhu, 1997). The network pinch limits the heat recovery for the existing heat exchanger network.

In practice, if the network pinch is being identified in a design study, then a practical minimum temperature difference of, say, 10 °C or 20 °C would be taken rather than the 0 °C used here for the sake of illustration. The principle is exactly the same. One or more matches in the existing network feature a practical minimum temperature difference for the heat exchangers when the network is pinched.

There are four ways in which the network pinch can be overcome and the performance of the existing heat exchanger network improved beyond that for the pinched condition (Asante and Zhu, 1997).

a) *Resequencing.* Figure 18.37 illustrates how *resequencing* can be used to overcome the network pinch. Resequencing moves

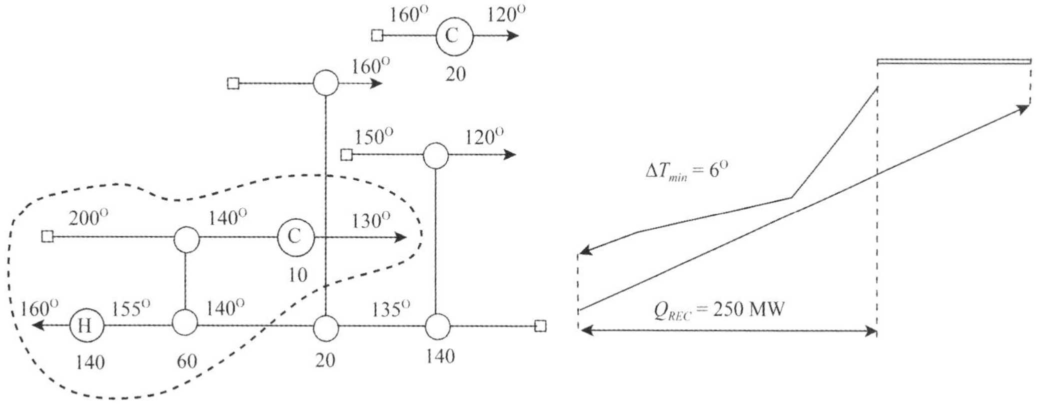

(a) Existing network with Q_{REC} increased to 220 MW.

(b) Composite curves for maximum energy recovery indicates Q_{REC} = 250 MW.

Figure 18.35

Maximizing the energy recovery of the existing heat echanger network even down to $\Delta T = 0$ °C gives an energy performance worse than the energy target. (Reproduced from Asante NDK and Zhu XX (1997) An Automated and Interactive Approach for Heat Exchanger Network Retrofit, *Trans IChemE*, **75**: 349, by permission of Elsevier.)

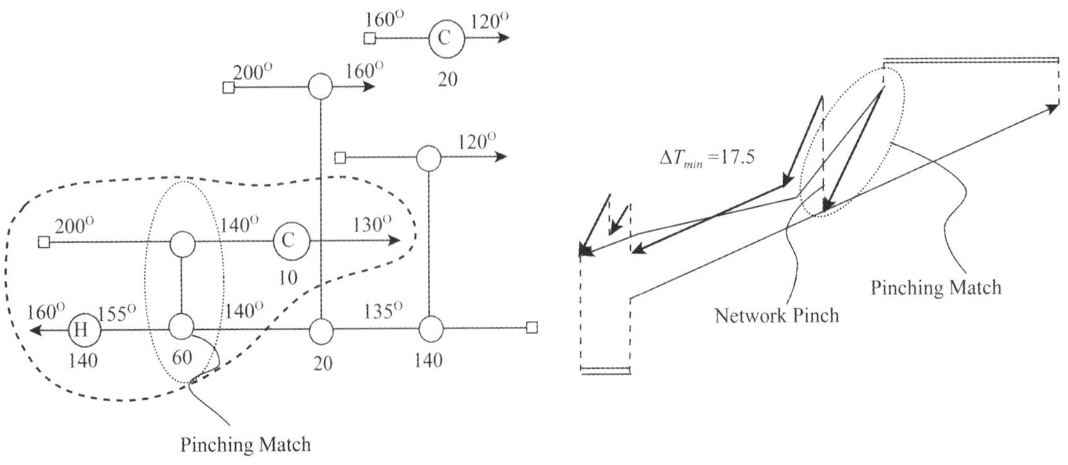

(a) Maximum heat recovery condition with Q_{REC} = 220 MW.

(b) Composite curves with Q_{REC} = 220 MW.

Figure 18.36

The network pinch limits the energy recovery in the existing heat exchanger network. (Reproduced from Asante NDK and Zhu XX (1997) An Automated and Interactive Approach for Heat Exchanger Network Retrofit, *Trans IChemE*, **75**: 349, by permission of Elsevier.)

the unit to a new location in the network, but between the same streams as the original match. Figure 18.37 shows a cold stream being heated by two hot streams. One of the hot stream profiles indicates that it is a pinching match and features a minimum temperature difference of 0 °C (again for the sake of illustration). In Figure 18.37, the pinching match is adjacent to another hot stream that has a finite temperature difference for the heat exchange. If the position of the two hot streams is swapped by a simple resequence, as shown in Figure 18.37, then the pinching match no longer limits. This means that there is now new scope to reduce the energy consumption of the network by exploiting a utility path, as shown in Figure 18.37. If the utility path is exploited to its limit, then a new network pinch is created, but now at a lower utility consumption for the network.

b) *Repiping*. *Repiping* is very similar to resequencing. Like resequencing, repiping moves the unit to a new location in the network. However, in repiping the unit can be moved to a location involving streams other than in the original location, rather than be restricted to operate between the same streams. Repiping is a more general case than resequencing, but might not be practical for a variety of reasons, for example, materials of construction or mechanical pressure rating being unsuitable

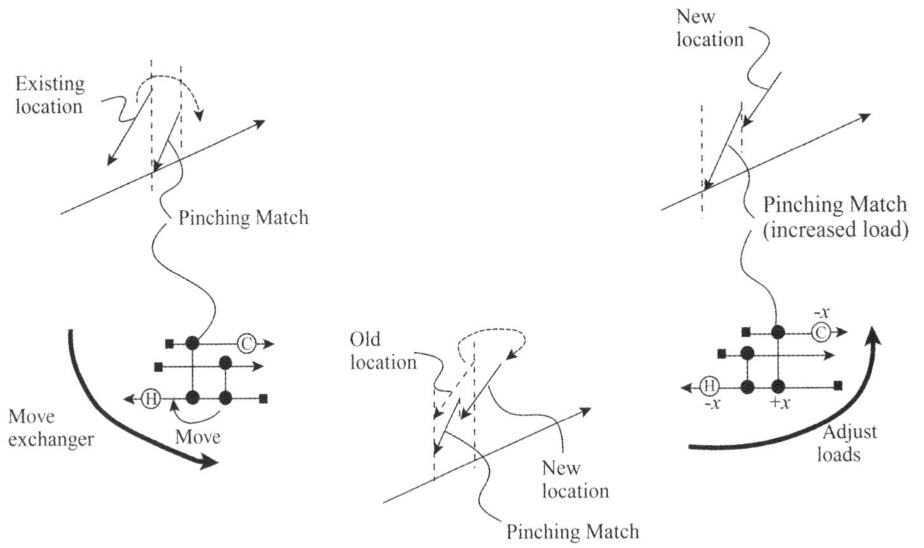

Figure 18.37

Resequencing a match can be used to overcome the network pinch. (Reproduced from Asante NDK and Zhu XX (1997) An Automated and Interactive Approach for Heat Exchanger Network Retrofit, *Trans IChemE*, **75**: 349, by permission of Elsevier.)

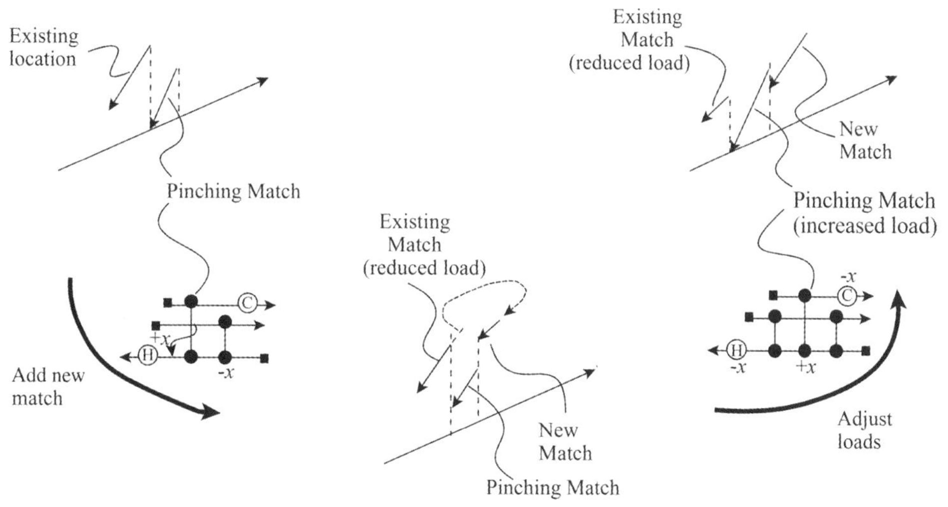

Figure 18.38

Adding a new heat exchanger can be used to overcome the heat network pinch. (Reproduced from Asante NDK and Zhu XX (1997) An Automated and Interactive Approach for Heat Exchanger Network Retrofit, *Trans IChemE*, **75**: 349, by permission of Elsevier.)

for other streams. The basic principle of repiping is the same as that for resequencing, but a distinction needs to be made for practical reasons.

c) *Inserting a new match.* Figure 18.38 shows how the network pinch can be overcome by inserting a new match. Again the principle is illustrated by two hot streams providing heat to a cold stream. One of the matches is pinched. If a new match is inserted such that the heat duty on the hot stream adjacent to the pinching match is decreased and replaced by the new match, then the position of the pinching match can be changed such that it is no longer pinching. This introduces scope to exploit the utility path to reduce the utility consumption of the network, until it is again pinched. The network is now pinched again, but at a lower utility consumption.

d) *Introduce additional stream splitting.* The fourth way to overcome the network pinch is by introducing additional stream splitting to the existing network, as illustrated in Figure 18.39. In this case, it can be seen that two matches are pinched simultaneously. By introducing a stream split, the cold stream profiles in the two pinched units are now such that one of the pinching matches is no longer pinched. This means that there is scope to exploit a utility path and reduce the energy consumption. This is being carried out to its limit in Figure 18.39 such that the network is again pinched, but at a lower utility consumption. Again, it should be emphasized that in practice a finite practical minimum temperature difference should be used rather than 0 °C. However, any assumption of minimum temperature difference to identify, and then overcome, the network pinch does not guarantee an optimum retrofit.

As an example, Figure 18.40a illustrates an existing heat exchanger network. The existing hot utility consumption is 14,453 kW. As a first step to retrofit, the existing network can be pinched by exploiting the degrees of freedom for energy reduction to the maximum without changing the network structure. This can be achieved by a linear programming (LP) optimization used to adjust the heat loads throughout the network with the objective of minimizing the energy consumption, subject to a constraint of a minimum temperature difference throughout the network. In this case the optimization will be constrained such that there is a practical minimum temperature difference of 10 °C in all of the

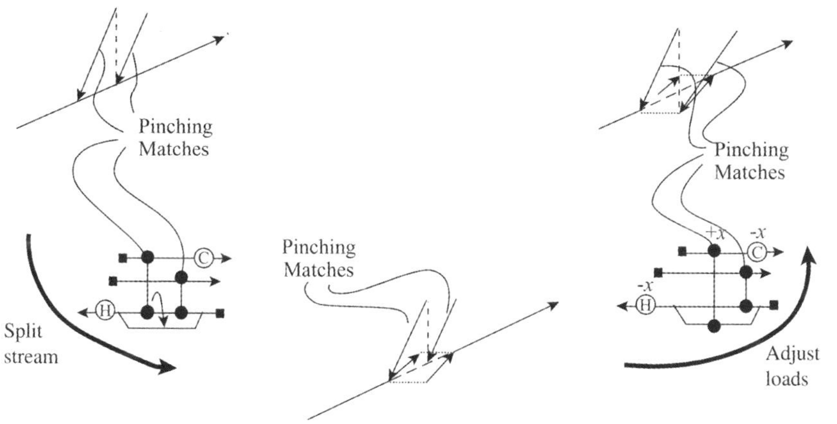

Figure 18.39

Changing the stream splitting arrangement can be used to overcome the network pinch. (Reproduced from Asante NDK and Zhu XX (1997) An Automated and Interactive Approach for Heat Exchanger Network Retrofit, *Trans IChemE*, **75**: 349, by permission of Elsevier.)

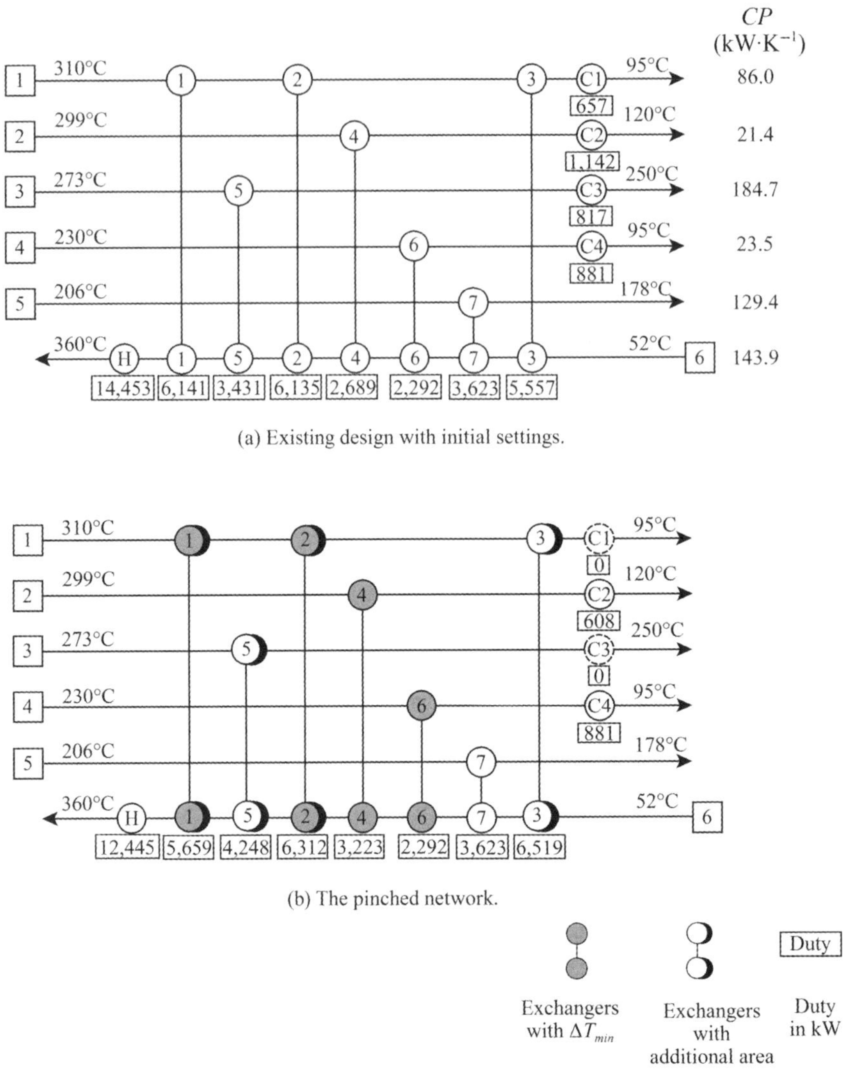

Figure 18.40
Existing network design.

matches. Any new network area can then be calculated from the new operating conditions for each match. The result is shown in Figure 18.40b. The hot utility consumption can in principle be reduced to 12,445 kW without changing the network structure, subject to a constraint that no heat exchanger uses a temperature difference less than 10 °C. It can be seen that, although a significant energy reduction is possible, additional heat transfer area needs to be added to Exchangers 1, 2, 3 and 5. The additional area can be added as new area or created indirectly using heat transfer enhancement. Whilst this shows a possible retrofit, it is not necessarily an economic retrofit. To make sure that the retrofit is economic, a capital energy trade-off would need to be carried out. The capital cost of any new heat transfer area, or the cost of heat transfer enhancement, would need to be estimated and its costs annualized over a specified period. Combining this annualized capital cost with the annual operation cost gives a total cost that can be minimized. This is a nonlinear programming (NLP) optimization problem. The result of this might well be to eliminate some of the additional heat transfer area or heat transfer enhancement that has been included in the pinched network, but at the penalty of decreasing the level of energy saving.

Any further reduction in the energy consumption requires a change in the network structure. Figure 18.41a illustrates the potential for a resequence. In this case, Exchanger 7 is moved to a new location and the new network structure is pinched through an LP optimization. This allows the energy consumption to be reduced to 11,715 kW. Unfortunately, this is at the penalty of additional heat transfer area or heat transfer enhancement required for most of the process-to-process heat exchangers. Again, although this shows a possible retrofit, it is unlikely to be an economic retrofit and should be subject to capital energy optimization using NLP optimization, as discussed above. This would remove some of the additional area or heat transfer enhancement, but at the expense of penalty in energy saving.

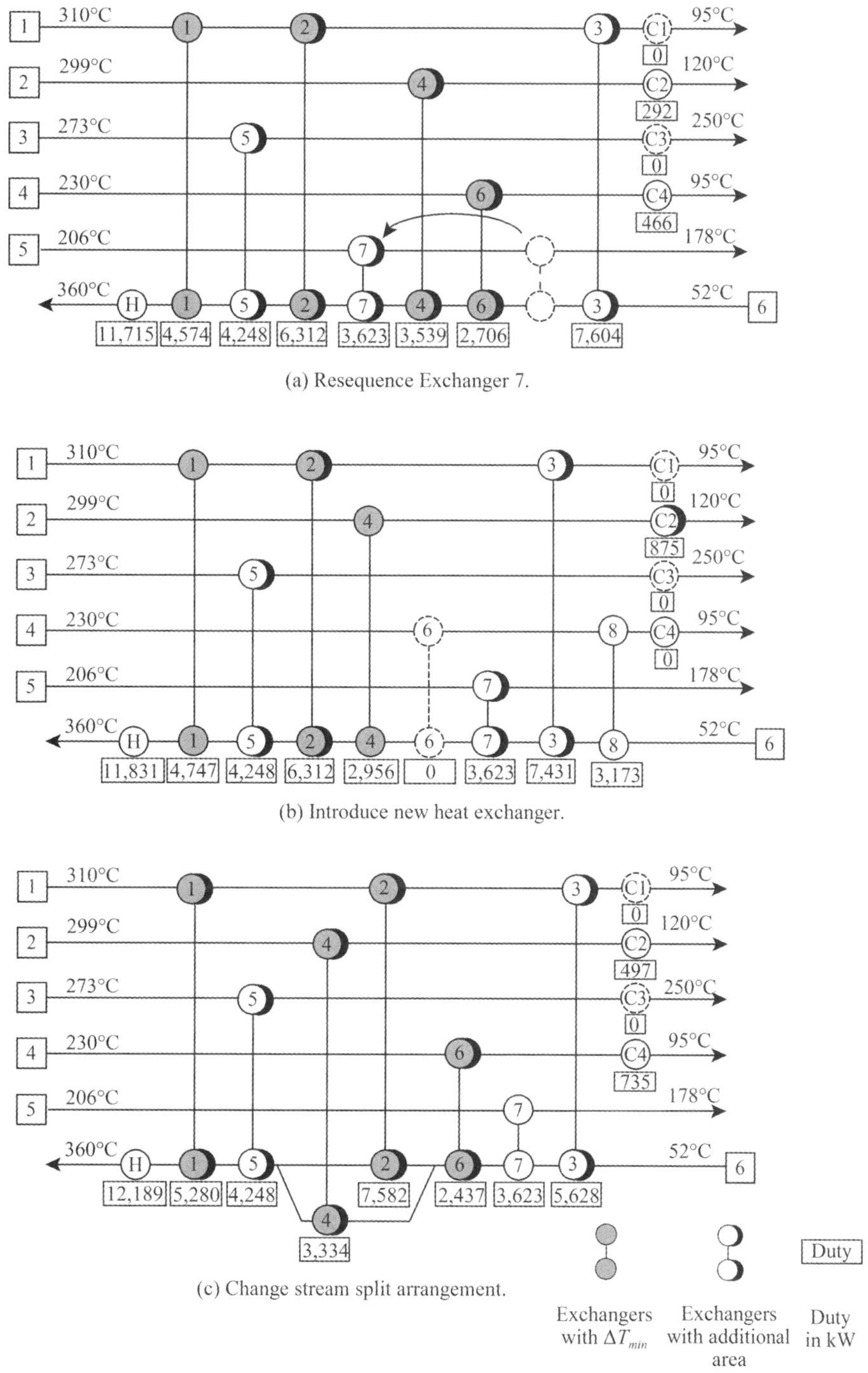

Figure 18.41
Network changes are required to overcome the network pinch.

Rather than resequence a heat exchanger, as illustrated in Figure 18.41a, a new heat exchanger can be introduced, as illustrated in Figure 18.41b. This shows a new Heat Exchanger 8 introduced to change the network structure and, after LP optimization, allowing the energy to be reduced to 11,831 kW, subject to a minimum temperature difference of 10 °C. New heat exchanger area or heat transfer enhancement is required in four of the existing process-to-process heat exchangers, together with additional area for one of the coolers. Again, this network change could be subject to a capital energy optimization.

Finally, Figure 18.41c illustrates the effect of introducing a new stream split arrangement into the network. As before, the network has been pinched using LP optimization, subject to a minimum temperature difference of 10 °C. Now the energy consumption is decreased to 12,189 kW with a penalty of additional area or heat transfer enhancement required for all process-to-process heat exchangers, except Heat Exchanger 7. This again should be subject to a capital energy trade-off to obtain an optimum retrofit.

In the example in Figures 18.40 and 18.41, the resequence option shown in Figure 18.41a seems to be the most attractive, but this is a superficial assessment and each option should to be subjected to a capital–energy trade-off optimization before screening.

The modified network should be subjected to cost optimization in order to obtain the correct setting for the capital–energy trade-off. The optimization of a heat exchanger network with a fixed structure is a nonlinear programming (NLP) optimization, as discussed previously. When dealing with the optimization of existing heat exchanger networks, it needs to be recognized that there is existing heat transfer area in place, but new area (or heat transfer enhancement) needs to be installed in some parts of the network to improve the network performance. The existing heat transfer area has zero capital cost and only new heat transfer area, heat transfer enhancement, pipework modifications and any civil engineering need to be included in the capital costs for the optimization. Care is required when specifying the form of the capital cost correlation in retrofit. Designers often like to use a cost per unit area, such as:

$$\text{Capital cost} = bA \tag{18.15}$$

where A = heat transfer area
b = cost coefficient

If the capital cost of new heat transfer area is expressed in the form of Equation 18.15, then this will lead to poor retrofit projects. The problem with Equation 18.15 is that the optimization is likely to spread new heat transfer area or heat transfer enhancement in the network in many locations, without incurring a cost penalty associated with the many modifications that would result. To ensure that new heat transfer area is not spread around throughout the existing heat exchanger network, a capital cost correlation should be used that is of the form:

$$\text{Capital cost} = a + bA^c \tag{18.16}$$

where a, b, c = cost coefficients

In Equation 18.16, the coefficient a is a threshold cost that is incurred even if a small amount of heat transfer area is installed. If a large threshold cost is used in the cost correlation, then this would lead the optimization to attempt to concentrate any new heat transfer area required in as few locations as possible in the network. This, in turn, will lead to fewer modifications in the retrofit project to accommodate the new heat transfer area requirements.

It should be noted that a resequence or repipe does not involve zero capital cost, even though no new heat exchanger equipment might be purchased. The pipework modifications for a resequence or repipe might be very expensive. Also, equipment might need to be relocated, perhaps incurring civil engineering costs. Methods for capital cost estimation for retrofit were discussed in Chapter 2.

Whilst the concept of the network pinch provides insights to the structural changes necessary to improve the network performance, it does not provide guidance as to how to choose the best network structural changes. The identification of which changes to make is not intuitive. For this, an automated approach needs to be introduced.

18.11 Automated Methods of Heat Exchanger Network Retrofit

Given that it is difficult to identify the key network changes in structural retrofit by intuition, the approach to heat exchanger network retrofit based on the concept of the network pinch can be automated (Zhu and Asante, 1999). A superstructure can be created for network resequencing (or repiping) and the best resequence (or repipe) identified by optimizing the superstructure of resequences (or repipes) for an assumed practical ΔT_{min} (Zhu and Asante, 1999). This can be formulated as an MILP optimization if the network is optimized for minimum energy consumption with a fixed ΔT_{min} (Zhu and Asante, 1999). Inclusion of area calculations to estimate heat exchanger capital costs would make the optimization nonlinear. Similarly, a superstructure can be created for positioning new matches or stream split modifications in the existing network and the superstructure optimized for minimum energy cost for an assumed practical ΔT_{min} (Zhu and Asante, 1999).

After each suggested modification has been identified using MILP, the network can be subjected to a detailed capital–energy trade-off requiring NLP optimization. Optimization of the capital–energy trade-off is illustrated in Figure 18.42. Structural modifications are first explored using MILP and then the setting of the capital–energy trade-off corrected using NLP. By decomposing the problem in this way, what is overall an MINLP problem, is carried out by MILP followed by NLP.

This approach is readily extended to carry out the structural and capital–energy optimization simultaneously. This requires a superstructure to be developed around the existing network structure by adding possible new features to the existing structure, including network resequencing, introducing new matches or stream-split modifications in the existing network. For the objective of minimum total cost, including the optimization of the capital–energy trade-off, the resulting optimization problem is a mixed integer nonlinear programming problem (MINLP). If the MINLP is constrained to allow a single modification, the retrofit strategy can follow an evolutionary approach, as illustrated in Figure 18.43. The network is first optimized within the existing structure, then one modification identified and optimized, then a second modification identified and optimized, and so on. This approach has a number of practical advantages. It allows a series of different network retrofit designs to be developed, each with increasing levels of energy saving corresponding with more complex and more expensive retrofits. Each can then be investigated in detail to explore the practicalities of implementation. At each stage, if necessary, the optimizations can be re-run with additional constraints if investigation of the

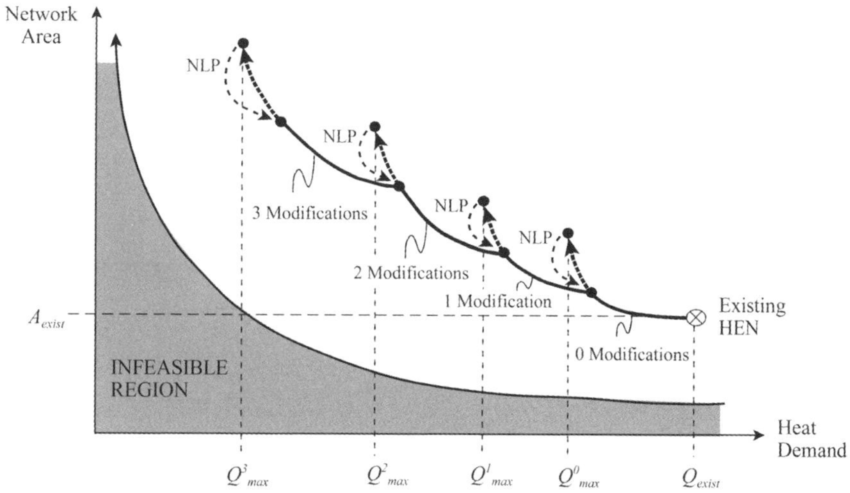

Figure 18.42
Retrofit based on the network pinch can proceed in a stepwise approach.

implementation identifies unanticipated problems. However, the approach also has disadvantages. It will not necessarily lead to the best retrofit, because different combinations of network changes can be chosen. To obtain the best retrofit requires all options to be considered simultaneously.

A better approach is to again follow an evolutionary strategy, but consider the modifications simultaneously. This can be done by first optimizing the existing network. Then the retrofit superstructure can be optimized by an MINLP optimization, constraining to a single network change. The original network is then reoptimized for retrofit by an MINLP optimization constraining to two network changes, then the original network is reoptimized for three network changes, and so on. This is illustrated in Figure 18.44. At each stage the optimizations can be re-run with additional constraints if investigation of the implementation identifies unanticipated problems.

Following this approach allows a series of potential retrofits to be identified, each with increasing energy saving, but each with increasing complexity and capital cost. This is necessary because it is impossible to include all practical details and issues regarding the layout, congestion and piping arrangements of an existing process in the optimization. Preparing a list of options gives greater flexibility to choose a retrofit package that is both economic and can be implemented without excessive modifications to the existing process.

Rather than using deterministic optimization of a superstructure to identify retrofit options, an approach based on stochastic search optimization can be used, such as simulated annealing in a similar approach to that for new design discussed earlier. The approach starts from the existing network design. Then simulated annealing is used to identify structural changes involving resequencing, repiping, new matches, or changes to stream splitting arrangements

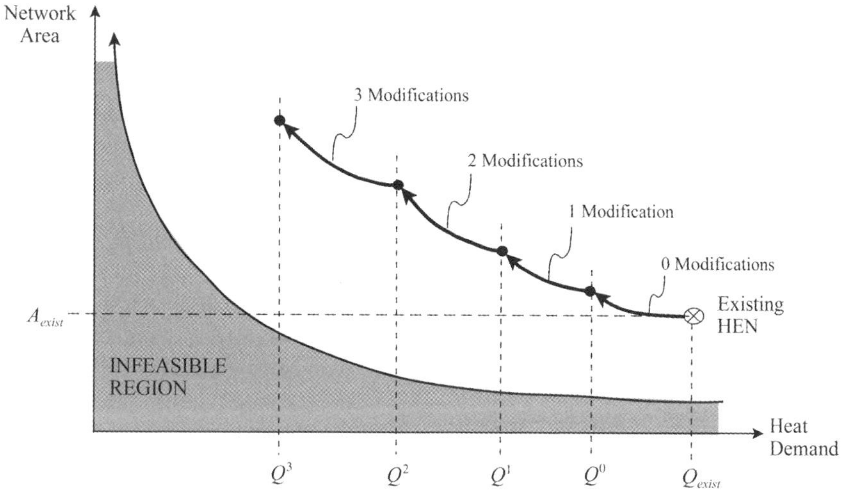

Figure 18.43
Using MINLP or stochastic optimization contstrained to one change at a time.

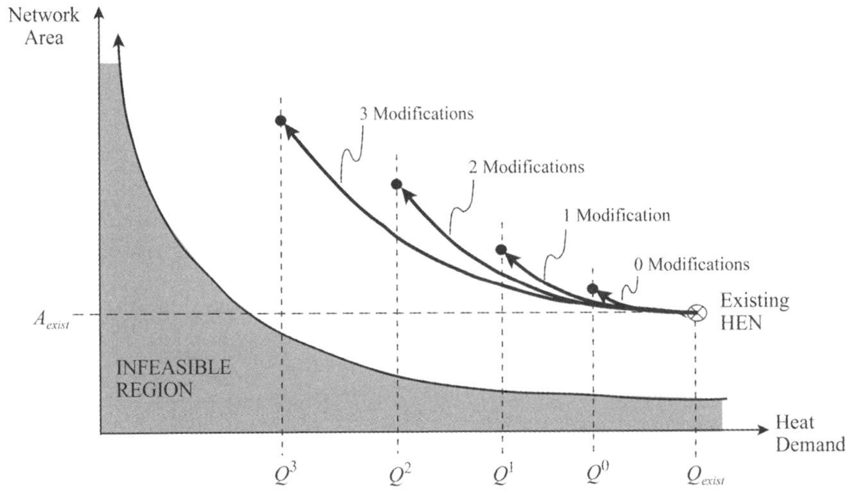

Figure 18.44
Using MINLP or stochastic optimization constrained by the number of modifications.

through a series of moves. After each move, the energy balance needs to be restored by some mechanism, the new network simulated and the objective function evaluated. As with the use of deterministic optimization, the best approach is to carry out a series of retrofit optimizations in which the number of network modifications is restricted by a constraint, gradually increasing the number of allowed modifications. The optimizations should include the capital–energy trade-off. Again at each stage, if necessary, the optimizations can be re-run with additional constraints if investigation of the implementation identifies unanticipated problems. The preferred retrofit can then be chosen from the list of possible options.

This approach guides the retrofit to simple and practical solutions and has the major advantage that it allows the designer to assess modifications and to keep control over the complexity of the retrofit as the retrofit design develops.

18.12 Heat Exchanger Network Design – Summary

A good initialization for heat exchanger network design is to assume that no individual exchanger should have a temperature difference smaller than ΔT_{min}. Having decided that no exchanger should have a temperature difference smaller than ΔT_{min} two rules were deduced in Chapter 17. If the energy target set by the composite curves (or the problem table algorithm) is to be achieved, there must be no transfer heat across the pinch by:

- process-to-process heat transfer,
- inappropriate use of utilities.

These rules are both necessary and sufficient for the design to achieve the energy target given that no individual exchanger should have a temperature difference smaller than ΔT_{min}.

The design of heat exchanger networks can be summarized in five steps.

1) Divide the problem at the pinch into separate problems.
2) The design for the separate problems is started at the pinch, moving away.
3) Temperature feasibility requires constraints on the *CP*s to be satisfied for matches between the streams at the pinch.
4) The loads on individual units are determined using the tick-off heuristic to minimize the number of units. Occasionally, the heuristic causes problems.
5) Away from the pinch, there is usually more freedom in the choice of matches. In this case, the designer can discriminate on the basis of operability, plant layout, and so on.

If the number of hot streams at the pinch above the pinch is greater than the number of cold streams, then the cold streams must be split to satisfy the ΔT_{min} constraint. If the number of cold streams at the pinch below the pinch is greater than the number of hot streams, then the hot streams must be split to satisfy the ΔT_{min} constraints. If the *CP* inequalities for all streams at the pinch cannot be satisfied, this also can necessitate stream splitting.

If the problem involves more than one pinch, then between pinches the design should be started from the most constrained pinch.

Remaining problem analysis can be used to make a quantitative assessment of matches in the context of the whole network without having to complete the network.

Once the initial network structure has been defined, then loops, utility paths and stream splits offer the degrees of freedom for manipulating network cost in multivariable continuous optimization. When the design is optimized, any constraint that temperature differences should be larger than ΔT_{min} or that there should not be heat transfer across the pinch no longer applies. The objective is simply to design for minimum total cost.

For more complex network designs, especially those involving many constraints, equipment specifications, and so on, design methods based on the optimization of a superstructure can be used. Approaches based on the deterministic optimization of a superstructure or stochastic search optimization can be used.

Network retrofit can be performed by resequencing and repiping of heat exchangers, the introduction of new exchangers and additional stream splitting. Deterministic optimization of a superstructure or stochastic search optimization can be used to identify network changes. A series of potential retrofits should be identified, each with increasing energy saving, but each with increasing complexity and capital cost. This is necessary because it is impossible to include all practical details and issues regarding the layout, congestion and piping arrangements of an existing process in the optimization. Preparing a list of options gives greater flexibility to choose an option that is both economic and can be implemented without excessive modifications to the existing process. New heat transfer area to existing matches can be added through the addition of new heat exchanger shells in series or parallel or through heat transfer enhancement to existing exchangers.

18.13 Exercises

1. The stream data for a process are given in the Table 18.9.

   a) For a ΔT_{min} of 10 °C, the pinch is at 90 °C for hot streams and 80 °C for cold streams. Design a heat exchanger network for maximum energy recovery that features the minimum number of units.

   b) By relaxing the constraint of $\Delta T_{min} = 10$ °C, shift the heat load through a utility path such that the network uses only hot utility and no cold utility at all.

2. The process stream data for a heat recovery network problem are given in Table 18.10.

   a) Determine the energy targets for hot and cold utility for $\Delta T_{min} = 20$ °C.

   b) Design a heat exchanger network to achieve the energy targets in the minimum number of units.

Table 18.9

Stream data for Exercise 1.

Stream		Supply temperature (°C)	Target temperature (°C)	Heat capacity flowrate (MW·K^{-1})
No.	Type			
1	Hot	200	100	0.4
2	Hot	200	100	0.2
3	Hot	150	60	1.2
4	Cold	50	140	1.1
5	Cold	80	120	2.4

Table 18.10

Stream data for Exercise 2.

Stream	T_S (°C)	T_T (°C)	CP (MW·K^{-1})
1. Hot	300	80	0.15
2. Hot	200	40	0.225
3. Cold	40	180	0.2
4. Cold	140	280	0.3

   c) Identify the scope to reduce the number of units by manipulating loops and utility paths by sacrificing energy consumption.

   d) Design a network to realize this scope and restore $\Delta T_{min} = 20$ °C.

3. The heat recovery stream data for a process are given in Table 18.11. A problem table analysis for $\Delta T_{min} = 10$ °C results in the heat recovery cascade given in Table 18.12.

   a) Design a heat recovery network in the minimum number of units. The number of units is below what might have been expected from the target for the number of units. Why?

   b) Low-pressure saturated steam can be generated from saturated boiler feedwater at an interval temperature of 115 °C. Determine how much low-pressure steam can be generated and design a network to achieve this duty.

Table 18.11

Stream data for Exercise 3.

Stream	T_S (°C)	T_T (°C)	CP (kW·K^{-1})
1. Hot	180	40	200
2. Hot	150	40	400
3. Cold	60	180	300
4. Cold	30	130	220

Table 18.12

Heat recovery cascade for Exercise 3.

Interval temperature (°C)	Heat flow (MW)
185	6.0
175	3.0
145	0
135	3.0
65	8.6
35	20.0

Table 18.13

Stream data for Exercise 4.

Stream	T_S (°C)	T_T (°C)	CP (MW·K^{-1})
1. Cold	18	123	0.0933
2. Cold	118	193	0.1961
3. Cold	189	286	0.1796
4. Hot	159	77	0.2285
5. Hot	267	80	0.0204
6. Hot	343	90	0.0538

c) The network design from Part b results in two steam generators. Evolve the network to remove one of these by suffering a penalty in steam generation, but maintain $\Delta T_{min} = 10\,°C$.

4. The process stream data for a heat recovery problem are given in Table 18.13. A problem table analysis reveals that $Q_{Hmin} = 13.95$ MW, $Q_{Cmin} = 8.19$ MW, hot stream pinch temperature is 159 °C and cold stream pinch temperature is 149 °C for $\Delta T_{min} = 10\,°C$.

a) Design a heat recovery network with the minimum number of units. (Hint: below the pinch it may be necessary to split a stream away from the pinch to achieve the minimum number of units.)

b) Evolve the design to eliminate stream splits below the pinch, while maintaining the same number of units, by allowing a temperature difference slightly smaller than 10 °C.

5. The data for a heat recovery problem are given in the Table 18.14. The problem table cascade is given in Table 18.15 for $\Delta T_{min} = 20\,°C$.

Table 18.15

Problem table cascade for Exercise 5.

Interval temperature (°C)	Heat flow (MW)
220	20.95
210	19.95
150	6.75
125	0
110	4.2
100	12
95	13.7
85	14.5
75	16.5
70	18.25
55	28.75
40	31.75

Table 18.14

Stream data for Exercise 5.

Stream		T_S (°C)	T_T (°C)	Heat capacity flowrate (MW·K^{-1})
No.	Type			
1	Hot	120	65	0.5
2	Hot	80	50	0.3
3	Hot	135	110	0.29
4	Hot	220	95	0.02
5	Hot	135	105	0.26
6	Cold	65	90	0.15
7	Cold	75	200	0.14
8	Cold	30	210	0.1
9	Cold	60	140	0.05
Steam		250	—	—
Cooling water		15	—	—

Table 18.16

Stream data for Exercise 6.

Stream		T_s (°C)	T_T (°C)	Stream heat duty (MW)
No.	Type			
1	Hot	150	50	−20
2	Hot	170	40	−13
3	Cold	50	120	21
4	Cold	80	110	15

Given these data:

a) Set out the stream grid.

b) Design a maximum energy recovery network.

6. The stream data for a process are given in the Table 18.16. Steam is available between 180 and 179 °C and cooling water between 20 and 40 °C. For $\Delta T_{min} = 10$ °C, the minimum hot and cold utility duties are 7 MW and 4 MW respectively. The pinch is at 90 °C on the hot streams and 80 °C on the cold streams.

a) Calculate the target for the minimum number of units for maximum energy recovery.

b) Develop two alternative maximum energy recovery designs, keeping units to a minimum.

c) Explain why the design below the pinch cannot achieve the target for the minimum number of units.

Table 18.18

Heat flow cascade for Exercise 7.

Interval temperature (°C)	Heat flow (kW)
349	1528
303	1758
273	1368
210	675
205	520
165	0
140	250
135	255
95	1095
85	1255
83	1283
35	851

Table 18.17

Stream data for Exercise 7.

Stream		T_s (°C)	T_T (°C)	CP (kW·K^{-1})
No.	Type			
1	Hot	170	88	23
2	Hot	278	90	2
3	Hot	354	100	5
4	Cold	30	135	9
5	Cold	130	205	20
6	Cold	200	298	18

d) How many degrees of freedom are available for network optimization?

7. The stream data for a process are given in the Table 18.17. A problem table heat cascade for $\Delta T_{min} = 10$ °C is given the Table 18.18. Hot utility is to be provided by a hot oil circuit with a supply temperature of 400 °C. Cooling water is available at 20 °C.

a) Calculate the minimum flowrate of hot oil if C_P for the hot oil is 2.1 kJ·kg^{-1}·K^{-1}, assuming ΔT_{min} process-to-process heat recovery is 10 °C and process to hot oil to be 20 °C.

b) Design a heat exchanger network for maximum energy recovery in the minimum number of units ensuring $\Delta T_{min} = 10$ °C for all process-to-process heat exchangers throughout the network.

c) Suggest an alternative network design below the pinch to eliminate stream splits by accepting a violation of ΔT_{min}.

d) What tools could have been used to develop the design below the pinch more systematically?

8. The stream data for a process are given in Table 18.19.

a) Sketch the composite curves for $\Delta T_{min} = 10$ °C.

b) Determine the target for hot and cold utility for $\Delta T_{min} = 10$ °C.

Table 18.19

Stream data for Exercise 8.

Stream		T_S (°C)	T_T (°C)	Heat duty (MW)
No.	Type			
1	Hot	150	30	7.2
2	Hot	40	40	10
3	Hot	130	100	3
4	Cold	150	150	10
5	Cold	50	140	3.6

Table 18.20

Stream data for Exercise 9.

Stream		T_S (°C)	T_T (°C)	Heat capacity flowrate ($MW \cdot K^{-1}$)
No.	Type			
1	Hot	500	100	4
2	Cold	50	450	1
3	Cold	60	400	1
4	Cold	40	420	0.75

c) Design a maximum energy recovery network in the minimum number of units for $\Delta T_{min} = 10\,°C$.

d) Can the number of units be reduced by evolution of the network?

9. The stream data for a process are given in Table 18.20. $\Delta T_{THRESHOLD}$ for the problem is 50 °C and ΔT_{min} is 20 °C. A problem table analysis on these data produces the cascade given in Table 18.21 for $\Delta T_{min} = 20\,°C$.

a) Design a heat exchanger network for this problem that achieves maximum energy recovery in the minimum number of units.

b) Determine how much steam at a condensing temperature of 180 °C can be generated by this process.

c) Sketch the composite curves for the process showing the maximum steam generation at 180 °C.

d) Design a network that achieves maximum energy recovery in the minimum number of units and that generates the maximum possible steam at 180 °C.

Table 18.21

Heat flow cascade for Exercise 9.

Interval temperature (°C)	Heat flow (MW)
490	0
460	120
430	210
410	255
90	655
70	600
60	582
50	575

References

Ahmad S, Linnhoff B and Smith R. (1990) Cost Optimum Heat Exchanger Networks: II. Targets and Design for Detailed Capital Cost Models, *Comp Chem Eng*, **14**: 751.

Asante NDK and Zhu XX. (1997) An Automated and Interactive Approach for Heat Exchanger Network Retrofit, *Trans IChemE*, **75**: 349.

Dolan WB, Cummings PT and Le Van MD (1990) Algorithm Efficiency of Simulated Annealing for Heat Exchanger Network Design, *Comp Chem Eng*, **14**: 1039.

Floudas CA, Ciric AR and Grossmann I.E. (1986) Automatic Synthesis of Optimum Heat Exchanger Network Configurations, *AIChEJ*, **32**: 276.

Gunderson T and Naess L. (1988) The Synthesis of Cost Optimal Heat Exchanger Networks: An Industrial Review of the State of the Art, *Comp Chem Eng*, **12**: 503.

Jiang N, Shelley JD, Doyle D and Smith R. (2014) Heat Exchanger Network Retrofit with a Fixed Network Structure, *Applied Energy*, **127**: 25.

Linnhoff B and Ahmad S. (1990) Cost Optimum Heat Exchanger Networks: I. Minimum Energy and Capital Using Simple Models for Capital Cost, *Comp Chem Eng*, **14**: 729.

Linnhoff B and Hindmarsh E. (1983) The Pinch Design Method of Heat Exchanger Networks, *Chem Eng Sci*, **38**: 745.

Linnhoff B, Townsend DW, Boland D, Hewitt GF, Thomas BEA, Guy AR and Marsland R.H. (1982) *A User Guide on Process Integration for the Efficient Use of Energy*, IChemE, Rugby, UK.

Press WH, Teukolsky SA, Vettering WT and Flannery BP (2007) *Numerical Recipes: The Art of Scientific Computing*, Cambridge University Press.

Saboo AK, Morari M and Colberg R.D. (1986) RESHEX – An Interactive Software Package for the Synthesis and Analysis of Resilient Heat Exchanger Networks – II. Discussion of Area Targeting and Network Synthesis Algorithms, *Comp Chem Eng*, **10**: 591.

Tjoe TN and Linnhoff B. (1986) Using Pinch Technology for Process Retrofit, *Chem Eng*, **April**: 47.

Yee TF, Grossmann IE and Kravanja Z. (1990) Simultaneous Optimization Models for Heat Integration – I. Area and Energy Targeting of Modeling of Multi-stream Exchangers, *Comp Chem Engg*, **14**: 1151.

Zhu XX and Asante NDK (1999) Diagnosis and Optimization Approach for Heat Exchanger Network Retrofit, *AIChE J*, **45**: 1488.

Zhu XX, O'Neill BK, Roach JR and Wood RM (1995) New Method for Heat Exchanger Network Synthesis Using Area Targeting Procedures, *Comp Chem Eng*, **19**: 197.

Chapter 19

Heat Exchanger Networks III – Stream Data

The heat exchanger network targeting and design methods presented in Chapters 17 and 18 maximize heat recovery for a given set of process conditions. However, before any analysis can be performed, the material and energy balance needs to be represented as a set of hot and cold streams. This is often not straightforward. First, the process conditions are often not fixed rigidly. Process conditions such as pressures, temperatures and flowrates might have the freedom to be changed within certain limits. The flexibility to change the processing conditions, where this is possible, can be exploited to improve the heat recovery further. But, even if the process conditions are fixed for a given process flowsheet and material and energy balance, it is still not straightforward to interpret the flowsheet as a set of hot and cold streams.

Consider first the issue of changing the process conditions and how those changes might be directed to improve the heat recovery.

19.1 Process Changes for Heat Integration

Consider the composite curves in Figure 19.1a. Any process change that (Umeda, Harada and Shiroko, 1979a; Umeda, Nidda and Shiroko, 1979b):

- increases the total hot stream heat duty above the pinch,
- decreases the total cold stream heat duty above the pinch,
- decreases the total hot stream heat duty below the pinch,
- increases the total cold stream heat duty below the pinch

will bring about a decrease in utility requirements. This is known as the *plus–minus principle* (Linnhoff and Parker, 1984; Linnhoff and Vredeveld, 1984). These simple guidelines provide a reference for appropriate design changes to improve the targets. The changes apply throughout the process to reactors, recycle flowrates, distillation columns, and so on.

If a process change, such as a change in distillation column pressure, allows shifting a hot stream from below the pinch to above, it has the effect of increasing the overall hot stream duty above the pinch and therefore decreasing hot utility. Simultaneously, it decreases the overall hot stream duty below the pinch and decreases cold utility. Shifting a cold stream from above the pinch to below decreases the overall cold stream duty above the pinch, decreasing hot utility, and increases the overall cold stream duty below the pinch, reducing cold utility. Thus, one way to implement the plus–minus principle, as illustrated in Figure 19.1b, is (Linnhoff *et al.*, 1982):

- shifting hot streams from below to above the pinch or
- shifting cold streams from above to below the pinch.

Another way to relate these principles is to remember that heat integration will always benefit by keeping hot streams hot and keeping cold streams cold (Linnhoff *et al.*, 1982).

19.2 The Trade-Offs Between Process Changes, Utility Selection, Energy Cost and Capital Cost

Although the plus–minus principle is the basic reference in guiding process changes to reduce utility costs, it takes no account of capital costs. Process changes to reduce utility consumption will normally bring about a reduction in temperature difference in the process, as indicated in Figure 19.1. Thus, the capital–energy trade-off (and hence ΔT_{min}) might need to be readjusted after process changes.

In addition, the decrease in driving forces in Figure 19.1 caused by the process changes also affects the potential for using multiple utilities. For example, as the driving forces above the pinch become smaller, the potential to switch duty from high-pressure to

Chemical Process Design and Integration, Second Edition. Robin Smith.

Companion Website: www.wiley.com/go/smith/chemical2e

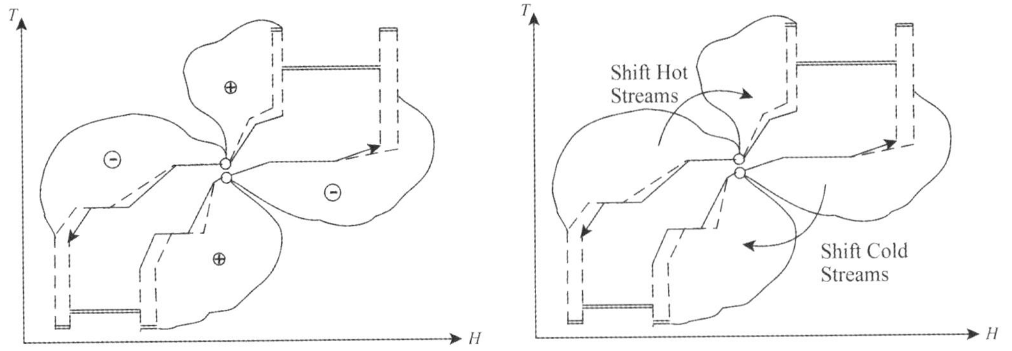

(a) The plus/minus principle.

(b) Shifting streams through the pinch in the right direction enacts the plus/minus principle.

Figure 19.1

The plus/minus principle guides process changes to reduce utility consumption. (Reproduced from Smith R and Linnhoff B, 1998, *Trans IChemE ChERD*, **66**: 195 by permission of the Institution of Chemical Engineers.)

low-pressure steam, as discussed in Section 17.7, decreases. Process changes are competing with better choice of utility levels, heat engines and heat pumps for available spare driving forces.

19.3 Data Extraction

Having discussed the way in which changes to the basic stream data can improve targets, an even more fundamental question now needs to be addressed. Before any heat integration analysis can be carried out, the basic stream data needs to be extracted from the material and energy balance. In some cases, the representation of the stream data from the material and energy balance is straightforward. However, there are a number of pitfalls that can lead to errors and missed opportunities. Missed opportunities can arise through extracting unnecessarily too many constraints.

Consider now the basic principles of data extraction for heat integration.

1) *Stream identification.* Figure 19.2a shows part of a flowsheet in which a feed stream is heated from 10 °C to 70 °C before being filtered. After the filter, it is heated from 70 °C to 135 °C and then from 135 °C to 200 °C before it is fed to a distillation column. A fundamental question for the representation of stream data for heat integration is "How many streams are there in this part of the flowsheet?" It could be assumed that there is a stream from 10 °C to 70 °C, another from 70 °C to 135 °C and a third from 135 °C to 200 °C. These are three cold streams. There are also three hot streams currently preheating the cold streams. One stream is cooled from 110 °C to 30 °C, another from 150 °C to 90 °C and a third from 210 °C to 170 °C. If the data are extracted in this way, then there will be some very neat matches between hot and cold streams: the ones that already exist. Extracting the data in this way does not seem to open up any opportunities for improvement. Fundamental questions need to be asked about what exactly is essential as far as the stream data are concerned. Heating the outlet of the filter from 70 °C to 135 °C is part of a previously suggested solution. It is not a constraint that the process heating should stop at 135 °C. The feed stream needs to be heated to 200 °C, but the midpoint at 135 °C is not a constraint. The feed is heated from 10 °C to 70 °C before entering a filter. It may be that the temperature at which the filtration can take place has some flexibility and does not need to be

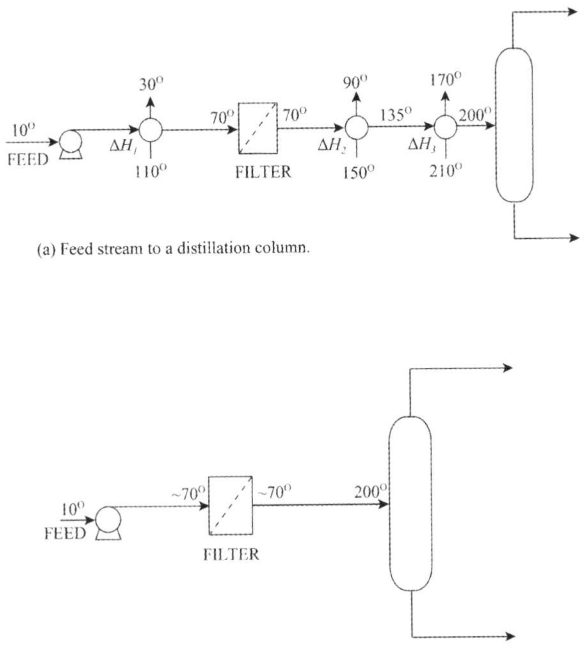

(a) Feed stream to a distillation column.

(b) Data extraction assuming the filter must operate around 70°C.

Figure 19.2

Stream identification.

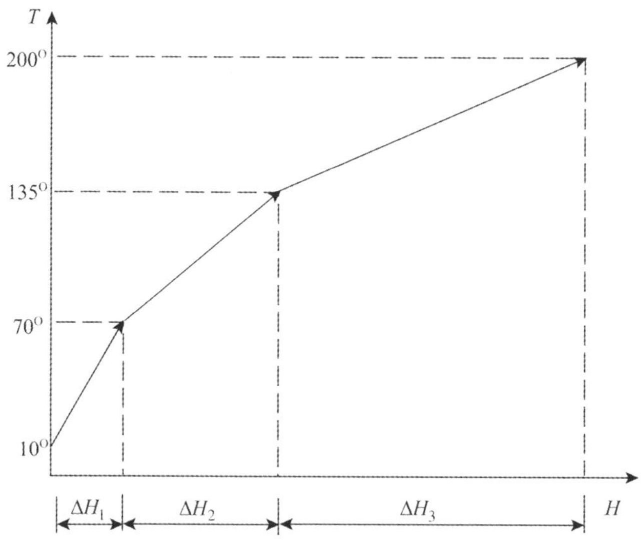

Figure 19.3

Estimates of temperature-enthalpy profiles from existing exchanger heat duties and temperatures.

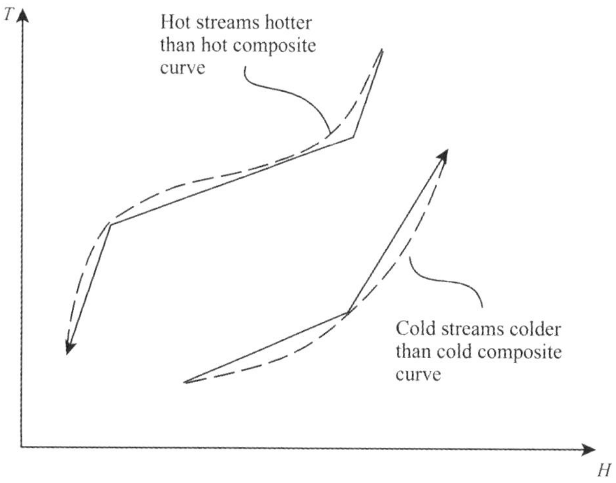

Figure 19.4

Linearization of nonlinear temperature-enthalpy profiles.

rigidly at 70 °C. Thus, for this problem, it would seem to be appropriate to extract the feed stream to the distillation as two streams, one from 10 °C to 70 °C and a second from 70 °C to 200 °C. The operation of the filter at 70 °C is kept flexible, as shown in Figure 19.2b.

2) *Temperature–enthalpy profiles.* Targeting of heat exchanger networks has assumed that the heat capacities of the streams are constant, leading to straight lines when plotted on a temperature–enthalpy plot. However, heat capacities are often not constant and this must somehow be represented, otherwise serious errors might occur. Consider again the feed preheat for the distillation shown in Figure 19.2a. If the physical properties are known for the stream, then the temperature–enthalpy profile of the feed stream can be obtained from physical property correlations. This is one possibility. Another possibility is to represent the temperature–enthalpy profiles approximately, as illustrated in Figure 19.3. The known temperatures and heat duties for the existing heat exchangers can be plotted on a temperature–enthalpy profile, as shown in Figure 19.3. This shows a nonlinear temperature–enthalpy profile taken directly from the flowsheet but represented as three linear *segments*. Such an approximation is good enough for many purposes. It is also a particularly convenient approach to adopt when dealing with retrofit of existing flowsheets.

Suppose that the actual behavior of temperature versus enthalpy is known and is highly nonlinear, as shown in Figure 19.4. How can the nonlinear data be linearized so that the construction of composite curves and the problem table algorithm can be performed? Figure 19.4 shows the nonlinear streams being represented by a series of linear segments. The linearization of the hot streams should be carried out on the under side (low-temperature side) of the curve and the linearization of the cold streams on the upper side (high-temperature side) of the curve. This is a conservative way to represent the nonlinear data, as the linearizations will come closer than the actual curves.

There is a great temptation to use a large number of linear segments to represent nonlinear stream data. This is very rarely necessary. Most nonlinear stream data can be represented reasonably well by two or three linear segments, as illustrated in Figure 19.4.

3) *Mixing.* Figure 19.5 illustrates a mixing junction in which a stream at 100 °C is mixed with a stream at 50 °C to produce a

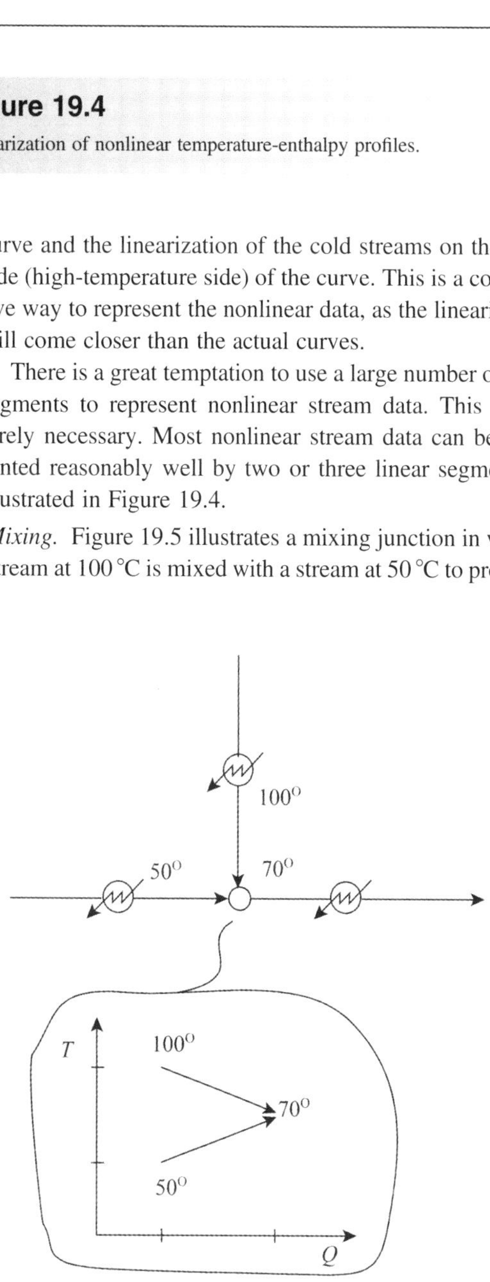

Figure 19.5

Nonisothermal mixing carries out direct contact heat transfer.

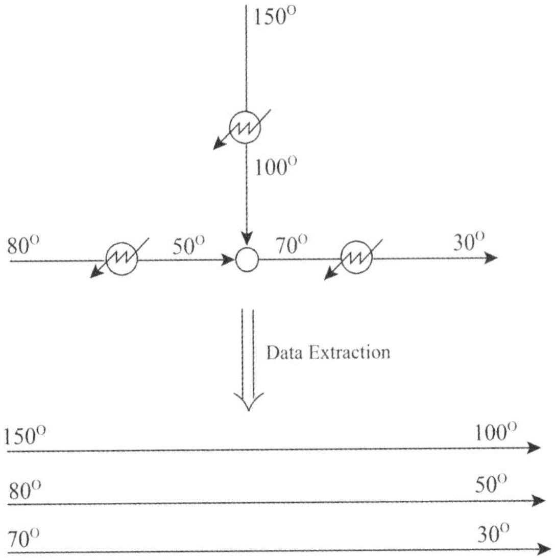

(a) Data for problem table analysis.

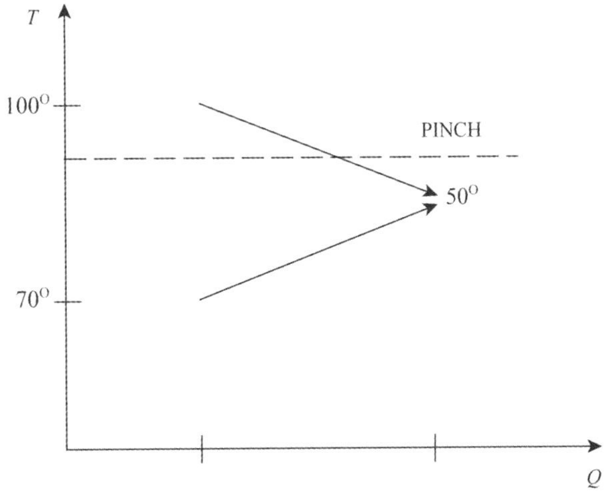

(b) Direct contact heat transfer might transfer heat across the pinch.

Figure 19.6

Nonisothermal mixing degrades temperature driving forces and might transfer heat across the pinch.

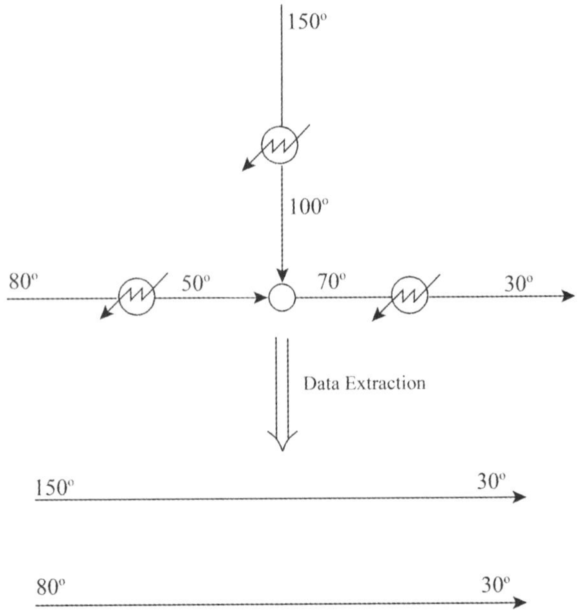

Figure 19.7

Streams should be assumed to mix isothermally for energy targeting.

combined stream with a temperature of 70 °C. Great caution must be exercised with such mixing points, as the mixing acts as a heat transfer unit. The mixing transfers heat directly rather than indirectly through a heat exchanger. This is illustrated in Figure 19.5.

The data for a problem table analysis could be taken as that for the mixer in Figure 19.6a where one stream is cooled from 150 °C to 100 °C, another is cooled from 80 °C to 50 °C and both of these are combined to produce a stream with 70 °C. This 70 °C stream is then cooled to 30 °C. If the stream data are extracted as illustrated in Figure 19.6a, then the heat transfer that takes place in the mixing junction is embedded within the stream data as being inevitable, and this might lead to lost opportunities. It might be, for example in Figure 19.6b, that the pinch temperature is somewhere between 100 °C and 50 °C. If this is the case, then some cross-pinch heat transfer will be embedded within the stream data. This will be a lost opportunity to improve the energy performance of the system.

The cross-pinch heat transfer that occurs in the mixing junction can never be corrected. Mixing streams non-isothermally therefore carries out heat transfer that should be avoided. To avoid this, streams need to be mixed isothermally. Figure 19.7 illustrates how the data should be extracted for this mixing junction. For there to be no prejudice in terms of heat transfer, it should be assumed that the streams are mixed at 30 °C in the first instance. This means extracting the streams as two separation streams, one from 150 °C to 30 °C and another from 80 °C to 30 °C. This assumes that the mixing takes place at 30 °C, for which there can be no heat transfer that takes place as a result of the mixing. Even if the mixing junction does not transfer heat across the pinch, it does degrade driving forces for the overall heat integration. In the example in Figure 19.6, high temperature heat at 100 °C is degraded to 70 °C. The overall heat exchange can never benefit by such degradation.

4) *Utilities.* In general, utilities should not be extracted from an existing flowsheet and included in the heat integration. For example, suppose a high-temperature heating duty could be carried out either with high-pressure steam or

with hot oil generated in a furnace. The cold stream that the utility is heating needs to be included. However, the high-pressure steam or hot oil should not be included. The targeting should be carried out, the grand composite curve constructed, and then the designer should make the decision as to the best choice of hot utility. Most uses of utilities, either hot or cold, fall into this category. However, there are other cases that are not so straightforward. Suppose steam is being used in a distillation, but injected as live steam directly into the distillation in order to reduce the partial pressure of the vaporizing components. This is common in refinery distillation, as discussed in Chapter 8. In this case, although the steam is generated in a boiler that is part of the utility system, the steam, when it enters the distillation, becomes part of the process. The distillation will not work without the injection of live steam. In this case, the steam that is injected into the process is effectively a process stream, rather than a utility. However, the steam will most often be generated in the utility system and should be left out of the stream data and considered separately in the utility system.

5) *Effective temperatures.* When extracting stream data to represent the heat sources and heat sinks for the heat exchanger network problem, care must be exercised so as to represent the availability of heat at its effective temperature. For example, consider the part of the process represented in Figure 19.8. The feed stream to a reactor is preheated from 20 °C to 95 °C before entering the reactor. The effluent from the reactor is at 120 °C and enters a quench that cools the reactor effluent from 120 °C to 100 °C. The vapor leaving the quench is at 100 °C and needs to be cooled to 40 °C. The quenched liquid also leaves at 100 °C but needs to be cooled to 30 °C. How should the data be extracted?

The fundamental question regarding representation of the part of the flowsheet in Figure 19.8 as heat sources and heat sinks is at what temperature the heat becomes available for heat recovery opportunities. Even though the reactor effluent is at 120 °C, the heat is not available at this temperature because the reactor effluent needs to be quenched. The heat only becomes available at 100 °C as a vapor stream that needs to be condensed and cooled to 40 °C and as a quench liquid at 100 °C that needs to be cooled to 30 °C.

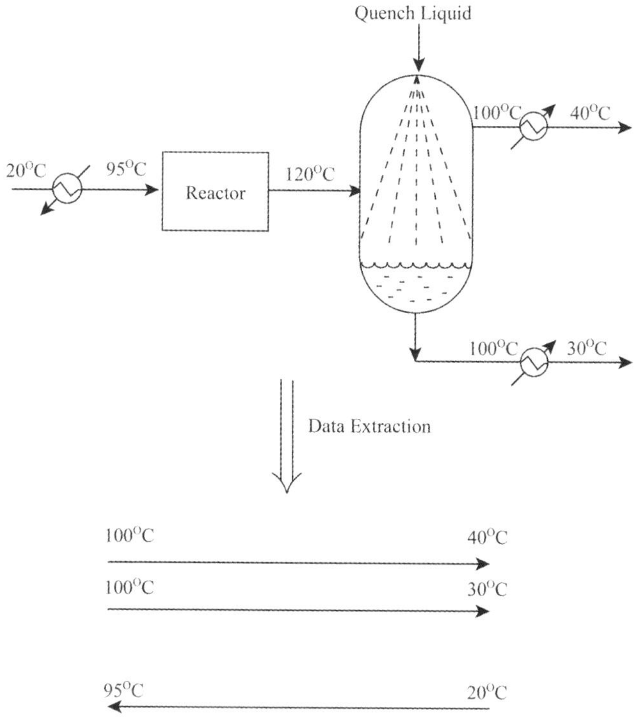

Figure 19.8

Data must be extracted at effective temperatures.

6) *Reboiling.* When reboiling a liquid there might be a significant difference between the bubble and dew points. If the data are extracted as a vaporization profile starting at the bubble point and ending at the dew point it would mean that any heat integration could in principle exploit this profile in countercurrent heat transfer. However, this is often not practical when the equipment to be used is taken into account. As discussed in Chapters 8 and 12, there are three types of reboiler used for most reboiling applications: kettle, vertical thermosyphon and horizontal thermosyphon. In kettle reboilers the vaporization takes place on the outside of the tube bundle and there is mixing around the tube bundle as the boiling occurs. In a vertical thermosyphon, vaporization takes place in the inside of tubes, but only a small fraction of the liquid entering the tubes is vaporized. In a horizontal thermosyphon, vaporization takes place on the outside of the tubes with only a small fraction of the entering liquid being vaporized and there is mixing around the tube bundle as the boiling occurs. Thus it is not possible in these types of reboiler to exploit the difference in temperature between the bubble and dew points. The safe way to extract reboiler data is to assume that the reboiler heating duty occurs at the dew point of the entering liquid being vaporized. Only if special equipment, such as plate heat exchangers, is to be used should the actual vaporization profile be included in the stream data.

7) *Condensing.* Similar to reboiling, for condensation there might be a significant difference in temperature between the dew point and bubble point of the condensing vapor. Again, if the data are extracted as a condensing profile starting at the dew point and ending at the bubble point it would mean that any heat integration could in principle exploit this profile in countercurrent heat transfer. However, like the common types of reboilers, the common types of shell-and-tube condensers, as discussed in Chapter 12, do not allow the temperature difference between dew and bubble points to be exploited. For example, condensers often use condensation on the shell

side of horizontal shell-and-tube heat exchangers in which the vapor and liquid mix as the condensation occurs. The safe way to extract data for condensers is to assume the condensation cooling duty occurs at the bubble point of the entering vapor being condensed. Only if special equipment such as plate heat exchangers is used should the actual condensing profile be included in the stream data.

8) *Heat losses.* For the majority of cases, heat losses to the environment from hot surfaces will be small compared to other heat duties and can be neglected. Occasionally, it might be necessary to accept a significant heat loss and to account for it in the energy balance. If the heat loss is from a cold stream, then it should be included with the process duty, as the heat loss must be serviced either from heat recovery or hot utility. If the loss is from a hot stream, the heat loss can be accounted for by splitting the stream from which the heat loss is occurring into two components: the process duty and the heat loss. The heat loss should then be left out of the analysis if the heat loss from the hot stream is to be accepted.

9) *Soft constraints.* In Figure 19.2b, there was a situation in which there was some flexibility to change the temperature at which a filtration takes place. These are termed *soft constraints*. There is not complete freedom to choose the conditions under which the operation takes place, but there is some flexibility to change the conditions. Another example of a soft constraint is the product storage temperature. There is sometimes flexibility to choose the temperature at which material is stored. How should such soft constraints be directed to benefit the overall heat integration problem?

When extracting soft constraints, the plus–minus principle needs to be invoked, such that the flexibility in choosing conditions is directed to reduce the utility consumption, as indicated in Figure 19.1a. Consider the example in Figure 19.9, in which a product stream needs to be cooled before entering the storage tank. There is some flexibility to choose the storage temperature, but what temperature should be chosen? Composite curves are shown in Figure 19.9, and these indicate that the hot stream pinch temperature is 65 °C. The product cooling is a hot stream and it is only of use down to a temperature of 65 °C. Thus, 65 °C would seem to be a sensible storage temperature, as far as the heat integration is concerned.

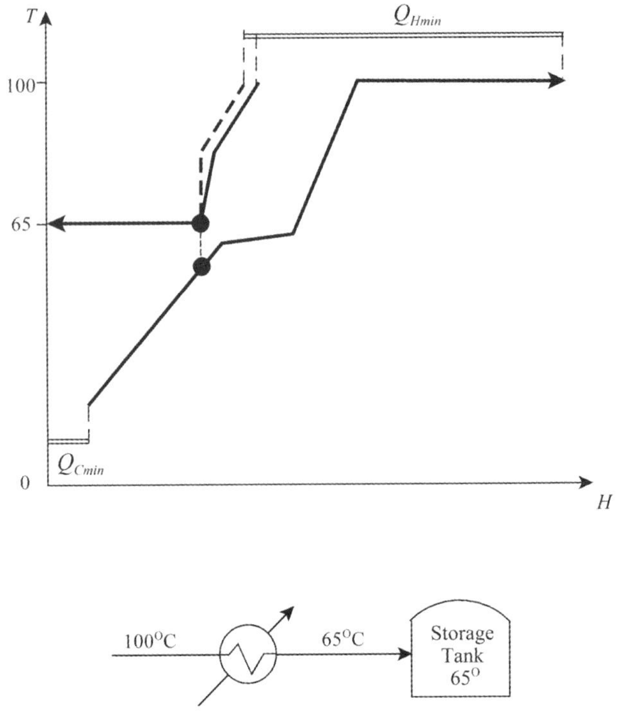

Figure 19.9
Soft temperatures.

10) *Enforced matches.* There are often some features of the heat exchanger network that the designer might wish to accept as fixed. This is often the case in retrofit. For example, a match between a hot and a cold stream might exist that is considered to be already appropriate or too expensive to change. If this is the case, then the part of the hot and cold streams involved in the enforced match should be left out of the analysis and added back in at the end of the analysis. Another typical case encountered in retrofit situations is where hot and cold streams with small heat duties are being serviced by utilities. Even though the temperatures might make the streams tempting to add into the analysis, if their heat duties are small, they will not make a significant difference, and changing them is unlikely to be economic. Again, such streams can be accepted as enforced matches and left out of the analysis. Only streams with large heat duties are likely to have a significant influence on the outcome.

11) *Data accuracy.* It is important to try and work with a set of data that is consistent when designing the heat exchanger network. Data inconsistencies are more likely to occur in retrofit situations than in new design. If there are inconsistencies in the data, it should be adjusted to make the energy balance consistent. Also, especially in retrofit situations, accurate data might not be available across the whole problem. If this is the case, then it is better to adjust the data to make it self-consistent and carry out a preliminary analysis. The most accurate data will be required where the problem is most constrained, in the region of the heat recovery pinch. Until a preliminary analysis is performed, the location of the heat recovery pinch is unknown. Having obtained insights into where data errors are likely to be a problem, then better data can be obtained for only those parts of the problem that are sensitive to data errors. Care should be taken to avoid spending considerable effort refining data that will have no significant influence on the analysis.

Example 19.1 Phthalic anhydride is an important intermediate for the plastics industry. Manufacture is by the controlled oxidation of *o*-xylene or naphthalene. The most common route uses *o*-xylene via the reaction:

$$\underset{o\text{-xylene}}{C_8H_{10}} + \underset{\text{oxygen}}{3O_2} \longrightarrow \underset{\text{phthalic anhydride}}{C_8H_4O_3} + \underset{\text{water}}{3H_2O}$$

A side reaction occurs in parallel:

$$\underset{o\text{-xylene}}{C_8H_{10}} + \underset{\text{oxygen}}{5/2O_2} \longrightarrow \underset{\text{carbon dioxide}}{8CO_2} + \underset{\text{water}}{5H_2O}$$

The reaction uses a fixed-bed vanadium pentoxide–titanium dioxide catalyst that gives good selectivity for phthalic anhydride, providing the temperature is controlled within relatively narrow limits. The reaction is carried out in the vapor phase with reactor temperatures typically in the range 380 to 400 °C.

The reaction is exothermic, and multitubular reactors are employed with direct cooling of the reactor via a heat transfer medium. A number of heat transfer media have been proposed to carry out the reactor cooling, such as hot oil circuits, water, sulfur, mercury, and so on. However, the favored heat transfer medium is usually a molten heat transfer salt, which is a eutectic mixture of sodium–potassium nitrate–nitrite.

Figure 19.10 shows a flowsheet for the manufacture of phthalic anhydride by the oxidation of *o*-xylene. Air and *o*-xylene are heated and mixed in a venturi, where the *o*-xylene vaporizes. The reaction mixture enters a tubular catalytic reactor. The heat of reaction is removed from the reactor by recirculation of molten salt. The temperature control in the reactor would be difficult to maintain by methods other than molten salt. Heat is removed from the molten salt by generating high pressure steam.

The gaseous reactor product is cooled first by boiler feedwater before entering a cooling water condenser. The cooling duty provided by the boiler feedwater has been fixed to avoid condensation. The phthalic anhydride in fact forms a solid on the tube walls in the cooling water condenser and is cooled to 70 °C. Periodically, the on-line condenser is taken off-line and the phthalic anhydride melted off the surfaces by recirculation of high-pressure hot water. Two condensers are used in parallel, one on-line performing the condensation duty and one off-line recovering the phthalic anhydride. The heat duties shown in Figure 19.10 are time-averaged values. The noncondensible gases contain small quantities of byproducts and traces of phthalic anhydride and are scrubbed before being vented to atmosphere.

The crude phthalic anhydride is heated and held at 260 °C to allow some byproduct reactions to go to completion. Purification is by continuous distillation in two columns. In the first column, maleic anhydride and benzoic and toluic acids are removed overhead. In the second column, pure phthalic anhydride is removed overhead. High-boiling residues are removed from the bottom of the second column. The reboilers of both distillation columns are serviced by a fired heater via a hot oil circuit.

There are two existing steam mains. These are high-pressure steam at 41 bar superheated to 270 °C and medium-pressure steam at 10 bar superheated at 180 °C. Boiler feedwater is available at 80 °C and cooling water at 25 °C to be heated to 30 °C.

a) Extract the data from the flowsheet.

b) Plot the composite curves and the grand composite curve.

Solution

a) From the flowsheet in Figure 19.10, the stream data for the heat recovery problem are presented in Table 19.1.

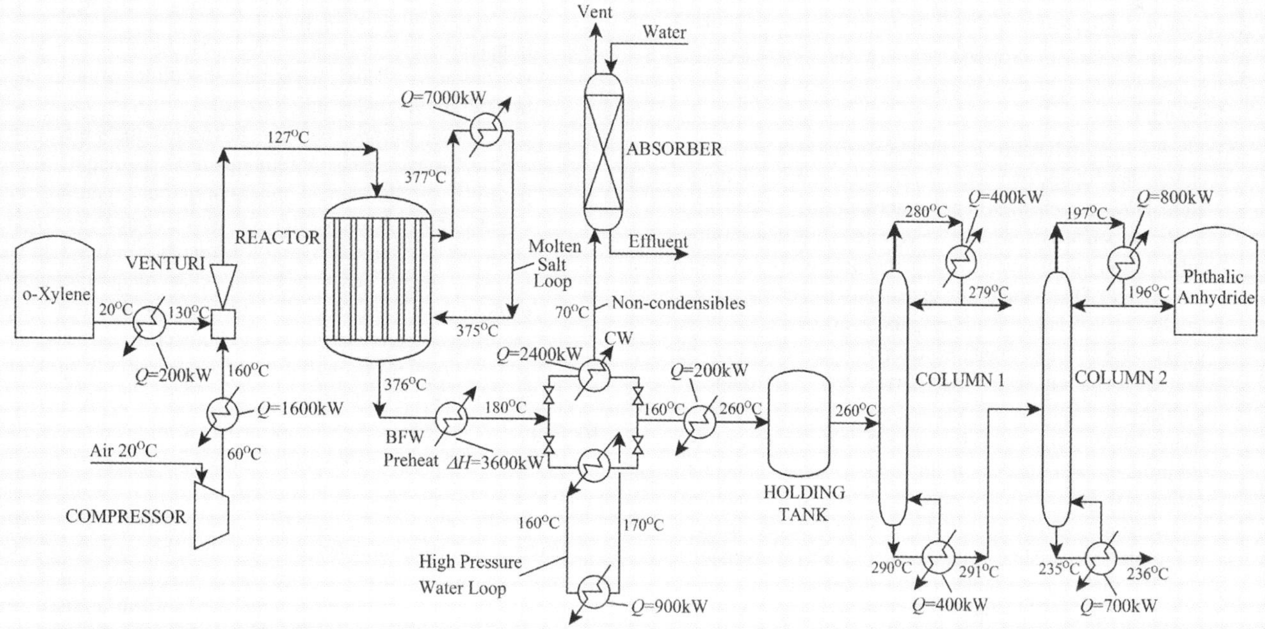

Figure 19.10

Outline of phthalic anhydride flowsheet.

Table 19.1

Stream data for the process in Figure 19.1.

Stream			T_s	T_T	ΔH
No.	Name	Type	(°C)	(°C)	(kW)
1	Reactor cooling	Hot	377	375	−7000
2	Reactor product cooling	Hot	376	180	−3600
3	Product sublimation	Hot	180	70	−2400
4	Column 1 condenser	Hot	280	279	−400
5	Column 2 condenser	Hot	197	196	−800
6	Air feed	Cold	60	160	1600
7	*o*-Xylene feed	Cold	20	130	200
8	Product melting	Cold	70	160	900
9	Holding tank feed	Cold	160	260	200
10	Column 1 reboiler	Cold	290	291	400
11	Column 2 reboiler	Cold	235	236	700

A number of points should be noted about the data extraction from the flowsheet:

1. The reactor is highly exothermic and the data have been extracted as the molten salt being a hot stream. The basis of this is that it is assumed that the molten salt circuit is an essential feature of the reactor design. Thereafter, there is freedom within reason to choose how the molten salt is cooled.
2. The product sublimation and melting are both carried out on a noncontinuous basis. Thus, time-averaged values have been taken.
3. The product sublimation and product melting imply a linear change in enthalpy over a relatively large change in temperature. However, changes of phase normally take place with a relatively small change in temperature. Thus, the product sublimation might involve desuperheating over a relatively large range of temperature, change of phase over a relatively small change in temperature and subcooling over a relatively large range in temperature. Product melting might involve heating to melting point over a relatively large range of temperature, followed by melting over a relatively small change in temperature. Thus, representation of the product sublimation and product melting as a linear change in enthalpy seems to be inappropriate. To overcome this, these two streams could be broken down into linear segments to represent this nonlinear temperature–enthalpy behavior. Here, for the sake of simplicity, the streams will be assumed to have a linear temperature–enthalpy behavior.
4. The air starts at 20 °C, but it is heated to 60 °C in the compressor by the increase in pressure. If the compressor is an essential feature of the process, then the heating between 20 and 60 °C is serviced by the compressor and should not be included in the heat recovery problem.
5. The air and *o*-xylene are mixed at unequal temperature in the venturi, where the *o*-xylene vaporizes. Mixing at unequal temperatures provides heat transfer by direct contact and might in principle be direct contact heat transfer across the pinch, the location of which is as yet unknown. Thus, accepting the direct contact heat transfer might lead to unnecessarily high energy targets if the mixing causes heat transfer across the pinch. The problem is avoided in targeting by mixing streams, where possible, at the same temperature, thus avoiding any direct contact heat transfer. Of course, once the targets have been established and the location of the pinch known, streams can then be mixed at unequal temperatures in the design away from the pinch in the knowledge that there is no cross-pinch heat transfer. In this case, the process conditions will be accepted, initially at least, because of the vaporization occurring in the mixing.

b) Figure 19.11a shows the composite curves for the process. The problem is clearly threshold in nature, requiring only cooling, with a threshold ΔT_{min} of 86 °C. Figure 19.11b shows the grand composite curves for $\Delta T_{min} = 10$ °C. The reason ΔT_{min} has been taken to be 10 °C and not 86 °C is that the cooling will be supplied by the introduction of steam generation, which will turn the threshold problem into a pinched problem, as discussed in Chapter 16, for which the value of ΔT_{min} is 10 °C for this problem.

19

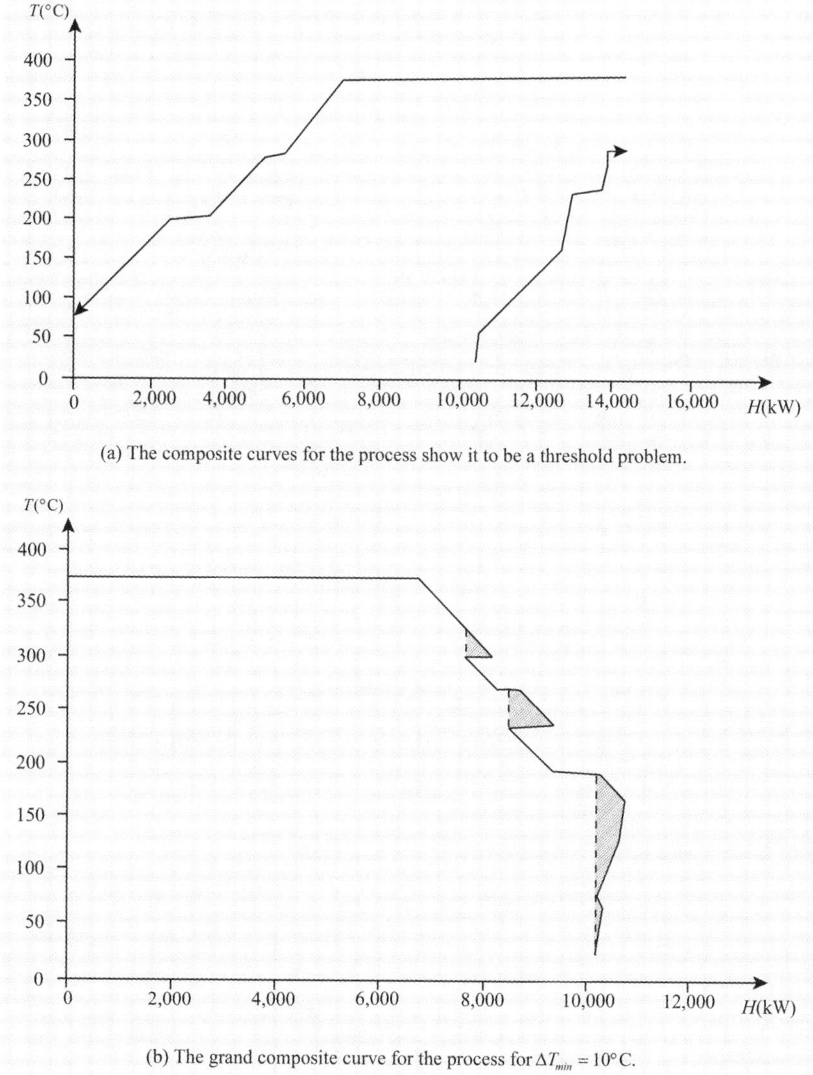

Figure 19.11
The composite and grand composite curves for the phthalic anhydride process.

19.4 Heat Exchanger Network Stream Data – Summary

The basic reference in guiding process changes to reduce utility costs is the plus–minus principle. However, process changes so identified can change the capital–energy trade-off and utility selection.

For data extraction from a flowsheet:

- Only essential constraints are included in the stream data.
- Stream data should be linearized on the safe side.
- Mixing should take place isothermally.
- Utilities should not be extracted with the stream data.
- Data should be extracted at effective temperatures.
- Reboiling and condensing data should not necessarily be extracted as countercurrent heat transfer.
- Soft constraints should exploit the plus–minus principle to improve the targets.
- Heat losses from a cold stream can be accounted for by the inclusion of a fictitious cold stream.
- Heat losses from a hot stream can be accounted for by splitting the hot stream to represent the loss.
- Accurate data are not always required throughout the whole problem.
- Enforced matches can be accounted for by leaving parts of the stream data out of the problem.

19.5 Exercises

1. The process flowsheet for a cellulose acetate fiber process is shown in Figure 19.12. Solvent is removed from the fibers in a dryer by recirculating air. The air is cooled before it enters an

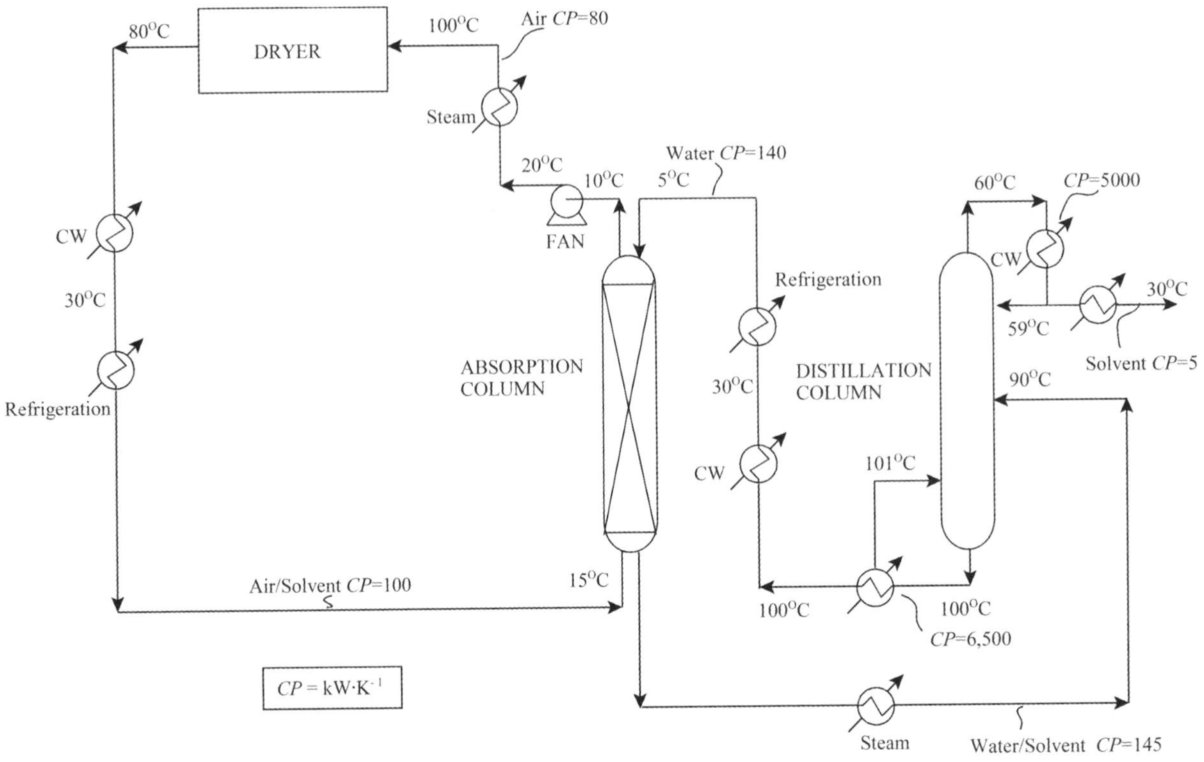

Figure 19.12

Flowsheet of a process for the manufacture of cellulose acetate fiber.

absorber where the solvent is absorbed in water. The solvent–water mixture is separated in a distillation column and the water recycled. The process is serviced by saturated steam at 150 °C, cooling water at 20 °C and refrigerant at −5 °C. The temperature rise of both the cooling water and refrigerant can be neglected.

a) Extract the stream data from the flowsheet and present them as hot and cold streams with supply and target temperatures and heat capacity flowrates.

b) Sketch the composite curves for the process for $\Delta T_{min} = 10\,°C$.

c) Determine the minimum number of units for a heat exchanger network that uses maximum cooling water.

d) Determine the scope for network simplification by using less than the maximum possible cooling water.

e) Design a network for the process that achieves maximum energy recovery in the minimum number of units when no cooling water is used (i.e. only steam and refrigeration).

f) Suggest an obvious use of cooling water instead of refrigeration for the network in Part e.

2. Figure 19.13 shows two situations where a feed stream is heated before entering an agitated reactor vessel. The temperature of the reactor needs to be controlled to be 100 °C. In the first case (Figure 19.13a), the feed is preheated

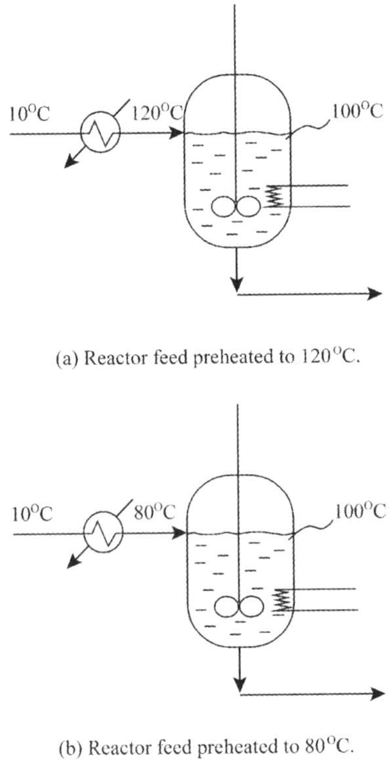

Figure 19.13

Data extraction for a reactor feed.

to a higher temperature than the reactor. In the second case (Figure 19.13b), the feed is preheated to a lower temperature than the reactor. In each case, how should the data for the feed stream be extracted if:

a) Heat of reaction is removed from the reactor by cooling water and it is considered essential to remove the heat of reaction by cooling water for safety reasons.

b) Heat of reaction is to be added to the reactor using steam and it is considered essential to add the heat of reaction by steam for control reasons.

References

Linnhoff B and Parker SJ (1984) Heat Exchanger Network with Process Modifications, *IChemE Annual Research Meeting*, Bath, April.

Linnhoff B and Vredeveld DR (1984) Pinch Technology Has Come of Age, *Chem Eng Prog*, **July**: 40.

Linnhoff B, Townsend DW, Boland D, Hewitt GF, Thomas BEA, Guy AR and Marsland RHA (1982) *User Guide on Process Integration for the Efficient Use of Energy*, IChemE, Rugby, UK.

Umeda T, Harada T and Shiroko K (1979) A Thermodynamic Approach to Synthesis of Heat Integration Systems in Chemical Processes, *Comp Chem Eng*, **3**: 273.

Umeda T, Nidda K and Shiroko K (1979) A Thermodynamic Approach to Heat Integration in Distillation Systems, *AIChE J*, **25**: 423.

Chapter 20

Heat Integration of Reactors

20.1 The Heat Integration Characteristics of Reactors

The heat integration characteristics of reactors depend both on the decisions that have been made for the removal or addition of heat and the reactor mixing characteristics. In the first instance, adiabatic operation should be considered since this gives the simplest design.

1) *Adiabatic operation.* If adiabatic operation leads to an acceptable temperature rise for exothermic reactors or an acceptable decrease for endothermic reactors, then this is the option that would normally be chosen. If so, then the feed stream to the reactor requires heating and the effluent stream requires cooling in most cases. The heat integration characteristics are thus in most cases a cold stream (the reactor feed) if the feed needs to be increased in temperature or vaporized and a hot stream (the reactor effluent) if the product needs to be decreased in temperature or condensed. The heat of reaction appears as increased temperature of the effluent stream in the case of an exothermic reaction or decreased temperature in the case of an endothermic reaction.
2) *Heat carriers.* If adiabatic operation produces an unacceptable rise or fall in temperature, then the option discussed in Chapters 6 and 14 is to introduce a heat carrier. The operation is still adiabatic, but an inert material is introduced with the reactor feed as a heat carrier. The heat integration characteristics are as before. The reactor feed is in most cases a cold stream and the reactor effluent a hot stream. The heat carrier serves to increase the heat capacity flowrate of both streams.
3) *Cold shot and hot shot.* Injection of cold fresh feed for exothermic reactions or preheated feed for endothermic reactions to intermediate points in the reactor can be used to control the temperature in the reactor. Again, the heat integration characteristics are similar to adiabatic operation. The feed is most often a cold stream if it needs to be increased in temperature or vaporized and the product a hot stream if it needs to be decreased in temperature or condensed. If heat is provided to the cold shot or hot shot streams, these are additional cold streams.
4) *Indirect heat transfer with the reactor.* Although indirect heat transfer with the reactor tends to bring about the most complex reactor design options, it is often preferable to the use of a heat carrier. A heat carrier creates complications elsewhere in the flowsheet. A number of options for indirect heat transfer were discussed earlier in Chapter 6.

The first distinction to be drawn, as far as heat transfer is concerned, is between the plug-flow and mixed-flow reactor. In the plug-flow reactor shown in Figure 20.1, the heat transfer can take place over a range of temperatures. The shape of the profile depends on the following:

- Inlet feed concentration
- Inlet temperature
- Inlet pressure and pressure drop (gas-phase reactions)
- Conversion
- Byproduct formation
- Heat of reaction
- Rate of cooling/heating
- Presence of catalyst diluents or changes in catalyst through the reactor.

Figure 20.1a shows two possible thermal profiles for exothermic plug-flow reactors. If the rate of heat removal is low and/or the heat of reaction if high, then the temperature of the reacting stream will increase along the length of the reactor. If the rate of heat removal is high and/or the heat of reaction is low, then the temperature will decrease. Under conditions between the two profiles shown in Figure 20.1a, a maximum can occur in the temperature at an intermediate point between the reactor inlet and exit.

Figure 20.1b shows two possible thermal profiles for endothermic plug-flow reactors. This time, the temperature decreases for low rates of heat addition and/or high heat of reaction. The temperature increases for the reverse conditions. Under conditions between the profiles shown in Figure 20.1b, a minimum can occur in the temperature profile at an intermediate point between the inlet and exit.

Chemical Process Design and Integration, Second Edition. Robin Smith.

Companion Website: www.wiley.com/go/smith/chemical2e

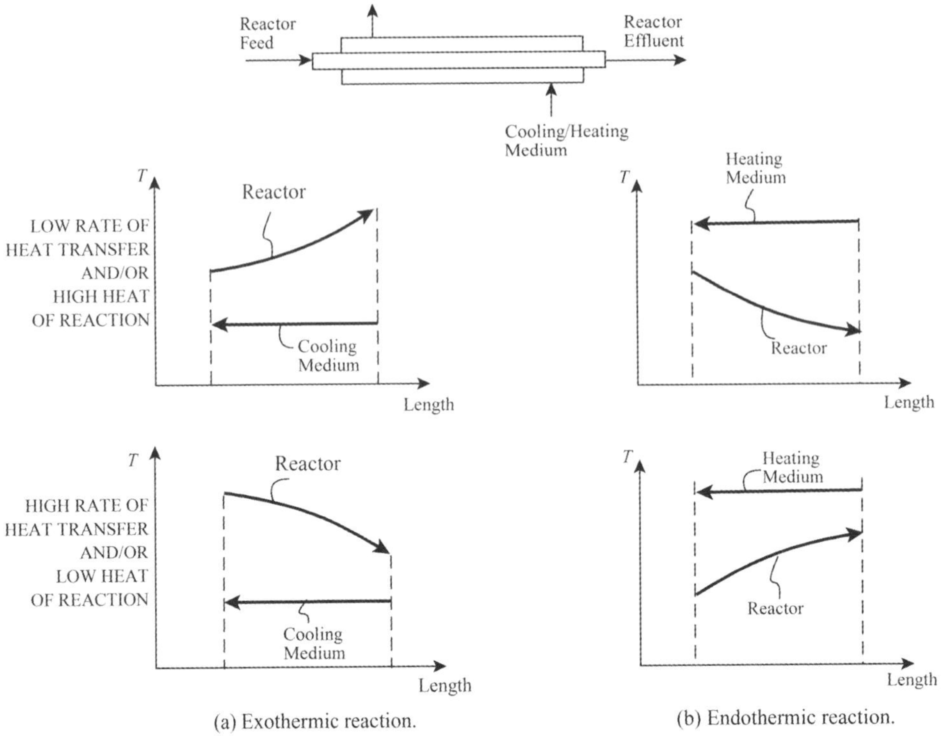

Figure 20.1
The heat transfer characteristics of plug-flow reactors.

The thermal profile through the reactor will, in most circumstances, be carefully optimized to maximize selectivity, extend catalyst life, and so on. Because of this, direct heat integration with other process streams is almost never carried out. The heat transfer to or from the reactor is instead usually carried out by a heat transfer intermediate. For example, in exothermic reactions, cooling might occur by boiling water to generate steam, which, in turn, can be used to heat cold streams elsewhere in the process or across the site.

By contrast, if the reactor is mixed-flow, then the reactor is isothermal. This behavior is typical of stirred tanks used for liquid-phase reactions or fluidized-bed reactors used for gas-phase reactions. The mixing causes the temperature in the reactor to be effectively uniform.

For indirect heat transfer, the heat integration characteristics of the reactor can be broken down into the following three cases:

a) If the reactor can be matched directly with other process streams (which is unlikely), then the reactor profile should be included in the heat integration problem. This would be a hot stream in the case of an exothermic reaction or a cold stream in the case of an endothermic reaction.

b) If a heat transfer intermediate is to be used and the cooling/heating medium is fixed, then the cooling/heating medium should be included and not the reactor profile itself. Once the cooling medium leaves an exothermic reactor, it is a hot stream requiring cooling. Similarly, once the heating medium leaves an endothermic reactor, it is a cold stream requiring heating. For cooling and heating media such as heat transfer oils and molten salts, these will be returned directly to the reactor once the heat has been removed or added.

c) If a heat transfer intermediate is to be used but the temperature of the cooling/heating medium is not fixed, then both the reactor profile and the cooling/heating medium should be included. The temperature of the heating/cooling medium can then be varied within the content of the overall heat integration problem to improve the targets, as described in Chapter 19.

In addition to the indirect cooling/heating within the reactor, the reactor feed is an additional cold stream, if it needs to be increased in temperature or vaporized, and the reactor product an additional hot stream, if it needs to be decreased in temperature or condensed.

For the ideal-batch reactor, the temperature can be assumed to be uniform throughout the reactor at any instant in time for the purposes of heat integration. Figure 20.2a shows typical variations in temperature with time for an exothermic reaction in a batch reactor. A family of curves illustrates the effect of increasing the rate of heat removal and/or decreasing heat of reaction. Each individual curve assumes the rate of heat transfer to the cooling medium to be constant for that curve throughout the batch cycle. Figure 20.2b shows typical curves for endothermic reactions. Again, each individual curve in Figure 20.2b assumes the rate of heat addition from the heating medium to be constant throughout the batch process.

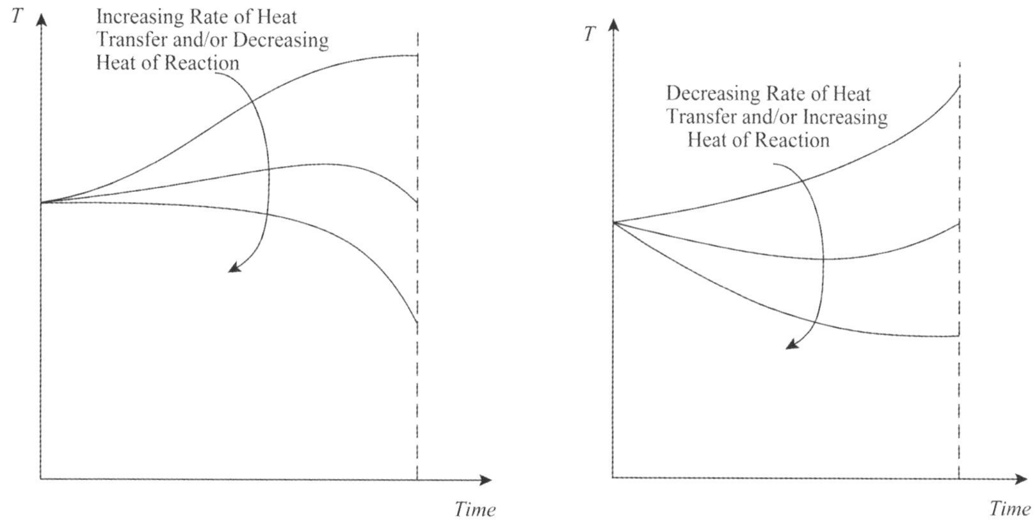

(a) Exothermic reaction. (b) Endothermic reaction.

Figure 20.2
The heat transfer characteristics of batch reactors.

Fixing the rate of heat transfer in a batch reactor is often not the best way to control the reaction. The heating or cooling characteristics can be varied with time to suit the characteristics of the reaction (see Chapter 16). Because of the complexity of batch operation and the fact that operation is usually on a small scale, it is rare for any attempt to be made to recover heat from a batch reactor, or supply heat by recovery. Instead, utilities are normally used.

The heat duty on the heating/cooling medium is given by

$$Q_{REACT} = -(\Delta H_{STREAMS} + \Delta H_{REACT}) \tag{20.1}$$

where

Q_{REACT} = reactor heating or cooling required
$\Delta H_{STREAMS}$ = enthalpy change between feed and product streams
ΔH_{REACT} = reaction enthalpy (negative in the case of exothermic reactions)

5) *Quench.* As discussed in Chapter 6, the reactor effluent may need to be cooled rapidly (quenched). This can be by indirect heat transfer using conventional heat transfer equipment or by direct heat transfer by mixing with another fluid.

If indirect heat transfer is used with a large temperature difference to promote high rates of cooling, then the cooling fluid (e.g. boiling water) is fixed by process requirements. In this case, the heat of reaction is not available at the temperature of the reactor effluent. Rather, the heat of reaction becomes available at the temperature of the quench fluid. Thus, the feed stream to the reactor is a cold stream, the quench fluid is a hot stream and the reactor effluent after the quench is also a hot stream. This was discussed under data extraction in Chapter 19.

The reactor effluent might require cooling by direct heat transfer because the reaction needs to be stopped quickly, or a conventional heat exchanger would foul, or the reactor products are too hot or corrosive to pass to a conventional heat exchanger. The reactor product is mixed with a liquid that can be recycled, cooled product or an inert material such as water. The liquid vaporizes partially or totally and cools the reactor effluent. Here, the reactor feed is a cold stream and the vapor and any liquid from the quench are hot streams.

Now consider the placement of the reactor in terms of the overall heat integration problem.

20.2 Appropriate Placement of Reactors

In Chapter 17, it was seen how the pinch takes on fundamental significance in improving heat integration. Now consider the consequences of placing reactors in different locations relative to the pinch.

Figure 20.3 shows the background process represented simply as a heat sink and heat source divided by the pinch. Figure 20.3a shows the process with an exothermic reactor integrated above the pinch. The minimum hot utility can be reduced by the heat released by reaction.

By comparison, Figure 20.3b shows an exothermic reactor integrated below the pinch. Although heat is being recovered, it is being recovered into part of the process, which is a heat source. The hot utility requirement cannot be reduced, since the process above the pinch needs at least Q_{Hmin} to satisfy its enthalpy imbalance.

There is no benefit by integrating an exothermic reactor below the pinch. The appropriate placement for exothermic reactors is above the pinch (Glavic, Kravanja and Homsak, 1988).

Figure 20.4a shows an endothermic reactor integrated above the pinch. The endothermic reactor removes Q_{REACT} from the process above the pinch. The process above the pinch needs at least Q_{Hmin} to satisfy its enthalpy imbalance. Thus, an extra Q_{REACT} must be

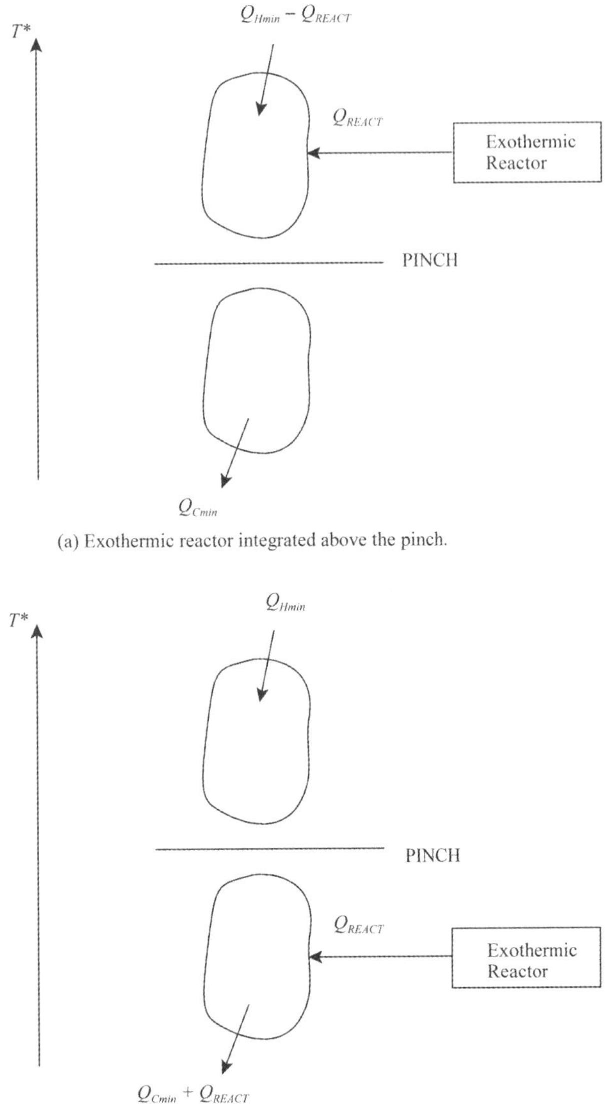

(a) Exothermic reactor integrated above the pinch.

(b) Exothermic reactor integrated below the pinch.

Figure 20.3

Appropriate placement of an exothermic reactor.

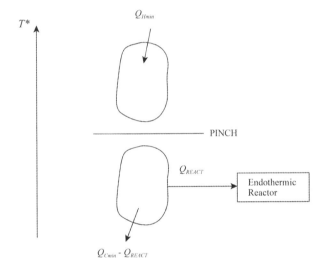

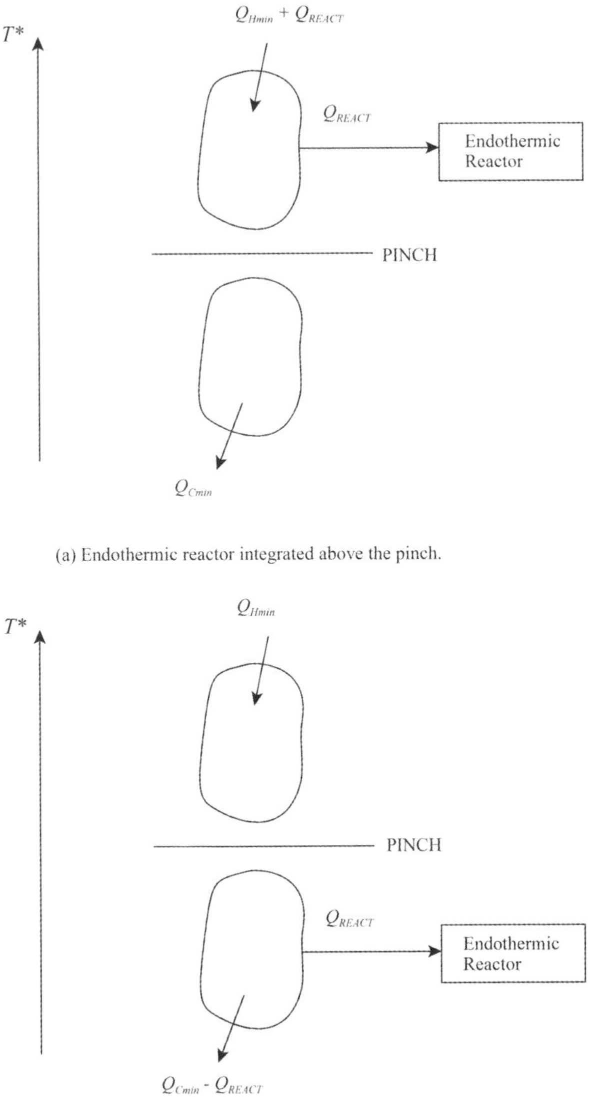

(a) Endothermic reactor integrated above the pinch.

(b) Endothermic reactor integrated below the pinch.

Figure 20.4

Appropriate placement of an endothermic reator.

imported from hot utility to compensate. There is no benefit in integrating an endothermic reactor above the pinch. Locally, it might seem that a benefit is being derived by running the reaction by recovery. However, additional hot utility must be imported elsewhere to compensate.

By contrast, Figure 20.4b shows an endothermic reactor integrated below the pinch. The reactor imports Q_{REACT} from part of the process that needs to reject heat anyway. Thus, integration of the reactor serves to reduce the cold utility consumption by Q_{REACT}. There is an overall reduction in hot utility because, without integration, the process and reactor would require $(Q_{Hmin} + Q_{REACT})$ from the utility.

There is no benefit in integrating an endothermic reactor above the pinch. The appropriate placement for endothermic reactors is below the pinch (Glavic, Kravanja and Homsak, 1988).

20.3 Use of the Grand Composite Curve for Heat Integration of Reactors

The above appropriate placement arguments assume that the process has the capacity to accept or give up the reactor heat duties at the given reactor temperature. A quantitative tool is needed to assess the capacity of the background process. For this purpose, the grand composite curve is used and the reactor profile treated as if it was a utility, as explained in Chapter 17.

The problem with representing a reactor profile is that, unlike utility profiles, the reactor profile might involve several streams. The reactor profile involves not only streams such as those for indirect heat transfer shown in Figure 20.1, but also the reactor feed and effluent streams that can be an important feature of the reactor heating and cooling characteristics. The various streams associated with the reactor can be combined to form a grand composite curve for the reactor. This can then be matched against the grand composite curve for the rest of the process. The following example illustrates the approach.

Example 20.1 Consider again the process for the manufacture of phthalic anhydride discussed in Example 19.1. The data was extracted from the flowsheet in Figure 19.10 and listed in Table 19.1. The composite curves and grand composite curve are shown in Figure 19.11.

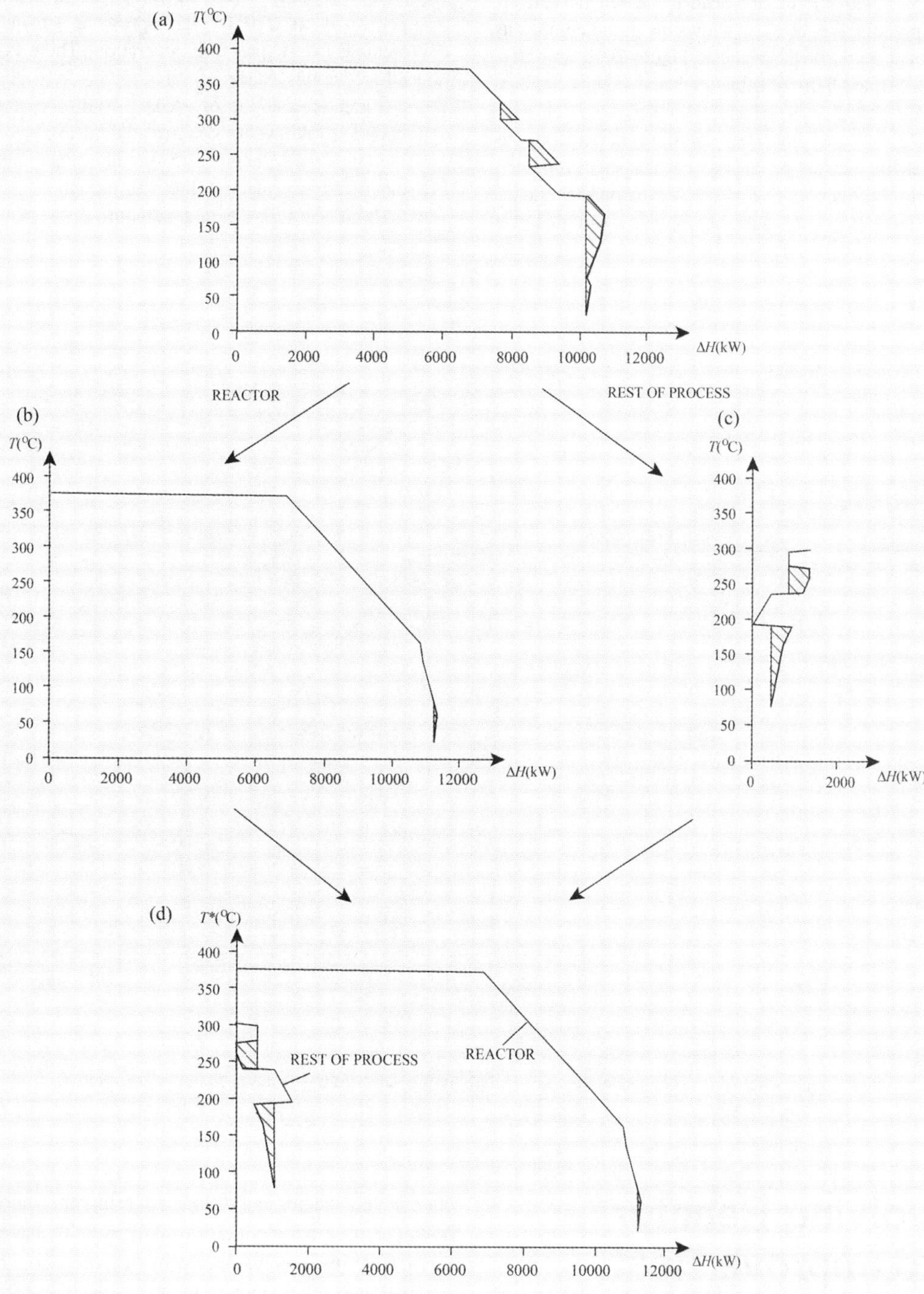

Figure 20.5

The problem can be divided into two parts, one associated with the reactor and the other with the rest of the process ($\Delta T_{min} = 10\,°C$) and then superimposed.

a) Examine the placement of the reactor relative to the rest of the process.
b) Determine the utility requirements of the process.

Solution

a) The stream data used to construct the grand composite curve in Figure 20.5a include those associated with the reactor and those for the rest of the process. If the placement of the reactor relative to the rest of the process is to be examined, those streams associated with the reactor need to be separated from the rest of the process. Figure 20.5b shows the grand composite curves for the two parts of the process. Figure 20.5b is based on Streams 1, 2, 6 and 7 from Table 19.1 and Figure 20.5c is based on Streams 3, 4, 5, 8, 9, 10 and 11.

In Figure 20.5d, the grand composite curves for the reactor and that for the rest of the process are superimposed. To obtain maximum overlap, one of the curves must be taken as a mirror image. It can be seen in Figure 20.5d that the reactor is appropriately placed relative to the rest of the process. Had the reactor not been appropriately placed, it would have been extremely unlikely that the reactor would have been changed to make it so. Rather, to obtain appropriate placement of the reactor, the rest of the process would more likely have been changed.

b) Figure 20.6 shows the grand composite curve for all the streams with a steam generation profile matched against it. The process cooling demand is satisfied by the generation of high-pressure (41 bar) steam from boiler feedwater, which is superheated to 270 °C. High-pressure steam generation is preferable to low-pressure generation. There is apparently no need for cooling water.

A greater amount of steam would be generated if the noncondensible vent was treated using catalytic thermal oxidation (see Chapter 25) rather than absorption. The exotherm from catalytic thermal oxidation would create an extra hot stream for steam generation.

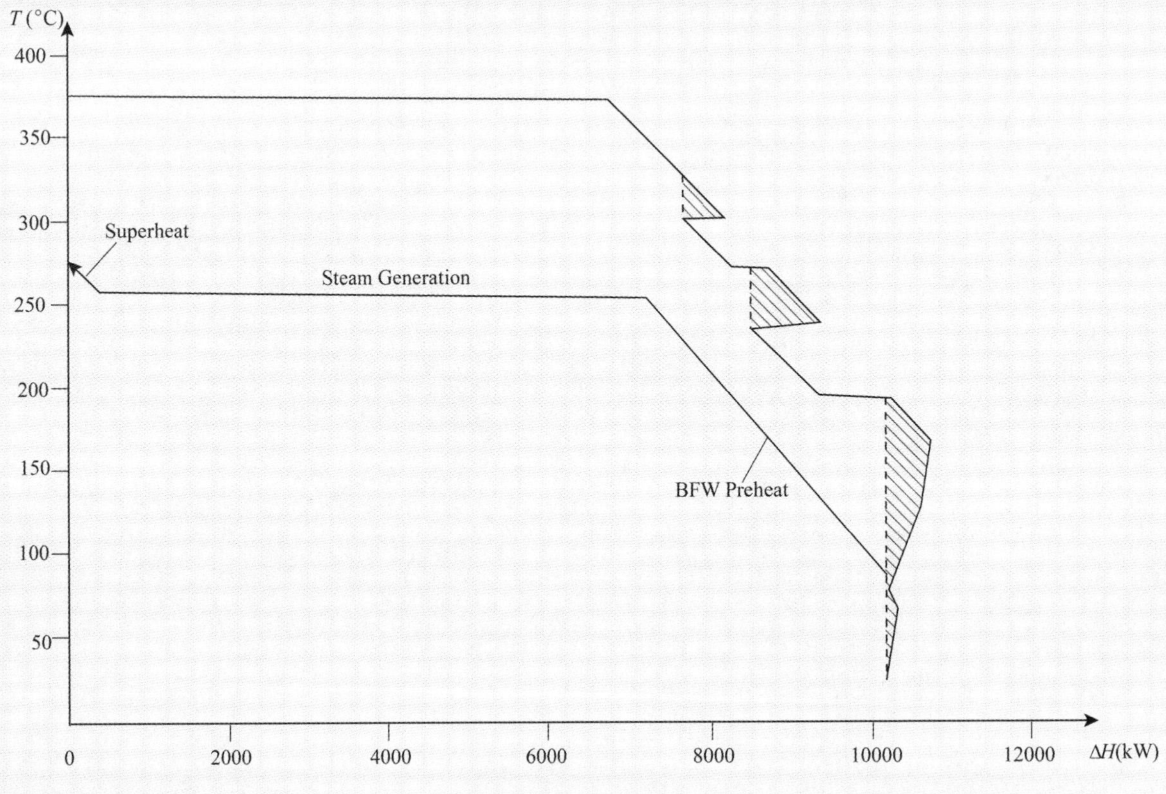

Figure 20.6
The grand composite curve for the whole process apparently requires only high-pressure steam generation from boiler feedwater.

20.4 Evolving Reactor Design to Improve Heat Integration

If the reactor is inappropriately placed, then the process changes might make it possible to correct this. One option would be to change the reactor conditions to bring this about. Most often, however, the reactor conditions will probably have been optimized for selectivity, catalyst performance, and so on, which, taken together with safety, materials-of-construction constraints, control, and so on, makes it unlikely that the reactor conditions would be changed to improve heat integration. Rather, to obtain appropriate placement of the reactor, the rest of the process would most likely be changed.

If changes to the reactor design are possible, then the simple criteria introduced in Chapter 19 can be used to direct those changes. Heat integration will always benefit by making hot streams hotter and cold streams colder. This applies whether the heat integration is carried out directly between process streams or through an intermediate such as steam. For example, consider the exothermic reactions in Figure 20.1a. Allowing the reactor to work at a higher temperature improves the heat integration potential if this does not interfere with selectivity or catalyst life or introduce safety and control problems, and so on. However, if the reactor must work with a fixed intermediate cooling fluid, such as steam generation, then the only benefit will be a reduced heat transfer area in the reactor. The steam becomes a hot stream available for heat integration after leaving the reactor. If the pressure of steam generation can be increased, then there may be energy or heat transfer area benefits when it is integrated with the rest of the process.

Care should be taken when preheating reactor feeds within the reactor using the heat of reaction. This is achieved in practice simply by passing the cold feeds directly to the reactor and allowing them to be preheated by mixing with hot materials within the reactor. However, if the exothermic reactor is appropriately placed above the pinch and the feeds start below the pinch, then the preheating within the reactor is cross-pinch heat transfer. In this case, feeds should be preheated by recovery using streams below the pinch before being fed to the reactor. This increases the heat generated within the reactor and heat integration will benefit from the increased heat available for recovery from the reactor.

20.5 Heat Integration of Reactors – Summary

The appropriate placement of reactors, as far as heat integration is concerned, is that exothermic reactors should be integrated above the pinch and endothermic reactors below the pinch. Care should be taken when reactor feeds are preheated by heat of reaction within the reactor for exothermic reactions. This can constitute cross-pinch heat transfer. The feeds should be preheated to pinch temperature by heat recovery before being fed to the reactor.

Appropriate placement can be assessed quantitatively using the grand composite curve. The streams associated with the reactor can be represented as a grand composite curve for the reactor and then matched against the grand composite curve for the rest of the process.

If the reactor is not appropriately placed, then it is more likely that the rest of the process would be changed to bring about appropriate placement rather than changing the reactor. If changes to the reactor design are possible, then the simple criterion of making hot streams hotter and cold streams colder can be used to bring about beneficial changes.

Table 20.1

Stream data for a process with an exothermic chemical reactor.

Stream		Enthalpy change (kW)	T_S (°C)	T_T (°C)
No.	Type			
1	Hot	7000	377	375
2	Hot	3600	376	180
3	Hot	2400	180	70
4	Cold	2400	60	160
5	Cold	200	20	130
6	Cold	200	160	260

20.6 Exercises

1. The stream data for a process involving a highly exothermic chemical reaction are given in Table 20.1.

a) Sketch the composite curves for $\Delta T_{min} = 10\,°C$, confirm that it is a threshold problem and determine the cold utility requirements.

b) It is proposed to use steam generation as cold utility. Assume saturated boiler feedwater is available and that the steam generated is saturated in order to calculate how much steam can be generated by the process at a pressure of 41 bar. The temperature of saturated steam at the pressure is 252 °C and the latent heat is $1706\,kJ \cdot kg^{-1}$.

c) If the steam is superheated to a temperature of 350 °C, calculate how much steam can be generated at 41 bar. Assume the heat capacity of steam is $4.0\,kJ \cdot kg^{-1} \cdot K^{-1}$.

d) Describe what would happen if the steam was generated from boiler feedwater at 100 °C with a heat capacity of $4.4\,kJ \cdot kg^{-1} \cdot K^{-1}$. How would the steam generation be calculated under these circumstances?

Reference

Glavic P, Kravanja Z and Homsak M (1988) Heat Integration of Reactors: I. Criteria for the Placement of Reactors into Process Flowsheet, *Chem Eng Sci*, **43**: 593.

20

Chapter 21

Heat Integration of Distillation

21.1 The Heat Integration Characteristics of Distillation

The dominant heating and cooling duties associated with a distillation column are the reboiler and condenser duties. In general, however, there will be other duties associated with heating and cooling of feed and product streams. These sensible heat duties usually will be small in comparison with the latent heat changes in reboilers and condensers.

Both the reboiling and condensing processes normally take place over a range of temperature. Practical considerations, however, usually dictate that the heat to the reboiler must be supplied at a temperature above the dew point of the vapor leaving the reboiler and that the heat removed in the condenser must be removed at a temperature lower than the bubble point of the liquid, as discussed in Chapter 19. Hence, in preliminary design at least, both reboiling and condensing can be assumed to take place at constant temperatures.

21.2 The Appropriate Placement of Distillation

Consider now the consequences of heat integrating simple distillation columns (i.e. one feed, two products, one reboiler and one condenser) in different locations relative to the heat recovery pinch. The distillation takes heat Q_{REB} into the reboiler at temperature T_{REB} and rejects heat Q_{COND} at a lower temperature T_{COND}. There are two possible ways in which the column can be heat integrated with the rest of the process. The reboiler and condenser can be integrated either across, or not across, the heat recovery pinch.

Chemical Process Design and Integration, Second Edition. Robin Smith.

Companion Website: www.wiley.com/go/smith/chemical2e

1) *Distillation across the pinch.* This arrangement is shown in Figure 21.1a. The background process (which does not include the reboiler and condenser) is represented simply as a heat sink and heat source divided by the pinch. Heat Q_{REB} is taken into the reboiler above the pinch temperature and heat Q_{COND} rejected from the condenser below the pinch temperature. Because the process sink above the pinch requires at least Q_{Hmin} to satisfy its enthalpy balance, the Q_{REB} removed by the reboiler must be compensated for by introducing an extra Q_{REB} from hot utility. Below the pinch, the process needs to reject Q_{Cmin} anyway, and an extra heat load Q_{COND} from the condenser has been introduced.

   By heat integrating the distillation column with the process and by considering only the reboiler, it might be concluded that energy has been saved. The reboiler has its heat requirements provided by heat recovery. However, the overall situation is that heat is being transferred across the heat recovery pinch through the distillation column and that the consumption of hot and cold utilities in the process must increase correspondingly. There are fundamentally no savings available from the integration of a separator across the pinch (Umeda, Niida and Shiroko, 1979; Linnhoff, Dunford and Smith, 1983).

2) *Distillation not across the pinch.* Here the situation is somewhat different. Figure 21.1b shows a distillation column entirely above the pinch. The distillation column takes heat Q_{REB} from the process and returns Q_{COND} at a temperature above the pinch. The hot utility consumption changes by $(Q_{REB} - Q_{COND})$. The cold utility consumption is unchanged. Usually, Q_{REB} and Q_{COND} have a similar magnitude. If $Q_{REB} \approx Q_{COND}$, then the hot utility consumption is Q_{Hmin} and there is no additional hot utility required to run the column. It takes a "free ride" from the process. Heat integration below the pinch is illustrated in Figure 21.1c. Now the hot utility is unchanged, but the cold utility consumption changes by $(Q_{COND} - Q_{REB})$. Again, given that Q_{REB} and Q_{COND} usually have similar magnitudes, the result is similar to heat integration above the pinch.

All these arguments can be summarized by a simple statement: the appropriate placement for distillation is not across the

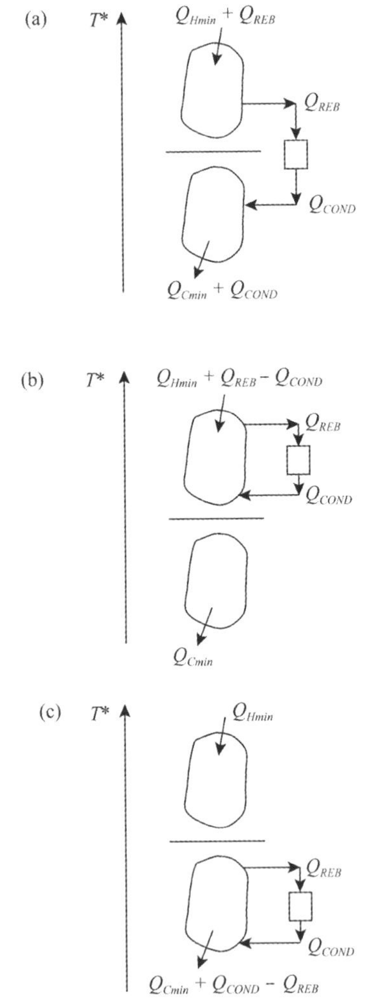

Figure 21.1

The appropriate placement of distillation columns. (Reproduced from Smith R and Linnhoff B, 1998, *Trans IChemE ChERD*, **66**: 195 by permission of the Institution of Chemical Engineers.)

pinch (Umeda, Niida and Shiroko, 1979; Linnhoff, Dunford and Smith, 1983). Although the principle has been developed with regard to distillation columns, it clearly applies to any separator that takes in heat at a higher temperature and rejects heat at a lower temperature.

If both the reboiler and condenser are integrated with the process, this can make the column difficult to start up and control. However, when the integration is considered more closely, it becomes clear that both the reboiler and condenser do not need to be integrated. Above the pinch, the reboiler can be serviced directly from the hot utility with the condenser integrated above the pinch. In this case, the overall utility consumption will be the same as that shown in Figure 21.1b. Below the pinch, the condenser can be serviced directly by cold utility with the reboiler integrated below the pinch. Now the overall utility consumption will be the same as that shown in Figure 21.1c.

21.3 Use of the Grand Composite Curve for Heat Integration of Distillation

The appropriate placement principle can only be applied if the process has the capacity to provide or accept the required heat duties. A quantitative tool is needed to assess the source and sink capacities of any given background process. For this purpose, the grand composite curve can be used. Given that the dominant heating and cooling duties associated with the distillation column are the reboiler and condenser duties, a convenient representation of the column is therefore a simple "box" representing the reboiler and condenser loads (Linnhoff, Dunford and Smith, 1983). This "box" can be matched with the grand composite representing the remainder of the process. The grand composite curve would include all heating and cooling duties for the process, including those associated with separator feed and product heating and cooling, but excluding reboiler and condenser loads.

Consider now a few examples of the use of this simple representation. A grand composite curve is shown in Figure 21.2a. The distillation column reboiler and condenser duties are shown separately and are matched against it. The reboiler and condenser duties are on opposite sides of the heat recovery pinch and the column does not fit. In Figure 21.2b, although the reboiler and condenser duties are both above the pinch, the heat duties prevent a fit. Part of the duties can be accommodated and, if heat integrated, that would be a saving, but less than the full reboiler and condenser duties.

The distillation columns shown in Figure 21.3 both fit. Figure 21.3a shows a case in which the reboiler duty can be supplied by hot utility. The condenser duty must be integrated with the rest of the process. Another example is shown in Figure 21.3b. This distillation column also fits. The reboiler duty must be supplied by integration with the process. Part of the condenser duty in Figure 21.3b must also be integrated, while the remainder of the condenser duty can be rejected to cold utility.

21.4 Evolving the Design of Simple Distillation Columns to Improve Heat Integration

If an inappropriately placed distillation column is shifted above the heat recovery pinch by changing its pressure, the condensing stream, which is a hot stream, is shifted from below to above the pinch. The reboiling stream, which is a cold stream, stays above the pinch. If the inappropriately placed distillation column is shifted below the pinch, then the reboiling stream, which is a cold stream, is shifted from above to below the pinch. The condensing stream stays below the pinch. Thus appropriate placement is a particular case of shifting streams across the pinch, which in turn is a particular case of the plus–minus principle (see Chapter 19).

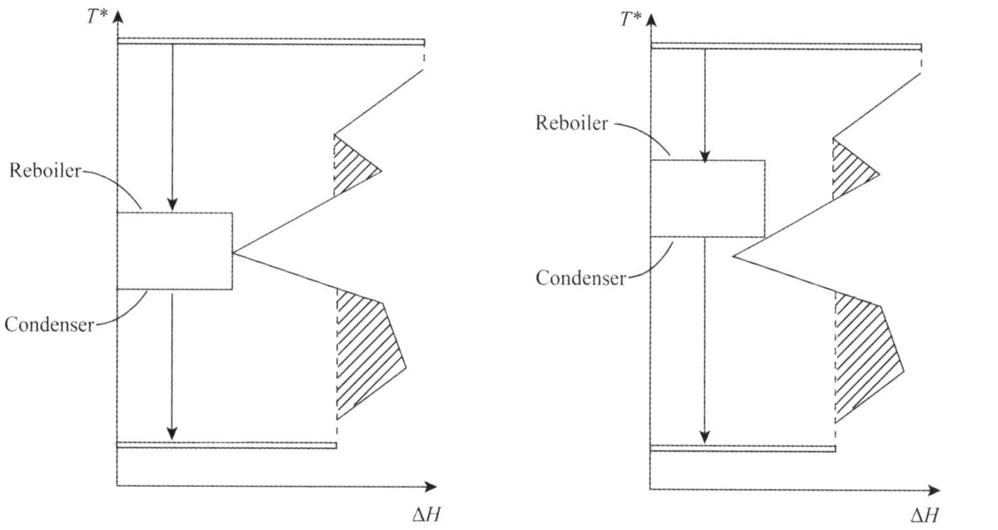

(a) An inappropriately placed distillation column.

(b) A distillation column that can only be partially integrated.

Figure 21.2

Distillation columns that do not fit against the grand composite curve. (Reproduced from Smith R and Linnhoff B, 1998, *Trans IChemE ChERD*, **66**: 195 by permission of the Institution of Chemical Engineers.)

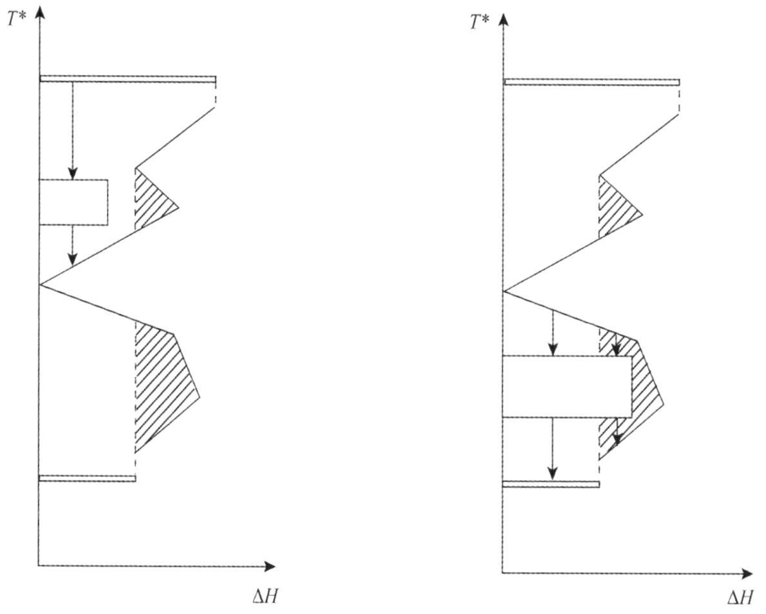

(a) A column appropriately placed above the pinch.

(b) A column appropriately placed below the pinch.

Figure 21.3

Distillation cloumns that fit against the grand composite curve. (Reproduced from Smith R and Linnhoff B, 1998, *Trans IChemE ChERD*, **66**: 195 by permission of the Institution of Chemical Engineers.)

If a distillation column is inappropriately placed across the pinch, it may be possible to change its pressure to achieve appropriate placement. Of course, as the pressure is changed, the shape of the "box" also changes, since not only do the reboiler and condenser temperatures change but also the difference between them. The relative volatility will also be affected, generally decreasing with increasing pressure. Thus, both the height and the width of the box will change as the pressure changes. Changes in pressure also affect the heating and cooling duties for column feed and products. These streams normally would be included in the background process. Hence, the shape of the grand composite curve will also change to some extent as the column pressure changes. However, as pointed out previously, it is likely that these effects will not be significant in most processes by comparison with the latent heat changes in condensers and reboilers, since the sensible heat loads involved will usually be small by comparison.

Constraints will often present themselves over which the pressure of the distillation columns will operate. For example, it

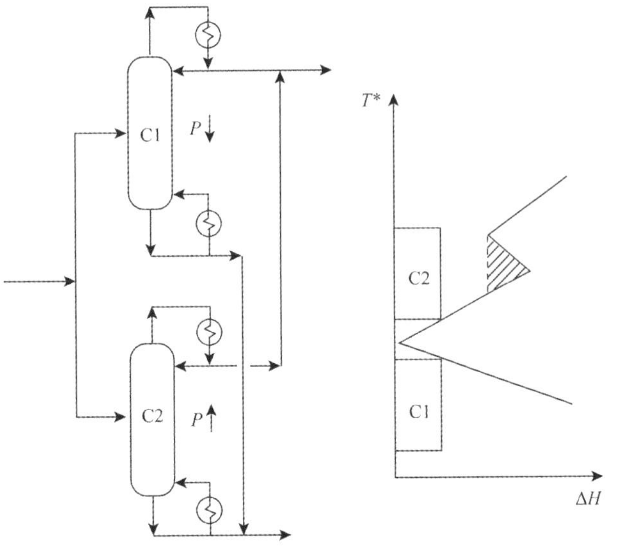

(a) Conventional double effect distillation.

(b) Integration of double effect columns separately.

Figure 21.4

Double-effect distillation. (Reproduced from Smith R and Linnhoff B, 1998, *Trans IChemE ChERD*, **66**: 195 by permission of the Institution of Chemical Engineers.)

is often the case that the maximum pressure of a distillation column is restricted to avoid decomposition of material in the reboiler. This is especially the case when reboiling high molar mass material. Distillation of high molar mass material is often constrained to operate under vacuum conditions. However, unless absolutely necessary, vacuum distillation is best avoided. Clearly, if the pressure of the distillation column is constrained, then this restricts the heat integration opportunities.

If the distillation column will not fit either above or below the pinch, either because of the heat duties or constraints, then other design options can be considered. One possibility is *double-effect distillation*, as shown in Figure 21.4a (Smith and Linnhoff, 1988). The column feed is split and fed to two separate parallel columns. The classical application of double-effect distillation is to choose the relative pressures of the columns such that the heat from the condenser of the high-pressure column can be used to provide the reboiler heat to the low-pressure column. In isolation, the scheme would save energy, approximately halving the energy consumption by using the same energy twice in different temperature ranges. The energy reduction will be at the expense of increased capital cost. However, used on a stand-alone basis in this way, in reducing the heat load on the system, the temperature difference over the distillation system increases. If an attempt is made to heat integrate this double-effect distillation with the rest of the process, the increased temperature difference across the system might create problems. The increased temperature difference might prevent integration above the pinch (perhaps because the required high temperature in the reboiler creates fouling) or below the pinch (perhaps because the required low temperature in the condenser would require expensive refrigeration).

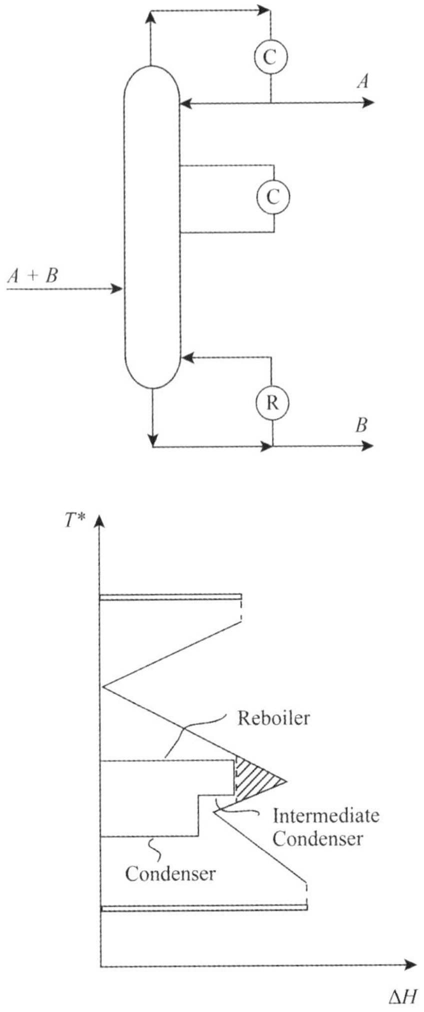

Figure 21.5

Distillation column with intermediate condenser. The profile can be designed to fit the background process. (Reproduced from Smith R and Linnhoff B, 1998, *Trans IChemE ChERD*, **66**: 195 by permission of the Institution of Chemical Engineers.)

However, there is no fundamental reason why these two columns must be linked together thermally. Figure 21.4b shows two columns that are not linked thermally and, as a result, each can be individually appropriately placed. Obviously, the capital cost of such a scheme will be higher than that of a single column, but it may be justified by favorable energy savings.

Another design option that can be considered if a column will not fit into the grand composite curve is the use of an intermediate condenser, as illustrated in Figure 21.5. The shape of the "box" is now altered, because the intermediate condenser changes the heat flow through the column with some of the heat being rejected at a higher temperature in the intermediate condenser. In the particular design shown in Figure 21.5, the match with the grand composite curve would require that at least part of the heat rejected from the intermediate condenser should be passed to the process. An analogous approach can be used to evaluate the possibilities for the use of intermediate reboilers. For intermediate reboilers, part

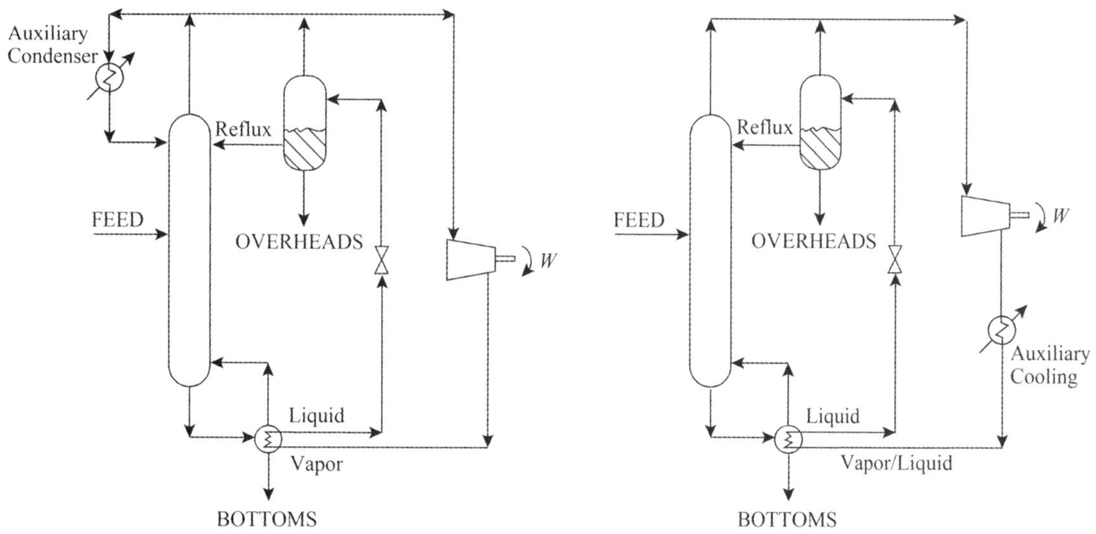

(a) A typical vapor recompression scheme for above ambient temperature distillation.

(b) A typical vapor recompression scheme for below ambient temperature distillation.

Figure 21.6
Heat pumping in distillation. Vapour recompression schemes.

of the reboiler heat is supplied at an intermediate point in the column, at a temperature lower than the reboiler temperature. Flower and Jackson (1964), Kayihan (1980) and Soares-Pinto *et al.* (2011) have presented procedures for the location of intermediate reboilers and condensers.

21.5 Heat Pumping in Distillation

Various heat-pumping schemes have been proposed as a means of saving energy in distillation. One approach is to use a heat pumping fluid in a *closed loop*, like the schemes discussed in Chapter 17. An alternative in distillation is to use the column overhead vapor as the heat pumping fluid in an *open loop*. Such open-loop schemes are known as *vapor recompression*. Many different arrangements are possible. Two possible schemes are illustrated in Figure 21.6. Figure 21.6a shows a scheme typical for above ambient temperature distillation. Some auxiliary cooling is normally required to satisfy the overall energy balance. Figure 21.6a features an auxiliary condenser. If the feed is significantly subcooled, some auxiliary reboiling might be required. Figure 21.6b shows a scheme typical for below ambient processes. If an auxiliary cooler is required, it is better to supply the cooling at a higher temperature to avoid the use of refrigeration. Compared with Figure 21.6a, the scheme in Figure 21.6b requires auxiliary cooling at a higher temperature. Again, if the feed is significantly subcooled, some auxiliary reboiling might be required.

Open-loop (vapor recompression) schemes have an advantage over closed-loop systems in terms of the temperature lift required. Closed-loop systems require indirect heat transfer in two heat exchanger operations for the temperature lift, each requiring a finite temperature difference. Open-loop systems require indirect heat transfer in one heat exchange operation for the temperature lift. Thus, the temperature lift, and hence power requirements, are generally smaller for open-loop systems compared with closed-loop systems. However, in closed-loop systems the heat pumping fluid can be chosen from a range of fluids to keep the power requirements potentially lower than for the same temperature lift using the distillation overhead vapor as the working fluid.

For heat pumping to be economic on a stand-alone basis across the distillation column, it must operate across a small temperature difference, which for distillation means close boiling mixtures. In addition, unless the column is constrained to operate either on a stand-alone basis or at a pressure that would mean it would be across the pinch, heat integration is likely to be preferred. Heat pumping schemes for distillation are most likely to be attractive for the distillation of close boiling mixtures in constrained situations (Smith and Linnhoff, 1988).

21.6 Capital Cost Considerations for the Integration of Distillation

The design changes suggested so far for distillation columns have been motivated by the incentive to reduce energy costs by more effective integration between the distillation column and the rest of the process. However, there are capital cost implications when the distillation design and the heat integration scheme are changed. These implications fall into two broad categories: changes in distillation capital cost and changes in heat exchanger network capital cost. Obviously, these capital cost changes should be

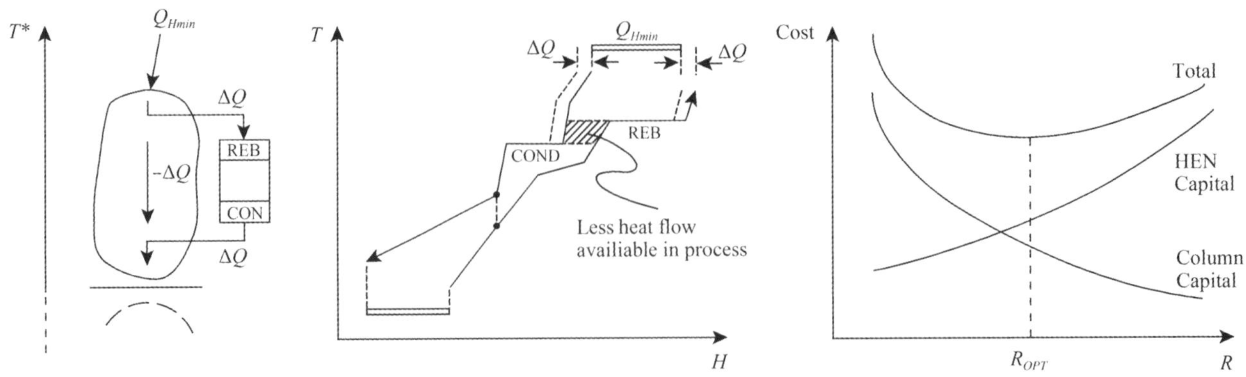

Figure 21.7

The capital–capital trade-off for an appropriately integrated distillation column. (Reproduced from Smith R and Linnhoff B, 1998, *Trans IChemE ChERD*, **66**: 195 by permission of the Institution of Chemical Engineers.)

considered together, along with the energy cost changes, in order to achieve an optimum trade-off between capital and energy costs.

1) *Distillation capital costs.* The classical optimization in distillation, as discussed in Chapter 8, is to trade off the capital cost of the column against the energy cost for the distillation by changing the reflux ratio. In Chapter 8, this was discussed for the situation of distillation columns operating on utilities and not integrated with the rest of the process. Experience gained with this traditional optimization has led to rules of thumb for the selection of reflux ratios. Typically, the optimum ratio of actual to minimum reflux ratio is usually around 1.1 or lower. Practical considerations often prevent a ratio of less than 1.1 being used except in special circumstances, as discussed in Chapter 8.

   If the column is inappropriately placed with the process, then an increase in reflux ratio causes a corresponding overall increase in energy and the trade-off rules apply. However, if the column is appropriately placed, then the reflux ratio can often be increased (up to a limit) without changing the overall energy consumption, as shown in Figure 21.7. Increasing the heat flow through the column decreases the requirement for distillation stages but increases the vapor rate. In designs initialized by traditional rules of thumb, this would have the effect of decreasing the capital cost of the column. However, the corresponding decrease in heat flow through the heat exchanger network will have the effect of decreasing temperature differences and increasing the capital cost of the heat exchanger network, as shown in Figure 21.7. Thus, the trade-off for an appropriately integrated distillation column becomes one between the capital cost of the column and the capital cost of the heat exchanger network (Smith and Linnhoff, 1988) (Figure 21.7).

   Consequently, the optimum reflux ratio for an appropriately integrated distillation column will be problem specific and is likely to be quite different from that of a stand-alone column operated from utilities.

2) *Heat exchanger network capital costs.* It is easy for the designer to become carried away with the elegance of "packing boxes" into space around the grand composite curve. However, the full implications of integration are only clear when the corresponding composite curves of the process with the distillation column are considered. Temperature differences become smaller throughout the process as a result of the integration. This means that the capital–energy trade-off should be readjusted, and a larger ΔT_{min} might be required. The optimization of the capital–energy trade-off might undo part of the savings achieved by appropriate integration.

   Unfortunately, the overall design problem is even more complex in practice. Large temperature differences in the process (i.e. space in the grand composite curve) could equally well be exploited to allow the use of moderate temperature utilities or the integration of heat engines, heat pumps, and so on, in preference to integration of distillation columns. There is thus a three-way trade-off between distillation design and integration, utility selection and the capital–energy trade-off (ΔT_{min} optimization).

21.7 Heat Integration Characteristics of Distillation Sequences

The problem of distillation sequencing was discussed in Chapter 10, where the distillation columns in the sequence were operated on a stand-alone basis using utilities for the reboilers and condensers. Following the approach in Chapter 10, the best few nonintegrated distillation sequences can be found. These sequences can then be heat integrated as discussed above. Figure 21.8 shows how heat integration can be applied within a two-column direct distillation sequence for the separation of three products. In Figure 21.8a, the first distillation column has been increased in pressure such that the condenser of the first column can provide the heat for the reboiler of the second column, sometimes known as *forward integration*. In Figure 21.8b, the pressure of the second column has been increased such that the condenser of the second column can provide the heat for the reboiler of the first,

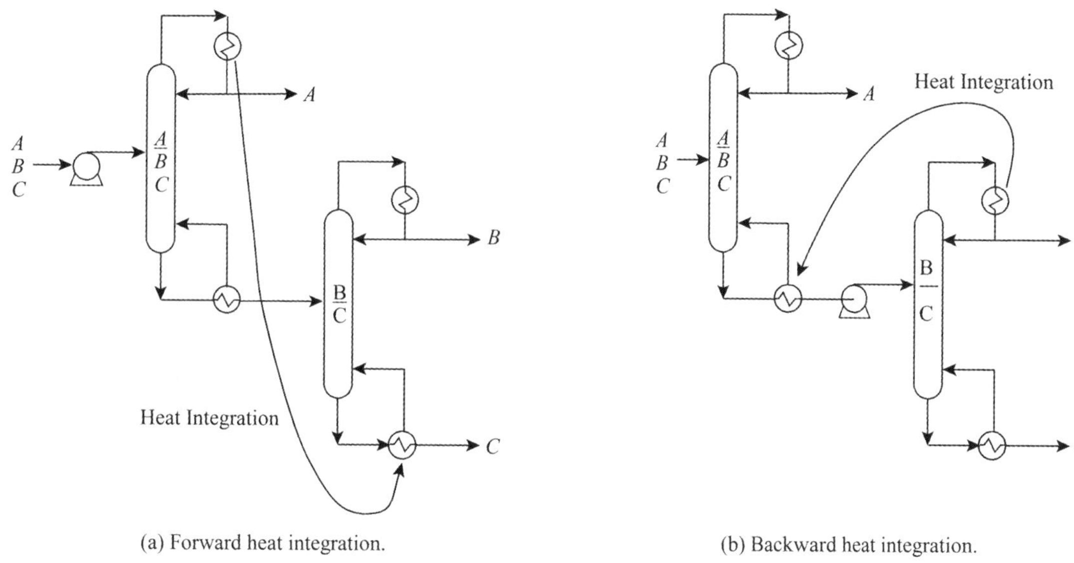

Figure 21.8

Heat integration of a sequence of two simple distillation columns.

sometimes known as *backward integration*. Both schemes will bring about a significant reduction in the energy requirement. However, these are generally five heating/cooling duties associated with a distillation column:

- feed preheating/cooling,
- cooling in the condenser,
- heating in the reboiler,
- overhead product heating/cooling,
- bottoms product heating/cooling.

Each of these duties in the distillation can be generally satisfied in one of the following ways:

- hot or cold utilities,
- heat integration with reboiling/condensing duties of other distillation columns,
- heat integration with background process heating and cooling utilities,
- heat pumping around the distillation using a closed loop,
- heat pumping around the distillation using an open loop,
- heat pumping from other distillation column condenser or process cooling duties using a closed loop,
- heat pumping from other distillation column condenser duties using an open loop,
- heat pumping from the distillation condenser duty to other distillation or process heating duties using a closed loop,
- heat pumping from the distillation condenser duty to other distillation or process heating duties using an open loop.

In addition, as will be discussed in Chapter 24, for low-temperature processes requiring complex refrigeration systems, energy integration can also be carried out with the refrigeration system. Further opportunities are presented by introducing intermediate reboiling and condensing to the distillation. One approach to the overall problem breaks down the design procedure into two steps of first determining the best nonintegrated sequence and then heat integration. This assumes that the two problems of distillation sequencing and heat integration can be decoupled.

However, there are complex interactions between the design of the distillation arrangement and the heat integration. Not only can different distillation sequences of simple columns be chosen, but the pressure of the columns can be varied between practical constraints. As the pressures of the columns are varied, the energy integration options discussed above vary. Another factor that can create problems as the pressure of each column in the sequence is varied, is the effect on the feed condition for downstream columns. If two columns are at the same pressure, then a saturated liquid leaving one column will be saturated as it enters the second column. However, if the pressure of the first column becomes higher than the second as a result of a pressure change, then the saturated liquid from the first column will become a subcooled feed to the second. If the pressure of the first column becomes lower than the second as a result of a pressure change, then the saturated liquid from the first column will become a partially vaporized or superheated feed to the second. The feed condition can have a significant influence on the column design.

Constraints might be applied for the sake of reducing the capital costs (e.g. to avoid long pipe runs). In addition, constraints might be applied to avoid complex heat integration arrangements for the sake of operability and control (e.g. to have heat recovery to a reboiler from a single source of heat, rather than two or three sources of heat).

In addition to these issues regarding simple columns, there are also issues associated with the introduction of complex columns into the sequence. Figure 21.9a illustrates the thermal characteristics of a direct sequence of two simple columns. Once the two columns are thermally coupled, as illustrated in Figure 21.9b, the

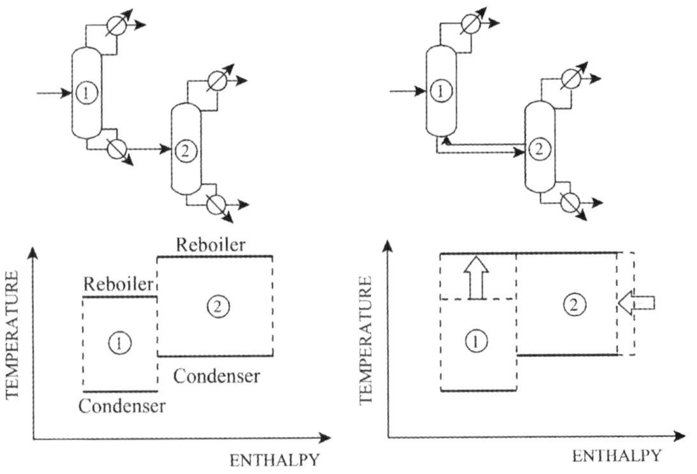

(a) The direct sequence. (b) Thermally coupled direct sequence.

Figure 21.9

Thermally coupling the direct sequence changes both the loads and levels.

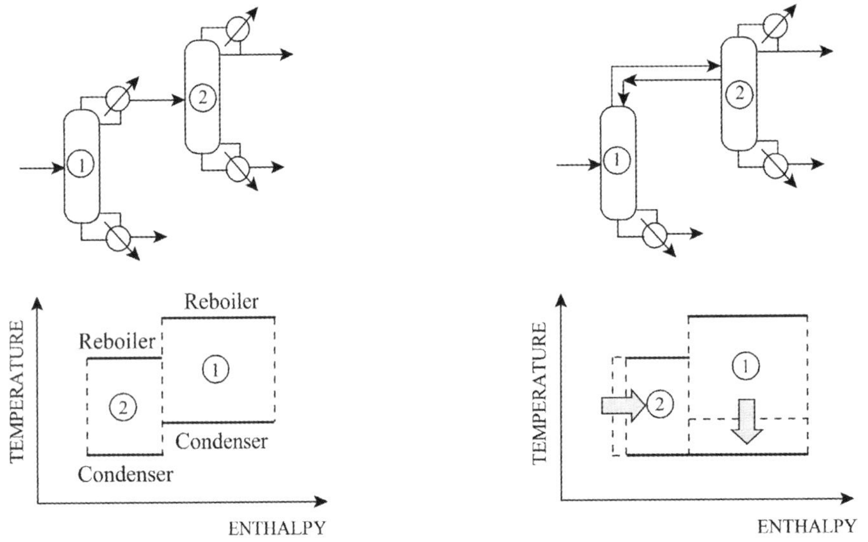

(a) The indirect sequence. (b) Thermally coupled indirect sequence.

Figure 21.10

Thermally coupling the indirect sequence changes both the loads and levels.

overall heat load is reduced. However, all of the heat must be supplied at the highest temperature for the system. Thus there is a trade-off in which the load is reduced, but the levels required to supply the heat become more extreme. The corresponding case for the indirect sequence is shown in Figure 21.10. As the indirect sequence is thermally coupled, the heat load is reduced, but now all of the heat must be rejected at the lowest temperature. Thus, there is a benefit of reduced load but a disadvantage of heat rejection at more extreme levels. The same problem occurs with thermally coupled prefractionators and partition (dividing wall) distillation columns. However, this time both heat supply and rejection must be carried out at the most extreme temperatures.

All of these arguments demand a more sophisticated approach than a sequential approach in which the sequence and heat integration (including complex columns) are explored simultaneously.

Example 21.1 Two distillation columns have been sequenced to be in the direct sequence (see Figure 21.8). Opportunities for heat integration between the two columns are to be explored. The operating pressures of the two columns need to be chosen to allow heat recovery. Data for Column 1 and Column 2 at various pressures are given in Tables 21.1 and 21.2.

Medium-pressure (MP) steam is available for reboiler heating at 200 °C. Cooling water is available for condensation, to be returned to the cooling tower at 30 °C. Assume a minimum permissible temperature difference for heat transfer of 10 °C. Determine the minimum utility requirements for:

a) both columns operating at 1 bar,
b) forward heat integration,
c) backward heat integration.

Table 21.1
Data for Column 1.

P (bar)	T_{COND} (°C)	T_{REB} (°C)	Q_{COND} (kW)	Q_{REB} (kW)
1	90	120	3000	3000
2	130	160	3600	3600
3	140	170	4000	4000
4	160	190	4300	4300

Solution

a) Heat integration between the condenser and reboiler is feasible if

$$T_{COND} \geq T_{REB} + \Delta T_{min}$$

From Tables 20.1 and 20.2, when both columns operate at 1 bar, no heat integration is possible. Thus,

$$\begin{aligned} Q_{Hmin} &= 3000 + 5500 \\ &= 8500\,\text{kW} \\ Q_{Cmin} &= 3000 + 5500 \\ &= 8500\,\text{kW} \end{aligned}$$

b) Consider forward heat integration by increasing the pressure of Column 1. Because heat duties will increase with increasing pressure, low operating pressures are preferred. Thus, Column 2 is kept at 1 bar with a reboiler temperature of 130 °C. This means that the minimum condensing temperature of Column 1 must be 140 °C, which corresponds with a pressure of 3 bar. From Tables 20.1 and 20.2:

$$\begin{aligned} Q_{Hmin} &= 4000 + (5500 - 4000) \\ &= 5500\,\text{kW} \\ Q_{Cmin} &= 0 + 5500 \\ &= 5500\,\text{kW} \end{aligned}$$

c) For backward heat integration, the appropriate operating pressures are 1 bar for Column 1 and 2 bar for Column 2.

Table 21.2
Data for Column 2.

P (bar)	T_{COND} (°C)	T_{REB} (°C)	Q_{COND} (kW)	Q_{REB} (kW)
1	110	130	5500	5500
2	130	153	6000	6000
3	150	175	6300	6300
4	163	190	6500	6500
5	170	200	6600	6600

Thus, from Tables 20.1 and 20.2:

$$\begin{aligned} Q_{Hmin} &= 0 + 6000 \\ &= 6000\,\text{kW} \\ Q_{Cmin} &= 3000 + (6000 - 3000) \\ &= 6000\,\text{kW} \end{aligned}$$

Thus, to minimize utility costs, forward heat integration should be used with Column 1 at 3 bar and Column 2 at 1 bar.

21.8 Design of Heat Integrated Distillation Sequences

In order to explore distillation sequencing and heat integration simultaneously, a more automated approach to the problem is required. Chapter 10 discussed how distillation sequences can be developed automatically. Figure 10.19 illustrates how all options for the separation of the mixture of five components using simple columns can be embedded in a superstructure that can be optimized to remove redundant features. The optimization involves not only the distillation configuration but also the operating pressure, reflux ratio and feed condition and choice of a partial or total condenser for each column. It should be again noted that there are interactions between columns in series as far as the feed condition is concerned. Figure 10.20 then shows that any two columns connected in series can in principle be replaced by different complex column arrangements. The appropriate complex column arrangement options depend on whether the two simple columns in series to be replaced are in the direct or indirect sequence. In Chapter 10 it was discussed how this structure can be subjected to structural optimization using stochastic search optimization (e.g. simulated annealing algorithm, see Chapter 3). Stochastic search optimization starts with any feasible distillation sequence (e.g. one of the simple column sequences). The design is then evolved in the optimization through a series of structural moves, as discussed in Chapter 10, to improve the design. In addition to the structural moves, changes in column pressure, reflux ratio, feed condition and condenser type of all columns need to be explored. After each move the design needs to be simulated and evaluated.

Also as discussed in Chapter 10, rather than using stochastic search optimization and moves from an initial design, a comprehensive superstructure can be developed by embedding all possible combinations of simple and complex columns into a single superstructure and optimized using deterministic optimization (MINLP).

Whichever approach is used, the approach so far assumes that all the heating and cooling will be satisfied by utilities. For most problems heat integration will have a significant effect on the final design, and for the reasons discussed above, the optimization of the distillation structure and heat integration need to be carried out simultaneously. The simplest approach, which applies to new design, is to use energy targeting from the problem table by assuming a ΔT_{min}. In this approach, whenever a structural change is made to the distillation and the design simulated and evaluated, energy costs are included from energy targeting. This is straightforward if only a single hot and cold utility is involved. If multiple utilities are involved, the energy targeting needs to optimize the utility mix for each distillation structural move when it is evaluated. A linear program can be used to optimize a mix of utilities matched against the grand composite curve. However, rather than use the energy targeting methods described in Chapter 17, linear programming can be used to provide a heat exchanger network design, as long as a ΔT_{min} is fixed. This network design can then be used to evaluate each change in the structure of the distillation system. Whichever method is used to provide the heat integration performance, the heat exchanger network evaluation is carried out as a subproblem to the basic distillation structural optimization.

If the heat integrated distillation problem is a retrofit, then an evolutionary move can be adopted from the existing distillation structure. This can be carried out using stochastic search optimization starting from the existing distillation structure and evolving in a series of moves from the existing structure. In comparison with the new design problem, retrofit cannot use energy targeting to adequately represent the retrofit potential for heat integration. For retrofit, the heat exchanger network should be evolved from the existing heat exchanger network structure using the methods described in Chapter 18. Each change in the distillation structure or operating conditions creates a new heat integration problem. This new heat exchanger network problem needs to be passed to a retrofit algorithm, as described in Chapter 18, to retrofit the existing heat exchanger network to the new conditions. Most retrofit problems will only be economic for a small number of equipment changes. Thus, it will normally be necessary to constrain the number of changes allowed in the retrofit optimization.

21.9 Heat Integration of Distillation – Summary

The appropriate placement of distillation columns when heat integrated is not across the heat recovery pinch. The grand composite curve can be used as a quantitative tool to assess integration opportunities. The scope for integrating conventional distillation columns into an overall process is often limited. Practical constraints often prevent integration of columns with the rest of the process. If the column cannot be integrated with the rest of the process, or if the potential for integration is limited by the heat flows in the background process, then attention must be turned back to the distillation operation itself and complex arrangements considered.

The use of complex columns (side-strippers, side-rectifiers and thermally coupled prefractionators) reduces the overall heat duties for the separation at the expense of more extreme temperatures for reboiling and condensing. Heat integration benefits from smaller duties, but more extreme temperatures make the heat integration more difficult).

Thus, the introduction of constraints and complex columns demands a simultaneous solution of the sequencing and heat recovery problems. This can be carried out on the basis of the deterministic optimization of a superstructure or stochastic search optimization.

21.10 Exercises

1. Table 21.3 represents a problem table cascade ($\Delta T_{min} = 10\,°C$). The utilities available are given in Table 21.4. The power required by refrigeration is given by

$$W = \frac{Q_C}{0.6}\left(\frac{T_H - T_C}{T_C}\right)$$

where W = power required for the refrigeration cycle
Q_C = the cooling duty
T_C = temperature at which heat is taken into the refrigeration cycle (K)
T_H = temperature at which heat is rejected from the refrigeration cycle (K)

Assume that only a single level of refrigeration can be used and heat rejection from refrigeration is to cooling water.

Table 21.3

Heat flow cascade for Exercise 1.

Interval temperature (°C)	Heat flow (kW)
160	1000
150	0
130	1100
110	1400
100	900
80	1300
40	1400
10	1800
–10	1900
–30	2200

Table 21.4

Utility data for Exercise 1.

Utility	Cost
Steam at 180 °C	$\$135\,kW^{-1}\cdot y^{-1}$
Cooling water from 20 °C to 40 °C	$\$4.5\,kW^{-1}\cdot y^{-1}$
Electricity	$\$620\,kW^{-1}\cdot y^{-1}$

The process also has a distillation column presently using steam and cooling water. The reboiler is at 120 °C, the condenser is at 90 °C and each has a duty of 1400 kW.

a) Calculate the utility costs without column integration.

b) Calculate the utility costs by integrating the column at the current pressure.

c) It is proposed to integrate the column by decreasing the column pressure. It can be assumed that the temperature difference across the column and the reboiler and condenser duties remain fixed as the column pressure is changed. Calculate the utility cost of an integrated column.

d) Determine the most appropriate reboiler and condenser temperatures to integrate the column. Calculate the utility cost.

e) What could be done to decrease the refrigeration cost?

2. A problem table analysis for a given process produces the heat flow cascade in Table 21.5 for $\Delta T_{min} = 10$ °C.

a) A distillation column, which separates a mixture of toluene and diphenyl into relatively pure products, is to be integrated with the process. The operating pressure of the column has been fixed initially to atmospheric (1.013 bara). At this pressure, the toluene condenses overhead at a constant temperature of 111 °C and diphenyl is reboiled at a constant temperature of 255 °C. What would be the consequence of integrating the distillation column with the process at a pressure of 1.013 bara?

b) Can you suggest a more appropriate operating pressure for the distillation column if it is to be integrated with the process? The reboiler and condenser loads are both 4.0 MW and can be assumed not to change significantly with pressure. The vapor pressures of toluene and diphenyl can be represented by:

$$\ln P_i = A_i - \frac{B_i}{T + C_i}$$

Table 21.5

Problem table cascade for Exercise 2.

Interval temperature (°C)	Cascade heat flow (MW)
295	18.3
285	19.8
185	4.8
145	0
85	10.8
45	12.0
35	14.3

Table 21.6

Vapor pressure constants.

Component	A_i	B_i	C_i
Toluene	9.3935	3096.52	−53.67
Diphenyl	10.0630	4602.23	−70.42

where P_i is the vapor pressure (bar), T is the absolute temperature (K) and A_i, B_i and C_i are constants, which are given in Table 21.6. Vacuum operation should be avoided and the reboiler temperature kept as low as possible to minimize fouling.

c) Sketch the shape of the grand composite curve after the distillation column has been integrated.

3. A direct sequence of two distillation columns is to separate a mixture at its bubble point into three products *A*, *B* and *C*. Steam is available at 200 °C and cooling water at 30 °C. The minimum temperature difference for heat transfer is 10 °C. Data for the two columns are given in Tables 21.7 and 21.8.

Assuming the feeds to both columns are both saturated liquid, calculate the minimum utility requirement for:

a) heat integration when both columns operate at 1 bar,

b) forward heat integration by selecting appropriate operating pressures,

c) backward heat integration by selecting appropriate operating pressures.

Table 21.7

Exercise 3 data for Column 1.

P (bar)	T_{COND} (°C)	T_{REB} (°C)	Q_{COND} (kW)	Q_{REB} (kW)
1	90	120	3000	3000
2	130	160	3600	3600
3	140	170	4000	4000
4	160	190	4300	4300

Table 21.8

Exercise 3 data for Column 2.

P (bar)	T_{COND} (°C)	T_{REB} (°C)	Q_{COND} (kW)	Q_{REB} (kW)
1	110	130	5500	5500
2	130	153	6000	6000
3	150	175	6300	6300
4	163	190	6500	6500
5	170	200	6600	6600

21

Table 21.9

Exercise 4 data for Column 1.

P (bar)	T_{COND} (°C)	T_{REB} (°C)	Saturated liquid feed		Saturated vapor feed	
			Q_{COND} (kW)	Q_{REB} (kW)	T_{FEED} (°C)	Q_{FEED} (kW)
1	90	130	3000	3000	110	2000
2	110	152	4000	4000	130	1800
3	130	173	5000	5000	150	1600
4	150	195	6000	6000	170	1500

Table 21.10

Exercise 4 data for Column 2.

P (bar)	T_{COND} (°C)	T_{REB} (°C)	Saturated liquid feed		Saturated vapor feed	
			Q_{COND} (kW)	Q_{REB} (kW)	T_{FEED} (°C)	Q_{FEED} (kW)
1	120	140	3000	3000	130	1500
2	140	165	4000	4000	150	1300
2.5	150	178	4500	4500	160	1200

4. A direct sequence of two distillation columns produces three products A, B and C. The feed condition and operating pressures are to be chosen to maximize heat recovery opportunities. To simplify the calculations, assume that condenser duties do not change when changing from saturated liquid to saturated vapor feed. This will not be true in practice, but simplifies the exercise. Assume also that the reboiler duty for saturated liquid feed is the sum of the reboiler duty for saturated vapor feed plus the heat duty to vaporize the feed. Data for the two columns are given in Tables 21.9 and 21.10.

Cooling water is available with a return temperature of 30 °C and a cost of \$4.5 $kW^{-1}{\cdot}y^{-1}$. Low-pressure steam is available at a temperature of 140 °C and a cost of \$90 $kW^{-1}{\cdot}y^{-1}$. Medium-pressure steam is available at a temperature of 200 °C and a cost of \$135 $kW^{-1}{\cdot}y^{-1}$. The minimum temperature difference allowed is 10 °C.

a) List the possible heat integration opportunities (include steam generation opportunities and feed heating opportunities).

b) Calculate the minimum cost for backward heat integration by optimizing the column pressures for saturated liquid feeds.

c) Calculate the minimum cost for backward heat integration by optimizing the column pressures, but disallow heat recovery between condensers and reboilers. Keep both feeds to be saturated liquids.

d) If the feeds to both columns are saturated vapor, calculate the minimum utility cost, but disallow heat recovery between condensers and reboilers. Use the pressures from Part b above.

e) Calculate the minimum cost for saturated vapor feeds but keeping the column pressures to 1 bar.

5. Consider the use of a side-rectifier or side-stripper for the separation of a three-product mixture. Assume that thermally coupled columns operate at the same pressure. Also, assume the feed to be saturated liquid. Data for the operation of the two arrangements are given in Tables 21.11 and 21.12.

Table 21.11

Side-rectifier data for Exercise 5.

P (bar)	T_{COND1} (°C)	T_{REB1} (°C)	Q_{COND1} (kW)	Q_{REB1} (kW)	T_{COND2} (°C)	Q_{COND2} (kW)
1	90	140	2000	4500	120	2500
2	110	165	2500	5500	140	3000
2.5	120	178	3000	6500	150	3500

Table 21.12

Side-stripper data for Exercise 5.

P (bar)	T_{COND1} (°C)	T_{REB1} (°C)	Q_{COND1} (kW)	Q_{REB1} (kW)	T_{REB2} (°C)	Q_{REB2} (kW)
1	90	140	5000	3000	120	2500
2	110	165	6000	3500	140	2500
2.5	120	178	7000	4000	150	3000

Using the utility data from Exercise 4:

a) When both columns operate at 1 bar, which of the two complex column arrangements will have lower utility costs?

b) Compare the results with the direct sequence of heat-integrated simple columns operating at 1 bar.

c) Optimize column pressure in both complex column arrangements to minimize utility costs.

References

Flower JR and Jackson MA (1964) Energy Requirements in the Separation of Mixture by Distillation, *Trans IChemE*, **42**: T249.

Kayihan F (1980) Optimum Distribution of Heat Load in Distillation Columns Using Intermediate Condensers and Reboilers, *AIChE Symp Ser*, **192**(76): 1.

Linnhoff B, Dunford H and Smith R (1983) Heat Integration of Distillation Columns into Overall Processes, *Chem Eng Sci*, **38**: 1175.

Smith R and Linnhoff B (1988) The Design of Separators in the Context of Overall Processes, *Trans IChemE ChERD*, **66**: 195.

Soares-Pinto F, Zemp R, Jobson M and Smith R (2011) Thermodynamic Optimization of Distillation Columns, *Chem Eng Sci*, **66**: 2920.

Umeda T, Niida K and Shiroko K (1979) A Thermodynamic Approach to Heat Integration in Distillation Systems, *AIChE J*, **25**: 423.

Chapter 22

Heat Integration of Evaporators and Dryers

22.1 The Heat Integration Characteristics of Evaporators

Evaporation processes usually separate a single component (typically water) from a nonvolatile material, as discussed in Chapter 9. As such, it is good enough in most cases to assume that the vaporization and condensation processes take place at constant temperatures.

As with distillation, the dominant heating and cooling duties associated with an evaporator are the vaporization and condensation duties. As with distillation, there will be other duties associated with the evaporator for heating or cooling of feed, product and condensate streams. These sensible heat duties will usually be small in comparison with the latent heat changes.

Figure 22.1a shows a single-stage evaporator represented on both actual and shifted temperature scales. Note that in the shifted temperature scale, the evaporation and condensation duties are shown at different temperatures even though they are at the same actual temperature. Figure 22.1b shows a similar plot for a three-stage evaporator.

Like distillation, evaporation can be represented as a box. This again assumes that any heating or cooling required by the feed and concentrate will be included with the other process streams in the grand composite curve. However, with evaporation, the temperature difference across the box can be manipulated by varying the heat transfer area. Increasing the heat transfer area between stages allows a smaller temperature difference between stages and hence a smaller overall temperature difference, and vice versa.

Chemical Process Design and Integration, Second Edition. Robin Smith.

Companion Website: www.wiley.com/go/smith/chemical2e

22.2 Appropriate Placement of Evaporators

The concept of the appropriate placement of distillation columns was developed in the preceding chapter. The principle also applies to evaporators. The heat integration characteristics of distillation columns and evaporators are very similar. Thus, evaporator placement should not be across the pinch (Smith and Linnhoff, 1988).

22.3 Evolving Evaporator Design to Improve Heat Integration

The thermodynamic profile of an evaporator can also be manipulated. The approach is similar to that used for distillation columns. The degrees of freedom are obviously different (Smith and Linnhoff, 1988; Smith and Jones, 1990).

Consider the three-stage evaporator against a background process, as shown in Figure 22.2a. At the chosen pressure, the evaporator will not fit against the grand composite curve. The most obvious possibility is to first try an increase in pressure to allow appropriate placement above the pinch (Figure 22.2b).

Now suppose that the required increase in pressure in Figure 22.2b would cause unacceptably high levels of decomposition and fouling in the evaporator as a result of the increase in temperature. The possibility of increasing the number of stages from three to, say, six could now be considered in order to allow a fit to the grand composite curve above the pinch (Figure 22.3a). The evaporator fits, but there still might be a problem of product degradation because of high temperatures. However, it is not necessary for all evaporator stages to be linked thermally with each other. Instead, Figure 22.3b shows a six-stage system with three effects appropriately placed above the pinch and three below. This could be either a conventional six-stage system in which the first three and last three stages are not linked thermally or,

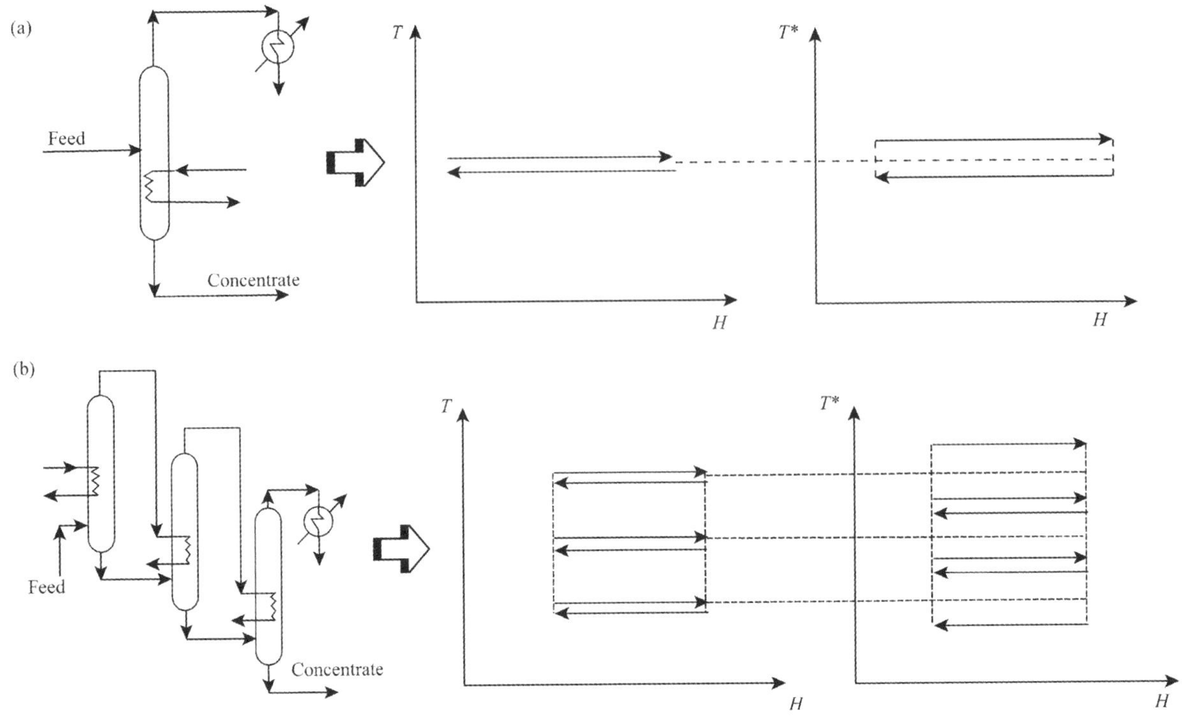

Figure 22.1

The representation of evaporators in shifted temperatures. (Reproduced from Smith R and Jones P.S (1990) The Optimal Design of Integrated Evaporation Systems, *J Heat Recovery Syst CHP*, **10**: 341, with permission from Elsevier.)

alternatively, two parallel three-stage systems, analogous to double effecting in distillation.

Yet another design option is shown in Figure 22.3c in which the heat flow (and hence mass flow) is changed between stages in the evaporator. Figure 22.3c shows an arrangement in which part of the vapor from the second stage is used for process heating rather than evaporation in the third stage. This means that more evaporation is taking place in the first two stages than the third and subsequent stages. It should be noted that even if the heat flow through the multistage evaporator is constant, the rate of evaporation will decrease because the latent heat increases as the pressure decreases.

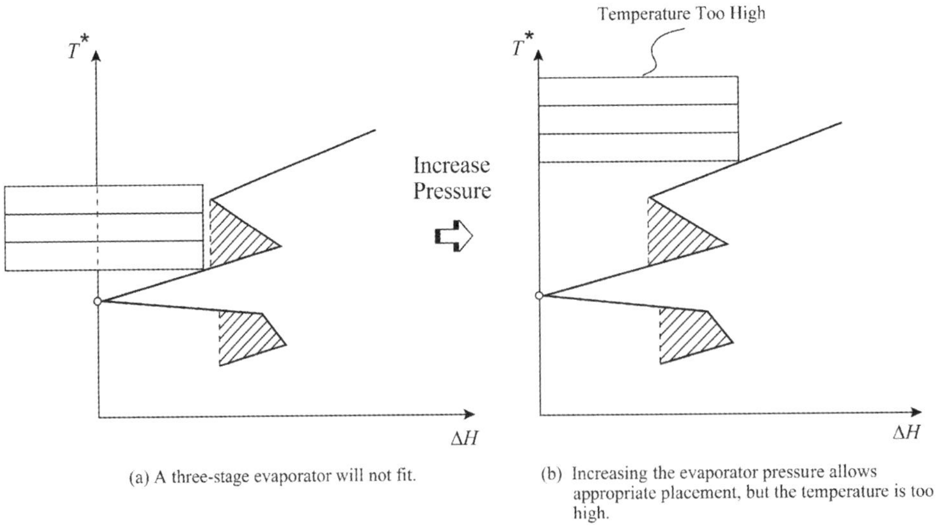

Figure 22.2

Integration of a three-stage evaporator.

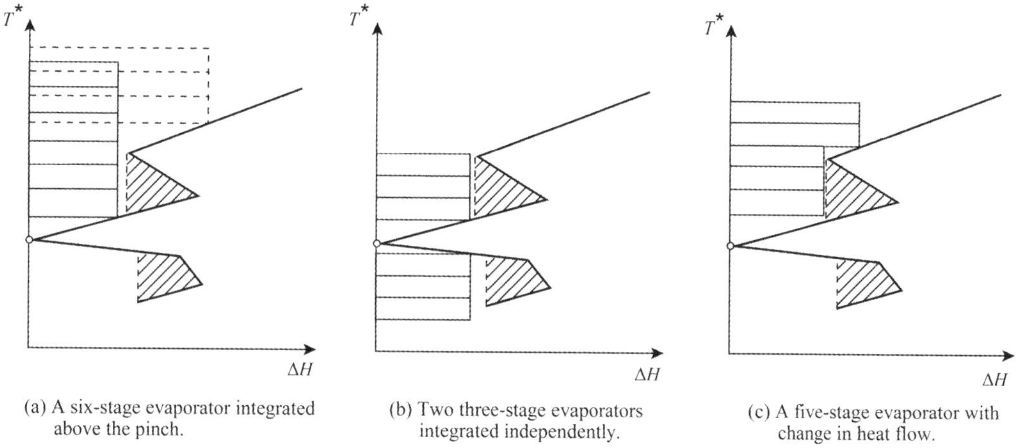

Figure 22.3

Evaporator design with the help of the grand composite curve. (Reproduced from Smith R and Linnhoff B, 1998, *Trans IChemE ChERD*, **66**: 195 by permission of the Institution of Chemical Engineers.)

If the evaporator cannot be integrated with the rest of the process because of the heat duty or constraints, another option is to use heat pumping. As with heat pumping in distillation, heat pumping of evaporators only makes sense if the evaporator cannot be integrated or must operate across the pinch. In practice, in many processes where there is a large evaporator duty, the pinch is caused by the evaporator vaporization and condensation duties. In such a situation heat pumping can make sense, because heat pumping around the evaporator also heat pumps around the pinch (see Case Study in Section 22.6). This will be most often vapor recompression, in basically the same arrangement as distillation. The evaporated vapor, which is very commonly water vapor, is compressed and used in the vaporization, as illustrated in Figure 22.4. Figure 22.4 illustrates a *mechanical vapor recompression*. For water evaporation, a steam ejector can also be used for the compression in *thermal vapor recompression*. Although Figure 22.4 shows a single-stage evaporator with a mechanical recompression, multiple stages can be combined with heat pumping and many different configurations are used.

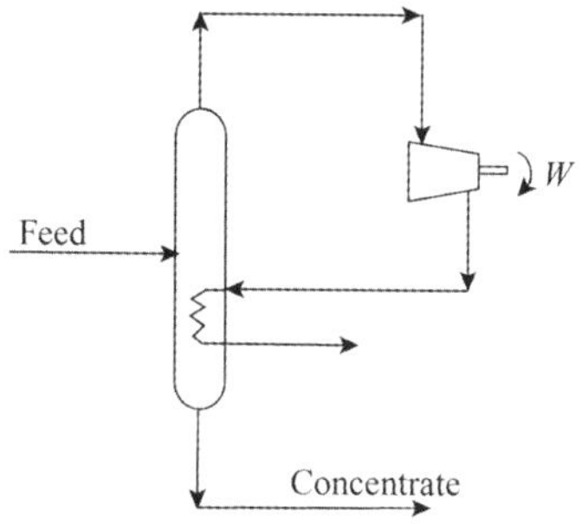

Figure 22.4

Evaporator mechanical vapor recompression.

22.4 The Heat Integration Characteristics of Dryers

The heat input to dryers, as discussed in Chapter 9, is to a gas and, as such, it takes place over a range of temperatures. Moreover, the gas is heated to a temperature higher than the boiling point of the liquid to be evaporated. The exhaust gases from the dryer will be at a lower temperature than the inlet, but again the heat available in the exhaust will be available over a range of temperatures. The thermal characteristics of dryers tend to be design specific and quite different in nature from both distillation and evaporation. Figure 22.5 illustrates the heat integration characteristics of a typical dryer.

22.5 Evolving Dryer Design to Improve Heat Integration

It was noted that dryers are quite different in character from both distillation and evaporation. However, heat is still taken in at a high temperature to be rejected in the dryer exhaust. The appropriate placement principle, as applied to distillation columns and evaporators, also applies to dryers. The plus–minus principle from Chapter 19 provides a general tool that can be used to understand the integration of dryers in the overall process context. If the designer has the freedom to manipulate drying temperature and gas flowrates, then these can be changed in accordance with the plus–minus principle in order to reduce overall utility costs.

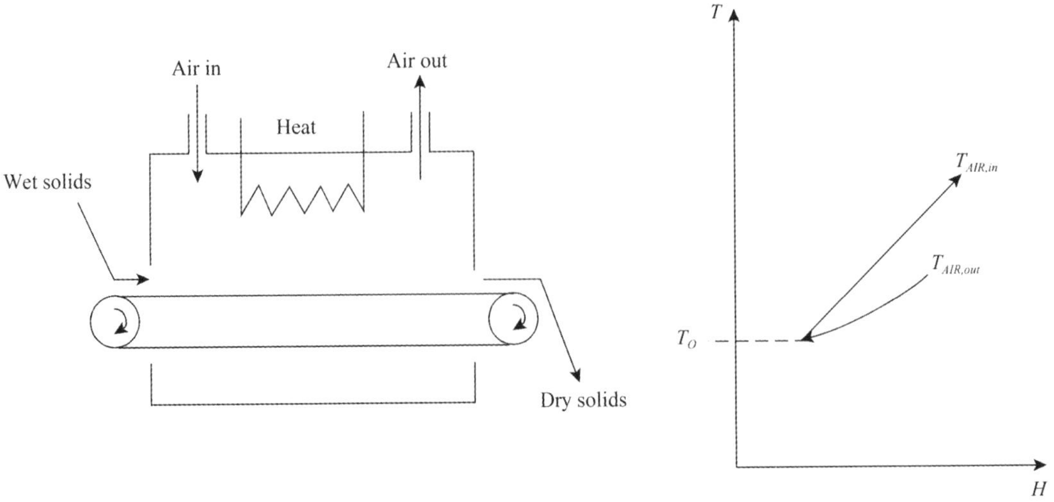

Figure 22.5

The heat integration characteristics of dryers.

Figure 22.6

A plant for the production of animal feed. The heat pump encroaches into a "pocket" in the grand composite curve. (Reproduced from Smith R and Linnhoff B, 1998, *Trans IChemE ChERD*, **66**: 195 by permission of the Institution of Chemical Engineers.)

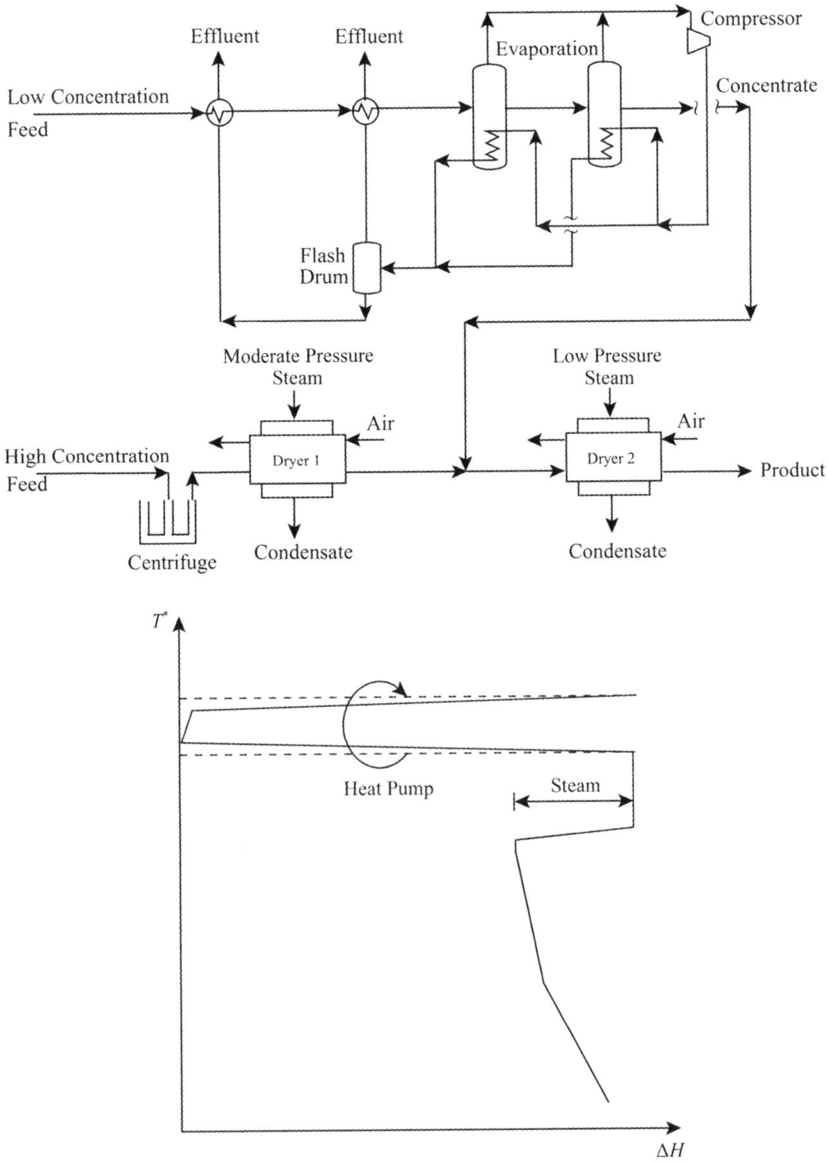

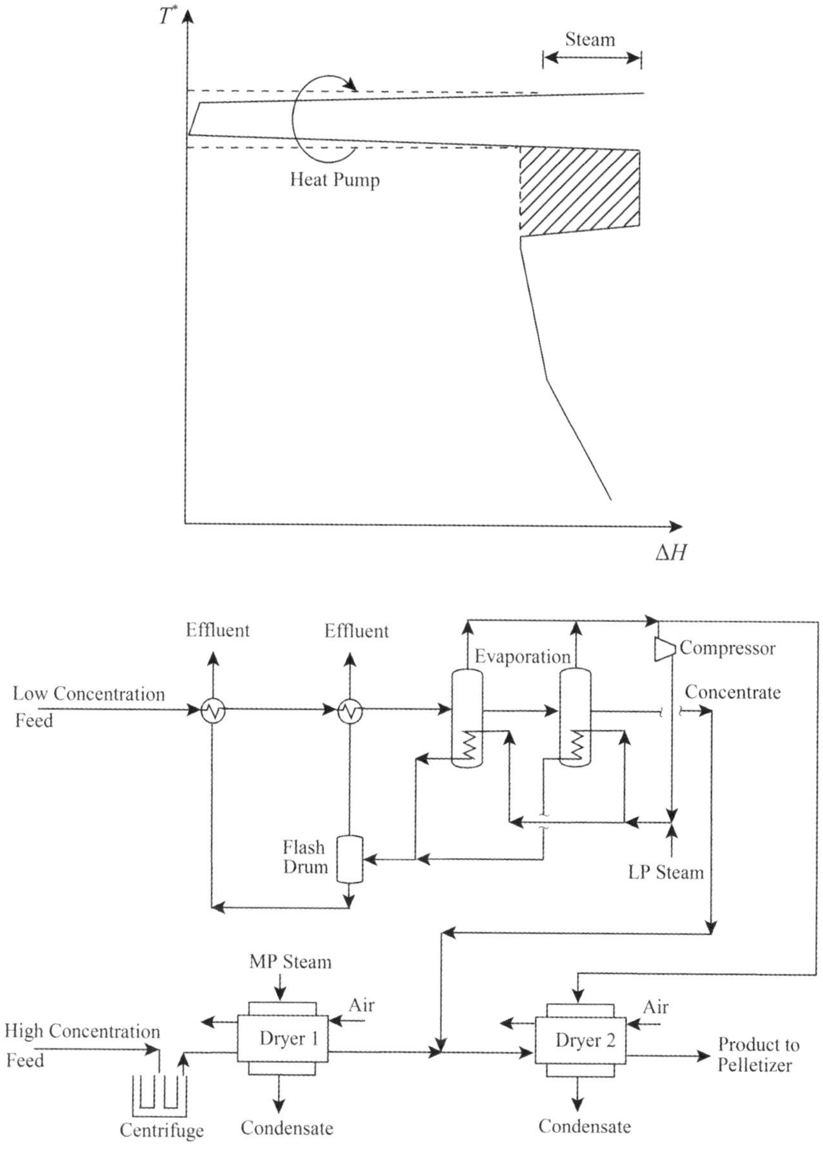

Figure 22.7

A simple modification reduces the load on the heat pump, saving electricity. (Reproduced from Smith R and Linnhoff B, 1998, *Trans IChemE ChERD*, **66**: 195 by permission of the Institution of Chemical Engineers.)

22.6 A Case Study

Figure 22.6 shows a plant for the production of animal feed from spent grains from a food process. The plant has two feeds, one of low and one of high concentration solids. Water is removed from the low concentration feed by an evaporator, followed by a rotary dryer. Water is removed from the high concentration feed by a centrifuge, followed by two stages of drying in rotary dryers. As is usual with this type of plant, the evaporators and dryers have been designed on a stand-alone basis without consideration of the process context. Optimization of the evaporator on a stand-alone basis has indicated that heat pumping using a mechanical vapor recompression system would be economic.

Figure 22.6 shows the grand composite curve for this process and the location of the evaporator heat pump. The heating duty for the first dryer has been omitted from the grand composite curve, since the required temperature is too high to allow integration with the rest of the process. The heat pump can be seen to be appropriately placed across the pinch. However, the cold side, below the pinch, encroaches into a pocket in the grand composite curve. If the design of the heat pump is changed so as not to encroach into the pocket, the result shown in Figure 22.7 is obtained. The resulting steam consumption is virtually unchanged, but energy costs will be lower. This results from the reduced load on the heat pump leading to a reduction in electricity demand.

22.7 Heat Integration of Evaporators and Dryers – Summary

Like distillation, the appropriate placement of evaporators and dryers is that they should be above the pinch, below the pinch, but not across the pinch. The grand composite curve can be used to assess appropriate placement quantitatively.

Also like distillation, the thermal profile of evaporators can be manipulated by changing the pressure. However, the degrees of freedom in evaporator design open up more options.

Table 22.1

Problem table cascade for the background process.

T^*(°C)	ΔH(kW)
211	0
191	2800
185	3280
171	2580
125	4880
105	2880
95	2680
80	4330
60	4930
51	4480
45	3580

Dryers are different in characteristic from distillation columns and evaporators in that the heat is added and rejected over a large range of temperatures. Changes to drier design can be directed by the plus–minus principle.

22.8 Exercises

1. Table 22.1 presents the problem table cascade data for a process for $\Delta T_{min} = 10\,°C$.

   An evaporation process is to be integrated with the process. The evaporator is required to evaporate $1.77\,kg{\cdot}s^{-1}$ of water. The latent heat of vaporization of the water can be assumed to be $2260\,kJ{\cdot}kg^{-1}$ and to be constant. Cooling water is available at 25 °C to be returned to the cooling tower at 35 °C. Suggest an outline evaporator configuration that will allow heat integration of the evaporator with the background process.

References

Smith R and Jones PS (1990) The Optimal Design of Integrated Evaporation Systems, *J Heat Recovery Syst CHP*, **10**: 341.

Smith R and Linnhoff B (1988) The Design of Separators in the Context of Overall Processes, *Trans IChemE ChERD*, **66**: 195.

Chapter 23

Steam Systems and Cogeneration

Most processes operate in the context of an existing site in which a number of processes are linked to the same utility system. The utility systems of most sites have evolved over a period of many years without fundamental questions being addressed as to the design and operation of the utility system. The picture is complicated by individual production processes on a site belonging to different business areas, each assessing investment proposals independently from one another and each planning for the future in terms of their own business. Yet, the efficiency of the site infrastructure and the required investment is of strategic importance and must be considered across the site as a whole, even if this crosses the boundaries of different business areas.

There are a number of reasons why site steam and power systems need to be studied in process design:

1) A grassroot utility system may need to be designed. However, this is rare, and most situations would involve modification of an existing system.
2) Modifications might be required to an existing utility system as a result of changes in the demand for steam and power on a site. This might result from a new process starting up, a process closing down, changes in process capacity or basic modifications to the process technology.
3) An old utility system might need a revamp to replace old equipment.
4) Modifications to the configuration might be required to an existing utility system to reduce its operating costs.
5) Changes to the operation of the utility system might allow reduced operating costs.
6) The choice of utility for a new process on the site requires an understanding of the costs and constraints associated with the utility system that is servicing it.
7) Before energy conservation projects are implemented in existing processes on a site, the true value of those savings needs to be established. This is usually not possible without studying the utility system.
8) The true energy costs associated with a production expansion require the true cost implications in the utility system to be established, even if there is no capital investment required in the utility system.

Figure 23.1 shows a typical site utility system. Various processes operate on the site and are connected to a common utility system. In Figure 23.1, a very high pressure steam is generated in utility stream boilers. A gas turbine generates power and exhausts its flue gas to a *heat recovery steam generator* (HRSG) to generate high-pressure steam. Steam is expanded in steam turbines to high-, medium- and low-pressure mains, producing power. The final exhaust steam from the steam turbines is expanded to vacuum conditions and condensed against cooling water. It may be that this power generation needs to be supplemented by the import of power from outside power generation via the grid. It might also be the case that excess power is generated on the site and exported.

In Figure 23.1, three levels of steam are distributed around the site, and the various processes are connected to the steam mains. Process *A* in Figure 23.1 has a local fired heater. It imports low-pressure steam and exports high- and medium-pressure steam via the site distribution system. Process *B* imports both high- and low-pressure steam. Process *C* imports high- and medium-pressure steam and exports low-pressure steam via the low-pressure steam main. Finally, Processes *A*, *B* and *C* all require the rejection of waste heat to the ambient, and this is achieved using a cooling water circuit. Some sites use other methods for cooling. These will be discussed in Chapter 24.

This is a complex system to analyze. Steam is available at different pressures and temperatures. Some processes use steam while others generate steam. There are interactions between the processes and the utility system via steam use and generation. There are also interactions between the processes on the site through the steam mains. Some processes export waste heat into the steam system, while other processes use this waste heat by drawing from the steam system. This is heat recovery between

Chemical Process Design and Integration, Second Edition. Robin Smith.

Companion Website: www.wiley.com/go/smith/chemical2e

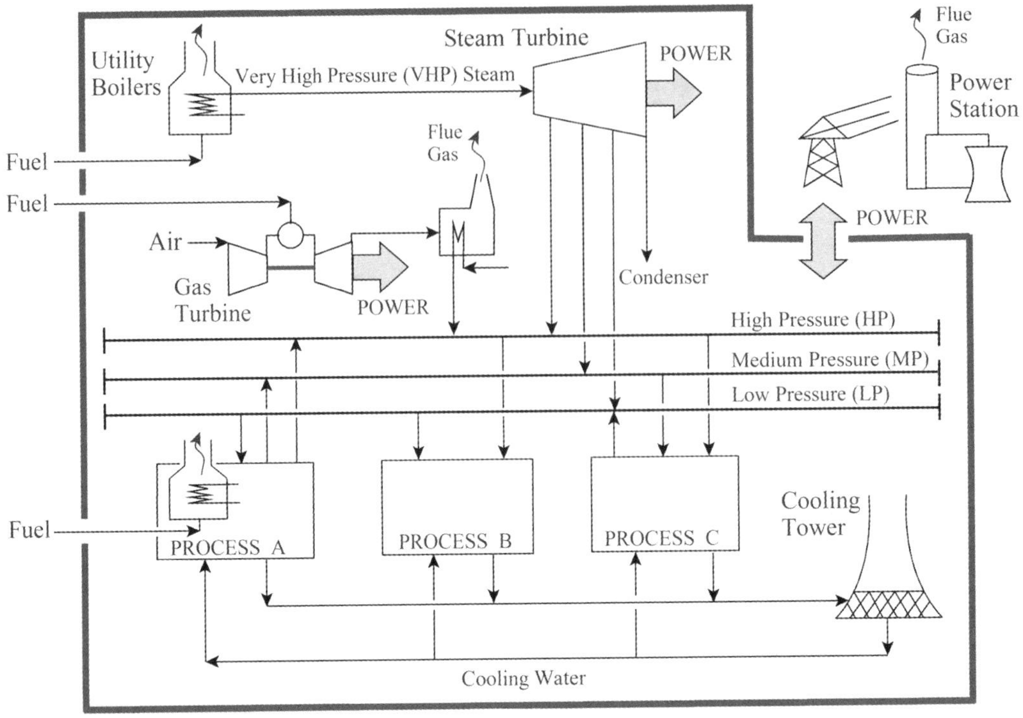

Figure 23.1

A typical site utility system.

processes on the site using steam as an intermediate for heat transfer. In Figure 23.1, there is *cogeneration* (combined heat and power generation) from steam turbines. Strictly, cogeneration is the production of useful heat and power from the same heat source. Generation of power might be through a steam turbine or gas turbine coupled to an electric generator for the production of electricity in a *turbogenerator*, or a steam turbine or gas turbine coupled directly to a machine on a *direct drive* (e.g. a steam turbine driving a process compressor directly). However, the byproduct heat must be useful to class the power generation as cogeneration. Cogeneration is the most efficient way to produce heat and power and lowers energy costs. It also lowers the overall emissions of combustion gases, although this can only be judged properly by including the emissions created by external power generation resulting from any imported power or the savings in emissions from external power generation resulting from the exported power (see Chapter 25).

Another important feature of steam and cogeneration systems is that the load on the system can vary significantly through time. Processes on the site will start up and shut down, and the capacity of the processes will vary according to market demands, maintenance requirements, and so on. There are likely to be changes in heating requirements between winter and summer. Also, the utility system itself requires equipment to be taken off-line and maintained and can suffer breakdown, which affects the capacity of the system. Yet, despite these variations, it is important that the steam and cogeneration system continues to satisfy the site demands under variable and even upset conditions. This requires considerable flexibility in the steam and cogeneration system to operate in different conditions. In turn, this means that there is a requirement for significant spare capacity or *contingency* or *redundancy* in the equipment. This might mean that discrete spare items of equipment are installed to cope with the variable conditions, or equipment is simply oversized.

Steam is distributed around the site for various purposes:

1) *Steam heaters.* Steam is by far the most commonly used heating medium. Shell-and-tube heat exchangers are the most common, but other heat exchanger designs and pipe coils are also used.
2) *Steam tracing.* Fluids in tanks, pipes and other process equipment often have characteristics that cause them to freeze, become too viscous, or condense undesirably at ambient temperature, even if the tanks, pipes and equipment are insulated. In order to prevent such problems, additional heat can be added, combined with insulation. A tube or small-diameter pipe carrying steam attached to the vessel or pipe for this purpose is referred to as *steam tracing*. Jackets can also be used instead of tubes.
3) *Space heating.* Production buildings, workshops, warehouses and offices can require space heating. Pipes, radiators and fan heaters can be used to create space heating from steam.
4) *Water heating through live steam injection.* Rather than use indirect steam heating in a heat exchange device, steam can be injected directly into water to provide heating. However, care must be exercised that any treatment chemicals in the steam do not cause unacceptable contamination of the water (e.g. in food processing operations).

5) *Boiler feedwater deaeration.* One of the operations required to prepare the water to be used for steam generation involves the use of steam to strip out gases dissolved in the water in a *deaeration* operation. Deaeration will be discussed in more detail later in this chapter.
6) *Power generation in steam turbines.* A steam turbine converts the energy of steam into power by expanding the steam across rows of *blades* to create rotational power. Steam turbines will be discussed in more detail later in this chapter.
7) *Steam ejectors.* As discussed in Chapter 13, and illustrated in Figures 13.16 and 13.17, steam can be used for the production of vacuum in steam ejector systems.
8) *Flaring.* Combustible waste gases are often disposed of by combustion in flares. Flares will be considered in more detail in Chapter 25. Mixing of the waste gases with combustion air at the flare tip is often assisted by the use of steam jets. Large amounts of steam can be consumed, especially under upset process conditions.
9) *Atomization of fuel oil in combustion operations.* The combustion of viscous fuel oil in burners often requires steam to atomize the fuel oil for combustion.
10) *Steam injection into combustion processes for NO_X abatement.* Steam can be injected into combustion processes to lower emission of oxides of nitrogen (NO_X). The steam acts to decrease the flame temperature. This will be discussed in Chapter 25.
11) *Steam distillation.* As pointed out in Chapter 8, steam is added to some distillation operations to assist the separation. This is particularly important in crude oil distillation.
12) *Reactor steam.* As pointed out in Chapters 4 to 6, steam might be required as a reactant and is sometimes fed to reactors along with the reactants in order to decrease the partial pressure of the reactants.
13) *Reactor decoking.* Reactors processing hydrocarbon feeds often suffer from the deposition of coke (carbon) on the reactor surfaces. Such deposits of coke can be removed by the application of steam. To achieve this, the reactor is taken off-line periodically and decoked by the application of steam jets.
14) *Soot blowing.* Furnace and boiler heat transfer surfaces exposed to the flue gas become fouled through the deposition of soot and ash. This fouling reduces the efficiency of the heat transfer. *Soot blowing* removes the deposits by applying jets of steam periodically.

Steam used in flaring, decoking of reactors and soot blowing will be short-term demands on the system, requiring flexibility in its operation.

It is no accident that steam is used extensively for process heating, as it provides a number of useful features that include:

- energy can be generated at one point and distributed;
- it is a convenient form of transferring energy around;
- it has a wide range of operating temperatures;
- it has a high heat content through the latent heat;
- the temperature is easy to control through control of the pressure;
- it can be used to generate power in steam turbines and generate vacuum in steam ejectors;
- it does not require expensive materials of construction;
- it is nontoxic and losses are easily replaced.

Power is also required across the site to drive machines and as electrical power. Shaft power is required to drive various process and utility machines:

- Electricity generators
- Process gas compressors
- Air compressors
- Refrigeration compressors
- Fans
- Pumps
- Mixers
- Solids conveyors (e.g. belt and screw conveyors)
- Solids processing equipment (e.g. crushing, grinding and pelletizing).

Electricity is required across the site (depending on the nature of the processes) for:

- Electrical process heating (e.g. immersion heaters and electric ovens)
- Electrical tracing (the electrical equivalent of steam tracing)
- Electrolysis operations (e.g. chlorine and aluminum production)
- Operations requiring an electrical field (e.g. electrocoagulation and electrodialysis)
- Instrumentation and control systems
- Lighting.

The major components of the steam system will be discussed next, before integrating the components into an efficient steam and cogeneration system.

23.1 Boiler Feedwater Treatment

Figure 23.2 shows a schematic representation of a boiler feedwater treatment system. Raw water from a reservoir, river, lake, borehole or a seawater desalination plant is fed to the steam system. However, it needs to be treated before it can be used for steam generation. The treatment required depends both on the quality of the raw water and the requirements of the utility system. The principal problems with raw water are (Kemmer, 1988; Betz, 1991):

- suspended solids,
- dissolved solids,
- dissolved salts,
- dissolved gases (particularly, oxygen and carbon dioxide).

Thus, the raw water entering the system in Figure 23.2 may need to be first filtered to remove suspended solids. If required, this would commonly be carried out using sand filtration. Figure 23.3

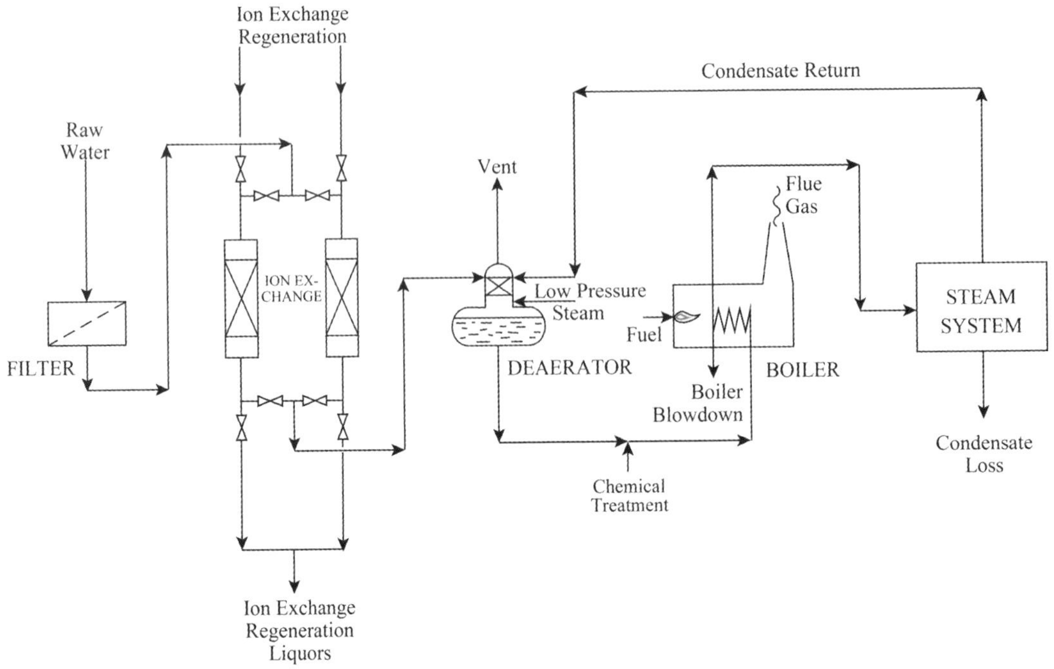

Figure 23.2

Boiler feedwater treatment.

illustrates the design of two different types of sand filter, one using gravity to provide the flow, the other using pressure from a pump. Sand filters are typically capable of removing 90% to 95% of suspended solids, removing solids down to around 10 to 20 μm. The sand beds need to be periodically taken off-line and back-flushed with clean water to remove the captured solids. Microfiltration using a membrane can also be used and goes further than sand filtration, removing particles down to around 0.5 μm.

Dissolved salts, which can cause fouling and corrosion in the steam boiler, then need to be removed. The principal problems are associated with calcium and magnesium ions, which would otherwise deposit as scale on heat transfer surfaces. The presence of silica is also a particular problem. Silica can form low-conductivity

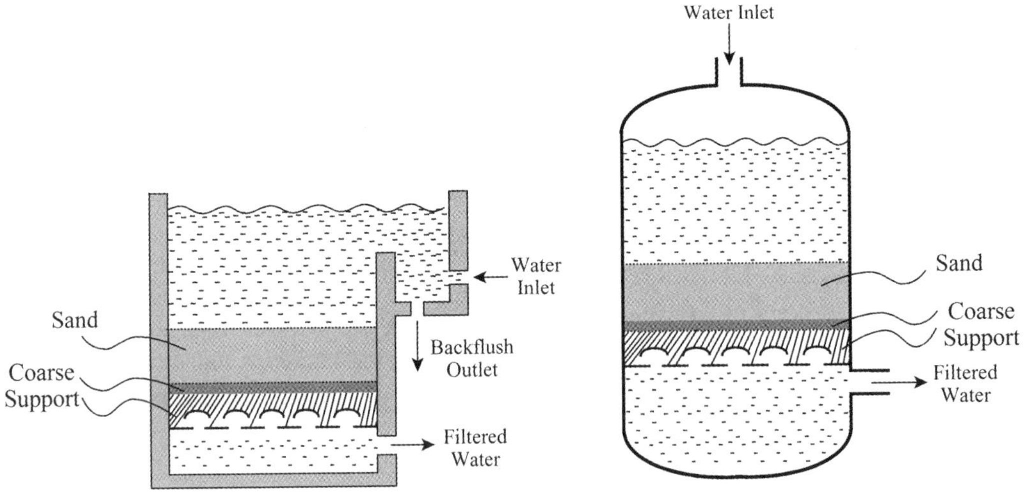

(a) Gravity filter. (b) Pressure filter.

Figure 23.3

Sand filters for the removal of suspended solids.

deposits in the boiler and if carried from the boiler in water droplets or a volatile form can cause damage to steam turbine blades, particularly the low-pressure section of steam turbines where some condensation can occur. The quality of feedwater required depends on the boiler operating pressure and boiler design. Sodium zeolite *softening* is the simplest process and is used to remove scale-forming ions, such as calcium and magnesium, sometimes referred to as *hardness*. The softened water produced can be used for low- to medium-pressure boilers. In this process, the water is passed through a bed of sodium zeolite, and the calcium and magnesium ions are removed according to (Kemmer, 1988; Betz, 1991):

$$\begin{bmatrix} Ca \\ Mg \end{bmatrix} \begin{bmatrix} SO_4 \\ 2Cl \\ 2HCO_3 \end{bmatrix} + Na_2Z \longrightarrow Z \begin{bmatrix} Ca \\ Mg \end{bmatrix} + \begin{bmatrix} Na_2SO_4 \\ 2NaCl \\ 2NaHCO_3 \end{bmatrix}$$

where Z refers to zeolite. The resin is regenerated by treatment with sodium chloride solution (Kemmer, 1988; Betz, 1991):

$$Z \begin{bmatrix} Ca \\ Mg \end{bmatrix} + 2NaCl \longrightarrow Na_2Z + \begin{bmatrix} Ca \\ Mg \end{bmatrix} Cl_2$$

The softening process removes the sparingly soluble calcium and magnesium ions and replaces them with less objectionable sodium ions. Softening alone is not sufficient for most high-pressure boiler feedwaters. Also, the processes on a site often have a requirement for *deionized water*. *Deionization*, in addition to removing hardness, removes other dissolved solids, such as sodium, silica, alkalinity and mineral ions such as chloride, sulfate and nitrate. Deionization is essentially the removal of all inorganic salts. In this process, a *strong acid cation resin* converts dissolved salts into their corresponding acids and a *strong base anion resin* in the hydroxide form removes these acids. The cation ion-exchange resin exchanges hydrogen for the raw-water ions according to (Kemmer, 1988; Betz, 1991):

$$\begin{bmatrix} Ca \\ Mg \\ 2Na \end{bmatrix} \begin{bmatrix} 2HCO_3 \\ SO_4 \\ 2Cl \\ 2NO_3 \end{bmatrix} + 2Z\,H \longrightarrow 2Z \begin{bmatrix} Ca \\ Mg \\ 2Na \end{bmatrix} + \begin{bmatrix} 2H_2CO_3 \\ H_2SO_4 \\ 2HCl \\ 2HNO_3 \end{bmatrix}$$

To complete the deionization process, water from the cation unit is passed through a strong base anion exchange resin in the hydroxide form. The resin exchanges hydrogen ions for both highly ionized mineral ions and the more weakly ionized carbonic and silicic acids according to (Kemmer, 1988; Betz, 1991):

$$\begin{bmatrix} H_2SO_4 \\ 2HCl \\ 2H_2SiO_3 \\ 2H_2CO_3 \\ 2HNO_3 \end{bmatrix} + 2Z\,OH \longrightarrow 2Z \begin{bmatrix} SO_4 \\ 2Cl \\ 2HSiO_3 \\ 2HCO_3 \\ 2NO_3 \end{bmatrix} + 2H_2O$$

As the ion-exchange beds approach exhaustion, they need to be taken off-line and regenerated. The cation resins are regenerated with an acid solution, which returns the exchange sites to the hydrogen form. Sulfuric acid is normally used for this. The anion exchange resin is regenerated using sodium hydroxide solution, which returns the exchange sites to the hydroxyl form.

Rather than use ion exchange, water can be deionized using membrane processes. Both nanofiltration and reverse osmosis can be used, but only reverse osmosis is capable of producing high-quality boiler feedwater. Nanofiltration is only capable of separating covalent ions and larger univalent ions such as heavy metals. Nanofiltration and reverse osmosis have been discussed in detail in Chapter 9. The main advantage of reverse osmosis over ion exchange is that it does not require the costly and hazardous chemicals required by ion exchange to regenerate the beds. The regeneration liquors also create an environmental burden. In addition, the regeneration of ion exchange beds can take many hours, with beds off-line for a significant period. On the other hand, reverse osmosis only requires periodic routine maintenance.

Having removed the suspended solids and dissolved salts, the water then needs to have the dissolved gases removed, principally oxygen and carbon dioxide, which would otherwise cause corrosion in the steam boiler. The usual method to achieve this is deaeration, which removes dissolved gases by raising the water temperature and stripping the dissolved gases (Kemmer, 1988; Betz, 1991). There are many different designs of deaerator. Figure 23.4 illustrates a typical design. The water to be deaerated enters at the top of the deaerator via a spray system. This water is contacted with injected steam. Some form of packing or plates is normally used to assist the contact between the boiler feedwater and steam. Boiler feedwater is heated to within a few degrees of saturation temperature of the steam. Most of the noncondensable gases (principally, oxygen and free carbon dioxide) are released into the steam. Deaerated water decends to a storage tank below,

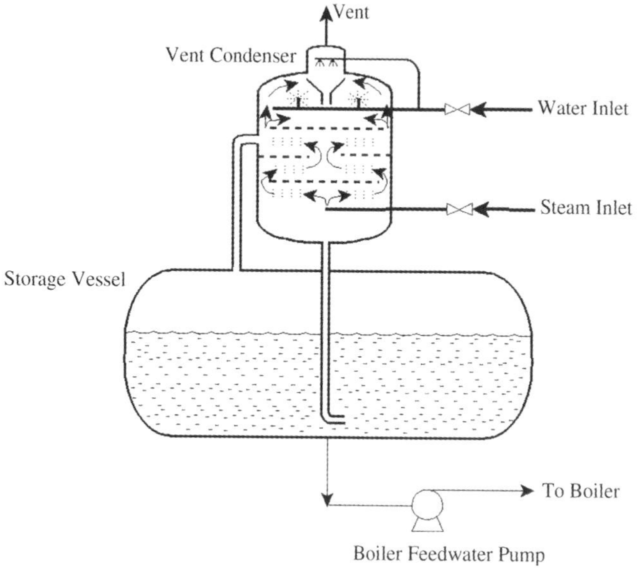

Figure 23.4
Boiler feedwater deaeration.

where a steam blanket protects it from recontamination. The deaeration steam flows up through the deaerator and most of it is condensed to become part of the deaerated water. A small portion of the steam, which contains the noncondensable gases released from water, is vented to atmosphere. There is a minimum temperature difference between the deaeration steam and the final boiler feedwater temperature to make the deaerator function effectively. A temperature difference of at least 10 °C is required. Some designs of deaerator include a vent condenser that sprays in a small portion of the boiler feedwater feed to a spray system prior to venting the steam and noncondensable gases in order to minimize the vented steam, so reducing the visible plume from the deaerator and improving energy efficiency.

Even after deaeration, there is some residual oxygen that needs to be removed by chemical treatment (Kemmer, 1988; Betz, 1991). After the deaerator, oxygen scavengers (e.g. hydroquinone) are added to react with the residual oxygen that would otherwise cause corrosion. Unfortunately, the boiler feedwater treatment does not remove all of the solids in the raw water, and the deposition of solids in the boiler is another problem. Phosphates can be added to precipitate any calcium and magnesium away from the heat transfer surfaces of the boiler. Polymer dispersant can also be added to help keep any precipitate dispersed. *Chelants* (weak organic acids) can be used. These have the ability to complex with many cations (calcium, magnesium and heavy metals, under boiler conditions). They accomplish this by locking the metals into a soluble organic ring structure. Another corrosion problem is in the condensate system. Steam condensate contains carbon dioxide, which is corrosive to steel. If left untreated, condensate can have a pH of 5.0 to 6.5, which is corrosive to the steel piping in the condensate return lines. Volatile amines (e.g. cyclohexylamine, diethylaminoethanol) can be added to the boiler feedwater to prevent corrosion in the condensate system. As the steam condenses, these amines neutralize the acid (H^+) generated by the dissolution of carbon dioxide or other acidic contaminants in the steam condensate system.

The purity of the boiler feedwater and the treatment required depend on the pressure of the boiler. The higher the pressure of the boiler, the purer the feed used.

The deaerated treated boiler feedwater then enters the boiler. Evaporation takes place in the boiler and the steam generated is fed to the steam system. Solids (either suspended or dissolved) not removed by the boiler feedwater treatment build up in the boiler, along with products of corrosion. If the solids are allowed to build up they can lead to foaming and carryover of solids into the steam. The solids will also lead to scale formation in the boiler, resulting in the deterioration of performance and possibly localized overheating and boiler failure. The control of the concentration of solids is achieved by purging, known as *blowdown*. The blowdown can be carried out continuously or intermittently. The total dissolved solids (TDS) are limited according to the boiler design and pressure. Table 23.1 presents some typical values.

Blowdown is normally controlled by measuring the conductivity of the water. Blowdown rates are usually quoted as the ratio of flowrate of blowdown to evaporation rate. The blowdown rate required depends on the boiler feedwater treatment system and the pressure of the boiler. Typical blowdown rates are:

- small, low-pressure boilers: 5 to 10%,
- large, high-pressure boilers: 2 to 5%,
- very large boilers with very pure feed: less than 2%.

The steam from the boiler goes to the steam system to perform various duties. The steam is ultimately condensed somewhere in the steam system. Some condensate is returned to the deaerator and some lost from the system to effluent. Sometimes, the returned condensate is subjected to deionization in a *condensate-polishing* step to remove any traces of dissolved solids that might have been picked up. Condensate return as high as 90% or greater is possible, but return rates are often significantly lower than this. Higher levels of condensate return than 90% might not be able to be justified because of the capital cost of the pipework required for condensate return or the possibility of some condensate being contaminated. This constant loss of condensate from the steam system means that there must be a constant make up with freshwater.

Table 23.1

Typical allowable solids levels for boilers at different pressures.

Pressure range (barg)	Total dissolved solids (ppm)
15–25	3000
25–35	2000
35–45	1500
45–60	500
60–75	300
75–100	100

Example 23.1 A small package fire-tube boiler has makeup water that contains 250 ppm total dissolved solids (TDS). The steam system operates with 50% condensate return. Estimate the blowdown rate. Assume that the maximum limit for the TDS in the boiler is 3000 ppm. Assume that there are no solids in the evaporation or the condensate return.

Solution Referring to Figure 23.5:

$$m_M C_M = m_F C_F = m_B C_B$$

where
m_M, C_M = flowrate and concentration of solids in makeup water
m_F, C_F = flowrate and concentration of solids in the boiler feed
m_B, C_B = flowrate and concentration of solids in the blowdown

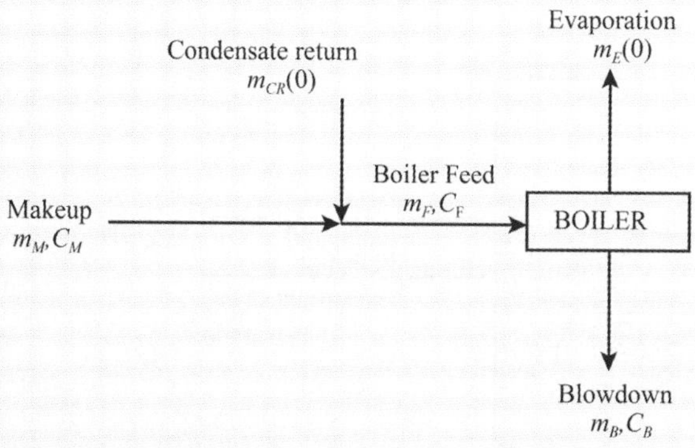

Figure 23.5

Boiler blowdown material balance.

Also:

$$m_F = m_M + m_{CR}$$

Combining these equations gives:

$$(m_F - m_{CR})C_M = m_B C_B$$

For a condensate return fraction of CR:

$$\begin{aligned}(m_F - CR\, m_E)C_M &= m_B C_B \\ (m_E + m_B - CR\, m_E)C_M &= m_B C_B\end{aligned}$$

Rearranging:

$$\begin{aligned}\frac{m_B}{m_E} &= \frac{C_M(1 - CR)}{C_B - C_M} \\ &= \frac{250(1 - 0.5)}{3000 - 250} \\ &= 0.038 \\ &= 3.8\%\end{aligned}$$

Example 23.2 The condensate return rate on a site is unknown. The boilers are high-pressure water-tube boilers having a makeup water that contains 5 ppm total dissolved solids (TDS). The TDS in the boilers are controlled to be 100 ppm. Flowrate measurements indicate that the blowdown rate based on the evaporation rate is 2%. Assuming that there are no solids in the evaporation or the condensate return, calculate the condensate return rate.

Solution Again referring to Figure 23.5, a mass balance for the total dissolved solids, as with Example 23.1 gives:

$$(m_E + m_B - CR \cdot m_E)C_M = m_B C_B$$

Rearranging:

$$\begin{aligned}CR &= 1 - \frac{m_B}{m_E}\left(\frac{C_B - C_M}{C_M}\right) \\ &= 1 - 0.02\left(\frac{100 - 5}{5}\right) \\ &= 0.62 \\ &= 62\%\end{aligned}$$

23.2 Steam Boilers

There are many types of steam boilers, depending on the steam pressure, steam output and fuel type (Dryden, 1982). Pressures are normally of the following order:

- 100 barg, used for power generation;
- 40 barg is the normal maximum pressure for distribution (higher pressure requires materials other than carbon steel);
- 10–40 barg are the conventional distribution pressure levels;
- 2–5 barg are typically the lowest pressures used for distribution.

Figure 23.6 illustrates a *fire-tube* (or *shell*) boiler. Water is preheated by the stack gas in an *economizer* and enters the boiler shell. The shell is a large metal cylinder housing a pool of boiling water that is vaporized by a hot flue gas flowing through the inside of tubes. The steam pressure is contained by the large cylindrical shell, which imposes certain mechanical limitations. A large-diameter shell requires a thicker wall to contain the same pressure as a small-diameter shell. The economic limit for such designs is around 20 barg (but they are available at higher pressures). Fire-tube boilers are normally used for small heating duties of low-pressure steam. Fuels are normally restricted to be gas or light fuel oil. One of the main disadvantages of the design is difficulty of providing superheat to the steam.

Higher duty boilers use a *water-tube* arrangement, as shown in Figure 23.7. The water is first preheated in an economizer from the stack gas and introduced into the *steam drum* behind a baffle. The steam drum is connected to a lower *water* or *mud drum* by the boiler tubes. Because the cold water is dense, it descends in the "downcomer" towards the water drum. This causes hot water to flow upwards from the water drum in the front tubes closer to the flame. Continued heating of the water in the front tubes causes partial vaporization of the water. Steam is separated from the hot water in the steam drum. The steam separated in the steam drum enters a superheating section before entering the steam system. Water-tube boilers are suitable for high pressures and high steam output. Fuels can be gas, fuel oil or pulverized solid fuel. In the case of solid fuels, the boiler must have some mechanism for the removal of ash produced by combustion from the base of the radiant chamber.

Another design of boiler is illustrated in Figure 23.8 that uses a fluidized bed combustion system. This is used for the combustion of solid fuels (coal, petroleum, coke, or biomass). The combustion air is typically preheated by the stack gas before entering the combustion zone. Fuel particles are suspended in a hot, bubbling

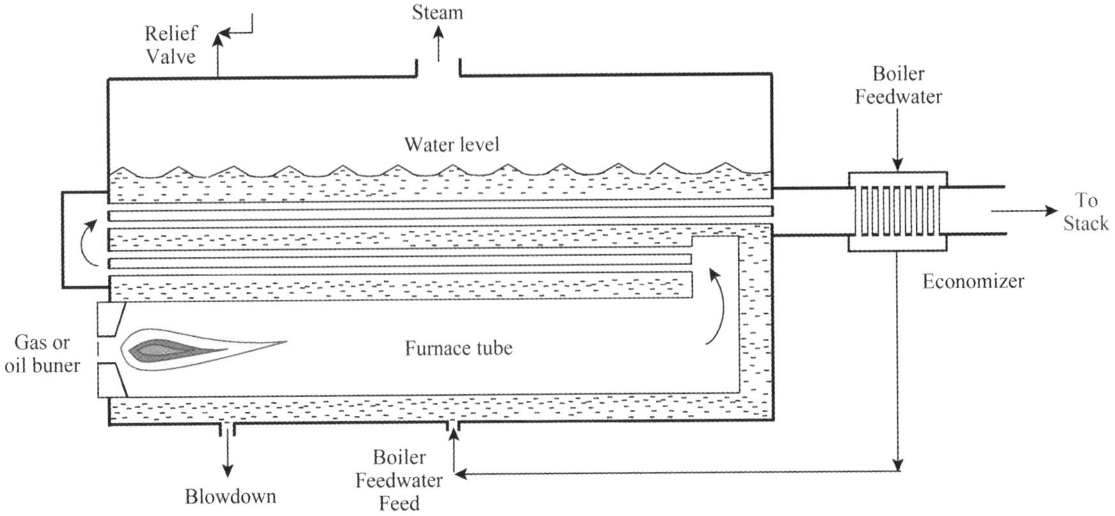

Figure 23.6
Fire-tube (or shell) boiler.

fluidized bed of fuel and ash. The fuel burns throughout its volume with no flame. Light ash particles leave the top of the bed, together with waste gases. The fluidized bed creates good contact between the fuel and combustion air and combustion takes place at temperatures lower than a conventional boiler from 800 to 900 °C, with a resulting reduced formation of oxides of nitrogen (see Chapter 25). The solid fuel can have a high moisture and ash content and can contain significant amounts of sulfur as impurity. Limestone can be added to the combustion, which reacts with any sulfur to form calcium sulfate and leaves with the spent ash, rather than the sulfur being discharged to the atmosphere with the flue gas as oxides of sulfur. Figure 23.8 illustrates a bubbling fluidized bed boiler. An alternative design uses a *circulating fluidized bed*. Circulating beds use a higher fluidizing velocity, causing the particles to pass through the main combustion chamber and into a cyclone, from which the larger particles are extracted and returned to the combustion chamber.

Many other designs of boiler are available. The draft in boilers can be *forced draft* in which a fan is used to blow air into the fired heater. The fired heater is thus at a pressure slightly above atmospheric. Alternatively, in *induced draft* designs a fan is located in the duct for the waste combustion gases after the boiler

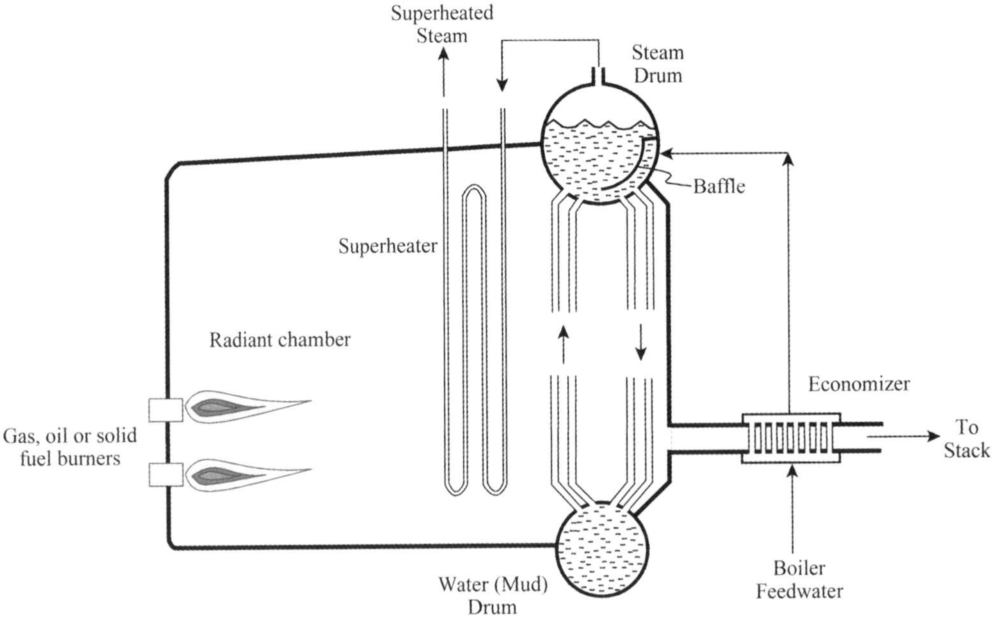

Figure 23.7
Water-tube boiler.

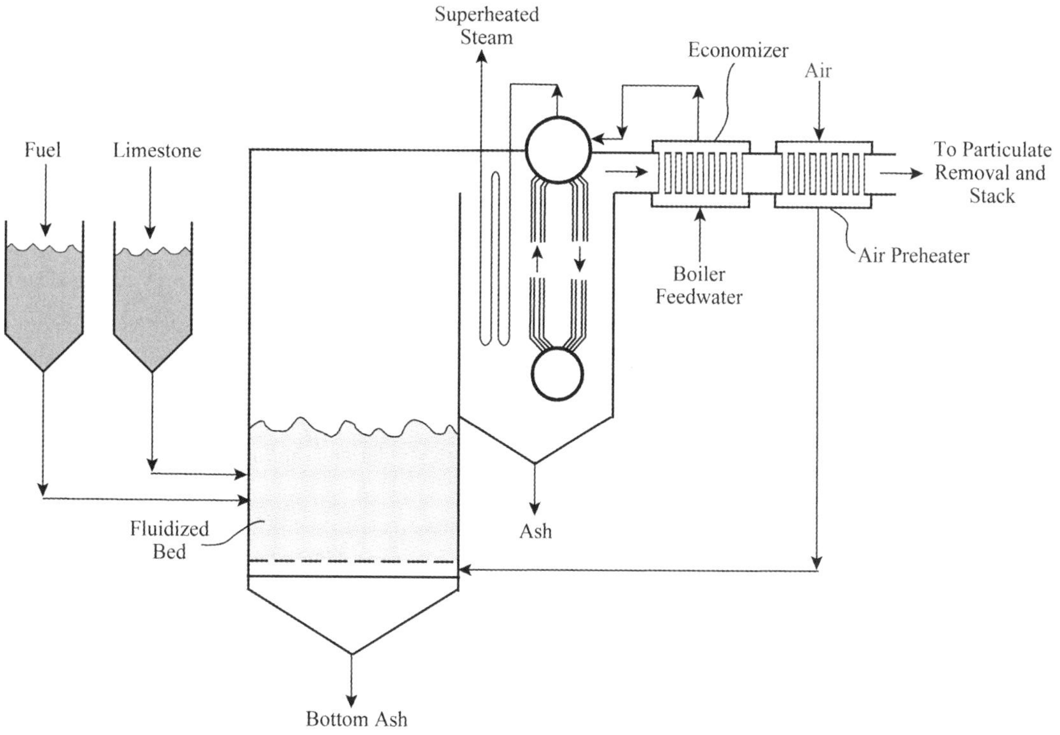

Figure 23.8
Fluidized-bed boiler.

23

and before entering the stack. This creates a pressure in the boiler slightly below atmospheric. For large boilers requiring a significant amount of equipment in the flue to treat the exhaust gases to remove oxides of sulfur and/or oxides of nitrogen, as well as equipment for heat recovery, a combination of forced and induced draft might need to be used in a *balanced draft* arrangement.

Quantifying the performance of steam boilers requires the heat input, heat output and losses to be quantified.

1) *Heat input.* Heat input can be based on the gross or net calorific value of the fuel. As discussed in Chapter 12, the *gross calorific value* of the fuel is the quantity of heat released from the complete combustion of a specified amount of the fuel at standard temperature (15 °C or 25 °C) and standard pressure (1.01325 bar) with water initially present in the fuel and that from the combustion products in the liquid state. This latent heat from the condensation of the water is rarely recovered in combustion equipment. The *net calorific value* of the fuel assumes that the latent heat of water vapor is not recovered and is defined as the quantity of heat released from the complete combustion of a specified amount of the fuel at standard temperature (15 °C or 25 °C) and standard pressure (1.01325 bar) with water initially present in the fuel and that from the combustion products in the vapor state at standard temperature. The relationship between net calorific value and gross calorific value is given by:

$$NCV = GCV - m_{COND}\Delta H_{VAP} \tag{23.1}$$

where NCV = net calorific value ($J{\cdot}m^{-3}$, $kJ{\cdot}m^{-3}$, $J{\cdot}kg^{-1}$, $kJ{\cdot}kg^{-1}$)
GCV = gross calorific value ($J{\cdot}m^{-3}$, $kJ{\cdot}m^{-3}$, $J{\cdot}kg^{-1}$, $kJ{\cdot}kg^{-1}$)
m_{COND} = mass of condensate formed per unit volume of gas or unit mass of solid/liquid fuel (kg)
ΔH_{VAP} = latent heat of vaporization at standard temperature ($J{\cdot}kg^{-1}$, $kJ{\cdot}kg^{-1}$)The net calorific value will be used as the basis here for heat input:

$$Q_{INPUT} = m_{FUEL}NCV \tag{23.2}$$

where Q_{INPUT} = heat input from fuel (W, kW)
m_{FUEL} = flowrate of fuel ($m^3{\cdot}s^{-1}$, $kg{\cdot}s^{-1}$)

2) *Heat output.* The heat output is the heat supplied from the boiler feedwater supply temperature to the final temperature of the steam. Some boiler standards also consider the heat in the blowdown to be heat output. In principle, heat can be recovered from the blowdown, but at least part of this heat, and sometimes all of it is lost. It is therefore most appropriate to consider the blowdown as part of the losses, but accounting for any heat recovery that might be carried out, for example for boiler feedwater preheat. The output is defined as:

$$Q_{OUTPUT} = m_{STEAM}(H_{STEAM} - H_{BFW}) \tag{23.3}$$

where Q_{OUTPUT} = heat output to steam generation (W, kW)
m_{STEAM} = flowrate of steam ($kg \cdot s^{-1}$)
H_{STEAM} = enthalpy of the steam ($J \cdot kg^{-1}$, $kJ \cdot kg^{-1}$)
H_{BFW} = enthalpy of the boiler feedwater ($J \cdot kg^{-1}$, $kJ \cdot kg^{-1}$)

3) *Heat losses.* The heat losses from the boiler are:

a) *Stack loss.* Flue gases leaving the furnace at a high temperature carry a significant amount of heat as they are cooled to ambient temperature in the environment. This is normally by far the largest loss from the boiler. To quantify the stack loss in principle requires knowledge of the flue gas flowrate, its composition and the enthalpy of the individual components. Equation 12.196 provides an approximate heat capacity as a function of temperature.

b) *Heat loss from unburnt combustible gases.* The most common unburnt combustible gases are carbon monoxide and hydrogen. It is usual to find only carbon monoxide in any significant amounts in flue gases, but in principle it should only be there in negligible quantities. The use of excess air should normally keep the carbon monoxide to very small amounts. To quantify the heat loss from unburnt combustible gases requires knowledge of the flowrates of the flue gases and their net calorific value.

c) *Heat loss from unburnt combustible fuel.* Unburnt solid fuel can leave the furnace with the bottom ash from the base of the combustion chamber or leave with fly ash that is carried with the flue gases. Unburnt liquid fuel can leave with the flue gases. To quantify the heat loss from unburnt solid fuel requires knowledge of the unburnt solid fuel flowrate in the bottom ash and fly ash and the net calorific value of the fuel. For a coal-fired boiler losses can be 2% or higher. To quantify the heat loss from unburnt liquid fuel requires knowledge of the unburnt liquid fuel flowrate carried with the flue gases and the net calorific value of the fuel. Such losses from an oil-fired boiler can be in the range 0.2 to 0.5%.

d) *Heat loss from the ash.* Sensible heat in the bottom ash and the fly ash leaving with the flue gases is another source of loss.

e) *Casing losses.* Heat is lost to the environment from the outer surface of the boiler by radiation and convection. Such losses can be estimated by assuming a heat transfer coefficient of the order of $250\ W \cdot m^{-2} \cdot K^{-1}$ and outside skin temperature of 55 °C. Casing losses as a percentage decrease with the size of the boiler. For gas- and oil-fired boilers casing losses range from 0.25% for large boilers to 1.25% for small boilers. For coal-fired boilers, casing losses range from 0.5% to 1.5% for the designs of boiler used in the process industries.

f) *Blowdown losses.* As pointed out above, the heat available in the blowdown water is taken in some standards to be part of the boiler output. However, even if heat recovery from the blowdown is used, at least part of the blowdown heat will be lost. Thus, blowdown loss can be taken to be the heat in the blowdown water from its discharge temperature down to the reference temperature. The discharge temperature might be that in the boiler, in other words the steam saturation temperature, or the blowdown temperature after heat recovery.

g) *Atomizing steam losses.* Steam is sometimes used to atomize fuel in burners using fuel oil. This can be accounted for from the flowrate of atomizing steam and the enthalpy of the atomizing steam.

The performance of steam boilers is measured by boiler efficiency. Two different methods are used to calculate the boiler efficiency: the direct method and the indirect method. In the direct method, the efficiency is calculated directly from the heat output from the boiler and the fuel input:

$$\eta_{BOILER} = \frac{Q_{OUTPUT}}{Q_{INTPUT}} = \frac{m_{STEAM}(H_{STEAM} - H_{BFW})}{m_{FUEL}\, NCV} \tag{23.4}$$

where η_{BOILER} = boiler efficiency (–)

In the indirect method, the efficiency is calculated from the fuel input and the losses:

$$\eta_{BOILER} = 1 - \frac{Q_{LOSS}}{Q_{INTPUT}} \tag{23.5}$$

where Q_{LOSS} = sum of the losses for the stack, unburnt combustible gas, unburnt combustible solid or liquid fuel, ash, casing, blowdown and atomizing steam (W, kW)

The definition used for boiler efficiency here is based on the net calorific value of the fuel. Efficiencies can be based on the gross calorific value or the net calorific value. Values based on the gross calorific value will be lower than those based on the net calorific value.

Figure 23.9 shows the steam generation profile in the boiler as it is matched against the boiler flue gas from the theoretical flame temperature as discussed in Chapter 12 (typically 1800 °C) to ambient temperature. Figure 23.9 shows the steam being preheated from subcooled boiler feedwater to superheated steam. Maximum steam temperature is normally below 600 °C. Higher temperatures require expensive materials of construction. In turn, the flue gas is cooled to the stack temperature before release to the atmosphere. The flue gases are normally kept above their dew point before release to atmosphere, particularly if there is any sulfur in the fuel. The heat loss in the stack Q_{STACK} is the major inefficiency in the use of the fuel.

The efficiency of boilers changes with the boiler load. Figure 23.10a illustrates the variation of boiler performance with load. The maximum load at which the boiler can be operated continuously is the *maximum continuous rating* (MCR). Operation in excess of MCR can be maintained for only short periods. The point of maximum boiler efficiency is known as the *economic continuous rating* (ECR). ECR is usually in the range of 60 to 80% of MCR, depending on the boiler design. The average working load of the boiler should correspond with ECR. Thus, for most applications it is best to have ECR at around 80% of MCR. Boilers can normally be operated as low as 20% of MCR. Boiler

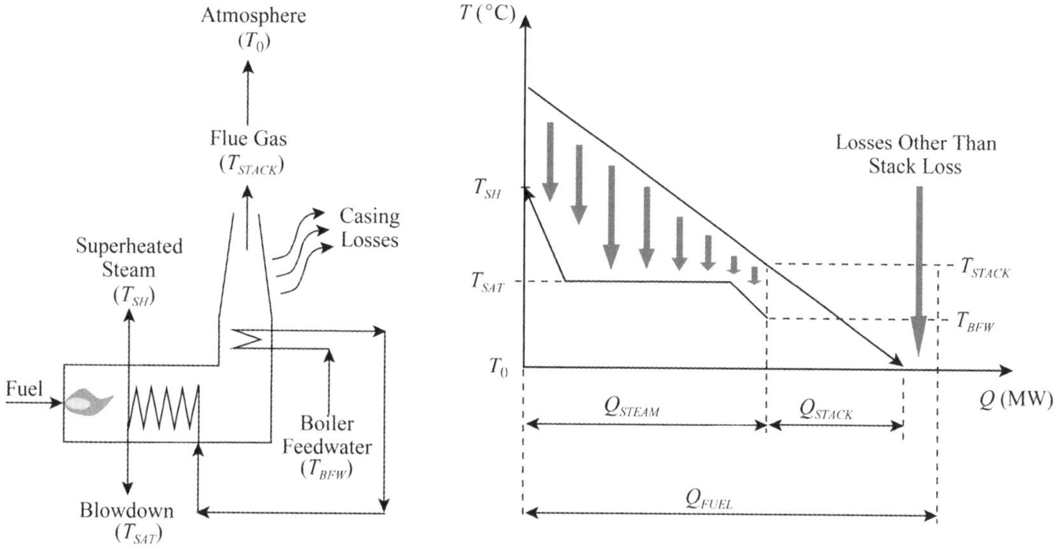

Figure 23.9
Model of steam boiler.

efficiencies can range from 75% to 90%, but can be as high as 93% or 94% for large water-tube boilers. There are two major factors that contribute to the variation of efficiency with load. The casing loss, principally from radiation, does not depend on the load. Thus, at low load proportionally more heat is lost as a result of the casing loss. This causes a decrease in efficiency as the load is decreased, and a corresponding increase as the load is increased. The other major effect causing the variation in efficiency is due to the flowrate of the flue gas. Increasing the flowrate of flue gas as the boiler load is increased results in a less than proportionate increase in the rate of heat transfer from the combustion gases to the heat transfer surfaces. The two effects combine to result in a maximum in the boiler efficiency, as illustrated in Figure 23.10a.

Modeling the variation of efficiency with a load across a wide range is difficult to achieve accurately using a single modeling equation. A polynomial equation could be used. However, the generally poor accuracy of the data available and the resulting problems the nonlinearity causes in optimization call for a simpler

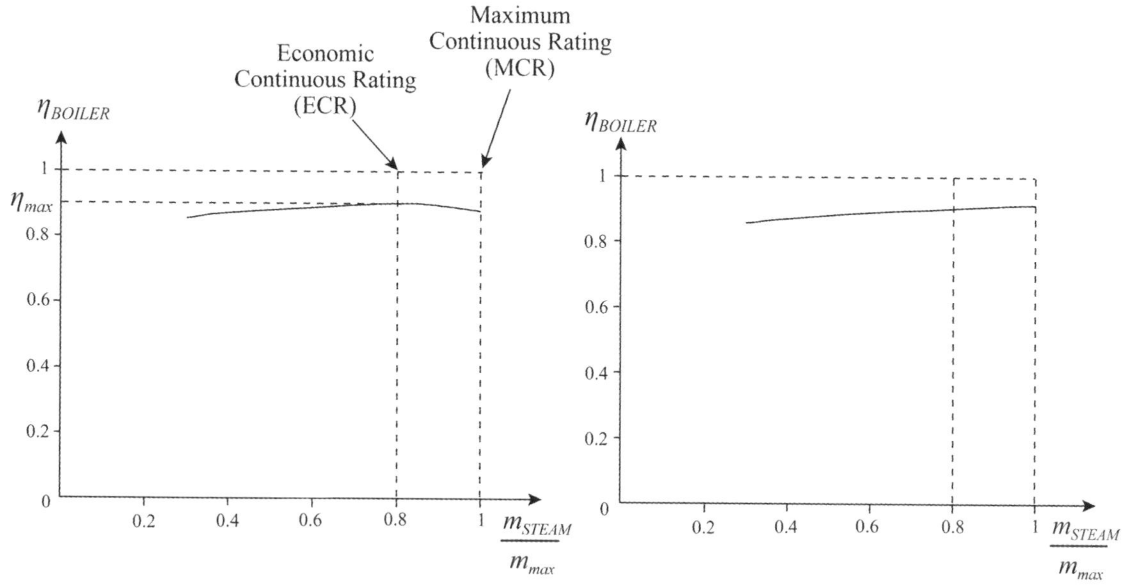

(a) The boiler efficiency as a function of load normally shows a maximum.

(b) Boiler efficiency showing only a monotonic increase in efficiency with increase in load.

Figure 23.10
Steam boiler performance at part-load.

approach. A simpler form of equation can be used over narrower ranges (Shang, 2000):

$$\eta_{BOILER} = \frac{1}{a + b\dfrac{m_{max}}{m_{STEAM}}} \tag{23.6}$$

where m_{max} = maximum flowrate of steam from the boiler ($kg{\cdot}s^{-1}$)

m_{STEAM} = mass flowrate of steam generated in the boiler ($kg{\cdot}s^{-1}$)

a, b = correlating parameters

For a given boiler, the correlating parameters a and b can be determined by fitting data points from part-load operation. The values of a and b depend on the design of boiler in new design and also on the age and maintenance of the boiler in existing equipment. From Equation 23.6, the fuel required for the boiler is given by:

$$\begin{aligned} Q_{FUEL} &= \frac{Q_{STEAM}}{\eta_{BOILER}} \\ &= \frac{\Delta H_{STEAM}\, m_{STEAM}}{\eta_{BOILER}} \\ &= \Delta H_{STEAM}[a m_{STEAM} + b m_{max}] \end{aligned} \tag{23.7}$$

where ΔH_{STEAM} = enthalpy difference between generated steam and boiler feedwater

This simpler form of Equation 23.6 has the advantage that it leads to a linear equation between fuel consumption and steam generation for use in optimization. Equation 23.7 is adequate for many modeling and optimization applications. It provides a linear relationship between Q_{FUEL} and m_{STEAM}, with ΔH_{STEAM}, m_{max}, a and b all being constant for a given boiler. However, the shape of the curve in Figure 23.10b only shows a monotonic increase of efficiency with load. It may be required to model an efficiency profile that goes through a maximum, in which case Equation 23.7 can be fitted to different ranges of m_{STEAM}/m_{max} with one on each side of ECR. This allows a linear model to be retained.

Given that the largest loss from the boiler is in the stack, reducing the stack loss has the biggest effect on the boiler efficiency. Every 22 °C reduction in the flue gas temperature increases the boiler efficiency by 1%. *Economizers* can be used to preheat the incoming boiler feedwater from the flue gas and decrease the stack temperature, as illustrated in Figures 23.6 to 23.8. These are heat exchangers between the incoming boiler feedwater and the hot flue gases before they are vented to atmosphere (Dryden, 1982). This increases the energy efficiency of the boiler. Another option to increase energy efficiency is to use an *air preheater* (Dryden, 1982), in which there is heat recovery from the hot flue gas to the incoming cold combustion air, as illustrated in Figure 23.8. This has the disadvantage that the combustion temperature is increased with a resulting increase in the formation of oxides of nitrogen (NO_X). The *acid dew point* limits the minimum temperature for flue gases containing oxides of sulfur (SO_X). If condensation occurs with flue gases containing SO_X, this will cause corrosion of metal surfaces. Metal heat exchange surfaces should normally be above 130 to 160 °C in flue gas containing any SO_X. The acid dew point depends on the sulfur content of the fuel and the amount of excess air used.

Excess air also has a major influence on the boiler efficiency. Excess air is essential to maintain the efficiency combustion process. However, too much excess leads to unnecessary stack loss. Boilers are typically controlled to have excess oxygen in the flue gas of around 3%. Every 1% above this will decrease the efficiency by 1%.

Example 23.3 A medium-sized boiler has the following specifications:

Steam output	$31.6\ kg{\cdot}s^{-1}$
Steam pressure	60 bar
Steam temperature	500 °C
BFW temperature	100 °C
Fuel, natural gas	96.5% CH_4, 0.5% C_2H_6, remainder noncombustible
GCV	$38{,}700\ kJ{\cdot}m^{-3}$ at 15 °C
Fuel consumption	$2.9\ m^3{\cdot}s^{-1}$

Determine the boiler efficiency based on the net calorific value of the fuel.

Solution

$$CH_4(vap) + 2O_2 \rightarrow CO_2 + 2H_2O(vap)$$

1 kmol CH_4 produces 2 kmol H_2O, 0.965 kmol CH_4 produces $2 \times 0.965 \times 18 = 34.74$ kg H_2O

$$C_2H_6(vap) + 3\tfrac{1}{2}O_2 \rightarrow 2CO_2 + 3H_2O(vap)$$

1 kmol C_2H_6 produces 3 kmol H_2O, 0.005 kmol C_2H_6 produces $3 \times 0.005 \times 18 = 0.27$ kg H_2O. Thus, 1 kmol of gas produces $34.74 + 0.27 = 35.01$ kg of H_2O

$$\begin{aligned} \text{Also, volume occupied by 1 kmol of gas} &= (RT/P) \\ &= (8314.5 \times 288/1.013 \times 10^5) \\ &= 1223.64\ m^3 \text{ at 1.013 bar and 15 °C} \end{aligned}$$

$$\begin{aligned} \text{Therefore, steam formed per } m^3 \text{ of gas} &= 35.01/23.64 \\ &= 1.481\ kg \end{aligned}$$

Then, using the relationship between GCV and NCV:

$$Q_{GCV} = Q_{NCV} + m_{COND}\Delta H_{VAP}$$

$$Q_{NCV} = 38{,}700 - (1.481 \times 2441.8) = 35{,}084\ kJ{\cdot}m^{-3}$$

Taking the steam properties from steam tables, the heat transferred to the working fluid is given by:

$$H_{STEAM} - H_{BFW} = 3421 - 419.1 = 3001.9\ kJ{\cdot}kg^{-1}$$

The boiler efficiency is given by:

$$\eta_B = (3001.9 \times 31.6) / (2.9 \times 35084)$$
$$= 0.93\,(93\%)$$

23.3 Gas Turbines

In its simplest form, the gas turbine consists of a *compressor* and a *turbine* (or *expander*). In a *single-shaft* gas turbine, these are mechanically connected and are rotating at the same speed, as shown in Figure 23.11a. Air from ambient is compressed and raised in temperature as a result of the compression. A portion of the compressed air enters a combustion chamber where fuel is fired to raise the temperature of the gases. Most of the air from the compressor provides cooling for the combustor walls. The hot, compressed mixture of air and combustion gases then flows to the inlet of the turbine. Combustor outlet temperatures range between 800 °C to over 1600 °C. In the turbine, the gas is expanded to develop power to drive the compressor. Around two-thirds of the power produced by the turbine is needed to drive the compressor. However, by virtue of the energy imported in the combustion process, excess power is produced. The higher the expander inlet temperature, the better is the performance of the machine. Temperatures in the turbine are limited by turbine blade materials.

Different machine configurations are possible with gas turbines. Figure 23.11b shows a *split-shaft* or *twin-shaft* arrangement. This is mechanically more complex. The first turbine provides the power necessary to drive the compressor. The second turbine provides the power for the external load. Figure 23.11c shows a *hollow-shaft* or *twin-spool* arrangement in which the compression and expansion are separated into two stages linked by concentric shafts. The high-temperature turbine drives a high-pressure air compressor and a low-pressure turbine is used to drive the low-pressure compressor and the load. Each shaft is run at the optimum speed, improving the overall efficiency. The expanded gas can be discharged directly to atmosphere, as shown in Figure 23.11a to c, or can be used to preheat the air at the exit of the compressor before entering the combustion chamber in a *regenerator*, as illustrated in Figure 23.11d. This increases the efficiency of the machine. However, the increased pressure drop on both the compressed air and turbine exhaust sides of the recuperator creates some inefficiency in the cycle.

The efficiency of the gas turbine can be defined as:

$$\eta_{GT} = \frac{W}{Q_{FUEL}} \tag{23.8}$$

where η_{GT} = gas turbine efficiency (–)
W = turbine shaft power (kW)
Q_{FUEL} = heat released from fuel (kW)

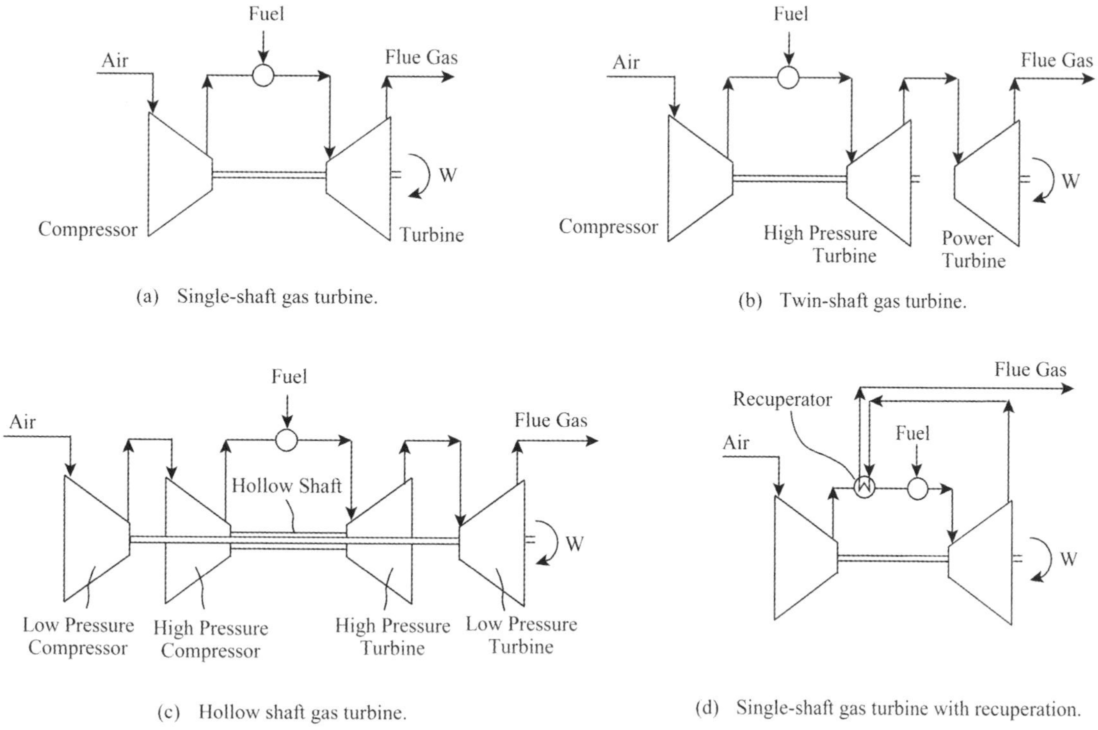

Figure 23.11
Gas turbine configurations.

23

Table 23.2
Gas turbine types.

Industrial	Aero-derivative
• Lower-pressure ratio (typically up to 15) • Robust • Lower efficiency (typically 30 to 35%) • Long periods of continuous operation without overhaul • Lower capital cost per unit power output • Sizes over 300 MW	• Higher-pressure ratio (typically up to 30) • Lightweight • Higher efficiency (typically up to 45%) • Require more frequent maintenance than industrial machines • Higher capital cost per unit power output (require advanced materials) • Sizes limited up to 65 MW

The efficiency of gas turbines can alternatively be expressed as a heat rate, which is simply the inverse of efficiency:

$$HR = \frac{Q_{FUEL}}{W} \tag{23.9}$$

where HR = gas turbine heat rate (–)

Depending on the definition of W and Q_{FUEL} in Equation 23.9, the heat rate often quoted is units dependent. The performance of the machine is normally specified at maximum load at International Standards Organization (ISO) conditions of 15 °C, 1.013 bar and 60% relative humidity. For families of gas turbines from a particular manufacturer, the performance at maximum load can be correlated according to:

$$Q_{FUEL,max} = a + bW_{max} \tag{23.10}$$

where $Q_{FUEL,max}$ = heat released from fuel at maximum load and ISO conditions (MW)
W_{max} = turbine shaft power at maximum load and ISO conditions (MW)
a, b = constants depending on the family of machine and turbine manufacturer

For example, $a = 2.668$, $b = 21.99$ MW for 25 MW < W_{max} < 250 MW (Brooks, 2001; Varbanov, Doyle and Smith, 2004). Equation 23.10 can be rearranged to give:

$$\eta_{GT,max} = \frac{1}{a + \frac{b}{W_{max}}} \tag{23.11}$$

Equations 23.10 and 23.11 can be used for preliminary sizing of machines. However, it should be emphasized that such correlations apply only to a class of machines from a particular manufacturer.

The basic characteristics of gas turbines are:

- Sizes are restricted to standard frame sizes ranging from under 500 kW to over 300 MW.
- *Microturbines* are available in the size range 25 kW to 500 kW.
- Electrical efficiency is 30 to 45% (increasing to over 50% for designs with cooled turbine blades).
- Exhaust temperatures are in the range 450 to 600 °C.
- Fuels need to be at a high pressure.
- Fuels must be free of particulates and sulfur.

The most common fuels used are gas (natural gas, methane, propane, synthesis gas) and light fuel oils. Contaminants such as ash, alkalis (sodium and potassium) and sulfur result in deposits, which degrade performance and cause corrosion in the hot section of the turbine. Total alkalis and total sulfur in the fuel should both be less than 10 ppm. Gas turbines can be equipped with dual firing to allow the machine to be switched between fuels (e.g. natural gas and light fuel oil).

Gas turbines can be classified as *industrial* or *frame* machines and *aero-derivative* machines, which are mechanically more complex. The aero-derivative machines are lighter weight units derived from aircraft engines. Table 23.2 compares the characteristics of the two broad classes of gas turbine machines.

Aero-derivative gas turbines are typically used for offshore oil exploration applications where weight and efficiency are a premium, to drive compressors for natural gas pipelines and in stand-alone power generation applications for peak periods of high power demand. For stand-alone applications, gas turbine efficiency becomes a critical issue. However, if heat is to be recovered from the gas turbine exhaust, the efficiency becomes less important as the waste heat is utilized.

Gas turbines require a start-up device, which is usually an electric motor. The power of the start-up device can be up to 15% of the gas turbine power, depending on the size and design of the machine. Once the gas turbine has been started, the motor can be switched off, run as a helper motor to boost the power from the gas turbine or switched to be a generator to generate electricity from the gas turbine.

The gas turbine performance is a function of a number of important parameters.

1) *Expander inlet temperature.* The power produced by the turbine (expander) is proportional to the absolute temperature of the inlet gases. An increase in the expander inlet temperature increases the power output and efficiency of the machine. The maximum temperature is constrained by the turbine blade materials. Expander inlet temperatures up to 1300 °C can be

achieved without cooling of the turbine blades. Air cooling of the turbine blades can be used in which 1 to 3% of the main air flow is directed through internal passages within the blades to provide cooling. This allows turbine inlet temperatures to be increased to 1600 °C and higher.

2) *Pressure ratio.* As the design pressure ratio across the compressor increases, the power output initially increases to a maximum and then starts to decrease. The optimum pressure ratio increases with increasing expander inlet temperature. Pressure ratios for industrial machines are typically in the range 10 to 15 but can be higher. For aero-derivative machines, pressure ratios are typically 20 to 30.
3) *Ambient conditions.* The performance of the machine is normally specified at International Standards Organization (ISO) conditions of 15 °C, 1.013 bar and 60% relative humidity. The power consumed by the compressor is proportional to the absolute temperature of the inlet air. Thus, a decrease in the ambient temperature increases both the power output and the efficiency for a given machine, and vice versa. For example, at 40 °C, the power output might typically drop to around 90% of the ISO-rated power. At 0 °C, the power output might increase to 106% of the ISO-rated power. In hot climates, inlet air cooling (e.g. by using a spray of water into the inlet air or an inlet air chiller) can be used to increase performance. However, care must be taken to avoid carryover of water into the compressor (e.g. by the installation of a coalescing pad), otherwise this can cause compressor fouling and degrade performance. Decreasing relative humidity causes the power output and efficiency to decrease. Increasing altitude causes the power output to decrease.
4) *Working load.* The efficiency drops off quickly as the load decreases. The more the load is decreased from 100% rated capacity, the steeper is the decline in the machine performance. The decline in performance depends on the machine and the control system. For example, at 70% of full load, the ratio of the part-load efficiency to the efficiency at full load might vary typically between 0.95 and 0.85, depending on the machine and the control system. Twin-shaft and hollow-shaft machines have a better part-load performance than single-shaft machines. Part-load performance depends on the size of the machine. However, in a gas turbine cogeneration system the machine is often designed such that the gas turbine is not routinely operated at full load in order to increase the reliability of the machine.
5) *Back-pressure.* Before the gases are vented to atmosphere, they might go through a device that has a pressure drop. This could be a heat recovery steam generator (HRSG), a furnace or a NO_X treatment unit. The back-pressure created by such a device decreases the power generation.

The combustion within the gas turbine creates emissions. The principal concern is usually NO_X. NO_X emissions will be dealt with in detail in Chapter 25, but it is worth reviewing the problems associated with gas turbines at this point. NO_X emissions from gas turbines can be dealt with in one of three ways as follows:

1) *Staged combustion.* In a standard combustion arrangement in a gas turbine, the fuel and air are mixed in the combustor. This leads to a high local peak flame temperature in which the nitrogen in the combustion air reacts with oxygen to produce NO_X. *Staged* or *premixed* combustion alleviates this problem by premixing the fuel and air in a substoichiometric mixture. This first stage of combustion involves a fuel-rich zone. Additional air is then mixed and the combustion completed. Such staged combustion lowers the peak flame temperature and the NO_X formation. NO_X emissions can be reduced to 10 ppm for gaseous fuels and 25 ppm for liquid fuels.
2) *Steam injection.* Steam can be injected into the combustion zone as an inert material with the purpose of reducing the peak flame temperature and thereby reducing NO_X formation. NO_X emissions can be reduced by typically 60% by steam injection. An obvious drawback is that the injected steam is lost to atmosphere and the latent heat in the steam cannot be recovered. A positive side effect of the steam injection is that it increases the power output due to the higher mass flowrate through the turbine. Indeed, steam injection over and above that required for NO_X suppression can be used to increase power production during times of peak power demand.
3) *Treatment of the exhaust gases.* If staged combustion or steam injection cannot satisfy the regulatory requirements for NO_X emissions, then the gas turbine exhaust must be treated to remove the NO_X. Alternatively, if an existing gas turbine installation requires the NO_X emissions to be decreased, then, again, the exhaust gases must be treated. The usual method to treat NO_X emissions in gas turbine exhausts is by *selective catalytic reduction*. This will be dealt with in more detail in Chapter 25, but involves the injection of ammonia upstream of a catalyst to chemically reduce the NO_X to nitrogen.

Gas turbines can be modeled as three separate sections:

1) The air compressor can be modeled by Equation 13.42 or 13.43 with an efficiency of 85 to 90% (Perry and Green, 1997).
2) The combustion chamber where the fuel is injected and reacts with the compressed air can be modeled as a combustion process as detailed in Chapter 12. However, there is also a significant pressure loss that must be accounted for. The pressure loss varies from 2 to 8% and the combustion temperature from 800 to over 1600 °C, depending on the machine design (Perry and Green, 1997).
3) The turbine expander can be modeled by Equation 13.72 or 13.73 with an efficiency from 88 to 92% (Perry and Green, 1997).

To model a specific machine, the compressor efficiency, combustion temperature, combustor pressure drop and turbine efficiency can be varied until the model agrees with the ISO-rated performance specified by the manufacturer.

Whilst this modeling approach is useful for conceptual design, it is not readily adapted for use in the modeling and optimization of an existing system if part-load operation is to be varied. This is because the parameters used to model the machine at ISO-rated conditions are likely to change with load. An alternative approach can be developed based on the Willans Line approach, named after the Victorian mechanical engineer Peter William Willans. To develop such an approach, the fuel-to-air ratio and steam-to-air ratio can be defined as:

$$f = \frac{m_{FUEL}}{m_{AIR}} \qquad s = \frac{m_{STEAM}}{m_{AIR}}$$

A mass balance gives:

$$\begin{aligned} m_{EX} &= m_{AIR} + m_{STEAM} + m_{FUEL} \\ &= \frac{f+s+1}{f} m_{FUEL} \end{aligned} \tag{23.12}$$

where m_{AIR}, m_{STEAM}, m_{FUEL}, m_{EX} = mass flowrates of the inlet air, steam injected (if any), fuel and exhaust respectively

An energy balance across the gas turbine gives (Shang, 2000):

$$\begin{aligned} m_{AIR}H_{AIR} + m_{STEAM}H_{STEAM} + m_{FUEL}H_{FUEL} + m_{FUEL}\Delta H_{COMB} \\ - m_{EX}H_{EX} - W - W_{LOSS} = 0 \end{aligned} \tag{23.13}$$

where H_{AIR}, H_{STEAM}, H_{FUEL}, H_{EX} = specific enthalpy of the inlet air, injected steam (if any), fuel and exhaust respectively
ΔH_{COMB} = net heat of combustion of the fuel
W = power production
W_{LOSS} = power lost due to mechanical inefficiencies

Combining Equations 23.12 and 23.13 and rearranging gives (Shang, 2000; Varbanov, Doyle and Smith, 2004):

$$W = n\, m_{FUEL} - W_{LOSS} \tag{23.14}$$

where $n = \frac{H_{AIR}}{f} + \frac{s}{f} H_{STEAM} + H_{FUEL} + \Delta H_{COMB} - \frac{f+s+1}{f} H_{EX}$

If the gas turbine has a fixed fuel-to-air ratio and fixed steam-to-air ratio at part load and the mechanical losses are assumed to be independent of load, then Equation 23.14 forms a linear relationship between machine output and mass flow of fuel. The relationship between shaft power and mass flow through a heat engine is sometimes referred to as a Willans Line (Shang, 2000; Varbanov, Doyle and Smith, 2004). Whether this is strictly true or not depends on the control system for the gas turbine. Thus, the Willans Line for the gas turbine can be written as:

$$W = n\, m_{FUEL} - W_{INT} \tag{23.15}$$

where W_{INT} = intercept of the Willans Line (kW, MW)

However, Equation 23.15 assumes that the gas turbine operates at ISO conditions (ambient conditions of 15 °C, 1.013 bar and 60% relative humidity). Most importantly, a correction is needed for the ambient temperature. Thus, Equation 23.15 can be modified to include a correction for ambient temperature:

$$W = (n\, m_{FUEL} - W_{INT})(a + bT_0) \tag{23.16}$$

where T_0 = ambient temperature (°C)
a, b = correlating parameters for power

Thus, to model the performance of a gas turbine at part load requires n, W_{INT}, a, and b to be fitted to gas turbine operating data by regression. The term $(a + bT_0)$ must be constrained to equal 1.0 at 15 °C. The correlating parameters for variation of power with ambient temperature are specific to a particular gas turbine model.

Figure 23.12 shows integrated gas turbine arrangements in which the exhaust from the gas turbine is used to generate steam in a *heat recovery steam generator* (HRSG) before being vented to atmosphere. It is possible to fire fuel after the gas turbine to increase the temperature of the gas turbine exhaust before entering the HRSG, as gas turbine exhaust is still rich in oxygen (typically 15% O_2). This is known as *supplementary* or *auxiliary* firing. The steam from the HRSG might be used directly for process heating or be expanded in a steam turbine system to generate additional power. Figure 23.12a shows an HRSG in which a single pressure level of steam is generated. Figure 23.12b shows a more complex arrangement where two levels of steam are generated in a *dual-pressure* HRSG. Figure 23.13a shows the corresponding representation of a single-pressure HRSG without supplementary firing on a temperature–enthalpy diagram. The steam profile includes boiler feedwater preheat, latent heat and superheat. Maximizing the heat recovery from the gas turbine exhaust into steam generation will be limited by a pinch between the gas turbine exhaust and the steam generation profile. Figure 23.13b shows a dual-pressure HRSG in a temperature–enthalpy diagram. This includes boiler feedwater preheat and superheat for both the low-pressure (LP) and high-pressure (HP) levels. The dual-pressure HRSG allows a better match between the steam profile and the gas turbine exhaust. This means that a greater amount of heat recovery is possible when compared with a single pressure HRSG. In the case of the dual pressure HRSG, there are degrees of freedom for how much LP and HP steam are generated, leading to two pinches that limit the heat recovery potential. The largest gas turbines sometimes use a triple-pressure HRSG, although these are normally reserved for power station applications.

Figure 23.13 shows the potential to generate steam from a gas turbine exhaust without supplementary firing. The potential to generate steam can be increased by introducing firing of fuel after the gas turbine. There are three firing modes for gas turbines:

1) *Unfired HRSG.* An unfired HRSG uses the sensible heat in the gas turbine exhaust to raise steam.
2) *Supplementary fired HRSG.* Supplementary firing raises the temperature by firing fuel to use a portion of the oxygen in the exhaust. Supplementary firing uses convective heat transfer and temperatures are limited to a maximum of 925 °C by ducting materials. However, the supplementary firing temperature can be increased to 1200 to 1300 °C if the walls of the HRSG are water cooled.
3) *Fully fired HRSG.* Fully fired HRSGs make full utilization of the excess oxygen to raise the maximum amount of steam in the HRSG. Full firing means reducing the excess oxygen to a minimum of around the 3% normally demanded by all

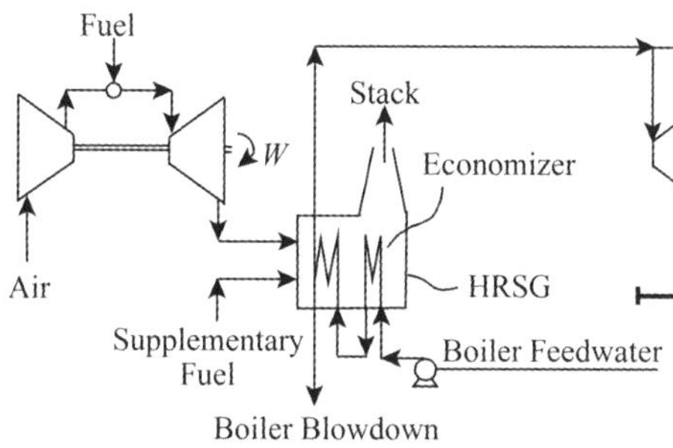

(a) Single-pressure HRSG.

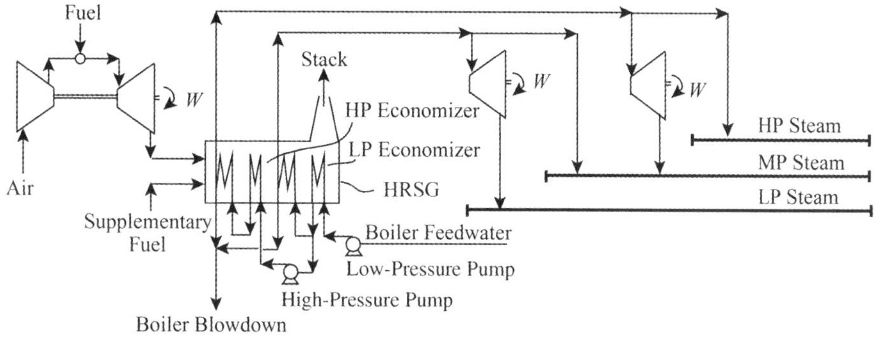

(b) Dual-pressure HRSG.

Figure 23.12

Gas turbine with heat recovery steam generator (HRSG).

combustion processes to ensure efficient combustion. However, this means that radiant heat transfer will result from full firing. Essentially, full firing means that the gas turbine exhaust is used as preheated combustion air to a steam boiler.

Figure 23.14 shows supplementary firing on a temperature–enthalpy diagram. The supplementary firing increases the temperature of the exhaust, thereby increasing the amount of available heat and the potential for steam generation. Supplementary firing

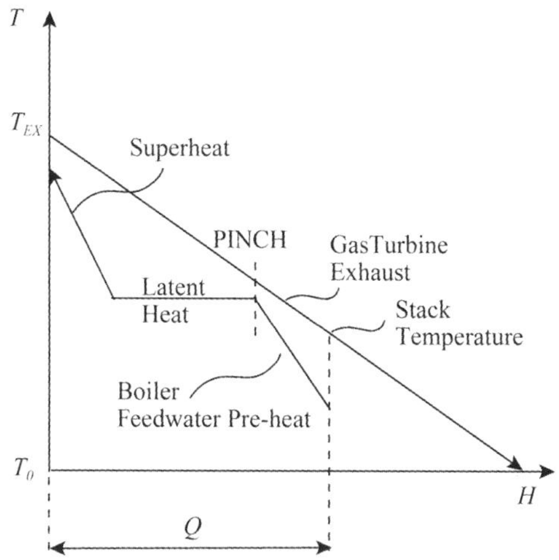

(a) Single Pressure HRSG.

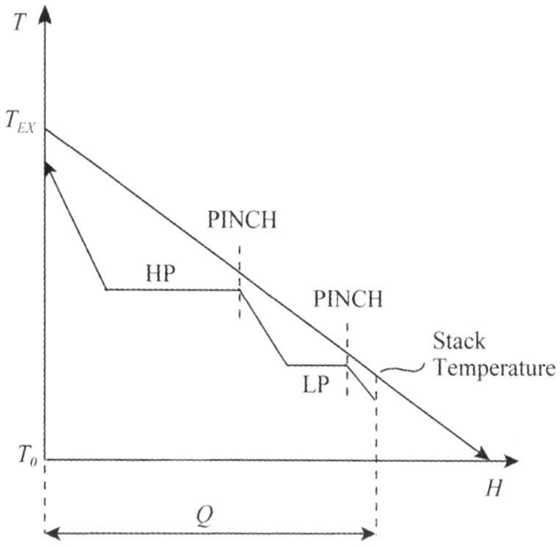

(b) Dual Pressure HRSG.

Figure 23.13

Temperature–enthalpy representation of HRSGs.

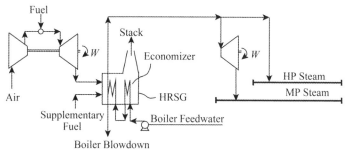

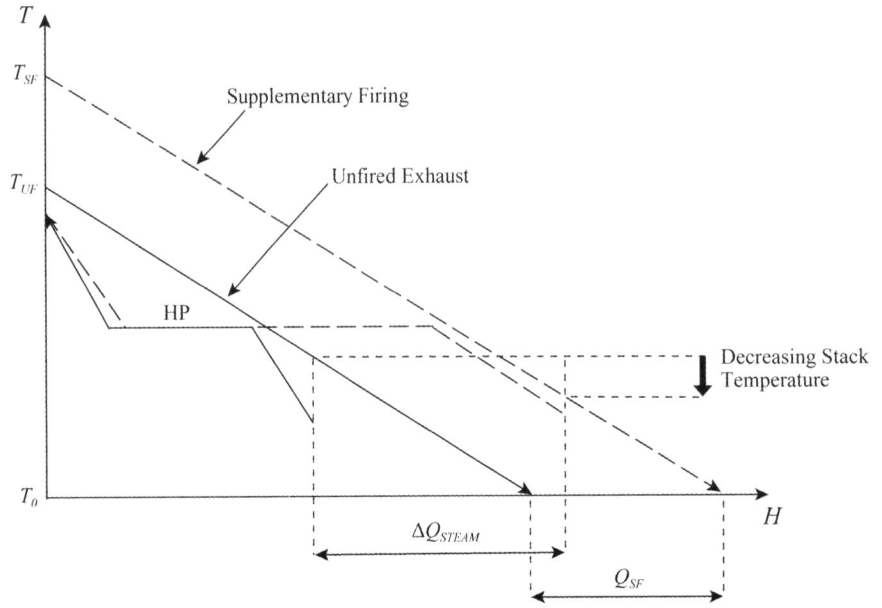

Figure 23.14
Supplementary fired HRSG.

allows greater amounts of steam to be generated with a high efficiency. Higher steam superheat temperatures are possible. Also, importantly, supplementary firing introduces more flexibility to the steam system. As already discussed, gas turbines are inflexible in terms of their part load performance. Supplementary firing allows the amount of steam generation to be varied without changing the load on the gas turbine. Figure 23.14 shows a situation in which the supplementary firing allows a closer match between the gas turbine exhaust and the steam generation profile when compared with the unfired case. The closer match for the supplementary fired case results in a lower stack temperature when compared with the unfired case. This can mean that the additional heat to steam generation from supplementary firing is greater than the heat input from fuel from the supplementary firing, as a result of the lower stack temperature.

Example 23.4 A gas turbine is to be modeled at ISO conditions. From the manufacturer's data, the air flowrate is 641 $kg \cdot s^{-1}$, the pressure ratio is 17 and the turbine inlet temperature is 1250 °C. It can be assumed that the gas has a compressibility factor of 1.0 and mean molar mass of 28.8 $kg \cdot kmol^{-1}$. Calculate:

a) The power required for the compressor and the outlet temperature from the compressor, assuming the isentropic efficiency to be 0.85 and $\gamma = 1.38$ for the gas.
b) The power generated by the turbine expansion and the exhaust temperature, assuming the pressure drop across the combustor to be 6% of the inlet pressure, exhaust pressure to be 1.013 bar, isentropic efficiency to be 0.88 and $\gamma = 1.32$ for the gas.
c) The shaft power generated by the gas turbine, neglecting mechanical losses.
d) The heat required from fuel, turbine efficiency and turbine heat rate in $kJ \cdot kWh^{-1}$, assuming the mean molal heat capacity of the gas to be 33.2 $kJ \cdot kmol^{-1} \cdot K^{-1}$.

Solution

a) The power required by the compressor can be calculated from Equation 13.43:

$$W = \left(\frac{\gamma}{\gamma - 1}\right)\frac{ZRT_{in}}{\eta_{IS}}\left[1 - \left(\frac{P_{out}}{P_{in}}\right)^{(\gamma-1)/\gamma}\right]$$

$$= \frac{641}{28.8}\left(\frac{1.38}{1.38 - 1}\right) \times \frac{1.0 \times 8314.5 \times 288.15}{0.85} \times \left[1 - \left(\frac{17 \times 1.013}{1.013}\right)^{(1.38-1)/1.38}\right]$$

$$= -2.693 \times 10^{8}\ W$$

The outlet temperature from the compressor can be calculated from Equation 13.47:

$$T_{out} = \left(\frac{T_{in}}{\eta_{IS}}\right)\left[\left(\frac{P_{out}}{P_{in}}\right)^{(\gamma-1)/\gamma} + \eta_{IS} - 1\right]$$

$$= \left(\frac{288.15}{0.85}\right)\left[(17)^{(1.38-1)/1.38} + 0.85 - 1\right]$$

$$= 688.8\ K$$

$$= 415.6\ °C$$

b) Power generated by the expander can be calculated from Equation 13.73:

$$W = \left(\frac{\gamma}{\gamma - 1}\right)\eta_{IS,E} ZRT_{in}\left[1 - \left(\frac{P_{out}}{P_{in}}\right)^{(\gamma-1)/\gamma}\right]$$
$$= \left(\frac{641}{28.8}\right)\left(\frac{1.32}{1.32 - 1}\right) \times 0.88 \times 1.0 \times 8314.5 \times 1523.15$$
$$\times \left[1 - \left(\frac{1.013}{0.94 \times 17 \times 1.013}\right)^{(1.32-1)/1.32}\right]$$
$$= 5.006 \times 10^8 \text{ W}$$

Exhaust temperature from the turbine can be calculated from Equation 13.80:

$$T_{out} = T_{in}\eta_{IS,E}\left[\left(\frac{P_{out}}{P_{in}}\right)^{(\gamma-1)/\gamma} + \frac{1}{\eta_{IS,E}} - 1\right]$$
$$= 1523.15 \times 0.88$$
$$\times \left[\left(\frac{1.013}{0.94 \times 17 \times 1.013}\right)^{(1.32-1)/1.32} + \frac{1}{0.88} - 1\right]$$
$$= 867.4 \text{ K}$$
$$= 594.2\,^{\circ}\text{C}$$

c) Shaft power generated by the gas turbine

$$= (5.006 - 2.693) \times 10^8$$
$$= 2.313 \times 10^8 \text{ W}$$
$$= 231.3 \text{ MW}$$

d) Heat required from fuel

$$= \left(\frac{641}{28.8}\right) \times 33.2 \times (1250 - 415.6)$$
$$= 6.165 \times 10^5 \text{ kJ·s}^{-1}$$
$$\text{Heat rate} = \frac{6.165 \times 10^5 \times 3600}{231.3 \times 10^3}$$
$$= 9595 \text{ kJ·kWh}^{-1}$$
$$\eta = \frac{231.3 \times 10^3}{6.165 \times 10^5}$$
$$= 0.375$$

Example 23.5 Data are given for a gas turbine firing natural gas at ISO conditions:

$$W = 26,070 \text{ kW}$$
$$HR = 12,650 \text{ kJ·kWh}^{-1}$$
$$m_{EX} = 446,000 \text{ kg·h}^{-1}$$
$$T_{EX} = 488\,^{\circ}\text{C}$$

Using these data calculate:

a) The fuel consumption ΔH_{FUEL} and efficiency η_{GT} at standard conditions.

b) Heat recovered from the exhaust if $T_{STACK} = 100\,^{\circ}$C and if C_P of the exhaust is 1.1 kJ·kg^{-1}·K^{-1}.

c) W_{max}, HR, η_{GT} and Q_{FUEL} at an ambient temperature of 0 °C. The change in performance can be estimated from

$$W_{max}(\text{kW}) = 29,036 - 189.4T_0$$
$$HR(\text{kJ·kWh}^{-1}) = 12,274 + 31.32T_0$$

where T_0 = ambient temperature (°C)

d) The heat available in the exhaust if it is supplementary fired to 850 °C for ISO conditions.

Solution

a) From Equation 23.9:

$$\Delta H_{FUEL} = W\,HR$$
$$= \frac{26,070(\text{kW}) \times 12,650(\text{kJ·kWh}^{-1})}{3600(\text{kJ·kWh}^{-1})}$$
$$= 91,607 \text{ kW}$$
$$\eta_{GT} = \frac{1}{HR}$$
$$= \frac{3600\,(\text{kJ·kWh}^{-1})}{12,650\,(\text{kJ·kWh}^{-1})}$$
$$= 0.28$$

b)

$$Q_{EX} = mC_P(T_{EX} - T_{STACK})$$
$$= \frac{446,000 \times 1.1 \times (488 - 100)}{3600}$$
$$= 52,876 \text{ kW}$$
$$= 52.9 \text{ MW}$$

c) At 0 °C:

$$W_{max} = 29,036 - 189.4T_0$$
$$= 29,036 - 189.4 \times 0$$
$$= 29,036 \text{ kW}$$
$$HR = 12,274 + 31.32T_0$$
$$= 12,274 \text{ kJ·kWh}^{-1}$$
$$\eta_{GT} = \frac{3600}{12,274}$$
$$= 0.29$$
$$Q_{FUEL} = W\,HR$$
$$= \frac{29,036 \times 12,274}{3600}$$
$$= 98,997 \text{ kW}$$

d)

$$Q_{SF} = mC_P(T_{SF} - T_{EXHAUST})$$
$$= \frac{446,000 \times 1.1 \times (850 - 488)}{3600}$$
$$= 49,333 \text{ kW}$$
$$= 49.3 \text{ MW}$$

Total heat in the exhaust with supplementary firing

$$= 52.9 + 49.3$$
$$= 102.2 \text{ MW}$$

23

The emphasis on gas turbine integration so far has been to utilize the exhaust for generation of steam in unfired, supplementary fired or fully fired systems. However, the exhaust can be used in other ways. In principle, it can be used directly for process heating. This would require a direct match between process fluids and the hot gas turbine exhaust gases. Safety issues might, however, prevent this. If the process fluid is flammable and a leak develops, then flammable material will leak into a duct of hot, oxygen-rich gas with all the potential for fire and explosion. The gas turbine exhaust can also be used for drying. Drying, as discussed in Chapter 7, usually involves the removal of water by evaporation into hot air. In some applications, a gas turbine can be ducted directly into a dryer to provide the hot air. However, care should be taken to ensure that the small amounts of combustion products (including NO_X) do not contaminate the process. Finally, the hot exhaust gases from a gas turbine can be used to supply preheated combustion air (albeit slightly depleted in oxygen) to furnaces. Examples of the application of this type of arrangement are in ethylene production with the furnace reactors and fired heaters in petroleum refinery applications.

23.4 Steam Turbines

A steam turbine converts the energy of steam into power by expanding the steam across rows of *blades* mounted on *wheels*, causing the wheels to rotate by extracting energy from the steam (Elliot, 1989). Steam turbine blades can be categorized as *impulse* or *reaction*, although most combine elements of the two (Elliot, 1989). An *impulse turbine* is rather like a water wheel. A single-wheel impulse turbine is illustrated in Figure 23.15a. Impulse nozzles orient the steam to flow in well-formed high-speed jets. Moving *blades* or *buckets* absorb the kinetic energy of the jet and convert it into mechanical work. As the steam strikes the moving blades, it suffers a change in direction and, therefore, momentum. This gives rise to an impulse on the blades. In a pure impulse turbine, no pressure drop occurs in the moving blades (except that caused by friction). The pressure drop occurs across the stationary blades only. Figure 23.15b illustrates an impulse turbine with two wheels.

The principle of a *reaction turbine* is somewhat different. A reaction turbine with a single wheel is illustrated in Figure 23.15c. Steam passing through the fixed blades undergoes a small decrease in pressure and its velocity increases. It then enters the row of moving blades and, just as in the impulse turbine, suffers a change in direction, and therefore momentum. However, during its passage through the blades, the steam undergoes a further drop in pressure. The resulting increase in velocity gives rise to a reactive force in the direction opposite to that of the added velocity. This is the same reaction principle used to propel rockets and to propel airplanes using jet engines. The gross propelling force in a steam turbine is a combination of impulse and reaction. Thus, in reality, the blades in a so-called reaction turbine are partly impulse and partly reaction. Figure 23.15c illustrates a reaction turbine with a single wheel. Figure 23.15d illustrates a reaction turbine with two wheels.

The blades are mounted on a rotating shaft. The rotating shaft is supported within a *casing*. The casing supports the bearings for the rotating shaft, the stationary blades and the steam inlet nozzles.

Steam turbine sizes vary from under 100 kW to over 250 MW. Impulse machines are used for small turbines, and the high-

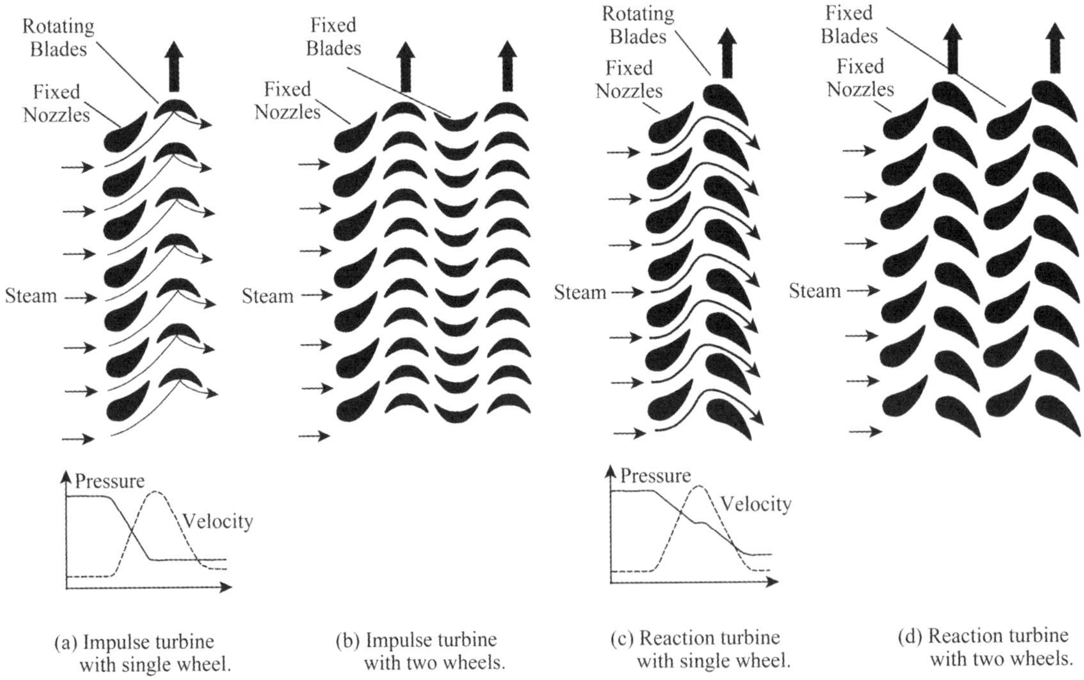

Figure 23.15

Steam turbine types.

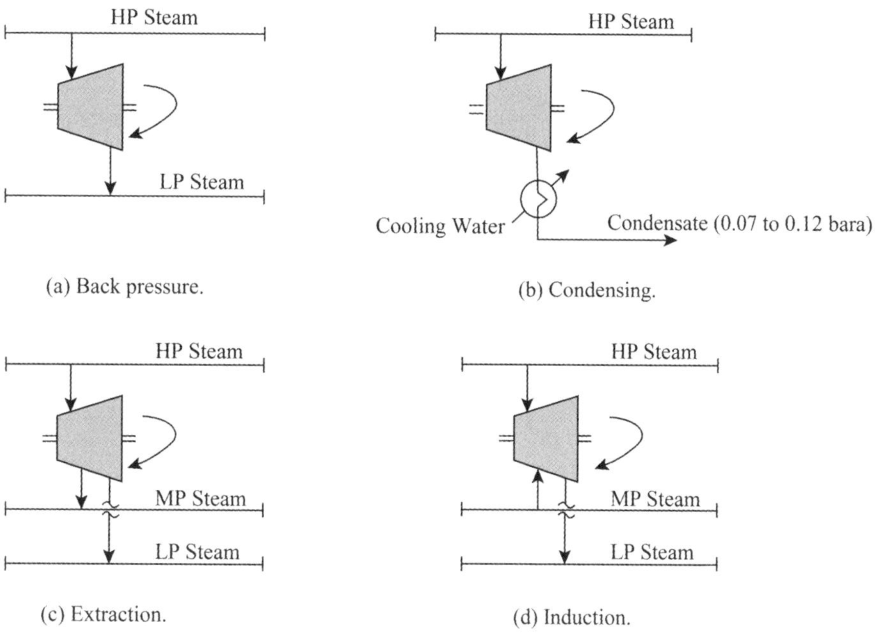

Figure 23.16
Configuration of steam turbines.

pressure section of large turbines use impulse blades. The low-pressure sections of large turbines tend to use reaction blades. The maximum steam inlet temperature depends on the machine design and is normally below 585 °C. Higher inlet temperatures require special materials of construction.

Figure 23.16 shows the basic ways in which a steam turbine can be configured. Steam turbines can be divided into two basic classes:

- *back-pressure* turbines, which exhaust steam at pressures higher than atmospheric pressure;
- *condensing* turbines, which exhaust steam at pressures lower than atmospheric pressure.

The exhaust pressure of a steam turbine is fixed by the operating pressure of the downstream equipment. Figure 23.16a shows a back-pressure turbine operating between high-pressure and low-pressure steam mains. The pressure of the low-pressure steam mains will be controlled elsewhere (see later).

Figure 23.16b shows a condensing turbine. Three types of condensers are used in practice as follows:

1) *Direct contact*, in which cooling water is sprayed directly into the exhaust steam.
2) *Surface condensers*, in which the cooling water and exhaust steam remain separate (see Chapter 12). Rather than using cooling water, boiler feedwater can be preheated to recover the waste heat.
3) *Air coolers* again use surface condensation but reject the heat directly to ambient air (see Chapter 24).

Of the three types, surface condensers using cooling water are by far the most common. When a volume filled with steam is condensed, the resulting condensate occupies a smaller volume and a vacuum is created. Any noncondensable gases remaining after the condensation are removed by steam ejectors or liquid ring pumps (see Chapter 13). The pressure at the exhaust of the condensing turbine is maintained at a reasonable temperature above cooling water temperature (e.g. 40 °C at 0.07 bara to 50 °C at 0.12 bara, depending on the temperature of the cooling water). The higher the pressure difference across the steam turbine, the more the energy that can be converted from the turbine inlet steam into power. For a given temperature difference between the condensing steam and the cooling medium, the lower the temperature of cooling, the more the power that can be generated for a given flowrate of steam with fixed inlet conditions.

Both back-pressure and condensing turbines can be categorized further by the steam that flows through the machine. Figure 23.16c shows an *extraction* machine. Extraction machines bleed off part of the main steam flow at one or more points. The extraction might be uncontrolled and the flows dictated by the pressures at the inlets and outlets and the pressure drops in the sections of the turbine. Alternatively, the extraction might be controlled by internal control valves. Figure 23.16c shows a single extraction with both the extraction and the exhaust being fed to steam mains. The exhaust could have been taken to vacuum conditions and condensed, as in Figure 23.16b.

Figure 23.16d shows an *induction* turbine. Induction turbines work like extraction machines, except in reverse. Steam at a higher pressure than the exhaust is injected into the turbine to increase the flow partway through the machine to increase the power production. In a situation like the one shown in Figure 23.16d, an excess of medium-pressure (MP) steam generation over and above that for process heating is used to produce power and exhaust into a low-pressure steam, where there is a demand for the low-pressure steam for process heating. Again, the exhaust could have been taken to vacuum conditions and condensed.

Any given machine will have minimum and maximum allowable steam flowrates. In the case of extraction and induction machines, there will be minimum and maximum flowrates allowable in each turbine section. These minimum and maximum flowrates are determined by the physical characteristics of individual turbines and specified by the turbine manufacturer.

The expansion process in a steam turbine transforms a portion of the energy of inlet steam to power. The turbine efficiency can be defined as:

$$\eta_{ST} = \frac{W}{W_{IS}} = \frac{W}{m\,\Delta H_{IS}} \tag{23.17}$$

where η_{ST} = steam turbine efficiency (–)
W = turbine shaft power (kW)
W_{IS} = power corresponding with an isentropic expansion (kW)
ΔH_{IS} = enthalpy change to the outlet pressure having the same entropy as the inlet steam ($kJ \cdot kg^{-1}$)
= $H_{in} - H_{IS}$
H_{in} = specific enthalpy of the inlet steam ($kJ \cdot kg^{-1}$)
H_{IS} = specific enthalpy of steam at outlet pressure having the same entropy as the inlet steam ($kJ \cdot kg^{-1}$)
m = turbine steam flow ($kg \cdot s^{-1}$)

The overall turbine efficiency can be represented by two components: the *isentropic efficiency* and the *mechanical efficiency*. The efficiency with which energy is extracted from steam is characterized by the isentropic efficiency, introduced in Figure 2.1 and Equation 2.3, defined as:

$$\eta_{IS} = \frac{H_{in} - H_{out}}{H_{in} - H_{IS}} = \frac{\Delta H}{\Delta H_{IS}} \tag{23.18}$$

where η_{IS} = turbine isentropic efficiency (–)
H_{in} = specific enthalpy of the inlet steam ($kJ \cdot kg^{-1}$)
H_{out} = specific enthalpy of the outlet steam ($kJ \cdot kg^{-1}$)
H_{IS} = enthalpy of steam at the outlet pressure having the same entropy as the inlet steam ($kJ \cdot kg^{-1}$)
ΔH = $H_{in} - H_{out}$

The isentropic efficiency reflects the inefficiency associated with the flow of the steam through the turbine and characterizes the efficiency at which energy is extracted from the steam. The mechanical efficiency reflects the efficiency with which the energy that is extracted from steam is transformed into useful power and accounts for machine frictional losses and heat losses. The mechanical efficiency is high (typically 0.97 to 0.99) (Siddhartha and Rajkumar, 1999). In addition to the mechanical efficiency being much higher than the isentropic efficiency, in most cases the mechanical efficiency does not change significantly with load. By contrast, the isentropic efficiency does change significantly with load. The overall steam turbine efficiency can be defined as:

$$\eta_{ST} = \eta_{IS}\,\eta_{MECH} \tag{23.19}$$

where η_{ST} = overall steam turbine efficiency
η_{MECH} = mechanical efficiency

Steam turbine efficiency is dictated by a number of factors. The most significant among them are:

- turbine size in terms of maximum power load,
- inlet pressure of the steam,
- outlet pressure of the steam,
- operating load (part-load conditions).

If the turbine is driving an electricity generating set, then there will also be an efficiency associated with electricity generation (typically 95 to 98%).

The steam turbine power output can in principle be modelled by Equations 13.72 or 13.75. However, the difficulties of defining physical properties that vary significantly through the turbine means that it is usual to calculate the power from an enthalpy balance across the machine and an efficiency, as defined in Equations 23.17 to 23.19. From an energy balance across the turbine:

$$W = \eta_{MECH}\, m(H_{in} - H_{out}) \tag{23.20}$$

Combining Equations 23.18 and 23.20 gives:

$$W = \eta_{MECH}\,\eta_{IS}\, m\,\Delta H_{IS} \tag{23.21}$$

This is useful for preliminary design, but η_{MECH} and η_{IS} must be known. Because $\eta_{MECH} \gg \eta_{IS}$, and η_{MECH} can be assumed to be a value of 0.97 to 0.99, the biggest problem in using Equation 23.21 is knowledge of η_{IS}. Peterson and Mann (1985) presented data to demonstrate that for steam turbines operating at maximum load, steam turbine efficiency increases with the size of the turbine and turbine inlet pressure for a fixed outlet pressure. Varbanov, Doyle and Smith (2004) demonstrated from turbine data that at maximum load, different sized turbines follow a linear relationship between shaft power and isentropic power. At maximum load, it can be assumed that (Varbanov, Doyle and Smith, 2004; Sun and Smith, 2015):

$$W_{IS,max} = aW_{max} + b \tag{23.22}$$

where $W_{IS,max}$ = power corresponding with an isentropic expansion at maximum load (kW)
W_{max} = turbine shaft power at maximum load (kW)
a, b = modeling coefficients

Writing the efficiency from Equation 23.21 at maximum load and combining with Equation 23.22 gives:

$$\eta_{ST,max} = \frac{1}{a + \dfrac{b}{W_{max}}} \tag{23.23}$$

where $\eta_{ST,max}$ = steam turbine efficiency at maximum load (–)

Equation 23.23 has the same form as the equation for gas turbine efficiency at maximum load (Equation 23.11).

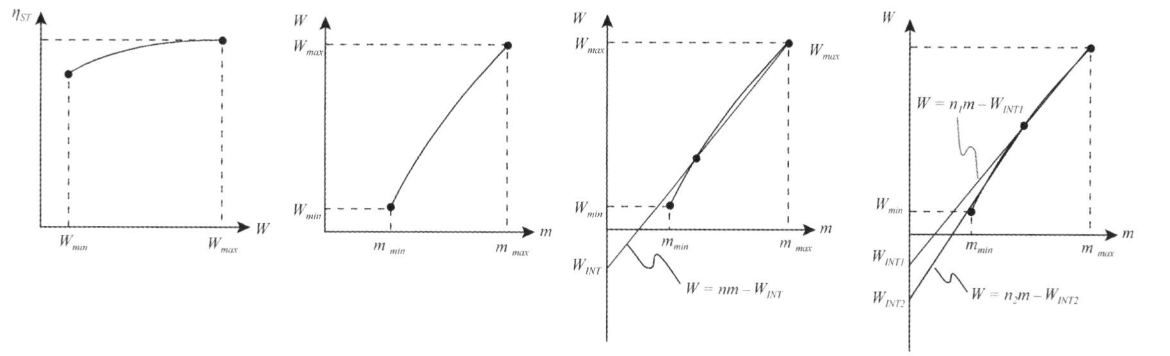

(a) Efficiency versus power output.
(b) Power output versus steam flowrate.
(c) A straight line can be used to represent turbine performance.
(d) Straight line segments can be used to improve the representation.

Figure 23.17
Steam turbine performance versus load.

Coefficients a and b are functions of the inlet and outlet pressures.

Equations 23.23 can be used to predict the efficiency of a steam turbine at maximum load if W_{max} is specified. This is useful for design. Alternatively, if the design flowrate of steam is known, rather than the power requirement, W_{max} can be eliminated by combining Equations 23.21 and 23.22 to give:

$$\eta_{ST,max} = \frac{1}{a}\left(1 - \frac{b}{m_{max}\Delta H_{IS}}\right) \quad (23.24)$$

Steam systems often need to be flexible in order to service site demands that can change significantly through time. For a steam turbine driving a machine such as a compressor on a direct drive, the load will be constant. However, for machines driving an electricity generator, the load is likely to change significantly through time. Generally, the efficiency of steam turbines decreases with decreasing load. Figure 23.17a illustrates the relationship between turbine efficiency and mass flow through the turbine. Because $\eta_{MECH} \gg \eta_{IS}$, the major contribution to the nonlinear trend of the overall efficiency with a part-load is from the isentropic efficiency η_{IS}. A turbine model needs to capture this behavior.

The relationship between shaft power and the mass flow of steam is again sometimes called the Willans Line, after Peter William Willans. A typical Willans Line for a steam turbine is illustrated in Figure 23.13b. For many machines this is almost a straight line. The power–steam flow relationship in Figure 23.13b can be represented over a reasonable range by a linear relationship, as shown in Figure 23.17c. The straight-line relationship is given by (Mavromatis, 1996; Mavromatis and Kokossis, 1998; Varbanov, Doyle and Smith, 2004; Sun and Smith, 2015):

$$W = n\,m - W_{INT} \quad (23.25)$$

where W = shaft power produced by the turbine, not including mechanical losses (kW)
m = mass flowrate of the steam through the turbine ($kg{\cdot}s^{-1}$)
n = slope of the linear Willans Line ($kJ{\cdot}kg^{-1}$)
W_{INT} = intercept of the linear Willans Line (kW)

If a single straight-line relationship, as shown in Figure 23.17c, is not adequate, then a series of linear segments can be used, as shown in Figure 23.17d. Each linear segment in Figure 23.17d is represented by an equation of the form of Equation 23.25, each with its own slope and intercept. It is worth preserving the linearity of the model rather than introducing a nonlinear model, as the linear model has advantages for use in optimization, as will be discussed later. However, most often in practice, a single straight line gives an adequate representation.

Whilst Equation 23.22 allows the maximum point in Equation 23.25 to be defined, the intercept point must somehow be defined. Assume:

$$W_{INT} = c\,W_{max} \quad (23.26)$$

where c = modeling coefficient

Writing Equation 23.25 at maximum load and combining with Equation 23.26 gives:

$$W_{max} = \frac{n\,m_{max}}{(1+c)} \quad (23.27)$$

Combining Equations 23.22 and 23.27 gives:

$$n = \frac{(1+c)}{a}\left(\Delta H_{IS} - \frac{b}{m_{max}}\right) \quad (23.28)$$

Combining Equations 23.22 and 23.26, assuming $W_{IS,max} = m_{max}\,\Delta H_{IS}$, gives:

$$W_{INT} = \frac{c}{a}(m_{max}\,\Delta H_{IS} - b) \quad (23.29)$$

Equations 23.28 and 23.29 define the slope and the intercept for Equation 23.25. The a, b and c coefficients in Equations 23.28 and

Table 23.3

Steam turbine modeling coefficients (Sun and Smith, 2015).

	Back-pressure turbines	Condensing turbines
a_1	1.1880	1.3150
a_2	-2.9564×10^{-4}	-1.6347×10^{-3}
a_3	4.6473×10^{-3}	-0.36798
b_1	449.98	-437.77
b_2	5.6702	29.007
b_3	-11.505	10.359
c_1	0.20515	7.8863×10^{-2}
c_2	-6.9517×10^{-4}	5.2833×10^{-4}
c_3	2.8446×10^{-3}	-0.70315

23.29 need to be fitted to actual turbine data. The form of the correlations for coefficients a, b and c is given by (Sun and Smith, 2015):

$$a = a_1 + a_2P_{in} + a_3P_{out} \quad (23.30)$$

$$b = b_1 + b_2P_{in} + b_3P_{out} \quad (23.31)$$

$$c = c_1 + c_2P_{in} + c_3P_{out} \quad (23.32)$$

where a_1 to a_3, b_1 to b_3 and c_1 to c_3 are modeling coefficients. These modeling coefficients have been fitted to data from 70 back-pressure turbines at 214 operating states and 104 condensing turbines at 335 operating states (Sun and Smith, 2015), giving a mean error of power prediction of 2%. The modeling coefficients are given in Table 23.3.

For any load of a given steam turbine with fixed inlet pressure, back-pressure and inlet temperature, the isentropic enthalpy change remains constant. As a result of changing isentropic efficiency, the actual enthalpy drop across the turbine changes with load. However, the change cannot exceed the isentropic enthalpy change. The model can be used to predict the variation of turbine efficiency with load:

$$\eta_{ST} = \frac{W}{W_{IS}} = \frac{W}{m\,\Delta H_{IS}} = \frac{n\,m - W_{INT}}{m\,\Delta H_{IS}} \quad (23.33)$$

The model can also be used to predict the variation isentropic efficiency turbine efficiency with load if the mechanical efficiency is known:

$$\eta_{IS} = \frac{\Delta H}{\Delta H_{IS}} = \frac{W}{m\,\eta_{MECH}\,\Delta H_{IS}} = \frac{n\,m - W_{INT}}{m\,\eta_{MECH}\,\Delta H_{IS}} \quad (23.34)$$

The outlet enthalpy from the turbine can be calculated by an energy balance. The outlet enthalpy is the inlet enthalpy minus both the useful power extracted from the steam and the mechanical losses:

$$H_{out} = H_{in} - \frac{W}{\eta_{MECH}\,m} \quad (23.35)$$

If the outlet enthalpy and pressure are known, then the outlet temperature can be calculated from steam properties.

This approach to steam turbine modeling has so far been restricted to simple turbines with a single feed and single exhaust. In principle, the approach can be extended to complex steam turbines by representing the sections of complex turbines as simple turbines on the same shaft (Chou and Shih, 1987). This decomposition approach is illustrated in Figure 23.18. Figure 23.18 shows two examples of complex turbines that have been decomposed into the corresponding basic components involving simple turbine units, splitters and mixers. Figure 23.18a shows how an extraction machine with a single extraction can be modeled by two simple turbine units with the appropriate connections. Figure 23.18a shows a controlled extraction. Other designs of machine have uncontrolled extraction. Figure 23.18b illustrates how an induction machine can be modeled by simple turbine units with the appropriate connections. It should be noted that the control features and internal flow patterns in complex turbines can create additional inefficiencies when compared with two simple turbines. Thus, the decomposition approach might over-predict the power generation for given steam flows. This can be compensated for by adding some additional pressure drop between the sections. Alternatively, some of the coefficients in the model can be regressed to fit a particular turbine.

It is often necessary to model an existing turbine as part of the modeling of an existing steam system, for optimization of the existing system or retrofit study. The data from Table 23.3 can be used if no operating data are available. If operating data are available, the a, b and c coefficients can be fitted to measured data by regression. However, it is often the case that only a limited number of operating points are available. In this case, some of the coefficients can be taken from Table 23.3 and others regressed from measured data. When fitting data for a complex turbine, such as those shown in Figure 23.18, a set of a coefficients and b coefficients are required for each section of the complex turbine. These can be partly set by Table 23.3 and partly regressed. One problem faced when regressing data for a complex turbine is that the breakdown of total power generated to that generated in each section of the turbine is not known, but only the total power. In this situation, the measured temperature of the extraction streams can be taken as an indirect measure of the power extracted in the turbine section. The regression for a complex turbine thus becomes (Sun and Smith, 2015):

$$\text{minimize} \quad \sum_i \left[\left(\frac{W_{CALCULATED} - W_{MEASURED}}{W_{MEASURED}} \right)_i^2 + \left(\frac{T_{out,CALCULATED} - T_{out,MEASURED}}{T_{out,MEASURED}} \right)_i^2 \right] \quad (23.36)$$

subject to $0.4 < \eta_{ST} < 0.9$

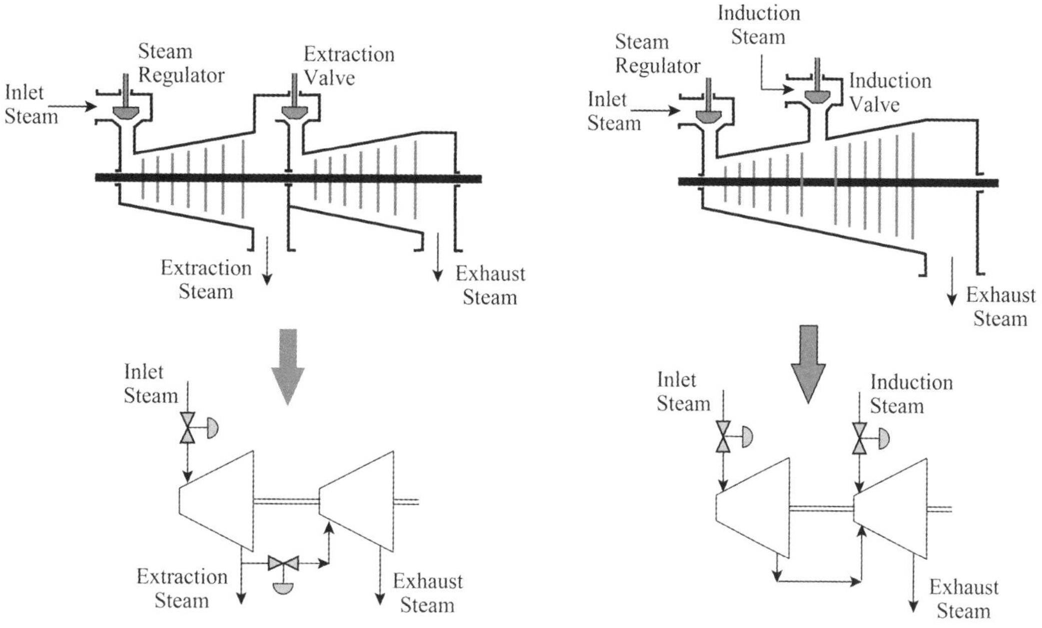

(a) Extraction steam turbines. (b) Induction steam turbines.

Figure 23.18
Modeling complex steam turbines.

where $(W_{CALCULATED} - W_{MEASURED})_i$ = difference between the calculated and measured total power for data point i

$\left(T_{out,CALCULATED} - T_{out,MEASURED}\right)_i$ = difference between the calculated and measured steam outlet temperature for data point i

Example 23.6 A process heating duty of 25 MW is to be supplied by the exhaust steam of a back-pressure turbine. High-pressure steam at 100 barg with a temperature of 485 °C is to be expanded to 20 barg for process heating. The heating duty of the 20 barg steam can be assumed to be the sum of the superheat and latent heat. Assume the mechanical efficiency to be 97%.

a) Determine the power production, steam flowrate and turbine efficiency for a fully loaded back-pressure turbine.

b) If the turbine is sized for an additional 20% extra flowrate, but operated at the same process heating duty, determine the power production

Solution

a) The conditions for the inlet steam are fixed, but the conditions of the outlet steam will depend on the performance of the turbine. First, estimate the steam flowrate from the process heating duty. A good approximation is that the sum of the heat content of the superheat and latent heat is constant from inlet to outlet. At the turbine inlet, the heat content of the superheat is higher than that of the outlet, but the latent heat is lower in the inlet than in the outlet. The two trends tend to cancel each other out, with the total heat content of superheat and latent heat being approximately constant across the turbine. From steam tables, the enthalpy of the superheated steam H_{SUP}, enthalpy of the saturated steam H_{SAT} and enthalpy of the saturated condensate H_L at 100 barg are:

$$H_{SUP} = 3335\ \text{kJ}\cdot\text{kg}^{-1} \text{ at } 485\ ^\circ\text{C}$$
$$H_{SAT} = 2726\ \text{kJ}\cdot\text{kg}^{-1}$$
$$H_L = 1412\ \text{kJ}\cdot\text{kg}^{-1}$$

Assuming the heat content of the superheat and latent heat of steam to be constant, the flowrate of steam can be estimated to be:

$$m = \frac{25 \times 10^3}{3335 - 1412} = 13.0\ \text{kg}\cdot\text{s}^{-1}$$

Now calculate the parameters for the steam turbine model:

$$\begin{aligned} a &= a_1 + a_2 P_{in} + a_3 P_{out} \\ &= 1.1880 - 2.9564 \times 10^{-4} \times 101.01 + 4.6473 \times 10^{-3} \times 21.01 \\ &= 1.2558 \end{aligned}$$

$$\begin{aligned} b &= b_1 + b_2 P_{in} + b_3 P_{out} \\ &= 449.98 + 5.6702 \times 101.01 - 11.505 \times 21.01 \\ &= 781.01 \end{aligned}$$

$$\begin{aligned} c &= c_1 + c_2 P_{in} + c_3 P_{out} \\ &= 0.20515 - 6.9517 \times 10^{-4} \times 101.01 + 2.8446 \times 10^{-3} \times 21.01 \\ &= 0.19470 \end{aligned}$$

Next, determine the isentropic enthalpy change ΔH_{IS}. From steam properties, the entropy of the inlet steam is:

$$S_{SUP} = 6.5432\ \mathrm{kJ \cdot kg^{-1} \cdot K^{-1}}$$

From steam properties, the enthalpy of steam at 20 barg with this entropy is:

$$H_{SUP} = 2912\ \mathrm{kJ \cdot kg^{-1}}$$

Thus:

$$\begin{aligned} \Delta H_{IS} &= 3335 - 2912 \\ &= 423\ \mathrm{kJ \cdot kg^{-1}} \end{aligned}$$

Calculate the turbine power at maximum flowrate from Equation 23.22:

$$\begin{aligned} W_{max} &= \frac{W_{IS,max} - b}{a} \\ &= \frac{m_{max}\,\Delta H_{IS} - b}{a} \\ &= \frac{13 \times 423 - 781.01}{1.2558} \\ &= 3757\ \mathrm{kW} \end{aligned}$$

The outlet steam enthalpy is given by an energy balance from Equation 23.35:

$$\begin{aligned} H_{out} &= H_{in} - \frac{W}{\eta_{MECH}\, m} \\ &= 3335 - \frac{3757}{0.97 \times 13.0} \\ &= 3037\ \mathrm{kJ \cdot kg^{-1}} \end{aligned}$$

Now revise the steam flow. From steam properties at 20 barg, $H_L = 920\ \mathrm{kJ \cdot kg^{-1}}$:

$$\begin{aligned} m &= \frac{25 \times 10^3}{3037 - 920} \\ &= 11.81\ \mathrm{kg \cdot s^{-1}} \end{aligned}$$

Compare this with the original estimate of $13.0\ \mathrm{kg \cdot s^{-1}}$. The calculation is now repeated with the revised steam flow:

$$\begin{aligned} W_{max} &= 3356\ \mathrm{W} \\ H_{out} &= 3042\ \mathrm{kJ \cdot kg^{-1}} \\ m &= 11.78\ \mathrm{kg \cdot s^{-1}} \end{aligned}$$

A further iteration with the revised steam flow gives:

$$\begin{aligned} W_{max} &= 3346\ \mathrm{W} \\ H_{out} &= 3042\ \mathrm{kJ \cdot kg^{-1}} \\ m &= 11.78\ \mathrm{kg \cdot s^{-1}} \end{aligned}$$

Further iteration will not bring any significant change. Calculate the steam turbine efficiency at maximum load from Equation 23.24:

$$\begin{aligned} \eta_{ST,max} &= \frac{1}{a}\left(1 - \frac{b}{m_{max}\,\Delta H_{IS}}\right) \\ &= \frac{1}{1.2558}\left(1 - \frac{781.01}{11.78 \times 423}\right) \\ &= 0.672 \end{aligned}$$

b) The turbine is to be sized with a spare capacity of an extra 20% flowrate:

$$m_{max} = 1.2 \times 11.78 = 14.14\ \mathrm{kg \cdot s^{-1}}$$

Part-load performance is predicted by the Willans Line from Equation 23.25. First calculate the slope and intercept point from Equations 23.28 and 23.29:

$$\begin{aligned} n &= \frac{(1 + c)}{a}\left(\Delta H_{IS} - \frac{b}{m_{max}}\right) \\ &= \frac{(1 + 0.19470)}{1.2558}\left(423 - \frac{781.01}{14.14}\right) \\ &= 349.9\ \mathrm{kJ \cdot kg^{-1}} \end{aligned}$$

$$\begin{aligned} W_{INT} &= \frac{c}{a}(m_{max}\,\Delta H_{IS} - b) \\ &= \frac{0.19470}{1.2558}(14.14 \times 423 - 781.01) \\ &= 806.2\ \mathrm{kW} \end{aligned}$$

From Equation 23.25:

$$\begin{aligned} W &= n\,m - W_{INT} \\ &= 349.9 \times 11.78 - 806.2 \\ &= 3315\ \mathrm{kW} \end{aligned}$$

The outlet steam enthalpy is given by an energy balance from Equation 23.35:

$$\begin{aligned} H_{out} &= H_{in} - \frac{W}{\eta_{MECH}\, m} \\ &= 3335 - \frac{3315}{0.97 \times 11.78} \\ &= 3045\ \mathrm{kJ \cdot kg^{-1}} \end{aligned}$$

Now revise the steam flow:

$$\begin{aligned} m &= \frac{25 \times 10^3}{3045 - 920} \\ &= 11.76\ \mathrm{kg \cdot s^{-1}} \end{aligned}$$

This is a slightly lower flowrate than a turbine sized for maximum load to supply 25 MW of heat at 20 barg, as slightly less power is produced by the larger turbine operating at part load.

23.5 Steam Distribution

Steam is distributed around the site for various purposes, as discussed earlier. When using steam in steam heaters, steam tracing and space heating devices, it is preferred to feed the steam close to saturation conditions. Superheated steam is normally not preferred for process heating, as desuperheating to saturation before condensation as part of the process duty involves a poor heat transfer coefficient. Also, the higher temperature from the superheat might cause fouling or product degradation on the process side. Steam is often desuperheated locally to within 3 °C to 5 °C of saturation before entering the steam heater by mixing the steam with boiler feedwater under temperature control, as illustrated in Figure 23.19.

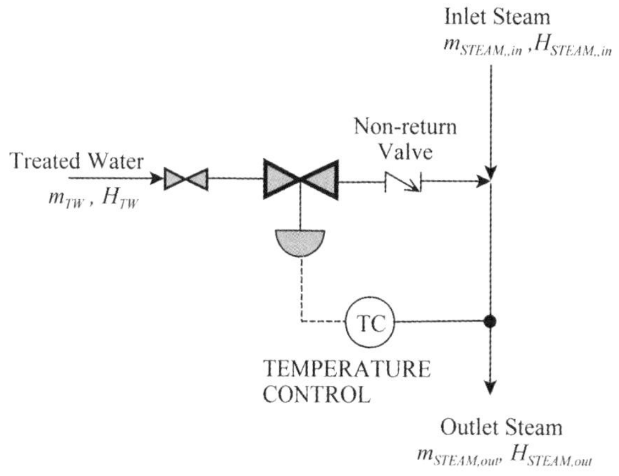

Figure 23.19
A desuperheater.

When using steam in steam heaters, steam tracing and space heating devices, some mechanism is required to allow the condensate to leave the heat transfer device, whilst preventing uncondensed steam from leaving. Also, condensation occurs in the pipes used for steam distribution around the site and this condensate must be prevented from building up. *Steam traps* are devices designed to allow condensate to pass, whilst not allowing steam to pass. It is also important that any air ingress before start-up is allowed to leave, as this can cause a significant deterioration in the heat transfer coefficient. Various designs of steam trap are used in the process industries:

a) Figure 23.20a shows a *float trap*. Such traps are commonly used on process equipment. A float-type trap operates from the difference in density between steam and condensate. In Figure 23.20a, as condensate builds up within the steam trap the float rises and lifts the valve off its seat, allowing condensate to leave. The design in Figure 23.20a requires air to be vented manually. A more sophisticated design can incorporate a thermostatic air vent that allows the initial air to pass whilst the trap is also handling condensate.

b) Figure 23.20b shows an *inverted bucket steam trap*. These are also used on process equipment. The inverted bucket is attached by a lever to a valve. The inverted bucket features a small vent hole in the top of the bucket. Condensate flows from the bottom of the bucket through to the outlet. Any steam entering the trap causes the bucket to become buoyant, which

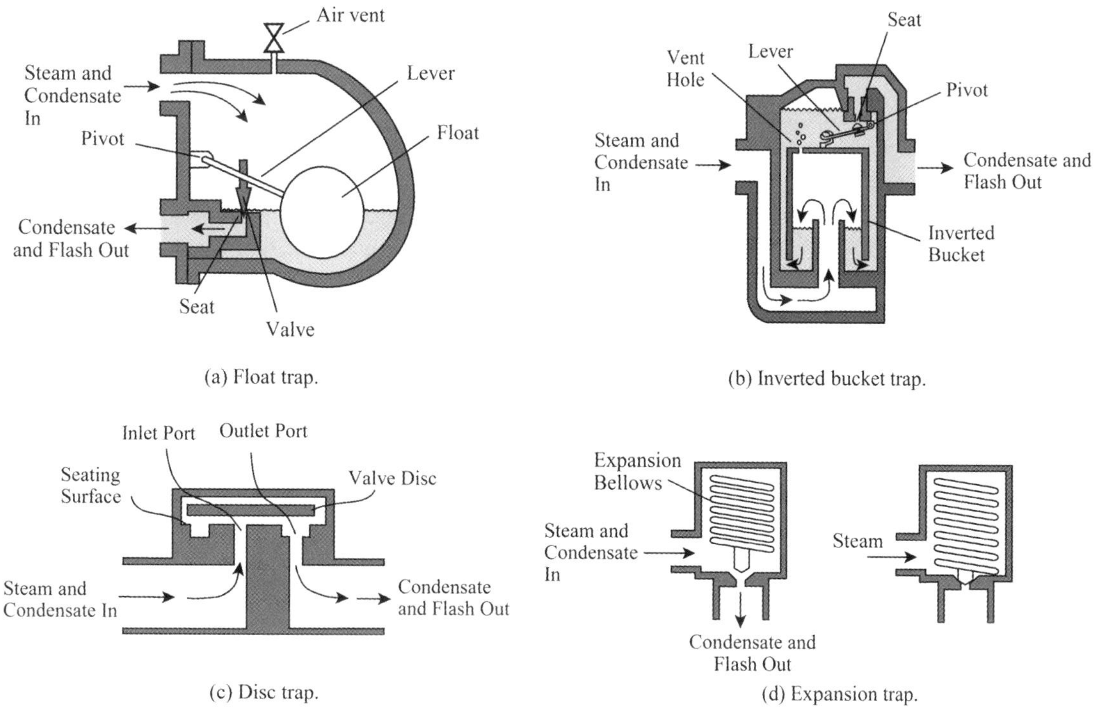

Figure 23.20
Steam traps.

rises and closes the outlet. The trap remains closed until the steam in the bucket has condensed or bubbled through the vent hole. The bucket then sinks, opening the trap. Any air entering the trap at start-up will also give the bucket buoyancy and close the valve. The vent hole allows air to escape.

c) Figure 23.20c shows a *disk steam trap*. At start-up, incoming pressure raises the disk and condensate and air are discharged under the disk to the outlet. When steam enters the trap, static pressure above the disk forces the disk against the valve seat and it closes. At the same time, steam entering at a high velocity creates a low-pressure area under the disk, also forcing it to close. When condensate enters, the pressure against the disk decreases and the trap opens. Disk traps are commonly used to allow condensate release from steam mains.

d) Figure 23.20d shows a *bellows steam trap*. The bellows trap uses a fluid-filled thermal element (bellows) that operates through thermal expansion and contraction. At start-up the trap is open, allowing air and condensate to be removed from the system. When steam enters the trap the fluid in the bellows vaporizes and expands as the temperature increases, causing the bellows to close the valve. Condensate entering the trap causes the temperature to decrease, the fluid in the bellows condenses and contracts, causing the bellows to open the valve. These traps are used on process equipment, steam mains and steam tracing.

Many other designs of steam trap are available. For large heat transfer duties, it is good practice to recover the steam that is flashed as the condensate reduces in pressure. Such an arrangement is shown in Figure 23.21. Steam enters the steam heater and condensate (in practice, with some steam) passes through the trap. Flashing occurs before the mixture enters a settling drum that allows the flash steam to be separated from the condensate. The flash steam would then be fed to a steam main at the appropriate pressure.

Steam condensate should be recovered and returned to the deaerator for boiler feedwater wherever possible. Return to the boiler might be via feedwater treatment for *polishing* and then deaeration if there is some danger of contamination. Steam condensate from steam heaters, steam tracing and space heating devices should normally be returned to the deaerator. For steam

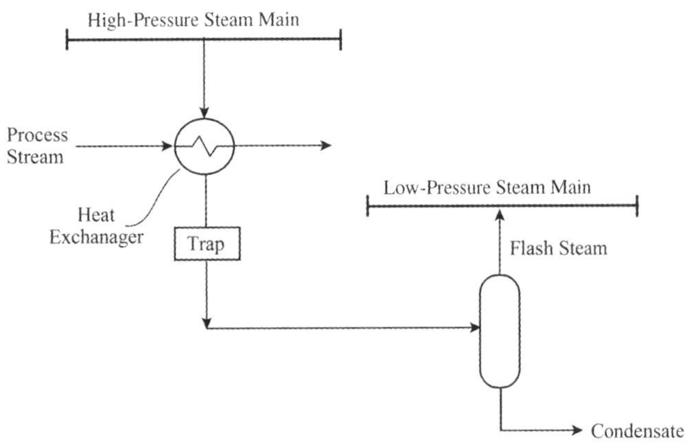

Figure 23.21
Flash steam recovery.

ejector systems, condensate collected in the hotwell, as illustrated in Figure 13.17, is in most cases too contaminated to be returned directly to the deaerator. The condensate is either treated and disposed of, or treated and reused (but not necessarily as boiler feedwater). The steam injected into distillation operations is ultimately condensed, but the condensate is highly contaminated and either treated and disposed of or treated and reused (not necessarily as boiler feedwater). Similarly, condensate from dilution steam used in reactors and recovered within the process will be too contaminated for direct reuse as boiler feedwater. In some cases, reactor dilution steam is generated within the process in a heat exchanger, rather than a steam boiler (e.g. from heat recovery), fed to the reactor and the condensate recycled directly in a closed loop. This is possible, because the dilution steam can be contaminated by process components without causing problems in the reactor. Any steam used for flaring, fuel oil atomization, NO_X abatement in combustion processes, reactor decoking and soot blowing operations is lost.

Given the various users of steam on the site, a distribution system is required. *Steam mains* or *headers* distribute steam around the site at various pressures. The number of steam headers and their pressures depend on the various processes requirements and the requirements to generate power from steam centrally or locally. Figure 23.22 shows a schematic of a typical steam distribution system. Raw water is treated before entering the boilers that fire fuel to generate high-pressure (HP) steam. Figure 23.22 illustrates the features of a typical steam system. It is common to have at least three levels of steam. On larger sites, steam may also be generated at a very high pressure (typically 100 bar), which will only be used for power generation in steam turbines in the boiler house. Steam would then be distributed around the site, which for larger sites would typically be at three pressures. On small sites there might be only a single pressure of steam distributed around the site. Back-pressure turbines let steam down from the high-pressure mains to the lower-pressure mains to generate power. Supplementary power may be generated using condensing turbines. Operating a condensing turbine from the highest-pressure inlet maximizes the power production. The system in Figure 23.22 shows flash steam recovery into the medium-pressure and low-pressure mains. Also, as shown in Figure 23.22, the boiler blowdown is flashed and flash steam recovered before being used to preheat incoming boiler feedwater and being sent to the effluent. Whether flash steam recovery is economic is a matter of economy of scale.

Also shown in Figure 23.22 are *letdown stations* between the steam mains to control the mains pressures via a pressure control system. It is desirable to have a low flow in the letdown stations. Instead, it is preferred to let down through steam turbines to generate power. However, zero flow in letdown stations is also undesirable. It is preferred to have some flow (typically a few tons per hour) for control and to avoid condensation in the pipework. In some cases, the letdown stations also have desuperheaters, as illustrated in Figure 23.22. When steam is let down from a high to a low pressure under adiabatic conditions in a valve the amount of superheat increases. Desuperheating, if carried out, is achieved by the injection of boiler feedwater under temperature control, which evaporates and reduces the superheat. There are two important factors determining the desirable amount of superheat in the steam mains.

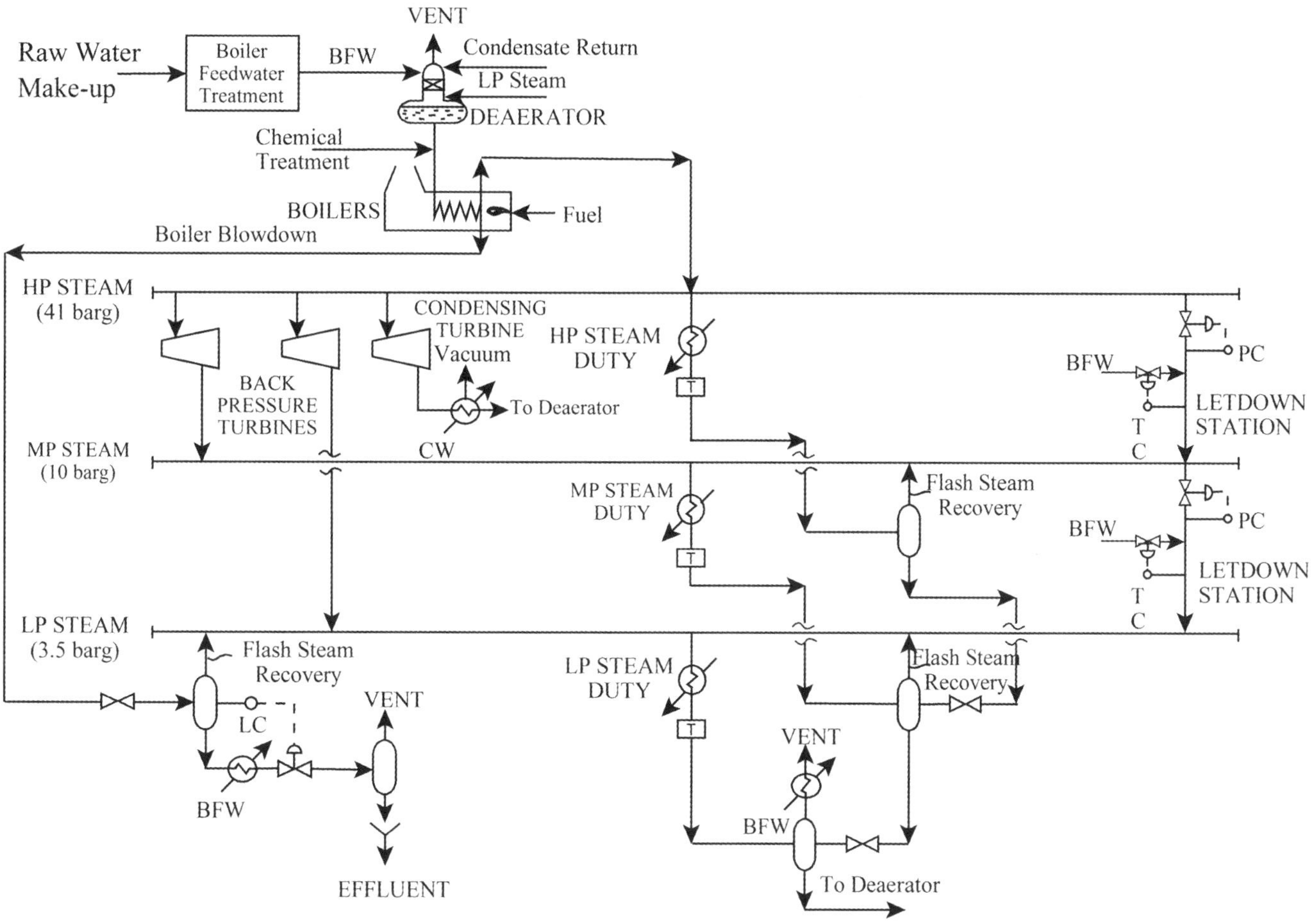

Figure 23.22

Features of a typical steam distribution system.

a) Steam heating is most efficiently carried out using the latent heat from condensation, rather than having to desuperheat before condensing the steam. Thus, the design of steam heaters benefits from having no superheat in the steam. However, having no superheat in the steam mains is undesirable, as this would lead to excessive condensation in the steam mains. It is desirable to have at least 10 to 20 °C superheat in steam mains to avoid excessive condensation in the mains.

b) In addition to using steam for steam heating, it is also used for power generation by expansion through steam turbines. Steam turbines might generate electricity centrally that is distributed to motors around the site. Alternatively, steam turbines might be used to drive machines directly. Expansion in steam turbines reduces the superheat in the steam as it is reduced in pressure. If there is not enough superheat in the inlet steam, then condensation can take place in the steam turbine. While a small amount of condensation in the machine is acceptable, excessive condensation can be damaging to the machine. Also, if the steam turbine is exhausting to a steam main, then it is desirable to have some degree of superheat in the outlet to maintain some superheat in the outlet low-pressure steam main.

Consider the simple steam turbine cascade in Figure 23.23a in which there are no inputs to the steam mains, apart from the utility boilers and steam turbines. The steam mains pressures and the vacuum steam pressure have been fixed. The isentropic efficiencies of all the steam turbines in the cascade have also been specified. Once the temperature of the steam generated in the utility boilers has been specified, then all the steam mains temperatures are also fixed as a result. This follows from the definition of isentropic efficiency. The temperature and pressure of the VHP steam fixes its enthalpy. Specifying $\eta_{IS,\text{VHP-HP}}$ fixes the HP steam enthalpy, and since the HP steam pressure is fixed, the temperature is also fixed. In turn, all the lower pressure mains conditions are also fixed down the cascade. As the steam passes through the steam turbine cascade, the amount of superheat in the steam decreases. The steam main with the lowest amount of steam superheat is the low-pressure main and there should be at least 10 to 20 °C of superheat. Also, if there is condensing power generation as in Figure 23.23a, then the superheat is further decreased to the point where the steam might start to condense. More than 12% wetness in the steam, or slightly higher in some circumstances, can lead to damage of the steam turbine blades. Thus, if there is not enough superheat in the low-pressure main or there is too much

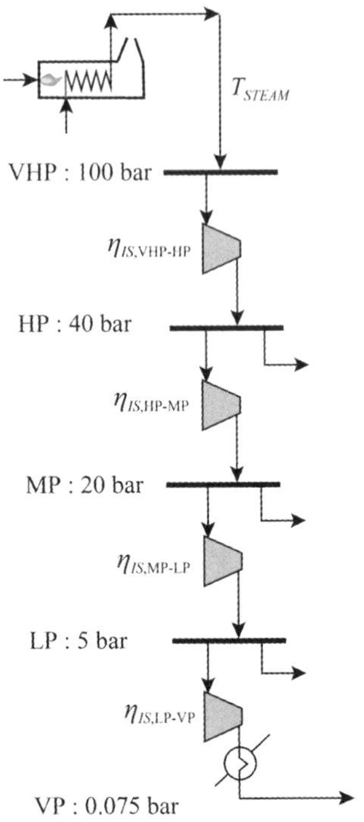

(a) Steam turbine cascade with no letdown flow.

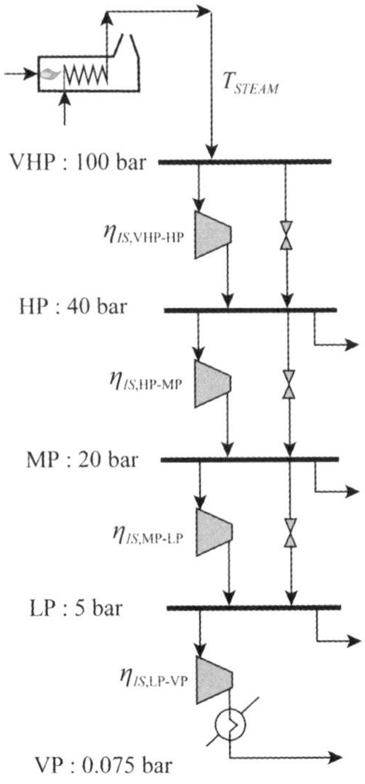

(b) Steam turbine cascade with letdown flow.

Figure 23.23
A simple steam turbine cascade.

wetness in the vacuum steam, then in Figure 23.23a the only way to increase the low-pressure steam superheat or decrease the vacuum steam wetness is to increase the temperature of the utility VHP steam. However, there is a maximum temperature that can be used with the VHP steam created by the limitations of the materials of construction of the pipework and steam turbines. Maximum steam turbine inlet temperatures are typically 500 °C to 585 °C, depending on the machine design. If the temperature required to satisfy the minimum superheat and maximum vacuum pressure wetness constraints is greater than the maximum of, say, 550 °C, then one way to satisfy the constraints and keep the temperature below 550 °C is to allow some flow in letdown stations between the steam mains, as shown in Figure 23.23b. As steam is expanded through a valve, because no energy is extracted, the superheat of the steam at lower pressure is increased. So in Figure 23.23b steam could be cascaded down through the letdown stations in parallel with the flow through the steam turbines in order to satisfy all of the constraints. However, any flow in letdown stations should be minimized, as this represents a missed opportunity to generate power in the steam turbines. In practice, there might be flow of steam into the various steam mains from process steam generation that will also affect the superheat in the steam mains. This will be examined in more detail in the following sections in the context of site targeting.

The general policy on steam usage for heating is that low-pressure steam should be used in preference to high-pressure steam. Using low-pressure steam for steam heating:

- allows power generation in steam turbines from the high-pressure steam;
- provides a higher latent heat in the steam for the steam heater;
- leads to lower capital cost heat transfer equipment due to the lower pressures.

It should be noted that there will be significant heat losses in the distribution system that might be typically 10% of the fuel fired in the boilers.

23.6 Site Composite Curves

Just as it is useful to have energy targets for individual processes, it is also useful to have energy targets for the site. This requires a thermodynamic analysis for the site to develop *site composite curves*. Site composite curves provide a temperature–enthalpy picture for the whole site, analogous to those for individual processes developed in Chapter 17. There are two ways in which such curves can be developed. The first relates to a new design situation.

A new design situation would start from the grand composite curves of each of the processes on the site and would combine them together to obtain a picture of the overall site utility system (Dhole and Linnhoff, 1992). This is illustrated in Figure 23.24, where two processes have their heat sink and heat source profiles from their grand composite curves combined to obtain a *site hot composite curve* and a *site cold composite curve*, using the procedure

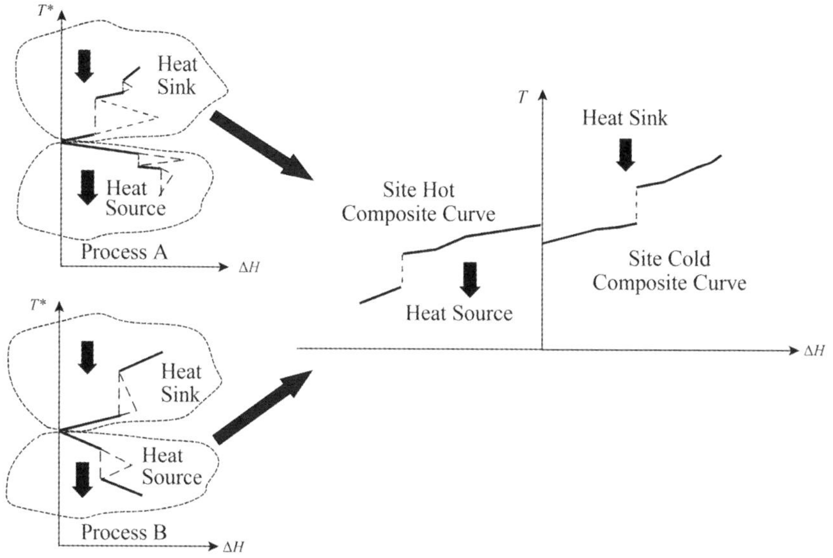

Figure 23.24

Site composite curves for new design can be produced by combining the grand opposite curves for the individual processes.

developed for composite curves in Chapter 17. Wherever there is an overlap in temperature between streams, the heat loads within the temperature intervals are combined together.

In Figure 23.24, the pockets of additional heat recovery in the grand composite curve have been left out of the site analysis. This assumes that this part of the heat recovery will take place within the processes (see Chapter 17). The remaining heat sink profiles and heat source profiles are combined to produce the site composite curves (Dhole and Linnhoff, 1992). While it is usually justified to leave out the pockets of additional heat recovery in a process and accept the in-process heat recovery, some processes demand that information from within the pockets should be included. Figure 23.25 shows a grand composite curve typical of processes involving highly exothermic chemical reactions. The grand composite curve shows only a cooling profile and no heating requirement, as a result of the reaction exothermic heat. There is a large pocket of additional heat recovery, which has not been isolated in Figure 23.25. The temperatures within the pocket are such that high-pressure steam can be generated within the pocket. If high-pressure steam is generated within the pocket, as indicated in Figure 23.25, then heat is used to generate steam where previously heat recovery would have satisfied the cooling requirements. This disturbs the energy balance within the pocket. To compensate for this, low-pressure steam can be used to provide heating within the pocket, where previously heat recovery would have satisfied the heating requirements within the pocket. In practice, exploiting the pocket in the way indicated in Figure 23.25, using a combination of high-pressure and low-pressure steam, is likely to be economic if there is a significant difference in value between the high- and low-pressure steam. Thus, the extraction of the data from the grand composite curves in such scenarios should include part of the profiles within the pocket, as shown in Figure 23.25. However, when should such opportunities be exploited? The situation arises when the temperature difference across the pocket spans two steam levels. If the pocket does not span two steam levels, then the pocket should be isolated, as shown in Figure 23.24. Also, even if the pocket spans a large temperature range, the heat duties within the pocket must be large enough to justify the complication of introducing an extra steam level into the design of that process.

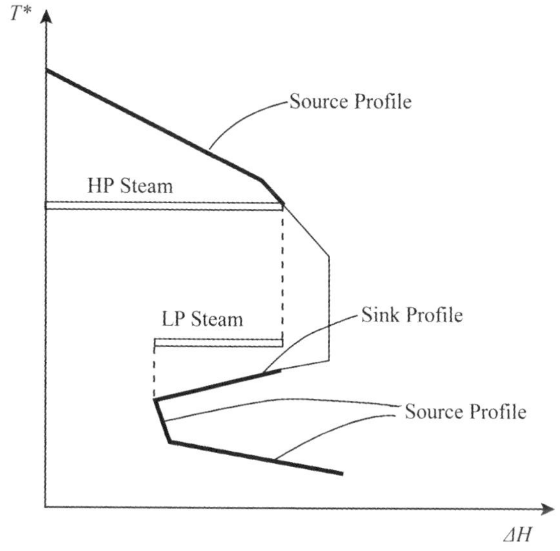

Figure 23.25

Sometimes the details of the grand composite curve pockets should be included in the site profiles.

One further point needs to be noted regarding the construction of the site composite curves. The temperatures are shifted over and above the shift included in the construction of the grand composite curve. If the original hot and cold streams were shifted by $\Delta T_{min}/2$ to produce the grand composite curve, then site composite curves require an additional shift of $\Delta T_{min}/2$ to give a total shift of ΔT_{min}, as illustrated in Figure 23.26 (Raissi, 1994). If different values of ΔT_{min} apply to different processes, then data for each grand composite curve are given their individual shift in ΔT_{min} before the steams are combined in the construction of the site composite curves. Even further, within a grand composite curve for an individual process, different streams might have different shifts

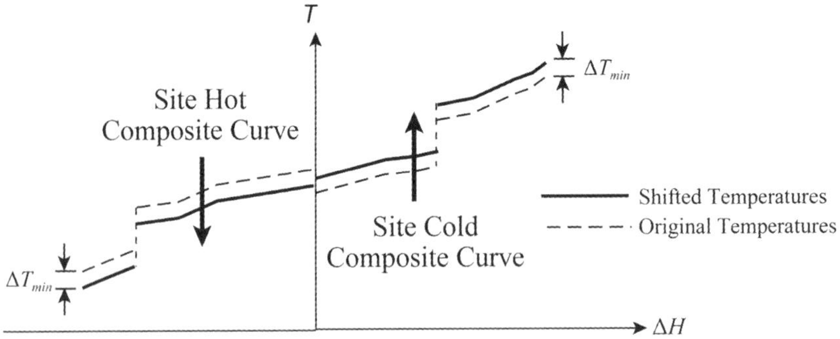

Figure 23.26

Site composite curves are required to have a total shift of ΔT_{min} from the original temperature.

in ΔT_{min} in the construction of the grand composite curve, as discussed in Chapter 17. Each stream must ultimately be shifted by the ΔT_{min} for that stream before the construction of the site composite curves.

The other way to construct the site composite curves relates more to a retrofit situation. In a retrofit situation, the plants and their heat recovery systems are already in place. The heat recovery might already be maximized and in agreement with the target set by the composite curves and the problem table algorithm (see Chapter 17). If this is the case, then the grand composite curve will give an accurate reflection of the process utility demand. However, in retrofit, the heat recovery might not be at its maximum. Thus in retrofit situations, the grand composite curve does not necessarily give an accurate picture of utility demands, as it assumes maximum heat recovery. If the existing amount of heat recovery is assumed to be fixed (whether maximized or not), the site profiles can be constructed from the individual process duties within each of the utility heat exchangers on the site. The temperature–enthalpy profiles of the process streams within each of the utility heat exchangers are used to construct the site composite curves in exactly the same way as discussed in Chapter 17 for process composite curves. This way of constructing the site composite curves thus accepts the existing heat recovery system within each process, good or bad. Again, the temperature needs to be shifted for the site composite curves. Starting from the individual process stream data, this needs to have a shift of ΔT_{min}, as indicated in Figure 23.26. One additional advantage of this approach relative to that based on grand composite curves is that fewer data are required for the construction. Construction of the grand composite curve for a process requires data for all heat sources and sinks in the process to be collected, whereas the alternate approach requires only data from the utility heat exchangers to be collected.

Following these procedures allows composite curves for the total site to be developed that give a picture of the heating and cooling requirements of the total site, both in terms of enthalpy and temperature (Dhole and Linnhoff, 1992). As an example, Table 23.4 presents the data for a site involving five processes. Steam is generated at very high pressure and distributed around the site at three lower pressures; ΔT_{min} is 10 °C for all processes (Sun, Doyle and Smith, 2015a).

Figure 23.27 shows the individual composite curves for the individual processes after the ΔT_{min} shift. Figure 23.28 shows the composite curves for the individual processes combined to give a *site hot composite curve* and a *site cold composite curve* by following the same approach used for the construction of process composite curves described in Chapter 17. Figure 23.29a shows the steam generation and steam use profiles matched against the site composite curves. For the sake of clarity, for now, only the latent heat part of the steam generation and use profiles is shown in Figure 23.29a. Later, the complexities associated with boiler feedwater preheat, steam superheat, steam desuperheating and condensate cooling will be considered. It should be noted, by contrast with the composite curves for individual processes, that direct heat recovery between the site composite curves is not allowed. All heating, cooling and recovery in the site analysis takes place through the utility system only. Figure 23.29a shows an ideal match between the steam generation and steam use and the site composite curves. The targets are set for site cooling by starting with the highest temperature cooling utility, in this case high-pressure (HP) steam generation. This is matched against the site hot composite curve and maximized. The second highest temperature cooling utility is medium-pressure (MP) steam generation. This is now maximized. The next highest temperature cooling utility is low-pressure (LP) steam generation. This is now maximized, with the residual cooling being satisfied by cooling water. To set the targets for steam use, the lowest temperature heating utility is first maximized. In this case, the lowest temperature heating utility is LP steam use. Having maximized the LP steam use, the next lowest heating utility is maximized, in this case MP steam. The residual high temperature heating is taken up by HP steam.

It should be noted that, in Figure 23.29a, the steam is represented at its actual temperature and when a steam profile touches a site composite curve, this implies ΔT_{min} between the steam and the process because of the ΔT shift built into the construction of the site composite curves. This is analogous to the match between utility profiles and grand composite curves discussed in Chapter 17. In practice, the streams in the construction of the site hot composite curve are ΔT_{min} hotter than plotted, and the streams in the site cold composite curve are ΔT_{min} colder than plotted. The steam profiles are plotted at their actual temperatures.

Figure 23.29b shows the site composite curves, but now relating to a retrofit situation. This time the loads on the various steam mains have been fixed to their existing duties (Dhole and Linnhoff, 1992). There is now a mismatch between the site composite curves and the steam profiles. In the example in Figure 23.29b, the steam generation matched against the site hot composite curve is on target. However, the steam use is poorly matched against the site cold composite curve. In Figure 23.29b, the LP steam duty should be higher, the MP steam duty higher and

Table 23.4

Process and steam data for a total site (Sun, Doyle and Smith, 2015a).

Process A					Process B				
Stream	T_S (°C)	T_T (°C)	ΔH (MW)	CP (MW·K^{-1})	Stream	T_S (°C)	T_T (°C)	ΔH (MW)	CP (MW·K^{-1})
1	300	280	30	1.5	1	270	260	10	1
2	148	135	10	0.769231	2	260	241	10	0.5263
3	135	110	20	0.8	3	241	240	20	20
4	110	100	10	1	4	240	220	10	0.5
					5	220	200	5	0.25
					6	200	150	5	0.1
					7	150	135	10	0.6667
					8	135	90	10	0.2222

Process C					Process D				
Stream	T_S (°C)	T_T (°C)	ΔH (MW)	CP (MW·K^{-1})	Stream	T_S (°C)	T_T (°C)	ΔH (MW)	CP (MW·K^{-1})
1	169	174	10	2	1	209	210	20	20
2	168	169	10	10	2	149	150	20	20
3	159	168	10	1.1111	3	104	105	30	30
4	179	160	5	0.2632	4	119	118	20	20
5	160	150	15	1.5	5	101	100	30	30
6	150	135	5	0.3333	6	95	94	20	20
7	135	90	5	0.1111					
8	90	85	8	1.6					
9	85	84	12	12					

Process E					Steam and utility data				
Stream	T_S (°C)	T_T (°C)	ΔH (MW)	CP (MW·K^{-1})	Main	T_S (°C)	T_T (°C)	T_{SAT} (°C)	Pressure (bar)
1	235	237	5.7143	2.8571	VHP	111	550	310.9	100
2	230	235	16.1039	3.2208	HP	105	270	250.3	40
3	180	230	18.1818	0.3636	MP	105	232	212.4	20
4	160	180	30	1.5	LP	105	172	157.8	5
5	110	160	20	0.4	CW	20	30		
6	95	110	5	0.3333					
7	90	95	25	5					
8	110	90	40	2					
9	90	80	20	2					

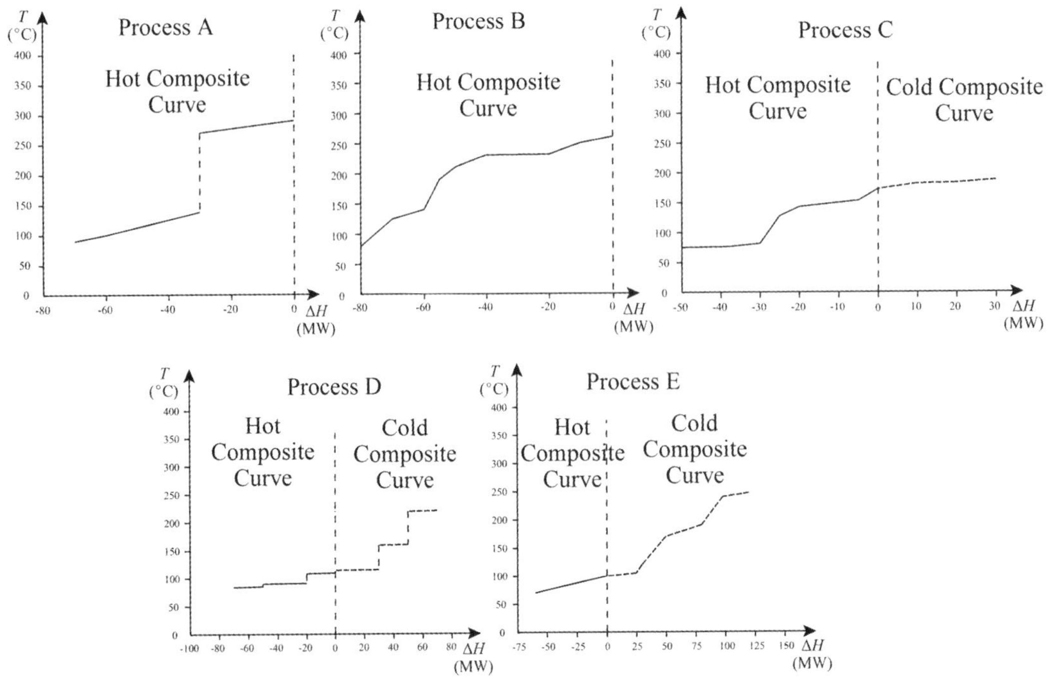

Figure 23.27

Individual composite curves after the ΔT_{min} shift for the example site with five processes. (Reproduced from Li Sun, Steve Doyle, Robin Smith, Heat Recovery and Power Targeting in Utility Systems. In Press. Ms. Ref. No.: EGY-D-14-03308, with permission from Elsevier.)

the HP duty lower. Ultimately, the penalty for such a mismatch is lost opportunity for cogeneration in steam turbines. A greater use of the lower-pressure steam for heating allows more steam to be expanded from the HP level through to the lower levels, and hence more power to be generated in steam turbines.

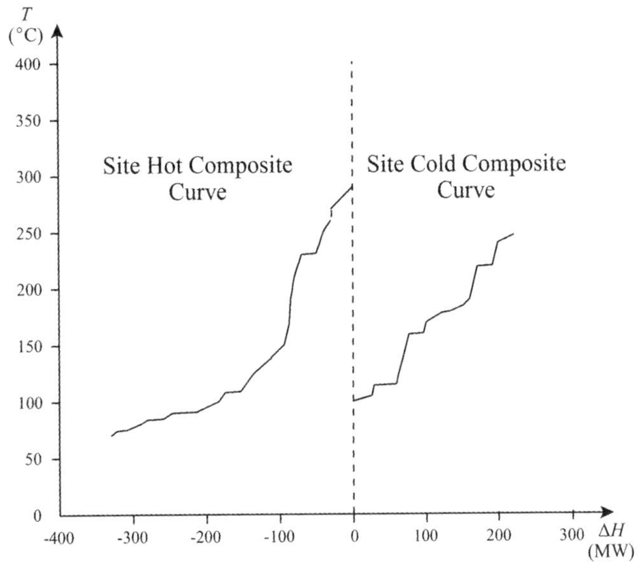

Figure 23.28

Individual process composite curves can be combined to give the site composite curves.

While Figure 23.29 is useful in being able to set targets for steam generation and steam use and identify missed opportunities in retrofit, the recovery of heat between processes through the steam system has not yet been addressed. Figure 23.30a shows a site in which HP steam, MP steam and LP steam are both generated and used on the site. Steam that is generated by waste heat does not need to be generated from the utility system from burning fuel in utility steam boilers. The steam generated by processes is fed into the steam mains, which is subsequently used by other processes on the site, as illustrated in Figure 23.30b. This heat recovery between processes through the steam system needs to be included in targeting for the site.

Figure 23.31a shows no overlap between the site composite curves. If this setting was used in practice, any HP, MP or LP steam generated from the site hot composite curve would have to be vented and all of the HP, MP and LP steam used by the site cold composite curve would need to be let down from utility VHP steam. In practice, venting steam like this would never be done. However, to obtain the whole picture requires consideration of power generation alongside heat recovery, which will be included in the next section. Figure 23.31b illustrates how heat recovery for the site can be targeted by overlapping the steam profiles. Figure 23.31b shows the region of overlap between the steam profiles to be a measure of the heat recovery across the site through the steam system (Raissi, 1994; Klemes *et al.*, 1997). For the setting in Figure 23.31b, some of the MP steam and all of the LP steam generated from the site composite would still need to be vented. Again this would not be done in practice, but the power generation implications will be considered in the next section.

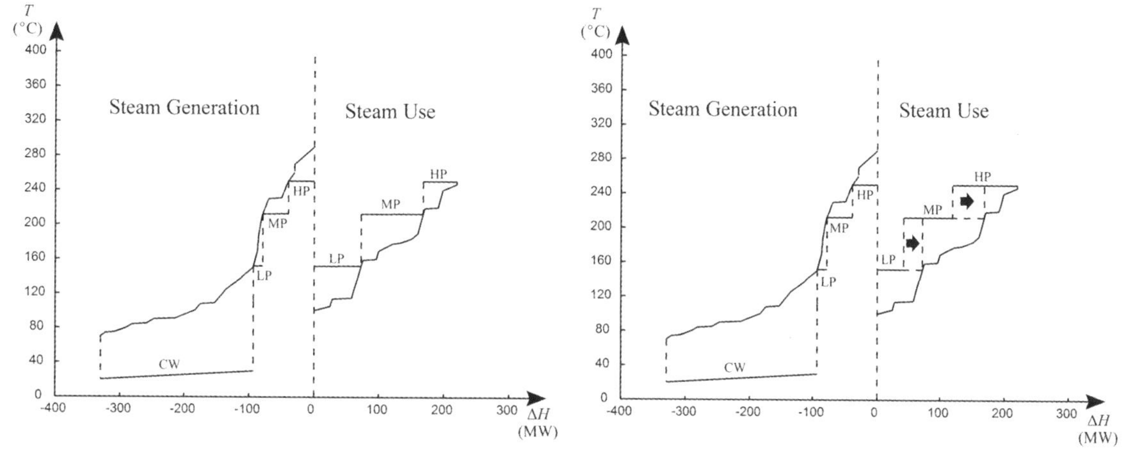

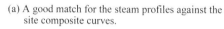

(a) A good match for the steam profiles against the site composite curves.

(b) A poor match between the steam profiles and the site composite curves.

Figure 23.29

Steam generation can be matched against the site hot composite curve and steam use against the site cold composite curve. (Reproduced from Li Sun, Steve Doyle, Robin Smith, Heat Recovery and Power Targeting in Utility Systems. In Press. Ms. Ref. No.: EGY-D-14-03308, with permission from Elsevier.)

Some of MP steam and all of LP steam use against the site composite curve would need to be let down from utility VHP steam. The amount of overlap between the steam profiles is a degree of freedom available to the designer.

Figure 23.32 shows that increasing the heat recovery between the site composite curves decreases the steam that must be generated in the utility boilers and decreases the site heat rejection. In other words, increasing the heat recovery decreases the heat flow through the system from utility steam generation through site cooling, and vice versa. If the overlap between the site steam profiles is maximized, as shown in Figure 23.32b, this minimizes the steam generation in the utility boilers and the site heat rejection. The limit is set by the *site pinch* (Raissi, 1994; Klemes *et al.*, 1997). Figure 23.33 illustrates the overall significance of the site pinch, dividing the site into a heat deficit above the site pinch and a heat surplus below the site pinch. This is analogous to the process pinch dividing individual processes into two parts, as discussed in Chapter 17.

So far, a number of issues have not been addressed for the steam profiles:

1) *Boiler feedwater preheating.* Boiler feedwater is fed to process steam generation at the deaeration temperature, which will be below saturation temperature. Preheating of the boiler feedwater prior to vaporization can be carried out by recovery from the site hot composite curve.
2) *Steam superheating.* Steam fed to the steam mains from process steam generation should be superheated and this can also be carried out by heat recovery from the site hot composite curve.
3) *Steam desuperheating.* Steam fed to process steam heaters, if superheated, involves a poor heat transfer coefficient until saturation conditions are attained. The design of steam heaters benefits from the desuperheating of steam prior to use. If this is carried out, boiler feedwater from the deaerator is injected into the steam under temperature control to typically bring it within

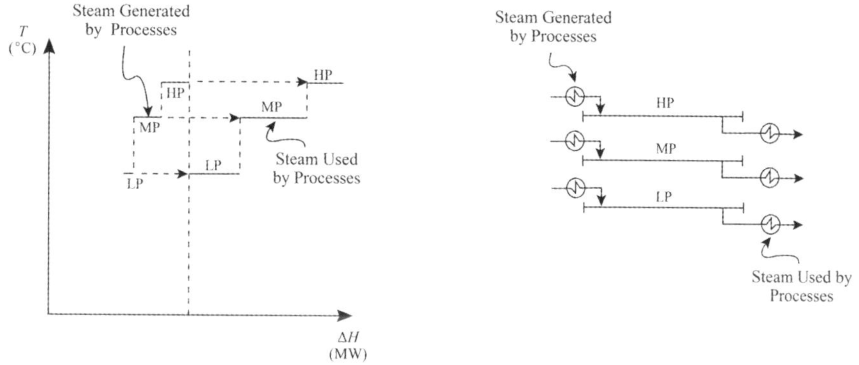

(a) Steam System Profiles.

(b) Steam flows between processes via the steam mains.

Figure 23.30

Site heat recovery through the steam system.

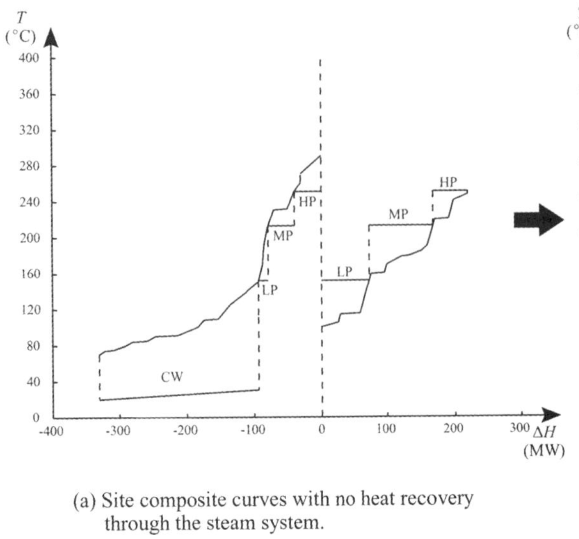

Figure 23.31

Overlapping the site composite curves gives the potential for heat recovery through the steam system. (Reproduced from Li Sun, Steve Doyle, Robin Smith, Heat Recovery and Power Targeting in Utility Systems. In Press. Ms. Ref. No.: EGY-D-14-03308, with permission from Elsevier.)

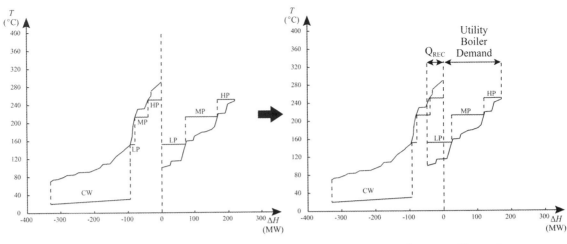

10 °C of saturation. The benefit is smaller and cheaper heat exchangers and in some cases less damage to sensitive process fluids. However, for the same process heating load, the mass flowrate of steam increases and additional boiler fuel is required to compensate for the desuperheating.

4) *Condensate heat recovery.* After condensation of the steam in process steam heaters, additional heat can be extracted from the condensate before returning the condensate to the boiler. This has drawbacks for the design of the heat transfer equipment, reducing overall heat transfer coefficients. However, less energy is lost in the condensate system, which possibly reduces boiler fuel.

These options need to be included in the targets so that they can be screened as to the benefit or penalty of different options. Figure 23.34 illustrates how this can be allowed for. Figure 23.34a shows as an example two steam generation profiles, each with boiler feedwater preheating and steam superheating. Figure 23.34b shows how these two profiles can be combined to produce a composite steam profile, in the same way as composite curves are constructed for the process streams. Following the same approach as illustrated in Figure 23.34 allows all additional features to be included in the steam profiles. Figure 23.35a shows the site hot composite curve matched against a composite steam profile that includes boiler feedwater preheating and steam

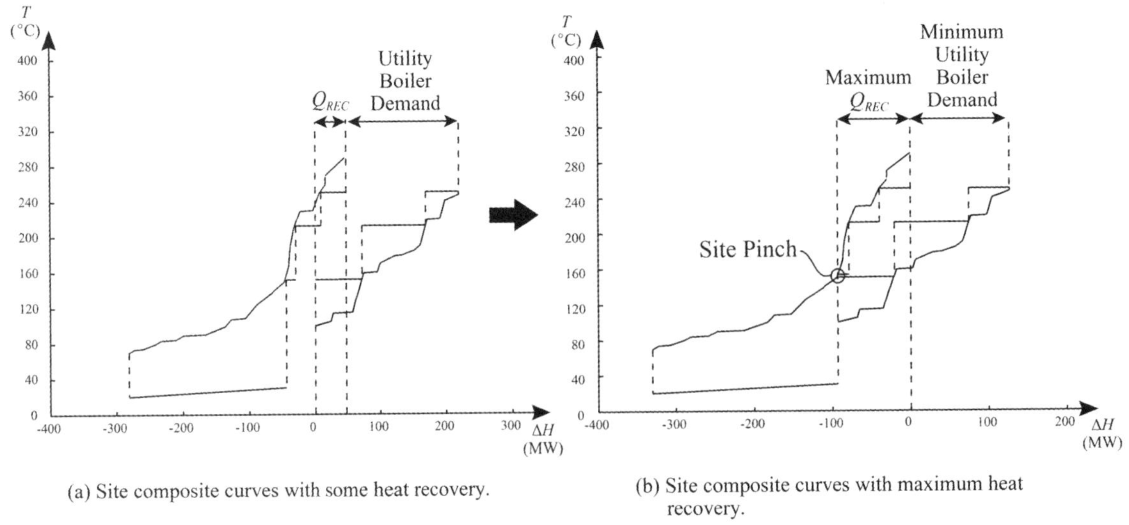

Figure 23.32

Maximizing the overlap minimizes the utility boiler demand. (From Sun L, Doyle S and Smith R, 2015, *Energy*, **84**: 196, reproduced by permission.)

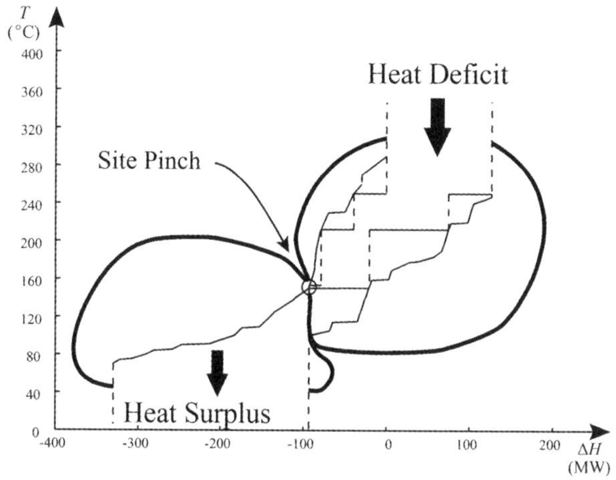

Figure 23.33

Maximizing the overlap creates a pinch for the site, dividing it into a heat sink and heat source, (analogous to individual process pinch).

superheating, along with vaporization, and a site cold composite curve matched against a composite steam profile that includes steam superheating and condensate cooling, along with condensation. Figure 23.35b shows the overlap between site composite curves maximized to create a site pinch and minimize the utility boiler demand (Sun, Doyle and Smith, 2015a).

It should be noted that in Figure 23.35 the superheat for both the steam generation and use have been assumed to be the same. The superheat for the generation is a matter of design of the heat transfer equipment. In principle, any degree of superheat could be designed for. On the other hand, steam for steam use will be drawn from the appropriate steam header with whatever superheat is available in the steam main. The temperature in the steam header will be a function of the superheat temperature for process steam generation, temperature of the utility steam, the efficiency of the steam turbines expanding between the headers, the amount of letdown flow between the steam headers and any desuperheating between the steam headers. Thus, in practice, the superheat temperature for the steam use is not as straightforward to set as with the generation.

Figure 23.36 shows a comparison of various options for the steam use profile to be matched against the site cold composite curve. Another option is shown in Figure 23.37, which features flash steam recovery in the steam use profile matched against the site cold composite curve. As an example, Figure 23.38 shows the site composite curves including boiler feedwater preheating, steam superheating, steam desuperheating to saturation prior to use, but no condensate cooling. Another example shown in Figure 23.39 includes boiler feedwater preheating, steam superheating, steam desuperheating to saturation prior to use and condensate cooling (Sun, Doyle and Smith, 2015a). 23

Important features that have been left out of the targeting so far are associated with the utility steam generation. The utility boiler feedwater preheating, vaporization and steam superheating could have been included in principle. This would have required the boiler flue gas profile to be included in the site hot composite curve. However, the utility boilers are normally kept as self-contained systems with their own economizers, rather than integrated with the site. Although this is normally the case, integration options can in principle be explored.

Whilst the targets developed so far are thermodynamically feasible, they might be difficult to achieve in practice because of the resulting design complexity. Consider the site composite curves shown in Figure 23.39. Where the boiler feedwater recovers heat from the site hot composite curve, might require the boiler feedwater to be circulated around a number of processes to achieve the temperature in Figure 23.39, because a number of different processes are brought together in the construction of the site hot composite curve. Also, the superheating of the steam might require the steam to be circulated around a number of processes to achieve

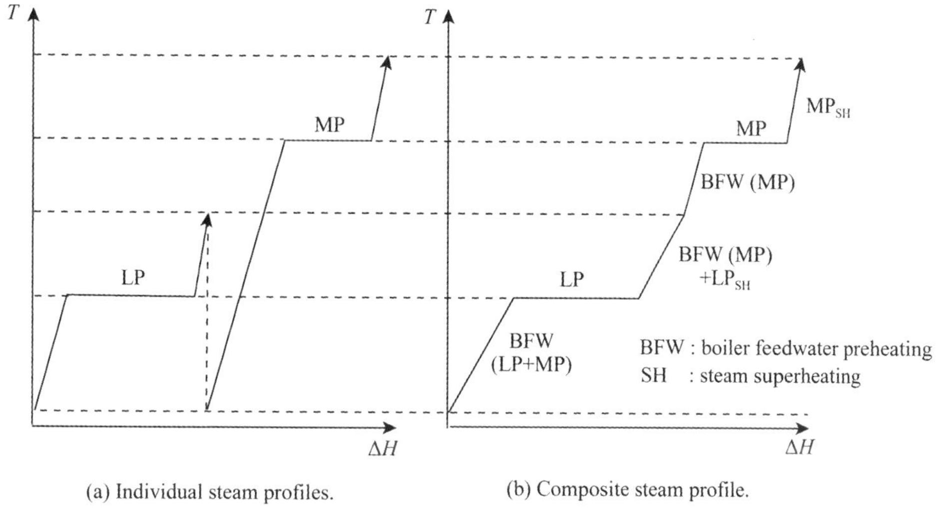

Figure 23.34

Boiler feedwater preheating (or boiler feed water cooling) and steam superheating (or desuperheating) can be included by creating a composite of the steam profiles. (Reproduced from Li Sun, Steve Doyle, Robin Smith, Heat Recovery and Power Targeting in Utility Systems. In Press. Ms. Ref. No.: EGY-D-14-03308, with permission from Elsevier.)

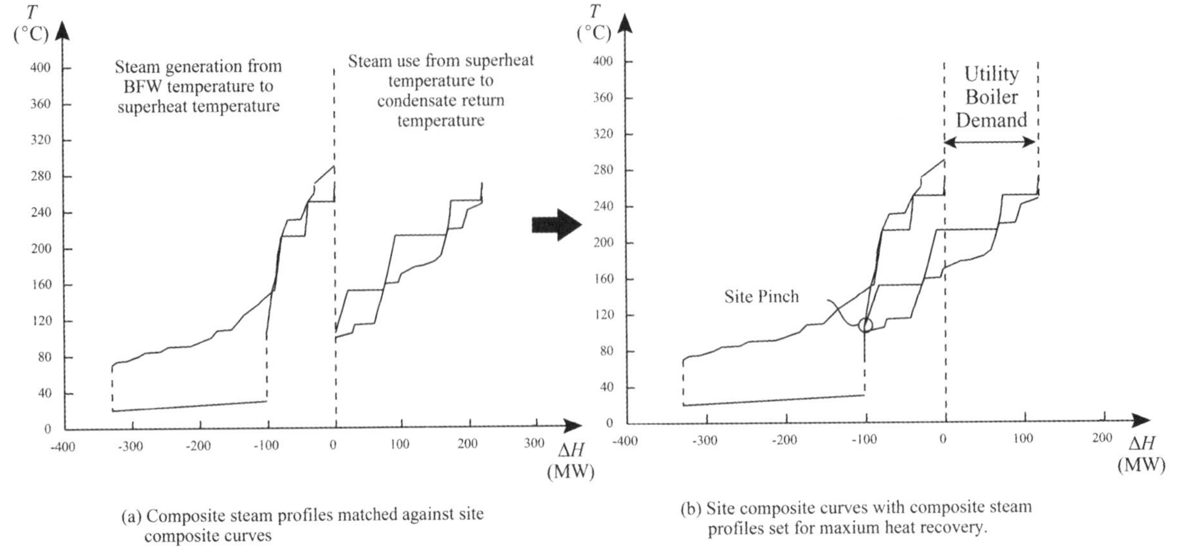

Figure 23.35

Site composite curves including BFW preheating, steam superheating, superheated steam use and condensate cooling. (Reproduced from Li Sun, Steve Doyle, Robin Smith, Heat Recovery and Power Targeting in Utility Systems. In Press. Ms. Ref. No.: EGY-D-14-03308, with permission from Elsevier.)

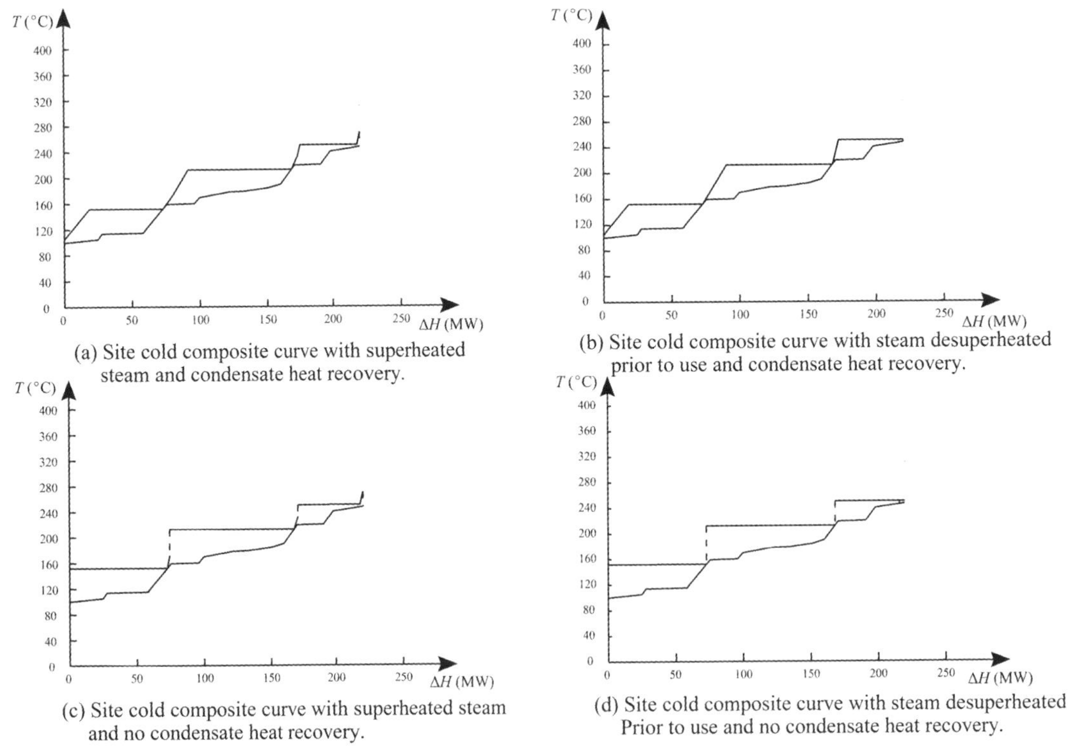

T (°C)
ΔH (MW)
(a) Site cold composite curve with superheated steam and condensate heat recovery.
(b) Site cold composite curve with steam desuperheated prior to use and condensate heat recovery.
(c) Site cold composite curve with superheated steam and no condensate heat recovery.
(d) Site cold composite curve with steam desuperheated Prior to use and no condensate heat recovery.

Figure 23.36

Various options are possible for the steam use profile to match against the site cold composite curve. (Reproduced from Li Sun, Steve Doyle, Robin Smith, Heat Recovery and Power Targeting in Utility Systems. In Press. Ms. Ref. No.: EGY-D-14-03308, with permission from Elsevier.)

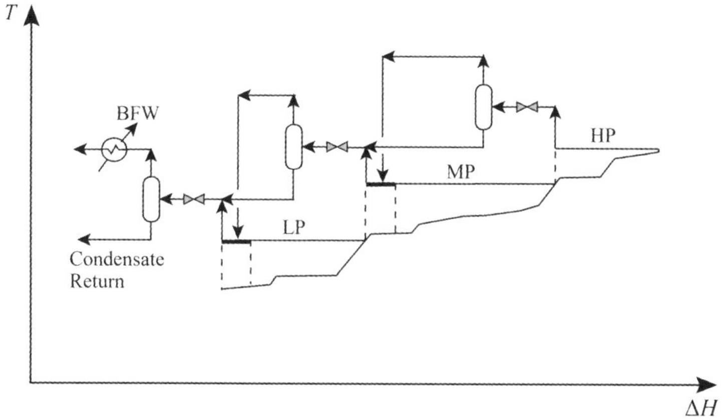

Figure 23.37

Site cold composite curve with flash steam recovery. (Reproduced from Li Sun, Steve Doyle, Robin Smith, Heat Recovery and Power Targeting in Utility Systems. In Press. Ms. Ref. No.: EGY-D-14-03308, with permission from Elsevier.)

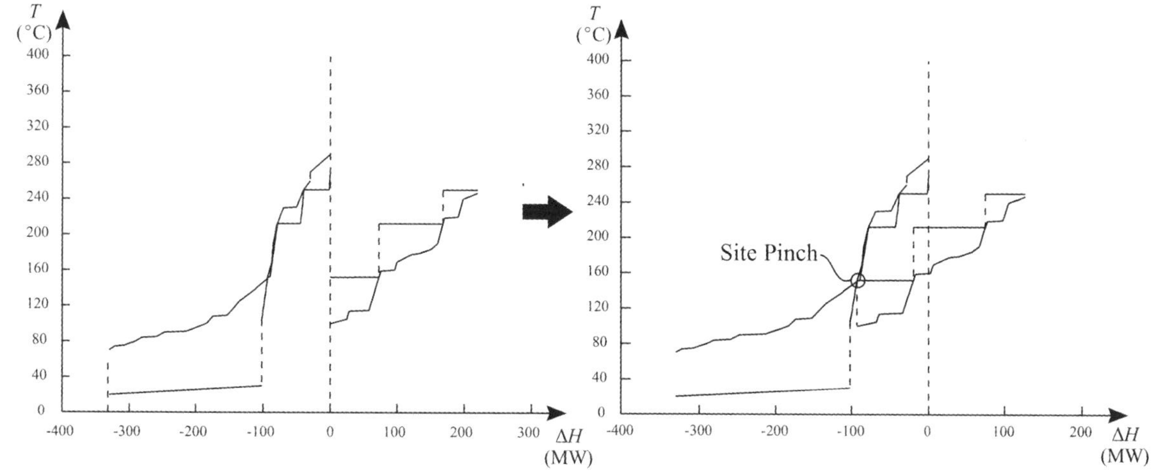

(a) Steam profiles matched against the site composite curves.

(b) Site composite curves with steam profiles set for maxium heat recovery.

Figure 23.38

Site composite curves including BFW preheating, steam superheating, steam desuperheating prior to use and no condensate cooling. (Reproduced from Li Sun, Steve Doyle, Robin Smith, Heat Recovery and Power Targeting in Utility Systems. In Press. Ms. Ref. No.: EGY-D-14-03308, with permission from Elsevier.)

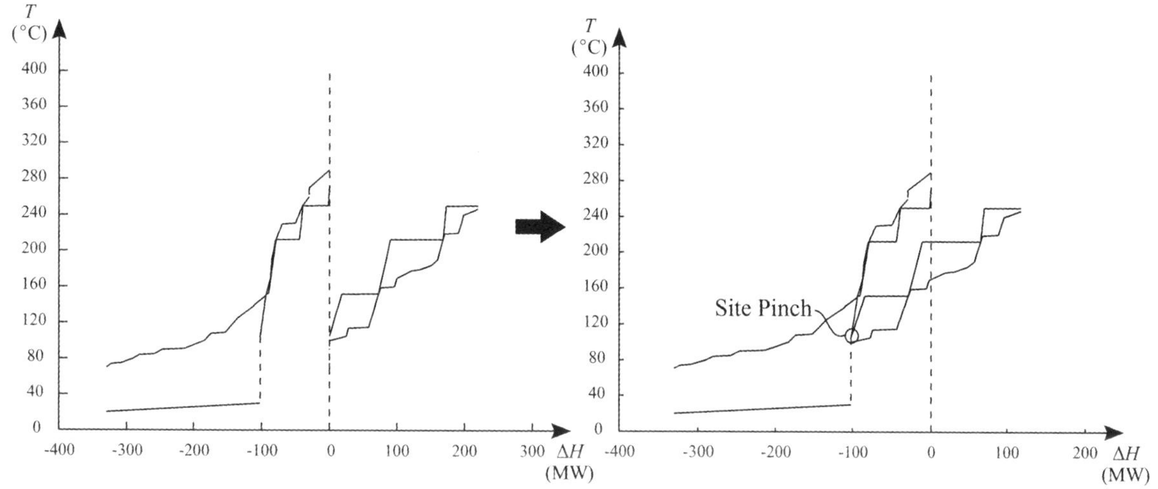

(a) Steam profiles matched against composite curves.

(b) Site composite curves with steam profiles set for maxium heat recovery.

Figure 23.39

Site composite curves including BFW preheating, steam superheating, steam desuperheating prior to use and condensate cooling. (Reproduced from Li Sun, Steve Doyle, Robin Smith, Heat Recovery and Power Targeting in Utility Systems. In Press. Ms. Ref. No.: EGY-D-14-03308, with permission from Elsevier.)

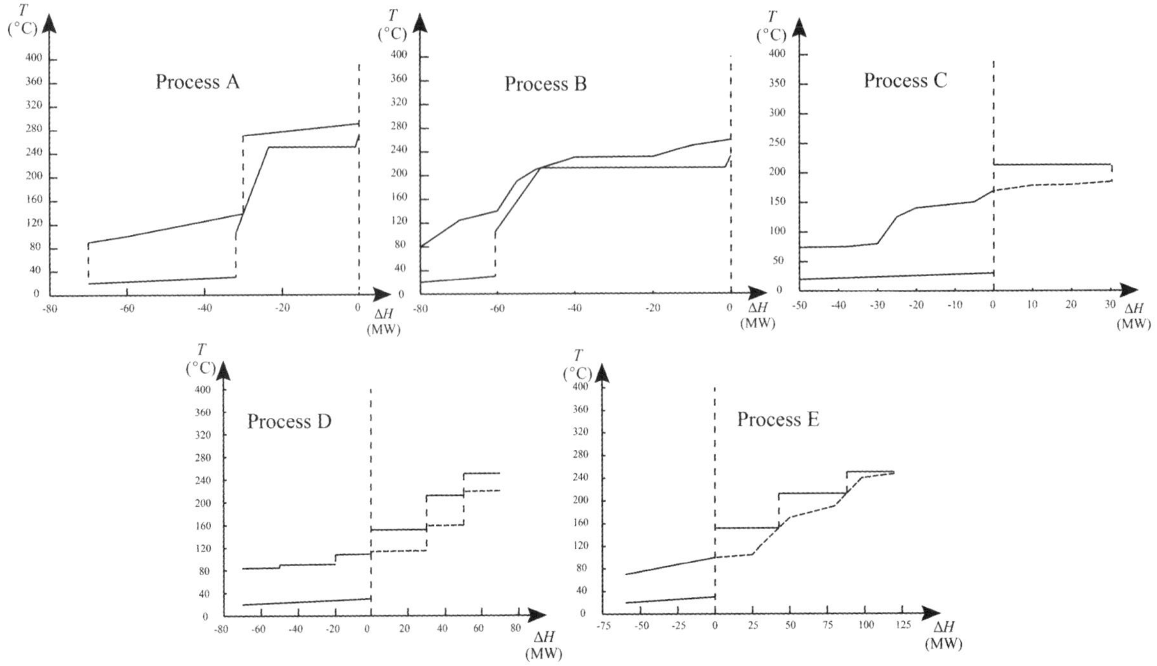

Figure 23.40

Targets can be set for the steam generation and use for the individual processes on the site. (Reproduced from Li Sun, Steve Doyle, Robin Smith, Heat Recovery and Power Targeting in Utility Systems. In Press. Ms. Ref. No.: EGY-D-14-03308, with permission from Elsevier.)

the temperature in Figure 23.39. Circulating steam in this way is even more problematic than circulating boiler feedwater. The same argument applies to the condensate cooling against the site cold composite curve shown in Figure 23.39. Thus, although the targets set in Figure 23.39 are thermodynamically feasible, they might involve significant undesirable design complexity (Sun, Doyle and Smith, 2015a).

This problem can be overcome by decomposing the site data (Sun, Doyle and Smith, 2015a). Figure 23.40 shows again the hot and cold composite curves for the individual processes on the site after the ΔT_{min} shift. Also in Figure 23.40, steam profile targets have been matched against the individual processes. Figure 23.41a shows the steam profiles of the individual processes combined to give steam profiles for the whole site. Figure 23.41b shows the

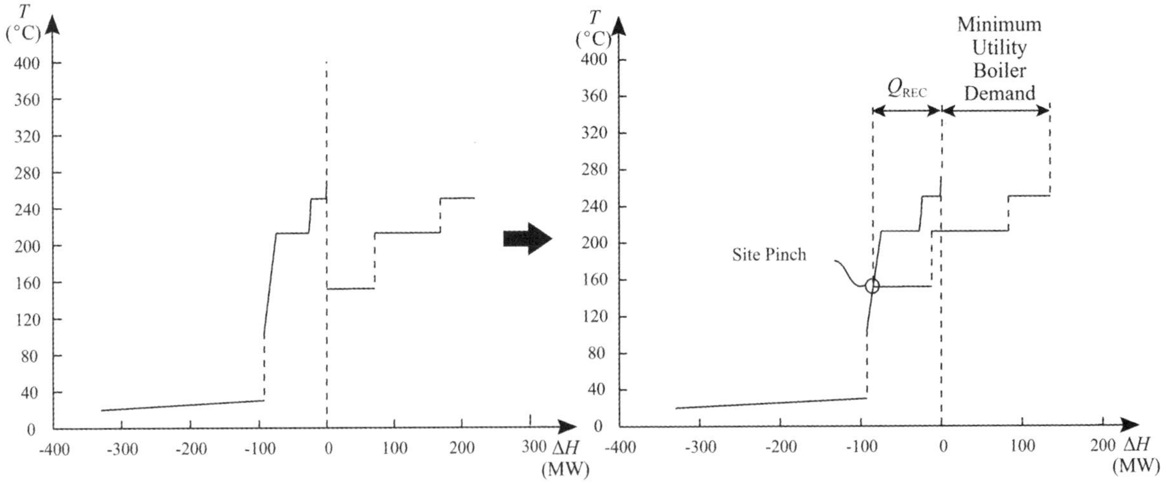

(a) Site steam profiles with no overlap.

(b) Site steam profiles with maximum overlap.

Figure 23.41

Site steam profiles based on decomposed site including boiler feedwater preheat, steam superheating, steam desuperheating to saturation prior to use and no condensate cooling. (Reproduced from Li Sun, Steve Doyle, Robin Smith, Heat Recovery and Power Targeting in Utility Systems. In Press. Ms. Ref. No.: EGY-D-14-03308, with permission from Elsevier.)

same profiles, but overlapped to maximize heat recovery through the steam system and minimize utility boiler demand. It should be noted that the targets set by the construction in Figure 23.41 are achieved without the need to pass boiler feedwater, steam or condensate around the various processes on the site to achieve the target heat recovery. The targets set in Figure 23.41 are based on the decomposed targets for the site. Because the data have been decomposed, the steam profiles shown in Figure 23.41 will not match against the site composite curves. Hence the site composite curves have not been shown in Figure 23.41. The targets set by the decomposed approach in Figure 23.41 (for the case of boiler feedwater preheat, steam superheating and steam desuperheating to saturation prior to use) will show a higher utility boiler demand and site cooling than the corresponding targets based on the site composite curves in Figure 23.38. However, both targets are useful in understanding the potential benefits of different options for the site utility system.

In setting the appropriate amount of heat recovery across the site through the steam system, various trade-offs need to be considered. Before this can be done, the cogeneration implications of the site heat recovery need to be quantified.

23.7 Cogeneration Targets

So far, targets for the site have been restricted to the heat duties for steam generation and steam use. However, this does not provide the complete picture, as the cogeneration potential from the expansion of steam in steam turbines needs to be included. This requires a simple model of power generation to be combined with the heating and cooling duties. A simple isentropic efficiency model can be used for this purpose based on Equation 23.23. The isentropic efficiency can be taken from typical values or calculated from Equation 23.23.

The cascade of steam in a steam turbine network was considered in Figure 23.23. It was noted that if there is no process steam recovery into the steam mains, once the steam mains pressures, the temperature of the utility steam and the isentropic efficiency of the turbines have been specified, then the temperatures of all the steam mains are fixed as a result. However, there are practical constraints that need to be maintained in the operation of the steam system. There is a minimum practical amount of superheat that needs to be maintained in the steam mains in order to prevent excessive condensation (typically 10 to 20 °C). Also, for the condensing turbine in Figure 23.23, the minimum dryness of the steam at the exit needs to be maintained to typically a minimum of 90%.

As an example, consider the case given in Figure 23.42a based on the data from Table 23.4 for the case of boiler feedwater preheat and superheat for steam generation and for the steam use desuperheating to saturation prior to use and no condensate recovery (Sun, Doyle and Smith, 2015a). Calculation of the steam turbine cascade requires the model of the steam expansion to be considered simultaneously with the target heating and cooling duties from the site composite curves (or from the decomposed steam profiles). The flowrate of steam generated at a given level is determined by matching the steam generation profile (given the assumed boiler feedwater feed condition and steam superheat temperature) against the corresponding cooling duty from the site hot composite curve (or from the decomposed steam profile). The flowrate of steam used at a given level is determined by matching the steam use

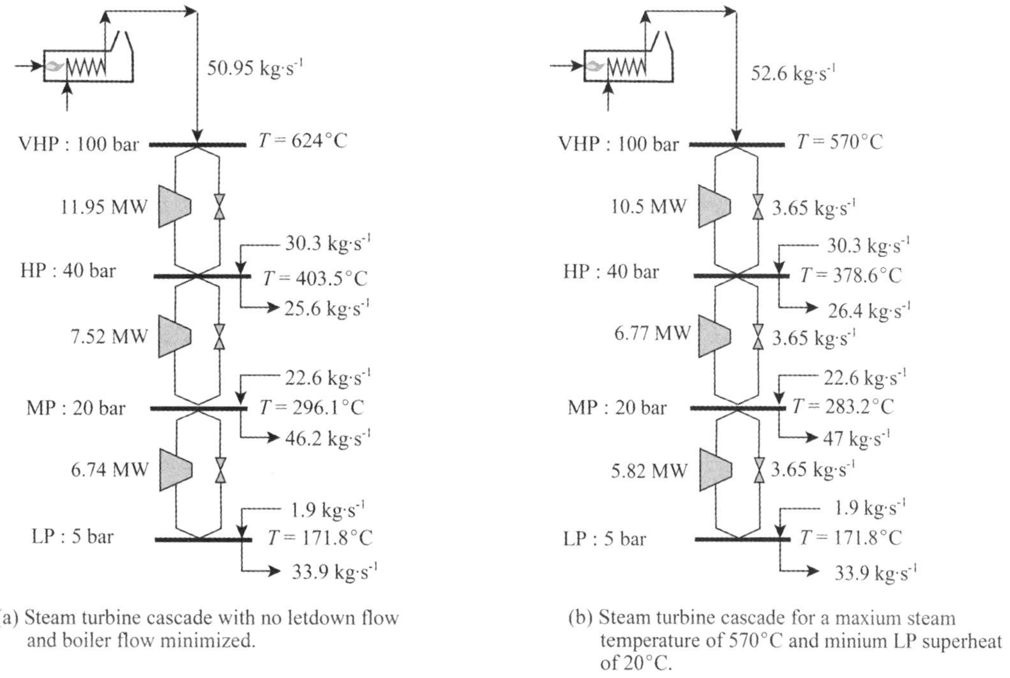

Figure 23.42
Steam turbine cascade for power generation targeting for the case with BFW preheating, steam superheating, steam desuperheating prior to use and no condensate cooling. (Reproduced from Li Sun, Steve Doyle, Robin Smith, Heat Recovery and Power Targeting in Utility Systems. In Press. Ms. Ref. No.: EGY-D-14-03308, with permission from Elsevier.)

profile (given the properties of the steam header at that level) against the heating duty for the site cold composite curve (or from the decomposed steam profile). The properties of the steam in each header are determined by a material and energy balance for that header. This is determined by the properties and flowrate of the steam generated that is feeding the header, the flowrate through, inlet steam conditions and isentropic efficiency of the steam turbine feeding the header, and the flowrate through and inlet steam conditions of the letdown station feeding the header. The calculation can start with the utility steam and work down through the cascade from high to low pressure simulating the steam turbine and letdown expansions, performing a material and energy balance for each header to determine the header conditions and the flowrates to and from each header to service the cooling and heating duties. Figure 23.42a shows the results assuming an isentropic efficiency for all steam turbines of 0.75, with the constraint that all the steam mains should have a superheat of at least 20 °C. It is also assumed that the letdown flow through the valves is zero, even though practical reasons, as discussed earlier, dictate that it would be better to maintain a small flow. To achieve the constraint of a minimum of 20 °C superheat requires the temperature of the utility steam to be varied by trial and error, with the steam cascade calculated for each iteration until the constraint for superheat is met. It can be seen in Figure 23.42a that this requires a utility steam temperature of 624 °C. This temperature is too high and should be below 600 °C because of limitations of the materials of construction.

If it is assumed that the maximum temperature is 570 °C, then at this temperature it will not be possible to maintain a minimum superheat of 20 °C in the LP main. One way to overcome this problem is to fix the temperature of the utility steam to its maximum of 570 °C and expand steam through the letdown stations to increase the superheat, in the steam mains to the required minimum value. Expanding steam through a steam turbine decreases the amount of superheat because of the energy extracted for power generation, whereas expanding steam in a valve increases the amount of superheat, as this is an adiabatic process. Figure 23.42b shows a steam turbine cascade in which the utility boiler steam has been set to a temperature of 570 °C and the flow of steam in the letdown stations has gradually been increased until the minimum superheat temperature in the mains has been achieved. In this case, to achieve the constraint of a minimum of 20 °C in the LP main with a maximum utility boiler steam temperature of 570 °C requires a flow of 3.65 $kg \cdot s^{-1}$ of steam through the letdown stations. An alternative way to overcome the constraint would be to increase the amount of superheat in the steam generated by recovery. Figure 23.42b gives the cogeneration target for the example based on the assumed maximum utility steam temperature, turbine isentropic efficiency and minimum steam main superheat.

Different options can now be compared on the basis of the targets for steam generation, steam use and cogeneration. Table 23.5 shows a comparison of the different options based on an assumed maximum temperature for the utility steam of 570 °C, turbine isentropic efficiency of 0.75 and minimum steam main superheat of 20 °C (Sun, Doyle and Smith, 2015a).

Table 23.5 lists the boiler flowrate, the letdown flow rate and power generation for various utility options based on a utility steam temperature of 570 °C and minimum mains superheat of 20 °C. Any desuperheating of steam prior to use is assumed to be carried out locally. Also shown in Table 23.5 is the power generation per unit of boiler steam flow as a basis of comparison between the different options. It can be seen that the decomposed options in each case perform worse than the integrated options, as expected. Table 23.5 shows that the option for superheated steam use and no condensate heat recovery is best for both the integrated and decomposed targets. Desuperheating increases the utility boiler demand. This is caused by the need to compensate for the injection of boiler feedwater, which must be heated from its supply temperature to the saturation temperature of the steam by direct heat transfer. In addition, desuperheating increases the flowrate of condensate after providing the heating duty, which creates additional losses from the condensate system. Table 23.5 shows that condensate heat recovery is not beneficial. This is because condensate heat recovery extracts heat at a high level, rather than allowing that heat to be cascaded down in pressure for power production. It should be noted that the model used here does not include the features of the condensate return system and the effect on the deaeration steam. Condensate subcooling will degrade the heat returned to the deaerator, increasing the deaeration steam. These targets can be used to explore different options, but ultimately each needs to be evaluated by a more detailed simulation of the whole utility system.

The steam turbine expansion zones can be represented graphically, as shown in Figure 23.43 (Sun, Doyle and Smith, 2015a). This shows both the site composite curves and the expansion zones for the steam in the steam turbine network. To identify the expansion zones, an energy balance must be performed between each of the steam mains. The heat used to generate steam from the site hot composite curve will be used to offset heat required by the site cold composite curve. If there is a deficit of steam, then this must be cascaded through a steam turbine from a level above. If there is a surplus of steam, then the surplus can be cascaded down to the next level.

So far, all of the settings for a heat recovery between the site composite curves have been set to a maximum, thereby minimizing the utility boiler load. Figure 23.44 shows a setting between the site composite curves that does not feature maximum heat recovery through the steam system, and therefore does not have a site pinch. This means that more heat needs to be imported from the utility boilers and more heat rejected from the site. However, in Figure 23.44 it can be seen that the additional heat rejected from the site is achieved by expanding steam through to vacuum conditions and then condensing against cooling water in condensing power generation. Thus, the steam turbine network in Figure 23.44 involves a combination of power generation from back-pressure steam turbines and condensing steam turbines. The more steam import from the utility boilers, the more condensing power generation there is. This is a trade-off that needs to be explored economically, based on the cost of the additional steam generation, profit from power generation and the capital cost required.

In Figures 23.43 and 23.44, a single steam turbine is shown expanding between the different steam levels. In practice, this might be simple steam turbines expanding between each level, simple steam turbines operating across levels, extraction steam turbines operating between and/or across multiple levels or

23

Table 23.5

Site targets for different utility options (Sun, Doyle and Smith, 2015a).

Integrated options					
Source profile	**Sink profile**	**Boiler flowrate ($kg \cdot s^{-1}$)**	**Letdown flowrate ($kg \cdot s^{-1}$)**	**Power generation (MW)**	**Power generation per unit boiler flow ($MW \cdot kg^{-1} \cdot s$)**
BFW preheat + steam superheat	Superheated steam use + condensate heat recovery	50.29	3.6	20.83	0.414
BFW preheat + steam superheat	Superheated steam use, no condensate heat recovery	49.95	3.9	22.68	0.454
BFW preheat + steam superheat	Desuperheated steam use + condensate heat recovery	52.66	3.3	21.35	0.405
BFW preheat + steam superheat	Desuperheated steam use, no condensate heat recovery	52.06	3.7	23.10	0.444
Decomposed options					
Source profile	**Sink profile**	**Boiler flowrate ($kg \cdot s^{-1}$)**	**Letdown flowrate ($kg \cdot s^{-1}$)**	**Power generation (MW)**	**Power generation per nnit boiler flow ($MW \cdot kg^{-1} \cdot s$)**
BFW preheat + steam superheat	Superheated steam use + condensate heat recovery	53.35	3.1	21.48	0.403
BFW preheat + steam superheat	Superheated steam use, no condensate heat recovery	53.25	2.8	23.18	0.435
BFW preheat + steam superheat	Desuperheated steam use + condensate heat recovery	55.95	2.8	21.84	0.390
BFW preheat + steam superheat	Desuperheated steam use, no condensate heat recovery	55.48	2.6	23.68	0.427

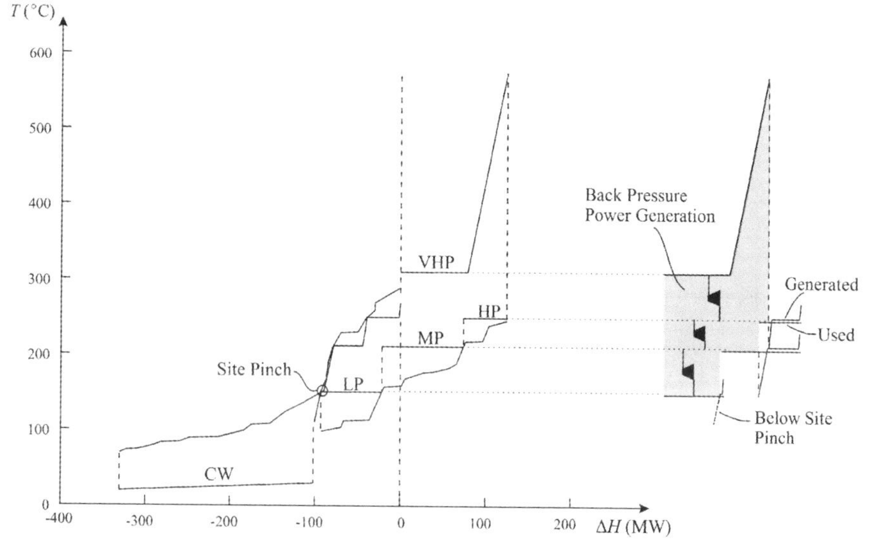

Figure 23.43

Power generation for the site for pinched conditions. (Reproduced from Li Sun, Steve Doyle, Robin Smith, Heat Recovery and Power Targeting in Utility Systems. In Press. Ms. Ref. No.: EGY-D-14-03308, with permission from Elsevier.)

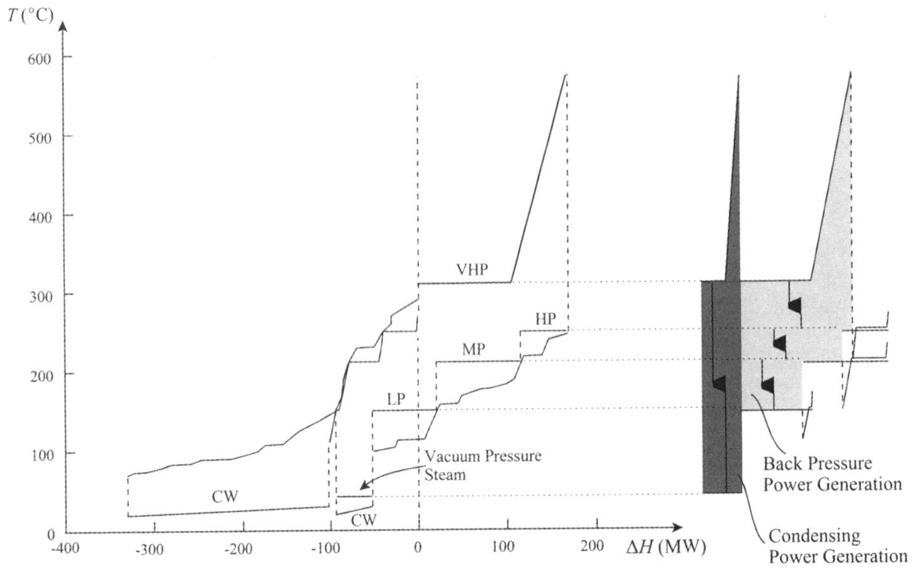

Figure 23.44

Power generation for the site for unpinched conditions showing condensing power generation. (Reproduced from Li Sun, Steve Doyle, Robin Smith, Heat Recovery and Power Targeting in Utility Systems. In Press. Ms. Ref. No.: EGY-D-14-03308, with permission from Elsevier.)

combinations of simple and extraction machines operating between and/or across different levels. Some of the many possible configurations are illustrated in Figure 23.45. Also, rather than have one steam turbine servicing each expansion, as implied by Figure 23.45, expansion duties can be split between several machines operating in parallel. If the only feed to each main is expanded across steam turbines, all with the same isentropic efficiency, then all turbine networks will perform the same, irrespective of the configuration. However, in practice there are different feeds and turbines do not all have the same isentropic efficiency, which means that there will be differences between the performances of different steam turbine network configurations. Final design of the steam turbine must consider driver allocation. The shaft power required by the process and utility machines might be provided by the provision of electricity for an electric motor or by a steam turbine in a direct drive. This means that steam turbines might drive turbogenerators to generate electricity or drive individual process machines on direct drives. Driver allocation will be considered in the next section.

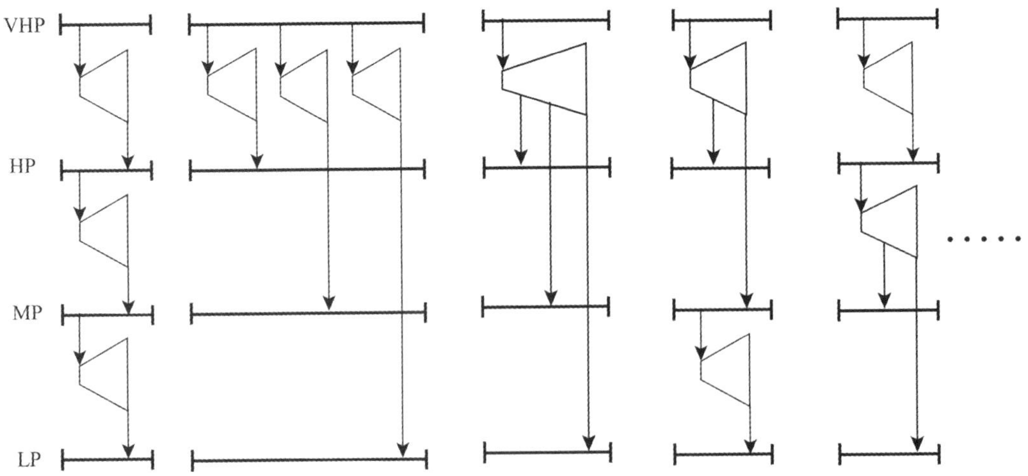

Figure 23.45

Some of the many possible steam turbine network configurations.

23.8 Power Generation and Machine Drives

The site requires both heat and power, and wherever possible this should be provided through cogeneration. Power is required across the site to drive machines and as electrical power. Electricity can be generated on site or imported from central electricity production. Electricity might be generated at a capacity above the site requirements and exported. The generation of electricity on site can be carried out through a machine driving an electricity generator. These include:

- Steam turbine
- Gas turbine
- Gas engine (reciprocating engine firing gas)
- Diesel engine (reciprocating engine firing diesel oil).

Once electricity has been generated, it can be used in electric motors to drive process machines. Alternatively, the above driver machines can be coupled directly to process machines in direct drives.

If cogeneration is to be the objective, then the temperature at which the byproduct heat from power generation becomes available is an important consideration. Gas turbines produce a hot exhaust gas with a temperature that depends on the particular machine, but high enough to generate high-pressure steam. Gas engines and diesel engines produce a hot exhaust gas at around 450 °C that can in principle be used to generate some steam. Also, the cylinders need to be cooled by jackets supplied with cooling water. This provides a source of heat up to 95 °C. Thus, in the case of gas engines and diesel engines, around half of the waste heat becomes available only at a low temperature. This is appropriate, for example, if large amounts of hot water are needed, but inappropriate if large amounts of steam are needed. Thus, for most sites in the process industries gas engines and diesel engines are not appropriate for cogeneration schemes. The discussion here will therefore concentrate on steam turbines and gas turbines.

Before considering the most appropriate drive for process machines, it is important to consider the type of cogeneration system appropriate for the site. To clarify this, two measures are introduced. The site power-to-heat ratio is defined as (Kenney, 1984; Kimura and Zhu, 2000):

$$R_{SITE} = \frac{W_{SITE}}{Q_{SITE}} \tag{23.37}$$

where R_{SITE} = site power-to-heat ratio (–)
W_{SITE} = power demand for the site, including both shaft power and electricity requirements (kW, MW)
Q_{SITE} = process heating demand for the site (kW, MW)

Site power to heat ratios can vary typically between 0.03 and 3 depending on the nature of the processes. For example petroleum refining has a power to heat ratio of typically around 0.5 based on steam requirements, but goes as low as typically 0.1 once the heat to process furnaces is included along with the steam heating. In this context, it is the steam heating that is most relevant.

Another useful measure of the utility system performance is the *cogeneration efficiency*. Of the fuel fired in the utility system, some of this energy produces power, some provides useful process heat and some is lost. The cogeneration efficiency recognizes the amount of fuel consumed to produce both power and useful process heat, and can be defined as (Kenney, 1984; Kimura and Zhu, 2000):

$$\eta_{COGEN} = \frac{W_{SITE} + Q_{SITE}}{Q_{FUEL}} \tag{23.38}$$

where η_{COGEN} = cogeneration efficiency (–)
Q_{FUEL} = fuel fired in the utility system (kW, MW)

Having established the basic definitions, a plot can be developed between cogeneration efficiency with the site power-to-heat ratio (R_{SITE}). It will be assumed initially that power is not to be imported to or exported from the site. Figure 23.46 shows the variation of η_{COGEN} with R_{SITE} for the case of power generated from steam turbines only. Steam turbines are best suited for sites with a low site power-to-heat ratio (R_{SITE}). The development of the curve assumes that the site heating demand is fixed. The power demand gradually increases, increasing the site power-to-heat ratio. Start with the limiting case assuming zero demand for power, $R_{SITE} = 0$, as shown in Figure 23.46. At this point, there is no attempt to generate power by expansion in steam turbines. Instead, all of the steam is expanded through letdown stations. As R_{SITE} increases from zero, the steam is expanded through steam turbines to meet the power demand, with the remainder being expanded in letdown stations. As R_{SITE} increases further in Figure 23.46, eventually all of the potential to expand steam through back-pressure steam turbines has been realized. This corresponds with full cogeneration and the site utility system being pinched at R_{PINCH}. As additional power is generated above R_{PINCH} in Figure 23.46, this requires a gradual increase in the amount of condensing power generation from utility steam. This curve asymptotically approaches the efficiency for a stand-alone steam generation cycle.

If cogeneration efficiency (rather than cost) is used to dictate the import and export policy for electricity from the site, then the regions where electricity should be imported and exported are illustrated in Figure 23.47 (Varbanov *et al.*, 2004). At low values of R_{SITE} below R_{PINCH}, the site heat demand means that the site has the potential to cogenerate more power than it requires in the steam turbines. Given that power generation in this situation is more efficient than stand-alone power generation, it would be beneficial from the point of view of the efficiency to realize the full potential for cogeneration between $R_{SITE} = 0$ and $R_{SITE} = R_{PINCH}$ and to export the electricity not required on the site. At high values of R_{SITE}, as shown in Figure 23.47, η_{COGEN} decreases and is likely to fall below that for centralized stand-alone power generation $\eta_{CENTRAL}$. This is because the power generation cycles used in centralized power generation tend to be more complex and more efficient than the simple steam generation cycles used on process sites. The maximum efficiency for conventional centralized stand-alone steam generation is around 40%, but can be significantly higher for the most advanced stand-alone cycles. Now, from the

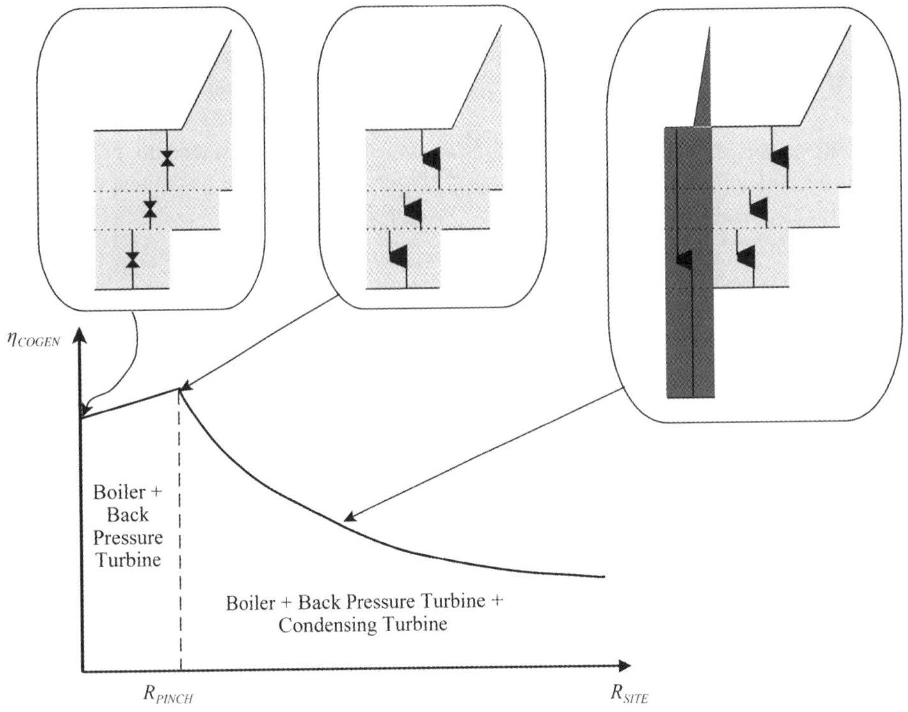

Figure 23.46

The site power-to-heat curve for complete on-site power generation in steam turbines.

point of view of efficiency, it would be sensible to stop power generation on the site when η_{COGEN} falls below the efficiency of centralized power generation $\eta_{CENTRAL}$ and to import the balance of power required by the site. Of course, these arguments are based on thermodynamic efficiency, rather than cost, and a full economic analysis is likely to change these thresholds. For example, the availability of large amounts of cheap fuel might make power export attractive over a wide range of values of

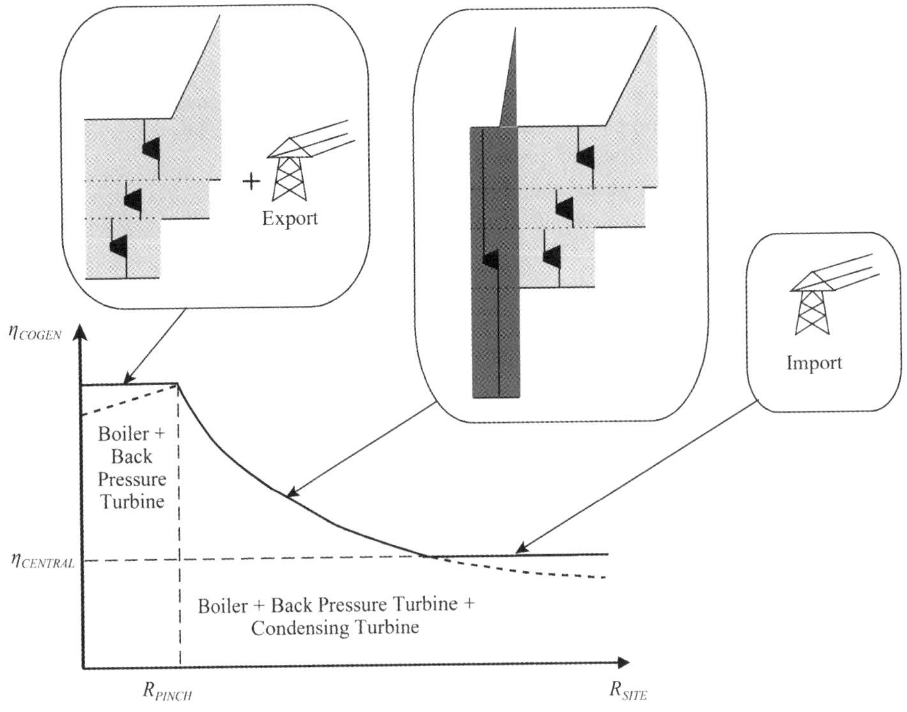

Figure 23.47

The site power-to-heat curve for complete on-site power generation in steam turbines.

R_{SITE}, despite poor cogeneration efficiency. It should also be recalled from Chapter 2 that power costs (both import and export) are usually subject to significant tariff variations according to the season of the year (winter versus summer), the time of the week (weekend versus weekday) and the time of day (night versus day). Even though transient economic factors might distort the picture, it is still important to understand the fundamentals of the trade-offs.

It has been noted that steam turbines are best suited to low values of R_{SITE}. As the site power demand increases, satisfying the site demands for cogenerated power is best suited to gas turbines, and combinations of gas turbines and steam turbines. The use of gas turbines will lead to cogeneration efficiencies higher than that for steam turbines for higher values of R_{SITE} (Kenney, 1984). Start by considering a gas turbine with an HRSG in isolation:

$$\begin{aligned} R_{GT} &= \frac{W_{GT}}{Q_{REC}} \\ &= \frac{W_{GT}}{\eta_{REC}(Q_{FUEL,GT} - W_{GT})} \\ &= \frac{1}{\eta_{REC}\left(\dfrac{Q_{FUEL,GT}}{W_{GT}} - 1\right)} \\ &= \frac{1}{\eta_{REC}\left(\dfrac{1}{\eta_{GT}} - 1\right)} \end{aligned} \quad (23.39)$$

where R_{GT} = gas turbine power-to-heat ratio (–)
W_{GT} = power generated by gas turbine (kW, MW)
$Q_{FUEL,GT}$ = fuel fired in the gas turbine (kW, MW)
Q_{REC} = heat recovered from the gas turbine exhaust (kW, MW)
η_{REC} = fractional heat recovery from the gas turbine exhaust in the HRSG (–)
η_{GT} = gas turbine efficiency (–)

Also:

$$\begin{aligned} \eta_{COGEN,GT} &= \frac{W_{GT} + Q_{REC}}{Q_{FUEL,GT}} \\ &= \frac{W_{GT} + Q_{REC}}{W_{GT}/\eta_{GT}} \\ &= \eta_{GT}\left(1 + \frac{1}{R_{GT}}\right) \end{aligned} \quad (23.40)$$

where $\eta_{COGEN,GT}$ = gas turbine cogeneration efficiency (–)

For example if $\eta_{REC}=0.75$ and fixed, then for $\eta_{GT}=0.3$, $R_{GT}=0.57$, $\eta_{COGEN,GT}=0.83$, and for $\eta_{GT}=0.5$, $R_{GT}=1.33$, $\eta_{COGEN,GT}=0.88$.

Figure 23.48 shows the variation of η_{COGEN} with R_{SITE} for the case of power generated from gas turbines together with steam turbines. The initial point for the curve at R_{GT} relates to power generation only from gas turbines that have been sized to satisfy the site heating demands from steam generation in HRSGs. Initially,

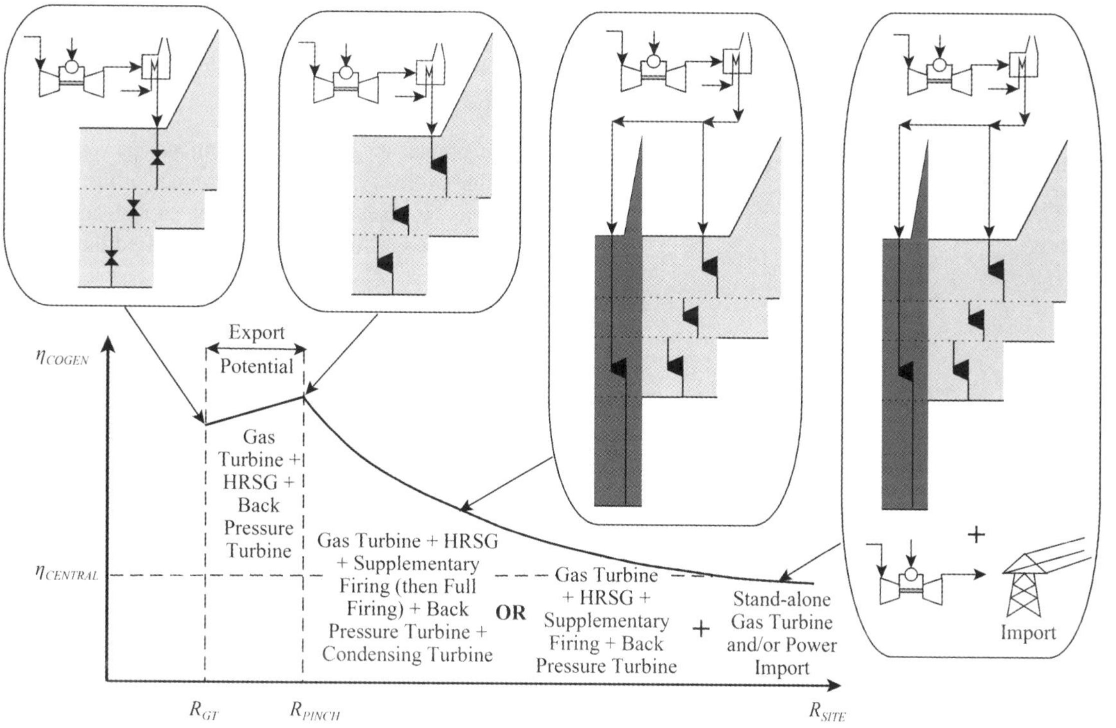

Figure 23.48

The site power-to-heat curve for complete on-site power generation in a combined cycle using gas and steam turbines.

there is no generation of power from steam turbines and all steam is expanded through letdown stations. As R_{SITE} is increased, letdown station expansion is gradually switched to steam turbine expansion, increasing η_{COGEN} until all of the capacity for back-pressure expansion has been exploited. R_{PINCH} corresponds with the site being pinched and full cogeneration capacity from steam turbines. Similar to the steam turbine only case in Figure 23.47, there is potential for power export between R_{GT} and R_{SITE} in Figure 23.48. Also, as with the steam generation only case, R_{SITE} can be increased by introducing condensing power generation through supplementary firing and then full firing. As R_{SITE} increases, the cogeneration efficiency decreases to the point where the efficiency would point to introducing additional gas turbines without heat recovery, or power import from centralized generation, or both.

The same basic approach discussed here for gas turbines can be used for gas engines and diesel engines if there is a demand for hot water rather than steam.

All R-curves will differ to some extent, depending on the site utility heating and cooling and equipment specifications. R-curve analysis can provide insights into efficiency improvements in utility systems. However, it has inherent limitations. Based purely on thermodynamics, it does not account for costs. For instance, it does not account for different fuels with different prices or, perhaps, low-efficiency boilers burning cheaper fuels and more efficient boilers burning expensive fuels. In this context, the most efficient operation of the utility system does not always mean the lowest cost operation. Consequently, there is also a need for an economic tool to understand the potential for cost saving in utility systems. However, R-curve analysis provides targets for the cogeneration potential of a site at different values of power requirement and is therefore useful in setting an overall framework that can be further optimized by considering economics.

Once the high-level decisions have been made as to whether boilers, steam turbines, gas turbines and HRSGs are to be used in the cogeneration system, important decisions need to be made regarding the major process drives on the site. The drivers used include:

- Electric motors
- Steam turbines
- Gas turbines
- Turboexpanders.

Electric motors are by far the most commonly used driver. Motors are available in a wide range of sizes from 1 kW to over 50 MW. The efficiency varies according to the size and design of the motor. A first approximation for the efficiency of the motor can be obtained from:

$$\eta_{MOTOR} = \frac{1}{1.039 + \dfrac{0.4235}{W}} \qquad (23.41)$$

where η_{MOTOR} = efficiency of the motor (–)
W = power produced by the motor at maximum load (kW)

Motors can be operated at part load, down to typically 50% of full load.

Steam turbines are also used quite commonly as direct drives, especially on relatively large shaft power duties. Gas turbines are used as direct drives, but only on the largest duties, such as those encountered in the liquefaction of natural gas. The use of turboexpanders is mostly restricted to situations where the expander is used to create a subambient temperature within the process, coupled with the recovery of power from the process streams (see Chapter 24).

Broadly, there are two basic options for the allocation of drivers to process machines. Power can be generated and distributed for use in electric motors or direct drives can be used. A direct drive (e.g. steam turbine driving a large pump) can often be the cheapest solution relative to a large steam turbine producing power, distribution of the power and use of the power in an electric motor to provide the driver. On the other hand, a large single (and efficient) generator can service many electric motor drives. Direct drives are inflexible as far as the utility system is concerned. For example, if the steam turbine is driving a large pump, then the flow through the steam turbine needs to be maintained to keep the pump running, even if the heat produced at the exhaust of the steam turbine might not be required. Electric generators are more flexible as far as the utility system is concerned. They can be turned up and down and can service many drives with electric motors. There are often many small drivers required on a site that are most economically serviced by electric motors, rather than direct drives. The best solution is usually a combination of electric generators producing electricity for electric motors and direct drives.

There are many issues to be considered for the selection of the most appropriate combination of drives:

- Economic analysis
- Fit to the existing infrastructure
- Process requirements
- Safety issues in the case of power failure
- Space limitations.

For example, suppose a new steam turbine is to be installed in an existing utility system. The economics of the new steam turbine need to be analyzed in terms of capital and operating costs. However, issues that need to be addressed include:

- Boiler capacity
- Fuel system capacity
- Steam mains capacity
- Water treatment capacity
- Cooling tower capacity
- Available space.

The process requirements include consideration of:

- Rotational speed
- Speed variability
- Operational duration without overhaul
- Size and weight
- Reliability.

Many issues need to be considered when choosing the most appropriate combination of drivers. Choosing the best combination

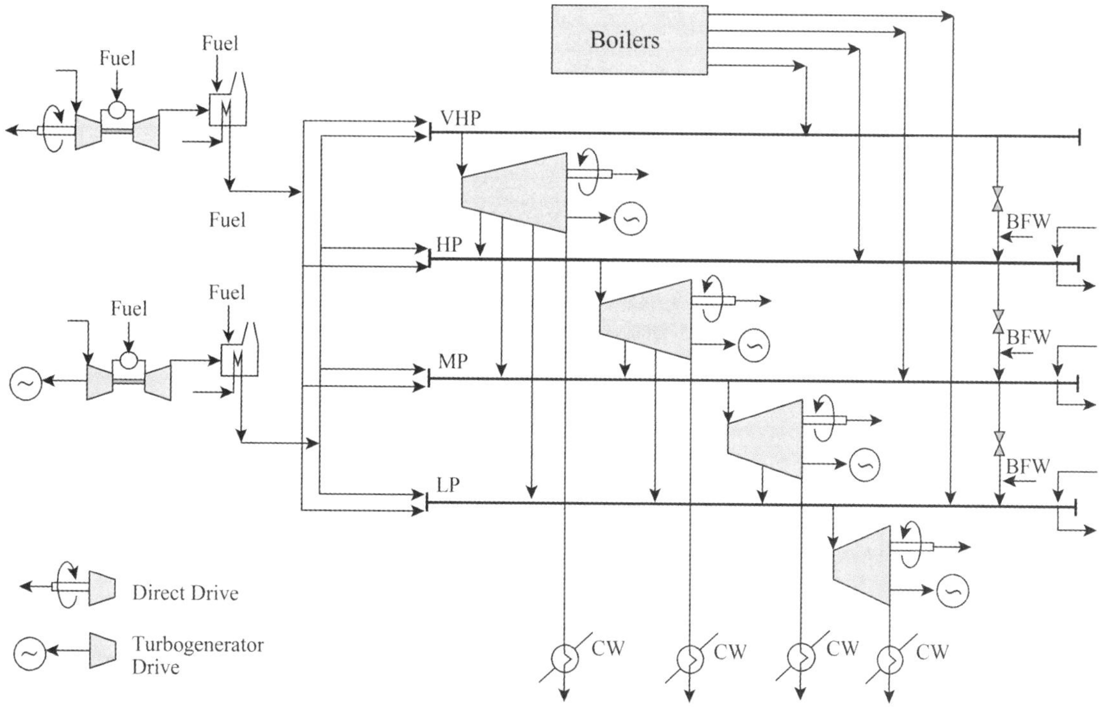

Figure 23.49

Superstructure for the optimization of utility system structure.

is a trade-off between capital costs, operating costs, system flexibility and reliability.

Because of the complexity of the problem in most cases, the best way to design the system configuration is through the optimization of a superstructure. Figure 23.49 illustrates the basic approach, without showing all of the details (Del Nogal *et al.*, 2010). All structural features that could be in the final design are included. Multiple boilers service the different steam headers. Multiple gas turbines with HRSGs with possible supplementary firing also service the different steam headers. Gas turbines might carry process shaft power loads on direct drives or drive turbogenerators. For direct drives, multiple process loads can be included on the same shaft. Multiple gas turbine frames can be incorporated to service both direct drives and turbogenerators. Stand-alone gas turbines can also be included. Figure 23.49 also features a steam turbine network with all possible steam turbine locations and connections. Each steam turbine might be a direct drive or drive a turbogenerator. A model needs to be developed for the superstructure that is capable of simultaneously optimizing the structure and parameters in the utility system (Del Nogal *et al.*, 2010). The optimization will remove the redundant features. Where possible, R-curve analysis should be used to screen out inappropriate features from the superstructure prior to the optimization.

23.9 Utility Simulation

Once the utility system structure has been fixed, or an existing system is to be studied, then a simulation model needs to be developed. The simulation model allows evaluation of the capacity and operation of the various system components, flowrates, pressures and temperatures of steam in various parts of the system, power generation and power import/export and operating costs.

Consider a simple example for an existing system shown in Figure 23.50 for which it is desired to establish a steam balance.

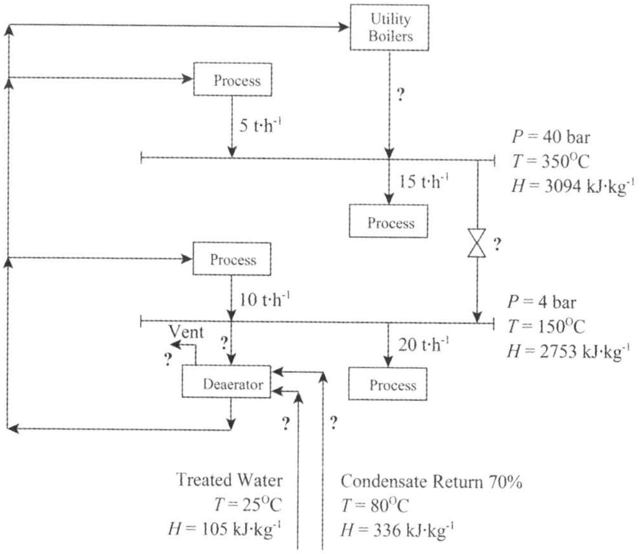

Figure 23.50

Example of a steam balance.

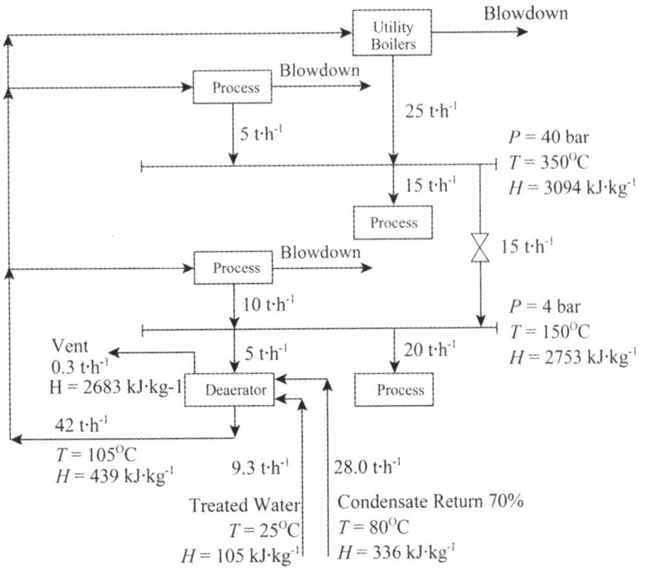

(Assume blowdown to be 5%)

Figure 23.51

Estimation of the deaeration steam allows the missing flows to be determined.

Figure 23.50 shows that there are two steam mains and no steam turbines. Steam is let down from the high- to low-pressure mains using a letdown station, which does not have desuperheating. Figure 23.50 shows the process steam use and the process steam generation from waste heat recovery. Treated water enters the deaerator at 25 °C. The site has a 70% condensate return (based on the rate of steam generation) at a temperature of 80 °C, as shown in Figure 23.50. Figure 23.50 shows the various flows around the system that are unknown because of a lack of instrumentation for measurements. The flow from the utility boilers, the letdown flow, deaerator steam, treated water and condensate return flowrates are all unknown. For this particular problem, a steam balance can be established if the flow of deaeration steam is known. Assume that the flow to the deaerator is 5 t·h^{-1}. Also assume that the blowdown rate is 5%. Having assumed the deaerator flow to be 5 t·h^{-1} fixes the letdown flow to be 15 t·h^{-1} and the flow from the utility boilers to be 25 t·h^{-1}, as illustrated in Figure 23.51. Assuming a 5% blowdown rate for the process and utility steam boilers now allows the boiler feedwater flow to be calculated as 42 t·h^{-1}. For a condensate return rate of 70%, the flowrate of condensate return to the deaerator is 28.0 t·h^{-1}. Assuming 5% of the deaeration steam is vented (0.3 t·h^{-1}) allows the treated water makeup to be calculated as 9.3 t·h^{-1}.

Thus, in this simple example, assumption of the deaeration steam allows the steam balance to be established. However, this is based on an assumed deaerator flow. The actual flow to the deaerator can be calculated from a heat balance around the deaerator. Figure 23.52 shows the flows into and out of the deaerator. If the boiler feedwater flow and condensate flows are known, together with an assumed value of the vent steam, then the

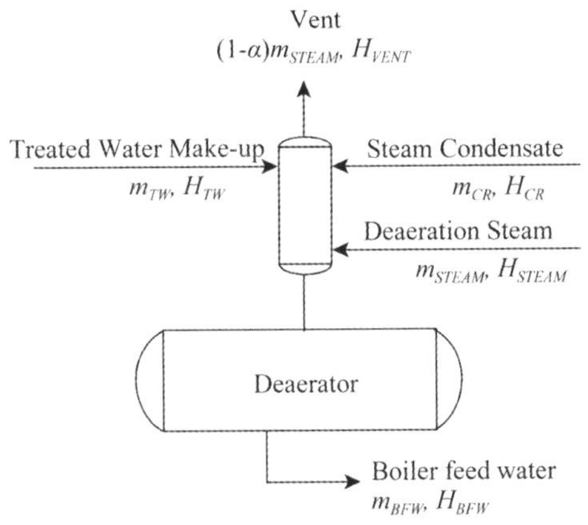

Figure 23.52

Deaerator material and energy balances.

flowrate of deaeration steam can be calculated from an energy balance around the deaerator.

A material balance around the deaerator gives:

$$m_{TW} = m_{BFW} - m_{CR} - m_{STEAM}(1 - \alpha) \tag{23.42}$$

where m_{TW}, m_{BFW}, m_{CR}, m_{STEAM} = flowrates of treated water, boiler feedwater, condensate return and deaeration steam respectively

α = proportion of deaeration steam vented

An energy balance around the deaerator assuming the enthalpy of the vented steam to be saturated at the deaerator pressure gives:

$$m_{TW}H_{TW} + m_{CR}H_{CR} + m_{STEAM}H_{STEAM} = m_{BFW}H_{BFW} + \alpha\, m_{STEAM}H_{VENT} \tag{23.43}$$

where H_{TW}, H_{BFW}, H_{CR}, H_{STEAM}, H_{VENT} = specific enthalpy of the treated water, boiler feedwater, condensate return deaeration steam and vented steam respectively

Combining Equations 23.42 and 23.43 and rearranging gives:

$$m_{STEAM} = \frac{m_{BFW}(H_{BFW} - H_{TW}) - m_{CR}(H_{CR} - H_{TW})}{(H_{STEAM} - H_{TW}) - \alpha(H_{VENT} - H_{TW})} \tag{23.44}$$

Substituting the values for the example and assuming $\alpha = 0.05$ gives:

$$m_{STEAM} = \frac{42(439 - 105) - 28.0(336 - 105)}{(2753 - 105) - 0.05(2683 - 105)} = 3.0\ \text{t·h}^{-1}$$

A new steam balance can now be calculated, the energy balance around the deaerator revised and the procedure repeated until convergence is achieved. A converged steam balance is shown in Figure 23.53.

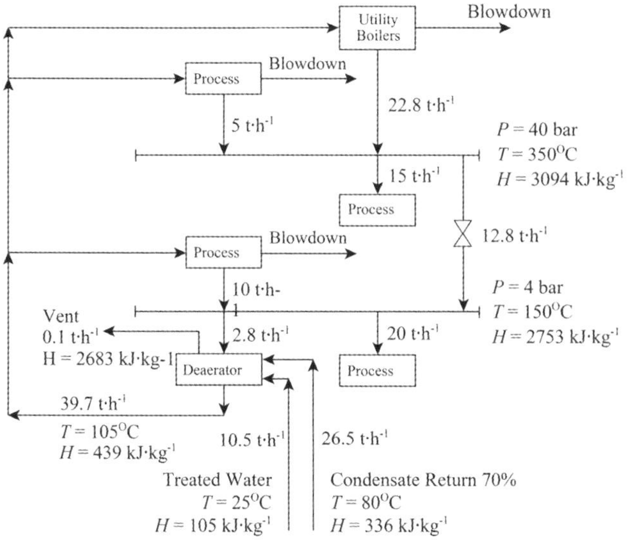

(Assume blowdown to be 5%)

Figure 23.53
Revising the deaerator steam flow allows the steam flows to be updated.

In other problems, the letdown stations might use desuperheating. Figure 23.19 illustrates a desuperheater. A material balance gives:

$$m_{STEAM,out} = m_{STEAM,in} + m_{TW} \tag{23.45}$$

An energy balance gives:

$$m_{STEAM,out}\, H_{STEAM,out} = m_{STEAM,in}\, H_{STEAM,in} + m_{TW}\, H_{TW} \tag{23.46}$$

Combining Equations 23.45 and 23.46 and rearranging gives:

$$m_{STEAM,out} = m_{STEAM,in} \frac{H_{STEAM,in} - H_{TW}}{H_{STEAM,out} - H_{TW}} \tag{23.47}$$

Steam flow and conditions of the inlet steam are known. Outlet conditions are usually fixed by the downstream main conditions, and hence Equation 23.47 can be solved.

Flash steam recovery might also feature. Condensate or blowdown is fed to the flash drum, as illustrated in Figure 23.54. A material balance gives:

$$m_{FS} = m_{COND,in} - m_{COND,out} \tag{23.48}$$

An energy balance gives:

$$m_{FS}\, H_{FS} = m_{COND,in}\, H_{COND,in} - m_{COND,out}\, H_{COND,out} \tag{23.49}$$

Combining Equations 23.47 and 23.48 and rearranging gives:

$$m_{FS} = m_{COND,in} \frac{H_{COND,in} - H_{COND,out}}{H_{FS} - H_{COND,out}} \tag{23.50}$$

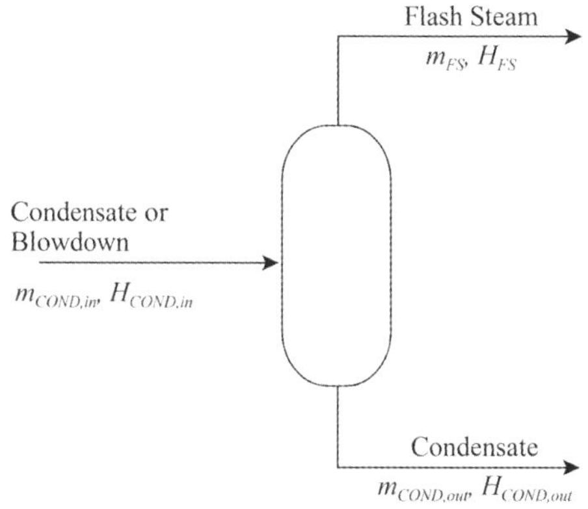

Figure 23.54
Flash steam recovery.

23

The flash drum pressure will be known as this will be set by the pressure of the main into which the flash steam will be recovered. The flash steam and condensate outlet enthalpies are set by saturation conditions.

When setting up simulation models using software, two basic approaches can be used, as discussed in Chapter 15. Writing the various mass and energy balance equations allows them to be solved simultaneously in an equation based approach. Alternatively, a sequential modular approach can be adopted, based on balances around each steam header in turn, starting from the highest pressure header and working down to the lower pressure headers.

23.10 Optimizing Steam Systems

Large and complex utility systems often have significant scope for optimization, even without changing the utility configuration. Consider now the optimization of a fixed utility configuration. First, the degrees of freedom that can be optimized in utility systems need to be identified.

1) *Multiple steam generation devices.* Consider the production of high-pressure (HP) steam in Figure 23.55. The HP steam needs to be expanded through a path from the very high pressure (VHP) mains. In turn, the VHP steam can be produced by Boiler 1, Boiler 2 or the gas turbine HRSG. If all of these steam generation devices are identical in terms of performance and operating cost, then there is nothing to choose between generating the steam in any of the three steam generation devices. However, this is not likely to be the case. In this example, the two boilers might have different fuels, with different fuel costs and different efficiencies, and the gas turbine (perhaps with supplementary firing in the HRSG) will have completely different characteristics from the steam boilers. Thus, there are degrees of freedom created by multiple steam generation devices with different costs of fuels, different boiler efficiencies

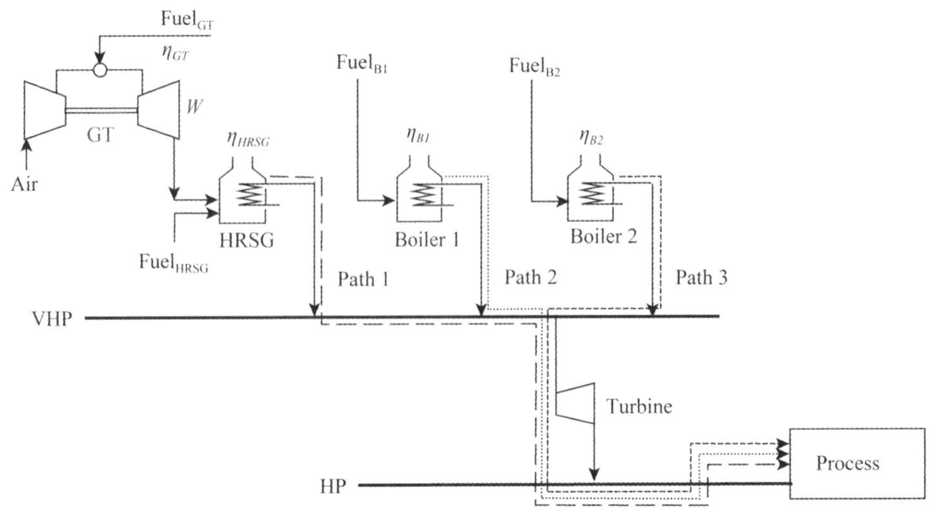

Figure 23.55

Multiple steam generation devices provide degrees of freedom for optimization. (Reproduced from Varbanov PS, Doyle S and Smith R (2004) Modeling and Optimisation of Utility Systems, *Trans IChemE*, **82A**: 561, with permission from Elsevier.)

and different power generation potential. Individual steam boilers and HRSGs will have minimum and maximum flows.

2) *Multiple steam turbines.* The HP steam required by the processes in Figure 23.56 can be generated by expanding through either Turbine *T*1 or *T*2, or through the letdown station, or a combination. Expanding the steam through the turbines is usually more economic than through the letdown station as there is a power credit resulting from expansion through the steam turbines. Moreover, Turbines *T*1 and *T*2 are likely to have different efficiencies for the expansion between VHP and HP steam. Thus, multiple expansion paths create degrees of freedom to choose the most appropriate path through which the steam is expanded.

If a steam turbine generates electricity, then the flow through the turbine can be varied within the minimum and maximum flows allowed by the machine. If a steam turbine is connected directly to a drive (e.g. a back-pressure turbine driving a large pump), then there is most likely to be no flexibility to change the flowrate through the steam turbine as this is fixed by the power requirements of the process machine. Process machines can, in some cases, be equipped with both a steam turbine and an electric motor, allowing the drive to be switched between the two

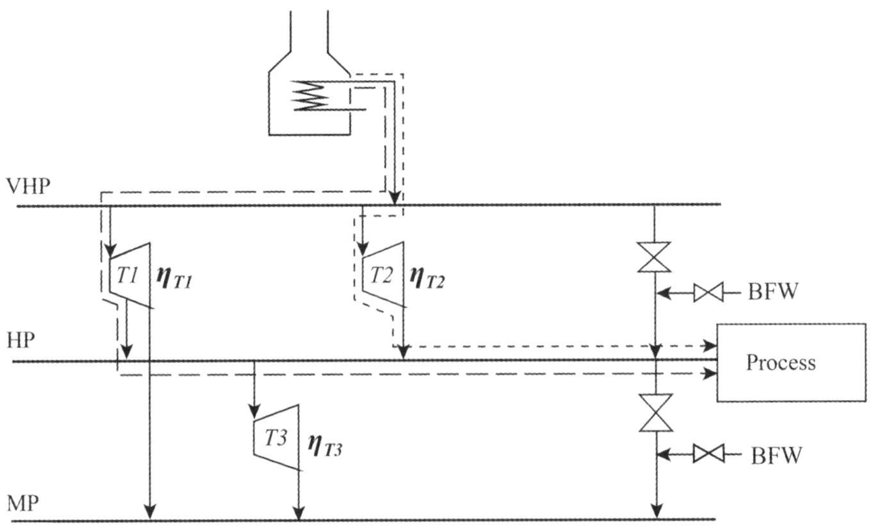

Figure 23.56

Steam turbine networks and let-down stations provide degrees of freedom for optimization. (Reproduced from Varbanov PS, Doyle S and Smith R (2004) Modeling and Optimisation of Utility Systems, *Trans IChemE*, **82A**: 561, with permission from Elsevier.)

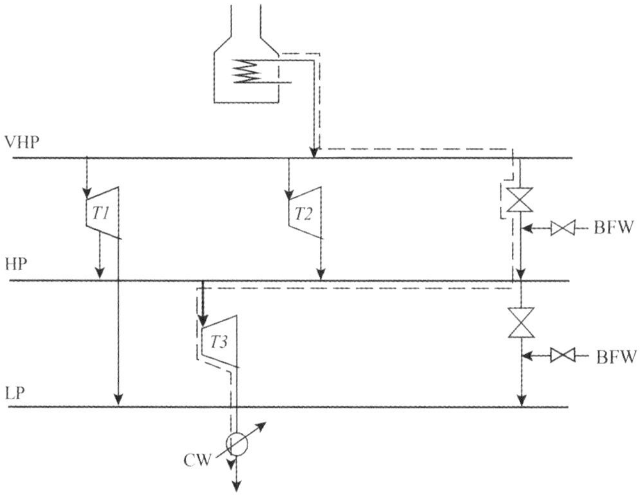

Figure 23.57

Let-down stations provide degrees of freedom to bypass constraints in the utility system.

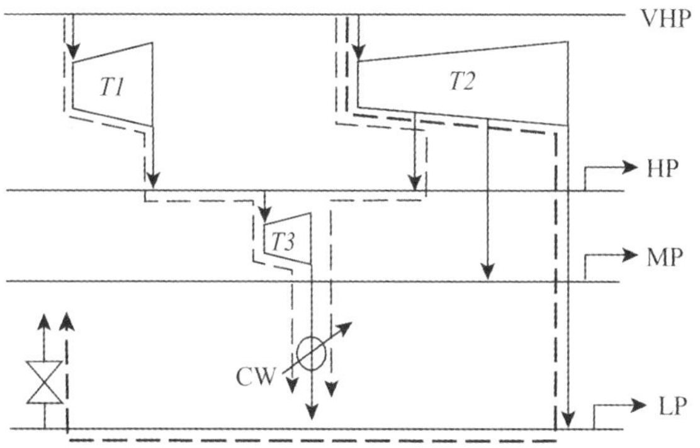

Figure 23.58

Condensing turbines and vents provide additional degrees of freedom for optimization.

according to steam demands in the utility system and operating costs. It should be remembered that operating costs might vary significantly as a result of different electricity tariffs, according to the time of day, the day of the week and the season of the year.

Extraction steam turbines will have minimum and maximum flows in each section of the turbine. This leads to minimum and maximum flows at each extraction level.

3) *Letdown stations.* If there is a requirement for steam to be expanded, then flow through the steam turbines should normally be maximized in order to maximize the power generation. However, a steam balance must be maintained, and letdown stations are still important degrees of freedom in the optimization, as shown in Figure 23.57. Steam expanding through a letdown station might be able to bypass a flow constraint in a steam turbine somewhere in the system and increase the power generation indirectly. Some steam flow through letdown stations might be required to maintain minimum superheat temperatures in steam mains, and a small flow is most often maintained for control and to avoid condensation in the pipes. In some instances, the letdown is desuperheated by injection of boiler feedwater after expansion, increasing the steam flow after expansion. This can replace steam generation in the utility boilers and at the same time increase the steam flows in the lower-pressure parts of the system. As discussed earlier, a minimum superheat is required in steam mains to avoid excessive condensation in the mains. Steam heaters benefit from having steam close to saturation, whereas if the steam is to be used in a steam turbine to generate power locally, then the steam needs to be superheated.

   Thus, large letdown flows are usually seen to be a missed cogeneration opportunity but might provide a degree of freedom to bypass bottlenecks in the steam system. Also, flow through letdown stations allows the level of superheat in the lower-pressure mains to be controlled independently of the flow through the steam turbines.

4) *Condensing turbines.* Condensing turbines provide extra degrees of freedom for power generation, as shown in Figure 23.58. Although the heat from the exhaust of a condensing turbine is not required, it provides flexibility in power generation to reduce the cost of imported power or to increase the credit from exporting surplus power.
5) *Vents.* Figure 23.58 also shows a vent on the LP steam main. At first sight, it might not seem sensible to vent steam directly to ambient. This steam is being generated at the VHP level, for which fuel costs are incurred. However, in expanding from the VHP main down to the LP main through steam turbines, power is generated. It might therefore be the case that for a high differential between fuel and power costs, it is economic to vent steam. However, it should always be better to expand the steam through a condensing turbine if there is one and there is spare capacity, rather than vent steam. Also, in some regulatory regimes steam venting might not be allowed.

Figure 23.59 shows an example of an existing site utility system. Steam is generated at high pressure (HP), which is distributed around the site, along with steam at medium pressure (MP) and low pressure (LP). Steam is expanded through a network of steam turbines from the HP main to the lower-pressure mains. Letdown stations are used to control the mains pressures. Steam is used for process heating from the HP, MP and LP mains. Turbines *T*1 to *T*6 generate electricity. Turbines *T*5 and *T*6 are condensing turbines. Turbines *DRV*1 and *DRV*2 are driver turbines connected directly to process machines.

To optimize the system in Figure 23.59 requires a model to be developed for the whole system that accounts for:

- cost of fuel per unit of steam generation in the boilers,
- cost of fuel per unit of power and steam generation in the gas turbine/HRSG,
- power generation characteristics of the gas turbine,
- power generation characteristics of the steam turbines,
- cost of imported power,
- credit for exported power,

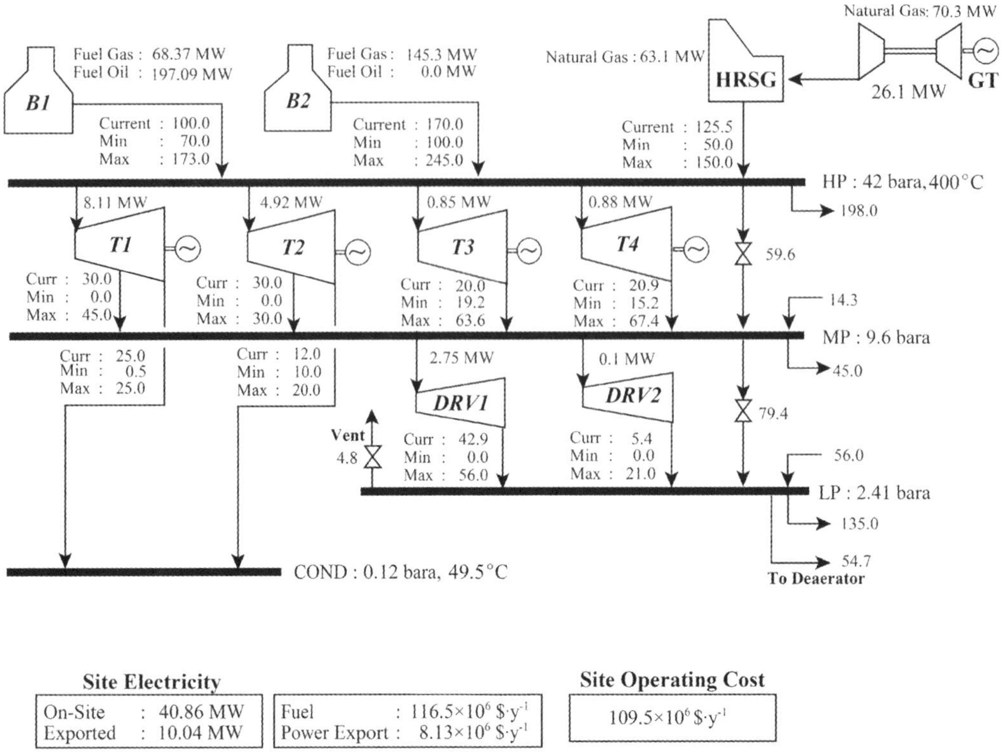

Figure 23.59

Current operating conditions for a site utility system (flows in t·h^{-1}). (Reproduced from Li Sun, Steve Doyle, Robin Smith, Heat Recovery and Power Targeting in Utility Systems. In Press. Ms. Ref. No.: EGY-D-14-03308, with permission from Elsevier.)

- costs for running the steam and power generation, other than fuel costs,
- flowrate constraints throughout the system,
- cost of demineralized water require to replace condensate, blowdown and vent losses,
- cooling water costs.

Figure 23.59 shows the maximum, minimum and current flows in each part of the system. Ancillary data for the utility system in Figure 23.59 are given in Table 23.6. Data for the fuels used by the utility system in Figure 23.59 are given in Table 23.7. Data for the steam turbines were given in Table 23.3.

If the conditions in the steam mains (temperature and pressure) are fixed and the steam boilers, steam turbines, gas turbines performance models and costs are assumed constant or are linear functions, then the optimization can be carried out with a linear program. In practice, even if the pressures of the steam mains are fixed by pressure control, the temperature (and enthalpy) will vary as the flow through the steam turbines and letdown stations vary. If this is taken into account, then the model becomes a nonlinear optimization. The optimization can be carried out using a nonlinear program or an iterative linear programming scheme. For an iterative linear programming approach, the steam system model is first simulated and the pressures and temperatures of the steam mains fixed. The model is then optimized using a linear program with the temperatures of the mains fixed. Resimulation of the model then allows the temperatures of the steam mains to be determined. These

Table 23.6

Ancillary data for the site utility system in Figure 23.59.

Ambient conditions	25 °C, 1.103 bar
Boiler feedwater temperature	105 °C
Boiler feedwater return temperature	30 °C
Condensate return rate	78.1%
Boiler efficiency	90%
HRSG efficiency	90%
Site power demand	30 MW
Minimum power import	0 MW
Maximum power import	30 MW
Minimum power export	0 MW
Maximum power export	50 MW
Power cost (import)	0.118 $·kWh^{-1}
Power value (export)	0.0942 $·kWh^{-1}
Power distribution losses	2%
Cooling water cost	0.005 $·kWh^{-1}
Demineralized water cost	0.005 $·kg^{-1}

Table 23.7

Fuel costs for the utility system in Figure 23.59.

Fuel	Net heating value (kJ·kg^{-1})	Density (kg·m^{-3})	Price	
			(\$·kg^{-1})	(\$·kWh^{-1})
Fuel oil	40,245	890	0.6260	0.0560
Fuel gas	32,502	0.668	0.3174	0.0352
Natural gas	46,151	0.668	0.4880	0.0381

Table 23.8

The performance of the optimized utility system in Figure 23.60.

Cost	Base case (\$·y^{-1})	Optimized system (\$·y^{-1})
Export power	8.13×10^6	6.07×10^6
Fuel	116.5×10^6	101.2×10^6
Total operating cost	109.5×10^6	92.27×10^6

temperatures are then fixed, the model reoptimized using a linear program and resimulated, and so on, until convergence is achieved.

For large sites, the conditions in a long steam main will not be uniform throughout the main because the steam main will not be well mixed and can also have a significant pressure drop along the main. Thus, the conditions in the same main can be different in different parts of a large site. If this turns out to be an important issue, then if the steam mains are assumed to be well mixed, additional steam mains must be created in the model to reflect the different conditions in different geographic locations for the same steam main.

Figure 23.60 shows the conditions of the optimized system. Table 23.8 compares the cost of the base case conditions in Figure 23.59 with those for the optimized system shown in Figure 23.60. Note that the settings might need to be adjusted slightly to allow a small flow through the HP-MP letdown station.

Thus, in this case, the optimization is capable of reducing the total operating cost by 13%. The scope for cost reduction through optimization will be case specific, but for large complex systems is most often around 5% or less.

Finally, all of the discussions so far have assumed that there is one set of operating conditions that need to be optimized. This is almost never the case. To begin with, shutdown and maintenance of process and utility plant dictate that there will be different steam flows and constraints during these periods, giving rise to different operating cases. To add to this complication, the cost of import power and the credit for export power will, in most cases, vary

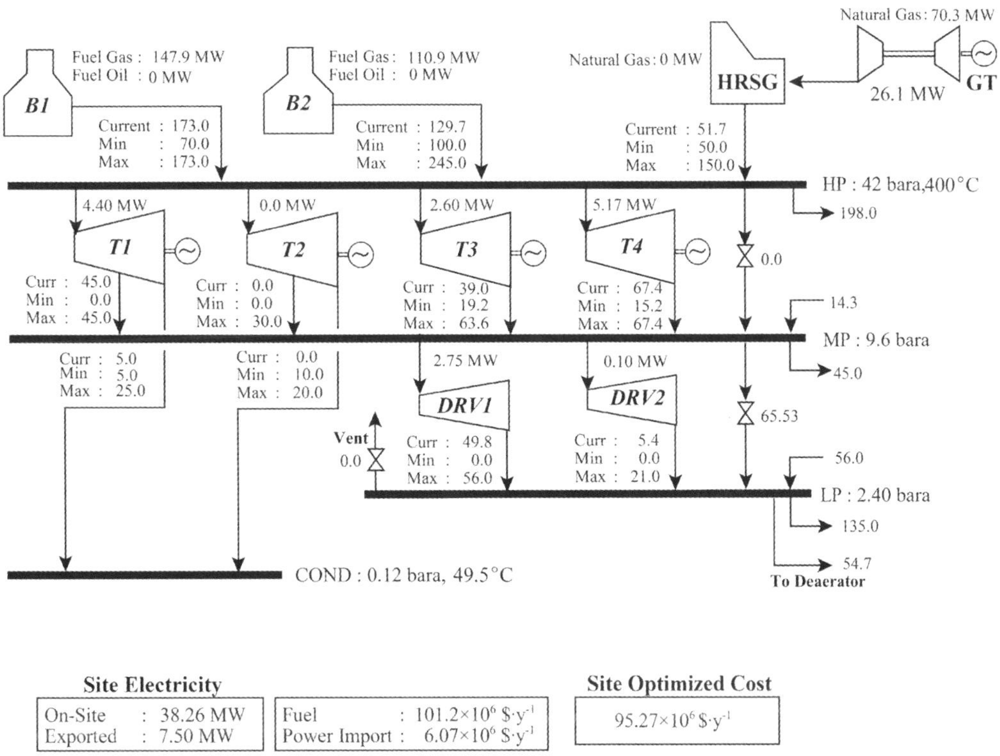

Figure 23.60

Optimized operating conditions for a site utility system (flows in t·h^{-1}). (Reproduced from Li Sun, Steve Doyle, Robin Smith, Heat Recovery and Power Targeting in Utility Systems. In Press. Ms. Ref. No.: EGY-D-14-03308, with permission from Elsevier.)

according to the time of day, time of week and season of the year. These tariffs are normally negotiated with the power supplier and form a fixed pattern, but often a complex one. Each of these different tariffs also presents a different operating case for optimization. Thus, many different optimizations are required to reflect the picture for each case throughout the year. The operational setup of the utility system should then be changed for each case to reflect the new conditions. On-line optimization is helpful in this respect.

If an optimization across the whole year is required, perhaps for a design modification, then the different operating cases need to be weighted according to the duration of each case and combined to give an annual picture (Iyer and Grossmann, 1998).

23.11 Steam Costs

Steam costs were considered in brief in Chapter 2. The philosophy of costing steam was on the basis of the fuel cost to raise the highest-pressure steam and the value of power generated in expanding to lower levels subtracted from this fuel cost. A simple isentropic efficiency model for the steam turbines was used to calculate the value of power generated as a result of the expansion through steam turbines. This represents a simplified picture of steam costs. It is adequate for new design and where full details of the steam system are not available. It does not take account of existing equipment, equipment performance and the constraints associated with both existing equipment and existing steam networks. For an existing steam system, the true cost of steam can be established using the modeling and optimization techniques discussed in this chapter. Two distinct cases need to be considered.

1) *Steam cost for fixed steam heat loads.* In the first case, the steam heat loads for the processes on a site are fixed and the objective is to calculate the cost of the steam for the allocation of costs to the processes and businesses on a site. To establish the true steam costs for an existing steam system requires that a model for the existing system needs to be first developed. The model should reflect the performance of the existing equipment and the constraints in the existing equipment and steam network. The model should also include the existing steam heating demands for the various processes. This model can then be optimized, as described in the previous section. The cost to generate the steam in the utility boilers at the highest pressure can be calculated from the optimized model. This will principally be the cost of fuel but can also include other costs, such as raw water, water treatment, labor, power for the boiler feedwater pumps and cost of deaeration steam. Having calculated the cost of the highest-pressure steam, the model will also provide the amount of power generated by expansion of the highest-pressure steam to the next highest pressure. The value of the power generated by the expansion can then be subtracted from the cost of generating the highest-pressure steam in the utility boilers to give the cost of the next-to-highest pressure steam. The calculation is repeated for the next pressure level lower, and so on. Thus, the cost of each steam level is the cost of the next highest level minus the value of the power generated from the expansion from the next highest level. If there are utility boilers generating steam into the lower-pressure steam mains, then the cost of operating these boilers must also be added to the cost of steam at that level.

   In extreme cases, this approach to costing steam might lead to steam at low pressure having a negative cost. This can happen if fuel is cheap and power is expensive. This is not a fundamental problem as it reflects the true economics. The site utility system is benefiting from the availability of a heat sink for the low-pressure steam. However, it should be emphasized that if the low-pressure steam has a negative cost this does not necessarily mean that it is economic to increase use of low-pressure steam significantly. As will be discussed later, the cost of steam is likely to change as its consumption changes.

   It should also be noted that variation in electricity tariffs that might change through the time of day, day of the week and time of the year will change the optimization and, therefore, the steam costs. An average can be taken according to the relative duration of the tariffs.

2) *Steam cost for changed steam heat loads.* In the second case, the steam costs need to be determined when heat loads are changed. This might be a project for reduction in energy demand (e.g. retrofit of a heat exchanger network for increased heat recovery). Alternatively, a project might involve an increase in heat demand as a result of commissioning of new plant or expansion of an existing plant. The starting case is again a model for the steam system, again optimizing for the existing steam heating loads. It might be suspected that the steam cost calculated from the model for the existing steam loads can be used to provide steam costs for larger or smaller steam loads for a given steam main. However, this is not the case. The optimum settings for the steam system change once the loads for the steam mains change. Constraints on the existing equipment will also be encountered and all of this needs to be accounted for.

Consider again the optimized steam system in Figure 23.60. Suppose that it is possible to reduce the HP steam demand for the processes on this site, for example by improving the heat recovery within the processes. How much is such a steam saving actually worth? The HP steam is being generated from the steam boilers and gas turbine HRSG. If steam is saved, the first obvious action to take is to reduce the steam generation in the utility boilers and accept a saving in fuel costs as a result. Unfortunately, the saving in fuel costs would also be accompanied by an increase in power costs. This results from the reduction in the flow of steam through the steam turbines, which in turn reduces the cogeneration and power export is decreased (or in other cases additional power imported). It is therefore not so straightforward to determine exactly what the cost benefits associated with a steam saving would be. Another way to deal with the surplus of steam at the HP level created by a steam saving would be to pass the heat through an alternative path, for example to the condensing turbines. This would allow additional power to be generated, with the resulting cost benefit. In a complex utility system, the heat can flow through the utility system through many paths. The flow through different paths will have different cost implications.

In assessing the true cost benefits associated with a steam saving, the steam and power balance for the site utility system must be considered, together with the costs of fuel and power (or power credit in an export situation). In general, a surplus of steam resulting from a steam saving in a process demand can be exploited by:

- saving fuel in the utility boilers;
- generating extra power by passing the steam to a condensing turbine;
- generating extra power by passing the surplus steam to a vent through back-pressure turbines;
- expanding the steam to a lower level than previously used by switching a steam heater from a higher-pressure steam to a lower-pressure steam.

If steam is being generated, rather than used, then the same basic arguments apply. For example, suppose additional HP steam was being generated by a process into the HP main in Figure 23.60 by improving process heat recovery. This leads to a surplus of HP steam and the same arguments apply as to what is the most efficient way to exploit the surplus of HP steam.

Steam demand for a given main might decrease as a result of improved energy recovery or from a change in production on the site. Alternatively, steam demand might increase as a result of an increase in the rate of production. In general, the steam balance of a main might change as a result of:

- increasing or decreasing process steam demand;
- increasing or decreasing process steam generation;
- switching a process demand from one steam main to another;
- changing the utility system, for example, shutdown of a boiler, steam turbine, and so on.

The only way to reconcile the true cost implications of a change in steam demand created by an energy reduction project or a change in production is to use the optimization techniques described in the previous section. An optimization model of the existing utility system must first be set up. Starting with the steam load on the main with the most expensive steam (generally the highest pressure), this is gradually reduced and the utility system reoptimized at each setting of the steam load. The steam load can only be reduced to the point where the flowrate constraints are not violated.

The concept of steam marginal cost is used to define the cost implications of a change in steam demand. Steam marginal cost is defined as the change in utility system operating cost for unit change in steam demand for a given steam main (Sun, Doyle and Smith, 2015b):

$$MC_{STEAM} = \frac{\Delta Cost}{\Delta m_{STEAM}} \qquad (23.51)$$

where MC_{STEAM} = marginal cost of steam
$\Delta Cost$ = change in cost
Δm_{STEAM} = change in steam flowrate

It is important to emphasize that the change in the operating cost is taken between the optimum operation before the steam demand change and the optimum operation after the steam demand change for the current step. Obviously, the result is context-specific and, for different operating conditions, each steam level will have a different marginal steam cost. Even further, the marginal steam cost can change for a given level, depending on how much steam is saved from the steam main.

Most often, the objective is to determine the value of steam saving resulting from potential energy reduction projects (Varbanov *et al.*, 2004, Sun, Doyle and Smith, 2015b). The first step is to optimize the operation of the utility system for the initial steam demands. The demand on the selected main is then gradually decreased. At each step in the reduction, the whole utility system is optimized. The procedure is then repeated until no further decrease in steam usage (or increased generation) is possible.

Figure 23.61 shows a plot of steam marginal cost versus the steam savings for the HP steam in Figure 23.60 (Sun, Doyle and Smith, 2015b). The pattern is complex and involves step changes up and down. Each step change in the marginal cost is caused by the optimization being constrained by a new constraint or a new combination of constraints. For example, in Figure 23.61, starting from the current HP steam load of 198 $t{\cdot}h^{-1}$ the marginal steam cost for the initial saving in HP steam with the other steam loads fixed is 32.9 $\${\cdot}t^{-1}$. Then there is a step increase at a saving in HP steam of 11.5 $t{\cdot}h^{-1}$ to a marginal cost of 33.9 $\${\cdot}t^{-1}$. This step is caused by the HP steam load being reduced by using a combination of a back-pressure and a condensing turbine. The condensing turbine was operating at the condition of maximum extraction flowrate to the MP main and minimum flowrate to condensation. The letdown between the HP and MP flowrate was zero. Once the HP saving is greater than 11.5 $t{\cdot}h^{-1}$ (HP load below 186.5 $t{\cdot}h^{-1}$), it is more economic to shut down the condensing turbine, set both the back-pressure turbine to maximum flowrate, with the balance of steam passed through the HP–MP letdown. Further steam saving continues to an HP saving of 85.6 $t{\cdot}h^{-1}$ (HP load below 112.4 $t{\cdot}h^{-1}$). At this point, both boilers are at minimum firing and the HRSG is not being fired. Because the gas turbine is assumed to be running constantly at full load, steam savings beyond 85.6 $t{\cdot}h^{-1}$ are achieved by reducing the steam generation in the HRSG. However,

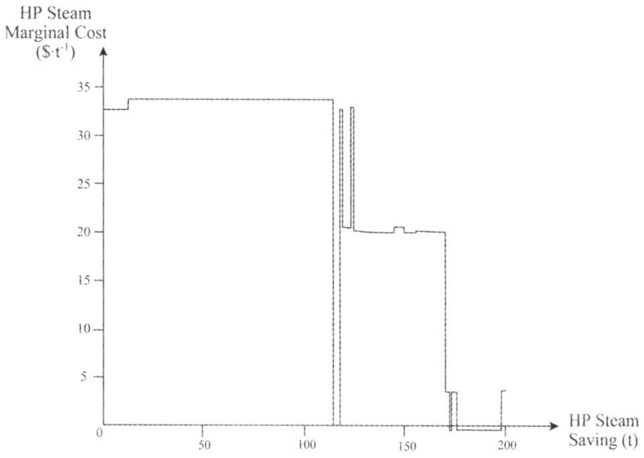

Figure 23.61

Marginal costs for the high-pressure steam. (Reproduced from Li Sun, Steve Doyle, Robin Smith, Heat Recovery and Power Targeting in Utility Systems. In Press. Ms. Ref. No.: EGY-D-14-03308, with permission from Elsevier.)

because the boilers are at minimum firing and no fuel is being fired in the HRSG, the increase in steam saving brings no cost saving and the marginal cost goes to zero. This continues for a further 3.5 $t{\cdot}h^{-1}$ savings when a condensing turbine is activated with maximum flowrate to the MP main and minimum flowrate to condensation, one of the back-pressure turbines reduces its flowrate, the HP–MP letdown flowrate goes to zero and Boiler 1 flowrate increases slightly above its minimum flowrate. The value of the additional power generation is greater than the value of the lost power generation from the back-pressure turbine and the increased cost of fuel for the boiler. Each further step change in the marginal cost of the HP steam in Figure 23.61 is caused by the effects of constraints on the system.

The pattern of marginal cost is difficult to interpret in terms of formulating an energy conservation strategy. The marginal cost can be accumulated according to (Sun, Doyle and Smith, 2015b):

$$CC_{STEAM} = \sum MC_{STEAM} \times \Delta m_{STEAM} \tag{23.52}$$

where CC_{STEAM} = cumulative cost of steam

The cumulative cost can then plotted against the steam saving, as shown in Figure 23.62. This shows that the overall benefit of saving HP steam decreases as the saving increases. There is no benefit from saving HP steam beyond 170 $t{\cdot}h^{-1}$.

The next question is whether the marginal and cumulative cost for HP steam is affected by changes in the steam demand for the other steam mains. Consider now the marginal cost of the MP steam in Figure 23.60. Initially, to calculate the marginal cost of the MP steam, the original flowrate of HP steam is reinstated to be 198 $t{\cdot}h^{-1}$. Figure 23.63 shows the marginal cost of the MP steam for an HP steam flowrate of 198 $t{\cdot}h^{-1}$. Also shown in Figure 23.63 are the MP marginal costs for HP steam flowrates of 86 $t{\cdot}h^{-1}$ and 31 $t{\cdot}h^{-1}$. It is clear that the marginal steam costs for different steam mains interact with each other. The marginal cost for MP steam depends on how much HP steam saving is taken. Again the marginal costs in Figure 23.63 show a pattern of step changes

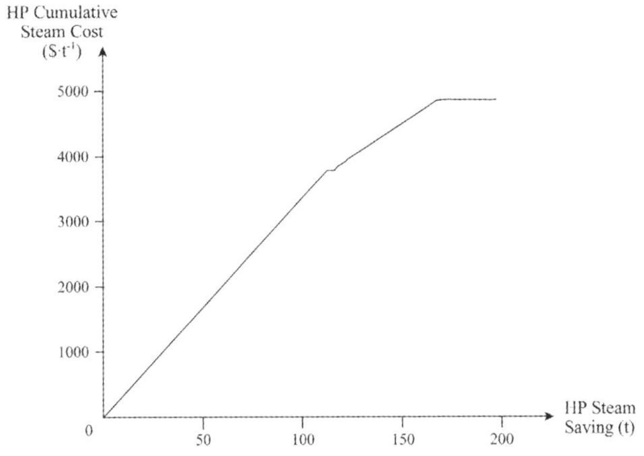

Figure 23.62

Cumulative cost for the high-pressure steam. (Reproduced from Li Sun, Steve Doyle, Robin Smith, Heat Recovery and Power Targeting in Utility Systems. In Press. Ms. Ref. No.: EGY-D-14-03308, with permission from Elsevier.)

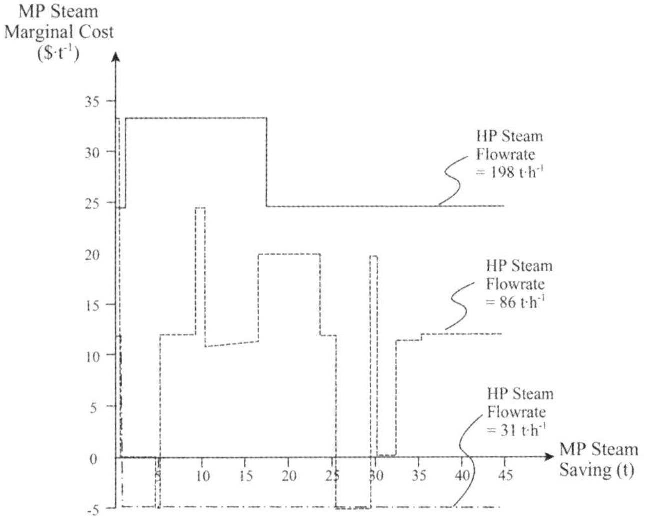

Figure 23.63

Marginal costs for the medium-pressure steam for different savings in high-pressure steam. (Reproduced from Li Sun, Steve Doyle, Robin Smith, Heat Recovery and Power Targeting in Utility Systems. In Press. Ms. Ref. No.: EGY-D-14-03308, with permission from Elsevier.)

caused by the constraints for the steam system. The cumulative steam costs for MP steam corresponding with HP steam savings of 198 $t{\cdot}h^{-1}$, 86 $t{\cdot}h^{-1}$ and 31 $t{\cdot}h^{-1}$ are shown in Figure 23.64. It is clear from Figure 23.64 that the greater the HP steam saving, the lower the benefit from saving MP steam. Indeed, beyond a certain point for HP steam saving, saving MP steam is counterproductive (Sun, Doyle and Smith, 2015b).

This example illustrates that simple steam costs cannot be used to evaluate the benefit from steam saving when processes are serviced by complex utility systems. A simulation and optimization model is needed and must be used to evaluate the true value of

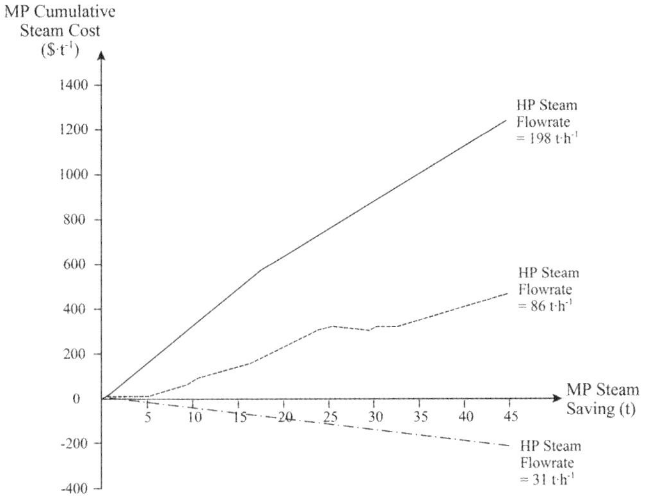

Figure 23.64

Cumulative cost for the medium-pressure steam for different savings in high-pressure steam. (Reproduced from Li Sun, Steve Doyle, Robin Smith, Heat Recovery and Power Targeting in Utility Systems. In Press. Ms. Ref. No.: EGY-D-14-03308, with permission from Elsevier.)

steam saving by exploring combinations of saving across different steam mains. Different combinations of savings from different steam mains can be tested. Different combinations will lead to different marginal costs and different cost benefits. Plotting marginal and cumulative steam costs can help to clarify a steam saving strategy. It must be remembered that the marginal and cumulative costs do not indicate that a certain level of steam saving is possible, but only the economic consequences of targeting a certain level of steam saving.

It should also be emphasized that variation in electricity and fuel tariffs will change the picture. Thus, graphs equivalent to Figure 23.61 to 23.64 could be produced for peak versus off-peak electricity costs, and so on. Therefore, to obtain a value of steam savings appropriate for a design study, an average needs to be taken across the different operating scenarios. The average can be taken by weighting the marginal costs for different tariff scenarios according to their relative duration.

The task for the designer now is to consider those processes using HP steam and to determine whether economic energy saving projects can be identified.

The arguments so far have focused on retrofitting existing processes on a site to reduce the steam consumption or increase the steam generation. On the other hand, it might be the case that a new process is to be added to the site, creating increased demands on the steam system. The same kind of analysis can be performed by increasing the consumption of steam, rather than decreasing it, and determining the marginal cost of steam for the new process. Again, steps are likely to be encountered in the steam marginal cost created by the constraints in the existing utility system. This might then call for the utility system to be modified to overcome the constraints in order to reduce the costs associated with provision of steam to the process.

Figures 23.61 to 23.64 show that it is in general incorrect to attribute a single value for steam at a given level when the load can change (increase or decrease). In general, the value of steam at a given level depends on how much is being consumed, as well as fuel cost, power cost, equipment performance, equipment constraints, and so on. The fact that the price of steam at a given level can vary contrasts with the common practice of attributing a single marginal cost for each level of steam, irrespective of load. A change in steam load might not only arise from an energy saving project or a new process being commissioned but also might arise from the day-to-day variations in steam load as a result of changes in the pattern of production for the existing processes. Using the approach here allows the true cost of steam to be established for any pattern of steam loads. Details of the existing utility system are included in the cost analysis and all of the costs and constraints associated with that system. This can give a very different picture as to the costs associated with changing steam load at any level, compared to an approach that does not take such detail into account.

23.12 Steam Systems and Cogeneration – Summary

Most process heating is provided from the distribution of steam around sites at various pressure levels. Normally, two or three steam mains will distribute steam at different pressures around the site. The steam system is not only important from the point of view of process heating but is also used to generate a significant amount of power on the site.

Boiler feedwater treatment is required for the raw water to remove suspended solids, dissolved solids, dissolved salts and dissolved gases (particularly oxygen and carbon dioxide). The raw water entering the site is commonly filtered and dissolved salts removed using softening or ion exchange. The principal problems are calcium and magnesium ions. The water is then stripped to remove the dissolved gases and treated chemically before being used for steam generation.

There are many types of steam boilers, depending on the steam pressure, steam output and fuel type. Blowdown is required to remove the dissolved solids not removed in the boiler feedwater treatment. The efficiency of the boiler depends on its load. 23

Gas turbines consist of a compressor and a turbine mechanically connected to each other. Release of energy to the compressed gases after the compressor, from the combustion of fuel, creates a net production of power. Gas turbines are available in a wide range of sizes but are restricted to standard frame sizes. The hot exhaust gas is useful for raising steam and the temperature of the exhaust gas can be increased by using supplementary firing. The performance of gas turbines can be modeled as a compressor and expander connected together with a pressure drop in the combustion chambers. Part load can be modeled by a simple correlation involving the mass flowrate of fuel.

Steam turbines are used to convert part of the energy of the steam into power and can be configured in different ways. Steam turbines can be divided into two basic classes: back-pressure turbines and condensing turbines. The efficiency of the turbine and its power output depend on the flowrate of steam to the turbine. The performance characteristics can be modeled by a simple linear relationship over a reasonable range of operation.

The steam system configuration normally generates most of the steam in boilers at the highest pressure, feeding to a high-pressure steam main. The highest-pressure steam main is normally used for power generation, rather than process heating. Steam is expanded from high to lower levels by either steam turbines or expansion valves. Steam expanded across letdown valves can be used to provide additional superheat to the low-pressure main or desuperheated by the injection of boiler feedwater. It is also good practice to recover the flash steam from large flowrates of high-pressure steam condensate.

Site composite curves can be used to represent the site heating and cooling requirements thermodynamically. This allows the analysis of thermal loads and levels on site. Using the models for steam turbines and gas turbines allows cogeneration targets for the site to be established. A cost trade-off needs to be carried out in order to establish the optimum trade-off between fuel requirements and cogeneration.

The site power-to-heat ratio is very important in determining the most appropriate cogeneration system for the site and the machine driver selection. Drivers are required for many different types of process machine on a site. Power can be generated and distributed to drive electric motors or direct drives can be used. Direct drives are inflexible as far as the utility system operation is concerned. A combination of direct drives and electric motors is usually the best solution for the site as a whole.

Complex steam systems usually feature many important degrees of freedom to be optimized. Multiple steam generation devices, multiple steam turbines, letdown stations, condensing turbines and vents all provide important degrees of freedom for optimization. To establish the steam costs for retrofit of site processes requires an optimization model to be developed. This allows the steam loads for process heating to be gradually decreased and the steam system reoptimized at each setting. The result in cost savings establishes the true cost of steam for retrofit projects aiming to reduce steam consumption on the site.

23.13 Exercises

1. A boiler with a capacity of 100,000 kg·h^{-1} is required to produce steam of 40 bar and 350 °C. The feedwater from the deaerator is at 100 °C and contains 100 ppm dissolved solids. Maximum dissolved solids allowed in the boiler is 2000 ppm. The steam system operates with 60% condensate return. Enthalpy data for steam are given in Table 23.9. Calculate:
   - **a)** the blowdown rate,
   - **b)** the energy consumption for a boiler efficiency of 88%.
2. A steam turbine operates with inlet steam conditions of 40 barg and 420 °C and can be assumed to operate with an isentropic efficiency of 80% and a mechanical efficiency of 95%. Using steam properties from Table 23.10 calculate the power production for a steam flowrate of 10 kg·s^{-1} and the heat available per kg in the exhaust steam (i.e. superheat plus latent heat) for outlet conditions of:
   - **a)** 20 barg,
   - **b)** 10 barg,
   - **c)** 5 barg.
   - **d)** What do you conclude about the heat available in the exhaust steam as the outlet pressure varies?
3. A steam turbine is operating between an inlet steam of 40 barg and 420 °C and an outlet steam of 5 barg. Using the Willans Line Model with parameters from Table 23.3 and steam properties from Table 23.10, calculate the power production for a turbine at full load with a flowrate of steam of 10 kg·s^{-1}.
4. The inlet steam to an extraction turbine is 30 bar and 400 °C and can be assumed to operate with a mechanical efficiency of 95%. The extraction steam is at 10 bar and the exhaust is condensed at 0.12 bar. The steam turbine is fully loaded with a throttle flow of 40 kg·s^{-1} to the inlet, 15 kg·s^{-1} going to extraction and the remaining 25 kg·s^{-1} going to condensation. Using the Willans Line Model with parameters from Table 23.3 and steam properties from Tables 23.11, calculate the power production.
5. A site steam system needs to supply steam at medium pressure (MP) for process heating purposes of 120 MW at 200 °C and at low pressure (LP) of 80 MW at 150 °C. High-pressure (HP) steam is generated in the site boilers at 40 bar and 420 °C and expanded through steam turbines to the MP and LP pressures. Saturation temperatures for the steam should be 10 °C above the temperature at which the heating is required. Assuming the isentropic efficiency of the turbine to be 70% and the mechanical efficiency to be 97% calculate the cogeneration target. Steam properties can be taken from Table 23.12.

Table 23.9

Enthalpy data for Exercise 1.

	Enthalpy (kJ·kg^{-1})
Steam at 40 bar, 350 °C	3095
Saturated steam at 40 bar	2800
Saturated condensate at 40 bar	1087
Water at 100 °C	422

Table 23.10

Steam properties for Exercise 2.

	H_{SUP} (kJ·kg^{-1})	H_L (kJ·kg^{-1})	S_{SUP} (kJ·kg^{-1}·K^{-1})	ΔH_{IS} (kJ·kg^{-1})
420 °C, 40 barg	3261	1095	6.828	0
20 barg	–	920.1	–	188.8
10 barg	–	781.4	–	346.1
5 barg	–	670.8	–	473.8

Table 23.11

Steam properties for Exercise 4.

	H_{SUP} (kJ·kg^{-1})	S_{SUP} (kJ·kg^{-1}·K^{-1})	ΔH_{IS} (kJ·kg^{-1})
30 bar, 400 °C	3232	6.925	0
10 bar	–	–	290.1
0.12 bar	–	–	1016.3

Table 23.12

Steam properties for Exercise 5.

Streams	P (bar)	H_{SUP} ($kJ \cdot kg^{-1}$)	S_{SUP} ($kJ \cdot kg^{-1} \cdot K^{-1}$)	T_{SAT} (°C)	H_{SAT} ($kJ \cdot kg^{-1}$)	H_L ($kJ \cdot kg^{-1}$)	ΔH_{IS} ($kJ \cdot kg^{-1}$)
HP	40	3262	6.841	250.3	2800	1087	0
MP	19.1	–	–	210	2796	897.7	207.6
LP	6.18	–	–	160	2757	675.5	464.0

6. The problem table cascade for a process is given in Table 23.13 for $\Delta T_{min} = 10\,°C$. This process is to be integrated with a steam turbine with a single outlet in a cogeneration scheme. The inlet steam is at 40 bar and 350 °C. The rate of power generation can be determined from the Willans Line Model with parameters from Table 23.3, assuming the turbine to be fully loaded and a mechanical efficiency of 97%. Calculate the power generation for a fully loaded turbine if the inlet steam flow to the turbine is fixed to satisfy process heating requirements. Steam properties can be obtained from Table 23.14.

7. The following data are given for a gas turbine:

Power generation	W	15 MW
Power efficiency	η_{GT}	32.5%
Exhaust flow	m_{EX}	$58.32\ kg \cdot s^{-1}$
Exhaust temperature	T_{EX}	488 °C
Exhaust heat capacity	$C_{P,EX}$	$1.1\ kJ \cdot kg^{-1} \cdot K^{-1}$
Stack temperature	T_{STACK}	100 °C

a) Calculate the fuel consumption Q_{FUEL}.

b) Assuming 72% of the heat available in the exhaust can be recovered for steam generation, how much steam (Q_{STEAM}) can be generated from an unfired HRSG in MW? What is the cogeneration efficiency η_{COGEN} for the system of gas turbine and HRSG?

c) With supplementary firing ($T_{SF} = 800\,°C$), calculate the fuel consumption Q_{SF} for the supplementary firing, available heat from the exhaust Q_{EX} and steam generation for the same HRSG with an 83% efficiency.

d) Calculate the power generation efficiency η_{POWER} and cogeneration efficiency η_{COGEN} for the system of gas turbine and HRSG with supplementary firing. Does supplementary firing improve η_{POWER} or η_{COGEN}?

Table 23.13

Problem table cascade for Exercise 6.

Interval temperature (°C)	Heat flow (MW)
400	15
400	20
170	20
115	21
115	11.5
100	12
100	0
80	1
80	17.5
50	18
50	5

8. A gas turbine has the following performance data:

Electricity output	13.5 MW
Heat rate	$10{,}810\ kJ \cdot kWh^{-1}$
Exhaust flow	$179{,}800\ kg \cdot h^{-1}$
Exhaust temperature	480 °C

The exhaust is to be used to generate steam in an unfired HRSG with a minimum stack temperature of 150 °C. The specific heat capacity of the exhaust is $1.1\ kJ \cdot kg^{-1} \cdot K^{-1}$. Enthalpy data for steam are given in Table 23.15. Calculate the following:

a) Fuel requirement in MW.

b) The efficiency of the gas turbine.

c) Amount of steam at 10 bara and 200 °C that can be produced from boiler feedwater at 100 °C in an unfired HRSG. Assume $\Delta T_{min} = 20\,°C$ for the heat recovery steam generator.

9. Complete the steam balance provided in Figure 23.65.

a) Calculate extraction flows through $T1$ and $T2$.

b) Calculate condensing flows of $T1$ and $T2$ from a power balance around each turbine, assuming the power generation is fixed, that the mechanical efficiency is 97% and the outlet enthalpy of the steam turbines is equal to the enthalpies of the steam mains. Steam properties for the inlet to steam turbines can be taken to be those of the steam mains, which are given in Table 23.16. Isentropic enthalpy changes between the steam mains in Figure 23.65 are given in Table 23.17.

c) Determine the steam production required from the utility boilers to meet the steam demand.

d) Calculate the isentropic and overall efficiency for Turbine $T4$ (load = 3.75 MW), assuming the inlet conditions are

23

Table 23.14

Enthalpies for Exercise 6.

Streams	P (bar)	H_{SUP} (kJ·kg^{-1})	S_{SUP} (kJ·kg^{-1}·K^{-1})	T_{SAT} (°C)	H_{SAT} (kJ·kg^{-1})	H_L (kJ·kg^{-1})	ΔH_{IS} (kJ·kg^{-1})
Inlet	40	3095	6.587	250.3	2800	1087	0
Outlet	1.985	–	–	120	2706	503.7	602.3

Table 23.15

Enthalpy data for Exercise 8.

	Enthalpy at 10 bar (kJ·kg^{-1})
Steam at 200 °C	2827
Saturated steam (179.9 °C)	2776
Saturated condensate (179.9 °C)	763
Water at 100 °C	420

given in Table 23.17 and the mechanical efficiency is 97%.

e) Turbine $T3$ is to be shut down and the driver switched to an electric motor. It can be assumed that the inlet conditions to the turbines do not change as a result of any changes to the steam balance. The steam flows through Turbines $T4$ and $T5$ are to be maintained at their current values. However, the flows through Turbines $T1$ and $T2$ can be changed to rebalance the steam system, whilst maintaining the current power output. How much steam is needed from the utility boilers if this is done? What are the cost savings from the change? Assume that the conditions of the steam in each main are fixed to the values in Table 23.16.

Boiler efficiency = 0.92
Fuel cost = 5.69 \$·GJ^{-1}
Power cost = 55 \$·MWh^{-1}

10. A company is considering implementing a cogeneration scheme. At the moment, the company is supplied with 50,000 MWh of electricity from the grid each year and the company produces steam of 80,000 MWh per year from a gas-fired boiler. It is proposed to replace this arrangement with a gas turbine. With the cogeneration scheme, the waste heat from the gas turbine can be used for the steam generation to satisfy the steam demand. Assume the power generation

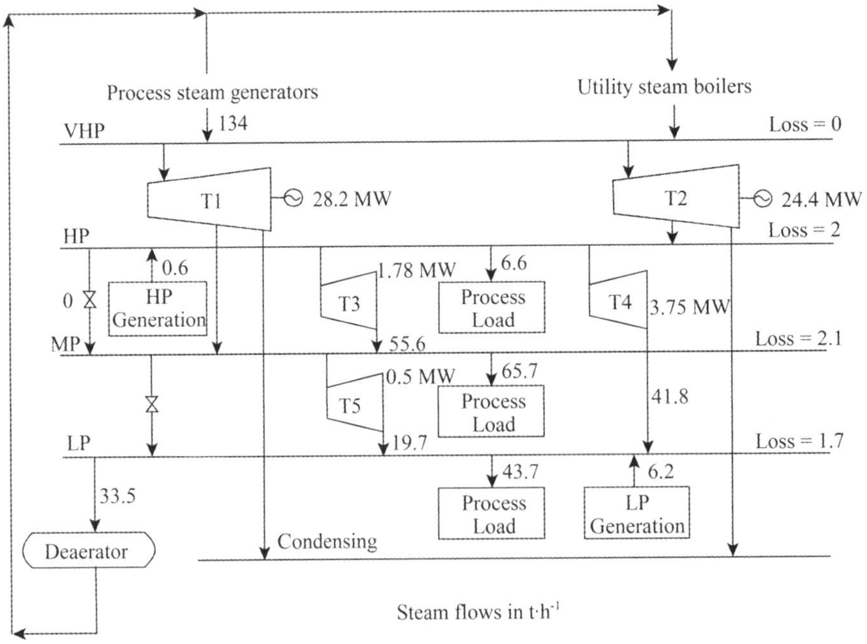

Figure 23.65

Steam balance for a site.

Table 23.16

Steam properties for the steam mains in Figure 23.65.

	Pressure (bar)	Superheat temperature (°C)	Enthalpy ($kJ \cdot kg^{-1}$)	Saturation temperature (°C)
VHP	100	500	3375	311
HP	40	400	3216	250
MP	15	300	3039	198
LP	4.5	200	2858	148
Condensation	0.12	–	2591	49.4
BFW	–	–	504	–

efficiency of the gas turbine is 0.35, average electricity tariff = 0.063 $\$ \cdot kWh^{-1}$, gas price = 0.015 $\$ \cdot kWh^{-1}$, boiler efficiency is 0.9, the installed cost of the gas turbine per kW electricity is 1000 $\$ \cdot kWh^{-1}$ and the company operates for 8000 hours annually.

a) Calculate the operating costs for both the current operating system and the proposed cogeneration scheme.

b) Determine the savings and payback time that can be achieved by implementing the cogeneration scheme.

c) What issues should be considered in recovering the exhaust heat from a gas turbine?

11. It is proposed to locate a thermal oxidizer adjacent to a chemical plant, with the purpose of supplying steam generated from the flue gas of the thermal oxidizer to the chemical plant for $\Delta T_{min} = 10\,°C$. The problem table cascades for the two processes are given in the Table 23.18. Note that T^* represents shifted temperatures for the process streams with $\Delta T_{min}/2 = 5\,°C$.

a) Plot the total site composite curves.

b) What are the minimum heat requirements for the total site assuming one intermediate steam main at 180 °C is installed and only latent heat of the steam is to be used?

c) Is there a site pinch and, if so, at what temperatures?

Table 23.17

Isentropic enthalpy changes in Figure 23.65.

Inlet	Outlet	ΔH_{IS} ($kJ \cdot kg^{-1}$)
VHP	HP	271.7
VHP	MP	507.6
HP	MP	235.8
HP	LP	467.5
HP	Condensation	991.6
MP	LP	231.6

Table 23.18

Data for Exercise 1 1.

Thermal oxidizer		Chemical plant	
T^* (°C)	Q (MW)	T^* (°C)	Q (MW)
605	0	245	17
155	18	125	11
		45	0

References

Betz (1991) *Handbook of Industrial Water Conditioning*, 9th Edition, Betz Laboratories Inc.

Brooks FJ (2001) GE Gas Turbine Performance Characteristics, GE Power Systems GER-3567H.

Chou CC and Shih YS (1987) A Thermodynamic Approach to the Design and Synthesis of Utility Plant, *Ind Eng Chem Res*, **26**: 1100.

Del Nogal FL, Kim J-J, Perry S and Smith R (2010) Synthesis of Mechanical Driver and Power Generation Configurations, Part 1: Optimization Framework, *AIChE Journal*, **56**: 2356.

Dhole VR and Linnhoff B (1992) *Total Site Targets for Fuel, Cogeneration, Emissions and Cooling, ESCAPE – II Conference*, Toulouse, France.

Dryden IGC (1982) *The Efficient Use of Energy*, Butterworth Scientific.

Elliot TC (1989) *Standard Handbook of Powerplant Engineering*, McGraw-Hill.

Iyer RR and Grossmann IE (1998) Synthesis and Operational Planning of Utility Systems for Multiperiod Operation, *Comp Chem Eng*, **22**: 979.

Kemmer FN (1988) *The Nalco Water Handbook*, 2nd Edition, McGraw-Hill.

Kenney WF (1984) *Energy Conservation in Process Industries*, Academic Press.

Kimura H and Zhu XX (2000) R-Curve Concept and Its Application for Industrial Energy Management, *Ind Eng Chem Res*, **39**: 2315.

Klemes J, Dhole VR, Raissi K, Perry SJ and Puigjaner L (1997) Targeting and Design Methodology for Reduction of Fuel, Power and CO_2 on Total Sites, *J Applied Thermal Eng*, **17**: 993.

Mavromatis SP (1996) Conceptual Design and Operation of Industrial Steam Turbine Networks, PhD Thesis, UMIST, UK.

Mavromatis SP and Kokossis AC (1998) Conceptual Optimisation of Utility Networks for Operational Variations – I Targets and Level Optimisation, *Chem Eng Sci*, **53**: 1585.

Perry RH and Green DW (1997) *Perry's Chemical Engineers' Handbook*, 7th Edition, McGraw-Hill.

Peterson JF and Mann WL (1985) Steam System Design: How it Evolves, *Chem Eng*, **92** (21): 62.

Raissi K (1994) Total Site Integration, PhD Thesis, UMIST, UK.

Shang Z (2000) Analysis and Optimisation of Total Site Utility Systems, PhD Thesis, UMIST, UK.

Siddhartha M and Rajkumar N (1999) Performance Enhancement in Coal Fired Thermal Power Plants. Part II: Steam Turbines, *Int J Energy Res*, **23**: 489.

Sun L and Smith R (2015) Performance Modelling of New and Existing Steam Turbines, *Ind Eng Chem Res*, **54**: 1905.

Sun L, Doyle S and Smith R (2015a) Heat Recovery and Power Targeting in Utility Systems, *Energy*, **84**: 196.

Sun L, Doyle S and Smith R (2015b) Understanding Steam Costs for Energy Conservation Projects, *Applied Energy*, **161**: 647.

Varbanov PS, Doyle S and Smith R (2004) Modeling and Optimisation of Utility Systems, *Trans IChemE*, **82A**: 561.

Varbanov PS, Perry SJ, Makwana Y, Zu XX and Smith R (2004) Top Level Analysis of Utility Systems, *Trans IChemE*, **82A**: 784.

Chapter 24

Cooling and Refrigeration Systems

24.1 Cooling Systems

Most industrial processes require an external heat sink for heat removal and temperature control. Waste heat rejection should be minimized as much as possible, both to conserve energy and to prevent damage to the environment. Priority should be given to recover waste heat in the first instance within the process. If waste heat is available at a high enough temperature and is not required for process heating within the process itself, then opportunities should be explored to pass the heat to other useful heat sinks on the site. Heat can be passed between processes on the site through the steam system, as discussed in Chapter 23. Other heat transfer fluids, such as hot oil, can be used to transfer heat around at high temperatures. Below steam temperature, heat can be passed around a site using hot water. Heat can also sometimes be rejected usefully into the utility system, for example, for boiler feed water preheating or combustion air preheating. Once the opportunity for heat recovery and rejection to useful external heat sinks has been exhausted, then heat must be rejected to the environment.

If heat rejection is required at temperatures above ambient temperature then three different methods can be used:

- Once through water cooling
- Recirculating cooling water systems
- Air coolers.

If heat rejection is required below that which can normally be achieved using cooling water or air cooling, then refrigeration is required. A refrigeration system is a heat pump with the purpose of providing low-temperature cooling. This means that heat must be rejected at a higher temperature, to ambient via an external cooling utility (e.g. cooling water), a heat sink within the process, or to another refrigeration system. Cooling systems require integration with the processes on the site, with other cooling systems and also the hot utility systems.

Start by considering the above-ambient temperature cooling.

24.2 Once-Through Water Cooling

Once-through cooling systems take water from a river, canal, lake or the sea, use it for cooling and then return it to its source. Heat rejection to the environment in this way does have environmental consequences and can have an impact on ecosystems. Any change of temperature in surface and groundwater resulting from waste heat rejection affects the chemical, biochemical and hydrological properties of the water and potentially has an impact on the overall ecosystem. Once-through cooling systems require large amounts of water to be used on a single-use basis. This method is simple and cheap but has environmental consequences and problems can be created in the process from fouling, unless the water is treated before use.

When the use of fresh cold water is limited, or environmental regulations limit the heat rejection to rivers, canals, lakes and the sea, then air cooling or recirculating cooling water systems must be used. Recirculating cooling water systems are the most common method used for heat rejection to the environment.

24.3 Recirculating Cooling Water Systems

Figure 24.1 illustrates the basic features of a recirculating cooling water system (Kröger, 1998). Cooling water from the cooling tower is pumped to heat exchangers where waste heat needs to be rejected from the process to the environment. The cooling water is in turn heated in the heat exchanger network and returned to the cooling tower. The hot water returned to the cooling tower flows down over packing and is contacted countercurrently or in cross-flow with air. The packing should provide a large interfacial area

Chemical Process Design and Integration, Second Edition. Robin Smith.

Companion Website: www.wiley.com/go/smith/chemical2e

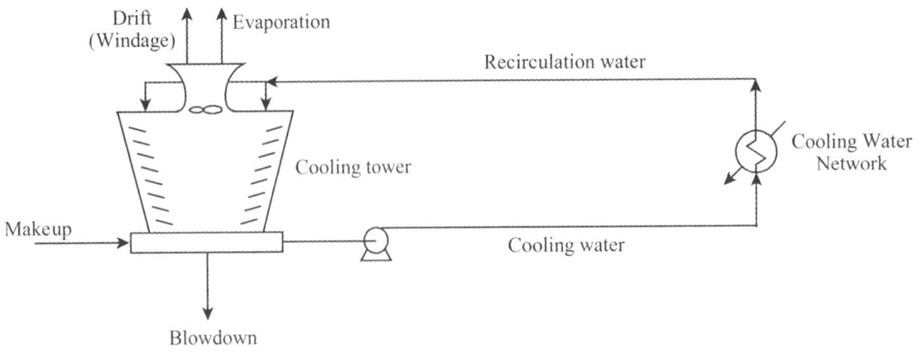

Figure 24.1
Cooling water systems.

for heat and mass transfer between the air and the water. The air is humidified and heated, and rises through the packing. The water is cooled mainly by evaporation as it flows down through the packing. The evaporated water leaving the top of the cooling tower reflects the cooling duty that is being performed. Water is also lost through *drift* or *windage*. Drift is droplets of water entrained in the air leaving the top of the tower. Drift has the same composition as the recirculating water and is different from evaporation. Drift should be minimized because it wastes water and can also cause staining of buildings, and so on, that are distant from the cooling tower. *Drift loss* is around 0.1 to 0.3% of the water circulation rate. *Blowdown*, as shown in Figure 24.1, is necessary to prevent the build-up of contamination in the recirculation. *Makeup water* is required to compensate for the loss of water from evaporation, drift and blowdown. The makeup water contains solids that build up in the recirculation as a result of the evaporation. The blowdown purges these, along with products of corrosion and microbiological growth. Both corrosion and microbiological growth need to be inhibited by chemical dosing of the recirculation system. *Dispersants* are added to prevent deposit of solids, *corrosion inhibitors* to prevent corrosion and *biocides* to inhibit biological growth. In Figure 24.1, the blowdown is shown to be taken from the cooling tower *basin*. It can be taken as cold blowdown, as shown in Figure 24.1. Alternatively, it can be taken from the hot recirculation water before it is returned to the cooling tower as *hot blowdown*. Taking hot blowdown might be helpful in increasing the heat rejection from the cooling system, but might not be acceptable environmentally from the point of view of the resulting increase in effluent temperature.

Many different designs of cooling tower are available. These can be broken down into two broad classes as follows.

1) *Natural draft.* Natural draft cooling towers consist of an empty shell, usually hyperbolic in shape and constructed in concrete. The upper, empty portion of the shell merely serves to increase the draft. The lower portion is fitted with the packing. The draft is created by the difference in density between the warm humid air within the tower and the denser ambient air.
2) *Mechanical draft cooling towers.* Mechanical draft cooling towers use fans to move the air through the cooling tower. In a *forced draft* design, fans push the air into the bottom of the tower. *Induced draft* cooling towers have a fan at the top of the cooling tower to draw air through the tower. The tower height for mechanical draft towers does not need to extend much beyond the depth of the packing. Mechanical draft cooling towers for large duties often comprise a series of rectangular *cells* constructed together, but operating in parallel, each with its own fan.

The type of packing used can be as simple as *splash bars* but is more likely to be packing similar in form to that used in absorption and distillation towers. The temperature limitation of the packing needs careful attention. Plastic packing has severe temperature limitations, as far as the cooling water return temperature is concerned. If the temperature is too high, the plastic packing will deform and this will result in a deterioration of cooling tower performance. Polyvinylchloride is limited to a maximum temperature of around 50 °C. Other types of plastic packing can withstand temperatures up to around 70 °C.

Figure 24.2 gives the basis of a cooling water system model. For the cooling tower:

$$T_0 = f(F_2, T_2, G, T_{WBT}) \tag{24.1}$$

$$F_0 = f(F_2, T_2, G, T_{WBT}) \tag{24.2}$$

$$F_E = f(F_2, T_2, G, T_{WBT}) \tag{24.3}$$

where
T_0 = outlet temperature of the cooling tower
T_1, T_2 = inlet temperature and outlet temperature for the heat exchanger network (T_2 = inlet temperature to the cooling tower)
T_{WBT} = wet bulk temperature of the inlet air (the equilibrium temperature attained by water that is vaporizing into the inlet air and is always less than the dry bulk temperature of the air into which vaporization is taking place)
F_0 = flowrate of water from the cooling tower
F_1, F_2 = flowrate of water to and from the heat exchanger network
F_E = evaporation rate
G = flowrate of air

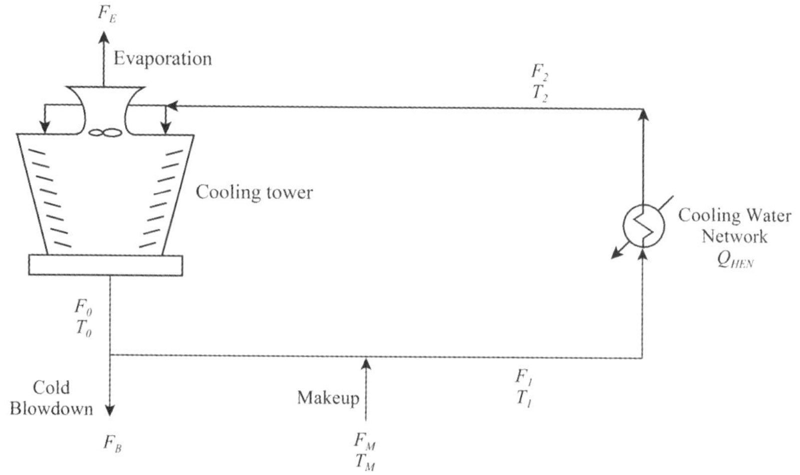

Figure 24.2
Cooling water system model.

The difference between T_0 and T_{WBT} measures the degree of unsaturation of the inlet air. If the air is initially saturated with water vapor, then neither vaporization of the liquid nor depression of the wet bulb temperature occurs. A simple cooling tower model that can be used in conceptual design is presented elsewhere (Kim and Smith, 2001).

There are some general trends that can be observed for the design of cooling towers in terms of the temperature and flowrate of the inlet cooling water to the tower. Increasing the temperature of the inlet to a given cooling tower design for a fixed flowrate increases the performance of the cooling tower and allows more heat to be removed (Kim and Smith, 2001). On the other hand, if the flowrate to the inlet of a given cooling tower is decreased for a fixed inlet temperature, then the performance of the cooling tower improves, allowing more heat to be removed (Kim and Smith, 2001). Thus, the performance of a cooling tower is maximized by maximizing the inlet temperature to the cooling tower, and minimizing the inlet flowrate (Kim and Smith, 2001). There will be constraints on the maximum return temperature determined by the maximum temperature allowable for the cooling tower packing, and fouling and corrosion issues, as discussed above.

Cooling water coolers are usually connected in parallel, as illustrated in Figure 24.3a. If process cooling duties allow and the coolers are connected in series, as illustrated in Figure 24.3b, this has the effect of decreasing the cooling water flowrate and increasing the return temperature. Parallel arrangements as in Figure 24.3a favor the design of the heat exchange coolers, whereas series arrangements as in Figure 24.3b favor the design of the cooling tower. For example, if an existing cooling tower is at maximum capacity, its capacity might be able to be increased by changing some of the coolers from parallel to series arrangements.

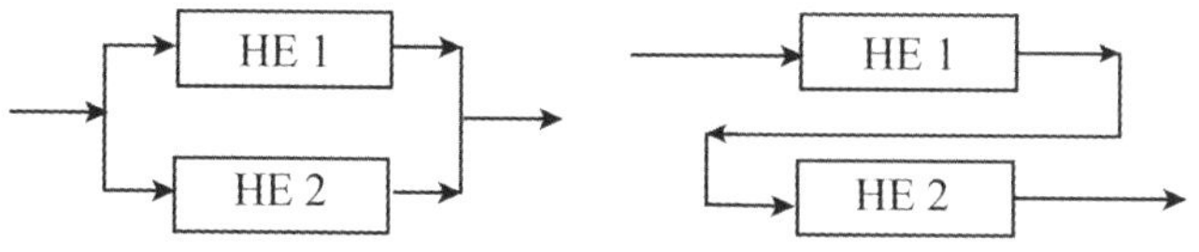

(a) Cooling water coolers in parallel. (b) Cooling water coolers in series.

Figure 24.3
Parallel and series arrangements for cooling water coolers. (Reproduced from Kim J-K and Smith R (2001) Cooling Water System Design, *Chem Eng Sci*, **56**: 3641, with permission from Elsevier.)

Referring back to Figure 24.2, the cooling tower makeup and blowdown form a mass balance according to:

$$F_1 = F_0 - F_B + F_M \tag{24.4}$$

where F_1 = flowrate of water to the heat exchanger network
F_B = flowrate of cooling tower blowdown
F_M = flowrate of makeup water

If the heat capacity of the water is assumed to be constant, then an energy balance can also be written:

$$F_1 T_1 = (F_0 - F_B)T_0 + F_M T_M \tag{24.5}$$

where T_1 = inlet temperature to the heat exchanger network (after addition of makeup)
T_0 = outlet temperature from the cooling tower (before addition of makeup)
T_M = temperature of the makeup water

The heat load on the heat exchanger network is given by

$$Q_{HEN} = F_2 C_P (T_2 - T_1) \tag{24.6}$$

where Q_{HEN} = heat exchanger network cooling duty
C_P = heat capacity of the water

As evaporated water is pure, solids are left behind in the recirculating water, making it more concentrated than the makeup water. The blowdown purges the solids from the system. Note that the blowdown has the same chemical composition as the recirculated water. *Cycles of concentration* is a comparison of the dissolved solids in the blowdown compared with those in the makeup water. For example, at three cycles of concentration, the blowdown has three times the solids concentration as the makeup water. For calculation purposes, blowdown is defined to be all nonevaporative water losses (drift, leaks and intentional blowdown). In principle, any soluble component in the makeup and blowdown can be used to define the concentration for the cycles. For example, chloride and sulfate being soluble at high concentrations can be used. The cycles of concentration are thus defined to be:

$$CC = \frac{C_B}{C_M} \tag{24.7}$$

where CC = cycles of concentration
C_B = concentration in blowdown
C_M = concentration in makeup

A mass balance for the solids entering with the makeup (assuming zero solids in the evaporation) gives:

$$F_M C_M = F_B C_B \tag{24.8}$$

Combining Equation 24.7 with Equation 24.8 gives:

$$CC = \frac{F_M}{F_B} \tag{24.9}$$

given that:

$$F_M = F_B + F_E \tag{24.10}$$

Combining Equations 24.9 and 24.10 gives:

$$F_B = \frac{F_E}{CC - 1} \tag{24.11}$$

$$F_M = \frac{F_E CC}{CC - 1} \tag{24.12}$$

Figure 24.4 shows the relationship between the makeup, blowdown, evaporation and drift versus the cycles of concentration. For a given design of cooling tower, fixed heat duty and fixed conditions, the evaporation and drift will be constant as the cycles of concentration increase. However, as the cycles of concentration increase, the blowdown decreases and hence the makeup water decreases. This can be an important consideration from the point of view of site water consumption, as discussed in Chapter 26.

The consequences of an increase in the cycles of concentration are that, as the level of dissolved solids increases, corrosion and deposition tendencies also increase. The result is that, although increasing the cycles of concentration decreases the water requirements of the cooling system, the required amount of chemical dosing also increases.

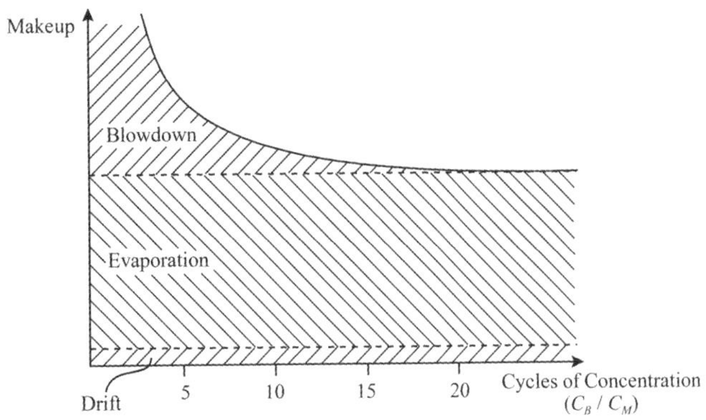

Figure 24.4
Relationship between makeup water and blowdown for cooling towers.

24.4 Air Coolers

Heat above ambient temperature can be rejected directly to the environment by the use of *air-cooled heat exchangers* or *fin-fan exchangers* (Kröger, 1998). Three designs are illustrated in Figure 24.5. In these designs, the fluid to be cooled passes through

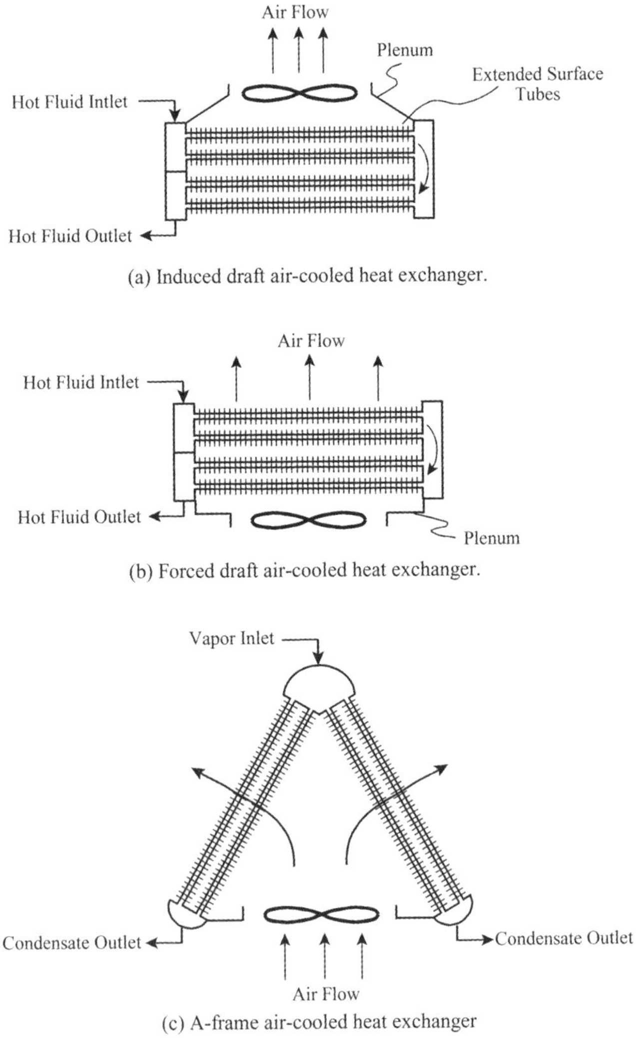

Figure 24.5
Air-cooled heat exchangers.

the inside of tubes and ambient air is blown across the outside by the use of fans. The outside surface most often features extended surfaces to enhance the heat transfer. High fin tubes are normally used, as discussed in Chapter 12. Fins can be engineered in one of three ways:

- Wrapping an 'L'-shaped aluminum strip that is footed on the tube.
- Helically wrapping a strip of aluminum into a pre-cut helical groove on the outside of the tube.
- Extruding fins from the wall of an aluminum tube. Bimetallic tubes consist of an inner tube and an outer tube. The outer tube is the aluminum tube with extruded fins. The inner tube can have the same materials and dimensions as standard heat exchanger tubes.

If integral fins are not used, then the bond between the tube wall and the fins creates an additional heat transfer resistance. Similarly, if bimetallic tubes are used there is an additional heat transfer resistance caused by the bond between the tubes. However, this is difficult to quantify and is neglected for preliminary design.

The enhancement in the heat transfer depends on the dimensions of the tube, the dimensions of the fins, the number of fins per unit length, the method by which the fins are formed on the tube, tube arrangement (square or triangular pitch), tube and fin materials of construction, as well as air and tube-side velocities, physical properties, temperatures and fouling characteristics. Figure 24.6 shows three typical air cooler layouts from the many possibilities. The tubes are grouped into rectangular *bundles*, typically 1 to 2 m wide. The bundles are factory assembled. Within each bundle, the tubes are arranged in 2 to 10 rows. Bundles are grouped into *bays* containing one or more bundles in parallel, with each bundle connected to inlet and outlet headers. Each bay can be serviced by one to three fans, depending on the geometry of the tube layout.

The headers can be fitted with pass partition plates, to create different tube pass arrangements, in a similar way to shell-and-tube heat exchangers. Figure 24.5a and b shows twopass arrangements, which are most common, but one- to four-pass arrangements can be used. Such *cross-flow* arrangements require F_T correction factors. As discussed in Chapter 12, it can be assumed that the air side is mixed but the tube side is unmixed. However, whilst there will be some mixing normal to the flow, the mixing will not be perfect across the whole cross-section of the gas flow. A conservative assumption is to consider the gas to be completely mixed normal to the flow. Equations 12.107 and 12.108 give the F_T for cross-flow in which the gas side is considered to be mixed normal to its flow, but the tube side flows through tubes in a single pass and is unmixed (Figure 12.18a). Equations 12.109 and 12.110 give the F_T for cross-flow in which the gas side is considered to be mixed normal to the flow, but the tube side flows through tubes in a two-pass countercurrent arrangement and is unmixed when flowing though the tubes, but mixed to a uniform temperature between the two passes (Figure 12.18a).

Typical outside film transfer coefficients are 50 to 70 $W \cdot m^{-2} \cdot K^{-1}$ and fouling coefficients typically 3000 $W \cdot m^{-2} \cdot K^{-1}$. Typical overall heat transfer coefficients are presented in Table 24.1. The values in Table 24.1 assume high fin tubes, but the heat transfer area is based on the bare tube area.

In Figure 24.5a the tubes are mounted horizontally and the air drawn across the tubes in an *induced draft* arrangement. In Figure 24.5b the tubes are also mounted horizontally but the air is driven across the tubes in a *forced draft* arrangement. The horizontal arrangements in Figure 24.5a and b are suitable if no condensation is taking place, as would be the case for cooling of liquids or cooling of gas without condensation. If the cooling duty is condensing, then the *A-frame* arrangement in Figure 24.5c is used, in which the tubes are sloped to allow separation of the condensed liquid and vapor.

For a given mass flowrate of air, an induced draft fan in principle requires slightly greater power than a forced draft fan, as the induced draft fan is compressing hot air with a lower density and high volume than a forced draft equivalent. However, induced draft

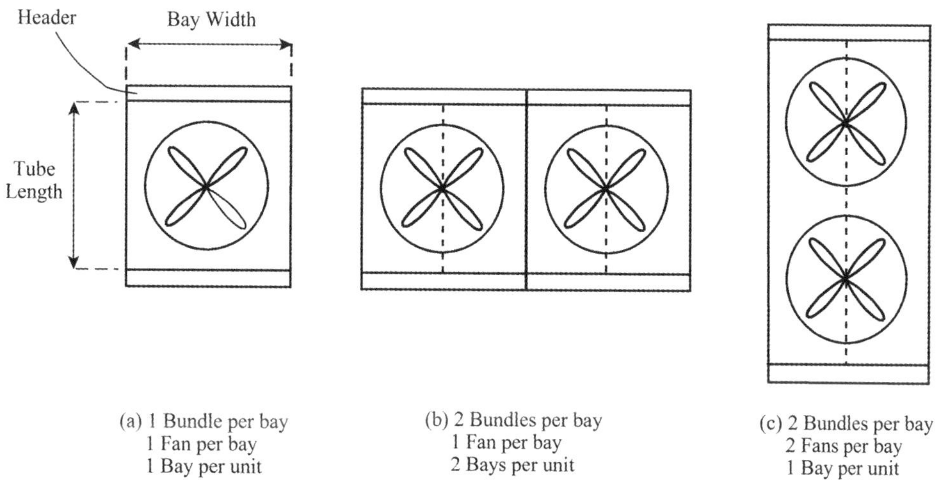

Figure 24.6

Some typical layouts of the air-cooled heat exchangers.

Table 24.1

Typical overall heat transfer coefficients for air-cooled heat exchangers (Hudson Products, 2015).

Duty	Overall heat transfer coefficient ($W \cdot m^{-2} \cdot K^{-1}$)
Liquid coolers	
Water	680–880
Light hydrocarbons	400–680
Heavy hydrocarbons	30–170
Gas coolers	
Air or flue gas	55–170
Hydrocarbon gasses	170–510
Condensers	
Steam	770–1100
Hydrocarbons	400–600

arrangements provide a more uniform air distribution over the tube bundle. Forced draft arrangements have the disadvantage that hot air can be recirculated back to the air inlet. The performance of the units is affected significantly both by ambient conditions and by fouling of the outside surfaces. The air flow across the tubes can be controlled by the use of variable pitch fan blades and variable speed electric motors to drive the fans. The outside surfaces should be cleaned regularly to maintain performance, typically by water sprays. Spraying water on the outside surface can also be used to temporarily enhance the heat transfer when the ambient temperature is high. If the exchangers have to work in very low temperatures under winter conditions, then they can be erected within enclosures equipped with louvers to allow some recirculation of hot air and prevent freezing.

General guidelines for the design of air-cooled heat exchangers are (Kröger, 1998; Serth, 2007; Hudson Products, 2015):

- Air temperature is normally taken to be the temperature only exceeded for 5% of the year. This means that the cooler will be overdesigned for 95% of the year and will require appropriate control systems to avoid overcooling. An air temperature near, but below, the maximum to be encountered needs to be chosen.
- Recirculation in congested areas can increase the effective inlet air temperature by 1 to 2 °C compared with ambient air temperatures in uncongested areas.
- Air velocity based on the face area is usually between 3 and $6\ m \cdot s^{-1}$.
- The fan projected area should normally be greater than 40% of the face area of the tubes.
- The pressure drop for the air across the tube bundle is normally between 100 and $200\ N \cdot m^{-2}$.
- Fan efficiency is typically assumed to be around 70% in preliminary design.
- Minimum temperature approach (the difference between the process fluid outlet temperature and the design air temperature) is not normally lower than 10 °C and often taken to be 20 °C. More than one tube pass might be used, in which case this applies for each tube pass.
- Tube-side liquid velocity is usually between 1 and $2\ m \cdot s^{-1}$.
- Logarithmic mean temperature difference correction F_T depends on the number of tube rows, number of tube passes and the tube-side flow pattern. A value of 0.9 is normally assumed in preliminary design for a single pass, 0.96 for two passes and 0.99 for three passes. Passes should be arranged for countercurrent flow, as in Figure 24.5a and b.
- Tube bundles are normally rectangular and consist typically of 2 to 10 rows of finned tubes arranged on a triangular pitch.
- Tubes can have diameters ranging from 1.59 cm to 15.2 cm, the most common being 2.54 cm.
- The tube pitch is usually between 2 and 2.5 tube diameters.
- Tube lengths can be between 2 m and 18 m.
- The fin pitch is 275 to $430\ fins \cdot m^{-1}$, the fin height is 8 to 25.4 mm, the fin thickness is 0.25 to 0.9 mm and clearance between adjacent fins is 3 to 10 mm.
- Tube bundles are normally factory assembled, which creates a width limit for transport of around 3.7 m. Of course greater widths can be created in the final assembly by installing bundles side by side.

The outside heat transfer coefficient for banks of finned tubes on a triangular pitch can be calculated from Equation 12.111 (Briggs and Young, 1963). The details of the calculation have been given in Section 12.8.

Air-cooled heat exchangers of the type in Figure 24.5 are very common in some industries, particularly when the plant is located in a region where water is scarce for cooling tower systems. Air-cooled heat exchangers also have the environmental advantage of eliminating cooling tower plumes.

Example 24.1 A hydrocarbon stream with a flowrate of $100{,}000\ kg \cdot h^{-1}$ is to be cooled in an air-cooled heat exchanger from 90 °C to 50 °C. The heat capacity of the hydrocarbon stream can be assumed to be $2.3\ kJ \cdot kg^{-1} \cdot K^{-1}$. Carry out a preliminary design assuming air temperature to be 25 °C (including a small allowance for recirculation). The heat capacity of air can be assumed to be $1.01\ kJ \cdot kg^{-1} \cdot K^{-1}$ and density $1.16\ kg \cdot m^{-3}$. For an assumed outlet air temperature of 40 °C:

a) Develop a preliminary design based for the tube bundle based on tubes with an outside diameter of 2.54 cm, on a triangular pitch of 2.5 tube diameters and 10 m tube length with an overall heat transfer coefficient of $500\ W \cdot m^{-2} \cdot K^{-1}$.

b) Estimate the power requirements of the fan based on an assumed pressure drop across the tubes for the air of $150\ N \cdot m^{-2}$, a fan efficiency of 70% and a motor efficiency of 90%.

Solution

a)

$$\text{Process heat duty} = \frac{100,000}{3600} \times 2.3 \times (90-50)$$
$$= 2555.6\ \text{kW}$$

$$\text{Flowrate of air} = \frac{2555.6}{1.01 \times (40-25)} \times \frac{1}{1.16}$$
$$= 145.42\ \text{m}^3 \cdot \text{s}^{-1}$$

$$\Delta T_{LM} = \frac{(90-40)-(50-25)}{\ln\left(\dfrac{90-40}{50-25}\right)}$$
$$= 36.07\ ^\circ\text{C}$$

$$R = \frac{T_{P1} - T_{P2}}{T_{G2} - T_{G1}} = \frac{90-50}{40-25} = 2.6667$$

$$P = \frac{T_{G2} - T_{G1}}{T_{P1} - T_{G1}} = \frac{40-25}{90-25} = 0.2308$$

Start by assuming one pass. From Equation 12.107:

$$F_T = \frac{\ln\left[\dfrac{1-P}{1-RP}\right]}{(1-R)\ln\left[1+\dfrac{1}{R}\ln(1-RP)\right]} \qquad R \neq 1$$

$$= \frac{\ln\left[\dfrac{1-0.2308}{1-2.6667\times 0.2308}\right]}{(1-2.6667)\ln\left[1+\dfrac{1}{2.6667}\ln(1-2.6667\times 0.2308)\right]}$$

$$= 0.9374$$

Assuming $U = 500\ \text{W}\cdot\text{m}^{-2}\cdot\text{K}^{-1}$, the heat transfer area based on plain tubes is given by:

$$A = \frac{Q}{U\Delta T_{LM} F_T}$$
$$= \frac{2555.6}{0.5 \times 36.07 \times 0.9374}$$
$$= 151.2\ \text{m}^2$$

$$\text{Area of a single tube} = \pi \times 0.0254 \times 10$$
$$= 0.7980\ \text{m}^2$$

$$\text{Number of tubes} = \frac{151.2}{0.7980} = 189.5$$
say 190

Try two tube rows. Width of bundle for Equation 12.114:

$$= N_{TR} p_T + \frac{3p_T}{2} - d_R$$
$$= \frac{190}{2} \times 0.0254 \times 2.5 + \frac{3 \times 0.0254 \times 2.5}{2} - 0.0254$$
$$= 6.102\ \text{m}$$

$$\text{Face area} = 10 \times 6.102 = 61.02\ \text{m}^2$$

Check air velocity.

$$\text{Flowrate of air} = \frac{2555.6}{1.01 \times (40-25)} \times \frac{1}{1.16}$$
$$= 145.42\ \text{m}^3 \cdot \text{s}^{-1}$$

$$\text{Air velocity based on face area} = \frac{145.42}{61.02} = 2.38\ \text{m} \cdot \text{s}^{-1}$$

This velocity is low. Also check the length-to-width ratio of the face:

$$= \frac{10}{6.102} = 1.6$$

A higher face velocity that can be arranged to be serviced by two fans would be preferred. Try four tube rows in two passes. From Equation 12.109:

$$F_T = \frac{\ln\left[\dfrac{1-P}{1-RP}\right]}{2(1-R)\ln\left[1-\dfrac{1}{R}\ln\left(\dfrac{\sqrt{\dfrac{1-P}{1-RP}}-\dfrac{1}{R}}{1-\dfrac{1}{R}}\right)\right]} \qquad R \neq 1$$

$$= \frac{\ln\left[\dfrac{1-0.2308}{1-2.6667\times 0.2308}\right]}{2(1-2.6667)\ln\left[1-\dfrac{1}{2.6667}\ln\left(\dfrac{\sqrt{\dfrac{1-0.2308}{1-2.6667\times 0.2308}}-\dfrac{1}{2.6667}}{1-\dfrac{1}{2.6667}}\right)\right]}$$

$$= 0.9829$$

$$A = \frac{Q}{U\Delta T_{LM} F_T}$$
$$= \frac{2555.6}{0.5 \times 36.07 \times 0.9829}$$
$$= 144.2\ \text{m}^2$$

$$\text{Number of tubes} = \frac{144.2}{0.7980} = 180.7$$
say 184

$$\text{Width of bundle} = N_{TR} p_T + \frac{3p_T}{2} - d_R$$
$$= \frac{184}{4} \times 0.0254 \times 2.5 + \frac{3 \times 0.0254 \times 2.5}{2}$$
$$- 0.0254$$
$$= 2.991\ \text{m}$$

$$\text{Face area} = 10 \times 2.991$$
$$= 29.91\ \text{m}^2$$

Air velocity based on face area

$$= \frac{145.42}{29.91}$$

$$= 4.86\ \text{m} \cdot \text{s}^{-1}$$

This velocity is more reasonable. Check the length-to-width ratio of the face

$$= \frac{10}{2.991} = 3.34$$

This could be arranged as two fans.

b) Assuming a pressure drop of air across the bundle of 150 $N{\cdot}m^{-2}$, power

$$= \frac{150 \times 145.42}{0.7 \times 0.9}$$

$$= 34,619\ \text{W}$$

This preliminary design would need to be followed by a design that accounted for the inside and outside heat transfer coefficients in detail.

Example 24.2 From the preliminary design in Example 24.1, carry out a more detailed design. The initial design in Exercise 24.1 estimated that 184 tubes of length 10 m would be needed in four rows with two passes. Tubes can be assumed to be steel with aluminum fins. Process data are given in Table 24.2 and geometry data in Table 24.3.

Table 24.2

Data for fluids.

	Hydrocarbon	Air
Flowrate ($kg{\cdot}h^{-1}$)	100,000	–
Initial temperature (°C)	90	25
Final temperature (°C)	50	40
Density ($kg{\cdot}m^{-3}$)	663	1.16
Heat capacity ($kJ{\cdot}kg^{-1}{\cdot}K^{-1}$)	2.3	1.01
Viscosity ($N{\cdot}s{\cdot}m^{-2}$)	2.45×10^{-4}	1.87×10^{-5}
Thermal conductivity ($W{\cdot}m^{-1}{\cdot}K^{-1}$)	0.10	0.0267
Fouling coefficient ($W{\cdot}m^{-2}{\cdot}K^{-1}$)	5000	3000

Table 24.3

Air cooler geometry data.

Tube outer diameter d_O (mm)	25.4
Tube inner diameter d_I (mm)	19.8
Tube pitch p_T (mm)	$2.5 \times d_O$
Tube length L (m)	10.0
Number of fins N_F (m^{-1})	350
Fin height H_F (m)	0.0127
Fin thickness δ_F (m)	0.0005
Tube thermal conductivity ($W{\cdot}m^{-1}{\cdot}K^{-1}$)	45
Fin thermal conductivity ($W{\cdot}m^{-1}{\cdot}K^{-1}$)	206

Solution The convective heat transfer coefficient can be calculated from Equation 12.113. First calculate the tube-side heat transfer coefficient. The tube-side velocity is given by:

$$v_T = \frac{m_T(N_P/N_T)}{\rho\left(\pi d_I^2/4\right)}$$

$$= \frac{(100,000/3600)(2/184)}{663\left(\pi \times 0.0198^2/4\right)}$$

$$= 1.479\ \text{m} \cdot \text{s}^{-1}$$

Now calculate the Reynolds number for the tube side, which is given by:

$$Re = \frac{\rho v_T d_I}{\mu}$$

$$= \frac{663 \times 1.479 \times 0.0198}{2.45 \times 10^{-4}}$$

$$= 79,247$$

This is a turbulent flow and the tube-side heat transfer coefficient can be calculated from Equation 12.59:

$$K_{hT4} = 0.023 \frac{k}{d_I} \text{Pr}^{0.4} \left(\frac{\rho d_I}{\mu}\right)^{0.8}$$

$$= 0.023 \times \frac{0.10}{0.0198} \times \left(\frac{2300 \times 2.45 \times 10^{-4}}{0.10}\right)^{0.4} \times \left(\frac{663 \times 0.0198}{2.45 \times 10^{-4}}\right)^{0.8}$$

$$= 1408.1$$

$$h_T = K_{hT4} v_T^{0.8}$$

$$= 1408.1 \times 1.479^{0.8}$$

$$= 1925.8\ \text{W} \cdot \text{m}^2 \cdot \text{K}^{-1}$$

To calculate the outside heat transfer coefficient, first calculate the maximum velocity of the flue gas from Equation 12.116. The flowrate and maximum velocity of the air are given by:

$$m_G = \frac{2555.6}{1.01 \times (40 - 25)}$$
$$= 168.7 \text{ kg} \cdot \text{s}^{-1}$$

The maximum gas velocity is given by Equation 12.116:

$$v_{max} = \frac{m_G}{\rho L \left[N_{TR}(p_T - d_R - 2H_F \delta_F N_F) + \dfrac{3p_T}{2} - d_R \right]}$$

$$= \frac{168.7}{1.16 \times 10.0 \left[\dfrac{184}{4}(2.5 \times 0.0254 - 0.0254 - 2 \times 0.0127 \times 0.0005 \times 350) + \dfrac{3 \times 2.5 \times 0.0254}{2} - 0.0254 \right]}$$

$$= 8.988 \text{ m} \cdot \text{s}^{-1}$$

$$Re = \frac{\rho d_R v_{max}}{\mu}$$
$$= \frac{1.16 \times 0.0254 \times 8.988}{1.87 \times 10^{-5}}$$
$$= 14,162$$

$$Pr = \frac{C_P \mu}{k}$$
$$= \frac{1.01 \times 10^3 \times 1.87 \times 10^{-5}}{0.0267}$$
$$= 0.7074$$

From Equation 12.113:

$$Nu = 0.134\, Re^{0.681}\, Pr^{1/3} \left(\frac{1}{H_F N_F} - \frac{\delta_F}{H_F} \right)^{0.2} \left(\frac{1}{\delta_F N_F} - 1 \right)^{0.1134}$$

The outside film coefficient is given by:

$$h_O = \frac{0.0267}{0.0254} \left[0.134 \times 14,162^{0.681} \times 0.7074^{1/3} \times \left(\frac{1}{0.0127 \times 350} - \frac{0.0005}{0.0127} \right)^{0.2} \times \left(\frac{1}{0.0005 \times 350} - 1 \right)^{0.1134} \right]$$

$$= 71.72 \text{ W} \cdot \text{m}^{-2} \cdot \text{K}^{-1}$$

The fin efficiency now needs to be calculated from Equation 12.117:

$$\kappa = \left(\frac{2h_O}{k_F \delta_F} \right)^{1/2}$$
$$= \left(\frac{2 \times 71.72}{206 \times 0.0005} \right)^{1/2}$$
$$= 37.318$$

$$\psi = \left(H_F + \frac{\delta_F}{2} \right) \left[1 + 0.35 \ln \left(\frac{d_R + 2H_F + \delta_F}{d_R} \right) \right]$$
$$= \left(0.0127 + \frac{0.0005}{2} \right) \times \left[1 + 0.35 \ln \left(\frac{0.0254 + 2 \times 0.0127 + 0.0005}{0.0254} \right) \right]$$
$$= 0.01614$$

$$\eta_F = \frac{\tanh(\kappa \psi)}{\kappa \psi}$$
$$= \frac{\tanh(37.318 \times 0.01614)}{37.318 \times 0.01614}$$
$$= 0.8944$$

Calculation of the weighted efficiency from Equation 12.119 requires A_{ROOT} and A_{FIN} to be calculated:

$$A_{ROOT} = \pi d_R L(1 - N_F \delta_F)$$
$$= \pi \times 0.0254 \times 10.0 \times (1 - 350 \times 0.0005)$$
$$= 0.6583 \text{ m}^2$$

$$A_{FIN} = \frac{\pi N_F L}{2} [(d_R + 2H_F)^2 - d_R^2 + 2\delta_F(d_R + 2H_F)]$$
$$= \frac{\pi \times 350 \times 10.0}{2} [(0.0254 + 2 \times 0.0127)^2 - 0.0254^2 + 2 \times 0.0005 \times (0.0254 + 2 \times 0.0127)]$$
$$= 10.920 \text{ m}^2$$

$$\eta_W = \frac{A_{ROOT} + \eta_F A_{FIN}}{A_{ROOT} + A_{FIN}}$$
$$= \frac{0.6583 + 0.8944 \times 10.920}{0.6583 + 10.920}$$
$$= 0.9004$$

The total outside area of each tube is given by:

$$A_O = 0.6583 + 10.920$$
$$= 11.578 \text{ m}^2$$

The overall heat transfer coefficient is now calculated from Equation 12.120:

$$\frac{1}{U}=\frac{1}{\eta_W h_O}+\frac{R_{OF}}{\eta_W}+\frac{A_O}{\pi d_R L 2k_W}\frac{d_R}{}\ln\left(\frac{d_R}{d_I}\right)+\left(\frac{A_O}{\pi d_I L}\right)\left(R_{IF}+\frac{1}{h_T}\right)$$

$$U=\left[\frac{1}{0.9004\times 71.72}+\frac{1/3000}{0.9004}+\frac{11.578}{\pi\times 0.0254\times 10.0}\times\frac{0.0254}{2\times 45}\ln\left(\frac{0.0254}{0.0198}\right)\right.$$

$$\left.+\left(\frac{11.578}{\pi\times 0.0254\times 10.0}\right)\left(\frac{1}{5000}+\frac{1}{1925.8}\right)\right]^{-1}$$

$$=36.602\,\mathrm{W\cdot m^{-2}\cdot K^{-1}}$$

From Example 24.1, $F_T = 0.9829$ and the heat transfer area required, based on finned tube area, is given by:

$$\begin{aligned}A_{REQUIRED} &= \frac{Q}{U\Delta T_{LM}F_T}\\ &= \frac{2555.6\times 10^3}{36.602\times 36.07\times 0.9829}\\ &= 1969.4\,\mathrm{m^2}\end{aligned}$$

Area available is given by:

$$\begin{aligned}A_{AVAILABLE} &= N_T\times A_O\\ &= 184\times 11.578\\ &= 2130.4\,\mathrm{m^2}\end{aligned}$$

The area available is higher than the area required by $(2130.4-1969.4)=26.1\,\mathrm{m^2}$. In Example 24.1 the overall heat transfer coefficient was assumed to be $500\,\mathrm{W{\cdot}m^{-2}{\cdot}K^{-1}}$. In this example, the overall heat transfer coefficient was calculated to be $36.602\,\mathrm{W{\cdot}m^{-2}{\cdot}K^{-1}}$ based on the finned area. Converting to bare tubes:

$$\begin{aligned}U_{BARE\ TUBE} &= U_{FINNED}\frac{A_{FINNED}}{A_{BARE\ TUBE}}\\ &= 36.602\times\frac{11.578}{\pi\times 0.0254\times 10}\\ &= 531.1\,\mathrm{W\cdot m^{-2}\cdot K^{-1}}\end{aligned}$$

This is slightly higher than the assumed overall heat transfer coefficient in Example 24.1, leading to an over design in area of 8%. This can be taken as contingency. Alternatively, the design can be adjusted. Any surplus or shortfall in area can be corrected by adjusting several parameters. The flowrate of air, tube dimensions and layout, fin dimensions and fin pitch could all be changed. The simplest solution is often to adjust the number of tubes. However, it should be noted that it would be unsafe to assume that simply adding extra tubes to an existing arrangement would solve a problem of a shortfall in area. This is because adding extra tubes decreases the tube-side velocity, decreasing the inside heat transfer coefficient, and increases the face area, decreasing the outside velocity and heat transfer coefficient. In this case, the over design might be corrected by removing tubes. However, this would increase the air velocity and outside heat transfer coefficient. It might be necessary to adjust both the number and length of the tubes to maintain a reasonable length to width ratio.

24.5 Refrigeration

There are many processes such as gas separation and natural gas liquefaction that operate below ambient temperature. If heat rejection is required below ambient temperature, then refrigeration is required. A refrigeration system is a heat pump that provides cooling at temperatures below the level that can normally be achieved using cooling water or air cooling (Gosney, 1982; Isalski, 1989; Dossat, 1991). Thus, the removal of heat using refrigeration leads to its rejection at a higher temperature. This higher temperature might be to an external cooling utility (e.g. cooling water), a heat sink within the process or to another refrigeration system. Refrigeration is important in chemical engineering operations where low temperatures are required to condense gases (Isalski, 1989), crystallize solids, control reactions, and so on, and for the preservation of foods and air conditioning. Generally, the lower the temperature of the cooling required to be serviced by the refrigeration system and the larger its duty, the more complex the refrigeration system. Processes involving the liquefaction and separation of gases (e.g. ethylene, liquefied natural gas, etc.) are serviced by complex refrigeration systems.

There are two broad classes of refrigeration:

- Compression refrigeration
- Absorption refrigeration.

Compression refrigeration cycles are by far the most common and absorption refrigeration is only applied in special circumstances.

Consider first a simple ideal compression refrigeration cycle using a pure component as the refrigerant fluid, as illustrated in Figure 24.7a. Process cooling is provided by a cold liquid refrigerant in the *evaporator*. A mixture of vapor and liquid refrigerant at Point 1 enters the evaporator where the liquid vaporizes and produces the cooling effect before exiting the evaporator at Point 2. The refrigerant vapor is then compressed to Point 3. At Point 3, the vapor is not only at a higher pressure, but is also superheated by the compression process. After the compressor, the vapor refrigerant enters a *condenser* where it is cooled and condensed to leave as a saturated liquid at Point 4 in the cycle. The liquid is then expanded to a lower pressure to Point 1 in the cycle. The expansion process partially vaporizes the liquid refrigerant across the expander, producing a cooling effect to provide refrigeration, and the cycle continues. Figure 24.7b shows the cycle on a temperature–entropy diagram. The diagram shows a two-phase envelope, inside of which the refrigerant is present as both vapor and liquid. To the left of the two-phase envelope, the refrigerant is liquid, and to the

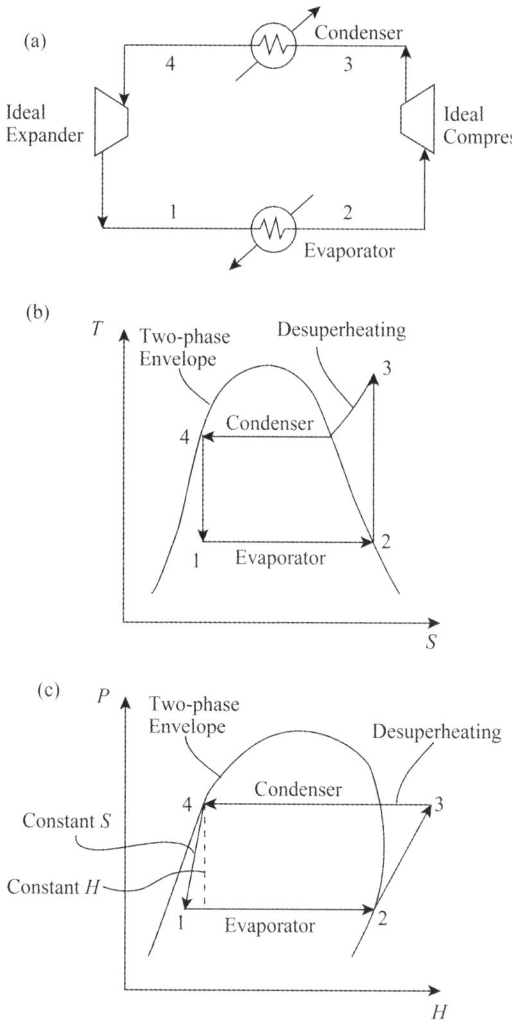

Figure 24.7
A simple (ideal) compression refrigeration cycle.

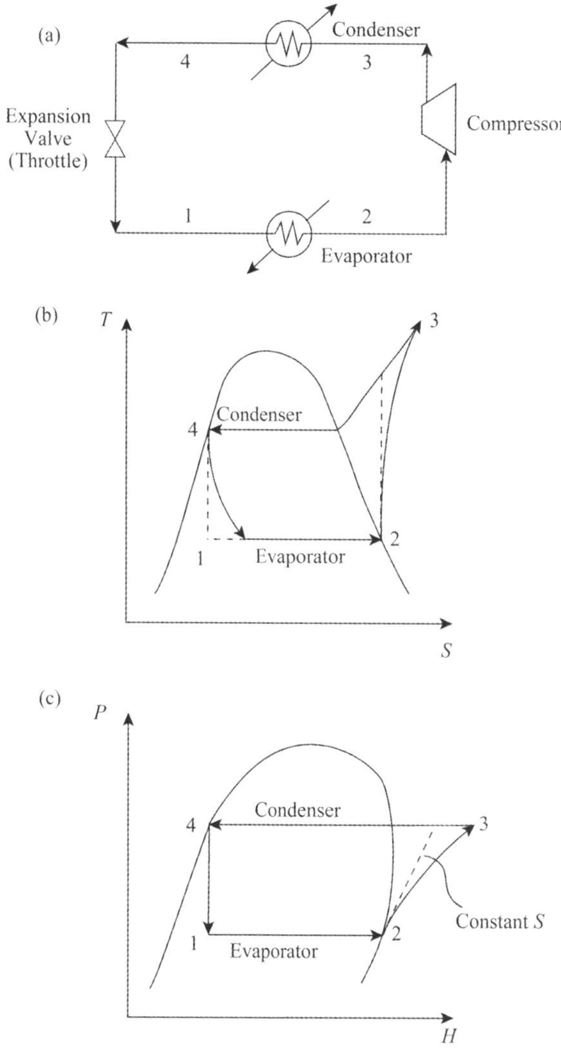

Figure 24.8
Simple compression cycle with expansion valve and nonideal compression.

right of the two-phase envelope, the refrigerant is vapor. Starting at Point 1 with a two-phase mixture, the refrigerant enters the evaporator, where there is an isothermal vaporization to Point 2. From Point 2 to Point 3, there is a compression process that, at this stage, is assumed to be ideal. This is an isentropic process finishing at Point 3. From Point 3, the refrigerant is first cooled to remove the superheat and is then condensed to Point 4, where there is a saturated liquid. The refrigerant is expanded from Point 4 to Point 1, which is assumed here to be an ideal expansion with no increase in entropy. In Figure 24.7c, the same cycle is represented on a pressure–enthalpy diagram. Again the two-phase envelope is evident. From Point 1 to Point 2, the enthalpy of the refrigerant increases in the evaporator. From Point 2 to Point 3, the pressure is increased in the compressor with a corresponding increase in enthalpy. From Point 3 to Point 4, there is no change in pressure but a decrease in the enthalpy due to the desuperheating and the condensation. Between Point 4 and Point 1, there is a decrease in pressure in the expander. The constant entropy assumption leads to a decrease in the enthalpy to Point 1. The cycle in Figure 24.7 assumed ideal isentropic compression and expansion processes. Next consider nonideal compression and expansion processes.

Figure 24.8a again shows a simple compression cycle involving a pure refrigerant fluid, but now with an *expansion valve* or *throttle* rather than an ideal expander. It also features a nonideal compression. On a temperature–entropy diagram, as shown in Figure 24.8b, the compression process now features an increase in entropy. The expansion also features an increase in entropy. On the pressure–enthalpy diagram in Figure 24.8c, the nonideal compression features a greater enthalpy change between Points 2 and 3 when compared with an isentropic compression. The expansion between Points 4 and 1 is now assumed to be isenthalpic, leading to a vertical line on the pressure–enthalpy diagram.

Figure 24.9a shows the corresponding simple compression cycle, but with subcooled condensate. The cycle is basically the same as before, except that the liquid leaving the condenser is now

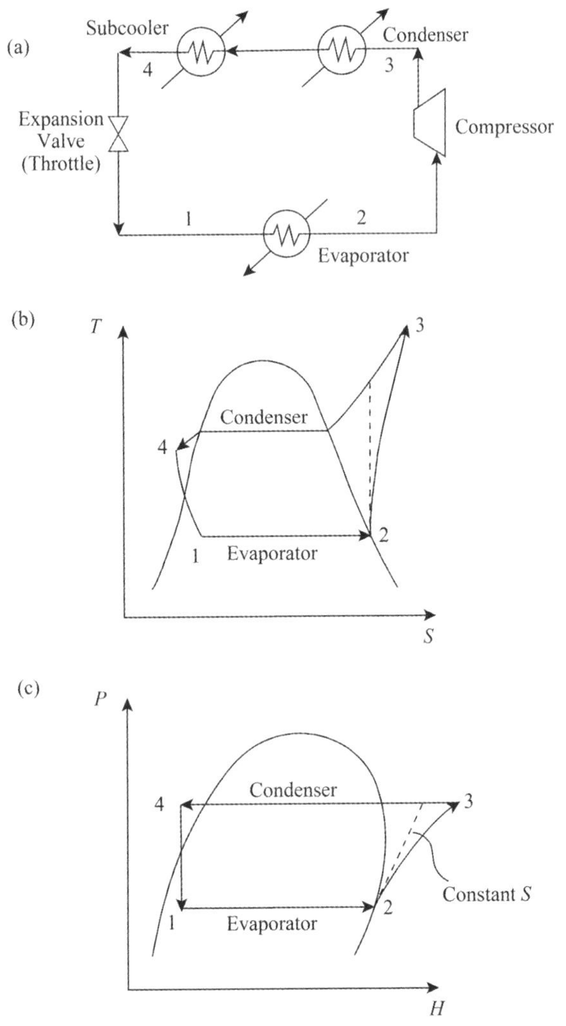

Figure 24.9

Simple compression cycle with subcooled condensate.

subcooled rather than being saturated. The subcooling effect can be seen on the temperature–entropy diagram in Figure 24.9b and the pressure–enthalpy diagram in Figure 24.9c. Subcooling the liquid in the way shown in Figure 24.9 brings a benefit to the cycle. The fact that the liquid is subcooled when entering the expansion process means that there is less vaporization in the expansion process to produce refrigerant at the required temperature. This leads to an increase in the proportion of the liquid refrigerant at Point 1 entering the evaporator. The liquid refrigerant provides the cooling and the vapor present does not provide a cooling effect as the vapor enters and leaves at the same temperature without change in enthalpy. If a greater proportion of the refrigerant entering the evaporator is liquid, it means that the flowrate around the cycle for a given refrigeration duty can be decreased. This, in turn, reduces the compression duty. Subcooling the liquid in this way therefore increases the overall efficiency of the cycle.

In Figures 24.7a, 24.8a and 24.9a, the expansion of the liquid refrigerant produces a two-phase vapor–liquid mixture that is fed directly to the evaporator to provide process cooling. For a vapor–liquid mixture of a pure refrigerant at saturated conditions, the vapor provides no cooling, as it enters and leaves the evaporator at the same temperature without change in enthalpy (neglecting any pressure drop). Only the vaporization of the liquid produces cooling. Some designs of refrigeration use a vapor–liquid separation drum after the expansion of the refrigerant. The vapor from the separation drum is then fed directly to the compressor. The liquid from the separation drum is fed to the evaporator and then to the compressor after vaporization. This arrangement with a vapor–liquid separator is thermodynamically equivalent to the design without a separator (neglecting pressure drops). The separation drum adds capital cost and complexity to the design without bringing any thermodynamic benefit. Whether such a separation drum is used largely depends on the type of heat exchanger used for the evaporator. For example, if the evaporation takes place on the shell side of a kettle reboiler, then feeding a two-phase vapor–liquid mixture does not present any problems and a vapor–liquid separator would not normally be justified. However, if the evaporation takes place in the channels of a plate-fin heat exchanger, then feeding a two-phase vapor–liquid mixture would present problems, as it is difficult to distribute a two-phase mixture evenly across a manifold of channels. Under such circumstances, a vapor–liquid separation drum would normally be used.

The performance of refrigeration cycles is measured as a *coefficient of performance* (COP_{REF}), as illustrated in Figure 24.10. The coefficient of performance is the ratio of cooling duty performed per unit of power required.

$$COP_{REF} = \frac{Q_C}{W} \tag{24.13}$$

where Q_C = cooling duty
W = refrigeration power required

The higher the coefficient of performance, the more efficient is the refrigeration cycle. An ideal coefficient of performance can be defined by:

$$\text{Ideal } COP_{REF} = \frac{Q_C}{W} = \frac{T_{EVAP}}{T_{COND} - T_{EVAP}} \tag{24.14}$$

where T_{EVAP} = evaporation temperature (K)
T_{COND} = condensing temperature (K)

Actual performance can be predicted by introducing a Carnot Efficiency into Equation 24.14:

$$COP_{REF} = \frac{Q_C}{W} = \eta_C \frac{T_{EVAP}}{T_{COND} - T_{EVAP}} \tag{24.15}$$

where η_C = Carnot Efficiency (–)

The Carnot Efficiency is a function of the physical properties of the working fluid, the cycle configuration, system pressures in the cycle and the compressor performance. The Carnot Efficiency of the refrigeration cycle can be determined by a simulation of the cycle. This will be discussed later. An approximate value of 0.6 can be used for a first estimate of cycle performance.

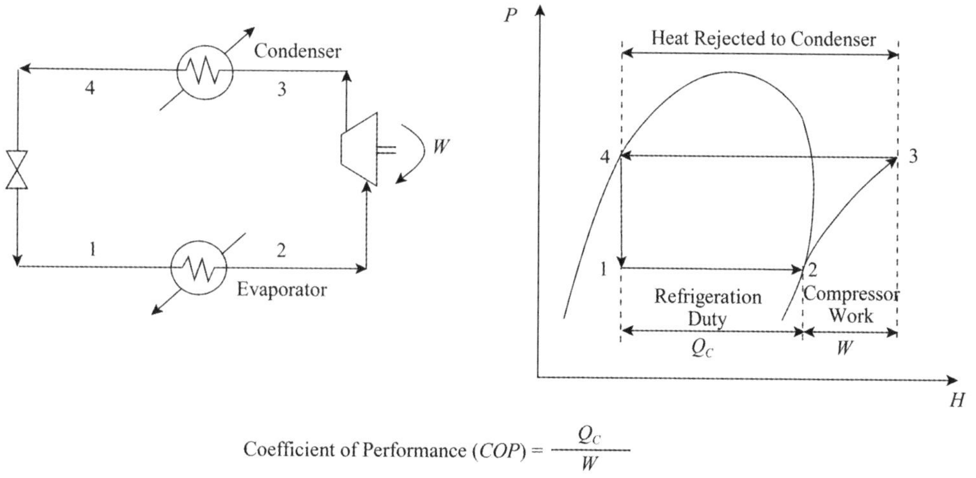

Figure 24.10
Performance of practical refrigeration cycles.

24

It is obvious from Equation 24.15 that the larger the temperature difference across the refrigeration cycle ($T_{COND} - T_{EVAP}$), the lower will be the coefficient of performance and the higher will be the power requirements for a given cooling duty.

Example 24.3 Estimate the coefficient of performance and power requirement for a refrigeration cycle with $T_{EVAP} = -30\,°C$, $T_{COND} = 40\,°C$ and $Q_{EVAP} = 3$ MW.

Solution

$$COP_{REF} \approx 0.6\frac{243}{(313-243)} = 2.08$$

$$W \approx \frac{3}{2.08} = 1.4\ \text{MW}$$

Simple cycles like the ones discussed so far can be used to provide cooling as low as typically −40 °C. For lower temperatures, complex cycles are normally used.

One way to reduce the overall power requirement of a refrigeration cycle is to introduce multistage compression and expansion, as shown in Figure 24.11a. The expansion is carried out in two stages with a vapor–liquid separator between the two stages, often called an *economizer*. Vapor from the economizer passes directly to the high-pressure compression stage. Liquid from the economizer passes to the second expansion stage. The introduction of an economizer reduces the vapor flow in the low-pressure compression stage. This reduces the overall power requirement. Figure 24.11a also shows an intercooler for the vapor between the low-pressure and the high-pressure compression stages. This intercooler reduces further the compressor power in the high-pressure compression stage, reducing the overall power requirement. The cycle is shown as a pressure–enthalpy diagram in Figure 24.11b.

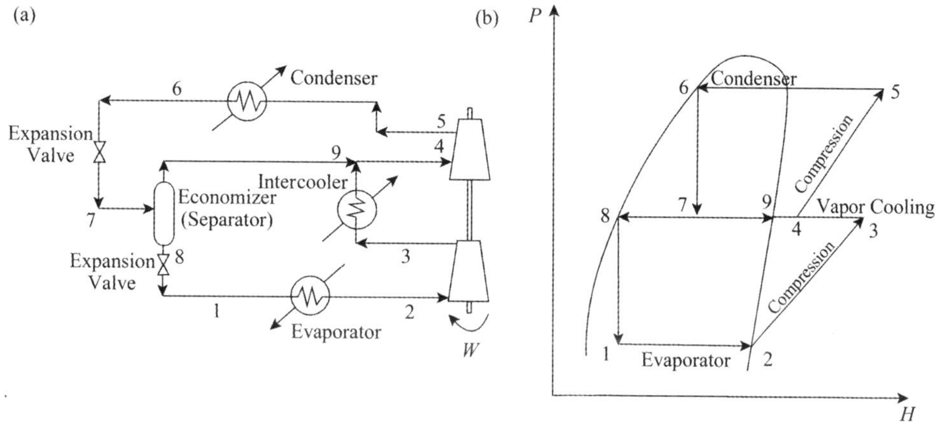

Figure 24.11
Multistage compression and expansion with an economizer.

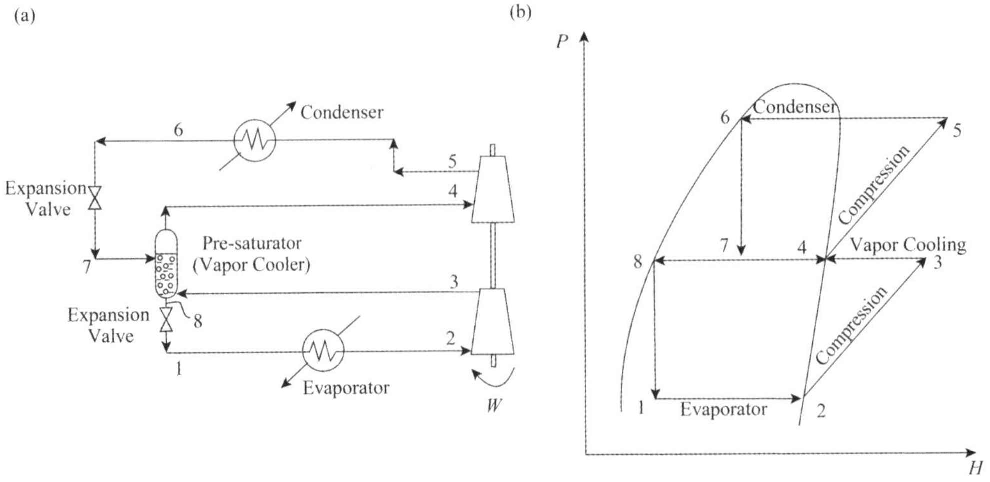

Figure 24.12
Multistage compression and expansion with a presaturator.

Figure 24.12a shows another way to reduce the overall power requirement of a refrigeration cycle by introducing multistage compression and expansion. Again the expansion is carried out in two stages with a vapor–liquid separator between the two. However, this time the cooled liquid and vapor from the first expansion stage is contacted directly with the compressed vapor from the low-pressure compression stage in a *presaturator*. The presaturator cools the vapor from the low-pressure compression stage by direct contact and acts as a vapor–liquid separator between the stages. The vapor leaving the presaturator being passed to the high-pressure compression stage is saturated. Liquid from the presaturator passes to the second expansion stage. Again, the introduction of a presaturator reduces the vapor flow in the low-pressure compression stage, reducing the overall power requirement. The cycle is shown in Figure 24.12b as a pressure–enthalpy diagram.

Figure 24.13a shows a *cascade cycle*. This involves two refrigerant cycles linked together. Each cycle will use a different refrigerant fluid. The low-temperature cycle provides the process cooling in the evaporator and rejects its heat to the other cycle, which rejects its heat in the condenser to cooling water. Cascade cycles are used to provide very low temperature refrigeration where a single refrigerant fluid would not be suitable to operate across such a wide temperature range. In the cascade in Figure 24.13a, each cycle comprises a simple cycle. The cascade cycle is represented in Figure 24.13b in a pressure–enthalpy diagram. The temperature

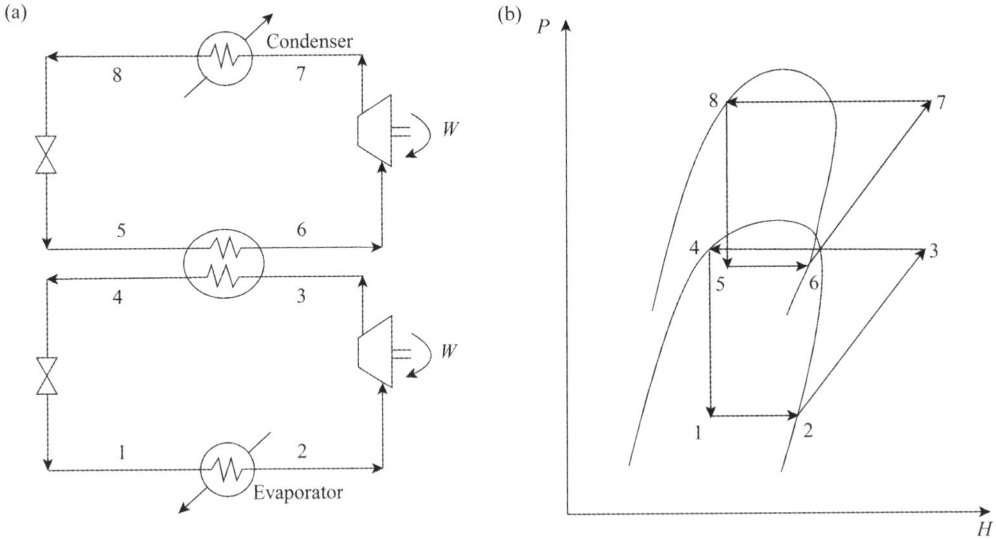

Figure 24.13
A cascade cycle.

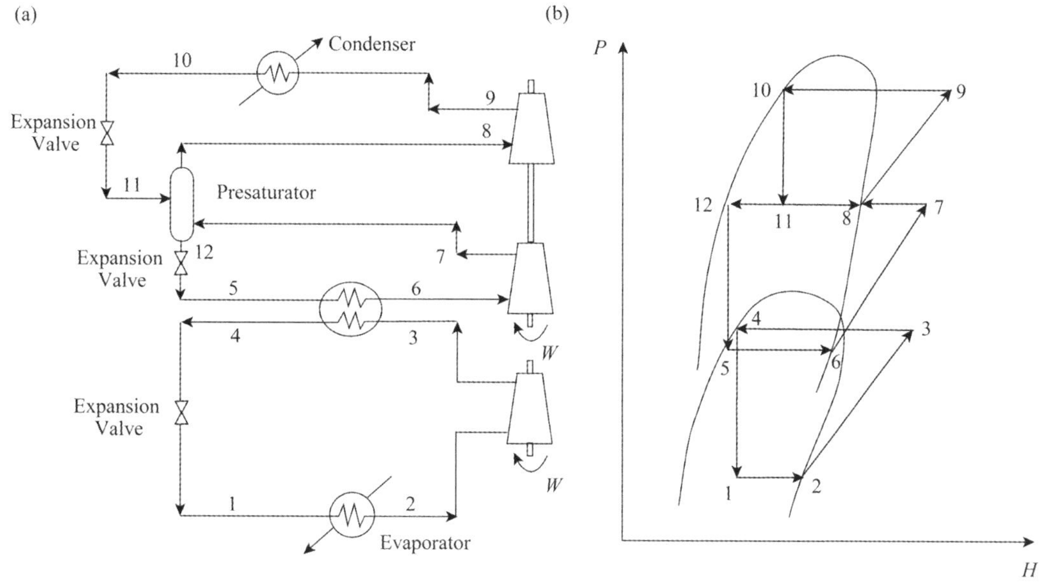

Figure 24.14
Cascade cycle with multistage compression and expansion.

of the interface between the two cycles is normally specified in terms of the temperature of the evaporator of the upper (high-temperature) cycle and is known as the *partition temperature*. The partition temperature is an important degree of freedom.

Complex refrigeration cycles can be built up in different ways. Figure 24.14 illustrates an example of a cascade cycle with multistage compression and expansion of the high-temperature cycle.

Figure 24.15a shows a *two-level* refrigeration cycle. Level 1 and Level 2 provide process cooling at different temperatures. This reduces the refrigerant flow through the low-pressure compression stage and reduces the overall power requirements. It is represented in Figure 24.15b in a pressure–enthalpy diagram.

To illustrate one of the many ways the different features can be put together, Figure 24.16a shows a two-level cascade system with presaturators. The evaporators in Level 1 and Level 2 provide process cooling at different temperatures. Two cycles are cascaded together, each separate cycle having multistage compression involving presaturators. The cycle is shown in a pressure–enthalpy diagram in Figure 24.16b. This is only one of many possible complex arrangements that combine the various features.

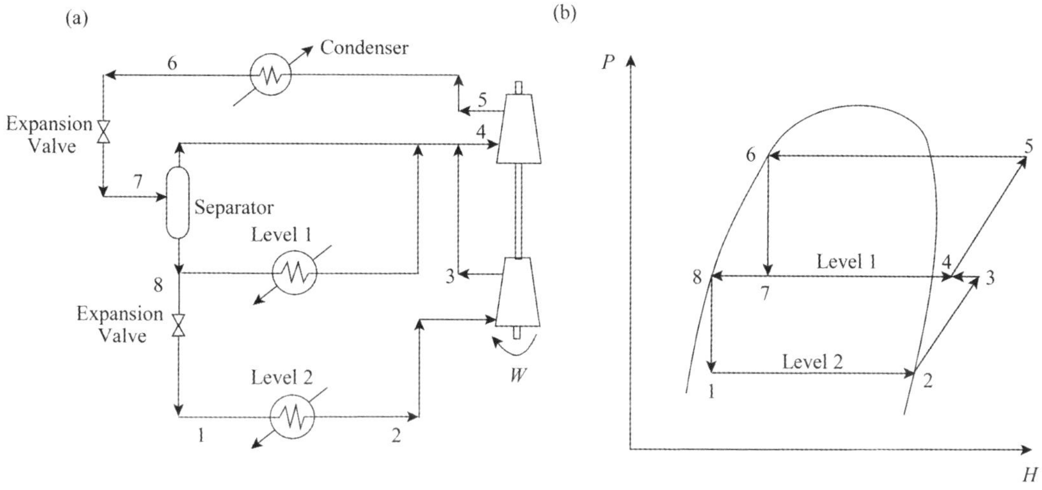

Figure 24.15
A two-level refrigeration cycle.

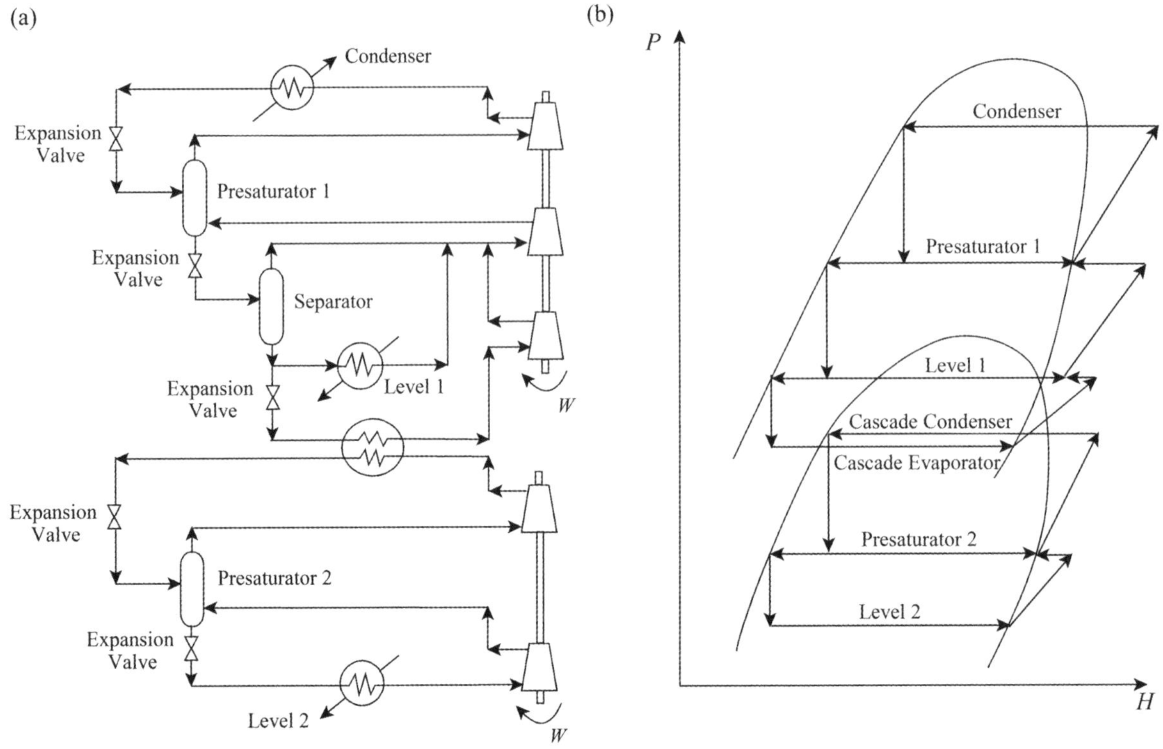

Figure 24.16
A two-level cascade system with presaturators.

24.6 Choice of a Single-Component Refrigerant for Compression Refrigeration

Now consider the main factors affecting the choice of refrigerant for compression refrigeration, but for now restricting consideration to single components.

1) *Freezing point.* First, the evaporator temperature should be well above the freezing temperature at the operating pressure. The freezing points of some common refrigerants at atmospheric pressure are given in Table 24.4.
2) *Vacuum operation.* The second consideration, illustrated in Figure 24.17, is that, at the evaporator temperature, evaporator pressure below atmospheric pressure should be avoided. An evaporator pressure above atmospheric avoids potential problems with the ingress of air into the cycle, which can cause performance and safety problems. However, special designs can use evaporator pressures below atmospheric. The boiling points of some common refrigerants are given in Table 24.4 at atmospheric pressure.
3) *Latent heat of vaporization.* It is desirable to have a refrigerant with a high latent heat. A high latent heat will lead to a lower flowrate of refrigerant around the loop and reduce the power requirements correspondingly. The shape of the two-phase envelope is such that as the temperature of the refrigerant increases, the latent heat decreases as the critical temperature is approached. It is therefore desirable to have evaporator and condenser temperatures significantly below the critical

Table 24.4
Freezing and normal boiling points for some common refrigerants.

Refrigerant	Freezing point at atmospheric pressure (°C)	Boiling point at atmospheric pressure (°C)
Ammonia	−78	−33
Chlorine	−101	−34
n-Butane	−138	0
i-Butane	−160	−12
Ethylene	−169	−104
Ethane	−183	−89
Methane	−182	−161
Propane	−182	−42
Propylene	−185	−48
Nitrogen	−210	−196

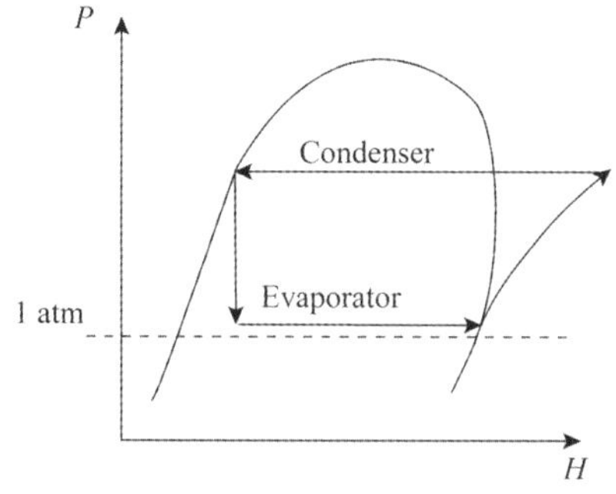

Figure 24.17

Choice of refrigerant for pressure.

temperature. Also, as the condenser temperature approaches the critical point, a greater portion of the heat is extracted in the desuperheating, relative to condensation, as illustrated in Figure 24.18. This reduces the coefficient of performance, as it increases the average heat rejection temperature. It also increases the heat transfer area in the condenser. Given that it is desirable to operate away from the critical point, where the latent heat of vaporization is high for a given refrigerant fluid, how close would it be desirable to go to the critical temperature? The change of latent heat with temperature can be correlated using the Watson Equation (Poling, Prausnitz and O'Connell, 2001):

$$\frac{\Delta H_{VAP2}}{\Delta H_{VAP1}} = \left(\frac{T_C - T_2}{T_C - T_1}\right)^{0.38} \tag{24.16}$$

where ΔH_{VAP1}, ΔH_{VAP2} = heat of vaporization at temperatures T_1 and T_2 respectively
T_1, T_2 = absolute temperature (K)
T_C = critical temperature (K)

Writing Equation 24.16 between the normal boiling point T_{BPT} and temperature T:

$$\frac{\Delta H_{VAP}}{\Delta H_{VAP,BPT}} = \left(\frac{T_C - T}{T_C - T_{BPT}}\right)^{0.38} \tag{24.17}$$

where $\Delta H_{VAP,BPT}$ = heat of vaporization at the normal boiling point

Let λ be the ratio of the heats of vaporization, such that from Equation 24.17

$$\lambda = \left(\frac{T_C - T}{T_C - T_{BPT}}\right)^{0.38} \tag{24.18}$$

Rearranging Equation 24.18 gives:

$$T = T_C - \lambda^{2.63}(T_C - T_{BPT}) \tag{24.19}$$

Thus, for a given refrigerant with normal boiling point T_{BPT} and critical temperature T_C, Equation 24.19 allows the maximum temperature to be fixed, if the minimum value of λ is specified. For example, for ethylene, $T_C = 282$ K and $T_{BPT} = 169$ K. If the minimum heat of vaporization in the evaporator is specified to be no lower than 50% of that at the normal boiling point, then the maximum temperature is given by:

$$\begin{aligned} T &= 282 - 0.5^{2.63}(282 - 169) \\ &= 264 \text{ K} \end{aligned}$$

The value of λ is at the discretion of the designer and more conservative values of 60% or 70% could be taken. A value of 50% is probably as low as would be desirable for many applications.

4) *Shape of the two-phase envelope.* Another factor affecting the choice of refrigerant relates to the shape of the two-phase region on a temperature–entropy diagram, as illustrated in Figure 24.19, which shows two different two-phase regions. The important difference is the slope of the saturated vapor line to the right of the two-phase region. Figure 24.19b shows a

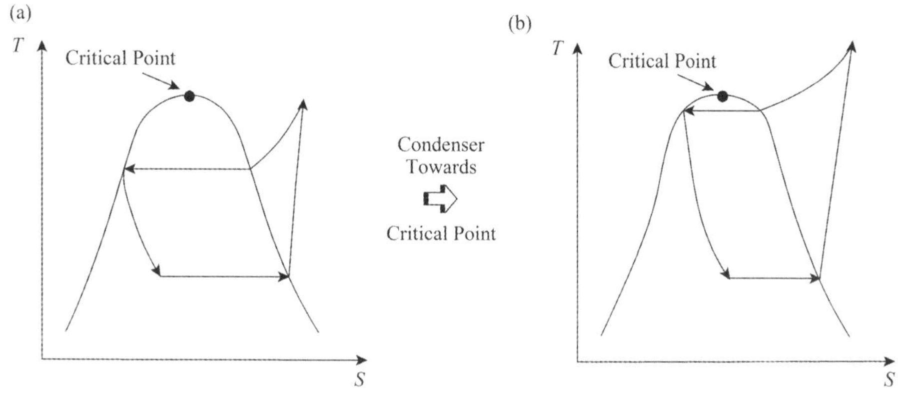

Figure 24.18

Choice of refrigerant-critical point.

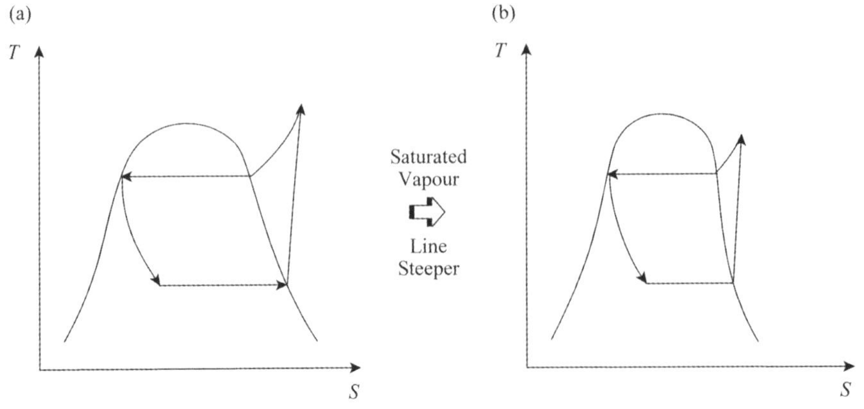

Figure 24.19

Choice of refrigerant-saturated vapor line.

two-phase region with a steep slope relative to the one in Figure 24.19a. If the slope is steep, as is the case with the two-phase region in Figure 24.19b, this reduces superheat and decreases heat extracted in desuperheating relative to condensation. In turn, this increases the coefficient of performance, as it decreases the average heat rejection temperature. It also decreases the heat transfer area.

5) *General considerations.* The refrigerant should, as much as possible, be nontoxic, nonflammable, noncorrosive and have low ozone depletion potential and a low global warming potential (see Chapter 25). Fluids that are already present in the process should be used where possible. Introducing new materials for refrigeration introduces new storage requirements and new safety and environmental problems.

Example 24.4 Figure 24.20 gives the operating ranges of a number of refrigerants. The upper limit of the operating range in Figure 24.20 has been set to ambient temperature if the critical temperature is well above ambient temperature or to a temperature corresponding with a heat of vaporization of 50% of that at atmospheric pressure. The lower temperature in Figure 24.20 has been limited by the normal boiling point to avoid vacuum conditions in the refrigeration loop. It should be noted that the operating range of the refrigerants can be extended at the lower limit by operation under vacuum conditions and at the upper limit by relaxing the restriction that the latent heat should not be lower than 50% of that at atmospheric pressure. The refrigerants in Figure 24.20 have been placed in approximate order of power requirement for a given refrigeration duty. Four process streams given in Table 24.5 require refrigeration cooling. Given the information in Figure 24.20, make an initial choice of refrigerants for each of the streams in Table 24.5.

Solution

Stream 1. At 175 K, the only suitable refrigerant from Figure 24.20 is ethylene. Methane is too close to its upper limit. If ethylene is chosen as a refrigerant and heat rejection is required to ambient at 313 K, then it needs to be cascaded with another refrigerant for heat rejection to ambient. In principle, ethylene could be cascaded with chlorine, ammonia, *i*-butane, propylene or propane. However, the overlap with *i*-butane is small and will require a small temperature difference in the heat exchanger linking the two cycles.

Table 24.5

Process streams to be cooled by refrigeration.

Stream no.	Temperature (K)
1	175
2	200
3	230
4	245

The choice of refrigerant for the cascade will depend on a number of factors. Choosing chlorine will minimize the power for the system, but introduces significant safety problems. Choosing a component already in the process would be desirable in order to avoid introducing new safety problems (e.g. propylene).

Stream 2. At 200 K, either ethane or ethylene would be suitable refrigerants, with ethane being slightly better from the point of view of the power requirements. As for Stream 1, a cascade system would be required for heat rejection to ambient temperature.

Stream 3. At 230 K, propylene, ethane and ethylene would all be suitable refrigerants. Given that propylene requires the lowest power and does not require a cascade for heat rejection to ambient, this would be a suitable choice.

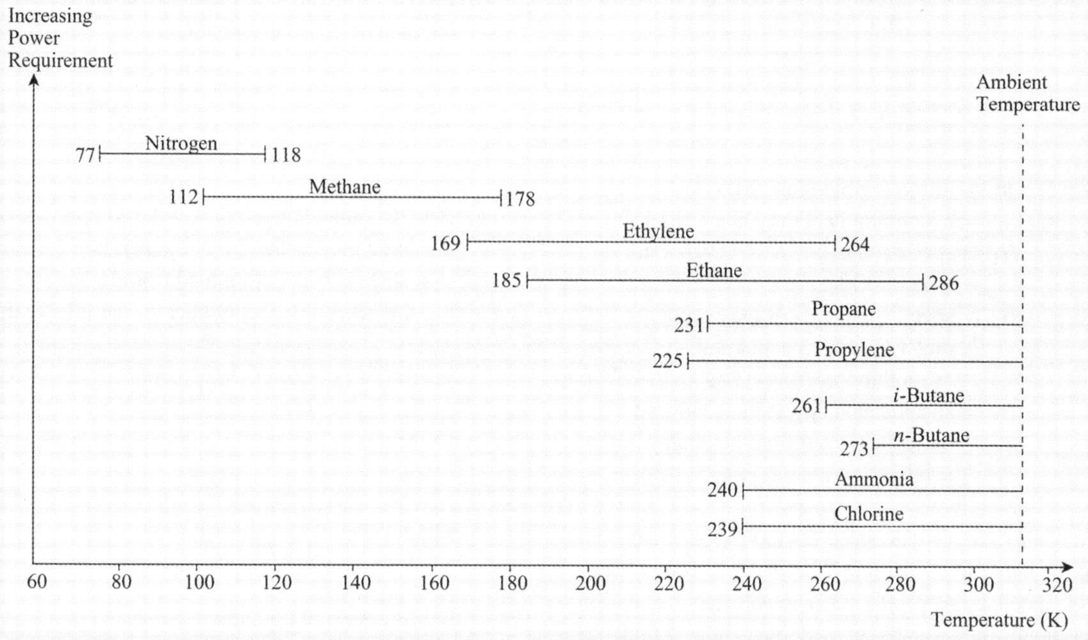

Figure 24.20

Operating ranges of refrigerants.

Stream 4. At 245 K, chlorine, ammonia, propylene and propane could all be chosen. In principle, ethane and ethylene could also have been included but at 245 K they are too close to their critical temperature and would require significantly higher refrigeration power than the other options. The safety problems associated with chlorine are likely to be greater than ammonia. Thus, ammonia might be a suitable choice of refrigerant. Choosing a component already in the process would be desirable.

24

24.7 Targeting Refrigeration Power for Pure Component Compression Refrigeration

To evaluate the impact of each design option, a quick and reliable method for estimating refrigeration power requirements can be very useful (Lee, Zhu and Smith, 2000; Lee, 2001). The benefits from setting targets for refrigeration power are to:

- evaluate refrigeration power requirements before complete design;
- assess the performance of the whole process prior to detailed design;
- allow many alternative design options to be screened quickly and estimated reliably;
- assess energy and capital costs.

1) *Simple cycles.* Consider a simple cycle as shown in Figure 24.8. The problem is how to estimate the actual power requirement, given the cooling duty (Q_{EVAP}), condensing temperature (T_{COND}) and evaporating temperature (T_{EVAP}).

The calculation procedure is performed as follows (Lee, Zhu and Smith, 2000; Lee, 2001):

a) Find the vapor pressure of the refrigerant in the condenser P_{COND}^{SAT} at T_{COND} and the vapor pressure in the evaporator P_{EVAP}^{SAT} at T_{EVAP} using a correlation for the saturated liquid vapor pressure, such as the Antoine equation (see Appendix A):

$$\ln P^{SAT} = A - \frac{B}{C + T} \tag{24.20}$$

where P^{SAT} = saturated liquid vapor pressure (bar)
A, B, C = constants determined experimentally for each refrigerant
T = temperature (K)

Once these pressures have been set, this sets the pressure difference across the compressor and the throttle if the pressure drops in the heat exchangers, flash drums and the pipework are neglected.

b) An energy balance around the evaporator gives the mass flowrate of refrigerant. Referring to Figure 24.8:

$$m = \frac{Q_{EVAP}}{H_2 - H_1} = \frac{Q_{EVAP}}{H_2 - H_4} \qquad (24.21)$$

where
m = refrigerant flowrate (kg·s^{-1})
Q_{EVAP} = evaporator heat duty (J·s^{-1}, kJ·s^{-1})
H_1 = specific enthalpy at the evaporator inlet (J·kg^{-1}, kJ·kg^{-1})
H_2 = specific enthalpy at the evaporator outlet (saturated vapor enthalpy at the evaporator pressure) (J·kg^{-1}, kJ·kg^{-1})
H_4 = specific enthalpy at the condenser outlet (saturated liquid enthalpy at the condenser pressure) (J·kg^{-1}, kJ·kg^{-1})

The enthalpy can be calculated for most refrigerant fluids using the departure function of the Peng–Robinson Equation of State (see Appendix A). The enthalpy calculations are complex and usually carried out using commercial physical property software packages. However, given the capability to determine the enthalpy, Equation 24.21 allows the refrigerant mass flowrate to be determined from a knowledge of the evaporator and condenser pressures. Once the mass flowrate is known, the volumetric flowrate into the compressor can be obtained from the vapor density, which can again be obtained from an equation of state such as the Peng–Robinson Equation. Thus:

$$F = \frac{m}{\rho_V} \qquad (24.22)$$

where
F = volumetric flowrate (m^3·s^{-1})
ρ_V = vapor density (kg·m^{-3})

c) Estimate the compressor efficiency. Compressor manufacturer's data should be used where possible. If this is not available, then for a reciprocating compressor, the isentropic efficiency can be estimated from Equation 13.69:

$$\eta_{IS} = 0.1091(\ln r)^3 - 0.5247(\ln r)^2 + 0.8577 \ln r + 0.3727 \quad 1.1 < r < 5 \qquad (13.69)$$

where
η_{IS} = isotropic efficiency
r = pressure ratio P_{out}/P_{in}

For a centrifugal compressor, the polytropic efficiency can be estimated from Equation 13.70:

$$\eta_P = 0.017 \ln F + 0.7 \qquad (13.70)$$

where
η_P = polytropic efficiency
F = inlet flowrate of gas (m^3·s^{-1})

d) For centrifugal compressors, estimate the polytropic coefficient n (see Chapter 13):

$$n = \frac{\gamma\eta_P}{\gamma\eta_P - \gamma + 1} \qquad (13.60)$$

e) Find the outlet temperature of the compressor (T_{out}). For an adiabatic model, the outlet temperature for each compression stage can be determined from Equation 13.47, assuming the intercoolers cool back to the inlet temperature:

$$T_{out} = \left(\frac{T_{EVAP}}{\eta_{IS}}\right)\left[(r)^{(\gamma-1)/\gamma} + \eta_{IS} - 1\right] \qquad (24.23)$$

where $r = \sqrt[N]{\left(\frac{P_{COND}^{SAT}}{P_{EVAP}^{SAT}}\right)}$

N = number of compressor stages

For a polytropic model, the outlet temperature can be determined from Equation 13.52:

$$T_{out} = T_{EVAP}(r)^{(n-1)/n} \qquad (24.24)$$

The polytropic coefficient n can be determined from Equation 13.60 if η_P is known.

f) An energy balance around the compressor gives:

$$W = m(H_2 - H_3) \qquad (24.25)$$

where
W = power required for compression (kJ·s^{-1} = kW)
m = refrigerant flowrate (kg·s^{-1})
H_2 = specific enthalpy at the compressor inlet P_{EVAP}, T_{EVAP} (kJ·kg^{-1})
H_3 = specific enthalpy at the compressor outlet P_{COND}, T_{out} (kJ·kg^{-1})

The compressor inlet and outlet enthalpies can be calculated using the departure function of an equation of state (see Appendix A) using a commercial physical property software package. Thus, the refrigeration power requirement of the compressor for this simple vapor-compression cycle is obtained. Alternatively, rather than perform an energy balance around the compressor, the refrigeration power can be calculated directly for a reciprocating compressor from Equation 13.65:

$$W = \left(\frac{\gamma}{\gamma - 1}\right)\frac{P_{EVAP}F_{in}N}{\eta_{IS}}\left[1 - (r)^{(\gamma-1)/\gamma}\right] \qquad (24.26)$$

where W = power required for compression (N·m·s^{-1} = J·s^{-1} = W)

P_{EVAP}, P_{COND} = inlet and outlet pressures for the compressor ($N{\cdot}m^{-2}$)
F_{in} = inlet volumetric flowrate ($m^3{\cdot}s^{-1}$)
γ = ratio of heat capacities C_P/C_V (–)
η_{IS} = isentropic efficiency (–)

$$r = \sqrt[N]{\left(\frac{P_{COND}^{SAT}}{P_{EVAP}^{SAT}}\right)}$$

N = number of compressor stages

For a centrifugal or axial compressor, the refrigeration power can be calculated directly from Equation 13.67:

$$W = \frac{n}{n-1}\frac{P_{EVAP}F_{in}N}{\eta_P}\left[1-(r)^{(n-1)/n}\right] \qquad (24.27)$$

where η_P = polytropic efficiency (ratio of polytropic power to actual power)

2) *Multistage cycles.* The calculation procedure for targeting refrigeration power for simple cycles can be extended to multistage cycles, such as the one shown in Figure 24.21a. In order to be able to apply the procedure, the temperature after the mixing junction between Levels 1 and 2 (T_{MIX} in Figure 24.21a) must be determined. T_{MIX} is not at saturated conditions, as a result of the superheat from the low-pressure (Level 2) compressor. The effect of the mixing between stages in a multistage cycle, leading to superheated conditions at the inlet to the high-pressure compression stage, can be estimated by correcting the temperature between the stages by weighting according to mass flows (Lee, Zhu and Smith, 2000; Lee, 2001). An energy balance around the mixing junction in Figure 24.21a assuming the heat capacity of the vapor to be constant gives (Lee, Zhu and Smith, 2000; Lee, 2001):

$$(m_1 + m_2)T_{MIX} = m_2T_{out} + m_1T_{EVAP1} \qquad (24.28)$$

where m_1, m_2 = mass flowrate for the refrigerant in Levels 1 and 2 ($kg{\cdot}s^{-1}$)
T_{MIX} = temperature after the mixing junction (K)
T_{out} = outlet temperature of the compressor for Level 2 (low-pressure compressor) (K)
T_{EVAP1} = evaporator temperature for Level 1 (K)

Rearranging Equation 24.28 gives:

$$T_{MIX} = \frac{m_2T_{out} + m_1T_{EVAP1}}{m_1 + m_2} \qquad (24.29)$$

The outlet temperature of the high-pressure compressor in Figure 24.21a is found by using Equation 24.23 or 24.24 with T_{EVAP} replaced by T_{MIX}. Thus, mass flow weighting of flows around mixing junctions can be used to extend the procedure for refrigeration power targeting to complex cycles.

However, attempting to target the power requirements for complex problems in this way is very restrictive. It requires the configuration of the refrigeration system to be specified. The more complex the problem, the more options available in the design of the refrigeration cycle (or cycles). What is required is an approach that allows the interactions between the design of

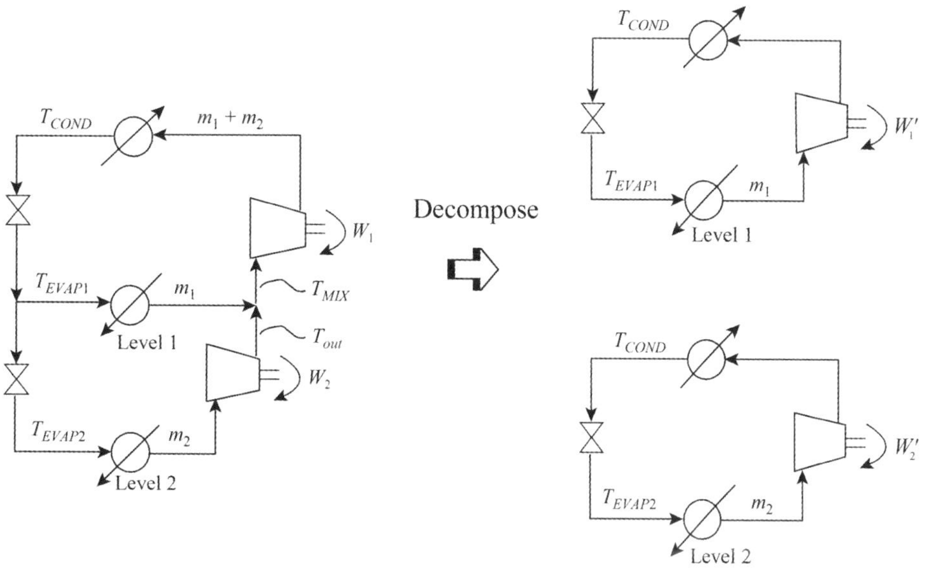

(a) A two-stage refrigeration cycle.

(b) Decomposition of a two-stage cycle into two simple cycles.

Figure 24.21
Targeting refrigeration work for a multistage cycle.

the process and the refrigeration to be explored without restricting options.

A multistage cycle, such as the one shown in Figure 24.21a, can be decomposed into an assembly of simple vapor-compression cycles in order to estimate the power requirements. Figure 24.21b shows the two-stage cycle being modeled as two simple cycles operating in parallel to estimate the power requirements. One simple cycle operates between Level 1 and ambient with cooling load given by the evaporator for Level 1. The other simple cycle operates between Level 2 and ambient with cooling load given by the evaporator for Level 2.

Thus, for the purposes of refrigeration power targeting for screening options, the power requirements can be estimated from an assembly of simple cycles. Rather than reject heat to ambient, heat rejection from the refrigeration cycle to the process or to another refrigeration cycle can also be allowed for, as will be discussed later. As the design evolves, the simple cycles used for targeting can be merged together to produce complex cycles. As the simple cycles are merged, a number of effects can change the power of the merged design to differ from that predicted by an assembly of simple cycles.

- Complex cycles can introduce nonisothermal mixing in the cycle, which increases the power requirements. In multistage cycles, such as the one shown in Figure 24.21a, the high-pressure compressor is subjected to superheated inlet conditions, which is a result of mixing of the saturated stream from the low-pressure (Level 1) evaporator with the superheated compressor outlet from the low-pressure (Level 2) compression stage. The superheated inlet conditions to the high-pressure compression stage lead to an overall increase in the power requirements when compared with two simple cycles operating in parallel, as in Figure 24.21b.
- The introduction of vapor–liquid separators (economizers) acts to reduce the power requirements. For example, if a vapor–liquid separator is introduced into the cycle in Figure 24.21a, as in Figure 24.15, vapor from the first (Level 1) throttle is not passed to the low-pressure (Level 2) compressor. This reduces the load on the low-pressure compressor and the overall power requirements.
- If compressor loads can be merged, such that a larger compressor is required, then the larger compressor is likely to have a higher efficiency than the smaller compressors used in simple cycles.

This means that the refrigeration power targets calculated by decomposing the refrigeration system into an assembly of simple cycles might be lower or higher than the final refrigeration design, depending on the problem and the options chosen by the designer. Complex cycles are likely to be cheaper in capital cost than a collection of simple cycles. Thus, in determining the final design for the refrigeration, there are difficult trade-offs to be considered, involving operating cost, capital cost, complexity, operability and safety.

Example 24.5 Following Example 24.3, calculate the target for refrigeration power for a cooling duty of 3 MW with an evaporator temperature of −30 °C and refrigeration condenser temperature of 40 °C using ammonia, propane and propylene. Assume that reciprocating compressors are to be used. Enthalpies can be calculated from the Peng–Robinson Equation of State.

Solution From saturated liquid–vapor pressure data, for an evaporator temperature of −30 °C and a condenser temperature of 40 °C, the following pressures are obtained.

	Pressure (bar)		
	Ammonia	**Propylene**	**Propane**
Evaporator	1.2009	2.1191	1.6815
Condenser	15.548	16.431	13.718
Pressure ratio	12.95	7.75	8.16

The values of the heat capacity ratio γ can be obtained from physical property data at average conditions.

	Ammonia	**Propylene**	**Propane**
γ	1.30	1.13	1.16

From the values of γ and the pressure ratios, it is clear that the pressure ratio for ammonia is very high for a single-stage compression. The guidelines for gas compression given in Chapter 13 indicate that propylene and propane compressions should be able to be achieved in a single compression stage, but the ammonia would seem to require two compression stages with intercooling. However, in this case, the gas entering the compressor is at a low temperature, which results in a correspondingly lower outlet temperature from the compressor. This would need to be examined in some detail. Indeed, if a two-stage compression is introduced, it would be sensible to consider using an economizer, together with intercooling, to reduce vapor flow in the low-pressure compression stage. In this case, the ammonia compression will be assumed to be carried out in a single stage so as to compare with the other refrigerants on a common basis.

Calculation of the power requirements for the three refrigerants requires the flowrate to be calculated for a duty of 3 MW. This can be calculated from the enthalpy difference across the evaporator $(H_2 - H_1)$. The enthalpy difference across the evaporator is assumed to be the difference between the saturated vapor enthalpy at the evaporator pressure and the saturated liquid enthalpy at the condenser pressure. This assumes no subcooling of the refrigerant.>

	Ammonia	**Propylene**	**Propane**
$H_2 - H_1\,(\text{kJ}\cdot\text{kg}^{-1})$	1047	245.8	229.33
$m = \dfrac{Q_{EVAP}}{(H_2 - H_1)}\,(\text{kg}\cdot\text{s}^{-1})$	2.865	12.21	13.10
$\rho_V\,(\text{kg}\cdot\text{m}^{-3})$	1.007	4.605	3.850
$F_{in} = \dfrac{m}{\rho_V}\,(\text{m}^3\cdot\text{s}^{-1})$	2.845	2.650	3.402

The compressor efficiency (in this case, isentropic efficiency) can now be estimated and in turn the outlet temperature and compressor power.

	Ammonia	Propylene	Propane
η_{IS} (−) (from Equation 13.69)	0.960	0.866	0.870
T_{out} (K) (from Equation 24.23)	447	318	337
W (MW) (from Equation 24.26)	1.24	1.50	1.60

From these results, the high temperature of the ammonia from the compressor outlet should be noted. This reinforces the points made earlier regarding ammonia compression.

In principle, ammonia is the best refrigerant fluid in terms of power requirement. However, this conclusion disregards the potential practical problems associated with compression. There is little to choose between propylene and propane in terms of the power requirements.

It is also interesting to compare the more detailed calculations performed here with the approximate calculation in Example 24.3. The power was estimated to be 1.4 MW on the basis of the *COP* for an ideal cycle with an assumed 60% efficiency.

24.8 Heat Integration of Pure Component Compression Refrigeration Processes

Figure 24.22a shows an example of a refrigeration cycle matched against a grand composite curve. Heat is rejected from the process into the refrigerant at the lowest temperature. In Figure 24.22a, the heat rejection from the refrigeration cycle is to cooling water. Closer inspection of the grand composite curve in Figure 24.22a shows that it has a large requirement for heat at a temperature below cooling water temperature. Thus, Figure 24.22b shows a refrigeration cycle that rejects its heat into the process to reduce the power requirement with a lower temperature lift and to save hot utility.

Figure 24.23 shows the same grand composite curve as in Figure 24.22, but cooled below ambient by a two-level refrigeration system. The grand composite curve is used to determine the heat rejection at each refrigeration level. Using a two-level refrigeration system, as shown in Figure 24.23, reduces the power requirement compared with the single-level arrangement shown in Figure 24.22. This results from the smaller temperature lift required by the higher temperature refrigeration level.

Figure 24.24 shows a much more complex refrigeration cycle matched against the same grand composite curve. Heat is rejected between Points 1 and 2 into the lowest level of refrigeration. Another level of refrigeration is placed between Points 3 and 4. So far, this is the same solution as shown in Figure 24.23. Figure 24.24 shows an additional feature in which some heat is rejected back into the process between Points 5 and 6. Rejecting the heat in the way shown in Figure 24.24 at an intermediate temperature saves power for the rejection of heat that would otherwise have to be lifted above the pinch temperature before rejection. However, the intermediate heat rejection between Points 5 and 6 in Figure 24.24 disturbs the energy balance in the pocket of the grand composite curve, requiring a third level of refrigeration to be introduced between Points 7 and 8 for heat rejection from the process. Going from the single-level refrigeration in Figure 24.22 to the two-level refrigeration in Figure 24.23 and on to the three-level refrigeration system with intermediate heat rejection in Figure 24.24 saves power with each additional level of complexity. However, the capital cost increases as the complexity increases. Thus, whether it is worthwhile to introduce a complex solution as the one shown in Figure 24.24 depends very much on the economy of scale and the trade-off between energy costs and capital costs.

When refrigeration profiles are matched against the grand composite curve, either heat removal from the process or heat rejection to the process, the selection of temperature is not always straightforward. Sometimes, the refrigeration level fits neatly

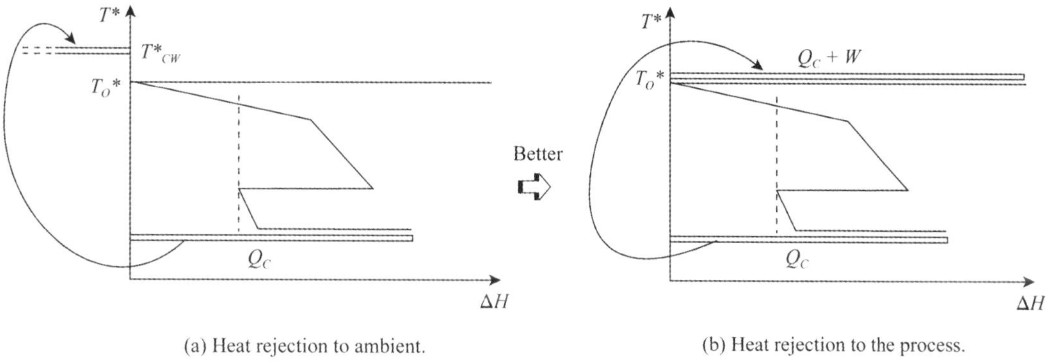

Figure 24.22
Heat integration of a single level refrigeration system.

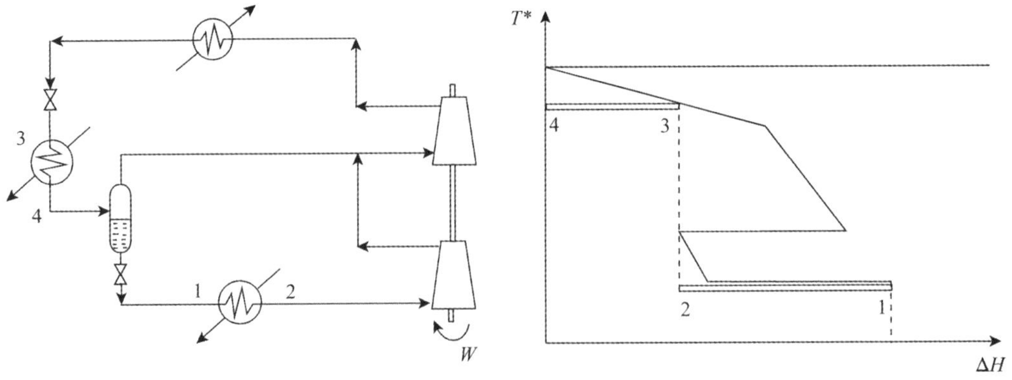

Figure 24.23
Heat integration of a two-level refrigeration system.

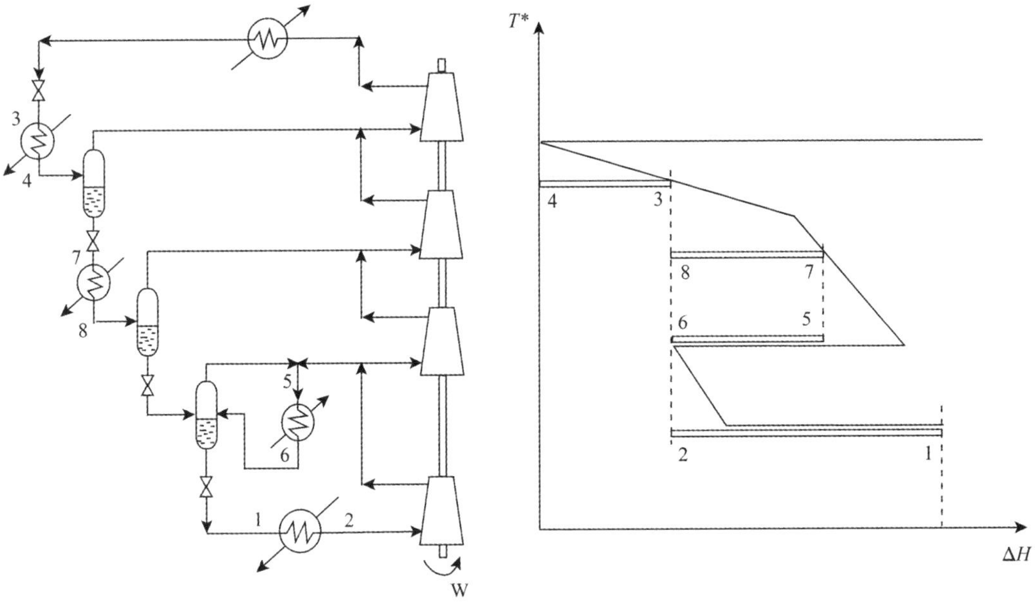

Figure 24.24
Exploiting the pockets in the grand composite curve reduces the overall temperature lift and the shaftwork requirements.

against a flat portion of the grand composite curve (e.g. Figure 24.23 between Points 1 and 2). Sometimes, the refrigeration level needs to be fitted against a sloping section of the grand composite curve, as shown in Figure 24.25. In this case, there is a degree of freedom to choose the level of the refrigeration. For the example in Figure 24.25, there are two levels of refrigeration. As the higher temperature refrigeration level is increased, the heat load and power requirement per unit of heat load both decrease. This will decrease the cost of the power requirements for the higher temperature refrigeration level. However, as the heat load on the higher temperature refrigeration level decreases, the heat load on the lower temperature refrigeration level increases, increasing its power requirement. As the temperature of the higher-level refrigeration decreases, these trends reverse. There is thus a degree of freedom that needs to be optimized to determine the heat duty allocated to each of the two levels of refrigeration. To analyze

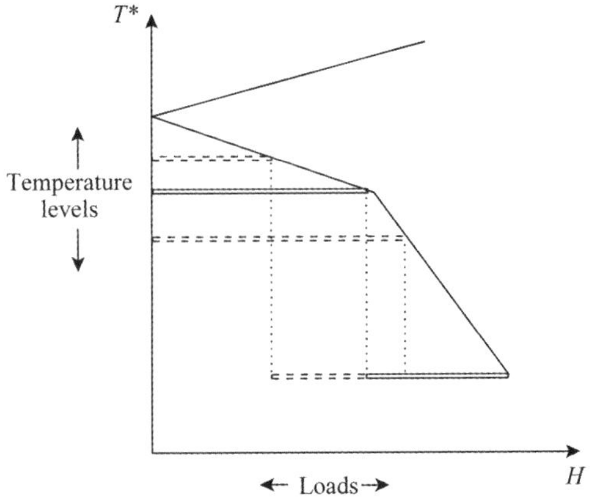

Figure 24.25
Optimization of refrigeration levels.

such a trade-off requires a quantitative prediction of the power required for given refrigeration evaporation and condensing temperatures.

If the refrigeration system involves a cascade cycle, the partition temperature is an important additional degree of freedom to be optimized.

Example 24.6 Determine the refrigeration requirements for the low-temperature distillation process from Example 17.1 shown in Figure 17.19 for $\Delta T_{min} = 5\,°C$.

a) Plot the grand composite curve from the heat cascade given in Figure 17.21b and determine the temperature and duties of the refrigeration if two levels of refrigeration are to be used. Assume that both vaporization and condensation of the refrigerant occur isothermally.

b) Calculate the power requirements for the refrigeration for heat rejection to cooling water operating between 20 and 25 °C approximated by Equation 24.15 with a Carnot Efficiency of 0.6.

c) Repeat the calculation from Part b using refrigeration power targeting, assuming propylene as the refrigerant and a reciprocating compressor.

d) Heat rejection from the refrigeration system into the process can be used to reduce the refrigeration power requirements. Calculate the power using Equation 24.15 with a Carnot Efficiency of 0.6.

e) Repeat the calculation from Part d using refrigeration power targeting, assuming propylene as the refrigerant and a reciprocating compressor.

Solution

a) Figure 24.26a shows a plot of the grand composite curve from the problem table cascade given in Figure 17.21b. Also shown in Figure 24.26a are two refrigeration profiles.

	T^* (°C)	T (°C)	Q_{EVAP} (MW)
Level 1	−24.5	−25	1.04
Level 2	−42.5	−45	(1.84 − 1.04) = 0.8

b) For heat rejection to cooling water:

$$T_H = 25 + 5 = 30\,°C = 303\,K$$

$$W_1 = \frac{1.04}{0.6}\left[\frac{303 - 248}{248}\right] = 0.38\,MW$$

$$W_2 = \frac{0.8}{0.6}\left[\frac{303 - 228}{228}\right] = 0.44\,MW$$

Total power for heat rejection to cooling water = 0.38 + 0.44 = 0.82 MW.

c) Repeat the calculation from Part b but now using refrigeration targeting, assuming propylene as the refrigerant. Designate Cycle 1 to operate between −25 °C and 30 °C and Cycle 2 to operate between −45 °C and 30 °C.

	Pressure (bar)	
	Cycle 1	Cycle 2
Evaporator	2.5239	1.1291
Condenser	13.0081	13.0081
Pressure ratio	5.15	11.52

The pressure ratio is a little high for Cycle 2 to be carried out in a single compression stage. However, single-stage compression will be assumed for the sake of comparison between different options. The heat capacity ratio γ will be assumed to be constant with a value of 1.13.

	Cycle 1	Cycle 2
Q_{EVAP} (MW)	1.04	0.8
$H_2 - H_1\,(kJ \cdot kg^{-1})$	279.4	258.2
$m = \frac{Q_{EVAP}}{(H_2 - H_1)}\,(kg \cdot s^{-1})$	3.722	3.098
$\rho_V\,(kg \cdot m^{-3})$	5.492	2.593
$F_{in} = \frac{m}{\rho_V}\,(m^3 \cdot s^{-1})$	0.6778	1.195

The compressor efficiency (in this case, isentropic efficiency) can now be estimated. This allows the outlet temperature to be calculated.

	Cycle 1	Cycle 2
η_{IS} (−) (from Equation 13.69)	0.849	0.928
T_{out} (K) (from Equation 24.23)	302	328
W (MW) (from Equation 24.26)	0.36	0.41

Total power for heat rejection to cooling water = 0.36 + 0.41 = 0.77 MW.

d) Now assume that part of Level 2 can be rejected to the process above the pinch as shown in Figure 24.26b. The heat exchangers represented in Figure 24.26b might be several exchangers in practice. The rejection load is fixed at 0.54 MW:

$$W = \frac{Q_{EVAP}}{0.6}\left[\frac{T_{COND} - T_{EVAP}}{T_{EVAP}}\right]$$

$$W = \frac{Q_{COND} - W}{0.6}\left[\frac{T_{COND} - T_{EVAP}}{T_{EVAP}}\right]$$

$$W = \frac{0.54 - W}{0.6}\left[\frac{(5 + 273) - 248}{248}\right]$$

$$W = 0.14\,MW$$

Process cooling by Level 2 by this arrangement across the pinch

$$= 0.54 - 0.14$$

$$= 0.4\,MW$$

24

Figure 24.26
Refrigeration system for Example 6.

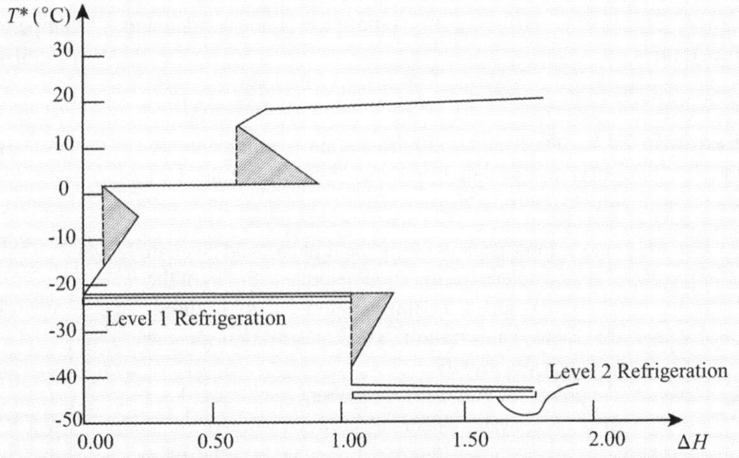

(a) Two refrigeration levels can be used to provide process cooling.

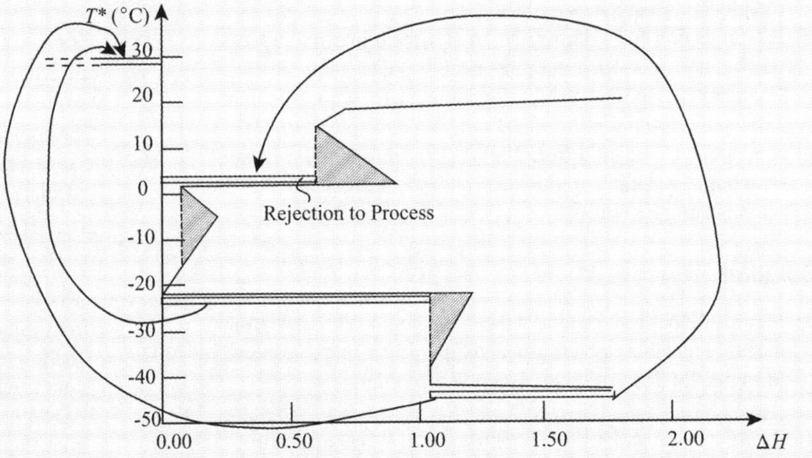

(b) Heat rejection to the process allows the power requirements to be reduced.

The balance of the cooling demand on Level 2 (0.8 – 0.4) = 0.4 MW, together with the load from Level 1 must be either rejected to the process at a higher temperature above the pinch or to cooling water. The process has a heating demand at 20 °C, which means that heat could be rejected at 25 °C. However, there seems little advantage in such an arrangement since the heat can be rejected to cooling water at 30 °C, and rejection to the process would add to the complexity of both design and operation. Assume that the rest of the rejection heat goes to cooling water:

$$W_1 = \frac{1.04}{0.6}\left[\frac{303 - 248}{248}\right] = 0.38\ \text{MW} \quad \text{(as before)}$$

$$W_2 = \frac{0.8 - 0.4}{0.6}\left[\frac{303 - 228}{228}\right] = 0.22\ \text{MW}$$

Total refrigeration power for part rejection of Level 2 to the process

$$= 0.38 + 0.22 + 0.14$$

$$= 0.74\ \text{MW}$$

Thus, the saving in refrigeration power by integration with the process

$$= 0.82 - 0.74$$

$$= 0.08\ \text{MW}$$

e) Repeat the above calculation using refrigeration power targeting. Designate the cycle operating between –45 °C and 5 °C to be Cycle 3.

	Pressure (bar) Cycle 3
Evaporator	1.1291
Condenser	6.7033
Pressure ratio	5.15

For Cycle 3, the refrigerant flowrate and power are dictated by the heat load on the condenser, $Q_{COND} = Q_{EVAP} + W$. Since W is unknown, some trial and error is required. Assume initially that $Q_{COND} = 0.4$ MW for Cycle 3.

Q_{EVAP} (MW)	0.4	0.38	0.388
$H_2 - H_1\,(\text{kJ}\cdot\text{kg}^{-1})$	258.2	258.2	258.2
$m = \dfrac{Q_{EVAP}}{(H_2 - H_1)}\,(\text{kg}\cdot\text{s}^{-1})$	1.549	1.472	1.503
$\rho_V\,(\text{kg}\cdot\text{m}^{-3})$	2.593	2.593	2.593
$F_{in} = \dfrac{m}{\rho_V}\,(\text{m}^3\cdot\text{s}^{-1})$	0.5974	0.5676	0.5795
η_{IS} (–) (from Equation 13.69)	0.852	0.852	0.852
T_{out} (K) (from Equation 24.23)	308	308	308
W (MW) (from Equation 24.26)	0.156	0.149	0.152
Q_{COND} (MW)	0.556	0.529	0.540

The duty on Cycle 1 remains unchanged from Part c, but the duty on Cycle 2 must be reduced from Part c to compensate for the duty on Cycle 3. For Cycle 2:

$$\begin{aligned} Q_{EVAP} &= 0.8 - 0.388 \\ &= 0.412\,\text{MW} \end{aligned}$$

Adjusting the calculation in Part c to reduce Q_{EVAP} to 0.412 MW:

$$m = 1.596\,\text{kg}\cdot\text{s}^{-1}$$

$$F_{in} = 0.6153\,\text{m}^3\cdot\text{s}^{-1}$$

$$W = 0.211\,\text{MW}$$

Total refrigeration power for part rejection of Level 2 to the process from refrigeration targeting

$$\begin{aligned} &= 0.36 + 0.21 + 0.15 \\ &= 0.72\,\text{MW} \end{aligned}$$

Thus, the saving in refrigeration power by integration with the process from refrigeration targeting

$$\begin{aligned} &= 0.77 - 0.72 \\ &= 0.05\,\text{MW} \end{aligned}$$

This is only a small benefit.

24

24.9 Mixed Refrigerants for Compression Refrigeration

It was discussed above how pure refrigerants have a restricted temperature range over which they can operate. The working range of refrigerant fluids can be extended and modified by using a mixture for the refrigerant rather than a pure component. Mixed refrigerants can then be applied in simple, multistage or cascade refrigeration systems. However, unlike pure refrigerants and azeotropic mixtures, the temperature and vapor and liquid compositions of nonazeotropic mixtures do not remain constant at constant pressure as the refrigerants evaporate or condense. Use of mixed refrigerants can be particularly effective if the cooling duty involves a significant change in temperature. The composition of the mixture is selected such that the liquid refrigerant evaporates over a temperature range similar to that of the process cooling demand. Figure 24.27 compares a pure refrigerant (or an azeotropic mixture) and a mixed refrigerant, both cooling a low-temperature heat source that changes temperature significantly. In Figure 24.27a, it can be seen that a pure refrigerant (or azeotropic mixture) evaporates at a constant temperature, leading to a small difference in temperature at one end of the heat exchanger, but a large difference in temperature at the other end. By contrast, a mixed refrigerant, as shown in Figure 24.27b, evaporates over a range of temperature and follows more closely the low-temperature cooling profile. Small temperature differences throughout the heat exchange in low-temperature systems lead to lower refrigeration power requirements. A higher average temperature for the

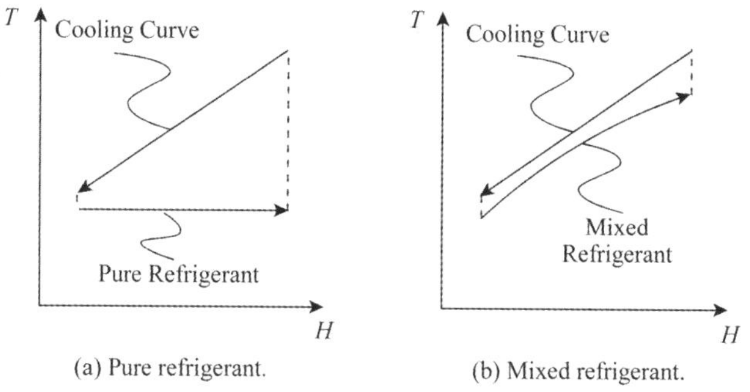

Figure 24.27
Pure and mixed refrigerants.

refrigeration evaporation means a lower temperature lift and a lower power requirement.

Figure 24.28 illustrates a simple refrigeration cycle using a mixed refrigerant. Figure 24.28 shows temperature enthalpy profiles for both the process cooling demand and the refrigerant evaporation profile. It can be seen that the refrigerant evaporation profile for the process cooling demand tends to follow the process cooling demand with a much smaller temperature difference than would be the case for a pure refrigerant. The match between the process cooling and the refrigerant evaporation would normally be carried out in plate-fin heat exchanger in a counter-current match with small temperature differences. It should be

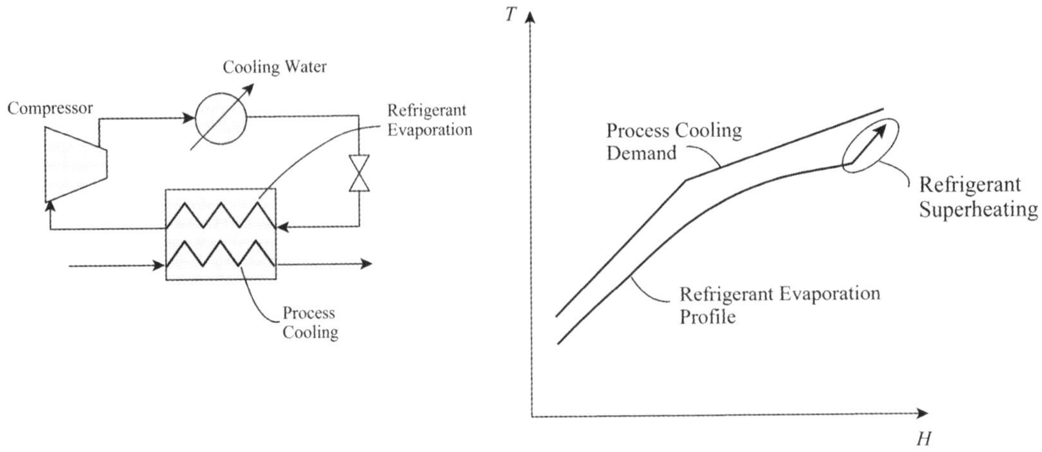

Figure 24.28
A simple refrigeration cycle using a mixed refrigerant.

noted that it is normal practice in such cycles to feature some superheating of the refrigerant, as illustrated in Figure 24.28. Superheating the refrigerant before it is returned to the compressor protects the compressor against potential damage from liquid entering the compressor.

The design of a mixed refrigerant system requires the degrees of freedom to be manipulated simultaneously. These are as follows (Lee, Smith and Zhu, 2003).

1) *Pressure level.* The choice of pressure level for the mixed refrigerant evaporation affects the temperature difference between the process cooling curve and refrigerant evaporation curve. This is illustrated in Figure 24.29, where a natural gas stream is cooled and liquefied using a mixed refrigerant. The mixed refrigerant evaporation profile shown in Figure 24.29a is infeasible as a result of a temperature cross. Increasing the compressor discharge pressure, as shown in Figure 24.29b creates feasible temperature differences throughout the natural gas liquefaction in this case. However, this is at the expense of increased compressor power. Increasing the overall temperature difference will increase the refrigeration power requirements.
2) *Refrigerant flowrate.* Increasing the refrigerant flowrate can widen the temperature difference between the process cooling curve and refrigerant evaporation curve, and vice versa. This is illustrated in Figure 24.30, again for a natural gas liquefaction process. The mixed refrigerant evaporation profile shown in Figure 24.30a is infeasible as a result of a temperature cross. Increasing the refrigerant flowrate, as illustrated in Figure 24.30b, creates feasible temperature differences

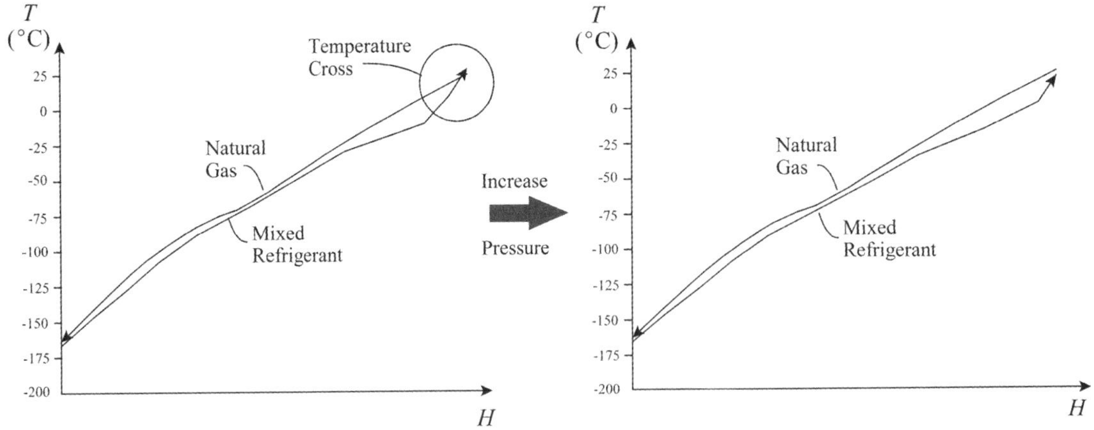

(a) Liquefaction of a natural gas stream exhibiting a temperature cross.

(b) Increasing the compressor discharge pressure with the same suction pressure creates feasible temperature differences throughout at the expense of increased compressor power.

Figure 24.29
Effect of pressure on temperature feasibility for liquefaction of a natural gas stream using a mixed refrigerant.

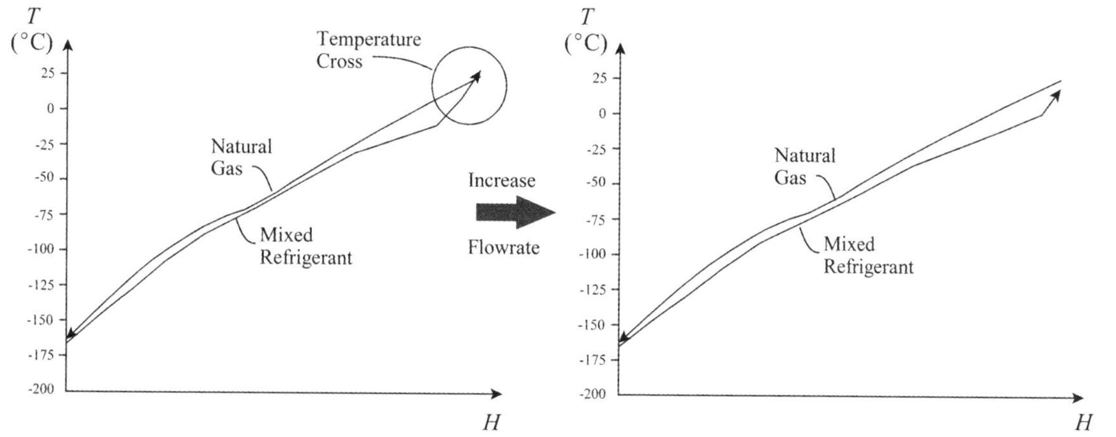

(a) Liquefaction of a natural gas stream exhibiting a temperature cross.

(b) Increasing the refrigerant flowrate creates feasible temperature differences throughout at the expense of increased compressor power.

Figure 24.30

Effect of refrigerant flowrate on temperature feasibility for liquefaction of a natural gas stream using a mixed refrigerant.

throughout the natural gas liquefaction in this case. However, increasing the refrigerant flowrate increases the compressor power. If the refrigerant flowrate is too high, there might be some wetness in the inlet stream to the compressor, which should be avoided. Therefore, the refrigerant flowrate can be changed only within a feasible range.

3) *Refrigerant composition.* The composition of the mixture can be varied to achieve the required characteristics. Introduction of new components or replacing existing components by new ones provides additional freedom to achieve better performance. Unlike adjusting the refrigerant pressure level and flowrate, optimization of the composition does not inevitably mean that increasing the temperature difference in some part of the heat exchange will result in higher refrigeration power requirements. This is illustrated in Figure 24.31 for the same natural gas liquefaction. Figure 24.31a shows a mixed refrigerant evaporation profile that features a temperature cross. In this case, the refrigerant composition can be changed to alter the mixed refrigerant evaporation profile. Figure 24.31b shows an alternative refrigerant evaporation profile through changing the composition that features feasible temperature differences throughout. In this case, change actually lowers the compressor power. Since pressure levels and refrigerant flowrate can only be changed within certain ranges, the refrigerant composition is the most flexible and significant variable when designing mixed refrigerant systems.

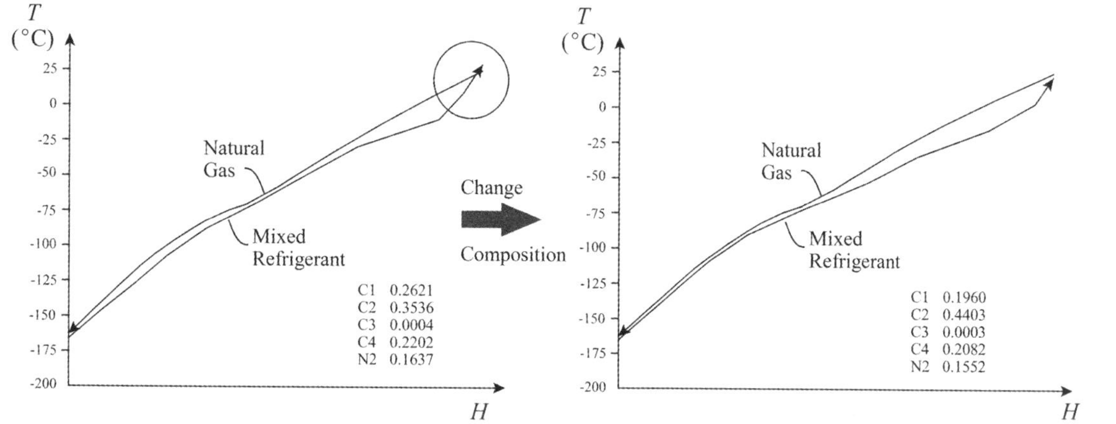

(a) Liquefaction of a natural gas stream exhibiting a temperature cross.

(b) Changing refrigerant composition creates feasible temperature differences throughout and allows a decrease in compressor power.

Figure 24.31

Effect of refrigerant composition on temperature feasibility for liquefaction of a natural gas stream using a mixed refrigerant.

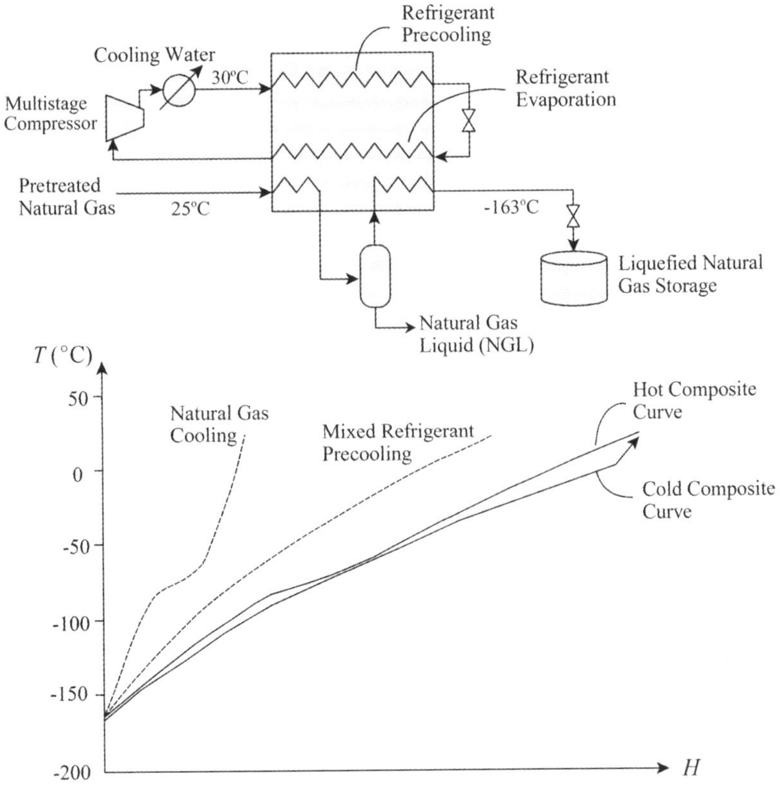

Figure 24.32

A simple liquefied natural gas (LNG) process. (Reproduced with permission from Lee G-C, Smith R and Zhu XX (2003) Optimal Synthesis of Mixed-refrigerant Systems for Low-temperature Processes, *Ind Eng Chem Res*, **41**: 5016. Copyright © 2003, American Chemical Society.)

The major difficulty in the selection of refrigerant composition for mixed refrigerants comes from the interactions between the variables and the small temperature difference between the process cooling curve and refrigerant evaporation curve. Any change in the refrigerant composition, evaporating pressure or the refrigerant flowrate will change the shape and position of the refrigerant evaporation curve.

Optimization can be used to vary the refrigerant composition, flowrate and pressure level to obtain the desired refrigerant properties and conditions. Ideal matching between the process cooling curve and refrigerant evaporation curve would lead to the evaporation curve tending to follow the general shape of the cooling curve, but with the two curves not necessarily being parallel with each other, and the refrigerant profile will normally feature some superheating.

To determine the optimum settings for the compressor suction and discharge pressures, the refrigerant flowrate and composition requires a simulation model that is subjected to optimization. The degrees of freedom can be manipulated using nonlinear programming (e.g. SQP, see Chapter 3) to perform the optimization by manipulating the composition, flowrate and pressures (Lee, Smith and Zhu, 2003). The objective is normally to minimize the compressor power. However, this approach is vulnerable to the highly nonlinear characteristics of the optimization. Alternatively, a stochastic search method can be used at the expense of the additional computational requirements. Genetic algorithms, as discussed in Chapter 3, have proven to be successful in developing good solutions (Wang, 2004; Del Nogal *et al.*, 2008). Great care must be taken to ensure temperature feasibility throughout the cooling process. The temperatures need to be checked at discrete points along the cooling profile to ensure that the temperature difference is greater than some practical minimum.

One major application of mixed refrigerant systems is in the liquefaction of natural gas (Kinard and Gaumer, 1973; Bellow, Ghazal and Silverman, 1997; Finn, Johnson and Tomlinson, 1999). A mixture of hydrocarbons (usually with carbon numbers in the C_1 to C_5 range) and nitrogen is normally used as refrigerant (see Figure 24.31). The refrigerant characteristics are chosen such that there is a close match between the cooling and heating profiles with small temperature differences throughout the whole temperature range for a specific refrigeration demand (but not necessarily parallel). The simplest form of process for liquefaction of natural gas using a mixed refrigerant is illustrated in Figure 24.32. The function of such a process is to convert natural gas to the liquid state for transportation and/or storage. The natural gas enters the heat exchanger at ambient temperature and high pressure and is to be liquefied by the mixed refrigerant flowing countercurrently through the main heat exchanger. The main heat exchanger could typically be a plate-fin design (see Chapter 12). The mixed refrigerant is then recompressed and partially condensed, normally by cooling water. Because there is no other heat sink in the process that can totally condense the mixed refrigerant, it must be condensed by the cold refrigerant itself. Thus, the mixed refrigerant after leaving the compressor is first cooled by cooling water and then passes through the main heat exchanger, where it is condensed by matching against the evaporating refrigerant. It is then expanded across a valve and returns countercurrently through the heat exchanger to the compressor.

Figure 24.32 shows the temperature profiles of the three streams in the main heat exchanger: the natural gas stream, the warm refrigerant stream before the expansion valve and the cold refrigerant stream after the expansion valve (Lee, Smith and Zhu, 2003). The natural gas stream and the warm refrigerant stream need to be combined as a hot composite curve, as described in Chapter 17, to give the total cooling demand. The mixed refrigerant evaporates and condenses at two constant pressure levels, while both the evaporating and condensing temperatures vary over a wide range. The choice of the two pressure levels affects the temperature difference between the hot composite curve and the refrigeration evaporation profile. It should be noted that, in this case, as the degrees of freedom are manipulated, this alters the shape of both the cooling curve and the refrigerant evaporation curve. This means that the cooling curve (hot composite curve) must be recalculated, as the optimization progresses (Lee, Smith and Zhu, 2003).

Many different structural variations are possible around the simple cycle shown in Figure 24.32. The cycle can be made more complex by introducing multiple expansion and compression stages. These structural variations can be screened using structural optimization, for example with a genetic algorithm (Del Nogal *et al.*, 2008). Beyond that, many different complex cycles can be used that exploit mixed refrigerants. Two different mixed refrigerant cycles can be cascaded or a complex multistage pure refrigerant cycle used as precooling before using a mixed refrigerant cycle.

24.10 Expanders

Expanders are machines that convert pressure energy into work. The inlet might in principle be a liquid under pressure that partially vaporizes, a vapor that cools and partially condenses as a result of the expansion or any gas that cools as a result of the expansion. The purpose of the machine might be to recover power, or create a cooling effect, or both.

Three types of expanders are used:

a) *Reciprocating expanders.* Reciprocating expanders are analogous to reciprocating compressors, but operating in reverse. Designs typically adopt two pistons in two cylinders with the pistons connected by a common shaft. These are used in low-temperature applications, but are uncommon devices and restricted to applications with a low flowrate and high pressure ratio.

b) *Radial flow turbines.* Radial flow turbines, sometimes referred to as *turboexpanders*, are analogous to centrifugal compressors, but operating in reverse. The flow is at an angle to the turbine shaft. Machines most often feature a single impellor. Radial flow turbines are the devices most commonly used in low-temperature applications.

c) *Axial flow turbines.* Axial flow turbines operate in an analogous way to axial compressors, but operating in reverse. Gas flows parallel to the turbine shaft. Axial flow turbines are normally used for high flowrates with high inlet temperatures where the prime function is to generate power from change in gas pressure and will not be considered further here.

There are three ways in which the power recovered by a turbine can be dealt with:

a) *Expander-compressor.* In this arrangement, an expander is coupled directly to a compressor and the power recovered in the expander used to satisfy fully or partially a compressor duty.

b) *Expander-generator.* In this arrangement, an expander is linked to an electricity generator with the electricity sent to a local electrical network.

c) *Expander-brake.* In some situations, it will be uneconomic to recover the power in small-scale applications. In this case, the recovered power is transformed into waste heat in a hydraulic braking arrangement. Such arrangements can be attractive in small-scale applications as a result of the effective cooling provided by the expanded gas.

Substituting the throttle valve in a compression refrigeration cycle by a process expander both increases the efficiency of the cycle and allows power recovery. For example, consider Figure 24.7 in which the liquid refrigerant is expanded isentropically. Compared with the expansion across a throttle valve in Figure 24.8, isentropic expansion will produce more liquid after the expansion when compared with a throttle valve. This reduces the compression duty, as the higher liquid fraction requires a lower refrigerant flowrate to provide the same cooling duty. Whilst the expansion in Figure 24.7 is not possible in practice, if the expansion can be made more isentropic than a throttle valve by using an expander, then this can bring the benefit of reducing the compression requirements.

Expansion of a process gas stream across an expander can also bring benefits when compared with expansion across a throttle valve. Such expanders will most often be radial flow turbines. Whilst the performance of centrifugal compressors is normally characterized in terms of a polytropic efficiency, the performance of radial flow turbines is normally characterized in terms of an isentropic efficiency. Isentropic efficiency is normally high, ranging between 70 and 90%, and in some cases higher than 90%. Such machines can typically operate with an outlet of over 40% liquid by mass without a significant loss of efficiency. Figure 24.33 illustrates how effective expanders can be in providing low-temperature cooling. Figure 24.33 shows a number of different expansion paths starting with propylene at 10 bar and 350 K. Isentropic expansion gives an outlet temperature of 256 K when expanded to 1 bar. As the isentropic efficiency decreases, the expansion path to the same outlet pressure leads to higher and higher outlet temperatures. The expansion path corresponding with $\eta_{IS} = 0\%$ corresponds with expansion across a valve. The gas after expansion across a valve has a much higher temperature of 340 K when compared with an isentropic expansion. This explains why it is sometimes desirable to use an expander for the gas expansion of a gas to produce low-temperature cooling and to not recover the power, but simply use a brake.

As an example in the use of expanders, consider Figure 24.34. This shows a process for the liquefaction of natural gas using nitrogen gas in a refrigeration cycle. Nitrogen from compression is

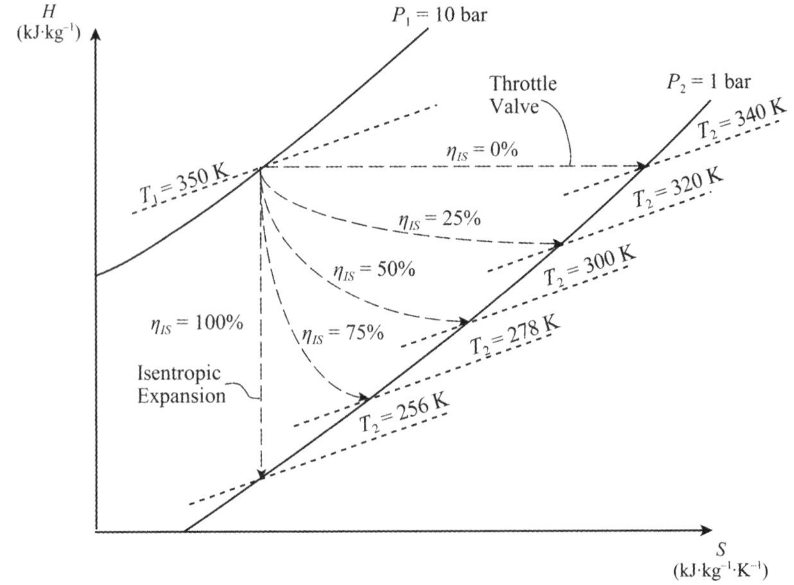

Figure 24.33

Expansion of a gas along different paths. An example using the expansion of propylene.

cooled to ambient temperature and then enters a multistream heat exchanger, typically a plate-fin heat exchanger (see Chapter 12). The nitrogen refrigerant is first precooled by heat transfer in the multistream heat exchanger before entering a process expander, where the nitrogen expands and cools to the minimum temperature required for the natural gas cooling. The nitrogen from the outlet of the expander enters the multistream heat exchanger, where it provides both the cooling for the natural gas and the precooling for the nitrogen refrigerant. The nitrogen at the exit of the multistream heat exchanger is then compressed. Power recovered from the expander provides part of the compression power. The remainder of the compressor power must be supplied by an electric motor, steam turbine or gas turbine. The nitrogen refrigerant cools the natural gas in the heat exchanger, after which the natural gas is

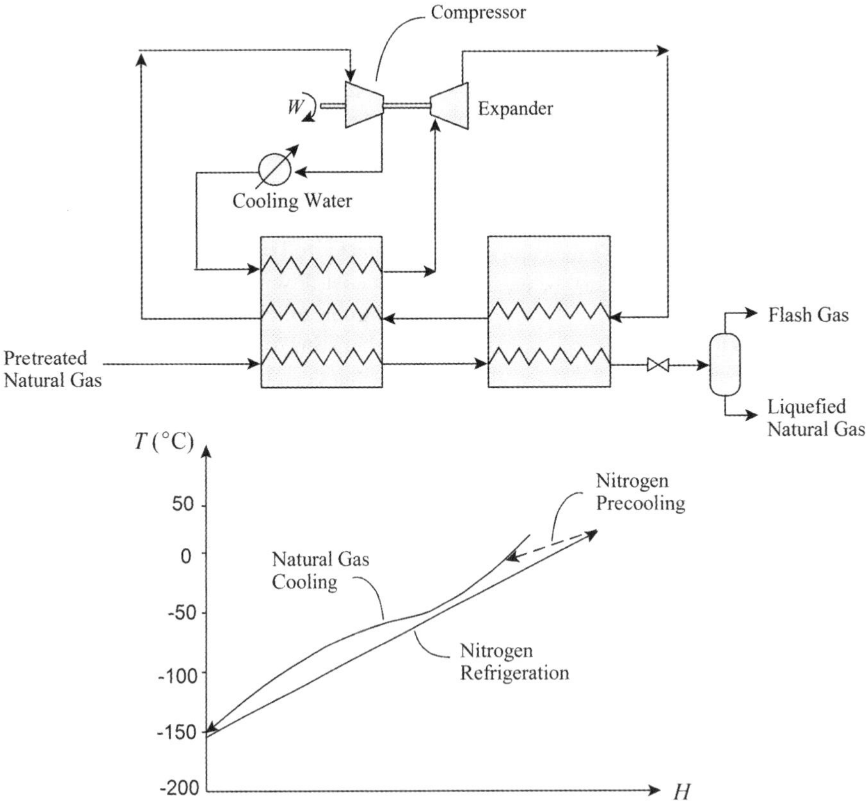

Figure 24.34

A nitrogen expander process for natural gas liquefaction.

flashed to a lower pressure to complete the cooling. Such refrigeration cycles using nitrogen gas are inefficient but are used for small-scale natural gas liquefaction processes where the safety of the nitrogen refrigerant is preferred to be hydrocarbon refrigerants normally used for liquefaction. The efficiency of the liquefaction can be improved by introducing two cycles heat integrated together. The warmer cycle might use nitrogen, methane or a mixture of nitrogen and methane and the colder cycle completes the cooling using nitrogen refrigerant. Both cycles use expanders.

The modelling of expanders has the same basis as compressors. Given that an ideal expansion is the reverse of an ideal compression, then an adiabatic ideal gas (isentropic) expansion can be calculated from Equation G.22:

$$W_S = \frac{\gamma}{\gamma - 1} P_{in} V_{in} \left[1 - \left(\frac{P_{out}}{P_{in}} \right)^{(\gamma-1)/\gamma} \right] \tag{24.30}$$

where W_S = ideal work produced (J)
P_{in} = inlet pressure ($N{\cdot}m^2$)
P_{out} = outlet pressure ($N{\cdot}m^2$)
V_{in} = inlet volume (m^3)
γ = heat capacity ratio C_P/C_V (–)

For a real expansion, this becomes:

$$W = \left(\frac{\gamma}{\gamma - 1} \right) \eta_{IS,E} P_{in} V_{in} \left[1 - \left(\frac{P_{out}}{P_{in}} \right)^{(\gamma-1)/\gamma} \right] \tag{24.31}$$

where W = actual work produced (J)
$\eta_{IS,E}$ = isentropic efficiency for the expansion process (–)

Expressing V_{in} as a volumetric flowrate returns a value of W in $J{\cdot}s^{-1}$. Alternatively, the work can be expressed in terms of the molar flowrate by introducing the compressibility factor, noting $PV = ZRT$ (see Appendix A):

$$W = \left(\frac{\gamma}{\gamma - 1} \right) \eta_{IS,E} Z R T_{in} \left[1 - \left(\frac{P_{out}}{P_{in}} \right)^{(\gamma-1)/\gamma} \right] \tag{24.32}$$

where W = actual work produced ($J{\cdot}kmol^{-1}$)
R = universal gas constant ($8314.5\ J{\cdot}K^{-1}{\cdot}kmol^{-1}$)
Z = compressibility factor (–)

The compressibility factor Z can be determined from an equation of state (see Appendix A). The outlet temperature can be calculated from Equation 13.80:

$$T_{out} = T_{in} \eta_{IS,E} \left[\left(\frac{P_{out}}{P_{in}} \right)^{(\gamma-1)/\gamma} + \frac{1}{\eta_{IS,E}} - 1 \right] \tag{24.33}$$

where T_{out} = temperature of the outlet stream for a real expansion
T_{in} = temperature of the inlet stream

Equations 24.31, 24.32 and 24.33 can be used to model gas expansion. However, these equations need to be used with caution, as they are based on ideal gas behavior and heat capacity is assumed to be constant. Providing the pressure is not high and the gas does not deviate significantly from ideal behavior, then the equations can be used to model the expansion. In addition, the value of γ can vary significantly through the expansion process.

An alternative way to calculate the work for an adiabatic expansion discussed in Chapter 13 is from the difference between the total enthalpy of the outlet and inlet flows:

$$\begin{aligned} W &= H_{in} - H_{out} \\ &= \eta_{IS,E}\left(H_{in} - H_{out,IS}\right) \end{aligned} \tag{24.34}$$

where W = power produced from expansion ($N{\cdot}m{\cdot}s^{-1} = J{\cdot}s^{-1} = W$)
H_{in} = total enthalpy of the inlet stream
$H_{out,IS}$ = total enthalpy of the outlet stream for an isentropic expansion
H_{out} = total enthalpy of the outlet stream for a real expansion
$\eta_{IS,E}$ = isentropic efficiency for the expansion process (–)

The calculation starts with the inlet enthalpy and, given the inlet enthalpy and the pressure, the entropy of the inlet is calculated. Then, assuming the outlet entropy is the same as the inlet (isentropic) and given the outlet pressure, the outlet enthalpy is calculated. From the isentropic enthalpy change, the real outlet enthalpy can then be calculated by multiplying the isentropic enthalpy change by the isentropic efficiency to get the real enthalpy change. Neglecting any mechanical losses, this gives the overall power production. Knowing the outlet enthalpy and pressure allows the outlet temperature to be calculated from physical properties. This is the method normally used in simulation software, but is inconvenient to carry out using hand calculations.

Example 24.7 A small-scale liquefied natural gas process is required to produce 210 tons per day of liquefied natural gas. The small-scale nature of the process makes it a possible candidate for a nitrogen expander cycle, as in Figure 24.34. The required flowrate of nitrogen to achieve the cooling is $0.43\ kmol{\cdot}s^{-1}$. The nitrogen expander can be modelled using an isentropic efficiency of 0.85. The large change in conditions of both temperature and pressure across the expander leads to a variation in the value of γ through the process. Given that the variation in γ is nonlinear with conditions, a simple average is not necessarily representative and a value of 1.42 can be taken to be representative of the overall expansion process. The compressor can be assumed to be a multistage centrifugal compressor to be modelled by a polytropic compression with a maximum pressure ratio per stage of 3. It can be assumed that the intercoolers cool the gas back to the compressor inlet temperature and there is no pressure drop across the intercoolers. The universal gas constant R can be taken to be $8314.5\ J{\cdot}kmol^{-1}{\cdot}K^{-1}$. Values of the compressibility of nitrogen can be determined from the Peng–Robinson Equation of State (see Appendix A).

a) The conditions at the expander inlet are 268 K, 59.9 bar with a compressibility of 0.966. The outlet pressure is 1.6 bar. Calculate the expander outlet temperature assuming the heat capacity is constant.

b) Calculate the power recovery from the expander.
c) The conditions at the compressor inlet are 303 K, 1.4 bar with a compressibility of 0.999. The required outlet pressure is 60.1 bar. Calculate the net power input for the process.

Solution

a) Assuming heat capacity and γ to be constant, the expander outlet temperature can be calculated from Equation 24.33:

$$T_{out} = T_{in}\eta_{IS,E}\left[\left(\frac{P_{out}}{P_{in}}\right)^{(\gamma-1)/\gamma} + \frac{1}{\eta_{IS,E}} - 1\right]$$

Substituting for the inlet conditions and outlet pressure:

$$T_{out} = 268 \times 0.85\left[\left(\frac{1.6}{59.9}\right)^{(1.42-1)/1.42} + \frac{1}{0.85} - 1\right]$$

$$= 118.2\text{ K}$$

b) Power recovery from the expander can be calculated from Equation 24.32:

$$W = m\left(\frac{\gamma}{\gamma - 1}\right)\eta_{IS,E}ZRT_{in}\left[1 - \left(\frac{P_{out}}{P_{in}}\right)^{(\gamma-1)/\gamma}\right]$$

$$= 0.43\left(\frac{1.42}{1.42 - 1}\right) \times 0.85 \times 0.966 \times 8314.5$$

$$\times 268\left[1 - \left(\frac{1.6}{59.9}\right)^{(1.42-1)/1.42}\right] = 1.75 \times 10^6\text{ W}$$

c) First determine the number of compression stages from Appendix G:

$$r = \sqrt[N]{\frac{P_{out}}{P_{in}}} \qquad \text{(G.32)}$$

Try $N=3$:

$$r = \sqrt[3]{\frac{60.1}{1.40}} = 3.50$$

This is too high. Try $N=4$:

$$r = \sqrt[4]{\frac{60.1}{1.4}} = 2.56$$

Thus, assuming that four stages with intercooling are needed, calculate the volumetric flowrate:

$$F_{in} = m \times \frac{ZRT_{in}}{P_{in}}$$

$$= \frac{0.43 \times 0.999 \times 8314.5 \times 303}{1.40 \times 10^5}$$

$$= 7.73\text{ m}^3 \cdot \text{s}^{-1}$$

Calculate the polytropic efficiency from Equation 13.70:

$$\eta_P = 0.017\ \ln F + 0.7$$

$$= 0.017\ \ln(7.73) + 0.7$$

$$= 0.73$$

The polytropic coefficient n can be calculated from Equation 13.60:

$$n = \frac{\gamma\eta_P}{\gamma\eta_P - \gamma + 1}$$

$$= \frac{1.42 \times 0.73}{1.42 \times 0.73 - 1.42 + 1}$$

$$= 1.68$$

The power required for compression can be calculated from Equation 13.67:

$$W = \frac{n}{n-1}\frac{P_{in}F_{in}N}{\eta_P}\left[1 - (r)^{(n-1)/n}\right]$$

$$= \left(\frac{1.68}{1.68 - 1}\right) \times \frac{1.40 \times 10^5 \times 7.73 \times 4}{0.73}\left[1 - (2.6)^{(1.68-1)/1.68}\right]$$

$$= -6.92 \times 10^6\text{ W}$$

The net power input

$$= (-6.92 + 1.75) \times 10^6$$

$$= -5.17 \times 10^6\text{ W}$$

Thus the net power requirement is 5.17 MW.

It should be noted that because of the large change in conditions of both temperature and pressure across the process, there is a change in both the heat capacity and γ. Greater accuracy would therefore be expected from the approach based on enthalpy given in Equation 24.34. However, this calculation requires simulation software.

A gas expanded in an expander might be a process gas. If a gas or vapor process stream is available at a high pressure and downstream conditions do not require this high pressure, the stream can be expanded to provide useful cooling, useful power or both. The cooling might allow partial condensation for recovery of the less volatile components in a mixture or to provide a cold stream for refrigeration purposes. Expansion across a throttle valve provides cooling only, whereas expansion across a process expander can provide more effective cooling as well as power recovery. As an example, suppose a distillation is required to be carried out at low temperature and high pressure for the separation of a mixture involving light gases. Uncondensed light gases from the condenser at high pressure, and not required to be at high pressure, can be expanded to cool the gases, which in turn can be used, for example, to precool the feed to the distillation.

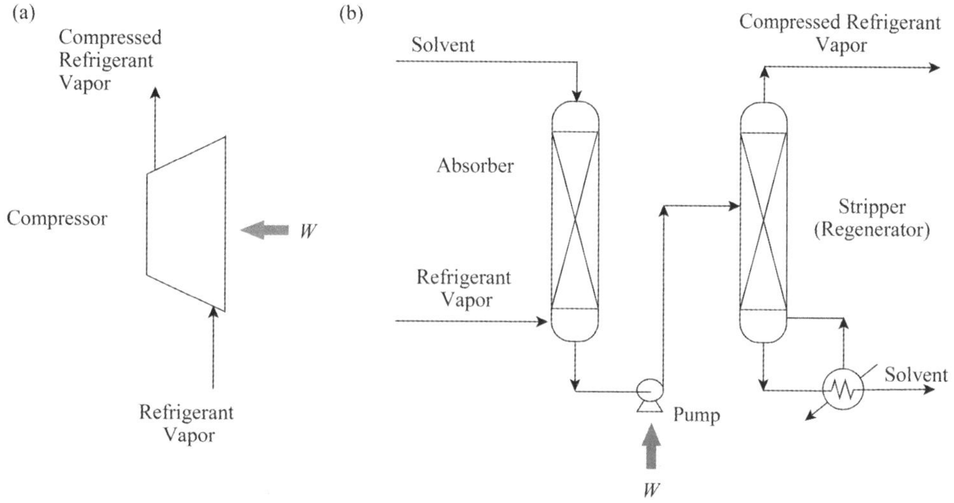

Figure 24.35
Compression versus absorption refrigeration.

24.11 Absorption Refrigeration

Consider now absorption refrigeration. Compression refrigeration is powered by a compressor compressing refrigerant vapor (Figure 24.35a). The basic problem with this is the high compression costs. Figure 24.35b illustrates an alternative way to bring about the compression. A refrigerant vapor is first absorbed in a solvent. The resulting liquid solution can have its pressure increased using a pump. The compressed refrigerant is then separated from the solvent in a stripper (regenerator). The pump requires significantly less power to bring about the increase in pressure compared with the corresponding gas compression. The overall effect is to increase the pressure of the refrigerant with far less power. The drawback is that a heat supply is needed for the stripper (regenerator).

Figure 24.36 shows a typical absorption refrigeration arrangement. To the left of the cycle in Figure 24.36 is the absorber and stripper (regenerator) arrangement. Low-pressure

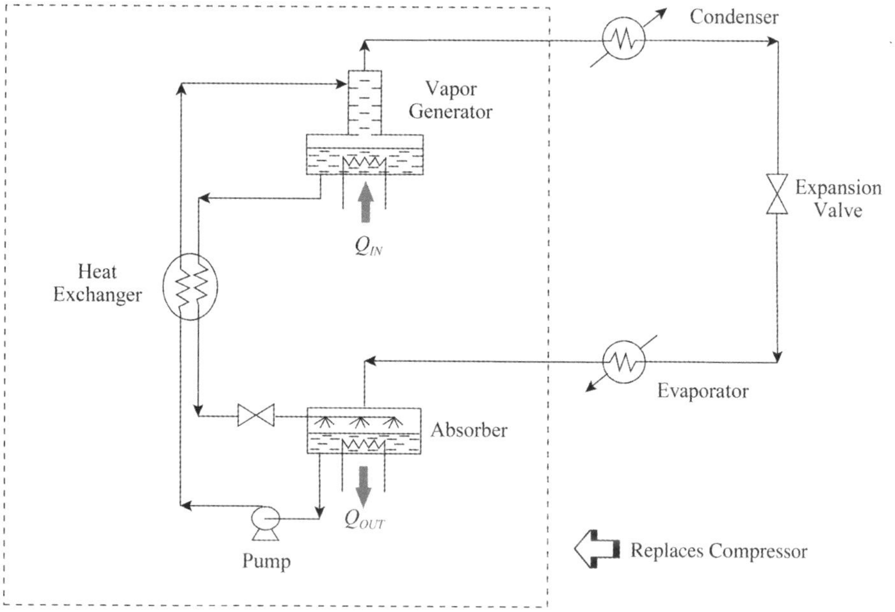

Figure 24.36
A typical absorption refrigeration arrangement.

Table 24.6
Common working fluids for absorption refrigeration.

Refrigerant	Solvent	Lower temperature limit (°C)
Water	Lithium bromide	5
Ammonia	Water	−40

refrigerant vapor from the evaporator is first absorbed in a solvent, which is then increased in pressure and then increased in temperature in a heat exchanger. The refrigerant then enters the vapor generator where the refrigerant is stripped from the solvent. Heat is input to the vapor generator and the solvent is cooled in the heat exchanger, decreased in pressure and returned to the absorber. The high-pressure vapor from the vapor generator is condensed in the condenser, expanded in the expansion valve to produce the cooling effect and then enters the evaporator to provide process cooling.

The features of absorption refrigeration are that there is a low power requirement relative to compression refrigeration, but heat input to the vapor generator (stripper) is required. Heat output from the absorber is required, usually to cooling water. Absorption refrigeration is only useful for moderate temperature refrigeration.

The most common working fluids for absorption refrigeration are given in Table 24.6, together with the working range.

When should absorption refrigeration be used rather than compression refrigeration? There are two important criteria. The first is that absorption refrigeration can only be used when moderate levels of refrigeration are required. Second, it should only be used when there is a large source of waste heat available for the vapor generator. This must be at a temperature greater than 90 °C, but the higher the better.

24.12 Indirect Refrigeration

Indirect refrigeration is often used. Figure 24.37 shows a liquid intermediate being recycled to provide the cooling to a process load. Heat is removed from the liquid intermediate by a refrigeration cycle. This arrangement is used for distribution of refrigeration across a number of process loads or when contact between refrigerant and process fluids is unacceptable for safety or product contamination reasons. The arrangement leads to higher power requirements than direct refrigeration because of the extra temperature lift required in the heat exchanger between the refrigerant fluid and intermediate liquid.

The liquid intermediates used are typically water solutions of various concentrations such as salts (e.g. calcium chloride, sodium chloride), glycols (e.g. ethylene glycol, propylene glycol) and alcohols (e.g. methanol, ethanol), or pure substances such as acetone, methanol and ethanol.

24.13 Cooling Water and Refrigeration Systems – Summary

Rejection of waste heat is a feature of most chemical processes. Once the opportunities for recovery of heat to other process streams and to the utilities system (e.g. steam generation) have been exhausted, then waste heat must be rejected to the environment. The most direct way to reject heat to the environment above ambient temperature is by the use of air-cooled heat exchangers. Cooling water using once-through cooling systems based on river

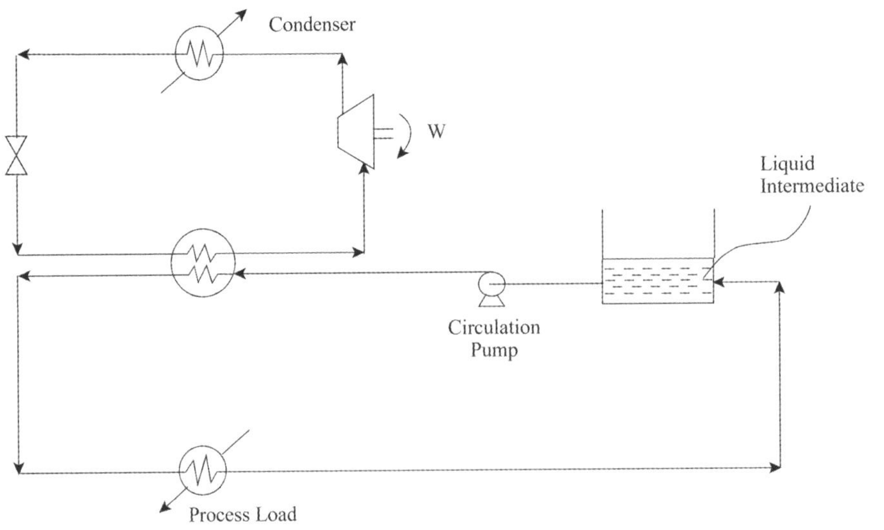

Figure 24.37
Indirect refrigeration.

water, and so on, can be used. However, such once-through systems require large amounts of water and the waste heat rejected to aquatic systems can create environmental problems.

The most common way to reject waste heat above ambient temperature is through recirculating cooling water systems. Both natural draft and mechanical draft cooling towers are used. Heat is rejected through the evaporation of water, but other losses of water result from the need for blowdown to prevent build-up of undesired contaminants in the recirculation and drift losses. Increasing the cycles of concentration reduces the blowdown losses, probably at the expense of increased chemical dosing to prevent fouling and corrosion.

The design of the cooling tower and the design of the cooling water network interact with each other. Cooling duties can be configured in parallel or series. Decreasing the flowrate of cooling water and increasing the cooling tower return temperature increase the effectiveness of the cooling tower. Placing coolers in series, rather than parallel, benefits the performance of the cooling tower through reduced flowrate and increasing return temperature. However, this is at the expense of reduced temperature differences in the coolers and increased pressure drop for the cooling water. Placing coolers in series for a few critical reuse opportunities might be extremely effective in improving the design.

Refrigeration is required to produce cooling below ambient temperature. There are two broad classes of refrigeration:

- Compression refrigeration
- Absorption refrigeration.

Compression refrigeration is by far the most common. Simple cycles can be employed to provide cooling typically as low as −40 °C. For lower temperatures, complex cycles are used. Economizers, presaturators and the introduction of multiple refrigeration levels allow multistage compression and expansion to be used to reduce the power requirements for refrigeration. Very low temperature systems require the use of cascade cycles with two refrigerant cycles linked together, with each cycle using a different refrigerant fluid.

The choice of refrigerant fluid depends on a number of issues. Use of an evaporator pressure below atmospheric pressure is normally avoided. It is desirable to have a high latent heat at the evaporator conditions in order to reduce the flowrate around the refrigeration loop to reduce the power requirements correspondingly. Various other factors relating to the shape of the two-phase region for the refrigerant fluid also affect the choice of refrigerant, as well as the toxicity, flammability, corrosivity and environmental impact.

Refrigeration cycles offer many opportunities for heat integration with the process. These can be explored using the grand composite curve.

Refrigeration power can be targeted for specific refrigerant fluids by performing an energy balance around the cycle. Simple, multistage and complex cycles can be targeted for minimum refrigeration power.

Mixed refrigerant systems use a mixture as refrigerant instead of pure refrigerants. Evaporation of the refrigerant mixture over a range of temperature allows a better match between the refrigerant duty and refrigerant evaporation if the refrigeration duty varies significantly in temperature. When designing mixed refrigerant systems, important degrees of freedom include:

- composition of the refrigerant mixture,
- pressure levels,
- refrigeration flowrate.

Expanders are more effective in producing low temperatures than expansion across a valve. Expanders can be used within the refrigeration cycle. Alternatively, if a process gas or vapor is at a high pressure and is not required at high pressure downstream, then it can be let down across an expander either to produce a low-temperature heat sink for the process or to recover power.

Absorption refrigeration is much less common than compression refrigeration. Absorption refrigeration reduces power requirements by compressing the refrigerant fluid dissolved in a solvent using a pump.

24.14 Exercises

1. Cooling water is being circulated at a rate of 20 m^3 min^{-1} to a cooling network. The cooling water from the cooling tower is at a temperature of 25 °C and is returned at 40 °C. Measurements on the concentrations of the feed and circulating water indicate that there are five cycles of concentration. Assuming that the heat capacity of the water is 4.2 kJ·kg^{-1}·K^{-1} and the makeup water is at a temperature of 10 °C, calculate:

   a) the rate of evaporation assuming the latent heat of vaporization to be 2423 kJ·kg^{-1} at the conditions in the cooling tower;

   b) the makeup water requirement, assuming drift losses are negligible.

2. What options might have been considered to accommodate the new cooling duty without having to invest in a new cooling tower if the cooling tower is near maximum capacity?

3. For cooling duties at

   a) −40 °C,

   b) −60 °C,

   estimate the power required to reject 1 MW of cooling duty to cooling water with a return temperature of 40 °C. Assume the Carnot Efficiency to be 0.6, $\Delta T_{min} = 5$ °C for the refrigeration evaporator and $\Delta T_{min} = 10$ °C for the condenser. Suggest suitable refrigeration fluids for these two duties using Figure 24.20.

4. A cascade refrigeration system is to be used to service a cooling duty with an evaporation temperature of −90 °C and rejecting heat ultimately to cooling water with a rejection temperature of 30 °C. The partition temperature of the interface between the two cycles is the temperature of the evaporator of the upper (high-temperature) cycle and is to be optimized. The minimum temperature allowed for the refrigerant in the upper cycle is 225 K and the maximum temperature allowed for the refrigerant in the lower cycle is 264 K. Assuming the cycles can be modelled by a Carnot Efficiency model with an efficiency of

Table 24.7

Stream data for Exercise 5.

Stream	T_S (°C)	T_T (°C)	ΔH (MW)
1. Hot	65	−35	−15
2. Hot	−54	−55	−4
3. Cold	−85	55	7
4. Cold	−6	−5	12

0.6, determine the partition temperature to minimize the total power input. Assume $\Delta T_{min} = 5\,°C$.

5. The stream data for a low-temperature process are given in Table 24.7.

Assume $\Delta T_{min} = 10\,°C$. Steam is available at 120 °C and cooling water at 25 °C to be returned at 35 °C. Refrigeration can be modeled using the Carnot Efficiency model with an efficiency of 0.6.

a) Carry out a problem table analysis and plot the grand composite curve.

b) Match a single-level refrigeration system using a pure refrigerant against the process. Determine the power required if heat is rejected from the refrigeration cycle to cooling water.

c) Match a two-level refrigeration system using a pure refrigerant against the process. Determine the power required if heat is rejected from the refrigeration cycle to the process.

d) For the two-level refrigeration system, devise a scheme to reduce the power requirement by heat rejection to the process rather than to cooling water. All of the heat from the refrigeration should be rejected to the process.

Table 24.8

Stream data for Exercise 6.

Stream		T_S (°C)	T_T (°C)	Stream heat duty (MW)
No	Type			
1	Hot	−20	−20	1.0
2	Hot	−40	−40	1.0
3	Hot	20	0	0.8
4	Cold	20	20	1.0
5	Cold	0	0	1.0
6	Cold	0	20	0.2
7	Cold	−40	20	0.6

6. The data for a heat recovery problem are given in Table 24.8. Carry out a problem table analysis on the data for $\Delta T_{min} = 5\,°C$ and plot the grand composite curve.

a) To reduce power requirements to a minimum, two levels of refrigeration are to be used. What are the temperatures and duties on the two levels?

b) Both the vaporization and condensation of the refrigerant occur isothermally in the refrigeration cycle. Assuming heat is to be rejected from the refrigeration cycle to cooling water, calculate the power requirements. Cooling water operates between 20 °C and 25 °C. The power required for the refrigeration system is given by Equation 24.15.

c) Rather than reject heat from the refrigeration cycle to cooling water, part can be rejected back into the process.

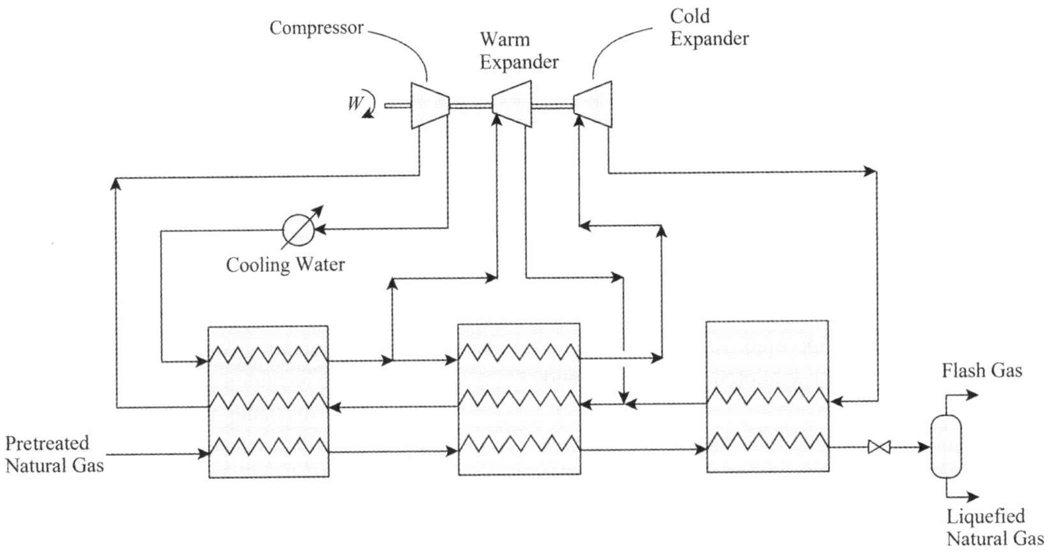

Figure 24.38

The dual nitrogen expander process in Exercise 7.

Table 24.9

Process data for Exercise 7.

	Warm cycle	Cold cycle
Flowrate ($kmol \cdot s^{-1}$)	0.60	0.26
Expander inlet temperature (K)	263	182
Expander inlet pressure (bar)	59.8	59.7
Expander outlet pressure (bar)	12.5	12.6
Heat capacity ratio γ	1.49	1.78
Expander inlet compressibility Z	0.962	0.806

Calculate the power requirement if the maximum heat possible is rejected to the process. The balance is to be rejected into cooling water.

7. A liquefied natural gas process is to use a dual nitrogen expander cycle, as shown in Figure 24.38. Table 24.9 gives the process conditions and fluid properties. The universal gas constant R can be taken to be $8314.5\ J \cdot kmol^{-1} \cdot K^{-1}$.

   a) The expanders can be modelled using an isentropic expansion with an isentropic efficiency of 0.85. Calculate the power recovery from the expander.

   b) A centrifugal compressor is to be used to compress the nitrogen with an inlet temperature of 300 K, inlet pressure of 12.3 bar, outlet pressure of 59.9 bar and compressibility of 0.995. The maximum pressure ratio per stage is 3. It can be assumed that intercoolers cool the nitrogen back to the inlet temperature and that there is no pressure drop across the intercoolers. The compressor can be modelled using a polytropic efficiency and γ for the compression is 1.4. Calculate the net power input for the process.

References

Bellow EG, Ghazal FP and Silverman AJ (1997) Technology Advances Keeping LNG Cost-Competitive, *Oil Gas J*, **June**: 74.

Briggs DE and Young EH (1963) Convection Heat Transfer and Pressure Drop of Air Flowing Across Triangular Pitch Banks of Finned Tubes, *Chem Eng Progr Syp Ser*, **59** (41): 1.

Del Nogal F, Kim JK, Perry S and Smith R (2008) Optimal Design of Mixed Refrigerants, *Ind Eng Chem Res*, **47**: 8724.

Dossat RJ (1991) *Principles of Refrigeration*, 3rd Edition, Prentice Hall.

Finn AJ, Johnson GL and Tomlinson T (1999) Developments in Natural Gas Liquefaction, *Hydrocarbon Process*, **April**: 47.

Gosney WB (1982) *Principles of Refrigeration*, Cambridge University Press.

Hudson Products (2015) Basics of Air-Cooled Heat Exchangers, www.hudsonproducts.com/products/finfan/.

Isalski WH (1989) *Separation of Gases*, Oxford Science Publications.

Kim JK and Smith R (2001) Cooling Water System Design, *Chem Eng Sci*, **56**: 3641.

Kinard GE and Gaumer LS (1973) Mixed Refrigerant Cascade Cycles for LNG, *Chem Eng Prog*, **69**: 56.

Kröger DG (1998) *Air-Cooled Heat Exchangers and Cooling Towers*, Department of Mechanical Engineering, University of Stellenbosch, South Africa.

Lee G-C (2001) *Optimal Design and Analysis of Refrigeration Systems for Low-Temperature Processes*, PhD Thesis, UMIST, UK.

Lee G-C, Smith R and Zhu XX (2003) Optimal Synthesis of Mixed-Refrigerant Systems for Low-Temperature Processes, *Ind Eng Chem Res*, **41**: 5016.

Lee G-C, Zhu XX and Smith R (2000) *Synthesis of Refrigeration Systems by Shaftwork Targeting and Mathematical Optimisation, ESCAPE 10, Florence, Italy*.

Poling BE, Prausnitz JM and O'Connell JP (2001) *The Properties of Gases and Liquids*, McGraw-Hill.

Serth RW (2007) *Process Heat Transfer: Principles and Applications*, Academic Press.

Wang J (2004) *Synthesis and Optimization of Low Temperature Gas Separation Processes*, PhD Thesis, UMIST, UK.

Chapter 25

Environmental Design for Atmospheric Emissions

25.1 Atmospheric Pollution

There are many types of emissions to atmosphere, and these can be characterized as particulate (solid or liquid), vapor and gaseous. Overall, the control of atmospheric emissions is difficult because the majority of emissions come from small sources that are difficult to regulate and control. Legislators therefore control emissions from sources that are large enough to justify monitoring and inspection. Industrial emissions of major concern are as follows.

1) PM_{10}. Particulate material less than 10 μm diameter is formed as a byproduct of combustion processes through incomplete combustion and through reactions between sulfur and nitrogen compounds in the atmosphere. PM_{10} is a particular problem as it causes damage to the human respiratory system.
2) $PM_{2.5}$. Particulate material less than 2.5 μm diameter forms in the same way as PM_{10}, but it can penetrate deeper into the respiratory system than PM_{10} and can therefore be more dangerous.
3) O_3. Ozone is a very reactive compound present in the upper atmosphere (stratosphere) and the lower atmosphere (troposphere). Whilst ozone is vital in the stratosphere, its presence at ground levels is a danger to human health and contributes to the formation of other pollutants.
4) *VOCs*. A volatile organic compound (VOC) is any compound of carbon, excluding carbon monoxide, carbon dioxide, carbonic acid, metal carbides or carbonates and ammonium carbonate, which participate in atmospheric photochemical reactions (US Environmental Protection Agency, 1992a). VOCs are precursors to ground-level ozone production and various photochemical pollutants. They are major components in the formation of smog through photochemical reactions (Harrison, 1992; De Nevers, 1995). There are many sources of VOCs, as will be discussed later.
5) SO_X. Oxides of sulfur (SO_2 and SO_3) are formed by the combustion of fuels containing sulfur and as a byproduct of chemical production.
6) NO_X. Oxides of nitrogen (principally NO and NO_2) are formed in combustion processes and as a byproduct of chemicals production.
7) *CO*. Carbon monoxide is formed by the incomplete combustion of fuel and as a byproduct of chemicals production.
8) CO_2. Carbon dioxide is formed principally by the combustion of fuel but also as a byproduct of chemicals production.
9) *Dioxins and furans*. Dioxins and furans are the abbreviated names for a family of over 200 different toxic substances that share similar chemical structures. They are not commercial chemical products but are trace-level unintentional byproducts from thermal oxidation, the processing of metals and paper manufacture. Dioxins and furans are also released naturally from forest fires and volcanoes. They are widely distributed throughout the environment at extremely low concentrations, but resist degradation and accumulate. They are considered to be human carcinogens and can promote adverse noncancer health effects.

Atmospheric emissions are controlled by legislation because of their damaging effect on the environment and human health. These effects can be categorized into their local effects and global effects. There are six main problems associated with atmospheric emissions.

1) *Urban smog*. Urban smog is commonly found in modern cities, especially where air is trapped in a basin. It is observable as a brownish colored air. The formation of urban smog is through complex photochemical reactions involving sunlight (*hf*) that can be characterized by:

$$VOCs + NO_X + O_2 \xrightarrow{hf} O_3 + \text{Other photochemical pollutants} \tag{25.1}$$

Photochemical pollutants such as ozone, aldehydes and peroxynitrates such as peroxyethanoyl (or peroxyacetyl

Chemical Process Design and Integration, Second Edition. Robin Smith.

Companion Website: www.wiley.com/go/smith/chemical2e

nitrate - PAN) are formed. Ozone and other photochemical pollutants have harmful effects on living organisms. High levels of these pollutants can cause breathing difficulties and bring on asthma attacks in humans. Warm weather and still air exacerbate the problem. In addition to VOCs and NO_X, the problem of urban smog is made worse by particulate emissions and carbon monoxide from incomplete combustion of fuel.

2) *Acid rain.* Natural (unpolluted) rain and other forms of precipitation are naturally acidic with a pH often in the range of 5 to 6 caused by carbonic acid from dissolved carbon dioxide and sulfurous and sulfuric acids from natural emissions of SO_X and H_2S. Human activity can reduce the pH very significantly down to the range of 2 to 4 in extreme cases, mainly caused by emissions of oxides of sulfur, but also oxides of nitrogen. Because atmospheric pollution and clouds travel over long distances, acid rain is not a local problem. The problem may manifest itself a long way from the source. Acid rain leads to the *acidification* of oceans, freshwater and soil. Problems associated with acid rain include:
   - damage to plant life, particularly in forests;
   - acidification of soil leading to a decline in crop and pasture production, as some nutrients become less available, while other components in the soil may reach toxic levels;
   - acidification of water, leading to dead lakes and streams, loss of aquatic life and possible damage to the human water supply;
   - corrosion of buildings, particularly those made of marble and sandstone.
3) *Ocean acidification.* Ocean acidification is the decrease in pH of the oceans, caused by the uptake of carbon dioxide from the atmosphere. A significant proportion of the anthropogenic carbon dioxide emissions dissolve in oceans. When carbon dioxide enters the ocean, it combines with seawater to produce carbonic acid, which increases the acidity of the water, lowering its pH. A consequence of the oceans becoming more acidic is that marine creatures cannot extract the carbonate ion they need to grow shells and skeletons. Even if creatures are able to build shells and skeletons in more acidic water, they spend more energy, decreasing reproduction. When shelled organisms are at risk, the entire food chain is also at risk.
4) *Eutrophication.* Eutrophication occurs when a body of terrestrial or marine water suffers an increase in nutrients that leads to enhanced growth of aquatic vegetation or *phytoplankton* and *algal blooms*. This disrupts normal functioning of the ecosystem, causing a variety of problems such as a lack of oxygen needed for aquatic life to survive, decreased biodiversity, changes in species composition and dominance, and toxicity effects. NO_X emissions contribute to eutrophication.
5) *Ozone layer destruction.* The upper atmosphere contains a layer rich in ozone. Whilst ozone in the lower levels of the atmosphere is harmful, ozone in the upper levels of the atmosphere is essential as it absorbs considerable amounts of ultraviolet light that would otherwise reach the Earth's surface. The destruction of ozone is catalyzed by oxides of nitrogen in the upper atmosphere:

$$NO^{\bullet} + O_3 \longrightarrow NO_2^{\bullet} + O_2$$
$$NO_2^{\bullet} + O \longrightarrow NO^{\bullet} + O_2$$
$$NO_2^{\bullet} \xrightarrow{hf} NO^{\bullet} + O$$

   Destruction is also initiated by certain halocarbon compounds such as:

$$CCl_2F_2 \xrightarrow{hf} {}^{\bullet}CClF_2 + Cl^{\bullet}$$
$$Cl^{\bullet} + O_3 \longrightarrow ClO^{\bullet} + O_2$$
$$ClO^{\bullet} + O \longrightarrow Cl^{\bullet} + O_2$$

   The $Cl^{\bullet}$ can then react with further ozone. Destruction of ozone has led to thinning of the ozone layer, especially over the North and South Poles where the ozone layer is thinnest. The result of ozone layer destruction is increased ultraviolet light reaching the Earth, potentially increasing skin cancer and endangering polar species. The thinning that leads to lower absorption of ultraviolet light also causes a cooling of the stratosphere and disruption of weather patterns.
6) *The Greenhouse Effect.* Gases such as CO_2, CH_4, N_2O and H_2O are present in low concentrations in the Earth's atmosphere. These gases reduce the Earth's emissivity and reflect some of the heat radiated by the Earth. Thus, the effect is to create a "blanket" to keep the Earth warmer than it would otherwise be. The problem arises mainly from burning fossil fuels and clearing forests by burning. The result is that global temperatures increase, leading to melting of the polar ice caps and glaciers, rising sea levels, desertification of areas, thawing of permafrost, increased weather disruptions and changes to ocean currents.

When applying legislation to atmospheric emissions, the regulatory authorities can either control emissions from individual points of release or combine all of the releases together as a combined release from an imaginary "bubble" around the whole manufacturing facility.

25.2 Sources of Atmospheric Pollution

As noted above, one of the major problems with atmospheric emissions is the number of potential sources. Solid emissions arise from:

- solids drying operations,
- kilns used for high-temperature treatment of solids,
- metal manufacture,
- crushing, grinding and screening operations for solids,
- any solids handling operation open to the atmosphere,
- incomplete combustion or fuel ash from furnaces, boilers and thermal oxidizers,
- incomplete combustion in flares, and so on.

Vapor emissions are even more difficult to deal with as they have more sources. Some are from single identifiable *point* sources, some are small, slow leakages from, for example, pipe flanges and valves and pumps seals, known as *fugitive* emissions. Vapor emission sources include:

- condenser vents.
- venting of pipes and vessels.
- inert gas purging of pipes and vessels.
- process purges to atmosphere.
- drying operations.
- application of solvent-based surface coatings.
- open operations such as filters, mixing vessels, and so on, leading to evaporation of VOCs.
- drum emptying and filling, leading to evaporation of VOCs.
- spillages of VOCs.
- process plant ventilation of buildings processing VOCs.
- storage tank loading and cleaning.
- road, rail and barge tank loading and cleaning.
- fugitive emissions through gaskets and shaft seals.
- fugitive emissions from sewers and effluent treatment.
- fugitive emissions from process sampling points.
- incomplete combustion of fuel in furnaces, boilers and thermal oxidizers.
- incomplete combustion in flares, and so on.

In larger plants, significant reductions in VOC emissions can usually be made by controlling major sources, such as tank venting, condensers and purges, and by inspection and maintenance of gaskets and shaft seals.

The largest volume of atmospheric emissions from process plants occurs from combustion gaseous emissions. Such emissions are created from:

- furnaces, boilers and thermal oxidizers,
- gas turbine exhausts,
- flares,
- process operations where coke needs to be removed from catalysts (e.g. fluid catalytic cracking regeneration in refineries), and so on.

In addition to the gaseous emissions from the combustion of fuel, gaseous emissions are also produced by chemical production, for example, SO_X from sulfuric acid production, NO_X from nitric acid production, HCl from chlorination reactions, and so on.

Example 25.1 A storage tank with a vent to atmosphere is to be filled at 25 °C with a mixture containing benzene with a mole fraction of 0.2 and toluene with a mole fraction of 0.8. Estimate the concentration of benzene and toluene in the tank vent:

a) at 25 °C,
b) corrected to standard conditions of 0 °C and 1 atm.

Assume that the mixture of benzene and toluene obeys Raoult's Law and the molar mass in kilograms occupies 22.4 m^3 in the vapor phase at standard conditions. The molar masses of benzene and toluene are 78 kg·kmol^{-1} and 92 kg·kmol^{-1} respectively. The vapor pressures of benzene and toluene at 25 °C are 0.126 bar and 0.0376 bar respectively.

Solution

a)
$$y_{BEN} = 0.2 \times \frac{0.126}{1.013} = 0.0249$$

$$y_{TOL} = 0.8 \times \frac{0.0376}{1.013} = 0.0297$$

The balance is inert gas (air). Thus, the concentration at 25 °C is given by:

$$C_{BEN} = y_{BEN} \times \frac{1}{22.4\left(\frac{T}{273}\right)} \times 78$$
$$= 0.0249 \times \frac{1}{22.4\left(\frac{298}{273}\right)} \times 78$$
$$= 0.0794 \text{ kg} \cdot \text{m}^{-3}$$
$$= 79,400 \text{ mg} \cdot \text{m}^{-3}$$

$$C_{TOL} = y_{TOL} \times \frac{1}{22.4\left(\frac{T}{273}\right)} \times 92$$
$$= 0.0297 \times \frac{1}{22.4\left(\frac{298}{273}\right)} \times 92$$
$$= 0.112 \text{ kg} \cdot \text{m}^{-3}$$
$$= 112,000 \text{ mg} \cdot \text{m}^{-3}$$

b) Corrected to standard conditions of 0 °C and 1 atm:

$$C_{BEN} = y_{BEN} \times \frac{1}{22.4} \times 78$$
$$= 0.0249 \times \frac{1}{22.4} \times 78$$
$$= 0.0867 \text{ kg} \cdot \text{m}^{-3}$$
$$= 86,700 \text{ mg} \cdot \text{m}^{-3}$$

$$C_{TOL} = y_{TOL} \times \frac{1}{22.4} \times 92$$
$$= 0.0297 \times \frac{1}{22.4} \times 92$$
$$= 0.122 \text{ kg} \cdot \text{m}^{-3}$$
$$= 122,000 \text{ mg} \cdot \text{m}^{-3}$$

These approximate calculations reveal that the concentrations are many orders of magnitude greater than those permitted for such compounds in most situations (permitted levels are typically less than 10 mg·m^{-3}). More precise calculations could be performed with an equation of state (e.g. the Peng–Robinson Equation of State, see Appendix A).

25.3 Control of Solid Particulate Emissions to Atmosphere

The selection of equipment for the treatment of solid particle emissions to atmosphere depends on a number of factors (Stenhouse, 1981; Svarovsky, 1981; Rousseau, 1987; Dullien, 1989; Schweitzer, 1997):

- size distribution of the particles to be separated,
- particle loading,
- gas throughput,
- permissible pressure drop,
- temperature.

There is a wide range of equipment available for the control of emissions of solid particles to atmosphere, as discussed in Chapter 7. These methods are classified in broad terms in Table 25.1 (Stenhouse, 1981).

1) *Gravity settlers.* Gravity settlers have already been discussed in Chapter 7 and illustrated in Figures 7.1 and 7.3. These function purely on the basis of density difference and are used to collect coarse particles. They may be used as prefilters. Only particles in excess of 100 μm can reasonably be removed (Stenhouse, 1981; Rousseau, 1987; Dullien, 1989; Schweitzer, 1997).
2) *Inertial separators.* Inertial separators were also discussed in Chapter 7 and illustrated in Figure 7.4. The particles are given a downward momentum to assist the settling. Only particles in excess of 50 μm can be reasonably removed (Stenhouse, 1981; Rousseau, 1987; Dullien, 1989; Schweitzer, 1997). Like gravity settlers, inertial collectors are widely used as prefilters.
3) *Cyclones.* Cyclones are also primarily used as prefilters. These have already been discussed in Chapter 7 and illustrated in Figure 7.6. The particle-laden gas enters tangentially and spins downwards and inwards, ultimately leaving the top of the unit. Particles are thrown radially outwards to the wall by the centrifugal force and leave at the bottom.

   Cyclones can be used under conditions of high particle loading. They are cheap, simple devices with low maintenance requirements. Problems occur when separating materials that have a tendency to stick to the cyclone walls.
4) *Scrubbers.* Scrubbers are designed to contact a liquid with the particle-laden gas and entrain the particles with the liquid. They offer the obvious advantage that they can be used to remove gaseous as well as particulate pollutants. The gas stream may need to be cooled before entering the scrubber. Some of the more common types of scrubbers are shown in Figure 7.11.

   Packed columns are widely used in gas absorption, but particulates are also removed in the process (Figure 7.11a). The main disadvantage of packed columns is that the solid particles will often accumulate in the packing and require frequent cleaning. Some designs of packed column allow the packing to self-clean through inducing movement of the bed from the up-flow of the gas. Spray scrubbers are less prone to fouling and use a tangential inlet to create a swirl, enhancing the separation, as illustrated in Figure 7.11b. An increase in the efficiency of particulate removal can be achieved at the expense of increased pressure drop using a venturi scrubber (Figure 7.11c).
5) *Bag filters.* Bag filters, as discussed in Chapter 7 and illustrated in Figure 7.10b, are probably the most common method of separating particulate materials from gases. A cloth or felt filter material is used, which is impervious to the particles. Bag filters are suitable for use in very high dust load conditions. They have a high efficiency but suffer from the disadvantage that the pressure drop across them may be high.
6) *Electrostatic precipitators.* Electrostatic precipitators are used where collection of fine particles at a high efficiency coupled with a low-pressure drop is necessary. The arrangement is illustrated in Figure 7.9. Particle-laden gas enters a number of tubes or passes between parallel plates. The particulates are charged and are deposited on the earthed plates or tube walls. The walls are mechanically *rapped* periodically to remove the accumulated dust layer. Electrostatic precipitators can be used to treat large gas flows, e.g. boilers, cement plants, waste thermal oxidizers.

Table 25.1

Methods of control of emissions of solid particles.

Equipment	Approximate particle size range (μm)
Gravity settlers	>100
Inertial separators	>50
Cyclones	>5
Scrubbers	>3
Venturi scrubbers	>0.3
Bag filters	>0.1
Electrostatic precipitators	>0.001

25.4 Control of VOC Emissions

Release limits for VOCs are set for either specific components (e.g. benzene, carbon tetrachloride) or as VOCs for organic compounds with a lower environmental impact and classed together and reported, for example as toluene.

The hierarchy appropriate for control of VOC emissions is in descending order of preference (Hui and Smith, 2001):

1) Eliminate or reduce VOC emissions at source through changes in design or operation.
2) Recover the VOC for reuse.
3) Recover the VOC for thermal oxidation with process heat recovery.

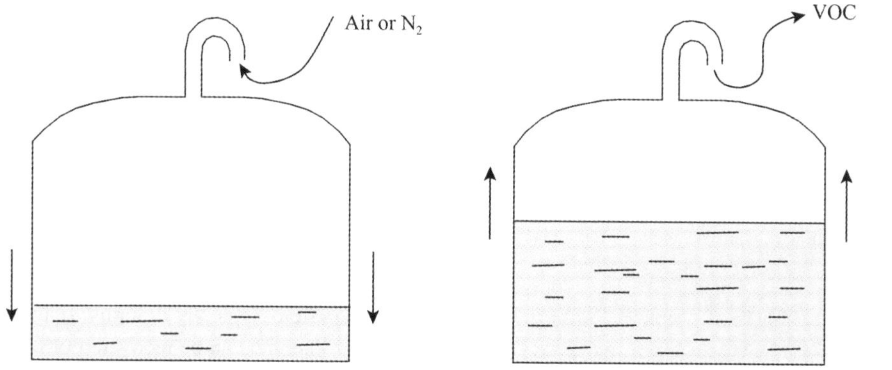

Figure 25.1

Tank loading causes VOC emissions to atmosphere.

4) Thermal oxidation of the VOC-laden gas stream with process heat recovery.
5) Oxidation of the VOC-laden gas stream without process heat recovery.

The problem with eliminating or reducing VOC emissions at source is that the sources of VOC emissions can be many and varied.

Tank loading can be a significant source of VOC emissions from atmospheric storage tanks used for the storage of organic liquids. Figure 25.1 shows an atmospheric storage tank first being emptied. As the liquid level falls, air or inert gas must be drawn into the tank to prevent the tank from collapsing from the pressure of the atmosphere. The air or nitrogen drawn into the vapor space will become saturated with vapor from the volatile liquid in the tank. As the tank is filled, the liquid level rises, displacing the air or nitrogen from the vapor space in the tank that is now saturated in VOC material. There are a number of ways in which such VOC emissions can be prevented from being emitted from storage tanks. Figure 25.2 shows a simple technique involving a balancing line. The atmospheric storage tank in Figure 25.2 is fitted with a vacuum/pressure relief system to guard against overpressure or underpressure of the storage tank. As the atmospheric storage tank is emptied and the road tanker filled in Figure 25.2, the displaced vapor from the road tanker is pushed back into the atmospheric storage tank. In this way, displaced vapor from the road tanker is prevented from being released to the atmosphere. The same system works if the road tanker is being emptied into the atmospheric storage tank, rather than being filled from the atmospheric storage tank as shown in Figure 25.2. If the atmospheric storage tank is being filled, then the displaced vapor from the atmospheric storage tank enters the road tanker vapor space and is not released to the atmosphere. Clearly, the same technique works for rail cars and barges.

25

Another way storage tanks can "breathe" to atmosphere results from the expansion and contraction of the tank contents as the temperature changes between day and night. However, such effects tend to be smaller than those associated with tank filling and emptying.

Floating roofs can be used, as illustrated in Figure 25.3:

1) *External floating roof.* In this arrangement, the roof floats directly on the surface of the stored liquid and is open to the atmosphere (Figure 25.3a). The floating roof has a seal system for the gap between the roof and the tank wall. In principle, this eliminates breathing of the tank to atmosphere, but maintaining

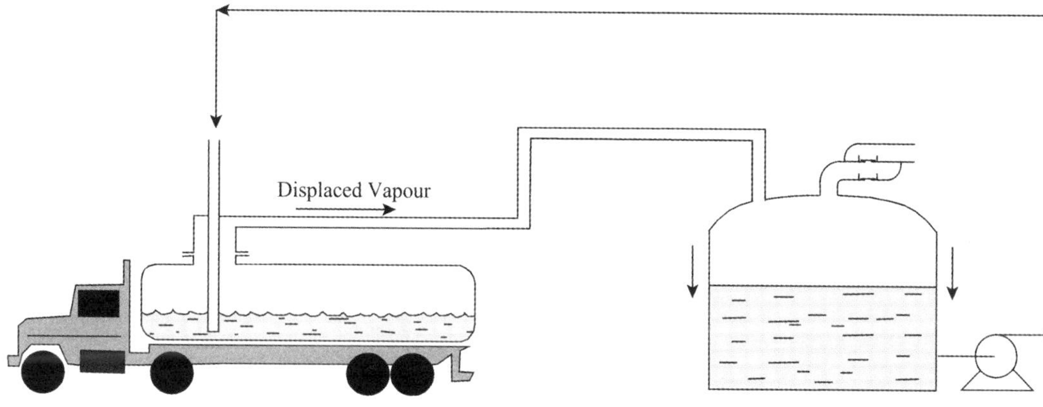

Figure 25.2

A balancing line can prevent VOC emissions when filling and emptying tankers.

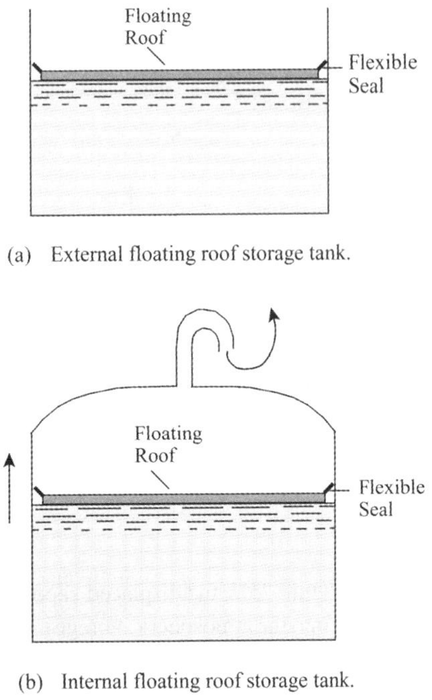

Figure 25.3

Floating roof and flexible membranes can be used to prevent the release of material. (Reproduced from Smith R and Petela EA (1992) Waste Minimization in the Process Industries Part 2 Reactors, *Chem Eng*, Dec(509–510): 17, by permission of the Institution of Chemical Engineers.)

a reliable seal between the edge of the floating roof and the tank wall can be problematic.

2) *Internal floating roof.* In this arrangement, the tank has a fixed roof in addition to the floating roof, and the floating roof is not open to the atmosphere (Figure 25.3b). The space between the floating and fixed roofs should normally be purged with an inert gas. The inert gas would be vented to atmosphere after treatment. Internal floating roofs are used when there is likely to be a heavy accumulation of rainwater or snow on the floating roof.

Liquid effluent systems can be significant sources of VOC emissions. If organic material is sent to effluent, along with aqueous effluent, then evaporation from open drains can create significant VOC emissions. If this is the case, then having separate sewers for organic and aqueous waste, as illustrated in Figure 25.4, can eliminate the problem. The organic waste is collected in a closed system, typically draining to a sump tank, from which the organic material can be recovered or treated before disposal. The segregation system shown in Figure 25.4 has many other benefits than simply eliminating VOC emissions from drains. Segregation of the effluents, as illustrated in Figure 25.4, is also useful due to reduced volume of liquid waste for treatment, as will be discussed in Chapter 26.

Sampling of organic liquids for laboratory analysis, usually for reasons of quality control, can also be a significant source of VOC emissions. Figure 25.5 shows a traditional sampling technique involving a sample point with a valve. Significant amounts of waste can be created in order to create a representative sample. The sample vessel would then be taken to the laboratory, the required sample taken and the remainder or the contents of the sample vessel sent to waste. The material going to drain from this procedure can create VOC emissions from evaporation from the sewers, as discussed above. For some chemical processes where a considerable amount of laboratory analysis is required, sampling waste can be a significant source of waste sent to liquid effluent. Figure 25.5 shows a method to eliminate sampling waste and, therefore, the VOC emissions associated with sampling waste. A sample vessel is connected directly to the pipe from which the sample needs to be taken by two tubes. The two tubes are connected to the pipe on either side of a restriction, such as an orifice plate or valve. If the valves in the tubes connecting the sample to the pipe are open, then the pressure drop across the restriction will create a flow through

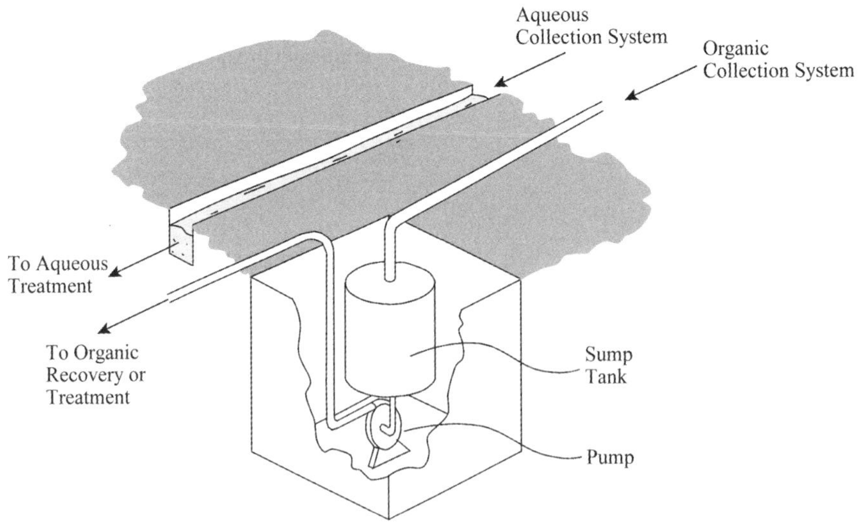

Figure 25.4

Segregation of aqueous and organic waste can prevent VOC emissions from open drains.

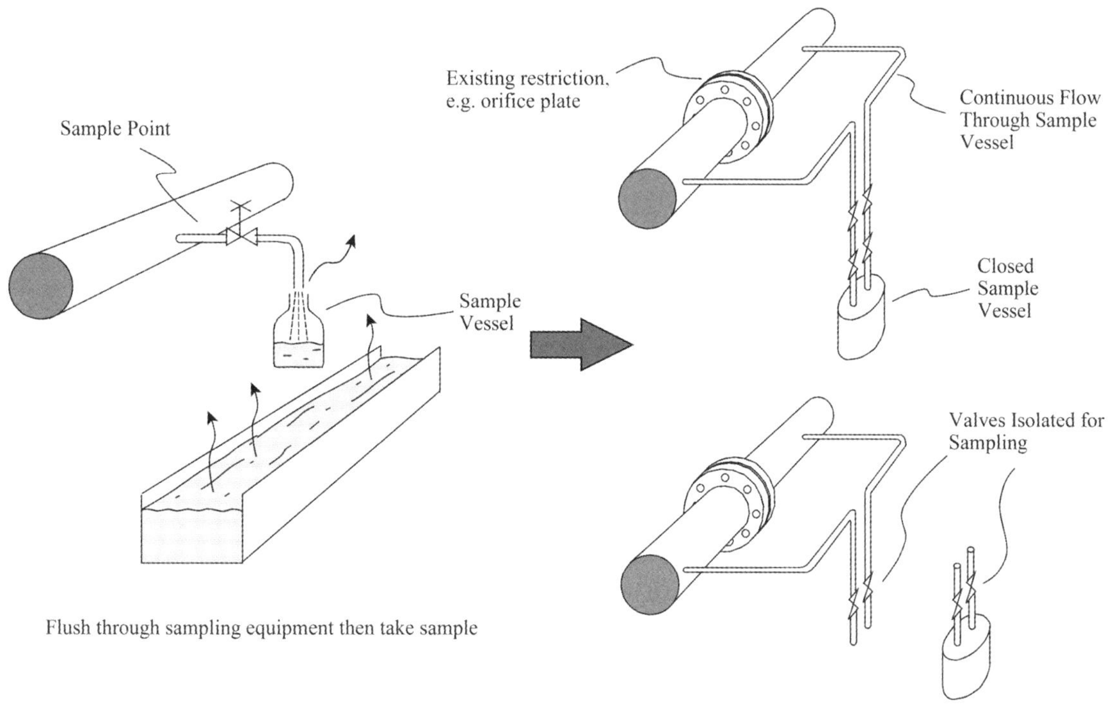

Figure 25.5
Prevention of emissions from sampling.

25

the sample vessel, ensuring a representative sample is obtained. When the sample needs to be taken, valves are closed to isolate the closed sampling vessel. This can then be taken to the laboratory for analysis. Any unused material in the sample vessel is returned to the plant and reconnected, eliminating the waste of sampling material from the laboratory.

Slow leaks from gaskets and from compressor, pump and valve seals can create significant fugitive emissions of VOCs. Such emissions can be reduced at source by better maintenance. However, more sophisticated sealing arrangements are desirable to eliminate the problem at source.

Once VOC emissions have been eliminated or reduced at source, then recovery of the VOC for reuse should be considered. Figure 25.6 shows a vapor recovery system associated with an atmospheric storage tank. The storage tank is fitted with a vacuum-pressure relief valve, which allows air (or nitrogen) into the vapor space when the liquid level falls (Figure 25.6a) but forces the vapor out through a recovery system when the tank is filled (Figure 25.6b).

Many processes involve open operations (e.g. filters, drum handling, etc.) that create VOC emissions. If this is the case, it is often not practical to enclose individual operations, but to install a ventilation system that draws a continuous flow across the operations into a duct and then a vapor recovery system, before being released to the atmosphere.

The methods of VOC recovery are:

- condensation,
- membranes,
- absorption,
- adsorption.

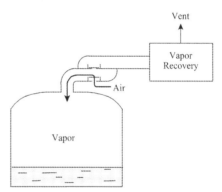

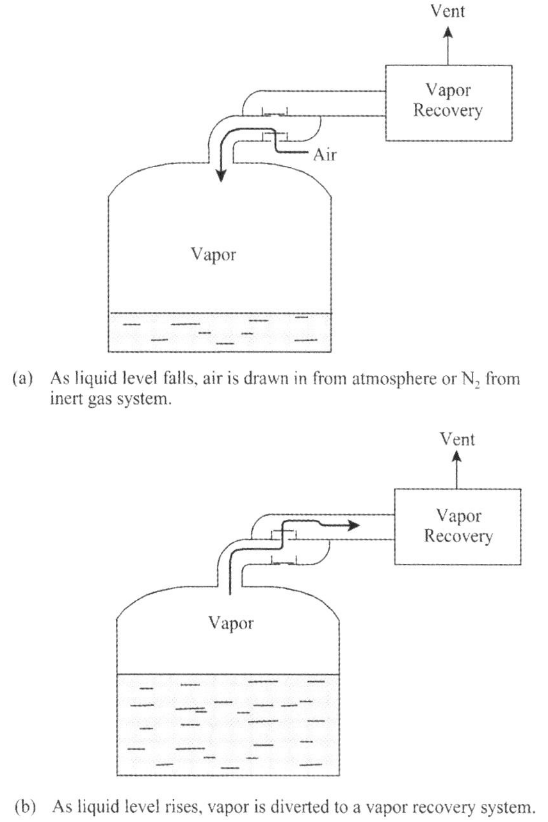

Figure 25.6
Storage tank fitted with a vapor treatment system. (Reproduced from Smith R and Petela EA (1992) Waste Minimization in the Process Industries Part 2 Reactors, *Chem Eng*, Dec(509–510): 17, by permission of the Institution of Chemical Engineers.)

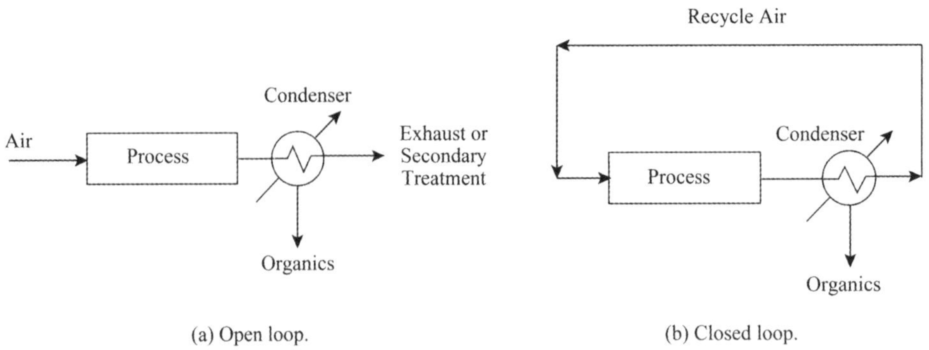

Figure 25.7

Condensation of volatile material.

1) *Condensation.* Condensation can be accomplished by an increase in pressure or a decrease in temperature. Most often, a decrease in temperature is preferred. Figure 25.7 illustrates two ways in which VOC recovery can be accomplished from an air stream used in a process. Figure 25.7a shows an open loop in which air enters the process, picks up VOCs and enters a condenser for recovery of the VOCs before being vented directly or passed to secondary treatment before venting. An alternative way is shown in Figure 25.7b. This illustrates a recirculation system in which the VOC is recovered from recycled air. If this is possible, it is a better arrangement than Figure 25.7a as the exhaust is eliminated.

Unfortunately, condensation of VOCs usually requires refrigeration. Figure 25.8a shows an arrangement with refrigerated condensation using a primary refrigerant in a simple refrigeration loop. Figure 25.8b shows an arrangement in which a secondary refrigerant is used in the VOC condenser. The use of a secondary refrigerant is less efficient than using the primary refrigerant directly. However, if a number of cooling duties are required, then a single refrigeration loop cooling a secondary refrigerant can be used for a number of condenser duties.

A further complication of refrigeration is that, if the gas stream contains water vapor or organic compounds with high freezing points, then these will freeze in the condenser, causing it to block up. It is therefore often necessary to precool the gas stream to remove water before condensing the VOCs. Figure 25.9a shows an arrangement in which a refrigeration loop is first used to precool the gas stream to the region of 2 to 4 °C to remove water vapor and high-boiling organics. A second condenser using a second refrigeration loop then condenses the VOCs. Figure 25.9b shows the same basic arrangement but with a two-level refrigeration system, instead of two separate loops. In both cases in Figure 25.9, it is likely that the VOC condenser will have to be taken off-line periodically to

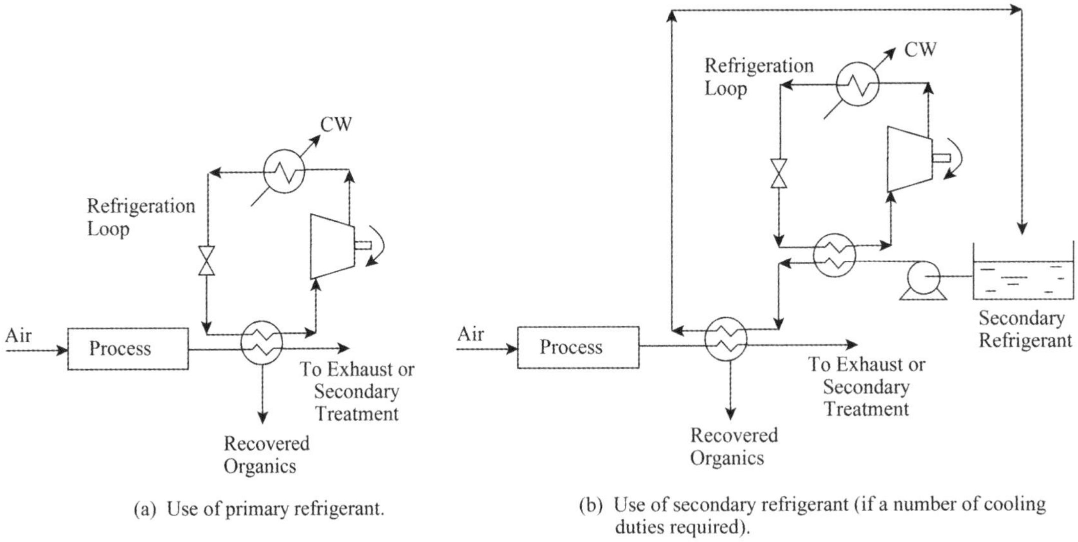

Figure 25.8

Refrigeration is usually required for the condensation of vapor emissions.

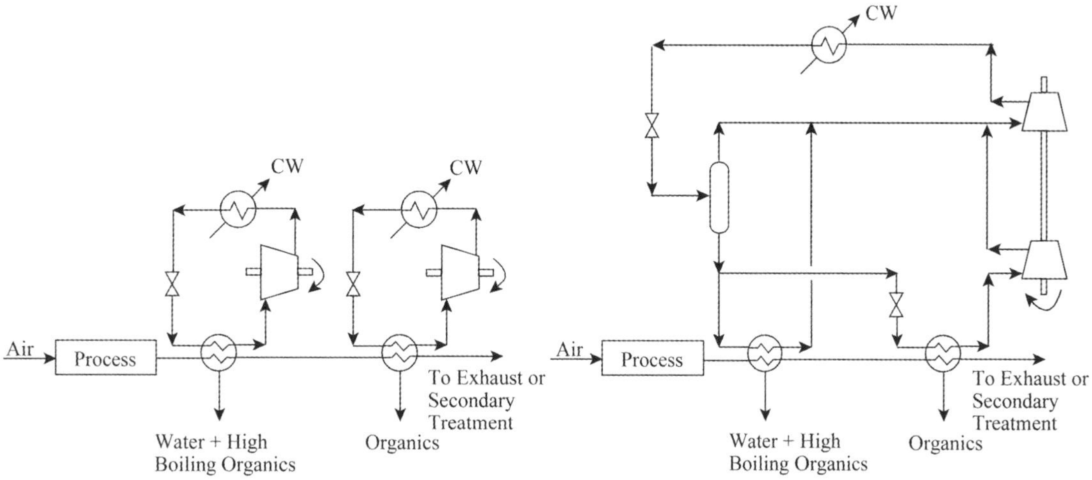

(a) Two independent refrigeration loops.

(b) Two level refrigeration system.

Figure 25.9

If a gas stream contains water vapor or organic compounds with high freezing points, then precooling is necessary.

remove frozen water and organic material that will accumulate over time.

If the site uses nitrogen as inert gas and stores the nitrogen as a liquid, then this can be used as a source of refrigeration. Rather than vaporizing the liquid nitrogen by low-pressure steam or an air heat exchanger, the liquid nitrogen can be used to cool a VOC-laden gas stream to condense the organic material.

The problem of freezing on the condenser surface can be avoided by using direct contact condensation, as shown in Figure 25.10. A secondary refrigerant is contacted directly with the vent stream containing VOC. A refrigeration loop is used to cool the secondary refrigerant. The mixture of secondary refrigerant and condensed VOC then needs to be separated and the secondary refrigerant recycled.

It is usually uneconomic to compress a gas in order to bring about the condensation of VOC. Figure 25.11 shows a flowsheet in which the VOC-laden vent gas is compressed in order that condensation can take place using cooling water rather than refrigeration. The arrangement in Figure 25.11 uses an expander on the same shaft as the high-pressure compression stage in order to recover part of the work of compression. Around 60% of the work of compression can be recovered in an arrangement like the one in Figure 25.11.

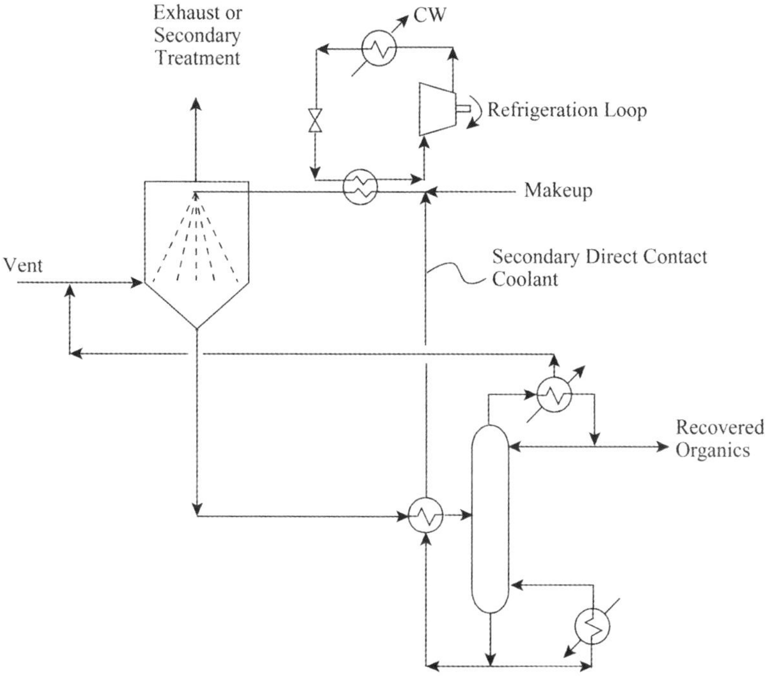

Figure 25.10

Direct contact condensation.

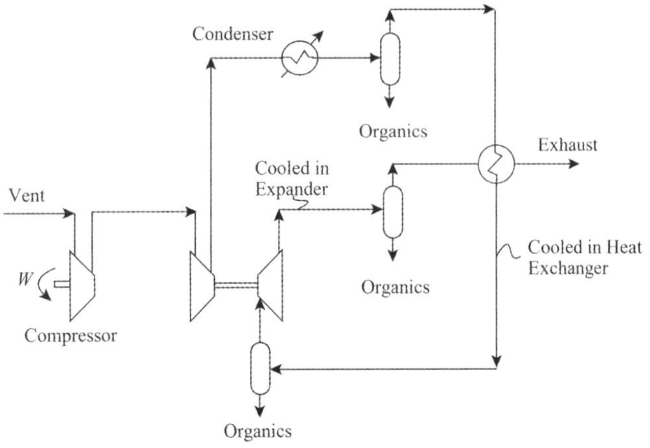

Figure 25.11
Compression of the vent with recovery of compression power.

Example 25.2 An air stream contains acetone vapor. Estimate the following.

a) The outlet concentration that can be achieved by cooling to 35 °C at 1 atm pressure (cooling water temperature).
b) The outlet concentration that can be achieved by cooling to –20 °C at 1 atm pressure.
c) The outlet temperature to which the air stream must be cooled to obtain an outlet concentration of 20 mg·m^{-3}.

The concentrations should be quoted at standard conditions of 0 °C and 1 atm. It can be assumed that the molar mass in kilograms occupies 22.4 m^3 at standard conditions. The molar mass of acetone can be taken to be 58 kg·kmol^{-1}. The vapor pressure of acetone is given by:

$$\ln P^{SAT} = 10.0310 - \frac{2940.46}{T - 35.93}$$

where P^{SAT} = saturated liquid vapor pressure (bar)
T = absolute temperature (K)

Solution

a) Acetone will start to condense when the partial pressure equals the vapor pressure. At 35 °C (308 K), the vapor pressure is given by:

$$P^{SAT} = \exp\left[10.0310 - \frac{2940.46}{308 - 35.93}\right]$$
$$= 0.460 \text{ bar}$$

For a partial pressure of 0.460 bar, the mole fraction is given by:

$$y = \frac{0.460}{1.013}$$
$$= 0.454$$
$$C = y \times \frac{1}{22.4} \times 58$$
$$= 0.454 \times \frac{1}{22.4} \times 58$$
$$= 1.18 \text{ kg} \cdot \text{m}^{-3}$$
$$= 1.18 \times 10^6 \text{ mg} \cdot \text{m}^{-3}$$

b) At –20 °C (253 K):

$$P^{SAT} = 0.0297 \text{ bar}$$
$$C = \frac{0.0297}{1.013} \times \frac{1}{22.4} \times 58$$
$$= 0.0759 \text{ kg} \cdot \text{m}^{-3}$$
$$= 75{,}900 \text{ mg} \cdot \text{m}^{-3}$$

c) For a concentration of 20 mg·m^{-3} (20 × 10^{-6} kg·m^{-3}):

$$y = \frac{20 \times 10^{-6}}{58} \times 22.4$$
$$= \frac{P^{SAT}}{1.013}$$
$$P^{SAT} = 1.013 \times \frac{20 \times 10^{-6}}{58} \times 22.4$$
$$= 7.825 \times 10^{-6} \text{ bar}$$

Rearranging the vapor pressure correlation:

$$T = 35.93 - \frac{2940.46}{\ln P^{SAT} - 10.310}$$
$$= 35.93 - \frac{2940.46}{\ln 7.825 \times 10^{-6} - 10.310}$$
$$= 169 \text{ K}$$
$$= -104 \text{ °C}$$

This temperature is outside of the range of the vapor pressure correlation and below the freezing point of acetone (–95 °C). However, it illustrates just how low temperatures must be to achieve very low concentrations of VOC.

Example 25.2 illustrates that low temperature is required for a high recovery rate of VOCs. However, extremely low temperatures are required to achieve high environmental standards. Therefore, condensation is often used in conjunction with another process (e.g. adsorption) to achieve high environmental standards.

When the VOC-laden gas stream contains a mixture of VOCs, then the calculations must be performed using the methods described for single-stage equilibrium calculations in Chapter 8. The temperature at the exit of the condenser must be assumed, together with a condenser pressure. The vapor fraction is then solved by trial and error using the methods described in Chapter 8, and the complete mass balance can be determined on the basis of the assumption of equilibrium.

The major advantage of condensation is that VOCs are recovered without contamination, in contrast with absorption and adsorption when water or other materials may be present after recovery. The significant disadvantage of refrigeration is that it has a relatively low efficiency of recovery (typically less than 95%) and the associated high operating costs.

2) *Membranes.* VOCs can be recovered using an organic-selective membrane that is more permeable to organic vapors than permanent gases. Figure 25.12 shows one possible arrangement for recovery of VOCs using a membrane (Hydrocarbon Processing's Environmental Processes, 1998). A vent gas is compressed and enters a condenser in which VOCs are

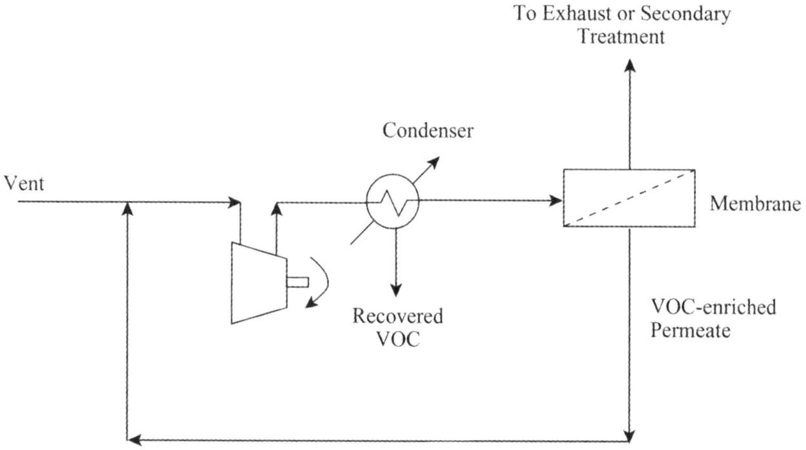

Figure 25.12
Recovery of VOCs using membranes.

recovered. The gases from the condensers then enter a membrane unit, in which the VOC permeates through the membrane. A VOC-enriched permeate is then recycled back to the compressor inlet. Recovery rates for such an arrangement can be as high as 90 to 99%. Other arrangements than the one shown in Figure 25.12 are possible, again using a combination of membrane and condensation.

3) *Absorption.* Physical absorption, as discussed in Chapter 9, can be used for the recovery of VOCs. If possible, the solvent used should be regenerated and recycled, with the VOC being recovered. If the VOC is water-soluble (e.g. formaldehyde), then water can be used as the solvent. However, high-boiling organic solvents are most often used for VOC absorption. Efficiency of recovery depends on the VOC, solvent and absorber design, but efficiencies of recovery can be as high as 95%. The efficiency of recovery increases with decreasing temperature and increases with increasing pressure. When using an organic solvent, care must be exercised in the selection of the solvent that vaporization of the solvent into the gas stream does not create a new environmental problem.

4) *Adsorption.* Adsorption of VOCs is most often carried out using activated carbon with in situ regeneration of the carbon using steam. The details have been discussed in Chapter 9. A number of different arrangements are used for adsorption. One possible arrangement is shown in Figure 25.13. This involves a three-bed system. The first bed encountered by the VOC-laden gas stream is the primary bed. The gas from the primary bed enters the secondary bed, which will have just been regenerated using steam and is now being cooled by the flow of the gas stream. The third bed is off-line, being regenerated using steam. The steam from the regenerating bed is condensed, along with the recovered organics, and the condensate separated to recover

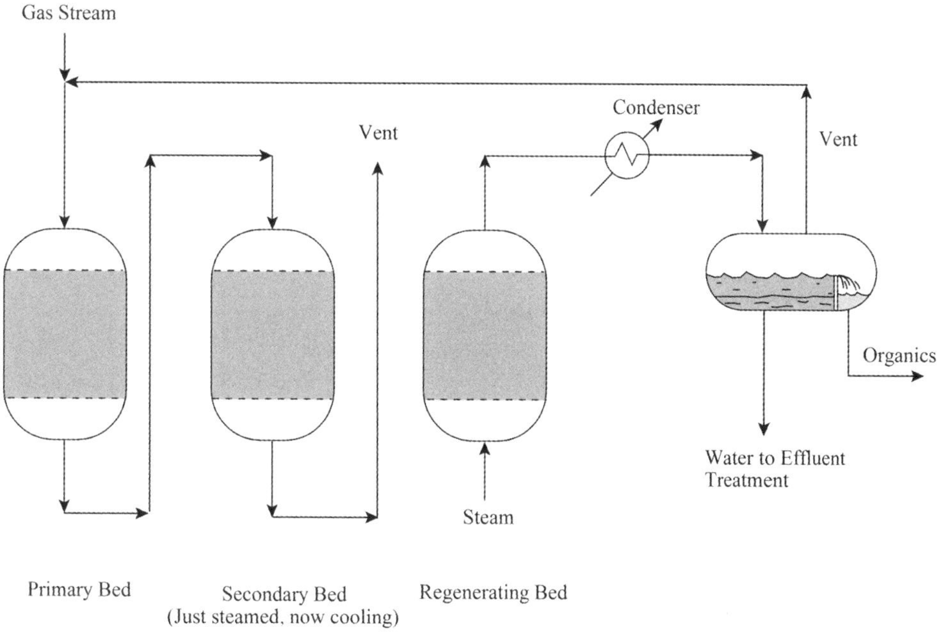

Figure 25.13
Adsorption with a three-bed system.

the organics. The vent from the condenser receiver will normally be too concentrated to vent directly to atmosphere and therefore will normally be connected back to the inlet of the primary bed. Once the primary bed has become saturated and breakthrough occurs (see Chapter 9), the beds are switched. The current secondary bed becomes the new primary bed. The regenerated bed becomes the secondary bed and the previous primary bed is now regenerated. The switching cycle is normally based on a timing arrangement.

Systems with one bed can be used for small duties (typically less than 5 kg per day discharge of organic material) as *disposable cartridges*. Two-bed arrangements can be used when environmental standards are not too demanding. One bed is absorbing whilst the other is being regenerated, with constant switching between the two. More complex cycles than the one in Figure 25.13 can be used involving four beds, instead of three, for more demanding duties.

The efficiency of recovery for adsorption:

- increases with molar mass of the VOC (carbon adsorbers are not good for controlling low-boiling VOCs with molar mass less than typically 40),
- increases with decreasing adsorption temperature,
- increases with increasing pressure,
- increases with decreasing concentration of water vapor in the gas stream due to competition of water for the adsorption sites.

In summary, condensation and absorption are usually the simplest methods of VOC recovery. Recovery methods can be effectively used in combination, but at a cost. Adsorption is usually the only method capable of achieving very low concentrations of VOC. If the gas stream contains a mixture of VOCs, then the recovered liquid might not be suitable for reuse and will need to be separated by distillation or destroyed by thermal oxidation.

Once the potential for minimizing VOC emissions at source has been exhausted and the recovery of VOCs exhausted, then any remaining VOC must be destroyed. There are two methods of VOC destruction (US Environmental Protection Agency, 1992a):

- thermal oxidation (including catalytic thermal oxidation),
- biological treatment.

Consider now the various methods of thermal oxidation.

1) *Flare stacks.* Flares use an open combustion process with oxygen for the combustion provided by air around the flame. Good combustion depends on the flame temperature, residence time in the combustion zone and good mixing to complete the combustion. The flare stack can be dedicated to a specific vent stream or for a combination of streams via a header. Flares can be categorized by:

   - the height of the flare tip (elevated versus ground flares),
   - the method of enhancing mixing at the flare tip (steam assisted, air assisted, unassisted).

   The most common type used in the process industries is the steam-assisted elevated flare. This is illustrated in Figure 25.14. Combustion takes place at the flare tip. The flare tip requires pilot burners, an ignition device and fuel gas to maintain the pilot flame. In steam-assisted flares, steam nozzles around the perimeter of the flare tip are used to provide the mixing. As shown in Figure 25.14, before the vent stream enters the flare stack, a knockout drum is required to remove any entrained liquid. Liquids must be removed as they can extinguish the flame or lead to irregular combustion in the flame. Also, there is a danger

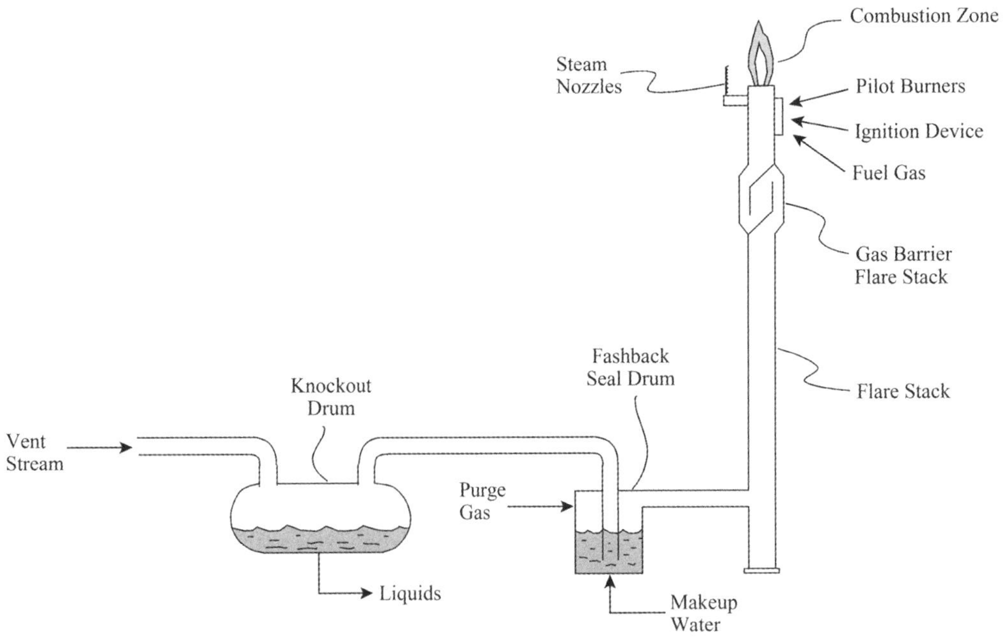

Figure 25.14
An elevated steam-assisted flare stack.

that the liquid might not be completely combusted, which can result in liquid reaching the ground and creating hazards. The vent header needs to be protected against the flame propagating back into the header. A *gas barrier* or a *flame arrester* prevents the flame passing down into the vent header when the vent stream flowrate is low. A *flashback seal drum* containing water provides additional protection. In some designs the flashback seal drum is incorporated into the base of the flare stack.

Flare performance can achieve a destruction efficiency of up to 98% with steam-assisted flares. Halogenated and sulfur-containing compounds should not be flared. Flares are applicable to almost any VOC-laden stream and can be used when there are large variations in the composition, heat content and flowrate of the vent stream.

However, flares should only be used for abnormal operation or emergency upsets when there is a short-term requirement to deal with an abnormal flowrate of VOC-laden gases. For normal operation, other methods should be used.

2) *Thermal oxidation.* In some cases, vent streams can be directed to be the inlet combustion air to steam boilers, such that any VOCs are oxidized to CO_2 and H_2O. However, more often it is necessary to use a *thermal oxidizer* specifically designed to deal with waste streams containing VOCs. Figure 25.15 illustrates three designs of thermal oxidizers. Figure 25.15a shows a vertical thermal oxidizer. The vent enters a refractory-lined combustion chamber fed with auxiliary air and auxiliary fuel. Figure 25.15b shows the corresponding arrangement mounted horizontally. Figure 25.15c shows a fluidized bed arrangement. In this arrangement, the vent enters a bed of fluidized inert material (e.g. sand), along with supplementary fuel. The bed is fluidized by an airstream. The fluidized bed provides good mixing in the combustion zone. It also provides some heat storage that can compensate for variations in the entering vent stream.

Conditions in the thermal oxidizer are typically 25% oxygen over and above stoichiometric requirements to ensure complete oxidation to CO_2 and H_2O. Minimum temperatures required for VOCs comprising carbon, hydrogen and oxygen are around 750 °C, but 850 to 900 °C are more typical. The residence time of the VOCs in the combustion zone is typically 0.5 to 1 second. When thermally oxidizing halogenated organic compounds, the minimum temperature used is in the range of 1100 to 1300 °C, with a residence time of up to two seconds. When oxidizing waste containing halogenated materials, it is important to avoid conditions that form dioxins and furans. Dioxins and furans can form as the gases from the thermal oxidation are cooled, especially if there is fly ash present. The greatest formation of dioxins and furans is as the gases are cooled between 500 °C and 200 °C. Gases can be cooled quickly by quenching (water injection) to reduce formation. Typical conditions for thermal oxidation of halogenated materials are greater than 1200 °C, cooling to 400 °C in a waste heat boiler and then quench to 70 °C. Permitted emission levels of dioxins and furans are typically less than 0.1 ng/m^3. Adsorption on powdered activated carbon or a carbon filter can be used for treatment of the exhaust gases for dioxins and furans. Alternatively, catalytic oxidation using a honeycomb catalyst can be used.

25

Supplementary fuel is required for start-up and required if the feed concentration of organic material is low or the feed concentration varies. Supplementary fuel is usually required for vent streams, as processed vents are normally designed to operate below the lower flammability limit in the case of destruction of VOCs in air or below the minimum oxygen concentration for VOCs in an inert gas mixture (see Chapter 28). The VOC should normally be less than 25 to 30% of the lower flammability limit for air mixtures or less than 40% of the minimum oxygen concentration for inert gas mixtures. These limits can be increased if on-line analysis equipment is installed to monitor VOC concentrations continuously. If the vent stream is above this limit, the airflow in the vent is normally increased accordingly. Otherwise, nitrogen

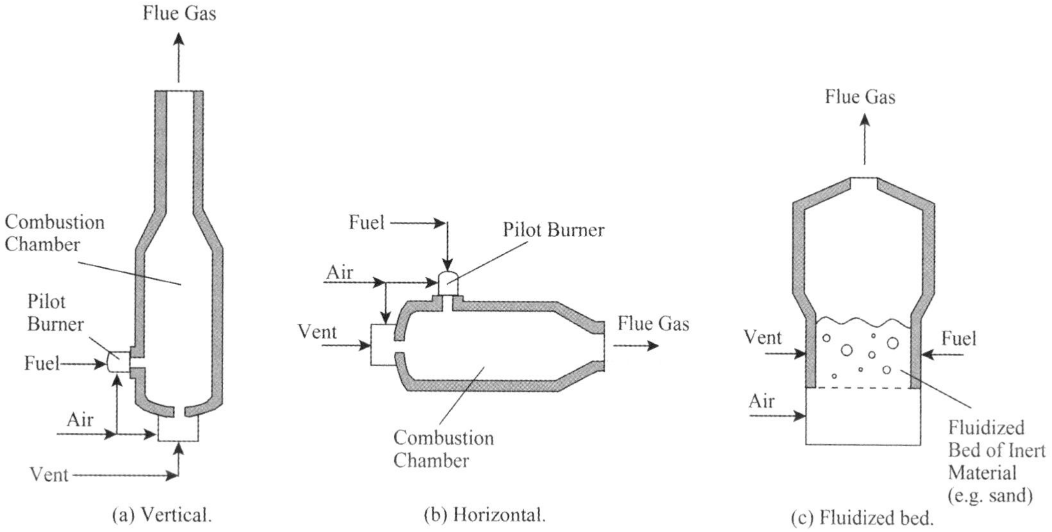

Figure 25.15
Thermal oxidation.

can be introduced, but this is more expensive. The arrangements in Figure 25.15 are not energy efficient, with no attempt to recover heat from the flue gas. There are two general ways in which the heat in the flue gas can be exploited. In the first, the heat in the flue gas is used to preheat the incoming feed stream to reduce the fuel requirements. In the second, the heat is used to generate steam, which is exported for process use.

Figure 25.16a shows a thermal oxidizer with *recuperative heat recovery* to preheat the incoming feed stream. In this arrangement, fuel and air enter directly into the combustion zone. However, the vent stream is preheated by heat transfer through the walls of the combustion chamber. Many different schemes for recuperative heat recovery through indirect exchange of heat in heat transfer equipment are possible. Figure 25.16b shows a thermal oxidizer with *regenerative heat recovery* to preheat the incoming feed stream. Fuel and air enter the combustion zone directly. However, the vent stream enters via a hot bed that preheats the vent stream prior to entering the combustion chamber. The scheme in Figure 25.16b shows two regenerative beds that are switched periodically. As the bed used to heat the vent stream cools, the other bed is being heated by the hot exhaust gases before being vented. Once the bed that is preheating the vent stream cools, then the beds are switched such that the cool bed is then preheated by the hot flue gas and the new hot bed used to preheat the incoming vent stream. The material used in the regenerative beds can be refractory material or metal. It should be noted that with a two-bed regenerative system, when one bed is switched from heating mode to cooling mode, that bed will contain vapor feed that has not yet been treated. This untreated vapor within the bed will thus be vented to atmosphere directly untreated. This problem can be overcome by the introduction of an additional bed in a three-bed system. However, this introduces additional capital cost and complexity of operation. Alternatively, the regeneration can be carried out continuously by a cylindrical rotating bed rather than having beds being switched. The rotating bed is in the form of a wheel that slowly rotates. The ducting arrangement is such that the incoming vent stream and the hot exhaust gas both flow axially across the wheel, with different parts of the wheel being exposed to each stream. Different parts of the wheel are thus in heating and cooling mode at the same time. This achieves the same purpose as the switching beds shown in Figure 25.16b but operates continuously. This continuous operation is preferable to switching beds, but it is mechanically more complex and introduces some challenging sealing requirements for the ducting of streams around the wheel.

If the vent is to be preheated prior to combustion, it should be lower than the flammability limit or the minimum oxygen concentration (see Chapter 28). Recuperative heat recovery above 70% is usually not economic for gas-to-gas heat exchange. Heat recovery increases to typically 80% if steam generation is used. Regenerative heat recovery can be as high as 95%.

If the vent stream contains halogenated material or sulfur, the exhaust gases must be scrubbed before release. Figure 25.16c shows a typical arrangement in which the cooled exhaust gases are scrubbed first with water and then with sodium hydroxide solution before being vented. The water

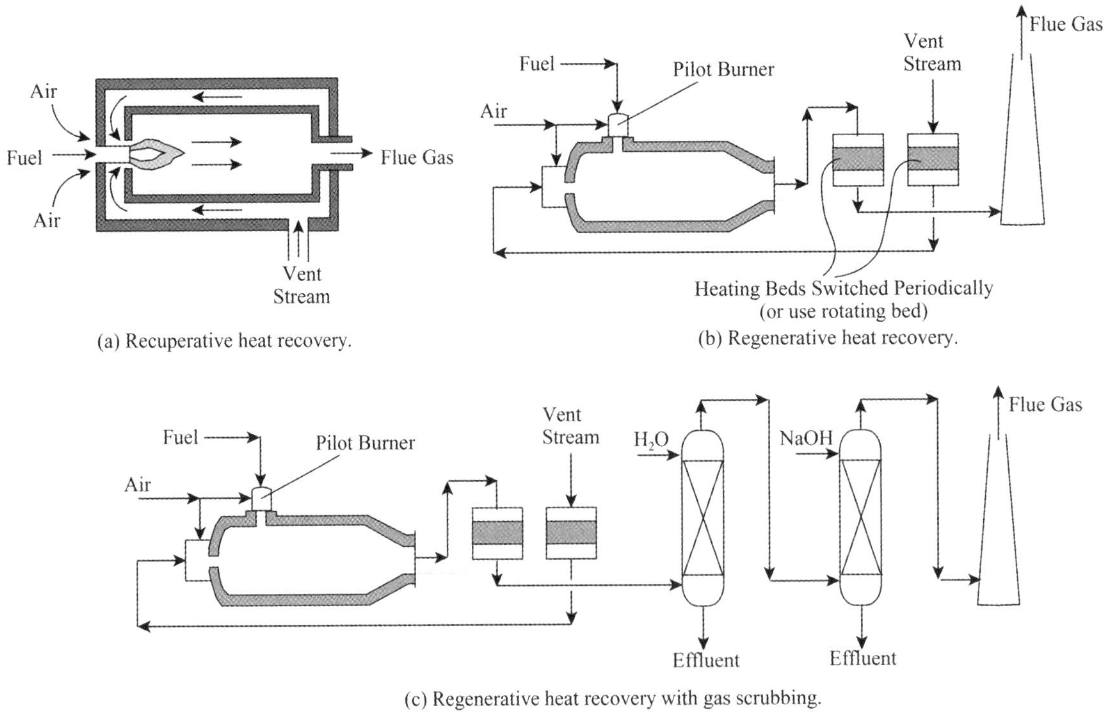

Figure 25.16

Recuperative and regenerative heat recovery from thermal oxidation.

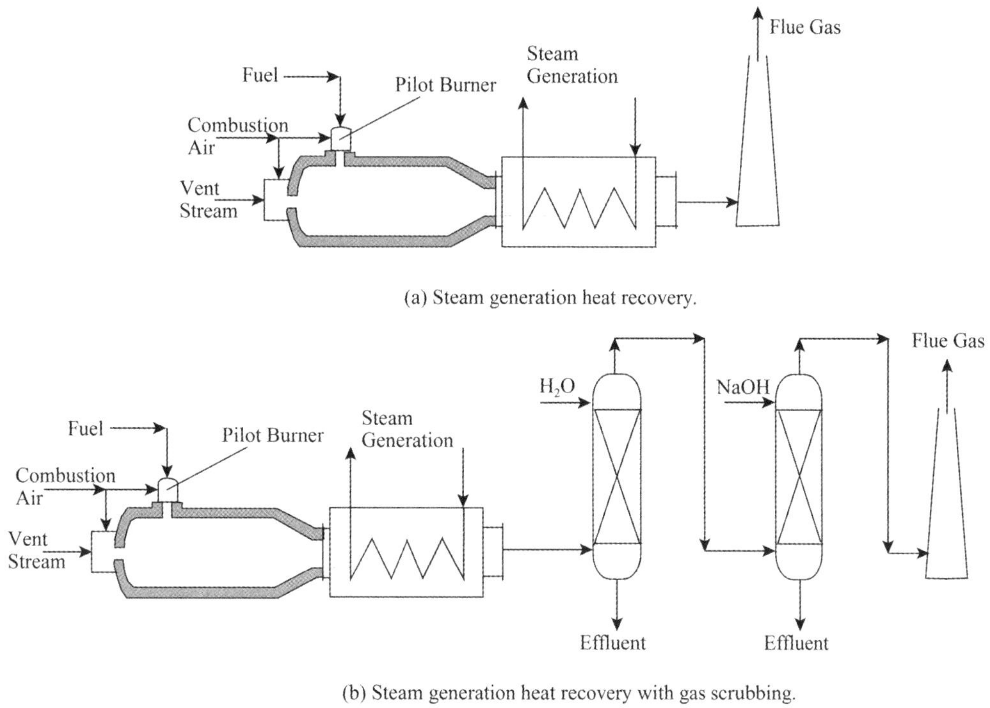

Figure 25.17
Steam generation heat recovery from thermal oxidation.

and sodium hydroxide scrubbers shown in Figure 25.16c are likely to be on a recirculation arrangement with a purge stream to liquid effluent. The arrangement in Figure 25.16c shows the exhaust gases from the thermal oxidation being cooled by regeneration before scrubbing. The hot flue gases from the thermal oxidizer must be cooled in one way or another before entering the scrubbers. This can be done by heat recovery, or by quenching with direct water injection, or a combination. Legislative requirements might dictate that there is no visible plume from the stack. If this is the case, the stack must be maintained above 80 °C, which might require the flue gas to be reheated before release.

Figure 25.17a shows an arrangement in which the hot gases from the thermal oxidizer are used to generate steam for process use. Figure 25.17b shows the corresponding arrangement with scrubbing before release.

The performance of thermal oxidizers usually gives destruction efficiency greater than 98%. When designed to do so, destruction efficiency can be virtually 100% (depending on temperature, residence time and mixing in the thermal oxidizer). Thermal oxidizers can be designed to handle minor fluctuations but are not suitable for large fluctuations in flowrate.

3) *Catalytic thermal oxidation.* A catalyst can be used to lower the combustion temperature for thermal oxidation and to save fuel. Figure 25.18 shows a schematic of a catalytic thermal oxidation arrangement. Combustion air and supplementary fuel, along with preheated VOC-laden stream, enter the preheater before the catalyst bed. Heat recovery, as illustrated in Figure 25.18, is not always used. The catalysts are typically platinum or palladium on an alumina support or metal oxides such as chromium, manganese or cobalt. The vent stream should be well below the lower flammability limit (see Chapter 28, typically 25 to

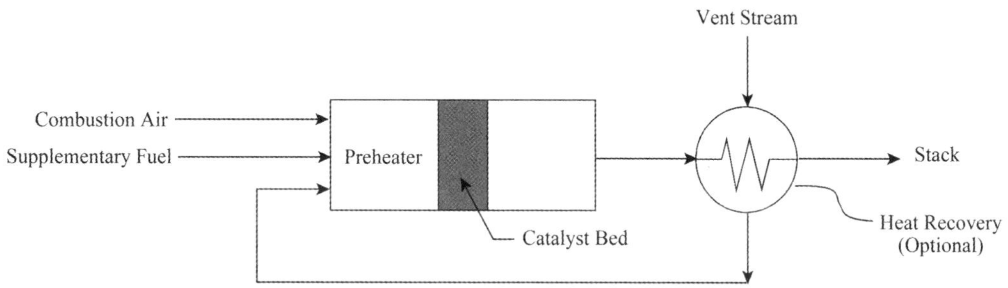

Figure 25.18
Catalytic thermal oxidation.

30% of the lower flammability limit). The catalyst can be deactivated by compounds containing sulfur, bismuth, phosphorus, arsenic, antimony, mercury, lead, zinc, tin or halogens. Catalysts are gradually deactivated over time, and their life is generally two to four years. Operating temperature for the catalyst should not normally exceed 500 °C to 600 °C; otherwise the catalyst can suffer from sintering. Typical operating conditions are between 200 °C and 480 °C with a residence time of 0.1 seconds. Catalytic thermal oxidation is not suitable for halogenated compounds because of the lower combustion temperature used. If the heat release in the combustion is such that the temperature rise in the catalyst is too high, then part of the cooled stack gas can be recycled and mixed with the incoming vent stream. Recuperative heat recovery is possible up to 70 to 80% and regenerative heat recovery up to 95%. The destruction efficiency is up to 99.9%, but the arrangement is not suitable for large fluctuations in flowrate.

The advantage of catalytic thermal oxidation is that the lower temperature of operation can lead to fuel savings (although effective heat recovery without a catalyst can offset this advantage). The major disadvantages of catalytic thermal oxidation are that the catalyst needs to be replaced every two to four years and the capital cost tends to be higher than thermal oxidation without a catalyst. Catalytic thermal oxidation also tends to increase the pressure drop through the system.

4) *Gas turbines.* If the vent flow is both constant and has a high flowrate of air with a significant VOC loading, then a gas turbine can be used for thermal oxidation. A few gas turbine models have been designed specifically for such applications. This allows power to be generated directly from the vent flow. Such applications should be restricted to nonhalogenated, sulfur-free VOCs. If the VOC loading is high, then the application can be achieved without the use of supplementary fuel; otherwise supplementary fuel will be required. Destruction efficiency can be up to 99.9%.

Example 25.3 A stream of air containing 0.4 vol % butane at 20 °C is to be thermally oxidized at 800 °C. A minimum excess of air of 25% is to be used. The heat of combustion of butane is 2.66×10^6 kJ·kmol^{-1} and the mean heat capacity of the combustion gases can be taken to be 37 kJ·kmol^{-1}·K^{-1}.

a) Is there sufficient oxygen in the inlet air for efficient combustion?
b) What efficiency of heat recovery would be necessary to sustain combustion without auxiliary fuel at steady state?

Solution

a) The combustion requirements are given by:

$$C_4H_{10} + 6\tfrac{1}{2}O_2 \rightarrow 4CO_2 + 5H_2O$$

Thus $6\tfrac{1}{2}$ kmol O_2 required for combustion per kmol of butane

$$= 0.004 \times 6\tfrac{1}{2}$$
$$= 0.026$$

O_2 in steam(kmol)

$$= 0.21 \times (1 - 0.004)$$
$$= 0.2092$$

Excess oxygen

$$= \frac{0.2092 - 0.026}{0.2092}$$
$$= 88\%$$

Thus, there is plenty of excess air for the combustion.

b) Temperature rise from combustion

$$= \frac{0.004 \times 2.66 \times 10^6}{37}$$
$$= 288\ ^\circ C$$

Inlet temperature required before combustion

$$= 800 - 288$$
$$= 512\ ^\circ C$$

Feed preheat duty

$$= (512 - 20) \times 37$$
$$= 492 \times 37 \text{ kJ} \cdot \text{kmol}^{-1} \text{ gas}$$

Heat available in exhaust gases

$$= (800 - 20) \times 37$$
$$= 780 \times 37 \text{ kJ} \cdot \text{kmol}^{-1} \text{ gas}$$

Efficiency of heat recovery to avoid supplementary fuel

$$= \frac{492}{780}$$
$$= 63\%$$

5) *Biological treatment.* Microorganisms can be used to oxidize VOCs (Hydrocarbon Processing's Environmental Processes, 1998). The VOC-laden stream is contacted with microorganisms. The organic VOC is food for the microorganisms. Carbon in the VOC is oxidized to CO_2, hydrogen to H_2O, nitrogen to nitrate and sulfur to sulfate. Biological growth requires an ample supply of oxygen and nutrients. Figure 25.19 shows a *bioscrubber* arrangement. The VOC-laden gas stream enters a chamber and rises up through a packing material, over which flows water. Nutrients (phosphates, nitrates, potassium and trace elements) need to be added to promote biological growth. The microorganisms grow on the surface of the packing material. The water from the exit of the tower needs to be adjusted for pH before recycle to prevent excessively low pH. As the film of microorganism grows on the surface of the packing through time, eventually it becomes too thick and detaches from the packing. This excess *sludge* must be removed from the recycle and disposed of.

An alternative arrangement, as illustrated in Figure 25.20, replaces the packing with soil or compost in a *biofilter*. One advantage of the biofilter is that the compost usually used for such systems contains a reservoir of nutrients to sustain biological growth and additional nutrients do not normally need to be added. However, biological activity eventually starts to decrease and the bed will need to be replaced after typically five years.

Destruction efficiency for biological treatment is typically up to 95%, but some VOCs are very difficult to degrade. Biological treatment is limited to low concentration streams (typically less than 1000 ppm) with flowrates typically less than 100,000 m^3·h^{-1} (US Environmental Protection Agency, 1992b).

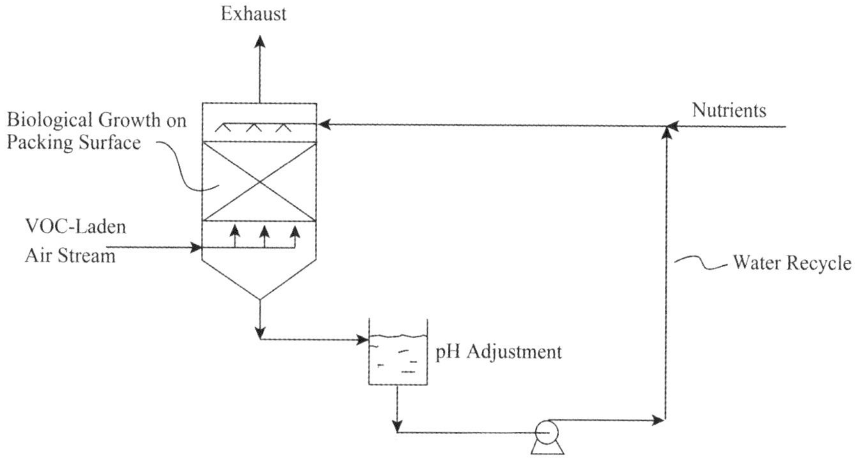

Figure 25.19
A bioscrubber for treatment of VOC.

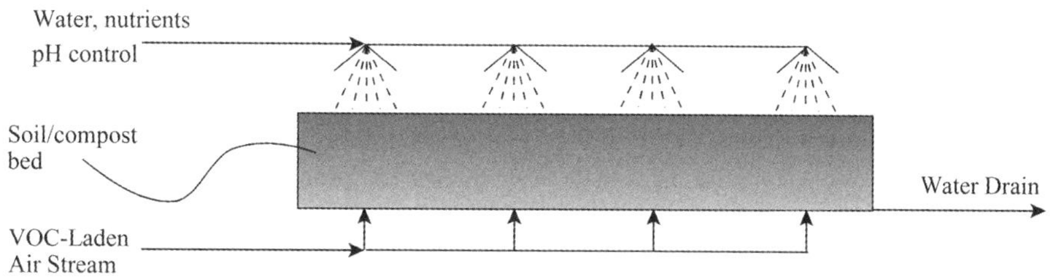

Figure 25.20
A biofilter for treatment of VOC.

25.5 Control of Sulfur Emissions

Sources of sulfur emissions of SO_2 and SO_3, collectively referred to as SO_X, from the process industries are:

- chemical production, for example, sulfuric acid production, sulfonation reactions, and so on;
- smelting processes, for example, production of copper;
- fuel processing operations, for example, fuel desulfurization;
- combustion of sulfur-bearing fuel.

When emissions of SO_2 and SO_3 are formed in combustion processes, chemical equilibrium would dictate that as the temperature is decreased, the equilibrium would favor SO_3 formation. However, in practice the mixture does not achieve chemical equilibrium and the mixture tends to be dominated by SO_2.

Before considering treatment of sulfur emissions to atmosphere, their minimization at source should be considered. Sulfur emissions can be minimized at source through:

- increase in process yield in chemical production,
- increase in energy efficiency leading to less fuel burnt,
- switch to a fuel with lower sulfur,
- desulfurizing the fuel prior to combustion.

Sulfur emissions from combustion processes can be reduced by switching to a fuel with lower sulfur. The sulfur content of fuels varies significantly. Generally, the sulfur content is gas < liquid < solid. This order arises mainly from the relative ease with which the fuels can be desulfurized. Desulfurizing coal prior to combustion is extremely difficult. It must be first ground to very fine particles (of the order of 100 μm) to *liberate* the mineral (inorganic) sulfur. The light coal can then be separated from the heavy mineral in a flotation process. Unfortunately, even this desulfurization only removes 30 to 60% of the sulfur, as the remaining sulfur is organic. Desulfurizing a liquid fuel is much more straightforward and is commonly practiced in refinery operations. The sulfur is present in the form of mercaptans, thiols, thiophenes, and so on. Desulfurizing the liquid fuel requires reaction with hydrogen over a catalyst at elevated temperature and pressure. The organic sulfur in the liquid fuel is reacted to H_2S. Desulfurizing gaseous fuel is the most straightforward. The sulfur in gaseous fuel is mostly present as H_2S. This can be removed straightforwardly, for example, in an amine separation absorption, as discussed in Chapter 9.

Removal of sulfur from gas streams is in general either removal of H_2S or removal of SO_2 (Crynes, 1977). Generally, removal of the sulfur in the form of H_2S is much more straightforward than in the form of SO_2. H_2S can be removed by absorption, as already discussed. Chemical absorption using amines is the most commonly used method. However, other solvents can be used for chemical absorption, for example, potassium carbonate. Physical absorption is also possible using solvents such as refrigerated methanol (the Rectisol Process) or a mixture of the dimethyl ethers of polyethylene glycol (the Selexol Process).

When it is necessary to combust a fuel containing a high sulfur content, gasification is the preferred route. Gasification is the partial combustion of the fuel using pure oxygen (or air) in the presence of steam. Temperatures up to 1600 °C and pressures up to 150 bar are used. The product of the gasification is a mixture of CO, CO_2, H_2, H_2O, CH_4, H_2S, COS, N_2, NH_3 and HCN. The sulfur is converted mostly to H_2S (typically 95%), with the balance to COS (carbonyl sulfide). The acid gases consisting of H_2S, COS, CO_2 are removed by countercurrent absorption with a regenerative solvent. Depending on the solvent used, COS might first need to be converted to H_2S in a COS hydrolysis unit. HCN is converted to NH_3 and CO in the hydrolysis unit. H_2S and CO_2 can be removed either simultaneously or selectively, depending on the synthesis gas composition from the gasifier and the final synthesis gas specification. The synthesis gas (H_2 and CO) can then be used as fuel, free of any sulfur. Hence, gasification has advantages over conventional combustion when dealing with fuels with high sulfur content.

Figure 25.21 shows a flowsheet for an integrated gasification combined cycle for power generation using a combination of a gas turbine and a steam turbine. Figure 25.21 also shows the option of using the synthesis gas as a chemical feedstock. Fuel containing the sulfur, along with steam and oxygen, enters the gasification process. The gasification products are then treated to remove particulate material. COS may then be converted to H_2S in a hydrolysis unit by reaction over a catalyst, followed by H_2S removal. This synthesis gas can then be used as a chemical feedstock or combusted in a gas turbine combined cycle, as shown in Figure 25.21. In Figure 25.21, the option of integrating the gas turbine with the process has been included. Integrating the gas turbine in Figure 25.21 involves using the compressor for the gas turbine to provide both the compressed gas for the gas turbine cycle and the compressed air for the air separation unit that produces the oxygen. The nitrogen from the air separation unit, under pressure, is fed to the combustion of the gas turbine for NO_X reduction through lowering the peak flame temperature (see later in Chapter 25) and then expanded in the turbine with the air from the compressor to recover the pressure energy from the nitrogen. The hot exhaust gases from the gas turbine are used to generate steam before being vented. The steam is used to generate further power in a steam turbine. Many variations around the basic theme in Figure 25.21 are possible. The important thing about gasification is that it is an effective method of producing chemical feedstock, power and heat from fuel that has high sulfur content as the sulfur is not oxidized all the way to SO_2 before being treated.

Having removed sulfur in the form of H_2S, it cannot be vented to atmosphere. The sulfur from H_2S is normally recovered as

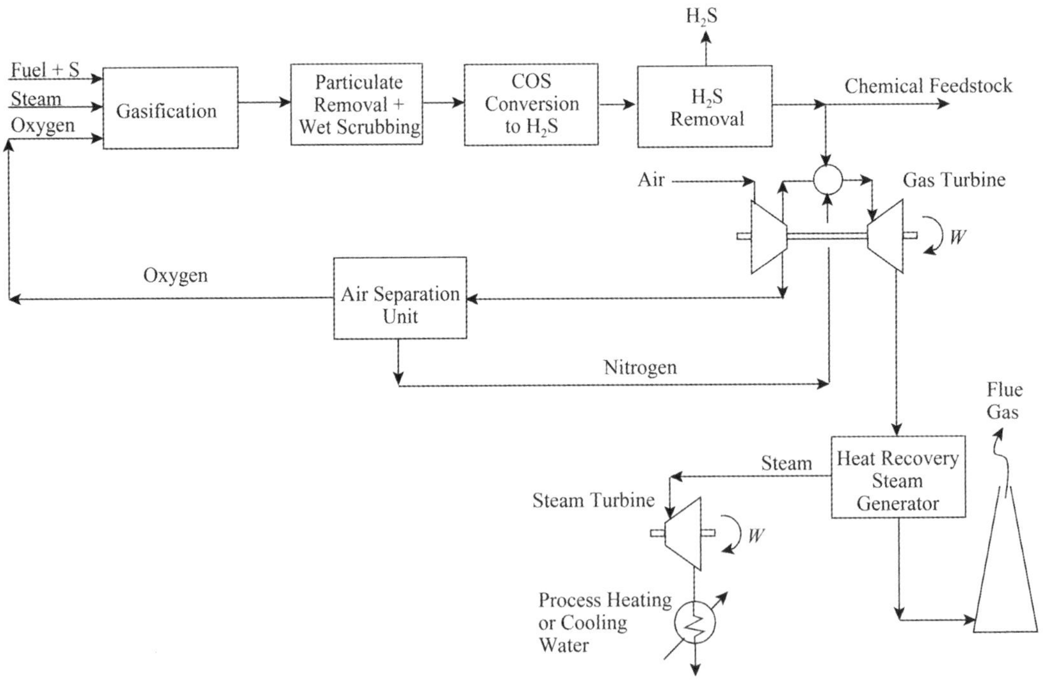

Figure 25.21

Integrated gasification combined cycle.

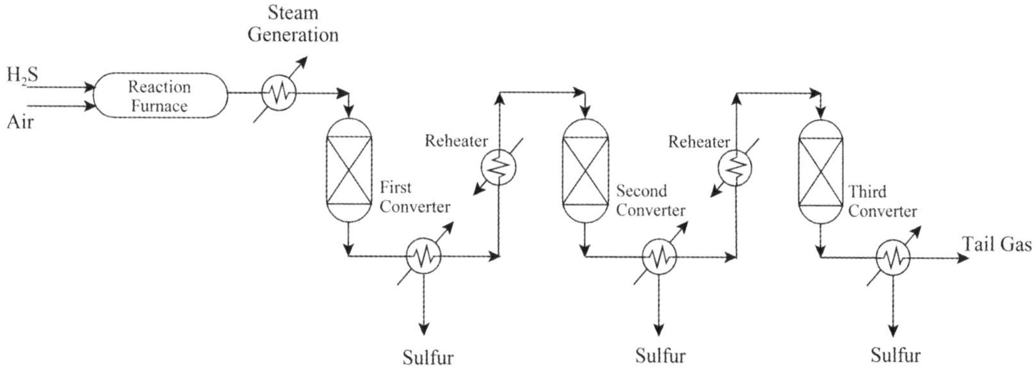

Figure 25.22

Removal of H_2S using the Claus Process.

elemental sulfur by partial oxidation in a Claus Process. Figure 25.22 shows the outline of a Claus Process. Hydrogen sulfide in air enters a reaction furnace where a partial combustion takes place. After cooling the gases by steam generation, the gases enter a converter. The chemistry is:

$$\begin{array}{lrcl} \text{Furnace} & 2H_2S + 3O_2 & \rightarrow & 2H_2O + 2SO_2 \\ \text{Converter} & 2H_2S + SO_2 & \rightarrow & 2H_2O + 3S \\ \text{Overall} & H_2S + \frac{1}{2}O_2 & \rightarrow & S + H_2O \end{array}$$

After the converter in Figure 25.22, the sulfur is recovered by condensation. Conversion is limited by reaction equilibrium to about 60%, so several (usually three) stages of conversion are used, as shown in Figure 25.22. After each conversion stage, the sulfur is condensed and the gas is reheated before going to the next conversion stage. The tail gas contains CO_2, H_2S, SO_2, CS_2, COS and H_2O and cannot be released directly to atmosphere. Various processes are available for the cleanup of Claus Process tail gas. Figure 25.23 shows a process in which the sulfur compounds are converted to H_2S by reaction with hydrogen. Hydrogen is generated by partial oxidation of fuel gas. Sulfur and sulfur compounds are hydrogenated to hydrogen sulfide in the hydrogenation reactor. Gases from the reactor are cooled by generating steam and then in a quench column by recirculating water (Figure 25.23). Hydrogen sulfide is then separated from the other gases by absorption (e.g. in an amine solution). The solvent from the absorber is regenerated in the stripper and the hydrogen sulfide recycled to the Claus Process.

25

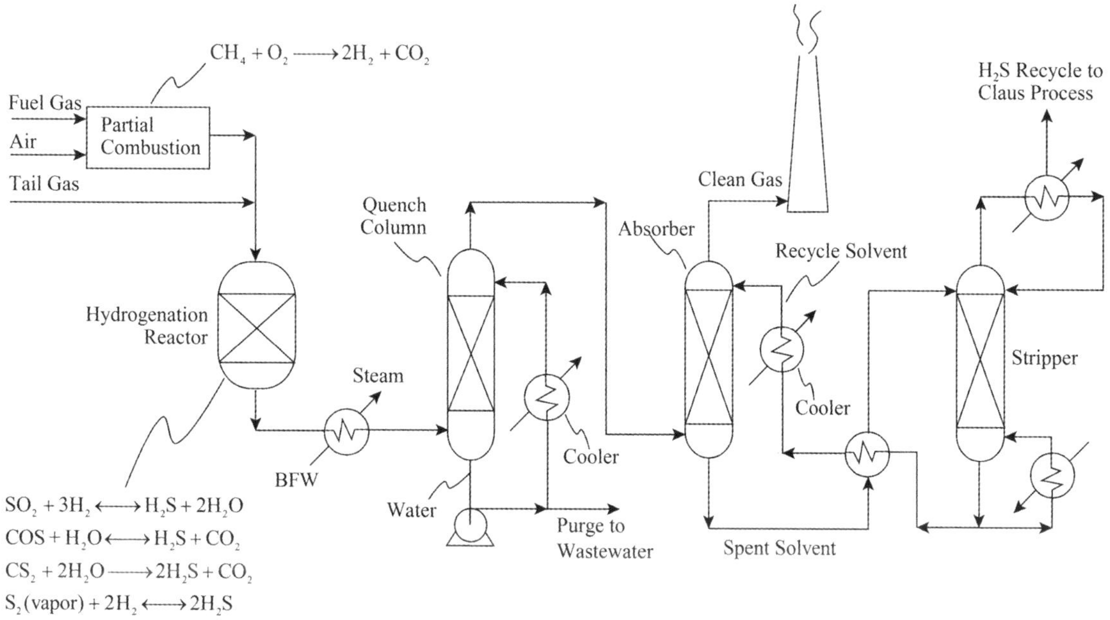

Figure 25.23

Claus Process tail gas clean-up.

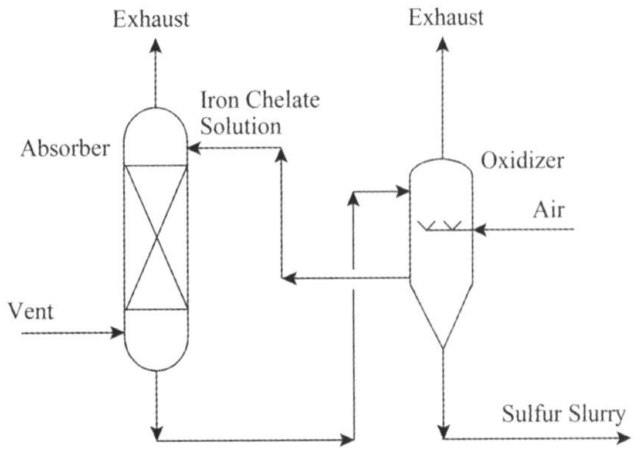

Figure 25.24

Removal of H_2S by partial oxidation of H_2S using iron chelate.

An alternative process for the removal of H_2S by partial oxidation is shown in Figure 25.24 (Hydrocarbon Processing's Environmental Processes, 1998). This uses an iron chelate to partially oxidize the H_2S. As shown in Figure 25.24, the gas is contacted in an absorber with the iron chelate solution. Fe^{3+} reacts with H_2S to produce Fe^{2+} and elemental sulfur. Reduced iron chelate solution is oxidized by sparging with air and returned to the absorber. This process can operate over a wide range of conditions.

So far, the techniques to recover sulfur have all related to sulfur in the form of H_2S. Consider now how sulfur in the form of SO_2 can be dealt with. SO_2 can be removed from gas streams using sodium hydroxide solution. In an oxidizing environment:

$$\underset{\text{sodium hydroxide}}{2NaOH} + SO_2 \rightarrow \underset{\text{sodium sulphite}}{Na_2SO_3} + H_2O$$

$$\underset{\text{sodium sulfite}}{Na_2SO_3} + \tfrac{1}{2}O_2 \rightarrow \underset{\text{sodium sulfate}}{Na_2SO_4}$$

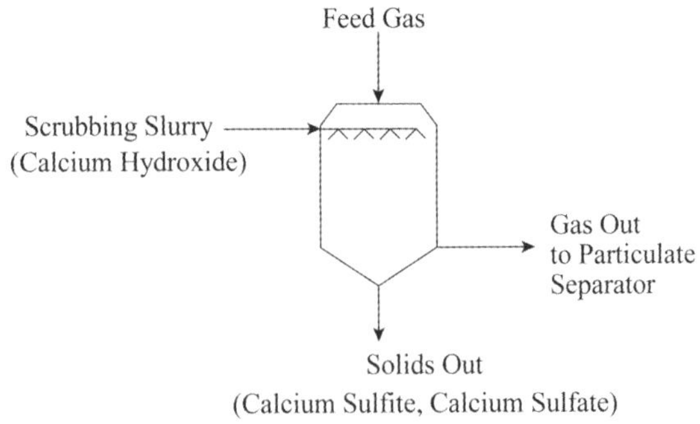

Figure 25.26

Removal of SO_2 using dry limestone scrubbing.

However, this is only economic for small-scale applications, as the sodium hydroxide is expensive relative to other reagents that can be used. A less expensive reagent to remove sulfur dioxide is limestone, which can be reacted to produce solid calcium sulfate (gypsum), according to (Crynes, 1977):

$$\underset{\text{Limestone}}{CaCO_3} + SO_2 \rightarrow \underset{\text{Calcium sulfite}}{CaSO_3} + CO_2$$

$$CaSO_3 + \tfrac{1}{2}O_2 + 2H_2O \rightarrow \underset{\text{Gypsum}}{CaSO_4 \cdot 2H_2O}$$

The flowsheet for the removal of SO_2 using wet limestone scrubbing is shown in Figure 25.25 (Crynes, 1977). The gas stream containing SO_2 enters a scrubbing chamber into which a limestone slurry is sprayed. The slurry is then collected, separated by thickening and filtration. This creates a solid material (gypsum) that does have some potential for use as building material.

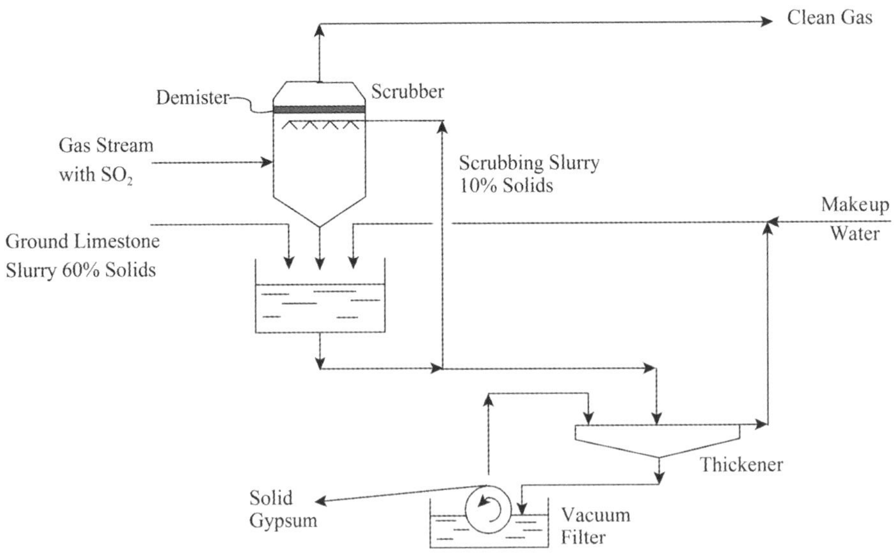

Figure 25.25

Removal of SO_2 using wet limestone scrubbing.

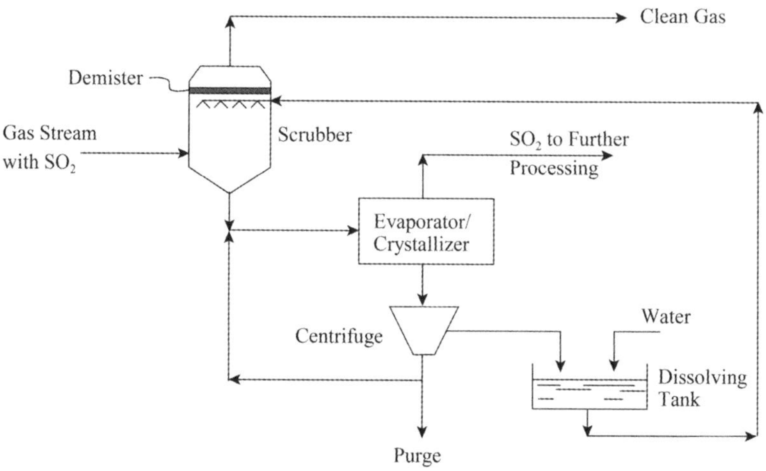

Figure 25.27

Removal of SO_2 using the Wellman–Lord Process.

However, the solid product is usually of low value and often needs to be disposed of.

Whilst wet limestone scrubbing can be effective for large-scale utility systems such as centralized power stations, it is not readily applied to smaller-scale processes, such as utility boilers on processing sites. Rather than use wet limestone scrubbing, as illustrated in Figure 25.25, "dry" scrubbing can be used. Dry scrubbing refers to the state of the waste byproduct being dry, rather than the wet byproduct from wet limestone scrubbing. The process is illustrated in Figure 25.26. An aqueous slurry of calcium hydroxide (or sometimes sodium carbonate) is sprayed into a flue gas. The reaction is:

$$\underset{\text{calcium hydroxide}}{Ca(OH)_2} + SO_2 \rightarrow \underset{\text{calcium sulfite}}{CaSO_3} + H_2O$$

$$CaSO_3 + {}^1\!/\!_2O_2 \rightarrow \underset{\text{calcium sulfate}}{CaSO_4}$$

Water evaporates and reaction products are removed as dry solids. Advantages over wet scrubbing include simplicity, cheaper construction, a dry waste product and no wastewater. Disadvantages relative to wet scrubbing include less efficient use of reagent and lower SO_2 removal efficiency. SO_2 removal efficiency is typically 70 to 90%. Around 2.2 tons dry waste are produced per ton of SO_2 recovered.

Another process for removal of SO_2 is the Wellman–Lord process (Crynes, 1977). The reactions are:

$$SO_2 + \underset{\text{sodium sulfite}}{Na_2SO_3} + H_2O \rightarrow \underset{\text{sodium bisulfite}}{2NaHSO_3}$$

$$2NaHSO_3 \xrightarrow{\text{crystallize}} Na_2SO_3 + H_2O + SO_2$$

A flowsheet for the Wellman–Lord process is shown in Figure 25.27. Again the gas stream with SO_2 enters a scrubber into which is sprayed a sodium sulfite solution. This is then fed to an evaporator/crystallizer to crystallize out the resulting sodium bisulfite, which converts the sodium bisulfite back to sodium sulfate, releasing the SO_2. The crystals are dissolved in water and recycled to the scrubber. The effect of the Wellman–Lord process is to produce a concentrated SO_2 stream from a dilute SO_2 stream. The resulting concentrated SO_2 still needs to be treated, typically by conversion to sulfuric acid. Compared with the limestone scrubbing process, the Wellman–Lord process treats the sulfur gas stream without producing a solid waste byproduct. On the other hand, it requires a sulfuric acid process to convert the concentrated SO_2 stream.

The SO_2 from the Wellman–Lord process or any other concentrated SO_2 stream (e.g. gases from copper smelting) can be oxidized to SO_3 to produce sulfuric acid, as discussed in Chapter 5.

Rather than oxidize the SO_2, it can be reduced in a process similar to the one illustrated in Figure 25.23. The SO_2 is reduced to H_2S. The resulting H_2S can then be separated and fed to a sulfur recovery process.

25

Example 25.4 A power plant produces 500 standard $m^3 \cdot s^{-1}$ of exhaust gas with 0.1% SO_2 by volume. It is required to remove 90% of the SO_2 before the gas is discharged to atmosphere. Three methods of removal of SO_2 are to be evaluated. Assume the molar mass of a gas in kilograms occupies 22.4 m^3 at standard conditions.

a) The first option is absorption at atmospheric pressure in water taken from a local river at a temperature of 10 °C. Assume the concentration of water at the exit of the absorber to be 80% of equilibrium. The equilibrium can be assumed to follow Henry's Law:

$$p_i = H_i x_i$$

where p_i = partial pressure of Component i
H_i = Henry's Law Constant (determined experimentally)
x_i = mole fraction of Component i in the liquid phase

Assuming ideal gas behavior ($p_i = y_i P$):

$$x_i = \frac{y_i^* P}{H_i}$$

where y_i^* = mole fraction of Component i in vapor phase in equilibrium with the liquid
P = total pressure
H_{SO_2} = 22 atm at 10 °C

How much water would be required for the removal of the SO_2?

b) The second option is absorption in sodium hydroxide solution. Assume that the sodium hydroxide and SO_2 react in an oxidizing environment to produce sodium sulfate. Calculate the quantity of material required for the absorption.

c) The third option is absorption in a slurry of calcium carbonate. Calculate the quantity of material required for the absorption. Assume the plant operates for 8600 $h \cdot y^{-1}$.

Solution

a)

$$x_{SO_2} = \frac{0.8 y^*_{SO_2} P}{H_{SO_2}} = \frac{0.8 \times 0.001 \times 1}{22} = 3.6 \times 10^{-5}$$

Now calculate the gas flowrate G:

$$G = 500 \times \frac{1}{22.4} = 22.32 \text{ kmol} \cdot s^{-1}$$

Assuming gas and liquid flowrates constant:

$$G(y_{in} - y_{out}) = L(x_{out} - x_{in})$$

$$L = \frac{G(y_{in} - y_{out})}{(x_{out} - x_{in})} = \frac{22.32(0.001 - 0.1 \times 0.001)}{(3.6 \times 10^{-5} - 0)} = 558 \text{ kmol} \cdot s^{-1} = 558 \times 18 \times \frac{1}{10^3} t \cdot s^{-1} = 10 \, t \cdot s^{-1}$$

which is an impractically large flowrate.

b)

$$2NaOH + SO_2 + {}^1\!/_2 O_2 \rightarrow Na_2SO_4 + H_2O$$

$$SO_2 \text{ in exhaust} = 500 \times \frac{1}{22.4} \times 0.001 = 0.0223 \text{ kmol} \cdot s^{-1}$$

NaOH required to remove 90%

$$= 2 \times 0.0223 \times 0.9 \text{ kmol} \cdot s^{-1}$$
$$= 2 \times 0.0223 \times 0.9 \times 40 \text{ kg} \cdot s^{-1}$$
$$= 1.606 \text{ kg} \cdot s^{-1}$$
$$= \frac{1.606 \times 3600 \times 8600}{10^3} t \cdot y^{-1}$$
$$= 49{,}722 \, t \cdot y^{-1}$$

c) $CaCO_3 + SO_2 + {}^1\!/_2 O_2 + 2H_2O \rightarrow CaSO_4 \cdot 2H_2O + CO_2$

SO_2 in exhaust $= 0.0223 \text{ kmol} \cdot s^{-1}$

$CaCO_3$ required to remove 90%

$$= 0.0223 \times 0.9 \text{ kmol} \cdot s^{-1}$$
$$= 0.0223 \times 0.9 \times 100 \text{ kg} \cdot s^{-1}$$
$$= 2.007 \text{ kg} \cdot s^{-1}$$
$$= \frac{2.007 \times 3600 \times 8600}{10^3}$$
$$= 62{,}137 \, t \cdot y^{-1}$$

25.6 Control of Oxides of Nitrogen Emissions

Nitrogen forms eight oxides, but the principal concern is with the two most common ones:

- nitric oxide (NO),
- nitrogen dioxide (NO_2).

These are collectively referred to as NO_X. NO_X emissions are produced from:

- chemicals production (e.g. nitric acid production, nitration reactions, etc.),
- use of nitric acid in metal and mineral processing,
- combustion of fuels.

NO_X from the combustion of fuels is formed initially as NO, which subsequently oxidizes to NO_2. There are three sources of NO production from the combustion of fuels (Wood, 1994).

1) *Fuel NO.* Organically bound nitrogen in fuel reacts with oxygen to form NO.

$$[\text{Fuel N}] + {}^1\!/_2 O_2 \rightleftharpoons NO$$

Organic nitrogen is more reactive than nitrogen in air. Fuel NO formation is weakly dependent on temperature.

2) *Thermal NO.* Thermal NO is formed when molecular nitrogen in the combustion air reacts with oxygen to form NO, according to the reaction:

$$N_2 + O_2 \rightleftharpoons 2NO$$

The reaction mechanism is via free radicals and is highly temperature dependent.

3) *Prompt (rapidly forming) NO.* Prompt or rapidly forming NO is formed when molecular nitrogen reacts with hydrocarbon radicals in a flame. The formation only occurs in the flame and is weakly dependent on temperature.

Figure 25.28 illustrates the contributions of the three mechanisms to NO formation. It can be seen from Figure 25.28 that both the fuel and prompt NO are weakly dependent on temperature. Below around 1300 °C thermal NO formation is negligible. However, at the highest temperatures thermal NO is the most important. Once NO has been formed, it can then oxidize to NO_2 according to:

$$NO + {}^1\!/_2 O_2 \rightleftharpoons NO_2$$

NO_X in flue gases is predominantly (of the order 90%) NO. NO oxidizes in the atmosphere to NO_2.

NO_X formation in combustion depends on the fuel and the type of combustion device. Generally, coal produces the highest NO_X concentrations, then fuel oil, with natural gas producing the lowest concentrations. Thus, one way of reducing NO_X formation is to change fuel. For example, a large steam boiler fired with fuel oil might produce a flue gas with 150 ppm NO_X, which might be reduced to 100 ppm by switching to natural gas. However, it should be noted that the actual NO_X concentration depends on both the fuel and the combustion device, and these are only illustrative figures.

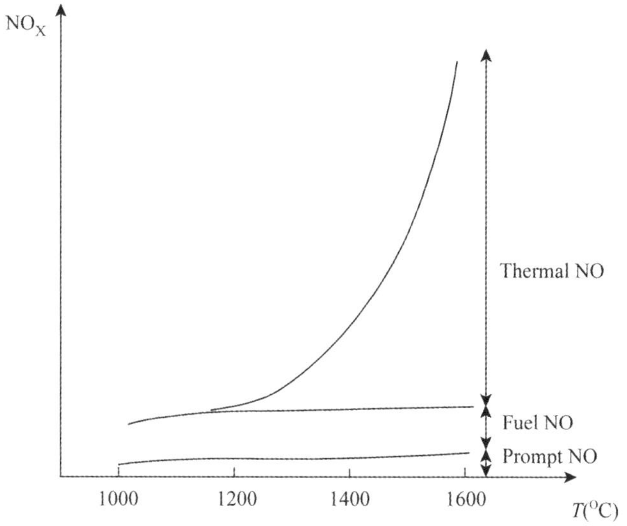

Figure 25.28
The contributions to NO_X formation.

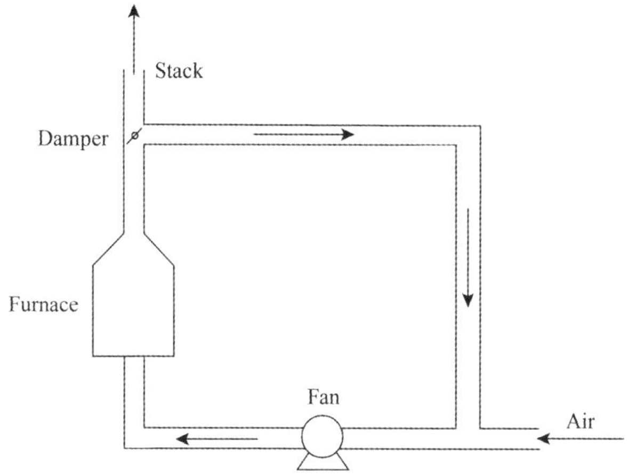

Figure 25.29
Flue gas recirculation.

Minimization of NO_X emissions at source means:

- increase the process yield in chemicals production,
- increase in energy efficiency leading to less fuel burnt,
- switch to a fuel with lower nitrogen content,
- minimize NO_X formation in combustion processes.

Switching to a fuel with lower nitrogen content typically means switching from coal to fuel oil, or fuel oil to natural gas. It should be noted that nitrogen can be a significant percentage of natural gas (e.g. 5%), but the nitrogen will be molecular nitrogen and, therefore, behaves like nitrogen in the combustion air and be relatively unreactive.

NO_X formation in combustion processes can be reduced by (Wood, 1994; Garg, 1994) the following, which all reduce the flame temperature:

1) *Reduced air preheat.* Preheating the air to combustion processes increases the flame temperature and furnace efficiency (see Chapter 12). However, the increase in flame temperature increases NO_X formation. Thus, reducing preheat lowers the thermal NO_X formation. However, this lowers the furnace efficiency at the same time.
2) *Flue gas recirculation.* Recirculating part of the flue gas, as shown in the Figure 25.29, reduces the peak flame temperature through the increased mass of inert material in the flame and reduces the thermal NO_X formation. Usually 10 to 20% of the combustion air is recirculated.
3) *Steam injection.* Steam injection reduces the combustion temperature by adding an inert to the combustion process. In principle, water or steam could be injected, but dry superheated steam is usually used. This reduces the furnace efficiency as energy is used to produce the steam, and the latent heat in the steam cannot be recovered. It is only used for moderate NO_X reduction. Steam injection can also be used in gas turbines.
4) *Reduced excess air.* Reducing the amount of excess air also reduces the NO_X formation.
5) *Air staging.* Air staging in the burner can be used to reduce NO_X formation. The principle is illustrated in Figure 25.30a. Fuel and primary air is injected into the flame with a substoichiometric amount of oxygen. Secondary air is then injected to complete the combustion to give overall greater than stoichiometric requirements. The effect is to reduce the fuel NO and the thermal NO by reducing the peak flame temperature. Air staging can be used both in furnaces and gas turbines.
6) *Fuel staging.* Fuel staging is illustrated in Figure 25.30b. The primary fuel and primary air are first combusted at greater than stoichiometric requirements. In the staging zone, secondary fuel (typically 10 to 20% of the primary fuel) is burned under substoichiometric (reducing) conditions. The temperature should be at least 1000 °C in the staging zone. In the final combustion zone, air is added to complete the combustion. Low temperatures (less than 1000 °C) are preferred in the final combustion zone to minimize NO formation.

Table 25.2 compares the performance of the NO_X minimization techniques (Wood, 1994).

Once NO_X formation has been minimized at source, if the NO_X levels still do not achieve the required environmental standards, then treatment must be considered.

The first option that might be considered is absorption into water. The problem is that NO_2 is soluble in water, but NO is only sparingly soluble. To absorb NO, a complexing agent or oxidizing agent must be added to the solvent. One method that can be used is to absorb into a solution of hydrogen peroxide according to:

$$2NO + 3H_2O_2 \rightarrow 2HNO_3 + 2H_2O$$

$$2NO_2 + H_2O_2 \rightarrow 2HNO_3$$

This allows nitric acid to be recovered from the oxides of nitrogen. It can be a useful technique, for example, in metal finishing operations where nitric acid is used. More commonly, NO_X is

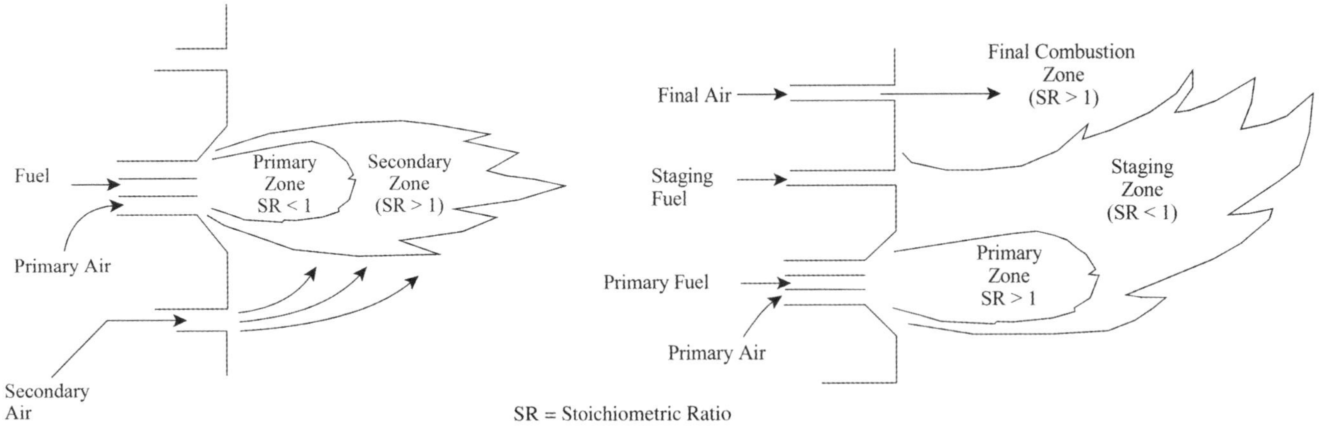

Figure 25.30

Low NO_X burners.

Table 25.2

Performance of the NO_X minimization techniques (Wood, 1994).

Technique	NO_X reduction
Reduced air preheat	25–65%
Flue gas recirculation	40–80%
Water/steam injection	40–70%
Reduced excess air	1–15%
Air staging	30–60%
Fuel staging	30–50%

removed using reduction. Ammonia is usually used as the reducing agent, according to the reactions:

$$6NO + 4NH_3 \rightarrow 5N_2 + 6H_2O$$

$$4NO + 4NH_3 + O_2 \rightarrow 4N_2 + 6H_2O$$

$$2NO_2 + 4NH_3 + O_2 \rightarrow 3N_2 + 6H_2O$$

This can be carried out without a catalyst in a narrow temperature range (850 to 1100 °C) and is known as *selective noncatalytic reduction* (Hydrocarbon Processing's Environmental Processes, 1998). Below 850 °C, the reaction rate is too slow. Above 1100 °C, the dominant reaction becomes:

$$4NH_3 + 5O_2 \rightarrow 4NO + 6H_2O$$

Thus, the technique can become counterproductive. A typical arrangement for selective non-catalytic reduction is shown in Figure 25.31. Aqueous ammonia is vaporized and mixed with a carrier gas (low-pressure steam or compressed air) and injected into nozzles located in the combustion device for optimum temperature and residence time (Hydrocarbon Processing's Environmental Processes, 1998). NO_X reduction of up to 75% can be achieved. However, *slippage* of excess ammonia must be controlled carefully.

Rather than selective noncatalytic reduction, the reduction can be carried out over a catalyst (typically vanadium pentoxide and tungsten trioxide dispersed on titanium dioxide) at 250 to 450 °C. This is known as selective catalytic reduction. Figure 25.32 shows a typical selective catalytic reduction arrangement. Either

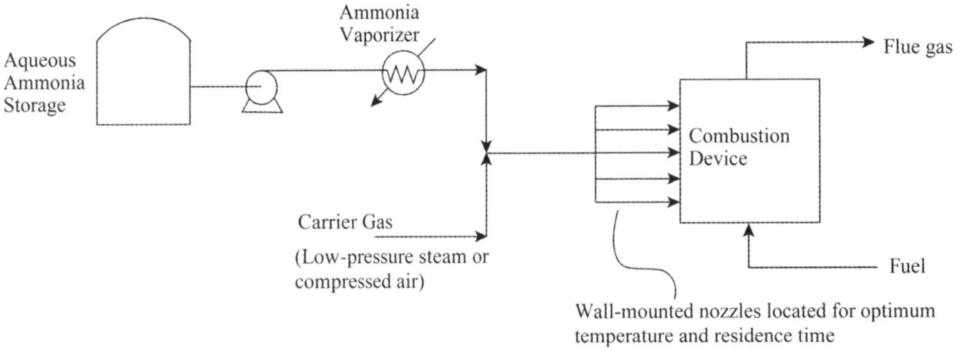

Figure 25.31

Removal of NO_X using selective noncatalytic reduction.

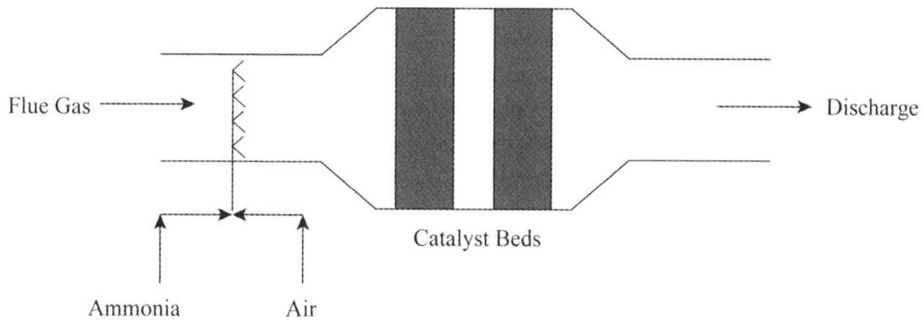

Figure 25.32

Removal of NO_X using selective catalytic reduction.

anhydrous or aqueous ammonia can be used. This is mixed with air and injected into the flue gas stream upstream of the catalyst. Removal efficiency of up to 95% is possible. Again, slippage of excess ammonia needs to be controlled.

Example 25.5 A gas turbine exhaust is currently operating with a flowrate of $41.6\,kg{\cdot}s^{-1}$. The exhaust contains 200 ppmv NO_X to be reduced to $60\,mg{\cdot}m^{-3}$ (expressed as NO_2) at 0 °C and 1 atm. The NO_X is to be treated in the exhaust using low-temperature selective catalytic reduction. Ammonia slippage must be restricted to be less than $10\,mg{\cdot}m^{-3}$, but a design basis of $5\,mg{\cdot}m^{-3}$ will be taken. Aqueous ammonia is to be used at a cost of $300\,\$\cdot t^{-1}$ (dry NH_3 basis). Estimate the cost of ammonia if the plant operates $8000\,h{\cdot}yr^{-1}$. Assume the exhaust gas composition to be similar to that of air. Molar mass data are given in Table 25.3. Assume the molar mass in kilograms occupies $22.4\,m^3$ at standard conditions.

Table 25.3

Molar mass data.

	Molar mass ($kg{\cdot}kmol^{-1}$)
Air	28.9
NO_2	46
NH_3	17

Solution Volume of exhaust at standard conditions

$$= 41.6 \times 22.4 \times \frac{1}{28.9}$$
$$= 32.2\ m^3 \cdot s^{-1}$$

Concentration of NO_2 at standard conditions

$$= y \times \frac{1}{22.4} \times 46$$
$$= 0.0002 \times \frac{1}{22.4} \times 46$$
$$= 0.0004107\ kg \cdot m^{-3}$$
$$= 410.7\ mg \cdot m^{-3}$$

NO_2 to be removed

$$= 32.2(410.7 - 60) \times 10^{-6}\ kg \cdot s^{-1}$$
$$= 0.01129\ kg \cdot s^{-1}$$

The reduction reaction is given by:

$$2NO_2 + 4NH_3 + O_2 \rightarrow 3N_2 + 6H_2O$$

$$NH_3 \text{ required for } NO_2 = 0.01129 \times \frac{17 \times 4}{46 \times 2}$$
$$= 8.345 \times 10^{-3}\ kg \cdot s^{-1}$$
$$NH_3 \text{ slippage} = 32.2 \times 5 \times 10^{-6}$$
$$= 0.161 \times 10^{-3}\ kg \cdot s^{-1}$$
$$\text{Cost of } NH_3 = (8.345 + 0.161) \times 10^{-3} \times 3600 \times 8000 \times 300 \times 10^{-3}$$
$$= \$73,500\ y^{-1}$$

25.7 Control of Combustion Emissions

The major emissions from the combustion of fuel are CO_2, SO_X, NO_X and particulates (Glassman, 1987). The products of combustion are best minimized by making the process efficient in its use of energy through efficient heat recovery and avoiding unnecessary thermal oxidation of waste through minimization of process waste. Flue gas emissions can be minimized at source by:

- increased energy efficiency at the point of use (e.g. better heat integration),
- increased energy efficiency of the utility system (e.g. increased cogeneration),
- improvements to combustion processes (e.g. low NO_X burners),
- changing fuel (e.g. changing from fuel oil to natural gas).

For a given energy consumption, fuel change is the only way to reduce CO_2 and SO_X emissions at source. For example, fuel switch from coal to natural gas reduces the CO_2 emissions for the same heat release because of the lower carbon content of natural gas. Fuel change can also be useful for reducing NO_X emissions. Once emissions have been minimized at source, then treatment can be considered to solve any residual problems.

1) *Treatment of SO_X.* The various techniques that can be considered for the treatment of SO_X are:
   - absorption into NaOH, $CaCO_3$, water, and so on,
   - oxidation and conversion to H_2SO_4,
   - reduction by conversion to H_2S and then sulfur.

2) *Treatment of NO_x.* The techniques used for treatment of NO_X are:
   - absorption into an NO-complexing agent or oxidizing agent,
   - reduction using selective noncatalytic or catalytic processes.
3) *Treatment of particulates.* The various methods for treatment of particulates were reviewed in Chapter 7. These include:
   - scrubbers,
   - inertial collectors,
   - cyclones,
   - bag filters,
   - electrostatic precipitators.
4) *Treatment of CO_2.* If CO_2 needs to be separated from the other combustion products for the purpose of reducing greenhouse gas emissions, there are different ways this can be done, depending on the combustion arrangement. There are two general combustion arrangements; postcombustion and precombustion separation (Steeneveldt, Berger and Torp, 2006).
   a) *Postcombustion CO_2 separation.* Figure 25.33 shows two arrangements for postcombustion CO_2 separation. The first arrangement in Figure 25.33a shows a standard combustion arrangement with air used for combustion of the fuel. The separation following the combustion is best suited to chemical absorption because of the low partial pressure of the CO_2 in the combustion gases. An amine or mixture of amines in an aqueous solution can be used, as discussed in Chapter 9. The amines used are monoethanolamine (MEA), diethanolamine (DEA) or methyldiethanolamine (MDEA). Rather than use an amine or mixture of amines, a hot aqueous solution of potassium carbonate (K_2CO_3) can be used.

      The other postcombustion separation arrangement shown in Figure 25.33b is oxy-combustion. The process starts with an air separation plant to provide pure oxygen for the combustion. This removes the nitrogen from the combustion process, giving products of combustion that are essentially CO_2 and water vapor. The CO_2 can then be separated by condensation of the water vapor. However, use of pure oxygen alone would create temperatures from combustion of hydrocarbon in excess of 3000 °C, which is problematic for the design of the combustion chamber. Thus, an inert dilution stream needs to be added to the combustion chamber to control the flame temperature. This can be a recycle of CO_2 from after the separation, or in principle steam can be added.
   b) *Precombustion CO_2 separation.* Rather than wait until after the combustion process to separate CO_2, various processes can be used to first produce synthesis gas (mixture of H_2 and CO), the CO converted to CO_2 and the CO_2 separated, leaving H_2 for combustion and/or chemical use. Figure 25.34 shows four arrangements for such precombustion. The four arrangements differ in the way the synthesis gas is produced. Figure 25.34 shows two different *catalytic reformer* processes. In a steam reformer the dominant reaction for a general hydrocarbon feed is:

$$C_nH_m + nH_2O \rightleftharpoons nCO + (n + m/2)H_2$$

The steam reforming reaction is strongly endothermic and takes place in a fired heater. Steam reforming is the preferred method for feeds with light hydrocarbons, but is difficult to scale down for small scales.

An alternative reforming process for light hydrocarbon feeds shown in Figure 25.34 is autothermal reforming. This utilizes the heat of a partial oxidation reaction to drive the steam reforming reaction and is better suited to smaller-scale applications than steam reforming. An autothermal reformer uses two reaction zones, a combustion zone and a catalytic zone. The overall reaction is exothermic. Fuel, steam and oxygen are mixed and combusted in a substoichiometric flame in the combustion zone according to:

$$C_nH_m + (n/2 + m/4)O_2 \rightleftharpoons nCO + m/2H_2O$$

In the catalytic zone, the reaction proceeds as the steam reforming reaction above.

The other two options illustrated in Figure 25.34 are partial oxidation reactions. Catalytic partial oxidation, which can also be used for light hydrocarbon feeds, uses oxygen (but can use air) according to:

$$C_nH_m + n/2O_2 \rightleftharpoons nCO + m/2H_2$$

The final option in Figure 25.34 is gasification, which has been described earlier. Although the reactions in gasification are complex, the dominant reaction is the partial oxidation reaction above. Gasification can be used for a wide range of feeds, solid, liquid or gas. It is normally used

Figure 25.33

Post-combustion arrangements for the separation of CO_2 with air and oxygen combustion.

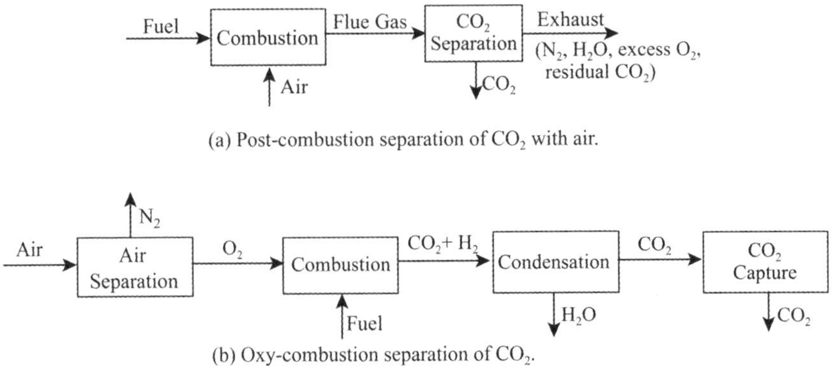

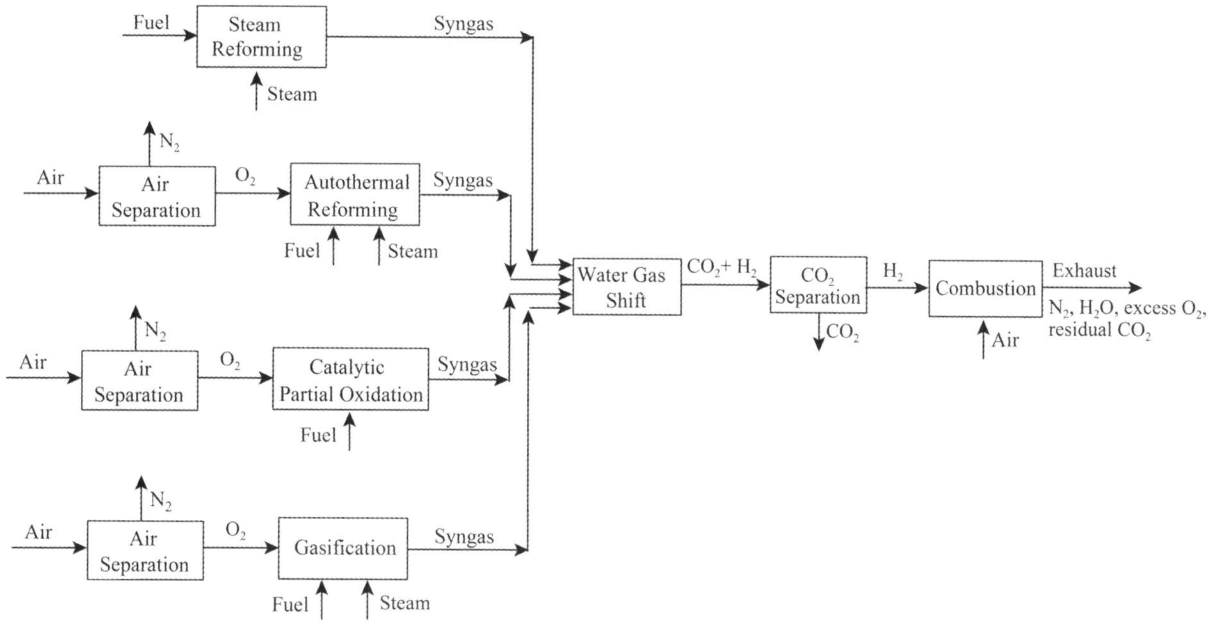

Figure 25.34

Pre-combustion arrangements for the separation of CO_2.

for the more difficult feeds of biomass, coal, petroleum residue and other heavy hydrocarbons. Because of the nature of the feeds normally used, particulates need to be removed from the gas stream after the gasification.

All of the precombustion arrangements in Figure 25.34 also feature the *water gas shift reaction*:

$$CO + H_2O \rightleftharpoons CO_2 + H_2$$

Whilst this reaction occurs during the reforming and partial oxidation reaction steps, it is forced to a high conversion in a catalytic water gas shift reactor, as shown in Figure 25.34. In principle, the CO can be forced to a high conversion. Whether a high conversion is justified depends on the ultimate use of the synthesis gas and the method of separation to be used after the water gas shift. If the synthesis gas is to be used for chemical conversion (as well as combustion) then a high conversion might be justified. Alternatively, if only a bulk separation of CO_2 is required before combustion to reduce environmental discharge, then a high conversion might not be justified. The conversion of CO in the water gas shift reaction is favored by low temperature. Thus, to obtain a high conversion requires two water gas shift reaction steps, one at high temperature and the second at low temperature with intermediate cooling. Different catalysts are used in the high and low temperature shifts. The high temperature shift operates in the range of 300 °C to 450 °C and takes advantage of high reaction rates at high temperature and can give conversion to below 3% carbon monoxide on a dry basis at the reactor exit. The low temperature shift operates in the temperature range 180 °C to 250 °C and can reduce the CO content in the product to as low as 0.2% on a dry basis at the reactor exit. Whether one or two shifts are used, the resulting gas is essentially a mixture of hydrogen, carbon dioxide, water vapor with small amounts of CO (and H_2S if there is sulfur in the feed).

The CO_2 then needs to be separated (and H_2S if there is sulfur in the feed), leaving essentially pure hydrogen for combustion and/or chemical use. Separation of the CO_2 in precombustion arrangements allows a wider range of techniques, as the CO_2 is available at a higher partial pressure than the case of postcombustion with air. Separation can be by:

- reactive absorption, e.g. using amines or hot potassium carbonate,
- physical absorption, e.g. using refrigerated methanol (the Rectisol Process) or a mixture of the dimethyl ethers of polyethylene glycol (the Selexol Process),
- pressure swing adsorption, e.g. using alumina or zeolite,
- membrane,
- cryogenic separation by condensing the CO_2 at low temperature.

Once the CO_2 is separated, the remaining gas is essentially pure hydrogen, which can be used for chemical purposes or combusted to provide heat, or power, or both in cogeneration systems. Subsequent combustion produces mainly water vapor.

The separated CO_2 can then be used or disposed of. The options are:

- Small-scale use in food and beverage products, fire extinguishers, enhanced growth in greenhouses, etc. However, such uses are too small to make a significant difference to the global effects of CO_2 emissions.

- Conversion of the CO_2 to chemical products (e.g. carbon monoxide, methane, methanol, or formic acid). However, conversion to chemical products requires energy input, which might create additional CO_2 emissions depending on the source of the energy.
- Separated CO_2 can be dehydrated, compressed to liquid and deposited in geological storage or used for enhanced gas/oil recovery. Such *sequestration* requires a stable location for long-term disposal and is an expensive way to deal with CO_2 emissions.

If it is important to significantly reduce CO_2 emissions to the atmosphere; overwhelmingly the best solution is demand reduction through increased energy efficiency.

Example 25.6 A medium fuel oil is to be burnt in a furnace with 10% excess air. Ambient temperature is 10 °C and 60% relative humidity. Saturated vapor pressure of water at 10 °C is 0.0123 bar. The analysis of the fuel is given in Table 25.4.

Estimate the composition of the flue gas.

Table 25.4

Fuel analysis for Example 25.6.

	Molar mass (kg·kmol^{-1})	Fuel analysis (mass%)
C	12	83.7
H	1	12.0
S	32	4.3

Solution First, calculate the mole fraction of water vapor in the combustion air. The partial pressure of water in the combustion air is given by:

$$p^{SAT}(10\,°\text{C}) = 0.0123 \text{ bar}$$

$$p_{H_2O} = 0.6 \times 0.0123 = 0.00738 \text{ bar}$$

Mole fraction water in combustion air:

$$y_{H_2O} = \frac{p_{H_2O}}{P} = 0.00738$$

	Fuel				Stoichiometric flue gas from fuel		
	Molar mass (kg·kmol^{-1})	m (kg)	N (kmol)	N_{O_2} (kmol)	N_{CO_2} (kmol)	N_{SO_2} (kmol)	N_{H_2O} (kmol)
C	12	83.7	6.975	6.975	6.975		
H	1	12.0	12.000	3.000			6.000
S	32	4.3	0.134	0.134		0.134	
		100.0		10.109	6.975	0.134	6.000

$$N_2 \text{ from combustion air} = 10.109 \times \frac{79}{21} = 38.03 \text{ kmol}$$

$$H_2O \text{ from combustion air} = (10.109 + 38.03)\left(\frac{y_{H_2O}}{1 - y_{H_2O}}\right) = 48.139\left(\frac{0.00738}{1 - 0.00738}\right) = 0.358 \text{ kmol}$$

$$\text{Stoichiometric moist air} = 10.109 + 38.03 + 0.358 = 48.497 \text{ kmol}$$

$$\text{Stoichiometric flue gas} = 6.975 + 0.134 + 6.000 + 38.03 + 0.358 = 51.497 \text{ kmol}$$

Assume 10% excess air:

$$N_{O_2} = 0.1 \times 10.109 = 1.011 \text{ kmol}$$

$$N_{N_2} = 0.1 \times 38.03 + 38.03 = 41.833 \text{ kmol}$$

$$N_{H_2O} = 0.1 \times 0.358 + 0.358 + 6.000 = 6.394 \text{ kmol}$$

$$\text{Total flue gas flowrate} = 1.011 + 41.833 + 6.394 + 6.975 + 0.134 = 56.347 \text{ kmol}$$

$$y_{CO_2} = \frac{6.975}{56.347} = 0.1238$$

$$y_{SO_2} = \frac{0.134}{56.347} = 0.0024$$

$$y_{N_2} = \frac{41.833}{56.347} = 0.7424$$

$$y_{H_2O} = \frac{6.394}{56.347} = 0.1135$$

$$y_{O_2} = \frac{1.011}{56.347} = 0.0179$$

In addition to these components, some NO_X will be formed. This depends on the combustion device, but assuming a typical value of 300 ppm:

$$y_{NO_x} \approx 0.0003$$

25.8 Atmospheric Dispersion

The objective must be to reduce atmospheric emissions to a minimum or at least below legislative requirements. However, there is inevitably some residual emission and this must be safely dispersed in the environment. The factors that affect the dispersion of gases to atmosphere are (De Nevers, 1995):

- temperature,
- wind,
- turbulence.

Temperature is a critical factor. Generally the temperature of the atmosphere decreases with height, and the actual change of

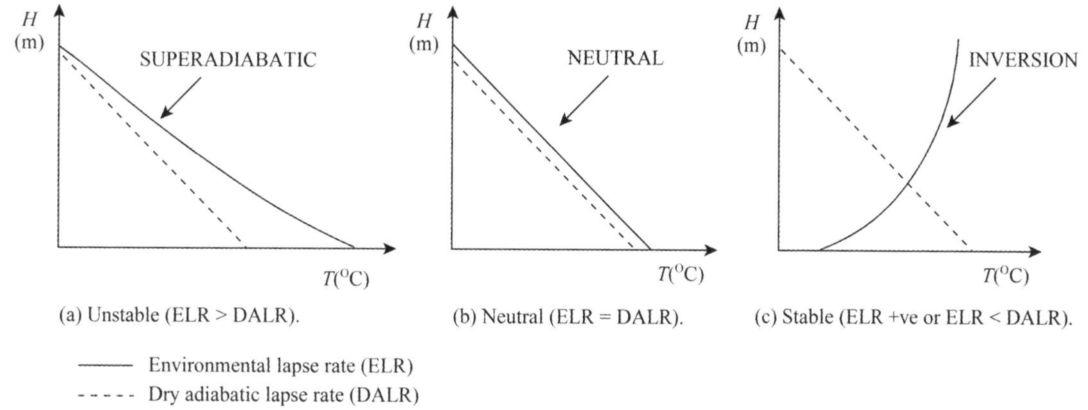

Figure 25.35
Temperature and atmospheric stability.

temperature with height is known as the *environmental lapse rate*. Air originating at the surface of the Earth, if rising, will cool owing to expansion as the pressure changes. The rate of cooling is known as the *dry adiabatic lapse rate* and is about 9.8 °C per kilometer, until condensation occurs. The environmental lapse rate (ELR) will determine what happens to *pockets* of air if they are forced to rise. Figure 25.35a shows the situation where the ELR has a greater change of temperature with height than the dry adiabatic lapse rate. This means that a small volume of air displaced upward will become less dense than the surroundings and would continue its upward motion. This is a desirable condition for atmospheric dispersion, termed *unstable conditions*. Figure 25.35b shows a situation in which the ELR and dry adiabatic lapse rate are roughly equal, giving *neutral conditions*. In this case, there is no tendency for a displaced volume to gain or lose buoyancy. The third situation is shown in Figure 25.35c, in which the ELR is such that the temperature increases with height, known as an *inversion*. These are termed *stable conditions* and provide a strong resistance to vertical motion of a displaced volume. Such stable conditions are problematic from the point of view of gas dispersion.

In the lower levels of the atmosphere, the ELR changes with the time of day. Figure 25.36 shows a typical daily variation of atmospheric stability. Starting before sunrise, the minimum temperature is at the surface. This is caused by heat loss from long-wave radiation. It creates an inversion (increase of temperature with height) up to perhaps 100 meters. Soon after sunrise in Figure 25.36, warming of the surface layer occurs, but the inversion remains at higher levels. Around midday in Figure 25.36, warming has extended from the ground, and now there are unstable conditions (decrease of temperature with height) throughout the lower atmosphere. Near sunset in Figure 25.36, there is radiation from the ground, and an inversion begins to extend from the surface.

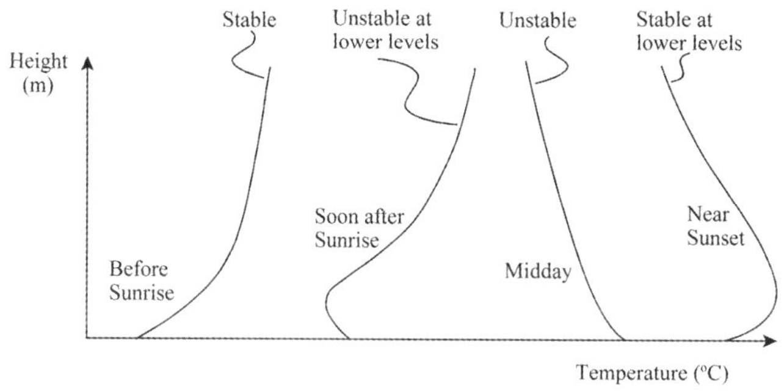

Figure 25.36
Typical daily variation of atmospheric stability.

Not only does wind direction change but also the wind speed increases with height above the ground to a maximum at a height at which speed is equal to the *free air* or *geostrophic* wind speed. The rate of change of wind speed with height is affected by the atmospheric conditions. It is also affected by the terrain. The buildings in urban areas, for example, slow the air close to the ground, meaning that the maximum speed occurs at a greater height than, for example, over level country.

The third factor affecting dispersion is turbulence. Mechanical turbulence is caused by the roughness of the Earth's surface. Away from the surface, convective turbulence (heated air rising and cooler air falling) becomes increasingly important. The amount of turbulence and the height to which it operates depend on the surface roughness, wind speed and atmospheric stability.

The problem for the designer is to determine the appropriate stack height. This is illustrated in Figure 25.37. This shows that the effective stack height is a combination of the *actual stack height* and the *plume rise*. The plume rise is a function of discharge velocity, temperature of emission and atmospheric stability (De Nevers, 1995).

The emissions from the stack itself must comply with environmental regulations relating to concentration and flowrate of pollutants. However, the stack must also be high enough such that any pollutant reaching the ground must be lower than ground level concentrations specified by the regulatory authorities. Pollution concentration at ground level depends on many factors, the most important being:

- height of the emitting stack,
- velocity and temperature of the emission from the stack,

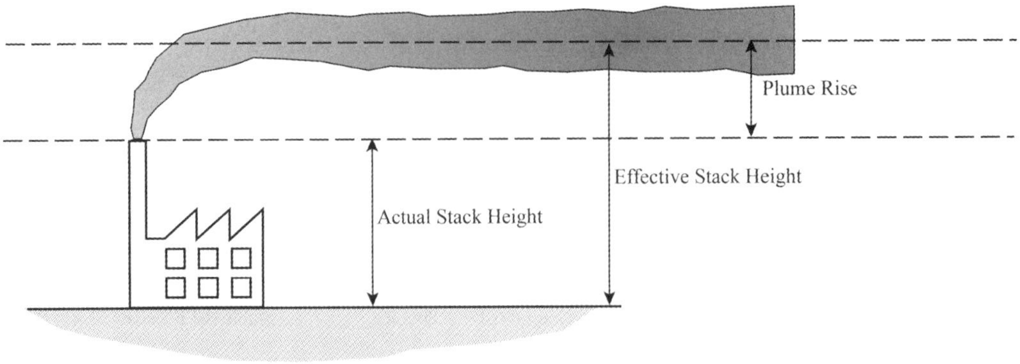

Figure 25.37
Stack height.

- wind speed,
- atmospheric stability,
- precipitation,
- topology of the surrounding terrain.

Calculation of pollution concentration at ground level requires specialized modeling techniques. These need to consider meteorological conditions over an extended period, often several years, to ensure that ground level concentrations remain below safe levels under all potential weather scenarios. These considerations are outside the scope of this text.

25.9 Environmental Design for Atmospheric Emissions – Summary

Emissions to atmosphere can be categorized according to their phase (gas, vapor, liquid and solid). Industrial emissions of major concern are:

- VOCs
- NO_X
- SO_X
- CO
- CO_2
- Dioxins and furans
- Particulates
- PM_{10}
- $PM_{2.5}$.

Urban smog, acid rain, ocean acidification, eutrophication, ozone layer destruction and the greenhouse effect are environmental problems caused by atmospheric emissions.

Solid (particulate) emissions are produced from incomplete combustion of fuels, drying operations, crushing and grinding operations, solids handling operations, and so on. The largest volume of emissions is from products of combustion (CO_2, CO, NO_X, SO_X and particulates). Acid gases are produced by chemical production (CO_2, NO_X, SO_X, H_2S, HCl). Vapor emissions are often more difficult to deal with because of the variety of sources.

Remediation processes to deal with atmospheric emissions include:

- removal of particulates,
- condensation,
- membranes,
- adsorption,
- physical absorption,
- chemical absorption,
- thermal oxidation,
- gas dispersion.

The hierarchy of control of VOC emissions is:

- eliminate or reduce VOC emissions at source,
- recover the VOC for reuse,
- recover the VOC for thermal oxidation with process heat recovery,
- thermal oxidation of the VOC-laden gas stream with process heat recovery,
- oxidation of the VOC-laden gas stream without process heat recovery.

Four methods can be used for VOC recovery:

- condensation,
- membranes,
- absorption,
- adsorption.

Once VOCs have been minimized at source and recovery possibilities have been exhausted, then any residual VOC needs to be destroyed using:

- flares (but only for abnormal and emergency release),
- thermal oxidation,

- catalytic thermal oxidation,
- gas turbines,
- bioscrubbers,
- biofiltration.

Atmospheric sulfur emissions can be minimized at source by improving the process yields and desulfurization of fuels prior to combustion. The difficulty of fuel desulfurization is solid > liquid > gas. Sulfur can be removed from emissions either as SO_2 or H_2S. Removal of the H_2S can be by:

- physical absorption,
- chemical absorption,
- gasification of material followed by H_2S removal,
- partial oxidation to elemental sulfur.

Removal of SO_2 can be by:

- absorption using sodium hydroxide,
- wet limestone scrubbing,
- dry limestone scrubbing,
- Wellman–Lord process,
- or many other methods.

NO_X (principally NO and NO_2) emissions are produced in chemical production, metal and mineral processing and combustion of fuels. NO_X formed in combustion processes is by three mechanisms (fuel, thermal and prompt NO). NO_X emissions can be minimized at source by increasing process yields in chemical production, switching to a fuel with lower nitrogen content or minimizing NO_X formation in combustion processes. After minimizing the production of NO_X, removal by either oxidation or reduction can be considered. NO_X can be absorbed in hydrogen peroxide, which forms nitric acid. Reduction is usually carried out using ammonia:

- without a catalyst in a narrow temperature range (850 to 1100 °C),
- with a catalyst with higher removal efficiency (up to 95%) at lower temperatures (250 to 450 °C).

Flue gas emissions can be minimized at source by:

- increased energy efficiency at the point of use,
- increased energy efficiency of the utility system,
- improvements to combustion processes,
- changing fuel.

Increased energy efficiency at the point of use can be achieved effectively by improved heat integration. Increased energy efficiency of the utility system can be improved through better matching between processes and the utility system, improved cogeneration, and so on. Improvements to combustion processes are effective for NO_X reduction.

25.10 Exercises

1. A storage tank is filled with toluene at 40 °C and has a vent that is open to the atmosphere:

   a) Estimate the concentration of toluene in the vent from the tank. Is this above a legislative limit of $80\ mg{\cdot}m^{-3}$?

   b) The concentration of toluene in the vent from the tank is to be reduced using a refrigerated condensation system. Is it possible to achieve the legislative requirement of $80\ mg{\cdot}m^{-3}$ by cooling the vent to −20 °C?

   c) Sketch a flowsheet to recover any volatile organic compounds from the vent by refrigerated condensation.

   d) What problems are likely to be encountered in using such a refrigerated condensation system?

   e) If the legislative limit of $80\ mg{\cdot}m^{-3}$ cannot be achieved by cooling to −20 °C, what would be the required pressure of the tank to achieve the limit, given the same refrigerated condensation system? Is this a realistic proposal?

   f) Suggest at least three alternatives to refrigerated condensation that would satisfy the environmental requirement.

   It can be assumed that the molar mass in kilograms at standard conditions occupies $22.4\ m^3$. The vapor pressures of toluene can be represented by:

   $$\ln P^{SAT} = 9.3935 - \frac{3096.52}{T - 53.67}$$

   where P^{SAT} = saturated liquid vapor pressure (bar)
   T = absolute temperature (K)

   The freezing point of toluene is −95 °C and its molar mass is $92\ kg{\cdot}kmol^{-1}$.

2. A covered tank with a volume of $50\ m^3$ is used for the storage of acetone. The tank has a vent to atmosphere. The temperature of the storage area changes from 5 °C during the night to 25 °C during the day. The tank breathes to atmosphere as a result of the change in temperature. Assume that the vapor space above the liquid reaches equilibrium with the temperature of the surrounding air, the vapor follows ideal gas behavior and that the change in volume of the liquid acetone can be neglected.

   a) Estimate the concentration of acetone in the tank vent (corrected to standard conditions of 0 °C and 1 atm) as the temperature begins to rise.

   b) Estimate the concentration of acetone in the tank vent (corrected to standard conditions of 0 °C and 1 atm) at the maximum temperature.

   c) Estimate the change in volume if the tank is on average 20% full. Assume ideal gas behavior.

   d) Estimate the mass of acetone vented to atmosphere from the expansion assuming a vapor pressure at the average temperature.

   e) Regulations dictate that the concentration and mass of volatile organic compounds (VOCs) should be related to the toluene equivalent. Express the concentration and mass released as toluene equivalent (according to the ratio of the molar masses). The regulations dictate that the vent should be a concentration of less than $80\ mg{\cdot}m^{-3}$ (toluene equivalent) if the mass vented is greater than $5000\ kg{\cdot}y^{-1}$ toluene equivalent. If the temperature change is assumed to take place 365 days per year, does this emission violate the regulations?

f) One way of reducing the mass released would be to operate the tank with a higher average level. What would the average level need to be in order to reduce the mass released by half?

g) Would you consider the estimate in Part d above to be high or low in terms of the concentration and the mass vented?

h) How would you estimate the mass vented more accurately than taking the vapor pressure at the average temperature?

i) Without using abatement equipment, how would you reduce the breathing losses resulting from temperature variations with fixed level in the tank?

It can be assumed that the molar mass in kilograms occupies $22.4\ m^3$ at standard conditions. The molar mass of acetone can be taken to be $58\ kg{\cdot}kmol^{-1}$ and toluene $92\ kg{\cdot}kmol^{-1}$. The vapor pressure of acetone is given by:

$$\ln P^{SAT} = 10.0310 - \frac{2940.46}{T - 35.93}$$

where P^{SAT} = saturated liquid vapor pressure (bar)
T = absolute temperature (K)

3. Toluene is used as a solvent for the application of surface coatings. The solvents evaporate as a result of the application, creating a problem for the emission of volatile organic compounds (VOCs). The legislative framework for the emission of VOCs requires that the mass load of VOC emissions allowed to be released to atmosphere should be less than 60% of the mass of solids deposited during the coating process. The coating operations are currently depositing $50\ t{\cdot}y^{-1}$ of solids. The concentration of the solids in the coating material is 20%, the remainder being toluene.

a) Calculate the mass of toluene released during the coating operations and the mass that needs to be recovered or destroyed as a result of the legislation.

b) It is proposed to solve the emissions problem by installing a ventilation system using air to collect the vapors and to destroy these using thermal oxidation. For safety reasons, the concentration of the flammable material in the ventilation system must be less than 30% of the lower flammability limit. The lower flammability limit for toluene in air is 1.2% by volume. The release of VOCs can be assumed to be evenly distributed over 8000 hours per year of operation. Calculate the air flowrate to the thermal oxidizer in $m^3{\cdot}h^{-1}$ required to collect the VOCs that need to be destroyed.

c) Toluene is also used for cleaning purposes and is stored on the site in a covered tank with a volume of $10\ m^3$ that is maintained on average to be half full. The tank has a vent to atmosphere. The temperature of the storage area changes from 5 °C during the night to 30 °C during the day. The tank breathes to atmosphere as a result of the change in temperature. Assume that the vapor space above the liquid reaches equilibrium with the temperature of the surrounding air, the vapor follows ideal gas behavior and the change in volume of the toluene can be neglected. Estimate the concentration of toluene in the tank vent (corrected to standard conditions of 0 °C and 1 atm) and the mass released as a result of the expansion, using the vapor pressure at the mean temperature.

d) Regulations dictate that the vent should be a concentration of less than $80\ mg{\cdot}m^{-3}$ if the mass vented is greater than $5000\ kg{\cdot}y^{-1}$. If the temperature change is assumed to take place 365 days per year, does this emission violate the regulations?

It can be assumed that the molar mass in kilograms at standard conditions occupies $22.4\ m^3$. The vapor pressures of toluene can be represented by:

$$\ln P^{SAT} = 9.3935 - \frac{3096.52}{T - 53.67}$$

where P^{SAT} = saturated liquid vapor pressure (bar)
T = absolute temperature (K)

The freezing point of toluene is −95 °C and its molar mass $92\ kg{\cdot}kmol^{-1}$.

4. A flue gas with a flow of $10\ Nm^3{\cdot}s^{-1}$ (Nm^3 = normal m^3) contains 0.1% vol NO_X (expressed as NO_2 at 0 °C and 1 atm) and 3% vol oxygen. It is proposed to remove NO_X by absorption in water to 100 ppmv for discharge. Whilst NO_2 is highly soluble, NO is only sparingly soluble and there is a reversible reaction in the gas phase between the two according to:

$$NO + {}^1\!/_2 O_2 \rightleftharpoons NO_2$$

The equilibrium relationship for the reaction is given by:

$$K_a = \frac{p_{NO_2}}{p_{NO}\, p_{O_2}^{0.5}}$$

where K_a is the equilibrium constant for the reaction and p the partial pressure. At 25 °C, the equilibrium constant is 1.4×10^6 and at 725 °C is 0.14.

a) Calculate the mole ratio of NO_2 to NO assuming chemical equilibrium at 25 °C and at 725 °C.

b) Calculate the flowrate of NO_2 and NO assuming chemical equilibrium at 25 °C and at 725 °C. Assume the kilogram molar mass occupies $22.4\ m^3$ at standard conditions.

c) It is proposed to remove the NO_X by absorption in water at 25 °C. Do you consider this to be a good option if chemical equilibrium is assumed?

d) In the worst case, if equilibrium is not attained, none of the NO would react to NO_2. Calculate the flowrate of water to remove the NO to a concentration of 100 ppmv by absorption in water at 25 °C and 1 atm. The solubility of NO in water can be assumed to follow Henry's Law:

$$x_i = \frac{y_i^* P}{H}$$

where x_i = mole fraction of Component i in the liquid phase

Table 25.5

Fuel analysis for Exercise 5.

	% by wt
C	78
O	7
H	3
S	1
Ash	6
H_2O	6

y_i^* = mole fraction of Component i in the vapor phase in equilibrium with the liquid
P = total pressure
H = Henry's Law constant

At 25 °C, $H = 28$ atm for NO. If the concentration in the gas phase at the exit of the absorber is assumed to be 80% of equilibrium and the gas phase is assumed to be ideal, calculate the flowrate of water required.

e) Suggest other methods to treat the NO_X.

The molar masses of NO and NO_2 are 30 kg·kmol^{-1} and 46 kg·kmol^{-1} respectively.

5. Coal with the analysis in Table 25.5 is to be burnt in 20% excess air in a boiler. The ash contains calcium (0.15% of the coal) and sodium (0.18% of the coal). Assume that the calcium reacts to calcium sulfate, $CaSO_4$, and the sodium to sodium sulfate, Na_2SO_4. Assume all the remaining sulfur reacts to SO_2. Regulations on the flue gas are to be based on a dry gas (free of water vapor) at 0 °C and 1 atm.

a) What proportion of the sulfur in the coal is trapped in the ash and what proportion is oxidized to SO_2?

b) What is the mass of the ash after combustion as a result of sulfate formation?

c) Calculate the flowrate of the flue gases per kilogram of fuel for 20% excess air on a dry basis, neglecting any ash carried from the boiler. Assume air is 21% O_2 and 79% N_2 and the kilogram molar mass occupies 22.4 m^3 at standard conditions.

d) Calculate the composition of the flue gas on a dry basis.

Atomic and molar masses are given in Table 25.6.

6. Coal with the analysis in Table 25.7 is to be burnt in 20% excess air in a boiler.

The ash contains calcium (0.2% of the coal) and sodium (0.24% of the coal). Assuming the calcium reacts to form solid calcium sulfate, $CaSO_4$, and the sodium to form solid sodium sulfate, Na_2SO_4:

a) What proportion of the sulfur in the coal is trapped in the ash and what proportion is oxidized to SO_2?

Table 25.6

Atomic and molar masses.

	Atomic/molar mass (kg·kmol^{-1})
H	1
C	12
O	16
N	14
Na	23
S	32
Ca	40
CO_2	44
H_2SO_4	98
$CaCO_3$	100
$CaSO_4$	136
Na_2SO_4	142

b) One method to prevent the emission of the remaining sulfur as gaseous SO_2 is to carry out the combustion in a fluidized bed with the addition of limestone ($CaCO_3$) to react the sulfur to calcium sulfate ($CaSO_4$). What mass of limestone must be added per mass of coal to prevent the emission of the remaining SO_2, assuming a 20% excess of limestone is added?

c) In what other way could limestone be used to prevent the emission of the SO_2? Write down the key reactions and steps in the process. What advantages could the alternative method have?

d) An alternative to the reaction of SO_2 with limestone is to recover the SO_2 in a Wellman–Lord process. If the resulting SO_2 is converted to sulfuric acid, how much sulfuric acid (as 100% pure) would be produced per mass of coal combusted?

Atomic and molar masses are given in Table 25.6.

Table 25.7

Fuel analysis for Exercise 6.

	% by wt
C	75
O	7
H	3
S	2
Ash	7
H_2O	5

7. A gas turbine with power output of 10.7 MW and an efficiency of 32.5% burns natural gas. In order to reduce the NO_X emissions to the environmental limits, 0.6 kg steam is injected into the combustion per kg of fuel. The airflow through the gas turbine is $41.6\,kg{\cdot}s^{-1}$. The composition of the natural gas can be assumed to be effectively 100% methane with a molar mass of $16\,kg{\cdot}kmol^{-1}$. The kilogram molecular volume can be assumed to occupy $22.4\,m^3$ at standard conditions.

a) Calculate the mass flowrate of natural gas. The fuel value of the natural gas is $34.9\,MJ{\cdot}m^{-3}$ at standard conditions.

b) Calculate the flowrate of steam for NO_X abatement.

c) Without the steam flow, the turbine produces 10.7 MW power. Assuming the power output is proportional to the mass flow through the turbine, estimate the power output with NO_X abatement.

d) Assume that the steam used for NO_X abatement was raised in a boiler burning natural gas (assumed to be 100% methane) with an efficiency of 90%. The condition of the steam is 20 bar and 350 °C with an enthalpy of $3138\,kJ{\cdot}kg^{-1}$ and is raised from boiler feedwater at 100 °C with an enthalpy of $420\,kJ{\cdot}kg^{-1}$. Estimate the total CO_2 emissions from the boiler and the gas turbine.

e) Instead of using a gas turbine, a steam turbine could have been used to generate power. For an ideal expansion of the 20 bar steam to 0.075 bar in a steam turbine, a flowrate of 3.7 kg of steam is required per kWh of power production. The steam turbine can be assumed to have an efficiency of 0.75. Calculate the CO_2 emissions using the power output calculated from Part c as the basis.

f) Compare the emissions from the gas turbine and steam turbine for power generation. Suggest why one process is better than the other.

References

Crynes BL (1977) *Chemical Reactions as a Means of Separation – Sulfur Removal*, Marcel Dekker Inc.

De Nevers N. (1995) *Air Pollution Control Engineering*, McGraw-Hill, New York.

Dullien FAL (1989) *Introduction to Industrial Gas Cleaning*, Academic Press.

Garg A (1994) Specify Better Low-NO_x Burners for Furnaces, *Chem Eng Prog*, **Jan**: 46.

Glassman J (1987) *Combustion*, 2nd Edition, Academic Press.

Harrison RM (1992) *Understanding Our Environment: An Introduction to Environmental Chemistry and Pollution*, Royal Society of Chemistry, Cambridge, UK.

Hui C-W and Smith R (2001) Targeting and Design for Minimum Treatment Flowrate for Vent Streams, *Trans IChemE*, **79A**: 13.

Hydrocarbon Processing's Environmental Processes '98 (1998) *Hydrocarbon Process*, **August**: 71.

Rousseau RW (1987) *Handbook of Separation Process Technology*, John Wiley & Sons, Inc., New York.

Schweitzer PA (1997) *Handbook of Separation Process Techniques for Chemical Engineers*, 3rd Edition, McGraw-Hill, New York.

Steeneveldt R, Berger B and Torp, TA (2006) CO_2 Capture and Storage Closing the Knowing – Doing Gap, Chemical Engineering Research and Design, **84** (A9): 739.

Stenhouse JIT (1981) Pollution Control, in Teja AS, *Chemical Engineering and the Environment*, Blackwell Scientific Publications.

Svarovsky L (1981) *Solid–Gas Separation*, Elsevier Scientific, New York.

US Environmental Protection Agency (1992a) *Hazardous Air Pollution Emissions from Units in the Synthetic Organic Chemical Industry – Background Information for Proposed Standards*, Vol. 1B, *Control Technologies*, US Department of Commerce, Springfield, VA.

US Environmental Protection Agency (1992b) *Control of Air Emissions at Superfund Sites*, Technical Report EPA/625/R - 92/012.

Wood SC (1994) Select the Right NO_x Control Technology, *Chem Eng Prog*, **Jan**: 32.

Chapter 26

Water System Design

In the past, water has been assumed to be a limitless low-cost resource. However, there is now increasing awareness of the danger to the environment caused by the overextraction of water. In some locations, future increases in water use are being restricted. At the same time, the imposition of ever-stricter discharge regulations has driven up effluent treatment costs, requiring capital expenditure with little or no productive return. Figure 26.1 shows a schematic of a greatly simplified water system for a processing site. Raw water enters the processing environment and might need some raw water treatment. This might be something as simple as sand filtration. In many instances, the raw water supply is good enough to enter the processes directly. Water is used in various operations as a reaction medium, solvent in extraction processes, for cleaning, and so on. Water becomes contaminated and is discharged to effluent. Also, as shown in Figure 26.1, some of the raw water is required for the steam system. It first requires upgrading in boiler feedwater treatment to remove suspended solids, dissolved salts and the dissolved gases before being fed to the steam boiler, as discussed in Chapter 23. Deionized water might also be required for process use. The steam from steam boilers is distributed and some of the steam condensate is not returned to boilers but is lost to effluent. The boiler requires blowdown to remove the build-up of solids, as discussed in Chapter 23. Also, as discussed in Chapter 23, the ion-exchange beds used to remove dissolved salts and soluble ions need to be regenerated by saline solutions or acids and alkalis and this goes to effluent. Finally, as shown in Figure 26.1, water is used in evaporative cooling systems to make up for the evaporative losses and blowdown from the cooling water circuit, as discussed in Chapter 24. All of the effluents tend to be mixed together, along with contaminated storm water, treated centrally in a wastewater treatment system and discharged to the environment. It is usual for most of the water entering the processing environment to leave as wastewater. If the use of water can be reduced, then this will reduce the cost of water supplied. However, it will also reduce the cost of effluent treatment, as a result of most of the water entering the processing environment leaving as effluent. There is thus considerable incentive to reduce both fresh water consumption and wastewater generation.

Figure 26.2a shows three operations, each requiring fresh water and producing wastewater. By contrast, Figure 26.2b shows an arrangement where there is a *reuse* of water from Operation 2 to Operation 1. Reusing water in this way reduces both the volume of the fresh water and the volume of wastewater, as the same water is used twice. However, for such an arrangement to be feasible, any contamination level at the outlet of Operation 2 must be acceptable at the inlet of Operation 1. Not all operations require the highest quality of water. For example, the extraction process described in Section 10.7 for desalting crude oil prior to distillation does not need the highest quality water. Some level of contamination is acceptable, but certain specific contaminants (e.g. hydrogen sulfide and ammonia in the case of crude oil desalting) can cause problems. Another example would be a multistage washing operation. Low-quality water could be used in the initial stages and high-quality water used in the final stages. There are many examples when water with some level of certain contaminants is acceptable for use rather than using the highest quality water.

Figures 26.2c and 26.2d both show arrangements involving *regeneration*. Regeneration is a term used to describe any treatment process that regenerates the quality of water such that it is acceptable for further use. Figure 26.2c shows *regeneration reuse*, where the outlet water from Operation 2 is too contaminated to be used directly in Operation 3. A regeneration process between the two allows reuse to take place. Regeneration reuse reduces both the volume of the fresh water and the volume of the wastewater, as with reuse, but also removes part of the *effluent load* (i.e. kilograms of contaminant). The regeneration, in addition to allowing a reduction in the water volume, also removes part of the contaminant load that would have to be otherwise removed in the final effluent treatment before discharge.

A third option is shown in Figure 26.2d, where a regeneration process is used on the outlet water from the operations and the water is *recycled*. The distinction between the regeneration reuse shown in Figure 26.2c and the *regeneration recycling* shown in Figure 26.2d is that in regeneration reuse the water only goes through any given operation once. Figure 26.2c shows that the water goes from Operation 2 to regeneration, then to Operation 3 and then discharge. By contrast, in Figure 26.2d, the water can go through the same operation many times. Regeneration recycling

Chemical Process Design and Integration, Second Edition. Robin Smith.

Companion Website: www.wiley.com/go/smith/chemical2e

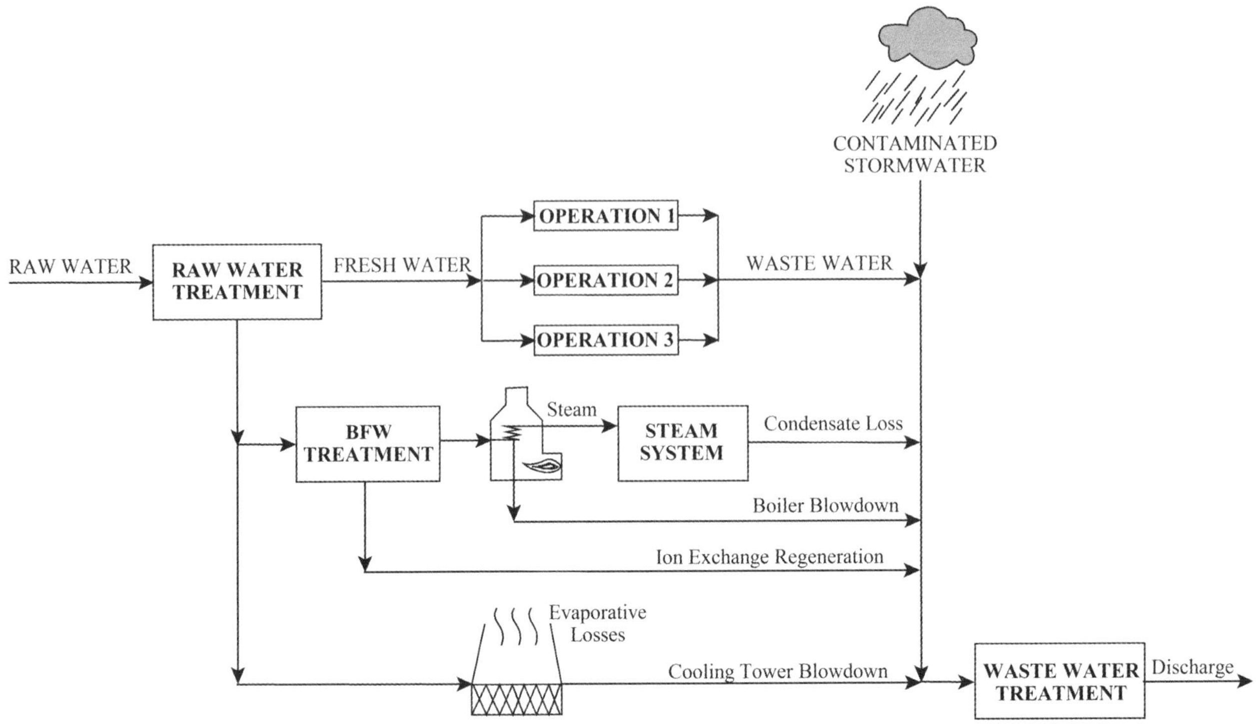

Figure 26.1
A typical water and effluent treatment system.

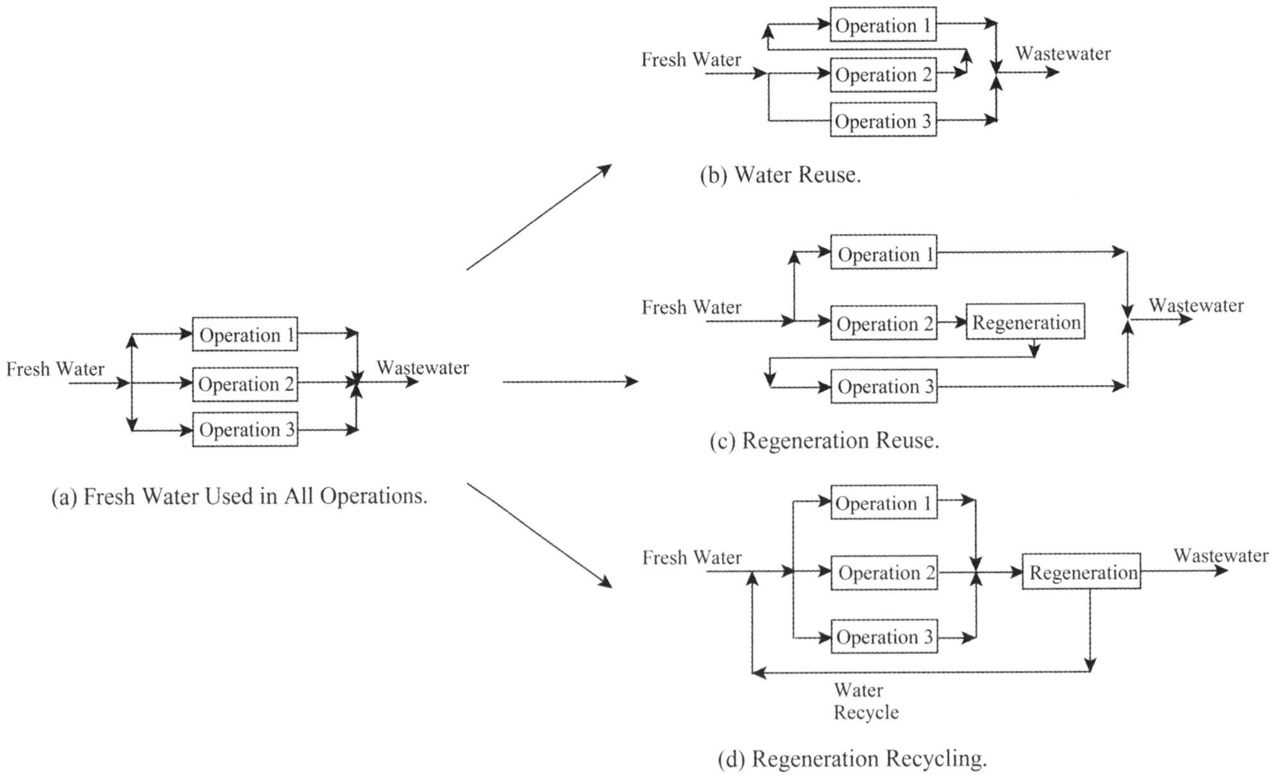

Figure 26.2
Water reuse and regeneration.

reduces the volume of fresh water and wastewater and also reduces the effluent load by virtue of the regeneration process taking up part of the required effluent treatment load.

Regeneration reuse and regeneration recycling are similar in terms of their outcomes. Regeneration recycling allows larger reductions in the fresh water use and wastewater generation than in the regeneration reuse. However, problems can be encountered with regeneration recycling. Regeneration costs can be high. What is usually a greater problem, however, is that the recycling can allow the build-up of undesired contaminants in the recycle, such as microorganisms or products of corrosion. These contaminants, if not removed in the regeneration, might build up to the extent of creating problems to the process.

As shown in Figure 26.1, all of the wastewater was collected together and treated centrally before discharge. Another way to deal with the effluent treatment is by *distributed effluent treatment* or *segregated effluent treatment*. The basic idea is illustrated in Figure 26.3. In addition to some reuse of water between Operation 2 and Operation 1, some local treatment (distributed treatment) is taking place on the outlet of Operation 1 and on the outlet of Operation 3 before going to final wastewater treatment and discharge. Another feature of the arrangement in Figure 26.3 is that a number of the effluents are no longer being treated before discharge. The arrangement in Figure 26.1 is such that the more contaminated effluents from the outlet of Operations 1, 2 and 3 need to be diluted by the less contaminated streams before being treated and discharged. By contrast, as shown in Figure 26.3, the pretreatment on the outlet of Operation 1 and Operation 3 is such that the dilution is no longer required before final effluent treatment and discharge. This means that the streams with only light contamination now no longer need to be treated.

The capital cost of aqueous waste treatment operations is generally proportional to the total flow of wastewater and the operating cost generally increases with decreasing concentration for a given mass of contaminant to be removed. Thus, if two streams require different treatment operations, it makes no sense to mix them and treat the two streams in both treatment operations. This will increase both capital and operating costs. The concept of distributed effluent treatment is one that tends to treat effluents before they are mixed together. Treatment is made specific to individual (or small numbers of) contaminants while still concentrated. The benefit is that, by avoiding mixing, this increases the potential to recover material, leading to less waste and lower cost of raw materials. However, the overriding benefit is usually that the effluent volume to be treated is reduced significantly, leading to lower effluent treatment costs overall.

Before developing more systematic ways of designing water systems, first consider the general issues of water contamination and treatment.

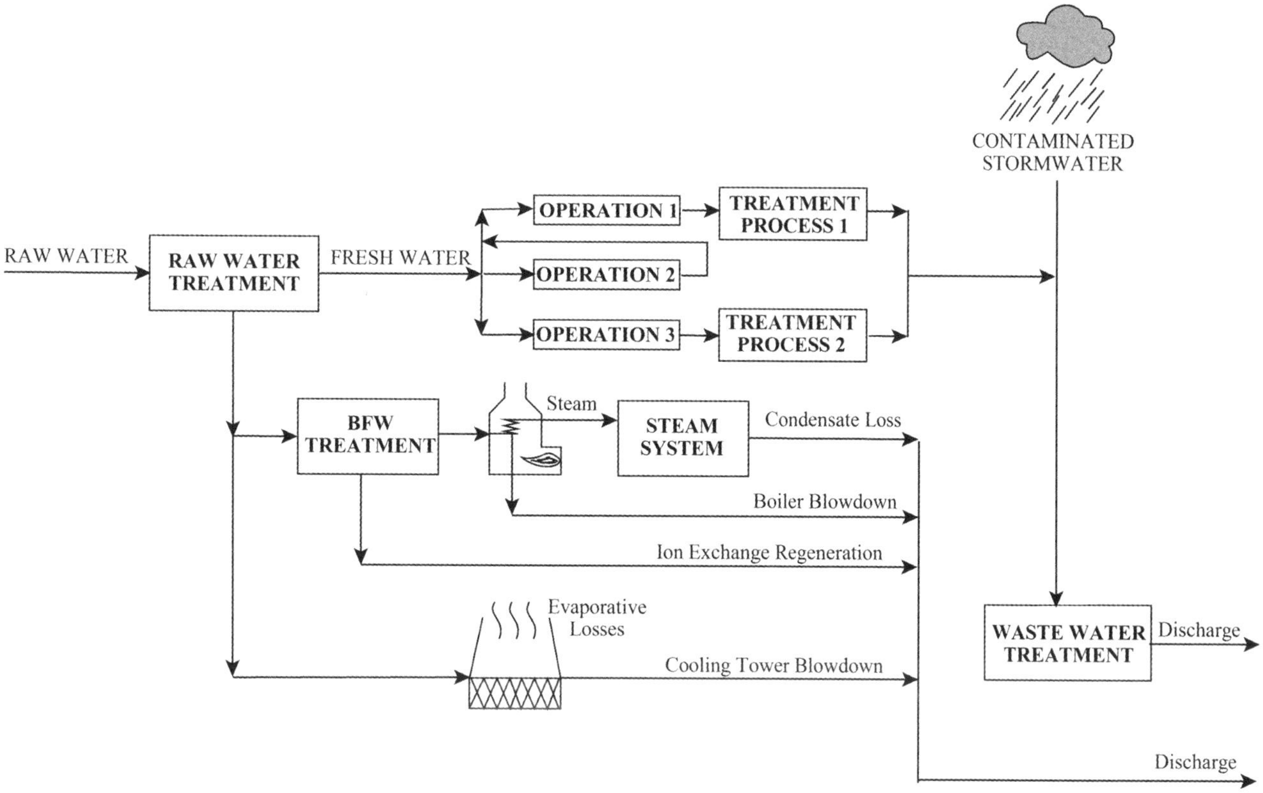

Figure 26.3
Distributed effluent treatment.

26.1 Aqueous Contamination

There are two significant reasons why water contamination needs to be considered. The first is that aqueous effluent must comply with environmental regulations before discharge. The environmental effects of pollution can be direct, such as toxic emissions providing a fatal dose of toxicant to fish, animal life and even human beings. The effects can also be indirect. Toxic materials that are non-biodegradable such as insecticides and pesticides, if released to the environment, are absorbed by bacteria and enter the food chain. These compounds can remain in the environment for long periods of time, slowly being concentrated at each stage in the food chain until ultimately they prove fatal, generally to predators at the top of the food chain, such as fish or birds. The concentration, and perhaps load, of contamination of various specified contaminants must be less than the regulatory requirements. The second reason is that contaminant levels will affect the feasibility of reuse and recycling of water, as shown in Figure 26.2. If water is to be reused or recycled, then the level of inlet contamination to the operation receiving reused or recycled water must be acceptable. What types of contamination need to be considered?

Consider first aqueous emissions of organic waste material. When this is discharged to the receiving water, bacteria feed on the organic material. This organic material will eventually be oxidized to stable end products. Carbon in the molecules will be converted to CO_2, hydrogen to H_2O, nitrogen to NO_3^-, sulfur to SO_4^{2-}, and so on. As an example, consider the degradation of urea:

$$\underset{\text{urea}}{CH_4N_2O} + \underset{\text{oxygen}}{9/2\,O_2} \rightarrow \underset{\substack{\text{carbon}\\\text{dioxide}}}{CO_2} + \underset{\text{water}}{2H_2O} + \underset{\text{nitrate}}{2NO_3^-}$$

This equation indicates that every molecule of urea requires 9/2 molecules of oxygen for complete oxidation. The oxygen required for the reactions depletes the receiving water of oxygen, causing the death of aquatic life. Standard tests have been developed to measure the amount of oxygen required to degrade a sample of wastewater (Tchobanoglous and Burton, 1991; Berne and Cordonnier, 1995; Eckenfelder and Musterman, 1995).

1) *Biochemical oxygen demand (BOD).* A standard test has been devised to measure biological oxygen demand (BOD) in which the oxygen utilized by microorganisms in contact with the wastewater over a five-day period at 20 °C is measured (usually termed BOD_5). The period of the test can be extended to a much longer period (in excess of 20 days) to measure the *ultimate demand*. While the BOD_5 test gives a good indication of the effect the effluent will have on the environment, it requires five days to carry out (or longer for the ultimate BOD). Other tests have been devised to accelerate the oxidation process.
2) *Chemical oxygen demand (COD).* In the chemical oxidation demand (COD) test, oxidation with acidic potassium dichromate is used. A catalyst (silver sulfate) is required to assist oxidation of certain classes of organic compounds. Chemical oxygen demand results are generally higher than BOD_5, since the COD test oxidizes materials that are only slowly biodegradable. Although the COD test provides a strong oxidizing environment, certain organic compounds are oxidized only slowly, or not at all.

   The ratio of BOD_5 to COD varies according to the contamination. The ratio can vary between 0.05 and 0.8, depending on the chemical species (Eckenfelder and Musterman, 1995). Domestic sewage has a value of typically around 0.37 (Wang and Smith, 1994a). An average value across all types of contaminants is around 0.35.
3) *Total oxygen demand (TOD).* The total oxygen demand (TOD) test measures the oxygen consumed when a sample of wastewater is oxidized in a stream of air at high temperature (up to 1200 °C) in a furnace. Under these harsh conditions, all the carbon is oxidized to CO_2. The oxygen demand is determined from the change in oxygen in the carrier gas. The resulting value of TOD embraces oxygen required to oxidize both organic and inorganic substances present.

   The relationship between BOD_5, COD and TOD for the same organic waste is in the order,

$$BOD_5 < COD < TOD$$

4) *Total organic carbon (TOC).* The total organic carbon (TOC) test measures the carbon dioxide produced when a sample of wastewater is subjected to a strongly oxidizing environment. One option is to oxidize the sample in a stream of air at a high temperature (around 1000 °C) in a furnace, similar to the TOD test, but measuring the change in CO_2 rather than the change in O_2. To allow complete oxidation of all carbon compounds also requires a catalyst such as copper oxide or platinum. Rather than using high temperature in a furnace, other strongly oxidizing environments can also be used. For example, UV light with sodium persulfate as a digesting reagent can be used. To obtain the TOC requires that inorganic carbon compounds be removed prior to test, or results corrected for their presence. The inorganic carbon can be removed prior to test by adding acid to convert the inorganic carbon to carbon dioxide that must be stripped from the sample by a sparge carrier gas.

A chemical process is designed from knowledge of physical concentrations, whereas aqueous effluent treatment systems are designed from knowledge of BOD_5, COD and TOD. Thus, the relationship between BOD_5, COD and TOD and the concentration of waste streams leaving the process needs to be established. Without measurements, relationships can only be established approximately. The relationships between BOD_5, COD and TOD are not easy to establish, since different materials will oxidize at different rates. To compound the problem, many wastes contain complex mixtures of oxidizable materials, perhaps together with chemicals that *inhibit* the oxidation reactions.

If the composition of the waste stream is known, then the *theoretical oxygen demand* (ThOD) can be calculated from the appropriate stoichiometric equations. As a first level of approximation, it can be assumed that the ThOD would be equal to the COD or TOD. The following example will help to clarify these relationships.

Example 26.1 A process produces an aqueous waste stream containing 0.1 mol % acetone. Estimate the COD and BOD_5 of the stream.

Solution First, calculate the ThOD from the equation that represents the overall oxidation of the acetone:

$$\underset{\text{acetone}}{(CH_3)_2CO} + \underset{\text{oxygen}}{4O_2} \rightarrow \underset{\substack{\text{carbon}\\\text{dioxide}}}{3CO_2} + \underset{\text{water}}{3H_2O}$$

Approximating the molar density of the waste stream to be that of pure water (i.e. 56 kmol·m^{-3}), then theoretical oxygen demand

$$\begin{aligned} &= 0.001 \times 56 \times 4 \quad \text{kmolO}_2 \cdot \text{m}^{-3} \\ &= 0.001 \times 56 \times 4 \times 32\ \text{kgO}_2 \cdot \text{m}^{-3} \\ &= 7.2\ \text{kg O}_2 \cdot \text{m}^{-3} \end{aligned}$$

Thus:

$$\begin{aligned} \text{COD} &\cong 7.2\ \text{kg} \cdot \text{m}^{-3} \\ \text{BOD}_5 &\cong 7.2 \times 0.35 \\ &\cong 2.5\ \text{kg} \cdot \text{m}^{-3} \end{aligned}$$

Effluent treatment regulations might specify a level of BOD_5, COD or both. Increasingly, the tendency is toward the specification of *toxicity*. This measures the toxicity of an effluent to some kind of living species. Other contaminants that might be specified are:

- specifically nominated contaminants (e.g. phenol, benzene, etc.),
- heavy metals (e.g. chromium, cobalt, vanadium, etc.),
- halogenated organic compounds,
- organic nitrogen,
- organic sulfur,
- nitrates,
- phosphates,
- suspended solids,
- pH, and so on.

Nitrates and phosphates in particular can cause *eutrophication* of the receiving water, either terrestrial or marine water. Nitrates and phosphates act as nutrients that lead to enhanced growth of aquatic vegetation or *phytoplankton* and *algal blooms*. This disrupts normal functioning of the ecosystem, causing a variety of problems such as a lack of oxygen needed for aquatic life to survive, decreased biodiversity, changes in species composition and dominance, and toxicity effects.

In addition to the levels of contamination, typically specified as parts per million (ppm) or milligrams per liter (mg·l^{-1}), there might be regulations for the total discharge of a contaminant in kilograms per day (kg·d^{-1}) or tons per year (t·y^{-1}). There might also be a specification for the maximum effluent temperature.

Water treatment processes can be classified in order as (Tchobanoglous and Burton, 1991):

- Primary (or pretreatment)
- Secondary (or biological)
- Tertiary (or polishing).

26.2 Primary Treatment Processes

Primary or *pretreatment* of wastewater can involve either physical or chemical treatment, depending on the nature of the contamination, and serves three purposes:

- Allows reuse or recycling of water.
- Recovers useful material from the wastewater where possible.
- Prepares the aqueous waste for biological treatment by removing excessive load or contaminants that will inhibit the biological processes in biological treatment.

The pretreatment processes may be most effective when applied to individual waste streams from particular processes or process steps before effluent streams are combined for biological treatment.

A brief review of the primary treatment methods will now be given (Tchobanoglous and Burton, 1991; Berne and Cordonnier, 1995; Eckenfelder and Musterman, 1995). Pretreatment usually starts with phase separation if the effluent is a heterogeneous mixture.

1) *Solids separation.* Methods used for solids separation include most of the commonly used techniques:
   - Sedimentation
   - Centrifugal separation
   - Filtration.

   Clarifiers can be used, as discussed in Chapter 7 and as illustrated in Figure 7.2. As an example of the effectiveness of clarifiers, in petroleum refinery applications they are capable of removing typically 50 to 80% of suspended solids, together with 60 to 95% of dispersed hydrocarbon (which rises to the surface), 30 to 60% of BOD_5 and 20 to 50% of COD (Betz Laboratories, 1991). The performance depends on the design and the effluent being treated. The performance can be enhanced by the use of chemical additives. The chemical additives neutralize the electric charges on the particles that cause them to repel each other and remain dispersed.

   Settling to remove solids from aqueous effluent can be enhanced by the use of centrifugal forces in hydrocyclones. Their design was discussed in Chapter 7. For solids removal, the solids are driven toward the wall of the hydrocyclone and removed from the base. Centrifugation can also be used for solids removal, but restricted to smaller volumes. Solids removal by centrifugation can typically be in the range 50 to 80%, increasing to typically 80 to 95% with chemical addition.

   Filtration can be used to remove solid particles down to around 10 μm. Both cake filtration and depth filtration can be used, as discussed in Chapter 7.
2) *Coalescence.* Coalescence by gravity in simple settling devices can often be used to separate immiscible liquid–liquid mixtures. The coalescence can be enhanced by the use of mesh pads and centrifugal forces, as discussed in Chapter 7.

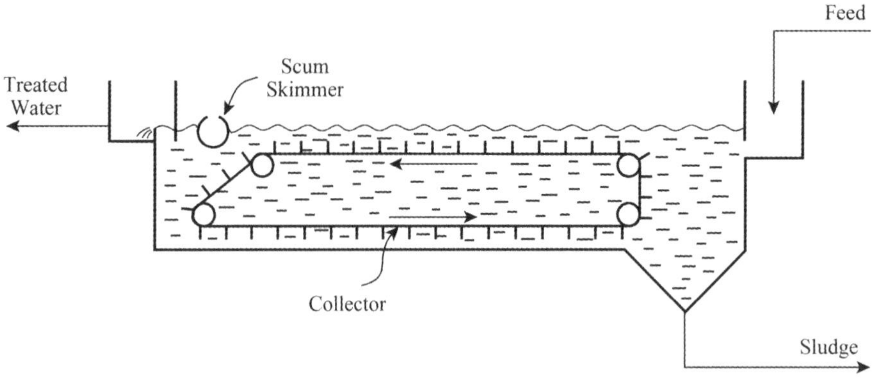

Figure 26.4

A typical API (American Peroleum Institute) separator.

In petroleum and petrochemical plants, a common device used for the separation of dispersed hydrocarbon liquids from aqueous effluent is the American Petroleum Institute (API) separator, as illustrated in Figure 26.4. This is a simple settling device in which the effluent enters a large volume. The resulting low velocity allows light particles of hydrocarbon to rise to the surface and any heavy solid particles to settle to the base of the device. Rakes might be employed to move the light material along the surface and the heavy material along the base for collection. The light material can be removed using a *scum skimmer*. This, for example, could be as simple as a horizontal pipe with a slot in its upper surface. Rotation of the device, such that the slot is just below the surface of the liquid, allows the light material (hydrocarbon) to overflow into the pipe for collection. In petroleum refinery applications, API separators typically remove 60 to 99% of dispersed hydrocarbon liquids, together with 10 to 50% of suspended solids, 5 to 40% of BOD_5 and 5 to 30% of COD (Betz Laboratories, 1991). It is a very simple device used for the first stage of treatment. The performance depends on the design and the effluent being treated, and the performance can be enhanced by the use of chemical additives.

Centrifugal forces can also be exploited in hydrocyclones. As discussed in Chapter 7, designs are different for the removal of solids and dispersed oil. In contrast to solids removal, for oil removal the water is driven to the wall of the hydrocyclone and the treated water removed from the base. Performance for the removal of oil from an aqueous effluent is typically in the range 70 to 90%.

3) *Flotation.* Flotation can be used to separate solid or immiscible liquid particles from aqueous effluent, as discussed in Chapter 7. These particles should have a lower density than water and be naturally hydrophobic. Typical applications include removal of dispersed hydrocarbon liquids from petroleum and petrochemical effluents and removal of fibers from pulp and paper effluents. Direct air injection is usually not the most effective method of flotation. Dissolved-air flotation is usually superior. In dissolved-air flotation, air is dissolved in the effluent under a pressure of several atmospheres and then liberated in the flotation cell by reducing the pressure, as discussed in Chapter 7. The fine air bubbles attach to the particles and rise to the surface. Once the particles have floated to the surface, they are collected by skimming. Flotation is particularly effective for the separation of very small and light particles. As an example of the performance of such units, when applied to effluent from a petroleum refinery, dissolved-air flotation can typically remove 70 to 85% of dispersed oil and 50 to 85% of suspended solids, together with 20 to 70% of BOD_5 and 10 to 60% of COD (Betz Laboratories, 1991). The performance depends on the equipment design and the effluent being treated. In such applications, it is normally used after an API separator has been used for preliminary treatment.

4) *Membrane processes.* Conventional filtration processes can separate particles down to a size of around 10 μm (micrometers). If smaller particles need to be separated, a porous polymer membrane can be used. Microfiltration retains particles down to a size of around 0.05 μm. A pressure difference across the membrane of up to 4 bar is used. The two most commonly used arrangements are spiral wound and hollow fiber, as discussed in Chapter 7.

In ultrafiltration, the effluent is passed across a semipermeable membrane (see Chapter 9). Water passes through the membrane, while submicron particles and large molecules are rejected from the membrane and concentrated. The membrane is supported on a porous medium for strength, as discussed in Chapter 9. Ultrafiltration is used to separate very fine particles (typically in the range 0.001 to 0.02 μm), microorganisms and organic components with molar mass down to 1000 $kg{\cdot}kmol^{-1}$. Pressure drops are usually in the range 1.5 to 10 bar.

Reverse osmosis and nanofiltration are high-pressure membrane separation processes (typically 10 to 50 bar for reverse osmosis and 5 to 20 bar for nanofiltration), which can be used to reject dissolved inorganic salt or heavy metals. The processes were discussed in Chapter 9 and are particularly useful for removal of ionic species, such as sodium, magnesium, nitrate, sulfate, chloride ions, and so on. Depending on the membrane system design and the ionic species, these can be removed with an efficiency of up to 85 to 99%.

Configurations used include tubes, plate-and-frame arrangements and spiral-wound modules. Spiral-wound modules should be treated to remove particles down to 20 to 50 μm, while hollow fiber modules require particles down to 5 μm to be removed. If necessary, pH should be adjusted to avoid extremes of pH. Also, oxidizing agents such as free chlorine must be removed. Because of these restrictions, reverse osmosis is only useful if the wastewater to be treated is free of heavy contamination. The concentrated waste material produced by membrane processes should be recycled if possible, but might require further treatment or disposal.

Electrodialysis can be an alternative to reverse osmosis, as discussed in Chapter 9.

5) *Stripping.* Volatile organic compounds and dissolved gases can be stripped from wastewater. The usual arrangement would involve wastewater being fed down through a column with packing or trays and the stripping agent (usually steam or air) fed to the bottom of the column.

If steam is used as a stripping agent, either live steam or a reboiler can be used. The use of live steam increases the effluent volume. The volatile organic materials are taken overhead, condensed and, if possible, recycled to the process. If recycling is not possible, then further treatment or disposal is necessary. A common application of steam stripping is the stripping of hydrogen sulfide and ammonia from water in petroleum refinery operations to regenerate contaminated water for reuse. If significant quantities of both H_2S and NH_3 need to be stripped, then this can be a problem. The ideal pH for stripping H_2S is below 5, since above 5 sulfide is primarily in the form of HS^- and S^{2-} ions that will not readily strip. On the other hand, at pH less than 7 the ammonia is present predominantly as NH_4^+ and it will not readily strip. At a pH above 10, the ammonia is present in solution predominantly as free NH_3, which will strip much more readily. The optimum strategy would be to use two separate strippers, one operating at low pH (e.g. by the addition of sulfuric acid before stripping) to strip the H_2S. The second stripping would operate at high pH (e.g. by the addition of sodium hydroxide before stripping) to strip the NH_3. However, the cost of this two-stage approach can be prohibitive, so a compromise of operation using a pH around 8 allows reasonable removal of both gases. Injecting sodium hydroxide near the bottom of the tower improves the ammonia stripping while still allowing hydrogen sulfide stripping at the top.

Steam stripping is capable of removing typically 90 to 99% H_2S, 90 to 97% NH_3 and 75 to 99% organic materials. It should be noted, though, that some organic materials are resistant to steam stripping, and it is thus not a universal solution to contamination with organic materials.

If air is used as a stripping agent, further treatment of the stripped material will be necessary. The gas might be fed to a thermal oxidizer or some attempt made to recover material by use of adsorption.

6) *Crystallization.* If contamination in wastewater has a solubility that varies significantly with temperature, then cooling might allow crystallization of a significant proportion of the contamination. Crystallization was discussed in Chapter 9. Unfortunately, cooling crystallization to treat wastewater streams might require the use of refrigeration to cool below cooling water temperature and this can be expensive to install and incur a high cost of power to operate. An alternative way to create supersaturation is to use evaporation, as discussed in Chapter 9. This can be expensive to operate in terms of the heat required for evaporation, unless the heat required can be supplied by heat recovery or the latent heat available in the evaporated water can be recovered. An additional benefit is the production of relatively clean water from the evaporated water.

7) *Evaporation.* An extreme case of the use of evaporative crystallization is to use evaporation to simply concentrate the contamination as a concentrated waste stream. This will generally only be useful if the wastewater is low in volume and the waste contamination is nonvolatile. The relatively pure evaporated water might still require treatment after condensation if it is to be disposed of. The concentrated waste can then be recycled or sent for further treatment or disposal. As with evaporative crystallization, the cost of energy for such operations can be prohibitively expensive, unless the heat required for evaporation can be supplied by heat recovery or the heat available in the evaporated water can be recovered.

8) *Liquid–liquid extraction.* With liquid–liquid extraction, wastewater containing organic waste is contacted with a solvent in which the organic waste is more soluble. The waste is then separated from the solvent by evaporation or distillation and the solvent recycled.

One common application of liquid–liquid extraction is the removal and recovery of phenol and compounds of phenol from wastewaters. Although phenol can be removed by biological treatment, only limited levels can be treated biologically. Variations in phenol concentration are also a problem with biological treatment, since the biological processes take time to adjust to the variations.

9) *Adsorption.* Adsorption can be used for the removal of organic compounds (including many toxic materials) and heavy metals (especially when complexed with organic compounds). Activated carbon is primarily used as the adsorbent, although synthetic resins are also used. Both fixed and moving beds can be used, but fixed beds are by far the most commonly used arrangement. For activated carbon, the removal of organic compounds depends, amongst other things, on the molar mass and the polarity of the molecule. A general trend is that nonpolar molecules (e.g. benzene) tend to adsorb more readily than polar molecules (e.g. methanol) on activated carbon. As an example, adsorption with activated carbon when applied to petroleum refinery effluent can remove typically 70 to 95% BOD_5 and 70 to 90% COD, depending on the effluent being treated (Betz Laboratories, 1991).

As the adsorbent becomes saturated, regeneration is required. When removing organic materials, activated carbon can be regenerated by steam stripping or heating in a furnace. Stripping allows recovery of material, whereas thermal regeneration destroys the organic material. Thermal regeneration requires a furnace with temperatures above 800 °C to oxidize the adsorbates. This causes a loss of carbon of around 5 to 10% per regeneration cycle.

10) *Ion exchange.* Ion exchange is used for selective ion removal and finds some application in the recovery of specific materials from wastewater, such as heavy metals. As with adsorption processes, regeneration of the medium is necessary. Resins are regenerated chemically, which produces a concentrated waste stream requiring further treatment or disposal.

11) *Wet oxidation.* In wet oxidation, an aqueous mixture is heated in the liquid phase under pressure in the presence of air or pure oxygen, which oxidizes the organic material. The efficiency of the oxidation process depends on reaction time and pressure. Temperatures of 150 to 300 °C are used, together with pressures of 3 to 200 bar, depending on the process and the nature of the waste being treated and whether a catalyst is used. Oxidation at low temperatures and pressures is only possible if a catalyst is used. Carbon is oxidized to CO_2, hydrogen to H_2O, chlorine to Cl^-, nitrogen to NH_3 or N_2, sulfur to SO_4^{2-}, phosphorus to PO_4^{2-}, and so on.

Wet oxidation is particularly effective in treating aqueous wastes containing organic contaminants with a COD up to 2% prior to biological treatment. Chemical oxygen demand can be reduced by up to 95% and organic halogen compounds by up to 95%. Organic halogen compounds are particularly resistant to biological degradation. If designed for a specific organic contaminant (e.g. phenol), removal can be 99% or greater. Wet oxidation is often used prior to biological treatment to pretreat wastes that would otherwise be resistant to biological oxidation.

A basic flowsheet for a wet oxidation process is shown in Figure 26.5. Although the oxidation reactions release heat, the process might still require a net input of heat from an external source (e.g. steam or hot oil). In some cases, the heat release can be high enough to avoid the need for an external source of heat. The largest cost associated with the process is the capital cost of the high-pressure reactor, which normally has an internal titanium cladding.

12) *Chemical oxidation.* Chemical oxidation can be used for the oxidation of organic contaminants that are difficult to treat biologically. It can be used to kill microorganisms by oxidation. When used before biological treatment, organic pollutants that are difficult to treat biologically can be oxidized to simpler, less refractory organic compounds. Chlorine (as gaseous chlorine or hypochlorite ion) can be used. In solution, chlorine reacts with water to form hypochlorous acid (HOCl) according to:

$$Cl_2 + H_2O \rightarrow HOCl + HCl$$
$$HOCl \rightleftharpoons H^+ + OCl^-$$

Then, for example, the reaction for nitrogen removal is:

$$2NH_3 + 3HOCl \rightarrow N_2 + 3H_2O + 3HCl$$

Heterogeneous solid catalysts can enhance the performance. Chlorination has the disadvantage that small quantities of undesirable halogenated organic compounds can be formed.

Hydrogen peroxide is another common oxidizing agent. It is used as a 30 to 70% solution for the treatment of:

- cyanides,
- formaldehyde,
- hydrogen sulfide,
- hydroquinone,
- mercaptans,
- phenol,
- sulfites, and so on.

Ozone can also be used as an oxidizing agent, but because of its instability, it must be generated on site. It is a powerful oxidant for some organic materials, but others are oxidized only slowly or not at all. Ozone is only suitable for low concentrations of oxidizable materials. A common use is for the sterilization of water.

The effectiveness of these chemical oxidizing agents is enhanced by the presence of ultraviolet (UV) light and solid catalysts. Chemical oxidation is also sometimes applied after biological treatment.

13) *Sterilization.* In some processes, such as food and beverage and pharmaceutical processes, water might need to be sterilized before it is reused or recycled. Chemical oxidation (e.g. ozonation) can be used. Ultraviolet light is an alternative for lightly contaminated water. Alternatively, a combination of chemical oxidation and UV light can be used. Also, heat treatment (*pasteurization*) can be used. Heat treatment of water to 80 °C provides sterilization that is often good enough for many purposes.

14) *pH adjustment.* The pH of the wastewater often needs adjustment prior to reuse, discharge or biological treatment.

Figure 26.5
A typical wet oxidation process.

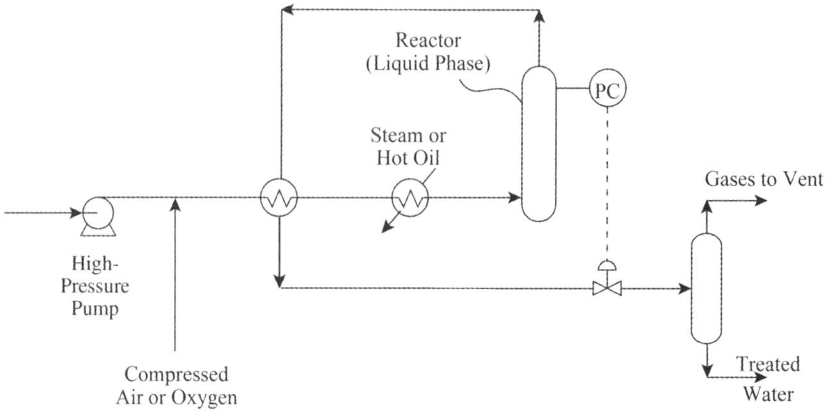

For biological treatment, the pH is normally adjusted to between 8 and 9. Bases used include sodium hydroxide, calcium oxide and calcium carbonate. Acids used include sulfuric acid and hydrochloric acid.

15) *Chemical precipitation.* Chemical precipitation followed by solids separation is particularly useful for separating heavy metals. The heavy metals of particular concern in the treatment of wastewaters include cadmium, chromium, copper, lead, mercury, nickel and zinc. This is a particular problem in the manufacture of dyes and textiles and in metal processes such as pickling, galvanizing and plating.

Heavy metals can often be removed effectively by chemical precipitation in the form of carbonates, hydroxides or sulfides. Sodium carbonate, sodium bisulfite, sodium hydroxide and calcium oxide can be used as precipitation agents. The solids precipitate as a *floc* containing a large amount of water in the structure. The precipitated solids need to be separated by thickening or filtration and recycled if possible. If recycling is not possible, then solids are usually disposed of to landfill.

The precipitation process tends to be complicated when a number of metals are present in solution. If this is the case, then the pH must be adjusted to precipitate out the individual metals, since the pH at which precipitation occurs depends on the metal concerned.

26.3 Biological Treatment Processes

In *secondary* or *biological treatment*, a concentrated mass of microorganisms is used to break down organic matter into stabilized wastes. The degradable organic matter in the wastewater is used as food by the microorganisms. Biological growth requires supplies of oxygen, carbon, nitrogen, phosphorus and inorganic ions such as calcium, magnesium and potassium. Domestic sewage satisfies the requirements, but industrial wastewaters may lack nutrients and this can inhibit biological growth. In such circumstances, nutrients may need to be added. As the waste treatment progresses, the microorganisms multiply, producing an excess of this *sludge*, which cannot be recycled.

There are three main types of biological process (Tchobanoglous and Burton, 1991; Berne and Cordonnier, 1995; Eckenfelder and Musterman, 1995):

1) *Aerobic.* Aerobic reactions take place only in the presence of free oxygen and produce stable, relatively inert end products, such as carbon dioxide and water. Aerobic reactions are by far the most widely used. The oxidation reactions are of the type:

$$\underset{\text{Organic matter}}{\begin{bmatrix} C \\ O \\ H \\ N \\ S \end{bmatrix}} + O_2 + [\text{Nutrients}] \rightarrow CO_2 + H_2O + NH_3 + [\text{Cells}] + [\text{Other end products}]$$

Endogenous respiration reactions also occur, which reduce the sludge formation:

$$[\text{Cells}] + O_2 \rightarrow CO_2 + H_2O + NH_3 + [\text{Energy}]$$

Nitrification reactions can occur, in which organic nitrogen and ammonia are converted to nitrate:

$$\text{Organic nitrogen} \rightarrow NH_4^+$$
$$NH_4^+ + CO_2 + O_2 \rightarrow NO_2^- + H_2O + [\text{Cells}] + [\text{Other end products}]$$
$$NO_2^- + CO_2 + O_2 \rightarrow NO_3^- + [\text{Cells}] + [\text{Other end products}]$$

The nitrification reactions are inhibited by high concentrations of ammonia.

2) *Anaerobic.* Anaerobic reactions function without the presence of free oxygen and derive their energy from organic compounds in the waste. Anaerobic reactions proceed relatively slowly and lead to end products that are unstable and contain considerable amounts of energy, such as methane and hydrogen sulfide:

$$\underset{\text{Organic matter}}{\begin{bmatrix} C \\ O \\ H \\ N \\ S \end{bmatrix}} + [\text{Nutrients}] \rightarrow CO_2 + CH_4 + [\text{Cells}] + [\text{Other end products}]$$

3) *Anoxic.* Anoxic reactions also function without the presence of free oxygen. However, the principal biochemical pathways are not the same as in anaerobic reactions, but are a modification of aerobic pathways and hence are termed anoxic. Anoxic reactions are used for *denitrification* to convert nitrate to nitrogen:

$$NO_3^- + BOD \rightarrow N_2 + CO_2 + H_2O + OH^- + [\text{Cells}]$$

Various methods are used to contact the microorganisms with the wastewater. In a completely mixed system, the hydraulic residence time of the wastewater and the solids residence time of the microorganisms would be the same. Thus, the minimum hydraulic residence time would be defined by the growth rate of the microorganisms. Since the crucial microorganisms can take a considerable time to grow, this would lead to hydraulic residence times that would be prohibitively long. To overcome this, a number of methods have been developed to decouple the hydraulic and solids residence times.

1) *Aerobic digestion.*

a) *Suspended growth* or *activated sludge.* This method is illustrated in Figure 26.6. Biological treatment takes place in a tank where the waste is mixed with a flocculated biological sludge. To maintain aerobic conditions, the tank must be aerated. Sludge separation from effluent is normally achieved by gravity sedimentation. Part of the sludge is recycled and excess sludge is removed. The hydraulic flow pattern in the aeration tank can vary between extremes of mixed flow and plug flow. For mixed-flow reactors, the wastewater is rapidly dispersed throughout the

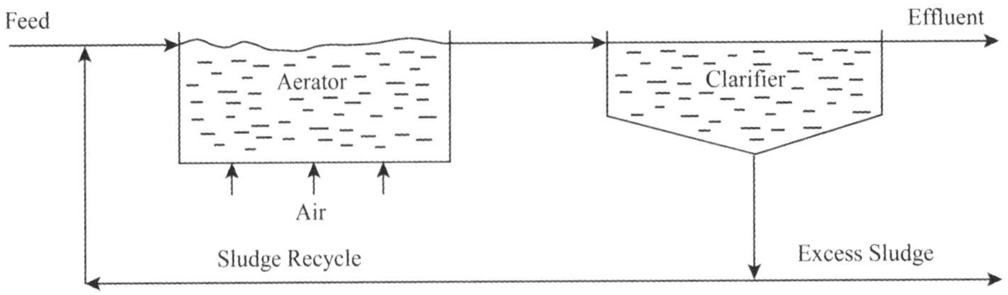

Figure 26.6

Suspended growth aerobic digestion.

reactor and its concentration is reduced. This feature is advantageous at sites where periodic discharges of more concentrated waste are received. The rapid dilution of the waste means that the concentration of any toxic compounds present will be reduced, and thus the microorganisms within the reactor may not be affected by the toxicant. Thus, mixed-flow reactors produce an effluent of uniform quality in response to fluctuations in the feed.

Plug-flow reactors have a decreasing concentration gradient from inlet to outlet, which means that toxic compounds in the feed remain undiluted during their passage along the reactor, and this may inhibit or kill many of the microorganisms within the reactor. The oxygen demand along the reactor will also vary. On the other hand, the increased concentration means that rates of reaction are increased, and for two reactors of identical volume and hydraulic retention time, a plug-flow reactor will show a greater degree of BOD_5 removal than a mixed-flow reactor.

The biochemical population can be specifically adapted to particular pollutants. However, in the majority of cases, a wide range of organic materials must be dealt with and mixed cultures are used.

Nitrification–denitrification reactions can be carried out in suspended growth by separating the reactor into different cells. A typical arrangement would control the first cell under anoxic conditions to carry out the denitrification reactions. These reactions require organic carbon and this will already be present in mixed effluents. If there is insufficient organic carbon in the feed, then this must be added (e.g. adding methanol). The water from the anoxic cell would then overflow into a cell maintained under aerobic conditions in which the nitrification reactions are carried out, along with the oxidation of organic carbon. The nitrification reactions produce nitrate and this requires a recycle back to the anoxic cell.

In aerobic processes, the mean sludge residence time is typically 5 to 15 days. The hydraulic residence time is typically 3 to 8 hours. However, this depends on the effluent to be treated and the design of reactor used. Suspended growth aerobic processes are capable of removing up to 95% of BOD. The inlet concentration is restricted to a maximum BOD of around $1\,kg{\cdot}m^{-3}$ ($1000\,mg{\cdot}liter^{-1} \approx 1000\,ppm$) or COD of around $3.5\,kg{\cdot}m^{-3}$ ($3500\,mg{\cdot}liter^{-1} \approx 3500\,ppm$). Chloride (as Cl^-) should be less than 8 to $15\,kg{\cdot}m^{-3}$. The performance of the process can be enhanced by the use of pure oxygen, rather than air. This reduces the plant size and increases the allowable inlet concentration of COD to around $10\,kg{\cdot}m^{-3}$, but does not have a significant effect on the overall removal. If powdered activated carbon is added, this increases the overall removal slightly. More importantly, though, it helps to treat refractory organic compounds by adsorption and evens out fluctuations in the performance caused by fluctuations in the feed concentration.

b) *Attached growth (film) methods.* Here the wastewater is trickled over a packed bed through which air is allowed to percolate. A *biological film* or *slime* builds up on the packing under aerobic conditions. Oxygen from the air and biological matter from the wastewater diffuses into the slime. As the biological film grows, it eventually breaks its contact with the packing and is carried away with the water. Packing material varies from pieces of stone to preformed plastic packing. Figure 26.7 shows a typical attached growth arrangement. Alternatively, the wastewater can be used to fluidize a bed of carbon or sand on to which the film is attached.

Attached growth processes are capable of removing up to 90% of BOD_5 and are thus less effective than suspended growth methods. Nitrification–denitrification reactions can also be carried out in attached growth processes.

2) *Anaerobic digestion.* With wastewaters containing a high organic content, the oxygen demand may be so high that it becomes extremely difficult and expensive to maintain aerobic conditions. In such circumstances, anaerobic processes can provide an efficient means of removing large quantities of organic material. Anaerobic processes tend to be used when BOD_5 levels exceed $1\,kg{\cdot}m^{-3}$ ($1000\,mg{\cdot}liter^{-1}$). However, they are not capable of producing very high quality effluents and further treatment is usually necessary.

The inability to produce high-quality effluents is one significant disadvantage. Another disadvantage is that anaerobic processes must be maintained at temperatures between 35 and 40 °C to get the best performance. If low-temperature waste heat is available from the production process, then this is not a problem. One advantage of anaerobic reactions is that the methane produced can be a useful source of energy. This can be fed to steam boilers or burnt in a heat engine to produce power.

a) *Suspended growth methods.* The contact type of anaerobic digester is similar to the activated sludge method of aerobic

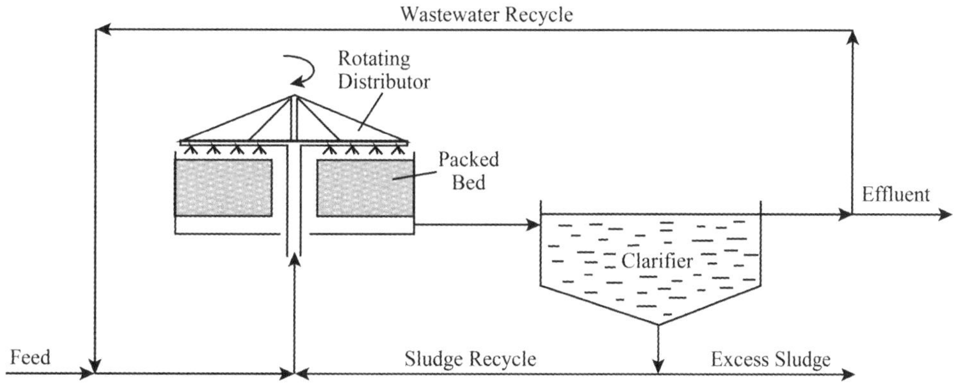

Figure 26.7
Attached growth aerobic digestion.

treatment. The feed and microorganisms are mixed in a tank (this time closed). Mechanical agitation is usually required since there is no air injection. The sludge is separated from the effluent by sedimentation or filtration, part of the sludge recycled and excess sludge removed.

Another method is the *upflow anaerobic sludge blanket*, illustrated in Figure 26.8. Here the sludge is contacted by upward flow of the feed at a velocity such that the sludge is not carried out of the top of the digester.

A third method of contact known as an *anaerobic filter* also uses upward flow but keeps the sludge in the digester by a physical barrier such as a grid.

b) *Attached growth (film) methods.* As with aerobic digestion, the microorganisms can be encouraged to grow attached to a support medium such as plastic packing or sand. In anaerobic attached growth digestion, the bed is usually fluidized rather than a fixed-bed arrangement, as shown in Figure 26.9.

Anaerobic processes typically remove 75 to 85% of COD (Tchobanoglous and Burton, 1991; Eckenfelder and Musterman, 1995).

3) *Reed beds.* Reed bed processes require reeds (phragmites) to be planted in soil, sand or gravel in the wastewater.

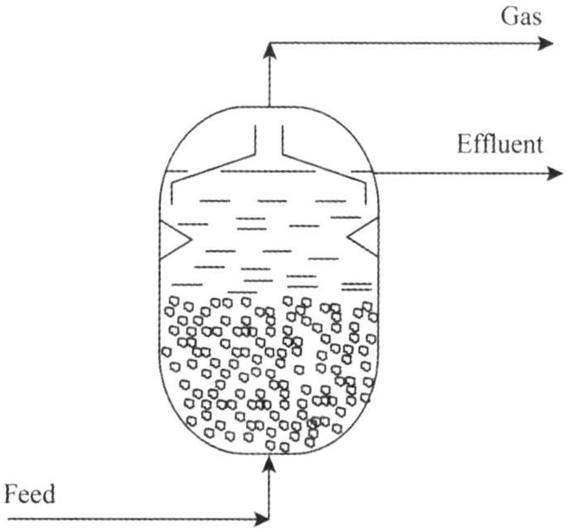

Figure 26.8
Suspended growth anaerobic digestion using an upward flow anaerobic sludge blanket.

Figure 26.10 shows a typical reed bed arrangement. Oxygen from air is transported through the leaves, stems and rhizomes to high concentrations of microorganisms in the root zone. Aerobic treatment takes place in the region of the rhizomes, with anoxic and anaerobic treatment in the surrounding soil.

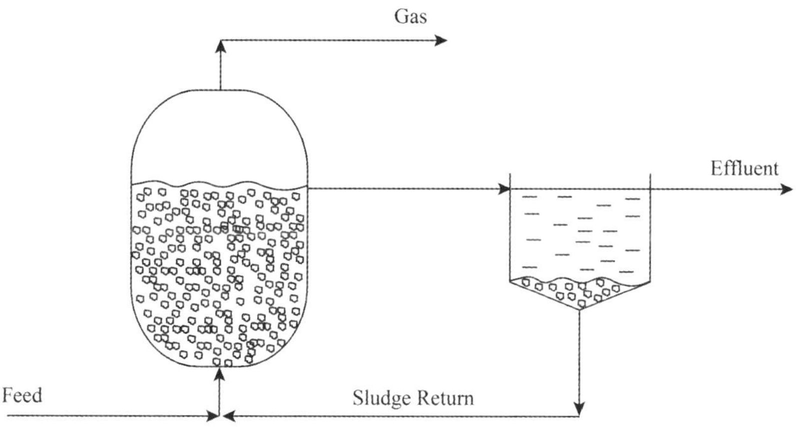

Figure 26.9
Attached growth anaerobic digestion using a fluidized anaerobic bed.

26

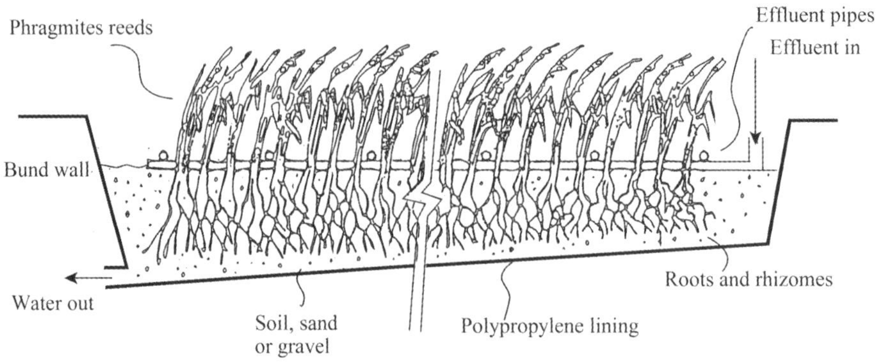

Figure 26.10
A reed bed.

Different flow arrangements other than that shown in Figure 26.10 can also be used. Reed beds are capable of removing 80% to 90% of BOD_5 (Schierup, Brix and Lorenzen, 1990). They are also capable of removing around 30% of total nitrogen and around 30% of total phosphorus (Schierup, Brix and Lorenzen, 1990). They have the advantage of no sludge disposal. The maximum inlet concentration can be as high as $3.5\ kg{\cdot}m^{-3}$ BOD_5. A significant disadvantage of reed beds is the time taken for them to become fully established and they need between six months and two years for this. The difference between winter and summer performance needs to be considered for such arrangements. Another significant disadvantage is that the reed beds need to be refurbished every 7 to 10 years.

Excess sludge is produced in most biological treatment processes, which must be disposed of. The treatment and disposal of sludge is a major problem that can be costly to deal with. Anaerobic processes have the advantage here, since they produce considerably less sludge than aerobic processes (of the order of 5% of aerobic processes for the same throughput). Sludge disposal can typically be responsible for 25 to 40% of the operating costs of an aerobic biological treatment system. Treatment of sludge is primarily aimed at reducing its volume. This is because the sludge is usually 95 to 99% water and the cost of disposal is closely linked to its volume. The water is partly free, partly trapped in the flocs and partly bound in the microorganisms. Anaerobic digestion of the sludge can be used, followed by dewatering. The dewatering can be by filtration or centrifugation. Alternatively, filtration or centrifugation can be used directly to carry out the dewatering. For centrifugation, the dewatering process can be enhanced by the addition of clay. Adding powdered activated carbon to aerobic suspended growth processes can also help with sludge dewatering. The resulting water content after these processes is reduced to typically 60 to 85%. The water content can be reduced to perhaps 10% by drying. The sludge may finally be used for agricultural purposes (albeit a poor fertilizer) or thermally oxidized.

Large sites might require their own biological treatment processes for final treatment before discharge. Smaller sites might rely on local municipal treatment processes, which treat a mixture of industrial and domestic effluent, for final effluent discharge.

Table 26.1
Comparison of biological wastewater treatments.

Aerobic	Anaerobic	Reed beds
$BOD_5 < 1\ kg{\cdot}m^{-3}$ (higher if O_2 used)	$BOD_5 > 1\ kg{\cdot}m^{-3}$	$BOD_5 < 3.5\ kg{\cdot}m^{-3}$
Stable end products (CO_2, H_2O, etc.)	Unstable end products (CH_4, H_2S, etc.)	Stable and unstable end products
BOD_5 removal up to 95%	BOD_5 removal 75–85%	BOD_5 removal 60–80%
High sludge formation	Low sludge formation	No sludge disposal

Table 26.1 provides a summary of the main features of biological wastewater treatment.

Table 26.2 summarizes the treatment processes that can be used for various types of contamination.

26.4 Tertiary Treatment Processes

Tertiary treatment or *polishing treatment* prepares the aqueous waste for final discharge. The final quality of the effluent depends on the nature and flow of the receiving water. Table 26.3 gives an indication of the final quality required (Tebbutt, 1990).

Aerobic digestion is normally capable of removing up to 95% of the BOD. Anaerobic digestion is capable of removing less, in the range 75 to 85%. With municipal treatment processes, which treat a mixture of domestic and industrial effluent, some disinfection of the effluent might be required to destroy any disease-causing organisms before discharge to the environment. Tertiary treatment

Table 26.2

Summary of treatment processes for some common contaminants.

Suspended solids	Dispersed oil	Dissolved organic
Gravity separation	Coalescence	Biological oxidation (aerobic, anaerobic, reed beds)
Centrifugal separation	Centrifugal separation	Chemical oxidation
Filtration	Flotation	Activated carbon
Membrane filtration	Wet oxidation	Wet oxidation
	Thermal oxidation	Thermal oxidation
Ammonia	**Phenol**	**Heavy metals**
Steam stripping	Solvent extraction	Chemical precipitation
Air stripping	Biological oxidation (aerobic)	Ion exchange
Biological nitrification	Wet oxidation	Adsorption
Chemical oxidation	Activated carbon	Nanofiltration
Ion exchange	Chemical oxidation	Reverse osmosis
		Electrodialysis
Dissolved solids	**Neutralization**	**Sterilization**
Ion exchange	Acid	Heat treatment
Reverse osmosis	Base	UV light
Nanofiltration		Chemical oxidation
Electrodialysis		
Crystallization		
Evaporation		

processes vary, but constitute the final stage of effluent treatment to ensure that the effluent meets specifications for disposal. Tertiary processes used include:

1) *Filtration.* Examples of such processes are microstrainers (a fine screen with openings) and sand filters. They are designed to improve effluents from biological treatment processes by removing suspended material and, with it, some of the remaining BOD_5. Sand filtration can remove effectively all of the remaining BOD_5 in many circumstances.
2) *Ultrafiltration.* Ultrafiltration was described under pretreatment methods. It is used to remove finely divided suspended solids and, when used as a tertiary treatment, can in many circumstances remove virtually all the BOD_5 remaining after biological treatment.

Table 26.3

Typical effluent quality for various receiving waters (Tebbutt, 1990).

Receiving water	Typical effluent	
	BOD_5 ($mg \cdot l^{-1}$)	Suspended solids ($mg \cdot l^{-1}$)
Tidal estuary	150	150
Lowland river	20	30
Upland river	10	10
High-quality river with low dilution	5	5

3) *Adsorption.* Some organic materials are not removed in biological systems operating under normal conditions. Removal of residual organic material can be achieved by adsorption. Both activated carbon and synthetic resins are used. As described earlier under pretreatment methods, regeneration of activated carbon in a furnace can cause carbon losses of perhaps 5 to 10%.
4) *Nitrogen and phosphorus removal.* Since nitrogen and phosphorus are essential for growth of the microorganisms, the effluent from secondary treatment will contain some nitrogen and phosphorus. The amount that is discharged to receiving waters can cause eutrophication and has a considerable effect on the growth of algae. If discharge is to a high-quality receiving water and/or dilution rates are low, then removal may be necessary. Nitrogen principally occurs as ammonium (NH_4^+), nitrate (NO_3^-) and nitrite (NO_2^-). Phosphorus principally occurs as orthophosphate (PO_4^{3-}). A variety of biological and chemical processes are available for the removal of nitrogen and phosphorus (Tchobanoglous and Burton, 1991; Eckenfelder and Musterman, 1995). These processes produce extra biological and inorganic sludge that requires disposal.
5) *Disinfection.* Chlorine, as gaseous chlorine or as the hypochlorite ion, is widely used as a disinfectant. However, its use in some cases can lead to the formation of toxic organic chlorides and the discharge of excess chlorine can be harmful. Hydrogen peroxide and ozone are alternative disinfectants that lead to products that have a lower toxic potential. Treatment is enhanced by ultraviolet light. Indeed, disinfection can be achieved by ultraviolet light on its own.

26.5 Water Use

Water is used for a wide variety of purposes in process operations:

- reaction medium (vapor or liquid),
- extraction processes,
- steam stripping,
- steam ejectors for production of vacuum,

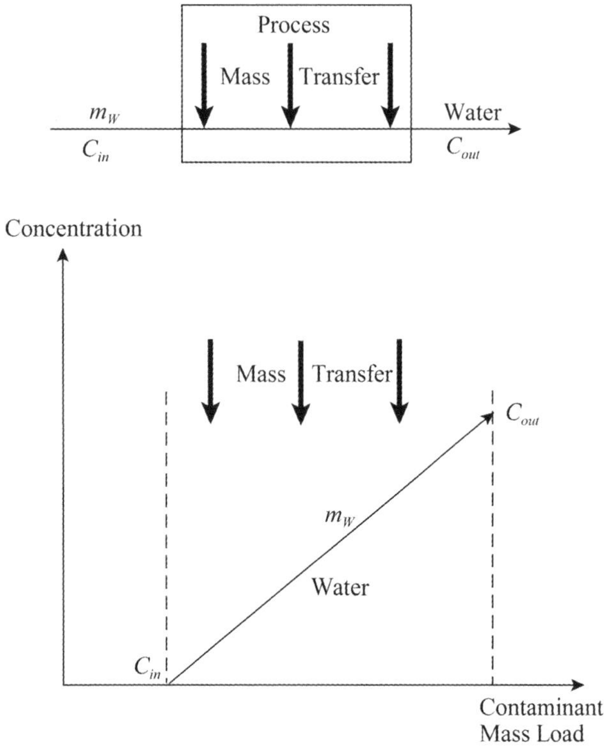

Figure 26.11

Water use representation.

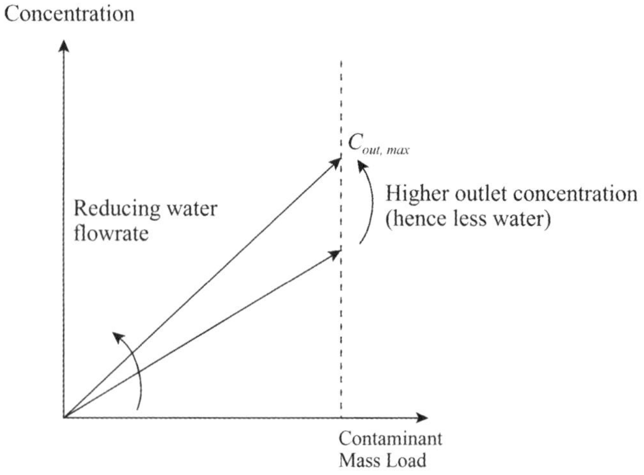

Figure 26.12

Reduction in the water flowrate is limited by minimum flowrate or maximum outlet concentration limits.

- equipment washing,
- hosing operations, and so on.

The one thing all of these operations have in common is that the water comes into contact with process materials and becomes contaminated. Figure 26.11 shows the quantitative representation of this on the plot of concentration versus mass load of contaminant. The water starts with no contamination and its level of contamination increases as a result of the mass transfer.

If the flowrate of water to an operation is decreased by some change to the operation, then for the same mass load transferred the reduction in the water flowrate will lead to a steeper line and higher outlet concentration, as shown in Figure 26.12. The reduction in the water usage will be limited by either the operation requiring some minimum flowrate, below which it cannot operate, or the outlet concentration from the operation goes to a maximum value. The maximum outlet concentration might be set by a number of considerations:

- maximum solubility of a contaminant,
- corrosion limitations,
- fouling limitations,
- minimum of mass transfer driving force,
- minimum flowrate requirements,
- maximum inlet concentration for downstream treatment.

If all operations use clean water, then reducing the flowrate to its minimum value can be used to minimize the water consumption, as illustrated in Figure 26.12. However, such an approach misses the opportunity to reuse water. To open up the opportunity to reuse water between operations, some level of inlet contamination must be accepted. Figure 26.13 shows a water-using profile such that both the inlet concentration and outlet concentration have been set to their maximum values. This particular setting, where both the inlet and outlet concentrations are set to their maximum values, can be used to define the *limiting water profile* (Takama *et al.*, 1980; Wang and Smith, 1994a). As shown in Figure 26.14, the limiting

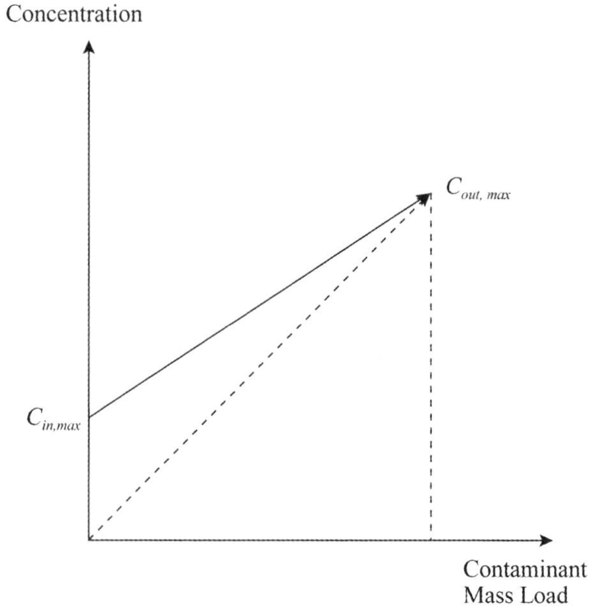

Figure 26.13

An alternative water profile uses more water but accepts slightly contaminated water.

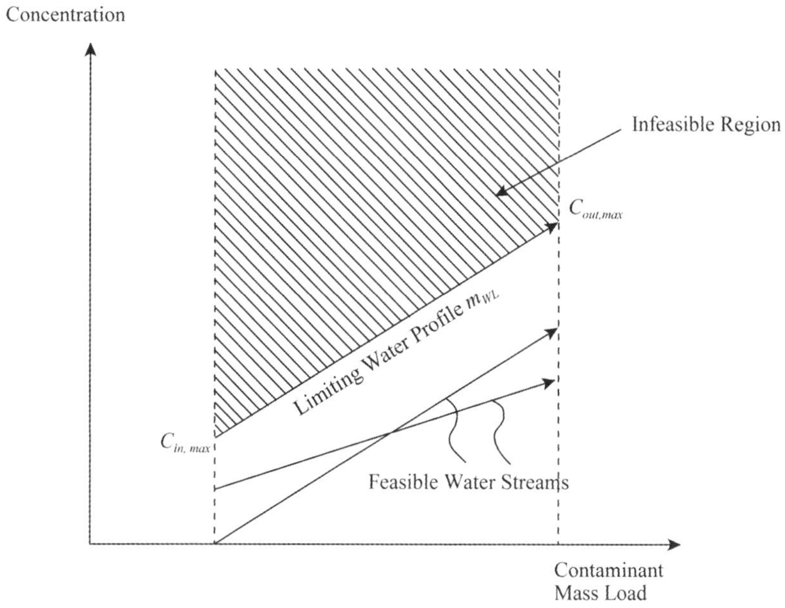

Figure 26.14

The limiting water profile.

water profile is used to define a boundary between feasible and infeasible concentrations. The concentration of a water profile is considered to be feasible as long as it is below the limiting water profile, as illustrated in Figure 26.14. This approach will be used later to identify reuse opportunities. The approach has a number of advantages:

- Operations with different characteristics can be compared on a common basis (e.g. water used in an extraction process versus a hosing operation).
- It does not require a model of the operation to represent the mass transfer.
- It does not depend on any particular flow pattern (countercurrent versus cocurrent).
- It works on any type of water-using operation (e.g. fire-water makeup, cooling tower makeup, and so on).

26.6 Targeting for Maximum Water Reuse for Single Contaminants for Operations with Fixed Mass Loads

As already noted, if fresh water is being used for all operations, then minimizing the flowrates to individual operations will minimize the water consumption. However, this misses opportunities for reuse. If some level of inlet contamination is allowed, then, in principle, water can be reused between operations. To illustrate how the overall minimum water consumption can be targeted when allowing reuse, consider the data for the simple problem given in Table 26.4. This specifies maximum inlet and outlet concentrations or *limiting concentrations* for a single contaminant. The single contaminant might be a specific component (e.g. phenol, acetone, starch) or an *aggregate property* (e.g. total organic material, suspended solids, dissolved solids, COD). Later the approach will be extended to systems where limiting concentrations for multiple contaminants are specified.

A number of points need to be noted regarding the data in Table 26.4. First, the concentration is specified on the basis of the mass flowrate of water and not the mass flowrate of the mixture:

$$C = \frac{m_C}{m_W} \quad \text{not} \quad C = \frac{m_C}{m_W + m_C} \tag{26.1}$$

where C = concentration (ppm)
m_C = mass flowrate of contaminant (g·h^{-1}, g·d^{-1})
m_W = mass flowrate of pure water (t·h^{-1}, t·d^{-1})

In most problems, the concentration of contaminant is so small that there is virtually no difference between the concentration based on the mass flowrate of water and the mass flowrate of the mixture. However, it is important to be consistent and follow the convention given in Equation 26.1. The other point to note is regarding the units. It is convenient to define the flowrate in terms of metric tons (typically tons per hour or tons per day). It is

Table 26.4

Data for a problem with four operations.

Operation number	Contaminant mass (g·h^{-1})	C_{in} (ppm)	C_{out} (ppm)	m_{WL} (t·h^{-1})
1	2000	0	100	20
2	5000	50	100	100
3	30,000	50	800	40
4	4000	400	800	10

also convenient to define the concentration in terms of parts per million (ppm). If the basic unit of flowrate is taken to be tons and concentration to be parts per million, then the mass load is measured in grams (typically grams per hour or grams per day).

The flowrate in Table 26.4 refers to the *limiting water flowrate* (m_{WL}). The limiting water flowrate is the flowrate required if the specified mass of contaminant is picked up by the water between the maximum inlet and outlet concentrations. If an operation has a maximum inlet contaminant concentration greater than zero and it is fed with water with zero concentration, then, for the specified mass load, a lower flowrate than the limiting water flowrate could be used. The case to be considered here is where the mass load of contaminant for each operation is fixed, but the flowrate is allowed to vary. Later the case for operations with fixed flowrates will be considered.

Analysis of the data in Table 26.4 can be started by calculating the flowrate that would be required for each of these operations if each operation was fed by fresh water with zero concentration. The relationship between mass pickup of contaminant, mass flowrate of water and concentration change is given by:

$$\Delta m_C = m_W \, \Delta C \tag{26.2}$$

where Δm_C = mass pickup of contaminant ($g \cdot h^{-1}$, $g \cdot d^{-1}$)
m_W = flowrate of pure water ($t \cdot h^{-1}$, $t \cdot d^{-1}$)
ΔC = concentration change (ppm)

For fresh water feed and minimum flowrate for the streams from Table 26.4:

$$m_{W1} = \frac{2000}{100 - 0} = 20\, t \cdot h^{-1}$$

$$m_{W2} = \frac{5000}{100 - 0} = 50\, t \cdot h^{-1}$$

$$m_{W3} = \frac{30,000}{800 - 0} = 37.5\, t \cdot h^{-1}$$

$$m_{W4} = \frac{4000}{800 - 0} = 5\, t \cdot h^{-1}$$

$$\text{Total flowrate of fresh water} = 20 + 50 + 37.5 + 5 = 112.5\, t \cdot h^{-1}$$

Consider now the possibility of reuse. To determine the maximum potential for reuse, Figure 26.15a shows the four operations plotted on axes of concentration versus mass load. Note that the concentrations are maximum inlet and outlet concentrations (limiting concentrations). Figure 26.15b shows a *limiting composite curve* of the four water streams (Wang and Smith, 1994a). The construction of the limiting composite curve is analogous to the composite curves for energy developed in Chapter 17. To construct the limiting composite curve in Figure 26.15b, the diagram is divided into concentration intervals and the mass load within each concentration interval is combined to obtain the limiting composite curve (Wang and Smith, 1994a). This represents a quantitative profile of the single-stream equivalent to the four separate streams. It is a combined boundary between feasible and infeasible

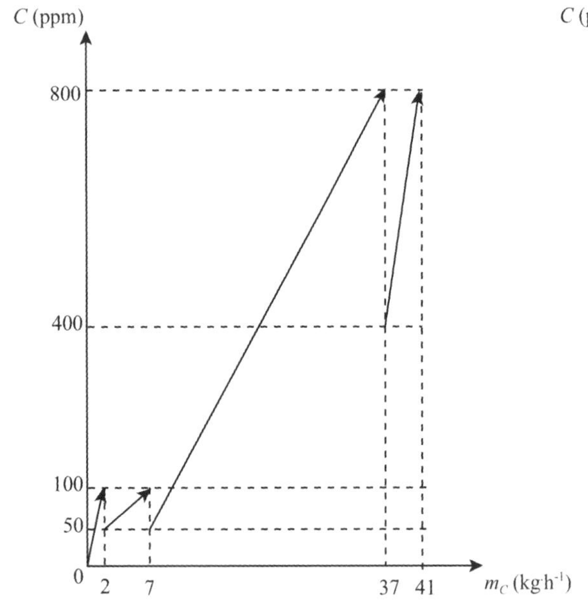

(a) Limiting water data for the four operations.

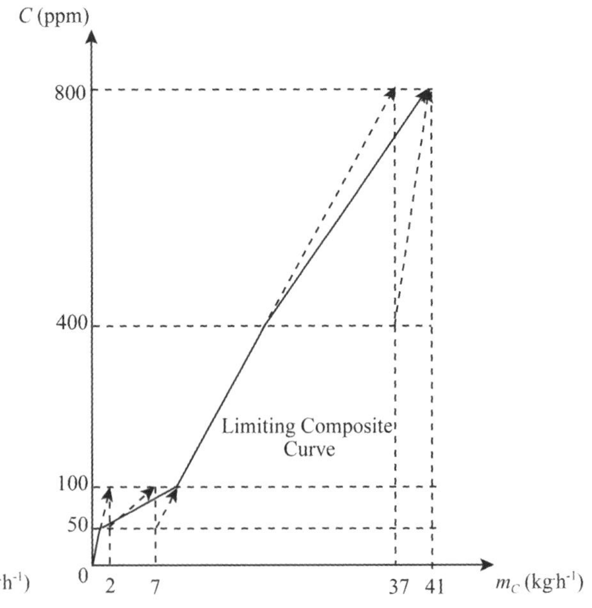

(b) The limiting composite curve.

Figure 26.15

Construction of the limiting composite curve for the simple example from Table 26.4. (Reproduced from Wang YP and Smith R (1994a) Wastewater Minimization, *Chem Eng Sci*, **49**: 981, with permission from Elsevier.)

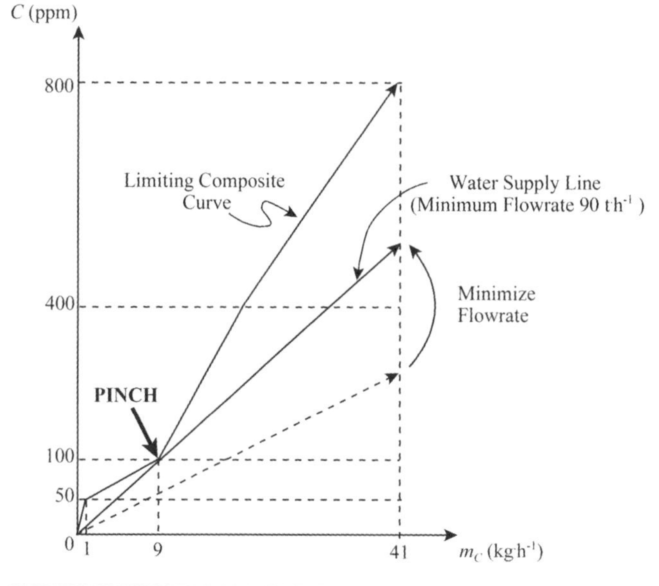

Figure 26.16

Targeting minimum water flowrate for a single contaminant.

concentrations. To target for the minimum water flowrate, a *water supply line* is drawn to represent the water supply, as shown in Figure 26.16 (Wang and Smith, 1994a). The water supply line starts at zero concentration in this case (as it must for this problem to satisfy the concentration requirements). In other cases, the water supply might not necessarily start from zero, depending on the quality of the water fed to the process. In principle, any slope can be drawn, as long as it is below the limiting composite curve. If the minimum water flowrate is to be obtained, the steepest line possible must be drawn for the water supply. This is shown in Figure 26.16, where the steepest slope corresponds with the water supply line touching the limiting composite curve at the *pinch point*. The pinch point corresponds with the concentration boundary that limits the slope of the water supply line. Knowing the mass load from zero concentration up to the pinch point (9000 g·h^{-1} in this case) and the concentration of the pinch (100 ppm in this case), the flowrate for the system can be calculated from Equation 26.2 to be 90 t·h^{-1}. This represents the minimum flowrate target for the case where the mass load is fixed and the flowrate allowed to vary in each operation, providing concentrations are kept below minimum concentrations.

A number of points should be noted regarding Figure 26.16. If a steeper water supply line was drawn than the one shown in Figure 26.16, it would cross the limiting composite curve and concentrations will be infeasible at some point. The fact that the water supply line touches the limiting composite curve does not imply, for example, a zero concentration difference in a mass exchange operation. It must be remembered that any allowances for driving forces have already been included in the limiting data. Where the water supply line touches the limiting composite curve implies that the water concentration goes to its maximum value at that point. This could correspond with a minimum mass exchange driving force, maximum solubility limit, and so on. At points in Figure 26.16 other than zero concentration and the pinch point, the concentrations of the water will all be below their maximum values. The point where the water concentration goes to its maximum at the pinch point dictates the minimum flowrate for the system.

An alternative graphical representation has been suggested involving a plot of concentration (or purity) versus volumetric flowrate of water (Dhole *et al.*, 1996). Using this approach, a plot is developed involving the water sinks (inlet concentration and flowrate) and water sources (outlet concentration and flowrate). While this plot is an alternative representation, it does not allow the minimum water flowrate to be targeted in a similar way to that developed in Figure 26.16 (Polley and Polley, 2000).

Rather than use a graphical approach, a cascade analysis can be performed, analogous to the problem table algorithm for heat exchanger networks discussed in Chapter 17 (Foo, 2007, 2012). Whilst such cascade analysis has the advantage of providing more of an algorithmic approach to targeting, it does not provide insights in the way that graphical methods do.

26.7 Design for Maximum Water Reuse for Single Contaminants for Operations with Fixed Mass Loads

Having seen how to set a target for the data in Figure 26.16 assuming operations have fixed mass loads and allowing the flowrates to vary, consider now how to achieve that target in design (Kuo and Smith, 1998a). Figure 26.17 illustrates the basis of the design strategy. First, the minimum water requirements in each region of the design are identified. In this simple example, there are two design regions: above the pinch and below the pinch. Below the pinch in Figure 26.17, by definition, the full amount of the target minimum flowrate is needed (90 t·h^{-1} in this example). Above the pinch, not all 90 t·h^{-1} are required and the process could, in principle, operate with a lower flowrate than the target. The minimum flowrate required above the pinch is determined by a simple mass balance (45.7 t·h^{-1} in this example). The design regions are identified by straight lines drawn between the convex points, to identify the *pockets* in the problem. The basis of the design strategy is to use the target flowrate below the pinch and then to use only the required amount above the pinch with the balance going to effluent. This design strategy serves two purposes:

a) The mass balance is tightly defined, allowing very specific rules to be applied in the design.

b) The strategy is compatible with minimizing effluent treatment costs overall through distributed effluent treatment, to be discussed later.

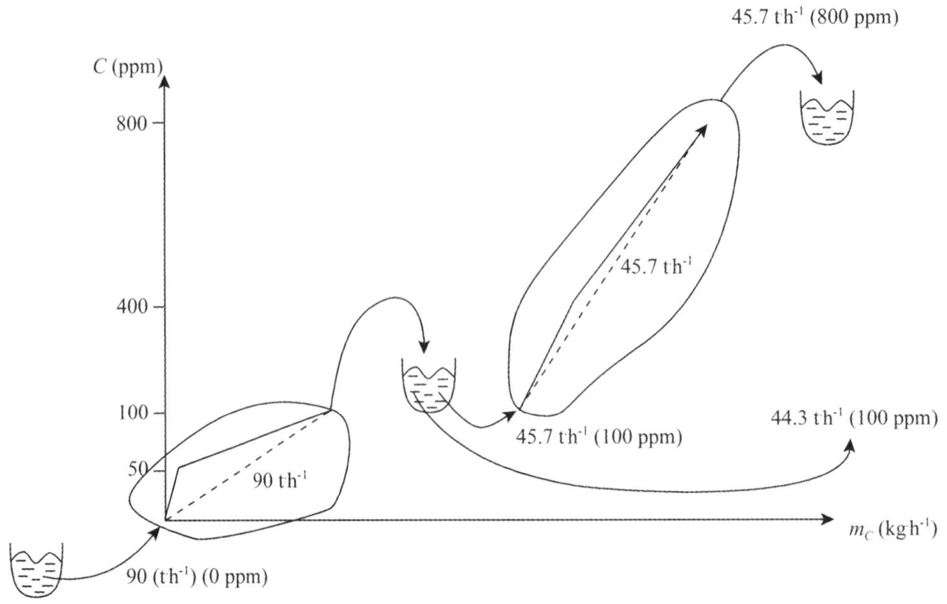

Figure 26.17

The basis of the design strategy.

Having formulated the basic design strategy, a grid can then be set up as shown in Figure 26.18 (Kuo and Smith, 1998a). The design grid starts by setting up three *water mains*, corresponding with fresh water concentration, pinch concentration and the maximum (effluent) concentration from Figure 26.17. The flowrate required by each water main is shown at the top of the main and the wastewater generated by the main at the bottom. Streams representing the individual operation requirements are superimposed on the water mains. Operation 1 has limiting data corresponding with fresh water at the inlet and 100 ppm at the outlet and is therefore shown between the fresh water main and the pinch concentration water main. Operation 2 starts at 50 ppm and terminates at 100 ppm

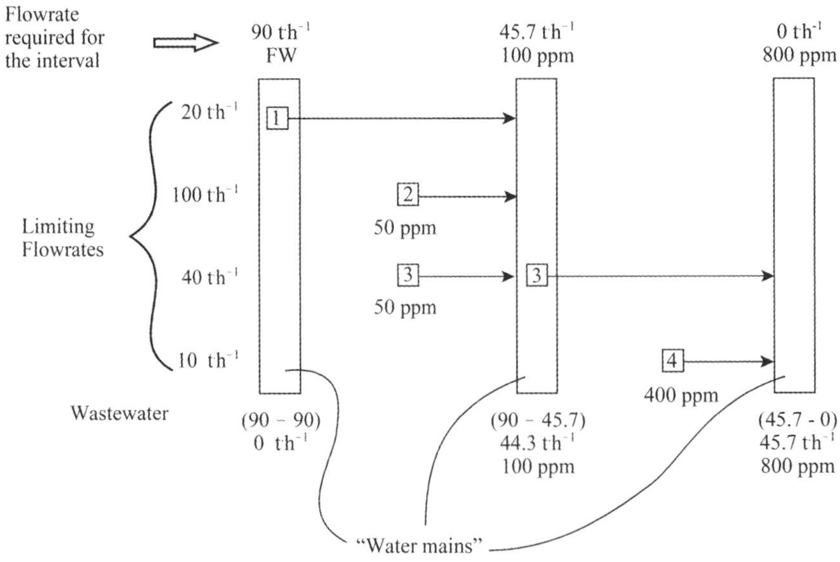

Figure 26.18

The design grid for the water system.

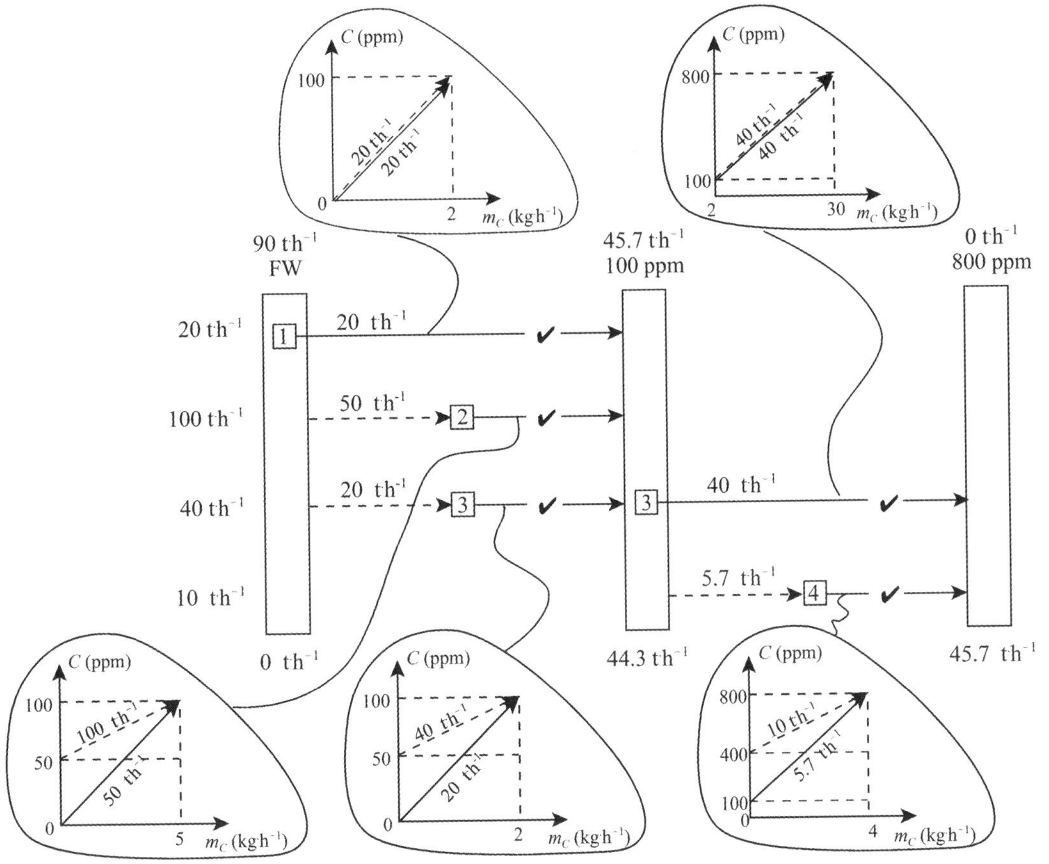

Figure 26.19

The streams are connected with the water mains.

and is therefore shown between the fresh water and pinch concentration mains. Operation 3 is divided into two parts as it features both below- and above-pinch concentrations. Operation 4 starts at 400 ppm and ends at 800 ppm and therefore features between the pinch concentration and final concentration mains. The operations are then connected to the appropriate water mains, as illustrated in Figure 26.19. Figure 26.19 also shows the concentrations and limiting concentrations versus mass load in the individual operations. This, in principle, is a working design. However, there is a problem created by Operation 3, in that below the pinch it receives a flowrate of $20\,t{\cdot}h^{-1}$ of fresh water, but at the pinch it receives a flowrate of $40\,t{\cdot}h^{-1}$ of water at 100 ppm. This change in flowrate in the middle of Operation 3 would be impractical for most operations. It could be practical if Operation 3 involved, for example, an operation with multiple stages of washing. If this was the case, then the different stages could be fed with different flowrates and qualities of water. However, for most situations this change in flowrate would be impractical.

The change in flowrate for Operation 3 is in fact readily corrected. Consider Figure 26.20, which shows the concentration versus mass load for an operation with a change in flowrate similar to Operation 3 in Figure 26.19. A mass balance around Part 1 in Figure 26.20 gives:

$$m_{W1}(C_{PINCH} - C_0) = m_{W2}(C_{PINCH} - C_{in,max}) \qquad (26.3)$$

Moving the mixing junction to the inlet of the operation, as shown in Figure 26.20, and carrying out a mass balance on the new mixing junction gives:

$$\begin{aligned} C_{in} &= \frac{(m_{W2} - m_{W1})C_{PINCH} + m_{W1}C_0}{m_{W2}} \\ &= \frac{m_{W2}C_{PINCH} - m_{W1}(C_{PINCH} - C_0)}{m_{W2}} \end{aligned} \qquad (26.4)$$

Substituting Equation 26.3 into Equation 26.4 gives:

$$\begin{aligned} C_{in} &= \frac{m_{W2}C_{PINCH} - m_{W2}(C_{PINCH} - C_{in,max})}{m_{W2}} \\ &= C_{in,max} \end{aligned} \qquad (26.5)$$

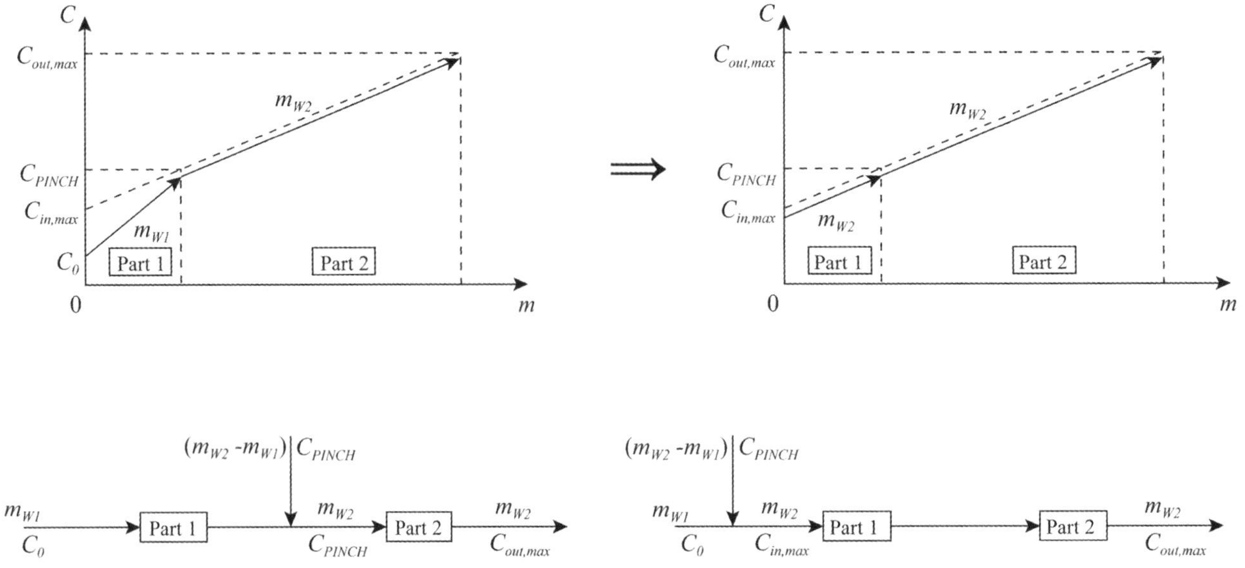

Figure 26.20

An operation involving a change in flowrate. (Reproduced from Kuo W-CJ and Smith R (1998a) Designing for the Interactions Between Water Use and Effluent Treatment, *Trans IChemE*, **A76**: 287301, with permission from Elsevier.)

In other words, if the mixing junction is moved from the middle of the operation to the beginning of the operation, then there is a constant flowrate throughout the operation corresponding with an inlet concentration after mixing of the maximum inlet concentration. Also, by definition, the flowrate will be the limiting water flowrate.

In Figure 26.21, the change in flowrate for Operation 3 that previously occurred at the pinch concentration water mains is now added at that concentration to the inlet of Operation 3. The design now features a constant flowrate in all of the operations and achieves the target minimum flowrate of $90\,t{\cdot}h^{-1}$. The arrangement shown in Figure 26.21 involves reuse of water

Figure 26.21

Correcting the change in flowrate in the design grid.

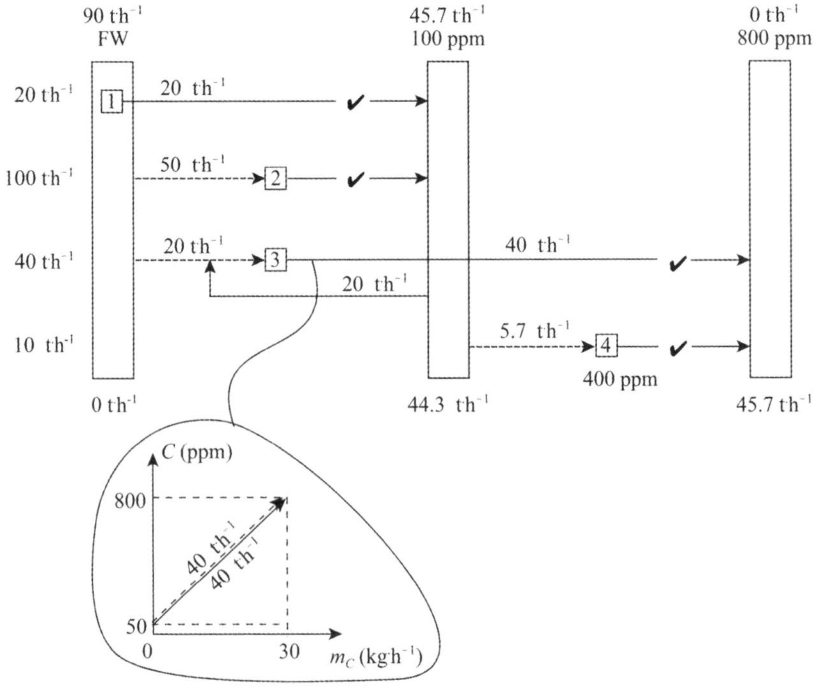

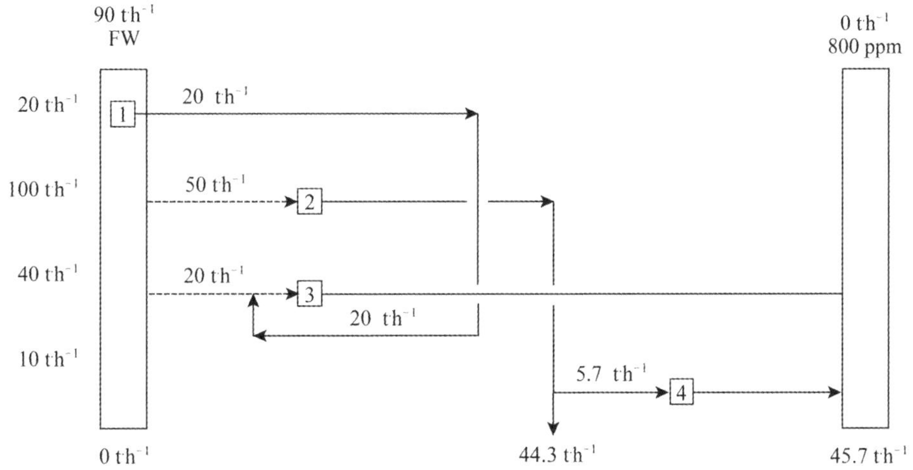

Figure 26.22
Water network without the intermediate water main.

from Operations 1 and 2 into Operations 3 and 4 via a water main at the pinch concentration of 100 ppm. An alternative way to arrange the design is to make the connections directly, rather than through an intermediate water main. Figure 26.22 shows a direct connection from Operation 1 to Operation 3 and another from Operation 2 to Operation 4 with $44.3\,t{\cdot}h^{-1}$ going to wastewater from Operation 2. The arrangement shown in Figure 26.22 is the only one possible arrangement of connections between the sources and the sinks. Figure 26.23 shows the final design as a conventional flowsheet. The design in Figure 26.23 can be evolved to produce alternative networks. For example, rather than splitting the flow of water at the inlet of Operation 1 with $20\,t{\cdot}h^{-1}$ going through the operation and $20\,t{\cdot}h^{-1}$ bypassing, all $40\,t{\cdot}h^{-1}$ could be put through Operation 1. Then, the outlet of Operation 1 could be fed directly to the inlet of Operation 3. Other options are possible.

This simple example illustrates the basic principles of water network design for maximum reuse for a single contaminant with a fixed mass load and varying flowrate. A number of issues need to be considered that would apply to more complex examples. Consider Figure 26.24 involving three water mains and three operations. Operation 2 above the pinch terminates at a concentration less than the concentration for the high-concentration water main. The outlet of Operation 2 must not be fed directly into this final water main. The basis of the mass balance from Figure 26.17 dictates that all streams must achieve the concentration of the water mains into which they are being fed; otherwise the mass balance will be violated. Thus, in Figure 26.24a, the outlet of Operation 2 must not be fed to the final water mains, but must be used again and brought to the concentration of the water main into which it is being fed (Figure 26.24b).

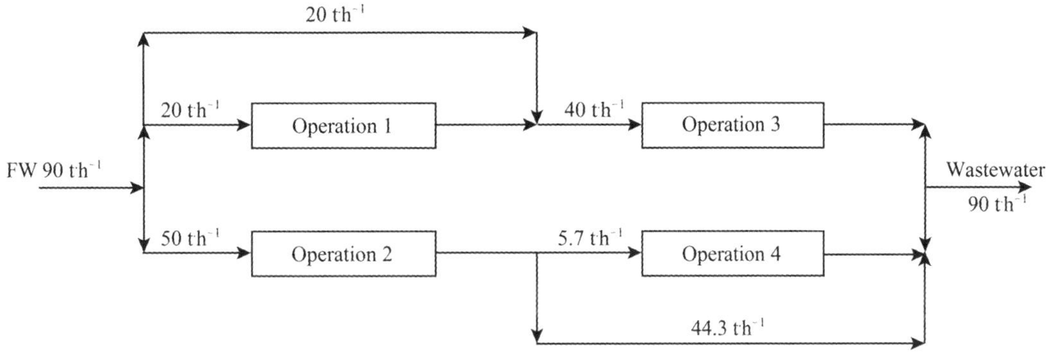

Figure 26.23
Flowsheet for the water network.

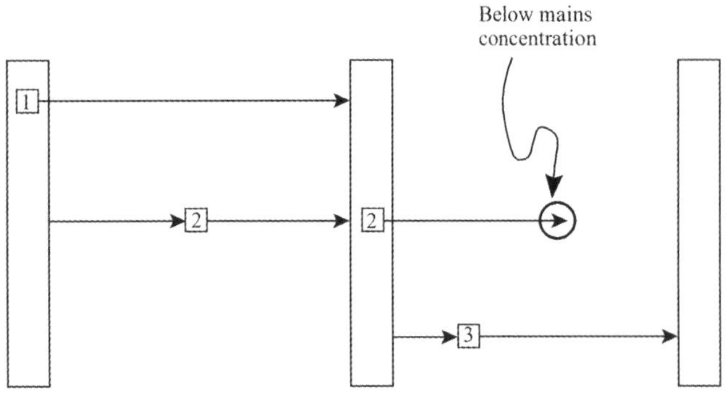

(a) The final concentration of a stream might be below that of a water main.

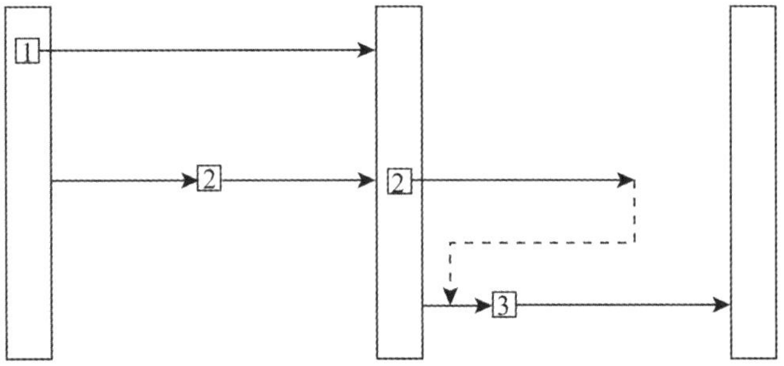

(b) Water should not be discharged to water mains until it has reached mains concentration.

Figure 26.24

Connecting to water mains if the final concentration of a steam is below mains concentration.

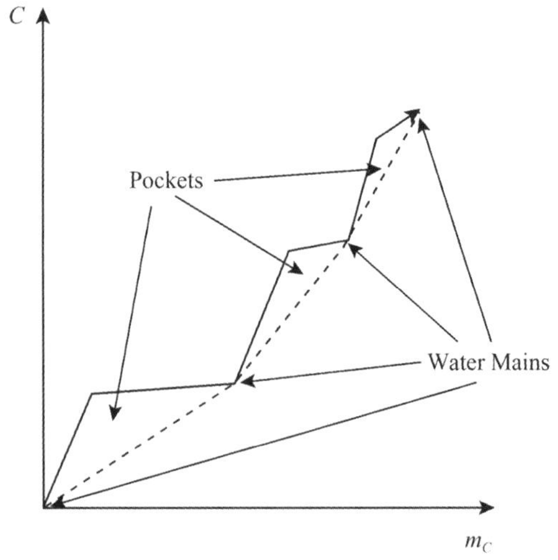

Figure 26.25

In more complex problems there might be more design intervals.

Figure 26.25 shows a more complex limiting composite curve that involves three design regions, rather than two. The design regions are identified by straight lines drawn between the convex points, to identify the pockets in the problem. A water main is required at each of the extreme points. In Figure 26.25, there would be four water mains required in the design grid. However, the basic principles are exactly the same as those described so far, but with the additional water main.

Example 26.2 Table 26.5 presents water-use data for a simple example involving three operations.

a) Target the minimum water consumption for the system through maximum reuse.
b) Design a network for the target water consumption.

Table 26.5

Water-use data for Example 26.2.

Operation number	Contaminant mass ($g \cdot h^{-1}$)	C_{in} (ppm)	C_{out} (ppm)	m_{WL} ($t \cdot h^{-1}$)
1	6000	0	150	40
2	14,000	100	800	20
3	24,000	700	1000	80

Solution

a) Figure 26.26 shows the limiting composite curve for Example 26.2. Figure 26.27 shows the limiting composite curve with the appropriate water supply line pinched at 150 ppm. From Table 26.5, the mass load up to 150 ppm can be calculated as:

$$\text{Mass load up to pinch concentration} = 40(150-0)+20(150-100) = 7000\ g \cdot h^{-1}$$

$$\text{Minimum flowrate} = \frac{\text{Mass load up to pinch concentration}}{\text{Concentration change of water to pinch}} = \frac{7000}{150-0} = 46.7\ t \cdot h^{-1}$$

b) The design to achieve the target must first identify the number of design regions. From the limiting composite curve in Figure 26.27, there are two design regions between zero and 150 ppm and 150 and 1000 ppm. Below the pinch, the first design region requires 46.7 $t \cdot h^{-1}$ by definition. Above the pinch, the concentration change extends from 150 to 1000 ppm with a mass load of 37,000 $g \cdot h^{-1}$. This corresponds with a minimum flowrate requirement above the pinch of 43.5 $t \cdot h^{-1}$. Figure 26.28 shows the design grid for the example with the streams in place. Figure 26.29 shows the completed network design.

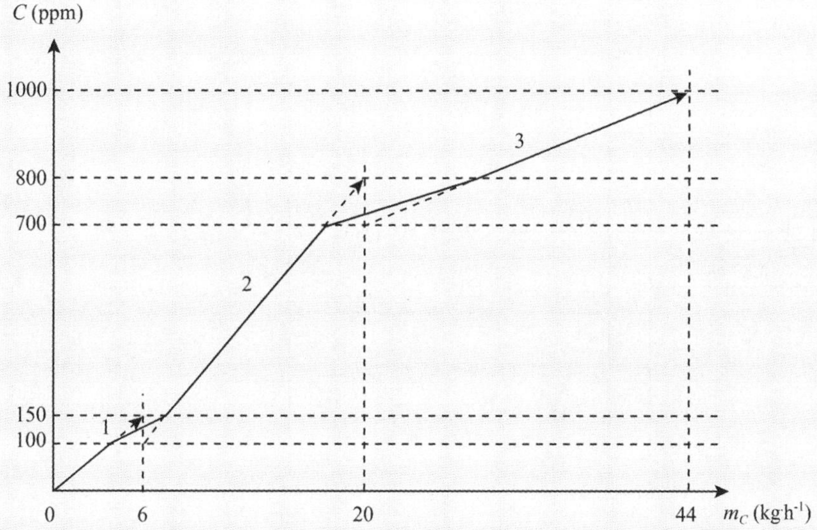

Figure 26.26

The limiting composite curve for Example 26.2.

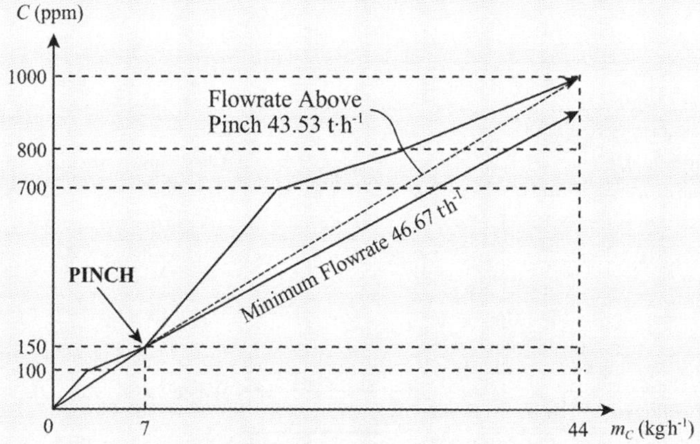

Figure 26.27

The minimum water target for Example 26.2.

The outlet from Operation 2 at 800 ppm has not been put directly into the water main at 1000 ppm. If this had been done, the mass balance would have been violated. Instead, further use of the water from the outlet of Operation 2 requires it to be fed to the inlet of Operation 3, finally discharging to the 1000 ppm main. Figure 26.30 shows a flowsheet for the completed water network.

Even for a small problem involving three operations this turns out to be a solution that would be difficult to be achieved by inspection. Of course, the practicality and operability of a network would have to be explored before this design was accepted. It might be that some design features are considered to be inoperable. In which case, the design would have to be simplified. However, the consequence might be an increase in the water consumption. The procedure simply allows the maximum potential for the reuse to be identified; thereafter the practicality must be explored and the design evolved if necessary.

26

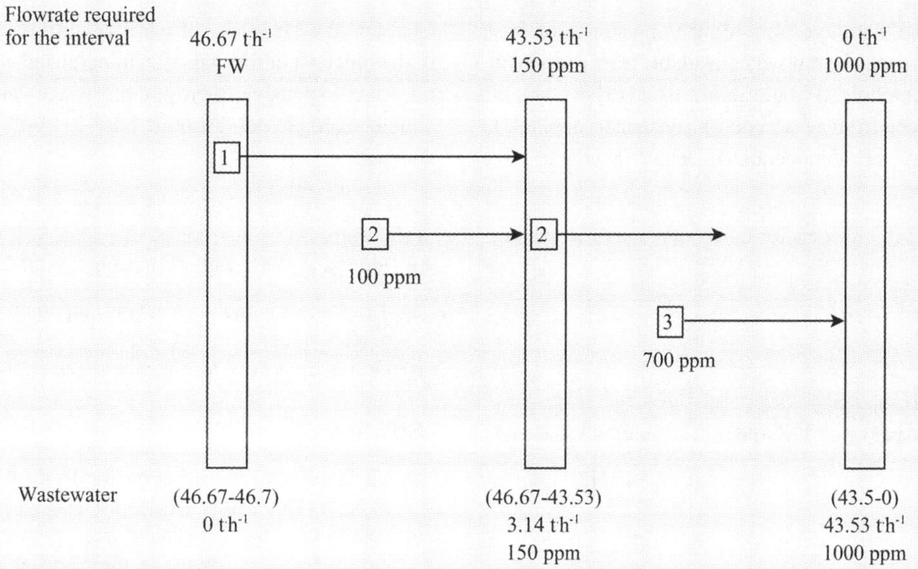

Figure 26.28

The design grid for Example 26.2.

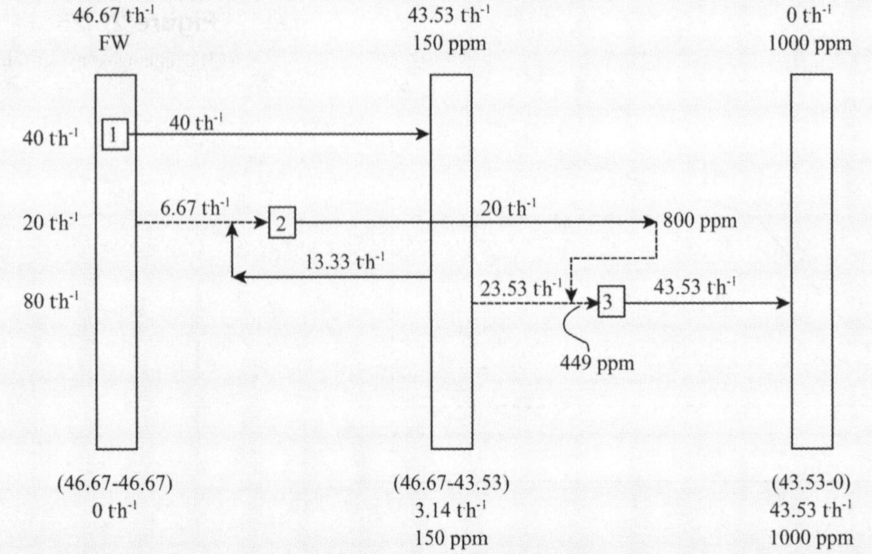

Figure 26.29
The completed network design for Example 26.2.

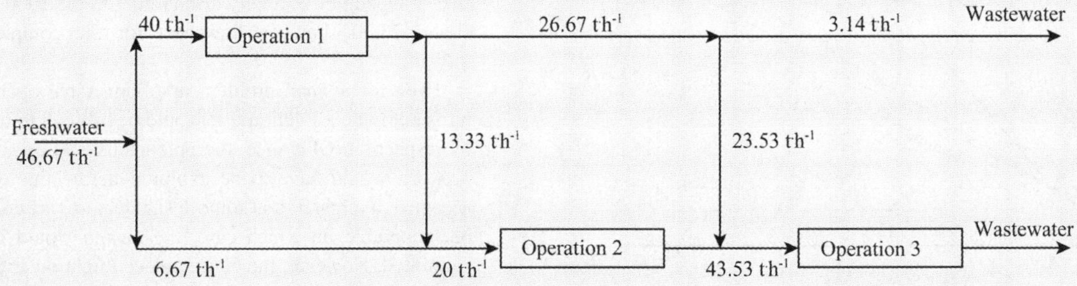

Figure 26.30
Flowsheet for the water network for Example 26.2.

It is often the case that there is more than one source of fresh water available. There may be raw water available from a reservoir, lake, river, canal or borehole. It would also be usual to have potable water available sourced from a reservoir or a desalination plant. In addition, there may be demineralized water available (see Chapter 23). The order of water quality is usually also the order of the unit cost of the water. Demineralized water will normally be the most expensive, then potable water, and so on. Consider a simple example to illustrate how to deal with multiple water sources.

Example 26.3 Limiting data for three operations are given in Table 26.6.

Table 26.6

Limiting data for three operations.

Operation	Contaminant mass (kg·h^{-1})	$C_{in,max}$ (ppm)	$C_{out,max}$ (ppm)	m_{WL} (t·h^{-1})
1	2	0	100	20
2	5	50	100	100
3	30	50	800	40

Two sources of fresh water are available, WSI with a concentration of 0 ppm and WSII with a concentration of 25 ppm:

a) Target the minimum water consumption for the system through maximum reuse.

b) Design a network for the target water consumption.

Solution

a) The limiting composite curve for the three operations is plotted in Figure 26.31. Matched against it is a water supply line that is a composite of the two water sources. Between 0 ppm and 25 ppm only WSI can satisfy the requirements of the process. Here the flowrate of WSI has been minimized to 20 t·h^{-1}. This is dictated

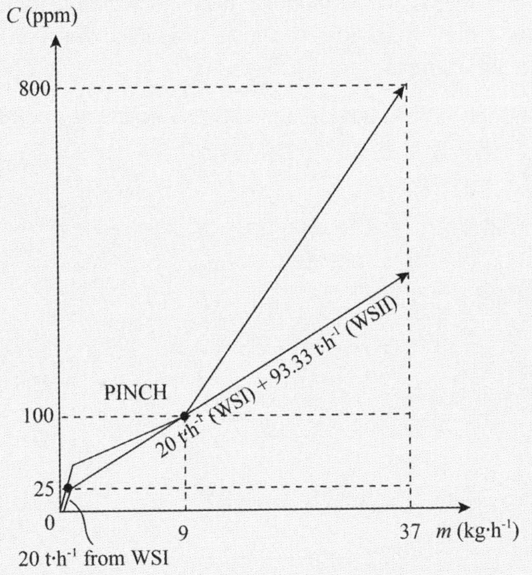

Figure 26.31

Targeting for multiple water sources for Example 26.3.

by the slope of the limiting composite curve between 0 ppm and 25 ppm. The requirements of the process above 25 ppm are being satisfied by the continued use of the 20 t·h^{-1} of WSI and the balance by the introduction of WSII at 25 ppm. The amount of WSII required can be determined by a mass balance below the pinch:

$$\begin{aligned} 9000 &= m_{WSI}\Delta C_{WSI} + m_{WSII}\Delta C_{WSII} \\ &= 20 \times (100 - 0) + m_{WSII}(100 - 25) \\ m_{WSII} &= 93.3\,\text{t}\cdot\text{h}^{-1} \end{aligned}$$

Rather than use a graphical approach to set targets in this way, a cascade analysis can be used as an alternative (Foo, 2007, 2012).

b) Figure 26.32a shows the corresponding design grid for the two water sources. An additional water main is introduced into the design grid to account for the second water source. Figure 26.32b shows the design as a flowsheet.

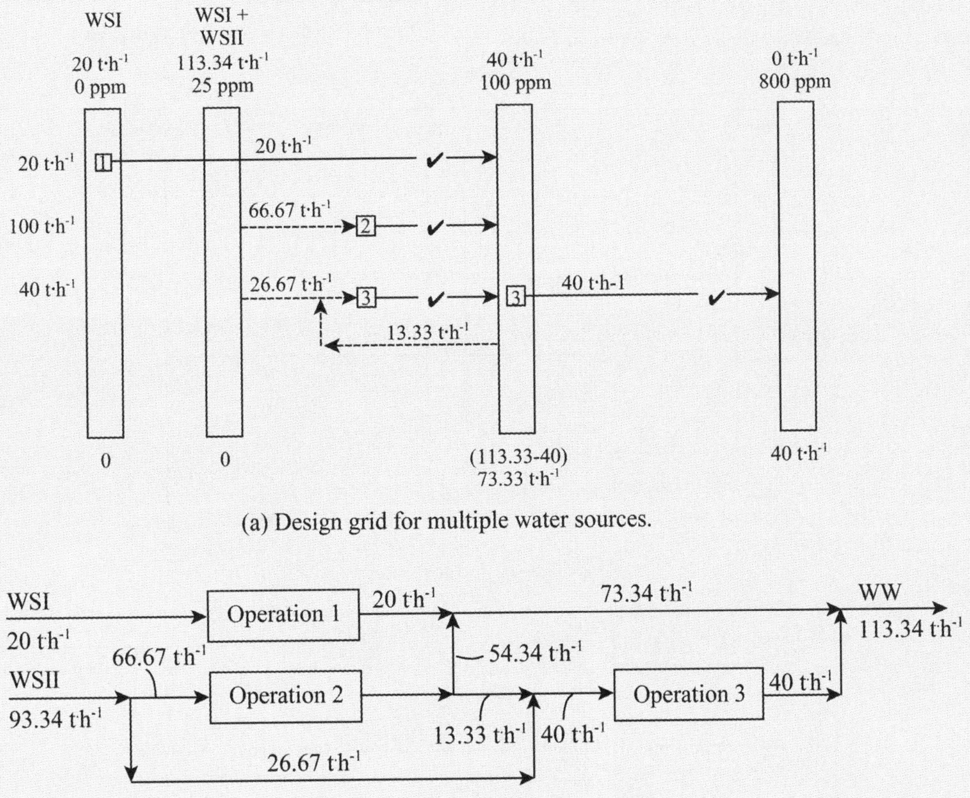

(a) Design grid for multiple water sources.

(b) Flowsheet for multiple water sources.

Figure 26.32

Design for multiple water sources for Example 26.3.

Another complication that often arises is loss of water from the system. This could be, for example, the loss of water to effluent from a hosing operation or the evaporative loss to atmosphere from a cooling tower, neither of which becomes available for reuse. To illustrate how water losses can be accounted for, consider the following example.

Example 26.4 An operation is to be added to those in Table 26.6 with a maximum inlet concentration of 80 ppm and a flowrate of $10\,t{\cdot}h^{-1}$, all of which is lost:

a) Target the minimum water consumption for the system through maximum reuse.

b) Design a network for the target water consumption.

Solution

a) Figure 26.33 shows the limiting composite curve for the three operations from Table 26.6 (i.e. excluding the flowrate loss). Matched against the limiting composite curve is a water supply line that shows a change in slope where the flowrate loss occurs. The target flowrate can be determined by a mass balance below the pinch.

$$\begin{aligned} 9000 &= m_W(\Delta C\, before\ loss) + (m_W - f_{LOSS}) \times (\Delta C\, after\ loss) \\ &= m_W(80 - 0) + (m_W - 10) \times (100 - 80) \\ m_W &= 92.0\,t \cdot h^{-1} \end{aligned}$$

b) Figure 26.34a shows the corresponding design grid for the loss. Note that no additional water main is required to account for the water loss. It should also be noted that the fresh water target for the three operations in Table 26.6 with a single source of fresh

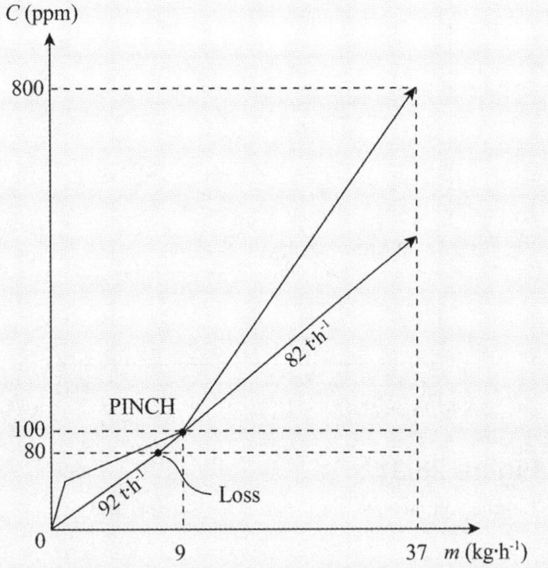

Figure 26.33

Targeting for water loss for Example 26.4.

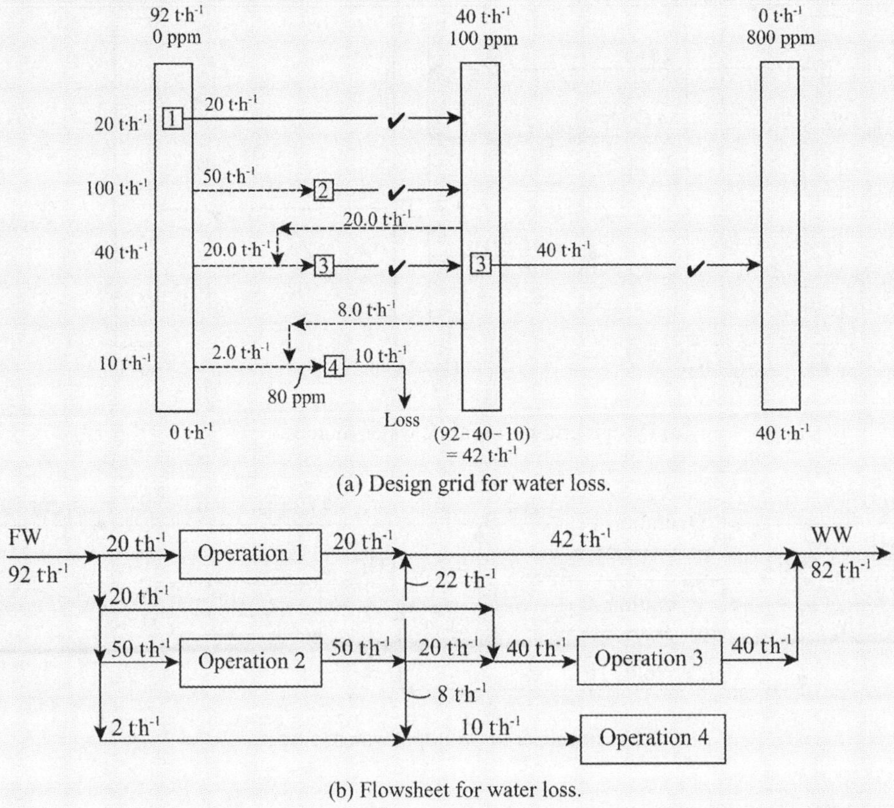

Figure 26.34

Design for water loss for Example 26.4.

water at 0 ppm is $90\ t \cdot h^{-1}$, which increases to $92\ t \cdot h^{-1}$ with a water loss of $10\ t \cdot h^{-1}$ at 80 ppm, rather than $100\ t \cdot h^{-1}$. This is because the water loss can be partially fed by reuse. Had the same loss occurred above the pinch concentration, then there would have been no increase in flowrate as a result of the introduction of the water loss. It should be noted that the target in Figure 26.34a is based on the assumption that the loss is being fed at its maximum inlet concentration. Figure 26.34b shows the design as a conventional flowsheet.

The design procedure can be summarized in four steps (Kuo and Smith, 1998a):

1) Set up the grid.
2) Connect operations with water mains.
3) Correct for changes in flowrate in individual operations.
4) Remove intermediate water mains and connect operations directly.

The final step in the procedure may or may not be appropriate. A design with an intermediate water main at the pinch concentration might in some circumstances be convenient. Having a water main on the plant corresponding with the one introduced for the design procedure might be a convenient way of distributing slightly contaminated water around the plant and provides flexibility that direct connections between operations do not provide. Also, in batch operation, the demands on the water system and the production of water from the outlet of the operations will vary through time. This normally requires buffering capacity to bridge between the times when the water becomes available and when it is required (Wang and Smith, 1995b; Gunaratnam and Smith, 2005). Thus, the intermediate water main in a batch environment could easily be envisaged to be a storage tank providing buffer capacity between operations producing water and consuming water during different time intervals.

26.8 Targeting for Maximum Water Reuse for Single Contaminants for Operations with Fixed Flowrates

One key feature of this approach developed so far is that the water flowrates to individual processes have been allowed to vary as the inlet concentration of contaminant in the water varied. However, many processes have a fixed flowrate requirement. For example, vessel cleaning, hosing operations, hydraulic transport, and so on, tend to require a fixed flowrate irrespective of the concentration of contaminant at the water inlet. An additional complication, as discussed above, is that in some processes there is a fixed flowrate that is lost and cannot be reused. Now suppose that each of the operations has a fixed flowrate specified to be that of the limiting water flowrate. In principle, the concentrations can vary up to the maximum inlet and outlet concentration, but the flowrate is fixed.

Consider the single operation in Figure 26.35a in which the maximum inlet concentration has been fixed to be $C_{in,max}$, the maximum outlet concentration has been fixed to be $C_{out,max}$ and the water flowrate fixed to be m_{WL}. Also shown in Figure 26.35a is the target minimum flowrate m_{Wmin} for an inlet concentration C_0. This seems to be a contradiction until it is considered how the fixed flowrate might be achieved in design. Figure 26.35b shows a design that imports m_{Wmin} and achieves the fixed flowrate through the process m_{WL} by *local recycling* of $(m_{WL} - m_{Wmin})$. A simple mass balance proves that the inlet concentration to the process is within the limit. The concentration after mixing the import m_{Wmin} with the local recycling of $(m_{WL} - m_{Wmin})$ is:

$$C_{in} = \frac{C_0 m_{Wmin} + C_{out,max}(m_{WL} - m_{Wmin})}{m_{WL}} \quad (26.6)$$

From targeting:

$$m_{Wmin} = \frac{C_{out,max} - C_{in,max}}{C_{out,max} - C_0} m_{WL} \quad (26.7)$$

Replacing m_{Wmin} in Equation 26.6 with Equation 26.7:

$$\begin{aligned} C_{in} &= \frac{1}{m_{WL}}\left[m_{WL}C_0\frac{C_{out,max} - C_{in,max}}{C_{out,max} - C_0} + m_{WL}C_{out,max}\right. \\ &\qquad \left. - m_{WL}C_{out,max}\frac{C_{out,max} - C_{in,max}}{C_{out,max} - C_0}\right] \\ &= C_{in,max} \end{aligned}$$

Therefore, the inlet concentration is within the maximum allowable limit for the operation (Wang and Smith, 1995a). Consider now the situation where multiple operations are involved, rather than a single operation. Suppose Figure 26.35a represents a mass load interval in a limiting composite curve. The stream in Figure 26.35a now represents several streams. The total flowrate requirements for the multiple processes exceed m_{Wmin}, but again local recycling can be used to satisfy the constraints. Different recycle arrangements are possible, as illustrated in Figure 26.36. A simple mass balance demonstrates that there is a recycle flowrate to satisfy the constraints. In Figure 26.36a the total flowrate for the three processes is $\sum m_{WLi}$. The recycle flowrate is then $(\sum m_{WLi} - m_{Wmin})$. The inlet concentration to the processes is given by:

$$C_{in} = \frac{m_{Wmin}C_0 + C_{out,max}\left[\left(\sum m_{WLi}\right) - m_{Wmin}\right]}{\sum m_{WLi}} \quad (26.8)$$

From targeting:

$$m_{Wmin}\left(C_{out,max} - C_0\right) = \left(\sum m_{WLi}\right)\left(C_{out,max} - C_{in,max}\right) \quad (26.9)$$

Replacing m_{Wmin} in Equation 26.8 with Equation 26.9:

$$\begin{aligned} C_{in} &= \frac{\left(\sum m_{WLi}\right)\left(C_{out,max}\right) - \left(\sum m_{WLi}\right)\left(C_{out,max} - C_{in,max}\right)}{\sum m_{WLi}} \\ &= C_{in,max} \end{aligned}$$

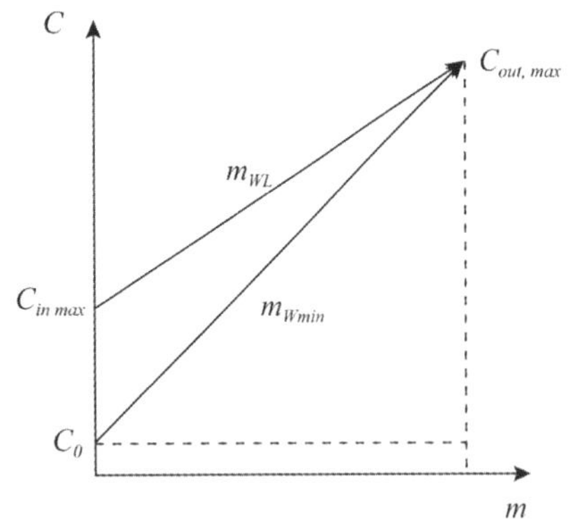

(a) Targeting for a single operation with a fixed flowrate allowing local recycling.

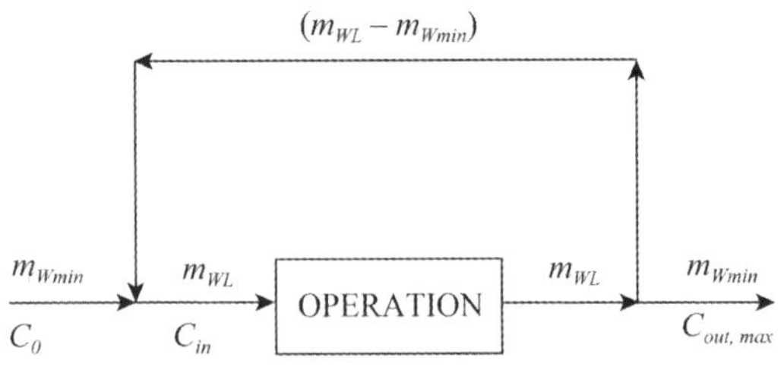

(b) Design for a single operation with a fixed flowrate allowing local recycling.

Figure 26.35

Local recycling to maintain flowrate constraints. (Reproduced from Wang YP and Smith R (1995a) Wastewater Minimization with Flowrate Constraints, *Trans IChemE*, **A73**: 889, by permission of the Institution of Chemical Engineers.)

Thus local recycling can be used to overcome flowrate constraints both with single and multiple operations (Wang and Smith, 1995a). However, local recycling is not always acceptable. Contaminants can build up in a recycle that can cause process problems. Such contaminants might be products of corrosion or microorganisms, and so on.

Consider now how a single operation, as illustrated in Figure 26.35a, can be designed using m_{Wmin}, but without local recycling and exploit reuse only (Wang and Smith, 1995a). In Figure 26.37a the operation has been split into two parts. Whether or not an operation can be split into parts depends very much on the nature of the operation. For example, a multistage washing operation can be readily split, whereas a steam stripping operation cannot be readily split. In Figure 26.37a the operation has been split such that each part has a flowrate requirement, which is less than or equal to m_{Wmin}, with Part 2 taking the balance of the flowrate. Figure 26.37 shows that if water is reused between these two parts, the design will be feasible, providing the part requiring the lower flowrate is fed first. On the other hand, Figure 26.37c shows the same split, but in a different order. Figure 26.37d shows water being reused between the two parts with the part requiring the larger flowrate being fed first. Now the outlet concentration from Part 1 exceeds the maximum inlet concentration for Part 2 and the design is infeasible. Had either of the part operations in Figure 26.37a exceeded m_{Wmin} in terms of its flowrate requirements, it could have been split further until all of the parts required less than or equal to m_{Wmin}. The arguments that apply to the order of matching would apply here also in that the water must be used in the part with the smallest flowrate requirement, then the part with the next largest flowrate, and so on. It can be concluded that the single operation in Figure 26.35a can be designed using m_{Wmin} without local recycling provided the operation can be split into parts such that each part has a flowrate requirement less than or equal to m_{Wmin}. The reuse design must then be such that the water is used in ascending order of flowrate requirements (Figure 26.38a). It can be demonstrated that the inlet concentration to the final process part in Figure 26.38a is $C_{in,max}$ by a simple mass balance. The inlet concentration C_{in} is given by:

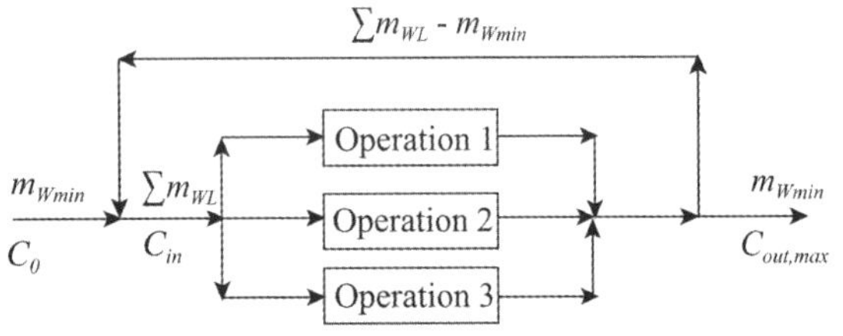

(a) Local recycling around several operations.

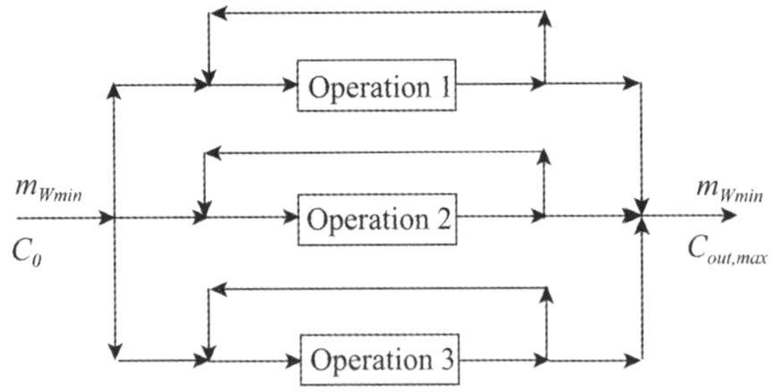

(b) Local recycling around individual operations.

Figure 26.36

Local recycling around several operations to maintain flowrate constraints can be implemented in different ways. (Reproduced from Wang YP and Smith R (1995a) Wastewater Minimization with Flowrate Constraints, *Trans IChemE*, **A73**: 889, by permission of the Institution of Chemical Engineers.)

$$C_{in} = C_0 + \frac{(m_{WL} - m_{W3})(C_{out,max} - C_{in,max})}{m_{Wmin}} \qquad (26.10)$$

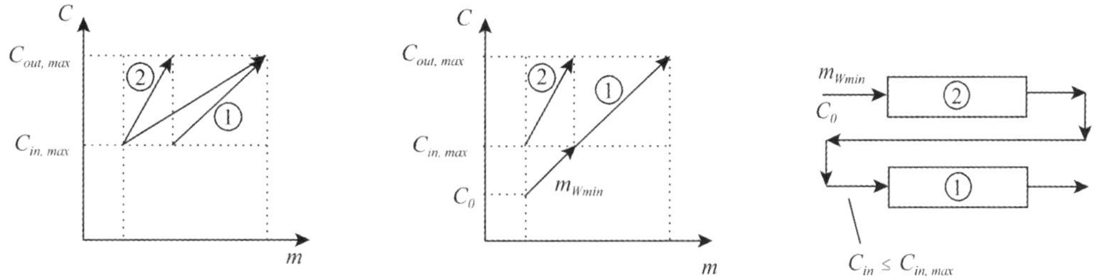

(a) Splitting an operation into two parts with Part 1 taking f_{min}.

(b) Reuse in which the part with the lower flowrate is fed first leads to a feasible design.

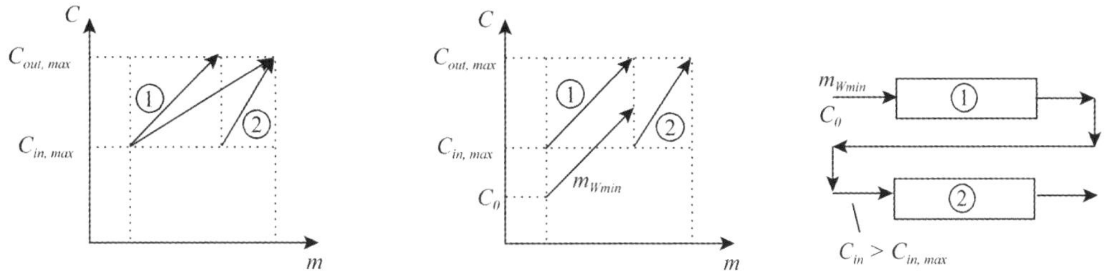

(c) Splitting an operation into two parts with Part 1 taking f_{min}, but used in a different order.

(d) Reuse in which the part with the larger flowrate is fed first leads to an infeasible design.

Figure 26.37

A single operation can be split into two parts to satisfy flowrate constraints, providing the part with the smallest flowrate is fed first. (Reproduced from Wang YP and Smith R (1995a) Wastewater Minimization with Flowrate Constraints, *Trans IChemE*, **A73**: 889, by permission of the Institution of Chemical Engineers.)

26

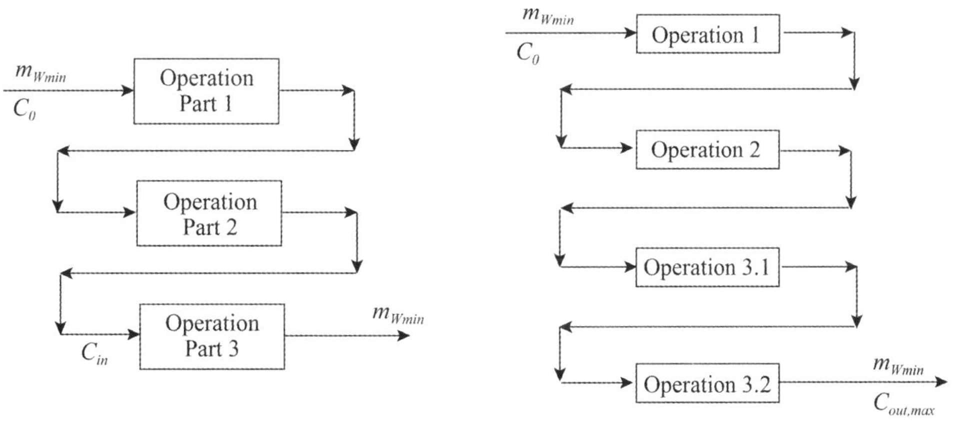

$m_{WL,Part1} \le m_{WL,Part2} \le m_{WL,Part3} = m_{Wmin}$

$m_{WL1} \le m_{WL2} \le m_{WL3.1} \le m_{WL3.2} = m_{Wmin}$

(a) Splitting a single operation into three parts with water used in ascending order of requirements.

(b) Multiple operations with water used in ascending order of requirements.

Figure 26.38

Flowrate requirements can be satisfied by reuse of water in ascending order of flowrate requirements. (Reproduced from Wang YP and Smith R (1995a) Wastewater Minimization with Flowrate Constraints, *Trans IChemE*, **A73**: 889, by permission of the Institution of Chemical Engineers.)

From targeting:

$$m_{WL}\left(C_{out,max} - C_{in,max}\right) = m_{Wmin}\left(C_{out,max} - C_0\right) \qquad (26.11)$$

Therefore, combining Equations 26.10 and 26.11 (assuming $m_{W3} = m_{Wmin}$):

$$\begin{aligned} C_{in} &= C_0 + \frac{1}{m_{Wmin}}\left[m_{WL}\left(C_{out,max} - C_{in,max}\right)\right] \\ &\quad - \frac{1}{m_{Wmin}}\left[m_{Wmin}\left(C_{out,max} - C_{in,max}\right)\right] \\ &= C_0 + C_{out,max} - C_0 - C_{out,max} + C_{in,max} \\ &= C_{in,max} \end{aligned}$$

Thus, for a single operation with fixed flowrate requirements defined by m_{WL} in Figure 26.35a, it is possible to design with a flowrate of m_{Wmin} by either local recycling or reuse with the operation split such that each part requires a flowrate smaller than or equal to m_{Wmin} and the parts are fed in ascending order of flowrate requirement.

Consider now the situation when multiple operations are involved. Suppose that, rather than a single operation, the stream in Figure 26.35a now represents several operations. If local recycling is not acceptable then Figure 26.38b shows a design for the interval with reuse. This follows the philosophy adopted for reuse with single operations. The difference now is that the operation is naturally split since the overall system already consists of several operations. As long as each operation requires a flowrate less than or equal to m_{Wmin}, there is no need to split individual operations. In Figure 26.38b, Operation 3 has been split in order to meet the requirement for being less than m_{Wmin}. In Figure 26.38 the order of reuse must be increasing order of flowrate requirement in order to satisfy all of the constraints.

If local recycling is not desirable and an operation cannot be split to be less than or equal to m_{Wmin}, then the minimum target for an interval can no longer be m_{Wmin}. Since each operation for reuse must have a flowrate requirement less than or equal to m_{Wmin} for reuse, if operation splitting is not allowed for the interval then the target flowrate m_{WT} is given by:

$$m_{WT} = \max\{m_{Wmin}, m_{WLi}\} \qquad (26.12)$$

where m_{Wmin} is given by the steepest water supply line that can be matched against the limiting composite curve and m_{WLi} are the flowrate requirements for each Operation *i*. Note that Equation 26.12 only guarantees that operation splitting within any individual interval is not necessary. It might still be necessary to split an operation between intervals. The worked examples later will illustrate this point.

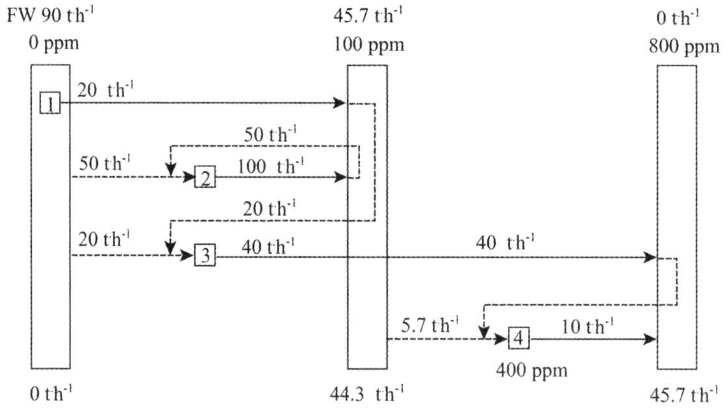

(a) Design grid for local recycling.

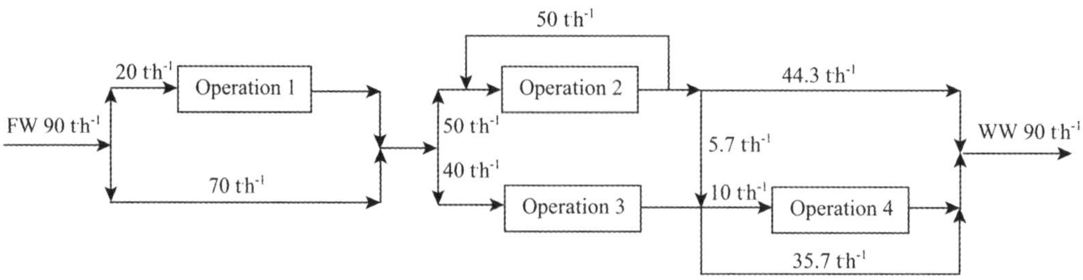

(b) Flowsheet for local recycling.

Figure 26.39
Design for local recycling.

26.9 Design for Maximum Water Reuse for Single Contaminants for Operations with Fixed Flowrates

The design procedure developed in Section 26.7 for fixed mass load can be readily adapted to design for fixed flowrates once the correct targets have been established. This can be illustrated by returning to the example data given in Table 26.4. Now it will be assumed that the concentration can vary up to the maximum, but the limiting water flowrate in Table 26.4 is fixed. First consider the case when local recycling is allowed. If local recycling is allowed, the target flowrate given in Figure 26.16 still applies. Also, for design the flowrate requirements above and below the pinch are still given by Figure 26.17. Figure 26.39a shows the design grid for local recycling. The flowsheet corresponding with this design grid is shown in Figure 26.39b.

If local recycling is not allowed, then a different target needs to be applied. Figure 26.40a shows the limiting composite curve, but with targets applied separately below and above the pinch. Below the pinch the target from Equation 26.12 is given by:

$$\begin{aligned} m_{WT} &= \max\{m_{Wmin}, m_{WLi}\} \\ &= \max\{90, 20, 100, 40\} \\ &= 100\,\mathrm{t \cdot h^{-1}} \end{aligned}$$

The pinch corresponds with the concentration interval boundary that limits the slope of the water supply line, which is again 100 ppm. The target above the pinch from Equation 26.12 is given by:

$$\begin{aligned} m_{WT} &= \max\{m_{Wmin}, m_{WLi}\} \\ &= \max\{45.93, 40, 10\} \\ &= 45.93\,\mathrm{t \cdot h^{-1}} \end{aligned}$$

Figure 26.40b shows the combined targets for no local recycling. Figure 26.41a shows the design grid for no local recycling corresponding with the targets in Figure 26.40b. The flowsheet corresponding with this design grid is shown in Figure 26.41b. It should be noted that in order to achieve the target from Equation 26.12, Operation 3 has been split into two parts. This results from the fact that the targeting equation has been applied over more than one interval below the pinch and above the pinch. The target equation only strictly applies for no operation splitting across a single interval. However, Figure 26.41 shows a design to achieve the targets. Most often, it will be undesirable to split an operation. In order to remove the split, the design in Figure 26.41 must be evolved. However, eliminating the split for Operation 3 will inevitably bring some penalty in the water consumption. Figure 26.42a shows a design grid for an evolution of the design to remove the operation split. It can be seen that the water consumption increases from $100\,\mathrm{t{\cdot}h^{-1}}$ to $111.43\,\mathrm{t{\cdot}h^{-1}}$. Figure 26.42b shows the corresponding flowsheet.

Consider again Examples 26.2 to 26.4 for a fixed mass load, but now constraining the flowrates to be fixed.

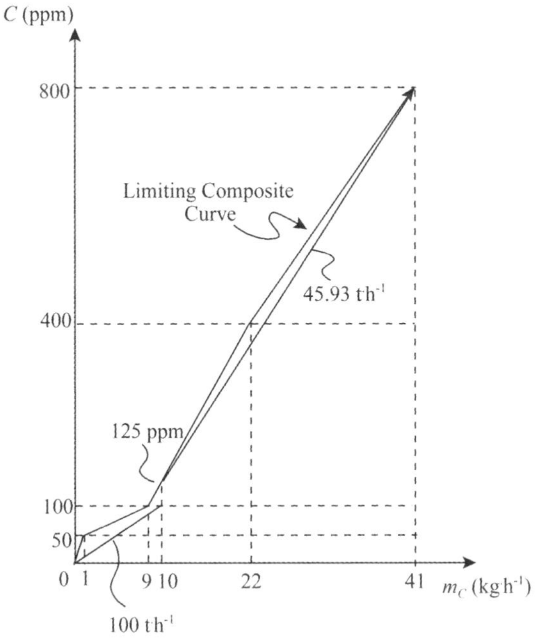

(a) Targeting below and above the pinch for no local recycling.

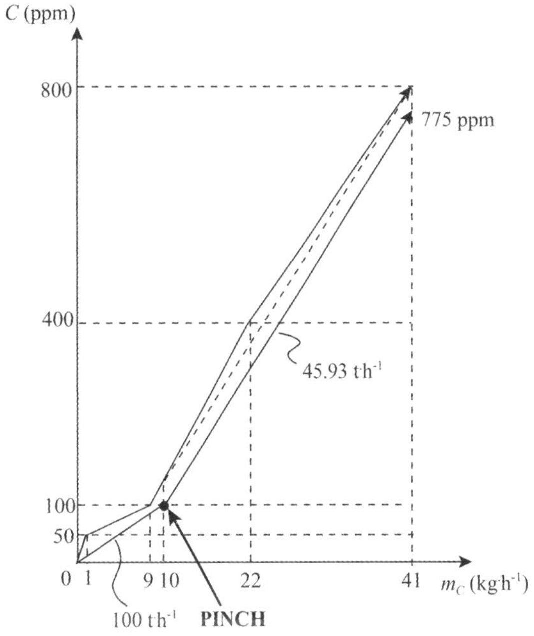

(b) Overall tagets for no local recycling.

Figure 26.40

Targets for no local recycling.

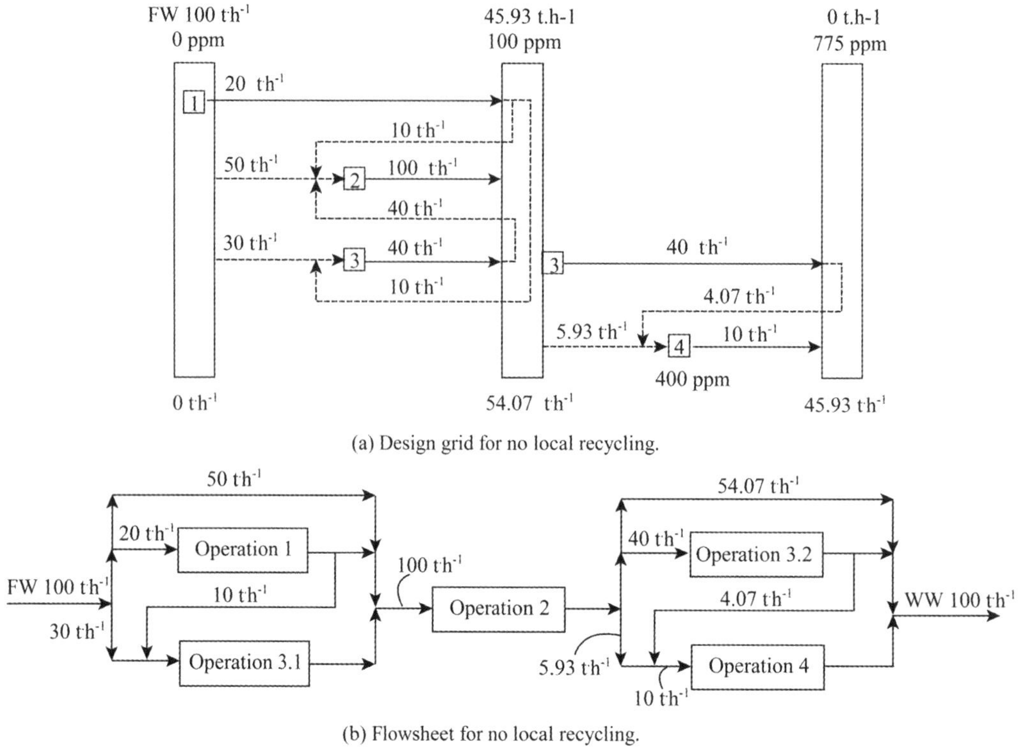

Figure 26.41

Design for no local recycling.

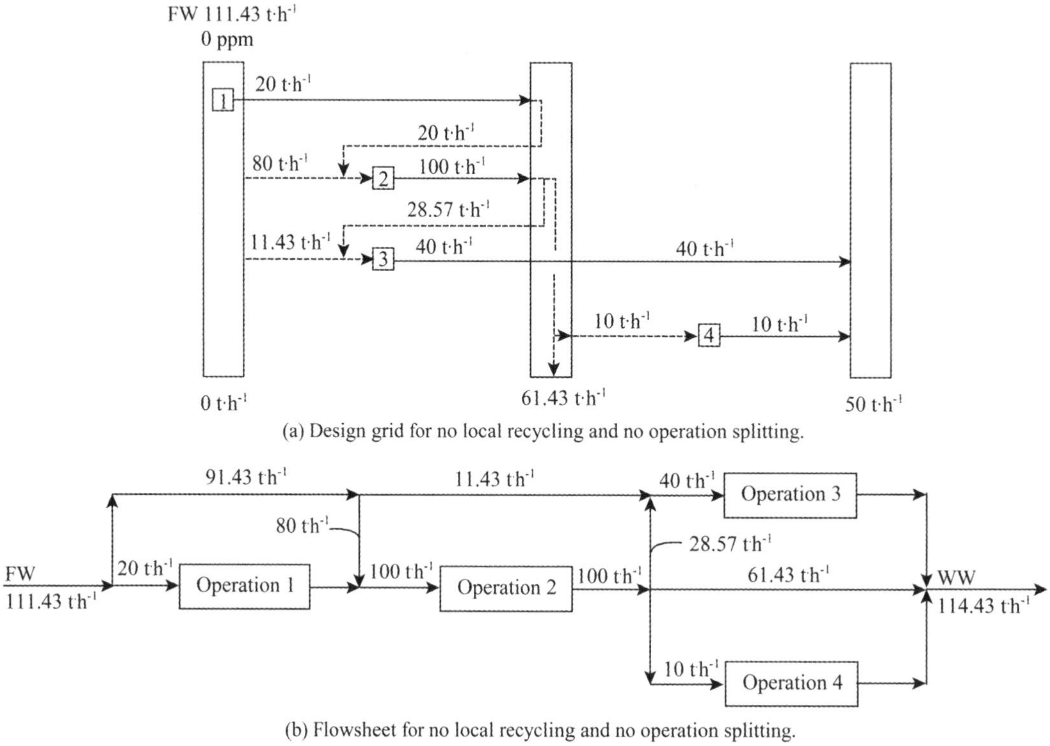

Figure 26.42

Evolution of the design for no local recycling and no operation splitting.

Example 26.5 Table 26.5 presents water-use data involving three operations.

a) Target and design the minimum water consumption for the system through maximum reuse with the flowrates fixed, allowing local recycling.
b) Target the minimum water consumption for the system through maximum reuse with the flowrates fixed, but no local recycling.
c) Design a network for the target water consumption with fixed flowrates and no local recycling.

Solution

a) When local recycling is allowed, the targets correspond with those given in Figure 26.27. Figure 26.43a shows the design grid to achieve the targets. Figure 26.43b shows the corresponding flowsheet.

b) When no local recycling is allowed, the target below the pinch is given by:

$$\begin{aligned} m_{WT} &= \max\{m_{Wmin}, m_{WLi}\} \\ &= \max\{46.67, 40, 20\} \\ &= 46.67\ \mathrm{t \cdot h^{-1}} \end{aligned}$$

The target above the pinch is given by:

$$\begin{aligned} m_{WT} &= \max\{m_{Wmin}, m_{WLi}\} \\ &= \max\{43.53, 20, 80\} \\ &= 80\ \mathrm{t \cdot h^{-1}} \end{aligned}$$

These targets are shown against the limiting composite curve in Figure 26.44. The target above the pinch is greater than that below in this case. Feeding 46.67 t·h^{-1} below the pinch up to a concentration of 150 ppm and then mixing with the balance of 33.33 t·h^{-1} of fresh water to be fed to the process above the pinch would lead to a feed concentration of 87.5 ppm. However, this mixing arrangement is not necessary in the design, as will be seen later. Overall the process must be fed with 80 t·h^{-1} of fresh water.

c) Figure 26.45a shows the design grid. It should be noted that the water mains are at concentrations of 0 ppm, 150 ppm and 550 ppm, corresponding with the water supply targets in Figure 26.44. Figure 26.45a shows the completed design grid to meet the targets and Figure 26.45b the corresponding flowsheet.

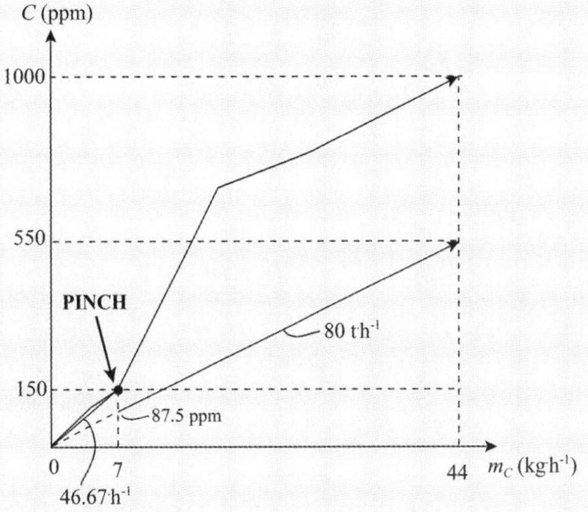

Figure 26.44

The minimum water target for no local recycling for Example 26.5

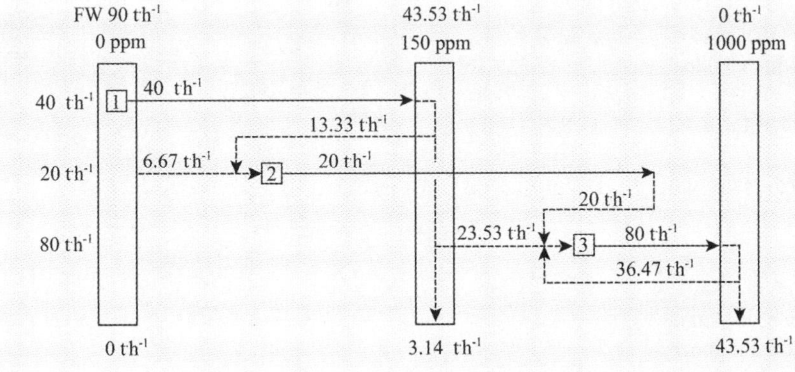

(a) Design grid for local recycling.

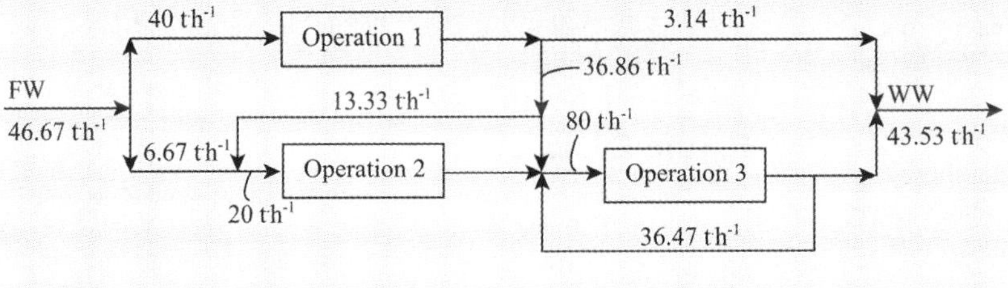

(b) Flowsheet for local recycling.

Figure 26.43

Design for local recycling for Example 26.5.

26

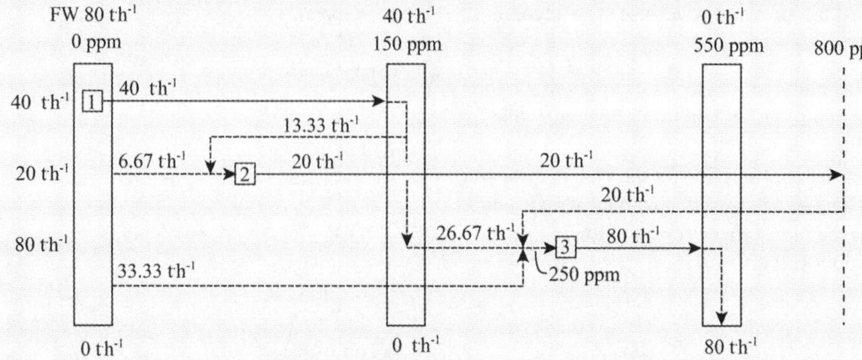

(a) Design grid for no local recycling.

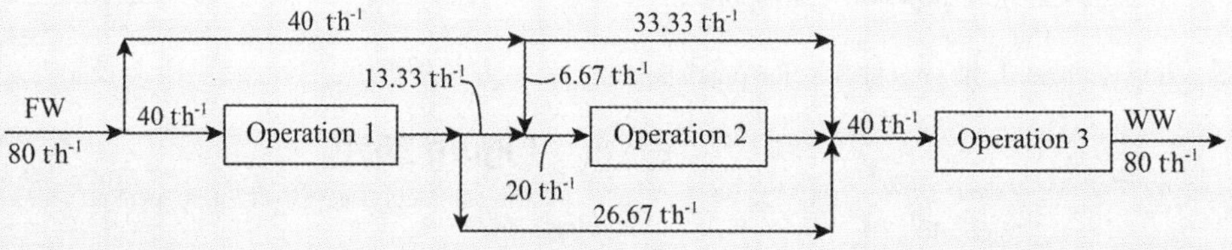

(b) Flowsheet for no recycling.

Figure 26.45
Design for no local recycling for Example 26.5.

Example 26.6 Limiting data for three operations are given in Table 26.6. Two sources of fresh water are available: WSI with a concentration of 0 ppm and WSII with a concentration of 25 ppm:

a) Target the minimum water consumption for the system through maximum reuse for fixed flowrates with no local recycling.
b) Design a network for the target water consumption for fixed flowrates with no local recycling.
c) Evolve the design from Part b to remove any operation splits.

Solution

a) The limiting composite curve for the three operations is plotted in Figure 26.46, with the targets above and below the pinch shown.
b) Figure 26.47a shows the design grid for the two water sources with water mains at 0 ppm, 25 ppm, 100 ppm and 800 ppm, corresponding with the targets in Figure 26.46. Figure 26.47a shows the completed design grid and Figure 26.47b the corresponding flowsheet.
c) Figure 26.48a shows an evolution of the design grid to remove the spit in Operation 3. It can be seen that there is a penalty of an extra 13.33 t·h^{-1} of WSII to remove the operation split. The corresponding flowsheet is shown in Figure 26.48b.

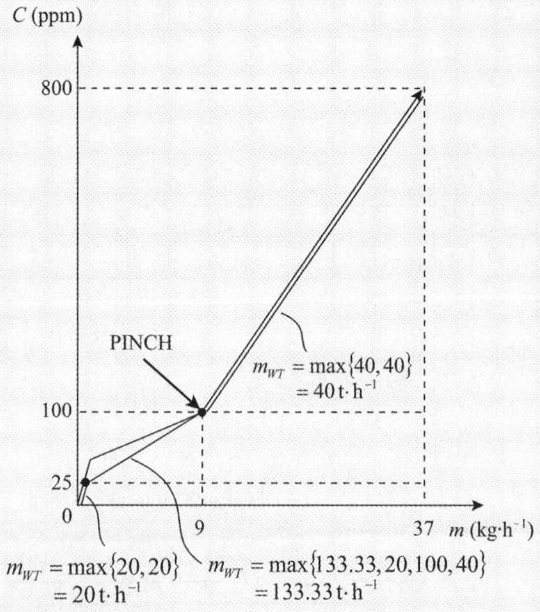

Figure 26.46
Targeting for multiple water sources with fixed flowrates without local recycling for Example 26.6.

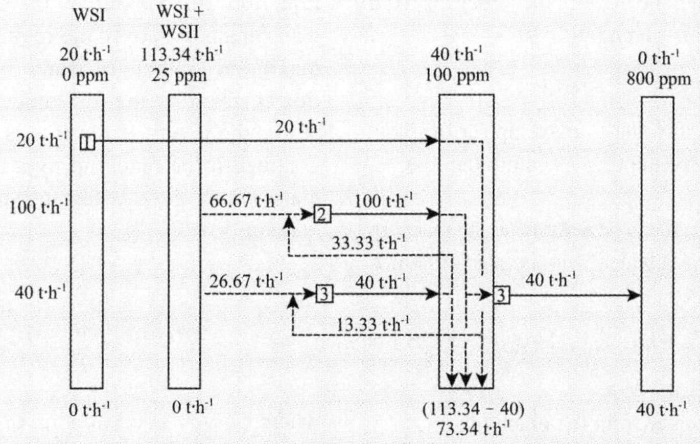

(a) Grid diagram for two water sources with process fixed flowrates to meet the targets.

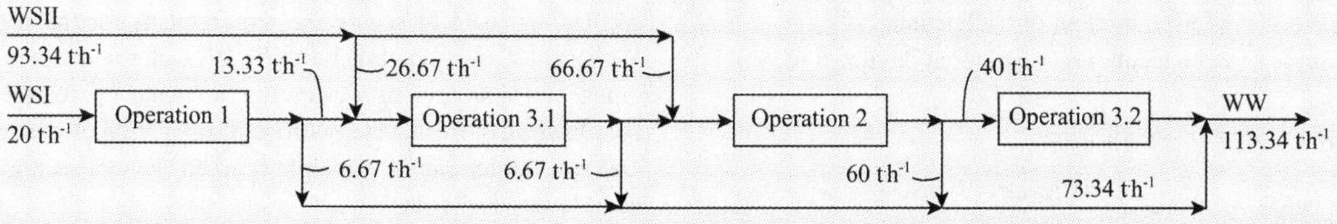

(b) Flowsheet for two water sources with process fixed flowrates to meet the targets.

Figure 26.47

Design to meet targets for multiple water sources with fixed flowates without local recycling for Example 26.6.

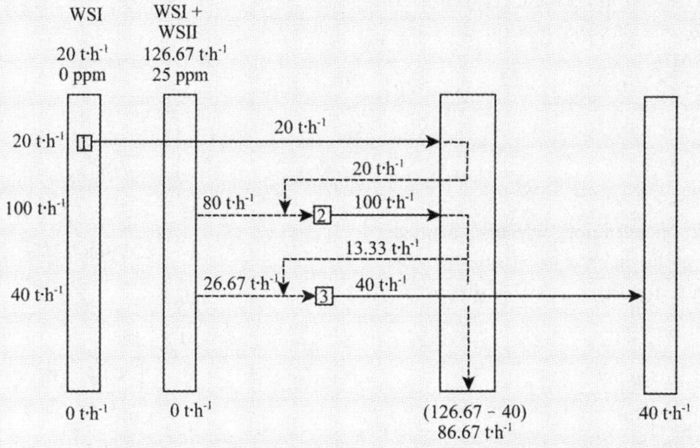

(a) Grid diagram for two water sources with process fixed flowrates with no operation splits.

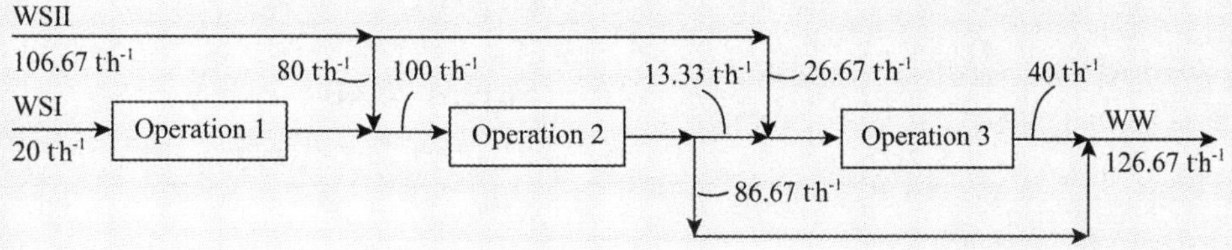

(b) Flowsheet for two water sources with process fixed flowrates with no operation splits.

Figure 26.48

Evolution of the design for multiple water sources without split operations for Example 26.6.

When there is a fixed loss from the system, the target for fixed flowrates with no local recycles is given by:

$$m_{WT} = \max\{m_{Wmin}, m_{WLi}\} + m_{WTLOSS} \qquad (26.13)$$

where m_{WTLOSS} = target for fresh water to make up for the loss over and above the target without the loss

If the loss is below the pinch, then it can be assumed to be made up by a combination of fresh water and water from the pinch main. A mass balance for the makeup gives:

$$m_{WLOSS}C_{LOSS} = m_{WTLOSS}C_0 + m_{WPINCH}C_{PINCH} \qquad (26.14)$$

where m_{WLOSS} = flowrate of water loss
m_{WPINCH} = flowrate of water from the pinch main to make up for the water loss
C_{LOSS} = concentration at which the water loss occurs
C_0 = concentration of fresh water
C_{PINCH} = concentration of water at the pinch

Rearranging Equation 26.14 gives:

$$m_{WTLOSS} = m_{WLOSS}\frac{(C_{PINCH} - C_{LOSS})}{(C_{PINCH} - C_0)} \qquad (26.15)$$

Equation 26.15 assumes there is enough water at the pinch water main to satisfy the requirements for the water loss. The maximum flowrate of water available at the pinch concentration is given by the difference between the targets below and above the pinch. If the loss is above the pinch then it can be assumed to be made up by a combination of fresh water and water from the wastewater main. A mass balance for the makeup gives:

$$m_{WTLOSS} = m_{WLOSS}\frac{(C_{WW} - C_{LOSS})}{(C_{WW} - C_0)} \qquad (26.16)$$

where C_{WW} = concentration of the wastewater

Again, Equation 26.16 assumes there is enough water at the wastewater main to satisfy the requirements for the water loss. Rather than use a combination of fresh water and either water at pinch concentration or wastewater to make up for the loss, a combination of fresh water with both water at pinch concentration and wastewater can in principle be used. However, this is more complex to deal with.

Example 26.7 An operation is to be added to those in Table 26.6 with a maximum inlet concentration of 80 ppm and a flowrate of $10\,t{\cdot}h^{-1}$, all of which is lost:

a) Target the minimum water consumption for the system through maximum reuse for fixed flowrates with no local recycling.
b) Design a network for the target water consumption for fixed flowrates with no local recycling.
c) For the design from Part b, evolve the design to remove any split operations.

Solution

a) If the water loss is below pinch concentration, and it is assumed that there will be enough water at the pinch water main, then the target for the makeup water for the loss is given by Equation 26.15:

$$m_{WTLOSS} = m_{WLOSS}\frac{(C_{PINCH} - C_{LOSS})}{(C_{PINCH} - C_0)}$$
$$= 10\frac{(100-80)}{(100-0)}$$
$$= 2\,t \cdot h^{-1}$$
$$m_{WT} = \max\{m_{Wmin}, m_{WLi}\} + m_{WTLOSS}$$
$$= \max\{90, 20, 100, 40\} + 2$$
$$= 102\,t \cdot h^{-1}$$

The target above the pinch is given by:

$$m_{WT} = \max\{m_{Wmin}, m_{WLi}\}$$
$$= \max\{40, 40\}$$
$$= 40\,t \cdot h^{-1}$$

These targets are shown in Figure 26.49.

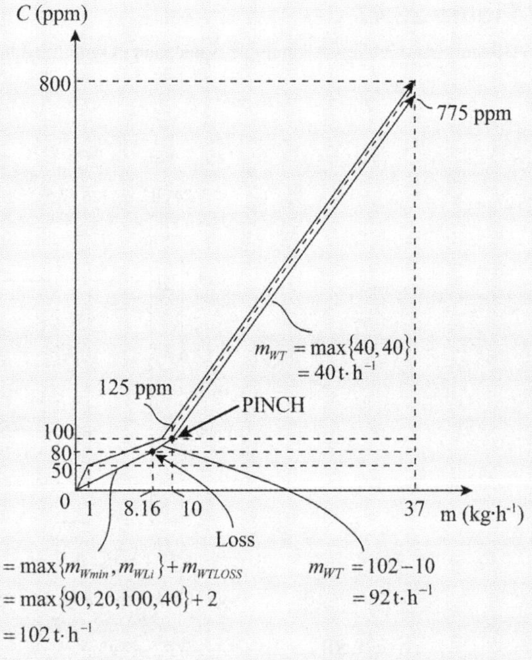

Figure 26.49
Targeting for no local recycling with a water loss for Example 26.7.

b) For the design, from Figure 26.49, there should be three water mains at 0 ppm, 100 ppm and 775 ppm. Figure 26.50a shows the design grid to meet the target featuring a water loss of $10\,t{\cdot}h^{-1}$. Figure 26.50b shows the corresponding flowsheet.
c) It can be observed in Figure 26.50 that Operation 3 has been split into two parts in order to achieve the target. Figure 26.51a shows

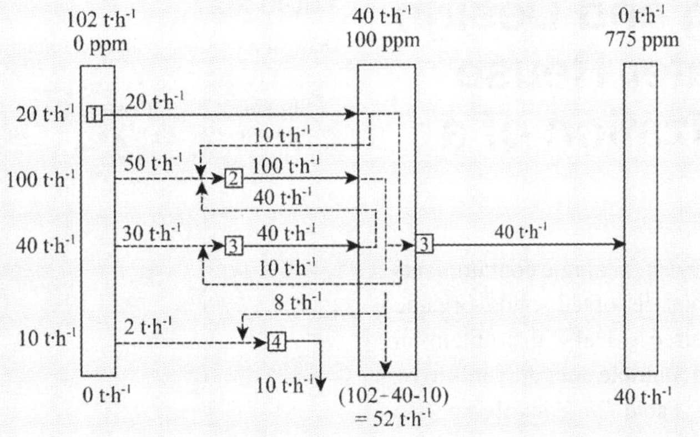

(a) Design grid for water loss with fixed flowrates.

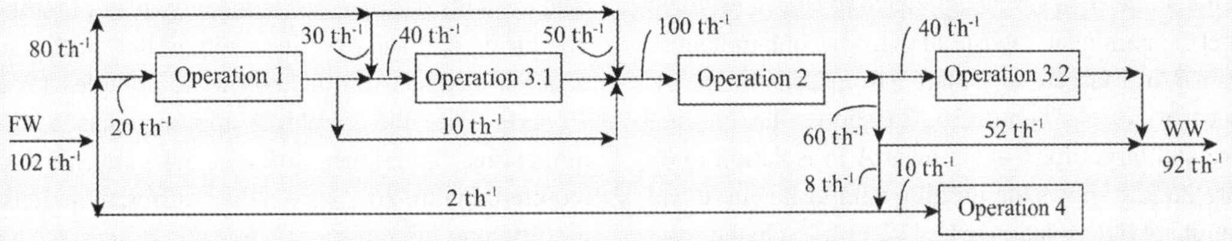

(b) Flowsheet for water loss with fixed flowrates.

Figure 26.50

Design to meet the targets for fixed flowrates and no local recycling with a water loss for Example 26.7.

the design grid with the design evolved to remove the split in Operation 3. Figure 26.51b shows the corresponding flowsheet. It can be seen that the penalty to remove the split in Operation 3 is an extra $20\,t{\cdot}h^{-1}$ of fresh water.

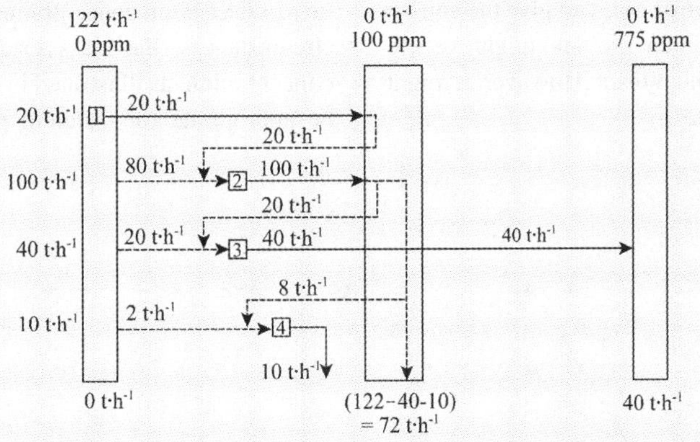

(a) Design grid for water loss with fixed flowrates with no operation split.

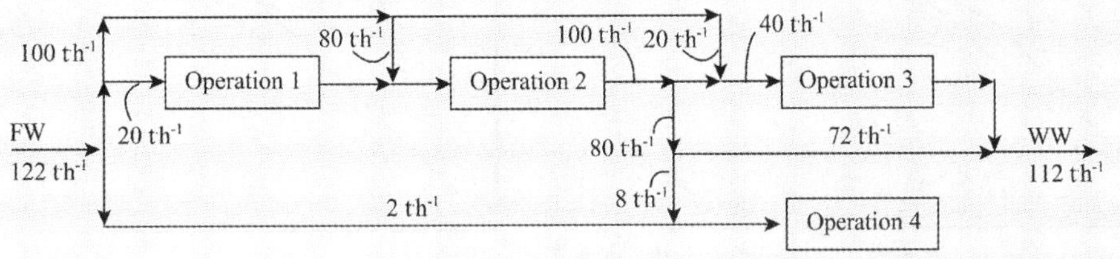

(b) Flowsheet for water loss with fixed flowrates with no operation split.

Figure 26.51

Evolution of the design for fixed flowrates and no local recycling with a water loss and no stream split for Example 26.7.

26.10 Targeting and Design for Maximum Water Reuse Based on Optimization of a Superstructure

The methods presented so far are adequate for single contaminants (e.g. total solids, suspended solids, total dissolved solids, organic concentration, etc.), but it is often required to deal with problems in which the concentration limits require multiple contaminants to be specified. Consider the problem in Table 26.7 involving two operations.

This time, maximum concentrations are specified separately for two Contaminants *A* and *B*. A simple approach to such a problem might be to target Contaminant *A* in isolation from Contaminant *B*, then Contaminant *B* to be targeted in isolation from Contaminant *A*, and the worst case chosen. Figure 26.52a shows the limiting composite curve and target for Contaminant *A* in isolation to be 115 t·h^{-1}. Figure 26.52b shows the limiting composite curve and target for Contaminant *B* in isolation to be 112.1 t·h^{-1}. On the basis of these calculations, it would therefore be suspected that the target would be the worst case and would be 115 t·h^{-1}. In fact, the true target for this problem is 126.5 t·h^{-1} by considering Contaminant *A* and Contaminant *B* simultaneously. The problem with considering each contaminant in isolation is that it does not allow for the fact that when Contaminant *A* is picked up, Contaminant *B* is picked up at the same time and vice versa.

In some problems, the simple approach of targeting the individual contaminants and taking the worst case can give the correct answer. The target for a multicontaminant problem might well be the largest target of the individual contaminant. However, it might be a larger value, as it is in this case.

Table 26.7

Data for a problem with two contaminants.

Operation number	m_{WL} (t·h^{-1})	Contaminant	C_{in} (ppm)	C_{out} (ppm)
1	90	*A*	0	120
		B	25	85
2	75	*A*	80	220
		B	30	100

A more sophisticated approach is therefore required when dealing with multiple contaminants. It is possible to extend the graphical approach described here to deal with multiple contaminants, but it gets significantly more complex (Wang and Smith, 1994a). Also, the graphical approach has a number of other limitations. Sometimes different mass transfer models might be required. Figure 26.53 gives three different models that are useful in different circumstances for the design of water networks. Figure 26.53a shows the model involving a fixed mass load, but allowing the flowrate to vary. Figure 26.53b shows the case that an operation requires a fixed flowrate, irrespective of the inlet concentration. Different inlet concentrations can be chosen, but the slope must be fixed. A third option is shown in Figure 26.53c where the outlet concentration from the operation is fixed, as is the flowrate. This last case can happen if water is coming into contact with a contaminant that is sparingly soluble and can only dissolve up to a maximum concentration, the maximum solubility. In these circumstances, the mass load will vary according to the inlet concentration, as illustrated in Figure 26.53c. Other models might be appropriate for different problems. As an example of how

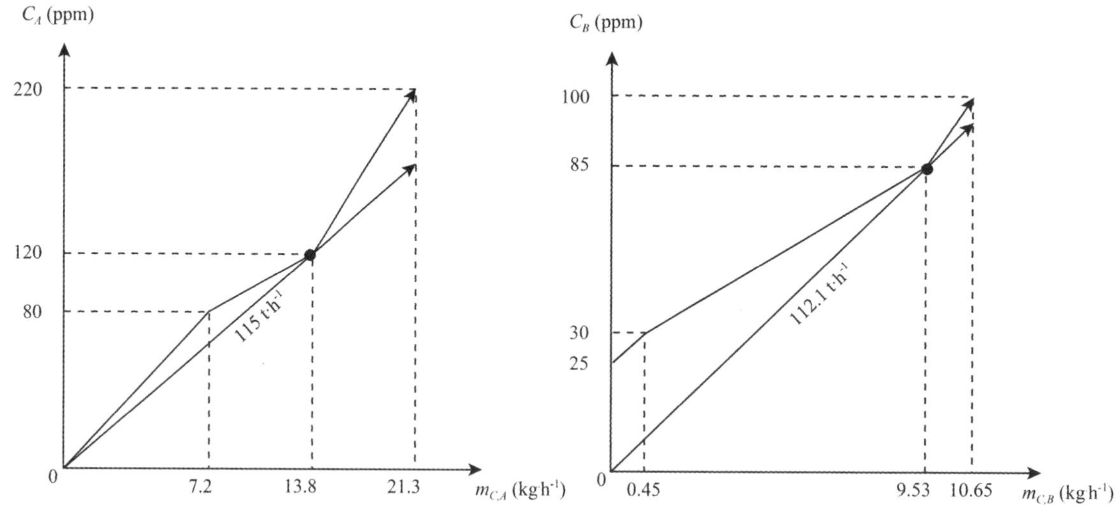

(a) Target for Contaminant *A* in isolation.

(b) Target for Contaminant *B* in isolation.

Figure 26.52

Targeting individual contaminants for a multiple contaminant problem from Table 26.7.

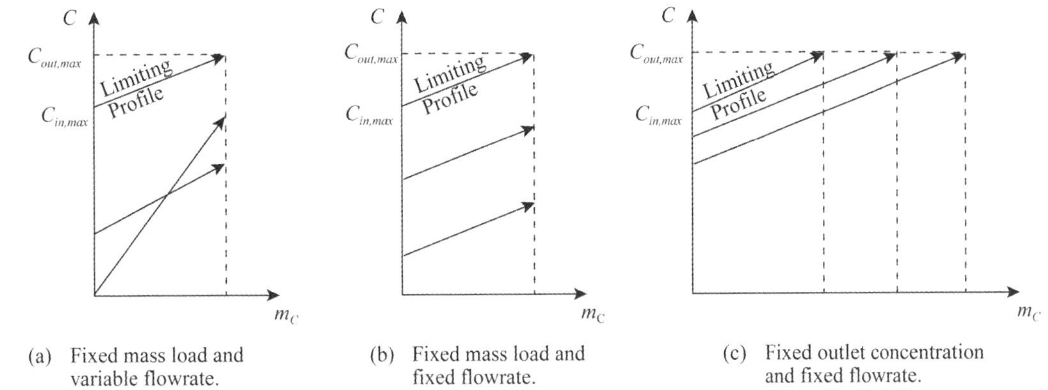

Figure 26.53
Different mass transfer models.

different mass transfer models might be required, consider again the example of the crude oil desalter described in Section 10.7. This is a simple extraction process in which crude oil is mixed with water to extract salt from the crude oil into the water. The function of the desalter is to extract salt. Thus, the load of salt is specified. The flowrate of water can be varied to some extent, but there will be limits on this. At the same time that the water extracts the salt from the crude oil, other contaminants will be transferred from the crude oil to the water. These might typically be hydrocarbons with a maximum solubility and thus a maximum concentration might wish to be specified for such contaminants. Thus, not only might it be necessary to specify multiple contaminants, it also might be necessary to specify different mass transfer models in an individual operation. Also, the flowrate of water might be fixed, or allowed to vary only within a specified range.

In addition to these complexities, there may be other issues that need to be included in the analysis that are not readily included using the graphical approach. These might include forbidden matches (e.g. because a long pipe run may be necessary), compulsory matches (e.g. for operability) and capital cost issues (e.g. the cost of running pipes between operations).

To include all of these complexities requires a different approach from the graphical approaches described so far. The design approach based on the optimization of a superstructure can be used to solve such problems. Figure 26.54 shows the superstructure for a problem involving two operations and a single source of fresh water (Doyle and Smith, 1997). The superstructure allows for reuse from Operation 1 into Operation 2, reuse from Operation 2 to Operation 1, local recycles around both operations, fresh water supply to both operations and wastewater discharge from both operations. All structural features that are candidates for the final design have been included in the superstructure. The mass balances need to be modeled, cost models set up and the superstructure can then be optimized, as illustrated in Figure 26.54.

The objective functions could be as simple as minimizing the supply of fresh water, or could be more sophisticated and minimize cost. Minimum cost could include features such as (Gunarantnam *et al.*, 2005):

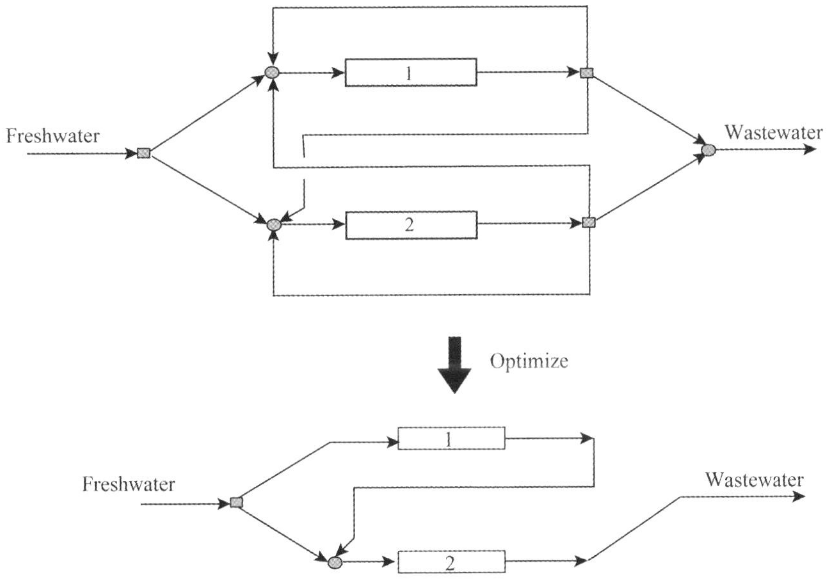

Figure 26.54
Water network design based on the optimization of a superstructure.

26

- Multiple water sources with different costs
- Effluent treatment costs
- Piping costs.

The cost of water from a given source will be proportional to its flowrate. Effluent treatment costs can also be formulated as a function of flowrate. The economics of water reuse is, however, often critically dependent on piping costs. Piping costs can be included in the optimization by providing information on the approximate length of pipe required to connect a water source and a water sink, together with cost information for the piping. During the course of the optimization, the flowrate of water will be calculated. Assuming a reasonable velocity of 1 to 2 $m \cdot s^{-1}$ allows the pipe diameter to be calculated. From knowledge of the length and diameter, the cost of a pipe between a source and sink can be estimated. Given this information, the optimization will tend to produce solutions avoiding long and expensive pipe connections (Gunarantnam *et al.*, 2005).

The nature of the optimization problem can turn out to be linear or nonlinear depending on the mass transfer model chosen (Doyle and Smith, 1997). If a model based on a fixed outlet concentration is chosen, the model turns out to be a linear model (assuming linear cost models are adopted). If the outlet concentration is allowed to vary, as in Figure 26.53a and Figure 26.53b, then the optimization turns out to be a nonlinear optimization with all the problems of local optima associated with such problems. The optimization is in fact not so difficult in practice as regards the nonlinearity, because it is possible to provide a good initialization to the nonlinear model. If the outlet concentrations from each operation are initially assumed to go to their maximum outlet concentrations, then this can then be solved by a linear optimization. This usually provides a good initialization for the nonlinear optimization, as the network design will tend, in most instances, to try to force concentrations to their maximum outlet concentrations to maximize water reuse (Doyle and Smith, 1997).

The advantages of an optimization-based approach are that it provides automation for complex problems, provides a design as well as a target and allows all kinds of constraints and costs to be included. One disadvantage is that there is no longer a graphical representation allowing conceptual insights. This shortcoming can be overcome by displaying the output from the optimization in graphical form. This is most conveniently done by using one contaminant at a time. The optimization provides a network design with inlet and outlet concentrations that can be taken and combined together in a composite plot, one contaminant at a time. By superimposing the flowrate predicted by the optimization on to this composite curve, conceptual insights can be gained.

26.11 Process Changes for Reduced Water Consumption

So far the discussion on water minimization has restricted consideration to identify opportunities for water reuse. Maximizing water reuse minimizes both fresh water consumption and wastewater generation. However, the process constraints for inlet concentrations, outlet concentrations and flowrates have so far been fixed. Often there is freedom to change the conditions within the operation. Typical process changes that might be contemplated include:

- Introducing a local recycle around a scrubber or washing operation.
- Removing water from the operation completely if it is extraneous (e.g. replace the scrubbing operation used to separate an organic contaminant from a vapor stream by condensation, or replacing an extraction process using water by cooling crystallization).
- Increasing the number of stages in an extraction process to reduce the water demand for a given separation.
- Increasing the number of stages in absorption and scrubbing operations that use water to reduce the water consumption for the same separation.
- Replacing live steam injection for distillation operations with a reboiler using indirect steam heating.
- Introducing multiple stages for equipment cleaning, such that the initial stages are carried out using slightly contaminated water (e.g. to remove residual solids), followed by high-quality water for the final stages.
- In multiproduct plants, scheduling operations to minimize product changeovers and thereby reduce the cleaning requirements between product changeovers.
- Introducing mechanical cleaning of vessels (e.g. wall wipers) and pipes (e.g. pipeline pigging) that process viscous materials and require periodic cleaning with water.
- For cleaning agitated vessels, rather than filling up the vessel and use agitation for the cleaning, introducing a local recycle using less water, but with a spray system inside the vessel to give effective cleaning with less water.
- Fixing triggers to hoses to prevent unattended running of water from the hoses when not in use, so that as soon as the trigger is released by the operator, the flow is stopped.
- Introducing local recycles around liquid ring pumps.
- Replacing direct contact condensation using a water spray by indirect condensation using cooling water.
- Introducing local recycles around quench operations that use a water spray with some indirect cooling in the recycle loop.
- For batch operations, introducing solenoid valves in water scrubbers, direct contact condensation and the flushes used to protect pump and agitator seals, such that if the operation is not in use, the flow of water is stopped.
- Improving the energy efficiency to reduce steam demand and hence reduce the wastewater generated by the steam system through boiler blowdown, boiler feed water treatment and condensate loss (see Chapter 23).
- Increasing condensate return for steam systems to reduce makeup water requirements, reduce aqueous waste from boiler feed water treatment and boiler blowdown (see Chapter 23).
- Improving control of cooling tower blowdown (see Chapter 24) for evaporative cooling water circuits to increase the cycles of concentration and reduce the cooling tower blowdown rate.

- Decreasing the heat duty on evaporative cooling tower circuits by better energy efficiency to decrease the evaporation from the cooling tower and, hence, decrease the makeup requirements.
- Introducing air cooling to replace evaporative cooling towers (see Chapter 24).
- Considering reducing the water mains pressure or isolating areas when not in use, if leaks and unaccountable losses are a problem from underground piping systems, and so on.

Listed above are just some of the many possible changes that can be made to operations to reduce their inherent water demand, but which changes are going to have an influence on the overall system? Consider the limiting composite curves in Figures 26.16 and 26.27. It is clear that only changes at or below the pinch for the system will reduce the demand for water for the overall system. A process located above the pinch should have more than enough water available in the system through the appropriate reuse opportunities. Thus, in general, the ways in which the overall system can benefit from process changes are:

- For a given volumetric flowrate for the processes below the pinch, shift the mass load of contaminant from below the pinch to above the pinch by increasing concentrations.
- For a fixed mass load of contaminant for the processes below the pinch, decrease the volumetric flowrate of the processes below the pinch, thus increasing concentrations.
- A combination of the above two measures.

The location of an operation relative to the pinch is one criterion for deriving a benefit from process changes. The other criterion is one of scale. The larger the flowrate fed to an operation, the larger is likely to be the overall effect of the process changes. To understand the resulting magnitude of suggested process changes on the overall water consumption, the targeting methods described above should be used to screen options.

26.12 Targeting for Minimum Wastewater Treatment Flowrate for Single Contaminants

Wastewater treatment in the process industries is most often carried out in a central treatment facility. Effluent streams are collected together in a common sewer system and passed to a central treatment facility, or, in some instances, given some pretreatment and sent off-site to a municipal treatment facility. The problem with centralized treatment is that combining waste streams that require different treatment technologies leads to a cost of treating the combined streams that is likely to be more expensive than the individual treatment of the separate streams. This situation results from the fact that the capital cost of most waste treatment processes is related to the total flowrate of wastewater. Also, operating costs for treatment usually increase with decreasing concentration for a given mass load of contaminant to be treated. On the other hand, if two waste streams require exactly the same treatment then it makes sense to combine them for treatment to obtain economies of scale for the equipment.

A philosophy of design is required that will allow the designer to combine effluents for treatment when it is appropriate, but also segregate them if it is appropriate. Consider now the targeting of the minimum treatment flowrate for a given set of effluent streams that need to be treated to bring the concentration to below an environmental discharge limit.

Figure 26.55a shows a system involving four effluent streams with different inlet concentrations that all need to be treated to remove mass load and bring down the concentration to an acceptable level for environmental discharge C_e. To obtain an overall picture, rather than deal with four separate effluent streams, the

26

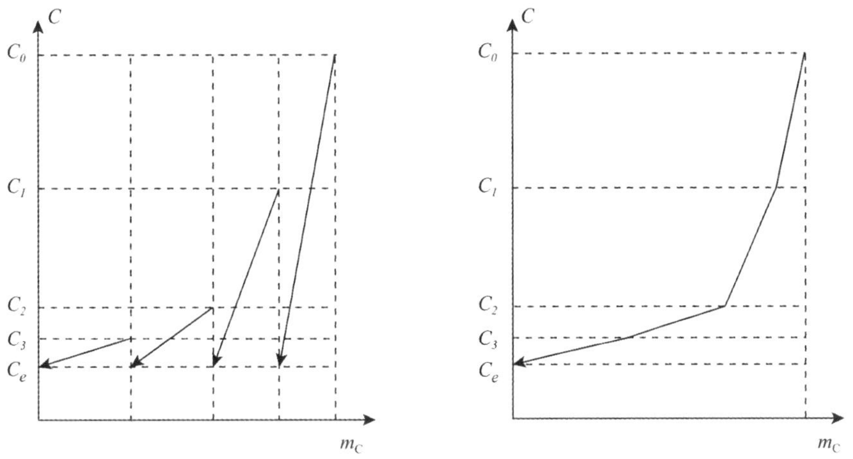

(a) Wastewater streams.

(b) Effluent composite curve.

Figure 26.55

A composite curve of wastewater streams. (Reproduced from Wang YP and Smith R (1994a) Wastewater Minimization, *Chem Eng Sci*, **49**: 3127, with permission from Elsevier.)

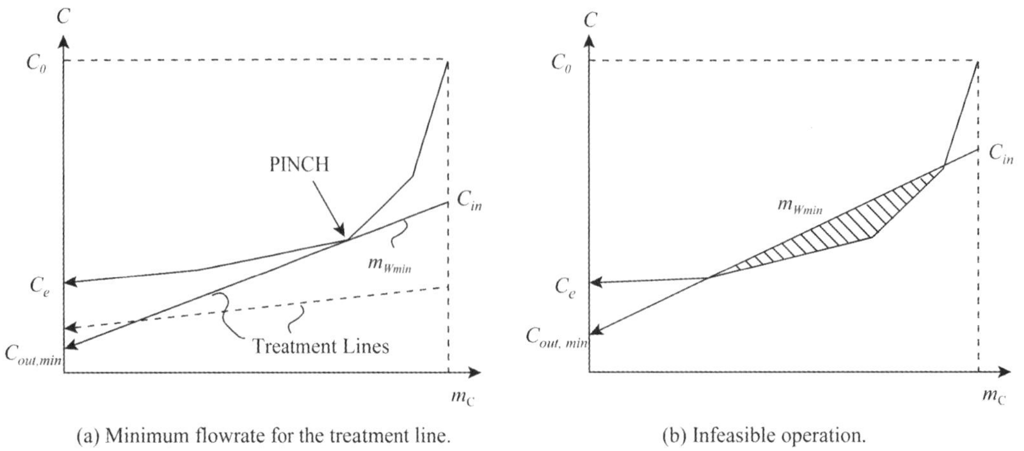

Figure 26.56

Targeting minimum wastewater treatment flowrate. (Reproduced from Wang YP and Smith R (1994a) Wastewater Minimization, *Chem Eng Sci*, **49**: 3127, with permission from Elsevier.)

streams can be combined together to produce a *composite effluent stream* (Wang and Smith, 1994b). The construction is analogous to that for the limiting composite curve. The diagram is divided into concentration intervals and the mass load of the streams within each concentration interval combined together (Figure 26.55b). This provides a picture of the overall effluent treatment problem and what is required to happen to the effluent streams.

The mass load can be removed in a number of different network arrangements. The objective in designing an effluent treatment network is to minimize its cost. In the first instance, this means minimizing the flowrate through the treatment process. Figure 26.56a shows a composite effluent curve. Superimposed are *treatment lines* representing the actual performance of the effluent treatment process (Wang and Smith, 1994b). The effluent treatment process is required to remove the mass load in order to bring down the effluent concentration to its environmental discharge limit. The steeper the treatment line, the lower will be the flowrate through the effluent treatment. The maximum slope of the treatment line in Figure 26.56a is dictated at the low concentration end by a minimum outlet concentration, below which the treatment process cannot operate, and the *pinch point* between the effluent composite curve and treatment line (Wang and Smith, 1994b). The treatment line cannot be made any steeper than shown in Figure 26.56a. If the slope is increased, as shown in Figure 26.56b, then the treatment line crosses the effluent composite curve. This is infeasible, as it requires mass to be treated in concentration ranges where it is not available. Any attempt to design this flowrate will inevitably lead to an infeasible mass balance in the design of the effluent treatment system.

In specifying effluent treatment processes, two specifications are used in conceptual design. The simplest case is where the concentration at the outlet of an effluent treatment process is specified to be a fixed value. Targeting with a fixed outlet concentration is illustrated in Figure 26.57a. The effluent composite curve is first constructed. The outlet concentration for the effluent treatment is specified and the slope of the treatment line pivoted around this value until the treatment line pinches with the effluent composite curve. The slope of the line will then dictate the minimum treatment flowrate.

The performance of the treatment process is more likely to be specified by a removal ratio:

$$R = \frac{\text{Mass of contaminant removed}}{\text{Mass of contaminant fed}} = \frac{m_{W,in}C_{in} - m_{W,out}C_{out}}{m_{W,in}C_{in}} \tag{26.17}$$

where R = removal ratio
$m_{W,in}$, $m_{W,out}$ = inlet and outlet flowrates
C_{in}, C_{out} = inlet and outlet concentrations

In many instances, the change in flowrate through the treatment process can be neglected, giving:

$$R = \frac{C_{in} - C_{out}}{C_{in}} \tag{26.18}$$

Figure 26.57b shows the construction for targeting the minimum wastewater treatment flowrate when the removal ratio has been specified. If the initial effluent treatment line shown dotted in Figure 26.57b is considered, then there is a point of origin corresponding with a specified removal ratio. Applying the geometry of similar triangles, as shown in Figure 26.57b and rearranging indicates that the ratio $\Delta m_T/m_T$ equals the removal ratio. Thus, a given removal ratio corresponds with a given ratio of $\Delta m_T/m_T$ and, therefore, a fixed point of origin for the treatment line. If the treatment line is then pivoted around this point of origin, as shown in Figure 26.57b, the ratio $\Delta m_T/m_T$ and, therefore, the removal ratio are both fixed. Finding the steepest slope where the treatment line is pinched against the composite effluent curve corresponds with the minimum treatment flowrate.

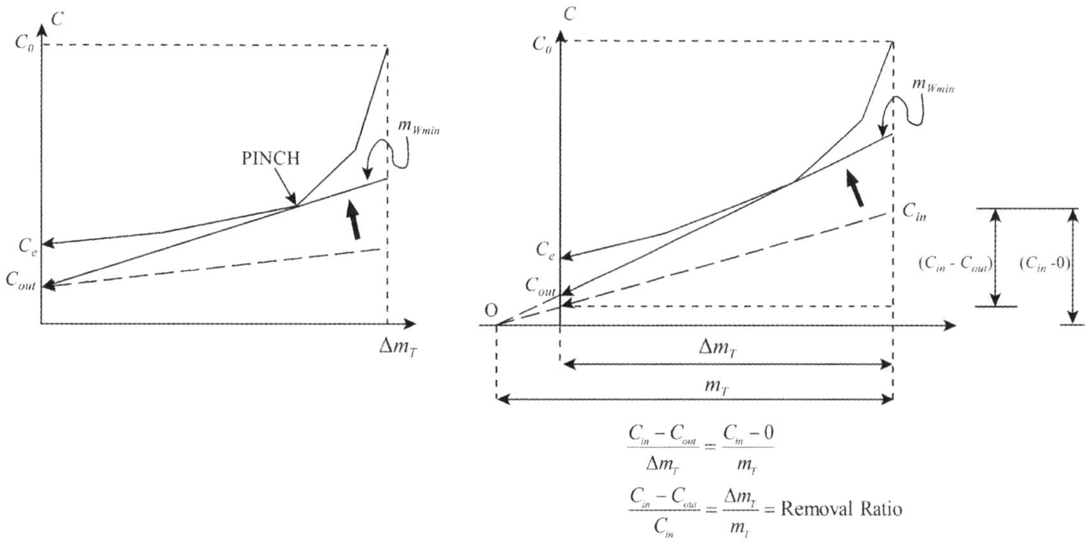

(a) Targeting with fixed outlet concentration.

(b) Targeting with fixed removal ratio.

Figure 26.57

Targeting minimum wastewater treatment flowrate. (Reproduced from Wang YP and Smith R (1994a) Wastewater Minimization, *Chem Eng Sci*, **49**: 3127, with permission from Elsevier.)

26

Example 26.8 The data for a wastewater treatment problem are given in Table 26.8. Centralized treatment would correspond with a flowrate to be treated of $75\,t{\cdot}h^{-1}$. For an environmental discharge limit of 20 ppm, determine the minimum treatment flowrate.

a) For a process with a fixed outlet concentration of 10 ppm.
b) For a treatment process with a removal ratio of 95%.

Table 26.8

Data for an effluent treatment problem.

Stream	C_{in} (ppm)	Water flowrate ($t{\cdot}h^{-1}$)
1	250	40
2	100	25
3	40	10

Solution

a) Figure 26.58 shows the effluent composite curve plotted for the three streams in Table 26.8. The end of the effluent treatment line is fixed with a concentration of 10 ppm and the steepest line drawn until it pinches with the effluent composite curve at 100 ppm. The removal flowrate is then given by:

$$m_{Wmin} = \frac{\Delta m_T}{\Delta C} = \frac{5400}{100 - 10} = 60\,t{\cdot}h^{-1}$$

b) For a fixed removal ratio with no change in flowrate of water:

$$R = 0.95 = \frac{C_{in} - C_{out}}{C_{in}}$$

Rearranging gives:

$$C_{in} = 20C_{out} \tag{26.19}$$

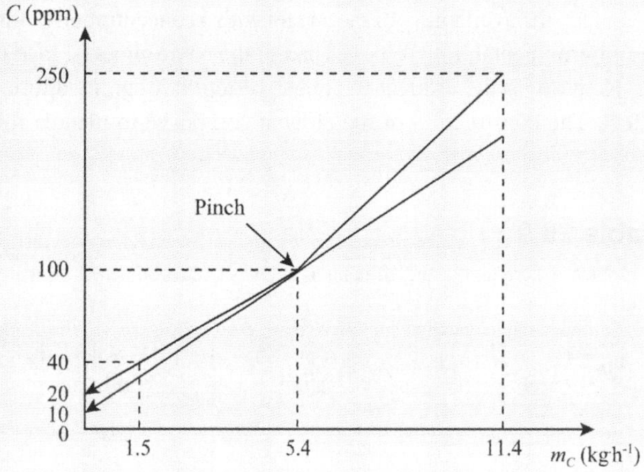

Figure 26.58

Targeting for a fixed outlet concentration of 10 ppm in Example 26.8.

The total mass load is given by:

$$\begin{aligned}\Delta m_T &= m_{Wmin}\Delta C \\ 11,400 &= m_{Wmin}(C_{in} - C_{out})\end{aligned} \quad (26.20)$$

For the minimum flowrate, the treatment line will pass through the pinch point at 100 ppm. Writing the mass balance below the pinch point:

$$5400 = m_{Wmin}(100 - C_{out}) \quad (26.21)$$

Dividing Equation 26.20 by Equation 26.21 gives:

$$\frac{11,400}{5400} = \frac{C_{in} - C_{out}}{100 - C_{out}}$$

Substituting Equation 26.19 gives:

$$C_{out} = 10 \text{ ppm}$$

Thus:

$$\begin{aligned}C_{in} &= 200 \text{ ppm} \\ m_{Wmin} &= 60 \text{ t·h}^{-1}\end{aligned}$$

By coincidence the answer is the same in both cases, that is, 10 ppm outlet concentration corresponds with a removal ratio of 0.95.

One complication not considered so far that is very commonly encountered in effluent systems is where some of the effluents are already below their environmental discharge limit. Table 26.9 presents a problem involving four effluent streams. The environmental discharge limit is 20 ppm.

Note from Table 26.9, that Stream 4 with a concentration of 10 ppm is already below the environmental limit of 20 ppm. Suppose that this is to be treated with an operation capable of producing 5 ppm outlet concentration. How can the target minimum flowrate be set under such circumstances?

In fact, the availability of the stream with a concentration below the environmental limit removes part of the environmental load to be taken up by the effluent treatment system through its dilution effect. The construction of the effluent composite to include this effect is shown in Figure 26.59. It can be seen that as the stream starting at 10 ppm increases in concentration to the specified environmental limit of 20 ppm, it takes up part of the effluent load from the other streams that would otherwise have to be removed by the effluent treatment process. The dilution effect of the stream below the environmental limit effectively creates a new environmental limit greater than 20 ppm. The target for the problem is shown in Figure 26.60. In this case, the specification for the treatment process was based on an outlet concentration. If the specification had been based on the removal ratio, then the basic approach would be the same. The only change would be the specification of the effluent treatment line.

Sometimes the effluent treatment cannot be accomplished in a single effluent treatment process. For example, different treatment processes might be required to treat contaminants over different

Table 26.9

A problem where one of the effluents is below the environmental discharge limit.

Stream	C_{in} (ppm)	Water flowrate (t·h^{-1})
1	200	40
2	100	30
3	30	20
4	10	80

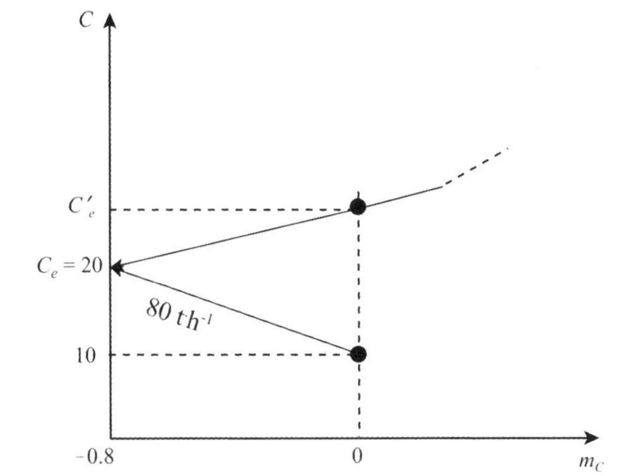

Figure 26.59

A stream below the environmental limit defines a new effective environmental limit.

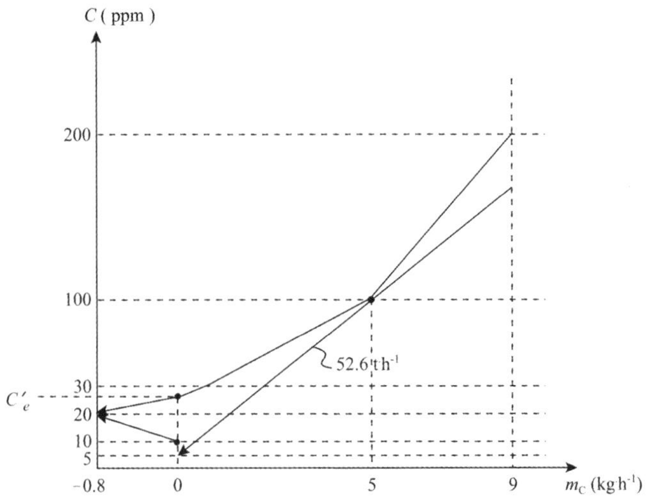

Figure 26.60

Target for the problem in Table 26.9.

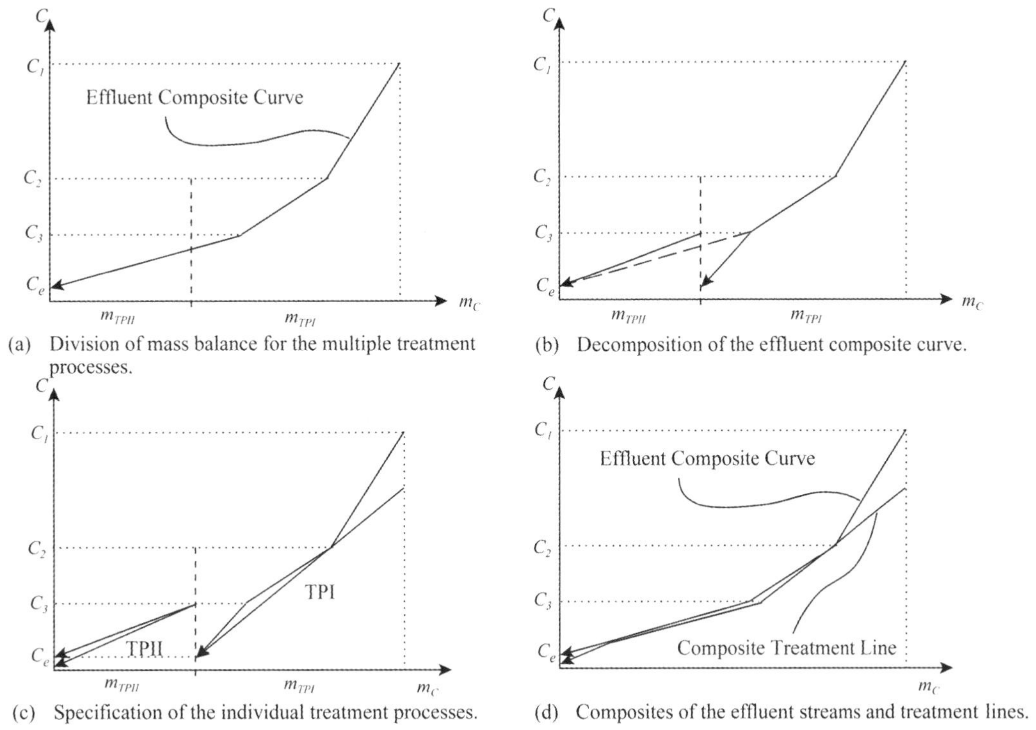

Figure 26.61

Targeting for multiple treatment processes.

concentration ranges. It might also be cheaper to use multiple treatment processes, rather than a single process. For example, when treating petroleum refinery effluent, it is usually cheap to start the treatment with an API Separator and then complete the treatment with other treatment processes, such as dissolved-air flotation (see Chapter 7) and biological treatment. The API Separator is a cheap process to remove part of the contamination before passing the effluent to more expensive treatment processes. Thus, limitations in the performance or the costs associated with effluent treatment processes might dictate the requirement for multiple treatment processes.

The graphical approach to targeting the minimum treatment flowrate can be extended to multiple treatment processes, as shown in Figure 26.61 (Kuo and Smith, 1998b). In Figure 26.61a, the effluent composite curve is divided into parts of the effluent mass load being treated by the different processes (m_{TPI} and m_{TPII} in Figure 26.61a). The effluent composite curve is then decomposed into two parts at this point of division in the mass load, as shown in Figure 26.61b. Decomposition is the opposite procedure from creating a composite. In Figure 26.61b, the mass balance division is such that the effluent composite curve between C_1 and C_e needs to be decomposed. In this way, two separate effluent composite curves are created for treatment by Treatment Process I (TPI) and Treatment Process II (TPII). Treatment lines for TPI and TPII are then matched against the respective decomposed effluent composite curves, as shown in Figure 26.61c. The matching of the treatment lines against the decomposed effluent composite curve is now the same as that for single treatment processes, as described previously. Specifications can be based on outlet concentration or removal ratio. Figure 26.61d shows that decomposed effluent composite curves and the treatment lines combined to obtain composites for each. It should be noted that the treatment line for TPII in Figure 26.61c would actually cross the original effluent composite curve and would appear to be infeasible. However, a composite of the two treatment lines does not cross the original effluent composite curve. The combination of the two treatment lines leads to a feasible solution, as seen in Figure 26.61d (Kuo and Smith, 1998b).

Finally, the relative contaminant mass load on the two treatment processes can usually be varied within reasonable limits. As load is shifted from TPI to TPII, and vice versa, their relative capital and operating costs will change. This is a degree of freedom that can be optimized.

26.13 Design for Minimum Wastewater Treatment Flowrate for Single Contaminants

Having established how to set the target for the minimum treatment flowrate, the next question is how to design to achieve the target. The design principles to achieve the target are based

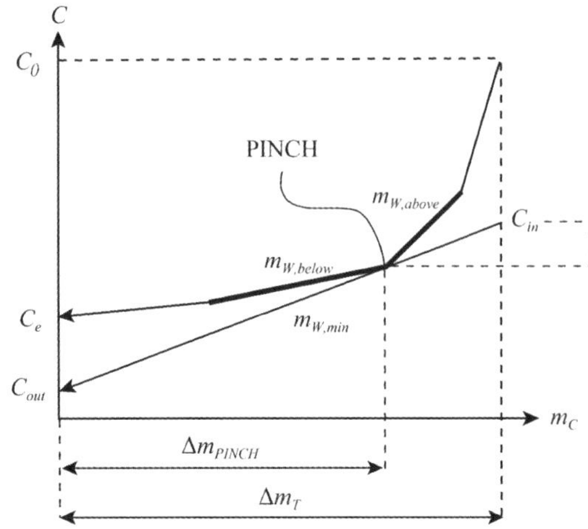

Figure 26.62

Achieving the treatment flowrate target. (Reproduced from Wang YP and Smith R (1994b) Design of Distributed Effluent Treatment Systems, *Chem Eng Sci*, **49**: 3127, with permission from Elsevier.)

on the starting concentrations of the streams relative to the pinch concentration:

$$\textbf{Group I.} \quad C_{I,j} > C_{PINCH} \tag{26.22}$$

$$\textbf{Group II.} \quad C_{II,j} = C_{PINCH} \tag{26.23}$$

$$\textbf{Group III.} \quad C_{III,j} < C_{PINCH} \tag{26.24}$$

where $C_{I,j}$, $C_{II,j}$, $C_{III,j}$ = concentration of Stream j in Groups I, II and III respectively.

If $m_{W,above}$ and $m_{W,below}$ are defined to be the total wastewater flowrates in the concentration intervals immediately above and below the pinch concentration respectively, as shown in Figure 26.62, then:

$$m_{W,above} = \sum_j^{S_I} m_{WI,j} \tag{26.25}$$

and

$$m_{W,below} = \sum_j^{S_I} m_{WI,j} + \sum_j^{S_I} m_{WII,j} \tag{26.26}$$

where $m_{W,above}$, $m_{W,below}$ = water flowrate immediately above and below the pinch

$m_{WI,j}$, $m_{WII,j}$ = water flowrate for Stream j in Group I and II respectively

S_I = total number of streams in Group I

S_{II} = total number of streams in Group II

Because the treatment line with a flowrate m_{WT} is pinched with the composite curve at this point, the following relation holds:

$$m_{W,above} \le m_{Wmin} \le m_{W,below} \tag{26.27}$$

which means that:

$$\sum_j^{S_I} m_{WI,j} \le m_{Wmin} \le \sum_j^{S_I} m_{WI,j} + \sum_j^{S_{II}} m_{WII,j} \tag{26.28}$$

Therefore, the minimum treatment flowrate target m_{Wmin} is given by:

$$m_{Wmin} = \sum_j^{S_I} m_{WI,j} + \theta \sum_j^{S_{II}} m_{WII,j} \tag{26.29}$$

where $0 \le \theta \le 1$

Equation 26.29 reveals that the treatment flowrate target can be achieved when all the wastewater streams in Group I are treated, together with some of the wastewater streams from Group II and none from Group III.

Having identified the streams to be treated in order to achieve the flowrate target, it is still necessary to prove that the mean inlet concentration of those streams, $C_{mean,in}$, is equal to the targeted inlet concentration for the treatment process, C_{in}. If the treatment flowrate target m_{Wmin} consists of the wastewater streams defined by Equation 26.29, then the mean inlet concentration of the streams treated, $C_{mean,in}$, is given by:

$$C_{mean,in} = \frac{\sum_j^{S_I} m_{WI,j} C_{I,j} + \theta \sum_j^{S_{II}} m_{WII,j} C_{PINCH}}{\sum_j^{S_I} m_{WI,j} + \theta \sum_j^{S_{II}} m_{WII,j}} \tag{26.30}$$

From Figure 26.62, the targeted inlet concentration, C_{in}, for the treatment flowrate target is given by:

$$C_{in} = C_{PINCH} + \Delta C = C_{PINCH} + \frac{\Delta m_T - \Delta m_{PINCH}}{m_{Wmin}} \tag{26.31}$$

where ΔC is the concentration change through the treatment process above the pinch. Then knowing that:

$$\Delta m_T - \Delta m_{PINCH} = \sum_j^{S_I} m_{WI,j}(C_{I,j} - C_{PINCH}) \tag{26.32}$$

where Δm_T = total mass treatment load

Δm_{PINCH} = mass treatment load below the pinch

Equations 26.32 and 26.29 can be substituted into Equation 26.31 to give the targeted inlet concentration, C_{in}:

$$\begin{aligned} C_{in} &= C_{PINCH} + \frac{\sum_j^{S_I} m_{WI,j}(C_{I,j} - C_{PINCH})}{\sum_j^{S_I} m_{WI,j} + \theta \sum_j^{S_{II}} m_{WII,j}} \\ &= \frac{\sum_j^{S_I} m_{WI,j} C_{I,j} - \sum_j^{S_I} m_{WI,j} C_{PINCH} + \sum_j^{S_I} m_{WI,j} C_{PINCH} + \theta \sum_j^{S_{II}} m_{WII,j} C_{PINCH}}{\sum_j^{S_I} m_{WI,j} + \theta \sum_j^{S_{II}} m_{WII,j}} \\ &= \frac{\sum_j^{S_I} m_{WI,j} C_{I,j} + \theta \sum_j^{S_{II}} m_{WII,j} C_{PINCH}}{\sum_j^{S_I} m_{WI,j} + \theta \sum_j^{S_{II}} m_{WII,j}} \\ &= C_{mean,in} \end{aligned} \tag{26.33}$$

Equation 26.33 proves that the mean inlet concentration of all the streams treated is equal to the targeted inlet concentration.

Summarizing the above discussion gives the following design rules, which must be observed to achieve the treatment flowrate target.

Design Rule I. If wastewater streams start above the pinch concentration, then all of the flowrate of all of these streams must be treated.

Design Rule II. If wastewater streams start at the pinch concentration, then these streams are partially treated and partially bypass the effluent treatment.

Design Rule III. If wastewater streams start below the pinch concentration, then all of these streams bypass the treatment process completely.

Design Rule II is dictated by the mass balance. The amount of the wastewater streams from Group II to be treated can be calculated from:

$$m_{WII} = \frac{\Delta m_T - \Delta m_{CI}}{C_{PINCH} - C_{out}} \qquad (26.34)$$

where

$$\Delta m_{CI} = \sum_{j}^{S_I} m_{WI,j}(C_{I,j} - C_e) \qquad (26.35)$$

These three basic rules create a segregation policy to ensure that the target is achieved in design.

If the effluent treatment involves multiple treatment processes, as shown in Figure 26.61, then the same basic approach can be followed as for single treatment processes. The problem is decomposed, as shown in Figure 26.61. This provides the target and the basis of the design. Each treatment process has its own pinch and the network design for each can be developed separately. The network designs are then simply connected together (Kuo and Smith, 1998b).

Example 26.9 Consider again Example 26.8. The target for the three streams in Table 26.8 was determined to be $60\,t{\cdot}h^{-1}$. This corresponded both with a treatment process achieving 10 ppm at its outlet and also one with a removal ratio of 95%. Design an effluent treatment network to achieve the target of $60\,t{\cdot}h^{-1}$.

Solution To achieve the target in practice requires that the starting concentrations relative to the pinch be first identified. The pinch is at 100 ppm and Stream 1 starts above the pinch. Stream 2 starts at the pinch and Stream 3 starts below the pinch. Thus, Stream 1 must be completely treated, Stream 2 partially treated and partially bypassed and Stream 3 totally bypasses the effluent treatment. The question remains as to how much of Stream 2 goes through the effluent treatment and how much bypasses. If Stream 1 with a flowrate of $40\,t{\cdot}h^{-1}$ must go through the effluent treatment and the target for effluent treatment is $60\,t{\cdot}h^{-1}$, then $20\,t{\cdot}h^{-1}$ from Stream 2 must go through the effluent treatment process. Figure 26.63 shows the design structure for the effluent treatment, together with the flowrates.

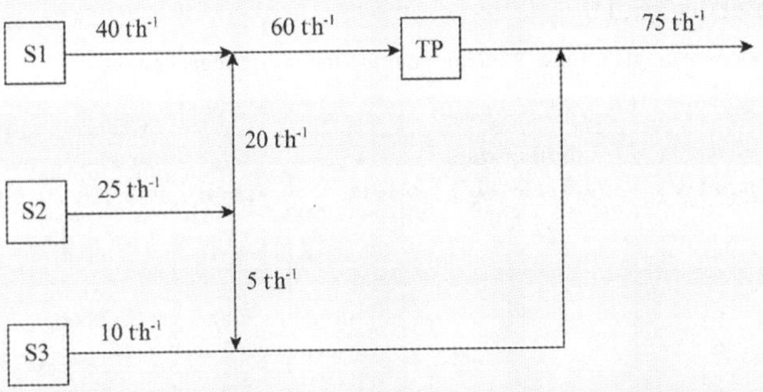

Figure 26.63

Design structure for Example 26.9.

The design to achieve the target is therefore a matter of following three simple rules. It is worth highlighting a few points though. First, streams starting at the pinch must be partially treated and partially bypassed in order to achieve the target. This splitting of the stream, while it is necessary to solve the problem as posed, is often not acceptable in practice. It provides design complexity in the need to split the stream. Also, regulatory authorities often find such an arrangement to be unacceptable, even though it solves the problem according to their regulations. Environmental regulations often have requirements for criteria such as *best practice* or *best environmental option*. It is therefore often considered to be not the best environmental practice to partially treat a stream. If part of the stream needs to be treated, then why not the whole stream? If the partial bypass is eliminated and all of the flow for the stream starting at the pinch is fed through the treatment process, then the performance of the system will exceed that of the regulatory requirements. While this might seem to be a drawback in small systems, in practice it is not such a problem. In practice, rather than dealing with three streams, it is likely that the number of streams from a site might be of the order of 100 or more. In this situation, changing perhaps one stream that is at the pinch to fully go through the treatment process, rather than partially bypass, will have a very small effect on the overall system performance (Smith, Petela and Howells, 1996).

Another point regarding the design process is the effect of the streams that are already below the environmental limit. In practice, for a large site, many of the streams will already be below the environmental limit. These are dealt with as shown in Figure 26.60. In design terms, these should not be treated, but bypassed along with all other streams below the pinch concentration (Smith, Petela and Howells, 1996).

26.14 Regeneration of Wastewater

The basic principle of regeneration was introduced in Figure 26.2. Regeneration is a treatment process but is applied with the

Table 26.10

Water-using data for a problem requiring the use of regeneration.

Stream number	Limiting water flowrate ($t \cdot h^{-1}$)	C_{in} (ppm)	C_{out} (ppm)	Contaminant mass flowrate ($g \cdot h^{-1}$)
1	20	0	200	4000
2	50	100	200	5000
3	30	100	400	9000

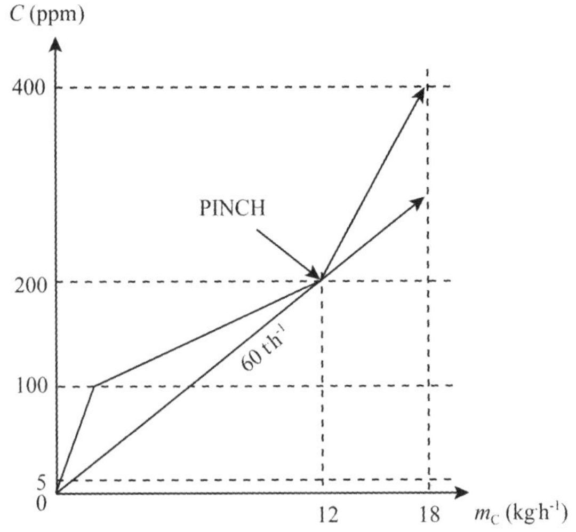

Figure 26.64

Target for the data in Table 26.10 for reuse only.

objective of reusing or recycling the water, rather than discharge to the environment. Regeneration of wastewater is most likely to be economic if:

- it allows process materials to be recovered;
- it substitutes investment in treatment for discharge by taking up part of the effluent load, as well as reducing the overall demand for water;
- limits are being imposed on fresh water consumption.

Consider the data in Table 26.10 for three water-using operations. It is desired to minimize the consumption of fresh water and generation of wastewater. A regeneration process is available with a performance that can achieve an outlet concentration of 5 ppm.

Start by considering the performance of the system for reuse only. Figure 26.64 shows that the target for reuse only turns out to be $60\ t \cdot h^{-1}$. For targeting regeneration reuse, Figure 26.65a shows a simplistic approach to targeting (Wang and Smith, 1995a). Fresh water is taken into the process, in this case at 0 ppm, and in Figure 26.65a used to a concentration of 200 ppm. Water with this concentration then enters a regeneration process that decreases its concentration down to 5 ppm. If it is assumed that the flowrate of water from the regeneration process is the same as fed to the regeneration process, then the water can be used after regeneration to complete the process. Because the flowrates of water before and after regeneration are the same, the slopes of the fresh water and regenerated water lines are the same. The target is obtained by moving the fresh water and regenerated water lines in parallel to achieve the steepest slope against the limiting composite curve. In this case, it corresponds with a flowrate of fresh water and regenerated water of $30.4\ t \cdot h^{-1}$. It should be noted in Figure 26.65a that the profile for the fresh water line crosses the limiting composite curve. However, this is still feasible as the

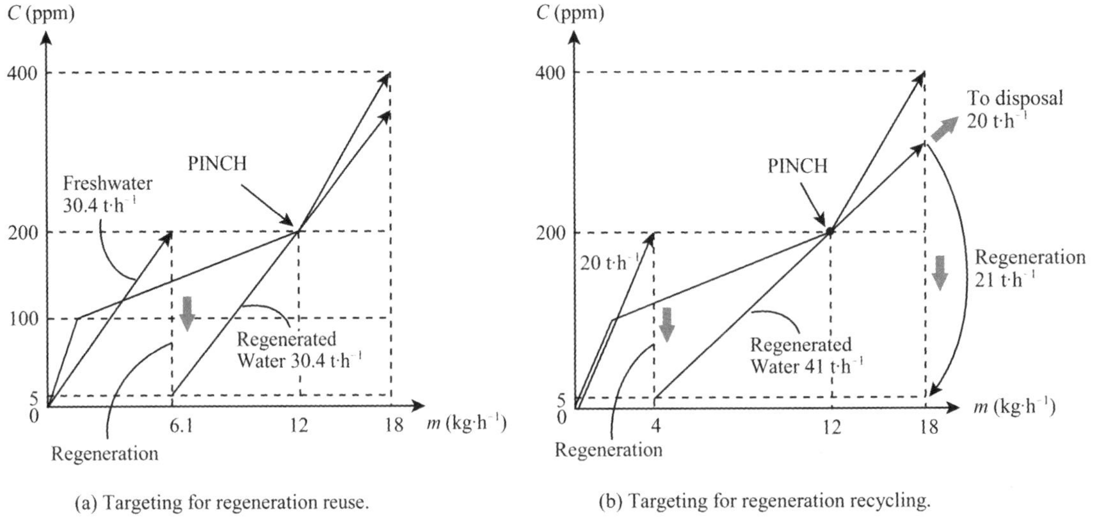

Figure 26.65

A simple approach to target fot regeration reuse and regeneration recycling.

composite of the fresh water and regenerated water lines does not cross the limiting composite curve. In Figure 26.65a the fresh water has been used up to 200 ppm, which is the pinch concentration, before being fed to the regeneration process. It can be shown that for many problems regeneration at the pinch concentration minimizes the consumption of fresh water (Wang and Smith, 1995a). However, feeding the regeneration at pinch concentration does not always lead to the minimum flowrate of fresh water for regeneration reuse (Bai, Feng and Deng, 2007).

If regeneration recycling is to be used, then a simplistic targeting is illustrated in Figure 26.65b (Wang and Smith, 1995a). In this case, if recycling is to be allowed, then the minimum flowrate of fresh water is fed to the system, dictated by the slope of the limiting composite curve at low concentrations. At 200 ppm this enters the regeneration process where it is reduced in concentration to 5 ppm. To match against the limiting composite curve, the regenerated water line needs to be increased in flowrate in order to match against the limiting composite curve, 20 $t \cdot h^{-1}$ for this example. The difference in flowrates between the fresh water line and the regenerated water line are reconciled by recycle of some of the water after regeneration, as illustrated in Figure 26.65b, 21 $t \cdot h^{-1}$ in this example. Again, it has been chosen to feed the regeneration process at the pinch concentration of 200 ppm in order to achieve the minimum flowrate. However, feeding the regeneration at pinch concentration does not always lead to the minimum flowrate of fresh water for regeneration recycling (Feng, Bai and Zheng, 2007).

The simplistic approach to targeting for regeneration reuse and regeneration cycling in Figure 26.65 has two fundamental drawbacks. First, the shape of the limiting composite curve dictates the optimum concentration at which to feed the regeneration process. In many cases, this is the pinch concentration, but it might be higher or lower than pinch concentration (Bai, Feng and Deng, 2007; Feng, Bai and Zheng, 2007). Second, any approach based on a construction like the one in Figure 26.65 will most often require operations to be split with one part being fed by fresh water and another by regenerated water in the design (Wang and Smith, 1995a). Such an operation splitting is most often impractical.

A better approach is to divide the operations into two groups. One group consists of the operations that must be fed by fresh water. The second group consists of operations that can be fed by regenerated water (Kuo and Smith, 1997). The grouping strategy depends on whether it is desirable to include or avoid recycling. Take first the case of regeneration reuse, in which no recycling is allowed.

Figure 26.66 gives a simple algorithm to provide an initial grouping for problems exploiting regeneration reuse in which recycling is to be avoided (Kuo and Smith, 1997). The initial grouping is such that the streams are first divided into those that require a lower concentration than the regeneration can achieve, and those that can accept a concentration greater than regeneration can achieve. In Figure 26.66, some of the operations that can accept a concentration greater than regeneration can achieve are also put to fresh water, such that the balance of the mass loads between fresh water and regeneration water below the reuse pinch in Figure 26.66 is as evenly distributed as possible. In this example, from the data given in Table 26.10, Operation 1 must be fed with fresh water. It is then a matter of deciding whether Operations 2 and 3 are fed by fresh water or regenerated water. The mass loads below the reuse pinch are: 26

Stream 1 4000 $g \cdot h^{-1}$
Stream 2 5000 $g \cdot h^{-1}$
Stream 3 3000 $g \cdot h^{-1}$

Given that Operation 1 belongs to the fresh water group, then an arrangement with Operations 1 and 3 in the fresh water group and Operation 2 in the regenerated water group balances the mass loads below the reuse pinch as evenly as possible.

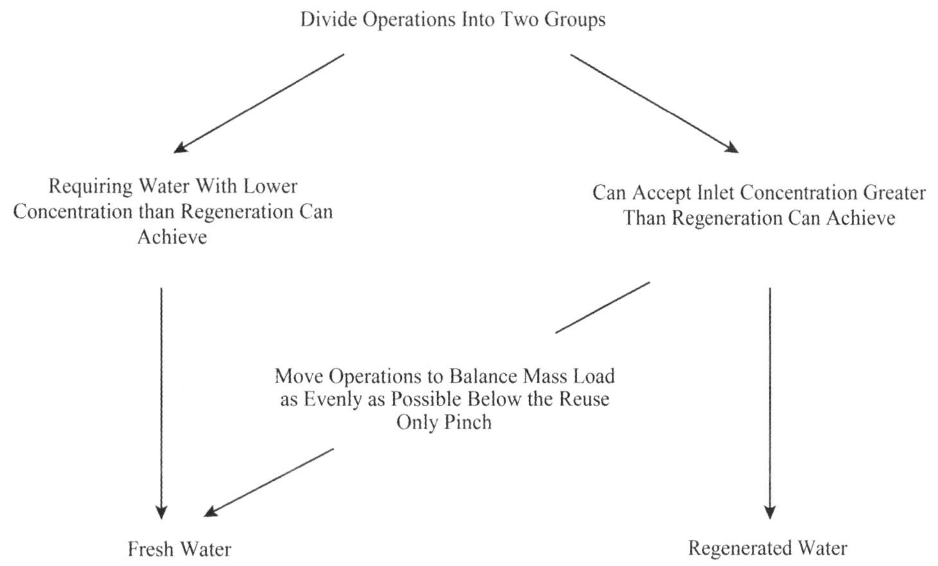
Divide Operations Into Two Groups
Requiring Water With Lower Concentration than Regeneration Can Achieve
Can Accept Inlet Concentration Greater Than Regeneration Can Achieve
Move Operations to Balance Mass Load as Evenly as Possible Below the Reuse Only Pinch
Fresh Water
Regenerated Water

Figure 26.66
Algorithm to provide an initial grouping for regeneration reuse.

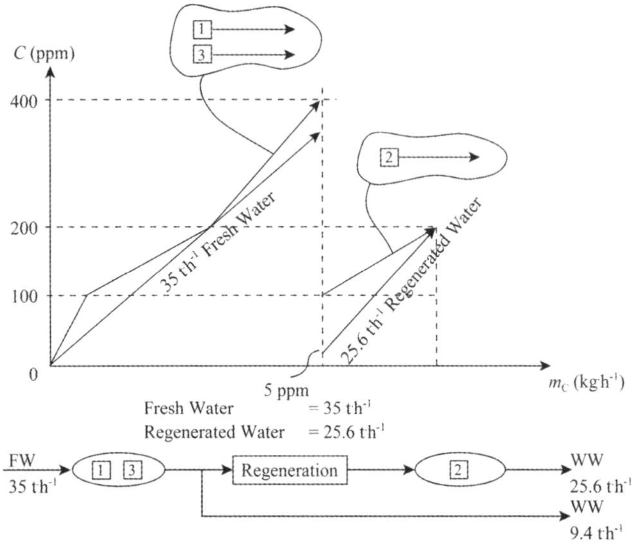

Figure 26.67

The targets for the data from Table 26.10 for regeneration reuse exploiting a regeneration process with an outlet concentration of 5 ppm.

This arrangement is shown in Figure 26.67. The streams being fed by fresh water (Operations 1 and 3) are used to construct a limiting composite curve. For these two streams the target fresh water is $35\,t\cdot h^{-1}$. Only Operation 2 is included in the group using regenerated water and the target for regenerated water starting at 5 ppm is $26.6\,t\cdot h^{-1}$. Given that the target for fresh water of $35\,t\cdot h^{-1}$ is greater than the target for regenerated water of $26.6\,t\cdot h^{-1}$, some of the water from the exit of Operations 1 and 3 goes directly to effluent, the remainder being regenerated and used in Operation 2. While Figure 26.67 provides an initial grouping, there is sometimes scope to improve the targets through migration of streams from the fresh water group into the regenerated water group and vice versa. This scope to improve arises from the fact that there are now two pinches for the system: the pinch for fresh water and the pinch for regenerated water. In some problems, exploiting the interactions between the pinches allows the targets to be improved. There are three basic approaches to refine the grouping of streams by such migration:

1) A conceptual approach based on the fresh water and regeneration pinches (Kuo and Smith, 1997)
2) A combinatorial search
3) Mixed integer programming (Gunarantnam *et al.*, 2005).

In fact, in this simple example there is no scope to improve the targets by migration and the final target is that given in Figure 26.67. Design to achieve the target uses exactly the same water network design procedures as described so far but carried out separately on the two groups of streams. The two designs are then connected together via a regeneration process to comply with the outline flow diagram shown in Figure 26.67. Figure 26.68 shows a design for regeneration reuse. Compared with a fresh water target for reuse only of $60\,t\cdot h^{-1}$, regeneration reuse allows the fresh water to be reduced to $35\,t\cdot h^{-1}$.

As with regeneration reuse, the objective of regeneration recycling is to divide the water-using operations into two groups, those requiring fresh water and those requiring regenerated water (Kuo and Smith, 1997). Figure 26.69 presents an algorithm to provide initial grouping for regeneration recycling (Kuo and Smith, 1997). This time the algorithm is as simple as dividing the streams into those that require water with a lower concentration than regeneration can achieve and those that can accept a concentration greater than regeneration can achieve. The targets based on this initial grouping are shown in Figure 26.70. Again, there is sometimes scope to improve the initial targets by migration. This again arises from the fact that there are two pinches in the problem, and the interactions between the two pinches can in some cases be exploited. In contrast to regeneration reuse, however, the migration will only move streams from regenerated water on to fresh water in order to reduce regenerated water costs (Kuo and Smith, 1997). This migration again can be based on a conceptual approach (Kuo and Smith, 1997), a combinatorial search or mixed integer programming (Gunarantnam *et al.*, 2005). Once the final targets have been accepted, design is a matter of designing the network for the two groups separately and then joining the two designs together via the regeneration process and the recycle. The final design to achieve the target is given in Figure 26.71. The target for fresh water consumption is now reduced to $20\,t\cdot h^{-1}$, in contrast with $35\,t\cdot h^{-1}$ for regeneration reuse and $60\,t\cdot h^{-1}$ for reuse only.

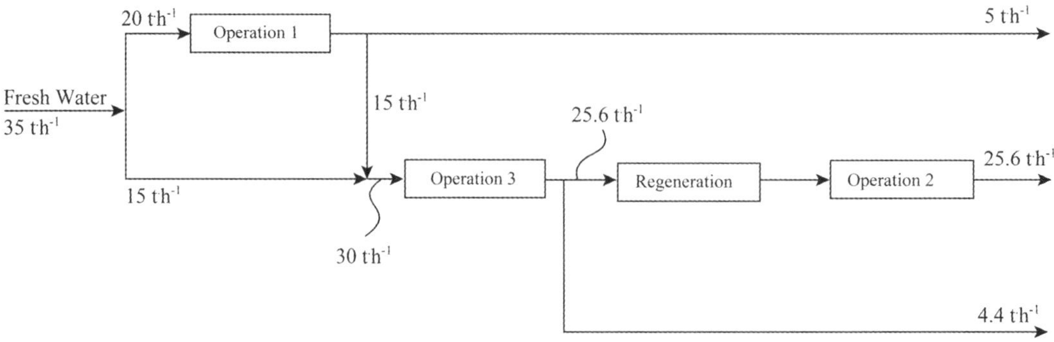

Figure 26.68

Design for regeneration reuse for the problem in Table 26.10.

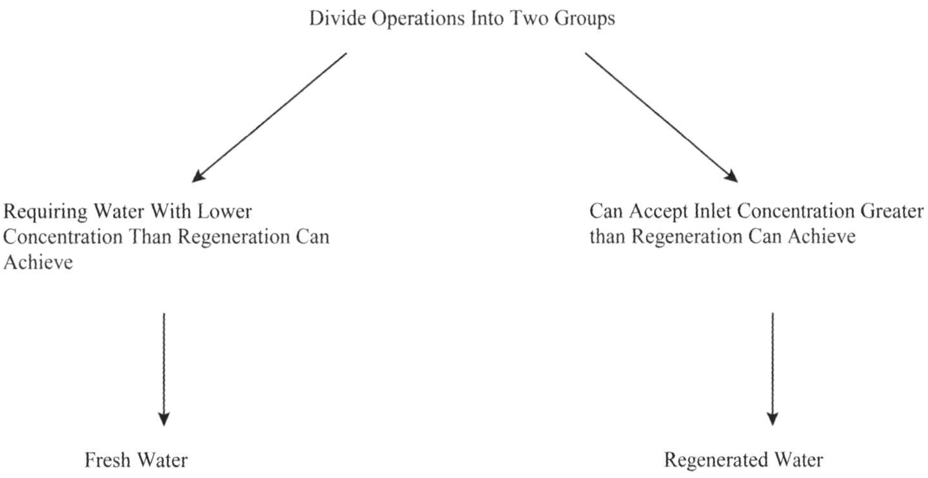

Figure 26.69
Algorithm to provide initial grouping for generation recycling.

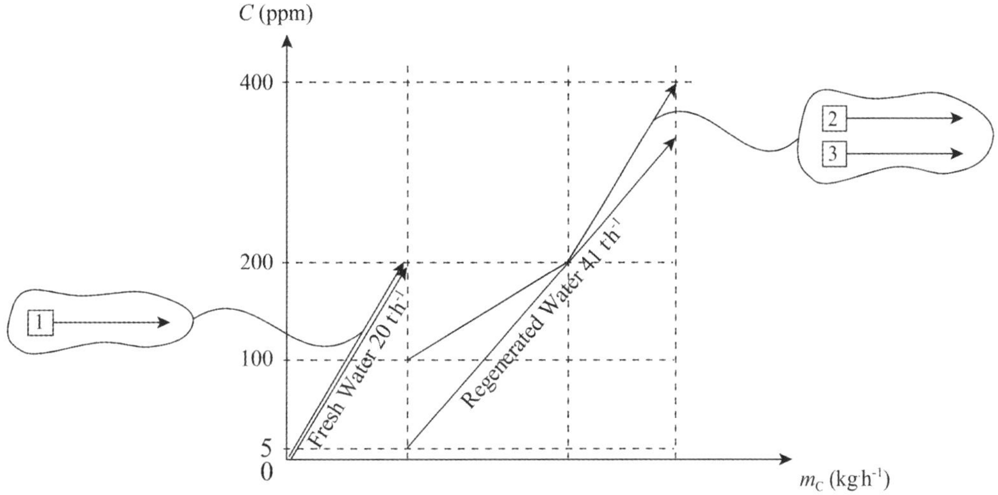

Figure 26.70
The targets for the data from Table 26.10 for regeration recycling exploiting a regeneration process with an outlet concentration of 5 ppm.

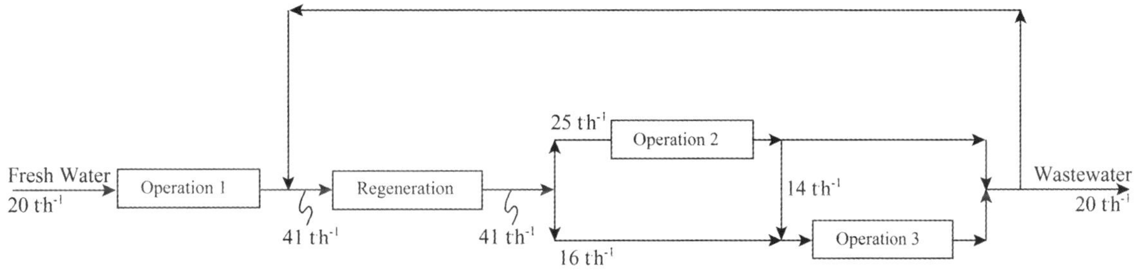

Figure 26.71
Design for regeneration recycling for the problem in Table 26.10.

26.15 Targeting and Design for Effluent Treatment and Regeneration Based on Optimization of a Superstructure

As with the case for water minimization, the graphical methods used for effluent treatment and regeneration have some severe limitations. As before, multiple contaminants, constraints, piping and sewer costs, multiple treatment processes and retrofit are all difficult to handle. To include all of these complications requires an approach based on the optimization of the superstructure.

Figure 26.72 presents a superstructure for the design of an effluent treatment system involving three effluent streams and three treatment processes (Kuo and Smith, 1998b). The superstructure allows for all possibilities. Any stream can go to any effluent process and potential bypassing options have been included. Also, the connections toward the bottom of the superstructure allow for the sequence of the treatment processes to be changed. To optimize such a superstructure requires a mathematical model to be developed for the various material balances for the system and costing correlations included. Such a model then allows the superstructure to be optimized and a final design obtained, as shown in Figure 26.72. The advantages of the superstructure approach are that targeting for large complex problems is facilitated, the design process is automated, it provides a design as well as a target, constraints can be readily included, piping and sewer costs can be included and the cost trade-offs explored (Galen and Grossmann, 1998; Gunarantnam *et al.*, 2005). The major disadvantage of the approach is that the optimization is more difficult than the corresponding optimization for water reuse. The general problem is nonlinear and the best approach is to provide an initial solution for the nonlinear optimization via a simplified linear model (Galen and Grossmann, 1998; Gunarantnam *et al.*, 2005).

A disadvantage of the automated approach based on the optimization of a superstructure is that conceptual insights are lost. However, as with the case of optimization of water reuse networks, graphical representations based on concentration versus mass load can be constructed from the output of the optimization. The actual concentrations from the design are used to plot the effluent composite curve and the composite of the effluent treatment processes. This can be done on a single component basis, as illustrated in Figure 26.73. Figure 26.73 shows the output from an optimization involving the treatment of two contaminants. The composite treatment line involves multiple treatment processes and pinches with the effluent composite curve for Contaminant *A*. Also shown in Figure 26.73 is the corresponding representation for Contaminant *B*. The effluent composite curve is different. Also, the composite treatment line shows some interesting features. First, it is not pinched. Second, the composite treatment line extends beyond the effluent composite curve. This indicates that Contaminant *B* has been overtreated. This presumably has arisen as a result of a necessity to fulfill the discharge requirements for Contaminant *A*. Contaminant *A* and *B* are treated simultaneously in the various treatment processes and Contaminant *A* was obviously limiting in this particular problem.

A further option for treatment processes is that, rather than used for effluent discharge, they can be used for regeneration of wastewater for reuse or recycling. In principle, any treatment process can

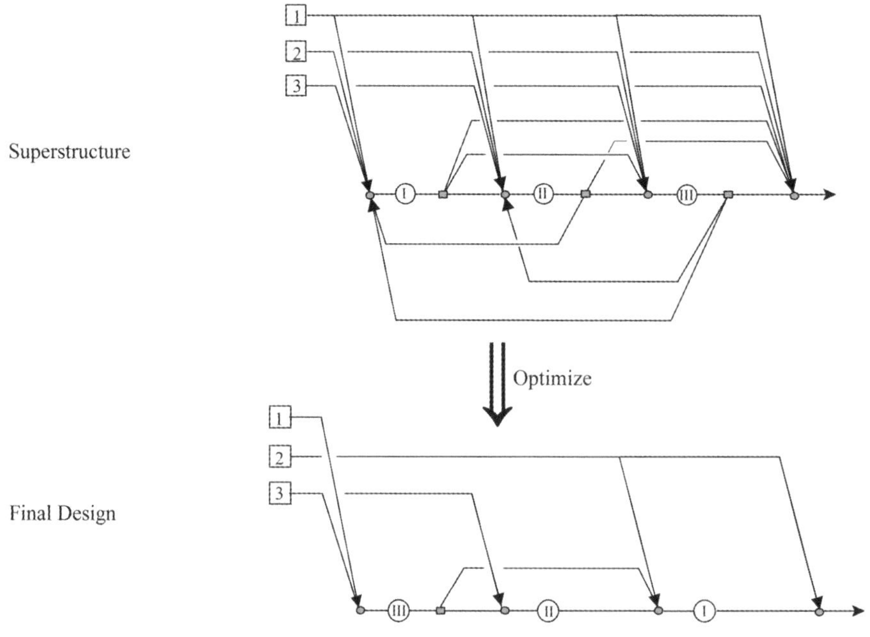

Figure 26.72
Superstructure for the design of an effluent treatment process with three streams and three treatment processes.

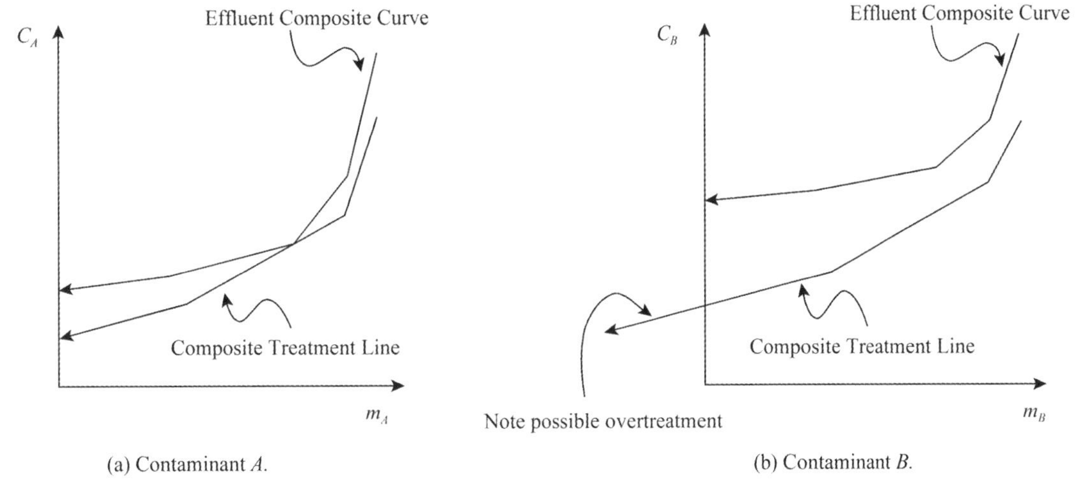

Figure 26.73
Plots of composition versus mass load can be developed from the output of an optimization calculation.

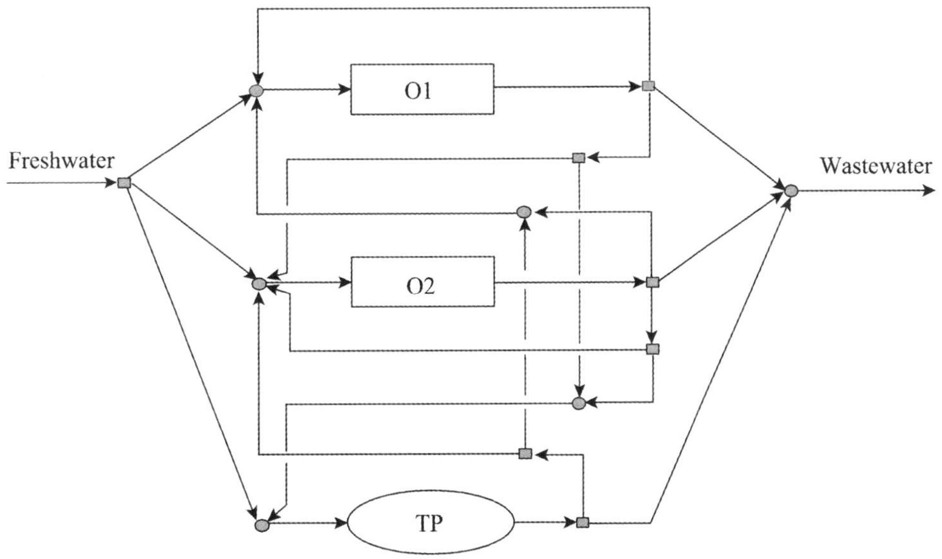

Figure 26.74
A superstructure allowing for a treatment process to be used for either regeneration or discharge.

be used for effluent discharge or wastewater regeneration. Figure 26.74 shows a superstructure for two operations and a single treatment process (Gunarantnam *et al.*, 2005). The superstructure allows for all of the basic reuse options between Operations 1 and 2, together with the use of the treatment process for regeneration reuse or regeneration recycling. Alternatively, the same treatment process can be used for treatment and discharge of effluent. In this way, the analyses for reuse only, regeneration reuse, regeneration recycling and effluent treatment for discharge can all be examined simultaneously by the optimization of the superstructure in Figure 26.74.

26.16 Data Extraction

When considering heat exchanger network design, the important issue of data extraction was highlighted, as lost opportunities can result from poor data extraction in heat exchanger network design. Similarly, there are fundamental issues associated with data extraction for the design of water systems.

1) *Water balance.* Before an existing system can be studied from the point of view of water consumption and effluent treatment, the first step must be to establish a water balance. For a new

design, this is a question simply of extracting the information relating to water streams from the flowsheets and material and energy balance for the processes on a site and the associated utility systems. However, it is rarely the case that the designer is faced with a completely new site. The most common situation is retrofit of an existing site, with all of the existing units in place, or the addition of new units to an existing site, or shutdown of units that modifies the basic site configuration. The objective could be to meet changes in the site load through addition or shutdown of units, improve the existing environmental and economic performance or to meet changes in environmental regulations. In retrofit situations, it is often very difficult to produce a water balance, yet it is a necessity. General lack of information, poor records and poor instrumentation mean that it is usually not possible to close a water balance to better than 90% accountability. This certainly applies to overall sites and will often apply to individual processes. A water balance that accounts for at least 90% of the water in a process or on a site is a necessary starting point.

2) *Contaminants.* The next issue to address is which contaminants should be included in any analysis? The answer is, as few contaminants as absolutely necessary. Contaminants should be lumped together where possible, for example, total organic material, salt, suspended solids, total solids, and so on. In some instances, it will be necessary to include individual components (e.g. phenol), if a particular constraint relating specifically to that component must be included. Care should be taken not to include contaminants into an analysis simply because analytical data are available from process outlets. For water reuse analysis, contaminants should only be included where there is an inlet restriction. A good strategy in developing an analysis is to start by assuming that the whole problem can be analyzed with a single contaminant, for example, dissolved solids. Additional contaminants should then only be added where necessary. It should be noted that the methods for studying water systems described here are not attempting to simulate the water system. All that is of concern at this stage in the design of water-using systems is whether the contaminant specifications reflect the constraints that allow or prevent reuse or recycling of water. For effluent treatment systems, the study can often be performed by putting in the most important contaminants related to discharge regulations. Of course, when the design is completed, it must be checked to make sure that all environmental regulations are complied with.

3) *Flowrate constraints.* Some processes require fixed flowrates, for example, cooling tower makeup, hosing operations, steam ejectors, and so on. In extracting the data, the designer needs to identify such flowrate constraints. If the objective of the study is to make a major impact in the consumption of fresh water and wastewater generation, then it makes sense not to include small streams in the analysis. Reuse or recycling of small streams is unlikely in most cases to be economic. Only the largest streams should be studied.

4) *Data accuracy.* Water systems are usually complex and there is usually a lack of data for a study. It is important to avoid collecting a lot of unnecessary data early in a study. The most accurate data requirements will be around the area of the pinch and below the pinch for the system. By definition, this is the data that is limiting. Therefore, a good approach is to collect approximate data first and to carry out a preliminary analysis. An initial target and analysis of the data will then reveal where the design is most sensitive to data errors. Collecting better data in these areas can be carried out to refine the target. This approach should avoid collecting too much unnecessary data.

5) *Limiting conditions.* When determining the opportunities for reuse, regeneration reuse and regeneration recycling, the starting point is the limiting water data that determine the most contaminated water acceptable to an operation. But how are limiting conditions determined? There is no straightforward answer to this and many factors can influence the setting of the limiting conditions:
   - judgment and experience,
   - corrosion limitations,
   - maximum solubility,
   - simulation studies,
   - laboratory studies,
   - plant trials,
   - sensitivity analysis.

   Sensitivity analysis is a particularly useful tool when determining the limiting concentrations. Starting with initial limiting conditions, that might well be the existing concentrations, an initial target can be set as illustrated in Figure 26.75. The data can then be adjusted to increase the inlet concentrations and retargeted. It might be that the result is sensitive or insensitive to changes in the inlet concentration. Operation 2 in Figure 26.75 shows that a small increase in the inlet concentration initially brings a large reduction in the water requirement. Further increases in the inlet concentration gradually bring diminishing returns. The point where flowrate becomes insensitive to changes in the inlet concentrations should then be examined more closely for practicality using simulation, laboratory studies, and so on. By contrast, large increases in the inlet concentration to Operation 1 only bring a modest reduction in the overall flowrate. Carrying out such a sensitivity analysis allows the critical design variables to be identified in order that they can be studied more closely.

6) *Treatment data.* When carrying out a study of an effluent treatment system, as with a water reuse study, as few contaminants as possible should be included. Again, contaminants should be lumped together where possible, for example, suspended solids, COD, BOD_5. When studying the performance of an existing biological treatment process, great care should be exercised if a mixed effluent from different processes is being treated. For example, suppose a site has an existing biological treatment unit that has an existing performance that removes 90% of the COD. Suppose now that this performance is no longer acceptable and an increase in the performance is required, either on COD removal or individually nominated contaminants. Additional capacity could be added to the biological treatment unit, distributed-treatment applied upstream, or waste minimization applied at source (see

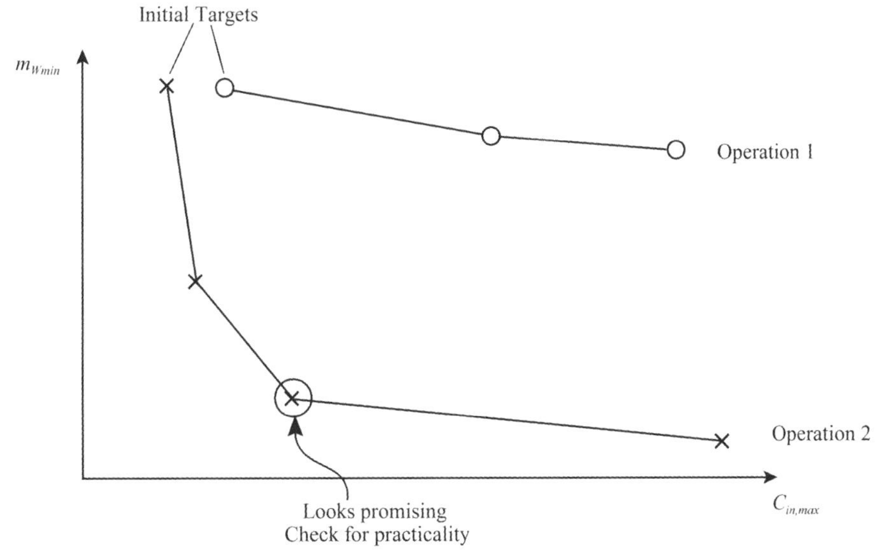

Figure 26.75
Sensitivity analysis to determine the limiting conditions.

Chapter 27). It might be extremely deceptive simply to assume that the existing performance of the biological treatment can be improved by adding extra capacity. It might be that when the performance of the treatment on the individual effluents from the different processes is examined more closely, the effect is quite different on the different effluent streams. It might be that some effluent streams, because the contaminants are easily digested, are treated almost completely in the biological treatment unit. Other effluents from other processes might be largely unchanged by the biological treatment, because the materials are refractory. In other words, some effluent streams are overtreated, others are undertreated and the overall effect is to give a removal ratio for COD of 90%. Such situations can occur on chemical processing sites involving a diverse range of chemical processes. To study an improvement in the system for such circumstances would require the effluent to be broken down into different categories of COD, depending on the origin, rather than just an overall COD. Each process would be given a COD category characterized by how effective the biological treatment was on that effluent. To identify how the individual streams are degraded requires biological digestion tests to be carried out on the individual streams (e.g. BOD_5). The relative rates of digestion can then be modeled to give the overall removal ratio of the mixed effluent in the existing biological treatment unit. The options of waste minimization at source, pretreatment or additional biological treatment capacity can then be evaluated with greater confidence.

7) *Environmental limits.* Each site will have environmental discharge limits specified by regulatory authorities. These might relate to concentrations, total load (i.e. total kilograms of contaminant discharged) or a combination of both. It is usual to have the major contaminants specified as concentrations (especially BOD_5 and COD). Given environmental *discharge limits* or *consent limits* from the regulatory authorities, it would be a bad practice to design precisely for those concentrations. The problem is, as with steam systems, water systems are almost never at steady state. Upset conditions cause surges in both the concentration and volume of effluent. For example, effluent is often collected in effluent pits until the pit is full, when the control system starts a pump and empties it. This causes both an increase in the flowrate and the concentration of the effluent. Given the dynamic nature of water systems, it is good practice to design for discharge concentrations significantly below the consent limits. Design for 60% of the consent limit is usually reasonable to allow for upset conditions.

26.17 Water System Design – Summary

Increasing awareness regarding the problems of overextraction of water and increasingly strict discharge regulations mean that water systems must be designed to be more efficient than in the past. Water consumption and wastewater generation can be reduced through reuse, regeneration reuse and regeneration recycling.

Distributed effluent treatment requires that a philosophy of design be adopted that segregates effluent for treatment, wherever appropriate, and combines it for treatment, where appropriate. Various primary, secondary and tertiary treatment processes are available to achieve the required discharge concentrations.

Maximum water reuse can be identified from limiting water profiles. These identify the most contaminated water that is acceptable in an operation. A composite curve of the limiting water profiles can be used to target the minimum water flowrate. While this approach is adequate for simple problems, it has some severe limitations. A more mathematical approach using the optimization of a superstructure allows all of the complexities of multiple contaminants, constraints, enforced matches, capital and operating costs to be included. A review of this area has been given by Mann and Liu (1999).

Creating a composite of the effluent streams and matching a treatment line against this composite curve, with an appropriate specification, can target the minimum flowrate of effluent to be treated. Targeting and design for simple problems is straightforward. Regeneration of wastewater requires that the streams for regeneration be separated into two groups: those that require fresh water and those that require regenerated water. Various methods are available to divide the operations into two groups. Targeting and design for effluent treatment and regeneration can be based on the optimization of a superstructure.

Water systems can be very complex to study and potentially require an enormous amount of data if a whole site is to be studied. Understanding the data requirements more fully can prevent collection of unnecessary data and avoid missed opportunities.

26.18 Exercises

1. A process involves four water-using operations as detailed in Table 26.11.
   a) Sketch the limiting water composite curve for the operations for fixed mass load.
   b) Target the water flowrate for reuse only for fixed mass load.
   c) Generate the grid diagram and design the water-using operation network to achieve the target for fixed mass load.
2. Four operations in a process are required to be supplied with water. Table 26.12 gives the maximum inlet and outlet concentrations for the water and the mass load of contamination in each operation.
   a) If freshwater with no inlet contamination is used in each of the four processes, calculate the total water requirements in t·h^{-1}.
   b) If water is reused up to its maximum potential, calculate the corresponding water requirements in t·h^{-1} for fixed mass load and generate the grid diagram and design the water-using operation network to achieve the target.
3. Consider a water-using network with a single contaminant. The limiting data for the problem are given in Table 26.13.
   a) Construct the limiting water composite curve and target minimum fresh water for fixed mass load without Operation 4.
   b) Target the minimum fresh water and wastewater flowrate for maximum reuse for fixed mass load considering all operations.
   c) Design a network for fixed mass load to achieve the target. There is one constraint to designing a water network. Water from only Operation 1 can be reused for other operations. Show the design steps and represent the water network as a conventional flowsheet.
4. Design a water network for the data in Table 26.13 to maximize water reuse, but constraining the operations to operate at fixed flowrates:
   a) When local recycling around operations is acceptable.
   b) When local recycling around operations in not acceptable.
   c) Suppose water for Operation 4 is a cooling tower makeup. There are several process changes that could reduce the water requirement for cooling tower operation. Suggest process changes.

Table 26.11

Limiting water data for Exercise 1.

Operation	Limiting water flowrate (t·h^{-1})	$C_{in,max}$ (ppm)	$C_{out,max}$ (ppm)
1	20	0	300
2	46	100	300
3	14	100	600
4	40	400	600

Table 26.12

Limiting water data for Exercise 2.

Operation	$C_{in,max}$ (ppm)	$C_{out,max}$ (ppm)	Mass load (kg·h^{-1})
1	0	400	16
2	200	400	16
3	200	1000	32
4	700	1000	6

Table 26.13

Limiting water data for Exercise 3.

Operation	$C_{in,max}$ (ppm)	$C_{out,max}$ (ppm)	Limiting water flowrate (t·h^{-1})	Flowrate loss (t·h^{-1})
1	0	100	30	–
2	25	100	50	–
3	50	200	40	–
4	70	–	10	10

Table 26.14

Limiting water data for Exercise 5.

Operation	$C_{in,max}$ (ppm)	$C_{out,max}$ (ppm)	m_C ($g \cdot h^{-1}$)	Limiting water flowrate ($t \cdot h^{-1}$)	Water loss ($t \cdot h^{-1}$)
1	0	100	1000	10	—
2	50	100	1000	20	—
3	100	400	6000	20	—
4	0	10	200	20	—
5	100	200	2000	20	—
6	100	—	2000	20	20

5. Consider a water-using network with a single contaminant, represented by the COD. The limiting data for the problem are given in Table 26.14.

a) Target the minimum fresh water and wastewater flowrates for maximum reuse only for a fixed mass load.

b) Design a network to achieve the target.

c) Show the design steps and represent the water network as a conventional flowsheet.

6. The effluent data for a plant are given in Table 26.15. For an environmental discharge limit of 50 ppm:

a) Sketch the composite curve for the effluent streams.

b) Target the treatment flowrate for a fixed outlet concentration $C_{OUT} = 20$ ppm.

c) Target for the treatment flowrate removal ratio $RR = (C_{in} - C_{out})/C_{in} = 0.97$.

d) Determine the stream groups for treatment system design.

e) Sketch the flow pattern.

7. A chemical process has three major water-using operations that are contaminated mainly by NH_3. The effluent data are given in Table 26.16.

Table 26.15

Effluent data for Exercise 6.

Stream	C (ppm)	Water flowrate ($t \cdot h^{-1}$)
1	1000	20
2	600	53.4
3	200	22.6

Table 26.16

Effluent data for Exercise 7.

Stream	C (ppm)	Water flowrate ($t \cdot h^{-1}$)
1	300	50
2	150	30
3	80	20

The effluent streams are currently combined and sent to a steam stripper to remove NH_3 from wastewater. The environmental discharge limit for NH_3 is 30 ppm.

a) Construct the effluent composite curve for this process.

b) What is the minimum treatment flowrate for the steam stripper (removal ratio of steam stripper = 90%)?

c) Design a distributed-treatment network that achieves the minimum treatment flowrate target in Part b. Draw the final design as a conventional flowsheet, giving flowrate and concentration levels for all wastewater streams.

d) The steam stripper is used to separate NH_3 from wastewater. If the wastewater stream contains NH_3 as well as H_2S, describe process modifications or operating changes in the steam stripper that can remove both contaminants.

8. A water network produces the six effluent streams in Table 26.17. The environmental discharge limit for the contaminant concentration is 50 ppm. There are two alternative treatment processes available on the site, which are capable of reducing the contaminant concentration respectively to 20 ppm and 40 ppm (Table 26.18).

a) Draw the effluent composite curve for the problem and determine the targets for each treatment process used separately.

b) Draw the conventional flowsheet for the treatment system for each treatment process and determine which option is the most economic.

c) Draw a superstructure for the problem.

Table 26.17

Effluent streams for Exercise 8.

Stream	Water flowrate ($t \cdot h^{-1}$)	C (ppm)
1	25	200
2	40	350
3	50	200
4	90	150
5	85	120
6	40	75

Table 26.18

Effluent treatment processes for Exercise 8.

Treatment process	C_{out} (ppm)	Unit treatment cost ($\$ \cdot t^{-1}$)	Maximum flowrate capacity ($t \cdot h^{-1}$)
TP1	20	2.25	700
TP2	40	1.5	500

Table 26.19

Water data for Exercise 9.

Operations	C_{in} (ppm)	C_{out} (ppm)	Water flowrate ($t \cdot h^{-1}$)
1	0	100	20
2	0	100	30
3	0	330	50

9. In a country with severe water scarcity, the government has placed limits on the amount of fresh water that may be used by industry. A company has three main water-using operations. In each operation, the water picks up contamination in the form of suspended solids (SS). The existing concentrations of SS are given in Table 26.19.

The company has now been told that it must reduce the total water demand by at least 10%. Carry out a water minimization study to determine the scope for reusing water. Discussions with the operations managers have concluded that not all the operations need to be fed with fresh water. The maximum concentrations that can be tolerated at the inlet of the operations are given in Table 26.20.

a) What is the contaminant mass load and limiting flowrate of each operation? Assume that the outlet concentrations in Table 26.19 are the maximum allowable ones.

b) What is the minimum fresh water flowrate that is required if water is reused for fixed mass loads?

c) Can the company achieve the required water reduction by reusing water?

d) Design a network that will achieve the minimum water target for fixed mass loads. Draw the final design as a conventional flowsheet, giving flows and concentrations for all water streams.

10. Table 26.21 lists the limiting data for a water network. The environmental discharge limit for the contaminant is 50 ppm. Fresh water is available with a concentration of 0 ppm. A treatment process is available that is capable of achieving a contaminant outlet concentration of 20 ppm.

a) Construct the limiting water composite curve and obtain the target for the minimum fresh water flowrate for reuse only for fixed mass load.

Table 26.20

Maximum inlet concentrations for Exercise 9.

Operation	$C_{in,max}$ (ppm)
1	0
2	50
3	30

Table 26.21

Water data for Exercise 10.

Operation	Limiting water flow ($t \cdot h^{-1}$)	m_C ($g \cdot h^{-1}$)	$C_{in,max}$ (ppm)	$C_{out,max}$ (ppm)
1	10	1000	0	100
2	55	8250	100	250
3	40	4000	300	400

b) Design a water-use network that achieves this target. Draw the network as a conventional flowsheet.

c) From the streams directed to effluent, devise a distributed treatment problem. Construct the effluent composite curve. Obtain the target for the minimum treatment flowrate.

d) Design a distributed effluent treatment network that achieves the minimum treatment target. Draw the network as a conventional flowsheet.

11. The data for five water-using operations are shown in Table 26.22. Targeting indicates that the process requires $80\,t \cdot h^{-1}$ for reuse only. It is proposed to reduce this target by introducing regeneration with a removal ratio of 0.95. Different stream groupings are to be explored for regeneration. In each case, sketch the flow pattern of how the water flows would be distributed between the two

Table 26.22

Water data for Exercise 11.

Operation	Mass load of contaminant ($kg \cdot h^{-1}$)	$C_{in,max}$ (ppm)	$C_{out,max}$ (ppm)	Limiting water flowrate ($t \cdot h^{-1}$)
1	8	0	200	40
2	5	100	200	50
3	6	100	400	20
4	6	300	400	60
5	8	400	600	40

groups. It is not necessary to determine the flow pattern within each group.

a) The streams are divided into two groups for regeneration. The first group contains Operations 1 and 2 and the second group Operations 3, 4 and 5. Determine the targets for fresh water and regenerated water.

b) An alternative grouping has only Operation 1 in the first group with the remainder in the second group. Again, determine the targets for fresh water and regenerated water.

c) A third grouping has Operations 1 and 4 in the first group with the remainder in the second group. Determine the targets for fresh water and regenerated water.

d) Which group would you recommend?

References

Bai J, Feng X and Deng C (2007) Graphical Based Optimisation of Single-Contaminant Regeneration Reduce Water Systems, *Chem Eng Res Dev*, **85**: 2127.

Berne F and Cordonnier J (1995) *Industrial Water Treatment*, Gulf Publishing.

Betz Laboratories (1991) *Betz Handbook of Industrial Water Conditioning*, 9th Edition.

Dhole VR, Ramchandani N, Tainsh R and Wasilewski M (1996) Make Your Process Water Pay for Itself, *Chem Eng*, **103**: 100.

Doyle SJ and Smith R (1997) Targeting Water Reuse with Multiple Contaminants, *Trans IChemE*, **B75**: 181.

Eckenfelder WW and Musterman JL (1995) *Activated Sludge Treatment of Industrial Wastewater*, Technomic Publishing.

Feng X, Bai J and Zheng X (2007) On the Use of Graphical Method to Determine the Targets of Single-Contaminant Regeneration Recycling Water Systems, *Chem Eng Sci*, **62**: 2127.

Foo DCY (2007) Water Cascade Analysis for Single and Multiple Impure Fresh Water Feed, *Trans IChemE Part A*, **85**: 1169.

Foo DCY (2012) *Process Integration for Resource Conservation*, CRC Press.

Galen B and Grossmann IE (1998) Optimal Design of Distributed Wastewater Treatment Networks, *Ind Eng Chem Res*, **37**: 4036.

Gunaratnam M and Smith R (2005) Design of Water Systems with Buffering Capacity for Operability, *Trans IChemE*, **A86**: 852–862.

Gunaratnam M, Alva-Argaez A, Kokossis A, Kim JK and Smith R (2005) Automated Design of Total Water Systems, *Ind Eng Chem Res*, **44**: 588.

Kuo W-CJ and Smith R (1997) Effluent Treatment System Design, *Chem Eng Sci*, **52**: 4273.

Kuo W-CJ and Smith R (1998a) Designing for the Interactions Between Water Use and Effluent Treatment, *Trans IChemE*, **A76**: 287–301.

Kuo W-CJ and Smith R (1998b) Design of Water-Using Systems Involving Regeneration, *Trans IChemE*, **B76**: 94.

Mann JG and Liu YA (1999) *Industrial Water Reuse and Wastewater Minimization*, McGraw-Hill.

Polley GT and Polley HL (2000) Design Better Water Networks, *Chem Eng Prog*, **Feb**: 47.

Schierup HH, Brix H and Lorenzen B (1990), Wastewater Treatment in Constructed Reed Beds in Denmark - State of the Art, in Cooper PF and Findlater BC, *The Use of Constructed Wetlands in Water Pollution Control*, Pergamon Press.

Smith R, Petela EA and Howells J (1996) Breaking a Design Philosophy, *Chem Eng*, **606**: 21.

Takama N, Kuriyama T, Shiroko K and Umeda T (1980) Optimal Water Allocation in a Petroleum Refinery, *Comp Chem Eng*, **4**: 251.

Tchobanoglous G and Burton FL (1991) *Metcalf & Eddy Wastewater Engineering – Treatment, Disposal and Reuse*, McGraw-Hill.

Tebbutt THY (1990) *BASIC Water and Wastewater Treatment*, Butterworths.

Wang YP and Smith R (1994a) Wastewater Minimization, *Chem Eng Sci*, **49**: 981.

Wang YP and Smith R (1994b) Design of Distributed Effluent Treatment Systems, *Chem Eng Sci*, **49**: 3127.

Wang YP and Smith R (1995a) Wastewater Minimization with Flowrate Constraints, *Trans IChemE*, **A73**: 889.

Wang YP and Smith R (1995b) Time Pinch Analysis, *Trans IChemE*, **A73**: 905.

Chapter 27

Environmental Sustainability in Chemical Production

Chemical processes should be designed to maximize the *sustainability* of industrial activity. Maximizing sustainability requires that industrial systems should strive to satisfy human needs in an economically viable, environmentally benign and socially beneficial way (Azapagic, 2014). Sustainability thus has economic, environmental and social dimensions. For process design, it is the environmental dimension that is of prime concern. However, when addressing the design from the environmental dimension, the outcome must still be economically viable.

Sustainability requires that the boundary of consideration should go beyond the immediate boundary of the manufacturing facility to maximize the benefit to society and to avoid adverse health effects, ecological damage, unnecessary use of materials, unnecessarily high burdens on transportation, and to avoid odor and noise nuisances.

One of the prime potential causes of ecological damage is environmental emissions from processes. When dealing with environmental emissions from processes, there are two approaches:

1) Treat the emission using thermal oxidation, biological digestion, and so on, to a form suitable for discharge to the environment, the so-called *end-of-pipe* treatment.
2) Reduce or eliminate the production of the emission by increasing the efficiency of raw materials, energy and water use, also referred to as *waste minimization*.

The problem with relying on end-of-pipe treatment is that once waste has been created, it cannot be destroyed. The waste can be concentrated or diluted, its physical or chemical form can be changed, but it cannot be destroyed. Thus, the problem with end-of-pipe effluent treatment systems is that they do not so much solve the problem but move it from one place to another. For example, aqueous solutions containing heavy metals can be treated by chemical precipitation to remove the metals. If the treatment system is designed and operated correctly, the aqueous stream can be passed on for further treatment or discharged to the receiving water. But what about the precipitated metallic sludge? This is usually disposed of to landfill (Smith and Petela, 1991a).

Environmental sustainability requires that emissions be reduced or eliminated by increasing the efficiency of use of raw materials, energy and water. However, environmental sustainability has broader dimensions than simply avoiding emissions. For chemical process design, processes should make use of materials of construction that depletes as little as practicable the reserves of scarce materials of construction. This applies particularly to the use of exotic materials of construction that make use of rare metals. The equipment installed should create as little waste as possible from maintenance. Process raw materials should be used as efficiently as is economic and practicable, both to prevent the production of waste that can be environmentally harmful and to preserve the reserves of manufacturing raw materials as much as possible. Processes should use as little energy as is economic and practicable, both to prevent the build-up of carbon dioxide in the atmosphere from burning fossil fuels and to preserve the reserves of fossil fuels. Water must also be consumed in sustainable quantities that do not cause deterioration in the quality of the water source and the long-term quantity of the reserves, particularly aquifers. Aqueous and atmospheric emissions must not be environmentally harmful and solid waste to landfill must be eliminated wherever feasible.

To direct a process design to maximize the sustainability of industrial activity requires that its location, construction and decommissioning, its feeds and products, its waste streams, and its connection to the transportation and wider industrial infrastructure be considered in terms of the whole *life cycle* of the process.

27.1 Life Cycle Assessment

Life cycle assessment (*LCA*) is a cradle-to-grave analysis of a product, process or service that evaluates all upstream and downstream resource requirements and environmental impacts (Curran,

Chemical Process Design and Integration, Second Edition. Robin Smith.

Companion Website: www.wiley.com/go/smith/chemical2e

1996; Guinee, 2002; Azapagic and Perdan, 2011). Whilst in principle LCA can consider a range of different product options, here it will be assumed that the product has been fixed and the manufacture of that product is the focus of the LCA. The assessment should start with the extraction of the raw materials from natural resources. In principle, different raw materials and process routes can be used for the same product. The various transformations of the raw materials are followed through to the manufacture of the final consumer product, the distribution and use of the product, recycling of the product (if this is possible) and finally to its eventual disposal. Each step in the life cycle uses resources and creates waste. Resource utilized and waste generated by transportation and the manufacture and maintenance of processing equipment should also be included.

LCA assesses the environmental aspects and potential impacts associated with a process by:

- Compiling an inventory of materials of construction, manufacturing raw materials, energy and water inputs and environmental releases
- Evaluating the potential environmental impacts associated with the inputs and releases
- Interpretation of the results to improve the sustainability of the process.

An LCA involves four components, known as *phases*:

1) *Goal and scope definition.* The first step is to define the purpose of the study and establish the system boundaries by defining what will be included in the analysis and what will be excluded. If the boundary is too wide, then the time and effort needed to collect information on the inputs and outputs and perform the analysis will be considerable. On the other hand, defining the boundary too narrowly can create an underestimate of the true life cycle impacts. It is necessary to keep the LCA manageable, but also comprehensive enough to capture the main environmental impacts.
2) *Inventory analysis.* A life cycle inventory quantifies requirements for materials of construction, process raw materials, energy and water, and quantifies atmospheric emissions, aqueous emissions, and solid wastes for the entire life cycle of the process, including plant construction and decommissioning. The life cycle inventory provides the basis for evaluation and comparison of environmental impacts or any potential improvements.

   Once the boundaries of the analysis have been established, a flow diagram can be developed. Unit processes inside of the system boundary link together to form a life cycle picture of the required inputs to and outputs from the whole system, including construction and decommissioning. Particular attention should be paid to the use of materials of construction that use rare elements. Decommissioning should as much as possible emphasize the use of process components or recycling of materials of construction. Decommissioning should also account for the reuse of, or disposal of, the process inventory after production has ceased.

   Each subsystem within the boundary requires inputs of materials, energy, water and should include transportation of raw materials and resources. Outputs from each subsystem are products, byproducts, atmospheric emissions, aqueous emissions and solid wastes and should include transportation of the outputs.

   a) *Raw materials and products.* Raw material inputs to each subsystem should account for energy, water and transportation associated with the extraction of the raw materials and delivery to the point of use. Products and byproducts are valuable outputs. Byproducts are process outputs that are not the main focus of the production, but have value and are not classified as wastes. The resource use, energy and water consumption, and emissions must be allocated to the products and byproducts. Waste byproducts must be assessed in terms of their environmental impacts, including the reuse or disposal of the process inventory as part of decommissioning.
   b) *Energy.* Energy input included in the inventory in the first instance considers *direct process energy* use and transportation. Transportation energy is the energy required for transportation such as pipelines, trucks, rail cars, barges and ships. Energy input beyond the primary combustion of the fuel needs also to be included, as the energy used in fuel combustion is only part of the total energy input. *Indirect energy* or *precombustion energy* accounts for the energy used to extract fuel (e.g. mining coal, drilling for oil), to process the raw fuel into usable fuels and to transport the fuels to the point of use. Indirect energy is the total energy necessary to deliver a usable fuel to the point of use. For example, indirect energy for fuel oil should account for oil exploration and production, refining and transporting the fuel oil to the point of use. Only through inclusion of the indirect energy contributions can the life cycle energy demand of the system be fully accounted for.

      Import of electricity from centralized power generation might involve coal, natural gas, oil, waste-to-energy, nuclear power, hydropower, solar, wind power or geothermal energy to generate electricity. The fuel mix associated with the various power generation sources feeding the electricity grid needs to be weighted to obtain a mean value for the sources feeding the power grid. Inefficiencies that must be accounted for include both the inefficiencies in the individual power generation cycles and the transmission losses. Some sources of power generation involve combustion of fossil fuels, others do not. The contribution from the non-fossil-fuel sources can be accounted for as their *fossil fuel equivalency*. For example, nuclear fuel indirect energy for nuclear power includes the energy for mining and fuel processing, as well as the increased energy associated with building the power plant relative to other power station designs.
   c) *Water.* Water input to the inventory measures the amount of water withdrawn from all sources. The input water incorporated into the products and byproducts, or evaporated, must also be accounted for in the input. Aqueous output to the environment should include those amounts still present in the waste stream after wastewater treatment, and

represent actual discharges into receiving waters. The most commonly reported aqueous pollutants are biological oxygen demand (BOD), chemical oxygen demand (COD), suspended solids, dissolved solids, hydrocarbon, phenol, nitrates and phosphates, but many other measures are relevant depending on the circumstances.

d) *Atmospheric emissions.* Output of emissions to atmosphere should be accounted for as discharges to the atmosphere after passing through emission control devices. Some emissions, such as fugitive emissions from pump, compressor and valve seals, pipe joints and storage areas, will not pass through control devices before release to the environment, but still must be included. Typical atmospheric emissions include particulates, volatile organic compounds (VOCs), oxides of sulfur, oxides of nitrogen, carbon monoxide and carbon dioxide. Atmospheric emissions from the production and combustion of fuel for process or transportation energy, as well as the process emissions, are included in the life cycle inventory.

e) *Solid waste.* For solid wastes a distinction needs to be made between industrial solid wastes and consumer solid wastes, as they are generally disposed of in different ways. Industrial solid waste is the solid waste generated during the production of a product (including its packaging if appropriate) and is typically divided into two categories: process solid waste and fuel-related solid waste. Fuel-related solid waste includes mineral extraction wastes, fuel processing solid wastes, residual ash from the combustion of solid fuels and solids collected from air pollution control devices. If included in the boundary of the analysis, consumer solid waste includes both the product and its packaging once it has met its intended use and is discarded to the municipal solid waste system.

3) *Impact assessment.* Impact assessment evaluates the potential human and ecological impacts and resource depletion from the use of materials, energy and water, and releases to the environment identified in the inventory analysis. Examples of impacts might be contribution to global warming from the release of greenhouse gases or eutrophication of receiving water from the discharge of nitrates and phosphates.

a) *Definition of impact categories.* Impacts are the consequences caused by the input and output streams from the inventory on human health, the environment or the future availability of natural resources. Table 27.1 lists impact categories commonly used in LCA. The impact categories are quantified in terms of *equivalents* (*eq.*) of the mass of a reference species (Azapagic and Perdan, 2011). For example, the global warming potential of an emission of greenhouse gases is quantified as a mass flow of kilograms of CO_2 equivalents.

b) *Classification.* Classification translates the resource use and emissions to the environment to the appropriate impact category. For example, carbon dioxide emissions might be classified to the global warming impact category. On the other hand, oxides of nitrogen might be classified to photochemical smog formation, acidification and eutrophication.

c) *Impact characterization.* Impact characterization uses factors to characterize the impact of inputs and outputs of different species from the inventory into a single representative indicator. Impact indicators are typically characterized by:

$$\begin{aligned}\text{Impact indicator} &= \text{Characterization factor}\\ &\quad \times \text{Quantity of input or output}\end{aligned}$$

In order to compare the impact of different species as a single factor, different species are expressed as the equivalent of a reference species. Table 27.1 gives examples of impact indicators (Azapagic and Perdan, 2011). Data for impact indicators can be obtained from various databases.

For example, all greenhouse gases can be expressed in terms of their global warming potential (GWP) using carbon dioxide as a reference. The GWP of a species compares the global warming of the species with that for the same mass of carbon dioxide. Carbon dioxide thus has a GWP factor of 1, and this is assumed not to change with time. The effect of other greenhouse gases on the environment changes through time and an average value is calculated over a specified time, usually 20, 100 or 500 years. For example, the GWP factor for methane is 86 for a 20-year time horizon and 34 for a 100-year time horizon. The reference period for the assessment of global warming environmental impact is normally taken to be 100 years. If the 100 year GWP factor for methane is 34, this means that 1 kilogram of methane has 34 times more potential as a global warming agent than 1 kilogram of CO_2 over a 100-year time horizon. A mixture of 1 kilogram of CO_2 and 1 kilogram of methane would then have a global warming potential of 35 kilograms of CO_2 equivalents for a 100-year horizon.

d) *Normalization.* Normalization expresses the different impact potentials on a common scale by dividing by a selected reference value to allow comparisons across impact categories. There are numerous ways a reference value can be selected for normalization. One common way to normalize is on a per capita basis per year in an area defined as global, regional or local. It should be noted that normalized data can only be compared within a given impact category and not across impact categories.

e) *Grouping.* Grouping involves bundling of impact categories to final damage groups to better facilitate the interpretation of the results. For example, grouping might be according to air, aqueous or solid emissions. Alternatively, grouping might be according to area (global, regional or local).

f) *Weighting.* Weighting assigns weights to the different impact categories and resources reflecting their perceived relative importance. However, there is no universally correct way to assign weights. Producing a single weighted impact factor assists the screening of options, but the interpretation of weighted results must always be viewed with caution. Only if one option is better than another option in all impact categories can it be interpreted as being clearly better in terms of its life cycle assessment.

4) *Improvement analysis.* The purpose of a life cycle assessment within the development of a process design is to select the preferred process option with a life cycle perspective of the impacts on the environment and human health. It should be

Table 27.1

Common impact categories (Azapagic and Perdan, 2011).

Impact category	Description	Reference species	Calculation
Abiotic resource depletion (ADP) of elements	Depletion of metals and minerals	Antimony	$ADP = \sum_i ADP_i B_i$ (kg Sb eq.) ADP_i = abiotic depletion of Resource i B_i = quantity of Resource i
Abiotic resource depletion (ADP) of energy	Depletion of fossil fuels	MJ energy	$ADP = \sum_i LHV_i B_i$ (MJ) LHV_i = lower heating value of Fossil Fuel i B_i = quantity of Fossil Fuel i
Primary energy demand (PED)	Total use of energy in the life cycle, including renewable and nonrenewable sources	MJ energy	$PED = \sum_i LHV_i B_i + \sum_i \frac{EF_i + Q_i}{\eta_{EFi}} + \sum_i \frac{ENC_i}{\eta_{ENC_i}}$ (MJ) LHV_i = lower heating value of Fuel i consumed for heating B_i = quantity of Fuel i consumed for heating EF_i = quantity of electricity consumed that is generated from combustion of Fuel i Q_i = useful heat from cogeneration derived from electricity generation from the combustion of Fuel i η_{EFi} = cogeneration efficiency of electricity generation from combustion of Fuel i ENC_i = quantity of electricity consumed that is generated from Non-combustible Source i (e.g. nuclear, hydro, wind, solar) η_{ENCi} = efficiency of power generation from Non-combustible Source i
Global warming potential (GWP)	Global warming potential of greenhouse gases, e.g. CO_2, CH_4 and N_2O	Carbon dioxide	$GWP = \sum_i GWP_i B_i$ (kg CO_2 eq.) GWP_i = global warming potential factor of Species i B_i = emissions of Greenhouse Gas i
Stratospheric ozone layer depletion potential (ODP)	Potential of emissions of chlorofluro-hydrocarbons and other halogenated hydrocarbons to deplete the ozone layer	Trichloro-fluoromethane (CFC-11)	$ODP = \sum_i ODP_i B_i$ (kg CFC-11 eq.) ODP_i = ozone depletion factor for Species i B_i = emissions of ozone depleting Species i
Human toxicity potential (HTP)	Toxicity to humans of species release to air, water and soil	1, 4-Dichloro-benzene (1,4-DB)	$HTP = \sum_i HTP_{Ai} B_{Ai} + \sum_i HTP_{Wi} B_{Wi} + \sum_i HTP_{Si} B_{Si}$ (kg 1, 4-DB eq.) HTP_{Ai}, HTP_{Wi}, HTP_{Si} = toxicological classification factors for air, water and soil B_{Ai}, B_{Wi}, B_{Si} = emissions of Species i to air, water and soil
Eco-toxicity potential (ETP)	Potential for freshwater aquatic, marine aquatic, freshwater sediment, marine sediment and terrestrial toxicity	1, 4-Dichloro-benzene (1,4-DB)	$ETP = \sum_i \sum_j ETP_{ij} B_{ij}$ (kg 1, 4-DB $eq.$) ETP = ecotoxicity classification factor of Species i released into Medium j (freshwater aquatic, marine aquatic, freshwater sediment, marine sediment, terrestrial) B_{ij} = emission of Species i into Medium j
Photochemical oxidant creation potential (POCP)	Potential of VOCs and NO_X to create photochemical smog	Ethylene	$POCP = \sum_i POCP_i B_i$ (kg C_2H_4 eq.) $POCP_i$ = photochemical oxidant classification factor for Species i B_i = emission of Species i participating in the formation of photochemical smog
Acidification potential (AP)	Potential for acid deposition, particularly due to emissions of SO_X, NO_X and NH_3	Sulfur dioxide	$AP = \sum_i AP_i B_i$ (kg SO_2 eq.) AP_i = acidification potential for Species i B_i = emission of acidification Species i
Eutrophication potential (EP)	Potential of nutrients to cause overfertilization of water and soil and the growth of algae, particularly through emissions of NO_X, NH_4^+, PO_4^{3-}, NO_3^-	Phosphate (PO_4^{3-})	$EP = \sum_i EP_i B_i$ (kg PO_4^{3-} eq.) EP_i = eutrophication potential for Species i B_i = emission of eutrophication Species i

used to evaluate opportunities to reduce material, energy and water inputs and environmental impacts of the emissions from a life cycle perspective. However, there must be a clear understanding of the uncertainty and the assumptions used to generate the assessment. Choosing different boundaries might change the conclusions. Characterization factors must be extracted from a database and there is uncertainty regarding the reliability of the factors. Thus, life cycle assessment cannot on its own be used to choose a design option, but provides additional information to be considered when choosing a design option.

To the process designer, life cycle assessment is useful because focusing exclusively on the performance of the process in isolation can sometimes create problems elsewhere in the life cycle. The designer can often obtain useful insights by changing the boundaries of the system under consideration to be wider than those of the process being designed. However, having identified a problem, the designer must focus mainly on improvement analysis.

Example 27.1 A gaseous vent stream with a flowrate of $1000\ Nm^3 \cdot h^{-1}$ needs to be released to the environment. The gas contains 47% methane, with the balance being an inert gas. The methane can be assumed to have a GWP factor of 34 for a time horizon of 100 years. The inert gas can be assumed not to make any contribution towards global warming. Calculate the global warming potential of the vent stream for a 100-year time horizon:

a) If it is released directly to the atmosphere.

b) If it is combusted in a steam-assisted elevated flare with a destruction efficiency of 98%. The steam required for the flare is at 2 bar and 140 °C with a flowrate of $500\ kg \cdot h^{-1}$. The energy required to generate the steam can be assumed to be $2601\ kJ \cdot kg^{-1}$. The overall steam generation and distribution efficiency can be assumed to be 85%. Steam generation can be assumed to be from natural gas. Natural gas is required for the flare pilot burner with a flowrate of $15\ Nm^3 \cdot h^{-1}$. The natural gas can be assumed to be effectively pure methane with a net calorific value of $50\ MJ \cdot kg^{-1}$. Any methane that is combusted in the pilot flame and the steam generation can be assumed to be completely combusted to CO_2. Other products of combustion will be formed in the flare, such as CO and NO_X (including N_2O), but these are difficult to estimate and extremely small and will be neglected. Any water vapor formed in combustion processes can be assumed not to make a contribution towards global warming.

It can be assumed that the molar mass of a gas in kilograms occupies $22.4\ m^3$ in the vapor phase at standard conditions. The molar masses of methane and carbon dioxide are $16\ kg \cdot kmol^{-1}$ and $44\ kg \cdot kmol^{-1}$ respectively.

Solution

a) Mass flowrate of CH_4 in the vent

$$= 1000 \times 0.47 \times \frac{1}{22.4} \times 16$$
$$= 335.7\ kg \cdot h^{-1}$$

Global warming potential of the vent stream for direct release to the atmosphere

$$= 335.7 \times 34$$
$$= 11,413\ kg\ CO_2 \text{ equivalents per hour}$$

b) If the vent is to be released to atmosphere via a steam-assisted flare, then there are four sources of contribution to global warming. These are the uncombusted methane, the products of the vent combustion, products of combustion of the pilot fuel and products of combustion from generation of steam for the flare steam. Emissions associated with the construction of the equipment will be neglected, as will be indirect energy emissions. First is the uncombusted methane.

Mass flowrate of uncombusted methane released for 98% destruction

$$= 335.7 \times 0.02$$
$$= 6.714\ kg \cdot h^{-1}$$

For combustion of the methane in the vent:

$$CH_4 + 2O_2 \rightarrow CO_2 + H_2O$$

Since 1 kmol of CO_2 is formed for each kmol of CH_4 combusted, CO_2 formed from combustion of the vent methane

$$= 1000 \times 0.47 \times 0.98 \times \frac{1}{22.4}$$
$$= 20.56\ kmol \cdot h^{-1}$$
$$= 20.56 \times 44\ kg \cdot h^{-1}$$
$$= 904.6\ kg \cdot h^{-1}$$

Pilot gas flowrate

$$= 15 \times \frac{1}{22.4}$$
$$= 0.6696\ kmol \cdot h^{-1}$$

If the pilot gas is assumed to be 100% methane and the methane is assumed to be completely combusted, then the flowrate of CO_2 formed from the pilot

$$= 0.6696 \times 44\ kg \cdot h^{-1}$$
$$= 29.46\ kg \cdot h^{-1}$$

If the fuel to generate the flare steam is assumed to be 100% methane, then the CH_4 required to generate the steam with an overall efficiency of 85%

$$= 2601 \times 500 \times \frac{1}{50 \times 10^3} \times \frac{1}{0.85}$$
$$= 30.6\ kg \cdot h^{-1}$$
$$= \frac{30.6}{16}\ kmol \cdot h^{-1}$$
$$= 1.913\ kmol \cdot h^{-1}$$

CO_2 formed from the generation of the flare steam

$$= 1.913 \times 44$$
$$= 84.17\ kg \cdot h^{-1}$$

Global warming potential for release to atmosphere from a flare stack

$$= 6.714 \times 34 + 904.6 \times 1 + 29.46 \times 1 + 84.17 \times 1$$
$$= 1,247\ kg\ CO_2 \text{ equivalents per hour}$$

27

Flaring of the vent gas reduces the global warming potential of the release by 89% relative to direct release. It should also be noted for the flare that the uncombusted methane and emissions from the pilot fuel and steam generation contribute 27% towards the total global warming potential. Flaring should only be used for abnormal operation and emergency release. Continuous flow of vent gas should be released to atmosphere after combustion in a thermal oxidizer. This should reduce the global warming potential of the release even further, but for the continuous flow only.

27.2 Efficient Use of Raw Materials Within Processes

Consider improvement analysis with respect to raw materials use for the process being designed or modified before considering upstream and downstream issues. The efficient use of process raw materials both prevents the production of waste that can be environmentally harmful and preserves the reserves of those raw materials. Consider how the efficiency of use of raw materials can be increased for individual processes.

1) *Reducing waste from chemical reactors when recycling is difficult*
   a) *Increasing conversion for single irreversible reactions.* If unreacted feed material is difficult to separate and recycle, it is necessary to force as high conversion as possible. If the reaction is irreversible, then the low conversion can be forced to higher conversion by a longer residence time in the reactor, a more effective catalyst, higher temperature, higher pressure or a combination of these.
   b) *Increasing conversion for single reversible reactions.* The situation becomes worse if the fact that unreacted feed material is difficult to separate and recycle coincides with the reaction being reversible. Chapter 5 considered what can be done to increase equilibrium conversion.
      - *Excess reactants.* An excess of one of the reactants can be used as shown in Figure 5.8a.
      - *Product removal during reaction.* Separation of the product before completion of the reaction can force a higher conversion, as discussed in Chapter 5. Figure 5.6 showed how this is done in sulfuric acid processes. Sometimes the product (or one of the products) can be removed continuously from the reactor as the reaction progresses, for example, by allowing it to vaporize from a liquid-phase reactor or incorporating a membrane in the reactor to allow it to permeate through the membrane.
      - *Inert material concentration.* The reaction might be carried out in the presence of an inert material. This could be a solvent in a liquid-phase reaction or an inert gas in a gas-phase reaction. Figure 5.8b shows that if the reaction involves an increase in the number of moles, then adding inert material will increase equilibrium conversion. On the other hand, if the reaction involves a decrease in the number of moles, then inert concentration should be decreased (Figure 5.8b). If there is no change in the number of moles during reaction, then inert material has no effect on equilibrium conversion.
      - *Reactor temperature.* For endothermic reactions, Figure 5.8c shows that the temperature should be set as high as possible, consistent with materials of construction limitations, catalyst life and safety. For exothermic reactions, the ideal temperature is continuously decreasing as conversion increases (Figure 5.8c).
      - *Reactor pressure.* In Chapter 5, it was deduced that vapor-phase reactions involving a decrease in the number of moles should be set to as high a pressure as practicable, taking into account that the high pressure might be expensive to obtain through compressor power, mechanical construction might be expensive and high pressure brings safety problems (Figure 5.8d). Reactions involving an increase in the number of moles should ideally have a pressure that is continuously decreasing as conversion increases (Figure 5.8d). Reduction in pressure can be brought about either by a reduction in the absolute pressure or by the introduction of an inert diluent.

   If the separation and recycle of unreacted feed material is not a problem, then squeezing extra conversion from the reactor is usually not necessary.
2) *Reducing waste from primary reactions that produce waste byproducts.* If a waste byproduct is formed from the reaction, as for example:

$$\text{FEED 1} + \text{FEED 2} \rightarrow \text{PRODUCT} + \text{WASTE BYPRODUCT} \tag{27.1}$$

   then the waste can only be avoided by different reaction chemistry, that is, a different reaction path and possibly different raw materials.
3) *Reducing waste from multiple reactions producing waste byproducts.* In addition to the losses described in Section 2 above for single reactions, multiple reaction systems lead to further waste through the formation of waste byproducts in secondary reactions. A brief review from Chapters 4 to 6 reveals what can be done to minimize waste byproduct formation.
   a) *Reactor type.* Firstly, make sure that the correct reactor type has been chosen to maximize selectivity for a given conversion in accordance with the arguments presented in Chapter 4.
   b) *Reactor concentration.* Selectivity can often be improved by one or more of the following actions (Smith and Petela, 1991b):
      - Use an excess of one of the feeds when more than one feed is involved.
      - Increase the concentration of inert material if the byproduct reaction is reversible and involves a decrease in the number of moles.
      - Decrease the concentration of inert material if the byproduct reaction is reversible and involves an increase in the number of moles.
      - Separate the product partway through the reaction before carrying out further reaction and separation.

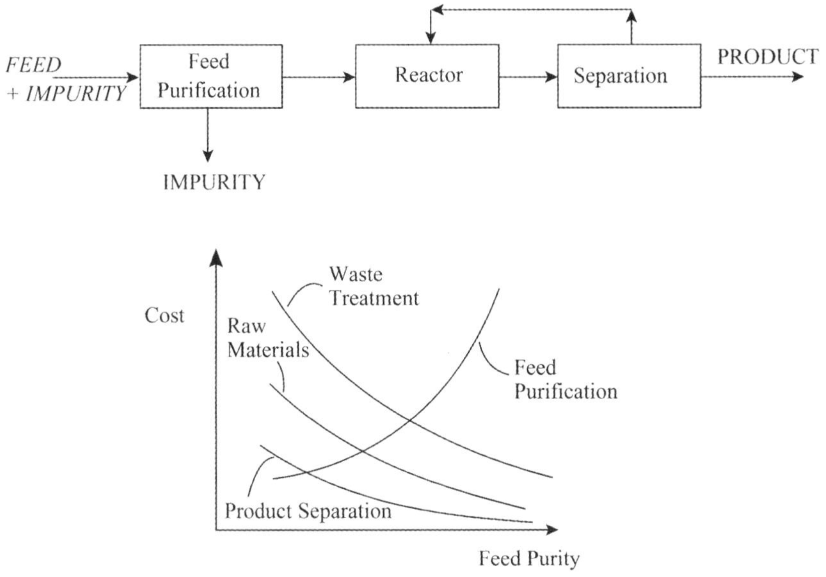

Figure 27.1

If feed impurity undergoes a reaction then there is an optimum feed purity. (Reproduced from Smith R and Petela E.A (1991c) Waste Minimization in the Process Industries Part 3 Separation and Recycle Systems, *Chem Eng*, **513**: 24, by permission of the Institution of Chemical Engineers.)

- Recycle waste byproducts to the reactor if by product reactions are reversible.

Each of these measures can, in the appropriate circumstances, minimize waste (Figure 5.9).

c) *Reactor temperature and pressure.* If there is a significant difference between the effects of temperature or pressure on primary and byproduct reactions, then temperature and pressure should be manipulated to improve selectivity and minimize the waste generated by byproduct formation.

d) *Catalysts.* Catalysts can have a major influence on selectivity. Changing the catalyst can change the relative influence on the primary and byproduct reactions. This might result directly from the reaction mechanisms at the active sites, or the relative rates of diffusion in the support material, or a combination of both.

4) *Reducing waste from feed impurities that undergo reaction.* If feed impurities undergo reaction, it causes waste of feed material, products or both. Avoiding such waste is most readily achieved by purifying the feed. Thus, increased feed purification costs are traded off against reduced raw materials, product separation and waste disposal costs (Figure 27.1).

5) *Reducing waste by upgrading waste byproducts.* Waste byproducts can sometimes be upgraded to useful materials by subjecting them to further reaction in a different reaction system. An example was quoted in Chapter 14 in which hydrogen chloride, which is a waste byproduct of chlorination reactions, can be upgraded to chlorine and then recycled to a chlorination reactor.

6) *Reducing catalyst waste.* Both homogeneous and heterogeneous catalysts are used. In general, heterogeneous catalysts should be used whenever possible, rather than homogeneous catalysts, since separation and recycling of homogeneous catalysts can be difficult, leading to waste. Heterogeneous catalysts are more common in large-scale processes. However, they degrade and need replacement. If contaminants in the feed material or recycle shorten catalyst life, then extra separation to remove those contaminants before entering the reactor might be justified. If the catalyst is sensitive to extreme conditions, such as high temperature, then some measures can help avoid local hot spots and extend catalyst life:

- better flow distribution,
- better heat transfer,
- introduction of catalyst diluent,
- better instrumentation and control.

Fluidized-bed catalytic reactors tend to generate loss of catalyst through attrition of the solid particles, causing fines to be generated, which are lost. More effective separation of catalyst fines from the reactor product and recycling the fines will reduce catalyst waste up to a point. Improving the mechanical strength of the catalyst is likely to be the best solution in the long term.

7) *Recycling waste streams directly.* Sometimes waste can be reduced by recycling waste streams directly. This is the first option in Figure 27.2. If this can be done, it is clearly the simplest way to reduce waste and should be considered first. Most often, the waste streams that can be recycled directly are aqueous streams, which, although contaminated, can substitute part of the freshwater feed to the process, as discussed in Chapter 26.

Figure 27.3a shows a simplified flowsheet for the production of isopropyl alcohol by the direct hydration of propylene (Smith and Petela, 1991c). Different reactor technologies are available for the process and separation and recycle systems vary, but Figure 27.3a is representative. Propylene containing propane as an impurity is reacted with water according to the reaction:

$$\underset{\text{propylene}}{C_3H_6} + \underset{\text{water}}{H_2O} \rightarrow \underset{\text{isopropyl alcohol}}{(CH_3)_2CHOH}$$

Some small amount of byproduct formation occurs. The principal byproduct is di-isopropyl ether. The reactor product is cooled and a phase separation of the resulting vapor–liquid

27

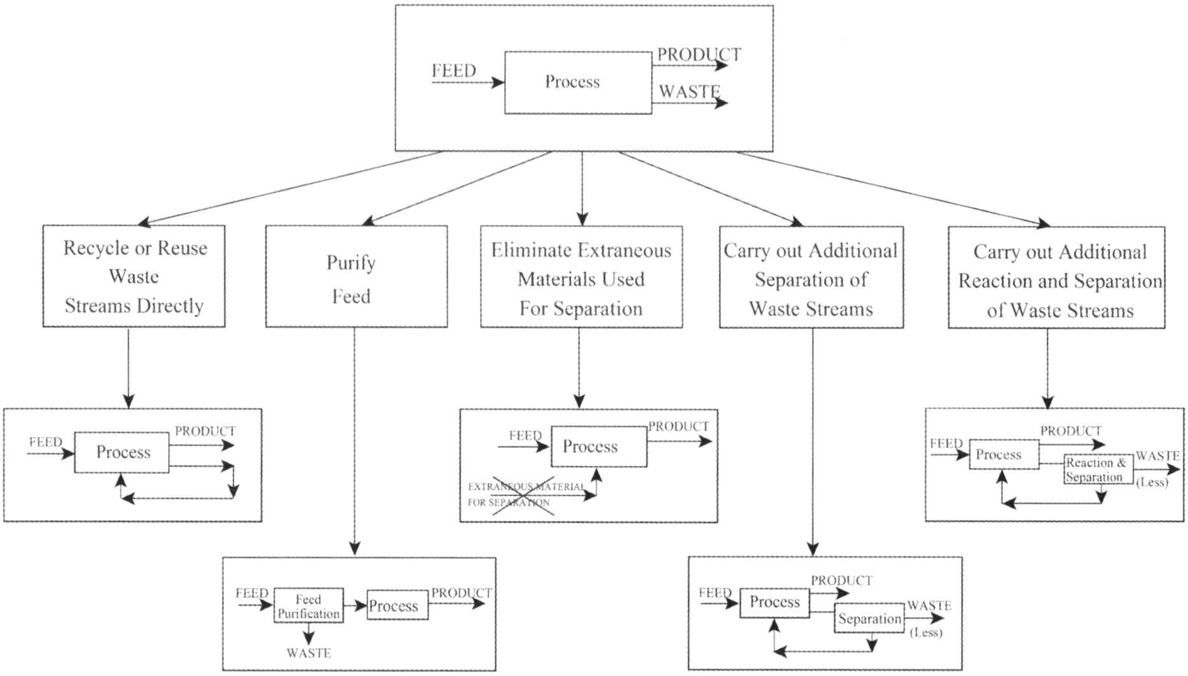

Figure 27.2
Waste minimization in separation and recycle systems.

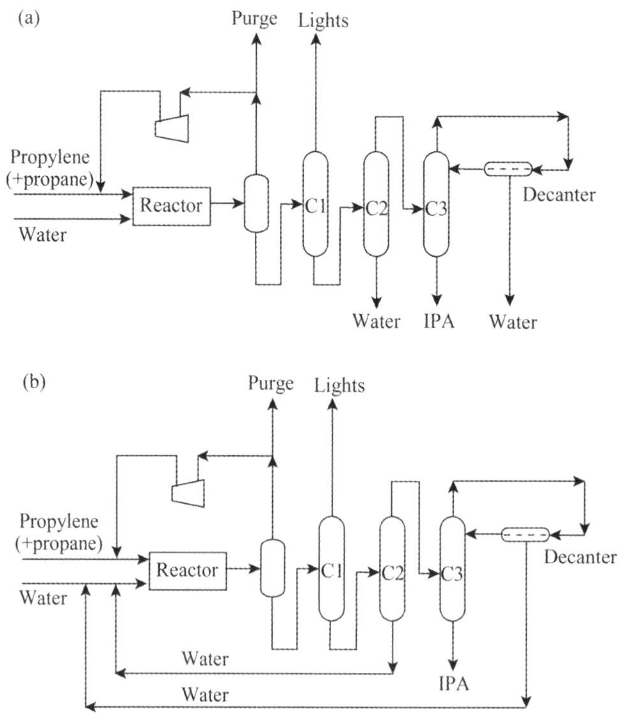

Figure 27.3
Outline flowsheet for the production of isopropyl alcohol by direct hydration of propylene. (Reproduced from Smith R and Petela E.A (1991c) Waste Minimization in the Process Industries Part 3 Separation and Recycle Systems, *Chem Eng*, **513**: 24, by permission of the Institution of Chemical Engineers.)

mixture produces a vapor containing predominantly propylene and propane and a liquid containing predominantly the other components. Unreacted propylene is recycled to the reactor and a purge taken to prevent the build-up of propane. The first distillation in Figure 27.3a (Column C1) removes light ends (including the di-isopropyl ether). The second (Column C2) removes as much water as possible to approach the azeotropic composition of the isopropyl alcohol–water mixture. The final column in Figure 27.3a (Column C3) is an azeotropic distillation using an entrainer. In this case, one of the materials already present in the process, di-isopropyl ether, can be used as entrainer.

Wastewater leaves the process from the bottom of the second column and the decanter of the azeotropic distillation column. Although both of these streams are essentially pure water, they will nevertheless contain small quantities of organics and must be treated before final discharge. This treatment can be avoided altogether by recycling the wastewater to the reactor inlet to substitute part of the freshwater feed (Figure 27.3b). If waste streams can be recycled directly, this is clearly the simplest method for reducing waste. However, most often, additional separation is required.

8) *Feed purification.* Impurities that enter with the feed inevitably cause waste. If feed impurities undergo reactions, then this causes waste from the reactor, as already discussed. If the feed impurity does not undergo reaction, then it can be separated out from the process in a number of ways, as discussed in Section 14.1. The greatest source of waste occurs when a purge is used. Impurity builds up in the recycle, and it is desirable for it to build up to a high concentration to minimize waste of feed materials and product in the purge. However, two factors limit the extent to which the feed impurity can be allowed to build up:

   a) High concentrations of inert material can have an adverse effect on reactor performance.

   b) As more and more feed impurity is recycled, the cost of the recycle increases (e.g. through increased recycle gas compression costs, etc.) to the point where that increase outweighs the savings in raw materials lost in the purge.

   In general, the best way to deal with a feed impurity is to purify the feed before it enters the process. This is the second option shown in Figure 27.2.

Returning to the isopropyl alcohol process from Figure 27.3, propylene is fed to the process containing propane as a feed impurity. In Figure 27.3, the propane is removed from the process using a purge. This causes waste of propylene, together with a small amount of isopropyl alcohol. The purge can be eliminated if the propylene is purified before entering the process. In this case, the purification can be achieved by distillation. Examples of where similar schemes can be implemented are plentiful.

Many processes are based on an oxidation step for which air would be the first obvious source of oxygen. A partial list would include acetic acid, acetylene, acrylic acid, acrylonitrile, carbon black, ethylene oxide, formaldehyde, maleic anhydride, nitric acid, phenol, phthalic anhydride, sulfuric acid, titanium dioxide, vinyl acetate and vinyl chloride (Chowdhury and Leward, 1984). Because the nitrogen in the air is not required by the reaction, it must be separated and leave the process at some point. Because gaseous separations are difficult, the nitrogen is normally removed from the process using a purge if a recycle is used. Alternatively, the recycle is eliminated from the design and the reactor forced to as high a conversion as possible to avoid recycling. The nitrogen will carry with it process materials, both feeds and products, and will probably require treatment before final discharge. If the air for the oxidation is substituted by pure oxygen, then, at worst, the purge will be very much smaller. At best it can be eliminated altogether. Of course, this requires an air separation plant upstream of the process to provide the pure oxygen. However, despite this disadvantage, very significant benefits can be obtained, as the following example shows.

Consider vinyl chloride production (see Example 4.1). In the "oxychlorination" reaction step of the process ethylene, hydrogen chloride and oxygen are reacted to form dichloroethane:

$$\underset{\text{ethylene}}{C_2H_4} + \underset{\text{hydrogen}}{2HCl} + \underset{\text{oxygen}}{\tfrac{1}{2}O_2} \longrightarrow \underset{\text{dichloroethane}}{C_2H_4Cl_2} + \underset{\text{water}}{H_2O}$$

If air is used, then a single pass with respect to each feedstock is used and no recycle to the reactor (Figure 27.4a). Thus, the process operates at near stoichiometric feedrates to achieve high conversions. Typically, between 0.7 and 1.0 kg vent, gases are emitted per kilogram of dichloroethane produced (Reich, 1976). If the air is substituted by pure oxygen, then the problem of the large flow of inert gas is eliminated (Figure 27.4b). Unreacted gases can be recycled to the reactor. This allows oxygen-based processes to be operated with an excess of ethylene, thereby enhancing the HCl conversion without sacrificing ethylene yield. Unfortunately, this introduces a safety problem downstream of the reactor. Unconverted ethylene can create explosive mixtures with the oxygen. To avoid explosive mixtures, a small amount of nitrogen can be introduced.

Since nitrogen is drastically reduced in the feed and essentially all ethylene is recycled, only a small purge is required to be vented. This results in a twenty- to hundredfold reduction in the size of the purge (Reich, 1976).

9) *Elimination of extraneous materials for separation.* The third option from Figure 27.2 is to eliminate extraneous materials added to the process to carry out separation. The most obvious example would be addition of a solvent, either organic or aqueous. Also, acids or alkalis are sometimes used to precipitate other materials from solution. If these extraneous materials used for separation can be recycled with a high efficiency, there is not a major problem. Sometimes, however, they cannot be recycled efficiently. If this is the case, then waste is created by the discharge of that material. To reduce this waste, alternative methods of

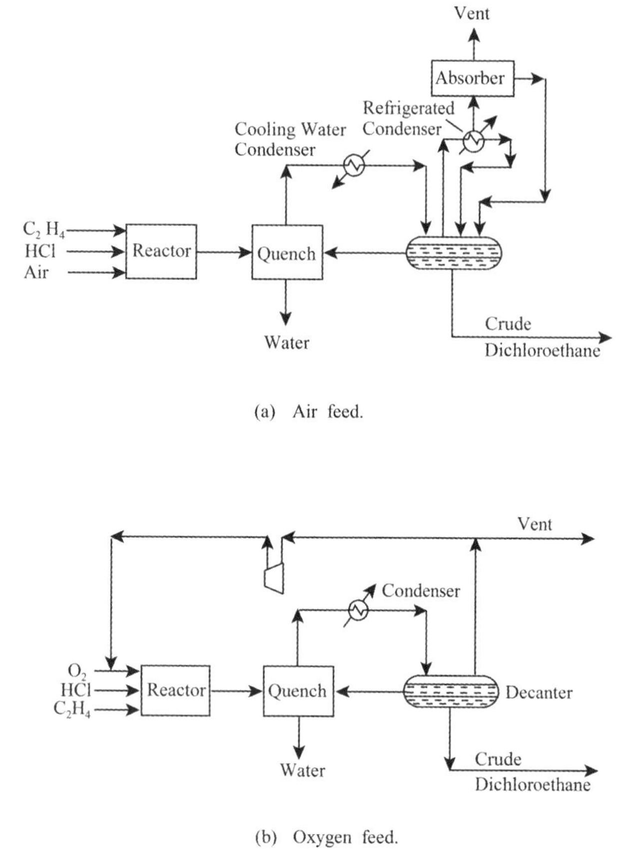

Figure 27.4

The oxychlorination step of the vinyl chloride process. (Reproduced from Smith R and Petela E.A (1991c) Waste Minimization in the Process Industries Part 3 Separation and Recycle Systems, *Chem Eng*, **513**: 24, by permission of the Institution of Chemical Engineers.)

separation are needed, such as use of evaporation instead of precipitation.

As an example, consider again the manufacture of vinyl chloride. In the first step of this process, ethylene and chlorine are reacted to form dichloroethane:

$$\underset{\text{ethylene}}{C_2H_4} + \underset{\text{chlorine}}{Cl_2} \rightarrow \underset{\text{dichloroethane}}{C_2H_4Cl_2}$$

A flowsheet for this part of the vinyl chloride process is shown in Figure 27.5 (McNaughton, 1983). The reactants (ethylene and chlorine) dissolve in circulating liquid dichloroethane and react in solution to form more dichloroethane. Temperature is maintained between 45 and 65 °C, and a small amount of ferric chloride is present to catalyze the reaction. The reaction generates considerable heat.

In early designs, the reaction heat was typically removed by cooling water. Crude dichloroethane was withdrawn from the reactor as a liquid, acid-washed to remove ferric chloride, then neutralized with dilute sodium hydroxide and purified by distillation. The material used for separation of the ferric chloride can be recycled up to a point, but a purge must be taken. This creates waste streams contaminated with chlorinated hydrocarbons that are environmentally very damaging and must be treated prior to disposal.

The problem with the flowsheet shown in Figure 27.5 is that the ferric chloride catalyst is carried from the reactor with the product. This is separated by washing. If a design of reactor can be found that prevents the ferric chloride leaving the reactor, the effluent problems created by the washing and neutralization are avoided. Because the ferric chloride is nonvolatile, one way to do this would be to allow the heat of reaction to raise the reaction mixture to boiling point and remove the product as a vapor, leaving the ferric chloride in

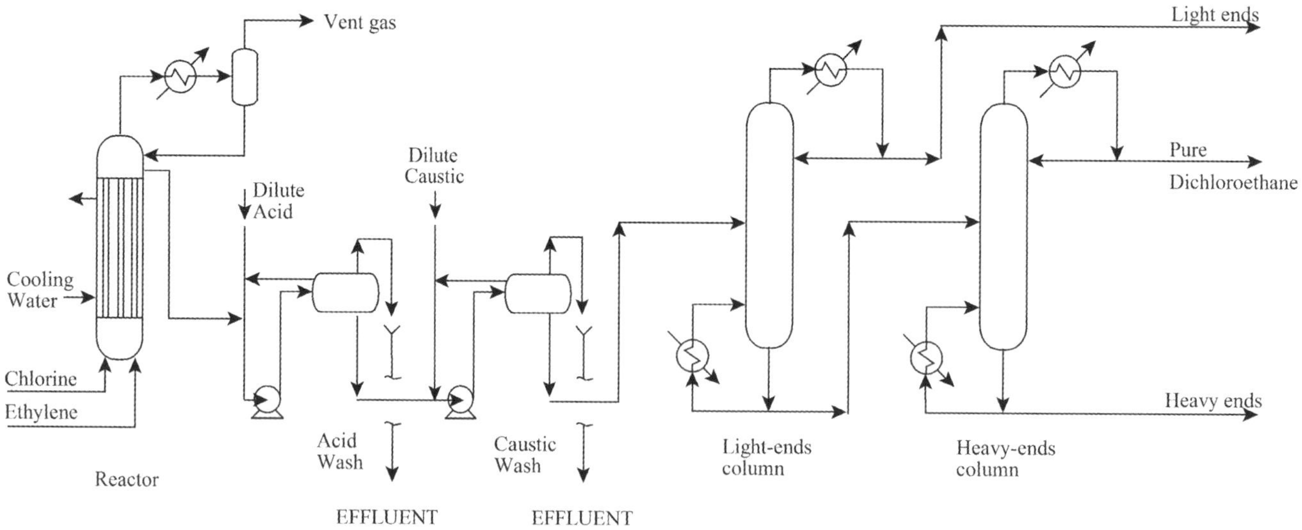

Figure 27.5

The direct chlorination step of the vinyl chloride process using a liquid phase reactor. (From McNaughton KJ, *Chem Engg*, 1983, Dec 12, 1983: 54, reproduced by permission.)

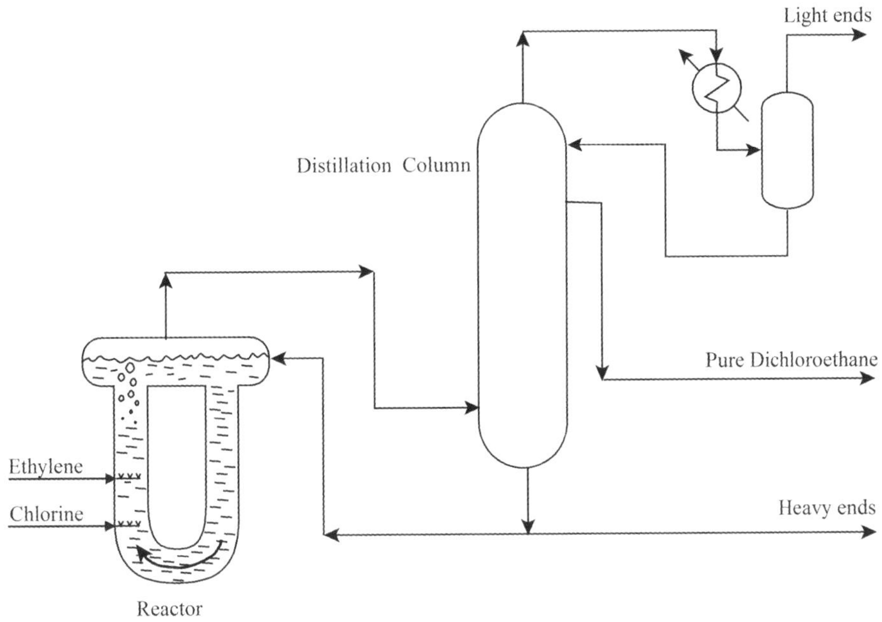

Figure 27.6

The direct chlorination step of the vinyl chloride process using a boiling reactor eliminates the washing and neutralization steps and the resulting effluents. (From McNaughton KJ, *Chem Engg,* 1983, Dec 12, 1983: 54, reproduced by permission.)

the reactor. Unfortunately, if the reaction mixture is allowed to boil, then there are two problems:

- ethylene and chlorine are stripped from the liquid phase, giving a low conversion;
- excessive byproduct formation occurs.

This problem is solved in the reactor shown in Figure 27.6 (McNaughton, 1983). Ethylene and chlorine are introduced into circulating liquid dichloroethane. They dissolve and react to form more dichloroethane. No boiling takes place in the zone where the reactants are introduced or in the zone of reaction. As shown in Figure 27.6, the reactor has a U-leg in which dichloroethane circulates as a result of gas lift and thermosyphon effects. Ethylene and chlorine are introduced at the bottom of the U-leg, which is under sufficient hydrostatic head to prevent boiling.

The reactants dissolve and immediately begin to react to form further dichloroethane. The reaction is essentially complete at a point only two-thirds up the rising leg. As the liquid continues to rise, boiling begins and finally the vapor–liquid mixture enters the disengagement drum. A slight excess of ethylene ensures essentially 100% conversion of chlorine.

As shown in Figure 27.6, the vapor from the reactor flows into the bottom of a distillation column and high-purity dichloroethane is withdrawn as a sidestream, several trays from the column top (McNaughton, 1983). The design shown in Figure 27.6 is elegant in that the heat of reaction is conserved to drive the separation and no washing of the reactor products is required. This eliminates two aqueous waste streams, which would inevitably carry organics with them, requiring treatment and causing loss of materials.

It is often possible to use the energy release inherent in the process to drive the separation system by improved heat recovery and in so doing carry out the separation at little or no increase in operating costs.

10) *Additional separation and recycling.* Once the possibilities for recycling streams directly have been exhausted, and feed purification and extraneous materials for separation eliminated that cannot be recycled efficiently, attention is turned to the fourth option in Figure 27.2, the degree of material recovery from the waste streams that are left. It should be emphasized that once the waste stream is rejected, any valuable material turns into a liability as an effluent material. The level of recovery for such situations needs careful consideration. It may be economic to carry out additional separation of valuable material with a view to recycling that additional recovered material, particularly when the cost of downstream effluent treatment is taken into consideration.

Perhaps the most extreme situation is encountered with purge streams. Purges are used to deal with both feed impurities and byproducts of reaction. In the previous section, reduction of the size of purges by feed purification was considered. However, if it is impractical or uneconomic to reduce the purge by feed purification, or the purge is required to remove a byproduct of reaction, then additional separation can be considered.

Figure 27.7 shows the basic trade-off to be considered, as additional feed and product material is recovered from waste streams and recycled. As the fractional recovery increases, the cost of the separation and recycle increases. On the other hand, the cost of the lost material decreases. It should be noted that

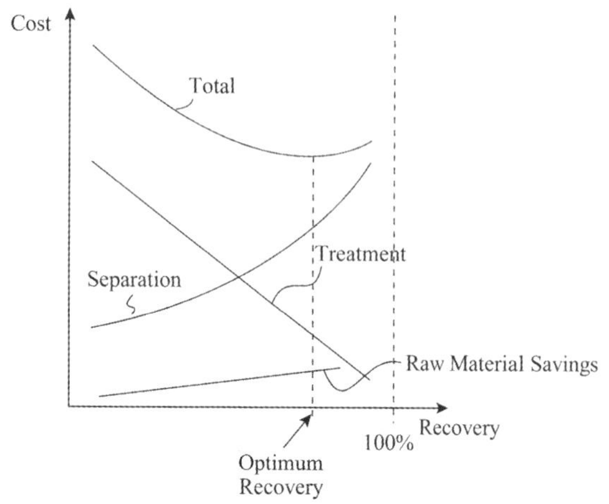

Figure 27.7
Effluent treatment costs should be included with raw materials costs when traded off against separation costs to obtain the optimum recovery. (Reproduced from Smith R and Petela E.A (1991c) Waste Minimization in the Process Industries Part 3 Separation and Recycle Systems, *Chem Eng*, **513**: 24, by permission of the Institution of Chemical Engineers.)

the raw materials cost is a *net* cost, which means that the cost of lost material should be adjusted to either:

a) add the cost of waste treatment for unrecovered material or

b) deduct the fuel value if the recovered material is to be burnt to provide useful heat in a furnace or boiler.

Figure 27.7 shows that the trade-off between separation and net raw materials cost gives an economic optimum recovery. It is possible that significant changes in the degree of recovery can have a significant effect on costs, other than those shown in Figure 27.7 (e.g. reactor costs). If this is the case, then these must also be included in the trade-offs.

It must be emphasized that any energy costs for the separation in the trade-offs shown in Figure 27.7 must be taken within the context of the overall heat integration problem. The separation might after all be driven by heat recovery.

11) *Additional reaction and separation of waste streams.* Sometimes it is possible to carry out further reaction as well as separation on waste streams. Some examples have already been discussed in Chapter 14.

12) *Process operations for raw materials efficiency.* Many of the problems associated with waste from process operations can be mitigated if the process is designed for low inventories of material in the process. This is also compatible with design for inherently safer processes (Chapter 28). Other ways to minimize waste from process operation are (Smith and Petela, 1992):

- Minimize the number of shutdowns by designing for high availability. Install more reliable equipment or standby equipment.
- Design continuous processes for flexible operation, for example, high turndown rate rather than shutdown.
- Consider changing from batch to continuous operation. Batch processes, by their very nature, are always at unsteady state, and thus difficult to maintain at optimum conditions.
- Install enough intermediate storage to allow reworking of off-specification material.
- Changeover between products causes waste since equipment must be cleaned. Such waste can be minimized by scheduling an operation to minimize product changeovers (see Chapter 16).
- Install a waste collection system for equipment cleaning and sampling waste, which allows waste to be segregated and recycled where possible. This normally requires separate sewers for organic and aqueous waste, collecting to sump tanks and recycle or separate and recycle if possible. This was discussed in Chapter 25. If equipment is steamed out during the cleaning process, the plant should allow collection and condensation of the vapors and recycling of materials where possible.
- Reduce losses from fugitive emissions and tank breathing, as discussed in Chapter 25.

There are many other sources of waste associated with process operations that can only be dealt with in the later stages of design, or after the plant has been built and became operational. For example, poor operating practice can mean that the process operates under conditions for which it was not designed, leading to waste.

27.3 Efficient Use of Raw Materials Between Processes

When considering raw materials efficiency within individual processes, it becomes clear that there are often limited options to reuse and recycle material. Practical and economic considerations constrain the performance or reactors and separation and recycle systems. Once the boundaries of the performance have been encountered there is little option other than to consider the material not used for products, and byproducts with no commercial value, to be waste. Yet a waste stream from one process can be a useful feed to another process. Thus, an improvement analysis from the perspective of raw materials must consider upstream and downstream processes.

Many of the products from the chemical industry, such as transportation fuels, detergents, soaps and cosmetics, are consumer end products. However, some three-quarters of chemicals manufactured are not consumer products, but are used to make other products by other parts of the chemical industry. The industry uses a wide range of raw materials from crude oil to minerals to air. The industry needs to form an integrated network of production plants linked by material flows. It is necessary as much as possible to use byproducts from an individual process that have no direct commercial value as raw material for other processes for the efficient use of raw materials and, at the same time, minimize the amount of waste and emissions. Maximizing

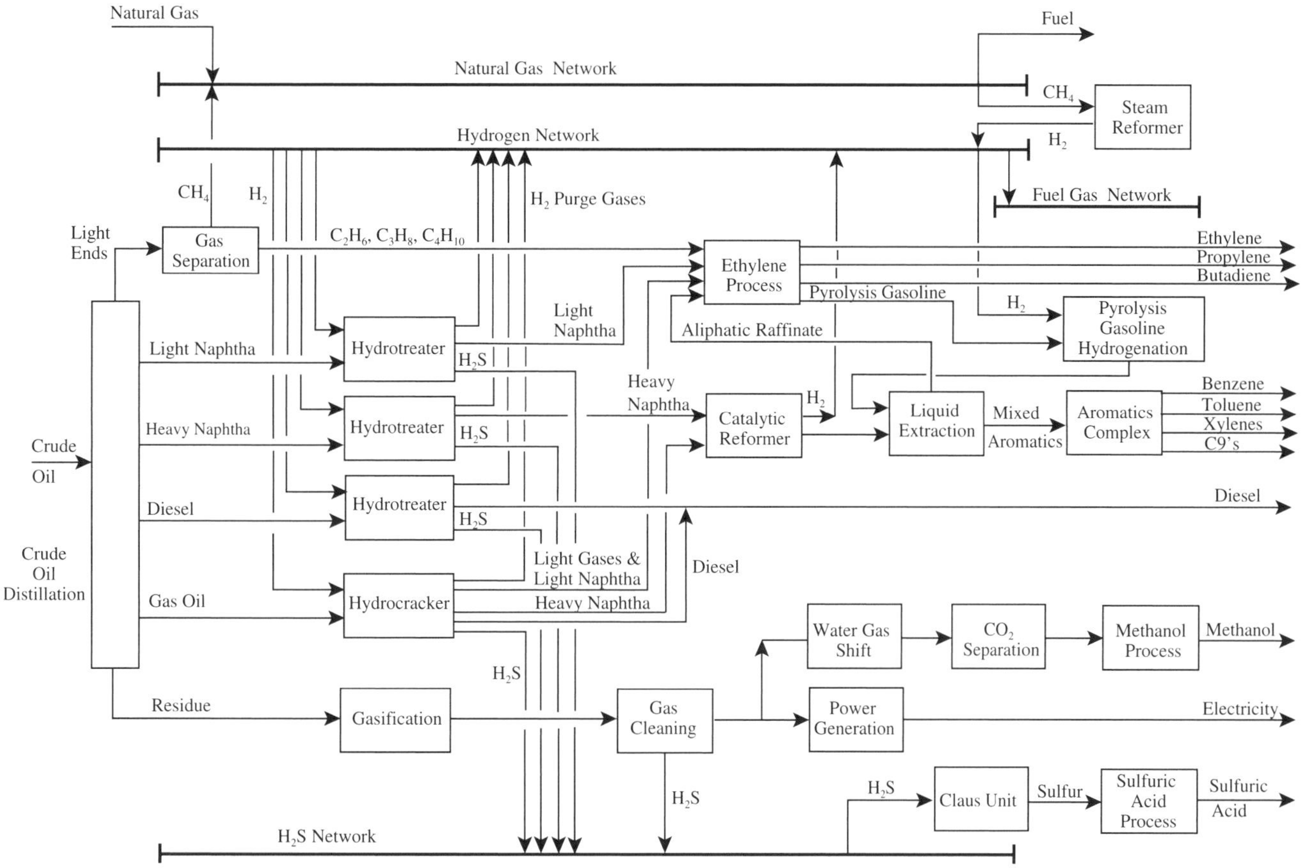

Figure 27.8

An example of an integrated process network.

the efficiency of raw materials use and, at the same time, minimizing the discharge of process effluents requires *process networks* or *integrated chemical clusters* to be created involving different production processes that reuse and recycle materials between the different processes.

Figure 27.8 shows an example of a process network. The process raw material is crude oil. This is first distilled into various fractions. However, before the distillation fractions can be used, each fraction (with the exception of the residue) must be *hydrotreated* by reaction with hydrogen to remove sulfur and nitrogen before being transformed into products. In the case of the gas oil, this is fed to a *hydrocracker*, which in addition to removing the sulfur and nitrogen also breaks up the larger hydrocarbon molecules into smaller molecules. The hydrogen feed required by the hydrotreaters and hydrocracker in Figure 27.8 is generated in a *steam reformer* (see Chapter 5) and is distributed in a hydrogen network. The hydrogen network is a distribution system involving different purities and pressures of hydrogen. In turn, the hydrogen production in the steam reformer requires a feed of natural gas supplied from a natural gas network that is also distributed for fuel. The nitrogen impurity in the hydrotreater and hydrocracker feed streams reacts to NH_3, which is readily absorbed in water. The sulfur impurity reacts to H_2S, which is collected in an H_2S network. The H_2S is transformed into elemental sulfur in a Claus Process (see Chapter 25) and then to sulfuric acid (see Chapter 5). The hydrotreaters and hydrocracker create purge gas steams that contain significant amounts of hydrogen, albeit at significantly lower hydrogen concentrations than the hydrogen feed. These purge streams are normally returned to the hydrogen network for reuse in other processes, or potentially upgrading through purification such as membranes or pressure swing adsorption (see Chapter 9). However, ultimately some gas with low quality needs to be rejected from the hydrogen network to a fuel gas system for use as fuel in fired heaters and boilers. In Figure 27.8 the light naphtha from the crude oil distillation, which is mainly straight-chain and branched alkanes, is transformed to alkene (olefinic) products in an ethylene process. The heavy naphtha, which contains aromatic compounds and cyclic alkanes as well as straight-chain and branched alkanes, is fed to a *continuous catalytic reformer*. The continuous catalytic reformer reacts with the feed to increase the aromatic portion of the feed. The reactor product from the continuous catalytic reformer is passed to a liquid extraction process (see Chapter 9) that separates the aliphatic material as raffinate from the aromatic compounds. The aliphatic raffinate from the extraction is fed to the ethylene process and the mixed aromatics from the extraction is fed to an aromatics complex to be transformed into various pure aromatic products. The *pyrolysis gasoline* (also known as *debutanized aromatic concentrate* or *pygas*) from the ethylene process is hydrogenated to saturate various alkene components (which would otherwise cause problems in the downstream processes) and is then fed to the reformer liquid extraction to separate the aliphatic and aromatic compounds. The diesel fraction from the crude oil distillation in Figure 27.8 is sold as diesel fuel. The gas oil fraction is transformed in a *hydrocracker* into lighter products that are fed to the ethylene process, a middle product that is fed to the continuous catalytic reformer and a heavy fraction that is sold as diesel. The residue from the crude oil distillation in Figure 27.8 is sent to a gasifier (see Chapter 25) where the residue is transformed into synthesis gas consisting mainly of H_2 and CO, but with quantities of CO_2, H_2O, CH_4, H_2S, COS, N_2, NH_3 and HCN. After cleaning of the gas to remove particulates and sulfur, in Figure 27.8 part is used to generate electricity in a combined cycle (see Chapter 25) and part has its composition adjusted before being transformed into methanol.

Many variations around the arrangement in Figure 27.8 are possible. For example, hydrogen for the hydrogen network could be produced by the gasification, rather than a steam reformer. Different liquid fuels (e.g. jet fuel) can be manufactured. Alternatively, no liquid fuels might be produced and all streams used for chemical production. Different chemicals can be manufactured. The most important point to be noted from the example in Figure 27.8 is that some processes produce byproducts that do not have a direct commercial value. These are reused or recycled in other processes to be transformed into useful products and byproducts with a commercial value. A network is created that uses materials in an overall efficient way by transforming as much of the raw material as possible into useful products and byproducts. Process networks need to be designed such that, as much as possible, byproducts from processes with no commercial value can be used as feeds to other processes. However, there are significant constraints that prevent reuse and recycling between different processes, particularly the reaction chemistry of the receiving process. The objective is to ideally transform all of the raw material into useful products and byproducts with nothing left over. It is clear that the wider the process network envelope, the more likely it is to be able to integrate the material flows efficiently.

Such complex networks as the one in Figure 27.8 need to navigate through volatile commercial markets with variations in the demand for, and price of, the products and valuable byproducts. The linking of process material flows links the operation of the various processes. Processes can breakdown and require shutdown for maintenance. This necessitates the strategic use of storage for the intermediate products. The design of process networks is therefore challenging from a number of perspectives. Raw materials efficiency should be maximized overall and the process network must be designed to be as flexible as possible, both to accommodate changes in the market and the operability of the various processes. Another problem in creating process networks is that the network might stretch across the boundaries of different companies with the resulting contractual complexity. Waste streams from one company finding use in another company is sometimes referred to as *industrial symbiosis* or *industrial ecology*.

To maximize the sustainability of a manufacturing system requires the material flows within and between processes to be exploited as efficiently as possible. The whole supply chain, both upstream and downstream, needs to be considered to maximize sustainability. However, this is not the whole picture and other parts of the manufacturing system need to be considered. Energy, water and waste treatment also contribute to both the economic and environmental impact. This will be considered later in this chapter.

27.4 Exploitation of Renewable Raw Materials

Rather than use crude oil, natural gas and coal as sources of raw materials for the production of chemical products, *renewable* sources of raw materials can be used as an alternative for the production of many chemicals. Renewable sources of raw materials can be defined to be resources that are replenished on a human timescale. Biomass as a renewable source of raw materials can be used for the production of many chemicals. Biomass is biological material derived from living or recently living organisms and includes the entire biological food chain from simple microorganisms to mammals and other animals. At the primary level, the production of biomass is driven by the capture of sunlight, which is transformed into chemical energy by photosynthesis. In the context of renewable sources of raw materials, biomass most often refers to plant-based materials and photosynthetic microorganisms such as some algae and fungi. To grow, these forms of biomass absorb carbon dioxide from the atmosphere. A non-inclusive list of sources of biomass is:

- wood from forestry or from wood processing,
- carbohydrate-based crops such as corn, sugarcane and wheat,
- oil-based crops such as palm, rapeseed and canola,
- high-yield crops grown specifically to be exploited for chemical or energy production, including woody crops such as willow or poplar and elephant grasses,
- agricultural residues from harvesting or processing,
- food waste from food and drink manufacture,
- consumer food waste,
- industrial waste from biochemical processing,
- algae or fungi harvested for chemical or energy production.

Various processes can be used to convert the biomass to intermediate products that can then be used for chemical production:

1) *Fermentation.* Fermentation is a biochemical process that converts sugars to acids, gases and alcohols. The acids and alcohols are useful chemicals. For example, ethanol produced from fermentation can be dehydrated to diethyl ether or ethylene, oxidized to acetaldehyde and on to acetic acid, reacted with carboxylic acids to form esters, and so on.
2) *Bioaccumulation.* Some microorganisms can exploit biomass by consuming it and converting it into oils and/or bioplastics, which accumulate inside them. The products can then be

harvested directly with some downstream processing. This can constitute an effective transformation of biomass to bioplastics or to oils.

3) *Anaerobic digestion.* As discussed in Chapter 26, anaerobic digestion can be used for biochemical conversion of biomass to a gas consisting mainly of methane and carbon dioxide.
4) *Photobiological production of hydrogen from algae.* Algae produce hydrogen under certain conditions. If certain types of algae are deprived of sulfur they will switch from the production of oxygen, as in normal photosynthesis, to the production of hydrogen.
5) *Gasification and synthesis gas conversion.* Like coal and petroleum residue, biomass can be transformed into synthesis gas consisting mainly of hydrogen and carbon monoxide in a gasification process. The hydrogen and carbon monoxide can then be used as the basis of the production of many chemicals. The synthesis gas can be reacted to methanol and then on to acetic acid, dimethyl ether, formaldehyde, olefins (alkenes), and so on. Alternatively, the synthesis gas can be reacted to ammonia and then on to ammonium phosphate, nitric acid, urea, and so on. Another option is to react the synthesis gas to liquid fuels (gasoline, jet fuel and diesel) using Fischer Tropsch technology. Thus, synthesis gas from gasification can be used as the starting point for an enormous variety of chemicals.
6) *Pyrolysis.* Pyrolysis is thermal decomposition occurring in the absence of oxygen at temperatures higher than 600 °C. The products from the pyrolysis are bio-oil, water, char (carbon) and a gas consisting mainly of hydrogen and carbon monoxide.
7) *Hydrothermal liquefaction.* Both gasification and pyrolysis require dried biomass as feed, and the processes occur at temperatures higher than 600 °C. Hydrothermal liquefaction involves direct liquefaction of biomass to bio-oil in the presence of water, and perhaps a catalyst. The reaction temperature is less than 400 °C.
8) *Transesterification.* Diesel fuel (biodiesel) can be produced by the transesterification of vegetable oils with short-chain alcohols (typically methanol or ethanol). Large quantities of impure byproduct glycerol are also produced.
9) *Biosynfining.* Biosynfining catalytically converts the triglycerides and/or fatty acids from fats, algae and vegetable oils to a synthetic kerosene and synthetic diesel that are rich in alkanes, a renewable naphtha and light hydrocarbon gases. The process involves three steps. First, the raw feedstocks are treated to remove catalyst poisons and water. In the second step, the fatty acid chains are deoxygenated and transformed into mainly alkanes in a hydrotreater by reaction with hydrogen. In the third step the long straight-chain alkanes are hydrocracked to shorter branched alkanes. The hydrocracked products fall mainly in the kerosene and naphtha boiling range. The naphtha stream and light gases can be used as feed to an ethylene process.

It would be a mistake to assume that just because a source of materials is renewable, that this leads to a process that is more sustainable. This can only be judged by a full life cycle assessment of alternative raw material sources. The production of biomass can compete with finite resources for the production of food and creates its own discharges to the environment.

27.5 Efficient Use of Energy

Consider now improvement analysis with respect to energy for the process being designed or modified. Later the wider energy network must be considered. The sources of emissions associated with energy use in processes are hot utilities (including cogeneration), cold utilities and the cooling water system. Furnaces, steam boilers, gas turbines and diesel engines all produce emissions as gaseous combustion products. These combustion products contain carbon dioxide, oxides of sulfur and nitrogen, and particulates (metal oxides, unburnt carbon and hydrocarbon), which contribute in various ways to the greenhouse effect, acidification and the formation of smog. As well as gaseous waste, the combustion of coal produces solid waste as ash from combustion and particulates collected from the flue gas. In addition to gaseous emissions, steam generation creates aqueous emissions from boiler feedwater treatment. The emissions created by energy use tend to be less environmentally harmful than process waste. However, the quantities of energy-related emissions tend to be larger than process waste. This sheer volume can then result in greater environmental impact than process emissions.

1) *Life cycle energy emissions.* When considering energy related emissions, it is tempting to consider only the local emissions from the process and its utility system (Figure 27.9a). However, this only gives part of the picture. The emissions generated from central power generation associated with any power import are just as much part of the process as those emissions generated on-site (Figure 27.9b). Precombustion emissions should also be included if data are available. These emissions should be included in the assessment of energy related emissions:

$$\begin{aligned}\text{Life cycle emissions} = {} & [\text{Emissions from on-site utilities}] \\ & + [\text{Emissions from central power generation corresponding with the amount of electricity imported}] \\ & - [\text{Emissions saved at central power generation corresponding with the amount of electricity exported from the site}] \\ & + [\text{Indirect energy emissions}]\end{aligned} \qquad (27.2)$$

Cogeneration can make a very significant contribution to energy related emissions. Assessing only the local effects of cogeneration is misleading. Cogeneration increases the local utility emissions. Besides the fuel burnt to supply the heating demand, additional fuel must be burnt to generate the power. It is only when the emissions are viewed on a life cycle basis and the emissions from central power generation are viewed that the true picture is obtained. Centralized power stations have a poor efficiency of power generation compared to a

Figure 27.9

Local and global emissions. (Reproduced from Smith R and Petela E.A (1992b) Waste Minimization in the Process Industries Part 5 Utility Waste, *Chem Eng*, **523**: 32, by permission of the Institution of Chemical Engineers.)

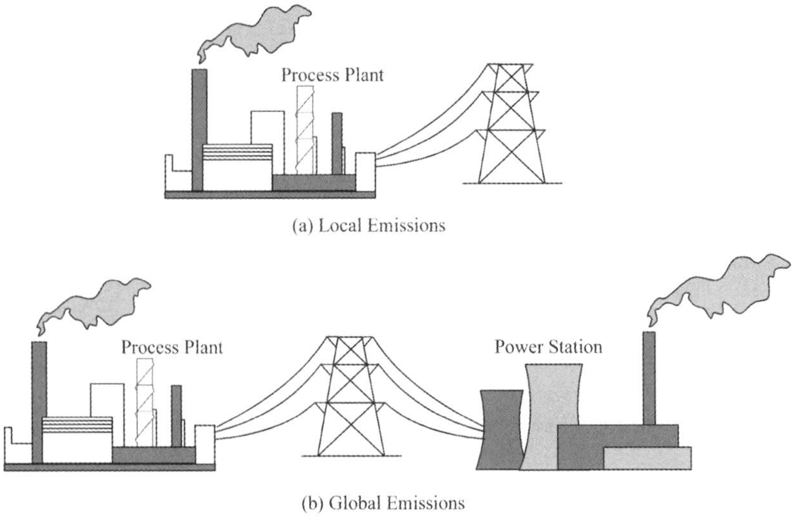

cogeneration system. Once the other inefficiencies associated with centralized power generation are taken into account, such as distribution losses, the gap between the efficiency of cogeneration systems and centralized power generation widens.

As an example, consider a process that requires a furnace to satisfy its hot utility requirements. Suppose it is a state-of-the-art furnace with a thermal efficiency of 90% producing $300\,kg\,CO_2 \cdot h^{-1}$ for each megawatt of heat delivered to the process. Power is being imported from centralized generation via the grid. If, instead of the furnace, a gas turbine is installed, this produces $500\,kg\,CO_2 \cdot h^{-1}$ for each megawatt of heat delivered to the process, an increase in local emissions of $200\,kg\,CO_2 \cdot h^{-1}$ per megawatt of heat. However, the gas turbine also generates 400 kW of power, replacing that much in centralized generation. If the same power was generated centrally to supplement the furnace, $450\,kg\,CO_2 \cdot h^{-1}$ would be released from centralized generation, giving a life cycle emission of $750\,kg\,CO_2 \cdot h^{-1}$ for the furnace plus power from the grid (Smith and Delaby, 1991).

2) *Fuel switch.* The choice of fuel used in furnaces and steam boilers has a major effect on the gaseous emissions from products of combustion. For example, a switch from coal to natural gas in a steam boiler can lead to a reduction in carbon dioxide emissions of typically 40% for the same heat released. This results from the lower carbon content of the natural gas. In addition, it is likely that the switch from coal to natural gas will also lead to a considerable reduction in both SO_X and NO_X emissions. Such a fuel switch, while being desirable in reducing emissions, might be expensive. If the problem is SO_X and NO_X emissions, there are other ways to combat these, as discussed in Chapter 25.

3) *Energy efficiency of the process.* If the process requires a furnace or steam to provide hot utility, then any excessive use of hot utility will produce excessive emissions from the generation of hot utilities through excessive generation of CO_2, NO_X, SO_X and particulates. Improved heat recovery will reduce the overall demand for utilities and hence reduce energy related emissions. Process energy efficiency has been discussed in detail in Chapters 17 to 22.

4) *Energy efficiency across processes.* Most process heating is provided from a steam network involving steam at various pressures. The processes on a site are linked to the steam network and, as discussed in Chapter 23, can exchange heat through the network. Utility steam boilers are normally required to supplement any steam generation from waste heat in the processes. Typically there are two or three steam mains to distribute steam at different pressures around the site. Local heating from a fired heater might be required in some processes. The steam network is also an integral part of the power generation system.

Site composite curves can be used to represent the site heating and cooling requirements thermodynamically, as discussed in Chapter 23. This allows the analysis of thermal loads and levels on the site. Using the models for steam turbines and gas turbines allows cogeneration targets for the site to be established (Chapter 23).

Once the energy recovery within and between processes has been maximized, there is inevitably low temperature waste heat rejected from the site. Petroleum, petrochemical and chemical processes cannot use large quantities of low temperature waste heat by the nature of their processes. Below 200 °C there is limited use for the heat. Below 100 °C there is even less use for the heat. However, this is not true of biochemical, food and beverage processes, which mostly operate at lower temperatures. To make efficient use of energy across processes requires process networks to be designed such that the recovery of energy across the network is maximized, in a similar way to the recovery of material flows is maximized. The priority must be to minimize the waste heat rejection. This has been discussed in detail in Chapter 23.

5) *Exploitation of low temperature waste heat.* Once waste heat rejection has been minimized, rather than reject heat to the

environment in cooling towers or air coolers at temperatures in excess of 100 °C, options to use the waste heat include:

a) *Boiler feedwater preheating.* Boiler feedwater makeup is around ambient temperature and low-temperature waste heat can be used to preheat the makeup water before being fed to the deaerator. This decreases the steam fed to the deaerator and in turn the fuel fired in the steam boilers. In addition to the preheating of boiler feedwater makeup, condensate return can also be preheated before entering the deaerator. If higher temperature waste heat is available, the boiler feedwater after the deaerator can be preheated before entering the steam boilers, which will decrease the fuel fired in the boilers.

b) *Space heating onsite.* The site might feature packaging areas, warehouses, workshops and offices that require space heating. Low-temperature waste heat can be used to substitute steam generated in the steam boilers and electric space heating.

c) *District heating offsite.* District heating supplies heat, typically, in the form of hot water to commercial and domestic users outside the factory fence. The economic performance depends on the capital cost of the distribution system, the distance that hot water must be transported and the pumping cost for the distribution system. Operation of the system is problematic, as the demand for heating will vary with the time of the day, day of the week and season of the year. Yet chemical manufacturing mostly operates at a relatively steady state. Storage of hot water can to some extent overcome this problem.

d) *Power generation using condensing turbines.* Chapter 23 considered power generation in steam turbines. Expansion in backpressure turbines is preferred, as long as the expanded steam has a use for further power generation or process heating. Ultimately, if expanded steam has no scope for process heating, additional power can be generated by expansion down to vacuum pressure to a temperature just above the cooling water temperature. However, the wetness of the steam after expansion must be no more than around 10%, otherwise it can lead to turbine damage (see Chapter 23). The heat from the steam turbine exhaust is then rejected to ambient at a temperature slightly above ambient.

e) *Power generation using Organic Rankine and Kalina Cycles.* Steam is the normal working fluid for the conventional power generation. Steam offers advantages of low cost, nonflammability and nontoxicity. However, in the case of low-temperature heat sources, steam as a working fluid operates at a relatively low efficiency. Replacing steam in a steam Rankine cycle as the working fluid with an organic fluid in an Organic Rankine Cycle can increase the power generation efficiency for low-temperature applications compared with steam. However, even with an organic working fluid, the efficiency is still low. The main challenge with the Organic Rankine Cycle is the choice of appropriate working fluid (Victor, Kim and Smith, 2013). Aside from the working fluid, a significant difference between a conventional Rankine cycle and an Organic Rankine Cycle is the machine employed for the expansion of the fluid to generate power. A conventional turbine cannot be used and a specialized expander device is required to cater for the properties of the organic working fluid. Another option is the Kalina Cycle. Rather than using steam or an organic fluid, the Kalina Cycle uses a binary mixture of ammonia and water as the working fluid. The binary properties of the mixture produce a nonisothermal evaporation and condensation in the cycle, which can contribute to a higher efficiency as compared with pure working fluids. The composition of the mixture of the working fluid in a Kalina Cycle can be optimized to suit the heat input and rejection temperatures. However, the power generation efficiencies of both Organic Rankine Cycles and Kalina Cycles are low for a low-temperature heat source (below 200 °C).

f) *Compression heat pumping.* Compression heat pumps have been described in Section 17.10 and a schematic of a simple compression heat pump is shown in Figure 17.36. The heat pump absorbs heat at a low temperature in the evaporator, consumes power when the *working fluid* is compressed and rejects heat at a higher temperature in the condenser. The condensed working fluid is expanded and partially vaporizes. The cycle then repeats. The performance of the compression heat pump is measured by the coefficient of performance:

$$COP_{CHP} = \frac{\text{Useful heat output}}{\text{Input energy}} = \frac{Q_{COND}}{W} \tag{27.3}$$

where COP_{CHP} = coefficient of performance of compression heat pump (–)
Q_{COND} = useful heat output from the heat pump condenser (W, kW)
W = power input for the compressor (W, kW)

Performance depends on the choice of working fluid, the efficiency of the compressor and the temperatures of the evaporator and condenser. The performance can be predicted by multiplying the ideal *COP* (*COP* of a reversible heat pump) by a Carnot efficiency:

$$\begin{aligned} COP_{CHP} &= \frac{Q_{COND}}{W} = \eta_{CHP} COP_{CHP}^{IDEAL} \\ &= \eta_{CHP} \frac{T_{COND}}{T_{COND} - T_{EVAP}} \end{aligned} \tag{27.4}$$

where η_{CHP} = Carnot efficiency of compression heat pump (–)
T_{COND} = condenser temperature (K)
T_{EVAP} = evaporator temperature (K)

The smaller the temperature lift required to make the waste heat useful, the more likely the heat pump is to be economic. The Carnot efficiency can be predicted by (Oluleye *et al.*, 2016):

$$\eta_{CHP} = A + \frac{B}{COP_{CHP}^{IDEAL}} = A + B\left(\frac{T_{COND} - T_{EVAP}}{T_{COND}}\right) \tag{27.5}$$

where A, B = correlating coefficients given in Table 27.2 assuming a compressor isentropic efficiency of 75%

Table 27.2

Correlating coefficients for mechanical heat pumps (Oluleye, Smith and Jobson, 2016).

Working fluid	T_{COND} (°C)	T_{EVAP} range (°C)	A	B
Propylene	50	10–40	0.6705	−0.4313
	60	10–50	0.6391	−0.4264
	70	10–60	0.5913	−0.3847
	80	10–70	0.5110	−0.2541
Propane	50	10–40	0.6811	−0.5107
	60	10–50	0.6550	−0.5290
	70	10–60	0.6165	−0.5254
	80	15–70	0.5554	−0.4697
i-Butane	50	10–40	0.7230	−0.5688
	60	10–50	0.7108	−0.5859
	70	10–60	0.7024	−0.6724
	80	15–70	0.6871	−0.7265
	90	20–80	0.6663	−0.7795
	100	25–90	0.6375	−0.8269
	110	25–100	0.5861	−0.7916
n-Butane	50	10–40	0.7319	−0.5154
	60	10–50	0.7259	−0.5602
	70	10–60	0.7181	−0.6077
	80	15–70	0.7081	−0.6582
	90	20–80	0.6952	−0.7107
	100	25–90	0.6781	−0.7639
	110	30–100	0.6551	−0.8149
	120	35–110	0.6217	−0.8504
	130	35–110	0.5586	−0.7767
Chlorine	50	10–40	0.7228	−0.3337
	60	10–50	0.7144	−0.3261
	70	10–60	0.7037	−0.3167
	80	10–70	0.6901	−0.3035
	90	20–80	0.6724	−0.2837
	100	25–90	0.6492	−0.2526
	110	30–100	0.6175	−0.2022

(continued)

Table 27.2 *(Continued)*

Working fluid	T_{COND} (°C)	T_{EVAP} range (°C)	A	B
Ammonia	50	10–40	0.7267	−0.4774
	60	10–50	0.7132	−0.4003
	70	15–60	0.7006	−0.3861
	80	25–70	0.6932	−0.4851
	90	25–80	0.6628	−0.3684
	100	25–90	0.6294	−0.282
	110	25–100	0.5732	−0.1195
	120	25–100	0.4971	0.0586

Equations 27.4 and 27.5 provide a simple model for the screening of options. However, once the screening has been completed, the model should be verified with a detailed simulation.

g) *Absorption heat pump.* The basic difference between a simple vapor compression refrigeration cycle and an absorption heat pump cycle is that the absorption heat pump cycle replaces the compressor with a combined unit consisting of a *generator*, an *absorber*, a *pump* and a *throttling valve* (Figure 27.10). Increase in the pressure of the working fluid is achieved by dissolving the working fluid in a solvent, pumping the solution to a higher pressure and then stripping the working fluid from the solvent by the input of heat. Replacement of the compressor with a pump leads to a significant reduction in the power requirements. However, the absorption cycle requires a source of heat for the stripping and is therefore suited to applications where a source of relatively high temperature waste heat is available. The basic flow arrangement is shown in Figure 27.10a and again in Figure 27.10b in terms of pressure and temperature. Low-temperature waste heat from the process is used to vaporize the *working fluid* (sometimes referred to as *refrigerant*) in the *evaporator* Q_{EVAP}. This heat input to the evaporator from the process is from a source of waste heat. The low-pressure vapor working fluid from the evaporator enters the absorber where it is absorbed at low pressure in a solution of the working fluid in a *solvent*. A quantity of heat Q_{ABS} is released at medium temperature as the working fluid vapor is absorbed. This heat should be recovered as useful heat. The solution from the absorber at a low pressure is then increased in pressure in a pump and increased in temperature in an *economizer*. The solution then enters the vapor generator where the working fluid is stripped from the solvent at high pressure from the input of Q_{GEN} from a higher temperature waste heat source. The high-pressure solution is cooled in the economizer, decreased in pressure and returned to the absorber. The working fluid vapor at high pressure and high temperature

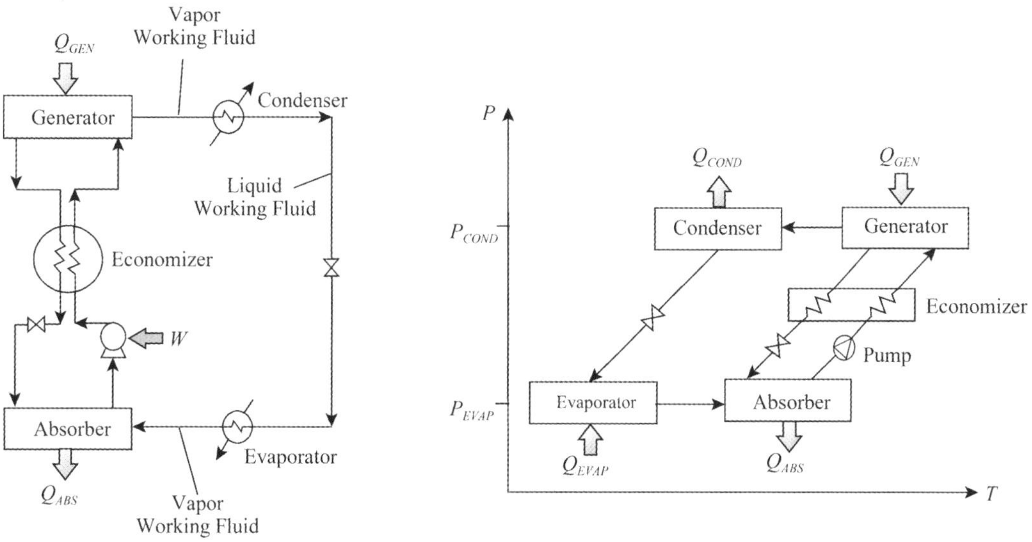

(a) Absorption heat pump cycle.

(b) Pressure–temperature representation of the absorption heat pump cycle.

Figure 27.10
Absorption heat pumping.

from the vapor generator is condensed in the *condenser* by the removal of heat Q_{COND}, which should be recovered as useful process heat. The liquid working fluid from the condenser is expanded in a valve and then enters the evaporator where low-temperature waste heat is absorbed from the process and the cycle repeats.

Water can be used as the working fluid (refrigerant) with lithium bromide as the solvent (absorbent). Alternatively, ammonia can be used as the working fluid (refrigerant) with water as the solvent (absorbent). When ammonia–water is used a *rectifier* may be required after the generator. The rectifier is a refluxed cooler, rejecting heat to a sink and increasing the concentration of the vapor. Other combinations of working fluid and solvent are possible. Water–lithium bromide systems are most commonly used. The phase equilibrium in the cycle can be represented using a Dűhring plot in which the boiling point of the solution is plotted against that of the refrigerant (water). Figure 27.11 shows a Dűhring plot for the water–lithium bromide system. The temperature of the solution in the generator can be represented on the x axis (abscissa) with the corresponding condenser temperature represented on the y axis (ordinate). The lines on the plot represent fixed concentration of the solution. Fixing the generator and condenser temperatures thereby fixes the concentration of the solution in the generator. Alternatively, fixing the generator temperature and the concentration fixes the condenser temperature or fixing the condenser temperature and the concentration fixes the generator temperature. Similarly, the temperature of the absorber can be represented on the x axis with the corresponding evaporator temperature on the y axis. To achieve the required temperatures requires the pressures in the cycle to be adjusted according to the phase equilibrium. The pressure in the condenser (or evaporator) can be fixed by the temperature and vapor pressure relation for water. This then also fixes the pressure in the generator (or absorber).

The performance of the absorption heat pump is again measured by the coefficient of performance:

$$COP_{AHP} = \frac{\text{Useful heat output}}{\text{Input energy}} = \frac{Q_{ABS} + Q_{COND}}{Q_{GEN} + W_{PUMP}} \tag{27.6}$$

where COP_{AHP} = coefficient of performance of absorption heat pump (–)
Q_{ABS} = useful heat output from the heat pump absorber (W, kW)
Q_{COND} = useful heat output from the heat pump condenser (W, kW)
Q_{GEN} = heat input from the heat pump generator (W, kW)
W_{PUMP} = power input for the liquid pump (W, kW)

In most cases the power input from liquid pumping can be neglected. A simplified model for the COP_{AHP} can again be developed on the basis of the ideal *COP*. For an ideal cycle, assuming expansions and pumping are carried out reversibly with no energy input or loss, the reversible heat pump becomes a reversible Carnot engine in which the entropy change in the evaporator and condenser are equal (see Dodge, 1944; Hougen, Watson and Ragatz, 1959; Smith and Van Ness, 2007):

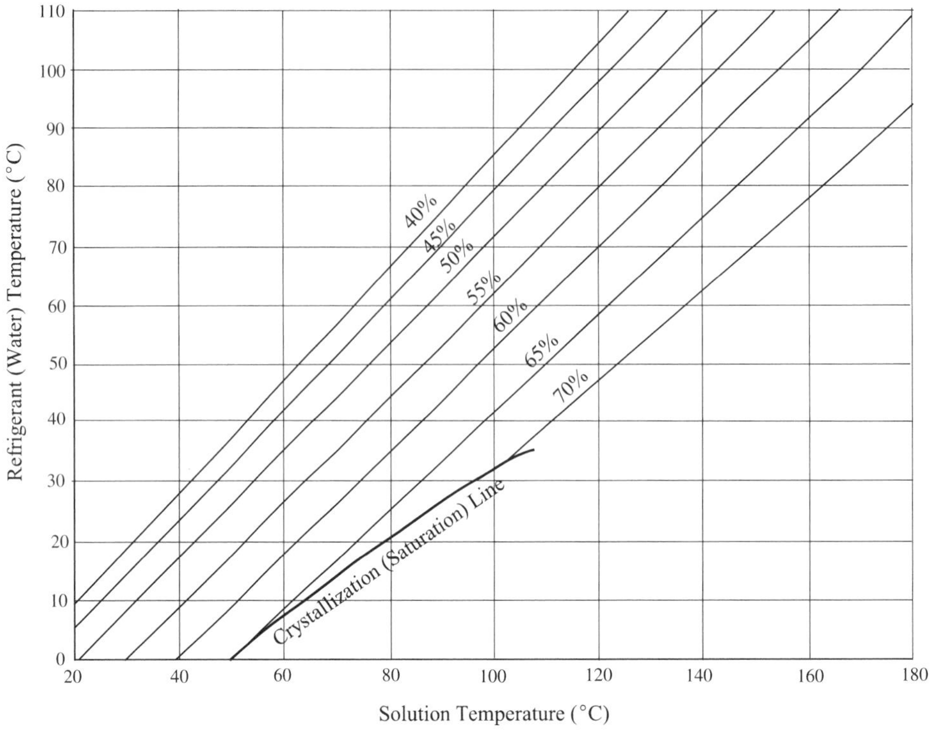

Figure 27.11
Dühring plot for the water–lithium bromide system.

$$\frac{Q_{EVAP}}{T_{EVAP}} = \frac{Q_{COND}}{T_{COND}} \quad (27.7)$$

where Q_{EVAP} = heat input to the evaporator (W, kW)
T_{EVAP} = temperature of the evaporator (K)
T_{COND} = temperature of the condenser (K)

For an ideal cycle, the change in entropy for the whole system is zero. Thus:

$$\frac{Q_{EVAP}}{T_{EVAP}} + \frac{Q_{GEN}}{T_{GEN}} = \frac{Q_{ABS}}{T_{ABS}} + \frac{Q_{COND}}{T_{COND}} \quad (27.8)$$

where T_{GEN} = temperature of the generator (K)
T_{ABS} = temperature of the absorber (K)

Combining Equations 27.7 and 27.8:

$$\frac{Q_{GEN}}{T_{GEN}} = \frac{Q_{ABS}}{T_{ABS}} \quad (27.9)$$

An overall energy balance, neglecting the energy input from the liquid pumping, gives:

$$Q_{EVAP} + Q_{GEN} = Q_{ABS} + Q_{COND} \quad (27.10)$$

Combining Equations 27.6, 27.7, 27.9 and 27.10 and neglecting the energy input from the pump:

$$COP_{AHP}^{IDEAL} = 1 + \left(1 - \frac{T_{ABS}}{T_{GEN}}\right)\left(\frac{T_{EVAP}}{T_{COND} - T_{EVAP}}\right) \quad (27.11)$$

where COP_{AHP}^{IDEAL} = ideal coefficient of performance of absorption heat pumps (–)

The actual coefficient of performance is given by:

$$\begin{aligned} COP_{AHP} &= \eta_{AHP} COP_{AHP}^{IDEAL} \\ &= \eta_{AHP}\left[1 + \left(1 - \frac{T_{ABS}}{T_{GEN}}\right)\left(\frac{T_{EVAP}}{T_{COND} - T_{EVAP}}\right)\right] \end{aligned} \quad (27.12)$$

Neglecting the energy input from liquid pumping, for a water–lithium bromide system the Carnot efficiency is given by (Oluleye, Smith and Jobson, 2015):

$$COP_{AHP} = A\,\eta_{AHP} + B \quad (27.13)$$

where A, B = correlating coefficients given in Table 27.3 for the water–lithium bromide system

Equations 27.12 and 27.13 must be solved simultaneously to obtain COP_{AHP}. These provide a simple model for the screening of options, but the model should be verified with a detailed simulation once the screening has been completed.

Table 27.3

Correlating coefficients for water–lithium bromide absorption heat pumps (Oluleye, Smith and Jobson, 2016).

T_{GEN} (°C)	T_{EVAP} (°C)	$T_{COND} = T_{ABS}$ (°C)	A	B
90	$20 < T_{EVAP} < 30$	50	−2.5064	3.4299
100	$20 < T_{EVAP} < 30$	50	−0.7448	2.2099
	$30 < T_{EVAP} < 40$	60	−2.9497	3.7592
110	$20 < T_{EVAP} < 30$	50	−0.5081	2.0366
	$30 < T_{EVAP} < 40$	60	−0.7478	2.2099
	$40 < T_{EVAP} < 50$	70	−2.4461	3.3795
140	$40 < T_{EVAP} < 50$	80	−1.7978	2.8816

h) *Heat transformers.* A heat transformer can also be used to increase the temperature of waste heat to a temperature high enough to make it useful. The basic flow arrangement is shown in Figure 27.12a and again in Figure 27.12b in terms of pressure and temperature. The arrangement has the same components as the absorption heat pump, but by contrast with the absorption heat pump, the condenser and generator work at low pressure and the evaporator and absorber work at high pressure. Waste heat at an intermediate temperature (between the process demand level and the environmental level) is supplied to the evaporator and the generator. Useful heat at a higher temperature is rejected from the absorber. In Figure 27.12, waste heat at an intermediate temperature Q_{GEN} to be upgraded is input to the generator, which vaporizes part of the working fluid from a solution of working fluid and solvent. The vaporized working fluid flows to the condenser where it is condensed, rejecting heat Q_{COND}. The heat rejected from the condenser might be recovered usefully as process heat or rejected to the environment via cooling water or air cooling. The liquid working fluid from the condenser is increased in pressure using a pump and enters the high-pressure evaporator. The working fluid is then vaporized in the evaporator by using a second quantity of intermediate temperature waste heat Q_{EVAP}. The vaporized working fluid enters the absorber, where it is absorbed by a solution of the working fluid and solvent. The heat of absorption increases the temperature of the solution. The absorber delivers useful heat Q_{ABS} at a higher temperature. The solution from the absorber is cooled in the economizer and reduced in pressure in a throttle valve before entering the generator. The cycle then repeats.

27

Overall, the heat transformer has the ability to increase the temperature of the solution to above the waste heat source temperature, which contrasts with the absorption

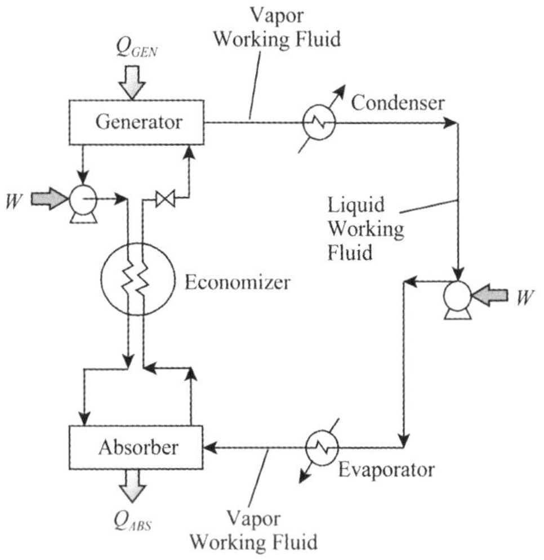

(a) Heat transformer cycle.

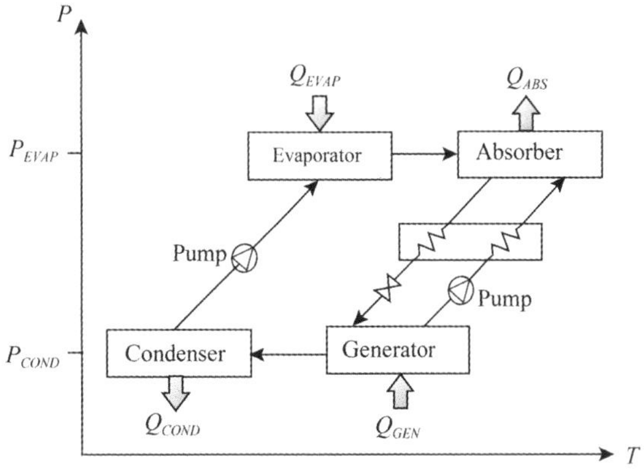

(b) Pressure–temperature representation of the heat transformer cycle

Figure 27.12

Heat transformer heat pumping.

heat pump where the highest temperature heat is input. The use of liquid pumps in the heat transformer means that it requires less power to operate than a compression heat pump. The input of waste heat is necessary to vaporize the working fluid for absorption and to release the working fluid from the solvent.

As with the absorption heat pump, water can be used as the working fluid with lithium bromide as the solvent. This is the most commonly used system. As with the absorption heat pump, the relationship between the absorber and evaporator temperatures and the generator and condenser temperatures can be represented on a Dűhring plot, as shown in Figure 27.11 for the water–lithium bromide system. Ammonia can also be used as working fluid with water as the solvent. Other combinations are possible.

The performance of the absorption heat transformer is measured by the coefficient of performance:

$$COP = \frac{\text{Useful heat output}}{\text{Input energy}} = \frac{Q_{ABS}}{Q_{GEN} + Q_{EVAP} + W_{PUMP}} \tag{27.14}$$

Equations 27.7 to 27.10 also apply to the absorption heat transformer. Combining Equations 27.14, 27.7, 27.9 and 27.10, neglecting the energy input from the liquid pump:

$$COP_{AHT}^{IDEAL} = \frac{T_{ABS}(T_{EVAP} - T_{COND})}{T_{ABS}(T_{EVAP} - T_{COND}) + T_{COND}(T_{ABS} - T_{COND})} \tag{27.15}$$

An alternative, but equivalent, form of Equation 27.15 is given by:

$$COP_{AHT}^{IDEAL} = \frac{T_{ABS}(T_{EVAP} - T_{COND})}{T_{GEN}(T_{EVAP} - T_{COND}) + T_{EVAP}(T_{ABS} - T_{GEN})} \tag{27.16}$$

Neglecting the energy input from liquid pumping, for a water–lithium bromide system the Carnot efficiency is given by (Oluleye, Smith and Jobson, 2016):

$$COP_{AHT} = \eta_{AHT} COP_{AHT}^{IDEAL} \tag{27.17}$$

$$COP_{AHT} = A\,\eta_{AHT} + B \tag{27.18}$$

where A, B = correlating coefficients given in Table 27.4 for the water–lithium bromide

Equations 27.17 and 27.18 need to be solved simultaneously to obtain COP_{AHT}. Again, the model should be verified with a detailed simulation once the screening has been completed.

i) *Absorption refrigeration.* Absorption refrigeration was discussed in Section 24.11. Absorption refrigeration uses the same cycle as the absorption heat pump, but operates across ambient temperature to provide refrigeration. This might be additional refrigeration capacity for the site or to substitute compression refrigeration by utilizing waste heat. Alternatively, cooling can be exported from the site. Like the absorption heat pump, the absorption refrigeration cycle is suited to applications where a source of relatively high temperature waste heat is available for the generator. The cycle is the same as that shown in Figure 27.10. Heat is absorbed from the process in the evaporator Q_{EVAP} to vaporize the working fluid, which is the refrigeration cooling duty. Heat Q_{ABS} is released at medium temperature as the working fluid vapor is absorbed in the solvent. Where possible this heat should be exploited as useful heat. Otherwise, the heat needs to be released to the environment via cooling water or air cooling. The working fluid is stripped from the solvent at high pressure from the input of Q_{GEN} from a high-temperature waste heat source. The heat Q_{COND} removed in the condenser might be recovered usefully as process heat or rejected to the environment via cooling water or air cooling.

The same working fluids as used in absorption heat pumps and heat transformers can also be used in absorption refrigeration cycles. Waste heat input should preferably be in excess of 90 °C. The water-lithium bromide system can be used for cooling down to 5 °C and ammonia-water

Table 27.4

Correlating coefficients for absorption heat transformers for T_{COND} = 30 °C (Oluleye, Smith and Jobson, 2016).

T_{EVAP} (°C)	T_{ABS} (°C)	T_{GEN} (°C)	A	B
40	$60 < T_{ABS} < 90$	$50 < T_{GEN} < 80$	0.6356	−0.0549
50	$70 < T_{ABS} < 100$	$50 < T_{GEN} < 80$	0.6303	−0.0461
60	$80 < T_{ABS} < 110$	$50 < T_{GEN} < 80$	0.6270	−0.0392
70	$90 < T_{ABS} < 120$	$50 < T_{GEN} < 80$	0.6190	−0.0305
80	$100 < T_{ABS} < 130$	$50 < T_{GEN} < 80$	0.5797	−0.00704
90	$120 < T_{ABS} < 140$	$60 < T_{GEN} < 80$	0.6568	−0.0407

system can be used down to −40 °C. The refrigeration produced from the waste heat in absorption refrigeration can be used for a number of purposes:

- service new refrigeration duties in a process;
- substitute a refrigeration duty being serviced by compression refrigeration to reduce the overall power input;
- substitute cooling water in a condenser with refrigeration to increase material recovery and decrease raw material losses;
- substitute cooling water to cool the inlet gas to a compressor and in the compressor interstage cooling to reduce the power input to compression;
- cool the inlet air to gas turbines to increase the efficiency (the exhaust heat from a gas turbine can be used to drive an absorption heat pump to cool the inlet air);
- provide air conditioning to onsite packaging areas, warehouses, workshops and offices in hot climates;
- export cooling offsite in the form of chilled water or chilled solutions (e.g. glycol–water, salt–water or alcohol–water) at temperatures below 0 °C to provide process cooling to neighboring businesses (e.g. food processing);
- export cooling in the form of chilled water to provide air conditioning offsite in hot climates.

The performance of the absorption refrigeration is measured by the coefficient of performance:

$$COP_{AR} = \frac{\text{Useful cooling}}{\text{Input energy}} = \frac{Q_{EVAP}}{Q_{GEN} + W_{PUMP}} \qquad (27.19)$$

Again neglecting the power input from liquid pumping:

$$COP_{AR} = \frac{Q_{EVAP}}{Q_{GEN}} = COP_{AHP} - 1 \qquad (27.20)$$

For waste heat below 100 °C, the most effective way to exploit the heat is to heat water for process use in processes that require hot water (e.g. food and beverage processes), boiler feedwater preheat, space heating onsite and district heating offsite. At a higher temperature, say in the range 100 °C to 200 °C, the waste heat can be used to generate additional power in an Organic Rankine Cycle or a Kalina Cycle. The cycle efficiency depends on the choice of working fluid, the temperature of the heat source and the heat rejection temperature. However, with waste heat of 200 °C, cycle efficiency is typically less than 20%. Heat pumping using vapor compression cycles is expensive, both from the point of view of capital cost and power requirements. The use of heat transformers reduces the power input relative to a vapor compression cycle, but the capital cost can be prohibitive. For heat pumping using the simple cycles described here, the economic temperature lift is normally less than 50 °C. More complex cycles can also be used to provide greater efficiency, but still the economic temperature lift is normally less than 80 °C. There is no simple solution to the exploitation of low temperature waste heat to increase the sustainability of operations. Heat recovery must be maximized in the higher-temperature operations in order to minimize the generation of low-temperature waste heat. Thereafter, the wider the horizon for the integration of waste heat, the more likely that an efficient use can be found for the waste heat (e.g. district heating systems).

Example 27.2 Figure 27.13a shows part of a site cold composite curve that is being serviced by heating from low-pressure steam. Figure 27.13b shows part of a site hot composite curve that is being serviced by cooling from cooling water. Cooling water is available at 20 °C, to be returned to the cooling tower at 25 °C. A source of higher temperature waste heat has also been identified in a fired heater flue gas with a temperature of 390 °C. Figure 27.13c shows the profile of the stack loss from the fired heater. The heat capacity flowrate of the flue gas is 15.974 kW · K^{-1}. The flue gas can be used to generate steam at 160 °C from boiler feedwater at 103 °C. The furnace is using a sulfur-free fuel gas, which allows the flue gas to be cooled to 100 °C. Assume the latent heat of steam generation is 2087 kJ · kg^{-1} and the enthalpy of water at 160 °C and 103 °C is 670 kJ · kg^{-1} and 432 kJ · kg^{-1} respectively. It should be noted in Figure 27.13 that the composite curves and the furnace stack loss profile have a ΔT_{min} adjustment of 10 °C incorporated. The utility streams are shown at their actual temperatures. It is proposed to substitute as much of the current low-temperature steam heat as possible with a combination of steam recovered from the flue gas and heat pumping part of the heat being rejected to cooling water.

a) Calculate how much steam can be generated from the furnace flue gas.
b) Using the steam generated from the flue gas to substitute the existing steam consumption, calculate the power required to provide the balance of the low-temperature process heating from the output of a compression heat pump, assuming ammonia to be used as the working fluid.
c) Assuming the power is supplied to the compression heat pump with a generation efficiency of 40%, calculate how much waste heat is created by the generation of the required power.
d) Rather than use a compression heat pump, it is proposed to use a water–lithium bromide absorption heat pump. In principle, the heat from the furnace flue gas can be used directly to provide the generator heat for the absorption heat pump. However, it is more practical to generate steam from the flue gas and use this to provide the generator heat. Calculate how much heat can be recovered using an absorption heat pump.

27

Solution

a) Assuming the steam generation is pinched at the steam generation temperature of 160 °C, heat available for steam vaporization

$$= 15.974 \times (380 - 160)$$
$$= 3514\,\text{kW}$$

Flowrate of steam

$$= \frac{3514}{2087}$$
$$= 1.684\,\text{kg} \cdot \text{s}^{-1}$$

Check whether the boiler feedwater preheat can be accommodated. Heat required for boiler feedwater preheat

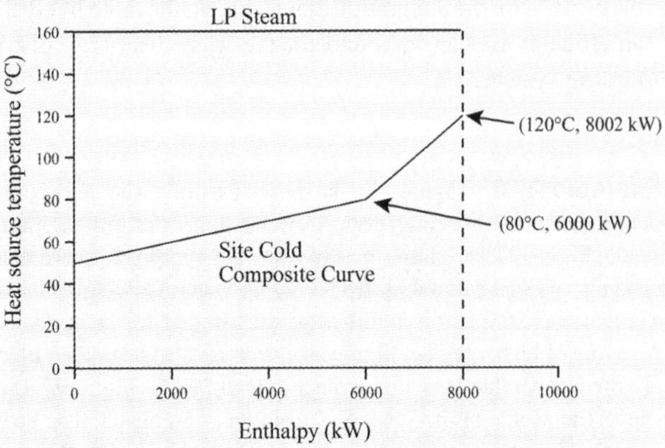

(a) Part of the site cold composite curve.

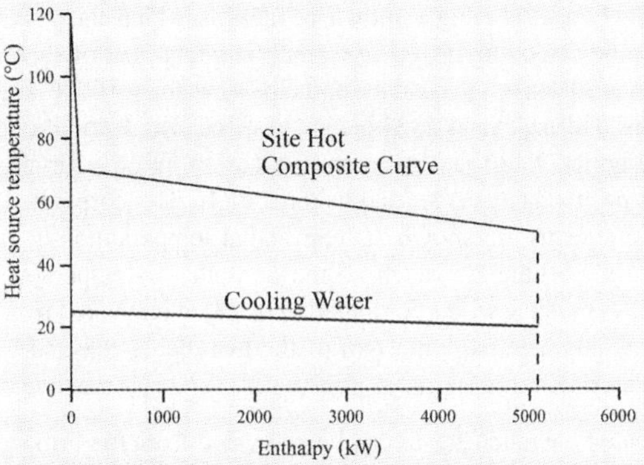

(b) Part of the site hot composite curve.

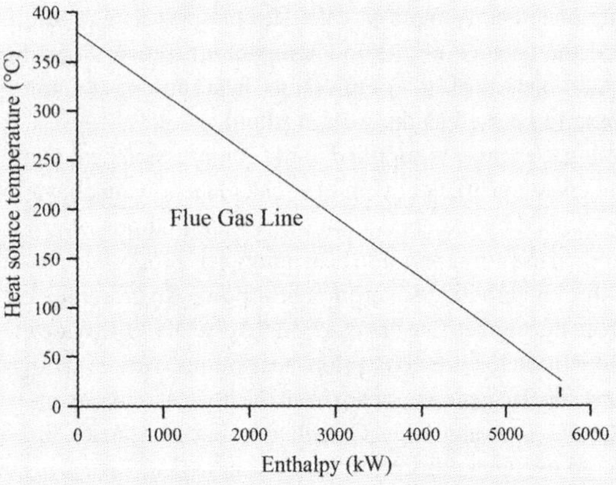

(c) Stack loss from a furnace that is a possible heat source.

Figure 27.13

Heat sink and heat sources for a site heat recovery problem.

$$= 1.684 \times (670 - 432)$$
$$= 401 \text{ kW}$$

Total heat for steam generation

$$= 3514 + 401$$
$$= 3915 \text{ kW}$$

Heat available in the flue gas cooling down to 103 °C

$$= 15.974 \times (380 - 103)$$
$$= 4425 \text{ kW}$$

Thus, the flue gas can generate 1.684 $kg \cdot s^{-1}$ saturated steam at 160 °C. This is shown in Figure 27.14a.

b) Assuming only the steam generated in the flue gas is used and only the latent heat of the generated steam is being used for heating, then heating required from the heat pump

$$= 8002 - 3514$$
$$= 4488 \text{ kW}$$

This is the heat required to be upgraded by the compression heat pump, as shown in Figure 27.14b. Calculate the *COP* for the heat pump initially assuming $T_{COND} = 80$ °C and $T_{EVAP} = 50$ °C. From Equation 27.5 and Table 27.2:

$$\eta_{CHP} = A + B\left(\frac{T_{COND} - T_{EVAP}}{T_{COND}}\right)$$
$$= 0.6932 - 0.4851 \times \left(\frac{353 - 323}{353}\right)$$
$$= 0.6520$$

From Equation 27.4:

$$COP_{CHP} = \eta_{CHP}\frac{T_{COND}}{T_{COND} - T_{EVAP}}$$
$$= 0.6520 \times \left(\frac{353}{353 - 323}\right)$$
$$= 7.672$$

From Equation 27.4:

$$W = \frac{Q_{COND}}{COP_{CHP}}$$
$$= \frac{4488}{7.672}$$
$$= 585 \text{ kW}$$
$$Q_{EVAP} = Q_{COND} - W$$
$$= 4488 - 585$$
$$= 3903 \text{ kW}$$

The cooling is shown in Figure 27.14c.

c) Assuming the power generation efficiency is 40%, the waste heat generated by the power for the compression heat pump

$$= 585 \times \frac{60}{40}$$
$$= 877.5 \text{ kW}$$

This illustrates the point that outside of the site boundary a significant amount of waste heat is generated as a result of the power imported for the heat pump to reduce the waste heat on the site.

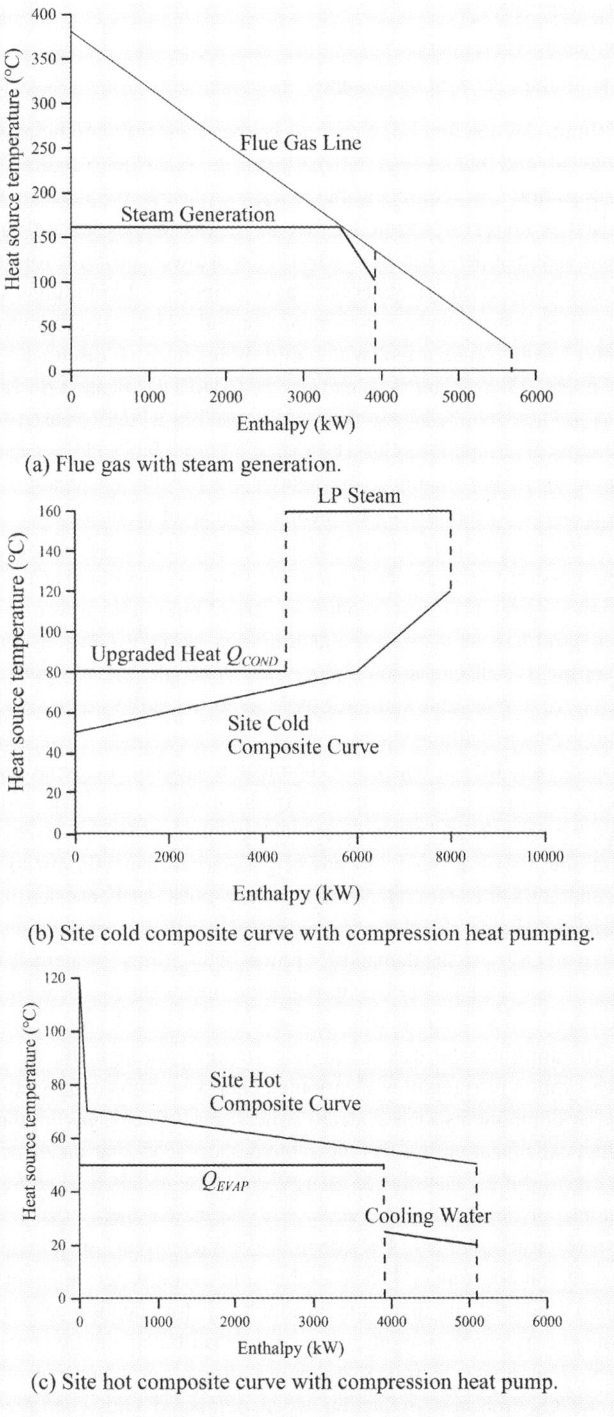

Figure 27.14

Heat sink and heat sources with compression heat pumping.

d) Initially assume $T_{COND} = T_{ABS} = 80\,°C$ and $T_{EVAP} = 50\,°C$. The generator temperature might be dictated by the heat available in the heat source. However, the operating temperatures are constrained by the Dűhring plot. Crystallization must be avoided. For a steam temperature of 160 °C, then $T_{GEN} = 140\,°C$. This allows $\Delta T_{min} = 10\,°C$ between both the flue gas and the steam generation and the steam use and generator. A combination of $T_{COND} = 80\,°C$ and $T_{GEN} = 140\,°C$ in Figure 27.11 implies a concentration of rich solution (absorbent) of 63% in the generator. A combination of $T_{EVAP} = 50\,°C$ and $T_{ABS} = 80\,°C$ in Figure 27.11 implies a concentration of weak solution of 52% in the absorber. Thus, from Figure 27.11 no crystallization will occur. For the absorption heat pump from Equation 27.12:

$$COP_{AHP}^{IDEAL} = 1 + \left(1 - \frac{T_{ABS}}{T_{GEN}}\right)\left(\frac{T_{EVAP}}{T_{COND} - T_{EVAP}}\right)$$

$$= 1 + \left(1 - \frac{353}{413}\right)\left(\frac{323}{353 - 323}\right)$$

$$= 2.564$$

From Table 27.3, $A = -1.7978$ and $B = 2.8816$ and Equations 27.12 and 27.13 are given by:

$$COP_{AHP} = \eta_{AHP} COP_{AHP}^{IDEAL}$$

$$= 2.564\eta_{AHP}$$

$$COP_{AHP} = A\,\eta_{AHP} + B$$

$$= -1.7978\eta_{AHP} + 2.8816$$

Solving these equations simultaneously gives:

$$\eta_{AHP} = 0.6606$$

Then:

$$COP_{AHP} = 0.6606 \times COP_{AHP}^{IDEAL}$$

$$= 0.6606 \times 2.564$$

$$= 1.694$$

From Equation 27.6, assuming the energy for liquid pumping is negligible:

$$Q_{ABS} + Q_{COND} = COP_{AHP} \times Q_{GEN}$$

$$= 1.694 \times 3514$$

$$= 5953\ \text{kW}$$

Thus, the heat upgraded is 5953 kW. This just fits against the site cold composite curve at 80 °C in Figure 27.13. This leaves (8002 – 5953) = 2049 kW from steam heating.

There are many trade-offs to be explored before the design is finalized. In particular, it is clear from Figure 27.14 that a higher temperature could have been taken for the evaporator. The temperature of the evaporator and condenser and the overall heat load in both the compression and absorption heat pumps should be varied in an optimization before the final design can be fixed.

27.6 Integration of Waste Treatment and Energy Sytems

The waste treatment and energy systems are most often considered separately. Yet integration of these systems can increase the sustainability of production systems.

For a number of waste streams, combustion in a thermal oxidizer is the only practical way to deal with them. This is particularly true of solid and concentrated waste and toxic wastes such as those containing halogenated hydrocarbons, pesticides, herbicides, and so on. Many of the toxic substances encountered resist biological degradation and persist in the natural environment for a long period. Unless they are in dilute aqueous solution, the most effective treatment is usually thermal oxidation. Thermal oxidizers are also used to dispose of excess sludge from aerobic wastewater treatment processes. The excess sludge is typically dewatered by filtration or centrifugation, then dried and finally thermally oxidized. Depending on the waste being treated, the thermal oxidizer may or may not require auxiliary firing from fuel oil or natural gas. The various designs of thermal oxidizer for gaseous waste were discussed in Chapter 25. The designs discussed in Chapter 25 are capable of oxidizing both gaseous and liquid waste, but are not suited to solid waste. If solid waste needs to be oxidized, then two classes are suited to such duties:

1) *Rotary kilns.* Rotary kilns involve a cylindrical refractory-lined shell mounted at a small angle to horizontal and rotated at low speed. Solid material, sludges and slurries are fed at the higher end and flow under gravity along the kiln. Liquids can also be oxidized. Rotary kilns are ideal for treating solid waste but have the disadvantage of high capital and maintenance costs.
2) *Hearth thermal oxidizers.* This type of thermal oxidizer is primarily designed to oxidize solid waste. Solids are moved through the combustion chamber mechanically using a rake.

Thermal oxidizers located on the waste producer's site can be fitted with waste heat recovery systems, usually steam generation, which can be fed into the site steam mains.

Gasification is a flexible process that can turn waste streams containing mainly carbon and hydrogen into synthesis gas that can be used for power generation (or chemical production). Waste from the process site might be supplemented by biomass imported to the site for gasification.

If waste material is concentrated and biodegradable, then the use of anaerobic digestion can be used to treat the waste, producing methane as a byproduct that can be used in the energy system.

Weak aqueous solutions of organic material that are still too strong to be treated by biological treatment can be treated by wet oxidation (see Section 26.2). The wet oxidation process can create excess heat that can be exploited within the energy system.

27.7 Renewable Energy

Renewable sources of energy are ones that are replenished on a human timescale. Renewable energy sources include:

- wind power,
- hydropower,
- wave power,
- tidal power,
- solar energy,
- geothermal energy,
- biomass, bio-oil and biogas.

Within the boundary of a chemical processing site only biomass, bio-oil and biogas can be exploited to make any significant substitution of nonrenewable energy sources. Biogas can be used as a direct substitute for natural gas in furnaces, gas turbines and steam boilers. Bio-oil can be used as a direct substitute for fuel oil in furnaces and steam boilers, as can biodiesel. Exploiting solid biomass for combustion is more problematic. The use of fluidized-bed steam boilers fed with solid biomass was discussed in Chapter 23. Solid biomass can also be fed to a gasification process to produce synthesis gas that can be used as a fuel (or used for chemical production). Solid biomass and fossil fuel sources can be combined. For example, the residue feed to the gasification in Figure 27.8 could be supplemented with solid biomass.

Renewable fuels can be used within the boundary of a chemical processing site to generate renewable power in combined heat and power systems. However, other sources of renewable power, such as wind and solar power generation, can normally only make a marginal contribution to on-site power generation. Power generation from renewable sources can be imported to the site to substitute nonrenewable sources.

Beyond the boundary of a chemical processing site, renewable energy can be exploited more fully and integrated with the wider energy system across a geographical region. *Distributed energy* is an alternative approach to power generation that refers to local generation of electricity, possibly at very much smaller scales than traditional centralized electricity production. Power production is moved closer to the demand for power. This allows greater use of combined heat and power through the use of district heating in colder climates. In hotter climates heat rejected from power generation can be used to produce chilling through the use of absorption refrigeration. Distributed energy systems mainly comprise components such as steam boilers, gas turbines, gas engines, fuel cells, heat pumps or renewable devices like solar panels and wind turbines. Distributed energy could be installed to serve a house or building, or gathered in a distributed energy center to serve large numbers of residential, commercial and in principle industrial consumers. If heat in the form of hot water or steam is also produced and distributed through a district heating network, the energy generation efficiency of the distributed energy centers could be greater than in 85%, representing a significant potential for CO_2 emissions reduction when compared with centralized power generation.

Distributed energy can be explored at domestic, commercial and industrial levels in an integrated approach to take advantage of industrial waste heat already available in a particular geographical area. The problem with residential and commercial energy requirements though is that they are variable through the time of the day, day of the week and month of the year. On the other hand, large-scale industrial systems are fairly constant in energy demand.

Due to varying heat and power demands as well as electricity tariffs, the distributed energy center must be flexible enough to provide the peak demands. This will depend on the geographic location and demand for heating in winter or cooling in summer.

This feature creates trade-offs with respect to the selection of combined heat and power units. For example, a single unit provides economy of scale but might generate electricity only half of the year because there is no demand for the heat rejected. Alternatively, several smaller units can be installed, some of which can keep running throughout the year.

27.8 Efficient Use of Water

The efficient use of water has two dimensions. The first is water use within and across processes. The second is related to energy use in steam systems and heat rejection in cooling water systems.

1) *Efficient use of process water.* Water is used within processes for a variety of purposes:
   - reaction medium in chemical reactors (vapor or liquid),
   - solvent in absorption operations and extraction processes,
   - energy and mass transfer stream in steam stripping,
   - motive fluid in steam ejectors,
   - cleaning and decontaminating medium in equipment washing and hosing operations,
   - etc.

   Reusing water between these various operations reduces both the volume of the freshwater and the volume of wastewater. Regeneration reuse increases the quality of water such that it is acceptable for further use. Regeneration reuse reduces both the volume of the freshwater and the volume of the wastewater, as with reuse, but also removes part of the effluent load by removing part of the contaminant load that would have to be otherwise removed in the final effluent treatment before discharge. A third option is where a regeneration process is used on the outlet water from the operations and the water is recycled. Methods to target and design for these three options have been discussed in detail in Chapter 26.
2) *Efficient use of water in steam systems.* Figure 23.2 shows a schematic representation of a boiler feedwater treatment system. If condensate is lost and not returned then this must be compensated by makeup with freshwater. This makeup results in aqueous emissions:
   a) Wastewater is generated in the deionization process when the ion-exchange beds are regenerated with saline, or acid and alkaline solutions.
   b) Wastewater is generated from *boiler blowdown*. The main problem with boiler blowdown is that it is contaminated with water treatment chemicals.
   c) The lost condensate does not create a direct problem since it is likely to be contaminated only with perhaps a few parts per million of amines added to prevent corrosion in the condensate system. The major problems are indirect. The heat loss caused by the condensate loss must ultimately be made up by burning extra fuel and the generation of extra products of combustion.

   In addition, the loss of condensate also creates a loss of energy in the hot condensate that must be compensated by combustion of additional fuel. These aqueous emissions and heat loss from the steam system can be reduced by increasing the percentage of condensate returned. The aqueous emissions can also be reduced by increasing the energy efficiency through increased heat recovery, reducing the load on the steam system.
3) *Efficient use of water in cooling systems.* Any additional heat imported into process energy systems as a result of inefficiencies in the heat recovery system will lead to that additional heat being rejected to cold utility. Such unnecessary load on cooling water systems requires additional heat to be rejected to the environment and also gives rise to additional wastewater generation. Most cooling water systems recirculate water rather than using "once-through" arrangements. Water is lost from recirculating systems in the cooling tower mainly through evaporation but also, to a much smaller extent, through drift (see Chapter 24). The build-up of solids is prevented by cooling tower blowdown. Cooling tower blowdown is the source of the largest volume of wastewater on many sites. The blowdown will contain corrosion inhibitors, polymers to prevent solid deposition and biocides to prevent the growth of microorganisms. Unnecessary load on cooling water systems creates unnecessary heat rejection to the environment, water makeup and blowdown. Cooling tower blowdown can be reduced by improving the energy efficiency of processes through increased heat recovery, thus reducing the thermal load on cooling towers or by increasing the cycles of concentration (see Chapter 24). Alternatively, cooling water systems can be switched to air-coolers, which eliminates the problem of makeup water load and emissions of blowdown, but requires additional power for the air cooler fans.

27.9 Sustainability in Chemical Production – Summary

Chemical processes should be designed to maximize the sustainability of industrial activity. Maximizing sustainability requires that industrial systems should strive to satisfy human needs in an economically viable, environmentally benign, and socially beneficial way (Azapagic, 2014). To direct a process design to maximize the sustainability of industrial activity requires its location, construction and decommissioning, its feeds and products, its waste streams, and its connection to the transportation and wider industrial infrastructure to be considered in terms of the whole life cycle of the process. Life cycle assessment (LCA) is a cradle-to-grave analysis of a process that evaluates all upstream and downstream resource requirements and environmental impacts. The boundary of consideration should go beyond the immediate boundary of the manufacturing facility.

Individual processes must as much as possible minimize waste production through the efficient use of raw materials. If waste can be minimized at source, not only are effluent treatment costs reduced but also raw materials costs.

Some processes produce byproducts that do not have a direct commercial value. A process network needs to be created that uses materials in an overall efficient way by transforming as much of the raw material as possible into useful products and byproducts.

Rather than use crude oil, natural gas and coal as sources of raw materials for the production of chemical products, renewable sources of raw materials can be used as an alternative for the production of many chemicals. However, it would be a mistake to assume that, just because a source of materials is renewable this leads to a process that is more sustainable. This can only be judged by a full life cycle assessment of alternative raw material sources.

When considering energy related emissions, it is tempting to consider only the local emissions from the process and its utility system. However, the emissions generated from central power generation associated with any power import are just as much part of the process as those emissions generated on-site. Precombustion emissions should also be included if data are available. Improved heat recovery will reduce the overall demand for utilities and hence reduce energy related emissions. This can be considered both at a process and site level.

The waste treatment and energy systems are most often considered separately. Yet integration of these systems can increase the sustainability of production systems.

Within the boundary of a chemical processing site only biomass, bio-oil and biogas can be exploited to make any significant substitution of nonrenewable energy sources.

The sustainability of water use can be increased by reuse, regeneration reuse and regeneration recycling both within processes and across processes.

27.10 Exercises

1. A reaction between organic compounds is carried out in the liquid phase in a stirred-tank reactor in the presence of excess formaldehyde. The organic reactants are nonvolatile in comparison with the formaldehyde. The reactor is vented to atmosphere via an absorber to scrub any organic material carried from the reactor. The absorber is fed with freshwater and the water from the absorber rejected to effluent. The major contaminant in the aqueous waste from the absorber is formaldehyde.

- **a)** If the absorption system is kept, how can the volume of aqueous waste be reduced from the system?
- **b)** How might the organic waste to effluent be eliminated?

2. A chemical manufacturing site has a large aqueous effluent flowrate that passes through biological treatment before discharge to a river. Although the outlet concentrations of pollutants from the biological treatment are within permitted limits, the temperature at the outlet is too high. The maximum temperature permitted for discharge is 30 °C, whereas the current outlet is 40 °C. The inlet temperature to biological treatment is 36 °C. A temperature rise of 4 °C occurs across biological treatment. Heat is generated within the biological treatment from the reactions but is also lost to the environment directly. The longer the residence time in biological treatment, the greater the heat loss. It has been proposed to solve the problem by installing a cooling tower downstream biological treatment. A better solution would be to solve the problem at source.

- **a)** What factors could be changed at the inlet to the biological treatment to alleviate the problem?
- **b)** The processes that create the effluent are batch in nature and involve various washing operations at different temperatures. What changes should be sought to solve the problem?

3. A chemical production site producing a variety of specialty chemicals has a problem with its aqueous effluent. The site produces aqueous effluent that is currently discharged without treatment. The effluent has a high load of organic material and has a low pH. The regulatory authorities have demanded a reduction in the organic load before discharge of 90%, together with neutralization. An effluent treatment system has been designed and costed. The treatment system consists of collecting all of the effluent steams together, followed by neutralization using lime and then biological treatment. The cost of both the neutralization and biological treatment operations are proportional to the volume of effluent to be treated. The cost of biological treatment also increases with the load of organic material. The cost of the treatment system is unacceptable and the company is prepared to consider an alternative solution based on waste minimization.

- **a)** The worst effluent stream, in terms of its organic load, comes from Operation 1 of Plant A. This effluent is created when an organic product is washed free of salts in an extraction operation. This is done by mixing the product with water in a tank followed by separation of the water from the organic product by settling in a decanter. The washing operation picks up organic product as well as the salts. The salts are extremely soluble, whereas the organic product is sparingly soluble. The effluent leaving this operation is saturated with organic contaminants, but well below saturation for the salts. Taking this operation in isolation, what can be done to reduce the effluent volume and organic load?
- **b)** Another operation, Operation 2, in Plant A uses water in a cooling circuit. The water is used for condensation of organic vapor by direct contact. In this operation, the organic vapor is passed through a vessel into which is sprayed the cooling water. The resulting two-phase mixture is again separated by settling in a decanter. The water becomes saturated with organic contaminants and is recirculated through a cooling tower. A purge must be taken from the cooling circuit, which is sent to effluent. This purge is another highly contaminated effluent stream from Plant A. What can be done to reduce the effluent from Plant A as a whole by integrating Operations 1 and 2? Explain what effect your suggestions are likely to have on the volume and the load.
- **c)** Plant B uses water to scrub hydrogen chloride from a vent. The resulting water is highly acidic but not contaminated with organic material. The other effluents on the site are

essentially neutral. Can anything be done to reduce the cost of treating the effluent?

d) Plant C produces an aqueous effluent contaminated with organic contaminants. In addition, there is no policy for recovery of steam condensate resulting from the use of steam for heating. Large quantities of steam condensate are sent to drain. There is also a large cooling tower on-site that requires a large water makeup to compensate for evaporative losses. The blowdown from the cooling tower is sent to drain and contains no organic contaminants. Taking the site as a whole, what, in addition to the measures suggested so far, can be done to reduce effluent treatment costs?

References

Azapagic A (2014) Sustainability Considerations for Integrated Biorefineries, *Trends in Biotechnology*, **32**: 1.

Azapagic A and Perdan S (2011) *Sustainable Development in Practice: Case Studies for Scientists and Engineers*, 2nd Edition, Wiley-Blackwell.

Chowdhury J and Leward R (1984) Oxygen Breathes More Life into CPI Processing, *Chem Eng*, **19**: 30.

Curran MA (1996) *Environmental Life Cycle Assessment*, McGraw-Hill, New York.

Dodge BF (1944) *Chemical Engineering Thermodynamics*, McGraw-Hill.

Guinee JB (2002) *Handbook on Life Cycle Assessment, Operational Guide to the ISO Standards*, Kluwer Academic Publishers.

Hougen OA, Watson KM and Ragatz RA (1959) *Chemical Process Principles. Part II Thermodynamics*, John Wiley & Sons.

McNaughton KJ (1983) Ethylene Dichloride Process, *Chem Eng*, **12**: 54.

Oluleye O, Smith R and Jobson MR (2016) Modelling and Screening Heat Pumping Options for the Exploitation of Low Temperature Waste Heat in Process Sites, *Applied Energy*, **169**: 267.

Reich P (1976) Air or Oxygen for VCM? *Hydrocarbon Process*, **March**: 85.

Smith R and Delaby O (1991) Targeting Flue Gas Emissions, *Trans IChemE*, **69A**: 492.

Smith R and Petela EA (1991a) Waste Minimization in the Process Industries. Part 1 The Problem, *Chem Eng*, **506**: 31.

Smith R and Petela EA (1991b) Waste Minimization in the Process Industries. Part 2 Reactors, *Chem Eng*, **509/510**: 12.

Smith R and Petela EA (1991c) Waste Minimization in the Process Industries. Part 3 Separation and Recycle Systems, *Chem Eng*, **513**: 24.

Smith R and Petela EA (1992) Waste Minimization in the Process Industries. Part 4 Process Operations, *Chem Eng*, **517**: 9.

Smith JM and Van Ness HC (2007) *Introduction to Chemical Engineering Thermodynamics*, 7th Edition, McGraw-Hill.

Victor RA, Kim JK and Smith R (2013) Composition Optimization of Working Fluids for Organic Rankine Cycles and Kalina Cycles, *Energy*, **55**: 114.

Chapter 28

Process Safety

Although safety considerations have been left until late in this text, they should not be left to the final stages of the design. Consideration has been left late here to allow the many issues in process design that have an impact on safety to be dealt with first. Safety considerations must be embedded from the very beginning of a process design. Early decisions made purely for process reasons can often lead later to problems of safety, health and environment that require complex solutions. It is far better to consider safety issues early as the design progresses when making the basic process decisions.

Start by considering the three major hazards in process plant, which are fire, explosion and toxic release (Lees, 1989).

28.1 Fire

The first major hazard in process plant is fire, which is usually regarded as having a disaster potential lower than both explosion and toxic release (Lees, 1989). However, fire is still a major hazard and can under the worst conditions approach explosion in its disaster potential. Fire requires a combustible material (gas or vapor, liquid, solid, solid in the form of a dust dispersed in a gas), an oxidant (usually oxygen in air) and usually, but not always, a source of ignition. Consider now important factors in assessing fire as a hazard.

1) *Flammability limits.* A flammable gas or vapor will burn in air only over a limited range of composition. Below a certain concentration of the flammable gas, the *lower flammability limit*, the mixture is too "lean" to burn. Above a certain concentration, the *upper flammability limit*, it is too "rich" to burn. Concentrations between these limits constitute the *flammable range*.

   Combustion of a flammable gas–air mixture occurs if the composition of the mixture lies in the flammable range *and* if there is a source of ignition. Alternatively, the combustion of the mixture can occur without a source of ignition if the mixture is in the flammable range and is heated to its *autoignition temperature*.

   The most flammable mixture is usually the stoichiometric mixture for combustion. It is often found that the concentrations of the lower and upper flammability limits are approximately one-half and twice that of the stoichiometric mixture respectively (Lees, 1989; Crowl and Louvar, 1990).

   Flammability limits are affected by pressure. The effect of pressure changes is specific to each mixture. In some cases, decreasing the pressure can narrow the flammable range by raising the lower flammability limit and reducing the upper flammability limit until eventually the two limits coincide and the mixture becomes nonflammable. Conversely, an increase in pressure can widen the flammable range. However, in other cases, increasing the pressure has the opposite effect of narrowing the flammable range (Lees, 1989; Crowl and Louvar, 1990).

   Flammability limits are also affected by temperature. An increase in temperature usually widens the flammable range (Lees, 1989; Crowl and Louvar, 1990).

2) *Limiting (minimum) oxygen concentration (LOC).* Whereas the lower flammability limit measures the lowest concentration that will allow combustion of a vapor–air mixture, sometimes inert gas (usually nitrogen, but sometimes carbon dioxide or steam) is added to the mixture to prevent combustion. The *limiting* (or *minimum*) *oxygen concentration* is the minimum concentration of oxygen below which the reaction cannot generate enough energy for the mixture (including inerts) to allow self-propagation of a flame, independent of the concentration of fuel. It is expressed in units of volume percent of oxygen in combustible material. The LOC varies with pressure and temperature. It is also dependent on the type of inert (nonflammable) gas.

3) *Minimum ignition energy.* The minimum ignition energy is the minimum energy of an ignition source, such as a spark, required to ignite a vapor. The value of the minimum ignition energy varies widely. A typical value for flammable vapors is 0.025 mJ (Crowl, 2012).

4) *Autoignition temperature.* The autoignition temperature of a gas or vapor is the temperature at which it will ignite spontaneously in air, without any external source of ignition.

5) *Flash point.* The *flash point* of a liquid is the lowest temperature at which it gives off enough vapor to form an

Chemical Process Design and Integration, Second Edition. Robin Smith.

Companion Website: www.wiley.com/go/smith/chemical2e

ignitable mixture with air. The flash point generally increases with increasing pressure.

6) *Limiting oxygen concentration of a dust.* The *limiting oxygen concentration* of a dust is the minimum concentration of oxygen (displaced by an inert gas such as nitrogen) capable of supporting combustion of dust that is dispersed in the form of a cloud. A mixture with oxygen concentration below the limiting oxygen concentration is not capable of supporting combustion and hence cannot support a subsequent dust explosion.
7) *Minimum ignition temperature of a dust.* The *minimum ignition temperature* of a dust is the lowest temperature at which dust that is dispersed in the form of a cloud can ignite. The minimum ignition temperature is an important factor in evaluating the sensitivity of dust to ignition sources such as hot surfaces. Increasing particle size of dust and increasing moisture content both increase the minimum ignition temperature.

28.2 Explosion

The second of the major hazards is explosion, which has a disaster potential usually considered to be greater than fire but lower than toxic release (Lees, 1989). Explosion is a sudden and violent release of energy. The energy released in an explosion on a process plant is either of the following.

1) *Chemical energy.* Chemical energy derives from a chemical reaction. The source of the chemical energy is exothermic chemical reactions or combustion of flammable material (dust, vapor or gas). Explosions based on chemical energy can be either *uniform* or *propagating*. An explosion in a vessel will tend to be a uniform explosion, while an explosion in a long pipe will tend to be a propagating explosion. For a dust, the *minimum explosible concentration* is the lowest concentration in $g{\cdot}m^{-3}$ in air that will give rise to flame propagation on ignition.
2) *Physical energy.* Physical energy may be pressure energy in gases, thermal energy, strain energy in metals or electrical energy. An example of an explosion caused by release of physical energy would be fracture of a vessel containing high-pressure gas.

Thermal energy is generally important in creating the conditions for explosions rather than a source of energy for the explosion itself. In particular, superheat in a liquid under pressure causes flashing of the liquid if it is accidentally released to the atmosphere.

There are two basic kinds of explosions involving the release of chemical energy:

1) *Deflagration.* In a *deflagration*, the flame front travels through the flammable mixture relatively slowly.
2) *Detonation.* In a *detonation*, the flame front travels as a shock wave followed closely by a combustion wave that releases the energy to sustain the shock wave. The detonation front travels with a velocity greater than the speed of sound in the unreacted medium.

A detonation generates greater pressures and is more destructive than a deflagration. The peak pressure caused by the deflagration of a hydrocarbon–air mixture or a dust mixture at atmospheric pressure is of the order of 8 to 10 bar. However, a detonation may give a peak pressure of the order of 20 bar. A deflagration may turn into a detonation, particularly if traveling down a long pipe.

Just as there are two basic kinds of explosions, they can occur in two different conditions:

1) *Confined explosions. Confined explosions* are those that occur within vessels, pipework or buildings. The explosion of a flammable mixture in a process vessel or pipework may be a deflagration or a detonation. The conditions for a deflagration to occur are that the gas mixture is within the flammable range *and* that there is a source of ignition. Alternatively, the deflagration can occur without a source of ignition if the mixture is heated to its autoignition temperature. An explosion starting as a deflagration can make the transition into a detonation. This transition can occur in a pipeline but is unlikely to happen in a vessel.
2) *Unconfined explosions.* Explosions that occur in the open air are *unconfined explosions*. An unconfined vapor cloud explosion is one of the most serious hazards in the process industries. Although a large toxic release may have a greater disaster potential, unconfined vapor explosions tend to occur more frequently (Lees, 1989). Most unconfined vapor cloud explosions have been the result of leaks of flashing flammable liquids. The problem of the explosion of an unconfined vapor cloud is not only that it is potentially very destructive, but also that it may occur some distance from the point of vapor release and may thus threaten a considerable area. The Flixborough disaster in the UK in 1974 was caused by the release of flashing cyclohexane liquid. This created a flammable vapor cloud that resulted in an unconfined vapor cloud explosion, killing 28 people and seriously injuring 36 out of a total of only 72 people on site.

   If the explosion occurs in an unconfined vapor cloud, the energy in the blast wave is generally only a small fraction of the energy theoretically available from the combustion of all the material that constitutes the cloud. The ratio of the actual energy released to that theoretically available from the heat of combustion is referred to as the *explosion efficiency*. Explosion efficiencies are typically in the range of 1 to 10%. A value of 3% is often assumed.

   The rupture of a vessel containing pressurized liquid can lead to a boiling liquid expanding vapor explosion (BLEVE). When the vessel ruptures, the pressure preventing the liquid from boiling is lost. If the rupture is catastrophic then a large mass of liquid will boil instantaneously, causing an extremely rapid expansion. Depending on the substance involved, the temperatures and pressure may lead to an explosion. The substance does not need to be flammable to create an explosion, as the stored energy might be enough to cause an explosion. However, if a flammable material is subject to a BLEVE and there is a source of ignition, a secondary explosion might be created by the primary BLEVE. BLEVEs can be caused by an external fire, or chemical reaction overpressurizing a vessel, or from any mechanical failure of the vessel.

When processing flammable materials, the hazard of an explosion should in general be minimized by avoiding flammable gas–air mixtures in the process. This can be done either by changing process conditions or by adding an inert material. It is bad practice to rely solely on elimination of sources of ignition.

28.3 Toxic Release

The third of the major hazards and the one with the greatest disaster potential is the release of toxic chemicals (Lees, 1989). The hazard posed by toxic release depends not only on the chemical species but also on the conditions of exposure. The high disaster potential from toxic release arises in situations in which large numbers of people are briefly exposed to high concentrations of toxic material. However, the long-term health risks associated with prolonged exposure at low concentrations over a working life also present serious hazards. In the Bhopal disaster in 1984, a release of toxic vapor (methyl isocyanate) to the air caused the death of approximately 3800 people in the immediate aftermath of the release. Subsequently it has been estimated that a further 10,000 to 20,000 people have since died from gas-related diseases. Even further, many thousands more have suffered disabling injuries.

For a chemical to affect health, a substance must come into contact with an exposed body surface. The three ways in which this happens are by inhalation, skin contact and ingestion.

In process design, the primary consideration is contact by inhalation. This happens either through accidental release of toxic material to the atmosphere or the *fugitive emissions* caused by slow leakage from pipe flanges, valve glands, pump and compressor seals. Tank filling also causes emissions when the rise in liquid level causes vapor in the tank to be released to atmosphere, as discussed in Chapter 24.

The acceptable limits for toxic exposure depend on whether the exposure is brief or prolonged. *Lethal concentration* for airborne materials and *lethal dose* for nonairborne materials are measured by tests on animals. The limits for brief exposure to toxic materials that are airborne are usually measured by the concentration of toxicant that is lethal to 50% of the test group over a given exposure period, usually four hours. It is written as LC_{50} (lethal concentration for 50% of the test group). The test gives a comparison of the absolute toxicity of a compound in a single concentrated dose, *acute exposure*. For nonairborne materials, lethal dose LD_{50} refers to the quantity of material (e.g. mg per 100g or kg body weight of the test animal) administered, which results in death of 50% of the test group. It should be emphasized that it is extremely difficult to extrapolate tests on animals to human beings. Non-animal test methods are being introduced.

The limits for prolonged exposure are expressed as the *threshold limit values*. These are essentially acceptable concentrations in the workplace. There are three categories of threshold limit values:

1) *Time weighed exposure.* This is the time weighted average concentration expressed in parts per million (ppm) or $mg{\cdot}m^{-3}$ for a specified period of time, normally an 8-hour workday or 40-hour workweek to which workers can be exposed, day after day, without adverse effects. Short excursions (typically less than 30 minutes during a workday) above the limit are allowed if compensated by other excursions below the limit.
2) *Short-term exposure.* This is the maximum concentration to which workers can be exposed for a short period of time, usually 15 minutes, without suffering from (1) intolerable irritation, (2) chronic or irreversible tissue change or (3) narcosis of sufficient degree to increase accident proneness, impair self-rescue or materially reduced efficiency. However, short-term exposure is only permitted as long as the time weighted average is not exceeded. Additional restrictions might demand no more than four excursions per day, with at least 60 min between exposure periods, and provided the daily time weighted value is not exceeded.
3) *Ceiling exposure.* This is the concentration that should not be exceeded, even instantaneously.

Low-level toxic emissions are most often fugitive in nature. Such emissions can be reduced by using leak-tight equipment (e.g. changing from packing to mechanical seals or even using sealless pumps, etc.) and regular maintenance checks. Equipment can also be enclosed and ventilated. Reducing vapor emissions from tank filling has been discussed in Chapter 24.

28.4 Hazard Identification

From the beginning of a process design the hazards need to be identified, such that where possible the design eliminates or mitigates the hazard when making basic design decisions. This applies both to the basic process decisions and the layout and location of the facility. Considerations of layout and location should not be left until late in the design. Only through an awareness of the hazards can decisions be made to eliminate or mitigate hazards.

28

Chemical processes by their nature are hazardous. Particular hazards are associated with the inventories of hazardous materials and the use of extreme operating conditions, but there are numerous others:

1) *Inventories of hazardous material.* Any inventory of flammable or toxic material can potentially bring hazards. However, the larger the inventory of flammable or toxic material, the greater the hazard. The inventories to be avoided most of all are flammable or toxic liquids held under pressure above their boiling points. Gases leak at a lower mass flowrate than liquids through an opening of a given size. Flashing liquids leak at about the same rate as a subcooled liquid but then turn into a mixture of vapor and spray on release. The spray, if fine, is just as hazardous as the vapor and can be spread as easily by the wind. Thus, the leak of a flashing liquid through a hole of a given size produces a much greater hazard than the corresponding leak of gas (Kletz and Amyotte, 2010).
2) *High pressure.* The use of high pressure greatly increases the stored energy in the plant. Although high pressures in themselves do not pose serious problems in materials of construction, the use of high temperatures, low temperatures or corrosive chemicals together with high

pressure does (Lees, 1989). With high-pressure operation, the problem of leaks becomes much more serious, since this increases the mass flowrate of fluid that can leak out through a given sized hole. This is particularly so when the fluid is a flashing liquid.

3) *Vacuum pressure*. Vacuum pressures are not in general as hazardous as the other extreme operating conditions. However, one particular hazard that does exist in vacuum-pressure plant handling flammable materials is the ingress of air with consequent formation of a flammable mixture.
4) *High temperature*. The use of high temperatures in combination with high pressures greatly increases stored energy in the plant. The heat required to obtain a high temperature is often provided by furnaces, which carry a number of hazards. Rupture of the tubes in the radiant zone carrying the process fluid can lead to explosions. If the fuel-to-air ratio is not controlled adequately, then substoichiometric combustion can lead to emissions of carbon monoxide and soot. The burners can be a source of ignition in the event of a release of flammable material elsewhere. There are materials-of-construction problems associated with high-temperature operation (see Appendix B). Equipment is subject to thermal stresses, especially during start-up and shutdown and creep is a major problem (see Appendix B).
5) *Subambient temperature*. Subambient temperature processes (below cooling water temperature) most often require a refrigeration loop to create the low temperature. This requires extra equipment that has the potential to fail and additional inventory of flammable (e.g. propane) or toxic (e.g. ammonia) material required for the refrigerant fluid. The process itself can contain large amounts of fluids kept in the liquid state by pressure and/or low temperature. If for any reason it is not possible to keep them under pressure or keep them cold, then the liquids will begin to vaporize. If this happens vapor needs to be vented and impurities in the boiling liquids are liable to precipitate from solution as solids, especially if equipment is allowed to boil dry. Deposited solids may not only be the cause of blockage but also in some cases the cause of explosions. It is necessary, therefore, to ensure that the fluids entering a low-temperature plant are purified. Another problem that can occur at low temperatures is hydrate formation. Gas hydrates are nonstoichiometric crystalline solids comprised of hydrocarbons such as methane, ethane or propane trapped within the cavities of a lattice of water molecules. Formation of gas hydrates requires low temperature, high pressure, free water and the presence of a hydrocarbon. If formed, gas hydrates can block pipelines including relief and venting systems. A severe materials-of-construction problem in low-temperature processes is low-temperature embrittlement (see Appendix B). Also, in low-temperature as in high-temperature operations, the equipment is subject to thermal stresses, especially during start-up and shutdown.
6) *Utility failure*. Failure of power, cooling systems (cooling water, air cooling or refrigeration), inert gas and instrument air can lead to extreme hazards. The safety of the process must, as much as possible, not be compromised by utility failure.
7) *Runaway reactions*. Often the worst safety problem that can occur with reactors is when an exothermic reaction generates heat at a faster rate than the cooling system can remove it. Such *runaway reactions* are usually caused by coolant failure, perhaps for a temporary period, or reduced cooling capacity due to perhaps a pump failure in the cooling water circuit. The runaway happens because the rate of reaction, and hence the rate of heat generation, increases exponentially with temperature, whereas the rate of cooling increases only linearly with temperature. Once heat generation exceeds available cooling capacity, the rate of temperature rise becomes progressively faster (Tharmalingam, 1989). If the energy release is large enough, liquids will vaporize, and overpressurization of the reactor can occur.
8) *Unreliability*. Historically, many of the major safety incidents that have happened in the past have been associated with equipment failure and maintenance. Designs should aim to be as reliable as possible and not subject to unexpected failure. Unexpected failure of equipment causes process upsets that can create safety problems and also the requirement for maintenance, which brings safety issues through the need to isolate, decontaminate and dismantle equipment. Maintenance might be *preventive* ahead of any failure by carrying out maintenance according to a schedule, or through monitoring of equipment performance, to avoid breakdown or failure. Maintenance might also be *corrective* to repair equipment that has already failed. A systematic analysis of reliability, availability and maintainability is required once the basic design has been developed. This might point to choosing more reliable equipment. Standby equipment is also commonly used to avoid unnecessary process upsets. The spare capacity or *redundancy* created by the standby equipment can be used in the event of failure of on-line equipment to avoid process upsets. The policies for equipment monitoring and preventive maintenance also impact the reliability. The maintenance strategy needs to be considered early in order to increase asset reliability. A reliability, availability and maintainability analysis is needed that considers the inherent reliability of equipment, the inclusion and operating policy for standby equipment, preventive maintenance and the inspection policy in a holistic analysis. Details of such considerations are outside the scope of this text.
9) *Maintenance hazards*. Process equipment, such as pumps, compressors, valves, vessels, instrumentation, and so on, needs to be inspected and maintained. Whenever equipment needs to be maintained, in most cases it needs to be isolated, decontaminated and dismantled. Appropriate design features need to be included to allow the safe inspection and maintenance of equipment. Safe access is required to equipment to conduct maintenance. It is important that reliable equipment is incorporated into the design to avoid unnecessary maintenance.

10) *Lifting hazards*. Some maintenance operations need heavy equipment to be lifted. Safe provision must be made for such lifting by the inclusion of cranes in the design or safe access for portable cranes and clear lifting pathways.

11) *Transportation*. Transportation of raw materials to the process, and products from the process, can create hazards both inside and outside of the facility. The transportation of materials and the storage of materials are intimately linked, as discussed in Sections 14.5 and 16.13. In particular, transportation of hazardous materials to and from the facility by road and rail cars should be avoided where possible.

12) *Access and evacuation hazards*. Personnel need safe access to, and safe evacuation from, process areas for both operations and maintenance personnel. Safe access to equipment must also be ensured for maintenance operations.

13) *Access for emergency vehicles*. In the event of an incident, emergency vehicles will need access for firefighting and evacuation of personnel.

14) *Location of occupied buildings*. Occupied buildings, such as control rooms, maintenance workshops, analytical laboratories and offices, should not be located in blast zones or where any release of toxic material would create risk to the personnel. In general, process plant should be designed to minimize occupancy within hazardous areas. This might mean centralizing control rooms at suitably remote locations.

15) *Wind direction*. The layout of a facility should, as much as possible, allow the prevailing wind to disperse the release of any hazardous material in the safest direction.

16) *Natural and environmental hazards*. Extreme weather (hurricanes, typhoons, tornados, extreme heat, extreme cold) and earthquakes need to be considered and measures taken to mitigate the consequences of such hazards.

17) *Other hazards*. The list of hazards above is by no means comprehensive. Different types of processes can create hazards that apply to only a small set of processes. For example, offshore oil and gas production brings many hazards that only apply to those operations through helicopter and marine transportation, and so on.

Formal managerial processes can be used for the identification of hazards, known as HAZID studies (Kletz, 1999). This is a structured brainstorming technique and typically involves process design, process operations, safety systems engineers, process control and instrumentation, mechanical engineering, commissioning and project management personnel. Such managerial tools should be used early in a project as soon as process flow diagrams, preliminary material and energy balances and preliminary plot layouts are available. The process design also needs to be considered in the context of the existing site infrastructure, transportation, weather and local conditions so that hazards *external* to the process can be identified as well as in the *internal* hazards. The HAZID study should not only identify potential hazards but also consider the consequences of hazards and identify safeguards for the elimination or mitigation of hazards.

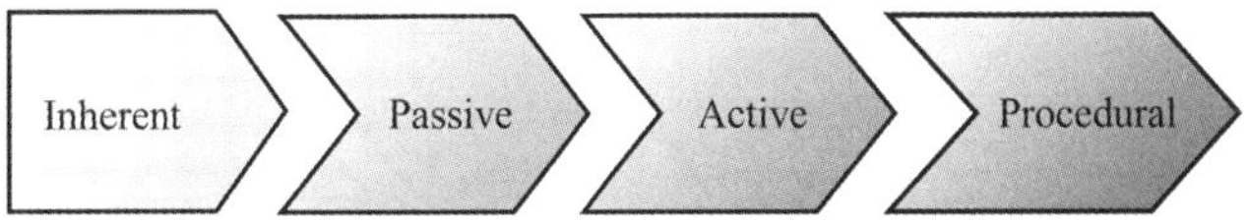

Figure 28.1

The hierarchy of process safety management. (Reproduced from Center for Chemical Process Safety (2009) *Inherently Safer Chemical Processes*, 2nd Edition, with permission from John Wiley & Sons.)

28.5 The Hierarchy of Safety Management

When managing safety, there is a hierarchy that should be followed, whether directed towards reducing the frequency or consequence of potential safety incidents, as illustrated in Figure 28.1 (Center for Chemical Process Safety, 2009).

a) *Inherent safety*. Make changes integral to the process to eliminate or reduce hazards at source; e.g. change from a process that uses a flammable solvent to a water-based process.

b) *Passive safety*. Incorporate design features that reduce the frequency or consequence of a hazard without the active functioning of a device; e.g. incorporate fire or blast protection walls.

c) *Active safety*. Using process control, safety instrumented systems and process alarms to detect and respond to a hazardous condition, and allow a mitigating action to be taken; e.g. detection of a high liquid level in a storage tank shuts off the feed pump. Human intervention is also active safety. However, human intervention is on the whole less reliable than process control and safety instrumentation systems.

d) *Procedural safety*. Using administrative controls and emergency response to prevent safety incidents or minimize the effects of an incident; e.g. use of safety permits for control of maintenance work.

All four levels in the hierarchy can contribute to the overall safety of the process. However, inherent safety differs in that it seeks to eliminate or reduce potential hazards at source.

28.6 Inherently Safer Design

Whilst no process can be made inherently safe, the design objective should be to strive for an *inherently safer design* throughout the evolution of a design. The intent of inherently safer design is to eliminate the hazard completely or reduce its magnitude sufficiently to eliminate the need for elaborate safety systems and procedures. Hazard elimination or reduction should be accomplished by means that are inherent in the process design. Designs that avoid the need for hazardous materials, use less of them, use them at less extreme temperatures and pressures or dilute them with

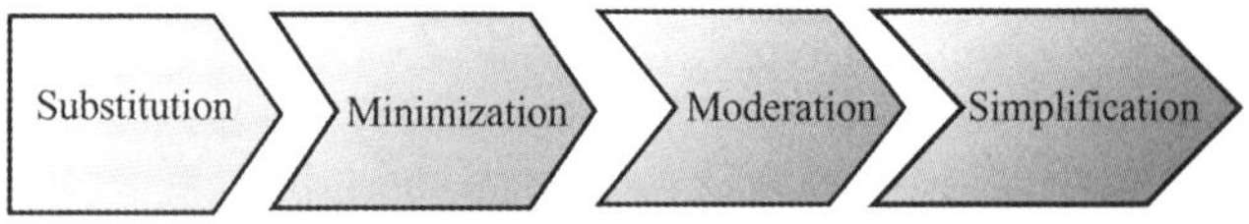

Figure 28.2
The hierarchy of inherently safer design.

inert materials will be *inherently safer* and will not require elaborate safety systems (Kletz and Amyotte, 2010). There is a hierarchy that applies to inherently safer design, as illustrated in Figure 28.2 (Center for Chemical Process Safety, 2009):

1) *Substitution.* The best way of dealing with a hazard in a process design is to remove it completely. Where possible, a hazardous material should be replaced with a less hazardous one. For example:
   - Change the reaction path to avoid hazardous material in the chemistry.
   - Replace an organic solvent with water.
   - Heat transfer media are sometimes liquid hydrocarbons used at high pressure. When possible, higher boiling liquids should be used. Better still, the flammable material should be substituted with a nonflammable medium such as water or molten salt.
   - Some operations need to be carried out at low temperature that needs refrigeration. The refrigeration fluid might be propylene, for example, and present a major hazard. Operation of the process at a higher pressure, on the one hand, brings increased hazards in the process equipment but, on the other hand, might eliminate the refrigeration or allow a less hazardous refrigeration fluid.
   - Etc.

   Often, hazardous materials are an integral part of a process and cannot be substituted, for example, flammable hydrocarbons in petroleum refining and petrochemicals processes.
2) *Minimization.* Once the possibilities to substitute hazardous materials have been exhausted, the inventory of hazardous material should be minimized. The inventories to be avoided most are flashing, flammable liquids or flashing, toxic liquids.
   a) *Reactors.* Reaction rates can be improved by the use of catalysts or alternative catalysts if a catalyst is already used. Alternatively, more extreme operating conditions can be used. More extreme conditions may reduce inventory appreciably. However, more extreme conditions bring their own problems. A very small reactor operating at a high temperature and pressure may be inherently safer than one operating at less extreme conditions because it contains a much lower inventory. A large reactor operating close to atmospheric temperature and pressure may be safe for different reasons. Leaks are less likely and, if they do happen, the leak will be small because of the low pressure. Also, little vapor is produced from the leaking liquid because of the low temperature. A compromise solution employing moderate pressure, temperature and medium inventory may combine the worst features of the extremes. The compromise solution may be such that the inventory is large enough for a serious explosion or a serious toxic release if a leak occurs; the pressure will ensure the leak is large and the high temperature results in the evaporation of a large proportion of the leaking liquid (Kletz and Amyotte, 2010).

      The potential hazard from runaway reactions is reduced by reducing the inventory of material in the reactor or reducing the operating temperature. Batch operation requires a larger inventory than the corresponding continuous reactor. Thus, there may be a safety incentive to change from batch to continuous operation. Alternatively, the batch operation can be changed to semibatch in which one (or more) of the reactants is added over a period. The advantage of semibatch operation is that the feed can be switched off in the event of a temperature (or pressure) excursion. This minimizes the chemical energy stored up for a subsequent exotherm. For continuous reactors, plug-flow designs require smaller volumes and hence smaller inventories than mixed-flow designs for the same conversion, as discussed in Chapter 4.
   b) *Distillation.* There is a large inventory of boiling liquid, sometimes under pressure, in a distillation column, both in the base and held up in the column. If a sequence of columns is involved, as discussed in Chapter 10, the sequence can be chosen to minimize the inventory of hazardous material. Use of the partition column or dividing-wall column shown in Figure 10.13c will reduce considerably the inventory relative to two simple columns. Partition columns or dividing-wall columns are inherently safer than conventional arrangements since they not only lower the inventory but also the number of items of equipment is fewer, and hence there is a lower potential for leaks.

      The column inventory can also be reduced by the use of low hold-up column internals. Packed columns have a lower inventory than tray columns. The column base can be designed with a smaller diameter than the rest of the column, but maintaining the same liquid height, to reduce the inventory in the base. Distillate receivers can be designed to be smaller or eliminated altogether. Thermosyphon reboilers have a lower inventory than kettle reboilers. Peripheral equipment such as reboilers can be located inside the column (Kletz and Amyotte, 2010).
   c) *Heat transfer operations.* Enhancing heat transfer with shell-and-tube heat exchangers makes them smaller and reduces the inventory. As discussed in Section 12.12, heat exchanger designs with a higher area density than shell-and-tube heat exchangers can be used, reducing the inventory.
   d) *Storage.* The largest inventories of hazardous materials are most often held up in the storage of raw materials and products and intermediate (buffer) storage. The most obvious way of reducing the inventory in storage is by locating producing and consuming plants near each other so that hazardous intermediates do not have to be stored or transported (Kletz and Amyotte, 2010). It may also be possible to reduce storage requirements by making the design more flexible. Adjusting the capacity could then

be used to cover for delays in the arrival of raw material, upsets in one part of the plant, and so on, and thus reduce the need for storage (Kletz and Amyotte, 2010).

Large quantities of toxic gases such as chlorine and ammonia and flammable gases such as propane and ethylene oxide can be stored either under pressure or at atmospheric pressure under refrigerated conditions. If there is a leak from atmospheric refrigerated storage, the quantity of hazardous material that is discharged will be less than the corresponding pressurized storage at atmospheric temperature. For large storage tanks, refrigeration is safer. However, this might not be the case with small-scale storage, since the refrigeration equipment provides sources for leaks. Thus, in small-scale storage, pressurized storage may be safer (Kletz and Amyotte, 2010; Lees, 1989).

The optimization of reactor conversion was considered in Chapter 15. As the conversion increased, the size (and cost) of the reactor increased but that of separation, recycle and heat exchanger network systems decreased. The same trends also occur with the inventory of material in these systems. The inventory in the reactor increases with increasing conversion, but the inventory in the other systems decreases. Thus, in some processes, it is possible to optimize for minimum overall inventory (Boccara, 1992). In the same way as reactor conversion can be varied to minimize the overall inventory, the recycle inert concentration can also be varied. It might be possible to reduce the inventory significantly by changing reactor conversion and recycle inert concentration without a large cost penalty if the cost optimization profiles are fairly flat.

3) *Moderation.* The use of an unnecessarily high-temperature hot utility or heating medium should be avoided. This may have been a major factor that led to the runaway reaction at Seveso in Italy in 1976, which released toxic material over a wide area. The reactor was liquid-phase and operated in a stirred tank (Figure 28.3). It was left containing an uncompleted batch at around 160 °C, well below the temperature at which a runaway reaction could start. The temperature required for a runaway reaction was around 230 °C (Cardillo and Girelli, 1981). The reaction was normally carried out under vacuum at about 160 °C in a reactor heated by steam at about 300 °C. The temperature of the liquid could not rise above its boiling point of 160 °C at the operating pressure. In this accident, the steam was isolated from the reactor containing the unfinished batch and the agitator was switched off. The reactor walls below the liquid level fell to the same temperature as the liquid, around 160 °C. The reactor walls above the liquid level remained hotter because of the high-temperature steam that had been used (but now isolated). Heat then passed by conduction and radiation from the walls to the top layer of the stagnant liquid, which became hot enough for a runaway reaction to start (Figure 28.3). Once started in the upper layer, the reaction then propagated throughout the reactor. If the steam had been cooler, say 170 °C, the runaway could not have occurred (Kletz and Amyotte, 2010).

The occurrence of flammable gas and dust mixtures should be avoided, rather than relying on the elimination of sources of ignition. This can be achieved in the first instance by changing the process conditions such that dust mixtures are below their limiting oxygen concentration and gas mixtures are outside their flammable range. If this is not possible, inert material such as nitrogen, carbon dioxide or steam should be introduced.

For the particular case of dusts, electrical charge can accumulate on a powder. The resulting spark discharges can lead to fires or dust explosions. When processing dusty materials, the use of electrically insulating materials of construction (e.g. plastic pipes) should be avoided, as this allows the accumulation of electrical charge. High relative humidity can reduce the resistivity of some powders, increasing the rate of charge decay, decreasing the accumulation of static charge.

Moderating the process conditions will result in a safer plant, providing the moderation does not increase the inventory of hazardous material. Also, moderating the conditions can reduce the reliance on protective equipment (Kletz and Amyotte, 2010).

4) *Simplification.* Simpler facilities are often inherently safer. For processes containing hazardous material, minimize:

- number of equipment items,
- pipe lengths,
- pipe joints,
- equipment penetrations for instrumentation (i.e. where possible use nonintrusive instruments).

This minimizes the potential for leaks. "What you don't have, can't leak" (Kletz and Amyotte, 2010). Eliminating unnecessary equipment and unnecessary complexity is also likely to make the process control simpler, process operation more straightforward and the process operation more tolerant of maloperation. It will also simplify maintenance and be generally more forgiving of poor maintenance practice.

Relief systems are expensive and bring considerable environmental problems. Sometimes it is possible to dispense with

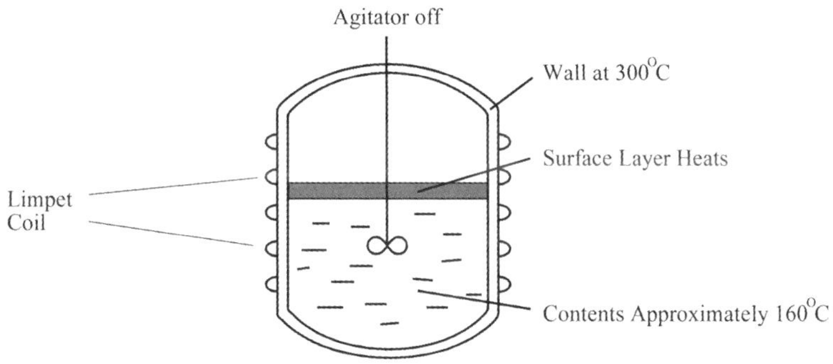

Figure 28.3
Schematic of the Seveso reaction system.

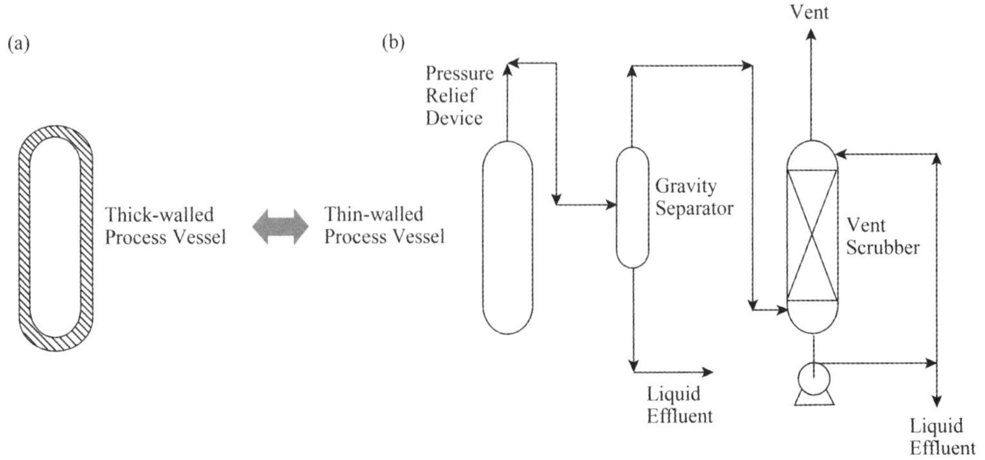

Figure 28.4
A thick-walled pressure vessel might be economic when compared with a thin-walled vessel with its relief and venting system.

relief systems and all that comes after them by using stronger vessels, strong enough to withstand the highest pressures that can be reached (Kletz and Amyotte, 2010). This is a form of passive safety. Regulations might dictate that a relief device must still be installed, for example to protect in case of an external fire. However, their size and relief capacity, as well as the hazards associated with the opening of relief devices might be reduced or eliminated. It might also be possible to eliminate catch pots, quench systems, scrubbers and flare stacks necessary to dispose of material from an emergency relief. Figure 28.4 shows as an example a comparison between a stronger vessel in Figure 28.4a and a weaker vessel with its relief and venting system in Figure 28.4b (Kletz and Amyotte, 2010). A stronger vessel may often be safer and cheaper (Kletz and Amyotte, 2010).

Example 28.1 A process involves the use of benzene as a liquid under pressure. The temperature can be varied over a range. Compare the fire and explosion hazard of operating with a liquid process inventory of 1000 kmol at 100 °C and 150 °C, on the basis of the theoretical combustion energy resulting from catastrophic failure of the equipment. The normal boiling point of benzene is 80 °C, latent heat of vaporization 31,000 kJ·kmol^{-1}, specific heat capacity 150 kJ·kmol^{-1}·K^{-1} and heat of combustion 3.2×10^6 kJ·kmol^{-1}. Assume that physical properties are constant over the temperature range.

Solution The fraction of liquid vaporized on release is calculated from a heat balance (Crowl and Louvar, 1990). The sensible heat above saturated conditions at atmospheric pressure provides the heat of vaporization. The excess heat in the superheated liquid is given by:

$$m\,C_P(T_{SUP} - T_{BPT})$$

where m = mass or moles of liquid
C_P = heat capacity
T_{SUP} = temperature of the superheated liquid
T_{BPT} = normal boiling point

If the mass of liquid vaporized is m_V, then:

$$m_V = \frac{mC_P(T_{SUP} - T_{BPT})}{\Delta H_{VAP}}$$

where m_V = mass or moles of liquid vaporized
ΔH_{VAP} = latent heat of vaporization

Thus, the vapor fraction (*VF*) is given by:

$$VF = \frac{m_v}{m} = \frac{C_p(T_{SUP} - T_{BPT})}{\Delta H_{VAP}}$$

For operation at 100 °C:

$$VF = \frac{150(100 - 80)}{31,000} = 0.097$$

$$m_V = 0.097 \times 1000 = 97 \text{ kmol}$$

$$\text{Theoretical combustion energy} = 97 \times 3.2 \times 10^6 = 310 \times 10^6 \text{ kJ}$$

For operation at 150 °C:

$$VF = 0.339$$

$$m_V = 339 \text{ kmol}$$

$$\text{Theoretical combustion energy} = 1085 \times 10^6 \text{ kJ}$$

Thus, against this measure of inherent safety the fire hazard is 3.5 times larger for operation at 150 °C compared to operation at 100 °C. In fact, the true potential fire load will be greater than the energy release calculated. In practice, such a release of superheated liquid generates large amounts of fine spray in addition to the vapor. This can double the energy release based purely on vaporization.

28.7 Layers of Protection

Once an inherently safer process design has been established, process safety requires multiple layers of protection to be added, as illustrated in Figure 28.5. Each layer of protection has its own level of risk reduction. It is important that each safety layer is independent of the others. If there is a failure in one safety layer then other layers should not be affected and still perform risk reduction.

1) *Layout and passive safety barriers.* The layout of the plant needs to consider safe access and evacuation for personnel, access for maintenance (including mobile cranes) and access for emergency vehicles. The layout also needs to separate personnel and, as much as possible, critical equipment from hazards during normal operations. The location of the plant control room is critical in this respect. Layout should also seek where possible to separate potential hazards from each other to avoid hazards from spreading should an incident occur.

   Containment dikes are installed around storage tanks to prevent leakages from flowing to other locations to spread the hazard caused by a release of liquid. Locations that might be susceptible to spillage should have sloped containment pads for the same purpose. The prime function is for process safety (Center for Chemical Process Safety, 2009):

   - limiting the spread of a fire and preventing exposure of other equipment if a flammable material spills and is ignited;
   - preventing contact of incompatible reactive materials in case of leaks or spills;
   - limiting the spread of spilled corrosive material and preventing contact with equipment that could be damaged by contact with the corrosive material.

   Sloped containment pads also provide environmental protection and prevent contamination of soil and surface water. Such pads are installed for pumps, process buildings and structures, road and rail car loading areas, and any other potential spillage location.

   Firewalls can be used to provide a fireproof barrier to prevent the spread of fire between or through structures to protect personnel and safety critical equipment. Blast walls carry out the same function and prevent the escalation of events due to explosions. Blast walls are expected to retain their integrity against any blast loading, as well as fire, that may subsequently follow.

   Fire-resistant cabling can be used to ensure critical equipment and emergency lighting continues to receive power during a local fire for a period of time.

2) *Process control.* The control system compensates for the influence of external *disturbances* such as changes in feed flowrate, feed composition, temperature and pressure and product demand. Ensuring safe operation within the normal operating envelope is the most important task of a control system. This is achieved by monitoring the process conditions and maintaining them within the normal operating envelope. A control mechanism is introduced that makes changes to the process in order to cancel out the negative impact of disturbances or to respond to changes in setpoints made by operations personnel. In order to achieve this, instruments must be installed to measure the operational performance of the plant. These *measured variables* could include temperature, pressure, flowrate, composition, level, pH, density and particle size. Having measured the variables that need to be controlled, other variables need to be *manipulated* in order to achieve the control objectives. A control system is then designed, which responds to variations in the measured variables and manipulates variables to control the process. Control valves must be designed such that the most probable failure mode results in the *fail safe* condition. The fail safe condition might be fully closed, fully open, maintaining the current valve position

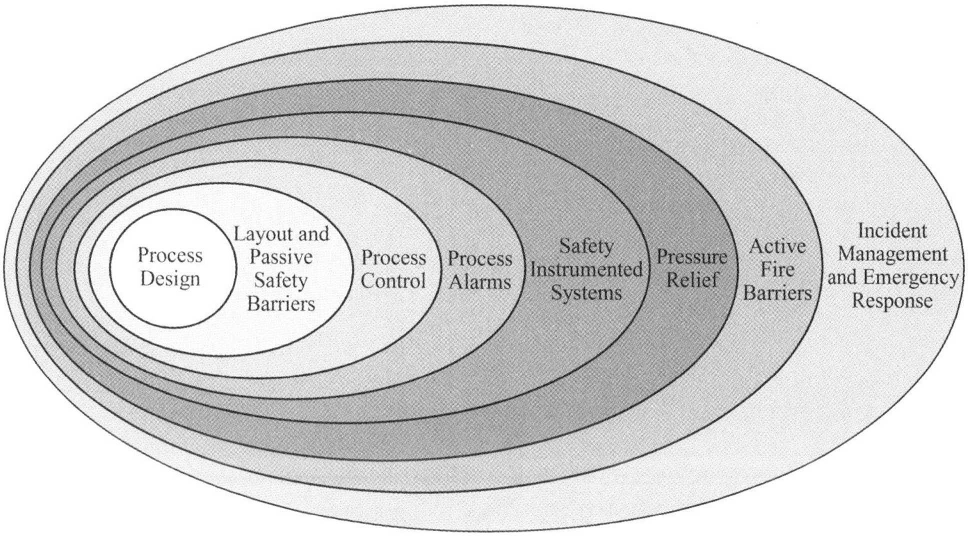

Figure 28.5

Layers of protection for the process design.

to maintain the current operation or to move to a predefined open position between 0 and 100% to keep the flow from going below a threshold.

3) *Process alarms.* Alarms are used to alert process operators to abnormal operating situations at the limits of the normal operating envelope. These are normally high or low levels of process settings, such as a high or low liquid level in a vessel. Operator intervention can then in principle prevent personnel and environmental hazards, and maintain equipment integrity and product quality. It is important that the operator can quickly and accurately identify the alarms that require immediate action. The usual requirement is that an alarm must have an associated operator action or actions, which are documented. The alarm system should be independent from the process control system and can be of different types. *Basic alarms* detect deviations for single process measurements. *Aggregated alarms* or *common alarms* combine the state of a number of basic alarms and together describe the state of a process or subsystem more precisely than a single alarm. *Model-based alarms* are generated by online simulation of the process. *Key alarms* are critical alarms presented in a way that makes them distinguishable even during situations where many alarms are activated. Safety-related alarms should be defined as key alarms, but also other alarms could be included if appropriate.

4) *Safety instrumented systems.* Safety instrumented systems are designed to prevent or mitigate hazardous events by taking the process to a safe state when there are violations in predetermined conditions, usually at or near the safe operating limits. A safety instrumented system is comprised of one or more *safety instrumented functions*. Safety instrumented functions consist of a combination of sensors, logic solvers and final control elements. Sensors collect information (e.g. temperature, pressure, flowrate) to determine whether a hazardous situation exists. Such sensors are dedicated to the safety instrumented function and not part of the control system. The purpose of the logic solvers of safety instrumented functions is to determine what action needs to be taken based on the information gathered. Logic solvers are controllers that read the signals from the sensors and execute preprogrammed actions with the final control elements to prevent a hazard. Logic solvers should provide both fail-safe and fault-tolerant operation. The final control elements are typically on–off valves. The safety instrumented function should safely isolate the hazardous part of the process in the event of a hazardous situation. Activation of the safety instrumented function may also trigger the control system to put the associated control valve into a safe state

5) *Pressure relief.* Any system that has the potential to have its pressure increased beyond safe levels requires safety devices to protect people and equipment. Pressure relief devices are relief valves and bursting disks as illustrated in Figure 28.6. Figure 28.6a shows a typical relief valve in which an adjustable spring is used to offset the internal pressure. The *set pressure* is the pressure at which the relief valve begins to operate. If the internal pressure exceeds the set pressure for the valve, then the valve opens. Different mechanisms can be used from the simple spring arrangement in Figure 26.6a. There are different categories of relief valves, depending on the application:

   a) *Safety valve.* A safety valve is characterized by a full-opening or "pop" action and recloses when the pressure drops to a safe level. Such designs are normally used on gas and steam applications.

   b) *Relief valve.* A relief valve has a gradual lift, generally proportional to the increase in pressure over the opening pressure. It is primarily used on liquid service applications.

   c) *Safety relief valve.* A safety relief valve can be used for compressible fluids or liquids depending on its application. It functions as a safety valve for compressible fluids with rapid opening and as a relief valve for liquid applications with an opening in proportion to the increase in pressure.

The other class of pressure relief device is *bursting disks* or *rupture disks*, as illustrated in Figure 28.6b. A bursting disk is a thin membrane (usually metal) that fails and ruptures at a critical

Figure 28.6
Pressure relief devices.

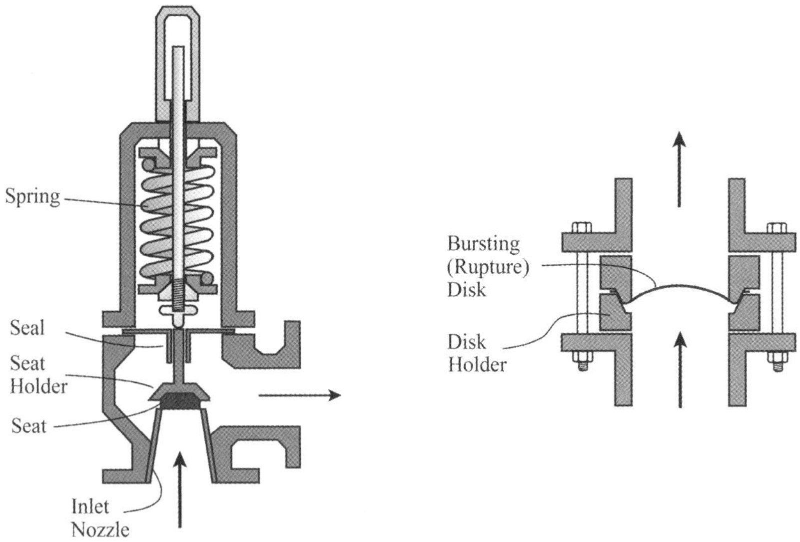

(a) Relief valve.

(b) Bursting (rupture) disk.

pressure. The bursting disk has a one-time use and, unlike a pressure relief valve, cannot close after the pressure has been relieved. There are different classes of bursting disks:

a) *Forward acting*. Figure 28.6b illustrates a forward acting disk in which the internal pressure acts against the concave surface. As the pressure increases beyond the allowable pressure the disk bulges and fails in tension. Some designs use scoring on the convex surface of the disk to allow more controlled failure.

b) *Reverse acting*. In a reverse acting disk, the internal pressure acts against the convex side of the disk. Compression loading causes it to reverse, snapping through the neutral position and causing it to fail by a scoring pattern on the surface of the disk on the low-pressure side or a knife blade penetration located on the low-pressure side.

The main distinction between bursting disks and pressure relief valves is that pressure relief valves will reclose, whereas bursting disks cannot and need to be replaced if activated, which might mean a plant shutdown. This would suggest that pressure relief valves are the best solution and they are more commonly used. However, pressure relief valves are exposed to process material that might corrode or cause plugging of the device, preventing it from operating properly after time in service. There is thus a reliability issue. This means that pressure relief valves will need to be inspected and maintained regularly. It might also mean that two pressure relief valves are installed in parallel in case one fails to operate. An alternative solution is to install a pressure relief valve and bursting disk in parallel with the pressures of the two devices set such that the pressure relief valve opens first and the bursting disk operates only if the pressure relief valve fails to relieve the excess pressure. Another problem with pressure relief valves is that they can leak process material if not properly closed or under vacuum conditions allow air ingress. Bursting disks can be installed upstream of pressure relief valves to protect the valve from process material and prevent leaks. If excess pressure is imposed on the system, the busting disk and pressure relief valve both operate, but when the pressure subsides the pressure relief valve recloses to allow the process to continue operation if appropriate. In some cases the transient pressures that occur in an upset condition mean that the pressure relief valve cannot act fast enough to relieve the pressure and a bursting disk must be used.

Process design specifies the *operating pressure* under normal operating conditions. The *maximum operating pressure* extends beyond normal operating conditions to include startup, shutdown and other off-design conditions. The *maximum allowable working pressure* or *maximum allowable operating pressure* is the maximum gage pressure that the weakest component of the system can safely withstand for a designated temperature. This is based on design codes and can vary from 10% to 25% above the maximum operating pressure of the system.

Pressures might exceed the maximum allowable working pressure for one or more of the following reasons:

- failure of the process control system,
- failure of a safety instrumented function,
- equipment failure,
- failure of a cooling medium (e.g. failure of condenser cooling in distillation whilst maintaining reboiling),
- outlet closed whilst maintaining the inlet pressure,
- runaway chemical reactions,
- heat from external fire,
- thermal expansion of liquids under shut-in conditions,
- operator error.

These problems can be made worse by poor equipment design, poor equipment sizing or poor maintenance. Vessels and pipes must be protected from pressures above the maximum allowable working pressure by pressure relief. The highest set pressure is equal to maximum allowable working pressure. *Overpressure* is the pressure increase over the set pressure of the relieving device as a percentage. Overpressure is limited to 10% above the maximum working pressure, except in the case of a fire. In the case of a fire, the overpressure must not exceed 21% above the maximum working pressure.

Emergency discharge from relief valves and bursting disks can be dealt with in a number of ways:

a) Direct discharge to atmosphere under conditions leading to rapid and safe dilution.

b) Total containment in a connected vessel, with ultimate disposal being deferred.

c) Partial containment, in which some of the discharge is separated either physically (e.g. gravitational or centrifugal means) or chemically (e.g. absorption).

d) Combustion in a flare. Flare systems normally include a catchpot that collects liquids and passes gases to flare (see Chapter 24).

Sizing of vent lines is far from straightforward and beyond the scope of this text. For example, if a vessel under pressure containing a level of liquid is to be relieved of pressure, it might be suspected that the vent line can be sized to remove enough vapor from above the liquid surface to relieve the pressure. In practice sudden loss of pressure can cause the liquid to boil and be carried into the vent along with the vapor. If this is the case, the relief device and the vent line must be sized to handle the required two-phase flow.

6) *Active fire barriers*. Sprinkler and deluge systems provide active fire protection to spray water on to a fire via a firewater distribution piping system. For *sprinkler systems* each sprinkler head is closed by a heat-sensitive plug that prevents water from flowing until the ambient temperature around the sprinkler head reaches the design activation temperature of the individual sprinkler head. In *deluge systems* all sprinklers connected to the firewater distribution system are open. Water is not present in the piping until the system operates. These systems are used for special hazards where rapid fire spread is a concern, as they provide application of water over the entire hazard.

7) *Incident management and emergency response*. The final layer of protection is incident management and emergency response.

If a large safety event occurs this layer responds in a way that minimizes ongoing damage, injury or loss of life. It includes evacuation plans and firefighting.

28.8 Hazard and Operability Studies

A hazard and operability study (HAZOP) is a structured, qualitative methodology that identifies potential safety and environmental hazards and major operability problems, assesses consequences and generates recommendations for action, but does not attempt to modify the design. A formal managerial process is used for the HAZOP studies (Kletz, 1999, Crawley and Tyler, 2015). This is a structured brainstorming carried out by a multidisciplinary team. The team has a *Chairman* leading the team and a *Scribe* to record the outcome of the HAZOP study. The makeup of the team depends on the process technology. Typically the makeup of the team includes one or more of process design engineers, process control engineers and process operations representatives. Depending on the nature of the process, other specialists may be involved.

For a HAZOP to be performed, the design should be at an advanced stage, but not to the stage where the design has been fixed. However, process and instrumentation diagrams (P&IDs) must have been developed. The P&IDs are analyzed by dividing them into *nodes*. A node is a section of equipment on the P&ID with clearly defined boundaries, within which process parameters are to be investigated for *deviations*. For example, a node could be:

- a tank with its associated equipment,
- a pump and pipe transferring liquid from one tank to another,
- a phase separator,
- a compressor, etc.

In each case, the node would include the associated control and instrumentation. Simple nodes can be combined where appropriate to give a *compound node*. For example, a pipe, pump and heat exchanger might be combined to a single node. The decision of how big a node should be is at the discretion of the team carrying out the HAZOP. If the node is too big and complex, the analysis might miss important problems. The intention of how the plant is expected to operate in the absence of deviations from the design intent needs to be defined before the deviations are explored.

The next step is for each node to generate meaningful deviations from the intended operation by coupling a *guideword* and a *parameter*. The guideword qualitatively defines a change and the parameter defines what parameter is to be changed. Table 28.1 lists standard guidewords and their generic meaning. Table 28.2 lists examples of parameters for process operations.

A deviation can be generated by choosing a parameter and combining it with each guideword in turn to see if a meaningful deviation results (the parameter-first approach). The alternative approach is to take a guideword and examine each parameter in turn (the guideword-first approach). The first seven guidewords in Table 28.1 are the ones normally used, with the others included if appropriate. The approach for continuous and batch processes is slightly different. Variations of the standard set can be used,

Table 28.1

Standard guidewords for HAZOP and their generic meanings (Crawley and Tyler, 2015).

Guideword	Meaning
No (not, none)	None of the design intent is achieved
More (more of, higher)	Quantitative increase in a parameter
Less (less of, lower)	Quantitative decrease in a parameter
As well as (more than)	An additional activity occurs
Part of	Only some of the design intention is achieved
Reverse	Logical opposite of the design intention occurs
Other than (other)	Complete substitution – another activity takes place OR an unusual activity occurs or uncommon condition exists
Other useful guidewords include:	
Where else	Applicable for flows, transfers, sources and destinations
Before/after	The step (or some part of it) is effected out of sequence
Early/late	The timing is different from the intention
Faster/slower	The step is done/not done with the right timing

Table 28.2

Examples of possible parameters for process operations (Crawley and Tyler, 2015).

Flow	Phase
Pressure	Speed
Temperature	Particle size
Mixing	Measure
Stirring	Control
Transfer	pH
Level	Sequence
Viscosity	Signal
Reaction	Start/stop
Composition	Operate
Addition	Maintain
Separation	Services
Time	Communication
Aging	

depending on the circumstances. The guidewords are used to search for meaningful deviations from the intended operation. The choice or parameters must be agreed by the HAZOP team and will depend on the process being studied. Parameters other than those listed in Table 28.1 might be appropriate depending on the process. However, many of the parameters listed in Table 28.1 will not apply to many problems. When seeking deviations, not every guideword will combine with a parameter to give a meaningful deviation. There is no point discussing combinations that do not have a physical meaning. Some examples of meaningful combinations are given in Table 28.3.

Each node is examined with this approach and the design examined for potential hazardous scenarios and problematic

Table 28.3

Examples of meaningful combinations of parameters and guidewords (Crawley and Tyler, 2015).

Parameter	Guidewords that can give a meaningful combination
Flow	None, more of, less of, reverse, elsewhere, as well as
Temperature	Higher, lower
Pressure	Higher, lower, reverse
Level	Higher, lower, none
Mixing	Less, more, none
Reaction	Higher (rate of), lower (rate of), none, reverse, as well as/other than, part of
Phase	Other, reverse, as well as
Composition	Part of, as well as; other than
Communication	None, part of, more of, less of, other, as well as

operability issues. Follow-up actions are recorded. It must be emphasized that HAZOP is not a design activity, but is intended for assurance of the design only. Corrections and modifications to the design are carried out outside of the HAZOP.

28.9 Layer of Protection Analysis

A hazard and operability study identifies hazardous scenarios, but should not deal with their prevention or mitigation. After hazardous scenarios have been identified, and outside of the hazard and operability study, various actions need to be taken to mitigate the hazardous scenarios identified. A *layer of protection analysis (LOPA)* can be carried out after the hazard and operability study has identified hazardous scenarios.

For a hazardous scenario, the *likelihood* of the hazardous scenario is determined from historic data, but without the benefit of protection layers. A specific hazardous scenario may have many initiating causes, all of which need to be examined from the perspective of layers of protection. For this, a failure frequency needs to be defined as failure events per year. That *unmitigated event likelihood* is then compared against a *target mitigated event likelihood*. If the unmitigated event likelihood is higher than the target mitigated event likelihood, this must be addressed through use of independent protection layers to reduce the likelihood in order to meet the target event likelihood. It should be emphasized that any gap between the unmitigated event likelihood and target mitigated event likelihood is best reduced or closed through the application of the principle of inherently safer design.

The probability that a system will fail to perform a specified function on demand requires the *probability of failure on demand* to be determined. Probability of failure on demand is a statistical measure of how likely it is that a system or device will operate and be ready to serve the function for which it was intended. It is based on historic data and is numerically between 0 and 1 (where 0 indicates impossibility and 1 indicates certainty). For example, a relief valve might have a probability of failure on demand of 1×10^{-2}. This means that is it likely to fail on demand once every hundred times it is required to function.

Safety instrumented functions can fail in two basic ways. The first way is a spurious trip, which usually results in an unplanned but safe process shutdown. In the second type of failure, the failure remains undetected, continuing unsafe process operation. If an emergency demand occurred, the safety instrumented function would fail to operate for a true process demand. Among other things, it is influenced by the reliability of the system, the interval at which it is tested, as well as how often it is required to function. To determine the probability of failure on demand for the safety instrumented function, historic failure data for the system components are required. This failure rate is used in conjunction with the test interval to calculate the probability of failure on demand. The test interval accounts for the length of time before an unknown fault is discovered through testing.

When safety instrumented functions are applied to reduce the mitigated event likelihood to below the target, each order of magnitude reduction required for the safety instrumented function corresponds with a discrete *safety integrity level*. The lower the safety integrity level, the higher the probability of failure on demand, where safety integrity level 1 (SIL-1) has the lowest safety integrity level and level 4 (SIL-4) has the highest level. However, as the safety integrity level increases, the cost and complexity of the system also increases. SIL-1 typically consists of a single sensor, single logic solver and single final element. Higher levels require multiple sensors (for fault tolerance), multiple channel logic solvers (for fault tolerance) and multiple final elements. Levels higher than SIL-3 tend not to be used in the process industries. The higher the level, the more complex testing and maintenance becomes throughout the life of the facility. If not tested and maintained adequately high SIL safety instrumented functions will not provide the protection required, leading potentially to very hazardous situations. The analysis should therefore strive not to use SIL-3 systems and to minimize as much as possible the use of SIL-2 systems. The reliance on such systems can be avoided by exploiting other protection layers. In particular, it is best avoided by inherently safer design.

28.10 Process Safety – Summary

The major hazards in the process industries are fire, explosion and toxic release. Potential hazards need to be identified early and, where possible, design changes incorporated to eliminate or mitigate hazardous scenarios. The hierarchy of process safety management is given in Figure 28.1. Various layers of protection can be applied to eliminate or mitigate hazards, as illustrated in Figure 28.5. These layers must be independent of each other.

The most significant improvements to process safety are achieved by incorporating inherently safer design features. The philosophy of inherently safer design should be applied from the very beginning of the design and continued through the life of the design activity, constantly reviewing the design to make it inherently safer. The hierarchy of inherently safer design is given in Figure 28.2. The following changes should be considered to improve safety (Kletz and Amyotte, 2010; Lees, 1989; Crowl and Louvar, 1990; Center for Chemical Process Safety, 2009):

Reactors
- Batch to continuous.
- Batch to semibatch.
- Mixed-flow reactors to plug-flow.
- Reduce the inventory in the reactor by increasing temperature or pressure, by changing catalyst or by better mixing.
- Lower the temperature of a liquid-phase reactor below the normal boiling point, or dilute it with a safe solvent.
- Lower the temperature to reduce the danger of runaway reactions.
- Substitute a hazardous solvent.
- Externally heated/cooled to internally heated/cooled.

Distillation
- Choose the distillation sequence to minimize the inventory of hazardous material.
- Use partition or dividing-wall columns to reduce the inventory relative to two simple columns and reduce the number of items of equipment, and hence lower the potential for leaks.
- Use of low hold-up internals.

Heat transfer operations
- Use water or other nonflammable heat transfer media.
- Use a lower temperature utility or heat transfer medium.
- Use a liquid heat transfer medium below its atmospheric boiling point if flammable or toxic.
- If refrigeration is required, consider higher pressure if this allows the refrigeration to be eliminated or a less hazardous refrigerant to be used.

Storage
- Locate producing and consuming plants near to each other so that hazardous intermediates do not have to be stored and transported.
- Reduce storage by increasing design flexibility.
- Store in a safer form (less extreme pressure, temperature or in a safer chemical form).

Relief systems
- Consider strengthening vessels rather than relief systems, but relief valves may still be needed for the case of fire relief.

Overall inventory
- Consider changes to reactor conversion and recycle inert concentration to reduce the overall inventory.

Designs that avoid the need for hazardous materials or use less of them or use them at lower temperatures and pressures or dilute them with inert materials will be inherently safer. When synthesizing a process, the occurrence of flammable gas mixtures should be avoided, rather than relying on the elimination of sources of ignition.

One of the principal approaches to making a process inherently safe is to limit the inventory of hazardous material. The inventories to be avoided most of all are flashing flammable or toxic liquids, that is, liquids under pressure above their atmospheric boiling points. If plants can be designed so that they use safer raw materials and intermediates or not so much of the hazardous ones or use the hazardous ones at lower temperatures and pressures or diluted with inert materials, then many problems later in the design can be avoided (Kletz and Amyotte, 2010).

As the design progresses, it is necessary to carry out Hazard and Operability Studies (Lees, 1989; Crawley and Tyler, 2000). HAZOP studies should not be seen as part of the design, but is an assurance. Having identified hazardous scenarios a layer of protection analysis (LOPA) needs to be carried out to determine which independent protection layers are needed to reduce the probability of the hazard occurring to a target mitigated event likelihood.

28.11 Exercises

1. When processing flammable and toxic materials, why are superheated liquids above their atmospheric boiling point to be most avoided?

2. **a)** Compare the flammability hazard of storing 1000 kmol of cyclohexane at 100 °C and 200 °C, using catastrophic failure of the vessel as a basis for comparison. The atmospheric boiling point of cyclohexane is 81 °C, its latent heat of vaporization is 30,000 kJ·kmol^{-1}, liquid heat capacity is 210 kJ·kmol^{-1}·K^{-1} and its heat of combustion is 3.95×10^6 kJ·kmol^{-1}.

   b) How much of the theoretical heat of combustion would you expect to be released in the event of an explosion?

3. A chemical process uses as raw material hazardous Component *A*. Component *A* is fed as liquid under pressure to a continuous stirred tank vessel (mixed-flow reactor) in which it performs a first-order reaction to useful Product *B*. The reaction in the vessel is isothermal first order with a rate of reaction given by $-r_A = k_0 \exp(-E/RT)$, where r_A is the rate of reaction (kmol·min^{-1}), $k_0 = 1.5 \times 10^6$ min^{-1}, $E = 67,000$ kJ·kmol^{-1}, $R = 8.3145$ kJ·K^{-1}·kmol^{-1}, T is the reactor temperature (K) and C_A the concentration of *A* (kmol·m^3). T is also the supply temperature of *A* whose yet unknown inventory m_A is in the form of a superheated liquid. The total amount of *B* to be produced is 100 kmol·h^{-1}. T and m_A are to be selected with the additional consideration of safety. For mixtures of *A* and *B* the normal boiling point can be assumed to be 70 °C, the latent heat of vaporization 25,000 kJ·kmol^{-1}, the liquid specific heat capacity 140 kJ·kmol^{-1}·K^{-1} and the heat of combustion 2.5×10^6 kJ·kmol^{-1}. The residence time of the reactor is 1 hour and the safety is measured in terms of fire and explosion hazards on the basis of the theoretical combustion energy resulting from catastrophic failure of the equipment.

   a) An initial choice of 80 °C has been made for T. Calculate the required inventory of *A*, the vapor fraction of *A* in the feed and the theoretical combustion energy at this temperature.

   b) Repeat the calculation at 130 °C and comment on whether or not this is a safer design.

References

Boccara K (1992) *Inherent Safety for Total Processes*, MSc Dissertation, UMIST, UK.

Cardillo P and Girelli A (1981) The Seveso Runaway Reaction: A Thermoanalytical Study, *IChemE Symp Ser*, **68**, 3/N: 1.

Center for Chemical Process Safety (2009) *Inherently Safer Chemical Processes*, 2nd Edition, John Wiley & Sons.

Crawley F and Tyler B (2015) *HAZOP: Guide to Best Practice*, 3rd Edition, IChemE, UK.

Crowl DA (2012) Minimise the Risks of Flammable Materials, *Chem Eng Progr*, **April**: 28.

Crowl DA and Louvar JF (1990) *Chemical Process Safety – Fundamentals with Applications*, Prentice Hall.

Kletz TA (1999) *Hazop and Hazan*, 4th Edition, IChemE, UK.

Kletz TA and Amyotte P (2010) *Process Plants: A Handbook of Inherently Safer Design*, 2nd Edition, CRC Press, Taylor & Francis Group.

Lees FP (1989) *Loss Prevention in the Process Industries*, Vol. 1, Butterworths.

Tharmalingam S (1989) Assessing Runaway Reactions and Sizing Vents, *Chem Eng*, **Aug**: 33.

Appendix A

Physical Properties in Process Design

Design calculations most often require knowledge of physical properties. This includes thermodynamic properties, phase equilibrium and transport properties. In practice, the designer often uses a commercial physical property or simulation software package to access such data. It is important to understand the basis and limitations of the methods and correlations, so that the most suitable methods can be chosen and not used inappropriately.

A.1 Equations of State

The relationship between pressure, volume and temperature for fluids is described by *equations of state*. For example, if a gas is initially at a specified pressure, volume and temperature and two of the three variables are changed, the third variable can be calculated from an equation of state.

Gases at low pressure tend towards *ideal gas* behavior. For a gas to be ideal:

- the volume of the molecules should be small compared with the total volume;
- there should be no intermolecular forces.

The behavior of ideal gases can be described by the ideal gas law (Hougen, Watson and Ragatz, 1954, 1959):

$$PV = NRT \tag{A.1}$$

where P = pressure ($N \cdot m^{-2}$)
V = volume occupied by N kmol of gas (m^3)
N = moles of gas (kmol)
R = gas constant ($8314\ N \cdot m \cdot kmol^{-1} \cdot K^{-1}$ or $J \cdot kmol^{-1} \cdot K^{-1}$)
T = absolute temperature (K)

The ideal gas law describes the actual behavior of most gases reasonably well at pressures below 5 bar.

If standard conditions are specified to be 1 atm ($101{,}325\ N \cdot m^{-2}$) and 0 °C (273.15 K), then from the ideal gas law, the volume occupied by 1 kmol of gas is 22.4 m^3.

For gas mixtures, the *partial pressure* is defined as the pressure that would be exerted if that component alone occupied the volume of the mixture. Thus, for an ideal gas:

$$p_i V = N_i RT \tag{A.2}$$

where p_i = partial pressure ($N \cdot m^{-2}$)
N_i = moles of Component i (kmol)

The mole fraction in the gas phase for an ideal gas is given by a combination of Equations A.1 and A.2:

$$y_i = \frac{N_i}{N} = \frac{p_i}{P} \tag{A.3}$$

where y_i = mole fraction of Component i

For a mixture of ideal gases, the sum of the partial pressures equals the total pressure (Dalton's Law):

$$\sum_i^{NC} p_i = P \tag{A.4}$$

where p_i = partial pressure of Component i ($N \cdot m^{-2}$)
NC = number of components (–)

The behavior of real gases and liquids can be accounted for by introducing a compressibility factor (Z), such that (Hougen, Watson and Ragatz, 1954, 1959; Poling, Prausnitz and O'Connell, 2001):

$$PV = ZRT \tag{A.5}$$

where Z = compressibility factor (–)
V = molar volume ($m^3 \cdot kmol^{-1}$)
Z = 1 for an ideal gas and is a function of temperature, pressure and composition for mixtures. A model for Z is needed.

Chemical Process Design and Integration, Second Edition. Robin Smith.

Companion Website: www.wiley.com/go/smith/chemical2e

In process design calculations, *cubic* equations of state are most commonly used. The most popular of these cubic equations is the Peng–Robinson equation of state given by (Peng and Robinson, 1976; Poling, Prausnitz and O'Connell, 2001):

$$Z = \frac{V}{V-b} - \frac{aV}{RT(V^2 + 2bV - b^2)} \tag{A.6}$$

where $Z = \dfrac{PV}{RT}$ (A.7)

V = molar volume

$a = 0.45724\dfrac{R^2T_c^2}{P_c}\alpha$

$b = 0.0778\dfrac{RT_c}{P_c}$

$\alpha = (1 + \kappa(1 - \sqrt{T_R}))^2$

$\kappa = 0.37464 + 1.54226\omega - 0.26992\omega^2$

T_C = critical temperature

P_C = critical pressure

ω = acentric factor

$= \left[-\log\left(\dfrac{P^{SAT}}{P_C}\right)_{T_R=0.7}\right] - 1$

R = gas constant

$T_R = \dfrac{T}{T_C}$

T = absolute temperature

The acentric factor is obtained experimentally. It accounts for differences in molecular shape, increasing with nonsphericity and polarity, and tabulated values are available (Poling, Prausnitz and O'Connell, 2001). Equation A.6 can be rearranged to give a cubic equation of the form:

$$Z^3 + \beta Z^2 + \gamma Z + \delta = 0 \tag{A.8}$$

where $Z = \dfrac{PV}{RT}$

$\beta = B - 1$

$\gamma = A - 3B^2 - 2B$

$\delta = B^3 + B^2 - AB$

$A = \dfrac{aP}{R^2T^2}$

$B = \dfrac{bP}{RT}$

This is a cubic equation in the compressibility factor Z, which can be solved analytically. The solution of this cubic equation may yield either one or three real roots (Press *et al.*, 1992). To obtain the roots, the following two values are first calculated (Press *et al.*, 1992):

$$q = \frac{\beta^2 - 3\gamma}{9} \tag{A.9}$$

$$r = \frac{2\beta^3 - 9\beta\gamma + 27\delta}{54} \tag{A.10}$$

If $q^3 - r^2 \geq 0$, then the cubic equation has three roots. These roots are calculated by first calculating:

$$\theta = \arccos\left(\frac{r}{q^{3/2}}\right) \tag{A.11}$$

Then, the three roots are given by:

$$Z_1 = -2q^{1/2}cos\left(\frac{\theta}{3}\right) - \frac{\beta}{3} \tag{A.12}$$

$$Z_2 = -2q^{1/2}cos\left(\frac{\theta + 2\pi}{3}\right) - \frac{\beta}{3} \tag{A.13}$$

$$Z_3 = -2q^{1/2}cos\left(\frac{\theta + 4\pi}{3}\right) - \frac{\beta}{3} \tag{A.14}$$

If $q^3 - r^2 < 0$, then the cubic equation has only one root, given by:

$$Z_1 = -sign(r)\left\{\left[\left(r^2 - q^3\right)^{1/2} + |r|\right]^{1/3} + \frac{q}{\left[\left(r^2 - q^3\right)^{1/2} + |r|\right]^{1/3}}\right\} - \frac{\beta}{3} \tag{A.15}$$

where sign(r) is the sign of r, such that sign(r) = 1 if $r > 0$ and sign (r) = −1 if $r < 0$.

If there is only one root, there is no choice but to take this as the compressibility factor at the specified conditions of temperature and pressure, but if three real values exist, then the largest corresponds to the vapor compressibility factor and the smallest is the liquid compressibility factor. The middle value has no physical meaning. A superheated vapor might provide only one root, corresponding to the compressibility factor of the vapor phase. A subcooled liquid might provide only one root, corresponding to the compressibility factor of the liquid phase. A vapor–liquid system should provide three roots, with only the largest and smallest being significant. Equations of state such as the Peng–Robinson equation are generally more reliable at predicting the vapor compressibility than the liquid compressibility.

For multicomponent systems, mixing rules are needed to determine the values of a and b (Poling, Prausnitz and O'Connell, 2001; Oellrich *et al.*, 1981):

$$b = \sum_i^{NC} x_i b_i \tag{A.16}$$

$$a = \sum_i^{NC}\sum_j^{NC} x_i x_j \sqrt{a_i a_j}(1 - k_{ij}) \tag{A.17}$$

where k_{ij} is a binary interaction parameter found by fitting experimental data (Oellrich *et al.*, 1981).

Hydrocarbons:	k_{ij} very small, nearly zero
Hydrocarbon/Light gas:	k_{ij} small and constant
Polar mixtures:	k_{ij} large and temperature dependent

Interaction parameters for a number of binary pairs have been given by Oellrich *et al.* (1981).

In addition to the Peng–Robinson (PR) equation of state (Peng and Robinson, 1976), another popular cubic equation of state is the Soave–Redlich–Kwong (SRK) equation (Soave, 1972). The two equations are similar, but the PR equation gives a better prediction of the liquid density. There are many variations on the basic forms of the PR and SRK equations. One significant weak point of the cubic equations is their reliability when predicting the behavior of mixtures containing significant amounts of hydrogen. If there are significant amounts of hydrogen, finding the correct root of the cubic equation is problematic. In situations where hydrogen is a problem, it is advisable to use the Chao–Seader–Grayson–Streed model (Chao and Seader, 1961; Grayson and Streed, 1963). This is a semi-empirical model, but it is tailored to predicting the behavior of hydrogen-rich hydrocarbon mixtures. Other forms of equations of state include those derived from statistical thermodynamics, such as the SAFT (Statistical Associating Fluid Theory) (Chapman *et al.*, 1989, Kontogeorgis and Folas, 2010). Many other equations of state are available for specialist applications. Table A.1 compares the characteristics of some of the most commonly used equations of state.

Table A.1

Characteristics of some of the most commonly used equations of state.

Equation of state	Characteristics	Reference
Peng–Robinson (PR)	Cubic equation of state used for hydrocarbons and gases. Cryogenic systems down to −250 °C. Pressures up to 1000 bar. Variations on the PR equation can be used for three phase flash calculations. Difficulties can arise obtaining the correct root to the cubic equation close to the critical point. Not recommended for heavy hydrocarbon mixtures at low pressures. Many variations available are derived from the original PR equation.	Peng and Robinson (1976)
Soave–Redlich–Kwong (SRK)	Cubic equation of state used for hydrocarbons and gases. Produces comparable results to PR, but the range of application is smaller. Less accurate than PR for the prediction of liquid density (molar volume). Cryogenic systems down to −140 °C. Pressures up to 350 bar. Not as reliable as PR for nonideal systems. Variations on SRK can be used for three phase flash calculations involving water as the second liquid phase. Difficulties can arise obtaining the correct root to the cubic equation close to the critical point. Not recommended for heavy hydrocarbon mixtures at low pressures. Many variations available are derived from the original SRK equation.	Soave (1972)
Chao–Seader–Grayson–Streed (CSGS)	Recommended for hydrocarbon mixtures with a high hydrogen content. Also recommended for mixtures containing mainly liquid or vapor H_2O (but use steam tables for pure H_2O). Can be used for three phase flash calculations involving water as the second liquid phase. Temperatures from −20 °C to 430 °C. Pressures up to 200 bar.	Chao and Seader (1961), Grayson and Streed (1963)
Benedict–Webb–Rubin (BWR)	Requires 8 specific coefficients, together with binary interaction parameters for mixtures. Accurate for hydrocarbon systems and gases, but not polar mixtures.	Benedict, Webb and Rubin (1940)
Benedict–Webb–Rubin–Starling (BWRS)	Requires 11 specific coefficients, together with binary interaction parameters for mixtures. Accurate for hydrocarbon systems and gases, but not polar mixtures. Commonly used for compression calculations for hydrocarbon systems.	Starling (1973)
Lee–Kesler–Plocker (LKP)	A modification of the BWR equation to improve prediction over a wider range of substances. Accurate general method for nonpolar substances and mixtures. Can be more accurate than PR and SRK for liquid density for hydrocarbon mixtures.	Lee and Kesler (1975), Plocker, Knapp and Prausnitz (1978)
SAFT	Use when strong hydrogen bonding is present and for mixtures involving polymers.	Chapman *et al.*, 1989, Kontogeorgis and Folas (2010)

Example A.1 Using the Peng–Robinson equation of state:

a) Determine the vapor compressibility of nitrogen at 273.15 K and 1.013 bar, 5 bar and 50 bar, and compare with an ideal gas. For nitrogen, $T_C = 126.2$ K, $P_C = 33.98$ bar and $\omega = 0.037$. Take $R = 0.08314$ bar·m^3·kmol^{-1}·K^{-1}.
b) Determine the liquid density of benzene at 293.15 K and compare this with the measured value of $\rho_L = 876.5$ kg·m^{-3}. For benzene, molar mass = 78.11 kg·kmol^{-1}, $T_C = 562.05$ K, $P_C = 48.95$ bar and $\omega = 0.210$.

Solution

a) For nitrogen, Equation A.8 must be solved for the vapor compressibility factor. In this case, $q^3 - r^2 < 0$ and there is one root given by Equation A.15. The solution is summarized in Table A.2.
From Table A.2, it can be seen that the nitrogen can be approximated by ideal gas behavior at moderate pressures.
b) For benzene, at 293.15 K and 1.013 bar (1 atm), Equation A.8 must be solved for the liquid compressibility factor. In this case, $q^3 - r^2 > 0$ and there are three roots given by Equations A.12 to A.14. The parameters for the Peng–Robinson equation are given in Table A.3.
From the three roots, only Z_1 and Z_2 are significant. The smallest root (Z_1) relates to the liquid and the largest root (Z_2) to the vapor. Thus, the density of liquid benzene is given by Equation A.5:

$$\rho_L = \frac{78.11}{Z_1 RT/P}$$
$$= \frac{78.11}{0.0036094 \times 24.0597}$$
$$= 899.5 \text{ kg} \cdot \text{m}^{-3}$$

This compares with an experimental value of $\rho_L = 876.5$ kg · m^{-3} (an error of 3%).

Table A.3

Solution of the Peng–Robinson equation of state for benzene.

	Pressure 1.013 bar
κ	0.68661
α	1.41786
a	28.920
b	7.4270×10^{-2}
A	4.9318×10^{-2}
B	3.0869×10^{-3}
β	−0.99691
γ	4.3116×10^{-2}
δ	-1.4268×10^{-4}
q	9.6054×10^{-2}
r	-2.9603×10^{-2}
$q^3 - r^2$	9.9193×10^{-6}
Z_1	0.0036094
Z_2	0.95177
Z_3	0.04153

Table A.2

Solution of the Peng–Robinson equation of state for nitrogen.

	Pressure 1.013 bar	Pressure 5 bar	Pressure 50 bar
κ	0.43133	0.43133	0.43133
α	0.63482	0.63482	0.63482
a	0.94040	0.94040	0.94040
b	2.4023×10^{-2}	2.4023×10^{-2}	2.4023×10^{-2}
A	1.8471×10^{-3}	9.1171×10^{-3}	9.1171×10^{-2}
B	1.0716×10^{-3}	5.2891×10^{-3}	5.2891×10^{-2}
β	−0.99893	0.99471	−0.94711
γ	-2.9947×10^{-4}	-1.5451×10^{-3}	-2.3004×10^{-2}
δ	-8.2984×10^{-7}	-2.0099×10^{-5}	-1.8767×10^{-3}
q	0.11097	0.11045	0.10734
r	-3.6968×10^{-2}	-3.6719×10^{-2}	-3.6035×10^{-2}
$q^3 - r^2$	-2.9848×10^{-8}	-7.1589×10^{-7}	-6.1901×10^{-5}
Z_1	0.9993	0.9963	0.9727
RT/P (m^3·kmol^{-1})	22.42	4.542	0.4542
Z_1RT/P (m^3·kmol^{-1})	22.40	4.525	0.4418

A.2 Phase Equilibrium for Single Components

The phase equilibrium for pure components is illustrated in Figure A.1. At low temperatures, the component forms a solid phase. At high temperatures and low pressures, the component forms a vapor phase. At high pressures and high temperatures, the component forms a liquid phase. The phase equilibrium boundaries between each of the phases are illustrated in Figure A.1. The point where the three phase equilibrium boundaries meet is the *triple point*, where solid, liquid and vapor coexist. The phase equilibrium boundary between liquid and vapor terminates at the *critical point*. Above the critical temperature, no liquid forms, no matter how high the pressure. The phase equilibrium boundary between liquid and vapor connects the triple point and the critical point, and marks the boundary where vapor and liquid coexist. For a given temperature on this boundary, the pressure is the *vapor pressure*. When the vapor pressure is 1 atm, the corresponding temperature is the *normal boiling point*. If, at any given vapor pressure, the component is at a temperature less than the phase equilibrium, it is *subcooled*. If it is at a temperature above the phase equilibrium, it is *superheated*. Various expressions can be used to represent the vapor pressure curve.

The simplest expression is the Clausius–Clapeyron equation (Hougen, Watson and Ragatz, 1954, 1959):

$$\ln P^{SAT} = A - \frac{B}{T} \tag{A.18}$$

where A and B are constants and T is the absolute temperature. This indicates that a plot of $\ln P^{SAT}$ versus $1/T$ should be a straight line. Equation A.18 gives a good correlation only over small temperature ranges. Various modifications have been suggested to extend the range of application, for example, the Antoine equation (Hougen, Watson and Ragatz, 1954, 1959: Poling, Prausnitz and O'Connell, 2001):

$$\ln P^{SAT} = A - \frac{B}{C+T} \tag{A.19}$$

where A, B and C are constants determined by correlating experimental data (Poling, Prausnitz and O'Connell, 2001). Extended forms of the Antoine equation have also been proposed. For example:

$$\ln P^{SAT} = A + \frac{B}{C+T} + DT + E\ln T + FT^G \tag{A.20}$$

where A, B, C, D, E, F, and G are constants determined by correlating experimental data. Extended forms of the Antoine equation allow prediction of the saturated liquid vapor pressure of a greater range of temperatures. Great care must be taken when using correlated vapor pressure data not to use the correlation coefficients outside the temperature range over which the data has been correlated; otherwise, serious errors can occur.

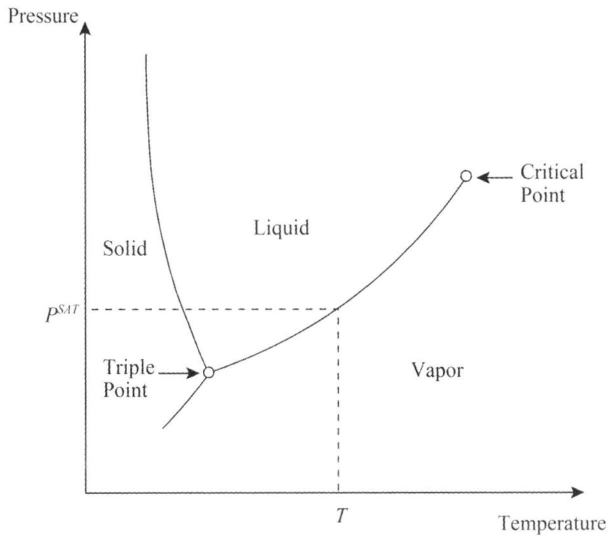

Figure A.1
Phase equilibrium for a pure component.

A.3 Fugacity and Phase Equilibrium

Having considered single-component systems, multicomponent systems now need to be addressed. If a closed system contains more than one phase, the equilibrium condition can be written as:

$$f_i^I = f_i^{II} = f_i^{III} \quad i = 1, 2, \ldots, NC \tag{A.21}$$

where f_i is the fugacity of Component i in Phases I, II and III and NC is the number of components. Fugacity is a thermodynamic pressure, but has no strict physical significance. It can be thought of as an "escaping tendency". Thus, Equation A.21 states that if a system of different phases is in equilibrium, then the "escaping tendency" of Component i from the different phases is equal.

A.4 Vapor–Liquid Equilibrium

Thermodynamic equilibrium in a vapor–liquid mixture is given by the condition that the vapor and liquid fugacities for each component are equal (Hougen, Watson and Ragatz, 1959):

$$f_i^V = f_i^L \tag{A.22}$$

where f_i^V = fugacity of Component i in the vapor phase
f_i^L = fugacity of Component i in the liquid phase

Thus, equilibrium is achieved when the "escaping tendency" from the vapor and liquid phases for Component i are equal.

The *vapor-phase fugacity coefficient*, ϕ_i^V, can be defined by the expression:

$$f_i^V = y_i \phi_i^V P \tag{A.23}$$

where y_i = mole fraction of Component i in the vapor phase
ϕ_i^V = vapor-phase fugacity coefficient
P = system pressure

The *liquid-phase fugacity coefficient* ϕ_i^L can be defined by the expression:

$$f_i^L = x_i \phi_i^L P \tag{A.24}$$

The *liquid-phase activity coefficient* γ_i can be defined by the expression:

$$f_i^L = x_i \gamma_i f_i^O \tag{A.25}$$

where x_i = mole fraction of Component i in the liquid phase
ϕ_i^L = liquid-phase fugacity coefficient
γ_i = liquid-phase activity coefficient
f_i^O = fugacity of Component i at standard state

For moderate pressures, f_i^O can be approximated by the saturated vapor pressure, P_i^{SAT}; thus, Equation A.25 becomes (Poling, Prausnitz and O'Connell, 2001):

$$f_i^L = x_i \gamma_i P_i^{SAT} \tag{A.26}$$

Equations A.22, A.23 and A.24 can be combined to give an expression for the K-value, K_i, which relates the vapor and liquid mole fractions:

$$K_i = \frac{y_i}{x_i} = \frac{\phi_i^L}{\phi_i^V} \tag{A.27}$$

Equation A.27 defines the relationship between the vapor and liquid mole fractions and provides the basis for vapor–liquid equilibrium calculations on the basis of equations of state. Thermodynamic models are required for Φ_i^V and Φ_i^L from an equation of state. Alternatively, Equations A.22, A.23 and A.26 can be combined to give:

$$K_i = \frac{y_i}{x_i} = \frac{\gamma_i P_i^{SAT}}{\phi_i^V P} \tag{A.28}$$

This expression provides the basis for vapor–liquid equilibrium calculations on the basis of *liquid-phase activity coefficient models*. In Equation A.28, thermodynamic models are required for ϕ_i^V (from an equation of state) and γ_i from a liquid-phase activity coefficient model. Some examples will be given later. At moderate pressures, the vapor phase becomes ideal, as discussed previously, and $\phi_i^V = 1$. For an ideal vapor phase, Equation A.28 simplifies to:

$$K_i = \frac{y_i}{x_i} = \frac{\gamma_i P_i^{SAT}}{P} \tag{A.29}$$

When the liquid phase behaves as an ideal solution $\gamma_i = 1$ and Equation A.29 simplifies to:

$$K_i = \frac{y_i}{x_i} = \frac{P_i^{SAT}}{P} \tag{A.30}$$

which is Raoult's law and represents both ideal vapor- and liquid-phase behavior. Correlations are available to relate component vapor pressure to temperature, as discussed in Section A.2.

By contrast, highly nonideal behavior in which $\gamma_i > 1$ (positive deviation from Raoult's Law) forms a *minimum-boiling azeotrope*. At the azeotropic composition, the vapor and liquid are both at the same composition for the mixture. The lowest boiling temperature is below that of either of the pure components and is at the minimum-boiling azeotrope (see Figure 8.4b). For highly nonideal behavior in which $\gamma_i < 1$ (negative deviation from Raoult's Law) a *maximum-boiling azeotrope* can be formed (see Figure 8.4c). This maximum-boiling azeotrope boils at a higher temperature than either of the pure components and would be the last fraction to be distilled, rather than the least volatile component, which would be the case with nonazeotropic behavior.

The vapor–liquid equilibrium for noncondensable gases in equilibrium with liquids can often be approximated by Henry's Law (Hougen, Watson and Ragatz, 1954, 1959; Poling, Prausnitz and O'Connell, 2001):

$$p_i = H_i x_i \tag{A.31}$$

where p_i = partial pressure of Component i
H_i = Henry's Law constant (determined experimentally)
x_i = mole fraction of Component i in the liquid phase

Assuming ideal gas behavior ($p_i = y_i P$):

$$y_i = \frac{H_i x_i}{P} \tag{A.32}$$

Thus, the K-value is given by:

$$K_i = \frac{y_i}{x_i} = \frac{H_i}{P} \tag{A.33}$$

A straight line would be expected from a plot of y_i against x_i.

The ratio of equilibrium K-values for two components measures their *relative volatility*:

$$\alpha_{ij} = \frac{K_i}{K_j} \tag{A.34}$$

where α_{ij} = volatility of Component i relative to Component j

These expressions form the basis for two alternative approaches to vapor–liquid equilibrium calculations:

a) $K_i = \phi_i^L / \phi_i^V$ forms the basis for calculations based entirely on equations of state. Using an equation of state for both the liquid and vapor phase has a number of advantages. First, f_i^O need not be specified. Also, in principle, continuity at the critical point can be guaranteed with all thermodynamic properties derived from the same model. The presence of noncondensable gases, in principle, causes no additional complications. However, the application of equations of state is largely restricted to nonpolar components.

b) $K_i = \gamma_i P_i^{SAT} / \phi_i^V P$ forms the basis for calculations based on liquid-phase activity coefficient models. It is used when polar molecules are present. For most systems at low pressures, ϕ_i^V can be assumed to be unity. If high pressures are involved, then ϕ_i^V must be calculated from an equation of state. However, care should be taken when mixing and matching different models for γ_i and ϕ_i^V for high-pressure systems to ensure that appropriate combinations are taken.

Example A.2 A gas from a combustion process has a flowrate of $10\ m^3 \cdot s^{-1}$ and contains 200 ppmv of oxides of nitrogen, expressed as nitric oxide (NO) at standard conditions of 0 °C and 1 atm. This concentration needs to be reduced to 50 ppmv (expressed at standard conditions) before being discharged to the environment. It can be assumed that all of the oxides of nitrogen are present in the form of NO. One option being considered to remove the NO is by absorption in water at 20 °C and 1 atm. The solubility of the NO in water follows Henry's Law, with $H_{NO} = 2.6 \times 10^4$ atm at 20 °C. The gas is to be contacted countercurrently with water such that the inlet gas contacts the outlet water. The concentration of the outlet water can be assumed to reach 90% of equilibrium. Estimate the flowrate of water required, assuming the molar mass of gas in kilograms occupies $22.4\ m^3$ at standard conditions of 0 °C and 1 atm.

Solution

$$\text{Molar flowrate of gas} = 10 \times \frac{1}{22.4}$$
$$= 0.446\ \text{kmol s}^{-1}$$

Assuming the molar flowrate of gas remains constant, the amount of NO to be removed

$$= 0.446(200 - 50) \times 10^{-6}$$
$$= 6.69 \times 10^{-5}\ \text{kmol} \cdot \text{s}^{-1}$$

Assuming the water achieves 90% of equilibrium and contacts countercurrently with the gas, from Henry's Law (Equation A.32):

$$x_{NO} = \frac{0.9 y_{NO}^* P}{H_{NO}}$$

where y_{NO}^* = equilibrium mole fraction in the gas phase

$$x_{NO} = \frac{0.9 \times 200 \times 10^{-6} \times 1}{2.6 \times 10^4}$$
$$= 6.9 \times 10^{-9}$$

Assuming the liquid flowrate is constant, water flowrate

$$= \frac{6.69 \times 10^{-5}}{6.9 \times 10^{-9} - 0}$$
$$= 9696\ \text{kmol} \cdot \text{s}^{-1}$$
$$= 9696 \times 18\ \text{kg} \cdot \text{s}^{-1}$$
$$= 174{,}500\ \text{kg} \cdot \text{s}^{-1}$$

This is an impractically large flowrate.

A.5 Vapor–Liquid Equilibrium Based on Activity Coefficient Models

In order to model liquid-phase nonideality at moderate pressures, the liquid activity coefficient γ_i must be known:

$$K_i = \frac{\gamma_i P_i^{SAT}}{P} \tag{A.35}$$

γ_i varies with composition and temperature. There are three popular activity coefficient models (Poling, Prausnitz and O'Connell, 2001):

a) Wilson
b) Nonrandom two liquid (NRTL)
c) Universal quasi-chemical (UNIQUAC)

These models are semi-empirical and are based on the concept that intermolecular forces will cause nonrandom arrangement of molecules in the mixture. The models account for the arrangement of molecules of different sizes and the preferred orientation of molecules. In each case, the models are fitted to experimental binary vapor–liquid equilibrium data. This gives binary interaction parameters that can be used to predict multicomponent vapor–liquid equilibrium. In the case of the UNIQUAC equation, if experimentally determined vapor–liquid equilibrium data are not available, the Universal Quasi-chemical Functional Group Activity Coefficients (UNIFAC) Method can be used to estimate UNIQUAC parameters from the molecular structures of the components in the mixture (Poling, Prausnitz and O'Connell, 2001).

a) *Wilson equation.* The Wilson equation activity coefficient model is given by (Wilson, 1964: Poling, Prausnitz and O'Connell, 2001):

$$\ln \gamma_i = -\ln\left[\sum_j^{NC} x_j \Lambda_{ij}\right] + 1 - \sum_k^{NC}\left[\frac{x_k \Lambda_{ki}}{\sum_j^{N} x_j \Lambda_{kj}}\right] \tag{A.36}$$

where

$$\Lambda_{ij} = \frac{V_j^L}{V_i^L} \exp\left[-\frac{\lambda_{ij} - \lambda_{ii}}{RT}\right] \tag{A.37}$$

where V_i^L = molar volume of pure Liquid i
λ_{ij} = energy parameter characterizing the interaction of Molecule i with Molecule j
R = gas constant
T = absolute temperature

$$\Lambda_{ii} = \Lambda_{jj} = \Lambda_{kk} = 1$$

$\Lambda_{ij} = 1$ for an ideal solution
$\Lambda_{ij} < 1$ for positive deviation from Raoult's Law
$\Lambda_{ij} > 1$ for negative deviation from Raoult's Law

For each binary pair, there are two adjustable parameters that must be determined from experimental data, that is, $(\lambda_{ij}-\lambda_{ii})$, which are temperature dependent. The ratio of molar volumes V_j^L/V_i^L is a weak function of temperature. For a binary system, the Wilson equation reduces to (Poling, Praunitz and O'Connell, 2001):

$$\ln \gamma_1 = -\ln[x_1 + x_2\Lambda_{12}] + x_2\left[\frac{\Lambda_{12}}{x_1+\Lambda_{12}x_2} - \frac{\Lambda_{21}}{x_1\Lambda_{21}+x_2}\right] \quad \text{(A.38)}$$

$$\ln \gamma_2 = -\ln[x_2 + x_1\Lambda_{21}] - x_1\left[\frac{\Lambda_{12}}{x_1+\Lambda_{12}x_2} - \frac{\Lambda_{21}}{x_1\Lambda_{21}+x_2}\right] \quad \text{(A.39)}$$

where

$$\Lambda_{12} = \frac{V_2^L}{V_1^L}\exp\left[-\frac{\lambda_{12}-\lambda_{11}}{RT}\right], \quad \Lambda_{21} = \frac{V_1^L}{V_2^L}\exp\left[-\frac{\lambda_{21}-\lambda_{22}}{RT}\right] \quad \text{(A.40)}$$

The two adjustable parameters, $(\lambda_{12}-\lambda_{11})$ and $(\lambda_{21}-\lambda_{22})$, must be determined experimentally (Gmehling, Onken and Arlt, 1977–1980).

b) *NRTL equation.* The NRTL equation is given by (Renon and Prausnitz, 1968; Poling, Prausnitz and O'Connell, 2001):

$$\ln \gamma_i = \frac{\sum_j^{NC} \tau_{ji}G_{ji}x_j}{\sum_k^{NC} G_{ki}x_k} + \sum_j^{NC} \frac{x_jG_{ij}}{\sum_k^{NC} G_{kj}x_k}\left(\tau_{ij} - \frac{\sum_k^{NC} x_k\tau_{kj}G_{kj}}{\sum_k^{NC} G_{kj}x_k}\right) \quad \text{(A.41)}$$

where $G_{ij} = \exp(-\alpha_{ij}\tau_{ij})$, $G_{ji} = \exp(-\alpha_{ij}\tau_{ji})$

$$\tau_{ij} = \frac{g_{ij}-g_{jj}}{RT}, \quad \tau_{ji} = \frac{g_{ji}-g_{ii}}{RT}$$

$G_{ij} \neq G_{ji}$, $\tau_{ij} \neq \tau_{ji}$, $G_{ii} = G_{jj} = 1$, $\tau_{ii} = \tau_{jj} = 0$

$\tau_{ij} = 0$ for ideal solutions

and g_{ij} and g_{ji} are the energies of interactions between Molecules i and j; α_{ij} characterizes the tendency of Molecule i and Molecule j to be distributed in a random fashion, depends on molecular properties and usually lies in the range 0.2 to 0.5. For each binary pair of components, there are three adjustable parameters, $(g_{ij}-g_{jj})$, $(g_{ji}-g_{ii})$ and $\alpha_{ij}(=\alpha_{ji})$, which are temperature dependent. For a binary system, the NRTL equation reduces to (Poling, Prausnitz and O'Connell, 2001):

$$\ln \gamma_1 = x_2^2\left[\tau_{21}\left(\frac{G_{21}}{x_1+x_2G_{21}}\right)^2 + \frac{\tau_{12}G_{12}}{(x_2+x_1G_{12})^2}\right] \quad \text{(A.42)}$$

$$\ln \gamma_2 = x_1^2\left[\tau_{12}\left(\frac{G_{12}}{x_2+x_1G_{12}}\right)^2 + \frac{\tau_{21}G_{21}}{(x_1+x_2G_{21})^2}\right] \quad \text{(A.43)}$$

where $G_{12} = \exp(-\alpha_{12}\tau_{12})$, $G_{21} = \exp(-\alpha_{21}\tau_{21})$,

$$\tau_{12} = \frac{g_{12}-g_{22}}{RT}, \quad \tau_{21} = \frac{g_{21}-g_{11}}{RT}$$

The three adjustable parameters, $(g_{12}-g_{22})$, $(g_{21}-g_{11})$ and $\alpha_{12}(=\alpha_{21})$, must be determined experimentally (Gmehling, Onken and Arlt, 1977–1980).

c) *UNIQUAC equation.* The UNIQUAC equation is given by (Anderson and Prausnitz, 1978a; Poling, Prausnitz and O'Connell, 2001):

$$\ln \gamma_i = \ln\left(\frac{\Phi_i}{x_i}\right) + \frac{z}{2}q_i\ln\left(\frac{\theta_i}{\Phi_i}\right) + l_i - \frac{\Phi_i}{x_i}\sum_j^{NC} x_jl_j + q_i\left[1 - \ln\left(\sum_j^{NC}\theta_j\tau_{ji}\right) - \sum_j^{NC}\frac{\theta_j\tau_{ij}}{\sum_k^{NC}\theta_k\tau_{kj}}\right] \quad \text{(A.44)}$$

where $\Phi_i = \dfrac{r_ix_i}{\sum_k^{NC} r_kx_k}$, $\theta_i = \dfrac{q_ix_i}{\sum_k^{NC} q_kx_k}$

$$l_i = \frac{z}{2}(r_i-q_i) - (r_i-1), \quad \tau_{ij} = \exp-\left(\frac{u_{ij}-u_{jj}}{RT}\right)$$

u_{ij} = interaction parameter between Molecule i and Molecule j ($u_{ij} = u_{ji}$)

z = coordination number ($z = 10$)

r_i = pure component property, measuring the molecular van der Waals volume for Molecule i

q_i = pure component property, measuring the molecular van der Waals surface area for Molecule i

R = gas constant

T = absolute temperature

$u_{ij} = u_{ji}$, $\tau_{ii} = \tau_{jj} = 1$

For each binary pair, there are two adjustable parameters that must be determined from experimental data, that is, $(u_{ij}-u_{jj})$, which are temperature dependent. Pure component properties r_i and q_i measure molecular van der Waals volumes and surface areas and have been tabulated (Gmehling, Onken and Alt, 1977–1980). For a binary system, the UNIQUAC equation reduces to (Poling, Prausnitz and O'Connell, 2001):

$$\ln \gamma_1 = \ln\left(\frac{\Phi_1}{x_1}\right) + \frac{z}{2}q_1\ln\left(\frac{\theta_1}{\Phi_1}\right) + \Phi_2\left(l_1 - l_2\frac{r_1}{r_2}\right) - q_1\ln(\theta_1+\theta_2\tau_{21}) + \theta_2q_1\left(\frac{\tau_{21}}{\theta_1+\theta_2\tau_{21}} - \frac{\tau_{21}}{\theta_2+\theta_1\tau_{12}}\right) \quad \text{(A.45)}$$

$$\ln \gamma_2 = \ln\left(\frac{\Phi_2}{x_2}\right) + \frac{z}{2}q_2\ln\left(\frac{\theta_2}{\Phi_2}\right) + \Phi_1\left(l_2 - l_1\frac{r_2}{r_1}\right) - q_2\ln(\theta_2+\theta_1\tau_{12}) + \theta_1q_2\left(\frac{\tau_{12}}{\theta_2+\theta_1\tau_{12}} - \frac{\tau_{21}}{\theta_1+\theta_2\tau_{21}}\right) \quad \text{(A.46)}$$

where $\Phi_1 = \dfrac{r_1 x_1}{r_1 x_1 + r_2 x_2}, \Phi_2 = \dfrac{r_2 x_2}{r_1 x_1 + r_2 x_2}$

$$\theta_1 = \frac{q_1 x_1}{q_1 x_1 + q_2 x_2}, \theta_2 = \frac{q_2 x_2}{q_1 x_1 + q_2 x_2}$$

$$l_1 = \frac{z}{2}(r_1 - q_1) - (r_1 - 1),$$

$$l_2 = \frac{z}{2}(r_2 - q_2) - (r_2 - 1)$$

$$\tau_{12} = \exp - \left(\frac{u_{12} - u_{22}}{RT}\right)$$

$$\tau_{21} = \exp - \left(\frac{u_{21} - u_{11}}{RT}\right)$$

The two adjustable parameters, $(u_{12} - u_{22})$ and $(u_{21} - u_{11})$, must be determined experimentally (Gmehling, Onken and Arlt, 1977–1980). Pure component properties r_1, r_2, q_1 and q_2 have been tabulated (Gmehling, Onken and Arlt, 1977–1980).

Since all experimental data for vapor–liquid equilibrium have some experimental uncertainty, it follows that the parameters obtained from data reduction are not unique (Poling, Prausnitz and O'Connell, 2001). There are many sets of parameters that can represent the experimental data equally well, within experimental uncertainty. The experimental data used in data reduction are not sufficient to fix a unique set of "best" parameters. Realistic data reduction can determine only a region of parameters (Hougen, Watson and Ragatz, 1959).

Published interaction parameters are available (Gmehling, Onken and Arlt, 1977–1980). However, when more than one set of parameters is available for a binary pair, which should be chosen?

a) Check if the experimental data is thermodynamically consistent. The Gibbs–Duhem equation (Hougen, Watson and Ragatz, 1954, 1959) can be applied to experimental binary data to check its thermodynamic consistency and it should be consistent with this equation.

b) Choose parameters fitted at the process pressure.

c) Choose data sets covering the composition range of interest.

d) For multicomponent systems, choose parameters fitted to ternary or higher systems, if possible.

A.6 Group Contribution Methods for Vapor–Liquid Equilibrium

The vapor–liquid equilibrium models discussed in Section A.5 depend on knowledge of binary interaction parameters for each binary pair in the mixture to have been fitted from experimental data. It is often the case that binary interaction parameters are not available for all of the binary pairs in the mixture being studied. If only fragmentary data or no data are available, then vapor–liquid equilibrium can be estimated using *group contribution methods*. Using this approach, a molecule is divided into functional groups. The division into functional groups can be somewhat arbitrary. Molecule-to-molecule interactions are considered to be weighted sums of the group-to-group interactions. Thus in a multicomponent system, group contribution methods assume that each functional group behaves in a manner independent of the molecule in which it appears. It is necessary to correlate group-to-group interactions from experimental data for binary systems. It is then possible to calculate molecule-to-molecule interactions, and therefore phase equilibrium for molecular pairs when no experimental data are available. The advantage of the group contribution approach is that, because the number of possible distinct functional groups is much smaller than the number of distinct possible molecules, then it is possible to correlate more group-to-group interactions than molecule-to-molecule interactions. However, there are also many disadvantages, to be discussed later.

The description of functional groups is illustrated in Figure A.2, which shows some examples of molecules divided into functional groups. Figure A.2a shows n-pentane, which consists of two CH_3 groups and three CH_2 groups. Figure A.2b shows ethanol, which

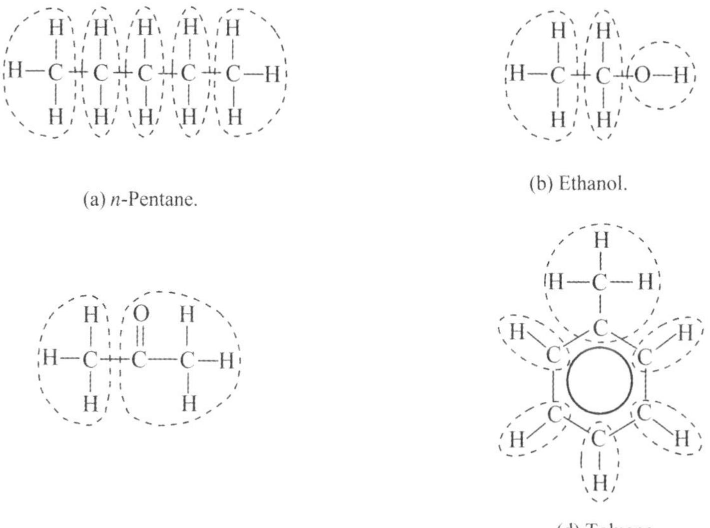

(a) n-Pentane.

(b) Ethanol.

(c) Acetone.

(d) Toluene.

Figure A.2

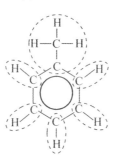

Molecules as functional groups.

consists of one CH_3 group, one CH_2 group and one OH group. Figure A.2c shows acetone, which consists of one CH_3 group and one CH_3CO group. Finally, Figure A.2d shows toluene, which consists of five ACH (aromatic-CH) groups and one $ACCH_3$ (aromatic-CCH_3) group.

To calculate the activity coefficient requires the group mole fraction X_k to be defined. This is analogous to the mole fraction of a molecule and is defined as:

$$X_k = \frac{\sum_i x_i \upsilon_k^i}{\sum_i x_i \left(\sum_k \upsilon_k^i\right)} \tag{A.47}$$

where X_k = mole fraction of interaction Group k in Molecule i
υ_k^i = number of interaction Groups k in Molecule i

For example, in a mixture with the mole fraction of acetone $x_{AC} = 0.4$ and the mole fraction of toluene $x_{TOL} = 0.6$, the group fraction of the CH_3 group in the mixture is given by:

$$X_{CH3} = \frac{x_{AC} \times 1}{x_{AC} \times 2 + x_{TOL} \times 6} = \frac{0.4 \times 1}{0.4 \times 2 + 0.6 \times 6} = 0.0909$$

The most commonly used group contribution method for vapor–liquid equilibrium is the UNIFAC (UNIQUAC Functional Group Activity Coefficient) method. The UNIFAC model represents the activity coefficient of Species i by a combinatorial component (C) and a residual component (R):

$$\ln \gamma_i = \ln \gamma_i^C + \ln \gamma_i^R \tag{A.48}$$

Both parts are based on the UNIQUAC equation above. The combinatorial component is given by:

$$\ln \gamma_i^C = \ln\left(\frac{\Phi_i}{x_i}\right) + \frac{z}{2} q_i \ln\left(\frac{\theta_i}{\Phi_i}\right) + l_i - \frac{\Phi_i}{x_i} \sum_j^{NC} x_j l_j \tag{A.49}$$

where $\Phi_i = \dfrac{r_i x_i}{\sum_j^{NC} r_j x_j} \qquad \theta_i = \dfrac{q_i x_i}{\sum_j^{NC} q_j x_j}$

$$l_i = \frac{z}{2}(r_i - q_i) - (r_i - 1), \quad z = 10$$

x_i = mole fraction of Component i
z = coordination number ($z = 10$)
r_i = pure component volume parameter for Molecule i
q_i = pure component surface area parameter for Molecule i

The pure component r_i and q_i parameters are calculated from the group volume R_k and surface area Q_k contributions:

$$r_i = \sum_k \upsilon_k^i R_k, \quad q_i = \sum_k \upsilon_k^i Q_k \tag{A.50}$$

The group volume and surface area contributions R_k and Q_k are normally obtained from tables (Poling, Prausnitz and O'Connell, 2001). The residual component of the activity coefficient γ_i^R is defined by:

$$\ln \gamma_i^R = \sum_k^{All\ groups} \upsilon_k^i \left(\ln \Gamma_k - \ln \Gamma_k^i\right) \tag{A.51}$$

where Γ_k = group residual activity coefficient
Γ_k^i = residual activity coefficient of group k in a reference solution containing only molecules of Type i

The group activity coefficient for both Γ_k and Γ_k^i is given by:

$$\ln \Gamma_k = Q_k \left[1 - \ln\left(\sum_m \theta_m \psi_{mk}\right) - \sum_m \frac{\theta_m \psi_{km}}{\sum_n \theta_n \psi_{nm}}\right] \tag{A.52}$$

where θ_m = summation of the area fraction of group m over all groups
$= \dfrac{Q_m X_m}{\sum_n Q_n X_n}$
X_m = mole fraction of Group m in the mixture (–)
$\psi_{nm} = \exp\left(-\dfrac{a_{mn}}{T}\right)$
a_{mn} = group interaction parameter (K)
T = absolute temperature (K)

The group interaction parameters a_{mn} must be correlated from experimental data. It should also be noted that $a_{mn} \neq a_{nm}$.

To apply the UNIFAC method requires two different classes of group to be defined: *subgroups* and *main groups*. Subgroups are the smallest building blocks and the main groups are used to group subgroups together. For example, subgroups CH_3, CH_2, CH and C all belong to the main group CH_2. The reason for this is that, although the subgroups have different volume and surface area parameters R_k and Q_k, the interaction parameters are the same for all subgroups within a main group. Table A.4 gives some examples of R_k and Q_k parameters and Table A.5 the corresponding interaction parameters.

There are a number of important restrictions for the UNIFAC method:

- The UNIFAC method does not distinguish between isomers.
- Noncondensable gases are excluded. However, such gases can be modeled separately using Henry's law.
- Polymers and electrolytes are excluded.
- It is limited to moderate pressures of less than 10 bar.
- Temperatures are limited approximately to the range 15 °C to 150 °C.
- Conditions must be remote from critical conditions.
- UNIFAC parameters based on vapor–liquid equilibrium data cannot be used for liquid–liquid equilibrium predictions. Variations from the original method allow the same parameters to be used for vapor–liquid and liquid–liquid equilibrium predictions.
- The group contribution method assumes that a given group will behave the same in any molecule. However, this is not the case.

An alternative group contribution method to UNIFAC is the ASOG (Analytical Solution Of Groups) Method (Poling, Prausnitz

Table A.4

Examples of group volume and surface area parameters (Poling, Prausnitz and O'Connell, 2001).

Main group		Subgroup		R_k	Q_k
Number	Group	Number	Group		
1	CH_2	1	CH_3	0.9011	0.848
		2	CH_2	0.6744	0.540
		3	CH	0.4469	0.228
		4	C	0.2195	0.000
2	C=C	5	CH_2=CH	1.3454	1.176
		6	CH=CH	1.1167	0.867
		7	CH_2=C	1.1173	0.988
		8	CH=C	0.8886	0.676
		70	C=C	0.6605	0.485
3	ACH	9	ACH	0.5313	0.400
		10	AC	0.3652	0.120
4	$ACCH_2$	11	$ACCH_3$	1.2663	0.968
		12	ACCH2	1.0396	0.660
		13	ACCH	0.8121	0.348

Table A.5

Examples of interaction parameters (Poling, Prausnitz and O'Connell, 2001).

Main Group	1	2	3	4
1	0	86.02	61.13	76.5
2	−35.36	0	38.81	74.15
3	−11.12	3.446	0	167.0
4	−69.7	−113.6	−146.8	0

and O'Connell, 2001). This is less commonly used than the UNIFAC Method, even though it was developed earlier.

Despite their convenience, group contribution methods must not be seen as the first choice for the prediction of phase equilibrium, but a choice of last resort when binary interaction parameters are not available for other methods.

A.7 Vapor–Liquid Equilibrium Based on Equations of State

Before an equation of state can be applied to calculate vapor–liquid equilibrium, the fugacity coefficient ϕ_i for each phase needs to be determined. The relationship between the fugacity coefficient and the volumetric properties can be written as:

$$\ln \phi_i = \frac{1}{RT}\int_V^{\infty}\left[\left(\frac{\partial P}{\partial N_i}\right)_{T,V,N_j} - \frac{RT}{V}\right] dV - RT\ln Z \qquad \text{(A.53)}$$

For example, the Peng–Robinson equation of state for this integral yields (Poling, Prausnitz and O'Conell, 2001):

$$\ln \phi_i = \frac{b_i}{b}(Z-1) - \ln(Z-B) - \frac{A}{2\sqrt{2}B}\left(\frac{2\sum_k x_k a_{ik}}{a} - \frac{b_i}{b}\right)\ln\left(\frac{Z+(1+\sqrt{2})B}{Z+(1-\sqrt{2})B}\right) \qquad \text{(A.54)}$$

A

where $A = \frac{aP}{R^2T^2}$

$B = \frac{bP}{RT}$

Thus, given critical temperatures, critical pressures and acentric factors for each component, as well as a phase composition, temperature and pressure, the compressibility factor can be determined and hence component fugacity coefficients for each phase can be calculated. Taking the ratio of liquid to vapor fugacity coefficients for each component gives the vapor–liquid equilibrium K-value for that component. This approach has the advantage of consistency between the vapor- and liquid-phase thermodynamic models. Such models are widely used to predict vapor–liquid equilibrium for hydrocarbon mixtures and mixtures involving light gases.

A vapor–liquid system should provide three roots from the cubic equation of state, with only the largest and smallest being significant. The largest root corresponds to the vapor compressibility factor and the smallest is the liquid compressibility factor. However, some vapor–liquid mixtures can present problems. This is particularly so for mixtures involving light hydrocarbons with significant amounts of hydrogen, which are common in petroleum and petrochemical processes. Under some conditions, such mixtures can provide only one root for vapor–liquid systems, when there should be three. This means that both the vapor and liquid fugacity coefficients cannot be calculated and is a limitation of such cubic equations of state.

If an activity coefficient model is to be used at high pressure (Equation A.28), then the vapor-phase fugacity coefficient can be predicted from Equation A.54. However, this approach has the disadvantage that the thermodynamic models for the vapor and liquid phases are inconsistent. Despite this inconsistency, it might be necessary to use an activity coefficient model if there is reasonable liquid-phase nonideality, particularly with polar mixtures.

A.8 Calculation of Vapor–Liquid Equilibrium

In the case of vapor–liquid equilibrium, the vapor and liquid fugacities are equal for all components at the same temperature and pressure, but how can this solution be found? In any phase equilibrium calculation, some of the conditions will be fixed. For example, the temperature, pressure and overall composition might be fixed. The task is to find values for the unknown conditions that satisfy the equilibrium relationships. However, this cannot be achieved directly. First, values of the unknown variables must be guessed and checked to see if the equilibrium relationships are satisfied. If not, then the estimates must be modified in the light of the discrepancy in the equilibrium, and iteration continued until the estimates of the unknown variables satisfy the requirements of equilibrium.

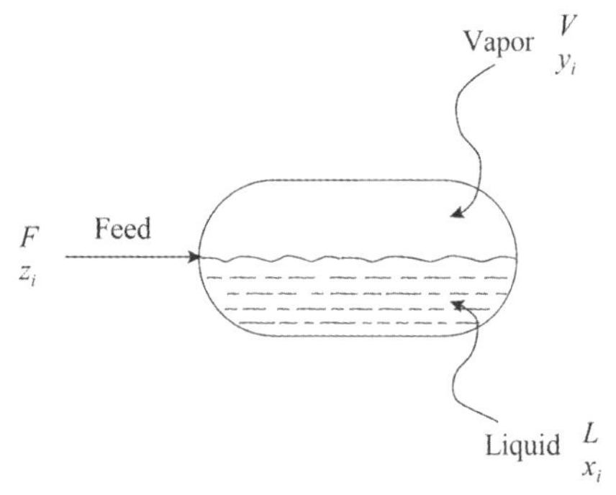

Figure A.3
Vapour–liquid equilibrium.

Consider a simple process in which a multicomponent feed is allowed to separate into a vapor and a liquid phase with the phases coming to equilibrium, as shown in Figure A.3. An overall material balance and component material balances gives (see Chapter 8):

$$y_i = \frac{z_i}{\frac{V}{F} + \left(1 - \frac{V}{F}\right)\frac{1}{K_i}} \tag{A.55}$$

$$x_i = \frac{z_i}{(K_i - 1)\frac{V}{F} + 1} \tag{A.56}$$

where
F = feed flowrate ($kmol \cdot s^{-1}$)
V = vapor flowrate from the separator ($kmol \cdot s^{-1}$)
z_i = mole fraction of Component i in the feed (–)
y_i = mole fraction of Component i in vapor (–)
x_i = mole fraction of Component i in liquid (–)

The vapor fraction (V/F) in Equations A.55 and A.56 lies in the range $0 < V/F < 1$.

For a specified temperature and pressure, Equations A.55 and A.56 need to be solved by trial and error. Given that:

$$\sum_i^{NC} y_i = \sum_i^{NC} x_i = 1 \tag{A.57}$$

where NC is the number of components, then:

$$\sum_i^{NC} y_i - \sum_i^{NC} x_i = 0 \tag{A.58}$$

Substituting Equations A.55 and A.56 into Equation A.58, after rearrangement gives (see Chapter 8):

$$\sum_i^{NC} \frac{z_i(K_i - 1)}{\frac{V}{F}(K_i - 1) + 1} = 0 = f(V/F) \tag{A.59}$$

Equation A.59 is known as the Rachford–Rice Equation (Rachford and Rice, 1952). To solve Equation A.59, start by assuming a value of V/F and calculate $f(V/F)$ and search for a value of V/F until the function equals zero. If it is necessary to calculate the *bubble point* (see Chapter 8):

$$\sum_i^{NC} z_i K_i = 1 \tag{A.60}$$

Thus, to calculate the bubble point for a given mixture and at a specified pressure, a search is made for a temperature to satisfy Equation A.60.

Alternatively, temperature can be specified and a search made for a pressure, the *bubble pressure*, to satisfy Equation A.60. Another special case is when it is necessary to calculate the *dew point*. In this case, $V/F = 1$ in Equation A.59, which simplifies to:

$$\sum_{i}^{NC} \frac{z_i}{K_i} = 1 \qquad \text{(A.61)}$$

Again, for a given mixture and pressure, temperature is searched to satisfy Equation A.61. Alternatively, temperature is specified and pressure searched for the *dew pressure*.

If the K-value requires the composition of both phases to be known, then to start the calculation, a temperature is assumed. Calculation of K-values requires knowledge of the vapor composition to calculate the vapor-phase fugacity coefficient and that of the liquid composition to calculate the liquid-phase fugacity coefficient. If the liquid composition is known, the vapor composition is unknown and an initial estimate is required. Once the K-value has been estimated from an initial estimate of the vapor composition, the composition of the vapor can be reestimated, and so on.

Example A.3 Calculate the vapor composition of an equimolar liquid mixture of methanol and water at 1 atm (1.013 bar):

a) assuming ideal vapor- and liquid-phase behavior, that is, using Raoult's Law and

b) using the Wilson equation.

Vapor pressure in bar can be predicted for temperature in Kelvin from the Antoine equation using coefficients in Table A.6 (Poling, Prausnitz and O'Connell, 2001). Data for the Wilson equation are given in Table A.7 (Gmehling, Onken and Arlt, 1977–1980). Assume the gas constant $R = 8.3145\ \text{kJ}\cdot\text{kmol}^{-1}\cdot\text{K}^{-1}$.

Table A.6

Antoine coefficients for methanol and water (Gmehling, Onken and Arlt, 1977–1980).

	A_i	B_i	C_i
Methanol	11.9869	3643.32	−33.434
Water	11.9647	3984.93	−39.734

Table A.7

Data for methanol (1) and water (2) for the Wilson equation at 1 atm (Gmehling, Onken and Arlt, 1977–1980).

V_1 ($\text{m}^3\cdot\text{kmol}^{-1}$)	V_2 ($\text{m}^3\cdot\text{kmol}^{-1}$)	$(\lambda_{12}-\lambda_{11})$ ($\text{kJ}\cdot\text{kmol}^{-1}$)	$(\lambda_{21}-\lambda_{22})$ ($\text{kJ}\cdot\text{kmol}^{-1}$)
0.04073	0.01807	347.4525	2179.8398

Table A.8

Bubble-point calculation for an ideal methanol–water mixture.

z_i	$T = 340$ K		$T = 360$ K		$T = 350$ K	
	K_i	z_iK_i	K_i	z_iK_i	K_i	z_iK_i
0.5	1.0938	0.5469	2.2649	1.1325	1.5906	0.7953
0.5	0.2673	0.1336	0.6122	0.3061	0.4094	0.2047
1.00		0.6805		1.4386		1.0000

Solution

a) K-values assuming ideal vapor and liquid behavior are given by Raoult's Law (Equation A.30). The composition of the liquid is specified to be $x_1 = 0.5$, $x_2 = 0.5$ and the pressure 1 atm, but the temperature is unknown. Therefore, a bubble-point calculation is required to determine the vapor composition. The bubble point can be calculated from Equation A.60 or from Equation A.59 by specifying $V/F = 0$. The bubble point is calculated from Equation A.60 in Table A.8. The procedure can be carried out in spreadsheet software. Table A.8 shows the results for Raoult's Law.

Thus, the composition of the vapor at 1 atm is $y_1 = 0.7953$, $y_2 = 0.2047$, assuming an ideal mixture.

b) The activity coefficients for the methanol and water can be calculated using the Wilson equation (Equations A.38 and A.39). The results are summarized in Table A.9.

Thus, the composition of the vapor phase at 1 atm is $y_1 = 0.7863$, $y_2 = 0.2136$ from the Wilson equation. For this mixture, at these conditions, there is not much difference between the predictions of Raoult's Law and the Wilson equation, indicating only moderate deviations from ideality at the chosen conditions.

Table A.9

Bubble-point calculation for a methanol–water mixture using the Wilson equation.

z_i	$T = 340$ K			$T = 350$ K			$T = 346.13$ K		
	γ_i	K_i	z_iK_i	γ_i	K_i	z_iK_i	γ_i	K_i	z_iK_i
0.5	1.1429	1.2501	0.6251	1.1363	1.8092	0.9046	1.1388	1.5727	0.7863
0.5	1.2307	0.3289	0.1645	1.2227	0.5012	0.2506	1.2258	0.4273	0.2136
1.00			0.7896			1.1552			0.9999

A

Example A.4 2-Propanol (isopropanol) and water form an azeotropic mixture at a particular liquid composition that results in the vapor and liquid compositions being equal. Vapor–liquid equilibrium for 2-propanol–water mixtures can be predicted by the Wilson equation. Vapor pressure coefficients in bar with temperature in Kelvin for the Antoine equation are given in Table A.10 (Gmehling, Onken and Arlt, 1977–1980). Data for the Wilson equation are given in Table A.11 (Gmehling. Onken and Arlt, 1977–1980). Assume the gas constant $R = 8.3145\ \text{kJ}\cdot\text{kmol}^{-1}\cdot\text{K}^{-1}$. Determine the azeotropic composition at 1 atm.

Solution To determine the location of the azeotrope for a specified pressure, the liquid composition has to be varied and a bubble-point calculation performed at each liquid composition until a composition is identified, whereby $x_i = y_i$. Alternatively, the vapor composition could be varied and a dew-point calculation performed at each vapor composition. Either way, this requires iteration. Figure A.4 shows the x–y diagram for the 2-propanol–water system. This was obtained by carrying out a bubble-point calculation at different values of the liquid composition. The point where the x–y plot crosses the diagonal line gives the azeotropic composition. A more direct search for the azeotropic composition can be carried out for such a binary system in a spreadsheet by varying T and x_1 simultaneously and by solving the objective function (see Section 3.8):

$$(x_1K_1 + x_2K_2 - 1)^2 + (x_1 - x_1K_1)^2 = 0$$

The first expression in parentheses in this equation ensures that the bubble-point criterion is satisfied. The second expression in parentheses ensures that the vapor and liquid compositions are equal. The solution of this is given when $x_1 = y_1 = 0.69$ and $x_2 = y_2 = 0.31$ for the system of 2-propanol–water at 1 atm.

Table A.10

Antoine equation coefficients for 2-propanol and water (Gmehling, Onken and Arlt, 1977–1980).

	A_i	B_i	C_i
2-Propanol	13.8228	4628.96	−20.524
Water	11.9647	3984.93	−39.734

Table A.11

Data for 2-propanol (1) and water (2) for the Wilson equation at 1 atm (Gmehling, Onken and Arlt, 1977–1980).

V_1 ($\text{m}^3\cdot\text{kmol}^{-1}$)	V_2 ($\text{m}^3\cdot\text{kmol}^{-1}$)	$(\lambda_{12} - \lambda_{11})$ ($\text{kJ}\cdot\text{kmol}^{-1}$)	$(\lambda_{21} - \lambda_{22})$ ($\text{kJ}\cdot\text{kmol}^{-1}$)
0.07692	0.01807	3716.4038	5163.0311

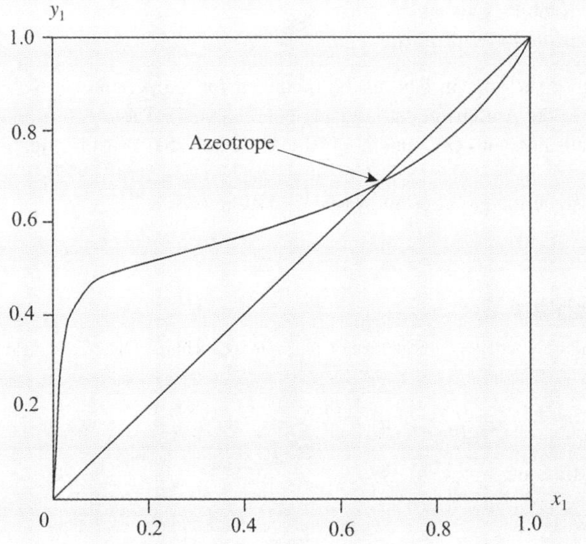

Figure A.4

An x–y plot for the system 2-propanol (1) water (2) from the Wilson Equation at 1 atm.

Example A.5 Repeat the calculation from Example A.4, but using the UNIFAC Model and compare the predictions of the Wilson equation with UNIFAC. 2-Propanol ($CH_3CHOHCH_3$) consists of two CH_3, one CH and one OH functional groups. Water is itself a functional group. Group volume and surface area parameters are given in Table A.12. Group interaction parameters are given in

Table A.12

Group volume and surface area parameters (Poling, Prausnitz and O'Connell, 2001).

Component	Subgroup	Main group number	R_k	Q_k
2-Propanol	CH_3	1	0.9011	0.848
	CH	1	0.4469	0.228
	OH	5	1.0000	1.200
Water	H_2O	7	0.9200	1.400

Table A.13

Interaction parameters (Poling, Prausnitz and O'Connell, 2001).

	CH_3	CH	OH	H_2O
CH_3	0	0	986.5	1318
CH	0	0	986.5	1318
OH	156.4	156.4	0	353.5
H_2O	300.0	300.0	−229.1	0

Table A.13. Note in Table A.13 that there are no interactions between subgroups in the same main group and that $a_{mn} \neq a_{nm}$. Vapor pressure coefficients for the Antoine equation are given in Table A.10.

Solution Figure A.5 shows a comparison between the *x–y* diagrams predicted by the UNIFAC Method and the Wilson equation. It can be seen that the UNIFAC Method predicts the formation of an azeotrope and gives a good prediction of the composition of the azeotrope in this case. However, at lower concentrations than the azeotrope there are significant errors in the prediction by the UNIFAC Method. For this system overall, the prediction of the vapor–liquid equilibrium is quite good. This is not surprising given that the conditions are moderate and there are only four functional groups involved. As the number of functional groups increases, the error in prediction is likely to increase. Also, the 2-propanol–water system is a well-studied system and the UNIFAC Method would be expected to give a good prediction. Despite the fact that the prediction in this case is quite good, other more complex systems can show serious errors with the UNIFAC Method and the method should only be used when no vapor–liquid equilibrium data are available.

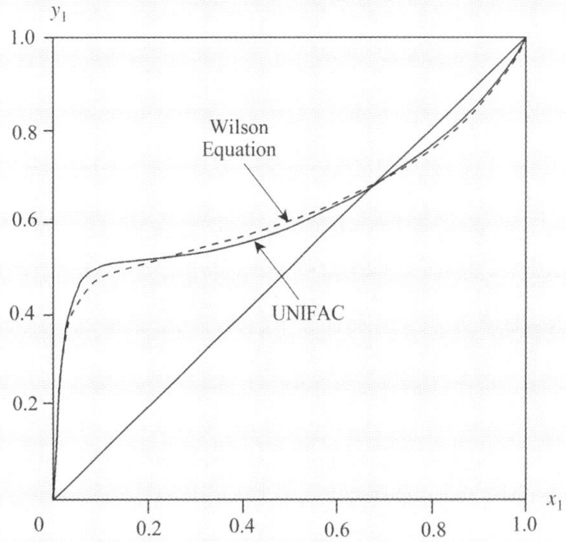

Figure A.5

An *x–y* plot for the system 2-propanol (1) water (2) from the UNIFAC Model at 1 atm.

A

A.9 Liquid–Liquid Equilibrium

As the components in a liquid mixture become more chemically dissimilar, their mutual solubility decreases. This is characterized by an increase in their activity coefficients (for positive deviation from Raoult's Law). If the chemical dissimilarity, and the corresponding increase in activity coefficients, become large enough, the solution can separate into two-liquid phases.

Figure A.6 shows the vapor–liquid equilibrium behavior of a system exhibiting two-liquid phase behavior. Two-liquid phases exist in the areas *abcd* in Figures A.6a and A.6b. Liquid mixtures outside of this two-phase region are homogeneous. In the two-liquid phase region, below *ab*, the two-liquid phases are subcooled. Along *ab*, the two-liquid phases are saturated. The area of the two-liquid phase region becomes narrower as the temperature increases. This is because the mutual solubility normally increases with increasing temperature. For a mixture within the two-phase region, say Point *Q* in Figure A.6a and A.6b, at equilibrium, two-liquid phases are formed at Points *P* and *R*. The line *PR* is the *tie line*. The analysis for vapor–liquid separation in Equations A.55 to A.59 also applies to a liquid–liquid separation. Thus, in Figures A.6a and A.6b, the relative amounts of the two-liquid phases formed from Point *Q* at *P* and *R* follow the Lever Rule (see Chapter 8 and Equation 8.24).

In Figure A.6a, the azeotropic composition at Point *e* lies outside the region of two-liquid phases. In Figure A.6b, the azeotropic composition lies inside the region of two-liquid phases. Any two-phase liquid mixture vaporizing along *ab*, in Figure A.6b, will vaporize at the same temperature and have a vapor composition corresponding with Point *e*. This results from the lines of vapor–liquid equilibrium being horizontal in the vapor–liquid region, as shown in Figure A.6b. A liquid mixture of composition *e*, in Figure A.6b, produces a vapor of the same composition and is known as a *heteroazeotrope*. The *x–y* diagrams in Figure A.6c and d exhibit a horizontal section, corresponding with the two-phase region.

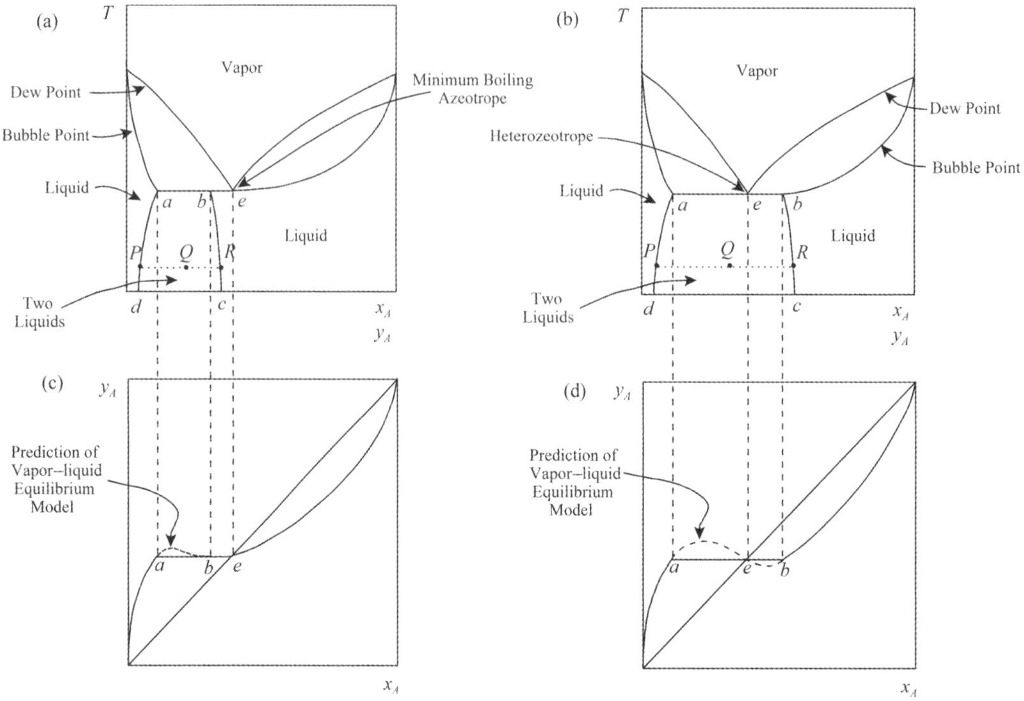

Figure A.6
Phase equilibrium featuring two liquid phases.

For liquid–liquid equilibrium, the fugacity of each component in each phase must be equal:

$$(x_i\gamma_i)^I = (x_i\gamma_i)^{II} \tag{A.62}$$

where *I* and *II* represent the two-liquid phases in equilibrium. The equilibrium K-value or *distribution coefficient* for Component *i* can be defined by

$$K_i = \frac{x_i^I}{x_i^{II}} = \frac{\gamma_i^{II}}{\gamma_i^I} \tag{A.63}$$

A.10 Liquid–Liquid Equilibrium Activity Coefficient Models

A model is needed to calculate liquid–liquid equilibrium for the activity coefficient from Equation A.62. Both the NRTL and UNIQUAC equations can be used to predict liquid–liquid equilibrium. Note that the Wilson equation is not applicable to liquid–liquid equilibrium and, therefore, also not applicable to vapor–liquid–liquid equilibrium. Parameters from the NRTL and UNIQUAC equations can be correlated from vapor–liquid equilibrium data (Gmehling, Onken and Arlt, 1977–1980) or liquid–liquid equilibrium data (Sorenson and Arlt, 1980; Macedo and Rasmussen, 1987). The UNIFAC Method can be used to predict liquid–liquid equilibrium from the molecular structures of the components in the mixture (Poling, Prausnitz and O'Connell, 2001).

A.11 Calculation of Liquid–Liquid Equilibrium

The vapor–liquid x–y diagram in Figure A.6c and A.6d can be calculated by setting a liquid composition and calculating the corresponding vapor composition in a bubble point calculation. Alternatively, vapor composition can be set and the liquid composition determined by a dew-point calculation. If the mixture forms two-liquid phases, the vapor–liquid equilibrium calculation predicts a maximum in the x–y diagram, as shown in Figure A.6c and A.6d. Note that such a maximum cannot appear with the Wilson equation.

To calculate the compositions of the two coexisting liquid phases for a binary system, the two equations for phase equilibrium need to be solved:

$$(x_1\gamma_1)^I = (x_1\gamma_1)^{II},\ (x_2\gamma_2)^I = (x_2\gamma_2)^{II} \tag{A.64}$$

where

$$x_1^I + x_2^I = 1,\ x_1^{II} + x_2^{II} = 1 \tag{A.65}$$

Given a prediction of the liquid-phase activity coefficients, from, say, the NRTL or UNIQUAC equations, then Equations A.64 and A.65 can be solved simultaneously for x_1^I and x_1^{II}. There are a

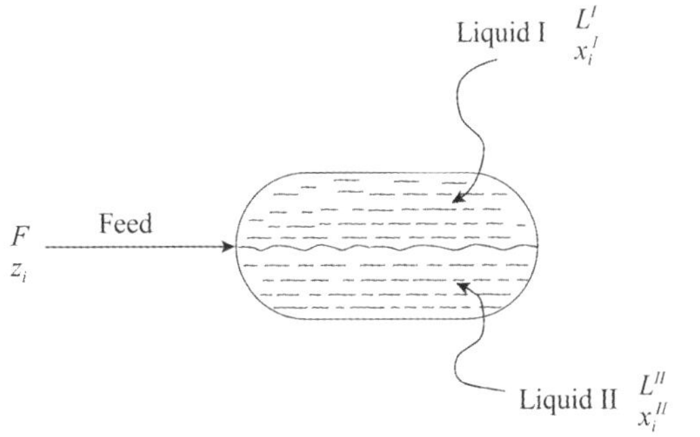

Figure A.7

Liquid–liquid equilibrium.

number of solutions to these equations, including a trivial solution corresponding with $x_1^I = x_1^{II}$. For a solution to be meaningful:

$$0 < x_1^I < 1, \quad 0 < x_1^{II} < 1, \quad x_1^I \neq x_1^{II} \qquad \text{(A.66)}$$

For a ternary system, the corresponding equations to be solved are:

$$\begin{aligned}(x_1\gamma_1)^I &= (x_1\gamma_1)^{II}, \\ (x_2\gamma_2)^I &= (x_2\gamma_2)^{II}, (x_3\gamma_3)^I = (x_3\gamma_3)^{II}\end{aligned} \qquad \text{(A.67)}$$

These equations can be solved simultaneously with the material balance equations to obtain x_1^I, x_2^I, x_1^{II} and x_2^{II}. For a multicomponent system, the liquid–liquid equilibrium is illustrated in Figure A.7. The mass balance is basically the same as that for vapor–liquid equilibrium, but is written for two-liquid phases. Liquid *I* in the liquid–liquid equilibrium corresponds with the vapor in vapor–liquid equilibrium and Liquid *II* corresponds with the liquid in vapor–liquid equilibrium. The corresponding mass balance is given by the equivalent to Equation A.59:

$$\sum_i^{NC} \frac{z_i(K_i - 1)}{\frac{L^I}{F}(K_i - 1) + 1} = 0 = f(L^I/F) \qquad \text{(A.68)}$$

where F = feed flowrate (kmol·s^{-1})
L^I = flowrate of Liquid *I* from the separator (kmol·s^{-1})
z_i = mole fraction of Component *i* in the feed (–)
K_i = K-value, or distribution coefficient, for Component *i* (–)

Also, the liquid–liquid equilibrium K-value needs to be defined for equilibrium to be:

$$K_i = \frac{x_i^I}{x_i^{II}} = \frac{\gamma_i^{II}}{\gamma_i^I} \qquad \text{(A.69)}$$

$x_1^I, x_2^I, \ldots, x_{NC-1}^I$ and $x_1^{II}, x_2^{II} \ldots, x_{NC-1}^{II}$, and L^I/F need to be varied simultaneously to solve Equations A.68 and A.69.

Example A.6 Mixtures of water and 1-butanol (*n*-butanol) form two-liquid phases. Vapor–liquid equilibrium and liquid–liquid equilibrium for the water–1-butanol system can be predicted by the NRTL equation. Vapor pressure coefficients in bar with temperature in Kelvin for the Antoine equation are given in Table A.14 (Gmehling, Onken and Arlt, 1977–1980). Data for the NRTL equation are given in Table A.15, for a pressure of 1 atm (Gmehling, Onken and Arlt, 1977–1980). Assume the gas constant $R = 8.3145$ kJ·kmol^{-1}·K^{-1}.

a) Plot the *x–y* diagram at 1 atm.
b) Determine the compositions of the two-liquid phase region for saturated vapor–liquid–liquid equilibrium at 1 atm.

Solution

a) For a binary system, the calculations can be performed in spreadsheet software. As with Example A.4, a series of bubble-point calculations can be performed at different liquid-phase compositions (or dew-point calculations at different vapor-phase compositions). The NRTL equation is modeled using Equations A.42 and A.43. The resulting *x–y* diagram is shown in Figure A.8. The *x–y* diagram displays the characteristic maximum for two-liquid phase behavior.
b) To determine the compositions of the two-liquid phase region, the NRTL equation is set up for two-liquid phases by writing Equations A.42 and A.43 for each phase. The constants are given in Table A.15. Note that if x_1^I and x_1^{II} are specified, then:

$$x_2^I = 1 - x_1^I \quad \text{and} \quad x_2^{II} = 1 - x_1^{II}$$

A search is then made by varying $x_1{}^I$ and $x_1{}^{II}$ simultaneously (e.g. using a spreadsheet solver) to solve the objective function (see Section 3.8):

$$\left(x_1^I\gamma_1^I - x_1^{II}\gamma_1^{II}\right)^2 + \left(x_2^I\gamma_2^I - x_2^{II}\gamma_2^{II}\right)^2 = 0$$

This ensures liquid–liquid equilibrium. Trivial solutions whereby $x_1^I = x_1^{II}$ need to be avoided. The results are shown

A

Table A.14

Antoine coefficient for water and 1-butanol (Gmehling, Onken and Arlt, 1977–1980).

	A_i	B_i	C_i
Water	11.9647	3984.93	–39.734
1-Butanol	10.3353	3005.33	–99.733

Table A.15

Data for water (1) and 1-butanol (2) for the NRTL equation at 1 atm (Gmehling, Onken and Arlt, 1977–1980).

$(g_{12} - g_{22})$ (kJ·kmol^{-1})	$(g_{21} - g_{11})$ (kJ·kmol^{-1})	α_{ij} (–)
11,184.9721	1649.2622	0.4362

in Figure A.8 to be $x_1^I = 0.59$, $x_1^{II} = 0.98$. The system forms a heteroazeotrope.

To ensure that the predicted two-phase region corresponds with that for the saturated vapor–liquid–liquid equilibrium, the temperature must be specified to be the bubble point predicted by the NRTL equation at either $x_1 = 0.59$ or $x_1 = 0.98$ (366.4 K in this case).

Care should be exercised in using the coefficients from Table A.15 to predict two-liquid phase behavior under subcooled conditions. The coefficients in Table A.15 were determined from vapor–liquid equilibrium data at saturated conditions.

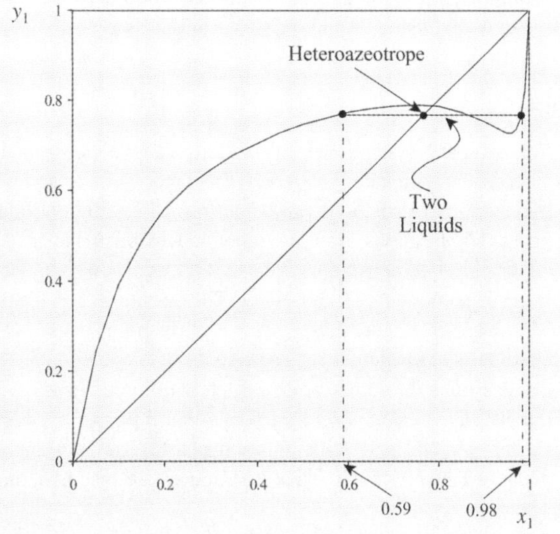

Figure A.8

An x–y plot for the system water (1) and 1-butanol (2) from the NRTL Equation at 1 atm.

Although the methods developed here can be used to predict liquid–liquid equilibrium, the predictions will only be as good as the coefficients used in the activity coefficient model. Such predictions can be critical when designing liquid–liquid separation systems. When predicting liquid–liquid equilibrium, it is always better to use coefficients correlated from liquid–liquid equilibrium data, rather than coefficients based on the correlation of vapor–liquid equilibrium data. Equally well, when predicting vapor–liquid equilibrium, it is always better to use coefficients correlated to vapor–liquid equilibrium data, rather than coefficients based on the correlation of liquid–liquid equilibrium data. Also, when calculating liquid–liquid equilibrium with multicomponent systems, it is better to use multicomponent experimental data, rather than binary data.

A.12 Choice of Method for Equilibrium Calculations

Choosing the most appropriate method from a number of options can be critical to obtaining a reliable design. Phase equilibrium (vapor–liquid, liquid–liquid, solid–liquid, etc.) is usually the most critical choice for physical properties. The choice of the most appropriate method for vapor–liquid, liquid–liquid and vapor–liquid–liquid equilibrium depends on a number of factors:

1) The nature of the components (polarity, electrolytic, polymeric, reactive)
2) Pressure
3) Temperature
4) The availability of experimental data.

The significant issues in choosing the phase equilibrium model are:

1) *Nonpolar mixtures.* For the prediction of vapor–liquid equilibrium for nonpolar mixtures of hydrocarbons and light gases, an equation of state method is normally used. Such mixtures are characterized by only moderate deviations from ideality in the liquid phase. Table A.1 lists the equations of state normally used for process design. By far the most commonly used methods are Peng–Robinson and Soave–Redlich–Kwong, with Chao–Seadar–Grayson–Streed being used when there are large amounts of hydrogen present. K-values are calculated using Equations A.27 and A.53. Limitations of the various equations of state are discussed in Table A.1.
2) *Polar mixtures.* For prediction of vapor–liquid equilibrium for mixtures of polar substances, or mixtures of polar and nonpolar substances, an activity coefficient model is normally used. Such mixtures are characterized by significant deviations from ideality in the liquid phase. K-values are calculated using Equation A.28 for high-pressure systems and Equation A.29 for low-pressure systems, when $\phi_i^V = 1$. Models are required for the liquid phase activity coefficient γ_i. Table A.16 compares the characteristics of the most commonly used models. These models should only be used when remote from the critical point. If supercritical gases are present in the mixture, then these can be modeled using Henry's Law. Each of the activity coefficient models can handle very strong nonideality. If the model requires adjustable parameters based on experimental data, then it is preferred if possible that the adjustable parameters be fitted from data in the temperature, pressure and composition range of the operation.
3) *Pressure.* For pressures below 5 bar, the vapor phase can be assumed to be ideal. For pressures above 5 bar, the vapor phase fugacity coefficient ϕ_i^V must be calculated from an equation of state. The calculation of vapor-liquid equilibrium based on equations of state from Equation A.27 naturally takes the vapor phase nonideality into account. However, activity coefficients based on Equation A.28 require an equation of state to calculate ϕ_i^V.
4) *Liquid–liquid systems.* Various methods can be used to predict liquid–liquid equilibrium. The most commonly used methods in process design are the NRTL, UNIQUAC and UNIFAC methods as discussed in Table A.16. Care should be taken not to use interaction parameters correlated for vapor–liquid equilibrium to predict liquid–liquid equilibrium. Also, care should be exercised when predicting

Table A.16

The commonly used activity coefficient models.

Activity coefficient model	Characteristics	Reference
Wilson equation	Requires two adjustable parameters for each binary pair that have been correlated from experimental data. Cannot predict the existence of two liquid phases.	Wilson (1964)
NRTL (non-random two liquid) equation	Requires three adjustable parameters for each binary pair that have been correlated from experimental data. Can be used for VLE and LLE. Parameters for VLE should not be used for LLE.	Renon and Prausnitz (1969)
UNIQUAC (universal quasi-chemical) equation	Requires two adjustable parameters for each binary pair that have been correlated from experimental data. Can be used for VLE and LLE. Parameters for VLE should not be used for LLE.	Anderson and Prausnitz (1978a, 1978b)
UNIFAC (universal quasi-chemical functional group activity coefficients)	Does not require interaction parameters fitted from experimental data. Molecules represented as functional groups. Interactions between molecules modelled as interactions between functional groups. Can be used for VLE and LLE. The original version required different parameters for VLE and LLE, but more recent derivatives of the method can model VLE and LLE with a single set of parameters. Temperatures restricted from 15 °C to 150 °C for VLE and from 15 °C to 40 °C for LLE. Should only be used in the absence of experimental data for binary interactions.	Fedenslund, Gmehling and Rasmussen (1977)

behavior under subcooled conditions (e.g. in a decanter) as binary interaction parameters are mostly determined at saturated conditions.

5) *Electrolytes.* Electrolytes are aqueous mixtures containing solutes that ionize either completely (e.g. sodium chloride) or partially (e.g. acetic acid). Phase equilibrium for such systems must account for the long-range interactions of the charges between ionic species. Special models must be used to predict phase equilibrium for electrolyte systems. Such models are available in commercial software.

6) *Polymers.* Solutions of polymers in solvents of relatively smaller molar mass exhibit highly nonideal phase equilibrium behavior. Special models are required for such mixtures, normally available in commercial software.

7) *Reactions.* Various separation processes that also involve reaction need to be modeled. The reaction might be a simple chemical association or *oligomerization* in the vapor phase to form dimers, tetramers and hexamers in the vapor phase. For example, formic and acetic acids can undergo dimerization in the vapor phase, acetic acid tetramerization and hydrogen fluoride hexamerization. The standard methods, such as the activity coefficient models in Table A.16, need to be modified to account for the partial pressure of the oligomer as well as the partial pressures of the monomer. This requires special models available in commercial software. More complex reactions involve systems such as the separation of H_2S and CO_2 using amines. An amine or mixture of amines in an aqueous solution can be used, as discussed in Chapter 9. The amines used are monoethanolamine (MEA), diethanolamine (DEA) or methyldiethanolamine (MDEA). These react with the H_2S and CO_2 in the absorption process, with the reaction being reversed in the regeneration process. Special models are available in commercial software to model simultaneously the reaction and separation.

8) *Petroleum and petroleum fractions.* Process design calculations with crude oil and petroleum fractions present a special problem. It is not possible to obtain a detailed analysis for crude oil and petroleum fractions, except for the lightest components. It is usually possible to obtain an analysis for gases and hydrocarbons up to C_6 compounds and possibly up to around C_{10}. For the heavier components it is not possible to characterize them individually. By contrast with detailed analysis, *bulk properties* can be more readily obtained. The more important bulk properties are the boiling temperature profile of crude oil (or petroleum fraction) versus the cumulative volume distilled and the corresponding density distribution. The crude oil (or petroleum fraction) is represented by pure components for the lighter components (where known) and a mixture of hypothetical hydrocarbon components known as *pseudo components*. Empirical correlations allow physical properties to be attributed to the pseudo components. Each pseudo component is characterized by a molar mass, critical properties and an acentric factor. The pseudo components can then be modeled by an equation of state. The properties of the pseudo components are adjusted to allow the predicted properties of the mixture to agree with the measured bulk properties of the mixture. This then allows process design calculations to be performed. The number of pseudo components used to represent the mixture is at the discretion of the designer. Cutting the petroleum streams into pseudo components requires specialized software available in commercial simulation packages.

9) *Sour water systems.* Sour water refers to streams in petroleum refineries that consist of water contaminated with various dissolved gases. The gases of greatest concern are hydrogen sulfide and ammonia. Such sour water needs to be subject to separation, which requires specialized software available in commercial simulation packages.
10) *Steam tables.* Steam tables are tabulated thermodynamic properties of water and steam that are the most accurate method of predicting the thermodynamic properties of pure water in various states. Interpolation must be carried out for conditions between the data points. However, the method can only be applied to pure water.
11) *Availability of experimental data.* For nonpolar systems, lack of availability of experimental data for interaction parameters does not present such a serious problem, as such mixtures are characterized by only moderate deviations from ideality in the liquid phase. However, the same cannot be said for polar mixtures. If experimental interaction parameters are not available for polar mixtures, then the UNIFAC method can be used. However, this must be used with extreme caution. If parameters are available for UNIQUAC, then those should be used and the UNIFAC method only used for the binary pairs where experimental data are not available.
12) *Sensitivity analysis.* There is always uncertainty regarding physical property data. Whether errors in physical property data are likely to be a problem or not can be tested by carrying out a sensitivity analysis. Perturbing the physical property parameters and repeating the process design calculation tests whether errors are likely to be a problem in the final design. For example, for vapor liquid equilibrium, the consequences of errors in the data are likely to be much more serious if the relative volatility is small. As an example, consider the separation by distillation of propylene and propane in ethylene production, which has a small relative volatility. An error of +1% in the prediction of the K-value for propane can lead to the requirement for 26% more trays or 25% more reflux (Streich and Kirstenmacher, 1979).

A.13 Calculation of Enthalpy

The calculation of enthalpy is required for energy balance calculations. It might also be required for the calculation of other derived thermodynamic properties. The absolute value of enthalpy for a substance cannot be measured and only changes in enthalpy are meaningful. Consider the change in enthalpy with pressure. At a given temperature T, the change in enthalpy of a fluid can be determined from the derivative of enthalpy with pressure at fixed temperature, $(\partial H/\partial P)_T$. Thus, the change in enthalpy relative to a reference is given by:

$$\left[H_P - H_{P_O}\right]_T = \int_{P_O}^{P} \left(\frac{\partial H}{\partial P}\right)_T dP \tag{A.70}$$

where H_P = enthalpy at pressure P (kJ·kmol^{-1})
H_{P_O} = enthalpy at reference pressure P_O (kJ·kmol^{-1})
P = pressure (bar)
P_O = reference pressure (bar)

The change in enthalpy with pressure is given by (Hougen, Watson and Ragatz, 1954):

$$\left(\frac{\partial H}{\partial P}\right)_T = V - T\left(\frac{\partial V}{\partial T}\right)_P \tag{A.71}$$

Also:

$$\left(\frac{\partial V}{\partial T}\right)_P = \left[\frac{\partial}{\partial T}\left(\frac{ZRT}{P}\right)\right]_P = \frac{R}{P}\left[Z + T\left(\frac{\partial Z}{\partial T}\right)_P\right] \tag{A.72}$$

where Z = compressibility (–)
R = universal gas constant

Combining Equations A.70 to A.72 gives the difference between the enthalpy at pressure P and that at the standard pressure P_O and is known as the *enthalpy departure*:

$$\left[H_P - H_{P_O}\right]_T = -\int_{P_O}^{P} \left[\frac{RT^2}{P}\left(\frac{\partial Z}{\partial T}\right)_P\right]_T dP \tag{A.73}$$

Equation A.73 defines the enthalpy departure from a reference state at temperature T and pressure P_O.

The value of $(\partial Z/\partial T)_P$ can be obtained from an equation of state, such as the Peng–Robinson equation of state, and the integral in Equation A.73 can be evaluated (Poling, Prausnitz and O'Connell, 2001). The enthalpy departure for the Peng–Robinson equation of state is given by (Poling, Prausnitz and O'Connell, 2001):

$$\left[H_P - H_{P_O}\right]_T = RT(Z-1) + \frac{T(da/dT) - a}{2\sqrt{2}\,b}\ln\left[\frac{Z + (1+\sqrt{2})B}{Z + (1-\sqrt{2})B}\right] \tag{A.74}$$

where

$$\frac{da}{dT} = -0.45724\frac{R^2T_C^2}{P_C}\kappa\sqrt{\frac{\alpha}{T\,T_C}}$$

$$B = \frac{bP}{RT}$$

where a, b, α and κ are defined in Equation A.6. Equations of state, such as the Peng–Robinson equation, are capable of predicting both liquid and vapor behavior. The appropriate root for the equation of a liquid or vapor must be taken, as discussed previously. Equation A.74 is therefore capable of predicting both liquid and vapor enthalpy (Poling, Prausnitz and O'Connell, 2001). Equations of state such as the Peng–Robinson equation are generally more reliable at predicting the vapor compressibility than the liquid compressibility, and hence are more reliable predicting vapor enthalpy than liquid enthalpy.

One problem remains. The reference enthalpy must be defined at temperature T and pressure P_O. The reference state for enthalpy can be taken as an ideal gas. At zero pressure, fluids are in their

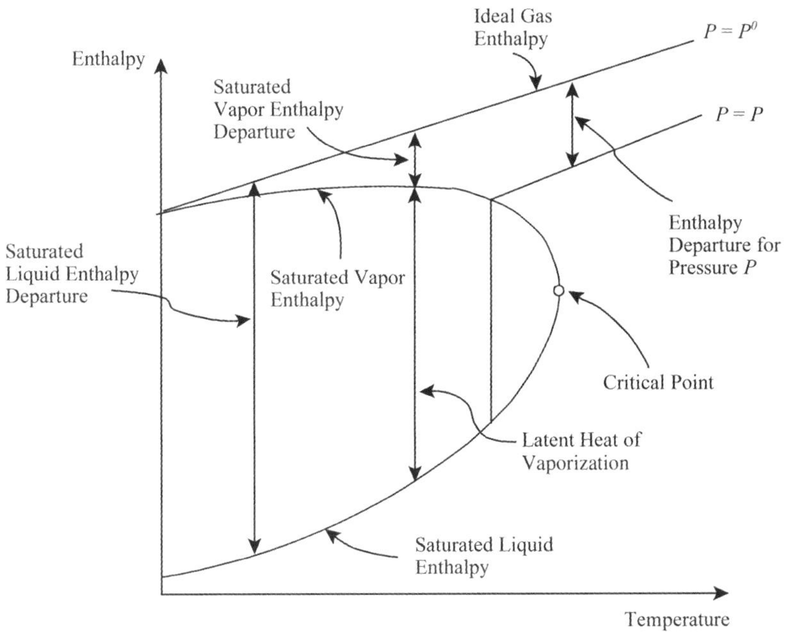

Figure A.9
The enthalpy departure function.

ideal gaseous state and the enthalpy is independent of pressure. The ideal gas enthalpy can be calculated from ideal gas heat capacity data (Poling, Prausnitz and O'Connell, 2001):

$$H_T^O = H_{T_O}^O + \int_{T_O}^{T} C_P^O dT \qquad (A.75)$$

where H_T^O = enthalpy at zero pressure and temperature T ($kJ \cdot kmol^{-1}$)
$H_{T_O}^O$ = enthalpy at zero pressure and temperature T_O, defined to be zero ($kJ \cdot kmol^{-1}$)
T_O = reference temperature (K)
C_P^O = ideal gas enthalpy ($kJ \cdot kmol^{-1} \cdot K^{-1}$)

The ideal gas enthalpy can be correlated as a function of temperature, for example (Poling, Prausnitz and O'Connell, 2001):

$$\frac{C_P^O}{R} = \alpha_0 + \alpha_1 T + \alpha_2 T^2 + \alpha_3 T^3 + \alpha_4 T^4 \qquad (A.76)$$

where $\alpha_0, \alpha_1, \alpha_2, \alpha_3, \alpha_4$ = constants determined by fitting experimental data

To calculate the enthalpy of a liquid or gas at temperature T and pressure P, the enthalpy departure function (Equation A.73) is evaluated from an equation of state (Hougen, Watson and Ragatz, 1959). The ideal gas enthalpy is calculated at temperature T from Equation A.76. The enthalpy departure is then added to the ideal gas enthalpy to obtain the required enthalpy. Note that the enthalpy departure function calculated from Equation A.73 will have a negative value. This is illustrated in Figure A.9. The calculations are complex and usually carried out using physical property or simulation software packages. However, it is important to understand the basis of the calculations and their limitations.

A.14 Calculation of Entropy

The calculation of entropy is required for compression and expansion calculations. Isentropic compression and expansion is often used as a reference for real compression and expansion processes. The calculation of entropy might also be required in order to calculate other derived thermodynamic properties. Like enthalpy, entropy can also be calculated from a departure function:

$$[S_P - S_{P_O}]_T = \int_{P_O}^{P} \left(\frac{\partial S}{\partial P}\right)_T dP \qquad (A.77)$$

where S_P = entropy at pressure P
S_{P_O} = entropy at pressure P_O

The change in entropy with pressure is given by (Hougen, Watson and Ragatz, 1959):

$$\left(\frac{\partial S}{\partial P}\right)_T = -\left(\frac{\partial V}{\partial T}\right)_P \qquad (A.78)$$

Combining Equations A.72 and A.77 with Equation A.78 gives:

$$[S_P - S_{P_O}]_T = -\int_{P_O}^{P} \left[\frac{RZ}{P} + \frac{RT}{P}\left(\frac{\partial Z}{\partial T}\right)_P\right]_T dP \qquad (A.79)$$

The integral in Equation A.79 can be evaluated from an equation of state (Poling, Prausnitz and O'Connell, 2001). However, before this *entropy departure function* can be applied to calculate entropy, the reference state must be defined. Unlike enthalpy, the reference state cannot be defined at zero pressure, as the entropy of a gas is infinite at zero pressure. To avoid this difficulty, the standard state can be defined as a reference state at low pressure P_O (usually chosen to be 1 bar or 1 atm) and at the temperature under

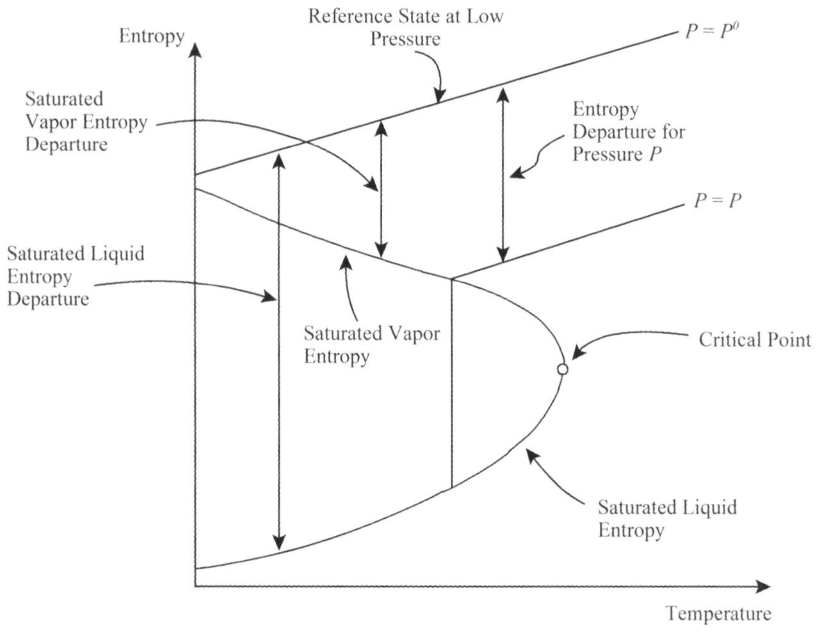

Figure A.10
The entropy departure function.

consideration. Thus,

$$S_T = S_{T_O} + \int_{T_O}^{T} \frac{C_P^O}{T} dT \qquad \text{(A.80)}$$

where S_T = entropy at temperature T and reference pressure P_O
S_{T_O} = entropy of gas at reference temperature T_O and reference pressure P_O

To calculate the entropy of a liquid or gas at temperature T and pressure P, the entropy departure function (Equation A.79) is evaluated from an equation of state (Poling, Prausnitz and O'Connell, 2001). The entropy at the reference state is calculated at temperature T from Equation A.80. The entropy at the reference state is then added to the entropy departure function to obtain the required entropy. The entropy departure function is illustrated in Figure A.10. As with enthalpy departure, the calculations are complex and are usually carried out in physical property or simulation software packages.

A.15 Other Physical Properties

In addition to the thermodynamic properties of molar volume, density, enthalpy, enthalpy of vaporization, heat capacity and entropy, transport properties (viscosity, thermal conductivity and diffusivity) are also required for design calculations. The accuracy required of physical property data depends on the use to which the data will be put. Physical property predictions can be based on:

- Correlated experimental data
- Estimation methods.

Correlated experimental data should be used whenever possible. If no such data are available, then the designer must resort to estimation methods.

1) *Correlated experimental data.* The prediction of the physical properties of mixtures of components based on correlated experimental data starts with the prediction of the physical properties of the individual components and then these are combined according to mixing rules. Such mixing rules introduce errors depending on the physical property and the reliability of the mixing rule. For most calculations, the variation of the property with temperature is the most important consideration. Some typical forms of correlations used are as follows:

a) *Liquid density.* Liquid density can be calculated from an equation of state. However, equations of state do not always provide a high enough accuracy for prediction of liquid density for process design calculations. Also, it is not always convenient to use an equation of state in a complex design calculation. Thus, for liquid density, correlated data are most often used. A form of correlation commonly used is:

$$\rho_L = \frac{A}{B^{\left(1+(1-T/C)^D\right)}} \qquad \text{(A.81)}$$

where ρ_L = molar liquid density (kmol·m^{-3})
A, B, C, D = correlating coefficients for liquid density from experimental data
T = absolute temperature (K)

Whereas Equation A.81 provides a reliable correlation over a wide temperature range, it requires four coefficients. A simpler equation that can be used over smaller temperature ranges is given by:

$$\rho_{LT} = \frac{\rho_{LT_O}}{1 + A(T - T_O)} \qquad \text{(A.82)}$$

where ρ_{LT} = liquid density at temperature T (kmol·m $^{-3}$)
ρ_{LT_0} = reference liquid density at temperature T_O (kmol·m $^{-3}$)
A = correlating coefficient for liquid density from experimental data
T_O = reference temperature (K)

The density of a mixture of NC Components i is given by:

$$\overline{\rho_L} = \frac{1}{\sum_i^{NC} \frac{x_i}{\rho_{Li}}} \qquad (A.83)$$

where $\overline{\rho_L}$ = molar density of the liquid mixture (kmol·m $^{-3}$)
ρ_{Li} = molar liquid density of Component i (kmol·m $^{-3}$)
x_i = mole fraction of Component i (–)

The mean molar mass can be used to convert molar to mass density:

$$\overline{M} = \sum_i^{NC} x_i M_i \qquad (A.84)$$

where $\overline{M}$ = molar mass of the mixture (kg·kmol^{-1})
M_i = molar mass of Component i (kg·kmol^{-1})

b) *Vapor density.* The calculation of vapor density from an equation of state is more reliable than liquid density. At moderate conditions, the ideal gas law can be used. Otherwise, a cubic equation of state can be used. If there are significant amounts of hydrogen in the mixture then the Chao–Seader–Grayson–Streed equation of state should be used.

c) *Liquid viscosity.* For liquid viscosity, the variation in temperature can be correlated by the equation:

$$\ln \mu_L = A + \frac{B}{T} + C \ln T + DT^E \qquad (A.85)$$

where μ_L = liquid viscosity (N·s·m^{-2})
A, B, C, D, E = correlating coefficients for liquid density from experimental data
T = absolute temperature (K)

Over relatively small ranges of temperature this can be simplified to:

$$\ln \mu_L = A + \frac{B}{T} \qquad (A.86)$$

The viscosity of a liquid mixture of NC Components i is given by:

$$\ln \overline{\mu_L} = \sum_i^{NC} x_i \ln \mu_{Li} \qquad (A.87)$$

where $\overline{\mu_L}$ = viscosity of the liquid mixture (N·s·m^{-2})
μ_{Li} = viscosity of Component i (N·s·m^{-2})

d) *Vapor viscosity.* For vapor viscosity, the variation in temperature can be correlated by the equation:

$$\mu_V = \frac{AT^B}{\left(1 + \frac{C}{T} + \frac{D}{T^2}\right)} \qquad (A.88)$$

where μ_V = vapor viscosity (N·s·m^{-2})
A, B, C, D = correlating coefficients for vapor viscosity from experimental data

Over relatively small ranges of temperature this can be simplified to:

$$\mu_V = AT^B \qquad (A.89)$$

The viscosity of a vapor mixture of NC components i is given by (Wilke, 1950):

$$\ln \overline{\mu_V} = \sum_{i=1}^{NC} \frac{\mu_{Vi}}{1 + \frac{1}{x_i} \sum_{\substack{j=1 \\ j \neq i}}^{NC} x_j \phi_{ij}} \qquad (A.90)$$

$$\phi_{ij} = \frac{\left[1 + \sqrt{\frac{\mu_{vi}}{\mu_{vj}}} \left(\frac{M_j}{M_i}\right)^{1/4}\right]^2}{\sqrt{8\left(1 + \frac{M_i}{M_j}\right)}} \qquad (A.91)$$

where $\overline{\mu_V}$ = viscosity of the vapor mixture (N·s·m^{-2})
μ_{Vi} = vapor viscosity of Component i (N·s·m^{-2})
M_i = molar mass of Component i (kg·kmol^{-1})

e) *Liquid thermal conductivity.* For liquid thermal conductivity, the variation in temperature can be correlated by the equation:

$$k_L = A + BT + CT^2 + DT^3 + ET^4 \qquad (A.92)$$

Over relatively small ranges of temperature this can be simplified to:

$$k_L = A + BT \qquad (A.93)$$

where k_L = liquid thermal conductivity (W·m^{-1}·K^{-1})
A, B, C, D, E = correlating coefficients for liquid thermal conductivity from experimental data

The thermal conductivity of a liquid mixture of NC components i is given by:

$$\overline{k_L} = \sum_i^{NC} x_i k_{Li} \qquad (A.94)$$

A

where $\overline{k_L}$ = viscosity of the liquid mixture ($W \cdot m^{-1} \cdot K^{-1}$)
k_{Li} = viscosity of Component i ($W \cdot m^{-1} \cdot K^{-1}$)

f) *Vapor thermal conductivity.* For vapor thermal conductivity, the variation in temperature can be correlated by the equation:

$$k_V = \frac{AT^B}{1 + \frac{C}{T} + \frac{D}{T^2}} \quad (A.95)$$

The thermal conductivity of a vapor mixture of NC Components i is given by:

$$\overline{k_V} = \sum_i^{NC} x_i k_{Vi} \quad (A.96)$$

where $\overline{k_V}$ = viscosity of the vapor mixture ($W \cdot m^{-1} \cdot K^{-1}$)
k_{Vi} = viscosity of Component i ($W \cdot m^{-1} \cdot K^{-1}$)

2) *Estimation methods.* There are two broad categories of estimation methods:

a) Methods whereby known properties of a compound are used to estimate the unknown properties. For example, the surface tension of a liquid can be estimated from the critical properties and the normal boiling point (Poling, Prausnitz and O'Connell, 2001):

$$\sigma = P_C^{2/3} T_C^{1/3} Q(1 - T_R)^{11/9} \quad (A.97)$$

where σ = surface tension ($dyn \cdot cm^{-1}$)
P_C = critical pressure (bar)
T_C = critical temperature (K)
T_R = reduced temperature (–)
$= \frac{T}{T_C}$
T = temperature (K)

$$Q = 0.1196\left[1 + \frac{T_{BR}\ln\left(\frac{P_C}{1.01325}\right)}{1 - T_{BR}}\right] - 0.279$$

T_{BR} = reduced normal boiling point (–)
$= \frac{T_B}{T_C}$
T_B = normal boiling point (K)

b) Group contribution methods, which are based on the concept that a particular physical property of a compound can be considered to be made up of the contributions from the constituent chemical groups and chemical bonds. This is the same basic approach used for the group contribution methods for vapor–liquid equilibrium. For example, liquid heat capacity can be estimated as a function of temperature from the equation (Ruzicka and Domalski, 1993):

$$C_{PL} = R\left[A + B\frac{T}{100} + D\left(\frac{T}{100}\right)^2\right] \quad (A.98)$$

where $A = \sum_{i=1}^{k} n_i a_i, B = \sum_{i=1}^{k} n_i b_i, D = \sum_{i=1}^{k} n_i d_i$

R = universal gas constant
T = absolute temperature (K)
n_i = number of groups of type i
k = total number of different group types
a_i, b_i, d_i = constants tabulated for different functional groups (Poling, Prausnitz and O'Connell, 2001)

A.16 Physical Properties in Process Design – Summary

Most design calculations require knowledge of physical properties. This includes thermodynamic properties, phase equilibrium and transport properties. In practice, the designer often uses a commercial physical property or simulation software package to access such data.

Vapor and liquid equilibrium can be calculated on the basis of activity coefficient models or equations of state. Activity coefficient models are required when the liquid phase shows a significant deviation from ideality. Equation of state models are required when the vapor phase shows a significant deviation from ideality. Equations of state are normally applied to vapor–liquid equilibrium for hydrocarbon systems and light gases under significant pressure. Such vapor–liquid equilibrium calculations are normally carried out using physical property or simulation software packages.

Prediction of liquid–liquid equilibrium also requires an activity coefficient model. The choice of models of liquid–liquid equilibrium is more restricted than that for vapor–liquid equilibrium, and predictions are particularly sensitive to the model parameters used.

It should be noted that the methods outlined in this appendix for phase equilibrium are not appropriate for systems in which the components exhibit chemical association in the vapor phase (e.g. acetic acid), reactions in the liquid phase, polymers or electrolytic systems.

Both enthalpy and entropy can be calculated from an equation of state to predict the deviation from ideal gas behavior. Having calculated the ideal gas enthalpy or entropy from experimentally correlated data, the enthalpy or entropy departure function from the reference state can then be calculated from an equation of state.

Other physical properties, especially particular transport properties, are correlated as a function of temperature. As an alternative to the correlation of experimental data, physical properties can be estimated from estimation methods, such as group contribution methods.

A.17 Exercises

1. For the following mixtures, suggest suitable models for both the liquid and vapor phases to predict vapor–liquid equilibrium.
   a) H_2S and water at 20 °C and 1.013 bar.
   b) Benzene and toluene (close to ideal liquid-phase behavior) at 1.013 bar.
   c) Benzene and toluene (close to ideal liquid-phase behavior) at 20 bar.
   d) A mixture of hydrogen, methane, ethylene, ethane, propylene and propane at 20 bar.
   e) Acetone and water (nonideal liquid phase) at 1.013 bar.
   f) A mixture of 2-propanol, water and diiospropyl ether (nonideal liquid phase forming two-liquid phases) at 1.013 bar.

2. Air needs to be dissolved in water under pressure at 20 °C for use in a dissolved-air flotation process (see Chapter 7). The vapor–liquid equilibrium between air and water can be predicted by Henry's Law with a constant of 6.7×10^4 bar. Estimate the mole fraction of air that can be dissolved at 20 °C, at a pressure of 10 bar.

3. The system methanol–cyclohexane can be modeled using the NRTL equation. Vapor pressure coefficients for the Antoine equation for pressure in bar and temperature in Kelvin are given in Table A.17 (Gmehling, Onken and Arlt, 1977–1980). Data for the NRTL equation at 1 atm are given in Table A.18 (Gmehling, Onken and Arlt, 1977–1980). Assume the gas constant $R = 8.3145$ kJ·kmol^{-1}·K^{-1}. Set up a spreadsheet to calculate the bubble point of liquid mixtures and plot the x–y diagram.

4. At an azeotrope $y_i = x_i$, Equation A.29 simplifies to give:

$$\gamma_i = \frac{P}{P_i^{SAT}} \qquad \text{(A.100)}$$

   Thus, if the saturated vapor pressure is known at the azeotropic composition, the activity coefficient can be calculated. If the composition of the azeotrope is known, then the compositions and activity of the coefficients at the azeotrope can be substituted into the Wilson equation to determine the interaction parameters. For the 2-propanol–water system, the azeotropic composition of 2-propanol can be assumed to be at a mole fraction of 0.69 and temperature of 353.4 K at 1 atm. By combining Equation A.100 with the Wilson equation for a binary system, set up two simultaneous equations and solve Λ_{12} and Λ_{21}. Vapor pressure data can be taken from Table A.10 and the universal gas constant can be taken to be 8.3145 kJ·kmol^{-1}·K^{-1}. Then, using the values of molar volume in Table A.11, calculate the interaction parameters for the Wilson equation and compare with the values in Table A.11.

5. Mixtures of 2-butanol (*sec*-butanol) and water form two-liquid phases. Vapor–liquid equilibrium and liquid–liquid equilibrium for the 2-butanol–water system can be predicted by the NRTL equation. Vapor pressure coefficients for the Antoine equation for pressure in bar and temperature in Kelvin are given in Table A.19 (Gmehling, Onken and Arlt, 1977–1980). Data for the NRTL equation at 1 atm are given in Table A.20 (Gmehling, Onken and Arlt, 1977–1980). Assume the gas constant $R = 8.3145$ kJ·kmol^{-1}·K^{-1}.
   a) Plot the x–y diagram for the system.
   b) Determine the compositions of the two-liquid phase region for saturated vapor–liquid–liquid equilibrium.
   c) Does the system form a heteroazeotrope?

A

Table A.17

Antoine coefficients for methanol and cyclohexane (Gmehling, Onken and Arlt, 1977–1980).

	A_i	B_i	C_i
Methanol	11.9869	3643.32	–33.434
Cyclohexane	9.1559	2778.00	–50.024

Table A.18

Data for methanol (1) and cyclohexane (2) for the NRTL equation at 1 atm (Gmehling, Onken and Arlt, 1977–1980).

$(g_{12} - g_{22})$ (kJ·kmol^{-1})	$(g_{21} - g_{11})$ (kJ·kmol^{-1})	α_{ij} (–)
5714.00	6415.36	0.4199

Table A.19

Antoine coefficients for 2-butanol and water (Gmehling, Onken and Arlt, 1977–1980).

	A_i	B_i	C_i
2-Butanol	9.9614	2664.0939	–104.881
Water	11.9647	3984.9273	–39.734

Table A.20

Data for 2-butanol (1) and water (2) for the NRTL equation at 1 atm (Gmehling, Onken and Arlt, 1977–1980).

$(g_{12} - g_{22})$ (kJ·kmol^{-1})	$(g_{21} - g_{11})$ (kJ·kmol^{-1})	α_{ij} (–)
1034.30	10,098.50	0.4118

References

Anderson TF and Prausnitz JM (1978a) Application of the UNIQUAC Equation to Calculation of Multicomponent Phase Equilibriums. I Vapor–Liquid Equilibrium, *Ind Eng Chem Proc Des Dev*, **17**: 552.

Anderson TF and Prausnitz JM (1978b) Application of the UNIQUAC Equation to Calculation of Multicomponent Phase Equilibriums. II Liquid–Liquid Equilibrium, *Ind Eng Chem Proc Des Dev*, **17**: 562.

Benedict M, Webb GB and Rubin LC (1940) An Empirical Equation for Thermodynamic Properties of Light Hydrocarbons and Their Mixtures. I. Methane, Ethane, Propane and *n*-Butane, *J Chem Phys*, **8**: 334–345.

Chao C and Seader JD (1961) A General Correlation of Vapor–Liquid Equilibria in Hydrocarbon Mixtures, *AIChEJ*, **7**: 598.

Chapman WG, Gubbins KE, Jackson G and Radosz M (1989) SAFT: Equation-of-State Solution Model for Associating Fluids, *Fluid Phase Equilibria*, **52**: 31.

Fedenslund A, Gmehling J and Rasmussen P (1977) *Vapor–Liquid Equilibrium Using UNIFAC: A Group Contribution Method*, Elsevier.

Gmehling J, Onken U and Arlt W (1977–1980) Vapor–Liquid Equilibrium Data Collection, Dechema Chemistry Data Series.

Grayson G and Streed CW (1963) Vapor–Liquid Equilibria for High Temperature, High Pressure Hydrogen–Hydrocarbon Systems, Paper 20-PD7, 6th World Petroleum Conference, Frankfurt, Germany.

Hougen OA, Watson KM and Ragatz RA (1954) *Chemical Process Principles. Part I Material and Energy Balances*, 2nd Edition, John Wiley & Sons.

Hougen OA, Watson KM and Ragatz RA (1959) *Chemical Process Principles. Part II Thermodynamics*, 2nd Edition, John Wiley & Sons.

Kontogeorgis GM and Folas GK (2010) *Thermodynamic Models for Industrial Applications*, John Wiley & Sons.

Lee BI and Kesler MG (1975) A Generalized Thermodynamic Correlation Based on Three-Parameter Corresponding States, *AIChE J*, **21**: 510.

Macedo EA and Rasmussen P (1987) Liquid–Liquid Equilibrium Data Collection, Dechema Chemistry Data Series.

Oellrich L, Plöcker U, Prausnitz JM and Knapp H (1981) Equation-of-State Methods for Computing Phase Equilibria and Enthalpies, *Int Chem Eng*, **21** (1): 1.

Peng D and Robinson DB (1976) A New Two Constant Equation of State, *Ind Eng Chem Fundam*, **16**: 59.

Plocker U, Knapp H and Prausnitz JM (1978) Calculation of High-Pressure Vapor–Liquid Equilibrium from a Corresponding-States Correlation with Emphasis on Asymmetric Mixtures, *Ind Eng Chem Proc Des Dev*, **17**: 243.

Poling BE, Prausnitz JM and O'Connell JP (2001) *The Properties of Gases and Liquids*, McGraw-Hill.

Press WH, Teukolsky SA, Vetterling WT and Flannery BP (1992) *Numerical Recipes in Fortran*, Cambridge University Press.

Rachford HH and Rice JD (1952) Procedure for Use in Electrical Digital Computers in Calculating Flash Vaporization Hydrocarbon Equilibrium, *Journal of Petroleum Technology*, Sec. 1, **Oct**: 19.

Renon H and Prausnitz JM (1968) Local Composition in Thermodynamic Excess Functions for Liquid Mixtures, *AIChE Journal*, **14**: 135.

Ruzicka V and Domalski ES (1993) Estimation of the Heat Capacities of Organic Liquids as a Function of Temperature Using Group Additivity, *J Phys Chem Ref Data*, **22**: **597**, 619.

Soave G (1972) Equilibrium Constants from a Modified Redlich–Kwong Equation of State, *Chem Eng Sci*, **27** (6): 1197.

Sorenson JM and Arlt W (1980) *Liquid–Liquid Equilibrium Data Collection*, Dechema Chemistry Data Series.

Starling, KE (1973) *Fluid Properties for Light Petroleum Systems*, Gulf Publishing Company.

Streich M and Kistenmacher H (1979) Property Inaccuaracies Influence Low Temperature Designs, *Hydrocarbon Process*, **58**: 237.

Wilke CR (1950) A Viscosity Equation for Gas Mixtures, *J Chemical Physics*, **18**: 517.

Wilson GM (1964) A New Expression for Excess Energy of Mixing, *J Am Chem Soc*, **86**: 127.

Appendix B

Materials of Construction

Choice of materials of construction affects both the mechanical design and the capital cost of equipment. Estimation of the capital cost and preliminary specification of equipment for the evaluation of performance requires decisions to be made regarding the materials of construction. Consider first the mechanical properties of materials.

B.1 Mechanical Properties

1) *Yield and tensile strength.* When a solid material is subjected to an increasing *stress* (force per unit area) normal to the stressed area, the material deforms as measured by the *strain* (the ratio of change in length of the material to the initial length). *Tensile stress* acts normal to the stressed area to lengthen the material. Figure B.1 shows a typical stress–strain profile for a metal specimen under tensile stress. Initially, stress is proportional to strain, as expressed by Hooke's law. Within this linear range, deformation is *elastic* and not permanent and the material will return to its original shape when the stress is removed. If stress is increased beyond the linear range up to the *yield strength*, deformation continues to be elastic and deformation is not permanent, even though stress is no longer proportional to strain. The yield strength marks the elastic limit beyond which the material will no longer go back to its original shape when the stress is removed. The yield strength is not a sharply defined point and is often defined as the stress that will cause a permanent deformation of 0.2% of the original dimension. As stress is increased beyond the yield strength, *plastic deformation* occurs and strains are only partially recoverable. The material is permanently deformed if the stress is completely released. Increased stress causes the material to elongate uniformly along its length to the point where the stress–strain curve in Figure B.1 reaches a maximum. The maximum point is known as the *tensile strength* or *ultimate strength*. As stress is increased beyond this point further, plastic deformation is concentrated in the region of the weakest point and the material begins to "neck" or thin down locally until fracture occurs. Since designers are interested in maximum loads that can be carried by the complete cross-section before necking, the stress at the maximum point is taken to be the design basis, rather than the point of fracture.

   Nondeformable materials, such as cast iron, do not show the same behavior as Figure B.1. Brittle materials typically fail while the deformation is elastic. The tensile strength and fracture strength are the same for brittle materials.

   If the stress is applied normal to the stressed area, but to compress (shorten) the material, it is a *compressive stress*. Compressive strength is not equal to tensile strength.

   For mechanical design of equipment, the allowable stress should be limited to be below the yield strength. However, since the yield strength is difficult to determine accurately, the allowable stress is taken as either the yield strength or tensile (ultimate) strength divided by a safety factor greater than one. The allowable stress for structural steel is typically one-half to two-thirds of the yield strength or one-quarter of the tensile strength at the working temperature. The use of tensile strength as the design basis is necessary for materials such as cast iron that do not exhibit a yield point. The design basis depends on the material of construction and the applicable design code being followed.

2) *Ductility and malleability. Ductility* is the ability of a material to deform plastically under tensile stress before fracturing. On the other hand, *malleability* measures the ability of a material to deform under compressive stress. Both of these mechanical properties measure the extent to which a solid material can be deformed plastically without fracture. A material with high ductility will be able to be drawn into long, thin wires without breaking. Ductility and malleability are especially important in metalworking, as materials that crack or break under stress cannot be manipulated using metal forming processes, such as rolling and drawing. Brittle materials cannot be formed by processes such as rolling and drawing and must be cast or thermoformed.

3) *Toughness. Toughness* is the ability of a metal to deform plastically and to absorb energy before fracture. Brittleness implies sudden failure; toughness is the opposite. Of prime importance is the ability to absorb energy before fracture. Toughness requires a combination of strength and ductility.

Chemical Process Design and Integration, Second Edition. Robin Smith.

Companion Website: www.wiley.com/go/smith/chemical2e

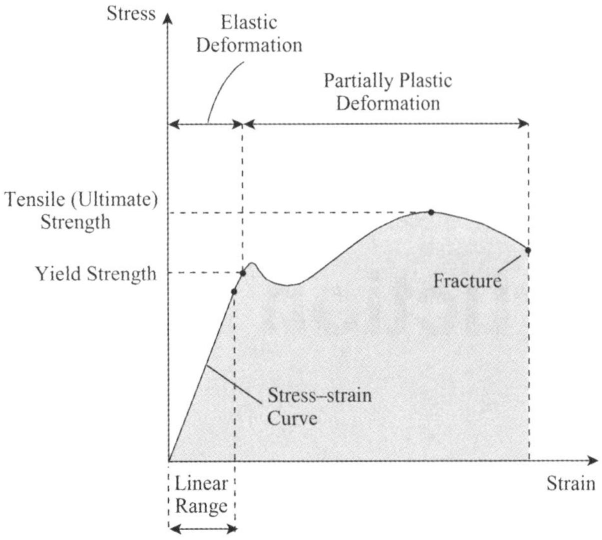

Figure B.1
Stress–strain curve for a test material.

One measure of toughness is the area under the stress–strain curve from a tensile test (see Figure B.1). This value is termed *material toughness* and has units of energy per volume. In order to be tough, a material must have both high strength and ductility. Brittle materials that are strong but with limited ductility are not tough. Ductile materials with low strengths are also not tough. To be tough, a material should withstand both high stresses and high strains.

4) *Hardness.* Hardness is a characteristic of material that determines the resistance of a material to wear and abrasion. This is important for materials of construction in duties that are exposed to abrasive material, particularly solids and slurries. Various tests can be used for hardness leading to different measures of hardness. Hardness can be increased in many metals by heat treatment or cold working of the metals.
5) *Fatigue.* Fatigue is progressive and localized structural damage that occurs when a material is subjected to cyclic loading. The stress that causes such damage can be much less than the yield or tensile strength of the material. If the loads are above a certain threshold, microscopic cracks begin to form. Eventually a crack can reach a critical size and propagate suddenly, causing structural fracture. The shape of the structure significantly affects the fatigue life. Square holes or sharp corners can lead to elevated local stresses where fatigue cracks can initiate. Round holes and smooth shapes increase the fatigue strength. *Fatigue limit* quantifies the amplitude of cyclic stress that can be applied to the material without causing fatigue failure. Ferrous alloys and titanium alloys have a distinct limit, i.e. a stress amplitude below which there appears to be no number of cycles that will cause failure. Other metals such as aluminum and copper do not have a distinct limit and will eventually fail even from small stress amplitudes.
6) *Creep.* Creep is the tendency of a material to deform slowly under the influence of mechanical stresses. The deformation may become large enough for a component to no longer be able to perform its function. It can occur as a result of long-term exposure to high levels of stress below the yield strength. Creep is more severe when the material under stress is exposed to elevated temperature for long periods, and generally increases near its melting point. Creep is of particular concern when materials are subjected to a combination of high stress and high temperature. As a general guideline, for metals the effects of creep deformation generally become noticeable at approximately 30% of the melting point (measured in absolute temperature). Creep creates initially a relatively high level of strain that slows with increasing time.

B.2 Corrosion

In chemical manufacture, corrosion resistance is most often the prime consideration in materials selection. Corrosion is the deterioration of materials (usually metals) as a result of chemical reactions between the materials and the surrounding environment, for example the oxidation of iron in the presence of water by an electrolytic process. Whilst the main concern is corrosion caused by process fluids, external corrosion in the surrounding environment can also be an issue. There are various mechanisms for corrosion that depend on the material of construction and the corrosive environment. The rate of corrosion is also highly temperature dependent. In broad terms, corrosion can be categorized as either *uniform* or *localized.* The main types of corrosion are:

1) *Uniform corrosion. Uniform* or *general* corrosion is attack to the material that is distributed more or less uniformly over a surface, and proceeds more or less uniformly at a uniform rate. It should be noted that uniform corrosion can also be accompanied by localized corrosion. An acceptable rate of corrosion of relatively low-cost material, such as carbon steel, is of the order of 0.25 $mm \cdot y^{-1}$ or less (Hunt, 2014). For more expensive materials, such as stainless steel, an acceptable rate of corrosion is of the order 0.1 $mm \cdot y^{-1}$ (Hunt, 2014).
2) *Galvanic corrosion. Galvanic* corrosion occurs when two different metals have contact, either physically or electrically, in a common electrolyte. A *galvanic couple* can be created in which the more *active metal* (the anode) corrodes at an accelerated rate and the more *noble metal* (the cathode) corrodes at a lower rate. The surface area of the two different metals affects the corrosion rates. If each material was immersed separately in the electrolyte, each metal would corrode at its own uniform rate. The same principle can be applied for corrosion protection through the use of *sacrificial electrodes*. For example, a zinc electrode can be connected electrically as a sacrificial anode to protect steel.
3) *Erosion corrosion. Erosion* corrosion is a corrosion process enhanced by the erosion action of flowing fluids. The erosion action might result from high velocity and turbulence of a fluid, or the presence of impinging solid particles or bubbles in a liquid stream, or impinging solid particles or droplets in a gas stream. *Cavitation*, which is the formation and sudden collapse of vapor bubbles in a liquid, can also facilitate erosion

corrosion. The erosion action removes protective (or passive) oxide layers from the surface of the material, enhancing the corrosion. The erosion corrosion may also be accompanied by mechanical erosion. Erosion corrosion can be mitigated through avoiding sudden changes in the direction in fluid flow, avoiding the formation of jets directed at surfaces in the flow, installation of impingement plates to protect vulnerable surfaces, or the use of more resistant materials and surface coatings.

4) *Crevice corrosion. Crevice* corrosion is localized corrosion in small crevices where fluid can stagnate. Such crevices can occur in a metal surface or at joints, screw threads, and so on. To function as a corrosion site, a crevice has to be wide enough to allow entry of the corrosive fluid, but narrow enough to ensure the corrosive fluid remains stagnant. Crevices can then develop a local chemistry that is different from the bulk fluid. Net anodic reactions can occur within the crevice and net cathodic reactions exterior to the crevice, causing corrosion similar to galvanic corrosion. Changing the design to avoid crevices and potential corrosion sites can mitigate crevice corrosion.
5) *Pitting corrosion. Pitting* corrosion is localized corrosion that leads to the creation of small holes in the metal surface. Pitting corrosion is usually found on metals and alloys such as aluminum alloys, stainless steels and stainless alloys when an ultrathin passive film (oxide film) protects the metal surface from corrosion. If this layer is chemically or mechanically damaged and does not immediately repassivate, it can become anodic, with the main area becoming cathodic, leading to localized galvanic corrosion.
6) *Stress corrosion cracking.* Stress corrosion cracking is a failure mechanism caused by a combination of a susceptible material (normally a ductile metal), tensile stress and a specific corrosive environment. Tensile stress is required to open up cracks in the material. The stress can be either directly applied or can be present in the form of residual stresses frozen into the material from fabrication. Cracks produced by the stress can grow rapidly in a specific corrosive environment. Stress corrosion cracking is highly chemically specific and only occurs when certain alloys are exposed to very specific chemical environments. For example, copper and its alloys are susceptible to ammonia compounds, mild steels are susceptible to alkalis and stainless steels are susceptible to chlorides. Temperature is also an important factor. Stress corrosion cracking can be mitigated by selecting combinations of materials and process environments that avoid the problem, stress relieving components by heat treatment after fabrication or welding.
7) *Hydrogen embrittlement.* Embrittlement is a phenomenon that causes loss of ductility in a material, making it brittle and reducing its load-bearing capacity. Hydrogen embrittlement involves the diffusion of hydrogen atoms into the metal. If steel is exposed to hydrogen at high temperatures, hydrogen atoms diffuse into the alloy and react with carbon in the alloy to form methane or combine with other hydrogen atoms to form hydrogen molecules in tiny pockets. Since these molecules are too large to diffuse through the metal, pressure builds at grain boundaries and voids within the metal, causing minute cracks to form. Copper alloys can be embrittled if exposed to hydrogen. The hydrogen diffuses through the copper and reacts with inclusions of Cu_2O to form water, which then forms pressurized bubbles at the grain boundaries and voids. Hydrogen embrittlement can be avoided by choosing a metal alloy that is resistant.

B.3 Corrosion Allowance

When equipment is mechanically designed, after determining the wall thickness needed to meet mechanical requirements, an extra thickness is added to compensate for the reduction in the wall thickness as a result of corrosion over the life of the equipment. This is referred to as the *corrosion allowance*. Thus, a corrosion allowance must be added to the wall thickness to meet the mechanical requirements over the life of the equipment. As noted above, an acceptable rate of uniform corrosion for relatively low-cost material, such as carbon steel, is of the order of 0.25 $mm{\cdot}y^{-1}$ and of the order of 0.1 $mm{\cdot}y^{-1}$ for more expensive materials, such as stainless steel (Hunt, 2014). However, this should be considered to be the maximum. The actual rate of uniform corrosion must be determined and used as the basis of the corrosion allowance. Then, because the penetration depth for the corrosion can vary across the surface, a corrosion allowance is normally assigned a safety factor of two.

B.4 Commonly used Materials of Construction

The most commonly used materials of construction used in chemical plant are:

1) *Carbon steel.* Carbon steel is an alloy formed from iron and carbon with nominal amounts of other elements. Steel is considered to be carbon steel when the percentages of other trace elements do not exceed certain values. The maximum content of manganese is 1.65%, silicon 0.60% and copper 0.60%. Steel that also contains higher or specified quantities of other elements such as nickel, chromium or vanadium is referred to as *alloy steel*. Changing the amount of carbon changes the properties of the steel. Lower carbon steels are softer and more easily formed and steels with higher carbon content are harder and stronger, but less ductile and more difficult to machine and weld. Carbon steels can be classified as:
   - Low carbon steel, also known as mild steel, which has a carbon content of typically 0.05% to 0.25%, with up to 0.4% manganese. It is a low-cost material that is easy to fabricate. It is not as hard as higher carbon steels.
   - Medium carbon steel has a carbon content of typically 0.25% to 0.55%, with 0.60% to 1.65% manganese. It is ductile and strong with good wear properties.
   - High carbon steel has a carbon content of typically 0.55% to 0.95%, with 0.30% to 0.90% manganese. It is very

strong and holds its shape well, making it ideal for springs and wire.

- Very high carbon steel has a carbon content of typically 0.96% to 2.1%. The high carbon content makes it an extremely strong but brittle material.

An alloy of iron with more than 2.1% carbon as the main alloying element is classed as cast iron. In addition to carbon, cast irons must also contain from 1% to 3% silicon. Cast iron cannot be used for pressure containment because of its brittleness, but can be used for some equipment components, such as pump casings. High silicon irons with carbon up to 1.1% and silicon up to 15% have excellent resistance to attack by sulfuric acid and most organic acids, but still suffer from brittleness.

2) *Stainless steel.* Stainless steels are the most commonly used materials of construction for corrosion resistance (Pitcher, 1976). Stainless steel differs from carbon steel mainly by the amount of chromium present, with a minimum of 10.5% chromium. Other alloying elements are added to enhance properties such as strength at high or cryogenic temperatures, ease of fabrication and weldability (ability to be welded). These additional elements include nickel, molybdenum, titanium, copper, carbon and nitrogen. Stainless steels protect against corrosion as they contain enough chromium to form a passive film of chromium oxide, which prevents further surface corrosion. The oxide bonds strongly to the metal surface and blocks corrosion from spreading to the internal structure of the metal. Thus, stainless steels are most effective in oxidizing environments.

There are many grades of stainless steels. The American Iron and Steel Institute (AISI) and the American Society of Testing and Materials (ASTM) have developed designations such as 304, 430, etc. These are not specifications, but relate to steel grade composition ranges only. Stainless steels have different metallurgical structures and can be divided into five types:

a) *Austenitic grades.* Austenitic grades are the most commonly used types in chemical plants. The most common austenitic grades are iron–chromium–nickel steels that form the 300 series. Austenitic grades cannot be hardened by heat treatment, but can be hardened by cold-working. The letter 'L' after a stainless steel type indicates low carbon (e.g. 304L), in which the carbon is kept to 0.03% or below. These are used to provide extra corrosion resistance after welding. However, L-grades are more expensive and more carbon at high temperatures imparts greater physical strength. High carbon H-grades contain between 0.04% and 0.10% carbon and retain strength at extreme temperatures. Some of the more commonly used grades are (Pitcher, 1976):

Type	Typical composition	Typical applications
304	18% to 20% Cr, 8% to 11% Ni, 0.08% C, balance Fe	The most common of the austenitic grades. Widely used for chemical processing equipment in lesser corrosive environments. It is also widely used for process equipment in the food and beverage industries.
304L	18% to 20% Cr, 8% to 11% Ni, 0.03% C, balance Fe	Low carbon version of Type 304. Used when welding and heat treatment are a problem with Type 304.
316	16% to 18% Cr, 10% to 14% Ni, 2% to 3% Mo, 0.08% C, balance Fe	Molybdenum used to control pitting corrosion. Widely used for chemical processing equipment in more corrosive environments. Also used for process equipment in the food and beverage and pulp and paper industries.
316L	16% to 18% Cr, 10% to 14% Ni, 2% to 3% Mo, 0.03% C, balance Fe	Low carbon version of Type 316. Used when welding and heat treatment are a problem with Type 316.
317	18% to 20% Cr, 11% to 15% Ni, 3% to 4% Mo, 0.08% C, balance Fe	Higher percentage of molybdenum than Type 316 for extra corrosion resistance. Suitable for highly corrosive environments.
317L	18% to 20% Cr, 11% to 15% Ni, 3% to 4% Mo, 0.03% C, balance Fe	Low carbon version of Type 317 when welding and heat treatment are a problem with Type 317.

b) *Martensitic grades.* Martensitic grades are stainless alloys that are both corrosion resistant and can be hardened by heat treatment. These grades are chromium steels with no nickel. The martensitic grades are used where strength, hardness and wear resistance are required. Some of the more commonly used grades are (Pitcher, 1976):

Type	Typical composition	Typical applications
410	11.5% to 13.5% Cr, 0.6% Ni, 1% Mn, 0.15% C, balance Fe	Basic martensitic grade. Low-cost, general-purpose stainless steel that can be heat treated and is used when corrosion is not severe. Typical applications are for highly stressed parts needing a combination of strength and corrosion resistance.
420	12% to 14% Cr, 0.6% Ni, 0.2% to 0.4% C, balance Fe	Increased carbon improves mechanical properties.
431	15% to 17% Cr, 1.25% to 2.5% Ni, 0.2% carbon maximum, balance Fe	Better corrosion resistance than type 410 and good mechanical properties. Typical applications are for high-strength parts, such as valves and pumps.
440	16% to 18% Cr, 0.6% Ni, 0.75% Mo maximum, 0.6% to 1.2% C, balance Fe	Further increases the chromium and carbon to improve strength and corrosion resistance.

c) *Ferritic grades.* Ferritic grades are suited to applications that need to resist corrosion and oxidation, but also require a high resistance to stress corrosion cracking. These grades cannot be heat treated. Ferritic grades are more corrosive resistant than martensitic grades, but generally not as good as the austenitic grades. Like martensitic grades, these grades are chromium steels with no nickel. Some of the more commonly used grades are (Pitcher, 1976):

Type	Typical composition	Typical applications
430	16% to 18% Cr, 0.12% C maximum, balance Fe	The basic ferritic grade, with slightly less corrosion resistance than Type 304. High resistance to nitric acid, sulfur gases, and many organic and food acids.
405	11.5% to 14.5% Cr, 0.8% C maximum, 0.1% to 0.3% Al, balance Fe	Lower chromium, but aluminum added to prevent hardening when cooled from high temperatures. Typical applications include heat exchangers.
442	18% to 23% Cr, 1% Mn, 0.2% carbon, balance Fe	Increased chromium to improve resistance to oxidation scaling. Typical applications include furnace and heater parts.
446	23% to 27% Cr, 1.5% Mn maximum, 0.2% carbon maximum, balance Fe	Contains even more chromium than Type 442 to offer high resistance to oxidation scaling at high temperatures and when sulfur may be present.

d) *Duplex grades.* These grades are a combination of austenitic and ferritic material. They have higher strength and better resistance to stress corrosion cracking.

e) *Precipitation hardening grades.* Precipitation hardening grades offer a combination of strength, ease of fabrication, ease of heat treatment, and corrosion resistance not found in the other classes. These grades are used for bar, rod, wire, forgings, sheet and strip products.

3) *Nickel and nickel alloys.* Nickel and its alloys have good resistance to many chloride-bearing and reducing environments that attack stainless steels. The resistance of nickel to reducing environments is enhanced by molybdenum and copper. Pure nickel is generally only used for sodium and potassium hydroxide services, with its alloys being preferred for most applications. The most common grades of nickel alloys are (Hughson, 1976):

a) *Hastelloy.* Hastelloy alloys offer high resistance to uniform corrosion attack, localized corrosion, stress corrosion cracking, and ease of welding and fabrication. Some of the more commonly used grades are:

Hastelloy	Typical composition	Typical applications
Type B	28% Mo, 2% Fe, 1% Cr, balance Ni	Hydrochloric acid, hydrogen chloride gas, sulfuric, acetic and phosphoric acids. Thermal oxidation of chlorinated wastes.
Type C	16% Mo, 5% Fe, 16% Cr, balance Ni	Mineral (inorganic) acids (e.g. sulfuric, nitric, phosphoric, etc), formic and acetic acids, acetic anhydride, chlorine, chlorine contaminated solutions (organic and inorganic), hypochlorite, brine solutions and seawater. Paper pulp processes.

(*Continued*)

Hastelloy	Typical composition	Typical applications
Type G	3% Mo, 30% Fe, 21% Cr, 2% Cu, balance Ni	Hot sulfuric and phosphoric acids, oxidizing and reducing agents, acid and alkaline solutions, mixed acids, sulfate compounds, contaminated nitric acid and hydrofluoric acid.

b) *Inconel.* Inconel is a family of austenitic nickel–chromium-based alloys. When heated, Inconel forms a stable, passive oxide layer protecting the surface from further attack. Inconel retains strength over a wide temperature range and is attractive for high-temperature applications where aluminum and steel would be vulnerable to creep. Some of the more commonly used grades are:

Inconel	Typical composition	Typical applications
Type 600	7% Fe, 16% Cr, balance Ni	Applications that require resistance to corrosion and heat. Reducing conditions for alkaline solutions. Resistant to chloride-ion stress corrosion cracking.
Type 625	9% Mo, 2.5% Fe, 22% Cr, balance Ni	Sulfuric acid at low temperatures. Phosphoric acid and sodium hydroxide solutions at high temperatures, but low concentrations.
Type 825	3% Mo, 30% Fe, 21% Cr, balance Ni	Reducing environments such as sulfuric and phosphoric acids. Oxidizing environments such as nitric acid solutions, nitrates and oxidizing salts. Most organic acids including boiling concentrated acetic acid, acetic-formic acid mixtures, maleic and phthalic acids. Most alkaline solutions.

c) *Monel.* Monel is a group of nickel alloys, primarily composed of nickel (up to 67%) and copper (typically 32%), iron (typically 2%), with small amounts of manganese, carbon and silicon. It is used for sulfuric acid, hydrochloric acid and reducing conditions. It can be used for many alkalis, but not at all concentrations. It has less resistance to alkalis than pure nickel. Monel can be used in seawater and brackish water services. It is not vulnerable to stress corrosion cracking by any of the chloride salts. Compared with steel, Monel is difficult to fabricate as it work-hardens very quickly.

4) *Aluminum and aluminum alloys.* Aluminum is principally used in heat transfer applications, because of its high thermal conductivity. Aluminum plate-fin heat exchangers are used extensively in cryogenic applications. Aluminum fins are widely used to enhance the heat transfer in high transverse fins for air-cooled heat exchangers. One of the main limitations of aluminum is that its strength declines significantly above 150 °C. The highest working temperature is normally taken to be 200 °C. However, it has excellent low-temperature properties and can be used down to −250 °C. Many mineral acids attack aluminum, but it can be used with concentrated nitric acid above 82%, concentrated sulfuric acid and most organic acids. Aluminum cannot be used with strong alkali solutions. As with stainless steels and nickel alloys, corrosion resistance is a result of the formation of a thin oxide layer, which means that it is most suited to use in oxidizing conditions.

A number of aluminum alloys are available. The alloys are mainly to improve the mechanical properties, and most have lower corrosion resistance than the pure metal.

5) *Copper and copper alloys.* Copper has the attraction that it has an extremely high thermal conductivity for use in heat transfer equipment. However, pure copper is rarely used in chemical plant equipment. Copper is generally attacked by mineral acids, but is generally resistant to caustic alkalis, organic acids and salts. Cupronickels containing 10 to 30% nickel have good corrosion resistance and are used for heat exchanger tubing. Resistance to seawater is particularly good for cupronickels and they are often used for heat exchangers where seawater is used as the coolant.
6) *Lead and lead alloys.* Lead was one of the traditional materials of construction in chemical plant. It was particularly important in the construction of sulfuric acid plants. Lead relies for its high resistance to corrosion on layers of insoluble lead salts that form on the surface. However, the use of lead and lead alloys has declined as other metals and plastics have replaced it.
7) *Titanium.* Titanium has good corrosion resistance in oxidizing and mild reducing environments. It is resistant to nitric acid at almost all concentrations. It is resistant to hot chloride solutions and performs better than stainless steel for seawater duties. However, fabrication with titanium is difficult.
8) *Zirconium.* Zirconium is similar to titanium in its corrosion properties. However, zirconium is more resistant to hydrochloric acid and resistance to chlorides, except ferric and cupric. Zirconium resembles titanium in its difficulty of fabrication. All welding must be carried out in an inert medium.
9) *Tantalum.* Tantalum is practically inert to many oxidizing and reducing acids. It is attacked by hot alkalis and hydrofluoric acid. The mechanical properties of tantalum are similar to mild steel, but it has a higher melting point.
10) *Glass.* Glass chemical plant equipment is available from specialist equipment manufacturers. Pipe joints use polytetrafluoroethylene (PTFE) gaskets. Glass has excellent resistance to all acids except hydrofluoric acid. There are two major drawbacks to its use. First, it is brittle and not suited to high pressures and can also be damaged by thermal shock. Second, the availability of equipment is limited in scope and restricted to relatively small scales. However, glass-lined steel combines the corrosion resistance of glass with the strength of steel. Thus, glass-lined stirred reactor vessels are quite common.
11) *Plastic materials.* Commonly used plastic materials of construction are polyvinyl chloride (PVC), acrylonitrile–butadiene–styrene (ABS) and polyethylene (PE). Generally, plastics have high resistance to weak mineral acids and inorganic salt solutions when metals might not be suitable. By comparison with metals, plastics are limited to use at relatively low temperatures. The use of plastic materials is generally restricted to tanks, pipes and valves. However, specialist designs of heat exchangers and pumps are also available in plastic materials. The most commonly used plastic materials are:

PVC	Most frequently specified of all thermoplastic piping materials. Resistant to corrosion and chemical attack by acids, alkalis, salt solutions and many other chemicals. However, it is attacked by polar solvents such as ketones and aromatics. The maximum service temperature for PVC is 60 °C under pressure and 80 °C in drainage.
ABS	Temperature range is −40 °C to 82 °C. ABS is resistant to a wide variety of process materials.
PE	Of the different grades of PE, process equipment is usually manufactured from high-density PE for strength and hardness. High-density PE tanks can be used up to 55 °C. Pipe is usually made from medium or high density PE. PE is generally used for gas distribution, water lines and slurry lines.

12) *Composite materials.* A composite material combines two or more materials, often ones that have very different properties. The properties of the two materials combine together to give the composite properties. The most commonly used composite material is fiberglass. Glass fiber reinforced plastics (GRP) or fiberglass reinforced plastics (FRP) use polyester resins reinforced with glass fiber. On its own, glass is strong but brittle and it will break if bent. The matrix holds the glass fibers together and protects them from damage by spreading the forces acting on them. Glass reinforced plastics are relatively strong and have a good resistance to dilute mineral acids, inorganic salts and many solvents, but are less resistant to alkalis. The use of glass reinforced plastics is mostly restricted to tanks and vessels that can be fabricated in sizes up to 20 m.

Some advanced composites are now made using carbon fibers instead of glass. These materials are lighter and stronger than fiberglass but more expensive to produce and therefore do not find wide application in chemical plant.
13) *Linings.* Many corrosion resistant materials are not suitable for the fabrication in large-scale equipment because they are difficult to weld, too brittle, too soft or too expensive. Rather than use an expensive alloy for the construction of large-scale equipment, a cheaper material such as carbon steel can be used, with a lining of corrosion resistant material to prevent damage. Glass can be used to line vessels and pipes. Plastic linings can be used. The most chemically resistant plastic for such applications is polytetrafluoroethylene (PTFE), which is resistant to all alkalis and acids except fluorine and chlorine gas at elevated temperatures. Rubber has been used extensively in the past to line tanks. Both natural and synthetic rubbers can be used. Natural rubber has good resistance to most alkalis and most acids, but not nitric acid. Synthetic

rubbers can be used with nitric acid and strongly oxidizing environments, but are generally not suited to chlorinated solvents.

In addition to the use of linings for corrosion resistance, linings are also necessary in high-temperature equipment to protect the metal shells of furnaces. Refractory bricks and cements are used as lining for fired heaters, boilers and the furnace reactors. Carbon steel shells must be maintained below 500 °C in operation.

B.5 Criteria for Selection

Many factors go into the selection of materials of construction for a particular application:

1) *Corrosion.* The primary consideration in many chemical process applications is corrosion. The general characteristics of corrosion and the use of different materials have been discussed earlier. Corrosion charts are available to give more detail in the selection process (Green and Perry, 2007). However, such corrosion charts should be used with great caution. Corrosion is very difficult to predict with precision. In addition to the bulk fluid major components, it is dependent on temperature, on low levels of components in the process materials and the preparation of the material of construction. It is also dependent on localized phenomena, as discussed earlier. The most reliable way to predict corrosion resistance is to use experience from working processes operating with the same or a similar process. Laboratory tests are useful, but are not as effective as tests carried out on working plant processing the same chemicals. Test samples can be located in working plant and tested for corrosion resistance. However, again the temperature is important in the tests, as well as the composition of the process materials.

   The primary choice for material for most applications is carbon steel, if corrosion permits. An alternative is to use plastic materials, but these are limited by their inability to withstand elevated temperatures and by their low strength. Thus, lack of corrosion resistance for carbon steel, or adverse temperature and pressure in the case of plastic material, force the use of more expensive materials. Most corrosion problems encountered in process design can be solved by using one of the various grades of stainless steel discussed earlier (Pitcher, 1976). However, sometimes the corrosion problems are too severe for stainless steels to handle. In such cases, it might be necessary to resort to more expensive metals, such as nickel alloys. As a cheaper alternative, it is sometimes possible to line equipment constructed with cheaper material, such as carbon steel, with glass, PTFE or some other corrosion resistant material.

2) *Temperature.* In addition to the corrosion phenomena being highly temperature dependent, temperature significantly affects the mechanical properties of the material. Plastic materials are generally restricted for use below 80 °C. Copper can be used up to 260 °C and aluminum up to 460 °C. Carbon steel is limited to use below 500 °C. Stainless steel grades 304 and 316 and nickel can be used up to 760 °C. Some other stainless steel grades, such as 321 and 347, and Monel can be used up to 815 °C. Type 446 stainless steel, Inconel and titanium can be used up to 1100 °C. Creep resistance is an important consideration for materials being exposed to high temperature, especially when in combination with high stress.

   However, high temperature is not the only problem. Metals can fail catastrophically when exposed to very low temperatures through *brittle fracture*. Carbon steel can be used down to −45 °C. Stainless steels can be engineered for lower temperatures, but this depends on the grade. Some austenitic grades of stainless steel can be used as low as −270 °C. Aluminum can be used down to −250 °C.

3) *Pressure.* For processes working under significant pressure, the main consideration is the strength of the material of construction, which will be temperature dependent.

4) *Abrasive environments.* Processing solids and slurries can create an abrasive environment for materials of construction. Under such conditions, materials of construction without the necessary hardness will quickly deteriorate.

5) *Ease of fabrication.* Fabrication involves cutting, bending, machining, joining and assembling to produce functional equipment. The mechanical properties of materials and their ease of fabrication are intimately linked. The mechanical properties can change as a result of the fabrication. Properties such as brittleness, ductility, malleability, and the work hardening and heat treatment properties of materials are all important factors in the ease of fabrication.

   Joining of materials is another important consideration. *Welding* joins metals by causing coalescence of the metals being joined. This is most often achieved by melting the metals to be joined and adding a filler material to form a pool of molten material that cools to become a strong joint. Many different energy sources can be used for welding. However, not all metals are easily welded and welding different metals together can present problems. Brazing is an alternative joining process for metals in which a filler metal is melted between the metals to be joined. The filler metals used in brazing usually have melting points between 450 °C and 1000 °C, but must have a lower melting point than the metals being joined. Unlike welding, the metals to be joined are not melted. The filler metal melts and reacts metallurgically with the metals to be joined, forming strong, permanent joints. Brazing has the advantage of producing less thermal stresses than welding.

   Various methods are available for joining plastics and composites. These can be classed into mechanical fastening, adhesive and solvent bonding, and welding. Welding is an effective method of permanently joining plastic components, but can only be applied to thermoplastics and thermoplastic elastomers.

6) *Availability of standard equipment.* When designing a chemical process, it is necessary to choose equipment (e.g. pumps, compressors, valves and pipes) from suppliers that have a limited range of specifications, including materials of construction. The choice of material of construction for such equipment might be dictated by the options available from equipment suppliers.

7) *Cost.* Cost effectiveness is one of the main considerations in process design. The choice of material of construction has major cost implications, as illustrated in Table 2.2, which shows that equipment constructed from stainless steels will be between 2.4 to 3.4 times more expensive than if constructed from carbon steel, between 3.6 and 4.4 times more expensive if constructed from nickel alloys and for titanium 5.8 times as expensive as carbon steel. There is therefore a great incentive to avoid exotic materials of construction. An alternative discussed above is to use low-cost material with a corrosion resistant lining.

B.6 Materials of Construction – Summary

Many factors enter into the choice of the materials of construction. Among the most important are:

- mechanical properties (particularly tensile and yield strength, compressive strength, ductility, toughness, hardness, fatigue limit and creep resistance),
- effect of temperature on mechanical properties (both low and high temperatures),
- ease of fabrication (machining, welding, and so on),
- corrosion resistance,
- availability of standard equipment in the material,
- cost.

A much more detailed discussion of selection of materials of construction has been given by Kirby (1980).

References

Green DW and Perry RH (2007) *Perry's Chemical Engineer's Handbook*, McGraw-Hill.

Hughson RV (1976) High-Nickel Alloys for Corrosion Resistance, *Chemical Engineering*, **Nov**: **125**.

Hunt MW (2014) Develop a Strategy for Material Selection, *Chem Eng Progr*, **May**: **42**.

Kirby GN (1980) How to Select Materials, *Chemical Engineering*, **Nov**: **86**.

Pitcher JH (1976) Stainless Steels: CPI Workhorses, *Chemical Engineering*, **Nov**: **119**.

Appendix C

Annualization of Capital Cost

Derivation of Equation 2.7 is as follows (Holland, Watson and Wilkinson, 1983). Let

P = present worth of estimated capital cost
F = future worth of estimated capital cost
i = fractional interest rate per year
n = number of years

After the first year, the future worth F of the capital cost present value P is given by:

$$F(1) = P + Pi = P(1+i) \tag{C.1}$$

After the second year, the worth is:

$$\begin{aligned} F(2) &= P(1+i) + P(1+i)i \\ &= P(1+i)^2 \end{aligned} \tag{C.2}$$

After the third year, the worth is:

$$\begin{aligned} F(3) &= P(1+i)^2 + P(1+i)^2 i \\ &= P(1+i)^3 \end{aligned} \tag{C.3}$$

After year n, the worth is:

$$F(n) = P(1+i)^n \tag{C.4}$$

Equation C.4 is normally written as:

$$F = P(1+i)^n \tag{C.5}$$

Take the capital cost and spread it as a series of equal annual payments A made at the end of each year, over n years. The first payment gains interest over $(n-1)$ years and its future value after $(n-1)$ years is:

$$F = A(1+i)^{n-1} \tag{C.6}$$

The future worth of the second annual payment after $(n-2)$ years is:

$$F = A(1+i)^{n-2} \tag{C.7}$$

The combined worth of all the annual payments is:

$$F = A[(1+i)^{n-1} + (1+i)^{n-2} + (1+i)^{n-3} + \cdots + (1+i)^{n-n}] \tag{C.8}$$

Multiplying both sides of this equation by $(1+i)$ gives:

$$F(1+i) = A[(1+i)^n + (1+i)^{n-1} + (1+i)^{n-2} + \cdots + (1+i)] \tag{C.9}$$

Subtracting Equations C.8 and A.9 gives:

$$F(1+i) - F = A[(1+i)^n - 1] \tag{C.10}$$

which on rearranging gives:

$$F = \frac{A[(1+i)^n - 1]}{i} \tag{C.11}$$

Combining Equation C.11 with Equation C.5 gives:

$$A = \frac{P[i(1+i)^n]}{(1+i)^n - 1} \tag{C.12}$$

Thus, Equation 2.7 is obtained.

Reference

Holland FA, Watson FA and Wilkinson JK (1983) *Introduction to Process Economics*, 2nd Edition, John Wiley & Sons, Inc., New York.

Appendix D

The Maximum Thermal Effectiveness for 1–2 Shell-and-Tube Heat Exchangers

The derivation of Equation 12.34 is as follows (Ahmad, 1985). From Bowman, Mueller and Nagle (1940):

When $R \neq 1$:

$$F_T = \frac{\sqrt{R^2+1}\ln\left(\frac{1-P}{1-RP}\right)}{(R-1)\ln\left(\frac{2-P\left(R+1-\sqrt{R^2+1}\right)}{2-P\left(R+1+\sqrt{R^2+1}\right)}\right)} \tag{D.1}$$

When $R = 1$:

$$F_T = \frac{\left(\frac{\sqrt{2}P}{1-P}\right)}{\ln\left(\frac{2-P\left(2-\sqrt{2}\right)}{2-P\left(2+\sqrt{2}\right)}\right)} \tag{D.2}$$

The maximum value of P, for any R, occurs as F_T tends to $-\infty$. From the F_T functions above, for F_T to be determinate:

1) $P < 1$

2) $RP < 1$

3) $\frac{2-P(R+1-\sqrt{R^2+1})}{2-P(R+1+\sqrt{R^2+1})} > 0$

Condition 3 applies to Equation D.2 when $R = 1$. Both Conditions 1 and 2 are always true for a feasible heat exchange with positive temperature differences.

Taking Condition 3, either:

a) $$P < \frac{2}{R+1-\sqrt{R^2+1}} \quad \text{and} \quad P < \frac{2}{R+1+\sqrt{R^2+1}} \tag{D.3}$$

or

b) $$P > \frac{2}{R+1-\sqrt{R^2+1}} \quad \text{and} \quad P > \frac{2}{R+1+\sqrt{R^2+1}} \tag{D.4}$$

but not both. Consider Condition b in more detail. For positive values of R, $R+1-\sqrt{R^2+1}$ is a continuously increasing function of R, and:

- as R tends to 0, $R+1-\sqrt{R^2+1}$ tends to 0;
- as R tends ∞, $R+1-\sqrt{R^2+1}$ tends to 1.

For Condition b to apply, for positive values of R, $P > 2$. However, $P < 1$ for a feasible heat exchange. Thus, Condition b does not apply.

Now consider Condition a. Because:

$$R+1+\sqrt{R^2+1} > R+1-\sqrt{R^2+1} \tag{D.5}$$

both inequalities for Condition a are satisfied when:

$$P < \frac{2}{R+1-\sqrt{R^2+1}} \tag{D.6}$$

Thus, the maximum value of P for any value of R, P_{max}, is given by:

$$P_{max} = \frac{2}{R+1+\sqrt{R^2+1}} \tag{D.7}$$

References

Ahmad S (1985) *Heat Exchanger Networks: Cost Trade-offs in Energy and Capital*, PhD Thesis, UMIST, UK.

Bowman RA, Mueller AC and Nagle WM (1940) Mean Temperature Differences in Design, *Trans ASME*, **62**: 283.

Chemical Process Design and Integration, Second Edition. Robin Smith.

Companion Website: www.wiley.com/go/smith/chemical2e

Appendix E

Expression for the Minimum Number of 1–2 Shell-and-Tube Heat Exchangers for a Given Unit

The derivation of Equations 12.45 to 12.47 is as follows (Ahmad, 1985). From Bowman (1936), the value of P over N_{SHELLS} number of 1–2 shells in series, P_{N-2N}, can be related to the value of P for each 1–2 shell, P_{1-2}, according to:

$R \neq 1$:

$$P = \frac{1 - \left(\dfrac{1 - P_{1-2}R}{1 - P_{1-2}}\right)^{N_{SHELLS}}}{R - \left(\dfrac{1 - P_{1-2}R}{1 - P_{1-2}}\right)^{N_{SHELLS}}} \tag{E.1}$$

$R = 1$:

$$P = \frac{P_{1-2}N_{SHELLS}}{P_{1-2}N_{SHELLS} - P_{1-2} + 1} \tag{E.2}$$

Now, the maximum possible value of P_{1-2} in a 1–2 shell is (see Appendix D):

$$P_{1-2max} = \frac{2}{R + 1 + \sqrt{R^2 + 1}} \tag{E.3}$$

The value of P_{1-2} required in each 1–2 shell to satisfy a chosen value of X_P is defined by:

$$P_{1-2} = X_P P_{1-2max} \tag{E.4}$$

This therefore requires that:

$R \neq 1$:

$$P = \frac{1 - \left(\dfrac{1 - \dfrac{2X_PR}{R + 1 + \sqrt{R^2 + 1}}}{1 - \dfrac{2X_P}{R + 1 + \sqrt{R^2 + 1}}}\right)^{N_{SHELLS}}}{R - \left(\dfrac{1 - \dfrac{2X_PR}{R + 1 + \sqrt{R^2 + 1}}}{1 - \dfrac{2X_P}{R + 1 + \sqrt{R^2 + 1}}}\right)^{N_{SHELLS}}} \tag{E.5}$$

$R = 1$:

$$P = \frac{\dfrac{2X_PN_{SHELLS}}{2 + \sqrt{2}}}{\dfrac{2X_PN_{SHELLS}}{2 + \sqrt{2}} - \dfrac{2X_P}{2 + \sqrt{2}} + 1} \tag{E.6}$$

These expressions define P_{N-2N} for N_{SHELLS} number of 1–2 shells in series in terms of R and X_P in each shell. The expressions can be used to define the number of 1–2 shells in series required to satisfy a specified value of X_P in each shell for a given R and P_{N-2N}. Hence, the relationship can be inverted to find that value of N that satisfies X_P exactly in each 1–2 shell in the series:

$R \neq 1$:

$$N_{SHELLS} = \frac{\ln\left[\dfrac{1 - RP}{1 - P}\right]}{\ln W} \tag{E.7}$$

Chemical Process Design and Integration, Second Edition. Robin Smith.

Companion Website: www.wiley.com/go/smith/chemical2e

where

$$W = \frac{R + 1 + \sqrt{R^2 + 1} - 2X_P R}{R + 1 + \sqrt{R^2 + 1} - 2X_P} \quad \text{(E.8)}$$

$R = 1$:

$$N_{SHELLS} = \left(\frac{P}{1-P}\right)\left(\frac{1 + \frac{\sqrt{2}}{2} - X_P}{X_P}\right) \quad \text{(E.9)}$$

Choosing the number of 1–2 shells in series to be the next largest integer above N_{SHELLS} ensures a practical exchanger design satisfying X_P.

References

Ahmad S (1985) *Heat Exchanger Networks: Cost Trade-offs in Energy and Capital*, PhD Thesis, UMIST, UK.

Bowman RA (1936) Mean Temperature Difference Correction in Multipass Exchangers, *Ind Eng Chem*, **28**: 541.

Appendix F

Heat Transfer Coefficient and Pressure Drop in Shell-and-Tube Heat Exchangers

Figure F.1 shows the basic layout of a shell-and-tube heat exchanger. The basic correlations used in the model are suitable for the following conditions (Wang *et al.*, 2012):

1) Single-phase heat transfer in a shell-and-tube heat exchanger
2) Plain tubes
3) Single segmental baffles with 20% to 50% cut
4) Physical properties are assumed constant, based on an average between the inlet and outlet conditions.

F.1 Heat Transfer and Pressure Drop Correlations for the Tube Side

1) *Tube-side heat transfer coefficient.* Three equations are used to calculate the tube-side heat transfer coefficient, depending on the flow regime. The Seider–Tate correlation (Equation F.1) is used for the laminar region (Serth, 2007; Kraus, Welty and Aziz, 2011). The Hausen correlation (Equation F.2) is used for the transition region and the Dittus–Boelter correlation (Equation F.3) is used for the turbulent region (Bhatti and Shah, 1987):

$$Nu = 1.86\left[RePr\left(\frac{d_I}{L}\right)\right]^{1/3} \quad \text{for } Re \leq 2100 \text{ and } L \leq 0.05RePr\,d_I \tag{F.1}$$

$$Nu = 0.116(Re^{2/3} - 125)Pr^{1/3}\left[1 + \left(\frac{d_I}{L}\right)^{2/3}\right] \quad \text{for} \quad 2100 < Re < 10^4 \tag{F.2}$$

$$Nu = CRe^{0.8}Pr^{0.4} \quad \text{for } Re \geq 10^4 \tag{F.3}$$

where
Nu = tube-side Nusselt number $= \dfrac{h_T d_I}{k}$
C = constant (0.024 for heating, 0.023 for cooling)
Re = tube-side Reynolds number $= \dfrac{\rho v_T d_I}{\mu}$
Pr = tube-side Prandtl number $= \dfrac{C_P \mu}{k}$
d_I = tube inner diameter
L = tube length
h_T = tube-side heat transfer coefficient
ρ = tube-side fluid density
C_P = tube-side heat capacity
μ = tube-side fluid viscosity
k = tube-side fluid thermal conductivity
v_T = mean fluid velocity inside the tubes $= \dfrac{m_T(N_P/N_T)}{\rho(\pi d_I^2/4)}$
m_T = mass flowrate on the tube-side
N_P = number of tube passes
N_T = number of tubes

Fluid physical properties are taken at the average bulk fluid temperature between the tube-side flow inlet and outlet.

Chemical Process Design and Integration, Second Edition. Robin Smith.

Companion Website: www.wiley.com/go/smith/chemical2e

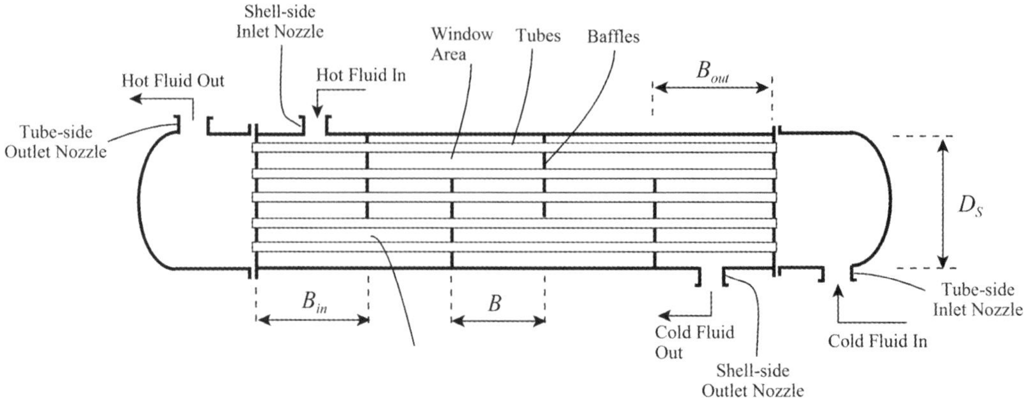

Figure F.1

Geometry of shell-and-tube heat exchangers.

The tube-side heat transfer coefficient can be calculated from Equations F.1 to F.3:

a) *For laminar flow* $Re \leq 2100$ and $L \leq 0.05 Re\, Pr\, d_I$:

$$\begin{aligned} h_T &= 1.86\frac{k}{d_I}\left[\left(\frac{\rho v_T d_I}{\mu}\right)Pr\left(\frac{d_I}{L}\right)\right]^{1/3} \\ &= 1.86\frac{k}{d_I}\left[\left(\frac{\rho d_I}{\mu}\right)Pr\left(\frac{d_I}{L}\right)\right]^{1/3} v_T^{1/3} \\ &= K_{hT1} v_T^{1/3} \end{aligned} \tag{F.4}$$

where

$$K_{hT1} = 1.86\frac{k}{d_I}\left[\left(\frac{\rho d_I}{\mu}\right)Pr\left(\frac{d_I}{L}\right)\right]^{1/3} \tag{F.5}$$

b) *For transition flow* $2100 < Re < 10^4$:

$$\begin{aligned} h_T &= \frac{k}{d_I}\times 0.116\left[\left(\frac{\rho v_T d_I}{\mu}\right)^{2/3} - 125\right]Pr^{1/3}\left[1+\left(\frac{d_I}{L}\right)^{2/3}\right] \\ &= 0.116\frac{k}{d_I}\left(\frac{\rho v_T d_I}{\mu_i}\right)^{2/3}Pr^{1/3}\left[1+\left(\frac{d_I}{L}\right)^{2/3}\right] - 0.116\times 125\frac{k}{d_I}Pr^{1/3}\left[1+\left(\frac{d_I}{L}\right)^{2/3}\right] \\ &= 0.116\frac{k}{d_I}\left(\frac{\rho d_I}{\mu}\right)^{2/3}Pr^{1/3}\left[1+\left(\frac{d_I}{L}\right)^{2/3}\right]v_T^{2/3} - 14.5\frac{k}{d_I}Pr^{1/3}\left[1+\left(\frac{d_I}{L}\right)^{2/3}\right] \\ &= K_{hT2}v_T^{2/3} - K_{hT3} \end{aligned} \tag{F.6}$$

where

$$K_{hT2} = 0.116\frac{k}{d_I}\left(\frac{\rho d_I}{\mu}\right)^{2/3}Pr^{1/3}\left[1+\left(\frac{d_I}{L}\right)^{2/3}\right] \tag{F.7}$$

$$K_{hT3} = 14.5\frac{k}{d_I}Pr^{1/3}\left[1+\left(\frac{d_I}{L}\right)^{2/3}\right] \tag{F.8}$$

c) *For fully developed turbulent flow* $Re \geq 10^4$:

$$\begin{aligned} h_T &= C\frac{k}{d_I}\left(\frac{\rho v_T d_I}{\mu}\right)^{0.8}Pr^{0.4} \\ &= K_{hT4}v_T^{0.8} \end{aligned} \tag{F.9}$$

where

$$K_{hT4} = C\frac{k}{d_I}Pr^{0.4}\left(\frac{\rho d_I}{\mu}\right)^{0.8} \tag{F.10}$$

2) *Tube-side pressure drop.* The total tube-side pressure drop ΔP_T for a single shell comprises the pressure drop in the straight tubes (ΔP_{TT}), pressure drop in the tube entrances, exits and reversals (ΔP_{TE}), and pressure drop in nozzles (ΔP_{TN}) (Serth, 2007).

---

a) *Friction loss in the straight section of tubes* ΔP_{TT} (Serth, 2007):

$$\Delta P_{TT} = \frac{2N_P c_{fT} L\rho v_T^2}{d_I} \tag{F.11}$$

where N_P = number of tube passes per shell
c_{fT} = tube-side friction factor

The Fanning friction factor is given by:

$$c_{fT} = F_C Re^{m_f} \tag{F.12}$$

where

$F_C = 16$, $m_f = -1$ for $Re \leq 2100$
$F_C = 5.36 \times 10^{-6}$, $m_f = 0.949$ for $2100 > Re > 3000$
$F_C = 0.0791$, $m_f = -0.25$ for $Re \geq 3000$

Substituting Equation F.12 into Equation F.11 gives:

$$\begin{aligned} \Delta P_{TT} &= \frac{2N_P F_C Re^{m_f} L\rho v_T^2}{d_I} \\ &= \frac{2N_P F_C \left(\frac{\rho v_T d_I}{\mu}\right)^{m_f} L\rho v_T^2}{d_I} \\ &= \frac{2N_P F_C \left(\frac{\rho d_I}{\mu}\right)^{m_f} L\rho v_T^{2+m_f}}{d_I} \end{aligned} \tag{F.13}$$

b) *Pressure loss due to the sudden contractions, expansions and flow reversals through the tube arrangement ΔP_{TE}* (Serth, 2007):

$$\Delta P_{TE} = 0.5\alpha_R \rho v_T^2 \rho \tag{F.14}$$

where

$\alpha_R = 3.25\,N_P - 1.5$ for $500 \leq Re \leq 2100$
$\alpha_R = 2N_P - 1.5$ for $Re > 2100$

c) *Pressure loss in the inlet and outlet nozzles ΔP_{TN} per shell* (Serth, 2007):

$$\begin{aligned} \Delta P_{TN} &= \Delta P_{TN,inlet} + \Delta P_{TN,outlet} \\ &= C_{TN,inlet}\rho v_{TN,inlet}^2 + C_{TN,outlet}\rho v_{TN,outlet}^2 \end{aligned} \tag{F.15}$$

where

$C_{TN,inlet} = 0.75$ for $100 \leq Re_{TN,inlet} \leq 2100$
$C_{TN,inlet} = 0.375$ for $Re_{TN,inlet} > 2100$

$$Re_{TN,inlet} = \frac{\rho v_{TN,inlet} d_{TN,inlet}}{\mu}$$

$$v_{TN,inlet} = \frac{m_T}{\rho(\pi d_{TN,inlet}^2/4)}$$

$d_{TN,inlet}$ = inner diameter of the inlet nozzle for the tube-side fluid

$C_{TN,outlet} = 0.75$ for $100 \leq Re_{TN,outlet} \leq 2100$
$C_{TN,outlet} = 0.375$ for $Re_{TN,outlet} > 2100$

$$Re_{TN,outlet} = \frac{\rho v_{TN,outlet} d_{TN,outlet}}{\mu}$$

$$v_{TN,outlet} = \frac{m_T}{\rho(\pi d_{TN,outlet}^2/4)}$$

$d_{TN,outlet}$ = inner diameter of the outlet nozzle for the tube-side fluid

Thus, the total pressure drop for the tube-side per shell is:

$$\begin{aligned} \Delta P_T &= \Delta P_{TT} + \Delta P_{TE} + \Delta P_{TN} \\ &= \frac{2N_P c_f L\rho v_T^2}{d_I} + 0.5\alpha_R \rho v_T^2 + C_{TN,inlet}\rho v_{TN,inlet}^2 \\ &\quad + C_{TN,outlet}\rho v_{TN,outlet}^2 \\ &= \frac{2N_P F_C \left(\frac{\rho d_I}{\mu}\right)^{m_f} L\rho v_T^{2+m_f}}{d_I} + 0.5\alpha_R \rho v_T^2 \\ &\quad + C_{TN,inlet}\rho v_{TN,inlet}^2 + C_{TN,outlet}\rho v_{TN,outlet}^2 \\ &= K_{PT1} N_P L v_T^{2+m_f} + K_{PT2} v_T^2 + K_{PT3} \end{aligned} \tag{F.16}$$

where

$$K_{PT1} = \frac{2F_C \left(\frac{\rho d_I}{\mu}\right)^{m_f} \rho}{d_I} \tag{F.17}$$

$$K_{PT2} = 0.5\alpha_R \rho \tag{F.18}$$

$$K_{PT3} = \rho\left(C_{TN,inlet} v_{TN,inlet}^2 + C_{TN,outlet} v_{TN,outlet}^2\right) \tag{F.19}$$

For heat exchangers with multiple shells connected in series, the total pressure drop on the tube-side fluid is estimated by the product of the pressure drop per shell and shell number in series:

$$\Delta P_{T,N_{SHELLS}} = N_{SHELLS} \Delta P_T \tag{F.20}$$

where $\Delta P_{T,N_{SHELLS}}$ = total pressure drop on the tube-side across N_{SHELLS}
ΔP_T = pressure drop per shell, given by Equation F.16
N_{SHELLS} = number of shells connected in series

For heat exchangers with multiple parallel shells, the total pressure drop on the tube-side fluid is equal to the pressure drop in a single shell ΔP_T.

F.2 Heat Transfer and Pressure Drop Correlations for the Shell Side

1) *Shell-side heat transfer coefficient.* The shell-side heat transfer coefficient is calculated from (Ayub, 2005; Wang *et al.*, 2012):

$$h_S = \frac{0.06207 F_S F_P F_L J_S k^{2/3}(c_P \mu)^{1/3}}{d_O} \tag{F.21}$$

where h_S = shell-side heat transfer coefficient
d_O = tube outer diameter
k = shell-side fluid thermal conductivity

c_P = shell-side fluid specific heat capacity
μ = shell-side fluid viscosity
F_s = shell-side geometry factor to allow for baffle cut, baffle arrangement and flow regime (Wang *et al.*, 2012):

$$F_S = -5.9969 \times 10^{-4} Re_S^2 + 0.6191 Re_S + 17.793 \quad Re_S \le 250 \tag{F.22}$$

$$F_S = 1.40915 Re_S^{0.6633} B_C^{-0.5053} \quad 250 < Re_S \le 125{,}000 \tag{F.23}$$

Re_S = Reynolds Number based on tube outside diameter
$= \dfrac{\rho d_O v_S}{\mu}$
ρ = shell-side fluid density
v_S = shell-side fluid velocity
B_C = baffle cut (–)
F_P = the pitch factor, which depends on the tube layout of the bundle
= 1.0 for triangular and diagonal square pitch
= 0.85 for in-line square pitch
F_L = the leakage factor to allow for all the stream leakage, which is a function of bundle configuration
= 0.9 for straight-tube bundle
= 0.85 for U-tube bundle
= 0.8 for floating-head bundle
J_S = correction factor for unequal baffle spacing (Taborek in Hewitt, 1998; Serth, 2007)

$$= \frac{(N_B - 1) + (B_{in}/B)^{2/3} + (B_{out}/B)^{2/3}}{(N_B - 1) + (B_{in}/B) + (B_{out}/B)} \quad \text{for } Re_S < 100$$

$$= \frac{(N_B - 1) + (B_{in}/B)^{0.4} + (B_{out}/B)^{0.4}}{(N_B - 1) + (B_{in}/B) + (B_{out}/B)} \quad \text{for } Re_S \ge 100 \tag{F.24}$$

B = central baffle spacing
B_{in} = inlet baffle spacing
B_{out} = outlet baffle spacing

The definition of the shell-side fluid velocity needs careful consideration. The velocity in cross flow will vary as the fluid flows from one baffle window (or the inlet nozzle) to the next window (or to the exit nozzle). The shell-side velocity is normally defined by reference to the velocity at the widest point, which is on the shell diameter and the slowest velocity across the shell. The velocity will depend on the area through which the fluid flows. Figure 12.15a shows that the cross-sectional area at the widest point is a combination of the area between the outside of the bundle and the inside of the shell and the area within the bundle between the tubes.

$$\begin{aligned} A_{CF} = {} & Baffle\ spacing \times [Area\ between\ the\ outside\ of\ the \\ & bundle\ and\ inside\ of\ the\ shell \\ & + Number\ of\ tubes\ across\ the\ widest\ point \\ & \times Space\ between\ tubes] \end{aligned} \tag{F.25}$$

where A_{CF} = shell-side cross-flow area

For the area between the tubes in the tube bundle, Figure F.2a shows flow across a square pitch (90°) layout. For a 90° layout,

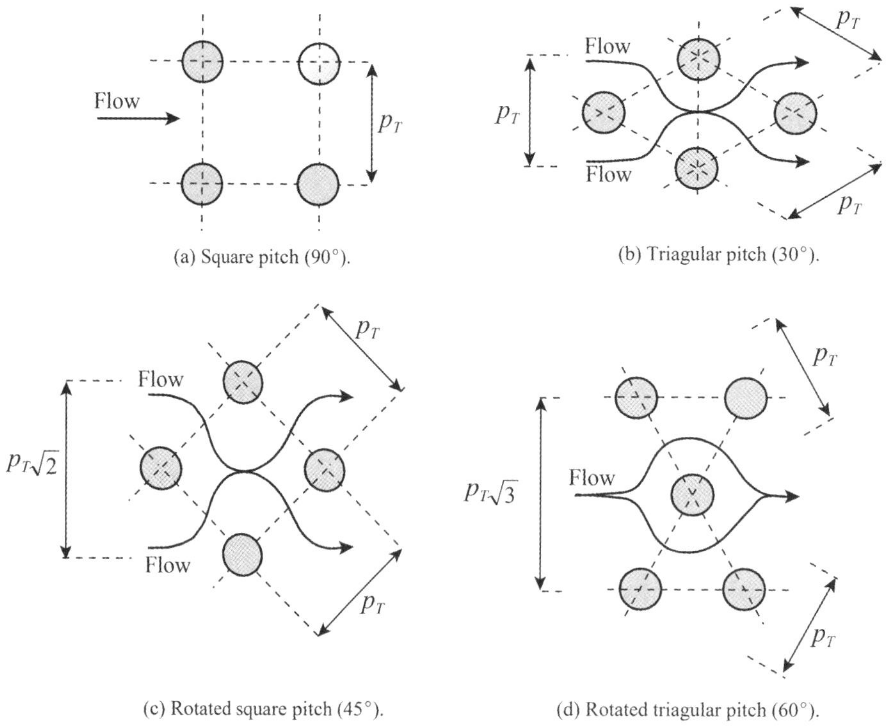

Figure F.2
Tube configurations.

the number of tubes across the cross-sectional area is $(D_B - d_O)/p_T$ and the gap between the tubes is $(p_T - d_O)$. Thus for a 90° layout:

$$A_{CF} = B\left[(D_S - D_B) + \frac{(D_B - d_O)}{p_T}(p_T - d_O)\right] \tag{F.26}$$

where A_{CF} = shell-side cross flow area
B = baffle spacing
D_S = shell inside diameter
D_B = outside diameter of the tube bundle
p_T = tube pitch
d_O = tube outside diameter

Figure F.2b shows the flow across a triangular (30°) layout The flow area is taken to be the minimum flow area, which is the minimum of $(p_T - d_O)$ and $2 \times (p_T - d_O)$, which is $(p_T - d_O)$. Thus the flow area for a triangular layout is also given by Equation F.26.

Figure F.2c shows flow across a rotated square (45°) layout. The minimum flow area is the minimum of $p_T\sqrt{2}$ and $2 \times (p_T - d_O)$. For the tube pitches used in practice (say $p_T = 1.25d_O$), the minimum is $2(p_T - d_O)$. Thus, for a 45° layout:

$$A_{CF} = B\left[(D_S - D_B) + \frac{(D_B - d_O)}{\sqrt{2}p_T}2(p_T - d_O)\right] \tag{F.27}$$

Finally, Figure F.2d shows flow across a rotated triangular (60°) layout. The minimum flow area is the minimum of $p_T\sqrt{3}$ and $2 \times (p_T - d_O)$. For the tube pitches use in practice the minimum is $2(p_T - d_O)$. Thus, for a 60° layout:

$$A_{CF} = B\left[(D_S - D_B) + \frac{(D_B - d_O)}{\sqrt{3}p_T}2(p_T - d_O)\right] \tag{F.28}$$

Equations F.26, F.27 and F.28 can be combined to give:

$$A_{CF} = B\left[(D_S - D_B) + \frac{(D_B - d_O)}{p_{CF}p_T}(p_T - d_O)\right] \tag{F.29}$$

where p_{CF} = pitch correction factor for flow direction
= 1 for 90° and 30° layouts
= $\sqrt{2}/2$ for 45° layouts
= $\sqrt{3}/2$ for 60° layouts

Assume:

$$v_S = \frac{m_S}{\rho A_{CF}} \tag{F.30}$$

where m_S = mass flowrate on the shell-side

Then combining Equations F.29 and F.30 gives:

$$v_S = \frac{m_S}{\rho B\left[(D_S - D_B) + \dfrac{(D_B - d_O)(p_T - d_O)}{p_{CF}p_T}\right]} \tag{F.31}$$

The shell-side pressure drop is a strong function of the baffle spacing B, since the shell-side fluid velocity v_S is inversely proportional to the baffle spacing B. From Equation F.21, the shell-side heat transfer coefficient h_S is:

$$h_S = K'_{hS}F_S \tag{F.32}$$

where

$$K'_{hS} = \frac{0.06207F_PF_LJ_Sk^{2/3}(c_P\mu)^{1/3}}{d_O}$$

a) *For $Re_S \le 250$.* Substituting Equation F.22 into Equation F.32 gives:

$$\begin{aligned} h_S &= K'_{hS}(-5.9969 \times 10^{-4}Re_S^2 + 0.6191Re_S + 17.793) \\ &= -5.9969 \times 10^{-4}K'_{hS}\left(\frac{\rho d_O v_S}{\mu}\right)^2 \\ &\quad + 0.6191K'_{hS}\frac{\rho d_O v_S}{\mu} + 17.793K'_{hS} \\ &= K_{hS1}v_S^2 + K_{hS2}v_S + K_{hS3} \end{aligned} \tag{F.33}$$

where

$$\begin{aligned} K_{hS1} &= -5.9969 \times 10^{-4}K'_{hS}\left(\frac{\rho d_O}{\mu}\right)^2 \\ &= -5.9969 \times 10^{-4} \times \frac{0.06207F_PF_LJ_Sk^{2/3}(c_P\mu)^{1/3}}{d_O} \times \left(\frac{\rho d_O}{\mu}\right)^2 \\ &= -3.722 \times 10^{-5}\frac{F_PF_LJ_Sk^{2/3}c_P^{1/3}\rho^2 d_O}{\mu^{5/3}} \end{aligned} \tag{C.34}$$

$$\begin{aligned} K_{hS2} &= 0.6191K'_{hS}\frac{\rho d_O}{\mu} \\ &= 0.6192 \times \frac{0.06207F_PF_LJ_Sk^{2/3}(c_P\mu)^{1/3}}{d_O} \times \frac{\rho d_O}{\mu} \\ &= 0.03843\frac{F_PF_LJ_Sk^{2/3}c_P^{1/3}\rho}{\mu^{2/3}} \end{aligned} \tag{F.35}$$

$$\begin{aligned} K_{hS3} &= 17.793K'_{hS} \\ &= 17.793 \times \frac{0.06207F_PF_LJ_Sk^{2/3}(c_P\mu)^{1/3}}{d_O} \\ &= 1.104\frac{F_PF_LJ_Sk^{2/3}(c_P\mu)^{1/3}}{d_O} \end{aligned} \tag{F.36}$$

b) *For* $250 < Re_S \le 125{,}000$. Substituting Equation F.23 into Equation F.32 gives:

$$\begin{aligned} h_S &= 1.40915K'_{hS}Re_S^{0.6633}B_C^{-0.5053} \\ &= 1.40915K'_{hS}\left(\frac{\rho d_O v_S}{\mu}\right)^{0.6633}B_C^{-0.5053} \\ &= K_{hS4}v_S^{0.6633} \end{aligned} \tag{F.37}$$

F

where

$$K_{hS4} = 1.40915 K'_{hS} \left(\frac{\rho d_O}{\mu}\right)^{0.6633} B_C^{-0.5053}$$

$$= 1.40915 \frac{0.06207 F_P F_L J_S k^{2/3} (c_P \mu)^{1/3}}{d_O} \left(\frac{\rho d_O}{\mu}\right)^{0.6633} B_C^{-0.5053}$$

$$= 0.08747 \frac{F_P F_L J_S k^{2/3} c_P^{1/3} \rho^{0.6633}}{\mu^{0.33} d_O^{0.3367} B_C^{0.5053}} \quad \text{(F.38)}$$

2) *Shell-side pressure drop.* The simplified Delaware method (Kern and Kraus, 1972) can be used for the calculation of the shell-side pressure drop. The total pressure drop ΔP_S for the shell-side in one shell includes the pressure drop in the straight section of shell (ΔP_{SS}) and pressure drop in nozzles (ΔP_{NS}) (Kern and Kraus, 1972).

a) *Pressure drop in the straight section of shell ΔP_{SS} per shell* (Wang *et al.*, 2012; Kern and Kraus, 1972):

$$\Delta P_{SS} = \Delta P_{SS,20\%} (B_C/0.2)^{m_{fo}} \quad \text{(F.39)}$$

where

$$\Delta P_{SS,20\%} = (N_B - 1)\Delta P_{SB,20\%} + R_S \Delta P_{SB,20\%} \quad \text{(F.40)}$$

$$\Delta P_{SB,20\%} = \frac{c_{fS} D_S \rho v_S^2}{2 d_e} \quad \text{(F.41)}$$

$$R_S = (B/B_{in})^{1.8} + (B/B_{out})^{1.8} \quad \text{(F.42)}$$

$\Delta P_{SS,20\%}$ = pressure drop in the straight section of shell with 20% baffle cut
$\Delta P_{SB,20\%}$ = pressure drop in one central baffle spacing zone with 20% baffle cut
m_{fo} = −0.26765 for $20\% < B_C \le 30\%$
m_{fo} = −0.36106 for $30\% \le B_C < 40\%$
m_{fo} = −0.58171 for $40\% \le B_C \le 50\%$
N_B = number of baffles
R_S = correction factor for unequal baffle spacing
B_C = baffle cut (%)
B = central baffle spacing
B_{in} = inlet baffle spacing
B_{out} = outlet baffle spacing
D_S = shell inner diameter
c_{fS} = shell-side friction factor

$$= 144[f_1 - 1.25(1 - B/D_s)(f_1 - f_2)] \quad \text{(F.43)}$$

$$f_1 = a Re_{Se}^{-0.125} \quad \text{(F.44)}$$

a = 0.008190 for $D_S \le 0.9$ m
= 0.01166 for $D_S > 0.9$ m

$$f_2 = b Re_{Se}^{-0.157} \quad \text{(F.45)}$$

b = 0.004049 for $D_S \le 0.9$ m
= 0.002935 for $D_S > 0.9$ m

$$Re_{Se} = \frac{\rho v_S d_e}{\mu}$$

v_S = shell-side fluid velocity
d_e = shell-side equivalent diameter

The equivalent diameter is used to calculate the flow in noncircular channels in the same way as a circular tube. The equivalent diameter is defined as:

$$d_e = \frac{4 \times \text{Flow Area}}{\text{Wetted Perimeter}} \quad \text{(F.46)}$$

For a square pitch (see Figure 12.14a):

$$d_e = \frac{4 \times \left(p_T^2 - \frac{\pi d_O^2}{4}\right)}{\pi d_O} = \frac{4}{\pi}\frac{p_T^2}{d_O} - d_O \quad \text{(F.47)}$$

For a triangular pitch (see Figure 12.14c):

$$d_e = \frac{4\left(\frac{1}{2}p_T^2 \frac{\sqrt{3}}{2} - \frac{1}{2}\frac{\pi d_O^2}{4}\right)}{\frac{\pi d_O}{2}} = \frac{2\sqrt{3}}{\pi}\frac{p_T^2}{d_O} - d_O \quad \text{(F.48)}$$

Equations F.47 and F.48 can be expressed as:

$$d_e = C_{De} \frac{p_T^2}{d_O} - d_O \quad \text{(F.49)}$$

where C_{De} = pitch configuration factor
= $4/\pi$ for square pitch
= $2\sqrt{3}/\pi$ for triangular pitch

Thus, Equation F.39 can be written as:

$$\Delta P_{SS} = \Delta P_{SS,20\%} (B_c/20\%)^{m_{fo}}$$
$$= \left[(N_B - 1)\Delta P_{SB,20\%} + R_S \Delta P_{SB,20\%}\right]\left(\frac{B_C}{0.2}\right)^{m_{fo}}$$
$$= (N_B - 1 + R_S)\frac{c_{fS} D_S \rho v_S^2}{2 d_e}\left(\frac{B_C}{0.2}\right)^{m_{fo}}$$
$$= (N_B - 1 + R_S)\left\{144\left[f_1 - 1.25(1 - B/D_S)(f_1 - f_2)\right]\right\} \times \frac{D_S \rho v_S^2}{2 d_e}\left(\frac{B_C}{0.2}\right)^{m_{fo}}$$
$$= (N_B - 1 + R_S)\left[36\left(5\frac{B}{D_S} - 1\right) a Re_{Se}^{-0.125} + 180\left(1 - \frac{B}{D_S}\right) b Re_{Se}^{-0.157}\right]\frac{D_S \rho v_S^2}{2 d_e}\left(\frac{B_C}{0.2}\right)^{m_{fo}}$$

$$= (N_B - 1 + R_S)\left[36\left(5\frac{B}{D_S}-1\right)a\left(\frac{\rho v_S d_e}{\mu}\right)^{-0.125}\right.$$

$$\left.+180\left(1-\frac{B}{D_S}\right)b\left(\frac{\rho v_S d_e}{\mu}\right)^{-0.157}\right]\frac{D_S \rho v_S^2}{2d_e}\left(\frac{B_C}{0.2}\right)^{m_{fo}}$$

$$= 36\left(5\frac{B}{D_S}-1\right)(N_B - 1 + R_S)\frac{aD_S\rho}{2d_e}\left(\frac{B_C}{0.2}\right)^{m_{fo}}\left(\frac{\rho d_e}{\mu}\right)^{-0.125} v_S^{1.875}$$

$$+180\left(1-\frac{B}{D_S}\right)(N_B - 1 + R_S)\frac{bD_S\rho}{2d_e}\left(\frac{B_C}{0.2}\right)^{m_{fo}}\left(\frac{\rho d_e}{\mu}\right)^{-0.157} v_S^{1.843}$$

$$= K_{PS1}v_S^{1.875} + K_{PS2}v_S^{1.843} \quad \text{(F.50)}$$

where

$$K_{PS1} = 18\left(5\frac{B}{D_S}-1\right)(N_B - 1 + R_S)\frac{aD_S\rho}{d_e}\left(\frac{B_C}{0.2}\right)^{m_{fo}}\left(\frac{\rho d_e}{\mu}\right)^{-0.125} \quad \text{(F.51)}$$

$$K_{PS2} = 90\left(1-\frac{B}{D_S}\right)(N_B - 1 + R_S)\frac{bD_S\rho}{d_e}\left(\frac{B_C}{0.2}\right)^{m_{fo}}\left(\frac{\rho d_e}{\mu}\right)^{-0.157} \quad \text{(F.52)}$$

where

$$\begin{aligned} a &= 0.008190 && \text{for } D_S \le 0.9\text{ m} \\ &= 0.01166 && \text{for } D_S > 0.9\text{ m} \\ b &= 0.004049 && \text{for } D_S \le 0.9\text{ m} \\ &= 0.002935 && \text{for } D_S > 0.9\text{ m} \end{aligned}$$

b) *Pressure loss in the inlet and outlet nozzles ΔP_{NS} per shell* (Serth, 2007):

$$\begin{aligned} \Delta P_{NS} &= \Delta P_{NS,inlet} + \Delta P_{NS,outlet} \\ &= C_{NS,inlet}\rho v_{NS,inlet}^2 + C_{NS,outlet}\rho v_{NS,outlet}^2 \end{aligned} \quad \text{(F.53)}$$

where

$C_{NS,inlet} = 0.75$ for $100 \le Re_{NS,inlet} \le 2100$

$C_{NS,inlet} = 0.375$ for $Re_{NS,inlet} > 2100$

$$Re_{NS,inlet} = \frac{\rho v_{NS,inlet} d_{NS,inlet}}{\mu}$$

$$v_{NS,inlet} = \frac{m_S}{\rho(\pi d_{NS,inlet}^2/4)}$$

$d_{NS,inlet}$ = inner diameter of the inlet nozzle for the shell-side fluid

$C_{NS,outlet} = 0.75$ for $100 \le Re_{NS,outlet} \le 2100$

$C_{NS,outlet} = 0.375$ for $Re_{NS,outlet} > 2100$

$$Re_{NS,outlet} = \frac{\rho v_{NS,outlet} d_{NS,outlet}}{\mu}$$

$$v_{NS,outlet} = \frac{m_S}{\rho\left(\pi d_{NS,outlet}^2/4\right)}$$

$d_{NS,outlet}$ = inner diameter of the outlet nozzle for the shell-side fluid.

Thus, the total pressure drop for the shell-side per shell is:

$$\begin{aligned} \Delta P_S &= \Delta P_{SS} + \Delta P_{NS} \\ &= K_{PS1}v_S^{1.875} + K_{PS2}v_S^{1.843} + C_{NS,inlet}\rho v_{NS,inlet}^2 \\ &\quad + C_{NS,outlet}\rho v_{NS,outlet}^2 \\ &= K_{PS1}v_S^{1.875} + K_{PS2}v_S^{1.843} + K_{PS3} \end{aligned} \quad \text{(F.54)}$$

where

$$K_{PS3} = \rho\left(C_{NS,inlet}v_{NS,inlet}^2 + C_{NS,outlet}v_{S,outlet}^2\right) \quad \text{(F.55)}$$

For heat exchangers with multiple shells connected in series, the total pressure drop on the shell-side fluid is estimated by the product of the pressure drop per shell and shell number in series:

$$\Delta P_{S,N_{SHELLS}} = N_{SHELLS}\Delta P_S \quad \text{(F.56)}$$

where

$\Delta P_{S,N_{SHELLS}}$ = total pressure drop on the shell-side

ΔP_S = pressure drop per shell, given by Equation F.54

N_{SHELLS} = number of shells connected in series

For heat exchangers with multi parallel shells, the total pressure drop on the shell-side fluid is equal to the pressure drop in a single shell ΔP_S.

References

Ayub ZH (2005) A New Chart Method for Evaluating Single-Phase Shell Side Heat Transfer Coefficient in a Single Segmental Shell and Tube Heat Exchanger, *Appl Therm Eng*, **25**: 2412.

Bhatti MS and Shah RK (1987) Turbulent and Transition Convective Heat Transfer in Ducts. In: Kakac S, Shah RK and Aung W (eds.), *Handbook of Single-Phase Convective Heat Transfer*, John Wiley & Sons, Chapter 4.

Hewitt GF (1998) *Handbook of Heat Exchanger Design*, Begell House Inc.

Kern DQ and Kraus AD (1972) *Extended Surface Heat Transfer*, McGraw-Hill, New York.

Kraus AD, Welty JR, Aziz A (2011) *Introduction to Thermal and Fluid Engineering*, CRC Press.

Serth RW (2007) Design of shell-and-tube heat exchangers. In: Serth RW (ed.), *Process Heat Transfer: Principles and Applications*, Burlington Academic Press, Chapter 5.

Wang YF, Pan M, Bulatov I, Smith R and Kim JK (2012) Application of intensified heat transfer for the retrofit of heat exchanger network, *Appl Energy*, **89**: 45.

Appendix G

Gas Compression Theory

G.1 Modeling Reciprocating Compressors

The mechanical work to compress a gas is the product of the external force acting on the gas and the distance through which the force moves. Consider a cylinder with cross-sectional area A containing a gas to be compressed by a piston. The force exerted on the gas is the product of the pressure (force per unit area) and the area A of the piston. The distance the piston travels is the volume V of the cylinder divided by the area A. Thus:

$$\begin{aligned} W &= \int_{V_1}^{V_2} PAd\left(\frac{V}{A}\right) \\ &= \int_{V_1}^{V_2} PdV \end{aligned} \tag{G.1}$$

where W = work for gas compression
P = pressure of the gas
V = volume of the gas
A = area of the cylinder and piston

For gas compression, the final volume V_2 is less than the initial volume V_1 and the work of compression is negative. For an expansion process, the work is positive. A compressor adds energy to the gas by doing work. A simple ideal compression process is shown in Figure G.1. In compressing the gas from pressure P_1 and volume V_1 to pressure P_2 and volume V_2, the work as defined by Equation G.1 is the area under the graph. The integral in Equation G.1 can be transformed from integration over V to integration over P by considering the areas in Figure G.1:

$$\begin{aligned} W &= \int_{V_1}^{V_2} PdV \\ &= -\int_{V_2}^{V_1} PdV \\ &= -\left[\int_{P_1}^{P_2} VdP - P_2V_2 + P_1V_1\right] \\ &= P_2V_2 - P_1V_1 - \int_{P_1}^{P_2} VdP \end{aligned} \tag{G.2}$$

Thus:

$$\int_{V_1}^{V_2} PdV + P_1V_1 - P_2V_2 = -\int_{P_1}^{P_2} VdP \tag{G.3}$$

At this stage, the compression process is considered to be frictionless.

To evaluate the integral in Equation G.1 requires the pressure to be known at each point along the compression path. In principle, compression could be carried out either at constant temperature or adiabatically. Most compression processes are carried out close to adiabatic conditions. Adiabatic compression of an ideal gas along a thermodynamically reversible (isentropic) path can be expressed as (Hougen, Watson and Ragatz, 1959, Coull and Stuart, 1964):

$$PV^{\gamma} = \text{constant} \tag{G.4}$$

where $\gamma = C_P/C_V$
C_P = heat capacity at constant pressure
C_V = heat capacity at constant volume

From Equation G.4 for an adiabatic ideal gas (isentropic) compression:

$$\frac{P_1}{P_2} = \left(\frac{V_2}{V_1}\right)^{\gamma} \tag{G.5}$$

where P_1, P_2 = initial and final pressures
V_1, V_2 = initial and final volumes

A general ideal gas adiabatic (isentropic) compression process is given by:

$$P = \frac{P_1V_1^{\gamma}}{V^{\gamma}} \tag{G.6}$$

where P and V are the pressure and volume, starting from initial conditions of P_1 and V_1. Substituting Equation G.6 into Equation G.1 gives:

$$\begin{aligned} W &= P_1V_1^{\gamma}\int_{V_1}^{V_2}\left(\frac{1}{V^{\gamma}}\right)dV \\ &= P_1V_1^{\gamma}\left[-\frac{1}{(\gamma-1)V^{\gamma-1}}\right]_{V_1}^{V_2} \\ &= \frac{P_1V_1}{\gamma-1}\left[1-\left(\frac{V_1}{V_2}\right)^{\gamma-1}\right] \end{aligned} \tag{G.7}$$

Chemical Process Design and Integration, Second Edition. Robin Smith.

Companion Website: www.wiley.com/go/smith/chemical2e

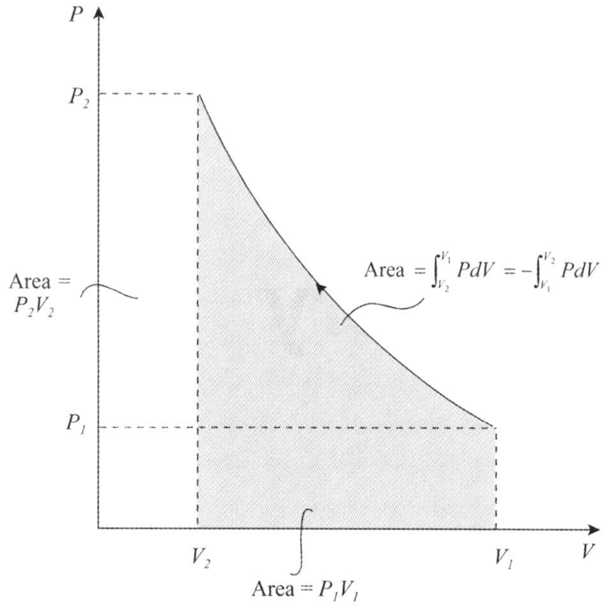

Figure G.1
A simple ideal compression process.

Combining Equations G.7 and G.5 gives:

$$W = \frac{P_1 V_1}{\gamma - 1}\left[1 - \left(\frac{P_2}{P_1}\right)^{(\gamma-1)/\gamma}\right] \quad \text{(G.8)}$$

Equation G.8 only considers the work accompanying a change of state. In a reciprocating compressor, these changes form only one step in a cycle of changes. Figure G.2 represents the pressure and volume changes that occur in the cylinder of an ideal reciprocating compressor.

Between Points 1 and 2 in Figure G.2, the intake and discharge valves are closed and the gas in the cylinder is compressed from P_1 to P_2. When the pressure reaches P_2, the discharge valve opens and the gas is pushed from the cylinder between Points 2 and 3 in Figure G.2. Between Points 3 and 4, the intake and discharge valves are closed and any gas remaining in the cylinder is expanded to the intake pressure of P_1. Between Points 4 and 1, the intake valve opens and the suction stroke draws gas into the cylinder at pressure P_1. The total work for the cycle is the sum of the work for the four steps. The work required by the compression is often termed *shaft work* W_S. Thus:

$$W_S = W_{12} + W_{23} + W_{34} + W_{41} \quad \text{(G.9)}$$

where
$$W_{12} = \int_{V_1}^{V_2} PdV$$
$$W_{23} = \int_{V_2}^{V_3} PdV = P_3V_3 - P_2V_2$$
$$W_{34} = \int_{V_3}^{V_4} PdV$$
$$W_{41} = \int_{V_4}^{V_1} PdV = P_1V_1 - P_4V_4$$

Substituting in Equation G.9 gives:

$$\begin{aligned} W_S &= \int_{V_1}^{V_2} PdV + P_3V_3 - P_2V_2 + \int_{V_3}^{V_4} PdV + P_1V_1 - P_4V_4 \\ &= \left[\int_{V_1}^{V_2} PdV + P_1V_1 - P_2V_2\right] + \left[\int_{V_3}^{V_4} PdV + P_3V_3 - P_4V_4\right] \end{aligned} \quad \text{(G.10)}$$

Combining Equations G.10 and G.3 gives:

$$\begin{aligned} W_S &= -\int_{P_1}^{P_2} VdP - \int_{P_3}^{P_4} VdP \\ &= -\int_{P_1}^{P_2} VdP + \int_{P_4}^{P_3} VdP \end{aligned} \quad \text{(G.11)}$$

For an adiabatic ideal gas (isentropic) compression or expansion:

$$\begin{aligned} \int_{P_1}^{P_2} VdP &= P_1^{1/\gamma} V_1 \int_{P_1}^{P_2} \left(\frac{1}{P^{1/\gamma}}\right) dP \\ &= P_1^{1/\gamma} V_1 \left[-\frac{1}{\left(\frac{1}{\gamma} - 1\right) P^{1/(\gamma-1)}}\right]_{P_1}^{P_2} \\ &= \frac{\gamma}{1-\gamma} P_1 V_1 \left[1 - \left(\frac{P_2}{P_1}\right)^{(\gamma-1)/\gamma}\right] \end{aligned} \quad \text{(G.12)}$$

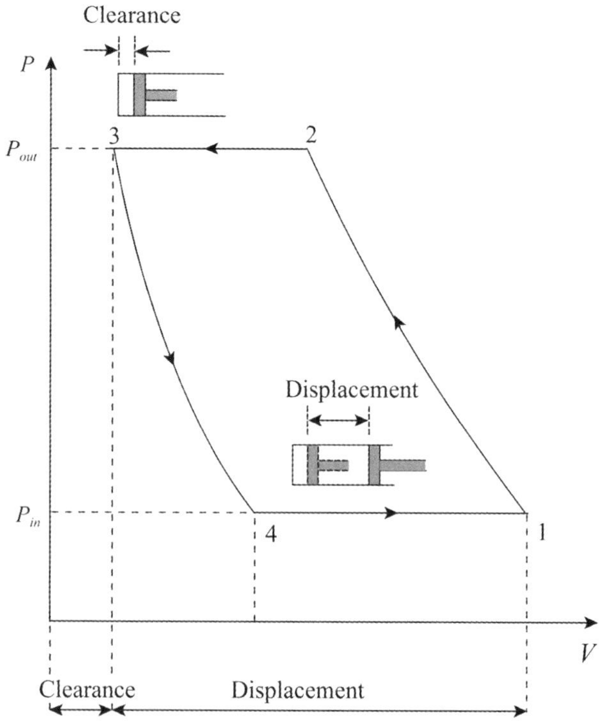

Figure G.2
Ideal single-stage reciprocating gas compressor.

Thus, from Equations G.11 and G.12:

$$W_S = -\frac{\gamma}{1-\gamma} P_1 V_1 \left[1 - \left(\frac{P_2}{P_1}\right)^{(\gamma-1)/\gamma}\right]$$

$$+\frac{\gamma}{1-\gamma} P_4 V_4 \left[1 - \left(\frac{P_3}{P_4}\right)^{(\gamma-1)/\gamma}\right] \tag{G.13}$$

Given that $P_1 = P_4 = P_{in}$, $P_2 = P_3 = P_{out}$ and $(V_1 - V_4) = V_{in}$:

$$W_S = \left(\frac{\gamma}{\gamma-1}\right) P_{in} V_{in} \left[1 - \left(\frac{P_{out}}{P_{in}}\right)^{(\gamma-1/\gamma)}\right] \tag{G.14}$$

Equation G.14 is the work required for an ideal adiabatic (isentropic) compression. To account for inefficiencies in the compression process, the isentropic compression efficiency is introduced (see Figure 13.18):

$$W = \left(\frac{\gamma}{\gamma-1}\right) \frac{P_{in} V_{in}}{\eta_{IS}} \left[1 - \left(\frac{P_{out}}{P_{in}}\right)^{(\gamma-1/\gamma)}\right] \tag{G.15}$$

where W = work required for gas compression (J)
P_{in}, P_{out} = inlet and outlet pressures ($N{\cdot}m^2$)
V_{in} = inlet gas volume (m^3)
η_{IS} = isentropic efficiency (–)
γ = heat capacity ratio C_P/C_V (–)

Expressing V_{in} as a volumetric flowrate returns a value of W in watts.

G.2 Modeling Dynamic Compressors

Unlike reciprocating compressors, dynamic compressors involve a constant flow through the compressor. Consider again adiabatic compression of an ideal gas in a dynamic compressor similar to a reciprocating compressor as above. A volume of gas V_1 flows into the compressor. Compression work W_1 is required to force the gas into the system. The constant force exerted on the gas is $P_1 A_1$, where A_1 is the cross-sectional area of the inlet duct. The distance through which the gas is forced as it enters the system is $-V_1/A_1$. The negative value of $-V_1/A_1$ results from the force acting from the surroundings on the system. Thus:

$$\begin{aligned} W_1 &= P_1 A_1 \left(-\frac{V_1}{A_1}\right) \\ &= -P_1 V_1 \end{aligned} \tag{G.16}$$

Similarly, for the outlet from the compressor, the system must do work on the surroundings to force the gas out, such that:

$$W_2 = P_2 V_2 \tag{G.17}$$

The work done by the compression is described by Equation G.1. Thus:

$$\int_{V_1}^{V_2} P dV = W_S + W_1 + W_2 \tag{G.18}$$

Substituting Equations G.16 and G.17 into Equation G.18:

$$\int_{V_1}^{V_2} P dV = W_S - P_1 V_1 + P_2 V_2 \tag{G.19}$$

Rearranging Equation G.19:

$$W_S = \int_{V_1}^{V_2} P dV + P_1 V_1 - P_2 V_2 \tag{G.20}$$

which from Equation G.3 gives:

$$W_S = -\int_{P_1}^{P_2} V dP \tag{G.21}$$

For an adiabatic ideal gas compression, from Equation G.12:

$$W_S = \frac{\gamma}{\gamma-1} P_{in} V_{in} \left[1 - \left(\frac{P_{out}}{P_{in}}\right)^{(\gamma-1)/\gamma}\right] \tag{G.22}$$

Introducing the isentropic compression efficiency gives:

$$W = \left(\frac{\gamma}{\gamma-1}\right) \frac{P_{in} V_{in}}{\eta_{IS}} \left[1 - \left(\frac{P_{out}}{P_{in}}\right)^{(\gamma-1)/\gamma}\right] \tag{G.23}$$

Equation G.23 is the same result as for a reciprocating compressor in Equation G.15. In a reciprocating compressor, the net effect of the cycle is a flow process, even though intermittent nonflow steps are involved. Expressing V_{in} as a volumetric flowrate returns a value of W in watts.

G.3 Staged Compression

If the temperature rise accompanying single-stage gas compression is unacceptably high the overall compression can be broken down into a number of stages with intermediate cooling.

Consider a two-stage compression in which the intermediate gas is cooled down to the initial temperature. The total work for a two-stage adiabatic gas compression of an ideal gas is given by:

$$W_S = \frac{\gamma}{\gamma-1} P_1 V_1 \left[1 - \left(\frac{P_2}{P_1}\right)^{(\gamma-1)/\gamma}\right]$$

$$+\frac{\gamma}{\gamma-1} P_2 V_2 \left[1 - \left(\frac{P_3}{P_2}\right)^{(\gamma-1)/\gamma}\right] \tag{G.24}$$

where P_1, P_2, P_3 = initial, intermediate and final pressures
V_1, V_2 = initial and intermediate gas volumes

For an ideal gas with intermediate cooling to the initial temperature:

$$P_1V_1 = P_2V_2 \qquad \text{(G.25)}$$

Combining Equations G.24 and G.25:

$$W_S = \frac{\gamma}{\gamma-1}P_1V_1\left[2 - \left(\frac{P_2}{P_1}\right)^{(\gamma-1)/\gamma} - \left(\frac{P_3}{P_2}\right)^{(\gamma-1)/\gamma}\right] \qquad \text{(G.26)}$$

The intermediate pressure P_2 can be chosen to minimize the overall work of compression. Thus:

$$\frac{dW_S}{dP_2} = 0 = \frac{\gamma}{\gamma-1}P_1V_1\left[-\left(\frac{1}{P_1}\right)^{(\gamma-1)/\gamma}\left(\frac{\gamma-1}{\gamma}\right)P_2^{-1/\gamma} + P_3^{(\gamma-1)/\gamma}\left(\frac{\gamma-1}{\gamma}\right)P_2^{(1-2\gamma)/\gamma}\right] \qquad \text{(G.27)}$$

Simplifying and rearranging Equation G.27 gives:

$$P_2^{(2\gamma-2)/\gamma} = (P_1P_3)^{(\gamma-1)/\gamma}$$

or

$$P_2 = \sqrt{P_1P_3} \qquad \text{(G.28)}$$

Rearranging Equation G.28 gives:

$$\frac{P_2}{P_1} = \frac{P_3}{P_2} = \left(\frac{P_3}{P_1}\right)^{1/2} \qquad \text{(G.29)}$$

Thus, for minimum shaft work, each stage should have the same compression ratio, which is equal to the square root of the overall compression ratio. The implication of this from Equation G.24 is that, given $P_1V_1 = P_2V_2$ for an ideal gas, if the pressure ratio is the same for each stage, then the power input is the same for each stage. In summary, for a staged compression, if the gas is assumed to be ideal, with intercooling to the inlet temperature and no pressure drop across the intercooler, then the pressure drop per stage is the same and the power input per stage is the same for the lowest overall power input. This result is readily extended to N stages. The minimum work is obtained when the compression ratio in each stage is equal:

$$\frac{P_2}{P_1} = \frac{P_3}{P_2} = \frac{P_4}{P_3} = \cdots = r \qquad \text{(G.30)}$$

where r = compression ratio

Since

$$\left(\frac{P_2}{P_1}\right)\left(\frac{P_3}{P_2}\right)\left(\frac{P_4}{P_3}\right) = \cdots = r^N = \frac{P_{N+1}}{P_1} \qquad \text{(G.31)}$$

The pressure ratio for minimum work for N stages is given by:

$$r = \sqrt[N]{\frac{P_{out}}{P_{in}}} \qquad \text{(G.32)}$$

The total shaft work for N stages is then given by:

$$W_S = \frac{\gamma}{\gamma-1}P_{in}V_{in}N\left[1 - (r)^{(\gamma-1)/\gamma}\right] \qquad \text{(G.33)}$$

Introducing the isentropic compression efficiency gives:

$$W = \frac{\gamma}{\gamma-1}\frac{P_{in}V_{in}N}{\eta_{IS}}\left[1 - (r)^{(\gamma-1)/\gamma}\right] \qquad \text{(G.34)}$$

The corresponding equation for a polytropic compression is given by:

$$W = \frac{n}{n-1}\frac{P_{in}V_{in}N}{\eta_P}\left[1 - (r)^{(n-1)/n}\right] \qquad \text{(G.35)}$$

Consider now a two-stage compression in which the intermediate gas is cooled to a defined temperature T_2, different from the inlet temperature T_1, and there is a pressure drop across the intercooler ΔP_{INT}. Let T_2 and P_2 now be the outlet temperature and pressure of the intercooler. For an ideal gas:

$$P_2V_2 = \frac{T_2}{T_1}P_1V_1 \qquad \text{(G.36)}$$

Substituting Equation G.36 into Equation G.24 and allowing for the pressure drop in the intercooler gives:

$$W_S = \frac{\gamma}{\gamma-1}P_1V_1\left[1 - \left(\frac{P_2+\Delta P_{INT}}{P_1}\right)^{(\gamma-1)/\gamma}\right] + \frac{\gamma}{\gamma-1}\left(\frac{T_2}{T_1}\right)P_1V_1\left[1 - \left(\frac{P_3}{P_2}\right)^{(\gamma-1)/\gamma}\right] \qquad \text{(G.37)}$$

Differentiating with respect to intermediate pressure P_2 (assuming T_2 to be constant) gives:

$$\frac{dW_S}{dP_2} = 0 = \frac{\gamma}{\gamma-1}P_1V_1\left[-\left(\frac{1}{P_1}\right)^{(\gamma-1)/\gamma}\left(\frac{\gamma-1}{\gamma}\right)(P_2+\Delta P_{INT})^{-1/\gamma} + \frac{T_2}{T_1}P_3^{(\gamma-1)/\gamma}\left(\frac{\gamma-1}{\gamma}\right)P_2^{(1-2\gamma)/\gamma}\right] \qquad \text{(G.38)}$$

Rearranging Equation G.38 gives:

$$\frac{P_2^{2\gamma-1}}{(P_2+\Delta P_{INT})} = \left(\frac{T_2}{T_1}\right)^{\gamma}(P_1P_3)^{\gamma-1} \qquad \text{(G.39)}$$

If $\Delta P_{INT} = 0$ and $T_2 = T_1$, Equation G.39 simplifies to Equation G.28. Equation G.39 predicts the intermediate pressure for minimum shaft work for compression of an ideal gas. The corresponding expression for a polytropic compression is given by replacing γ by n in Equation G.39. The important thing to note is that if the intercooling is not back to the original inlet temperature and there is a pressure drop in the intercooler, then Equation G.32 does not predict the pressure ratio for minimum power consumption. In practice, this is normally allowed for by breaking the calculation down into different compressor stages with the appropriate allowance for change in temperature and pressure drop between stages. However, the effect on the overall shaft work for compression is often insensitive to these effects.

References

Coull J and Stuart EB (1964) *Equilibrium Thermodynamics*, John Wiley & Sons.

Hougen OA, Watson KM and Ragatz RA (1959) *Chemical Process Principles. Part II Thermodynamics*, John Wiley & Sons.

Appendix H

Algorithm for the Heat Exchanger Network Area Target

Figure H.1 shows a pair of composite curves divided into vertical enthalpy intervals. Also shown in Figure H.1 is a heat exchanger network for one of the enthalpy intervals, which will satisfy all of the heating and cooling requirements. The network shown in Figure H.1 for the enthalpy interval is in grid diagram form. Hot streams are at the top, running left to right. Cold streams are at the bottom, running right to left. A heat exchange match is represented by a vertical line joining two circles on the streams being matched. The network arrangement in Figure H.1 has been placed such that each match experiences the ΔT_{LM} of the interval. The network also uses the minimum number of matches $(S-1)$. Such a network can be developed for any interval, provided each match within the interval:

a) completely satisfies the enthalpy change of a stream in the interval and

b) achieves the same ratio of *CP*s as exists between the composite curves (by stream splitting if necessary).

As each such match is successively placed in the interval, the minimum number of matches can be achieved because there is one fewer stream to match *and* the *CP* ratio of the remaining streams (that is, the ratio of ΣCP_H and ΣCP_C of the remaining streams) in the interval still satisfies the *CP* ratio between the composite curves.

It is, thus, always possible to achieve the interval design with $S-1$ matches, with each match operating with the log mean temperature differences of the interval.

Now consider the heat transfer area required by enthalpy interval *k*, in which the overall heat transfer coefficient is allowed to vary between individual matches:

$$A_k = \frac{1}{\Delta T_{LM_k}} \sum_{ij} \frac{Q_{ij,k}}{U_{ijk}} \tag{H.1}$$

where
A_k = network area based on vertical heat exchange in enthalpy interval *k*
ΔT_{LM_k} = log mean temperature difference for enthalpy interval *k*
$Q_{ij,k}$ = heat load on match between hot stream *i* and cold stream *j* in enthalpy interval *k*
$U_{ij,k}$ = overall heat transfer coefficient between hot stream *i* and cold stream *j* in enthalpy interval *k*

Introducing individual film transfer coefficients:

$$A_k = \frac{1}{\Delta T_{LM_k}} \sum_{ij} Q_{ij,k} \left(\frac{1}{h_i} + \frac{1}{h_j} \right) \tag{H.2}$$

where h_i, h_j are film transfer coefficients for hot stream *i* and cold stream *j* (including wall and fouling resistances). From Equation H.2:

$$A_k = \frac{1}{\Delta T_{LM_k}} \left(\sum_{ij} \frac{Q_{ij,k}}{h_i} + \sum_{ij} \frac{Q_{ij,k}}{h_j} \right) \tag{H.3}$$

Since enthalpy interval *k* is in heat balance, then summing overall cold stream matches with hot stream *i* gives the stream duty on hot stream *i*:

$$\sum_{j}^{J} Q_{ij,k} = q_{i,k} \tag{H.4}$$

where $q_{i,k}$ = stream duty on hot stream *i* in enthalpy interval *k*
J = total number of cold streams in enthalpy interval *k*

Similarly, summing overall hot stream matches with cold *j* gives the cold stream duty on cold stream *j*:

$$\sum_{i}^{I} Q_{ij,k} = q_{j,k} \tag{H.5}$$

Chemical Process Design and Integration, Second Edition. Robin Smith.

Companion Website: www.wiley.com/go/smith/chemical2e

Figure H.1

Within each enthalpy interval it is possible to design a network with $(S-1)$ matches. (Reproduced from Ahmad S and Smith R, 1989, *IChemE ChERD*, **67**: 48, by permission of the Institution of Chemical Engineers.)

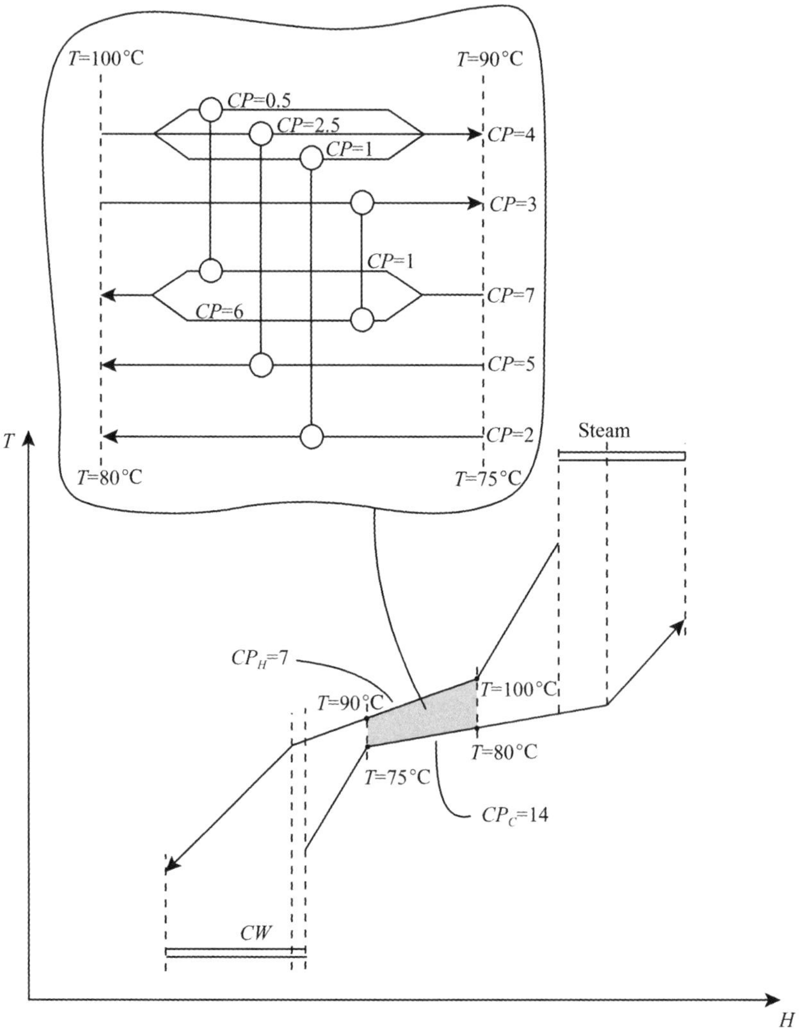

where $q_{j,k}$ = stream duty on cold stream j in enthalpy interval k
I = total number of hot streams in enthalpy interval k

Thus, from Equation H.4:

$$\sum_{ij} \frac{Q_{ij,k}}{h_i} = \sum_{i}^{I} \frac{q_{i,k}}{h_i} \tag{H.6}$$

and from Equation H.5:

$$\sum_{ij} \frac{Q_{ij,k}}{h_j} = \sum_{j}^{J} \frac{q_{j,k}}{h_j} \tag{H.7}$$

Substituting these expressions in Equation H.3 gives:

$$A_k = \frac{1}{\Delta T_{LM_k}} \left(\sum_{i}^{I} \frac{q_{i,k}}{h_i} + \sum_{j}^{J} \frac{q_{j,k}}{h_j} \right) \tag{H.8}$$

Extending this equation to all enthalpy intervals in the composite curves gives:

$$A_{NETWORK} = \sum_{k}^{INTERVALS\ K} \frac{1}{\Delta T_{LM_k}} \left(\sum_{i}^{HOT\ STREAMS\ I} \frac{q_{i,k}}{h_i} + \sum_{j}^{COLD\ STREAMS\ J} \frac{q_{j,k}}{h_j} \right) \tag{H.9}$$

Index

References to figures are given in italic type. References to tables are given in bold type.

Chemical Process Design and Integration, Second Edition. Robin Smith

Companion Website: www.wiley.com/go/smith/chemical2e